33rd Polymer Processing Society Annual Meeting 2017 (PPS-33)

Cancun, Mexico
10 - 14 December 2017

Volume 1 of 2

ISBN: 978-1-5108-6330-9

Printed from e-media with permission by:

Curran Associates, Inc.
57 Morehouse Lane
Red Hook, NY 12571

Some format issues inherent in the e-media version may also appear in this print version.

Printed by Curran Associates, Inc. (2018)

For permission requests, please contact Polymer Processing Society
at the address below.

Polymer Processing Society
810 E. 10th St.
Lawrence, KS 66044
USA

Phone: +1-800-627-0326
+1-785-865-9403
Fax: +1-785-843-6153

pps@allenpress.com

Additional copies of this publication are available from:

Curran Associates, Inc.
57 Morehouse Lane
Red Hook, NY 12571 USA
Phone: 845-758-0400
Fax: 845-758-2633
Email: curran@proceedings.com
Web: www.proceedings.com

TABLE OF CONTENTS

VOLUME 1

S01 – EXTRUSION PROCESSES

S02 – INJECTION MOLDING

S03 – PROCESS MODELING

S04 – BLOW MOLDING, THERMOFORMING AND ROTOMOLDING

S05 - RECYCLING

S06 – FIBER, MEMBRANES AND BARRIER

S07 – MIXING AND COMPOUNDING

S08 – POLYMER BLENDS

S09 – POLYMER NANOCOMPOSITES

S10 - COMPOSITES

VOLUME 2

S11 – BIOPOLYMERS & NATURAL FIBERS

S12 – POLYMER FOAMS AND FILMS

S13 – RUBBER AND ELASTOMERS

S14 – RHEOLOGY AND RHEOMETRY

S16 – REACTIVE PROCESSING

S17 – ADVANCED PROCESSING TECHNOLOGY

S18 – AUTOMOTIVE AND MEDICAL

S19 – ADDITIVE TECHNOLOGY

S20 – SOFT MATTER APPLICATIONS

Nano-Extrusion Process for Production of Organic Photovoltaic Cells: Proof of Concept

Ali Nourdine,[a] Lionel Flandin,[a] Nicole Albérola,[a] Lara Perrin,[a] Emilie Planès,[a] Anne Hiltner,[b,c] and Eric Baer,[b,c]

[a] *Université Savoie Mont-Blanc, LEPMI, F-73000 Chambéry*
[b] *Center for Layered Polymeric Systems, Case Western Reserve University, Cleveland, OH 44106-7202, USA*
[c] *Department of Macromolecular Science and Engineering, Case Western Reserve University, Cleveland, OH 44106-7202, USA*

Abstract. For the first time, the extrusion process is used to produce a perfectly nanostructured organic photoactive layer. Extrusion is an interesting inexpensive and environment-friendly (solventless) option allowing an industrial production scale for the polymer solar cell technology. In this work, this plausible alternative to usual large scale wet-processing techniques (inkjet printing, roll-to-roll coating…) was considered in order to elaborate bulk lamellar ordered heterojunctions employing polythiophene and fullerene C_{60} based materials, respectively electron donor and acceptor, which were specifically formulated for the preparation of extrudable solar cells. An alternating electron donor/acceptor (D/A) lamellar structure close to the ideal interpenetrating system was targeted and successfully realized. Continuous D and A domains of about 24nm were obtained, i.e. size close to the exciton diffusion length (10-20nm) for π-conjugated organic materials; and a photovoltaic response was evidenced under AM1.5 light simulator conditions. This work open the way to a new promising organic photovoltaic processing method. This work demonstrates the successful use of extrusion as technique for the fabrication of lamellar nanostructured photo-active polymeric films, and suggests a potential application for large scale fabrication of organic photovoltaic solar cells. The originality and novelty of this work first consist in using a one-step continuous extrusion process in order to produce a photo-active nano-multilayer film based on 1024 D(24nm)/A(24nm) alternating layers. After this, the total number of D and A nano-layers was increased to 38000 (19000 D/A interfaces) by consolidation of 37 extruded nano-multilayered films. And a photovoltaic response was evidenced under light, which confirms the achievement of this functional perfect nano-architectured structure by extrusion process.

Keywords: Organic photovoltaic cell; extrusion; rheological properties; poly(3-octylthiophene); fullerene-grafted polystyrene.

In this conference paper, the groundbreaking proposed concept consists in the realization of an ideal nanoscaled lamellar morphology realized by alternatively layering D and A polymeric materials via the "coextrusion forced assembly technique" presented in scheme 1.[1] This process can be used for large scale production and this is a plausible solventless and environment-friendly processing alternative compared to usual wet-processing techniques, using for instance printing [2 3 4 5] or roll-to-roll coating. [2 6 7, 8] This "forced assembly technique", which has been developed by Layered Polymeric Systems (CLiPS) at Case Western Reserve University (CWRU), allows for the layering of two polymers using layer multiplication coextrusion techniques.[9-16] This original process has been used for structuring usual polymers, and the thickness of these layers has reached nanometer sizes, allowing for unique optical, mechanical, transport, and structure properties.[9-16]

With this ideal architecture applied to a photo-active layer, the photovoltaic devices should thus lead to high and stable photovoltaic efficiencies. Nevertheless, in order to maximize the layers uniformity and overall D/A heterojunction quality, D and A polymeric materials must be coextruded at a viscosity-match temperature [17-19] and within a viscosity window determined by the extrusion pressure (upper limit of 0.1MPa.s) and the melt strength (lower limit of 0.01MPa.s). [17] Indeed, two main defects related to the difference of rheological behaviors and viscoelastic characteristics of polymers can be obtained: layer-to-layer non-uniformity of thickness [19-24] and phase's discontinuity due to interfacial instability. [18, 19, 23, 24] Accordingly, to study the feasibility of co-extrusion as new concept for production of OPV layers, the principal criteria which we firstly considered in the presented work is the processing ability of D and A materials, keeping in mind that their intrinsic functional properties could be adjusted subsequently in order to optimize performances.

Scheme 1. Presentation of the extrusion process used and photographs taken during the fabrication of the nano-multilayers film [1]

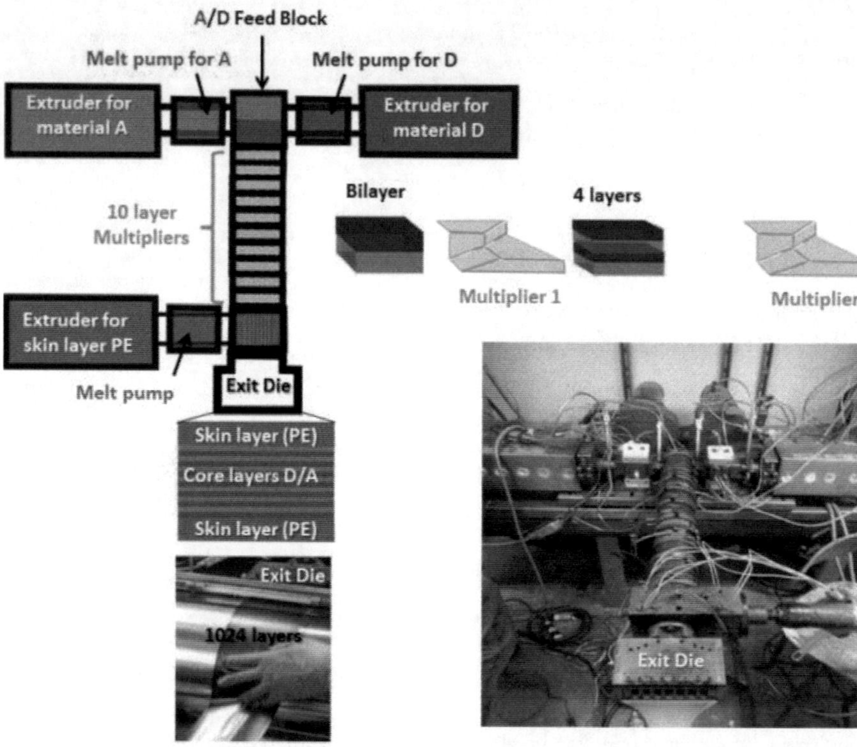

D and A materials were specifically synthetized or formulated as demonstration materials to investigate the melt-processability by extrusion process [1] of the active layers for photovoltaic application, taking into account several needed criteria -firstly- for the special processing conditions and -secondly- for the photovoltaic properties. Then, the fabrication of "multi nano-layers" photo-active films using the forced assembly coextrusion process and the final realization of "D/A nanolayers solar cells" will be both expounded.

Formulation of the D material

In order to validate the melt-processability of the chosen commercial P3OT polymer ($M_{w\ eq.PS}$ = 92kg.mol^{-1}), rheological tests were carried out using a melt flow indexer. The minimal temperature for coextrusion of a semi-crystalline polymer is limited to the temperature corresponding to its total melt (about 195°C for P3OT). At a temperature of 200°C, P3OT exhibits melt viscosity of 0.0081MPa.s. This melt viscosity value is slightly below the range required for our coextrusion process (i.e. [0.01-0.1MPa.s]); an adjustment of the rheological behavior is thus essential. A blend of P3OT with another semi-crystalline polymer having similar processing temperature and higher viscosity was chosen.

To achieve this latter approach, a semi crystalline polypropylene PP polymer was chosen as diluent polymer. The P3OT/PP polymers blends were elaborated using a twin screw extruder at temperature close to 195°C. This temperature was chosen above the melting temperature of all constituents, and well below their TGA decomposition temperatures (higher than 300°C under air, and 400°C under nitrogen). A viscous

P3OT/PP system was obtained; its composition and homogeneity were controlled by TGA and DSC. The real composition of the D material after melt blending by extrusion can be checked using TGA analysis under inert atmosphere. A good accordance was found with the initial composition introduced in the extruder, which indicates thermal stability of the P3OT/PP system during the extrusion process. A blend composition of 65wt.% and 35wt.% for respectively P3OT and PP, leading to a melt viscosity of 0.15MPa.s at 195°C, was selected in order to both be inside the compatible process viscosity range and remain close to the viscosity of the acceptor system (see Figure 1).

Synthesis and formulation of the A material

A polystyrene grafted by 12 vol.% of fullerene C_{60}, which is above the C_{60} percolation threshold (4vol.%) [25] and below the C_{60} aggregate formation threshold, [26] was chosen [27] and synthesized in large scale. In order to validate the melt-processability of this $PSCN_{60}MS12$ polymer, rheological tests were thus carried out using a melt flow indexer. The processing temperature for amorphous polymers is from 50°C above T_g, which is about 100°C for $PSCN_{60}MS12$. Consequently, the temperature range chose for the processing of D material is also adequate for the extrusion of A material. When controlling the melt viscosity of $PSCN_{60}MS12$ using MFI, an unexpected behaviour was observed. Indeed, if the first melt viscosity value measured is well inside the coextrusion process compatibility range (i.e. [0.01-0.1MPa.s]), this value is not stable and increases drastically by two orders of magnitude after 3 MFI cycles as shown in Figure 3 (0.015, 0.068 and 0.71 MPa.s for respectively cycles 1, 2 and 3 at a temperature of 200°C). This behaviour could be explained by possible chemical cross-linking (formation of fullerene-fullerene covalent bonds by ageing) and/or by phase separation resulting in the formation of fullerene aggregates as previously evidenced for higher fullerene contents in our previous work[26]. It was thus decided to reduce the fullerene content by adding a slight ratio of polystyrene polymer with close viscosity, and it was found that the initial rheological behaviour of the $PSNC_{60}MS12$/PS blend can be stabilized and maintained constant during the extrusion process. The addition of 30wt.% of PS into $PSNC_{60}MS12$ enables to adjust the melt viscosity to 0.02MPa.s at 195°C, which is both compatible with the process and close to the one of the chosen D material. Moreover, the resulting C_{60} content (8vol.% or 13wt.%) remains over the percolation threshold. Thus, the semiconductor character and the applicability in photovoltaic solar cells are preserved.

A $PSNC_{60}MS12$/PS acceptor material with a blend composition of 70wt.% for $PSNC_{60}MS12$ and 30wt.% for PS was elaborated using a twin screw extruder at the temperature of 195°C. This temperature is well below the decomposition temperature of the A selected material (close to 300°C for all constituents, determined by TGA under air). In addition, a TGA analysis performed under nitrogen on $PSNC_{60}MS12$/PS blend after extrusion indicates that the PS/C_{60} composition remains stable during extrusion (the C_{60} content being represented by the degradation residue at 550°C).Amorphous PS polymer was chosen as diluent polymer because of its presumed physico-chemical compatibility with $PSNC_{60}MS$ in order to obtain a miscible and homogeneous system which was confirmed by optical microscopy analysis.

Compatibility of A and D materials with the extrusion process

Figure 1 summarizes the logarithmic viscosities measured at different temperatures for D (P3OT/PP 65/35wt.%) and A ($PSNC_{60}MS12$/PS 70/30wt.%) materials and their constitutive polymers (P3OT, PP, $PSNC_{60}MS12$, PS). Upper and lower extrusion processing limits are also identified on the graph.

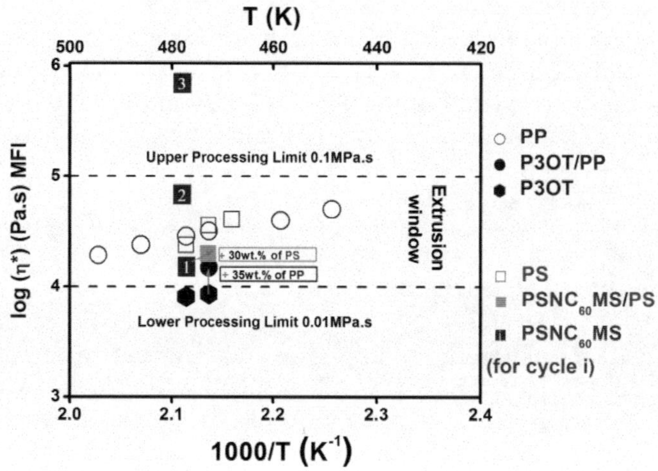

Figure 1. Logarithmic viscosity (MFI) of constitutive polymers materials as function of temperature [28]

A reasonable processing temperature close to 195°C was chosen. At this temperature, pure P3OT exhibits melt viscosity below the lower limit. The resulting readjusted viscosity by addition of 35 wt.% of PP matches the required range of viscosity. In contrast, the melt viscosity of pure $PSNC_{60}MS12$ at this range of temperature is adequate with extrusion process but its rheological behaviour is instable. The addition of 30 wt.% of PS allowed a slight modulation of the viscosity while obtaining a stable rheological behaviour. Another important parameter for the nano-layers realization success using our process is the close rheological behaviour of the two selected materials at chosen temperature. Indeed, the ratio of their viscosity should not exceed 3. This is fully the case of selected materials at 195°C: $\eta_A / \eta_D = 1.3$.

Photoactive layer realization: A/D multi-layers

A nanolayered film with 1024 alternating D/A layers was successfully extruded (see the photograph of the film in Scheme 1), and its continuous and uniform layered structure was revealed by atomic force microscopy observation (see phase image presented in **Erreur ! Source du renvoi introuvable.**). D/A nanolayers are well-resolved with an average layer thickness of 24nm, and the total thickness of the obtained film is close to 55μm (PE10μm-multiD/A 25μm-PE10μm).

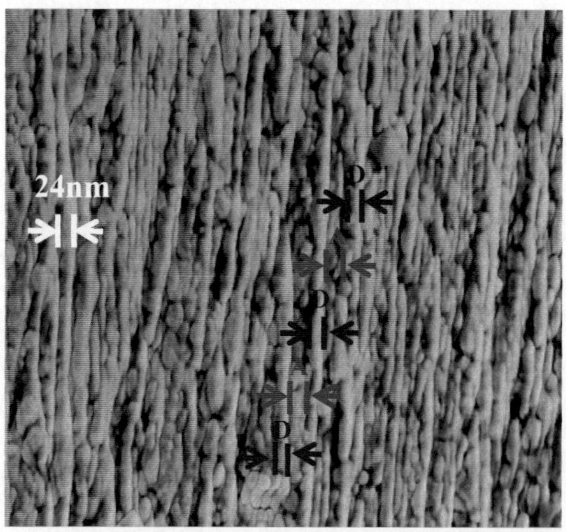

Figure 2. AFM phase image of D/A extruded multi nano-layers (tapping mode) [1]

Conclusion

This work demonstrates the successful use of extrusion as technique for the fabrication of lamellar nanostructured photo-active polymeric films, and suggests a potential application for large scale fabrication of organic photovoltaic solar cells. The originality and novelty of this work first consist in using a one-step continuous extrusion process in order to produce a photo-active nano-multilayer film based on 1024 D(24nm)/A(24nm) alternating layers. After this, the total number of D and A nano-layers was increased to 38000 (19000 D/A interfaces) by consolidation of 37 extruded nano-multilayered films. And a photovoltaic response was evidenced under light, which confirms the achievement of this functional perfect nano-architectured structure by extrusion process. The next step, consisting in the use of pure D and A extrudable polymers, should thus led to high and durable photovoltaic performances and make the extrusion method even competitive.

ACKNOWLEDGMENTS

This work was performed within the framework of the Centre of Excellence of Multifunctional Architectured Materials "CEMAM" n° ANR-10-LABX-44-01. The Institut National de l'Energie Solaire (INES) and the financial supports from Assemblée des Pays de Savoie, Université de Savoie are gratefully acknowledged.

The authors also thank J. CARR and S. ARMSTRONG for their fruitful help and valuable scientific exchange.

REFERENCES

1.A. Nourdine, L. Flandin, N. Alberola, L. Perrin, E. Planes, A. Hiltner and E. Baer, *Sustainable Energy & Fuels*, 2017, **1**, 2016-2027.
2.F. C. Krebs, *Solar Energy Materials and Solar Cells*, 2009, **93**, 394-412.
3.C. N. Hoth, P. Schilinsky, S. A. Choulis and C. J. Brabec, *Nano Letters*, 2008, **8**, 2806-2813.
4.A. Lange, M. Wegener, C. Boeffel, B. Fischer, A. Wedel and D. Neher, *Solar Energy Materials and Solar Cells*, 2010, **94**, 1816-1821.
5.J. Yang, D. Vak, N. Clark, J. Subbiah, W. W. H. Wong, D. J. Jones, S. E. Watkins and G. Wilson, *Solar Energy Materials and Solar Cells*, 2013, **109**, 47-55.
6.R. Søndergaard, M. Hösel, D. Angmo, T. T. Larsen-Olsen and F. C. Krebs, *Materials Today*, 2012, **15**, 36-49.
7.C. Koidis, S. Logothetidis, A. Ioakeimidis, A. Laskarakis and C. Kapnopoulos, *Organic Electronics*, 2013, **14**, 1744-1748.
8.Y. Liu, T. T. Larsen-Olsen, X. Zhao, B. Andreasen, R. R. Søndergaard, M. Helgesen, K. Norrman, M. Jørgensen, F. C. Krebs and X. Zhan, *Solar Energy Materials and Solar Cells*, 2013, **112**, 157-162.
9.R. Y. F. Liu, T. E. Bernal-Lara, A. Hiltner and E. Baer, *Macromolecules*, 2004, **37**, 6972-6979.
10. T. E. Bernal-Lara, R. Y. F. Liu, A. Hiltner and E. Baer, *Polymer*, 2005, **46**, 3043-3055.
11. R. Y. F. Liu, T. E. Bernal-Lara, A. Hiltner and E. Baer, *Macromolecules*, 2005, **38**, 4819-4827.
12. F. Ania, F. J. BaltÃ¡-Calleja, S. Henning, D. Khariwala, A. Hiltner and E. Baer, *Polymer*, **51**, 1805-1811.
13. Y. Jin, A. Hiltner and E. Baer, *Journal of Polymer Science Part B: Polymer Physics*, 2007, **45**, 1138-1151.
14. Y. Jin, A. Hiltner, E. Baer, R. Masirek, E. Piorkowska and A. Galeski, *Journal of Polymer Science Part B: Polymer Physics*, 2006, **44**, 1795-1803.
15. R. Y. F. Liu, A. P. Ranade, H. P. Wang, T. E. Bernal-Lara, A. Hiltner and E. Baer, *Macromolecules*, 2005, **38**, 10721-10727.
16. R. Y. F. Liu, Y. Jin, A. Hiltner and E. Baer, *Macromolecular Rapid Communications*, 2003, **24**, 943-948.
17. M. Ponting, A. Hiltner and E. Baer, *Macromolecular Symposia*, 2010, **294**, 19-32.
18. D. R. Paul, *Polymer Blends*, Elsevier Science, 1978.
19. J. Dooley, *Viscoelastic Flow Effects in Multilayer Polymer Coextrusion*, Technische Universiteit Eindhoven, 2002.
20. C. D. Han, *Journal of Applied Polymer Science*, 1973, **17**, 1289-1303.
21. B. L. Lee and J. L. White, *Transactions of The Society of Rheology*, 1974, **18**, 467-492.
22. El Kissi N., Piau J. M. and Chaigneau R. G., *Journal*, 2003, **4**, 17.
23. C. D. Han and R. Shetty, *Polymer Engineering & Science*, 1978, **18**, 180-186.
24. H. Mavridis and R. N. Shroff, *Polymer Engineering & Science*, 1994, **34**, 559-569.
25. A. Nourdine, L. Perrin, R. m. d. Bettignies, S. p. Guillerez, L. Flandin and N. Alberola, *Polymer*, 2011, **52**, 6066-6073.
26. L. Perrin, A. Nourdine, E. Planes, C. Carrot, N. Alberola and L. Flandin, *Journal of Polymer Science Part B: Polymer Physics*, 2013, **51**, 291-302.
27. A. Nourdine, L. Perrin, C. Carrot, E. Baer, L. Flandin and N. Alberola, *European Polymer Journal*, 2017, **96**, 1-9.
28. A. Nourdine, L. Flandin, N. Albérola, L. Perrin, E. Planès, A. Hiltnerb and E. Baer, *Sustainable Energy and Fuels*, 2017.

Modeling of the solid throughput and pressure build-up of single screw smooth barrel extruders under consideration of the backpressure and high screw speeds

Jan Trippe and Volker Schöppner

University of Paderborn, Warburger Straße 100, 33098 Paderborn, Germany

jan.trippe@ktp.upb.de

Abstract. The task of the solid conveying zone is to collect the entering material and transfer it to the compression zone. Therefore, its upstream interface is the hopper and its downstream interface is the melting zone where the first melting takes place. Through the screw rotation the material is conveyed forward. Common known models for describing the solid conveying in single screw extruders assume that the screw channel is always completely filled so that a block flow induced. At high screw speeds however, these assumptions are no longer valid. In previous investigations, simulations based on the Discrete Element Method have been executed to analyze and describe the throughput behavior at screws speeds up to 2000 rpm. The model for describing the solid conveying behavior allows a variation of general screw and hopper geometry parameters but neglects the backpressure that inhibits the solid movement. For that reason, further investigations are presented in this paper, which allow an extension of the known model by considering the pressure build-up in the feeding zone. Astonishingly, at high screw speeds a low filling degree in the screw channel has been observed which didn´t meet the expectations that a backpressure always leads to a fully filled channel due to a particle backlog. Afterwards the conducted simulations have been validated by experimental investigations. Therefore, a test stand has been designed, which allows solid conveying analysis with adjustable backpressures. The results showed a good agreement with the formulated model which now allows the calculation of solid conveying zones with consideration of the pressure build-up and filling degree.

Keywords: solid conveying, discrete element method, dissipation

PACS: 89.20.Bb, 89.20.Kk, 89.90.+n

INTRODUCTION

The task of the solid conveying section is to collect the entering material and transfer it to the melt dominated sections whereby a certain pressure built up is necessary. Therefore, the upstream interface is the hopper and the downstream interface is the section in which the first melting occurs. Beneath the hopper opening, the screw rotation collects and conveys the material forward. Commonly known models for describing the solid conveying in single screw extruders assume that the screw channel is always completely filled so that a block flow is induced because the friction forces among the particles are higher than the forces acting on the screw and barrel. At high screw speeds, however, these assumptions are no longer valid. To analyze and mathematically describe the solid conveying behavior simulations based on the Discrete Element Method have been executed in previous investigations. Based on this, a model for describing the solid conveying behavior was developed which enables a variation of general screw and hopper geometry parameters but therefore neglects the backpressure that inhibits the solid movement. For that reason, further investigations based on the Discrete Element Method are presented in this paper. Varied backpressures are now considered in the simulations which enable pressure build-up in the feeding zone. Astonishingly, at high screw speeds a low filling degree in the screw channel was observed which did not fulfill the expectations of an always fully filled channel due to a backpressure forced particle backlog. Based on the investigations, a mathematical model has been developed which now fully comprehends the effects occurring in the solid conveying section. The conducted simulations were subsequently validated by experimental investigations. Therefore, a test bench was designed which enables solid conveying analysis with adjustable backpressures. The results indicate a good agreement with the formulated model which now enables the calculation of solid conveying zones with consideration of the pressure build-up and filling degrees.

SIMULATIONS

The simulations have been executed with the DEM software EDEM (Software Company: DEM Solutions Ltd.) which has already been successfully used in other works [1-5] for analyzing solid conveying processes. Simulation set-ups with a screw diameter between 30 mm and 90 mm were built up via a CAD model import (Figure 1). The Hysteretic Spring model based on the investigations by Walter and Braun [6-9] was used which considers particle deformation when two particles are in collision. For the simulator a Rayleigh time step was set to $1.05x10^{-5}$ s, which is automatically calculated from the shear modulus. However, to create a stable simulation a fixed time step of 15% of the Rayleigh time was used. The granules were not varied in size and have a fixed diameter of 4 mm. The Poisson´s ration, shear modulus and density were averaged to represent a high range of polymers. This results in a higher density as it is typical for polymers like PP or PS, but as will be described later the density in the final modeling is taken into account again. The used plastic properties are shown in Table a. In addition, the used process and geometric parameters are listed:

TABLE a). Design of Experiments (CCD-Design)

Parameter	Unit	Values
Diameter D	mm	30…90
Channel Depth h/D	-	0.1…0.2
Pitch t/D	-	0.7…1.3
Flight width e/D	-	0.1
Backpressure p	bar	0…200
Screw Speed N	1/min	50…2000
Granule Size d	mm	4
Solid Density ρ	kg/m³	1500
Shear Modulus G	Mpa	600
COF µi,a	-	0.1…0.6
Rolling Friction	-	0.01
Poisson Ratio	-	0.4
COR	-	0.5

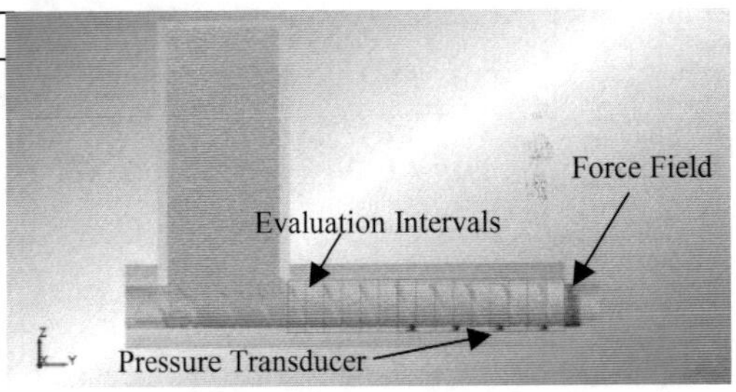

FIGURE 1. Simulation Set-Up

$$p = \frac{N_p \cdot F_p}{A_q} \tag{1}$$

The values for the coefficient friction (COF) between steel-plastic μ_a and plastic-plastic μ_i are within the range of typical polymers. The Coefficient of Restitution (COR) which is typically measured in a drop test is not varied because investigations by Leßmann and Weddige [10-11] show that the influence on the solid conveying throughput is negligible. A high range for the screw speed was set to analyze whether a low filling degree occurs regardless of particle backlog effects which result from the backpressure. Compared to the investigations implemented by Moysey and Thompson [1-5], the backpressure simulations presented in this paper are not based on a leak gap. It is implemented via a force field at the end of the conveying section. If a particle enters the force field a normal force acting on the particle against the extrusion direction is automatically applied. The nominal pressure in the axial direction is then to be obtained over the sum of the particles N_p in the force field multiplied by the set particle force F_p. In Equation 1, A_q is the ring cross-sectional area of the channel in the force field. An excellent agreement is obtained with the pressure calculated in accordance with equation 1. The average deviation is 0.4% and the maximum deviation found is 3%. For all simulated combinations, the mass flow and axial pressure profile over the screw length were analyzed. For evaluating the axial pressure profile, the barrel is segmented in the axial direction into sections of 0.3 D length (see Figure 1). Within the segments, the cumulative forces which act between the barrel and the particles located in the screw channel are detected.

RESULTS AND DISCUSSION

For the modeling of the throughput behavior, the mass flows obtained from the simulations have been detected and compared with the respective screw speed. If simply the simulated throughputs were plotted against the simulated screw speeds, a vertical shift of the data points along an axis would result, since different back pressures were simulated for the same screw speed. Accordingly, partial filling effects cannot be analyzed and the curve cannot be mathematically described. Moreover, it would not be possible to distinguish between a reduced throughput coming from an underfeeding or coming from a higher backpressure set at a similar screw speed. Therefore, dimensionless values are used in the following. Due to the dimensionless plotting, the data points are described in their slope independently of the backpressure. That is, for mathematic detection, e.g. according to Schneider, which takes into account the influence of different back pressures, but does not calculate with an under-feeding, the dimensionless representation of a linear relationship must result (see Figure 2). If, however, the simulation data points are plotted directly, a deviation of the linearity can be detected, which indicates an under-feeding independent of the backpressure.

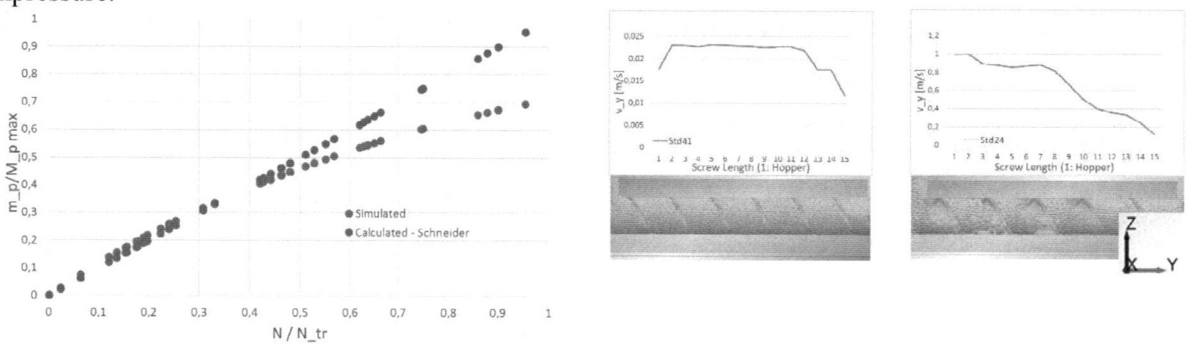

FIGURE 2. Simulation results
(a) Simulated data in comparison with calculated data (Schneider model), (b) Average particle velocity in conveying direction (Y); left: 50 rpm, bp=100 bar; right: 1650 rpm, bp=163 bar

For the description of the dimensionless throughput, the mass flow is divided by the maximum throughput which can flow from the hopper into the screw channel. The maximum throughput rate is determined as follows [10]:

$$\dot{m}_{max} = 3{,}615 * \rho_s * g^{0,5} * L_T * D^{1,5} * \frac{h}{D} * \left(\frac{t}{D}\right)^{-0,27} * \left(\frac{D}{D_{max}}\right)^{0,75} \tag{2}$$

The maximum throughput depends on the hopper opening length as well as the screw geometry, because it reduces the free volume beneath the hopper. The screw speed is considered to be dimensionless via the transition screw speed, which represents the intersection point from the linear throughput behavior of the solids conveying channel and the maximum throughput coming from the hopper (see figure 3). To describe the linear function range, the model according to Schneider [12] has been modified (see section Modeling). The simulation results plotted in figure 3 firm that a degressive throughput screw speed behavior also occurs with backpressure. This behavior can be confirmed visually by using selected simulations (see Figure 2, right). The left section in Figure 2 (b) shows a data point located in the linear region of the plotted data points. It can be seen that for low screw speeds the backpressure ensures a backlog of the particles from which a full filling of the channel results. However, with increasing screw speed, the total screw length is no longer needed to overcome the backpressure, which then results in a partial filling of the first screw flight. Conversely, this results in different particle velocities between the fully filled and partially filled areas, as the number of particles per volume element varies. Furthermore, the axial pressure profile has been evaluated which indicate a pronounced exponential course as expected. Schneider also calculates this function with a compacted solid´s bed [12]. However, he always assumes a full filling of the screw channel resulting from the backpressure, so that a potential shift in the pressure increase during partially filled processes is not taken into account. However, as seen before, it can be anticipated that a partial filling can also be observed for backpressure-dependent conveying processes. Therefore, a mathematical adaptation of the pronounced exponential course is necessary for partially filled screw channels. The location of the pressure increase, which correlates with the location of the full filling, is decisive.

MODELING

In this section a model will be presented which takes into account the throughput limitation at high screw speeds under consideration of the back pressure. The observed saturation function (Figure 2) is limited by the two characteristic asymptotes for the maximum input throughput $\dot{m}_{max}$ as well as the linear function for the description of the conveying in the fully filled channel. The following approach is therefore used to describe this characteristic:

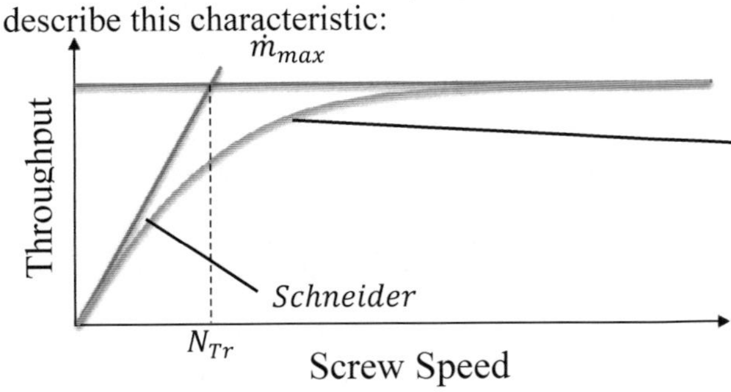

$$\frac{\dot{m}}{\dot{m}_{max}} = 1 - e^{-(\frac{N}{N_{tr}})*1.239} \tag{3}$$

$$with\ N_{tr} = \frac{\dot{m}_{max}}{\rho_s * B * H * \pi * D} * \frac{\sin(\alpha + \varphi)}{\sin(\alpha)} \tag{4}$$

FIGURE 3. Solid conveying model considering underfeeding at high screws speeds and backpressure

$\dot{m}$ is the simulated mass flow at a given screw speed N. $\dot{m}_{max}$ is the geometry-dependent maximum throughput, described by equation 2 and N_{Tr} is the transition screw speed, which is calculated by inserting $\dot{m}_{max}$ in the throughput calculation by Schneider [12] (see equation 4). A known disadvantage of Schneider´s model is that it is too sensitive regarding the coefficient of friction. Therefore, a fixed coefficient of friction has been approximated using an EXCEL solver and the existing simulation database. The target value has thereby been determined by maximizing the coefficient of determination which results in calculating the data points by means of Equation 3. The result is a coefficient of friction of 0.327 with a coefficient of determination of 0.968. This leads to significantly more stable calculation results, compared to the variable-coefficient modeling, where the coefficient of determination is only 0.643. Since the coefficient of friction also corresponds approximately to the mean value of common polymer friction values and is thus physically plausible, the specifying of the coefficients of friction is well suited for imaging the processes present in practice. An experimental confirmation of this situation can be found in the next section. In a second step, the mathematical presentation of the pressure development will be introduced. In order to specifically detect a displacement of the pressure build-up in partially filled regions, the following approach has been selected:

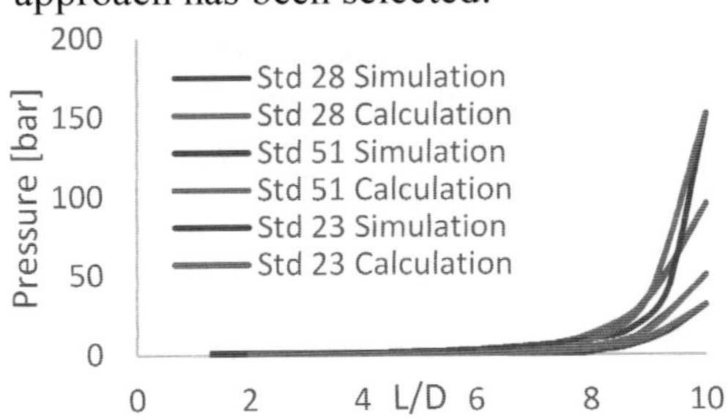

$$p(x) = p_{Hopper} + p_{FMP} \cdot e^{\left[(\frac{x_{FMP}-x}{x_{FMP}-x_{FF}})\cdot \log(\frac{p_{Hopper}}{p_{FMP}})\right]} \tag{5}$$

FIGURE 4. Simulated and calculated pressure profiles

The solid conveying section ends at the location of the first melt pool, so that the pressure p_{FMP} which is applied here takes place as a backpressure in the function at the local coordinate x_{FMP}. The pressure decreases over the length difference from x_{FMP} and x_{FF} to the pressure p_{Hopper} below the hopper. X_{FF} describes the coordinate from which a backlog of the granulate bed caused by the backpressure leads to a full filling of the screw channel. The coordinate can be determined by means of a regression function based on the simulations. As shown in Figure 4, the calculation with the equation 5 leads to a good agreement with the simulated pressure profiles.

VALIDATION

The validation of the found model was carried out with a specially designed test bench. For this purpose, the extended feed section of a 30 mm extruder was used (Figure 5). The backpressure is created by a spring which can be adjusted via a screw. The counterforce is transmitted via a bar construction to an axially and rotatable mounted cone. Between this cone and the barrel flange, the conveyed granules pass through the discharge gap. The backpressure results from the opposing force which is set in the process and the annular cross-sectional area of the cone at the outlet gap. Different channel depths, screw speeds and materials were tested and demonstrate a good agreement with the found equation 3. In a next step, this will be validated on larger machines.

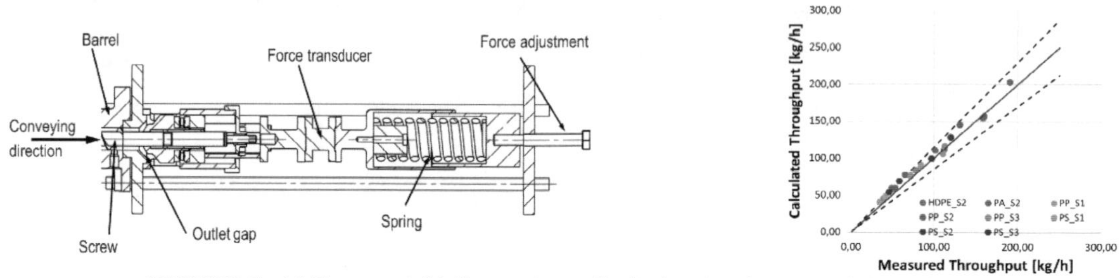

FIGURE 5. (a) Test stand (b) Comparison of calculated and measured throughputs

ACKNOWLEDGMENTS

We like to thank the Deutsche Forschungsgemeinschaft (DFG), which has made these studies possible by financing the SCHO 551/23-1 project.

REFERENCES

1. P. A. Moysey and M. R. Thompson, Polym. Eng. Sci 44, 2203-2215 (2004).
2. P. A. Moysey and M. R. Thompson, Powder Techn. 153, 95-107 (2005).
3. P. A. Moysey and M. R. Thompson, Chem. Eng. Sci. 62, 3699-3709 (2007).
4. P. A. Moysey and M. R. Thompson, Polym. Eng. Sci. 48, 62-73 (2008).
5. P. A. Moysey et al., Int. Polym. Proc. 23, 301-311 (2008).
6. O. R. Walton, Chapter 25: Numerical Simulation of Inelastic Frictional Particle-Particle Interactions", Particulate Two-Phase Flow, Ed. M. C. Roco, Butterworth-Heinemann, Boston (1993)
7. O. R. Walton, „(Linearized) Elastic-Plastic contact model", Firmenschrift, DEM Solutions (2006)
8. O. R. Walton and R.L. Braun, „Stress Calculations for Assemblies of Inelastic Spheres in Uniform Shear", Acta Mechanica 63, S. 73-86 (1986)
9. O. R. Walton and R.L. Braun, „Viscosity and Temperature Calculations for Assemblies of Inelastic Frictional Disks", Journal of Rheology 30, S. 949-980 (1986)
10. J.-S. Leßmann, „Berechnung und Simulation von Feststoffförderprozessen in Einschneckenextrudern bis in den Hochgeschwindigkeitsbereich", Dissertation, Paderborn University (2016)
11. R. Weddige, „Analyse und Optimierung des Prozessverhaltens von Einschneckenextrudern im Bereich der Feststoffförderung und der druckabhängigen Viskosität", Dissertation, Paderborn University (2015)
12. K. Schneider, „Der Fördervorgang in der Einzugszone eines Extruders", Dissertation, RWTH Aachen (1968)

pps proceeding
Symposium: S01 - Extrusion processes
Oral Presentation

Comparison Flow Rate Calculations Using Power Law Fluid and Carreau Fluid

rauwendaal chris [(1)*]

(1) ree - California - USA

For several polymers the Power Law Fluid approximation does not accurately describe the melt flow behavior. In these cases the Carreau Fluid generally leads to a more accurate description of the flow behavior. This paper will compare results of flow rate calculations using the Power Law fluid and the Carreau fluid using analytical solutions for both these fluids.

Online Elongational Viscosity Measurement for an Improved Evaluation of Melt Quality – A Smart Sensor for Continuous Plastic Processes

Dipl.-Ing. Sven Wolf [1, a]
Dipl.-Ing. Hans-Jürgen Luger [2]
Univ.-Prof. Dipl.-Ing. Dr. Jürgen Miethlinger MBA [2]

[1]*Leistritz Extrusionstechnik GmbH, Markgrafenstrasse 29-39, 90459 Nürnberg, Germany*
[2]*Institute of Polymer Extrusion and Compounding, Johannes Kepler University, Altenberger Straße 69, 4040 Linz, Austria*

[a)] *Corresponding author: swolf@leistritz.com*

Abstract. Rheological behavior of plastic melt is rather complex as it depends on various interactions. In single phase systems, it depends mainly on macromolecular structures of the polymer. In multi-phase systems, it depends on macromolecular properties of polymers, filler properties, interactions between polymers, interaction between fillers and the polymer(s) as well as interaction between the fillers itself. Inline monitoring of rheological characteristics for continuous plastic processes are realized mostly based on relative values – e.g. to monitor pressure values – or with inline viscosity measurement systems. Such viscosity measurement systems are meanwhile somehow established monitor devices for complex or expansive continuous plastic processes in the extrusion or compounding industry [e.g. Dynisco/Göttfert]. Due to the measuring principle of those inline viscosity measuring systems only shear viscosities are measured. Simplified systems - with a fixed throughput rate and capillary geometry – are also called inline MFI/MFR sensors, whereas some others can measure entire viscosity curves over a certain shear rate range. Further elongational viscosity characterizations of polymers or polymer systems are mainly conducted offline by using equipment like the Rheotens system [Göttfert et al.]. Beside time delay between production and measuring, also measurements in a molten state or at certain temperatures are somehow difficult. Within the scope of this contribution an online elongational viscosity measuring instrument and its advantages will be presented.

Keywords: Online Rheometer, Elongational Viscosity, Smart Sensor, Hyperbolical Slit Die.
PACS: 83.85 Jn

INTRODUCTION

As online shear rheometers (either in a by-pass version or direct into the stream) are more or less state of the art, the measurement of elongation behavior of plastic melt – especially inline - is rather seldom. Based on works such as Pabedinskas et al. [1], Thermo Scientific HAAKE introduced the X-Die as a first device to determine the elongation viscosity in a converging slit based on inlet pressure losses. For Newtonian fluids Olivera et al. [2] developed a very small (micrometer size) elongation slit rheometer, where a hyperbolical contraction of the flow channel at constant channel height is used to measure at constant deformation rates. To use enhanced rheological data for quality monitoring of plastic melts in single and multi-phase systems for continuous processes, the Institute of Polymer Extrusion and Compounding at the Johannes Kepler University, Linz/Austria together with the company Leistritz Extrusionstechnik, Nuremberg/Germany developed a new online elongational as well as shear rheometer.

The heart of the new rheometer is the special designed and meanwhile patented slit die [3] for viscous fluids which allows the measurement of elongational viscosity based on a constant average elongation rate with standard pressure transducers avoiding dead spots. The slit die is divided into four sections (see Fig. 1), whereas in the second section ($0 \leq x2 \leq L2$) the width is reducing on a hyperbolical function (Equ. 1) and the height on a linear function (Equ. 2).

$$y_2(x_2) = \frac{C}{a+k_1 \times x_2} \times \frac{1}{z(x_2)} \qquad (1) \qquad\qquad z_2(x_2) = \frac{H_1}{2} - k_2 \times x_2 \qquad (2)$$

In the following third section ($0 \leq x3 \leq L3$) either the height is reduced H3(x) at constant width B3 (case A) or the width is reduced B3(x) at constant height H3 (case B). For case A Equation 3 is valid for the width.

$$y_2(x_3) = y_2(x_2 = L_2) \qquad (3) \qquad\qquad z_3(x_3) = \frac{\tilde{C}}{\tilde{a}+\tilde{k} \times x_3} \qquad (4)$$

To ensure a constant elongation rate the height is described in Equation 4. This section 3 avoids expanding after contraction and therefore prevents dead zones. The constants of the above-mentioned Equations are determined based on geometrically boundary conditions as well as desired elongation rates.

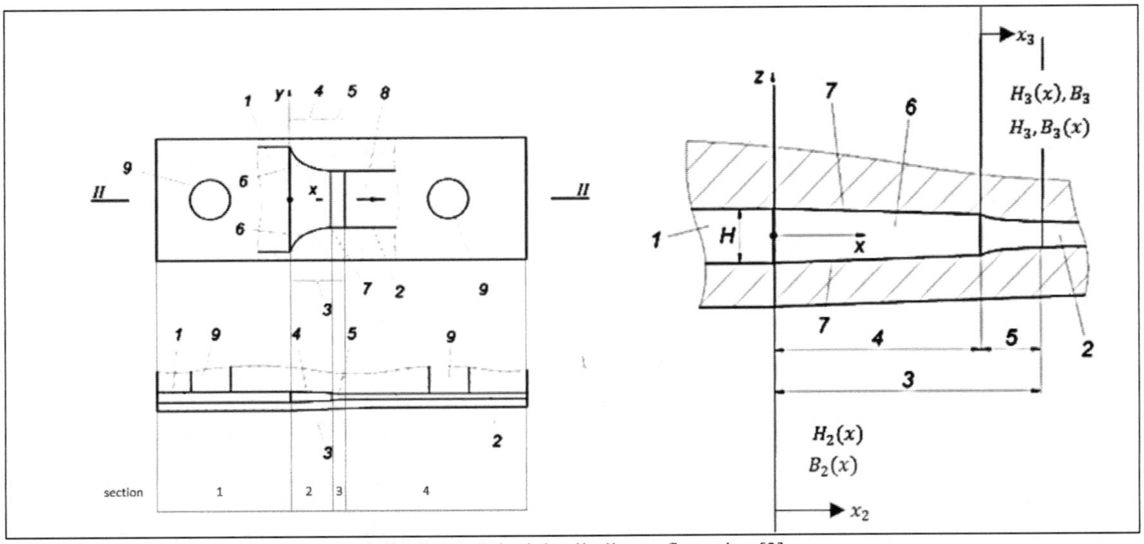

FIGURE 1. Principle slit die configuration [3]

The new developed semi-hyperbolical slit die (see Fig. 2a) has been incorporated into an online rheometer (Fig. 2b) adjusted with a very precise electrically controlled melt-pump to ensure different shear and elongational rates. The principle flow arrangement can be seen in Fig. 2c

FIGURE 2. a) Semi-hyperbolic slit die, b) Online elongation & shear rheometer, c) principle arrangement

Based on two slit sections and a hyperbolical contraction part the rheometer is designed for constant monitoring of shear viscosities with shear rates between 10 up to 10.000 s-1 and elongation rates between 5 up to 75 s-1. Adjustable throughput rates and designated flow channels also guarantee the application for a wide range of polymers (low and high viscous) and compounds (including pigments/fillers and fibers).

EXPERIMENTAL

Conformity with offline measurements

First studies have shown that good accordance to offline measurements can be found. To ensure good conformity within a certain spectrum the principle tests have been carried out with a flexible LDPE grade used for blow molding (Purell1840H, LyondellBasell) and a high molecular weight, low melt flow rate PP pipe grade (PP-RA 130E, Borealis). The reference measurements where done on the one hand with a SER-elongational tool [4] on an industrial rheometer (Anton Paar GmbH, Austria), where also the shear viscosity based on a plate-plate rheometer has been determined. On the other hand, high pressure capillary rheometer data were measured and based on Cogswell's models investigated to get representative elongational data at higher deformation rates. A further estimation for elongational data has been created by fitting storage module (G') and loss module (G'') measurements from plate-plate Rheometer by K-BKZ constitutive model [5]. Finally, also the elongation viscosity was measured with the Thermo HAAKE X-Die. All this data is presented in Figure 3.

Fig. 3a) and Fig. 3b) show good conformity between offline and online shear viscosity measurements of "softer" LDPE and "harder" PP starting at typical shear rates > 10 s-1. For online elongational viscosity data the congruence against offline measurements is good enough within the measuring range of the online Rheometer, to get significant inline material data. The crosscheck with HPCR data evaluated according to Cogswell [6] and K-

BKZ-elongational models even support the measurement principle of the On-Line rheometer. Fig. 3b) shows also the advantage of the developed slit die, as probably negative influences of dead swirls (elongational viscosity measurements from X-Die) for high viscous materials can be avoided and headed to more accurate data.

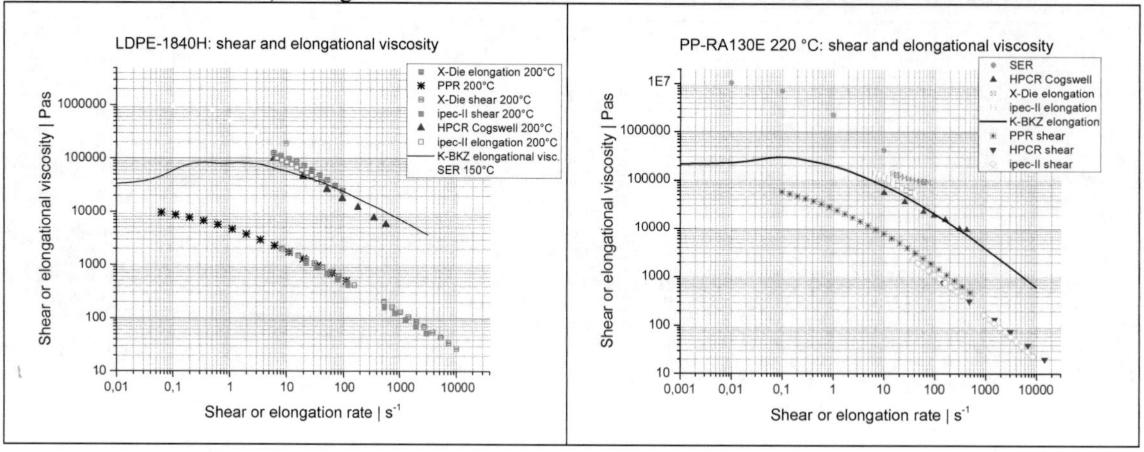

FIGURE 3. Comparison of different shear and elongational measurements a) LDPE-1840H b) PP-RA 130E

Although it is wise for every rheological data to crosscheck or compare measurements, it is shown that the figures of the new developed slit die will be in conformity with typical online as well as offline measurements and can be therefore used for characterization of rheological behavior of plastic melt.

Exemplary Measurements and Results

The realization of inline measurements of shear viscosity as well as elongational viscosity provides a wide range of possibilities. For example, it can be used for an easy and more accurate inline melt quality related measurement of material data during lab trials. As seen in Fig. 4 degradation can be easily seen with online shear as well as elongational viscosity measurements.

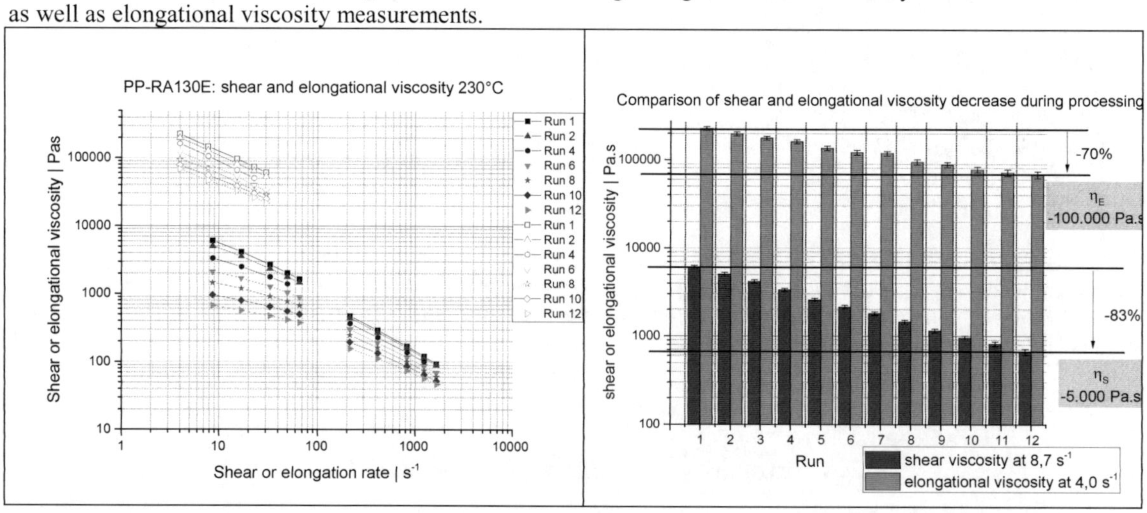

FIGURE 4. Elongational and shear viscosity by degradation of polymer through multiple processing

This high viscous-resolution of the melt quality inline measured with the elongational/shear rheometer allows also to look quite closer into material properties than usually. E.g. while producing 24% fiber reinforced PP-compounds with a different mixture of short (max. length: 1.5 mm) and long (max. length: 4.5 mm) glass fibers typically no differences can be seen in shear viscosity in terms of different fiber length distributions (See Fig. 5). However, at elongational viscosity, differences in "shorter" and "longer" glass fibers at the same amount can be detected. Due to shear stress an orientation of the glass fibers will happen in the slit die where shear viscosity is measured. This fiber orientation won't create any viscosity differences in flow direction and therefore no significant viscosity data change can be seen. Contrariwise the length of fibers will play a significant role in strain processes. These differences were detected by the measured elongation viscosity at the online Rheometer.

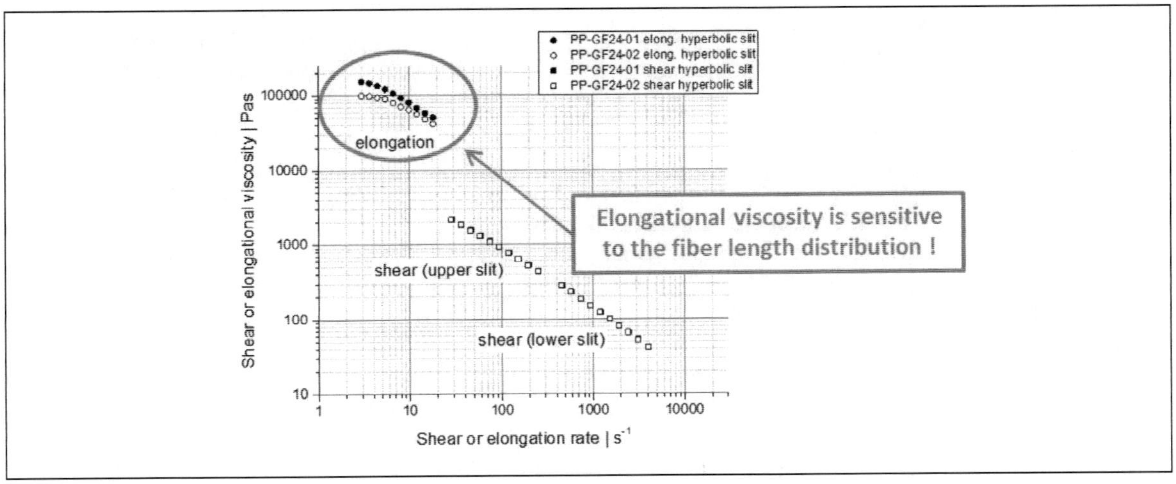

FIGURE 5. Inline shear and elongational viscosity of 24% glass fiber reinforced compounds with different ("shorter" and "longer") glass fiber length; PP-GF24-01 is 20:80 whereas PP-GF24-02 is 70:30

But not only the fiber length also the impact of the type of fiber can be detected and/or measured much more accurate with the elongational viscosity. As seen in Figure 6., the shear viscosity on four different fiber reinforced PP compounds – two with natural cellulose fibers and two with glass fibers – show nearly no difference in the shear viscosity data. By contrast, the elongational viscosity varies significantly and the differences of the amount of fibers as well as the kind of fiber can be easily seen.

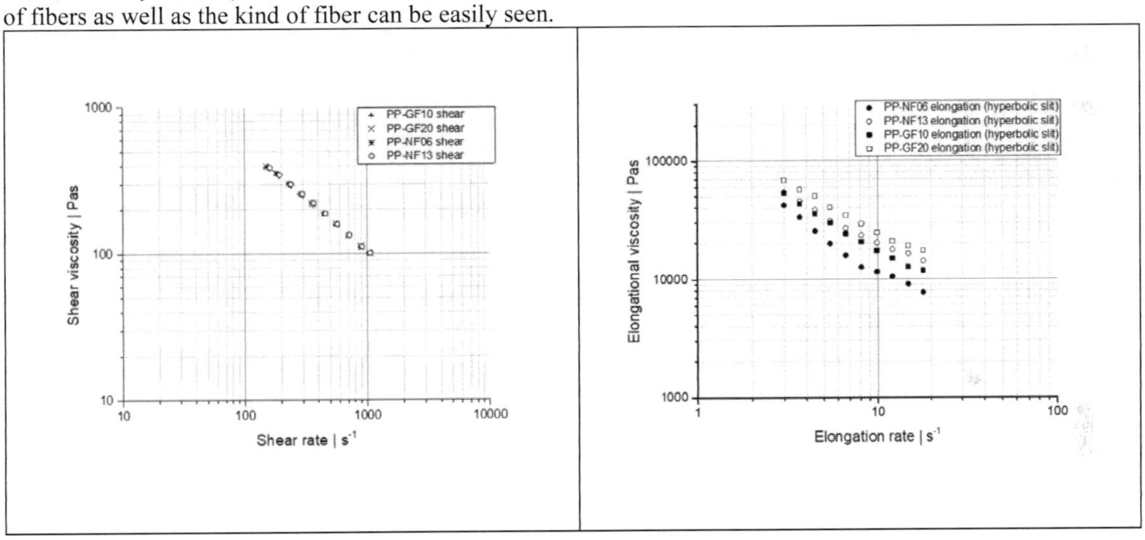

FIGURE 6. Inline shear and elongational viscosity of PP compounds with 10% and 20% glass fibers as well as 6% and 13% of natural cellulosic fibers

Smart Sensor for Continuous Plastic Processes

As shown above online information will lead to a better understanding of compounding and/or extrusion processes. Those data will improve the classification of the material as well as the design work e.g. for the formulation of compounds. In addition, the use of this rheometer is not limited to lab-scale trials or continuous R&D work, but also to evaluate production processes. Obviously during a production run the focus is not on viscosity curves but on changing of viscosity related data. Therefore, simplified production values like MFI or IV-values can be predicted easily and much more accurate as not only one viscosity data is measured at once. The measurement of two shear viscosity data as well as one elongational viscosity data improves the accuracy of the inline MFI or IV measurement.

To ensure the safe use of the new online rheometer - called i4LDS RHEO - a sturdy housing with a measuring flange has been designed (See Fig. 7 a). The rheometer is applicable to the extruder or extrusion line independent from operation side. A hand switch ensures that it can be operated as a real-time measurement sensor either in a bleeding version for thermal sensitive materials or as a refeed version, where the melt is pushed back to the melt stream (zero waste sensor). A possible arrangement within an extrusion line is seen in Fig. 7 b).

FIGURE 7. a) Online Rheometer i^{4}LDS-RHEO as smart sensor; b) extrusion line arrangement of i^{4}LDS-RHEO

Two software executions - either process or research - and the possibility to have the Rheometer i4LSD RHEO as a stand-alone or an integrated sensor ensure a versatile application spectrum within the continuous plastic processing industry. In addition, real time measured melt quality data will be capable to integrate into relevant process control strategies. First results [7] have been proven that inline measuring of viscous data combined with predicted control systems led to very precise polymer compound production. When such inline measurement systems are combined with active ingredients (like peroxide for PP) MFI variations during compounding of PP can be minimized.

SUMMARY AND OUTLOOK

The newly developed two stage slit die and the integration into the online Rheometer, called i4LDS RHEO, has approved its versatility for different plastic materials. Its high viscous-resolution of the measured melt quality with elongational and shear viscosity can be used easily in Lab or R&D extrusion processes as well as for monitoring continuous plastic processes in production. Implementation of this smart sensor into extrusion lines in combination with intelligent process control strategies will ensure a continuous and sustainable production within quality limits.

Especially the "higher" viscous resolution due to the combination of elongation and shear viscosity will probably enlarge the application of the i4LDS RHEO towards to food and pharma processes and will support the continuous production within that industry. The use as a control sensor for reactive extrusion, viscous measurements of melt strength where elongational viscosity will play an important role but due to process conditions the measurement is not easy (e.g. foam extrusion) and much more assume that the measuring principle of the i4LDS RHEO will have a bright future. I am quite curious about further developments and measurement results done with the i4LDS RHEO.

ACKNOWLEDGMENTS

This research was partially founded by the FFG (Austrian Research Promotion Agency) and the entire development work of the slit die will be published in [8]. The authors would like to thank the fruitful discussions within the project group as well as with the colleagues at the Johannes Kepler University Linz, Austria. A special gratitude is given to the author's colleagues from the company Leistritz Extrusionstechnik GmbH as well as for the support and the material provided for the study.

REFERENCES

[1] A. Pabedinskas, W.R. Cluett and S.T. Balke, "Development for an In-Line Rheometer Suitable for Reactive Extrusion", Polym. Eng. Sci., vol. 31, no. 5, pp. 365-375, 1991

[2] M.S.N. Olivera, M.A. Alves, F.T. Pinho and G.H. McKinley, "Viscous flow through microfabricated hyperbolic contractions", Exp. Fluids, vol. 43, pp. 657-669, 2007

[3] J. Miethlinger et al, WO 2016/197169 A1, "Measuring Nozzle for Determining the Extensional Viscosity of Polymer Melts"

[4] M.L. Sentmanat, "Miniature universal testing platform: from extensional melt rheology to solid-state deformation behavior", Rheol. Acta, vol. 43 pp. 657-669, 2004

[5] E. Mitsoulis, "50 years of the K-BKZ Constitutive Relation for Polymers", ISRN Polym. Sci., vol. 2013, Article ID 952379, 22 pages, 2013

[6] F.N. Cogswell, "Converging flow of polymer melts in extrusion dies", Polym. Eng. Sci., vol. 12, no. 1 pp. 64-73, 1972

[7] S. Wolf, "Industry 4.0 for PP-Large Scale Compounding", presentation at the "AMI Polyolefin Additives 2015" conference, October 2015 Cologne/Germany

[8] H.J. Luger, Ph.D.-Thesis in progress, Institute of Polymer Extrusion and Compounding, Linz/Austria.

Extrusion Die Design by using Network Theory with Cross Flows Behavior

B. Jacobkersting[a] and V. Schöppner[b]

[a] *Faculty of Mechanical Engineering, University of Paderborn, Germany – bianka.jacobkersting@uni-paderborn.de*
[b] *Kunststofftechnik Paderborn, University of Paderborn, Germany – volker.schoeppner@ktp.upb.de*

Abstract. Extrusion dies exert influence on later final product quality. Therefore it´ll make a point the dimensional and die design by using programs for calculation and simulation frequently. For the implementation of product design, it is significant to understand the flow conditions and to be able to predict the flow behavior accurately. Special rheological flow phenomena in the plastic melt as cross flows, which flow perpendicular to the main flow, should be taken into account. This phenomenon is caused by pressure gradients transverse to the direction of extrusion and both the flow distribution and the pressure consumption are influenced in the die. Network theory is a simple numerical method for a holistic one-dimensional representation (GEB) in a spreadsheet program (p.e. Excel), which can design an optimal uniform flow rate distribution and low pressure drop. The cross flow behavior can´t be described with this method as yet. Therefor a linear equation system according to the Gauss algorhythm was developed, which can calculate the cross flows in the die with rectangle cross-section. In this network crosslinks are implemented to take into account the cross flows. The equation system is set up from the network, which corresponds to the number of the desired partial volume flows in the number of established equations. Furthermore the technical measurement entry of cross flows was conduced about the evaluation of ellipsoid shape according to the flow direction and the alignment according to the flow direction. Dead-stop experiments were performed by adding a blowing agent to the extrusion process. Negatives of die with gas-filled bubbles were prepared and evaluated with image analysis software across the half width of die. Afterwards the network theory was validated by Computational Fluid Dynamics (CFD).

Keywords: flat die, extrusion, network theory, coathanger geometry, rectangular cross-section, calculation, cross flows
PACS: 02.10 Dc, 02.10 Kn, 02.10 Pn, 02.10 Ud, 02.10 Yn, 47.85 md, 81.05 Lg, 83.50 Ha, 83.50 Uv, 83.60 a, 83.80 S

INTRDUCTION

The plastics industry is calling for films of ever higher-quality, especially with a multi-layer structure [1]. Increasing competitive and cost pressure as well as growing demands on the resource-efficient production of plastic films have led to the constant improvement and optimization of film extrusion lines [2]. Particular importance is attached to the dimensioning and design of extrusion dies. This is a complex process for which calculation and/or simulation programs are being more and more frequently deployed. One simple method of numerical calculation is the one-dimensional study of the flow path using the network theory. For carrying out the product design by means of virtual die design, it is necessary to be fully acquainted with the flow processes. In extrusion dies with a slit-shaped orifice, cross flows also occur in addition to the main flows in the extrusion direction [3]. The calculation method of the network theory has been further developed to take into account the cross flows in pressure-throughput relationships by modelling with a Gauß algorithm. Besides the analytical and holistic numerical calculation method, another calculation method for designing extrusion dies is the holistic one-dimensional representation (GEB) technique, which is based on the network theory and is predominantly used by applying simplifications. The material behavior is taken to be Newtonian and a detailed flow velocity distribution in the radial direction is neglected, which means that a spatial direction for the calculation can be eliminated [4, 5]. Here, the system of the network theory is transferred from electrotechnology to flow technology. In electrotechnology, in the case of an electrical circuit with a branch, reference is made to a network in which the voltages and currents can be calculated via Kirchhoff's laws (mesh and node law). To be able to describe and calculate the die geometry using the network theory, an exact simulation of the die geometry with flow resistances is needed. The die geometry is broken down into a network of individual segments. The methodology is based on the parallel and series connection of linear flow resistances, whereby non-linearity of resistances is taken into account by the iterative solution. By applying Kirchhoff's laws, the network of the simulated die geometry can be quickly solved by the GEB in a simple spreadsheet program (e.g. Excel) [4]. The existing program architecture [4, 5] was developed using the representative values. By dividing up the die zones (manifold, preland, die, restrictor bar and adjustable lip section) into individual segments, it is possible to approximate the die contours with the aid of tubular and rectangular geometries. With a holistic examination of the flow paths, it is then possible to calculate the volume flow distribution, pressure loss and optimum die geometry.

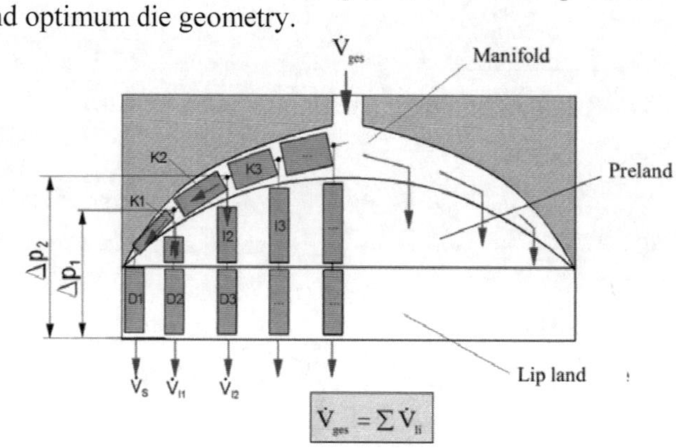

FIGURE 1. Calculation method for the network theory [5]

MODELING OF A SIMPLE NUMERICAL CALCULATION METHODOLOGY TO IMPLEMENT CROSS FLOWS

Cross flows occur in the extrusion direction alongside the main flows and are characterized by a perpendicular flow behavior to the main flow direction. Pressure gradients transverse to the extrusion direction cause this special flow characteristic, which has an influence on the volume flow distribution and pressure consumption in the die. For this, in analogy to Kirchhoff's laws of electrotechnology, a network was created that also contains crosslinks. With the network, it is possible to study any desired number of downstream zones in addition to the manifold and the preland. When modeling the flow channel, the slit or flat film die (Figure 2) is divided up into segments of simple geometry, which are linked to form a network. Calculation of the volume flows for the individual segments is carried out as described in the previous section. To take account of cross flows, the segmentation of the preland is further refined whereas the die zone and manifold correspond to the subdivision of the GEB. The alignment of the resistances is carried out in the longitudinal and transverse directions of the flow in order to realistically illustrate the flow process. With the aid of the network theory, a resistance network can be generated as shown in Figure 2. Nodes were consecutively numbered and meshes were marked (blue arrows represent the volume flow of a sub-segment, whereby the arrows perpendicular to the main direction of flow include the cross flow). The incoming volume flow reflects half the volume flow provided by the extruder, whereas the outgoing volume flow is made up of the partial volume flows from the die segments. The prerequisite for calculating the resistance network is the application of the Visual Basic for Applications (VBA). Via the program code according to [6], a linear equation system with n equations and n unknowns (volume flows) was defined which is then solved with the help of the Gauß algorithm. The requirement made on the solution algorithm is that an equation system with n volume flows with n equations is created for an unambiguous solution, so that the input volume flow is fixed as an unknown variable and the exit volume flow is assumed to be constant. If the same result is calculated for the input volume flow as for the given output volume flow, the equation system has been correctly solved. To solve the mesh equations, knowledge of the die resistance and viscosity at certain points in the die is required. The connection between pressure and volume flow can be determined via the die resistance W and the representative viscosity.

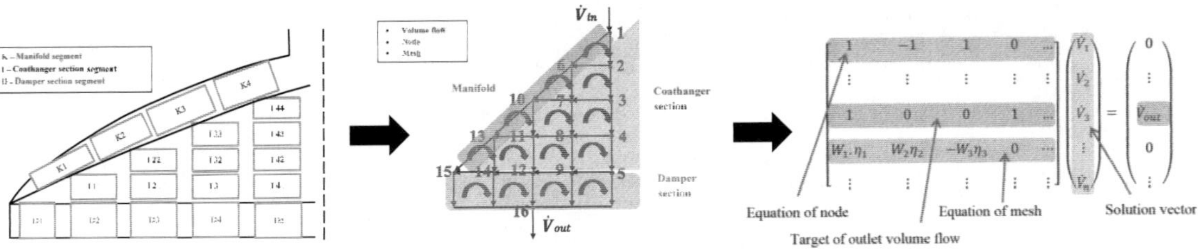

FIGURE 2. Procedure for modeling the solution matrix

Subsequently, an automatic calculation is carried out of the products from viscosity and resistance (Figure 3). With the transfer of the total volume flow, the volume flows at the die orifice are added together and compared with the total volume flow. Via the "Calculate" button, the solutions can be generated in the application. The individual partial volume flows (x_1 to x_n) are assigned with the aid of the network. Here, transverse volume flows in the network directed towards the outer side of the die imply a positive value.

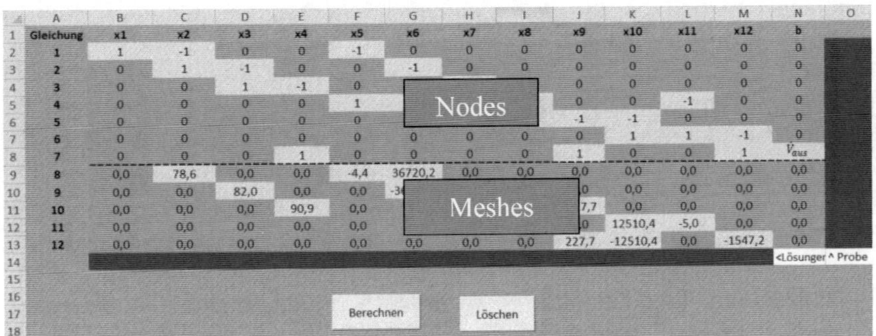

	A	B	C	D	E	F	G	H	I	J	K	L	M	N	O
1	Gleichung	x1	x2	x3	x4	x5	x6	x7	x8	x9	x10	x11	x12	b	
2	1	1	-1	0	0	-1	0	0	0	0	0	0	0	0	
3	2	0	1	-1	0	0	-1	0	0	0	0	0	0	0	
4	3	0	0	1	-1	0				0	0	0	0	0	
5	4	0	0	0	0	1				0	0	-1	0	0	
6	5	0	0	0	0	0				-1	-1	0	0	0	
7	6	0	0	0	0	0	0	0	0	0	1	1	-1	0	
8	7	0	0	0	1	0	0	0	0	1	0	0	1	$\dot{V}_{aus}$	
9	8	0,0	78,6	0,0	0,0	-4,4	36720,2	0,0	0,0	0,0	0,0	0,0	0,0	0,0	
10	9	0,0	0,0	82,0	0,0	0,0	-3			0	0,0	0,0	0,0	0,0	
11	10	0,0	0,0	0,0	90,9	0,0				7,7	0,0	0,0	0,0	0,0	
12	11	0,0	0,0	0,0	0,0	0,0				0	12510,4	-5,0	0,0	0,0	
13	12	0,0	0,0	0,0	0,0	0,0	0,0	0,0	0,0	227,7	-12510,4	0,0	-1547,2	0,0	
14														<Lösunger ^ Probe	

FIGURE 3. Excel spreadsheet with input equation system [7]

EXPERIMENTAL INVESTIGATION

To check the suggested solution approach, a dead-stop examination was performed. A blowing agent was added to the extrusion process to specifically induce the formation of gas-filled bubbles. 0.2 % sodium hydrogen carbonate was added to the polystyrene starting material. During the stable process, the machine was halted and the extrusion process abruptly interrupted. The extruded die negatives from the flat die showed small gas-filled bubbles distributed evenly over all zones of the die negative. Towards the die orifice, they became smaller. By varying the speed, the aim was to derive tendencies with regard to the flow behavior through the arrangement, alignment, size or number of individual bubbles. Using a single-lens reflex camera, digital images were made of the extruded negatives. The image data were processed using the Fiji software. With the help of corresponding plug-ins, the arrangement, alignment, size and number of the individual bubbles over the individual zones in the flat die can be evaluated. A striking point was the alignment of the individual bubbles which, because of their main direction, assumed an ellipsoid form. The connection was examined between the alignment of the ellipsoids with the flow direction and thus with the cross flow, and the arrangement of several bubbles in the flow direction in the relaxation zone (lip area) vs. the angle geometry in degrees of the bubble:

$$\tan \varphi = \frac{\vartheta_{0z}}{\vartheta_{0x}}$$

RESULTS AND VERIFICATION

An alignment of the bubbles is observed towards the side of the die. The angle of the evaluated bubbles of the extruded negative is smaller or larger than 90°, which implies cross flows in addition to the main flow. Furthermore, it was found that the alignment of the ellipsoids in the direction of extrusion increases towards the edge of the die (Figure 4).

Another approach for examining the cross flow was selected via the arrangement of several bubbles in the flow direction in the relaxation zone (lip area). The lip area was selected for the examination because previous findings had shown that no further cross flows should occur in this area.

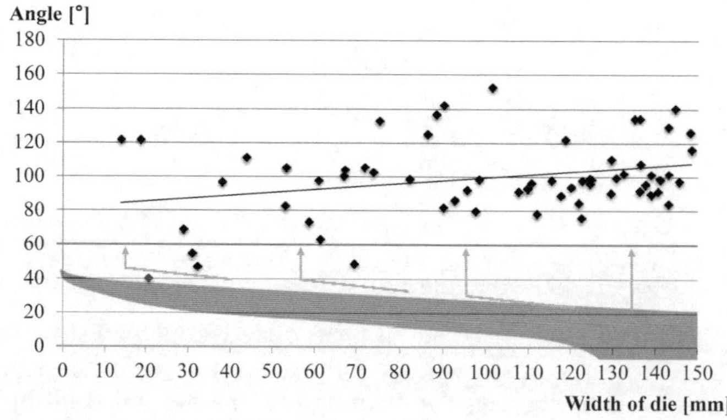

FIGURE 4. Alignment of the ellipsoids in the extrusion direction

An arrangement of the bubbles in the flow direction was postulated because of the collapse of the droplets and was used as a possible method for further evaluation. The calculation of the droplet collapse was carried out with the aid of the functional relationship via the Weber number, which is characterized by the ratio of the inertial forces to the surface tension forces. For polystyrene, a Weber number of 1.64 was measured at a temperature of 220° C, a density of 1,028 kg/m, a flow velocity of 0.15 m/s, a mean cell radius of 0.16 cm and a surface tension of 43.65 N/m. This figure gives a clue as to the droplet collapse mechanism; it is below 12. The bubbles collapse according to the vibrational breakup. The evaluation of the bubble arrangement in the lip area shows a uniform distribution over half the die width. The average for the bubble orientation is 90.2°, which corresponds to an alignment in the flow direction. The objective observation is confirmed; the bubbles become arranged in rows. From this it can be derived that there are no cross flows in the lip area of the die. It should be noted that the selected approach for determining cross flows via the arrangement of several bubbles in the flow direction in the relaxation zone (lip area) can be used, as can the approach via the alignment of an individual bubble in the preland in the flow direction. For the experimental verification of the generated model, pressures were measured at various speeds and with various materials. Through the operating point-dependent design of the laboratory die, the tooling die was subsequently modeled for the selected materials with the GEB, and the solution equation system (LGS) of the calculation methodology according to Gauß. A comparison of the results with the measurements shows that, overall, a more uniform volume flow distribution is obtained with the LGS. This becomes clear in the following table. The distribution of the LGS is more uniform at all points in the die than the original solution.

TABLE a). Summary measurement results – Volume flow distribution over the segments in a comparison of GEB and LGS

Operation point Segment	GEB [mm³/s]	GEB $\pi_{\dot{V}}$	LGS [mm³/s]]	LGS $\pi_{\dot{V}}$
#D1	1363	0.72	1634	0.97
#D2	1957	1.04	1918	1.02
#D3	1601	0.85	1766	0.93
#D4	2227	1.18	2045	1.08
#D5	2290	1.21	2074	1.1
$\dot{V}_{average} = 1888\ mm^3/s$				

To validate the characterization of the cross flow behavior in the flat die, in addition to the dead-stop tests, Finite Element Method simulations (FEM) were carried out for alignment of the individual bubbles and for the arrangement of several bubbles in the extruded negative. A comparison between the experimental results and the numerical calculation options aimed to generate a description for characterizing the cross flow behavior in the flat die. After determining the material-specific parameters, the process-related and geometric data of the die were entered. Evaluation of the numerical simulations was carried out via two aspects the flow velocities (or the volume flow) and the pressure. For further evaluation, a determination of the relationship between the flow velocities and the angles (Figure 5) is required:

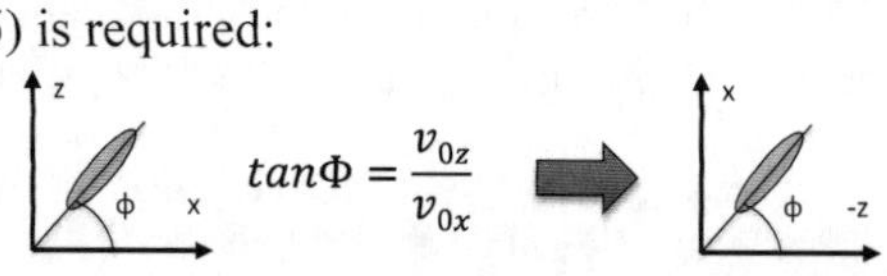

$$tan\Phi = \frac{v_{0z}}{v_{0x}}$$

$$z < 0:\ \Phi = tan^{-1}\frac{v_x}{v_z} + 180°$$

$$z > 0:\ \Phi = tan^{-1}\frac{v_x}{v_z} - 180°$$

FIGURE 5. Angle function of the flow velocity in the die [7]

The evaluation of the mold contours simulated by FEM shows a distinct accumulation of cross flows in the area of the coathanger. Overall, a flow alignment is recognizable from the center of the die to the edge. The alignment decreases as in the previous test results over the zones and, up until the die orifice, becomes the same as the alignment in the flow direction: A laminar flow is present. With the FEM calculation, it can be confirmed that, in the preland, in contrast to the die zone, cross flows are present; conformity exists between simulations and the experimental tests. In all processes, cross flows were found in the area of the coathanger. Nevertheless, not all deviations in the main flow pointed in the same direction. A comparison of the experimental test results with the simulated results was also targeted via the bubble arrangement. Figure 7 shows a comparison of the processes via the flow angle at the die orifice. The bubble orientation of the experimental test was very close to the main flow direction of 90°, so that the minor differences indicate small cross flows in the area of the lip zone. The comparison of the simulation processes with the experimental result via the bubble arrangement provides a similar result (Figure 6).

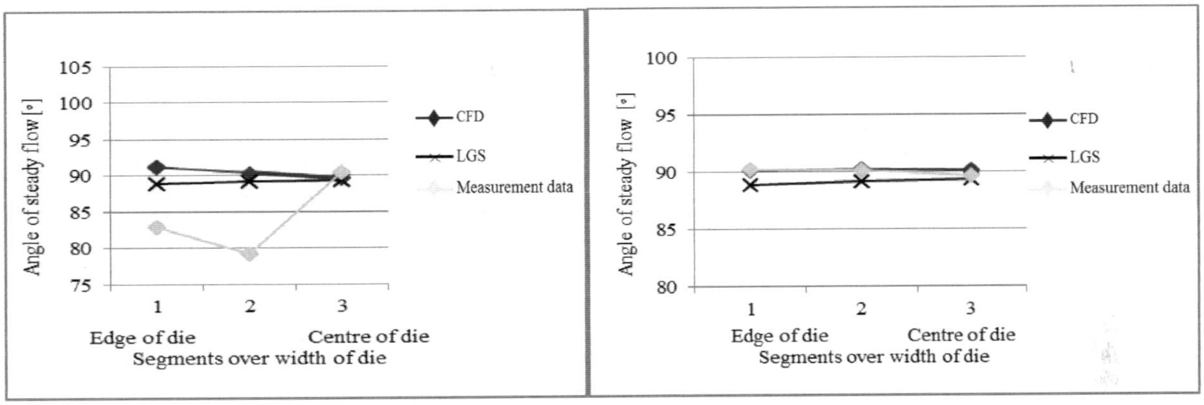

FIGURE 6. Verification – LGS vs. CFD

It was found that the LGS can provide information about the cross flows in the extrusion die because the results are approximately the same as with the three-dimensional simulation process. The experimental tests also tend to deviations from the main flow direction, but in a more marked form than the numerical simulations. The methods for visualizing the cross flow and the calculation principle for taking account of the cross flow show a uniform distribution at the mold die orifice. The alignment of the arranged bubbles virtually corresponds to that of the main flow direction. Cross flows provide a balancing-out of flow differences across the zones and become correspondingly smaller in the extrusion direction, which is primarily reflected by the evaluation across the angle of the evaluated processes. A comparison of the test results and the subsequent evaluation shows that, in the preland, a deviation from the main flow direction exists, which is balanced out by flows perpendicular to the extrusion direction and that the flow angles in the subsequent zones approach a 90° alignment. At the die orifice, both the numerical processes and the experimental tests show an alignment of the flow angles in the main flow direction. A calculation method based on the network theory has thus been found that takes account of cross flows and allows a more accurate calculation of volume flows and pressure drops in the extrusion die.

REFERENCES

1. H. Seidel, S. Wenigmann, A. Tondorf and T, Schmitz, *Immer mehr Schichten,* München: Carl Hanser Verlag, 2006, pp. 139.
2. M. Bussmann, *Ein kalibrierbares integratives Modell zur Beschreibung des Schlauchbildungsprozesses in der Blasfolienextrusio*n, Dissertation an der Universität Duisburg-Essen, 2010.
3. S. Seibel, *Konzeptionelle Aspekte bei der Gestaltung und Auslegung von Extrusionswerkzeugen*, Dissertation Universität Paderborn, Shaker Aachen, 2007.
4. A. Limper, *Verfahrenstechnik der Thermoplastextrusion*, München, Carl Hanser Verlag, 2013.
5. H. Stieglitz, *Ganzheitliche Betrachtung der Fließvorgänge in Breitschlitz- und Wendelverteilerwerkzeugen*, Dissertation Universität Paderborn, Janus Druck Borchen, 2001.
6. R. Hirte, *Komplexe Berechnung: Lineares Gleichungssystem beliebigen Grades*, Internet resource: http://rhirte.de/informatik/pdf/rechnen, date of access: 31th July 2012.
7. B. Jacobkersting, Weiterentwicklung Netzwerktheorie basierender Werkzeugauslegung, Dissertation Universität Paderborn, Shaker Verlag Aachen, 2017.

Mixing Characteristics of Different Kneading Elements: an Experimental Study

Yasuya Nakayama[a*], Satoshi Esaki[a], Kenta Nakao[a], Toshihisa Kajiwara[a], Takahide Takeuchi[b], Koichi Kimura[b], and Hideki. Tomiyama[b]

[a]*Department of Chemical Engineering, Kyushu University, 744 Motooka, Nishi-ku, Fukuoka 819-0395, Japan*
[b]*Hiroshima Plant, The Japan Steel Works Ltd. 1-6-1 Funakoshi-minami, Hiroshima 736-8602, Japan*

kajiwara@chem-eng.kyushu-u.ac.jp
nakayama@chem-eng.kyushu-u.ac.jp

Abstract. We discuss the melt-mixing characteristics of different geometries of pitched-tip kneading disks (ptKD). Extrusion experiment was performed under the fully-filled state in the kneading zone by controlling the fill lengths for different KDs. We study the variation of the particle size distribution and the rheological response of ABS (acrylonitrile-butadiene-styrene copolymer) caused by melt-mixing. We found that the increase or decrease of the fine and coarse particles depends on the type of KD. Furthermore, the change of the particle size distribution affects the storage modulus at low frequencies. From these results, different mixing characteristics of different KDs are evaluated.

Keywords: Twin-screw extrusion, Mixing element, ABS
PACS: 83.50.Xa Mixing and blending

INTRODUCTION

Mixing of highly viscous materials, such as polymer blends, polymer composites, and rubber compounds, is commonly performed in twin-screw devices, such as twin-screw extruders and twin-screw kneaders, because the material exchange between neighboring chambers is effective at mixing. The channel geometries in twin-screw devices, which is mainly determined by screw elements normally have the narrow gap regions, such as tip-barrel clearance and the intermeshing regions, which is mainly responsible for generating a high level of shear rate and shear stress. In contrast, the other regions from the narrow gaps are responsible for mixing, so that different designs of the mixing elements have been developed [1, 4, 5, 6, 7, 8]. Understanding the relation between the element geometries and the mixing characteristics is a key issue in optimizing the mixing process [1, 2, 3].

Mixture quality of a product is characterized in the two aspects that are the homogeneity in the spatial concentration distribution, and the size distribution of dispersed phases. The characteristics of a mixing process can be assessed through the mixture quality of the extrudate. Mixing process is mainly performed by the flow caused by rotation of the elements, and is most promoted when the material fully fills the melt-mixing zone. However, the channel inside a twin-screw extruder is not necessarily fully filled. Since degree of filling or fill length depends on the geometries of elements under a fixed throughput rate and screw rotation speed. When experimentally comparing the mixture characteristics for different elements under a fixed operation condition, difficulty comes from the difference of the fill lengths which directly affects the mixture quality of the extrudates. Intrinsic mixing characteristics of an element need to be measured at fully-filled state.

In this paper, we experimentally investigate the difference of the mixing characteristics of the pitched-tip kneading disks and the conventional one. Extrusion experiments with different elements under a fully-filled state in the melt-mixing zone and a given operation condition are performed. By analyzing the mixture quality of the extrudates, we discuss the mixing characteristics of the different elements.

EXPERIMENTAL

ABS (acrylonitrile butadiene styrene) resin (UMG-ABS: R-80) which contains 60 wt% polybutadiene rubber was diluted by adding SAN (styrene-acrylonitrile copolymer, UMG-ABS: S101N) resin to be 12 wt% of rubber. By this dilution, single rubber particles are easily identified from SEM image. Twin-screw extruder used in this study is TEX28V from Japan Steel Works Ltd. Melt-mixing zone was arranged to L/D=4.5 (equivalent to three segments of one piece in Fig. 2) just before the die (Fig. 1). Five different elements shown in Fig. 2 are tested for melt-mixing zone: one is conventional forward kneading disks (FKD), and the others are four different types of pitched-tip kneading disks (ptKD) which are denoted as Fs-Ft, Fs-Bt, Bs-Ft, and Bs-Bt where Fs/Bs indicates forward/backward disk stagger and Ft/Bt indicates the forward/backward pitched tips.

The difference in the pumping abilities of different elements results in difference in the fill length in the melt-mixing zone under a fixed throughput rate and screw rotation speed, leading to insufficient output of the intrinsic mixing characteristics which makes it difficult to compare different kneading elements. To overcome this difficulty, we explicitly set the melt-mixing zone to be fully-filled under a throughput rate and a screw rotation speed by adjusting the channel width in the die. The fully-filled state in the melt-mixing zone is monitored by the two pressure sensors equipped at the inlet of the melt-mixing zone. Pressure values as a function of the screw speed under 10 kg/h extrusion is shown in Fig. 3 where we can identify the fully-filled state at the change of the slope. Keeping the fully-filled state for different elements, ABS resin was melt-mixed under a throughput rate of 10kg/h and the screw rotation speed of 350 rpm with the barrel temperature of 200°C. The size distribution of the rubber particle in the extrudates as well as the melt viscoelasticity are analyzed to discuss the mixing characteristics of the different elements.

Exrudates of ABS are compression molded into plates at a temperature above softening point. After that, ABS was stained with osmium tetraoxide to get a contrast between SAN matrix and polybutadiene rubber in SEM image. The preprocessed samples were observed by SEM (Zeiss ULTRA55, located at The Ultramicroscopy Research Center Kyushu University). Obtained SEM images were processed to be binarized using a software ImageJ to identify the single rubber particles (Fig. 4), from which the area-equivalent diameter is extracted to calculate the distribution of the rubber particle sizes.

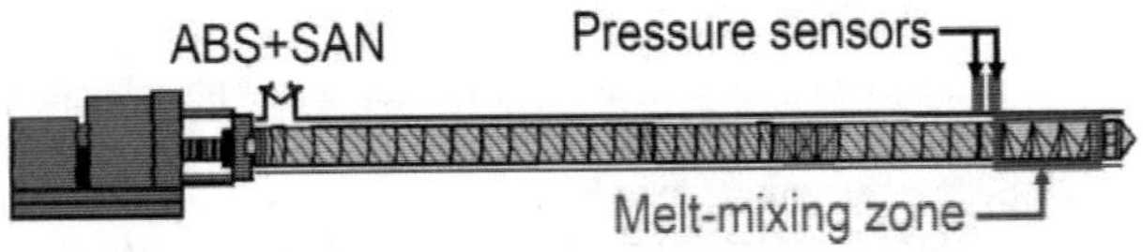

FIGURE 1. Configuration of the twin-screw extruder. Melt-mixing zone of L/D = 4.5 is set just before the die, at which different kneading elements are configured. Two pressure sensors are equipped adjacent to the inlet of the melt-mixing zone in order to detect the fully filled state of the melt-mixing zone.

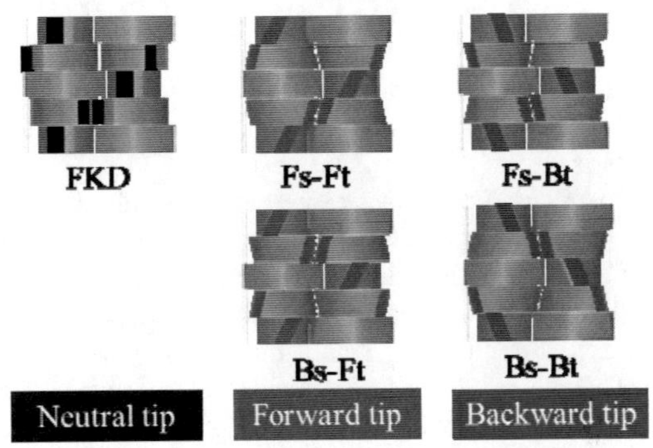

FIGURE 2. Top view of the five types of the kneading elements discussed in this paper. Each element has a dimension of L/D = 1.5.

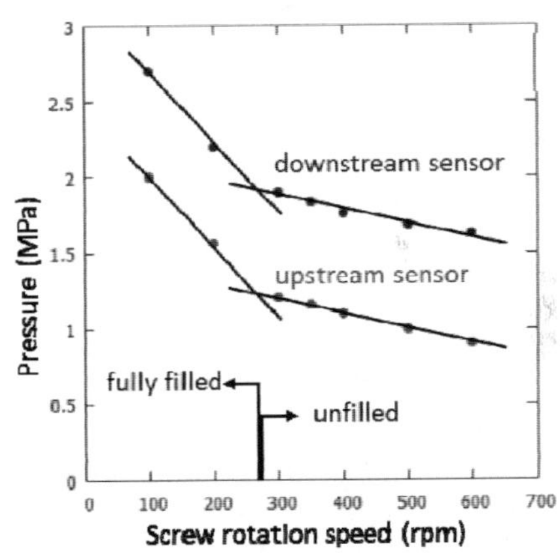

FIGURE 3. Pressure values from the two sensors at the inlet of the melt-mixing zone as a function of screw rotation speed under a constant throughput of 10 kg/h.

The particle size in the polymer composites is known to affect the modulus amplitude as well as the long-time response [9, 10, 11]. By analyzing the melt viscoelasticity of the extrudates, the mixing characteristics of different elements can be evaluated. The dynamic shear moduli of the molten extrudates were measured at different temperatures under a nitrogen atmosphere using a rheometer (Rheosol-G1000, UBM, Kyoto, Japan) in parallel-plate geometry with a plate diameter of 25 mm. The strain amplitude was set at 0.1, which falls within the linear response regime.

RESULTS AND DISCUSSION

Figure 4 shows the binarized SEM images of the extrudates with different elements as well as that of the unprocessed sample for comparison. In all the cases in Fig. 4, the particles of polybutadiene rubber (black) are almost evenly distributed in the matrix of SAN (white), and no significant aggregation is observed. The mean diameter of the particles in the unprocessed sample is 259 nm consistent with the value reported by the supplier, while those after melt-mixing lie between 259 and 277 nm. The average size of the rubber phase rarely changes by melt-mixing since the mean size in the original ABS is already small. In contrast, distribution in particle size is observed in Fig. 4 and varies between the melt-mixing elements.

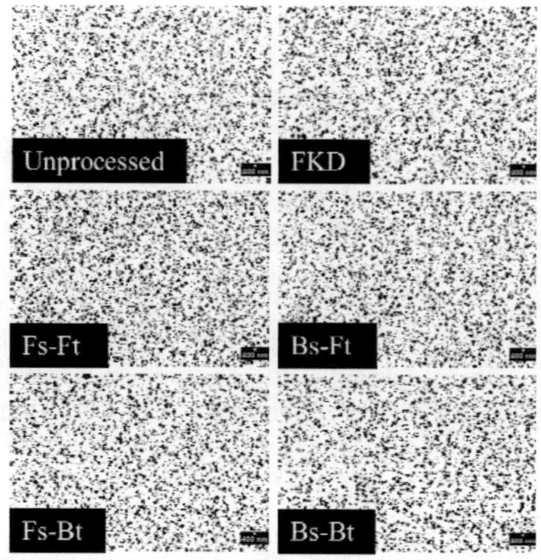

FIGURE 4. Binarized SEM images of the extrudates melt-mixed with different kneading elements. Black region is the rubber particles of polybutadiene while the white region is styrene-acrylonitrile copolymer which consists of the matrix of ABS resin.

Fig. 5 shows the probability density functions of particle diameter in the extrudates, and clearly demonstrates that the change in the particle size distribution depends on the melt-mixing elements. For the unprocessed sample, two peaks at around 200 nm and 120 nm, which are less than the mean value, are observed. For further quantification of the change in particle size distribution, we divide the size domain into four intervals with three quartile points in the distribution of the unprocessed sample, and calculate the relative change in the fractions of the four intervals.

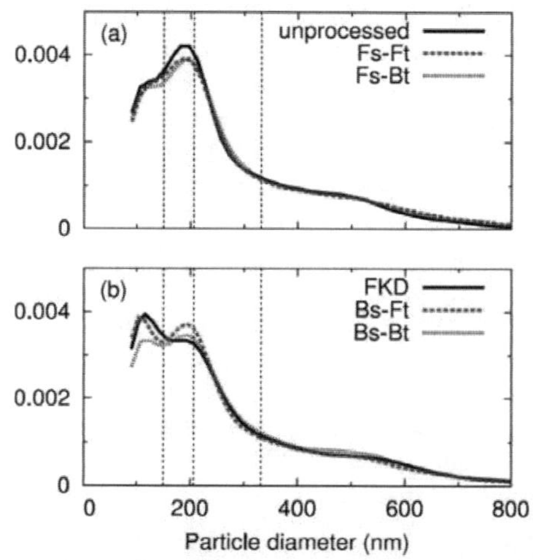

FIGURE 5. Probability density function of particle diameter for the extrudates melt-mixed with different kneading elements. The vertical dotted lines indicate the quartiles in the distribution of the unprocessed sample.

Figure 6 shows the relative changes in the fractions of the four intervals from those of the unprocessed sample. For Fs-Ft and Fs-Bt, the change in the size distribution is relatively small (Figs. 5(a) and 6), but we observe the decrease in the fractions of fine particles smaller than 206.8 nm and the increase in the fractions of coarse particles. For Bs-Bt, the increase in the coarsest-particle fraction (larger than 331.7 nm) is the highest. In contrast, for FKD and Bs-Ft, the increase in the fine-particle fraction (smaller than 149.7 nm) is noticeable (see also Fig. 5(b)). Especially for Bs-Ft, only the finest-particle fraction (smaller than 149.7 nm) increases, and the other three fractions of the particle decrease, showing a remarkable size reduction with Bs-Ft.

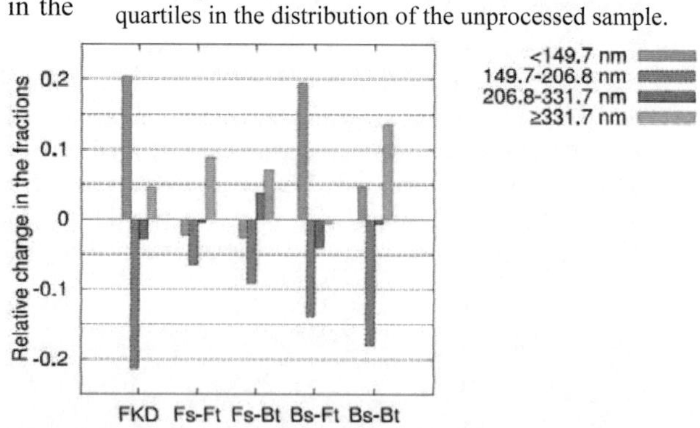

FIGURE 6. Relative changes in the fractions in the four intervals, which is defined by the quartiles of the unprocessed sample shown in Fig. 5, after melt-mixing with the different kneading elements.

The increase of the first fraction (smaller than 149.7 nm) reflects a dispersive ability. The results in Fig. 6 suggest a certain level of dispersive ability of FKD and Bs-Ft. On the other hand, the increase of the

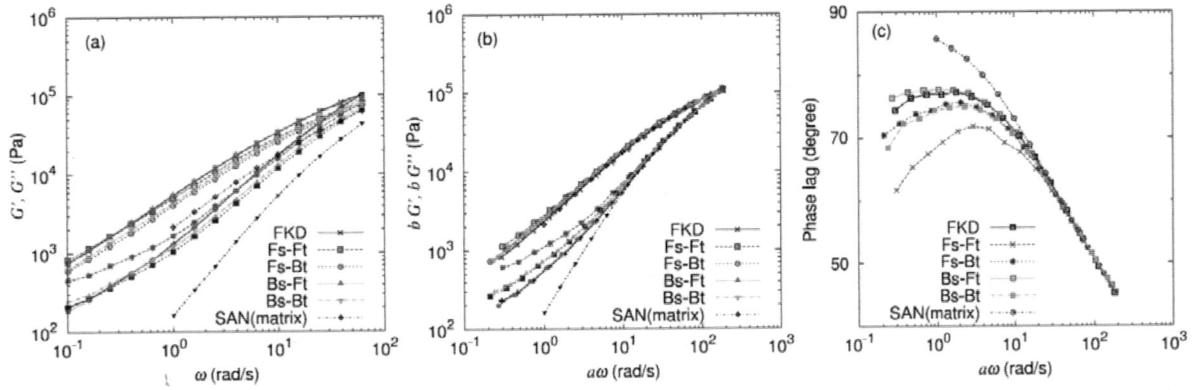

FIGURE 7. Linear modulus of the extrudates melt-mixed by different kneading elements: (a) raw data at 200◦C, (b) the data shifted to that of SAN, and (c) the phase lag after the frequency shift to SAN data.

fourth fraction (larger than 331.7 nm) reflects a degree of weakness of mixing, supposing that the aggregation of fine particles induces the coarsening of particles due to weak mixing. The results in Fig. 6 suggest a high mixing ability of Bs-Ft as well as a modest mixing ability of Fs-Ft, Fs-Bt, and Bs-Bt. From the change in the size distribution, the difference of dispersive and mixing abilities for different kneading elements were characterized.

The change in the particle size distribution caused by the melt-mixing modifies the rheological response of the extrudates. Fig. 7(a) shows the storage and loss moduli at 200°C of the extrudates by the different kneading elements as well as those of SAN, the matrix of the ABS, for comparison. At high frequencies, we observe that the modulus has a similar frequency-dependence as for SAN, but the amplitude of the modulus depends on the kneading elements. Furthermore, at low frequencies where the terminal region for SAN, the storage modulus, G', for the extrudates starts to increase. Prior works reported, as the rubber content increases, a plateau of G' develops at low frequencies, which is called the second plateau [9, 10, 11]. Since the rubber content of 20 wt% used in this repor is relatively low, only the onset of the second plateau was observed.

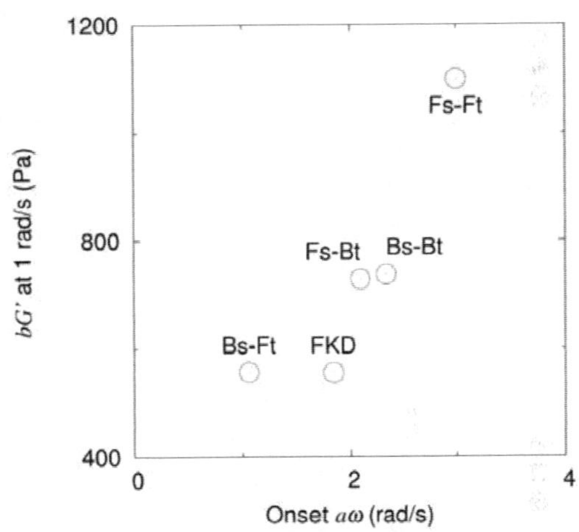

FIGURE 8. Storage modulus at 1 rad/s from the low-frequency elastic regime and onset frequency of the second plateau, at which frequency the phase lag have a maximum.

In Fig. 7(b), $G'(\omega)$ and $G''(\omega)$ for the extrudates are shifted to those for SAN, showing that the loss modulus $bG''(a\omega)$ almost coincides irrespective of the mixing elements, but the onset of the second plateau in the storage modulus $bG'(a\omega)$ depends on the mixing elements. Fig. 7(c) shows the phase lag, G''/G', after shifting to that for SAN, from which we observe that maxima in the phase lag clearly indicate the onset of the second plateau.

In Fig. 8, we plot the onset frequency of the second plateau determined from Fig. 7(c) and bG' value at 1 rad/s from Fig. 7(b) as a measure for the second plateau. Fig. 8 demonstrates a correlation between a value of bG' at a low frequency and the onset frequency, indicating that the onset frequency can characterize the low-frequency elastic property of ABS resin. Moreover, the shift factor for the amplitude of modulus, b, for high frequencies has a correlation with the onset frequency (Figure not shown). From these results, the effects of the size distribution induced by the melt-mixing are supposed to be reflected primarily in the onset frequency of the second plateau.

The onset frequency of the second plateau is low for Bs-Ft and FKD, for which cases the increase in the fine-particle fraction in the extrudates is a high level. The onset frequency is high for other cases where the coarse-particle fraction increases. Although the onset frequency itself describes only a limited aspect of the quality of the extrudates in comparison with the size distribution, but it surely shows some dependence on the size distribution.

CONCLUSIONS

Mixing characteristics of four kinds of pitched-tip kneading disks and a conventional kneading disk were analyzed experimentally. In order to evaluate intrinsic mixing characteristics of the kneading elements with different pumping abilities, we maintained the melt-mixing zone to be fully filled under a given throughput rate and a screw rotation speed by adjusting the pressure at the downstream of the melt-mixing zone. The melt-mixing of an ABS resin resulted in the change in the distribution of the rubber particle size. The particle size distributions of the extrudates with different elements reflect the mixing characteristics of the elements. It turns out that Bs-Ft ptKD and a conventional FKD have a high level of dispersive ability while Fs-Ft and Bs-Bt ptKDs have a suppressed mixing ability, and Fs-Bt ptKD have a moderate level of mixing. Furthermore, the changes in the particle size distributions were reflected in the melt rheology of the extrudates. Figure 9 shows the relative changes in the fractions of the intervals of the finest and coarsest particles, qualitatively indicating different mixing characteristics of the different elements.

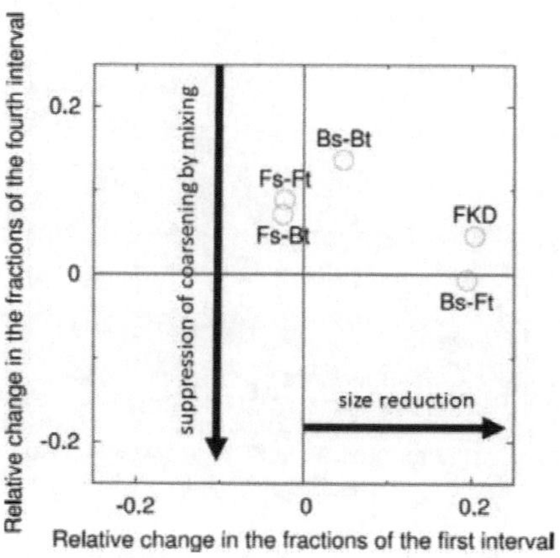

FIGURE 9. Relative changes in the fractions of the first (finest particles) and fourth (coarsest particles) intervals of the particle size distributions. The increase in the fine-particle fraction reflects the dispersive ability while the decrease in the coarse-particle fraction reflects the mixing ability.

ACKNOWLEDGMENTS

This work has been supported by Grants-in-Aid for Scientific Research (JSPS KAKENHI) under Grants Nos. 26400433, and 15H04175. Financial support from Hosokawa Powder Technology Foundation is also greatfully acknowledged. The numerical calculations were partly carried out using the computer facilities at the Research Institute for Information Technology at Kyushu University. The SEM observation was carried out at the Ultramicroscopy Research Center, Kyushu University.

REFERENCES

1. K. Kohlgruber, *Co-Rotating Twin Screw Extruder*, Munich, Hanser, 2007.
2. C. Rauwendaal, *Polymer Extrusion, 5 ed.*, Munich, Hanser, Munich, 2014.
3. Y. Nakayama, T. Kajiwara, and T. Masaki, *AIChE J.* **62**, 2563 (2016).
4. Y. Nakayama *et al.*, *Chem. Eng. Sci.* **66**, 103 (2011).
5. Y. Nakayama *et al.*, *Nihon Reoroji Gakkaishi (J. Soc. Rheol. Jpn.)*, **44**, 281 (2016).
6. K. Hirata *et al.*, *Intern. Polym. Process.*, **28**, 368 (2013).
7. K. Hirata *et al.*, *Polym. Eng. Sci.*, **54**, 2005 (2014).
8. S. Yamada *et al.*, *Intern. Polym. Process.*, **30**, 451 (2015).
9. Y. Aoki, *J. Soc. Rheol. Jpn.*, **7**, 20 (1979).
10. L. Castellani and P. Lomellini, *Rheol. Acta*, **33**, 446 (1994).
11. J. J. Cai and R. Salovey, *Polym. Eng. Sci.*, **39**, 1696 (1999).

Extrusion melt flow characterized in real time using in-line rheo-optical techniques

Canevarolo Sebastião V. ***(1)*, Bicalho Luciana A. (2), Gasparini Thiago M. (3)***

(1) Universidade Federal de São Carlos, Departamento de Engenharia de Materiais - São Paulo - Brazil, (2) Universidade Federal de São Carlos, Programa de Pós-Graduação em Ciencia e Engenharia de Materiais - SP - Brazil, (3) Universidade Federal de São Carlos, Programa de Pos-Graduação em Ciencia e Engenharia de Materiais - SP - Brazil

The polymer melt flow can be studied in-line using simultaneous rheo-optical techniques. The rheological behavior is obtained recording the pressure drop along the slit-die. Optical effects like light scattering and structure anisotropy are measured as turbidity, birefringence and low angle laser light scattering (LALLS). For this purpose we have designed and constructed an instrumented slit-die with three ports to measure pressure drop along the slit and two transparent windows for the optical measurements. It includes measuring turbidity and birefringence in the first window and LALLS at the second. The morphology of flowing polymer blends and composites are evaluated in the molten flowing field using, i) a known flowing field produced by a Cambridge Shearing System and ii) during extrusion in a Werner & Pfleiderer ZSK 30. Using different polymer blends and processing conditions the rheo-optical variables can be obtained and the morphological characteristic of the polymer flow be studied. The results can be compared with any conventional off-line measurement like, thermal (DSC), spectroscopic (FTIR) and morphology (POM, SEM, TEM), taken from samples collected during extrusion. Examples of various set of data will be presented during the conference, including diluted polymer blends and composites.

A New Open-Source Methodology to Balance the Flow in Complex Extrusion Dies

Carneiro O. S. (1)*, Rajkumar A. (1), Ferrás L. L. (1), Fernandes C. P. (1), Sacramento A. (2), Nóbrega J. M. (1)

(1) IPC-i3N - Institute for Polymers and Composites, University of Minho - Minho - Portugal, (2) Soprefa - Componentes Industriais SA. - Porto - Portugal

In this work we present a detailed description of a die design methodology aimed to improve the flow balance of complex extrusion dies, totally based on open source computer codes. The work encompasses the description of the overall design methodology, the implementation of the energy conservation equation in an existing OpenFOAM® solver, which then will be capable of simulating the steady non-isothermal flow of incompressible generalized Newtonian fluids, and its application to two industrial case studies. The new design methodology, is composed by a sequence of stages involving the use of different open-source software packages, namely: i) Free CAD – for the generation of the flow channel geometry; ii) Salome – the previous geometry is imported to this package where its boundary is divided in different groups to allow applying different boundary conditions. This boundary regions are exported as STL ASCII files from Salome; iii) SnappyHexMesh – based on the previous STL ASCII files from Salome, this package is used to generate, in a semi-automatic manner, the corresponding unstructured mesh; iv) Developed solver (OpenFOAM®) – is used to perform the numerical simulation in the tentative flow channel geometry and is followed by the post-processing, in which the predicted results are used to access the flow distribution quality. The methodology encompasses two alternative approaches to improve the flow distribution: modification of the flow channel geometry and/or boundary conditions. For each approach, an industrial case study is presented in detail. The results obtained with these case studies evidence the capabilities and practical usefulness of the developed methodology, and demonstrate that the computational design aid is an excellent alternative to the experimental trial-and-error procedure commonly used in industry.

Juxtaposing Technological Progress and Modeling of Blown Film Extrusion

VLACHOPOULOS JOHN (1)*

(1) McMaster University - ON - Canada

In blown film extrusion, the polymer is melted in an extruder and the hot melt is pumped through a die to form a thin-walled tube, which is simultaneously axially drawn and radially expanded. Industrial installations can produce more than a ton an hour of film, ranging in thickness usually from 0.01 to 0.5mm. Computer simulation had an enormous impact in die design for reducing film thickness variation and for eliminating stagnant flow areas which could produce local polymer degradation. Such simulations are usually based on some sort of Hele-Shaw flow approximation of a Generalized Newtonian Fluid through the spiral die channels and occasional 3-D simulations Mathematical modeling of blown film bubble formation is carried out on the basis of the thin membrane approximation for the prediction of bubble shape, velocity, thickness, temperature, strain rates and stresses. The standard bubble shape and the high-neck wine glass shape can be predicted as explained in J. Vlachopoulos and V. Sidiropoulos "Die Flow Analysis and Mathematical Modeling of Film Blowing" Chapter 4, T. Kanai and G.A. Campbell (Editors) "Film Processing Advances" Hanser (2014). The rheological properties influence the bubble shape, however, in industrial installations, the bubble shape is controlled by mechanical means and cooling air jets impinging on the bubble, externally and internally. Instabilities are significantly reduced by devices that change the opening to suit an appropriate blow up ratio, bubble cages and bubble guides. There is a dichotomy between academic research and technological progress. While most modern blown film lines are multilayer (frequently more than 7), academic papers are nearly always restricted to modeling of monolayer bubble formation to predict shape and any associated instabilities. In industry, the bubble shape is really controlled by mechanical means.A proposal will be made on how approach the modeling problem by assuming the bubble shape as a parameter.

Analysis of Neck-like Deformation Behavior in High-Speed Melt Spinning Process of PET by using High-Resolution On-Line Diameter Measurement Apparatus

Choi Young-Chan [1], Nam In-Woo [1], An Hyung-Joo [1], Jong Il [2], Kikutani Takeshi [3], Hahm Wan-Gyu [1]*

[1] Korea Institute of Industrial Technology - Ansan - Korea, [2] Kolon Industries, Inc. - Gumi - Korea, [3] Tokyo Institute of Technology - Tokyo - Japan

In this study, to verify the characteristic behavior of fiber deformation in high-speed melt spinning process of high molecular weight poly(ethylene terephthalate) (IV 1.2), on-line measurements of fiber diameter were performed for the melt spinning line at the take-up velocities of 1 to 5 km/min by using 2 way (XY-axis) high-resolution optical measurement apparatus with sampling rate of maximum 16 kHz. Overall diameter profiles obtained in the spinning line showed that neck-like deformation accompanying strain-induced crystallization started to appear from around 4km/min and its position shifted toward the spinneret with an increase in the take-up velocity. Detailed analysis of neck-like deformation through high-resolution simultaneous diameter measurements at two XY positions of close distance confirmed that the position of necking deformation fluctuates along the spinning line and maximum strain rate varies significantly depending on the take-up velocity and the direction of its movement.

High Speed Extrusion of Modified Polycarbonate

V. Schöppner [a], M. Pohl

[a] *Kunststofftechnik Paderborn, Paderborn University – volker.schoeppner@ktp.upb.de; max.pohl@ktp.upb.de*

Abstract. An important topic in the area of research and development in extrusion technology is to increase the economical efficiency. In general, an increase in mass flow rate is done by designing bigger machines, but this also increases costs for maintenance and batch changes. Another possibility to increase throughput is to use smooth barrel extruders with high screw speeds (from 2 m/s on). In further investigations several issues were detected in the field of high speed extrusion and the processing of polycarbonate. Because of the high shear sensitivity of this polymer, at higher screw speeds temperature increases massively. The reason for the lack of melting is the enthalpy of amorphous polymers. To handle this issue we decided to analyze the behavior of a modified polycarbonate with a high shear thinning range and the same zero viscosity as a typical extrusion material. In this essay, the manufacturing of a special blended polycarbonate is presented as well as first results of processing this material at high screw speeds. These investigations are based on the evaluation of various speed ranges on an extruder (D = 30 mm) with vacuum degassing and a direct torque drive. In addition, several pressure sensors along the axial screw length and a temperature sensor in the area of the screw tip were installed. With this arrangement, it is possible to make a statement on temperature homogeneity, pressure development and the melting process. The overall goal is to achieve optimization strategies for the high speed extrusion of polycarbonate.

Keywords: Single Screw Extrusion, High Speed Extrusion, Smooth Barrel, Polycarbonate
PACS: 89.20.Bb, 89.20.Kk, 89.90.+n, 81.20.Hy

INTRODUCTION

To increase the throughput of a single screw extruder one possibility is to increase the screw speed. Researches have shown that for circumferential speeds over 2 m/s on a small extruder lead to the same throughput as on bigger machines [1]. Thereby, the specific energy can be reduced, because of the smaller screws size and the operation with a direct power engine [2]. In general, a grooved bushing extruder can be used to increase throughput, but this method is nearly pushed to the limit [3]. Next to standard extrusion concepts there are some special concepts to overcome the limits in throughput. For example the “S-truder” witch separates the melt and solid material or the “Helibar” in which a barrier screw is combined with a grooved bushing that is located along the whole screw. Both machines allow higher screw speeds on smaller units [3,4]. In further investigations, different researches have been made with semi-crystalline polymers like PE or PP on a smooth barrel extruder [1-8]. A result of this scientific publications was that the melt temperature is difficult to control in high speed extrusion but very important for the melt quality. However, most of the examined materials have a high shear shining ability. This leads to a better management of the temperature at higher screw speeds because of the lower dissipation entry.

First insights into extrusion of polycarbonate (amorphous nearly newtonian flow behavior) showed that a transferability from semi-crystalline polymers is hard to find [9]. For polycarbonate the limits in temperature and pressure are reached too fast to enable high speed extrusion. One opportunity to solve this issue is to create a special screw geometry for exactly this material type. However, for the work presented in this paper a modified polycarbonate with the same zero viscosity as the initial material and higher shear thinning rate is investigated. Thereby, it is possible to produce different types of polycarbonate at high screw speeds. Investigations were done with screw speeds from 750 rpm to 1000 rpm on a smooth barrel 30 mm extruder. Here relevant parameters like melt temperature and pressure have been analyzed.

FIRST RESULTS ON HIGH SPEED EXTRUSION OF POLYCARBONATE

In general, high screw speeds on small machines can be energy-efficient due to the machine size and the advantage of a direct torque engine.. Hereby, the melt quality is a target quantity, which is limited by the mechanical and the thermal stress. Massive heating is generated by the dissipative shearing resulting from the screw speed and the material behavior. Studies have shown that an increasing of the screw speed in combination with the processed polycarbonate leads to the anticipated difficulties and damage to the polymer due to high temperatures (critical value is 360 °C). This was confirmed in own investigations which have been executed in advance of this publication. Therefore, on a 30 mm smooth barrel extrusion line a throughput of 170 kg/h was generated at 1000 rpm with a melt temperature about 360 °C witch is over the critical value of 340 °C. However, in comparison with slower screw speeds, the chain damage is lower due to a slighter influence of thermal damage resulting from a shorter residence time. For the described investigations, a polycarbonate with an MVR of 6 cm^3/10 min (300 °C; 1.2 kg) was used. A high strength, impact strength and stiffness characterizes this polymer. Also the viscosity is nearly of an Newtonian behavior [9]. Based in this results a concept will be presented in this paper to overcome these challenges. All machine parts and settings for the process are the same as in past investigations.[9]. In detail a material modification with a pronounced shear thinning should decrease the damaging issues by pressure and temperature.

MATERIAL DEVELOPMENT

The processing of polycarbonate on a degassing screw is the state of the art. Typically, a cylindrical shearing section in front of the degassing section is used to prevent flooding. However, at high screw speeds the thin shear gap leads to high shear rates. At a throughput of 170 kg/h with 1000 rpm, which is currently the target of optimization, shear rates about 800 1/s were reached. (Figure 1)

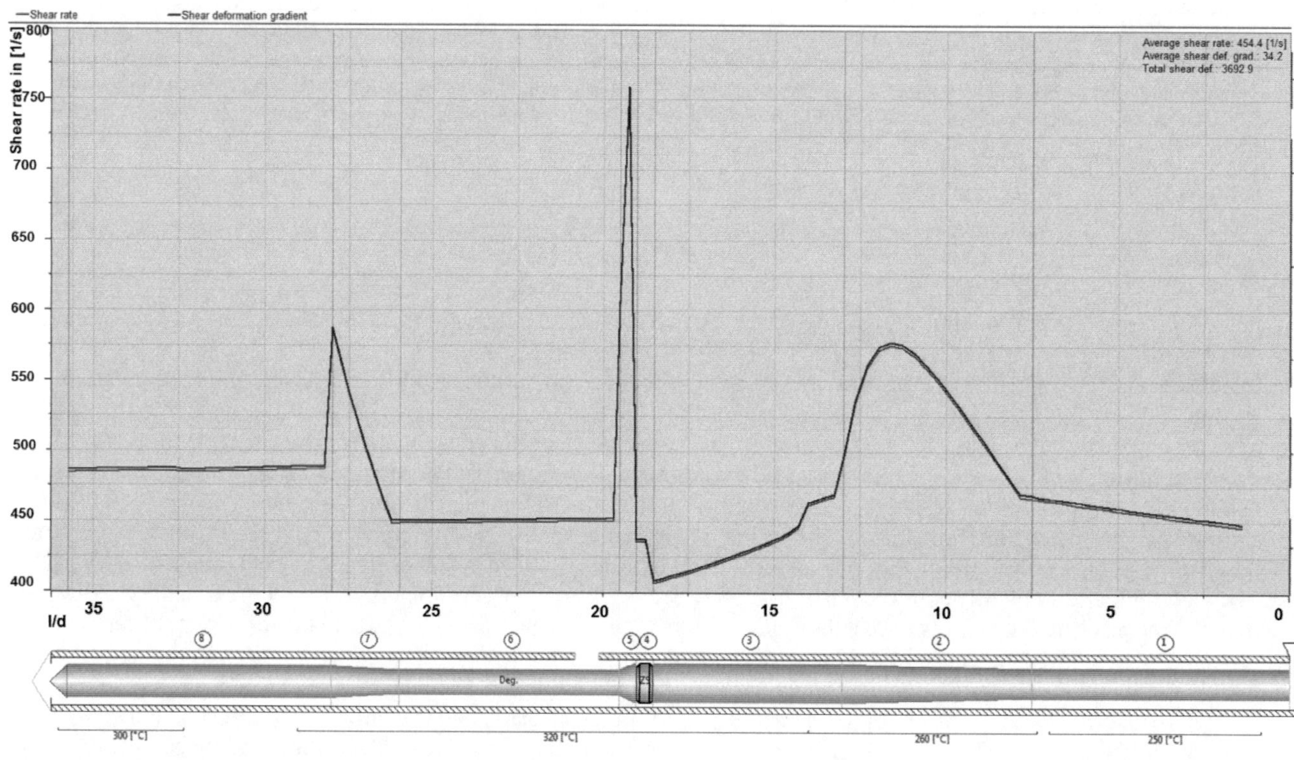

FIGURE 1. Axial course of shear stress at 1000 rpm

In order to facilitate the flow through the screw channels, a low viscosity is much better. Thereby for the research results presented in this paper a special PC Blend has been created. Here it is important to obtain a zero viscosity like the initial polycarbonte (PC standard), because a specific viscosity is needed at ambiente pressure for the downstream equipment(cast film extrusion, chill rolls). For the work presented in this paper, a very high viscous and a low viscous resin were mixed. In order to determine the different proportions the following equation was used in which viscosity of polymer is described by η and mass by w [10].

$$log(\eta_M) = w_A log(\eta_A) + w_B log(\eta_B) \quad (1)$$

Based on this equation a mixture for a blend was calculated and compounded on a twin screw extruder. Figure 2 shows curves for viscosity and shear rate for the initial material, both mixture resins and the final blend.

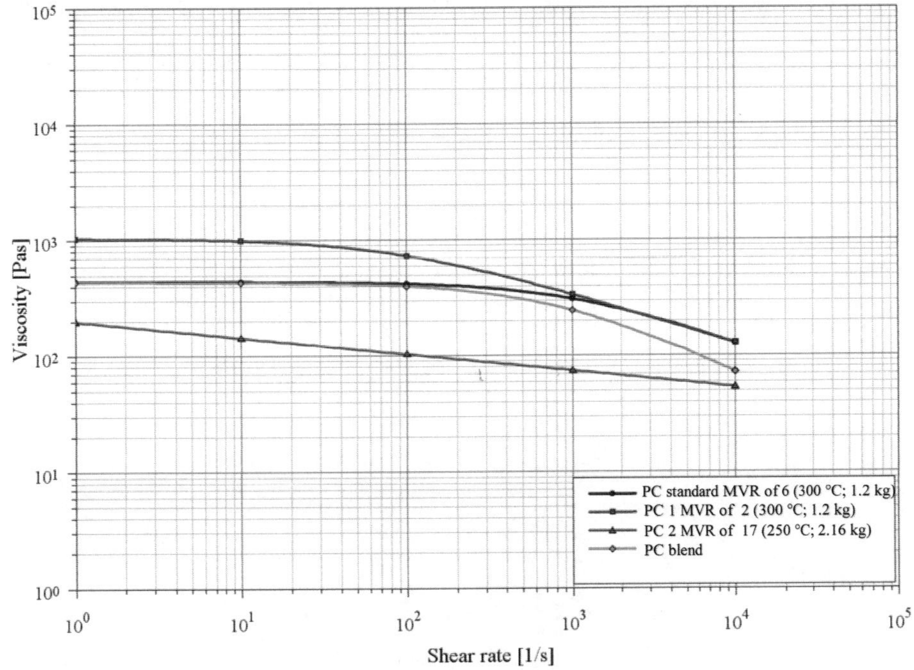

FIGURE 2. Rheology – Viscosity and Shear rate of the different polycarbonates used for the investigations

For the PC blend 84% of PC1 was mixed with 16% of PC2. At higher shear rates a significant increase of shear thinning is recognizable. With this material an extensive test series was carried out.

EXPERIMENTAL SETUP AND EXECUTION

Based on further investigations the same high-speed extruder from ESDE GmbH with a 30 mm degassing screw and a smooth barrel is used [9]. The barrel has a total length of 44.3 L/D and several pressure measure units as well as thermocouples have been used. To remove the moisture of the melt a degassing section was installed. In figure 3 the barrel temperature profile and the measurement setup is shown.

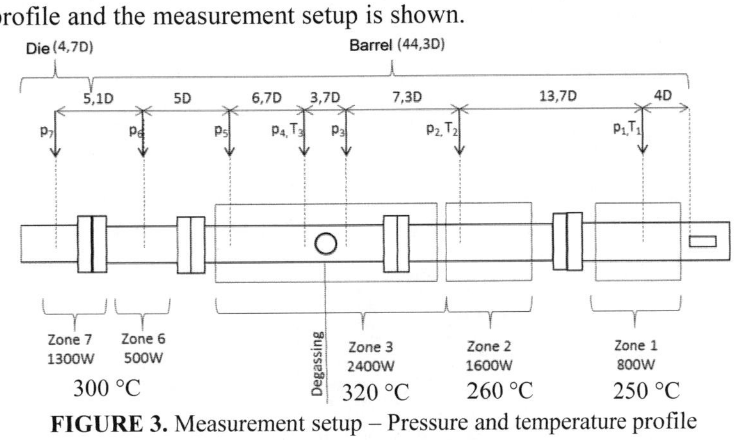

FIGURE 3. Measurement setup – Pressure and temperature profile

The pressure was measured at the transition of compression section 1 to metering section 1 (p2) because here the highest pressure takes place. In addition, the temperatures are measured at the die manually. The goal is to reach a temperature around 290 - 320 °C for an optimal processing. With the standard material and at a screw speed over 750 rpm a value over 340 °C was measured. A backpressure was used to keep the investigations close to industry standards where the pressure drop in a film extrusion die is around 35 -80 bar for PC.

For this paper as well as in further investigations the primary goal was a throughput rate of about 180 kg/h and a good melty quality. Therefore, to compare the results, the same settings and operations points at 750 rpm and 1000 rpm were used. The following table indicates the speed, torque, throughput and the temperature at the screw tip.

TABLE a). Summary measurement results – Screw speed, torque, throughput and melt temperature

Operation point	Screw speed [rpm]	Torque [Nm]	Throughput [kg/h]	Melt temperature [°C]
#1 standard	750	355	126	343
#2 standard	1000	360	170	357
#3 blend	750	355	134	298
#4 blend	1000	343	182	316

Table a) shows that with higher screws speeds for the standard PC the torque, throughput and melt temperature reaches a critical limit. On the contrary, the blend at high screw speeds leads to lower melt temperature compared to the standard type which also leads to a smaller torque. The following chart shows the melt temperature at the screw tip.

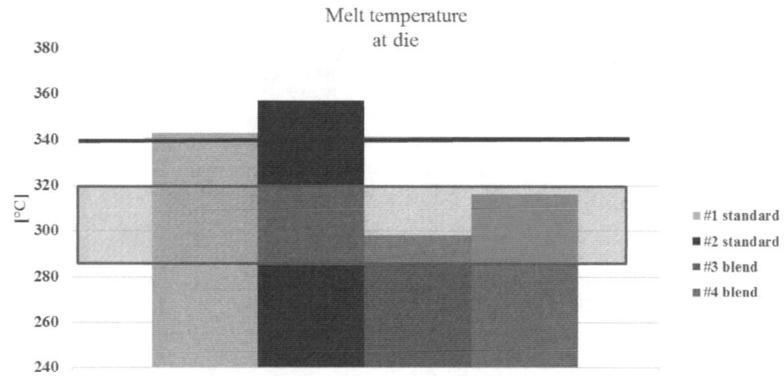

FIGURE 4. Melt temperature at die – manually measured

A known limitation for the material degradation without oxygen is described in the literature [11] and drawn in this graphic as a red line, which means the melt temperature should not pass 340°C. The green field is the standard process range for creating polycarbonate products. In Figure 4 the two columns on the left side shows the temperature issue from the PC standard because of the dissipative heat input due to the Newtonian behavior at high screw speeds. The other two bars represent the new PC blend, which has a higher shear thing behavior. Because of this, the heat input of the screw is lower and a good melt temperature for processing is generated. An analogous behavior can be seen for the pressure in the following figure 5.

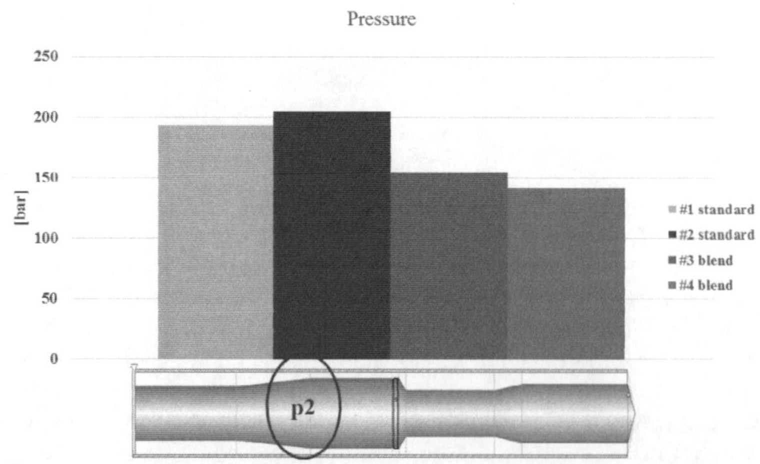

FIGURE 5. Pressure at p2 (crossing of compression section to metering section)

At Figure 5 the PC standard tends to higher pressures and the PC blend has a lower pressure at this position. Furthermore higher screw speeds leads at the blended material to a decreased pressure due to the higher shear velocity and the shear thinning behavior. Therefore, the desired effect making the melt thinner has been enabled. By this procedure it is possible to reduce the stress by pressure of the molten material for about 30%.

SUMMARY AND OUTLOOK

In this paper, it is shown that a modification of polycarbonate helps to prevent the anticipated difficulties of damaged polymer due to high temperatures at high-speed extrusion. In detail, the zero viscosity of a standard material for the film cast extrusion has been used as scale for the new PC blend. The blend was created by using the logarithmic mixing rule to determine the needed ratios of the both components. To create comparable results with the results from the foregone investigations in [9], the same test setup for screw size, barrel temperature and so on have been chosen. The generated blends show a decreased melt temperature and pressure at the desired throughput of 182 kg/h on a 30 mm extruder at 1000 rpm. Now further investigations are necessary to analyze the mechanic behavior of the blend. Moreover, studies are planned in which a cast film is processed and tensile tests are carried out as well as Yellowness index is to be measured.

ACKNOWLEDGMENTS

We like to thank the Deutsche Forschungsgemeinschaft (DFG) which has made these studies possible by financing the SCHO 551/19-1 project and Covestro AG for material donation.

REFERENCES

1. T. Pohl, Entwicklung schnelldrehender Einschneckensysteme für die Kunststoffverarbeitung auf Basis theoretischer Grundlagenuntersuchungen, Dissertation, University of Paderborn, 2003.
1. H. Hörmann, *Theoretische und experimentelle Betrachtung schnelllaufender Einschneckenextruder*, Dissertation, University of Paderborn, 2014.
2. M. Großmann, *Leistungs- und Effizienzsteigerung in der Einschneckenextrusion durch alternative Plastifiziertechnik*, Dissertation, University of Duisburg, 2011.
3. J. Wortberg and G. Karrenberg, *Alternative Plastifizierkonzept für die High-Speed-Extrusion – Der High-Speed-S-Truder mit rotierender Schneckenhülse*, Journal of Plastics Technology 11, Munich, 2015.
4. E. Grünschloß, *HELIBAR® - AN UNIVERSAL PLASTICATING SYSTEM FOR SINGLE-SCREW-EXTRUDERS AND INJECTION-MOULDING-MACHINES*, Stuttgarter Kunststoff-Kolloquium, Institut für Kunststofftechnologie der University of Stuttgart, 2013.
5. H. Rahal, *Alternative Methoden zur Feststoffförderung und Plastifizerung in der Extrusion*, Dissertation, University of Duisburg, 2007.
6. P. Gorczyca, *Analyse und Optimierung von Einschneckenextrudern mit schnelldrehenden Schnecken*, Dissertation, University of Duisburg, 2011.
7. C. Bonten and O. Kast, *Untersuchung eines schnelllaufenden Einschneckenextruders mit wendelgenuteter Plastifizierzone*, Journal of Plastics Technology 11, Munich, 2015.
8. K. Abbas, *Degradation of PC during recycling in a Capillary Rheometer*, PES 7/1980, 1980.
9. V. Schoeppner and M. Pohl, First insight into the High Speed Extrusion of Polycarbonate, PPS 32, 2016.
10. L. A. Utracki and M. R. Kamal, *Melt Rheology of Polymer Blends*, Polym. Eng. Sci., 1982
11. B. Dolezel, *Die Beständigkeit von Kunststoffen und Gummi*, CHV, 1978.

1- Dimensional Computer Simulations to understand the internal behavior of a Twin Screw Extruder

***Balakrishnan Nitish* [(1)]*, *Bigio David* [(2)]**

[(1)] University of Maryland, College Park - Maryland - USA, [(2)] University of Maryland- College Park - Maryland - USA

Polymer processing using extrusion has been widely used in industries because of the compounding capabilities of the extruders. The properties of the existing base polymers could be improvised by compounding with additives rather than the need to fabricate new base polymers. The mixing of input materials in the extruder is extremely critical to produce superior quality products. Quantification of stress history in the extruder could be used to understand the mixing behavior within the extruder. Residence Stress Distribution (RSD) is a methodology that was implemented to quantify the amount of material that undergoes a particular amount of stress within the Twin Screw Extruder (TSE). Percent Break-Up is a metric that was developed using RSD Methodology to compare the stress history experienced by polymer melts within the TSE with varying operating conditions (screw speed (N), specific throughput (Q/N). Experiments carried out using the UMD RSD methodology suggested that percent break-up (%BU) is a function of N and Q/N. In order to validate the findings of the RSD Experiments, 1-Dimensional Computer Simulation called Ludovic is used. Ludovic provides an insight into the behavior observed on performing the RSD experiment, as it helps correlate the internal behavior of the extruder i.e. shear rate, temperature, viscosity profile within the extruder, to the RSD results. The results obtained from the experiments are statistically analyzed using JMP to predict the influence of the operating conditions (N, Q/N) on the properties of the polymer melt (temperature, pressure, residence time etc.). Similarly, properties such as degree of fill in the screw channel are analyzed using the computer simulations and used to explain the internal behavior of the extruder.

Micro and Nano-layered Tubes and Pipes via Continuous Multilayer Co-Extrusion

***Schneider Tyler* [(1)]*, *Maia Joao* [(1)]**

[(1)] Case Western Reserve University - Ohio - USA

Continuous multilayered co-extrusion is a relatively mature technique that goes back to the 1970s, when it was first developed. Historically, two of the main limitations of the technology have been its limitation to sheet or film extrusion of materials of matched viscosities. Our group has recently overcome the Relating rheology and simulations to latter via the introduction of novel, non-square layering dies, which allowed systems with viscosity ratios up to 10 and elasticity ratios up to 100 to be micro-layered. In this work, we strive to overcome the former, by designing, manufacturing and validating a new system designed to multilayer tubes and pipes. Rheology and simulations are utilized to design and validate the multiplication system at research scale and to provide suggestions for upscale to industry. In particular, ANSYS® POLYFLOW® is utilized to evaluate shear, pressure, velocity profiles, and residence times through the designed flow path using commercially available commodity and engineering thermoplastics. Viscosity matched or mismatched layer systems are then evaluated using moving interface simulations to gain insight on layer structure during extrusion. Finally, the layer structure tracked using simulations was then compared with experimental results produced from a manufactured die. The presence of a weld line in this flow path required the outer mandrel to be rotated thereby superimposing a rotational deformation to the extrudate. The effect of shear rate and shear orientation on the elimination of the weld line and on the mechanical strength of the extrudate is also demonstrated and suggestions are made to produce viable multilayer tubing systems with high layer numbers and thin layer structures.

Investigation of the Barrel Temperature Profile on the Process Behavior of Single Screw Extruders and Strategies to Determine the Optimal Temperature Control

Volker Schöppner[a], Verena Resonnek[a]

[a]*Paderborn University, Warburger Straße 100, 33098 Paderborn, Germany*

verena.resonnek@ktp.upb.de

Abstract. At present, polymer extrusion is one of the most important manufacturing processes due to the increasing demand of polymer products. In order to reach the desired amounts of polymer, the melt quality of the polymer plays a crucial role. The melt quality is not solely dependent on the chemical composition of the base polymer but also on the condition of preparing and processing the polymer. Indeed, the resulting melt quality is highly dependent on the parameter settings of the extruder.
Since polymer materials have a temperature-dependent behavior, the temperature progress over time is one significant aspect. Hence, the thermal stress of the polymer particles during the extrusion process is essential to preserve a good melt quality. In the wake of this, the determination of a decent barrel temperature along the extruder is one of the primary challenges in the field of extrusion. Besides, generating a thermal homogeneous polymer melt, a progression of the throughput, consequently an increase of the economic efficiency, is possible. In return, an inauspicious barrel temperature can lead to a deterioration of the material properties and in this manner, it will lead to a decrease in the economic efficiency. In addition to that, the thermal radial melting profile and the pressure fluctuations of the melt during the process is significantly influenced by the barrel temperature. Furthermore, with respect to high pressure fluctuations, an inopportune barrel temperature setting can lead to a substantially higher wear of the screw and the barrel.
In this article the effect of different barrel temperature profiles on the process behavior is in the focus. Firstly, different criteria for defining or respectively measuring the melt quality are considered. After that, two different strategies to determine a convenient barrel temperature setting are taken into account. The collected target figures are utilized to find correlations between the barrel temperature setting of each barrel zone and the polymer melt quality.

Keywords: **Single Screw Extrusion, Temperature Profile, Barrel Temperature, Temperature Settings**

INTRODUCTION

Nowadays, in many industrial fields, polymers have clearly illustrated their technological and economic benefits. Under this aspect, single screw extruders are the most common. For the conveying of the polymer resin, the material is transferred into the extruder barrel. Here, the resin is positioned in the annular space between the extruder screw and the barrel. As a matter of fact, the rotation of the extruder screw is converted into frictional force which is converted into heat that acts on the material. On the one hand, the polymer is melted due to the frictional heat generation. On the other hand the heat is conducted from the barrel heaters. Approximately 80 % of the energy input results from the dissipation whereas 20 % come from the thermal input of the barrel heater [1].

The focal point of many researches is the improvement of extrusion processes regarding different parameter settings. Taking into account finding a decent barrel temperature setting in practice, so called "trial and error" approaches are applied, whereas the melting process is not sufficiently abetted. Since polymers have a distinct temperature-time dependent behavior, specifically the barrel temperature setting exerts a great influence on the extruder performance. Thus, finding an optimal setting of the barrel heaters is beneficial for the resulting polymer melt quality. Indeed, the barrel temperature influences the discharge temperature, pressure fluctuation and the throughput behavior.

The issue of applying these trial and error methods is that they are usually based on the longstanding expertise of the operating process engineer. However, in many cases, the suggested barrel temperature profile does not meet the optimal barrel temperature setting. Moreover, in case of changes in the process setting, i.e. the rpm or the resin, the barrel temperature needs to be adjusted. Already little changes in the barrel temperature can lead to defective melt quality. In this respect, it appears that the determination of an optimal barrel temperature setting, not least because of the tremendous interference, offers a large potential for process optimization. This is the reason why, in this paper, the focus lays on correlations between the barrel temperature setting and the extruder performance.

INVESTIGATION OF THE BARREL TEMPERATURE

The properties of polymer products strongly depend on the process conditions, especially on the thermal process settings during the extrusion process. In this article, the barrel temperature setting is called into question. Here, the axial temperature for each barrel zone is examined. The intention is to find out in what manner the local extruder barrel temperature is affecting the extruder performance. Moreover, the aim is to detect process rules that can be applied in order to set a decent barrel temperature setting to obtain a good melt quality. For this purpose, two different studies have been taken place.

The first was an experimental investigation concentrating on the axial temperature profile. The aim was to find significant changes in predefined objective criterions i.g. the discharge fluctuations of local pressure and temporal temperature as well as the overall performance of the extruder and the discharge temperature. These extrusion trials have been done at steady state conditions.

The second investigation focused on single barrel zones in accordance to the screw section. Here, each sectional barrel was varied in order to see correlation between the set point temperature and the criterions. These extrusion studies were performed at an unsteady state conditions. In the following both investigation are presented.

EQUIPMENT, MEASUREMENT TECHNIQUES AND MATERIALS

All examinations were carried out on a 45 mm diameter single screw extruder (Battenfeld BEX1-45-30B) with a length of 32 L/D. The extruder had a maximum driving power of 125 kW and a maximum revolution speed of 585 rpm. A three sectional and a barrier screw with no mixing or shearing elements were used. The extruder had a smooth bushing in the feed zone machine and was equipped with six separate temperature zones and the set temperature can be set individually. The feed zone (zone1) of the extruder is thermally decoupled from the barrel and had two heater bands with a max. performance of 1640 W and 650W. Here the feed zone was cooled with a water temperature control with an inlet temperature of 15 °C. The extruder had four heating/cooling combinations for the temperature control of the barrel with a max. performance of 2850 W (zone 2-3), respectively 3500 W (zone 5) which are equipped with ceramic resistance heater bands and cooling fans. A throttle die with two heat bands with a max. power of 650 W was used. In order to maintain constant process conditions, the screw of the throttle die which is responsible for changing the back pressure had not been moved for the entire examinations.

Furthermore, each barrel zone was equipped with a different sensor. In general each barrel provided a spring loaded temperature sensor which was located with a depth about 2/3 of the barrel wall thus to minimize measurement errors (see [2]). Additionally, each well depth was filled with thermo-lubrication to sustain an enhanced heat transfer between the sensor and the heated barrel wall. Moreover, since solid conveying is dominant in the feeding section, zones 2 and 3 were equipped with optical pressure sensors (FOS, Type EDS 2C). These optical sensors are very rugged designed thus the membrane will not be deformed from the solid granule [2]. Zone 2-4 are assigned with wall mounted thermocouples (Thermoexpert, type J). In zone 6 (throttle die), a strain gauge pressure transducer (Gneuß, DA-250, type J) as well as a bridge with multiple probes (customized production by FOS, type J) is mounted to measure the radial temperature across the melt channel. Therefore six thermocouple junctions (in a distance of 0, 4.5, 9, 13.5, 18 and 22.5 mm) were placed across the melt flow along the diameter. The installed measurement techniques, the power of the barrel heater bands and the throttle die are depicted in figure 1. A data acquisition program (DASYLab, National Instruments) was used to communicate between the experimental instruments and the computer device.

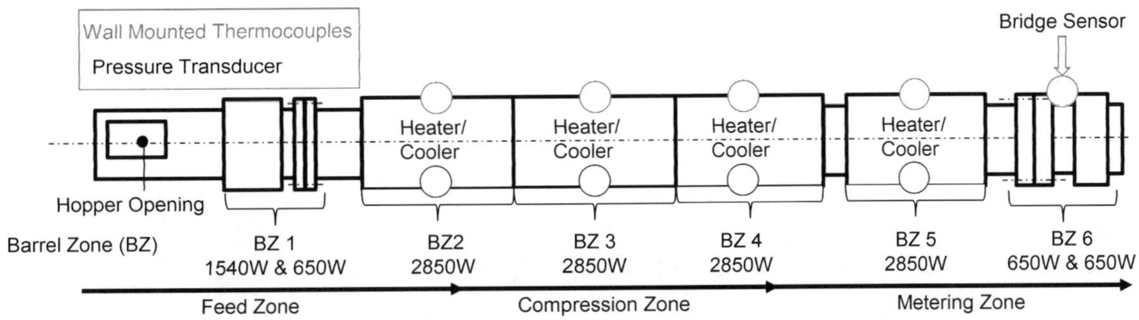

Figure 1 Battenfeld Extruder BEX1-45-30B with Depicted Temperature Zones and Measurement Techniques

Polypropylene (PP) and polystyrene (PS) resins were used for the extrusion studies. The PP material was a homopolymer with a solid density of 0.9 g/cm3 and a melt flow rate of 8 dg/min (230°C/2.16 kg), the

PS resin had a solid density of 1.04 g/cm3 and a melt flow rate of 9.5 dg/min (200°C/5 kg). The melting temperature for the resins were measured and constituted 167.69 °C for PP and 98.78 °C for PS.

Examination of the Axial Barrel Temperature Profile

The experimental setup consists of the extruder described above using a three-sectional screw with a compression ratio of 2:1 and a barrier screw with a compression ratio of 2.5:1. The barrel temperature zones were assigned to the underlying section of the screw. Therefore the feed zone consists of BZ 1 and BZ 2, the compression zone of BZ 3 and BZ 4 and the metering zone of BZ5 and BZ6. The temperature and pressure sensors were mounted to a connector box and the data acquisition program acquired all signals with a frequency of 100 Hz (barrel temperature, melt temperature, melt pressure). The other signals (i.e. power performance, rotational speed, engine load, actual vs. target barrel temperature and throughput) were also recorded.

The two resins were run under four different barrel temperature profile settings with four different temperature levels (from low to high temperatures and one in between) and four different rotational speeds (25, 50, 75, 100 rpm). The barrel temperature profiles are illustrated in figure 2 and include four different types:

- A flat profile; meaning that all barrel zones were set on the same temperature.
- A hump profile, which means that the feeding and the metering zone are much cooler than the compression zone.
- A ramp profile with an increasing barrel temperature from the feeding to the metering zone.
- A decrease profile which is reversed order to the ramp profile. The feed zone is set on high temperatures whereas the metering zone is set much higher.

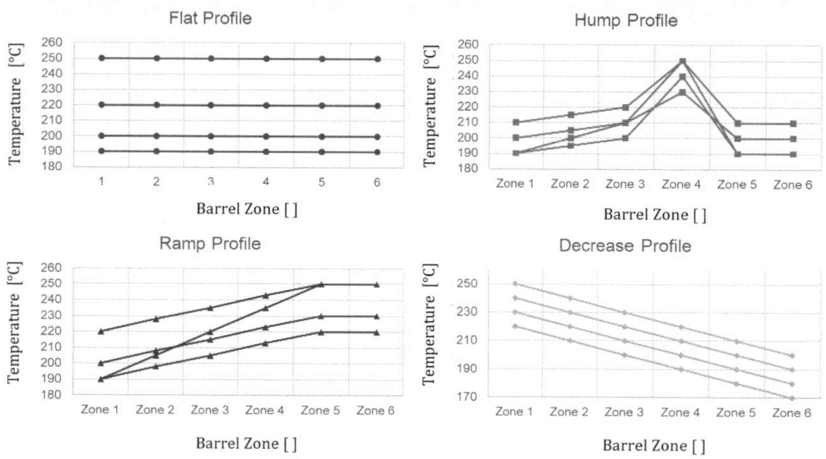

Figure 2 Graphical Illustration of the Experimental Design of the Barrel Temperature Setting for each Screw Geometry, Resin and Rotational Speed and Varying Temperature Levels

The statistical analysis of the temperature profiles for each resin and screw geometry in accordance to the objective criterion clearly showed interference. Figure 3 illustrates an excerpt of the analysis of PP with a barrier screw and a rotational speed of 75. The results show that the target criterions (throughput, performance, discharge temperature, pressure variation) are affected by the barrel temperature. It becomes obvious that the throughput, for instance in figure 3, can be increased by 27 % when setting a ramp barrel profile with a high temperature level. Regarding the total performance, it becomes evident that the hump profile had the least energy consumption for all temperature levels,

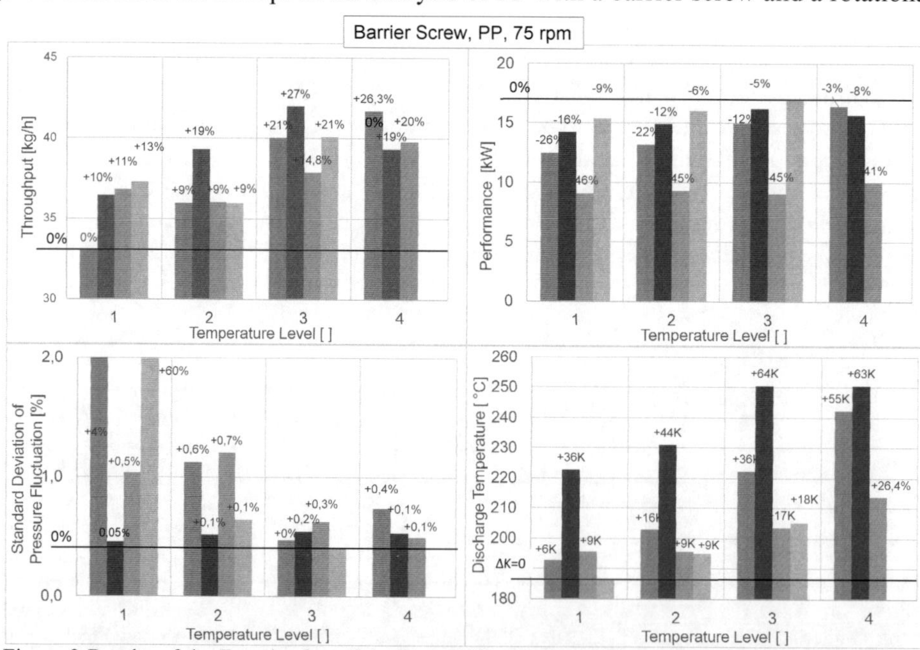

Figure 3 Results of the Examination of the Four Different Barrel Temperature Profiles for PP with a Barrier Screw and 75 rpm Regarding the Target Criterions Throughput, Total Performance, Pressure Fluctuations and Discharge Temperature

whereas the decrease profile had the most. Taking into account the pressure fluctuation (given as the standard deviation in percent), the lowest temperature level in general lead to higher pressure fluctuations. When taking a closer look at the temperature setting it becomes apparent that a lower barrel temperature in the feed zone for three-zone screws and in comparison to this a higher temperature setting for barrier screws lead to lower pressure fluctuations. The discharge temperature is highly affected by the barrel temperature as well. As seen in figure 3 the ramp profile and also the flat profile showed much higher discharge temperatures (up to 55 K) in comparison to the hump and decrease profile. Similar results could be observed with each screw design and resin.

The results of the examination of the axial temperature profile lead to the assumption that each sectional barrel temperature is responsible for different effects. In this regard, the coefficient of friction is strongly dependent on the temperature. Therefore the barrel temperature can be set to such an extent that the friction coefficient increases. By increasing the friction a more stable conveying will occur and thus the pressure fluctuations will decrease. The temperature setting at the compression zone respectively, is suspected to support the melting process whereas the temperature at the metering zone might affect the discharge temperature. In order to demonstrate these hypothesis dynamic examinations of each barrel zone were carried out.

DYNAMIC INVESTIGATIONS ON THE BARREL TEMPERATURE ZONE

In [3] RAUWENDAAL uses a dynamic optimization methods to determine the correct barrel temperature. In his examinations he changes the barrel temperature in the feed section from a low to a high temperature and records the dynamic pressure response of the extruder. However, a graph which shows the pressure variation over the barrel temperature can be plotted. In this manner RAUWENDAAL finds the correct barrel temperature by detecting the location (barrel temperature) of the lowest pressure variation.

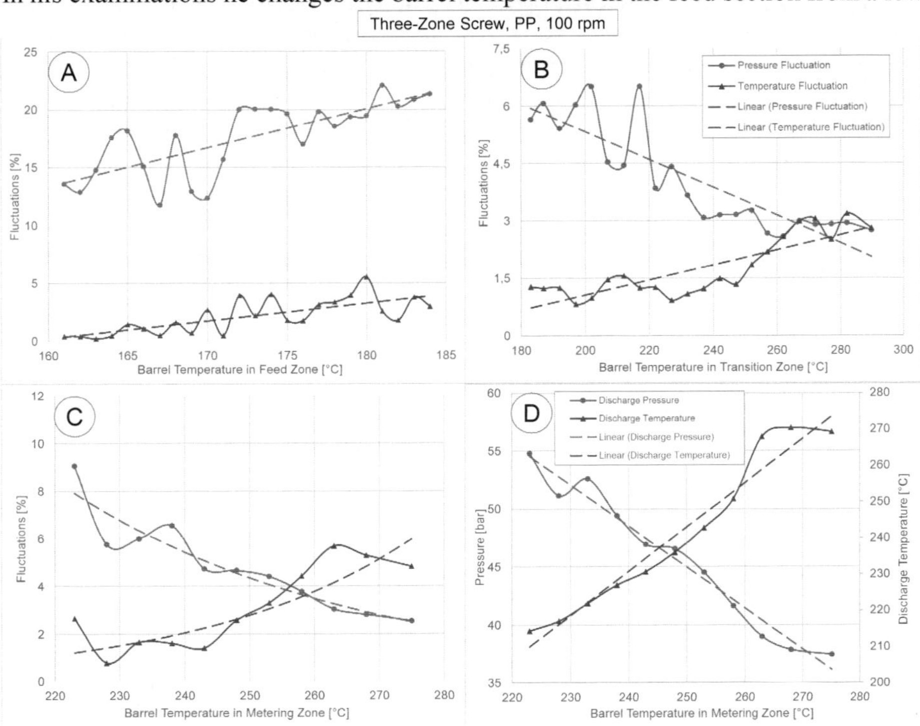

Figure 4 Results of the Dynamic Optimization Experiments with a Varying Barrel Temperature in the Feed, Transition and Metering Section with a Three-Zone Screw, PP resin at 100 rpm

The following examinations have been carried out in accordance to RAUWENDAALs methods with different rotational speeds (25, 50, 75, 100 rpm), the PP and PS resins and the two screw designs. Additionally, every sectional barrel zone was varied. The following graphs (see figure 4) show an excerpt of the recorded fluctuations of the temporal temperature and pressure as well as the actual discharge temperature and pressure for PP with a three-zone screw for the three barrel sections (feed A, transition B and metering zone C). Graph D in figure 4 shows the discharge pressure and temperature when changing the barrel of the metering section. All the results of the dynamic investigations clearly show that the pressure and temporary temperature fluctuations can be controlled by choosing a decent barrel temperature setting.

In figure 4 for the barrel temperature in the feed zone (A) it can said that the pressure and the temporal temperature fluctuations at the discharge increase with an increasing barrel temperature. In figure 5 the throughput over the barrel temperature is depicted. Graph A shows that the throughput decreases with an increasing barrel temperature in the feed zone. These results prove that a cooler temperature in the feed zone leads to a higher friction coefficient and thus stabilizes the conveying process which can be observed by the decreasing fluctuations and furthermore, it proves that a better melting process takes place.

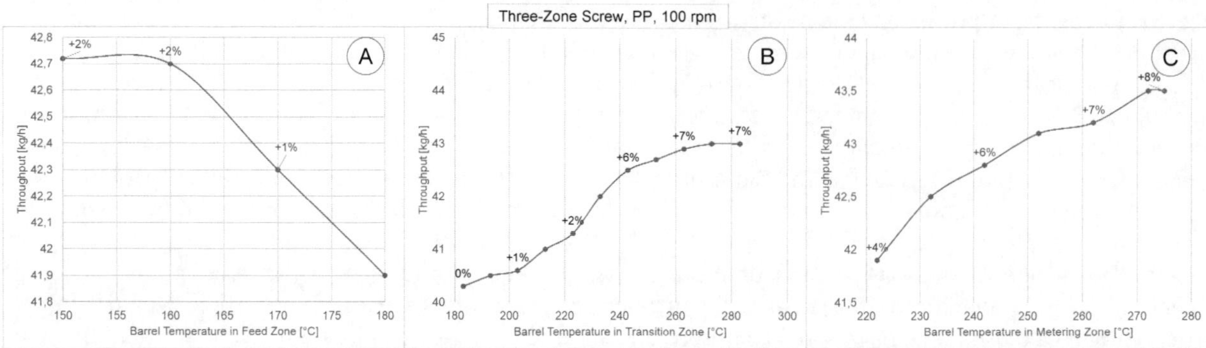

Figure 5 Throughput Behavior over the Barrel Temperature in: A: Feed Zone, B: Transition Zone and C: Metering Zone for PP and a Three-Zone Screw at 100 rpm

Regarding the transition zone (figure 4, B) it is noticeable that the temporal temperature fluctuation operates vice versa to the pressure fluctuations. The higher the barrel temperature got, the lower the pressure fluctuations but the higher the temporal temperature fluctuations were. Taking a look at the throughput behavior in figure 5, graph B the throughput increases with an intensifying barrel temperature. However, this incident verifies the assumption that the temperature in the transition zone supports the melting process.

Taking into account the temperature in the metering zone, graph C and D in figure 4 show that the pressure and temporal temperature fluctuations behave similarly in comparison to the fluctuations in the transition zone since the temporal temperature fluctuations rise with an increasing barrel temperature, whereas the pressure fluctuation reduce. Graph C in figure 5 demonstrates that the throughput certainly increases but when comparing to graph D in figure 4, the discharge temperature, logically, also rises with an increasing barrel temperature. Hence, the temperature in the metering section affects the discharge temperature and should be set close to the desired melt temperature.

CONCLUSION AND FUTURE WORK

In this paper the extruder barrel temperature was examined in order to find correlations between the set barrel temperature and the extruder performance. The carried out experiments clearly showed effects of the barrel temperature control on the extrusion performance. As shown from the experimental results, the effects on the discharge pressure variation, temporal temperature variation and throughput behavior are highly significant. For the purpose of producing a good melt quality, the barrel temperature setting needs to be set correctly. However, the examination of the material behavior of the PP and the PS resin in combination with a three-zone scree and a barrier screw demonstrate that those materials can be generalized referring to the barrel temperature setting. This means that for those materials it is possible to develop recommended actions regarding the barrel temperature. In this manner this incident can be realized by using the virtual signs of the extruder like the pressure and temperature fluctuations over time. The aim of this project is the development an intelligent barrel temperature control on the foundation of basic settings in conjunction with a regulation algorithm that ensures a focused process control.

ACKNOWLEDGEMENT

We are grateful for the support of the project by the Deutsche Forschungsgemeinschaft (DFG-Project SCHO 551/29-1).

REFERENCES

[1] C. Hopmann, W. Michaeli, Einführung in die Kunststoffverarbeitung, 7th Edition, Carl Hanser Verlag Munich ; 2015; p.119

[2] C. Rauwendaal; Polymer Extrusion, 4th edition, Carl Hanser Verlag, Munich, 2001, p. 78-82, p. 81-82

[3] C. Rauwendaal, Time to Learn about Dynamic Optimization of Extruder Barrel Temperatures, Plastics Technology, 2008

FRICTIONAL OR SLIP HEATING IN THE EXTRUSION OF LINEAR POLYETHYLENE MELTS

PEREZ-GONZALEZ JOSE (1)*, MARIN-SANTIBAÑEZ BENJAMIN MARCOS (2), RODRIGUEZ-GONZALEZ FRANCISCO (3)

(1) INSTITUTO POLITECNICO NACIONAL, ESFM - CIUDAD DE MEXICO - Mexico, (2) INSTITUTO POLITECNICO NACIONAL, ESIQIE - CIUDAD DE MEXICO - Mexico, (3) INSTITUTO POLITECNICO NACIONAL, CEPROBI - YAUTEPEC, MORELOS - Mexico

Frictional or slip heating has been recognized as the heating produced by slip flow of polymer melts. This type of heating must be distinguished from the well-known viscous heating since they have different physical origin. In particular, extrusion of polymer melts with slip at the die generates frictional heating. The relationship between slip flow and frictional heating during the continuous extrusion of a non-slipping linear low-density (LLDPE) and a slipping high-density polyethylene (HDPE), respectively, both pure as well as blended with a fluoropolymer processing aid (PA), was investigated in this work by rheo-particle image velocimetry and thermal imaging. Significant rises in temperature were measured under slip and no slip conditions, being these much higher than the values predicted by the adiabatic flow assumption. Clear difference was made between viscous and frictional heating before the stick-slip regime for the LLDPE, even though they could not be distinguished from one another at higher stresses. Such a difference, however, could not be made for the slipping HDPE, since overall in the presence of slip, frictional and viscous heating act synergistically to increase the melt temperature.

Compounding of Polyethylene Composites Using High Speed Twin and Quad Screw Extruders

Mansour M. Albareeki[a], Stephen B. Driscoll[a], and Carol Forance Barry[a]

[a]*Department of Plastics Engineering, University of Massachusetts Lowell, Lowell, MA, USA*

Abstract. The effect of high screw speed twin and quad screw extruders on mixing was investigated using polyethylene microcomposites with 1 wt% calcium carbonate. Using screws with similar designs extruder type influenced the effects that screw speeds of 300 to 200 rpm had on extruder residence time, melt temperature, drive torque, and head pressure. Parallel plate rheology indicated significant chain scission of the polymers, but better filler dispersion at screw speeds of 900 and 1500 rpm and better dispersion in the quad screw extruder. In the quad screw extruder, the lower melt temperatures and greater shear allowed better mixing at higher screw speeds than the twin screw extruder. The dispersion was confirmed with transmission electron microscopy.

Keywords: Extrusion, Twin screw extrusion.

INTRODUCTION

Improved dispersive mixing of fillers with higher viscosity melts, greater shear and elongational flows, and longer residence times has been shown for mixing of thermoset rubber in Banbury mixers and thermoplastics in twin screw extruders[1]. In traditional co-rotating twin screw extruders where screw speeds have reached 1200 rpm, no issues have been reported during the mixing process[2-4]. Ultra-high speed twin screw extruders, operated at screw speeds as high as 4000 rpm, have produced degradation of even relatively stable polymers[5-8]. Changes in screw design and selections of screw speeds have been a trade-off between the greater shear/elongation and longer residence times that improve mixing and reducing these parameters, particularly shear to prevent or reduce polymer degradation[6-8]. These compromise has often lead for filler agglomeration and the resultant defects and brittle failure in parts[7].

In contrast, quad screw extruders contain four co-rotating, parallel, fully intermeshing screws. These extruders have greater free volume, but greater number of intermeshing regions ("bite points") which produce high shear stresses compared to twin screw extruders. Prior work on mixing with these quad screw extruders has been reported for polymer blends at relatively low speeds (< 1000 rpm)[9-10] and color concentrates at higher speeds (~1500 rpm)[11]. When compared to the high speed twin screw extruders, the quad extruder has shown better mixing of color concentrates[11]. Given the limited work on quad screw extruders, this study was an investigation of mixing a particulate-filled polymer composite with high speed twin and quad screw extruders. One objective was to understand the effects of screw speed on the drive torque, residence time, head pressure, and melt temperature in these extruders. The second objective was to determine the effects of the residence time and number of bite points (i.e., kneading blocks) on the viscosity and dispersion in the resultant composites.

METHODOLOGY

Materials

For this work, polyethylene was selected as the matrix polymer because its relatively high stability during melt processing. The five commercially-available polymers were three low density polyethylenes (LDPE), one linear low density polyethylene (LLDPE), and one medium density polyethylene (MDPE). The complex viscosity of the neat resins was measured at temperatures of 190, 205, and 220°C using parallel plate rheometry. The reported densities melt flow indices (MFI) as well as the calculated melt indices (n) and activation energies for flow (E_a) are listed in Table 1. The LDPEs exhibited viscosities that corresponded to their melt indices and had n values of 0.42-0.59. LLDPE and LDPE 1 had similar viscosities at low frequencies, but LLDPE showed less shear thinning. At low frequencies, MDPE had a similar viscosity to LDPE 2, but had a power law index of 0.51. The filler was calcium carbonate ($CaCO_3$), which due to its high surface area, is difficult to disperse and stabilize within a polymer matrix[13-16]; the calcium carbonate filler (Omya, grade: Omyafilm-SY) had a reported particle diameter of 0.5-50 μm.

TABLE 1. Materials Properties

Material	Density (kg/m^3)	MFI (g/ 10 min)	n	E_a (kDa)
LDPE 1	920	0.18	0.42	15.9
LDPE 2	920	0.65	0.50	32.2
LDPE 3	920	1.85	0.59	39.9
LLDPE	921	0.60	0.68	14.8
MDPE	936	0.30	0.51	11.6

Compounding

Composites of polyethylene containing 1 wt. % calcium carbonate were prepared by melt blending in (1) an ultra high speed co-rotating 15-mm-diameter twin-screw-extruder with a length-to-diameter (L/D) ratio of 60:1 (Technovel, model: KZW15TW-45/60MG-NH (-4400)) and (2) an ultra high speed co-rotating 15-mm-diameter quad-screw extruder with an L/D ratio of 45:1 (Technovel, model: WDR15QD-45MG-NH (-2200)). Screws for the quad screw extruders contained four mixing sections with respective lengths of 15, 25, 15, and 55 mm (i.e., about 16% of the total screw length). For the twin screw extruder, the four mixing sections had the same composition, but were twice as long as those in the quad screw extruder; (the mixing sections were about 25% of the total screw length). Therefore, both the twin and quad screw extruders had the same number and type of mixing elements. Melt exiting the extruders passed through a single strand die for twin screw extruder and a double strand die for the quad screw extruder; when corrected for the number of screws, the dimensions of the die openings provided similar pressure drops in both extruders. The extruded strands were cooled in a water bath and then pelletized.

For compounding, the barrel and die temperatures were 205°C for all zones. The $CaCO_3$ and polymers were pre-mixed and then the premix was fed through the first feed port at a constant feed rate of 1.68 kg/h; this feed rate was limited by the capacity of the twin screw extruder and the measured error was ± 5%. The screw speeds were 300, 900, 1500, and 2000 rpm. During each extrusion trial, drive torque was recorded from control panel; the resin temperature and pressure were determined by using 'K' type thermocouple and Dynisco pressure transducer, respectively, located in the die adaptor; and the residence time was measured as the time from the introduction of a colored pellet to the extruder until that color change was observed in the strands.

Characterization

The rheology of the composites was determined using a parallel plate rheometer (TA Instruments, model: ARES). Samples were prepared by placing a ring and shim on the bottom plate of the rheometer, adding and preheating the sample pellets, and removing the extra melt to achieve a 2-mm gap between the two plates. Measurements were performed at a temperature of 205°C. Strain sweeps were used to identify the linear viscoelastic region and to study the breakdown of the composites' structure. These sweeps were performed at strains of 0.1 to 100% and at a frequency of 0.16 Hz (1 rad/s) for the linear viscoelastic region and 3 Hz (18.85 rad/s) for structure breakdown. Frequency sweep measurements were performed between 0.02 and 15.92 Hz (0.01-100 rad/s) at a strain extent of 10%.

Dispersion of the calcium carbonate in the polyethylene was evaluated using transmission electron microscopy (TEM). The extruded pellets were cryomicrotomed to 60-nm-thick slices.

RESULTS AND DISCUSSION

Impact on Extrusion Parameters

As shown in Figure 1a, extruder residence time decreased with increasing screw speed and was shorter for the quad screw extruder (L/D = 40:1) than for the twin screw extruder (L/D = 60:1). The difference in residence time between the two extruders depended on the screw length, the greater travel distance with the four parallel screws of the quad extruder, and the quad screw extruder's greater free volume which decreased residence time in the kneading blocks. Prior work for twin screw extruders[16] has shown that the residence time for partially-filled screw channels depends on screw speed, whereas with completely-filled channels, the residence time is proportional to the ratio of the free volume to the volumetric flow rate. Determination of the free volume for the twin screw extruder followed Rauwendaal[17] and the quad screw extruder was treated as a twin screw extruder plus two additional screws without screw-barrel clearances at the sides of the screws. The calculated free volumes for the twin and quad screw extruders were 165 cm^3 and 273 cm^3, respectively. With these free volume values, the relative residence time in the twin and quad screw extruders (t_{TSE}/t_{QSE}) was proportional to $L_{TSE}V_{QSE}/L_{QSE}V_{TSE}$, where L and V are the extruder length and free volume in filled channels, respectively, in the twin and quad screw extruders.

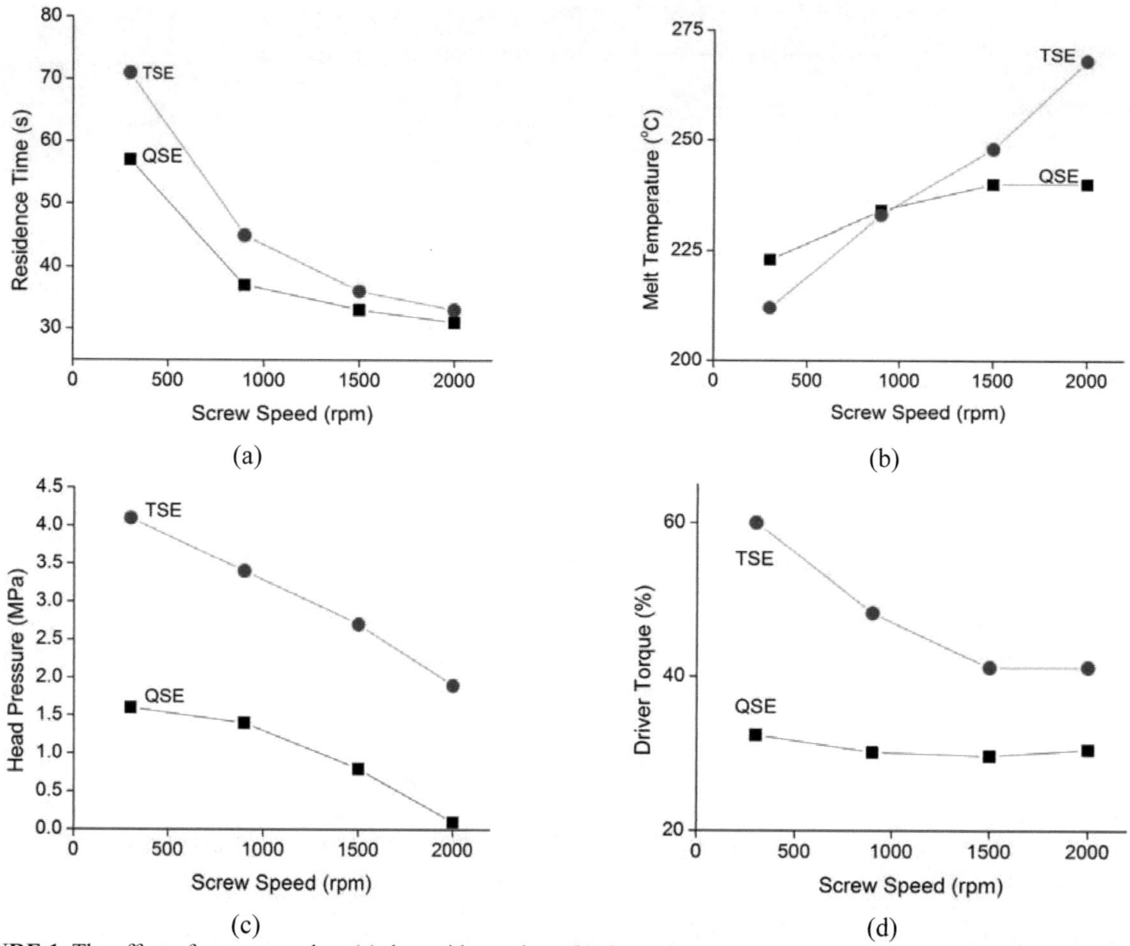

FIGURE 1. The effect of screw speed on (a) the residence time, (b) the melt temperature, (c) head pressure, and (d) drive torque of LDPE 2 in the twin and quad screw extruders.

Increasing the screw speed of the twin screw extruder from 300 to 2000 rpm produced a a linear 60°C increase in melt temperature (Figure 1b). With the quad screw extruder, however, there was 25°C increase in melt temperature, with the temperature leveling off when the screw speed exceeded 1500 rpm. These temperature increases were consistent with previous observations[11]. In the twin screw extruder, the melt flows in a tight figure-8 pattern, which creates significant friction at polymer-barrel wall and polymer-screw interfaces. With kneading blocks, this flow pattern produced significant increases in polymer temperature (which were observed as spikes in measured barrel temperature). Although the twin and quad screw extruders had the same clearance between the screw and barrel walls, the open area in the extruder cross-sections was 1.69 cm^2 and 4.03 cm^2 for the twin and quad screw extruders, respectively. This design reduced friction as polymer flowed along the quad screw extruder.

As shown in Figure 1c, (1) the head pressure for both extruders decreased with increasing screw speed and (2) the head pressure was significantly lower in the quad screw extruder compared to the twin screw extruder. These head pressures reflected the levels of shear in the extruders, with the quad screw extruder imparting less shear because of its greater free volume. In addition, with a screw speed of 2000 rpm, the head pressure in the quad screw extruder varied from -0.1 to 0.1 MPa due to the relatively low feed rate used in this work; other work (not reported) showed stable head pressures at high screw speeds when higher feed rates were used for the quad screw extruder.

As illustrated in Figure 1d, the drive torque for quad screw extruder was lower than that for the twin screw extruder. Again, this difference was attributed to the greater free volume and lower shear rates in the quad screw extruder. The lower shear rates produced less shear thinning and reduced melt temperature increases in the quad screw extruder. As a result, less power was needed to rotate the screws and the drive torque for the quad screw extruder did not decrease as much with increasing screw speed as it did with the twin screw extruder.

Impact on Mixing

The "degradation" of unfilled LDPE 2 was evaluated using the ratio of the zero-shear-rate viscosity of the processing and unprocessed resin (η/η_o). With the twin screw extruder, η/η_o increased at 900 rpm and then decreased below 1 for higher screw speeds. For the quad screw extruder, the η/η_o was about 0.5 for screw speeds of 300 and 900 rpm and also decreased for the two highest screw speeds. The higher screw speeds increased the stress applied to the polymer, producing chain scission in both extruders, whereas the lower melt temperatures and increased bite points in the quad screw extruder increased the shear stress. The increase in η/η_o with the twin screw extruder may have been due to the physical cross-linking of the polymer chains[18].

Figure 2a presents η/η_o for LDPE 2 composites; an η/η_o value greater than 1 indicated better dispersion of the $CaCO_3$ in the polymer melt. The quad screw extruder enhanced mixing at screw speeds of 300-1500 rpm, whereas the twin screw extruder only provided this performance for speeds of 300 and 900 rpm. This improvement in dispersion using the quad screw extruder was attributed to greater shear/elongation and higher viscosity (due to lower melt temperatures) in the quad screw extruder. The better dispersion was confirmed with TEM images (Figure 3).

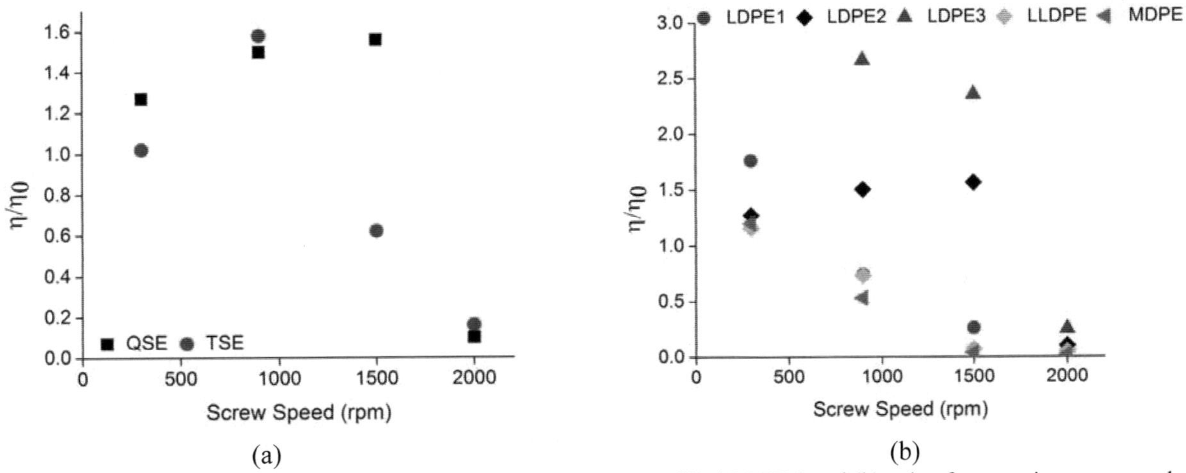

FIGURE 2. The effect of screw speed on (a) η/η_o of calcium carbonate-filled LDPE 2 and (b) η/η_o of composites compounded in the quad screw extruder

As shown in Figure 2b, mixing performance in the quad depended on the polymer's melt viscosity. For all materials mixed at N = 300 rpm, adding $CaCO_3$ improved η/η_o. With the higher screw speeds, the η/η_o seemed to depend on material viscosity, with the higher viscosity resins (LDPE1, LLDPE, and MDPE) exhibiting poorer mixing than the lower viscosity resins (LDPE2 and LDPE3). The cause for these results is still under investigation. Strain sweeps indicated that, in the linear viscoelastic region (strain of about 0-60%), the complex viscosity (η^*) of the filled materials changed only slightly compared to η^* of the neat materials. At higher strain rates, however, screw speed produced a difference in the complex viscosity, with a sharp decrease in viscosity at 300 rpm correlating with poor $CaCO_3$ dispersion. When the screw speed was increased to 1500 rpm, the complex viscosities of the neat and filled materials were similar, indicating better dispersion of $CaCO_3$. The complex viscosity-strain curves of $CaCO_3$-filled LDPE2 also were affected by extruder type; for both extruders, increasing the screw speed increased the linear viscosity region (to higher strains), indicating better dispersion of the $CaCO_3$. With the quad screw extruder, the complex viscosity-strain curve became nearly linear at high strains, which suggested better dispersion of the $CaCO_3$.

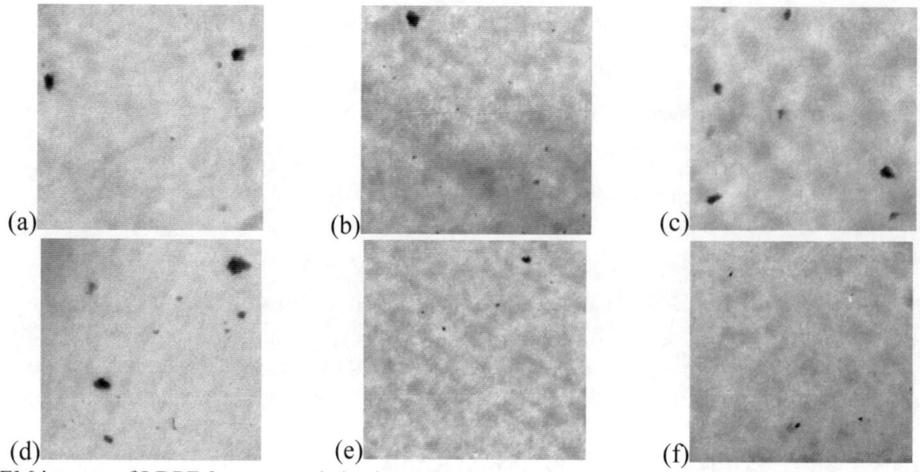

FIGURE 3. TEM images of LDPE 2 compounded using (a-c) the twin screw extruder with speeds of 300, 600, and 900 rpm, respectively, and (d-f) the quad screw extruder with speeds of 300, 600, and 900 rpm, respectively.

CONCLUSIONS

When mixing polyethylene and $CaCO_3$using ultra high speed co-rotating twin and quad screw extruders with equivalent screw designs, the quad screw extruder showed lower residence time, drive torque, head pressure, and melt temperature increases than the twin screw extruder. This difference was attributed to the greater free volume in the quad screw extruder. Rheology indicated significant chain scission of the polymers at higher screw speeds as well as with the quad screw extruder. While higher screw speeds produced better dispersion of the $CaCO_3$, the lower melt temperatures and greater shear in the quad screw extruder provided better mixing than the twin screw extruder. Better dispersion occurred with the lower viscosity resins.

ACKNOWLEDGMENTS

The authors would like to thank Technovel Corporation (Osaka, Japan) for the use of their extruders..

REFERENCES

1. D. Kim, J. S. Lee, C. F. Barry, and J. L. *Journal of Applied Polymer Science* **109**, 2924-2934 (2008).
2. K. Kohlgrüber, *Co-Rotating Twin-Screw Extruders: Fundamentals, Technology, and Applications*, Cincinnati, OH: Hanser Publications, 2012.
3. P. Peltola, E. Välipakka, J. Vuorinen, S. Syrjälä, and K. Hanhi, *Polym. Eng. Sci*, **46**, 995-1000 (2006).
4. T. Villmow, B. Kretzschmar, and P. Pötschke, *Compos. Sci. Technol.*, **70**, 2045-2055 (2010).
5. J. J. Boone, G. Pigeon, C. Thellen, J. A. Ratto, and R. Whitehouse, *Proc. Ann. Tech. Conf. Soc. Plast. Eng.*, **71**, 1591699 (2013).
6. A. Farahanchi, R. Malloy, and M. J. Sobkowicz Kline, *Polym. Eng. Sci.*, **56**, 743-751 (2016).
7. A. Farahanchi, E. Boehm, N. Orbey, and R. Malloy, *Polym. Eng. Sci.*, **57**, 60-68 (2017)
8. A. Farahanchi and M. J. Sobkowicz Kline, *Polymer Degradation and Stability* **138**, 40-46 (2017).
9. K. Matsumoto, M. Nagai, K. Hamakawa, S. Nishitsuji, and T. Inoue, *J. Appl. Polym. Sci.*, **129**(1), 443-448 (2013).
10. S. Suzuki, S. Nishitsuji, and T. Inoue, T., *Polym. Eng. Sci.*, **52**, 958-1963 (2012).
11. M. M. Linares Rodriguez, "Study of the Compounding Characteristics of Twin-Screw and Quad-Screw Extruders on Pigment Dispersion," M.S. Thesis, University of Massachusetts Lowell, 2013.
12. J. Lei and R. Zhou, *Polym. Eng. Sci.*,**40**, 1529-1533 (2000).
13. T. Xie, H. Liu, Y Ou, and G. Yang, *J. Polym. Sci. Part B: Polym. Phys.*, **43**, 3213-3221 (2005).
14. S. Sahebian, S. M. Zebarjad, J. V. Khaki, and S. A. Sajjadi, J. *Materials Processing Technology*, **209**, 1310-1317 (2009).
15. R. H. Elleithy, I. Ali, I., M. A. Ali, and S. M. Al-Zahrani, *J. Appl. Polym. Sci.*, **117**, 2413-2421 (2010).
16. K. Kohlgrüber, *Co-Rotating Twin-Screw Extruders: Fundamentals, Technology, and Applications*, Cincinnati, OH: Hanser Publications, 2012, p. 166.
17. C. Rauwendal, *Polymer Extrusion*, 5th ed., Munich, Germany: Carl Hanser Verlag, 2014, pp. 705-714.
18. A. E. Hamielec, P. E. Gloor, and S. Zhu, *Canadian J. Chem. Eng.*, **69**, 611-618 (1991).

Analysis of Shrink Void Formation in Extrusion

Rauwendaal Chris [1]*

[1] REE - CA - USA

Voids in extrusion can be caused by air entrapment, moisture, poor venting, and non-uniform cooling. Voids caused by non-uniform cooling are typically referred to as shrink voids; they are common in thick-walled products (2+ mm). They result from non-uniform shrinkage that occurs in the cooling of the extrudate. The shrinkage occurs as a result of the increase in density as the polymer cools. This paper presents an experimental and theoretical study of void formation. The cooling process is analyzed using the simulation software chillWARE. The paper will outline measures that can be taken to eliminate shrink voids.

Modeling and Experimental Studies on Single Screw Extrusion of Wood-Polymer Composites

Wilczyński Krzysztof [1]*, Buziak Kamila [1], Lewandowski Adrian [1], Wilczyński Krzysztof J. [1], Nastaj Andrzej [1]

[1] Warsaw University of Technology - Warsaw - Poland

Studies on rheology and extrusion of wood-polymer composites WPC are very limited. Computer models currently available do not allow to simulate WPC extrusion correctly. Extensive experimental studies have been performed for single screw extrusion of PP/wood flour composites of WF concentrations (from 25% to 70%). Screw Pulling-out Technique was used to study material melting, screw filling, and pressure was measured. An effect of screw speed and flow rate (starvation level) was studied. It was concluded that WPC extrusion is totally different from extrusion of thermoplastics. Material transport and melting are strongly dependent on the composition and operating conditions. Contiguous solid melting mechanism CSM which is typical for flood fed single screw extrusion of thermoplastics was not observed for high content of WF. In this case, one-dimensional melting mechanism was seen, and melting occured at the solid bed/melt film interface, and the molten material penetrated the solid bed such that a melt pool was not formed at the active flight of the screw. With an increase of PP content, more than 50%, CSM mechanism appeared, and a melt pool at the active flight was formed which increased in width when melting progressed. WPC single screw extrusion has been simulated using GSEM program. This is a global model which includes solid conveying, melting and melt flow. The model allows to predict pressure and temperature profiles, melting of the polymer, etc. A new, one-dimensional melting model has been developed and implemented into the program code. Three-dimensional, non-Newtonian FEM simulations using Ansys-Polyflow have been performed for material flow with yield stress and slipping at the walls. Dimensionless screw pumping characteristics (flow rate vs. pressure gradient) have been developed and implemented into the program code to model the material flow in the extruder. FEM modeling for material melting has been discussed. Simulations have been validated experimentally. Acknowledgments: The authors would like to acknowledge support from National Science Center, Poland (DEC-2015/19/B/ST8/00948).

The Finite Element Simulation of Alternating Multilayer Co-extrusion Process

Yu Huaning (1)*

(1) Sichuan University - Sichuan - China

Alternating multilayered co-extrusion technology is a simultaneous extrusion of two heterogeneous polymer melts, which used to produce films with thousands and thousands layers, with special laminating-multiplying elements (LMEs). In this paper, the three-dimensional, isothermal, generalized Newtonian polymeric fluid flow (PP) in LMEs was simulated using the finite elements. In addition, the geometry of LME was also changed as a reference factor. Simulation results, especially velocity profiles, shear stress fields, slip profiles and extrusion deformations were discussed. The analysis shows adding a postpositional rectangle channel or increasing wall slip helps to improve extruded profiles. At last, the effects of fluid's flow rate ratio and viscosity ratio on the interface of PP/PP are further examined.

New Potentials for the Design of Extrusion Dies by Using Additive Manufacturing: Thermo-Rheological Optimization of a Blown Film Extrusion Die

Bastian Neubrech[1, a)], Johannes Wortberg[1], Stefan Kleszczynski[2] and Gerd Witt[2]

[1] *University of Duisburg-Essen, Institute of Product Engineering, Engineering Design and Plastics Machinery, Lotharstraße 1, 47057 Duisburg, Germany*
[2] *University of Duisburg-Essen, Institute of Product Engineering, Manufacturing Technology, Lotharstraße 1, 47057 Duisburg, Germany*

[a)] bastian.neubrech@uni-due.de

Abstract. In polymer processing, the blown film extrusion is one of the most important processes for the production of films. Thereby, the extrusion die can be seen as the form-giving component, which transforms the melt into a tubular film at the die exit. State of the art extrusion dies are designed as spiral mandrel dies and they are credited with positive characteristics regarding weld lines respectively material and thickness distribution at the outlet. Nevertheless, the temperature distribution shows potentials for improvement, as the layer of melt affects the heat transfer away from the spiral mandrel. It has been simulated, that in operation a temperature delta of up to 20°C between the housing and the spiral mandrel is established, which influences the cooling behavior of the inner and outer product surface. For conventional production methods, such a die consists of several components, which have to be assembled.

The aim of this paper is the identification of potentials for a novel die design for blown film extrusion processes, originated by a different manufacturing method for such dies. The chosen procedure will be the additive manufacturing method of selective laser melting. Thereby, a thermo-rheological optimization is achieved by realizing new geometrical structures, which cannot be manufactured by conventional procedures. Focus of the design process is an improved temperature household of the new die geometry. In a first step, the die requirements and the geometric restrictions are defined and the restrictions of the manufacturing process are gathered and will be implemented in a 3D-CAD master model. Different geometrical structure elements are investigated and will lead to a basis model for the extrusion die.

Keywords: blown film extrusion, additive manufacturing, extrusion die design, automatic optimization
PACS: 81.20.Hy, 83.50.Uv, 07.05.Tp, 87.55.de

INTRODUCTION

Within the field of polymer processing, the extrusion die is described to be the form-giving component of an extrusion line. Apart from the general type of the product, the product quality is greatly influenced by the extrusion die. This includes mechanical properties as well as the surface topology and the wall thickness distribution. The latter is particularly interesting for the production of blown film semi-finished products and affects the downstream processes. Consequently, the correct design and dimensioning is highly relevant. Since increasing requirements are placed on the product, the die design is becoming more complex to ensure product quality. This results in the superimposition of multiple influencing factors that needs to be considered. Even though computer-aided design is gaining importance in the design process, there is a discrepancy between research and practice, based on empirical knowledge garnered iteratively. The conventional manufacturing process still restricts die designs and flow channels.

Against this backdrop, certain synergy effects arise, because additive manufacturing offers potentials for optimizing the design process of extrusion dies. In this context, its feasibility has been investigated for blow molding dies [1], as well as for flat sheet dies [2]. This manufacturing method holds certain degrees of freedom, allowing novel, complex interior structures and flow channel geometries that cannot be realized by conventional manufacturing methods. In particular, the melt conveying channels and their designs are not restricted by the accessibility of the machine tools. This allows a new way of thinking for the design process to adjust the flow channels for the flow situation and to optimize the extrusion die thermo-rheologically.

This paper aims to outline the potentials which result from implementing additive manufacturing into the design process of extrusion dies exemplified by a small-scale laboratory blown film extrusion plant. At first, the conventional manufacturing process will be described. This section will focus on the computer-aided design process as the basis for the future tasks. Afterwards, the potentials for the novel design will be clarified and the general approach will be described. The outlook summarizes further tasks in order to present a thermo-rheologically optimized die design in the future.

CONVENTIONAL MANUFACTURING

The conventional manufacturing process employs analytical calculations for the occurring cross-sections based on material properties, mass flow rates and the idealized basic geometry. Those calculations are performed with the representative viscosity and are sufficiently accurate for designing simple flow channel geometries. The results can yield information about pressure loss, wall shear stress and shear rates. Nevertheless, for more complex geometries, the flow situation is assessed with the help of network theory, which originated in the field of electronic engineering [3, 4].

By using modern Computational Fluid Dynamics (CFD) software, the automatic optimization of extrusion dies is gaining importance in their design. The main advantage is that information of the design process is secured and this knowledge is transferred into the software environment, making it quantifiable with the help of quality criteria. The need for an experienced engineer to analyze the results is becoming obsolete, and the outcome does not have to be optimized in an iterative, manual process. Thus, the computer-aided design process leads to substantial time-savings, as indicated in Figure 1.

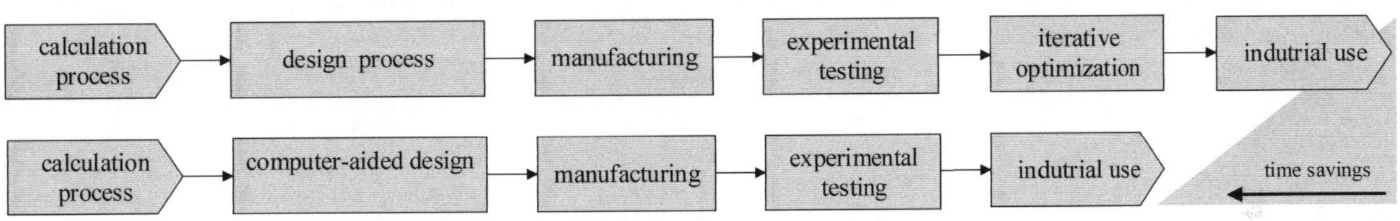

Figure 1. Schematic comparison of the conventional and the computer-aided design process

There are certain approaches to optimize extrusion dies automatically, but here the holistic approach of te Heesen [5] is cited as the basis for future works. It can be described as a coupled approach combining the network theory, generating an initial geometry based on the defined process conditions, and iterative 3D flow simulations, optimizing the geometry with a bandwidth of design parameters. The evaluation is achieved by certain design criteria, which quantify the flow situation [5, 6].

PROCEDURE

The current approach focuses on improving the extrusion die design with regard to melt homogeneity, temperature distribution and pressure consumption. This is to be achieved through a concept, which is not restricted by the limits of conventional manufacturing processes. Figure 2 presents the structure of the approach. The general optimization procedure can be divided into two modules that interact with each other. Since the thematic field of the manufacturing process is covered by the Chair for Manufacturing Technology at the University of Duisburg-Essen, it is only briefly introduced in this paper. The Chair for Engineering Design and Plastic Machinery covers the computer-aided optimization of the die design. The first module deals with customizing the additive manufacturing process to meet the requirements, which are defined for extrusion dies in polymer processing. The chosen procedure is the selective laser melting. In the context of the processing method, the main task is to optimize the surface topology in order to ensure the required product qualities.

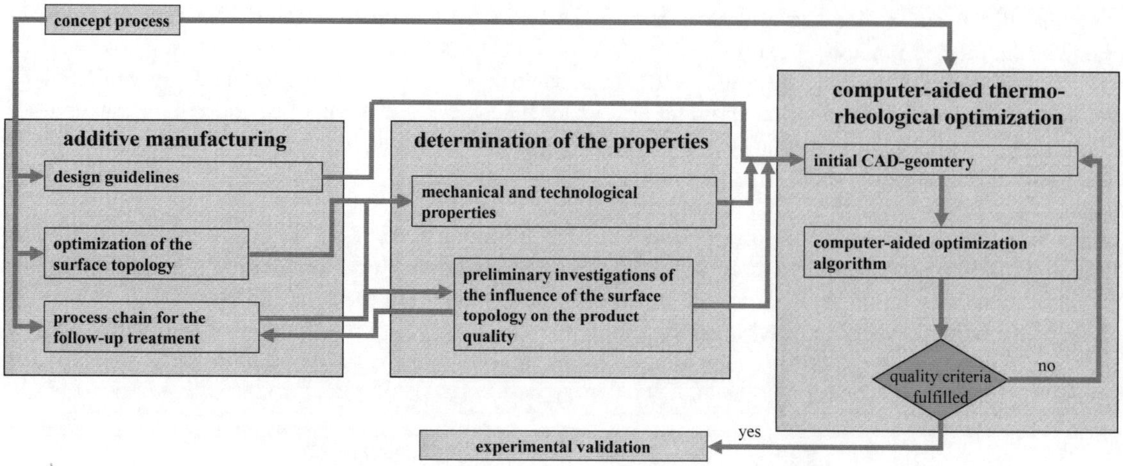

Figure 2. Procedure for the development of the new die design

Plastic products are subject to high optical and haptic requirements that have to be guaranteed. Even so, whereas conventional manufacturing methods achieve a surface roughness of $R_z = 1\ \mu m$, commercial laser melting machines can obtain R_z values of between 20 – 50 μm, depending on the part orientation and the follow-up treatment. Consequently, within the scope of the manufacturing module, the general surface topology has to be optimized. On the one hand, this is achieved by a series of experiments determining the optimal exposure parameters, and on the other hand, by developing novel exposure strategies. Apart from that, an appropriate process chain for the follow-up treatment has to be developed. Challenging for the surface treatment is the fact that the quality-determining surfaces are located inside the die geometry, which makes them inaccessible for most of the follow-up treatments. The present investigations for a suitable process chain comprise mechanical-abrasive and electro-chemical treatments as well as a combination of both. Previous investigations by Szöke [7] on the effect of surface topology of extrusion dies with respect to color and material change processes found that the surface topology is relevant, not the global roughness of the surface.. Examinations showed that supposedly rougher, but undulating surfaces lead to lower changeover times because of the positive flow effects of the microscopic surface topology. The results will be evaluated in preliminary investigations employing test samples in a flow channel.

Finally, design guidelines have to be defined; that will be handed over to the second module. Even though the design process is not subject to restrictions known from the conventional manufacturing process, new restrictions do emerge. These new restrictions are mainly characterized by the accuracy and feasibility of the manufacturing technology. In general, the investigations cover the presentability of geometric elements and determine the manufacturing limits. Table 1 depicts a short overview of the classification of the manufacturing process and the assigned parameters. These will be examined within the production of test samples and the respective results will be transferred to a computer-aided design (CAD) master model to save the knowledge and keep it accessible for the optimization algorithm.

Table 1: Classification of the manufacturing process

Material parameters	Geometrical parameters	Process parameters
-mech.- technological properties	- min. presentability	- layer thickness
-particle size distribution	- crit. overhang angles	- scan strategies
-residual stresses	- scaling factor	- exposure strategies
-...	- ...	- ...

The second module contains the design of the melt-conveying flow channels and the computer-aided thermo-rheological optimization. The global goal is to develop a new die design with flow-optimized flow channels. A blown-film extrusion die has been chosen for the evaluation, because this processing method — due to the biaxial stretching — is well suited to reveal failures in the melt homogeneity.

The basis for the optimization is an initial CAD model that offers a bandwidth of parameters to adjust the geometry for the flow situation. The restrictions for those design parameters are gained from the aforementioned module. In order to define the flow situation, the process has to be classified. The division is analogous to the manufacturing process and is summarized in Table 2. These parameters will form the boundary conditions for the operating point.

Table 2: Classification of the extrusion process

Material parameters	Geometrical parameters	Process parameters
-viscosity	-flow channel cross-sections	- mass flow rate
-spec. volume	-geometric structures	- pressure demand
-thermal conductivity	-dimensioning	- temperature profiles
-...	-...	- ...

The focus of the current approach is the automatized geometry optimization against the background of the manufacturing method. The general procedure (Fig. 3) is derived from the approach of [5] and is adjusted for the new demands. The starting point is an initial design concept that is completely parameterized. The manufacturing limits will be implemented into the CAD design. The quantitative rating of the design is based on quality criteria, which have been introduced in [5] and are adjusted to the present flow situation. Key factors for the quantification are the mass flow distribution and wall shear stress ratios.

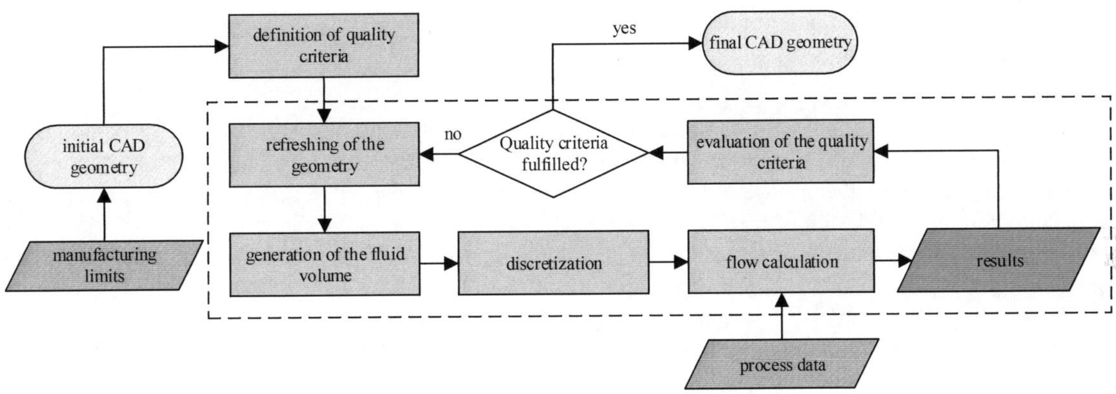

Figure 3. Optimization sequence

The central component is the optimization algorithm to automatically generate the fluid volume and prepare it for the flow calculation. The results will be evaluated without the need for human interference, and the geometry is adjusted until the criteria are fulfilled.

POTENTIALS FOR THE DEVELOPMENT

In a first step, the temperature household of a spiral mandrel die has been investigated. Therefore, simulations have been performed to account for the material heating of the die. Figure 4 shows the results for a conventional extrusion die on the left side. There, an effect becomes noticeable, which is not considered in regular simulations. The inner die heats up significantly in contrast to the housing. This can be attributed to the insulating effect of the melt between inner and outer die. This effect takes place while processing and leads to a temperature gradient of up to 20°C in the 1 mm melt film. Thus, preliminary simulations show the effect of cooling channels within the inner die. The right figure clarifies this. A helical cooling channel with a diameter of 4 mm has been integrated, whereby the cooling fluid is air with an inlet temperature of 25°C. This is already sufficient to cool down the spiral mandrel and balance the temperature gradient in the melt film. Further investigations will integrate the interior die cooling into the novel die design.

Currently, the primary pre-distribution is being reconsidered. The spiral mandrel die is chosen for blown film extrusion because of its good characteristics of avoiding weld lines. Nevertheless, a mandrel support with conical flow surface seems to be the most efficient design to transfer a tube flow into a ring shaped flow. Problematic is merely the support geometry to fix the mandrel. Here, the advantages of additive manufacturing offer opportunities. Complex support structures with, e.g., drop-shaped profiles can be manufactured, and they combine the advantages of the pre-distribution by eliminating the disadvantages. Further flow simulations will validate this concept.

Mass flow rate	200 kg/h
Initial wall temperature	210°C
Material	HDPE

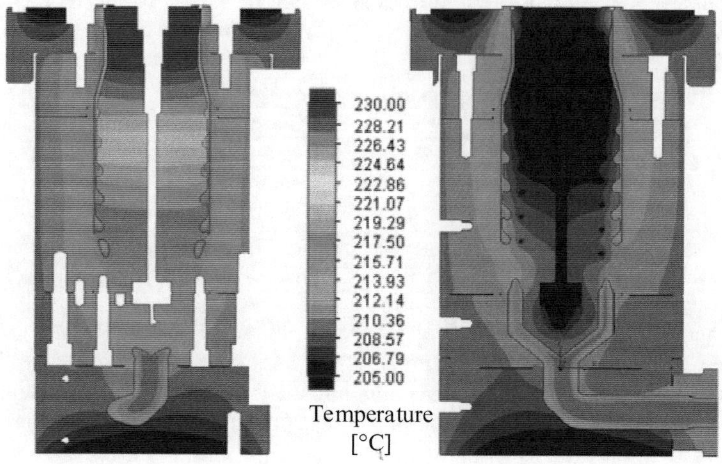

Figure 4. Comparison of CFD results of a spiral mandrel die without (l.) and with inner die cooling (r.)

OUTLOOK

It can be shown that the additive manufacturing of an extrusion die is promising—offering the potential for further optimizations in the processing of blown film plastic products. This new treatment enables a die design, no longer subjected to the restrictions of conventional manufacturing technology. Instead, additive manufacturing allows the design of melt-conveying areas to be completely adjusted to the flow situation. This approach entails an automatic optimization of the design with regard to the process conditions and manufacturing limits.

In a first step, the potential for an improved temperature household could be demonstrated; this will be integrated in the design process in future research. The introduced general optimization procedure will be further developed in the project.

The next step is the final definition of the die design as the initial geometry for the optimization. Therefore, considerations of the ideal primary pre-distribution are made and validated with the help of CFD - simulations. First results show the concept of mandrel support dies with a conical flow surface to be promising, because the bridge geometries can now be adjusted to reduce weld lines.

ACKNOWLEDGMENTS

We would like to thank the German Research Foundation (Deutsche Forschungsgemeinschaft, DFG, Project-No.: WO 302/61-1) for supporting this research project entitled "Thermo-rheological optimization of additive manufactured extrusion dies".

REFERENCES

[1] Groß, H.: *Neuartige Mischdüsen vermindern Schmelzeinhomogenitäten*, Extrusion, 06/2015, S. 36-37.

[2] Hopmann, C., Yesildag, N., Bremen, S.: *Surface quality of profile extrusion dies manufactured by Selective Laser Melting.* RTeJournal – Forum für Rapid Technologie, Vol. 2015.

[3] Michaeli, W.: *Extrusion Dies for Plastics and Rubber*. Carl Hanser Verlag, 2003, ISBN: 978-3-446-22561-9.

[4] Hopmann, C.: Eilbracht, S.: *Towards automatic optimization of flow channel geometries in complex multi-level dies for film extrusion.* Proceedings: EngOpt 2013 - 3rd International Conference on Engineering Optimization, 1.-5. July 2012, Rio de Janeiro.

[5] te Heesen, O.; *Ganzheitlicher Ansatz zur Dimensionierung und Optimierung von Extrusionswerkzeugen am Beispiel von Wendelverteilern.* Dissertation Universität Duisburg-Essen, 2015

[6] te Heesen, O.; Wortberg, J.; et. al: *Automatic Optimization of Spiral Mandrel Dies by a Coupled Approach Using the Methods of Flow Analysis Networks and Three Dimensional CFD-Simulations.* Proceedings of the Polymer Processing Society 31th Annual Meeting (PPS-31), 07.-11. June 2015, Jeju Island, Korea.

[7] Szöke, L.; *Modellbildung von Produktwechselprozessen in Extrusionswerkzeugen.* Dissertation Universität Duisburg-Essen, 2014.

Simulative Evaluation Of The Optimization Potential Of Additively Manufactured Static Mixing Elements For Extrusion

Christian Hopmann[a], Malte Schön[a*], Matthias Theunissen[a], Wilhelm Meiners[b] and Liyaowei Shen[b]

[a] *Institute of Plastics Processing (IKV) at RWTH Aachen University, Seffenter Weg 201, 52074 Aachen, Germany*
[b] *Fraunhofer Institute for Laser Technology (ILT), Steinbachstr. 15, 52074 Aachen, Germany*

Corresponding author: malte.schoen@ikv.rwth-aachen.de

Abstract. : In plastics extrusion, the homogeneity of the melt is an essential pre-requisite for uniform product quality. Static mixing elements for the extrusion of plastics serve to homogenize the melt by splitting and recombining the flow of polymer melt, thus removing e.g. hot streaks in the melt. However, the geometric complexity of static mixers, which typically consist of crossed bars of metal joined together, makes both the design and the manufacture of specialized mixer geometries difficult. Against this background, both simulation of melt flow in static mixers and additive manufacturing of static mixers by means of Selective Laser Melting are investigated. In order to accurately model the flow through the multitude of small gaps between the bars, the open-source CFD software OpenFOAM is employed. While OpenFOAM already supports very high resolution meshes and parallel computing, the interaction between melt temperature and melt viscosity and the effect of viscous heating are added to the simulation model. By means of the new simulation model, thermal mixing occurring during flow of a highly viscous HPDE through a reference static mixer is evaluated numerically. Selective Laser Melting not only allows an economic production of small and very small lots, but also offers design freedoms that have not been available before. This allows the production of static mixer geometries that cannot be manufactured with conventional processes. A framework for the exploitation of this advantage is shown. As an example, the potential thermal benefits of tempering channels located within each bar of the static mixer are explored.

Keywords: Additive manufacturing, Selective Laser Melting, Static Mixers, thermal mixing

INTRODUCTION

Extruded plastics can be found in many important applications in modern society, such as films providing hygienic packaging of both foodstuffs and medical instruments. In the extrusion process, the plastics is molten (typically within in a single screw extruder) and pushed through a die that gives the melt the form required by the product (e.g. thin film). The homogeneity of the product depends heavily on the homogeneity of the melt. If fillers or additives are poorly distributed in the melt, the film might fail to provide the required attributes in its application, and if the temperature of the melt shows local differences, the melt's viscosity is similarly affected. Lower viscosities lead to higher throughputs in a corresponding local section of the die's outlet, resulting in a thinner film [1]. For this reason, static mixers can be employed to homogenize the melt both thermally and materially. These static mixers consist of multiple sets of bars, arranged crosswise. As the melt is split up, moved around and recombined numerous times, the melt's properties across a given cross section level out. However, the mixers cause a considerable pressure loss which in turn reduces the productivity of a given extrusion line [2]. Multiple avenues exist to reduce pressure loss in polymer-bearing equipment, the premier ones being an optimization of the flow channel geometry and local changes of the melt temperature [3]. Geometry optimization can be achieved by a series of practical experiments or - more economically - by means of Computational Fluid Dynamics (CFD) simulations. Using CFD, multiple geometries can be compared without the expenses associated with manufacturing each geometry and running a battery of tests on them. Optimization using CFD tends to generate the best results when as few restrictions as possible are placed on the geometries [4]. On the other hand, local changes to the melt temperature are frequently employed, primarily in injection moulding, where variothermal processes dynamically change the mould's surface temperature, achieving a reduction in moulding pressure. Multiple technologies are available for this kind of temperature manipulation, but most of them are not suitable for extrusion processes in which no opportunity to affect the walls of the plastics-filled die or mould from the 'inside presents itself. Conformal tempering channels, however, could be positioned in an extrusion die just as well as in a mould. The large contact area of a static mixer and the polymer processed and the fact that the mixer sits right in the middle of the material stream make changes to melt temperature feasible in extrusion in spite of the polymer's poor thermal conductivity [5]. However, the high geometric complexity of conformal tempering channels poses a difficult task for conventional manufacturing processes. Additionally, conventional processes typically used to produce static mixers, such as milling, stamping and joining or casting, decrease the degree of design freedom. A high degree of design freedom is necessary to achieve gains when optimizing the flow channel with CFD. For this reason, an additive manufacturing (AM) process, which manufactures the mixer without the use of a shape-giving form for casting or any assembly or post-processing, is employed [6]. One main advantage of this approach to the manufacture of static mixers is the greatly increased design freedom: The component bars of the static mixer no longer must be be designed in a way that allows either assembly and joining or casting and tempering channels can be positioned freely. One promising AM process is Selective Laser Melting (SLM). This process produces metal parts with mechanical properties almost identical to those of conventional metal processing techniques and needs neither tooling specific to a certain part nor a secondary sintering treatment [7]. Against this background, the requirements and opportunities offered by using SLM to produce static mixers are investigated. This article focuses on two facets of the investigations: Firstly, creating a system of construction that enables a quick design of static mixers geometries while considering both the geometric complexity and the suitability for additive manufacturing and secondly, setting up a simulation environment providing flow information for the evaluation of the quality of the mixing.

SELECTIVE LASER MELTING

Selective Laser Melting (SLM) has been developed by the Fraunhofer Institute for Laser Technology (ILT) as a technology for additive manufacturing. It enables the layer-by-layer manufacture of metal components based on a 3D model. After breaking down the 3D model provided as an input into layers (slicing), the vectors of the laser are calculated and the actual production process begins. Three repeating steps can be identified (Figure 1): First, powder material is deposited on the substrate of the building platform as a thin layer by the help of a brush or a reservoir containing the powder. Second, the laser beam rasters the build area, irradiating the regions in which solid material is to be contained. The powder melts and resolidifies. Finally, the building platform is lowered by the thickness of a single layer and the SLM machine is ready to begin anew with the first step. In comparison to other AM processes for metal parts, SLM produces almost identical mechanical properties and densities compared to conventionally produced parts. This is achieved by completely melting the powder's particles [7].

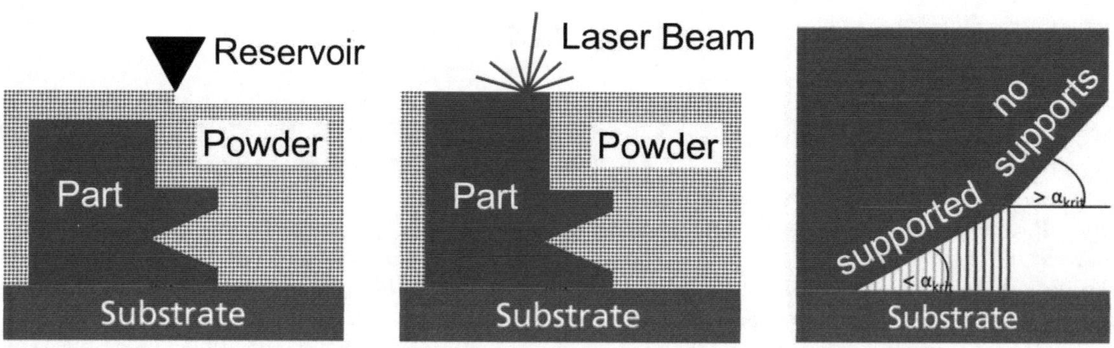

FIGURE 1. (a) first and (b) second second step of the SLM process, (c) Angle α indicating where supports are needed for a functioning SLM build-up

REQUIREMENTS OF THE SLM PROCESS

As the SLM process builds up a part layer by layer, two superimposed layers are joined by a quasi-weld as the solid metal particles turns into a liquid melt. Whether this molten and now solidifying bead attaches to the rest of the structure is dependant on two factors: how far the molten metal has been solidified and how much solid material is available below the bead. In a worst-case scenario, the bead is surrounded by powder in all directions, leaving no solid path on which heat can be dissipated and no point of attachment. Consequently, an area composed of multiple 'isolated' beads will overheat, attaching to powder that was not supposed to be molten and leave a discoloured surface. Also, such a bead is bound to be dislocated once the powder-carrying reservoir is moved across it. For this reason, it is necessary to design the volume to be built up in a way that ensures a good connection of the beads to an existing body of solid material during the entirety of the build process. One of the ways to achieve this goal is to provide small disposable support structures than are removed once manufacturing is complete. However, having to both reserve space for these structures and a path to access and remove them, restricts the freedom of design. Whether supports are necessary for certain geometry depends on the parameters of the material and the process. Considering a sloping overhang (Figure 1 (c)), the connection between bead and solid is the stronger the steeper the overhang slopes [8]. The angle α between overhang and build plane is used to characterise this. For the given process, in which a powder of 1.4404 steel is processed on a SLM 280 Selective Laser Melting machine (SLM Solutions Group AG, Luebeck, Germany) build-up tests have revealed that angle of up to 45° can be realised without having to resort to supports. This result must obviously inform the design of the static mixers.

CONSTRUCTION SYSTEMATIC FOR STATIC MIXERS

For reference, a static mixer with a diameter D of 41 mm, of the type SMB plus 40/3 (Promix Solutions Gmbh, Wetzlar, Germany) is analysed. A reconstructed CAD 3D model of the mixer shows that over 60 dimensions (excluding chamfers and radii) are necessary to capture all details of its geometry. This situation it difficult to rapidly generate changed static mixers for an iterative optimization using computational fluid dynamics. For this reason, the new 3D model of the mixer is not composed from individual bars, but rather from the mixer's sub-elements. The given mixer essentially consists of three building blocks, offset from each other by a 90 degree rotation around the extrusion axis. A pattern of repetition can be observed in each building block. This pattern consists of three-dimensional elements shaped like an H on one plane and shaped like a cross on the plane orthogonal to it (Figure 2, (a)). For this reason, the basic building block of the static mixers will be called 'H-cross'. A three-dimensional grid of these 'H-crosses' forms the framework from which a cylindrical part can be cut away. This part already mirrors the structure of static mixers (Figure 2, (b)). As explained earlier, the build-up phase of the SLM process requires structures to either support themselves or to be propped up with dedicated supports. Seeing that these supports diminish the design space, the H-cross is optimised to be self-supporting. The new shape is shown on the right in Figure 2 (c). A new mixer block is constructed from these new shapes (Figure 2 (d)).

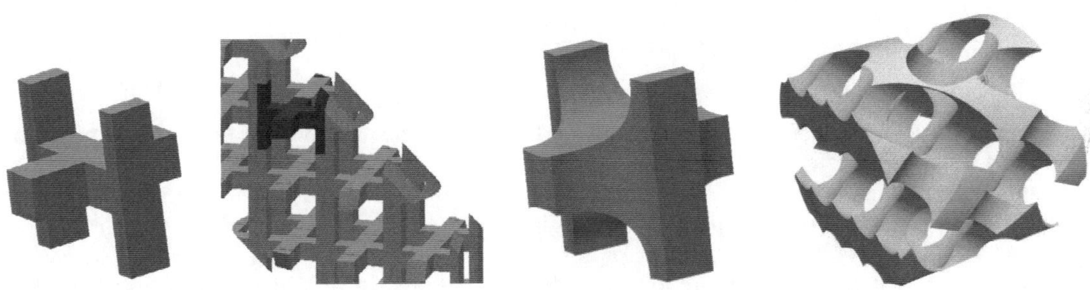

FIGURE 2. (a) basic repetition unit of the mixer, (b) mixer composed of these elements, (c) 'H-cross'optimized to support itself, (d) mixer element composed of optimised 'H-Crosses'

SIMULATING FLOW IN STATIC MIXERS WITH OPENFOAM

Broadly speaking, realistic simulation results can only be achieved when the modelling accounts for all relevant phenomena occurring in the static mixer (such as shear thinning of the fluid and viscous dissipation) and the mesh resolution is high enough to capture the correct shape of the temperature, velocity and pressure fields. The reference mixer's complex geometry splits the flow into 12 segments during the first pass of the melt through the bars alone, requiring 144 times as many cells as before. The open-source software package OpenFOAM (OpenFOAM Foundation Ltd., London, United Kingdom) is employed as it handles meshes composed of many cells and parallel processing well.

Out of the numerous flow models contained within OpenFOAM, the finite volume method solver *simpleFoam* is identified as suitable for the given extrusion problem, since the this solver captures both the incompressible and steady-state nature of flows in extrusion. However, *simpleFoam* does not model heat transfer by default and the models for shear-thinning fluids available do not account for the temperature dependence of the viscosities. Having access to the source code and taking into account the modular design of OpenFOAM, these features can be added easily. Consequently, a new solver is implemented, based on [9]. This solver solves the differential equation of temperature advection and diffusion with viscous heating (also known as shear heating) as a source term in addition to the equations for the conservation of mass and impulse. The Carreau model of viscosity combined with a temperature-based viscosity shift according to Williams, Landel and Ferry is implemented in OpenFOAM. For pipe flow, these models are compared to analytical solutions based on the concept of representative viscosity and found to be in good accordance, with average differences around 0.2 % (Figure 3 (a)).

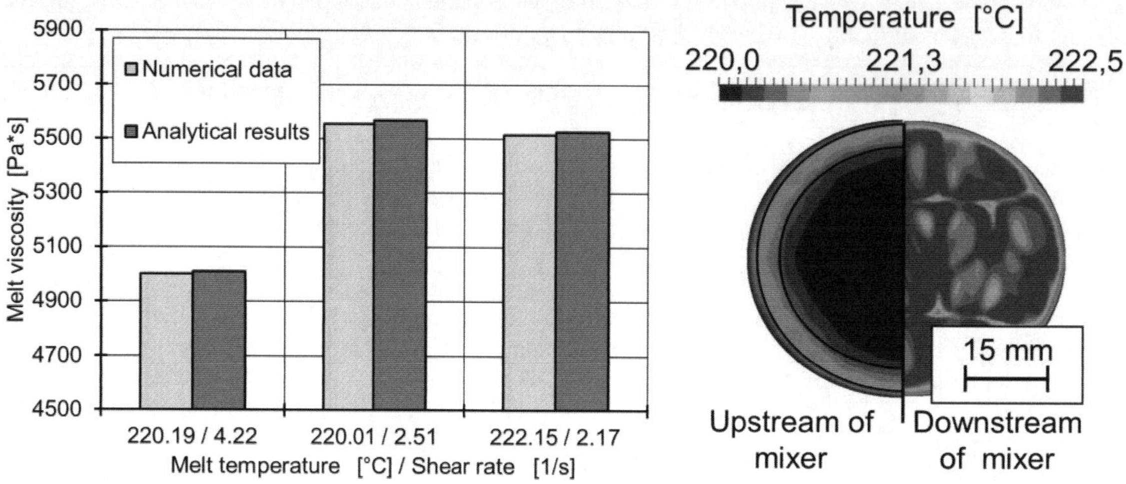

FIGURE 3. (a) Comparison of numerical and analytical results for melt viscosity at typical temperatures and shear rates, (b) Temperature field ahead and after the reference mixer (HDPE, 220° C, 100 kg/h)

SIMULATION RESULTS

Using the newly developed models, laminar, non-isothermal, incompressible, steady state flow of polymer through the reference static mixer is investigated. The material considered is a highly viscous HPDE (Hostalen 9550F, Basell Polyolefine GmbH, Frankfurt am Main, Germany) at a throughput of 100 kg/h. The process temperature considered is 220 °C. Initially, the temperature of the mixers's bars is left to be determined numerically (Neumann boundary condition). The mesh encompasses two sections of pipe flow (Diameter 41 mm, length 123 mm) located up- and downstream of the mixer. These sections ensure that flow through the mixer is not distorted by effects such as entry length. Independence of results and mesh resolution can be shown for a mesh with 2.9 million cells. As an initial thermal inhomogeneity, the effect of viscous heating in the upstream pipe flow is considered. As figure 3 (b) shows, the melt roughly 1 mm away from the outer diameter (indicated by the annulus-shaped marker) the inlet of the mixer is overheated by roughly 0.7 °C, while the melt leaving the mixer has an average temperature of 221.7 °C. This indicates that shear heating in the mixer is stronger than the shear heating causing the actual thermal inhomogeneity. Thermal homogeneity $e_{thermal}$ of the melt in the two cross sections is calculated with following formulae:

$$e_{thermal} = \frac{\delta_{inlet} - \delta_{outlet}}{\delta_{inlet}} \quad \text{wherein}$$

$$\delta_{thermal} = \sum | T_i - T_{average} |$$

T_i being the temperature at a given node point of the mesh and $T_{average}$ the temperature averaged across all nodes of either the mixer's inlet or outlet. Consequently, high values of δ mean that melt temperatures diverge strongly from the average.An $e_{thermal}$ of 1 would indicate no thermal deviation at the outlet, which in turn represents perfect thermal mixing. Additionally, an $e_{thermal}$ of -1 indicates that the thermal inhomogeneity at the outlet is twice the thermal inhomogeneity at the inlet. As described above, shear heating caused by the mixer is greater than the shear heating that was supposed to be 'homogenized away'. Accordingly, $e_{thermal}$ takes a value of -2.55 for the given scenario, clearly showing negative thermal mixing. As there is the option of conformal tempering channels when using additively manufactured static mixers, it is investigated whether varying the temperature of the surface of the mixer's bars can improve this situation. In two follow-up simulations, the temperature of the mixer bars is set to 218 °C and 222 °C respectively. For heated bars, the pressure loss increases by 5.51 bar, while thermal homogeneity drops to -2.58. For cooled bars, the pressure loss also increases by 5.72 bar. However, thermal homogeneity improves to a value of -1.42. The increase in pressure loss in both scenarios involving tempered mixers can be explained by that fact that the temperatures present at the mixer's bars are higher than 222 °C if not controlled by tempering. Accordingly, tempering the bar surface to 222 °C or even 218 °C causes the local viscosities - and with them the pressure loss - to increase. A surface temperature of 218 °C, however, helps to combat viscous dissipation and therefore curbs negative thermal mixing.

CONCLUSION

Static mixers play an important role in the thermal and material homogenization of polymer melts during extrusion, since this homogeneity directly affect product quality. In order to reduce the pressure loss of static mixers, conformal tempering channels and CFD-based geometry optimisation are considered. Both approaches require the mixers to be manufactured additively using Selective Laser Melting (SLM) in order to either integrate the channels or provide a sufficient degree of design freedom. In order to streamline the creation of 3D models of the highly complicated mixer geometry and to accommodate the SLM process' requirements, a method for composing the geometry from very basic repetition units is introduced. Using this method a mixer geometry that is self-supporting and can be built up using SLM without the need for disposable supports is created. The open-source software OpenFOAM is enhanced in order to capture non-isothermal incompressible flows and temperature-dependent viscosities. In a test case featuring a mild thermal inhomogeneity at the mixer's inlet, viscous dissipation inside the mixer is found to cause an even greater thermal inhomogeneity, leading to a negative thermal mixing performance. Reducing the surface temperature of the mixer bars slightly increases pressure loss, but pushes thermal mixing towards a positive performance. In future works, other combinations of local temperatures and mixer geometries will be considered.

ACKNOWLEDGMENTS

The research project 18712N of the Forschungsvereinigung Kunststoffverarbeitung depicted in this article is sponsored as part of the "Industrielle Gemeinschaftsforschung und -entwicklung (IGF)" by the German Bundesministerium für Wirtschaft und Energie (BMWi) due to an enactment of the German Bundestag through the AiF. We would like to extend our thanks to all organizations mentioned

REFERENCES

1. W. Michaeli, Extrusion Dies for Plastics and Rubber, Munich, Vienna: Carl Hanser Verlag, 2016.
2. W. Tauscher, G. Schneider, Static mixing device for viscous melts. Patent US 4692030 A, 1987.
3. H.T. Kim, J.P. Darby, G.F. Wilson, *Polymer Engineering and Science* **13**: 372 - 381, 2004.
4. I. Szarvasy, J. Sienz, J. Pittman and E. Hinton, *Int Polym Process* **1**, 28-39, 2000.
5. C. Hopmann, N. Lammert, Y. Zhang, *Cellular Polymers* **36**, 151 - 166, 2017.
6. J.H. Schleifenbaum, "Verfahren und Maschine zur individualisierten Produktion mit High Power Selective Laser Melting", Ph.D. Thesis, RWTH Aachen University, 2012.
7. W. Meiners, "Selective Laser Melting – Additive Manufacturing for series production of the future?", Intermat 2011, Luxembourg, 2011.
8. F. Calignano, *Materials and Designs* **64**, 203 - 213, 2015.
9. M. Becker, *viscousHeatingSolver - Adding temperature transport and viscous heating to simpleFoam.* http://www.dhcae-tools.com/docs/visousHeatingSolver.pdf, 2012.

Highly endurable hydrostatic pressure polyethylene pipe prepared by rotation extrusion

Pi Lin (1)*, Nie Min (1), Wang Qi (1)

(1) Sichuan University - Sichuan - China, (2) Sichuan University - Sichuan - China

A small amount of cross-linked polyethylene (XPE) was added in polyethylene (PE) matrix and then was processed into the pipe by self-designed rotation extrusion system. The experimental results showed that during rotation extrusion of the PE pipe, the polymer melt helically moved forward in an off-axis direction to induce molecular orientation and shish-kebab alignment deviate from the axial direction. However, due to sufficient relaxation of oriented molecules caused by high temperature, numerous spherulites arose in the pure PE pipe. With the addition of XPE, the flow character of PE matrix was tailored so the relaxation was suppressed and the flow effect on the formation of shish kebab was amplified, bringing about denser shish kebabs off the axial direction in the PE pipe. As a result, the PE pipe with excellent endurance to hydrostatic pressure was prepared. The hydrostatic pressure endurance limit of PE pipe was up to 11.7MPa, which was 130% of the standard one.

Modelling the Contamination Behaviour of Polymer Melt Filters

Volker Schöppner, Peter Meilwes

Kunststofftechnik Paderborn, Paderborn University, Warburger Str. 100, 33098 Paderborn, Germany
volker.schoeppner@ktp.upb.de; peter.meilwes@ktp.upb.de

Abstract. In the field of polymer processing, extrusion is one of the most common processing methods. Not only in the processing of recycled materials, but also when using virgin polymers, contaminations can occur during the storage or processing of the material. This may adversely affect the melt quality and by this also the quality of the final extrusion product. Examples of possible contaminations are metal particles which are caused by wear and tear of the extruder, or degradation products of the processed material.

As a result of the increasing demands on the quality of extrusion products, especially in the field of fiber and film extrusion, the filtration of polymer melts is widely used and state of the art. To remove unwanted debris out of the melt, different metal filter media are used. Some examples of typical filter media are different kinds of wire mesh, filters made of nonwoven metal fibers or sintered metal powder discs. By using this filter media, it is possible to remove foreign particles like solid particles as well as soft components so called "gels" from the melt. This ensures a high quality extrusion result.

In order to calculate the lifetime of a filter in an extrusion process, a new possibility of modelling the filter contamination is presented in this work as a part of the PET processing. At first, a method for reproducible rapid contamination of filter materials was developed in pre-investigations. After this, a high number of experimental investigations have been done in order to generate a wide data basis for the modelling of the filter contamination. This soiling behavior is then mathematically modelled, in that manner that the filter lifetime can be calculated.

Keywords: Polymer Melt Filtration, Extrusion
PACS: 47.56.+r; 81.20HY; 83.80Sg

INTRODUCTION

As a result of increasing quality requirements on extrusion products, especially in the field of fiber and film extrusion, filtration of polymer melts is widely used today. An example of the need of a melt filtration is the fiber extrusion of polyethylene terephthalate (PET). Even the smallest contaminations of the melt in the micron range can lead to a weakening of the fiber strand [1].

During extrusion processes, there are multiple possibilities of contaminating the polymer material. On one hand the contamination can be brought in by the raw material itself. During storage or transportation of the material there are lots of chances of contaminations. Also the process itself can contaminate the polymer melt. Examples of occurring contaminations are metal particles which are caused by wear and tear of the extruder, or degradation products of the processed material. Degradation processes in the extrusion process can cause the formation of so called gel particles in the melt. In extrusion products, for example films, gels can result in so called fish-eyes or discolorations of the material. Basically there are two kinds of contaminations. On one hand the contamination can consist of inorganic particles like metal abrasion. This contamination leads to an accretion of particles in the filtration media. Also in addition, there may be a structure of a filter cake on the filtration media. On the other hand the contamination can consist of organic gels. These can consist of the processed polymer itself or different polymers with a different viscosity. These "soft" contaminations lead to an effective clogging of the filter media [1, 3].

In order to model the contamination behavior of polymer melt filters it is necessary to develop a method for a reproducible rapid contamination of filter media. In industrial extrusion processes, the filter lifetime lasts several days up to weeks depending on the degree of contamination. However, in a laboratory investigation, this contamination should ideally be done in a few minutes or hours. This rapid test contamination was already developed and presented in further works [2].

The designed test contamination consists of 30 % calcium carbonate powder and 70 % crosslinking polyethylene powder. This enables the simulation of gels and also solid Contaminations. This allows a realistically fast-contamination of the polymer melt. The defined contamination is absolutely necessary for the laboratory investigations in order to ensure a constant, defined load on the melt with foreign particles for the modeling. The contamination must also guarantee the reduction of the filter lifetime to a level that can be realized in the laboratory. In industrial plants the lifetime is several hours to weeks depending on the degree of contamination in the polymer melt. In laboratory investigations, however, this soiling must be carried out in just a few minutes. Based on the compilation of a test contamination, experimental investigations have been carried out on the contamination behavior of industrially relevant filter discs. In this work, the contamination behavior of nonwoven polymer melt filters are supposed to be modelled, to calculate the lifetime of filter media in a process with a defined material contamination.

BRINGING IN A DEFINED MELT-CONTAMINATION

The experimental investigations have been carried out on an extruder D45 33 L/D using PET. A filter housing for 12" filter discs has been used. Three filter discs have been installed in the process for each extrusion cycle. The tests have been carried out on nonwoven filter discs with a filter fineness of 20 µm, 10 µm and 5 µm, which are also often used in real industrial processes. To decouple pressure increases in the extruder due to the increasing filter contamination, a melt pump has been installed between the extruder and the filter housing. Because of this, the extruder was able to work on a stationary operating point. The increasing pressure at the filter did not show any influence on the throughput behavior of the extruder. This is due to the melt pump.
The entire assembly can be rinsed by a bypass directly in front of the filter housing. This enables to guarantee the highest possible reproducibility of the test results because the soiling process can always be started at a defined operating point. The throughput was continuously set at 50 kg/h in the experimental investigations, in order to achieve a realistic flow rate in the filter. The PET melt thereby was contaminated with 1 % of the contamination of calcium carbonate and crosslinking polyethylene powder. Immediately in front and behind the filter housing, the melt pressure has been measured and recorded. This allowed the detection of the increasing pressure during the contamination. The process was operated until the pressure drop above the filter medium was 100 bar above the initial pressure drop. This is shown in Fig. 1.

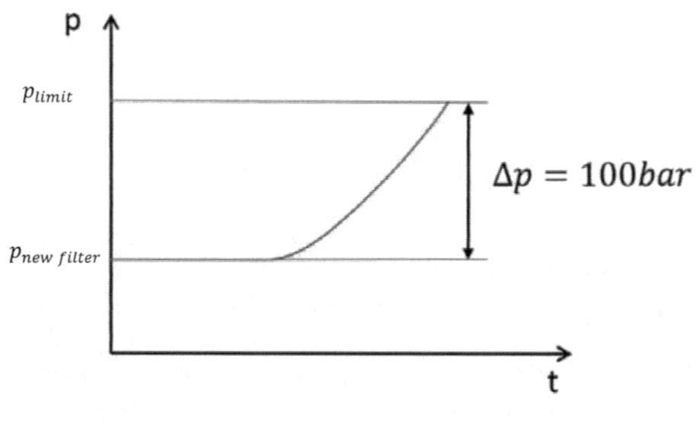

Figure 1. Forced pressure loss at the filtration media

100 bar is a typical limit value where a filter has to be replaced. Otherwise the structure of the filter discs could be damaged. To examinate the influence of the filter fineness, investigations with the same amount of dirt have been carried out with different filter fineness.

PRESSURE INCREASE BEHAVIOUR OF NONWOVEN FILTER DISCS

As already shown in the validation tests of the test contamination, a reproducible soiling behavior of the filter discs [2] is shown in Fig. 2.

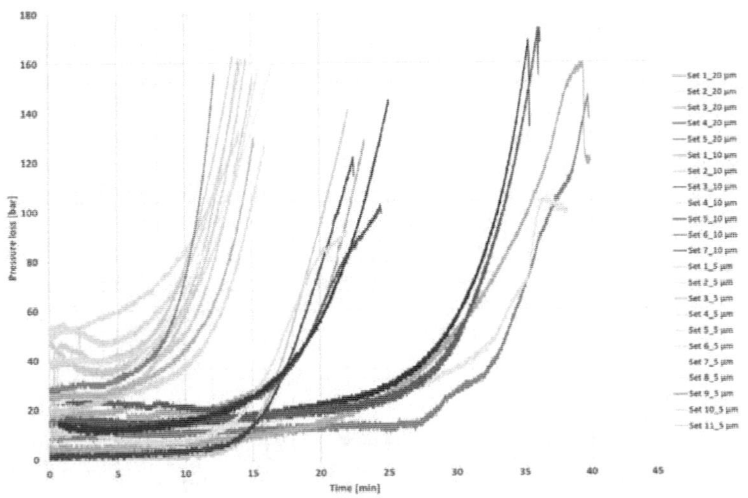

Figure 2. Experimental results for the pressure drop of the filtration media

Figure 2 shows that the pressure always increases with nearly the same profile. The curves with a filter fineness of 10 µm and 5 µm show a similar characteristic.

The finer the filter medium, the shorter the soiling times. This is attributed to inter alia of the constantly maintained dirt size. In this case, the tendency is that a halving of the filter fineness from, for example, 20 µm to 10 µm with the same soiling approximately halves the soiling time. A similar behavior can also be observed when viewing the curves for a filter fineness of 10 µm and 5 µm.

CONSIDERATION OF THE PRESSURE DROP OF THE UNLOADED FILTER HOUSING

For the modeling, the pressure profiles must still be cleaned by the pressure loss of the filter housing. This guarantees that only the pressure loss on the filter element is modeled. Due to the complex geometry of the filter housing, CFD simulations have been used. This simulations have been carried out using the software ANSYS-Fluent in order to determine the pressure loss. Here a pressure loss of about 3.2 bar took place.

This is subtracted from the initial pressure loss of the uncontaminated filter disc in order to allow a correct modeling. The simulation results are shown in Fig.4. A mathematical modeling of the soiling behavior can now be carried out with the pressure increase along the soiling time. Fig. 3 shows the simulation results for the filter housing.

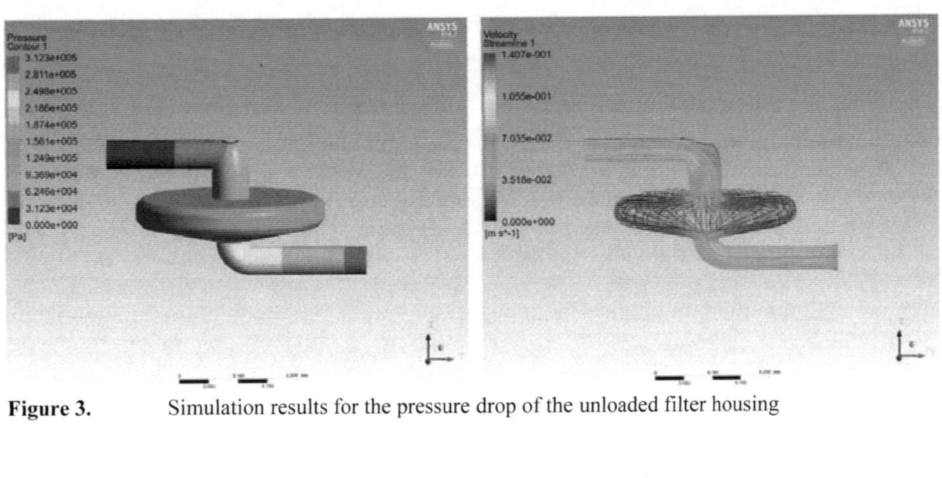

Figure 3. Simulation results for the pressure drop of the unloaded filter housing

PREVIOUS CALCULATION POSSIBILITIES

Currently, there are quite a few approaches to calculate the pressure increase on a filter medium. Some of them are presented here. The simplest way to calculate a filter contamination is to carry out the pressure loss calculation of the filter medium several times with a constant increasing flow resistance. The increasing flow resistance has to be proportionally to the previously flowed melt stream. This corresponds to a constant pollution of the volumetric flow [4]. However, this does not take into account the type and size of the contamination. In [5], the increasing pressure on the filter medium is calculated according to the Darcy law. Here, also an increasing flow resistance is used that is due to a filter cake which builds up at the filtration media.

An approach to include the particle size into the calculation of the increasing pressure drop of the filtration media is presented in [6]. In this model, only sieve filtration and a filter cake are used. In this modeling, the number of pores on the surface of the filter medium is calculated. With a given particle size distribution and throughput, it is then possible to determine how much material is needed to close these pores. All described models so far have the limitation, that they are designed for newtonian fluids. In [7] and [8] pressure increases are simulated in similar form by contamination.

Beside the restriction of newtonian flow behavior, all these calculation methods have the restriction, that there is no easy way to use process characteristics for the pressure calculation. This should be changed by the new modeling

MATHEMATICAL MODELLING OF THE FILTER CONTAMINATION

The increase in the pressure due to contamination is recorded in the experimental investigations and can be described by an exponential function. The goal of the modeling is the mathematical description of contamination behavior in order to calculate the filter lifetime. For this purpose the measured curves for each filter fineness is combined to form a mean value curve (Fig. 4). This can then be described mathematically.

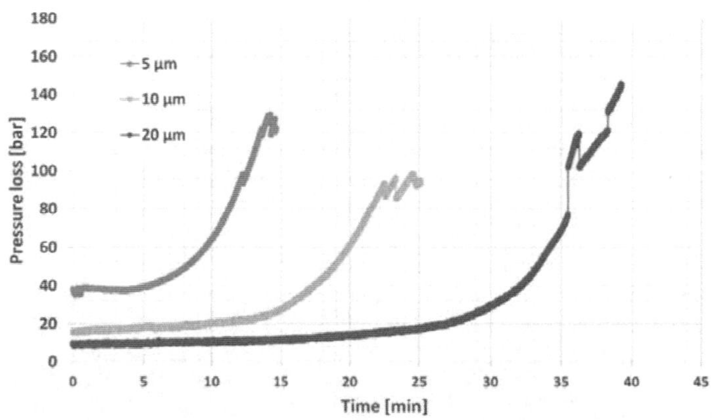

Figure 4. Mean value curves for the different filter fineness's

The modeling contains some characteristic values. The pressure loss of the uncontaminated filter is a variable value in order to be able to consider different filter fines with different pressure losses. In addition, a limiting pressure has to be specified in which the filter is considered to be completely soiled. In order to calculate the filter lifetime, the time for reaching the limit pressure has also to be an influencing variable. For the curves shown in Fig. 4, the filter contamination behavior can be calculated with the equation

$$p(t) = p_{Start} + p_{limit} * e^{\left(\frac{t_{End}-t}{t_{End}-t_{increase}} * \alpha\right)} \quad \alpha = -3{,}08 \qquad \text{Eq. 1}$$

The pressure increase behavior is parameterized by the following variables: p_{Start} represents the initial pressure loss of an unpolluted filter. p_{limit} is the maximum permissible pressure loss which is to be regarded as the time of complete clogging of the filtration media. t_{End} is the time of reaching this maximum pressure. $t_{increase}$ defines the time from which the filter experiences an increase in pressure drop. By applying this function to the pressure loss profile for 10µm, the curve shown in Fig. 5 is obtained.

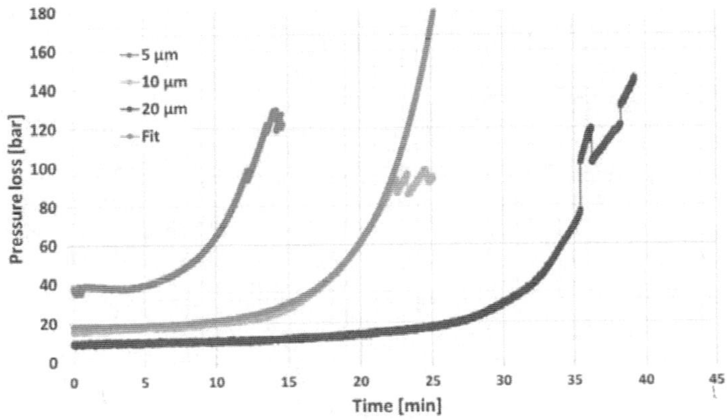

Figure 5. Mean value curves for the different filter fineness's with curve fitting

It becomes clear that the modeled exponential function provides a good description of the contamination behavior. There is also a good agreement with the other curve fittings. In order to make the function easier to handle, the variables t_{End} and $t_{increase}$ should be made describable by predetermined influencing factors, ideally the filter fineness or the dirt particle size. By this, the pressure loss could be made predictable on the basis of known parameters independently of preliminary tests. From the pressure curves it can be assumed that there is be a linear relationship. If the respective variable is plotted as a function of the ratio of filter fineness to dirt particle size, a linear behavior is actually the result.

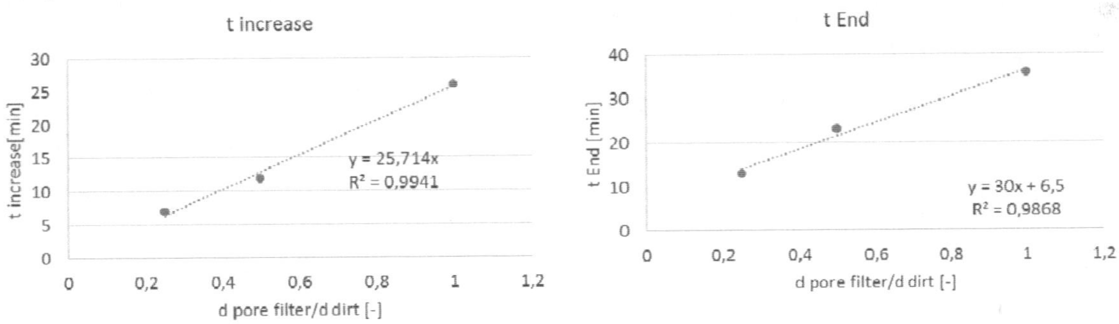

Figure 6. Modelling of t_{End} and $t_{increase}$

The values t_{End} *and* $t_{increase}$ can be described by the following equations:

$$t_{increase} = 1542{,}84\ s * \frac{d_{Pore\ Filter}}{d_{dirt}} \qquad \text{Eq. 2}$$

$$t_{End} = 1800\ s * \frac{d_{Pore\ Filter}}{d_{dirt}} + 390\ s \qquad \text{Eq. 3}$$

$d_{Pore\ Filter}$ describes the filter fineness, d_{dirt} describes the average diameter of a dirt particle.

When equations 2 and 3 are used in equation 1, there is a formal relation which allows to calculate the pressure increase at the filter medium as a function of the starting pressure, the filter fineness, the dirt particle size and the permissible limit pressure. The formal relation is then:

$$p(t) = p_{Start} + p_{limit} * e^{\left(\frac{(1800\ s*\frac{d_{Pore\ Filter}}{d_{dirt}}+390\ s)-t}{\left(1800\ s*\frac{d_{Pore\ Filter}}{d_{dirt}}+390\ s\right)-(1542{,}84\ s*\frac{d_{Pore\ Filter}}{d_{dirt}})}*\alpha\right)} \qquad \text{Eq.4}$$

With this equation it is now possible to calculate the pressure increase within a constant melt contamination. When the equation is transformed to the time t, the filter lifetime can be calculated to:

$$t = -\frac{257*\ln\left(\frac{p(t)}{p_{limit}}\right)*\frac{d_{Pore\,Filter}}{d_{dirt}}-1800*\alpha*\frac{d_{Pore\,Filter}}{d_{dirt}}+390*\ln\left(\frac{p(t)}{p_{limit}}\right)-390*\alpha}{\alpha} \qquad \text{Eq. 5}$$

Filter discs with a filter fineness of 20 µm have been used. The dirt particles had an average diameter of 20 µm. According to the manufacturer, the filter discs are to be exposed to a maximum pressure increase of 120 bar due to soiling, in order to avoid damages of the filtration media. If, for example, the time at which a pressure increase of 100 bar is reached has to be calculated, the following parameters are obtained for the calculated process:

$$\alpha = -3{,}08 \qquad d_{Pore\,Filter} = 20\mu\text{m} \qquad d_{dirt} = 20\mu m$$

$$p(t) = 100bar \quad p_{limit} = 120bar$$

If these parameters are used in the equation 5, a filter lifetime of 2151 seconds (35.9 minutes) is obtained until a pressure difference of 100 bar is reached at the filter medium. This corresponds to the values that have been determined experimentally. With this result it is possible to calculate the filter lifetime up to any pressure difference.

CONCLUSION AND OUTLOOK

With the modeling presented in this work, a first step towards the calculation of the contamination behavior of polymer melt filters is made. In order to identify influencing factors further investigations have to be carried out. In this way, the model presented can, for example, be expanded by the influence of the dirt concentration in the polymer melt. The pressure loss of uncontaminated filter elements are an important input factor. This will be further investigated by using CFD flow simulations. Furthermore, the existing calculation models, such as Darcy's law should be validated for their applicability to the polymer melt filtration.

ACKNOWLEDGEMENTS

The research project (KF2363833SL3) described here has been sponsored as part of the "Zentrale Innovationsprogramm Mittelstand (ZIM)" by the German Bundesministerium für Wirtschaft und Energie (BMWi). The research Project is conducted in cooperation with "SCHWING Fluidtechnik GmbH".

REFERENCES

1. VDI Verlag, *Filtrieren von Kunststoffschmelze,* Düsseldorf: VDI-Verlag 1981, pp. 1-6, 57-58, 129-133, 141-146
2. Schöppner, V.; Meilwes, P.; *Realitätsnah und definiert verschmutzen*, Zeitschrift Kunststoffe 08-2016; Carl Hanser Verlag, München 2016
3. J. Scheirs; *Compositional and Failure Analysis of Polymers – A Practical Approach*, Chichester: John Wiley & Sons Ltd, 2000, pp. 270-273, 275-278, 288-297
4. Schlottmann, R.; Potente, H.; *Berechnungsverfahren für Großflächenschmelzefilter*, Zeitschrift Kunststofftechnik 4(2008)2; Carl Hanser Verlag, München 2008
5. Koch, M.; Kramer, G.; *Permeability distribution for filter media characterization in solid liquid filtration*, Powder Technologie, Elsevier; 2014
6. Meireles, M.; Prat, M.; Estachy, G.; *Analytical modeling of steady-state filtration process in an automatic self-cleaning filter,* Chemical Engineering Research and Design 100(2015) S. 15-26; Elsevier; 2015
7. Steffens, J.; Coury, J.R.; *Collection efficiency of fibre filters operating on the removal of nano-sized aerosol particles II. Heterogeneous fibers*, Separation and Purification Technologie 58(2007) S. 106-112; Elsevier; 2007
8. Giuliani, J.; *Particle arrestance modeling within fibrous porous media,* Journal of fluids Engineering Vol 121, 1 S. 155-162; ASME New York; 1999

STUDY OF BULK AND CONFINED CRYSTALLIZATION OF ETHYLENE-VINYL ALCOHOL

FREYMOND Clément (1)*, GUINAULT Alain (1), ROLLAND Sébastien (1), GERVAIS Matthieu (1), SOLLOGOUB Cyrille (1)

(1) Laboratoire PIMM, Arts et Métiers ParisTech-CNRS-CNAM - PARIS - France

Ethylene-vinyl alcohol (EVOH) copolymer is commonly combined in the food packaging industry with polyolefins because it offers very good barrier properties to gas and organic solvents. These properties come from the crystalline phase (degree of crystallinity and morphology) depending on copolymer composition and crystallization conditions [1]. It is well known that confining a polymer at the nanoscale modifies crystallization habit, possibly leading to a great variety of crystalline morphologies exhibiting unique preferential orientations [2]. The aim of this study is to investigate EVOH bulk crystallization phenomenon and its modification when confining the polymer in nanoscale space. After having investigated the bulk crystallization of EVOH, two techniques have been used to confine the polymer: the classical spin coating [3] and the forced-assembly coextrusion [4]. Both techniques can induce spatial confinement in a different way. In the first technique, the film is spin-coated on a substrate, while in the second, EVOH is coextruded with another polymer chosen to impose hard confinement. Different techniques (AFM, X-ray diffraction) have been used to characterize the crystalline morphology in the bulk and lamellas orientation in thin layers. The effect of the layer thickness, the crystallization temperature, and the interface has been investigated. Acknowledgment This research was supported by FUI NOXY II project. References [1] A. Alvarez, K. Kenny, et V. Vazquez, « Isothermal Crystallization of Poly(vinyl alcohol–co–ethylene) », J. Appl. Polym. Sci., vol. 89, 2003. [2] R. E. Prud'homme, « Crystallization and morphology of ultrathin films of homopolymers and polymer blends », Prog. Polym. Sci., vol. 54 55, p. 214 231, mars 2016. [3] O. K. Muratoğlu, A. S. Argon, et R. E. Cohen, « Crystalline morphology of polyamide-6 near planar surfaces », Polymer, vol. 36, no 11, p. 2143 2152, janv. 1995. [4] J. M. Carr, D. S. Langhe, M. T. Ponting, A. Hiltner, et E. Baer, « Confined crystallization in polymer nanolayered films: A review », J. Mater. Res., vol. 27, no 10, p. 1326 1350, mai 2012.

Enhanced Neutron Shielding Ability of Polyethylene Composites through Alternating Multi-layered Structure

***Zhang Xianlong* [1]*, *Zhang Xiaomeng* [2], *Wu Hong* [2], *Guo Shoayun* [2], *Wang Yuzhong* [2]**

[1] Sichuan university - Sihcuan - China, [2] Sichuan university - Sichuan - China

The neutron radiation was often encountered in a wide range of the industries, such as aerospace, healthcare industry and nuclear power plants, and it was an arduous challenge to shield the neutron radiation to protect the human health and equipment safety. In order to fabricate effective neutrons shielding material, the alternating multilayered (high density polyethylene)/(high density polyethylene/ boron nitride) (HDPE/(HDPE/BN)) and (HDPE/BN)/(HDPE/Barium sulfate (BaSO4)) composites were fabricated through a multilayered co-extrusion system. The HDPE/BN layers in the alternating multilayered HDPE/(HDPE/BN) and (HDPE/BN)/(HDPE/BaSO4) composites had a continuous and layered distribution structure, and the BN particles oriented along the extrusion direction. Moreover, the collision probability between incident photons and flake-shaped particles will be enhanced through alignment of the oriented BN particles. Therefore, the neutron transmittance was obviously decreased with increasing the layer number. Meanwhile, Compared to the traditional polymer blended materials, the alternating multilayered composites showed an excellent shielding efficiencỹIn addition, the results of dynamic rheological analysis showed the alternating multilayered composites with high layer number can weaken the cross-linking effects induced by secondary radiation. Furthermore, the BaSO4 was effective to shield the secondary radiation so that average melting point in nanoscale (Nano- Tm) ((HDPE/BN)/(HDPE/BaSO4)) > Nano- Tm (HDPE/(HDPE/BN)) that was obtained through Nano-TA analysis.

Polyethylene Injection Molding for IR System Lenses

Hiroshi Ito[a*], Ryo Kaneda[b], Toshihiro Takahashi[b], Masayasu Takiguchi[c], and Motoharu Hijikata[c]

[a] *Graduate School of Organic Materials Science, Yamagata University, Japan*
[b] *Yamagata Research Institute of Technology, Japan*
[c] *Yamagata Factory, Chino Corporation, Japan*

Abstract. An experimental investigation of injection molding changed the molding conditions and mold surface roughness was conducted to assess the high infrared radiation (IR) transmittance with an opaque state (low visibility ray (VIS) transmittance) necessary for IR system lenses as a target high-density polyethylene (HDPE) IR transmission material. Data for IR and VIS transmittance of molded parts were evaluated using measurements or observation results obtained for surface roughness, thickness, crystallinity, and the internal structure. Results shows that the IR transmittance rises with deccreased skin layer thickness and the VIS transmittance decreased because of the rise of crystallinity. These transmittances were obtained at high mold temperature simultaneously, and it was confirmed that the influence of the surface roughness became small.

Keywords: HDPE, injection molding, optical properties, crystallinity, skin-core structure, molecular chain orientation.
PACS: 83.50.Uv

1. INTRODUCTION

High-density polyethylene (HDPE) is a general purpose type polymer used in materials such as food containers, buckets, and so on. In addition, HDPE is used as a condenser lens for some infrared radiation (IR) systems such as radiation thermometers to transmit IR of the 8–12 μm region [1, 2]. Such lenses are produced by injection molding. When these HDPE lenses are used for IR systems, it is necessary to make them sufficiently thin to obtain high IR transmittance. However, thinning them makes them transparent, which makes it possible to view the inside of the device. Such lenses would be useful for other IR systems such as thermal imaging systems if it were possible to make them sufficiently thin for high IR transmittance in an opaque state.

In our earlier study [3-5], an experimental investigation of injection molding changed the molding conditions and mold surface roughness was conducted to obtain the optical properties necessary for an IR system lens as described above using a cavity of disk-shape. Data for IR and VIS transmittance of molded parts were evaluated using measurements or observation results obtained for surface roughness, thickness, crystallinity, and the internal structure.

In this report, these investigation results are introduced.

2. EXPERIMENTAL

2.1 Material and Mold

Commercially available HDPE (Suntec-HD J240; Asahi Kasei Corp.) was used for this study. The melt flow rate (MFR) was 5 g/ 10 min at 190°C. Predrying was not done before molding.

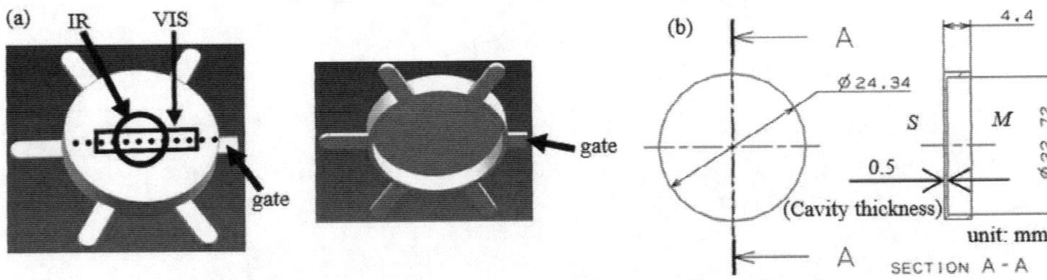

FIGURE 1. 3D-CAD model (a) and cavity dimensions (b)

A three-dimensional CAD model that includes the cavity shape and dimensions is presented in Fig. 1. The thickness of cavity is 0.5 mm. In addition, the mold having three kinds of different surface roughness *Ra* (arithmetical mean roughness) was used, as shown in Table 1.

TABLE 1. Mold surface roughness *Ra* (μm).

0.420	0.258	0.085

2.3 Injection Molding Machine and Molding Conditions

All electric servo drive injection molding machines (J110ELII; The Japan Steel Works, Ltd., Japan) used to conduct this experiment had 160 mm/s maximum injection speed, 188 MPa maximum injection pressure, 1078 KN clamping force, and 35 mm screw diameter.

Regarding the injection molding conditions, only the mold temperature was changed: 20, 40, 65, and 80°C. The cylinder temperature, injection time, and cooling time were, respectively, 200°C, 0.3 s, and 15 s.

2.4 Measurement and Evaluation

2.4.1 IR and VIS Transmittance

The IR and VIS transmittance were measured, respectively, using an FTIR imaging system (Spotlight 400; PerkinElmer Inc., USA) and a UV-VIS spectrophotometer (UV-2450; Shimadzu Corp., Japan). The measured areas of IR and VIS transmittance were, respectively, ϕ8 mm and 1 mm × 12 mm of the central part in the molded parts, as shown in Fig. 1(a). The measured wavelengths of IR and VIS were, respectively, 8–12 μm and 380–780 nm. Transmittance was evaluated as an average of measured values. Furthermore, an evaluation of surface roughness, thickness, crystallinity and skin-core structure in molded parts was carried out.

2.4.2 Surface Roughness and Thickness of Molded Parts

Surface roughness *Ra* of the molded parts was measured using a 3D optical surface profiler (New View 7300; Zygo Corp., USA). The measured length of the central part in the molded parts was 0.7 mm. Furthermore, the molded part thickness was measured using a digimatic indicator (ID-C150B; Mitutoyo Corp., Japan). The measured location was a central part in the molded parts.

2.4.3 Crystallinity

Regarding the central part ϕ4 mm of these molded parts, the heat of fusion ΔHm was measured using a differential scanning calorimeter (DSC, DSC8500; PerkinElmer Inc., USA) that had been calibrated with indium under a flowing nitrogen atmosphere at a heating rate of 10°C/min. Measurements were taken using the same heating rate of 10°C/min from 50°C to 150°C. Crystallinity (Xc) was calculated as

$$Xc = \left(\Delta Hm / \Delta Hc\right) \times 100. \quad (1)$$

where ΔHc stands for the heat of fusion values for 100% crystalline polymers. It becomes 293 J/g in HDPE.

Furthermore, a sample sliced every 0.01 mm was made from the surface of molded parts of the stationary side using an ultraprecision machine (AHN60-3D; JTEKT Corp., Japan). Then the distribution of the crystallinity in the thickness direction was evaluated using the method described above.

2.4.4 Internal Structure Observations

To investigate the internal structure of molded parts through the cross-section, thin slices of 60 μm were cut using a microtome (RM2235; Leica Microsystems GmbH, Germany). The cutting location is a broken line part presented in Fig. 1(a). Slices were observed using a polarized light microscope (PLM, Eclipse LV100ND; Nikon Corp., Japan) with 90° cross-polarized light. For this study, the skin–core structure was evaluated as three layers: a skin layer, a shear layer, and a core layer. Each layer thickness was measured from a cross-sectional observation photograph.

3. RESULTS AND DISCUSSION

3.1 Effects of Surface Roughness of Molded parts on IR and VIS Transmittance

Results of having measured the surface roughness of molded parts, it was not depended on the mold temperature, and it was obtained about the same value with the surface roughness of used mold. Figure 2 shows the IR and VIS transmittance for the surface roughness *Ra* of molded parts. When the surface roughness of the molded parts became large, the IR and VIS transmittance decreased. Furthermore, when the mold temperature is low, then the line slope becomes higher. Results show that transmittance more strongly receives the influence of the surface roughness. This tendency was stronger for VIS transmittance. In other words, the VIS of a short wavelength is affected more strongly by the surface roughness. However, when the mold temperature becomes higher, the line slope decreases. Results show that the surface roughness influence becomes slight, probably because of the influence of the thickness, the internal structure, or the increased crystallinity of molded parts.

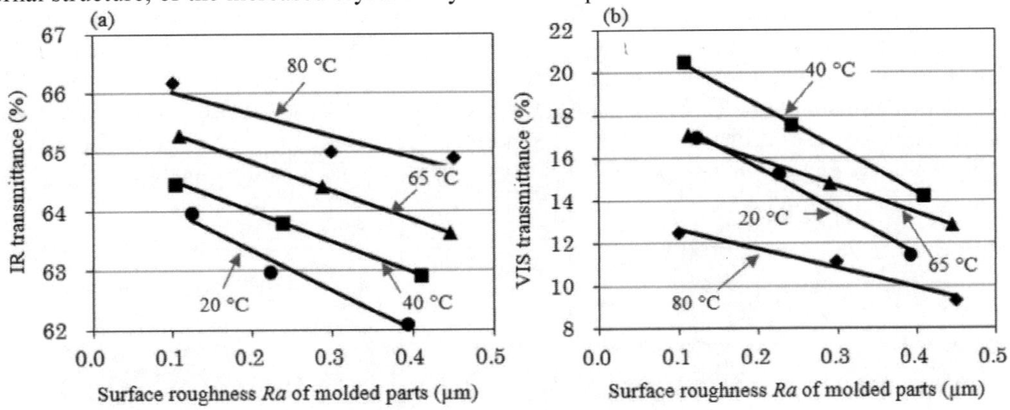

FIGURE 2. Effects of surface roughness of molded parts and mold temperature on transmittance of IR (a) and VIS (b).

3.2 Effects of Molded Part Thickness and Crystallinity on IR and VIS Transmittance

Results of having measured the thickness of molded parts, it was obtained the tendency to be thin, but there was not change in mold temperature of 65°C and 80°C when the surface roughness was small. The IR transmittance increased by the higher mold temperature, as shown in Fig. 2(a). Therefore, it was thought that the difference in internal structure affected it other than molded parts thickness. On the other hand, the decrease in VIS transmittance by a higher mold temperature had no relation to the molded part thickness, and it was affected by the crystallinity of molded parts portrayed in Fig. 3. However, the VIS transmittance decreased even though the crystallinity is low at a mold temperature of 20°C because the transmittance in the long wavelength side decreased from the VIS transmission spectrum presented in Fig. 4. We infer that this factor is not influenced by the crystallinity, but it is influenced by the difference in internal structure as with the IR transmittance.

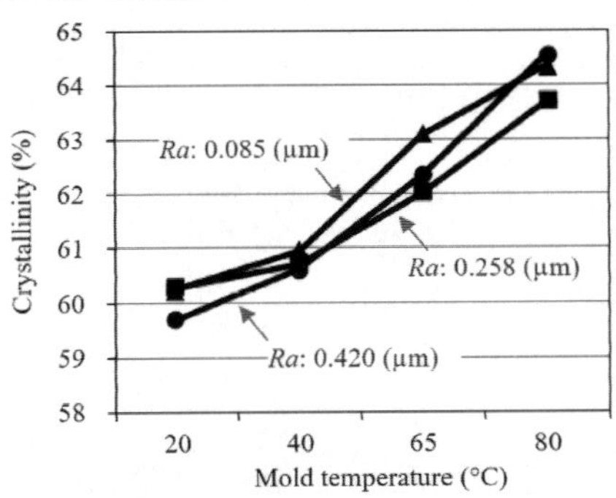

FIGURE 3. Variation of crystallinity of molded parts obtained at different mold temperatures.

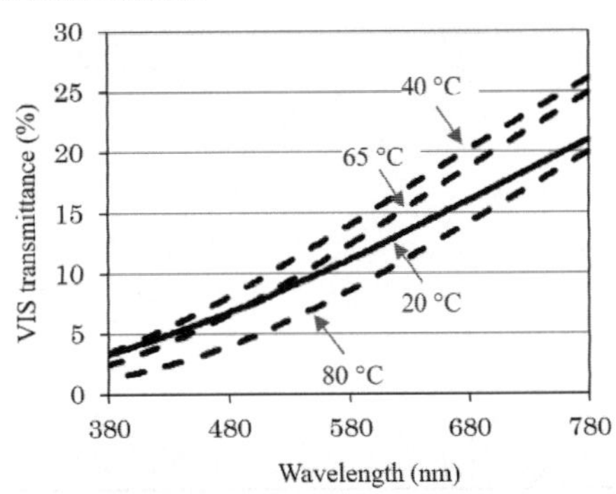

FIGURE 4. Effects of wavelength and mold temperature on VIS transmittance of molded parts.

3.3 Effects of Internal Structure and Crystallinity on IR and VIS Transmittance

Figure 5 portrays a cross-sectional observation photograph of molded parts used mold of surface roughness *Ra* 0.420 μm obtained by changing the mold temperature. For all mold temperatures, because the orientation occurred to the center of the molded parts, differences in the skin layer, shear layer, and core layer were not clear. However, the boundary of the layer with the knife mark and without the knife mark was judged clearly when the thin slices were produced using a microtome, as presented by the broken line in Fig. 5. Therefore, in this study, it was evaluated as the skin layer from the surface of molded parts to the boundary and as the shear layer inside from the boundary. The measurement results for each layer are presented in Table 2. The value shown in parentheses of this table is the depth of the skin layer from the molded part surface. This skin layer depth is half of the skin layer thickness. To assess the validity of this position as a boundary, in addition to the crystallinity of the center of the molded parts, the distribution of the crystallinity of the thickness direction was measured. The result is presented in Fig. 6 for every mold temperature. Between the surface and the center of the molded parts, the position at which the crystallinity decreased was found. This position is shown by an arrow with a value in Fig. 6. When the mold temperature was high, it became a small value. The skin layer depth shown in Table 2 was compared with the decreased position of crystallinity presented in Fig. 6. The result is presented in Fig. 7. It was an almost identical value, except for the mold temperature of 20°C. Therefore, we infer that the different internal structure is formed by a boundary line depicted in Fig. 5 and that the skin layer and the shear layer are proper. Figure 8 presents the IR transmittance for the skin-layer depth. The IR transmittance rises with decreased skin-layer depth.

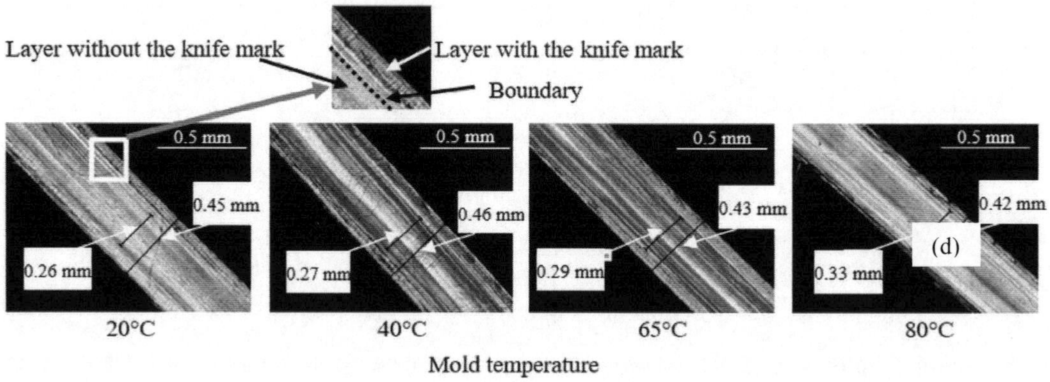

FIGURE 5. Cross-sectional photograph of molded parts obtained at different mold temperatures, dimensions shown in the photograph are the layer thickness evaluated as the shear-layer and the whole thickness of molded parts.

TABLE 2. Comparison of shear-layer and skin-layer thicknesses under different mold temperatures.

Mold temperature (°C)	Shear-layer thickness (mm)	Skin-layer thickness (mm) (Depth from surface)
20	0.26	0.19 (0.095)
40	0.27	0.19 (0.095)
65	0.29	0.14 (0.070)
80	0.33	0.09 (0.045)

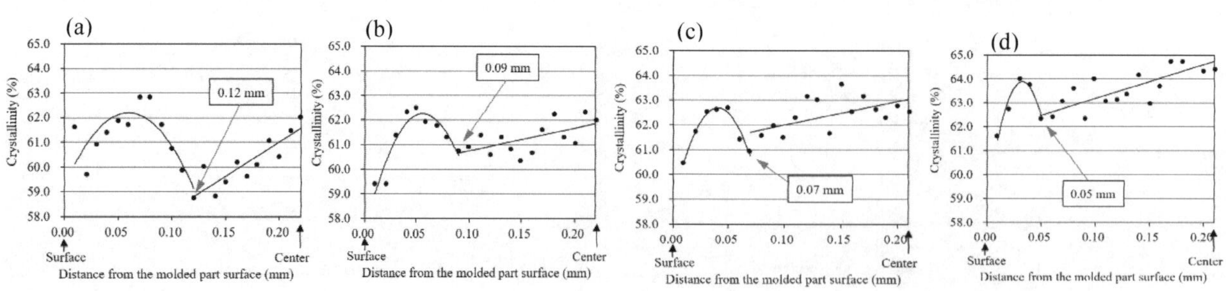

FIGURE 6. Distribution of crystallinity in the thickness direction of molded parts with different mold temperatures: 20°C (a), 40°C (b), 65°C (c), and 80°C (d).

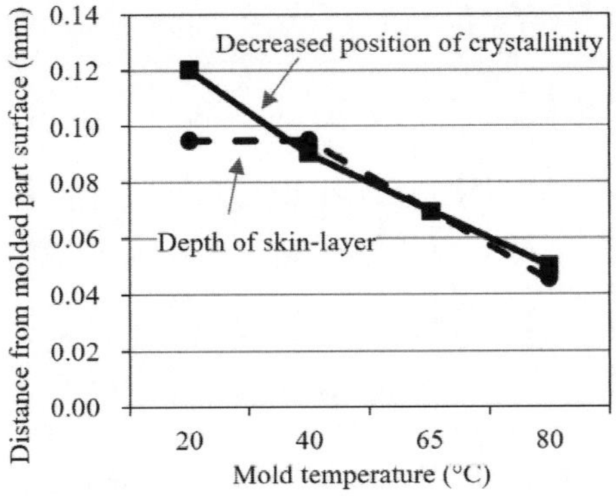

FIGURE 7. Comparison of skin-layer depth and decreased position of crystallinity under different mold temperatures.

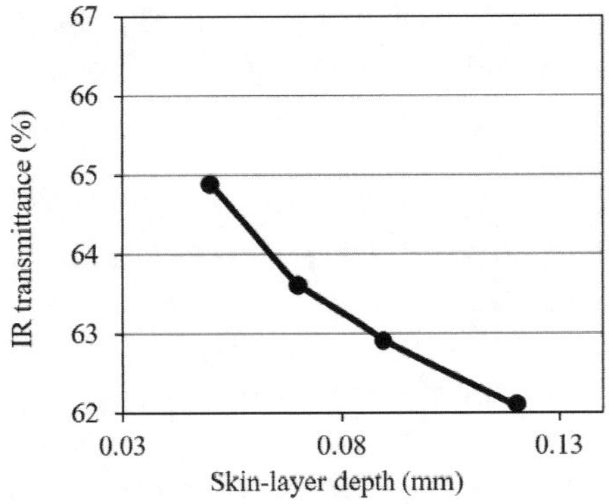

FIGURE 8. Effects of skin-layer depth from surface of molded parts on IR transmittance.

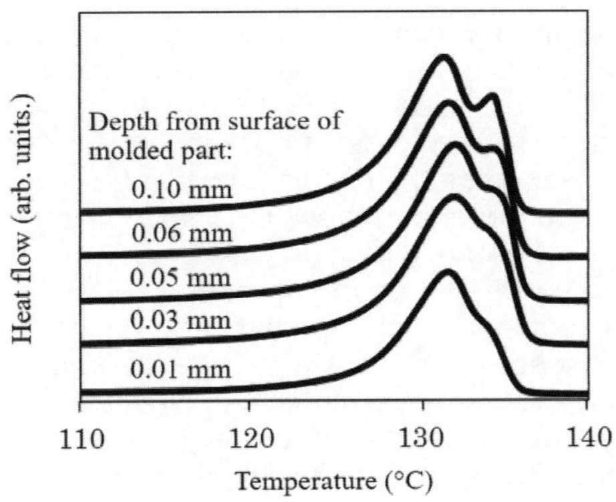

FIGURE 9. DSC curve obtained by each layer sliced on thickness of 0.01 mm in molded part produced under mold temperature of 80°C.

In addition, Fig. 9 shows the DSC curves obtained when crystallinity of Fig. 6(d) was evaluated. From this Fig. 9, it is revealed that orientation is small in the surface layer of the molded parts. Therefore, results of this investigation suggest that a factor of the increase in the IR transmittance was decreased in the skin layer having small orientation. Regarding the decrease of VIS transmittance at a mold temperature of 20°C, we investigate it.

4. CONCLUSIONS

Experimental investigations of injection molding changed the molding conditions and mold surface roughness was conducted to obtain the high IR transmittance and low VIS transmittance necessary for an IR system lens produced by HDPE. Data for IR and VIS transmittance of molded parts were evaluated using measurements or observation results obtained for surface roughness, thickness, crystallinity, and the internal structure. Results shows that the IR transmittance rises with deccreased skin layer thickness and the VIS transmittance decreased because of the rise of crystallinity. These transmittances were obtained at high mold temperature simultaneously, and it was confirmed that the influence of the surface roughness became small.

REFERENCES

1. F. Erismann, H. Fong, G.L. Graef, and S.H. Risbud, *J. Appl. Polym. Sci.*, **65**, 2727-2732 (1997).
2. A.R. Abdul Manaf and J. Yan, *Precis. Eng.*, **43**, 429-438 (2016).
3. R. Kaneda, T. Takahashi, M. Takiguchi, M. Hijikata, and H. Ito, *Int. Polym. Proc.*, **31**, 385-392 (2016).
4. R. Kaneda, T. Takahashi, M. Takiguchi, M. Hijikata, and H. Ito, *Int. Polym. Proc.*, **32**, 237-244 (2017).
5. R. Kaneda, T. Takahashi, M. Takiguchi, M. Hijikata, and H. Ito, *Seikei Kakou*, **29**, 90-92 (2017).

The effect of processing conditions on the microstructure of LCP and PBT micro-moldings, and its influence on the optimization of LDS process

Sepulveda Alexandra T. (1), Pitães Alexandre (2), Viana Julio C. (3), Pontes Antonio J. (1)*

(1) Institute for Polymers and Composites,I3N - University of Minho - Guimarães - Portugal, (2) bPólo de Inovação em Engenharia de Polímeros, University of Minho - Guimarães - Portugal, (3) Institute for Polymers and Composites,I3N - University of Minho - Guimarães - Portugal - Guimarães - Portugal

The fabrication of micro-LDS (Laser Direct Structuring) components by micro-injection molding requires adjustment of polymeric materials, process and parts design. LDS process is based on the selective structuring/activation of a surface, by laser, of a component previously injection molded, using appropriate materials for the application. The process of activation results in a physicochemical reaction of the material constituents, forming catalytic elements for the subsequent process of superficial deposition of a conductive material. When the polymer is exposed to laser, the metal complex decomposes into elemental metal and residual organic groups. Then, the laser engraves the circuit tracks in the surface of the component, leaving a roughened surface that contains embedded metal particles [1]. These particles serve as nuclei for growing of crystals during subsequent copper deposition step. Injection moulding is one of the most important and versatile technique to produce polymeric parts with complex geometries. The rescale of the conventional injection molding process to obtain miniaturized components is not valid [2-3]. Most components obtained by the conventional injection moulding process present a typical skin-core morphology that influences the final properties of polymeric parts. Generally, a frozen layer is formed near the mold wall while the core is slowly cooling down until the ejection temperature. Nevertheless, the high aspect ratios and small dimensions of micro-componentes in micro-injection molding, results in a rapid cooling of material and consequently higher molecular orientation in the surface of micro-parts. This work aims at studying the influence of processing conditions on the microstructure and distribution of seed metals of two different LDS-compatible materials and to optimize the selective metallization of micro-moldings by using optimal processing conditions. The metallization of parts at macro-scale does not cause major constrains regarding the interfacial mixing. However, considering that in micro-parts the available width can be in the range of few to several hundred of micrometers, the selective metallization may be conditioned.

Dimensional accuracy of Acrylonitrile Butadiene Styrene injection molded parts produced in a pilot production with an additively manufactured insert

Michael Mischkot[a], Ali Davoudinejad[a], Alessandro Charalambis[a], Thomas Hofstätter[a], Guido Tosello[a*], David Bue Pedersen[a] and Hans Nørgaard Hansen[a]

[a]*DTU Mechanical Engineering, Produktionstorvet 427A/323B, DK-2800 Kgs. Lyngby, Denmark*

guto@mek.dtu.dk

Abstract. Injection molding inserts manufactured additively by vat photopolymerization have become a serious option for significantly faster and more economical prototyping and pilot production due to technological progress and advancements in photopolymer materials in the recent years. 10 000 parts of a geometry including micro-features have been injection-molded in Acrylonitrile Butadiene Styrene (ABS) with a single 20x20x2.5 mm^3 injection molding insert manufactured in a photopolymer composite material. This research investigates the dimensional accuracy of the injection molded parts as a function of inserts wearing and deformation with increasing shot number.

Keywords: Additive manufacturing; Micro injection molding; Soft tooling
PACS: 06.20.-f Metrology

INTRODUCTION

In recent years, additive manufacturing soft tooling has become a considerable option for the manufacturing of injection molding inserts in prototyping and pilot production, as demonstrated in numerous publications, e.g., [1] and [2]. Its key advantages include low costs and fast manufacturing. Also, the process chain via injection molding enables a larger selection of polymer materials for the final part, thereby fulfilling a broader range of material requirements. Compared to conventional tooling, usually a low number of parts in the one to three-digit range can be produced with one insert. [3]

With technological progress, also longer tool lifetimes have become reality, as reported by [4], using a fiber-reinforced insert manufactured with vat photopolymerisaton for the production of more than 2 500 parts in polyethylene low-density. [5] presented the production of 2 800 parts injection molded in acrylonitrile butadiene styrene (ABS) as part of a long-term experiment. This research presents the continuation of that experiment.

MATERIALS AND METHODS

Insert Geometry

The test geometry consists of a plate containing two cuboid structures with 800 µm pillars attached as well as two heart-shaped elements. The consequent insert design is presented in Figure 1. A draft angle of 2° was added to the vertical walls to facilitate ejection of the injection molded parts.

Tooling

A stereolithography printer (3D Systems, SLA 3500, Solid State Nd:YVO_4 laser, 354.7 nm wavelength) was used to manufacture the injection molding insert in a ceramic composite photopolymer (SOMOS® Perform). A heat treatment was performed after the printing to increase the photopolymer's heat deflection temperature (268°C HDT @ 0.46 MPa, ASTM Method D648).

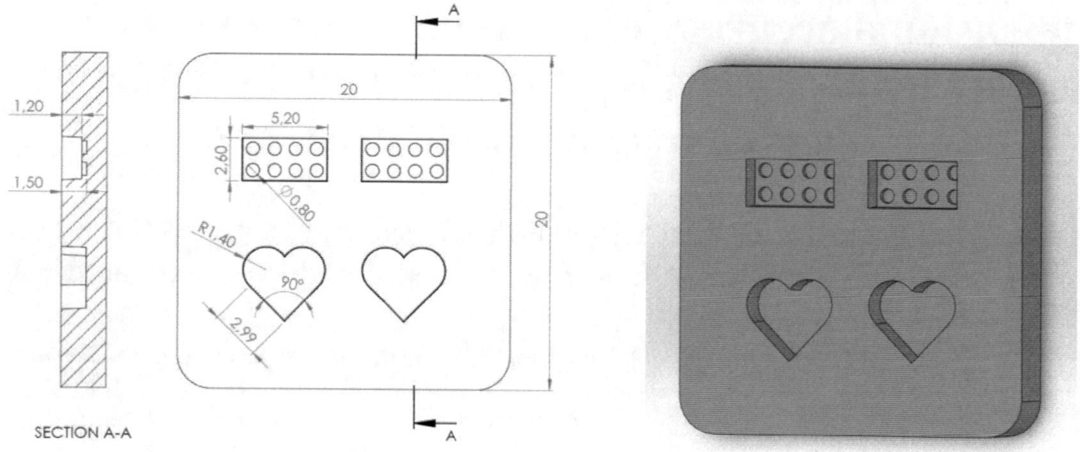

FIGURE 1. Sketch (a, left) of the insert geometry (20 x 20 x 2.5 mm^3) comprising two cuboid structures with pillars (diameter 800 μm, height 0.3 mm) and two heart-shaped elements as well the corresponding 3D-model (b, right).

Injection Molding

The injection molding was performed with a packing pressure of 950 bar using an Engel® e-motion 170/110 injection molding machine. The cycle time was 21 seconds with a cooling time of 10 seconds, the injection speed 15 mm/s, and the melt temperature 235°C.

Metrological Assessment

For the metrological assessment, a laser-scanning microscope (OLS4100 Lext from Olympus) was used to take images of selected samples of the injection molded parts. SPIP™ from Image Metrology was used to perform the dimensional analysis of the images.

RESULTS

Lifetime

More than 10 000 parts were injection molded successfully with one insert after which the experiment was stopped. This is significantly longer than other sources report [1-5]. A possible explanation for the long lifetime of the insert is the mold design. Only the part of the cavity on the injection side is manufactured in the photopolymer material. The cavity part on the ejector side is made of steel. Given the photopolymer's very low heat conductivity (0.178 – 0.187 W/mK, [6]) compared to steel, most of the heat is transported through the metal parts of the cavity, reducing the thermal energy load for the insert and, consequently, its thermal ageing.

Dimensional Accuracy

Pillar Diameter

Figure 2 shows the development of the pillar diameter. The diameters are deviating significantly from the nominal diameter (500-600 μm vs. 800 μm). This is a result of the printing process, which does not produce sharp edges (Figure 3).

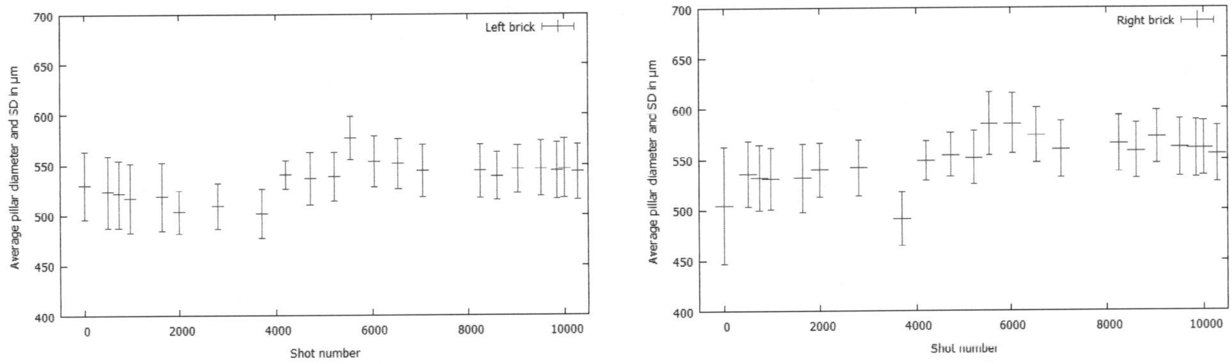

FIGURE 2. Average pillar diameter (left brick, right brick, in µm) and standard deviation over 10 000 shots.

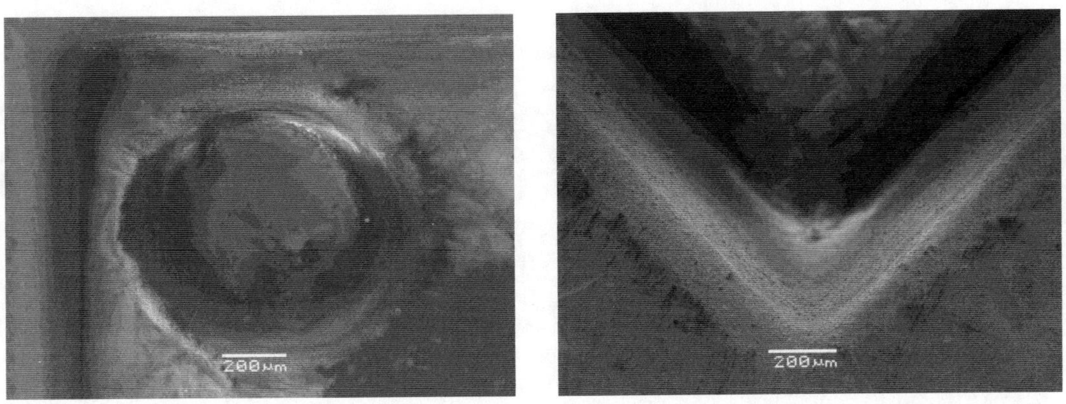

FIGURE 3. SEM images of the insert. Left: Top left pillar in the left cuboid. Right: Bottom of the bottom left heart.

Heart Corners

Figure 4 shows the angle of the 90° corners of the hearts. For the left heart, all values are between 88° and 92°, apart from three measurements between 500 and 1 000 shots at the left heart. For the right heart, all values are in the range 90±2°.

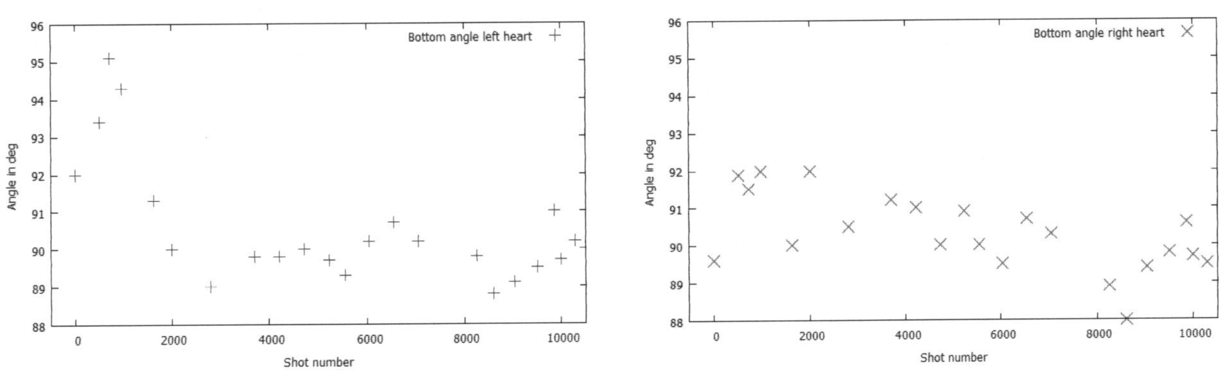

FIGURE 4. Angles at the bottom of the hearts (nominal value 90°) over 10 000 shots.

Depth of Cuboid Elements

The depth of the cuboids with and without the pillars (Section A-A in Figure 1, nominal values $h_1 = 1.2$ mm and $h_2 = 1.5$ mm, respectively) were constant within a range of ±20 μm throughout the experiment (Figure 5).

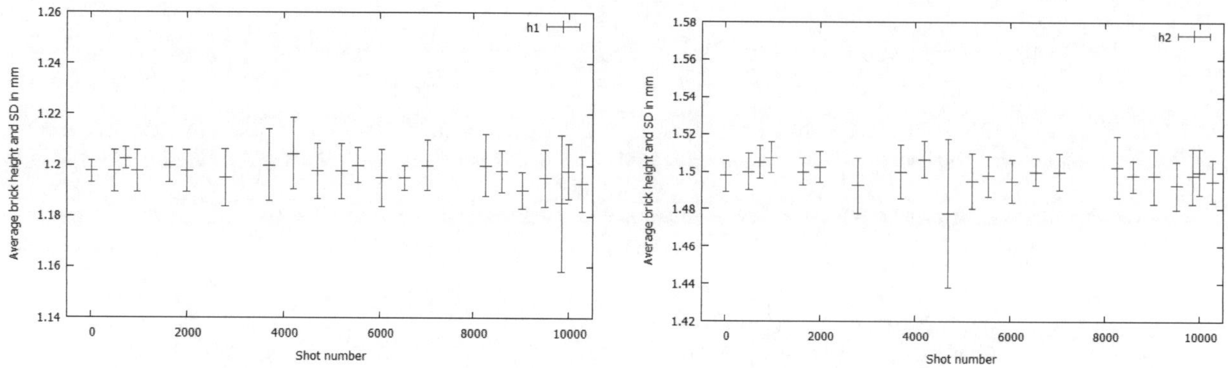

FIGURE 5. Average depth of cuboids and standard deviation over 10 000 shots (left: without pillars, right: with pillars).

CONCLUSIONS

With more than 10 000 parts being produced with a single insert, it has been shown that soft tooling based on vat photopolymerization can be used to injection mold five-digit part numbers, given appropriate conditions.

Concerning the dimensional accuracy, the measured diameters of the cylindrical elements attached to the cuboids were significantly smaller than the nominal values. An explanation for this phenomenon is the undesired curing of photopolymer material remaining in corners and edges after the printing process.

The measurements of the 90° corners at the bottoms of the hearts were within 90±2°, apart from three measurements between at the left heart 500 and 1 000 shots.

The depth of the cuboids remained constant within ±20 μm, showing the negligible insert wear during this long-term experiment.

Suggestions for Future Research

It is suggested to investigate the thermal aging process of the photopolymer material to determine the factors leading to the reported differences in insert lifetime. Further, the modification of the liquid photopolymer before printing (adding particles or fibers) or the printed part (coating) seems a promising direction for future investigations. Finally, thermal simulations have been found to be a useful and accurate tool for additive manufacturing soft tooling [5]. Consequently, the reader is encouraged to engage in in-depth simulations with both multiphysics as well as injection molding simulation software.

ACKNOWLEDGEMENTS

The research presented in this paper was carried out and financed in the framework of a project between the Technical University of Denmark (DTU) and the Manufacturing Academy of Denmark (MADE), (http://en.made.dk). Funding from Innovation Fund Denmark (http://innovationsfonden.dk/en) is greatly acknowledged.

REFERENCES

1. A.D. Lantada et al., "Toward mass production of microtextured microdevices: linking rapid prototyping with microinjection molding", *Int. J. Adv. Manufact. Technol.* **76**(5–8), 1011–1020 (2015).
2. J. Rajaguru et al., "Development of rapid tooling by rapid prototyping technology and electroless nickel plating for low-volume production of plastic parts", *Int. J. Adv. Manufact. Technol.* **78**(1–4), 31–40 (2015).
3. I. Gibson et al., "Additive Manufacturing Technologies", Springer, DOI 10.1007/978-1-4939-2113-3.
4. T. Hofstätter et al., "Evolution of surface texture and cracks during injection molding of fiber-reinforced, additively-manufactured, injection molding inserts", *Proceedings of ASPE Summer Topical Meeting 2016: Dimensional Accuracy and Surface Finish in Additive Manufacturing.*
5. M. Mischkot et al., "Performance Simulation and Verification of Vat Photopolymerization Based, Additively Manufactured Injection Molding Inserts with Micro-Features", *Industrializing Additive Manufacturing – Proceedings of Additive Manufacturing in Products and Applications – AMPA2017*, DOI 10.1007/978-3-319-66866-6_16.
6. J. G. Kovacs et al., „Evaluation of measured and calculated thermal parameters of a photopolymer", *International Communications in Heat and Mass Transfer* **38**(2011) 863-867.

A Novel Microcellular Foam Injection Molding Technology Using Non-Supercritical CO_2, N_2 or Air

Long Wang[*,a], Shota Ishihara[a], Yuta Hikima[a], Masahiro Ohshima[**,a], Atsushi Yusa[b], and Hiromasa Uezono[c]

[a]*Department of Chemical Engineering, Kyoto University, Kyoto 615-8510, Japan*
[b]*Technology Development Department, Hitachi Maxell, Ltd., Ibaraki, Osaka, 567-8567, Japan*
[c]*Injection Molding Machine Division, Japan Steel Works Ltd., Hiroshima 736-8602, Japan*

Abstract. A new foam injection molding technology was developed to produce microcellular foams without supercritical fluids and with gas directly delivered from its cylinder. As for the conventional foam injection molding process, the blowing agents, such as nitrogen (N_2) and carbon dioxide (CO_2), are pressurized to a level higher than their critical pressures using a supercritical fluid pump unit (SCF) to introduce the blowing agents into molten polymer. Behaving differently, we have designed a resilient & innovative cellular foam injection molding (RIC-FIM) technology to fabricate microcellular foams without using the SCF pump. This new system can directly deliver N_2 or CO_2 through a pressure vessel and a venting hole into the barrel from their gas cylinders. Microcellular foaming can be achieved with this delivery system, a specially designed screw configuration and operation scheme. Interestingly, owing to the performance of the developed machine setup, air can be used as a blowing agent for microcellular foaming. The concentration of gas dissolved in the polymer is controlled by the secondary pressure of the reducing valve of the gas cylinder and the pressure vessel of the delivery system. As a vital industrial material, polypropylene foams with various expansion ratios were fabricated here using the developed FIM technology with either N_2, CO_2, or air as the blowing agents. It is clearly revealed that the cell morphology, cell size and cell density can be controlled by manipulating the delivery pressure.

Keywords: Non-Supercritical Gas, Foam Injection Molding, Microcellular Foam, Polypropylene
PACS: 82.35.-x, 61.41.+e, 61.20.Lc, 61.25.Hq, 81.05.Rm

INTRODUCTION

The microcellular foaming technology was developed since the 1980s, and it was first proposed by Professor Suh et al. of the Massachusetts Institute of Technology (MIT) [1]. The first commercial application of the microcellular technology was the MuCell extrusion and injection molding processes, which was licensed by Trexel, Inc. The MuCell process configuration mainly consisted of a specifically configured screw and a metering system for the supercritical fluid [2]. In its foaming process, supercritical carbon dioxide (sc-CO_2) or nitrogen (sc-N_2) was used as the physical blowing agent (PBA). With the great commercial success of the MuCell technology, various microcellular foam injection molding (FIM) technologies, such as the Optifoam system from the Institute of Plastics Processing (IKV) and the Ergocell process licensed by Demag, were successively developed [3].

Recently, we introduced a novel FIM technology named resilient and innovative cellular foam injection molding, i.e., RIC-FIM I, at FOAMS2015 [4]. This developed RIC-FIM I process demonstrated that the pressurization of N_2 or CO_2 to a supercritical state was unnecessary for obtaining injection-molded foams [4,5]. The foamed products showed that microcellular polypropylene (PP) foams could be obtained using a non-supercritical fluid, such as 8 MPa N_2 and 5 MPa CO_2 [4,5]. The PBA was directly introduced from a gas cylinder into the molten polymer through an injection valve, which was controlled through a specifically designed operation sequence and screw configuration.

To distinctly differentiate the new process from the conventional MuCell technology, our RIC-FIM technology was further upgraded and simplified to deliver the PBA solely from the vent hole while removing the injector valve. This newly developed FIM technology, namely, RIC-FIM II, was applied to the foaming of PP resins. Moreover, due to limited void fraction in the regular FIM process, an FIM technique with core-back operation was used here, which

[*]*E-main address*: wangl.kevin@cheme.kyoto-u.ac.jp (Long Wang) ohshima@cheme.kyoto-u.ac.jp (Masahiro Ohshima)

greatly enlarged the void fraction of the injection-molded products [6]. To check the flexibility of our developed FIM process, N_2 with various delivery pressures was directly introduced from a gas cylinder and used for the foaming process. The use of CO_2 and air as the PBAs to control the cell morphology were also investigated.

DESIGN OF THE NOVEL INJECTION-MOLDING MACHINE

Figure 1 shows a schematic diagram of the first version of the RIC-FIM machine. As shown in Figure 1, the injector valve for PBA delivery was directedly connected to a gas cylinder, which enabled the PBA to be dosed into the molten polymer in a non-supercritical state [4,5]. The amount of injected gas was regulated by the injector valve opening time and the set point of the reducing pressure regulator on the cylinder. Using a reducing pressure regulator, the pressure of the supplied PBA could be controlled. Moreover, an auxiliary vent hole with a venting vessel was installed between the injector valve and the nozzle. Through this venting hole, excess gas that remained un-dissolved in the molten polymer could be released into the atmosphere, or extra PBA could be supplemented into the polymer melt.

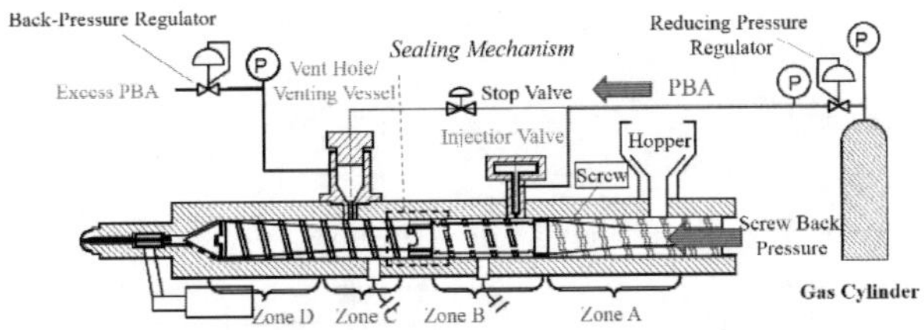

FIGURE 1. Schematic diagram of the RIC-FIM I machine.

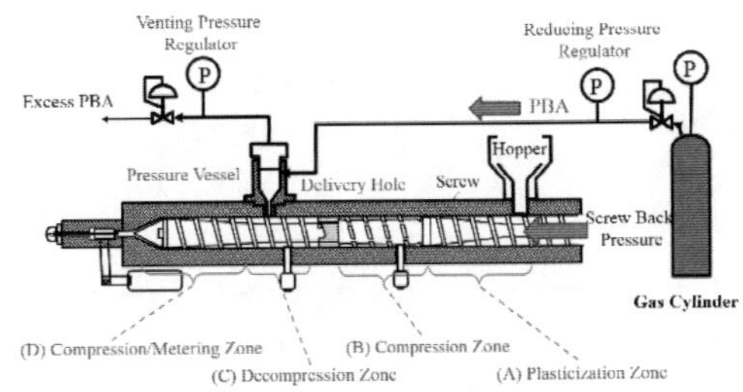

FIGURE 2. Schematic diagram of the RIC-FIM II machine.

To completely distinguish the new method from the regular MuCell technology, we proposed the removal of the injector valve from the RIC-FIM I machine and used the previously auxiliary venting vessel as the dominant gas-dosing component. Figure 2 illustrates a schematic diagram of the new version of our developed RIC-FIM machine (RIC-FIM II). In this upgraded version, the PBA is directly provided from a gas cylinder and introduced into the polymer melt via a delivery hole in a pressure vessel. Thus, the auxiliary vent hole and venting vessel in the RIC-FIM I machine were separately transformed into a delivery hole and pressure vessel in the new FIM machine.

EXPERIMENTAL

Materials

Isotactic polypropylene (iPP, F133A) with a weight-average molecular mass of 379 kg/mol and a melt flow rate (230 °C/2.16 kg) of 3.0 g/10 min was provided by the Prime Polymer Co., Ltd., Tokyo, Japan. A crystal nucleating agent, 1,3:2,4 bis-O-(4-methylbenzyliden)-D-sorbitol gelling agent (Gel-all MD), was used as a bubble nucleating agent and 0.5 wt % of MD was used [7]. N_2 with 0.999 purity and CO_2 with 0.99 purity, which were supplied from the Showa Denko Gas Products Co., Ltd., were independently used as the physical blowing agent. Synthetic air, which was also obtained from Showa Denko Gas Products Co., Ltd., with 80% N_2 and 20% O_2 was used as the blowing agent.

Foam Injection Molding Experiments

Foaming experiments were carried out on a 35-ton clamping force electric injection molding machine with a screw diameter of 22 mm (J35AD-AD30H, Japan Steel Work, Ltd. Hiroshima, Japan). Excluding the injector valve unit and its operation procedure, the machine setup and operating procedure of this new FIM technology was the same as those of the previous RIC-FIM I machine. The mold consisted of a rectangular cavity (70 mm × 50 mm) with an initial thickness of 2 mm, which was able to increase to 10 mm. Additionally, a data logging system (Mold Marshaling system EPD-001, Futaba Corp., Chiba Prefecture, Japan) was used to monitor the real temperature and pressure of the injected polymer in the cavity. The operating parameters of the FIM experiments are summarized in Table 1.

TABLE (1). Processing conditions for the foam injection molding process.

Parameters	Values
Melt temperature (°C)	200 (N_2, CO_2), 190 (Air)
Mold temperature (°C)	40
Pressure of gas cylinder (MPa)	12 (N_2, Air) 6 (CO_2)
Pressure of delivery vessel (MPa)	4, 5, and 6 (N_2), 5 (CO_2), 5 and 6 (air)
Injection speed (mm/s)	200
Injection pressure (MPa)	160
Dwelling time (s)	6.2–8.0
Shot size (mm)	45
Core-back rate (mm/s)	20
Core-back distance (mm)	2, 4, 6, 8

Characterizations

Cell morphology of the foamed samples was investigated using a scanning electron microscope (Tiny-SEM Mighty-8, Technex, Japan) operating at an accelerating voltage of 17 kV. Prior to observation, fractured surfaces of the prepared samples were gold-coated using a VPS-020 Quick Coater (ULVAC KIKO, Ltd., Japan). The obtained SEM images were further analyzed and the cell density, N_0, is usually calculated using the following expression:

$$N_0 = (\frac{n}{A})^{3/2} \tag{1}$$

where n is the number of cells in the SEM micrographs; A is the area of the micrograph. Similarly, the average cell size is also directly obtained from the SEM micrographs, assuming that the cell shape is spherical. A gas pycnometer (AccuPycII, Shimadzu, Kyoto, Japan) was used to detect the open cell content (OCC) of the foamed samples.

RESULTS AND DISCUSSION

Cell Structures with N_2 Delivery

Figure 3 shows the cell morphology in core region of PP foams with a fixed void fraction of 50%. SEM images were taken from both perpendicular and parallel views to the core-back direction. As shown in Figure 3, it revealed that fine cellular structure of injection-molded foams was achieved by using a very low delivery pressure approximately 4 MPa. This indicated the feasibility of only using vent hole as gas dosing points and the possibility of using low-pressure non-supercritical gas as a PBA. With the increase of delivery pressure, smaller cell sizes and finer cellular structures were achieved in the PP foams, which was related to the increased concentration of N_2 dissolved in PP melt because of enhanced gas pressure. This signified that the gas content of PBA dissolved in polymer could be controlled by delivery pressure regulator.

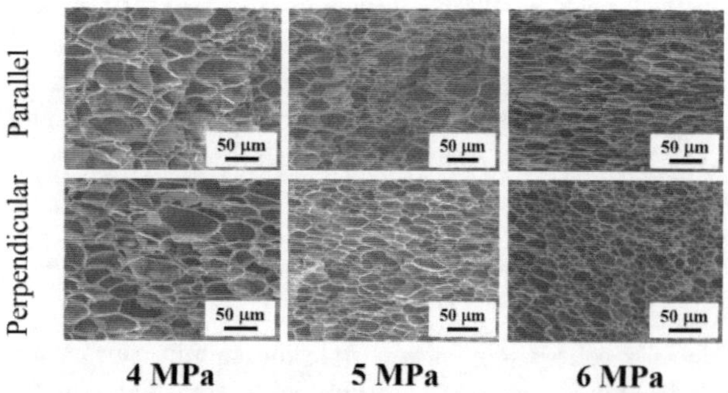

FIGURE 3. Effect of N_2 pressure on cell morphology.

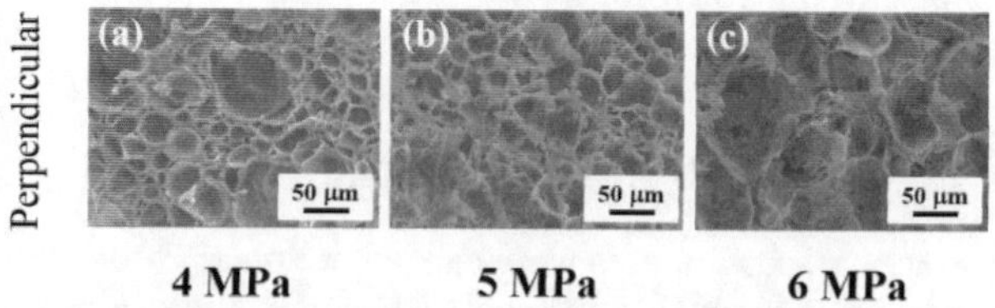

FIGURE 4. Effect of N_2 pressure on maximum void fraction.

To further compare the effect of different delivery pressure, we have examined the highest-achieved void fraction and is shown in Figure 4. The optimum cellular structures in core region observed perpendicular to the core-back direction were presented for each gas pressure. It was found that the highest void fraction of PP foams under 4, 5, and

6 MPa of N_2 were 66 (3-folds), 75 (4-folds), and 80% (5-folds), respectively. This increased maximum void fraction with enhanced gas pressure additionally verified the flexibility of using vent hole controls gas concentration dissolved in the polymer melt. Moreover, as shown in Figure 4, microcellular FIM foams with a void fraction as high as 80 % was achieved for PP foams under 6 MPa of N_2.

Cell Structures with CO_2 Delivery

Figure 5 shows the cell morphology of PP foams with various void fraction prepared under the delivery pressure of 5 MPa. As discussed above, the highest void fraction for the foams prepared at the same pressure of N_2 was only 75%, while a higher void fraction reaching 80 % was achieved by using CO_2. This signified that a better foaming performance was achieved in PP foams by utilizing CO_2 as the PBA at the same delivery pressure. Generally, Figure 5 reveals that bubble sizes of the PP foams increased with the rising of void fraction.

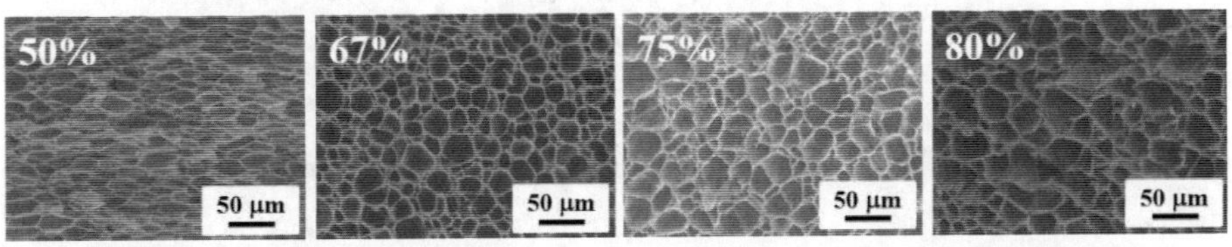

FIGURE 5. Effect of void fraction on the cell structures of CO_2 foaming samples.

Figure 6 illustrates the cell density and average cell size of PP foams as a function of void fraction by using CO_2 as the PBA. For comparison, the results of N_2 foaming under the same delivering pressure were also presented. It manifested that microcellular injection-molded foams, whose cell size was 15–25 μm and cell density ranging of 4.8 $\times 10^7$– 2.4 $\times 10^8$ cells/cm^3, were successfully prepared using CO_2 as a PBA. Cell size of PP foams slightly increased with the enlargement of void fraction in CO_2 foaming, while the change was more severe in N_2 foaming. As shown in Figure 6a, compared with the N_2 foaming, using of CO_2 produced a higher cell density for PP sample at the same delivery pressure and void fraction. This promotion in cell density was more substantial for PP foams fabricated at a higher void fraction. Moreover, regarding the N_2 delivery, the 80% void fraction injection-molded foams were difficulty to be prepared. This might be due to the solubility difference between N_2 and CO_2 in PP resins. It is well-known that CO_2 exhibited a much higher solubility than N_2 in PP melt under the same pressure and temperature. Larger void fraction might need a high concentration of PBA, and thus the use of CO_2 was favorable in preparing for high void fraction FIM foams. Likewise, cell sizes of the PP foams prepared using CO_2 were smaller than that of N_2, and the distinction in cell size was more obvious for high void fraction PP foams.

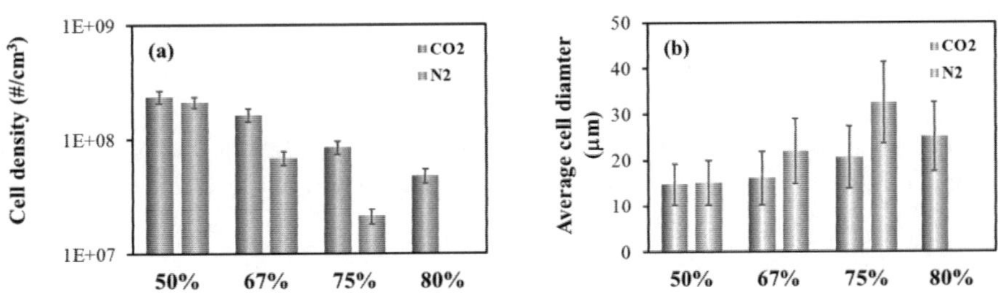

Figure 6. Variations of (a) cell density and (b) average cell diameter for PP foams under the delivery pressure of 5 MPa.

Cell Structures with Air Delivery

Figure 7 presents the effect of air pressure on the cell morphology of PP foams prepared under various void fraction. The optimum cellular structures were selected for each condition. It was found that microcellular injection-molded PP foams were successfully prepared using a low-pressure air as the foaming agent. The enhanced gas pressures endowed the prepared PP foams with finer cellular structures and smaller bubble sizes, which was attributed to the introduction of a greater degree of thermodynamic instability with enhanced gas content. For instance, with a slight increase of 1

MPa of air pressure, cell density of a 50% void fraction PP foams was enhanced from 4.04×10^8 cells/cm^3 to 1.63×10^9 cells/cm^3. Similarly, cell sizes of the obtained PP foams were correspondingly reduced with an enhancement of gas pressure, demonstrating a tiny cell size about 7 μm under the pressure of 6 MPa. This demonstrates that the air concentration dissolved in the molten polymer can be regulated by the delivering pressure and simultaneously influence the final cellular structures.

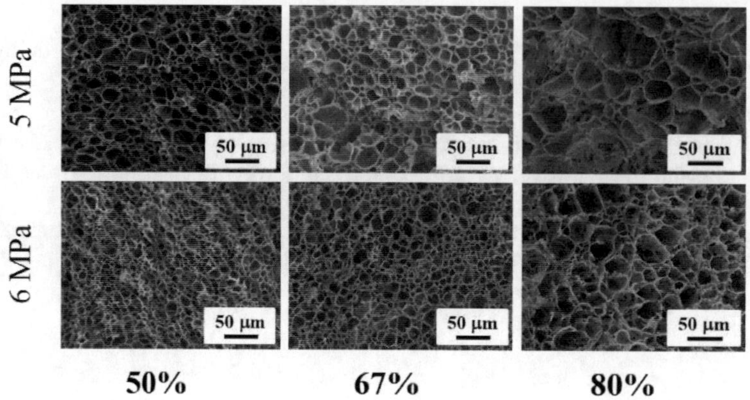

FIGURE 7. Effect of air pressure on the cell morphology of PP foams.

CONCLUSIONS

In this study, our new microcellular FIM technology (RIC-FIM) was further simplified to reduce the machine cost while still keeping the foaming performance and producing fine cellular structure. The key feature of this novel and viable technology is that it does not need a high-pressure pump or injector valve, as the injector valve was still required in the previous RIC-FIM I machine. In this newly-developed RIC-FIM II, PBAs could be delivered straightly from their gas cylinders to molten polymer through the delivery vessel and hole. Controllability of the cell morphology of PP foams was comparatively investigated by using CO_2, N_2 and air as the PBAs. We revealed that the delivery of PBA through the vessel and hole could produce high void fraction PP foams. Specifically, with the increase of delivery pressure, cell size of PP foams was decreased and its cell density was correspondingly increased, which exhibited that the concentration of PBA dissolved in molten polymer was well-adjusted by the vessel pressure. Interestingly, compared with N_2 foaming, using of CO_2 produced finer cellular structures in PP.

ACKNOWLEDGMENTS

The research was conducted in association with the Special Priority Research Areas (White Biotechnology) of the Advanced Low Carbon Technology Research and Development Program (ALCA) of the Japan Science and Technology Agency.

REFERENCES

1. J. E. Martini, F. A. Waldman, N. P. Suh, *SPE ANTEC Tech. Pap.* **43**, 674–676 (1982).
2. N. P. Suh, Innovation in Polymer Processing, ed. Stevenson, J. F. Hanser Publication (Cincinnati), Chapter 3, 1996.
3. J. Y. Xu, Microcellular Injection Molding; John Wiley and Sons Ltd.: New York, Chapter 1, 2010.
4. A. Yusa, S. Yamamoto, H. Goto, M. Ohshima, Proc. of FOAMS2015, Kyoto, 2015.
5. A. Yusa, S. Yamamoto, H. Goto, H. Uezono, F. Asaoka, L. Wang, M. Ando, S. Ishihara, M. Ohshima, *Polym. Eng. Sci.* **57**, 105-113 (2017).
6. L. Wang, S. Ishihara, M. Ando, A. Minato, Y. Hikima, M. Ohshima, *Ind. Eng. Chem. Res.* **55** 11970-11982 (2016).
7. R. Miyamoto, S. Yasuhara, H. Shikuma and M. Ohshima, *Polym. Eng. Sci.* **54**, 2075–2085 (2014).

Surface Structure Analysis for Injection Molded Direct Joining using Abrasive Blasting

Fuminobu Kimura[a], Yuta Tamura[a], Eiji Yamaguchi[b], and Yusuke Kajihara[a]

[a] *Institute of Industrial Science, the University of Tokyo, 4-6-1, Komaba, Meguro-ku, Tokyo, 153-8505, Japan*
[b] *SINTOKOGIO, LTD., 3-1, Honohara, Toyokawa, Aichi, 442-8505, Japan*

Abstract. This paper describes that surface structure analyses for injection molded direct joining (IMDJ) process. The IMDJ process is one of metal/polymer direct joining processes and consists of a surface treatment and insert injection molding. The surface treatment produces a fine structure on a surface of a metal workpiece. By insert injection molding, the molded polymer is joined with the metal workpiece via the treated surface. Since the feature of surface structure could affect joining characteristics, this study investigated the surface structure and joining strength. We two- and three dimensionally measured surface profiles produced by micro blasting using various sized alumina media. Using the measured profiles, we analyzed them by using the roughness parameters in two dimensional and by a particle analysis in three dimensional. Results show there is a trade-off relationship.

Keywords: Metal/polymer direct joining; Injection molding; Abrasive blasting; Surface analysis
PACS: 81.20.Vj; 68.35.Ct; 81.20.Hy

INTRODUCTION

Replacing parts of structural materials of automobile bodies from metals to light-weight materials, such as polymers, reduces greenhouse gas emissions and is one of effective and typical environmental measures. To use both the metal and polymer materials in one product, we need joining of metal and polymer materials, which usually uses some joining parts, e.g. bolt screws and adhesives. A direct joining technique, which does not require the joining parts, has recently attracted attention in industrial fields since the direct joining has many advantages.

Injection molded direct joining (IMDJ) process is one of the direct joining between metal and polymer materials. The IMDJ process is a combination of two basic processes; one is a surface treatment of a metal workpiece and the other is insert injection molding. The surface treatment produces a fine structure on a surface of the metal workpiece by using several techniques; abrasive blasting [1-3], laser processing [4, 5], chemical processing [6-8]. A size of the fine structure ranges from tens nanometer scale to micrometer scale. In the molding process, a polymer material is injected to a mold where the treated metal workpiece is previously placed. The melted polymer material flows into the surface fine structure during the molding, and then cools and hardens. After being ejected from the mold, the molded polymer is joined with the metal workpiece. Figure 1 shows an overview of the IMDJ process (this case uses abrasive blasting as the surface treatment).

The features of the surface structure, such as shape, size, and condition, could affect joining characteristics. In this study, we investigated geometric properties of the surface structure produced by micro blasting using various types of media. In the previous studies [2, 3], effects of surface roughness have been investigated. The effects of surface structure, however, still remains unclear since the investigations were limited and uncompleted. To improve understanding of the surface structure, this study deals with surface profile measurements and analyses in two and

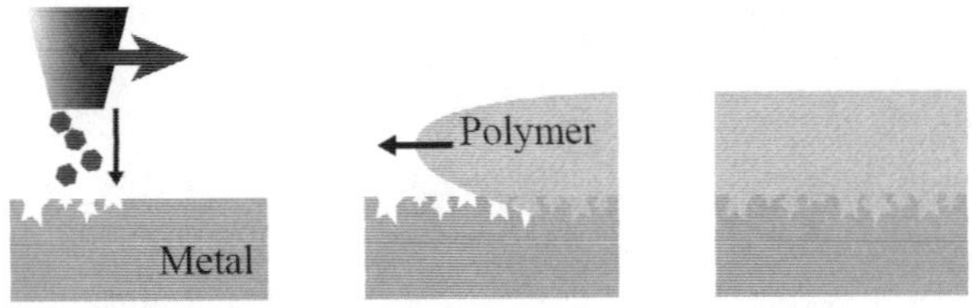

(a) (b) (c)

FIGURE 1. Overview of injection molded direct joining assisted by abrasive blasting. (a) A surface of a metal workpiece is fine-structured by abrasive blasting. (b) A polymer material is injection molded on the fine-structured metal workpiece. (c) Metal and polymer materials are joined directly.

three-dimensional. The results show that there is not difference of roughness parameters between media types, alumina particle and glass beads, although the strength is completely different. Considering only alumina particles, we show the different size of particles has no correlation to the strength since both dimples depth and dimple density is affected by the media size.

INJECTION MOLDED DIRECT JOINING PROCESS

Micro Blasting

We used a micro blasting technique to form a micro-scale structure on a surface of a metal workpiece. The micro blasting machine propels blasting media by compressed air (1.0 MPa) from a scanning nozzle (Fig. 2). The propelled media strike against the workpiece, and then the surface structure is formed. In this study, two types of the blasting media were used; one is alumina particles and the other is glass beads. Figure 3 shows photo images of the media. We prepared five different sizes of alumina particles and one size of glass beads as shown in TABLE 1. In this study, we used aluminum alloy A5052 as a metal workpiece.

Injection Molding

The molding process employed a commercial injection molding machine (ROBOSHOT α-S100iA, FANUC) and an original mold. A pressure sensor (6158A, Kistler Japan) and a temperature sensor (EPSSZT-04.0×030, Futaba Corporation) were installed in the mold to measure states of the mold cavity. Based on the cavity measurement, we controlled parameters of the molding machine. As a polymer material, we used polybutylene terephthalate (PBT. Toraycon® 1101G-X54, Toray), which is reinforced by containing short glass fibers of 30wt%. TABLE 2 shows a molding condition used in the experiment.

A shape of the mold cavity was designed to produce a single-lap joint as shown in Fig. 4. The size or shape of this joint follows ISO 19095; the joining area is 5 mm × 10 mm. A photo image of produced IMDJ sample is shown in Fig. 5. A typical evaluation for the single-lap joint is to measure shear strength of joining by a tensile test. We performed the tensile test by using an original tensile tester developed for the single-lap joint [9]. The tensile test applies strain to a sample with constant speed (1 mm/min) until the sample is broken, and measures applied load. The shear strength, s, is calculated as: $s = l_{max} / a$, where l_{max} is maximum load during the test, and a is joining area (50 mm^2).

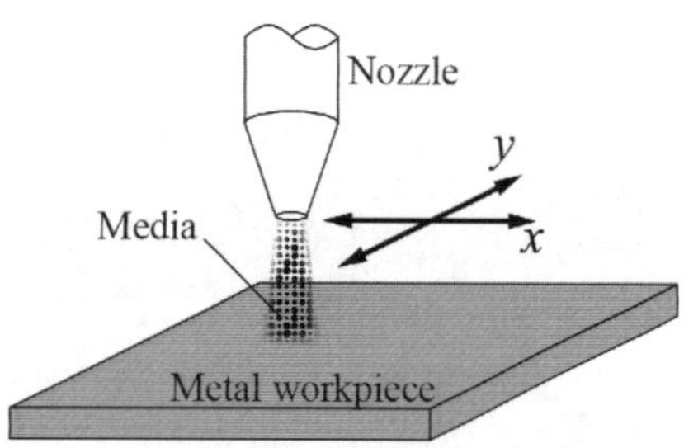

FIGURE 2. Schematic illustration of a micro blasting.

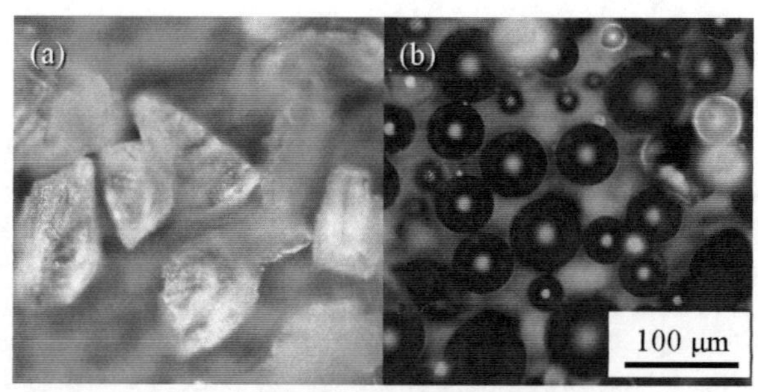

FIGURE 3. Optical microscope image of blasting media; (a) alumina particles, (b) glass beads.

TABLE 1. Specification of blasting media.

label	Model	Avg. size (μm)	Type
a	WA#400	47.5	
b	WAF180	71.5	
c	WAF150	84.5	Alumina
d	WAF120	108	
e	WAF80	181	
F	GB-AH	68.0	Glass

TABLE 2. Molding condition.

Condition	value
Packing pressure	90 MPa
Holding pressure	60 MPa
Holding time	8 s
Injection speed	300 mm/s
Melt temperature	270°C
Mold temperature	140°C

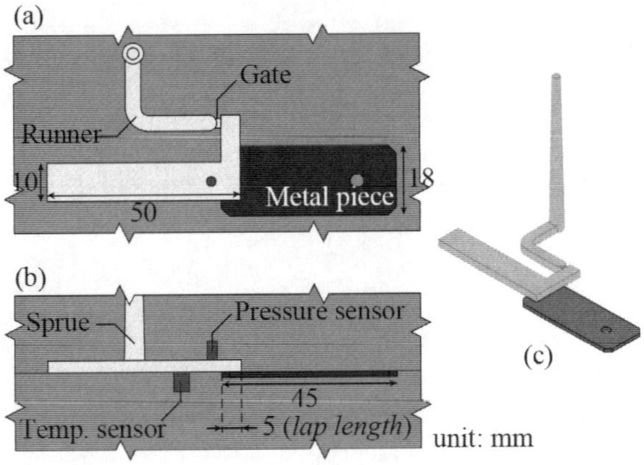

FIGURE 4. Schematic illustration of a mold; (a) top-view, (b) front-view, (c) single-lap joint produce by the mold.

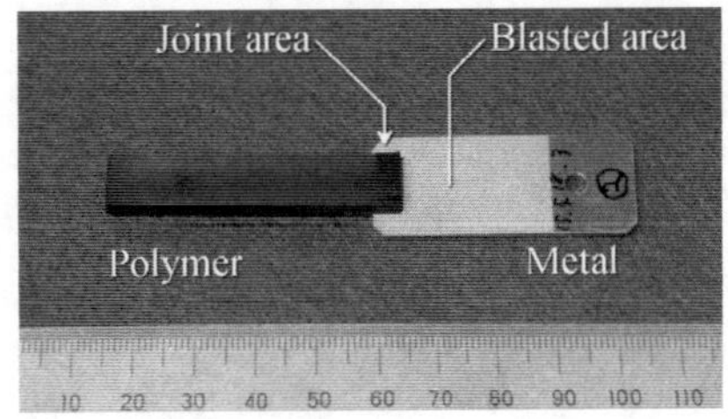

FIGURE 5. Photo image of IMDJ sample.

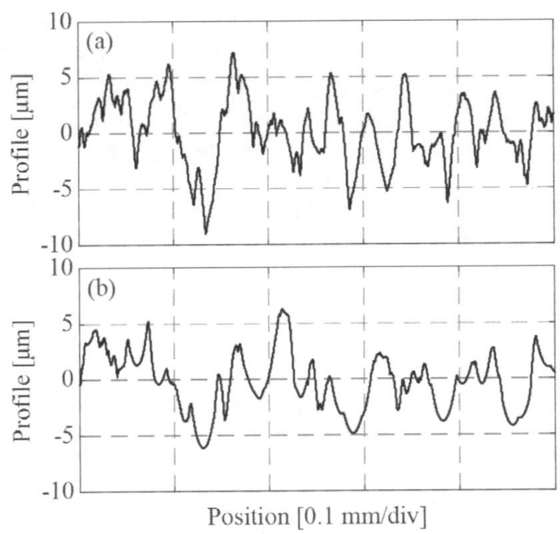

FIGURE 6. Measurement examples of surface profiles; (a) blasted by alumina particles and (b) by glass beads.

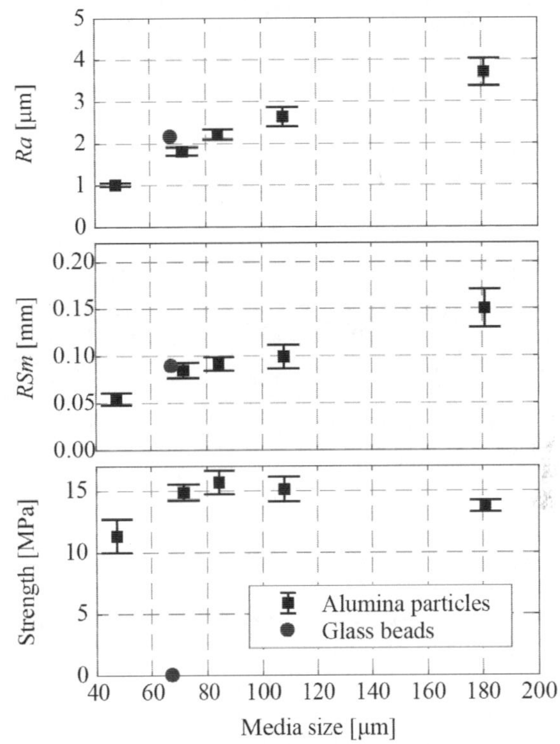

FIGURE 7. Calculated roughness parameters, *Ra* and *RSm*, and shear strength as functions of media size.

EVALUATION

Surface Profile Measurement and Roughness Calculation

To evaluate effects of the difference of the media, we measured surface profiles on the blasted surfaces. We used a two-dimensional (2D) surface roughness meter (SE300, Kosaka laboratory) that equips a diamond contact needle (apex radius: 2.0 μm. angle 60 degree) and a three-dimensional (3D) surface roughness meter (ET4000, Kosaka laboratory) that equips a diamond contact needle (apex radius: 0.5 μm. angle 60 degree).

Figure 6 shows measurement examples of 2D surface profiles made by (a) alumina particles and (b) glass beads. In the profile (a), both the peaks and valleys have similar shape, which is a sharp edge. On the other hand, the shapes of the peaks and valleys are different in the profile (b); most of valleys are downward arc shape curves (large radii), while the peaks are sharp shapes (small radii). The arc shape curves were made by replication of the shape of the glass beads.

From the measured 2D profiles, we calculated roughness parameters that were arithmetical mean deviation of the assessed profile, *Ra*, and mean width of the profile elements, *RSm*. Figure 7 shows the roughness parameters, *Ra* and *RSm*, and the shear strength of the joining samples under different media conditions. Markers and error bars show the mean values and the standard deviations, respectively.

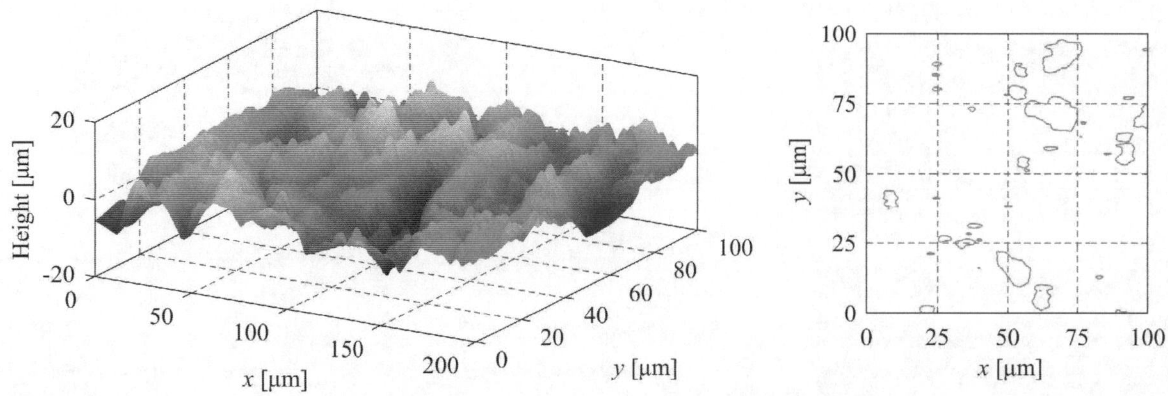

FIGURE 8. (Left) Example of three-dimensional surface profile that was provided by alumina particle. (Right) Contour image (100 μm x 100 μm) obtained by slicing the 3D profile at $z = 5$ μm. For the particle analysis, we counts the closed loop, which represents a filled part of a section of a workpiece.

Both the roughness parameters have positive correlations to the media size regardless of the difference of media types (shapes and materials). However, the strength of IMDJ provided by the glass beads is zero, which means that the metal workpiece blasted by the glass beads could not joined with the polymer part. This is because the profile made by the glass beads has the smooth arc valleys and then provides a small anchoring effect. Regarding the alumina particles, we cannot see any tendencies of relationship between the media size and the strength. This is because both the parameters increase with increasing of the media size. The larger *Ra*, which means the deeper dimples, could make the higher strength, on the other hand, the larger *RSm*, which means the lower density of dimples, could make the lower strength. Since increasing *Ra* and *RSm* provides a trade-off relationship for the strength, there was no tendency.

Particle Analysis

Figure 8 (Left) shows an example of three-dimensional (3D) profile measurements. From the 3D profiles, we carried out a particle analysis. The procedure of the particle analysis is as follows. First, the 3D profile is sliced in the *x-y* plane at a particular *z* level to get a contour image of the profile (Fig. 8 (Right)). The second step counts the number of closed particles (loops) from the contour image. Since the particle represents a filled part of a section of a workpiece, the number of the particles means the number of convex parts of the workpiece surface at the *z* level. Next, the slicing *x-y* plane is shifted by Δz (=1 μm), then it counts the particles again. Repeating this sequence from top to bottom of the surface makes a distribution of the convex parts. Figure 9 shows two-dimensional illustration of counted profile to simplify the concept of the particle analysis.

Figure 10 shows results of the particle analysis for each alumina particle. The difference of the markers means the model number (count number of the media). To easily see the data, several offsets were added to the result; offsets for the result of #80, #120, #150, #180, and #400 are 0, 100, 200, 300, and 400, respectively. From this plot, the maximum of the numbers (the peak value) is larger if the workpiece blasted by the smaller media. The profile made by the smaller media has higher density of dimples and could provide higher strength of joining. On the other hand, the range of *z* level, which is the distribution range of data, is larger if the media size is larger. The large range means the deep dimples that could provide the strong joining. Therefore, the result of the particle analysis shows a trade-off relationship between media size and strength as well, which is reasonable since it is same as the 2D roughness analysis.

CONCLUSION

This study dealt with surface structure analyses to evaluate the injection molded direct joining (IMDJ). Since strength of IMDJ could be affected by features of a surface structure of a metal workpiece, we investigated relationship between the joining strength and the geometric properties of the surface structure produced by the micro blasting using various media. The surface profiles were analyzed by using surface roughness parameters, *Ra* and *RSm*, that relates depth and density of dimples on the surface, respectively. In addition, we carried out a particle analysis to evaluate the surface in three-dimension. The results are summarized as follows:

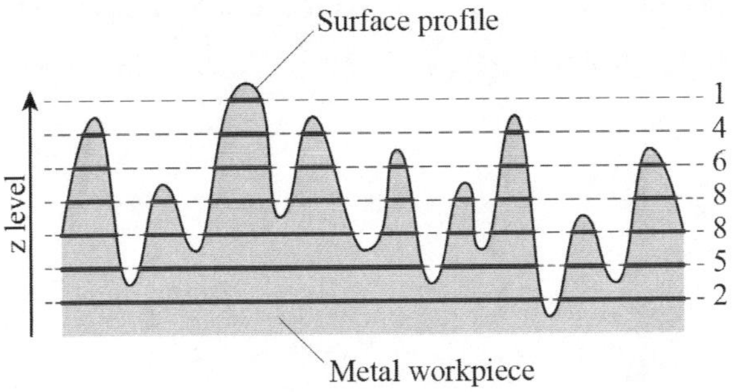

FIGURE 9. Concept of procedure of a particle analysis. To simplify the explanation, the concept image drawn in two-dimensional. The procedure counts the crossed parts of the workpiece surface at each z-level.

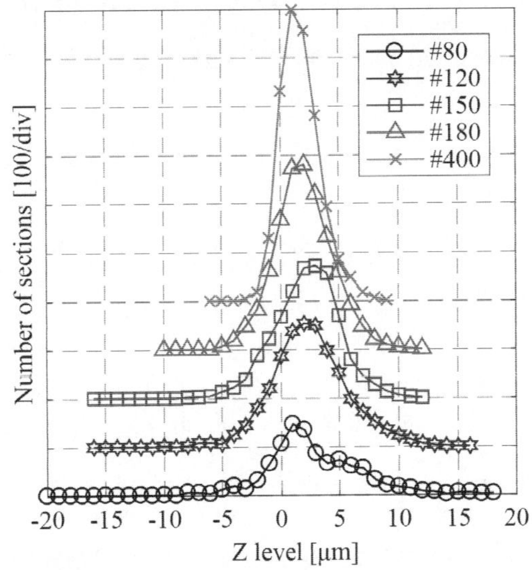

FIGURE 10. Result of a particle analysis. The data on each line are added offsets to see it easily.

- The micro blasting using glass beads as media cannot provide a surface strongly joined with molded polymer although the roughness parameters are in the same range with the surface made by alumina particles.
- The larger media in size provide the higher *Ra* and *RSm*.
- The strength has no correlation to the media size.
  - This could be due to a trade-off relationship between effects of *Ra* and *RSm* on the strength.

In this study, we analyzed the surface structures in the viewpoint of difference of media. Therefore, the roughness parameters, *Ra* and *RSm*, could not be controlled independently. To evaluate the surface by the independent parameters, we will investigate the effects of other blasting conditions as well as media in future work.

ACKNOWLEDGMENTS

This study was supported by JSPS KAKENHI (#15K17946) and The Foundation for the Promotion of Industrial Science. A part of the experiments was supported by Yokoi group, the University of Tokyo, Japan.

REFERENCES

1. K. Ramani and B. Moriarty, Thermoplastic bonding to metals via injection molding for macro-composite manufacture. Polym Eng Sci, 38(5), pp. 870-877, 1998.
2. G. Lucchetta et al., Aluminum sheet surface roughness correlation with adhesion in polymer metal hybrid overmolding. CIRP Ann – Manuf Technol, 60(1), pp. 559-562, 2011.
3. Y. Tamura, et al. Metal-polymer direct joining with abrasive blasted metals. In proc. of ICPE 2016, C309-8173, 2016.
4. K. Taki, et al., Direct joining of a laser-ablated metal surface and polymers by precise injection molding. Microsyst Technol, 22, pp. 31-38, 2016.
5. B. Huang et al., Experimental investigation of the strength of polymer-steel direct adhesion (PSDA) joints with micro-structures ablated by laser. Journal of Materials Processing Technology, 249, pp. 407-414, 2017.
6. N. Andoh and K. Osumi, "Nano Molding": permanent bond between aluminum and polymer. Kunststoffe international, 9, pp. 106-111, 2005.
7. P. Fabrin, et al., Adhesion of thermoplastic elastomer on surface treated aluminum by injection molding. Polymer Engineering & Science, 47(8), pp. 1187-1191, 2007
8. F. Kimura et al., Effects of molding conditions on injection molded direct joining using a metal with nano-structured surface. Precision Engineering, 45, pp. 203–208, 2016.
9. S. Kadoya et al., Development of a tensile test machine for metal-polymer direct joining. In proc. of ISMTII 2015, 1112, 2015.

Full surface sink mark inspection on large high gloss plastic components by robot vision

D. P. Gruber[a*], M. Haselmann[b]

Polymer Competence Center Leoben GmbH, Roseggerstraße 12, 8700 Leoben, Austria

[a)] dieter.gruber@pccl.at
[b)] matthias.haselmann @pccl.at

Abstract. In this contribution a new methodology for automated sink mark assessment by machine vision is presented. The methodology is based on phase measuring deflectometry in combination with discrete orthogonal polynomials to render curvature representations of the surface to be measured. It is capable to provide objective and repeatable results for very large components (>1000mm) within 5 seconds for the first time.

INTRODUCTION

Sink marks as a visible surface defect form a persistent problem especially in the automotive component industry. Even a height variation of only 1 to 3 micrometers causes a visible sink mark on a glossy surface. This is particularly apparent in case of experienced examiners. Such small height-changes are frequently caused by back side construction elements of polymer components and they are not completely avoidable in most cases. In view of constantly increasing quality requirements for surface components in premium cars, the immaculateness of glossy parts has to be ensured by inspection of every produced part. Unfortunately, the occurrence of sink marks is not a binary "good" / "bad" event. The transition between an acceptable and an unacceptable sink mark is smooth for the naked human eye. Therefore, human assessment of sink marks suffers from and subjectivity and non-reproducibility, which causes major problems between manufacturers and customers. In this paper a new methodology for automated sink mark detection with fast and repeatable results also on large surfaces by machine vision is presented.

METHODOLOGY AND RESULTS

The methodology is based on phase measuring deflectometry in combination with discrete orthogonal polynomials to render curvature representations of the surface to be measured (compare Fig. 1). It is capable to provide objective and repeatable results even for very large components. With the new methodology sink mark measurement can be performed within 5 seconds at parts of 1200mm length. Therefore, the methodology is suitable for automated in-line sink mark quality assessment by machine vision during production and hereby solves a huge problem in the field of high quality component production.

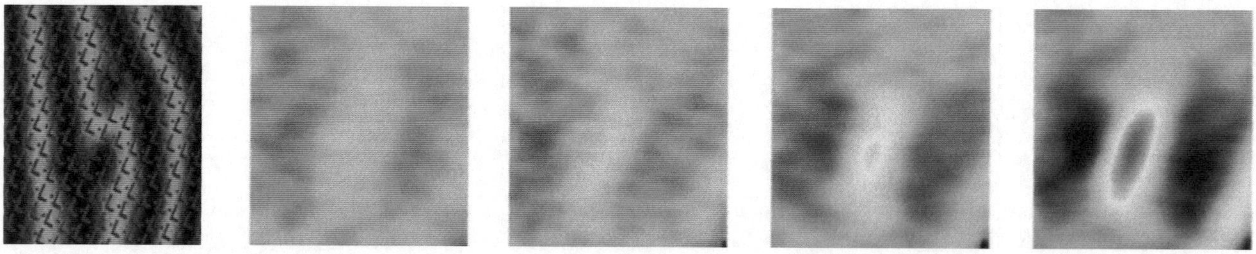

FIGURE 1. a) Example of a glossy part surface with a decorative pattern that shows a critical sink mark. The white stripes origin projected patterns for phase measuring deflectometry. b) to d) show calculated curvature representations showing the same surface region with increasing visible sink mark intensity.

pps proceeding
Symposium: S02 - Injection molding
Oral Presentation

A virtual study on cooling a hotspot in injection molding by laser-metal-deposited mold insert cores

Berger Gerald Roman [1]*, Zorn David [2], Bevc Franz [3], Brandstätter Elmar [4], Görgl Richard [4]

[1] Montanuniversitaet Leoben, Dpt. Polymer Engineering & Science - Styria - Austria, [2] Montanuniversitaet Leoben, Institute of Injection Molding - Styria - Austria, [3] MAHLE Filtersysteme Austria GmbH - Carinthia - Austria, [4] Joanneum Research Materials - Styria - Austria

The aim of this study was to optimize the manufacturing process of an oil filter housing for automotive applications. To improve the cooling of a mass accumulation, a copper-beryllium alloy mold insert with drilled cooling channels was installed, which reduced the hotspot significantly; however, its wear resistance was too weak for the polymer used (PA66GF35). Thus we searched for a solution where cooling performance is equal and wear resistance is superior. By using simulation software Sigmasoft v.5.1, we analyzed two first approaches: A) We filled up the insert core by copper and shielded it by a 3mm steel skin. B) We designed conformal cooling channels (Ø3.5 mm) in a Selective Laser Sinter (SLS)-able steel grade insert. In the insert socket region, the cooling was improved significantly by both designs. As a consequence, we had to switch off the cooling pulse earlier not to let fall the mold insert surface temperature below the minimal recommended 60°C, even though the (cooling time determining) finger-like insert tips were cooled worse, as neither copper nor the conformal channels reached that region. Thus we had to improve the cooling in the critical fingers and weaken the cooling in the socket region by splitting the mold insert cooling in 2 separate circuits, where the critical fingers were cooled by C) slim conformal copper cores in a copper heat sink, D) cylindrical copper pins growing from a copper heat sink, and E) cylindrical copper pins surrounded by cooling water flow at their free ends, all in a steel mold insert. At the moment of ejection (after 75s), where the recommended ejection temperature is below 210°C, the relevant hotspot temperature at the molded part surface was 141°C, 203°C, 210°C, and 211°C, for the original design, C), D), and E), respectively. To conclude: Conformal copper cores were able to substitute a copper-beryllium mold insert without raising the cycle time. Future studies will try to manufacture those by Laser Metal Deposition.

Direct flow visualization of hesitation during injection molding of thermoplastic polymers

Francesco Regi[a], and Guido Tosello[a]

[a]*Department of Mechanical Engineering, Technical University of Denmark, Produktionstorvet, Building 427A, 2800 Kgs. Lyngby, Denmark*

Abstract. A special mold provided with a glass window has been used in order to directly evaluate the flow progression during the filling phase of the injection molding process. The flow of the polymer was recorded at 500 frames per second using a high-speed camera. Two unfilled thermoplastic polymers, acrylonitrile butadiene styrene (ABS) and polypropylene (PP), were used to fill a 50 mm x 18 mm staircase geometry cavity, which was specifically designed to evaluate the hesitation effects. Complete and incomplete (short shot) filling of the cavity was performed with two different velocities and in the case of two geometry thicknesses of the same design. Two different methods for the reproduction of the short shots were used. In the first, the maximum applied pressure was limited and progressively raised (MP Method), in the second the desired positions were obtained with a consecutive increase of the stroke length (standard SS method). The analysis of the video recordings highlighted that flow progression and hesitation were mainly influenced not only by the geometry, but also by the velocity and the material.

Keywords: Glass Mold, High Speed Camera, Injection Molding, Afterflow, Hesitation, Thermoplastic.
PACS: 83.50.Uv.

INTRODUCTION

In recent years, glass inserted molds have received great attention as a direct flow visualization methodology in injection molding and other related manufacturing processes. Although their use presents some limitations, several authors have chosen glass inserted mold coupled with various camera setups in order to analyze the flow behavior [1-9].

Murata et al. [1,2] have obtained simultaneous visualization from two directions thanks to a setup consisting of a camera on the glass side, projected light through the back side glass and mirrors. Yang et al. [3] have characterized a microinjection-molding tool and investigated the flow into micro parts with high injection speed. They used a quartz glass window in combination with a reflection mirror and captured videos at 1000 frames per second with high-speed camera.

Nian [4] has chosen the same setup but using different pigments in the polymer blend, he obtained further information about the flow. Han and Yokoi [5] used a higher speed camera capable of recording at 13,500 frames per second and a microscope in order to analyze the filling of micro grooves on a glass prism insert.

Layser and Coulter [6], investigated the hesitation of the melt flow in the mold cavity. Their results showed that the flow hesitation is primarily dependent on the material, but also on the melt temperature and packing pressure. Similarly, in the present paper, we observed a strong influence of the material, but also of the velocity and geometry used.

Spares et al. [7] have analyzed the thermal field generated during microinjection molding with a thermal imaging camera. Whiteside et al. [8] investigated micro molding with a high-speed camera and sapphire glass insert, which allowed them to study different effects in microinjection molding.

Recently, Sorgato et al. [9] have integrated a sapphire window and a mirror into a system originally designed for microinjection molding venting. Thanks to this new setup, they could evaluate the effect of vacuum venting on the replication and definition micro/nano features.

Flow visualization during injection molding has also been performed with mold designs different from the ones described. In other solutions, a transparent insert was directly applied to the mold cavity parallel to the injection direction, which allowed the direct filming of the flow from the side without the use of a reflective mirror.

Yanev et al. [10,11] added a sapphire insert to this mold design and performed a simulation comparison to verify the filling performance of the software. Similarly, Guerrier et al. [12] used a mold with a glass window and an integrated multi-turn induction heating to capture the effect of induction heating with a high-speed camera. Guerrier

[13] used the same mold to directly compare simulation and experimental moldings from an injection molding machine.

The purpose of the present study was to characterize the hesitation effect in thin cavities using the experimental setup developed in [12],[13] where a turned insert parallel to the injection screw allows the direct visualization of the flow in the cavity.

EXPERIMENTAL

Glass mold and Machine

A special mold was used where the injection plane was approx. perpendicular to the opening and closing plane of the glass mold, not to hamper the opening/closing procedure. The glass was tilted of 8°, i.e. 82° with respect to the opening plane, as shown in Figure 1.

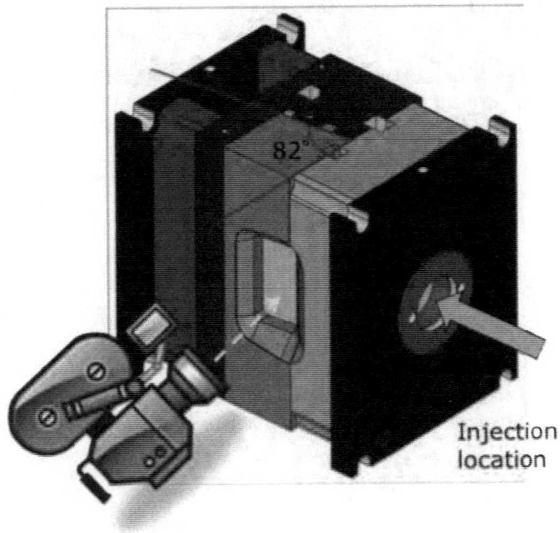

FIGURE 1. Schematic of the mold and of the tilting angle of the glass insert

This angle was chosen in order to avoid a too tight fit and therefore a possible breakage of the glass when the holding force was applied. The glass window was made of a 55 mm thick Borosilicate glass (width 60 and height 140 mm), and it was capable of withstanding at least 130 MPa of machine pressure during injection, and 50 MPa during packing. In Figure 2 the mold and the position of the camera are depicted.

An AOS Technologies AG (Fislisbach, Switzerland) X-PRI high-speed camera able to record 1280 x 1024 pixels images at 500 frames per second was used. A spotlight and two light bulbs were installed to increase the brightness of the video. A signaling diode turned on when the filling phase began and turned off at the switch over point. In addition, the mold was provided by three pressure sensors, two thermocouples, two interchangeable inserts, a cooling system and an ejector system installed on an Engel e-motion 110 with a screw diameter of 25 mm.

Test Part and Materials

Two inserts with the same geometry were designed to match the constraints imposed by the glass mold geometry and the objectives of the study. The staircase geometry was chosen in order to evaluate the extent of the hesitation and the afterflow in steps of different thickness, particularly in very thin sections where those effects become preponderant. The geometry comprised of a later channel and several lateral steps. The only difference between the two inserts was the thickness of the lateral steps. The common dimensions of the inserts are presented in Figure 3, while in Figure 4 the dimensions of the steps of the first (Cavity 1) and of the second (Cavity 2) insert are shown in black and blue, respectively. Cavity 1 had a thickness of 0.3 mm at the smallest step, it raised by 0.3 mm at each step to a final value of 1.5 mm; Cavity 2, had a thickness of 0.5 mm at the smallest step and it increased by 0.25 mm at each step to a final 1.5 mm.

Acrylonitrile butadiene styrene (ABS), a thermoplastic amorphous plastic material with a melt flow index (MFI) of 34 cm^3/10min and Polypropylene (PP), a thermoplastic semi-crystalline opaque polymer with a MFI of XX were used to characterize the influence of the material on the hesitation.

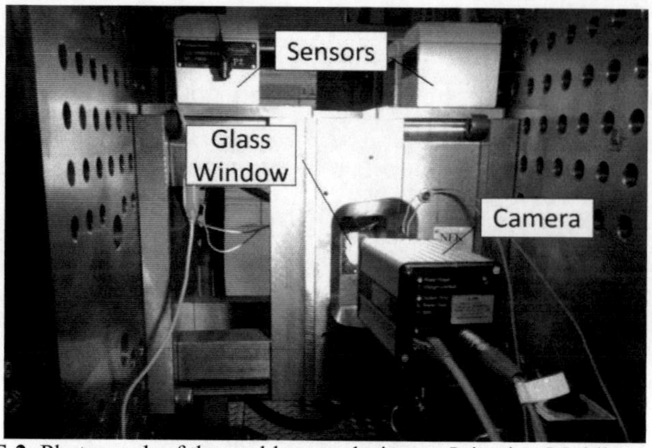

FIGURE 2. Photograph of the mold setup during an Injection Molding cycle. It comprises HSC, Glass Window and Sensors.

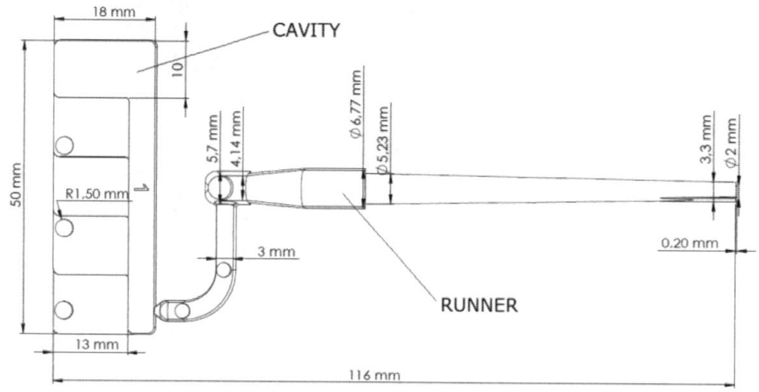

FIGURE 3. Technical drawing of the insert Cavity 1, comprehensive of gate and runner.

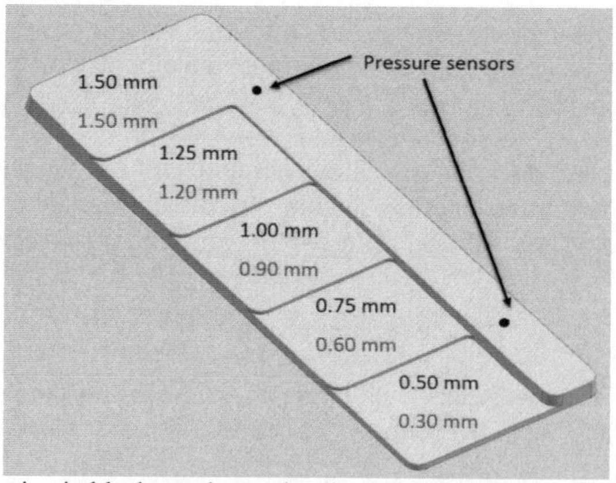

FIGURE 4. Schematic of the cavity: in black are shown the dimensions of cavity 1, in blue the dimensions of cavity 2. The positions of two sensors are also shown.

Experiments

For each material and each cavity the same setup of experiments was performed, for a total 20 runs. Table 1 summarizes the experimental combinations consisting of two complete fillings with a constant injection velocity of 40 mm/s (ComF) and 10 mm/s (ComM) and three short shots experiments. The short shots were obtained using two different methods.

In the maximum pressure (MP) method, the different filling percentages of the cavity were obtained limiting the maximum applied pressure and progressively raising the limit. MP short shots were produced with a very low injection speed.

In the second method, the desired positions were obtained with consecutive increments of the stroke length at a certain injection velocity. They were addressed as SS. Similarly to the complete experiments, SSF runs with an injection velocity of 40 mm/s, while SSM with 10 mm/s. Those velocities were chosen to allow a direct comparison between the complete and the short shots experiments.

The rest of the processing parameters were selected following the recommendations of the material suppliers and kept constant throughout the experiments.

To summarize:

- ComF refers to the complete filling experiments with an injection velocity of 40 mm/s;
- ComM refers to the complete filling experiments with an injection velocity of 10 mm/s;
- SSF refers to the short shots obtained varying the stroke length with injection velocity of 40 mm/s;
- SSM refers to the short shots obtained varying the stroke length with injection velocity of 10 mm/s;
- MP refers to the short shots achieved limiting the maximum applied pressure.

TABLE 1. Experimental Summary.

Cavity	Material	Velocity[mm/s]	Type
Cavity 1	ABS	40	Complete Filling
Cavity 2	PP	10	Short Shot (SS)
			Short Shot (MP)

Hesitation

Hesitation occurs wherever there are variations in wall thickness of injection molded parts. When the melt advances through a cavity and reaches a location where both a thin and thick region are present, the thinner region provides higher flow resistance slowing down the incoming fluid. On the other hand, the melt flowing through the thicker region continues at a faster flow rate.

In the thin region the melt will cool down more rapidly as it moves forward, increasing its viscosity. Besides, the reduced velocity will decrease the shear rate and increase the viscosity due to the non-Newtonian characteristics of polymer melts, therefore the melt will quickly become more viscous and the frozen layer will rapidly grow.

These combined factors can exaggerate the hesitation to the point at the material will completely freeze off.

It is possible to correlate the hesitation effects to the pressure drop through (1):

$$\Delta P = \frac{12 \cdot Q \cdot l \cdot \eta}{w \cdot h^3} \quad (1)$$

where Q is the flow rate, h is the thickness of the flow, l is the length, w is the width and η is the viscosity in a rectangular section.

Viscosity increases due to the non-Newtonian shear thinning behavior of the material and as consequence of the rapid cooling in the thin section. When the flow channel thickness is reduced, the pressure raises and therefore the viscosity increases due to the shear thinning behavior. Consequently, the velocity reduces and thus the frozen layer thickness rapidly increases. This causes a further increase of the viscosity which provokes an additional pressure rise. In the worst scenario, this combination of effects results in the premature freeze of the incomplete part, which causes a short shots. If the melt is instead able to fill the cavity, the hesitation effects can result in retained internal stresses of the part, as consequence of the high pressure, and in surface defects such as hesitation marks and weld lines.

HIGH SPEED VIDEO ANALYSIS

Method

A reference system was set in order to accurately follow the video records of the cavity fillings. As presented in Figure 5, the zero point was at the intersection between the thin step's external wall and the lateral channel internal wall. Y axis was parallel to the channel and the X axis was perpendicular along the step direction.

Measurements had been performed using "ImageJ" software [14]. The pixel distances were converted in millimeters with reference to the total length Y of the cavity.

The penetration degree X in the side steps was calculated as an average between the maximum penetration and the average penetration on each step. This value gives an estimation of the hesitation for a certain position Y in the lateral channel, which was then used as the reference for the evaluation. Therefore, the results are relative to a certain Y length traveled by the melt flow in the lateral channel. In the case of the short shots, this was the position where the melt stopped moving (before the retraction caused by the cooling), while in the case of the complete experiments, the same position was extracted by the video analysis. Although a total of three positions were chosen, only the results relative to Position 3 (Figure 5 and 6), are shown in this paper since these were the most significant.

The length of the cavity has been measured 12 times in order to evaluate the variability of the measurement method. The resulting standard deviation was 0.3 mm. The related uncertainty was calculated following the guidelines of GUM [15] and was equal to 0.098 mm paragraph.

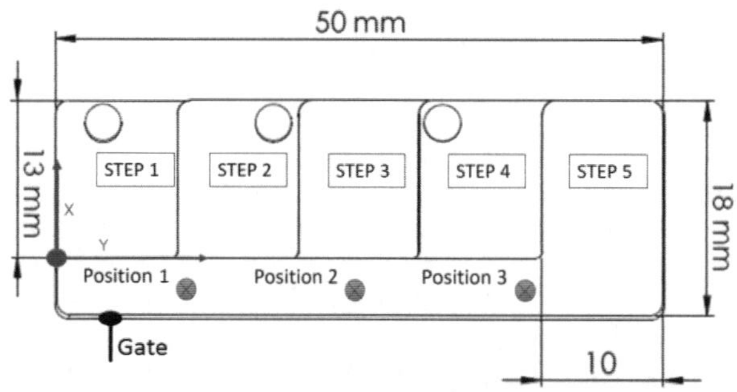

FIGURE 5. Reference system applied in the measurements and the melt positions in the lateral channel considered in the study. These positions are also the final length of the short shots.

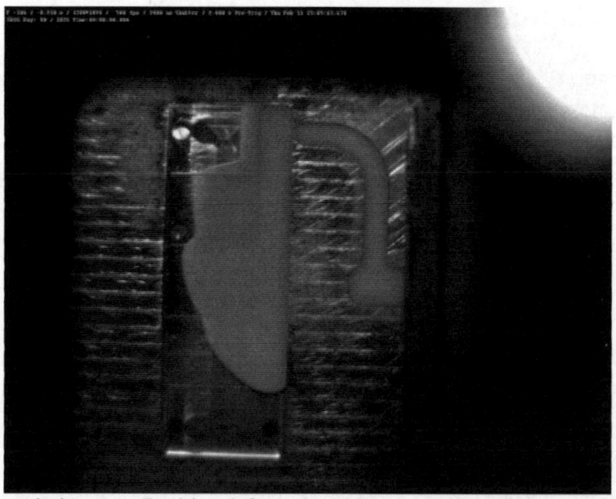

FIGURE 6. Extracted picture at Position 3 from the video recordings of a complete experiment.

Results

In all the experiments, the length of the melt flow in the lateral channel was chosen to match as accurately as possible the selected position. Therefore, the length Y was a measure of the accuracy of the position selection. Figure 7a shows the main effects plots for Y relatively to Material, Cavity and Type of experiment.

It can be noticed that the traveling distance in the lateral channel is lower in the case of Cavity 2 and PP experiments than in the case of Cavity 1 and ABS experiments. This suggests an overestimation of their relative hesitation. A notable difference of the length Y is also present between the short shots SS experiments and other experiments. Hence, on the opposite, the hesitation for the experiments SS will be underestimated.

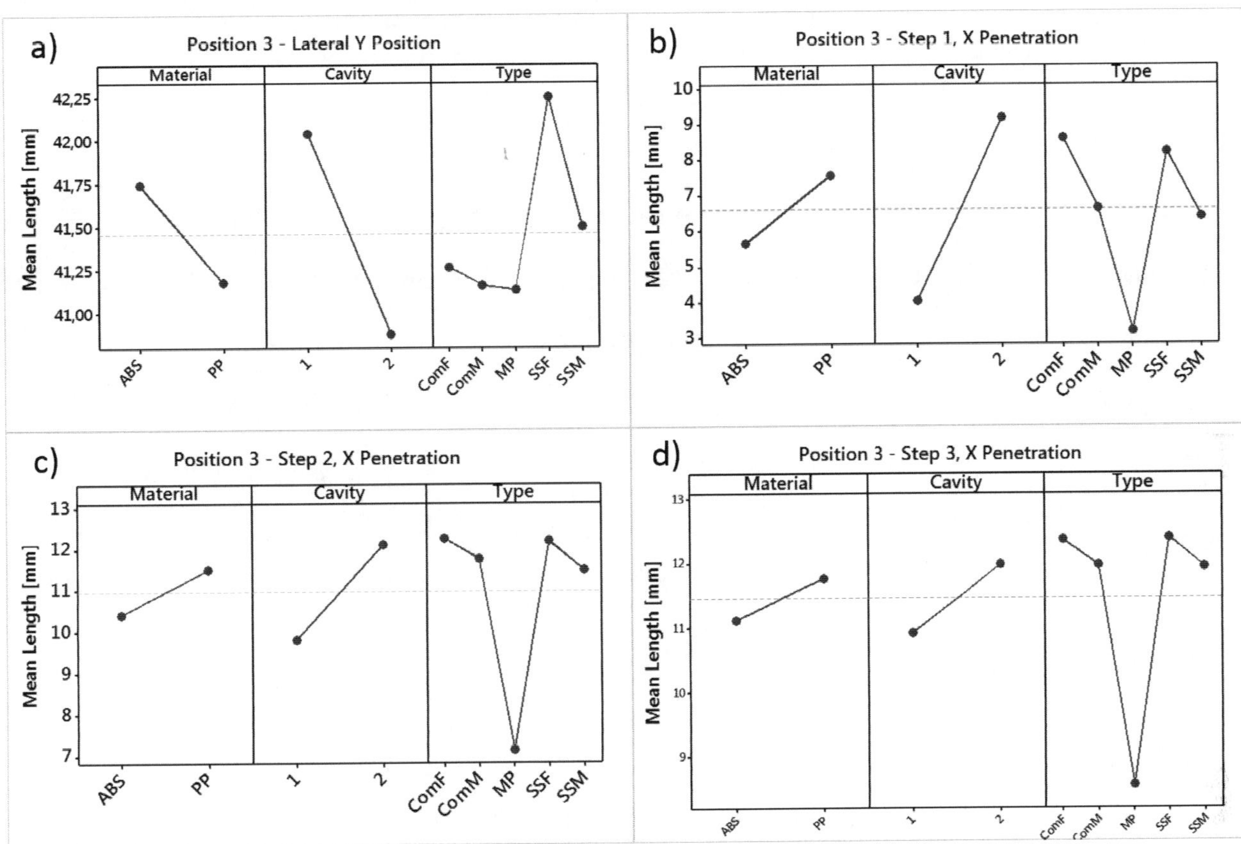

FIGURE 7. a) Main effects plots for Y length, measure of the accuracy of the position selection, relatively to Material, Cavity and Type of experiment. b) Main effects plots of the degree of penetration X in the Step 1, c), Step 2, d) Step 3.

Nevertheless, the main effect plots depicted in Figures 7b, 7c and 7d highlight that the hesitation is higher in the ABS experiments, and it is mostly affected by the thickness of the cavity, i.e. higher hesitation in thinner sections. In the case of the MP experiments, the flow was not able to penetrate in none of the step due the very low velocities applied; therefore, it will not be taken in consideration further in.

SS and Complete experiments were in close agreement, but a slight underestimation of the hesitation in short shots SS could be noticed. This needs certainly to be considered during the evaluation of filling in molds without glass windows.

It is also worth noticing how the effect of material and cavity diminishes for thicker steps, as expected by Equation 1. Similarly, the velocity (type of experiment) is more important than the material type for Step 1 (0.3 for cavity 1 and 0.5 mm for cavity 2) hesitation, while the opposite is true for thicker steps.

In conclusion, it was possible to draw a few summarizing observations:

- Thickness is the most important factor;
- The velocity of filling and the thickness are not linearly related;
- PP showed less hesitation effect than ABS;

- MP short shot method is not suitable for the characterization of the flow in the cavity;
- Short Shots underestimate the hesitation occurring during the filling phase.

CONCLUSION

The present study investigated the hesitation of unfilled thermoplastics through direct flow visualization. The setup included a transparent insert directly applied to the mold cavity. The mold cavity was parallel to the injection direction, so the flow was directly filmed from the side of the injection-molding machine with a high speed camera. The cavity, consisting of a staircase geometry specifically design to investigate hesitation effects, has been filled with an amorphous (ABS) and a, semi-crystalline (PP) polymers. The cavity thickness had a strong influence on the hesitation of the plastic melt, whereas velocity and type of material affected it less. When the thicknesses decreases to 0.3-0.5 mm, the velocity becomes mainly dependent on the type of thermoplastic.

In addition, the short shots at very low injection velocity, varying the max pressure, namely MP, did not provide a suitable characterization of the filling of the cavity. On the other hand, the short shots at consecutive increments of the stroke length underestimated the hesitation of a complete filling but provided a more accurate representation.

ACKNOWLEDGMENTS

This research work was part of the project 5163-00001B funded by Innovation Fund Denmark.

REFERENCES

1. Y. Murata et al., "Visual Analysis of Short-shot Front Behaviors by Glass-Inserted Mold" in *Seikei-kakou 6,* 1994, pp 703–711.
2. H. Yokoi et al., "Visualization Analysis of Undesired Phenomena on Molded Surface in Injection Molding Process." In *Journal of The Surface Finishing Society of Japan 51*, 2000, pp 1182–1186.
3. S. Yang et al., "Flow visualization of filling process during micro-injection molding." in *Intern. Polymer processing 4*, 2002, pp. 354–360.
4. S. Nian et al., "Flow visualization of filling with aid of colored billets during impact micro-injection molding." in *Intern. Polymer processing 4,* 2004, pp. 402–407.
5. X. Han et al., "Visualization analysis of the filling behavior of melt into microscale v-grooves during the filling stage of injection molding." in *Polym Eng Sci 46*, 2006, pp. 1590–1597.
6. G.S. Layser et al., "Effect of processing parameters on flow hesitation with glass-filled reinforced thermoplastics: Experimental investigation with direct visualization.", ANTEC Conference Proceedings, 2009, pp. 342-348.
7. R. Spares et al., "High speed thermal imaging of micromoulding.", ANTEC Conference Proceedings, 2009, pp. 2570-2573.
8. B. Whiteside et al., "Rheo-optical measurements in micromoulding.", ANTEC Conference Proceedings, 2009, pp. 2803-2810.
9. M. Sorgato et al., "Investigation of the influence of Vacuum Venting on Mould Surface Temperature" in *Micro Injection Moulding. International Journal of Advanced Manufacturing Technology 88*, 2017, pp. 547-555.
10. A.S.Yanev et al., "Direct visualization of conventional injection molding.", ANTEC Conference Proceedings, 2007, pp. 1097-1101.
11. A.S.Yanev et al., "Visualization of injection molding process." in *Materials Science Forum*, 2008, pp. 716-720.
12. P. Guerrier et al., "Three-Dimensional Numerical Modeling of an Induction Heated Injection Molding Tool with Flow Visualization." in *International Journal of Advanced Manufacturing Technology 85,* 2016, pp. 643-660.
13. P. Guerrier et al., "Flow Visualization and Simulation of the Filling Process during Injection Molding." in *CIRP – Journal of Manufacturing Science and Technology 16,* 2017, pp. 12-20.
14. A.C.Schneider et al., "NIH Image to ImageJ: 25 years of image analysis." in *Nature methods 9(7)*, 2012, pp. 671-675
15. Evaluation of measurement data — Guide to the expression of uncertainty in measurement. JCGM 100:2008

Effect of Laser-structured Metal Surface on Strength of Metal-Polymer Direct Joining

Kakeru Enami[a], Fuminobu Kimura[b], Keisuke Yokoyama[a], Takeshi Murakami[a] and Yusuke Kajihara[b]

[a]*Core Technology R & D Center, NSK Ltd., 1-5-50 Kugenumashinmei, Fujisawa-Shi, Kanagawa 251-8501, Japan*
[b]*Institute of Industrial Science, The University of Tokyo, 4-6-1 Komaba, Meguro-ku, Tokyo 153-8505, Japan*

Abstract. This paper experimentally and theoretically investigates the effect of metal surface structures on metal-polymer direct joining. The direct joining technique is based on a combination of a surface roughening process of metal and an insert molding. In this study, we applied a laser machine to form periodic micro-dimples with a different aspect ratio and periodicity. We manufactured lap joint specimens under a constant molding condition and carried out tensile-shear tests. Finite Element Method was used to investigate the effect of the aspect ratio. Test results showed that the joining strength depended on the aspect ratio and periodicity of the dimple. FEM results showed that the principal stress generated in polymer decreased and saturated with increasing the aspect ratio. We discussed the effect of the surface structures on the joining strength from experimental results and FEM results.

Keywords: Metal-Polymer Direct Joining, Insert Molding, Laser-structuring, Finite Element Analysis
PACS: 81.20.Vj, 83.50.Uv, 42.62.Cf, 02.70.Dh

1. INTRODUCTION

Composite materials of metal and polymer are now strongly required as a way of reducing weight in automotive industries. Conventional methods to obtain composite materials are mechanical fastening such as bolts or using adhesives. However, these methods often lead to an increase in manufacturing processes. A solution to the problem is to directly join metal and polymer structures, in other words, to join them without any extra part. Recently, many researchers have been studying direct joining techniques and proposed various types of joining methods. In this study, we focus on one of the joining techniques, based on a combination of a surface treatment of metal and an injection molding [1-7]. The direct joining technique enables to join metal and polymer structures at the same time when the polymer structure is produced. This method could be a promising candidate of obtaining composite materials. However, there are some challenges to be overcome toward practical use. The joining mechanism and optimum joining conditions still remain unclear. Recently, a lot of effort has been made to solve the problems using different ways of metal surface treatments. The surface treatment methods are chemical etching [1, 2], abrasive blasting [3-5] and laser machining [6, 7]. To reveal the joining mechanism, two important factors which affect the joining properties have been mainly focused: surface structures of metal and molding conditions. However, details of the joining mechanism remain to be elucidated.

In this study, we applied a laser machining technique to form micro-dimples on the metal surface. Laser machining enables to control surface structures geometrically. The controlled structures promote a better understanding of the joining mechanism. The aim of this study is to experimentally analyze the relationship between surface structures and joining strength. We focus on two important geometries of the structure. One is the periodicity of each dimple and the other is the aspect ratio of the dimple. The effect of the aspect ratio is also examined with finite element method (FEM).

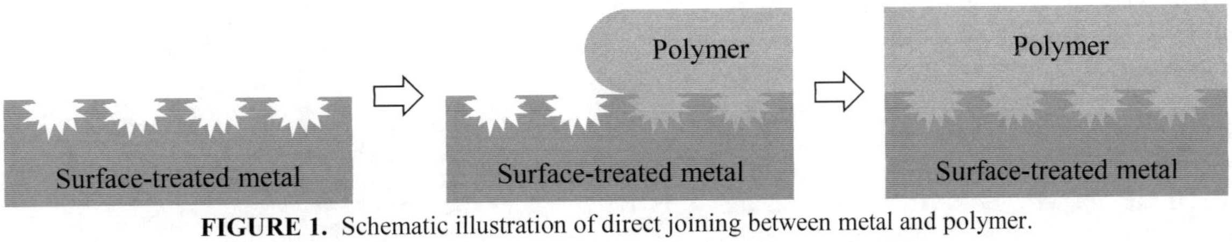

FIGURE 1. Schematic illustration of direct joining between metal and polymer.

TABLE (1). Laser machining conditions.

Pulse width	120 ns	Laser spot diameter	48 μm
Wavelength	1064 nm	Laser shots	1, 5, 10, 15
Laser power	50 W	Periodicity	75 ,112.5, 150 , 187.5 μm
Pulse repetition rate	50 kHz		

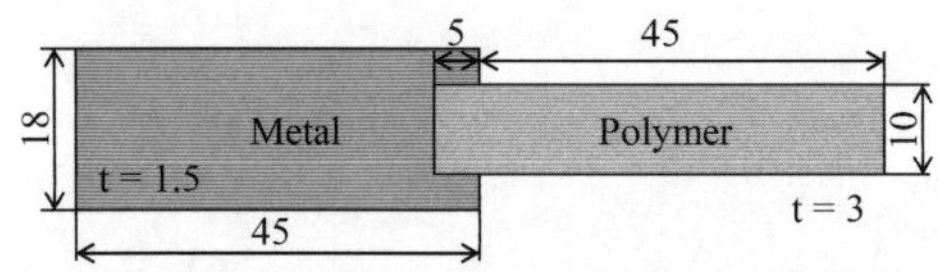

FIGURE 2. Lap joint specimen.

TABLE (2). Molding conditions.

Pack pressure	90 MPa
Holding pressure	60 MPa × 8 s
Injection speed	100 mm/s (38.0 cm^3/s)
Cylinder temperature	270 °C
Mold temperature	140 °C

2. EXPERIMENTAL

2.1 Joining Process

Figure 1 shows a schematic illustration of direct joining between metal and polymer. The joining process is as follows: (i) forming micro/nano structures on a metal part and (ii) insert molding (getting melted polymer into the surface structures). Details of each process is described below.

We utilized a nanosecond pulse laser to roughen a metal piece. The pulse width and wavelength of laser beam were 120 ns and 1064 nm, respectively. The metal piece was aluminum alloy A5052. We formed dimples on square lattice vertexes. Two parameters for this surface structure were controlled. One was the periodicity of each dimple and the other is the aspect ratio of the dimple. Table 1 shows laser machining conditions. A maximum laser power was 50 W. A reputation rate of laser beam was 50 kHz. We irradiated laser beam to make dimples in a depth direction. We controlled the depth of the dimple with the number of laser shots. The number to make one dimple was 1, 5, 10 and 15. The periodicity of each dimple was 75 μm, 113 μm, 150 μm and 188 μm.

In this study, we manufactured a lap joint specimen as shown in Fig. 2 to evaluate the joining strength between metal and polymer. We utilized an injection molding machine (ROBOSHOT α-S100iA, FANUC). The polymer material was polybutylene terephthalate containing 30 wt% of glass fibers (Toraycon® 1101G-X54, Toray). Table 2 shows molding conditions. Molding conditions were set to be constant because we focused on the effect of surface structures on the joining strength.

2.2 Evaluation of Joining Strength

To evaluate the joining strength of the lap joint specimens, we carried out tensile shear tests with an original tensile tester [8]. The tensile tester applied a displacement to the lap joint specimen at a constant speed of 1 mm/min and recorded a load during the tensile test. We defined the joining strength as the maximum load divided by the joining area (5 mm × 10 mm).

3. RESULTS AND DISCUSSION

3.1 Observation of Laser-structured Surface

Figure 3 shows a micrograph of laser-structured dimples on a metal surface. We can see that the diameter and periodicity of the dimple are well controlled. The diameter and depth of the dimple were evaluated with an optical microscope. The depth is defined as the distance between the upper surface and the lower surface of dimple at the respective focal points. Figure 4 is measurement results of the dimple dimensions. Figure 4(a) demonstrates that the diameter of the dimple

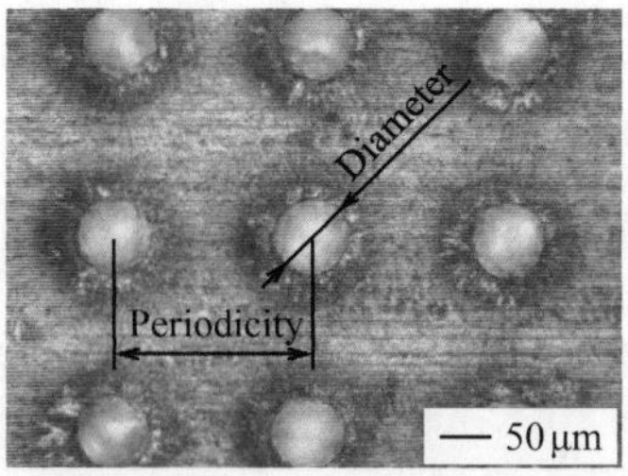

FIGURE 3. Micrograph of periodic dimples.

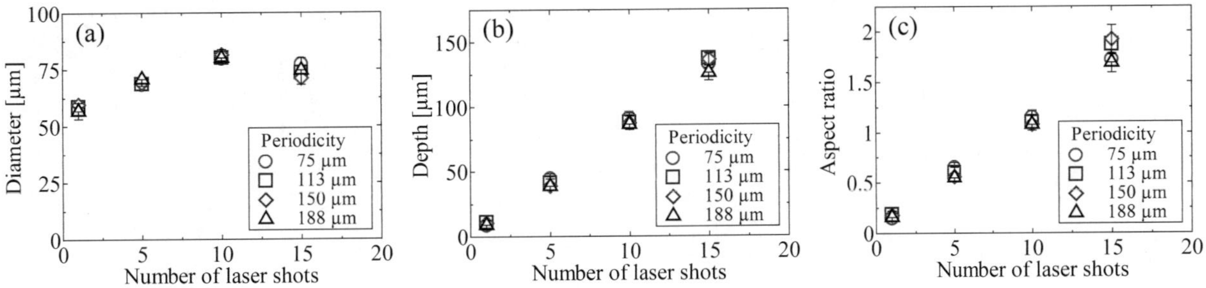

FIGURE 4. Measurement results of dimple dimensions.

increases and tends to saturate with increasing the number of laser shots regardless of the periodicity. From Fig. 4(b), there is a linear relationship between the depth of the dimple and the number of laser shots. These results lead to the aspect ratio which is the depth divided by the diameter as shown in Fig. 4(c). We formed the dimples with the aspect ratio in a range from 0.15 to 1.8.

3.2 Results of Tensile Test

Figure 5 shows a relationship between joining strength and aspect ratio. The specimens with the small aspect ratio and the large periodicity were broken while the specimens were being set to the tensile tester. In this case, we defined the joining strength as 0 MPa. When the aspect ratio is over a certain value, we can see that there is no clear correlation between joining strength and aspect ratio. The aspect ratio shown in the Fig. 5 is based on the dimensions of the dimple formed on the metal piece. In order to discuss the effect of the aspect ratio in detail, we need to evaluate an amount of polymer filling into the dimple, or a replication ratio of injected polymer to the dimple. Therefore, we conducted cross sectional observations of the joining specimens with scanning electron microscope, SEM, to reveal the replication ratio. Figure 6 shows SEM images of cross sections of the specimens with the aspect ratio of 0.6 and 1.8 and the periodicity of 113 μm. The polymer fills in the dimple even when the aspect ratio is 1.8. Thus, the horizontal axis in Fig. 5 (aspect ratio) would reflect the replication ratio of the polymer. These results strongly suggests that the aspect ratio would not affect the joining strength when the aspect ratio is over a certain threshold. The effect of the aspect ratio is discussed in section 3.3.

Figure 7 shows a relationship between joining strength and laser-treated area. The laser-treated area is a summation of the project area of each dimple on the joining surface. In Fig. 7, the larger laser-treated area causes the higher joining strength. The reason could be explained as follows. The laser-treated area defined above is determined by the number of dimples on the joining area. Thus, the large laser-treated area means that there are a large number of dimples on the joining area. Therefore, we can expect that the anchor effect by dimples is enhanced under the large laser-treated area, which leads to the higher joining strength.

In order to clarify how surface structures affect a fracture mode, we observed fracture surfaces of the

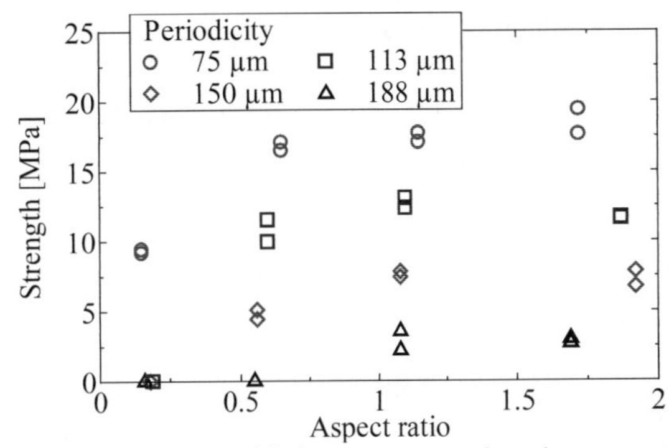

FIGURE 5. Relationship between strength and aspect ratio.

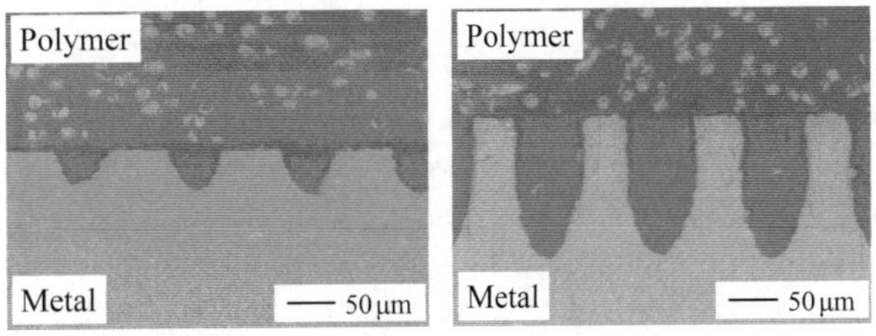

FIGURE 6. SEM images of cross sections of joining specimens: the aspect ratio is 0.6 (left) and 1.8 (right).

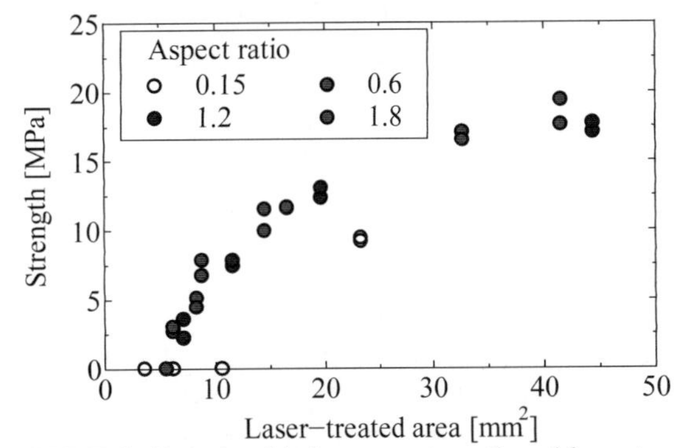

FIGURE 7. Relationship between strength and laser treated area.

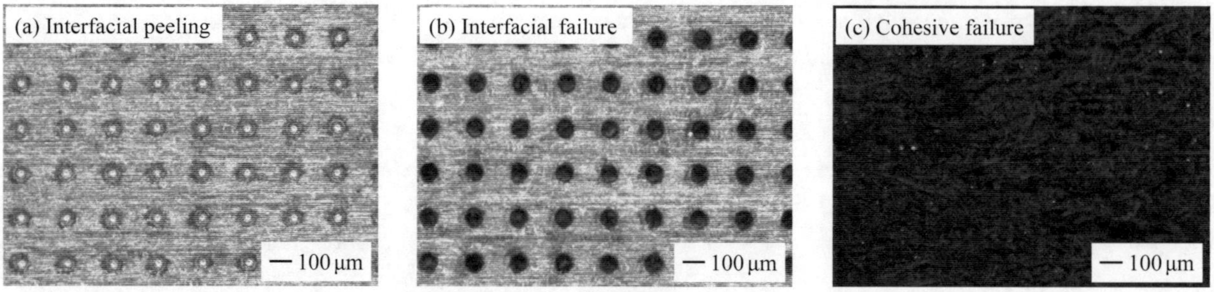

FIGURE 8. Fracture surfaces of metal pieces.

specimens. Typical fracture modes are shown in Fig. 8. The specimen with the low aspect ratio as 0.15, which showed the joining strength of 0 MPa, exhibited an interfacial peeling between metal part and polymer part. By increasing the aspect ratio, the fracture mode of the specimen changed to an interfacial failure. The specimen was broken with polymer remaining in the dimples. The specimens show the interfacial failures whose joining strength increases with increasing the laser-treated area. From Fig. 7, the joining strength saturates around 17.5 MPa when the laser-treated area is over 30 mm^2. In this case, a cohesive failure occurred on a part of joining area. Polymer slightly remained on the joining area and polymer part was broken. This should be one of the reasons why the joining strength is saturated. The aspect ratio and periodicity would affect the distribution of the stress generated in polymer injected into dimples. Therefore, depending on the surface structures, the joining strength and fracture mode might change.

3.3 Finite Element Analysis on the Effect of Aspect Ratio

3.3.1 Analysis Model

We theoretically analyzed the relationship between joining strength and aspect ratio of the dimple with finite element method, FEM, using an analysis software ANSYS 16.2. Figure 9 shows an analysis model of a joining specimen. A joining area was 1 mm ×1 mm. A model for the dimple was a truncated cone. The diameter on the joining area and the slope angle of the dimple were 0.5 mm and 75 °. To investigate the effect of the aspect ratio of the dimple, we varied the depth of the dimple. A displacement at the speed of 1 mm/min was applied to a side surface of a polymer model in X direction until a tensile load reached 1 N. As the boundary condition, nodes of side and bottom surfaces of a metal model were constrained. Friction was applied to the interface between metal and polymer models. Friction coefficient was set to 0.5. Young's modulus and poisson's ratio of the metal model were 71 GPa and 0.33, respectively. Young's modulus and poisson's ratio of the polymer model were 2.7 GPa and 0.35, respectively.

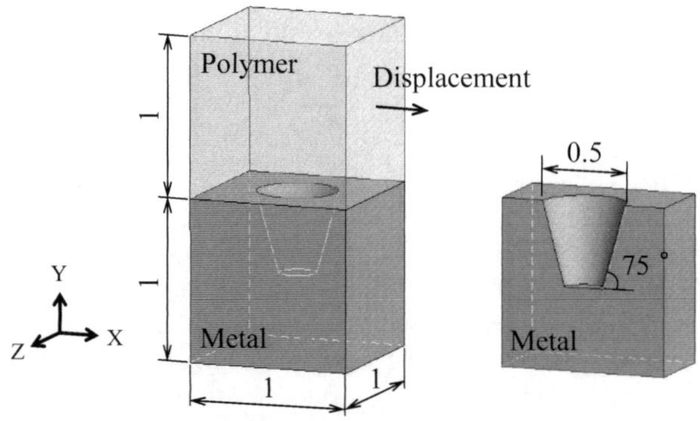

FIGURE 9. 3D FEM model and its cross section.

3.3.2 Analysis Results

We investigated how the aspect ratio of the dimple affect joining properties. Figure 10 shows the maximum principal stress generated in polymer against the aspect ratio. The maximum principal stress occurred at the edge of the dimple due to stress concentration. The maximum principal stress decreases and tends to saturate at the aspect ratio of 0.4 or more. The same load is applied to each FEM model. The principal stress differs depending on the aspect ratio. The polymer parts are broken when the principal stress is over the strength of polymer. The result suggests that the joining specimen with lower aspect ratio is broken at lower load compared to the specimen with higher aspect ratio. Under the assumption that the slope angle of dimple is constant, the joining strength might saturate with increasing the aspect ratio.

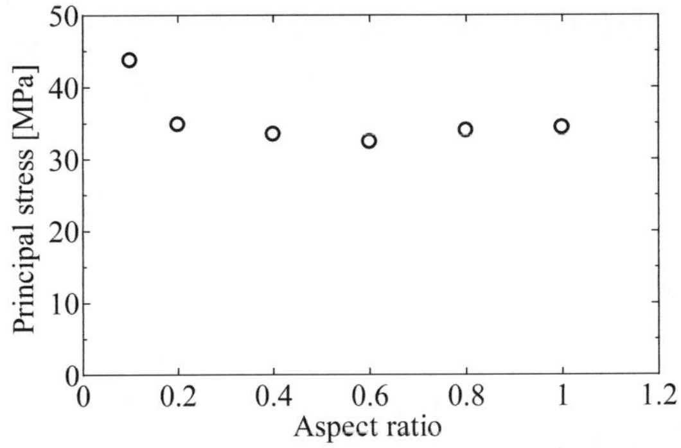

FIGURE 10. Principal stress generated in polymer against aspect ratio.

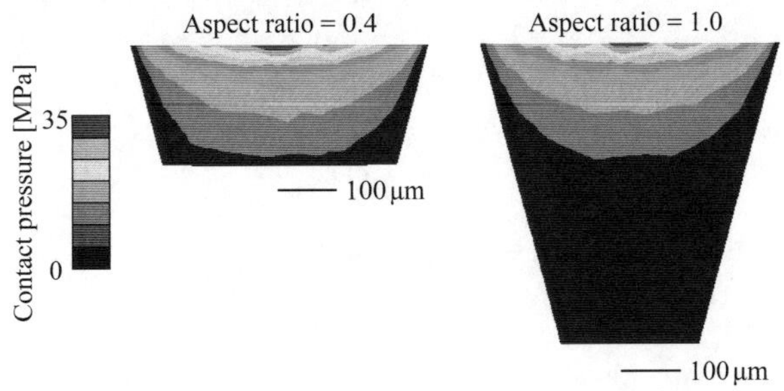

FIGURE 11. Contact pressure between metal and polymer: polymer in the dimple is viewed from the tensile direction.

Here we discuss the magnitude of the principal stress from the view point of a contact pressure between metal and polymer. When the tensile load is applied to the surface of the polymer model, the load is transferred to the metal model via polymer filled into the dimple. This causes the contact pressure on the interface. Figure 11 shows the distribution of the contact pressure. The distribution is almost the same when the aspect ratio is 0.4 or more. This implies that the bottom of the dimple does not support the load when the aspect ratio is high and that there is a threshold of the aspect ratio against a load capacity. The principal stress might saturate because the contact pressure is almost the same over a certain threshold of the aspect ratio.

The slope angle of the dimple is considered to also affect the fracture mode and joining strength. It is easily estimated that the small slope angle of the dimple would bring the interfacial peeling and low joining strength. This is because the anchor effect of the dimple is small.

4. CONCLUSIONS

This study experimentally and theoretically investigated the effect of surface structures of the metal piece on the joining strength. We focused two important geometries of surface structures. One is the aspect ratio of the dimple and the other is periodicity of the dimple. The following results were obtained.

The aspect ratio and periodicity of the dimple affects the joining strength and fracture modes. The joining strength increases with increasing the aspect ratio and saturates at the aspect ratio of 0.5 or more. The joining strength is enhanced by forming dimples with small periodicity. The low aspect ratio causes interfacial peelings. The high aspect ratio and small periodicity causes cohesive failures. FEM results shows that the principal stress generated in polymer injected into the dimple decreases and saturates with increasing the aspect ratio. The findings of this study will lead to the optimization of surface structures of metal parts for the metal-polymer direct joining.

ACKNOWLEDGMENTS

This work was supported by the Foundation for the Promotion of Industrial Science. A part of the experiments was supported by Yokoi group, the University of Tokyo, Japan.

REFERENCES

1. F. Kimura et al., Precision Engineering, 45, 203-208 (2016).
2. M. Naritomi and N. Ando, U. S. Patent Application Publication, 12/669,143 (2010).
3. K. Ramani and B. Moriarty, Polymer Engineering and Science, 38, 870-877 (1998).
4. G. Lucchetta et al., CIRP Annals - Manufacturing Technology, 60, 559-562 (2011).
5. X. Li et al., Journal of Materials Processing Technology, 249, 386-393 (2017).
6. K. Taki et al., Microsystem Technologies, 22, 31-38 (2016).
7. M. Seto et al., Journal of the Japan Society of Polymer Processing, 27(2), 68-74 (2015) (in Japanese).
8. S. Kadoya et al., Proceedings of 12th ISMTII, 1112 (2015).

Heat Transfer Analytical Models for the Rapid Determination of Cooling Time in Thermoplastic Injection Molding and Experimental Validation

Pignon Baptiste, Delaunay Didier, Boyard Nicolas, Sobotka Vincent

Laboratoire de Thermique et Energie de Nantes, UMR CNRS 6607, Polytech'Nantes, Rue Christian Pauc, CS 50609, 44306 Nantes Cedex 3, France

Abstract. Heat transfer during the cooling of a thermoplastic injected part directly affects the solidification of the polymer and consequently the quality of the part in term of mechanical properties, geometric tolerance and surface aspect. This paper proposes to mold designers a methodology based on analytical models to provide quickly the time to reach the ejection temperature depending of the temperature and the position of cooling channels. The obtained cooling time is the first step of the thermal conception of the mold. The presented methodology is dedicated to the determination of solidification time of a semi-crystalline polymer slab. It allows the calculation of the crystallization time of the part and is based on the analytical solution of the Stefan problem in a semi-infinite medium. The crystallization is then considered as a phase change with an effective crystallization temperature, which is obtained from Fast Scanning Calorimetry (FSC) results. The crystallization time is then corrected to take the finite thickness of the part into account. To check the accuracy of such approach, the solidification time is calculated by solving the heat conduction equation coupled to the crystallization kinetics of the polymer. The impact of the nature of the contact between the polymer and the mold is evaluated. The thermal contact resistance (TCR) appears as significant parameter that needs to be taken into account in the cooling time calculation. The results of the simplified model including or not TCR are compared in the case of a polypropylene (PP) with experiments carried out with an instrumented mold. Then, the methodology is applied for a part made with PolyEtherEtherKetone (PEEK).

Keywords: injection molding, crystalline material, cooling time.

PRELEMINARY DISCUSSION

Injection molding is one of the most common process in polymers forming. This is a cyclic process, in which each cycle can be decomposed in several steps: The filling step consists in injecting the melted polymer in the molding cavity during a short time. During the packing phase, a pressure is applied to compensate the thermal and crystallization shrinkages until the freezing of the gate. The cooling of the part continues until it is solidified and cooled enough to be ejected. The cooling time being the most significant one of the process, its determination is of primary importance to estimate production capabilities and costs. It includes the solidification time and depends on the expected ejection temperature. Heat transfer in the process can be described in a first approach by designers' experience and the use of simple formulas [2-4] even for the design of conformal cooling channel. This approach can be completed using simulations from a commercial software, especially for complex geometry parts. However, their use may be time-consuming in terms of computation and also requires input data especially boundary conditions and material properties, which are sometimes difficult to obtain. In this context, we have developed new analytical models suitable for semi-crystalline polymers.

Usual simplified models based on classical analytical solutions of conduction in slabs of finite thickness cannot be used for semi-crystalline polymers since they do not consider the crystallization. Furthermore, heat transfer and crystallization are strongly coupled: the crystallization depends on the temperature, pressure and mechanical stresses, while the temperature field is influenced by the crystallization enthalpy released and by the dependence of the thermophysical properties according to the crystallinity and temperature. The coupling can be simplified if the crystallization time is small compared to the characteristic conduction time, as proposed by Bénard and Advani [5]. In this case, the phase change can be considered as a moving boundary problem with a constant phase change temperature. This temperature is defined as the "effective phase change temperature of the polymer", which is highlighted by the crystallization quasi-plateau observed during the cooling of parts [6-8]. An analytical solution is the well-known Stefan one [9] valid only for semi-infinite medium. We extended this result [10], allowing to compute from the Stefan solution the solidification time of a slab with a finite thickness.

All the analytical models published consider a perfect contact between the polymer and the wall of the molding cavity. However, heat transfer at the interface between the polymer and the mold is impeded by the non-perfect contact mainly due to the roughness of the mold surface. Between asperities, an interstitial gas is trapped and its volume evolves during the cooling of the part due to the pressure variations. A thermal contact resistance (TCR) allows to model this phenomenon by a temperature drop between the surface temperatures of the polymer T_p and the mold surface temperature. The heat flux density φ crossing the interface and the TCR were measured using specific instrumentations [11,13]. We developed models to compute TCR during the process [14], considering the surface roughness, the process parameters and the polymer properties. However, this parameter is never considered for the rapid estimation of cooling time in injection molding of semi-crystalline polymers and there is no analytical solution available in the literature. A parametric model was developed by Merlin [15] to take into account the effect of TCR on the solidification time of slabs.

In this context, the present paper proposes an original method to determine analytically the solidification times, mold temperature variations and heat flux densities at the polymer/mold interface for semi-crystalline polymers. The influence of the thermal contact on the cooling time is evaluated. The results of the simplified model are compared with experimental ones using an instrumented mold, and with a model coupling heat transfer and crystallization [13].

DESCRIPTION OF THE MODEL

Analytical Model Considering a Perfect Contact

For the modeling of the cooling of semi-crystalline parts, the crystallization kinetics and the associated released enthalpy ΔH have to be taken into account. The solidification of a slab with a half thickness e/2 with a constant phase change temperature T_f under the hypothesis of [5] is obtained by solving the heat transfer equation in both amorphous and semi-crystalline phases, the thermophysical properties being considered as independent of the temperature in each phase. The location of the moving interface $x_f(t)$ is established in detail in [9-10].
The evolution of the crystallization front x_f for a semi-infinite medium is given by Eq. (1), introducing ξ solution of the transcendental equation Eq. (2).

$$x_f = 2\xi\sqrt{a_{p_l} t} \tag{1}$$

$$Ste_s \nu \frac{e^{-\xi^2/\nu^2}}{\mathrm{erf}\left(\frac{\xi}{\nu}\right)} - \frac{Ste_l e^{-\xi^2}}{erfc(\xi)} = \sqrt{\pi}\xi \tag{2}$$

Ste_s and Ste_l are the Stefan numbers for the solid and the melt, defined by the properties of the polymer: the average specific heat of theses phases, the crystallization latent heat ΔH, and ν the square root of the ratio between the thermal diffusivities of the solid and of the melt. Their exact expression is given in ref. [10]. They introduce also the injection temperature T_{inj}, the equivalent crystallization temperature T_f, and the contact temperature T_c at the polymer/mold interface during the initial phase of the cooling, just after the end of the filling step. T_c is given by the Eq. (3). b_{ps} and b_m are respectively the effusivities of the solid polymer and of the mold. T_{mi} is the temperature of the mold at the beginning of the filling. Eqs. (1, 2, and 3) must be solved iteratively, as described in [10]. To estimate the solidification time of a finite slab of half thickness e/2, we introduce a parameter constant c, defined as the ratio between the time required to solidify a phase change medium of finite thickness and the time necessary to solidify the same thickness in a semi-infinite medium obtained from Eq. (1).

$$T_c = \frac{b_{ps} T_f + b_m \,\mathrm{erf}\left(\frac{\xi}{\nu}\right) T_{mi}}{b_{ps} + b_m \mathrm{erf}\left(\frac{\xi}{\nu}\right)} \tag{3}$$

This parametric constant c was shown to be a function of only the three parameters Ste_s, Ste_l and ν. It is given by Eq. (4):

$$c = \left[0.39\nu^{-1.36} Ste_s^{-(0.028\nu+0.70)}\right].Ste_l + 1 \tag{4}$$

The solidification time t_{solid} is then given by the equation (5) [10]. The semi-infinite model provided also the contact temperature T_c given by Eq. (3) and the heat flux density φ_{msi} given by Eq. (6) just after the injection [16]:

$$t_{solid} = \frac{1}{a_{p_l}} \left(\frac{\frac{e}{2}}{2\xi c} \right)^2 \tag{5}$$

$$\varphi_{msi}(t) = \frac{b_{p_s}(T_f - T_c)}{\sqrt{\pi t}\ \mathrm{erf}(\xi)} \tag{6}$$

Analytical Model Considering a Polymer-Mold Thermal Contact Resistance (TCR)

Heat transfer in the part is also impeded by the thermal contact resistance between the polymer and the mold during injection molding. It consequently affects the solidification time of the part. Merlin [15] highlighted a relationship given by Eq. (7) describing the ratio of solidification times obtained for a finite Biot number (related to a given value of *TCR*) and an infinite Biot number (perfect contact):

$$\frac{t_{solid}(\mathrm{B_i})}{t_{solid}(\mathrm{B_i} = \infty)} = 1 + g(Ste_s, Ste_l, \nu).\frac{1}{B_i} \tag{7}$$

The parametric coefficient g was function of only three parameters Ste_s, Ste_l and ν. It could be expressed by a polynomial law, whose coefficients obtained by "numerical experiments", are given in [15]. The *TCR* is computed during the cycle by a model described into [14], which considers at the microscopic scale the surface roughness, the surface tension of the polymer and the contact angle between the polymer and the metal of the mold. This model takes also into account the equilibrium between the pressure into the molding cavity, the gas pressure inside the roughness "valleys" and the surface tension of the polymer. The solidification time becomes sensitive to *TCR* for values higher than 1.10^{-4} Km²/W and increase of 25% for *TCR* of 1.10^{-3} Km²/W and thickness of 5mm. For the same value of *TCR* it is doubled for thickness of 1mm.

EXPERIMENTAL STUDY

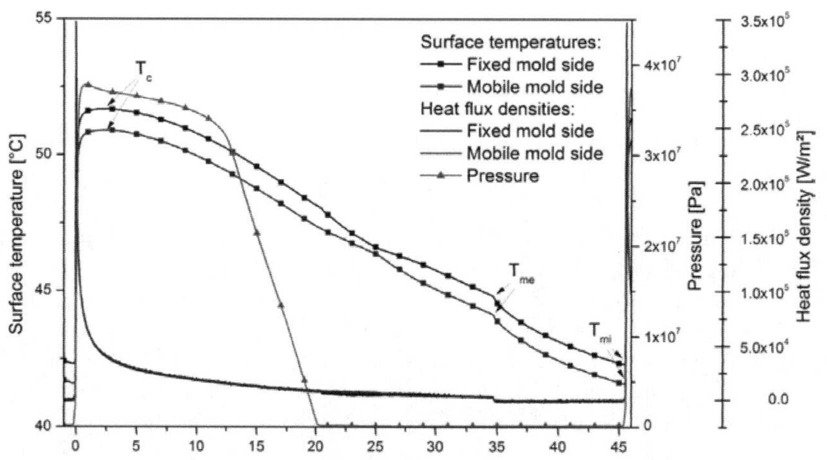

FIGURE 1. Experimental mold temperatures, heat flux densities and pressure during the processing of the PP slab for a packing pressure of 56 MPa

The mold used in this study has been developed several years ago within the framework of the SWIM project (Shrinkage and Warpage in Injection Molding) [7,17].The part is a square slab made with isotactic polypropylene (PP), which dimensions are 60 mm x 60 mm x 3 mm. A special feeding system was designed to ensure unidirectional flow through a 1.5mm thick gate. The molding cavity is instrumented with a pressure sensor Kistler© (type 6157B) positioned at 17 mm from the gate in the fixed mold side and with two heat flux sensors located at 47 mm from the gate and opposite to each other to measure heat flux exchanges on both sides of the molding cavity. A typical result is shown on figure 1. The Table 1 shows the comparison between the computed (Eq. (3)) and experimental contact temperatures. The comparison between the computed and experimental average heat flux densities are also presented. We see an excellent agreement, proving the ability of the model to correctly describe the cooling of a part.

TABLE 1: Comparison between experiments and analytical model results for contact temperatures and average heat flux densities for PP molding

	Contact temperature [°C]		Average heat flux density [W/m²]	
Packing pressure	Experimental	Calculated	Experimental	Calculated
2.8 10^7 Pa	50.7	52.2	15818	15838
5.6 10^7 Pa	51.7	52.9	16789	16933
8.4 10^7 Pa	52.4	53.5	17553	17739

However, a difficulty arises to solve the equations since the knowledge of T_f appearing in Eqs. (2-3) is required. This value depends on the crystallization kinetics of the material and of the thickness of the part. A method can consist in using DSC data to evaluate T_f as a function of the average cooling rate v of the part. A simple way is to estimate it through the relation $v = (T_{inj}-T_f) / t_{solid}$. Note that T_{inj} is the injection temperature. Equations (1-5) allow to calculate t_{solid}, so v and thus T_f. A few iterations are sufficient to converge to the value of T_f. To compute precisely the solidification time it is necessary to use a model coupling heat transfer and crystallization kinetics equations. It is done in reference [13]. We show on figure 2 the comparison between the solidification time obtained with the analytical model and the full numerical coupled model for polypropylene injected parts of several thickness, for several mold temperatures. We can observe a very good agreement.

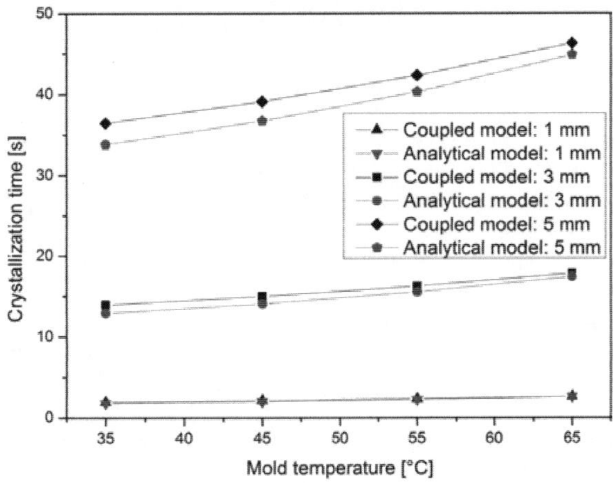

FIGURE 2: Comparison between the computed solidification time given by a coupled model and the simplified analytical model for PP solidification. Pressure is 40 MPa.

The same work was done with PEEK. The comparison of the solidification times obtained by the two models are presented on the figure 3. When the mold is too cold with thin parts, the model is not valid, because the part remains in the amorphous state, due to the quenching of the polymer as a consequence of the very rapid cooling.

The cooling time can be defined by the sum of the solidification time and the time elapsed between the end of the crystallization and the ejection of the part. This cooling time after the solidification depends on the temperature at which the ejection is expected and is fixed by the manufacturer. It can be easily computed, as explained in [10].

CONCLUDING REMARKS

A method is proposed for semi-crystalline polymers to determine the solidification time and the heat flux density at the interface between the polymer and the mold. The method is based on the modified analytical solution of the Stefan problem in semi-infinite medium to take into account the finite thickness of the part. The thermal contact resistance was introduced, which has a significant impact for solidification time determination. Results are in good agreement with experiments, which validated the simplified methodology to compute the cooling time of thermoplastics in injection molding. The crystallization of the slab was modeled by a constant effective phase change temperature, which is valid for low Deborah number. A protocol to determine this temperature was proposed and validated numerically.

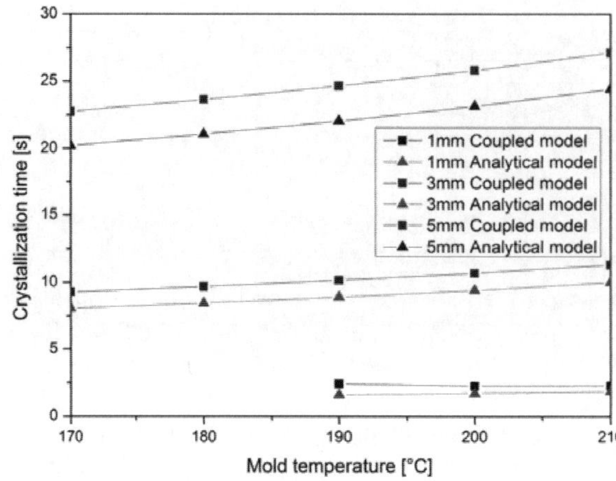

FIGURE 3: Comparison between the computed solidification time given by a coupled model and the simplified analytical model for PEEK solidification. Pressure is 40 MPa.

REFERENCES

1. O. Zöllner, Optimised mould temperature control, ATI 1104 (Application Technol. Information), Plast. Bus. Group, Bayer AG, Leverkusen. 1999.
2. X. Xu, E. Sachs, S. Allen, The design of conformal cooling channels in injection molding tooling, Polym. Eng. Sci. 41, 2001 pp. 1265–1279.
3. C.J. Yu, J.E. Sunderland, Determination of ejection temperature and cooling time in injection molding, Polym. Eng. Sci. 32 (1992) 191–197. doi:10.1002/pen.760320305.
4. D.M. Zarkadas, M. Xanthos, Prediction of cooling time in injection molding by means of a simplified semianalytical equation, Adv. Polym. Technol. 22 (2003) 188–208. doi:10.1002/adv.10048.
5. A. Bénard, S.G. Advani, Energy equation and the crystallization kinetics of semi-crystalline polymers: regimes of coupling,
6. D. Delaunay, P. Le Bot, R. Fulchiron, J.F. Luye, G. Regnier, Nature of contact between polymer and mold in injection molding. Part II: Influence of mold deflection on pressure history and shrinkage, Polym. Eng. Sci. 40 (2000) 1692–1700. doi:10.1002/pen.11301.
7. R. Le Goff, G. Poutot, D. Delaunay, R. Fulchiron, E. Koscher, Study and modeling of heat transfer during the solidification of semi-crystalline polymers, Int. J. Heat Mass Transf. 48 (2005) 5417–5430.
8. B. Pignon, X. Tardif, N. Lefèvre, V. Sobotka, N. Boyard, D. Delaunay, A new PvT device for high performance thermoplastics: heat transfer analysis and crystallization kinetics identification, Polym. Test. 45 (2015) 152–160.
9. J. Stefan, Über einige Probleme der Theorie der Wärmeleitung, Sitzungsberichte Der Österreichischen, Akad. Der Wissenschaften Math. Naturwissenschaftliche Klasse. 98 (1889) 473–484.
10. V. Sobotka, A. Agazzi, N. Boyard, D. Delaunay, Parametric model for the analytical determination of the solidification and cooling times of semi-crystalline polymers, Appl. Therm. Eng. 50 (2013) 416–421.
11. H. Massé, É. Arquis, D. Delaunay, S. Quilliet, P.H. Le Bot, Heat transfer with mechanically driven thermal contact resistance at the polymer-mold interface in injection molding of polymers, Int. J. Heat Mass Transf. 47 (2004) 2015–2027.
12. A. Bendada, A. Derdouri, M. Lamontagne, Y. Simard, Analysis of thermal contact resistance between polymer and mold in injection molding, Appl. Therm. Eng. 24 (2004) 2029–2040.
13. B. Pignon, N. Boyard, V. Sobotka, D. Delaunay, Heat transfer analysis at high cooling rate on the surface of thermoplastic parts, Int. J. Heat Mass Transf. 106 (2017) 253–262.
14. S.C. Somé, D. Delaunay, J. Faraj, J.-L. Bailleul, N. Boyard, S. Quilliet, Modeling of the thermal contact resistance time evolution at polymer–mold interface during injection molding: Effect of polymers' solidification, Appl. Therm. Eng. 84 (2015) 150–157.
15. Kevin Merlin, Caractérisation thermique d'un matériau à changement de phase dans une structure conductrice, PhD Thesis, University of Nantes, 2016
16. T. Loulou, D. Delaunay, The interface temperature of two suddenly contacting bodies , one of them undergoing phase change, Int. J. Heat Mass Transf. 40 (1997) 1713–1716.
17. D. Delaunay, P. Le Bot, R. Fulchiron, J.F. Luye, G. Regnier, Nature of contact between polymer and mold in injection molding. Part I: Influence of a non-perfect thermal contact, Polym. Eng. Sci. 40 (2000) 1682–1691.

Automatic Cooling Channel Design for Injection Moulds

Ch. Hopmann, M. Theunissen[a)], T. Schneppe, M. Schmitz

[1]Institute of Plastics Processing (IKV) in Industry and the Skilled Crafts at RWTH Aachen University, Seffenter Weg 201, D-52074 Aachen, Germany

[a)]Corresponding author: matthias.theunissen@ikv.rwth-aachen.de
URL: http://www.ikv-aachen.de

Abstract. During the injection moulding process, local differences in pressure and temperature lead to different shrinkage over the part. This inhomogeneous shrinkage can lead to internal stresses, which are compensated by an undesirable warpage of the part. A sophisticated design of the cooling channels can avoid most of these problems. In previous research, it was shown that a locally adjusted temperature distribution on the cavity surface leads to less warpage of the part. In order to achieve this temperature distribution, an adapted cooling channel design is necessary. This design process is not possible with an analytical calculation but is the result of an iterative and numerical optimisation process. In this paper, a simulative approach for an automatic cooling channel layout via an inverse heat conduction problem is presented. Based on previous results, the methodology is adapted to three-dimensional problems. This approach brings the local heat and cooling demand of the part in equilibrium based on the postulate of a locally homogeneous material density. By this methodology, a proper cooling channel design close to the mould surface is determined. Simulations show a significant reduction of warpage by at least 22 % which will be validated by experimental results.

INTRODUCTION

Cycle time and achievable part quality during injection moulding are influenced to a great extent by the mould tempering system. These two aspects provoke a target conflict. By using low mould temperatures, short cycle times can be achieved with the acceptance of poor part quality. Improved part quality, on the other hand, can often be achieved by slow cooling rates at high mould temperatures. Part quality in general is defined by several quality criteria, which can include optical, mechanical and geometrical properties.

Therefore, the overall objective of this paper is to introduce a paradigm shift in the classical mould design by placing the focus on the part and away from the cavity during the thermal design of injection moulds. In contrast to conventional thermal mould design, the thermally optimised cooling channel design is based on the assumption of a homogeneous temperature and density distribution in the part. In the course of previous investigations, a new methodology for the automatic design of the tempering layout in injection moulds has been investigated and expanded. The fundamental approach is solving an inverse problem for the calculation of the resulting temperature distribution in the mould based on a homogeneous temperature distribution in the part. This is solved iteratively by optimisation. Based on the calculated optimised temperature distribution, a cooling channel system can then be derived. [1-4]

In consequence of the material's dependence on temperature and density during the cooling phase, the need for a precise design of the tempering system of the injection mould becomes obvious. This material characteristic has to be taken into account during the thermal mould design. The aim is to find an optimum between a low cycle time and the best possible part properties, which in this case means the most homogeneous shrinkage possible. With increasing numerical possibilities, the appropriate material models and the progressing rise of computing power, the wish for an automated solution to locate the position and dimensions of the tempering system becomes realistic.

Different approaches to improve the thermal mould design phase were developed in recent years. They reach from ray tracing methods, to cluster algorithms and artificial neuronal networks, to find a proper solution for the moulds tempering system [5-7]. Common to all of these approaches is the forward headed nature, which emphasises the temperature control, but the evaluation of the solution is only possible after the simulation.

In contrast, several algorithms for iteratively optimising different thermal problems have been developed. Particle swarm optimisation is one possibility in order to investigate the variothermal operation of a mould in extrusion blow moulding [8]. In addition, simple geometrical considerations of the part and the mould can be used to determine suitable positions for temperature control channels. A route-finding algorithm connects the individual cooling channels and generates a plurality of realisable temperature control systems. [9]. Finally, a topology optimisation for the multimaterial design of mould cores for blow moulding shows an interesting optimisation approach. In this case, for a 2D problem, the optimum distribution for a high-temperature-conducting material within a low-conducting material is first determined based on various boundary conditions, e.g. the dissipated energy. Results are two-dimensional material distributions, which are strongly based on the chosen boundary conditions. [10]

All of the presented research results show interesting approaches to the field of optimisation but either they do not adapt the algorithms to the process of injection moulding or do not transfer the results consequently. Therefore, a methodology is presented in the following starting from the individual cooling demand of the part and calculating the resulting mould temperature [2].

OPTIMISATION OF TEMPERATURE DISTRIBUTION IN PART AND MOULD

Within the framework of the methodology presented in this article, an efficient tempering system can be developed in order to obtain economic cycle times. However, geometrical, mechanical as well as optical quality features of the component are to be maintained and improved. To realise dimensional accuracy, a locally uniform shrinkage needs to be achieved by appropriately designing tempering systems in order to minimise component stresses and resulting warpage. These requirements need to be translated into an objective assessment of a tempering layout and in the long term into an automated construction of the tempering system. The methodology used here is thus based on the principle that a shrinkage potential, which is the same over the whole part, leads to a uniform shrinkage and thus to no part warpage.

An equal shrinkage potential of a plastic component can be achieved by means of a homogeneous distribution of temperature and density at the time of demoulding. This consideration leads to an inversed problem in which a part with a homogeneous distribution of density and temperature is calculated. A resulting temperature distribution is then determined in the mould. The methodology developed is illustrated in a schematic flow chart in Figure 1 and described in detail below.

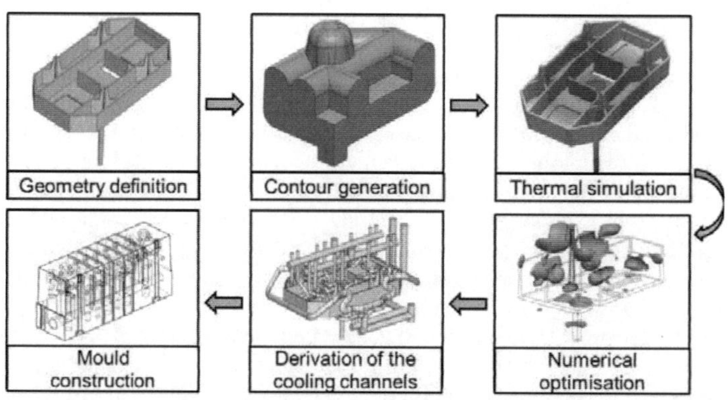

Figure 1: Flow chart of the methodology for the inverse three-dimensional mould construction

In this methodology, a surrounding contour is placed around the part, which defines the area to be optimised in the mould. A symmetrical test specimen is developed, which minimises the resulting calculation effort by cutting at the symmetry planes. On the outer contour, a temperature distribution is calculated which leads to a homogeneous distribution of the specific volume in the part. To convert this theoretically derived temperature distribution into reality, a tempering system can be placed in the area of the heat sinks. In contrast to the two-dimensional problem, the cooling channels now have to be adapted to two-dimensional isotherms, so-called morphological surfaces. These cooling channels are to have a water temperature corresponding to the morphological surfaces, which realise the optimised temperature distribution. Subsequently, a mould design can be derived from this.

By means of numerical methods, the temperature compensation process in the mould can be solved with the Fourier heat conduction equations for both the part and the surrounding mould. Since a thermal optimisation of an entire mould would considerably exceed the existing computing capacities, a contour is laid around the moulded part at a constant distance, which characterises the thermally relevant region to be optimised and in which a temperature control layout is subsequently designed.

Through an injection moulding simulation using the simulation software Sigmasoft, Sigma Engineering GmbH, Germany, a starting temperature distribution is determined for the part and the mould after the end of the filling phase. The simulation can take into account the relevant plastic-specific properties, e.g. shear heating in the filling phase and thus is able to calculate a realistic temperature distribution. This approach also prevents the subsequent optimisation from running into a local minimum due to unrealistic starting values. The temperature distribution is calculated in a mould without cooling channels and the local cooling demand of the melt is identified. The data fields of temperature and pressure over the entire part at the end of the filling phase, as well as the temperature in the mould, are exported and specified as starting parameters in the optimisation.

The optimisation is then carried out using the FEM software COMSOL from Comsol AB, Stockholm, Sweden. The aim is to find an optimised temperature distribution in the mould. For this purpose, a heat conduction problem must be calculated, which is solved over the cooling phase of the moulded part. A quality function determines and sums up local differences between a required demoulding temperature and the local temperature, as well as a local density and an averaged global density. At the end of the cooling phase, the homogeneity of density and temperature in the moulded part is evaluated. The optimisation parameter is the temperature distribution on the contour of the surrounding mould (T_C). By varying the temperature on the contour, the local temperature differences can be minimised in order to reach the target of a homogeneously tempered component at the end of the cooling phase. The quality function consists of two terms and is described in equation (1):

$$Q(T_C)=\sum_{i=1}^{m}\int_{\Gamma_1}\left(\frac{T_{Ejec}-T_{loc}(x_i,t_c;T_C)}{\omega_i}\right)^2 d\Gamma_1 \quad + \quad \sum_{j=1}^{k}\int_{\Gamma_2}\left(\frac{\bar{\rho}_{Ejec}-\rho_{loc}(x_j,t_c;T_C)}{\omega_j}\right)^2 d\Gamma_2 \tag{1}$$

In this proposed quality function $Q(T_C)$, the temperature differences are evaluated over the part surface (Γ_1), whereas the density is evaluated on a surface inside the part (Γ_2). Considering the first term, a demoulding temperature (T_{Ejec}) is set, which needs to be reached on the surface of the part at the end of cooling. The cooling time (t_c), at the end of which the demoulding temperature is to be reached, is initially estimated over a one-dimensional, analytical heat conduction approach for the thickest wall of the part and defined as a boundary condition. The quality function sums up the local deviations of the melt temperature (T_{loc}) to the ejection temperature (T_{Ejec}). Furthermore, the influence of temperature or density can be locally weighted ($\omega_{i/j}$) in order to favour cycle time or dimensional stability. The quality goal of dimensional accuracy is modeled via the density term of the quality function. Therefore, the difference of the local density (ρ_{loc}) to the averaged part density at ejection ($\bar{\rho}_{Ejec}$) is evaluated. This density is not determined on the surface of the mould but is calculated on a surface that is between the outer contour and the center of the part, since the moulding temperature on the wall of the cavity – and thus the density – deviates from the interior of the part. By eliminating differences in the local density at this surface located in the part, a homogeneous shrinkage potential of the part is achieved, which leads to a minimisation of part warpage. The now required temperature distribution on the surrounding volume can be determined numerically [11, 12]. In this approach, a conjugated gradient method is used for this purpose. The advantage of this method is to achieve faster convergence compared to the gradient method [13]. After successfully determining an optimised temperature distribution, a cooling channel system can be derived based on the temperature distribution, as described in the following chapters.

APPLICATION TO A THREE-DIMENSIONAL TEST GEOMETRY

The presented methodology was initially applied to a two-dimensional model at the Institute of plastics processing [2]. The algorithm was implemented with these simplifications and the possibilities for setting the weighting factors were investigated in various test series. Based on a symmetrical exemplary geometry with ribs, it was shown that the results so far are very promising to minimise part warpage. As the next step in the investigation of this methodology, the algorithms are transferred from the second to the third dimension, so that a full consideration of volumetric effects is now possible.

In this research project, the geometry shown in Figure 1 was developed in order to test the strengths of the methodology. The geometry has plastic-specific elements, such as openings, tabs and ribs, which are potentially prone to severe warping. In particular, the area of the centre tips is complicated and difficult to temper. Based on this exemplary geometry, the algorithm as described in Figure 1 was performed and an optimised cooling channel layout was derived manually. Further, a conventional design was made by means of drilled channels. For the conventional design, a best possible thermal design without consideration of the costs was designed to ensure comparability with the cost-intensive Selective Laser Melting (SLM) process. The two studied designs are shown in Figure 2. For the conventional design, tempering is only possible with the aid of riser holes and baffle plates, for which the optimised design allows continuous temperature control based on the local cooling demand.

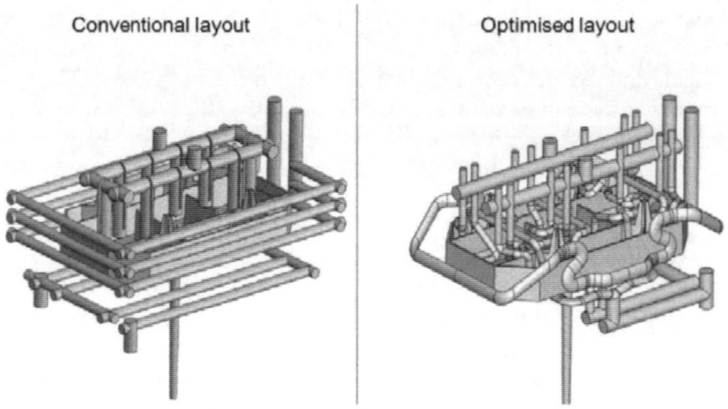

Figure 2. Comparison of a conventional tempering system with an optimised system

In order to evaluate the warpage results of the simulations, appropriate measurement points need to be defined for the warpage (see Figure 3). Two characteristic displacements are examined. First, the deflection is characterised. For this purpose, six measurement points on the back of the plate geometry are defined and the displacement of the points relative to each other in the z-direction defines the deflection. The bending of a flat geometry is a classic challenge in plastic manufacturing. Furthermore, the displacement of the four side walls is measured at the visualised measurement points.

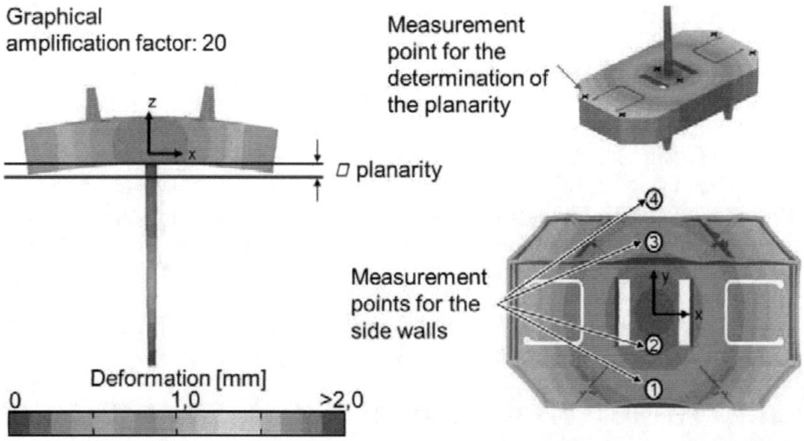

Figure 3: Comparison of the warpage for a conventional and an inverse constructed tempering system

For both tempering layouts, injection moulding simulations with the same process parameters are carried out and the resulting warpage of the parts produced with two different cooling channel layouts is calculated. The results of these simulations for cooling time and part warpage are summarised in Figure 4.

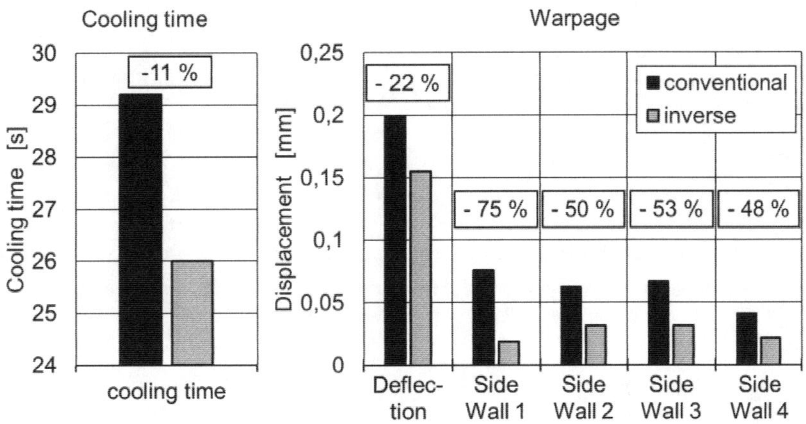

Figure 4. Results of the improved tempering system

The sprue was not taken into account for the calculation of the cycle time since this is determined by the cooling time. Because of its thickness due to limitations in the base mould, the sprue becomes the main impact factor for the cycle time. Considering the geometry without the sprue, a reduction of the cooling time by 11% compared to the conventional design was achieved. Furthermore, a considerable reduction of the warpage could be realized. The displacement of the side walls was reduced by at least 47 % in the simulation. Further, the flatness of the plate-like part could be improved by 22 % as compared with the conventional cooling. The described tempering systems are brought into a modular moulding concept for practical validation.

CONCLUSION AND OUTLOOK

The methodology for the construction of a thermally optimised, automated tempering channel design for a three-dimensional application case was presented in this article and its effectiveness demonstrated in simulation. Based on the assumption that a homogeneously tempered component undergoes a homogeneous shrinkage at the end of the cooling time, an optimum temperature distribution has been derived from the mould. A temperature control layout has been derived from the resulting temperature distribution. This shows that the automatic inverted solution leads to the best results. At the same time, however, it can be seen that part warpage cannot be avoided completely.

In order to implement the calculated temperature distribution in reality, a great challenge is to provide an accurate representation of the morphological surfaces within a mould insert through a cooling channel layout. Due to a large number of influencing factors on the quality and resulting solutions, a large number of possible tempering systems has to be considered. The available morphological surfaces are also to be connected to form a cooling channel layout that respects flow limitations. For the presented test case, the moulding distortion can be reduced at least by 22 % and the required cooling time by 10.9 %.

In the next step, it must now be demonstrated in practical experiments that the methodology presented can also be validated in reality. For this purpose, a modular mould concept is used in which inserts with different cooling systems can be used. Based on precise measurements of the produced parts, a statement can be made about the validity of the methodology in reality.

Furthermore, in the future, the sensitivity of the results of the design with different input parameters and designs will be examined. In addition, the cooling intensity of the cooling channels has until now been based on a constant coolant flow and thus also constant thermal boundary conditions, which is not always true and should be investigated further in the course of the research.

ACKNOWLEDGMENTS

The depicted research has been funded by the Deutsche Forschungsgemeinschaft (DFG) as part of the special research area "Sonderforschungsbereich 1120 Precision Manufacturing by Controlling Melt Dynamics and Solidification in Production Processes". We would like to extend our thanks to the DFG.

REFERENCES

1. A. Agazzi, V. Sobotka, R. LeGoff and Y. Jarny. Optimal cooling design in injection moulding process – A new approach based on morphological surfaces. In: *Applied Thermal Engineering* **52**: 170-178, 2013.
2. Ch. Hopmann and P. Nikoleizig. Inverse thermal mould design - Addressing the local cooling demand as quality function for an inverse heat transfer problem. In: *International Journal of Material Forming.* [Published online]. p. 1-12.
3. G. Menges, E. Haberstroh, W. Michaeli and E. Schmachtenberg. *Werkstoffe der Kunststoffe.* (Carl Hanser Verlag, Munich, Vienna, 2011).
4. G. Menges, W. Michaeli and P. Mohren. *Spritzgießwerkzeuge. Auslegung, Bau, Anwendung*. Carl Hanser Verlag, 2007.
5. P. Faßnacht, J. Kerkeling and R. Nickel. Künstlich intelligent statt manuell. *Plastverarbeiter* **62**:72-73, 2011.
6. V. Maag and K.-H. Küfer. Optimal cooling in injection molding and die casting. In: *Proceedings of the International Conference on Engineering Optimization*, Rio de Janeiro, Brazil, 01-05 June 2008.
7. J. Mehnen, T. Michelitsch, C. Lasarczyk and T. Bartz-Beielstein. Multi-objective evolutionary design of mold temperature control using DACE for parameter optimization. *International Journal of Applied Electromagnetics and Mechanics* **25**: 661-667, 2007.
8. C.-L. Xiao and H.-X. Huang. Optimal design of heating system for rapid thermal cycling mold using particle swarm optimization and finite element method. In: *Applied Thermal Engineering* **64:** 462-470, 2014.
9. J. M. Jauregui-Becker, G. Tosello, F. J. A. M. Van Houten and H. N. Hansen. Performance Evaluation of a Software Engineering Tool for Automated Design of Cooling Systems in Injection Moulding. In: *Procedia* **46**: 270-275, 2013.
10. A. Agazzi, R. Legoff and C. Truc-Vu. On the use of topology optimization for improving heat transfer in molding process. In: *AIP Conference Proceedings* **1769**: 40002, 2016.
11. W. Polifke and J. Kopitz. *Wärmeübertragung: Grundlagen, analytische und numerische Methoden*. (Pearson Deutschland GmbH, München, 2009).
12. N. Zabaras. Inverse Problems in Heat Transfer, in *Handbook of Numerical Heat Transfer,* edited by W. J. Minkowycz (John Wiley & Sons, Hoboken, 2009). p. 523-557.
13. M. Papageorgiou, M. Leibold and M. Buss. *Optimierung. Statische, dynamische, stochastische Verfahren für die Anwendung*. (Springer Vieweg, Berlin, Heidelberg, 2015).

Influence of Insert Component Position and Geometry on Shrinkage in Thermoplastic Insert Molding

Matthias Zeppenfeld*, Benedikt Müller, Sophia Heyl

Institute of Medical and Polymer Engineering, Technical University of Munich
Boltzmannstr. 15, 85748 Garching, Germany
matthias.zeppenfeld@tum.de

Abstract. In modern thermoplastic injection molding the overmolding of inserts allows a high level of functional integration. An important quality criterion for the manufacturing of those technical thermoplastic parts is the detailed and comprehensive understanding of the factors that influence shrinkage. Often local shrinkage differences result in loads on the inserts or cause dimensional tolerance inaccuracies. This study investigates the effect of different insert part geometries and positions along the flow path on the shrinkage of the overmolded part. A mold was designed to overmold square and cylindrically shaped metal inserts at six different mold positions. Overmold materials were polypropylene (PP), polybutyleneterephthalate (PBT), and polyamide (PA). The result of simulation and experimental measurements indicated significant lower shrinkage of the overmold with inserts parallel and normal to the flow direction. Square shaped inserts in most cases induced greater impact on shrinkage behavior compared to the cylindrical inserts. The insert position caused reduced shrinkage with inserts placed at the end of the flow path compared to positions near the feeding point. Additionally, normal to flow shrinkage decreased between central and lateral insert positions. Asymmetrical insert positions especially caused a partially shrinkage onto or rather off the insert resulting in mechanical load, critical for e.g. sensitive insert parts like RFID-transponders. Overall, these findings contribute to a more precise estimation of the achievable dimensional part accuracy to improve overmolding processes in injection molding.

Keywords: Insert Molding, Insert Shape, Insert Position, Shrinkage, Part Accuracy
PACS: 81.05.Lg, 81.20.Hy, 81.70.Fy, 82.35.Lr, 83.50.Uv, 83.80.Sg

INTRODUCTION

Injection molding, one of the most important polymer processing technologies has been particularly economically successful through the mass production of thermoplastic parts [1]. However, manufacturing of complex and high-quality technical plastic parts with tight tolerance requirements demands a detailed process knowledge and remains a challenge in injection molding [2]. This is due to the unique properties of polymers compared to classic structural materials (e.g. metals). For example is the viscoelastic material behavior of polymers which creates deformation effects during manufacturing, especially shrinkage and warpage (differential shrinkage). These deformation effects, resulting from the solidifying polymer melt, are the key parameters for quality characteristics like sink mark or voids as well as for the final dimensions and their tolerances [3-5]. The linear shrinkage, related to a specific part dimension results from the volumetric shrinkage and the shrinkage anisotropy. The anisotropy results from external shrinkage resistance, like geometrical constraints by the mold or insert parts, as well as internal shrinkage resistance (e.g. derived from molecule orientation or filler materials). As the parts complexity increases with the trend for enhancing the level of functional integration of plastic parts by overmolding bushings, nuts, pins or sensitive electronics via insert technique, the molding process similarly gets even more complicated [6-10]. For plastic parts with integrated inserts, especially with sensitive mechatronic inserts, knowledge about the specific shrinkage influences is important to ensure a media tight bonding and avoid stress on the insert part [11-13].

Annicchiarico and Alcock [4] have identified four main influence factors affecting shrinkage involving material behavior, technical issues like mold and specimen design as well as processing parameters. However, there is still a need for a detailed investigation of the influence of external shrinkage resistance caused by mold and specimen design, i.e. the geometry of the insert part and its position in the mold, on the dimensions of high integrative plastic parts. The obtained results of this study will provide indications how the insert part position and geometry are effecting the final part dimension. Additionally, this study improves upon the knowledge how to assess shrinkage related problems like deformation of the insert parts.

MATERIAL & METHODS

To evaluate the molding shrinkage in relation to insert position and geometry, a mold was designed to produce 40 x 40 x 2 mm shaped plates inspired by DIN EN ISO 294-3 [14]. The mold was filled via a pinpoint gate to avoid post-processing influences by cutting the gate and to allow a sufficient tactile measurement at different positions on the gate side of the test specimen. Steel inserts (square shaped (10 x 10 mm) and cylindrical (Ø = 10 mm), 2 mm thick) were fixed with exchangeable core pins in the cavity wall and subsequently overmolded at six different mold positions (see Figure 1). Position 3, 4 and 7 were not evaluated due to their symmetrical configuration. A standard washer (Misumi Group, Japan) was used as a cylindrical insert and measured with the following dimensions (n = 15): 9.91 ± 0.01 mm in outer diameter, 2.01 ± 0.01 mm in thickness, and 0.7245 ± 0.0019 g in weight. The square shaped insert parts were custom manufactured and measured (n = 15) with 10.17 ± 0.05 mm and 10.15 ± 0.08 mm side dimensions, 2.02 ± 0.02 mm in thickness and 1.5266 ± 0.0319 g in weight. To set-up the process and to collect additional information about the injection molding process a CoMo Injection process monitoring unit (Kistler Group, Switzerland) and a Kistler type 6183A pressure sensor, placed interchangeably at insert position 8 near the gate, was used. In order to investigate shrinkage effects of different technical thermoplastic polymers (see Table 1), 15 test specimen were manufactured at all positions and with both insert shapes out of polypropylene (PP, Sabic, Saudi Arabia), polybutyleneterephthalate (PBT, BASF, Germany), and polyamide (PA66, Albis Plastic, Germany). Parts (n = 15) without inserts were manufactured and measured as reference.

TABLE 1. Used thermoplastic materials.

Material	Type	Molding shrinkage [%]		Young`s modulus [MPa]
		parallel	normal	
PP	Sabic® PHC31	1.5	1.5	1300
PBT	Ultradur® B4520	1.5	1.7	2500
PA66	Altech® A 1000/109	0.6-2.2	0.6-2.2	3300

All 810 parts were manufactured on a KM 50-55 EX (KrausMaffei, Germany) molding machine and molding process parameters were adapted for each material ensuring an appropriate mold filling and cavity pressure curve. The molding shrinkage was calculated out of tool contour and molded part dimension parallel (P) or normal (N) to polymer flow (see Figure 1) and in thickness direction after 16-24h storage according to DIN EN ISO 294-4 [15]:

$$\text{Molding shrinkage} = \left(1 - \frac{\text{moulded part dimension}}{\text{tool contour dimension}}\right) * 100 \quad [\%] \tag{1}$$

These dimensional measurements were performed in or near (in case of sprue) the axis of the insert part with a METRO MT 60M length gauge (Heidenhain, Germany) indicated with a precision of 0.1 ± 0.05 µm (at 1N testing force) and a repeatability of 0.06 µm. Specimen weight was recorded with a LA230S high-precision balance (Sartorius, Germany). Finally, all obtained results were checked for normal distribution (Shapiro-Wilk) and evaluated statistically either by T-test or Mann-Whitney U-test (level of significance, e.g. in Figure 2, is indicated with asterisks (* ≙ 0.05, ** ≙ 0.01, *** ≙ 0.001).

Furthermore, insert induced residual stress distribution was analyzed optically by aid of the birefringence phenomenon after overmolding with polycarbonate (PC, Makrolon® 2858, Covestro, Germany). Injection molding simulation (Moldflow, Autodesk, USA) was additionally used, considering the experimentally derived settings, to evaluate shrinkage predictability in order to accelerate future mold design processes and experiments.

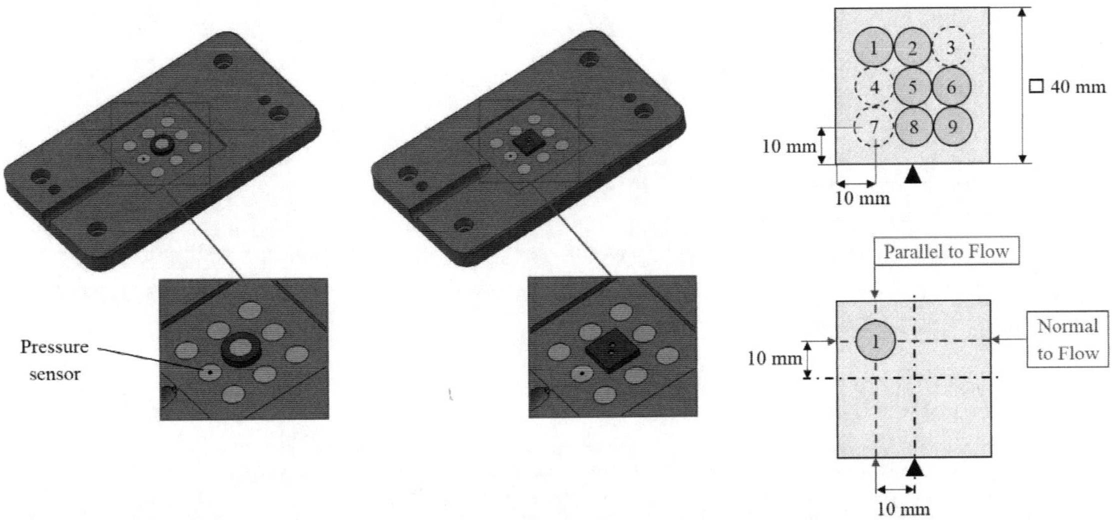

FIGURE 1. Cavity part of the injection mold with exchangeable core pins (left) to attach square shaped or cylindrical insert parts at 6 different mold positions (top right). Resulting influences on part shrinkage were measured parallel and normal to flow direction in the axis of the insert part depending on the different positions (bottom right).

RESULTS & DISCUSSION

Experimental data showed a significant ($p < 0.001$) shrinkage difference for the molded parts with inserts parallel as well as normal to flow direction compared to the reference without inserts (e.g. for PBT in detail see Figure 2). This effect was observable for the cylindrical as well as the square shaped insert parts, whereby maximum blocking of shrinkage occurred at positions near the edges (see Table 2).

Measurements of part dimensions parallel to flow direction indicated a lower shrinkage with inserts placed at position 1, 6 and 9 compared to more central inserts (Pos. 2, 5 or 8). The same effect was observed normal to flow direction where shrinkage decreased between central and lateral insert positions. Further analysis showed that shrinkage increased in parallel and normal direction with greater distance to the insert, but was still reduced even in a distance of 15 mm compared to a part without any constraint. Inserts placed in the middle mostly showed the smallest additional influence on the shrinkage of the test specimen because the part could shrink uniformly in all directions. Near the edges the shrinkage decreased especially by use of square shaped inserts due to the reduced polymer material between insert and cavity wall (see Figure 3). These results suggests that effort towards dimensional part accuracy should focus if insert parts are placed near the edges due to their enhanced impact on shrinkage. Further, the variation of the insert positions showed lower shrinkage for parts with inserts placed at the end of the flow path compared to positions near the feeding point (difference between pos. 2 and 8 parallel and normal to flow). This means shrinkage resistance increases with distance from the gate, at which especially the square shaped inserts can constrain shrinkage additionally.

TABLE 2. Maximum shrinkage constraint parallel and normal to flow direction between parts with inserts (angular: □ and round: O) compared to the reference for all tested materials. The position at which the maximum occurred is indicated in brackets.

Max. shrinkage constraint		PP	PBT	PA66
Parallel to flow [%]	O	25.4 (Pos. 1)	16.3 (9)	16.2 (1)
	□	29.0 (9)	21.5 (1)	23.7 (1)
Normal to flow [%]	O	24.5 (1, 2)	15.4 (1, 2)	14.6 (1)
	□	26.5 (1)	20.8 (6)	16.0 (1)

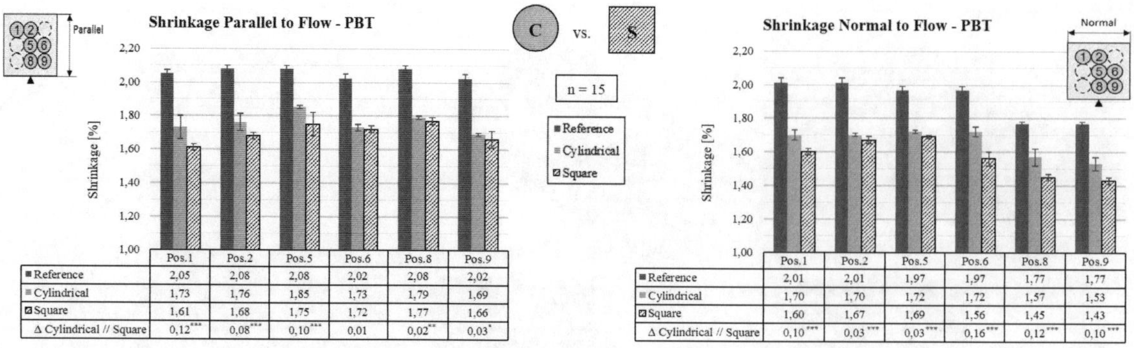

	Pos.1	Pos.2	Pos.5	Pos.6	Pos.8	Pos.9
Reference	2,05	2,08	2,08	2,02	2,08	2,02
Cylindrical	1,73	1,76	1,85	1,73	1,79	1,69
Square	1,61	1,68	1,75	1,72	1,77	1,66
Δ Cylindrical // Square	0,12***	0,08***	0,10***	0,01	0,02**	0,03*

	Pos.1	Pos.2	Pos.5	Pos.6	Pos.8	Pos.9
Reference	2,01	2,01	1,97	1,97	1,77	1,77
Cylindrical	1,70	1,70	1,72	1,72	1,57	1,53
Square	1,60	1,67	1,69	1,56	1,45	1,43
Δ Cylindrical // Square	0,10***	0,03***	0,03***	0,16***	0,12***	0,10***

FIGURE 2. Comparison of shrinkage for PBT parallel (left) and normal (right) to flow direction depending on different insert part geometries (cylindrical (C), square (S)), positions in the mold and reference parts without additional shrinkage constraints.

Regarding the insert geometry, most parts with square shaped inserts caused a lower shrinkage compared to the cylindrical inserts parallel to flow (except for PP (Pos. 1, 2, 6, 8), PA66 (Pos. 6, 8)) and normal to flow (except for PP (Pos. 2), PA66 (Pos. 5, 6, 8)). This can be explained by the greater surface area of the square, resulting in a greater ratio between insert and surrounding part and leads to a higher shrinkage resistance by the insert part (see Figure 3). Furthermore, in direction of the parts wall thickness (2 mm) the inserts increased shrinkage near the edges of the test specimen compared to the reference without inserts, especially for PBT / PA material.

Concerning insert induced residual stresses, a more angular isochromatic fringe pattern could be observed for the square shaped part compared to the cylindrical part (see Figure 4). Additionally the fringe pattern was shifted of about 1000 µm away from the insert compared to the cylindrical insert at the same position. Thus, a postponed and more unsteady stress concentration is caused by the sharp-edged insert.

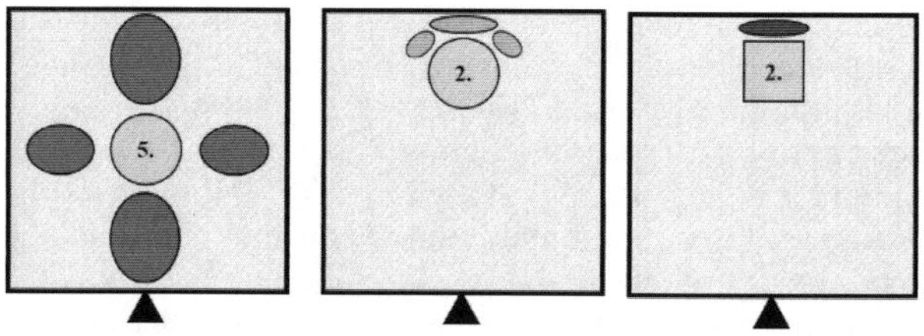

FIGURE 3. Schema of shrinkage behavior between different insert part geometries and positions. A uniform shrinkage (green) of polymer material can take place by a symmetrical part layout (e.g. at Position 5). Insert positions near the edges are significantly blocking shrinkage (orange), which is even increased by use of angular insert parts (red).

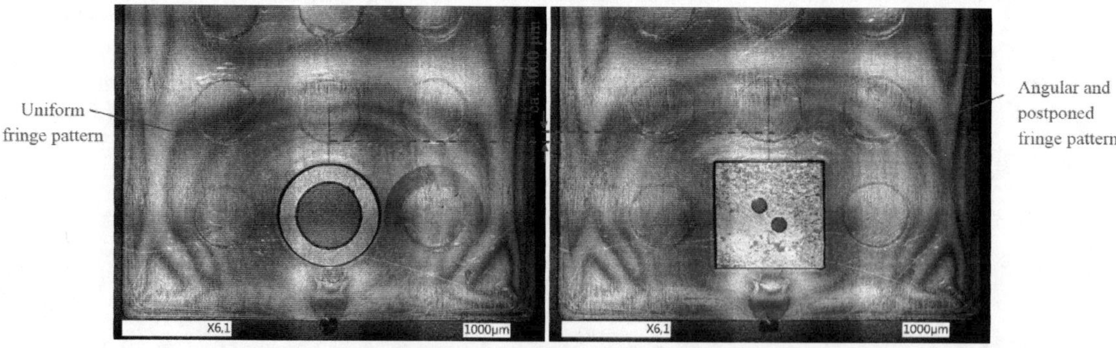

FIGURE 4. Optical analysis of insert induced stress. The square shaped insert parts at mold position 8 near the gate show a more angular and postponed profile of isochromatic fringe patterns compared to the round inserts at the same position.

Injection molding simulation helped to visualize a partial shrinkage onto or rather off the insert at asymmetrical insert positions (e.g. see Figure 5). Besides a resulting mechanical load, potentially critical for e.g. sensitive insert parts like RFID-transponder, asymmetrical shrinkage can be problematic for adhesion and media tightness due to the formation of spaces between insert and polymer. The injection molding simulation showed a sufficient predictability regarding the shrinkage tendencies with included insert parts parallel and normal to flow direction. Higher shrinkage resistance by square shaped parts compared to cylindrical inserts could also be predicted. Nevertheless, the absolute shrinkage in thickness direction was underestimated by the simulation. Also the experimentally observed decreasing shrinkage with rising distance from the gate could not be proven by the simulation. This may be caused by a delayed solidification and followed by an overestimated time of effective holding pressure. Regarding welding lines, the simulation showed a longer welding line right after the square shaped part compared to the round insert. This could be explained by a higher flow resistance of the angular insert, a prolonged flow path, and thus a colder melt front at the meeting point of the melt behind the insert.

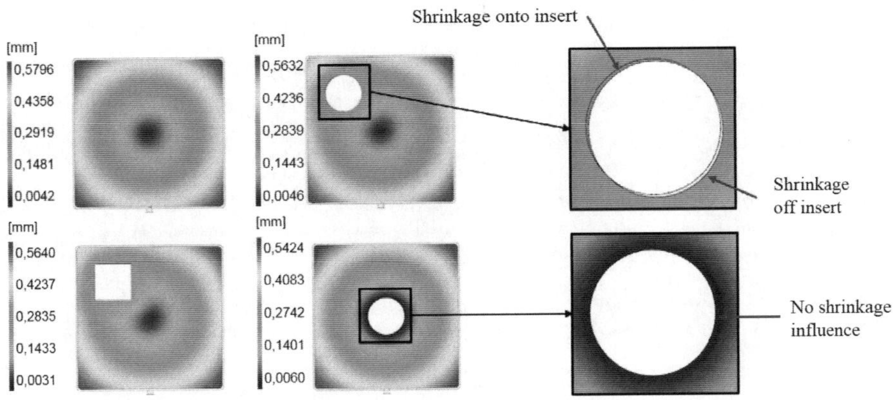

FIGURE 5. Injection molding simulation derived warpage of PBT test specimen without insert (top left), with angular insert (bottom left) and round insert (middle). An asymmetrical insert configuration can cause partial shrinkage onto or rather off the insert resulting in mechanical load on the insert (red areas indicate increased warpage, scaling factor = 1.0).

CONCLUSION

One of the main targets of this study was to investigate how the insert part geometry influences shrinkage of the surrounding polymer and consequently the dimensional accuracy of the injection molding part. The influences on shrinkage by inserts occurred in all three materials tested. The molded parts with two different kind of inserts (angular and round) showed significantly lower shrinkage parallel and normal to flow direction compared to parts without inserts, whereas angular shaped inserts mostly induced higher shrinkage resistance than round shaped inserts.

The second aspect evaluated was the effect of the insert position in the mold on shrinkage, including positions in the center, near the edges and near or apart the gate. It was observed that the insert position has a significant influence on shrinkage and dimensional accuracy of the injection molding part. Compared to a test specimen without inserts and to specimen with inserts at a central, symmetrical part position, a lower amount of shrinkage was observed for molded parts with inserts placed near the edges. This effect, assumed to result from asymmetrical shrinkage conditions and little material between cavity wall and insert, increases with distance from the gate and results in a lower shrinkage at the end of the flow path. A validation of the injection molding simulation with the experimental results showed a good alignment suggesting that simulation can help to save effort during the part and mold design process. However, due to remaining error, absolute and precise shrinkage values, especially in direction of parts wall thickness, should be checked experimentally.

Overall, this study aids the understanding of insert geometry and position induced shrinkage resistance in combination with different technical thermoplastic materials. Furthermore, shrinkage resulting in thermo-mechanical loads on inserts, critical for e.g. mechatronic inserts like RFID-transponders, can be assessed earlier. The investigated influences concerning round or angular shaped inserts, asymmetrical configurations and positions far or near the sprue can contribute to a more precise estimation of the achievable dimensional part accuracy.

REFERENCES

1. F. Johannaber, *Injection Molding Machines, A User's Guide*, Hanser Publications, Cincinnati, OH, 2007.
2. D. Kazmer, *Injection mold design engineering*, Hanser Publications, Cincinnati, OH, 2016.
3. F. de Santis, R. Pantani; V., Speranza, G. Titomanlio, *Analysis of Shrinkage Development of a Semicrystalline Polymer during Injection Molding*, Industrial & Engineering Chemistry Research, 2010. **5**(49), pp. 2469–2476.
4. D. Annicchiarico and J. R. Alcock, *Review of Factors that Affect Shrinkage of Molded Part in Injection Molding*, Materials and Manufacturing Processes, 29, 2014, pp. 662-682.
5. B.-R. Meyer and D. Falke, *Maßhaltige Kunststoff-Formteile, Toleranzen und Formteilengineering*, Hanser, Munich, 2013.
6. M. Grujicic, *Injection over molding of polymer-metal hybrid structures*, American Journal of Science and Technology, 2014. **1**(4), pp. 168-181.
7. T. Alajoki et al., *In-mould Integration of Electronics into Mechanics and Reliability of Overmoulded Electronic and Optoelectronic Components*, Microelectronics and Packaging Conference, Rimini, Italy, 2009, pp.1-6.
8. S.T. Amancio-Filho and J.F. dos Santos, *Joining of Polymers and Polymer–Metal Hybrid Structures: Recent Developments and Trends*, Journal of Polymer Engineerieng and Science, 2009. **49**(8), pp. 1461-1476.
9. M.-Y. Lyu and T. G. Choi, *Research Trends in Polymer Materials for Use in Lightweight Vehicles*, International Journal of Precision Engineering and Manufacturing, 2015. **16**(1), pp. 213-220.
10. M. Zeppenfeld et al., *Communication with Passive RFID Sensor Tags during Injection Molding of Medical Plastic Parts*, Smart SysTech 2017 - European Conf. on Smart Objects, Systems and Technologies, VDE Verlag, Berlin, Germany, 2017, pp.1-6.
11. M. Heinle and D. Drummer, *Measuring mechanical stresses on inserts during injection molding*, Advances in Mechanical Engineering, 2015. **7**(5), pp. 1–6.
12. P. Frey et al., *Influence of metal inserts with microformed edges on subsequent injection assembly moulding for media tight electronic systems*, 4th Int. Conf. on New Forming Technology, EDP sciences, Les Ulis, France, 2015, pp.1-7.
13. C. Heinle et al., *Assembly Molding of Tight Mechatronic System - A Model Approach*, Journal of Manufacturing Science and Engineering, 2010. **132**, pp. 1-4.
14. DIN EN ISO 294-3, *Plastics - Injection moulding of test specimens of thermoplastic materials, Part 3: Small plates*, 2003.
15. DIN EN ISO 294-4, *Plastics - Injection moulding of test specimens of thermoplastic materials, Part 4: Determination of moulding shrinkage*, 2003.

Optimizing the replication quality of a micro component via injection molding process simulation

Luca Antonio [(1)]*, Riemer Oltmann [(1)]

[(1)] University of Bremen - Bremen - Germany

Micro injection molding processes have to be optimized in order to ensure the required product quality at the micrometer level. Simulation programs in polymer micro replication technologies are used to optimize the mold design, to avoid the risks of costly re-engineering and to shorten the run-in time. Therefore, software simulation tools adapted from conventional injection molding can provide useful assistance for the optimization of the molding tools, mold inserts, micro component designs, and process parameters. The aim of this work is to investigate a medical micro component by simulating the micro injection molding process. The component has dimensions in the micrometer range and micro features that could lead to a challenging filling of the micro cavity. The process encompasses a large number of process parameters to be set properly in order to achieve a satisfactory replication quality of the micro component. The influence of mold design will be investigated by considering different runner and gate systems in order to find an optimal configuration. The filling, packing and warpage aspects of the micro component will be analyzed.

Micro Injection Molding Process Optimization Using Flash as Fingerprint Quality Indicator

Federico Baruffi[a*], Matteo Calaon[a] and Guido Tosello[a]

[a]*Technical University of Denmark, Department of Mechanical Engineering, Produktionstorvet, Building 427A, 2800 Kgs. Lyngby, Denmark*

febaru@mek.dtu.dk

Abstract. Replication-based manufacturing technologies provide a cost effective solution for producing complex and net-shaped components. Among them, micro injection molding holds a prominent place for its capability of accurately and precisely produce micro plastic parts. In this study, the effects of the micro injection molding process on the geometrical quality of an ultra-small (volume = 0.07 mm^3, mass = 0.1 mg) three-dimensional micro Polyoxymethylene (POM) component for medical applications were investigated. Preliminary experiments highlighted the need for venting channels in order to facilitate the evacuation of air from the micro cavity, thus achieving complete filling. If, on one hand, the implemented venting channel helped with this regard, on the other, it caused the generation of a micro-scaled flash on the part. In order to optimize the part geometry with respect to design specifications, the flash areal size was utilized as quality indicator along with the size of another defect, the gate mark. A design of the experiments approach was carried out in order to study the effects of melt temperature, mold temperature, holding pressure and injection speed. For this task, a two-level full factorial design was adopted. The flash and gate mark sizes of each molded part were characterized and measured using a state-of-the-art 3D focus variation microscope featuring sub-micrometric resolution. The results of the analysis showed that flash size can be successfully utilized to highlight the most significant process parameters with respect to the part quality. Being the flash area also measurable using an appropriate 2D camera, this opens the door to future in-line and cost-effective dimensional quality assurance.

Keywords: Micro injection molding, POM, Process optimization, Focus variation microscope
PACS: 83.50.Uv

INTRODUCTION

Micro injection molding (µIM) is the only technology enabling the mass production of polymer micro components. µIM was conceived as the miniaturized counterpart of conventional injection molding (IM) process with the aim of transfer to the micro scale the enormous advantages of IM in terms of repeatability and productivity [1]. µIM components are currently categorized in three classes [2]. The first lists components with outer dimensions in the millimeter range and locally featuring structures with size in the micrometer range. The second class contains parts having larger outer dimensions but tolerances in the micrometer range. Finally, the last class covers polymer components with outer dimensions in the micrometer range and mass in the milligram order.

Although IM and µIM share the same process cycle, downscaling IM is not straightforward since it involves many challenges that need to be dealt with to develop an accurate and precise process [3]. Dedicated micro machines, different modeling approaches, new measuring solutions and micro tooling processes have to be designed and optimized to efficiently support the µIM production. The recent research faced many of such issues, creating a consistent knowledge of the µIM process [4], [5].

The new trend of µIM research is related to process and product optimization. High temperature, injection flowrates and pressures are in fact needed to achieve the target replication at the micro scale [6], and this makes the process window more restricted when compare to IM. Therefore, process optimization is the key to ensure the required quality for micro polymer product manufactured by µIM. In literature, examples of µIM optimization aiming at enhancing cavity filling, process repeatability, venting conditions and demolding performances [2], [7], [8], [9] are reported.

One of the main challenges of μIM optimization is that, when the geometrical characteristics of the component are considered, demanding and time consuming experimental investigations based on off-line dimensional measurements are needed to tune the process [10], [11]. This experimental effort could be consistently diminished by the application of in-line dimensional measurement systems. Such systems, however, must be extremely fast in order to be successfully applied to a fast process such as μIM, in which cycle times are usually in the order of few seconds. Therefore, 2D high-resolution cameras are one of the best candidate for such a task, even though a 2D representation of a part geometry may not be enough to acquire all the functional geometries of 3D micro molded parts.

A possible approach to face this issue is the identification of a μIM fingerprint. This fingerprint, which can be any measurable characteristic of the molded part, has to be very sensitive to the process parameters that need to be optimized. This condition is essential since the aim is to carry out the product/process optimization by only acquiring and/or measuring the fingerprint. Therefore, a quantity that is strongly influenced by the process, and is somehow linked to part functionality, is needed. Another condition that the fingerprint has to fulfill is that it should be measurable by an in-line instrument with the aim of carrying out the optimization without relying on time-consuming and expensive off-line measurement systems.

The present paper studied the effect of μIM process parameters on the geometrical quality of an ultra-small 3D Polyoxymethylene (POM) micro medical component with the aim of identifying possible μIM fingerprint candidates. In order to accomplish this, the molded parts were measured and the relation between dimensional quality and investigated process parameters was quantified using a statistical DOE approach.

MATERIALS AND METHODS

The micro component object of the study is a three-dimensional POM component for medical applications. Figure 1 shows the main geometrical characteristics of the part. Being the volume equal to 0.07 mm^3 and the mass to 0.1 mg, it is, according to literature, the smallest component manufactured by μIM.

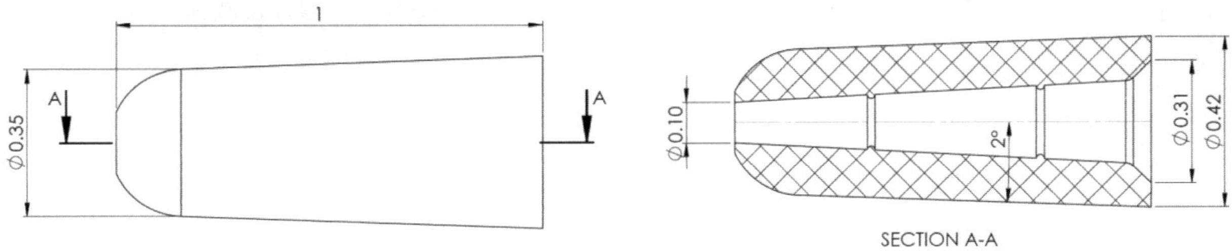

(a) (b)

FIGURE 1. (a) Geometry and outer dimensions in mm; and (b) section view.

The component has a full three-dimensional shape, being both the hole and the outer wall tapered and structured.

The material used was an unfilled POM from Celanese, Hostaform® C 27021. Such polymer grade was selected for its peculiar properties, namely self-lubrication and enhanced flowability. Injection molding experiments were carried out using a Wittmann-Battenfeld MicroPower 15 μIM machine. This machine has an injection unit consisting of a 14 mm plasticization screw and a 5 mm injection plunger. The one-cavity insert was machined by μEDM. A three-plate mold was used, ensuring automatic detachment of the ring gate that was positioned on the left side of the part according to Figure 1. The central hole of the micro part was created by means of a tapered micro pin protruding from the ejection plate. The features on the internal hole were designed to keep the part on the ejection side after the molding cycle was completed. A vacuum gripper acting in the direction parallel to the component axis allowed to automatically perform demolding.

Preliminary experiments highlighted the need of a venting channel to achieve a consistent filling of the part. The first shots without this modification, in fact, always provided parts showing defects caused by the presence of entrapped air in the cavity. Therefore, a circular 5 μm-deep venting channel was machined on the opposite side of the gate, that is the right side of the part according to Figure 1. Such a modification, which allowed performing a more stable process, generated flash around the outer diameter of the part, since the polymer melt was allowed to flow inside the venting channel.

Another defect that affected the part quality was a gate mark generated by the automatic detachment of the feeding system from the part. Such a defect, which may be negligible for conventional injection molded parts, becomes very relevant in the micro scale and therefore must be minimized to ensure a perfect functionality.

A design of experiment (DOE) approach was used to evaluate the effects of the process on the size of these two defects. Four parameters were varied: holding pressure, melt temperature, mold temperature and injection speed. A 2^4 full factorial design with five replications was performed. The parts molded during the first ten cycles of each process combinations were discarded to draw conclusion on a stable process. Table 1 shows the details of the experimental design.

TABLE 1. Experimental settings.

Process Parameters	Experimental Levels
Holding pressure p_{hold} [bar]	250, 500
Melt temperature T_{melt} [°C]	200, 220
Mold temperature T_{mold} [°C]	100,110
Injection speed v_{inj} [mm/s]	150, 350

The dimensional outputs were selected as indicators of the defects size. The flash size was characterized by measuring the area of the part, named A_{flash}, correspondent to the venting channel (see Figure 2). A larger measured area thus corresponds to a bigger flash. The gate mark was characterized by its length L_{mark} defined as distance between the extremity of the gate mark and the plane corresponding to the ideal detachment of the gate from the part (see Figure 2). Both these dimensional outputs were extrapolated from 3D acquisitions of the parts carried out with a 3D focus variation optical instrument (Alicona InfiniteFocus) with 0.44 μm lateral resolution. Each acquisition was repeated three times with the aim to performing robust measurements.

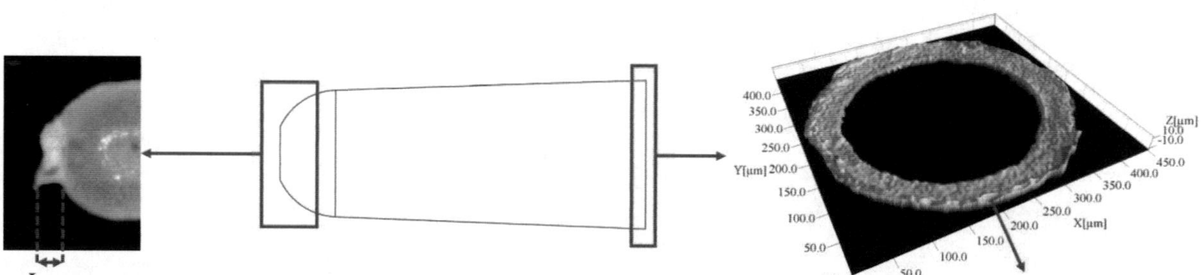

FIGURE 2. Details of the dimensional outputs for the analysis: length of the gate mark L_{mark} (left) and flash area A_{flash} (right)

RESULTS

Figure 3 shows the main effects plots for the two dimensional outputs under analysis.

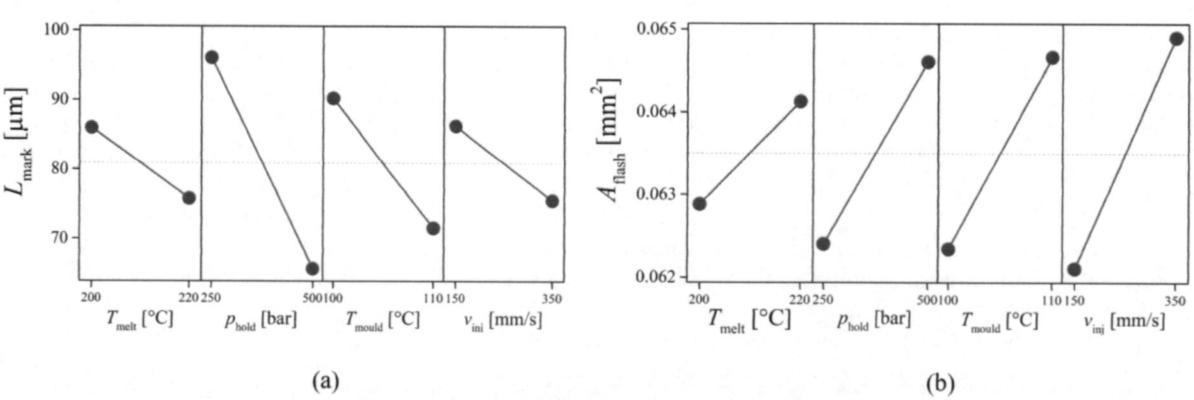

FIGURE 3. Main effects plot for L_{mark} (a) and A_{flash} (b)

The length of the gate mark on the parts was on average equal to 82 μm. Such a size is of course very relevant if the overall dimensions of the micro component are considered (see Figure 1). L_{mark} was generally lower when the high levels of the parameters were used. In particular, holding pressure showed the largest influence, since the use of 500 bar provided parts having on average a 30 μm shorter gate mark than those molded with p_{hold} equal to 250 bar. In order to further inspect the effect of the process on such defect, SEM images of parts molded with different parameters combinations were taken. When molding with all the variables at their low levels, the defect was much more evident and it almost occluded the upper hole, making the component not functional (see Figure 4(a)). A zone of deformation was clearly visible around the defect, meaning that the mechanism that generated the defect was indeed the breakage caused by detachment of the gate from the part. Molding with some parameters at high level improved the quality of the components, as shown in Figure 4(b) and 4(c). In these cases, the upper hole was more visible, but the gate mark still protruded greatly from the surface. Finally, molding with most the parameters at high levels (see Figure 4(d)) allowed to produce part almost free of gate mark. This difference of gate mark could be caused by a change of crystallinity and/or residual stresses when molding with different parameters. Finally, it can be concluded that, for the sake of minimizing the gate mark size, molding with the four process parameters at their high levels was necessary.

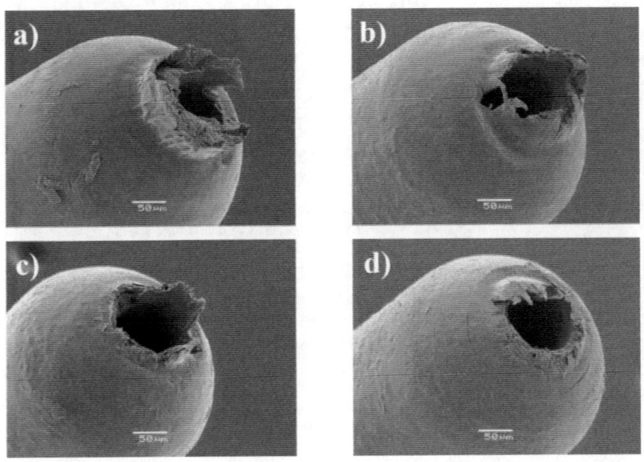

FIGURE 4. Appearance of the gate mark for different process parameters combinations: a) was molded with low T_{melt}, p_{hold}, T_{mold} and v_{inj}; b) was molded with low p_{hold} and v_{inj}; c) was molded with low T_{melt} and T_{mold} and d) was molded with low v_{inj}

The area on the back of the component was equal on average to 0.0635 mm^2. To check the actual size of the flash, the diameters of the mold corresponding to the inner and outer diameter of the part were measured. It was therefore calculated that a component without flash and perfectly replicated would have a 0.047 mm^2 area. Being all the measured areas greater than this value, it was proven the actual presence of flash on the molded parts. As for L_{mark}, the process variables showed a great effect on A_{flash}. In particular, the increase of all the four parameters resulted in an increase of the flash size. This observation was congruent with the expectations. An increase of T_{melt} and T_{mold} both facilitated the filling of the cavity because of the reduced viscosity of the polymer melt at high temperatures. The better filling, in this particular case, generated a larger flash since the circular venting channel was more easily filled. A higher level of v_{inj} also led to a decrease of viscosity thanks to the shear thinning behavior of polymers and hence to a larger flash. Increasing p_{hold} allowed more material to enter the cavity during the holding phase. This of course generated a larger flash on the part. In conclusion, molding with the low levels of the selected process parameters is advisable to minimize the size of this defect on the micro parts.

When comparing the results for the two outputs, it is easy to note that the effects of the process was opposite. This means that it was not possible to achieve the same effect on L_{mark} and A_{flash} simultaneously. Being the objective to minimize both the defects, it is a great challenge to optimize the part dimensional quality. A further optimization shall therefore be carried out by investigating the behavior of the parts within the experimental ranges of this study.

With regard to the product fingerprint identification, both the measured outputs showed a pronounced dependence on the μIM process. They are also both linked to the part quality, being a direct quantification of defects affecting it. However, the flash size is more easily measurable by an in-line system. In fact, once the molded component is ejected from the mold using a vacuum system, a proper 2D system could acquire the portion of the part correspondent to the flash and easily calculate A_{flash} during a single μIM cycle. A high-resolution camera would be needed, since the deviations of flash size caused by the process were in the order of 10^{-3} mm^2. On the other hand, L_{mark} is less easily measurable, being the gate mark length parallel to the demolding direction. Therefore, either a 3D measuring system or a different mold design would be needed to carry out in-line measurements.

CONCLUSIONS

In this study, the effects of the μIM process on the dimensional quality of an ultra-small three-dimensional micro POM component were investigated with the aim of identifying a suitable product fingerprint to use as a tool to carry out a fast and efficient dimensional quality assurance.

The two main defects affecting part quality, namely gate mark and flash, were characterized using dimensional indicators and measured with a focus variation microscope. Four process parameters (melt temperature, holding pressure, mold temperature and injection speed) were varied within a 2^4 full factorial design. The results of the measurements showed that both the defects were largely affected by process variations. In particular, increasing the process variables led to a decrease of gate mark length and, simultaneously, to an increase of flash size. This opposite behavior of the two defects poses a great challenge with respect to μIM optimization.

Finally, the flash area was selected as best fingerprint candidate, since it was proven a process indicator and, at the same time, can be measured in-line more easily than the gate mark length.

Future studies will be dedicated to a further optimization of the part defects within the experimental ranges investigated in this paper. The fingerprint concept will be also tested using actual in-line measurement of the flash area.

ACKNOWLEDGMENTS

This research work was undertaken in the context of MICROMAN project ("Process Fingerprint for Zero-defect Net-shape MICROMANufacturing", http://www.microman.mek.dtu.dk/). MICROMAN is a European Training Network supported by Horizon 2020, the EU Framework Programme for Research and Innovation (Project ID: 674801).

REFERENCES

1. J. Giboz, T. Copponnex, and P. Mélé, *J. Micromechanics Microengineering* **17**, R96-R109 (2007).
2. B. Whiteside, M. T. Martyn, P. D. Coates, G. Greenway, P. S. Allen, and P. Hornsby, *Plast. Rubber Compos.* **33**, 11-17 (2004).
3. C. Yang, X.-H. Yin, and G.-M. Cheng, *J. Micromechanics Microengineering* **23**, 1-21 (2013).
4. C. H. Wu and W. J. Liang, *Polym. Eng. Sci.* **45**, 1021-1030, (2011).
5. M. Heckele and W. K. Schomburg, *J. Micromechanics Microengineering* **14**, R1-R14 (2003).
6. V. Bellantone, R. Surace, G. Trotta, and I. Fassi, *Int. J. Adv. Manuf. Technol.* **67**, 1407-1421 (2012).
7. B. Sha, S. Dimov, C. Griffiths, and M. S. Packianather, *J. Mater. Process. Technol.* **183**, 284-296 (2007).
8. U. M. Attia and J. R. Alcock, *Int. J. Adv. Manuf. Technol.* **52**, 183-194 (2011).
9. C. A. Griffiths, G. Tosello, S. S. Dimov, S. G. Scholz, A. Rees, and B. Whiteside, *Microsyst. Technol.* **21**, 1677-1690 (2014).
10. D. Masato, M. Sorgato, and G. Lucchetta, *Mater. Des.* **95,** 219-224 (2016).
11. R. Surace, V. Bellantone, G. Trotta, V. Basile, F. Modica, and I. Fassi, *J. Micro Nano-Manufacturing* **4**, (2015).

Predicting Adhesive Strength for Two-Component Injection Molded Products of Thermoplastics and Thermoset Rubbers

W. Six[1, a)], G.J. Bex[2, b)], J. De Keyzer[3], F. Desplentere[1], A. Van Bael[2]

[1] *Materials Technology TC, KU Leuven Campus Bruges, Spoorwegstraat 12, 8200 Bruges, Belgium*
[2] *Materials Technology TC, KU Leuven Technology Campus Diepenbeek, Wetenschapspark 27, 3590 Diepenbeek, Belgium*
[3] *Sustainable Chemical Process Technology TC, KU Leuven Technology Campus Diepenbeek, Wetenschapspark 27, 3590 Diepenbeek, Belgium*

[a)]Corresponding authors: wim.six@kuleuven.be
[b)]gertjan.bex@kuleuven.be

Abstract. Two component injection molding is a widespread technique to produce products with a hard soft material combination. In most cases, the stiff component is overmolded with the soft component. Most commonly, the overmolded material is a thermoplastic elastomer (TPE) which is compatibilised for optimal adhesion to a certain thermoplastic material. However, TPE's have limited properties for chemical and thermal properties. For this applications, a thermoset rubber as EPDM, NBR, XBR, etc. can be used to replace the TPE. Although recommendations for good adhesion between certain material combinations can be found, overmolding thermoplastics with a thermoset rubber is not that common, and final properties of the injection molded product are mostly unknown. One of the most important properties is the adhesion between the two injection molded products. To predict the final adhesion quality, a tool was developed, inspired by experimental evidence for a simple interface geometry and using existing simulation software to predict relevant local physical quantities. This paper focuses on the optimization of the process settings in order to maximize the adhesion in a two component injection molded wheel-like product, manufactured with thermoplastic core and a thermoset rubber outer layer. To optimize the adhesion, the main focus is put on the temperature and the degree of cure at the interface.

Keywords: Overmolding, Rubbers, adhesion.

PACS: 68.35 Gy, 81.70-q, 81.050 Lg

INTRODUCTION

Numerical simulations are a common tool to predict final properties of injection molded products in function of the process and material parameters, this as well for single as two stage injection molded products. Examples are fiber orientation, fiber length, warpage, localized tensile modulus, etc. Next to predicting the final properties, those numerical simulations are also used to optimize the product properties. Although difficult, some software packages also start to implement predictions on strength of weld lines and the adhesion between two or multi component thermoplastic products [1][2]. To the best of our knowledge, for two component thermoplastic, thermoset rubber products, there is still a lack of knowledge for interfacial strength. This makes it difficult to predict the quality of adhesion between the two components. Moreover, in commercial available numerical simulation tools as Autodesk Moldflow, there even is no straight way to simulate this material combination in overmolding [7]. In previous research conducted by Bex et al [8], a specialized two-component injection mold was developed to produce test samples. These samples make it possible to monitor the influences of process settings, such as mold temperature, curing time, injection speed, packing-pressure etc. on the final adhesion for various material combinations such as HDPE/EPDM and PC/NBR. As follow-up to this experimental information, Six et al [3], developed a tool to predict and to visualize a factor of adhesion (between 0, no adhesion and 1, maximal realized adhesion during experiments) calculated from temperature, degree of cure and pressure at the interface between the two components. In this paper, the process settings for optimal adhesion in a wheel-like product for which the mold shall be produced, will be optimized.

PRODUCT AND MOLD GEOMETRY

Figure 1 shows the cross section of the wheel-like product (ø 56mm) and the channels to control the mold temperature. In the mold, there are two types of cooling channels, the blue channels are used to cool the thermoplastic product while the red channels are used to cure the thermoset rubber. For both channels, the temperature can be regulated between 20 and 180°C. In the mold design, it is important to avoid severe heat flow between those two channels since they generally are at a different temperature.. In the shown product, the core of the product (wall thickness of 3mm) will be molded first. The outer shell will be overmolded in a second phase. In the presented mold, it will be possible to implement an insert to change the thickness of the thermoset rubber ring ranging from 2 to 8 mm. In the presented study, the simulations are performed with HDPE M80064S, Sabic and EPDM 005K, provided by Hercorub, of which the most important data was measured and presented by Bex et al. [4] and Six et al. [5].

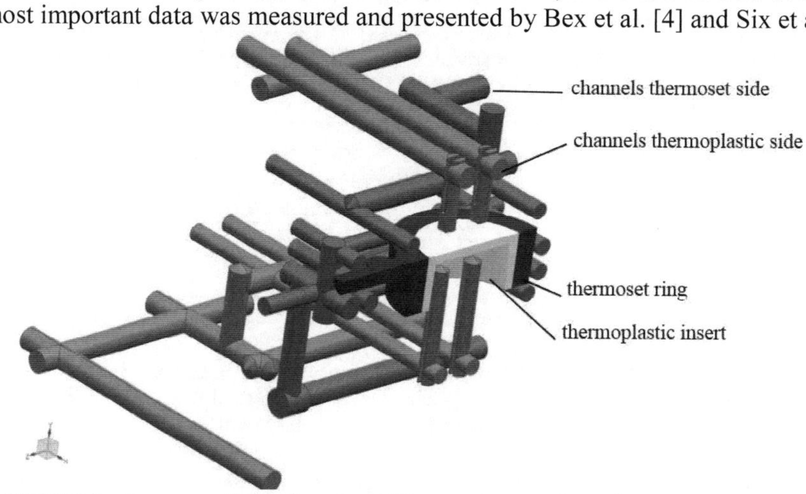

FIGURE 1. Cross section of a wheel-like product and cooling channels

SIMULATION

To predict the possible strength of adhesion between the thermoplastic and thermoset material, data is used presented by Bex et al. [4]. This data, is normalized and shown in table 1. As can be noted (grey zone), only two types of data are available, adhesion vs. degree of cure at an optimal interface temperature and adhesion vs. interface temperature at an optimal interface cure. Data points between those two curves are not available at this moment, The other values in this table are an estimation, test to confirm this values will be performed in the nearby future. This is also the case in the presented results, influence of temperature is only shown as if curing would be optimal, and influence of curing is if temperature would be optimal. The presented table shows interface temperatures ranging from 126°C to 138°C. Measurements on higher temperatures showed a large deformation of the interface surface, and thus resulting in a defective product. Furthermore, it can be noted that optimal temperature range for adhesion is very narrow and lays between 132 and 138°C (between T_{vicat} and T_{melt} [6]) . For curing, the optimal zone lays between 60 and 100% degree of cure. It should be stated that this is only valid for the presented material combination. Other material combinations have been measured but are not included in this paper.

The presented results show the influence of the temperature of the channels on the resulting adhesion, as they affect curing and interfacial temperature. The presented results exist out of two parts, a color plot, which shows the local value for the specified parameter, a local adhesion factor, this shows amount of adhesion which can be expected for the given parameter, and a histogram showing a summary of the results for adhesion as function of temperature and as function of cure. The initial presented temperature setting for the hot and cold side is 180°C-85°C. After an optimization step, an optimal temperature combination of 170°C-55°C is found. In the result section, the resulting surface temperatures, cure and effect on the adhesion are shown.

TABLE (1). Relative strength of adhesion in function of interface temperature and interface cure. Grey values are measured, white are estimates .

	degree of cure							
interface temperature	**0,09**	**0,18**	**0,38**	**0,57**	**0,73**	**0,83**	**0,89**	**1**
126	0,06	0,12	0,12	0,17	0,19	0,21	0,22	0,21
129	0,10	0,19	0,20	0,27	0,32	0,35	0,36	0,36
132	0,21	0,43	0,44	0,60	0,69	0,78	0,79	0,79
135	0,29	0,58	0,61	0,82	0,95	1,06	1,08	1,07
138	0,27	0,54	0,56	0,77	0,88	0,99	1,00	1,00

RESULTS AND DISCUSSION

The initial settings were derived from the mold settings used by Bex et al [4]. The temperature has been set on 180°C for the thermoset side and 85°C for the thermoplastic side. The simulated interface temperature is shown in Figure 2 (a) and ranges between 77 and 154°C. Taking into account the measured strength of adhesion as function of temperature, a new result which shows the expected value for strength of adhesion is created, this is shown in Figure 2 (b). As can be noted, almost the complete interface shows no adhesion. The temperature on the interface is too high which will cause a deformed interface. Combining all nodal data, a histogram of the simulated values on the interface can be created, which shows relevant information of the temperature distribution on the interface (Figure 2). On this histogram, Figure 2 (c) two blue vertical lines show the optimal interface temperature (adhesion factor > 0.8). The histogram in Figure 2 (c) shows a skew distribution with an average 11°C out of the optimal zone.

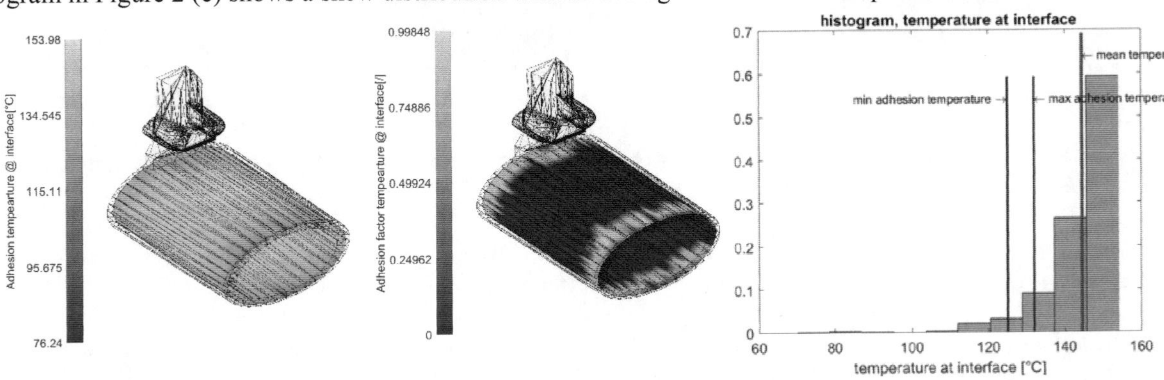

FIGURE 2. Interface temperature (a), factor of adhesion for temperature (b). Histogram of temperature at interface (c) for the 180°C-85°C setting

High interface temperatures result in a good curing at this interface. This is shown in Figure 3 (a). Although curing degrees range between 0 and 1, almost the complete interface is cured at the proposed cycle time of 1000s. As can be noted, a few small zones are not cured completely due to low interface temperatures. This is mostly caused by the difference in distance between the cooling channel and product interface. The histogram in Figure 3 (c) shows the distribution of adhesion factor. The simulation shows a skew distribution, with a high adhesion factor as function of cure. The average value lays on 0.95, an almost optimal value for adhesion as function of curing.

Evaluating those two results, it is clear that reduction of the temperature at the interface is necessary. This can be achieved by reducing the temperature at the rubber and/or the thermoplastic side. To avoid an increase in curing time or a lower curing at the interface, the most optimal way to reduce interface temperature is reducing the water temperature at the thermoplastic side. However, this will increase the temperature difference between the thermoset and the thermoplastic side, which increases an unwanted heat flow between those two sides. This heat flow can be limited through the use of insulating techniques.

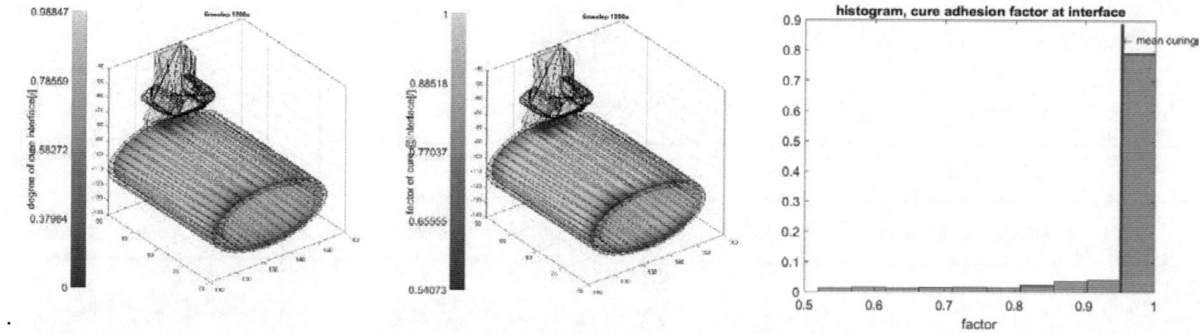

FIGURE 3. Degree of cure at interface (a) , factor of adhesion (b) for curing. Histogram of temperature at interface (c) for the 180°C-85°C setting

Via the presented tool by Six et al. [3], the temperature settings were optimized towards the optimal settings. An optimal temperature setting for this mold and product geometry was found for a temperature of 170°C for the thermoset side and 55°C for the thermoplastic side. In the example the presented optimal setting was found through trial and error. This of course can be automated with an optimization routine. Figure 4 left shows the temperature at the interface. It can be noted that the maximum temperature has been reduced to 140°C. The histogram in Figure 4 right also shows a normal distributed temperature around the average, which is within the zone for optimal adhesion.

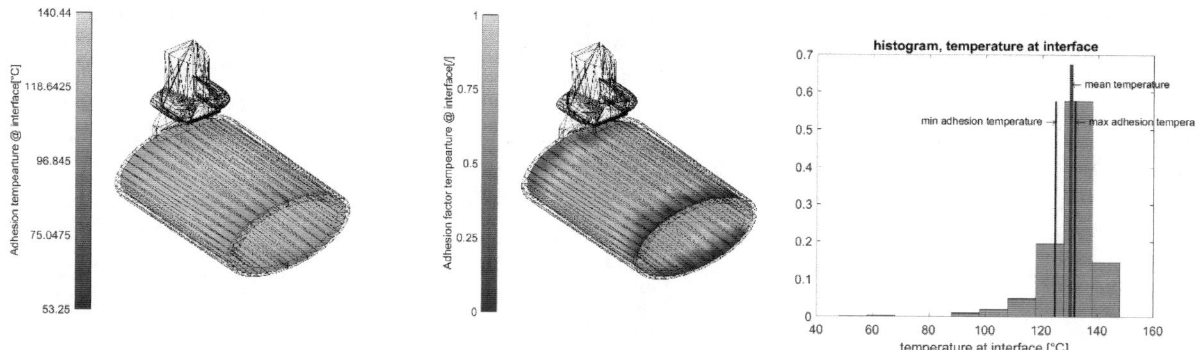

FIGURE 4. Interface temperature (a), factor of adhesion for temperature (b). Histogram of temperature at interface (c) for the 170°C-55°C setting

The resulting cure at the interface, shown in Figure 5 (a), is lower compared to the 180°C-85°C setting. This is expected as curing speed is directly correlated with the interface temperature. The average curing for this settings at the interface is 0.62, this results in an average factor of adhesion of 0.79. Although lower than in the 180°C-85°C setting, the reduction in average adhesion factor is only around 15% while the reduction in curing is around 30%. A longer curing time can increase the degree of cure, without influencing the interface temperature.

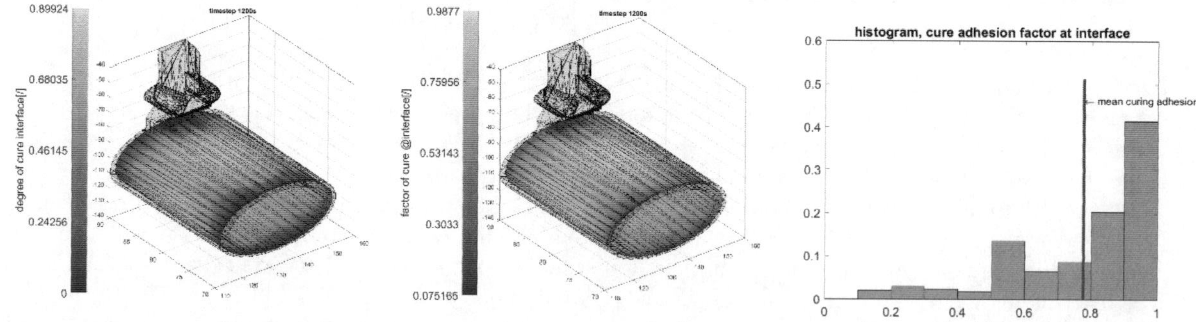

FIGURE 5. Degree of cure at interface (a) , factor of adhesion (b) for curing. histogram of temperature at interface (c) for the 170°C-55°C setting

CONCLUSION

A tool is presented in this paper that allows to optimize the adhesion quality between a rubber and a thermoplastic in two component injection molded products. For a test case, optimal temperature settings were found. Experimental validations are planned in the near future. The tool also can be used to optimize cooling channel geometry to reduce the deviation in temperature and curing degree. Another improvement would be automating the tool with an optimization routine to find the optimal process settings.

ACKNOWLEDGEMENT

The authors acknowledge the KU Leuven Impulsfonds for the financial support and the Company Hercorub NV for providing the uncured rubber material. G.J. Bex acknowledges the Flanders Research Foundation (FWO) for his PhD grant strategic basic research.

REFERENCES

[1] M.M. Bouwman T.G Donderwinkel, S. Wijskamp 3rd ITHEC procedings , Bremen, 2016, pp 37-40

[2] J. Onken and C. Hopmann, GAK, 69, No. 9, 2016, pp.574-580

[3] Six, W., Desplentere, F., Bex, G., De Keyzer, J., Van Bael, A. (2017). PPS Europe Africa procedings, Dresden, 2017, art.nr. S06-224

[4] Bex, G., Six W., De Keyzer, J., Desplentere, F., Van Bael, A., PPS Europe Africa procedings, Dresden, 2017

[5] Six, W., Desplentere, F., Bex, G., De Keyzer, J., Van Bael, A, 7th international PMI conference proceedings: Ghent, 2016, pp. 195-199.

[6] Sabic (2017,06 14) www.sabic-europe.com. cosulted on https://www.sabic.com/en/products/polymer/polyethylene-pe/sabic-hdpe

[7] Moldflow (2017,07 08) https://knowledge.autodesk.com consulted on https://knowledge.autodesk.com/support/moldflow-insight/learn-explore/caas/CloudHelp/cloudhelp/2018/ENU/MoldflowInsight-WhatsNew/files/GUID-352A6BCD-9F36-42B2-9896-46F632116F9C-htm.html

[8] Bex, G., De Keyzer, J., Desplentere, F., Van Bael, A.,.JAMT ,2017, No.- 92,pp. 2599-2607

On the Applicability of Micro Injection Molding Simulations to Multivariate Integrated Process/Product Optimization

Federico Baruffi[a*], Matteo Calaon[a], Guido Tosello[a] and René Elsborg[b]

[a]*Technical University of Denmark, Department of Mechanical Engineering, Produktionstorvet, Building 427A, 2800 Kgs. Lyngby, Denmark*
[b]*Ortofon A/S, Nakskov, Denmark*

febaru@mek.dtu.dk

Abstract. As in conventional injection molding, process simulations of micro injection molding are a powerful tool for the optimization of the design of mold, parts and process. However, the miniaturized scale of the micro products introduces relevant challenges in terms of both validation and accuracy of the simulations. In the present work, a case study based on the micro injection molding process of thermoplastic elastomer micro rings for sensors application is treated. Injection molding process simulations using Autodesk Moldflow Insight 2016® were applied with the aim of accomplishing two main tasks. The first concerns the prediction of the main parts defects (weld lines and air traps). The second is related to the accuracy of the simulations results in forecasting the effects of the main injection molding process parameters, namely mold temperature, melt temperature, injection speed and holding pressure, on the part geometry. The outcomes of the simulations were directly compared to real parts based on SEM inspections and optical measurements. The results showed that the model was capable of accurately capturing the position and the relevance of the micro ring weld lines and air traps. It was therefore proved that micro scaled defects can be successfully predicted using a suitable model. Finally, process simulations correctly predicted the effects of the four investigated process parameters on the part functional dimensions, opening the door to future micro injection molding virtual experiments.

Keywords: Process simulation, Micro injection molding, TPE
PACS: 83.50.Uv

INTRODUCTION

In recent days, the demand of miniaturized, complex shaped components has increased in many engineering fields such as biotechnology, avionics, communication, automotive, medicine, etc. [1]. The need of meeting this fast growing led to the development of new manufacturing processes [2]. Micro injection molding (μIM) is one of them: it is a process that effectively enables the manufacturing of micro components in the large volumes typical of replication-based processes [3].

As for conventional injection molding, process simulations are a powerful tool also for μIM. They are used to shorten micro products time-to-market by avoiding or reducing development and re-engineering time [4], leading to a consistent cost reduction.

Different commercial software packages such as Autodesk Moldflow®, Sigmasoft® and Moldex3D® are available on the market to simulate injection molding. However, such software tools are specifically made for conventional injection molding. When dealing with μIM, such commercial software packages can still be applied, but the results usually lack of the needed quantitative accuracy [5]. This happens because the polymer flow inside micro-cavities is characterized by peculiar phenomena that are not modeled inside commercial software. Firstly, when flowing inside a micro channel (i.e. a channel whose section has the main dimension in the micrometric range), a polymer has a strong tendency to slip [6], invalidating the no-slip conditions that is assumed as true for conventional injection molding models. Secondly, the rheological data at high shear rates, which are typical of μIM [7], are usually not measurable by means of common testing equipment such as shear of capillary rheometers [8]. This lack of information, which is caused by the extreme difficulty of measuring viscosity of a polymer melt flowing inside micro cavities, can seriously affect the simulation results. Other approximations made by commercial software that do not usually apply to micro molding are constant heat transfer coefficient [9], absence of the elongation component of viscosity and of surface tension [10]. Another typical complication is related to the validation phase. Pressure and temperature sensors cannot be easily placed inside micro cavities, and thus measured and simulated data are not usually comparable. Short-shots are also difficult to carry out when very small injection volumes are involved, preventing from performing a validation based on the filling pattern.

However, a proper modeling strategy can help to achieve more precise results. Marhöfer et al. [11] used a multi-scale mesh for modeling the micro injection molding process of a microfluidic device using Autodesk Molflow 2013®. By including mold blocks and barrel in the model, the authors showed how simulations results were largely improved. Costa et al. [12] studied the best strategies for carrying out a precise simulation of the injection molding of a flat thin cavity and of a miniaturized dog-bone shaped part using Autodesk Moldflow 2010®. Their results showed that the modeling strategy for the nozzle tip strongly influenced the accuracy of the simulation. Marhöfer et al. [13] successfully used process simulation in the gate design of two microfluidic components. The validation phase was based on SEM images of short shots that were compared with the correspondent filling steps.

The present work aims at applying injection molding process simulations as a virtual process/product optimization tool for thermoplastic elastomer (TPE) micro rings. To validate the numerical model, a functional geometry of the part was chosen as term of comparison between simulation results and real measurements. The effects of four different process parameters were then studied. The numerical model was also tested for the prediction of the main part defects.

MATERIALS AND METHODS

Experimental Details

The micro component object of the study was a TPE micro ring used in sensor applications. Figure 1 shows the main geometrical characteristics of the part.

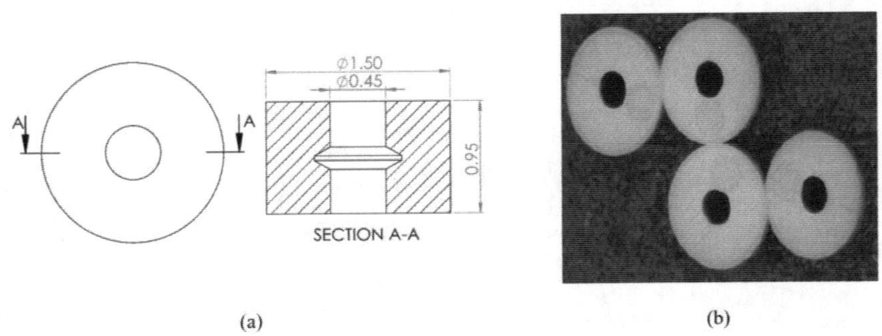

FIGURE 1. TPE micro rings. (a) Geometry and nominal dimensions in mm; and (b) appearance of molded rings.

The outer diameter (OD) was chosen as output for the validation of simulation results based on experimental results, being a geometry that has a decisive impact on the component functionality.

The material used was a styrene-ethylene-butylene-styrene (SEBS). Injection molding experiments were carried out using a Wittmann-Battenfeld MicroPower 15 μIM machine. The machine has an injection unit consisting of a 14 mm plasticization screw and a 5 mm injection plunger. Four cavities were machined on the mold insert. The central hole of the micro rings was created by means of structured micro pins.

A design of experiment (DOE) approach was used to evaluate the effects of the process on OD. Four parameters were varied: holding pressure, melt temperature, mold temperature and injection speed. A general full factorial design with five replications was performed. Mold and melt temperatures were varied on two levels, while holding pressure and injection speed on three, resulting in 36 different process combinations. The parts molded during the first ten cycles of each process combination were discarded to draw conclusion on a stable process. Table 1 shows the details of the experimental design.

TABLE 1. Experimental settings.

Process Parameters	Experimental Levels
Holding pressure p_{hold} [bar]	300, 500, 700
Melt temperature T_{melt} [°C]	210, 225
Mold temperature T_{mold} [°C]	30, 40
Injection speed v_{inj} [mm/s]	50, 70, 90

The diameters of the molded micro rings were measured with a focus variation optical instrument (Alicona InfiniteFocus) with 1.75 μm lateral resolution. A preliminary uncertainty assessment procedure allowed the calculation of the expanded uncertainty U, which was equal to 2.0 μm.

Simulation Setup

The commercially available software Autodesk Moldflow Insight (ASMI) 2016® was used for the numerical analysis. The simulation model comprised the four parts and the feed system (see Figure 2(a)). A 3D meshing was applied to such geometry. The commonly used 2.5 D Hele-Shaw approximation is in fact not suitable for μIM simulations since the hypothesis of constant pressure along the thickness of the micro geometries cannot be made [12], [13]. To reduce the computation time but, at the same time, achieve accurate results, a multi-scale mesh was used. Element sizes ranging from 500 μm to 50 μm size were set (see Figure 2(b)). To increase the accuracy, the region of interest (i.e. ring geometry and the gate) was more finely discretized than the sprue, where the polymer flow followed a relatively simple pattern. The model contained approximately 1.4×10^6 tetrahedrons.

In ASMI, the flow is modeled by Navier-Stokes equations in 3D. As regard the material data, the Cross-WLF and the Tait models are used to describe the rheological and thermodynamic behavior of the plastics. Material data provided by the material supplier were used to fit the two models and then imported in ASMI. A "Fill+Pack+Warp" analysis sequence was selected in order to obtain the final dimensions of the parts after the injection process is finished. As boundary conditions, the process settings employed in the actual experiments were selected to reproduce the real process conditions.

36 simulations were run in order to replicate the experimental DOE design. After the analyses were completed, the spatial coordinates of the nodes standing on the outer diameters were extracted (see Figure 2(c)) and a circle was mathematically fitted to calculate the OD value. For each process condition, the four ODs of the four cavities were averaged in order to eliminate any deviation introduced by non-perfect mesh symmetry.

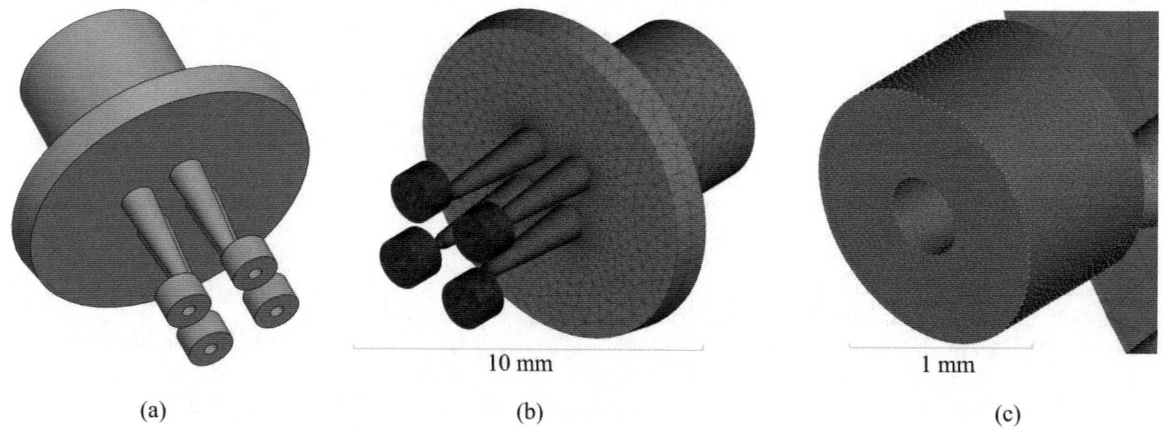

FIGURE 2. (a) CAD model of the parts with feed system as imported in ASMI; (b) Meshed model; and (c) detail of the meshed micro ring.

RESULTS

Prediction of Part Defects

Air evacuation from the mold is a very important condition for achieving optimal cavity replication. If air is not properly evacuated from the mold by means of parting lines and/or venting channels, it can prevent the polymer melt from completely filling the cavity. This typically causes defects on the surface of the part. In the present work, simulation results indicated the presence of five air traps on both outer and inner surface of the ring on the opposite side of the gate position (see Figure 3). This was due to the filling pattern: being the gate asymmetrical with respect to the ring axis, the polymer flow separates and then converges on the other side of the ring, sealing the cavity. Another defect caused by the asymmetrical gate was the presence of weld lines (see Figure 4) on the top and bottom ring surfaces. SEM images of real parts were taken and directly compared to simulation results. The process simulations accurately predicted both air traps and weld line position, proving to be a powerful tool for defects prediction in micro injection molded parts.

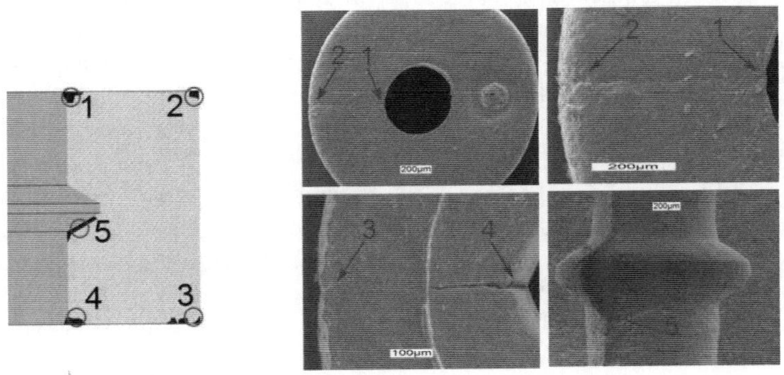

(a) (b)

FIGURE 3. (a) Air traps as predicted by process simulations; and (b) SEM images showing the same defects on real parts.

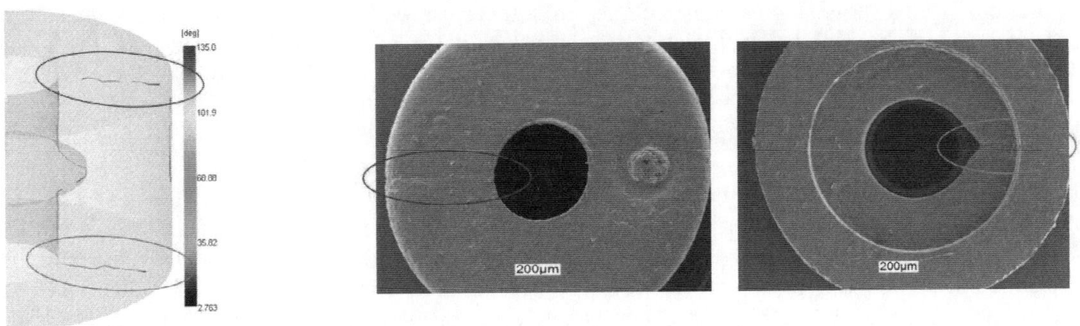

(a) (b)

FIGURE 4. (a) Weld lines as predicted by process simulations; and (b) weld lines on real parts. The weld line on the upper surface is indicated in red, while the one on the lower surface in blue.

Validation Based on DOE Results

To validate the use of the µIM model as a tool for product/process optimization, the OD measurements were compared to the numerical simulations. The effects of the four process parameters were quantified for both real and virtual measurements. Figure 5 shows the results in terms of main effects plots.

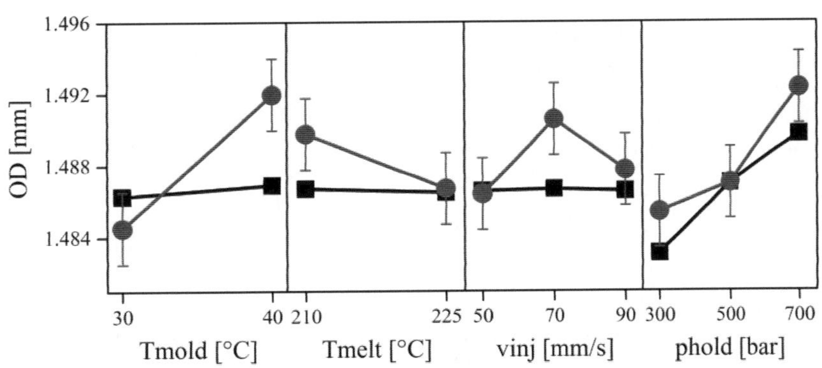

FIGURE 5. Main effects plots for OD. In black, the simulation results. In red, the measurements results of real parts. The error bars indicate the expanded uncertainty *U*.

On average, the deviations between experiments and simulation results was 1.6 μm. This value was completely covered by the uncertainty U, and therefore the simulation were able to accurately predict the outer diameter of the micro molded rings. As regard the effects of the process parameters on OD, a good agreement was also observed. The use of the high level of mold temperature provided an increase of OD for both simulations and real parts. This happened because, by increasing T_{mold}, a premature cooling of the polymer melt inside the cavity was prevented. In this case, the effect (i.e. the slope of the main effect plot) of the real process was larger than the simulated one. On the other hand, an increase of melt temperature resulted in a reduction of the outer diameter. This happening was predicted by simulations too. As for T_{mold}, the effect of the T_{melt} was underestimated by the numerical model. The effect of the injection speed was strongly non-linear, since the central level produced higher results than the low and high ones. This behavior, which was also predicted by the simulations, could be due to larger residual stresses originated at high injection speed because of the more drastic cooling rate. The effect of the real process was once again underestimated by the numerical analysis results. The effect of the holding pressure was the one that was best predicted by simulations. The direct relation between OD and p_{hold} was accurately predicted by the numerical model: the deviations of experimental and numerical results of the three p_{hold} levels ranged between 0.0 μm and 2.6 μm.

CONCLUSIONS

In this work, the use of μIM simulations as tool for process/product optimization was investigated.

The simulations results showed very high accuracy in comparison with experimental observations. An average deviation of 1.6 μm was observed between real parts measurements and numerical results, proving the adequacy of the model for μIM simulations. Simulations were also capable of accurately predicting the effects of the four process parameters: the slopes of the main effects plots for real and simulated results had the same sign in all the investigated ranges. The effect of the holding pressure was predicted particularly well, whereas the effects of other parameters were underestimated by the simulations. This could be caused by the fact that the effect of the holding pressure is better modeled by ASMI. Simulations were also capable to effectively predicting the position of the main defects affecting the micro rings quality. Air entrapments location and weld lines were in fact observed in real parts and simulation results in the same position.

It can be concluded that, in the present case study, μIM process simulations could be used as an effective tool for process and product optimization, leading to a consistent reduction of the experimental and metrological efforts needed to set up a production that is both defect-free and within tolerances.

ACKNOWLEDGMENTS

This research work was undertaken in the context of MICROMAN project ("Process Fingerprint for Zero-defect Net-shape MICROMANufacturing", http://www.microman.mek.dtu.dk/). MICROMAN is a European Training Network supported by Horizon 2020, the EU Framework Programme for Research and Innovation (Project ID: 674801).

REFERENCES

1. E. B. Brousseau, S. S. Dimov, and D. T. Pham, *Int. J. Adv. Manuf. Technol.***47**, 161-180 (2010).
2. L. Alting, F. Kimura, H. N. Hansen, and G. Bissacco, *CIRP Ann.* **52**, 635-657 (2003).
3. J. Giboz, T. Copponnex, and P. Mélé, *J. Micromechanics Microengineering* **17**, R96-R109 (2007).
4. G. Tosello, F. S. Costa, and H. N. Hansen, *Proc. Polym. Process Eng. 11,* 1-14, 2011.
5. G. Tosello, A. Gava, H. N. Hansen, H. Reinecke, G. Lucchetta, and A. Schoth, *67th Annual Technical Conference (ANTEC) of the Society of Plastic Engineers,* Chicago, 1787-1793, 2009.
6. L. Wang, Q. Li, W. Zhu, and C. Shen, *Microsyst. Technol.* **18**, 2085-2091 (2012).
7. A. Gava and G. Lucchetta, *Express Polym. Lett.* **6**, 417-426 (2012).
8. C. S. Chen, S. C. Chen, W. L. Liaw, and R. Der Chien, *Eur. Polym. J.* **44**, 1891-1898 (2008).
9. M. Heinle and D. Drummer, *Int. Polym. Process.* **30**, 434-441 (2015).
10. S. A. Khalilian, S. S. Park, and T. I. Freiheit, *Proc. 8th Int. Conf. Micro Manuf. (ICOMM 2013),* 50-56, 2013.
11. D. M. Marhöfer, G. Tosello, H. N. Hansen, and A. Islam, *Proc. 10th Int. Conf. Multi-material Micro Manuf (4M 2013),* 2013.
12. F. Costa, G. Tosello, and B. Whiteside, *Proc. Ppe'09,* 2009.
13. D. M. Marhöfer, G. Tosello, A. Islam, and H. N. Hansen, *J. Micro Nano-Manufacturing* **4** (2016).

Effect of an Inline-Plasma Treatment during Injection Molding Process on the Bonding Strength of Adhesive Bonded Joints

Elmar Moritzer[a], Timo Nordmeyer[a]

[a]*KTP at Paderborn University, Warburger Straße 100, 33098 Paderborn, Germany*

timo.nordmeyer@ktp.upb.de

Abstract. For technical applications, the bonding of thermoplastic materials becomes more and more important. To ensure good mechanical properties a good adhesion behavior is essential. The atmospheric-pressure (AP) plasma treatment is a common method to improve the surface characteristics, but it leads to further process steps between the injection molding and the adhesive bonding process. An entirely new process type is the Direct Injection Plasma (DIP) treatment, which combines the injection molding and the surface activation to a full-inline process. During the cooling phase of the injection molding cycle a pre-defined area of the molded part is activated by sending a plasma afterglow through a channel. In previous publication it was shown that the surface activation leads to high surface energies for a treatment length of more than 500 mm [1].

In this study the impact of the integrated plasma treatment on the injection molding process was investigated and compared with a conventional AD-Plasma treatment. The injection molding tool was investigated in relation to the specific shrinkage and warpage behavior for the plastic component. The effect of DIP treatment and an AD plasma treatment was determined. Additionally, the surface activation and bonding strength was analyzed.

Keywords: Injection molding, mold technology, adhesive bonding, adhesion, atmospheric plasma.
PACS: 81.65.Cf, 82.33.Xj, 82.35.Gh, 88.20.rb

INTRODUCTION

The need for surface activation is driven by lightweight construction as the use of thermoplastic materials becomes more and more important. Due to the low surface energies of many plastics, the plastics-processing industry frequently utilizes plasma technology to achieve the modification of part surfaces and to prepare them for subsequent process steps such as bonding with adhesives or multi component injection molding. When activating polymers, chemical functional groups produced on the surface, causing an increase in the surface energy, which leads to significant improvement in surface wettability and high adhesion strength for varnishes or adhesives. [2]
One typical surface treatment to increase the surface polarity of plastics and to improve adhesion properties is atmospheric pressure (AP) plasma treatment. For such treatments, jet nozzle concepts are often employed. When using surface treatments with AP plasma, the effectiveness of the treatment depends on multiple factors, such as the distance between substrate and jet and the speed at which the sample is treated. These and other influencing factors affect the treatment intensity and the homogeneity of the activated polymer surface. Typical applications of this method use CNC-automated machinery, which enables homogenous surface activation, but extend the chain of production, which is economically undesirable. Prior developments to integrate plasma treatments into the injection molding process have continued to use CNC-supported systems, with which the plasma treatment takes place in situ: the molded part is plasma treated at the end of the production cycle while resting in the opened tool [3, 4]. Such treatments often lead to the extension of the process time and to no complete integration into the process. It is only during the injection molding cycle itself that an in-situ and cycle-time-neutral plasma treatment is possible. The use of CNC-supported plasma treatment in that case is no longer a viable option. In previous publications, the Direct Injection Plasma (DIP) treatment process has been presented [1, 5]. With the DIP injection molding tool, an activation process has been developed for injection molding, which enables to activate a partial plastic surface inline, i.e. in a cycle time neutral manner.

The development of the DIP tool is based on application fields in which an adhesive or an expanding foam seal is applied in a circumferential groove, e.g. for products such as fuse box covers or housings of automobile headlamps. The test specimen is therefore a sheet with an internal 8 mm wide groove. The length of the treated region is 530 mm. The design includes two positions with in- and outlets for the plasma in the treated space. The plasma treatment can be carried out as single or dual mode (cf. Figure 1).

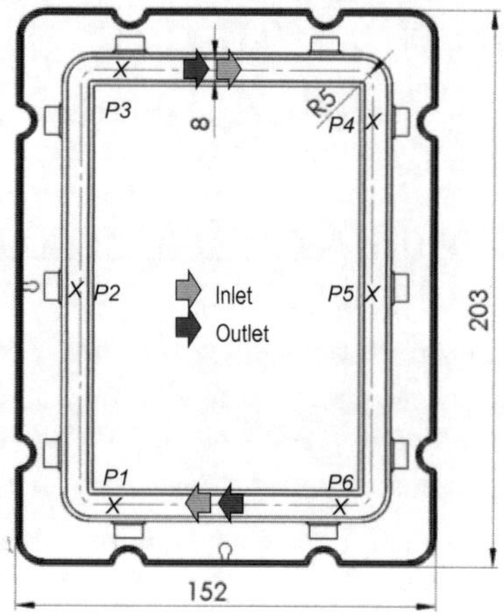

FIGURE 1. Test specimen with visualization of the in- and outlet for the plasma jet

The DIP treatment starts with completion of the holding pressure phase by pulling out a core and releasing a cavity between the tool and component, which is then used as channel for the plasma treatment. In the second step, plasma nozzles are moved into the cavity. The plasma used for the activation is ignited within the tool by two PFW-30 plasma nozzles (manufacturer Plasmatreat) and passes through the atmospherically insulated channel. The plasma flows through the channel and simultaneously activates the surface. Within the usual cooling time the surface activation ends and the part is ejected (cf. figure 2). There are high requirements for tool making and process design due to the necessary atmospherically insulation of the plasma within the channel, in order to eliminate the speed of the recombination of the plasma species with oxygen and having a negative effect on the surface treatment.

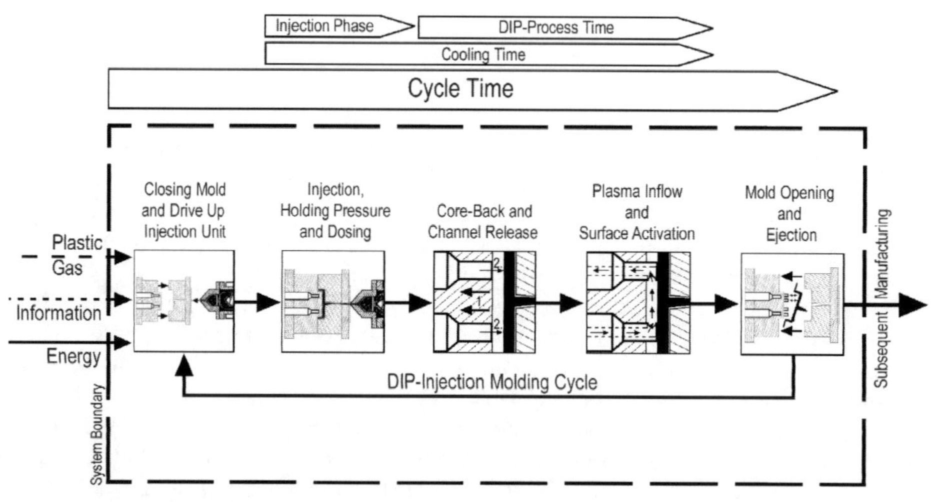

FIGURE 2. Illustration of the DIP injection molding cycle

Experimental Setup

The examinations were carried out with a non-reinforced polypropylene (PP-H, Moplen HP 500N, manufacturer: LyondellBasell Industries). The shrinkage behavior was determined by measuring the width and the length of the component 1 hour after the injection molding process. For the warpage behavior, the distance between the component and the reference surface has been measured at eight points and has been subtracted with the nominal value (cf. figure 3). The dependent variable was the arithmetic mean of all eight measurement locations.

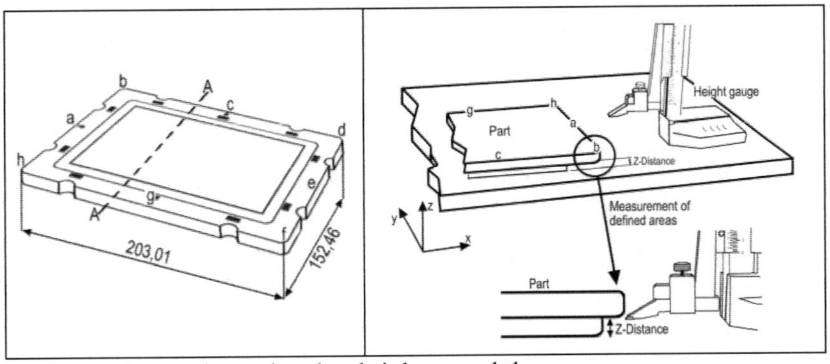

FIGURE 3. Measurement points to determine the shrinkage and the warpage

For the investigation of the surface activation and in accordance to DIN 55660-2 standard, the surface energy was determined along the channel using a static contact angle measurement (cf. figure 1). To characterize the surface energy, three test liquids were selected varying from nonpolar (diiodomethane) to strongly polar (completely de-ionized (DI) water). The degree of purity of the thiodiglycol used was > 99 %, that of diiodomethane 99.0 %. For all liquids, a dropping volume of 2.5 µl was selected. Before the measurement of the contact angle, a delay time of 5 s was set for the creation of the thermodynamic equilibrium at the three-phase point. The contact angles of the three test liquids were evaluated using the Owens-Wendt-Rabel-Kaelble model (OWRK model). The polar and disperse shares of the surface tensions of the liquids used are shown in table 1.

TABLE (1). Surface tensions of the different liquids

Testing liquid	Surface tension (total) [mN/m]	Surface tension (disperse) [mN/m]	Surface tension (polar) [mN/m]
Diiodemethane	50.8	50.8	0
Thioldiglycol	54	39.2	14.8
Water	72.1	19.9	52.2

The examinations aimed at the determination of the maximum bond strength between plasma activated surface and glue. To reach this aim, the tensile lap-shear strength of bonded assemblies was examined based on the standard DIN EN 1465. Due to the process principle, the treatment intensity depends on the treatment distance. Therefore, the bond strength is determined at six defined locations along the treatment path (cf. figure 1).

As adhesive system a 2-component epoxy resin adhesive (type: Araldite 2011, manufacturer: Huntsman) and a 2-component polyurethane adhesive (type: Araldite 2029-1, manufacturer: Huntsman) was used. The adhesive was applied within 6 h after the plasma treatment. The tensile lap-shear samples were hardened with a pressure of about 0.45 MPa and an adhesive layer thickness of about 0.05 mm for 48 h and subsequently tested. Due to the given component geometry, the sample geometry of the tensile lap-shear samples with a length of 48 mm, a width of 6 mm and a thickness of 2 mm was adjusted. The adhesive layer area was 6 mm x 6 mm and clamping length was 18 mm. The plasma treatment were carried out with the single mode. As process gas, nitrogen plasma (99.999% N_2) was selected. With a constant generator performance the volume flow rate was 120 slm.

Results

The shrinkage and warping behavior is an important target for industrial use, which have to be known. Due to the changed cooling conditions and the inflowing plasma, a modified shrinkage and delay behavior was assumed. In the first step, the aim was to find suitable operating settings and the optimal cooling time at which the molded part exhibited the minimal warpage. For this purpose, a filling and holding-pressure study was initially prepared at three different cavity temperatures (30-60°C). The lowest possible shrinkage potential was the combination of T_{cav} = 30 °C and a holding pressure of 300 bar. The cooling time was then gradually increased and the mean warpage was calculated. Figure 4 shows the mean warpage as a function of the cooling time. It can be shown that a cooling time of 20 s leads to the smallest warpage values (cf. figure 4).

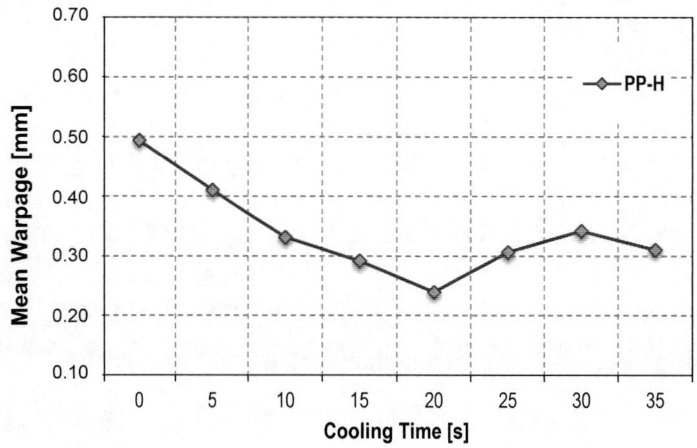

FIGURE 4. Variation of the cooling time

Furthermore, it was investigated whether the plasma treatment changes the warpage behavior during the cooling time. For this study, the DIP treatment time was gradually increased up to a total treatment time of 20 s. For comparison, the resulting warpage of the specimens after an AD plasma treatment was also determined. For this purpose, the specimens were prepared at optimal injection molding settings and then treated with a conventional AD plasma jet (RD1004 rotary nozzle, manufacturer Plasmatreat). The process gas was nitrogen and the volume flowrate 35 slm. Figure 5 shows that both plasma treatments lead to larger warpage values. However, the warpage behavior after a DIP treatment does no longer change after a treatment period of 10 s. In addition, the warpage behavior for the observed application case is significantly smaller compared to the AD plasma treatment. In summary, it can be concluded that the DIP treatment does not lead to larger warpage values than an AD plasma treatment.

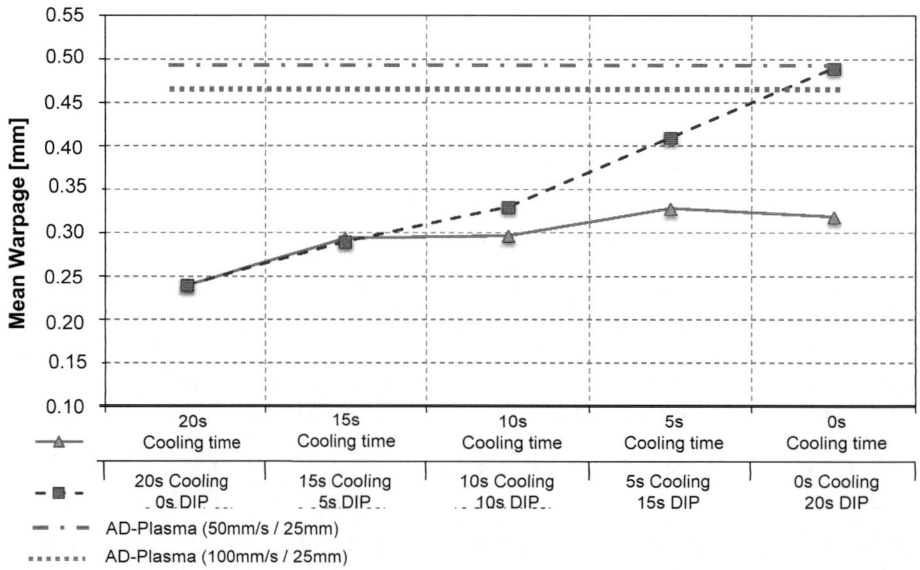

FIGURE 5. Effect of the cooling time, the DIP treatment time and the AD plasma treatment on the mean warpage

In [5] we show that the main influence factor of the DIP treatment is the treatment time. Homogenous surface activation is achieved with longer treatment times. For the observed process duration of 20 s, it can be seen that the surface activation decreases at a length of 300 mm (cf. figure 6). If the surface activation is to be carried out uniformly, the treatment time must be extended or two nozzles have to be used simultaneously. However, the bond strength analysis shows that the previous nonpolar material also leads to very high strengths with a treatment duration of 20 s. A correlation between bonding strength and the surface energy is not possible. The interaction between the adhesives and the incorporated polar functional groups at the surface is strong enough even for low activation to ensure high bonding strengths (cf. figure 6).

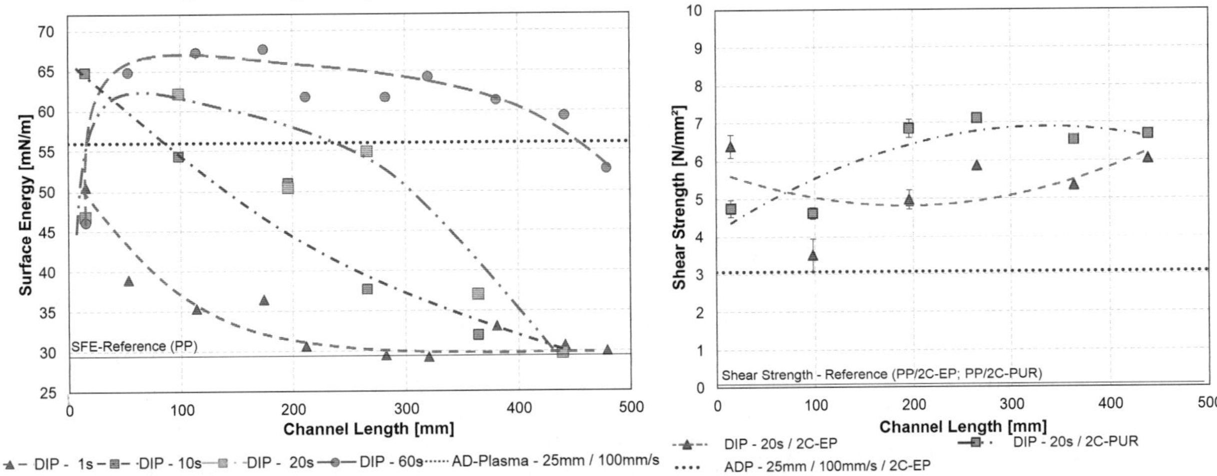

FIGURE 6. Left: Surface energy for 1 s, 10 s, 20 s and 60 s treatment time [5]. Right: Bonding strength for 10 s and 20 s treatment time for 2C-Epoxy and 2C-PUR

SUMMARY

In this paper, the influence of the DIP treatment is described on the warpage, surface activation and shear strength behavior for an unreinforced PP. It was found that the substitution of a conventional AD plasma treatment by the DIP method is possible. Both surface activation processes increase the warpage. However, the DIP treatment leads in comparison to the AD plasma treatment due to the fixation of the part in the mold to lower warpage values. The DIP treatment leads to high bonding strength, even if the surface activation is inhomogeneous, which shows the strong interaction between the adhesives and the incorporated polar functional groups.

ACKNOWLEDGMENTS

This research and development project was funded by the European Union, the European Font for Regional Development (EFRE) and by the State of North Rhine- Westphalia. We would like to thank Plasmatreat for their close and trustful cooperation.

Ziel2.NRW
Regionale Wettbewerbsfähigkeit und Beschäftigung

EUROPÄISCHE UNION
Investition in unsere Zukunft
Europäischer Fonds
für regionale Entwicklung

REFERENCES

1. E. Moritzer; T. Nordmeyer, C. Leister, M. A. Schmidt, A. Grishin, A. Knospe, *Investigations of the Surface Activation of Thermoplastic Polymers by Atmospheric Pressure Plasma Treatment with a Stationary Plasma Jet,* Proceedings of Polymer Processing Society 31th (PPS-31), Jeju Island (Korea), 2015
2. F. Awaja, M. Gilbert, G. Kelly, B. Fox, P.J. Pigram, Adhesion of polymers, Progress in Polymer Science, 34, 2009, 948-968.
3. G. W. Ehrenstein, S. Amesöder, M. Schuck, *In-Line-Activation of Polymer Surfaces with Atmospheric-Plasma leads to a Strong Bond in Multi-Material Molding*, Fifteenth International Conference & Exhibit Molding, New Orlreans (USA), 2005.
4. C. Schmachtenberg, M. Schuck, I. Kühnert, *Forming and Assembly in One Process*, Kunststoffe International, H. 4, S. 8-13, 2004.
5. E. Moritzer; T. Nordmeyer; L. Enneking; A. Grishin; A. Knospe; C. Buske: *Development of an Inline Plasma Treatment During Injection Molding Process*. 74th Annual Technical Conference of the Society of Plastics Engineers (ANTEC), Indianapolis (USA), 2016.

Comparison of cold and hot runner system concerning part accuracy in thermoplastic injection molding

Zeppenfeld Matthias [(1)]*, Trescher Matthias [(1)], Robeck Andreas [(1)], Gattinger Johannes [(1)]

[(1)] Technical University of Munich, Institute of Medical and Polymer Engineering - Bavaria - Germany

In modern thermoplastic injection molding cold runner (CR) and hot runner (HR) systems are used for the production of dimensionally accurate parts. This study aims for a direct comparison of both gate systems attachable to the same mold cavity and hence for the production of the same part geometry. Influences of the injection molding process related to both systems on part dimensions, weight and cavity pressure have been evaluated. For this purpose a mold with two cavities each containing cavity pressure/temperature sensors (Kistler Instrumente GmbH, Switzerland) has been designed to produce eight-sided, beaker-shaped parts (70x60mm) with 1mm wall thickness. By maintaining the cavity unchanged the gate system has been modified by a CR as well as a HR system. A design of experiments (n=14, i=15) has been performed for both setups comparing different process parameters to evaluate optimal operating points. Test specimen have been manufactured out of polybutylene terephthalate (PBT, Ultradur® B4520, BASF SE, Germany) on a KM125-520C2 (KrausMaffei Technologies GmbH, Germany) molding machine, stored and measured in dimensions and weight after 16 to 24h according to DIN 16742 and DIN EN ISO 294. Results show that the difference in standard deviation regarding dimensions (e.g. height up to -42%) and weight (up to -78%) is lower for the HR compared to the CR system indicating a more stable process. Due to a lower pressure drop in the HR gate system a higher injection speed (CR 80mm/s, HR 110mm/s) and a lower holding pressure (CR 850bar, HR 750bar) could be set for similar cavity pressures compared to the CR system. The influence of the holding pressure time and cooling time on the measured quality parameters was less with the CR system. To sum up the cavity filling was more robust with the HR system and the parts could be filled in a broader range of process parameters, e.g. the adjustable process window for the mass temperature has been 10°C in contrast to 5°C (CR setup).

Fast mold surface temperature evolution: challenges and opportunities

S. Liparoti[a], A. Sorrentino[b], V. Speranza[a], R. Pantani[a], and G. Titomanlio[a]

[a] *Department of Industrial Engineering, University of Salerno, Italy*

[b] *Institute for Polymers, Composites and Biomaterials IPCB-CNR - P.le E. Fermi 1, 80055 Portici - NA - Italy*

**gtitomanlio@unisa.it*

Abstract

Fast mold temperature evolution allows obtaining injection molded parts with lower filling pressure, better surface finishing, more accurate surface replication and reduced frozen-in orientation. Recently proposed, thin heating devices, with very low thermal inertia, located just below the mold cavity surface, allow obtaining high surface temperature during filling and low surface temperature during cooling stage. Several tests have been carried out injecting isotactic polypropylene into a rectangular cavity with the aim of analyzing the effects of heating temperature and heating time (equal or larger to the filling time) on both part quality and process evolution. As far as the part quality is concerned, a relevant improvement of the surface micro features replication and a lower orientation level have been attained by increasing heating temperature and time. From the point of view of the process, a significant reduction of filling pressure and pressure drops have been reached by increasing the mold surface temperature. The cycle time increases when the surface heating time increases. However, this increase has been found not much different from the surface heating time, if thin heating devices located at a short distance from the mold surface are adopted. Some secondary effects on the pressure evolutions related to the in-mold cooling and solidification have been also observed during the process and have been discussed with reference to the heating elements activity.

INTRODUCTION

The injection molding process is a widespread and reliable process for the production of polymeric parts with complex shapes, since it assures accuracy of geometry replication [1]. The process can be split into three stages: injection, packing and cooling. The processing time mainly depends on the cooling time thus a low mold temperature has to be adopted to assure a small processing time. The necessity of adopting temperature largely below the glass transition temperature (for the most part of the polymers) is a limit in the production of very thin parts (in the micro-metrical range) where the rapid cooling of the part induces a significant increase in viscosity preventing the complete filling of the cavity. To overcome the limitations due to the low mold temperature several technologies have been proposed to modulate mold surface temperature. Induction coils [2], radiation [3] or proximity heating [4] were proposed to increase mold surface temperature during the filling, however, high additional design and tool costs make the process not economically attractive and the small temperature increase/decrease rate makes very long the processing times. More promising technology is based on the use of a thin electric heater based on a poly(amide-imide) loaded carbon black layered below the cavity surface. This system allows increasing temperature in few seconds (1-3 s) during the screw movement [5] and its reduced thickness allows a fast decrease of temperature during the cooling of the part, therefore, the whole cycle time remains comparable with the cycle time of the conventional injection molding process. The rapid modulation of mold surface temperature makes the modified injection molding process more economically attractive [6,7].

The possibility to modulate mold surface temperature in short times allows to produce parts with reduced thickness [8] and also to replicate micro a nano-structures located on the mold surface [9]. This opens the possibility to produce micro-electromechanical systems (MEMS) applicable to the life science fields, such as DNA sequencing and clinical diagnostics devices, by injection molding assisted by the system for the fast control of mold surface temperature.

The potentially of this system have been discussed in this paper. In particular, the effect of mold surface temperature evolution on morphology, mechanical properties and replication of micro and nano-structures have been presented and discussed. The possibility to modulate optical properties and wettability is also proposed.

EXPERIMENTAL

Molding experiments were carried out using a polypropylene (i-PP) grade supplied by Montell (now Basell), with commercial name T30G (non-nucleated, Mw=376 000, Mw/Mn=6.7, tacticity=87.6%). A conductive polymeric film, polyimide loaded carbon black (ENSACO 260G, 30% w), with thickness of 50 μm and 40 Ω/square as sheet resistance [5], sandwiched between two insulating layers of polyimide was layered on both sides of the cavity giving rise to the mold surface heating by Joule effect. A steel layer was located on the heating device to protect it from the melt.

A 70-ton Negri-Bossi reciprocating screw, injection molding machine was used for the experiments. The polymer was injected into a line-gated rectangular cavity having length L=107 mm, width W=12.7 mm, and thicknesses S=1.5 mm. Fig 1 shows the cavity adopted for the injection experiments. The two heaters, 70 mm long, are red marked in the figure. Where the nickel insert is not present a steel layer of the same thickness is located. The molding machine and the mold were equipped with five piezoelectric transducers: one in the injection chamber (P0), one just before the gate (P1) and three in the cavity (P2, P3 and P4), located in the non-moving part of the mold (15, 60 and 105 mm downstream from the gate position). Moreover, a temperature sensor was located in the cavity, on the steel layer, in positions P2.

The nickel insert, 19x19 mm^2, with micro and nano-features, is eventually located downstream the gate, instead of the steel layer, to analyze replication accuracy.

All the experiments were carried out keeping constant the flow rate at 2.8 cm^3/s, the holding time at 8 s, the melt temperature at 220°C, and the mold temperature, except on the heated surface, at 28°C. TABLE 1 summarizes all the others experimental conditions.

The molded samples were analyzed by optical and Atomic Force microscopy. To analyze the effect of the processing conditions on the replicability, the micro-crosses marks, detected on the sample surface, have been observed by Atomic Force Microscope (AFM) in tapping mode (commercial probe tips with nominal spring constants of 42 N m^{-1}, resonance frequencies of 300 kHz, and tip radius of 7 nm have been used).

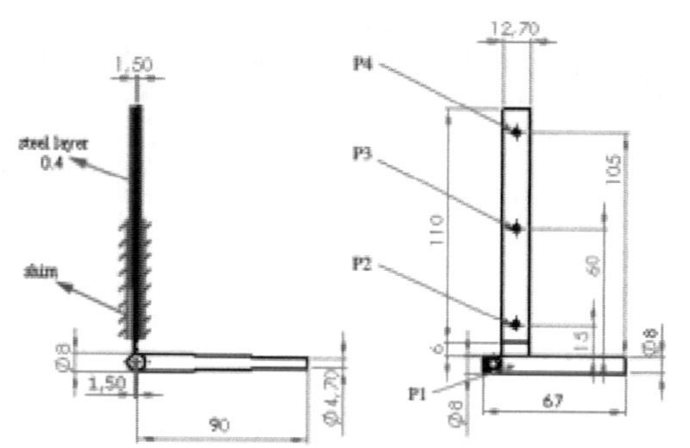

FIGURE 1 Geometry of the cavity (P1-P4 indicate the pressure transducers positions) and sketch of the heating device.

Test run	*P (W/cm^2)*	*T_{level} (°C)*	*t_h (s)*	*t_a (s)*	*P_{hold}*
T_off	0	28	0	0	360
120-1	7	120	1.3	2	360
120-13	7	120	13	2	360
120-20	7	120	20	2	360
150-1	10	150	1.3	2	360
150-6	10	150	6	2	360
150-12	10	150	13	2	360
150-20	10	150	20	2	360

TABLE 1 Operating conditions adopted for the injection molding experiments. (P=electrical power applied; T_{level}= temperature reached on the steel layer; t_h=time that the heater is activated after the polymer reaches position P2. t_a = time that the heater was activated before the contact with the polymer in P2: P_{hold}=holding pressure).

RESULTS AND DISCUSSION

FIGURES 2a show temperature evolutions in position P2 for the experiment T_off, performed without activating the heater, and for the experiment 150-13. During the experiment 150-13, the heater was activated 2 s before the melt reached position P2 (t=0); during such a time the temperature on the surface raised from the mold temperature to 150°C. When the polymer reaches position P2, the temperature reaches the maximum value and soon after it starts to decrease to the selected surface temperature.

The presence of the system for the control of mold surface temperature induces significant differences in pressure evolutions. Generally, the whole pressure drop is reduced. Furthermore, when the heater is active pressures in the cavity go to zero into two steps, the first one is due to the decrease of the average temperature from the melt temperature to the surface temperature and the second takes place at the heater deactivation, from the high temperature reached by the heating device to ejection temperature. This suggests that the crystallization/solidification take place only at the heater deactivation. Furthermore, the filling pressure is reduced with high mold surface temperature. This aspect is important especially in micro injection molding process, in which, due to the reduced thickness of the cavity (in the micrometrical range) high filling pressures are required. However, despite the high filling pressures, the solidification of the melt could prevent the filling of the whole cavity. FIGURE 2b shows the dependence of the filling length, obtained by injecting the iPP in a cavity with 0.5 mm thickness and 50 mm length, on the filling pressures for different mold surface temperatures. The results show that a temperature increase from 110°C to 140°C determines an increase of the filling lengths.

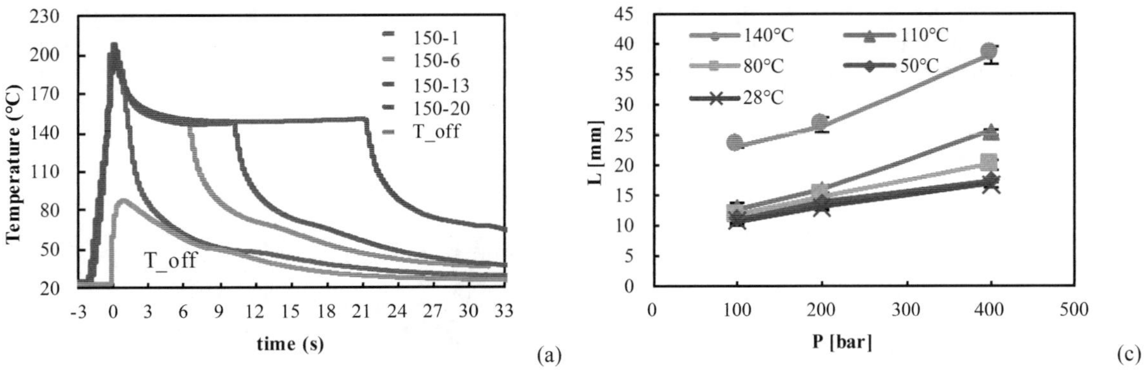

FIGURE 2 (a) Temperature evolutions for the experiments T_off, 150-1, 150-6, 150-13 and 150-20. (b) Filling length dependence on the filling pressure for different mold surface temperature.

The possibility to keep constant the surface temperature at the desired value for the desired time allows to modify the morphology developed along the sample thickness. FIGURE 3 shows the morphology developed with 150°C mold surface temperature and 360 bar holding pressure. The different layers developed along the sample thickness, namely the skin, the shear and the spherulitic layer, from the wall to the core, are strongly influenced by the temperature of the cavity surface. In particular, the skin layer disappears already with 1.3 s heating time. The thickness of the shear layer decreases by increasing the heating time. The shear layer completely disappears with longer heating time and the spherulitic layer impinges on the sample wall. Evidently, in that case, the stretch due to the flow has sufficient time to relax over all the sample thickness, the relaxed molecules can subsequently crystalize into a spherulitic morphology during cooling to ambient temperature. Mechanical analyses have been performed along the sample thickness on each morphological structure by AFM HarmoniX [10]. Cross sections elastic modulus distributions of the four samples, T_off, 120-1, 120-13 and 120-20 are reported in FIGURE 4. The results clarify that the oriented fibrils are characterized by the highest value of elastic modulus, the skin and the spherulitic layer show the lowest and an intermediate value of the modulus, respectively. The cross section of the samples obtained with high mold surface temperature and long heating time (120-20) is completely spherulitic, however, the modulus is not constant over the cross section; nevertheless, it appears smaller close to the surface, probably the higher cooling rate does not allow an efficient structuring within or among the polymer chains.

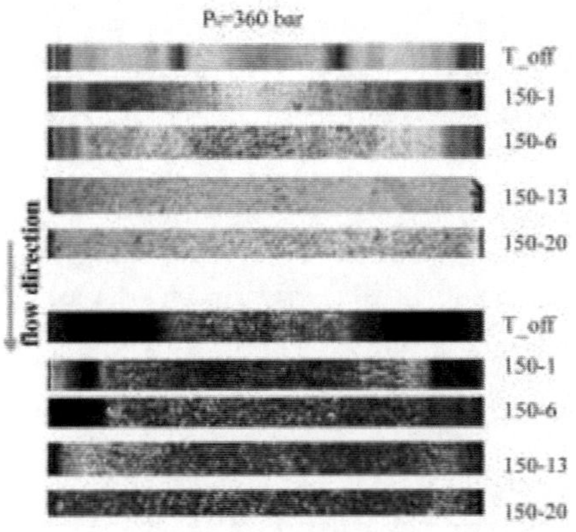

FIGURE 3 Optical micrographs of the slices cut close to position P2 from the samples produced with 150°C surface temperature held for different heating times and with packing pressures of 360bar.

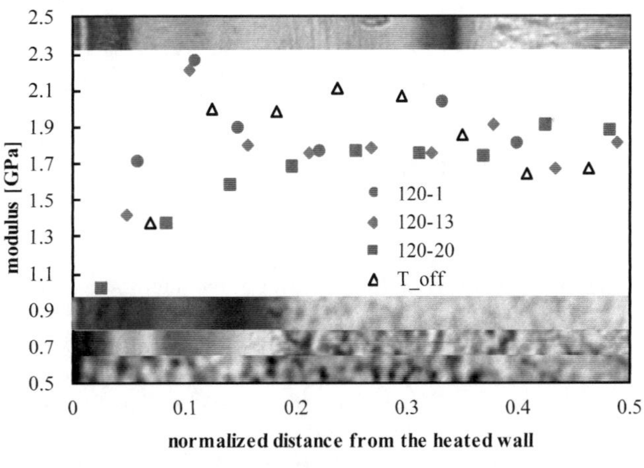

FIGURE 4 Elastic modulus distribution along the sample thickness obtained for the samples T_off, 120-05, 120-8 and 120-18.

The injection molding process coupled with the system for the fast evolution of mold surface temperature has been already adopted to obtain molding with micro and nano-metric channels replicated on the surface, which are of interest in microfluidic field. To this purpose, a specifically designed mold with micro and nano-crosses on the surface has been adopted [9]. FIGURE 5 shows the AFM height maps of the crosses located on the mold and of the crosses replicated on polymeric molded objects.

The AFM height images show that mold surface temperature is a very important parameter in achieving good microstructure replication. Indeed, the micro-cross replicated with 150°C mold surface temperature appears to have an edge much sharper than the micro-cross replicated without activating the heater, and this can be observed especially in the corners. The reduction of viscosity induced by the increase of cavity surface temperature is the main responsible for the increase in the replication accuracy. The replicability efficiency has been analyzed also for nano-crosses, as shown in the FIGURE 5. The increase of mold surface temperature is even more important in the replication of nano-features. Indeed, the replication of nano-feature with a suitable accuracy is possible only if high temperatures are adopted. In particular, the sample obtained with 150°C mold surface temperature shows more accurate replication of nano-crosses with respect to the sample obtained with 120°C. Probably, a wall temperature higher than the crystallization one favors the replication, since the replication proceeds after the filling of the whole cavity.

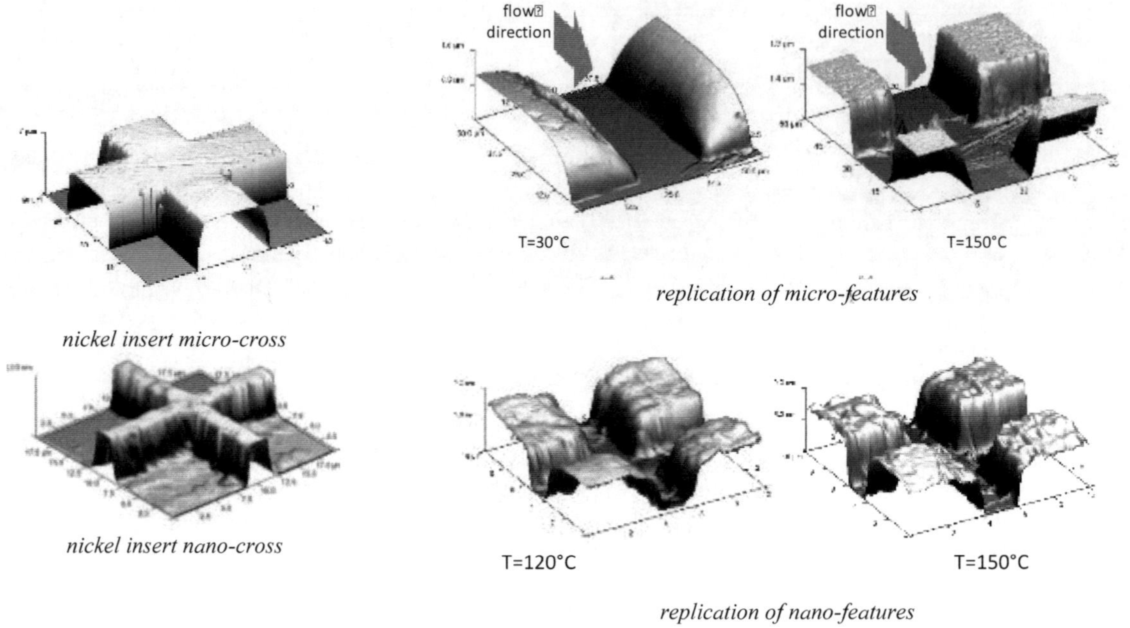

FIGURE 5 AFM height images of the micro-crosses (5 μm height) having a width of 20 μm and nano-crosses (60 nm height) having a width of 2 μm on the nickel insert and on the samples of iPP obtained without activating the heating device and activating the heating device.

CONCLUSIONS AND PERSPECTIVES

In this work, some advantages of coupling the injection molding process with fast control and evolution of mold surface temperature are pointed out showing that relevant characteristics of the moldings can be achieved by modulating the mold surface temperature evolution. It is important to point out that the technology allows fast control of the mold surface temperature, both during the heating (within a few seconds) and during the sample cooling (cooling time is comparable with the cooling time of a standard injection molding process).

The morphology distribution inside the molding can be calibrated by modulation of both the mold surface temperature and heating duration. Morphology distribution determines all moldings properties, such as mechanical properties and optical properties. An example of the effect of morphology distribution on local values of the modulus is reported in the paper. Morphology affects also the optical transmission properties through the sample thickness, which are directly related to the birefringence. This aspect is a key factor in the production of optical lens made of amorphous polymers.

Fast control and evolution of mold surface temperature influence also the replication accuracy of details on molding surface. This aspect is very intriguing in the biomedical field, since the control of replication accuracy allows to produce devices suitable for microfluidic applications and also to obtain surfaces with tunable wettability without using coating. Significant improvements of replication accuracy were achieved on iPP samples, by the use of the technology proposed in this work for the fast control of mold surface temperature.

REFERENCES

1 M.R. Kamal, A.I. Isayev, S.-J. Liu, Injection molding : technology and fundamentals, in: Hanser (Ed.), Progress in Polymer Processing, Carl Hanser Verlag GmbH & Co. KG, 2009: p. 737.

2 H.L. Lin, S.C. Chen, M.C. Jeng, P.S. Minh, J.A. Chang, J.R. Hwang, *International Communications in Heat and Mass Transfer. 39 514–522 (2012).*

3 S.-C. Chen, Y. Chang, Y.-P. Chang, Y.-C. Chen, C.-Y. Tseng, *International Communications in Heat and Mass Transfer. 36 1030–1035 (2009).*

4 D. Yao, T.E. Kimerling, B. Kim, *Polymer Engineering & Science. 46 938–945 (2006).*

5 S. Liparoti, G. Landi, A. Sorrentino, V. Speranza, M. Cakmak, H.C. Neitzert, *Advanced Electronic Materials. 2*

1600126 (2016).
6 S. Liparoti, A. Sorrentino, G. Titomanlio, *AiChE Journal. 62 2699 (2016).*
7 S. Liparoti, A. Sorrentino, G. Titomanlio, *RSC Advances. 6 99274–99281 (2016).*
8 F. De Santis, R. Pantani, *Journal of Materials Processing Technology. 237 1–11 (2016).*
9 V. Speranza, S. Liparoti, M. Calaon, G. Tosello, R. Pantani, G. Titomanlio, *Materials & Design. 133 559–569 (2017).*
10 S. Liparoti, A. Sorrentino, V. Speranza, G. Titomanlio, *European Polymer Journal. 90 79–91 (2017).*

Chemically adsorbed anti-adhesive organo silane coating of mold and dies for polymer processing

Kaynak Baris [(1)]*******, ***Kern Wolfgang*** [(1)]

[(1)] Chair in Chemistry of Polymeric Materials, Montanuniversität Leoben - Styria - Austria

In this work, metallic tools are surface modified with new synthetized dipodale silanes from the liquid phase. In particular, these molecular coatings are investigated with respect to adhesive properties, durability and long-term stability. Tool steel surfaces are treated with acid (dilute HNO3) and corona discharges to create hydroxyl groups on the surface ("activation"). The modification of the metal surface is achieved by subsequent reaction of the surface hydroxyl groups with organo silanes. The characterization of the modified surfaces is performed by infrared reflection absorption spectroscopy (IRRAS), X-ray photoelectron spectroscopy (XPS), atomic force microscopy (AFM), and contact angle measurements. It is shown that the surface energy of steel surfaces can be reduced significantly by application of the organo silanes. Using physic-chemical methods, the thermal stability of the silane layers is studied at 290°C in order to evaluate the applicability of such coatings for molds and dies used in the processing of thermoplastics (e.g. PC and PA). The stability is examined as a function of the structure of the modifying organosilane, layer thickness, and absence / presence of oxygen during the ageing test. The use of the chemically adsorbed dipodal silane films for injection molding greatly reduced the demolding forces and the static coefficient of friction. When appropriate molecular structures and application techniques are selected, anti-adhesive silane layers exhibit a surprising thermal stability which makes them suitable candidates for the coating of tool steel. Most important, a renewal of silane coatings can be done in a convenient process without employing critical or toxic chemicals. The results are discussed with respect to applications in polymer processing.

A study on replication and quality correlation of on-part, on-runner micro features and process conditions in polymer Injection molding

Giannekas Nikolaos (1)*

(1) Technical University of Denmark - Lyngby - Denmark

Quality control and assurance methods play a major role in the world of modern manufacturing. The evolution of manufacturing processes on efficiency and output has instigated the development of advanced and more flexible quality control and assurance methods. Especially for injection molded components where the precision and quality of process and parts requires verification for a high volume production. Quality control in a traditional metrological framework has been known to be a lengthy and time-consuming process, setting limitations to the number of parts that are assessed and accepted. The delayed detection of out-of-tolerance production due to time-consuming metrological inspection and short cycle time in injection molding can increase the scrap rate of the production. This paper presents a concept approach of a quality control system based on multiple inputs, such as signals from process regulating sensors (process conditions data) and metrology applied to dedicated micro features positioned on the runners, equal to those in the part. All inputs are used in order to access the quality of the produced plastic parts and establish part and process quality indicators ("Process and Product Fingerprints"). A designed experiment was employed to assess the micro feature replication on the part and runner, in correlation to the process condition's machine data signals and assess the effectiveness of the two types of fingerprints. It is presented that the process and part quality indicators derived from the machine process sensor signals and metrological inspection of on-part and on-runner micro features follow a similar trend, with the main factorials effects showing similar trends in correlation to process quality indicators changes . As such, process and part quality indicators can be calibrated to support a fast quality part quality monitoring system with the potential decrease the metrological efforts for the approval of injection-moulded parts.

Combined learning processes for injection moulding based on simulation and experimental data

Christian Hopmann[1], Sabina Jeschke[2], Tobias Meisen[2], Thomas Thiele[2], Hasan Tercan[2], Martin Liebenberg[3], Julian Heinisch[1] and Matthias Theunissen[1, a)]

[1]*Institute of Plastics Processing (IKV) at RWTH Aachen University, Seffenter Weg 201, D-52074 Aachen, Germany*
[2]*Institute of Information Management in Mechanical Engineering (IMA) - RWTH Aachen University, Dennewart-straße 27, D-52068 Aachen, Germany*
[3]*Knowledge-based Systems Group - RWTH Aachen University, Ahornstraße 55, D-52056 Aachen*

[a)]Corresponding author: Matthias.Theunissen@ikv.rwth-aachen.de

Abstract. Injection moulding enables the production of complex formed high-quality plastics parts in a single production step. To achieve a high and constant product quality, an appropriate process set-up with regards to product quality and process robustness is essential. A conventional process set-up requires expensive and time consuming experiments and know-how from the machine operator. One way to overcome this challenge is to make use of machine learning methods for process set-up. These methods can model the relationship between setting parameters and quality values and thus enable the identification of optimal working points. However, the training required for accurate modelling needs experimental data from extensive experiments for each process, too. Numerical simulation can predict quality values based on setting parameters without practical experiments. Whereas trends and the general dependencies between the parameters can be predicted with a satisfying accuracy, a certain discrepancy between the prediction of the simulation and the real process cannot be excluded. A combined approach using data from injection moulding simulations as well as experimental data appears promising to overcome the detriments of the solitary use of simulation for the training of machine learning algorithms. General dependencies could be gained from simulations without practical experiments and fine-tuning could be achieved by experimental trials with a minimal scope.

In this paper, data obtained from a 2.5 D injection moulding simulation is compared with experimental data from a plate specimen and a complex formed injection moulded part. Therefore, central composed designs of experiments are used to identify differences in the effects and interdependencies of six setting parameters on quality values like part weight and dimensions. Furthermore, differences in the absolute values and the functionality of the effects are considered. On this basis, a combined machine learning concept using simulation and experimental data is presented.

INTRODUCTION

A high and consistent product quality is of paramount importance for injection moulders. For the part quality and its reproducibility, the interactions between material properties, mould and machine equipment as well as the chosen setting parameters of the process are the main influencing factors. With the set-up of the machine parameters, the final product quality and the process efficiency can be adjusted. Therefore, the fast identification of a parameter set that is as proficient as possible is essential for an efficient production. In a conventional set-up process, the set-up relies primarily on the experience and the intuition of the moulding personnel. Since experienced moulding personnel is rare, standardised approaches to determine the parameters become more important [1]. They enable a more purposeful optimisation regarding defined target values than trial and error methods [2].

On the one hand, an appropriate design of experiments in combination with machine learning (ML) methods can help to systematise and optimise the process set-up [1-3]. However, the effort for the necessary practical experiments, which have to be conducted for every new mould and process anew, prevents a wide application of these methods in the industry. On the other hand numerical simulation approaches can be used to predict the effects of different setting parameters and to support the definition of suitable parameter settings. Though, the assumptions and simplifications for the numerical description of the phenomena during the injection moulding process likely cause differences between the simulation and the actual process [3]. According to this, long practical set-up times and experienced moulders are still necessary in the industrial environment. This paper proposes a combined approach based on ML from practical experiments and simulation data to overcome the disadvantages of both domains. The goal is to produce reliable predictions for the product quality with few experimental data in order to set up machine parameters.

RELATED WORK

In the following, different approaches in order to find and optimise suitable injection moulding parameters are presented. In the past, mainly trial and error methods relying on the experience and intuition of the machine operators have been used [2]. Later on in the 1980s and 1990s, process window approaches [4], case based reasoning approaches [5] and expert system based approaches [6] have been developed. The major issue with these systems is that they need a huge library of cases and knowledge, which has to be updated with changing equipment and advancements in technology [2]. Furthermore, the Taguchi method has been thoroughly analysed in research and is an effective way to determine process settings which obtain the desired result and which are robust against disturbances [1, 2]. However, if not combined with ML techniques, the optimum can only be chosen from predefined parameter levels [2]. Regarding ML methods, especially artificial neural networks (ANN), have been applied for a prediction of the relationships between setting parameters of the injection moulding process and single our multiple quality criteria, such as warpage, part weight and part dimensions. For a detailed overview refer to [2] and [3]. To find optimized setting parameters, the following steps are necessary: appropriate design of experiments, training of the chosen model (ML or regression models) and optimisation with a suitable method such as generic algorithms (GA) [7].

Besides the optimisation with ML methods based on experimental data, numerical simulations can be applied to generate a data basis [2, 3]. Thereby, optimised parameter settings can be found without the need for practical experiments and the simulation software can automate the data generation [8-10]. However, differences between the predicted quality and the optimised parameter settings are unavoidable and the found parameters can only be a starting point for the optimisation at the machine in the real world. An ML approach to optimise parameter settings for the injection moulding using a model which is originally based on simulation data and consecutively refined by practical experiments does not exist yet. In this study the differences between simulation and experimental data of two injection moulded parts are analysed in order to derive an appropriate combined ML approach which can significantly reduce the necessary amount of experimental data for an ML model which represents the actual moulding process. This could enable a practically feasible optimisation based on ML for injection moulding for the first time.

EXPERIMENTAL SETUP

For the analysis of the differences between simulation and experimental injection moulding data in order to design a combined learning process, a simple plate specimen and a more complex box specimen were analysed (see Fig. 1). Polypropylene of type PP 579S (Sabic Deutschland GmbH, Duesseldorf, Germany) was used for both parts. In a central composite design (CCD) with six factors consisting of a full factorial design with additional star points with the level of square root two, 77 different test points were analysed (see Table 1). For the injection moulding simulations, the software Cadmould 3DF v8.0.0.27 of the supplier Simcon kunststofftechnische Software GmbH, Wuerselen, Germany, was used. The simulations were conducted with packing pressure and warpage and a thermal wall temperature analysis for the injection and cooling phase has been performed in advance. The practical experiments for the plate specimen have been conducted on an injection moulding machine of type Allrounder 370A by the manufacturer Arburg GmbH & Co KG (Loßburg, Germany) and the box specimens on a 160-1000 CX by the manufacturer Kraussmaffei Technologies GmbH (Munich, Germany). Ten parts were produced for each test point after the process had reached a thermal equilibrium.

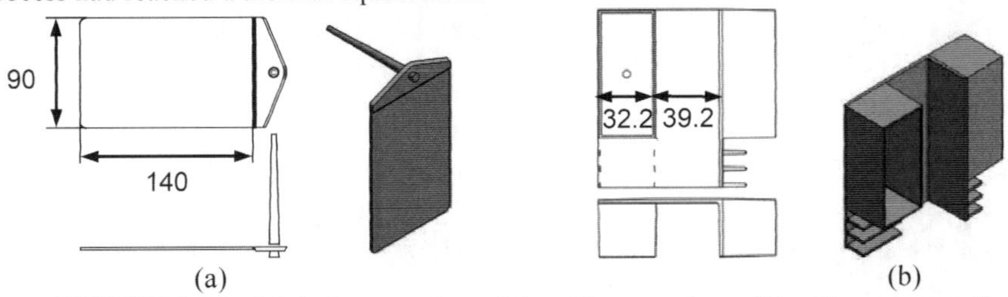

FIGURE 1. Moulded plate specimen (a) and box specimen (b) with measured dimensions

TABLE 1. Factors and test points used in a central composite design

		Plate specimen			Box specimen		
		0	**+/-1**	**+/-1.414**	**0**	**+/-1**	**+/-1.414**
A	Mould temperature [°C]	40	14	20	40	14	20
B	Melt temperature [°C]	240	14	20	225	14	20
C	Cooling time [s]	12	2.1	3	13	2.8	4
D	Packing pressure time [s]	6	1.4	2	4.5	1.4	2
E	Packing pressure [bar]	250	71	100	275	88	125
F	Injection speed [cm³/s]	20.7	5.3	7.5	63.5	9	12.7

RESULTS AND DISCUSSION

In Fig. 2 the resulting part weights of the simulated and averaged part weights moulded specimens are shown during the CCD. Without going into detail about the influence of the setting parameters, a similar general course of part weight of the simulation and the experimental part weights can be observed. However, the overall value of the part weights and the effect of certain parameters differ due to the assumptions and simplifications in the simulation.

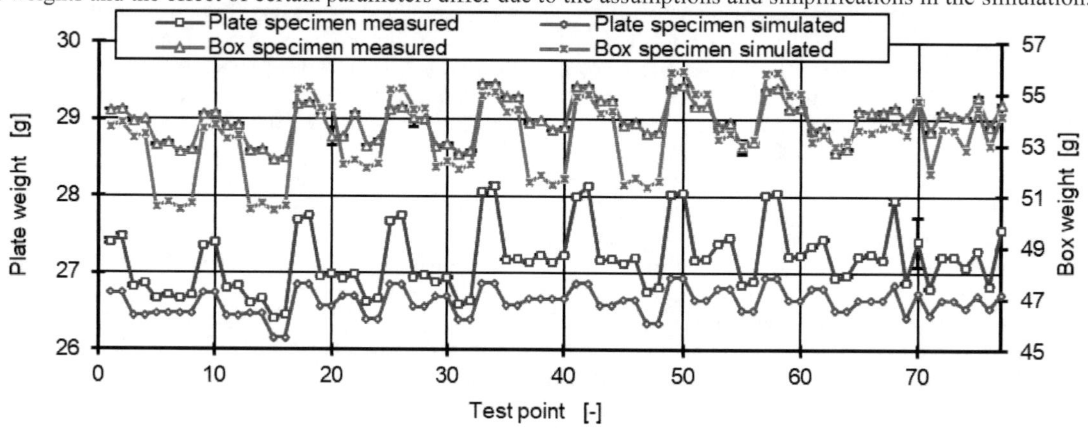

FIGURE 2. Part weight development over all test points

In Fig. 3 the standardised effects of the factors on the part weight (a) and the dimensions of the box specimen (b) are shown. In spite of the standardisation, factors with a negative effect are plotted. The 5 % significance levels for the standardised effects are smaller than 2 for all experiments and cannot be shown in the diagrams in Fig. 3.

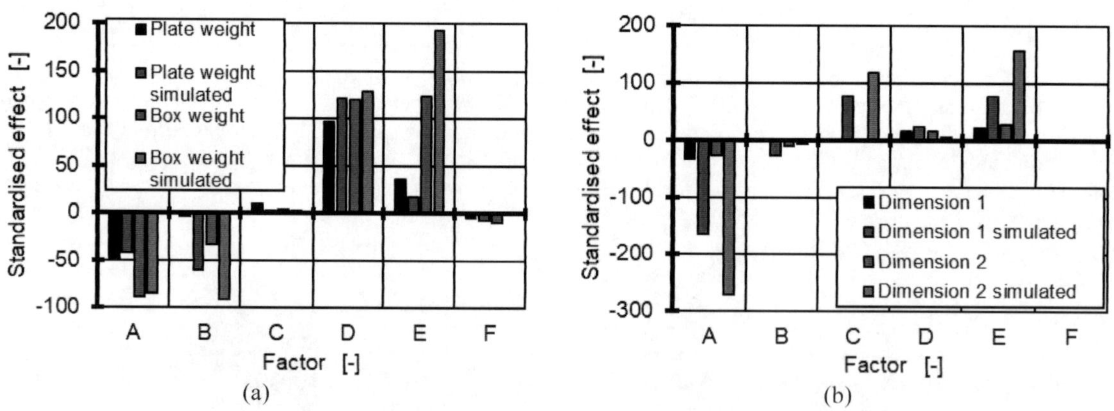

FIGURE 3. Standardised effects on part weights of both specimens (a) and dimensions of the box specimen (b)

Mould temperature (A) and melt temperature (B) in general have a negative effect on part weight since melt with higher temperatures has a larger specific volume. Packing time (D) and pressure (E) have a positive influence on the part weight as expected. Whereas the simulation predicts the direction of the influence of different parameters correctly, the overall values between simulation and experiments differ. Regarding the dimensions of the box specimens, the predicted effects of the simulation are higher because of the higher scattering between the different specimens in a test point in the experiments. The directions are predicted correctly. However, not all factor interactions are modelled properly.

In Fig. 4 the predicted effects of the setting parameters to the plate weight of a quadratic regression model with a coefficient of determination of 98 % are shown. It can be seen that the effects in injection moulding are not linear. In summary, a combined learning model using numerical and experimental data needs to be able to cope with differences in overall values and differences in the overall values of effects. Besides, interdependencies between the factors, the nonlinear behaviour and the scattering of the quality values in the experiments in contrast to single value simulations has to be considered.

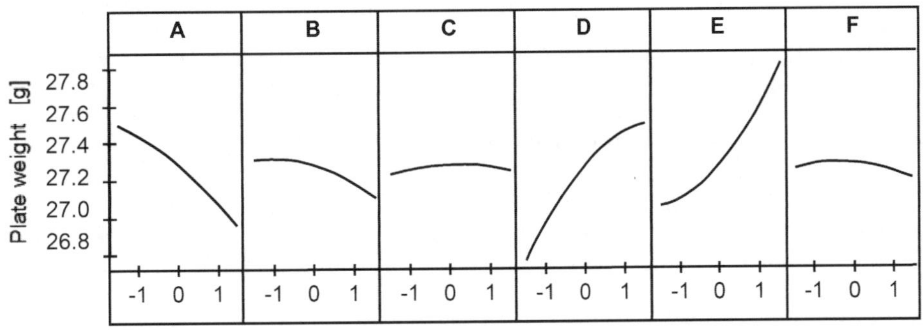

FIGURE 4. Effects (see Table 1) to (experimental) plate weight according to a quadratic regression model

COMBINED LEARNING CONCEPTS

The results show that the simulation predicts the influence of machine parameters in the correct directions. Thus, it is convenient to use observations (i.e. data) from the simulation and to build a ML-based approach that "transfers" the knowledge from these observations to the practical experiments. There are two major strategies to realise such an approach. On the one hand, ML can be used to learn the deviations of part weight and dimensions between the simulated values and the practical experiments. After that, the model can predict deviations for new test points (i.e. new experiments) and "add" them to numerical simulation outputs in order to gain predictions without the need for further practical experiments. On the other hand, a ML model can be trained on data from the simulation and, after that, be adjusted with few data from practical experiments. This is based on the idea that the model can extract useful knowledge from the simulation data concerning the overall interdependencies between process parameters and criteria. This relatively inaccurate knowledge (concerning the real world) is then refined based on few real data to be able to perform accurate and reliable predictions for the real moulding process. Fig. 5 depicts the proposed approach by pre-training a learning model and adjusting it in a second stage.

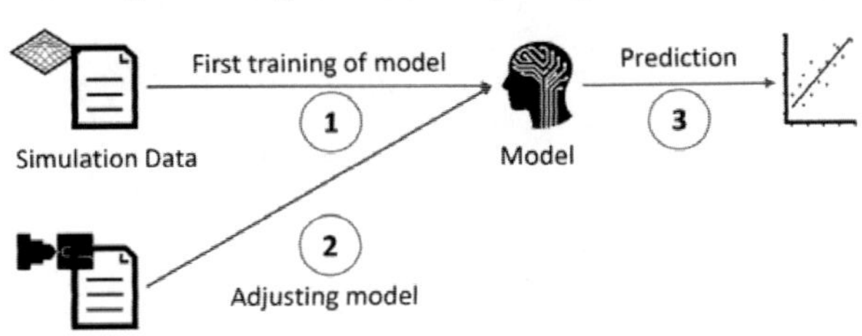

FIGURE 5. Combined approach for using machine learning on simulation (pre-training) and real data (refinement)

There exist different ML and regression techniques that can be used for this approach. The two key requirements are the capability for additional training on new data and – as seen in the previous results – the ability to identify non-linear relationships in the data. Both requirements are met by polynomial regression techniques (of higher degrees) and artificial neural networks [11]. Thus, we propose the combined approaches on the basis of one of these techniques.

CONCLUSION AND OUTLOOK

Simulation and experimental data from two injection moulded parts in a six factor central composite experiment design were analysed. Simulations and experiments were in good agreement regarding the directions of the important factors on part weight and dimensions. Even though, the overall effect values and the interdependencies between the varied setting parameters show deviations, it seems reasonable to use both data types in order to create a model between setting parameters and quality values in injection moulding. Based on the gained data, strategies for a combined learning model were presented.

The depicted combined learning methods will be implemented and analysed. Furthermore, especially for the part weight, it is conceivable to use data from known parts to reduce the learning effort for new moulds and parts.

ACKNOWLEDGMENTS

The depicted research has been funded by the Deutsche Forschungsgemeinschaft DFG as part of the Cluster of Excellence 'Integrative Production Technology for High-Wage Countries'. We would like to extend our thanks to the DFG. We give thanks to Arburg GmbH & Co KG (Loßburg, Germany), Kraussmaffei Technologies GmbH (Munich, Germany) and Sabic Deutschland GmbH (Duesseldorf, Germany) for supplying the injection moulding machines and plastics material used in the experiments.

REFERENCES

1. Fei, N.C., N.M. Mehat, and S. Kamaruddin, *Practical Applications of Taguchi Method for Optimization of Processing Parameters for Plastic Injection Moulding: A Retrospective Review.* ISRN Industrial Engineering, 2013. **2013**(DOI: 10.1155/2013/462174): p. 1-11.
2. Kashyap, S. and D. Datta, *Process parameter optimization of plastic injection molding: a review.* International Journal of Plastics Technology, 2015. **19**(1): p. 1-18.
3. Fernandes, C., et al., *Modeling and Optimization of the Injection-Molding Process: A Review.* Advances in Polymer Technology, 2016. **21683**(DOI: 10.1002/adv.21683): p. 1-21.
4. Pandelidis, I. and Q. Zou, *Optimization of injection molding design. Part II: Molding conditions optimization.* Polymer Engineering & Science, 1990. **30**(15): p. 883-892.
5. Kwong, C.K. and G.F. Smith, *A computational system for process design of injection moulding: Combining blackboard-based expert system and case-based reasoning approach.* The International Journal of Advanced Manufacturing Technology, 1998. **14**(4): p. 239-246.
6. Vagelatos, G.A., G.G. Rigatos, and S.G. Tzafestas, *Incremental fuzzy supervisory controller design for optimizing the injection molding process.* Expert Systems with Applications, 2001. **20**(2): p. 207-216.
7. Chen, W.-C., et al., *Process parameter optimization for MIMO plastic injection molding via soft computing.* Expert Systems with Applications, 2009. **36**(2): p. 1114-1122.
8. Mathivanan, D., M. Nouby, and R. Vidhya, *Minimization of sink mark defects in injection molding process – Taguchi approach* International Journal of Engineering, Science and Technology, 2010. **2**(2): p. 13-22.
9. Fasching, M., et al., *Robust process control for rubber injection moulding with use of systematic simulations and improved material data.* International Polymer Science and Technology, 2014. **41**(11): p. 640-644.
10. Shi, F., et al., *Optimisation of Plastic Injection Moulding Process with Soft Computing.* The International Journal of Advanced Manufacturing Technology, 2003. **21**(9): p. 656-661.
11. Engelbrecht, A. P., *Computational intelligence: an introduction.* (John Wiley & Sons, Chichester, England, Hoboken, NJ, 2007).

Modification of PP with CNT as painting agent: influence on performance of injected pieces

Costantino Maria Alejandra (1)*

(1) Universidad Nacional de Mar del Plata - Buenos Aires - Argentina

From an industrial point of view, final plastic products should meet both certain structural requirements and adequate aesthetic quality. Manufactured pieces are commonly subjected to different surface finishing techniques to achieve desired aesthetic requirements, being the electrostatic painting the most efficient procedure. It offers great economic, environmental and cosmetic advantages, among the great reduction of used paint and volatile emission as the most relevant. However, this technique cannot be directly applied in polymers due to their insulating properties. Plastic pieces need to be previously prepared by painting them with a primer, adding a further stage to the production cycle. To avoid this extra step, a painting agent could be added to the polymer to modify its electrical properties and make it suitable to be directly electrostatic painted. In this sense, carbon nanotubes (CNT) were proposed in this work as an electrostatic painting assistant for polypropylene (PP) injected pieces. The present work aims to evaluate how the addition of CNT to increase the panting transfer efficiency, influences the mechanical performance of intricate PP injected moldings. Then a complete characterization, including morphological and mechanical characterization – and their relationship – of double gated injection molded PP/CNT parts is presented. Differences in CNT orientation were found along the pieces, these dissimilar orientations induce a difference of an order of magnitude in the electrical properties of the part that would result in distinct paint transfer efficiency in the same piece. Moreover, mechanical performance of pieces was degraded by the addition of only 3% of CNT, especially in the weld line zone. It was then concluded that this modification would not lead to final high quality (mechanical and aesthetical) injected PP parts

Metal-Polymer Direct Joining using Nano Structures on Metal Surface: Effect of Surface Structure Size on Polymer Replication

Shotaro Kadoya, Fuminobu Kimura, and Yusuke Kajihara

The University of Tokyo, 4-6-1, Komaba, Meguro-ku, Tokyo, Japan

Abstract. We investigate the joint shear strength of the single lap joints processed by Injection Molded Direct Joining (IMDJ). In IMDJ method, roughened metal piece and thermoplastic are joined directly by usual injection molding. In this work, we utilized anodizing method to form nano structures on aluminum surface. Thermoplastic and anodized metal were successfully joined, whose shear strengths were more than 10 MPa. The results showed that the etchant concentration of the anodizing process affected the joint strength. For our future work, we need to change the structure size by anodizing and to carry out the insert molding process under various conditions.

Keywords: Direct joining, insert molding, metal surface treatment.
PACS: 81.20.Hy, 81.20.Vj, 81.65.Cf

INTRODUCTION

Polymer is widely used in many fields due to its lightness and design flexibility. To achieve both lightness and strength for industrial use, polymer is used together with metal in many cases. Conventionally, metal and polymer have been joined by using screws or adhesives. These days direct joining techniques, which needs no screws or no adhesives, is strongly required with the increase of the importance to energy saving. Direct joining can reduce weight of extra parts like screws or rivets and can bring higher productivity. In spite of its advantages, direct joining has not been applied to industries so much mainly because the reliability is not sufficient. Therefore, we need to elucidate the joining mechanism and to optimize the joining conditions for proper applications.

Various types of metal polymer direct joining have been proposed; (i) heat welding [1, 2], (ii) chemical bonding [3, 4], and (iii) micro/nano anchoring [5, 6]. In the welding method, metal and polymer are heated by some ways such as laser radiation, supersonic wave, and friction. Chemical bonding utilizes very thin layer on metal surface, which can bond metal and polymer by chemical effects. In our research, we focus on micro/nano anchoring which is called Injection Molded Direct Joining (IMDJ). IMDJ consists of two core techniques; metal surface roughening and polymer injection molding. Metal and polymer are joined directly as melted polymer flows into metal surface structures.

Our previous work revealed that the metal surface structure size could affect the behavior of melted polymer [7]. Particularly, the viscous resistance of melted polymer should be dominant when the surface structure is nano scale. We need to control nano structure size more finely and to analyze the polymer behavior. The purpose of our study is to control surface structure size from 100 nm to 10 μm scale and to investigate the dependence of the joint strength on the structure size in more detail. In this work, we proposed anodizing method to form and control metal surface structures investigated the joining potential with anodizing. We processed joint product and measured the joint strength as a basic verification.

INJECTION MOLDED DIRECT JOINING

IMDJ process is divided into three steps; (i) metal surface roughening, (ii) insert molding, and (iii) post-treatment. Figure 1 shows the overview of IMDJ process. First, micro/nano structures are formed on metal surface by laser processing, chemical etching, or other methods. In this work we utilized anodizing method due to its controllability of structure size. Second, polymer is formed with a treated metal piece by general injection molding method. Melted polymer flows into metal surface structures. After cooling, metal and polymer are joined hardly and directly. Finally, extra parts like sprue and runner are removed.

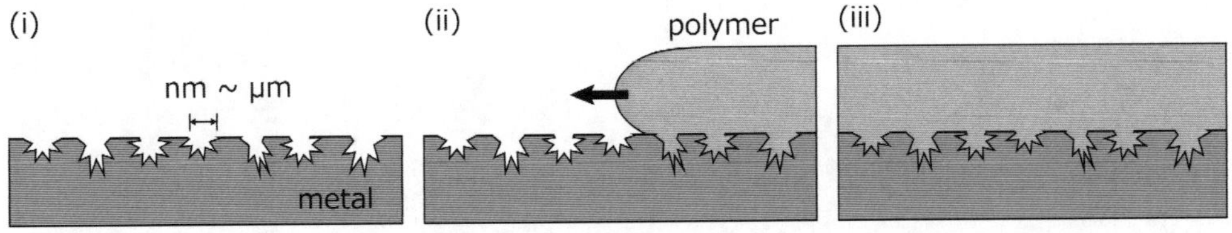

FIGURE 1. IMDJ process overview; (i) micro/nano structures are formed on metal surface, (ii) melted polymer flows into surface structures, and (iii) metal and polymer are joined directly.

EXPERIMENT

In this work, we processed joint products under several molding conditions, and measured their tensile shear strengths.

Materials and joint product

We joined aluminum alloy A5052 and PBT (Poly-butylene terephtalate; Toraycon 1101G-X54, Toray Industries, Inc.). The shape of lap joint product confirms to ISO19095. Metal part is 45 mm × 18 mm × 1.5 mm and polymer part is 50 mm × 10 mm × 3 mm. Metal and polymer parts are joined at 5 mm × 10 mm area. The schematic illustration and photo image of the single lap joint are shown in figure 2.

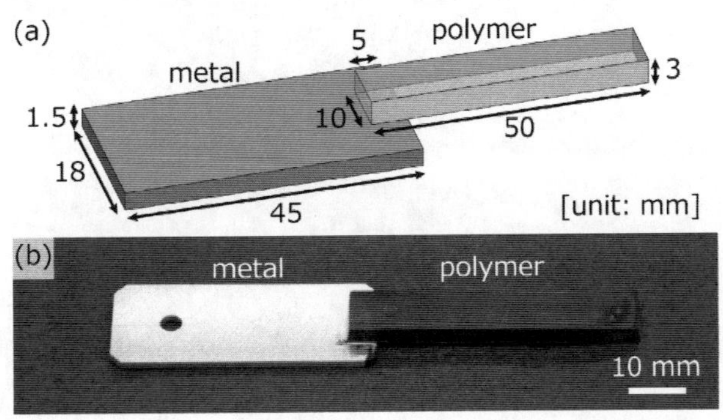

FIGURE 2. Single lap joint product; (a) a schematic illustration and (b) a photo image.

Metal surface treatment by anodizing process

We utilized anodizing process [8] to form nano structures on metal surface. In anodizing process, an oxidation layer grows on metal surface. Meanwhile, the layer is dissolved by the etchant at the bottom of metal surface irregularities. Both the layer processing and the solution make deep pore structures as shown in figure 3. We can control pore size by controlling anodizing conditions such as current, etchant concentration, and treating time length. In this work, we used phosphoric acid as etchant and carbon as cathode. We controlled current value with a constant-current power supply (PWR400H, Kikusui electronics corporation). The anodizing conditions are shown in table 1. Six aluminum pieces are treated at once with each condition.

Figure 4 shows the SEM images of anodized aluminum surface. The higher concentration brings the smaller pores. This result should be caused by the lower current density, which means the anodizing process proceeded slowly.

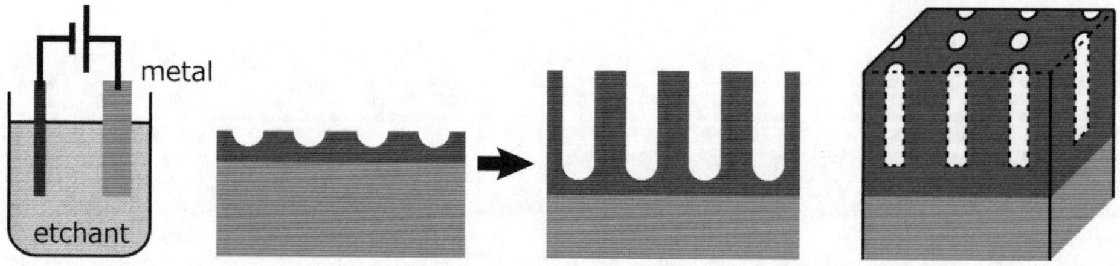

FIGURE 3. Schematic illustration of anodizing process.

TABLE 1. Anodizing conditions.

	Current [mA] (constant)	Etchant concentration [wt%]	treating time length [min]
A	10	1	60
B	30	1	60
C	100	1	60
D	10	2	60
E	30	2	60
F	50	2	60
G	100	2	60

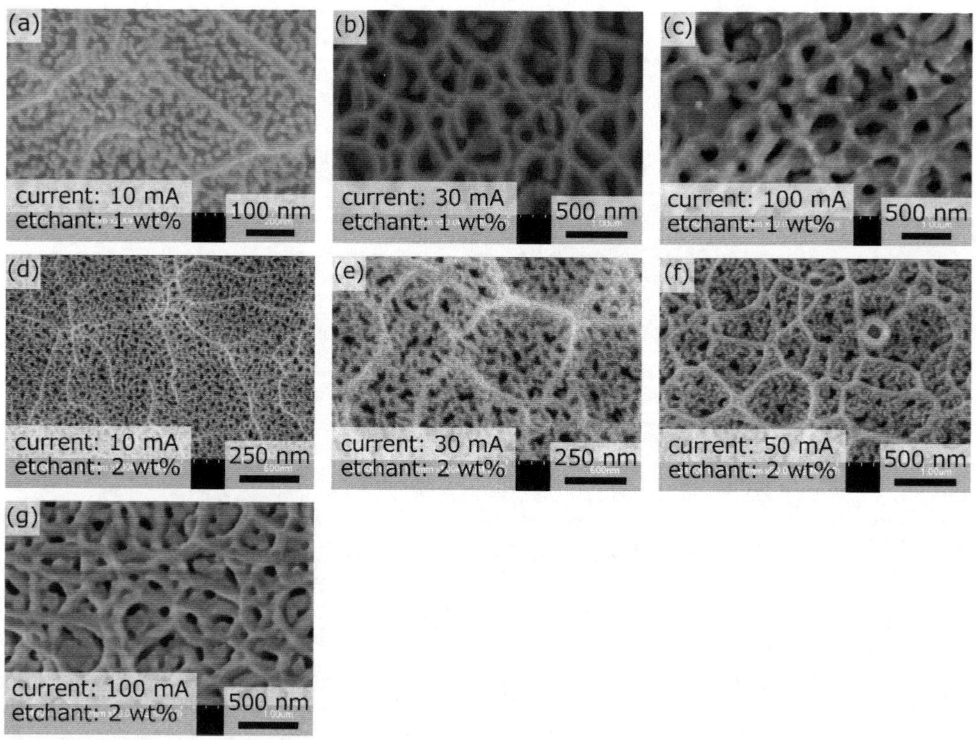

FIGURE 4. SEM images of anodized metal surface; (a) – (c) are anodized with 1 wt% etchant, and (d) – (g) are anodized with 2 wt% etchant.

Insert molding process

We used a commercial injection molding machine (ROBOSHOT α-S100iA, FANUC corporation), and original mold cores to form single lap joints. A schematic illustration of the mold is shown in figure 5. In this mold, a pressure sensor (6158A, Kistler Japan) and a thermo sensor (EPSSZT-04, Futaba corporation) are installed to measure cavity pressure and temperature during the molding process. The molding conditions are shown in table 2. We processed the joint samples under different injection speed conditions.

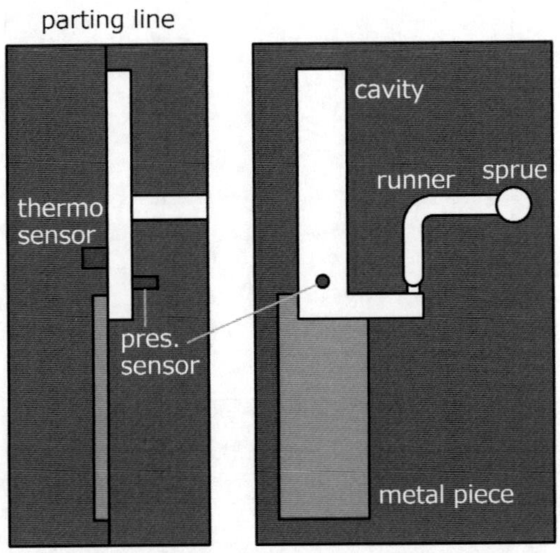

FIGURE 5. Schematic illustration of molding cores for single lap joint. A pressure sensor and a thermo sensor are installed.

TABLE 2. Molding conditions.

Injection speed	100 mm/s
(injection rate)	38.0 cm^3/s
Cylinder temperature	270 °C
Mold temperature	140 °C
Pack pressure	80 MPa
Holding pressure	40 MPa × 8 s

Tensile-shear test

We carried out the tensile-shear test with an original tensile tester [9]. This tester is optimized to test lap joints. The joint product was pulled with a constant speed (1 mm/min) to apply shear stress on the joint interface. We recorded the maximum force and calculated the joint strength with the joint area (50 mm^2).

RESULTS AND DISCUSSIONS

Figure 6 shows the tensile test result. The markers and error bars mean values and standard deviations of the tensile shear strength, respectively. The blue circle and red triangle markers mean the results for 1 wt% and 2 wt% etchant, respectively. Note that under condition C (current: 100 mA and etchant: 1 wt%) metal and polymer were not joined. First, the lower current caused the stronger joint, especially in 1 wt% etchant. Second, the higher etchant concentration brings the stronger joint in all current conditions.

The reason of negative correlation of the anodizing current should be caused by the pore size difference. With the higher current, the anodizing process proceeds faster and thus the pores could become larger. The reason of the difference between etchant concentrations could be explained as follows: with high concentration, the applied voltage changes because the current value is constant. Therefore, the solution process becomes slower and then fine structures remain inside the pores, as shown in figure 4 (b) and (e). These fine structure inside the pores could cause stronger anchoring.

The maximum strength in this work is over 10 MPa (current: 10 mA and etchant: 2 wt%), whereas our previous result study using other chemical etching process achieved over 20 MPa. To apply the industrial products widely, the joint strength should be over 20 MPa. Therefore, we need to optimize the anodizing conditions to process stronger joint.

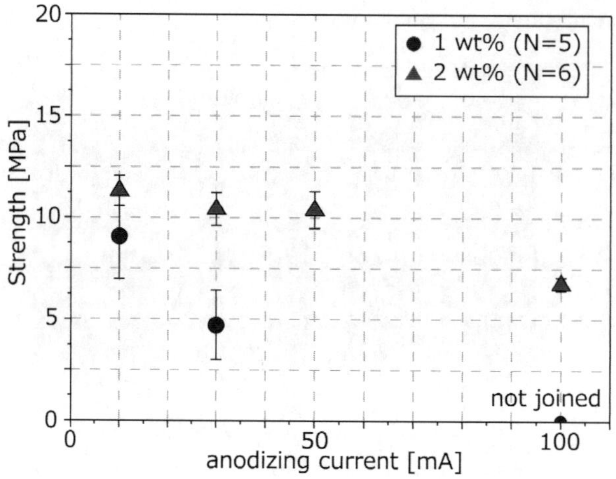

FIGURE 6. Tensile test result; anodizing current [mA] – joint strength [MPa]. The blue circles show the results of 1 wt% etchant and the red triangles show those of 2 wt% etchant.

SUMMARY

This work experimentally investigated the joint strength of single lap joints which are formed by anodizing and insert molding process. We formed several tenth pore structures on aluminum surface by anodizing method with various current and etchant concentration. The anodized metal and polymer were successfully joined for almost all conditions. The tensile test result shows that the smaller structure inside each pores could cause stronger anchoring, which leads to the higher joint strength. However the strongest joint in this work did not reached 20 MPa, which many industrial product needs. We need to optimize the anodizing process to stronger joint.

For our future work, we will control pore diameter or pore depth and process joint strength under various molding conditions. We need to carry out TEM analysis of joint interface except for the tensile test.

ACKNOWLEDGMENTS

This work was supported by The Foundation of the Promotion of Industrial Science and JSPS KAKENHI (#15K17946 and #17J00345). The experiments were partially supported by Yokoi group, The Univ. of Tokyo, Japan.

REFERENCES

1. Katayama S. and Kawahito Y., *Scripta Materiaria*, **59**, 12, 1247-1250 (2008).
2. Fortunato A. et al, *Intl. J. of Material Forming*, **3**, 1, 1131-1134 (2010).
3. Sasaki H. et al, *J. of Adhesion Science and Technology*, **13**, 4, 523-539 (1999).
4. Kang Z. X., Mori K., and Oishi Y., *Surface and Coating Technology*, **195**, 2-3, 162-167 (2005).
5. Ramani K. and Moriarty B., *Polymer Engineering & Science*, **38**, 5, 870-877 (1998).
6. Fabrin P. A., Hoikkanen M. E., and Vuarinen J. E., *Polymer Engineering & Science*, **47**, 8, 1187-1191 (2007).
7. Kadoya S., Kimura F., and Kajihara Y., *euspen's 16th International Conference & Exhibition* (2016).
8. Sulka G., *Nanostructured Materials in Electrochemistry*, 1-97 (2008).
9. Kadoya S., Kimura F., and Kajihara Y., In Proc. of 12th Intl. Symp. on Measurement Technology and Intelligent Instruments (2015).

In-line surface inspection of decorated film insert molding products

M. Haselmann[a], D. P. Gruber[b]

Polymer Competence Center Leoben GmbH, Roseggerstraße 12, 8700 Leoben, Austria

[a)]Matthias.Haselmann@pccl.at
[b)]Dieter.Gruber@pccl.at

Abstract. The decorations of film insert molding products are versatile and can show a high morphological complexity. Moreover, the decoration patterns of such plastic parts often reveal significant but allowed part-to-part variations. A classical machine vision system based on comparing a test part with a constant golden sample fails in this case. In this contribution a deep convolutional neural network is presented that is able to detect aesthetical defects within a varying decoration pattern.

INTRODUCTION

Film insert molding (FIM) is an injection molding process that enables free-form plastic parts with decorated and functional surfaces within one processing step. Imprinted plastic films are thermoformed according to the plastic part's surface geometry, inserted into the mold and back-injected with a thermoplastic. Possible decorative patterns are versatile and can show a high morphological complexity. A full-area surface inspection of several FIM products in a row is monotonous and exhausting for humans. A machine vision based inspection system as used for many other kinds of plastic parts would be very beneficial. However, FIM products have one very unique feature which almost makes it impossible to use traditional machine vision techniques. The thermoforming process, which brings the plastic film to the desired 3D-shape, suffers from slight variations depending on the position and distortion of decorative pattern elements. This leads to a unique appearance of every single manufactured part (compare Fig. 1a and 1b). Methodologies that are based on comparing a test part with a constant golden sample fail in this case.

METHODOLOGY AND RESULTS

In this contribution we present a deep convolutional neural network (DCNN) that is able to detect aesthetic defects on decorated plastic parts. The corresponding images were acquired in an in-line quality testing site. The total number of images required for the network's training is drastically reduced by a defect synthetization algorithm that is based on a stochastic process. The presented algorithm superimposes artificial defects on fault-free surface images so that the DCNN is able to learn the difference between flawless and defective surface sections. Thereby, the manual and time consuming labeling process of real defects in the training images is avoided. After training, the presented 7-layer DCNN is able to detect 33 out of 35 real defects on the test parts (see Fig. 1d). Vice versa 99.7% of flawless surface sections are correctly classified as fault-free by the DCNN (see Fig. 1c).

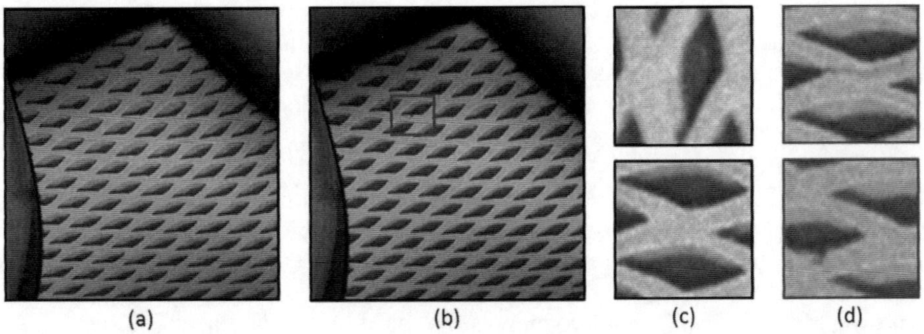

FIGURE 1. a-b) Surface images showing the same region of two different FIM parts with clearly visible part-to-part variations. **c-d)** Two fault-free and two defective surface sections properly detected by the DCNN.

Characterization of Injection Compression Molding (ICM) in Fresnel Lenses Production Using Amorphous Polymer Materials

Federico Baruffi[a*], D. Loaldi[a], Guido Tosello[a], Matteo Calaon[a], Danilo Quagliotti[a], Paolo Parenti[b], Massimiliano Annoni[b] and Igor Di Vora[c]

[a]*Technical University of Denmark, Department of Mechanical Engineering, Produktionstorvet, Building 427A, 2800 Kgs. Lyngby, Denmark*
[b]*Politecnico di Milano, Department of Mechanical Engineering, via La Masa 1, 20156 Milan, Italy*
[c]*Automotive Lighting Italia S.p.A., Via Industria 17, 33028, Tolmezzo (UD), Italy*

febaru@mek.dtu.dk

Abstract. Injection compression molding (ICM) is the leading process technology for the mass manufacturing of high precision polymer optics. To characterize the process, the effects of process factors on the part geometrical accuracy must be investigated. This study aims at quantifying the effect of the injection compression molding process on the replication accuracy in Fresnel lenses production. The replication fidelity was assessed at different distances from the gate using step height measurements of the Fresnel lens structured profile as indicator. Two different optical grade polymer materials were compared: Polymethil-Metacrylate (PMMA) and Cyclo-Olephin Polymer (COP). Three different process parameters, namely compression starting point, compression gap and packing pressure, were tested for the COP grade while only two for the PMMA grade. The influence of packing and compression was experimentally demonstrated, highlighting the importance of the packing phase to ensure full replication of the part. A significant interaction was detected between packing pressure and compression gap. The replication fidelity showed a great dependence on the distance from the gate. In particular, worse replication was observed far from the gate due to the presence of entrapped air, which was also influenced by the compression gap size.

Keywords: Injection Compression Molding, Process characterization, Fresnel lenses
PACS: 83.50.Uv

INTRODUCTION

Fresnel lenses are optical tools with enhanced light gathering properties and compact design [1], [2]. Such devices have many advantages on standard aspheric lenses: firstly, they allow to achieve better optical performances by means of reduced aspherical aberration [2]. Fresnel lens design also introduce valuable cost reduction due to the decreased thickness and consequent material saving in the production phase. This product is nowadays used in many different industrial sectors such as automotive, electronic, indoor and outdoor lightning, solar energy, etc. [3], [4].

As most other precision optical components, the geometry of Fresnel lenses demand particular high precision and accuracy to guarantee the functionality. At the moment, there are multiple solutions to manufacture Fresnel lenses. Guduru et al. [5] produced micro lenses with nano-features with laser ablation micro machining. Macro-scale lenses (i.e. the ones used in lighthouses) are usually produced by direct glass or polymer machining and subsequent mechanical or/and chemical polishing. Diamond turning is also reported as a feasible solution for manufacturing of radial Fresnel lenses [6]. However, replication processes are the preferred solution to produce meso- and micro-scale optics because of the opportunity of achieving throughputs typical of mass manufacturing [7].

Among replication-based processes, injection compression molding (ICM) is the leading technology for mass manufacturing of high precision optical lenses [8]. This process can be considered as an extension of the conventional injection molding (IM) process for thermoplastic materials. The first phase of ICM is the same as the filling in conventional injection molding. The main difference of ICM is that the mold is kept partially open to facilitate the flow of plastic inside the cavity. After injection is completed, the gap left during injection is closed by the movement of the movable plate. This phase, called compression, is peculiar of ICM and allows to achieve lower volumetric shrinkage, improved replication capability and more homogeneous material properties in comparison with IM [7]. These advantages make the process particularly suitable for manufacturing optical components.

In the recent decade, numerous researchers studied the replication capability of ICM in the production of structured surfaces. Rohde et al. [9] investigated the effect of ICM process parameters on the replication of micro structures on a flat disk. It was found that an increase of compression gap had a negative effect on the replication fidelity. Kuo et al. [10] compared conventional injection molding and injection compression molding when molding ultra-high molecular weight polyethylene (UHMWPE) inside micro grooves with width ranging from 50 µm to 5000 µm. The authors proved that both the technologies were capable of filling the mold micro geometries. However, it was clear that ICM provided a better replication than IM. Chao et al. [11] developed a two-stage ICM process for the production of fiber diffractive optic elements. Their results showed that the process allowed to achieve a very high replication of the lenses grooves.

The present study aims at investigating the effect of the ICM process on the replication accuracy in Fresnel lenses production. Two different optical polymer grades, namely Polymethil-Metacrylate (PMMA) and Cyclo-Olephin Polymer (COP), were molded. The replication performance was assessed by measuring the step height of the lens grooves using a laser confocal scanning microscope.

MATERIALS AND METHODS

The Fresnel lens object of this paper features a constant groove pitch of 748.0 µm and a variable step height, which is equal to 17.3 µm in the middle of the lens and 346.6 µm at its border (see Figure 1).

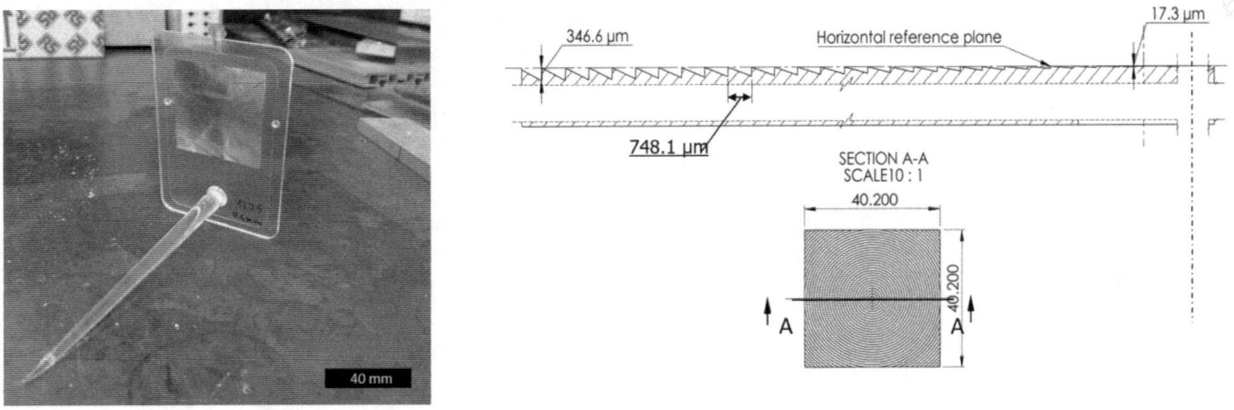

(a) (b)

FIGURE 1. ICM Fresnel lens. (a) Molded lens and feed system; and (b) design of the Fresnel lens.

ICM experiments were carried out using a Negri Bossi© v70-180 injection molding machine. The parts were molded on removable inserts manufactured with a diamond turning process.

The two tested material were a COP (commercial name: ZEONEX® E48R) and a PMMA (commercial name: ALTUGLAS® V 825T). Both these amorphous polymers are optical grades and therefore are suitable for Fresnel lenses manufacturing. Table 2 shows the main characteristics of such polymers.

TABLE 1. Main properties of the two materials.

Polymer	Density [g/cm^3]	MFI [g/10 min]	T_g [°C]
ZEONEX® E48R	1.01	25.0 (load 2.18 kg; temperature 280 °C)	139
ALTUGLAS® V 825T	1.19	2.8 (load 3.8 kg; temperature 230 °C)	108

A design of experiment approach (DOE) was used to evaluate the effect of the ICM process on the Fresnel lens replication fidelity. Two different experimental plans were designed for the two materials. For the COP materials, a

2^3 full factorial design with two extra screening conditions was performed. The three varied factors were compression gap, compression starting point (i.e. the stroke coordinate at which the compression phase start) and packing pressure. Moreover, two extra screening conditions, correspondent to pure injection (i.e. both packing and compression were off) and conventional injection molding (only compression was off) were run. For the PMMA polymer, a 2^2 full factorial design was performed by considering packing pressure and presence of compression (i.e. molding with or without compression) as factors. All the other process parameters were kept as constant. For both the materials, two lenses per process parameters combination were collected and measured. Table 2 and 3 show the details of the experimental design for the two materials.

TABLE 2. Experimental settings for COP material.

Process Parameters	Experimental Levels
Compression gap [mm]	13, 14
Compression start [mm]	0.5, 1
Packing pressure [bar]	0, 450

TABLE 3. Experimental settings for PMMA material.

Process Parameters	Experimental Levels
Compression	ON, OFF
Packing pressure [bar]	0, 450

For both the experimental plans, it is worth noticing that the low level of packing pressure was set at 0 bar, meaning that the possibility of performing ICM without the packing phase was investigated. The aim was therefore to verify if the replication enhancement introduced by the compression phase was enough to justify a process without the packing phase. The advantage of performing a packing-less ICM would be a consistent raise in productivity due to the elimination of the packing phase from the process cycle.

As response of the experimental analysis, the groove step height was selected. This quantity in fact indicates how well the polymer replicated the mold micro features. A laser confocal scanning microscope (Olympus© Lext OLS400) was used to perform areal measurements. This instrument is particularly suitable for transparent samples. A 20× objective lens was used, having a digital lateral resolution of 0.625 μm a vertical resolution of 10 nm. Being the step height variable on the lens profile (see Figure 1), measurements were taken in different positions around the lens central point: eight topography acquisitions were carried out, as shown in Figure 2. By doing this, the replication symmetry could be evaluated comparing "left" and "right" areas as well as the effect of the distance from the gate ("very near", "near", "central", "far" and "very far").

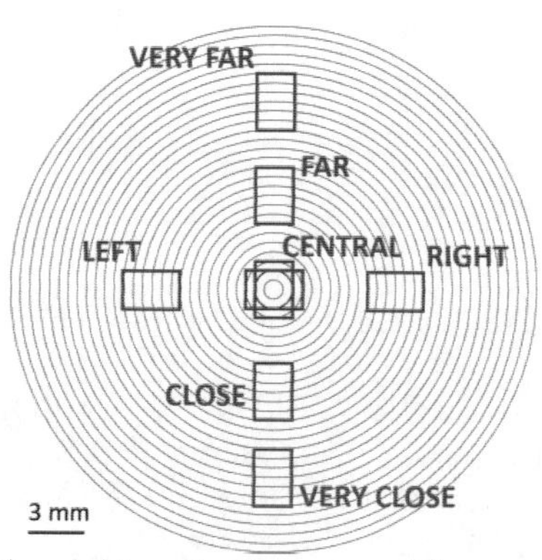

FIGURE 2. Map of the Fresnel lens and the eight evaluation areas. The gate position is placed down on the figure.

RESULTS

As response for the analysis of the results, the percentage replication with respect to the mold dimensions was used. By considering such an output, the level of replication can directly be assessed.

First, it was observed that pure injection and conventional injection molding did not allow to achieve a full replication of the mold geometry for both the materials (see Figure 3). In fact, the profiles extracted from the areal measurements showed a different morphology due to the incomplete replication. This clearly proved that packing was necessary to achieve the desired surface structure, and therefore the effect of compression was not sufficient to compensate the absence of a proper packing phase.

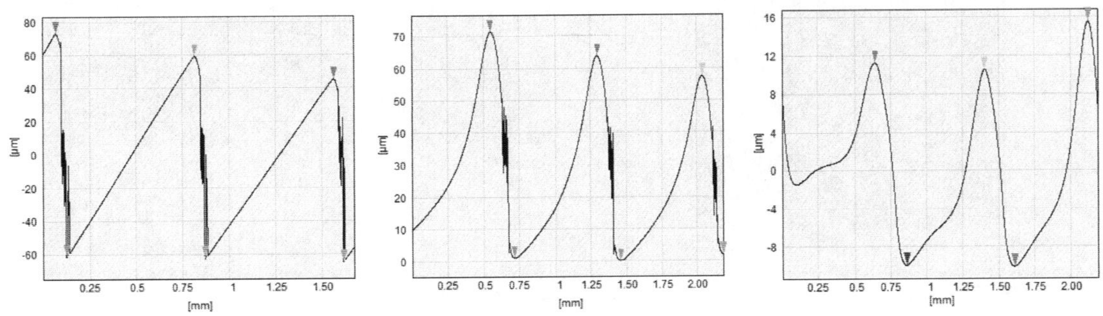

(a) (b) (c)

FIGURE 3. Molded lens profile with compression and packing (a), compression without packing (b) and pure injection (c)

Figure 4(a) shows the main effects plot of the step height replication for the COP material. The biggest effect was the distance from the gate. The central position provided the best replication, while the measurements far from the gate revealed the worst performance. This observation, also illustrated in Figure 5, was most probably caused by the presence of trapped air inside the cavity. This influenced mainly the features far from the gate because of the particular micro grooves geometry and the direction of the flow. The presence of air in the cavity depended on the ICM process parameters. In the compression stroke was large enough (i.e. the high value of 1 mm), the replication was generally enhanced since the air could evacuated more easily. On the other hand, an increase of the compression gap did not favor the air evacuation, leading to a worse replication of the mold surface: the larger the compression gap, larger the volume of air that needed to be evacuated from the cavity.

Figure 4(b) shows the main effects plot for the PMMA material. As for the previous case, the use of packing pressure was necessary to achieve a better replication. The distance from the gate played the same role, since the same phenomenon explained in Figure 5 happened. This justifies the hypothesis that the worse replication far from the gate was due to the improper venting of the mold, being the output equal for the two different materials.

For both materials, a lower replication level was observed for the central position. This was unexpected, since lower aspect ratio features, such as the ones in the middle of the lens (see Figure 1) are usually more easily replicated. The lenses molded with PMMA resulted in a less symmetric replication, being the features on the left better replicated than those on the right. A higher symmetry was achieved for COP. When comparing the two materials, it was observed that the COP provided a better overall replication performance, being the average equal to 91% against the 90% of the PMMA. However, PMMA allowed to achieve a higher maximum replication level in correspondence with the high packing pressure value. This also proves that the effect of the packing pressure was more relevant for the PMMA and that, by opportunely tuning the ICM process, the PMMA was capable of providing a more accurate replication of the mold micro features.

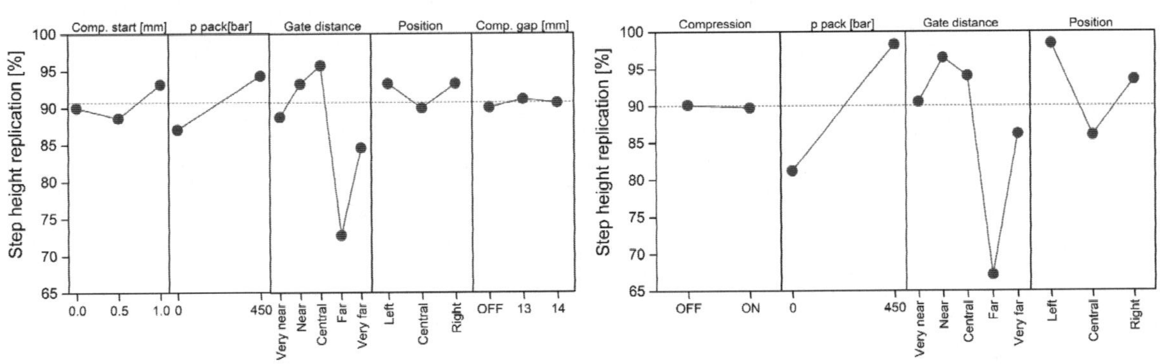

(a) (b)

FIGURE 4. Main effects plot for COP (a) and PMMA (b) materials

(a) (b)

FIGURE 5. Top view (a) and illustration (b) of the entrapped air in the position "far" from the gate

CONCLUSIONS

In this study, the effects of the ICM process on the replication of Fresnel lenses was studied on two polymer optical grades.

Measurements carried out with a laser scanning confocal microscope showed that pure injection and conventional injection molding did not allow to achieve a full replication of the mold geometry for both the materials. In general, the replication performance of the process deteriorated far from the gate position. This was caused by the presence of air trapped in the cavity that prevented to optimally fill the mold micro features. An increase of the compression gap helped this phenomenon, increasing the volume of air that remained in the cavity during the process. When comparing COP and PMMA, it was found that better replication could be achieved with the second one since it was more dependent on the variation of the investigated process parameters.

It can be concluded that performing an ICM cycle that comprises both packing and compression phase is necessary to achieve a satisfying replication level when molding Fresnel lenses. Being also the replication strongly dependent on the distance from the gate position, different gating solution should be analyzed in order to optimize the ICM dimensional output.

ACKNOWLEDGMENTS

This research work was undertaken in the context of MICROMAN project ("Process Fingerprint for Zero-defect Net-shape MICROMANufacturing", http://www.microman.mek.dtu.dk/). MICROMAN is a European Training Network supported by Horizon 2020, the EU Framework Programme for Research and Innovation (Project ID: 674801).

REFERENCES

1. B. A. Aničin, V. M. Babović, and D. M. Davidović, *Am. J. Phys.* **57**, 312-316 (1989).
2. A. Davis and F. Kühnlenz, *Opt. Photonik* **2**, 52-55 (2007).
3. H. P. Annen, L. Fu, R. Leutz, L. González, and J. Mbakop, *Proceedings of Spie--the International Society for Optical Engineering* **8112** (2011).
4. J. R. Egger, *Proc. Soc. Photo-optical Instrum. Eng.* 1979.
5. S. S. K. Guduru, F. Scotognella, L. Criante, R. M. Vazquez, R. Ramponi, and K. C. Vishnubhatla, *Opt. Eng.* **53**
6. R. Huang, X. Zhang, M. Rahman, A. S. Kumar, and K. Liu, *CIRP Ann. - Manuf. Technol.* **64**, 121-124 (2015).
7. H. N. Hansen, R. J. Hocken, and G. Tosello, *CIRP Ann. - Manuf. Technol.* **64**, 695-714 (2011).
8. W. Michaeli, S. Heßner, F. Klaiber, and J. Forster, *CIRP Ann. - Manuf. Technol.* **56**, 545-548 (2007).
9. M. Rohde, A. Derdouri, and M. R. Kamal, *Int. Polym. Process.* **24**, 288-297 (2009).
10. H.-C. Kuo and M.-C. Jeng, *Int. Polym. Process.* **26**, 508-516 (2011)
11. C. C. A. Chen, *Key Eng. Mater.* (2008)

Zero defect manufacturing in injection compression moulding of polymer Fresnel lenses

Baruffi Federico [1], Loaldi Dario [2]*, Tosello Guido [2], Calaon Matteo [2], Quagliotti Danilo [2]

[1] Technical University of Denmark - Denmark - Denmark, [2] Denmark technical University - Kgns. Lyngby - Denmark

Fresnel lenses are polymer optics with reduced dimensions and higher illumination properties, in comparison with conventional lenses. However, their structured profile demands high precision replication techniques when industrial scale manufacturing is concerned. The opportunity to perform a compression phase on an injection moulding machine allows to introduce a compression force on the polymer while it is injected in the cavity. This means that at the beginning of injection, the moulds are not closed and a volume of air stays inside the cavity. In this context, challenges during replication, related to geometrical design of the Fresnel lens are presented. By performing different experimental plans, in some process conditions, incomplete replication of the part is observed. The presence of air trap in the cavity is investigated and a geometrical dependent justification is proposed. The importance of gate location and distance, combined with insert design represents the most critical aspect to take into account when replication quality requires to be investigated. The lens aspect ratio plays a sensible role on the replication of the part.

Zirconia-based thin films via MOCVD for thermal barrier and protective applications in injection molding

Fornalczyk Gregor [(1)]*, Mumme Frank [(1)], Sommer Michaela [(1)], Frettloeh Vanessa [(1)]

[(1)] Gemeinnuetzige KIMW Forschungs-GmbH - North Rhine-Westphalia - Germany

Steel tools, which are used in industrial high-throughput processes like injection molding, are susceptible for wear and corrosion due to rapid cyclic temperature and pressure fluctuation as well as the use of abrasive polymers. For the protection of tool surfaces zirconia-based thin films can be applied by metal-organic chemical vapor deposition (MOCVD). Deposited zirconia provides mechanical strength and chemical inertness resulting in enhanced resistance to both wear and corrosion. In addition to these protective properties high-performance ceramics like yttria-stabilized zirconia (YSZ) are able to thermally insulate tool surfaces providing a more precise temperature regulation in injection molding with intent to avoid the formation of surface flaws, e.g. weld lines, in the later plastic parts. At the same time it enables the shortening of cycle times as well as the decrease of energy demands during the molding process. In this work we demonstrate the fabrication of zirconia-based thin films and multilayer systems on steel tools with complex 3D surfaces via MOCVD using metal acetylacetonates as precursor materials. Coating development was carried out by measuring the film thicknesses at different parameter settings. The usage of autonomous liquid flow controllers enables the formation of multilayer systems as well as the control of crystallinity by addition of different dopants to the material. Alternating crystalline and amorphous layers lead to an increase in flexibility and diminishment of local stress in the joint material. For process development substrates were engineered according to tool geometry containing typical cavities and defined cracks aspect ratios up to 1:60. That application enables the proof of conformity and the verification of homogeneous film thickness distribution. Exploitation of these results offers the coating of tools, which are tested regarding their desired properties by industrial project partners under production conditions.

Improving the shape accuracy of a micro component manufactured by injection molding

Luca Antonio [(1)]*

[(1)] University of Bremen - Bremen - Germany

Polymer micro injection molding is one of the most common technologies for cost-effective mass production of micro components. The part to be analyzed is a medical micro component that has a flat surface and characteristic micro features. The presence of these features causes a not uniform filling of the micro cavity, due to separation of the polymer melt during the filling stage. The aim of this investigation is to evaluate the quality of the manufactured micro component in terms of form deviation from the nominal geometry, considering the influence of the micro injection molding process parameters. The shape accuracy of the manufactured micro component and its features is not only affected by the complex filling process but also by the mold design and machining. The gate and runner configuration, as well as the surface finish of the cavity manufactured by micro milling play an important role in achieving a satisfactory replication quality.

The Simulation Method for Injection Molding Process of Short-fiber/Thermoplastic Composites Considering Fiber Interactions

Thanh Binh Nguyen Thi[a], Atsushi Yokoyama[b,*], Senji Hamanaka[c], Katsuhisa Yamashita[c], and Chisato Nonomura[c]

[a]*National Institute of Advanced Industrial Science and Technology, Central 5, 1-1-1 Higashi, Tsukuba, Ibaraki 305-8565, Japan*
[b] *Kyoto Institute of Technology, Matsugasaki, Sakyo-ku, Kyoto 606-8585, Japan*
[c] *Research Center, Toyobo Co., LTD, 2-1-1 Katata, Otsu, Shiga 520-0292, Japan*

Abstract. The theoretical fiber-fiber interaction model including the fiber dynamics simulation for calculating the fiber orientation was proposed in our previous research. Then, the three-dimensional fiber orientation distribution was predicted by combining our developed fiber interaction model and improved Anisotropic Rotary Diffusion - Retarding Principle Rate model. In this study, we proposed another theoretical model that has been introduced a fiber dynamics simulation in order to obtain a relation for fiber-matrix interaction coefficient C_M. This supposed that the coefficient is a exponential function of the fiber parameters, and strain rate. Simultaneously, the Micro-computed tomography system YXLON International Y.Cheetah µHD was employed to observe and measure the three-dimensional fiber orientation in short-glass fiber-reinforced polyamide 6 specimens, which were produced using plate-shaped cavity with the fiber contents ranging from 30 wt% to 65 wt%. The experimental and numerical results proved that this theory should be used for the 3D fiber orientation prediction with considering fountain flow.

Keywords: Injection molding, Fiber orientation, Short-fiber reinforced plastic, X-ray computed tomography
PACS: 81.20.Hy

INTRODUCTION

Properties of fiber-reinforced composite are dominated by the microstructure of the fabricated part rather than the properties of constituent materials. And it varies throughout the part due to local variations in the orientation of fibers. Thus, flow processing is the key to tailor the orientation microstructure of fiber reinforced composites to suit the desired application with minimum cost and maximum reliability. Moreover, due to the widespread use in many industrial applications of these composites, there is a considerable interest in development an accurate estimate the fiber orientation kinematic [1]. Thus, many researcher have studied on the fiber orientation using numerical methods. Mathematical models need to be developed with the emphasis on accurate predicting the final orientation pattern.

Jeffery's Hydrodynamic (HD) model and Folgar and Tucker's Isotropic Rotary Diffusion (IRD) model are two representative theoretical models over the past two decades. In general, the standard F-T orientation equation is a combination between HD and IRD model [2, 3]. Our previous research presented a developed fiber interaction model, which modified the rotary diffusion term in the Folgar-Tucker model in order to obtain a global fiber interaction coefficient [4, 5]. The fiber interaction model is introduced a fiber dynamics simulation in order to obtain accurate estimates of fiber interacts at all orientation states. This advance should permit software developers to proceed confidently as they build fiber orientation into their mold filling simulations. The fiber-fiber interaction in our proposed model is represented by a global fiber interaction coefficient. The steps to derive the equation for this coefficient in short fiber suspension as a function of the fiber aspect ratio, volume fraction and general shear rate was delineated. Then the relation of C_I was proposed, which is written as follows:

$$C_I = 0.214\left|\dot{\gamma}\right|\left[1 - e^{-0.224\frac{L}{D}\phi_f}\right], (1)$$

A new theoretical model - 2017KIT model for fiber suspension in injection molding is proposed. The fiber-fiber interaction, fiber-matrix interaction, fiber degradation mechanism, and anisotropic fiber diffusion in molding processes will be considered in this model. The 3D fiber orientation predictions of short/long-fiber-reinforced thermoplastics will be performed and compared with the experimental results.

FIBER ORIENTATION THEORY

A theoretical 2017KIT model for the detailed behavior of fiber suspension in molding processes is proposed regarding anisotropic fiber diffusion, fiber-fiber interaction (FFI model), fiber-matrix interaction, and fiber degradation mechanism (FDM model). The model will be expected to encounter the problems in its implementation in both injection and compression molding, as shown below:

$$\dot{A} = \dot{A}_{HD} + \dot{A}_{FFI}(C_I^*) + \dot{A}_{FDM}(C_D, C_M), (2)$$

The isotropic rotary diffusion model (IRD) developed C_I as the dimensionless interaction coefficient describes the strength of fiber-fiber interactions. In this research, FFI model is proposed based on fiber interaction model in our previous research. The spatial tensor describing fiber-fiber interaction is used. The model includes both anisotropic fiber diffusion and fiber-fiber interaction as shown in Eq. 3.

$$\dot{A}_{FFI} = \beta_1\left|\dot{\gamma}\right|\left[1 - \exp(-\beta_2\frac{L}{D}\phi_f)\right]\left[2C_I^* - 2(trC_I^*)A - 3(C_I^*.A + A.C_I^*) + 6A_4 : C_I^*\right], (3)$$

Furthermore, we introduce a fiber dynamics simulation in order to obtain fiber-matrix interaction coefficient C_M. In this simulation, the total number of the fibers in a suspension is distributed into uniform volumes and in which the fibers are curved by randomizing effects of stress flow. The steps to derive the equation of C_M in fiber suspension as a function of the fiber aspect ratio, volume fraction and strain rate will be delineated, as shown in Eq. 3. There are three common and likely mechanisms for fiber breakage: fiber-fiber/fiber-matrix interaction, and higher fiber loading. Similarly, the steps to derive the equation of C_D as a function of volume fraction and strain rate will be delineated (see Eq. 4).

$$C_M = \beta_3\left|\dot{\varepsilon}\right|\exp\left[\beta_4\phi_f\left(\frac{L}{D}\right)^2\right] \quad ; C_D = \beta_5\left|\dot{\varepsilon}\right|\exp(-\beta_6\phi_f), (4)$$

Where ϕ_f is fiber volume fraction, L/D is fiber aspect ratio, parameters β_1, β_2, β_3, β_4, β_5, β_6 are contants, C_I^* is a spatial tensor of fiber-fiber interaction, and $\left|\dot{\varepsilon}\right|$ is train rate.

RESULTS

Short-glass fiber reinforced polyamide 6 composite was prepared using a procedure described previously [3]. The composites with 30wt%, 50wt%, and 65wt% fiber concentration were used in this study. Experimental mold cavity is 100 x 100 x 2 [mm^3] injection-molded plate. The X-ray CT test specimens, 2 x 4 x 2 [mm^3], were cut from injection-molded sample. During sample preparation, great care was taken to ensure that the flow direction (X direction). The Y direction was determined according to the right-hand rule, and the Z direction is thickness direction. Their fiber orientation is measured Micro-Computed Tomography (CT) system Y.Cheetah µHD.

Numerical algorithm of this model for predicting the 3D fiber orientation in injection-molded parts is then described and incorporated into the simulation program of mold filling Moldflow. The three-dimensional finite element solver of Moldflow software is used. The parts are molded with tetrahedral elements. The fiber orientation calculation started from the gate, and a random orientation is applied at that location as the inlet boundary condition. Fiber orientation data in plated-shape cavity exhibits the typical shell-core-shell structure through the thickness.

The 3D fiber orientation predictions show a quantitative agreement with experimental and confirm the full understanding of fiber orientation in molding processes. Consequently, the microstructure of composite can be tailored in order to achieving high-performance composites be exercising control over the flow processing.

CONCLUSIONS

As the microstructure of composite is related to the flow-processing route of the fiber-reinforced suspensions and the geometry of the mold, and the microstructure of composite can be tailored in order to achieving high-performance composites by exercising control over the flow processing. Thus, detailed microstructure of composites may exploit new applications for prediction of physical and mechanical properties of short- and long-fiber-reinforced composites.

REFERENCES

1. S. G. Advani, "Molding Short-Fiber Composites, Flow Processing", in *Wiley Encyclopedia of Composites*, Second Edition, edited by L. Nicolais and A. Borzacchiello, John Wiley & Sons, Inc., 2012, pp. 1-12.
2. F. Folgar; C. L. Tucker J. Reinf. Plast. Compos. 1984, 3, 98.
3. T. B. Nguyen Thi, M. Morioka, A. Yokoyama, S. Hamanaka, K. Yamashita, and C. Nonomura, *J. Mater. Process. Tech* **129**, 1-9 (2015).
4. T. B. Nguyen Thi, A. Yokoyama, K. Ota, K. Kodama, K. Yamashita, Y. Isogai, K. Furuichi, and C. Nonomura, "Numerical Approach of the Injection Molding Process of Fiber-reinforced Composite with Considering Fiber Orientation" in *The 29th International Conference of the Polymer Processing Society-2013*, edited by V. Altstadt, AIP Conference Proceeding **1593**, American Institute of Physics, Melville, NY, 2014, pp. 571-577.
5. T. B. Nguyen Thi, A. Yokoyama, S. Hamanaka, K. Yamashita, and C. Nonomura, "Numerical Simulation Of Fiber Interaction In Short-Fiber Injection-Molded Composite Using Different Cavity Geometries" in *31st International Conference of the Polymer Processing Society-2015*, edited by Byungohk Rhee, AIP Conference Proceeding **1713**, American Institute of Physics, Melville, NY, 2016, pp. 040006.

Investigating the effect of the mold surface roughness and the process parameters on the quality of a micro component

Luca Antonio [(1)*]

(1) University of Bremen - Bremen - Germany

Polymer micro injection molding is one of the most common technologies for cost-effective mass production of micro components. The stringent requirements of part dimensions lead to an intricate design of the mold and the process itself. The mold surface roughness plays an important role on the quality of the manufactured micro parts. The part to be analyzed is a medical micro component that has a flat surface and characteristic micro features. The presence of these features causes a not uniform filling of the cavity, due to separation of the polymer melt during the filling stage and consequently to the formation of structures that increase the surface roughness of the manufactured micro components. The surface roughness of the micro part is not only affected by its features but also by the mold surface roughness and the selected process parameters. The aim of this investigation is to relate the influence of the mold surface roughness and the process parameters with the achievable surface roughness of the manufactured micro part. Mold cavities with different surface roughness but the same geometry will be manufactured and different process parameters will be employed in order to establish this relationship. It is expected to minimize the surface roughness of the micro part through the selection of the best process window.

RFID technology integration into injection molded plastic parts

***Frettloeh Vanessa* (1)*, *Beck Christopher* (1), *Hinzpeter Udo* (1)**

(1) Gemeinnuetzige KIMW Forschungs-GmbH - North Rhine-Westphalia - Germany

At times, when the cross-linking of manufacturing gains importance, technologies in which produced parts can be traced and additional information can be allocated to the products, are of enormous economic and technological interest. Since the RFID (radio frequency identification) technology enables all this, it is currently in the focus of the industries. At the current state of the art the RFID transponder are predominantly integrated manually or semi-automatically into the product or they are applied on the products. Therefore, the product costs reach a level which makes the technology uneconomical for low-priced mass products. Moreover, the RFID tags are exposed to environmental conditions and therefore not suitable for a long-term usage in most applications. An automatable technique for the integration of RFID tags into injection molded plastic parts was developed. The tool technique and the technical design of the RFID tags were adapted to each other and the essential construction indications were established. Due to the extreme conditions during the injection molding process of thermoplastics (pressure, temperature and shear), the RFID tags had to have a design retaining the functionality after integration into the final product. A carrier module was used to stabilize the geometry of the filigree coils so that even the sensitive NFC capable RFID tags could be integrated under conservation of functionality in glass fiber filled technical plastics like PPS. Furthermore, the carrier module enables an automatic delivery of the tags. With injection molded thermoset materials (lower viscosity of the material, higher temperatures, lower pressure in the cavity), it was possible to integrate and completely encapsulate RFID tags consisting only of the coil and the chip under conservation of its functionality into the plastic parts.

Effects of processing parameters on the real part surface structure of ABS and PC/ABS parts using simulation and experimental data

Siepmann Jens Peter (1)*, Janßen Mirco (1), Wortberg Johannes (1)

(1) University of Duisburg-Essen - Institute of Product Engineering - Northrhine-Westphalia - Germany

For the evaluation of 2D SEM and 3D confocal microscopy measurements of injection molded part surfaces 3D injection molding simulation is used. A design of experiment was implemented to the simulation which was the basis for the surface measurements. The focus was put on the reconstruction of the real process. In addition to that meshing was used to generate a sufficient dimension of the boundary layer. After the simulations were performed, local simulation parameters (shear rate, temperature, viscosity) were related to the 2D and 3D key figures (degree of orientation, roundness of caverns, cavern area, etc.) of the previous investigations. Hereby it could be stated that especially the deformation tendency of the butadiene particles is related to the simulated shear rates. Furthermore, the influence of local temperatures to the relaxation of the butadiene can be detected. Four different positions on the part were investigated and differences caused by geometry can be seen within the simulation results and also the surface measurements. The simulated shrinkage potential was identified to be a good indicator for the residual stresses within the part. It can be shown that higher simulated shrinkage potential effects the etching process and the resulting cavern structures, which are quality defining for the layer system.

Influences of processing and geometry on the 3D surface structure of injection molded parts in context of electroplating of plastics

Siepmann Jens Peter (1)*, Janßen Mirco (1), Wortberg Johannes (1)

(1) University of Duisburg-Essen - Institute of Product Engineering - Northrhine-Westphalia - Germany

By processing technical polymers like acrylonitrile butadiene styrene or polycarbonate/ acrylonitrile butadiene styrene blends with injection molding machines for a following electroplating coating there is, next to the electroplating parameters, the surface of the injection molded part which is responsible for the conjunction of the polymer and the metal and influenced by processing parameters. The influences of processing parameters and mold geometry on the surface structure of injection molded ABS and PC/ABS parts after etching within the electroplating process chain are investigated. For this purpose relevant parameters (screw advance speed, barrel temperature and cooling temperature) are varied in a design of experiment. All parts are prepared for the follow up electroplating by etching the surface with identical parameters (time, temperature). The part surface is measured at different positions using a confocal microscope, which was identified to offer sufficient resolution for micro- and nano-scale depth information of the surface structures. Surface parameters from DIN EN ISO 25178 (height parameters and functions parameters) are investigated to describe the surface properties depending und position on the part and processing parameters. By the connection of already published methods of the investigation of 2D shape and the results of these surface cavern structures and the 3D information findings regarding differences between ABS and PC/ABS were made.

Development of Micro Piezo Actuator System for In-Situ Injection-Embossing Molding Process of Structured Molding Optics.

Hou Chien-Yu (1), Chen Chao-Chang Arthur (1)*, Lai Chang-Shun (1)

(1) National Taiwan University of Science and Technology, Mechanical Engineering Department - Taipei city - Taiwan

This study aims to fabricate an aspherical reflective optical element(AROE) with micro-structure by closed-loop injection molding process. This process combines with the machine signal, molding data acquisition, micro vibratile (MV) and varithermo mold temperature (VMT) methods. In this research, the form error of AROE can be calculate in the macro scale, the groove filling ratio can be obtained in micro scale. From the experimental result, the aspheric form error (Rt) can be lowered to 0.92μm after optimization. Experimental results also show that with different piezo actuator waveform the micro groove filling ratio (GFR) is varied. The GFR of triangle wave is 57.14%, sine wave is 85.71%; and square wave is 87.75%. It is obviously to see that the ROE with MV-VMT is better than that by conventional injection molding process. In control loop system, the best parameters are as following, piezoelectric actuator frequency 200Hz and amplitude 40V. In this study, the best stray light of samples is 5.42%. Further study can focus on form error between aspherical design value and measurement result of AROE in order to augment the axial aspherical core design. Keywords: Aspherical Reflective Optical Element (AROE), Structured Optics, Micro Vibratile Injection Embossing Molding (MV-IEM), Varithermo Mold Temperature (VMT)

Research of Residual Birefringence and Optical Properties Measured by Injection Molding

***Chen Chao-Chang Arthur* [(1)*]*, Lee Jui-Yang* [(1)]*, Yu Chia-Wei* [(1)]

[(1)] National Taiwan University of Science and Technology, Mechanical Engineering Department - Taipei city - Taiwan

White light photoelastic theory is used to design residual birefringence analysis software and also used polariscope to measure the aspheric lens fabricated by injection molding. Moreover, the residual birefringence, modulation transfer function, effective focal length and radius error of aspheric lens have been investigated and discussed. Through the two-stage injection molding experiments, it can achieve the residual birefringence optimization. The main parameters include V/P position, packing pressure and filling velocity. Setting V/P position and two stage filling with lower packing pressure, it can reduce fringe order amount 32.7%. Results of self-assemble MTF measurement shows that the fringe order in image quality is less than 2.76N and MTF50 can be promoted to 13.6%, even MTF10 can be promoted to 40%. On the other hand, the commercial MTF measurement in 110 lp/mm the MTF can be improved to 15.4%. Through this result, residual birefringence is one of the factors which can affect lens image quality. Radius error and residual birefringence can also affect EFL and MTF. The larger residual birefringence causes less EFL error and decreases image quality. A white light photoelastic software can been modified and used to quantify fringe order less than 1N and maximize fringe order error approximately 0.0298N. Finally this study can adapt to measure difference material and geometry product. Future study can focus on high-resolution plastic lens for microscopic phone camera. Keywords: Residual birefringence, White light photoelastic, Injection molding, Aspheric lens, Modulation transfer function

Construction of a pipe support with overmoulded metal insert

***Anger Karsten* (1)*, *Lakemeyer Christoph* (1), *Lochbaum Ella* (1), *Haksteter Mark* (1)**

(1) Hadi-Plast GmbH & Co. KG - Hövelhof - Germany

Pipe supports contribute a big proportion to safe energy in modern buildings. Beside their function as the fastening of the pipeline, they have to insulate it from the environment. Conventional solutions consist of numerous components, of which every component fulfills one function. Because of this, conventional solutions are expensive and uncomfortable to install. The new developed pipe support provides a construction, that combines all functions in one component. The connection to the wall is given by an overmoulded metal insert. During the developing process, a with flame retardants filled polyethylene, that is foamed up with a chemical foaming agent, has been chosen as the material. It fulfills the mechanical requirements as well as the requirements to the insulation and fire prevention. Beside to the material, the focus of the developing process was on the construction of the pipe support. It was developed to the effect that the handling during the installation process is much easier and the pipe is well insulated. Moreover, the costs should decrease by being able to fasten different sizes of pipes with one pipe support. To reach this effect, the pipe support has a complex inner structure and pipe fastening, that promotes a good insulation by minimizing the contact surface between pipe and pipe support and using hollow chambers around the pipe. The pipeline is held by a minimal number of contact points, what leads to a contact surface, that is around 80% lower compared to conventional solutions. Different pipe sizes can be fastened with slats, that work as bending springs to fasten the pipeline. For installation or maintenance of the pipeline, the pipe support has snap-in hooks to open or close it as often as needed without additional tools. Finally, the injection molding tool and the process parameters are developed, whereby the challenge is to build a foam structure in some parts, and solid elements in other parts of the pipe support.

Total Solution from Injection Molding Modeling to Anisotropic Structural Analysis for Tensile Specimens

Kim Changsung Sean (1)*

(1) Corning Inc. - New York - USA

A total solution from Injection Molding Modeling to Anisotropic Structural Analysis has been presented for tensile specimens. A commercial polymer material with 20% weight glass fibers is used for molten fluid of injection molding. Viscoelastic modeling results of fiber orientation show a good agreement with X-ray images. It is noticed that fiber orientation is crucial to accurate prediction of warpage for dual-gated tensile specimen. Fiber orientation model shows a curved shrinkage of 0.68 mm. Dual-gated model gives a slightly shorter length than single-gated one by 80 microns which is identical to the experimental measurement. In addition, a typical feature of stress concentration along the weld line is successfully demonstrated through structural analysis using anisotropic material properties and fiber orientation data from viscoelastic injection molding modeling.

Scalar Parameters of the Anisotropic Rotary Diffusion Model for Fiber Orientation in Long-Fiber Composites

Jin Wang[a], Alex Bakharev[b], Franco Costa[b], and Shishir Ray[b]

[a]*Moldflow R&D Center, Autodesk, Inc., 2353 N. Triphammer Rd, Ithaca, NY, 14850, USA*
[b]*Moldflow R&D Center, Autodesk Australia Pty Ltd, 259-261 Colchester Road, Kilsyth, VIC, 3137, Australia*

jin.wang@autodesk.com

Abstract. The anisotropic rotary diffusion (ARD) model and its modification are widely used in commercial software packages to predict fiber orientation in long-fiber composites. The five scalar phenomenological parameters for the rotary diffusion tensor in the original ARD model are carefully selected by fitting its steady-state solution to experimental orientation data and requiring stable and physical orientation solutions under some typical flow conditions. In this paper, those parameters were determined for a wide range of typical fiber orientations in injection-molded parts. No obvious trend of those parameter values with respect to orientation was observed. Next, the rotary diffusion tensor was simplified so that it contains only two parameters and is independent of the flow. The parameters were determined by fitting to the diagonal components of orientation data. The original five-parameter and simplified two-parameter ARD models were tested in injection-molding simulations, and the fiber orientation predictions from the two models were found to be similar. Further, the two parameters were easily fitted to quartic polynomial functions of the diagonal components of orientation data, which makes the simplified ARD model more practical in engineering applications because it allows a straightforward selection of the orientation model parameters according to the expected orientation levels.

Keywords: Fiber Orientation, Long Fiber, Injection Molding, Anisotropic Rotary Diffusion
PACS: 47.85.md, 81.05.Ni, 83.50.Uv

The anisotropic rotary diffusion (ARD) model provides the general form for fiber interaction term, and the rotary diffusion tensor **C** is constructed from the orientation tensor **A** and the rate-of-deformation tensor **D**, as $\mathbf{C} = b_1 + b_2\mathbf{A} + b_3\mathbf{A}^2 + b_4\mathbf{D}/\|\mathbf{D}\| + b_5\mathbf{D}^2/\|\mathbf{D}\|^2$. The five b_i parameters were first identified over a wide range of typical fiber orientations, but no valid parameters are available for certain orientations and no obvious trend with respect to orientation was observed. The rotary diffusion tensor was further simplified as $\mathbf{C} = b_1 + b_3\mathbf{A}^2$. The valid values of b_1 and b_3 were determined for all typical orientations where $0.6 \le A_{11} \le 0.9$ and $A_{22} > A_{33}$. The 5-parameter and 2-paremeter ARD models, combined with the reduced strain closure (RSC) model, were used in three-dimensional simulations to predict the orientation in an injection-molded long-carbon-fiber composite, and the results are almost identical, as shown in Fig. 1. The predictions also show good agreement with experimental data. The b_1 and b_3 parameters were further fitted to quartic polynomial functions, as shown in Fig. 2. The polynomial functions can be easily used to select the two b_i parameters and so makes the ARD model a more practical engineering tool.

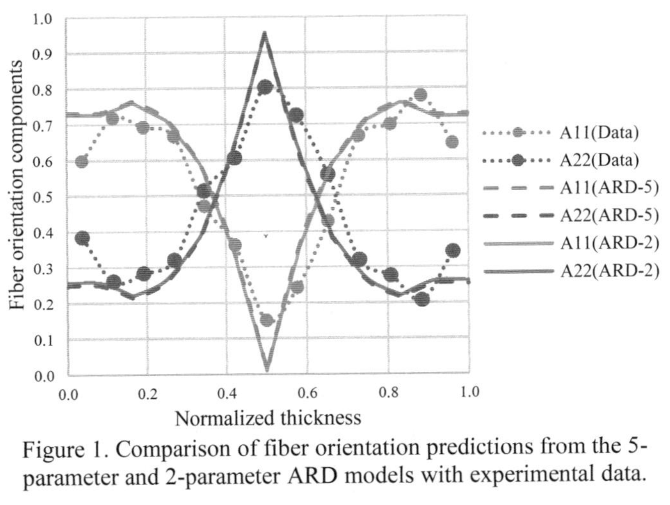

Figure 1. Comparison of fiber orientation predictions from the 5-parameter and 2-parameter ARD models with experimental data.

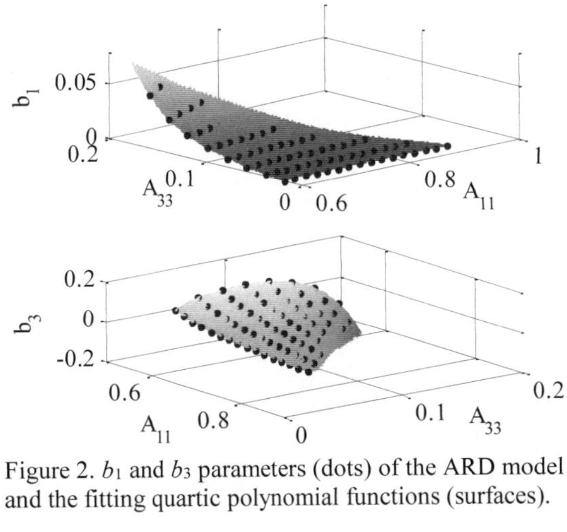

Figure 2. b_1 and b_3 parameters (dots) of the ARD model and the fitting quartic polynomial functions (surfaces).

Achievements and Challenges of Predicting the Mechanical Performance of Weld Lines in Injection-molded Short Fiber-reinforced Thermoplastics

Cruz Camilo [1]***, Baradi Mohamed Besher*** [1]***, Régnier Gilles*** [2]

[1] Robert Bosch GmbH, Corporate Sector Research and Advance Engineering, Plastics Engineering - Renningen - Germany, [2] PIMM, ENSAM, CNRS, CNAM - Paris - France

Weld Lines (WLs) appear frequently in the manufacturing of injection-molded thermoplastic parts and have a direct impact on their mechanical performance. In the literature, we can find an extensive list of papers characterizing the effect of WLs on the mechanical properties of reinforced thermoplastics, but the prediction of the mechanical properties in the vicinity of the WLs is still challenging. The formation of a WL during the injection molding of a short fiber-reinforced thermoplastic can induce locally several morphologies and microstructures, which strongly affect the local loading-dependent mechanical properties at the WL. In our vision, the most important local morphological/microstructural factors contributing to the mechanical performance at the WL are: matrix healing, fiber orientation and molecular orientation. In such context, we propose a two-step simulation strategy for predicting the mechanical properties at the WL: 1) the modeling of the locally induced morphologies/microstructures at the WL, and 2) the correlation of the last ones with the mechanical properties of the composite material. This paper deals basically with the first step of the integrative simulation scheme. We present firstly the current situation of the prediction of the final position of a WL using standard injection molding software and we emphasize on the remaining challenges. In terms of the prediction of the degree of the matrix healing, we propose a physical model based on the reptation theory and we show its numerical implementation using standard commercial software. By comparing the fiber orientation measurements by computed tomography with the fiber orientation predictions at the WL using commercial software, we evidence that the macroscopic fiber orientation models still have room for improvement. Finally, we outlook the modelling of the process-induced molecular orientation and we describe the particularities of its implementation using standard commercial software.

pps proceeding
Symposium: S03 - Process modeling
Oral Presentation

Flow Behavior of Polypropylene Melt in Capillary Dies

Mitsoulis Evan [(1)]*, Luger Hans-Jürgen [(2)], Neunhäuserer Andreas [(3)], Friesenbichler Walter [(3)], Miethlinger Jürgen [(4)]

[(1)] NTUA - Athens - Greece, [(2)] Johannes Kepler Universität Linz - Linz - Austria, [(3)] Department Polymer Engineering and Science, Montanuniversität Leoben - Styria - Austria, [(4)] Institute for Polymer Extrusion and Compounding Johannes Kepler Universität Linz - Linz - Austria

The flow of a highly viscous polypropylene (PP) melt in converging dies is studied both experimentally and computationally. A series of capillary experiments are carried out in order to study pressure drop and temperature rise induced by the polymer flow. Additional rheological measurements using a plate-and-plate rheometer and a high-pressure capillary rheometer were carried out to ascertain the material data necessary for calculations. Slip at the wall was also measured. The flow simulations are performed for an axisymmetric extrusion (conical dies) using the purely viscous Cross model and the integral K-BKZ viscoelastic model. The effect of rheological properties is investigated, and it becomes obvious that the integral K-BKZ model is able to capture the viscoelastic behavior of this PP melt provided that a pressure-dependence of viscosity is properly accounted for.

Morphology Development during polymer processing: comparison with iPP injection molding experimental results

V. Speranza, R.Pantani and G. Titomanlio[a)]

Department of Industrial Engineering, University of Salerno – ITALY

[a)] Corresponding author: gtitomanlio@unisa.it

The effect of temperature, pressure and flow on viscosity, relaxation time, nucleation density, crystallites growth rate, the distributions of deformation rate and cooling time during the process determine the distribution of the molecular stretch at the solidification which give rise to the morphology distribution in the final object. The start of crystallization has a dramatic effect on viscosity, relaxation time and, consequently, on the evolution of molecular stretch which in its turn affects growth and nucleation rates.

Each of the effects mentioned above was experimentally analyzed and described by a model for the iPP considered in this work. The ensemble of all specific models becomes a model for the morphology evolution during polymer processing and its application to the injection molding process was numerically developed into the UNISA code. The code was previously validated for predictions of evolutions during the process of the pressure distributions along the flow path and for the crystallinity at the sample surface.

In this work, we propose a criterion for fibrillar crystallization. The criterion refers to local values of both the work done and the molecular stretch; at the crystallization they have to be higher than critical values and, in addition, the whole amount of work has to be done while the molecular stretch is above a critical value. thickness of both the shear and spherulite layers into injection molded samples are satisfactorily described by the criterion.

INTRODUCTION

The description of morphology evolution during polymer processing is very complex and it is determined by the effect of temperature and pressure on viscosity, relaxation time, crystalline nuclei density and crystalline growth rate and, furthermore, by the effect of crystallinity on viscosity and relaxation time and of the flow on nuclei density and growth rate. The constitutive equation describing all these effects requires wide and delicate experimental campaigns. Obviously, a model of morphology evolution during each polymer processing operation is obtained by coupling the material constitutive equations with heat and momentum balance equations describing the specific polymer processing .

It is well known that crystalline polymers may crystallize into both fibrillar and spherulitic morphology and that the molecular stretch is high where the morphology is fibrillar and that the morphology is spherulitic where the molecular stretch is sufficiently small [1]–[5].

As a result final morphology distribution along the thickness direction of semi-crystalline injection molded objects is characterized by a sequence of different morphological layers, which from the surface to the mid-plane are: a thin skin layer; a fibrillar shear layer; a spherulitic layer through the mid-plane [6]–[9].

Several authors suggested that the achievement of a critical work at crystallization is related to the fibrillar crystallization, as long as the shear rate is larger than a limiting value and it is related also to a critical viscous work [1][5]. More recently, an additional attempt was made to clarify the criterion which induces the formation of the fibrillar structures. In particular a series of experiments were carried by continuously shearing the material, an PP (T30G), in a Linkam Cell within a narrow temperature range around 140°C [1]. It was found that the material has a limiting shear rate below which the crystallization takes place with spherulitic character; above that limiting shear rate the crystallization took place with fibrillar character when the shearing was held for a time sufficient to achieve about the same viscous work, of 20Mpa.

In this work the model for fibrillar crystallization mentioned above is adopted with the following additional simultaneity specification: the whole amount of work has to be done while the molecular stretch is already above the critical value; furthermore, if the molecular stretch is let to relax below that critical value, all previous work has to be cancelled in the accounting for the critical work. The injection molding process seems specifically designed to verify the simultaneity specification clarified above, since soon after filling the polymer not yet solidified relaxes considerable part of its stretch and afterwards, during the packing step, it undergoes further stretching.

Literature Experimental data of thickness of skin, shear and spherulitic layers along the cross section of 4 injection molded samples of iPP obtained changing flow rate, mold surface temperature and holding pressure are analyzed in this work, on the basis of the criterion reported above with the simultaneity specification.

The calculations for the viscous work and molecular stretch evolutions required for the criterion application have been done with the UNISA code of the injection molding process, developed at the University of Salerno on the basis of constitutive equation experimentally verified on the polymer, the iPP adopted for the four experiments mentioned above. The model implemented in the UNISA code and the values of the parameters of the constitutive equations of the iPP T30G are reported in [10].

MATERIAL AND INJECTION MOLDING

Four injection molding tests carried out by using an iPP (T30G, Basell) and already reported in the literature [7], [9] are considered in this work. The material was injected into a rectangular cavity of 120mm x 30mm x 2 mm. The evolution of the pressure distribution was recorded by five transducers (P0-P4) located along the flow path: P0 in the injection chamber, P1 just before the gate and P2,P3,P4 in the cavity (at 15, 60 and 105mm downstream from the gate);. Further details on geometry and adopted experimental procedure are available in [7], [9]. Main processing conditions adopted for the four tests are reported in Table 1.

Tab. 1. Processing conditions.

	Standard	Slow	High T	High P
Flow rate [cm^3/s]	15	**5**	15	15
Mold temperature [K]	303	303	**343**	303
Holding pressure [bar]	400	400	400	**700**

The effect of molding conditions on the multilayer morphology of injection molded part is evidenced in Fig. 1, where pictures of the samples taken in crossed polarized light are shown. The pictures refer to slices taken in position P3 and report just one half of the sample thickness, with the skin layer at the top, and the midplane at the bottom of each image. In all cases, the flow is parallel to the plane of the slices. The skin layer is evident in all cases except for the Slow condition. The shear layer appears as the darkest part of each picture, the spherulitic layer is instead the lightest one. The effect of the processing condition on the relative thickness of all the layers is quite relevant.

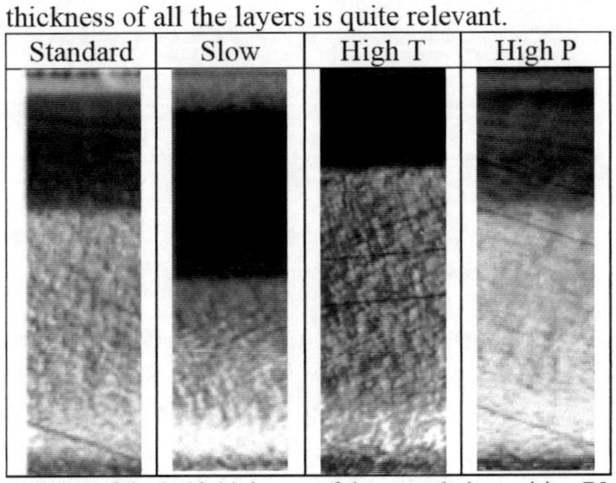

Fig. 1. Optical images, in cross polarized light, of the half thickness of the sample in position P3 for all the conditions reported in Table 1. The skin layer at the top, and the midplane at the bottom of the images.

Injection molding simulation

In order to simulate rheology, crystallinity and molecular stretch evolutions during the injection molding process, all the constitutive equations had to be extrapolated well beyond the experimental ranges, within which each equation had been identified. All the constitutive equations of that iPP resin (T30G) were recently revised [11] on the basis of data published after the publication of the paper which analyzed those injection moulding tests for the first time [7], [9].

The field equations adopted for the numerical simulation are reported in [10] and the capability of the overall model of reproducing main trends of experimental results for both pressure evolutions and spherulite dimension distribution on the cross section was already verified in a previous paper [7], [9].

The evolution of molecular stretch of the sub-chain ensemble was analyzed through the evolution of the molecular conformation tensor $\underset{\sim}{A}$ defined as

$$\underline{\underline{A}} = 3\,\frac{(\langle RR\rangle - \langle RR\rangle_0)}{\langle R_0^2\rangle} \qquad (1)$$

where $\underline{R}$ represents the end to end vector of a sub-chain. The evolution of the conformation tensor was described by a Covariant Maxwell-type equation with a single dominant relaxation time [7], [12], [13]. In this work, the difference, Δ, between the two main eigenvalues of the tensor $\underline{\underline{A}}$ was taken as an adequate measure of the molecular stretch.

The simultaneity specification and the injection molding process

The *simultaneity* specification, which clarifies that the work performed before the stretch reaches its critical value is not relevant to the purpose of achieving fibrillar crystallization, is physically grounded. Indeed, as it happens also during processing conditions (for instance the injection molding process), stretched polymer melt may have the time to (partially) relax before being stretched again. If the relaxation were complete, it is obvious that the whole previous history does not have any effect on what happens next. The *simultaneity* specification, stated above, starting from such a situations, defines a general approach still based on two critical values of the dissipated viscous work, W_c, and molecular stretch Δ_c.

Numerical calculations for the evolution of the viscous work integral and of the molecular stretch during the Standard test at several distances from the surface are reported in fig. 2a and fig. 2b respectively.

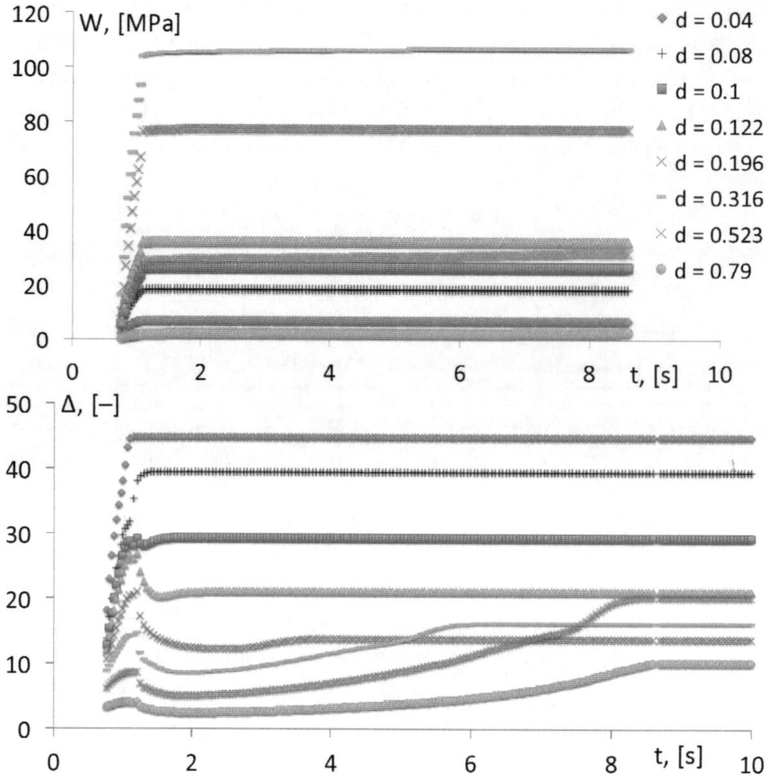

Fig. 2. Viscous work integral (a) and molecular stretch (b) vs time for several distances from the mold surface for the Standard test.

It is clear from the fig. 2b that the stretch grows during filling except where meanwhile there is a solidification; this is what happens for distances from the mold surface smaller than 0.1mm. For distances from the mold surfaces equal or larger than 0.1 mm, at the end of the filling step the polymer is not yet solidified, thus it undergoes some stretch relaxation and, afterwards, the stretch increases again by effect of the packing flow. All layers finally solidify at times increasing with the distance from the surface and solidification takes place by effect of crystallization which develops either during the filling step or during the relaxation or during the packing flow or also after the packing. The integral of the viscous work as shown in fig. 2a grows especially during the filling stage and at the very beginning of the packing stage when the flow is more intense. During the packing, only for the layers not yet solidified, there is a small increase of the viscous work due to the weak intensity of the flow.

The integral of the viscous work and the final stretch distributions of the Standard sample both up to the end of the filling steps (lines) and at the end of the injection molding process (symbols) are plotted in fig. 3 versus the distance from the surface. Moreover, the thickness of the shear layer, as observed in the optical image of Standard sample in fig.1, is indicated in fig. 3.

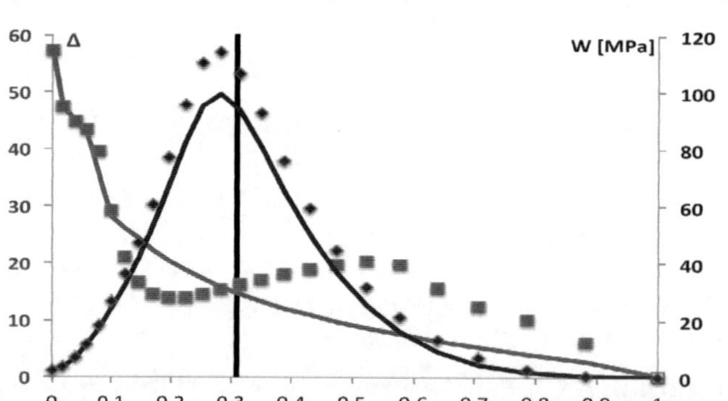

Fig. 3. Integral of the viscous power density (right axis) and molecular stretch parameter Δ (left axis) in position P3, as function of the distance from the surface as calculated for the "Standard" condition. The distributions of both variables at the end of the filling stage (lines) and at the end of the process (symbols) are reported. The experimental thickness of the shear layer is also reported as a vertical black line.

The two viscous work curves, reported in fig 3, have regular shapes: the viscous work increases from the surface toward a maximum (reached at about 0.3 mm) and then it decreases toward zero, at the mid-plane. The maximum viscous work is about two orders of magnitude larger than the value at the surface. The viscous work, being dissipated, grows continuously with time, thus, the viscous work at the end of the process is everywhere equal or larger (due to the packing flow) than the viscous work at the end of filling. However, because the packing flow is very slow, the difference between the two viscous work curves is quite small except in a narrow zone close to the maximum.

As can be verified from the work and stretch final distributions reported in fig. 3 for the standard sample, whatever reasonable choice would be made for the two critical values, W_c and D_c, without applying the simultancity specification, a reasonable comparison with the thicknesses and the positions of the shear layers experimentally observed in the position P3 can not be obtained. Thus, at the best, the achievement at the solidification of the two critical values, W_c and D_c, without the simultaneity specification, could be considered a necessary criterion for the crystallization into fibrillar morphology of the polymer, whereas a sufficient criterion needs also the simultaneity specification. Similar conclusion are obtained applying the same analysis to the other molding conditions reported in tab. 1.

If both the final viscous work and the final stretch are well above their critical values (W_c and D_c), the *simultaneity* specification denies the possibility to have fibrillar morphology if, after the stretch had reached its critical value Δ_c, only a quantity of work smaller than the critical one was spent, although the viscous work spent in that position during the whole process exceeded W_c. Similar situations take place very far from the mold surface, where the stretch undergoes a relevant increase by effect of the packing flow, whereas the viscous work, already grown during the filling step, does not undergo enough additional increase after the stretch reached its critical value. This is what happens during the standard test, considered in fig 3, for instance at d=0.52 mm from the surface where most of the viscous work is accumulated during the filling step with only a moderate increase of the stretch (see for instance fig. 2a and fig 2b for d=0.52) and, on the contrary, during the packing step the stretch undergoes a relevant increase with only a marginal increase of the viscous work (the difference between the dotted and the continuous curve of fig. 3 at d=0.52), not sufficiently to verify the *simultaneity* specification if the critical work is larger than about 4MPa. Vice versa, close to the wall, with final values of both the stretch and the integral of work similar to those of d=0.52 (d=about 0.1 mm in fig.3) the simultaneity specification is satisfied because both the molecular stretch and the viscous work grow simultaneously during a filling followed by a fast solidification, which assures that most of the work is done on the melt only when the critical stretch, Δ_c. has been already reached.

Comparison with experimental results

The cross section morphology of injection molded samples tipically reported includes three different layers, whose thickness is determined by the processing conditions: from sample surface to its mid-plane the skin layer, the shear (fibrillary) layer, the spherulitic layer. Thus the correct identification of the shear layer borders indentifies also the correct thickness of the other two layers, namely the complete thickness morphology.

Since the molecular stretch at the sample wall is typically very large, see for instance fig. 3, certainly larger than any reasonable value of Δ_c, the thickness of the skin layer is identified by the critical work W_c. On the contrary, it has to be pointed out that the value of Δ_c plays a relevant role on the other side of the shear layer, far from the wall. Far from the wall, after filling, a relevant fraction of the molecular stretch relaxes, and only a fraction of the work done during the packing satisfies the *simultaneity* with an adequate stretch; thus the value for Δ_c can substantially affect the internal border of the shear layer, the one closer to the midplane .

As a consequence, the critical values of W_c and Δ_c were determined as follows: the value of W_c was identified by comparison with the thickness of the skin layers; the value of Δ_c was determined by comparison with the internal borders of the shear layers.

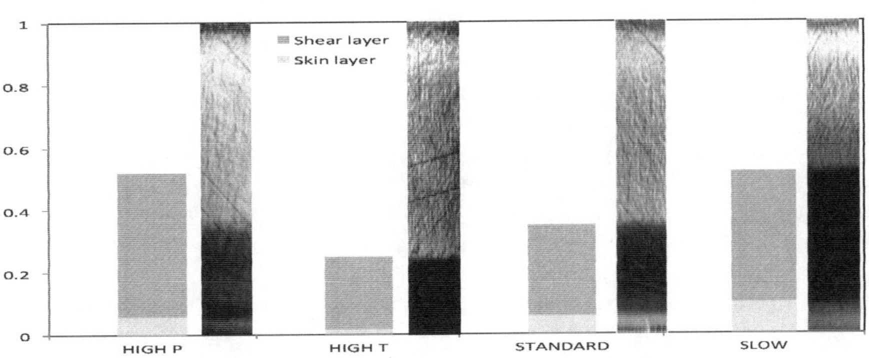

Fig. 4. Comparison between experiments and the results of a simulation conducted by using $\Delta_c = 7$ and $W_c = 10$ MPa.

The results of simulations carried out by using $\Delta_c = 7$ and $W_c = 10$ MPa are reported in fig. 4 as colored bars which represent the thickness of the skin and of the shear layers. In order to assure an easy comparison with experimental results, the images taken by a polarized optical microscope (already show in fig.1), are reported also in fig. 4 on the right of the thicknesses predicted by the simulation with the values of Δ_c and W_c just mentioned.

It can be noticed that the thicknesses of the skin and shear layers are correctly described for all the samples except for the shear layer thickness of the sample molded with holding pressure of 70 MPa (High P), for which the thickness is overestimated by the simulation. The reasons of this discrepancy can be found in the effect of pressure on relaxation which obviously becomes more significant when the pressure levels increase. A slightly smaller dependence on the relaxation time upon pressure can certainly be a cause for a thinner shear layer for sample obtained with higher holding pressure and therefore would give rise to a much better comparison also for that sample

CONCLUSIONS

In this work, a novel criterion is adopted to simulate the occurrence of oriented, fibrillar crystals in semicrystalline polymers. The criterion is based on the achievement of a critical viscous work, W_c, and of a critical molecular stretch , Δ_c, provided the following *simultaneity* specification is satisfied: the viscous work is spent only when the molecular stretch is already higher than its critical value.

The criterion is applied to injection molded samples of a well characterized polypropylene in four different molding conditions. In the injection molding process, the fibrillar layer (also called shear layer) is found at intermediate distances from the sample skin (between the skin and the spherulitic layers). Thus the geometrical limits of the shear layer on the cross section identify the thickness of all three layers on the injection moulded samples.

In order to predict the shear layer thickness, the injection molding tests were simulated by a software developed at the University of Salerno which is able to consider all the effects needed to describe the phenomena involved: the molecular stretch, the crystallinity evolution, and the mutual effects of crystallinity and flow-fields.

Adopting the *simultaneity* specification with appropriate values of Δ_c and W_c, a satisfactory description of both the skin and the shear layer thickness for all the conditions in which the holding pressure is 40MPa was reached. Some discrepancy in the thickness of the shear layer was found for the sample molded with a higher holding pressure (70MPa). In this case, the shear layer was predicted to be thicker than the real one. The discrepancy is consistent with the fact that the effect of pressure on relaxation time is slightly over-estimated. This is an aspect to be improved for the simulation of this specific condition.

REFERENCES

[1] R. Pantani, V. Nappo, F. De Santis, and G. Titomanlio, "Fibrillar morphology in shear-induced crystallization of polypropylene," *Macromol. Mater. Eng.*, vol. 299, no. 12, pp. 1465–1473, 2014.

[2] O. O. Mykhaylyk, P. Chambon, C. Impradice, J. P. A. Fairclough, N. J. Terrill, and A. J. Ryan, "Control of Structural Morphology in Shear-Induced Crystallization of Polymers," *Macromolecules*, vol. 43, no. 5, pp. 2389–2405, Mar. 2010.

[3] O. O. Mykhaylyk, P. Chambon, R. S. Graham, J. P. A. Fairclough, P. D. Olmsted, and A. J. Ryan, "The Specific Work of Flow as a Criterion for Orientation in Polymer Crystallization," *Macromolecules*, vol. 41, no. 6, pp. 1901–1904, Mar. 2008.

[4] J. F. Vega, D. G. Hristova, and G. W. M. Peters, "Flow-induced crystallization regimes and rheology of isotactic polypropylene," *J. Therm. Anal. Calorim.*, vol. 98, no. 3, pp. 655–666, Dec. 2009.

[5] H. Fang, Y. Zhang, J. Bai, and Z. Wang, "Shear-Induced Nucleation and Morphological Evolution for Bimodal Long Chain Branched Polylactide," *Macromolecules*, vol. 46, no. 16, pp. 6555–6565, Aug. 2013.

[6] P. C. Roozemond, M. van Drongelen, and G. W. M. Peters, "Modeling Flow-Induced Crystallization," in *Advances in Polymer Science*, Springer, Cham, 2016, pp. 243–294.

[7] R. Pantani, I. Coccorullo, V. Speranza, and G. Titomanlio, "Modeling of morphology evolution in the injection molding process of thermoplastic polymers," *Prog. Polym. Sci.*, vol. 30, no. 12, pp. 1185–1222, Dec. 2005.

[8] J. Karger-Kocsis and I. Csikai, "Skin-Core morphology and failure of injection-molded specimens of impact-modified polypropylene blends," *Polym. Eng. Sci.*, vol. 27, no. 4, pp. 241–253, Feb. 1987.

[9] R. Pantani, I. Coccorullo, V. Speranza, and G. Titomanlio, "Morphology evolution during injection molding: Effect of packing pressure," *Polymer (Guildf).*, vol. 48, no. 9, pp. 2778–2790, Apr. 2007.

[10] R. Pantani, V. Speranza, and G. Titomanlio, "Thirty Years of Modeling of Injection Molding . A Brief Review of the Contribution of UNISA Code to the Field," *Int. Polym. Process.*, vol. XXXI, no. 5, pp. 655–663, 2016.

[11] R. Pantani, V. Speranza, and G. Titomanlio, "Effect of Flow-Induced Crystallization on the Description of Morphology in Injection Molded Parts," *Eur. Polym. J. Submitt.*, 2017.

[12] G. Marrucci and G. Ianniruberto, "Constitutive equations for polymeric solutions close to the overlap concentration," *Chem. Eng. Sci.*, vol. 56, no. 19, pp. 5539–5544, Oct. 2001.

[13] R. G. Larson, *Constitutive equations for polymer melts and solutions*. Boston: Butterworths, 1988.

Multiscale simulation of polymer melt spinning process --- Analysis on states of well-entangled polymer chains ---

Taniguchi Takashi **(1)*, *Sato Takeshi*** **(1)**

(1) Dept. of Chemical Engineering, Kyoto University - Kyoto - Japan

We successfully applied a multiscale simulation method to a spinning process of a well-entangled polymer melt. In the spinning process, the macroscopic dynamics of the polymer melt and microscopic states of polymer chains are tightly connected through the strain rate and stress fields. A macroscopic model for the spinning process is constructed by using the standard and simplified 1D approach, thus ignoring the complex flow conditions near the die exit. In our multiscale simulation method, to evaluate the flow history of the stress, we employ a particle method in the Lagrangian manner at the macroscopic level and use the slip-link model to address entangled polymer dynamics at the microscopic level. Therefore, we can take into account the history effect of polymer melts and quantitatively evaluate the polymer dynamics, such as the entanglements. From our multiscale simulations, the number of entanglements decreases with flow down the spinning line because of the convective constraint release (CCR) effect. Moreover, we found that the number of entanglements generally decreases in the middle section of a polymer chain. This result shows that our multiscale simulation makes it possible to design a polymer melt that has desirable physical properties for producing a sophisticated fiber.

Simulation of the flow behavior in injection molding of polymer bonded ferrite microparticles based on rheological and thermal analysis

Tamara Florian[1], Gerhard Ziegmann[1, 2]

[1] Institute of Polymer Materials and Plastics Engineering, Clausthal University of Technology, Clausthal-Zellerfeld, Germany

[2]Clausthal Centre of Material Technology, Clausthal University of Technology, Clausthal-Zellerfeld, Germany

ABSTRACT

Injection molding is an important technology for fabrication of many industrial products, since it offers the possibility to produce complex components very efficiently with a small amount of finishing work. Therefore, this technology is quite attractive in mass production of highly filled, functional plastic parts e.g. polymer-bonded magnets (PBMs), which serve as a clever substitution for many metallic components. Due to the high particle content, the processability in injection molding is a critical aspect. Therefore, simulation is indispensable to assure the quality in the production of functional high-end products.

In this study, the rheological and thermal properties of polyphenylensulfide (PPS) filled with milled ferrite grains (150μm) up to particle contents of 50, 60, 65vol.-% are investigated. The viscosity-, pvT- and thermal conductivity measurements are carried out at a wide range of shear rate, temperature and pressure. The viscosity of the PPS melt increases more than ten times by adding 65-vol.% of particles due to the particle-particle and polymer-particle interactions. The specific volume of PPS increases with temperature but decreases with pressure and filler content. Therefore, the filler leads to a negligible shrinkage behavior and thus – complete mold filling– precondition to a good dimensional accuracy. Thermal conductivity of the filled material enhances remarkably.

The determined material properties (viscosity and pvT) are modelled according to Cross-WLF equation and using the Tait-pvT-Model, respectively. The modeled material properties serve as basis for simulation of spiral flow test with commercial software Moldflow®. Comparison of injection molded spiral flow test parts at different processing parameters show a good accordance to the simulation results.

1 INTRODUCTION

The functionalization of polymeric materials by particle modification offers a number of interesting new possibilities in the field of research and for industrial applications [1, 2, 3, 4, 5]. A major advantage of these kinds of composites is that the property profile can be tailored to the special requirements for different applications due to the characteristics and volume fraction of particles and polymer [6]. A significant advantage of these composites is that they can be processed with the high efficient manufacturing technologies of the plastics industry e. g. injection molding. This enables a cost-effective production of complex components with tailored material properties, which can be used without additional finishing work. In the development of these new composites, it must be noted that the use of the particles affects not only the property profile of the final product, but also the processing properties of the polymer melt. Due to the particle modification, the flow and solidification behavior is influenced significantly. The simulation of the flow behavior of particle-modified systems is a challenge for research because of the complex fluid dynamics caused by the particle-particle and polymer-particle interactions. However, knowledge of this behavior is necessary for the production of complex functional parts by injection molding processes in industrial applications. The simulation software Autodesk MoldFlow® was developed for the prediction of the processing behavior of polymers in industrial injection molding applications. It allows the simulation of the flow behavior of polymer melts. In order to use this tool, the present multi-phase material must be reduced to a single-phase material model. For standard materials, a database with all necessary material properties e.g. viscosity, pvt-behavior, thermal conductivity, specific heat capacity is available in the simulation software. For proprietary developed composites, a material-specific database must be determined. Thus, rheological and thermal properties of polymer-bonded composites have to be analyzed and modelled to provide a material-specific database for simulation and injection molding, which has not been investigated in literature up to now.

2 THEORY

In recent years, varieties of phenomenological and empirical models have been developed to describe the flow behavior of polymeric melts a function of temperature and shear rate. Based on the well-known Carreau-model [7, 8] several enhancements were carried out by Williams et. al. [9] and Cross [10] to describe viscosity η (T, p) of polymer melts:

$$\eta(\mathrm{T},\dot{\gamma}) = \frac{\eta_0}{1+\left(\frac{\eta_0 \cdot \dot{\gamma}}{\tau^*}\right)^{(1-n)}} \tag{1}$$

with

$$\eta_0 = D_1 \cdot \exp\left(\frac{-A_1 \cdot (T - D_2)}{A_2 + (T - D_2)}\right) \tag{2}$$

This approach for describing the temperature and shear rate dependency of viscosity is implemented in the Autodesk's Moldflow software.

The zero shear viscosity η_0 describes the Newtonian plateau in which the viscosity approaches a constant at very low shear rates. Furthermore, $\dot{\gamma}$ is the shear rate (1/s) , τ* is the critical stress level at the transition between Newtonian and the pseudoplastic behavior (Pa), n is the power law index in the high shear rate regime. D_1, A_1 and A_2 are data-fitted coefficients and D_2 is the reference temperature without considering the pressure dependency.

The first pvT model for polymeric materials was developed by Spencer and Gilmore (1949) by modification of the van der Waal equation [11]. The experimental data correlate with the property profile of the polymer melt in a temperature range well above the glass transition temperature. Among the models, the 2-domain Tait-pvT model is the most widely used one possibility for the description of the pressure-volume-temperature (pvT) behavior of polymeric materials. It describes the specific volume V(p,T) as a function of temperature T and pressure p for polymers.

$$v(p,T) = v_0(p,T) \cdot (1 - 0.0894 \cdot \ln[1 + p / B(T)]) + v_t(p,T) \tag{3}$$

The temperature region above the glass transition temperature ($T>T_t$) can be described by modification of equation 3 with:

$$v_0 = b_{1m} + b_{2m}(T - b_5) \tag{4}$$

$$B(T) = b_{3m} e^{-b_{4m}(T-b_5)} \tag{5}$$

$$v_t(p,T) = 0 \tag{6}$$

to

$$v(p,T) = \left(b_{1m} + \left(b_{2m} \times (T - b_5)\right)\right) \times \left(1 - \left(0.0894 \times \ln\left(1 + \left(\frac{p}{b_{3m}\exp(b_{4m} \times T - b_5)}\right)\right)\right)\right) \tag{7}$$

For the description of the temperature region below glass transition temperature ($T<T_t$) can be by modification of equation 3 with:

$$v_0 = b_{1s} + b_{2s}(T - b_5) \tag{8}$$

$$B(T) = b_{3s} e^{-b_{4s}(T-b_5)} \tag{9}$$

$$v_t(p,T) = b_7 \exp((b_8(T - b_5)) - (b_9 p)) \tag{10}$$

to

$$v(p,T) = \left(b_{1s} + \left(b_{2s} \times (T - b_5)\right)\right) \times \left(1 - \left(0.0894 \times \ln\left(1 + \left(\frac{p}{b_{3s} \times \exp(b_{42} \times T - b_5)}\right)\right)\right)\right) + b_7\exp\left(\left(b_8 \times (T - b_5)\right) - (b_9 \times T)\right) \tag{11}$$

The pressure dependence of T_t can be described by the following equation:

$$T_t = b_5 + b_6 \times p \tag{12}$$

The 13 parameters can be divided into three groups. The parameters b_{1m}-b_{4m} describe the curve progression in the region of the melt state of the polymer, b_{1s}-b_{4s} in the region of the solid state. The crystalline transition is described by b_5-b_7. B_5 is the transition of low pressure, b_6 is the slope of the transition change with pressure and b_7-b_9 specify the shape of the crystalline transition (Figure 1).

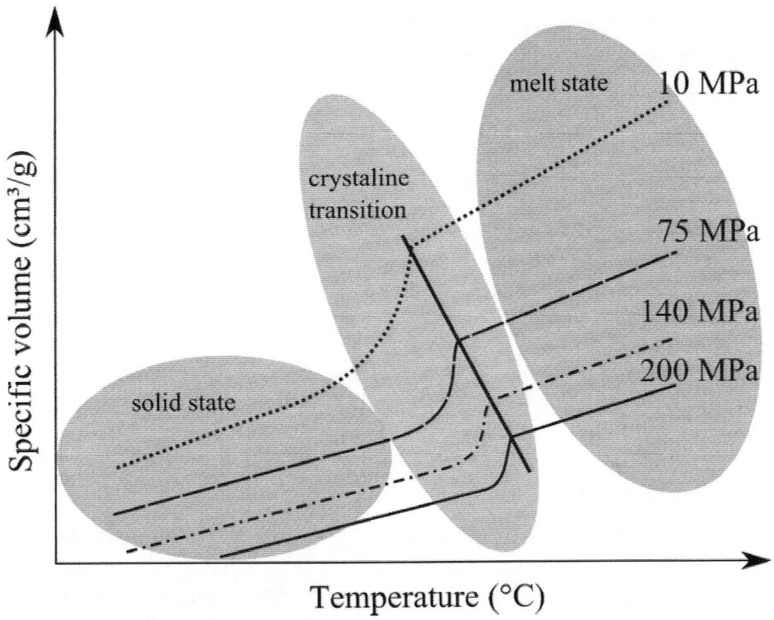

Figure 1. Schematic diagram of a pvT-diagram of a semi crystalline polymer.

The physical meaning of the parameters used in the equations 3-12 is summarized in Table 1 [12].

Table 1. Physical meaning of the parameters used in the 2-domain Tait-pvt model [11,12].

Parameter	Physical meaning	Parameter	Physical meaning
b_{1m}	Intercept of melt-state v-T line at p=0	b_{4s}	Pressure sensitivity below T_t (slope change)
b_{2m}	Slope of melt-state v-T line at p=0	b_5	Glass transition temperature at p=0
b_{3m}	Pressure sensitivity above T_t (line drop)	b_6	Rate of change of T_t with pressure
b_{4m}	Pressure sensitivity above T_t (slope change)	b_7	Radius of curvature below T_t
b_{1s}	Intercept of solid-state v_{sp}-T line at p=0	b_8	Radius of curvature below T_t
b_{2s}	Slope of solid-state v-T line at p=0	b_9	Change of curvature radius with pressure below T_t
b_{3s}	Pressure sensitivity below T_t (line drop)		

3 MATERIALS AND METHODS

As polymeric matrix, a high temperature thermoplastic, semi-crystalline polyphenylene sulfide (Fortron 0205, Ticona, Germany) with a specific density of 1.35 g/cm³ was chosen.

Irregular shaped iron oxide based magnetic metal powder (FSM150 supplied by Ferroxcube, Elmshorn, Germany) with a particle size of 150µm and a bulk density of 1,9 g/cm³ was selected as filler particles.

A summary of the properties of the filler and polymer can be found in Table 2.

Table 2. Material properties.

	Thermoplastic		Filler
Manufacturer	Ticona	Manufacturer	Ferroxcube
Designation	Fortron 0205	Designation	FSM 150
Water absorption in %	0.02	Water absorption in %	<0.5
Processing temperature in °C	290 - 320	Particle size in µm	50
Mechanical properties Young's modulus in MPa Yield strain in %	 4000 2	Chemical composition	Fe_2O_3 71.0 ± 3% MnO 20.7 ± 2% ZnO 8.3 ± 2%

Polymer and microparticles were compounded for 20 min at 290 °C and 40 rpm using a co-rotating kneader (PolyLab Rheomix 600p, ThermoHaake, Thermo Fisher Scientific, Karlsruhe, Germany). As filler contents 50, 60, 65 vol.% were chosen. The material properties Viscosity, pvt-behavior (p=pressure, v=specific volume, T=temperature) and thermal conductivity of the composites were characterized using high shear capillary rheometer (Rheograph75, Goettfert Werkstoff-Pruefmaschinen GmbH, Germany). Viscosity was determined from apparent viscosity using Bagley- and Rabinowitsch-Weissenberg corrections. PvT-behavior and thermal conductivity were analyzed in script-controlled methods. Thermal conductivity was measured at a temperature range from 50-200°C in steps of 25 °C by heat flow meter method (TCA 200-LT-A, NETZSCH-Gerätebau GmbH, Selb, Germany). The test sample is placed between two heated surfaces controlled at different temperatures with a heat flow from the hot to the cold surface. The thermal resistance of the interfaces between the sample and adjacent surfaces is reduced by applying a coupling agent, e.g., silicone heat sink compound [13]. Autodesk® MoldFlow® (Insight) and OriginLab® software were used to perform simulation and data fitting.

The composites were granulated and further injection molded in flow spirals according to ASTM D3123-09 using a hydraulic injection molding machine (Arburg Allrounder 220S, Arburg, Lossburg, Germany).

4 RESULTS AND DISCUSSION

Figure 2 shows the influence of the filler content on the viscosity at a constant temperature of 320 °C. The measurement was carried out at a shear rate range of 10 to 10^4 1 / s. Based on the curve, two effects can be determined. On one hand, the viscosity increases more than ten times by adding 65 vol% of particles. On the other hand, the composites show a much more pronounced shear thinning behavior. Similar behavior was measured on the complex viscosity of PP+FeSi composites at filler content up to 70 vol.-% [14].

The viscosity behavior was fitted using Cross-WLF equation (equation 1-2) and OriginLab® software. The parameters are mentioned exemplary for PPS FSM 150-60 in Table 3. The accuracy of the viscosity fitting was 99.97%.

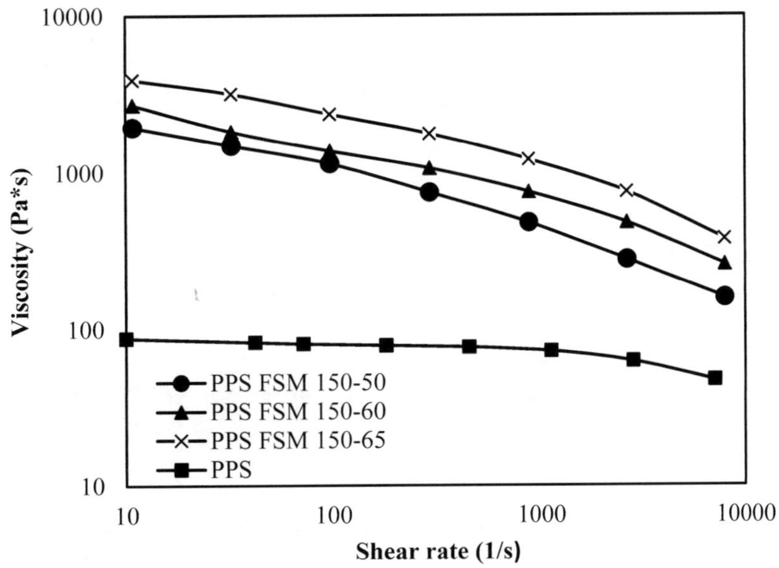

Figure 2. Effect of FSM filler content and shear rate viscosity of PPS and PPS + 65 vol.-% 150 FSM composite.

Table 3. Cross-WLF model parameters for PPS FSM 150-60.

Cross-WLF parameter	PPS FSM 150-60
D_1	$3.17*10^{12}$
D_2	273.39
A_1	171.75
A_2	$2.07*10^{3}$
η	0.26
τ*	$1.23*10^{5}$

Figure 4 shows pvT (p=pressure, v=specific volume, T=temperature) behavior of PPS and PPS FSM 150-60 composite. Specific volume of PPS decreases with increasing pressure and decreasing temperature. Considering the pressure-dependent volume change over the temperature of highly filled composites using the example of PPS FSM 150-60, it can be seen that the volume increase is significantly reduced compared to the unmodified PPS. During the temperature range of 30-330 °C the volume of pure PPS increases by approx. 150 mm³. In the same temperature range, only a rise of approx. 15 mm³ can be measured with the composite material. The reason for this is the lower thermal expansion coefficient of the metallic filler, which can be neglected compared to the thermoplastic matrix. In melting temperature range of PPS (approx. 285 °C) the pvT-behavior of PPS is characterized by declining the specific volume compared to PPS FSM 150-60 composite, caused by change of the free volume due to the crystallization characteristics of PPS. Since specific density is defined as the inverse specific volume, specific density of PPS and PPS FSM 150-60 composite can be directly calculated from Figure 3.

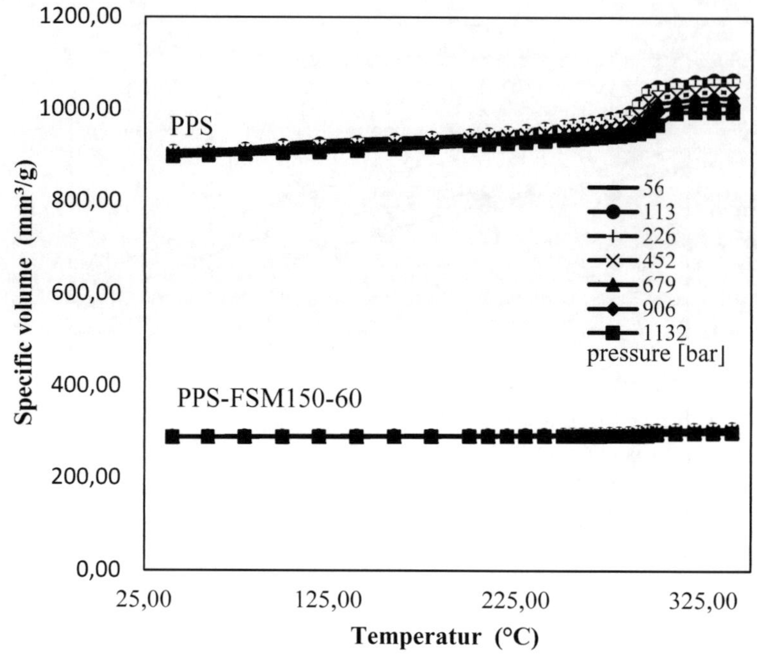

Figure 3. Effect of metal particle on pvT-behavior of PPS.

The pvT-behavior of PPS FSM 150-60 composite in Figure 5 was fitted using 2-domain Tait-pvT model described in Equations 3 - 12. The parameters of the Tait-pvT model were determined by global fitting in OriginLab® software and listed in Table 4 for PPS FSM 150-60 composite. The fitting of the pvT-behavior was performed with an accuracy of 99.4%.

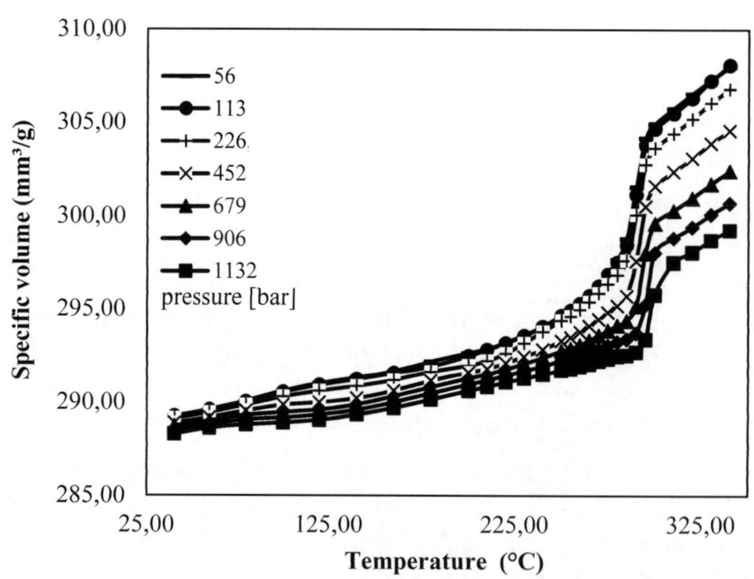

Figure 4. pvT-behavior of PPS FSM 150-60 composite.

Table 4. Tait-pvT parameter of PPS FSM 150-60.

Tait-pvT parameter	PPS FSM 150-60	Tait-pvT parameter	PPS FSM 150-60
b_{1m}	0.00095001	b_{4s}	$1.00*10^{-2}$
b_{2m}	$8.5405*10^{-8}$	b_5	362.2
b_{3m}	$5.99*10^{8}$	b_6	$5.82*10^{-8}$
b_{4m}	$1.00*10^{2}$	b_7	$4.05*10^{-8}$
b_{1s}	0.00095001	b_8	0.06
b_{2s}	$8.5405*10^{-08}$	b_9	0
b_{3s}	$5.99*10^{8}$		

The influence of the filler content on the thermal conductivity example is shown in Table 5. The comparison of the unfilled and filled material (PPS FSM 150-60) shows a significant increase in the thermal conductivity from about 0.21 W / (mK) to 0.67 W / (mK) at 50 ° C. by 60 vol.-% filler addition.

Table 5. Thermal conductivity of PPS and PPS FSM 150-60 in temperature range of 30-200 °C.

Temperature	50	75	125	150	175	200
PPS thermal conductivity in W/mK	0.22	0.21	0.21	0.22	0.23	0.23
PPS FSM 150-60 thermal conductivity in W/mK	0.68	0.71	0.87	0.93	0.96	0.94

Based on the experimental data and the Cross-WLF viscosity and Tait-pvT data modeling the flow behavior of polyphenylensulfide (PPS) filled with milled ferrite cores (FSM) with particle contents of 50, 60, 65 vol.-% can be simulated. Simulation results were created by Autodesk® Moldflow® software. Furthermore, melt temperature of 320 °C, flow rate of 22 cm³/s and demolding temperature (140 °C) were kept constant regarding to the material requirements. The simulation was performed at three different injection pressure steps (1250, 1500, 1750 bar).

Table 6 shows the comparison of the real injection molding results and the simulation of the flow length of PPS FSM 150-60 as a function of the injection pressure (1250 bar, 1500 bar, 1740 bar). The flow path length of the simulation corresponds with the real values. However, the expected significant increase in the flow length with increasing pressure is less pronounced than expected (1250bar = 48.66mm, 1500bar = 48.66mm, 1740bar = 50.66mm). Increasing injection pressure induces higher shear rate, whereby viscosity of PPS FSM 150-60-vol.% composite is reduced (see Table 6). The color sequence from blue to red shows the filling time and is not further discussed.

Table 6. Flow length of PPS FSM 150-60 as a function of the injection pressure (1250 bar, 1500 bar, 1740 bar). Reality vs. simulation.

PPS FSM 150-60, 1250bar	PPS FSM 150-60, 1500bar	PPS FSM 150-60, 1740bar

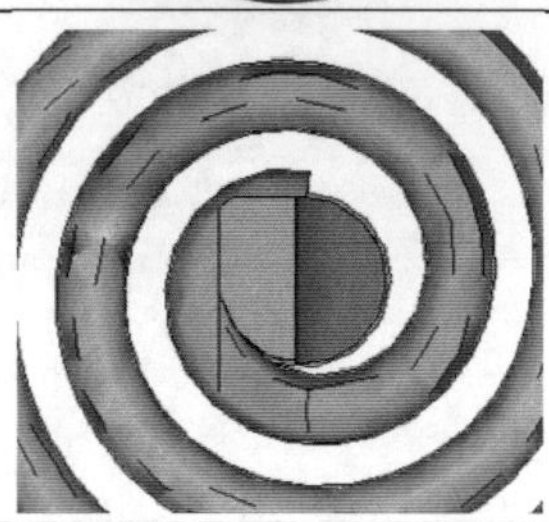	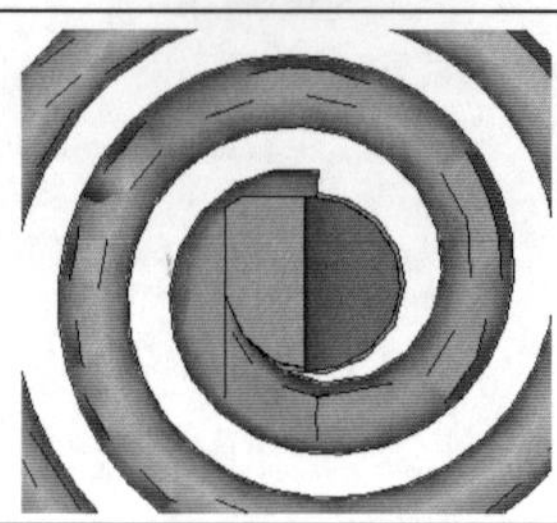	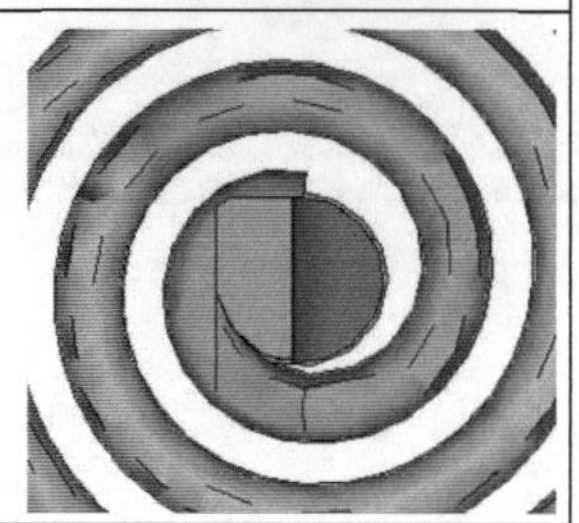

If the influence of the filler content is considered, the composites behave according to the theoretical principles as modelled in Table 4). At a constant pressure (1500 bar), the flow length decreases with increasing filler content due to the influence of the particles on the viscosity (see Table 7) caused by the particle-particle and particle-polymer interaction.

However, the simulation of the composite materials shows in all cases reduced flow properties compared to the neat polymer. Besides the increase of the viscosity, the changed thermal behavior of the material leads to a rapid freezing of the flow front in the mold due to the high thermal conductivity of the melt and the temperature difference between the mold wall and the melt. This is a challenge for the processing of these materials. Especially for components with long flow paths, many weld lines and narrow flow channels, an optimization of the processing parameters must be carried out in order to ensure a complete form filling of complex geometry.

Table 7. Flow length as a function of the filler content (50, 60, 65 vol.%) of PPS FSM 150 (injection pressure 1500 bar) Reality vs. simulation.

PPS FSM 150-50, 1500bar	PPS FSM 150-60, 1500bar	PPS FSM 150-65, 1500bar

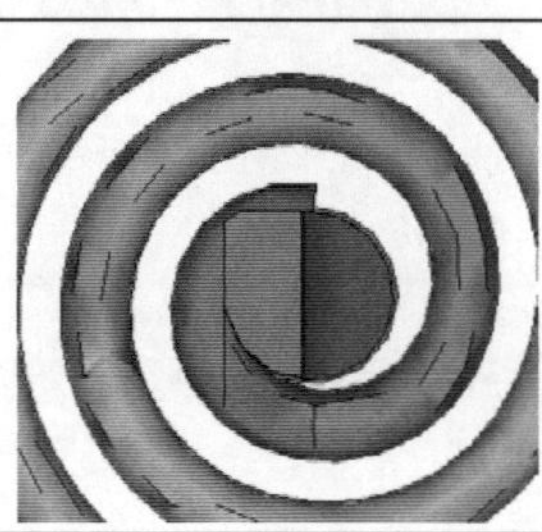	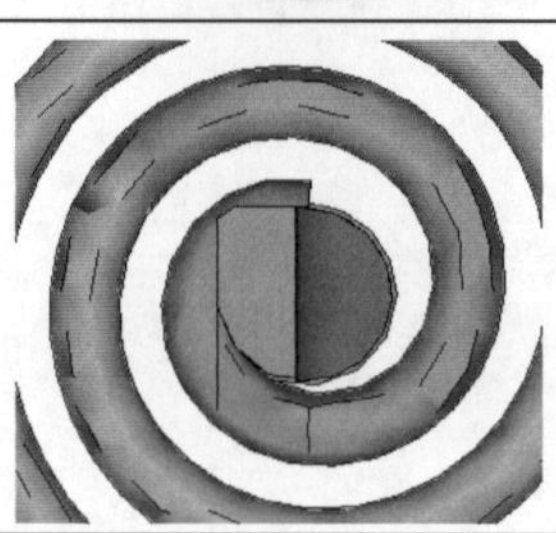	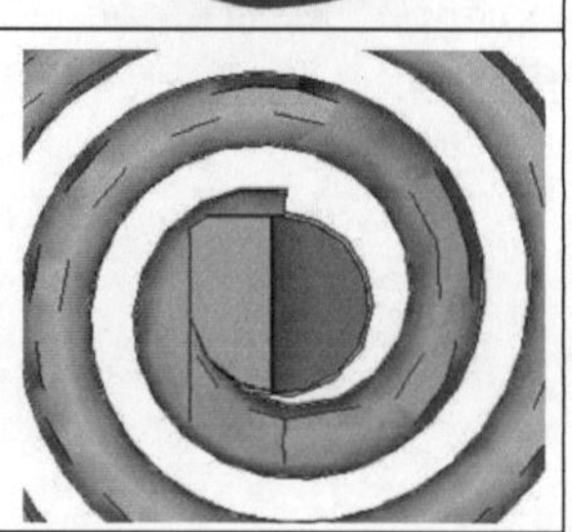

5 CONCLUSION

In this study, the viscosity, pvT-behavior and thermal conductivity of highly filled PPS-composites with soft magnetic fillers (milled ferrite coares, 150µm) were investigated in the process-specific parameter range. The characteristic values were determined for three different filler content levels (50, 60, 65 vol.-%) and serve with the thermal properties as a basis for the simulation of the material behavior. The viscosity of the PPS melt increases more than ten times by adding 65-vol.% of particles due to the particle-particle and polymer-particle interactions. The specific volume of PPS increases with temperature but decreases with pressure and filler content. Therefore, the filler leads to a negligible shrinkage behavior and thus complete mold filling– precondition to a good dimensional accuracy. Thermal conductivity of the filled material enhances remarkably.

The viscosity and the pvT-behavior were modeled by using cross-WLF equation and 2-domain Tait pvT-model. The modeled material properties serve as basis for simulation of spiral flow test with commercial software Moldflow®. Comparison of injection molded spiral flow test parts at different processing parameters show a good accordance to the simulation results. In order to use a conventional simulation tool, the used multi-phase material must be reduced to a single-phase material model. Despite the acceptable accuracy between the simulation and the real injection molding results, it is not possible with this simulation basis to make statements about the particle distribution and the homogeneity of the melt and thus the property profile of subsequent components.

REFERENCES

[1] D. M. Bigg, "Electrical properties of metal-filled polymer composites," In: Bhattacharya SK (ed), Metal-filled poly-mers: properties and applications. New York: Dekker, 1986, pp. 165–226.

[2] S. T. Tan, M. Q. Zhang, M. Z. Rong and H. M. Zeng, "Effect of interfacial modification on metal fiber filled polypropylene composites and property balance," Polymer Composites, Vol. 20, No. 3, 1999, pp. 406–412.

[3] T. Katsura, M. R. Kamal and L. A. Utracki, "Some properties of polypropylene filled with metal fibers," Polymer Composites, Vol. 6, No. 4, 1985, pp. 282–285.

[4] M. Y. Razzaq, M. Anhalt, B. Weidenfeller and L. Frormann, "Thermal, electrical and magnetic studies of magnetite filled polyurethane shape memory polymers," Material Science and Engineering: Part A, Vol. 444, 2007, pp. 227–235.

[5] M. Sarasa, D. Gerling, G. Kastinger and A. Schumacher, "Soft magnetic materials for electrical machines," In: Joint Czech Polish Conference on project GACR430813, Low voltage electrical machines. Brünn: Czech Republic, 2004, pp. 91–98.

[6] S. Kirchberg, U. Holländer, K. Möhwald, G. Ziegmann and F.-W. Bach, "Processing and characterization of injection moldable polymer-particle composites applicable in brazing processes," Journal of Applied Polymer Science, 2012, doi:10.1002/app.38862.

[7] R. B. Bird and P. J. Carreau, "A nonlinear viscoelastic model for polymer solutions and melts: part I," Chemical Engineering Science, Vol. 23, 1968, pp. 427–434.

[8] P. J. Carreau, I. F. MacDonald and R. B. Bird, "A nonlinear viscoelastic model for polymer solutions and melts: part II," Chemical Engineering Science, Vol. 23, 1968, pp. 901–911.

[9] M. L. Williams, R. F. Landel and J. D. Ferry, "The temperature dependence of relaxation mechanisms in amorphous polymers and other glass-forming liquids," Journal of the American Chemical Society, Vol. 77, 1955, pp. 3701–3707.

[10] M. M. Cross, "Relation between viscoelasticity and shear-thinning behaviour in liquids," Rheologica Acta, Vol. 18, 1979, pp. 609-614.

[11] J. Wang, "PVT Properties of Polymers for Injection Moulding, Some Critical Issues for Injection Molding", ISBN: 987-51-0297-7, InTech, China

[12] R. Y. Chang, C. H. Chen and K. S. Su, "Modifying the Tait equation with cooling-rate effects to predict the pressure-volume-temperature behaviors of amorphous poly-mers: modeling and experiments," Polymer Engineering and Science, Vol. 36, No. 13, 1996, pp. 1789-1795.

[13] Saleem, A., Frormann, L. and Iqbal, A. (2007). Mechanical, Thermal and Electrical Resisitivity Properties of Thermoplastic Composites Filled with Carbon Fibers and Carbon Particles, J. Polym. Res., 14:121-127.

[14] S. Kirchberg, "Simulation and Injection Molding of Ring-Shaped Polymer Bonded Nickel Braze Metal Composite Preforms Based on Rheological and Thermal Analyses", Open Journal of Composite Materials, 2013, doi:10.4236/ojcm.2013

pps proceeding
Symposium: S03 - Process modeling
Oral Presentation

Numerical Analysis Method for Deformation Behavior of Anisotropic Polymer shee

Yokoyama Atsushi (1)*, Kita Yui (1)

(1) Kyoto Institute of Technology - Kyoto - Japan

One stretching process is a process of molding a polymer film. In particular, since the productivity is good, longitudinal-lateral stretching method is often adopted. At this time, a tenter apparatus is generally used in the transverse stretching step, but transverse stretching is carried out while heat treatment is performed in order to efficiently stretch the film in the tenter apparatus. During stretching, the stress-strain curve of the film varies with temperature. It is considered that these conditions greatly affect the quality of the film after molding. However, the deformation behavior of the film in the tenter apparatus is not clearly elucidated. Therefore, in this research, numerical analysis is performed considering the conditions considered to influence the state of the film after molding, and the deformation behavior of the film in the tenter apparatus is considered.

Comparison of the Flow and Rheological Behavior of Two Semi-Structural Sheet-Molding-Compound (SMC) Based on a Hybrid Resin and Glass or Carbon Fibers

Hohberg, Martin[a]; Kärger, Luise[a]; Henning, Frank[a,b]; Hrymak, Andrew[c]

[a]*Karlsruhe Institute of Technology (KIT), Institute of Vehicle System Technology, Karlsruhe, Germany*
[b]*Fraunhofer Institute for Chemical Technology (ICT), Polymer Engineering, Pfinztal, Germany*
[c]*University of Western Ontario, Chemical and Biochemical Engineering, London, Canada*

Abstract. For a realistic process simulation of sheet molding compound (SMC), comprehensive and reliable viscosity data are essential. As the mean fiber-length in SMC is similar to the sample size of shear rheometers, an alternative rheology tool and method was developed. Therefore, a molding tool for the compression molding process was equipped with seven pressure sensors. By means of this tool, the pressure distribution was measured and thus the rheological behavior of two semi-structural SMC systems consisting of a hybrid UPPH resin (Unsaturated Polyester Polyurethane Hybrid) with glass or carbon fibers were estimated. The characterization of the carbon SMC is performed by using a previously developed compressible rheological shell-model, which was extended to describe a vinyl ester (VE) glass filled SMC [1]. This shell model describes the SMC rheological behavior during 1D flow in the compression molding process by considering the elongation flow in the core and the thin resin rich outer lubrication layer. Owing to the heat transfer at the mold surface, this lubrication layer has a relatively low viscosity leading to a local high shear rate. In contrast, the glass fiber SMC shows a significantly different pressure distribution and thus different flow behavior over the flow-length. It shows a significant pressure peak at the flow-front, which decreases after a certain flow-length and then follows the classical SMC rheological behavior. This cannot be modeled by the compressible shell-model, nor any other available SMC related model. Therefore, a new approach in addition to the classical SMC-rheological has to be developed. By comparing these two SMCs with their different fiber types, concepts to model the flow behavior of the hybrid glass SMC will be presented.

Keywords: Polymer-matrix composites (PMCs), Rheological properties, Process Modeling, Compression molding

PACS: 47.80.Fg Pressure and temperature measurements
82.35.−x Polymers: properties; reactions; polymerization

INTRODUCTION

Sheet Molding Compound (SMC) is a widely used composite material in the automotive and commercial vehicle industry. Owing to its Class-A surface quality, relatively low price and a density of c. 1.9 g cm^{-3}, the typical SMC is used in exterior applications. As SMC parts are manufactured in a compression molding process with a flow of the material, it has a great design freedom with nearly no clipping. Due to improvements in the compression molding process and new resin developments, investigations on applications with a combination of SMC and unidirectional fiber reinforcements were initiated. For these semi-structural applications the developed SMC has a higher fiber volume content than Class-A SMC, fillers are omitted, and a hybrid resin system is used to ensure crosslinking between the SMC and the unidirectional fiber reinforcements, which can be preformed and B-staged to freeze the shape [1].

Beside the material development, the simulation of such a material system is crucial to develop such hybrid parts. With the use of simulation it is possible to exploit the outstanding properties of both materials and their combination. Therefore, the combination and the transfer of results between mold filling and part finite element simulations is done through CAE-Chains [2, 3]. In such a CAE-Chain, process simulation has an important role as the resulting fiber orientation is directly linked to the mechanical properties of the part. By changing the position of the initial charge or by changing process parameters, the fiber orientation distribution could be influenced within a certain range. To predict the resulting fiber orientation distribution, SMC specific material models and corresponding material parameters are crucial [4].

The typical fiber length of SMC is 1 inch (approx. 25 mm), which is similar to the sample size of shear rheometers. Therefore, an alternative rheology tool and method was developed and implemented to characterize the rheological properties of SMCs. A tool for the compression molding process was equipped with seven pressure sensors. By linking these pressure sensors with the press data system, it is possible to trace the flow front, measure the compressibility and determine the viscosity data according to the SMC specific material models by post processing the data [5].

EXPERIMENTAL SETTING

Materials and equipment

The experiments were performed with two different SMCs using the same paste mixture and the addition of glass or carbon fibers. The paste was based on an unsaturated polyester polyurethane hybrid resin (UPPH) from DSM (Daron ZW 14142). This UPPH resin was then mixed with adherent, impregnation and flow aids as well as peroxide and isocyanate to achieve the designated properties during the preforming and the flow phase within the mold.

Glass fiber SMC consists of 12k glass fiber provided by Johns Manville (MultiStar® 272) and loaded to a fiber content of 41 wt.% (23 vol.%) of 1 inch (25mm) long chopped fibers. A fiber content of 55 wt.% (42 vol.%) was reached for carbon fiber SMC of 50k carbon fiber 1 inch long chopped fibers provided by Zoltek PX 35. These two materials were compounded at Fraunhofer Institute for Chemical Technology (ICT) in Pfinztal, Germany, on a modified HM-LB-800 Schmidt & Heinzmann SMC line [1].

The cavity of the tool is (length x width) 800 mm x 250 mm and the cavity thickness is adjustable in to produce plates between 1 mm and 5 mm of thickness. This tool was specially designed to measure rheological properties of SMC, with 7 pressure sensors aligned in the middle of the tool along the flow direction (see FIGURE 1). These pressure sensors were connected to a data logging system, which was also connected to the press system to simultaneously record data such as, the position of the parallelization-equalizing cylinder and the current working force.

The experiments were performed at the Fraunhofer ICT in Pfinztal, Germany on a Dieffenbacher COMPRESS PLUS DCP-G 3600/3200 AS. This press has a maximum working pressure of 36 000 kN and a maximum closing speed of 80 mm s-1.

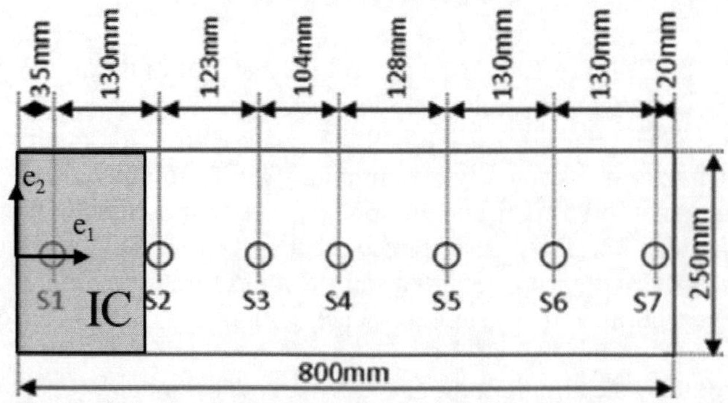

FIGURE 1. Position of the SMC initial charge and the pressure sensors along the lower mold part of the rheology tool [5].

Process parameter and conditions

For the experiments, the tool temperature was set to 140°C for the upper tool and to 145°C for the lower tool. These temperature setting helps to balance temperature differences due to convection of the mold. For the closing speed two different profiles were chosen. First, to compare the flow behavior under standard conditions, a constant closing speed of $\dot{h} = 1\ \mathrm{mm\ s^{-1}}$ was chosen. Second, a closing speed profile with a constant nominal deformation rate of $D = 0.1\ \mathrm{s^{-1}}$ was used to determine the rheological material parameter. The nominal deformation rate is defined in e_3-direction by the quotient of the closing speed $\dot{h}(t)$ and the gap height h(t):

$$D(t) = \frac{|\dot{h}(t)|}{h(t)} \quad (1)$$

To ensure the nominal deformation rate, the closing speed profile over the current height was predefined in the press settings. The maximum closing speed was limited to 3 000 kN for both materials to prevent too high pressures in the tool.

To enable a long flow length, the initial charge coverage of the tool was 20 % (160 mm x 250 mm). The initial charge had a thickness of 18 mm, which contains of 16 to 18 single SMC sheets. The average initial charge weight was 996 g for the glass fiber SMC and 970 g for the carbon fiber SMC. The room-tempered initial charge was placed at the left end of the tool, where it covered sensor S1. By using this sensor, the delay time from placing the material into the mold until the upper mold side is touching the material was determined to be 5s.

EXPERIMENTAL RESULTS

Pressure distributions

The pressure distributions for the compression experiments with a constant nominal deformation rate of $D = 0.1\ \mathrm{s^{-1}}$ for the carbon and the glass fiber UPPH SMCs are plotted in FIGURE 2 and FIGURE 3. The pressure and press working force are plotted as a function of the relative gap height ($h - h_f$), where h is the current height and h_f is the final height of the plate. By using this type of diagram, it is easier to compare different materials with different flow behaviors [5].

The pressure distribution of the carbon fiber SMC follows the typical SMC pattern. At the beginning, when the upper mold side is touching the material for the first time at a relative gap height of approximately 13.5 mm, a transverse compression (e3) takes place. In this short compression or squish [6] phase, no radial flow occurs, only a compaction of the charge, partly caused by a release of styrene and trapped air, can be observed. After reaching a threshold, the flow phase starts at a relative gap height of 12.5 mm. During this phase, the pressure is increasing constantly due to the hydrodynamic friction stress [5, 7–9]. At a relative gap height of approx. 0.5 mm, the press working force reaches its predefined limit. Owing to this fact, the pressure in the mold is not increasing anymore, as the closing speed is reduced and therefore the deformation rate decreased.

At a relative gap height of 4.4 mm, an abrupt decrease of the pressure at Sensor 1 can be observed. This is not a SMC specific effect, but can be attributed to a damaged short pegboard. Due to this damage, some material can flow through the gap and therefore the pressure is decreasing. After a short time the material is curing in this gap and so pressure is increasing again.

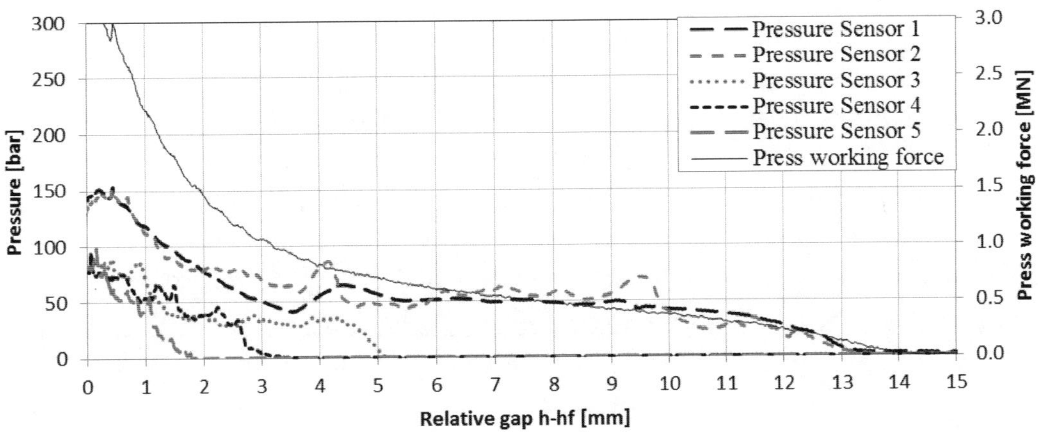

FIGURE 2. *Pressure distribution for the carbon fiber UPPH SMC for a nominal deformation rate of* $D = 0.1\ s^{-1}$.

The pressure distribution for the glass fiber SMC shows a significantly different behavior. The compression phase is similar to the one of the carbon fiber SMC. At the beginning of the flow phase, a high pressure at Sensor 1 and Sensor 2 is recorded until the pressure reaches a threshold at a relative gap height of 9 mm. This threshold is at a 1.5 times higher level than the pressure at the end of the molding process. Compared to the carbon fiber SMC, this threshold is approx. 5 times higher. As the press working pressure is increasing at the same time, this is not just a local effect at the locations of the pressure sensors, but a global effect throughout the whole material. After the threshold, the pressure is decreasing until it proceed in the same way as the carbon SMC. From a relative gap height of approx. 3 mm, the pressure values of the carbon and glass fiber SMC are nearly the same. The lower pressure values of Sensor 1 compared to Sensor 2 can be explained once again by the damaged pegboard. As the total pressure is higher for this case, the material is not curing in the gap and so the pressure at Sensor 1 is lower over a larger range. Still the amount of material flown through is so small, it can be neglected.

The effect of this local pressure peak is not only appearing at the beginning of the flow phase, but also at the flow front along the hole molding process. Sensor 3 to Sensor 5 also show this behavior when the flow front is reaching them, but with a lower amplitude.

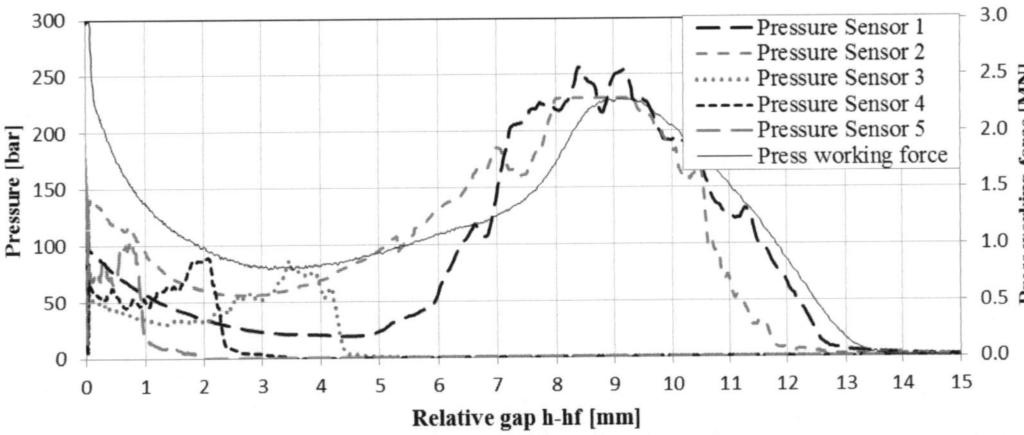

FIGURE 3. *Pressure distribution for the glass fiber UPPH SMC for a nominal deformation rate of* $D = 0.1\ s^{-1}$.

To identify the reason for the pressure peaks at the flow front, a series of short shots were made. During the first evaluation of these plates, three different areas were observed by optical assessment. At the flow front and at the opposite end of the plate from the initial charge, areas with a randomized oriented fiber distribution and complex curvature of the fibers at the surface are observed (see FIGURE 4). In the middle area, a more aligned fiber orientation at the surface is observed.

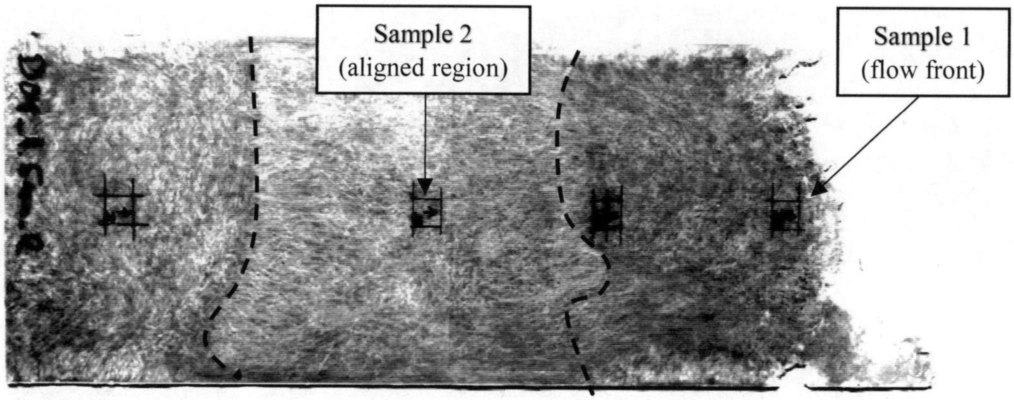

FIGURE 4. Gray-scale picture with high contrast of a glass fiber UPPH SMC short shot plate.

To identify which of these regions is responsible for the pressure peaks, the length along the plate of the peaks were evaluated. Therefore, the point with the first significant pressure decrease after the pressure peak is used. For this point, the current flow front is calculated [5] and the current sensor position is subtracted from this. Furthermore, the plates were sized by using visual criteria. The results from the pressure evaluation and the length of the swirled flow front region are plotted in FIGURE 5 for both plates of the short-shots study with 4.5mm thickness. The optical method leads only to one measuring point, but with the pressure evaluation method, a tracking of the flow front region could be done. As the correlation between the optical and the pressure evaluation seem to correspond, the flow front region with its highly swirled fiber orientation at the surface seems to be responsible for the pressure peak.

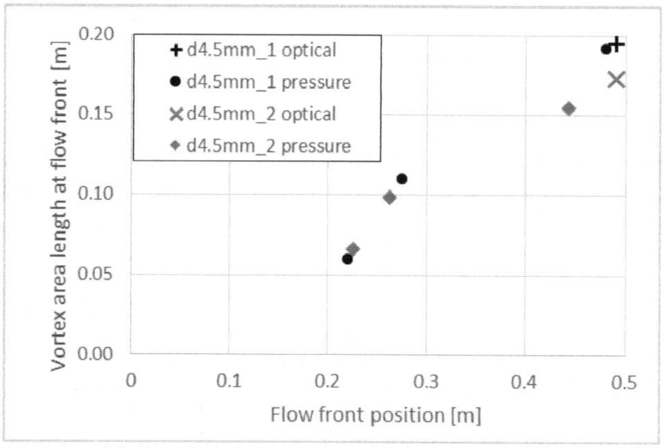

FIGURE 5. Vortex area at the flow front: optical measured and extracted from pressure distribution

Micro tomography scans

To understand what is happening during the flow phase and why the three regions develop, a series of micro tomography scans were done by the Institute for Applied Materials (IAM-WK) at the Karlsruhe Institute of Technology (KIT). Therefore, four samples of 20mm x 200mm, two of the flow front region and one each of the middle and end region, were excised from each 4.5mm thick plates. The positions of these samples are indicated in FIGURE 4.

Comparing the scans from the flow front (FIGURE 6) and the aligned region (FIGURE 7), the visual differences between these two regions can be confirmed. The flow front sample shows fibers which are highly swirled. Furthermore, the many fiber bundles have been separated into smaller bundles or single fibers. These effects were also detected by [10]. In contrast, the aligned region shows the typical fiber orientation originating from the manufacturing process. Especially in the cross-section, the randomly orientated fiber bundles can be recognized. At the surface, some separation and reorientation effect of the fiber bundles appeared.

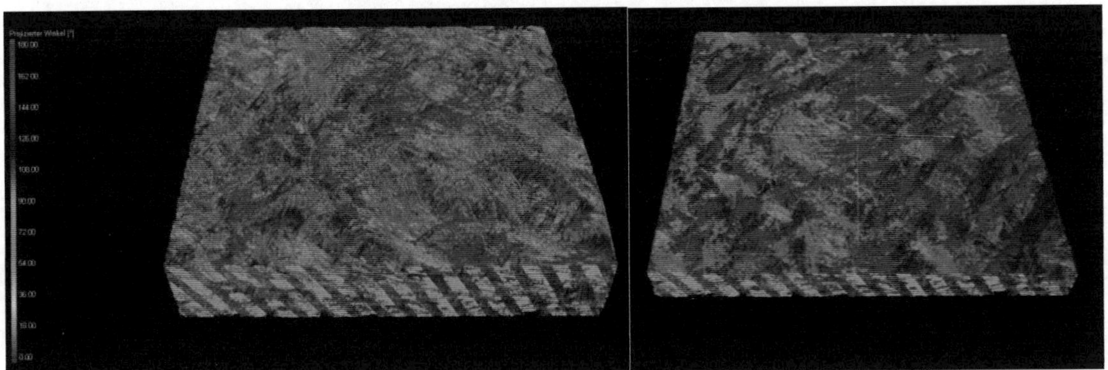

FIGURE 6. Micro tomography scan of the flow front (sample 1): view on top of the sample (left) and sliced in the middle (right).

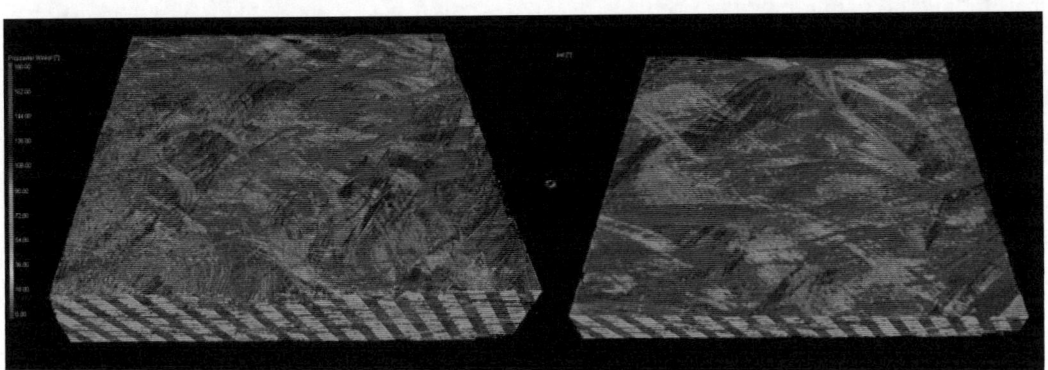

FIGURE 7. Micro tomography scan of the sample from the aligned region (sample 2): view on top of the sample (left) and sliced in the middle (right).

MODEL DISCUSSION

The typical flow model, which is also used for the rheological characterization of the carbon fiber SMC [5] divides the material in two regions. As the SMC initial charge is placed room-tempered into the hot mold, the outer layer of the material is heating up rapidly. Owing to the relatively low thermal conductivity, this layer has a changing thickness in the area of only a few ηm [11]. This layer contains nearly no fibers and because of the high temperature, the viscosity decreases dramatically thus functioning as a lubrication layer. The charge core region stays relatively cold during the molding process. Due to these two different regions, the lubrication layer is exposed to a high shear rate, while the core charge is loaded by normal forces. As SMC is a compressible non-Newtonian fluid [12, 13], the shear and the elongation viscosity are not coupled by the Trouton ratio and have to be determined separately.

The elongation based deformation of the core region leads to a material flow, where the core charge stays in the middle of the plate thickness and no mixing in the thickness direction occurs. Therefore, the measured pressure can be divided into a rheological stress (from the core charge region) and the hydrodynamic friction stress (from the lubrication layer) [5, 7–9].

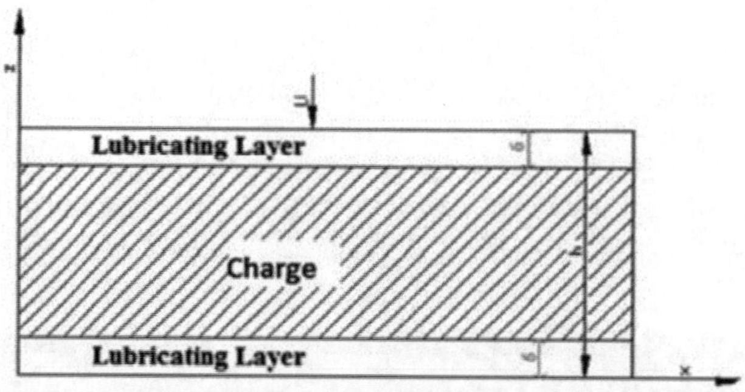

FIGURE 8. *Schematically cross-section of SMC material during the molding process [11].*

In contrast to this typical behavior of the carbon fiber SMC, the glass fiber SMC must have a different flow behavior to explain the pressure distribution. To identify this flow behavior, the evolution of the three regions (vortex region at both ends and aligned region in the middle) over the molding process were analyzed, by using the short-shot study. Taking into account the micro tomography scans, it seems as the core charge is moving in flow direction. Therefore, the lubrication layer or the complete sheet at the mold surface is flowing into the vortex regions. Furthermore, the core charge region is mainly transported in flow direction, while showing low elongation. In such a system, the main flow of the material would be at the mold surfaces, where the higher temperature leads to a lower viscosity.

The reason for the pressure peak could not be identified yet. It can be eliminated, that the material is speeding up so much, that the hydrodynamic pressure becomes relevant. With a flow front speed of 0.1 m s^{-1} after 4 s of the molding process and a 10 times higher speed in the lubrication layer, this would only lead to an increase of 1 bar. It is more realistic, that the fiber-fiber-interaction in the vortex regions leads to an increase of the homogenized shear viscosity and therefore to higher stresses. Furthermore, the fiber bundle separation and the curved fibers could cause an anisotropic viscosity [14, 15], which is higher than the isotropic viscosity. New methods to investigate this approach will be designed and conducted.

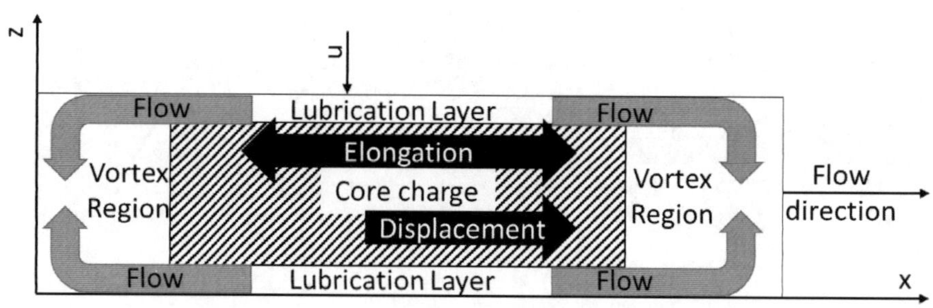

FIGURE 9. *Schematically cross-section of the probable glass fiber UPPH SMC flow during the molding process*

These two different flow behaviors are caused by the fiber type, as the paste is the same for post materials. Therefore, the thermal conductivity and consequently the temperature distribution seem to be relevant. As the UPPH is still a new resin under research, the material parameters, and the material property effects during the compression molding process are still to be investigated. Furthermore the ratio of the fiber stiffness to the matrix viscosity might have a significant effect [16]. Therefore, no reason for these two different flow regimes could be presented right now. The UPPH resin will be further investigated within the International Research Training Group "Integrated engineering of continuous-discontinuous long fiber-reinforced polymer structures" and so new results will be available within the next years.

CONCLUSION

In this work, a rheological tool was presented, which can be used to characterize rheological properties of SMC. Therefore, seven pressure sensors record the pressure distribution over the flow length over the compression molding process. This tool was used to investigate two SMC with glass or carbon fibers, based on the same UPPH resin. The carbon fiber UPPH SMC shows the typical flow behavior, which could be modelled by using a rheological stress term for the core region and a hydrodynamic friction stress term for the lubrication layer. In contrast to this typical pressure distribution, the glass fiber UPPH SMC shows pressure peaks at the flow front. After the vortex region at the flow front passed the sensor, the pressure is decreasing to the same level as the carbon fiber SMC and then follows the typical pressure trend.

To investigate and model this behavior, a series of short-shots were made. By visual examination and evaluation of the pressure data, the three different flow regions could be detected. Samples of these regions were prepared to place them into the micro tomography scanner to correlate the optical examination with the fiber orientations in the plate.

Taking all experimental results into account, a new flow behavior for such a SMC was proposed, which could explain the basic effects. As not all effects are completely understood right now, further investigations have to be done within the project.

ACKNOWLEDGMENTS

These investigations were carried out through R&D activities within the Technology Cluster Composites TC2 "InnoSMC" until 2014 and within the International Research Training Group "Integrated engineering of continuous-discontinuous long fiber-reinforced polymer structures" (GRK 2078) since 2015. The "TC2 InnoSMC" project was funded by the European Union through the program "European Funds for Regional Development" as well as state government of Baden-Wuerttemberg in Germany. Administrative agency of the program RWB-EFRE is the Ministry of Rural Development, Food and Consumer Protection. The International Research Training Group is funded by the German Research Foundation (DFG).

REFERENCES

1. D. Bücheler and F. Henning, Hybrid resin improves position and alignment of continuously reinforced prepreg during compression co-molding with sheet molding compound. ECCM17 - 17th European Conference on Composite Materials, München, Germany (2016).
2. L. Kärger, A. Bernath, F. Fritz, S. Galkin, D. Magagnato, A. Oeckerath, A. Schön, and F. Henning, Composite Structures, 132, 350 (2015).
3. M. Hohberg, L. Kärger, F. Henning, and A. Hrymak, "Process Simulation of Sheet Molding Compound (SMC) as key for the integrated Simulation Chain," in Simulation of Composites - Ready for Industry 4.0?, NAFEMS, Ed. (2016).
4. M. Hohberg, Process simulation of hybrid SMC structures with local continuous fiber reinforcements, Frankfurt (2015).
5. M. Hohberg, L. Kärger, F. Henning, and A. Hrymak, Composites Part A: Applied Science and Manufacturing, 95, 110 (2017).
6. P.T. Odenberger, H.M. Andersson, and T.S. Lundström, Composites Part A: Applied Science and Manufacturing, 35, 1125 (2004).
7. P. Dumont, L. Orgéas, S. Le Corre, and D. Favier, International Journal of Plasticity, 19, 625 (2003).
8. P. Dumont, L. Orgéas, D. Favier, P. Pizette, and C. Venet, Composites Part A: Applied Science and Manufacturing, 38, 353 (2007).
9. O. Guiraud, P.J.J. Dumont, L. Orgéas, and D. Favier, Composites Part A: Applied Science and Manufacturing, 43, 2107 (2012).
10. A. Motaghi and A.N. Hrymak, Polymer Composites (2017).
11. M.M. Shokrieh and R. Mosalmani, Journal of Reinforced Plastics and Composites, 33, 1183 (2014).
12. G. Kotsikos and A.G. Gibson, Composites Part A: Applied Science and Manufacturing, 29, 1569 (1998).
13. A.G. Gibson and S. Toll, Journal of Non-Newtonian Fluid Mechanics, 82, 1 (1999).
14. T. Tsigkopoulos, T. Osswald, R. Feulner, G. Hülder, and D. Drummer, Consideration of elongation effects for both fiber reinforced and unfilled fluids by means of an invariant constitutive model. The Society of Rheology 81st Annual Meeting, Madison, Wisconsin, USA (2009).
15. R. Bertóti and T. Böhlke, Mech Adv Mater Mod Process, 3, 751 (2017).
16. C. Pérez, D. Ramírez, and T.A. Osswald, "Mechanistic model simulation of a compression molding process: Fiber orientation and fiber-matrix separation," in Annual Technical Conference - ANTEC, Conference Proceedings (2015).

Development of a new coupled viscoelastic solver in OpenFOAM®

Fernandes Célio [1], Vukčević Vuko [2], Uroić Tessa [2], Ferrás Luís Lima [1], Carneiro Olga Sousa [1], Simões Ricardo [3], Jasak Hrvoje [4], Nobrega Joao Miguel [1]*

[1] University of Minho - Guimarães - Portugal, [2] Faculty of Mechanical Engineering and Naval Architecture - Zagreb - Croatia, [3] Institute for Polymer and Composites - Guimarães - Portugal, [4] Faculty of Mechanical Engineering and Naval Architecture, Wikki - Zagreb - Croatia

: Effective design approaches require the use of numerical modelling tools, to avoid long and high resource demanding experimental trial-and-error processes. The design tasks are even more difficult to undertake when dealing with complex rheology materials, as happens with the viscoelastic fluids, which present some counter intuitive phenomena. This framework has been motivating, during the last decades, the development of numerical modelling codes able to deal with viscoelastic fluids. The usual approaches employed, are based on segregated procedures, in which the mass, momentum and constitutive governing equations and solved iterative and sequentially. This sequential process is quite sensitive and prone to divergence, thus is often reported to require high relaxation factors, to assure the achievement of a converged solution. Moreover, the segregated approach is also know to limit substantially the Deborah numbers that can be achieved on the numerical calculation. This work reports the development of a coupled viscoelastic solver in OpenFOAM® computational library, in which the governing equations are solved simultaneously. The presented case studies clear evidence the advantages of the proposed approach, both in terms of calculation time and stability.

A New Scale-up / Scale-down Method for the Blown Film Extrusion

Benedikt Neubert[1], Christoph Dohm[1, a)] and Johannes Wortberg[1]

[1]*University of Duisburg-Essen, Institute of Product Engineering, Engineering Design and Plastics Machinery, Lotharstraße 1, 47057 Duisburg, Germany*

[a)]christoph.dohm@uni-due.de

Abstract. The development of innovative film products is a two-step process: New composite films or material recipes are first experimentally tested on a laboratory-scale blown film line, and then the results regarding process conditions and final film properties are transferred to a large-scale production plant. This procedure is generally known as scale-up. However, the transferability of results is not always ensured because of processing- and plant-specific restrictions. Thus, this paper presents a process-oriented scale-up / scale-down strategy (POSS), that should provide higher flexibility in practical use compared to existing scale-up strategies. To prove the feasibility of the process-oriented strategy, a series of experimental runs was carried out. An industrial production process on a high-capacity blown film line (> 750 kg/h) was selected as the starting point for a scale-down to a small-scale laboratory plant (scale-down factor of ten). Finally, mechanical properties of the film samples taken from the production process and the laboratory runs were measured and compared. The comparison showed that films with similar properties could be produced under process-oriented scale-down conditions.

Keywords: blown film extrusion, process-oriented scale-up / scale-down strategy, film properties, process time
PACS: 81.20.Hy, 68.60.Bs

INTRODUCTION

The blown film process is characterized by simultaneous stretching of the molten polymer in machine direction (MD) and circumferential direction (CD), which generally leads to anisotropic mechanical properties of the solid film. A cooling ring provides an air stream to cool the bubble while an additional system for internal bubble cooling (IBC) is optionally installed on large-scale blown film lines. Primarily influenced by the cooling conditions, dynamic processes, such as crystallization and relaxation, take place in the bubble formation zone, too. The period for one small element of the film to move from the die exit to the frost line is defined as the process time (t_p).

For film-producing companies, it is important to offer innovative and high-quality products in order to stay competitive on international markets. The development of new composite films or material recipes is usually performed on a laboratory-scale blown film line, because this saves materials, time and, thus, costs. Subsequently, the process conditions tested on the small-scale plant and the resulting final film properties need to be transferred to a large-scale production plant. This transfer of results between two differently scaled blown film lines is called scale-up [1] or, conversely, scale-down. Obtaining equal film properties independent of the plant's scale is the main objective during scale-up / scale-down. It is well known that blown film properties correlate with the underlying process parameters that, in turn, strongly interact with each other. Furthermore, the technical equipment of a production plant differs from that in a laboratory; hence, the adjustability of certain process parameters might be limited due to plant-specific or processing restrictions. Taking all these relations into account, the demand for a systematic and practicable scale-up / scale-down strategy arises.

Therefore, this paper presents the theoretical background and an experimental validation of a process-oriented scale-up / scale-down strategy (POSS), which, on the one hand, is based on existing scale-up strategies. On the other hand, since a degree of freedom is introduced without neglecting the aforementioned interdependences of major process influences, the POSS ought to be more flexible in practice.

PROCESS-ORIENTED SCALE-UP / SCALE-DOWN STRATEGY

In a scale-up / scale-down scenario, the difference in size between the laboratory plant (subscript: lab) and the production plant (subscript: pro) is generally described by the scaling factors k and s. They are defined as the ratios between the die diameters d_0 and the die gap widths h_0, respectively [2].

$$k=\frac{d_{0,pro}}{d_{0,lab}} \tag{1}$$

$$s=\frac{h_{0,pro}}{h_{0,lab}} \tag{2}$$

The basic requirement for maintaining mechanical film properties, independently of the plant's scale, is equal stretching in MD and CD, which is ensured by keeping the draw-down ratio (DDR) and the blow-up ratio (BUR) constant. The DDR equals the haul-off speed divided by the melt delivery rate, and the BUR is calculated as the quotient of the bubble diameter at the frost line and the die diameter. In accordance with Pearson [3], this requirement is called 'geometrical similarity'. Here, it has to be noted that Pearson, as well as Kanai et al. [4], also suggested to leave the ratio between the frost line height (x_f) and the die radius (r_0) unchanged. In contrast, this condition is not included in the POSS, because it is too restrictive with regard to possible restrictions in practical use. Since the film thickness (h_f) is an important product characteristic and, moreover, affects other film properties, it should be the same for both scales. Consequently, the scaling factor s equals one in the POSS. This corresponds to the scale-up condition presented by Butler et al. [5], whereas Kanai et al. allow the factor s to deviate from one.

Another essential requirement of the POSS is called 'dynamic similarity', which is achieved by keeping processing conditions similar in terms of process time. Thus, time- and temperature-dependent processes within the bubble formation zone, affecting final film properties, are considered. For a precise approximation of the process time, its determination should be based on the real shape of the bubble and the actual profile of the film velocity. Assuming that measured data for the bubble contour and the velocity profile as functions of height above the die exit are available, the process time is calculated according to Eq. (3). Accordingly, the bubble formation zone is divided into n small segments, and the residence time for each segment is calculated as the quotient of its length l_i and a local average film velocity $\bar{v}_i$ [6].

$$t_p=\sum_{i=1}^{n} \bar{v}_i^{-1} \cdot l_i \tag{3}$$

Butler et al. introduced the dimensionless fabrication time ratio (FTR) as a scale-up variable. It is the ratio between the polymer-specific relaxation time and the process time [5]. They suggested to estimate the process time on the basis of a linear film velocity profile. Since this simplification is not considered to be universally reasonable in practice, the approximation with real data for the bubble contour and the velocity profile is preferred in the context of the POSS.

Provided that the material composition, the layer structure as well as the melt temperature at the die exit (T_0) are comparable, similar material characteristics, such as density, viscosity or relaxation time, and similar material behavior during bubble formation can be expected. Here, it has to be assumed that processing influences due to thermal and mechanical stresses in the extruder and the die are negligible or, at least, comparable for the considered plants.

Finally, an operating point can be transferred between two blown film lines adjusting the mass flow rate ($\dot{m}$) and the frost line according to the following process-oriented scale-up / scale-down conditions (presented for scale-up):

$$\dot{m}_{pro}=w\cdot\dot{m}_{lab} \tag{4}$$

$$x_{f,pro}=\frac{w}{k}\cdot t_{p,lab}\cdot C_{lab} \tag{5}$$

$$C_{lab}=\dot{m}\cdot\left(\frac{1}{d_f\cdot h_f\cdot\rho_f}-\frac{1}{d_0\cdot h_0\cdot\rho_0}\right)\cdot\left[\pi\cdot \ln\left(\frac{d_0\cdot h_0\cdot\rho_0}{d_f\cdot h_f\cdot\rho_f}\right)\right]^{-1} \tag{6}$$

The flexible scaling of the mass flow rate with the variable w (Eq. 4) represents the aforementioned degree of freedom, which is a key element of the POSS. This enables the determination of suitable operating points within a process window. Theoretical boundaries for the value range of w are suggested to be k as the lower limit and k^2 as the upper limit. These boundaries represent the scale-up conditions for $\dot{m}$ according to the strategies by Butler et al. (k) and Kanai et al. (k^2). Under the condition of an equal process time for both scales, x_f is a function of the variable w and the parameters k, t_p and C (Eq. 5). In practice, the frost line height needs to be regulated, adjusting bubble cooling parameters such as the volumetric flow rate of the cooling air and/or its temperature. The constant C depends on the process state that is used as the starting point in the scale-up / scale-down scenario (laboratory process in case of scale-up). In Eq. (6), the variables d, h and ρ represent the bubble diameter, its thickness and the density of the polymer. The subscripts 0 and f indicate that it is either the corresponding value at the die exit or at the frost line.

As a result, the POSS allows to set the target process parameters $\dot{m}$ and x_f in accordance with existing processing- and plant-specific restrictions, for example, the maximum mass flow rate or the cooling capacity. Therefore, performing a scale-up / scale-down becomes more feasible, even for high values of k, which occur when high-capacity production plants are considered. In contrast to the POSS, the aforementioned scale-up strategies developed by Kanai et al. and Butler et al. calculate just one single, fully determined operating point for a given scale-up / scale-down scenario. Consequently, they do not provide sufficient flexibility to respond to possible constraints due to the production line or the laboratory plant.

EXPERIMENTAL PROCEDURE

Applying the process-oriented scale-up / scale-down strategy, an experimental scale-down was performed. A production process of a three-layer polyethylene film on a high-capacity blown film line ($d_{0,pro}$ > 600 mm) was used as the reference process (run 0). The mass flow rate exceeded 750 kg/h. For cooling the tubular film, a single-lip cooling ring and an IBC were installed. The large-scale process was transferred to a laboratory-scale plant, which was equipped with a single-lip cooling ring only. Six operating points with different mass flow rates were calculated and experimentally carried out on the small-scale plant (Run 1-6). Table 1 presents the corresponding process parameters.

TABLE 1. Process parameters for the reference process (run 0) and the laboratory-scale operating points (run 1-6).

Operating Point	w -	DSO (kg/hmm)	BUR -	DDR -	T_0 (°C)	x_f (m)	t_p (s)
Run 0	1	1.1875		20.4		0.930	
Run 1	0.011	0.125		16.7		0.110	
Run 2	0.013	0.150		16.7		0.135	
Run 3	0.016	0.188	1.66	16.6	195	0.167	5.1
Run 4	0.019	0.225		17.3		0.186	
Run 5	0.022	0.263		17.5		0.238	
Run 6	0.026	0.313		16.9		0.276	

Furthermore, the scale-down factors k and s were ten and one, respectively. The film thickness, the layer structure and the materials used were identical for both scales. Regarding the DDRs, runs 1-6 seem to deviate systematically from run 0. Although the DDR should have been the same for all runs, achieving the target film thickness had higher priority during the experiments. Measured data for the velocity profile and the bubble contour of the reference process were used to determine the process time (5.1 s) according to Eq. 3. Applying the POSS, the same process time was expected for the laboratory-scale runs. This assumption was experimentally confirmed by means of Run 2. For this purpose, the bubble was marked with a viscous paint at the die exit and the position of the marked spot, passing the bubble formation zone, was recorded with a video camera. Thus, a measured process time of 5.4 seconds resulted, which is in very good accordance with the calculation. Finally, film samples were taken from every run, including the reference process, and their mechanical properties were measured according to DIN EN ISO 527-3 (tensile properties) and DIN EN 14477 (puncture resistance).

EXPERIMENTAL RESULTS

The following figures contain the measurement results of the tensile test (Fig. 1 and Fig. 2) and of the puncture resistance test (Fig. 3). Each run is characterized by the mean value of ten samples. Moreover, the dotted line indicates the average of the laboratory-scale runs (Run 1-6) so that the comparison with the reference process (Run 0) is more intuitive.

Figure 1 shows the tensile stress at yield in machine direction (Fig. 1a) and circumferential direction (Fig. 1b). The stresses in CD are somewhat higher than in MD or are nearly comparable. This does not meet the expectation, since the stretching in CD is considerably lower than in MD. It is possible that the BUR, however, is high enough to overcompensate for the molecular orientation induced by stretching in MD. The deviation of the average stress (Run 1-6) from Run 0 in MD (Fig. 1a) can be explained by the difference between the corresponding DDRs (Table 1). Nevertheless, the laboratory-scale runs demonstrate a very good reproducibility of the tensile stress at yield under process-oriented scale-up / scale-down conditions, and the results are similar to those of the reference process.

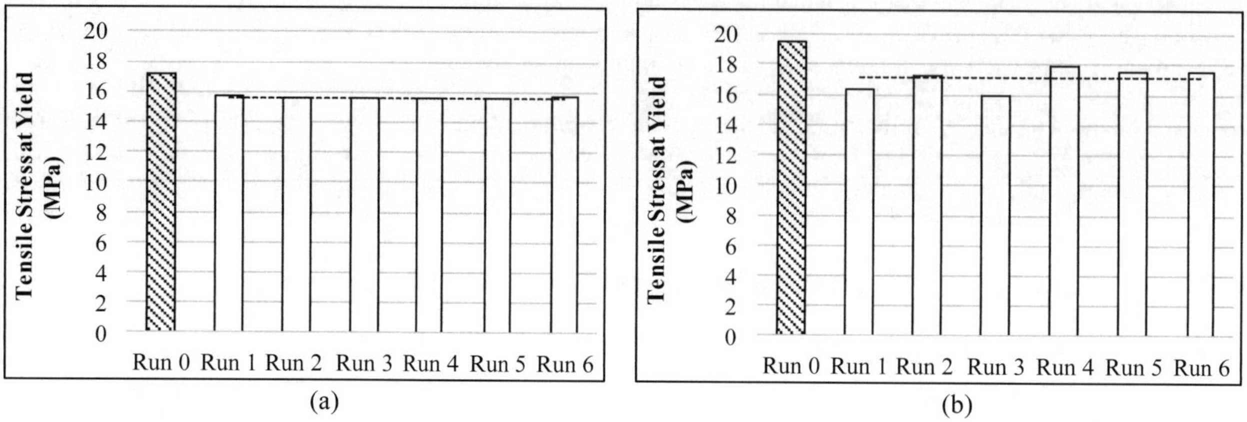

FIGURE 1. Tensile stress at yield (a) in machine direction and (b) in circumferential direction for runs 0-6.

The modulus of elasticity was measured as a secant modulus between 0.75 % and 1.25 % of elongation. Figure 2 shows the test results in machine direction (Fig. 2a) and circumferential direction (Fig. 2b). For all runs, the modulus measurements in MD are lower than in CD, which is in accordance with the typical findings in the literature [7, 8]. The average of Run 1-6 in MD is nearly 20 % higher than that of Run 0, but the reproducibility of the POSS is still observed (Fig. 2a). Furthermore, there are noticeable variations in the results for runs 1-6 in CD (Fig. 2b). They might be caused by variations in the film thickness due to the absence of a thickness control system in the laboratory plant. It is possible that these irregularities lead to uneven film stretching and, hence, affect the modulus of elasticity.

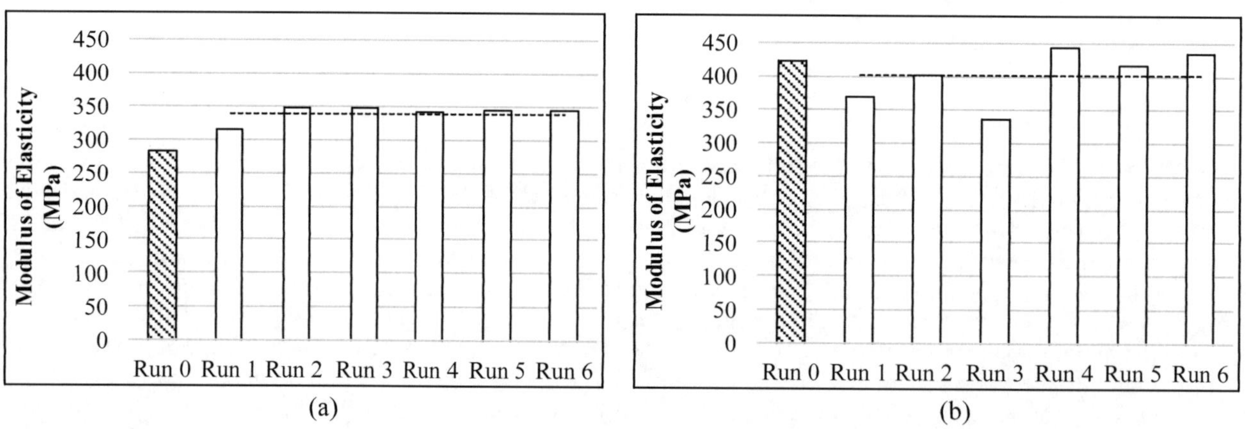

FIGURE 2. Modulus of elasticity (a) in machine direction and (b) in circumferential direction for runs 0-6.

Figure 3 shows the puncture resistance, the elongation and the penetration energy, resulting from the puncture resistance test. In general, the puncture resistance can be related to the ratio of anisotropy regarding the molecular orientation and, hence, the film's strength. It is assumed that an increasing ratio of anisotropy leads to decreasing puncture resistance. Considering the underlying BURs and DDRs (Table 1), the puncture resistance of Run 0 was expected to be higher than that of Run 1-6. The contrary finding (Fig. 3) could be explained by the different film thicknesses. It has to be noted that the mean thicknesses of the samples from Run 1-6 were slightly thinner than those from Run 0. The values for elongation and penetration energy are remarkably similar, for the laboratory-scale runs in particular.

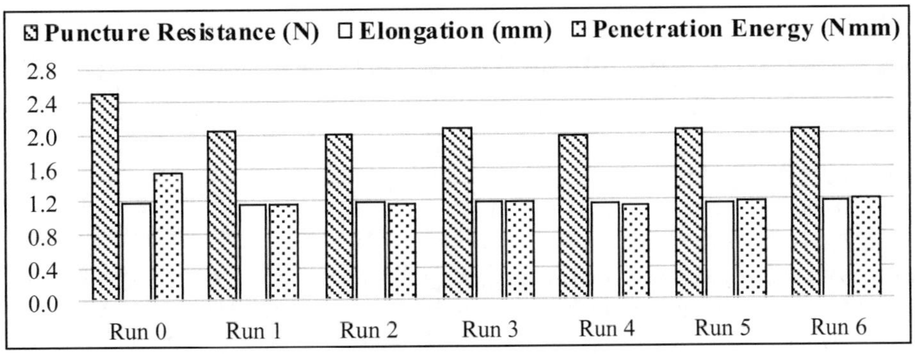

FIGURE 3. Puncture resistance, penetration energy and elongation for runs 0-6.

All in all, the results confirm a very good reproducibility of the considered mechanical properties, applying the POSS. Variations are generally wider in circumferential direction than in machine direction. Since the findings do not always correspond to the expected outcome, it has to be noted that the material composition used seems to be relatively robust. Therefore, the material characteristics might have a big impact on the final film properties and, thus, superimpose the processing influences to some extent.

CONCLUSION AND OUTLOOK

The process-oriented scale-up / scale-down strategy (POSS) for blown film extrusion allows a flexible adjustment of the mass flow rate as a target process parameter. Consequently, possible processing- and plant-specific restrictions can be considered in practice. Performing an experimental scale-down with a scale-down factor of ten, the feasibility of the POSS could be confirmed. The results demonstrate a good reproducibility of the mechanical properties.

Nevertheless, further studies validating a general applicability of the POSS should be carried out. In this context, more complex products, e.g. barrier films, and rather sensitive materials need to be regarded. Moreover, it would be interesting to use the POSS in the more common direction, performing a scale-up.

ACKNOWLEDGMENTS

We would like to thank the German National Science Foundation (DFG) for funding our research.

REFERENCES

1. W. Michaeli and J. Hauck, "Up-Scaling experimenteller Untersuchungsergebnisse von Labor- auf Produktionsanlagen," in *Wertschöpfung bei der Folienextrusion*, edited by VDI-Gesellschaft Kunststofftechnik (VDI-Verlag, Düsseldorf, 1999), pp. 165-199.
2. A. M. Sukhadia, J. Plast. Film Sheet. **10**(3), 213-234 (1994).
3. J. R. A. Pearson, *Mechanics of Polymer Processing* (Elsevier Applied Science Publishers, London, 1985), pp. 508-510.
4. T. Kanai, M. Kimura and Y. Asano, J. Plast. Film Sheet. **2**(3), 224-241 (1986).
5. T. I. Butler, S. Y. Lai and R. Patel, "Scale-up Factors Effecting Crystallization in Polyolefin Blown Film," in *1994 Polymer, Laminations & Coatings Conference* (TAPPI PRESS, Atlanta, GA, 1994), pp. 289-305.
6. G. J. Schmitz, "Experimentelle Analyse und numerische Modellierung der Deformationsvorgänge bei der Schlauchfolienextrusion," Ph.D. thesis, RWTH Aachen University, Germany, 1996.
7. F. Ohlendorf, "Vorhersage der mechanischen Folieneigenschaften bei der Schlauchfolienextrusion," Ph.D. thesis, RWTH Aachen University, Germany, 2004.
8. P. P. Tas, "Film Blowing: From Polymer to Product," Ph.D. thesis, Eindhoven University of Technology, Netherlands, 1994.

MODELING OF GAS DIFFUSION IN NANOCOMPOSITES FOR BARRIER APPLICATIONS

***ZID Sarra* [(1)]*, *ESPUCHE Eliane* [(2)]**

[(1)] PhD - Lyon - France, [(2)] HDR - LYON - France

The optimization of polymer barrier properties is currently of crucial importance for a wide range of applications from packaging to building or even energy applications.To meet the requirements of these applications, polymer matrices are often combined with impermeable (nano) fillers. Different nanofiller natures, shapes and contents have been experimentally used and a large range of barrier materials has been obtained. In the meantime,several numerical approaches have been developed to model gas diffusion properties of nanocomposite materials.However, these approaches only considered bidimensional systems.Nowadays, finite element modeling (FEM) approaches are being used increasingly as they allow the description of the materials in 3 dimensions.The aim of our work is to develop and validate a new Finite Element model allowing to predict the gas diffusion properties of nanocomposites for different nanofiller shapes, contents and dispersion states. In this approach, the mathematical formulation describing the mass transfer is based on Fick's partial differential equation and the model was solved numerically using the commercial package COMSOL Multiphysics.As a first step, a 3D geometry consisting of disk-shaped nanofillers dispersed in the polymer matrix was considered. A parallelepiped domain representing the nanocomposite periodic structure was built and discretized into finite elements using an adaptive mesh in order to get the most accurate results. The model was solved numerically for various boundary conditions and the decline in effective diffusivity due to a tortuous gas transport path was analyzed. Then, the effect of adding various shaped nanofillers for a volume fraction range from 0 to 10% was studied, with emphasizing on the influence of several parameters such as spacing between layers. The results obtained from our Finite Element modeling approach were discussed with respect to the nanofiller shape, aspect ratio and content.

Effect of Morphological Structure of Polymeric Foams on the Mitigation of Organic Contaminated Water

Pavani Cherukupally, Krishna Prasad Nagarajan Ramani, Amy M. Bilton, and Chul B. Park*
University of Toronto, Toronto, Canada
*Corresponding author: park@mie.utoronto.ca

Background and Motivation

The global water demand for agriculture, industrial processes, and domestic use is growing rapidly. Unfortunately, the available water resources are becoming increasingly scarce [1]. As a result, much research is currently focused on developing energy-efficient treatment of industrial wastewater such as oil sands tailings for reuse, energy generation, and product recovery [2]. For wastewater treatment, adsorbents in a fixed-bed configuration are widely used due to their efficacy, simplicity, and energy efficiency [3]. To advance the bed performance, this research is investigating a new class of adsorbent materials polymer foams for treatment of difficult-to-remove emulsified organic compounds from oil field water.

For practical implementation, the adsorbent has been configured in a lab-scale fixed-bed adsorbent filter (AF). However, this lab-scale system far away from commercial implementation. A key challenge is understanding of the complex phenomena governing the oil adsorption and accurately describing it in a computational model to enable system design. This paper describes a computational approach using COMSOL Multiphysics v. 5.2 to model and predict the performance of the foam based AF.

Approach

The goal of this work is to produce a generic model to enable an understanding of the physical, chemical, and electrical parameters of involved in the adsorption process. This model will be used to be able to identify parameters of a AF system shown in Fig. 1. In addition, the model will be used in design optimization studies and to extend the AF concept for other applications. This paper describes the preliminary AF model that has been developed.

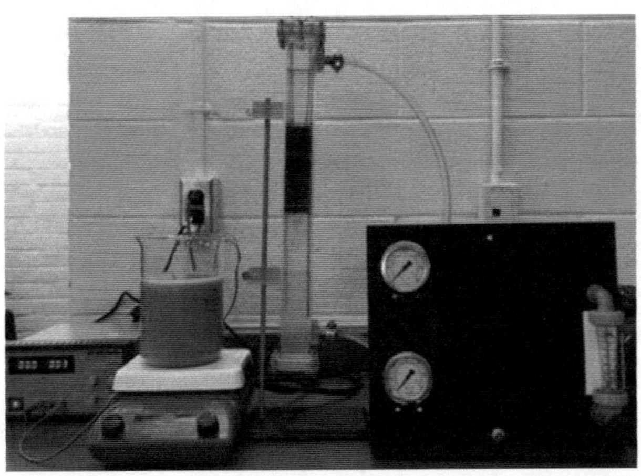

Figure 1. Hybrid Electrochemical and adsorbent filtration lab-scale system for emulsified oil-water separation

The AF model shown in Fig. 1 developed in COMSOL consists of stationary porous foam fixed-bed and mobile emulsified oily water. To model the adsorption process, the COMSOL-based framework outlined in Fig.2 was utilized. The coupled mass, energy, and momentum equations were solved numerically using the finite element method in COMSOL Multiphysics. In the preliminary simulations, the COMSOL built in adsorption model for mass transfer in porous adsorbent was used assuming instantaneous equilibrium between foam and oil, thereby neglecting mass transfer resistance. In future models, this resistance will be incorporated into the mass transport model. The mass transport model is coupled with built-in energy and momentum models predict the performance of the AF.

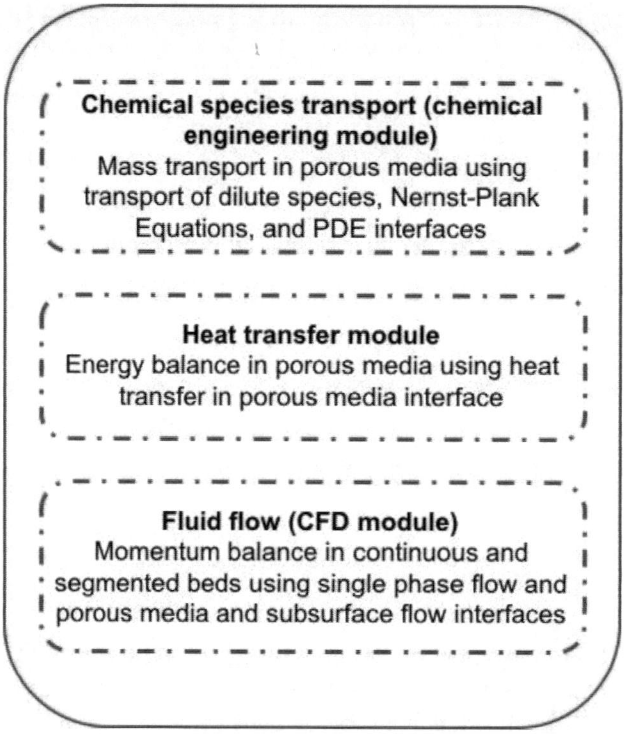

Figure 2. Model framework for investigating foam adsorbent filtration system using COMSOL Multiphysics

Preliminary Results

Fig. 3 and 4 shows preliminary model results evaluating the pressure drop for different AF bed configurations and the oil adsorption isotherms, respectively. For the lab-scale system, the segmented bed leads to lower pressure drop of 0.8 PSI and a higher oil mass uptake due to lower resistance forces. These preliminary results have already provided insight in design and the more detailed future studies will use the model to optimize the overall AF configuration.

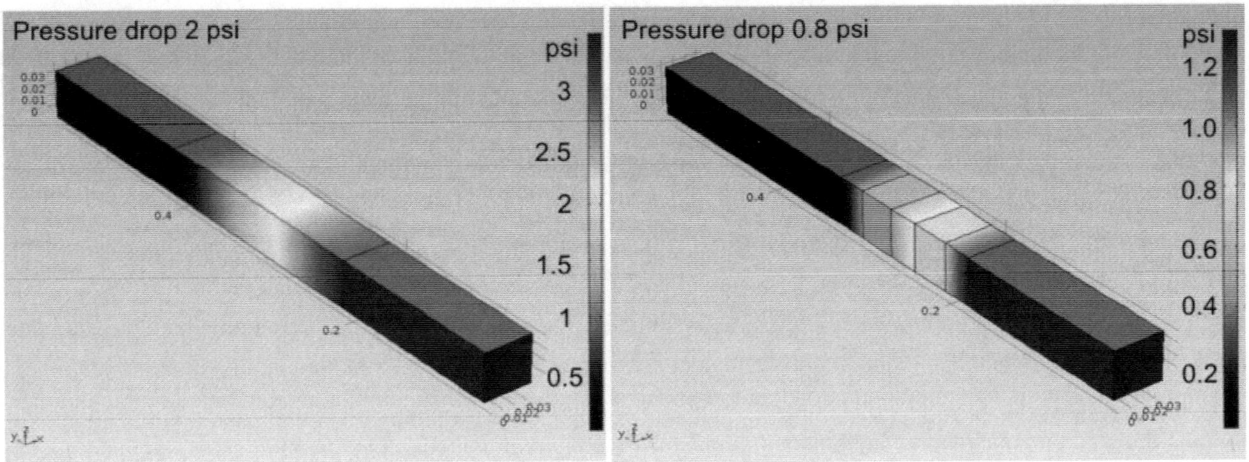

Figure 3. Pressure drops in continuous bed (left) and segmented bed (right)

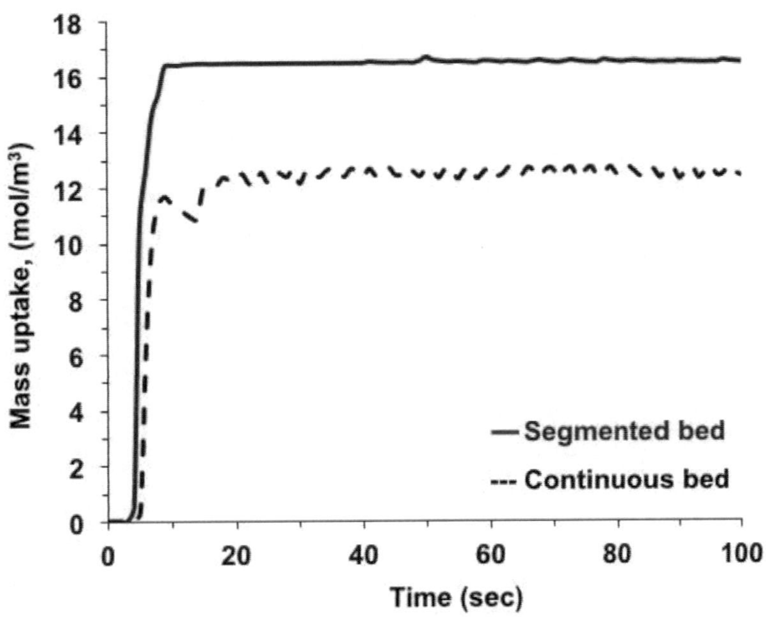

Figure 4. Oil droplets mass uptake in continuous and segmented beds

Conclusions

This paper presents preliminary models and lays the foundation towards design and development of foam based adsorbent filters. The approach can be used to evaluate AF for many applications including oily wastewater treatment, heavy metals removal, produced water treatment, and other separation processes.

References

1. M.A. Shannon et.al., Science and technology for water purification in the coming decades, Nature, 452, 301–310 (2008).
2. C.W. King and M.E. Webber, Water intensity of transportation. Environmental Science & Technology, 42, 7866–7872 (2008).
3. A. Fakhru'l-Razi et.al., Review of technologies for oil and gas produced water treatment, Journal of Hazardous Materials, 170, 530-551 (2009).
4. Tefera, Dereje Tamiru, et al., Two-dimensional modeling of volatile organic compounds adsorption onto beaded activated carbon, Environmental science & technology 47.20, 11700-11710 (2013).

A Combination of the eXtended Pom-Pom Model and the Stress-Thermal Rule to Predict Anisotropy in Thermal Conductivity in Non-Linear Polymeric Flows

David Nieto Simavilla[a*] and Wilco M. H. Verbeeten[a]

[a]*University of Burgos, Avda. Cantabria, s/n, 09006 Burgos Spain*
[*]*davidnietosimavilla@gmail.com*

Abstract. Advances in polymer processing enables more affordable technologies and significantly impacts the global plastics market which is expected to reach 654 billion USD by 2020. The cost/energy required to manufacture, recycle and dispose polymers is strongly affected by the thermo-physical properties and their dependence on state variables such as temperature and stress. The viscoelastic behavior of polymeric flows under isothermal conditions has been extensively researched. However, most of the processing of polymeric materials occurs under non-isothermal conditions and their implementation in simulations remains a challenge. In particular, flow-induced anisotropy in the thermal conductivity of polymers has been previously omitted in the study of industrially relevant flows. Our work combines evidence of a universal relationship between thermal conductivity and stress tensors (i.e. the stress-thermal rule) with differential constitutive equations for the viscoelastic behavior of polymers to provide predictions for the anisotropy in thermal conductivity in uniaxial, planar, equibiaxial and shear flow in commercial polymers. Special focus is placed on the eXtended Pom-Pom model which captures the non-linear behavior in both shear and elongation flows. The predictions provided by this approach are easily implemented in finite elements packages since the viscoelastic and thermal behavior is described by a single equation. Our results include predictions for the flow-induced anisotropy in thermal conductivity for low density polyethylene as well as validation of the method by comparison with available measurements. Remarkably, this approach allows universal predictions of anisotropy in thermal conductivity to be used in simulations of complex flows in which only the most fundamental rheological behavior of the material has been previously characterized (i.e. no adjusting parameters are needed in addition to those in the constitutive model).

Keywords: Anisotropy, Differential Constitutive Models, Flow Simulations in Polymers, Thermal Conductivity.
PACS: 65.60+a, 83.10.Gr, 83.50.Ax, 83.50.Jf

INTRODUCTION

The anisotropic nature of thermal transport in polymers plays an important role in their processing and final properties. Manufacturing processes like extrusion and injection molding are designed based on physical properties that are strongly dependent on temperature, such as viscosity. At the same time, the temperature distribution in the material being processed is characterized by thermo-physical properties (i.e. thermal conductivity), which are dependent not only on temperature but also on deformation[1-5]. These effects become significant when large temperature gradients are present, which is often the case in the processing of polymers due to their rather low thermal conductivity. While there has been a substantial amount of research characterizing the viscoelastic behavior of polymer systems during isothermal flows, the study of non-isothermal flows and the coupling between the thermo-physical properties and the process state variables remains a challenge. In this paper, we first introduce the concept of anisotropic thermal transport and its relation to deformation and stress through the *stress-thermal rule*. Then, we lay out the *eXtended Pom-Pom* model details used to describe the response of viscoelastic materials to flow. Finally, we combine experimental data, the *eXtended Pom-Pom* model and the *stress-thermal rule* to predict the anisotropy in thermal conductivity in steady-state and transient shear and uniaxial flows. The approach presented herein has important implications for the simulation of industrially relevant flows. Universality of the thermal transport response to stress in polymers allows for the use of a viscoelastic model to describe and predict both the rheological behavior and thermal transport. The minimal material characterization required results in reduced time and costs in the design and tuning of polymer manufacturing processes.

ANISOTROPY IN THERMAL CONDUCTIVTY AND THE STRESS THERMAL RULE

Thermal conductivity is a thermo-physical property governing the diffusive transport of energy. In oriented materials such as quartz crystals and polymers subjected to deformation, the thermal conductivity is an anisotropic property with different magnitude in different directions (See **¡Error! No se encuentra el origen de la referencia.**a). In such cases, Fourier's law is given by

$$\boldsymbol{q} = -\boldsymbol{k} \cdot \boldsymbol{\nabla T}, \tag{1}$$

where **k** is a generalized thermal conductivity tensor, **q** the energy flux, and **∇**T the temperature gradient. Experimental techniques, such as Forced Rayleigh Scattering (FRS)[2,4,5] and Infrared Thermography (IRT)[3], have characterized the anisotropy in thermal conductivity of polymers subjected to uniaxial extension and shear flows. These studies suggest a linear relationship between thermal conductivity and stress tensors. If energy is transferred more effectively along the backbone of the polymer chains than between them, then anisotropy in thermal conductivity should correlate with polymer chain orientation or strain. Based on this argument, it was proposed that the thermal conductivity $\boldsymbol{k}$ and stress τ tensors are linearly related by the *stress-thermal rule*[1] (See Figure 1b):

$$\boldsymbol{k} - \tfrac{1}{3}\delta \mathrm{tr}(\boldsymbol{k}) = k_{\mathrm{eq}} C_t \left(\boldsymbol{\tau} - \tfrac{1}{3}\delta \mathrm{tr}(\boldsymbol{\tau})\right), \tag{2}$$

where C_t is the stress thermal coefficient, and k_{eq} is the thermal conductivity of an unreformed sample that can be used to obtain the normalized thermal conductivity tensor $\boldsymbol{\alpha} = \boldsymbol{k}/k_{\mathrm{eq}}$. For uniaxial extension, the *stress-thermal rule* is expressed in its dimensionless form as (See Figure 1b),

$$\frac{k_{11}-k_{22}}{k_{eq}} = \alpha_1 - \alpha_2 = C_t G_N \frac{\tau_{11}-\tau_{22}}{G_N}, \tag{3}$$

where G_N is the plateau modulus characteristic of the material. A surprising result is that a universal (i.e. independent of polymer chemistry) value of ~0.03 is found for the normalized stress-thermal coefficient $C_t G_N$ for a wide variety of chemistries subjected to both uniaxial extension and shear flows[2-5] (See Table 1). Furthermore, the *stress-thermal rule* has been shown to hold beyond the limit imposed by finite extensibility in contrast with the *stress-optic rule* which fails at large stress[2] (i.e. thermal conductivity truly tracks stress as opposed to orientation or chain alignment). An example of the deviations from the equilibrium thermal conductivity in the directions parallel and perpendicular to the stretch in uniaxial extension is shown in Figure 1a.

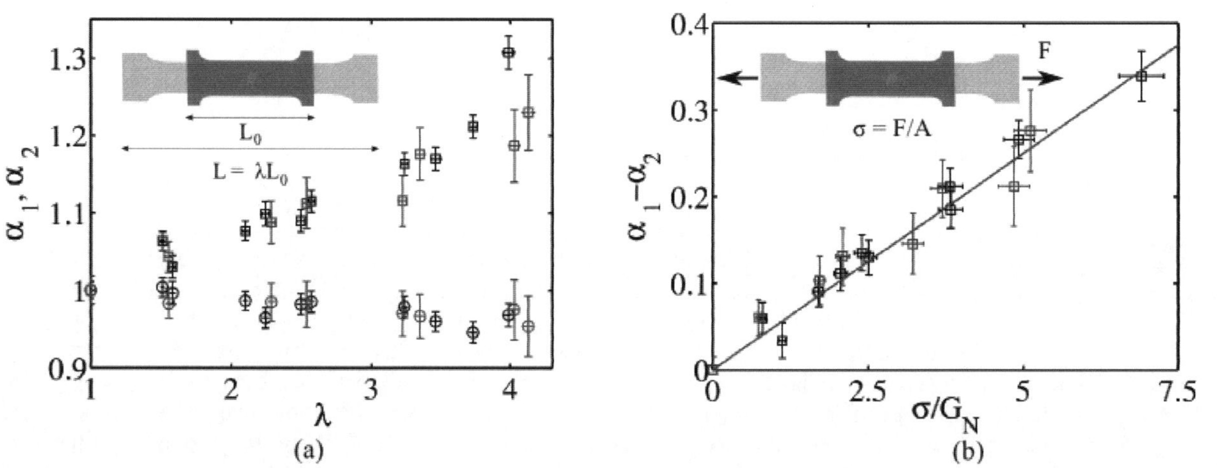

FIGURE 1. (a) Anisotropy in thermal conductivity as a function of stretch in a cross-linked polybutadiene elastomer subjected to uniaxial elongation obtained through FRS[2] (black) and IRT[3] (red) experiments. The thermal conductivity ratios α_1=k_{11}/k_{eq} and α_2=k_{22}/k_{eq} show an increase (decrease) in the thermal conductivity in the directions parallel (perpendicular) to the stretch with respect to the equilibrium value k_{eq}. **(b)** Test of the stress-thermal rule via IRT (red) and FRS (black).

Finally, for shear deformation we take into account two different components: 1) the anisotropy in thermal conductivity resulting from the first normal stress, which is equivalent to Eq. (3). And 2) an extra component to the thermal conductivity tensor due to the shear stress given by[4]:

$$\frac{k_{12}}{k_{eq}} = \alpha_{12} = C_t G_N \frac{\tau_{12}}{G_N}. \tag{4}$$

The extra component of the thermal conductivity tensor, which has not yet been measured experimentally, has important implications when applying Eq. (1) in energy balance (i.e. the temperature equation) used to study thermal transport since it results in "cross-flow" components $k_{12} \frac{\partial^2 T}{\partial x_1 \partial x_2}$ (i.e. a temperature gradient in the 2-direction will drive heat flow in the 1-direction).

TABLE 1. Summary of Stress-Thermal Coefficients and Plateau Modulus for Several Polymeric Materials[2-7].

Material (flow)	$C_t \times 10^{-4}$ [kPa^{-1}]	G_N [kPa]	$C_t G_N$
Polystyrene (uniaxial extension u.e.)	1.65±0.1	200	0.033±0.007
Poly (methyl methacrylate) (u.e.)	1.7±0.1	310	0.054±0.011
Polyisoprene (u.e.)	0.37±0.1	370	0.014±0.005
Polydimethylsiloxane (u.e.)	1.3±0.3	200	0.026±0.008
Polybutadiene (u.e.)	0.73±0.15	760	0.055±0.011
Polyisobutilene (shear)	1.2±0.67	320	0.038±0.022

VISCOELASTICITY MODEL: EXTENDED POM-POM

The eXtended Pom-Pom (XPP) model employed to describe the viscoelastic behavior of polymer melts was introduced by Verbeeten *et al.*[8] The original Pom-Pom model is based on tube theory combined with a simplified topology for branched molecules. The eXtended version of the Pom-Pom model is able to simultaneously describe the transient non-linear shear and extensional behavior in a quantitative manner and avoids the unrealistic discontinuities due to the finite extensibility condition that were present in the original version. Moreover, a second normal stress difference, that was not present in the original model, has been included. The linear viscoelasticity parameters, the plateau modulus $G_{0,i}$ and the linear relaxation time $\lambda_{0b,i}$ both of the ith mode, are determined from dynamic measurements, while the non-linear parameters (number of arms at the end of a backbone q_i, stretch relaxation time $\lambda_{0s,i}$, anisotropic drag parameter α_i) are fitted manually to the uniaxial extensional data only. In this study, we have used the same set of fitting parameters proposed by Verbeeten *et al.*[8] One of the advantages of the XPP model is that it is relatively easy to implement in finite element methods to simulate industrially relevant flows.[9-10]

PREDICTIONS OF THE COMBINED RHEOLOGICAL AND THERMAL TRANSPORT MODELS

By combining the eXtended Pom-Pom model stress predictions with the *stress-thermal rule* in Eqns. (2-4), we are able to predict the changes in the different components of the thermal conductivity tensor when a material is subjected to steady-state and transient deformations. Here, we focus on steady-state and transient start-up flows in a LDPE melt (IUPAC A) at 150 °C for which rheological data is available. However, once the model parameters have been determined, prediction for any number of flows (i.e. equibiaxial, planar…) inside as well as outside the deformation rates range that can be achieved in experiments is possible. An important parameter in the application of the *stress-thermal rule* is the plateau modulus $G_N \sim 250$kPa, which we have estimated as the sum of the plateau moduli $G_{0,i}$ of the generalized Maxwell model from dynamic measurements. The anisotropy in thermal conductivity predictions found via this approach are in good agreement with those observed for Polystyrene and Poly(methyl methacrylate) subjected to uniaxial extension[5] and polyisobutylene subjected to step shear[4] deformations.

Steady-State Shear and Uniaxial Deformation

Figure 2a presents the model predictions for shear viscosity and first normal stress coefficient as a function of the shear rate. Figure 2b shows the equivalent extensional viscosity as a function of the strain rate. Model predictions for the different moments of the stress tensor (insets in Figures 2c and 2d) are used to estimate the anisotropy in thermal conductivity under these two flow conditions. We confirm proper model rheological predictions for shear and uniaxial steady-state flows. In steady-state shear, we predict an increase in thermal conductivity of ~5% for shear component α_{12} and ~15% for the anisotropy $\alpha_{11} - \alpha_{22}$ due to the first normal stress. Steady-state uniaxial extension prediction shows a more significant increase in the anisotropy $\alpha_{11} - \alpha_{22}$ up to ~40% for a deformation rate $\dot{\varepsilon} = 20\ \mathrm{s}^{-1}$.

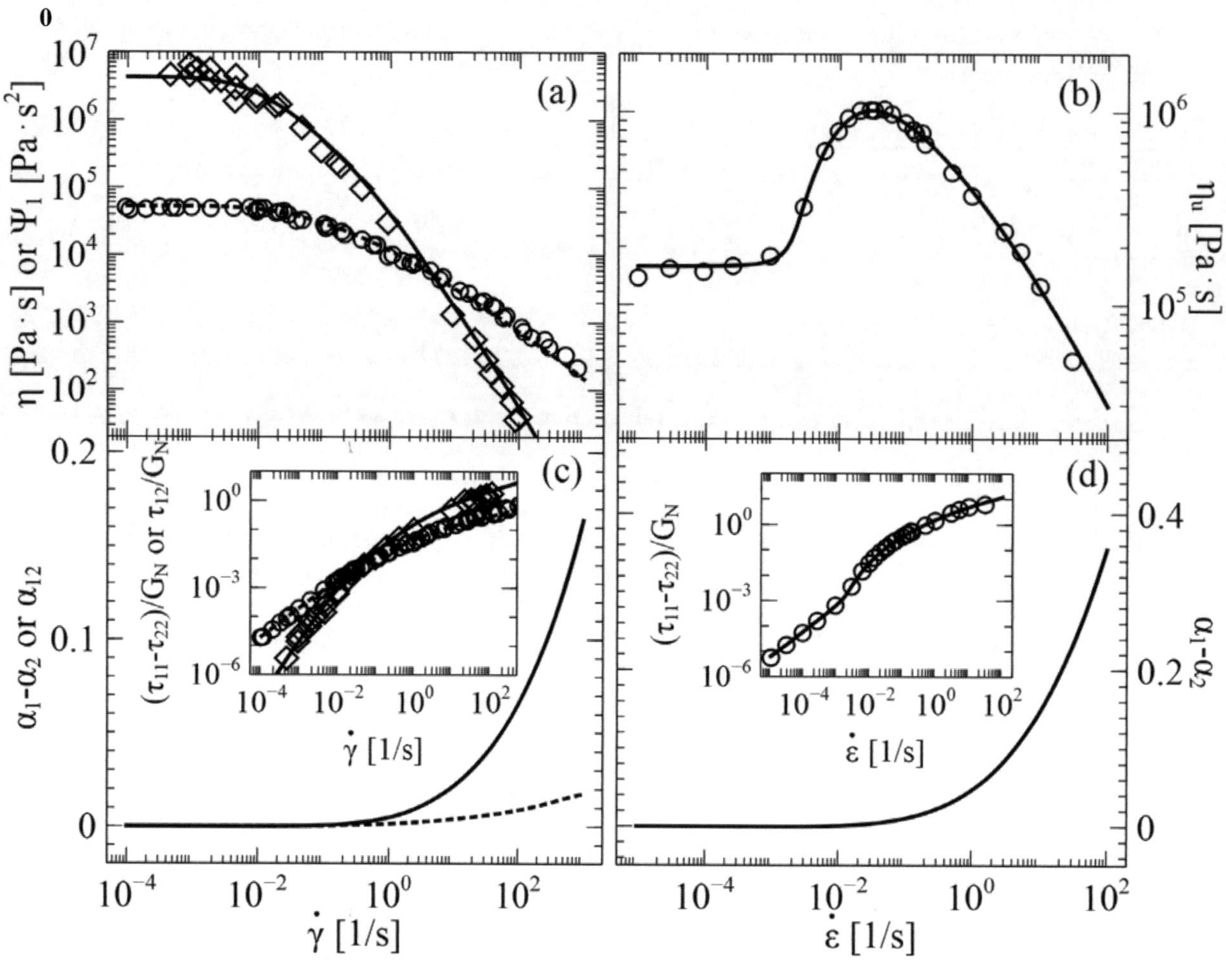

FIGURE 2. Steady-state shear and uniaxial results for LDPE melt (IUPAC A) at 150°C: **(a)** Shear viscosity (data: circles, model: dashed line) and first normal stress difference coefficient (data: diamonds, model: solid line) as a function of the shear rate. **(b)** Uniaxial viscosity (data: circles, model: solid line) as a function of the strain rate. **(c)** Predictions for the anisotropy $(k_{11} - k_{22})/k_{\mathrm{eq}} = \alpha_1 - \alpha_2$ (solid line) and the extra component $k_{12}/k_{\mathrm{eq}} - 1 = \alpha_{12}$ (dashed line) of the thermal conductivity tensor as a function of the shear rate. The inset shows the normalized normal stress difference (diamonds) and shear stress (circles) as a function of the shear rate. (**d**) Predictions for the anisotropy in thermal conductivity as a function of the strain rate. The inset shows the normalized normal stress as a function of the strain rate.

Transient Start-Up Shear and Uniaxial Deformation

Figures 3a and 3b present the transient start-up shear results for the IUPAC A LDPE melt at 150°C. We find good agreement between experiments and the model predictions. In Figures 3c and 3d, we include the shear and first normal stresses as a function of the shear strain. From the stress, we calculate the relevant components of the thermal conductivity tensor using Eqns. (2-4) (See Figures 3e and 3f). The stress goes through a maximum at $\gamma \sim 10$ which results on a non-monotonic behavior for the thermal conductivity tensor. In general, much lower values for the anisotropy in thermal conductivity (i.e. deviations from the equilibrium value) are observed due to a level of stress also much lower than in the steady-state flows or in the transient start-up for uniaxial extension (i.e. $(\tau_{11} - \tau_{22})/G_{\mathrm{N}}$ and τ_{12}/G_{N} <1).

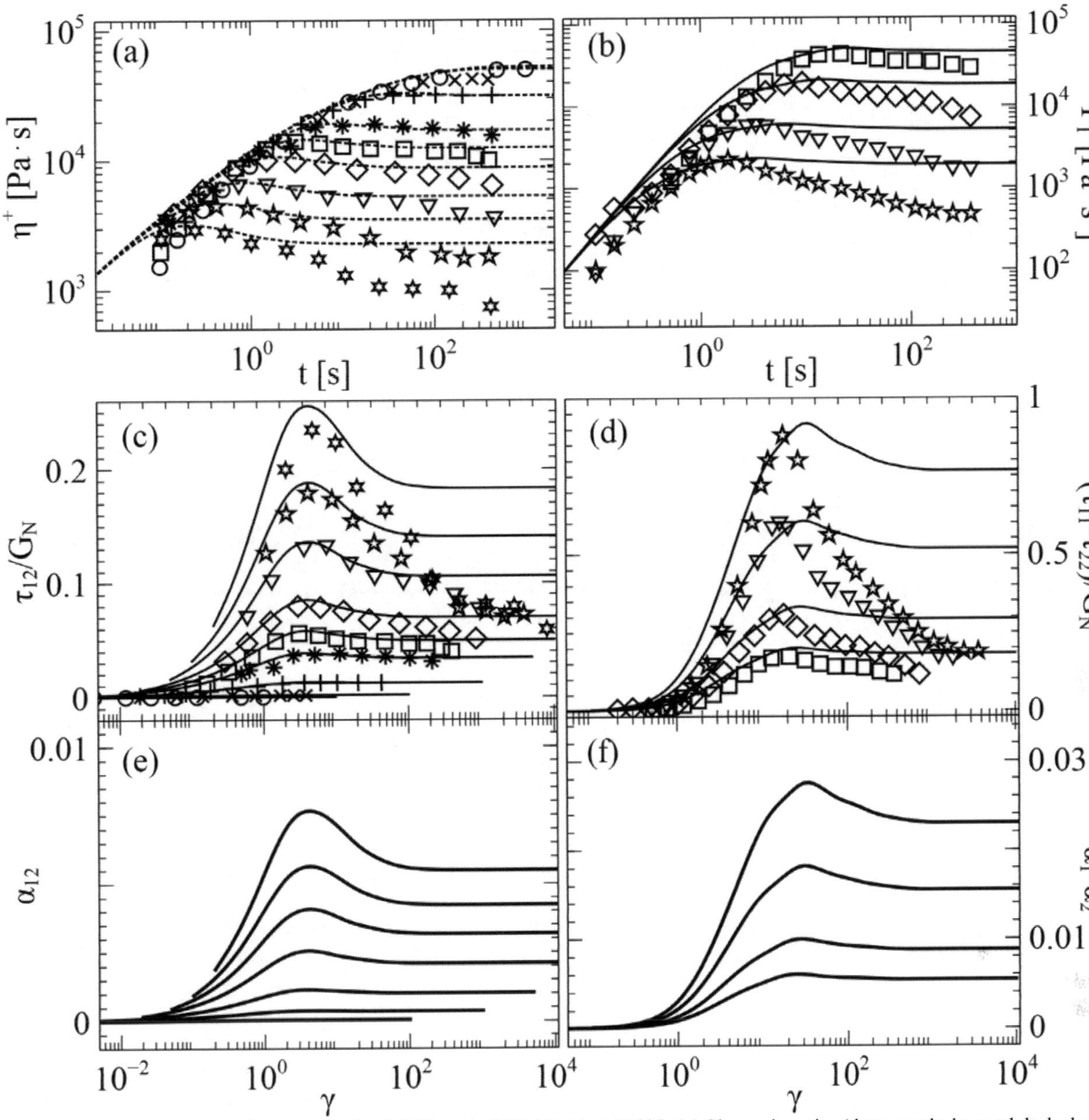

FIGURE 3. Transient start-up shear results for LDPE melt (IUPAC A) at 150°C: **(a)** Shear viscosity (data: symbols, model: dashed lines) as a function of time for shear rates $\dot{\gamma} = 0.001\ s^{-1}$ (circle), $0.01\ s^{-1}$ (cross), $0.1\ s^{-1}$ (plus), $0.5\ s^{-1}$ (asterisk), $1\ s^{-1}$ (square), $2\ s^{-1}$ (diamond), $5\ s^{-1}$ (triangle), $10\ s^{-1}$ (star-5), $20\ s^{-1}$ (star-6). **(b)** First normal stress coefficient (data: same symbols as in **(a)**, model: solid line) as a function of time. **(c)** Normalized shear stress as a function of shear strain (same symbols). **(d)** Normalized first normal stress as a function of shear strain (same symbols). **(e)** Normalized extra component of the thermal conductivity tensor $k_{12}/k_{\mathrm{eq}} - 1 = \alpha_{12}$ predicted from stress in **(c)** as a function of the shear strain. **(f)** Anisotropy $(k_{11} - k_{22})/k_{\mathrm{eq}} = \alpha_1 - \alpha_2$ in thermal conductivity predicted from stress in **(d)** as a function of the shear strain.

Finally, Figure 4a shows the transient start-up uniaxial extension viscosity data and model results. The stress-strain behavior is included in the inset in Figure 4a. In contrast to the transient start-up shear results, the stress increases monotonically with strain up to a maximum stress ~$6G_{\mathrm{N}}$ for the highest strain rate ($\dot{\varepsilon} = 30\ \mathrm{s}^{-1}$). The significantly higher stress results on high anisotropy in thermal conductivity (~25%) which is in good accordance with results for polystyrene and poly(methyl methacrylate) melts subjected to similar strains[5]. Interestingly the model predicts a plateau in the stress that results in a plateau in thermal conductivity (i.e. after a strain ε~5 the growth in anisotropy in thermal conductivity with strain stops once steady-state is reached).

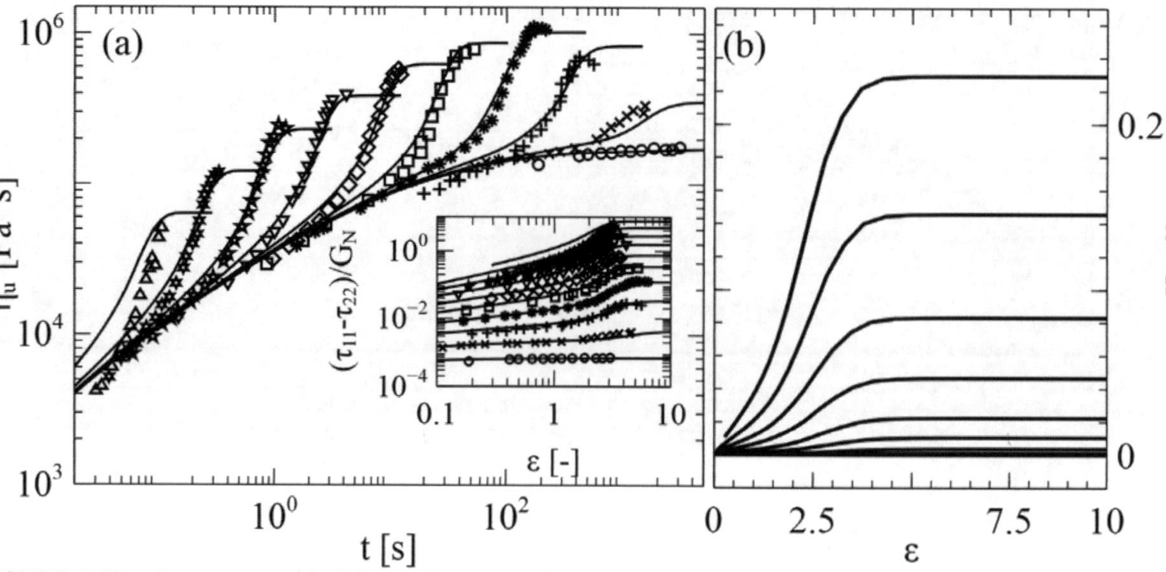

FIGURE 4. Transient start-up uniaxial extension results for LDPE melt (IUPAC A) at 150°C: **(a)** Extensional viscosity (data: symbols, model: solid lines) as a function of time for strain rates $\dot{\varepsilon}$ = 0.001 s^{-1} (circle), 0.003 s^{-1} (cross), 0.01 s^{-1} (plus), 0.03 s^{-1} (asterisk), 0.1 s^{-1} (square), 0.3 s^{-1} (diamond), 1 s^{-1} (down-triangle), 3s^{-1} (star-5), 10 s^{-1} (star-6) and 30 s^{-1} (up-triangle). The inset shows the normalized normal stress as a function of strain (same symbols). **(b)** Anisotropy $(k_{11} - k_{22})/k_{\text{eq}} = \alpha_1 - \alpha_2$ in thermal conductivity predicted from stress in **(a)** as a function of the strain.

CONCLUSIONS

We have combined the eXtended Pom-Pom model, which describes well the non-linear rheology of viscoelastic materials, with the *stress-thermal rule* to predict evolution of the thermal conductivity tensor when polymeric materials are subjected to flow. We observe higher deviations from the equilibrium thermal conductivity for both steady-state and transient start-up in uniaxial extension (up to ~40%) as a result of lower relaxation of the chain conformation giving rise to higher stress levels. For steady-state and transient start-up shear, deviations from the equilibrium thermal conductivity are lower. However, it is important to highlight that deformation rates in polymer manufacturing processes are significantly higher. While in polymer melt extrusion shear rates range typically between 1-100 s^{-1}, in processes such as injection molding and lubrication shear rates are typically 2-3 orders of magnitude higher[11] and, therefore, one would expect a more significant effect in the different components of the thermal conductivity tensor. Interestingly, the *stress-thermal rule* predicts an additional component (non-diagonal) k_{12} in the thermal conductivity tensor for shear flow. Although k_{12} is smaller in value than the diagonal counterparts, it could significantly affect the temperature equation during the simulation of non-isothermal flows. Our approach to model thermal transport in polymers has the advantage of requiring only a basic rheological characterization of the material to fit the model parameters and provide predictions of anisotropy in thermal conductivity for more complex flows. Since the model is amenable for its use in finite elements methods, future work will focus on the simulation of industrially relevant flows to study the effects of anisotropic thermal transport during and after polymer processing.

REFERENCES

1. van den Brule, B.H.A.A. Rheological Acta, 28 (1989).
2. Nieto Simavilla, D., Schieber, J.D. and Venerus, D. C. *Journal of Polymer Science Part B: Polymer Physics*, 50 (2012).
3. Nieto Simavilla, D. and Venerus, D. C. *Journal of Heat Transfer*, 136(11) (2014).
4. Venerus, D.C., Schieber, J.D., Balasubramanian, V., Bush, K. and Smoukov, S. *Physics Review Letters*, 93 (2004).
5. Gupta, S., Schieber, J.D. and Venerus, D.C. Journal of Rheology, 57 (2013)
6. Fetters, L. J., Lohse, D. J., Richter, D., Witten, T. A., and Zirkel, A. *Macromolecules*, 27 (1994).
7. Abdel-Goad, M., Pyckhou-Hitzen, W., Kahle, S., Allgaier, J., Richter, D. *Macromolecules*, 37 (2004).
8. Verbeeten, W.M.H., Peters, G.W.M. and Baaijens, F.P.T. *Journal of Rheology*, 45 (2001).
9. Verbeeten, W.M.H., Peters, G.W.M. and Baaijens, F.P.T. *Journal of Non-Newtonian Fluids Mechanics*, 108 (2002).
10. Verbeeten, W.M.H., Peters, G.W.M. and Baaijens, F.P.T. *Journal of Non-Newtonian Fluids Mechanics*, 117 (2004).
11. Barnes, H.A.; Hutton, J.F.; Walters, K., *"An Introduction to Rheology"*. Elsevier, Amsterdam, (1989).

Axisymmetric plastic flow of polymers near very rough walls

Alexandrov Sergey [(1)*]

[(1)] Institute for Problems in Mechanics - not applicable - Russia

There are a great number of experiments showing that skin layers are generated in the vicinity of frictional surfaces in polymer processing. Temperature is one of the main contributory mechanisms responsible for skin layer generation. In turn, friction is responsible for the temperature field in the vicinity of friction surfaces. In many cases, the effect of friction on temperature is studied assuming a friction law in terms of stress. However, such friction laws are only applicable in the regime of sliding whereas typical friction laws include sliding and sticking regimes. The relative velocity between the workpiece and tool vanishes in the regime of sticking and no heat is generated at the surface. Therefore, it is of importance to study qualitative solution behavior in the vicinity of frictional surfaces. In particular, it is known that the constitutive equations that are usually adopted for metals may or may not be compatible with the regime of sticking. The present paper extends this result to constitutive equations that can be used for polymers. A class of polymers can be treated as rigid plastic materials under certain conditions. The yield criterion of isotropic materials may include three stress invariants. However, in most cases yield criteria for polymers are assumed to be independent of the cubic invariant. On the other hand, a sensible yield criterion must depend of the linear invariant. The normality rule is often adopted for polymers. However, for several polymers the volume change is similar to that observed in metals. Therefore, the normality rule used in conjunction with a pressure-dependent yield criterion is not valid for such polymers. The present paper deals with such polymers. The plastic potential function is taken in the form of Mises. An axisymmetric boundary value problem is solved to study solution behavior in the vicinity of the friction surface for two yield criteria. Qualitative differences in solution behavior are discussed.

EFFECT OF FIBER LENGTH DISTRIBUTION ON LONG SEMI-FLEXIBLE FIBER ORIENTATION DURING INJECTION MOLDING

Hongyu Chen[1], Peter Wapperom[2] and Donald G. Baird[1]

[1]*Department of Chemical Engineering, Virginia Tech, Blacksburg, VA 24061*
[2] *Department of Mathematics, Virginia Tech, Blacksburg, VA 24061*

Abstract. When considering the mechanical performance of polymeric composites, Long-fiber (lengths > 1 mm) thermoplastic composites (LFTs) have attracted significant attention over short fiber (< 1 mm) composites. The fiber orientation distributions are crucial to the mechanical properties. This paper aims to study the length effect on fiber orientation inherent in long fiber systems. In this study, the Bead-Rod fiber orientation model was used which takes into account the flexibility of semi-flexible fibers that show small bending angles. Different averaging techniques were used to represent the average fiber length for the 30 wt% glass fiber composites which give different fiber length parameters for the Bead-Rod model. The sensitivity of the Bead-Rod model was evaluated about the fiber flexibility parameter k and length parameter l_B. Fiber orientation simulation was conducted for the Center-Gated-Disk (CGD) geometry and compared with experimental data. To further study the length effect on fiber orientation, short pellets (4 mm) and long pellets (12 mm) were used to generate injection molded parts with different fiber length distribution under the same fiber content (40 wt% and 50 wt%). The experimental orientation results between different fiber length distribution with both 40 wt% and 50 wt% fiber concentration are compared in this study. The simulations are in progress and will be presented at the conference.

INTRODUCTION

As reinforcing constituents in a polymer matrix, fibers attract tremendous interest, because the strength and stiffness imparted to the composite materials[1]. Fiber reinforced polymer composites are also in a wide variety of forms, ranging from discontinuous short fibers to continuous oriented fibers. Thermosetting resins are the most widely used matrix for continuous fiber polymer composites, while thermoplastic matrices are often used in discontinuous fiber reinforced polymer composites. If the fibers are aligned and if they are sufficiently long, discontinuous fiber composites can approach the performance of continuous fiber composites. Important microstructural variables for discontinuous fiber reinforced composites include fiber concentration, fiber length distribution, and fiber orientation distribution which is induced by typical polymer processing procedures especially injection molding[2]. Long-fiber (lengths > 1 mm) thermoplastic composites (LFTs) possess significant advantages over short fiber (< 1 mm) composites regarding their mechanical properties while retaining their ability to be injection molded. During the plasticating procedure, fiber lengths are prone to experience significant attrition, and as a result, before entering the mold cavity, the fiber suspension ends up with a broad fiber length distribution[3]. There is a lack of knowledge about the effects of fiber length and fiber length distribution on fiber orientation kinetics. This lack of information provides an opportunity to understand the length effect inherent in long fiber systems.

Background

It is customary to use a unit vector **p** to specify the orientation of a rigid axially symmetric particle while a long and flexible fiber is represented as three beads connected by a pair of orientation vectors, **p,** and **q** as shown in Figure 1[4]. To describe the orientation state of a suspension of rigid fibers, a tensor **A** defined as the second moment of the orientation vector **p** is given by:

$$\mathbf{A} = \langle \mathbf{pp} \rangle = \int \mathbf{pp}\psi(\mathbf{p}, t)\, d\mathbf{p} \qquad (1)$$

where ψ(**p**) is the probability distribution function statistically describing the orientation distribution the orientation state for a complete population of particles in a suspension.

(a) (b)

FIGURE 1. Rigid-rod representation of rigid fiber (a) and Bead-rod representation of flexible fiber.

To predict short (rigid) fiber orientation induced by flow, the strain reduction factor (SRF) model is widely used. This model has the advantage to capture slow orientation kinetics by multiplying the right-hand side of the Folgar-Tucker model by a factor α as below[5]:

$$\frac{D\mathbf{A}}{Dt} = \alpha[\mathbf{W}\cdot\mathbf{A} - \mathbf{A}\cdot\mathbf{W} + \xi[\mathbf{D}\cdot\mathbf{A} + \mathbf{A}\cdot\mathbf{D} - 2\mathbf{D}{:}\mathbf{A}_4] + 2C_I\dot{\gamma}(\mathbf{I} - 3\mathbf{A})], \tag{2}$$

where **W** is the vorticity tensor, and **D** the rate of deformation, C_I the phenomenological constant which is used to include the effect of fiber-fiber interactions on the motion of fibers, and **I** the unit tensor.

For long and flexible fibers, in the Bead-Rod model, the end to end tensor $\boldsymbol{r}$, is defined as the second moment of the end-to-end vector, with respect to ψ as:

$$\mathbf{r} = \iint (\mathbf{p} - \mathbf{q})\,(\mathbf{p} - \mathbf{q})\psi(\mathbf{p}, \mathbf{q}, t)d\mathbf{p}d\mathbf{q}. \tag{3}$$

The semi-flexible system[6] uses the similar moments of orientation notation as the rigid fiber model, but because of the two-rod system, two more moments rather than the **A** tensor occur and are defined by:

$$\mathbf{B} = \iint \mathbf{pq}\psi(\mathbf{p}, \mathbf{q}, t)d\mathbf{p}d\mathbf{q} \tag{4}$$

$$\mathbf{C} = \iint \mathbf{p}\psi(\mathbf{p}, \mathbf{q}, t)d\mathbf{p}d\mathbf{q}. \tag{5}$$

The end-to-end tensor, r, can also be normalized to obtain a tensor R regarding A and B:

$$\mathbf{R} = \frac{\mathbf{r}}{\mathrm{tr}(\mathbf{r})} = \frac{\mathbf{A} - \mathbf{B}}{1 - \mathrm{tr}(\mathbf{B})}. \tag{6}$$

The three moments of the Bead-Rod model give rise to a semi-flexible orientation model for use in a dilute solution, which has been extended by Ortman et al[7]. to concentrated solutions by adding the isotropic rotary diffusion term given as:

$$\frac{D\mathbf{A}}{Dt} = \alpha\left(\mathbf{W}\cdot\mathbf{A} - \mathbf{A}\cdot\mathbf{W} + \mathbf{D}\cdot\mathbf{A} + \mathbf{A}\cdot\mathbf{D} - 2\mathbf{D}{:}\mathbf{A}_4 - 2C_I\dot{\gamma}(3\mathbf{A} - \mathbf{I}) + \frac{l_B}{2}(\mathbf{Cm} + \mathbf{mC} - 2(\mathbf{m}\cdot\mathbf{C})\mathbf{A}) - 2k\big(\mathbf{B} - \mathbf{A}\mathrm{tr}(\mathbf{B})\big)\right) \tag{7}$$

$$\frac{\mathbf{DB}}{\mathrm{Dt}} = \alpha\Bigg(\mathbf{W}\cdot\mathbf{B} - \mathbf{B}\cdot\mathbf{W} + \mathbf{D}\cdot\mathbf{B} + \mathbf{B}\cdot\mathbf{D} - 2(\mathbf{D}:\mathbf{A})\mathbf{B} - 4C_{\mathrm{I}}\dot{\gamma} \quad + \frac{l_{\mathrm{B}}}{2}(\mathbf{Cm} + \mathbf{mC} - 2(\mathbf{m}\cdot\mathbf{C})\mathbf{B}) - 2k\big(\mathbf{A} - \mathbf{B}\mathrm{tr}(\mathbf{B})\big)\Bigg) \tag{8}$$

$$\frac{\mathbf{DC}}{\mathrm{Dt}} = \alpha\left(\nabla\mathbf{v}^{\mathrm{t}}\cdot\mathbf{C} - (\mathbf{A}:\nabla\mathbf{v}^{\mathrm{t}})\mathbf{C} - 2C_{\mathrm{I}}\dot{\gamma}\mathbf{C} + \frac{l_{\mathrm{B}}}{2}\big(\mathbf{m} - \mathbf{C}(\mathbf{m}\cdot\mathbf{C})\big) - \mathrm{k}\mathbf{C}\big(1 - \mathrm{tr}(\mathbf{B})\big)\right) \tag{9}$$

$$\mathbf{m} = \sum_{i=1}^{3}\sum_{j=1}^{3}\sum_{k=1}^{3}\frac{\partial^2 v_i}{\partial x_j \partial x_k}A_{\mathrm{jk}}\boldsymbol{\delta}_i \tag{10}$$

$$k_i = \frac{E_f}{8\eta_m}\left(\frac{d}{l_i}\right)^3 \tag{11}$$

where l_B is the half fiber length, and k_i the resistivity to bending for a single fiber which are directly related to fiber length and thus might provide an opportunity to understand the length effect inherent in long fiber systems. Strautins and Latz[6] have assumed fibers are in a non-uniform flow, where the velocity is twice differentiable. The flow effect on fiber bending is taken into account through the m vector. Ortman et al[7]. used a simply supported beam with a point load in the center to estimate the flexibility parameter, k_i, given by Eq.(11) for a fiber with length l_i. It is necessary to associate a single k_i value to the population of fibers, and in this paper the averaged value of k is calculated through the use of Eq.(11) by plugging in an averaged fiber length L_{av}.

Using different averaging techniques to calculate the average fiber length, L_{av} will lead to different values of k. The widely used number average and weight average fiber length are defined by Eq.(12) and Eq.(13), respectively:

$$L_n = \frac{\sum N_i l_i}{\sum N_i} \tag{12}$$

$$L_w = \frac{\sum N_i l_i^2}{\sum N_i l_i} \tag{13}$$

where, N_i is the number of fibers with length l_i in the system. In the expression for L_n, all fiber sizes contribute equally to the final average. The definition of L_w implies that longer fibers contribute more to the average than shorter fibers. A similar way to represent the distribution of fiber lengths is the z-average length, which includes a greater contribution with respect to fiber length relative to the weight average method. Huq and Azaiez compared the steady shear viscosity of one-model, bi-model, and tri-model samples obtained by mixing fibers of uniform lengths. The results revealed that L_n, L_w and L_z fail to accurately represent contributions from long and short fibers to the suspension rheology. Their study also suggested that the modified average fiber length, L_{HA}, was more appropriate. In this study, the length dependent parameters in the Bead-Rod model are evaluated by the L_n, L_w, L_z and L_{HA} average fiber length.

EXPERIMENTS

To evaluate the performance of Bead-Rod model, 12 mm long, 30 wt% glass fiber reinforced polypropylene pellets were injection molded into center gated disks and the experimental orientation results were compared with the simulation results. To further study the length effect on fiber orientation, short pellets (4 mm) and long pellets (12 mm) were used to generate injection molded parts with different fiber length distribution under the same fiber content (40 wt% and 50 wt%). The radial distance filled inside the mold corresponded to about 90% of the total radial length inside the mold and the disk geometry is shown in FIGURE 2.

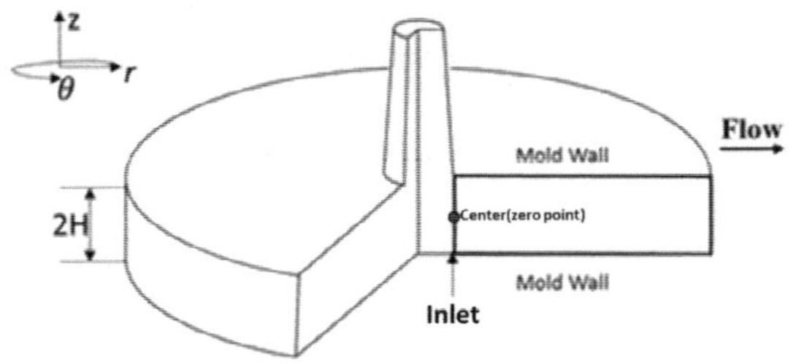

FIGURE 2. Center-gated disk geometry in this study.

The micrographic technique of García et al[8]. is used to obtain the fiber orientation at different radial locations along the mold cavity of the center-gated disk. The measurements of the elliptical footprints of the fibers in the r-z plane were used to determine the components of the orientation tensor.

RESULTS AND DISCUSSION

Evaluation of Bead-Rod Model

For both the rigid Folgar-Tucker and the Bead-Rod fiber orientation models the phenomenal parameter C_I and α are determined by fitting the orientation predictions to experimental data under startup of steady shear in a sliding plate rheometer. The parameter values are $(C_I, \alpha) = (0.005, 0.25)$ for the Folgar-Tucker model and $(C_I, \alpha) = (0.05, 0.1)$ for the Bead-Rod model. The fiber length distribution was experimentally determined and as shown in FIG.3.

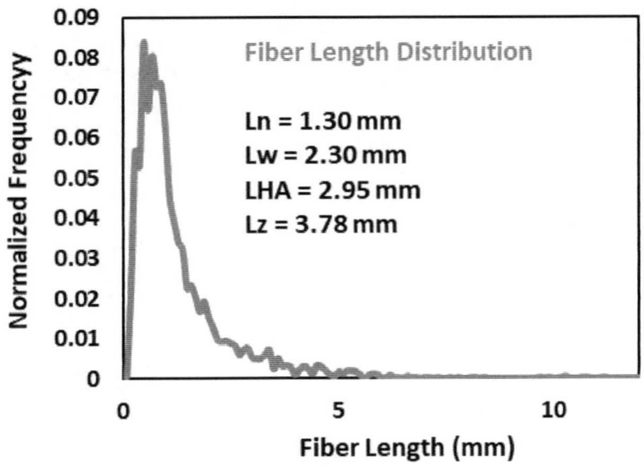

FIGURE 3. Fiber length distribution for glass fiber at a concentration of 30 wt%.

Orientation predictions with the Bead-Rod model were assessed by comparing model predictions with different fiber length related parameters. To verify the predictions of fiber orientation near the advancing front, experimental orientation values are compared to model predictions at 80% and 90% of the mold fill. Figure 4 (a) shows the comparison at r/H = 31.694. The results of the Rigid Folgar Tucker model show the characteristic shell-core-shell layer structures. The Bead Rod model predicted higher flow direction orientation compared with the rigid fiber model, which matches better with the experimental data. At the shell region, the predicted result of the rigid Folgar Tucker model is very similar to that of the Bead Rod model using the parameter values kL_n & L_n (the most rigid flexibility parameter and the shortest average fiber length in this study). As the value of L_{av} increases which is determined by the averaging technique and the corresponding flexibility parameter value decreases, the value of R_{11} gradually decreases near the wall. It seems the Bead-Rod model with the parameter set kL_z & L_z performs encouragingly well with both predicting the qualitative trends through the mold cavity as well as agreeing quantitatively with observed orientation for this LGF system. The exceptions are around z/H = 0.53 and 0.68 where the orientation is under predicted. Figure 4 (b) shows the simulation results at r/H = 45.994 (90 % flow length). It turns out all the simulations over predict the flow direction orientation from z/H = -0.54 to 0.53. Again the Bead-Rod model with parameters kL_z & L_z is observed to qualitatively better capture the relative broader orientation profile through the mold thickness.

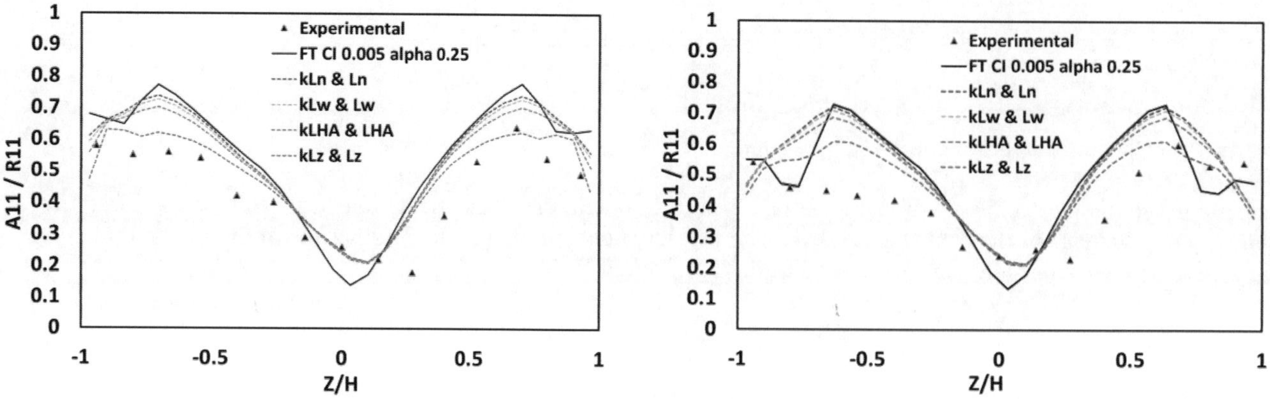

(a) (b)

FIGURE 4. Comparison of the simulation results and the experimental data at r/H = 31.694 (a) and r/H = 41.18 (b).

Length Effect on Fiber Orientation

To further study the length effect on fiber orientation, short pellets (4 mm) and long pellets (12 mm) were used to generate injection molded parts with different fiber length distribution under the same fiber content (40 wt% and 50 wt%). The experimental orientation results between different fiber length distribution with both 40 wt% and 50 wt% fiber concentration are compared in this study. Table 1 shows the final fiber lengths at the center of the disks starting with short and long pellets at different fiber concentration. It is clear that 12 mm pellets deliver longer fibers to the final disks in comparison to 4 mm pellets generated final parts. In both cases, the average fiber length appears to decrease with increasing fiber content. The average fiber lengths of 50 wt% composites of both 4 mm and 12 mm pellets created disks are significantly shorter than the corresponding lower content composites. This means that the fiber-fiber interaction mechanism might dominate at 50 wt% concentration.

Table 2. Final fiber lengths start with short and long pellets at different fiber concentration.

	40 wt%		**50 wt%**	
	4 mm pellets	**12 mm pellets**	**4 mm pellets**	**12 mm pellets**
L_n **(mm)**	0.84	1.02	0.475	0.81
L_w **(mm)**	1.48	2.7	0.726	1.58

FIG. 5 compares the flow direction orientations between short and long pellets generated disks with 40 wt% glass fiber content. Noticeable differences are observed at the inlet and the advancing front.). At the inlet and advancing front, the flow pattern changes quickly from the fully developed region within a very short range. At the inlet, the long pellets generated disk has a lower degree of flow direction orientation at the top half compared to the short pellets. At the advancing front, the short pellets have a broader fiber orientation distribution than the short ones. These phenomena might arise becasue shorter fibers are more prone to adjusting the alignment to the change of flow pattern. FIG. 6 compares the flow direction orientations between short and long pellets generated disks with 50 wt% glass fiber content. Again there are noticeable differences at the inlet and advancing front. The general trends are very similar to those of the 40 wt% fiber content. However, the orientation distribution of the short pellets generated disks at the advancing front is very asymmetric.

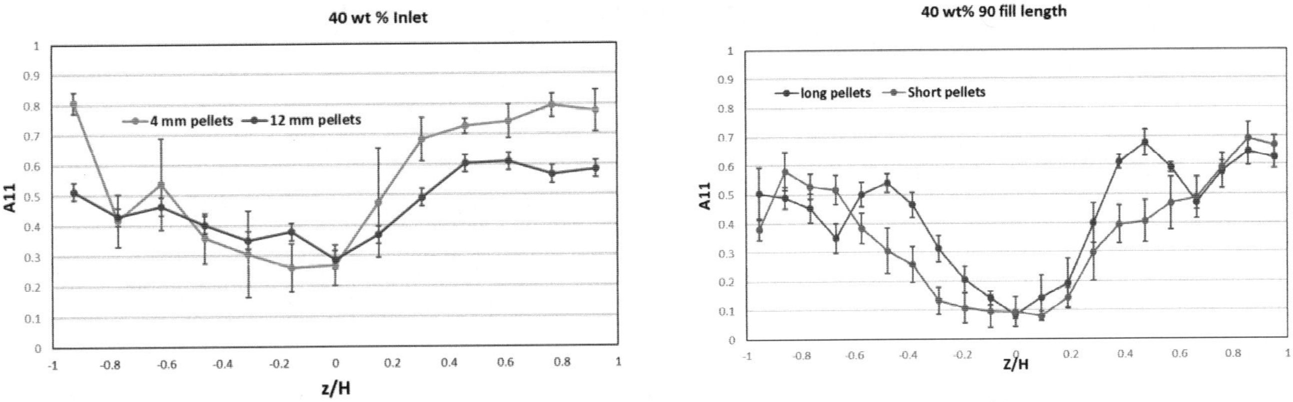

(a) (b)

FIGURE 6. Comparison of the flow direction fiber orientation between short and long pellets generated disks at the inlet (a) and the advancing front (b) under 40 wt% glass fiber content.

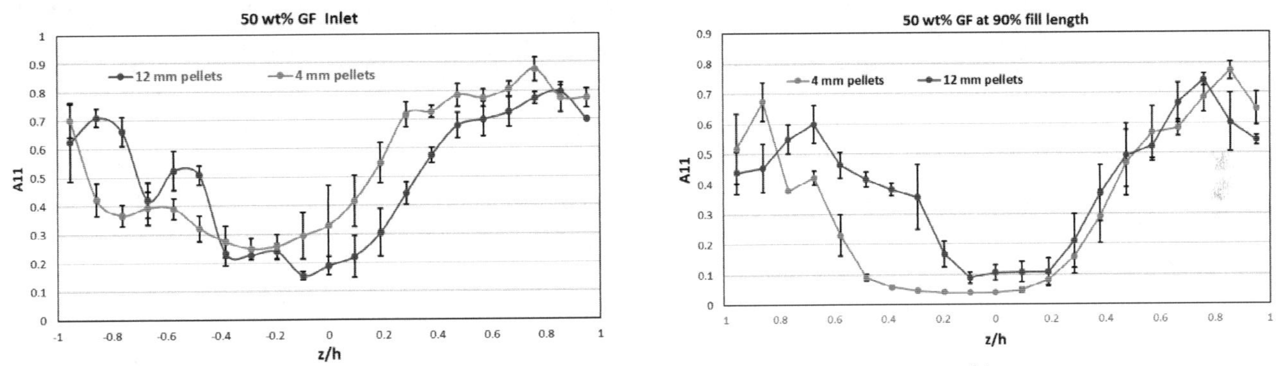

(a) (b)

FIGURE 6. Comparison of the flow direction fiber orientation between short and long pellets generated disks at the inlet (a) and the advancing front (b) under 50 wt% glass fiber content.

CONCLUSION

Performance of the Folgar-Tucker model and the Bead-Rod model in injection molding is assessed against experimental long fiber orientation data obtained from a center-gated disk. Along z/H profiles the rigid Folgar-Tucker model predicts the characteristic shell-core shell layer structure. However, at radial locations close to the front, the experimental data shows noticeable higher orientation values at the core region than the predicted results by the Folgar-Tucker model. The Bead-Rod model performs encouragingly more accurately predicting experimental fiber orientation at most positions investigated. The fiber length effect on the performance of the Bead-Rod model has also been investigated by using different fiber length averaging values. As the value of average fiber length L_{av} increases and the corresponding flexibility parameter value decreases, the core regions become wider and the flow direction orientation gradually decreases especially near the walls for the Bead-Rod model predictions. The fiber length distribution effect on fiber orientation has also been studied at different fiber concentration. There are noticeable differences at the inlet and advancing front. It might be reason that at the inlet and advancing front, the flow pattern changes quickly from the fully developed region within a very short range and shorter fibers are more prone to adjusting the alignment to the change of flow pattern.

REFERENCES

1. Phelps, J. H., *Processing-microstructure models for short-and long-fiber thermoplastic composites*. ProQuest: 2009.
2. Thomason, J. L.; Vlug, M. A., Influence of fibre length and concentration on the properties of glass fibre-reinforced polypropylene: 1. Tensile and flexural modulus. *Composites Part A: Applied Science and Manufacturing* **1996,** *27* (6), 477-484.
3. Phelps, J. H.; Abd El-Rahman, A. I.; Kunc, V.; Tucker Iii, C. L., A model for fiber length attrition in injection-molded long-fiber composites. *Composites Part A: Applied Science and Manufacturing* **2013,** *51* (0), 11-21.
4. Strautins, U. Flow-driven orientation dynamics in two classes of fibre suspensions. Ph. D. thesis, Technical University Kaiserslautern, 2008.
5. Nguyen, B. N.; Bapanapalli, S. K.; Holbery, J. D.; Smith, M. T.; Kunc, V.; Frame, B. J.; Phelps, J. H.; Tucker, C. L., Fiber Length and Orientation in Long-Fiber Injection-Molded Thermoplastics — Part I: Modeling of Microstructure and Elastic Properties. *Journal of Composite Materials* **2008,** *42* (10), 1003-1029.
6. Strautins, U.; Latz, A., Flow-driven orientation dynamics of semiflexible fiber systems. *Rheologica Acta* **2007,** *46* (8), 1057-1064.
7. Ortman, K.; Baird, D.; Wapperom, P.; Whittington, A., Using startup of steady shear flow in a sliding plate rheometer to determine material parameters for the purpose of predicting long fiber orientation. *Journal of Rheology (1978-present)* **2012,** *56* (4), 955-981.
8. Vélez-García, G. M.; Wapperom, P.; Baird, D. G.; Aning, A. O.; Kunc, V., Unambiguous orientation in short fiber composites over small sampling area in a center-gated disk. *Composites Part A: Applied Science and Manufacturing* **2012,** *43* (1), 104-113.

EXTENSION OF THE 3D-CFD SIMULATION TECHNIQUE TO VISUALIZE BLOCKAGE PHENOMENA IN MELTING PROCESSES IN POLYMER PROCESSING BASED ON A CUSTOM MATERIAL MODEL

Neubrech Bastian [1]*, Karrenberg Gregor [1], Wortberg Johannes [1]

[1] University of Duisburg-Essen - Duisburg - Germany

The extruder is the central component for any polymer processing procedure and the economic design benefits from the possibilities given by numerical flow simulations. Nevertheless, commercial software does not offer sufficient possibilities for calculating plastification processes. Therefore, an approach based on a custom material model for robust 3D-CFD simulations of the plastification of polymeric materials has been developed. Temperature-dependent material properties allow a differentiation between the solid and the liquid phase and make presupposed melting mechanisms obsolete. The numerical calculations have been performed for models with a screw diameter from 35 mm up to 100 mm and validated under high speed conditions. Nevertheless, the model shows deficits in dealing with blockage phenomena occurring in front of barrier elements or in the present case of a custom extruder concept with solid-melt separation. These deficits are reasoned by the simplification of simulating melting processes in one single fluid domain. This paper aims to extend the simulation techniques to depict the melting process in polymer processing more correctly. Therefore, the implementation of porous media is realized to visualize blockage phenomena. In a next step, the model is calibrated to guarantee reliable results. Finally, a reference simulation state is compared to the results with the extended approach to evaluated the gained knowledge.

A Bubble-Scale Continuum Approach to Model Foam Expansion and Bubble Pressurization

Rao Rekha R (1)*, Long Kevin (1), Roberts Chirstine C (1), Voskuilen Tyler (1), Noble David (1), Celina Mathew (1)

(1) Sandia National Laboratories - NM - USA

In this presentation, we discuss our new, experimentally driven, approach to foam expansion and polymerization modeling. Our foam, PMDI, is chemically blown foam, where carbon dioxide is produced via the reaction of water and isocyanate. The isocyanate also reacts with polyol in a competing reaction, which produces the polymer. The foam expansion kinetics are determined by tracking the molar concentration of water and carbon dioxide in both the gas and polymer phases, where the gas is quite soluble. Bubble-expansion is driven by a number of factors including gas generation rate, vapor-liquid equilibrium between the dissolved carbon dioxide and the gas pressure in the bubbles, and the polymer's evolving viscosity, gelation and vitrification. The evolving viscosity is linked to the polymerization reaction, which is modeling via condensation chemistry, where vitrification and glass transition temperature evolution must be included to correctly predict this quantity. A bubble-scale mathematical model has been developed that looks at the super saturation of carbon dioxide in the polymer and the concomitant bubble pressurization. The model is calibrated with bubble-expansion data from bubble growth experiments. For validation, the model is used to predict the density of a foam bar and compared to X–Ray CT data. Foaming, curing and density predictions in a complex mold is also discussed.

ACKNOWLEDGMENTS Sandia National Laboratories is a multi-program laboratory managed and operated by Sandia Corporation, a wholly owned subsidiary of Lockheed Martin Corporation, for the U.S. Department of Energy's National Nuclear Security Administration under contract DE-AC04-94AL85000.

Injection Moulding Simulation and Experimental Validation of Hearing Aid Shells

A. Islam (presenting author)[a, b*], X. Li[b]

[a]Centre for Acoustic-Mechanical Micro Systems, Technical University of Denmark, Ørsteds Plads, Building 352, DK-2800 Kgs. Lyngby, Denmark
[b]Department of Mechanical Engineering, Technical University of Denmark, Produktionstorvet, Building 427, Dk-2800 Kgs Lyngby, Denmark

mais@mek.dtu.dk

Abstract. Moldex3D with Boundary Layer Meshing (BLM) mode was adopted in this work to simulate the injection molding process of a hearing aid shell made of Polybutylene Terephthalate (PBT) with 30% glass fiber filled and predict the part warpage. Injection molding experiment was conducted to validate the prediction from Moldex3D. Injection time, injection pressure; pressure loss and warpage were treated as the main comparison criteria. Different parameters setting in Moldex3D were investigated to research their influence on the accuracy of simulation. Results showed that the injection molding process prediction from the simulation was relatively precise by considering the nozzle geometry and pressure effect on the viscosity in the simulation model and determine a proper heat transfer coefficient (HTC) value. But good agreement was not observed between the warpage of the experiment molded part and simulated parts. Warpage were dominated by the fiber orientation. Predicted warpage was found to be extremely dependent on the filling HTC (Heat Transfer Coefficient) and the fiber orientation model used in Moldex3D, both of which had significant influence on the fiber orientation.

Keywords: Moldex3D, Injection molding, Warpage, Fiber orientation.

INTRODUCTION

The parts of the hearing aid are held together in a protective case consists usually of upper and lower shells which are assembled together at the final stage of the production [1]. Fig 1 shows the picture of a hearing aid and its shells. The outer shells of hearing aids are made by injection molding. As they are usually thin walled structures, highly sophisticated molds and a lot process optimizations by trial and errors are used to make the successful parts. So, the development time and expense of the hearing aid are increased significantly. To reduce the time and cost for developing a new hearing aid, Computer Aided Engineering (CAE) software like Moldex3D, can be employed to simulate the process of injection molding to optimize the part and mold design and to optimize injection molding process parameters in advance. With the advances on the process simulation, it is highly likely that a lot of mold making efforts and many trial and errors can be reduced or even eliminated [2]. But this practice is not in vogue in the hearing aid industries due to the lack of validation of the simulation software for thin walled parts like hearing aid shells. The aim of this paper is validating and improving the prediction accuracy of one of the widely used injection moulding simulation software which is Moldex3D. Different parameters setting in Moldex3D were investigated to check their influence on the simulation accuracy. This paper will provide in-depth knowledge about the simulation process for highly demanding thin wall parts and show the way to improve the simulation accuracy.

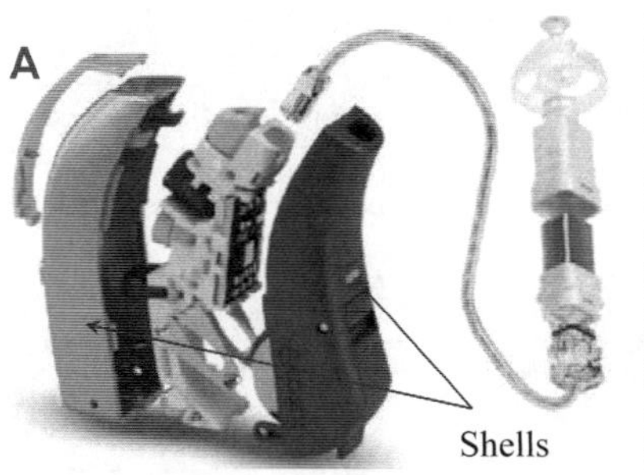

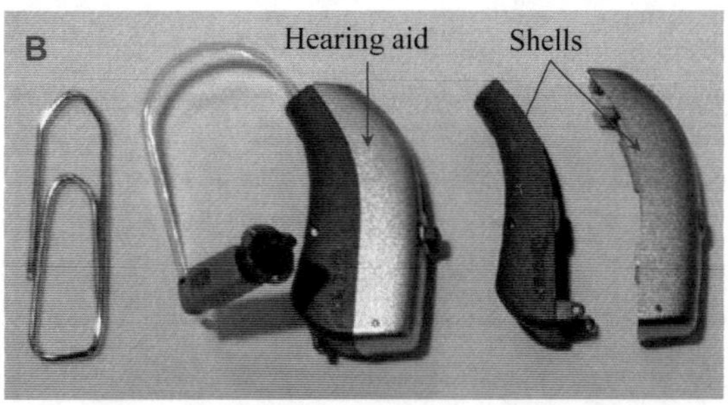

FIGURE 1. Exploded view of a hearing aid showing shells (A); picture of a hearing aid and its shells (B). (Image curtesy: Widex A/S, Denmark).

MATERIALS AND METHODS

The test geometry was a simplified hearing aid shell. The part is shown in Fig 2 (A) which had outer dimensions of about 32 mm × 8 mm × 24 mm (length × width × height) and wall thickness of 0.8 mm. The material selected for the experiment was 30% glass fiber reinforced PBT (Arnite TV4 261), a semi-crystalline thermoplastic produced by DSM Engineering Plastics. This material is widely used for the production of hearing aid shells and other thin walled structures. The injection molding experiments were performed on a combined injection machine from two different systems. The clamping unit was from Arburg ALLROUNDER 270A and injection unit was from Arburg ALLROUNDER 170S, taking the advantages of small shot size but large clamping force.

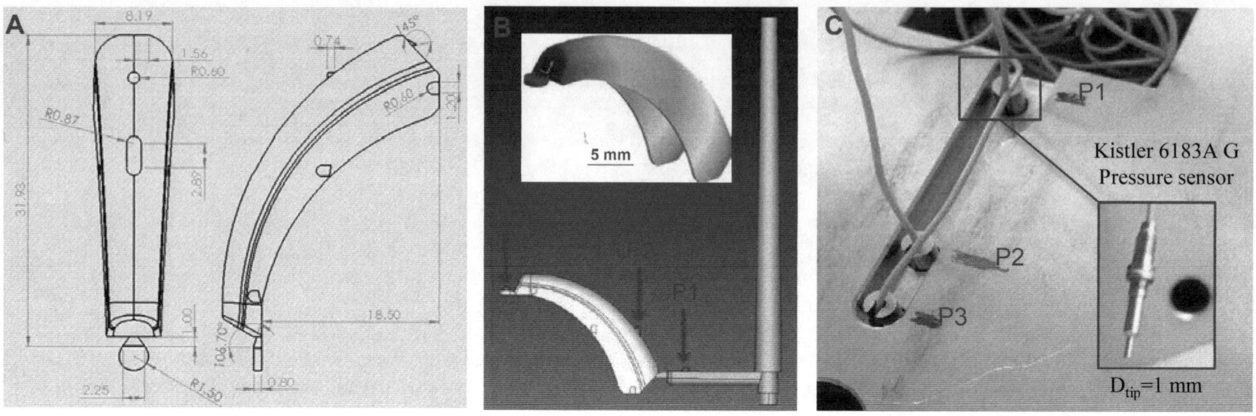

FIGURE 2. Drawing of the hearing aid shell (A); simulated and CAD model of the shell (B)- CAD model showing the position of the sensors; assembly of the sensor in the mould (C).

Cavity pressure measurement system: Three pressure sensors (Type 6183A G) from Kistler were inserted inside the mold to record the cavity pressure during injection molding process. As sketch shown in Fig 2(B), the first sensor denoted as P1 was located on the runner, the second one P2 was placed in the cavity, the last one P3 was at the end of the cavity, on an overflow location. All of these sensors were embedded perpendicularly in the upper half of the mold and were fixed in the mold. The sensitivity of the sensors was 2.2 pC/psi. Pressure sensors were connected to CoMo Injection system for collecting data, which was then transferred to Como Data Center software installed on the laptop via a cable. By placing a magnet with inductive sensor nearby the mold, the Como injection system was triggered when mold was closing.

Injection molding processing parameters: Injection molding experiments were performed with the molding parameters listed in Table 1. The same moulding parameters were used both for the simulation and experiment. Experiments consisted of a batch of full shots with packing, a batch of shots stopping at V/P switch over time without packing and 3 batches of short shots. Molding of each production batch included a temperature stabilization period of about 10 minutes, then a molding series of 20 parts followed by an actual series of 20 parts for analysis.

TABLE 1. Injection molding parameters.

Moulding parameters	Values
Melt temperature [°C]	270
Mold temperature [°C]	90.5
Back pressure [MPa]	12.5
Injection speed [mm/s]	37.5
V/P switch over time [s]	0.4
Packing pressure [MPa]	99.0
Packing time [s]	2.8
Cooling time [s]	8

TABLE 2. Meshing parameters for the simulation.

Simulation parameters	Setting
Cavity mesh type	2 layer BLM with offset ratio 0.3
Runner mesh type	2 washer layers
Global mesh size	0.4
Refinements base	Curvature and proximity
Mesh size on the gate	0.2
Cavity mesh element no.	59432
Runner mesh element no.	695832
Project settings	Machine mode-1

Simulation experiment: The real geometries of the hearing aid shell cavity, gate, cold runner as well as the cooling channels were directly imported into the Moldex3D Designer to build the model. In addition, three sensor nodes, i.e. the pink points in Fig 3 (A), with the same location as in the real mold were constructed to record the pressure change and then compare with the sensor pressure profiles in the experiment. In the simulation, the hot runner was also included in the model to resemble the machine geometry which consisted of nozzle and used part of the barrel. The internal geometry of the real nozzle was obtained by measuring the real nozzle part. The model constructed can be seen in Fig 3 (B). The models were meshed with the settings listed in Table 2.

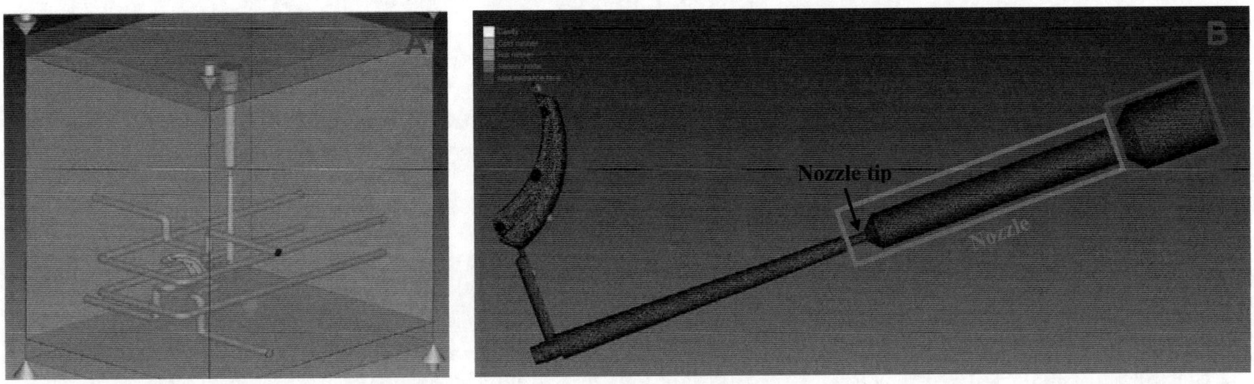

FIGURE 3. Simulation model in Moldex3D Designer (A); modelling of the machine geometry (B).

The material data for the simulation was obtained from the direct characterization of the material. There are 3 project setting methods in Moldex3D, namely CAE mode, machine mode-1 and machine mode-2. CAE mode is easier to use and usually for quick analysis while the Machine mode provides more realistic characteristic settings as actual injection molding machine. In this project, machine mode-1 was used for reliable results. The injection machine used in the simulations was created with the clamping unit and injection unit information from Arburg 270A and 170S injection machine respectively. Then the simulation was run. To make a straightforward comparison the CAD file, the molded CT scanned part and the simulated part were imported into GOM Inspect with preliminary fitting and best fit approach.

RESULTS AND DISCUSSION

Figure 4 shows the short shots comparison from the experiment with the simulation. The shape of the short shot and profile of the flow front have good agreement between the experiment and simulation.

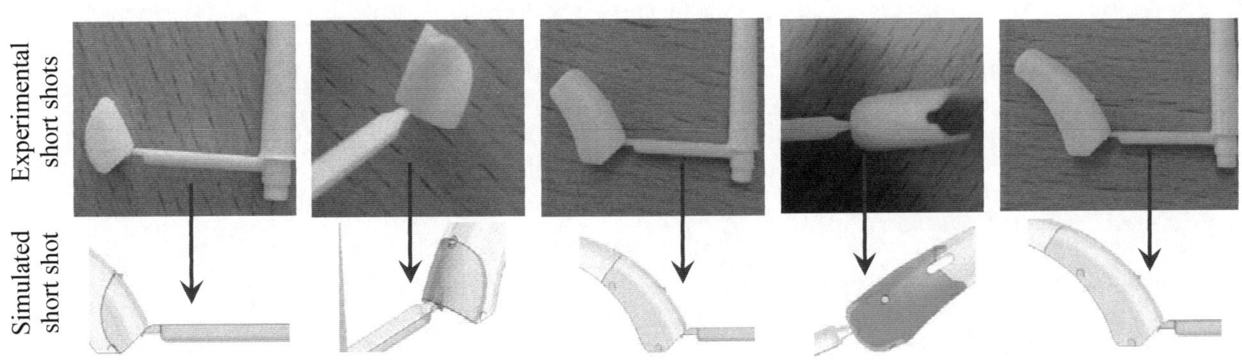

FIGURE 4. Short shot comparison between the experiment and simulation.

The injection pressure of the experiment and the simulation were 99.0 and 81.5 MPa respectively as shown in Fig 5 (A). Fig 5 (B) and (C) show the cavity pressure profiles from three pressure sensors from experiment and simulation respectively, where the pressure reached the peak point at the transition from the end of filling to the packing stage. It can be observed that the simulated pressure curves during the holding stage have more fluctuation compared with the experiment.

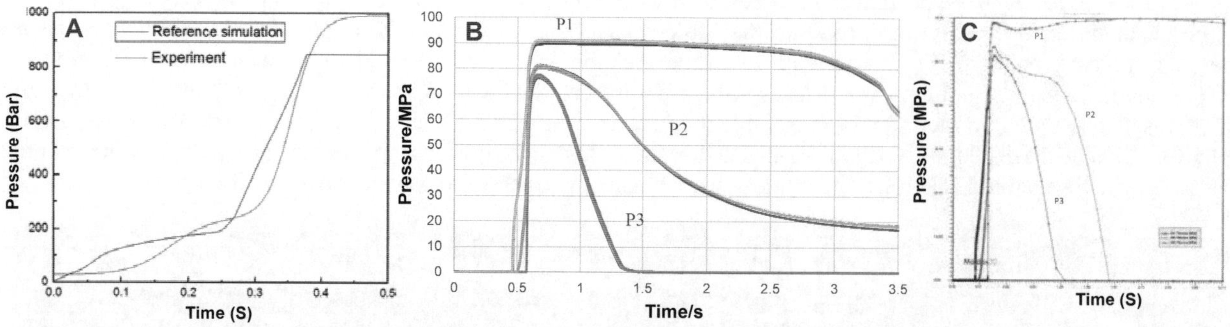

FIGURE 5. Simulated and experimental injection pressure profiles (A); Cavity pressure curves of 20 injection cycles recorded by three sensors inserted in the mold (B); Cavity pressure curves from simulation (C).

From the comparison between the experiment and simulation, following points can be observed:

- Melt front shapes are in good agreement between simulation and experiment. From the perspective of filling behavior, injection time and pressure distribution, Moldex3D R13 with BLM meshing mode has the ability to predict the injection molding process of a thin walled part correctly, even though it can't be 100% identical.
- Injection time, ram stroke, injection pressure and cavity pressure loss in the simulation are all smaller than the experimental values. Different factors are responsible for this and will be discussed later.
- The deformation modes of the simulated part and the molded part are different, simulated part shrank in the flow direction while molded part extended in this direction. By observing the displacement caused by fiber orientation effect and random fiber orientation effect in the simulation, speculation can be made that the difference of warpage is perhaps caused by the underestimation of fiber orientation effect in the simulation.

Effects of the factors influencing the accuracy of the simulation: Based on the discrepancy of the reference simulation and experiment results, a detail investigation on the influence of following factors in the simulation software were made to improve the accuracy of simulation prediction:

- Flow solver efficiency factor
- Pressure dependent viscosity
- Machine geometry
- Heat transfer coefficient (HTC)
- Fiber parameters settings (RPR)
- Crystalline effect
- Geometry boundary
- Viscoelasticity
- Flow rate of coolant

The investigation showed that the higher efficiency factor increased the time from V/P switch over point to the end of filling greatly, i.e. filling time minus injection time. For example, the simulation with efficiency factor 5, showed an increased time difference of 45.45% compared with the simulation done with efficiency factor 5. The pressure dependent viscosity parameter D has large influence on injection time and the pressure. The injection time, injection pressure as well as the pressure loss were all increased with rising pressure dependent viscosity parameter D. It was frequently reported that the modeling of nozzle and the barrel of injection machine as hot runner in the simulation increases the filling time and injection pressure, making the simulation more accurate and closer to the experiment result [3, 4]. But according to the current investigation, the hot runner in the simulation model didn't make much difference on warpage compared with those without hot runner. As the Moldex3D machine mode-1 already considers the compressibility in the barrel, the model only with nozzle geometry as hot runner in the simulation is suggested, which increases the V/P switch over time by 9.6%, ram stroke length by 9.4% and pressure loss from machine to the first sensor (P0-P1) by 76.7%.

Higher filling HTC can increase the injection time, injection pressure as well as the pressure loss in the cavity. But it also has the tendency to cause more shrinkage in the flow direction but less shrinkage in the transverse direction. The warpage in the part is dominated by the fiber orientation. The fiber orientation calculation is done at the filling phase in Moldex3D, so different filling HTC can cause different fiber orientation as well as warpage. The automatically determined HTC suggested by Moldex3D is not reliable; it seems to overestimate the HTC values. Due to the larger surface-to-volume ratio and the smaller thickness, the cooling and heat loss happens much faster for micro and thin walled parts, which indicates a better heat transfer condition for the parts with small dimension and thin walled structures. So from this perspective, HTC value should not be too small. And from the aspect of warpage, HTC should not be too large. For the hearing aid shell, a filling/packing/detached HTC of 3000/2500/1500 is recommended. When the material, geometry and process conditions change, HTC values should be optimized.

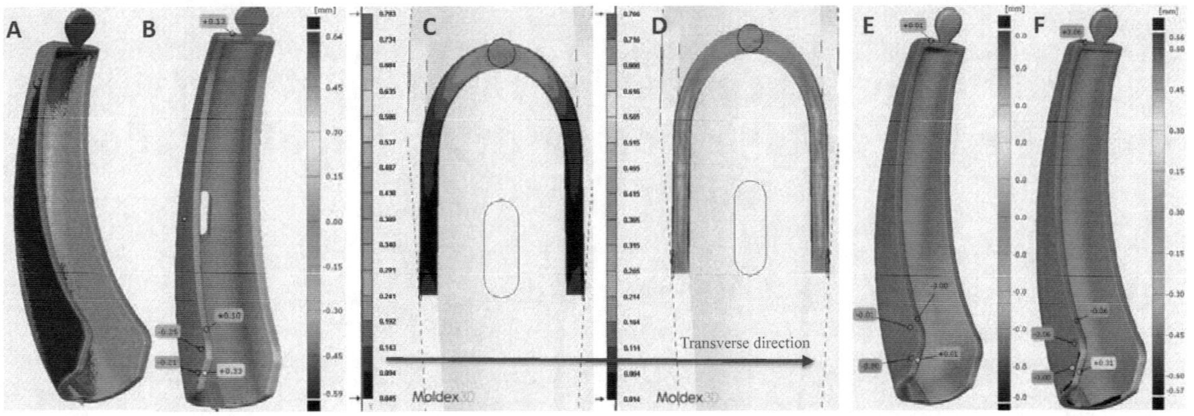

FIGURE 6. Geometrical comparison of CT scanned part (grey) and simulated part (blue)-(A), color mapping inspection of the CT scan part showing its deviation from simulated part (B); Fiber orientation in transverse direction of part with packing HTC 1500 (C) and with packing HTC 6000 (D); Simulated part geometry comparison with α=0.1 and 0.7 (E); α=0.1 and 0.99 (F).

Retarding Principal Rate (RPR) model parameter- α influences the warpage (see picture E and F of Fig 6) by changing the fiber orientation. Too large α will decrease the effect of fiber orientation, and thus cause a more isotropic deformation in the flow and transverse direction. The investigation shows that the α-value smaller or higher than 0.7, doesn't make any big effect. So the filler parameter values of short fibers, i.e. Ci=0.005, Cm=0 and α=0.7, suggested by Moldex3D, are fine for thin walled structure. The crystallization effect of the material is not recommended to consider for thin walled part due to the faster cooling, even for semi-crystalline material like PBT. When the part is thick enough, the crystalline effect should be considered and the packing HTC should be decreased to reflect the worse thermal contact between the mold and part due to the high shrinkage (coming from the crystallization effect).

CONCLUSION

With proper parameter settings, Moldex3D is able to predict the injection molding process in a relatively precise way. It is necessary to consider the effect of pressure on the viscosity in the simulation as it has large influence on injection time, pressure and pressure loss. Larger filling HTC can increase the injection time, pressure as well as the pressure loss. Larger packing HTC can increase the pressure loss but also bring more fluctuation to the pressure curves. To make an accurate prediction, it is necessary to consider the effect of pressure on viscosity, use proper heat transfer coefficient and include the nozzle geometry but not consider crystallization effect for small and thin walled part in the simulation. Currently, Moldex3D can't accurately predict the warpage for complex part made of fiber filled plastics. Predicted warpage of part made of the fiber filled plastics relies a lot on the fiber orientation [5], which is influenced significantly by the filling HTC and RPR fiber orientation model in Moldex3D.

ACKNOWLEDGMENTS

The authors acknowledge Widex A/S of Denmark for their support to the experimental and simulation works.

REFERENCES

1. A. Islam; H.N. Hansen; D. M. Marhöfer; et al., "Two-component micro injection moulding for hearing aid applications", International Journal of Advanced Manufacturing Technology, vol 62 (5), pp. 605-615, 2012.
2. R. Zheng, R. I. Tanner, X J. Fan, "Injection molding: integration of theory and modeling methods", Springer Science & Business Media, 2011.
3. F. S. Costa, G. Tosello, B. R. Whiteside, "Best practice strategies for validation of micro moulding process simulation," Proc. Polym. Process Eng. 2009 Conf., no. 1, pp. 259–261, 2009.
4. P. Guerrier, "Analysis of warpage of thin walled injection molded parts: Simulation and experiments,"Master thesis, Department of Mechanical Engineering, Technical University of Denmark, 2012.
5. S. Kleindel, R. Eder, H. Schretter, et al. Experimental and Numerical Investigation of Shrinkage and Warpage of a U-Shaped Injection Molded Part, International Polymer Processing, 31(1): 68-81, 2016.

Chemical Blowing Agent-based Processing of Integral-skin Polypropylene Cellular Composites in Rapid Rotational Foam Molding

Remon Pop-Iliev

UOIT-University of Ontario Institute of Technology, 2000 Simcoe St. N., Oshawa, ON, L1H7K4, Canada

Abstract. The traditional rotational molding process is the most widely implemented technology for low-cost fabrication of low-volumes of hollow/double-walled large-sized, seamless and virtually stress-free, single-piece plastic articles with complex shapes and/or moderate undercuts. However, demanding industrial end-use applications often require lightweight rotationally molded articles that could resist harsh environments while simultaneously demonstrating buoyancy, enhanced mechanical strength, thermal and/or acoustic insulative properties, as well as improved strength-to-weight ratios. In response to this need, over the years, several design modifications of the traditional rotational molding process advantageously enabled the manufacture of rotationally foam molded cellular composites having a distinct integral solid-skin that is fully encapsulating a foamed inner core or layer. Although the resulting cellular composite moldings were attributed with significantly enhanced properties compared to their hollow non-foamed predecessors, the inherent disadvantages of the traditional parent process, such as the very lengthy and energy-intensive processing cycle times and the very limited means for real-time process control, were only further aggravated. This paper focuses on a recently patented process that alleviates these drawbacks by replacing the close interactions between foamable and non-foamable polymeric resins to their effective independence. It is referred to as Rapid Rotational Foam Molding (RRFM). An experimental study in the domain of processing polypropylene (PP)-based integral-skin cellular composites using the RRFM technology is presented in details. In this context, by successfully decoupling (segregating) the skin-generating/article-shaping processing phase from the foamed-core-processing segment, RRFM proved to provide a viable solution to the common previously faced challenges in the manufacture of PP foams surrounded by an integral-solid-PP-skin with a desired shape.

Keywords: Rotational molding, Rotomolding, Extrusion, Foaming, Blowing agents, Integral-skin cellular polymeric composites.
PACS: 89.20.Kk 81.05.Qk

INTRODUCTION

Foam-reinforced integral-skin cellular polymeric composites form a very desirable class of rotationally molded plastic products because they can satisfy severe service requirements while demonstrating very high strength-to-weight ratios. Their prime applications include floating devices, navigational buoys, and lightweight, ultra-strong and insulative articles demanded by the marine, automotive, and the aerospace industries. Until now, applying chemical blowing agents (CBAs) was a mandatory routine in rotational foam molding operations that was imposed due to the intrinsic atmospheric nature of the process that uses vented molds [1,2].

BACKGROUND

Traditional rotational molding (or rotomolding) is a four-step cyclic process as depicted in Figure 1. It includes subjecting a pre-charged (with plastic), separable (cast or fabricated), vented, metal mold to biaxial rotation (or at least a rocking motion) first into a heated environment (e.g., using convection-based heath transfer within an oven), and subsequently, into a cooled environment (e.g., air and/or water sprinklers), after which the mold is opened, the part is removed, and the cycle repeated. It should be noted that, rotational molding production cycles are, inherently, and undesirably, lengthy because the plastic material charged into the bi-axially rotating mold has to be indirectly heated from room temperature to beyond its melting point and then indirectly cooled back until material solidification [1,2].

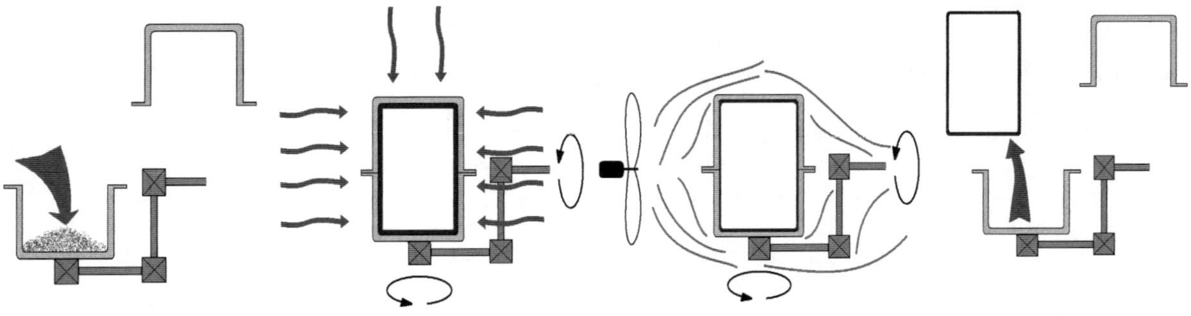

FIGURE 1. The Traditional Four-step Rotational Molding Process.

Conventional rotational foam molding represents a modification of the traditional rotational molding process that can serve for the manufacture of foam-reinforced rotational moldings that possess a distinct non-foamed outer skin that encapsulates entirely a foamed core or layer. Both non-foamable and foamable plastic resins are required for this process. They can be charged either subsequently or simultaneously into the mold within an identical molding cycle. This can be achieved either by interrupting the molding process, or continuously, in a single-charge fashion. Single-charge concepts, as depicted in Figure 2, are based on charging the mold with a mixture of predetermined quantities of non-foamable and foamable resins simultaneously at the outset of the cycle, so that the use of drop boxes or plastic bags becomes unnecessary. In this context, the conventional rotational foam molding process is culminating with the single-charge fabrication of integral-skin polymeric composite rotational moldings that include a layer of solid polymeric skin with a desired thickness, which encapsulates entirely the foamed core or layer. Importantly, these integral-skin rotational moldings are characterized with adjacent, but clearly distinct, layers of non-cellular (encapsulating) and cellular polymeric structures, normally consisting of identical or compatible polyolefin grades [1-4].

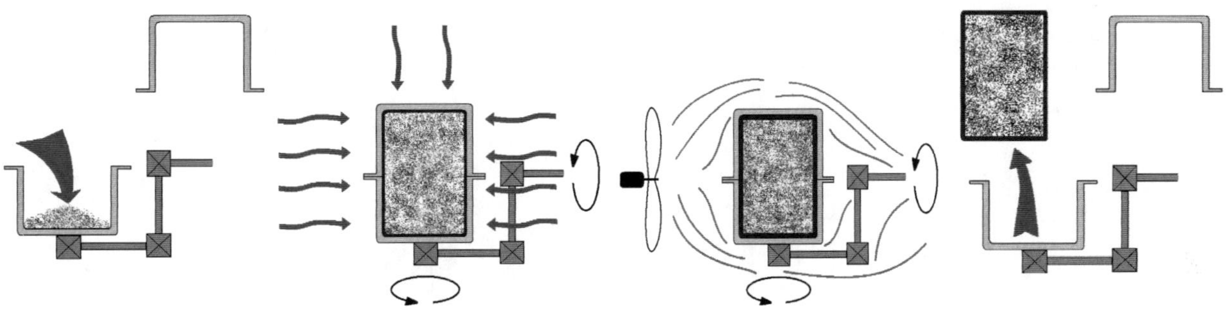

FIGURE 2. The Conventional Four-step Rotational Foam Molding Process.

Despite its obvious advantage: the ability to deliver integral-skin cellular composites in a single processing step, it has serious limitations because it suffers from inherently aggravating the fulfillment of crucial processing goals such as: (i) the execution of the adhesion of the non-foamable thermoplastic resin to the internal surface of the mold that should always take place prior to the thermal activation of the foaming resin (thereby avoiding skin protrusions), and (ii) obtaining a solid skin layer with a uniform thickness [3-7].

For example, the control of the cell size of rotationally foam molded foams crated by using a chemical blowing agent (CBA) is often aggravated by some inherent limitations that are unique to the rotational molding process such as the lengthy processing cycles. As a result, coarser-celled final cellular structures are often being yielded. In addition, inherently, the polymer close to the internal mold surface continues to be heated even after its foaming is completed, simply because the interior polymer has yet to reach the decomposition temperature of the blowing agent. Even when cooling is applied to the mold, the melt reaches its recrystallization temperature with a significant delay, during which time bubble coalescence and bubble collapse phenomena set out and accelerate. [2,5].

It is also inherent to the rotomolding process that, as the melt front progresses, the air pockets that have been entrapped inside the melt eventually become bubbles that will be subjected to diffusion controlled shrinkage and eventual disappearance. Prior fundamental research of the lifespan of chemical blowing agent-blown bubbles in non-pressurized non-isothermal polymer melts indicated that the lifespan of fine-celled bubbles is significantly shorter than the inherently lengthy heating portion of the rotational molding process. Therefore, fine-celled bubbles seldom reach the solidification stage of the cycle, which implies that only coarser-celled bubbles live long enough to participate in the final cellular structure. Thus, it was proved that it is quite difficult to generate the preferred kind of fine-celled bubbles in rotational foam molding unless the duration of the processing cycle would somehow be dramatically reduced [5].

Consequently, solutions for reducing the processing cycle time and the energy consumption of the rotational foam molding process to a fraction of its current levels without compromising product quality are very necessary.

RAPID ROTATIONAL FOAM MOLDING

Rapid Rotational Foam Molding (RRFM) is an innovative and patented [8,9] processing concept that is based on exploiting the unique synergistic effects resulting from the deliberate conjunction of extrusion with the rotational molding process. It offers a design solution that completely decouples the shaping of the part's solid skin from the formation of the skin-encapsulated foamed layer or core. Thereby, improved process controllability, significantly shorter processing cycle duration and reduced energy consumption can be readily achieved.

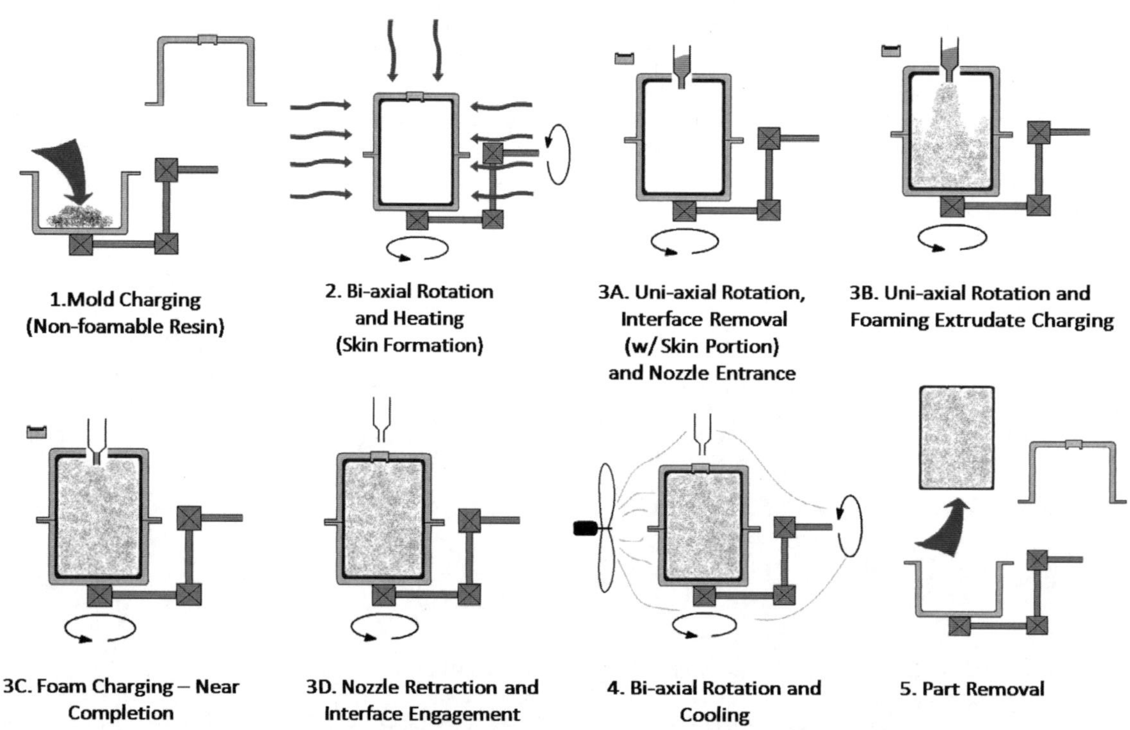

FIGURE 3. The Conventional Four-step Rotational Foam Molding Process.

The RRFM technology as depicted in Figure 3 can be summarized in four unique processing steps, as follows:

Step 1: At the inception of the process, the mold is manually charged with a predetermined amount of non-foamable plastic resin in powder form, which will be used to create the part's solid skin [8,9].

Step 2: The arm and the mold assembly are inserted into the oven by means of a carriage and translational mechanism. The mold rotates in a biaxial manner within the oven, utilizing the arm. This step is accomplished at a material-dependent elevated temperature for an accordingly set period of time to accomplish the creation of the skin. Simultaneously, the extruder is charged with predetermined amounts of dry blended polymer resin with a chemical blowing agent, suitable for achieving polymeric foam that exhibits a desired volume expansion ratio at a set melt temperature [8,9].

Step 3: The arm and mold assembly are then translated towards the extruder, with the arm rotation switched to a uni-axial mode and the mold aligned with the extrusion port. The extruder subsequently fills the hot mold with extrudate comprised of the desired foam for a period of time dependent on the melt flow rate and screw RPM. The filling process occurs at the mold-extruder interface gate, which has been designed so as to seal the mold during the conventional rotational molding cycle, facilitate the introduction of foam to it during the foam filling stage, and allow the skin to self-heal after the filling process is accomplished thereby allowing the skin to settle back at its original location, undamaged [8,9].

Step 4: The mold is then bi-axially rotated and cooled. Subsequently, the solidified part is removed [8,9].

By design, in RRFM, the shaping of the solid skin of the integral-skin cellular composite article is completely decoupled and independent from the formation of its foamed layer or core, which is eventually encapsulated by the skin. Unlike prior processing approaches, RRFM features direct introduction of non-chilled extruded foam onto the non-chilled skin within a uni-axially rotating mold via a specially designed "self skin-repairing" injection port. This makes it capable of achieving ~35% reductions of the processing cycle time and energy consumption compared to conventional rotational molding of analog parts, thereby setting a new, greener, processing state-of-the-art in the field. RRFM also provides a dramatic improvement of the controllability of the process. As a result, the experimentally obtained RRFM foam morphologies and skin-foam interfaces were adequate for both PE and PP grades as well as their combinations. However, most importantly, the successful inclusion of melt-extrusion in the processing equation of RRFM is enabling for the very first time implementing PBAs instead of CBAs, or even using their combination, in rotational foam molding operations. In addition to its improved processing controllability, the RRFM technology has the potential to be equipped with the necessary capabilities to produce new classes of ultra low-density rotationally molded integral-skin cellular composites with unique properties and within wider processing and materials windows, which is currently not achievable [8,9].

EXPERIMENTAL

Experimental Setup

The advantages of RRFM versus the conventional rotational foam molding technique were assessed and experimentally verified through the use of a custom design and development of a fully functional lab-scale experimental setup that efficiently combines melt extrusion with rotational molding, as depicted in Figure 4. The comprising components include an arm and mold assembly designed to achieve both uni-axial and bi-axial rotation modes, an arm-mold assembly carriage integrated with the main frame through a translational mechanism, an extruder with a modified extrusion port, and a retrofitted oven.

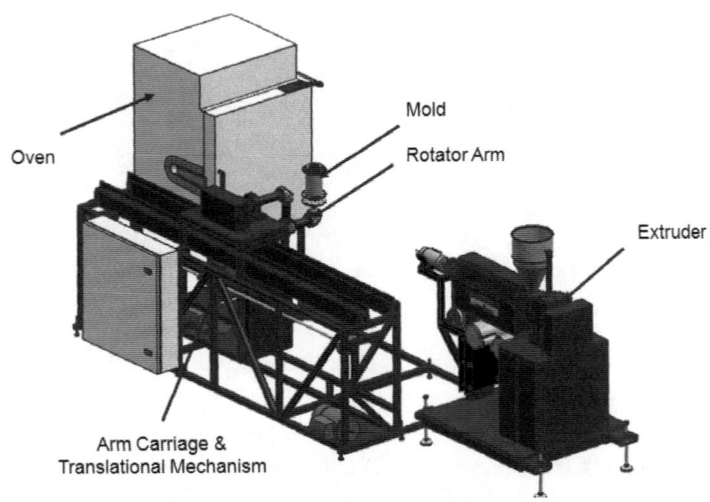

FIGURE 4. The Experimental Setup for Rotational Foam Molding.

The rotation of the mold that is mounted onto an offset rotating arm is powered by two variable geared motor systems that are contained by the carriage. The translational mechanism is capable of repositioning the arm-mold assembly between the oven and extruder through a cart-track setup that is actuated by a powered lead screw engaged with a ball nut fixed to the carriage frame. The lab-scale oven has been custom fit with a rotator arm opening to allow the mold and arm to be contained inside it during operation, and an additional pneumatic cylinder to allow automatic door operation.

During the foam-filling step, the filling of the mold is facilitated by using a specially designed insulated mold interface assembly, which, as demonstrated in Figure 5, consists of three main components: the insulation cap, the mold interface, and the mold connection adaptor. The mold connection adaptor is mounted onto the mold by means of bolts. The mold interface includes a channel that is filled with high temperature ceramic fiber insulation, which in turn facilitates the parting of the skin portion under the mold interface from the overall skin structure. The insulation cap holds the ceramic fiber securely within this channel during processing. The mold interface and insulation cap assembly are completely detachable from the mold body during filling by means of a "quick release" T-shaped handle that locks into a designated compartment within the mold interface. After the foam-filling step is completed, the interface is reattached to the mold to allow the foam to adhere to the semi-molten polymer attached to the interface. Figure 5(a) shows the cylindrical mold approaching the extruder nozzle while rotating uni-axially about its central axis; the quick release T-shaped handle is shown engaged with the interface assembly to instigate its removal process. Figure 5(b) illustrates the removal of the mold interface with the skin portion adhering to its lower surface at an instant prior to foam filling. Figure 5(c) demonstrates that the interface assembly is kept at the elevated temperature of the oven to preserve the molten state of the skin portion attached to interface while foam filling is in process. Finally, Figure 5(d) shows a cutout from a resulting experimental polypropylene (PP)-based integral-skin cellular composite sample demonstrating the achieved high quality morphology of the skin-foam interface.

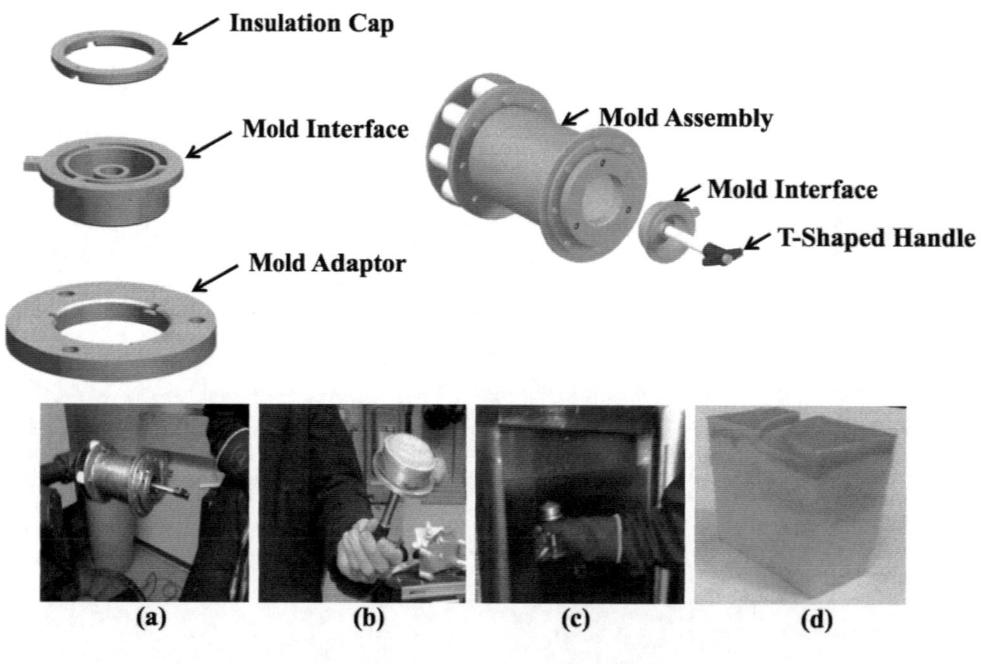

FIGURE 5. The Mold Interface and its Role in Enabling Foam-filling.

Experimental Materials

Four PP grades from LyondellBasell were used for experimentation purposes. The respective resin codes along with their supplied and measured commercial and thermal characteristics are listed in Table 1 [10]. The selected PP grades reflect differences in melt flow indexes (MFIs) as well as variation in material nature from PP homopolymers to PP copolymers. The resins were received in pellet form, were criogenically pulverized to mesh-35 sized powders, to enhance their processability as solid skin constituents in RRFM, and when used for foaming purposes, to enhance and facilitate their dispersion and distribution characteristics with the chemical blowing agent (CBA).

Table 1: PP resins used throughout the experimental batteries associated with the RRFM process

	PP Resin Code	PFHL451H	PFSR257M	PF6523	PFSB786
Characteristics	MFI [g/10min]	2.0	2.0	4.0	8.0
	Density [g/cm^3]	0.902	0.902	0.900	0.902
	Melt T [°C] (DSC result)	165.78	149.41	166.38	167.64
	Crystallization T [°C] (DSC result)	116.85	123.99	112.30	114.46
	Homopolymer (H) or Copolymer (C)	H	Random C	H	Impact C
	PP Tacticity	Isotactic	Isotactic	Isotactic	Isotactic

The CBA used to formulate the foamable resin was Celogen AZ-120 (exothermic CBA) and was obtained from Chrompton Chemicals. The molecular constituents of this thermally activated CBA are azodicarbonamide with amorphous silica. Celogen AZ-120 was selected because of its decomposition temperature range that is suitable to the properties of the experimental PP grades as well as for its small particle size, which assists the dispersion and distribution properties during pre-extrusion dry blending. Table 2 highlights the most important characteristics of Celogen AZ-120 [11].

Table 2: Properties of Celogen AZ-120

CBA: Celogen AZ -120 (exothermic)	
Chemical Composition	Azidocarbonamide, Silica
Bulk Density [g/cm^3]	7.961 – 9.755
Gas Yield φ [cm^3/g]	210 – 220
Decomposition range [°C]	205 – 213
Decomposition T [°C] (TGA result)	207.64
Gas Composition	N_2, CO_2, CO, NH_3
Particle size [μm]	2.0 – 2.4

RESULTS AND DISCUSSION

Out of three series of experimental trials (A,B and C) designed using Design of Experiments (DOE) software, Series A involved processing integral-skin cellular composites having a solid skin made of PFSB786 (≈5.5 [mm] thickness) encapsulating a completely foamed core of FHL451H revealed the best results obtained. Table 3 presents the respective experimental parameters implemented. The best resulting experimental samples from experimental scenarios A3 to A5 are shown in Figure 6. When visually comparing the generated moldings, it is apparent that experimental scenario A4 yielded the most uniform cell-sized foam core. Closer visual inspection of molding A4 revealed minimal cell coarsening present within the cellular community near the absolute center of the molding. This can be attributed to the probable incomplete evacuation of the processing gasses prior to the stabilization and recrystallization phase of the foam structure.

Table 3: Experimental parameters for experimental scenarios of Series A

Experiment Code	Processing T_{melt} Range [^{o}C]	Processing Melt Pressure Range [PSI]	Skin Weight [g]	Overall Part Weight [g]
A1	171 - 183	195 - 0	280	615
A2	170 - 183	200 - 0	280	618
A3	172 - 181	200 - 0	280	613
A4	178 - 185	180 - 0	280	618
A5	171 - 176	225 - 0	280	620
A6	175 - 181	200 - 0	280	460
A7	181 - 187	150 - 0	280	483
A8	176 - 182	150 - 0	280	481
A9	172 - 183	170 - 0	280	533

Processing scenario A5 created the coarsest cellular structure at the molding's foam core. Evidence of incomplete CBA decomposition is present when inspecting the top left corner of the presented illustration, which is the actual bottom right corner of the molding and the furthest point from the extruder nozzle. The reasons behind the displayed phenomenon can be attributed to the relatively lower melt temperature endured by the foamable formulation during the foam filling segment of this experimental run (171-176 [^{o}C]), which was not high enough to decompose the CBA, even though the extruder's screw RPM was at its maximum value. Molding A3 appears to demonstrate the finest skin thickness uniformity among the generated moldings. There is no evidence of significant cell coarsening or cell coalescence throughout the foam core of this sample. Altogether, moldings A3-A5 displayed a successful skin-foam bond at the insulated interface location associated with the mold filling gate. Consecutive successful moldings were also achieved as a result of conducting processing scenarios A1 and A2. It was thereby established that a high degree of control over the critical processing parameters characterizing RRFM is possible. Moreover, no significant filling gaps or protrusions at any location throughout the skin-foam boundary or within the distinct skin and foam layers were observed throughout the generated moldings.

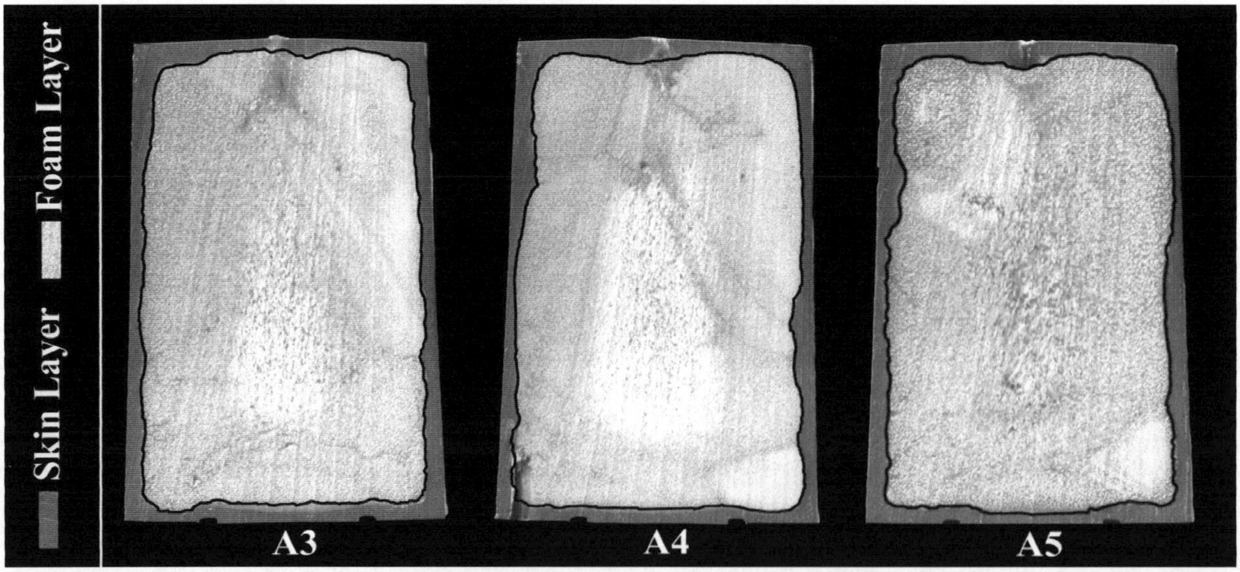

FIGURE 6. Specimens Associated with Experimental Scenarios A3 to A5.

Random samples were obtained from moldings A3-A5 for inspection using Scanning Electron Microscope (SEM). As illustrated in Figure 7, the SEM micrographs revealed a strong bond between the distinct skin and foam layers; this is justified by the fact that the foaming extrudate is introduced within the skin boundaries as the skin is still in a semi-molten/soft state, which allows the adherence of both layers together before the cooling process is commenced. Thus, PFHL451H as a foam-core comprising PP grade in RRFM, combined with a skin material made of PFSB786, was validated as successful.

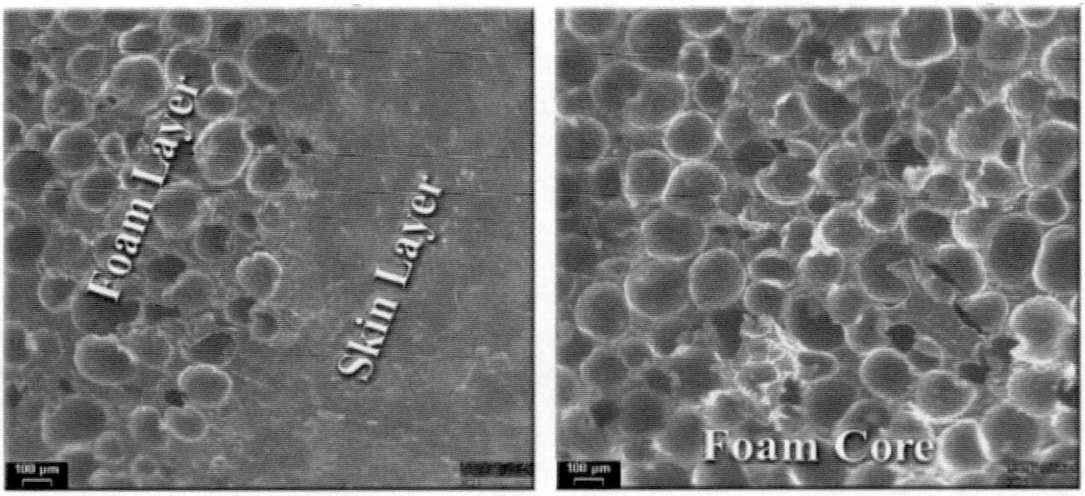

FIGURE 7. SEM micrographs of the skin-foam interface and the foam core morphologies of the formulation: PFSB786-skin & PFHL451H-foam core.

The obtained foam morphologies were investigated with respect to their cell population, which is represented by Equation (1) [12-14]:

$$N = (n_{cells})^{3/2} \cdot VER \tag{1}$$

In Equation (2), n_{cells} is the number of cells per unit area of $1 cm^2$. The value of n_{cells} for a selected sample within this series was found to be 1140 [cells/cm^2], and the corresponding cell population density was determined to be 153,963.3 [cells/cm^3]. Subsequently, the computation of the average cell size ($D_{average}$) was conducted using Equation (2) [12-14]:

$$D_{average} = \sqrt[3]{\left[(VER - 1) \cdot \frac{6}{\pi \cdot N}\right]} \tag{2}$$

The calculated average cell size for the presented 4-fold case was 334 [µm]. This indicates that the achieved foams are particularly close to a fine-cell classification. A similar approach was used to obtain an average cell size value of 401 [µm] for the 6-fold generated moldings. A bar chart, presented in Figure 8(a) was used to plot the achieved average cell size values.

Samples within this experimental series were cut further to extract cubes from the respective foam cores, exhibiting a minimum volume of 1 [in3], in order to calculate and quantitatively assess the particular foam densities according to ASTM D1622 – 08. Based on this principle, four cubes were obtained from each sample, two of which were extracted from the top and bottom ends of the molding, and two from around the central volume of the molding. Figure 8(b) presents a bar chart comparing an averaged value of the different average foam densities achieved across the 4-fold and 6-fold foam cores generated within this experimental series.

Samples A3 to A5 were also characterized in terms of skin thickness uniformity to investigate the effects of varying the processing parameters on the overall skin quality. Thickness uniformity is commonly mathematically interpreted by calculating the statistical root mean square (RMS or Rq) utilizing Equation (3) for a preset theoretical thickness of 5.5 [mm] [12-14]:

$$R_q = \sqrt{\frac{1}{N}\sum(a - 5.5)^2} \tag{3}$$

Where N is the number of skin thickness measurements carried out at different locations throughout the part (set at 10), and "a" is the value obtained for each measurement in [mm]. The accuracy in determining the level of skin uniformity increases with the conducted amount of skin thickness measurements. The calculated R_q values pertaining to the solid-skin uniformity associated with samples A3 to A5 are presented and compared in Figure 9. From the displayed results it is evident that the solid-skin thickness was most uniform throughout sample A3, as only a variation of about 0.8 [mm] was demonstrated across the skin structure (≈14.5% variation). Additionally, sample A4 had the least uniform skin thickness throughout its cross-section (33.6% variation). This result essentially implies that the increased RPM of the mold during the foam-filling step enhanced the uniformity of the resulting skin.

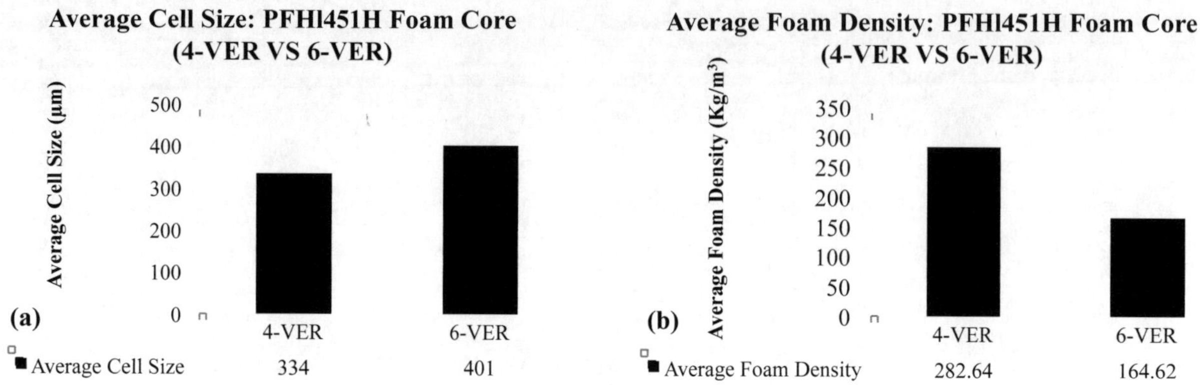

FIGURE 8. (a) Typical average cell size values obtained at 4-fold and 6-fold PP-PFHL451H foam core respectively; (b) Typical average foam density values obtained at 4-fold and 6-fold PP-PFHL451H foam core.

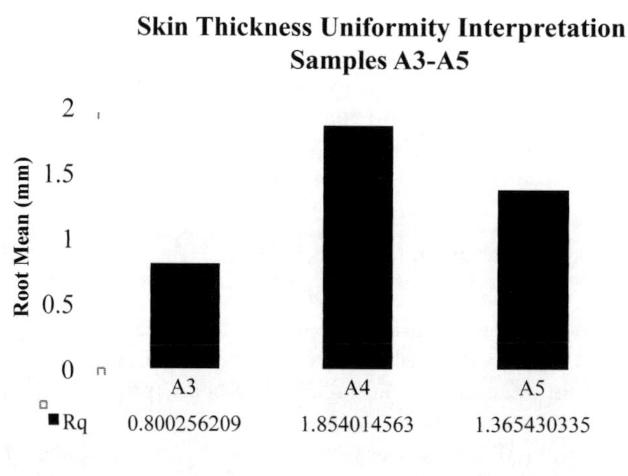

FIGURE 9. Skin Thickness Uniformity Analysis of Samples A3-A5.

General Effect of Resin MFI: Two PP homopolymers, PFHL451H (MFI 2 [g/10 min]) and PF6523 (MFI 4 [g/10 min]), with the second sustaining a MFI that is twice as high as the first, were proved to achieve successful moldings through the experimental runs. The achieved average cell densities, when utilizing both resins for foaming purposes, were rather comparable, with PF6523 exhibiting a slightly higher average foam density when foamed with respect to both, the 4-fold and a 6-fold formulation. The distribution of the foam structure associated with PFHL451H appeared to be more uniform than that of PF6523. Also, the lowest average cell size calculated for the foam obtained using PFHL451H proved to be smaller than the one obtained for PF65234, particularly at the higher volume expansion ratio (VER). This result allowed concluding that the PP grade with the lower MFI preserved its melt strength over the one with higher MFI during foaming in RRFM.

General Effect of Extruder Die Temperature: The extruder's die zone temperature is an essential parameter for controlling the nucleation and growth of foam cells in RRFM as well as the preservation of the blowing gasses within the foaming extrudate. It was observed through experiments that the melt strength of the PP grades studied in this paper improved significantly with decreasing the die zone temperature, while the cell size and the degree of cell rupture associated with the foamed extrudate increased gradually with increasing the die temperature.

CONCLUSION

The fundamental drawbacks of the conventional single-charge rotational foam molding process in manufacturing integral-skin plastic parts entailing partially or completely foamed cores include its intrinsically lengthy processing cycle times in addition to the variable quality of the respectively manufactured parts. In response to the indicated problem, a remedial process, referred to as Rapid Rotational Foam Molding, was designed, developed, and evaluated. This patented processing technology treats the aforementioned drawbacks by decoupling the foam fabrication segment from the entirety of the conventional practice and achieving it through the most efficient continuous plastic fabrication process: extrusion. The devised practice accomplishes up to 35% savings in processing cycle times by eliminating the need to utilize extensive and delicate compounding, palletizing, then heating procedures to achieve skin and foam layers with the desired attributes. Additionally, RRFM demonstrated improvements in the overall quality of manufactured articles compared to their analogs achieved using the conventional process.

In total, three distinct PP-based materials formulations, of different morphological, compositional, and rheological natures, were investigated throughout the concluding experimental work. The thoroughly analyzed results proved the success of the RRFM concept in processing integral-skin foamed core PP moldings for all three formulations. In terms of integral-skin quality, a bubble-free non-degraded solid integral skin was achieved all along the accomplished experimental series with calculated variability in skin thickness uniformity as low as 8.9% and as high as 33.7 %. In terms of foam characteristics, complete foam filling has been achieved in all cases, with no significant gaps, venting issues, or foam quality deterioration due to poor processing melt strength. An average foam density of as low as 152.75 [kg/m3] and as high as 308.51 [kg/m3] was achieved for the 6-VER and the 4-VER moldings, respectively. The SEM analysis performed to examine the morphologies pertaining to the achieved moldings revealed that an average cell size of as low as 334 [μm] was achieved, this indicates that with further refinement of the processing parameters it would be feasible to produce fine-celled PP foams with a more uniform cell-size distribution in RRFM. RRFM proved to provide a viable solution to the challenges faced in processing PP foams exhibiting an integral-solid-PP-skin by imposing a large degree of control over the entities of (i) the foam processing segment, by utilizing the astutely regulated extrusion process and (ii) the skin processing segment, by enforcing optimal conventional rotational molding parameters.

ACKNOWLEDGMENTS

The author is grateful to NSERC's Discovery Development Grant and NSERC's Chairs in Design Engineering Program for the financial support of this research.

REFERENCES

1. Beall, G.L., and Throne, J.L., 2004, Hanser/Gardner, Cincinnati
2. Crawford, R.J., and Throne, J.L., 2002, William Andrew Publishing.
3. Pop-Iliev, R. and Park, C.B., 2003, Journal of Cellular Plastics, Vol. 39, No. 1, pp. 49-58, 2003.
4. Pop-Iliev, R., Lee, K-H., and Park, C.B., 2006, Journal of Cellular Plastics, Vol. 42, No. 2, pp. 139-152.
5. Xu, D., Pop-Iliev, R., Park, C.B., and Fenton, R.G., 2005, Journal of Cellular Plastics, Vol. 41, No. 6, pp. 519-538.
6. Pop-Iliev, R., Xu, D., and Park, C.B., 2004, Journal of Cellular Plastics, Vol. 40, No. 1, pp. 13-25.
7. Pop-Iliev, R., Rizvi, G.M., and Park, C.B., 2003, Polymer Eng. and Sci., Vol. 43, No. 1, pp. 40-54.
8. Pop-Iliev, R., 2011, Acta Physica Polonica A, Vol. 120, No. 2, Aug. 2011, pp. 292-297.
9. Pop-Iliev, R, et. al., 2014, US Patent No: 8,628,704 B2.
10. LyondellBasell, Product Data and Technical Information Sheets, PF HL451H, PF SR257M, PF6523, PFSB786 (2009)
11. Chrompton Chemicals, Materials Safety Data Sheet, Celogen AZ (2009)
12. R. Pop-Iliev, G. M. Rizvi, and C.B. Park, SPE ANTEC Tech. Papers, 908, 1604 (2001)
13. R. Pop-Iliev, D. Xu and C.B. Park, Journal of Cellular Plastics, 40, 1, pp. 13-25 (2004)
14. R. Pop-Iliev and C.B. Park, Journal of Cellular Plastics, 39, 1, pp. 49-58 (2003)

A Combined Experimental And Numerical Study On Pulsed Heating Of Thermoplastic Sheets In Thermoforming

Bart Buffel[a] and Frederik Desplentere[a]

[a]KU Leuven Campus Brugge Spoorwegstraat 12, 8200 Brugge

Presenting author: bart.buffel@kuleuven.be
Reference author: bart.buffel@kuleuven.be

Abstract. In the present study, the potential of pulsed heating in the thermoforming process of thermoplastic sheets is investigated by means of both experimental measurements and numerical simulations. A specific test was developed to measure the through thickness temperature increase of an instrumented silicon sheet which was heated with an electronically controlled halogen heater element. The experimental data indicates the importance and positive effect of residual thermal radiation when the heater is switched off during pulsed heating. High "on/off" ratios and lower heating powers favor a more uniform through thickness temperature. In industry an optimum has to be found in these parameters to avoid unacceptable cycle time increases. The developed numerical model describes fast heating curves very well. For lower heating rates, the importance of convective heat losses when the heater elements are switched off, is shown.

Keywords: Thermoforming, Finite Difference, Heat Transfer
PACS: 44.10, 44.27, 44.40

INTRODUCTION

Thermoforming of thermoplastic sheets is typically done in three consecutive steps: (1) heating of the sheet, (2) deformation of the sheet on the mold and (3) cooling and demolding of the product. In a lot of cases some post production finishing steps, e.g. trimming, drilling and sawing functional openings are required as well. Despite the widespread application of this production technique, the level of automation in industrial production facilities is still rather limited. Especially thick gauge thermoforming, still requires a lot of manual actions. The main current goal for the thermoforming industry is a higher level of process control and automation [1].

The heating of the thermoplastic sheet is the most important step in the thermoforming process. It determines the cycle time, part quality and cost and is responsible for the major part of the total energy consumption [2]. Although some specific applications use convective hot air ovens, the heating is mainly done using infrared radiation. Different types of infrared heating elements exist. Traditionally, ceramic and quartz heater elements are commonly used heater elements in industry. These heating elements operate in the long wavelength spectrum (3-5μm). More recently, halogen and metal foil heaters were implemented in the thermoforming processes. These heaters emit at shorter wavelengths (0,76-2μm) and have a much faster response time compared to the long wavelength elements [3]. The shorter wavelength of these heaters results in deeper penetration of the infrared radiation into the polymer sheet and thus in more volumetric heating instead of a pure surface heating of the sheet [4]. In a comparative study, Schmidt et al. [3] have proven that halogen heater elements equipped with aluminum reflectors have the highest energy efficiency to heat up PS sheets for thermoforming applications.

On top of this, the very short response time to a change in parameter setting of the halogen heaters allows faster heating process changes and use of different heating patterns. Moreover halogen heater elements have the potential to be used in pulsed heating methods. In this method, the setting of the heater elements is not anymore limited to one or two fixed values during a thermoforming cycle, but is switched on and off at a controlled frequency with a certain power amplitude. Although this type of heating was first described in 1973 by Progelhof et al., the method is seldom applied in industry or further explored in literature. The high potential of the method was demonstrated by means of numerical simulations by Buffel et al. [5]. In the present paper, an experimental approach was adopted to characterize the pulsed heating of an instrumented silicon sheet and the most import parameters involved in the heating process. Next to this, the numerical results are validated against the experimental measurements.

EXPERIMENTAL TEMPERATURE MEASUREMENTS

An experimental test was developed to measure the through thickness temperature distribution of a sheet as function of time. The complete set up is schematically illustrated in figure 1. The core of this set up is an instrumented silicon sheet with embedded thermocouples. There are three thermocouples through thickness of the sheet at five different locations. In the present paper, the time-temperature data from the three thermocouples at the center of the sheet will be discussed. The experimental data were captured by a NI LabVIEW Signal express data acquisition system.

The halogen heater element (Philips 650W) was taken from a Geiss U8 thermoforming machine and is placed at a fixed position with respect to the silicon sheet and its emitted power as function of time determined by an Arduino controller. At the other side of the sheet, an IR camera was used in order to visualize the in plane temperature distribution. During the development of the experimental set up, the camera proved to be very helpful in optimizing the shape of the aluminum reflector and positioning the heater element.

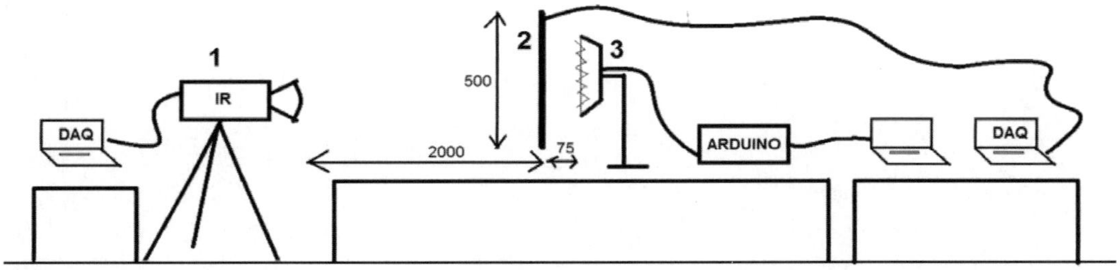

FIGURE 1. Schematic illustration of the complete experimental set up. (1) infrared camera to visualize in plane temperature distribution, (2) Instrumented silicon sheet to measure through thickness temperature distribution and (3) PHILIPS 650W Halogen heater element with an aluminum reflector linked to a Arduino controller.

Applying the pulsed heating method introduces a number of extra parameters in the heating process. In addition to the traditional setting of the preheating time and power, pulsed heating allows more flexibility in heating by using different power amplitudes, frequencies, delay times,….

During the experimental measurements, the sheet was heated from 30°C up to 90°C. This upper temperature was selected to avoid sagging and deformation of the sheet caused by viscoelastic effects at higher temperatures. To avoid initial start-up phenomena a heating cycle was applied five times to the halogen heater to reach constant and reproducible reference conditions between different test runs. In all experiments the halogen lamp was set to the maximum power for 30s before the pulsed heating phase is started.

In the present paper the influence of three different parameters on the through thickness temperature evolution will be discussed: first the influence of residual thermal radiation during the period of time in which the halogen lamp is switched off (= "time off") will be presented. Secondly the influence of different on/off time ratios during the pulsed heating phase is described. The last parameter which is discussed is the power of the halogen lamp during the pulsed heating phase.

MODELLING THE PULSED HEATING METHOD

The modelling of the pulsed heating of thermoplastic sheets is done by means of explicit finite difference. The model used in the present paper is based on the model described by Buffel et al. [5 & 6]. For an in depth description of the model, the authors refer to this paper.

In order to take pulsed heating into account the model was extended with different parameters: the moment at which the pulsed cycles are started, the "time on" and "time off" of the heater elements and the applied power during the "time on". Based on experimental observations, a parameter which refers to the residual heat radiation during the "time off" of the heaters was added. This parameter applies a certain amount of heating power during the "time off" and allows to quantify the residual heat transfer with respect to the applied power during the "time on".

In the present study, experimental data was used to validate the numerical model and the different physical processes which are taken into account. The results of the comparison between experimental and numerical data will identify bottlenecks which have to be resolved before the model can be used in predictive way.

RESULTS & DISCUSSION

In this section the results of the experimental temperature measurements are presented and compared to numerical simulations. This section is divided in subsections, each discussing the influence of a single parameter on the pulsed heating characteristics.

Residual Thermal Radiation of Switched off Halogen Heater Elements

When applying the pulsed heating method, the halogen heater element is electrically switched on and off at predefined time intervals. Since the halogen element is heated up to high temperatures (filament temperature up to 1700K) during the "time on", a substantial amount of thermal energy is still present during the "time off" in each cycle. This residual thermal radiation causes the surface temperature of the heated side to maintain a higher temperature compared to the case where no residual heat of the halogen element would be present. This was experimentally tested by placing a cardboard plate between the heater and sheet during the "time off". The influence of the residual thermal radiation is illustrated in figure 2.

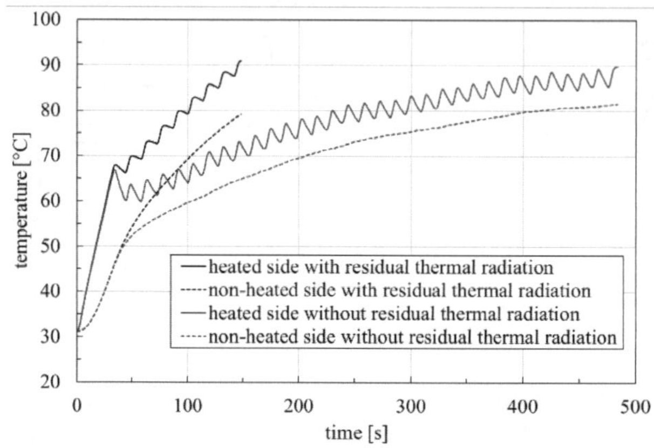

FIGURE 2. Experimental temperature data at the heated and non-heated side of the instrumented silicon sheet with residual thermal radiation of the heater element (black) and blocking of the residual thermal radiation (red).

Both situations were modelled by means of the finite difference model. Figure 3 shows a fairly good agreement between the experimental and calculated values in case the residual heat is taken into account. This allows to quantify the amount of residual thermal power being transmitted during the "time off". In this case, an average power of 24% with respect to the power emitted during the "time on" was still present during the "time off". When the residual heat is taken out of the experiments, the finite difference model starts to deviate from the numerical data after a certain amount of time. This is caused by an underestimation of the convective heat losses by the numerical model. At his point, the model does not allow variable ambient air conditions which would represent the real conditions in a better way.

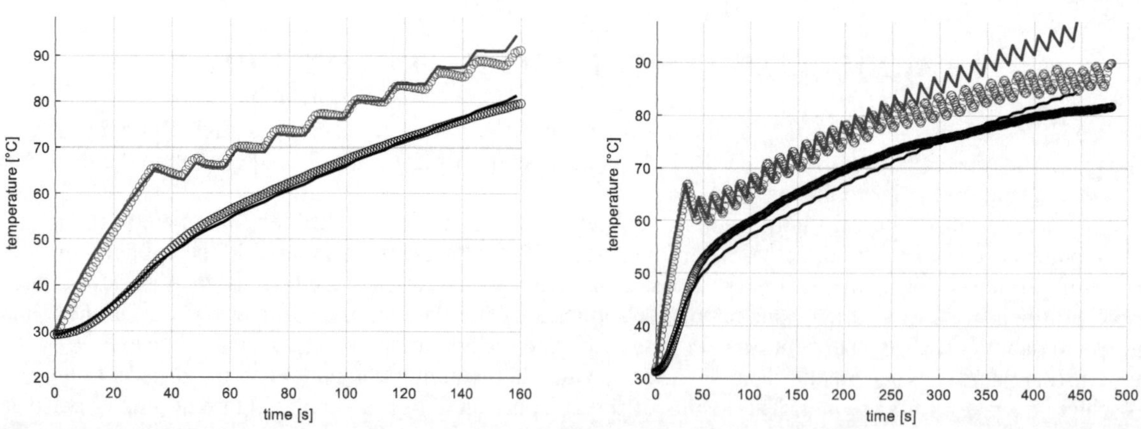

FIGURE 3. experimental(open dots) and numerically modelled (solid lines) time-temperature data at the heated (red) and non-heated (black) side of the instrumented sheet. With residual thermal radiation (left) and without (right)

Influence of Different on/off Ratios During the Pulsed Heating Phase

The left graph in figure 4 illustrates the influence of the ratio between the "time on" and "time off" on the temperature evolution of the instrumented silicon sheet. In this study, the "time on" was kept constant at 10s. During "time off" residual thermal radiation is present. The higher the on/off ratio is, the faster the heated side of the instrumented sheet reaches the targeted 90°C and the higher the through thickness temperature difference is. When compared to simple one step heating (not illustrated in the graph), the heating time to reach 90°C increases by 20% while the temperature difference between the heated and non-heated side decrease by 30%. These results were obtained without process setting optimization and indicates the potential of pulsed heating to obtain a more uniform through thickness temperature distribution without an excessive increase in heating time.

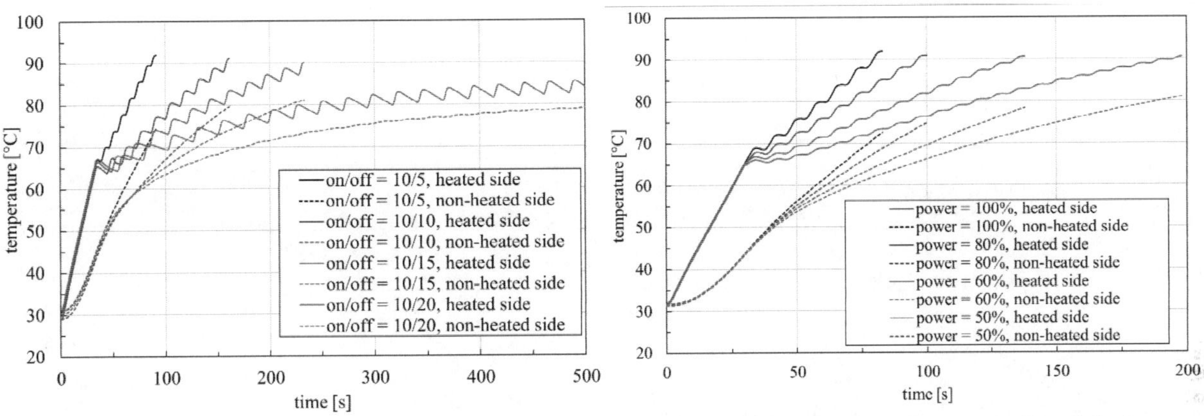

FIGURE 4. Experimentally measured temperature evolution at the heated and non-heated side of the instrumented silicon sheet illustrating the influence of the on/off time ratio (left) and the power applied during the pulsed heating phase (right).

A comparison between numerical and experimental data for two on/off ratios is made in figure 5. The developed model describes the experimental data very well for high "on/off" ratios. However due to an underestimation of the convective heat losses which cannot be neglected for lower "on/off" ratios, the numerical model overestimates the heating rate of the sheet at low "on/off" ratios.

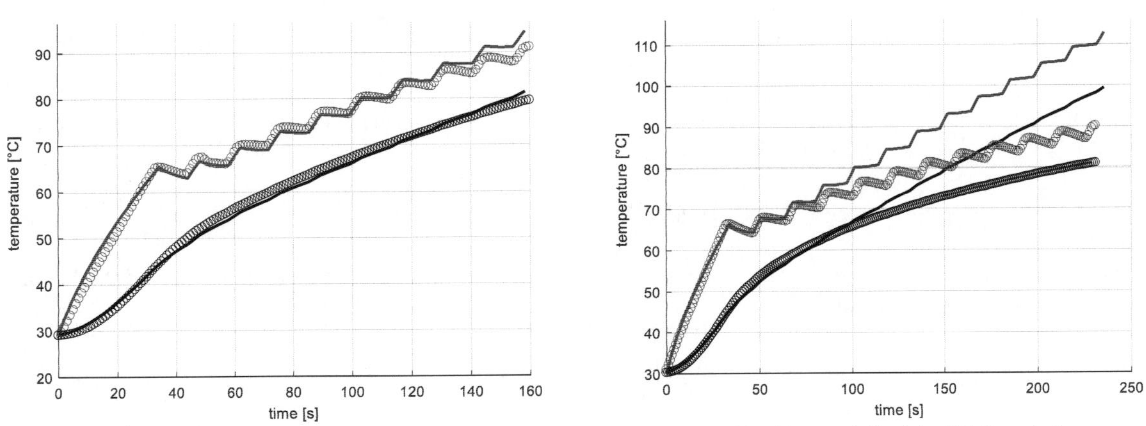

FIGURE 5. Comparison between experimental(open dots) and numerically modelled (solid lines) time-temperature data at the heated (red) and non-heated (black) side of the instrumented sheet. With on/off ratio equal to 10/10 (left) and 10/15 (right)

Influence of the Power of the Halogen Heater During the Pulsed Heating Phase

The influence of the applied heater power during the pulsed heating phase on the measured heating curves is presented in the right graph of figure 4. In these experiments, the "time on" was 5s and was equal to the "time off". The higher the heating power is, the faster the temperature of the heated surface increases and the larger the through thickness temperature difference is. By decreasing the power, the through thickness temperature distribution can be significantly improved. As in the previous comparisons between experimental and numerically modelled data, the model describes fast heating rates, caused by a high thermal radiative power, better than slow heating of the sheet. Again the convective heat transfer processes during "time off" have to be taken into account in order to improve the numerical model.

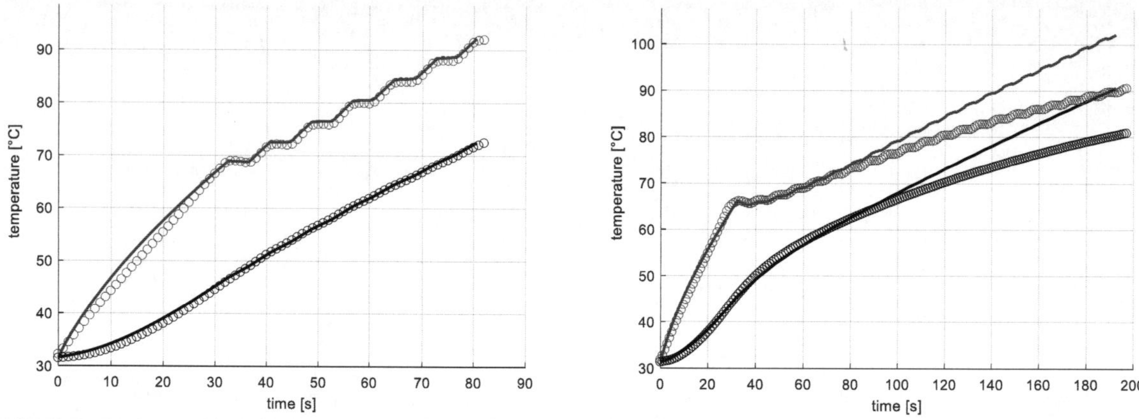

FIGURE 6. Comparison between experimental(open dots) and numerically modelled (solid lines) time-temperature data at the heated (red) and non-heated (black) side of the instrumented sheet. With heating power equal to 100% (left) and 50% (right)

CONCLUSIONS

The present study clearly indicates the potential of pulsed heating methods in the heating step of thermoforming thermoplastic sheets. Fast responding halogen heater elements allow different settings within a single process. The study has identified the positive effect of residual heat of the halogen elements in a pulsed heating sequence. In order to decrease through thickness temperature distributions without excessively increasing the heating time, the on/off ratio of the elements and heater element power have to be optimized. Future work will therefore focus on further process parameter optimization and improving the numerical model by adding variable convective heat losses.

ACKNOWLEDGEMENTS

The authors which to thank Wolfgang Daum from the Geiss AG thermoforming equipment company to provide sample reflectors for the experimental heater set up.

REFERENCES

1. E. Gilham, *Lean manufacturing*, SPE- European Thermoforming Division conference, Barcelona (Sitges), 2016.
2. S.A. Khan, P. Girard, N. Bhuiyan and V. Thomson, *Polym. Eng. Sci.* ***52*** ,625-636 (2012).
3. F.M. Schmidt, Y. Le Maoult and S. Monteix, *J. Mater. Process. Technol.* **143-144**, 225-231 (2003).
4. J.L. Throne, *Thermoforming Quarterly* **35**,22-27 (2016).
5. B. Buffel and F. Desplentere, PMI conference, Gent, 2016, pp. 221-226.
6. B. Buffel, B. Van Mieghem, A. van Bael and F. Desplentere, *Int. Polym. Proc.* **32**, 378-386 (2017).

pps proceeding
Symposium: S04 - Blow Molding, thermoforming and rotomolding
Oral Presentation

Friction Measurement Techniques for Thermoforming

Wuren Tuya (1), Martin Peter (1)*

(1) Queen's University Belfast - Belfast - UnitedKingdom

Plugs are a common feature of many thermoforming processes and are often essential for ensuring that the wall thickness distribution in a deep-draw thermoformed part is controlled and balanced. Generally, they are designed to locally capture the initial sheet thickness and greatly reduce its deformation in key areas prior to the final blowing stage that completes most thermoforming processes. Contact friction between the soft polymer sheet and the rigid plug surface therefore plays a fundamental role in controlling the magnitude of deformation that may occur locally and it is therefore critical in determining the overall wall thickness distribution. Over many years researchers have sought to measure and understand the nature of the frictional contact during thermoforming. Basic friction tests such as the sled test have been employed but often measurements have been difficult due to the soft nature of the polymer surface involved at thermoforming conditions. In addition, the test methods used for rigid bodies are unable to replicate the actual contact conditions, which involve the soft polymer surface stretching or elongating across the contacting surface. In this work the results from previous studies are compared to those of a new test in which heated polymer specimens are drawn across a contacting surface. This technique uses a rotational rheometer to stretch the polymer and thereby create a form of contact that more closely resembles the real process. The results obtained to date are in broad agreement with those previously obtained, but with some important differences. These are discussed as part of a wider review of the current state of the art in friction testing for thermoforming.

Effect of Surface Treatment on the Mechanical Properties of Hybrid Composite with Natural Fibers Produced by Rotomolding

López Gonzaleznúñez Rosa Gabriela[a], Ortega-Gudiño Pedro[a], González-Núñez Rubén[a], Robledo-Ortíz Jorge Ramón[b]

[a] *Departamento de Ingeniería Química , Universidad de Guadalajara, Blvd. Gral. Marcelino García Barragán # 1451, Guadalajara, Jalisco, 44430, México.*
[b] *Departamento de Madera, Celulosa y Papel, Universidad de Guadalajara, Carretera Guadalajara-Nogales Km 15.5, Las Agujas, Zapopan, Jalisco, 45510, México.*

Abstract. Hybrid composites are materials produced by combination of two or more types of fiber in a common matrix, the correct combination of these fibers may have certain advantages over materials produced with a single fiber type. Rotational molding is one of the fastest growing technologies in the area of polymer processing due to it is many advantages like the ability to produce large hollow parts without weld lines, its low residual stress and low equipment and manufacturing costs. In this work, polyethylene were processed with natural fibers (agave-coir) to obtain hybrid composites produced by dry-blending and rotational molding with a total fiber content of 20% wt. using different weight ratios of agave and coir fibers (100-0, 70-30, 50-50, 30-70 0-100). The fibers (both agave and coir) were chemically treated with maleated polyethylene (MAPE) in solution, in order to improve the fiber-matrix compatibility. The hybrid composites were characterized in terms of morphology and mechanical properties (tensile, flexion and impact). Hybrid composites reinforced with treated fibers, showed no significantly changes in their tensile and flexural modulus nor in its impact strength compared to composites produced with untreated fibers; however, tensile and flexural strength were considerably improved.

Keywords: Hybrid composites, reinforced thermoplastics, rotational molding.
PACS: 81.05.Mh, 81.05.Ni, 81.05.Pj, 81.05.Qk

INTRODUCTION

The use of natural fibers to obtain composites has been extensively studied by various polymer processing techniques such as compression or injection molding. When two or more different types of fibers are incorporated in the same polymeric matrix, the resultant materials called hybrid composite. In order to achieve the effect of "hybridization" a careful selection of fibers must be carried out considering its physical, chemical and mechanical properties. The correct combination of different fibers in a hybrid composite may have certain advantages over a material produced with a single fiber type [1]. Aji et al.[2] obtained hybrid composite with natural fibers by compression molding. They found that while pineapple leaf fibers increased flexural and tensile properties of the composite, the kenaf fiber improve impact strength and reduce water absorption.

Despite the fact that natural fibers have many advantages, their hydrophilic character results in poor interfacial bonding with most polymer matrices, resulting in poor mechanical behavior which limiting their applications. To solve this issue, a widely used solution is the incorporation of coupling agents during the processing step of composites [3]. Several authors refers that the maleated compounds are a good option for this purpose, for example Cisneros et al. [4] tested different types of fiber treatments to produce natural fiber composites by compression molding. The mechanical properties of composites based on fibers that were treated with maleated polyethylene (MAPE), showed substantially higher impact, tensile, and flexural properties up to 160, 400, and 100% respectively, compared to untreated fiber composites. Verdaguer et al. [5] treated wood flour with maleated polyethylene to produce composites by compressing molding. The results showed that surface treatment increased up to 110% the stiffness of composites.

Rotational molding (rotomolding) is a polymer processing method to produce hollow parts with different sizes, with the advantages of making plastic products with low residual stress and manufacturing costs and no weld lines [6]. Different types of polyethylene (PE) and natural fibers (agave, coir, pine, abaca banana, etc.) have been used to produce rotomolded natural fiber composites (NFC) [7]. Lopez-Bañuelos et al. [8] were the first to propose rotomolded PE-agave composites, and they found that the optimal mechanical properties of prototypes were obtained with a 10% wt. of agave fiber.

The aim of this work is to analyze the effect of the incorporation of two different natural fibers with and without surface treatment into a linear medium density polyethylene (LMDPE) matrix on the final properties of rotomolded hybrid composites. Natural fibers (agave and coir) were used with different weight ratio and a constant total content of 20%wt. Complete characterization in terms of morphology, density and mechanical properties (tensile, flexural and impact) are reported.

EXPERIMENTAL

Materials

The polymer matrix was rotomolding grade linear medium density polyethylene (LMDPE) RO 936 50 supplied by "Polimeros nacionales" that has a melt flow index of 5.0 g/10 min at 190°C and density of 0.936 g/cm^3. Agave fiber (*A. tequilana* Weber var. Azul) was obtained from residues of a local tequila company, coir fiber was obtained by GERMINAZA, S.A. DE C.V. Agave and coir fibers were grinded and sieved to keep particles between 354 and 297 µm corresponding to 50 mesh. For the surface treatment: sodium hydroxide from Analytyka, (Mexico), maleated polyethylene (MAPE) Epone C-26 from Westlake Chemicals (The United States) and xylene technical grade from Golden Bell reactivos (Mexico) were used.

Methodology

Fiber Treatment

The treatment was performed for both agave and coir fibers and was done in two steps as proposed by Cisneros López et al. [9]: mercerization and maleated polyethylene treatment. For mercerization the fibers was submerged for 15 min in a solution of 2% NaOH at room temperature, after this time fibers were rinsed several times with water to remove alkali residual and dried one week on room temperature, in this step hydroxyl groups on the fibers have been activate for the next stage due to alkali treatment increase an amorphous region on fibers also extracts a portion of hemicelluloses, lignin, pectin, and extractives [10]. For maleated polyethylene treatment 1% wt. MAPE was dissolved in xylene at 90°C, then pre-treated fibers were immersed for 30 min in the solution at the same temperature with high intensity mixing. Finally treated fibers were dried other week in room temperature for complete evaporation of xylene.

Hybrid Composites Preparation by Rotational Molding

Hybrid composites were produced with constant 20% wt. of fiber content with different agave-coir fiber weight ratios (1-0, 0.7-0.3, 0.5-0.5, 0.3-0.7, and 0-1). To produce rotomolded parts, a total mass of 350g of each composition of polymer and fiber was dispersed by dry blending using an industrial blender (JR Torrey PL2) for 6 min at 1750 rpm. Composites were produced in a laboratory-scale rotational machine with a cubic stainless steel mold of (16x16x15) cm^3. To easy extraction of part before load the material, a demolding agent (CP-500) was applied on internal surface of the mold. Then, the charged mold was closed, mounted on the rotating arm, and introduced into the oven which was previously heated at 280°C. The mold was kept rotating for 30 min at 270°C (±5). Later, the mold was pulled out from the oven and cooled by forced air until the internal air temperature reached 40°C. Finally, the mold was opened and the part was demolded. During all the rotomolding cycle (heating and cooling), a speed ratio of 1:4 was used.

Physical Characterization

Morphology and Density

Samples were fractured at cryogenic conditions (submerged in liquid nitrogen). Micrographs of hybrid composites were obtained by a scanning electron microscope HITACHI TM-1000 in order to characterize the structure and adhesion of composites. Density was measured by a gas pycnometer ULTRAPYC 1200e (Quantachrome Instruments) using nitrogen.

Mechanical properties

All the specimens for mechanical test were cut from rotomolded parts and tested at room temperature. Tensile properties were tested according to ASTM D-638, using an Instron 3345 universal testing machine with a 1KN load, the gage length and crosshead speed were fixed at 9.53 mm and 5 mm/min, respectively for type V dog-bone specimens. Three points flexural test were carried out using an Instron 3345 universal testing machine with a 1KN load, according to ASTM D-790, the speed for the flexural properties was set at 2 mm/min. Sample dimensions were (90x12.7xthickness) mm^3. For both tests (tensile and flexural), a minimum of 5 samples were used to report the average and standard deviation for modulus and strength. Impact strength was determined using a Charpy impact tester Ceast 9050 according to ASTM D-6110; the results were taken as the average value of 10 samples.

RESULTS AND DISCUSSIONS

Morphology

Figure 1 shows the cryogenic fracture of hybrid composites with a 50-50 composition. As well-known, rotomolding is a shear free polymer processing method, whereby the material is not required to move in one direction. Figure 1 shows that fibers are distributed randomly along piece. For untreated fiber hybrid composites (Figure 1 (a)) its possible to confirm the incompatibility between fibers and polymeric matrix, observations includes interfacial voids, fiber-fiber contact and domains with fiber agglomeration. In the other hand, hybrid composites prepared with treated fibers (Figure (b)) showed a better compatibility between fibers and matrix. It was possible to observe a reduced space between fiber and matrix, a better distribution, as well the absence or reduced incidence of voids, defects, and fiber pullout. . Similar results for this surface treatment of fiber were reported by Chimeni et al. [11], Verdaguer and Rodrigue [5] and Cisneros-López [9] for hemp, maple and agave fibers, respectively.

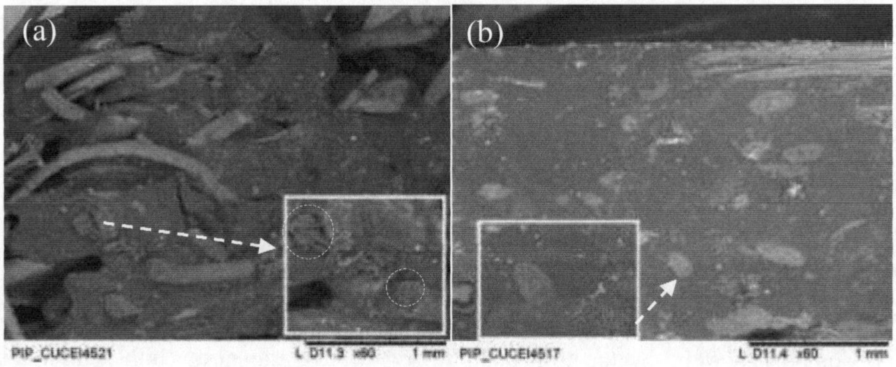

FIGURE 1. SEM Images of Hybrid Composite with 50-50 (agave-coir) composition. (a) Without Treatment, (b) Maleated Treatment.

Density

The density of hybrid composites increased with the incorporation of fibers, respect to neat PE (Table 1). It can be noticed that the density of the materials prepared with treated fibers decreases when compared with those with untreated fibers. This is a result of grafting lower density polymer chains on the fiber surface: polyethylene from MAPE has a lower real or skeletal density (the density of solids including only the pores not accessible to the gas) than agave and coir fibers. Raymond and Rodrigue [12]also reported lower density for MAPE treated maple fibers compared with untreated ones.

TABLE 1. Density of Hybrid Composite with Untreated and Treated fibers.

Sample (Agave-Coir)	Density (g/cm3)
PE	0.933 (±0.0018)
100-0	0.999 (± 0.0030)
100-0 (T)	0.999 (±0.0074)
70-30	0.976 (±0.006)
70-30 (T)	0.960 (±0.0084)
50-50	1.017 (±0.0070)
50-50 (T)	1.008 (±0.0060)
30-70	0.977 (± 0.0034)
30-70 (T)	0.955 (±0.0020)
0-100	1.016 (±0.0200)
0-100 (T)	0.988 (±0.0041)

Mechanical Properties

Impact strength

Figure 2 shows the results obtained from the Charpy impact tests of the treated and untreated hybrid composites. Impact strength decrease with the addition of fiber, because fiber-fiber contact induced stress concentrations points , providing sites for crack initiation and potential composite failure [13], Treated fibers composites showed significative changes, specially for agave and coir composites when compared to the untreated fibers, these represents an increased impact resistance: up to 20% (8.33 kJ / mm^2 at 10 kJ / mm^2) and 37% (6 kJ / mm^2 at 8.33 kJ / mm^2), respectively. This behavior corresponds to observations discussed in morphology section: SEM micrographics of treated fiber composites showed less interfacial voids and fiber agglomerations, which leads to a reduction in failure points. Cisneros-López et al.[4] reported similar results for MAPE-treated agave fiber composites prepared by compression at high fiber concentrations (30 and 40%): impact strength was increased due to an improved fiber distribution and fiber-matrix adhesion, leading to more efficient load distributions during test.

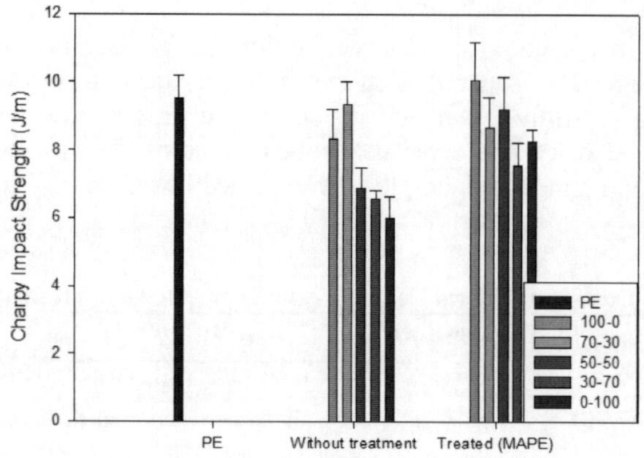

FIGURE 2. Charpy Impact Strength of Hybrid Composite with fiber untreated and maleated treated.

Flexural properties

Flexural moduli of hybrid composites are presented in figure 3 (a). Flexural moduli of treated fiber composites do not show significant changes when compared with untreated ones. On the other hand, it is observed that the maximum value of modulus is presented by 70-30 (agave-coir) composition, no matter if the fiber was treated or not. For untreated fiber composites the modulus is even greater than modulus obtained by the composites prepared with a single type of fiber, (hybrid effect). Ramezani et al.[14] and Perez-Fonseca et al. [15], mention that the properties of hybrid composites with the combination of fibers 70-30 "long-short fibers" are which usually have higher properties, because the "short fibers" occupy the spaces left by the "long fibers", better distributing the load when an effort is applied.

In figure 3 (b), contrary to what was shown by the moduli, the use of treated fibers presents an improvement in flexural strength, increasing by 103% for the composite of coir fiber (from 5.8 MPa to 11.8 MPa) compared to untreated composite. Similar behavior was presented for Cisneros-Lopez et al. [16] for composites of agave and coir treated with maleated PE, which showed an improvement in flexural strength at high fiber concentrations when compared to untreated ones, because decreases the space of interface between fiber and matrix, improving load distribution, as was showed in figure 1 (b).

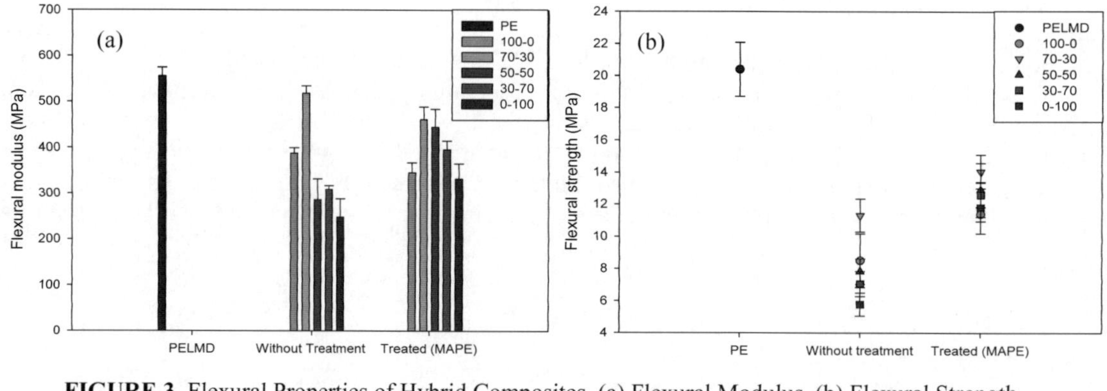

FIGURE 3. Flexural Properties of Hybrid Composites. (a) Flexural Modulus, (b) Flexural Strength.

Tensile properties

Tensile modulus of hybrid treated composites did not present significative changes (figure 4 (a)) when compared with untreated fiber composites, similarly of flexural properties, it is observed that the composition 70-30 shows the best performance, getting values even superior than neat PE, exceeding considerably those prepared with a single type of fiber (hybridization effect, as discussed in flexural properties).Tensile strength showed a significantly increase for treated composite presenting improves around 100% in comparison with untreated fiber composites. This improvements are related to surface treatment of natural fibers, and the consequently increased in fiber-matrix adhesion, which leads to an improved stress transfer between phases, as reported by several authors [5],[7],[13], and was confirmed in morphology section (figure 1 (b)): the treatment reduced the interfacial space between matrix and fiber, allowing a better fiber distribution.

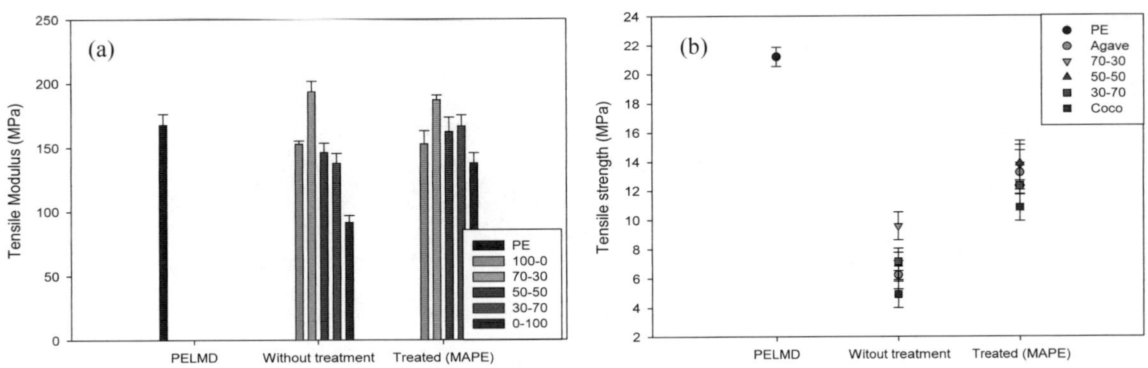

FIGURE 4. Tensile Properties of Hybrid Composites. (a) Tensile Modulus, (b) Tensile Strength.

CONCLUSIONS

Hybrid composites of agave-coir fiber were successfully produced by rotational molding, a composition 70-30 (agave-coir) presented the best mechanical performance. The use of treated fibers improves the compatibility of fibers with polymeric matrix: micrographs indicated decreases defects, voids, porosity and interfacial space. Tensile and flexion modulus did not change significantly with surface treatment of fibers, however, the impact, tensile and flexural strength, were considerably increased.

REFERENCES

1. Ashori, "Hybrid composites from waste materials," *J. Polym. Environ.*, **18**, 2012, pp. 65–70.
2. I. Aji, E. Zainudin, K. Abdan, S. Sapuan, and M. Khairul, "Mechanical properties and water absorption behavior of hybridized kenaf/pineapple leaf fibre-reinforced high-density polyethylene composite," *J. Compos. Mater.*, **47**, 2012, pp. 979–990.
3. G. M. Arifuzzaman Khan, M. S. Alam Shams, M. R. Kabir, M. a. Gafur, M. Terano, and M. S. Alam, "Influence of chemical treatment on the properties of banana stem fiber and banana stem fiber/coir hybrid fiber reinforced maleic anhydride grafted polypropylene/low-density polyethylene composites," *J. Appl. Polym. Sci.*, **128**, 2013, pp. 1020–1029.
4. O. Cisneros-López, J. Anzaldo, F. J. Fuentes-Talavera, R. González-Núñez, J. R. Robledo-Ortíz, and D. Rodrigue, "Effect of agave fiber surface treatment on the properties of polyethylene composites produced by dry-blending and compression molding," *Polym. Compos.*, 2015.
5. Verdaguer and D. Rodrigue, "Effect of surface treatment on the mechanical properties of wood-plastics composites produced by dry-blending," 2014, pp. 2021–2025.
6. Perot, K. Lamnawar, and A. Maazouz, "Optimization and modelling of rotational molding process," *Int. J. Mater. Form.*, **1**, 2008, pp. 783–786.
7. G. Torres and C. L. Aragon, "Final product testing of rotational moulded natural fibre-reinforced polyethylene," Polym. Test., vol. 25, no. 4, pp. 568–577.
8. R. H. López-Bañuelos, F. J. Moscoso, P. Ortega-Gudiño, E. Mendizabal, D. Rodrigue, and R. González-Núñez, "Rotational molding of polyethylene composites based on agave fibers," *Polym. Eng. Sci.*, **52**, 2012, pp. 2489–2497.

9. E. O. Cisneros-López, M. E. González-López, A. A. Pérez-Fonseca, R. González-Núñez, D. Rodrigue, and J. R. Robledo-Ortíz, "Effect of fiber content and surface treatment on the mechanical properties of natural fiber composites produced by rotomolding," *Compos. Interfaces*, **6440**, 2016, pp. 1–19,.
10. M. M. Kabir, H. Wang, K. T. Lau, and F. Cardona, "Chemical treatments on plant-based natural fibre reinforced polymer composites: An overview," *Compos. Part B Eng.*, **43**, 2012, pp. 2883–2892.
11. D. Y. Chimeni, J. L. Toupe, C. Dubois, and D. Rodrigue, "Effect of hemp surface modification on the morphological and tensile properties of linear medium density polyethylene (LMDPE) composites," *Compos. Interfaces*, **23**, 2016, pp. 405–421.
12. A. Raymond and D. Rodrigue, "Effect of Surface Treatment of Wood-Plastics Composites Produced By Rotomolding," 2014, pp. 2367–2371.
13. A. Nourbakhsh, A. Karegarfard, and A. Ashori, "Effects of Particle Size and Coupling Agent Concentration on Mechanical Properties of Particulate-filled Polymer Composites," *J. Thermoplast. Compos. Mater.*, **23**, 2009, pp. 169–174.
14. Pérez-Fonseca, J. R. Robledo-Ortíz, F. J. Moscoso-Sánchez, D. Rodrigue, and R. González-Núñez, "Injection molded self-hybrid composites based on polypropylene and natural fibers," *Polym. Compos.*, **35**, 2014, pp. 1798–1806.
15. Ramezani Kakroodi, S. Leduc, and D. Rodrigue, "Effect of hybridization and compatibilization on the mechanical properties of recycled polypropylene-hemp composites," *J. Appl. Polym.* Sci.,**124**, 2012, pp. 2494–2500.
16. E. O. Cisneros-López, A. A. Pérez-Fonseca, F. J. Fuentes-Talavera, J. Anzaldo, R. González-Núñez, D. Rodrigue, and J. R. Robledo-Ortíz, "Rotomolded polyethylene-agave fiber composites: Effect of fiber surface treatment on the mechanical properties," *Polym. Eng. Sci.*, **56**, 2016, pp. 856

ROTATIONAL MOLDING OF POLYLACTIC ACID BIOCOMPOSITES: USE OF AGAVE FIBERS AS AN APPROACH TO OPTIMIZE THE PROCESS

Cisneros López Erick Omar (1)*, Pérez Fonseca Aida Alejandra (1), Ramírez Arreola Daniel Edén (2), González García Yolanda (3), Robledo Ortíz Jorge Ramón (3)

(1) Departamento de Ingeniería Química, CUCEI, Universidad de Guadalajara - Jalisco - Mexico, (2) Departamento de Ingenierías, CUCSUR, Universidad de Guadalajara - Jalisco - Mexico, (3) Departamento de Madera, Celulosa y Papel, CUCEI, Universidad de Guadalajara - Jalisco - Mexico

In this study, biocomposites of polylactic acid (PLA) and agave fibers (Agave tequilana Weber var. Azul) were produced by rotational molding. In particular, a simple dry blending technique was used to disperse the agave fibers in the biodegradable polymer matrix. Different fiber contents were used between 0 and 40% wt. To compare the efficiency of the process, constant mass as well as constant thickness samples were produced. The samples were then characterized in terms of morphology and mechanical properties (tensile and flexion). The biocomposites were successfully obtained by both processing strategies (constant mass and constant thickness). In general, it was possible to observe that low fiber contents (10% wt.) produced the best morphology and stiffness (i.e. tensile modulus up to 11% higher than PLA) indicating that there is an optimum fiber content to get well-distributed fibers in the matrix. The constant mass parts were obtained with a thickness between 2 mm (PLA) and 5 mm (40% wt. agave fiber), and produced savings in materials cost of 10, 19, 29 and 39% for biocomposites with 10, 20, 30 and 40% of agave fiber respectively, compared to PLA. In the case of constant thickness parts, a dimension of 2 mm was obtained for all fiber contents by using a variable mass of material. Materials cost savings of 11, 31, 60 and 75% were achieved with 10, 20, 30 and 40% of agave fiber respectively, when compared to PLA. Also, a reduction up to 6 minutes on cycling time were produced with constant thickness parts, which represents 16% less processing time, at 40% of agave fiber.

Rapid Fabrication of Microspeaker Diaphragms with the Aid of Ultrasonic Thermoforming

Keun Park[a], Hyun-Joong Lee[a], Hojin Bae[a,b] and Andreas Gomer[c]

[a]*Seoul National University of Science and Technology, 232 Gongneung-ro, Nowon-gu, Seoul, 01811, Korea*
[b]*CAE Team, CADians Systems, 212 Gasan Digital 1-ro, Geumchun-gu, Seoul, 08052, Korea*
[c]*RWTH Aachen University, Steinbachstraße 53 B, 5207,4 Aachen, Germany*

Abstract. Owing to its rapid and energy-efficient heating capability, ultrasonic vibration energy has been used in various polymer fabrication processes for the purpose of polymer heating and softening. The purpose of this study is to use ultrasonic vibration energy in thermoforming, *so called ultrasonic thermoforming*, for efficient fabrication of microspeaker diaphragms. The ultrasonic thermoforming process was applied to fabrications of two types of diaphragms, a small microspeaker diaphragm for a mobile phone and a large diaphragm for a TV speaker. For the microspeaker diaphragm fabrication, the effect of the vibration transmission direction was investigated in terms of the forming quality. For the TV speaker diaphragm, infrared heating with additionally used to improve the forming quality of the ultrasonic thermoforming. In both process, effects of forming parameters on the forming quality were investigated. By applying the optimal forming condition, both diaphragms could be successfully fabricated with an excellent forming ratio within 10 seconds cycle time, which was more than 100 seconds in the traditional thermoforming process.

Keywords: microfabrication, microspeaker diaphragm, ultrasonic thermoforming, infrared heating.
PACS: 47.57.Ng

INTRODUCTION

As an efficient heating and softening method of thermoplastic polymer, ultrasonic vibration energy has been used in various polymer fabrication processes, including welding, injection molding, hot embossing and thermoforming [1]. This study aims to apply ultrasonic vibration energy to the thermoforming process, instead of the conventional thermal energy. Thermoforming is a process in which a thin polymer film is softened by appropriate heating, and is then deformed into the desired shape over a mold [2]. For the appropriate heating, the polymer film and mold should be heated higher than the polymer softening temperature before the forming stage. Accordingly, relatively long cycle time is required as long as several minutes, for proper heating and cooling of the mold [3].

Recently, ultrasonic vibration energy has been applied for thermoforming of thin polymer films [4-6]. In these studies, various speaker diaphragms were developed by applying ultrasonic thermoforming. A microspeaker diaphragm for a mobile phone (13.8 × 9.8 × 0.47 mm) was formed out of 32 μm films, where a soft elastomer mold was used instead of a metal mold in order to prevent damage of such a thin film [4]. In this process, the effect of vibration transmission direction was investigates to improve the forming accuracy of microspeaker diaphragms; direct and indirect ultrasonic thermoforming processes were compared thereby [5]. The ultrasonic thermoforming was further extended to the fabrication of a large diaphragm for a TV speaker (80.1 × 18.4 × 1.7 mm). The film thickness was 300 μm, which was more difficult to be formed than the previous microspeaker diaphragm [6]. To improve forming quality of such a large diaphragm, infrared heating was additionally used to preheat the polymer film before the forming stage. This article introduces a review of the previous results related to the ultrasonic thermoforming, and a new finding for further improvement is also discussed.

ULTRASONIC THERMOFORMING OF MICROSPEAKER DIAPHRAGMS

Ultrasonic thermoforming uses ultrasonic vibration energy to soften thermoplastic polymer films. In ultrasonic thermoforming a polymer film is placed between an ultrasonic horn and a mold. Ultrasonic vibration is transferred from the horn to the film, resulting in repetitive deformation and frictional heat on the polymer surfaces [4]. The softened film is then formed over a mold under the given pressure. This ultrasonic thermoforming was used to replace the conventional thermoforming process for two types of microspeaker diaphragms, a small diaphragm for a mobile phone and a large diaphragm for a TV speaker.

Ultrasonic Thermoforming of a Microspeaker Diaphragm for Mobile Phone

Figure 1 shows dimensional configuration of the microspeaker diaphragm for a mobile phone. This diaphragm is formed out of thin polyetheretherketones (PEEK) films with 32 μm thickness. Two small rectangular bands were formed in the inner region, and a large rectangular band was formed in the outer region. The width and height of the outer band were 1.4 and 0.47 mm, and a number of micro-corrugations were formed at four corners of the outer band. **Figure 2** shows schematic description of the ultrasonic thermoforming process. A steel mold was fabricated to have the bands with micro-corrugations, as a positive shape. A negative elastomer mold was prepared and was attached to an aluminum horn. A polymer film was placed between the negative elastomer mold and the positive metal mold, and ultrasonic waves were transferred to the polymer film through the elastomer mold. This process is advantageous in its heating capability because the elastomer mold acts as a thermal insulator [5].

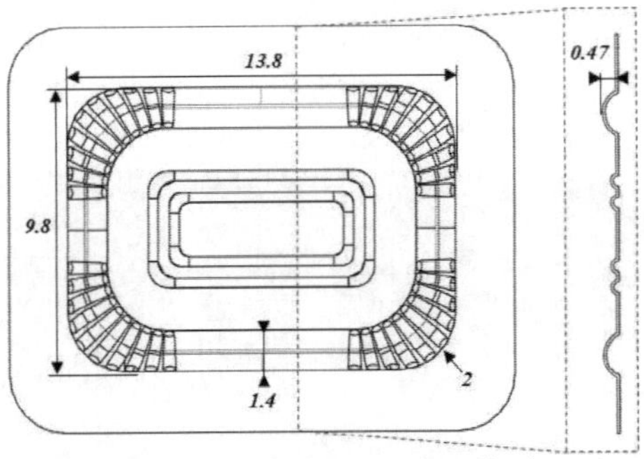

FIGURE 1. Configuration of microspeaker diaphragm

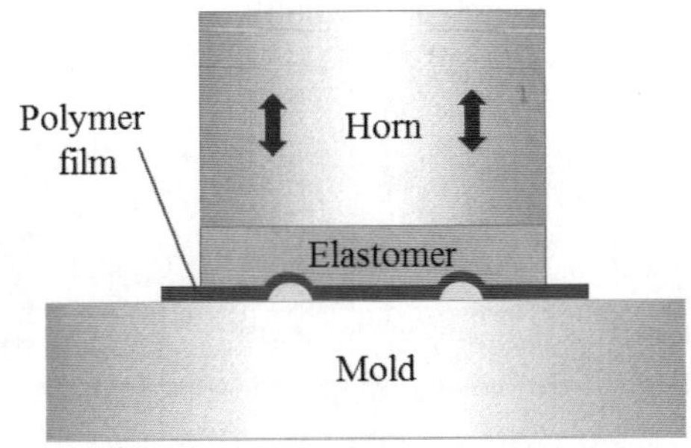

FIGURE 2. Schematic description of ultrasonic thermoforming

Ultrasonic thermoforming were conducted under the conditions of 0.4 MPa forming pressure, 0.3 s vibration time and 3.0 s holding time. To evaluate the heating capability, temperature distribution was measured during the thermoforming experiments, using a thermal imaging system (FLIR E50, FLIR Systems Inc.). **Figure 3** shows the temperature distributions after, which indicates that the temperature increased to 102.9°C, just after 0.3 s vibration. This rapid heating capability is superior to the conventional thermoforming process which requires long heating time (> 1 min) to heat the entire mold.

Figure 4 shows the measured profiles of the diaphragm across the outer band (Section A-A') and the micro-corrugation (Section B-B'). These profiles were measured using a high resolution 3D scanner (Rexcan DS2 5.0, Solutionix Co). The forming ratios were measured to 86.7% and 73.1% for the outer band and micro-corrugations, respectively. To improve forming ratio, ultrasonic thermoforming experiments were performed with variations in the three forming parameters: forming pressure, vibration time and holding time. **Table 1** summarizes the resulting forming ratios both in the band (FR_b) and corrugation (FR_c) regions according to the forming parameters. Based on these results, statistical analysis was performed to obtain optimal processing conditions; FR_c could be improved to 89.0% under the condition of 0.20 MPa forming pressure, 0.33 s vibration time, and 5.0 s holding time.

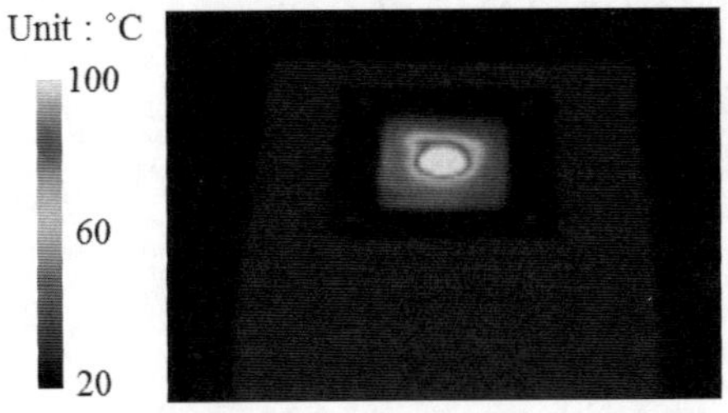

FIGURE 3. Temperature distribution after vibration

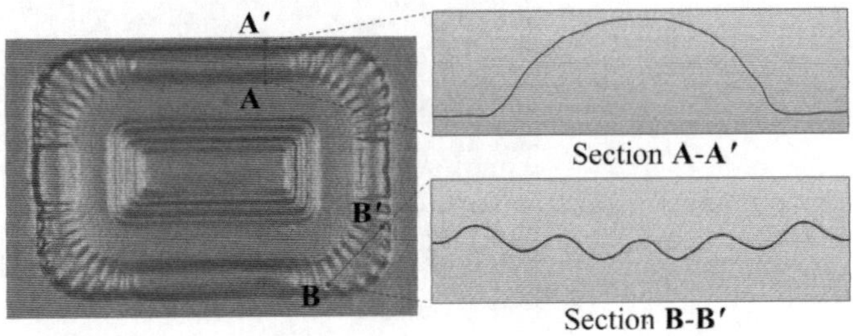

FIGURE 4. Surface profiles for the thermoformed diaphragm

TABLE 1. Comparison of forming ratio in the band (FR_b) and corrugation (FR_c) according to forming parameters

Parameters	Forming pressure (MPa)			Vibration time (s)			Holding time (s)		
Values	0.2	0.4	0.6	0.1	0.3	0.5	1.0	3.0	5.0
FR_b (%)	91.6	86.7	83.1	90.8	86.7	89.0	83.1	86.7	93.6
FR_c (%)	82.1	73.1	59.0	63.2	73.1	70.1	66.2	73.1	75.3

Ultrasonic Thermoforming of an TV Speaker Diaphragm with Infrared Heating

Figure 5 shows the dimensional configuration of the TV speaker diaphragm with a size of 80.1 × 18.4 mm. A rectangular-round band was designed to have a depth of 1.5 mm, and a number of micro-corrugations were to be formed on the band surface. The material was thermoplastic polyurethane (TPU) film of 300 μm thickness, which is more difficult to be formed than the previous microspeaker diaphragm if we consider its larger thickness and higher flexibility.

Ultrasonic thermoforming experiments were performed under the following conditions: 5.0 s vibration time, 0.3 MPa forming pressure and 4.0 s holding time. **Figure 6** shows photographs of the fabricated diaphragm from the top and sectional views, showing that the height of the protruded band was only 0.40 mm, which corresponds to a 26.7% forming ratio [6]. Although this forming ratio could be improved up to 60% when high forming conditions were applied (i.e. 9.0 s vibration time), this low forming ratio was not sufficient to obtain desired sound quality. Moreover, micro-corrugations were not replicated at all even in this condition, so that further study was required to improve the forming quality not only in the band region but also in the corrugation region.

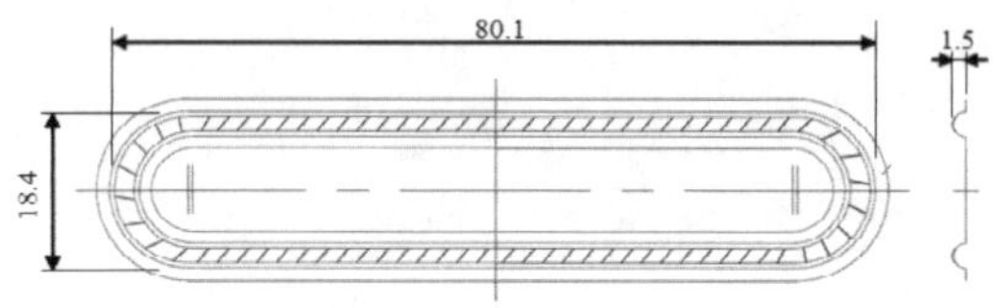

FIGURE 5. Configuration of TV speaker diaphragm

FIGURE 6. Top and sectional views of the fabricated diaphragm

To improve the forming quality, infrared heating was additionally applied to preheat the TPU film before the forming stage. **Figure 7** shows a schematic description of the infrared heating equipment. Two infrared lamps (850 nm wavelength, 500 W power) were installed above the TPU film. Infrared light was radiated on the upper face of the TPU film before for ultrasonic excitation, and the TPU film became heated and softened. Although previous study with one infrared lamp showed that the infrared heating had an effect to raise the film temperature higher than its softening temperature (150 °C) in 2 second, there was the temperature difference on the target area due to its unsymmetrical radiation [6].

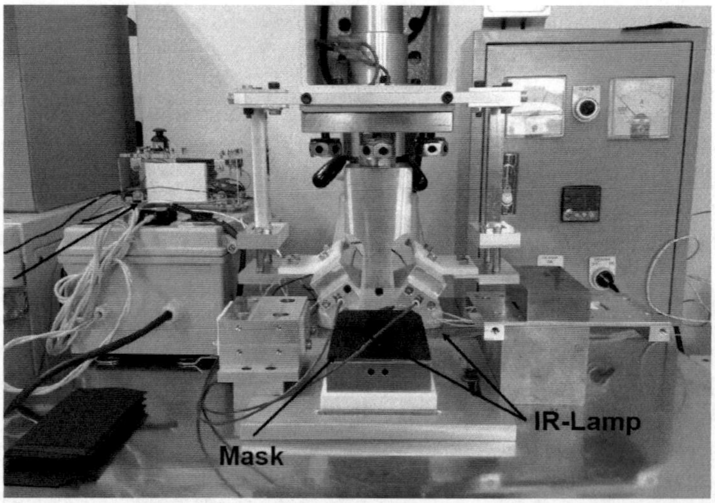

FIGURE 7. Experimental setup for ultrasonic thermoforming with infrared heating

In the proposed approach with two infrared lamps, however, the radiation showed symmetric characteristics. Various processing parameters (forming pressure, vibration time, holding time and infrared heating time) were investigated in terms of the forming quality. As a result, the best forming condition was determined to be 0.5 MPa forming pressure, 5.0 s vibration time, 10 s holding time and 5.0 s heating time. As a result, high forming ratios could be obtained as shown in **Figure 8**, which was higher than 95% in the band region and between 78.5 and 94.1 % in the micro-corrugation regions, respectively.

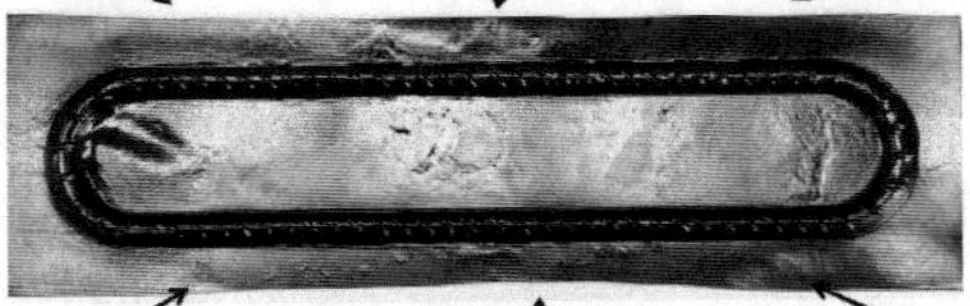

FIGURE 8. Thermoformed diaphragm with forming ratio at various locations

ACKNOWLEDGMENTS

This study was financially supported by the Commercialization Promotion Agency for R&D Outcomes, Ministry of Science, ICT and Future Planning, Republic of Korea.

REFERENCES

1. J. Sackmann, K. Burlage, C. Gerhardy, B. Memering, S. Liao and W. K. Schomburg, *Ultrasonics* **56**, 189-200 (2015).
2. J. L. Throne, *Technology of thermoforming*, Munich: Hanser Verlag GmbH & Co (1996).
3. H. Hosseini, B. B. Vasilivich, A. Mehrabani-Zeinabad, *Polym. Plast. Technol. Eng.* **45**, 1357-1362 (2006).
4. H. Bae, H. J. Lee and K. Park, *Microsyst. Technol.***23**, 1677-1686 (2017).
5. H. Bae, H. J. Lee and K. Park, *Int. J. Precis. Engng. Manuf.* **18**, 697-703 (2017).
6. H. J. Lee, D. J. Shin and K. Park, *J. Mech. Sci. Tech.***31** (in press).

FUNDAMENTALS OF ABUSE PERFORMANCE OF LLDPE/LDPE BLENDS IN BLOWN FILM APPLICATIONS

Rajen Patel, Teresa Karjala, Nilesh Savargaonkar^, Philip Salibi, and Lizhi Liu+*
The Dow Chemical Company, Freeport, Texas
^ Printpack Inc., 2800 Overlook Parkway, Atlanta, GA 30339
**4136 Lemoyne Way, Campbell, CA 95008; +Beijing Research Institute of Chemical Industry, Beijing, China*

Abstract
Blends of LLDPE and LDPE are used in many film applications. This study was undertaken to understand the behavior of LLDPE/LDPE blends over the entire composition range in terms of Elmendorf tear, dart impact, and puncture strengths in order to understand the relationships between these film properties and orientation effects and intrinsic properties of the blends.

Introduction
Blends of linear low density polyethylene (LLDPE) and high pressure low density polyethylene (LDPE) are used very commonly, particularly for films used in food and specialty packaging (F&SP) applications and also in some industrial and consumer packaging applications such as heavy duty shipping sacks (HDSS), green house films, and collation shrink films. There have been some studies performed to understand the effect of the addition of small amounts of LDPE, *i.e.* less than 20% to an octene LLDPE [1-2] to understand the impact of the LDPE on film physical properties such as Elmendorf tear strength, dart impact strength, etc. However, the effect of the addition of small amounts of LLDPE to LDPE (LDPE-rich blends) has not been studied as much [3], and hence is not as well understood. As a result, this study was undertaken to understand the behavior of LLDPE/LDPE blends over the entire composition range in terms of physical properties such as Elmendorf tear strength, dart impact strength, and puncture strength in order to understand the relationships between these film properties and orientation effects and intrinsic properties of the blends. Blends of three different LDPEs were studied with a LLDPE resin.

Experimental
Melt index (MI or I_2) was measured by ASTM D 1238, at 190 °C/2.16 kg and is reported in grams per 10 minutes. Density was measured as a quick density according to ASTM D 1928. Measurements were made within one hour of sample pressing using ASTM D792, Method B.

Melt strength was measured at 190 °C using a Göettfert Rheotens 71.97 melt fed with a Göettfert Rheotester 2000 capillary rheometer equipped with a flat entrance angle (180 degrees) of length of 30 mm and diameter of 2 mm. The pellets were fed into the barrel (length = 300 mm, diameter = 12 mm), compressed and allowed to melt for 10 minutes before being extruded at a constant piston speed of 0.265 mm/s. The extrudate passed through the wheels of the Rheotens located 100 mm below the die exit and was pulled by the wheels downward at an acceleration rate of 2.4 mm/s^2. Melt strength is reported as the plateau force (cN) before the strand broke.

Differential Scanning Calorimetry (DSC) was run on a TA Instruments Q1000 DSC, equipped with an RCS (refrigerated cooling system) and an autosampler. The sample was rapidly heated to 180 °C and held isothermal for 3 minutes in order to remove its thermal history. Next, the sample was cooled to -40 °C at a 10 °C/minute cooling rate and held isothermal at -40 °C for 3 minutes. The sample was then heated to 150 °C at 10 °C/min.

Monolayer films were produced on a blown film line in Dow's Film Application Development Center in B-1607, Freeport, TX. The films were 2 mil thick and produced at a rate of 300 lb/hr on a 250 mm die or 11.9 lb/hr/inch die circumference. The blow up ratio was 2.5. The melt temperature was between 380 and 420°F, and internal bubble cooling was utilized. A die gap of 70 mil was used for all the films, including the 100% LDPE films. The frost line height was measured between 36 and 38 inches for all the films.

Elmendorf tear testing in both the machine and cross directions (MD and CD) was done in accordance with ASTM D1922, type B – constant radius. Dart A testing was performed via ASTM D1709. The puncture test was performed using a modified ASTM D 5748 with a 0.5" diameter stainless steel probe. For intrinsic tear, measurements were done per the method described in [4].

Small Angle X-ray Scattering (SAXS) data were collected with a Rigaku instrument with a micro X-ray source. The sample to detector distance was 150mm, the data were calibrated with Silver Behenate, and the air background was corrected before data analysis.

Materials
The resins are listed in Table 1 along with their melt index and density. Blown film formulations were made with

10%, 20%, 40%, and 80% LDPE added to the octene LLDPE. In addition, blown films were made with 100% LLDPE and 100% LDPE.

Results and Discussion

Resin Characterization

The melt strength overlay is shown in Figure 1. LDPE-A and LDPE-B both have similar melt strength values. The higher MI LDPE-C has substantially lower melt strength, followed by LLDPE-1. The melt strength of the LLDPE/LDPE blends was also measured as shown in Figure 2. It is well known that high pressure LDPE resins have higher melt strength than Ziegler-Natta (Z-N) LLDPE resins due to the presence of long chain branching in the LDPE resins. Hence it was not surprising that addition of LDPE to a Z-N LLDPE resulted in higher melt strength. However, the melt strength of the blends of LLDPE/LDPE was significantly higher than the linear rule of mixtures at all blend ratios. This synergy in melt strength of LLDPE/LDPE has been reported earlier as well [5]. For LDPE-C it was observed that the melt strength of the blend with 40% LDPE was about the same as the melt strength of the blend with 80% LDPE and the melt strength for the 100% LDPE was significantly lower than the 20% LLDPE/80% LDPE blend. For LDPE-B and LDPE-A, the melt strength increased from 40% to 80% LDPE and then it dropped from 80% LDPE to 100% LDPE. In all three cases, the melt strength of the 100% LDPE resin was lower than that at 80% LDPE.

The DSC cooling curves are shown in Figure 3. The DSC curves within a given family, either LLDPE or LDPE, generally have higher crystallization temperatures with increase in density. At the same density, the crystallization point is lower for the LDPE as compared to the LLDPE due to narrow short chain branching distribution (SCBD) of the LDPE resins.

Film Results

Elmendorf Tear Strength

The films were tested in the machine direction (MD) as well as the cross direction (CD). The MD and CD tear strengths for the blends of LLDPE-1 with LDPE-A, LDPE-B, and LDPE-C as a function of the % LDPE are in Figures 4-5. The MD tear strength gradually decreased with the addition of LDPE up to 40% loading, almost reached a plateau value up to 80% LDPE, and then actually increased up to 100% LDPE. Interestingly, the same trend was observed for all blend films with the three LDPEs. It is possible that the MD tear strength actually went through a minimum at a loading between 40 and 80% LDPE since blends were not included in this LDPE range. However, more noteworthy was the observation of the plateau reached for the MD tear strength between 40 and 80% LDPE and then the increase in MD tear strength between 80 and 100% LDPE.

It is well known that Z-N catalyzed octene LLDPEs such as LLDPE-1 exhibit high tear strength (generally > 400 g/mil) because of presence of a high density fraction having high molecular weight and they also exhibit a relatively high tie-chain concentration [6]. High pressure LDPE resins do not have as much tie chain concentration due to the presence of high levels of long chain branching, hence the addition of such LDPE resins to octene LLDPE resins should result in lower tie chain concentration for the blend resulting in lower MD tear strength for the blend as compared to the pure LLDPE resin. Further, the addition of LDPE to Z-N LLDPE would be expected to result in increased amorphous phase orientation in the MD resulting in reduced MD tear strength for the LLDPE/LDPE blends. This was observed in this study as well up to 40% LDPE loading. Somewhat unexpectedly, the MD tear strength did not continue to decrease at higher LDPE loadings for such blends in blown films; but actually increased from 80 to 100% LDPE. It is believed that this unusual behavior is related to the orientation locked in the blown films for LLDPE/LDPE blends containing more than 40% LDPE.

Hence it was decided to measure intrinsic tear strength of such blends by making compression molded plaques, in order to remove the orientation effects on tear strength. The intrinsic tear data is plotted in Figure 6. In addition, it was decided to measure the melt strength of the LLDPE/LDPE blends and to analyze the blown films by small angle X-ray light scattering in order to understand the potential effects of orientation in the films on the MD tear strength of the blown films. The CD tear strength went through a maximum around 20% LDPE for all three LDPE resins used in this study as shown in Figure 5. The intrinsic tear strength of LLDPE/LDPE blends decreased with increasing LDPE content as expected, due to a decrease in the tie chain concentration.

Film Orientation by Small Angle X-Ray Scattering

It is known that tear strength closely relates to the lamellar orientation induced during the blown film process. The lamellar orientation can be quantitatively evaluated using a Herman's orientation function [7] defined as:

$$f_c = \frac{3\langle \cos^2 \phi \rangle - 1}{2}$$

In which ϕ is the angle between the chain axis (approximately the normal direction of the lamellar stack) and a reference axis. The $<\cos^2\phi>$ is defined as:

$$\langle \cos^2 \phi \rangle = \frac{\int_0^{\pi/2} I(\phi)\cos^2\phi \sin\phi d\phi}{\int_0^{\pi/2} I(\phi)\sin\phi d\phi}$$

in which I(ϕ) is the scattered intensity along the angle ϕ. For this SAXS work, the MD direction was used as the reference direction. The f_c represents the degree of lamellar orientation along the MD direction. A value of 1 represents perfect orientation, 0 represents random orientation, and -0.5 represents perfectly perpendicular orientation. As the SAXS peak is relatively broad for polyethylene, a band with a width of 5 pixels, which is crossing the scattering maxima, is used for the evaluation of lamellar orientation.

The scattering patterns are shown in Figure 7; all of the scattering patterns are preferably oriented in the MD. As shown in Figure 8, as the LDPE content is increased up to 20 wt%, there is a sharp increase in lamellar orientation and then a slight increase with additional LDPE up to 80 wt%. The neat LDPE shows a lower orientation. It is known from rheology that the relaxation of LDPE is slow; with the addition of LDPE to LLDPE, more orientation can be locked in the MD during crystallization of the film, due to a much higher draw down ratio in the MD as compared to the blow up ratio (BUR) in the CD. This is what is observed in Figure 8. It is also noted that the blend with 80 wt% of LDPE shows higher lamellar orientation than the neat LDPE. This can be interpreted from crystallization dynamics as well as from melt strength data. The LLDPE has a higher crystallization temperature than LDPE during a cooling process as shown in Figure 3. Therefore, for the blend with 20 wt% of LLDPE in the LDPE, the blend could crystallize at a higher temperature than LDPE. Also, blending 20 wt% LLDPE in LDPE resins increased the melt strength of the blend which also would result in higher MD orientation. Therefore, more MD orientation can be locked in the blend than for the neat LDPE. Typically, when polymer is crystallized at a higher temperature, the long period increases.

As seen in Figure 9, the scattering profiles of LDPE and its blend with 20 wt% of LLDPE looks very similar except that the blend shows a larger long period (16.4 nm) than that of the LDPE (15.5 nm) presumably due to a faster crystallization process in the blend. The scattering profiles of other blends with less LDPE look more similar to neat LLDPE which shows a significantly larger long period of 21 nm. With increasing LDPE content, the long period decreases. The change in lamellar orientation with the addition of different amounts of LDPE shows a similar trend as with melt strength (Figure 3) but the opposite trend as compared with MD tear strength (Figure 4). It is well known that typically the higher orientation in the MD direction leads to lower tear strength in the MD and higher tear strength in the CD. The change in MD tear strength with LDPE content is consistent with the lamellar orientation observed for the blends. However, it needs to be kept in mind that the Hermann orientation factor takes into account only the average orientation and does not consider localized orientation effects. It is generally known that the fracture during tear testing takes the path of least resistance which generally would be the path of most orientation and lowest tie chain concentration. Hence a combination of both these factors would be expected to have an influence on MD and CD tear strengths of the films.

Dart Impact Strength

The variation in dart impact strength with increase in LDPE loading of the LLDPE/LDPE blends is shown in Figure 10. The addition of any of these LDPEs to LLDPE-1 results in a significant reduction in the dart impact strength. The decrease in dart with increasing LDPE is also believed to be due to an increase in MD orientation (more unbalanced MD/CD orientation) and the decrease in intrinsic impact strength (decrease in tie chain concentration) with increasing LDPE. The error bars in Figure 10 represent the range of dart weights tested for a given film with a particular LLDPE/LDPE blend.

Puncture Strength

The variation in puncture strength with increase in LDPE loading of the LLDPE/LDPE blends is plotted in Figure 11. It was observed that the addition of any LDPE to LLDPE-1 resulted in a significant reduction in puncture strength. The decrease in puncture strength of the blends appeared to be almost linear with the LDPE content. The decrease in the puncture strength of the films with increasing LDPE content of the blends is believed to be due to an increase in the MD orientation (more unbalanced MD/CD orientation) and a decrease in the intrinsic toughness (decrease in the tie chain concentration) as the LDPE content was increased.

Conclusions

The MD tear strengths of LLDPE/LDPE blend blown films went through a minimum as the LDPE loading in the blends was increased. The minimum in MD tear strength was observed at LDPE loadings between 40 and 80%. This trend in MD tear strength was observed for all three LDPE resins studied. This observation was found to be related to the MD orientation effects in a blown film and seemed to be independent of the type of LDPE used in these blends. In contrast, the CD tear strengths of the blend blown films went through a maximum at about 20% LDPE and then showed a continuous decrease with increasing LDPE loading. This trend in CD tear strength was observed for all LDPEs. The intrinsic tear strength of the LLDPE/LDPE blends showed a continuous decrease with increasing LDPE loading for all three LDPE resins. It is believed that the intrinsic tear strength of the LLDPE/LDPE blends decreased continuously due to decreasing tie chain concentration with increasing LDPE content of the blends. Dart impact strength and puncture strength of blown films containing LLDPE/LDPE blends showed a continuous decrease with increasing LDPE loading for all three LDPE resins studied. Similar trends were observed for the decrease in intrinsic dart impact strength with increasing LDPE content of the blends. The decrease in the dart impact and puncture strengths of the LLDPE/LDPE films was thought to be occurring due to the decrease in tie chain concentration and increased MD orientation in the films with increasing LDPE content of the blends.

References

[1] P. Tas, L. Nguyen, S. Marshal, and M. Kashanian, *TAPPI Place Division Conference* (2004).

[2] P. Tas, L. Nguyen, and J. Teh, *J. Plast. Film Sheeting*, **21**, 265 (2005).

[3] D. J. Falla, *TAPPI Conference Proceedings* (1996).

[4] T. A. Plumley, K. Sehanobish, R. M. Patel, S. Y. Lai, S. P. Chum, and G. W. Knight, *J. Plast. Film Sheeting*, **11**, 269 (1995).

[5] K. Ho, L. Kale, and S. Montgomery, *J. Appl. Polym. Sci.*, **85**, 1408 (2002).

[6] R.M. Patel, K. Sehanobish, P. Jain, S.P. Chum, and G.W. Knight, *J. Appl. Polym. Sci.*, **60**, 749 (1996).

[7] K. Hong and G. Strobl, *Macromolecules*, **39**, 268 (2006).

Tables and Figures

Table 1: Melt index and density of LDPE and LLDPE.

Description	I_2	Density (g/cm^3)
LDPE-A	0.31	0.9248
LDPE-B	0.64	0.9206
LDPE-C	1.63	0.9191
LLDPE-1	0.95	0.9202

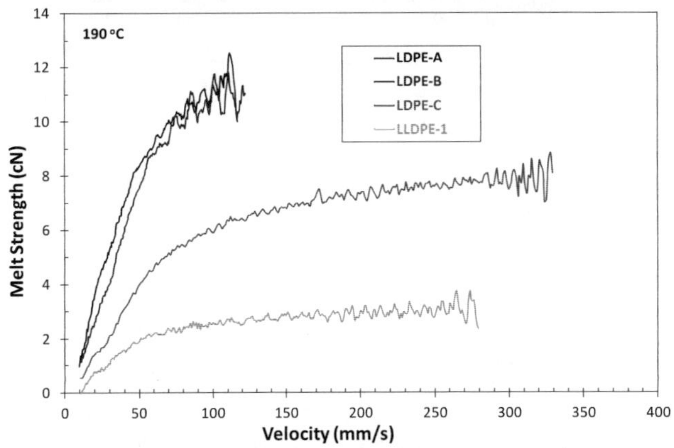

Figure 1: Melt strength of LDPE and LLDPE resins.

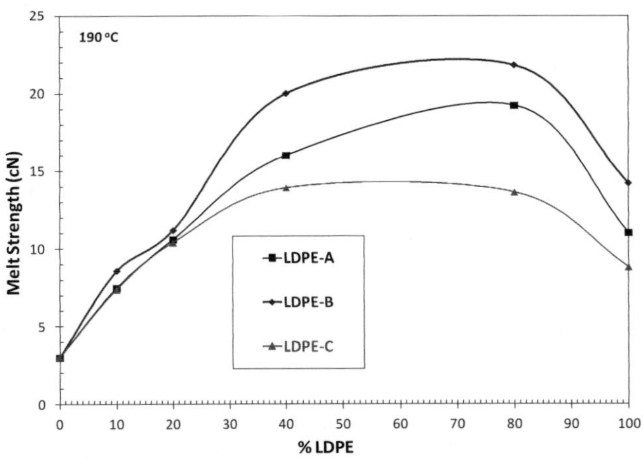

Figure 2: Melt strength of LLDPE/LDPE blends.

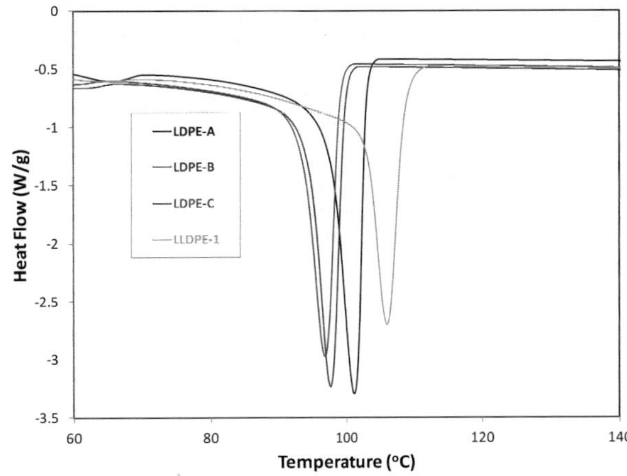

Figure 3: DSC cooling curves of LLDPE and LDPE resins.

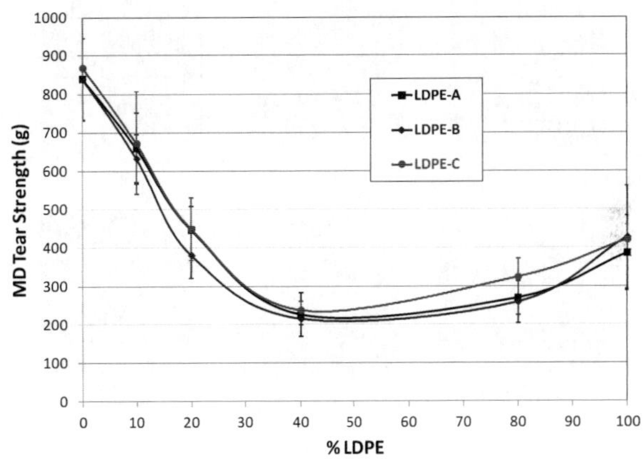

Figure 4. MD tear strength of 2 mil blown films of LLDPE/LDPE blends.

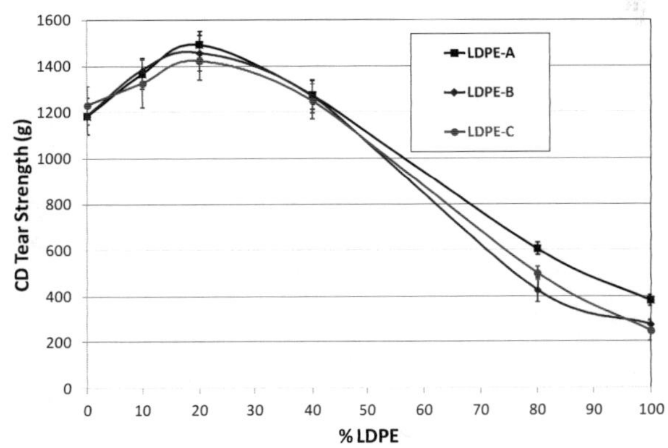

Figure 5. CD tear strength of 2 mil blown films of LLDPE/LDPE blends.

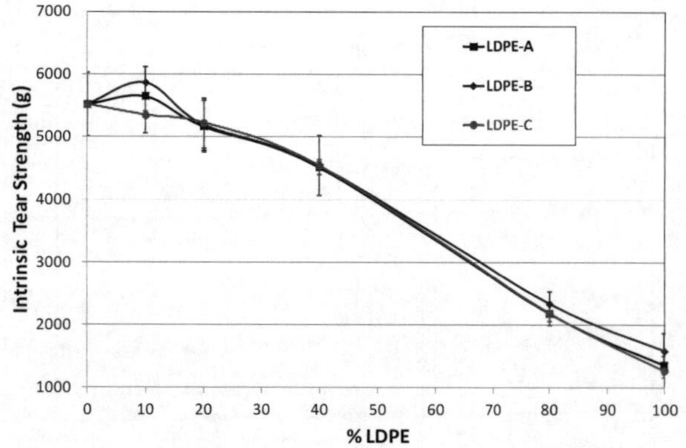

Figure 6. Intrinsic tear strength of LLDPE/LDPE blends.

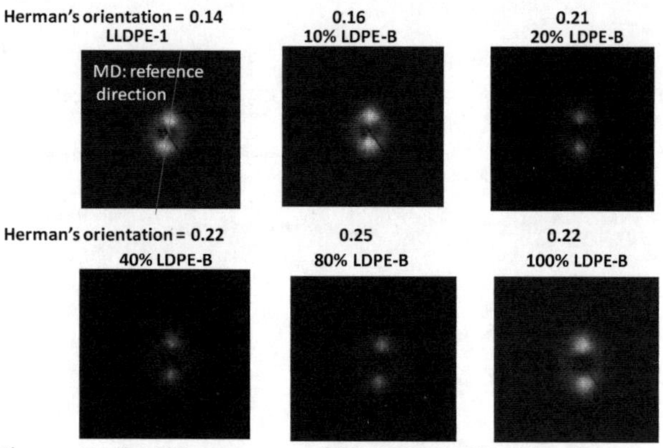

Figure 7: SAXS patterns of LLDPE/LDPE blends.

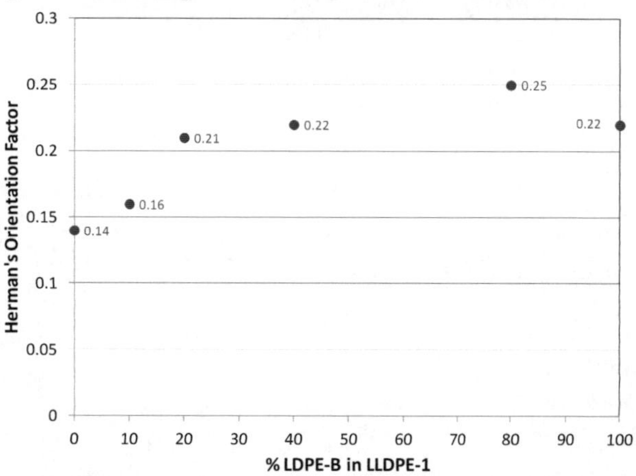

Figure 8: Herman's orientation values of LLDPE-1/LDPE-B blends.

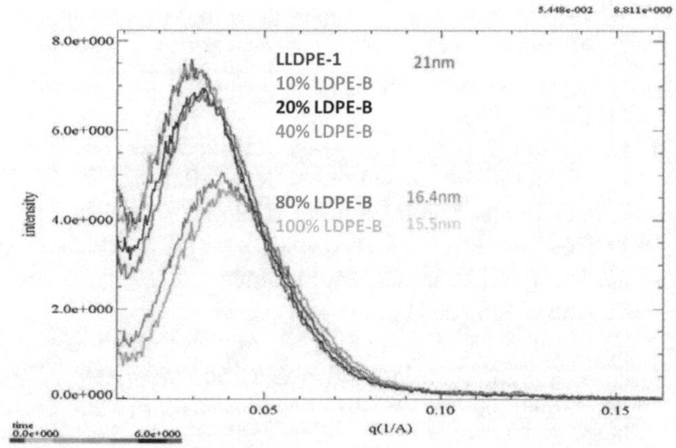

Figure 9: SAXS scattering profiles of LLDPE-1/LDPE-B blends.

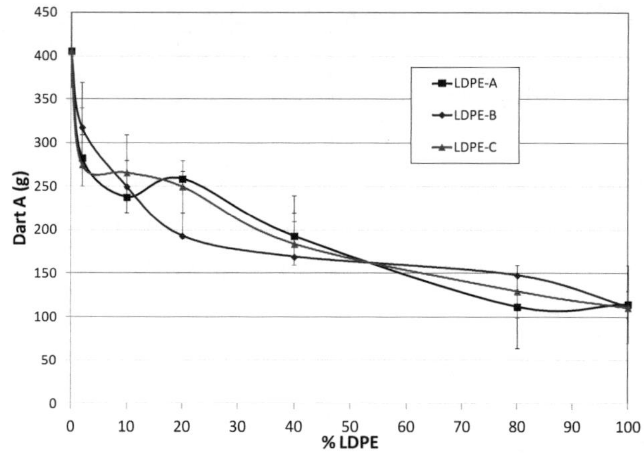

Figure 10: Dart impact strength of LLDPE/LDPE blends in blown films.

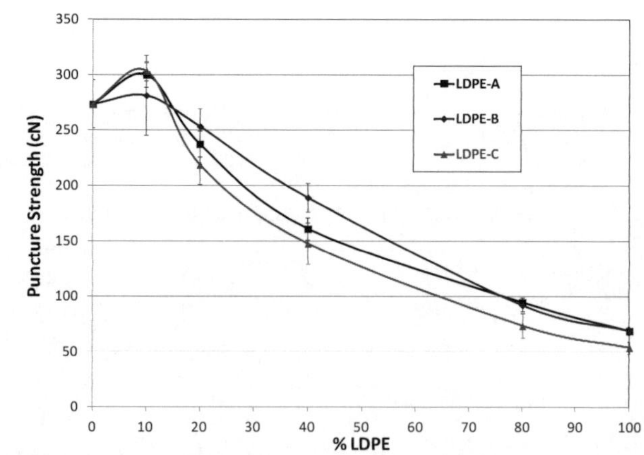

Figure 11: Puncture strength of blown films with LLDPE/LDPE blends.

Investigations on the Production of Hybrid Structures Using the Two-Stage GITBlow Process

Elmar Moritzer[a], Björn Landgräber[a*]

[a]*Paderborn University, Kunststofftechnik Paderborn (KTP), Germany*

bjoern.landgraeber@ktp.upb.de

Abstract. The Two-Stage GITBlow process was developed by the Kunststofftechnik Paderborn and combines the processes of gas assisted injection molding and blow molding. The parts feature large, thin-walled hollow spaces and additional functional elements. This is realized by using two consecutive gas injections and a mold with two cavities of different sizes. The first gas injection creates a hollow space within the part. In a second larger cavity the hollow space is inflated further by a second gas injection. The aim of the current research project is to manufacture hybrid structures consisting of GITBlow structures and continuous fiber-reinforced thermoplastic sheets. The new approach is to join the parts with the molten sheet material during the final phase of inflation in order to strengthen the parts locally. The parts produced with this process are characterized by their low weight, large volume and high compressive strength. To achieve the highest possible bond strength between the two components, it is necessary to maximize the temperature of the GITBlow part within the joining area. In a case study with Polycarbonate (Covesto Makolon OD 2015) and Styrene-Butadiene copolymer (Styrolution Styrolux 656C) the influence of the melt temperature and various process times on these factors can be shown. For PC a maximum temperature in the joining area of 165°C can be achieved while still maintaining optimal part quality (SBC: 120°C). Further investigations show that heat transfer within the part is of major importance for the final temperature distribution. With this knowledge an optimization of the process concerning the production of hybrid structures can be achieved.
Keywords: Gas assisted injection molding, lightweight, hybrid structures, weld strength, temperature distribution
PACS: 82.35.Gh, 81.05.Lg, 81.05.Qk

DESCRIPTION OF THE TWO-STAGE GITBLOW-PROCESS

The special injection molding process GITBlow, developed by the Kunststofftechnik Paderborn (KTP), combines the respective advantages of the established processes of gas-assisted injection molding ("GIT") and blow molding ("Blow"). The basic idea of GITBlow is to execute a preform production using conventional GAIM-technology followed by an enlargement of the cavity and a second gas injection to inflate the preform [1]. The parts produced with this technology feature large hollow spaces, which are typical for blow molding and additional functional elements such as reinforcing rips and connector elements, which are characteristic of injection-molded parts.

The process cycle of the Two-Stage GITBlow-process is characterized by a mold with two cavities of different size. The smaller one is used to produce the preform geometry, which is then transferred to the larger cavity by a rotary table. Inside this cavity the second gas injection takes place resulting in the inflation of the preform to the final part shape. As there is no reheating of the preform prior to the second gas injection, the heat distribution within the part is crucial for the inflation behavior. To ensure evenly distributed temperatures within the preform, the time for the transfer to the larger cavity (also called handling time) is utilized. During the transfer the preform is exposed to the surrounding air with relatively good thermal isolation properties. This leads to thermal equalization within the preform cross-section.

In the current project the focus is set on the production of hybrid structures using the Two-Stage GITBlow process. The goal is to reinforce the GITBlow parts partially with continuous fiber-reinforced thermoplastic sheet materials (also known as organo sheets). To achieve this, the process is modified and the thermoplastic sheets are placed as an insert during the handling time. As the sheets are heated to forming temperature before being placed inside the cavity it is intended to achieve a welding process when the preform is inflated by the second gas injection. As a result parts with local reinforcement due to the welding with the organo sheets are achieved. One major point of interest of the conducted investigations is the achievable weld strength at the bonding area between the GITBlow part and the organo sheet.

INVESTIGATED PARAMETER INFLUENCES ON THE WELD STRENGTH

The welding process within this modified GITBlow process is not completely comparable with conventional welding processes but has some similarities with hot plate welding. The geometry of the weld zone in this process is not even but has the shape of a semicircle. As seen in Figure 1 the contact area between the GITBlow part and the reinforcing organo sheet is nonplanar and it is mentionable that the temperatures within the welding area are lower compared to conventional welding processes. This is because the GITBlow part's temperature needs to be within the thermoelastic temperature range to enable the inflation of the part.

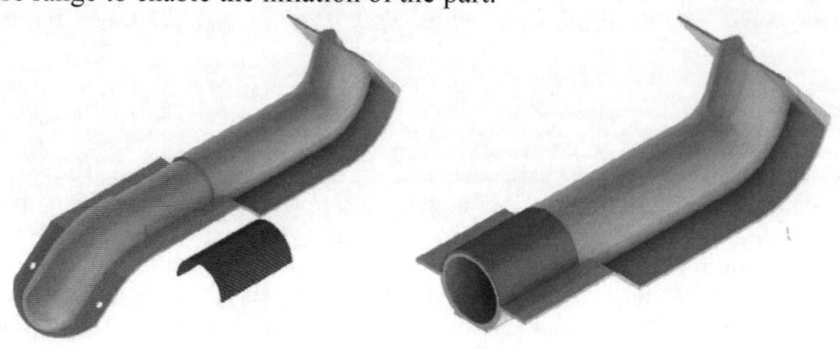

FIGURE 1. GITBlow base part with reinforcing structure alongside (left) and geometrical conditions in the joining area (right)

Choice of Materials and Surface Properties

The aim of all joining processes is to achieve a maximum weld strength within the joining area. The adhesion between the two joining partners can be described with several models focusing on individual aspects of the complex mechanisms of adhesion [2]. Generally adhesion can be categorized into mechanical and specific adhesion. Mechanical adhesion however can be neglected for the bonding process in the modified GITBlow process as there is no mechanical keying involved. The most advanced and most widely used theory to describe specific adhesion is the adsorption (or thermodynamic) theory based on the work of Sharpe and Schonhorn [3], [4]. The basis of this theory is that adhesive forces are the result of an intimate contact between the two substrates utilizing interatomic and intermolecular forces. To ensure close contact between the two substrates, an adequate wetting is necessary.

The wetting of a (liquid) substrate on the surface of a different substrate is characterized by the contact angle θ. The wetting behavior between two substrates and the resulting contact angle are affected by the surface energies of the different joining partners as they result in an interface energy. The correlation between the surface energies, the interface energy and the contact angle can be described by Young's equation [5]. The contact angle itself is only one aspect to be considered when optimizing adhesive bonding. -The Dupre equation (eqn 1) is derived from the Young equation and takes the resulting work of adhesion into account [6]. With this equation an analysis of the expected level of adhesion can be achieved.

$$W_A = \gamma_S + \gamma_{LV} - \gamma_{SL} \quad (1)$$

From eqn (1) can be concluded, that for a maximization of the adhesive bonding the interface energy γ_{SL} needs to tend against 0. If eqn (2) is taken into account it can be concluded, that this can be achieved if the disperse and polar portions of the surface energies of the two substrates are equal.

$$\gamma_{SL} = \left(\sqrt{\gamma_S^P} - \sqrt{\gamma_{LV}^P}\right)^2 + \left(\sqrt{\gamma_S^D} - \sqrt{\gamma_{LV}^D}\right)^2 \quad (2)$$

But as the disperse portions of the surface energy for the investigated materials are roughly the same, the focus is set on the polar portions. If the aspects mentioned above are concluded it can be seen, that for a maximization of the adhesion the ratio of the polar portions of the surface energies of the two substrates need to be equal. Additionally the total surface energy of the two substrates should be as high as possible [6]. Table (1) contains the available materials for the organo sheet matrices and for the GITBlow parts. The surface energies mentioned in this table were determined using the method of Owens Wendt Rabel und Kaelble and the Sessile Drop Technique.

TABLE (1). Ratio of the polar surface energy portions and expected work of adhesion for different material combinations

Organo sheet matrix	GITBlow part material	Work of adhesion / mN/m	Ratio of the polar surface energy portions
PC (Lanxess Tepex Dynalite)	PC (Covestro Makrolon OD2015)	76.46	1.49
	PC-ABS (Covestro Bayblend T85)	75.21	3.2
	SBS (Styrolution Styrolux 656C)	56.84	12.58
PP (Lanxess Tepex Dynalite)	PC (Covestro Makrolon OD2015)	76.15	0.73
	PC-ABS (Covestro Bayblend T85)	75.51	1.59
	SBS (Styrolution Styrolux 656C)	57.42	6.26

Even though the combination of both organo sheet matrices with SBS show a relatively high ratio of the polar surface energy portions, all combinations mentioned in Table (1) are tested for their welding capabilities.

Temperatures and Pressures in the Joining Area

In order to test the material combinations two boundary conditions are necessary. This includes the temperature of the GITBlow part and the organo sheet in the welding area and the welding pressure that is present in the GITBlow hybrid process. To investigate the temperatures an infrared camera is positioned atop of the mold to monitor the temperature distribution of the preform after the rotation of the rotary table and before the inflation. In a first step the time before the opening of the mold (cooling time) and the handling time of the GITBlow process are varied in order to determine the influence on the temperature level of the GITBlow part. Due to restrictions in the process behavior the cooling time can only be varied in fairly limited range between 10 and 15 seconds. A maximum in the temperature can be detected at a cooling time of 10 s and a handling time of about 21 s for all materials. Figure 2 shows the time-dependent temperature development of the SBS material, which is given as an example.

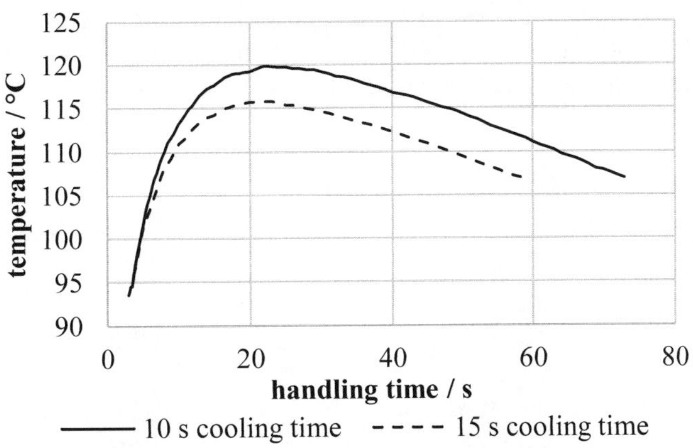

FIGURE 2. Influence of the cooling time and the handling time on the average temperature of the preform within the welding area (material shown: SBS Styrolux 656C, $T_{melt} = 245°C$)

While it is obvious, that a shorter cooling time leads to a higher temperature level, the optimum handling time of 21 s can be explained with the thermal equalization processes within the preform geometry mentioned above. The fact that this handling time is independent of the material type can be ascribed to the almost identical thermal conductivity of the three investigated materials. The following maximum temperatures can be measured: SBS Styrolux 656C: 120°C, PC Makrolon OD2015: 167°C, PC-ABS Bayblend T85XF: 155°C. In order to achieve realistic welding performances, the temperatures of the organo sheets are taken from the manufacturer datasheets. For the two matrix materials the manufacturer recommends the following forming temperatures: PP: 200°C, PC: 300°C

Besides the measured temperatures, the pressure level, which is present in the joining area, is of high importance as it is considered to be one of the two parameters with major influence on the weld strength. The second gas injection in the Two-Stage GITBlow process is set to pressures of 10 – 20 bar (0.1 – 0.2 N/mm²). As there can be no loss in pressure between the inside of the preform and the outer wall pressing against the organo sheet, these pressures are chosen for the investigation of the weld strength.

RESULTS OF THE DETERMINATION OF THE WELD STRENGTH

In order to determine the mechanical properties of the joining area a series of investigations on the welding process is conducted. As the geometry of the joining area of the GITBlow part and the organo sheet forms a semicircle the real parts are unsuitable for the determination of weld strengths. To encounter this, the investigations are conducted on a hot plate welding machine and simplified geometries are chosen for the test specimens. Combinations of all materials listed in Table (1) are tested for their welding capabilities. Besides the material combination the temperature of the unreinforced welding partner representing the GITBlow part and the welding pressure are varied. To provide a better understanding of the process the temperature of the unreinforced materials is set to the value determined in the experimental investigations with the infrared camera and to values 20°C above and below this. The welding pressure is set to three values in a range between 0.1 and 0.2 N/mm² as derived from the gas pressures in the GITBlow process.

The geometry of the test specimen is designed based on the European standard DIN EN 1465 and partly determined by the limitations of the welding machine. As the weld strength shall be measured in a tensile shear test, the welding zone is a rectangular area of (7 x 40 mm). The testing conditions for the lap welded specimens are taken from the standard mentioned above as well. It is recommended to achieve a failure of the welding zone within a time range of 45 – 85 s and set the testing speed accordingly. In a pre-investigation a testing speed of 2 mm/min is determined to fit these requirements.

For the material combinations with the PP based organo sheets there is no weld strength detectable. Even though the ratio of the polar portions of the surface energy and the calculated work of adhesion indicates good adhesive properties, it is impossible to weld the selected material combination.

Based on the analysis of the surface properties the combination PC-PC is the most promising as two nearly identical materials are brought into contact. At a stress of about 25 MPa the limits of the unreinforced PC material are reached. Independent of the welding parameters there is always cohesive failure detectable. For the production of hybrid structures using the GITBlow process this is a positive result as it shows the possibility to achieve good joining properties even though the temperatures of one of the welding partners is far below its melting temperature.

For the two remaining material combinations of PC organo sheets with PC-ABS and SBS the results are given in Figure 3. An increase in the temperature of the welding partner representing the GITBlow base part has only minor influence on the resulting shear strength and the results are all situated within the standard deviation. Therefore the impact of the temperature can almost be neglected.

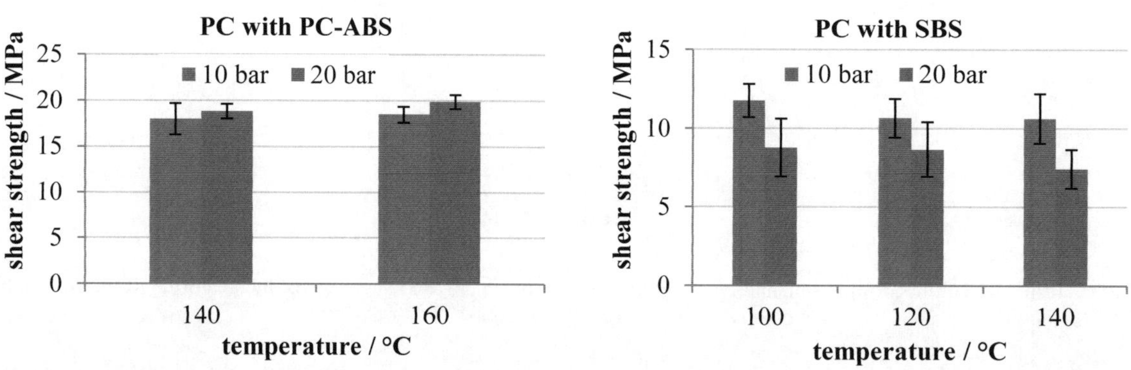

FIGURE 3. Influence of the temperature of the GITBlow base material and the welding pressure on the weld strength for the material combinations PC – PC-ABS (left) and PC – SBS (right). Temperature of the PC organo sheets: 300°C

The impact of the welding pressure is much more distinctive as there are significant decreases detectable for the PC – SBS combination at all temperature levels. For the PC – PC-ABS combination an increase in the weld strength of about 7% at a pressure increase from 10 to 20 bar (0.1 to 0.2 N/mm²) is measured. These opposing trends for both material combinations and their causes are subject of currently conducted further investigations. First results lead to the conclusion that in case of the highly viscous SBS material higher welding pressures lead to an increased displacement of the low viscous PC matrix material. The result is that the thickness of the matrix covering the fibers diminishes leading to a worse welding behavior compared to the lower pressure levels. For the PC-ABS material this impact is less distinctive.

The results for the temperatures of the unreinforced welding partner can be explained with the temperature distribution and development within the welding area. The contact temperature between the two substrates is a function of the materials' thermal conductivity, thermal capacity, density and initial temperature. Using the finite difference method (FDM) the temperaturc development within the two joining partners can be calculated with sufficient accuracy. In Figure 4 the temperatures of the material combination PC and PP organo sheet (thickness 2 mm) with unreinforced SBS (thickness 3.5 mm) are shown after 1s and 10s of contact which is a representative contact time expected for the hybrid GITBlow process. The initial temperatures of the organo sheets are 200°C (PP) and 300°C (PC) and 140°C for the SBS base part.

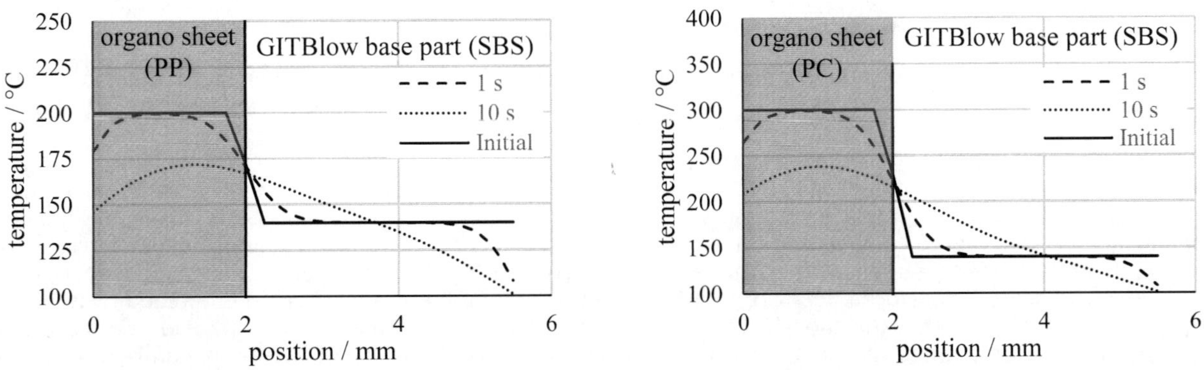

FIGURE 4. Calculated temperature distribution during the welding of PP organo sheets with SBS (left) and PC organo sheets with SBS (right)

Within the 10s contact time of the SBS with the PC organo sheets the initially very low temperature of 140°C rises to values above the SBS melting temperature of 160°C in a section of about 1 mm thickness. This leads to the conclusion that even though the temperatures of the GITBlow base part are much lower than the melting temperature of the material, a good welding result can be achieved due to the attainment of a melt layer caused by heat conduction from the organo sheet. Even with base part temperatures lower than the 140°C mentioned above, the heat conduction leads to a melting of the base plate leading to a good weld strength between the two materials. With the same calculation method it is possible to show, that when using the PP based organo sheets the temperatures in the unreinforced joining partner do not exceed the melting temperature of the three materials. This explains why the test samples with PP matrix show no appreciable weld strength.

In the presented paper the necessary basics for the production of hybrid structures using the Two-Stage GITBlow process are investigated. This includes the determination of possible material combination based on the surface properties. In a series of experimental investigations on the process using thermal imaging the temperature levels of the GITBlow part can be determined. Using lap welded samples the weld strength of the material combinations can be determined under representative process conditions. It can be concluded that high weld strengths can be achieved even though the temperatures of the GITBlow base part remain underneath their specific melt temperature.

ACKNOWLEDGMENTS

The authors would like to thank the Ministry of Innovation, Science and Research of the State of North Rhine-Westphalia for the financial support of the scientific work within the scope of the project "Light, Efficient, Mobile."

REFERENCES

1. M. Schäfers, Beitrag zur Entwicklung des Spritzgießsonderverfahrens GITBlow, Ph.D. Thesis, Paderborn University, 2009.
2. G. Habenicht: Kleben: Grundlagen, Technologien, Anwendungen, Springer-Verlag, Heidelberg, 2002
3. A. Pizzi , K. L. Mittal: Handbook of Adhesive Technology, Revised and Expanded, Marcel Dekker Inc., New York, 2003
4. L. H. Sharpe and H. Schonhorn, "Surface energetics, adhesion, and adhesive joints", Advances in Chemistry, 43, pp. 189-201, 1964
5. H.-J. Butt, K.-H. Graf, M. Kappl: Physics and Chemistry of Interfaces, Wiley-Vch-Verlag, Weinheim, 2003
6. A. J. Kinloch: Adhesion and Adhesives: Science and Technology, Chapman and Hall, London, 1987

The Effect of Internal Mould Water Spray Cooling on Rotationally Moulded Polyethylene Parts

Kearns Mark (1), McCourt Mark (1), Martin Peter (1)*

(1) Queen's University Belfast - Belfast - UnitedKingdom

The conventional method of cooling during the rotational moulding process is through the use of forced air. During the cooling phase of a typical rotomoulding cycle, large volumes of high velocity room temperature air are forced across the outside of the rotating rotomoulding tool to encourage cooling of the metal mould and molten polymer. Since no cooling is applied to the inside of the mould, the inner surface of the polymer (polyethylene) cools more slowly and will have a tendency to be more crystalline and the polyethylene will have a higher density in this region. The side that cools more quickly (in contact with the inside mould wall) will be less crystalline, and will therefore have a lower density. The major consequence of this difference in crystallinity is a buildup of internal stresses producing warpage and excessive shrinkage of the part with subsequent increased levels of scrap. Therefore excessive cooling on the outside of the mould should be avoided. One consequence of this effect is that the cooling time for a standard rotationally moulded part can be relatively long and this has an effect on the overall economics of the process in terms of part manufacture. A number of devices are currently on the market to enhance rotational moulding by introducing a water spray to the inside of the rotomoulding during cooling. Internal mould water spray cooling is shown to significantly reduce manufacturing cycle times whilst improving mechanical properties. This paper will highlight the positive effects of internal mould water spray cooling during rotational moulding on the mechanical properties, shrinkage and warpage of the part as well as demonstrate the reduction in cycle time possible.

Effect of hybridization on the physical and mechanical properties of polyethylene-(coir/agave) composites produced by rotomolded

López-GonzalezNuñez Rosa Gabriela* (1), *Pérez-Fonseca Aida Alejandra* (1), *Ortega-Gudiño Pedro* (1), *Robledo-Ortíz Jorge R.* (1), *Rodrigue Denis* (2), *González-Núñez Rubén* (1)

(1) Universidad de Guadalajara - Jalisco - Mexico, (2) Université Laval - Quebec - Canada

This work reports on the preparation of linear medium density polyethylene (LMDPE) based hybrid composites made with two natural fiber; coir and agave. The composites were produced by rotational molding. The efect of hybridization was analysed via morphological and mechanical test for three total fiber content (5, 10 and 15 wt%) and different agave-coir fiber ratios (100-0, 70-30, 50-50, 30-70 and 0-100). Moreover, the effect of coupling agent (maleated polyethylene) in the hybrid composite formulation was evaluated. The results showed a synergistic effect on the flexural and impact strength for the composition of 10 wt% of fiber with 70-30 coir-agave fiber ratio. As expected, the addition of a coupling agent improves substantially the quality of the polymer–fiber interface as well as the mechanical properties.

Decision-making tools for Design from Recycling

***Ragaert Kim* [(1)]*, *Hubo Sara* [(1)], *Veelaert Lore* [(2)], *Dubois Els* [(2)]**

[(1)] Ghent University - CPMT - Ghent - Belgium, [(2)] University of Antwerp - Antwerp - Belgium

Design from Recycling is a material-driven design methodology for the development of products, manufactured with existing recycled (polymer) materials. These materials are often quite complex (mixed plastics, contaminations...) and not so straightforward to use efficiently. Design from Recycling involves identifying the recycled polymer's strengths and weaknesses through extensive characterization, as well as identifying acceptable strategies for possible upgrading of the material quality where necessary. During the stage of product design, these characteristics are taken into account and matches are found between (new) products and recycled polymers suitable for producing them. The method is complimentary to Design for Recycling in order to come to a full circularity of polymer materials. Within this paper, decision-making tools developed for Design from Recycling are presented. This includes a decision tree based on polymer properties versus product demands, a hands-on tactile evaluation tool of the injection moulded recycled polymer and concise rules of thumb, based in industrial case studies.

Post-consuming Extruded Polystyrene Food Packaging Recycling

Schmidt, Albano[a], Zimmermann, Regina[a], Oro, Dieizon[a], Valerio F., De Felice[b]
Oliveira, Patrícia F.[b]; Bolzan, Ariovaldo[b] , Mara G.N. [b] and Machado, Ricardo A.

[a]*Termotécnica Ltda. Rua Albano Schmidt, 2750, Boa Vista, 89206-002. Joinville – SC – Brasil.*
[b]*Universidade Federal de Santa Catarina – Departamento de Engenharia Química e Engenharia de Alimentos – Campus Reitor João David Ferreira Lima, s/n - Trindade, Florianópolis - SC, 88040-970, Brasil*

Abstract. The vast application and high consuming of low density polymeric materials, such as expanded and extruded polystyrene, emphasizes the need for solutions concerning the disposal and recycling procedures for plastic materials. Extruded polystyrene (XPS), or polystyrene foam, has been largely applied as packaging material. A great amount of XPS is frequently found in landfills, despite environmental laws, governmental and institutional efforts all over the world. This survey presents a recycling method for food trays made from XPS, by using it in polystyrene production. Up to 10% of XPS food trays were added as part of the organic phase in the styrene suspension polymerization. Kinetic behavior of styrene free radical polymerization with added XPS was evaluated by bulk polymerization process and applying a mathematical model. On both procedures, benzoyl peroxide was used as the initiator in a concentration of 1.61×10^{-5} mol/g, in relation to the monomer mass. Results demonstrated that the final polystyrene obtained is suitable to the main applications in which polystyrene (PS) is used, regarding average molecular weight and residual monomer. Recycling of post-consuming XPS trays represents a feasible method, featuring as a singular alternative in the destination of this solid waste. Particles size distribution was in accordance to what is expected for this product. Styrene concentration limit in the final product agreed to what is stated in the Brazilian food packaging regulation, which is based on FDA guidelines. Particles size diameter and its distribution were favored by the addition of the polymer. Additionally, the mathematical model applied is suitable for exploratory studies, promoting materials and time saving, also making it possible to evaluate polymer molecular weight and conversion rate, which can be used to define its final application.

Keywords: Food packaging, Polystyrene, Recycling.
PACS: 81.30.Hd

INTRODUCTION

The increasing consumption and diversity of food and beverages, with different presentation packages and sizes, as well as of products from other areas, demand packaging produced in polystyrene foams, whether for convenience or need of the consumer market. Although there is a global growing movement towards sustainability, conscious consumption, and concern for environmental preservation policies in evidence, the production of petroleum-based polymers remains high. Polystyrene is a non-biodegradable polymer and neither are its semi rigid foams (expanded polystyrene – EPS, extruded polystyrene – XPS). Extruded polystyrene (XPS) is widely used as thermoformed packaging material by the food and beverages sectors. This material has an oriented structure and a low density [1]. Due to the fact that it is not reusable and to the great demand for this kind of package, a large amount of XPS is frequently found in landfills, as well as other low density polymeric materials [2]. This poses as an environmental challenge, due to the need for huge areas of disposal as well as the negative role in other materials degradation. The recycling of these materials is possible either by dissolution in solvents or by a well-known mechanical process. The greatest difficulty imposed by polystyrene recycling processes is that they must compensate transport costs, since polystyrene foams have low density and occupy large volumes.

However, any form of recycling should be related to the environment and not just from an economic point of view. That is, by breaking the production chain of a product by the insertion of recycled raw material, it is usually a reduction of part of the manufacturing process of this product which, consequently, implies a reduction of direct costs, in the conservation of energy and, of course, preservation of the environment.

The need to adopt methods for recycling is therefore imminent in order to reduce the amount of solid waste that can be benefited and also to preserve non-renewable natural resources.

Studies conducted by [3, 4] demonstrate the feasibility of the addition of expanded polystyrene (EPS) in the polymerization process of styrene in suspension as a means of recycling this material. Recycling XPS, like EPS, is based on the fact that when the monomer is polymer solvent, it can be easily incorporated into the reaction without compromising the final product.

In this sense, this survey presents a recycling method for post-consumer XPS food packaging trays by using this material in the styrene polymerization free radicals suspension reaction.

EXPERIMENTAL

The *in situ* recycling process that occurs by the addition of XPS-made food trays in the free radicals styrene polymerization reaction was carried out in suspension and also in bulk mode, as further described. All reagents were used as received. The styrene (Innova) is 99.95% pure, containing the inhibitor p-tert-butylcatechol at a concentration of 15 ppm. The initiators used in the reaction were benzoyl peroxide (BPO) and tert-butyl peroxide (PTB), both of technical grade (Atofina). Poly (vinyl pyrrolidone), PVP K-90, of technical grade was used as suspension polymerization stabilizing agent (Sul Polímeros). Distilled water with neutral pH and mean conductivity of 12.6 μS was used as the continuous phase in the suspension system. Gaseous nitrogen (AGA) grade 2.0 was used in order to guarantee an inert atmosphere to the polymerization reaction. The recyclable material used was extruded polystyrene, XPS, obtained in the form of post-consumer food trays, from the collection of waste from Florianópolis, SC. In order to remove excess dirt from the material, it was washed with commercial detergent, rinsed with abundant water, and then dried in an oven with air circulation at 50°C for 48h. The trays were then fragmented into a hammer grinder (TRF) with a 25cv and 1700rpm motor to facilitate the recycling process.

Bulk and Suspension Polymerization

Kinetic behavior of styrene free radical polymerization having XPS trays incorporated (2.5%, 5% and 10% w/w) was evaluated by bulk polymerization process. A mathematical model proposed by [8, 9] was applied. A standard styrene polymerization was performed for comparison. Benzoyl peroxide was used as the initiator at 90°C in a concentration of 1.61×10^{-5} mol/g and Terc-butyl peroxide as the initiator at 120°C in a concentration of 8.01×10^{-6} mol/g, both in relation to the monomer mass. The reaction time was of 9h.

Suspension polymerization took place in a stirred reactor, by adding 2.5%, 5% and 10% of XPS food trays as part of the organic phase. A standard styrene polymerization was also performed for comparison. Peroxides concentration followed that used for bulk polymerization. Agitation was set to 600rpm and the reaction ran for 6h at 90°C. Figure 1 illustrates the experimental unit. Reactions used to produce material for residual styrene analysis took place in a jacketed stainless steel reaction vessel, which can be pressurized so temperature can be risen up to 120°C. Reaction parameters remained the same as those described for the experimental unit shown in Figure 1. However, reaction was carried out for 9h and temperature risen up to 120°C after 6h for monomer depletion.

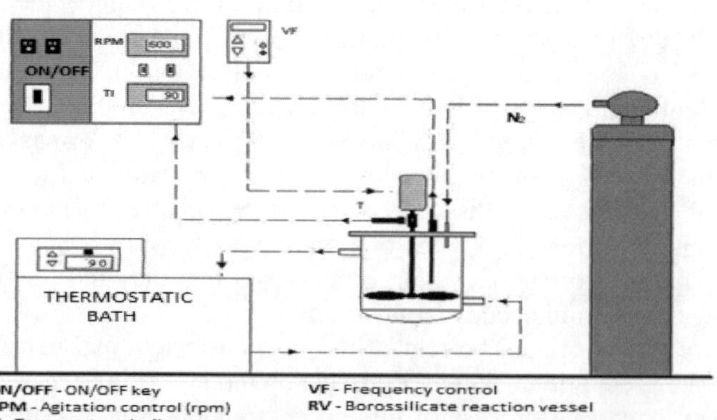

FIGURE 1. Suspension polymerization experimental unit scheme.

The produced polymer was characterized according to its molecular weight, residual monomer, particle size and its distribution. Molecular weight was measured by viscosimetry, previously described by [10] and [11]. Residual styrene was evaluated by gas chromatography. Particles size and its distribution were determined by gravimetric analysis using a set of Tyler series sieves. Kinetic behavior was evaluated by applying a mathematical model [8, 9] and comparing to experimental data.

RESULTS AND DISCUSSION

Results demonstrate that the final polystyrene is suitable to the main applications in which polystyrene (PS) is used, regarding average molecular weight (Mw) and residual monomer (styrene), as can be seen in Table 1. The use of this polystyrene as food packaging may pose health concern as residual styrene can migrated from PS, but the results obtained are considered safe. According to regulations from Brazil [5] and EU [6] there is no specific composition limit for the styrene monomer. In EU in order to safeguard food quality there are limits for the maximum amount of additives and volatile components that could migrate into food. Although no limit is stated for styrene, a maximum Specific Migration Limit of 10 mg/dm² of any substance among all that could be present is specified. Previous Brazilian regulatory act [7] had determined a Concentration Limit of styrene as 0.25%. This resolution has changed in 2012 [5]. However, regulatory acts also states that the taste of the food being packaged should not be negatively impacted by the plastic package, and the manner in which the package is made should comply with good manufacturing practices.

This study aimed to reach final products with an average molecular weight of more than 140,000 g/mol, according to what is adopted for some applications of polystyrene, thus proving the suitability of the recycling process to obtain a polymer that can be reincorporated into its industrial cycle.

It can be observed by Figure 2 that as the amount of XPS incorporated increases, particles size diameter is favored into regions of larger diameters, also narrowing its distribution curves. This behavior can be explained by the increase in the viscosity of the organic phase, leading to a reduction in the rate of breakup and favoring coalescence of droplets. Similar findings can be found elsewhere [3, 4].

TABLE 1. Average molecular weight (Mw) and residual styrene in polystyrene.

XPS in the Reaction (%)	Average Mw (g/mol)	Residual styrene (%)
0.0	190,524	< 0.0035
2.5	168,127	< 0.0035
5.0	174,722	< 0.0035
10.0	174,730	< 0.0035

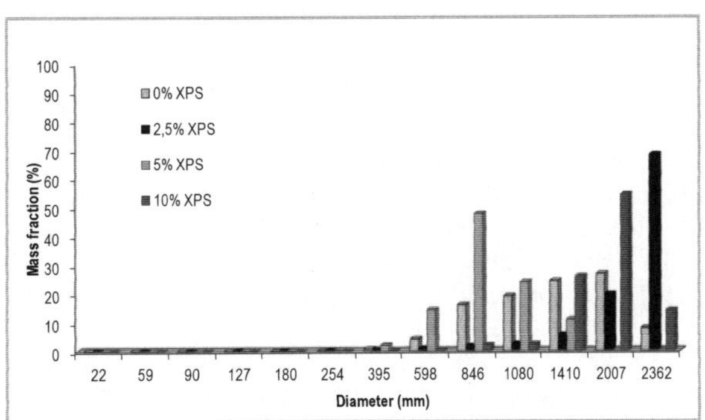

FIGURE 2. Particle size distribution results for suspension polymerization reactions.

The kinetic behavior for the free radical polymerization reaction in which were added different amounts of XPS was evaluated using experimental data and comparing it to simulated reactions, according to literature [8]. The referred mathematical model was modified concerning the initial organic phase. Since polystyrene conversion does not start from 0.0% in this work, simulations took into consideration the ratio of XPS dissolved in each reaction as the initial conversion. By the analysis of the results presented in Figure 3, it can be concluded that there is a good agreement between the experimental values and those obtained with the model. As the concentration of XPS increases it is possible to observe a deviation of the model from the experimental results in the phase of the gel effect. The correlation used to describe the gel effect is well known and widely used. However, it was obtained under optimum conditions. Thus, deviation between the experimental and simulated curves could be attributed to the prediction and correction that the model applies to the gel effect, inherent to suspension polymerization systems. As it can also be observed from data presented, higher conversion rates are obtained faster as the amount of XPS increases, since it promotes acceleration of the reaction kinetics.

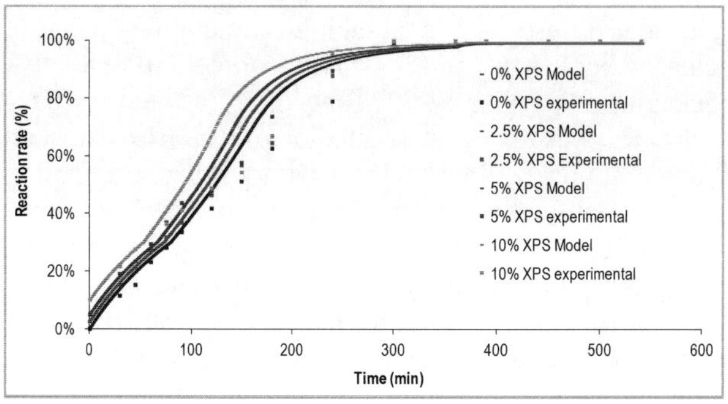

FIGURE 3. Kinetic of the free radical styrene polymerization, experimental and modeled results.

CONCLUSION

The reaction kinetics changed with the addition of extruded polystyrene, obtaining higher conversion rates as the percentage of added material increased. The addition of XPS favored PSD, narrowing the size range. The recycling of post-consumer foamed XPS food trays was then considered appropriate and represents a technically feasible method, featuring as an alternative for the destination of this solid waste.

ACKNOWLEDGMENTS

The authors acknowledge The Federal University of Santa Catarina for providing the infrastructure necessary to the development of this research, and also Termotécnica Ltda. for providing reagents and financial support to researchers involved in this work.

REFERENCES

1. G.C. Welsh, *Modern Styrenic Polymers: Polystyrenes and Styrenic Copolymers*, John Wiley & Sons, Inc., 2003, 233-246.
2. S. Calderoni, *Os bilhões perdidos no lixo*, 3rd Edition, São Paulo, 1999, edited by Humanitas.
3. R. Schlischting. "Influência da adição de poliestireno expandido no processo de polimerização do estireno em suspensão", M.Sc. Dissertation, The Federal University of Santa Catarina, 2003.
4. A. Cancelier. "Otimização e controle do processo de produção do poliestireno em suspensão". Ph.D. Thesis, The Federal University of Santa Catarina, 2004.
5. ANVISA. Regulamento Técnico sobre a lista positiva de monômeros, outras substâncias iniciadoras e polímeros autorizados para a elaboração de embalagens e equipamentos plásticos em contato com alimentos, RDC nº 56, 16 november 2012.
6. Commission Regulation (EU) nº10/2011, 14 January 2011, Plastic materials and articles intended to come into contact with food.
7. ANVISA. Disposições gerais para embalagens e equipamentos plásticos em contato com alimentos. Resolução n.105, 19 May 1999.
8. R.A.F. Machado and A. Bolzan, *Control of batch polymerization reactor, Chemical Engineering Journal* ***70***, 1-8 (1998).
9. A.W. Hui and A.E. Hamielec, *Thermal polymerization of styrene at high conversion and temperatures, Journal of Applied Polymer Science* **16**, 33, 749-769 (1972).
10. R.B. Bishop. *Practical Polymerization for styrene*. EUA, edited by Cahners Books, 469 p. (1971).
11. F.N. Billmeyer, *Textbook of polymer science*. New York: John Wiley & Sons, Inc., 3rd Ed. (1984).

pps proceeding
Symposium: S05 - Recycling
Oral Presentation

Preparation of composites with pinewood residues and recycled HDPE added with teak wood: Evaluation of resistance to accelerated weathering and subterranean termites (Nasutitermes corniger)

Zacarías-Calderón Edgar [1]*, Cruz-Estrada Ricardo Herbé [1], Cupul-Manzano Carlos Vidal [1]

[1] Centro de Investigación Científica de Yucatán, Unidad de Materiales - Yucatán - Mexico

The preparation of a composite with pinewood residues (Pinus sp.), and recycled high-density polyethylene (r-HDPE) is reported. The composite was added with teak wood particles (Tectona grandis) with potential to repel the attack of subterranean termites. Both pine and teak wood are residues of the forest industry, so their use in this work, as well as that of r-HDPE proposes a solution to the problem of solid residues accumulation. In addition, it helps to reduce the use of virgin raw materials such as wood and petroleum derivatives, which is a non-renewable resource. Thus, the originality of the work and its contribution to the care of the environment are clear, since it also proposes to prove the effectiveness of a forest industry waste as a biopreservative of wood-plastic composites (WPCs) against the attack of termites. The amount and formulations of WPCs were based on a 2^3 factorial experimental design, where the factors to be studied were 1) hours of UV irradiation in accelerated weathering, 2) days of exposure to termites, and 3) concentration of teak wood in the composite. In order to evaluate the impact that each factor could have on the composites, they were subjected to mechanical tests under tension and flexion. In addition, SEM and FTIR were used to analyze the fracture surfaces, and to determine the chemical modification of the composites components due to accelerated weathering and termites attack. In general, preliminary results indicate that teak wood was effective in repelling termite attack on the WPCs.

Spent coffee grounds composites with PVA aqueous solutions

***Duarte Fernando* (1)*, *Prata Alexandra* (1), *Nunes Hector* (1)**

(1) University of Minho - Guimaraes - Portugal

Coffee is one of the most popular and appreciated beverages around the world, being consumed for its stimulating and refreshing properties. As a consequence of this big market, the coffee industry is responsible for generating large quantities of residues, being spent coffee grounds (SCG) one of the most abundant bio residues. Due to its composition, SCG has a great potential for use in various applications, including the production of green materials composites. The aim of this work is to develop and characterize environmentally friendly composites using SCG with aqueous solutions of Poly(vinyl alcohol), PVA. The SCG/PVA composites, previously mixed manually with different amounts of SCG and with two different PVA aqueous solutions, were produced through thermo compression in a 150mm square mold, at 150ºC. Samples were evaluated in terms of physical properties (density) and mechanical properties (tensile strength, flexural strength, impact resistance) The results showed that it is possible to obtain with success composite materials using SCG and aqueous solutions of PVA, at least with 50% of the waste in weight. It was further concluded that the final composite mechanical behavior is influenced by the amount of coffee waste and mainly by the type of aqueous solution used. Regardless of the PVA solution be rigid or flexible, it was observed that increasing the concentration of the coffee residues increases the rigidity, hardness and density, reducing the impact strength.

Production of sheets based on pinewood residues and recycled HDPE: Preliminary experimentation to manufacture a roofing prototype

Cruz-Estrada Ricardo Herbé (1)*, Cupul-Manzano Carlos Vidal (1), Guillén-Mallette Javier (1), Rivero-Ayala Miguel Ángel (1), Casas-Beltrán Diego Armando (2)

(1) CICY, A.C. - Mérida, Yucatán - Mexico, (2) CICY, A.C. - Cancún, Quintana Roo - Mexico

About 60 cubic meters of pinewood residues (PR) are generated daily In Merida, Mexico. Most of them are a by-product of activities carried out in carpenters, sawmills and furniture factories. These residues are generally stored outdoors in the form of chips or sawdust without being exploited, allowing them to decompose. About 200 tons of plastic waste are produced per month, of which about 27% are containers and articles made of high-density polyethylene (HDPE-REC). It is therefore important to find an appropriate use for these types of residues to create new materials with practical applications, which would help to solve the problem of excessive accumulation and associated environmental problems. One of many potential uses is as building elements in housing, such as roofs. PRs were dried after collection, particle size (PS) was reduced, and were classified into three different PS ranges. The HDPE-REC was ground to obtain the proper PS. PR/HDPE-REC composites with PR contents of 5 to 70% by weight were prepared using a twin-screw laboratory extruder. Mechanical properties of the resulting composites and the effect of PR PS and the composition were evaluated. A coupling agent (CA) and a process aid (PA) were used. It was observed that flexural and tensile strength decreased with increasing PR percentage. In general, the decrease in PR PS increased the tensile and flexural strength. On the other hand, the impact resistance decreased by increasing the PR percentage and decreasing PR PS. A laboratory prototype was built for the production of corrugated sheets. It was experimented with various conditions, from which it is concluded that the best results in terms of appearance and processability were obtained using 40% by weight of PR with the highest PS. The intention is to scale up the process to pilot level to evaluate the feasibility of using the material for roofing

Ternary Blends of Poly(lactic acid), Waste Cellulose Fibers and Waste Cross-linked Polyurethanes Prepared by High-shear Thermo-kinetic Mixer

Oguzhan Oguz[a,b] and Yusuf Z. Menceloglu[a,b,*]

[a]*Faculty of Engineering and Natural Sciences, Materials Science and Nano Engineering, Sabanci University, 34956, Orhanli, Tuzla, Istanbul, Turkey*
[b]*Sabanci University Integrated Manufacturing Technologies Research and Application Center & Composite Technologies Center of Excellence, Teknopark Istanbul, 34906 Pendik, Istanbul, Turkey*
**Corresponding Author: yusufm@sabanciuniv.edu (Y. Z. Menceloglu)*

Abstract. Critical environmental and economical issues have been stimulating research in the mass production of sustainable materials for plastics market that favours low costs and high production rates for decades. Recently, we have presented a modular approach to produce Poly(lactic acid)/waste cellulose fibers (PLA/WCF) composites and Poly(lactic acid)/waste crosslinked polyurethanes (PLA/WCPU) binary blends using high-shear thermo-kinetic mixer. For PLA/WCF composites, we have achieved one-fold and 50% increase in elastic modulus and tensile strength of PLA, respectively, by the incorporation of 30 wt.% WCF. For PLA/WCPU blends, 5-fold increase in impact resistance of PLA have been achieved by the addition of 30 wt.% WCPU. As presented in PPS-32 (Lyon/France), these two binary systems (PLA/WCF and PLA/WCPU) have been demonstrated as practical guidelines for the mass production of high performance "green" composites and super-tough "green" polymer blends, respectively. Here, as part of our ongoing efforts, we mainly focused on the sustainable production of ternary blends of PLA, WCF and WCPU via high-shear mixing. This technique facilitates the ease of processing of composite/blend materials and ensures the homogeneous dispersion of natural wastes and/or synthetic thermoset additives in thermoplastics due to shear rates as high as 5000 rpm. Key parameters, such as optimal concentrations, homogeneous dispersion, interfacial interactions and crystallinity were examined for PLA/WCF/WCPU ternary blends with enhanced mechanical properties. The main conclusion of the study is that the extensive blending technology allows for the production of high performance "green" ternary blends with tunable mechanical properties as well as addressing the world-wide waste disposal problem by reusing of waste additives in biodegradable thermoplastics.

Keywords: PLA, waste cross-linked polyurethane, waste cellulose fiber, high-shear mixing, tensile and impact properties
PACS: 81.05.Lg

INTRODUCTION

Critical environmental and economical issues have been stimulating research in the mass production of sustainable materials for plastics market that favours low costs and high production rates for decades. Along with new industrial regulations and growing technological needs, biocompatible and biodegradable polymers have attracted particular interest as promising alternatives to petrochemical based plastics. In this regard, poly(lactic acid) (PLA) is one of the most attractive (bio)polymers due to its high stiffness and strength combined with biodegradability, recyclability and ease of processing that are vital requirements in the rational design and mass production of sustainable materials.[1–2] However, inherent brittleness of PLA, which is evidenced by poor impact resistance and fairly low elongation at break, is one of its main limitations in substituting commodity plastics.[1–2] Thus, various physical and/or chemical modifications of PLA are required to obtain decent properties needed for targeted applications. Various design strategies, such as reactive extrusion, dynamic vulcanization, physical blending and copolymerization, have been applied to enhance the impact strength of PLA. In a previous study[3], which has presented in PPS-32 (Lyon/France), we have demonstrated that WCPU can be used to tailor the impact strength of PLA. In this binary system (PLA/WCPU), 5-fold increase in impact resistance of PLA have been achieved by the addition of 30 wt.% WCPU.[3] In line with this, we have also shown that the impact strength of PLA gradually increases with increasing WCPU content.[3] However, incorporation of WCPU results in a gradual decrease in the elastic modulus and tensile strength of PLA. In another previous investigation[2–3], we have reported that waste cellulose fibers can be used to enhance the elastic modulus and tensile strength of PLA. Inspired by these two studies, here we mainly focus on tailoring stiffness and strength of super-tough PLA/WCPU blends via incorporation of WCF.

EXPERIMENTAL

Materials

A commercial grade of PLA (Ingeo 4043D) with a weight average molecular weight of 150 kDa was purchased from NatureWorks LLC (Minnetonka, MN). Waste cross-linked polyurethane (WCPU) powders and waste cellulose fibers (WCF) with an average diameter of 7-20 μm were kindly supplied by local polyurethane and textile companies in Istanbul, Turkey. Respective details can be found in References.[2–3]

Fabrication of Ternary Blends

All masterbatches were prepared by a Gelimat G1 laboratory scale high-shear thermo-kinetic mixer (Draiswerke, USA). In this technique, the particles are generally accelerated by the blades on a high-speed shaft. This acceleration basically imparts a fairly large amount of kinetic energy to the particles. This kinetic energy is mainly converted to the thermal energy when the particles collide with the chamber wall. The mixing of compounds most likely takes place between the tip of the blade and the wall mainly because of the centrifugal forces generated by the rotating blade.[2, 4–6] The shear rate calculations were previously reported in detail by Gopakumar and Page.[5] In accordance with these calculations by the authors, the dimensions of the Gelimat mixing chamber/shaft, and the rotational speed, imply that significantly high shear rates (about 10^4 s^{-1}) can be generated between the tip of the blade and the walls in this system.[5] For instance, these authors calculated the shear rates of 16400 and 13900 s^{-1} at the tip of the blade and at the wall, respectively, at a blade speed of 6000 rpm.[5] In our study, the mixer was operated with a shaft speed of 5200 rpm until the compounds temperature reached to 190 °C. Using the same approach reported by Gopakumar and Page,[5] the shear rates at the tip of the blade and at the wall, respectively, were calculated as 14200 and 12060 s^{-1} at a blade speed of 5200 rpm. Details related to thermokinetic mixing can also be found elsewhere.[2, 4–6] All batches were prepared with an identical load of 100 g. The compounds were then granulated with a typical lab-scale granulator under ambient conditions, and subsequently injection molded using an Xplore Instruments 12 mL injection molding machine. The injection molding temperature was 190 °C. The injection pressure was 10 bars and the mold temperature was 30 °C. The cycle time was totally 10 seconds. The code of PLA/WCPU-X/WCF-Y was used in the sample notation where the X and Y represent the weight fractions of waste cross-linked polyurethane powders and waste cellulose fibers in the blends.

Characterization Techniques

The tensile tests were performed in accordance with ASTM D3039 standard at a cross-head displacement rate of 2 mm/min on a Zwick Z100 model universal tester. The impact tests were conducted according to ISO 179 unnotched Charpy standard for injection molded samples using a CEAST impact tester. At least five specimens were tested for both analyses. All the tests were carried out under ambient conditions. The materials were also characterized by several structural, microscopic, spectroscopic, rheological and thermal analyses techniques. The data were not shown in here.

RESULTS

Tensile Properties

Elastic modulus (E) and ultimate tensile strength (σ_{max}) values of the samples were presented in Table 1. E and σ_{max} values of pure PLA decreased from 3079 MPa to 2175 MPa and from 60.8 MPa to 43.1 MPa by the addition of 15 wt.% WCPU. This material (PLA/WCPU-15) was employed as "control" sample to understand the effect of WCF content on the tensile properties. As listed in Table 1, E and σ_{max} values gradually increased by increasing WCF content.

Impact Properties

Results obtained from the unnotched Charpy impact tests of the materials were also listed in Table 1. The impact strength of PLA significantly increased with the incorporation of 15 wt.% WCPU. The impact strength of PLA/WCPU-15 is three times higher than that of pure PLA. No notable change was observed in the impact strength of PLA/WCPU-15 by the addition of WCF. The impact strengths of the ternary blends with 5, 10, 15 and 20 wt.% WCF (PLA/WCPU-15/WCF-5, PLA/WCPU-15/WCF-10, PLA/WCPU-15/WCF-15, PLA/WCPU-15/WCF-20) are quite comparable with that of PLA/WCPU-15.

TABLE 1. Tensile and impact properties of the samples.

Sample	*E* (MPa)	σ_{max} (MPa)	Charpy Impact Strength (kJ/m^2)
PLA	3079±154	60.8±3.04	15.9±0.795
PLA/WCPU-15	2175±112	43.1±1.28	68.8±1.823
PLA/WCPU-15/WCF-5	2550±104	48.4±1.43	67.9±1.973
PLA/WCPU-15/WCF-10	3153±87	52.7±1.12	68.3±1.715
PLA/WCPU-15/WCF-15	3350±98	56.9±1.03	67.8±1.697
PLA/WCPU-15/WCF-20	3589±66	61.2±1.32	68.2±1.803

DISCUSSION

Incorporation of WCPU into pure PLA results in a significant increase in the impact strength of PLA. In contrast, WCPU addition gives rise to a notable decrease in stiffness and strength of PLA. This limits the use of super-tough PLA/WCPU blends in a wide range of applications. In this study, we demonstrate that this problem can be solved by the addition of WCF into super-tough PLA/WCPU blends. Results obtained from the tensile and impact tests mainly suggest that waste cellulose fibers can be used to tailor the stiffness and strength of super-tough PLA/WCPU blends while maintaining impact resistance.

CONCLUSION

Ternary blends of PLA, WCPU and WCF were successfully produced by a high-shear thermo-kinetic mixer. This approach mainly suggests an economical way for the sustainable production of high performance blend materials, as it comprises the reuse of waste cross-linked polyurethane powders, waste cellulose fibers and the implementation of a facile technique, which is available to scale up, ensuring the ease of processing. In addition, this approach creates significantly high added-value for the mass production of blends as the reuse of WCPU and WCF also corresponds to the reduced consumption of thermoplastics during processing. The main conclusion of the study is that the extensive blending technology allows for the production of high performance “green” ternary blends with tunable mechanical properties as well as addressing the world-wide waste disposal problem by reusing of waste additives in biodegradable thermoplastics.

ACKNOWLEDGEMENTS

We sincerely thank Mr. Burak Yurt and Mr. Sezgin Sahin for their valuable contributions during the production.

REFERENCES

1. G. Koronis, A. Silva, M. Fontul, *Composites, Part B* **44**, 120−127 (2013).
2. O. Oguz, K. Bilge, E. Simsek, M. K. Citak, A. Wis, G. Ozkoc, Y. Z. Menceloglu; *Ind. Eng. Chem. Res.* **56**, 8568–8579 (2017).
3. O. Oguz and Y. Z. Menceloglu; “Green composite materials with enhanced stiffness and impact resistance”, In Proc. of 32nd International Conference of the Polymer Processing Society (PPS) Lyon, France, July 25–29, 2016.
4. O. Oguz, E. Simsek, K. Bilge, Y. Z. Menceloglu, *Int. Polym. Proc.* XXXII **5**, 1–7 (2017).
5. T. Gopakumar, D. Page, *J. Appl. Polym. Sci.* **96**, 1557−1563 (2005)
6. T. Gopakumar, D. Page, *Polym. Eng. Sci.* **44**, 1162−1169 (2004).

Combining polymeric waste streams to improve functional properties of post-consumer mixed polyolefines

Delva Laurens [(1)]*, De Tandt Ellen [(1)], Ragaert Kim [(1)]

[(1)] CPMT, Ghent University - Ghent Zwijnaarde - Belgium

Polymeric waste streams can be re-used as secondary material sources for polymer processing by extrusion or injection moulding. One of the major commercially available waste streams is the so called mixed polyolefines (MPO). These MPO mainly consist of a mixture of different grades of polyethylene and polypropylene. An important practical hurdle for the direct implementation of this waste stream is the inherent immiscibility of polyethylene and polypropylene in the melt, which leads to segregation within the final structure and adversely affects the reproducibility and mechanical properties of the manufactured parts. Melt mixing this primary waste stream with other polymeric waste streams could mitigate these drawbacks. Within the current research, a commercially available MPO waste stream was combined with post-industrial polyethylene-polyamide multilayer films and post-consumer recycled polyester fibers. The mechanical (tensile, flexural and impact) properties, physicochemical properties and morphology of the composites were characterized to assess the suitability of these combinations.

Functional evaluation of compatibilization systems for recycled PP-PET blends

L. Delva[1], M. Kuzmanovic[1], T. Wieme[1], N. Mys[1], K. Ragaert[1, a)]

[1]*CPMT Research Group, Department of Materials, Textiles and Chemical Engineering, Faculty of Engineering and Architecture, Ghent University, Technologiepark 915, 9052 Zwijnaarde, Ghent, Belgium*
[a)] Corresponding author: Kim.Ragaert@ugent.be

Abstract. The main goal of this study is to investigate the effect of different compatibilizers on mechanical properties in blends of polypropylene (PP) and poly(ethylene terephthalate) (PET). The blends have been prepared by twin screw extrusion followed by injection molding at a weight ratio of 80/20. The compatibilizers, PP grafted with maleic anhydride (PP-g-MA), polyolefin based elastomer grafted with MA (POE-g-MA), poly[styrene-b-(ethylene-co-butylene)-b-styrene] grafted MA (SEBS-g-MA), SEBS grafted glycidyl methacrylate (SEBS-g-GMA) and non-functionalized SEBS were added in a fixed weight percentage. Their effect on the mechanical performances of the blend was measured using tensile, flexural and impact tests. These tests revealed that, for this specific blend, improving the impact properties is most successful with POE-g-MA, while PP-g-MA performs better at improving the tensile and flexural properties. In comparison to the polyolefin-based compatibilizers, the SEBS-based compatibilizers displayed little or even a negative effect on the mechanical properties of the PP/PET blends.

INTRODUCTION

Today, the plastic waste disposal represents one of the biggest challenges for the environment. There is a great amount of plastic waste, which mostly comes from packaging applications, as many of these plastic products are not designed for long-term usage. One of the ways of reducing landfilling is by mechanically recycling these waste streams. Often, polypropylene (PP) and poly(ethylene terephthalate) (PET) are found in these packaging products but are very hard to separate. This implies that a lot of recycled plastic waste comes as a mixture of different types of plastics. As known, many polymers are immiscible in the melt phase, causing distinct phase morphologies and poor mechanical properties in the final recycled products. Hence, there are intensive studies of researchers on the development of new methods, which could overcome the immiscibility of polymer waste and by this improve the final product performance [1-5]. In order to mitigate the disadvantages of immiscibility, compatibilization is known as a promising method to improve the compatibility of PP/PET blends [3,5-10]. Conventional compatibilizers are polyolefins or rubber based polymers grafted with reactive functional groups, such as acrylic acid (AA), maleic anhydride (MA) or glycidyl methacrylate (GMA) [7]. The compatibilization of the blends can have a substantial influence on the interfacial properties between phases as the compatibilizer concentrates itself at the interface between the different phases during blending, preventing coalescence and resulting in smaller and finer dispersions as well as a better adhesion between the polymer phases [11,12,13]. Many researchers reported that PP grafted with maleic anhydride (PP-g-MA) could improve the compatibility of PP/PET blends. Heino et al. [14] found that poly[styrene-b-(ethylene-co-butylene)-b-styrene] (SEBS) grafted with MA or GMA could act as a compatibilizer to yield a finer dispersion by improving the interfacial adhesion in the mixture PET/PP. Ample studies reported results on the effect of adding different amounts of compatibilizers. This research focuses on adding different compatibilizers with a fixed weight percentage. A variety of compatibilizers exist for improvement of blend properties and in this article an extensive study on the comparison of the effectiveness of five different compatibilizers in the PP/PET blend is reported: PP-g-MA, polyolefin based elastomer grafted with MA (POE-g-MA), SEBS, SEBS-g-MA and SEBS-g-GMA. In this regard, the mechanical properties in terms of tensile, flexural and impact testing are investigated and compared to the virgin blend in order to determine the influence of the compatibilizers.

EXPERIMENTAL

Materials

The PP was kindly donated by ExxonMobil (PP6272NE1) with a melt flow rate (MFR) of 2.8 g/10 min (2.16 kg, 230 °C), and the used PET was LIGHTER C93 from Equipolymers, which is a bottle-grade material with an intrinsic viscosity of 0.80±0.02 dL/g. PET was dried in a vacuum oven for 24 h at 80 °C before processing, while PP was used as received.

Five different compatibilizers were used in this research. PP-g-MA (SCONA TPPP 2003 GB) and SEBS-g-GMA (SCONA TPKD 8304 PCC) were kindly donated by BYK Additives & Instruments. The POE-g-MA used was Acti-Tech 16MA13, kindly donated by Nordic Grafting Company (NGC). SEBS-g-MA (FG1901 G) and SEBS (G1657M) were kindly donated by Kraton. The reported grafting percentages of the functional groups are listed in Table 1. All compatibilizers were dried at 80 °C for 24 h before processing.

TABLE 1. Grafting percentage of MA and GMA functional groups in compatibilizers. Data in table was found in technical data sheets, provided by the supplier.

Material	Grafting percentage [wt%]
PP-g-MA	0.9-1.3
SEBS-g-GMA	3.5
POE-g-MA	1.3
SEBS-g-MA	1.4
SEBS	0

Sample Preparation

The polymers were mixed in a weight ratio of 80/20 PP/PET and compatibilizers were added in 6 wt%, while maintaining the same PP to PET ratio. Blends were prepared by extrusion followed by injection moulding. The PP and dried PET pellets, as well as dried compatibilizer were dry mixed in the correct weight ratio before being melt mixed. The melt mixing of the polymers (with and without compatibilizer) was done using a twin-screw extruder (Coperion ZSK18) with two co-rotating screws of 18 mm diameter. The rotation speed of the screw was set at 120 rpm, while the temperature profile was fixed at 205 °C, 245 °C, 250 °C, 250 °C, 255 °C, 255 °C, 260 °C, 260 °C, 260 °C, from hopper to die. The extrudate has been obtained as a sheet with dimensions of 25x1 mm. The next step consisted of shredding the sheets, followed by injection moulding with an Engel 80T using a temperature profile of 180 °C, 190 °C, 200 °C and 210 °C in a standard mould with a temperature of 30 °C, obtaining both ISO 527 tensile and ISO 178 impact and flexural bars.

Characterization

All mechanical characterization was conducted under controlled conditions, being at 23 °C and 50% humidity, after letting the samples rest for a minimum of 48 hours within this controlled environment.

The notched Charpy impact test was used to evaluate the toughness of the samples. The impact bars were analysed by using the Tinius Olsen IT 503 Pendulum Impact Tester according to ISO 179.The specimens were notched in the middle of the sample with a depth of 2 mm and then placed horizontally with the notch oriented away from the pendulum with energy of 2 J. At least 10 measurements were done for each sample.

The standard tensile bars were tested with an Instron 5565 tensile device according to ISO 527. During the tests, different test speeds were used before and after removing the extensometer (gauge length 12.5 mm), 3 mm/min and 10 mm/min respectively.

The three-point flexural test was performed with an Instron 4464 testing machine (Norwood, MA USA). The test was done according to the ISO 178 standard. The distance between both support spans was set at 48 mm, while the deflection rate was set at 5 mm/min. Analysis was performed using the Bluehill software.

At least 5 specimens were tested during flexural and tensile tests. Results are reported as mean ± standard deviation. Statistical analysis of all specimens was performed by software package SPSS Statistics 23 (Armonk, NY, USA) (t-independent test, p=0.05).

RESULTS AND DISCUSSION

Impact properties

It is well known that uncompatibilized polymer blends have inferior properties due to immiscibility between the phases [2,3]. The goal of this research was to investigate the influence of different compatibilizers on the mechanical properties of a PP/PET blend.

Figure 1 presents impact strength values of the different samples. A significant increase can be seen in toughness for the compatibilized blends PP-g-MA, POE-g-MA and SEBS-g-MA compared to the neat blend, and no significant increase for SEBS-g-GMA and SEBS. The samples containing the POE-g-MA compatibilizer showed the highest value for impact. This could be explained by the fact that POE contains randomly ordered propylene and ethylene blocks in an elastomeric structure compared to PP-g-MA containing only relatively rigid propylene. SEBS based compatibilizers are rubber-like as well but more rigid compared to POE backbones due to the styrene end blocks. POE-g-MA is as such more effective in improving the impact resistance by better dissipation of the impact energy [15,16]. In theory, the POE phase could initiate crazes contributing to the absorption of the impact energy to block the crack propagation [16]. PP-g-MA improves the compatibility of both phases as well, resulting in an improved impact strength compared with the non-compatibilized sample by better transferring of the stress over the two phases. However, when comparing with the POE-g-MA, the influence of the rubber-like backbone and the lack of it in the PP-g-MA is clearly visible by a decrease in impact energy [15]. Furthermore, the effect of the rubber behaviour of POE backbone is also confirmed by the results of the strain at break (Table 2), which are considerably higher compared to the more rigid SEBS and PP backbones.

PP-g-MA compatibilized blends show better results compared to blends containing SEBS based compatibilizers, because of better compatibility of the PP backbone with the PP matrix. On the other hand, SEBS-g-MA showed an increase relative to the other two SEBS based compatibilizers.

The polar carboxyl groups of MA grafted onto the compatibilizer's backbone improve adhesion with the PET phase by lowering the interfacial tension. In the case of PP-g-MA, adhesion between the PP and PET phases is further improved because of the same backbone structure. In this regard, the SEBS-compatibilizers do not benefit the same compatibility because of a difference in backbone structure, resulting in a decrease in adhesion between the two phases for these blends. This is reflected in the lower impact values of the SEBS-blends where SEBS-g-GMA and SEBS did not show any significant improvement compared to the uncompatibilized sample.

Different researchers have compared the differences between GMA and MA groups, where the GMA functional group was found to be more reactive than MA due to its bifunctional character [5,6,17]. Nevertheless, the difference in grafting percentage needs to be taken into account, where the GMA percentage is 3.5wt% in SEBS-g-GMA and 1.4wt% of MA content in SEBS-g-MA. Tanrattanakul et al. [18] have investigated the effect of different contents of grafted MA in compatibilizers for PET-rich blends and they reported that the increased percentage of grafting groups could have a negative effect such as a decrease in fracture energy.. However, for such a statement it would be necessary to verify the effect of grafting percentage of anhydride for our system, which is a PP-rich blend.

The blend with non-functionalized SEBS displays lower impact strength compared with neat blend PP/PET. In this case, it could be considered as a ternary blend causing a decrease in impact energy. The rubber phase is too low to be effective without grafted functional groups, however the SEBS could still be located at the interface and act as a compatibilizer, increasing the physical interactions. Nevertheless, this ternary structure makes it more complicated because it can contain cavities, core-shell and individual particles simultaneously, which do not effectively absorb the impact energy [6,19].

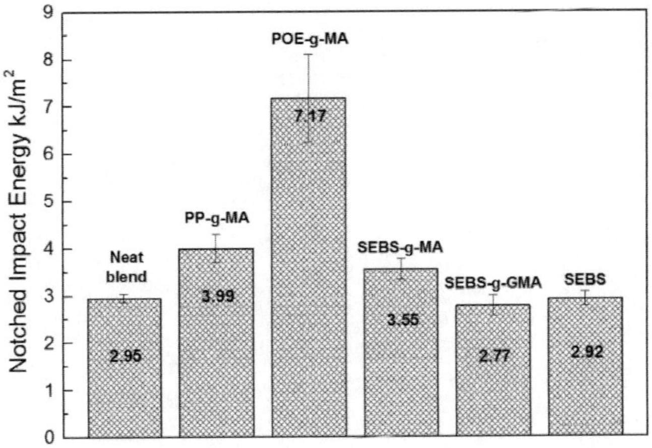

FIGURE 1. Comparison of impact strengths of the different blends.

Tensile and flexural properties

The tensile modulus and strength for all samples are given in Figure 2. These bar graphs show that PP-g-MA is the only one succeeding in maintaining the stiffness of the original blend. This can be explained by the higher stiffness of the PP backbone in the compatibilizer compared to other blends containing the softer backbones such as SEBS and POE. The PP-g-MA greatly improves the compatibility between PP and PET, reducing the interfacial energy, and forming an interphase layer between the matrix and particles, increasing interfacial bonding. Combined with the lower interfacial tension between the MA group and PET, a better interfacial adhesion was achieved when compared to SEBS-g-GMA and POE-g-MA.

It is also noted that the tensile strength for POE-g-MA is lowered considerably because of the aforementioned presence of elastomeric chains in the polymer backbone [15]. G'Sell et al. [15] explained that several mechanisms could contribute to the general deformation of the sample under tension: interface decohesion, cavitation at the PP/PET interface and the cavitation in POE-g-MA particles. Blends with SEBS, SEBS-g-MA and SEBS-g-GMA have shown a slight decrease in modulus compared POE-g-MA and a significant decrease compared to PP-g-MA and neat PP/PET.

TABLE 2. Strain at break of the blends.

Material	Strain at break%
Neat blend	9.83
PP-g-MA	102.16
POE-g-MA	401.07
SEBS-g-MA	13.58
SEBS-g-GMA	7.31
SEBS	13.08

Additionally, all tested blends have shown a mostly brittle breakage during the tensile tests, except for the PP-g-MA and POE-g-MA where necking was observed. This can also be seen in Table 2, as the strain at break is much higher for the blends containing POE-g-MA when compared to all other blends. This is because of the rubber blocks in the POE-g-MA and the improved compatibility between both phases of the blend. PP-g-MA improves the adhesion as well, but will show less strain due to its more rigid backbone. It was observed that non-compatibilized blend shows a low elongation at break, due to the low interfacial adhesion between phases. The strain was slightly improved with the SEBS and SEBS-g-MA addition to blend, likely because of their rubbery-structure. However, due to their low improvement in compatibility of the PP matrix with the PET this effect is barely noticeable.

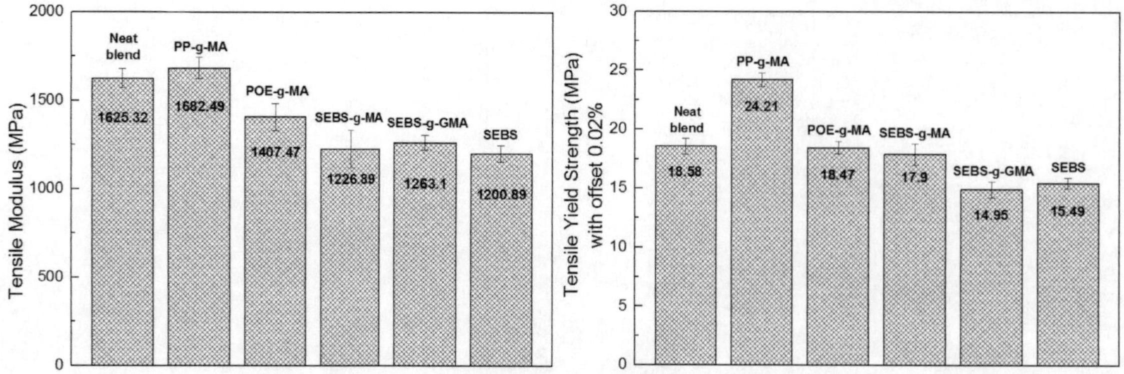

FIGURE 2. Comparison of tensile modulus and strength of the different blends

Figure 3 presents the results of the flexural modulus and strength of the tested samples. Depending on the polymer backbones and their functionality, existing research [7] has demonstrated that the addition of a rubber-based compatibilizer could lower the values of flexural properties. In our experiments, the highest flexural modulus and strength is achieved for PP-g-MA blends, which are consistent with the results obtained during the tensile testing. A similar improvement can be noticed in flexural strength compared to other samples. This improvement can be ascribed to the PP-g-MA compatibilizer because of the possibility that there occurs no decohesion between the PP matrix and the PET particles due to improved compatibility and stiff backbone. Hence, PP-g-MA has the optimal flexural properties when comparing with other compatibilizers. Further, there are no remarkable differences in flexural modulus and strength between other blends. The lower flexural modulus and strength are caused by the elastomeric nature of POE and SEBS backbones, which are not resistant to the applied flexural force.

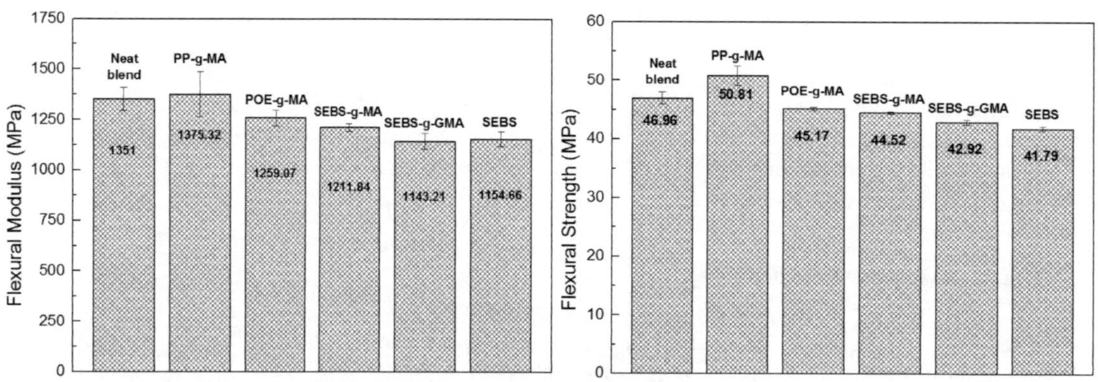

FIGURE 3. Comparison of flexural modulus and strength of the different blends

CONCLUSION

The effect of different compatibilizers on the mechanical properties of blends consisting of a 80/20 ratio PP/PET was studied. The comparison between different backbones and functional groups was made. Five different compatibilizers were investigated: PP-g-MA, POE-g-MA, SEBS-g-MA, SEBS-g-GMA and non-functionalized SEBS were used in a fixed percentage being 6 wt%.

The POE-g-MA compatibilizer exhibited a great improvement in impact strength due to the elastomeric nature of its backbone structure and improved compatibility between the PP and PET phases. PP-g-MA showed a slight improvement in impact strength due to its good compatibility with the PP phase because of its PP backbone. In the case of SEBS-g-MA, the slightly enhanced impact strength can be ascribed to the rubber blocks in SEBS backbone. However, it causes a smaller improvement in compatibility compared with PP-g-MA and POE-g-MA. In this regard, the lowest value was obtained by SEBS-g-GMA, i.e. the compatibilizer with the highest grafting percentage of anhydride. This could be the reason for the decrease in impact strength regarding to the literature, nevertheless these assumptions need experimental verification for our system. In the case of SEBS, lower impact strength was observed relative to the neat PP/PET blend which could be ascribed to the formation of ternary blends denoting a complete absence of compatibility between the phases as the rubber content is too low to react effectively without functional groups.

The tensile tests revealed that PP-g-MA improved the tensile modulus and strength the best due to the stiff nature of the backbone, while the other compatibilizers contain rubber-like blocks resulting in a decreased tensile modulus and strength. The same conclusion can be drawn for the flexural properties. The PP-g-MA shows a similar improvement due to good compatibility between the PP chains and the PP backbone of the compatibilizing agents, increasing adhesion between PP and PET. The other compatibilizers have not shown an improvement, due to the content of elastomeric groups. The MA-based compatibilizers (PP-g-MA, POE-g-MA and SEBS-g-MA) generally performed better than the non-MA based compatibilizers.

It can be concluded that improving the impact properties of the PP/PET blends is most successful with POE-g-MA, while PP-g-MA is better at improving the tensile and flexural properties. SEBS-based compatibilizers seem to have little or even a negative effect on the mechanical properties of the blends.

In future research the complementary SEM images could provide more insight in the effectiveness of the different compatibilizers.

ACKNOWLEDGMENTS

The authors gratefully acknowledge the experimental work of the Master students during the Master course Polymer and Composite Materials at Ghent University, Belgium.

REFERENCES

1. W.K. Pek. and U.K Ghosh, Environ Prog Sustain Energy, 34(4),1113-1119 (2015).
2. M. Kuzmanovic, L. Delva, L. Cardon and K. Ragaert, Polymers, 8(10), 355 (2016).
3. M. Asgari and M. Masoomi, Compos. Part B Eng. 43, 1164–1170 (2012).
4. K. Ragaert, L. Delva and K. Van Geem, Waste Manag, In press, (2017).
5. H. T. Chiu, and Y.K. Hsiao, J. Polym. Res. 13(2), 153-160 (2006).
6. M. Heino, J. Kirjava and P. Hietaoja, J. App Polym. Sci. 65(2), 241-249 (1997).
7. K. Jayanarayanan, S. Thomas and K. Joseph, Int. J. Plastics Tech. 19(1),84-105 (2015).
8. M. F. Champagne, M. A. Huneault, C. Roux and W. Peyrel, Polym. Eng. Sci. 39, 976–984 (1999).
9. Y. L. Zhu, C. Liang, Y. Bo and S. Xu, J. Polyme. Res. 22(3), 12 (2015).
10. M. Evstatiev, S. Fakirov and K. Friedrich. Polym. Compos. 149-167 (2005)
11. X. Zhang, B. Li, K. Wang, Q. Zhang and Q. Fu, Polymer 50(19), 4737-4744 (2009).
12. M. Entezam, H. A. Khonakdar, A. A. Yousefi, S. H. Jafari, U. Wagenknecht, G. Heinrich, and B. Kretzschmar, Macromol. Mater. Eng. 297(4), 312-328 (2012).
13. Y. X. Pang, D. M. Jia, H. J. Hu, D. J. Hourston and M. Song, Polymer 41, 1, 357-365 (2000)
14. M. Heino, J. Kirjava and P. Hietaoja, J. App Polym. Sci. 65(2), 241-249 (1997).
15. C. G'Sell, S.L. Bai and J.M. Hiver, Polymer, 45(17), 5785-5792 (2004).
16. F. Z. Wang, L. T. Drzal, Y. Qin and Z. Huang, Composites Part A,. 87,10-22 (2016)
17. E. P. A.van Bruggen, R. P. Koster, S. J. Picken and K. Ragaert, Int. Polym. Proc. 31(2): p. 179-187 (2016).
18. V. Tanrattanakul, W. G. Perkins, F. L. Massey, A. Moet, A. Hiltner and E. Baer, J. Mater. Sci., 32(18), 4749-4758 (1997).
19. O. Moini Jazani, M. A. Khoramabadi, M. M. Salehi and H. Riazi, J. Part. Sci. Tech. 1(1), 39-48 (2015).

Assessing the Environmental and Economic Performance of Alternative Car Chassis

Carla L. Simões[b], Carlos J. Ribeiro[b], Pedro Bernardo[c], António J. Pontes[a,b], C.A. Bernardo[a,b,*]

[a]*Institute for Polymers and Composites - IPC/I3N, Minho University, Azurém, 4800-058 Guimarães, Portugal*
[b]*Innovation in Polymer Engineering - PIEP, Minho University, Azurém, 4800-058 Guimarães, Portugal*
[c]*Car Multimedia Portugal, S.A., Rua Max Grundig, 35, Lomar, 4705-820 Braga, Portugal*

* Corresponding author: cbernardo@dep.uminho.pt

Abstract. The main objective of this study is to evaluate the life cycle environmental and economic performance of a car multimedia chassis containing metallic parts, and compare it with new, totally plastic, chassis designs. The Life Cycle Assessment and Life Cycle Costing methodologies were applied. All systems boundaries consider material and parts production, and the use and End of Life (EoL) phases of the chassis. The results showed that the former system has a higher environmental impact, the material production being the main contributor followed by the use phase, and Fossil depletion the most burdensome impact category. All total plastic scenarios enable approximately 40% weight reduction, mitigating both the Global Warming Potential and the Cumulative Energy Demand environmental impacts until the end of the use phase. However, this result is inverted including the EoL phase, as recycling the metal is more favourable than incinerating the polymer and recovering energy. All TPC scenarios present a higher cost. Although their assembly and use phases costs are lower than the corresponding BSL ones, this does not mitigate the higher material and production costs. Again, at EoL, recycling the metal is more cost favourable. The present work evidences that to make sustainable decisions environmental and economic considerations should be concurrently contemplated in product development.

Keywords: Life Cycle Assessment; economic analysis; plastic car chassis; sustainability; automotive parts; design for environment
PACS: 81; 88

INTRODUCTION

Together with safety and performance, environmental impact reduction and cost containment throughout the complete life cycle (LC) of a vehicle are the major issues for today's automotive industry. Consequently, decisions on product development must involve technological requirements as well as economic and environmental considerations [1]–[4]. Such considerations should be assisted by adequate LC thinking-based tools, such as Life Cycle Assessment (LCA) [5], [6] and Life Cycle Costing (LCC) [7], [8].

In this context, the main objective of this study, continuing a research programme focussed on the same subject, is to evaluate the LC environmental and economic performance of a car multimedia chassis containing metallic parts (thereafter called "BSL"), and compare it with new, totally plastic, chassis designs (thereafter called "TPC").

The present work focuses on how Life Cycle Thinking, LCA, and economic assessment methodologies can be used to evaluate environmental and economic impacts of alternative production systems.

METHODOLOGY

The LCA study was done according to the ISO 14040 series procedures [5], [6] and the LCC study according to the Ciroth et al. [7] and Swarr et al. guidelines [8]. The estimation of the polymer composite components production cost was done via a Process Based Cost Model (PBCM) [9], [10].

LCA and LCC results

The target product of this study is a car multimedia chassis whose function is to protect the electrical/electronic components that compose the car multimedia solution. Its main objective is to evaluate the LC environmental and economic performance of the selected car chassis that will constitute the baseline platform product, and compare it with a new car multimedia chassis design (total plastic chassis). The functional unit is a car multimedia chassis unit, which meets the mechanical, thermal and electrical requirements, as defined by the client, and has a service life of 200,000 km (maximum distance driven in the expected use life of a vehicle [11], [12]). All solutions under study have the essential function of mechanically protecting the multimedia contents, while guaranteeing adequate thermal and electromagnetic shielding. All systems boundaries consider material and parts production, and the use and end of life (EoL) phases of the chassis.

Thc selected LC impact assessment (LCIA) method was the ReCipe Endpoint (H/A) [13]. The Energy consumption, according to the Cumulative Energy Demand (CED) method [14], and the Global Warming Potential (GWP) impact categories [14] were analysed. The goal and scope steps of the LCC study were consistent with those of the LCA. Therefore, the functional unit, system boundary and other fundamental assumptions were the same.

The new car multimedia fully polymer chassis design (TPC) is still in development by the project team (Bosch, UM, PIEP) and, thus, not yet in real production. Therefore, its LC characterization is not a final version, but rather a first approach that will be updated in the future. Hence, the definition of the LC characterization of the TPC parts and product require the formulation of some assumptions. In this study the first version, subsequently called TPC01, will be studied. Additionally, four scenarios were developed with four possible alternative materials.

All the systems have been modelled by means of the commercial ecoinvent database [15] and, whenever possible, using field data from the companies involved in the study, which were ultimately summarized in the LC inventory that was performed in SimaPro 8.04 [16]. Some data did not exist in the ecoinvent database and, hence, their inventory was collected from other sources. The carbon fibres inventory data were collected from Schmidt and Watson [17], and the glass fibres and waste treatment by incineration with energy recovery inventory data from the European LC Database (ELCD) [18]. The car multimedia chassis EoL scenarios were modelled under the following assumptions: (i) incineration, considering environmental credits due to energy recovery; and (ii) recycling, considering that production of primary metal was avoided.

An environmental LCC was selected in the present work, which considers all direct costs of the different actors intervening in the product LC. The system was modelled using market prices. In the LCC inventory the data collected were adjusted, where appropriate, to Euros 2016.

Figure 1 a) compares the BSL and the 4 TPC01 possible material scenarios, on a functional unit basis. The BSL system has a higher impact in almost all environmental impact categories, except Climate change human health, Climate change ecosystems and Terrestrial ecotoxicity. The TPC01A system depicts the worse performance in these impact categories, in the case of the latter, together with TPC01C system. Considering the 4 TPC01 possible material scenarios, the TPC01D system shows the best environmental performances in all impact categories. These results are a consequence of the different environmental impacts of the various composites reinforcements. The carbon fibre has the highest environmental impact of all the reinforcement materials, followed by the stainless steel and the glass fibres. Graphite is the reinforcement that shows the smallest environmental impact.

The LCIA normalization results of the 5 systems are shown in Figure 1b), on a functional unit basis. The Fossil depletion, Climate change human health and Climate change ecosystems impact categories are the most significant environmental burdens of all systems. Fossil depletion is more significant for the BSL system, The two other ecosystems are more significant for the TPC01A and TPC01C systems, respectively.

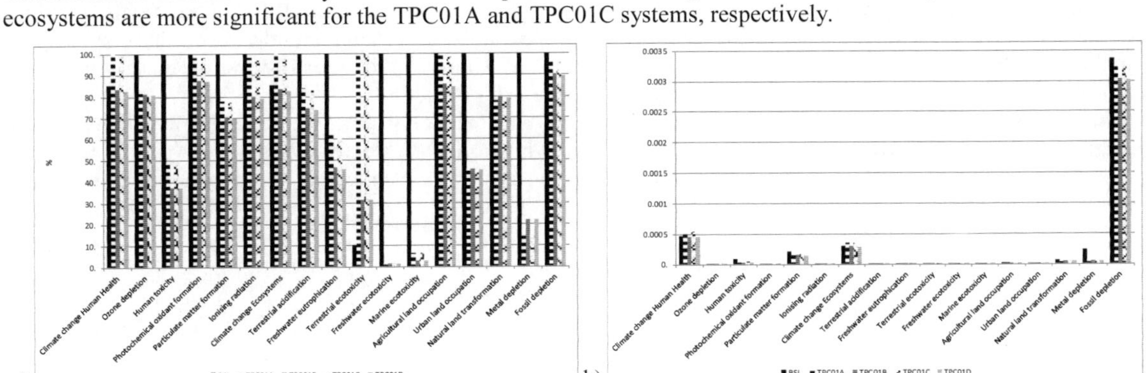

FIGURE 1. LCIA a) characterization and b) normalization results of the BSL versus TPC01 possible material scenarios, on a functional unit basis.

The single score results obtained (expressed in Eco-Indicator points (Pt)), on a functional unit basis, were 1.19 Pt, 1.13 Pt, 1.02 Pt, 1.12 Pt and 1.01 Pt, for the BSL, TPC01A, TPC01B, TPC01C and TPC01D, respectively. They are relatively close, in any case they show that, all caveats considered, the TPC01B/TPC01D systems are environmentally preferable (in circa of 15%).

Figure 2 a) shows the systems GWP characterization results, expressed as the differences between the BSL and the four TPC01 possible material scenarios, throughout their LC. These differences are presented (in the ordinate axis) as a function of the several LC phases (in the abscissa axis), each phase being represented by a line segment. All TPC01 possible material scenarios mitigate the GWP environmental impacts until the use phase, since the results have a negative value until that phase. The new concept designs (TPC01) allow achieving approximately 40% weight reduction, therefore enabling these results. These results are inverted, however, when the EoL phase is included. At EoL, recycling the metal (steel and aluminium) is more favourable regarding GWP than incinerating with energy recovery the composite, since an increase of this environmental impact category is observed (to overall positive values).

Figure 2 b) shows the systems CED characterization results, expressed as the differences between the BSL and the four TPC01 possible material scenarios, throughout their LC. Again all TPC01 possible material scenarios mitigate CED until the use phase, since the results have a negative value until that phase. However, these results are, once more, inverted when the EoL phase is included. The new concept designs (TPC01) allow achieving approximately 40% weight reduction, therefore enabling these results. It should be noted that the CED of the material production in the new designs TPC01A/TPC01C was higher than in the BSL. This was due to the fact that polymer matrix composites are produced through a more energy intensive (by unit weight) process than steel/aluminium, and the weight reduction doesn't mitigate this energy increase. At the EoL phase it was again observed that recycling the metal (steel and aluminium) is more favourable regarding CED than incinerating the composite and recovering energy, since a significant increase of this environmental impact category is observed.

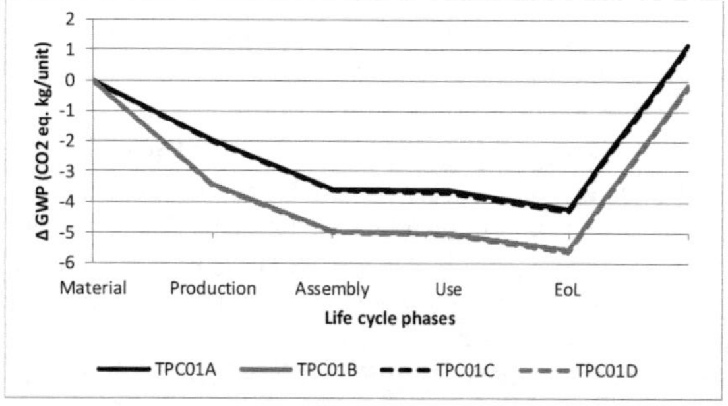

a)

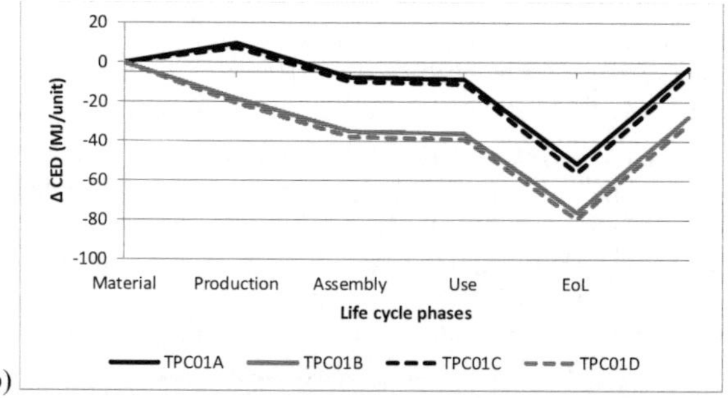

b)

FIGURE 2. Characterization a) GWP (CO2 eq. kg/unit) and b) CED (MJ/unit) results for the difference between the BSL and the TPC01 possible material scenarios (two series are hidden by the data of the others).

Figure 3 shows the systems cost results, expressed as the differences between the BSL and the TPC01 possible material scenarios, throughout their LC. Globally, the TPC01 scenarios do not mitigate the cost impact, since their cumulative differences along the LC have a final positive value. Although the assembly and use phases costs are lower than in the BSL, this does not mitigate the higher material and production costs.

In conclusion, the new concept designs allow achieving approximately 40% weight reduction, but not cost reduction, since the polymeric composite materials have a very high price. At EoL it is again observed that recycling the metal (steel and aluminium) is more favourable regarding cost than incinerating the polymer composite and recovering energy.

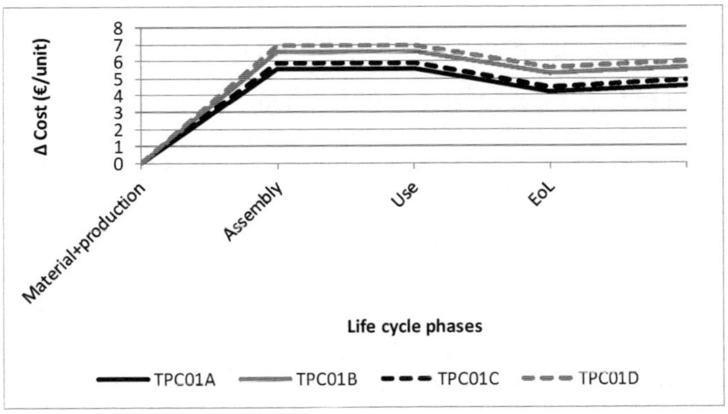

FIGURE 3. Cost results (€/unit) for the difference between the BSL and the TPC01 possible material scenarios.

CONCLUSIONS

In the present work it was concluded that the metallic chasis has the highest environmental impact of all the systems considered, the material production being the main contributor, followed by the use phase. The TPC01B/TPC01D system shows the best environmental performance (in circa of 15%), being, in this case, the use phase the main contributor, followed by the material production. The Fossil depletion, Climate change human health and Climate change ecosystems impact categories are the most significant environmental burdens in all systems.

All TPC01 possible material scenarios, which enable approximately 40% weight reduction, mitigate the GWP and CED environmental impacts until the use phase. However, these results are inverted when the EoL phase is included in the assessment. It should be noted that the CED of the material production in the new designs was higher than in the case of BSL. This is due to the fact that polymer matrix composites are produced through a more energy intensive (by unit weight) process than steel/aluminium, and the weight reduction doesn't mitigate this energy increase. At EoL, it was observed that recycling the metal (steel and aluminium) is more favourable regarding GWP and CED than incinerating the polymer composite and recovering energy.

All TPC01 possible material scenarios present a higher assembly cost. None of these scenarios mitigate the cost impacts, since the results have a positive value. It was observed that the TPC01 assembly and use phases costs are lower than the corresponding ones of the BSL. However this does not mitigate the higher material and production costs. The new concept designs allow achieving approximately 40% weight reduction, which does not lead to a cost reduction, since the polymeric composite materials have a very high price. Again, at EoL it was observed that recycling the metal (steel and aluminium) is more favourable regarding cost than incinerating the polymer composite and recovering energy.

ACKNOWLEDGMENTS

The present work was partially financed by the Portuguese Incentive System for Research and Technological Development, as co-promotion Project nº 36265/2013 (Project HMIExcel – 2013-2015). The authors thank the support of the Bosch Car Multimedia Portugal, S.A. company throughout the inventory phase of the study. Two of the authors (CAB, AJP) acknowledge the funding received from FCT, the Portuguese Foundation for Science and Technology, through project UID/CTM/ 50025/2013 and from the COMPETE 2020 Programme under project POCI-01-0145-FEDER-00768.

REFERENCES

1. N. S. Ermolaeva, M. B. G. Castro, and P. V. Kandachar, "Materials selection for an automotive structure by integrating structural optimization with environmental impact assessment," *Mater. Des.* **25**(8), 689–698 (2004).
2. P. Peças, I. Ribeiro, A. Silva, and E. Henriques, "Comprehensive approach for informed life cycle-based materials selection," *Mater. Des.* **43**, 220–232 (2013).

3. I. Ribeiro, P. Peças, A. Silva, and E. Henriques, "Life cycle engineering methodology applied to material selection, a fender case study," *J. Clean. Prod.* **16**(17), 1887–1899 (2008).

4. C. A. Ungureanu, S. Das, and I. S. Jawahir, "Subodh K. Das and Weimin Yi,n, eds., Life-cycle cost analysis: aluminium versus steel in passenger cars," in *Aluminium alloys for transportation, packaging, aerospace, and other applications*, edited by b Subodh K. Das, Weimin Yin, The minerals, Metals & Materials Society, Pittsburgh, PA, 2007.

5. International Organization for Standardization (ISO), "ISO 14040: 2006 - Environmental management – Life cycle assessment – Principles and framework," in *ISO 14000 International Standards Compendium*, Genève, Switzerland, 2006.

6. International Organization for Standardization (ISO), "ISO 14044: 2006 - Environmental management – Life cycle assessment – Requirements and guidelines," in *ISO 14000 International Standards Compendium*, Genève, Switzerland, 2006.

7. A. Ciroth, G. Huppes, W. Klöpffer, I. Rüdenauer, B. Steen, and T. Swarr, *Environmental life cycle costing*, 1st ed. Pensacola, FL: CRC Press, Publishing House Taylor and Francis, SETAC Press, 2008.

8. T. Swarr *et al.*, *Environmental life cycle costing: a code of practice*, 1st ed. New York, NY: Society of Environmental Toxicology and Chemistry, 2011.

[9. R. Kirchain, "Cost modeling of materials and manufacturing processes," in *Encyclopedia of Materials: Science and Technology*, edited by *K.* H. Jurgen Buschow, Robert Cahn, Merton Flemings, Bernhard Ilschner, edward Kramer, Subhash Mahajan and Patrick Veyssiere, vol. 2, Elsevier B.V., 2001, pp. 1718–1727.

10. C. L. Simões, R. Figueirêdo de Sá, C. J. Ribeiro, P. Bernardo, A. J. Pontes, and C. A. Bernardo, "Environmental and economic performance of a car component: assessing new materials, processes and designs," *J. Clean. Prod.* **118**, 105–117 (2016).

11. C. Koffler, "Life cycle assessment of automotive lightweighting through polymers under US boundary conditions," *Int. J. Life Cycle Assess.* **19**(3), 538–545 (2014).

12. R. A. Witik, J. Payet, V. Michaud, C. Ludwig, and J.-A. E. Månson, "Assessing the life cycle costs and environmental performance of lightweight materials in automobile applications," *Compos. Part Appl. Sci. Manuf.*, **42**(11) 1694–1709 (2011).

13. M. Goedkoop, R. Heijungs, M. Huijbregts, A. D. Schryver, J. Struijs, and R. van Zelm, *ReCipe 2008: A life cycle impact assessment method which comprises harmonised category indicators at the midpoint and the endpoint level - Report I Characterisation*, First edition. Amersfoort, The Netherlands: Ministry of Housing, Spatial Planning and the Environment, 2009.

14. M. Goedkoop, M. Oele, and S. Effting, *SimaPro Database Manual: Methods library*. The Netherlands: PRé consultants B. V., 2004.

15. The Swiss Centre for Life Cycle Inventories, *Ecoinvent database version 3.1*. Zürich, Switzerland, 2013.

16. *SimaPro 8.04*. The Netherlands: PRé consultants B. V., 2014.

17. J. H. Schmidt and J. Watson, "Eco Island Ferry: Comparative LCA of island ferry with carbon fibre composite based and steel based structures," 2.0 LCA consultants, 2014.

[18. European Commission - DG Joint Research Centre - Institute for Environment and Sustainability, *ELCD v2.0 database*. Brussels, Belgium, 2010.

Factors' evaluation and limiting properties in the reprocessability of the Polyethylene by renewable source

Douglas D. Oliveira[1], Giovanna A. D. Rosa[1], Luiz H. C. Damacena[1], Anderson Maia[1], Rondes F. S. Torin [1,2].

1 Núcleo de Pesquisa, Polímeros, Faculdade de Tecnologia de Mauá, Mauá, São Paulo, Brasil;

2 Centro de Engenharia, Modelagem e Ciências Sociais Aplicadas – CECS, Universidade de Federal do ABC, Santo André, São Paulo, Mauá.

daartdouglas@hotmail.com, giamabile@gmail.com.br, luizhcdamacena@gmail.com, maia.anderson@uol.com.br, rondesf@gmail.com

Abstract: Renewable source polyethylene is a bioplastic obtained through the Brazilian sugar cane developed by Braskem. The objective of this work is to evaluate limiting factors of processability of high density polyethylene (HDPE) from renewable sources without additives of oxidants, verifying the application of technical performance, aesthetics and reuse potential. The samples HDPE were submitted to successive process steps by means of a monoroscopic extruder, which is very common in plastic transformation industries, and showed a change in the visual appearance from the eighth reprocessed sample, being more pronounced only in the sixteenth reprocessing. The results obtained from colorimetric analyzes, mechanical tests (ASTM D 638 and ASTM D 256), DSC and rheological analyzes showed degradation of HDPE from the eighth reprocessing, without significantly affecting its mechanical properties, thus acquiring a single alteration in the aspect of the material, related to its aesthetic property, where it is possible to observe the significant yellowing of the samples. In relation to the sixteenth reprocessing, effective degradation occurred, when compared to the eighth reprocessing, it found a higher yellowing level. When evaluating the mechanical properties, it was observed that there was an increase in the elongation of the material, and a decrease in the tensile and impact strengths.

Keywords: HDPE, bioplastic, reprocessability, degradation, reuse.

1. Introduction

The use of polymer materials has grown a lot in the last decades, as we can observe in the amount of plastic utilities that surround us during the day-to-day. Research shows that the world market for thermoplastics has surpassed the margin of 180 million tons / year, among them Polyethylene (PE), due to its expressive use in the manufacture of disposable packaging [1]

To meet such demand the use of petroleum has become limiting, since there is a high cost and a greater attention to sustainability, becoming the crucial development of new plastic materials from renewable sources, which are called biopolymers, examples of these materials are polylactics, thermoplastics starch and green polyethylene. [9] Among them is the green PE, due to the representativeness it has in the market and easy accessibility that the consumer has. However, it is still important to preserve the environment, because plastic waste is the great villain of environmental problems, for the time it takes to undergo spontaneous degradation.

Recycling is a more viable solution for the treatment of plastic waste and there is a wide range of developments and technological studies focused on this subject, because some materials lose properties when reprocessed, for such reasons the present work aims to evaluate the behavior of green PE when reprocessed several times.

The PE used to carry out this work is from sugarcane, produced by Braskem, is not a biodegradable material and has the same chemical, physical and thermal characteristics of conventional PE, obtained from petroleum, so there is no restriction in the application. [1]

2. Materials and Methods

2.1. Materials

For this work the green HDPE SGM9450F provided by Braskem was used, it was processed sixteen times in a monorous NZ Extruder, in a temperature range of 190 to 220 ºC.

2.2. Colorimetric Assay

In a Delta Color Colorium 2 colorimeter, virgin HDPE was used as the comparative standard for the 8th and 16th reprocessing samples.

2.3. Tensile Testing

For the tensile test, the standard ASTM D638 was used, and the test bodies were made following the standard, 5 samples were made, at room temperature, under a tension rate of 50 mm / min. The universal machine used was manufactured by Kratos ® model KE 2000 MP, the software for data acquisition was Tracomp - Windows.

2.4. Impact Test

The impact tests were carried out based on ASTM D6110 - Charpy, by method A and test bodies made according to the same. Assays of the reprocessed HDPE samples were conducted at room temperature. The equipment used was the Zwick Roell® HIT25 / 50P, the pendulum was endowed with a nominal energy of 1J.

2.5. DSC Analysis

For the analyzes a DSC model Q20 / TA Instruments® was used. Two sweeps were performed in the range of 35 ° C to 200 ° C at 20 ° C / min, the first scan to erase the thermal history of the sample and the second heating ramp was collected.

2.6. Rheological Analysis

To analyze the rheological properties in a linear regime, the rheometer TA instruments AR 2000 Ex / TA instruments was used. After the identification of the linear viscoelastic region of the HDPE from a renewable source, oscillatory tests were performed to observe variations in the storage (G ') and loss (G' ') modules, as well as in the viscosity as a function of the shear rate.

3. Results and Discussions

3.1. Colorimetric

The colorimetric analysis is used to identify changes in the coloration of polymeric materials, and in this research, it served as a tool base, helping in the possible degradation by the difference of tone.

According to the colorimetric results concerning the virgin HDPE samples compared to HDPE 8 and HDPE 16, Figure 1 and 2, a reddish-yellow color was obtained in both reprocessed samples, HDPE 8 and HDPE 16, but with different degrees of yellowing. In the HDPE 8 sample the degree of yellowing according to Db * of 3.71 was lower than in the sample HDPE 16 which has Db * of 4.77. These results indicate that there was degradation from the 8th reprocessing, relative to the HDPE 8 sample, which was more pronounced in the 16th processing. Variations of this type are characteristic of thermo oxidative degradation. [3]

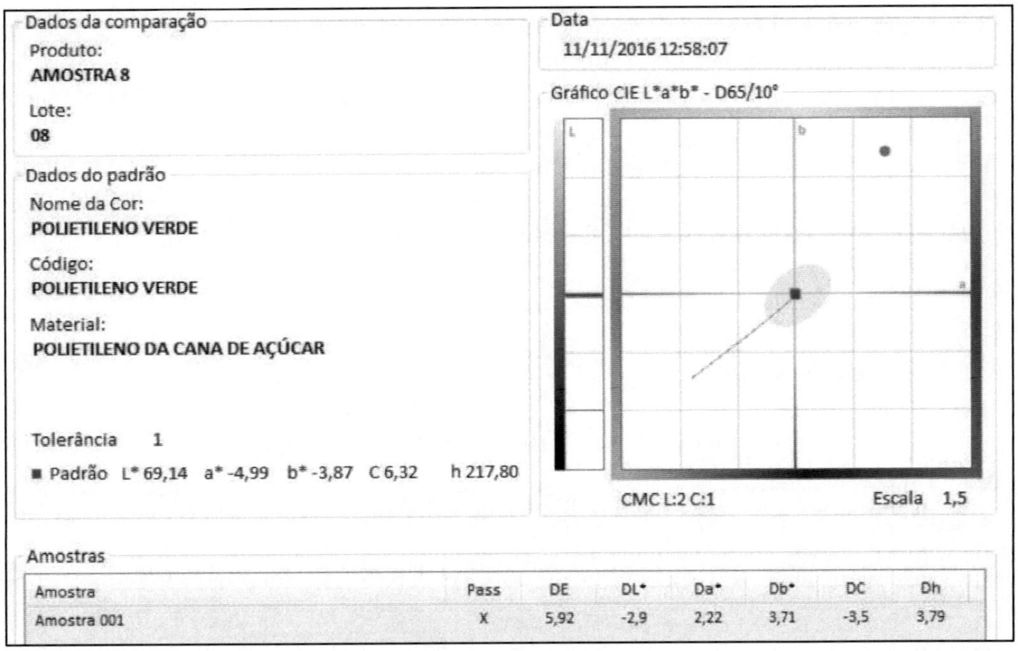

Figure 1: Comparison degree of yellowing between virgin HDPE and HDPE 8.

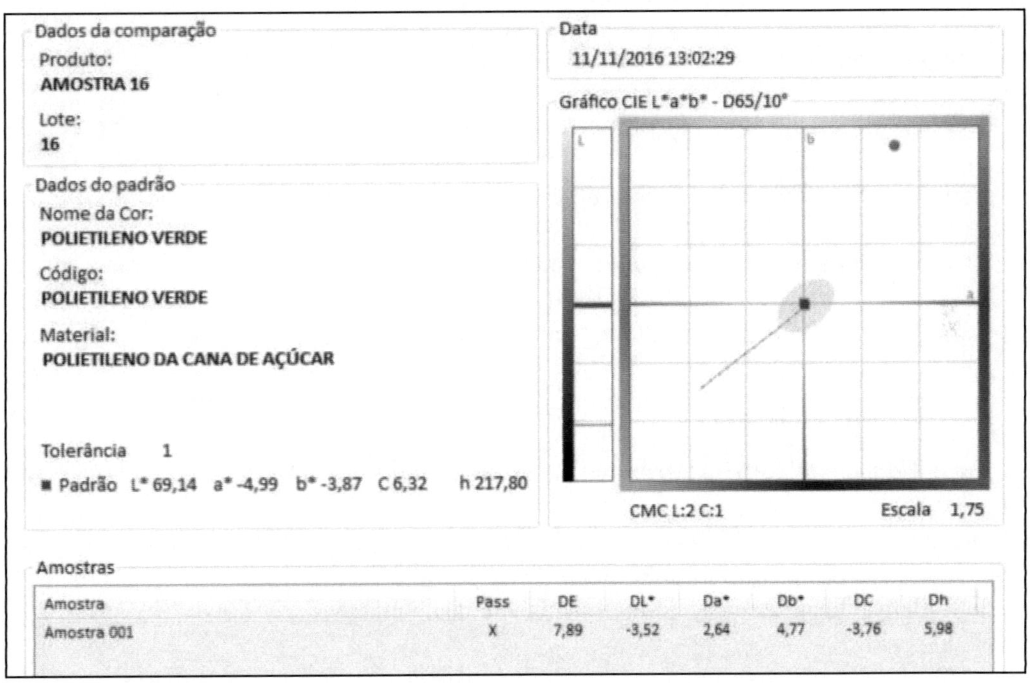

Figure 2: Comparison degree of yellowing between virgin HDPE and HDPE 16.

3.2. Mechanical Test

The mechanical tests are important for determining changes in the physical behavior of the analyzed material. In this way, mechanical tests were used to monitor structural changes related to the reprocessability of the HDPE from a renewable source.

3.2.1. Tensile Testing

The data obtained from the Traction Test for samples of virgin HDPE, HDPE 8th and 16th processing, referred to as HDPE 8 and HDPE 16, respectively, are listed in Table 1. All samples were analyzed following the ASTM D638 standard using five specimens, the data shown in Table 1 are the mean values found. The HDPE sample from the 8th processing shows a slight increase in the tensile strength limit, probably due to orientation of the crosslinked molecular structure of the material. The HDPE samples of the 16th reprocessing showed a decrease in tensile strength properties, the maximum strength reduced by 894.38, the tensile strength limit reduced by 22.01 compared to the virgin HDPE, and the maximum strength 949, 89 and the tensile strength limit 23,38. These results indicate that there was a process of thermal degradation resulting from intermolecular cleavage in the HDPE polymer chains.

On the other hand, the elongation at break was higher in the sample of the 16th reprocessing, since it is a characteristic attributed to the degradation by the processing. This result agrees with the literature, research done by Alcântara et. al. describes that the polyethylene has the degradation characteristic of increasing the molecular weight due to the split of the chains and the formation of cross-links that cause the greater stretching of the material. [2]

Samples	Maximum Force (N)	Resistence Limit (Mpa)	Stretching (%)	Linear Stretching (mm)
HDPE	949,89	23,38	14,28	16,42
HDPE8	971,66	23,92	15,57	17,90
HDPE16	894,38	22,01	32,45	37,32

Tabela 1: Traction Test Data of the HDPE, HDPE 8 and HDPE 16º reprocessing data.

3.2.1. Impact Test

The results obtained from the impact test for the samples of virgin HDPE, HDPE 8 and HDPE 16, summarized in Table 2, indicate minimum variations, that is, there were no large losses in the impact properties of the HDPE samples even after 16 reprocessing. These results may give the impression that even after 16 reprocessing the HDPE did not suffer degradation, however this characteristic is common of the polyethylene, being a semi crystalline polymer the dissipation of energy between the molecules is easier, with this the loss of physical properties are smaller. The tendency is that with the increase of the reprocessing by extrusion the variations in the mechanical properties are more remarkable. Since the increase of reprocessing directly impacts the formation of reticulations, re-folding causes a reduction in the mobility of the molecules and consequently, making it difficult to dissipate energy, which facilitates the degradative process by radical scission of the polymer chains. [2]

Samples	Impact Energy (J)	Result (J/m)
HDPE	0,8678	260,3
HDPE8	0,88	269,9
HDPE16	0,836	256,4

Tabela 2: Impact Test Data of the HDPE, HDPE 8 and HDPE 16 reprocessing data.

3.3. DSC Analysis

Differential Scanning Calorimetry (DSC) analyzes were conducted with the objective of investigating the thermal behavior, from the monitoring of melt (Tm) and crystallization (Tc) temperatures, according to the degree of reprocessing of HDPE.

DSC curves of the virgin HDPE samples were performed as a comparative standard for the HDPE 8 and HDPE 16 samples, Figure 3, 4 and 5, respectively. The DSC curves do not show significant variations in the endothermic event Tm, this may have occurred because there were no changes that could alter the melting temperature. [5]

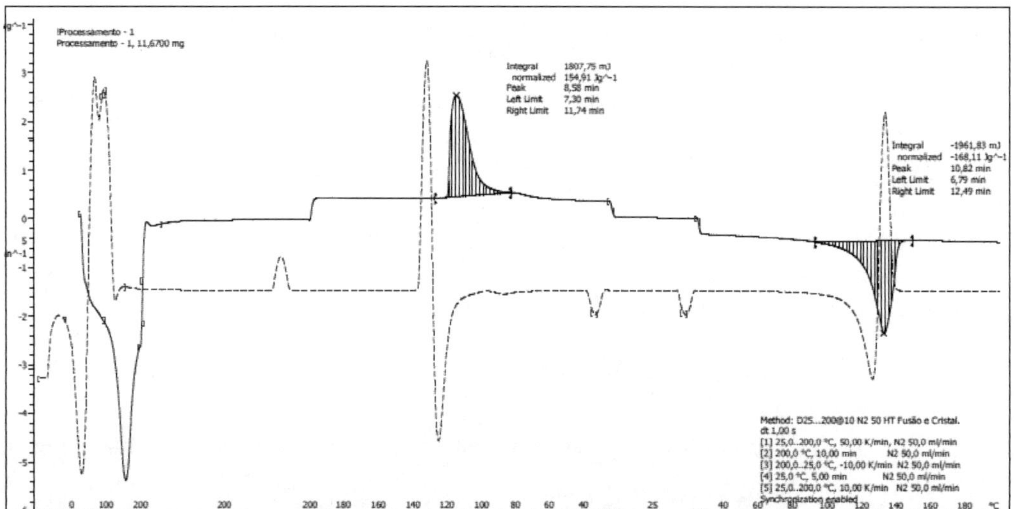

Figure 3: DSC curve of virgin HDPE.

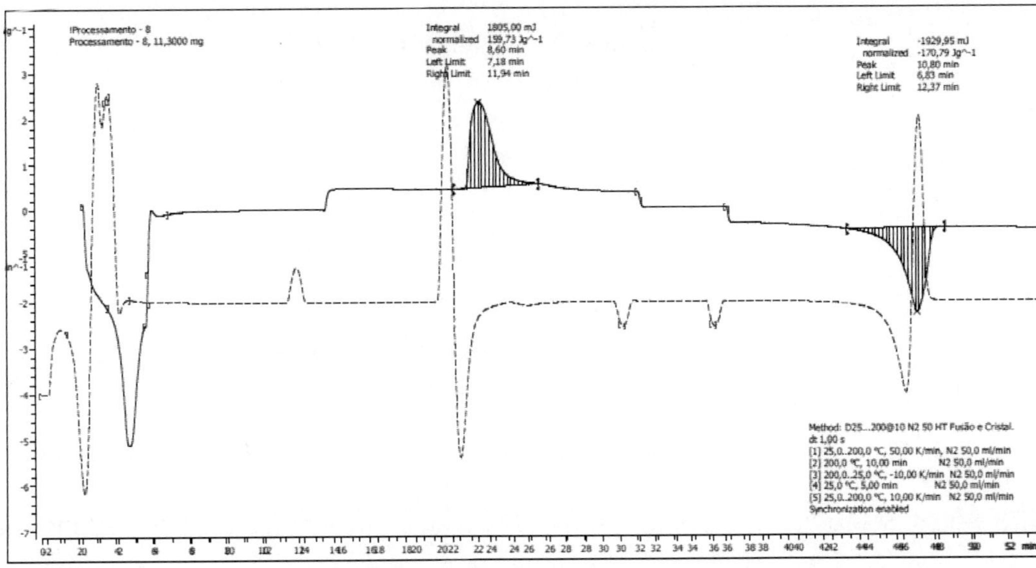

Figure 4: DSC curve HDPE 8.

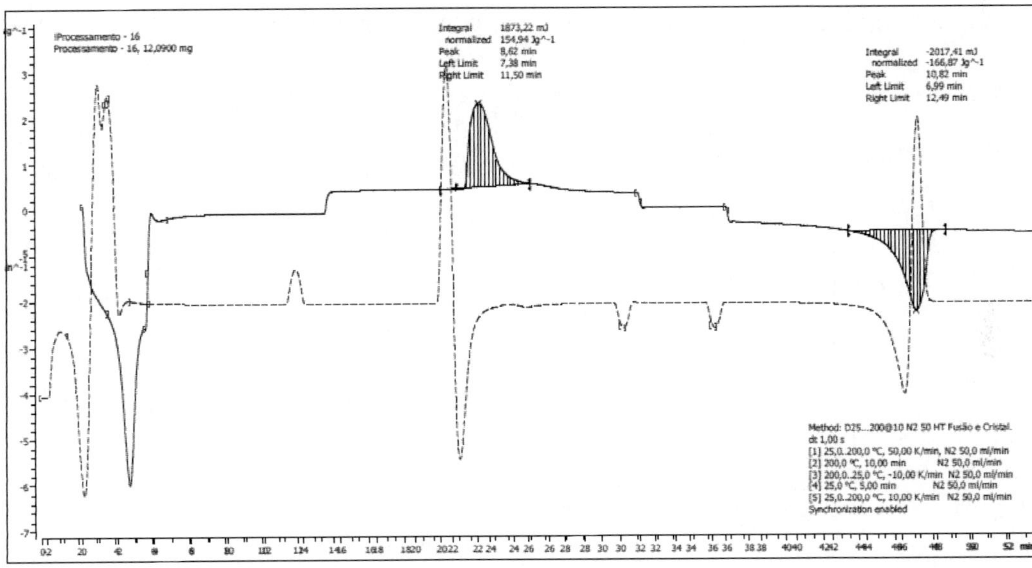

Figure 5: DSC curve HDPE 16.

Samples	Degree of Crystallinities
PE	57,97%
HDPE8	58,89%
HDPE16	57,54%

Table 3: Percentage of Degree Crystallinity.

In Table 3, demonstrates percentage values of the degree of crystallinity, calculated from the DSC curves, Figures 2, 3, and 4. The results show a slight increase of the crystallinity in the sample HDPE 8 with respect to the virgin HDPE, such increase may be associated to the compliance of the chains due to processing, but in the HDPE 16 sample the degradation effect due to the decline in crystallinity is remarkable, proving that the sample has crosslinks. [5] [6].

3.4. Rheological Tests

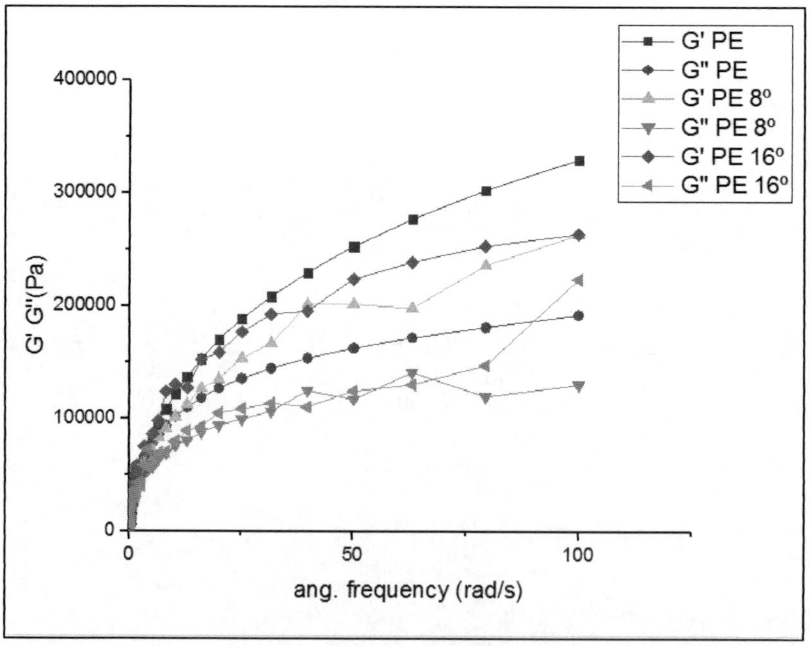

Figure 6: Graphs G 'and G "as a function of the frequency angle.

The oscillatory rheological analysis relates the linear viscoelastic properties of the polymeric materials and contributes to identify changes in the molecular structure of polymers in the molten state. The Figure 6 shows the storage modulus (G ') and loss modulus (G') as a function of the angular frequency of the reprocessed virgin HDPE and PEAD8 and PEAD16 samples. The curves of the G 'and G' modules, Figure 6, generally show an increase as a function of the applied frequency and variations in the elastomeric plateau, which are related to the molecular weight and molecular weight distribution, molecular chain and crosslinking of the side chains, which increases the volume of the polymer chain that alter the modules that storage and loss. [4], [10]

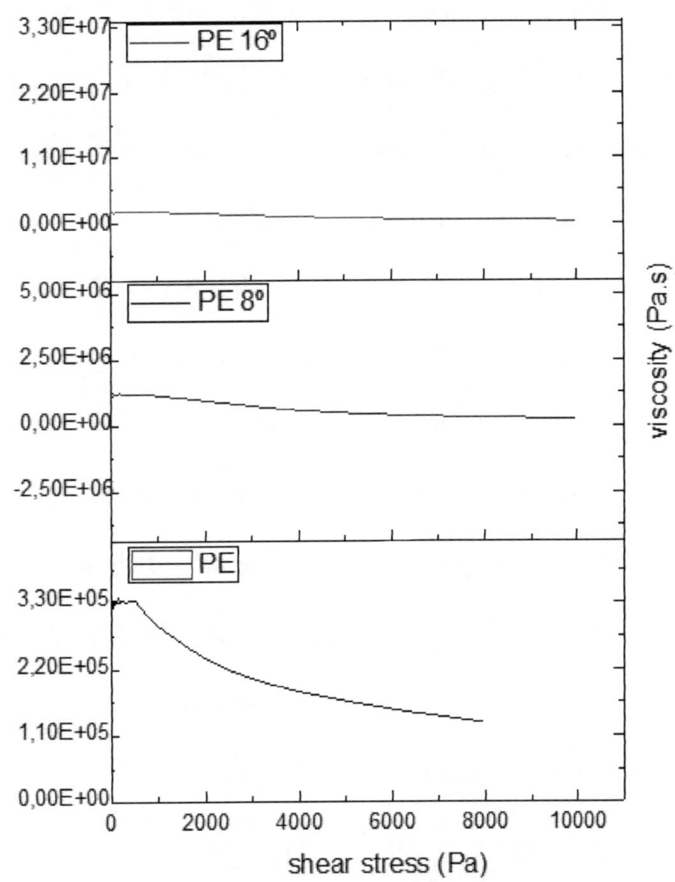

Figura 7: Viscosity in shear stress function of samples HDPE, HDPE 8 and HDPE 16

The Figure 7 show viscosity graphs in function of shear rate stress. When comparing them, it is evident that there is a variation in the viscosity between them, which is related to the decrease of the molecular mass, associated with the polymer chain degradation process, which infuences the incrase in crystallinity, due to the restriction of the amorphous phase. [8]

Conclusion

The results of the tensile test indicated a reduction in the viscoelastic properties of the sample HDPE 8 in relation to virgin HDPE. When comparing HDPE 8 with HDPE 16, there is an increase in the elastic region due to the decrease in the degree of crystallinity obtained by the DSC analysis. According to the rheological results, it shows changes in the storage and loss modulus, being related to the changes in the molecular structure of the polymer pertinent to the crosslinking and unsaturation of the polymer chain, from the various reprocessing. With the colorimetric results, it shows in the samples HDPE 8 and HDPE 16 yellowing, caused by the thermooxidative degradation. The viscosity results in the shear stress function of the virgin HDPE, HDPE 8 and HDPE 16 samples show that there was a reduction in viscosity as a function of shear stress, related to chain stricture, molecular weight decrease and amorphous phase restriction. Regarding the impact analyzes, there are no significant changes in the molecular structure in the HDPE 8 and HDPE 16 samples that influenced this property. Thus, there were no significant changes that could influence the mechanical and processing properties that prevent the industrial reuse of this high density polyethylene from renewable source.

Bibliography References

[1] BELLOLI, Rodrigo. Polietileno Verde do etanol da cana-de-açúcar brasileira: Biopolímeros de classe mundial. Tese (Bacharel em Engenharia química), Universidade Federal do Rio Grande do Sul, Porto Alegre, 2010.

[2] ALCÂNTARA, R. L.; CARVALHO, L. H.; RAMOS, S. M. L. S. Propriedades Mecânicas de Resíduos Plásticos Urbanos da Região Nordeste: Influência das Condições de Processamento. Revista: Polímeros, n.5, p.42, 1995.

[3] PAOLI, M. A. Degradação e estabilização de polímeros. 4. ed. São Paulo: Artlibler Editora. 2008.

[4] BRETAS, E. S. R.; d' Ávila, A. M. Reologia de Polimeros Fundidos. 2. Ed. São Carlos: EdUFSCar, 2010.

[5] LUCAS, F. E.; SOARES, G. B.; MONTEIRO, E. C. E. Caracterização de Polímeros: Determinação de Peso Molecular e Análise Térmica. 1. Ed. Rio de Janeiro: E-PAPERS, 2001.

[6] CANEVAROLO, V. S. Jr.; Ciência dos Polímeros: Um tesxto básico para tecnólogos e engenheiros. 2. Ed. São Paulo: Artiliber. 2006.

[7] GUIMARÃES, O. C. M.; COUTINHO, M. B. F.; ROCHA, C. G. M.; BRETAS, E. S. Reologia de Polietileno de Alta Densidade Tenacificado com Polietileno Elastomérico. Polímeros: Ciência e Tecnologia, vol. 13, nº 2, p. 135-140, 2003.

[8] BATALIOTTI, D, M.; Avaliação da degradação do polipropileno sob múltiplas extrusões com e sem antioxidante. Universidade Tecnológica Federal do Paraná, Londrina, 2016.

[9] ANDREW, W.; Handbook of Biopolymers and Biodegradable Plastics. 1. Ed. Waltham, USA: Sina Ebnesajjad. 2006

[10] Kealy, T.; Rheological Analysis of the Degradation of HDPE During Consecutive Processing Steps and for Different Processing Conditions, Journal of Applied Polymer Science 112 (2): (2009) 639-648.

[11] Mittal, V.; Renewable Polymers: Synthesis, Processing and Technology, 1. Ed. Salem, Massachusetts. USA: Scrivener Publishing LLC, 2012.

Evaluation of the recyclability potencial of Poly (Lactic Acid) via multiple extrusions

Maia Frenhe Bianca** [1], **Garcia Gonçalves Livia Maria** [1], **Rocha Rigolin Talita** [1], **Prado Bettini Silvia Helena** [1]*

[1] Federal University of São Carlos - São Paulo - Brazil

Large scale production allied to the advance in polymer usage has created an increasing problem regarding the destination of polymer waste. Polylactide (PLA) is a biopolymer, which, in specific conditions, can be considered biodegradable. The need for specific conditions to biodegradation combined to the expensive price of this resin, lead to mechanical recycling as an economically viable alternative to PLA waste. This work consists in an analysis of the viability of the reuse of PLA through reprocessing, via multiple extrusions. Degradation of a polyester as PLA is highly influenced by the presence of humidity. Therefore, the presence of humidity during reprocessing was also considered in the development of experiments, as well as the influence of temperature and shear rate inherent to the extrusion process. PLA was submitted to five cycles of extrusion, with or without a prior step of drying. The products of each cycle had its molecular weight distribution analyzed via SEC and quantification of end acid groups via titration. It was observed from the results, that the presence of humidity has a more significant effect to degradation, resulting in a more steep reduction of molecular weight between cycles. The results also have shown that controlling parameters as humidity, temperature and shear rate, it is possible to recycle PLA without loss in its properties.

Study of processing conditions on the reuse of p-aramid fibers of industrial waste in PA 6,6/p-aramid composites.

Scarassati Bello Thômas (1), Costa Lidiane Cristina (1), Prado Bettini Silvia Helena (1)*

(1) Universidade Federal de São Carlos - São Paulo - Brazil

The p-aramid fibers, commercially known as Kevlar®, have been studied due to its excellent mechanical properties, low densities, thermal stability, high flexibility, corrosion resistance, and other advantages when in comparison with traditional materials. However, due to the fact that the p-aramid fibers have melting point so high, the conventional recycle methods is uneconomic. In this way, the p-aramid fibers need to be discarded in specific landfills or incinerated. This issue in addition with the current demand for more efficiency in the use of the energetic available resources, the reuse of post-consumption products, and the high mechanical reinforcement potential that p-aramid could provide in polymeric composites have became p-aramid fibers in an interesting reinforcement material for the production of new thermoplastic products. However, the reuse of p-aramid fiber of industrial waste has several difficult issues on the processing due to its low density, high flexibility, and water absorption. Due to lack on the literature about processing of p-aramid fibers and as an alternative option to the recyclability issues, this work evaluated the reuse p-aramid fibers as mechanical reinforcement in polyamide 6,6(90%)/p-aramid(10%) composites through different processing conditions. The shear sensibility of the fibers on the presence or not of different surface treatments was evaluated through the measurements of the average length of the fibers using optical microscopy. The mechanical properties of tensile (ASTM D638) and flexure (ASTM D790) of the composites were also measured.

Polymer hollow fiber membranes for gas separation: A comparison between three commercial resins

Xiao Yuan Chen[ab], Serge Kaliaguine[a], Denis Rodrigue[a*] (presenting author)

[a] *Department of Chemical Engineering, 1065 Avenue de la Médecine, Université Laval, Quebec City, G1V 0A6, Canada*

[b] *Centre National en Électrochimie et en Technologies Environnementales, Collège de Shawinigan, Shawinigan, QC, G9N 6V8, Canada.*

**Denis.Rodrigue@gch.ulaval.ca*

Abstract. Polymer hollow fibre membranes are becoming more and more used for gas separation. In this work, asymmetric hollow fibre membranes were prepared by a phase inversion technique with three commonly used commercial polymers: polyethersulfone (PES), polyetherimide (Ultem® 1000), and polyimides (Matrimid® 5218). The effect of spinning parameters (composition of the dope and bore solution, bore flow rate, air gap distance, temperature of the spinneret and coagulation bath, as well as take-up speed) on the membrane morphology and gas permeation properties was investigated. The membrane separation performances were characterized in terms of their gas transport properties (permeance and selectivity) for different gases (H_2, CO_2, O_2, N_2, and CH_4) to relate this information with their cross-section morphology studied by scanning electron microscopy (SEM). Furthermore, dense flat membranes made from the same materials were also prepared by solvent casting to investigate the relationships between the gas separation performance and membrane configuration (hollow fibers and compact flat membrane). Finally, a comparison between the apparent skin layer thickness obtained from oxygen permeability/permeance and SEM image analyses gave good agreement between both methods.

Keywords: Polymer hollow fiber, flat membrane gas separation, permeance, selectivity skin layer, membrane configuration.
PACS: 81

INTRODUCTION

Membrane technology plays a major role in gas purification because of its high efficiency, low power consumption, easy control, simple maintenance and low investment cost, compared to conventional separation processes such as cryogenic distillation, adsorption and absorption [1]. This technology can be applied in several industrial sectors requiring gas purification units such as [2]: separation of the main component of air (O_2/N_2); separation of hydrogen from a synthesis gas (H_2/CH_4, H_2/CO_2); hydrogen recovery in ammonia plants (H_2/N_2, H_2/CH_4); methane concentration from biogas (CO_2/CH_4); natural gas purification (CO_2/CH_4), and CO_2 capture (CO_2/N_2).

For gas purification at an industrial scale, a large membrane surface area is needed. Currently, three types of configurations are used for industrial applications: hollow fiber module, spiral wound and envelope. Compared with other membrane configurations, hollow fibers provide higher surface area per unit volume (module) leading to higher separation capacity [2]. Actually, hollow fiber membrane modules have values on the order of 10000 m^2/m^3, which is significantly higher than spiral wound (200-1000 m^2/m^3) and envelope type (30-500 m^2/m^3) modules [2]. In addition, hollow fiber modules provide mechanical support, are easier to handle and the process is also simpler [3]. Membrane modules can be used for large or small-scale operations in several industries. However, most of the scientific literature describing the production and characterization of membranes for gas phase separations deals with flat membranes. Since hollow fibers are more appropriate for commercial/industrial implementation, the question as to whether the data obtained for flat membranes can be used to predict the separation performance of hollow fibers is of high interest, when both configurations are made from the same material.

There are several objectives in this work. The first one is to compare polyethersulfone, polyetherimide, and polyimide, three asymmetric hollow fiber membranes working under optimized conditions by measuring their separation performances for different gases (H_2, CO_2, O_2, N_2, and CH_4). These commercial polymers were selected to produce the membranes using the same method. From the samples obtained, the relationships between the gas separation performance of the hollow fibers and the intrinsic gas properties of the dense flat membranes are also discussed. The second objective is to compare the gas transport properties of a single fiber and modules (up to 20 fibers). Finally, an attempt at correlating the available literature on permeability-selectivity data between both membrane configurations is presented and a comparison between the apparent skin layer thickness obtained from oxygen permeability/permeance and SEM image analyses is made.

EXPERIMENTAL

Materials

Ultem® 1000 pellets and Matrimid® 5218 powders were purchased from GE Plastics and Huntsman Advanced Materials Americas Inc., respectively. Polyethersulfone (Ultrason E6020P) was graciously provided by BASF. All the polymers were dried overnight at 120°C before use. Their glass transition temperature and chemical structure are listed in Table 1. N-methyl-2-pyrrolidone (NMP, bp = 204°C, purity >99%), ethanol, methanol and hexanes were purchased from Caledon Laboratories Ltd. Tetrahydrofuran (THF, bp = 66°C) was purchased from Fisher Chemicals. Lithium nitrate ($LiNO_3$) was purchased from Sigma-Aldrich. Polydimethylsiloxane (PDMS, Sylgard 184) was supplied by Dow Chemicals.

TABLE 1. Glass Transition Temperature and Chemical Structure of the Polymers Used.

Materials	Tg (°C)	Chemical structure
Ultem® 1000 (PEI)	217	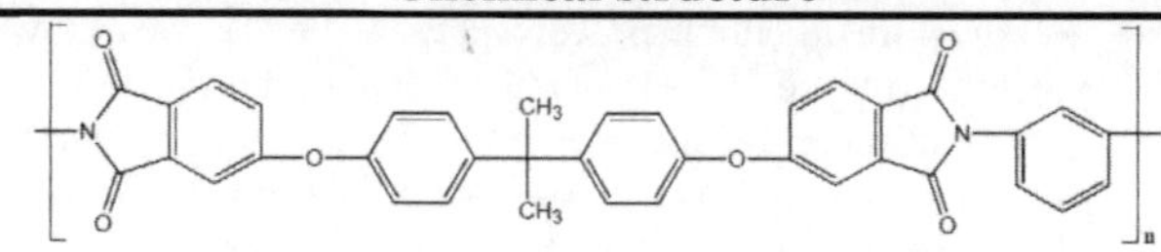
Matrimid® 5218 (PI)	319	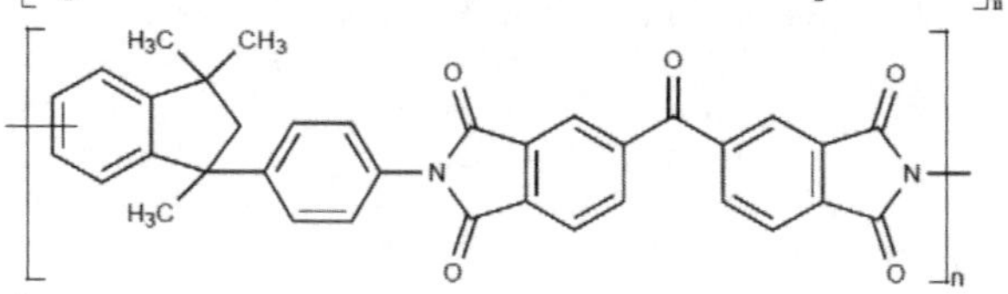
Ultrason E6020P (PES)	225	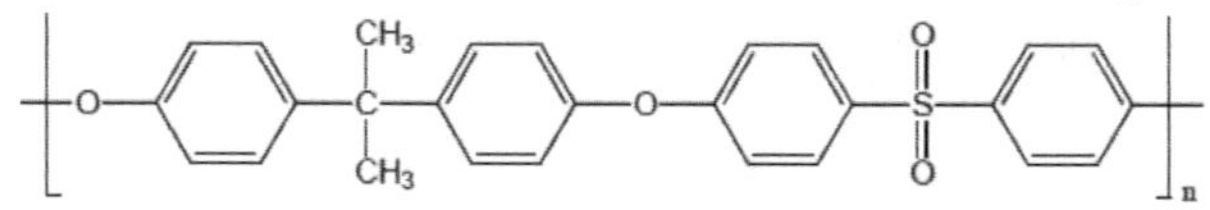

Sample Preparation

Flat membrane preparation: Flat membranes were made by 10-12% wt. solution casting with NMP as the solvent [4-5]. The nascent films were heated in a vacuum oven at 200°C to evaporate the residual solvent for 24 h.

Hollow fiber membrane preparation: Firstly, the three polymer powders were dried in a vacuum oven at 120°C for 3 h to remove any moisture and prepared three polymer dope solution. Then, the asymmetric structure of the hollow fiber membranes was produced by a dry-wet spinning system (phase inversion) using a complete set-up from EMI (European Membrane Institute). The spinning parameters for the polymers are present in Table 2. Finally, a post-treatment procedure was applied. The collected fibers were then soaked in fresh tap water for 72 h (water changed every day) to remove the residual solvent. The wet fibers were then immersed in three methanol baths for 20 min each, followed by immersion in three hexane baths for 20 min each before drying in air at room temperature for two days. Then, the fibers were immersed in a 3% wt. silicone rubber solution (PDMS) for 5 min to seal the defects on the outer surface and dried at 60°C for 4 h under vacuum [6].

TABLE 2. The Spinning Parameters for Ultem® 1000, Matrimid® 5218 and PES (E6020P).

Spinning conditions	Ultem 1000	Matrimid 5218	PES
Dope composition	30% Ultem + 35% NMP + 34% THF + 1% $LiNO_3$	58.8% NMP + 15% ethanol	28% PES +72% NMP
Bore liquid composition	80/20 NMP/water	80/20 NMP/water	80/20 NMP/water
Dope flow rate (ml/min)	3	3	3.75
Bore flow rate (ml/min)	3	3	3.75
Air gap (cm)	10	4.5	3
Spinning temperature (°C)	23	45	23
Take-up speed (m/min)	20.3	10.4	10.4

Characterizations

The weight curves (TGA) and their derivatives (DTG) were recorded using a TA Instruments model Q5000IR from 100 to 850°C at a heating rate of 10°C/min under air.
Tensile properties were measured on an Instron model 5565. From the stress-strain curves, Young's modulus, tensile strength and elongation at break were determined using ASTM D882. The initial grip separation was 100 mm and a deformation rate of 50 mm/min was applied.
Scanning electron microscopy was performed to determine the dimensions of the hollow fibers in terms of skin layer thickness and fiber diameters using a JEOL JSM-840A operated at 15-20 kV. The cross-sections were exposed by cryogenic fracture after immersion in liquid nitrogen.
Membrane porosity and density: The membrane porosity (ε) was calculated as [7]:

$$\varepsilon = \frac{\frac{1}{\rho m} - \frac{1}{\rho p}}{\frac{1}{\rho m}} \quad (1)$$

where ρ_p is the polymer density, while ρ_m is the membrane density determined by the ratio of weight over volume.

The gas transport properties were obtained using a variable pressure (constant volume) method [5-6]. The permeability (1 Barrer=10^{-10} cm^3 (STP) cm / cm^2 s^1 cm Hg) for flat membranes was determined by Equation (2), while the permeance (1 GPU=10^{-6} cm^3 (STP) / cm^2 s^1 cm Hg) for hollow fiber membranes was calculated by Equation (3).

$$P = \frac{22414}{A} \times \frac{V}{RT} \times \frac{l}{\Delta p} \times \frac{d_p}{d_t} \quad (2)$$

$$Q = \frac{22414}{A} \times \frac{V}{RT} \times \frac{1}{\Delta p} \times \frac{d_p}{d_t} \quad (3)$$

where *A* is the membrane area (cm^2); *p* is the pressure (psi); *V* is the downstream volume (cm^3); *R* is the universal gas constant (6236.56 cm^3 cmHg/mol K); *l* is the thickness of a flat membrane; *T* is the absolute temperature (K); and *dp/dt* is the permeation rate (psi/s). Therefore, the skin layer thickness evaluations were made from gas permeation measurements for membranes made of the same material:

$$l = \frac{Permeability}{Permeance} = \frac{P}{Q} \quad (4)$$

RESULTS AND DISCUSSION

Thermal properties

Thermogravimetric analysis (TGA) is providing information about the thermal stability of the hollow fibers compared to their neat state (Figure 1). The TGA results indicate that the weight is almost constant (plateau) up to 440°C which reflects the high thermal stability of these membranes. They also show the absence of residual NMP (solvent with a boiling point of 204.5°C). But in all cases, the $T_{d5\%}$ of the hollow fibers are lower than the neat polymers [6]. This is likely associated to the cross-linked silicone rubber (PDMS) coating on the membranes. The TGA traces reported in Figure 1 show that the PES membrane has the same thermal behavior as the neat PES. However, the TGA curves of Matrimid and Ultem membrane display slightly lower stability than their corresponding neat polymer which can be associated to the different porosity and the absence of additive in the PES membrane preparation. As expected, materials with higher porosity have a larger available surface area leading to better contact with the gases. So better diffusion of the gas molecules leads to lower thermal stability via faster reaction kinetics and less mass transfer resistance. Nevertheless, the results indicate that these three materials could be used at high temperature.

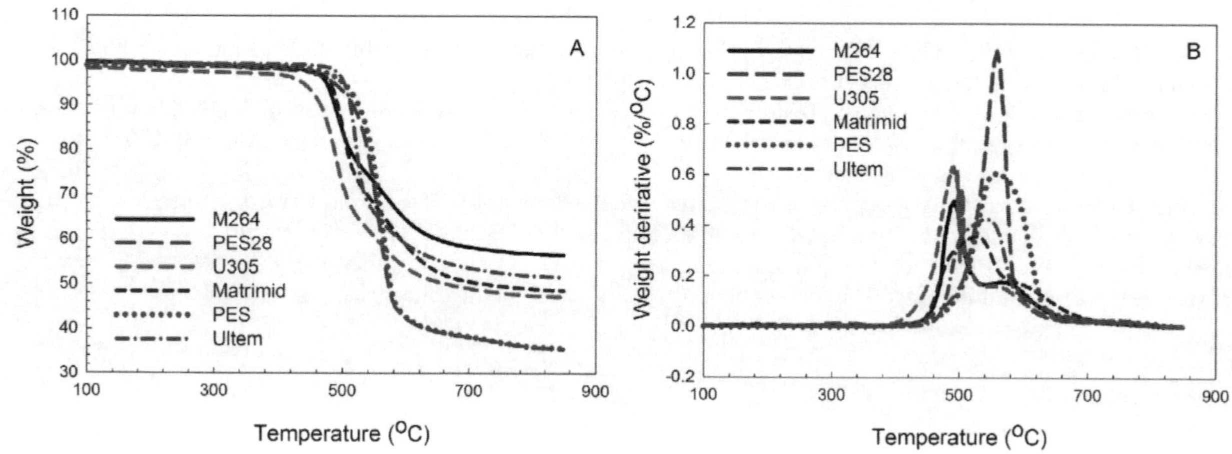

Figure 1: Thermal stability of the neat polymers and their hollow fiber membranes: (A) TGA and (B) DTG.

Morphology

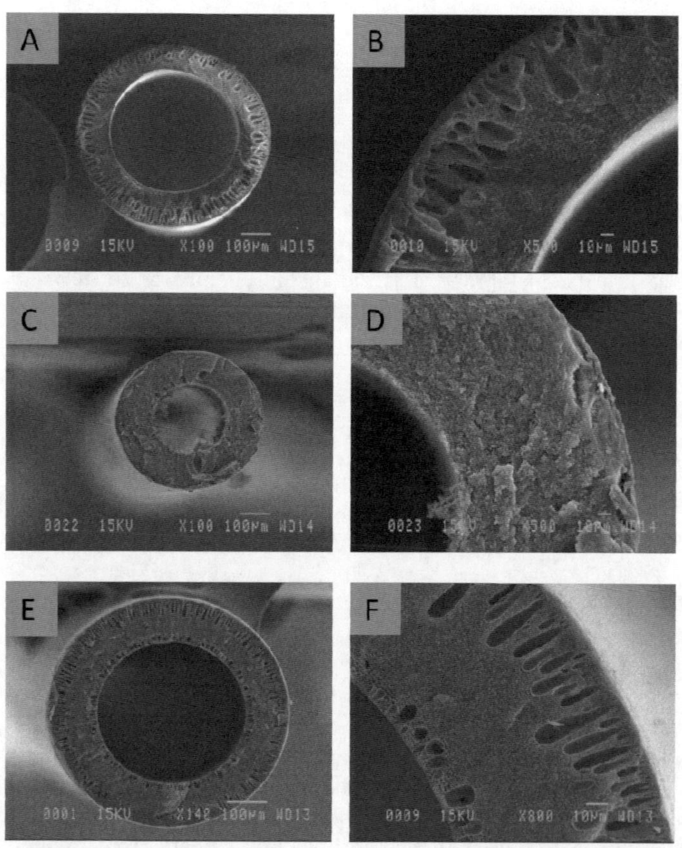

The fiber morphology was examined by SEM and the image are presented in Figure 2. It can be seen that a concentric cylinder morphology was obtained for all three hollow fibers. The good concentricity indicates a relatively uniform morphology, which potentially allows using a high feed pressure across the hollow fiber membrane. The dense skin layer, transition layer and open porous substructure of the hollow fibers are obviously observed in the images. Two long and wide finger-like structure closes to the inner and outer side of the membrane are observed in PES (Figure 2E and F), while only one finger-like structure is formed in Ultem (Figure 2A and B). The morphology of Matrimid is different without finger-like, but mainly a sponge-like structure. Additionally, a nodular structure can be seen in Figure 2C and D. As reported by Kesting [8], large finger-like macrovoids is generally formed when the coagulation process is fast, while a slow coagulation rate results in a porous sponge-like structure. In PES and Ultem, no non-solvent was used while in the case of Matrimid, 15% ethanol as non-solvent was added in the polymer solution. Therefore, the addition of a non-solvent in the polymer solution played an important role in controlling the morphology.

Figure 2. SEM images of U305 (A,B), M264 (C,D) and PES28 (E,F) hollow fiber asymmetric structure. (A,C,E) Overall cross-section of a fiber (100X) and (B,D,F) cross-section at higher magnification (500X).

Porosity is representative of the hollow fibers morphology as higher porosity (lower density) indicates a higher amount of free volume in the membrane. Here, Matrimid has higher porosity, followed by Ultem and PES (Table 3). However, the porosities reported in Table 3 are lower than the values for Ultem and PES reported elsewhere [9]. In the latter, asymmetric membranes made from 15% Ultem or 15% PES in 85% NMP with 0.05% water were produced leading to 79.8% and 83.4% of porosity, respectively.

TABLE 3. Density and Porosity of the three Hollow Fiber Membranes

Materials	Membrane density (g/cm^3)	Polymer density (g/cm^3)	Membranes porosity (%)
Ultem® 1000 (PEI)	0.528	1.27	55.9
Matrimid® 5218 (PI)	0.528	1.20	58.4
Ultrason E6020P (PES)	0.744	1.38	46.1

Gas Transport Properties

Table 4 presents the gas separation performance of the flat membranes measured at 35°C and 10 atm and hollow fiber membrane at 30°C and 2 atm. The O_2 permeability of Ultem, Matrimid and PES are 0.4, 1.32 and 0.44 Barrer respectively, leading to O_2/N_2 selectivity of 7.6, 7.2 and 6.3, respectively. Each measurement was made in triplicate with three different membranes of the same material. Comparison of the gas transport results shows that the highest permeance for all the gases was obtained for PES. However, all the selectivities for H_2/N_2, H_2/CH_4, CO_2/CH_4, CO_2/N_2 and O_2/N_2 are lower than for the other two polymers. Matrimid was the best material for gas separation, especially for CO_2/CH_4 and CO_2/N_2. Ultem is a lower cost material compared to Matrimid, and other commercial polyimides. Ultem also has higher mechanical properties and performs in continuous use up to 170°C and even up to its glass transition temperature (217°C). The selectivity of H_2/N_2 and H_2/CH_4 can reach 117 and 91.7 which are quite high. So, the final material selection must be made depending on the industrial application and which gas separation must be performed.

TABLE 4. Gas Separation Performance of the Flat (35°C and 10 atm) and Hollow Fiber Membranes (30°C and 2 atm) for the Three Polymers studied.

Membranes	Unit	Gases	U-flat	U-hollow fiber (U305)	M-flat	M-hollow fiber (M264)	PES-flat	PES-hollow fiber (PES28)
Permeability/ Permeance	Barrer/ GPU	H_2	6.53	38.5	18.0	56.4	6.80	58.7
		CO_2	1.46	7.53	6.50	17.7	3.22	29.6
		O_2	0.40	2.89	1.30	38.8	0.44	8.36
		CH_4	0.04	0.42	0.19	0.73	0.10	2.66
		N_2	0.05	0.33	0.18	0.52	0.07	1.59
Selectivity	-	H_2/N_2	130	117	94.2	109	67.6	37.1
		H_2/CH_4	163	91.7	98.2	77.3	94.7	22.1
		CO_2/N_2	29.2	22.8	35.0	34.0	46.1	18.6
		CO_2/CH_4	36.5	17.9	34.0	24.3	32.0	11.1
		O_2/N_2	7.60	8.80	7.20	7.46	6.30	5.29

The O_2/N_2 selectivities of the hollow fiber membranes of Matrimid and Ultem are higher than the dense films of known thickness (the intrinsic O_2/N_2 selectivities). This is due to a high shear and high chain orientation in the thin layer induced by the spinning process in the polymer matrix. This may have changed the packing density of polymer chains and free volume resulting in a reduction of gas mobility in the dense selective layer and a change in selectivity compared to the flat membrane. In our case, the take-up speed for Ultem and Matrimid hollow fiber membranes are 20 and 10 m/min (see Table 2). The high shear allows improving the O_2/N_2 selectivities. The average selectivities of PES asymmetric membrane samples were within 84% of the intrinsic O_2/N_2 selectivities of PES. The skin layer thickness was calculated by Eq.(4) and the values from SEM image analyses are listed in Table 5. The values obtained by both methods gave good agreement. Matrimid hollow fiber has the thickest skin layer, and PES has a very thin one. Because for each polymer different spinning conditions were used, it is difficult to fully compare the results. In general, longer air gap leads to thicker skin layer, while higher take-up speed produces thinner skin layer. U305 being prepared with an air gap of 10 cm with the presence of a volatile solvent in the dope has thicker dense skin layer. On the other hand, a higher take-up speed of 20 cm/min was used in this case. Therefore, the final skin layer thickness obtained was only of nm. M264 fiber was made with an air gap of 4 cm and a lower take-up speed of 10 cm/min, so the skin thickness was thicker (340 nm). For the PES28 membrane prepared with an air gap of 4 cm and a take-up speed of 10 cm/min, and no evaporation of a volatile solvent from the dope, produced a skin layer of only 52.5 nm. More details and analysis on these data can be found in our previous study [10].

TABLE 5. Skin Layer Thickness Evaluated by from O_2 Permeability/Permeance and SEM Image Analyses.

Hollow fiber Membrane	U305	M264	PES28
Skin layer thickness from O_2 permeation (nm)	138	340	52.5
Skin layer thickness from SEM images (nm)	150	400	50.0

CONCLUSION

In this work, three different commercial polymers (Ultem, Matrimid and Ultrason) were used to prepare asymmetric hollow fiber membranes. Their permeances were obtained for different gases (H_2, CO_2, O_2, N_2 and CH_4) under different conditions (pressure and temperature). The results showed that no single polymer can be universally used whatever the gas separation to be performed. A careful selection must be made based on the type of gas separation to perform and the operating conditions. In our case, based on selectivity values, Ultem performed best for H_2/N_2 (117), H_2/CH_4 (91.7) and O_2/N_2 (8.8) separation, while Matrimid performed better for CO_2/N_2 (34.0) and CO_2/CH_4 (24.3). Nevertheless, permeance values are very important for industrial applications. In this case, PES would be the best material for O_2 enrichment due to higher O_2 permeance. For CO_2/N_2 and CO_2/CH_4 separation, Matrimid would be the best membrane material due to its higher CO_2 permeance. Finally, different results might be obtained for other conditions and this is currently being investigated.

ACKNOWLEDGMENTS

The authors acknowledge the financial support of the Natural Sciences and Engineering Research Council of Canada (NSERC) and the Mitacs Acceleration program for a postdoctoral scholarship.

References

1. W. J. Koros, G. K. Fleming, Membrane-based gas separation. *J. Membr. Sci.* **83**, 1-80 (1993).
2. R. W. Baker in Advanced membrane technology and applications. N. N. Li, A. G. Fane, W. S. Winston Ho, T. Matsuura. John Wiley & Sons, Inc., New Jersey, 2008; pp. 557-80.
3. E. Drioli, L. Giorno, in Comprehensive membrane science and engineering, ITM-CNR, University of Calabria, Rende (CS), Italy: Elsevier Science 2010.
4. L. Wang, Y. M. Cao, M. Q. Zhou, S. J. Zhou, Q. Yuan, Novel copolyimide membranes for gas separation. *J. Membr. Sci.* **305**, 338-46 (2007)
5. X. Y. Chen, D. Rodrigue, S. Kaliaguine, Diamino-organosilicone APTMDS: A new cross-linking agent for polyimides membranes. *Sep. Pur. Techno.* **86**, 221-33 (2012).
6. X.Y. Chen, D. Rodrigue, S. Kaliaguine, A Comparison between Several Commercial Polymer Hollow Fiber Membranes for Gas Separation, *J. Membr. Sep. Techno.* **6**, 1-15 (2017)
7. G. H. Bakeri, A. F. Ismail, M. Rahimnejad, T. Matsuura, D. Rana, The effect of bore fluid type on the structure and performance of polyetherimide hollow fiber membrane in gas–liquid contacting processes. *Sep. Pur. Techno.* **98**, 262-69 (2012).
8. R. E. Kesting, Synthetic Polymeric Membranes, second ed.,Wiley, New York 1985.
9. G. H. Bakeri, A. F. Ismail, M. Rezaei, D. Arzhandi, T. Matsuura. Porous PES and PEI hollow fiber membranes in a gas–liquid contacting process—A comparative study. *J. Membr. Sci.* **475**, 57-64 (2015).
10. X.Y. Chen, S. Kaliaguine, D. Rodrigue, Correlation between performances of hollow fibers and flat membranes for gas separation, *Sep. Purif. Rev.* http://dx.doi.org/10.1080/15422119.2017.1324490.

High gas barrier properties of nanostructured multilayer thin films

Larocca Nelson [(1)]***, Pessan Luiz*** [(1)*]

[(1)] Federal University of São Carlos - São Paulo - Brazil

High gas barrier properties are usually required from polymer films in applications in food packaging. A strategy largely used to meet the high barrier requirements is to apply a thin layer of metallic film on the polymer substrate; however, this metalized structure does not meet other requirements in packaging such as transparency and adequacy to microwave ovens applications. An alternative technique that can easily produce a very thin coating on polymer films is the layer by layer technique (lbl), where the coating is assembled by alternating the deposition of positively and negatively charged materials on a substrate. If at least one of these materials is comprised of inorganic nanoplatelets, the coating can achieve high gas barrier values since these nanoparticles are virtually impermeable to gases and the lbl technique allow to be assembled in a highly aligned and orderly manner. In this study, the lbl technique was utilized to produce thin layers of multilayers films containing layered double hydroxides (LDH) inorganic nanoplatelets, which are deposited alternated with polyelectrolyte layers onto a polyethylene terephthalate (PET) substrate. PET covered with only 10 layers of LDH nanoparticles present a permeability reduction of 10x and 80x for oxygen and carbon dioxide, respectively, combined with high transparency to visible light. This combination of high gas barrier and transparency can be ascribed to the high spatial alignment of the LDH platelets and the absence of agglomeration, as shown by microscopy analysis.

Atmospheric pressure plasma treatment of carbon fibers for thermoplastic composites

Robeck Andreas [(1)]*, ***Fischer Theresa*** [(1)], ***Martin Judith*** [(1)]

[(1)] TU München - Bayern - Germany

The strength of short fiber reinforced plastic parts depends on the bonding strength between fiber and matrix. However, short fibers are normally delivered with suitable sizing for the used matrix. In other sectors of composite engineering Atmospheric pressure (AP) plasma treatment is already used to increase the strength of the laminate. In this thesis carbon fibers are activated by AP plasma treatment and impregnated in thermoplastic matrix. The aim of this activation is to increase the fiber-matrix-adhesion and thus to extend the application area of fiber-reinforced components produced by injection molding. Short fibers with two different sizes (PA and epoxy) are employed. PA 6 and PA 12 serve as matrix. To examine whether the activation of the fibers was successful, tension rods were produced using injection molding. To verify the improvement of mechanical properties tensile tests, the contact angle measurement and the microscopic analysis of the fracture surface were conducted. Furthermore, the fiber content and the fiber length were determined. The results of the fracture surface analysis show that plasma activated fibers are covered with matrix, whereas the surface of reference fibers is smooth. This implies a higher fiber-matrix-adhesion for the plasma activated fibers. The measurements of the contact angles yield a degradation of the contact angle on the plasma treated samples. This is due to the insertion of functional groups on the fiber surface which can generate chemical bonds to the functional groups of the matrix. These additional polar groups yield to a lower contact angle and to a higher fiber-matrix-adhesion. The analysis of the surface fracture and the contact angle measurements indicate the incorporation of functional groups and thereby the improvement of the interphase. The results can also be confirmed by the tensile test to some extent. The tensile tests of short fiber samples show an increase of the tensile strength.

Development of PAN/silazane Hybrid Polymer for Processing of Carbon-based Fibers with Remarkable Oxidation Resistance

Ribeiro Luiz Fernando Belchior (1), Marangoni Cintia (1), Motz Günter (2), MACHADO RICARDO ANTONIO FRANCISCO (1)*

(1) Universidade Federal de Santa Catarina - Santa Catarina - Brazil, (2) Universität Bayreuth - Bayern - Germany

Carbon fibers (CFs) are currently the most important reinforcement component used for high performance composites, representing a key material for the modern economy. The biggest drawback of CFs is their insufficient oxidation stability at temperatures >400°C. In this work, we proposed the modification of polyacrylonitrile (PAN), the most widely used carbon precursor, with a commercially available oligosilazane (ML33), as precursor for SiCN ceramics, which are well known to provide a remarkable resistance to oxidation (>1300°C). To improve the interaction of both educts (acrylonitrile and oligosilazanes) and to avoid phase separation, the in-situ free-radical polymerization of acrylonitrile in the presence of the oligosilazane was performed in order to obtain a hybrid precursor instead of a simple mixture. Thermogravimetric analysis demonstrated that depending on the AN/ML33 ratio the weight loss during pyrolysis is drastically reduced up to 63% due to crosslinking reactions between both components, confirmed by NMR spectroscopy investigations. Unexpectedly, with increasing AN content in the starting composition the thermal treatment up to 1000°C leads to a change from amorphous C/SixCyNz to C/Si3N4 nanocomposites confirmed by NMR, XRD and TEM measurements. Despite the high carbon content, the amorphous C/Si3N4 material possesses a remarkable oxidation resistance, due to the well distributed phases. The excellent oxidation resistance of the Si3N4 phase protects the material very efficiently from decomposition up to 800°C, which is 300 K more than that for pure carbon. The homogeneously distributed ceramic phase within the carbon matrix is responsible of the extraordinary intrinsic oxidation stability of this new type of carbon-based fibre.

Polymer-derived Mesoporous Silicon Carbide Membranes

Acosta Emanoelle Diz[a*], Mallmann Maíra Debarba[a], Bernard Samuel[b], Machado Ricardo Antonio Francisco[a]

[a]*Universidade Federal de Santa Catarina - Santa Catarina – Brazil*
[b]*Institut Europeen des Membranes - Universite Montpellier 2 -Montpellier – France*

Abstract. The synthesis of ceramic materials, originated from preceramic polymers, with multifunctional characteristics and potential to be applied in several areas (such as the aeronautic, automotive, chemical, electrical and food industries), becomes an object of great interest. This methodology enables the development of ceramics with particular compositions, textures, structures and shapes which are difficult to obtain by conventional powder-based ceramic processes. In this work it is presented some results and an overview of the processes and materials required for the production of inorganic layers to produce membranes, the synthesis methods, as well as other relevant concepts of the techniques used to obtain ordered structures. Here, the idea is to combine the properties of silicon carbide based ceramics with gas separation applications through the Polymer Derived Ceramics (PDCs) route combined with a suitable shaping process.

Keywords: Silicon Carbide, Block Copolymers, Polymer Derived Ceramic, Inorganic Membranes, Ceramics, Mesoporous layers, Inorganic layers.

PACS: 81.05.Je, 81.05.Qk, 81.05.Rm, 82.35.-x

INTRODUCTION

Polymer derived ceramic materials have garnered a lot of interest due to its great potential for the production of high performance materials with controlled composition, shape and porosity. Controlled porosity, in specific, is hard attribute to be achieved in PDCs. Some organic polymers may be used as a sacrificial material, allowing for uniformly distributed pores on the micro (< 2 nm), meso (2-50 nm) and macro (> 50 nm) scales, obtaining then a product with optimal characteristics for the production task-specific ceramics, such as: separation membranes, biomaterials, catalytic materials, lightweight structural materials, among others.

Under this context, an interesting strategy is to use the block copolymers (BCPs), applying styrene and butadiene (SBS) as the sacrificial materials. BCPs become appealing due to their capacity to "self-assemble" in very thermodynamically stable domains, controlling the shape and size of the domain and resulting a material with controlled pore size and porosity in the nanometric scale. The use of block copolymers as structure direct agents in porous non-oxide ceramics allows for the construction of high temperature nanostructure materials through easier methods, with less steps when compared, for example, with infiltration processes via templating (nanocasting). Besides SBS as the organic portion of the nano/mesostructure porogenic agent, Allylhydridopolycarbosilane (AHPCS) was used in this work as Silicon Carbide ceramic precursor. To design structured porous ceramic, an easy to mould ceramic precursor is necessary, AHPCS is a commercially available copolymer with higher ceramic yield leading to a SiC with free carbon[1] which results in a lower volume shrinkage during the polymer-to-ceramic conversion. SiC provenient ceramics possess unique characteristics, such as outstanding high temperature stability, better oxidation and corrosion resistance, high thermal conductivity and chemical and mechanical resistance[2].

To potentialize the ceramic precursor's processability, as well as optimize the final properties of the ceramic, Boron was incorporated to the AHPCS precursor at the molecular scales, using the synthesis path with borane dimethylsulfide (BDMS). Some authors investigated silane hydroboration, more recently, the group of Samuel Bernard studied the evolution of the incorporation of Boron with a high increase in the crosslinking degree of the AHPCS backbone. Moreover, the same authors reported the increasing of the ceramic yield using Boron modified AHPCS when compared with standard AHPCS, a consequence of Boron preventing the extraction of small polymeric fragments at low temperatures.

This previously described methodology is based on polymer derived ceramic (PDC) route used to prepare porous ceramics and has several advantages: the possibility to change the material shaping as films/layers, powders, foams to use in different applications; the porosity of the materials can be easily adjusted by changing the volume ratio between the organic polymer and the hydroborated ceramic precursor; pore size control according to the calcination temperature. In this work we show initial, albeit representative, results on the formation of a typical lamellar structure from SBS following chemical modification of the standard AHPCS using BDMS as a Boron source. Finally we demonstrate the shaping of the assembled material.

SBS Block Copolymer as Structured Directing Agents

Block Copolymers are defined as polymers containing two or more chains (homopolymers) with differents molecular composition linked at one end. The two different domains of block copolymers are covalently[3] linked in general with a nanometric size scale. Several morphologies have already been studied. This is dependent on the molecular mass of the constituent polymeric species and their composition, as well as the solvent type and the thermotropic/lyotropic processing characteristics.

EXPERIMENTAL PART

General Comments

All syntheses were carried out in a purified argon atmosphere by means of standard Schlenk manipulations and vacuum/argon-line techniques. AHPCS (SMP-10) structural formula $(SiH_2CH_2)_{0,9}(Si(allyl)HCH_2)_{0,1}$ was acquired from Starfire Systems®. Borane dimethylsulfide $BH_3{\cdot}SMe_2$ (2 M solution in Toluene) was obtained from Sigma–Aldrich and used without further purification. Toluene and Tetrahydrofuran (THF) (99.85 %, Extra Dry over Molecular Sieve, AcroSeal®) was obtained from Acros Organics. Block copolymers based on styrene and butadiene were donated by Kraton Polymers and used after dried in oven at 30 °C for 48 h. Manipulation of the chemical products was made inside an argon-filled glove box (Jacomex BS521) dried with phosphorus pentoxide.

Synthesis of Borane-modified Polycarbosilane Precursor (SMPB)

The Schlenk flask was dried at 120 °C overnight before pumping under vacuum and filling with Argon for synthesis. The flask equipped with a Teflon stirrbar and high-vacuum stopcock was charged with 5 mL of AHPCS corresponding to 5 %mol allyl groups and 50 mL of Toluene were introduced into a dropping funnel connected with the Schlenk. 1.7 mL of $BH_3{\cdot}SMe_2$ (2 M solution in Toluene) were added via syringe. The Schlenk was cooled down to 0 °C to allow slowly dropping the $BH_3{\cdot}SMe_2$ into the solution. Some gas evolution was observed. The reaction mixture was stirred under the same conditions for 3 h, allowed to reach room temperature. The synthesis was stopped after 72h of reaction and the Boron-modified polycarbosilane was kept in a flask under argon in the fridge.

SBS and SMPB@SBS films/layers

To produce the films of the block copolymers they were separately dissolved in a solvent. Toluene and THF were used to compare the different polymer morphology during work-up and isolation using these solvents. The solvent and the SBS were mixed together to form a precipitation-free solution with 5 wt% total polymer content, for an additional analysis was also used 10 %wt. After stirring overnight, the solution was cast in a Teflon® mold, and allowed to evaporate for 24 h at room temperature. After then, the polymer film was peeled from the Teflon® mold. This operation produces the so-called "as-assembled" materials that later were analyzed by AFM.

To finalize the analysis of behavior of structure using thermal treatment and swelling agent as the Silane compound, solutions were done using the block copolymers mixed with the borane-modified polycarbosilane precursor (SMPB) in solution. Different SBS concentrations were used keeping constant the quantity of SMPB as 0.2 g. In this paper we will show the results for the best sample using 43 %wt of block copolymer. Then the films were casted into the Teflon molds and kept for 24 h drying in glovebox. These films were directly pyrolysed at a rate of 60 ◦C.h^{-1} to 800 ◦C. The dwell time was 2 h. In order to examine the micro/nanostructures SEM analysis were performed in all of the ratios. The thickness of the prepared specimen was about 0.5 mm. All the operations cited here were done in a glove box with moisture and oxygen control.

Analysis Method

For microscopic structural evaluations, the structures of samples were observed with scanning electron microscope (SEM). The chemical structure of polymers was determined by transmission FT-IR spectroscopy using a

Nicolet Magna 550 Fourier transform-infrared spectrometer. The liquid samples were deposited between KBr windows whereas the solid samples were mixed with KBr (dried at 120 °C in air) and pressed into pellets. The structure of SBS films was determined by Atomic Force Microscopy using a Nanoscope III, quadrex controller.

Results

The polymeric structures may be visualized through specific techniques, such as Atomic Force Microscopy (AFM) or scanning electron Microscopy (SEM). In Figure 1 are shown the structures for 4 different types of SBS for which the molecular weight (obtained by Gel Permeation Chromatography-GPC and from the Manufacturer), as well as the polydispersivity index, are indicated in Table 1.

TABLE a). Results for the diferents Block Copolymers

SBS	Mw	Polydispersivity (Mw/Mn)	Mw (Manufacter)
A	116618	1,12778	125000
B	115408	1,14145	120000
C	158299	1,1469	170000
D	14225	1,1048	15000

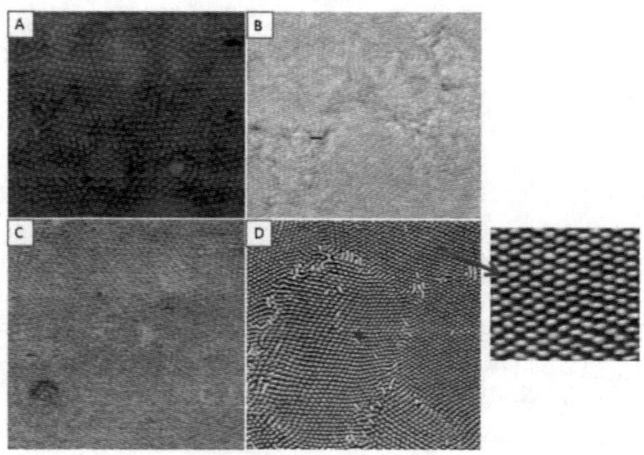

FIGURE 1. Ordered microstructure of polymer blocks A, B, C, D dissolved in solvent Toluene (5 wt% of SBS) and a zoom in the structure of SBS D.

It may be observed in Figure 1 the typical morphology of copolymer in styrene and butadiene block after evaporation of Toluene solvent at room temperature. It is clear that the structures consist of PS cylinders in an hexagonal arrangement in a continuous PB matrix, just as studied by Kim and Libera[4] when using the as-cast microstructure of SBS triblock copolymer (30 wt % PS, *M*w 112 000). The same authors obtained similar results for annealed microstructure: a mixed morphology with regions of vertical cylinders and regions in the process of transforming to the more stable in-plane cylinder morphology. This was also visualized in this work after a heat treatment at 120 °C for 4 hours. The morphology is also dependent upon the solvents in which the block polymer is prepared because of the effect of the solvent on the polymer morphology during work-up and isolation from solutions in different solvents. The two images from Force Atomic Macroscopy in Figure 2 for the same SBS, but one was prepared in Toluene and the other was prepared in THF. The difference in morphology results from Toluene being a good solvent for the polystyrene terminal blocks, while THF is a good solvent for the central block and a rather poor solvent for the terminal blocks.

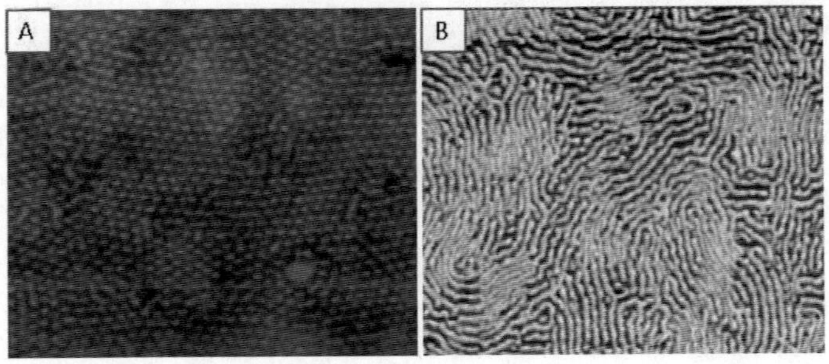

FIGURE 2. Differents morphologies using the same SBS but in differents solvents, A shows the SBS prepared in Toluene like Figure 1 and B the same SBS prepared in THF.

The ordered nano-scale structure created in the block copolymer/precursor hybrids can be preserved through the pyrolysis process, thereby leading to ordered nanostructures in the final ceramic. The versatility in producing a multitude of phases with varied length scale makes block copolymers ideal scaffolds for forming inorganic nanostructures, Figure 3 shows the structure produced after pyrolysis with a mesoporous layer from SMPB@SBS coated on a macroporous layer (Macroporous support from the work of C. Durif [5]).

FIGURE 3. The zoom of an open cel coated mesoporous layer from SMPB@SBS on a macroporous layer (Macroporous support from the work of C. Durif et al. [5])

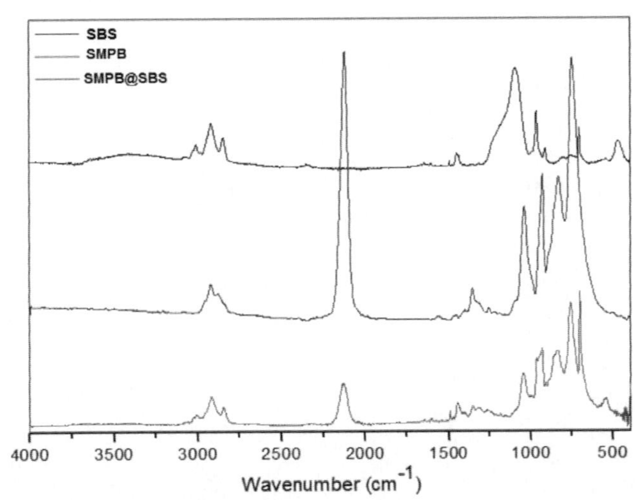

FIGURE 4. FTIR spectra of SMPB, SMPB@SBS and SBS.

All prepared samples (before pyrolisys) were shown to exhibit the characteristic bands related to AHPCS and to the BCP bands. Thus, the band at ca. ~2121cm-1 in the FTIR spectra was assigned to Si-H groups, whereas the bands at 1631 and 3070 cm-1 were assigned to the allyl-groups their absence indicate the full. the FT-IR spectrum of AHPCS shows successive peaks in the wavenumber range from 770 to 1049 cm^{-1} attributed to the stretching and deformation vibrations involving $SiCH_3$ groups (770 cm^{-1}), Si-C bonds (854.3 cm^{-1}), Si-H bonds (941.1 cm^{-1}) and Si-CH_2-Si units (1049 cm^{-1}).

CONCLUSIONS

The preliminary results of this work demonstrate that pre-ceramic polymers using in-block copolymer matrixes allow for the obtention of hybrid films of controlled porosity which are able to be converted into high performance ceramic materials.

AKNOWLEDEGMENTS

The authors would like to thank the National Council for Scientific and Technological Development (CNPq) and the National Council for the Improvement of Higher Education (CAPES) for the financial support, and the *Institut Europeen des Membranes.*

REFERENCES

1. Starfire Systems, Inc., 877 25th St., Watervliet, NY, 12189
2. Yu Yuxi et al. *Synthesis of nanostructured silicon carbide at ultralow temperature using self-assembled polymer micelles as a precursor*. JF- Journal of Materials Chemistry, 2011.
3. Gu X et al. Pattern transfer using block copolymers. Phil Trans R Soc, 2011.
4. G. Kim and and M. Libera. *Kinetic Constraints on the Development of Surface Microstructure in SBS Thin Films. Macromolecules* 1998 *31* (8), 2670-2672.
5. C. Durif. B. R. "Elaboration de membranes non-oxydes de type SiBC pour la maîtrise de la contamination dans les éacteurs à eau pressurisée", Ph.D. Thesis. Montpellier University, 2017.

Synthesis and Characterization of PVDF Nanofibers

AZZAZ Mohamed Cherif (a), DEMARQUETTE Nicole (a), ZEDNIK Ricardo (a)*

(a) École de Technologie Supérieure, Université du Québec,
1100 rue Notre-Dame Ouest, Montréal, Québec, Canada

Abstract. Piezoelectric materials have important applications as sensors and actuators, as they can reversibly convert between mechanical deformation and electric charge. PVDF (polyvinylidene fluoride) is a semi-crystalline polymer that presents piezoelectric characteristics. It also presents good flexibility, thermal properties, biocompatibility, and chemical inertness, that turn this material a promising candidate for microelectronic mechanical systems (MEMS) and biomedical applications. In this work, uniform PVDF fibers with tunable diameters ranging from 100nm to 1 μm were obtained by electrospinning, which is a processing method that enables the fabrication of low-cost nanofibers from polymeric solutions. The effect of processing conditions and material characteristics on the morphology and properties of PVDF fibers was studied thoroughly. In particular, we found that feed rate, applied electric field, solution viscosity, and solvent type, can be used to control the fiber diameter and crystalline structure of the polymer and hence its piezoelectric capacity.

Keywords: PVDF, electrospinning, nanofibers, polar phase, piezoelectricity.

PACS: 61.Materials science , 77.Dielectrics, piezoelectrics, and ferroelectrics and their properties, 81. Materials science

INTRODUCTION

Piezoelectric polymers are able to convert between mechanical deformation and electric charge, and are therefore used in sensors and actuators. PVDF is a widely used piezoelectric polymer because of its high flexibility, good thermal properties, low cost, and relatively high piezoelectric coefficients compared to other polymers [1, 2] . It commonly exists in four different crystalline structures: α, β, γ and δ, of which the polar β phase exhibits the largest piezoelectric effect [3]. Unfortunately the piezoelectric properties of PVDF are less attractive than that of many piezoelectric ceramics [4, 5].

Fortuitously, the piezoelectric properties of PVDF can be enhanced during its processing. In particular, electrospinning, which is a technique that enables the production of polymeric nanofibers has been shown to improve the functional properties of the polymer [4-7]: electrospinning of PVDF favors the formation of the β phase [6]. However, significant uncertainty exists as to the underlying mechanistic reasons of the enhanced piezoelectric character of the electrospan fibers: He and al. attribute the formation of the β phase to the choice of solvent used to prepare the polymeric solution [7]. Baji and al. argue that the addition of the solvent tetramethylammonium chloride (TMAC) increases the electrical conductivity of the solution and makes it possible to have fibers with uniform diameter. They also conject that this solvent promotes the formation of the polar β phase [8]. Zheng et al. studied the influence of different electrospinning parameters on the structure of PVDF nanofibers, and found that solvents with high evaporation rates can also improve the formation of the polar β phase [9]. Other authors argue that the extension force created by the electrical field between the needle and collector is the main cause for the formation of the β phase. Further, Yee et al. suggest that the use of a rotatory collector imposes a secondary mechanical force that increases the fiber alignment [10].

In our study, we report the effect of processing parameters on the crystalline structure and appearance of the β phase in PVDF nanofibers produced by electrospinning. We discuss the effect of electric field and solution viscosity on the morphology, diameter, and crystalline structure of PVDF nanofibers.

Experimental Methods

Polyvinylidene difluoride (PVDF) powder with the trade name Kynar 741 were supplied by Arkema and were used as received. A mixture solvent was prepared from 95% aq. N,N-Dimethylformamide (DMF) and acetone with a weight ratio 7:3. The concentration of polymer (PVDF powder) was chosen as 16% and 22%. During electrospinning, an electric field of 1.25×10^5 V/m was applied to the needle and a solution flow rate of 5μl/min was used. All tests were performed at room temperature.

Results and Discussion

Figure 1 shows representative electrospun PVDF fibers of different diameters.

Figure 1. SEM images of 750 nm and 150 nm diameter PVDF nanofibers.

The fiber diameters were calculated by image analysis. We find that decreasing the polymer concentration in the solution decreases the average diameter of fibers. This decrease is consistent with the large amount of solvent evaporated during electrospinning [11].

Crystalline phase identification was performed using X-ray diffraction. Figure 2 shows representative diffractograms of PVDF nanofiber samples and powder.

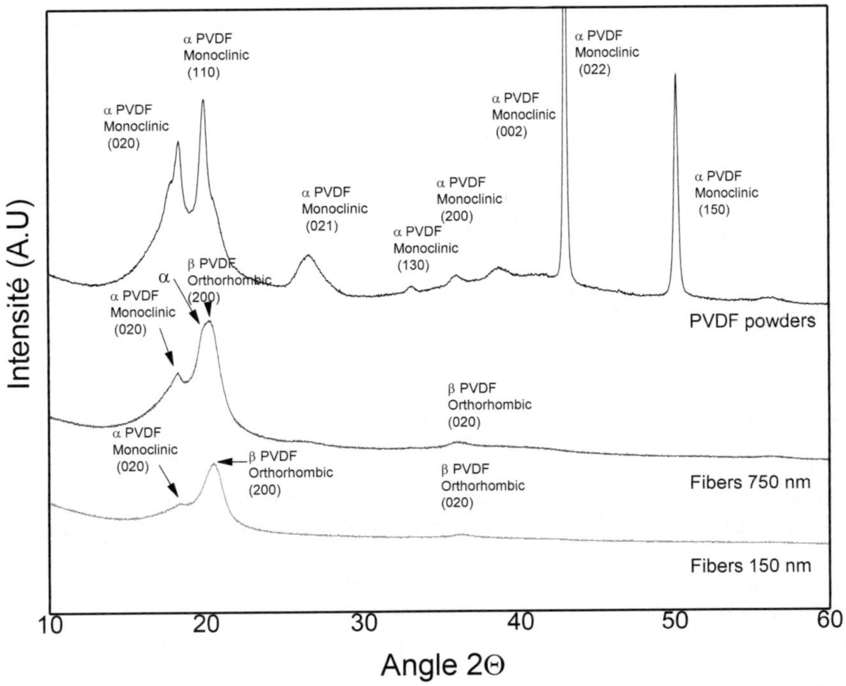

Figure 2. Representative X-ray diffraction patterns of powder and different nanofibers: 750 nm and 150 nm.

The initial raw material used is semi-crystalline and the non-polar α phase is predominant. After electrospinning, we notice the appearance of a new peak located at the 20.6° position, corresponding to the polar β phase [8]. We find that the relative volume fraction of the β phase increases with decreasing fiber diameter.

Thermal analysis was carried out by DSC to study the thermal properties of PVDF powder and electrospun fibers. In our analysis, each sample was heated to 240°C, followed by cooling to room temperature. Enthalpy of fusion and melting points are shown in Table 1. The data in table 1 shows an increase in the crystallinity of electrospun PVDF fibers with decreasing fiber diameter.

TABLE 1. Calorimetric measurements for PVDF powder and nanofibers.

Samples	Melting Point (°C)	Enthalpy of Fusion (J/g)	Crystallinity (%)
PVDF virgin powder	165	41	40
PVDF Fibers (750 nm)	165	51	50
PVDF Fibers (150 nm)	170	67	65

The increase in crystallinity is likely due to the molecular alignment encouraged by the electric field and mechanical stress imposed during electrospinning.

Conclusion

PVDF nanofibers were prepared using electrospinning, and characterized by SEM, XRD and DSC. The effect of electrospinning parameters (viscosity and electrical field) was studied by analyzing the fibers morphology, crystalline structure, and thermal properties. XRD results showed that the crystalline structure of the fibers was dominated by the polar β phase, without the need for any annealing or electric poling treatment: we demonstrate that electrospinning is a promising technique to produce piezoelectric PVDF nanofibers.

REFERENCES

[1] E. S. Cozza, O. Monticelli, E. Marsano, and P. Cebe, "On the electrospinning of PVDF: influence of the experimental conditions on the nanofiber properties," *Polymer International,* vol. 62, pp. 41-48, 2013.

[2] C.-T. Pan, C.-K. Yen, S.-Y. Wang, Y.-C. Lai, L. Lin, J. C. Huang, *et al.*, "Near-field electrospinning enhances the energy harvesting of hollow PVDF piezoelectric fibers," *RSC Adv.,* vol. 5, pp. 85073-85081, 2015.

[3] W. A. Y. Shu Huang, Wuiwui Chauhari Tjiu, Ye Liu,Masaya Kotaki,| and Y. C. F. B. J. M. T. L. a. X. Lu, "Electrospinning of Polyvinylidene Difluoride with Carbon Nanotubes:
Synergistic Effects of Extensional Force and Interfacial Interaction on
Crystalline Structures," *Langmuir,* vol. 24, pp. 13621-13626, 2008.

[4] A. Gheibi, M. Latifi, A. A. Merati, and R. Bagherzadeh, "Piezoelectric electrospun nanofibrous materials for self-powering wearable electronic textiles applications," *Journal of Polymer Research,* vol. 21, 2014.

[5] A. Sarvi, V. Chimello, A. B. Silva, R. E. S. Bretas, and U. Sundararaj, "Coaxial electrospun nanofibers of poly(vinylidene fluoride)/polyaniline filled with multi-walled carbon nanotubes," *Polymer Composites,* pp. n/a-n/a, 2013.

[6] J.-M. Park, G.-Y. Gu, Z.-J. Wang, D.-J. Kwon, P.-S. Shin, J.-Y. Choi, *et al.*, "Mechanical and electrical properties of electrospun CNT/PVDF nanofiber for micro-actuator applications," *Advanced Composite Materials,* vol. 25, pp. 305-316, 2015.

[7] L. He, J. Sun, X. Zheng, Q. Xu, and R. Song, "Effect of multiwalled carbon nanotubes on crystallization behavior of poly(vinylidene fluoride) in different solvents," *Journal of Applied Polymer Science,* vol. 119, pp. 1905-1913, 2011.

[8] A. Baji, Y. W. Mai, Q. Li, and Y. Liu, "Electrospinning induced ferroelectricity in poly(vinylidene fluoride) fibers," *Nanoscale,* vol. 3, pp. 3068-71, Aug 2011.

[9] J. Zheng, A. He, J. Li, and C. C. Han, "Polymorphism Control of Poly(vinylidene fluoride) through Electrospinning," *Macromolecular Rapid Communications,* vol. 28, pp. 2159-2162, 2007.

[10] W. A. Yee, J. Kong, C. Zhang, T. Liu, M. Kotaki, and X. Lu, "Polymorphism of electrospun polyvinylidene difluoride/carbon nanotube (CNT) nanocomposites: Synergistic effects of CNT surface chemistry, extensional force and supercritical carbon dioxide treatment," *Polymer,* vol. 53, pp. 5097-5102, 2012.

[11] K. F. Seeram Ramakrishna and W.-E. T. T.-C. L. Z. Ma, *An Introduction to Electrospinning and Nanofibers*, World Scientific ed., 2005.

Polyolefin Pipes With Crosslinked Barrier Layer Based On Water Soluble Polymers

G. Riess[a*], J. Schauberger[a], K. Berger[a], R. Kramer[a], H. Mattausch[b], C. Keimel[a], B. Haar[b]

[a] *Montanuniversitaet Leoben, Department of Polymer Engineering and Science, Chair of Chemistry of Polymeric Material, Leoben, Austria*
[b] *Montanuniversitaet Leoben, Department of Polymer Engineering and Science, Chair of Polymer Processing, Leoben, Austria*

corresponding and presenting author: riess@unileoben.ac.at

Abstract. This paper deals with the modification of silicate fillers (i.e. nanoscale silica particles) and with the properties of corresponding polymer composites. These composites are based on water-soluble polymers such as polyvinyl alcohol (PVA). These PVA composites can be cross-linked by UV light because the modified filler has a layer of photo initiator on the surface. The crosslinking effectivity is tested by sol gel analysis. Some applications of such polymer composites as barrier layer for polyethylene pipes are shown in the reference section.

Keywords: Barrier coatings, Polyvinylalcohol, Modified filler, Crosslinking
PACS: 82.35.Np

INTRODUCTION

High density polyethylene (HDPE) has been the preferred material for lining applications for crude-oil onshore and inland transport at higher temperature and pressure. The material shows a high chemical resistance to various metal-aggressive substances in the oil [1]. But one special aspect for materials in onshore pipe oilfield applications is the requirement of a high permeation resistance against low molecular weight hydrocarbons [2]. However, polyethylene material shows high permeation rate for hydrocarbons [3, 4]. It is also possible to use other high polar materials such as polyvinyl alcohol (PVA) as a coating to prevent permeation of hydrocarbon [5].

For the use of PVA as barrier material, it is also necessary to crosslink the PVA. Otherwise PVA can be solved by crude oil, which contains water. It is known that be PVA can be crosslink by UV-light [6], but in most case low molecular Initiator are used. These initiators can be leached by water. Aromatic diazosulfonates, which are described in the literature as photoreactive materials, are used as photoinitiator [7]. It is already known from the literature to incorporate such diazosulfonate groups in a polymer in order to obtain a chemical bond between a polymer and the surface of a filler [8, 9]. Our research deals with the attach of azosulfonate to filler surface and to use these as photo initiator for crosslinking of PVA.

EXPERIMENTAL

Synthesis of Particles with Azosulfonate Group

All chemicals are supplied form sigma-aldrich and wacker and used without purification. The details of synthesis and characterization of modified fillers are described in detail by Jörg Schauberger [10].

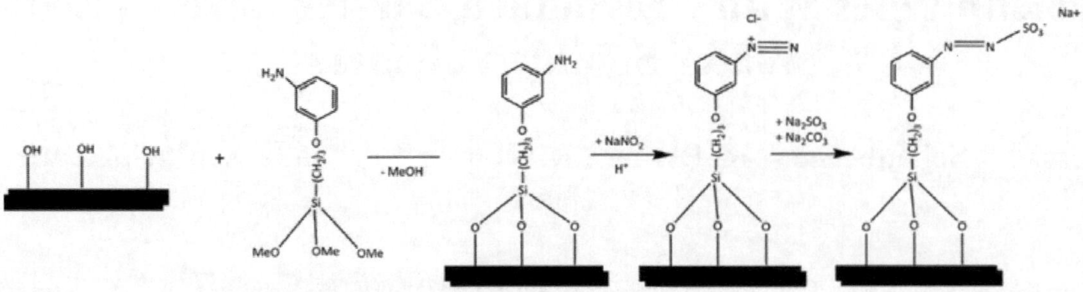

FIGURE 1. Modification of silica filler with azosulfonate gropus [10]

With this method, different fillers are modified. P1 to P3 are based on a fumed silica from Wacker (Wacker HDK® N20). NP are Silicon dioxide nanoparticles (10 – 20 nm) from Sigma Aldrich.

TABLE 1. Modified silica filler

notation	particle type	particle/silane (weight ratio)
P1	fumed silica	1:1
P2	fumed silica	1:2
P3	fumed silica	1:4
NP	SiO2 nanoparticles	1:1

Thin Solid Films with PVA and Particles with Azosulfonate Groups

Thin solid films were obtained by the gravity settling method. A PVA solution (5 wt.% in water) together with the dispersed particles was cast onto polystyrene petri dishes for sol gel analysis and irradiation experiments and onto a CaF_2 platelets for spectroscopic measurement. Both are dried at room temperature with a constant air flow. The absorption spectra were recorded at wavelengths ranging from 800 to 200 nm applying a medium scanning speed. The samples (thin films and coated CaF_2 platelets) were exposed to UV light. An iron doped medium pressure mercury vapour lamp was used. Thin PVA films were cut into 1 x 1 cm^2 pieces, weighed to obtain the initial sample weight and immersed in deionised water at 23 °C for 48 h to obtain equilibrium swelling. The swelling ratio was calculated according to equation [11]. The produced PVA composite films are listed in Table 2:

TABLE 2. PVA - modified particle composite

notation	particle	PVA (wt.-%)	particle (wt.-%)
P1.1	P1	97.5	2.5
P1.2	P1	95	5
P2.1	P2	97.5	2.5
P2.2	P2	95	5
P2.1	P3	97.5	2.5
P3.2	P3	95	5
NP.1	NP	97.5	2.5
NP.2	NP	95	5

RESULT AND DISCUSSION

The azosulfonate group as coating was used because this functional group is photoreactive and decomposes under irradiation according to the two different reaction mechanisms (see fig. 3). The first decomposition mechanism occurs in the presence of water. The second one is a radical mechanism; this photodecomposition mechanism takes place in the presence of alcohol.

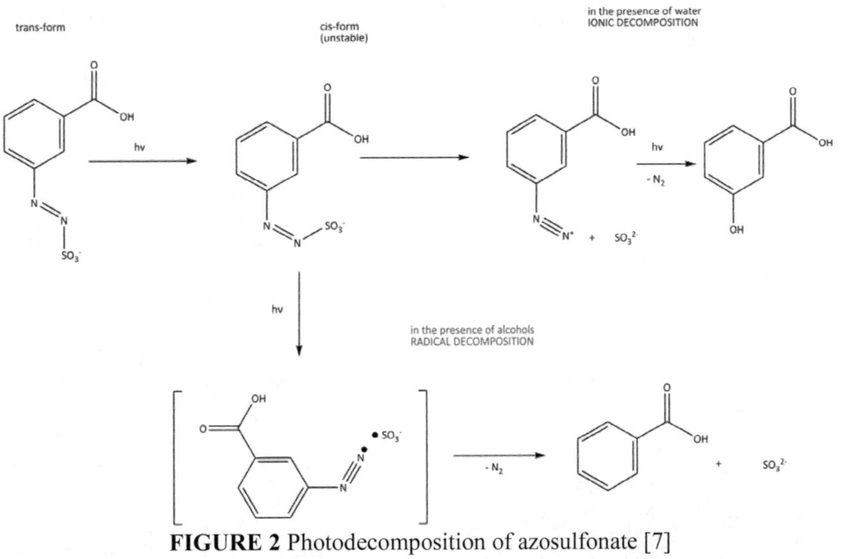

FIGURE 2 Photodecomposition of azosulfonate [7]

In an attempt to investigate UV induced decomposition of the aryl azosulphonate groups, UV-Vis absorption spectra of aqueous PVA-particle dispersions are recorded. Composite dispersions containing fumed silica based modified particles exhibit strong UV response, as a result of photoinduced decomposition of the azo groups. Additional photolysis kinetics for dispersions containing 5 wt.-% of photoactive particles are collected in the figure 3. Due to changes of the electron density of the aromatic system, the absorption peak at 290 nm shifts to lower values, which is a clue that phenol groups are formed [10]. The obtained PVA-particle dispersions exhibit two isobestic points at wavelengths of 251 nm and 263 nm due to the photodecomposition of the chromophore.

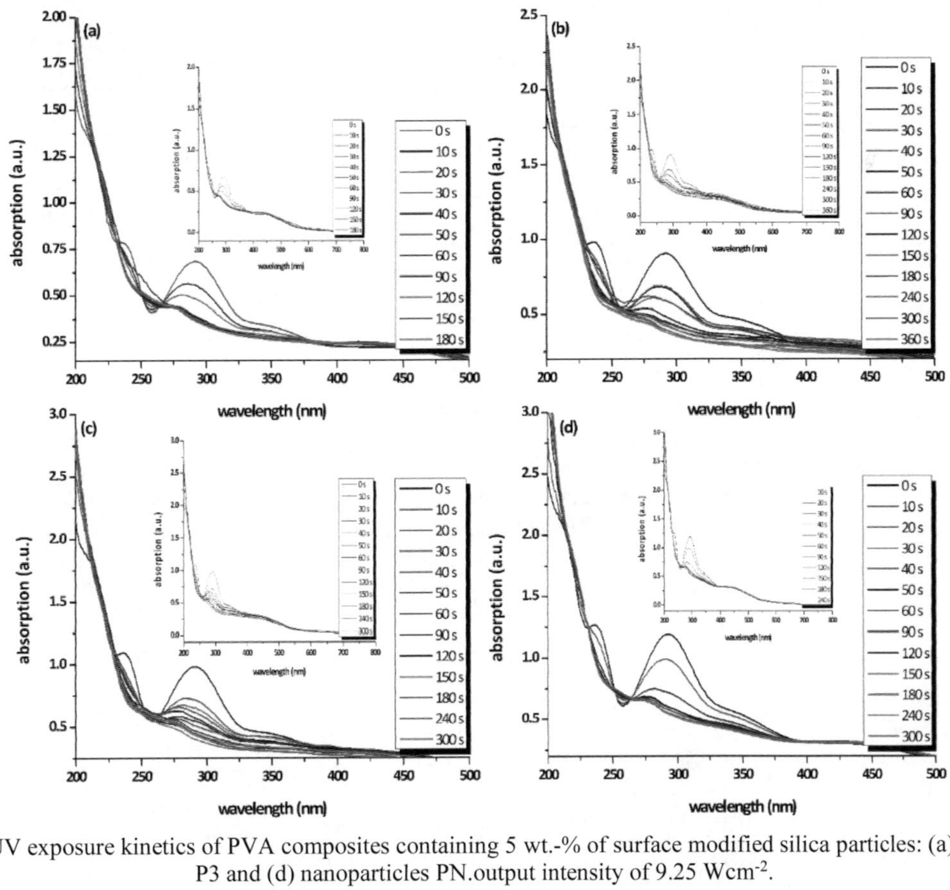

FIGURE 3 UV exposure kinetics of PVA composites containing 5 wt.-% of surface modified silica particles: (a) P1; (b) P2; (c) P3 and (d) nanoparticles PN.output intensity of 9.25 Wcm^{-2}.

UV exposure of PVA composite films were performed that were treated 60 min with radiation doses of 22.2 and 44.4 Jcm^{-2}.

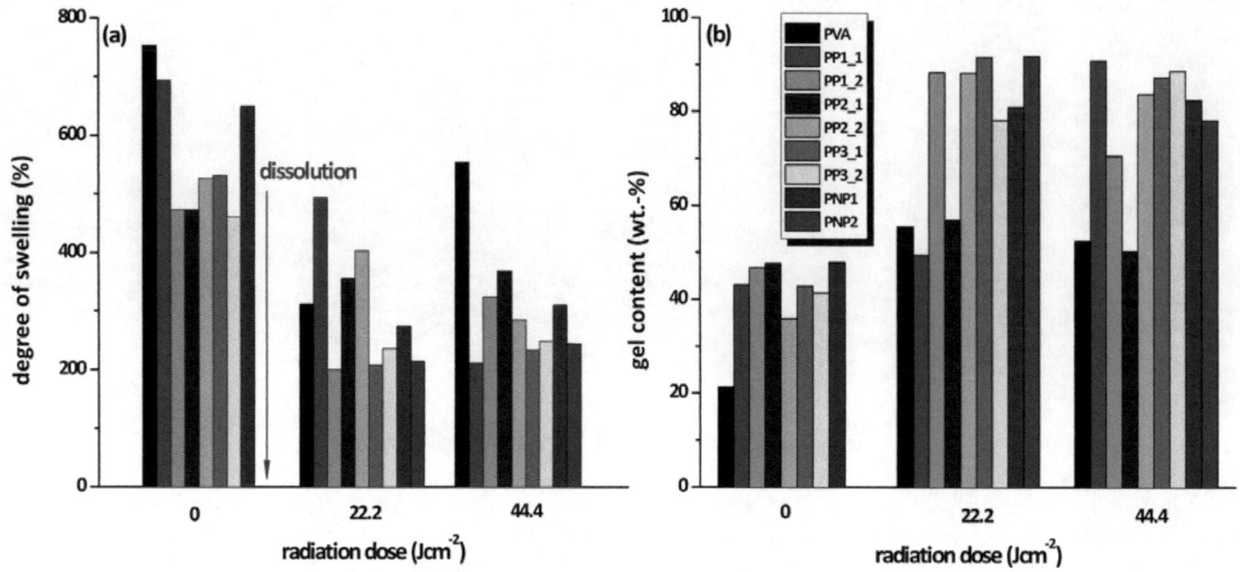

FIGURE 4 Swelling behaviour of PVA composites comprising photoreactive silica particles; UV exposure is performed with a medium pressure mercury vapour lamp

The UV exposure of composite samples containing of PVA and UV reactive silica particles results in a general decrease of the swelling behaviour as well as a strong increase of the insoluble fraction. While neat PVA exhibits a gel content of 55 wt.-%, silica composite samples and nanoparticle composites show significantly higher values (see Fig. 4). This is either resulted by the photodecomposition of the reactive groups, which leads to formation of crosslinks, as well as an increase of the crystalline fraction of the polymer due to heating. Whilst a composite comprising 5 wt.-% nanoparticles is dissolved prior UV exposure, the swelling behaviour is improved after a radiation dose of 22 Jcm^{-2}. Further irradiation can also lead to slight deterioration of the gel content, which is probably caused by chain scission

In summary, it can be stated that both developed systems with UV initiators coupled to fillers are suitable for crosslinking a water-soluble polymer, such as polyvinyl alcohol, under irradiation with UV light.

CONCLUSION

As the photoactive species is immobilised on particle surfaces, neither leaching of the chromophore out of composite nor migration onto the surface can take place. This would be crucial for the fabrication of coatings for applications with a contact to water. As the composite materials exhibit improved stability against dissolution by water, the application as advanced barrier materials with phototuneable properties may be feasible. Furthermore the application as resist material or protective water based coating should be considered, because a dual crosslinking mechanism consisting of heat treatment and UV light can be employed. Some application based on this PVA materials are developed [12, 13]. The behavior as barrier layer for polyethylene pipes is shown in the lecture.

REFERENCES

1. J. E. Ritums, A. Mattozzi, U. W. Gedde, M. S. Hedenqvist, G. Bergman, M. J. Palmlöf, *Polym. Sci. Part B Polym. Phys*. **44**, 641–648 (2006)
2. W. Rühl, *Fachzeitschrift für sichere und effiziente Rohrleitungssysteme* **2013**, 77–84 (2013)
3. K. Berger, C. Keimel, E. Helfer, B. Haar, A. Eder, H. Mattausch, G. Riess, W. Kern, *J. Appl. Polym. Sci. Vol.* **134,** DOI: 10.1002/APP.44968 (2017).
4. K. Berger, C. Keimel, E. Helfer, B. Haar, A. Eder, H. Mattausch, G. Riess, *J. Appl. Polym. Sci.* **134**, DOI: 10.1002/APP.45374 (2017).
5. R. Kramer, "Entwicklung von Compositematerialien mit Barriereeigenschaften", Ph.D. Thesis, Montanuniversität Leoben, 2016.
6. J. Schauberger, G. Riess, W. Kern, *J. Appl. Polym. Sci.* **129**, 3623–3628 (2013).
7. O. Nuyken, B. Voit, *Macromol. Chem. Phys.* **198**, 2337-2372 (1997)
8. B. Voit, F. Braun, L. Eng, C. Loppacher, S. Trogisch, *Macromol.Chem. Phys.* **203**, 1781-1789 (2002)
9. B. Voit, F. Braun, M. Gernert, B. Sieczkowska, M. Millaruelo, M. Messerschmidt, M. Mertig, J. Opitz, *Polym. Adv. Technol.* **17**, 691-693 (2006)
10. J. Schauberger, "Alternative methods for the crosslinking of poly(vinyl alcohol) by the use of functionalised inorganic particles", Ph.D. Thesis, Montanuniversität Leoben, 2013.
11. S.Y. Nam, Y.C. Nho, S.H. Hong, G.T. Chae,. H.S. Jang,. T.S. Suh, W.S.; Ahn, K.E. Ryu, H.J. Chun, *Macromol. Res.* **12**, 219–224 (2004)
12. A. Brandstätter, S. Schlögl, R. Kramer, G. Riess, EP3098070 A1 (25. Mai 2016)
13. S. Schlögl, R. Kramer, G. Riess, EP3057788 A1 (8. Okt. 2014)

A rheological study to predict the ability of Polyethylene oxide solutions to be processed by electrospinning.

Bourdon Laura [(1)], Courty Laura [(1)], Fulchiron Rene [(2)*], Salles Vincent [(1)], Brioude Arnaud [(1)]

[(1)] Laboratoire des Multimatériaux et Interfaces, Université Claude Bernard Lyon 1 - Lyon - France, [(2)] Ingénierie des Matériaux Polymères, Université Claude Bernard Lyon 1 - Lyon - France

The textiles and scaffolds designed for biomedical applications request smaller and smaller filaments diameter (submicronic) for helping drug delivery or resorbability. On the other hand, preserving the continuity of the filaments and controlling their trajectory can be requested for further processing (precise deposition, yarn fabrication ...). Combining electrospinning process with a winding system to fabricate yarns can be a solution. For usual electrospinning process, the filament is accelerated by the electric field and deposited on the target to form a mat and the possible jet turbulence is not really a process limit. However, in the case of winding, the filament must keep its straight shape all along the thread line. Obviously, the spinnability of the solution is governed by its rheological behavior. In the present work, the spinnability of different solutions of Polyethylene oxide (PEO) in water was tested with fixed electrospinning conditions. The solutions differed either by the POE concentration or by the molar mass of the POE (from 200 to 5000 kg/mol). They were classified in "spray", "Filament" or "Stable Filament" according to the spinning behavior: respectively, formation of droplets, formation of a filament but with bending instability and formation of a straight filament. Concurrently, the rheological behaviors of the solutions were characterized in steady mode or, when possible, in dynamic mode. The results were first analyzed in terms of specific viscosity (Etasp) versus the product concentration-intrinsic viscosity c[Eta]. The transition between semi-dilute particle solution and semi-dilute network solution is clearly obtained and very well correlated with the change in the spinning behavior from spray to filament. Furthermore, for a given product c[Eta], the polymer of highest molar mass leads to a better stability of the filament. This could be linked to the mean relaxation time of the solution.

Effect of Processing Conditions on Heat Seal Strength for Peelable Heat Sealing of Multilayered Polyethylene Films with Different Sealant Layers

N. Iwasaki Toshiharu [a], Takarada Wataru [b], Kikutani Takeshi [b*]

[a]*Research and Development Laboratory, Hosokawa Yoko Co., Ltd.*
[b]*Department of Materials Science and Engineering, Tokyo Institute of Technology,*
[*] *2-12-1, O-okayama, Meguro-Ku, Tokyo 152-8550, Japan, Tel +81-3-5734-2468*

Abstract. The effect of heat sealing temperature on peelable heat seal strength (HSS) before and after the sterilization was investigated for the polyethylene (PE) multilayered films with various sealant layers consisting of two high density PEs (HDPEs), linear low density PE (LLDPE) or low density PE (LDPE). To attain similar HSS, heat sealing at a temperature close to the melting peak temperature (Tm-p) was needed only in the case of LDPE. Heat sealing behavior at higher temperatures, to attain higher HSS, was also investigated for HDPE and LDPE. High HSS as well as suppressed reduction of HSS after the sterilization was achieved for HDPE at a heat sealing temperature well below its Tm-p. Increase of HSS with the increase of dwell time was also observed. On the other hand, for LDPE, high HSS and its stability after the sterilization were achieved only when the heat sealing temperature was increased to the value equivalent to its Tm-p. Improvement of the processing conditions for the shortening of the processing period was also investigated. For HDPE, higher sealing bar temperature was needed to gain similar HSS when the dwell time of the heat sealing was shortened from 4 to 2 s, whereas the attained maximum temperature at the interface was found to be similar. The characteristics of the heat sealed portion judged from the relation between HSSs before and after the sterilization were not affected by the dwell time of heat sealing, indicating that the shortening of the processing time by appropriate raising of the sealing bar temperature is possible.

Keywords: Peelable heat sealing; Polyethylene; Heat seal strength, Sterilization, Dwell time
PACS:47.85.md

INTRODUCTION

Flexible packages are widely used for packaging foods, medicals, etc. Those packages are prepared by the heat sealing of plastic films. There are two types of heat sealing with different characteristics, i.e. weld seal and peelable seal. For the weld seal, it is required to guarantee the sealing performance for the packages, while the peelable seal is applied for the easy-opening of packages by hand without using a cutter. Another important application of peelable seals is for the fabrication of "double-bag" packages for medical use, in that two different contents in two chambers, separated by a peelable seal, can be mixed by opening only the peelable seal of the bag but without breaking the entire package. The sealant layers applied for peelable seal are mostly composed of blends of two or more incompatible polymers, such as combinations of low density polyethylene and isotactic polybutene-1. In the cases of "double-bag" packages, however, both weld seal and peelable seal need to be fabricated for the same sealant material through the control of heat sealing conditions.

On the other hand, the flexible packages for cooked foods and medicals are sterilized by hot water under high pressure following packaging to maintain the hygiene of the filled contents. The sterilization process should affect the properties of the heat sealing portion; however, we know of only a few reports on the alteration of the heat seal strength (HSS) after the sterilization. It should be noted that sterilization process may affect the HSS more significantly in case of the peelable seals. Therefore, the aim of this research was set to investigate the characteristics of PE multi-layered films with different sealant layers through the investigation of HSS before and after the sterilization.

EXPERIMENTAL

The polymers used in this study were polyethylenes of various types, including Ziegler-Natta type high density polyethylene of high and low molecular weights (HDPE-H and HDPE-L), metallocene catalyzed linear low density polyethylene of low and high molecular weights (mLLDPE-1 and mLLDPE-2), and high pressure low density polyethylene (LDPE). The coextruded multilayered PE films were produced by the water-cooled tubular film blowing process. The prepared tubular films consisted of three layers. The outer layer was HDPE-H, and the middle layer was mLLDPE-2/HDPE-H blend with blend ratio of 9/1. For the inner layer (sealant layer), four different polymers, i.e. HDPE-H, HDPE-L, mLLDPE-1 and LDPE, were used. The thicknesses of the outer, intermediate and inner layers were 10, 230 and 10 μm, respectively. The total thickness was 250 μm. The heat sealing equipment adopted in this experiment consisted of heat sealing and cooling stages as shown in Fig.1. For the heat sealing stage, two sets of heat sealing bars, i.e. knurled shape bars and flat shape bars were used. For the cooling stage, only a set of flat shape bars was used. During the process, temperature at the interface of the two films was measured using a plate type thermocouple of thickness 70 μm. Regarding the conditions for the heat sealing stage, a pressure of 0.4 MPa was applied. The interval between the opening of the heat sealing bars and the closing of the cooling bars was about 6 s. In the cooling stage, pressure and temperature of the cooling bars were 0.4 MPa and 20 °C, respectively, and the same dwell times as for the heat sealing were adopted.

The heat sealed samples were cut to the size for the HSS measurement, put in a plastic bag filled with water and heat treated under sterilization conditions of 70, 90 or 110 °C for 30 min at 250 kPa using a high temperature and high steam pressure sterilization equipment. T-peel tests were carried out using a tensile testing machine. The relation between displacement and load was recorded at a loading speed of 300 mm/min, and the maximum peeling load was adopted as the HSS. In the HSS measurement, the T-peel test was stopped immediately after the starting of peeling for the observation of the cross-section of the starting point of peeling. It was revealed that the peeling progressed at the interface between the two sealant layers, as shown in Fig. 2. Thermal behaviors of the various PE samples were analyzed by differential scanning calorimetry. Three dimensional topography images of the surface of the sealant layer of the original films were obtained by using an atomic force microscope.

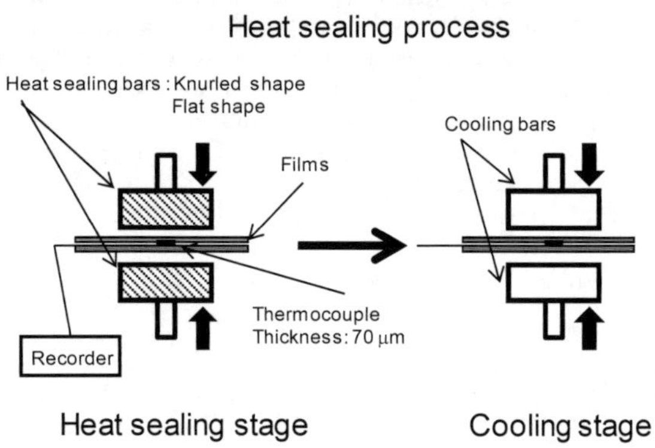

Figure 1. Schematic of heat sealing equipment.

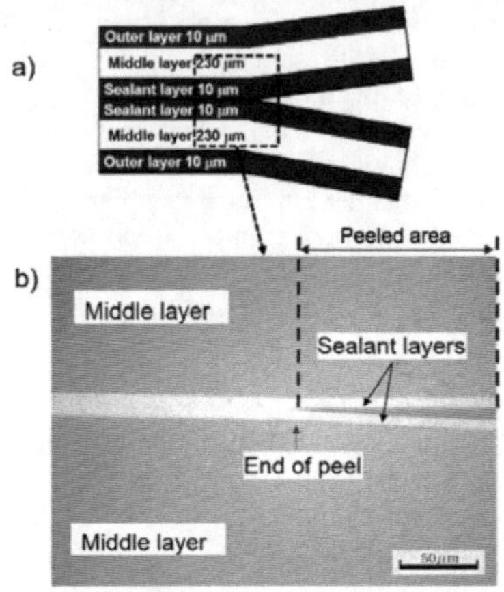

Figure 2. (a) Schematic of the stacked multilayered polyethylene films, and (b) optical micrograph of the cross-section of peeled portion.

RESULTS AND DISCUSSION

Effect of heat seal temperature on heat seal strength

HSSs for the films with four different sealant layers were plotted against heat sealing bar temperature, as shown in Figs. 3-a-d. In the figure for each sealant layer, the data before and after sterilization for the two types of heat sealing bars are shown. The heat sealing bar temperatures were selected so as to realize the HSS before sterilization of 1 to 20 N/15mm. The adopted heat seal temperature for the knurled shape sealing bars was generally higher than those for the flat shape sealing bars. For the HSS of 2 N/15mm before sterilization, temperature differences between flat and knurled shape bars for the HDPE-L, HDPE-H, LDPE and m-LLDPE-1 were 3.1, 3.1, 1.4 and 3.3 °C, respectively. It was considered that these differences originated from the difference in the contact area. More specifically, thermal conduction was more effective for the flat shape sealing bars, with larger contact area, in comparison with the knurled shape sealing bars with smaller contact areas.

The temperatures for realizing the HSS of 2 N/15 mm using the knurled shape bars were 127.3, 128.3, 112.9 and 111.7 °C for HDPE-L, HDPE-H, LDPE and m-LLDPE-1, respectively. If these temperatures are compared with the melting peak temperatures (*T*m-p) of the individual polymers, the differences were 6.7, 8.1, 2.0 and 9.5 °C, respectively. As described above, the temperature difference between the flat and knurled shape bars was around 3 degrees for HDPE-L, HDPE-H and mLLDPE1, while that was only 1.4 degrees for LDPE. Therefore, it can be said that the heat sealing at a temperature close to the *T*m-p was necessary to achieve reasonably high HSS only in case of LDPE. In order to investigate the reason for this difference, surface profiles of the four different sealant layers of the original films were investigated using an AFM. It was found that the surface roughness for the HDPEs was relatively high and that for the LDPE was low, while mLLDPE-1 showed the lowest degree of surface roughness. These results suggested that the difference between the melting temperature and the heat sealing temperature was not directly related with the surface characteristics, i.e. the state of contact between the two sealant layers. Therefore, it was speculated that the difference in the mobility of molecular chain for the formation of entanglements across the interface was the origin for such difference.

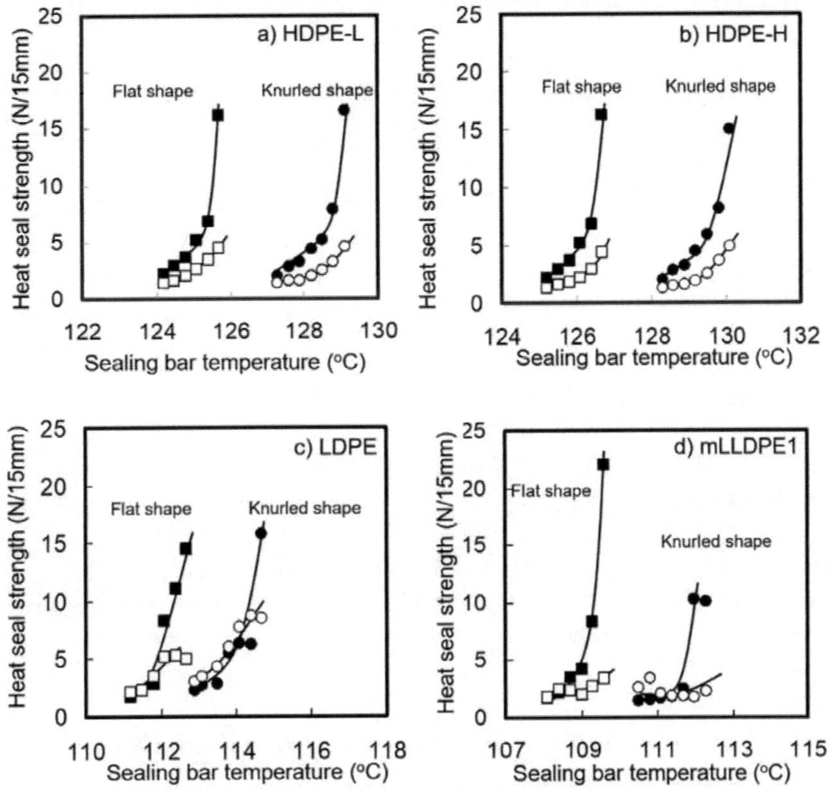

Figure 3. Relation between sealing bar temperature and heat seal strength before and after sterilization (closed and open symbols) at 110 °C for the films with sealant layers of a) HDPE-L, b) HDPE-H, c) mLLDPE-1 and d) LDPE. Melting peak temperature of these polymers 134.0, 136.4, 121.2, and 114.9 °C, respectively. Results obtained using flat shape bars and knurled shape bars are shown.

Relationship between heat sealing condition and sterilization process

In the above experiment, it was revealed that the HSS generally decreased after sterilization when heat sealing was conducted to realize relatively low HSS of up to 20 N/15mm for the peelable seal application. Therefor effect of the sterilization for the heat sealing of relatively high HSS was also investigated comparing the results for two different sealant layer materials, HDPE-H and LDPE.

Variation of HSS after the sterilization for the sealant layer of HDPE-H were investigated for various heat sealing temperatures with a dwell time of 4 s and for heat sealing temperature of 127.3 °C with dwell times of 4 or 15 s as shown in Fig. 4-a. For the dwell time of 4 s, the HSS before the sterilization increased from 5.5 to 50 N/15mm when the heat sealing temperature was increased from 126.1 to 130.0 °C, with the latter being 6.4 °C below *T*m-p. With the increase of dwell time from 4 to 15 s, the HSS for the heat sealing temperature of 127.3 °C increased from 38 to 49 N/15mm. A series of experiment revealed that for the samples with HSS less than 40 N/15mm, the HSS clearly decreased with the increase of sterilization temperature.

At the heat sealing temperature of 126.9 °C, HSS decreased dramatically even at the sterilization temperature of 70 °C. When the heat sealing temperature was increased to 127.3 °C, reduction of the HSS started when sterilization temperature exceeded 70 °C. When the dwell time was increased from 4 to 15 s, the HSS did not show a reduction, but rather showed a noticeable increase after the sterilization; however, the reason for this increase is not clear at this moment.

For the sealant layer of LDPE, the effect of sterilization temperature on HSS was investigated for the samples prepared with dwell times of 4 and 15 s and heat sealing temperatures of 113 and 115 °C, as shown in Fig.4-b. The HSS decreased significantly from 34 to around 10 N/15mm with the increase of sterilization temperature for the heat sealing temperature of 113 °C, which is two degrees lower than the *T*m-p. Extension of dwell time from 4 to 15 s did not bring about any significant change. On the other hand, when the heat sealing temperature was increased to 115 °C, i.e. *T*m-p, HSS before the sterilization increased to 40 N/15mm, and there was no reduction or even a slight increase of HSS with the increase of sterilization temperature. It was speculated that the different characteristics of these two polymers, in terms of the diffusivity of molecular chain and formability of entanglement among the molecular chains across the interface of heat sealing layers, was the origin of this different behavior.

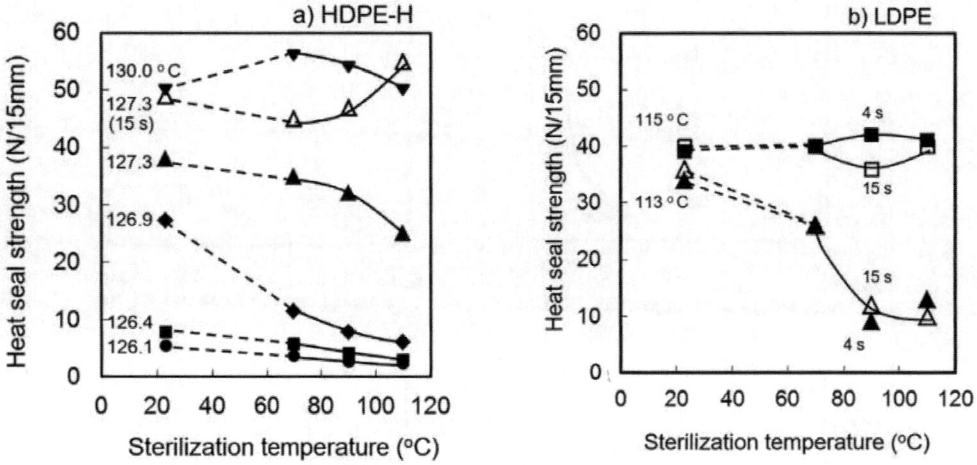

Figure 4. Variation of heat seal strength with sterilization temperature for a) HDPE-H and b) LDPE. The values before sterilization were plotted at the sterilization temperature of 23 °C. The flat shape sealing bar was used, and sealing pressure was 0.4 MPa. Two different dwell times of 4 s (closed symbols) and 15 s (open symbols) were adopted.

Analysis of heat seal strength for different dwell times

With the aim of improving the processing conditions by shortening the period of the heat sealing process, the effect of dwell time on the heat sealing behavior, as well as the characteristics of the heat sealed portion, was investigated. The effect of sealing bar temperature on the HSS of HDPE-H for the different dwell times of 2 and 4 s is shown in Fig. 5-a. To attain similar HSS after changing the dwell time from 4 to 2 s, the sealing bar temperature needed to be increased about 3 °C. To elucidate the reason for the necessity of changing the sealing bar temperature, the variations of the interface temperature with time for dwell times of 2 and 4 s were measured, as shown in Fig. 6. The heat sealing temperature was selected so as to attain the HSS of around 5 N/15mm. At the dwell time of 4 s, the interface temperature almost reached the maximum temperature at the end of the heat sealing, whereas significant overshooting of temperature was observed following the dwell time of 2 s. The maximum temperature for the dwell time of 4 s and HSS of 5.3 N/15 mm was only about 0.3 °C lower than those of 2 s and 4.4 N/15mm. The HSS plotted against the sealing bar temperature in Fig. 5-a was replotted against the attained maximum temperature at the interface in Fig. 5-b. To attain similar HSS for the dwell times of 2 and 4 s, the interface temperature for the dwell time of 2 s was only about 0.3 - 0.6 °C higher in comparison with that for the dwell time of 4 s. This result suggested that the HSS is mainly controlled by the attained maximum temperature at the interface. Necessity of reaching slightly higher temperature to gain similar HSS for the dwell time of 2 s may be attributable to the difference in the rate of temperature increase to reach the maximum temperature.

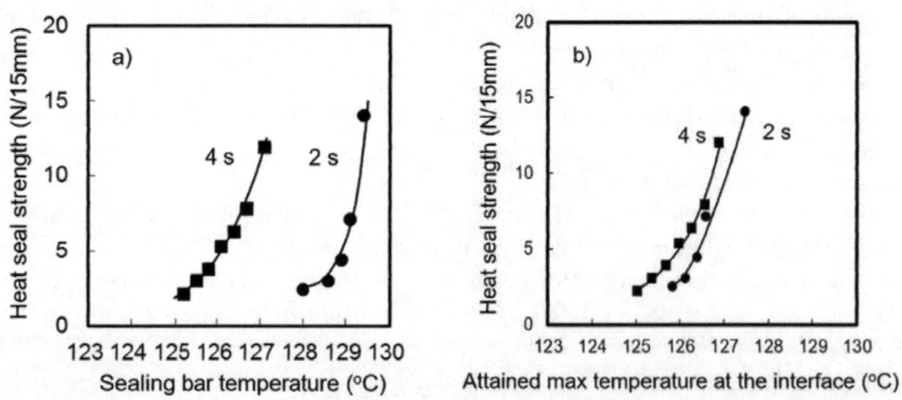

Figure 5. Variation of HSS with a) sealing bar temperature and b) attained maximum temperature at the interface for dwell times of 2 and 4 s for HDPE-H. The flat shape sealing bar was used and the sealing pressure was 0.4 MPa.

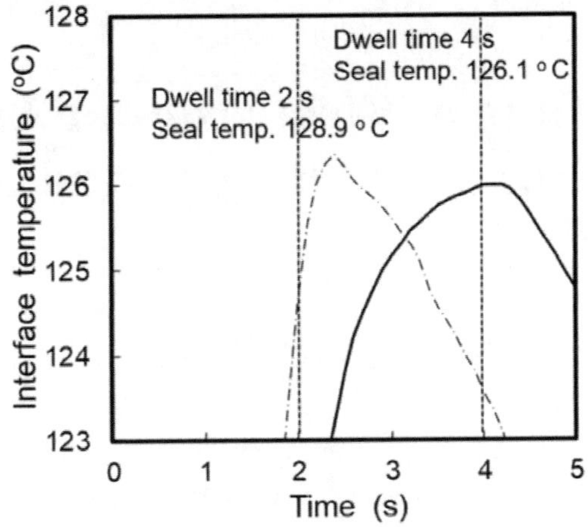

Figure 6. Variation of interface temperature with time for dwell times of 2 and 4 s for HDPE-H. The heat sealing temperature was selected so as to attain the heat seal strength of around 5 N/15mm. The flat shape sealing bar was used.

The characteristics of the heat sealed portion, judged from the relation between HSSs before and after the sterilization, were not affected by the dwell time, indicating that the shortening of the processing time by appropriate raising of the sealing bar temperature is practically feasible.

CONCLUSIONS

The effect of processing conditions on the HSS for peelable heat sealing of multilayered PE films with different sealant layers was investigated. Investigation on the effect of heat sealing temperature on HSS revealed that heat sealing at a temperature closer to the melting temperature was needed in the case of LDPE to attain similar HSS as HDPEs and mLLPE. The difference in the mobility of molecular chain for the formation of entanglements across the interface, which is known to be affected by the molecular chain branching, was speculated to be the origin for such difference. For more detailed investigation regarding the different characteristics of HDPE-H and LDPE, the heat sealing behavior at higher heat sealing temperatures to attain higher HSS were investigated. In the case of HDPE, high HSS as well as suppression of its reduction after the sterilization was achieved at a heat sealing temperature well below its *T*m-p. Increase of HSS with the increase of dwell time was also observed. On the other hand, in the case of LDPE, high HSS and stable HSS through the sterilization were achieved only when the heat sealing temperature was increased to a value equivalent to its *T*m-p. In addition, the effect of dwell time of heat sealing on the HSS before and after the sterilization was minimal. Possible improvement of the processing conditions for the shortening of processing period was also investigated for HDPE-H. Higher sealing bar temperature was needed to gain similar HSS when the dwell time was shortened from 4 to 2 s, whereas the attained maximum temperature at the interface was found to be similar. The shortening of the processing time by appropriate raising of the sealing bar temperature was found to be practically feasible.

REFERENCES

1. T. Iwasaki, W. Takarada, and T. Kikutani, *J. Polym. Eng.*, **36**(9), 909-916 (2016)
2. T. Iwasaki, W. Takarada, and T. Kikutani, *Seikei-Kakou*, **28**(1), 34-40 (2016)
3. T. Iwasaki, W. Takarada, and T. Kikutani, *SJ. Macromol. Sci., Pt-B,* http://dx.doi.org/10.1080/00222348.2017.1369250 (2017)

Chemical Grafting of Electrospun PAN Nanofibers For The Adsorption Of Dyes And Metal Ions

Fahad A. AlAbduljabbar[a], Sajjad Haider[b]*, Waheed. A.Almasry[b], Iqbal Mujtaba[a]

[a] School of Engineering, Department of chemical Engineering, Room Cl. 16, Universityof Bradford Bradford, West Yodshire, BD7 lDP, UK
[b]Department of Chemical Engineering, College of Engineering, King Saud University, P.O. Box 800, Riyadh 11421, Saudi Arabia, Saudi Arabia
(*Corresponding author e-mail: shaider@ksu.edu.sa)

Abstract. In present study, detail experiments were performed to prepare Polyacrylonitrile (PAN) nanofibers (NFs) membrane via electrospinning technique, graft the NFs membrane with ethanolamine (EA) functional group and utilize the grafted NFs membrane for the removal of dyes and metal ions from synthetic solution. After grafting the membrane was characterized with field emission scanning electron microscopy (FE-SEM) and fourier transforms infrared (FT-IR) spectroscopy. FESEM showed smooth morphology for NFs before and after grating whereas FT-IR confirmed the grafting of EA to nitrile group of PAN. The % grafting value with no change in the physical nature (remained soft) for EA was 12.18 %. The adsorption for all the dyes on to the grafted PAN NFs membranes were found to increase in the order PAN<EA whereas the dye adsorption affinity followed the order: methylene blue (MB) < safranin T (ST)< rhodamine b (RB). The adsorption of Cu (II) and Pb(II) ions also showed good adsorption onto the grafted membrane. Both (PAN and grafted) the membranes showed more affinity to Cu (II) due to its small size (atomic radii 0.128 Å) as compare to Pb (II) ions (atomic radii 0.175 Å); Cu (II) ions were easily diffused into the pores and adsorbed onto the internal surface of the adsorbent.

Keywords: Enter Keywords here. Nanofibers; Grafting; pollutants; Adsorption
PACS: 81.05.Lg

INTRODUCTION

Many industries, such as electrical, paper, plastics, dyestuffs and textile consume a substantial volume of water and also use chemicals of inorganic (such as gold (Au), palladium (Pd), and platinum (Pt) copper (Cu) and iron (Fe)) and organic nature (such as dye stuff, etc) during manufacturing. These industries generate a considerable amount of polluted wastewaters, which is a grave issue because metal ions causes various diseases such as ailments, dehydration, stomach ache, nausea, dizziness and/or lack of coordination in muscles, destruction of young children nervous systems, lung irritation, eye irritation, vomiting, abdominal pain, lung insufficiency, liver damage, etc. On the other hand dye stuff and pigments are mostly considered either inert or non-toxic, even though some are not totally harmless. Dyes causes aesthetic problems, reduce photosynthetic activity in the receiving waters (toxic to aquatic organism) and in some cases carcinogenicity. Many dyes are designed for their chemical stability and do not undergo biochemical degradation readily [1]. Interest in the environmental behavior of dyes and metal ions is prompted primarily due to their biological and chemical stability. To eliminate these pollutants from the wastewaters a number of conventional water treatment methods/processes such as coagulation [2], ultrafiltration [3], oxidization [4], electrochemical [5], photocatalytic degradation [6] and adsorption [7] have been widely investigated. However, most of these are complex, insufficient, and expensive and time consuming. Therefore adsorption, a simple technique, is considered as a favorable alternative for the removal of dyes and metal ions from wastewater streams. In recent years, a wide variety of materials such as fly ash [8], natural phosphate [9], bentonite [10], activated carbon [11], chitosan [12], and cyclodextrine polymer [13] are used for the removal of various dyes and metal ions.

Electrospun NFs with fiber diameter 10 to 500 nanometers (nm), large surface area per unit mass, high porosity, high gas permeability, and small interfibrous pore size has changed the whole picture of wastewater treatment. Polymers functionalities [14] plus the unique characteristics of NFs originated from being engineered in various forms have also allowed NFs to be used in advance applications such as filtration, multifunctional membranes [15], composite reinforcement [16], tissue engineering scaffolds [17], wound dressings [18], drug delivery [19], artificial organs [20], and vascular grafts, etc [21].

In our current work, which is a continuation of our previous works, PAN was electrospun and grafted with EA. The EA grafted PAN (EA-g-PAN) was then applied for the removal of dyes and metal ions in batch experiments.

EXPERIMENTAL

Materials

PAN (average molecular weight (Mw) 150,000), dimethyl formamide (DMF($_3H_7NO$)), EA (C_2H_7NO), MB ($C_{16}H_{18}ClN_3S.3H_2O$), RB ($C_{28}H_{31}ClN_2O_3$), ST ($C_{20}H_{19}ClN_4$), Lead(II) nitrate ($Pb(NO_3)_2$) and Copper(II) nitrate ($Cu(NO_3)_2$) were purchased from Sigma-Aldrich. Distilled water was used as solvent for the preparation of dyes and metal ions solutions. All the chemicals were of analytical grade and were used without further purification. Teflon frame were prepared locally.

Preparation Of The NFs Membrane And EA Grafting

The PAN NFs membranes sheets were prepared at optimal electrospinning conditions [22], dried and stored. EA-g-PAN membrane was prepared by immersing PAN-NFs membranes in EA solution (4.97 M) in a 250 mL sealed beaker. The mixture was heated while stirred on a water bath at 85-90 °C for different time durations (1, 2, 3 and 4 h). After reaction, the NFs membrane were washed several time with distilled water, dried at 80°C in an oven and stored for characterization. The % grafting was calculated according to eq.1.

$$C_n = \frac{W_1 - W_0}{W_0} \frac{M_0}{M_0} \times 100 \dots\dots\dots\dots\dots\dots 1$$

Where Cn is % conversion of nitrile group in PAN into an amine group, W_0 is the weight of the PAN NFs membrane before reaction, W_1 is the weight of the PAN NFs membrane after reaction, M_1 is the M_w of EA (61.08 g/mol) and M_o is the molecular weight of acrylonitrile monomer (53.06 g/mol)

Morphology Study

The morphologies of the as spun and EA-g-PAN NFs were studied by using a field emission electron microscope (FE-SEM). To study the surface morphologies of the EA-g-PAN NFs via FE-SEM, the NFs samples were fixed onto a holder with aid of a carbon tape and then placed in the sputtering machine for platinum coating (to their increase electrical conductivity). After platinum coating the NFs were examined by FE-SEM under high vacuum.

FT-IR Study

The changes occurred in chemical structure of PAN were studied by taking infrared spectra of the as spun and EA-g-PAN NFs membrane using FT-IR spectrometer (Bruker Vertex 70). For the FT-IR characterization, KBr discs of the samples were prepared by mixing and grounding the samples with KBr powder in mortar with a pestle. The mixture was then shaped into discs under hydraulic pressure. The samples discs were put into FT-IR and the spectral measurements were recorded in the wavenumber range of 600-2200 cm^{-1}. The data was processed by using Software OPUS 6.0 (Bruker), which was baseline corrected by rubber band method with CO_2 and H_2O bands excluded.

Adsorption Studies

Dried samples of the PAN and EA-g-PAN membranes were added separately to 10 mL synthetic solutions of dyes and metal ions (400 ppm) and shaked (in a shaker bath) by a batch technique as a function of time until 240 min and 24 h, respectively at 25 °C. Equilibrium time was determined from the saturation point of the adsorption kinetics data. Adsorption equilibrium isotherm was also studied as a function of the concentration of the dyes and metal ions at 25 °C. The concentrations of dyes in solution (after adsorption experiment) were determined with UV/VIS Spectrometer (Perkin Elmer, Lambda 35) whereas in case of metal ions an inductively coupled plasma mass (ICP MS (PerkinElmer)) spectrophotometer was used. The amount adsorbed was calculated using the following formula;

$$q = \frac{(C_0 - C_f)V}{M} \qquad \text{.......................2}$$

Where **q** is the amount adsorbed (mg·g−1), $\mathbf{C_0}$ and $\mathbf{C_f}$ are the initial and final concentrations (mg· L^{-1}), respectively for dyes and metal ions, **V** is the solution volume (L) and **M** is the amount of adsorbent (g) used.

RESULTS AND DISCUSSIONS

EA Grafting And Morphology Study

Table 1 shows the data for EA grafting to PAN. The % grafting increased with time and reached maximum (12%) until 2 h, further, increase in time reduced the % grafting. The increase in grafting with time until 2 h might be attributed to the increase in the interactions of the EA with the nitril group of the PAN to a critical value beyond which the interactions decreased. During the reaction the color of the EA grafted PAN NFs membrane changed from white to off white whereas the physical nature of NFs membrane remained soft. Fig. 1a-b shows the FE-SEM micrographs of PAN and EA-g-PAN NFs. The micrographs of EA-g-PAN NFs showed almost similar morphology to that of the PAN membrane without any cracks and degradation (Fig.1a-b).

Table (1). EA grafting to PAN NFs membrane**.**

Exp. No	Time (h)	Temp.(OC)	Conv. %	Coloure	Softness
1	1	85-90	6.820347	White	Soft
2	2		12.17799	Off White	Soft
3	3		8.568778	White	Soft
4	4		4.075368	White	Soft

FIGURE 1. FE-SEM micrographs of the surface of the NFs membranes: (a) PAN and (b) EA-g-PAN of the EA-g-PAN.

FTIR Study

Fig. 2 shows the FTIR spectra of the PAN and EA-g-PAN NFs membranes at various reaction times. PAN exhibited the absorption bands in region of 1700 cm^{-1} (C=O stretching), 1000-1300 cm^{-1} (C-O stretching), 2938 cm^{-1}

(CH stretching), 1453 cm^{-1} (CH_2 bending), and a stretching vibration at band at 2241cm^{-1} (C≡N) which suggested that the PAN was a copolymer of acrylonitrile and methylacrylate. All the characteristic bands of the PAN were observed in spectrum of EA-g-PAN. The sharp band at 2241cm^{-1}contineouly decreased in intensity, until its intensity decreased to maximum around 2 h. The conversion at this intensity was 12 % (Table 1) and the NFs membrane was soft. Besides, decreased intensities of the old, some new bands, whose intensities increased continuously with increase the reaction time, were also observed, e.g. broad bands in regions of 3300-3550 cm^{-1}(N–H/ O-H stretching) and 1630-1670 cm^{-1} ((C=N) and some traces of amide C=O stretching vibration), and 1440-1395 cm^{-1}(O–H bend) [23]. The data presented in the FTIR clearly shows that EA is grafted to PAN. The reduced broad band in the region of 1630 -1670 cm^{-1} could also be attributed to the formation of traces of hydroxamic acid [24].

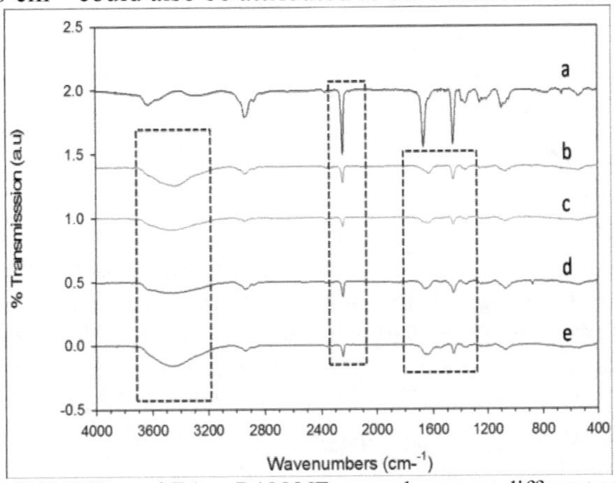

FIGURE 2. FT-IR spectra of the (a) PAN NFs and EA-g-PAN NFs membranes at different reaction time; (b) 1 h, (c) 2 h, (d) 3 h and (e) 4 h.

Adsorption Study

The adsorption of MB, RB and ST onto the PAN and EA-g-PAN NFs membranes from 400 ppm synthetic solution as a function of time was studied. The amount of adsorption for all the dyes was found to increase sharply until 30 min and leveled off at 50 min.(Figures are not shown here). At this time there is equilibrium between adsorption and desorption of dyes. Fig. 3a shows the equilibrium adsorption of MB, RB and ST membranes as a function of equilibrium concentration (Ce) onto the PAN and EA- g- PAN NFs membranes. Initially the adsorption of the dyes increased rapidly, however as the concentration increased, the rates of increase were reduced. The adsorption of all the dyes on to the EA- g- PAN NFs membranes was found to increase in the order PAN<EA. This increase in the dyes adsorption with grafting could be attributed to the increase in the surface and number of active sites onto the membranes. To understand the phenomenon of the adsorption onto the grafted membranes, Langmuir isotherm model (eqn.3) was used.

$$\frac{C_e}{q_e} = \frac{1}{K_L \cdot q_m \cdot} + \frac{C_e}{q_m} \quad ....3$$

Fig.3b shows that the adsorption data fitted to Langmuir model. Langmuir theory suggest that adsorption takes place at specific homogeneous sites within the adsorbent and once a dye occupied a reaction site, then no further adsorption occurred at that location (chemical adsorption) [25]. Table 2 shows the values of q_{max}, K_L and correlation coefficient (r^2) for Langmuir. The r^2 values varied from 0.94-0.99. .The q_{max} values calculated from the slope of the Langmuir plots for the adsorption of RB, ST and MB onto our grafted membranes were found reasonably high. The increased values suggested that our grafted membranes could find potential industrial application.

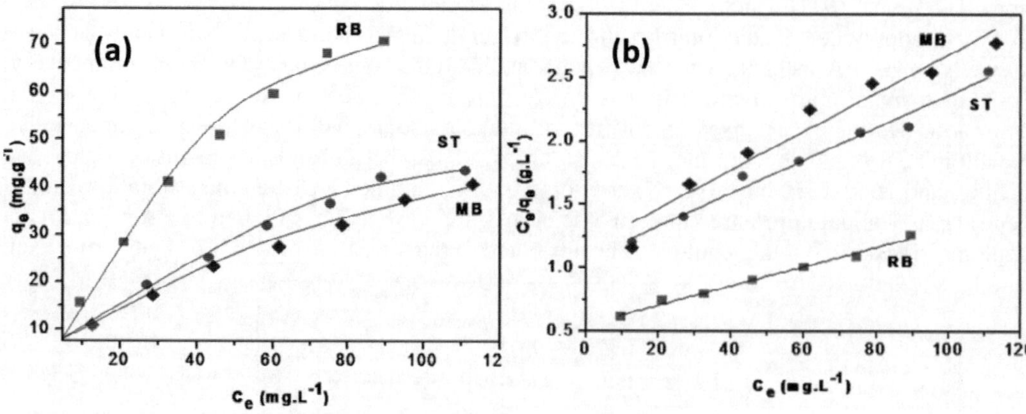

FIGURE 3. Adsorption of dyes (a) adsorption equilibrium and (b) Langmuir isotherm.

The adsorption of Cu (II) and Pb (II) metal ions was only studied onto the grafted PAN NFs membranes from 400 ppm synthetic solution as a function of time. The author has already reported the adsorption of these metal ions onto PAN NFs membrane [14]. The amount of adsorption for both metal ions increase sharply until 4 h and leveled off at 8 h (Figure is not shown here). Fig.4a shows the equilibrium adsorption of Cu (II) and Pb (II) ions as a function of equilibrium concentration (Ce) onto the grafted PAN NFs membranes. Initially the adsorption of metal ions increased rapidly, however as the concentration increased, the rates of increase were reduced. To understand the phenomenon of the adsorption onto the grafted membranes, Langmuir isotherm equation was used. Fig.4 b shows that the adsorption data fitted to Langmuir. The value of correlation coefficient (r^2) for Langmuir varied from 0.951-0.949. q_{max} and K_L calculated from the slope and intercept of the plots are shown in Table 2.

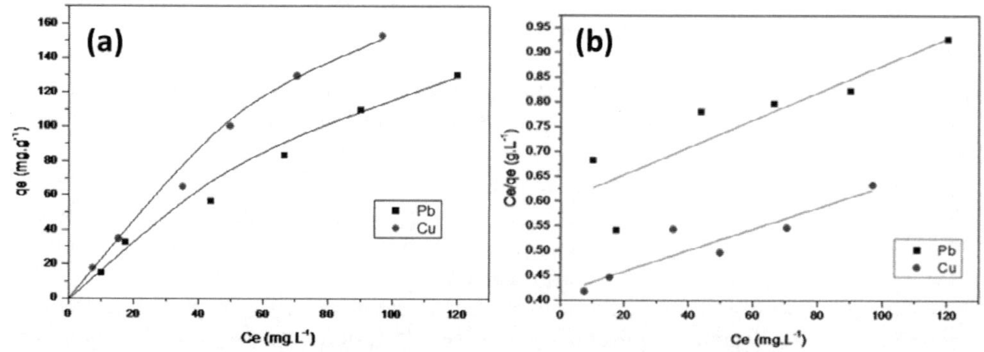

FIGURE4. Adsorption of metal ions (a) adsorption equilibrium and(b) Langmuir isotherm.

Table (2). Langmuir constants for the adsorption of dyes and metal ions onto grafted PAN NFs membranes.

Mem. Type	Dyes								
	MB			RB			ST		
	q_{max}	K_L	r^2	q_{max}	K_L	r^2	q_{max}	K_L	r^2
PAN	42.66	1.16	0.99	99.31	2.31	0.97	72.47	0.93	0.99
EA	65.62	0.86	0.94	130.72	1.80	0.99	77.16	0.93	0.98
	Metal ions								
	Cu(II)			Pb (II)					
	q_{max}	K_L	r^2	q_{max}	K_L	r^2			
EA	518.00	2.452	0.968	344.00	1.746	0.801			

CONCLUSION

EA-g-PAN NFs were successfully prepared using electrospinning technique and chemical grating. The morphology and chemical grafting of NFs were confirmed by FESEM and FTIR. Thc EA-g-PN NFs were applied to the removal of dyes and metal ions. The adsorption for all the dyes on to the grafted PAN NFs membranes were found to increase in the order PAN<EA. whereas dye adsorption affinity followed the order: MB < ST< RB. The adsorption of Cu (II) and Pb(II) ions also showed good adsorption onto the grafted membrane. Both (PAN and grafted) the membranes showed more affinity to Cu (II) due to its small size (atomic radii 0.128 Å) compared to Pb (II) ions (atomic radii 0.175 Å); Cu (II) ions were easily diffused into the pores and adsorbed onto the internal surface of the adsorbent.

ACKNOWLEDGEMENT

Financial support from the National Plan for Science and Technology, King Saud University, Saudi Arabia under the Project number 10NAN1017-2 is greatly acknowledged.

REFERENCES

1. R. Jain, V.K. Gupta, S. Sikarwar, J. Hazard. Mater. 182, 749-756 (2010).
2. J.W. Lee, S.P. Choi, R. Thiruvenkatachari, W.G. Shim, H. Moon, Water Res. 40, 435-444 (2006).
3. K. Majewska-Nowak, Desalination 71, 127-135 (1989).
4. I.A. Salem, M. El-maazawi, Chemosphere 41, 1173-1180 (2000).
5. S. Song, J. Fan, Z. He, L. Zhan, Z. Lin, J. Chen, X.Xu, Electrochim. Acta. 55, 3606-3613 (2010).
6. F. Xia, E. Ou, L. Wang, J. Wang, Dyes Pigm. 76, 76-81 (2008).
7. D. Suna, X. Zhang, Y. Wu, X. Liu, J. Hazard. Mater. 181, 335-342 (2010).
8. I.D. Mall, V.C. Srivastava, N.K. Agarwal, Dyes Pigm. 69, 210-223 (2006).
9. N. Barka, A. Assabbane, A. Nounah, L. Laanab, Y.A. Ichou, Desalination 235, 264-275 (2009).
10. B. Benguella, A. Yacouta-Nour, Desalination 235, 276-292 (2009).
11. V. Gomez, M.S. Larrechi, M.P. Callao, Chemosphere 69, 1151-1158 (2007).
12. A.R. Cestari, E.F.S. Vierira, A.M.G. Tavares, R.E. Bruns, J. Hazard. Mater. 153, 566-574 (2008).
13. G. Erini, Dyes Pigm. 77, 415-426 (2008).
14. K. Saeed, S. Haider, T. J. Oh, S. Y. Park, J. Membr. Sci. 322, 400-405 (2008).
15. Z. Ma, M. Kotaki, S. Ramakrishna, J. Membr. Sci. 265, 115-123 (2005).
16. G.M. Kim, G.H. Michler, P. Potschke, Polymer 46, 7346-7351 (2005).
17. Z. Ma, M. Kotaki, R. Inai, S. Ramakrishna, Tissue Eng. 11, 101-109 (2005).
18. H. Ueno, T. Mori, T. Fujinaga, Adv. Drug Deliv. Rev. 52, 105-115 (2001).
19. D.S. Katti, K.W. Robinson, F.K. Ko, C.T. Laurencin, J. Biomed, Materials Research 70, 286-296 (2004).
20. Z.M. Huang, Y.Z. Zhang, M. Kotaki, S. Ramakrishna, Compos. Sci. Technol. 63, 2223-2253 (2003).
21. J. Stitzel, L. Liu, M. Komura, J. Berry, S. Soker, G. Lim, M. Van Dyke, R. Czerw, J.J. Yoo, A. Atala, Biomaterials 27, 1088-1094 (2006) .
22. S.Haider, F. F. Binagag, A. Haider, W. A. Al-Masry, J. Polym. Res. 21, 371(2014).
23. IR Spectroscopy Tutorial: Carboxylic Acids; http://orgchem.colorado.edu/Spectroscopy/irtutor/carbacidsir.html.
24. N. Bilba, D. Bilba, G. Moroi, J. Appl. Polym. Sci. 92, 3730-3735 (2004).
25. I.D. Mall, V.C. Srivatava, N.K. Agarwal, Dyes Pigment. 69, 210223 (2006).

Reinforced Composite Structures with In-situ Shrinking Microfibers

Kim Eric S. [(1)], Lee Patrick C. [(1)]*, Huston Dryver [(1)]

[(1)] Department of Mechanical Engineering Univ of Vermont - VT - USA

This study describes an innovative fiber reinforcement technology with various applications including cementitious composite structures and fiber reinforced plastics. Unlike incumbent passive reinforcing microfiber technology, in situ shrinking microfibers that respond to an external stimulus such as heat or pH variations can induce pre-compression to matrix and make additional resistance from external loads, creating stronger composite structures. In this paper, heat activated shrinking (HAS) microfibers made from polyolefin, and pH activated shrinking (pHAS) microfibers made from chitosan biopolymer were used to study effects of shrinking microfiber reinforcement in concrete and epoxy laminate. Specimens with active shrinking microfibers, passive non-shrinking microfibers, as well as control samples were fabricated. For cementitious composite specimens, the mechanical properties of the samples were investigated with the results of compression and three-point bending tests. Compared to heat passive (HP) microfibers specimens, 45 % increase in the maximum compression strengths at specimens with 0.5 wt% fiber ratio, and 124 % increase in the maximum bending strengths at specimens with 1.0 wt% fiber ratio were achieved. With specimens with pHAS microfibers, 145 % increase in the maximum bending strengths were resulted, compared to the pH passive (pHP) microfiber reinforced specimens at 0.5 wt% fiber ratio. In addition, with the epoxy laminate composite specimens, T-peel tests were conducted to investigate the interlaminar reinforcing effect with HAS microfibers. Compared to HP microfibers specimens, up to 57.0 % increases in peel strengths were resulted.

Fiber Structure Development in High-Speed Melt Spinning of Core/Sheath Bicomponent Fibers of Poly(L-lactic acid)s with Non-equimolar D-lactide Contents

N. Roungpaisan [a], W.Takarada[a] and T.Kikutani[a]

[a] *Department of Materials Science and Engineering, Tokyo Institute of Technology, 2-12-1, O-okayama, Meguro-Ku, Tokyo 152-8550, Japan, Tel +81-3-5734-2468*

Abstract: Processablity, crystallizability and other properties of poly(lactic acid) depends strongly on the composition of L- and D-isomers. To gain more understanding regarding the effect of D-lactide content on the structure development behavior of poly(L-lactic acid) (PLLA) in the melt spinning process, bicomponent fibers of core/sheath (CS) cross-sectional configuration was selected, and the effect of take-up velocity on the structure and properties of resultant fibers was investigated. Applying the bicomponent melt spinning system for the PLLA of low and high D-lactide content (LowD, HighD), four kinds of CS fibers i.e. core/sheath= LowD/LowD, LowD/HighD, HighD/LowD and HighD/HighD, with equal compositions of the core and sheath components were extruded. In the spinning process, on-line measurement of the diameter profile of the spin-line was performed to acquire information on the structure development behavior. Birefringence, wide-angle X-ray diffraction and differential scanning calorimetry measurements were also carried out to analyze the structure of as-spun fibers prepared at various take-up velocities. In all cases, orientation-induced crystallization started above certain take-up velocities, in that crystallizability could be judged from the position of neck-like deformation in the spin-line. The neck-like deformation was closer to the spinneret in the LowD/LowD spin-line than the HighD/HighD spin-line, whereas the position of neck-like deformation was closest to the spinneret for the HighD/LowD bicomponent spinning. Birefringence of the as-spun fibers increased with the take-up velocity, and the LowD/LowD fibers showed higher birefringence than the HighD/HighD fibers at the same take-up velocity, whereas the birefringence of the LowD component in the HighD/LowD bicomponent fiber exhibited the highest birefringence. These results suggested that the orientation-induced crystallization of LowD in the bicomponent fiber was promoted because of the existence of HighD in the fiber cross-section.

Keywords: Poly(L-lactic acid); D-lactide content; Bicomponent fiber; Birefringence, Neck-like deformation

INTRODUCTION

Numerous plastic products as well as textiles are now fabricated utilizing bioplastics, which is known as the green and sustainable materials. Among variety of bio-based polymers, poly(lactic acid) (PLA) has been attracting the highest attention owing to its high processability with its glass transition temperature (Tg) being higher than the room temperature. Accordingly, various melt processing technologies have been applied for the production of fibers, films, bottles, moldings etc. [1]. The physical and mechanical properties of the PLA is known to depend strongly on the molecular structure consisting of L- and D-isomers. Takasaki et al [2] conducted a research on the high-speed melt spinning of poly(L-lactic acid)s (PLLAs) with different D-lactide contents and reported that the higher D-lactide content in PLLA brings about the suppression of crystallizability. On the other hand, high-speed melt spinning process is a technique for promotion of crystallization through the mechanism of orientation-induced crystallization. It has been reported that the investigation on the thinning behavior of the spin-line can be utilized for the estimation of structure development behavior knowing that the solidification of the spin-line occurs at Tg when the polymer is in an amorphous state, while starting of crystallization leads to the increase of solidification temperature. In the high-speed melt spinning process, crystallization is accompanied by the so-called neck-like deformation of the spin-line [2-4].

Core/sheath fiber is one of unique cross-sectional configurations of bicomponent fibers. In this characteristic design, one component (core) is fully covered by the second component (sheath). Various functional materials can be blended into either core or sheath polymer to introduce functionalities to the fibers. Furthermore, especially in the case of high-speed spinning, mutual interaction between the two polymers from the view point of spin-line dynamics brings about unique structure formation behavior in the spinning process. In previous studies, bicomponent spinnings of various combinations of polymers such as PET/PP, PBAT/PBT, LMPET/HMPET and so forth were attempted to introduce characteristic features to the structure formation behavior such as molecular orientation, orientation-induced crystallization etc. [3-5].

In this research, high-speed melt spinning of core/sheath bicomponent fibers consisting of two types PLLA with different D-lactide content was attempted with the aim of investigating the effect of mutual interaction between the polymers with similar fundamental characteristics except for the crystallizability.

EXPERIMENTAL

Two types of poly(lactic acid) pellet with different D-lactide contents (<0.5% and 4% labelled as LowD and HighD) and with the same MFI of 6 g/cm^3 kindly supplied by Corbion were used in this research. Melting temperature of these polymers were 175 and 155°C, respectively. The polymers were dried at 100°C for 24 hours under vacuum before the melt spinning experiment. The core-sheath (CS) bicomponent fibers were extruded using an extrusion system consisting of two sets of a screw extruder and a gear pump, attached with a spinning pack with a 0.5 mm diameter single-hole spinneret. Four types of CS fibers of PLAs were co-extruded feeding the core/sheath components of LowD/LowD, LowD/HighD, HighD/LowD and HighD/HighD. The throughput rate of 2.5 g/min was set for the core and sheath components. In other words, two types of single-component fiber of LowD and HighD with the throughput rate of 5.0 g/min, and two types of bicomponent fibers with the core/sheath components of LowD/HighD and HighD/LowD with the throughput rate for the core and sheath components of 2.5 g/min were prepared. The spinning temperature at the spin-pack was set at 230 °C. In the high-speed melt spinning, attainable maximum take-up velocity for the LowD/LowD was 8 km/min, whereas the maximum take-up velocity for the other three fibers was 7 km/min. In order to evaluate the thinning behavior of the fibers, on-line spin-line diameter measurement was carried out using a back-illumination type diameter monitor at the distances of every 10 cm up to 330 cm from the spinneret. To characterize the structure and properties of the as-spun fibers, measurements of refractive index, birefringence, wide-angle X-ray diffraction (WAXD) patterns and differential scanning calorimeter (DSC) thermograms were conducted.

RESULTS AND DISCUSSION

Thinning behavior along the spin-line

It was suggested that the use of D-lactide caused the deterioration of spinnability probably because of the suppression of the crystallizability. On-line measurement of diameter profile could be accomplished up to the take-up velocities of 7 km/min for the LowD single component fiber whereas measurement was possible up to 6 km/min for the bicomponent fibers and single component fiber of HighD. As shown in Fig. 1, increase in take-up velocities caused the reduction of final diameter because the throughput rate was kept constant in these experiments. At low take-up velocities of up to around 3 km/min, the spin-line diameter decreased smoothly with the increase of distance from the spinneret. From around 3 or 4 km/min, the neck-like deformation started to appear, indicating the staring of orientation-induced crystallization in the spinning process. The neck-like deformation was milder for the spin-line of single component HighD, whereas in the case of HighD/LowD bicomponent fibers, the neck-like deformation started to appear clearly even from 3 km/min, and accordingly the neck-like deformation was steeper and occurred at positions closer to the spinneret in comparison with the LowD single component and LowD/HighD bicomponent spinnings. These results suggested that the crystallizability of the spin-line was higher in the order of HighD/LowD > LowD/HighD ≅ LowD > HighD.

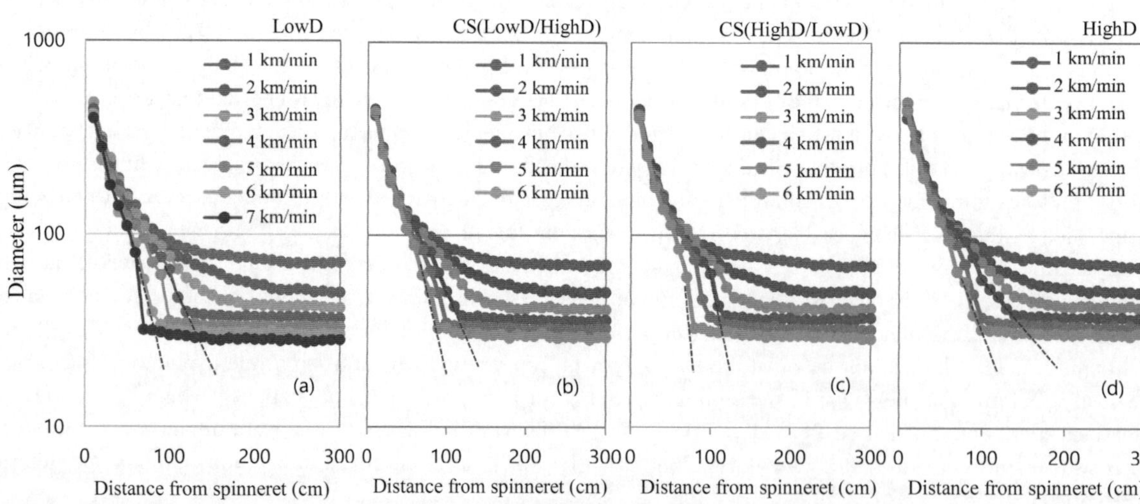

Figure 1. Diameter profiles along the spin-line for the spinning of LowD (a), CS(LowD/HighD) (b), CS(HighD/LowD) (c) and HighD (d) measured at various take-up velocities.

WAXD

The WAXD patterns of single- and bi-component PLA fibers obtained at various take-up velocities are shown in Fig. 2. The fibers prepared at the take-up velocity of 1 km/min only showed the amorphous halo of virtually no orientation, whereas at 3 km/min and above, all fibers exhibited the clear crystalline reflections of highly oriented α-form crystals of PLA. The strongest reflection on the equator can be assigned to the overlapping of (110) and (200) reflections at $2\theta = 17°$, and the second strongest reflection off the equator can be assigned to the (203) reflections at $2\theta = 19°$ [2]. Comparing the WAXD patterns at the same take-up velocities, the HighD fibers exhibited the broadest crystalline reflection with the strongest amorphous halo. This is because of the low crystallizability of this polymer. The amorphous halo in the WAXD patterns of bicomponent fibers were slightly stronger than that for LowD fibers. It is noteworthy that the intensity of amorphous halo for the bicomponent fibers and HighD fibers increased at the take-up velocity of 7 km/min.

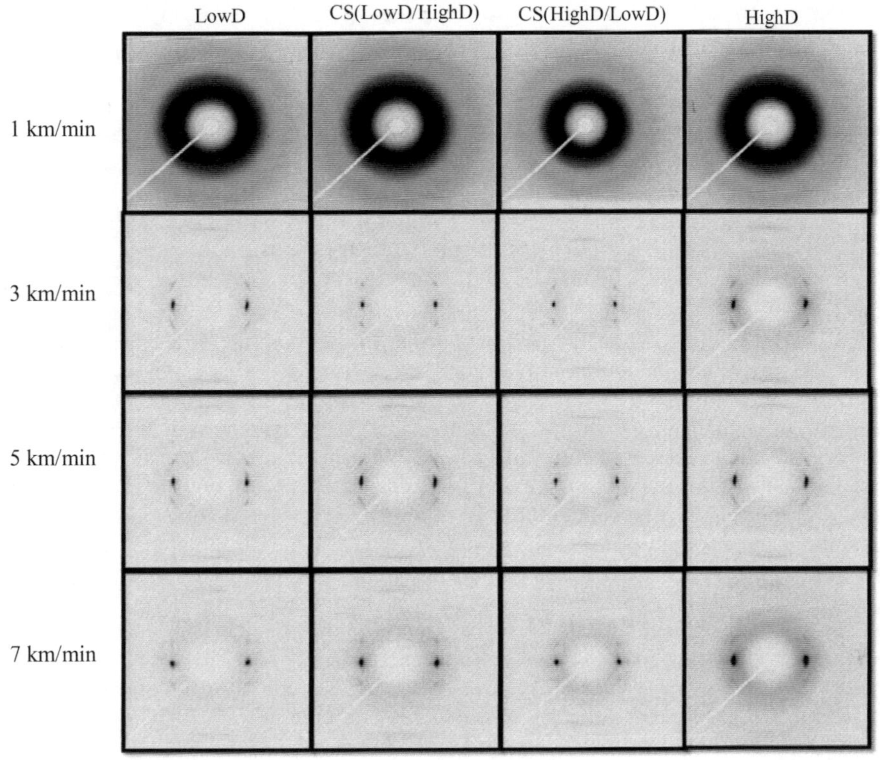

Figure 2. WAXD pattern of as-spun fibers at various take-up velocities.

Birefringence

Refractive indices of the core and sheath components in the as-spun PLA bicomponent fibers were evaluated separately utilizing an interference microscope. Measurements were conducted applying the polarizing filter in the directions parallel and perpendicular to the fiber axis. Birefirngence of the core and sheath components were obtained as the difference between the parallel and perpendicular refractive indices. Variation of birefringence with take-up velocity for the LowD and HighD components are shown in Fig. 3. The birefringence for the LowD component in the single- and bi-component fibers increased gradually with the increase of take-up velocity up to the take-up velocity of 5 or 6 km/min, and then started to decrease slightly with the further increase of the take-up velocity. The birefringence of the LowD in the sheath part of bicomponent fibers showed the highest values at high take-up velocities, whereas the single-component LowD fibers showed the lowest birefringence. Regarding the HighD component, birefringence of the single-component HighD fiber showed a similar trend with the change of take-up velocity in comparison with the single-component fiber of LowD, however the values were significantly lower. The birefringence of the HighD component in the bicomponent fibers exhibited much lower values in comparison with the single-component HigD, especially when the HighD was placed at the core. The birefringence decreased with the increase of take-up velocity.

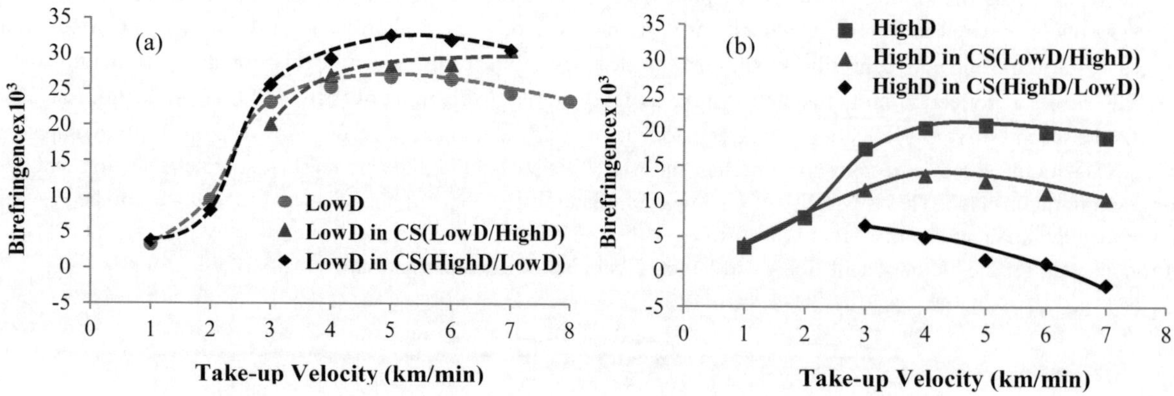

Figure 3. Change of birefringence with take-up velocity for the LowD PLA (a) and the HighD PLA (b) in the single- and bi-component PLA fibers.

Thermal properties

Thermal properties of the high-speed spun PLA fibers were investigated through the DSC analysis. The DSC thermograms of the HighD/LowD bicomponent fibers obtained at various take-up velocities are shown in Fig.4. With the increase of temperature, glass transition, cold crystallization and melting were observed consecutively. At low take-up velocities of 1 and 2 km/min, double cold crystallization exothermic peaks were observed. The peaks of lower and higher temperatures obviously correspond to the cold crystallization of LowD and HighD. With the increase of take-up velocity, temperature of cold crystallization decreased gradually because of the increase of molecular orientation. Above 4 km/min, the cold crystallization temperature started to increase again. This result corresponds to the reduction of birefringence with take-up velocity for the HighD component in the bicomponent fibers. On the other hand, three melting peaks were observed except for the 1 and 2 km/min fibers. The lowest melting peak (Peak L) should correspond to the melting of HighD. In the case of high-speed spun PLA fibers, Takasaki et. al. reported that the two melting peaks could be assigned to the melting of crystals formed in the spinning process, the peak at lower temperature, and the melting peak appearing as the result of melting and recrystallization in the DSC heating process [6]. In their case, the PLLA with D-lactide content of 1.3 % was used. Probably, the same concept can be applied for the melting peaks M and H in Fig. 4. Distinct appearance of high temperature melting peak may be caused because the D-lactide content of the PLLA used in this research was lower. Variation of temperature of melting peaks with the increase of take-up velocity is shown in Fig.5. It appears that the lower cold crystallization temperature yielded the higher melting peak temperature for HighD. For the higher temperature double melting peaks, which probably correspond to the melting of LowD, the temperatures of peaks at lower and higher temperatures gradually decreased and increased, respectively, with the increase of take-up velocity. This behavior was also similar to the one reported by Takasaki.

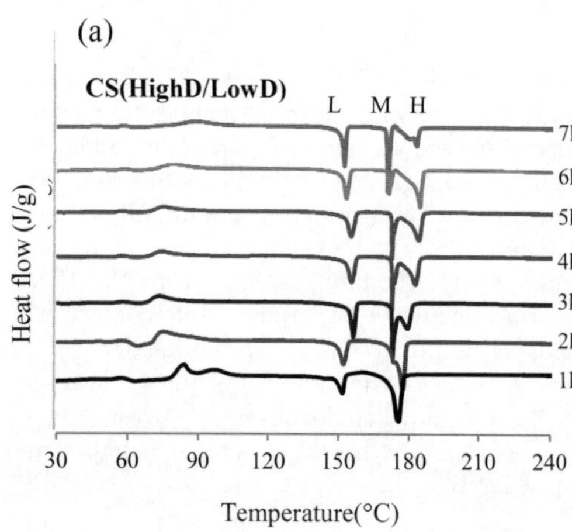

Figure 4. DSC thermograms of CS(HighD/LowD) as-spun fiber prepared at various take-up velocities.

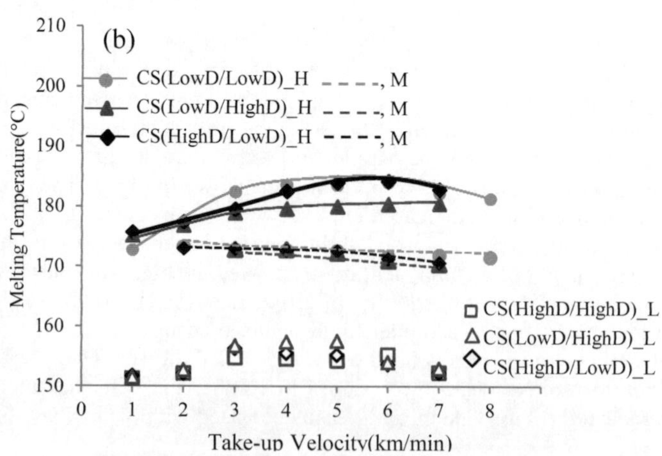

Figure 5. Variation of temperatures of multiple melting peaks with take-up velocity for PLA single- and bi-component fibers.

Mechanism of Structure Development in High-Speed Bicomponent Spinning

Summarizing the above, it was revealed that the orientation-induced crystallization started to occur from around 2 km/min accompanying the neck-like deformation in the spin-line. Crystallizability was higher in the order of HighD/LowD > LowD/HighD ≅ LowD > HighD, and the higher crystallizability corresponded to the occurrence of distinct neck-like deformation at the position closer to the spinneret, i.e. at higher temperature. In bicomponent spinning, no mutual interaction of two components was observed before the starting of crystallization at low take-up velocities, while once the orientation-induced crystallization of LowD started, orientation and crystallization of HighD was significantly suppressed. This tendency was more distinct when the LowD was used as the sheath component.

From the view point of spin-line dynamics, similar thinning behavior of LowD and HighD single component spinning at low speeds corresponds to the no mutual interaction of two components in bicomponent spinning of low speeds. Steeper neck-like deformation occurring at a position closer to the spinneret for the LowD in comparison with the HighD is the origin for the mutual interaction of two components in the bicomponenst spinning process, and eventually, molecular orientation of LowD was enhanced and that of HighD was suppressed in comparison with the individual single component spinning. Once orientation-induced crystallization of LowD in bicomponent fiber started to occur selectively, increase of take-up velocity leads to the occurrence of neck-like deformation at higher temperatures. This leads to the further orientation relaxation of the HighD. Regarding the exchange of the sheath and core components in the bicomponent spinning, crystallization of LowD could be promoted when it was used as the sheath component because of the enhanced cooling of the outer part of spin-line cross-section, and eventually difference in structure development of LowD and HighD could be enhanced.

CONCLUSIONS

The four types of core/sheath fibers consisting of low and high D-lactide content PLLAs were prepared by high-speed melt spinning process. In all cases, orientation-induced crystallization started above certain take-up velocities, in that crystallizability could be judged from the position of neck-like deformation in the spin-line. For the single-component spinning, the neck-like deformation was closer to the spinneret in the LowD spin-line than that in the HighD spin-line, whereas the position of neck-like deformation was closest to the spinneret for the HighD/LowD bicomponent spin-line. Birefringence of the as-spun fibers increased with the take-up velocity, and the LowD single-component fiber showed higher birefringence than the HighD single-component fibers at the same take-up velocity, whereas the birefringence of LowD component in the HighD/LowD bicomponent fiber exhibited the highest birefringence. These results suggested that the orientation-induced crystallization of LowD in the bicomponent fiber was promoted because of the existence of HighD in the fiber cross-section.

ACKNOWLEDGEMENT

The poly(lactic acid) pellets used in our experiments were supplied by Corbion. Our gratitude to Corbion for supporting our research.

REFERENCES

1. O. Avinc and A. Khoddami, Fibres. *Fibre Chemistry* **41**, 391-401(2009)
2. M. Takasaki, H. Ito, T. Kikutani, *Journal of Macromolecular Science Part-B,* **42**, 57-73 (2003)
3. T. Kikutani, J. Radhakrishnan, S. Arikawa, A.Takaku, N. Okui, X.Jin, F. Niwa , and Y.Kudo, *Journal of applied Polymer Science* **62**, 1913-1924 (1996)
4. J. Radhakrishnan, T. Kikutani, and N. Okui, *Textile Research Journal* **67(9),** 684-694 (1997)
5. X.Q. Shi, H. Ito and T. Kikutani, *Polymer* **47**, 611-616 (2006)
6. M. Takasaki, N. Fukushi, M. Yoshizawa, S. Onosato, M. Hanada, W. Takarada, Y. Kawahara, T. Kikutani, H. Kobayashi, and K. Tanaka, *Journal of Macromolecular Science Part-B,* **56**, 143-160 (2017)

Analysis of Fiber-thinning Behavior in Melt-blowing Process by using Numerical Analysis based on a Particle Method

Wataru Takarada, Sho Hatano and Takeshi Kikutani

Department of Materials Science and Engineering, Tokyo Institute of Technology
2-12-1-S8-31, Ookayama, Meguro-ku, Tokyo 152-8552 Japan

Abstract. Melt-blowing experiments were conducted in various spinning conditions and fiber formation behavior was investigated using high-speed camera recording and on-line velocity measurements. In addition, fiber diameter of resultant web was measured by SEM. Numerical model and calculation method based on a particle method was developed and fiber formation behavior in melt-blowing process was investigated using this model by comparison with experimental results. The model could represent fiber formation behavior in melt-blowing process by adjusting surface tension. Blown spin-line became unstable at low thorough-put conditions both in experiment and in numerical analysis. Instability of spin-line caused broadening of fiber diameter distribution in resultant webs. Numerical analysis suggested that almost half of total mass was occupied by a few thick fibers. The rest of mass formed huge numbers of fine fibers. In SEM observation, thick fiber were easily overlooked because its number was so small. It underestimate average fiber diameter. In other words, melt-blowing process can produce fine fibers in spite of relatively slow spinning speed because of instability of spin-line.

Keywords: Enter Keywords here.
PACS: 83.50.Uv

INTRODUCTION

The non-woven materials with fine fibers are strongly required in this field to improve both of high filtration accuracy and low pressure drop. Melt-blowing process is well developed process for produce non-woven materials with fine fibers. The non-woven materials which has a mean fiber diameter 1 micron or finer has already been developed commercially. However, fiber formation behavior in melt-blowing process has not been fully understood yet.

Basic numerical model of melt-spinning process and calculation method for that model in the 1960's. This model could describe spin-line deformation very well. The model was described in Eulerian coordinate system, so it was simple and powerful model for calculate steady-state spin-line. On the other hand, application of this model on melt-blowing process caused difficulty in calculation stability. In this study, numerical model and calculation method based on a particle method was developed and fiber formation behavior in melt-blowing process was investigated using this model.

EXPERIMENTAL

Melt-blowing Experiments

Lab-scale twin-screw extruder with metering pump were used with melt-blow die and hot air supply. Melt-blowing conditions are listed in Table.1. Stability of spin-line was observed by high-speed camera (Phantom V, Vison Research Inc.). Air velocity was measured by Pitot tube. Fiber diameter was observed by SEM photograph of resultant web.

TABLE 1. Melt-blowing conditions

Material	Polypropylene (MFR 1550 g/10min)
Spinneret	0.12 mm ϕ, 10 holes
Spinneret temperature	250 °C
Air heater temperature	290 °C
Throughput rate	8 – 500 mg/min/hole

Air supply	100 – 300 NL/min
Die to Collector Distance (DCD)	10 – 70 cm

Modeling for numerical analysis

Spin-line was described by series of cylindrical elements (particles) connecting each other. Fig.1 shows schematic illustration of the model and acting forces on an element. Force balance equation on the element i is described as:

$$F_i = -(F_g + F_{air} + F_{rheo} + F_{surface}) \qquad (1)$$

Here, F_i is the inertia force, F_g is the gravitational force, F_{air} is air friction force, F_{rheo} is the rheorogical force and $F_{surface}$ is the surface tension, respectively. Using the position of the element $x_{i,t}$, this equation can be described as:

$$\begin{aligned} ma_i = mg &+ \pi D_i \tau_{fi}(x_{i,t} - x_{i+1,t}) \\ &- \frac{A_i\eta_i}{dt}\left(\frac{x_{i,t+\Delta t}-x_{i+1,t+\Delta t}}{x_{i,t}-x_{i+1,t}} - 1\right) + \frac{A_{i-1}\eta_i}{dt}\left(\frac{x_{i-1,t+\Delta t}-x_{i,t+\Delta t}}{x_{i-1,t}-x_{i,t}} - 1\right) \\ &- \frac{\pi\gamma_i}{2}\left\{\left(D_i - \frac{\pi D_i{}^4}{4U}\right) - \left(D_{i-1} - \frac{\pi D_{i-1}{}^4}{4U}\right)\right\} \end{aligned} \qquad (2)$$

Here, a_i, D_i, A_i, η_i, γ_i and τ_{fi} is acceleration, diameter, cross-sectional area, viscosity, surface tension and air friction on the element i, respectively g is acceleration of gravity. In this study, temperature dependence on polymer density was neglected.

When the element moves on a constant acceleration a_i during a short period Δt, position could be described as:

$$x_{i,t+\Delta t} = x_{i,t} + v_i\Delta t + \frac{1}{2}a_i\Delta t^2 \qquad (3)$$

Here, v_i is the velocity of element i. Using equation (2) and (3), $x_{i-1,t+\Delta t}$, $x_{i,t+\Delta t}$ and $x_{i+1,t+\Delta t}$can be described using $x_{i-1,t}$, $x_{i,t}$ and $x_{i+1,t}$. Solving simultaneous equation for the all elements, motion of the elements could be analyzed.

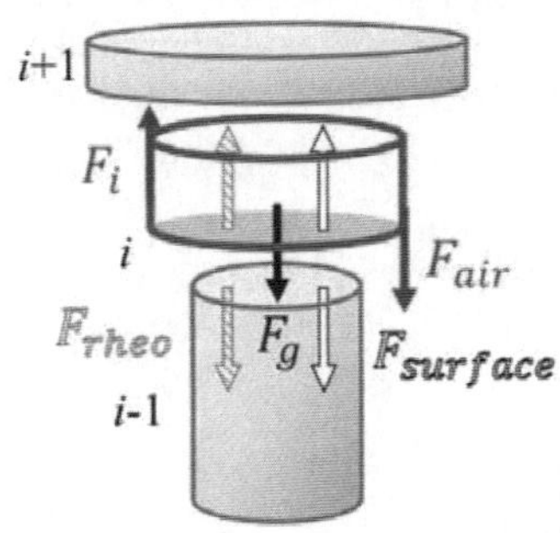

FIGURE 1. Schematic illustration of the model and acting forces on an element.

RESULTS AND DISCUSSIONS

Fig.2 shows the consecutive photograph of melt-blowing spin-line captured from high-speed video. Spin-line became unstable with low throughput and high air supply conditions. Spin-line diameter under the spinneret varied periodically. It likely to intermittent dripping in extremely low throughput condition. Numerical analysis could represent similar periodic instability shown in Fig.3 by adjusting the effect of surface tension.

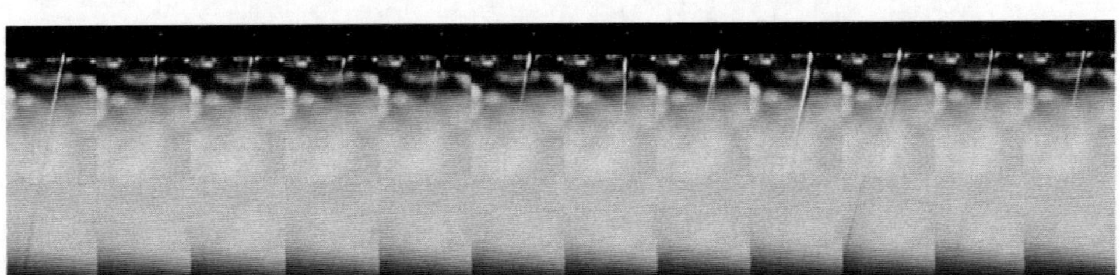

FIGURE 2. Consecutive photograph of melt-blowing spin-line. Interval of each photograph is 2 ms. Through-put: 13.5 mg/min/hole, Air supply: 300 NL/min

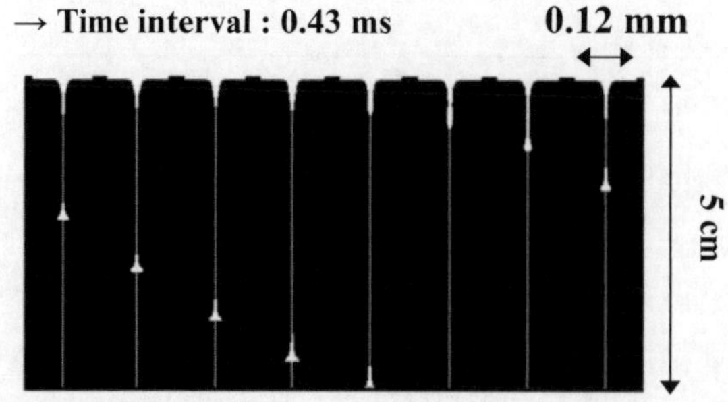

FIGURE 3. Calculated motion of melt-blowing spin-line by numerical analysis
Through-put: 12.5 mg/min/hole, Air supply: 300 NL/min, Adjustment factor of surface tension: 19.44

Fiber diameter distribution based on the number of fibers in each bin by numerical analysis and SEM measurement were shown in Fig. 4 a). General trend of diameter distribution was similar in numerical analysis and experiment except for the existence of very thick fiber which diameter was larger than 20 μm. Number of this thick fiber was samll, but it had huge weight because of large diameter. Fig. 4 b) shows probability profiles based on the weight in each bin. Probability of the fibers which diameter larger than 5 μm was 0.08 in number based, on the other hand, that was 0.83 in weight based. This thick fiber rarely observed under SEM measurement and sometimes people exclude those fibers from average calculation. This exclusion affected a little on the number-average diameter, but significant affected on weight-average diameter.

Results of numerical analysis in various conditions indicated that when the spin-line was stable, fiber diameter decreased with decreasing throughput linearly. In contrast, when the spin-line became unstable, number-average diameter decreased but weight-average diameter increased. So there was an optimum throughput to observe minimum weight-average diameter in the web. Some web properties, for example, adsorption properties are more sensitive for weight-average diameter than number-average diameter. If you requires such properties, you should adjust throughput closer to the optimum throughput.

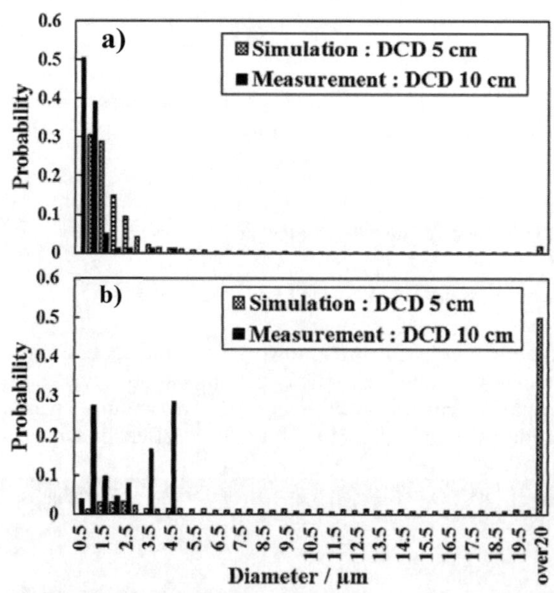

FIGURE 4. Diameter distribution in the resultant webs by numerical analysis (Simulation) and SEM measurements (Measurement). a) probability profiles based on number b) probability profiles based on weight

Experimental Study of a Multi-Row Meltblowing Process and Web Structures

Stephen Barilovits[a], Eunkyoung Shim[a], Saad Khan[a], and Benham Pourdeyhimi[a,b]

[a]*North Carolina State University*
[b]*Nonwovens Institute*

Abstract. The Biax Spunblown® process is a unique variant of the meltblowing process. Molten polymer, often polypropylene, is extruded from an array of outlets, each of which is individually surrounded by a concentric high velocity air stream. To provide the first thorough understanding of this unique process, we have experimentally investigated the effects of critical processing parameters on the fiber and web formation process. Air temperature, air speed, and polymer throughput have been related to fiber diameter distribution and distance to solidification. Mean fiber diameters ranged from less than 1 μm to almost 14 μm. Air temperature effects were investigated using high-resolution infrared thermography and indicated that substantial heat transfer occurs between the air and molten polymer prior to spinning. A detailed three-dimensional air temperature and velocity profile is presented for an array of supplied temperatures and air cavity pressures. Ultra-high-speed videography was used to observe the onset and propagation of bending instabilities, showing that for some conditions, complex bending perturbations likely occur when the polymer is still molten.

Keywords: Polypropylene, Meltblowing, Heat Transfer, Rheology
PACS: 61.41.+e, 83.50.Ax, 83.50.Jf, 83.60.Jk, 83.80.Sg, 07.07.Df, 07.070Hj, 07.07.Tw, 07.20.Dt, 07.57.Ty, 07.61.+m, 07.79.-v

---

Meltblowing is one of the preferred commercial methods of producing fine fibrous webs and fabrics. Meltblown fibers are produced by injecting a molten thermoplastic polymer into a high velocity heated air stream. Fibers are accelerated, stretched and cooled before landing on a collection belt. The traditional "slot die" design contains a single row of fibers that are attenuated by two converging air curtains. The fiber formation process using this die design is well understood since it has been the subject of much experimental and theoretical work. [1]–[4] Other designs have been patented and commercialized [5]–[7], but none have been as thoroughly studied as the slot die.

One such alternative meltblowing die design contains multiple rows of spinning fibers, each of which is surrounded by an annular air stream.[8] This design may have benefits in overall throughput, fiber structure produced, and may provide polymer processing versatility. However, little is known about multi-row meltblowing since published information is limited to its patents. The purpose of this investigation is to give the first thorough understanding of the fiber formation process in a multi-row meltlblowing design. A two-row meltblowing design (Spunblown® system from Biax Fiberfilm Corporation) was selected to examine the unique spinneret arrangement and aerodynamics.

(a)

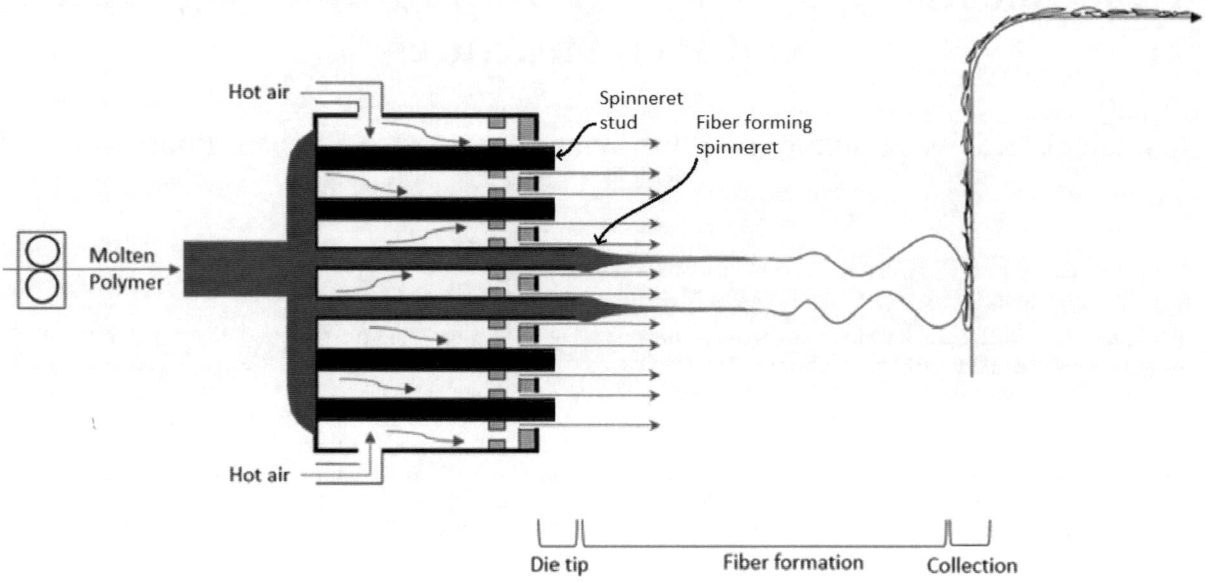

(b)

FIGURE 1. Two-row meltblowing schematic side view; indicated spinneret studs are non-fiber-forming spinnerets that emit only air and insulate the spinning fibers form entrained room air. (b) Photo of an eight fiber-forming spinneret Spunblown® die (photo courtesy Biax Fiberfilm Corporation)

To gain a fundamental understanding of the multi-row meltblowing process, polymer properties and processing conditions have been related to web properties for the die geometry shown in Figure 1. Polypropylene with a Melt Flow Rate (MFR) of 500 was chosen as the model polymer for this study since it is common industrially for this die, and is spinnable over a wide range of conditions. Thermal properties and flow behavior of the molten polymer have been examined to provide a window of optimal processability. Differential Scanning Calorimetry (DSC) gave a melting curve and indicated a minimum processing temperature while Thermogravimetric Analysis (TGA) gave a degradation curve and indicated a maximum processing temperature of about 250 °C. Melt rheology has been examined at low and high shear rates and viscosity is presented as a function of temperature and shear rate. These viscosities may be used to establish a viscosity-process-fiber diameter relationship (as opposed to a polymer specific relationship) to generalize the conclusions from this work.

In order to estimate process-relevant viscosities, polymer temperature must be known during the spinning process. An infrared (IR) thermography approach was adopted to measure fiber temperature during the spinning process as a function of distance from the spinneret. The IR emissivity of fibers was determined through experimental calibration of monofilament polypropylene and a previous report where emissivity of fibers was measured as a function of diameter for very small fibers. [9] An IR monitored fiber forming trial was performed spanning 47 conditions varying air temperature, air speed, polymer throughput, and die-collector-distance (DCD).

IR images were taken at each condition of the fibers forming near the die face. The processing window was established based on visual observations of the stability of the spinning fibers. IR results indicate that air temperature has a significant impact on polymer extrusion temperature and that larger fibers cool more slowly than smaller fibers.

In addition to process-relevant viscosities provided in part by the IR measurements, air speed and temperature must be quantified at various positions to explain the diameter attenuation. To do this, air speed and temperature were measured simultaneously for a three dimensional array of positions in front of the die face. Air speed and temperature was recorded for 62 positions for nine die pressure and air cavity temperature combinations (identical air conditions to the IR monitored fiber spinning trial). Air cavity pressures and temperatures were correlated to external air speed and temperatures, respectively. Examination of the air profile reveals that air heat always disperses more rapidly than does air momentum; this is especially obvious very close to the die face. These results indicate that non-fiber-forming air rows on the perimeter of the fiber forming rows may assist in reducing fiber diameter distribution.

Fiber diameter distributions for all webs formed were measured by Scanning Electron Microscopy (SEM). Average diameters ranged from 0.76 µm to 13.4 µm and depended on processing parameters in descending order of importance: polymer throughput, air temperature, air speed. A bar chart shown in Figure 2 displays the range, median, and mean fiber diameters for each process condition.

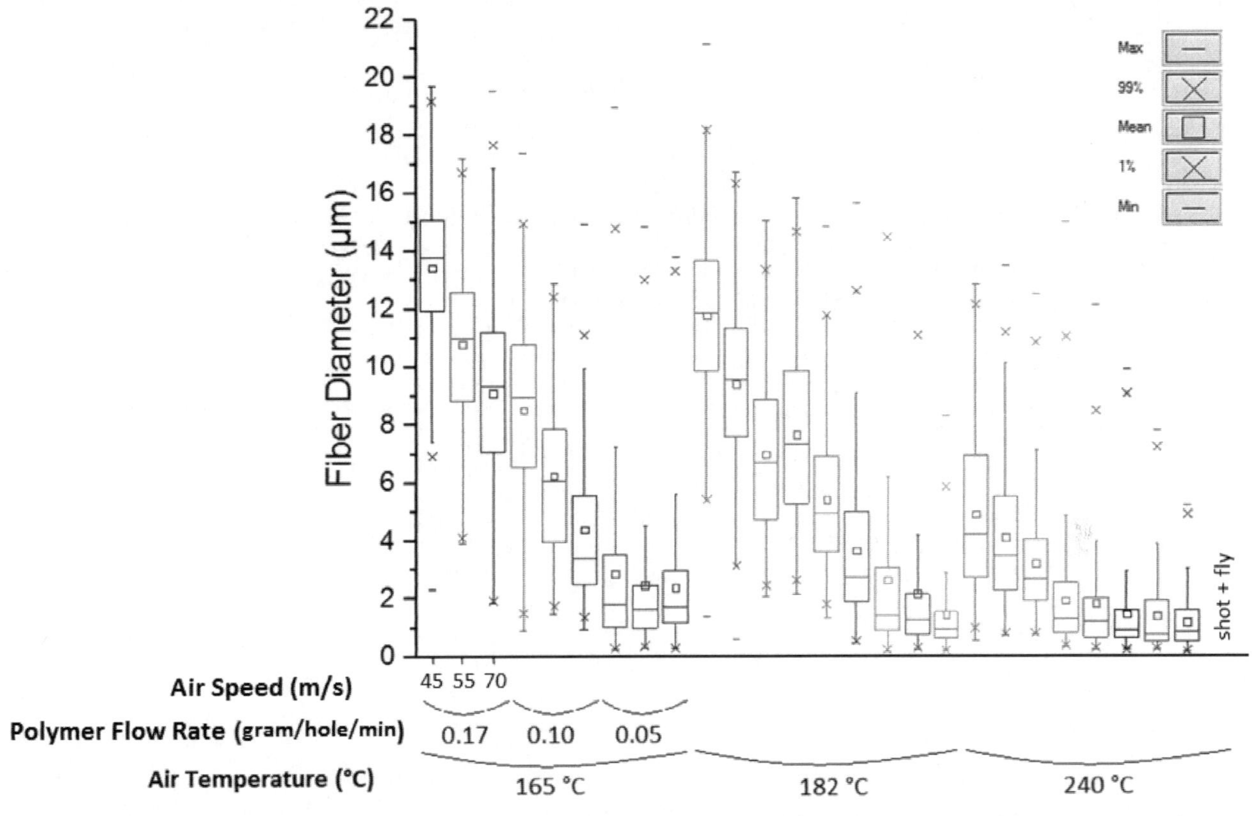

Figure 2: Fiber diameter distribution box-plot for air temperature, pressure, and polymer throughput conditions. DCD set to 40 cm, air speed and temperature values shown were measured 1 cm in front of spinnerets.

Samples with average fiber diameter below about 1.3 µm began to form shot. Resultant diameter distributions have been related to air temperature and velocity, as well as to the polymer viscosity deduced from the temperature as determined by the IR measurements.

Although the most fiber diameter attenuation occurs in the first centimeter past the spinnerets, rapid flapping motions - termed bending instabilities – farther from the die may play an important role in reaching the minimum fiber size possible. Ultra-high-speed videography (Phantom v1612) has been used to examine the extent to which bending instability action in the turbulent region affects fiber diameter, roping, and comingling of fibers before collection. A single fiber's lateral motion was tracked at several distances from the spinneret to observe and quantify the magnitude of bending instabilities. Unfortunately, resolution limitations prevent such direct tracking of a single fiber except for processing conditions producing the largest fibers spun. Bending instabilities close to the spinnerets however, are visible for all processing conditions and are present to a much greater extent when initial draw down rates are higher. Smaller fibers have a much lower bending stiffness and can begin flapping much closer to the die. Indeed, the distance between the spinneret and the point at which a fiber begins to fold back on itself was measurable for all conditions. When this distance was compared to the carefully measured air temperature profile, it is clear that for some conditions fibers are backfolding when the air surrounding them is at a higher temperature than the DSC determined crystallization temperature of this polypropylene. Because the fibers temperature, at points in front of the spinnerets, is expected to be greater than the air temperature, complex bending instabilities are likely occurring while the fiber is still molten. This suggests that the bending instabilities further reduce the fiber diameter when air temperature and initial attenuation rate are high.

ACKNOWLEDGMENTS

The authors kindly acknowledge the Nonwovens Institute and its member companies for the support of this research. Also special thanks to the Chemical and Biomolecular Engineering Department, and to North Carolina State University.

REFERENCES

1. A. L. Yarin, S. Sinha-ray, and B. Pourdeyhimi, "Meltblowing : Multiple polymer jets and fi ber-size distribution and lay-down patterns," *Polymer (Guildf).*, vol. 52, no. 13, pp. 2929–2938, 2011.
2. R. Butin, J. P. Keller, and J. Harding, "Non-Woven Mats by Melt Blowing," 38492411974.
3. M. Uyttendaele and R. L. Shambaugh, "Melt Blowing : General Equation Development and Experimental Verification," *AICHE J.*, vol. 36, no. 2, pp. 175–186, 1990.
4. D. H. Tan, C. Zhou, C. J. Ellison, S. Kumar, C. W. Macosko, and F. S. Bates, "Meltblown fibers: Influence of viscosity and elasticity on diameter distribution," *J. Nonnewton. Fluid Mech.*, vol. 165, no. 15–16, pp. 892–900, 2010.
5. B. Haynes and M. C. Cook, "Die for Producing Meltblown Multicomponent Fibers and Meltblown Nonwoven Fabrics," 7150616, 2006.
6. A. S. . Fabbricante, G. F. Ward, and T. J. Fabbrincante, "Micro-Denier Nonwoven Materials Made Using Modular Die Units," 6,114,017, 2000.
7. V. T. Marla, R. L. Shambaugh, and D. V Papavassiliou, "Using Swirl Dies To Spin Solid and Hollow Fibers," *Ind. Eng. Chem. Res.*, vol. 45, pp. 2331–2340, 2006.
8. E. C. A. Schwarz, "Apparatus and Process of Melt-Blowing A Fiberforming Thermoplastic Polymer and Product Produced Thereby," 4380570, 1983.
9. V. T. Marla, R. L. Shambaugh, and D. V Papavassiliou, "Use of an Infrared Camera for Accurate Determination of the Temperature of Polymer Filaments," *Ind. Eng. Chem. Res.*, vol. 46, pp. 336–344, 2007.

Poly(arylene ether ketone) ion exchange membranes grafted with long aliphatic pendant sulfonated groups for vanadium redox flow batteries

Kim Dukjoon* (1)

(1) Sungkyunkwan University - Kyunggi - Korea

As an energy storage system (ESS), all vanadium redox flow battery (VRFB) has drawn much attention for its long life time, flexible design, low maintenance cost, and high energy capacity. The proton exchange membrane, one of the core components of VRFB, is prepared from poly(arylene ether ketone) (PAEK) with long pendant sulfoalkyl groups. We identify the chemical structure of PAEK using 1H-NMR and FT-IR. The essential properties as a membrane in VRFB application such as proton conductivity, ion exchange capacity, and vanadium ion permeability along with the thermal, mechanical and chemical stability are tested for all membranes synthesized, and are compared with those of a commercial Nafion 117® membrane. As the proton conductivity increases with the degree of sulfonation (DS), the sulfonated PAEK membrane with DS of 40 % (S-PAEK-40) shows even higher proton conductivity than Nafion 117® membrane. All of the S-PAEK membranes are thermally stable at up to 200℃ and no chemical degradation is observed in NaOH solution in the present experiment. The long aliphatic side chains prohibit the approach (attack) of vanadium ion to the polymer backbone (phenyl rings) and thus it results in lower VO+2 permeability than Nafion 117. While the ion cluster dimensions of S-PAEK membranes are larger than that of Nafion 117® membrane, S-PAEK membranes display higher energy efficiency than Nafion 117® membrane. In addition, S-PAEK membranes show excellent stability during the cycling test with 100 cycles.

Effect Of Potassium Sorbate (KS) On Antibacterial Smart Packaging

Atabaki-Pasdar Pouria (1), Mehranpour Milad (1)*, Oromiehie Abdulrasoul (2), Bahari-Aban Babak (1)

(1) Islamic Azad University Science & Research branch - Tehran - Iran, (2) Islamic Azad University South Tehran branch - Tehran - Iran

In this article, antibacterial film based on low density polyethylene (LDPE), linear low density polyethylene (LLDPE), potassium sorbate (KS) and ethylene vinyl alcohol (EVOH) copolymer were prepared and characterized using field emission scanning electron microscopy(FESEM), immigration test and oxygen permeability. The oxygen barrier properties of the LDPE/LLDPE/EVOH/KS film were significantly better than those of the LDPE film. Moreover, the antibacterial film containing potassium sorbate decreased the growth rate and maximum growth of yeast, and extended the lag period before mold growth was apparent. Therefore, it can prevent or reduce the rate of microbial spoilage in low viscosity liquids and on the contacted surface area of solid food products and as the result may prolong the shelf life of food products when it is used as a packaging material. Furthermore, Total Plate Count (TPC) results demonstrate could decrease bacterial growth by adding KS in comparison with LDPE/LLDPE film. a novel sticker sensor has been fabricated based on methyl red, and tests have been conducted to detect the freshness of meat by its color change in real time (when the red methyl red paper changed to yellow). These are prepared by entrapping within a pH sensitive dye that responds, through visible color changes to the spoilage volatile compounds that contribute to a quantity known as total volatile basic nitrogen (TVBN). Finally, the methyl red was successfully used as a sticker sensor for the real-time monitoring of meat freshness in chiller conditions.

Water filtration by green electrospinning membranes

Ribba Laura (1)*, Cimadoro Jonathan Daniel (1), Goyanes Silvia (1)

(1) Universidad de Buenos Aires. FCEyN, Depto. de Física, LP&MC, IFIBA-CONICET - Buenos Aires - Argentina

The availability of affordable clean water is one of the technological, social, and economic main issues of the 21st century. The removal of solids through filtration is a fundamental step on the path to secure clean water from natural water sources as well as to recycle waste water for reuse. Different filtration techniques are currently applied, including micro and nanofiltration, reverse osmosis and diffusion dialysis among others. All these methods use a porous medium capable of retaining solids larger than its porosity size and allowing the passage of fluid and smaller particles. PVA electrospun meshes have been widely studied for many applications, in particular as filtering media. However, previous treatments are needed in order to enhance its low water stability. In this work we use a green method based in the addition of citric acid and a subsequent heat treatment to crosslink PVA. The aim of the current work was to develop PVA electrospun membranes with enhanced water stability and characterize its morphological changes as it is applied as nano-filters of aqueous systems. In this sense, the filtering performance of the mesh was characterized using TiO_2 nanoparticles suspensions and a complete study on the morphology of the mats was carried out through SEM and AFM techniques, following the changes with heat treatment and water exposure. Besides hydrophilicity and solubility of the membranes were also studied. Morphological changes were related to changes in the filtering efficiency.

Investigation of Fiber Diameter Distribution in a Multi-Row Meltblowing Process

Stephen Barilovits[a], Eunkyoung Shim[a], Saad Khan[a], and Benham Pourdeyhimi[a,b]

[a]*North Carolina State University*
[b]*Nonwovens Institute*

Abstract. Meltblowing is one of the most popular high-throughput methods of producing fine polymeric fibrous webs. The traditional design contains a single row of spinnerets between two converging high-velocity air curtains. A multi-row variant to the meltblowing process contains spinnerets that are individually surrounded by concentric air streams. The latter system produces fibrous webs with a unique fiber diameter distribution. We have begun an investigation of the origins of the fiber diameter distribution of a multi-row meltblowing system. Observations from ultra-high-speed videography of a two row system indicate that, for some conditions, fiber-forming throughput oscillates with time for each spinneret. The fact that a single spinneret can produce both large and small fibers contradicts the prevailing intuition that the fiber diameter distribution can be explained entirely by the two dimensional thermal profile characteristic of an array of spinnerets. A similar phenomenon has been observed in melt spinning processes, where it has been termed "draw resonance". We hypothesize that we are observing draw resonance in a meltblowing system and are experimentally and theoretically investigating the matter.

Keywords: Polypropylene, Meltblowing, Draw Resonance
PACS: 61.41.+e, 83.50.Ax, 83.50.Jf, 83.60.Jk, 83.80.Sg, 07.07.Df, 07.070Hj, 07.20.Dt, 07.61.+m, 07.79.-v

---

Meltblown fibers are produced by injecting a molten thermoplastic polymer into a high velocity heated air stream. Fibers are accelerated, stretch and cool before landing on a collection belt. The traditional "slot die" design contains a single row of fibers that are attenuated by two converging air curtains. The fiber formation process using this die design is well understood since it has been the subject of much experimental and theoretical work. [1]–[4] Other designs have been patented and commercialized [5]–[7], but none have been as thoroughly studied as the slot die.

One such alternative meltblowing die design contains multiple rows of spinning fibers, each of which is surrounded by an annular air stream.[8] This design may have benefits in overall throughput, fiber structure produced, and may provide polymer processing versatility. Although a comprehensive investigation of the process is lacking from the public literature, there is substantial collective knowledge about the process from users and designers in the industry. One of the known characteristic trends of multi-row meltblown fibers is a slightly wider fiber diameter distribution as compared to fibers formed using traditional meltblowing. A two dimensional array of spinning fibers has interior and perimeter spinnerets; fibers originating from interior spinnerets are more insulated from cool entraining air and may therefore draw down to a finer fiber than a perimeter fiber. The resulting web could therefore have a wider fiber diameter distribution than a web formed from a slot die design, in which all fibers are exposed to approximately the same drawing conditions.

While this intuitive explanation of the origins of the multi-row meltblowing fiber diameter distribution may be partly correct, there is reason to believe other factors play a role. A similarly wide fiber diameter distribution has been produced using just a two-row meltblowing die. This geometry is symmetric with all spinnerets on the perimeter. The fiber diameter distribution is therefore likely influenced by a phenomenon occurring at each spinneret.

To investigate the behavior of fast-moving fibers, ultra-high-speed videography (Phantom v1612) was used to observe the fibers in the region of highest attenuation rate, very close to the spinnerets. At high throughputs and low air temperatures, fibers near the spinneret have a nearly constant diameter with time. At reduced throughputs and higher air temperatures however, the fiber diameter at the spinneret oscillates with time. In the latter situation, although the flow of polymer through the spinneret is continuous, a bead or reservoir of polymer forms at the spinneret, causing the *fiber forming throughput* to fluctuate with time. In modest cases, the drawing action downstream of the spinneret pulls the fiber out from the bead faster than the spinneret can resupply the polymer, causing a temporary bead depletion. After the bead has been depleted, a very small fiber is still formed, but with less fiber forming throughput than the spinneret supplies. The bead can then grow again until it is swept away into a larger fiber. When the polymer throughput is lowered further, this pulsation at the spinnerets becomes more pronounced and the fiber may even break, causing shot. The observations made can explain the fiber diameter distribution and are worthy of this detailed description. The situation for a single fiber and spinneret over time is shown in Figure 1.

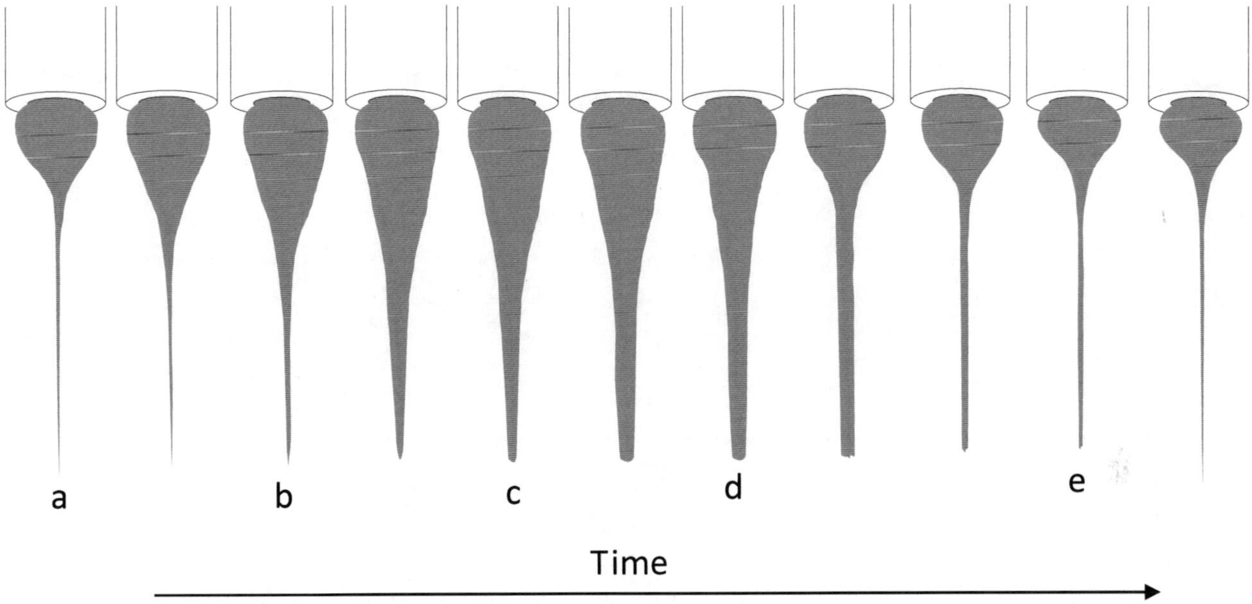

FIGURE 1. Schematic snapshots in time of the oscillatory behavior of a single fiber issuing from a single spinneret. At time (a) the bead size is at a minimum and the fiber formed is very narrow; at (b) the bead grows but the fiber formed remains small; at (c) the bead grows and extends as the fiber formed also enlarges; at (d) the bead size quickly shrinks as the fiber formed is large; at (e) the bead is depleted to a minimum and the fiber formed is narrow again.

The buildup of this bead of polymer occurs at the center of each spinneret, where it may be shielded from the high speed air stream flowing between spinnerets. Eventually it becomes large enough to encounter sufficient air drag to be drawn into a fiber, which then pulls more polymer out of the bead.

These oscillations are not invisible to the naked eye, to which they appear as pulsations or "spitting" from the spinnerets. Such periodic fluctuations of flow have been observed in other polymer processing applications such as melt spinning and film casting and are known as draw resonance.[9] The similarity of the effect and time scales involved indicate that draw resonance could help explain the fiber diameter distribution characteristic of the multi-row meltblowing process. To better understand the effect, the period of the pulsations was monitored as a function of processing condition, see Figure 2.

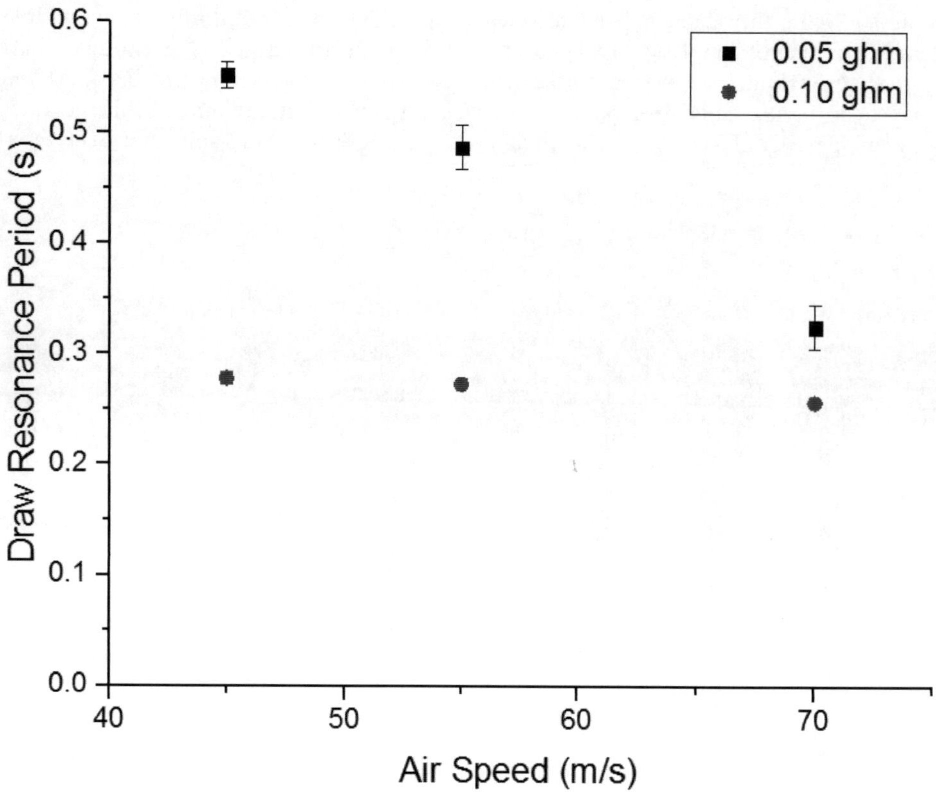

Figure 2: Draw resonance period versus air speed for two polymer throughputs (units of grams of polymer per hole per minute, ghm). Air temperature was 240 °C, and resonance was not observable at 0.17 ghm.

The resonance periods were quite repeatable for each condition so error bars were small. Throughput has the greatest impact on resonance period, with higher throughputs causing faster pulsations because the bead can refill more quickly (continuously at even higher throughputs). The dependence on air speed is more complex, with greater dependence at the low throughput. While the air flow field should be simulated to confirm these notions, higher air speed will be more turbulent and may draw the bead down to a fine fiber at a point closer to the spinneret. Thus turbulent air may limit the volume of region in front of the spinneret that is protected from high velocity air. Low throughputs may be more sensitive to air speed because the fibers completely sever and allow a higher percentage of the spinneret flow to fill the bead.

Draw resonance has not been used before to explain fiber diameter distribution in a meltblown web, but this die geometry is quite different from the well-studied design. Indeed, the spinneret annulus both protects a region in front of the outlet and causes the primary drawing force to be somewhat downstream of the outlet, a situation similar to melt spinning. In the traditional die, polymer may be exposed to a high velocity jet closer to the outlet, preventing bead formation and drawing the fiber immediately. Efforts are ongoing to determine if this observed pulsation is a caused by the same mechanism that causes draw resonance in melt spinning and other polymer processing applications.

ACKNOWLEDGMENTS

The authors kindly acknowledge the Nonwovens Institute and its member companies for the support of this research. Also special thanks to the Chemical and Biomolecular Engineering Department, and to North Carolina State University.

REFERENCES

1. A. L. Yarin, S. Sinha-ray, and B. Pourdeyhimi, "Meltblowing: Multiple polymer jets and fi ber-size distribution and lay-down patterns," *Polymer (Guildf).*, vol. 52, no. 13, pp. 2929–2938, 2011.
2. R. Butin, J. P. Keller, and J. Harding, "Non-Woven Mats by Melt Blowing," 38492411974.
3. M. Uyttendaele and R. L. Shambaugh, "Melt Blowing: General Equation Development and Experimental Verification," *AICHE J.*, vol. 36, no. 2, pp. 175–186, 1990.
4. D. H. Tan, C. Zhou, C. J. Ellison, S. Kumar, C. W. Macosko, and F. S. Bates, "Meltblown fibers: Influence of viscosity and elasticity on diameter distribution," *J. Nonnewton. Fluid Mech.*, vol. 165, no. 15–16, pp. 892–900, 2010.
5. B. Haynes and M. C. Cook, "Die for Producing Meltblown Multicomponent Fibers and Meltblown Nonwoven Fabrics," 71506162006.
6. A. S. . Fabbricante, G. F. Ward, and T. J. Fabbrincante, "Micro-Denier Nonwoven Materials Made Using Modular Die Units," 6,114,0172000.
7. V. T. Marla, R. L. Shambaugh, and D. V Papavassiliou, "Using Swirl Dies To Spin Solid and Hollow Fibers," *Ind. Eng. Chem. Res.*, vol. 45, pp. 2331–2340, 2006.
8. E. C. A. Schwarz, "Apparatus and Process of Melt-Blowing A Fiberforming Thermoplastic Polymer and Product Produced Thereby," 43805701983.
9. R. J. Fisher and M. M. Denn, "A Theory of Isothermal Melt Spinning and Draw Resonance," *AICHE J.*, vol. 22, no. 2, pp. 236–246, 1976.

Influence of Adjacent Defects on the Permeability of Plasmapolymer Coated Plastics

Christian Hopmann[a], Stefan Wilski[a], Jens Wipperfuerth[a], Felix Mitschker[b], Peter Awakowicz[b], Rainer Dahlmann[a]

[a]*Institute of Plastics Processing (IKV) at RWTH Aachen University*
Seffenter Weg 201, D-52074 Aachen, Germany

[b]*Electrical Engineering and Plasma Technology, Ruhr-University Bochum,*
Universitätsstraße 150, D-44780 Bochum, Germany

Abstract. Permeation through plastics is an important factor in food and medical packaging, or in solar technology. The cause for usage of plastics is the variety of advantages, but on the other hand, the high permeability for gases or vapours has to be taken into account to reduce the mass transport. For this reason, the material is often coated with high barrier layers, e.g. SiO_x-layer, that effectively reduces the permeation of gases and vapours. One of the established technologies to realise barrier coatings is the PECVD technique, in which a thin film of a few ten nanometres is generated by depositing the barrier film from the gas state to the solid state on the polymer. In theory, the permeation reduction should be much higher than typically measured. This is caused by defects in the barrier layer that arise by reasons of the film growing process during PECVD or contaminations on the polymer. A lot of research has been done to describe the mechanisms of permeation through thin films under consideration of the influence of adjacent defects. Typically a distance was defined under which an influence occur, but the influence itself was not described. In this paper, the focus is on the influence of adjacent defects. Assuming an amorphous polymer the permeation can be described by the Fick's-Law. This requires knowledge of the defect and defect-size distributions, which are predetermined by etching methods and subsequent SEM imaging. A critical defect spacing parameter and an attenuation function will be derived from the results to describe the influence of adjacent defects. The results show that a superposition principle can be applied. Depending on the number of adjacent defects and their size the permeability of an individually defect can be described. In the future, the permeation through defects will be described by a molecular dynamic approach. The output of the first step of the simulation will be a time and defect size dependent concentration function, which serves as input for here performed macroscopic simulation.

Keywords: Permeation, PECVD, Barrier Coatings, OTR, defects , Permeation Model
PACS: 51.20.+d, 52.25.Fi, 68.35.Fx

INTRODUCTION

Polymer packaging is widely used in many fields, especially in food packaging [1]. It is also important to have polymers with good barrier properties against gases like O_2/N_2. Conventional polymer films are polyethylene (PE), polypropylene (PP) or polyethylene terephthalate (PET) [2-6]. Durability and taste are increased with decrease of permeation of these gases. For this reason, coating the polymer film with e.g. a silicon oxide (SiO_x) layer increases the barrier characteristics but also maintain the economical low prices for food packaging. A lot of research has been done and published in order to describe the mechanisms of permeation [7-11] and many researchers stated that gas permeation takes place only through defects in the coating [12-14].

In the century of computation just a few works have made reasonable simulations describing gas permeation through barrier coatings. One reason is the lack of knowledge about the defect distribution, another reason is the absence of a general solution of the diffusion equation. Here, the influence of adjacent defects (also called neighbouring defects) is investigated to establish a model for further experiments. Barrier defects are often clustered [12], so simulating each defect by itself and adding them up will not be representing an accurate picture.

GENERAL THEORY AND ASSUMPTIONS

Permeation of gases through SiO_x in comparison to polymers is very slow. So a SiO_x-layer can sufficiently hinder permeation. Thus the assumption permeation only occurs through defects in the coating seems valid. If the defect distribution and the respective sizes are known it is possible to simulate the region of interest. Here a self-made program inside MATLAB 2016a is used to simulate the gas flow. Due to the low solubility of gases like oxygen or nitrogen in PP and PET the solution of gas molecules in the polymer matrix can be described by means of Henry's law:

$$\Delta c = S \cdot \Delta p, \tag{1}$$

where p is the partial pressure of the gas near the film surface, S is the solubility coefficient of the gas in the polymer and c is the concentration of the gas inside the polymer. By relating the diffusive flow to the concentration, the transport of the solved gas in the polymer is described by Fick's first law:

$$j = -D \cdot \nabla c, \tag{2}$$

where j is the gas flow and D the diffusion coefficient of the polymer. By means of Fick's second law the time dependency of the gas transport inside the film is included. The 3-dimensional case can be described with the following equation:

$$\frac{dc}{dt} = \nabla\left(D \cdot \nabla c\right) = D \cdot \left(\frac{\partial^2 c}{\partial x^2} + \frac{\partial^2 c}{\partial y^2} + \frac{\partial^2 c}{\partial z^2} \right) \tag{3}$$

The diffusion coefficient is assumed to be concentration-independent and isotropic. Otherwise, Equation 3 cannot be used and the diffusion becomes non-Fickian. In this case equation 3 is solved with the MATLAB program using finite difference method (FDM). In FDM the first and second derivative approximations are used to bring equation 3 into the form

$$\underline{\underline{M}} \cdot \underline{c}(x,t) = \underline{f(x,t+dt)}, \tag{4}$$

where M is the dimensionless discretization matrix, c being the concentration vector which contains values for all lattice points and f is the vector containing the concentration one time step further and the gas source and boundary conditions. For 3D the process is done for all three directions ($\Delta x, \Delta y, \Delta z$) and takes more computation time, accordingly. Boundary conditions (BC) are applied in the form

$$\underline{n} \cdot \left(D \cdot \nabla c\right) + q \cdot c = g, \tag{5}$$

where n is the outward unit normal, g and q are the Dirichlet and Neumann BC functions, respectively. The more elements are used, the more accurate is the resolution. Here an 200 x 200 x 200 nm element and a time step of 2 ms is used. In the simulated volume (Figure 1) the calculations of several billions of elements had to be considered, limited by the computational power. Also some assumptions are made to make this possible:

- The simulation of a symmetrical part of the simulated volume is sufficient to describe area
- The polymer is homogenous and swelling or chemical reactions can be neglected
- The gas flow out of the polymer is unhindered and equivalent to the flow into the edge layer
- The concentration at the edge layer is mirrored
- The surrounding environment is under normal pressure and 23 °C
- Fick's laws are valid

The source concentrations ($C_{i,in}$) through the coating are considered to be cylindrical and have the same concentration throughout the hole defect ($C_{i,out}=C_{i,in}$).

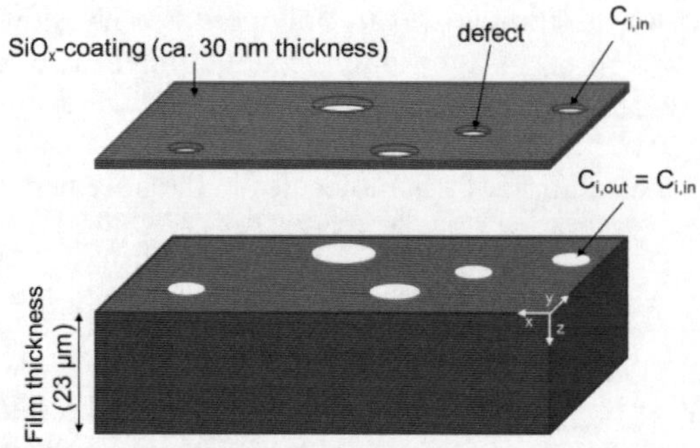

FIGURE 1. Simulated volume separated in polymeric film and barrier coating. C represents the normalised gas concentration.

SIMULATION OF ADJACENT DEFECTS

The defects in the coating were visualised by an etching process of the polymer surface underneath the coating and subsequent SEM imaging, which was developed by Da Silva Sobrinho et al. [12]. The statistical defect distribution is shown in Figure 2. The average defect size of a SiO_x coating on PET is about 1.0 µm in diameter. To investigate the influence of nearby defects, we considered varying defect pairs and compared them to two single defects.

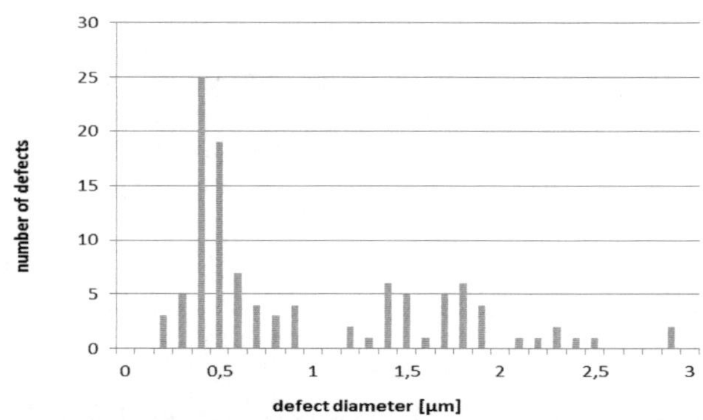

FIGURE 2. Statistical distribution of initial defect diameters in a 25 nm SiO_x coating (O2/HMDSO = 50:1) deposited on PET

Two defects in close vicinity to each other interact and the combined flow will be reduced compared to two defects far away from each other [12, 15]. In order to describe the influence of the defect spacing da Silva Sobrinho et al. [12] defined the "equivalent defect radius":

$$R_g = \sqrt{2 \cdot R \cdot d + R^2}\ , \tag{6}$$

where R is the defect radius and d the film thickness. Furthermore, he considers the defects to be independent when the distance "significantly exceeds $2R_g$" [12]. Zajec et al. [15] established a relation for no interaction when the defect spacing exceeds $6R_g$. By inserting the film thickness of 23 µm and median defect radius of 0.5 µm in Equation 6 the equivalent defect radius R_g is 4.8 µm. This results in a condition for independent defects with a distance larger than 9.6 µm or 28.9 µm. Additionally, the condition from Hanika et al. [13] considering an interaction between defects occurs if

$$\frac{L}{d} \leq 3 \tag{7}$$

is taken into account, where L is the characteristic defect spacing and d the film thickness. With a defect spacing $L > 69$ µm and a film thickness of $d = 23$ µm this condition is not fulfilled and no interaction between the defects should occur. Based on the above described findings from previous papers the flow rate through two defects at a distance of > 80 µm is assumed to have no interaction and thereby is normalised to 100 % for each defect.

By adding a second defect to the simulation with different defect spacings from 5 to 80 µm and the declared defect radii (0.3, 0.5, 0.7 µm), the influence on the oxygen concentration in the polymer and the resulting flow rate can be examined. An exemplary presentation of two defects with a radius of R2 (0.5 µm) as well as a pairing of R1 (0.3 µm) and R2 is displayed in Figure 3. The defect spacing is set to 40 µm and the oxygen concentration is monitored at a level ($z = 12$ µm) in the polymer as shown in Figure 3c (white dashed line). Two defects with the same size result in an eight-shaped isoline of concentration that develops into an oval with further distance to the centre (Figure 3a). As expected, the highest oxygen concentration is located directly beneath the defects. The same findings can be seen in Figure 3b with the smaller defect (R1) combined with a defect of R2. However, in this case the circular concentration lines of the smaller upper defect result in a smaller expanse of concentration and the image shows no symmetric concentration distribution. With this method the concentration in the solid material can be visualised to any given time step and at any level.

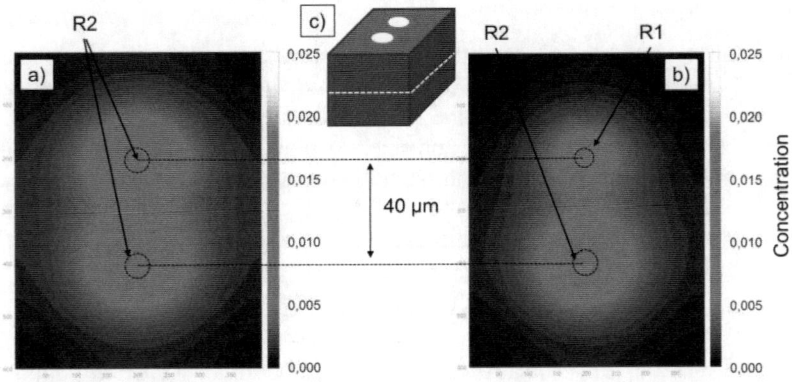

FIGURE 3. Oxygen concentration in PET at $z = 12$ µm. a) Two defects with radius R2 and 40 µm spacing. b) Two defects with radius R1, R2 and 40 µm spacing. c) Simulated volume.

In Figure 4 the results for the simulation of two adjacent defects is shown. It should be mentioned that the reduction in the normalized flow rate is given for each defect, so the percentages have to be doubled for real comparison. Generally all curves increase rapidly for short distances ($L < 25$ µm). For very close defects ($L < 15$ µm) the combined flow is 3-6% lower than for large distanced without any influence ($L > 60$ µm). As can be seen, the flow reduction is greater for larger defect areas, comparing the flow rate for the 0.3/0.3 µm and the 0.7/0.7 µm pair. The larger holes exhibit an increased influence, so the combined flow drops more significantly when brought together.

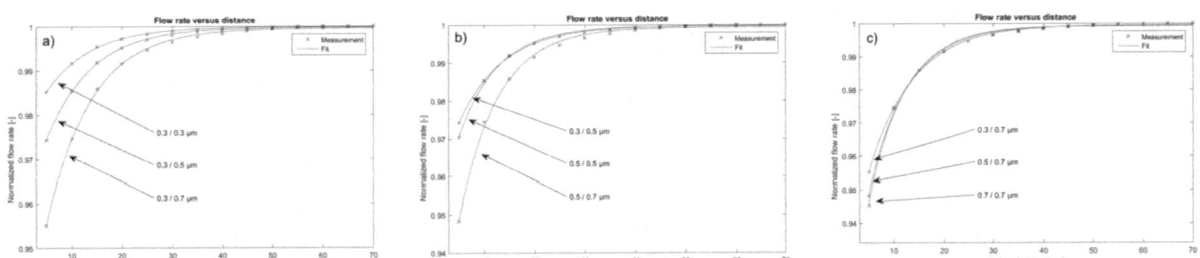

FIGURE 4. Reduction function depending on the defect spacing. a) Defect pairings 0.3/0.3 µm, 0.3/0.5 µm and 0.3/0.7 µm. b) Defect pairings 0.3/0.5 µm, 0.5/0.5 µm and 0.5/0.7 µm. c) Defect pairings 0.3/0.7 µm, 0.5/0.7 µm and 0.7/0.7 µm.

In order to validate the calculations of the influence of adjacent defects on mass transport with experimental data, the barrier layer is substituted by adhesive aluminium tape and drilled by means of ultrashort pulse lasers as seen in Figure 5. Thus, any arrangement of defects and distances with high reproducibility in the single-digit micrometre range is possible. Subsequently, the barrier properties (O_2 and H_2O) are to be calibrated with precise knowledge of the number of defects and surface area, but especially in the absence of non-visible nano defects, using the above calculations.

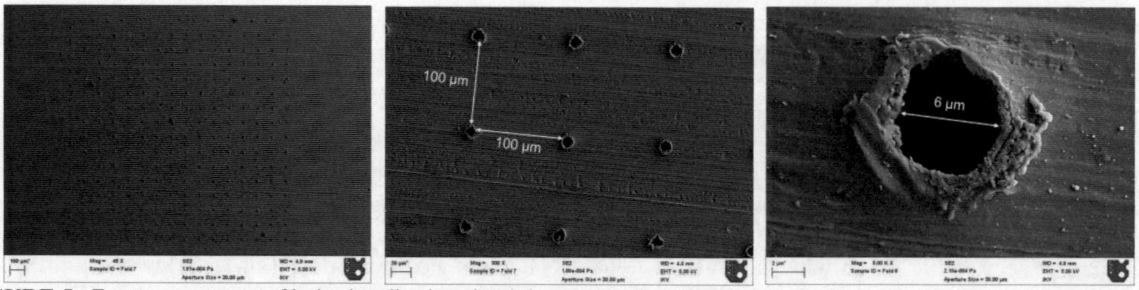

FIGURE 5. Free arrangement of holes in adhesive aluminium tape using ultrashort pulse lasers to investigate the influence of adjacent defects in the absence of nanoscale defects

CONCLUSION AND OUTLOOK

It is shown that adjacent defects hinder each other's permeation depending on their spacing and defect area. In case of clustered defect distributions, the influence of the flow reduction is expected to be significantly greater. The results of this paper can be used to better describe permeation through defective barrier layers and should be evaluated with regard to critical defect distributions and sizes. A reduction of the computing times is carried out with discretization methods such as the octree or multigrid method and thus enables the implementation of large, arbitrary defect distributions. Furthermore, the evaluation of the SEM images is to be automated by means of digital image processing methods. Thus, it is possible to implement real defect distributions directly into the presented simulation.

ACKNOWLEDGMENTS

Some of the work shown in this paper received support of the German Research Foundation (DPG) within the framework of the Transregional Collaborative Research Centre TRR 87/1 (SFB-TR 87) "Pulsed high power plasmas for the synthesis of nanostructured functional layers".

REFERENCES

1. L. Körner, "Diffusion barrier coatings for polymer containers processed by plasma enhanced chemical vapor deposition", ETH Zürich, Ph.D. Thesis, 2010
2. A. Bieder, A. Gruniger, R. von Rohr, *Surface and Coatings Technology*, 2005, 200 (1-4), 928-931
3. H. Chatham, *Surface and Coatings Technology*, 1996, 78 (1-3), 1-9
4. T.N. Chen, D.S. Wuu, C.C. Wu, C.C. Chiang, Y.P. Chen, R.H. Horng, *Journal of the Electrochemical Society*, 2006, 153 (10), F244–F248
5. J. Fahlteich, M. Fahland, W. Schönberger, N. Schiller, *Thin Solid Films*, 2009, 517 (10), 3075-3080
6. J. Lange, Y. Wyser, *Packaging Technology and Science*, 2003, 16 (4), 149-158
7. Malladi Sairam , S. K. Nataraj , Tejraj M. Aminabhavi , Sukumar Roy & C. D. Madhusoodana (2006), 35:4, 249-283
8. Tejraj M. Aminabhavi & U. Shanthamurthy Aithal (1991), *Journal of Macromolecular Science*, Part C, 31:2-3, 117-163
9. U. Shanthamurthy Aithal , Ramachandra H. Balundgi , Tejraj M. Aminabhavi & Shyam S. Shukla (1991), *Polymer-Plastics Technology and Engineering*, 30:4, 299-342
10. D. Kirchheim, M. Jaritz, F. Mitschker, M. Gebhard, M. Brochhagen, M. Boeke, A. Devie, P. Awakowicz, Ch. Hopmann, R. Dahlmann: Transport Mechanisms through PE-CVD Coatings: Influence of Temperature, Coating Properties and Defects on Permeation of Water Vapour. *Journal of Applied Physics* 50 (2017)
11. F. Mitschker, S. Steves, M. Gebhard, M. Rudolph, L. Schnücke, D. Kirchheim, M. Jaritz, M. Brochhagen, Ch. Hoppe, R. Dahlmann, M. Böke, J. Benedikt, I. Giner, T. de los Arcos, Ch. Hopmann, G. Grundmeier, A. Devi and P. Awakowicz: Influence of PE-CVD and PE-ALD on defect formation in permeation barrier films on PET and correlation to atomic oxygen fluence. *Journal of Applied Physics* 50 (2017)
12. A.S. da Silva Sobrinho, G. Czeremuszkin, M. Latreche, M.R. Wertheimer, *Journal of Vacuum Science & Technology*, 2000, 18, 149
13. M. Hanika, H.-C. Langowski, U. Moosheimer, W. Peukert, *Chemical Engineering Technology*, 2003, 26, 605-614
14. E. Schmachtenberg, F.R. Costa, S. Göbel, *Journal of Applied Polymer Science*, 2006, 99, 1485-1495
15. A. S. da Silva Sobrinho, G. Czeremuszkin, M. Latrèche, and M. R. Wertheimer, *Appl. Phys. A* 1999, 68, 103-105

Effect of Nanoclay on Intelligence Barrier Single-layer Films

Bahari-Aban Babak [1], Mehranpour Milad [1]*, Oromiehie Abdulrasoul [2], Atabaki-Pasdar Pouria [1]

[1] Islamic Azad University Science & Research branch - Tehran - Iran, [2] Islamic Azad University South Tehran branch - Tehran - Iran

In this article, nanocomposite films based on low density polyethylene (LDPE), Nano Clay (NC) and ethylene vinyl alcohol (EVOH) copolymer were prepared and characterized through using field emission scanning electron microscopy(FESEM), X-ray diffraction (XRD), oxygen permeability and immigration test. The oxygen barrier properties of the LDPE/EVOH/NC film were better than those of the LDPE/EVOH film. In addition to barrier properties. Furthermore, from the FESEM micrographs, the O2 permeability through the films was evaluated, which illustrated that adding both NC and EVOH to LDPE leads to a remarkable increase in the barrier properties of Polyethylene films. Total plate count(TPC) and total volatile basic nitrogen (TVBN) released from red meat during storage which has been fabricated for real-time monitoring of meat quality. Moreover, the amount of TVBN and TPC were remarkably decrease by using EVOH and Nanoclay in order to enhancing shelf life. Also, a new colorimetric indicator sensor based on methyl red was sensitive to pH. The methyl red as a freshness sensor worked based on pH increase as the basic spoilage volatile amines and microbial growth produced gradually in the package headspace, and subsequently, the color of the sensor will change for spoilage indication, which is easily visible to the naked eye. The results reveal that the sticker sensor could be used to determine the degree of red meat cut freshness, as the relationship between the color change of methyl red as a sensor response and the meat freshness follows a similar trend.

Effect of Process Engineering on the Morphology and Crystallization of Biobased Poly(Trimethylene Terephthalate) – PTT and Nylon 6 Blends

Amandine Codou[1,] Amar K. Mohanty[1, 2,*] and Manjusri Misra[1, 2,*]

[1]*Bioproducts Discovery and Development Centre, Department of Plant Agriculture, Crop Science Building, University of Guelph, Guelph, N1G 2W1, Ontario, Canada*
[2]*School of Engineering, Thornbrough Building, University of Guelph, Guelph, N1G 2W1, Ontario, Canada*
*Corresponding Author(s) - E mail: mohanty@uoguelph.ca

Abstract. Poly(trimethylene terephthalate) (PTT), a partially biobased polymer, was blended with nylon 6 in order to develop a new bio-material. PTT and nylon 6 blends were prepared in a twin-screw extruder while varying the temperature and speed of melt mixing. The effects of the processing conditions on blend morphology were highlighted by scanning electron microscopy (SEM) and polarized optical microscopy (POM). Both the domain size and the density of nucleation are strongly influenced by the processing temperature. Differential scanning calorimetry (DSC) was used to measure the influence of the operatory conditions on the thermal stability of the crystals formed.

Keywords: poly(trimethylene terephthalate), nylon 6, blends, crystallization
PACS: 83.80.Tc

INTRODUCTION

Polyamides, commercially known as nylons, are of particular interest due to their high engineering benefits (high range temperature and easy processability, good mechanical and thermal properties, etc.) and high application benefits (low density, durability, etc.). Poly(trimethylene terephthalate), from the polyester family, is a linear aromatic thermoplastic nowadays commercialized by DuPont from renewable sources, containing between 20 and 37% of renewable material (35% in this study). PTT is particularly suitable for carpet fibers application due to its low sensitivity to moisture [1], low modulus, better bending and tensile elastic recovery in comparison to PET and PBT; providing soft and stretchable fabrics [2,3]. Studying PA6-PTT blends is particularly relevant from a carpet recycling perspective. For fiber applications, while PA6.6 has a greater strain resistance than PA6, the latter has a better resistance to light, dyeability, elastic recovery, wear resistance and thermal stability [4]. The recovery of individual polymers is a long and costly process, for this reason, most of the time recycled fibers are extruded and re-used for car applications where the mechanical properties of the blends can be improved by the addition of chemicals or fillers [5].A special focus on the 50/50 wt.% blend of PTT/PA6 was selected since any type of morphology is possible at this ratio and where the morphology influences the properties the most. Then, in addition to being the most challenging ratio, the 50/50 wt.% blend additionally presents a very relevant case for the carpet recycling of PTT/PA6 blends as presented before. By varying the processing conditions, a preliminary investigation on the morphological development of the 50/50 wt.% PTT/PA6 blends was carried on scanning electron microscopy (SEM) and polarized optical microscopy (POM). Differential scanning calorimetry (DSC) was employed to explain the thermal properties of the blends.

EXPERIMENTAL

Polyamide 6 (PA6) and partially bio-based poly(trimethylene terephthalate) (PTT) were used for processing. Polyamide 6, product name Ultramid B27E was supplied by BASF. Poly(trimethylene terephthalate), PTT product name SORONA 3301 BK001 was supplied by Dupont. PTT/PA6 blends at the weight ratio 50/50 wt.% were melt-processed in a twin screw extruder (Leistritz-Micro-27 in co-rotation mode) at three different temperatures: 240, 250 and 265 °C. The pure polymers, PTT and PA6, were processed at 265 °C. The extruder screw speed was set at 70 rpm for all the samples and one test at 265 °C and 200 rpm was done. At the nozzle, the extrudates were quenched in a water bath before being re-pelletized. A scanning electron microscope Phenom ProX (Phenom World BV, the Netherlands) equipped with a back-scattered electron (BSE) detector was used to observe the morphology of the fractured surface of the blends. A polarized optical microscope (POM) (Nikon Instruments) equipped with a hot stage (Linkam Scientific Instruments) was used in this work. An ultramicrotome was used to prepare homogeneous samples of 1 μm thickness from the extruded pellets. After cutting, the polymer slices (processed at 240 and 265 °C) were taken and melted for 2 min on the hot stage of the microscope at their processing temperature, between two slices of glass. A cooling rate of 10 K/min was applied, in accordance with the DSC method described downstream. Standard DSC runs were carried out on a

DSC Q200 (TA Instruments). Temperature and enthalpy calibration were done using an indium standard. The samples were prepared in 40 μL aluminum pans of about 6 mg. The experiments were conducted under N_2 atmosphere (60 mL/min). The method applied consisted in a heating at 10 K/min from 0 °C to 250 °C, in order to observe the influence of the processing conditions, before melting the samples at 250 °C for 5 min and cooling them down at 10 K/min to record the inherent physical transitions. By integrating the total area of the endothermic and exothermic peaks, crystallization (ΔH_c) and melting enthalpies (ΔH_m) were determined. The crystallization and melting temperature recorded were measured at the maximum of the peak. Crystallinity of the extruded neat resins were calculated considering the equilibrium melting enthalpy of fully crystalline PTT $\Delta H^{\circ}_m \sim 146$ J/g [3] and PA6 $\Delta H^{\circ}_m \sim 230$ J/g [6] (selected because both α and γ polymorphs are present).

RESULTS AND DISCUSSION

Blends Morphology

The blends morphology was investigated by means of scanning electron microscopy (SEM), in order to observe the phase distribution of the polymers and get an estimation of the size of the domains, and by means of polarized optical microscopy (POM) giving insights into the morphology of the spherulites.

Scanning Electron Microscopy

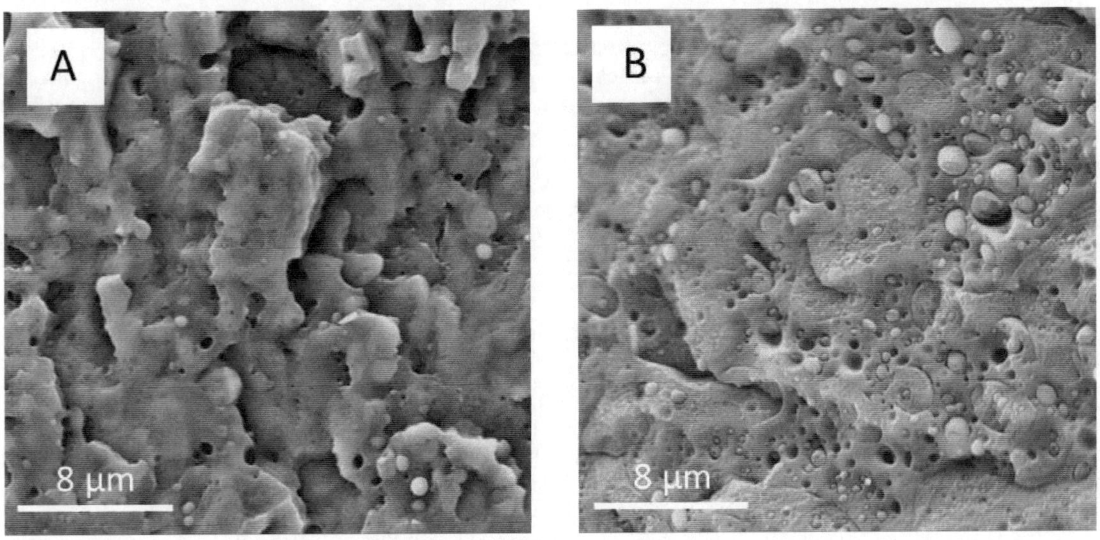

FIGURE 1. SEM (10000x) of PTT/PA6 50/50 wt.% blends after processing at A) 240 °C and B) 265 °C.

The observation of the images obtained by scanning electron microscopy of the processed blends at 240 °C FIGURE 1 A) and 265 °C B) highlights the immiscibility of PTT/PA6 blends for both processing temperatures investigated. On the other hand, the domain size is strongly influenced by the processing temperature applied and the higher the temperature and the smaller the droplets. Very small occlusions (typically below 0.5 μm) of PTT in the PA6 dispersed phase are distinctly observed after a process at 265 °C, see FIGURE 1 B). This partial encapsulation was also obtained in other polyester/amide blends by varying the polyamide/polyester ratio [7].

Polarized Optical Microscopy

The observation of the samples by polarized optical microscopy (POM) gives insights into the spherulitic morphologies by measuring the crystal size and the density of nucleation. Pictures were captured on cooling after melt crystallization (*i.e.* ~ 110 °C for both process at 240°C and 265 °C), see FIGURE 2.

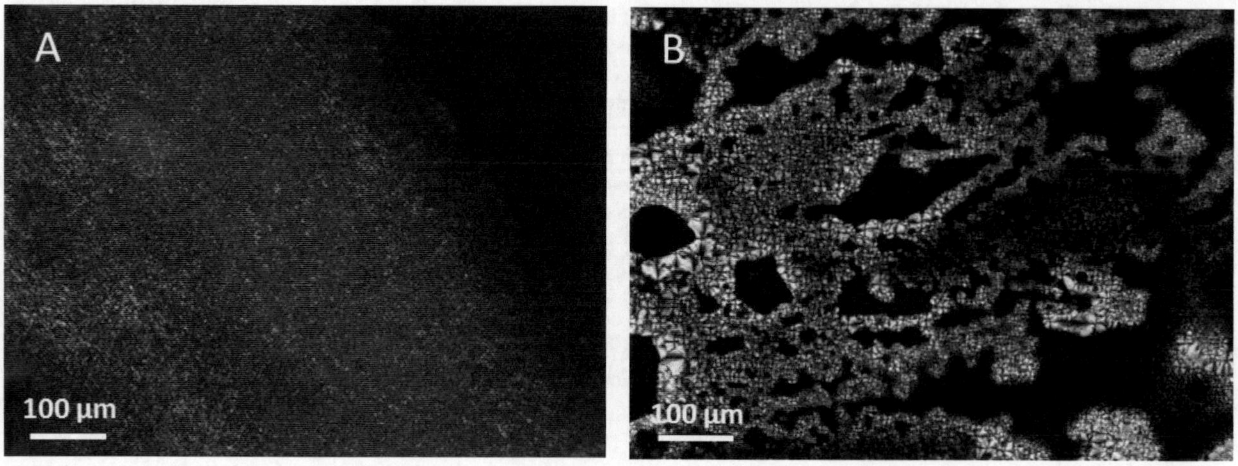

FIGURE 2. Polarized optical microscopy pictures of the PTT/PA6 50/50 wt.% blends after processing at A) 240 °C and B) 265 °C.

As we can see, the temperature applied on the sample is significantly impacting the crystallization of the blends. Indeed, the POM images show that a melting at 240 °C (FIGURE 2 A) results in densely packed, small, and unmeasurable spherulites impinged by one another. In comparison, the blend processed at 265 °C (FIGURE 2 B) clearly shows Maltese crosses of colorful birefringence, whose spherulites' average diameter is reaching ~ 20-50 µm. After recording the pictures, the cooling was applied up to room temperature, without showing any significant difference. From the morphology obtained we concluded that a melting at 240 °C resulted in formation of additional nuclei in comparison to a melting applied at 265 °C. Moreover, an increase of the brightness of the crystals can be observed, attributed to an increase of order in the crystals. As shown in the literature, both individual PA6 [8] and PTT [9] are melt crystallizing under spherulitic organization, showing Maltese crosses. Contrary to our expectations, at 265 °C a phase does not seem to crystallize as shown by the regions of dark contrast underlining area remnant in the amorphous state. In addition, the well dispersed PA6 droplets, previously observed by SEM, seems to have coalesced after a melting at 265 °C which is probably due to a decrease of the viscosity.

Crystallization

SEM revealed that the selection of an adapted processing temperature is essential since it is directly influencing both the blend morphology and crystallization. Based on the preliminary blend morphology investigation, a process at higher temperatures appeared to be favorable to smaller dispersed phase droplets and bigger crystals. DSC was used to elucidate the origin of the rate of crystallization decrease.

Differential scanning calorimetry

Differential scanning calorimetry was used to make a link between the morphology observed by SEM and the change of crystallization rate highlighted by POM. The method applied and the thermograms obtained after processing the neat polymers and blends at 240, 250 and 265 °C (the latter processed at 70 and 200 rpm) are displayed in FIGURE 3. The temperatures of transition and enthalpy of crystallization/melting are referred in *TABLE 1.*

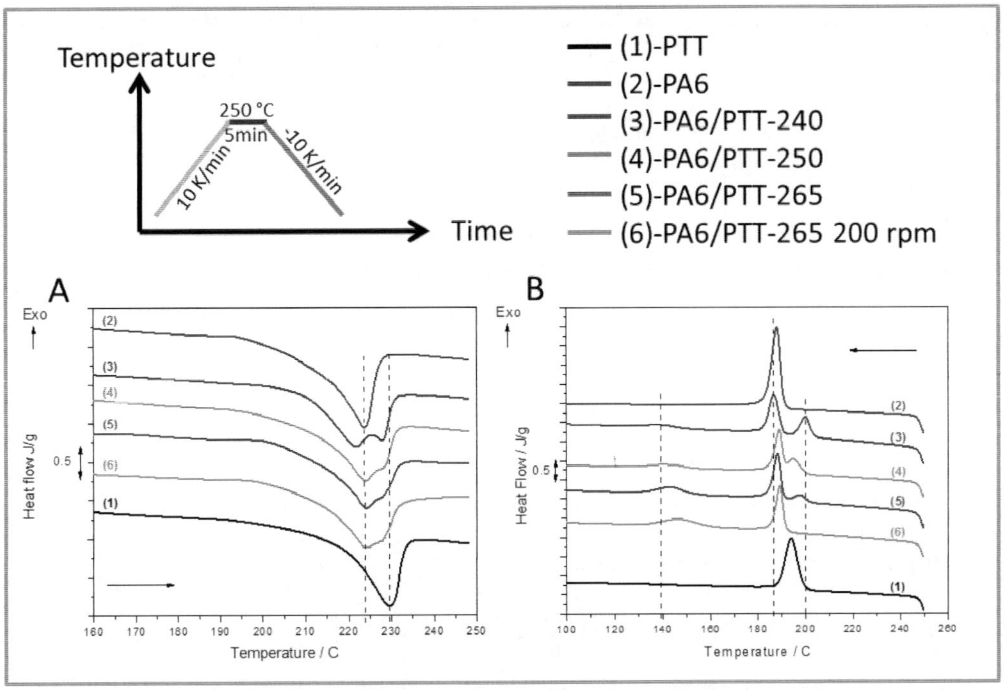

FIGURE 3. Method and thermograms corresponding to the neat polymers and their PTT/PA6 50/50 wt.% blends processed at 240, 250 and 265 °C on A) heating and B) cooling.

The extruded materials were used for DSC investigation in order to observe the influence of the processing conditions on the crystals stability, see FIGURE 3-A. Neither the α-relaxation nor the crystallization of the neat polymers and blends appeared on heating given that the materials already crystallized during the water bath cooling. The processing conditions have a significant influence on the melting transition of the blends. A process at low temperature shows that PA6 and PTT crystals are melting at two distinct temperatures (T_m) while an increase of the processing temperature shows a convolution of the peaks and an overall melting enthalpy decrease. In addition, a process at low temperature is detrimental to the thermal stability of the crystals formed as shown by the shift to lower temperature observed for both endothermic events. The melting of PTT crystals seems to be hindered by the process at higher temperature, stressing that the operatory conditions would be influencing the crystalline structure of the individual polymers. After integration of the melting peaks, a degree of crystallinity (X_c) equal to 30% was measured for PA6 while 40% was obtained for PTT, see TABLE 1. The peak convolution of the blends makes it impossible to assess their degree of crystallinity.

TABLE 1. Thermal properties of PA6 and PTT and their blends.

Sample	T_m °C		T_{mc} °C			ΔH_m J/g	X_c %
	1st peak	2nd peak	1st peak	2nd peak	3rd peak		
(1)-PTT	/	229	194	/	/	58	40
(2)-PA6	224	/	/	188	/	69	30
(3)-PA6/PTT-240	222	228	200	187	140	59	/
(4)-PA6/PTT-250	224	228	195	189	142	61	/
(5)-PA6/PTT-265	223	228	197	188	144	56	/
(6)-PA6/PTT-265-fast	224	228	/	189	147	54	/

T_m: Melting temperature; T_{mc}: Temperature of melt crystallization; ΔH_m: Enthalpy of fusion; X_c: Degree of crystallinity, $\Delta H°_m$ PTT ~ 146 J/g & $\Delta H°_m$ PA6 ~ 230 J/g

Focusing specifically on the cooling, the melt crystallization temperatures (T_{mc}) of the different samples are spread over a wide range of temperatures (**FIGURE 3 B**). The lower the processing temperature and shearing, the higher the temperature difference between the different crystallization exotherms, see TABLE 1. The first small exotherm corresponds to PTT melt crystallization at ~200 °C, followed by PA6 at ~187 °C and then a broader peak arising at ~140 °C assigned to the fractionated crystallization of PA6 dispersed phase [10,11,12,13]. While the bigger PA6 droplets crystallize at the bulk temperature, the smaller ones would crystallize at higher supercooling which is correlated to the lack of active nuclei able to start the crystallization growth [10]. This retarded melt crystallization is energetically dependent, higher the processing temperature (and shearing) and faster the crystallization, as shown by the shift to higher temperature. Furthermore, the first crystallization peak at ~200 °C clearly disappears with the increase of processing temperature and shearing. The disappearance of the PTT crystallization observed at higher temperature, by DSC, can be correlated to the absence of crystallization highlighted earlier by POM. Consequently, in the same fashion as the hindered PA6 phase, the droplets of PTT would crystallize at higher supercooling. No distinct exotherm of crystallization can be observed neither on cooling nor on the subsequent melting (not shown here), meaning that the rate of crystallization is significantly decreased. This result would suggest that the phase remnant in the amorphous state observed on POM pictures would be the PTT. In consequence, it seems that the presence of PA6 is hindering PTT crystallization.

CONCLUSIONS

After the process of 50/50 wt.% blends at varied temperatures and shearing, a significant reduction of the PA6 droplet size in the PTT matrix was highlighted by SEM pictures. This increase of phase compatibility in the blends was associated to a significant crystallization rate decrease, observed by POM and corroborated to DSC measurements. In the 50/50 wt.% blends, the hindered crystallization was associated to the PTT phase.

ACKNOWLEDGMENTS

The authors are thankful to the Ontario Ministry of Agriculture, Food and Rural Affairs (OMAFRA), Canada/University of Guelph-Bioeconomy for Industrial Uses Research Program Theme (Project # 030176); the Natural Sciences and Engineering Research Council (NSERC), Canada–Discovery Grants (Project # 401111) and Agriculture and Agri-Food Canada and Competitive Green Technologies (CGTech) through AgriInnovation Program (Project # TBD); Natural Sciences and Engineering Research Council, Canada for the Discovery Grants (Project #401111); and the Ontario Research Fund, Research Excellence Program; Round-7 (ORF-RE07) from the Ontario Ministry of Research and Innovation (MRI), currently known as the Ontario Ministry of Research, Innovation and Science (MRIS) (Project # 052665 and #052644).

REFERENCES

1. Khonakdar, H. A.; Jafari, S. H.; Asadinezhad, A. A Review on Homopolymer, Blends, and Nanocomposites of Poly(trimethylene terephthalate) As a New Addition to the Aromatic Polyesters Class, *Iran. Polym. J.*, **17**, 19-38 (2008).

2. Ward, I. M.; Wilding, M. A. The Mechanical Properties and Structure of Poly(m-methylene Terephthalate) Fibers, *J. Polymer Sci. Polymer Phys. Ed.*, **14**, 263-274 (1976).

3. Chuah, H. H.; Synthesis, Properties and Applications of Poly(Trimethylene Terephthalate), in Modern Polyesters: Chemistry and Technology of Polyesters and Copolyesters, Scheirs, J. and Long, T. E., John Wiley & Sons Ltd., 2003, chap.11, pp. 361-397.

4. Dubinskaite, K.; Milašius R. Investigation of Dynamic Properties of PA6 and PA6.6 Carpet Pile Yarns, *Mater. Sci.*, **11**, 288-291 (2005).

5. Wang, Y.; Zhang, Y.; Polk, M.; Kumar, S.; Muzzy, J. Recycling of Carpet and Textile Fibers, in Plastics and the Environment: A Handbook, Andrady, A. L., John Wiley & Sons, New York, 2003, chap. 16, pp. 697-725.

6. Campoy, I.; Gómez, M. A.; Marco, C. Structure and Thermal Properties of Blends of Nylon 6 and a Liquid Crystal Copolyester, *Polymer*, **39**, 6279-6288 (1998).

7. Retolaza, A.; Eguiazabal, J. I.; Nazabal, J. Structure and Mechanical Properties of Polyamide-6,6/Poly(Ethylene Terephthalate) Blends, *Polym. Eng. Sci.*, **44**, 1405-1413 (2004).

8. Kolesov, I.; Mileva, D.; Androsch, R. Mechanical Behavior and Optical Transparency of Polyamide 6 of Different Morphology Formed by Variation of the Pathway of Crystallization, *Polym. Bull.*, **71**, 581-593 (2014).

9. Xue, M.-L.; Sheng, J.; Yu, Y.-L.; Chuah, H. H. Nonisothermal Crystallization Kinetics and Spherulites Morphology of Poly(trimethylene terephthalate), *Eur. Polym. J.*, **40**, 811-818 (2004).

10. Tol, R. T.; Mathot, V. B. F.; Groeninckx, G. Confined crystallization phenomena in immiscible polymer blends with dispersed micro- and nanometer sized PA6 droplets, part 1: uncompatibilized PS/PA6, (PPE/PS)/PA6 and PPE/PA6 blends, *Polymer*, **46**, 369-382 (2005).

11. Tol, R. T.; Mathot, V. B. F.; Groeninckx, G. Confined crystallization phenomena in immiscible polymer blends with dispersed micro- and nanometer sized PA6 droplets, part 2: reactively compatibilized PS/PA6 and (PPE/PS)/PA6 blends, *Polymer*, **46**, 383-396 (2005).

12. Tol, R. T.; Mathot, V. B. F.; Groeninckx, G. Confined crystallization phenomena in immiscible polymer blends with dispersed micro- and nanometer sized PA6 droplets, part 3: crystallization kinetics and crystallinity of micro- and nanometer sized PA6 droplets crystallizing at high supercoolings, *Polymer*, **46**, 2955-2965 (2005).
13. Tol, R. T.; Mathot, V. B. F.; Groeninckx, G. Confined crystallization phenomena in immiscible polymer blends with dispersed micro- and nanometer sized PA6 droplets, part 4: polymorphous structure and (meta)-stability of PA6 crystals formed in different temperature regions, *Polymer*, *46*, 2966-2977 (2005).

Effect of the Amino-carboxylate as Volatile Compatibilizer in EVA/clay Nanocomposites

Julio Caceres[a,c], Jose Pilataxi[a], Miguel Vera[a], and David Vera[a], Antonio Diaz-Barrios[b], Andres Rigail-Cedeño[a,c],*

[a]*Faculty of Mechanical Engineering and Production Science,*
[c]*Center of Research and Development in Nanotechnology (CIDNA)*
ESPOL Polythenic University Escuela Superior Politecnica del Litoral,
Campus Gustavo Galindo Km 30.5 vía Perimetral, P.O. Box 09-01-5863, Guayaquil, Ecuador
[b]*School of Chemical Science and Engineering, Yachay Tech, Yachay City of Knowledge, Urcuqui, Ecuador*
*Corresponding author: arigail@espol.edu.ec

Abstract. The scope of this study involves the evaluation of the effect of amino-carboxylate (AC) compound on the structure, thermomechanical and rheological properties of EVA (19 wt. % vinyl acetate)-clay nanocomposites. Two commercial organoclays were pre-swollen in AC compound at 1, 2.5 and 5 wt. %, in a solvent free procedure. Later, a torque rheometer dispersed the organoclays/AC mixtures in EVA matrix. The X-ray diffraction (XRD) and Fourier transfer infrared (FTIR) spectroscopy carried out the volatilization and interaction of AC with the organoclays as well as the structure of nanocomposites. The shear flow in a parallel plate rheometer followed the change in the viscosity of the nanocomposites. These changes depended on the organoclay and the synergy between the organoclays and AC concentration. Thermal analysis was evaluated using a Thermogravimetric analysis (TGA) and Dynamic scanning calorimetry (DSC). The AC compound decreases the catalization effect of organoclay in the deacetylation reaction of EVA. Mechanical properties were also evaluated. The increase in the Young modulus and elongation (as measured via tensile testing), provided evidence of a good level of dispersion and surface interaction in the EVA/clay/AC nanocomposites. These results demonstrate the feasibility to use amine-carboxylate compounds as compatibilizer in polymer/clay nanocomposites to enhance their performance.

Keywords: vapor phase corrosion inhibitor, Amine carboxylate, EVA, clay, nanocomposite.
PACS: 81.05.Qk

INTRODUCTION

Poly(ethylene-co-vinyl acetate), EVA, has been used widely in several applications [1] such as packing of food or parts, photovoltaic films, shoes industry, cable jacketing, drug carrier and electrical insulation due its low-cost and low-energy manufacturing. Most of these applications require a balance in the mechanical, thermal, dielectric or permeability properties.

In fact, to accomplish the referred balance of properties and meet the corresponding industrial requirements, researchers have studied certain issues for improving the use of EVA in several applications. Thus, authors have analyzed potential problems in food packing, [2] drug delivery [3] and corrosion in photovoltaic films [4] probably due to acid acetic formation during EVA degradation [5]. In terms of EVA/clay nanocomposite, several variables including different types of organoclays [6], acetate content in EVA [7], degradation of EVA-clay [8] or compatibilization [9] have been studied. The aim was to improve the dispersion of the organoclays and enhance physical-chemical properties. Nevertheless, there are unexplored issues in degradation and dispersion of the organoclays in EVA nanocomposites and their performance.

In this research, an Amino-carboxylate (AC) compound was incorporated in EVA/clay nanocomposite processing. These compounds are environmental friendly and have been used extensively as vapor phase corrosion inhibitor (VpCI) for industrial applications [10]. Amino and carboxylate ions dissociate and interact with anionic and cationic metal surfaces [11]. Therefore, AC can potentially interact with all organoclay surfaces. Besides, AC would promote a green nanocomposite processing approach, since it avoids the use of solvents, acids or high temperatures during the mixing procedures. This is relevant in terms of development a new kind of non-toxic and easy processing compatibilizer. For this reason, EVA/organoclay nanocomposites with different concentration of AC were studied to determine the AC ability to disperse the organoclays. The objective was to improve physical-chemical and processing properties for obtaining high performance applications of these nanocomposites.

METHODS

Materials

Ethylene vinyl acetate (EVA), Braskem 3019 PE, was kindly provided by Nutec America. The material has 18% vinyl acetate content, melt flow of 2.5 gr/10 min as stated by the manufacturer. Two commercial organoclays, purchased from Southern Clay products, were: Cloisite 30B (C30B) and Cloisite 93A (C93A). The Amino-carboxylate (AC) compound, RM-210, was kindly provided by Cortec Corporation. The chemical structures of the materials are illustrated in Figure 1.

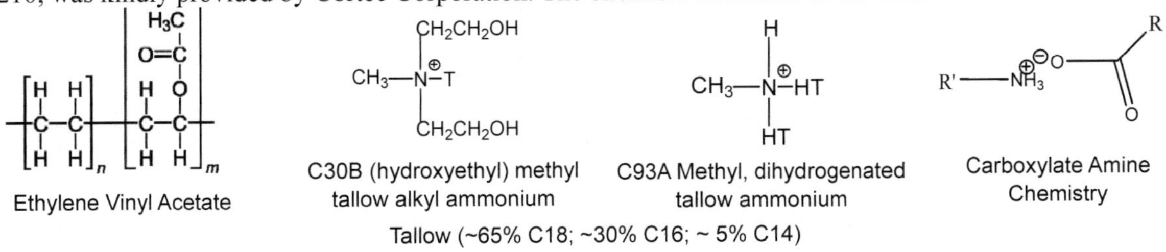

FIGURE 1. Chemical structures of materials used in this research.

Preparation

For organoclay pre-swelling, AC was mixed with organoclays using a Flacktek DAC 400.1 FVZ speed-mixer operating at 2500 RPM for 1 minute, and then heated at 60ºC during 10 min to induce AC volatilization. This procedure was repeated twice. For mixing composition of organoclays, AC parts were 5:1, 2:1 and 1:1, mixtures were kept in a dessicater for 24 hours until processing with EVA. For nanocomposite preparation, the mixtures EVA/Organoclay (5%wt)/AC (Content of AC: 1, 2.5 and 5% wt.) were melt mixing using Brabender Torque Rheometer at 60 r.p.m., from 90ºC for 10 minutes to 80ºC for 10 minutes. The transition time between both temperatures was about 10 minutes. The melted mixture was pressed in a compressing molding machine from 1000 psi to 3000 psi for about 95 °C for about 5 minutes, depending on the appropriate size and shape for further characterizations and evaluation.

Characterization and Evaluation

X-ray diffraction patterns were obtained using a Phillips PANalytical XPert-Pro model diffractometer equipped with Cu Kα (λ=1.54060 Å) radiation operated at 45 kV and 40 mA with a 1/8incident slit and 1/16" detector slit, scanning from 1° to 10° 2θ with a 0.5° step size a 20 second count time per step. In the case of morphology of the nanocomposites, similar operating condition were used, except for the incident and detector slit 1/4" and 1/8", respectively. FTIR absorption spectra of 50 µm films (32 scans) were analyzed in the mid-range using a Perkin Elmer FTIR Model 100. TGA characterization was performed using TA instrument model Q600 SDT. Approximately 10 mg of the sample was placed in alumina crucible and heated from room temperature to 600ºC, at a heating rate of 10ºC/min in nitrogen atmosphere. Glass transition, T_g , was obtained using DSC 200 operating from -50 ºC to 110 ºC at heating/cooling rate of 10ºC/min. Viscosity measurements were accessed using a Malvern Kinexus Pro+ Rotational Rheometer at 100 °C. A parallel plate (20 mm /20 mm) arrangement with a 2.5 mm gap setting was used. Tensile tests were based on ASTM D 638. A samples Typo IV were performed at room temperature at a tensile rate 1.96 in/min using a Shimdazu AGS 10 N load frame.

RESULTS AND DISCUSSION

Figure 2 illustrates the XRD patterns of organoclays/AC. The organoclays showed primary reflections at about 5° (~18 Å) and 3.54° (~25 Å) for C30B and C93A, respectively. The increase in the AC concentration in the organoclay mixture revealed different intercalation and dispersion behaviors. A well-defined peak at about 1.95° (~45 Å) was observed from lowest C93A:AC ratio in the case of C93A modification, while XRD patterns of C30B:AC ratios showed a broad shoulder reflection at same basal spacing. Carboxylate part and amino part developed polar and dispersive interactions with the surfactant and hydroxyl groups of the organoclays. Peaks at about 4.9 ° (~18 Å) and 6° (~14 Å) showed intercalations effects due to AC interaction via re-arrangements of the tallow and adsorption on organoclays surfaces. However, these peaks did not show the usual harmonic series peaks associated with the main reflection illustrated previously. Thus, mixed layering in clay occurred [12]. For the case of nanocomposites, EVA/C30B/AC also showed peaks at lower basal spacing due to regularly packed silica tatoids with collapse d-spacing of C30B, as reported in other EVA/clay studies [13]. It was also observed the presence of AC for the peak at about 8° (~11 Å) in its pure state. For the EVA/C93A, a peak of about 2.9° (~30 Å) was caused by intercalation from EVA into the galleries of C93A. Nevertheless, the XRD patterns of EVA/C93A/AC series showed an interlayer spacing collapse illustrated in the peak at about 6.8° (~ 13 Å). It seems that AC increases the polarity in the system. Hence, tertiary ammonium modifier from C93A is probably deprotonated by AC in the melted mixture EVA/AC. Similar deprotonation has also been observed with low molecular weight amines molecules [14].

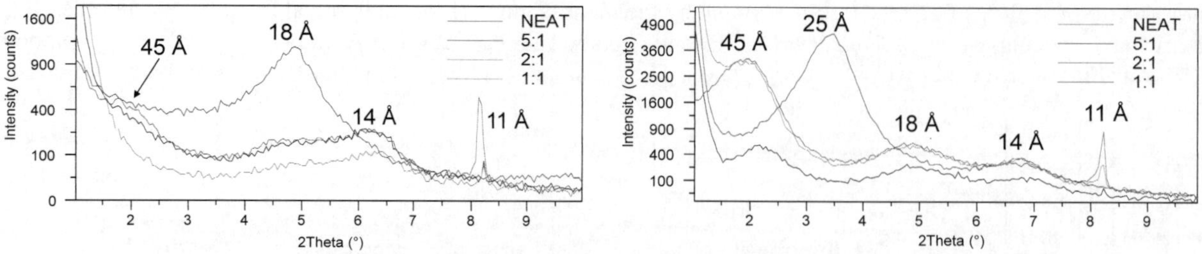

(a) (b)

FIGURE 2. X-ray of Cloisite 30B (a) and Cloisite 93A (b) a different organoclay: AC ratio.

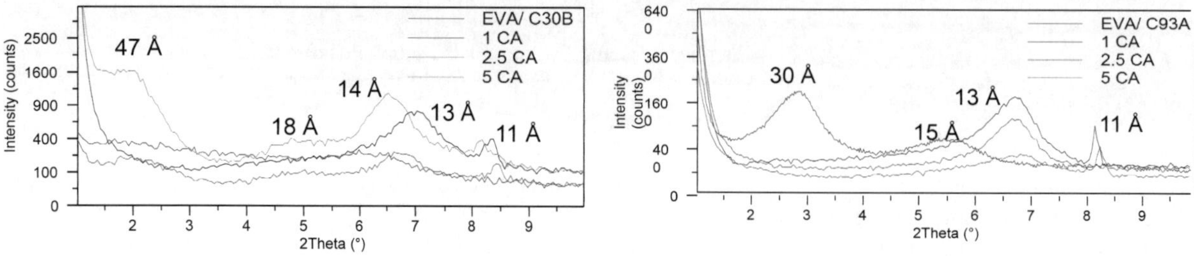

(a) (b)

FIGURE 3. X-ray of EVA/organoclay/Clay with Cloisite 30B (a) and Cloisite 93A (c) different %weight AC concentrations.

Figure 4 shows the FTIR spectrum of AC, EVA and EVA/Clay and the addition of 5 wt. % AC. Asymmetric COO^- stretching frequency was found at 1548 cm^{-1}, while those corresponding to symmetric COO^- stretching were found at 1364 cm^{-1} and 1338 cm^{-1} as a result of carboxylate ion from AC [15,16]. A peak at about 1555 cm^{-1} due to antisymmetric COO^- stretching mode from carboxylate ion of AC was only observed in EVA/C93A/5AC. It was also noted in other AC concentrations incorporated in EVA/C93A, but not noticed in EVA/C30B/AC. This peak confirmed the interaction of carboxylate ion in the nanocomposite. A peak at about 3190 cm^{-1} resulted from NH stretching vibration of the amine ion from AC. Moreover, it was observed a shoulder at about 3170 cm^{-1} due to amine ion interaction of AC in nanocomposites for both clays. The other peaks of AC were overlapped with EVA. However, the peaks at about 1236 cm^{-1} assigned to C-O-C stretching vibration and those at about 1740 cm^{-1} due to C=O stretching vibration of EVA [17] , shifted to shorter wave number for about 25 cm^{-1} and 17 cm^{-1},respectively as polarity increased with the addition of AC in all nanocomposites. However, it might also be the consequence of the interaction of AC, acetate groups with hydroxyl and ammonium functional groups in the C30B and C93A. It is well known that the high sensibility of C-O-C and C=O stretching vibration changes in the surrounding polarity is generally transformed into infrared band shifts [18].

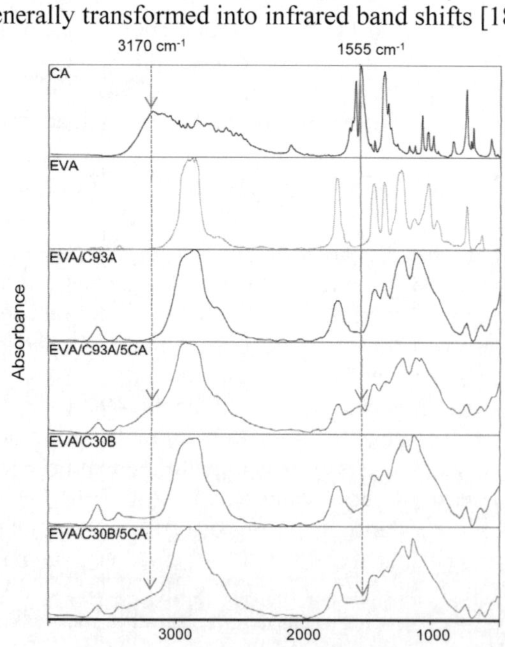

FIGURE 4. FTIR spectrum of AC, EVA and EVA nanocomposites including 5 wt. % of AC.

Figure 5 shows the flow curves of EVA/C30B series (a) and EVA/C93A series including 1, 2.5 and 5 wt. % AC. The presence of the both organoclays increased the viscosity at low shear rates due to dispersion and interaction of the organoclays surfaces and EVA matrix. The presence of 1 and 2.5 wt. % AC in EVA/C30B decreased its viscosity compared to EVA at lower shear rates. Nevertheless, its viscosity was slightly greater than EVA at 5wt% AC. Moreover, the AC decreased the viscosity in all EVA/C93A nanocomposites at lower and higher shear rates. The deprotonation of C93A surfactant decreased the interaction of this organoclay and EVA matrix. Also, the flow curves were close to EVA. Therefore, flow curves revealed a Newtonian behavior at low shear rates due to dispersion, interaction effects and power low behavior at high shear rates as a result of alignment of clay parallel to the flow, as reported [19]. Thus, AC can potentially improve the processing behavior in EVA/organoclays nanocomposites.

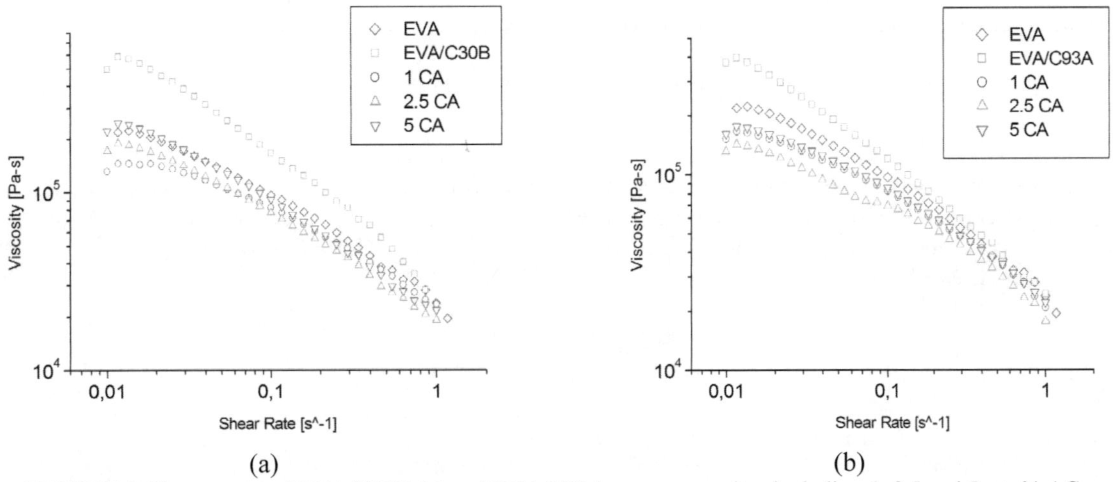

FIGURE 5. Flow curves of EVA/C30B (a) and EVA/C93A nanocomposites including 1, 2.5 and 5 wt. % AC.

Table (1) illustrates the thermal and mechanical properties of theses materials. The addition of both organoclays improved considerably the Young modulus (>70%) compared to EVA matrix. As demonstrated previously, platelet strongly interacted with soft EVA matrix. Then, the presence of rigid clay particles increased the rigidity of the system. Elongation at 10 Kg_f decreased considerably in EVA/C30B nanocomposite, due to the strong interaction of the hydroxyl groups from either clay or surfactant to polar groups of EVA. AC decreased the Young Modulus compared to EVA/Clays, but it was greater compared to EVA matrix. AC improved the elongation related to EVA nanocomposites. Moreover, there is a progressive increase either in elastic modulus or in elongation as Clay: AC ratios increased in the system.

In terms of thermal stability, there are two states of degradation. The first occurred as a result of de-acetylation. The second was due to a break-down of carbon in EVA backbone [8]. It is possible that hydroxyl groups of the C30B or ammonium group in C93A and acetate of EVA interacted with AC compounds. Consequently, scission of the acetate group was vanished from EVA. The weight loss temperature at 5% of EVA/clay/AC was similar to EVA, particularly at lowest Clay:AC ratio. In the second step, there was also an increase in the weight loss temperature at 30% related to EVA nanocomposites. This might result from a previous interaction of AC with unsaturated chains or previous formation of carbon-carbon double bonds in EVA backbone. There were slight changes in EVA T_g values from -1 to -3 °C. This could probably occur due to AC plastification effect on EVA polymer. These results indicated that AC can improve the thermal stability compared to EVA/Clay nanocomposites.

Table (1). Mechanical and Thermal properties of EVA nanocomposites systems.

	Young Modulus (MPa)	**Elongation, % @ 10 kgf**	**Tg (°C)**	**Weight Loss temperature at 5% (°C)**	**Weight Loss temperature at 30% (°C)**
EVA	25.0 ± 0.2	343.2 ± 4.5	-29	340	453
EVA/C93A	43.6 ± 2.3	240.0 ± 16.7	-29	316	439
1 AC	33.4 ± 1.6	291.9 ± 15.3	-30	337	442
2.5 AC	35.7 ± 2.2	293.8 ± 17.9	-30	333	450
5 AC	41.4 ± 0.8	197.6 ± 4.2	-30	296	448
EVA/C30B	45.8 ± 1.1	46.5 ± 7.2	-28	314	439
1 AC	34.7 ± 1.1	263.7 ± 15.0	-31	332	448
2.5 AC	34.9 ± 1.2	278.1 ± 17.5	-30	326	452
5 AC	36.0 ± 1.5	307.2 ± 5.3	-31	173	433

CONCLUSIONS

In this work, the effects of an amino-carboxylate compound in the morphology, thermal and mechanical properties of EVA/Clay nanocomposites have been investigated. XRD showed that AC enhanced the intercalated structure in both organoclays and EVA nanocomposite. C-O-C and C=O stretching vibration in FTIR showed AC, EVA and Clay interactions and also the presence of AC after processing. The change in viscosity at low and high shear rate depended on the organoclay and the synergy between the organoclays and AC concentration. It was observed improved thermal and mechanical properties in EVA/Clay/AC compared to the neat nanocomposite. AC showed a potential material for replacing conventional, functionalized or reactive compatibilizer for EVA nanocomposite applications. In addition, it promoted the green processing nanocomposite approach. Further studies will have to develop potential applications of these findings in protection packing for some applications such as photovoltaic or electronic encapsulation.

ACKNOWLEDGEMENTS

The authors thank the technical staff of the Laboratory of Testing Materials (LEMAT) and the Center of Nanotechnology Research and Development (CIDNA, acronym in Spanish) at ESPOL Polythecnic University. We are grateful for the support of the School of Mechanical Engineering and Production Science and the Research Dean at ESPOL Polythecnic University. The authors also acknowledge Nutec America Inc. and Cortec Corporation for kindly supplying the materials used in this research.

REFERENCES

1. A. Henerson. IEEE Electrical Insulation Magazine, **9,** 1, 30-38 (1993)
2. R. Boonnattakorn, V Chonhenchob, M. Siddiq, S. Singh. Packaging Technology and Science, **28**, 3, 241-252 (2015)
3. C. Schneider, R. Langer, D. Loveday, D. Hair. Journal of Controlled Release, **262**, 284-295 (2017)
4. M. Kempe, G. Jorgensen, K. Terwilliger, T.J. McMahon and C.E.Kennedy. IEEE 4th World Conference on Photovoltaic Energy May 7-12 (2006)
5. B. Rimez, H. Rahier, G.Van Assche, T. Artoos, M. Biesemans, B. Van Mele. Polymer Degradation and Stability, **91**, 800-810 (2008)
6. S. Peeterbroeck, M. Alexandre, R. Jerome, Ph. Dubois. Polymer Degradation and Stability **90**, 288-294 (2005)
7. W. Zhang, D. Chen, Q. Zhao, Y Fand. Polymer, **44**, 7953-7961 (2003)
8. G. Griffin, M. Pannirselvam, B. Martin. Macromolecular Symposium, **361**, 101-105 (2016)
9. R. Lujan-Acosta, S. Sanchez-Valdes, E. Ramirez-Vargas, L. Ramos-DeValle, A. Espinoza-Martinez, O. Rodriguez-Fernandez, T. Lozano-Ramirez, P. Lafleur. Material Chemistry and Physics, **46**, 3, 437-445 (2014)
10. B. Miksic. Anti-Corrosion Methods and Materials, **22,** 5-8 (1975)
11. B. Miksic and R. Miller. Fundamental Principles of Corrosion Protection with vapor phase corrosion inhibitors. 5th European Symposium on Corrosion Inhibitors, University of Ferrara, September Italy (1980)
12. D. Eckel, M. Balogh, P. Fasulo, W. Rodgers. Journal of Applied Polymer Science, **93**, 1110-1117 (2004)
13. S. Filippi, M. Paci, G. Polacco, N.T. Dintcheva, P. Magagnini. Polymer Degradation and Stability, **96,** 5, 823-832 (2011)
14. A.Rigail-Cedeño, D. Schmidt. Bio-based Epoxy Clay Nanocomposites, 32th Conference of Polymer Processing Society, July , Lyon, France (2016)
15. J. Oomens, J., and J. Steill. The Journal of Physical Chemistry A Letters. 112: 3281-3283 (2008)
16. M. Nara, H. Torii, and M. Tasumi. Journal of Physical Chemistry 100: 19812-19817 (1996)
17. R. Koopmans, R. Van Der Linden and E. Vansant. Polymer Engineering and Science, **22**, 4 879-882 (1982)
18. A. Barth. Progress in Biophysics & Molecular Biology, **74**, 141-173 (2000)
19. V. Pasanovic-Zujo, R. KumarGupta, S. Bhattacharya. Rheology Acta , **43**, 99-108 (2004)

Sub-Critical Gas-Assisted Processing of Ethylene Vinyl Alcohol + Nanoclay Composites

Thomas Ellingham[a,b], Galip Yilmaz[a,b], Lukas Duddleston[a], Zhutong Li[a,b] and Lih-Sheng Turng[a,b*]

[a]*Department of Mechanical Engineering, University of Wisconsin–Madison Madison, Wisconsin, 53706, USA*
[b]*Wisconsin Institute for Discovery, University of Wisconsin–Madison Madison, Wisconsin, 53715, USA*

ellingham@wisc.edu
turng@engr.wisc.edu

Abstract. Ethylene vinyl alcohol (EVOH) was melt compounded with an organo-modified nanoclay (NC) at a loading level of 5% by weight in a twin-screw extruder (TSE) using a sub-critical gas-assisted process (SGAP). Carbon dioxide (CO_2) and nitrogen (N_2) were used as blowing agents to induce foaming inside of the barrel of the extruder and as the melt exits the die. Samples processed with SGAP demonstrated better dispersion of NC versus the traditional extrusion process due to the additional stresses that bubble expansion provides during foaming.

Keywords: Nanoclay, nanocomposite, sub-critical gas, EVOH, twin-screw extrusion.
PACS: 83.50.Jf, 81.05.Qk

INTRODUCTION

Methods for dispersing clay nanoparticles include solvent-based methods, *in situ* polymerization, and melt processing (1–3). Melt processing is the most economically viable technique for scale-up and is the method used in this paper—specifically, twin-screw extrusion (TSE).

While nanofillers boast large increases in properties, these benefits are only obtainable if the particles are well mixed throughout the host polymer matrix (2,3). There are two types of mixing of concern: distributive and dispersive. Distributive mixing refers to how well the filler is evenly spaced throughout the bulk of the matrix with no attention paid to the size or existence of particle agglomerates. Dispersive mixing pertains to how well separated the individual particles so as not to form agglomerates (4). For 2-D fillers such as nanoclay (NC), the methods of dispersion are the breakup of large agglomerates into smaller ones and the exfoliation of individual platelets from a stack of platelets, sometimes preceded by intercalation, whereby the polymer or a solvent enters and increases the gallery spacing of the NC, thus easing exfoliation (1). Both of these methods can be accomplished when the host polymer matrix exerts a sufficient force on the particle during deformation in the molten state, where the amount of force applied to a particle agglomerate is proportional to the viscosity of the host polymer (5).

Most of the existing literature which utilize gas-assisted processing involves the use of supercritical CO_2 (SC-CO_2), which has several interesting properties. It has the viscosity of a gas, but the density of a liquid (which makes it able to be metered and pumped accurately). It also behaves as a polar solvent, which is ideal for interacting with polar group-containing polymers and organo-modified NC (1).

As SC-CO_2 dissolves into a polymer such as EVOH, it increases the free volume between the molecules, which decreases the viscosity. This makes processing of the polymer and SCF mixture easier, but it is detrimental to the ability of the polymer to break up agglomerates of fillers. As noted earlier, this is because the stresses a molten polymer can exert on a filler particle are directly proportional to the viscosity of that polymer (5). It has been shown in the literature that over-saturating a polymer with SC-CO_2 during SCF-assisted extrusion can be detrimental to the amount of exfoliation and dispersion of NC; it was proposed that the decrease in viscosity was the reason for this drawback (6). To avoid this decrease in viscosity, in this study work, gases below supercritical pressure were used in a so-called 'Sub-Critical Gas-Assisted' extrusion process (SGAP). Because the viscosity decrease associated with small molecules like CO_2 or N_2 is linear with concentration, and concentration scales linearly with pressure, sub-critical gases do not decrease the viscosity appreciably (7,8).

In the SGAP process, the sub-critical pressure N_2 or CO_2 introduced during extrusion causes the polymer melt to foam inside of the barrel and upon exiting the die. While in the barrel, as the gas-containing melt experiences high and low pressure zones, the foaming process is repeated several times. Cell growth or bubble expansion is an equibiaxial extensional flow generally not found in TSEs; therefore, any additional flows of this type add to the amount of stress in the polymer melt available to break up NC agglomerates. Equibiaxial extensional flows are especially efficient for this purpose because the associated material viscosity is six times greater than the shear case (9–11).

In this study, EVOH was compounded with NC at a loading of 5% by weight using TSE with sub-critical gas-assisted processing (SGAP) with N_2 or CO_2 as the blowing agent. Thermal, rheological, and morphological properties were studied. SGAP was found to be beneficial to dispersion and exfoliation of NC while the addition of inert blowing agents decreased thermo-oxidative degradation. This novel and facile method has a myriad of applications in the plastics industry for the cost-effective creation of high-performance, lightweight nanocomposites.

EXPERIMENTAL METHODS

Soarnol DT2904RB (Nippon Gohsei, Japan), a film blowing grade of ethylene vinyl alcohol copolymer (EVOH), in pellet form was provided by Bemis Company, Inc. (Neenah, WI). Cloisite 30B, an organo-modified NC, was donated by Southern Clay Products (now BYK Additives, Gonzales, Texas). Compressed CO_2 and N_2 were obtained and used as-received from Airgas, Inc. (Madison, WI).

Mixing was performed in a Leistritz ZSE-18 twin-screw extruder at 200°C and 125 rpm. The extruder was equipped with a standard CO_2 gas cylinder and regulator purchased from Airgas. A metal hose (1/8" diameter) with Swagelok fittings was used to connect the regulator to the barrel of the extruder. When SGAP was employed, the gas was set to ~0.35 MPa (50 psi) greater than the measured melt pressure at the die exit. Only forwarding elements and one reverse element just before the gas injector are used. The extrudate foams until solidification and subsequent pelletization. Pelletized strand samples were melt-pressed into films at 200°C to a thickness of about 200 μm using a Carver Auto M press.

SEM images were obtained on a LEO 1530 FESEM with a 3kV accelerating voltage using an in-lens detector on gold coated samples. TGA was performed on a TA Instruments Q50 ramping from 20°C to 600°C at a rate of 10°C/min under inert gas. Storage modulus (G') measurements were made on a TA Instruments AR 2000ex rheometer. A 25 mm parallel plate fixture at 190°C was used to test the materials with a gap of 500 μm and a strain of 1%. X-ray diffraction spectra were obtained on a Bruker D8 Discovery using a $Cu_{k\alpha}$ (1.54Å) emitter. An industrial micro computed tomography (μCT) system, Metrotom 800 from Carl Zeiss AG, was used to image the samples with a 50 kV accelerating voltage and 80 μA of current. The resolution was 5 μm.

RESULTS AND DISCUSSION

In Figure 1, the TGA weight loss curves are shown for the samples with and without NC. Several studies show that EVOH is easily degraded during thermal processing, like extrusion (12,13). The method of degradation is thermally induced chain-scission via oxidation. Another related process that alters the molecular structure of EVOH is cross-linking of the vinyl alcohol units, which may or may not change the molecular weight, but do change the microstructure—which affects further thermal degradation via volatile transport and free volume changes (among other things, like the rheological behavior of the melt) (14,15).

In the case of the samples without NC, the neat EVOH has the most delayed thermal degradation because it has undergone one less thermal processing step. In comparison to the samples processed with N_2 or CO_2, earlier onset of thermal degradation (T_{onset}) and a lower maximum degradation temperature (T_{max}) can be observed for the 0% NC samples. It appears that using either N_2 or CO_2 can prevent some of the thermo-oxidative reactions that take place during extrusion. This accounts for a change in T_{onset} of about 10°C for N_2 and 14°C for CO_2, and a difference in T_{max} of about 10°C for N_2 and 8°C for CO_2. It is likely that both gases displace environmental oxygen during extrusion in a so-called blanket effect, thus lowering the chance that a thermo-oxidative event will occur (13,16). For samples containing 5% NC, similar changes due to thermal processing and the blanket effect can be observed as well. The difference between T_{onset} for 5% NC gas-assisted and traditional processed samples is 11°C for N_2 and 8°C for CO_2; the difference in T_{max} is 17°C for N_2 and 16°C for CO_2.

The rheology of a polymer is closely related to the microstructure present in the material. In TGA, it was shown that there were changes that occurred during extrusion that contributed to a change in the properties of the polymer. These changes could be seen in the rheological behavior as well. In Figure 2, the changes in dynamic rheological response can be observed for the 0% NC samples, which manifest as a slight shift to larger values for the storage modulus (*G*'). The storage modulus refers to the solid-like component of the response of a fluid undergoing oscillatory deformation. *G*' is sensitive to changes in molecular weight (5,17). The overall trend is the same as in TGA.

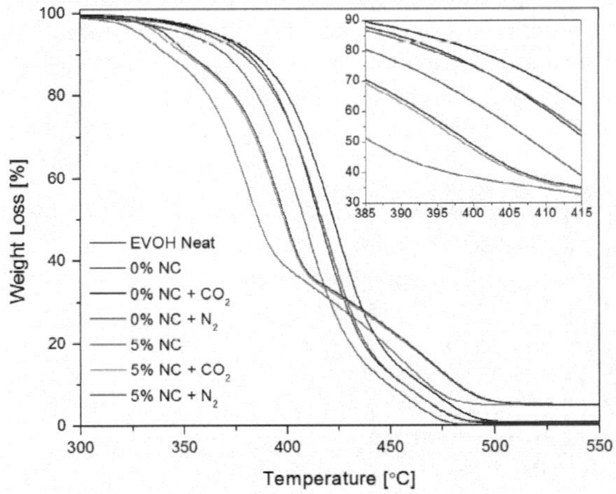

FIGURE 1. TGA curves

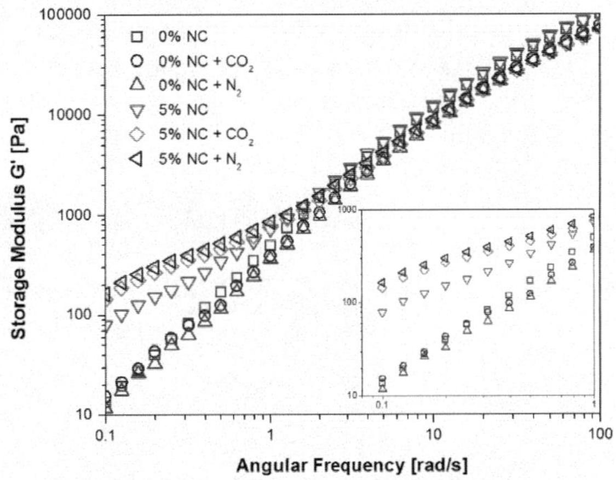

FIGURE 2. Storage Modulus (G') vs. Angular Frequency (rad/s).

In the samples containing NC, the trend in high frequency response is essentially the same as in the 0% NC samples, albeit with larger values of *G*' due to the addition of the filler. An interesting effect of the filler is the low-frequency tail behavior of NC-containing samples, as shown in Figure 2. In systems with a well dispersed nanofiller, the low-frequency end of the *G*' curve displays a fundamentally different slope than the rest of the curve. This is due to the formation of a pseudo-solid network, which is disrupted at higher frequencies and their resulting larger stresses. This network is a result of increased polymer–filler interaction and is indicative of better dispersion (1–3,6,18).

Another measure of dispersion of plate-like fillers such as NC is the x-ray diffraction pattern of the nanocomposite, specifically the peak resulting from the stacked nature of NC. Figure 3 shows the WAXS spectra of the NC-containing samples. The observable peaks correspond to the interlayer spacing of the NC in the melt. As this peak shifts to lower 2θ angles, this signifies a larger interlayer or gallery spacing and thus intercalation. From Bragg's law, for 5% NC without gas, this gives a gallery height of 3.20 nm (2θ = 2.76°) which is larger than NC alone (not shown), which has a gallery height of 1.79 nm (2θ = 4.93°) (19).

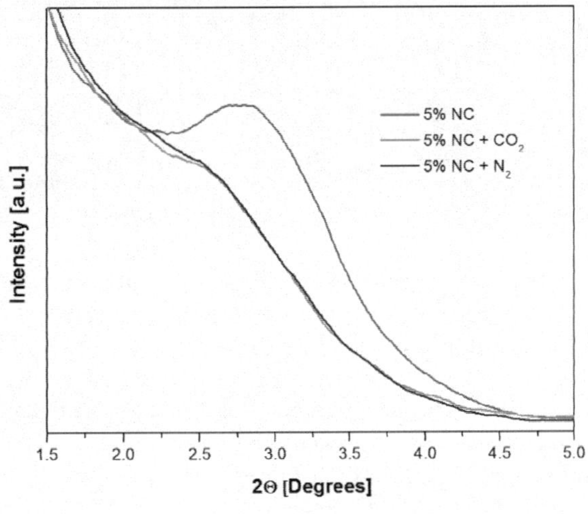

FIGURE 3. WAXS spectra of samples with nanoclay. The peaks correspond to the interlayer spacing or gallery height of the nanoclay in the EVOH.

The peaks for the gas-processed samples are shifted even further lower angles (and intensity), and the peaks are at too low of an angle to accurately detect. This diminishing peak angle, resulting from the smaller amount of stacked layers (for a given total amount of NC) available to interact with the x-ray light, suggest an increase in exfoliation of NC in SGAP samples (1–3,6,13).

In this study, both intercalation and exfoliation were obtained without the use of supercritical pressures for CO_2 and N_2, the latter of which was not expected to favor an interaction with either the polymer or the NC as it does not have a polar nature. As the curves for either of the gas-processed samples were very similar and had a lower intensity and 2θ angle as compared to the non-gas-processed sample, it would seem that gas-assisted processing is effective in exfoliating the nanoclay and that the added stress during bubble expansion plays a large role in the exfoliation process as opposed to the interaction of the gas with NC (infiltrating the gallery spacing via a CO_2- or N_2-phillic interaction) or the interaction between the gas with the polymer, which would reduce the viscosity at higher pressures.

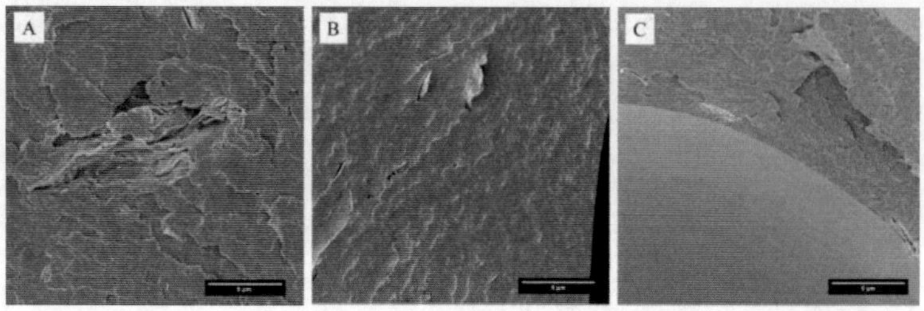

FIGURE 4. SEM of extruded samples with nanoclay: (A) 5% NC, (B) 5% NC + CO_2, and (C) 5% NC + N_2.

While rheology and x-ray diffraction can investigate the microstructure of the bulk material, SEM and μCT are useful to probe the local morphology of the samples. Figure 4 displays SEM images of the nanocomposites. Figure 4 (A) was extruded without gas and shows an agglomerate of NC. Conversely, Figures 4 (B) and (C) display samples processed with CO_2 and N_2, respectively. Here the NC particles become aligned perpendicular to the direction of bubble or cell growth in the polymer between the growing cells. As the cells expand, the elongational stresses add to the total stress the polymer melt exerts on the agglomerates, thus aiding in the effort to break them up.

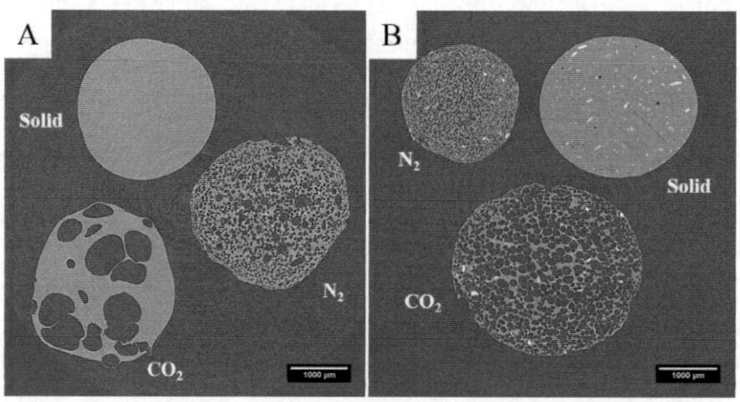

FIGURE 5. μCT of extruded pellets: (A) with 0% NC, (B) with 5% NC.

Cell morphology can be obtained via μCT, as presented in Figure 5. The samples without NC are shown in Figure 5 (A), where the CO_2 sample has much larger cells than the N_2 sample due to the difference in solubility and super-saturation levels between CO_2 and N_2, resulting in faster and more uniform cell nucleation for N_2 (20). The bright white particles in Figure 5 (B) are NC agglomerates. They are smaller and fewer in number with SGAP. From the μCT images, it can be observed that samples with NC have smaller and more numerous cells due to the cell nucleation effect of the nanofiller (3,21). As these cells grow, they impart elongational stresses that help break up NC agglomerates, exfoliating and dispersing the platelets. Because TSEs are starve-fed, having areas of low and high pressure, this process occurs in the barrel of the extruder several times with the cells collapsing, the gas re-solubilizing, and then foaming again (22,23). This process introduces a synergy that creates a final dispersion not otherwise attainable without foaming.

CONCLUSIONS

Sub-critical gas-assisted processing with either CO_2 or N_2 was found to be effective in increasing the dispersion of nanoclay (NC) in EVOH. It was shown via TGA that the gas can reduce the amount of thermal degradation that occurs in the final product. When NC is introduced into the melt during extrusion with sub-critical gas, it was observed through rheology, WAXS, SEM, and μCT that dispersion was increased. The better dispersion is due to the additional stresses incurred during bubble expansion inside of the barrel and upon die-exit. This effect is synergistic as the NC particles nucleate bubble growth.

ACKNOWLEDGMENTS

The authors would like to acknowledge the support of the Kuo K. And Cindy F. Wang Professorship, the Vilas Distinguished Achievement Professorship, Wisconsin Distinguished Graduate Fellowship, 3M Fellowship, Bemis Company, Nanocor, and the Wisconsin Institute for Discovery.

REFERENCES

1. Horsch S, Serhatkulu G, Gulari E, Kannan RM. Supercritical CO2 dispersion of nano-clays and clay/polymer nanocomposites. Polymer, 2006,47(21):7485–96.
2. Pavlidou S, Papaspyrides CD. A review on polymer-layered silicate nanocomposites. Progress in Polymer Science (Oxford), 2008,33(12):1119–98.
3. Sinha Ray S, Okamoto M. Polymer/layered silicate nanocomposites: a review from preparation to processing. Progress in Polymer Science, 2003,28(11):1539–641.
4. Alemaskin K, Manas-Zloczower I, Kaufman M. Simultaneous characterization of dispersive and distributive mixing in a single screw extruder. In: ANTEC 2003, 2003.
5. Osswald TA, Menges G. Material Science of Polymers for Engineers. Third Edit. 2012.
6. Zhao Y, Huang HX. Dynamic rheology and microstructure of polypropylene/clay nanocomposites prepared under Sc-CO2 by melt compounding. Polymer Testing, 2008,27:129–34.
7. Nalawade SP, Picchioni F, Janssen LPBM. Supercritical carbon dioxide as a green solvent for processing polymer melts: Processing aspects and applications. Progress in Polymer Science (Oxford), 2006,31(1):19–43.
8. Pyda M. High-pressure rheology and viscoelastic scaling predictions of polymer melts containing liquid and supercritical carbon dioxide. Journal of Polymer Science, Part B: Polymer Physics, 2001,39(23):3055–66.
9. Tadmor Z. Forces in Dispersive Mixing. Fundamentals of Industrial Chemical Engineering, 1976,15(4):346–7.
10. Meissner J. Polymer melt elongation - methods, results, and recent developments. Polymer Engineering and Science, 1987,27(8):537–46.
11. Ellingham T, Duddleston L, Turng L-S. Sub-critical gas-assisted processing using CO2 foaming to enhance the exfoliation of graphene in polypropylene + graphene nanocomposites. Polymer, 2017,.
12. Lagaron JM, Giménez E, Saura JJ. Degradation of high barrier ethylene-vinyl alcohol copolymer under mild thermal-oxidative conditions studied by thermal-oxidative conditions studied by thermal analysis and infrared spectroscopy. Polymer International, 2011,50(6):635–42.
13. Artzi N, Tzur a., Narkis M, Siegmann a. The effect of extrusion processing conditions on EVOH/clay nanocomposites at low organo-clay contents. Polymer Composites, 2005,26(3):343–51.
14. Garcia-Leiner M. and Lesser A. J. Intercalated Polymer Nanocomposites Prepared in Supercritical Carbon Dioxide. In: Material Research Society, 2004. p. 1528–32.
15. Hasan MM, Li YG, Li G, Park CB, Chen P. Determination of solubilities of CO2 in linear and branched polypropylene using a magnetic suspension balance and a PVT apparatus. Journal of Chemical and Engineering Data, 2010,55(11):4885–95.
16. Franco-Urquiza E, Santana OO, Gámez-Pérez J, Martínez AB, Maspoch ML. Influence of processing on the ethylene-vinyl alcohol (EVOH) properties: Application of the successive self-nucleation and annealing (SSA) technique. Express Polymer Letters, 2010,4(3):153–60.
17. Osswald T, Rudolph N. Polymer Rheology. 2014.
18. Nguyen QT, Baird DG. An improved technique for exfoliating and dispersing nanoclay particles into polymer matrices using supercritical carbon dioxide. Polymer, 2007,48(23):6923–33.
19. Jeong H, Kim BC, Kim EH. Structure and properties of EVOH/organoclay nanocomposites. Journal of Materials Science, 2005,40:3783–7.
20. Sun X, Kharbas H, Peng J, Turng L-S. A novel method of producing lightweight microcellular injection molded parts with improved ductility and toughness. Polymer, 2015,56:102–10.
21. Okamoto M, Nam PH, Maiti P, Kotaka T, Nakayama T, Takada M, et al. Biaxial Flow-Induced Alignment of Silicate Layers in Polypropylene/Clay Nanocomposite Foam. Nano Letters, 2001,1(9):503–5.
22. White JL, Coran AY, Moet A. Polymer mixing : technology and engineering. 2001.
23. White JL, Potente H. Screw Extrusion: Science and Technology. White JL, Potente H, editors. 2002. 35, p.

Development of Partially Filled Flow Simulation of High-Viscosity Fluid inside Mixer

Kazuhide Sekiyama, Sayaka Yamada and Tomokazu Nakagawa

Fluid Thermal & Chemical Technology Research Section, Mechanical Engineering Research Laboratory, Technical Development Group, Kobe Steel Ltd.,

Abstract. A simulation technique has been developed which is more accurate and suitable for the simulation of highly viscous fluid flow in partially filled mixers. This new technique is based on the Element-free Galerkin method (EFGM). To improve the accuracy, a "rearrangement" scheme has been employed which eliminates errors caused by the irregular distribution of data points. Through numerical tests, the new technique has been verified to be applicable to high viscosity fluid flow with sufficient accuracy. Furthermore, simulation examples for realistic mixers were carried out using the developed technique. From the results, it was found that the developed technique is capable to investigate the relationship between operating conditions and flow behavior inside realistic mixer.

Keywords: Plastic mixing, Continuous mixer, Simulation, Mesh-free method.
PACS: 46.15.-x

INTRODUCTION

Twin screw extruders and mixers have been used in polymer processing, because it has good mixing performance and provide mass-production. There are many studies about flow characteristics inside extruders and mixers. In particular, flow simulations are useful to obtain detailed information such as velocity and stress distributions. Yang et al.[1], for example, conducted a three-dimensional non-Newtonian flow analysis of the kneading disk region in a co-rotating twin-screw extruder on the basis of finite element method (hereinafter "FEM"). Kajiwara et al.[2], on the other hand, conducted a three-dimensional non-Newtonian flow analysis on a melt-conveying zone in an intermeshed counter-rotating twin-screw extruder by FEM. Many authors have been studying flow phenomena inside mixers using fully filled simulations[3][4]. On the other hand, there are a few studies handling partially filled state inside extruders and mixers[5][6].

In this study, Element-free Galarkin Method, one of the mesh-free methods, is applied to partially filled flow of melt plastic. Furthermore, a new scheme keeping distances among data points is introduced to improve simulation accuracy. As a validation, the simulation results are compared with the experimental results with a triangular-shaped rotor. This method enables accurate predictions of free surface shapes and stable simulations. Moreover, partially filled flow simulations in realistic plastic mixers are carried out using the developed simulation method.

SIMULATION METHOD

We applied Mesh-free method for present issue. The one of the characteristic features of a mesh-free method is it doesn't need "mesh" or "element". The feature enables to simulate large deformation flow and combining or separating flow. We selected Element-free Galarkin method[7].

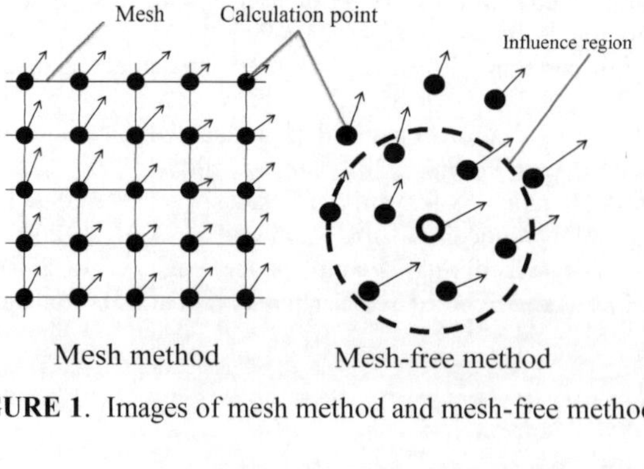

FIGURE 1. Images of mesh method and mesh-free method

This study has been conducted on the basis of the following assumptions. The material is a viscous body that is almost incompressible in isothermal conditions, in which the viscosity only depends on strain rate. The equation of motion is given as follows:

$$\rho\dot{\mathbf{u}} = \nabla \cdot \boldsymbol{\sigma} + \mathbf{b} \qquad \text{in } \Omega \tag{1}$$

$$\mathbf{n} \cdot \boldsymbol{\sigma} = \bar{\mathbf{t}} \qquad \text{in } \Gamma_t \tag{2}$$

$$\mathbf{u} = \bar{\mathbf{u}} \qquad \text{on } \Gamma_u \tag{3}$$

$$\sigma_{ij} = 2\mu\dot{\varepsilon}'_{ij} + \lambda\dot{\varepsilon}_v\delta_{ij} \tag{4}$$

where, Ω, object domain; Γ_t, boundary prescribing surface force; Γ_u; $\boldsymbol{\rho}$, density of fluid; $\dot{\mathbf{u}}$, time differential velocity vector; boundary prescribing velocity; $\boldsymbol{\sigma}$, stress tensor; $\mathbf{b}$, body force vector; ∇, vector differential operator; $\mathbf{u}$, velocity vector; $\mathbf{n}$, vector normal to boundary; $\bar{\mathbf{u}}$, boundary velocity vector; and $\bar{\mathbf{t}}$, surface force vector. σ_{ij}, $\dot{\varepsilon}'_{ij}$, $\dot{\varepsilon}_v$: stress, strain deviator, volumetric strain respectively, μ : viscosity of fluid. λ is a resistance coefficient to volume change and enough large against fluid viscosity. The boundary velocity is defined by the penalty method.

When the data point coordinates are altered in accordance with a time marching method, inter-particle distances become uneven with time, deteriorating the accuracy. Then, the original "rearrangement" scheme is employed. In the "rearrangement" scheme, a repulsive force works on the points when the inter-particle distance becomes smaller than setting distance, making the points to be arranged homogeneously.

VALIDATION OF DEVELOPED METHOD

In this section, we performed simulations and experiments for validation of accuracy of our simulation with simple shape mixer under partially filled condition. The mixer has a triangular-shaped rotor. Silicon oil (KF-96H-1 Mcs by Shin-Etsu Chemical Co., Ltd.) was used as a fluid. In the simulation, two-dimensional simulations were conducted by the model showed in figure 2 (a). The filling ratio was set at 80%. In experiment, the rotor has an axial length of 89 mm, and the barrel has an axial length of 90 mm. The rotor has no torsion in the axial direction so as to avoid the flow in the axial direction as much as possible.

The results of distributions of fluid inside the mixer are showed in figure 2 (b). The red solid lines in figure 2 (b) are the contours, retrieved from the experimental results, of the free surface. This figure indicates that the simulation result agreed well with experimental one qualitatively.

Figure 2 (c) shows the relation between the number of rotation and the torque acting on the rotor. This figure indicates that the simulation result ($T_{\text{cal.}}$) was a bit smaller than the experimental one ($T_{\text{exp.}}$). A possible reasons of this difference are as follows; 1) extra torque is also generated because the experiment was performed on an apparatus having gap between the rotor and the barrel, 2) in the experiment, an axial flow occurs although a little, which may cause the torque to be changed. On the basis of these results, it can be mentioned that the developed simulation enabled to predict the partially filled flow with high accuracy.

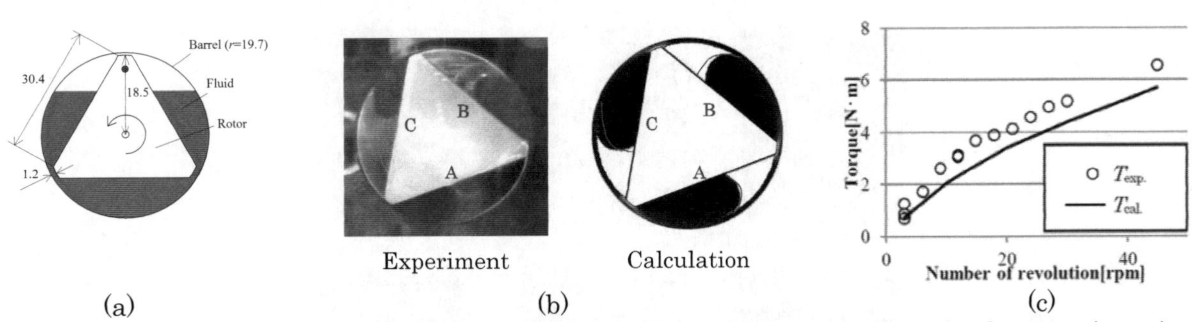

(a) (b) (c)

FIGURE 2. (a) Section shape of simple mixer, (b) Steady state flow condition at 6 rpm, (c) results of torque acting on inner rotor.

EXAMPLE OF SIMULATION INSIDE ACTUAL MIXERS

In this section, the developed method was applied to simulate realistic mixers. We investigated the flow inside continuous mixers. For the continuous mixers, a residence time is one of the most important mixing indices. The residence time depends on a filling ratio inside mixer. The filling ratio varies depending on an operating condition. However, the relationship between the filling ratio and operating condition is not clear. Therefore, in this section, we investigated the effect of the operating conditions to the filling ratios or the average residence times.

Numerical Simulation condition

We carried out partially filled simulations for melt mixing zones of counter-rotating continuous mixers. Calculation models and conditions of the simulations were shown below. There are both screw zone and mixing zone in each calculation model. Regarding screw zones, the both models have same screw shapes. Regarding mixing zone, they have same cross-sectional shape. The configurations are Model-1) only forward rotor; Model-2) forward and backward rotor. Nominal barrel diameter is 100 mm, mixing zone length is 144 mm, and barrel length is 254 mm. Viscosity of the fluid is 1000 Pa·s. The density is 900 kg/m^3. Rotation speeds were 200, 400, 600 rpm. Feed rates were 185, 544, 817 kg/hr. We investigated all combinations of two rotor shapes, three rotation speeds and three feed rates. The surface boundary condition is no slip condition. The outlet cross-sectional boundary condition is free outflow. The inlet cross-sectional boundary condition is constant velocity. In the initial states of the simulations, mixers are empty. Fluids were fed from the upstream ends of the screw zones. The simulations were finished when the flow behaviors were constant.

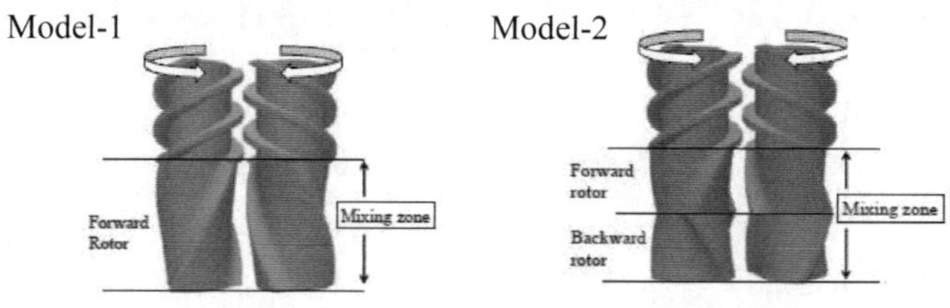

FIGURE 3. Calculation models

Results and discussion

Filling state and filling ratio inside mixer

We observed filling state inside mixers. The filling states at rotation speed 200 rpm obtained by the simulations are shown in figure 4 (the results of Model-1s are shown at upper; those of Model-2s are shown at bottom).

In the cases of Model-1s, polymer exists just only around the tips of the rotors. On the other hand, the mixing zones in Model-2s are almost fully filled with the polymer. This feature about filling state was appeared at other rotation speeds.

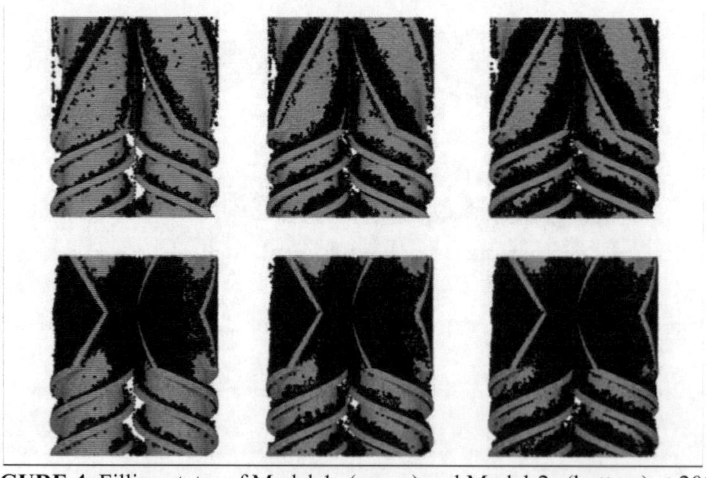

FIGURE 4. Filling states of Model-1s (upper) and Model-2s (bottom) at 200 rpm.

The relationship between the filling ratios and feed rates for all rotation speed cases were summarized in figure 5. Filling ratios at feed rates 185, 544, 819 kg/m^3 are shown in figure 5 (left). It is found that the filling ratios of Model-2s are larger than those of Model-1s in the all feed rates cases. Regarding the results of Model-1s, the larger feed rate is, the higher filling ratio is for all rotation speeds. The filling ratio is larger when the filling ratio is higher in the all cases of Model-1s. On the other hand, the filling ratios of Model-2s which have a backward rotor are around 85%.

From the results, it can be mentioned that torsion directions of rotors strongly affect to the filling rates. It can be also mentioned that the feed rate strongly affect to the filling rate in the case without backward rotors. On the other hand, the feed rate doesn't affect to the filling rate for the case with backward rotors.

The filling ratios at rotation speeds 200, 400, 600 rpm are shown in figure 5 (right). The filling ratios of Model-2s are also higher than those of Model-1s in the all rotation speeds. Regarding Model-1s, the higher rotation speed is, the lower the filling ratio is for all feed rates. The filling ratios of Model-2s are around 85% in the all cases.

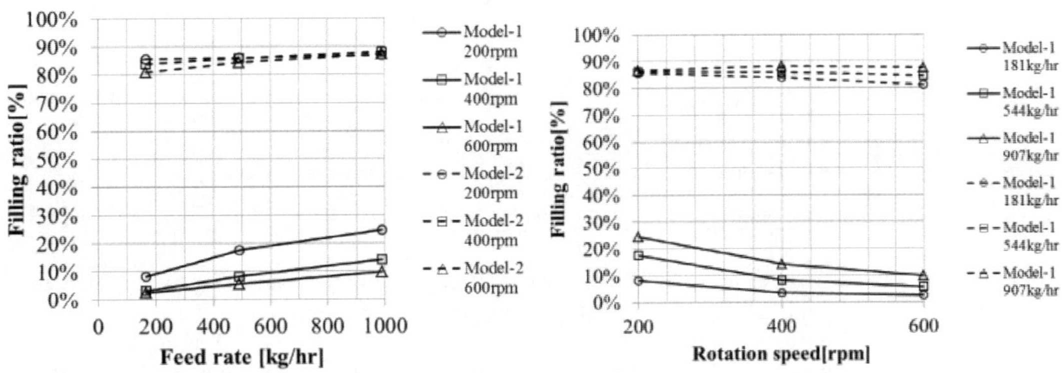

FIGURE 5. Relationship between filling ratio and feed rate (left), rotation speed (right)

Figure 6 shows the relationship between the filling ratios and the ratio of feed rate and rotation speed *Q/n*. Regarding Model-1s, there is strong correlation between filling ratio and *Q/n*. The larger *Q/n* is the higher filling ratios. In other words, the filling ratios of Model-1s can be controlled by *Q/n*. On the other hand, the filling ratio of Model-2s is almost constant in all region of *Q/n* because the filling ratios of Model-2s are almost constant.

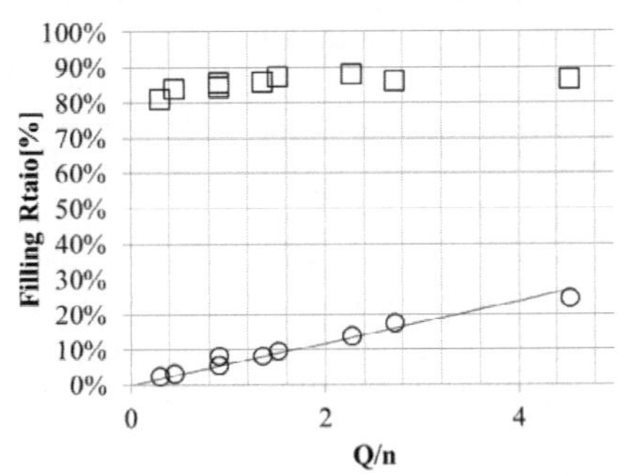

FIGURE 6. Relationship between filling ratio and *Q/n*

Average residence time

Average residence time(t_{ave}) is calculated by the equation shown below.

$$t_{ave} = \frac{\text{fluid volume inside mixing zone(filling ratio)}}{\text{feed rate}} \tag{5}$$

Average residence times at feed rates 185, 544, 819 kg/m^3 are shown in figure 7 (left). In both configurations, when the feed rate is larger, average residence time is shorter. The average residence times of Model-2s are longer than those of Model-1s in all feed rates. From these results, it can be mentioned that the reduction of feed rate or the configuration which has backward rotor shape is one of the ways to obtain good mixing performance.

Average residence times at rotation speeds 200, 400, 600 rpm are shown in figure 7 (right). Regarding Model-2s, the average residence time is constant in each feed rate. From the results shown in figure 7, the average residence times of Model-2 that have backward rotor are controllable by only feed rate.

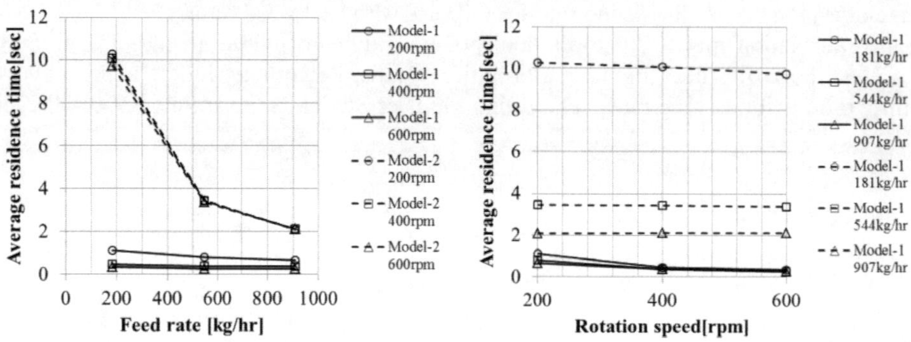

FIGURE 7. Relationship between average residence time and feed rate (left), rotation speed (right)

From the results, we can say the rotors with backward rotor can be appropriate for a mixing that needs longer residence time. It would lead strong mixing. On the other hand, we can say the rotors with only forward rotor can be appropriate for a mixing that needs shorter residence time. We can say that the configuration with only forward rotors strongly related with feed rate and rotation speed. We conclude that the mixer shape and operating conditions are strongly related with degree of mixing and we predict the degree of mixing by using our simulation technique.

CONCLUSION

In the present work, a new simulation technique for partially filled flow has been developed. The mesh-free method has been employed and new scheme called "rearrangement" has been proposed for partially filled simulations of plastic mixers. It is found that the scheme is effective to improve simulation accuracy and the developed simulation technique can predict partially filled flow phenomena with high accuracy.

Furthermore, the developed simulation technique has been used to investigate the effects of rotation speed and feed rate on the average residence time. The investigation was carried out for two different mixing rotor shapes. From the results, it is found that our developed simulation technique is capable to reveal flow behaviors inside counter-rotating continuous mixers and the operating condition that has strong correlation with the average residence time for each configuration. As a conclusion, it is found that our technique is useful to know flow phenomena inside mixer that is difficult to obtain by conventional simulation technique.

In the future, we will clarify the stress distribution or the residence time distribution of partially filled state to investigate more detailed mixing condition.

REFERENCE

1. Yang, H. and Manas-Zloczower I., *Polym. Eng. Sci.,* 32, 1411 (1992)
2. Kajiwara, T., Nakano, Y., Ninomiya, S., Funatsu, K., *Seikei-Kakou,* 5 (8) 557 (1993)
3. Nakayama Y. Nishihira N., Kajiwara T., Tomiyama H., Takeuchi T., Kimura K., *Nihon rheology gakkaishi*, 44, 5 (2016)
4. Ishikawa T., Kihara S., Funatsu K. *Polym. Eng. Sci.,* 40, 2 (2000)
5. Eitzlmayr A., Khinast J., *Chem. Eng. Sci.*, 134 (2015)
6. Sun X., Sakai M., *Chem. Eng. Sci.*, 139(2016)
7. Belytschko, T., Lu, Y. Y. and Gu, L., *Int. J. Numer. Methods Eng.*, 37, 229 (1994)

Effective Process Parameters of Twin Screw Extrusion for dispersion of Polymer

Dryer Ben [1], Ema Masuma Khatoon [1]*, Bigio David I. [1]

[1] Department of Mechanical Engineering, University of Maryland, College Park - Maryland - USA

In Pharmaceutical industries, dispersion of Active pharmaceutical ingredients (API) into polymer still remains challenging because of its crystalline structure. Twin screw extrusion (TSE) is very effective technique to dissolve polymer. But till now, prediction of dispersive mixing of polymer is difficult due to complex flows developing in TSE. In this paper, we proposed a new approach to quantify the percentage of breakup(%BU) of polymer in TSE. Residence stress distribution (RSD) method was used to quantify the stress history in a twin screw melt at any set of operating conditions. Optimization of process parameter for proper mixing is the main goal for commercial production. In this paper, a prediction of equation for process parameter has been made with the combination of RSD method and statistical data analysis. Screw speed, percentage of throughput, barrel temperature- Three different parameters were considered to predict this equation. Also, the effect of %BU on pharmaceutical properties such as – degradation and percentage of water content into polymer were also discussed.

Largely Enhanced Stretching Sensitivity of Polyurethane/CNTs Nanocomposites via Incorporation of Cellulose Nanofiber

Xu Shuman [1], Zhang Qin [2], Fu Qiang [2]*

[1] Sichuan University - Sichuan - China, [2] Sichuan University - Sichuan - China

Abstract. There is a growing demand to investigate stretchable sensor in recent years due to its importance and necessity in hi-tech areas. However, it's still a challenge to fabricate sensors with high sensitivity companied high tenacity. In this work, nano-fibrillated cellulose (NFC) was used as compatilizer of carbon nanotubes and thermoplastic polyurethane (TPU) matrix to fabricate high stretchable TPU/NFC@CNTs sensor. Then the mechanical and electrical properties, particularly the stretching sensitivity of the prepared TPU/NFC@CNTs nanocomposites were investigated. The result shows that compared with TPU/CNTs composites, the prepared TPU/NFC@CNTs has much better mechanical properties and electrical conductivity. What's more, when subjected a uniform uniaxial tension, the resistance of the TPU/NFC@CNTs samples exhibit a linear change as function of strain which is at least up to 300% and a very high sensitivity whose gauge factor close to 50. The excellent properties of TPU/NFC@CNTs nanocomposites should attribute to the addition of NFC which improved the dispersion state of CNTs as well as the interaction between CNTs and TPU matrix resulting in the easy load transfer from TPU matrix to the CNTs conductive network which is beneficial for the sensitivity of composites.
Keywords: carbon nanotubes, cellulose nanofiber, dispersion, tenacity, stretchable sensitivity

Twin Screw Extruders as Continuous Mixers for Pharmaceutical Dosage Forms: A Technical and Historical Perspective Compared to Plastics

Charles Martin

[1]*Leistritz Extrusion, 175 Meister Ave, Somerville NJ 08876 USA*

Abstract. Almost every plastic has been processed at some stage in its manufacturing past on a twin-screw extruder, a device that continuously mixes materials to impart desired properties into a product. Developed approximately 100 years ago for natural rubber/plastics applications, twin-screw extruders (TSEs) are now the most prevalent compounding device in the plastics industry. It is only in the last 25 years or so that TSEs have been utilized in pharmaceutical environments, about 50 years behind plastics! Technologies developed to make high-tech plastics products via TSEs are now being applied to advanced drug delivery systems to facilitate commodity, targeted, and alternative drug delivery systems. Interestingly, the evolution of twin screw extrusion for pharmaceutical applications has followed a somewhat similar path as previously trodden by plastics processing pioneers. The similarities are striking. Just like in the 1960s in plastics, it has become evident in the last 10 years why TSE processing offers significant advantages as compared to other manufacturing techniques. The history of processing plastics via TSEs was thoroughly documented by Dr. James White. As before, the superior mixing characteristics inherent with a TSE have allowed this device to dominate both batch and other continuous mixers for a variety of new dosage forms, including multi-functional drugs and devices. The historical development path and application of TSEs for plastics and pharmaceuticals will be compared and presented.

Keywords: amorphous solid dispersion; continuous manufacturing; continuous mixing; twin screw extrusion
PACS: Extrusion, 81.20.Hy

INTRODUCTION

If the old saying "the best indication of future performance is past performance" holds true, then the future for twin screw extrusion to develop and manufacture dosage forms is bright. Developed almost 100 years ago for food and natural rubber/plastics applications, twin screw extrusion (TSE) now generates some of the most cutting-edge drug delivery systems available.

Polymers are specified to function as thermal binders and act as drug depots and/or drug release retardants upon cooling and solidification. This means active pharmaceutical ingredients (APIs) need to be intimately mixed with a wide range of excipients. (any material in the dosage form that's not the drug) There are many mixing devices, but it will become evident that TSE processing offers significant advantages as compared to other techniques. An obvious advantage for pharmaceutical scientists is that solvents are generally not necessary for processing, which reduces the number of process steps as expensive drying equipment and time-consuming drying steps can be eliminated.

The well-characterized nature of the TSE process lends itself to ease of scale-up and process optimization while also affording the benefits of continuous manufacturing and adaptability to strenuous process requirements in an ever-changing regulatory and fiscal environment. What's occurring today for pharmaceuticals with regard to the implementation of extrusion technology to increase efficiencies and save costs is analog to what occurred 80 or so years ago in the plastics and food sectors of industry, where batch processes were replaced by continuous manufacturing alternatives for reasons that are now obvious.

This ability to mix materials to customize product performance has caused visionary pharmaceutical scientists to consider extrusion to enable therapies utilizing poorly soluble compounds through the generation of amorphous solid dispersions. The embrace by the pharmaceutical scientific community of melt extrusion has led to further research efforts and understanding of how the technology can be applied, and resulted in traditional plastics process techniques being transferred to manufacture novel dosage forms and unique multifunctional medical devices. It is now a fact that TSEs can produce reliable and robust drug product manufacturing alternatives and the technology shows unlimited potential for new, novel dosage forms.

TWIN SCREW EXTRUSION FROM A HISTORICAL PERSPECTIVE

Evolution of Batch to Continuous Processing in the Plastics Industry

The earliest commercial machines for mixing food and polymer melts (natural rubbers) were batch mixers. The first major mixer, a two-roll mill, was described in an 1836 patent. Processing technologies that used single screw pumping devices began to commonly appear in the mid to late 1800s for products such as pasta, soap and ceramic materials and reflected the transition that was underway from batch to continuous processing. Why did this transition take place? Early issues were primarily relating to "product quality", as evidenced by a patent on screw extrusion by Gray (1879) that involved extruding thermoplastic products for wire coating which argues the quality of the insulation properties was better than produced by a ram (batch) extruder. Interestingly, the issue of higher production rates was not raised, suggesting that improved pumping consistencies and quality drove the invention.

In the mid-1800s continuous mixing machines were beginning to be used, but seemingly were not widely commercialized. Continuous mixers began appearing in patent literature in the 1920s-1960s, however it was only during the second half of the 20th century that twin screw extruders became the accepted continuous mixer of choice for most applications. The first in, first out nature with a controlled residence time distribution of the TSE meant the materials experienced a more uniform residence time, heat and shear history as compared to other available continuous and batch type mixing devices.

In the 1950s and 60s a lot of work was performed using single screw extruders, with some success. A single screw extruder, as the name implies, uses a one-piece screw inside a one-piece barrel driven by a motor/gearbox- a much simpler design as compared to a twin-screw extruder. In the 1970s a tipping point occurred and much of the food/plastic industries transitioned from batch to continuous processing, primarily via twin screw extrusion.

In the early 1980s the TSE became the most preferred continuous mixer used to compound plastic formulations, primarily because TSEs are better mixers than SSEs due to several features relating to inter-screw interactions. That being said, the SSE maintains its' place for low-intensity mixing applications and high-pressure pumping, which is why this device makes most of the extruded items we use every day, i.e. drinking straws and packaging film.

During this time frame (1980s and 1990s) research activities on polymer mixing intensified at universities and research institutions such as University of Akron, Polymer Processing Institute and Ecole Polytechnique, to name a few. Exhaustive mixing studies were performed, thousands of papers were published and millions of dollars were invested in developing (and marketing) the "ultimate" continuous mixer. By the mid/late 1990s the co-rotating intermeshing twin screw extruder was generally deemed the device of choice for most mixing and many devolatilization processes.

Today, the TSE continues to dominate the market for continuous mixing and devolatilization, and the SSE dominates for high-pressure pumping for the extrusion of parts such as film, sheet, profiles, tubes, fibers, coatings, etc. Interestingly, in terms of the number of units currently installed each year, there are approximately ten times as many SSEs as compared to TSEs. For dedicated products there are opportunities for usage of SSEs for some mixing applications, particularly for multi-functional medical devices.

History of Twin Screw Extrusion for Pharmaceutical Applications

Interest in extrusion for pharmaceutical applications began in the 1980s. Some early work was performed by Goedecke GmbH in Germany that led to the installation of a 50 mm class co-rotating TSE for a melt granulation process. The machine supplier, Leistritz, struggled with the paperwork and wondered why Goedecke chose such a funny color for the equipment, but not much more and was unaware a new market was beginning to unfold. The extrusion equipment remains in production today.

The case study of Rezulin™, the first drug manufactured via melt extrusion, provides enlightening history. In 1989, Isaac Ghebre-Sellassie of Parke-Davis/Warner-Lambert (PD/WL), was asked by management to convert a spray dried process for the solubilization of a poorly soluble drug substance, troglitazone, to an alternative organic solvent-free manufacturing process. The spray drying process had been initially selected after all commonly used methods (fusing method, hot melt/solvent evaporation and solvent evaporation process) were evaluated to improve the bioavailability of troglitazone. However, the drug substance and processes that included melting had to be abandoned, which prompted non-traditional methodologies to be considered. This chain of events caused a laboratory scale twin screw extruder being installed in the PD/WL laboratories.

As a result, the PD/WL staff began to explore extrusion as an alternative process that generated an amorphous dispersion of the drug in a polymeric matrix and simultaneously ensured the stability of the drug substance during processing. Since the TSE was available and recognized to offer superior mixing capabilities with a short residence time, a decision was made to assess the melt extrusion (ME) process. The formulation and process were optimized using a Leistritz 34 mm TSE without degradation of the drug substance, even when the formulation was processed above the melting temperature of the drug substance, due to the short residence time in the TSE of less than 1 minute.

The Rezulin™ process, as it became known, was then transferred to a Leistritz 50 mm production scale TSE, and the product was approved by the FDA and introduced in 1997, becoming the first commercial product that utilized extrusion as a solubility enhancement manufacturing process. Interestingly, the initial batch size was 250 kilograms. Due to the consistency and reproducibility of the product the batch size was eventually increased to 5000 kilograms. During the time the product was in the market, there were no batch failures. A USA patent for the formulation and process was issued in 2004 (USA Patent 6,677,362). Somewhat similar work was performed in Europe by companies such as Knoll (later Abbott and Soliqs), Mundiepharma and Napp.

Simultaneously in the early 1990s, under the auspices of Jim McGinity and the University of Texas (UT), vigorous research efforts were exerted to develop alternative processes to solvent evaporation for the preparation of transdermal and transmucosal film delivery systems. During this timeframe, UT graduates (John Koleng, Feng Zhang, Michael Crowley and others) became the nucleus of the management team at PharmaForm, a UT incubator company that led pioneering efforts using extrusion as a manufacturing platform. During this time, extrusion technology became recognized as a viable manufacturing platform.

In 2003 the first book on the subject matter, Pharmaceutical Extrusion Technology, edited by Isaac Ghebre Selassie and Charles Martin, was published by Informa Healthcare. In addition to some of the early pioneers from the pharmaceutical world, many of the chapters were provided by extrusion experts from the plastics industry, focusing upon extrusion engineering principles.

Then a spark! The FDA's 2004 Process Analytical Technology (PAT) initiative provided drug makers a framework for pharmaceutical development, manufacturing and quality assurance through in-line monitoring. In a nutshell, the PAT initiative encouraged that continuous processes be considered for the manufacture of new dosage forms- the PAT initiative could have, literally, been composed by an extrusion equipment supplier since it denotes attributes inherent with the continuous extrusion mixing process.

During this time, another solubility enhanced product that was approved by the FDA was Kaletra™, which was a reformulation of a drug to treat HIV that provided a robust dosage form that did not require refrigeration to maintain stability. Kalletra™ applied melt-extrusion technology to improve the overall therapeutic efficacy of a drug, and currently represents the most commercially successful melt-extruded dosage form to enhance oral bioavailability of poor water-soluble drugs.

Now the activities relating to melt extrusion activity really began to accelerate. The seeds had been planted from the early development work at Parke-Davis, the ground had been fertilized at research institutions like University of Texas to develop an understanding of melt extrusion, and the FDA's 2004 PAT initiative had provided the germination effect that allowed melt extrusion usage to expand.

In the early 1990s there were only a handful TSE pharmaceutical class installations worldwide. In the next 10 years this number would dramatically increase into all niches of the pharmaceutical landscape, ranging from universities to large pharmaceutical companies, with smaller contract manufacturers and generic drug companies all becoming users of melt extrusion.

By 2010 almost every major pharmaceutical company had installed TSEs into their R&D equipment portfolios. In the last dozen years, Merck has become a leading proponent and has evolved into a leading-edge innovator introducing, developing and promoting various extrusion related technologies in the areas of foaming, devolatilization, shape extrusion and computer modelling. Novartis, GSK, BASF, Evonik, Dow Chemical, Grunenthal and Ashland have all been active in developing and marketing extrusion processes. Bend Research, PharmaForm (now Formex), ExxPharma, Rottendorf and Foster Delivery Science are examples of companies that offered contract services early on, with the list growing to 40 plus worldwide. In addition to the University of Texas, early research efforts began at University of Mississippi and Massachusetts Institute of Technology, followed by St. Johns, New Jersey Institute of Technology, University of Pittsburgh, Instituto de Capacitación del Plástico y del Caucho (ICIPC), NOLL GmbH, Halle Plastic GmbH and others.

While the majority of melt extrusion applications have focused on oral delivery, numerous commercially marketed extruded products are available which have changed the drug delivery paradigm. The most well-known is NuvaRing (Merck-Schering Plough) which uses a coextrusion process to imbed the API into an ethylene vinyl acetate (EVA) core matrix. Another melt extruded shaped system is Lacrisert™, a rod inserted into the pocket of the patient's lower eyelid to minimize symptoms associated with dry eyes. There are currently development efforts being expended for transdermal patches, dissolvable films, foamed, and co-extruded products that are made by the extrusion process. These products represent the most novel of melt extrusion developments.

Interestingly, the evaluation of co-rotating versus counter-rotating twin screw extrusion technologies has resurfaced in the pharmaceutical research arena. For instance, in 2012 a study was performed at the University of Texas (Austin, TX) that investigated a comparison between counter-rotating and co-rotating TSEs with a focus on the impact of the critical attributes of the solid dispersion. Both configurations produced visually and X-ray amorphous solid dispersions with up to 30% griseofulvin in copovidone having similar supersaturating capability. However, under the parameters of this study, the counter-rotating configuration produced a slightly better solid dispersion at a reduced temperature and with a residence time distribution that minimized the mean residence time, variance and skewness when compared to the co-rotating configuration. It became evident that using these attributes of counter-rotating extruders it may be possible to address challenging poorly soluble drug candidates where uniformity and/or minimization of thermal exposure are critical to the processing strategy.

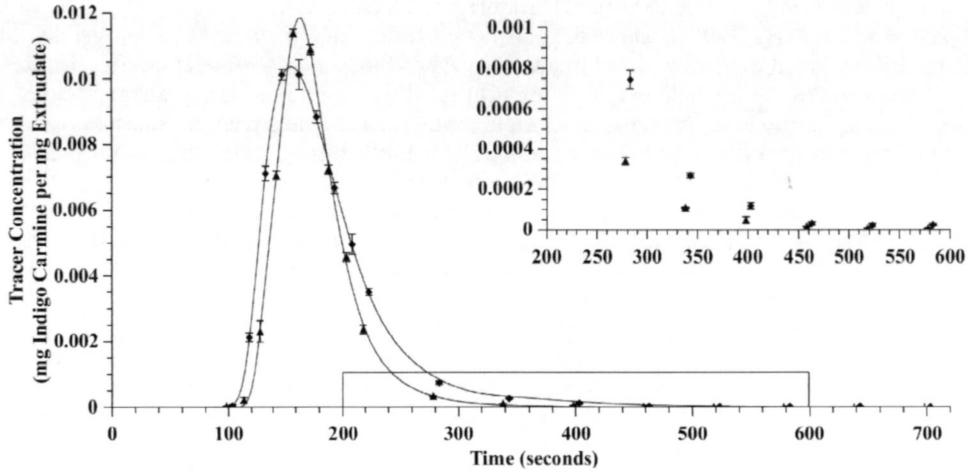

FIGURE 1. Residence time distribution results co-rotating versus counter-rotation TSEs.

EVOLUTION OF DESIGN FEATURES FOR TWIN SCREW EXTRUDERS

Controls and Instrumentation

In the 1940s control systems used manual devices such as hardwired push buttons with analog meters that used electro-mechanical relays to handle all the logic and interlocks. Direct Current (DC) drives and motors were used to vary the speed of the screw rpms and other devices. In the 1990s Alternating Current (AC) drives began replacing DC because AC motors were less complicated and maintenance intensive. The usage of AC motors/drives steadily increased and now accounts for 98% motors/drives below 1000 KW.

Analog temperature controllers were used in early extruders with PID (Proportional, Integral, Derivative) control loops to control the temperature of each heating zone. The analog controller eventually evolved into digital temperature controllers employing microprocessors using software for PID algorithms. Temperature control algorithms are now imbedded in the programmable logic controller (PLC) code. In the early 1980s the new digital gauges were introduced and today many of the monitoring signals are brought into the PLCs so that logic can be programmed to provide different control features associated with these signals.

The PLC started to replace electro-mechanical relays in the 1960s, and became commonly used on TSEs in the 1980s, when Machine Interface (MMI) with graphics began to be used. In the early 2000s the term Man Machine Interface became politically incorrect and evolved into Human Machine Interface (HMI) with alarming, recipes, data logging, trending and communications to other devices. In the last 10 years, the ability to link and integrate PLCs, referred to as distributive processing, has become available to integrate multiple sub-systems into a fully functioning manufacturing cell. Remote access, troubleshooting and service is now commonplace.

In pharmaceutical environments, TSE's often are specified to adhere to FDA part 11 of Title 21 of the Code of Federal Regulations which defines the criteria under which electronic records are deemed trustworthy. Practically speaking, drug makers must implement controls, audits and system validation for software and systems involved in processing electronic data. Protocols must be followed to limit system access to authorized individuals, operational checks, device checks, and controls over systems documentation. Strict adherence with regard to copies of records and record retention is part of the guideline.

Screws: Shafts Design

The motor transmits energy to the gearbox and then to shafts, screw elements and rotating screws the materials being processed. Screw shafts, upon which the elements are assembled, are typically the torque limiting factor to process materials. Torque is determined by the cross-sectional area of the shaft, the geometry of the shaft, the metallurgy and hardening of the shaft, and the shaft geometry. Shaft technologies have evolved to allow smaller diameter shafts to transmit higher torques, facilitating a higher OD/ID ratio and therefore more volume. The current state of the art shaft design is an asymmetrical splined shaft manufactured of 17-4 stainless steel. The geometry of each "tooth" is such that the resulting tangential force vector is isolated, allowing a smaller diameter shaft to transmit higher torque as compared to previous designs. One-piece screws (without shafts) are also possible for higher attainable torques.

Barrels, Heating/Cooling

In the 1950s TSE barrels were generally made from one-piece of steel and nitride hardened with external blowers for cooling. During this era screw rpms were much lower and external cooling blowers were used for cooling. In the 1970s and 1980s screw rpms increased and segmented barrels became the industry standard.

Modular barrels with replaceable liners, internal cartridge heaters and cooling bores are currently deemed state of the art. Internal cooling bores are close to the liner for maximum cooling effect and facilitate higher screw rpms without over-heating.

For pharmaceutical applications metallurgies are specified not to be additive, reactive or absorptive with the process, and are typically hardenable stainless steel or nickel based alloys. The same criteria applies to screws and barrels and any part the contacts the materials being processed.

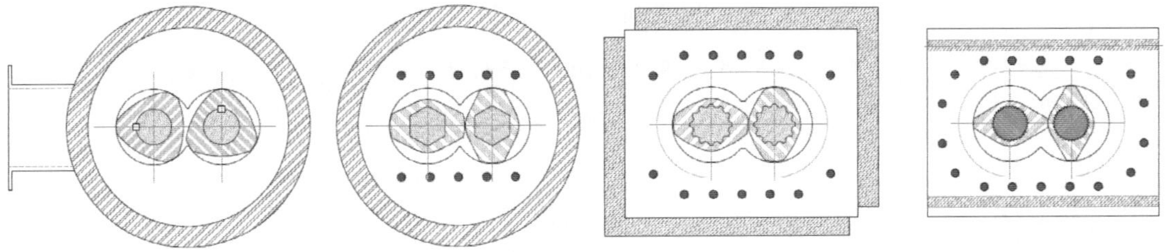

FIGURE 2: Evolution of OD/ID ratios, shafts and TSE barrel designs.

CONCLUSION

Before the twin screw extruder was deemed the best continuous mixer by the plastics community, other devices such as single screw extruders batch mixers were utilized. The superior mixing characteristics inherent with a TSE have allowed this device to dominate other continuous mixers and spawned intensive development efforts that resulted highly engineered formulations for the high-tech plastic products we use every day. Today, twin screw extrusion is a battle hardened, well proven, manufacturing process that has been validated in 24 hour/day industrial settings for more than half a century.

The evolution of extrusion in the pharmaceutical industry resembles what occurred in plastics. TSE designs for pharmaceutical applications replicate the unit operations and mixing mechanisms used for plastics, and the extensive production theory emanating from the plastics industry has allowed quick implementation of proven technologies for pharmaceutical processes. And there's much more to follow!

With so many drug candidates having solubility issues, it is clear extrusion will continue to play a significant role in producing amorphous molecular dispersions of these new chemical entities. As pharmaceutical scientists become more familiar with this technology, the number of extrusion related publications, patents, and multi-functional products continues to expand. The future is bright and the path is well proven for previously proven extrusion principles in pharmaceutical manufacturing environments. The best indication of the future is the past, so it is a certainty that twin screw extrusion will play a prominent role in the future of drug delivery systems.

REFERENCES

1. J. L. White, S. H. Bumm, Perspectives on the Transition from Batch to Continuous Mixing Technologies in the Compounding Industry. In: Polymer Processing XXV. Munich, Carl Hanser Verlag; 2010. p. 322-326
2. 2016 Year-end Machinery Reports. Society of Plastics Industry (SPI). Compiled by Veris Consulting. [cited 2015 Jan]
3. J. DiNunzio, C, Martin, J. McGinity, F. Zhang, Melt Extrusion. In: D. A. Miller, A. B. Watts, R. O. Williams III, editors. Formulating Poorly Water Soluble Drugs. New York: Springer; 2012. p. 311-362
4. M. Szycher, Intelligent Antimicrobial Products via Twin Screw Extrusion. Proceedings of the Leistritz Pharmaceutical Extrusion Seminar; 2015 Jun 17-18; Clinton, NJ.
5. FDA Process Analytical Technology Initiative, (2004)
6. I. Ghebre-Sellassie, C. Martin, Future Trends. In: I. Ghebre-Sellassie, C. Martin, editors. Pharmaceutical Extrusion Technology. Boca Raton, FL: CRC Press; 2003.p. 383-392
7. J. M. Keen, C. Martin, A. Machado, S. Harpreet, J. W. McGinity, J. DiNunzio, Research Paper: Investigation of Process Temperature and Screw Speed on Properties of a Pharmaceutical Solid Dispersion Using Corotating and Counter-rotating Twin Screw Extruders. In: Journal of Pharmacy and Pharmacology. Royal Pharmaceutical Society; 2013.p. 204-217
8. J. DiNunzio, C. Martin, F. Zhang. Melt Extrusion – Shaping Drug Delivery in the 21st Century. In: Drug Delivery Magazine. Pharmaceutical Technology; 2010.p. 30-37
9. I. Ghebre-Sellassie (personal communication)
10. R. Steiner (personal communication)
11. J. McGinity (personal communication)
12. P. Palmer (personal communication)

Three-dimensional numerical analysis method of particle size distribution in unsteady shear flow

Ryotaro Shimada[a] and Tsutomu Kono[a]

[a]*Processing Innovation Research Dept., Manufacturing technology Research Center, Hitachi, Ltd., 292, Yoshida-cho, Totsuka-ku, Yokohama-shi, Kanagawa-ken, 244-0817 Japan*

Abstract. We propose a numerical analysis method of the filler aggregation/dispersion behavior in resin composite that contributes to the optimization of the resin molding process. We calculated the flow of a resin composite during molding using computational fluid dynamics (CFD) and estimated the particle size distribution (PSD) of the cluster for each computational domain of CFD by solving the population balance equation (PBE). Our numerical analysis method is based on the thixotropy model of Usui. and the two-dimensional numerical simulation method of Masuda et al. In the thixotropy model of Usui, the PSD is calculated by taking into account the aggregation/dispersion rate of the cluster by Brownian coagulation, shear coagulation and shear breakage. The shear breakage rate of the cluster is evaluated by solving the energy balance of the bonding energy of the primary particles cut at the breakage and the drag applied to the cluster by the flow of the resin composite. We also calculated the composite viscosity using Krieger and Dougherty's model using the apparent solid volume fraction estimated from the calculated PSD. To solve the PBE at low calculation cost, the PBE was discretized using the fixed pivot technique of Kumar and Ramkrishna. We incorporated our proposed method into the general-purpose CFD software FLOW-3D®, and proved its accuracy in a numerical simulation of the backward-step flow.

Keywords: Fluid analysis, population balance, suspension, aggregation/dispersion, resin composite material.
PACS: 83.10.Rs, 83.50.Ha, 83.50.Uv, 83.60.Pq, 83.80.Hj

INTRODUCTION

To improve the performance of a resin composite product such as a semiconductor encapsulation material, electric wire coating, or insulating composite for high voltage equipment, the filler needs to become finer. As the filler becomes finer, the surface area to volume ratio becomes larger, so it easily aggregates due to the surface effect. This can greatly affect the performance of the resin composite material in terms of heat resistance, insulation, and thermal conductivity. Therefore, it is necessary to optimize the molding process at the design stage to maximize functionality. To carry out this optimization in a short period, numerical analysis of the particle dispersion during the molding process is effective.

There are various numerical analysis methods for particle aggregation/dispersion behavior such as fluid particle dynamics (FPD)[1], smoothed profile method (SPM)[2], computational fluid dynamics - discrete element method (CFD-DEM)[3], and CFD-aggregation kinetics method[4–8]. The CFD-aggregation kinetics analysis method is superior in analysis of large-scale molding processes due to its low calculation cost. With this method, the population balance equation (PBE) of the particle cluster in the flow of the suspension calculated by CFD without modeling individual particles is solved. For accurate evaluation with this method[6], it is necessary to accurately model the particle aggregation/dispersion kinetics. Masuda *et al.* proposed the two dimensional numerical simulation method, which calculates the aggregation/dispersion behaviors of clusters in the microchannel flow, based on the thixotropy model established by Usui[9,10]. In the thixotropy model of Usui, the particle aggregation/dispersion kinetics are modeled by Brownian coagulation, shear coagulation, and shear breakage. The shear breakage rate is evaluated by solving the energy balance of the bonding energy of the primary particles in the cluster and the shear force applied to the cluster by the flow of suspension. Although Masuda *et al.*'s method is limited to two-dimension and cannot calculate large clusters due to the calculation cost of higher-order simultaneous equations, it is effective in estimating the particle dispersion in the suspension during molding.

In this study, Masuda *et al.*'s method was extended to three-dimensional space to develop our numerical analysis method of particle dispersion contributing to optimization of the molding process of resin composites. In three-dimensionalization, the PBE was discretized using the fixed pivot technique of Kumar and Ramkrishna[11] to solve the PBE at low calculation cost. The proposed mehthod was incorporated into the general-purpose CFD software FLOW-3D®, and proved its accuracy in a numerical simulation of the backward-step flow.

METHODS

In Masuda *et al.*'s method, the PBE was not discretized because their target cluster size was small[6]. In this study, to extend to three-dimensional space with large cluster size, the PBE was discretized using the fixed pivot technique of Kumar and Ramkrishna using the following equations[12].

$$\frac{dN_i}{dt} = -(\boldsymbol{u_r}\cdot\nabla_r)N_i + \sum_{\substack{j\geq k \\ x_{i-1}\leq(x_j+x_k)<x_i}} \left(1-\frac{1}{2}\delta_{jk}\right)\left[\gamma_i^{(i-1)}(x_j+x_k)\beta(k,j)N_jN_k\right]$$
$$+\sum_{\substack{j\geq k \\ x_i\leq(x_j+x_k)<x_{i+1}}} \left(1-\frac{1}{2}\delta_{jk}\right)\left[\gamma_i^{(i)}(x_j+x_k)\beta(k,j)N_jN_k\right] - N_i\sum_{j=0}^{i_{max}}\beta(i,j)N_j \qquad (1)$$
$$+\sum_{j=i}^{i_{max}} N_j\nu(j)b(j)\pi_{0,i,j} - b(i)N_i$$

Here, t [s] is time. i [-], j [-], and k [-] are the classes of the discrete PBE, i_{max} [-] is the maximum class number, N_i [m^{-3}] and x_i [-] are the number density and number of primary particles per aggregate of the ith class cluster, respectively, $\boldsymbol{u_r}$ [m·s^{-1}] is the suspension flow velocity vector, $\beta(i,j)$ [m·s^{-1}] and $b(i)$ [m·s^{-1}] are the aggregation and breakage frequency functions, respectively, $\nu(i)$ [-] is the average number of clusters formed by breakage, and δ_{ij} [-] is the Kronecker delta function. The mass conservation functions $\gamma_i^{(i)}$ [-] and $\pi_{0,i,j}$ [-] are given by

$$\gamma_i^{(i)}(v) = \frac{x_{i+1}-v}{x_{i+1}-x_i},\quad \gamma_{i+1}^{(i)}(v) = \frac{v\ -x_i}{x_{i+1}-x_i},\qquad x_i\leq v<x_{i+1} \qquad (2)$$

$$\pi_{r,i,j} \equiv \int_{x_{i-1}}^{x_i} v^r\gamma_i^{(i-1)}(v)P(v,x_j)dv + \int_{x_i}^{x_{i+1}} v^r\gamma_i^{(i)}(v)P(v,x_j)dv\,, \qquad (3)$$

where $P(i,j)$ is the breakage probability function. In this study, the sectional spacing factor for geometric grid $R_p = 2$ was used ($x_1 = 1, x_{i+1} = 2x_i$). In Usui's thixotropy model, we assumed that only two fragments of equal size are obtained by breakage[9,10]. In this assumption, $v(i) = 2$ and $P(i, j) = \delta_{i+1, j}$ hold. In aggregation of particles in resin, assuming Brownian and shear aggregation are dominant in highly viscous resin, $\beta(i,j)$ is given by

$$\beta(i,j) = \alpha_b\frac{2k_bT}{3\eta_0}\left(\frac{d_{c,i}}{2}+\frac{d_{c,j}}{2}\right)\left(\frac{2}{d_{c,i}}+\frac{2}{d_{c,j}}\right) + \alpha_s\frac{4}{3}\left(\frac{d_{c,i}}{2}+\frac{d_{c,j}}{2}\right)^3\dot{\gamma}\,, \qquad (4)$$

where k_b [J·K^{-1}], T [K], η_0 [Pa·s], and $\dot{\gamma}$ [s^{-1}] are the Bolzmann constant, temperature, viscosity of the dispersion medium, and shear rate, respectively, and α_b (= 0.58) and α_s (= 0.60) are coefficients for Brownian coagulation and shear coagulation obtained by Higashitani *et al*[13]. The collision diameter of the ith class cluster $d_{c,i}$ [m] is defined as $d_{c,i} = \{x_i/(1-\varepsilon_i)\}^{1/3}d_p$, where d_p [m] and ε_i [-] are primary particle diameter and void fraction, respectively. It is known that ε_i increases as the number of primary particles in the cluster increases. Usui used the following empirical formula of ε_i assuming monomodal primary particles[9,10].

$$\varepsilon_i = \left(1-\frac{\phi_p}{\phi_{p,max}}\right)(1-x_i{}^{-0.4}) \qquad (5)$$

Here, ϕ_p [-] and $\phi_{p,max}$ [-] are actual solid volume fraction and maximum packing fraction. In the case of monomodal primary particles and rigid spheres, $\phi_{p,max} = 0.63$ holds. In Masuda *et al.*'s method, $b(i)$ was derived from solving the energy balance of the bonding energy of all primary particles in the cluster and the shear force[6]. However, when applying the fixed pivot technique to calculate large clusters, this estimated a smaller $b(i)$ than the actual. Therefore, by solving the energy balance of the bonding energy of the primary particles cut at the breakage, the following equation was used to determine $b(i)$.

$$b(i) = \begin{cases} 0 \quad , & i = 1 \\ \dfrac{3\pi\eta_s\dot{\gamma}^2}{4F_pN_{b,i}} d_{c,i}{}^3, & i > 1 \end{cases} \tag{6}$$

Here, η_s [Pa·s] and F_p [J] are the suspension viscosity and bonding energy of primary particles, respectively. In Usui's model, number of bonds to cut at the breakage of the ith class cluster $N_{b,i}$ [-] is defined as $N_{b,i} = x_i d_p / 2d_{c,i}$. Suspension viscosity is calculated by the following Krieger and Dougherty's model.

$$\frac{\eta_s}{\eta_0} = \left(1 - \frac{\phi}{\phi_{p,max}}\right)^{-[\eta]\phi_{p,max}} \tag{7}$$

Here, $[\eta]$ is inherent viscosity and is determined by the shape of the particle and as 2.5 for a rigid sphere. This model does not take into account the aggregation of particles. In Usui's model, suspension viscosity was estimated using apparent solid volume fraction ϕ [-] with the void fraction in the cluster [9,10] and ϕ was defined as

$$\phi = \sum_{i=1}^{i_{max}} \frac{\pi}{6} d_{c,i}{}^3 N_i \,. \tag{8}$$

The proposed method was incorporated into the general-purpose CFD software FLOW-3D® (ver.11.0.4) as a user subroutine. In the first step, fluid analysis was carried out to calculate mass, momentum, and energy conservation law by using FLOW-3D® standard function. The advection term of the PBE (the first term of Eq.(1)) was calculated in the step before PBE calculation by using the standard function of FLOW-3D®. In the next step, the PBE was calculated to evaluate the PSD by a user subroutine. The strain rate applied to the cluster in each computational domain by the flow of suspension were calculated from using the standard function of FLOW-3D®. In the last step, the local suspension viscosity was evaluated by the apparent solid volume fraction calculated from the PSD by a user subroutine. By repeating the above calculations at each time step, the particle aggregation/dispersion behavior was calculated.

Our proposed analysis method was applied to three-dimensional microchannel flow of Masuda *et al.*'s experiment[6] shown in **FIGURE 1**. Channel and backward-step size were set to 0.75 × 0.4 × 0.1 mm and 0.3 × 0.2 × 0.1 mm, respectively. A 50×26×5 grid was assigned in the computational domain, and a half-symmetric model was used with respect to the z-axis direction. The primary particle diameter d_p, temperature of the fluid T, dispersed media viscosity η_0, and bonding energy of primary particles F_0 were set to 3.5 μm, 293K, 0.5 Pa·s, and 8.0×10^{-14} J, respectively. The maximum class number i_{max} was set to 15 to decrease the calculation cost. At the inlet boundary, the velocity profile and number density of the ith class cluster N_i were set at the steady state value obtained by preliminary analysis assuming a flow path at infinity. At the outlet boundary, the suspension flowed out including the clusters.

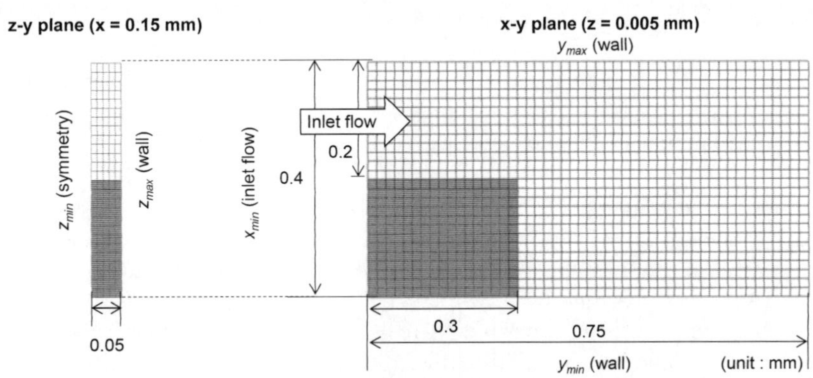

FIGURE 1. Computational domain and boundary conditions.

RESULTS AND DISCUSSION

FIGURES 2(a)~(c) show the contours of the strain rate of particles in cluster and the flow velocity vector at steady state at the x-y section of z = 0.005 and 0.045 mm and z-y section of x = 0.15 mm. **FIGURES 3(a)~(c)** show the contours of the mean number of particles in cluster at the same cross sections as above. In any cross sections, the fluid velocity became slower near the wall. The velocity and shear rate magnitude of the narrow area were greater than that of broad area. This is consistent with the experimental results of Masuda *et al*[6]. However, in their experiment, they did not observe the wall effect in the z-axis direction, and could not taking into account this effect from their two-dimensional simulation method. Our present three-dimensional simulation could express this effect. Near the wall region, the mean particle number tended to decrease. This is because the shear rate becomes larger near the wall. When the shear rate is high, the particle breakage becomes dominant. On the contrary, the particle aggregation becomes dominant at the low-shear-rate area. Specifically, a completely different particle-dispersion state was observed near the wall in the z-axis direction compared to the central area observed by Masuda *et al*. The mean particle number became smallest near the wall surface in the z-axis direction. In addition, at the corners of the channel, the mean particle number became slightly larger because the shear rate was low.

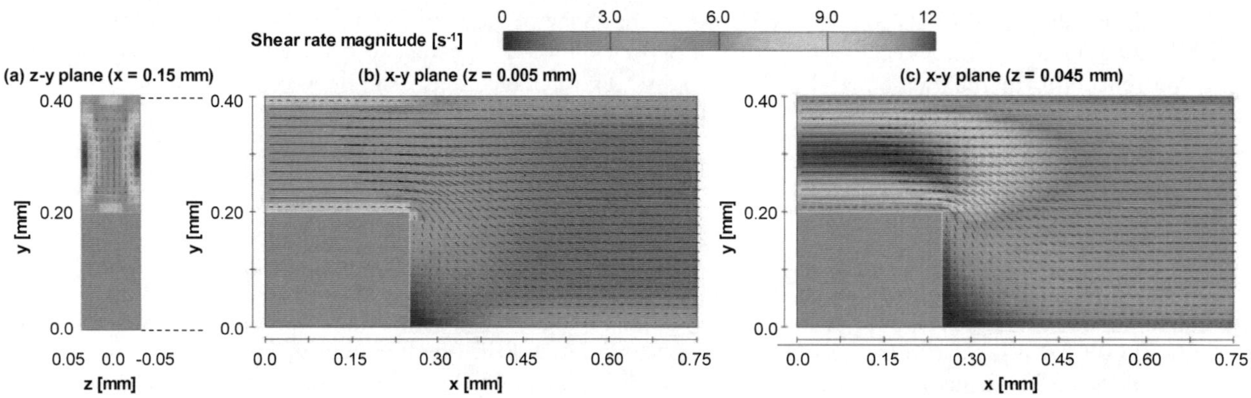

FIGURE 2. Strain rate magnitude contour and fluid velocity vector at steady state at (a) z-y plane (x = 0.15 mm), (b) x-y plane (z = 0.005 mm), and (c) x-y plane (z = 0.045 mm).

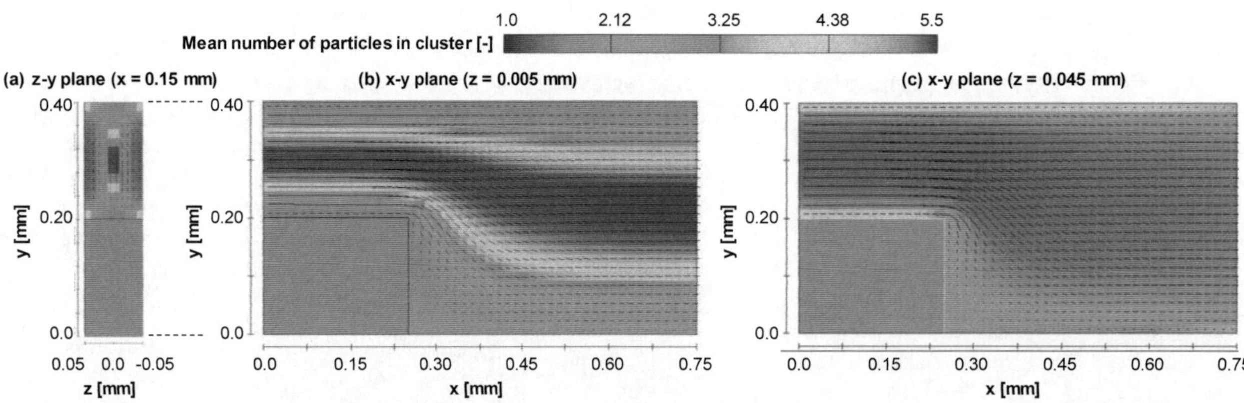

FIGURE 3. Mean number of particles in cluster contour and fluid velocity vector at steady state at (a) z-y plane (x = 0.15 mm), (b) x-y plane (z = 0.005 mm), and (c) x-y plane (z = 0.045 mm).

FIGURES 4(a)~(c) show the distributions of the mean number of particles in the cluster at x = 0.13, 0.37, and 0.49 mm, respectively. Experimental values were cited from Masuda *et al.*'s results. In the upstream region (x = 0.13 mm, z = 0.005 mm), the numerical results at the center area agree well with experimental results. The high peak was observed at y = 0.3 mm, where the shear rate was lowest. At x = 0.37 mm and z = 0.005 mm, the high peak was observed at y = 0.25 mm in both numerical and experimental results. However, the numerical results in the downstream region (x = 0.37 and 0.49 mm, z = 0.005 mm) slightly differed with the experiment results, as with the results of the two-dimensional analysis by Masuda *et al*. The difference became larger in the region near the y-axis lower wall. In our present simulation, the fluid and particle z-axis velocity was almost zero because the lift effect by the cluster rotation, gravity, and buoyancy were not taken into account. In Masuda *et al.*'s experiment, the fluid and particles in the low velocity regions moved in the z-axis direction[6]. Near the z-axis wall (x = 0.37 and 0.49 mm, z = 0.045 mm), the mean number of particles near the lower y-axis wall became high in the numerical results. Thus, in Masuda *et al.*'s experiment, it was considered that the z-axis direction flow caused the increase in the mean number of particles at z-axis center regions. The incorporation of this effect is for future work. However, the three-dimensional distribution of the mean number of particles were well estimated using our proposed analysis method. Our proposed analysis method can be a suitable tool for estimating the three-dimensional distribution of the PSD and correlating the physical properties during the resin molding process.

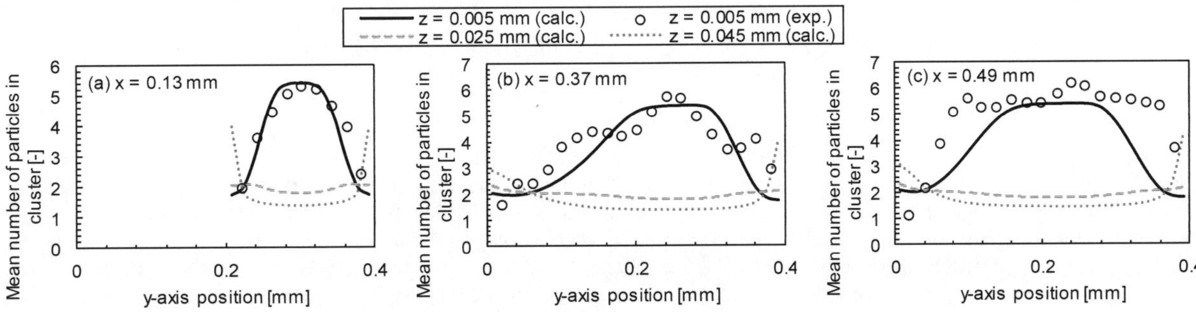

FIGURE 4. Mean number of particles in cluster at (a) x = 0.13, (b) x = 0.37, and (c) x = 0.49 mm.

Conclusion

We proposed a method of analyzing the three-dimensional distribution of the PSD in a non-uniform shear flow. The PSD is calculated by taking into account the aggregation/dispersion rate of the cluster by Brownian coagulation, shear coagulation and shear breakage. The proposed method was incorporated into the general-purpose CFD software FLOW-3D®, and proved its accuracy in a numerical simulation of the backward-step flow. In the downstream region (x = 0.37 and 0.49 mm, z = 0.005 mm), there was small disagreement between the numerical and experiment results. This might have been caused by the z-axis flow of fluid and particles, which was not taken into account our proposed analysis method. However, the calculated mean number of particles in clusters agree well with the experimental data, and the three-dimensional distribution of the PSD could be estimated. Our proposed analysis method can be a promising tool for estimating the PSD during the molding process.

Extension from the present monodisperse system to a polydispersity system (multiple types of primary particles) and evaluating the physical properties of composite materials due to the PSD are subjects for future work.

ACKNOWLEDGMENTS

The authors wish to acknowledge Associate Professors Yoshiyuki Komoda and Koji Masuda for their advice on the numerical model construction and interpretation of obtained data.

REFERENCES

1. H. Tanaka and T. Araki, Phys. Rev. Lett. **85**, 1338 (2000).
2. Y. Nakayama, K. Kim, and R. Yamamoto, Adv. Powder Technol. **21**, 206 (2010).
3. R. Sun and H. Xiao, Int. J. Multiph. Flow **72**, 233 (2015).
4. A.P. Kumar, D. Depan, N. Singh Tomer, and R.P. Singh, Prog. Polym. Sci. **34**, 479 (2009).
5. S. Rollié, H. Briesen, and K. Sundmacher, J. Colloid Interface Sci. **336**, 551 (2009).
6. K. Masuda, H. Suzuki, Y. Komoda, and R. Hidema, Nihon Reoroji Gakkaishi **41**, 75 (2013).
7. K. Masuda, H. Suzuki, Y. Komoda, and R. Hidema, J. Chem. Eng. Jpn **46**, 524 (2013).
8. K. Masuda, H. Suzuki, Y. Komoda, and R. Hidema, Nihon Reoroji Gakkaishi **43**, 85 (2015).
9. H. Usui, Kagaku Kogaku Ronbunshu **25**, 459 (1999).
10. H. Usui, Nihon Reoroji Gakkaishi **29**, 71 (2001).
11. S. Kumar and D. Ramkrishna, Chem. Eng. Sci. **51**, 1311 (1996).
12. K. Higashitani, R. Ogawa, G. Hosokawa, and Y. Matsumoto, J. Chem. Eng. Jpn **V 15**, 299 (1982).

Toughness Enhancement of PLA/Natural Rubber Blend by Melt Compounding with Clays

Jung Hyun Ahn[a], Hyun Geun Ock[c], Joung Sook Hong [b], Seung Jong Lee[a] and Kyung Hyun Ahn[a]

[a]*School of Chemical and Biological Engineering, Seoul National University, Korea*
[b] *Institute of Chemical Processes, Seoul National University, Korea*
[c]*Central Research Center, LG Hausys, Korea*

Abstract. Poly(lactic acid) (PLA) is an aliphatic polyester that possesses various advantageous physical and thermal properties compared to other commercial biopolymers. It is thermally stable and biodegradable which promises a great processability for industrial applications. Meanwhile, it also has restrictions for industrial applications due to its brittleness, long degradation time, and low melt strength. In order to overcome these drawbacks, in this study, natural rubber and mineral powders were added to PLA. We expected the addition of natural rubber featuring high elongational properties and impact strength to PLA would make up its brittleness, while natural clays would enhance melt strength as well as biodegradability. Nevertheless of outstanding properties of each polymer, the immiscibility between PLA and natural rubber (NR) consisting of polyisoprene is a challenge to obtain synergistic effect out of mixing. Depending on the distribution of the rubber phase and the compatibility between rubber and PLA, the brittleness is able to be limitedly improved. On the other hand, organically modified clays were added in order to induce compatibilization and to enhance mechanical properties. Clays were introduced to PLA/NR blend with a weight fraction, 0-10 wt%. With a small amount of clays, the size of the dispersed Natural rubber phase was effectively decreased to sub-micronsize which seems compatibilize between the matrix PLA and the rubber phase. Then, the extendibility was improved (With 3wt% clays, elongation at break of 10% NR/PLA increased 4 times larger than that of blend). Furthermore, the content of NR was broadly varied to observe the toughening effect. Mechanical properties including tensile strength and elongation breakage, rheological properties including storage and loss modulus as well as morphology were observed.

Keywords: Poly(lactic acid), Natural rubber, Clays, Compatibilization

INTRODUCTION

Poly(lactic acid) (PLA) has good performances in biodegradability showing reliable thermal stability and processability compared to other biopolymers. Because it can be decomposed by bacteria or microorganisms, PLA has gained much attention as an alternative to non-biodegradable plastics [1]. However, it still has restrictions for industrial applications due to its brittleness, long degradation time, and low melt strength. One possible approach to overcome these drawbacks is to use plasticizers [2]. However, this causes additional problems in melt compounding as well as sacrifices biodegradability. Instead, PLA has been studied to toughen by blending it with a highly ductile elastomer. Natural rubber is a thermoplastic elastomer consisting of polyisoprene featuring high elongational properties and good impact strength. Moreover, as it comes from the rubber tree in nature, it can be used to make eco-friendly materials with biodegradable PLA [3]. Immiscible blends system of PLA and rubbers is not thermodynamically stable. The phase separation as well as morphology change can reduce the profits received from two polymers. Thus, in order to toughen PLA to make up brittleness, it is important to add a compatibilizer to stabilize blend system. As a compatibilizer, organically modified clays (organoclays) have been investigated for immiscible polymer blend systems. When organoclays are added to a polymer blend, the clay particles at the interface of the two polymers reduce the drop size of the dispersed phase by lowering the interfacial tension [4-6]. For PLA/natural rubber systems, it is important to figure out the exact mechanism for the role of organoclays in order to toughen PLA and to get the desirable end-use properties for industrial applications. In this study, we investigated the toughening of PLA by addition of natural rubber as a minor phase and also the compatibilizing of PLA/natural rubber by organoclays.

EXPERIMENTALS

PLA (grade 4032D, Mn = 90,000 g/mol, Mw = 181,000 g/mol, PDI = 2.01) was purchased from Natureworks (USA). Natural Rubber (grade ISNR5, Mn = 260,000 g/mol, Mw = 780,000 g/mol, PDI = 3.00) was obtained from Rubber Board (India). Organoclays (Cloisite 20A, density = 1.77 g/cc) was provided by Southern Clay Product(USA). The modifier of the organoclay consists of two hydrogenated tallow (HT) and two methyl groups (CH3–). Melt compounding was carried out with an intensive mixer (Rheocomp mixer 600, MKE, Korea) for 8 min at 200 °C with a rotor speed of 100 rpm. The ratio of PLA and NR was broadly varied from 90/10 weight percent through 50/50 and the amount of clay varied from 0.5 to 5 wt%. Rheological measurements were carried out with RMS800 (Rheometrics, USA) under dynamic test mode with a 25 mm parallel plate fixture at 190 °C. The tensile test was conducted following the ASTM D638 type V with UTM (LF plus, Lloyd instruments Ltd). The experiments were repeated 10 times, and the data were averaged. The morphology was observed with Transmission Electron Microscopy (JEM-2100F, Jeol, accelerating voltage 200 kV).

RESULTS AND DISCUSSION

Figure 1 shows the TEM images of three representative blends (PLR, PLRC3 and PLRC10). The images clearly show that the NR droplets are uniformly distributed and stabilized inside the PLA matrix. The size of the dispersed NR phase dramatically decreases with the increase in clay concentration. With TEM images, the variation of drop size was quantitatively analyzed as a function of clay content. As the clay content increases, the D_n decreases from 1.5 um to 0.4 um and it saturated to the drop size of 0.4 um above 3 wt% clays. At clay concentrations lower than 3 wt%, the clay particles are mostly observed near the droplet interface. In addition to 10%NR/PLA blend, clays played a role as a compatibilizer for the NR/PLA blend system.

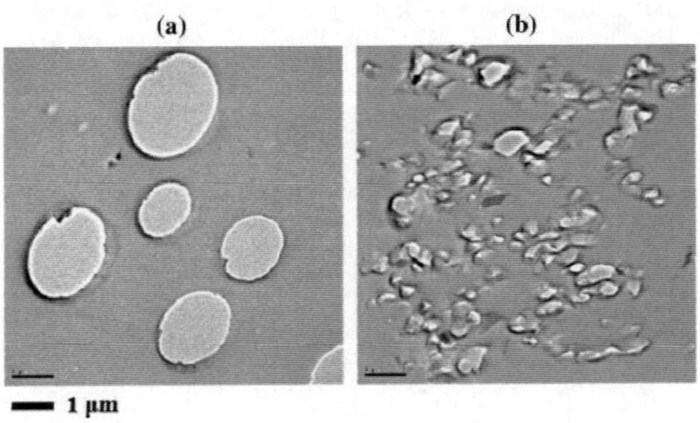

Figure 1. TEM micrographs for (a) 10wt% NR/PLA and (b) 10wt% NR/PLA with 3wt%organoclays.

The mechanical properties of the NR/PLA blends are listed in Table 1. The elongation at break of the blends with a clay content of 0.5–3 wt% is much larger than that of NR/PLA without clays. However, when the content is higher than 5 wt%, the extendibility is rather reduced. This transition of the tensile strain with respect to the clay content can be attributed to the different role of the clay on the mechanical properties according to complex contribution of clays to the blend depending on selective localization. Meanwhile, the tensile strength of the blend decreases with the addition of clays up to 1 wt%, and changed little with the further addition of clays. Below the concentration of 3 wt%, the elongational strain increases following the rubber-toughening.

TABLE (1). Elongation at break and tensile strength of the PLA/natural rubber blend depending on the content of clays

Blend	Clays	Elongation at break (%)	Tensile strength (MPa)
10wt%NR/PLA	0	17	45
	0.5	24	42
	1	27	32
	2	60	32
	3	78	34
	5	34	32

ACKNOWLEDGMENT

This work was supported by the National Research Foundation of Korea(NRF) grant funded by the Korea government(MSIP) (No. 2016R1E1A1A01942362)

REFERENCES

1. R.E. Drumright, P.R. Gruber, D.E. Henton, Adv. Mater. 12 (2000) 1841–1846.
2. F. Hassouna, J.-M. Raquez, F. Addiego, V. Toniazzo, P. Dubois, D. Ruch, Eur. Polym. J. 48 (2012) 404–415.
3. S. Ishida, R. Nagasaki, K. Chino, T. Dong, Y. Inoue, J. Appl. Polym. Sci. 113 (2009) 558–566.
4. J.S. Hong, N. Han, K.H. Ahn, S.J. Lee, C. Kim, Polymer 47 (2006) 3967–3975.
5. J.S. Hong, Y.K. Kim, K.H. Ahn, S.J. Lee, C. Kim, Rheol. Acta 46 (2007) 469–478.
6. J.S. Hong, Y.K. Kim, K.H. Ahn, S.J. Lee, J. Appl. Polym. Sci. 108 (1) (2008) 565-575.

Polyethylene Cellulose Composites and Nanocomposites: effect of fiber size, composition and processing

Maia Thiago Henrique Sousa* (1)*, Calazans Marília* (1)*, Lima Vitor* (2)*, Moreira Francys Kley Vieira* (1)*, Dufresne Alain* (3)*, Lucas Alessandra DE ALMEIDA* (4)

(1) DEMa-UFSCar - SP - Brazil, (2) PPGCM-UFSCar - SP - Brazil, (3) Pagora-INP-Grenoble - Grenoble - France, (4) DEMa-UFSCar - Select One - Brazil

This study investigates the use of an aqueous dispersion of polyethylene (PE) copolymer with a relatively high content of acrylic acid as a compatibilizing agent and as an alternative medium to obtain polyethylene cellulose composites and nanocomposites. Different scales were evaluated: bleached microfibers and cellulose nanofibrils. Another nanofibril containing 3% of lignin was also studied. Fibers content was varied from 5 to 30 wt.% and the appearance, optical, thermal, mechanical and the morphology of the films were evaluated. The dispersions containing the desired amounts of PE and fibers were stirred for 5 minutes and casted in a Teflon plate. The samples were dried at 60oC for about 24h and then hot pressed at 100oC for 5 min, followed by water cooling the films up to ambient temperature. The same reinforcement efficiency was achieved by cellulose micro and nanofibrils. But, The PE/CNF films are transparent up to 20 wt.% of NFC indicating a good dispersion of CNF, but a poor distribution, with PE-rich and CNF-rich regions observed by scanning electron microscopy, SEM. Cellulose microfibers composites presented low transparency and high haze. Lignin containing nanofibrils also caused a decrease on transparency, as well as a brown color effect, but the best mechanical properties were observed for this system. The effect of melt processing was evaluated by extruding the sample in a twin screw extruder, at 100 rpm and 100oC. Films were obtained by hot pressing. After melt processing, a decrease in mechanical properties were observed, what can be credited to an agglomeration of cellulose nanofibrils into micro fibrils, as observed by SEM. The presence of lignin was not enough to avoid the decreasing of properties and did not alter the thermal stability of the samples.

pps proceeding
Symposium: S07 - Mixing and compounding
Poster Presentation

Effect of Mixing Screw on Mechanical Properties of Injection Molded Polylactic Acid / Organic-Montmorillonite Composites

Huang Jhih-Hong (1)*, Chen Chao-Chang (1)

(1) National Taiwan University of Science and Technology, Mechanical Engineering Department - Taipei City - Taiwan

This study is devoted to design a new type of mixing screw in injection molding to investigate mechanical properties influence of polymer composites by injection molding process. Polylactic Acid (PLA) and PLA composites with Organic Montmorillonite (PLA/M) have been selected to be investigated by regular and mixing screw performed by injection molding. Relation between tensile strength, impact strength, hardness of PLA and PLA/M composites have been tested and discussed. The amino silane coupling agent is selected as interface improving agent of PLA and Organic Montmorillonite (PLA/S/M). Design of injection molding experiment is adopted to find the most significant factor for tensile strength and impact strength of composites sample. Design of such mixing screw has been Flow2000 / ANSYS-CFD simulation for mixing characteristics. Results of shear rate and flow velocity simulation with three feature designs of mixing characteristics. For the best to be fabricated. From experimental results, the better mechanical properties of PLA composites have been obtained by the PLA composites of PLA/M3.0 group. The regression analysis has been used to find the significant factor as melt temperature of injection molding parameter. The processing with low level melt temperature can achieve better mechanical results. With such two types of screw tests, the mechanical properties of PLA composite including tensile strength and impact strength by mixing screw are 0.9~12.7% lower than that of regular screw. Furthermore, process in low level melt temperature, the mechanical properties of tensile strength and impact strength of PLA/S/M composites by mixing screw are higher 7.0~70.6% than those of the single pitch screw. Result of this study can be futher applied on food tableware or medical aids in the one-time molding process in future. Keyword: Mixing screw of injection molding, Bio-degradable materials, Polylactic Acid, Organic Montmorillonite, Mechanical properties.

Combined effect of interfacial strength and fiber orientation on mechanical performance of short Kevlar fiber reinforced olefin block copolymer

FU SIRUI [1], YU BOWEN [1], CHEN FENG [1]*, FU QIANG [1]

[1] sichuan university - sichuan - China

As a novel thermoplastic elastomer, olefin block copolymers (OBCs) show a tremendous application potential in many areas. However, the practical use of OBCs is mainly restricted to the low load-bearing environments because of its poor low-strain mechanical strength and stiffness. In this work, short kevlar fiber was introduced into OBC matrix to improve its mechanical properties. The results show that introduction of pristine Kevlar fiber only gives rise to a slight increase in tensile strength of OBC owing to the poor interfacial strength, regardless of the presence of compatibilizer maleic anhydride-grafted polypropylene (MA-g-PP). However, a largely enhanced tensile strength can be easily obtained when hydrolyzed or polydopamine-coated Kevlar fiber was incorporated into the OBC matrix using the MA-g-PP as the reactive compatibilizer. More importantly, the tensile strength can be further enhanced with the orientation of kevlar fiber in the composites. As compared with compression molded composites with randomly orientated fibers, injection molded composites with highly oriented fibers exhibit a significantly higher tensile strength at the same composition. Moreover, it is interesting to observe that kevlar fiber can be easily exfoliated into several microfibers during melt-mixing with OBC, but the exfoliation has no evident effect on the reinforcement effect possibly because of the poor interfacial strength between the newly-formed microfibers and the OBC matrix. Our work demonstrates that mechanical performance of short fiber reinforced elastomer composites could be significantly improved by enhancing interfacial strength and fiber orientation using a facile surface modification strategy and conventional melt-processing technology.

Composite performance and durability of injection molded engineering compounds of polyamide 6 with glass fiber, talc, and biochar

Jubinville Dylan [1], Misra Manju [1], MOHANTY Amar K. [1]*

[1] University of Guelph - Ontario - Canada

The melt blending and durability of polyamide 6 with various fillers resulted in a comprehensive study of how properties can vary drastically depending on filler type. The formulations are targeted, ultimately, to be used for the fabrication of lightweight bio-composite materials for automotive usages. The compounds were prepared with a 70 %wt. polyamide 6 matrix and with a filler content of 30 %wt. The formulations were twin-screw extruded and injection molded with diversified fillers such as: talc, glass fiber, and biochar. The composites were also subjected to thermooxidative aging in order to observe their behaviour after being exposed to 140 ºC for 1000 hours. Differential scanning calorimetry (DSC) and Dynamic mechanical analyzer (DMA) were utilized to observe the evolution of the physical transitions (i.e. glass transition, melting and crystallinity) before and after aging. The morphological structure was investigated via scanning electron microscopy (SEM) of the impact fracture surface to better understand the filler dispersion and adhesion of the systems. Lastly, a rheological study was performed on the neat polymer and designed compounds to follow the progression of the visco-elasticity over frequencies at the processing temperature. Conventional fillers are typically dense materials, resulting in high density when applied in composite systems (~ 1.36 g/cm3). However, with the introduction of bio-based fillers, like biochar, the density value is lowered (~ 1.20 g/cm3) by around 12 % while still maintaining marginal performance compared to talc and glass fibers.

Rheological properties of thermoplastic starch/clay nanocomposites

Guarás, M. P. [a], Ludueña, L. N. [a] and Alvarez, V. A.[a]

[a] *Thermoplastic Composite Materials Group (CoMP) - Research Institute of Material Science and Technology (INTEMA) - Engineering Faculty - National University of Mar del Plata - Av. Colón 10890 - (B7608FDQ) Mar del Plata - Argentina*

Abstract. The use of conventional polymers in the past decades grew exponentially and it represents a wide field of oil consumption. These materials are not biodegradable, which causes a serious problem regarding the disposal of waste. Plastic packaging materials are also often contaminated by food products and biological substances which, in many cases, makes the recycling of these materials impractical and economically not convenient. In the last decade, the use of biodegradable polymers has had a growing interest. Thermoplastic starch (TPS) is one of the candidates for the manufacture of biodegradable polymer based products for packaging applications. Native starch (NS) is the term used to refer to starch found in plants. Under these conditions NS is in the form of granules composed of amylose and amylopectin in different quantities depending on the source. The main advantages of NS include its abundance, low cost, renewability and biodegradability. On the other hand, compared to synthetic polymers, NS has two main disadvantages: a) contains hydroxyl groups which impart hydrophilic properties; so, NS disintegrates in water and loses its properties when exposed to moisture and b) degradation starts at a temperature lower than its melting point. So, to modify NS became relevant. Modification is done by the breakdown of the NS granule when it is processed in the presence of a specific amount of plasticizer, at a given extrusion conditions, obtaining TPS. TPS is a material that generally has low stability at high humidity conditions and fragile nature. In this work cost-effective, biodegradable nanocomposites based on chemically modified TPS reinforced with nanoclays were prepared by reactive extrusion. Their rheological behavior using capillary rheometry was studied. Rheological characterization was focused in the processing shear rate range in order to give an initial screening about the adaptability of the TPS formulation to different and industrially available polymer processing techniques.

Keywords: Thermoplastic Starch, Nanocomposites, Melt Rheology.

INTRODUCTION

Until now petrochemical based polymers such as polyethylene (PE), polypropylene (PP), polystyrene (PS), among others; have been increasingly used as packaging materials, owing to relatively low cost, good mechanical properties and good vapor and oxygen barrier [1]. These materials are not fully recyclable and are not biodegradable, which caused a major environmental problem regarding to final disposal of waste. Eventually, oil-based polymer materials will be replaced in many applications by inexpensive and natural-based products. For this reason there is a growing interest in the used of biodegradable polymers which can help to minimize the environmental impact of "commodities" [2]. Among natural polymers, starch has been considered as one of the most promising candidates because of an attractive combination of availability, price and performance [3].

Starch is a completely biodegradable polymer of the group of polysaccharides derived from agricultural plants and it is susceptible to biological attack. In its granular form, starch is mostly composed of lineal amylose and highly branched amylopectin. To improve starch processability, the molecular order within the granules must be destroyed. When mixed with a limited amount of plasticizer and subjected to heat and shear, starch undergoes spontaneous destructurization. The result is a homogeneous melt known as "thermoplastic starch" (TPS) which can be processed by conventional plastics technologies [4, 5].

Starch may be chemically modified in order to improve properties such as mechanical properties, crystallinity and moisture absorption because it is very fragile and sensitive to moisture due to its hydrophilic nature [4]. Chemical modification by esterification of starch plays a role of internal plasticization and reduces the degree of crystallinity, so that the plasticizer molecule can penetrate deeper within the starch molecule for subsequent lamination. It was shown that maleic anhydride (MAH) has high reactivity in the preparation of esterified starch [6, 7].

One of the main strategies to improve competitiveness of this material with a high efficiency/cost ratio is the incorporation of nanofillers in the polymer matrix. The addition of nanofillers has been shown to be an effective way of improving both mechanical and barrier properties. Nanoclays have been extensively studied, due to their high availability and low cost [8]. Bentonite is one of the most popular clay mainly composed of layered aluminosilicates. Due to its laminar structure, clays are able to incorporate a variety of molecules in their interlamellar spaces. In order to compatibilize with different polymeric matrices, metallic cations present in interlamellar spaces are replaced by cations compatible (generally organic modifiers) with the polymer. Modifiers can change both physical and chemical properties of clay, such as hydrophilicity, interlamellar spacing and thermal resistance among others, allowing improving the intercalation of the polymer chains between their sheets [9, 10].

Polymers are complex rheological materials, that exhibit both viscous and elastic (viscoelastic) properties under conditions of strain, stress and temperature [11]. Processing techniques employed to obtain conformed pieces include injection, blow molding, profile extrusion and sheet extrusion followed by stamping or thermoforming. These processes, involve molten polymer subjecting to shearing flow under a wide range of share rates. The study of rheological behavior of polymers is an important area since it determines the processing technique for which the polymer will be applicable. It also affects the quality of finished products by aspects like: morphological structure obtained during processing, surface aspects, properties of finished products and material anisotropy.

The aim of this work is to study the rheological behavior of different TPS formulations by capillary rheometry to give an initial screening of the processability focusing on techniques such as injection molding, blow molding, profile extrusion and sheet extrusion followed by stamping or thermoforming.

EXPERIMENTAL

Materials

Corn starch provided by "Distribuidora Dos Hermanos" (Yin Yang) was used in powder form. The plasticizer used was glycerol (99% purity, DEM) and distilled water. Stearic acid (SA Shuchardt Merck OHG) was used as a lubricant for processing. Maleic anhydride (analytical grade, Anhedra) was used for chemical modification of starch (esterification). The clay used was a commercial bentonite with a cation exchange capacity (CEC) of 105 meq/100g supplied by Minarmco S.A. (Argentina). The raw material (Bentonite) used in this study consists predominantly of montmorillonite, as evidenced by X-ray diffraction (XRD). It contained quartz and feldspar as major impurities, as well as traces of gypsum and sepiolite. The clay modifier, benzalkonium chloride (CBK), whose chemical composition is $C_{17}H_{30}ClN$ and purity of 80%, was supplied by DEM Chemistry.

Preparation of thermoplastic starch (TPS)

Native starch films were prepared with 30% plasticizer by extrusion. For this, a pre-blend of 69.5% w / w of native starch, 30% w / w of plasticizer (90 % glycerol and 10% water) and 0.5% stearic acid. The mixtures were introduced into an extruder Extrualex Doble Argentina co-rotating twin screw 25mm in diameter and length / diameter ratio (L / D) = 38 with six heating zones (**FIGURE 1**). The rotation speed of screws was maintained at 50 rpm thus obtaining flow rates of between 1 and 2 kg / h depending on the composition of the material and the temperature profile used. The temperature profile used was 80/90/100/110/100/110ºC. Finally extrudate blend was pelletized.

Preparation of modified thermoplastic starch (MTPS)

Starch (400g) was introduced into a chemical reactor containing a solution of maleic anhydride (4% w / starch) and acetone (4lt) at 25 ° C for 24 h. It was then centrifuged and washed several times with distilled water, dried in an oven at 50 ° C, and finally milled in mortar. Once modified starch was obtained, it was processed under the same conditions as TPS.

Preparation of nanocomposites (NS)

Nanocomposites films were prepared by reactive twin screw extrusion. A single twin extrusion step was used for both, starch gelatinization and nanocomposite preparation. For this porpoise, a pre-mixture of 69.5% w/w of NS or MNS, 30% w/w of plasticizer (90 % glycerol and 10% water), 0.5% of stearic acid was used as matrix. Then, 3%

w/w of bentonite (BENT) or modified bentonite (CBK) was incorporated to the mixture and stirred with an industrial kitchen mixer until obtaining a homogeneous paste. The final paste was introduced at constant flow rate into the co-rotating twin screw at the same process conditions as TPS. Finally, the strands out of the extruder were pelletized.

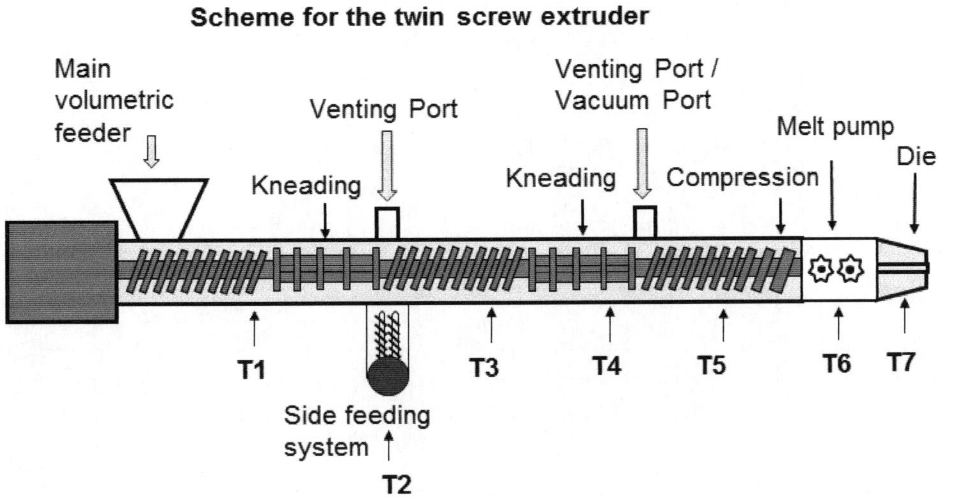

FIGURE 1. Scheme of co-rotating twin screw extruder Extrualex Doble Argentina available in CoMP.

Characterization of blends

Rheological properties were studied using a capillary rheometer, CEAST SR50, Instron. Bagley Test were performed with two capillary dies of L/D = 20 and L/D = 5, D = 1 mm at 170 °C.

RESULTS AND DISCUSSION

Melt viscosity al 170 °C and at two different share rates were obtained (**TABLE 1**). It is possible to observe that TPS and MTPS matrices have higher viscosity than nanocomposites. In all cases, viscosity decreases with the addition of nano-clay. On the other hand, MTPS had lower viscosity than TPS.

TABLE 1. Power Law constants and share viscosity of blends and nanocomposites.

Sample	η (Pa.s) (Share rate = 1000 1/s)	η (Pa.s) (Share rate = 2000 1/s)	Power Law Parameters	
			n	K
TPS	46.53	25.66	0.142	4.153
MTPS	8.02	2.83	0.505	5.319
TPS/BENT	28.16	21.35	0.600	2.630
TPS/CBK	30.45	14.09	0.159	4.173
MTPS/BENT	3.32	2.682	0.691	1.439
MTPS/CBK	3.28	1.97	0.446	2.310

Power Law parameters were calculated by the equation 1. The results are shown in **TABLE 1**. MTPS/BENT showed the highest Power Law exponent *n*, which indicates that the viscosity is less dependent on the shear rate. Lower *n* values indicate that polymers had lower entanglement density so are easier to disentangle under shear and more pseudoplastic.

$$\eta = K.\gamma^{(n-1)} \tag{1}$$

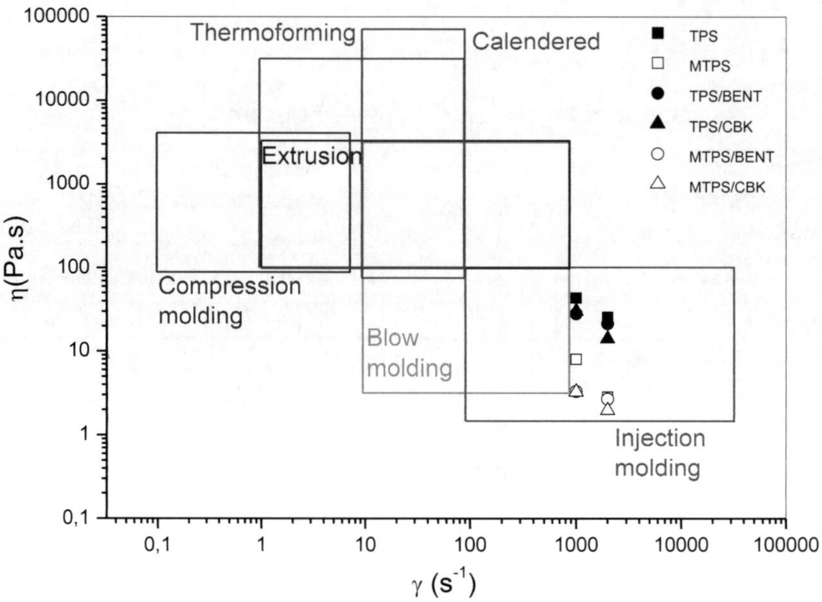

FIGURE 2. Intervals of shear rates usually achieved by different melt processing techniques.

FIGURE 2 shows the range of shear rates and viscosity that are commonly present in different melt processing methods. When the obtained values were plotted in that graphic we observed that blends can be processed by injection molding and blow molding at 170 ºC [12-14].

CONCLUSION

The TPS flow behavior is characterized by a significant pseudoplasticity and viscous function, which depend on starch modification and the incorporation of nanofillers. Nanocomposites showed lower viscosity than their neat matrix. At the same time, chemical modification decreases the viscosity of TPS. Power Law model was applied to fit the experimental values. According the viscosities obtained for the given share rates and at 170 ° C, it is possible to process the material by blow molding and injection molding.

ACKNOWLEDGMENTS

The authors acknowledge to UNMdP, CONICET and ANCyT for the financing to carry out the project.

REFERENCES

1. Siracusa, V., et al., *Biodegradable polymers for food packaging: a review.* Trends in Food Science & Technology, 2008. **19**: p. 634-643.
2. Firwan, M.J. and J.W. Strawbridge, *7: Plastics in food packaging*, in *Food Packaging Technology*. 2003. p. 174-240.
3. Avérous, L. and E. Pollet, *Chapter 2: Biodegradable polymers*, in *Environmental silicate nano-biocomposites*. 2012, Springer-Verlag: London.
4. Nafchi, A.M., et al., *Thermoplastic starches: Properties, challengers and prospects.* Starch Journal: Starch/Stärke, 2013. **65**: p. 61-72.
5. Ojedat, T., *Polymers and the Environmen*. Polymer Science. 2013, Croatia.

6. Miladinov, V.D. and M.A. Hanna, *Starch esterification by reactive extrusion.* Industrial crops and products, 2000. **11**: p. 51-57.
7. Zuo, Y., et al., *Study on the preparation of maleated thermoplastic starch by reactive extrusion.* Journal of Thermoplastic Composite Materials, 2014: p. 1-13.
8. Okamoto, M., *Polymer/Clay Nanocomposites*. Encyclopedia of Nanoscience and Nanotechnology, ed. H.S. Nalwa. Vol. 8. 791-843.
9. Wilhelm, H.M., et al., *Starch films reinforced with mineral clay.* Carbohydrate Polymers, 2003. **52**: p. 101-110.
10. Muller, C.M.O., J.B. Laurindo, and F. Yamashita, *Effect of nanoclay incorporation method on mechanical and water vapor barrier properties of starch-based films.* international Crops and Products, 2011. **33**: p. 605-610.
11. Chiou, B.-S., et al., *Rheology of starch-clay nanocomposites.* Carbohydrate Polymers, 2005. **59**: p. 467-475.
12. Franck, A., *Understandig Rheology of Thermoplastic Polymers*. TA Instruments.
13. RM, C., et al., *Some aspects concerning the rheology of biodegradable starch based materials.* Universitatea Politehnica Bucoresti, Scientific Bulletin Series B, 2010.
14. *High & Low Shear Rate Rheology*. 2013, Innovative Plastics, Sabic.

High cellulose content composites produced by melt processing

Lo Re Giada (1)*, Larsson Per (1), Wågberg Lars (1), Berglund Lars (1)

(1) KTH - Royal Institute of Technology - Stockholm - Sweden

Being the structural component of plants, cellulose is the most ubiquitous and abundant organic compound on the planet. For this reason, it has played a key role in the development of the human society. However, materials based on cellulose are still foreseen to be of significant importance in a sustainable society. In order to replace materials with higher carbon footprints, the cellulosic ones need to overcome their intrinsic limitations, especially related to strength, fragility, permeability to gases, water vapour in particular, and three-dimensional shaping. Improved properties together with new ones are required to expand the traditional application fields of cellulosic materials, as well as facilitating new market opportunities. This study contributes in overcoming the well-known restrictions in generating high cellulose content composites by melt-processing and open new routes in the processing of these biomaterials. In particular, cellulose fibers partially derivatized to dialcohol cellulose can be used to manufacture three-dimensional high-density materials by using conventional melt processing techniques, with or without the addition of a thermoplastic polymer. This work shows how the use in the melt-processing of chemically modified cellulose fibers can lead to the preparation of a new generation of highly sustainable materials with tunable properties that can be tailored for specific applications, providing bio-based materials with tremendous potential to replace less sustainable ones. Reference: [1] P. A. Larsson, L. Wågberg. Green Chem., 2016, 18, 3324

A Mesoscale Computational Study of the Effect of Shear Flow on Janus Rods Organization in Polymer Blends

Khani Shaghayegh* [1], *Maia Joao* [1]

[1] Case Western Reserve University - Ohio - USA

Janus particles are highly active at the interface of immiscible fluids and thus they can self-assemble into different structures when incorporated into multiphase systems. In the case of Janus rods, their anisotropy and amphiphilic nature result in superior properties and their self-assembly can be manipulated for fabrication of engineered materials. The structures formed by the Janus rods can be tuned by varying different parameters in the system. Therefore, understanding the effect of the deterministic factors on Janus rod self-assembly is of crucial importance and has the potential to pave the way for designing new functional materials. Herein, we perform mesoscale simulations in order to investigate the influence of individual parameters on the structures formed by Janus rods at the interface of two immiscible liquids. In particular, Dissipative Particle Dynamics simulations are performed to explore the effect of Janus structure and interface properties on the self-assembled structures. Janus rods exhibit varying orientational preferences at the interface and ultimately, these particles are incorporated into binary blends of immiscible polymers. The interfacial orientation and self-assembly of Janus rods are shown to be highly influential in controlling the phase separation process when added to a polymer blend. Moreover, flow can dramatically alter the localization of these particles within the polymer blend. Therefore, we not only monitor the microstructures formed by these systems at rest, but we also do so under flow conditions and upon relaxation after flow cessation. The results of this study can be used for designing new approaches for directing nano-particles into desired morphologies, which will subsequently tune the final characteristics and properties of nano-composites.

Transversal molecular orientation in an extruded sheet of polypropylene/rubber blend

Masayuki YAMAGUCHI,[1*] Yoshiaki FUNAHASHI,[1] Panitha PHULKERD,[1] Shohei IWASAKI,[2] and Yohei UCHIYAMA[2]

[1] *School of Materials Science, Japan Advanced Institute of Science and Technology, Japan – m_yama@jaist.ac.jp*
[2] *New Japan Chemical Co., Ltd.*

Abstract – The novel polymer blend with phase separation, in which the deformation direction of dispersed droplets is perpendicular to the molecular orientation of a matrix, was prepared using an extruded sheet composed of isotactic polypropylene (PP) and rubbery ethylene/butene-1 copolymer (EB) containing a nucleating agent, *N,N'*-dicyclohexyl-2,6-naphthalenedicarboxamide. It was found that the epitaxial crystallization of PP occurs on the nucleating agent with needle shape, in which PP chains are oriented perpendicular to the long axis of the needle crystals of the nucleating agent. Because of the hydrodynamic force at extrusion, the nucleating agents orient to the flow direction, leading to anomalous molecular orientation of PP chains, i.e., perpendicular to the flow direction. Moreover, dispersed EB droplets are deformed into the flow direction. As a result, the dispersed phase is deformed perpendicular to the orientation direction of PP chains. We further performed the measurements of mechanical properties considering the anisotropy. It was clarified that the rubber deformation greatly affects the crack propagation at the tear test, while the molecular orientation of PP matrix is responsible for the anisotropy in the dynamic tensile modulus greatly.

Keywords: Polypropylene; Nucleating agent; Molecular orientation; Polymer Blend

Introduction

For polymorphic crystalline polymers like isotactic polypropylene (PP), the crystalline form, as well as the crystalinity, plays an important role on the material performance. In general, crystalline form is determined by flow-field, cooling condition, applied pressure, and the addition of a nucleating agent. In the case of PP, it is well known that a product containing a large amount of β-form crystals exhibits marked toughness due to energy dissipation mechanism at the crystalline transformation process from β- to α-form during deformation, although the melting point is lower than that of α-form crystals [1,2]. Therefore, new types of efficient nucleating agents for β-modification are being developed in industry [3-6]. When adding a nucleating agent, the molecular orientation should be also considered to understand the mechanical properties, because the molecular orientation is enhanced due to rapid solidification [7]. Although molecular chaines always orient to the flow direction in general, it has been found that PP chains orient perpendicular to the flow direction by using the specific nucleating agent *N,N'*-dicyclohexyl-2,6-naphthalenedicarboxamide. This nucleating agent has been known as a high performance β-nucleating agent for PP. Furthermore, PP chains orient perpendicular to the long axis of the needle-shaped crystals of the nucleating agent [8-15]. Because the needle crystals are aligned to the flow direction due to the hydrodynamic force, PP chains show unusual orientation. Up to now, various PP products have been proposed using this phenomenon, such as transversal oriented sheet, injection-molded plate with plywood structure, and extruded sheet with woven cloth structure. However, it has not been applied for rubber-toughened PP. In this study, we controlled the chain orientation of PP using an immiscible blend of PP and a rubbery material.

Rubber-toughened PP materials are employed in various applications these days, because they show both high rigidity and mechanical toughness. Ethylene-α-olefin copolymer is usually employed as rubber, i.e., impact modifier. It has been well known that ethylene-α-olefin copolymers having a large amount of α-olefin show low interfacial tension and small interaction parameter with PP [16-20], because the chain rigidity, that is expressed by statistical segment length [21] or packing length [20], becomes similar to that of PP. Moreover, copolymers with butene-1, hexene-1, and octene-1 show good compatibility with PP as compared with ethylene-propylene copolymer, leading to fine dispersion in PP matrix. Here, the structure and properties of binary blends composed of PP and ethylene-butene-1 copolymer are demonstrated. In particular, orientation control of PP chains as well as the deformation of dispersed droplets of ethylene-butene-1 copolymer was performed using the specific nucleating agent, *N,N'*-dicyclohexyl-2,6-naphthalenedicarboxamide.

Experimental

Materials

A commercially available isotactic polypropylene (PP) (SunAllomer, PM600A, Melt Flow Rate (MFR) 7.5 [g/10 min at 230°C], M_n 63,000 and M_w 360,000) was employed. The stereoregularity is 96% [*mmmm*%, a meso-pentad value], and it contains no comonomer (propylene homopolymer). Moreover, ethylene-butene-1 copolymer (EB) (Mitsui Chemicals, TAFMER DF610, MFR 2.2 [g/10 min at 230°C], density 860 kg/m^3, and ethylene content 54 wt.%) was used as an impact modifier. *N,N'*-dicyclohexyl-2,6-naphthalenedicarboxyamide, as a nucleating agent (New Japan Chemical, NJ Star^{TR} NU-100) was used without further purification.

Both materials, i.e., PP and EB (70/30 in weight ratio), were melt-mixed with a small amount of thermal stabilizers and calcium stearate by a co-rotating twin-screw extruder at 260 °C. The amount of the nucleating agent was 1,000 ppm. The obtained pellets were fed into a single-screw extruder with a T-die at 200 °C to obtain a sheet with a thickness of 200 μm. The die lip was 0.5 mm. The sheet was stretched in the air gap (10 mm) between the die and chill roll. The chill-roll was controlled at 105 °C to promote β-form crystals. Extruded sheets of PP with/without the nucleating agent and PP/EB (70/30) without the nucleating agent were also prepared using the same conditions as reference samples.

Measurements

Shape of the nucleating agent in the molten PP and crystallization behavior of PP were observed using a polarized optical microscope equipped with a hot stage. A full-wave plate was inserted to determine the direction of molecular orientation. The polarized microscope was also used to investigate the distribution of the orientation birefringence in the thickness direction with a tilting compensator.

Thermal properties were measured using a differential scanning calorimeter (DSC)(Perkin Elmer, DSC 8000). Heating and cooling rates were 10 °C/min.

Crystalline form and molecular orientation of PP were characterized using an X-ray diffractmeter (Rigaku, PILATUS 3R 100K).

The morphology of the crystalline lamellae was observed by means of a transmission electron microscope (TEM) (JEOL, JEM-2000FX, Japan) with 160 kV acceleration voltage. The sample, embedded in epoxy resin, was trimmed with an ultra-microtome (RMC-Boeckeler, Ultramicrotome MT-XL) equipped with a diamond knife after exposure to the vapor of ruthenium tetraoxide at 40 °C for a day. Thin slices, with a thickness of 50 nm, were cut out from the stained samples at room temperature. Furthermore, the deformation of dispersed phase was characterized by a scanning electron microscope (SEM) (Hitachi, S4100). The cut surface was immersed in xylene to remove EB. After drying, the sample surface was coated with Pt/Pd using an ion-sputtering machine.

Tear strength of the extruded sheets was evaluated using a tensile test following ISO 6383-1 Trouser tear method using a tensile machine. Two types of sample specimens were cut out from the extruded sheet; one has a notch parallel to the flow direction (MD), and the other has a perpendicular notch to the flow, i.e., the transverse direction (TD). The stretching rate was 200 mm/min with 20 mm of an initial distance between crossheads. The tear-propagation resistance, defined as the force divided by the sample thickness, was measured as a function of the displacement of the crossheads.

Results and Discussion

Crystallization

The crystallization behavior of PP as well as the structure development of the nucleating agent was observed under crossed polars. Figure 1 shows the optical micrographs of the sample at the cooling process from 260 °C at a rate of 5 °C/min. The nucleating agent appears around at 220 °C as needle crystals, which were clearly observed in the left figure. Then, the crystallization of PP occurs around at 135 °C from the surface of the needle crystals. It should be noted that the chain-axis of PP crystals orients perpendicular to the long-axis of the needle crystals of the nucleating agent, which was confirmed by inserting a full-wave plate.

The effect of the EB addition on the crystallization behavior of PP was studied using DSC. It was found that the additon of EB does not affect the crystallization behavior of PP irrespective of the addition of the nucleating agent. The sample containing the nucleating agent showed higher crystallization temperature of PP due to the nucleating effect, which was not affected by EB. Because EB is immiscible with PP, this is a reasonable result as demonstrated in our previous study [22].

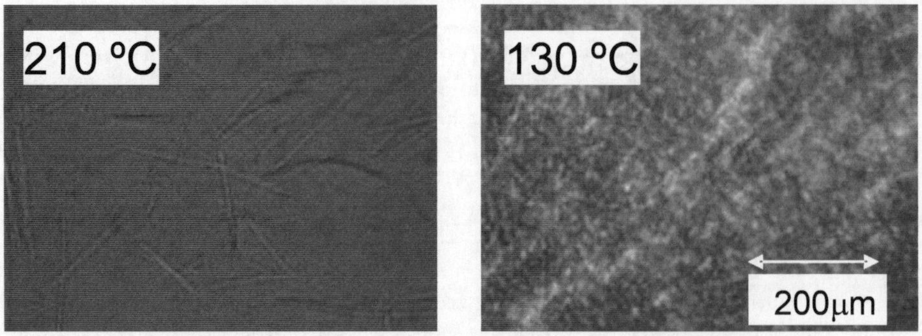

Fig.1 Optical micrographs of PP with 1000 ppm of the nucleating agent.
A full-wave plate was inserted between the crossed polars.

Structure of Extruded Sheet

The crystalline form and chain orientation of the extruded sheets were characterized by 2D wide-angle X-ray diffraction measurements (WAXD). It was found that the β-form crystals dominate in the sheet sample with a small amount of α-form crystals for the sheets containing the nucleating agent. In contrast, α-form is dominant in the sheets without the nucleating agent. The results demonstrate that the nucleating agent generates β-form crystals efficiently. The diffractions ascribed to polyethylene crystals were not detected, suggesting that EB shows a low level of crystallinity due to a large amount of butene-1 content. Furthermore, the orientation direction of PP chains in the extruded sheet containing the nucleating agent was found to be transversal direction (TD), i.e., perpendicular to the flow direction in the extruded sheet. In the case of the sheets without the nucleating agent, PP chains orient to the machine/flow direction, i.e., MD, irrespective of the EB addition.

TEM observation revealed that the needle-shaped nucleating agents orient to MD. The diameter is around 10-20 nm and the length is 100-200 nm. Moreover, most PP lamellae were found to orient perpendicular to the sheet plane and flow direction, i.e., ND orientation. Therefore, PP chains orient to TD, which corresponds with WAXD patterns. This unique orientation is attributed to the epitaxial crystallization of PP chains on the needle-shaped nucleating agents which align to the flow direction by the hydrodynamic force. The applied compressive stress at the chill-roll leads to the TD orientation of PP chains not ND [10].

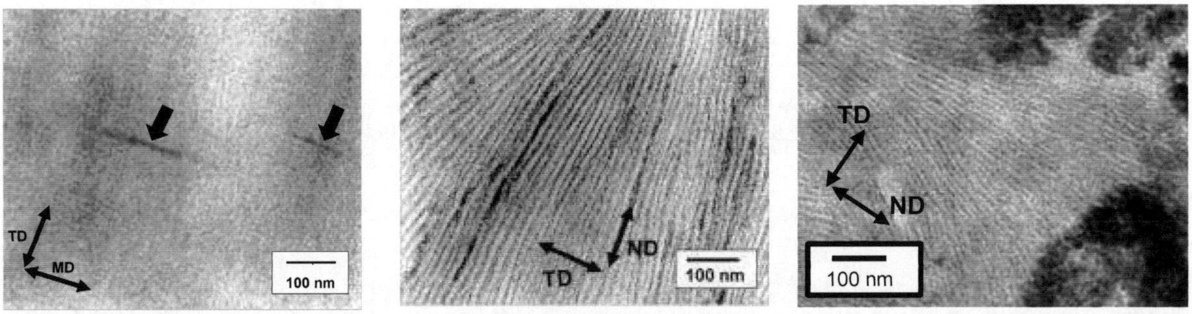

Fig.2 TEM pictures (left) the MD-TD plane of PP with the nucleating agent, (center) the TD-ND plane of PP with the nucleating agent, and (right) the TD-ND plane of PP/EB with the nucleating agent.

The deformation direction of EB dispersed phase was characterized by SEM using the sample treated with xylene to remove EB droplets. It was found that the dispersed phase was deformed to the flow direction greatly as similar to the nucleating agents. This is reasonable because the sheet was extruded through the T-die and stretched in the airgap. The aspect ratio of EB droplets was larger than 10.

Mechanical Properties of Extruded Sheet

Tear test was performed to confirm the role of molecular orientation of PP and the deformation of EB droplets. The result was shown in the following table.

Table I Tear-tests

Sample		Fracture behavior	Tear Strength (N/mm)	Sample		Fracture behavior	Tear Strength (N/mm)
PP	MD	Brittle	249	PP/EB	MD	Ductile	166
	TD	Brittle	237		TD	Ductile	193
PP with NA	MD	Ductile	221	PP/EB with NA	MD	Ductile	151
	TD	Brittle	284		TD	Ductile	199

* NA represents the nucleating agent.

As seen in the table, the strength and its anisotropy were strongly dependent upon the recipe. Without EB, only PP with the nucleating agent shows ductile behavior in MD. Furthermore, the tear strength in MD is almost similar to that in TD for the pure PP sheet, whereas the PP containing the nucleating agent shows a large value in the TD sample [11]. Consequently, the mechanical anisotropy was pronounced by the addition of the nucleating agent. In the case of the PP/EB blend, the tear strength in the TD sample was significantly larger that that in the MD sample, irrespective of the nucleating agent. The results demonstrate that PP orientation has no marked impact on the tear property. The deformation of EB particles plays a dominant role on the tearing. Such information has never been reported at the best of our knowledge, because the deformation direction of dispersion is always the same with the molecular orientation direction of a matrix in general. Regarding the effect of the nucleating agent, the tear strength of the MD sample is slightly enhanced, although the effect is not so obvious compared with pure PP.

Conclusions

The extruded sheet having unique structure, in which the matrix PP molecules are oriented in the transversal direction whereas the dispersed EB droplets are deformed to the flow direction greatly, was prepared using epitaxial crystallization of PP chains on the surface of the needle-shaped nucleating agent. It was found by the tear test that the deformed droplets are found to be more important for the crack propagation than the orientation of matrix chains.

References

1. J. Kargar-Kocsis, J. Varga, *J. Appl. Polym. Sci.*, **62**, 291 (1996).
2. J. Karger-Kocsis, *Polym. Eng. Sci.*, **36**, 203 (1996).
3. J. Varga, *J. Macromol. Sci.*, **41**, 1121 (2002).
4. X. Li, K. Hu, M. Ji, Y. Huang, G. Ahou, *J. Appl. Polym. Sci.*, **86**, 633 (2002).
5. J. Kotek, I. Kelnar, J. Baldrian, M. Raab, *Eur. Polym. J.*, **40**, 679 (2004).
6. B. Shentu, J. Li, G. Tan, Z. Weng, *Eur. Polym. J.*, **43**, 3036 (2007).
7. M. Tenma, M. Yamaguchi, *Polym. Eng. Sci.*, **47**, 1441 (2007)
8. M. Yamaguchi, T. Fukui, K. Okamoto, S. Sasaki, Y. Uchiyama, C. Ueoka, *Polymer*, **50**, 1497 (2009).
9. Y. Uchiyama, S. Iwasaki, C. Ueoka, T. Fukui, K. Okamoto, M. Yamaguchi, *J. Polym. Sci., Polym. Phys. Ed.*, **47**, 424 (2009).
10. M. Yamaguchi, Y. Irie, P. Phulkerd, H. Hagihara, S. Hirayama, S. Sasaki, *Polymer*, **51**, 5983 (2010).
11. P. Phulkerd, S. Nobukawa, Y. Uchiyama, M. Yamaguhci, *Polymer*, **52**, 4867 (2011).
12. P. Phulkerd, H. Hagihara, S. Nobukawa, Y. Uchiyama, M. Yamaguhci, *J. Polym. Sci., Polym. Phys. Ed.*, **51**, 897 (2013).
13. P. Phulkerd, S. Hirayama, S. Nobukawa, T. Inoue, M. Yamaguchi, *Polym. J.*, **46**, 226 (2014).
14. P. Phulkerd, S. Arayachukeat, T. Huang, T. Inoue, S. Nobukawa, M. Yamaguchi, *J. Macromol. Sci., Part B, Phys.*, **53**, 1222 (2014).
15. S. Sato, T. Maeda, M. Yamaguchi, *Macromol. Mater. Eng.*, **302**, 1600413 (2017).
16. M. Yamaguchi, H. Miyata, K. Nitta, *J. Appl. Polym. Sci.*, **62**, 87 (1996).
17. P. A. Weimann, T. D. Jones, M. A. Hillmyer, F. S. Bates, J. D. Londono, Y. Melnichenko, G. D. Wignall, K. Almdal, *Macromolecules,* **30**, 3650 (1997).
18. Y. Thomann, J. Suhm, R. Thomann, G. Bar, R. D. Maier, R. Mulhaupt, *Macromolecules,* **31**, 5441 (1998).
19. M. Yamaguchi, K. Suzuki, H. Miyata, *J. Polym. Sci., Polym.Phys. Ed.,* **37**, 701 (1999).
20. D. J. Lohse, *J. Macromol. Sci. Polym. Rev.*, **C45**, 289 (2005
21. F. Bates, G. H. Fredrickson, *Macromolecules*, **27**, 1065 (1994).
22. M. Yamaguchi, H. Miyata, K. Nitta, *J. Polym. Sci., Polym. Phys. Ed.*, **35**, 953 (1997).

Binary and Ternary PLA-based Bioblends with PBAT and PBSA

Ciftci Umit [(1)]***, Oguz Hazal*** [(1)]***, Yargici Ceren*** [(1)]***, Babayev Kemal*** [(1)]***, NOFAR Mohammadreza*** [(1)]*

[(1)] Istanbul Technical University - Maslak, Istanbul - Turkey

In this study, binary and ternary blends of polylactic acid (PLA), poly[(butylene adipate)-co terephlate] (PBAT) and poly[(butylene succinate)-co adipate] (PBSA) were prepared through injection molding (IM). Two different PLAs with low and high D-lactide contents (semi-crystalline and amorphous, respectively), as well as their mixture were used as the matrix phase in both binary and ternary blend systems. PLA phase, was fixed at 75 wt % or 50 wt % whereas the corresponding remaining 25 wt% or 50 wt% were either PBAT, PBSA, or equal mixture of PBAT and PBSA. To predict the morphology of the ternary blends, interfacial tension between each two polymers and spreading coefficient values were calculated by using Harkins equation. Negative values of spreading coefficient indicated a partial wetting morphology where droplets of one of the phases locate at the interface and demonstrate a three-phase contact. The effects of PBAT, PBSA, or their mixture, semi-crystalline, amorphous PLAs, or their mixtures, and varying blending ratios on the morphology of the blends were explored through using a scanning electron microscopy (SEM). Subsequently, the materials and morphological variation influences on the mechanical properties of blends were investigated and compared using tensile and three-point bending tests. Mechanical properties of the blends exhibited a dramatic improvement in elongation at break with increasing PBAT content (25 wt % to 50 wt %). Ternary blend containing 12.5 wt % PBAT and 12.5 wt % PBSA were shown to have superior flexural strength in comparison with binary blends.

The Processing of Alternating Multi-layered Functional Polymer Composites Through High Speed Thin-wall Injection Molding

Meirui Fu, Hua Deng*, Rongni Du, Mingfeng Fan, Yi Zhou and Qiang Fu

College of Polymer Science and Engineering, State Key Laboratory of Polymer Materials Engineering, Sichuan University, Chengdu 610065, China

Abstract. Various functions of polymer composites, including electrical conductive, thermal conductive, dielectric, EMI shielding, strain sensing, etc, has been extensively investigated. These functions need to be delivered by functional filler or structure. Therefore, the morphological control of these functional filler or structure should play vital role on the final properties of these functional polymer composites. In recent years, we have carried out a series of studies focusing on "The relationship between different morphological control methods--filler network morphology & property—functionality". Efforts have been made on investigating various morphology control methods including: construction and destruction of filler network during processing, multi-layered structure through high speed thin-wall injection molding and macroscopic segregated network structure. Herein, the formation of multi-layered structure during high speed thin-wall injection molding process will be introduced. The thermodynamic and kinetic issues influencing the formation of such structure and the application of such structure for the preparation of functional polymer composites will be systematically discussed.

Keywords: Alternating multi-layered structure, high speed thin-wall injection molding, confinement, polymer processing

INTRODUCTION

Multi-layered structure is thought as the core structure for many objects in nature: shell, the wing of butterfly and tree, etc., and responsible for many of their interesting properties. It is thought to be able to provide many useful properties for different materials, such as: robust mechanical properties, dielectric properties, barrier property, sound absorbing properties to name a few. Thus, much attention has been paid on the preparation of materials with multi-layered structure. Several methods have been used to prepare materials with such structure: layer-by-layer assembly, layer multiplying coextrusion, pressure-induced flow (PIF) processing, uncontrolled casting of polymer/clay mixtures, ice-templating and sintering of ceramics, bidirectional freezing, etc. Various functionalities or improved mechanical performance have been obtained thanks to such structure. Nevertheless, above methods are often quite complex or require complicated equipment.

In our recent studies, high speed thin-wall injection molding (HSTWIM) was used as an effective method to construct alternating multi-layered structure for polymer blends consisting of polypropylene (PP) and polyethylene (PE). Viscosity match, small interfacial tension, high shear rate and fast cooling rate are considered as the main mechanism responsible for the structural evolution from dispersed droplets to alternating multi-layered structure during injection molding. Such method provides an easy route for the preparation of alternating multi-layer functional polymer materials. Selective distribution of functional filler in one of the polymer phases can result in confinement of filler network, and thus enhanced functionalities, such as: anisotropic conductivity, barrier property.

EXPERIMENTAL SECTION

PE (Fushun Petro-chemical Corp., 2911), PP (Lanzhou Petrochemical Corp., T30S), æ ł1-æ łßŁº ł1Ł˘ (MMT), (Nanocor Corp., Nanomer I.44P: a dimethyl, dihydrogenated tallow ammonium salt modified montmorillonite) and multi-wall carbon nanotubes (MWNTs) (Nanocyl, S.A., NC7000) are commercial available materials purchased from respective producers. These materials were used as received.

First, MMT or MWNTs was blended with PE to prepare masterbatch consisting of 10 wt % filler in twin screw extruder with temperature in the range of 140–200 °C from hopper to die. After that, the masterbatch was further diluted into binary PE-MMT or MWNTs composites with different filler contents using twin screw extruder, respectively. Subsequently, single screw extruder was utilized to blend PP with PE-MMT or MWNTs pellets at weight proportion 5:5 to fabricate ternary PE-MMT (or MWNTs)/PP composites. Possible migration of filler to PP phase is hindered because of its relative weak shear. The obtained PE-MMT (or MWNTs)/PP pellets were then subjected to HSTWIM at an injection speed of 600 mm/s and a melt temperature of 200 °C to obtain naocomposite sheets with the size of 80 mm ×60 mm and a thickness of 0.4 mm. For different process condition study, no filler was added into the PE phase. And different process conditions was used as indicated.

RESULTS AND DISCUSSION

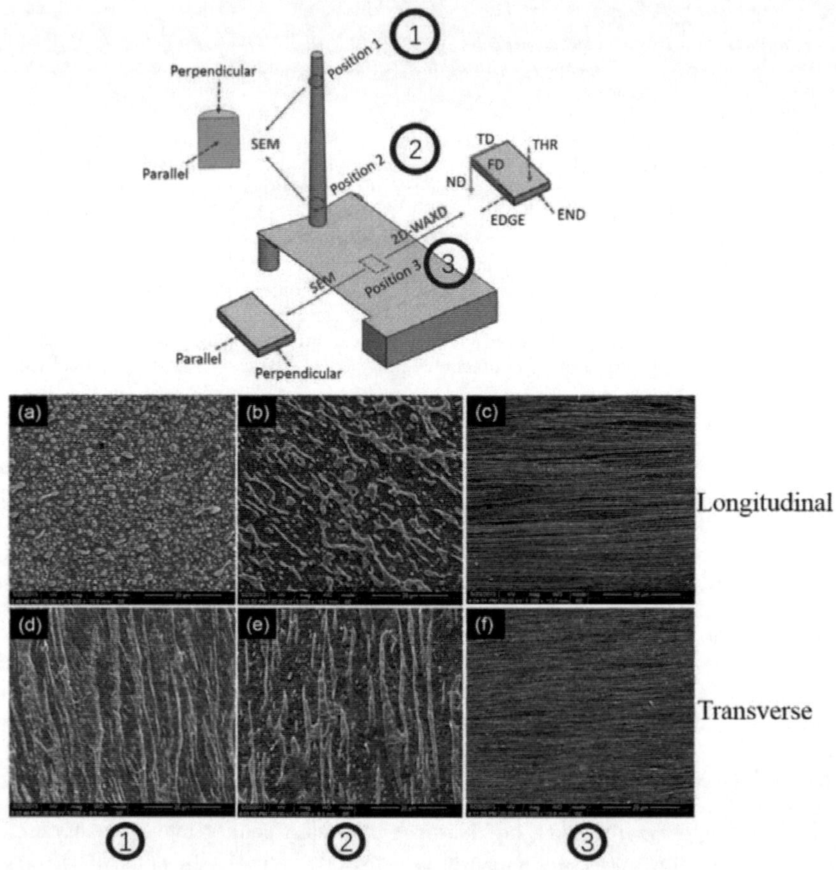

FIGURE 1. SEM morphology for the evolution of PP/PE blends phase morphology in the mold for high speed thin-wall injection molding at position 1, 2, 3. The PP phase was slightly etched with mixed acid to study the phase morphology.

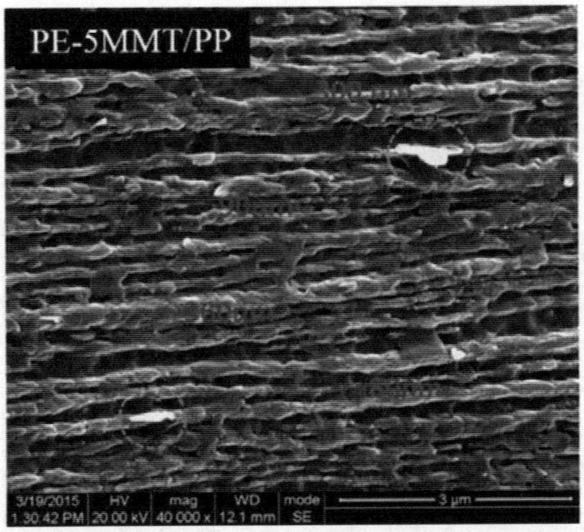

FIGURE 2. SEM image for etched PP/PE (50:50) blends containing 5 wt.% MMT in PE phase.

PP/PE blends could be deformed into alternating layered structure during high speed thin-wall injection molding process due to strong confinement on the polymer phases during processing, and small viscosity ratio, interfacial tension as well as melt elasticity (as shown in Figure 1). The phase morphology evolution in the mold could be observed. It can be noted that the phase morphology involves from cylinder-like morphology to small platelets, and to alternating multi-layered structure in the mold. The strong confinement effect from the mold to polymer blends is thought to be responsible for such phase morphology evolution. Thus, alternating layered functional filler networks could be constructed by selectively localize MMT (as shown in Figure 2) or MWNTs (as shown in Figure 3) into the PE phase of such blends. The confinement from the processing field as well as blends morphology has strongly limited the orientation or rotation of these inorganic fillers, resulting in highly oriented and anisotropic filler networks.

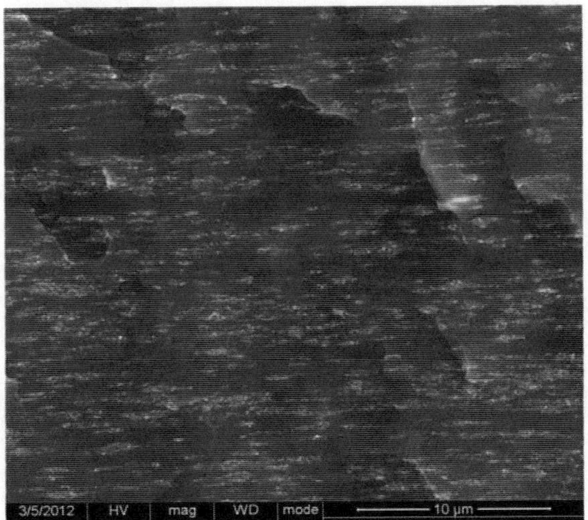

FIGURE 3. SEM image for PP/PE (50:50) blends containing 5 wt.% MWNTs in PE phase. Image is taken with SEM operated under contrast charge mode without gold coating, thus, MWNTs buried under polymer matrix could be observed.

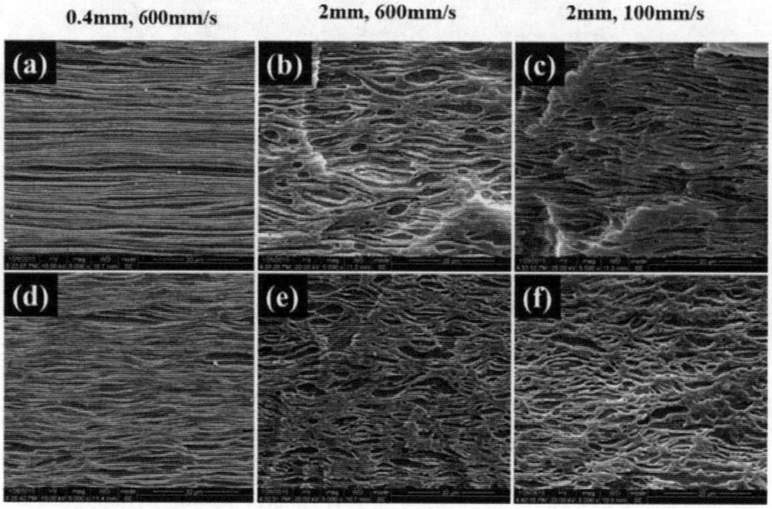

FIGURE 4. SEM image at two different directions (above: parallel direction, below: perpendicular direction) for etched PP/PE blends processed at different mold thickness and injection speed.

Figure 4 illustrates the phase morphology of PP/PE blends (50:50) processed under different mold thickness and injection speed. It is demonstrated that thinner mold thickness provides much stronger confinement, thus, much more organized structure comparing with blends processed at large thickness or lower injection speed.

CONCLUSION

It is demonstrated that multi-layered structure could be obtained by process PP/PE blends using high speed thin-wall injection molding. The phase morphology evolution could be observed in the mold, which illustrate an evolution from cylinder to small platelets, and to multi-layered structure. Additional MMT into the PE results in layered filler network in the injection molded parts. Furthermore, processing condition is shown to have important influence on the phase morphology. Such study provide an efficient method for the preparation of multi-layered structure.

ACKNOWLEDGMENTS

We would like to thank ExxonMobil Asia Pacific Research & Development Co., Ltd. for funding (LAW-2014-0184) and Dr. Robert Wittenbrink and Dr. Peite Bao from ExxonMobil, STC for fruitful discussion.

REFERENCES

[1] K. Jiang, F. L. Yu, H. W. Bai, J. Gao, H. Deng, Q. Zhang, Q. Fu, Journal of Applied Polymer Science 2012, 124, 4452.
[2] F. L. Yu, H. Deng, Q. Zhang, K. Wang, F. Chen, Q. Fu, Polymer, 2013, 54, 6425.
[3] F. Yu, H. Deng, H. Bai, Q. Zhang, K. Wang, F. Chen, Q. Fu, ACS Appl. Mater. Interfaces, 2015, 7, 10178.
[4] Y. Zhou, H. Deng, F. L. Yu, H. W. Bai, Q. Zhang, F. Chen, K. Wang, Q. Fu, Polymer, 2016, 99, 49.
[5] Y. Zhou, F. Yu, H. Deng, Y. Huang, G. Li, Q. Fu, The Journal of Physical Chemistry B, 2017.

Understanding modification of impact strength and energy absorbing properties in Novatein thermoplastic protein

Smith Matthew J. [(1)*], Verbeek Casparus J. R. [(1)]

(1) University of Waikato - Hamilton - NewZealand

Novatein thermoplastic protein, produced from bloodmeal, embrittles after production due to plasticiser loss, causing poor energy absorbing properties (impact strength = 0.5 kJ/m2). The addition of a low Tg second phase in Novatein means energy absorption during fracture can be improved. Several avenues were used to explore impact modification; namely addition of nano-scale core-shell particles, reactive blending of Novatein and functionalised PE, and bio-based ternary blends of Novatein, PBAT and epoxy functionalised chain extender. It was found that the greatest increase in energy absorption occurred in blends containing core-shell particles (impact strength at 30 wt. % particles = 3 kJ/m2). This was attributed to good particle dispersion, efficient stress transfer to the elastomeric particle and decrease in interparticle distance, allowing the matrix to elongate freely during fracture. In contrast, in-situ formation of morphologies conducive to impact modification through reactive extrusion of Novatein and modified PE proved difficult due to viscosity ratio and interfacial tension that weren't favourable in producing stable, dispersed morphologies. However, blends containing zinc ionomer performed well in mechanical testing, with high elongation and greatly increased impact strength (impact strength at 20 wt. % PE = 2.3 kJ/m2). This was attributed to strong ionic interactions between the zinc carboxylate salt and charged amino acids in the protein phase. Similar morphologies were displayed in Novatein/PBAT ternary compatibilized blends. The introduction of epoxy functionalities caused a fine dispersion of PBAT and a maintenance of tensile properties along with a threefold increase in notched impact energy (1.5 kJ/m2) was observed. In conclusion, the study shows that it is possible to significantly increase energy absorption of protein based thermoplastics through reactive blending of synthetic or biobased polymers, or incorporation of pre-synthesised particles.

Properties of blends based on Ethylene-Vinyl Acetate and Ethylene 1-Octene Copolymers and investigation of their crosslinking behaviour

Ehrich Christian [(1)]*, Ramezani Farzad [(1)], Klehm Jessica [(1)], Schulze Stefan H. [(1)]

[(1)] Fraunhofer IMWS - Sachsen Anhalt - Germany

In spite of its commercial importance as encapsulant film for solar cells, ethylene-vinyl acetate (EVA) shows some drawbacks. One of the most discussed is its susceptibility to the potential induced degradation (PID) effect which causes a solar module's power output capacity to degrade. As a solution, a polyolefin having low polarity and high volume resistivity can be blended with EVA to improve encapsulant performance. However, the requirements in regard to the mechanical and aging properties of materials applied as solar cell encapsulants make cross linking of the blended materials an utmost necessity. An Ethylene 1-Octene (EO) copolymer was blended with EVA in varying proportions. The EO and EVA (33% VA) were characterized through a high comonomer content and thus low crystallinity and density. Furthermore, a peroxidic cross linking agent was added to the blends in varying quantities and the cross linking reaction as well as the resulting mechanical, thermal and optical properties of films were studied. Emphasis was placed on extrusion at lower temperatures then described before to allow for blending and additivation with cross linking agent in one step. Techniques used for characterization included differential scanning calorimetry (DSC), dynamic shear rheometer (DSR) and dynamic mechanical analysis (DMA). Morphology was investigated by means of scanning electron microscope (SEM) and atomic force microscopy (AFM). Optical spectroscopy was applied to investigate transparency and scattering. It could be shown that blending at a low extrusion temperature of 100°C resulted in blend morphologies comparable to those of EVA-EO blends extruded at significantly higher temperatures reported earlier as viable blending conditions. Peroxidic crosslinking of the blends led to high gel contents, improved viscoelastic properties and improved morphology while crosslinking intensity was reduced with increasing EO content. Optical transmission improved with increasing EVA content and was shown to be stabilized against physical aging through the crosslinking process.

Development of Co-Continuous Morphology in Blends of Thermoplastic Starch and Low-Density Polyethylene

T. Mazerolles[a], M. C. Heuzey[b], M. Soliman[c], H. Martens[c], R. Kleppinger[c], M. A. Huneault[a]

[a]*Department of chemical and biotechnological engineering, Université de Sherbooke, Canada*
[b]*Department of chemical engineering, École Polytechnique de Montréal, Canada*
[c]*SABIC Global technologies B.V., T&I STC Geleen, The Netherlands.*

Abstract. Because of its low cost and availability, Thermoplastic Starch (TPS) has become an important feedstock in bioplastics. This study focuses on blends of potato starch-based TPS with Low Density Polyethylene (LDPE) and the morphology development of such blends around the phase inversion point. The effect of glycerol and water content on TPS rheology and the subsequent morphology development with LDPE of various viscosity levels has been investigated.

Keywords: Polyethylene, thermoplastic starch, blends.
PACS: 60

INTRODUCTION

Thermoplastic starch (TPS) is a destructured and plasticized version of natural starch. Its rheological and mechanical properties can be finely tuned depending on the plasticizer content and on the processing sequence used to properly destructure it, a process referred to as gelatinization. Thermoplastic starch is however highly hydrophilic and is typically blended as a minor phase into synthetic materials to lower its water sensitivity. In this work, our aim is to produce co-continuous structures as this is the first step in a global effort to produce fibrillary or lamellar TPS morphologies for specific packaging application. In this first effort, we will focus on the morphology development at the end of the twin-screw extrusion process used to produce the blends and we will assess the necessary conditions required to produce co-continuous structures.

METHODOLOGY

Blends have been mixed in one-step using a twin-screw compounder and extruded into strands following a process described elsewhere[1]. The investigated compositions included plasticizer levels of 30 to 40 wt% glycerol and 0 to 10 wt% water based on the TPS phase and of TPS fractions in LDPE ranging from 40 to 80 wt%. The TPS/LDPE viscosity ratio was modified either by increasing the TPS plasticization or by proper selection of the LDPE grade. Morphology assessment was carried out by Scanning Electron Microscopy on microtomed surfaces after the TPS phase had been extracted by acid etching in order to increase the contrast. For formulations where TPS formed the matrix, Energy Dispersive Spectroscopy has been coupled with SEM to identify components. Rheology measurements were carried out on pure TPS and on TPE/LDPE blends using capillary rheometry. Particular attention was given to the effect of TPS humidity on viscosity and therefore humidity levels were carefully measured by Karl Fisher titration prior to rheometry. Different types of blend morphologies were achieved from droplets to co-continuity and the relation between the phase transitions and the viscosity ratio was established.

RESULTS

The presentation will first review the main results obtained in terms of TPS and TPS/LDPE rheology, especially as a function of plasticizer content and temperature. While it is possible to match the viscosity of LDPE to that of TPS on selected shear rate regions and for selected temperature, it will be shown that the activation energy of the two materials are very different leading to a strong temperature dependency of the viscosity ratio. Nonetheless, it was possible to produce a mapping of the required LDPE melt flow index and of the TPS plasticizer content that lead to either dispersed or co-continuous morphologies. A typical example of the morphology evolution is summarized in Figure 1 where the TPS content was increased progressively while all other parameters were kept constant. The TPS was extracted in Figure 1a to c and thus the TPS phase is revealed by the cavities. The continuity of the TPS phase increased with TPS content from a low value at 50%TPS (i.e. 9%) to full continuity at 60 wt% and up. The blend with 80 wt% TPS was phase inverted and therefore selective removal of the TPS could not be used to reveal the structure. The LDPE dispersed phases can however be clearly observed on the micrograph 1d. Other results that will be described in the oral presentation include the effects on the blend morphology of LDPE viscosity, of starch's botanical source (potato vs corn) and the effect of plasticizer level and composition.

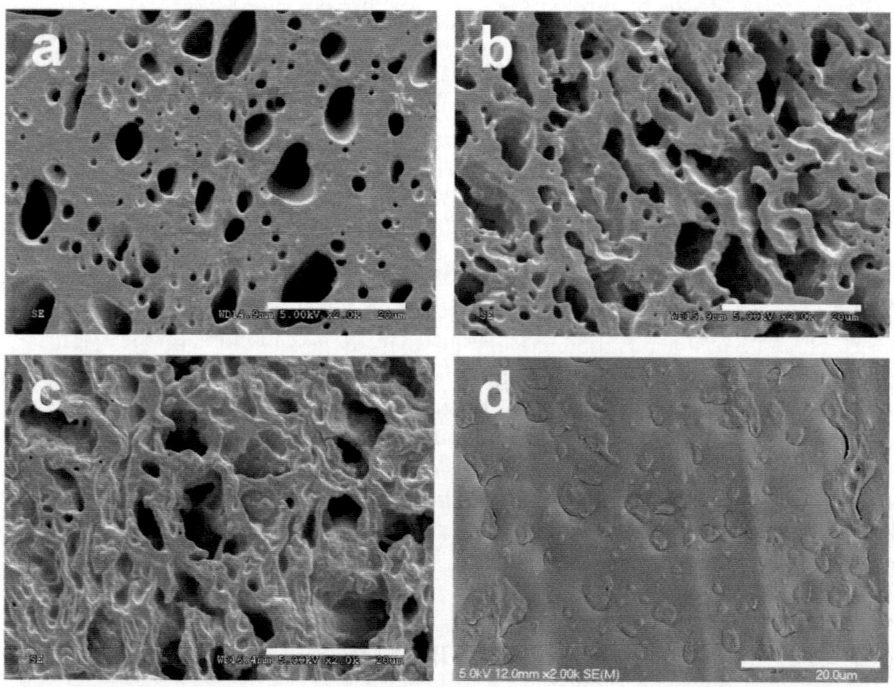

FIGURE 1. Effect of increasing TPS content on compatibilized blend morphology: (a) 50 wt% TPS c=19%, (b) 60 wt% TPS c=95%, (c) 70 wt% TPS c=89% and (d) 80 wt% TPS c=100%.

CONCLUSION

Co-continuous TPS/LDPE blends can be produced reproducibly by selecting materials compositions leading to a proper viscosity match. The mapping of the relationship has been achieved and will be disclosed in the oral presentation.

REFERENCES

1. M. A. Huneault and H. Li, *Journal of Applied Polym. Sci.*, 126, Issue SUPPL. 1, 25 October 2012, Pages E96-E108 (2012)

Developing conducting immiscible PP/PS blends with a percolated polyaniline/PA filler by tuning their specific interactions with the compatibilizer

Hejmady Prakhyat [1], Barathi Avanish [2], Cardinaels Ruth [3], Moldenaers Paula [2]*

[1] TU Eindhoven - Den Dolech 2, 5612 AZ, P.B. 513 - Netherlands, [2] KU Leuven - Celestijnenlaan 200F, 3001 Leuven - Belgium, [3] TU Eindhoven - Den Dolech 2, 5612 AZ, P.B. 513 - Netherlands

We have investigated the effect of master batch preparation techniques such as by extrusion and solution casting to develop a double-percolated network of conducting Polyaniline (PANI) in immiscible PP/PS blends. The strategy adopted to develop conducting blends was to concomitantly drive the morphology of the immiscible blend to a cocontinuous structure and control the localization of the PANI using specific interactions with an interfacially segregated SEBS-g-MA compatibilizer. A percolated composite of PANI in Polyamide (PA) is used as the conducting filler, and we envisage that the hydrogen bonding of PA with MA of the copolymer drives the localization of the PANI to the copolymer. Hence, master batches of the conducting PANI and copolymer with increasing PANI concentration were prepared by extrusion and solution casting. The master batches were subsequently added to the polymer constituents (PS or PP/PS) using different sequences of solution and melt mixing to localize the PANI either in the polar PS phase or partially at the interphase. The thus prepared blends led to a synergistic increase in the conductivity (6 decades) and dynamic moduli with increasing PANI and copolymer concentrations, whose effect was investigated for various blend compositions. Our findings were warranted by STEM micrographs, in which for blends with solution casted master batches, PANI-PA selectively localized at the interphase or in the minority matrix phase of the PP/PS blend. The latter was attained due to phase inversion with increased copolymer concentration. In addition, we established the essential role of interactions between the conductive PANI-PA composite and the MA functionality of the SEBS-g-MA in the localization process. In summary, various master batch techniques were employed to optimize the percolation of the conductive polymer PANI in PP/PS blends, allowing to generate easily processable conductive PP-PS blends.

Effect of Nano-fillers on Mechanical and Rheological Properties of Polyphenylene ether/High Impact Polystyrene/Styrene-Butadiene-Styrene Blends

López Martínez Erika I. (1)*, Mendoza Duarte Mónica E. (1), Estrada Moreno Iván A. (1), Zaragoza Contreras E. Armado (1), Vega Rios Alejandro (1), Flores Gallardo Sergio G. (1)

(1) CIMAV - Chihuahua - Mexico

Blends of polyphenylene ether (PPE), high impact polystyrene (HIPS) and styrene-butadiene-styrene (SBS) triblock copolymer with different concentrations of nano-fillers (nano-SiO2 or nano-TiO2) have been compounding by melt mixing using an internal mixer. The rheological, dynamic mechanical and mechanical properties of PPE/HIPS/SBS blends were investigated by means of oscillatory rheometry, dynamic mechanical analysis (DMA) and Izod impact measurements to determine the effect of the nano-fillers on the blends properties. The test results showed that the blends properties are affected when the processing conditions (order of addition of the components to the internal mixer and addition time of nano-fillers) are modified.

Developments in PLA-based blends over last two decades

Mohammadreza Nofar*[1,2] and Dilara Sacligil[2]

1: Metallurgical & Materials Engineering Department, Faculty of Chemical and Metallurgical Engineering, Istanbul Technical University, Maslak, Istanbul, 34469, Turkey
2: Polymer Science and Technology, Istanbul Technical University, Maslak, Istanbul, 34469, Turkey
**: Corresponding author: nofar@itu.edu.tr*

Abstract. Polylactide (PLA) is a well-developed commercial biobased, biodegradable, biocompatible, and non-toxic polymer that has competitive material and processing costs and high mechanical properties. Thereby, it can highly be considered for biomedical applications and as the most promising substitute for a wide range of commodity and engineering applications where non-compostable petroleum-based polymers are being used. PLA also possesses series of serious shortcomings such as low melt strength, slow crystallization kinetics, poor processability, formability, and foamability, high brittleness, low toughness, and low service temperature which limit its production and usage. To overcome these limitations, blending PLA with other polymers has been known as an inexpensive breakthrough which could also tailor the final properties of the PLA-based products. During the last two decades, researchers have investigated the synthesis, processing, characterization, and development of various PLA-based blend systems. This study reviews these investigations and compares the syntheses/processing-morphology-properties interrelationships in PLA-based blends developed thus far for various applications.

Keywords: PLA, multiphase, blend, PBAT, PBSA, nanocomposite, foam

INTRODUCTION

Poly (lactic acid) or polylactide (PLA), an aliphatic thermoplastic polyester, is one of the most extensively developed commercial biobased, biodegradable, and biocompatible polymers. It is derived from resources such as cornstarch, sugarcane, and other renewable biomass products and wastes. PLA can be synthesized through either direct condensation polymerization of lactic acid monomer or ring-opening polymerization of cyclic lactide dimer. In general, the ring-opening approach is used commercially to produce solvent-free high molecular weight PLA, since in the other approach water is also produced as an undesirable coproduct during polymerization [1-2].

PLA could be considered as a promising substitute for the petroleum based polymers in various commodity and engineering applications due to the comparable cost and properties of PLA when compared with those of petroleum-based polymers. Furthermore, due to its non-toxic features, PLA is an appropriate candidate for food packaging and other consumer products. It is also a biocompatible polymer which could also be applied in biomedical applications such as bone fixtures, drug delivery and tissue engineering [3-4].

Despite the noted advantages of PLA, there are several limitations such as brittleness, low toughness, low service temperature, narrow processing window, low melt strength, poor foamability and slow crystallization kinetics that PLA possess [2-3, 5-8]. One of the most common and practical approaches to overcome the above limitations is to blend PLA with other polymers. Thus, the blending polymer should exhibit a superior behavior in the area(s) where PLA shows weaknesses, in order to compensate for the weaknesses of PLA [9-10]. Moreover, development of PLA blend systems also open new avenues to extend PLA applications [11]. This will be possible through controlling/tailoring the ratio of PLA to the secondary blending polymer, the use of a third blending polymer or introducing functional nanoparticles, and controlling the morphology and localization of blending components, as well as benefiting from the generated synergies of interfacial interactions among the blending elements [9-10, 12-13].

REVIEW INVESTIGATIONS

Statistics

This report reviews the investigations related to PLA-based blends and compares the composition/processing-morphology-properties relationships developed so far for various applications. This includes the overview of the investigations that have been conducted on (i) miscible blends of poly L-lactide (PLLA) and poly D-lactide (PDLA), which generate stereo-complex crystals; (ii) binary immiscible/miscible blends of PLA with other thermoplastics; (iii) multifunctional ternary PLA blends using a third polymer or hybrid systems with fillers such as nanoparticles; and

(iv) PLA-based blend foam systems. Figure 1 provides detailed information regarding the number of research reports on various PLA-based blend systems over the last two decades, including binary PLA blends with biobased polymers, biodegradable polymers, biobased-biodegradable polymers, and synthetic-nondegradable polymers (non-biopolymers), ternary polymer blends of PLA with two other polymers, ternary blend nanocomposites of PLA with a polymer and a nanofiller, and PLA-based blend foams. It is clear that research on PLA-based blend systems grown dramatically in different areas during these decades. Figure 2 also shows the distribution and total number of PLA-based blend publications relating to the various blend systems noted above, as well as the incorporated type of nanofiller in PLA-based ternary blend nanocomposites.

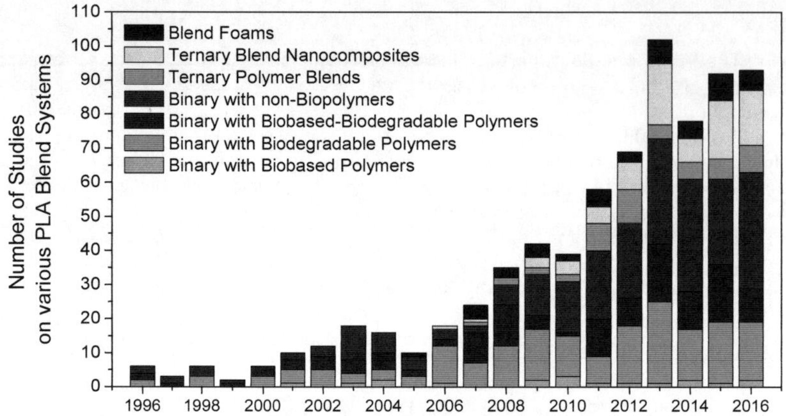

FIGURE 1. Number of studies reported on various PLA-based blend systems over the last two decades

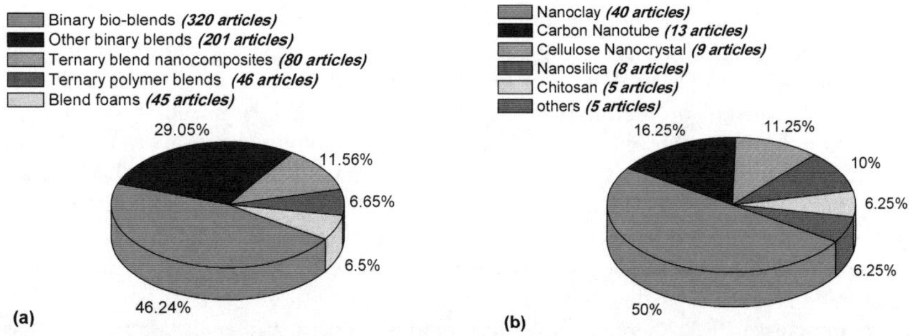

FIGURE 2. (a) Distribution and total number of studies reported on various PLA-based blends, and (b) distribution and numbers of articles on PLA-based ternary blend nanocomposites for various nanofillers.

PLA is a biobased/biodegradable biopolymer. Other biopolymers could be biobased but not biodegradable (normally called biobased) and others could be non-biobased but biodegradable (normally called biodegradable). Research activities on blends involving the above three categories of biopolymers have been growing at a significant rate. Figure 3 shows the frequency of studies conducted on binary PLA blends with other biopolymers as categorized above.

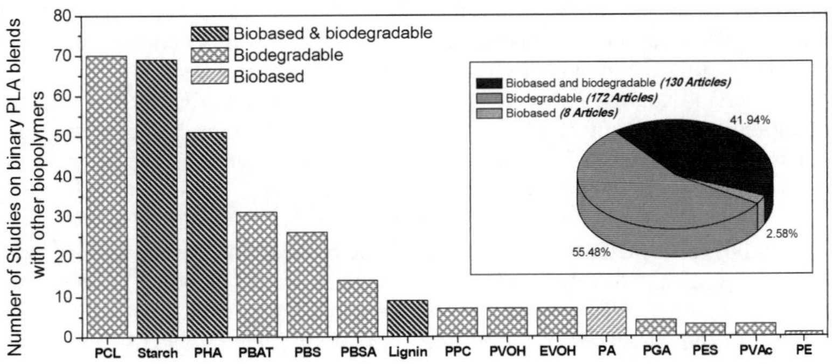

FIGURE 3. Number of studies conducted on binary PLA blends with other biopolymers

Manufacturing of PLA-based binary blend systems with synthetic/noncompostable thermoplastics and elastomers has also been explored because the development of such partial bioblends could also be beneficial from environmental and energy considerations. The majority of the investigations on these topics have been carried out since a decade ago, whereas the oldest studies belong to less than two decades ago. Figure 4 presents the number of these studies as a functions of the blending polymer. As seen, a large number of these investigations relates to PLA/rubber and PLA/polyurethane (PU) blends, as rubber and PU may provide high ductility and toughness to PLA.

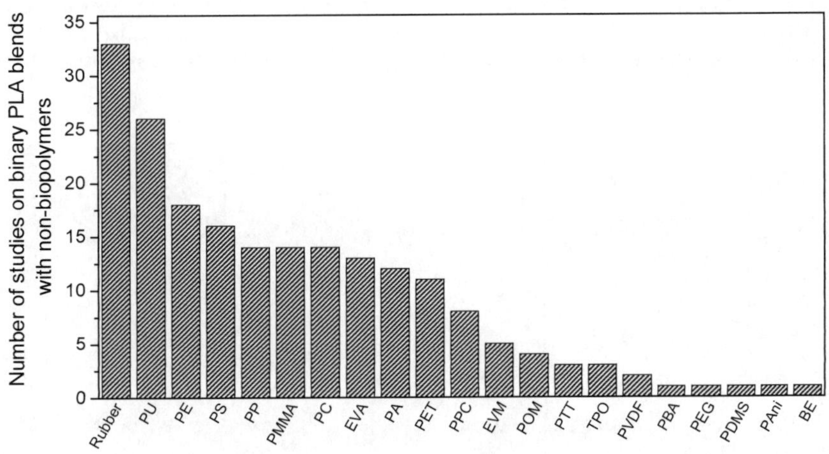

FIGURE 4. Number of studies on PLA blends with synthetic/nonbiodegradable thermoplastics/elastomers

CASE STUDY

Bioblends of PLA with PBAT and PBSA

While maintaining the biodegradable nature of the product, among the PLA-based blend systems with droplet-like morphology, poly[(butylene adipate)-co-terephthalate] (PBAT) and poly[(butylene succinate)-co-adipate] (PBSA) have contributed to significantly overcome some of the PLA's drawbacks [14-17]. It has been showed that the addition of PBAT and PBSA improves the ductility and toughness of PLA. However, they tend to reduce the tensile strength and stiffness of the final blend. We showed compared the interfacial and rheological properties of PLA-PBAT and PLA-PBSA blend systems and showed how their morphological stability depends on shear flow history [14-15].

In more specific case, we investigated the crystallization, rheological, and morphological properties and subsequently their influence on the bead foaming behavior of PLA-PBAT and PLA-PBSA blends with a given w/w of

75/25. The effects of PLA matrices with two different molecular weights (low and high molecular weight PLAs, LPLA and HPLA) and using two different processing techniques (internal mixer, BB, and twin-screw extruder, TSE) on the crystallization, rheological, and morphological properties of PLA-PBAT were examined; and subsequently their influence on the mechanical and bead foaming behavior was analyzed.

Figure 5 shows the blend and foam morphologies of various obtained blends as well as the ductility of the blends in differently processed blends. The tensile results show that similar blend morphologies of LPLA-PBAT (BB) and LPLA-PBSA (BB) resulted in a similar tensile behavior. In both cases, the strain at break increased to around 150 %, which represents a huge improvement over the PLA ductility with strain at break of 5 %. The energy at break was also improved from around 0.2 J/m in PLA to around 5.2 J/m in both blends. The ductility and energy at break increased in HPLA-PBAT (BB) with larger droplets (1.2 μm). Although the blend with a larger droplet morphology in HPLA-PBAT (BB) is expected to suppress the ductility and toughness, the existence of a PLA matrix with a larger molecular weight seems to have a more dominant effect on ductility and toughness improvements. HPLA-PBAT (TSE) revealed much higher ductility (265 %) and toughness (9.2 J/m) in comparison with HPLA-PBAT (BB) due to the presence of a much finer droplet size (0.5 μm).

The bead foaming results also showed that HPLA-PBAT (TSE) with finer PBAT droplet morphology exhibited an open-cell structure and a higher expansion ratio, whereas HPLA-PBAT (BB) with larger droplets (1.2 μm) revealed a more closed cell structure and a lower expansion ratio. The PBAT droplets in HPLA-PBAT (BB) favored the heterogeneous cell nucleation in HPLA but the expansion ratio was small. In HPLA-PBAT (TSE), however, the PBAT droplet size was too small to nucleate cells in the given processing condition. Similar to HPLA-PBAT (TSE), the LPLA-PBAT (BB) blend with a fine droplet morphology (0.7 μm) resulted in an open-cell foam structure. The lower molecular weight matrix could be another factor that caused cell rupture during expansion due to the less molecular entanglements in LPLA. In LPLA-PBSA (BB), however, uniform closed-cell foams with a high expansion ratio were obtained. The generation of high melting temperature crystals during the saturation increased the stiffness of the PBSA droplets. Therefore, PBSA droplets could more effectively enhance the cell nucleation and stabilize cell growth.

After the foaming, it was found that, except in HPLA-PBAT (BB) that showed the lowest expansion ratio, the foaming of the blends introduced a small degree of less perfected crystals within the amorphous PLA matrices. Around 1.5 % of PLA crystallinity was induced in HPLA-PBAT (TSE) and LPLA-PBAT (BB) systems; whereas this amount was increased to around 4 % in LPLA-PBSA (BB).

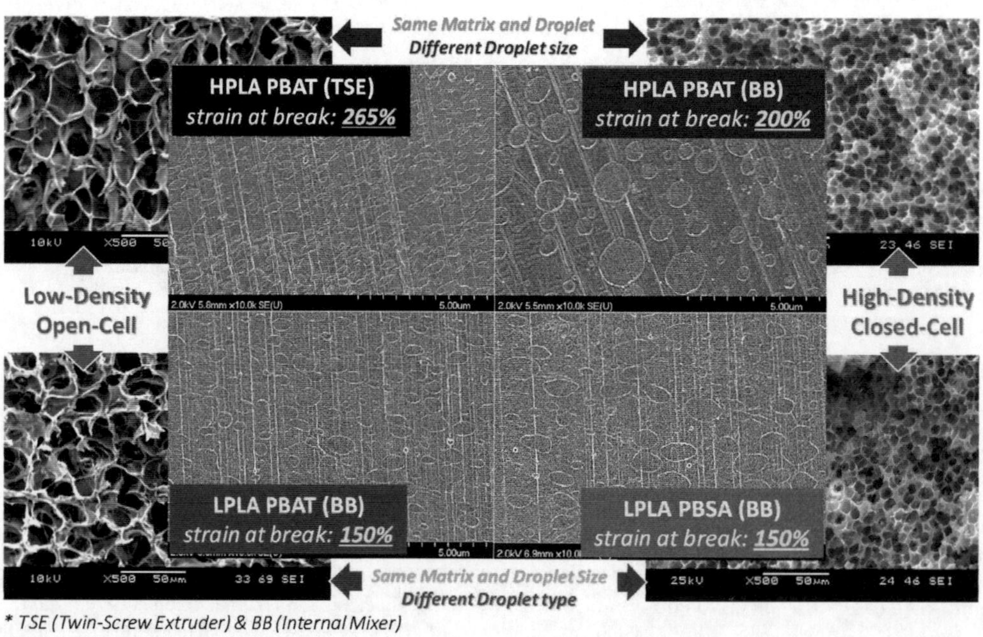

FIGURE 5. The blend and foam morphologies and ductility values of the blends for differently prepared blends systems

SUMMARY AND PERSPECTIVES

In this study, we reviewed the attempts to develop PLA-based blends. According to the literature, the interest of developing PLA-based blends with better performance and lower cost is growing every year and researchers are eagerly seeking more efficient and multifunctional products based on PLA. This is driven by the global concerns regarding energy and environmental issues which are growing with time. The lack of biodegradability of conventional plastics which are based on non-renewably petroleum makes the development of bio-based PLA products with desired properties a priority not only for academicians but also for plastics manufacturing companies. Also, the biocompatibility and non-toxic features of PLA render its compounds promising candidates for biomedical applications. According to the statistics shown in this report, the research on PLA-based blends has grown dramatically in different areas during the last few decades. Among binary systems, blending PLA with biopolymers has been of greater interest to researchers due to the full bio-feature of the final products. Amongst these biopolymers, blends of PLA with PCL, starch, PHA, and PBAT have been investigated more than blends with other biopolymers. PLA blends with non-biopolymers such as elastomeric thermoplastics, rubber and TPU have also been largely investigated. Moreover, the development of ternary and hybrid blends is growing more aggressively due to the possible introduction of further functional properties to the blend. These functional properties could be obtained due to the synergy resulting from interfacial interactions of different phases in the blend; hence, properties could differently be influenced. For blend nanocomposites, the use of nanoclay has also been explored more extensively due to the high performance of clay nanoplatelets, its low cost, and its easier dispersion in polymers according to the existence of various modified clays. Layered clay could also improve the barrier properties of PLA, which is important for packaging applications. The development of PLA blend cellular and microcellular foams has also turned to be an important research area to produce lighter products at lower costs and biodegradation features.

Several petroleum based plastics such as PS, PET, PE, and PP could be replaced within years by PLA compounds for various commodity and engineering applications. Besides energy and environmental concerns, this trend is due to price reductions of commercial PLAs, efforts to solve PLA's shortcomings via blending or other possible routes, and the introduction of PLA-based products with better performance and desired properties. Nevertheless, extensive research still needs to be carried out to develop PLA-based compounds/blends with superior performance, less shortcomings, lower cost and desirable multifunctional properties.

REFERENCES

1. Sinclair RG. *J. Macromol. Sci. Pure Appl. Chem.* **33**, 585-597 (1996).
2. Garlotta DJ. *J. Polym. Environ.* **9**, 63-84 (2001).
3. Nofar M, Park C.B, *Polylactide Foams, Fundamentals, Manufacturing, and Applications*, Elsevier, ISBN: 9780128139912
4. Nofar M, Ameli A, Park CB. *Macromol. Mater. Eng.* **299 (10)**, 1232-1239 (2014)
5. Nofar M, Park C.B, *Prog. Polym. Sci*, **39**, 1721-1741 (2014).
6. Saeidlou S, Huneault M, Li H, Park C.B, *Prog. Polym. Sci*, **37**, 1657-1677 (2012).
7. Rasal R.M, Janorkar A.V, Hirt D.E, *Prog. Polym. Sci*, **35**, 338-356 (2010).
8. Dorgan J, Williams J, *J. of Rheol*, **43**, 1141-1155 (1999).
9. Macosko CW, *Macromol Symp* **149**, 171-184 (2000).
10. Favis BD, *Can J Chem Eng*, **69**, 619-625 (1991).
11. Yu L, Dean K, Li L, *Prog Polym Sci*, **31**, 576-602 (2006).
12. Sarazin P, Lia G, Orts WJ, Favis BD, *Polymer* **49**, 599-609 (2008)
13. Liu H, Song W, Chen F, Guo L, Zhang J. *Macromolecules*, **44**, 1513-1522 (2011).
14. Nofar M, Maani A, Sojoudi H, Heuzey M-C, Carreau P.J, *J. of Rheol*, **59 (2)**, 317-333 (2015).
15. Nofar M, Huzey M-C, Carreau P.J, Kamal M.R Randall J, *J. of Rheol*, **60 (4)**, 637-648 (2015).
16. Nofar M, Tabatabaei A, Sojoudiasli H, Park C.B, Huzey M-C, Carreau P.J, Kamal M.R, *Eur. Polym. J.*, **90**, 231-244 (2017).
17. Nofar M, Huzey M-C, Carreau P.J, Kamal M.R, *Polymer*, **98**, 353-364 (2016).

Phase Behavior, Properties and Hydrolitic Stability of PMMA/PLA Blends

E. Rodriguez, M. Gonzalez, S. Shahbikian, B. Marcos, M.A. Huneault

Departement of Chemical and Biotechnological Engineering, Université de Sherbrooke, Canada.

Abstract. This presentation summarizes findings on the phase behaviour of the PMMA/PLA blends in the melt and solid-state, on the basic mechanical performance, on the temperature resistance of the material and finally on the hydrolytic stability enhancement observed when adding PMMA to PLA. Some key unexpected findings were highlighted. In particular, the phase behavior was process-history dependent indicating that attaining the equilibrium thermodynamic state requires sufficient mixing and diffusion time. The complex phase structure of the blend was investigated by classical thermal analysis (DSC and DMA) but also by monitoring the structure during accelerated hydrolysis experiments. These experiments led to the creation of nanometric pore structure that give insight on the phase structure of the blend. The presentation will aim at presenting our current understanding of this interesting blend system and to assess the usefulness and potential of this blend for engineering applications.

Keywords: PLA; PMMA; Blends; Phase behavior.
PACS: 60

INTRODUCTION

Blends of PLA and PMMA have been reported to be miscible or partially miscible depending on blending procedures and composition. Nonetheless, the PLA within this blend has the ability to phase separate by crystallization thus concentrating PMMA in the amorphous regions. Already from this point of view, this blend system offers a very interesting scientific case study. The blend however may also offer interesting promises from an application point of view: the PLA fraction offers a biobased component, the blend remains transparent, the temperature resistance is improved relative to pure PLA and the final cost of the material may be reduced since PMMA is currently more expensive than PLA. This work therefore aims at thoroughly investigating the phase behavior and properties of these blends.

EXPERIMENTAL

Amorphous and semi-crystalline PLA grades were used to highlight the role of crystalline development in the blends. For PMMA, two homopolymers, of different molecular weight were selected to highlight the effect of chain mobility on the phase transitions. The blends were prepared in the melt state using a twin-screw mixing device. The blend rheology, tensile properties, thermal properties by DMA and DSC were measured as a function of blend composition. These measurements were used to monitor the phase transitions of the blend during processing and over time in the quiescent conditions. Finally, the effect of PMMA content on the hydrolytic resistance of PLA/PMMA blends was assessed using accelerated tests in which the blends are subjected to aqueous media of various pH at temperatures above glass transition[1].

RESULTS

PLA and PMMA can be turned into a homogenous mixture but a sufficiently long mixing time was necessary to reach this equilibrium state. The importance of mixing dynamics may explain conflicting observations reported in the literature based different blending protocols. Nearly additive rheological behavior as function of the blend ratio and single-relaxation time in Cole-Cole plots indicated that the blends were miscible in the melt state. Wide thermal transitions hinted however at a wide distribution of composition that could infer that the system is at the verge of phase separation. Young modulus and tensile strength of the materials were always very close to the values predicted by a simple linear mixing rule and thus indicate that the properties are not affected by interfacial issues. The melt processing of the blends was much easier than the processing of the components. The viscosity of the blends was greatly reduced by the presence of the lower viscosity PLA through a dilution effect. The presence of PMMA, on the other hand, conferred some shear thinning that is not present in PLA and which is often a desired property in terms of processability. The presence of PMMA also increased the temperature resistance relative to that of pure PLA. In terms of hydrolytic stability, the presence of PMMA was shown to greatly help the hydrolytic stability. As little as 5%PMMA in PLA enabled retention of part integrity during hydrolysis. It was also effective at slowing surface erosion of the samples in aggressive alkaline media. Figure 1 shows a very interesting microstructure observed in the bulk of a hydrolyzing sample. The hydrolysis in this case has generated a phase separation and selective removal of PLA leading to a very fine porous structure. The remaining polymeric material seems to be agglomerated sphere. It could be postulated that composition fluctuation are intensified by the selective removal of PLA revealing morphological information on the blend initial state. It will be shown in the presentation that the two main erosion mechanisms, i.e. bulk and surface erosion, vary in relative intensity depending on the PMMA content and of the hydrolyzing media alkalinity. This was shown by plotting the sample porosity as a function of weight loss. For samples that lose weight exclusively by a bulk erosion mechanism, porosity should be proportional to weight loss while for surface erosion, porosity should not develop as material is removed only from the surface. Actual materials fall between these idealized cases depending on the PMMA concentration and on the aggressiveness of the hydrolyzing media.

FIGURE 1. SEM Micrograph taken form a fractured surface in the bulk of a 20%PMMA/80%PLA blend after 4 days in an alkaline solution. .

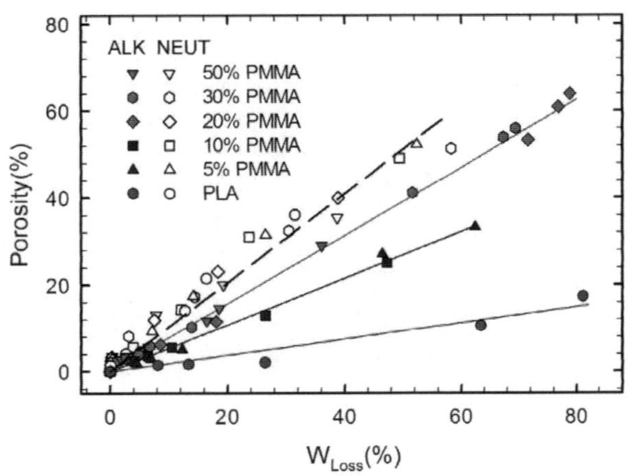

FIGURE 2. Porosity as a function of weight loss for blends of various compositions in neutral and alkaline (pH=12) media.

REFERENCES

1. E. Rodriguez et al, Journal of *Applied Polym. Sci.*, ***133*** (44), article number 44152.

Effect of blending styrene-methyl methacrylate-maleic anhydride copolymer on rheology and foaming of poly(methyl metacrylate)

Sugimoto Masataka (1)*, Kawahara Yuki (1), Sathish Sukumaran K. (1), Shindo Yuichi (2), Nishino Kohei (2)

(1) Yamagata University - Yamagata - Japan, (2) Denka K.K. - Chiba - Japan

We studied the effect of blending styrene-methyl methacrylate-maleic anhydride copolymer (SMA) on rheology and foaming of poly(methyl metacrylate) (PMMA). The blends are transparent over a wide range of blending ratio and can be miscible. Although the blends showed a slight increase of storage modulus in low frequency region at the molten state, the apparent change in non-linear viscoelasticity was not observed. At the identical carbon dioxide adsorbed amount, the blends showed very unusual foaming behaviors depending on the blend ratio.

Mechanical enhancement of olefin-block copolymer by using biodegradable Polylactic acid

YU BOWEN [1], WANG KE [1], FU QIANG [1]*

[1] sichuan university - sichuan - China

Olefin block copolymers (OBC) is a novel thermoplastic elastomer with excellent properties. However, the insufficiency in mechanical strength limits its application. Polylactic acid (PLA) is a biodegradable and renewable material with high strength and modulus. Hence by blending PLA with OBC can not only improve the mechanical properties of OBC but also reduce the consumption of oil resources. In this work, two kinds of PLA, poly(L-lactide) (PLLA) and sc-PLA powder, which is the stereocomplex of PLLA and its optical isomer poly(D-lactide) (PDLA), were used to enhance the property of OBC and ethylene-methyl acrylate-glycidyl methacrylate (EGMA) was used as a compatibilizer. It is found that as the PLLA contents vary form 5%wt to 30%wt, the tensile strength of OBC/PLLA composites will increase continuously and the elongation at break can be well preserved at the same time. However, the introduction of the sc-PLA have a different influence on the mechanical properties of OBC. With the variation of sc-PLA content, the tensile strength of the OBC/sc-PLA composites increase apparently at 10%wt and then remain unchanged. Meanwhile, the elongation at break decreased appreciably with the sc-PLA content. This can be attributed to the poor dispersion of sc-PLA particles at high loading content.

Dynamics of Hierarchical Morphology Landscape in Crystalline Polymer Blends

Kyu Thein* (1)

(1) University of Akron - OH - USA

The present talk is aimed at theoretical and experimental determination on thermodynamic phase diagrams of binary crystalline polymer blends and dynamics of phase transitions involving the competition between liquid-liquid phase separation and liquid-solid phase transition, specifically solidification of polymer crystals driven by thermal-quenching. We will discuss i) thermodynamic phase equilibria of crystalline polymer blends, ii) development of diverse crystal polymorphs during isothermal quenching, and iii) dynamics of hierarchical morphology subjected to thermal gradient driven by liquid carbon dioxide (CO2) spraying. To elucidate formation of diverse crystal polymorphs, phase diagrams of binary azobenzene/diacrylate mixture as well as azobenzene/nematic liquid crystalline solution were first established experimentally and tested theoretically in the context of combined free energy of Flory-Huggins theory for liquid-liquid demixing and phase field theory for crystallization. In an effort to understand the effect of local thermal gradients on hierarchical organization of single crystal polymorphs, the azobenzene/nematic liquid crystal blend sample was sprayed with compressed CO2. The liquid CO2 spraying generates spatial variations of local temperatures leading to local thermal gradients, which in turn drives the faceted single crystals to undergo hierarchical organization into a so-called 'spherulite'. The present paper is the first to demonstrate the development of single crystal polymorph upon spraying with liquid CO2, thereby creating localized thermal gradients. Of particular interest is the formation of a hierarchical spherulitic structure composed of lamellar aggregates, whereby discrete single crystal polymorphs grow on the nematic defect (i.e., disclination), fanning out from a common core. It should be emphasized that the present aggregate model of discrete single crystals is an alternative account for spherulitic growth.Supported by NSF-DMR1502543

pps proceeding
Symposium: S08 - Polymer blends
Oral Presentation

Processing/microstructure/mechanical performance relationship of LLDPE/PP double gated injected blends

Rosales Caren Soledad [(1)]*

[(1)] Universidad Nacional Mar del Plata, Argentina - Buenos Aires - Argentina

Polyethylene and propylene are two of the most abundant polymers in wastes. Their separation in common waste recovery operations is very difficult and a complete separation is sometimes not possible. This is the reason why polymer waste recovery is providing an increasing motivation for improving PE/PP properties as a blend. In this work, microstructure and fracture performance of blends of commercial polypropylene (PP) and linear low density polyethylene (LLDPE) is reported. Blends were prepared in a twin screw extruder, and then processed via injection molding into double-gated plaques that induced weld lines. The morphology of melt-crystallized polymers is known to be related to the fabrication technique. Then, microstructure and morphology of blends were analyzed by differential scanning calorimetry (DSC), X-ray diffraction (XRD) and polarized optical microscopy (POM) after each processing step, i.e. extrusion and injection molding. The aim was to evaluate the influence of processing in microstructure and morphology, based in the knowledge that they will influence final parts performance. Fracture tests were carried out on mode I double edge-notched tensile specimens at quasi-static conditions. Different Fracture Mechanics approaches were applied depending on the materials´ fracture behavior. The Essential Work of Fracture methodology was adopted for blends that exhibited ductile fracture behavior, whereas the stress intensity factor (K1c) was chosen to characterize the resistance to unstable fracture. It was also observed that fracture behavior was dependent on the type of blend and the location of the specimen in the injected piece, and the loading direction. Moreover, impact tensile tests were performed in dog-bone samples. Several behaviors were found, ranging from ductile with plastic deformation prior to rupture, to completely brittle.

Increasing thermal conductivity by interface modification

Seyed Esfahani Seyed Armin [1], ghahramani Nikoo [2], Mehranpour Milad [1]*, Nazockdast Hossein [2]

[1] Islamic Azad university,Science and Research Branch - Tehran - Iran, [2] Polymer Engineering Department, Amirkabir University Of Technology - Tehran - Iran

Nanoscale thermal transport has attracted considerable attention because of both fundamental scientific interest and important engineering applications. In ceramic based composites, phonon is responsible for thermal conduction. Phonon (lattice vibration) is affected by phonon scattering due to high thermal boundary resistance caused phonon scattering across the interface between the components which is lead to reduce thermal conductivity. Double percolation has been widely used to reduce the filler volume fraction. In this method interface between two phases is where the probability of phonon scattering is high. One way to reduce phonon scattering is localization of conductive nano fillers at the interface of the polymer blends. In this work localization behavior of reduce graphene oxide (rGO) and Boron nitride (BN) in melt mixed blend of polypropylene (PP) and polyamide 6 (PA6) were investigated. The blends were prepared with different mixing sequence to investigate different selective localization of fillers By trapping r-GO in the interface of polymer blends. therefore, interfacial resistance significantly reduced, it means that phonon scattering reduced and thermal conductivity increased in comparison when the interface is not filled by any conductive nano fillers. The results are verified by rheological measurement and electron microscopy images.

New method of blending polyurethane-urea with rubber: selection of processing parameters and effect of blend composition

Tahir Muhammad (1)*, Stöckelhuber Klaus Werner (1), Wießner Sven (1), Heinrich Gert (2)

(1) Leibniz-Institut für Polymerforschung Dresden - Dresden - Germany, (2) Technische Universität Dresden, Institut für Werkstoffwissenschaft - Dresden - Germany

The blending of various types of polyurethanes with rubbers has extensively been attempted to obtain new materials with an improved property spectrum. However, the opted solution-and melt-blending methodologies are highly time- and energy-intensive, which tends to forfeit the commercial realization of the reported blends offering limited performance benefits. This study realizes the in-situ synthesis of structurally different polyurethane-ureas during reactive blending with polar (CR, NBR) and nonpolar (NR, EPDM, sSBR) rubbers [1-2]. The critical processing parameters of the proposed reactive blending method are determined from the chemo-rheological investigations of the polyurethane-urea (PUU) precursors. The compositional, structural and morphological aspects influence the static- and dynamic-mechanical response of the prepared highly reinforced blends. Importantly, the scanning electron microscopy of the cryofractured specimens reveals strong interfacial adhesion for an effective stress transfer between phases of heterogeneous blends. The blends are always compatibilized irrespective of the kind of polyurethane-urea and rubber polymers. References: Tahir, M.; Stöckelhuber, K.W; Mahmood, N., Komber, H.; Heinrich G. Macromolecular Materials & Engineering 300 (2015) 242-250. 2 Tahir, M; Stöckelhuber, K.W.; Mahmood, N.; Komber, H.;Formanek, P.; Wießner, S.; Heinrich, G. European Polymer Journal 73 (2015) 75-87

Blending Decoloured Novatein® with Modified PLA

1.1 Abstract

Blends of semi-crystalline poly (lactic acid) (PLA), a thermoplastic polyester and decoloured Novatein® (DNTP), a thermoplastic protein material, were prepared using reactive extrusion. Free radical grafting was used to graft itaconic anhydride unto PLA, creating a reactive side group. The compatibility between DNTP and PLA blends was investigated by mechanical testing, scanning electron microscopy (SEM), differential scanning calorimetry (DSC) and wide-angle X-ray scattering (WAXS). SEM revealed a co-continuous phase structure existed in the blends. The addition of itaconic anhydride improved the dispersion of DNTP in the PLA matrix. DNTP accelerated cold crystallization of PLA in the blend and increase crystallinity of PLA in the blends. Tensile strength exceeding that of DNTP was observed for the blends. Secant modulus, elongation at break and impact strength were also improved.

1.2 Introduction

Renewed interest in bio-based polymer has been due to their environmentally friendly nature and potential use in packaging and agricultural industries [1-4]. Biopolymers have been considered an attractive alternative for petroleum based polymers because they are abundant, inexpensive, biodegradable, renewable and environmental friendly. Biopolymers such as protein, cellulose and starch are obtained from animal and plant sources.

Bloodmeal is one of the animal sources of protein, containing 90 to 95 % dry weight of protein [5-7]. Bloodmeal can potentially be used as an alternative resource for bioplastics in agricultural and horticultural applications such as weasand clip, weed pegs, biodegradable plant pots and biopolymer foams [8-10]. There are some limitations to wider application of bloodmeal based polymer due to their offensive odour and colour. However, the colour and odour were successfully eliminated through a pre-treatment with peracetic acid (PAA) [7] resulting to a bio-feed stock referred to as decoloured bloodmeal (DBM). Decoloured bloodmeal has been processed into thermoplastic known as Decoloured Novatein® (DNTP), using Triethylene glycol (TEG), water and sodium dodecyl sulphate (SDS) [11]. However, its properties are relatively poor compared to other polymers used in sheet production due to its hydrophilicity [12], also processibility is difficult due to strong intra- and inter-molecular interactions between the protein chains. Therefore, the improvement of its properties and processibility becomes imperative.

Studies have concentrated on property improvement of protein based polymers through association with other hydrophobic polyesters with desirable properties [13-18] . Association with other polymers can be achieved through blending, coating or lamination. Blending is the most effective and easier way of modifying polymer properties, although poor interfacial adhesion between the polymer phases will result in a material with poor properties [19]. Poor interfacial adhesion in polymer-polymer blends is as a result of incompatibility between the two phases. However, this can be addressed using compatibilizers to create reactive functional groups capable of reacting with both polymer phases resulting in improved properties [13, 15, 17, 18, 20].

PLA is one of the suitable polymers for blend with other protein bio-polymers because of its biodegradability, high strength and hydrophobic properties. PLA is a thermoplastic polyester synthesized from lactic acid, which is derived from corn starch and sugar beets [21]. Most polyesters are immiscible with proteins because of the different polarities [16, 17, 20, 22]. DNTP and PLA blends are thermodynamically immiscible having weak interfacial adhesion, resulting in inferior mechanical properties because of the large difference in their hydrophilicities. Interfacial modification plays an important role in manipulating the solid-state adhesion between the components of two incompatible material blend [23]. Therefore, compatibilization of DNTP and PLA blend is essential.

Poly(2-ethyl-2-oxazoline) (PEOX) and methylene diphenyl diisocyanate (MDI) have been used widely as compatibilizers for natural polymers and PLA blends [18, 22, 24, 25]. Recent developments in interfacial modification of PLA and natural polymer blends has been explored. This involved grafting a reactive moiety, such as maleic anhydride [10, 20, 23, 26-29], or itaconic anhydride [9, 30] onto PLA matrix to create functional reactive groups for the natural polymers.

The amide and amino groups of DNTP protein components would possibly react with the anhydride group of itaconic anhydride. The aim of this study was to identify the quantity of additives for DNTP formulation and processing conditions required to produce a homogenous blend of DNTP/PLA with an improved property. This was achieved by investigating varying combination of water, TEG and SDS to determine the best formulation for blending with PLA.

1.3 Material and Methods

1.3.1 Materials

Bloodmeal was obtained from Wallace Corporation Limited, New Zealand and used as received. Analytical grade itaconic anhydride (IA), dicumyl peroxide (DCP), acetone, 50 wt% hydrogen peroxide, technical grade sodium dodecyl sulphate (SDS) and triethylene glycol (TEG) were purchased from Sigma Aldrich NSW, Australia. Peracetic acid (Peraclean 5) was purchased from Evonik Industries, Morrinsville, New Zealand. Poly (lactic acid) (PLA) grade 3051D was purchased from NatureWorks Ltd in pellet form. Distilled water was produced onsite.

1.3.2 Sample preparation

1.3.2.1 Interfacial modification of PLA

PLA was modified through free radical grafting of itaconic anhydride [30] to create reactive side-chain groups. PLA was dried at 80 °C for 4 hours to control moisture. 4.2 parts per hundred PLA (pph) Itaconic anhydride and 0.8 pph dicumyl peroxide were dissolved in 30 mL acetone. The preformed solution was poured over the oven dried PLA and was kept in the fume hood for about 2 hours. The solution was decanted before oven drying the PLA for 3 hours at 50 °C. The modified PLA (PLA-g-IA) was extruded using a LabTech twin screw co-rotating extruder having a screw diameter of 20 mm and L/D of 44:1, at temperature profile of 145 (feed zone), 145, 165, 165, 180, 180,180, 180, 160, 160, 155 °C (die zone). A constant screw speed was maintained at 150 rpm. A vacuum pump was attached on the 7th heating zone of the extruder to get rid of vapour generated during extrusion. The pelletized PLA-g-IA was oven dried for 12 hours prior to blending with Decoloured Novatein (DNTP) to minimise hydrolysis during melt processing.

1.3.2.2 Bloodmeal decolouring and decoloured Novatein preparation

Bloodmeal was decoloured using a solution of peracetic acid (PAA) according to previous methods [7, 31]. 4 wt% PAA solution was prepared by diluting 5 wt% PAA stock solutions with distilled water at a constant ratio of 80:20 respectively. 150 g bloodmeal was decoloured by adding 450 g of 4 wt% PAA in a high-speed mixer. The mixture was allowed to mix continuously for 5 min to ensure homogenous decolouring of bloodmeal. 450 g of distilled water was added and mixed for another 5 min to ensure complete dilution of the slurry. The slurry was neutralized by adjusting to pH = 7 with sodium hydroxide solution. The neutralized slurry was filtered using a wire mesh sieve with aperture size 60 and subsequently washed by adding another 450 g of distilled water. The decoloured bloodmeal was dried approximately 15 hours in a 75 °C oven.

Decoloured Novatein (DNTP) was formulated by dissolving SDS in water heated to 60 °C while stirring. The solution was added to decoloured bloodmeal powder in a high-speed mixer and mixed for 5 min. TEG was added to the mixture and mixed for another 5 min to ensure homogeneous mixture is obtained. The mixed material was stored in an air tight bag overnight in a 2 °C fridge to equilibrate.

1.3.2.3 Blend formulation

Different DNTP formulations are shown in Table 1 and PLA grafting was performed prior to blending. All blends contained 50 wt% DNTP and 50 wt% PLA or PLA-g-IA. Blends were compounded using the same extruder as for grafted. The extrusion temperature varied from 100 to 180 °C (having the lowest temperature at the feed zone and the highest at die zone) depending on the DNTP content. The extrudate were granulated using a Tri-blade granulator from Castin Manufacturing Limited.

Table 1: Varying formulations contents of DNTP, dried bloodmeal and TEG plasticized used for PLA blends.

components	formulations				
	Dried decoloured bloodmeal (D)	TEG plasticized (T)	F2	F3	F4
Water	0	0	40	30	40
TEG	0	30	20	30	30
SDS	0	0	3	6	6

Note: all formulations contained 100 parts decoloured bloodmeal

1.3.2.4 Test specimens preparation

ASTM D638-14 Standard tensile test samples [32] and ISO 179-1:2010 impact test specimens [33] were injection moulded using a BOY 35A injection moulding machine. The samples were injected through a cold runner into a 60 °C water heated mould. The injection moulder has five heating zones including the feed and the die zone. The feed temperature remains constant at 100 °C, the barrel temperature varied between 100 to 140, and the die temperature varied from 120 to 140 depending on the formulation. The screw speed was constant at 150 rpm. The sample specimens produced were also used for thermal, moisture content and morphology testings.

1.3.3 Sample Analysis

All samples were conditioned for 7 days at 23 °C and 50% relative humidity before testing except otherwise stated.

1.3.3.1 Mechanical properties

Mechanical testing was performed according to ASTM D638 using Instron Universal Testing machine (model 33R4204) at a crosshead speed of 5 mm/min and an extensometer gauge length of 50 mm. 10 replicates were tested for each sample type to obtain an average value.
The impact testing bars produced from the injection moulder has diameter of 80 x 10 x 4 mm^3. Charpy edgewise impact strength was performed according to ISO 179-1:2010 using a RAY-RAN Pendulum Impact System. The bar tested was notched according to standard. 10 bars were tested to obtain average impact strength of the material.

1.3.3.2 Thermal analysis

Differential scanning calorimetry was conducted using a Perkin Elmer DSC 8500. About 5 mg of sample was crimp sealed in 30μL DSC aluminium pans. All samples were heated from 25 to 200 °C at 10 °C/min and kept on isothermal for 5mins before cooling to 25 °C at 10 °C/min. The data collected was analysed using Pyris software version 11.1.1.0492. The reported values were average of three replicates.

1.3.3.3 Wide angle X-ray scattering measurement (WAXS)

WAXS was used to measure the XRD pattern of the blends. WAXS was performed with a philips X-ray diffractometer operating at 40 Kv and 40 mA using CuKα radiation. The diffraction data was collected from 2θ values of 4° to 40° with a step size of 0.013°. A fixed 7.5 mm anti-scatter slit, fixed incidence beam mask of 10mm and a soller slit of 0.04 rad was used. The data collected was baseline corrected from 5-40° and amorphous halo was fitted to this region to determine crystallinity of the blends.

1.3.3.4 Phase Morphology

The phase structure of the blends was investigated using Hitachi S-4700 field emission scanning electron microscope (SEM). The injection moulded specimens were cryo-fractured using liquid nitrogen. The specimens were sputter coated with platinum using Hitachi E-1030 Ion sputter before scanning. For the digested surface, the samples were extracted with chloroform and then rinsed with hot water to remove PLA phase as DNTP is not soluble in chloroform. The extracted surface was dried and sputter coated prior to examination.

1.4 Results and Discussion

1.4.1.1 Morphological Characterisation

The facture and digested morphologies of the DNTP/PLA or PLA-g-IA blends samples with different DNTP formulations are shown in Figure 1 and Figure 2 respectively.

The cryo-fractured surface of blends without itaconic anhydride (IA) showed a relatively large and unevenly distributed DNTP agglomerates because of the inherent immiscibility. However, addition of PLA-g-IA revealed a much finer and more homogenously dispersed matrix indicating improved mixing and good interfacial adhesion as the DNTP agglomerate observed in the blends without IA are not distinct. Interstices were clearly observed in the blends without IA, indicating poor interfacial adhesion. Some interstices were still visible in the blends with PLA-g-IA but they are much smaller and fewer compared to the blends without IA. PLA is hydrophobic with poor polarity while DNTP has strong polarity and is hydrophilic. Therefore, the poor interfacial adhesion between the two phases led to coarse phase structure for both phases in the material matrix.

The clear phase separation observed in the blends without IA, is an indication of poor blend compatibility. Blending different formulation of DNTP with PLA-g-IA improved the morphology of the blend significantly as the reduction of the DNTP rich phase size was observed. In the modified PLA blend, the two polymers are proposed to form co-continuous phases, one that is rich in DNTP and another that is rich in PLA with sufficient adhesion between both phases. This is further investigated through SEM digested surface morphology of the blends.

Larger voids were observed in blends without IA (Figure 2) indicating a poor dispersion of PLA in the blends, corresponding to the DNTP agglomerates observed in Figure 1 for blends without IA. Blends with PLA-g-IA exhibited a much finer phase structure with relatively uniform voids having small dimension. This fine structure confirms an improved interfacial adhesion between the two phases are suggested in Figure 1 with PLA-g-IA. This suggests that blending immiscible polymer without compatibilization will result in a material with undesired properties due to the high interfacial tension between both material phases. F3P, DP and DgP disintegrated in chloroform suggesting that DNTP was the dispersed phase while PLA formed a continuous phase in the blends matrix.

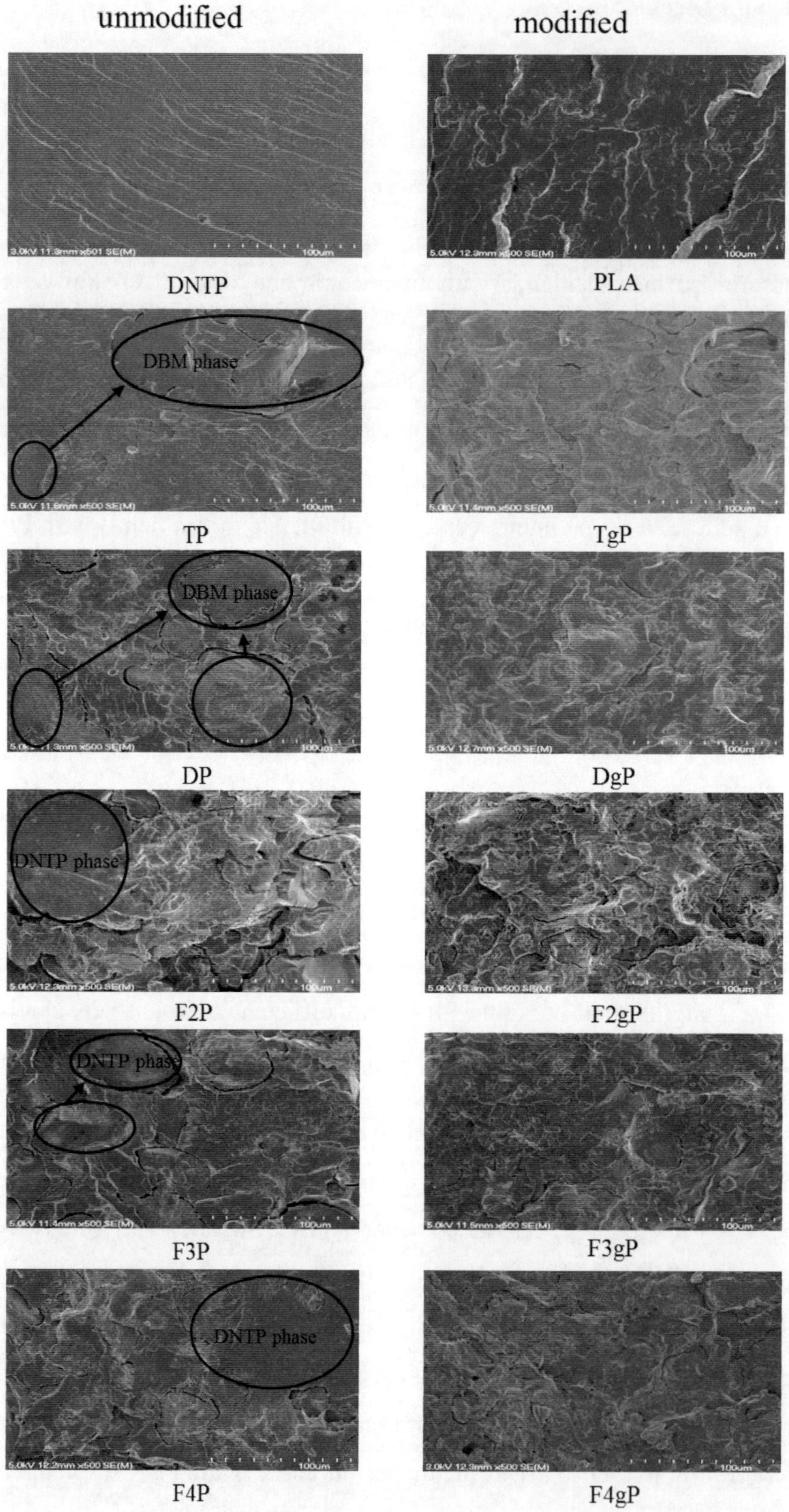

Figure 1: SEM cryo-fracture surface of DNTP/PLA blends with and without Itaconic anhydride. (DNTP) Decoloured Novatein, (TP) TEG plasticized decoloured bloodmeal, (DP) dried decoloured bloodmeal, (F2, F3 and F4) varied additives for DNTP

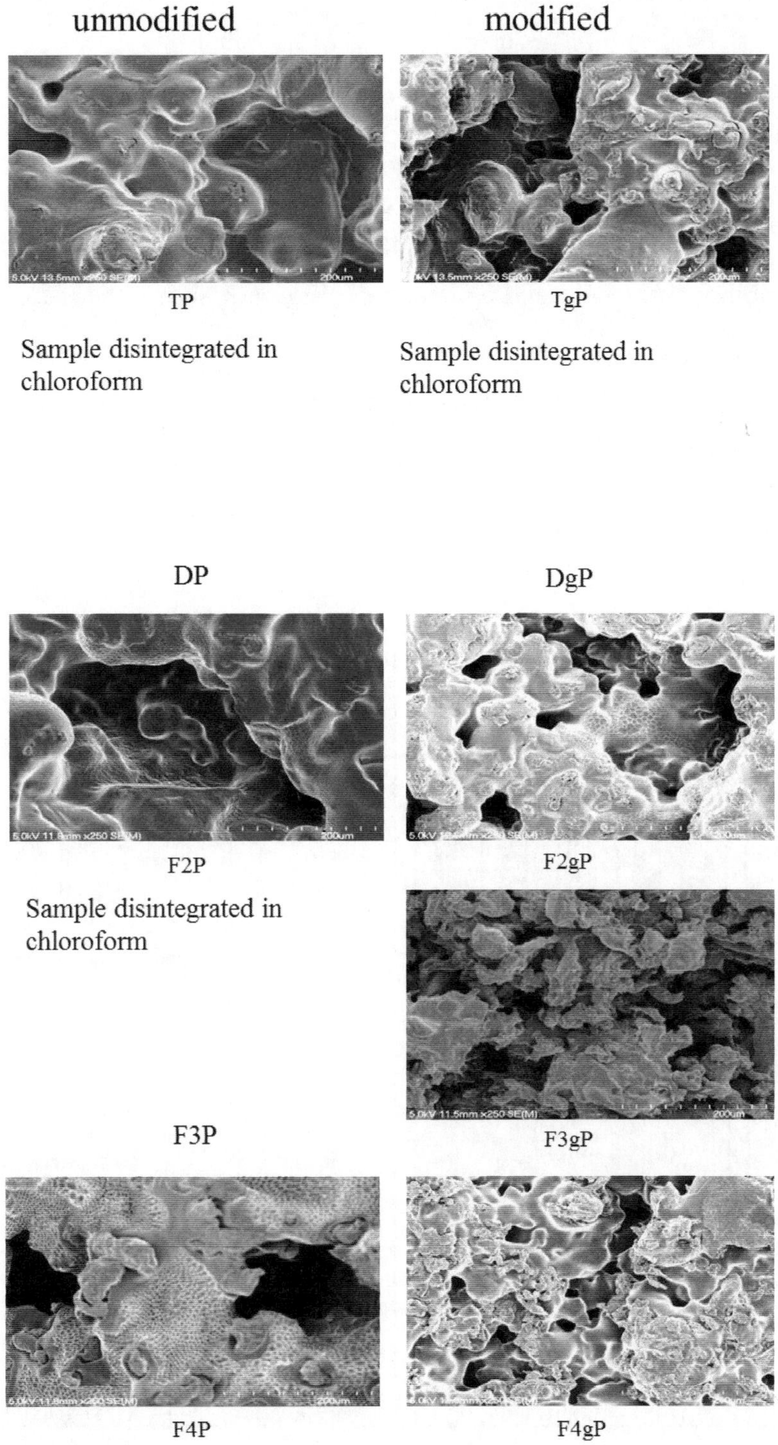

Figure 2: SEM Digested surface micrograph of DNTP/PLA blends with and without Itaconic anhydride. (DNTP) Decoloured Novatein, (TP) TEG plasticized decoloured bloodmeal, (DP) dried decoloured bloodmeal, (F2, F3 and F4) varied additives for DNTP

1.4.1.2 Mechanical properties

Properties of polymer blends are strongly influenced by its morphology. Materials performance during processing, storage and handling can be predicted through its mechanical properties as a function of flexibility, toughness and elongation. Figure **3** shows the mechanical properties of neat materials and blends (with and without PLA-g-IA). DNTP and PLA are known to be rigid polymers and show brittle behaviour which was reflected in the blend's elongation and modulus. The mechanical properties of blends were compared with that of the neat DNTP.

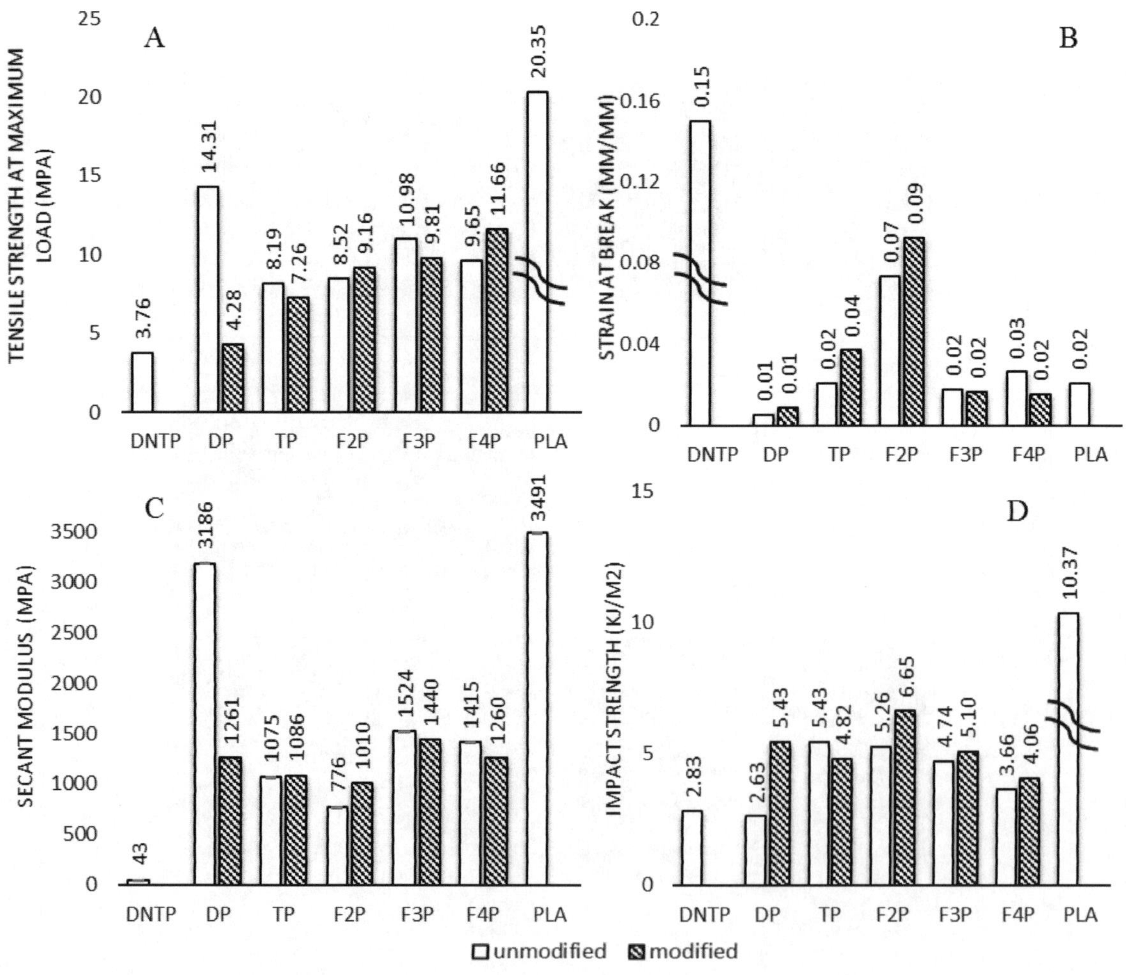

Figure 3: Mechanical property of DNTP/PLA blends with and with Itaconic anhydride

Tensile strength and modulus of both blends with and without PLA-g-IA increased while the elongation decreased when compared to neat DNTP. However, there was no significant difference observed in the mechanical properties of both compatibilized blends compared to the uncompatibilized blends except for DP blend which showed a drastic decrease in the tensile strength of the modified blends. This can be attributed to the interfacial adhesion as affected by the addition of Itaconic anhydride. It is thought that itaconic anhydride improved the interfacial adhesion between the both phases as observed in Figure 1 (modified). However, the interaction between both phases was weak resulting in slightly low or insignificant increase in tensile strength compared to uncompatibilized blends. The agglomerates of dispersed DNTP phase (Figure 1 unmodified) in DP blend are believed to act as stress concentrations, absorbing a similar level of stress as the PLA phase (i.e. stress was effectively transferred between both phases) resulting in high strength. Blending of DNTP and PLA led to a reduction in elongation compared to DNTP. This is due to the addition of the rigid PLA which restricted the movement of polymer chains, thereby decreasing the elongation at break. The interactions between the phases may also have caused a reduction in elongation. Marsilla et al, suggested same with the blends of protein thermoplastic and polybutylene succinate blends [17].

The secant modulus increased compared to neat DNTP suggested that interfacial adhesion was not sufficient to maintain the stiffness of each blend's components. An increase in impact strength of blends compared to neat DNTP was observed suggesting the blends ability to absorb energy before rupturing.
F2P modified blend had considerable increase in elongation at break exceeding that of all blends, an increase in tensile strength and impact strength more than double fold compared to neat DNTP with a lower modulus more than all the blends. This suggests better compatibility between PLA and this DNTP formulation.

1.4.1.3 Differential scanning calorimetry (DSC)

DSC thermograms of neat PLA, DNTP and their blends are shown in Figure 4, Figure 5 and there results are listed in Table 2. The glass transition, cold crystallization and melting endotherm can be observed in the DSC curve of neat PLA and each blend. However, there was no exothermic peak in the DSC curve for DNTP. Small water molecules act as plasticizers in the protein system reducing protein exothermic temperature [22, 34]. A single peak is observed for all blends suggesting a degree of miscibility. Double Tg have been reported as evidence of immiscibility by other researchers [35, 36]. The presence of a single peak in the cold crystallization and double-melting peaks for DNTP/PLA blends suggests that DNTP acted as a heterogenous nucleating agent for the PLA matrix, inducing and accelerating PLA crystallization. A glass transition temperature of 49.8 °C and cold crystallization temperature of 102.86 °C were observed for DP (dried bloodmeal/PLA blend) suggesting slower crystallization kinetics and the involvement of less flexible macromolecules. Similar Tg and cold crystallization temperature have been observed for other protein/ PLA blends [22]. However, with the addition of itaconic anhydride both Tg and cold crystallization shifted to lower temperature of 47.94 °C and 95.81 °C respectively. The blends of DNTP/PLA showed a slight increase in both Tg and cold crystallization temperature with the addition of Itaconic anhydride, except for F4P with an insignificant decrease in T_g. The increase observed is attributed to the restricted slippage and mobility of the PLA macromolecular segment due to an increased interaction between PLA and DNTP phases.

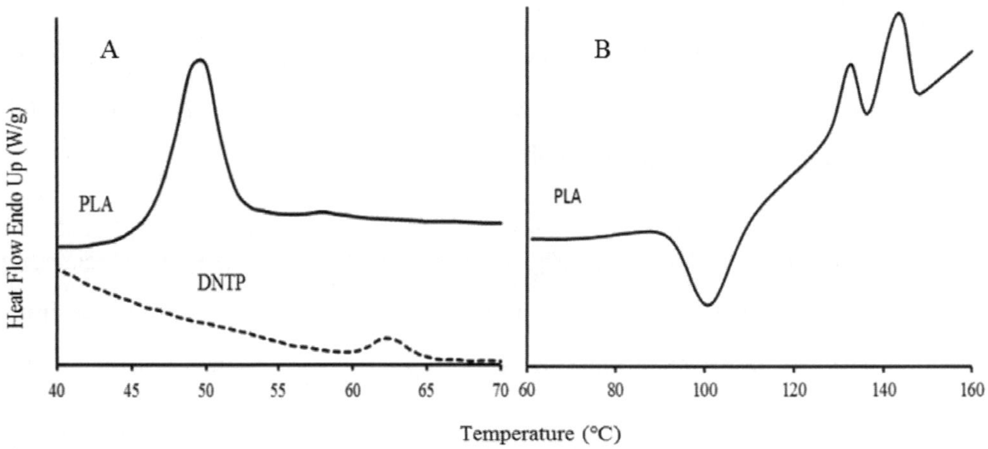

Figure 4: Glass transition temperature (T_g), crystallization peak (T_{cc}) and melting endotherms (T_m) of Pure PLA and DNTP

The PLA in the compatibilized blends exhibited cold crystallization at lower temperature compared to the neat PLA and uncompatibilized blends. This suggested that PLA-g-IA accelerated the crystallization of PLA and substantially increased PLA crystallinity. The melting of the PLA in the blend was reflected by the bimodal endotherm transition observed at ~ 120 – 145 °C. The peak at lower temperature reflects the melting of less paragon crystal in the boundary regions followed by the melting of recrystallized PLA at higher temperature [13, 35]. The ΔH_m values observed in the blends (Table 2) are larger than the ΔH_{cc} values. This indicates that the addition of DNTP promoted a certain degree of PLA crystallinity. Crystallization and melting temperatures of all blends showed a decrease compared to neat PLA except for DP with a T_g slightly higher than neat PLA. This could be attributed to the residual moisture in the blends which plasticizes the PLA molecules and increases its flexibility resulting in lower crystallization and melting temperatures. Zhang et al, observed a continuous decrease in T_g, T_{cc} and T_m with an increase in moisture content of SPI/PLA containing 3 phr PEOX conditioned at different relative humidity [13]. The blends showed a decrease in ΔH_{cc} with the addition of IA suggesting that Itaconic anhydride probably reduced the interfacial tension as a result of interaction between phases such as hydrogen bonding promoting the crystallization of PLA.

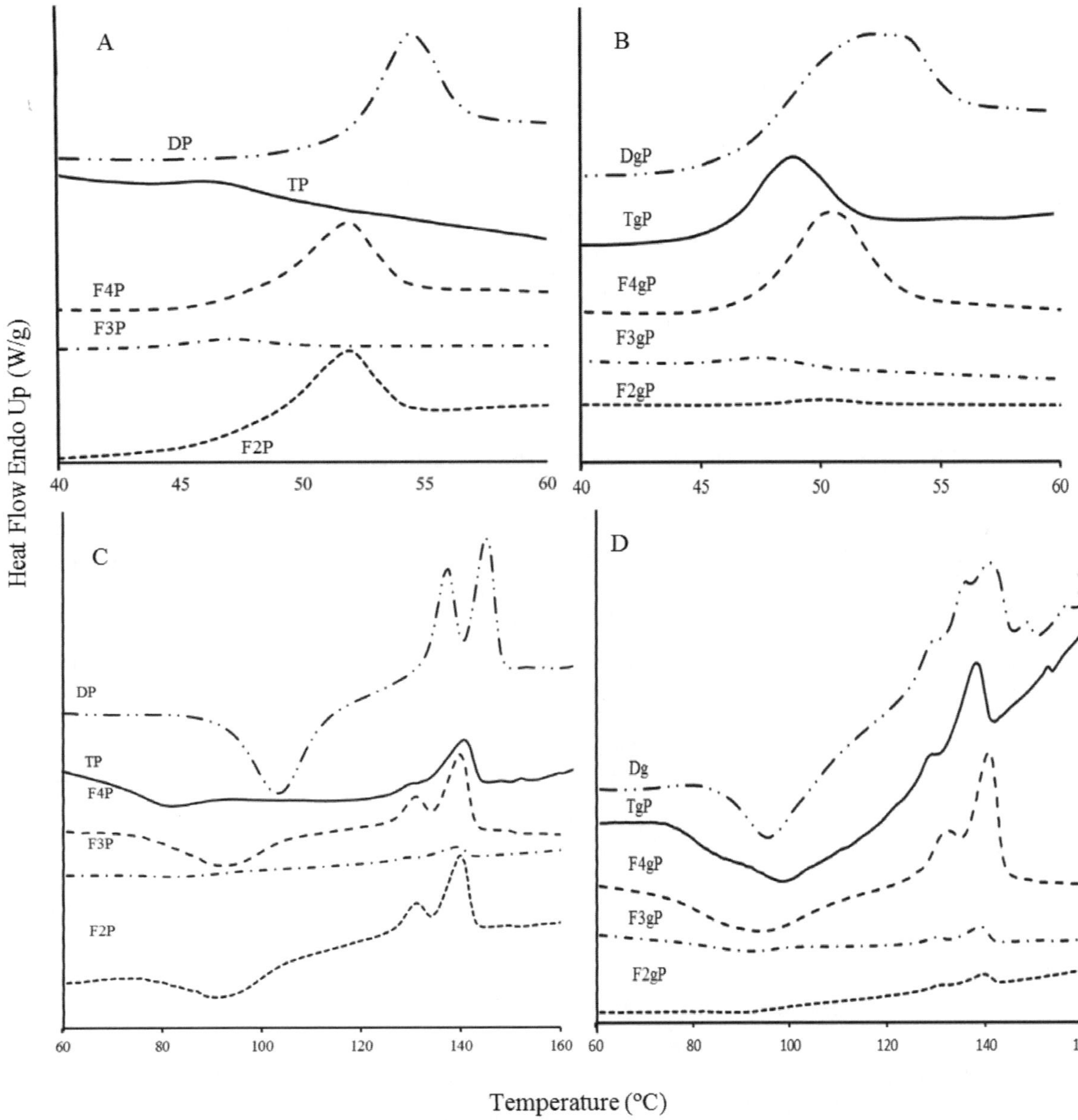

Figure 5: DSC first heat scan thermograms of different DNTP/PLA blends showing glass transition temperature (Tg), crystallization peak (Tcc) and melting endotherms (Tm) of the blends. (DNTP) Decoloured Novatein, (TP) TEG plasticized bloodmeal, (DP) dried bloodmeal, (F2, F3 and F4) DNTP with varying additives, (P) Neat PLA and (gPLA) modified PLA

Table 2: Glass transition temperature, cold crystallization peak and melting endotherms of blends

		cold crystallization		melting T_m (°C)		
Sample	T_g (°C)	T_{cc} (°C)	ΔH_{cc}^{a}(J/g)	1	2	ΔH_m(J/g)
PLA[a]	41.78	99.48	-2.736	131.81	142.89	2.4303
DNTP	60.5			nd	nd	
DP	49.8	102.86	-3.122	135.6	142.27	3.4074
DgP	47.94	95.81	-2.456	135.55	140.62	1.6384
TP	39.53	81.32	-0.6967	124	138.78	1.0073
TgP	45.51	99.38	-3.09	128.49	138.19	0.8297
F2P	47.37	92.08	-10.133	130.97	140.06	10.8316
F2gP	47.43	92.2	-0.6146	130.18	139.32	0.7878
F3P	43.59	83.98	-0.9813	129.29	139.21	1.601
F3gP	44.14	91.42	-0.6898	129.79	138.92	1.1314
F4P	47.51	91.58	-11.068	130.97	140.06	11.0577
F4gP	47.47	91.75	-5.3824	131.8	141.19	10.4048

[a] Data collected on the second heat scan of samples, T_g: glass transition temperature, T_{cc}: cold crystallization temperature, T_m: melting endotherm

1.4.1.4 Wide angle X-ray scattering (WAXS) measurement

WAXS analysis has been performed on the blends for better insight into the blend morphology. Data obtained for the XRD of a material can be used to study the material's phase change. Figure 6 presents the wide-angle X-ray diffractogram for neat PLA and DNTP, and blends of DNTP/PLA with and without compatibilizer. It can be observed the neat PLA shows only amorphous halo.

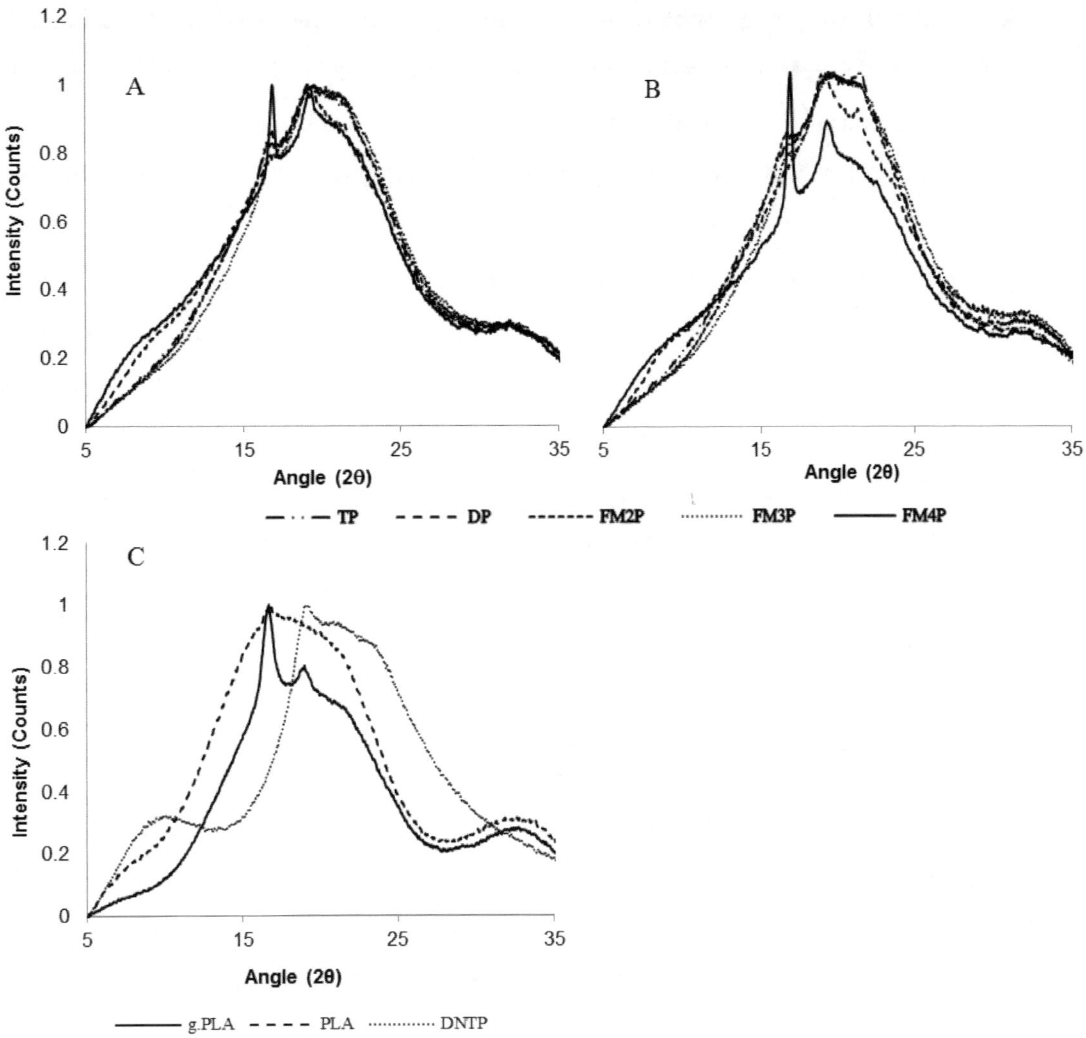

Figure 6: WAXS diffraction patterns for DNTP/PLA blends with and without Itaconic anhydride

Clear crystalline peaks were observed at 16° and 20° for PLA-g-IA, the peak at 2θ = 16° has been reported to be associated with (110)/(020) which is a reflection of α-form homo-crystal structure and 2θ = 22° was indexed as (100)/(-120)/(-210) which is a reflection of stereo-complex crystals [37-42]. The formation of stereo-complex crystal as well as α-form homo-crystal structure observed in PLA-g-IA suggests that significant chain orientations occurred during the grafting of PLA because of hydrolytic degradation probably due to decrease in molecular weight. These peaks were very strong and sharp in blends with formulation 4. This indicates the presence of significant crystallinity in the blends. The presence of stable crystal structure observed has been suggested to be evidence of some degree of PLA hydrolytic degradation during modification which is probably, reflected in the low tensile properties observed for the blends modified with itaconic anhydride Figure 3.

The peaks at 2θ = 9° observed in the diffractogram of DNTP corresponds to helical spacing and inter ß-sheet while the peak at 2θ = 20° corresponds to the repeated distance within each structure. The amorphous region of the compatibilized blends showed a slight reduction probably due to the integration of both PLA and DNTP amorphous regions as confirmed by the reduction of DNTP agglomerate in SEM thermogram Figure 1.

Peaks were observed at 2θ, 19° and 20° with an appearance of a shoulder at 22° for F2P, F3P and TP blends. However, with the addition of Itaconic anhydride, the peak at 19° became slightly broader for F3P and TP, and the shoulder at 22° for TP became clearer. There was a slight shift in the 2θ of both peaks for TP blend. A sharp peak at 2θ 19° and shoulders at 16° and 20° were observed for DP blends. The shoulder at 16° became broader and an appearance of a shoulder at 9° was observed while the shoulder observed at 2θ 22° became clearer. F4P showed a sharp peak at 2 16° and another clear peak at 19°. The peaks became sharper and had reduced intensity with the addition of itaconic anhydride.

1.5 Conclusion

Blends of PLA/DNTP with varying additives were successfully prepared through reactive extrusion. The processibility and flowability of DNTP increased with the addition of PLA and was injection mouldable without processing aids. Co-continuous phase morphology was observed in the blends. The mechanical properties of DNTP were greatly enhanced by the addition of PLA. The mechanical properties, SEM and DSC indicated compatibility of DNTP and PLA-g-IA. Adding itaconic anhydride led to improved mixing of the two phases, which resulted in finer phase structure. However, no significant effect was observed in the blend's mechanical properties with the addition of itaconic anhydride. This was attributed to itaconic anhydride, improving the dispersion of DNTP as observed in SEM thereby enhancing the interfacial adhesion. However, the interfacial adhesion is thought to be weak resulting in low mechanical properties compared to the uncompatibilized blends. All blends showed low elongation at break and brittle fracture failure in tensile testing, suggesting a gap for toughening and plasticization in future investigation. However, the data obtained from mechanical properties F2P modified with Itaconic anhydride seems to be the best DNTP formulation for blend with PLA as it showed better tensile strength, elongation at break and flexibility favourable for sheet processing. The crystallization temperature of compatibilized blends occurred at higher temperature indicating reduction of chain mobility.

DNTP /PLA blends showed an improved thermal stability compared to neat DNTP. The addition of Itaconic anhydride showed a further improvement in thermal stability of DNTP/PLA blends. WAXS of DNTP/PLA blends showed the formation of stereo-complex crystal as well as α-form homo-crystal structure on the PLA in the blends, suggesting the occurrence of significant chain orientations due to PLA hydrolysis. With the addition of Itaconic anhydride, the presence of these crystal structure became clearer which is thought to be responsible for the insignificant effect of Itaconic anhydride on the mechanical properties of DNTP/PLA blends. This suggests the manipulation of processing and blend conditions to better control for hydrolysis of PLA in the blends.

1.6 References

1. Rhim, J.-W., et al., *Preparation and properties of biodegradable multilayer films based on soy protein isolate and poly (lactide).* Industrial & engineering chemistry research, 2006. **45**(9): p. 3059-3066.
2. Rhim, J.-W., J.H. Lee, and P.K. Ng, *Mechanical and barrier properties of biodegradable soy protein isolate-based films coated with polylactic acid.* LWT-Food Science and Technology, 2007. **40**(2): p. 232-238.
3. Briassoulis, D., *An overview on the mechanical behaviour of biodegradable agricultural films.* Journal of Polymers and the Environment, 2004. **12**(2): p. 65-81.
4. Petersen, K., et al., *Potential of biobased materials for food packaging.* Trends in food science & technology, 1999. **10**(2): p. 52-68.
5. Grant, R., *Applied protein chemistry*. 1980: Applied Science Publishers.
6. Pickering, K.L., et al., *Plastics material*. 2012, Google Patents.
7. Low, A., C.J.R. Verbeek, and M.C. Lay, *Treating Bloodmeal with Peracetic Acid to Produce a Bioplastic Feedstock.* Macromolecular Materials and Engineering, 2014. **299**(1): p. 75-84.
8. van der Merwe, D.W., *The Environmental and Economic Impacts for Producing the Port Jackson from Novatein®*. 2014, University of Waikato.
9. Walallavita, A.S., C.J.R. Verbeek, and M.C. Lay, *Biopolymer foams from Novatein thermoplastic protein and poly (lactic acid).* Journal of Applied Polymer Science.
10. Gavin, C., M.C. Lay, and C.J. Verbeek. *Extrusion foaming of protein-based thermoplastic and polyethylene blends*. in *AIP Conference Proceedings*. 2016. AIP Publishing.
11. Low, A., C.J.R. Verbeek, and M.C. Lay, *Processing peracetic acid treated bloodmeal into bioplastic.* Chemeca 2012: Quality of life through chemical engineering: 23-26 September 2012, Wellington, New Zealand, 2012: p. 1433.
12. Low, A., *Decoloured bloodmeal based bioplastic*. 2012, University of Waikato.
13. Zhang, J., et al., *Morphology and properties of soy protein and polylactide blends.* Biomacromolecules, 2006. **7**(5): p. 1551-1561.
14. Zhong, Z. and S.X. Sun, *Properties of soy protein isolate/poly (ethylene ‐ co ‐ ethyl acrylate ‐ co ‐ maleic anhydride) blends.* Journal of applied polymer science, 2003. **88**(2): p. 407-413.
15. Li, Y.-D., et al., *Structure and properties of soy protein/poly (butylene succinate) blends with improved compatibility.* Biomacromolecules, 2008. **9**(11): p. 3157-3164.
16. Mungara, P., et al., *Processing and physical properties of plastics made from soy protein polyester blends.* Journal of Polymers and the Environment, 2002. **10**(1): p. 31-37.
17. Ku ‐ Marsilla, K. and C. Verbeek, *Compatibilization of Protein Thermoplastics and Polybutylene Succinate Blends.* Macromolecular Materials and Engineering, 2015. **300**(2): p. 161-171.

18. Marsilla, K. and C.J.R. Verbeek, *Properties of Bloodmeal/Linear Low‐density Polyethylene Blends Compatibilized with Maleic Anhydride Grafted Polyethylene.* Journal of Applied Polymer Science, 2013. **130**(3): p. 1890-1897.
19. Wang, L., R. Shogren, and C. Carriere, *Preparation and properties of thermoplastic starch‐polyester laminate sheets by coextrusion.* Polymer Engineering & Science, 2000. **40**(2): p. 499-506.
20. Zhu, R., H. Liu, and J. Zhang, *Compatibilizing effects of maleated poly (lactic acid)(PLA) on properties of PLA/Soy protein composites.* Industrial & Engineering Chemistry Research, 2012. **51**(22): p. 7786-7792.
21. Garlotta, D., *A literature review of poly (lactic acid).* Journal of Polymers and the Environment, 2001. **9**(2): p. 63-84.
22. Fang, K., et al., *Properties and morphology of poly (lactic acid)/soy protein isolate blends.* Journal of applied polymer science, 2009. **114**(2): p. 754-759.
23. Huneault, M.A. and H. Li, *Morphology and properties of compatibilized polylactide/thermoplastic starch blends.* Polymer, 2007. **48**(1): p. 270-280.
24. Aoi, K. and M. Okada, *Polymerization of oxazolines.* Progress in polymer science, 1996. **21**(1): p. 151-208.
25. Wang, H., X. Sun, and P. Seib, *Strengthening blends of poly (lactic acid) and starch with methylenediphenyl diisocyanate.* Journal of Applied Polymer Science, 2001. **82**(7): p. 1761-1767.
26. Carlson, D., et al., *Maleation of polylactide (PLA) by reactive extrusion.* Journal of applied polymer science, 1999. **72**(4): p. 477-485.
27. Nyambo, C., A.K. Mohanty, and M. Misra, *Effect of maleated compatibilizer on performance of PLA/wheat Straw‐Based green composites.* Macromolecular Materials and Engineering, 2011. **296**(8): p. 710-718.
28. Zhang, J.-F. and X. Sun, *Mechanical properties of poly (lactic acid)/starch composites compatibilized by maleic anhydride.* Biomacromolecules, 2004. **5**(4): p. 1446-1451.
29. Orozco, V.H., et al. *Preparation and characterization of poly (Lactic acid)‐g‐maleic anhydride+ starch blends.* in *Macromolecular symposia*. 2009. Wiley Online Library.
30. Marsilla, K.K. and C. Verbeek, *Modification of poly (lactic acid) using itaconic anhydride by reactive extrusion.* European Polymer Journal, 2015. **67**: p. 213-223.
31. Hicks, T.M., et al., *The role of peracetic acid in bloodmeal decoloring.* Journal of the American Oil Chemists' Society, 2013. **90**(10): p. 1577-1587.
32. Properties, A.S.D.o.M. *Standard test method for tensile properties of plastics*. 1996. American Society for Testing and Materials.
33. ISO, E., *179-1 (2010)."* Plastics. Determination of Charpy impact properties. Part. **1**.
34. Mo, X., X.S. Sun, and Y. Wang, *Effects of molding temperature and pressure on properties of soy protein polymers.* Journal of applied polymer science, 1999. **73**(13): p. 2595-2602.
35. Jiang, L., M.P. Wolcott, and J. Zhang, *Study of biodegradable polylactide/poly (butylene adipate-co-terephthalate) blends.* Biomacromolecules, 2006. **7**(1): p. 199-207.
36. Signori, F., M.-B. Coltelli, and S. Bronco, *Thermal degradation of poly (lactic acid)(PLA) and poly (butylene adipate-co-terephthalate)(PBAT) and their blends upon melt processing.* Polymer degradation and stability, 2009. **94**(1): p. 74-82.
37. Mihai, M., et al., *Extrusion Foaming of Semi‐Crystalline PLA and PLA/Thermoplastic Starch Blends.* Macromolecular bioscience, 2007. **7**(7): p. 907-920.
38. Gorrasi, G. and R. Pantani, *Effect of PLA grades and morphologies on hydrolytic degradation at composting temperature: assessment of structural modification and kinetic parameters.* Polymer degradation and stability, 2013. **98**(5): p. 1006-1014.
39. Zhang, X., et al., *Morphological behaviour of poly (lactic acid) during hydrolytic degradation.* Polymer Degradation and Stability, 2008. **93**(10): p. 1964-1970.
40. Furuhashi, Y., Y. Kimura, and H. Yamane, *Higher order structural analysis of stereocomplex‐type poly (lactic acid) melt‐spun fibers.* Journal of Polymer Science Part B: Polymer Physics, 2007. **45**(2): p. 218-228.
41. Tsuji, H., H. Takai, and S.K. Saha, *Isothermal and non-isothermal crystallization behavior of poly (l-lactic acid): effects of stereocomplex as nucleating agent.* Polymer, 2006. **47**(11): p. 3826-3837.

42. Mittal, V., et al., *PLA, TPS and PCL binary and ternary blends: structural characterization and time-dependent morphological changes.* Colloid and Polymer Science, 2015. **293**(2): p. 573-585.

Thermoplastic Starch/PBAT Blends Prepared by Solid-State Shear Pulverization

Lopes Henrique Solowej Medeiros [(1)*], Lucas Alessandra Almeida [(1)], Ambrósio José Donato [(2)], Oliveira Gean Henrique Marcatto [(1)], Ferreira Caio Moreira [(1)]

[(1)] UFSCAR - São Carlos SP - Brazil, [(2)] CCDM - UFSCAR - São Carlos SP - Brazil

Starches have being investigated as renewable resource alternative material for several materials applications, mainly packaging, because of its properties as biodegradability, low cost and abundance. However, starch has low mechanical properties and is highly hydrophilic. Thermoplastic starch blends are a key for these issues, but the amount of starch is limited due to the high viscosity of starch. A novel processing technique were employed in this work aiming a better blend processing, with improved compatibilization and easier processing at high starch concentrations. Blends of thermoplastic starch (TPS) and poly(butylene adipate-co-terephthalate) (PBAT) were prepared by Solid-State Shear Pulverization (SSSP or S3P), at room temperature, and conventional twin screw extrusion in different compositions (80, 70, 60 and 50% w/w of starch), both in the same twin screw extruder. Native cassava starch plasticization process was performed during 12 hours in an oven at 90 °C with 25% of glycerol and 10% of water (w/w) with further plasticization in twin screw extruder to obtain TPS, process known as Dry Blend. The processability and visual appearance of films processed by S3P showed to be better, easier to handle and more stable. Preliminary results of mechanical properties showed S3P presented a greater elastic modulus in materials containing higher amounts of starch (80% w/w), but a smaller tensile strength and elongation at break. Materials prepared by S3P showed a higher or equal water absorption compared to those processed directly by conventional twin screw extruder. Retrogradation of both materials showed very close results. Nevertheless, conventionally extruded films are a little less crystalline when compared to those processed by S3P. Analysis of retrogradation as a function of time, morphology and thermal-mechanical properties, are programmed for a better comparison between this two processing techniques.

Modification of Poly (lactic acid) with amine functional reactive elastomer via solvent-cast 3D printing (SC-3DP)

***Rissato Dognani de Oliveira Lucas* [1]*, *Morales Ana Rita* [1], *Carajelescov Nonato Renato* [1]**

[1] UNICAMP - São Paulo - Brazil

The effects of the addiction of amine functional reactive elastomer on Poly (lactic acid) (PLA) were investigated. The blends were produced varying the elastomer concentrations (8%, 14% and 20%) using chloroform as solvent, via solvent-cast 3D printing (SC-3DP) as films. Chemical interactions, morphology, thermal and mechanical properties of the PLA/elastomer blends were evaluated by Fourier transform infrared spectroscopy (FTIR), scanning electronic microscopy (SEM), differential scanning calorimetry (DSC), termogravimetry (TGA), dynamic mechanical analysis (DMA) and tensile strength test. IR spectra of the blends showed interactions between PLA carboxyl groups and elastomer amine groups that lead to the formation of amide groups. The increase of the vibration bands intensity of the carbonyl and hydroxyl groups suggests that hydrolysis effect was strong throughout the reaction. Thus, the water formed as a reaction subproduct acted like an expansion agent creating a porous morphology on the blends, as observed in the SEM images. According to the thermal properties analyses, the blends showed less thermal stability and higher degree of crystallinity than the neat PLA. The 8% PLA/Elastomer blend showed the higher crystallinity value, 4.2 times greater than that for neat PLA. However, the crystallinity of the blends reduced as the elastomer content was increased. The increase of elastomer content shifted the glass transition (Tg) of the PLA rich phase to lower temperatures while the Tg of the elastomer rich phase tended to be at higher temperatures. The cold crystallization temperature of PLA changed also to lower temperatures that could be due to the nucleation effect of the elastomer. Mechanical properties analysis showed no significant changes, except for the elongation that reduced by 50% compared to the neat PLA film, which could be explained by the foam structure formed during the 3D printing.

Properties of ABS / PVC blend expanded with azodicarbonamide and its combination with sodium bicabornate and oxibis(benzene-sulfonyl-hidrazine)

da Silveira Marcio [1], Bischoff Eveline [2], Simon Douglas [2], Francisquetti Edson [2]*

[1] IFRS- Grendene - Farroupilha-RS - Brazil, [2] IFRS - Farroupilha-RS - Brazil

The expansion of a polymer is the process in which gas cells are introduced into a product during processing, been consider an alternative to reduce the density of the material. This work evaluates samples injected with a blend of ABS/PVC in the ratio 60/40 w/w, using azodicarbonamide as its blowing agent in combination with sodium bicarbonate and oxybis(benzene-sulfonyl-hydrazine) as a lightweight alternative to ABS for industrial applications. Scanning electron microscopy evaluation shows that the samples have skin thicknesses and cell size distribution highly dependent of composition, with the results indicating that azodicarbonamide in combination with oxybis (benzenesulfonylhydrazine) has demonstrated better skin uniformity and uniformity in cell size distribution. Furthermore, the compression test also has shown better properties for expansion with azodicarbonamide combined with oxybis(benzene-sulfonyl-hydrazine).

Influence of Devulcanization and Revulcanization of Ground Tire Rubber in Dynamic Mechanical Properties of Blends Ground Tire Rubber/High Density Polyethylene

Fabiula Danielli Bastos de Sousa[a,b], Guo-Hua Hu[b], Sandrine Hoppe[b] and Carlos Henrique Scuracchio[c]

[a]Technology Development Center - CDTec, Universidade Federal de Pelotas - UFPel, Rua Gomes Carneiro, 1, 96010-610 Pelotas, RS, Brazil
[b]CNRS - Université de Lorraine, Laboratoire Réactions et Génie des Procédés - LRGP, 1 Rue Grandville, Nancy BP20451, 54001, France
[b]Materials Engineering Department - DEMa, Universidade Federal de São Carlos - UFSCar, Washington Luís (SP-310), Km 235, 13565-905 São Carlos, SP,Brazil

Abstract. *Ground tire rubber (GTR) was devulcanized in a system comprised of a conventional microwave oven adapted with a motorized stirring system with speed control by 5.5 minutes. Blends composed of GTR/high density polyethylene (HDPE) were prepared in a twin-screw extruder with two feeders at 180°C. The blends were produced in different concentrations, and both the effect of devulcanization and the dynamic revulcanization of GTR phase in the dynamic mechanical properties of the blends were analyzed. As main results, the devulcanization of GTR phase changed the crystallinity of HDPE phase of some blends, possibly because of changes in sizes of particles of devulcanized GTR and its distribution in the matrix. Concerning the revulcanization of the elastomeric phase, generally, it was observed the shift of the glass transition temperature (T_g) peak of the GTR to higher temperatures, and reduction in the area under the peak of tan δ for the T_g of the same phase, as a result of the restriction on the mobility of the GTR by dynamic revulcanization.*

Keywords: Devulcanization, Revulcanization, Ground Tire Ruber, Dynamic Mechanical Properties.

INTRODUCTION

In no other phase of human development was produced such amount of waste as currently. Its composition and quantity are directly related to the way the population lives, socioeconomic condition and the ease of access to consumer goods. The irregular disposition of such waste may cause harmful environmental impacts. One of the most dangerous solid wastes is the vulcanized rubber, which besides having high natural degradation time, has many chemical additives on its formulation and the possibility to store rain water when disposed in landfills, and may become breeding of vectors. So, recycling comes against this problem and devulcanization a way of recycling that restores the fluidity of the rubber. One of the applications of this devulcanized rubber is in the production of polymeric blends, like dynamically revulcanized ones.

EXPERIMENTAL

Materials and Methods

Materials

Ground waste truck tire (GTR) separated from non-elastomeric components was kindly supplied by Pirelli Pneus Ltda. As vulcanization additives, were used rubber accelerator TBBS (N-tert-Butyl-2-benzothiazole sulfenamide) and sulphur, both kindly donated by Pirelli Pneus Ltda. The TBBS is agent accelerant and sulphur the vulcanization agent. The HDPE used was a commercial product of Braskem with the trade name IA-59, used for injection molding applications.

Devulcanization of GTR

GTR was devulcanized in a system comprised of a conventional microwave oven adapted with a motorized stirring system with a speed control. The whole devulcanization process was done by using the maximum power of the oven, i.e., 820 W. About 65 g of the GTR sample was put into a 250 mL glass beaker with a stirring speed of 100 rpm. The time at which the material was exposed to microwaves was 5.5 min.

After the devulcanization process, all the samples were homogenized in a laboratory two roll mill and were added 1 phr of TBBS and 1 phr of sulphur, respectively. Shortly after the process, the resulting tacky rubber sheets were cut manually (dimensions of approximately 3 mm x 3 mm) for use in the production of blends in twin screw extruder.

Production of the blends in twin screw extruder

Blends were prepared in a twin-screw extruder of type Process 11 Thermo Scientific with 2 feeders. The temperatures of the barrel zones were fixed at 180°C, except the third zone for the second feeding mode whose temperature was fixed at 120°C to facilitate the introduction of the rubber phase. The screw speed was 250 rpm.

The nomenclature adopted is type X GTR Y+ad/Z HDPE, where Y represents the time of exposure of the GTR to the microwaves, '+ad' the presence of vulcanization additives in the sample, X and Z the amount of GTR and HDPE in the sample (in mass), respectively.

DMA

The dynamic-mechanical properties of polymer blends were analyzed by using a dynamic-mechanical analyzer DMA Q800 of TA Instruments. The analyses were carried out in Single Cantilever mode, 1 Hz of frequency, in heating ramp from -100 to 140°C and heating rate of 3°C/min. Glass transition temperatures of the rubbers were extracted from the maximum value of the tan δ peak.

The samples were previously pressed in a hydraulic press in a rectangular mould with the dimensions of 17,5x12,5x2 mm to 180°C for 5 minutes.

RESULTS AND DISCUSSION

Effect of devulcanization of elastomeric phase: Nonrevulcanized blends

Tan δ curves as function on the temperature of the blends containing GTR0 and GTR5.5 for verification of effect of GTR devulcanization on dynamic-mechanical properties of the nonrevulcanized blends are presented in Figure 1.

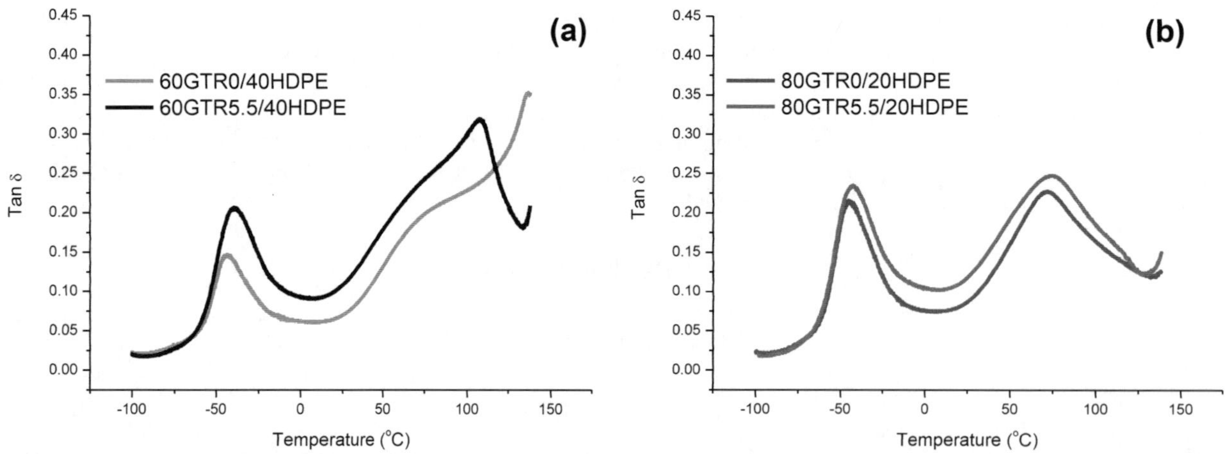

FIGURE 1: Tan δ curves as function of the temperature of the blends: (a) 60GTR/40HDPE, (b) 80GTR/20HDPE, containing GTR0 and GTR5.5 for verification of the effect of devulcanization in dynamic-mechanical properties of the nonrevulcanized blends.

According to Figure 1, there are two transitions on the phases of the blends: around -40° C refers to the T_g of the GTR and the other refers to the α transition of HDPE (T_α), which occurs around 100° C. The existence of two distinct transitions proves the character of immiscible blend. The temperature at which the transition α occurs is

related to the crystallinity degree and the average thickness of the crystallites. It is also related to the occurrence of rotations of small molecular sequences in the crystallites [1], which can be subdivided into smaller subdivisions known as pre-melting temperature [2]. The transition refers to the relaxation of the limits of the grains, a subdivision of α transition known as α ' [3].
According to Molefi et al. [4], the increase of the crystallinity of the thermoplastic phase causes the increase in the intensity of this peak, which is often checked in the case of HDPE. The increase in the intensity of the peaks relating to the transition of the thermoplastic phase was verified in some blends containing GTR5.5 because the devulcanization of the elastomeric phase resulted in particle size smaller and well distributed in the matrix, which may, possibly, have changed the crystallinity of matrix phase. Similarly, variations in the region of curves were also verified related to the thermoplastic phase of blends containing clay, which possibly acted as nucleation points to HDPE [5].
According to Cao et al. [6], the decrease of peak widths of tan δ in the region on the elastomeric phase transition is a result of the reduction of mobility of polymer chains. In the case of blends comprising of devulcanized GTR, extending the width of the peak in temperatures around -50° C as a result of the greater mobility of the GTR due to breaking of three-dimensional structure by devulcanization by microwaves. Also, in relation to the T_g values, were not observed large variations among the blends containing GTR0 and GTR5.5.

Effect of devulcanization and revulcanization of elastomeric phase: Revulcanized blends

Tan δ curves as function on the temperature of the blends containing GTR0 and GTR5.5 for verification of effect of GTR devulcanization in dynamic-mechanical properties of the revulcanized blends are presented in Figure 87.

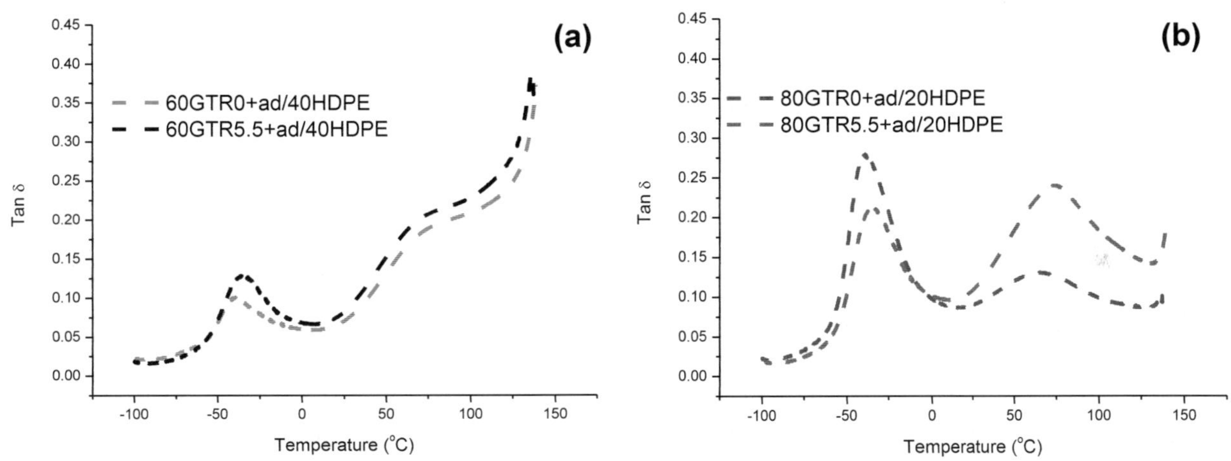

FIGURE 2. Tan δ curves as function of the temperature of the blends: (a) 60GTR+ad/40HDPE, (b) 80GTR+ad/20HDPE, containing GTR0 and GTR5.5 for verification of effect of devulcanization and revulcanization in dynamic-mechanical properties of revulcanized blends.

According to the Figure 2, it was observed a reduction in the area under the T_g peak of GTR phase of blends containing GTR0 due to the restriction on the mobility of the chains by the increase of cross-linking density [7]. Such behavior is due to the probable coexistence of two three-dimensional networks, the first from the vulcanization of the GTR and the second of dynamic revulcanization, what does not mean that it has higher cross-linkings density. This is also due to the increase in the E' values of GTR0 phase, which probably hindered the breaking of the GTR0 particles during processing; resulting in larger rubber particles.

T_g values of GTR phase of the analyzed blends were determined from the maximum points from the first peak of tan δ curves as function of temperature, and they are presented in Table 1.

In general, in the blends containing revulcanized GTR0, T_g values are smaller in comparison to revulcanized blends containing the same concentration of GTR5.5. Due to the devulcanization, the polymeric chains displayed greater freedom during the reactive processing, which probably has increased the number of effective shocks and resulted in increased cross-linkings density of elastomeric phase of blends containing GTR5.5.

Whereas in relation to nonrevulcanized blends, they showed a slight difference among the T_g values of the blends containing or not clay, which is considered to be in the range of the experimental error of the analysis.

TABLE 1. T_g values of the elastomeric phase of revulcanized and non revulcanized blends GTR0/HDPE and GTR5.5/HDPE.

Sample	Tg (°C)
60GTR0/40HDPE	-44.17
60GTR0+ad/40HDPE	-40.40
80GTR0/20HDPE	-45.46
80GTR0+ad/20HDPE	-39.63
60GTR5.5/40HDPE	-39.39
60GTR5.5+ad/40HDPE	-35.65
80GTR5.5/20HDPE	-43.19
80GTR5.5+ad/20HDPE	-34.83

CONCLUSIONS

In general, the devulcanization of GTR phase changed the crystallinity of HDPE phase of some blends, possibly due to changes in sizes of particles of devulcanized GTR and its distribution in the matrix. Concerning the revulcanization of the elastomeric phase, generally, it was observed the shift of the glass transition temperature (T_g) peak of the GTR to higher temperatures, and reduction in the area under the peak of tan δ for the T_g of the same phase, as a result of the restriction on the mobility of the GTR by dynamic revulcanization.

ACKNOWLEDGMENTS

The authors would like to thank Pirelli Pneus for the material donation. CAPES, FAPESP (process number 2010/15799-6) and CNPq (process number 201891/2011-5) for the financial support.

REFERENCES

1. Peacock AJ (2000) Handbook of polyethylene: Structures, properties and applications. Marcel Dekker, United States
2. <www.thermalsupport.com>, access in 19/9/13.
3. <http://www.hitachi-hitec-science.com/en/documents/technology/thermal_analysis/application_DMS_005e.pdf>, access in 03/7/14.
4. Molefi JA, Luyt AS, Krupa I (2010) Comparison of the influence of copper micro- and nano-particles on the mechanical properties of polyethylene/copper composites. J Mater Sci 45:82–88.
5. Tsai Y, Wu JH, Wu YT, et al (2008) Reinforcement of dynamically vulcanized EPDM/PP elastomers using organoclay fillers. Sci Technol Adv Mater.
6. Cao LM, Cao XD, Jiang XJ, et al (2013) In situ reactive compatibilization and reinforcement of peroxide dynamically vulcanized polypropylene/ethylene-propylene-diene monomer tpv by zinc dimethacrylate. Polym Compos 34:1357–1366.
7. Babu RR, Singha NK, Naskar K (2011) Effects of mixing sequence on peroxide cured polypropylene (PP)/ethylene octene copolymer (EOC) thermoplastic vulcanizates (TPVs). Part. II. Viscoelastic characteristics. J Polym Res 18:31–39.

SYNTHESIS AND CHARACTERIZATION OF NANOPARTICLES OF PVDF/PMMA BLENDS FOR PHOTOVOLTAIC BACKSHEET COATING

HA Jong Wook (1), PARK In Jun (1), LEE Soo Bok (1)*, KIM Kyung Jae (2)

(1) Korea Research Institute of Chemical Technology - Daejeon - Korea, (2) NICCA Korea - Daegoo - Korea

The nanoparticles of PVDF/PMMA blends with interpenetrating polymer networks were synthesized using the two-step polymerization method. In the first step, PVDF nanoparticles were prepared by emulsion polymerization using a fluorinated surfactant, and used as a seed for the following polymerization. In the second step, MMA emulsion was introduced to the emulsion of PVDF nanoparticles, and polymerized with a hydrocarbon surfactant. The nanoparticles of PVDF seed and PVDF/PMMA blend synthesized in this study were characterized by transmission electron microscopy (TEM) and differential scanning calorimetry (DSC) and contact angle measurements. The sizes of the nanoparticles of PVDF seed and PVDF/PMMA blend were approximately 55~75 and 70~90 nm, respectively. It could be confirmed by comparing the nanoparticle sizes and DSC data that the nanoparticles of PVDF/PMMA blends with interpenetrating polymer networks were well synthesized. The coating layers prepared using the nanoparticles of PVDF/PMMA blends with interpenetrating polymer networks show good compatibility with PET and low surface energy. We prepared and characterized a coating formulation of PVDF/PMMA nanoparticles for use in backsheet constructions for photovoltaic modules. The resulting backsheets coated with PVDF/PMMA nanoparticles with IPN show excellent weatherability, good adhesion strength between PVDF/PMMA coating and PET, and excellent gas-barrier properties for oxygen and water vapor.

Slip and Roughness at the Interface between Immiscible Polymer Liquids

***Sukumaran Sathish K.* [(1)*]*, Ito Takuya* [(1)]*, Komuro Ryohei* [(1)]*, Sugimoto Masataka* [(1)]**

[(1)] Yamagata University - Yamagata - Japan

In order to understand slip at the interface between two immiscible polymer liquids in the context of polymer extrusion, we have investigated coaxial two-phase flow using a capillary rheometer. To measure the slip velocity at the polymer/polymer interface we have modified and adapted the Mooney method. Usually, the Mooney method is used to investigate slip at the interface between a polymer and a solid wall. Using the modified Mooney method, we measured the dependence of the interfacial slip velocity on the interfacial shear stress for cylindrical core-sheath samples of two polymer pairs: polypropylene/polystyrene and poly(ethene-co-tetrafluoroethene)/polystyrene. When compared to previous work, we were able to investigate slip over a wider range of shear stress values using the new approach. In agreement with prior work on multilayer sandwiches undergoing drag flow and flow visualization of capillary flow, we found two distinct power-law regimes in the relationship between the interfacial slip velocity and the interfacial shear stress. The power-law exponent at low shear stress values and the critical stress for the transition between the power-law regimes varies with the polymer pair. However, the power-law exponent at high shear stresses appears to be independent of the particular polymer pair. We also investigated the flow conditions under which interfacial roughness was observed. The interfacial roughness increases with interfacial shear stress but only beyond a certain critical stress. Finally, the connection, if any, between interfacial slip and the roughness observed at the polymer/polymer interface will be explored.

Novel compatibilized polyamide ternary blends with improved hydrophobicity and mechanical properties

Anstey Andrew (1), Codou Amandine (1), MOHANTY Amar K. (1), Misra Manju (1)*

(1) University of Guelph - Ontario - Canada

Polyamide blending is a thoroughly investigated area, with aims of improving various aspects including density, toughness, moisture absorption, and cost. Due to the immiscibility of polyamides with many commodity polymers such as polypropylene (PP), a compatibilization approach is required to obtain a blend with desirable mechanical properties. However, even in well-compatibilized blends of polyamides with commodity polymers, the addition of the second polymer tends to significantly reduce the tensile strength and modulus of the blend. In this study, a ternary polymer blending approach was taken by blending nylon 6 (PA6) systems with PP and bio-based poly(lactic acid) (PLA). This approach was targeted to improve the hydrophobicity and price, while maintaining high tensile strength and modulus via the incorporation of PLA. Binary systems (PA6-PP) and ternary systems (PA6-PP-PLA) were compared at different blending ratios, with and without compatibilizing agents. The mechanical properties of the resulting blends were characterized via tensile and flexural analysis, impact testing, and moisture absorption testing. The phase morphology and dispersion was thoroughly investigated using scanning electron microscopy as well as atomic force microscopy. The crystallization behaviour of the blends was studied using differential scanning calorimetry. Overall, through a reactive extrusion approach, compatibilized ternary blends of PA6-PP-PLA were achieved with excellent properties comparable to those of neat PA6, with reduced density and improved hydrophobicity."

A nylon 6/PBT matrix for bio-based composites for automotive industry

Jubinville Dylan [1], Misra Manju [1], MOHANTY Amar K. [1]*

[1] University of Guelph - Ontario - Canada

The use of nylon in manufacturing is largely due to its high heat deflection temperature with good continuous-use temperatures, moderate chemical resistance, high strength even at elevated temperatures, and high stiffness. Poly (butylene terephthalate) (PBT) is a crystalline polymer with high molecular weight and good balanced properties. In some areas, PBT is replacing other engineering plastics due to its good chemical resistance, low water absorption, good processibility and its strength. The nylon 6/PBT blend was prepared with the formulation of 50:50 wt% through melt extrusion and injection molding with various mold temperatures in order to optimize the blending process. Through studying the mechanical properties a comparison in mold temperatures was explored and revealed slight changes in the Young's modulus. A differential scanning calorimeter (DSC) was utilized to study both the purity/miscibility, melting and crystallization behaviour of the blend as well as a dynamic mechanical analyzer (DMA) to observe the heat deflection temperature (HDT). A thermogravimetric analyzer (TGA) was also used to understand the blends degradation temperatures and purity/miscibility. A scanning electron microscope (SEM) of the impact fracture surface of the nylon/PBT was studied to look at the morphological structure. The blending of nylon 6 with PBT resulted in balanced properties where the formulation is targeted ultimately be used for the fabrication of lightweight bio-composite materials for automotive parts uses.

STUDY OF BARRIER PROPERTIES AND CRYSTALLIZATION OF PLA/PBT NANOBLENDS

Santos Luiza Gouvea [1], Costa Lidiane Cristina [1], Pessan Luiz Antonio [1]*

[1] UFSCar - Sao Paulo - Brazil

The inappropriate use and disposal of polymeric materials, combined with their resistance to degradation, produces considerable environmental problems, which encourages the pursuit for a balance between the durability for application and degradation after final disposal. An alternative is the use of biodegradable polymers which, for some applications, present a great potential in non-biodegradable polymers replacements. Poly (lactic acid) (PLA) has been widely applied as an environmentally friendly thermoplastic. PLA meets many requirements as a packing thermoplastic, however its crystallization rate PLA is relatively low and its barrier properties are not adequate such as high water vapour transmission rate for food packing application. In an attempt to obtain improved water vapour barrier properties and increase the crystallization rate, immiscible blends with poly (butylene terephthalate) (PBT) were produced in a co-rotating twin-screw extruder. Significantly improved properties can be achieved in immiscible polymer blends with polymer nanoparticles well disperse in a polymer matrix (Nanoblends). Compatibilized blends with 3, 5 and 10 wt% of disperse phase (PBT) in the matrix (PLA) were produced, as well as an uncompatibilized blend with 5wt% PBT. The compatibilizer used in this study was LOTADER AX 8840, with glycidyl methacrylate reactive groups. Preliminary results indicates that PBT increased the crystallization rate of PLA during cooling of the binary blend. Pure PLA did not crystalize at the same cooling rate. For the compatibilized blends, the cold crystallization exothermic peak during the first heating was displaced to lower temperatures, given the same condition, in accordance with the observation of an increase in the crystallization rate.

The Influence of Pre-processing and Processing Conditions on the Properties of HDPE and LLDPE Nanocomposites Incorporating Nanocrystalline Cellulose (CNC)

Musa R Kamal., Fatma Mahrous, Thomas Peret, Victor Tan, Wissam Abdallah
CREPEC, Department of Chemical Engineering, McGill University, Montreal - Quebec H3A 0C5– Canada

Abstract. There is growing interest in using cellulose nanocrystals (CNC) as nanofillers in the preparation of polymer nanocomposites. Since biodegradable, biocompatible CNC is derived from renewable biosources, much of the effort has been directed toward using it as the nanofiller with other bio-based polymers such as poly(lactic acid) (PLA). However, it would also of interest to explore the potential for incorporating CNC with other polymers such as polyethylene to produce viable polyolefin/CNC nanocomposites. While other methods such as solvent or water assisted processing techniques might be considered, it is commonly preferred to employ melt processing methods for economic and environmental reasons. The presentation will report on the status of our efforts to prepare nanocomposites by incorporating CNC in both high density polyethylene (HDPE) and linear low density polyethylene (LLDPE). Initial results indicated the need for reducing the size of feed polyethylene pellets to insure large surface contact areas between PE and CNC in the early stages of mixing. Moreover, effective premixing in the solid state was helpful in enhancing the uniformity of the nanocomposite. Morphology examination using scanning electron microscopy (SEM) of the various components and the CNC suggested that the porosity and network structure of the spray freeze dried CNC (SFDCNC) were effective in producing an interconnected reinforced network of polyethylene and CNC. This was most prominent at concentrations above the percolation level, where various rheological and mechanical properties showed significant changes. The results of WAXD and DSC showed that the crystallinity and crystallization kinetics of the two polymers were influenced to different levels by the presence of CNC. However, the optical transmittance of the nanocomposites was similar to that of the matrix resins. Other observations will be discussed. .

Keywords: CNC; LLDPE: HDP, Nanocomposites.
PACS: Replace this text with PACS numbers; choose from this list: http://www.aip.org/pacs/index.html

INTRODUCTION

There is growing interest in using cellulose nanocrystals (CNC) as nanofillers in the preparation of polymer nanocomposites (1-3). Since biodegradable, biocompatible CNC is derived from renewable biosources, much of the effort has been directed toward using it as the nanofiller with other bio-based polymers such as poly(lactic acid) (PLA) or in the presence of other more compatible polymers or compatibilizers. However, it would also be of interest to explore the potential for incorporating CNC with other important commercial polymers, such as polyethylene, to produce viable polyolefin/CNC nanocomposites. While other methods such as solvent or water assisted processing techniques might be considered, it is commonly preferred to employ melt processing methods for economic and environmental reasons. This manuscript provides a brief report on the status of our efforts to prepare nanocomposites by incorporating CNC in both high density polyethylene (HDPE) and linear low density polyethylene (LLDPE).

Our work and results with a variety of CNC-containing polymers, including polypropylene (PP), suggested the advantages of employing spray freeze dried cellulose nanocrystals in nanocomposite systems (2). Morphology examination using scanning electron microscopy (SEM) indicated that the porosity and network structure of the spray freeze dried CNC (SFDCNC) was more effective than spray dried CNC (SDCNC) or freeze dried CNC (SDCNC) in producing well dispersed CNC in a variety of polymers. This was supported by rheological, crystallinity and crystallization kinetics, mechanical properties and other analytical techniques

CRYSTALLIZATION KINETICS & CRYSTALLINE GROWTH

Two techniques were employed to evaluate the crystallization kinetics and the crystalline growth behavior of the polyethylene resins. Differential scanning calorimetry (Perkin Elmer Pyris-1) was used to study crystallization kinetics, and an Olympus Polarized light Microscope (PLM) Model BXP in conjunction with a Linkam hot stage Model THMS60 (equipped with CI93 Temperature Controller) was used to study crystallite growth..

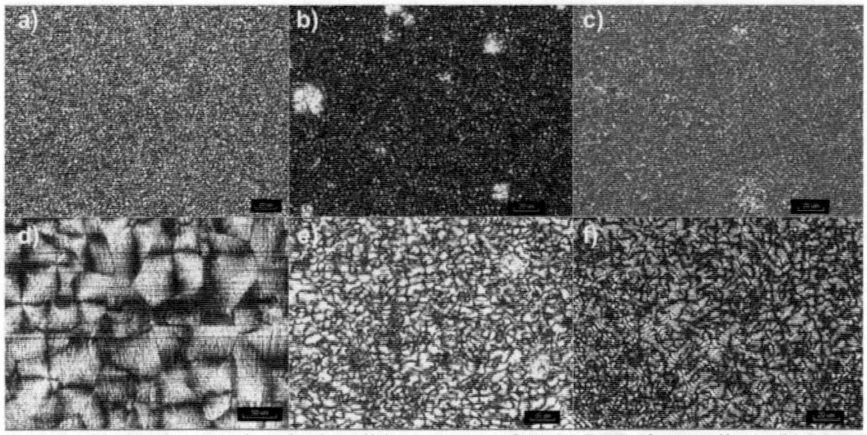

Figure 1. Photomicrographs of spherulitic structure of (a) LLDPE (fast cooling); (b) LLDPE/2.0SDCNC; (c) LLDPE/2.0SFDCNC; (d) HDPE (fast cooling); (e) HDPE/2.0SDCNC; (f) HDPE/2.0SFDCNC (band spherulites)

Both the Spray-freeze dried CNC filled LLDPE and HDPE showed that 4 to 5 micron small spherulites were formed surrounding CNC clusters. The large clusters in SFDCNC-HDPE appeared to be smaller than for SDCNC-HDPE, while the SFDCNC-LLDPE clusters appeared to be larger than the SDCNC-LLDPE. This may be due to the higher viscosity of the LLDPE, which tends to collapse the open-structure of the SFDCNC. The main observation, however, was that for both LLDPE and HDPE, spray freeze dried CNC exhibited at least twice the count of fine particles accumulated in the small particle range of 0.5 and 2.0 μm compared to spray dried CNC.

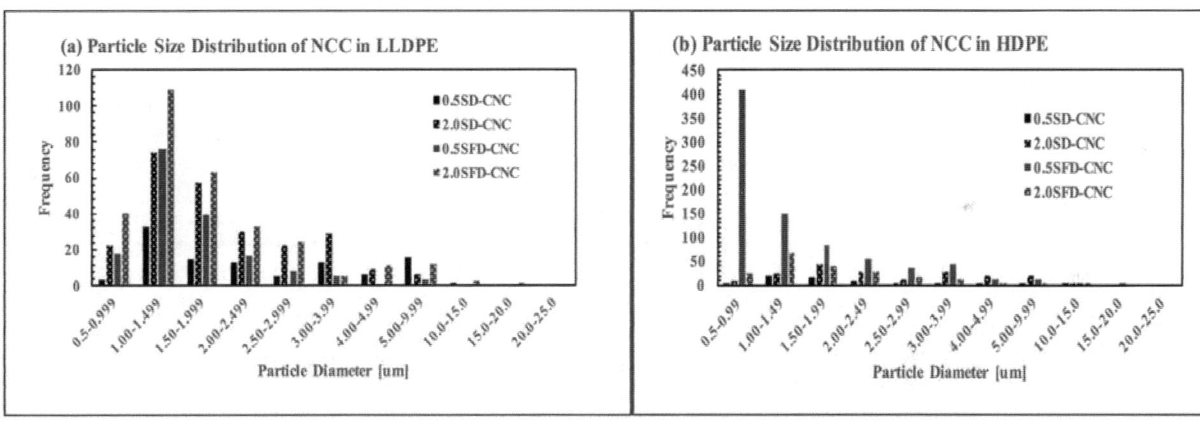

Figure 2. Particle size distribution of CNC in (a) LLDPE and (b) HDPE from PLM photomicrographs

Crystalline Growth Kinetics of CNC in LLDPE and HDPE

The crystallization half-times obtained experimentally using DSC were used to analyze the crystalline growth kinetics for each material by using the Lauritzen-Hoffman Theory for Regime II (4-6). Higher edge surface energy for LLDPE signifies a more highly strained or imperfect surface in the case of LLDPE as compared to that of HDPE. The magnitudes of edge surface energy, σ_e, obtained for the LLDPE and HDPE and their nanocomposites are summarized in Figure 3. The energy of work for chain folding can be estimated using the relationship: $q=2a_o b_o \sigma_e$ where $a_o = 4.55(10)^{-10}$ [m] and $b_\theta = 4.15(10)^{-10}$ [m] (4) for both LLDPE and HDPE. Comparisons of q values for LLDPE and HDPE are summarized in Figure 3.

Small aggregates of CNC aligned in a parallel manner and formed a bundle which produced a Maltese cross pattern when viewed under cross polarized microscope. Such patterns were observed in HDPE/SDCNC at about 4-5 microns in diameter, while the large clusters were very bright and in the order of 26-40 microns. Yet, the high viscosity of LLDPE destroyed the smaller clusters, while producing some larger clusters in the case of SFDCNC. For the SDCNC, higher viscosity of LLDPE produced smaller CNC clusters.

The higher edge surface energy for both the LLDPE with 0.5% SDCNC and SFDCNC was due to the combined effect of high viscosity of LLDPE and hindrance effect of the CNC. Increasing the CNC concentration to 2.0% in LLDPE enhanced the nucleation rate which counterbalanced the negative hindrance effect of CNC in both SDCNC and SFDCNC. In the case of HDPE, the edge surface energy increased with the CNC concentration.

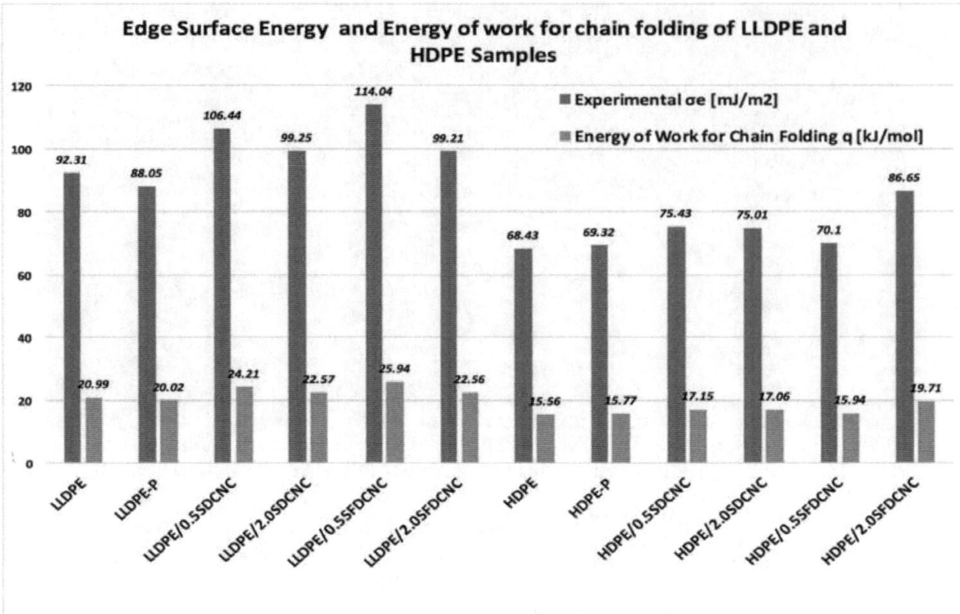

Figure 3. Edge Surface Energy and Energy of Work for Chain Folding Calculated from Experimental half time for LLDP and HDPE and their CNC nanocomposites

DISPERSION, RHEOLOGY & MECHANICAL PROPERTIES

It is well established that cellulose is thermally stable until approximately 300 °C. Also, our laboratory tests have demonstrated that the thermal stability especially during processing, rheological and crystallization measurements. Figure 4 shows the results of measurements of complex viscosity for both LLDPE and HDPE. As expected, the viscosity of LLDPE is higher than that of LLDPE. Both of them are significantly influenced by the incorporation of CNC. Both appear to experience a jump in the range of 2-3 % CNC concentration.

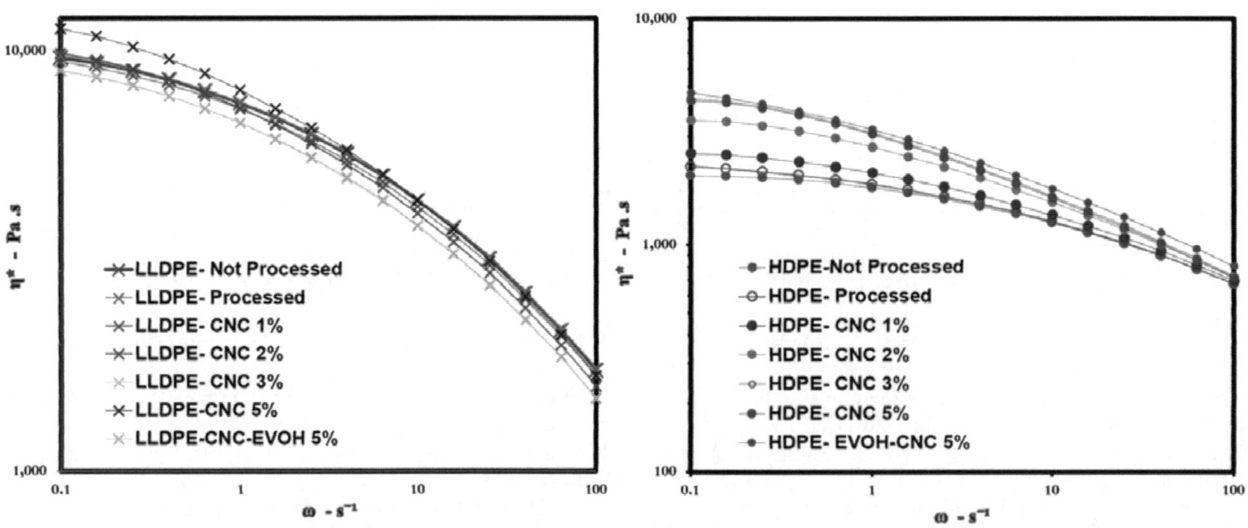

Figure 4: Complex viscosity of LLDPE & HDPE and their nanocomposites with various levels of CNC

Figure 5 indicates that the tensile modulus of HDPE is increased by around 80% upon incorporation of 5% SFDCNC. It should be noted that the tensile strength remains virtually unchanged, while the elongation at break is reduced significantly. Moreover, Figure 6, obtained by transmission electron microscopy (TEM), shows clearly that the quality of CNC dispersion at the nanoscale is very good. This is true for both LLDPE and HDPE.

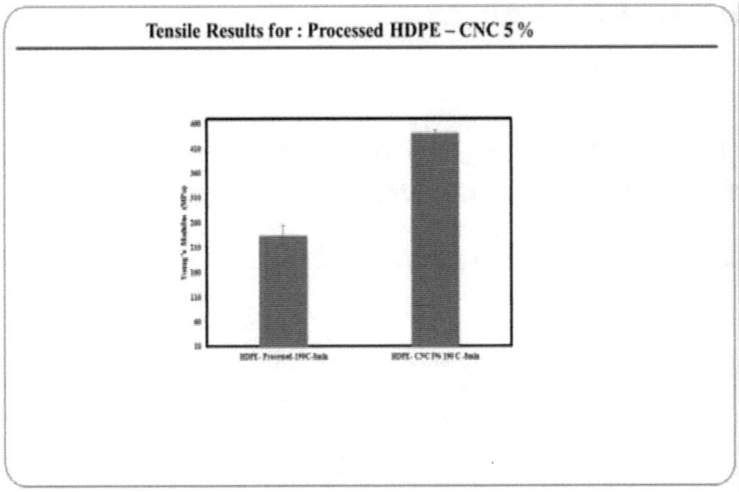

Figure 5: Tensile Strength of HDPE & HDPE –CNC 5%

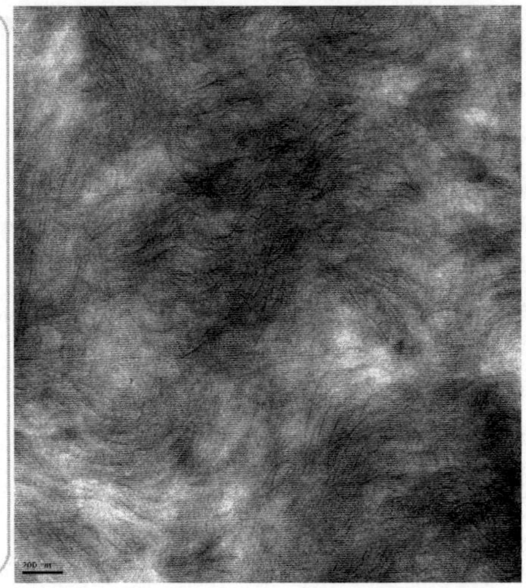

Figure 6: TEM Image of LLDPE 5% CNC

ACKNOWLEDGMENTS

This work was supported by the Natural Sciences and Engineering Research Council of Canada (NSERC) the NSERC Network for Innovative Plastic Materials and Manufacturing Processes (NIPMMP). The authors also wish to thank Professor Paula Wood-Adams & Professor Suong Hoa (Concordia University) and Dr. Hajatallah Vali and Dr. S. Kelly Sears from McGill University.

REFERENCES

1. Khoshkava, V. and Kamal, M. R.(2014). Effect of drying conditions on cellulose nanocrystal (CNC) agglomerate porosity and dispersibility in polymer nanocomposites. Powder Technol.261: 288-298..
2. Khoshkava, Vahid and Kamal, Musa R.(2014). Effect of Cellulose Nanocrystals (CNC) Particle Morphology on Dispersion and Rheological and Mechanical Properties of Polypropylene/CNC Nanocomposites. ACS Appl. Mater. Interfaces. 6(11): 8146-8157.
3. Khoshkava, V. and Kamal, M. R.(2013). Effect of surface energy on dispersion and mechanical properties of polymer/nanocrystalline cellulose nanocomposites. Biomacromolecules. 14(9): 3155-63.
4. Hoffman, J. D. and Miller, R. L., "Kinetic of crystallization from the melt and chain folding in polyethylene fractions revisited: theory and experiment", Polym., **38**, 3151-3212 (1997)
5. Hoffman, J.D. and Weeks, J.J., "Melting process and the equilibrium melting temperature of polychlorotrifluoroethylene", J. Res. Natl. Bur. Stand. A., Sec. A: Phys. Chem., **66A**, 13-28 (1962)
6. Lauritzen J. I. and Hoffman, J. D., "Extension of theory of growth of chain-folded polymer crystals to large undercoolings", J. Appl. Phys., **44**, 4340-4352 (1973)

Custom Nanomaterials for use in Hybrid Polymer Blend Nanocomposites for Electrical and EMI Shielding Applications

***Sundararaj Uttandaraman* (1)*, *Otero Ivonne* (1), *Shayesteh Ali* (1), *Arjmand Mohammad* (1)**

(1) University of Calgary - Alberta - Canada

Our group has a unique ability to span the entire spectrum of novel nanocomposite development starting from synthesizing custom carbon and metal nanomaterials, to mixing with polymer to create novel nanocomposites, and then post-processing to make parts, which can be tested for a variety of properties. Polymer nanocomposites have unique multifunctional properties resulting from the size and shape of the fillers and display superior electrical, thermal and mechanical properties to conventional polymer composites due to the nanoscale size of the filler. Multiwall carbon nanotubes (MWCNT) with and without doping, graphene-like materials like Graphene nanoribbons (GNR), MnO2 and ZnO nano wires, Copper and Silver nanowires (CuNW and AgNW) were synthesized and thenmelt-mixed with polymers via solution mixing and in miniature mixers. The electrical, electromagnetic interference (EMI) shielding and mechanical properties of the resulting nano composites were tested. The electrical, mechanical and rheological properties of these new materials were studied and evaluated for industrial applications such as personal electronics enclosures, shielding for aerospace, batteries and anti-static packaging. The morphology of the polymer blend nanocomposites was characterized by transmission electron microscopy (TEM) and scanning electron microscopy (SEM) and revealed unique network structures. Hybrid filler and polymer blend systems indicate that there is an optimum morphology and nanofiller ratio for best electrical properties. Hybrid nanocomposites displayed extremely low electrical percolation thresholds lower than 0.0005 volume fraction. The outstanding performance of these materials in X-band opens up opportunities for security and military applications. By controlling the dispersion of the hybrid filler system, we found that there is an excellent opportunity to use these materials for charge storage applications by properly dispersing a conductive and a semi-conductive filler.

Graphene Oxide-Based Polymer Nanocomposites: Processing and Characterization

Fechine Guilhermino [1], Andrade Ricardo [1]*, Domingues Sergio [1], Souza Eunézio [1], Tavares Maria Inês [2], Nagaoka Danilo [1], Amorin Leice [1], Oliveira Camila [1], Muñoz Pablo [1]

[1] Mackenzie Presbyterian University - São Paulo - Brazil, [2] IMA - Instituto de Macromoléculas Profa. Eloísa Mano - Rio de Janeiro - Brazil

When the Nobel Prize of Physics was announced in 2010 the superlatives properties of graphene have taken attention for the academic and industrial purposes. Improvements in strength, electrical and thermal conductive and also impermeability are characteristics that can be observed in polymers by the insertion of graphene. Graphene obtained by liquid exfoliation of graphite and graphene oxide (GO) and/or reduced graphene oxide (rGO) obtained by chemical exfoliation are the best candidates to be used in polymer nanocomposites since the products obtained by these methods are in a large quantity. In both techniques, liquid and chemical, few layers of graphene are dispersed in a solution and the big challenge is to keep the graphene exfoliated when it is added to the polymer matrix during the manufacturing of nanocomposite. Patents and manuscripts have been filed and published, respectively, showing several methods to prepare polymer-graphene nanocomposites (solution mixing, in situ polymerization and melt mixing), however, the scale-up to industrial manufacturing is limited. Melt mixing by using extruders is the one that can be easily moved for industries purposes. Here, we are introducing two different strategies to disperse GO into polymer matrix by using a twin screw extruder and polystyrene as a polymer template. The influence of the methods of GO insertion, processing parameters and GO concentration on the level of particles dispersion and polymer nanocomposite properties were evaluated by mechanical and thermal tests, molecular weight data, rheological measurements, Raman and FTIR spectroscopies, low-field nuclear magnetic resonance and X-ray microtomography. The results showed that the two strategies used here to prepare polymer-GO nanocomposite are very efficient in order to obtain a good GO dispersion, and consequently, printing to the polymer excellent properties at very low content of GO (less than 0.5% w/w).

Thermally Conductive Microcellular Polymer Nanocomposites of Graphene Nanoplatelets (GnP)

S. Mahdi Hamidinejad (presenting author) [a,b], Raymond K.M. Chu [a], Tobin Filleter [b], Chul B. Park [a]*(reference author)

[a] *Microcellular Plastics Manufacturing Laboratory, Department of Mechanical and Industrial Engineering, University of Toronto, Toronto, Ontario, Canada M5S 3G8*

[b] *NanoMechanics and Materials Laboratory, Department of Mechanical and Industrial Engineering, University of Toronto, Toronto, Ontario, Canada M5S 3G8*

hamidi@mie.utoronto.ca
park@mie.utoronto.ca

Abstract. A new class of thermally conductive microcellular polymer nanocomposites of graphene nanoplatelets (GnP) is reported. Foamed and solid high-density-polyethylene (HDPE)-GnP composites containing different GnP contents (0-18 vol.%) were injection-molded. Foamed composites were fabricated using a facile technique of melt mixing followed by supercritical fluid-treatment and physical foaming in an injection molding process. The effects of foaming on the dispersion, exfoliation, orientation and inter-connectivity of platelets, and heat dissipation functionality were investigated. The introduction of microcellular foaming, significantly changed the orientation of platelets, enhanced their inter-connectivity and further exfoliated GnPs in the polymer. Hence, foaming reduced the density of the injection-molded samples and enhanced the thermal conductivity of product up to 800% (3.75 W/m.k). The results revealed that lightweight, highly thermally conductive products with lower filler loading, can be fabricated using foam injection molding for various applications in miniaturized electronic devices, and electric motor systems.

Keywords: Thermal conductivity, polymer nanocomposites, graphene nanoplatelets, high-density-polyethylene (HDPE), foam injection molding, microcellular foaming.
PACS: 81.05.Qk

INTRODUCTION

The development of new generations of smaller, lighter and more powerful electronic devices requires more compact and lighter heat sinks. Multifunctional, highly thermally conductive polymer composites show promise as candidates to be employed as heat dissipation components in the electronic packaging technology. A polymer composite's thermal conductivity controls the heat transfer rate and the product's direction. The thermal conductivity of non-metallic materials is governed by the transfer of phonons, and this greatly depends on the type of fillers, their percolation, dispersion, and orientation. Depending on the fillers' dispersion/orientation and the processing method, the thermal conductivity or heat dissipation property of the polymer composite can vary in different directions (e.g., in-plane direction rather than through-plane direction). In other words, the effectiveness of highly thermal conductive fillers in a product depends on the ability the fillers to form a thermally conductive network in the desired direction [1]. The thermal conductivity of polymer composites is also greatly affected by the interaction of filler/polymer composites, which are governed by a chain structure and the polymer matrix's chemistry [2,3]. Conventionally, thermally conductive polymer composites, which are used as heat dissipation components, are usually loaded with a large amount of thermally conductive filler particles (> 60 wt.%) such as graphite and carbon fiber. This is done to achieve a thermal conductivity up to 10 W/mK [1]. With such a high filler loading, the amount of polymer matrix left to support the fillers and the composite's structural integrity is insufficient. As a result, these fabricated composites are extremely brittle, and this very much limits their utilization prospects. Hence, there is much motivation to develop polymer composites that are highly thermally conductive and which require a much less filler content. One promising way to overcome this drawback is to incorporate nanomaterials in the manufacturing of these composites. In this regard, graphene nanomaterials with their extraordinary thermal conductivity, higher aspect ratios, and mechanical properties that excel those of their micro-size counterparts are highly promising filler candidates in this context. The thermal and electrical conductivities of single-layer graphene have been reported, respectively, as ~5000 W/ (m.K) [4–6] and ~6000 S/cm [7]. However, the practical underpinning needed to economically manufacture graphene-based polymer composites is missing.

Herein, for the first time a facile SCF-assisted technique for manufacturing the GnP-polymer composites with enhanced thermal conductivity is presented. In this method GnP-HDPE nanocomposites are fabricated by using foam injection molding to create heat dissipation components. The microcellular GnP-HDPE nanocomposites showed heat transport properties that were superior to those reported in the literature [1,8–16]. This was due to the tailored microstructure within the fabricated GnP-HDPE nanocomposites.

EXPERIMENTAL SETUP

A 50-ton Arburg Allrounder 270/320C injection molding machine (Lossburg, Germany) with a 30-mm diameter screw, equipped with MuCell® technology (Trexel, Inc., Woburn, Massachusetts) was used for fabricating HDPE-GnP nanocomposites samples. The mold contained a rectangular cavity with a fan gate after the sprue. The mold cavity dimensions were 132 × 108 × 3 mm. More details are found in [17].

Two different types of HDPE-GnP nanocomposites, namely injection-molded solid (IMS) and injection-molded foam (IMF) treated with supercritical N_2 (Sc-N_2), are prepared. The processing parameters are summarized in **Table 1**.

TABLE (1). Processing parameters used in injection molding of solid and foamed composites.

Parameter	IMS	IMF
Melt temperature (°C)	210	210
Barrel pressure (MPa)	16	16
Screw speed (rpm)	300	300
Metering time (s)	12	12
Injection flow rate (cm^3s^{-1})	90	90
Mold temperature (°C)	75	75
Pack/hold pressure (bar)	300	N/A
Pack/hold time (s)	15	N/A
Gas injection pressure (MPa)	N/A	24
N_2 content (wt.%)	N/A	0.4
Nominal void fraction (%)	N/A	10

WAXD analyses were conducted on IMS and IMF nanocomposites using a Rigaku MiniFlex 600 X-ray diffractometer. The samples were frozen in liquid nitrogen, cryofractured and sputter-coated to examine the microstructure and morphology using Scanning Electron Microscopy (SEM; JEOL JSM-606). The total thermal conductivities of GnP-polymer nanocomposites and foams are measured via transient hot disk method. A transient plane source (TPS) hot disk thermal constants analyzer (Therm Test Inc., TPS 2500, Sweden) was used to measure the thermal conductivity of samples under ambient conditions using a Kapton sensor.

RESULTS AND DISCUSSION

Microstructure and morphology of GnP-polymer nanocomposites

Figure 1 demonstrates the core and skin microstructure of the IMS and IMF HDPE-9 vol.% GnP. The IMS samples' core and skin layers had completely solid structures. In the IMS samples' skin layers (about 500 µm on each side), the GnPs were aligned in the machine direction. This preferential fillers' alignment in the composites fabricated via injection molding has been well described and reported in literature [18–21]. However, in the core layer of IMS samples, the GnPs follow the fountain flow orientation and are randomly oriented throughout the entire part. Conversely, in the IMF samples, the skin-core morphology is hardly identified and the GnPs are highly random-oriented. The IMF samples have a microcellular structure with relatively random cell morphology.

The growth of cells causes different degree of rotation and displacement of GnPs leading to random orientation and further dispersion of GnPs and thus higher chance of interconnectivity. This leads to a particular morphology which is greatly different from that of IMS samples.

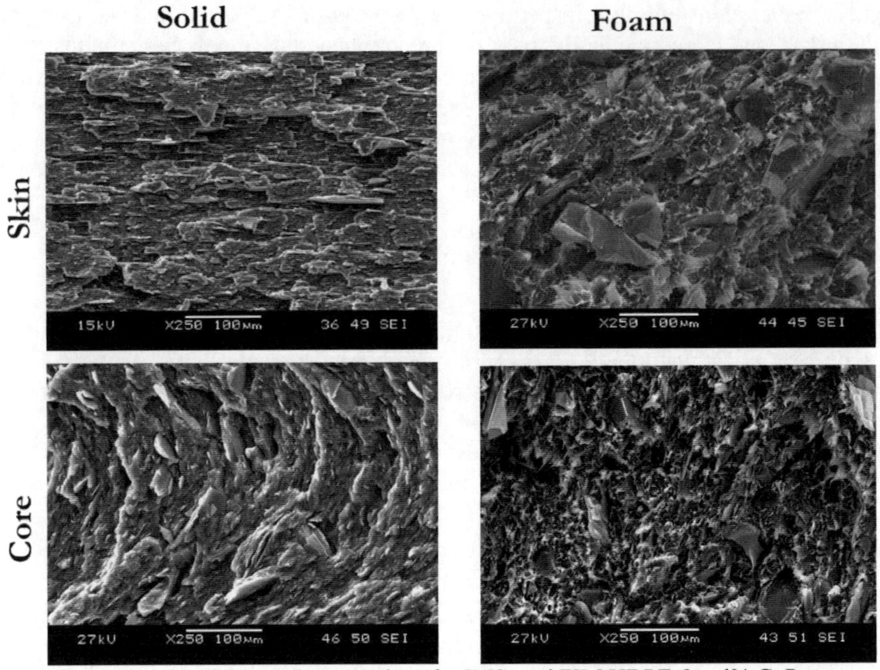

FIGURE 1. SEM micrographs of skin and core regions for IMS, and FIM HDPE-9 vol% GnP nanocomposites

To quantify the exfoliation level of GnPs after foaming, WAXD analyses are conducted. **Figure 2** shows the WAXD patterns of neat HDPE, GnP powder, IMS of HDPE-9 vol.% GnP samples its IMF counterparts. The diffraction peak at $2\theta = 26.5°$, is characteristic of the 002 reflection of graphite (I_{002}) resulting from the d-spacing between monolayer graphene sheets. The foaming of the GnP-HDPE nanocomposites results in 94% decrease in the intensity of I_{002} diffraction of initial value corresponding to unfoamed nanocomposites (Figure 3b). This suggests the occurrence of efficient exfoliation and structural disordering of the GnPs which is in agreement with literature [22–24]. However, there are still layered GnP structure retained in each stage, explaining the presence of the slight diffraction (I_{002}) even after physical foaming of the GnP-HDPE nanocomposites.

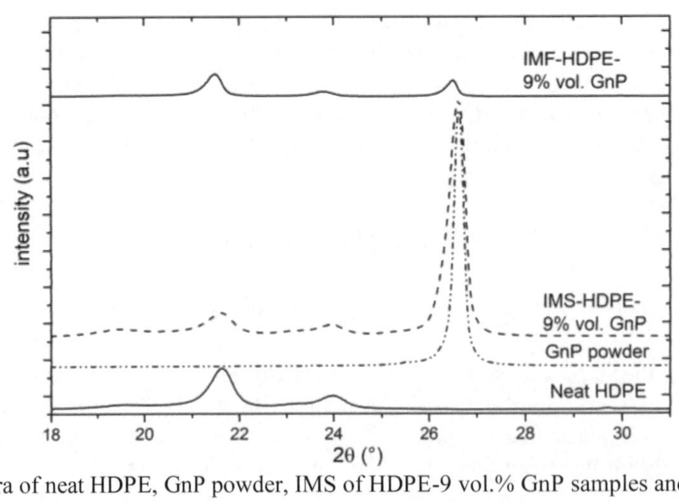

FIGURE 2. XRD spectra of neat HDPE, GnP powder, IMS of HDPE-9 vol.% GnP samples and its IMF counterparts

Thermal conductivity of the GnP-polymer nanocomposites

Figure 3 shows the total thermal conductivity of IMS and IMF samples as a function of GnP content. The thermal conductivity is reported as a function of final content of GnPs. As expected, in all samples (including IMS and IMF) the thermal conductivities of HDPE-GnP nanocomposites are highly dependent on the GnP content.

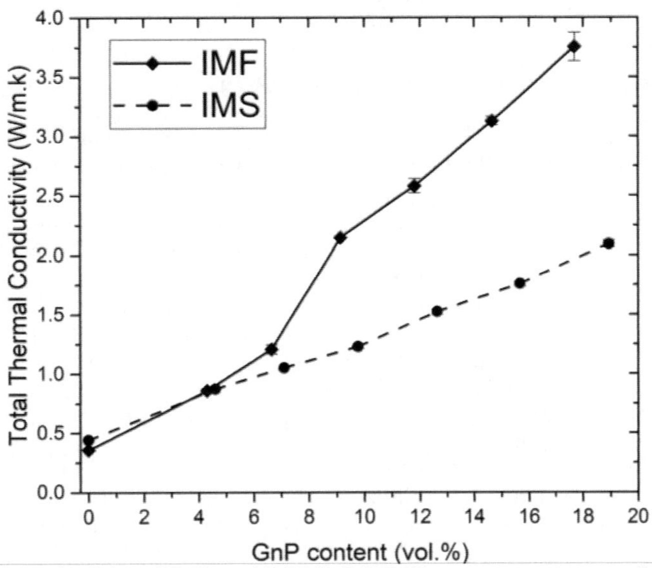

FIGURE 3. Total thermal conductivity of IMS, and IMF GnP-HDPE nanocomposites as a function of GnP content

In the IMS samples, the thermal conductivity corresponds to 404% enhancement over the neat HDPE samples at 18 vol.% of GnP and an enhancement of 21.3% per 1 vol.% loading. However, the thermal conductivity of the IMF counterparts exhibits about 800% enhancement over neat HDPE samples and enhancement of 46% per 1 vol.% loading. The thermal conductivities of IMF nanocomposites start excelling those of IMS at GnP loading more than 7 vol. %. This can be attributed to presence of enough GnPs in the polymer nanocomposites to form thermally conductive paths. In other word, below 7 vol. % GnP loading, the polymer matrix mediates between GnPs resulting in a polymeric gap which breaks the direct GnP-GnP contact.

The reason for this outstanding improvement in the thermal conductivity of IMF samples compared with IMS, stems from influential factors such as: (i) further exfoliation and dispersion of GnPs in the polymer matrix and; (ii) high level of orientation and interconnectivity of GnPs; and (iii) highly attenuated preferential orientation GnPs in IMF samples compared with IMS samples.

The developed process enhanced the thermal conductivity through tailoring a three-dimensional network with increased number of direct GnP-GnP contact, having much lower interfacial thermal resistance as compared to those with GnP-polymer contact due to lower phonon scattering [16].

CONCLUSION

A new class of highly thermally conductive GnP-polymer nanocomposites foam was introduced. Such nanocomposites foams were developed by a facile and industrially viable method of melt mixing followed by foam injection molding process. The HDPE-17.6 vol.% GnP nanocomposites foams showed a total thermal conductivity of 3.75 W/m.k, highly superior to the values of conventionally injection-molded counterparts. The reasons for this drastic difference include (i) further exfoliation and dispersion of GnPs in the polymer matrix; (ii) high level of interconnectivity among GnPs (direct GnP-GnP contact) and; (iii) high level of GnPs' random orientation. Therefore, physical foaming of GnP-polymer composites provides a novel and large-scale production method for developing tailored structures to enhance the heat dissipation functionality.

ACKNOWLEDGMENTS

The authors gratefully acknowledge Group NanoXplore Inc.'s donation of GnP powder and GnP-HDPE masterbatch, and the financial support of the Natural Sciences and Engineering Research Council of Canada (NSERC).

REFERENCES

1. Mosanenzadeh, S. G., Khalid, S., Cui, Y. & Naguib, H. E. High Thermally Conductive PLA-Based Composites With Tailored Hybrid Network of Hexagonal Boron Nitride and Graphene Nanoplatelets. *Polym. Compos.* **37,** 2196–2205 (2016).
2. Balandin, A. A. Thermal properties of graphene and nanostructured carbon materials. *Nat. Mater.* **10,** 569–581 (2011).
3. Yang, S.-Y. *et al.* Effect of functionalized carbon nanotubes on the thermal conductivity of epoxy composites. *Carbon N. Y.* **48,** 592–603
4. Balandin, A. A. *et al.* Superior Thermal Conductivity of Single-Layer Graphene. *Nano Lett.* **8,** 902–907 (2008).
5. Ghosh, S. *et al.* Extremely high thermal conductivity of graphene: Prospects for thermal management applications in nanoelectronic circuits. *Appl. Phys. Lett.* **92,** 151911 (2008).
6. Seol, J. H. *et al.* Two-dimensional phonon transport in supported graphene. *Science* **328,** 213–6 (2010).
7. Kim, H., Abdala, A. A. & MacOsko, C. W. Graphene/polymer nanocomposites. *Macromolecules* **43,** 6515–6530 (2010).
8. Kalaitzidou, K., Fukushima, H. & Drzal, L. T. Multifunctional polypropylene composites produced by incorporation of exfoliated graphite nanoplatelets. *Carbon N. Y.* **45,** 1446–1452 (2007).
9. Kim, H. S., Bae, H. S., Yu, J., Kim, S. Y. & Regev, O. Thermal conductivity of polymer composites with the geometrical characteristics of graphene nanoplatelets. *Sci. Rep.* **6,** 26825 (2016).
10. Song, S. H. *et al.* Enhanced thermal conductivity of epoxy-graphene composites by using non-oxidized graphene flakes with non-covalent functionalization. *Adv. Mater.* **25,** 732–737 (2013).
11. Yu, J., Cha, J. E. & Kim, S. Y. Thermally conductive composite film filled with highly dispersed graphene nanoplatelets via solvent-free one-step fabrication. *Compos. Part B Eng.* **110,** 171–177 (2017).
12. Guo, H., Li, X., Li, B., Wang, J. & Wang, S. Thermal conductivity of graphene/poly(vinylidene fluoride) nanocomposite membrane. *Mater. Des.* **114,** 355–363 (2017).
13. Ding, P. *et al.* Highly thermal conductive composites with polyamide-6 covalently-grafted graphene by an in situ polymerization and thermal reduction process. *Carbon N. Y.* **66,** 576–584 (2014).
14. Meng, F. *et al.* In situ intercalation polymerization approach to polyamide-6/graphite nanoflakes for enhanced thermal conductivity. *Compos. Part B Eng.* **117,** 165–173 (2017).
15. Araby, S. *et al.* Electrically and thermally conductive elastomer/graphene nanocomposites by solution mixing. *Polymer (Guildf).* **55,** 201–210 (2014).
16. Shtein, M., Nadiv, R., Buzaglo, M., Kahil, K. & Regev, O. Thermally conductive graphene-polymer composites: Size, percolation, and synergy effects. *Chem. Mater.* **27,** 2100–2106 (2015).
17. Lee, J. W. S., Wang, J., Yoon, J. D. & Park, C. B. Strategies to Achieve a Uniform Cell Structure with a High Void Fraction in Advanced Structural Foam Molding. *Ind. Eng. Chem. Res.* **47,** 9457–9464 (2008).
18. Ameli, A., Jung, P. U. & Park, C. B. Through-plane electrical conductivity of injection-molded polypropylene/carbon-fiber composite foams. *Compos. Sci. Technol.* **76,** 37–44 (2013).
19. Ameli, A., Jung, P. U. & Park, C. B. Electrical properties and electromagnetic interference shielding effectiveness of polypropylene/carbon fiber composite foams. *Carbon N. Y.* **60,** 379–391 (2013).
20. Motlagh, G. H., Hrymak, A. N. & Thompson, M. R. Improved through-plane electrical conductivity in a carbon-filled thermoplastic via foaming. *Polym. Eng. Sci.* **48,** 687–696 (2008).
21. Ameli, A., Wang, S., Kazemi, Y., Park, C. B. & Pötschke, P. A facile method to increase the charge storage capability of polymer nanocomposites. *Nano Energy* **15,** 54–65 (2015).
22. Matsumoto, M., Saito, Y., Park, C., Fukushima, T. & Aida, T. Ultrahigh-throughput exfoliation of graphite into pristine 'single-layer' graphene using microwaves and molecularly engineered ionic liquids. *Nat. Chem.* **7,** 730–736 (2015).
23. Ellingham, T., Duddleston, L. & Turng, L. S. Sub-critical gas-assisted processing using CO2 foaming to enhance the exfoliation of graphene in polypropylene + graphene nanocomposites. *Polym. (United Kingdom)* **117,** 132–139 (2017).
24. Yasmin, A., Luo, J. J. & Daniel, I. M. Processing of expanded graphite reinforced polymer nanocomposites. *Compos. Sci. Technol.* **66,** 1179–1186 (2006).

Development of a highly nucleated and dimensionally stable isotactic polypropylene/nanoclay composite using reactive blending

Ray Suprakas Sinha [(1)*], ***Bandyopadhyay Jayita*** [(2)]

[(1)] CSIR and UJ - Gauteng - SouthAfrica, [(2)] CSIR - Gauteng - SouthAfrica

Polypropylene (PP) is one of the most widely used commodity polymer. However, it has a low upper service temperature. Addressing this challenge, we introduced reactive blending of PP, maleic anhydride-grafted PP (PP-g-MA) and (3-aminopropyl)triethoxysilane (3APTES) in the presence of nanoclay, which results in a highly nucleated PP composite with improved dimensional stability. At the processing temperature, the reaction between maleic anhydride, which is already grafted on the PP chain, and 3APTES forms N-substituted maleimide-grafted PP. The results indicated that the introduction of branched structures on PP chains increased the chain bulkiness and led to the formation of stable nuclei that initiate crystallization at high temperatures. By offering nucleating sites, the dispersed nanoclay particles facilitated the nucleation process to a greater extent. Compared with the neat polymer matrix, the crystallization temperature of the composite was 15.5 °C higher with moderate improvements in the heat distortion and Vicat softening temperatures. This is most likely the first report where such an improvement in crystallization temperature has been achieved for isotactic PP homopolymers. In summary, reactive processing of the PP/nanoclay composite in the presence of PP-g-MA and 3APTES has potential for the development of high-performance PP-based advanced materials.

Effects of VOH or VA Co-monomer Contents on the Morphology, Barrier and Mechanical Properties of Polypropylene/Ethylene Copolymer/ Montmorillonite Nanocomposites

Castro Juan Felipe [1], Medina Jorge Alberto [1]*, Salcedo Felipe [1]

[1] Universidad de los Andes - Bogota - Colombia

Control on the morphology-processing-properties relationships of polyolefin/organoclay nanocomposites is paramount for food packaging applications. This research reports the influence of the intermolecular interactions between ethylene vinyl acetate copolymer (EVA) and ethylene vinyl alcohol copolymer (EVOH) each with organically modified montmorillonite (MMT) on the morphology, mechanical and barrier properties of polypropylene (PP) based films. PP/EVA and PP/EVOH (75/25%wt) blends were reinforced with 1, 3 and 5%wt MMT (introduced via a polypropylene grafted maleic anhydride) in a double screw internal mixer. Four different copolymers (EVA12%, EVA28% co-monomer content of vinyl acetate (VA), EVOH32%, and EVOH44% molar content of vinyl alcohol (OH)) were used to prove that a higher amount of favorable intermolecular interactions with MMT is obtained with higher content of VA or OH functional groups of the copolymer. Samples were compression-molded into films of 300µm. The structural characterization was made using X-ray diffraction, mechanical properties were characterized with simple tension essay, and oxygen permeation by oxygen transmission rate (OTR). XRD patterns showed a combination of intercalated/exfoliated morphologies for the MMT, with higher d-001 interplanar distance increments for the blends with higher content of polar functional groups. Mechanical properties and oxygen permeability of composites exhibited a higher improvement by the addition of MMT for higher intermolecular interactions in the composites. The results show that the exfoliated morphology can be obtained by maximizing the magnitude and number of interactions between MMT and polymer with the use of polar character copolymers. The higher the amount of interactions, given by the VA or OH polar functional groups, implies a more exfoliated morphology of the nanocomposites and in consequence a higher improvement on the mechanical and barrier properties of the composite material.

pps proceeding
Symposium: S09 - Polymer nanocomposites
Oral Presentation

Graphene Oxide-Based Thermoplastic Polyurethane Nanocomposites. Part 2: Influence of GO on Nanocomposite Structure and Properties

Andrade Ricardo [(1)]*, Amurin Leice [(1)], Danda Chaitanya [(2)], Munoz Pablo [(1)], Nagaoka Danilo [(1)], Domingues Sergio [(1)], De Souza Eunezio [(1)], Maia Joao [(2)], Fechine Guilhermino [(1)]

[(1)] Mackenzie Presbyterian University - Sao Paulo - Brazil, [(2)] Case Western Reserve University - Cleveland - USA

Thermoplastic polyurethanes (TPUs) are multi-block copolymers usually consisting of hard and soft segments. The hard segments (HSs), which are composed of diisocyanate and short-chain diols as chain extenders, are most often thermodynamically incompatible with soft segments (SSs) that are made of polyethers or polyesters. Therefore, phase separation occurs with HSs forming domains that consist in either glassy or semi-crystalline regions, which act as physical cross-links and provide stiffness and reinforcement. The reversibility of these regions enables melt processing of these materials at temperatures above the highest melting or glass transition temperature of the domains. The nanostructure of TPUs, and consequently their mechanical properties, depends on the chemical structure and processing conditions, namely temperature and flow histories. Moreover, at low temperatures, i.e., below the lower glass transition or melt temperature associated with the HSs, strain also originates significant changes in the nanostructure of the material. The graphene oxide (GO) has been studied in the context of many applications, such as polymeric nanocomposites. These particles 2D are able to tailor the excellent mechanical and thermal properties of the thermoplastic polyurethane. In this work, GO was added to the thermoplastics polyurethanes, via melt mixing, processed under fundamentally different types of flow (see Part I). The GO obtained by Hummers method, possesses functional groups, which was anticipated to link to the TPU structure improving its final properties. The presence of GO in TPU nanocomposites is possible to observe by the signature in the intensity of the Raman spectrum. The mechanical properties show a significant improvement in composites prepared via aggressive extensional mixing by comparison with mild extension or shear-dominated conditions.

ZnO@Carbon Nanotube Hybrid Nanocomposites with Excellent Dielectric Properties

***Shayesteh Zeraati Ali* (1)*, *Sundararaj Uttandaraman* (1)**

(1) University of Calgary - AB - Canada

This study investigates the effect of uniformly coated CNT with ZnO nanoparticles (ZnO@CNT) addition on the dielectric properties of ZnO@CNT/polyvinylidene fluoride (PVDF) hybrid nanocomposites. As a dielectric material, the ZnO@CNT/PVDF nanocomposites were prepared at various loadings of ZnO@CNTs by melt mixing, and they exhibited excellent dielectric properties in X-band frequency (8.2-12.4 GHz). The dielectric study showed that introducing ZnO nanoparticles on the surface of CNT improved dielectric properties by decreasing the dielectric loss tremendously. We attribute the low dielectric loss of the hybrid nanocomposites to the role of ZnO on hindering the CNT direct contact. The mechanism for dielectric properties of these nanocomposites has been explained in detail in the manuscript. This excellent dielectric properties of ZnO@CNT/PVDF would be beneficial for various flexible capacitors, such as embedded capacitors in printed circuit boards.

Hybrid Nanocolourants for Polymeric Films and Fibres

Joshi Mangala (1)*

(1) IIT Delhi - Delhi - India

Colour plays an important role in plastic manufacturing industry. Colour Master batches are made in several forms depending on the aesthetic required for the type of plastic it is being made for. For colouration of any plastic product either dyes or pigments are used. Both have its own advantages and disadvantages. A possible solution of this problem has been found by an easy process of combining organic dyes with layered natural or synthetic clays. This paper presents use of a new class of nanocolourants or nanopigment or PlanoColours for colouration of polymeric films and fibers. These combine the advantages of dyes with those of pigments and are more stable without increase in the particle size of dyes. The inspiration for this class of nanocolorants are the Maya Blue (MB) pigments that have attracted considerable interest in the past few years. This opens the way to the low-cost engineering of heavy metal free nonhazardous pigment powders based on indigoid and other dyes. Maya Blue is a famous artificial pigment widely used in murals, pottery and sculptures by the ancient Mayas and other people in Mesoamerica. The pigment, whose hue ranges from bright turquoise to dark greenish blue, exhibits a remarkable stability, being unaffected by the attack of acids, alkalis, oxidants, reducing agents, organic solvents, as well as painting ability, good UV, oxygen and thermal stability. Clay-dye nanocolourant pigment was prepared by using Rhodamine dyes and MMT clay. Nanopigment was characterised by TGA, computer colour matching system, XRD, SEM. It was further incorporated in polymer for preparation of coloured master batches on a twin screw extruder. The coloured master batch was used to prepare films and fibers and different properties such as thermal stability, UV stability, mechanical properties etc. of these samples were evaluated. Keywords: Nanocolourant, Maya Blue Pigment, Colour Masterbatch, Film, Fibre

Influence of boehmite nanoparticles on the curing kinetics and rheology of an epoxy matrix for liquid composite molding process

Dilmurat Abliz[1,2], Tobias Juergens[1], Tatjana Artys[1,2], Gerhard Ziegmann[1,2]
[1]Institute of Polymer Materials and Plastics Engineering, TU Clausthal, Agricolastr.6, 38678 Clausthal-Zellerfeld, Germany
[2]Clausthal Center for Materials Technology, TU Clausthal, Leibnizstraße 9, 38678 Clausthal-Zellerfeld, Germany

ABSTRACT

Nanoparticles can effectively increase the resin matrix properties, which is especially attractive for fabrication of high-performance fiber composites by liquid composite molding (LCM) processes. However, the cure kinetics and rheological behavior of the matrix, which is the most critical aspects in LCM processes, could be affected by the introduction of nanoparticles, depending on the surface functionalization and concentration. In this paper, the cure kinetics and rheology of boehmite nanoparticle modified epoxy matrix is characterized and modelled. Influence of boehmite nanoparticles on cure and rheological behavior is analyzed combining model parameter studies, and a generalized model is put forward to describe and predict the cure kinetics and rheology depending on time, temperature, and concentration of nanoparticles.

1 INTRODUCTION

Nanoparticles (NPs) can effectively improve the critical mechanical and thermal properties of fiber reinforced polymer (FRP) structures [1-7]. This is especially attractive for liquid composite molding (LCM) processes, which allows for low-cost production of high-performance FRPs. The effectiveness of NPs is dependent on the chemical (composition, surface modification) and physical (size, morphology, concentration) characteristics. However, depending on these characteristics, NPs affect the critical impregnation processing aspects such as matrix cure kinetics and rheology. By LCM manufacturing of FRPs, there is a strict criterion on viscosity limit – normally up to 1 Pa·s – on the applicable epoxy matrix. When the viscosity exceeds this value, the potential failure increases due to the decreased impregnation capabilities of the resin, by which the effective process window is defined.

There are plenty of studies that are concentrated on cure kinetics and rheological characterization and modelling of neat epoxy resin systems [8-13]. However, with the addition of NPs, the curing and rheological behaviors get more complex due to the interaction of the NPs with the matrix. The influence of NPs could be quite complex since both chemical as well as physical effects are involved. According to the earlier studies based on silica or alumina NP modified resin systems, both inhibiting and acceleration effects of NPs on the kinetics have been reported. NPs without surface treatment or with hydrophobic surface characteristics – as for the case in mostly investigated silica particles [14, 15][14, 15][14, 15][14, 15][14, 15][14, 15] – are reported to be inert to the epoxy reaction with no significant change on the activation energy and the reaction profile [14-16]. In comparison, NPs with special surface modification or hydrophilic surface characteristics showed mostly an accelerating effect on the reaction kinetics of the original epoxy resin, which is typically the case for alumina NPs [17, 18].

Boehmite is an aluminum hydroxide oxide – Al(OH)O, which is often used as a precursor for many aluminum products. Boehmite NP is price-efficient and the hydroxyl groups on the surface could be directly used as functional groups, or easily be modified to different functional groups (amine, acetic acid, milk acid etc.) to generate a good chemical stability and improved particle-matrix adhesion by possible covalent connection. According to earlier studies, boehmite NPs is shown to be quite efficient for enhancing the mechanical properties of FRPs [19-21]. Especially, the studies from Exner et al.[21] on boehmite NPs have shown an certain dependency of epoxy matrix crosslink density against the NP concentration and surface functionalities. However, there are only few reported studies with respect to the processing aspects of boehmite-modified epoxy matrix, especially regarding modelling of cure kinetics and rheology of boehmite NP systems. Of course, with the special surface characteristics and higher concentrations of the boehmite NPs concentrations – up to 15 wt% as targeted in this paper, the interactions could be much more complex with a remarkable influence on the process and final properties.

2 THEORIOTICAL BACKGROUND

2.1 Curing Kinetics Characterization and Modelling

By the curing characterization of the epoxy with Differential Scanning Calometry (DSC), the degree of cure at each point of time ($\alpha(t)$) is determined from the heat released up to that time ($\Delta H(t)$), the heat released during heat-up and equilibration (H_{init}) and the total heat released during the reaction (H_{total}):

$$\alpha(t) = \frac{\Delta H(t)}{H_{total}} \qquad \text{Eq. 1}$$

The reaction rate of a single step process is normally described by a rate function $K(T)$ which is dependent on the temperature (T) and a conversion function $f(\alpha)$ as following:

$$\frac{d\alpha}{dt} = K(T)f(\alpha) \qquad \text{Eq. 2}$$

The temperature dependent reaction rate function $K(T)$ can be expressed by an Arrhenius relationship:

$$K_i(T) = A_i e^{(-\frac{E_i}{RT})} \quad i = 1,2 \ldots. \qquad \text{Eq. 3}$$

Therein, A is the pre-exponential factor, E is the activation energy, R is the ideal gas constant and T is the absolute temperature. So , the development of the model is to determine the sets of reaction triplet – A: pre-exponential factor which is the frequency of vibrations of the activated complex[22], E: activation energy for the reaction and $f(\alpha)$: reaction mechanism function.

The reaction models can be principally categorized into two major types: n^{th} order (accelerating or decelerating) and autocatalytic. According to earlier investigations [8, 12, 23], the epoxy resin system shows an autocatalytic reaction, and the cure behavior of the matrix system could be well described by Kamal-Sourour model that is modified by Fournier's diffusion factor $f_d(\alpha)$ as follow:

$$\frac{d\alpha}{dt} = (K_1 + K_2\alpha^m)(1-\alpha)^n f_d(\alpha)$$
$$f_d(\alpha) = \frac{2}{1+\exp(\alpha-\alpha_c)/C} - 1 \qquad \text{Eq. 4}$$

Therein, m and n are reaction orders, α_c is the critical cure degree and C is a fit parameter.

2.2 Rheology Modelling

The development of the viscosity is dominated by the curing kinetics – the increased molecular size due to the curing. Besides, it is also affected by pressure, temperature, time, shear rate and NPs. The effects of these parameters and models are discussed and overviewed in paper [24]. In this work, the pressure and shear rate is approximated to be constant, therefore the viscosity could be considered only as a function of temperature (T) and cure (α):

$$\eta = \eta(T, \alpha) \qquad \text{Eq. 5}$$

Generally, the dependence of viscosity on temperature for uncured resin system is mostly described by William–Landel–Ferry (WLF) equation, derived from the free volume theory [25]. For the change of viscosity due to cure (time & temperature), a modified WLF model [26, 27], Castro-Macosko model [28] or a combined model of the both [26, 29] were studied in several papers. The Castro-Macosko model was used in several studies for modelling the rheological behavior of epoxy resin systems [30] or films [31] and showed promising results, so in this work the Castro-Macosko model is selected to model the rheological behavior of the matrix system.

$$\eta = \eta_\infty exp(\frac{E_\eta}{RT})(\frac{\alpha_{gel}}{\alpha_{gel} - \alpha})^{(A+B\alpha)} \qquad \text{Eq. 6}$$

Therein, E_η: flow activation energy, η_∞: viscosity at T_∞, α_{gel}: cure degree at gelation, A and B: fit parameters.

3 EXPERIMENTAL

3.1 Boehmite NP

In this paper, boehmite NPs that are functionalized with acetic acid (CH_3COOH), as shown in the following Figure 1, are investigated. The special functionalized boehmit NP are called as KE later for the simplification. The boehmite NPs with a primary particle size of 20nm are used in the study are provided by SASOL (DISPERAL®. HP 14).

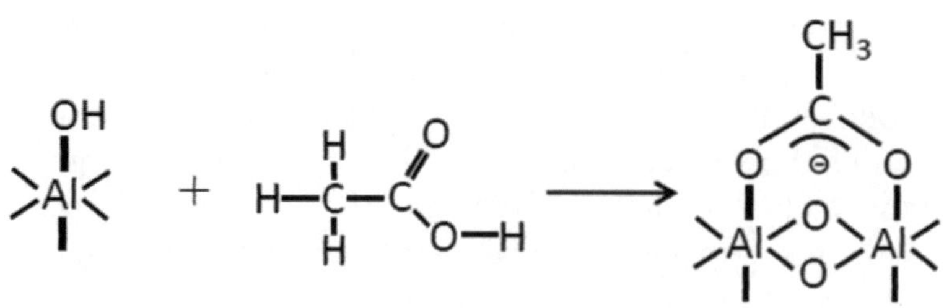

Figure 1: Schematic illustration of boehmite NP functionalized with Acetic acid (CH_3COOH)

3.2 NP-epoxy resin matrix

Epoxy matrix: a commercial epoxy resin system LY 556 (DGEBA-type epoxy) with corresponding anhydride hardener HY917 (methyltetrahydrophthalic) and accelerator DY070 (1-methyl imidazole) from Huntsman Advanced Materials (Switzerland) GmbH is used. The resin, curing agent and accelerator are mixed at the recommended mass ratio of 100:90:1.

Boehmite NPs are dispersed into the epoxy matrix to produce masterbatches with 40 wt% boehmite NPs, at the Institute for Particle Technology (IPAT)- TU Braunschweig. The x10%, x50% and x90% values of dispersed particle size within the produced masterbatch are separately 81 nm, 104 nm and 134 nm, as shown in Figure 3.

Figure 2: Morphology of the boehmite NPs (dry aggregate particle structure)

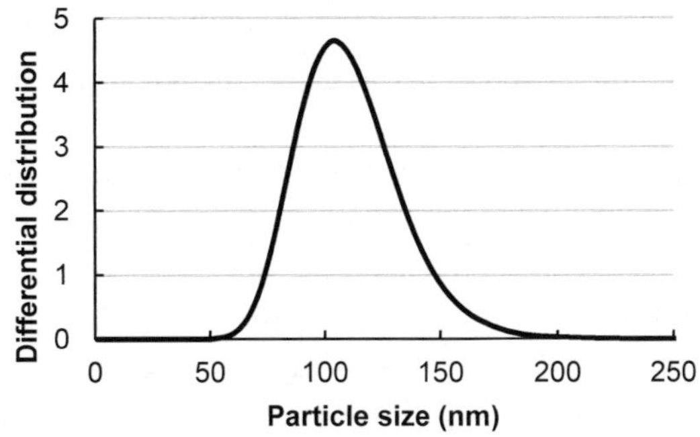

Figure 3: Particle size distribution of boehmite-epoxy masterbatch

For the experiments, the masterbatch is diluted to the desired particle concentration with a planetary mixing machine (Thinky Mixer ARV-310) twice with the following parameters: 1000 rpm under vacuum of 0.3 bar absolute pressure for 1.5 minutes, then at 2000 rpm under vacuum of 0.1 bar absolute pressure for another 1.5 minutes at room temperature. The particle concentration is assured by thermal gravimetric measurements (TGA) after delusion.

3.3 DSC Measurements

A TA Q-2000 DSC from TA Instrument is used to measure the reaction heat and the glass transition temperature (T_g) of the samples. The attached cooling system RCS90 allows for cooling the samples down quickly to −90 °C. Resin samples of approximately 6–8 mg are placed in aluminium hermetic pans.

Dynamic measurements are performed from −80 °C to 250 °C by 2 and 10 K/min to characterize the dynamic curing behavior and to obtain the total reaction enthalpy. The cured specimens are checked for the maximum obtainable glass transition temperature ($T_{g\infty}$) using a subsequent dynamic scan from 20 °C to 250 °C. For the isothermal measurements, several samples are prepared in one batch and stored at −18 °C to stop the reaction (proved feasible up to 5 days). The samples are tested isothermally at 100, 110, 120 and 130 °C, respectively. Before the isothermal tests, they are cooled down to −20 ° C, heated up quickly to the desired cure temperature, and cured at this temperature until the heat flow rate remained constant. At this point, the T_g exceeds the T_{cure} by which the sample vitrifies due to the glass transition, therefore the reaction only proceeds at negligible rate due to the diffusion controlled reaction. The samples are subsequently cooled down to 40 °C and heated up, until it reaches the aging peak (also called as annealing peak) due to the thermal aging effect [32, 33]. Directly after the aging peak the samples are quenched again to 40 °C to overcome the disturbance of the aging peak to determine the Tg and rest enthalpy. At the third cycle, the samples are heated again using 10 K/min dynamic scans up to 280 °C until the samples are fully cured and the final T_g as well as the residual enthalpy (H_{res}) are measured. The heat flow and temperature are recorded over time. The heat flow data is adjusted using a horizontal baseline, which extends from heat flow values at the end of reaction to initial values.

3.4 Rheological Measurement

The rheological measurements are conducted by a TA AR-2000 instrument from TA Instrument, with a 40 mm diameter aluminum-plate geometry by plate-plate mode. A platin temperature controller is used as the basic temperature control element to accurately control the temperature. Dynamical viscosity measurement is carried out with 5K/min, from 20 °C to about 140°C, until the viscosity reaches the limit of 1.5 Pa.s. Isothermal viscosity measurements are carried out with a shear rate of 50 1/s, also until the viscosity reaches the limit of 1.5 Pa.s. Besides, oscillation viscosity measurements are conducted to determine the gel time and conversion.

4 RESULTS AND DISCUSSION

4.1 Dynamic DSC

The total reaction enthalpy and T_g depending on the concentration of boehmite NPs are shown in the following Table 1. As the total enthalpy is dependent on the unit mass of the sample, with the increase of the boehmite NP concentration, the proportion of the reactive resin and curing agent will be decreased, which inevitably leads to the decrease of the total enthalpy. Therefore, the total enthalpy is normalized by the by subtracting the weight concentration of the boehmite NP on the enthalpy.

Table 1: Total reaction enthalpy and Tg by different mixtures

Mixture	Average enthalpy (J/g)	Normalized average enthalpy (J/g)	T_{g0} (°C): uncured sample	$T_{g\infty}$ (°C): cured sample
0% (Pure resin)	354.4 ± 3.9	354.4 ± 3.9	- 37.99 ± 0.08	136.32 ± 3.48
KE 5%	313.4 ± 3.65	329.8 ± 3.84	- 35.99 ± 1.13	127.18 ± 0.89
KE 10%	293.4 ± 1.65	325.9 ± 1.83	- 34.87 ± 1.71	121.83 ± 4.61
KE 15%	266.3 ± 1.2	313.3 ± 1.41	- 34.43 ± 0.74	118.68 ± 0.17

According to the normalized average enthalpy, there is still a gradual decrease in the total enthalpy. The glass transition temperature before the reaction – T_{g0} – showed a slight increase, which is presumably because that the increased amount of boehmite NPs could restrict the mobility of the epoxy molecules, so that a higher energy is needed for the initial glass transition. Conversely, the glass transition temperature after the full reaction – $T_{g\infty}$ – showed a remarkable decrease, indicating a reduced stability of the epoxy chains due to the NPs. Figure 4 shows the normalized heat flow for different mixtures by the dynamic DSC process (10 K/min). With addition of the boehmite NPs up to 15 wt%, the evolution of heat stayed similar without any extra peak, and the reaction peak temperature is almost the same, indicating no substantial change of the reaction by addition of the boehmite NPs. However, dynamic DSC investigations (heat-cool-heat) on just boehmite-epoxy suspensions (without any curing agent) as in the Figure 5 showed a remarkable reaction exothermic heat by the first heating cycle and an increase of the glass transition temperature at the second heating. This shows a certain reactivity of the boehmite NP with the epoxy, even without curing agent. Combining these aspects, it is reasonable to assume that the boehmite NPs could directly react with the epoxy to build covalent connections, nevertheless it seems that the reaction density between the boehmite NP and the epoxy seems to be weaker than that between the epoxy and cure agent, which leads to the decrease of the total enthalpy and $T_{g\infty}$.

The influence of the NPs to the T_g is closely related to the specific surface of the particles and surface functionality – the reactivity with the epoxy matrix. If the particles play a catalytic role participating in the epoxy reaction resulting a larger portion of immobilized polymer chain, an increase in the T_g is to be expected, or vice versa. The boehmite NP that is used in the study is proven by dynamic DSC studies to be reactive with the epoxy groups, presumably due to the excessive hydroxyl groups on the surface. The hydroxyl groups can spend hydrogen atoms, which are essential for the opening of oxirane rings of the resin or epoxy. However, based on the decrease in the final T_g, it is reasonable to assume that even if the boehmite NPs accelerate the curing in a certain degree at the initial stage of the curing, but the crosslinking density between the particle and epoxy interphase is lower than that with the anhydride curing agent, which inevitably leads to a decreased total crosslinking density. The influence of the boehmite NPs will be further analyzed and discussed later combining the cure kinetics modelling.

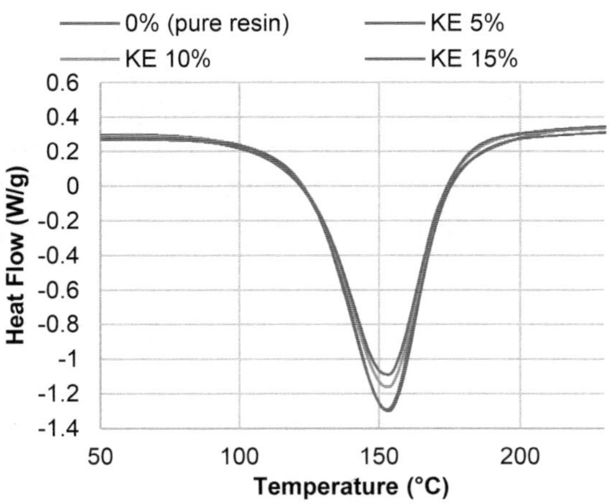

Figure 4: Normalized heat flow for for KE-suspension by dynamic DSC

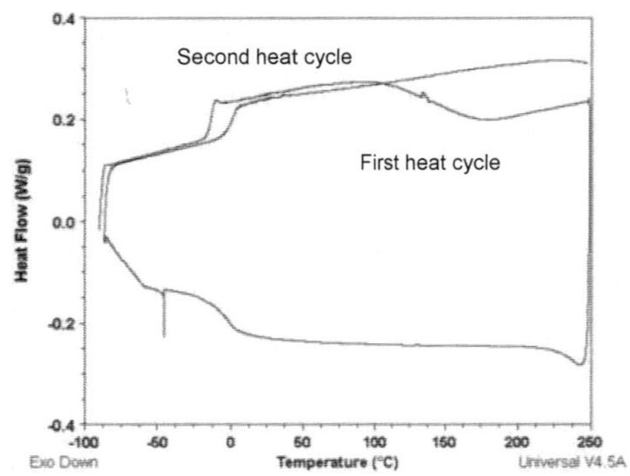

Figure 5: Dynamic DSC measurement on 15wt% boehmite-epoxy suspension without curing agent

4.2 Cure characterization and modelling

The kinetic parameters are estimated by Sestak-Berggren method [34] directly by the isothermal DSC process, as shown below in the Figure 6. The estimated kinetic parameters boehmite-filled epoxy suspensions are provided in According to the estimated model parameters, both pre-exponential factor and activation energy parameters showed a generally decreasing trend, indicating an acceleration effect of the boehmite NPs on reaction. In comparison, the reaction order *m* and *n* are almost constant around 0.5 and 1, respectively.

Table 2.

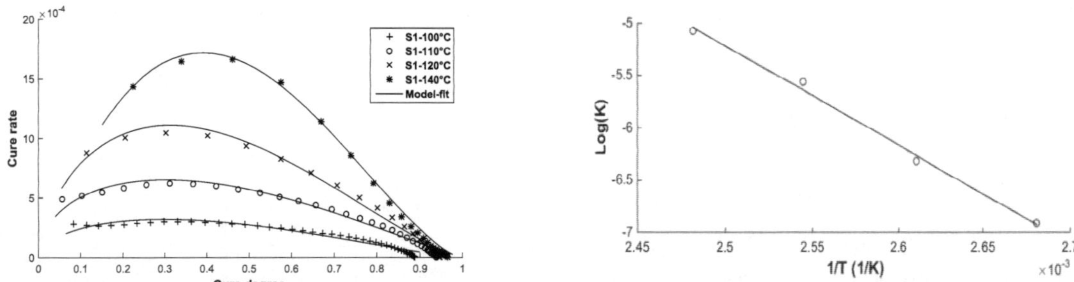

Figure 6:; left: determination of K, m and n by Sestak-Berggren plot; right: determination of A and E based Arrhenius linear fitting

According to the estimated model parameters, both pre-exponential factor and activation energy parameters showed a generally decreasing trend, indicating an acceleration effect of the boehmite

NPs on reaction. In comparison, the reaction order *m* and *n* are almost constant around 0.5 and 1, respectively.

Table 2: Estimated kinetic parameter start values S-K method for different mixtures

Mixture \ Kinetic parameter	A_1 /A_2 (1/sec)	E_2 /E_1 (J/mol)	m	n
0% (neat resin)	1.47E+07	72857.9	0.52	1.12
KE 5%	6.86E+07	77481.4	0.57	1.07
KE 10%	3.74E+07	76019.6	0.48	1.05
KE 15%	2.46E+05	59419.3	0.63	1.12

The final values for the model coefficients are determined by the regression analysis based on the non-linear least-square fitting of the cure curves – cure rate versus cure degree. It can be shown by Figure 7 that the curing behavior of the boehmite-modified epoxy matrix could also be well described by the modified Kamal-Sourour model with Fournier's model as diffusion factor. The final developed model parameters for the matrix systems based on the model with S-K start value and multi-target regression is provided in the

Table 3: Model parameters by Kamal-Sourour model modified with diffusion factor for different mixtures

Mixture \ Kinetic parameter	A_1 (1/sec)	A_2 (1/sec)	E_1(J/mol)	E_2 (J/mol)	m	n	R^2
0% (Pure resin)	1.99E+09	1.11E+06	93264.2	64411.6	1.00	1.26	0.997
KE 5%	1.97E+09	1.33E+06	93619.1	64433.1	1.00	1.21	0.998
KE 10%	1.99E+09	1.57E+06	93275.0	58165.0	1.00	1.19	0.995
KE 15%	1.05E+08	3.97E+06	83653.8	68784.4	1.00	1.18	0.992

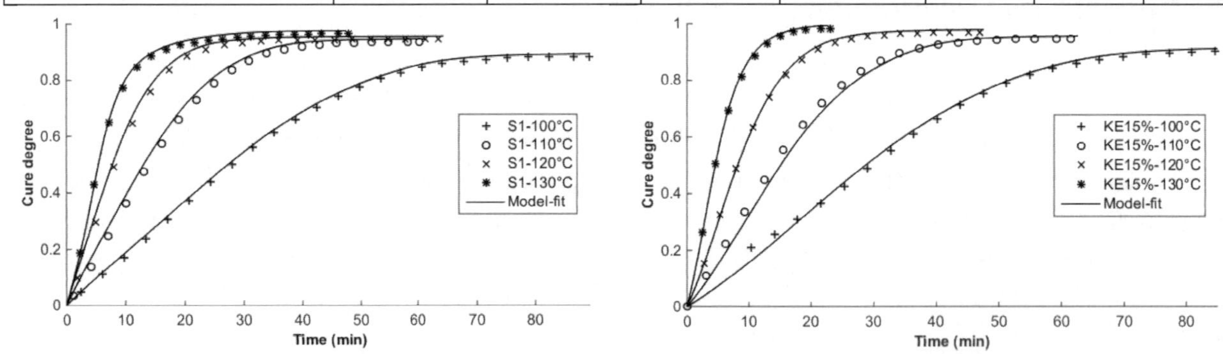

Figure 7: Cure kinetics modelling; Left: neat resin; Right: 15 wt% NP-modified resin

According to the model parameters, with the increase of the boehmite NPs, reaction orders *m* and *n* stayed around 1 and 1.2, indicating that the boehmite NPs don't change the reaction mechanism dramatically. If the change of pre-exponential factor (A_1, A_2) and activation energy (E_1, E_2) is plotted against the NP concentration, as shown in Figure 8, on one side, a remarkable influence of NPs on the model parameters could be seen, but on the other side, it is not possible to determine a clear tendency in the change of the model parameters with respect to NP concentration.

In the earlier studies, the influence of different model parameters that are based model-free and model-fitting methods, and different non-linear or linear regression algorithms are investigated, regarding the neat matrix system. It was found that the Kamal-Sourour model shows an ill-posed characteristic – it was not possible to develop a unique set of model parameters to describe the cure behavior based on the experimental data [23]. The ill-posed characteristic of the model can well explain why it is not possible to determine model parameter correlation regarding the influence of NPs. Therefore, some further model parameter sensitivity and correlation studies are needed.

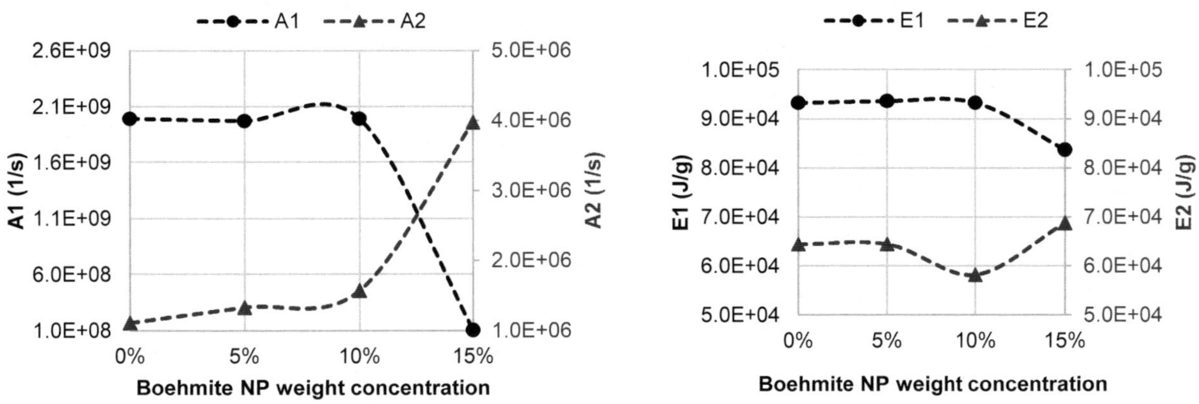

Figure 8: Change of pre-exponential factor (A_1, A_2) and activation energy (E_1, E_2) regarding the NP concentration

When the cure degree is set to zero in the Kamal-Sourour model as shown in Eq. 4, it is obvious that the model describes an initial cure rate at the very beginning which is determined by rate function K_1 as follows,

$$\frac{d\alpha}{dt} = K_1 \qquad \text{Eq. 7}$$

$$K_i(T) = A_i e^{\left(-\frac{E_i}{RT}\right)}$$

The rate function K_1 describes some secondary reactions that could be initiated by some side aspect, e.g., moisture in certain systems, which is mostly the case for amine cured epoxy systems [35]. As the target system in this paper is an anhydride cured epoxy resin, by which the influence of moisture on the cure kinetics can be reasonably assumed much smaller than compared to the main reaction, so that the two parameters could be set to zero. However, from the numerical point of view, the Kamal-Sourour model describes only the dependency of cure rate to cure degree, in order to develop the dependency of cure degree against time, an integral calculation based on Euler method is applied. If the initial cure rate – K_1 – is set to zero, it leads to a dead cycle by which the calculation could not be initiated. By combining the two aspects, it is reasonable to assume that the K_1 is a number that satisfies: $0<K_1<exp(-\infty)$. Therefore, it can physically be considered as zero, nevertheless it is bigger than zero so that the integral calculation can be initiated. Based on the assumption, the model parameter could be new determined as follows,

According to the new model parameters, with the increase of the boehmite NPs, the pre-exponential factor A_1 and activation energy E_1 that describes the initial cure rate kept constant at 2.2E-14 and 1E-10, respectively. The value is extremely small, so that physically could be seen as zero. The reaction orders m and n stayed around 0.5 and 1. The A_2 and E2 showed a very good correlation regarding the NP concentration, as shown in the following Figure 9.

Table 4: New model parameters by Kamal-Sourour model

Kinetic parameter / Mixture	A_1 (1/sec)	E_1 (J/mol)	A_2 (1/sec)	E_2 (J/mol)	m	n	R^2
0% (Pure resin)	2.20E-14	1.00E-10	5.42E+06	70106.66	0.4631	1.07	0.99
KE 5%	2.20E-14	1.00E-10	1.20E+06	64368.72	0.54	1.10	0.99
KE 10%	2.20E-14	1.00E-10	6.50E+05	62999.98	0.42	1.03	0.99
KE 15%	2.20E-14	1.00E-10	9.53E+04	56684.20	0.45	1.00	0.95

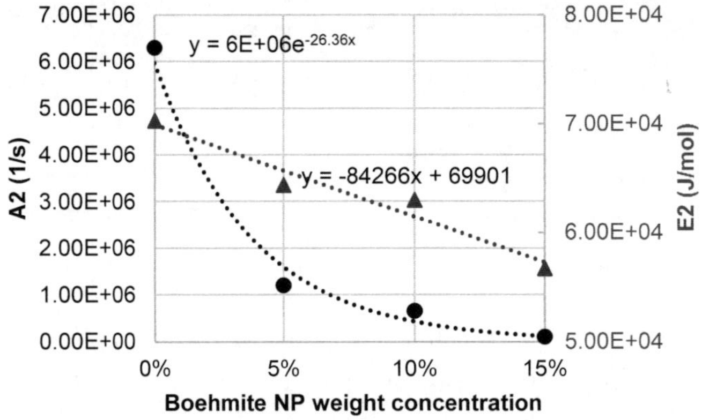

Figure 9: Change of A_2 and E_2 regarding the NP concentration by the new modelling

Based on the new modelling assumption and methods, it can be seen that the A_2 shows an exponential decrease and E_2 showed a linear decrease with increase of NP concentration. Based on the observation, the cure kinetics model for the NP-epoxy matrix could be generalized against NP weight concentration (x), to enable to describe and predict the change in the cure kinetics with respect to different NP concentrations.

Table 5: Generalized Kamal-Sourour model parameters against NP concentration

Kinetic parameter	A_1 (1/sec)	E_1 (J/mol)	A_2 (1/sec)	E_2 (J/mol)	m	n
Model parameters regarding NP concentration wt% (x)	2.2E-14	1.0E-10	6E+6exp(-26.36x)	-84266x+69901	0.47	1.04

4.3 Rheology characterization and modelling

The initial values of E_η (activation energy for flow) and η_∞(viscosity at T_∞) by Castro-Makasco model could be obtained from dynamic viscosity plots of $ln\eta$ versus $1/RT$. As the assumption only holds when the curing degree is zero, therefore, only viscosity data in the temperature range from 20 °C to 50 °C is considered, where the curing rate is very low to be presumed to be zero. This assumption could be also verified by the linear development of the curve as in Figure 10. Based on the fitting curve, the start values for the rheology model could be calculated.

Besides these two parameters, the critical curing degree at the gel point is needed for the rheological model. The gel point is measured by oscillation method in rheometer, and is determined by the time where the storage modulus (G') exceeds the loss modulus (G"), as shown in the Figure 11.

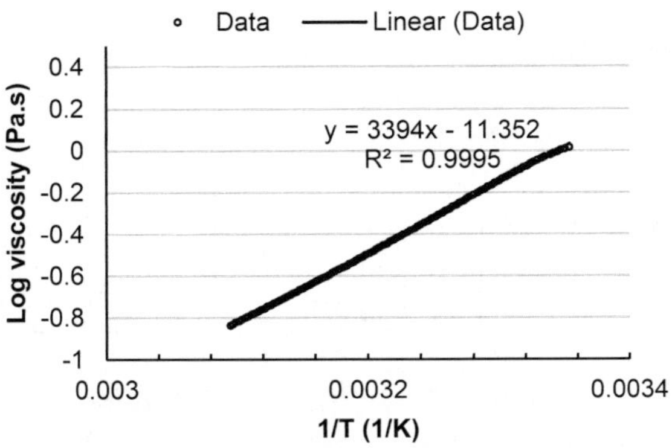

Figure 10: Log viscosity versus Temperature plot (pure resin)

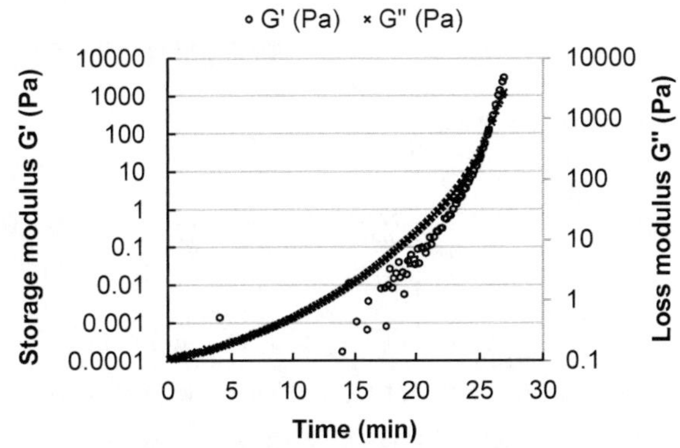

Figure 11: Oscillation viscosity test and determination of gel point (pure resin)

Based on the cure kinetics and get time, the gel conversion can be calculated. Similarly, the start values and cure degree at gel point for the boehmite particle-filled reactive suspensions are calculated as follow:

Table 6 : Estimated start kinetic parameters for rheological modelling by Castro-Macosko model

Kinetic parameter / Mixture	η_∞ (Pa.s)	E_η (J/mol)	Gel convertion
0% (Pure resin)	1.1746E-05	28219.3	0.5039
KE 5%	7.89722E-06	29949.5	0.5441
KE 10%	5.38443E-06	31514.3	0.5766
KE 15%	3.29864E-06	33291.8	0.6490

As indicated in the Castro-Macosko model, the development of the viscosity is dominated by the cure kinetics – the increased molecular size due to the curing. Therefore, the modelling of cure kinetics is necessary to predict the change of the curing degree regarding the temperature and time during the viscosity measurements, which is essential for the viscosity modelling. The rheological modelling is conducted by multi-target regression analysis with the determined start parameters, based on four isothermal measurements.

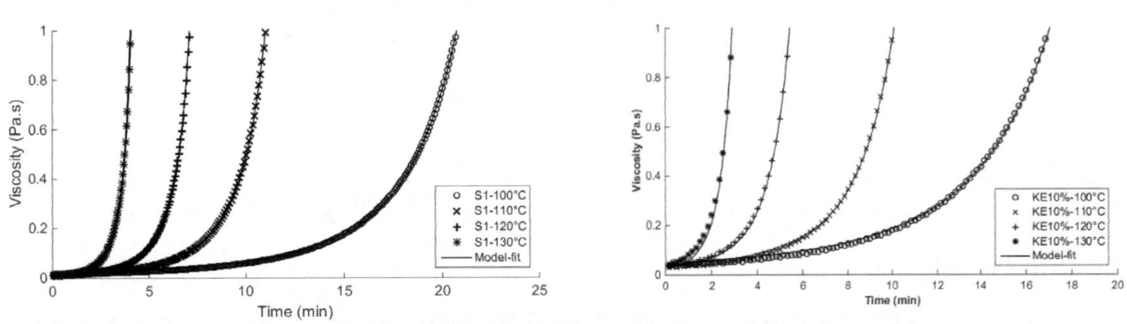

Figure 12: Rheological modelling with Castro-Macosko model (Left: neat epoxy matrix; Right: 10 wt% boehmite –epoxy matrix)

It could be seen that the rheological behavior of both the neat epoxy and boehmite-modified system could be described by the model quite well, and the model parameters are provided in the following Table 7.

Table 7 : Castro-Macosko model parameters for different mixtures

Model parameter / Mixure	η_∞ (Pa.s)	E_η (J/mol)	A	B	Gel convertion	R^2
0% (Pure resin)	7.73E-06	25327.36	−0.069T + 29.89	−0.105T + 42.09	0.5039	0.99
KE 5%	2.11E-06	28859.09	0.042T – 12.01	−0.105T + 42.65	0.5441	0.99
KE 10%	4.51E-06	29017.05	0.038T – 11.07	−0.0667T + 27.22	0.5766	0.99
KE 15%	7.1E-06	31085.82	0.169T – 58.46	−0.042T + 18.60	0.6490	0.99

If the model parameters are plotted against the NP concentration, it could be seen that the activation energy for flow E_η shows a linear increase and the η_∞ shows an exponential decrease with respect the NP concentration. The parameters A and B are temperature dependent fit parameters, by an isothermal temperature of 100 °C, they also show a linear correlation with respect to the NP concentration. Based on the modelling and correlation studies, it is now also possible to predict the rheological behavior depending on the NP concentration.

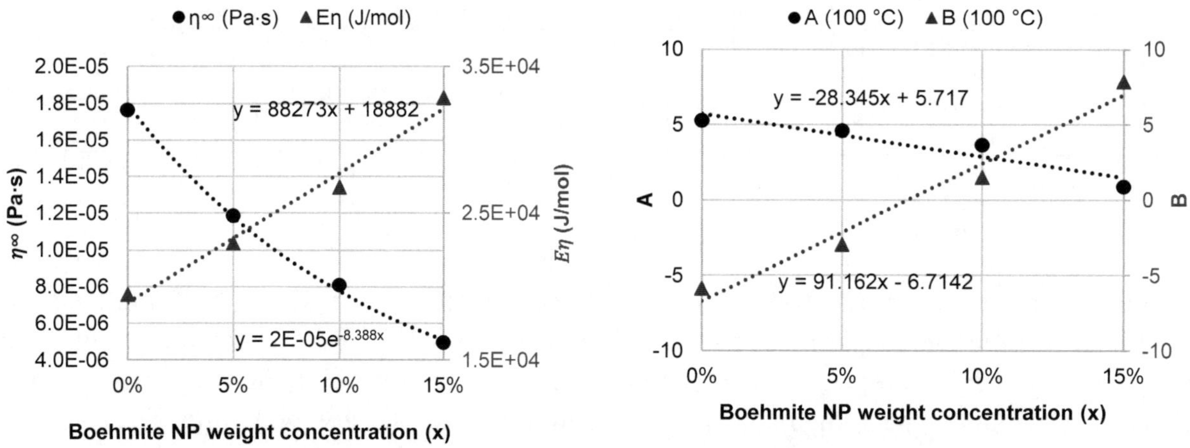

Figure 13: Change of rheological model parameters regarding the NP concentration

Table 8 : Generalized Castro-Macosko model parameters against NP concentration

Model parameters	Model parameters regarding NP concentration wt% (x)
η_∞ (Pa.s)	2E-5exp(-8.388x)
E_η (J/mol)	88273x+18882
A (100 °C)	−28.345x+5.717
B (100 °C)	91.162x-6.7142
Gel convertion	0.9356x+0.4982

5 CONCLUSTOINS

Curing kinetics and rheological behaviors are the most dominating material parameters for process design, control and final properties of the epoxy systems and their composites. Therefore, modelling the curing kinetic and rheology provides an indispensable basis for the process design, optimization and property enhancement.

In this paper, the influence of boehmite NPs on the cure kinetics and rheology of an epoxy resin system is systematically investigated. It could be shown that the boehmite NPs could build certain covalent connection with the epoxy, which inevitably influences the reaction kinetics of the matrix. It also could be shown a slight accelerating effect at the beginning until a certain cure degree that presumably due to the reactivity of the boehmite NPs with the epoxy, but after then it reversely inhibits the cure reaction – probably due to the spatial hindrance by the NPs, leading to a decreased polymerization network density of the matrix.

The standard Kamal-Sourour model showed an ill-posed characteristic due to which it was not possible to determine a clear correlation of the model parameters against the NP concentration. However, by assuming the parameter K_1 being infinitive small, it was possible to develop a good correlation between the model parameters with respect to NP concentration. At the end, generalized models for describing the cure kinetics and rheology of the NP modified epoxy resin system were proposed, which enables to describe and predict the cure and rheological behaviors depending on the time, temperature and the concentration of NPs.

ACKNOWLEDGEMETNS

This work originates from the Research Group Program FOR 2021"Acting Principles of Nano-Scaled Matrix Additives for Composite Structures" funded by the German Research Foundation (DFG: ZI 648/42-1; ZI 648/43-1).

References:

[1]. Mahrholz, T., J. Stangle and M. Sinapius, Quantitation of the Reinforcement Effect of Silica Nanoparticles in Epoxy Resins Used in Liquid Composite Moulding Processes. Composites Part A-Applied Science and Manufacturing, 2009. 40(3): p. 235-243.

[2]. Drescher, P., et al., Strengthening Fibre/Matrix Interphase by Fibre Surface Modification and Nanoparticle Incorporation Into the Matrix. Composites Science and Technology, 2013. 74: p. 60-66.

[3]. Sprenger, S., Epoxy Resin Composites with Surface-Modified Silicon Dioxide Nanoparticles: A Review. Journal of Applied Polymer Science, 2013. 130(3): p. 1421-1428.

[4]. Sprenger, S., Epoxy Resins Modified with Elastomers and Surface-Modified Silica Nanoparticles. Polymer, 2013. 54(18): p. 4790-4797.

[5]. Hsieh, T.H., et al., The Mechanisms and Mechanics of the Toughening of Epoxy Polymers Modified with Silica Nanoparticles. Polymer, 2010. 51(26): p. 6284-6294.

[6]. Abliz, D., et al., Curing Methods for Advanced Polymer Composites - a Review. Polymers & Polymer Composites, 2013. 21(6): p. 341-348.

[7]. Lee, G., et al., Enhanced Thermal Conductivity of Polymer Composites Filled with Hybrid Filler. Composites Part a: Applied Science and Manufacturing, 2006. 37(5): p. 727-734.

[8]. Rabearison, N., C. Jochum and J.C. Grandidier, A Cure Kinetics, Diffusion Controlled and Temperature Dependent, Identification of the Araldite LY556 Epoxy. Journal of Materials Science, 2011. 46(3): p. 787-796.

[9]. Garschke, C., et al., Cure Kinetics and Viscosity Modelling of a High-Performance Epoxy Resin Film. Polymer Testing, 2013. 32(1): p. 150-157.

[10]. Hardis, R., et al., Cure Kinetics Characterization and Monitoring of an Epoxy Resin Using DSC, Raman Spectroscopy, and DEA. Composites Part A-Applied Science and Manufacturing, 2013. 49: p. 100-108.

[11]. Ivankovic, M., et al., Curing Kinetics and Chemorheology of Epoxy/Anhydride System. Journal of Applied Polymer Science, 2003. 90(11): p. 3012-3019.

[12]. Huang, X. and B. Patham, Experimental Characterization of a Curing Thermoset Epoxy-Anhydride System—Isothermal and Nonisothermal Cure Kinetics. Journal of Applied Polymer Science, 2013. 127(3): p. 1959-1966.

[13]. Zhao, L. and X. Hu, Autocatalytic Curing Kinetics of Thermosetting Polymers: A New Model Based On Temperature Dependent Reaction Orders. Polymer, 2010. 51(16): p. 3814-3820.
[14]. Harsch, M., J. Karger-Kocsis and M. Holst, Influence of Fillers and Additives On the Cure Kinetics of an Epoxy/Anhydride Resin. European Polymer Journal, 2007. 43(4): p. 1168-1178.
[15]. Baller, J., et al., Interactions Between Silica Nanoparticles and an Epoxy Resin Before and During Network Formation. Polymer, 2009. 50(14): p. 3211-3219.
[16]. Louis, B.M., F. Klunker and P.A. Ermanni, Evaluation of Nanoalumina and Nanosilica Particle Toughened High Glass-Transition Temperature Epoxy for Liquid Composite Molding Processes. Journal of Composite Materials, 2016. 50(11): p. 1533-1545.
[17]. Sanctuary, R., et al., Influence of Al2O3 Nanoparticles On the Isothermal Cure of an Epoxy Resin. Journal of Physics-Condensed Matter, 2009. 21(0351183).
[18]. Baller, J., et al., The Catalytic Influence of Alumina Nanoparticles On Epoxy Curing. Thermochimica Acta, 2011. 517(1-2): p. 34-39.
[19]. Jux, M., et al., Effects of Al(OH)O Nanoparticle Agglomerate Size in Epoxy Resin On Tension, Bending, and Fracture Properties. Journal of Nanoparticle Research, 2017. 19(4).
[20]. Wu, Z.J., et al., Mechanical Properties of Epoxy Resins Reinforced with Synthetic Boehmite (AlOOH) Nanosheets. Journal of Applied Polymer Science, 2015. 132(414095).
[21]. Exner, W., et al., Nanoparticles with Various Surface Modifications as Functionalized Cross-Linking Agents for Composite Resin Materials. Composites Science and Technology, 2012. 72(10): p. 1153-1159.
[22]. Vyazovkin, S., et al., ICTAC Kinetics Committee Recommendations for Performing Kinetic Computations On Thermal Analysis Data. Thermochimica Acta, 2011. 520(1-2): p. 1-19.
[23]. Abliz, D., T. Artys and G. Ziegmann, Influence of Model Parameter Estimation Methods and Regression Algorithms On Curing Kinetics and Rheological Modelling. Journal of Applied Polymer Science, 2017. 134(30): p. 45137.
[24]. Halley, P.J. and M.E. Mackay, Chemorheology of Thermosets—an Overview. Polymer Engineering & Science, 1996. 36(5): p. 593-609.
[25]. Williams, M.L., R.F. Landel and J.D. Ferry, Mechanical Properties of Substances of High Molecular Weight .19. The Temperature Dependence of Relaxation Mechanisms in Amorphous Polymers and Other Glass-Forming Liquids. Journal of the American Chemical Society, 1955. 77(14): p. 3701-3707.
[26]. Ivankovic, M., et al., Curing Kinetics and Chemorheology of Epoxy/Anhydride System. Journal of Applied Polymer Science, 2003. 90(11): p. 3012-3019.
[27]. Mijovic, J. and C.H. Lee, A Comparison of Chemorheological Models for Thermoset Cure. Journal of Applied Polymer Science, 1989. 38(12): p. 2155-2170.
[28]. Castro, J.M. and C.W. Macosko, Studies of Mold Filling and Curing in the Reaction Injection-Molding Process. Aiche Journal, 1982. 28(2): p. 250-260.
[29]. Faria, H., F.M.A. Pires and A.T. Marques, Modeling the Rheology of SR1500 and LY556 Epoxies Under Manufacturer's Recommended Cure Cycles After Viscosimetry and Rheometry Characterization. Polymer Engineering & Science, 2014. 54(4): p. 831-839.
[30]. Hickey, C.M.D. and S. Bickerton, Cure Kinetics and Rheology Characterisation and Modelling of Ambient Temperature Curing Epoxy Resins for Resin infusion/VARTM and Wet Layup Applications. Journal of Materials Science, 2013. 48(2): p. 690-701.
[31]. Garschke, C., et al., Cure Kinetics and Viscosity Modelling of a High-Performance Epoxy Resin Film. Polymer Testing, 2013. 32(1): p. 150-157.
[32]. Cheng, S.Z.D., Handbook of Thermal Analysis and Calorimetry. Vol. 3, Vol. 3. 2002, Amsterdam; London: Elsevier.
[33]. Karkanas, P.I., I.K. Partridge and D. Attwood, Modelling the Cure of a Commercial Epoxy Resin for Applications in Resin Transfer Moulding. Polymer International, 1996. 41(2): p. 183-191.
[34]. ASTM E2070-13, Standard Test Method for Kinetic Parameters by Differential Scanning Calorimetry Using Isothermal Methods, ASTM International, West Conshohocken, PA, 2013, Www.Astm.Org.
[35]. Hülder, G., Reaktionskinetik Von Verbundmörtelsystemen Für Tragende Anwendungen Im Bauwesen. Journal of Plastics Technology, 2010. 3(6).

Graphene Oxide-Based Thermoplastic Polyurethane Nanocomposites. Part 1: Extensional vs Shear-Dominated Continuous Mixing and Rheological Behavior

Danda Chaitanya (1)*, Amurin Leice (2), Andrade Ricardo (2), Schneider Tyler (1), Fechine Guilhermino (2), Maia Joao (1)

(1) Case Western Reserve University - Ohio - USA, (2) MackGraphe - Graphene and Nanomaterials Research Center, Mackenzie Presbyterian University - Sao Paulo - Brazil

Polymer nanocomposites of Graphene Oxide (GO) have been known to impart enhanced mechanical behavior. While dispersing nanocomposites via conventional extrusion remains to be the quickest way producing polymer nanocomposites, attaining the desired dispersion has been a challenge. Extensional mixing elements embedded in twin-screw extruders (EME) are capable of dispersing nanocomposites in thermoplastics. EMEs generate an extension-dominated flow as opposed to kneading blocks, wherein flow is shear dominated. Polymer melt going through an EME passes through alternating converging and diverging zones. Polymer nanocomposites of 0.25 wt. % GO in Thermoplastic polyurethane (TPU) are processed in a 24 mm twin screw extruder. Four mixing elements, three EMEs with varying contraction ratios and channel length were compared with industry standard kneading block (KB) configurations. Optical microscopy followed by optical analysis processed by image J® reveals a superior dispersion with an aggressive mixing section to normal industry standard kneading element. Scanning electron microscopy (SEM) and X-ray diffraction studies were performed to qualitatively and quantitatively study the dispersion of GO in the TPU matrix. SEM images gathered on nanocomposites processed using the KB- screw configuration displayed larger GO platelets/ agglomerates, while materials processed through an aggressive EME displayed much smaller agglomerates. XRD data showed a shift of peak to lower d-spacing indicating platelets to be farther exfoliated in the case of nanocomposites processed employing the EME by comparison to those processed via aggressive KBs.

Thermoplastic polyurethane nanocomposites with hybrid fillers

Bonab Vahab Solouki (1), Manas-Zloczower Ica (1)*

(1) Case Western Reserve University - Ohio - USA

Thermoplastic polyurethanes (TPUs) are among the most versatile engineering polymers. The presence of hard and soft segments on their backbone and specific hydrogen bond interactions between the hard segments provide TPUs with outstanding engineering properties. Nanofillers are incorporated in the TPU matrix in order to compensate for the mechanical property loss due to weakening of the hydrogen bonding between hard segments at higher temperatures and also to enhance other properties such as thermal and electrical conductivity, fatigue life, tribological properties. In this study, hybrid carbon nanofillers with different geometric shapes, namely rod like (1D) carbon nanotubes and sheet like (2D) graphene nanoplatelets were incorporated in the TPUs. TPU nanocomposites were prepared via a sustainable in-situ polymerization of Polycaprolactonediol, 1,4-Phenyldiisocyanate (PPDI) and Butanediol (BDO) in the presence of hybrid carbon nanofillers using a two step method. Mechanical and thermal properties of TPU nanocomposites with different ratios of carbon nanotube/graphene nanofillers were investigated. Differences in mechanical behavior were attributed to the formation of different types of 3D percolated networks when using hybrid nanofillers at different ratios.

Using zinc oxide nanoparticles to improve the thermal stability of a high-performance benzoxazine resin

J. Horion[a,*], C. van Tieghem de Ten Berghe[a], L. Bonnaud[b], C. Bailly[a]
[a] Institute of Condensed Matter and Nanosciences (IMCN) – Bio & Soft Matter (BSMA), Université catholique de Louvain, Croix du Sud, box L7.04.02., Boltzmann building 1348 Louvain-la-Neuve, Belgium
[b] Laboratory of Polymeric and Composite Materials, Center of Innovation and Research in Materials and Polymers (CIRMAP), Materia Nova Research Center & University of Mons, 23 Place du Parc, B-7000 Mons, Belgium
* corresponding author : jeremy.horion@uclouvain.be

Introduction

Benzoxazine resins have attracted an increasing interest in the last two decades, due to their easy synthesis, very high molecular design flexibility and exceptional combination of properties [1]. Benzoxazine resins (BZO) can self-polymerize through an autocatalytic ring opening polymerization (ROP), they exhibit a near-zero volume shrinkage upon polymerization and present outstanding thermal and mechanical properties. These features make them suitable for a wide panel of applications [2].
Zinc oxide (ZnO) nanoparticles, on the other hand, have drawn attention in recent years due to their physical and chemical properties, such as chemical stability, low dielectric constant, high luminous transmittance, high catalysis activity, intensive ultraviolet and infrared absorption [3, 4]. ZnO nanoparticles have also proven to effectively improve the thermal properties of thermosets [5].
In this work, nanocomposites with a high thermal stability were developed using T-DDM, a new benzoxazine resin based on 4,4'-diaminodiphenylmethane (DDM) and tyrosol, and zinc oxide (ZnO) nanoparticles. First, some of the resin's kinetic parameters are investigated. Then, a synergetic effect of the ZnO nanoparticles on the ROP's enthalpy is evidenced, giving rise to a significant improvement of the degradation temperature and char yield. The interaction between the benzoxazine matrix and the ZnO nanoparticles is also investigated by FTIR spectroscopy.

Materials and methods

1. Benzoxazine matrix

The benzoxazine used is a T-DDM was synthetized by Materia Nova in Mons, Belgium. This benzoxazine precursor is based on tyrosol and DDM. The uncured resin was provided as a light yellow powder. Because the resin is produced at lab scale, small differences can arise between different batches, which can affect some of the materials properties. In order to avoid such issues, all the systems of the present study were made from the same batch.

2. Zinc oxide

The zinc oxide used in this study is the *Zano®20* from the Zinc Chemicals division of Umicore and is provided in form of a powder.

3. Nanocomposites preparation

After dissolving the T-DDM into chloroform, the desired amount of ZnO was added. The mixture was then tip-sonicated for a total of 90 seconds, divided in 3 pulses of 30 seconds separated by 30 seconds resting time. The tip-sonication was performed with an Omni-Ruptor 400, from OMNI International Inc.

4. Electron microscopy

Scanning electron microscopy (SEM) was performed on a JEOL 7600F device operating at 15.0 kV. Before SEM observation, the samples were prepared by cold breakage at liquid nitrogen temperature, and metallized with a 20nm thick chromium layer. This metallization was performed with a Sputter Coater 208HR and a Thickness Controller mtm20 from Cressington.

Transmission electron microscopy (TEM) was performed on a LEO 922 device operating at 120 kV. Prior to observation, the samples were cut to ultrathin sections under ambient conditions, by means of a Zeiss microtome, using a Diatome histo diamond knife with cutting angle of 35°. The sections were then collected and placed on 200 mesh carbonated copper grids.

5. Thermal analysis

Differential scanning calorimetry (DSC) experiments were conducted on a robotic DSC 821e from Mettler Toledo. The samples were put into 40 µL aluminum pans, and experiments were conducted under a nitrogen flow of 50 mL/min. Previously to each set of experiments, the machine was calibrated through a standard calibration test with Indium.

Thermogravimetric analysis (TGA) of the nanocomposite blends was performed with a TGA 851e from Mettler Toledo. The samples were put into alumina crucibles, and heated from room temperature to 800°C at a rate of 10°C/min and under a nitrogen flow of 100 mL/min.

6. Infrared spectroscopy

FTIR was done on a *Spectra Tech Continuum* FTIR device from the Thermo Electron Corporation. The nanocomposites were crushed to a powder, which was then mixed with KBr and pressed to small pads. The spectra were then acquired in transmission mode, with 128 scans and a resolution of 4 cm^{-1}.

Results and discussion

1. Kinetics of the T-DDM

Neat T-DDM samples were brought from 30°C to 300°C in a DSC furnace, with different heating rates. The ROP of benzoxazine resins is known to be an autocatalytic process [1]. Equation 1 gives the general expression for the reaction rate of autocatalytic systems.

$$\frac{d\alpha}{dt} = k \cdot (1-\alpha)^n \cdot \alpha^m \qquad (1)$$

where $\alpha = \alpha(t)$ is the fractional conversion, k is the kinetic rate constant and *(n+m)* is the overall order of the reaction. Under the assumptions that the specific heat is constant or follows at linear variation with the heating rate of the DSC experiments and that the reaction rate follows an Arrhenius temperature dependence, equation 1 can be transformed to :

$$\ln\left(\beta \frac{d\alpha}{dT}\right) = \ln(A) - \left(\frac{E_a}{RT}\right) + n \cdot \ln(1-\alpha) + m \cdot \ln(\alpha) \qquad (2)$$

with $\beta = dT/dt$ the heating rate, A the pre-exponential factor of the Arrhenius expression for the temperature dependence of the reaction rate, E_a the activation energy and R the gas constant. Equation 2 eventually allow us to determine, through a multi-linear regression on our DSC results, the kinetic parameters of the ROP of T-DDM, which are summarized in table 1.

β [°C/min]	A	E_a [kJ/mol]	n	m
5	3.99×10^7	90.25	1.63	0.59
10	4.01×10^7	89.10	1.57	0.69
15	4.00×10^7	88.00	1.69	0.78

Table 1. Summary of the kinetic parameters for the ROP of T-DDM

From the multi-linear regression, an activation energy around 89 kj/mol is found, while the order of the reaction is approximately 2.3. Nevertheless, one should remain careful with all these values, for several sources of inaccuracy can be easily identified. First, it is strongly expected that an endothermic reaction, due to the condensation polymerization of the hydroxyl functionalities of T-DDM, would occur simultaneously to the ROP. Second, it was assumed that the specific heat, c_p, was constant or followed a linear variation with the heating rate, and that the reaction rate followed an Arrhenius temperature dependence. These hypothesis would need to be verified. Finally, errors resulting from the measuring limits of the DSC apparatus or from the inaccuracy of the multi-linear regression should be considered.

2. Impact of ZnO on the thermal properties of T-DDM

As can be seen from the TEM micrographs presented on Figure 1, the dispersions of ZnO in T-DDM are quite good. Some individualized ZnO nanoparticles can be observed, however most of the particles remain located into small clusters which in turn are well dispersed in the matrix. These clusters average a micrometer in diameter.

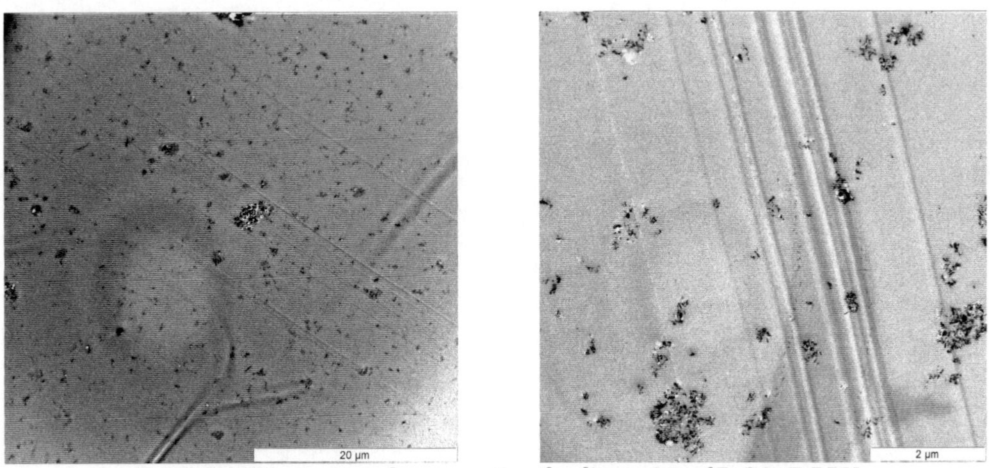

Figure 1. TEM micrographs presenting the dispersion of ZnO in T-DDM

The ROP's enthalpy was then measured for systems containing various amounts of ZnO nanoparticles. As can be seen on Figure 2, the measured enthalpy decreases, which can be attributed to the replacement of a part of the reactive species by the ZnO, which is not expected to react. Hence, the measured enthalpy should follow a simple rule of mixture, and plotting the enthalpy versus ZnO loading should give a straight line.

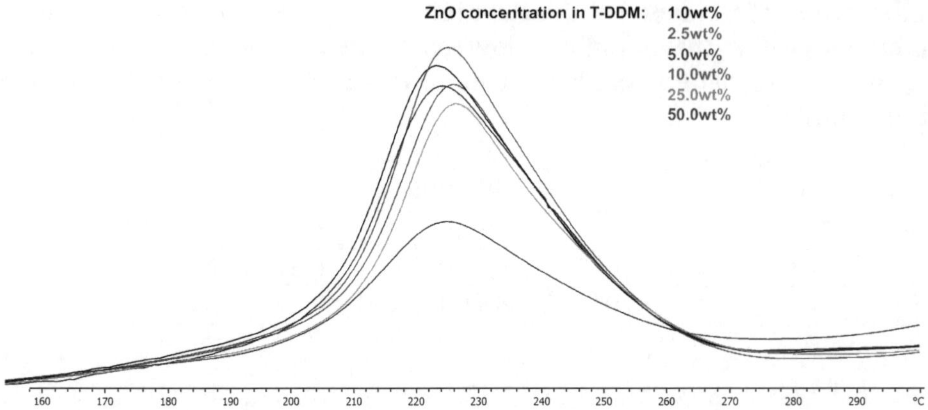

Figure 2. ROP's exotherm for blends of T-DDM with various loadings of ZnO nanoparticles

The measured enthalpies are plotted on Figure 3, and a polynomial fit was drawn. From this plot, one can clearly see a deviation from the expected straight line, which cannot be explained by the presence of T-DDM nor of zinc oxide alone. It strongly suggests an interaction between the ZnO nanoparticles and the T-DDM matrix, leading to a modification of the ROP's mechanism, or to additional bonds being formed.

An FTIR analysis was then performed on cured systems, both neat and loaded with ZnO, in order to search for the presence of additional peaks in the latter, indicating the formation of new bonds. On Figure 4, which presents the spectra obtained between 1800 and 650cm^{-1}, one can observe that the peak at 1480cm^{-1}, characteristic of the tetra-subsituted benzene formed during the ROP, disappears in the presence of ZnO nanoparticles, although a shouldering remains around 1490cm^{-1}. Moreover, a very intense peak appears with zinc oxide at 1678cm^{-1}. The nature of the bond related to this peak is not clear yet, but it may be related to the presence of some additional hydroxyl groups forming at the surface of ZnO nanoparticles.

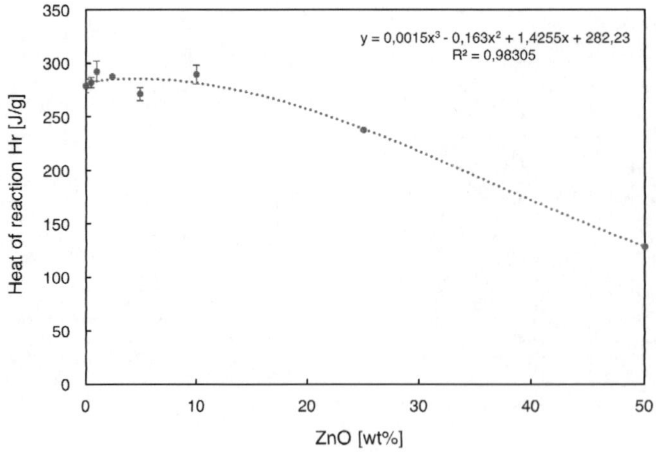

Figure 3. ΔH_r versus ZnO loading.

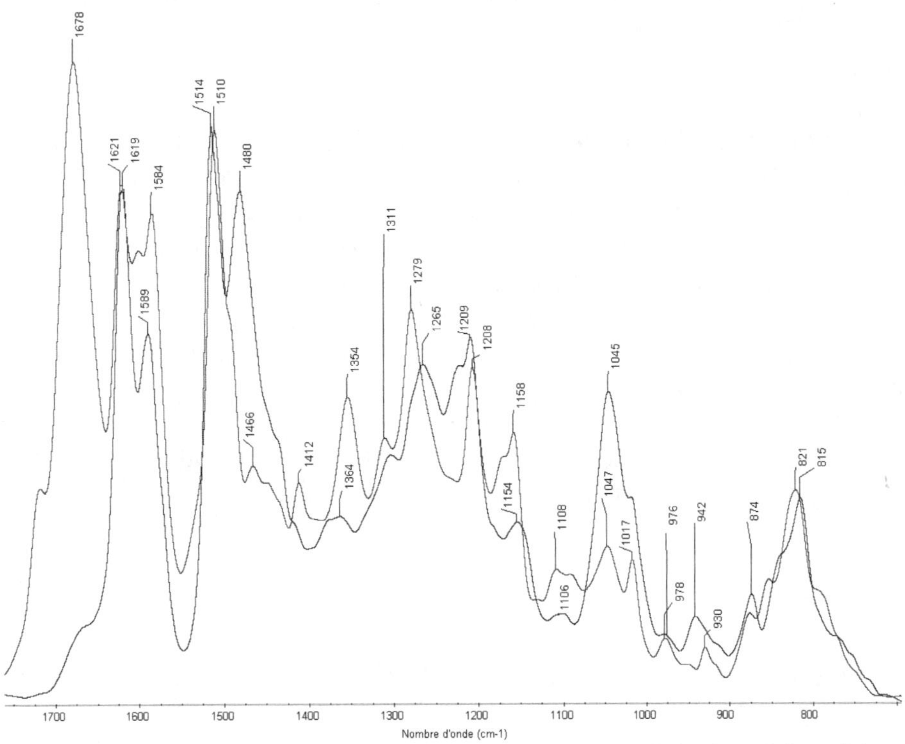

Figure 4. FTIR spectra of cured systems, neat (blue) and with 5wt% ZnO (purple)

Eventually, the thermogravimetric analysis performed on the nanocomposite systems lead to the results presented on figure 5. As one can observe, the interaction between the ZnO nanoparticles and the matrix, evidenced by DSC and FTIR, is responsible for an impressive synergetic effect on the systems thermal degradation. Indeed, the addition of only 0.5wt% ZnO leads to a spectacular 7% increase of the char mass at 800°C. Although this effect rapidly softens, with 10% additional char yield for 1wt% ZnO, a net char yield improvement of 17% for 5wt% nanoparticles added remains an impressive result. What's more, the zinc oxide also seem have an effect on the system degradation, pushing its initiation towards higher temperatures.

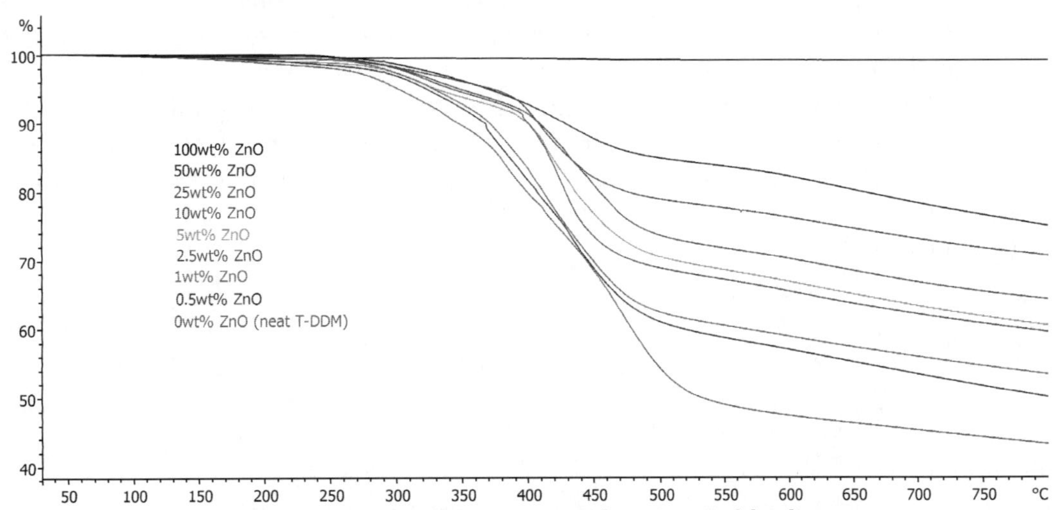

Figure 5. TGA of T-DDM systems with various ZnO loadings

Conclusions

The kinetic study on neat T-DDM allowed us to determine an activation energy around 89 kJ/mol for the ROP and an order of the reaction around 2.3. However, condensation reactions are expected to occur between the hydroxyl groups present in the network, thus introducing a bias which may be non-negligible. What's more, the hypothesis we made in our calculations may introduce additional errors.
The addition of zinc oxide nanoparticles in the T-DDM give rise to an interaction between the nanoparticles and the matrix, responsible for an increase of the reaction enthalpy, associated with the formation of new chemical bonds in the T-DDM network. Altough the nature of those bonds could not be ascertained for sure, the presence of the ZnO is responsible for an increase of the degradation temperatures, together with a spectacular synergetic effect on the char yield.

Acknowledgments

The authors are thankful to the Flycoat project and Walloon Region for financially supporting this research, and to the Materia Nova research center for providing the benzoxazine precursors.

Bibliography

1. H. Ishida, T. Agag. "Overview and Historical Background of Polybenzoxazine Research". *Handbook of Benzoxazine Resins* (2011), pp. 3–81
2. S. Rimdusit, C. Jubslip, S. Tiptipakorn, "Introduction to Commercial Benzoxazine and Their Unique Properties". *Alloys and Composites of Polybenzoxazines* (2013), pp. 1–23
3. AZoNano. *Zinc Oxide (ZnO) Nanoparticles – Properties , Applications.* 2013. URL: http://www.azonano.com/article.aspx?ArticleID=3348 (visited on 09/10/2017)
4. S. S. Kumar, P. Venkateswarlu, V. R. Rao, and G. N. Rao. "Synthesis, characterization and optical properties of zinc oxide nanoparticles". *International Nano Letters* 3.30 (2013), pp. 1–6
5. O. Zabihi, S. Mojtaba Mostafavi, Fatemeh Ravari, Aminreza Khodabandeh, Amin Hooshafza, Karim Zare and Mehrab Shahizadeh. The effect of zinc oxide nanoparticles on thermo-physical properties of diglycidyl ether of bisphenol A/2,2 -Diamino-1,1 -binaphthalene nanocomposites. *Thermochimica Acta*, 521 (2011) 49–58

Characterization of the nucleation efficiency of filler at the crystallization of polypropylene at high supercooling

Schawe Juergen E.K. (1)*

(1) Mettler-Toledo - Switzerland - Switzerland

The influence of filler on the crystallization kinetics of isotactic polypropylene is studied on the example of different carbon nanotubes (CNT) in a wide concentration range. Isothermal and non-isothermal crystallization was measured in a temperature range between 0 °C and 130 °C by fast scanning DSC measurements. The acceleration of the crystallization process is described by an efficiency parameter which is derived from the classical crystallization kinetics. This parameter is independent on the filler content and characterizes the ability of the filer to generate additional nuclei. We recommend the efficiency parameter for characterization the influence of fillers and nucleating agents on the crystallization kinetics.

Functionalized Graphene Oxide As a Compatibilizer for Polymer Blends

Kenig Samuel [1]*, Naveh Naum [1], Kol Adi [1], Bercovich Nofar [1]

[1] Shenkar College - Israel - Israel

With the emergence of cost effective processes for preparation of Graphene Oxide (GO), new routes have become feasible for functionalization of GO and its application as tailored made compatibilizers for polymer blends. The present study deals with oxidation of commercial stacked graphene to various degrees (5-50% oxygen content) followed by silanization in aqueous media to yield controlled grafting of graphene possessing a wide range of electrical conductivities. The resultant functionalized graphenes were subsequently used as compatibilizers for polyamide (PA)/polypropylene (PP) blends. The experimental program was accompanied by thermodynamic and dynamic analysis for investigating the localization of the functionalized GO moieties. Experimental results and the associated analysis indicated that the functionalized GO can be tailored to control its localization in the hydrophilic (PA) or the hydrophobic (PP) blend's phases or even at the interfacial region, with significant consequences as far as mechanical and electrical properties.

Permation properties of PA6/GnP-nanocomposites

Boldt Regine [(1)]*, Leuteritz Andreas [(1)], Langner Patrick [(1)], Wagenknecht Udo [(1)]

[(1)] Leibniz-Institut für Polymerforschung Dresden e.V. - Saxony - Germany

Graphene nanoplatelets (GnP) offer extraordinary good possibilities in order to enhance the permeation properties of nanocomposites. Due to their layered structure and the high aspect ratio paths of gas diffusion through polymer films can be elongated if the nanoplatelets are well dispersed and best exfoliated. But since size and morphology of GnP affects the degree of dispersion different morphology types of GnP have been used for preparation of nanocomposites based on semi-aromatic and aliphatic polyamide. Different compositions of both matrix and GnP materials were melt mixed in a small scale DSM X-Plore microcompounder and subsequently compression moulded. Rheological properties and permeation rate of He have been measured. Furthermore, morphology was investigated using electron microscopy. First results show, that the permeation rate of GnP/PA6-nanocomposites does not significantly correlate with the filling level of GnP but is influenced by the type of polymer. This effect is caused by different interaction between GnP and polymeric matrix and could be confirmed by rheological and morphological investigations.

Effect of Hydrophobic-Modification of Cellulose Nanofiber (CNF) on the Crystallization Kinetics of Polypropylene / CNF Composites

Yuta Hikima*, Megumi Ando, Minami Sodenaga, Fumiya Taguchi and Masahiro Ohshima

Dept. of Chem. Eng., Kyoto University, Katsura, Kyoto 615-8510, JAPAN

hikima@cheme.kyoto-u.ac.jp

Abstract. Cellulose nanofibers (CNFs) have been recently attracting a great deal of attention because of their superior characteristics in regard to mechanical strength, light weight and bio-renewable resource, and they are often compound with polymers to improve the mechanical property. To obtain the high-performance nanocomposites with hydrophobic polymers, such as polypropylene (PP), the surface modification of CNFs with the hydrophobic functional groups is needed to ensure the good dispersity of CNF in the polymer. In this study, PP / hydrophobic modified CNF (HM-CNF) composites were prepared and, with the fast scanning chip-calorimetry, the crystallization behaviors of the composites were measured under the practical processing conditions, such as the fast cooling rates conducted by injection molding machines. The effect of the chemical modification on the kinetics of PP crystallization was investigated by changing the length of the alkyl chain of modifier as C2, C8 and C14. The overall crystallization rate of PP/HM-CNF composites was increased at temperature above 40ºC by the presence of HM-CNF. This was because the HM-CNF acted as a nucleating agent and promoted the heterogeneous nucleation of PP crystal. The effect was strongly dependent on the kind of hydrophobic group and the temperature range. HM-CNFs' dispersity in PP increased with the alkyl chain length of the modifier, in contrast, the crystal nucleating effect of CNFs became weaker. The crystallization behaviors at the fast cooling rates were also influenced by hydrophobic modification of CNFs. At the cooling rates from 1 K/s to 1000 K/s, the crystallization temperature of the PP/C2-CNF composite rose at least 15ºC compared with pure PP.

Keywords: Polypropylene, Cellulose Nanofiber, Crystallization, Thermal Analysis
PACS: Solidification, 81.30.Fb

INTRODUCTION

Nano-sized cellulose fibrils called cellulose nanofibers (CNFs) have been recently attracting a great deal of attention because of their superior characteristics in regard to mechanical strength, light weight and bio-renewable resource. The use of CNFs as a nano-filler for polymer composites is promising to improve the rheological properties of the molten polymers as well as the product properties, such as a specific strength and a low coefficient of linear thermal expansion. The most important factor for bringing out the performance of polymer / CNF composites is the dispersity of CNFs in the polymer. The chemical modification of CNFs with hydrophobic functional group is effective in improving the dispersity of CNFs in hydrophobic polymers, for example, polypropylene (PP), with the result that, the mechanical property of polymer / hydrophobic-modified CNF (HM-CNF) composites increased [1]. On the other hand, the crystal nucleating effect of cellulose fibrils in the crystalline polymers has been reported [2-5]. Jenicek et al. investigated the effect of hydrophobic modification of cellulose surface on the polymer crystallization kinetics and dispersity in a PP matrix [2]. They revealed that the ability in crystal nucleation of cellulose fibrils was strongly affected by the kind of hydrophobic functional group. However, this research focused on the phenomena occurring at much slower cooling rates than the practical polymer processing processes due to the limitation of analytical instrument: a conventional DSC. Therefore, the crystallization behavior of PP / HM-CNFs composites under the practical processing conditions, such as the fast cooling rates conducted by injection molding machines, has not been investigated.

Recently, an advanced DSC method called fast scanning chip-calorimetry (FSC) was developed and has been applied for kinetic analysis of the polymer crystallization under fast cooling conditions [6, 7]. The FSC overcomes the difficulty of precise temperature control with fast heating / cooling rates by employing the micro-sized heater and temperature sensors fabricated on a membrane by the MEMS technique. The FSC is useful for evaluating the crystal nucleating effect of additives dispersed in polymer matrix as well as investigating the temperature range of polymer crystallization at the various cooling rates corresponding to the practical polymer processing conditions [7]. For example, by comparing the isothermal crystallization behavior of PP added the additive with that of pure PP in the wide range of crystallization temperature, the crystal nucleating effect of additives can be detected as the decrease of the half crystallization time observed above 40°C.

In this study, the crystallization behavior of PP / HM-CNF composites was investigated with FSC method in order to reveal the effect of the hydrophobic modification on the PP crystallization behavior under the fast cooling conditions. The bacterial CNFs hydrophobized by esterification with four different carboxylic acids were compounded with PP and evaluated their crystal nucleating effect from the crystallization kinetic analysis under isothermal conditions.

EXPERIMENTAL

Sample Preparation

The hydrophobic-modified CNFs were prepared according to the procedure reported by Agustin et al. [8]. The bacterial cellulose was esterified with acetyl (C2), octanoyl (C8), tetradecanoyl (14) and benzoyl (BNZ) chrolides. Esterification was carried out in an acyl chloride-pyridine system. The degrees of substitution (DS) were controlled as to be 0.4 for all HM-CNFs. As the length of alkyl chain in hydrophobic functional group increased, the hydrophobicity of HM-CNFs increased [9].

The provided HM-CNFs were crushed with a tabletop crusher (Wonder Crush, Osaka Chemical Co., Ltd.) and dried at 80°C for 5 hours in a hot oven. The dried HM-CNFs were mixed with PP powder (H50000, Prime Polymer Co., Ltd.) as the cellulose content became 5 wt. %. The mixtures of PP powder and HM-CNF were compounded at 180°C with 120 rpm for 20 minutes using a rotating twin screw mixer (Labo Plastomil, KF15V model, Toyoseiki).

Measurement

FSC measurements were performed using a Flash DSC1 (Mettler Toledo). The measurements were performed with about 50 ml/min nitrogen flow. The obtained PP / HM-CNF compounds were sliced with the thickness of 10 micrometer by a microtome. The sliced specimens were cut into the area sizes of approximately 100 μm x 100 μm, and then, the sample was directly placed on a micro thermal sensor. A new sensor was used for each sample. The FSC analyses were conducted with two methods: the isothermal crystallization analysis and the non-isothermal crystallization analysis. The measurement and analytical procedures are explained as follow sections. Each sample was successively measured twice for both methods to confirm the influence of the sample degradation.

Isothermal Crystallization Analysis

The isothermal crystallization analysis was performed to evaluate the crystal nucleating effect of HM-CNFs on PP crystallization. The sample was heated up from -50°C to 200°C with the heating rate of 1000 K/s, and then, melted for 1 second. The molten sample was quenched with the cooling rate of 4000 K/s to the crystallization temperature (T_c) programmed before. The crystallization behavior at an isothermal condition was recorded as change of the heat flow in the time domain for 10 seconds. The half crystallization time was calculated from this DSC curve for each T_c. After the isothermal crystallization at a T_c, the sample was quenched to -50°C with 4000 K/s. The measurements were repeated according to the cycle explained above with varying the T_c from 0°C to 120°C.

Non-Isothermal Crystallization Analysis

The non-isothermal crystallization analysis was performed to investigate the crystallization behavior of PP/HM-CNF composites during rapid cooling process with various constant cooling rates from 1 K/s to 1000 K/s. The sample was heated up from -50°C to 200°C with the heating rate of 1000 K/s, melted for 1 second, and cooled to -50°C with a constant cooling rate. The heat flow in the cooling DSC curve was divided by the cooling rate. The temperature range of crystallization was evaluated with three characteristic temperatures, the peak-top temperature (T_{cp}), the extrapolated onset temperature ($T_{c\ on}$) and the extrapolated end temperature ($T_{c\ end}$), determined from exothermic peak in the cooling DSC curve. Here, $T_{c\ on}$ and $T_{c\ end}$ means the temperatures at which PP crystallization begins and ends.

The effect of hydrophobic modification of CNFs on the temperature range of PP crystallization at various cooling rates was investigated.

RESULTS AND DISCUSSION

Isothermal Crystallization Analysis

Figure 1 shows the DSC curves of isothermal crystallization behavior of pure PP and PP/HM-CNF composites at the temperatures of 80, 90 and 100°C. The heat flow was normalized by the value at the peak top for comparing crystallization kinetics between the samples. The crystallization of PP/HM-CNF composites began earlier than that of pure PP sample. This fact suggests that the HM-CNFs accelerate the nucleation of PP crystal. The crystal nucleating effect of C2-CNF was much larger than that of C8- and C14 CNFs. To compare the dependence of crystallization kinetics on the crystallization temperature between the samples, the half crystallization time was calculated for each isothermal DSC curve.

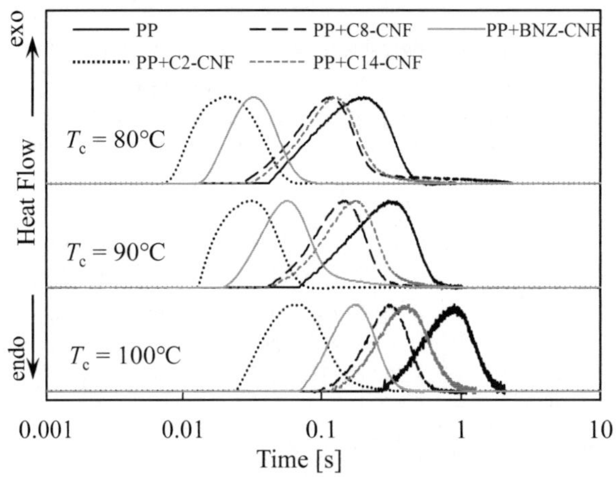

FIGURE 1. Normalized heat flow curves of isothermal crystallization of pure PP and PP/HM-CNF composites. The crystallization temperatures are 80, 90 and 100°C. The crystallization rates of PP increased by adding HM-CNFs.

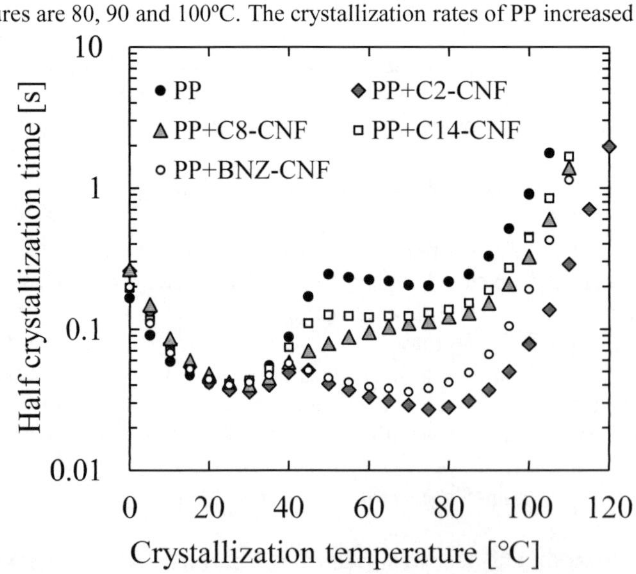

FIGURE 2. Half crystallization time as a function of the crystallization temperature for the pure PP and PP/HM-CNF composites. The crystal nucleating effect of HM-CNFs are observed above 40°C.

The half crystallization time is plotted against the crystallization temperature for each sample in Figure 2. The plot for pure PP shows the bimodal temperature dependence. This is a typical result for FSC isothermal crystallization analysis of pure PP [10]. The bimodal curve is due to the meso-phase formation dominated by the homogeneous nucleation below 50°C and the α-phase crystallization dominated by the heterogeneous nucleation above 50°C [10,

11]. The crystal nucleating effect of each HM-CNF was observed above 40°C in Fig. 2. This is due to the increase of the heterogeneous nucleation of PP crystal on the surface of HM-CNFs. In the temperature range between 40°C and 80°C, C2- and BNZ-CNF indicated great nucleating effect, and the effect increased as the alkyl chain in hydrophobic functional group became shorter. On the other hand, from 80°C to 120°C, the PP/C2-CNF composite indicated much faster crystallization rate than the other samples. In contrast, the crystal nucleating effect of BNZ-CNF decreased as well as those of C8- and C14-CNFs with the increase of crystallization temperatures. These results suggest that the hydrophobic modification significantly affects the temperature dependence of the nucleating effect of CNFs in the PP crystallization. The crystal nucleating effect of HM-CNF increases with the weak hydrophobic modification such as acetyl group (C2).

Non-Isothermal Crystallization Analysis

The crystallization kinetics of PP/HM-CNFs composites varied with the kind of hydrophobic functional group below 120°C. PP crystallizes below 120°C with faster cooling rates than the common condition in the conventional DSC measurement, so that the crystallization behavior of PP/HM-CNF composites should be investigated at fast cooling conditions as well as conducted in the practical manufacturing processes. Figure 3 shows the crystallization behavior of each PP/HM-CNF composites at the cooling rates of 10, 50, 100 and 500 K/s. Corresponding to the results in isothermal crystallization analysis, the temperature ranges of PP crystallization at cooling rates of 10, 50 and 100 K/s were highest for compounded with C2-CNF, followed by BNZ-CNF, C8-CNF and C14-CNF in that order. The crystallization peaks of PP/C2-CNF and PP/BNZ-CNF clearly appeared despite a fast cooling rate of 500 K/s.

Changes of crystallization temperature were plotted against the cooling rate for each sample in Figure 4. The crystallization temperatures, T_{cp}, $T_{c\ on}$, $T_{c\ end}$ of the PP/C2-CNF composite rose more than 15°C compared with pure PP at cooling rates faster than 5 K/s. This is due to the great crystal nucleating effect of C2-CNFs above 40 °C demonstrated in Figure 2. The crystallization onset temperatures of PP/C8-CNF, PP/C14-CNF and PP/BNZ-CNF indicated almost same dependence on the cooling rate. On the other hand, T_{cp} and $T_{c\ end}$ for the PP/BNZ-CNF composite were much higher than the others at the cooling rates faster than 50 K/s. At the cooling rate of 50 K/s, the $T_{c\ on}$ of these composites are almost 100°C as shown in Figure 4 (b), so that the crystallization proceeds below 100°C at the cooling rates faster than 50 K/s. Therefore, the differences in T_{cp} and $T_{c\ end}$ observed at the cooling rate faster than 50 K/s were dominated by the difference in the crystallization rates in the temperature range from 40°C to 100°C as shown in Figure 2. This result also means that the BNZ-CNF acts well as the crystal nucleating agent at the fast cooling conditions. As demonstrated above, the difference of crystallization behavior between the PP/HM-CNF composites at fast cooling conditions is strongly affected by the crystal nucleating effect evaluated in the isothermal crystallization analysis.

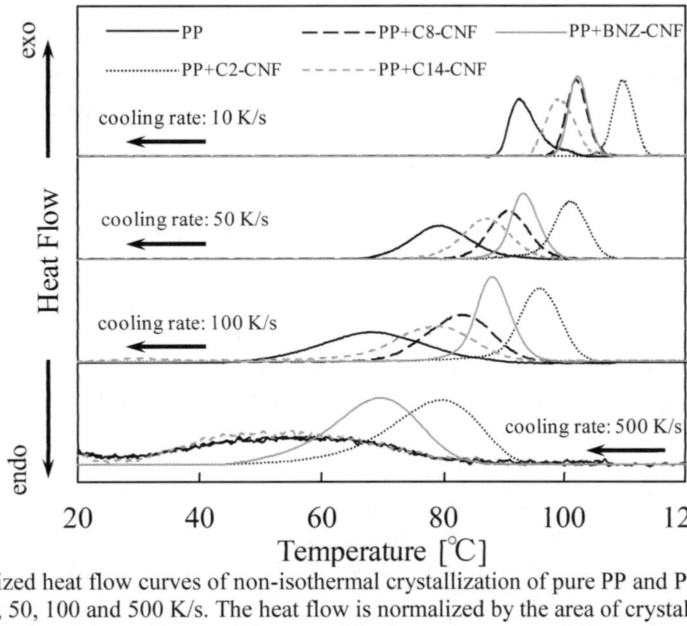

FIGURE 3. Normalized heat flow curves of non-isothermal crystallization of pure PP and PP/HM-CNF composites. The cooling rates are 10, 50, 100 and 500 K/s. The heat flow is normalized by the area of crystallization peak for comparing among the curves at the same cooling rate. The crystallization temperature of PP rose by adding HM-CNFs.

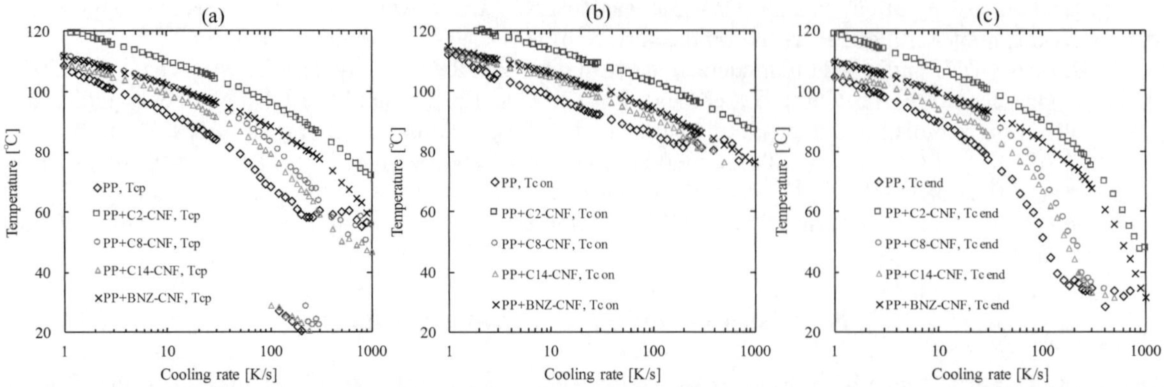

FIGURE 4. Crystallization temperatures of (a) peak-top (T_{cp}), (b) onset ($T_{c\ on}$) and (c) endset ($T_{c\ end}$) as a function of cooling rate. T_{cp}, $T_{c\ on}$ and $T_{c\ end}$ increased all over the range of cooling rate from 1 to 1000 K/s by adding C2-CNF. By adding BNZ-CNF, T_{cp} and $T_{c\ end}$ increased at the cooling rates above 50 K/s.

CONCLUSION

The crystallization behaviors of PP/HM-CNF composites were measured using the fast scanning chip-calorimetry. The crystal nucleating effect of HM-CNFs was observed in the temperature range above 40°C in which the heterogeneous nucleation is dominant. The acetylated (C2) CNFs indicated the strongest nucleating effect on the PP crystallization among the HM-CNFs which has the different length of alkyl chain (C2, C8 and C14) in the functional group. Thus, the nucleating effect increased as the hydrophobicity of the functional group became weak. By adding the C2-CNFs, the temperature range of PP crystallization rose at least 15°C compared with pure PP at the cooling rates from 1 K/s to 1000 K/s. The crystallization end temperature particularly increased at the cooling rates faster than 50 K/s by adding C2- and BNZ-CNFs due to their strong nucleating effect in the temperature range between 100°C and 40°C. In conclusion, the hydrophobicity of functional groups on the CNFs' surface influences significantly the crystal nucleating effect of CNFs, and consequently, the crystallization behavior at fast cooling rates.

ACKNOWLEDGMENTS

The authors would like to thank Professor Hiroyuki Yano of Research Institute for Sustainable Humanosphere in Kyoto University for providing the hydrophobic modified CNFs samples and helpful advices. We also thank Dr. Takeshi Semba and the members in the Polymer Materials Lab. of Kyoto Municipal Institute of Industrial Technology and Culture for the technical supports in the compounding process. This research was financially supported by the Advanced Low Carbon Technology Research and Development Program (ALCA) of the Japan Science and Technology Agency (JST).

REFERENCES

1. A. Sato, D. Kabusaki, H. Okumura, T. Nakatani, F. Nakatsubo, and H. Yano, Composites Part A: Applied Science and Manufacturing **83**, 72 (2016).
2. M. Janicek, M. Polaskova, R. Holubar, and R. Cermak, Cellulose **21**, 4039 (2014).
3. W. Ding, R. Chu, L. Mark, C. Park, and M. Sain, European Polymer Journal **71**, 231 (2015).
4. J. Chen, D. Wu, K. Tam, K. Pan, and Z. Zheng, Carbohydrate Polymers **157**, 1821 (2017).
5. V. Khoshkava, H. Ghasemi, and M. Kamal, Thermochim Acta **608**, 30 (2015).
6. V. Mathot, M. Pyda, T. Pijpers, G. Poel, E. van de Kerkhof, S. van Herwaarden, F. van Herwaarden, and A. Leenaers, Thermochimica Acta **522**, 36 (2011).
7. J. Schawe, P. Pötschke, and I. Alig, Polymer **116**, 160 (2017).
8. M. Agustin, F. Nakatsubo, and H. Yano, Cellulose **23**, 2887 (2016).
9. L. Tomé, M. Freire, L. Rebelo, A. Silvestre, C. Neto, I. Marrucho, and C. Freire, Green Chem 13, 2464 (2011).
10. J. Schawe, P. Vermeulen, and M. van Drongelen, Thermochimica Acta **616**, 87 (2015).
11. R. Androsch, A. Monami, and J. Kucera, Journal of Crystal Growth **408**, 91 (2014).

Processing challenges in using graphene platelets in polyamide for enhanced barrier properties

Wagenknecht Udo [1]*, Leuteritz Andreas [1], Boldt Regine [1], Kuehnert Ines [1], Schult Michael [1], Trampe Markus [1]

[1] Leibniz-Institut fuer Polymerforschung Dresden e. V. - Dresden - Germany

In order to reduce the gas permeation through polymer films layered particles are a suitable method by elongating the path of diffusion according to the Nielson model. High aspect ratio and orientation of fillers are the key parameters for this effect. Graphene offers in this respect extraordinary good possibilities, if exfoliated and well dispersed. Therefore, different commercially available types of graphene were compounded with semi-aromatic and aliphatic polyamide grades in laboratory scale and extruded to flat film. Composition and process parameters of compounding as well as of flat film extrusion were varied. Extruded samples were investigated using microscopy, rheology and permeation rate of Helium (He) as a very small probe molecule. The results show that there is no obvious correlation between permeation rate and filling level and processing conditions, whereas rheological experiments emphasize better interaction in case of semi-aromatic than aliphatic polyamide. It will be shown that besides the usual critical parameters like dispersion, distribution and platelet orientation the degassing of the matrix is one of the critical points. This phenomenon will be discussed and ways to avoid a non acceptable gas bubble concentration will be shown.

Banana Starch Biodegradable Films Reinforced with Cellulose Nanofibers Isolated from Banana Peel by Enzymatic Treatment

H. Tibolla[a], F.M. Pelissari[b], F.C. Menegalli[a]

1 Department of Food Engineering, School of Food Engineering, University of Campinas, Campinas – SP, Brazil
2 Institute of Science and Technology, Food Engineering, University of Jequitinhonha and Mucuri, Diamantina – MG, Brazil

Abstract. Using nanotechnology to develop films based on polymer nanocomposites containing nanometric fillers might be a new way to improve the mechanical and barrier properties of a given polymer. Because of their renewability, agricultural crop residues can be a valuable source of cellulose nanofibers. The edible part of the banana fruit constitutes only 12 wt% of the plant; the remaining parts become agricultural waste and cause environmental problems. Therefore, cellulose nanofibers (CNFs) were isolated from banana peel using enzymatic treatment (Xylanase enzyme). New nanocomposites were then prepared from a mixed suspension of banana starch and cellulose nanofibers using the casting method and the effect of the addition of these nanofibers on the properties of the resulting nanocomposites (CNFF) was investigated. Atomic force microscopy (AFM) topography shows that the structure of film was modified with CNFs incorporation. The roughness level of the control film (CF) and CNFF films were 4.10 nm and 16.58 nm, respectively. The CNFs dispersed well in the starch matrix and did not agglomerate. The CF film showed higher density, moisture and solubility in water when compared with CNFF film. CNFF film presented a reduction of water content at equilibrium, sorption coefficient, water diffusivity and water vapor permeability compared with CF. Films added with CNFs had the greatest values of tensile strength showing a tendency to form more resistant and rigid films. On the other hand, films without CNFs were flexible as a consequence of their higher elongation capacity. Produced films showed a complex structure formed between starch and CNFs resulting in a remarkable improvement on mechanical and barrier properties. CNFs offer potential for reinforcing biodegradable composites, producing a material with added value, being indicated for use on food packaging.

Keywords: natural fibers, nanotechnology, enzymatic hydrolysis
PACS: S09-442

INTRODUCTION

The interest in biodegradable films is a global trend that has increased significantly in recent years. This fact has motivated studies related to the development and characterization of films produced from biopolymers such as polysaccharides, which are capable to producing continuous matrices. Because starch is abundant, inexpensive, and degradable, several researchers have been used to produce biodegradable films. Simultaneously to studies of biodegradable composites production, cellulose nanofibers (CNFs) have been explored as an alternative to reinforce polymeric matrices which showed several advantages, for instance, CNFs have potential to improve the mechanical properties, thermal stability and barrier properties (water vapor and oxygen permeability) [1,2]. The vegetable fibers (cellulose source) of nanometer size have been explored to reinforce the polymer matrix, because the use of cellulose nanoparticles provides numerous advantages when incorporated into composites, besides presenting renewable and biodegradable characteristics [3]. The banana peel contains a considerable amount of cellulose (12%) and this material can be potentially applicable as a reinforcing component in high-performance composites [1]. Nowadays, different methods for the isolation of cellulose crystals on nanometer scale have been developed. The hydrolysis method is performed to individualize the cellulose fibrils of the cell wall. Enzymatic hydrolysis is a method that gained prominence in the last decade, which has several advantages compared to the chemical methods, such as milder reaction conditions and no need to use chemical reagents. Thus, it can be considered an environmentally friendly process [4,5]. Our previous studies showed that enzymatic treatment followed by high-pressure homogenization or not, successfully isolates cellulose nanofibers from the banana peel [6, 7].

This work aimed to study an agricultural waste (banana peel) as a raw material for preparing higher added value products. The potential of CNFs, obtained from banana peel by enzymatic treatment, was investigated when applied as reinforcing agent in polymer composites.

MATERIALS AND METHODS

Nanocomposites Production

The banana starch films were produced by the casting method according to the methodology adapted from Pelissari et al. [8], as well the isolation of the banana starch. The fruit, unripe banana from variety "Terra" (*Musa paradisiaca*), was obtained from the southeastern region of Brazil, and was not subjected to any postharvest treatment. CNFs were isolated by enzymatic hydrolysis using xylanase, according to the method described by Tibolla et al. [6]. The authors performed previous experimental tests to choose the best enzymatic hydrolysis conditions, in order to improve CNF properties that will be use as reinforcing agent of the polymeric matrices. Table 1 summarizes the characteristics of the CNFs used in this study. The first step of the nanocomposites production involved the dispersion of CNFs suspension (5 g of nanofibers/100 g of starch) by magnetic stirring for 15 min, followed by homogenization with an Ultra-Turrax (5.000 rpm for 15 min). The CNFs suspension was added to a water solution of 4 % (w:w, dry basis) of banana starch previously stirred for 30 min. The mixture was homogenized by mechanical stirring for 30 min, followed by heating up to the process temperature (81 °C) under gentle stirring. Glycerol (25 g of glycerol/100 g of starch) was added at this point, and the solution was maintained at this temperature for 15 min. After that, the film-forming suspension (FFS) was sonicated for 10 min and 70.4 g of the FFS was poured onto acrylic plates (18 cm x 21 cm) to controlling the thickness of the nanocomposites. The films were dried in a chamber with air circulation under controlled temperature (54 °C) and relative humidity (RH) (48 %) for 4 h. A film without CNFs (control) was produced for comparison purposes. Before characterization, the films were conditioned in desiccators under 58% RH, at 25 °C, for 48 h. The produced films were designated as: CF = control film without CNFs and CNFF = films containing 5% (w/w starch) CNFs.

TABLE (1). Characteristics of the CNFs added to the films, according to Tibolla et al. [7].

CNFs	Length (L) (nm)	Diameter (d) (nm)	Aspect ratio (L\d)	Zeta potential (mV)	Crystallinity index (%)
	1490.0	3.7	404.5	-29.1	61.5

Nanocomposites Characterization

Topography and Roughness by Atomic Force Microscopy (AFM)

The topography of the films was analyzed by AFM. The analysis was performed in an Nx-10 Atomic Force Microscope (Park systems, Suwon, Korea). To perform the analysis, a piece of film (10 x 10 mm), was placed on a grid with mica surface. The digital images were acquired with a camera, under controlled parameters (RH ≤ 10% RH and 25 ºC). AFM analyses were performed at the Laboratory for Surface Science (LCS) of the National Nanotechnology Laboratory (LNNano) (Campinas, Brazil). The morphology and roughness of the films were analyzed by image analysis using the Gwiddion software.

Thickness, Density and Mechanical Properties

Film thickness was measured using a digital micrometer (Digimatic Micrometer Series 293 MDC-Lite, Mitutoyo Corporation, Japan). The mean thickness of each film was determined from an average of 10 random measurements. To determine film density, 20 x 20 mm squares samples were cut and their thickness was measured (three random measurements). The film samples were dried at 105 ºC for 24 h and weighted. The density of the films was calculated as the ratio between the film weight and film volume (film thickness x film area). The density experiments were accomplished in triplicate, and the data are reported as mean values.

The film mechanical properties were investigated using a texture analyzer (Stable Micro Systems, model TA.TXplus, Surrey, UK, England), according to the standard method D882-02 [9], by taking an average of six determinations for each film sample. The samples were cut into 20 x 80 mm strips by means of a scalpel. The initial grip separation and crosshead speed were set at 40 mm and 1.0 mm/s, respectively.

Moisture Content and Water-Soluble Fraction

The film moisture content was analyzed by gravimetry, in triplicate, by drying the samples at 105 °C for 24 h [10]. The film water-soluble fraction (S) was determined by employing the methodology described by Gontard et al. [11]. Three discs (diameter= 20mm) of each film were stored in a desiccator containing silica gel (~0% RH) for 48 h. The film samples were immersed into 50 mL of water containing sodium azide (0.2 g/L) at 25 °C for 24 h, under sporadic agitation. The solution containing the samples was filtered, and insoluble matter was dried at 105 °C for 24 h. Analyses were carried out in quadruplicate, and the film water-soluble fraction (*S*, %) was calculated by means of difference of the initial dry sample weight (g) and the dry resulting material weight (g).

Water Vapor Permeability (WVP)

WVP was determined by gravimetry [12], with modifications. The film samples (diameter = 0.06 m) was placed on the circular opening (area = 0.00196 m^2) of a permeation cell and was sealed with sealant ring, to ensure that humidity migration would occur only through the film. The interior of the cell was filled with saturated $MgCl^2$ saline solution (33% RH) and the system was stored in a desiccator containing saturated $NaNO^2$ solution at 25 °C (64% RH). The weight gain was monitored every 30 min, for 8 h. The analyses were conducted in triplicate and, WVP was calculated on the basis of Equation 2 and expressed in g/(m.s.Pa). Where: w/t is the slope of the line of weight gain (w) as a function of time (t) (g/s); δ is the mean sample thickness (m); A is the sample permeation area (m^2); and ΔP is the difference in water vapor pressure through the sample for pure water at 25 °C (Pa).

$$WVP = \left(w.t^{-1}\right)\left[\delta.\left(A.\Delta P\right)^{-1}\right] \quad (1)$$

Water Uptake (WU)

The water absorption kinetics was determined according to the method described by Dufresne et al. [13]. Films with the dimensions of 20 x 20 mm and, with known thickness, were stored in a desiccator containing silica gel (~0% RH) for 48 h. After this period, the films were weighted (w_0) and conditioned at 25 °C in hermetically sealed flasks containing Na_2SO_4 saturated solution, to ensure a RH of 95%. The samples were removed at desired intervals and weighted (w_t) until equilibrium state was reached. WU was calculated by means of difference of the sample weight after time "t" of exposure to 95% RH (g) and the initial weight of the sample (g).

Statistical analysis

One-way analysis of variance (ANOVA) and a Tukey test of multiple comparisons with a significance level of 5% ($p \leq 0.05$) were run using a Statistic software (StatSoft Inc, Tulsa, Oklahoma, USA).

RESULTS AND DISCUSSION

Nanocomposites characterization

The AFM images of the sample films CF (1), CNFF (2) obtained from AFM (A) topography, (B) amplitude and (C) phase are shown in Figure 4. Incorporation of cellulose nanofibers significantly changed the microstructure of the nanocomposites (CNFF) as compared with the control film (CF). The CNFs had specifics characteristics, such as aspect ratio, surface charge and crystallinity (Table 1), that may account for the type of interactions between cellulose and polymer matrix components (starch, glycerol, and water) and culminate in distinct microstructures. The control film displayed a more uniform and smoother surface compared to the nanocomposites, which presented non-homogenous structures; i.e., irregular surfaces. Also, the roughness value of CF was 4.10 nm, which was increased when CNFs was incorporated into the starch matrix to 16.58 nm (CNFF). Despite the different microstructures, results demonstrated that the nanoparticles were well incorporated in the polymeric matrix which indicates that starch chemically attached to the nanoparticles.

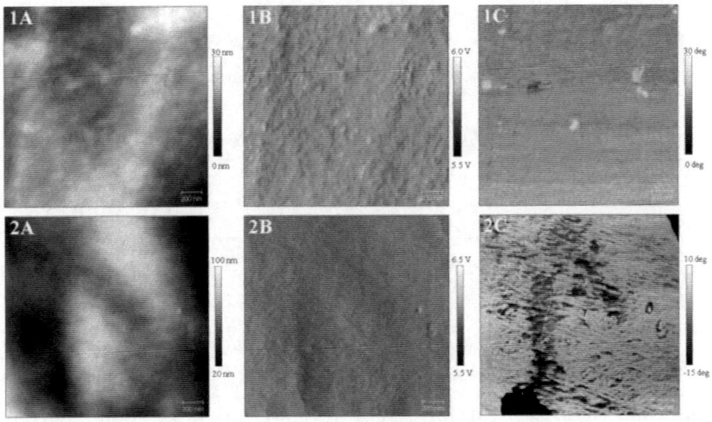

FIGURE 1. AFM images of topography (A) amplitude (B) and phase (C) images (scanning area 2.0 μm x 2.0 μm, scale bar = 200 nm) of the films: CF (1) and CNFF (2).

Table 2 shows the results of thickness, density, moisture and solubility in water of the produced films. The thickness values of the films was increased with the CNFs incorporation into the polymeric matrix ($p \leq 0.05$), even with the control of the FFS content per unit area of plate during the casting procedure. The control film presented higher density than the nanocomposites (no statistical difference; $p > 0.05$), as a result of the type of microstructure observed in the AFM analysis (Fig. 1). The incorporation of CNFs also reduced the moisture content in the films, showing that the presence of CNFs reduces the contact of the water with hydrophilic groups of the starch [14]. The CNFF was less water-soluble (36% reduction) as compared with the CF ($p < 0.05$). This may be due to formation of the nanofibers network, which hinders water diffusion through the matrix. Films slightly soluble in water are more indicated for use on food with high water activity due to their resistance [11].

TABLE (2). Results of thickness, density, moisture and solubility of control film (CF) and cellulose nanofiber film (CNFF)

Film	Thickness (μm)	Density (g/cm^3)	Moisture (%)	Solubility (%)
CF	76 ± 0.0^a	1.18 ± 0.08^a	18.61 ± 0.03^b	31.61 ± 1.02^b
CNFF	80 ± 0.5^b	1.11 ± 0.00^a	13.41 ± 1.49^a	20.23 ± 1.03^a

The mechanical properties (tensile strength, elongation at break and Young's modulus) of films are presented in Table 3. The produced films showed statistical difference for all properties studied ($p < 0.05$). CNFF presents lower values of elongation at break than CF. However, CNFF presents higher values of tensile strength and Young's modulus than CF. The evaluation of mechanical properties showed that reinforced film with cellulose nanofibers is more rigid and present greater mechanical resistance than control starch film. This means that incorporating cellulose crystallites into the starch matrix results in strong interactions between cellulose crystallites and the starch matrix, which restrict the chain motion of the starch matrix [15]. The Young's modulus was affected the most; this parameter increased >100% upon incorporation of the CNFs. The high zeta potential of the nanofiber suspensions (-29,1 mV) prevented the formation of aggregates, giving more stable colloidal suspensions with individualized cellulose fibers, and promoting a strong network of nanofibers inside the material.

TABLE (3). Mechanical properties of control film (CF) and cellulose nanofiber film (CNFF)

Film	Tensile strength (MPa)	Elongation at break (%)	Young's modulus (MPa)
CF	8.9 ± 0.1^a	78.3 ± 6.9^b	283.7 ± 7.3^a
CNFF	16.1 ± 0.9^b	12.6 ± 0.3^a	759.6 ± 46.3^b

Table 4 summarizes the water vapor permeability (WVP) of films in gradient of relative humidity of the 33%-64%, and the typical kinetic parameters for water uptake: water content at equilibrium (M_∞), diffusion coefficient (D), and sorption coefficient (S). The CNFF film presented a lower water diffusivity (6.68 x 10^{-8} mm^2/s) and lower water uptake at equilibrium (2.75%) when compared with CF film (4.91 x 10^{-8} mm^2/s and 3.28%, respectively), however there was no statistical difference ($p < 0.05$). The sorption coefficient (S) decreased with the CNFs incorporation into the starch matrix ($p < 0.05$). These results are in accordance with the values of solubility in water.

These results can be explained by the strong hydrogen bonding interactions between starch and cellulose crystallites, since these interactions tend to stabilize the starch matrix when the nanocomposites are submitted to high humidity [16]. Reduced water absorption by the film is strongly related to the diminished diffusion coefficient imposed by the nanofibers and to decreased sorption of the penetrant [3].

TABLE (4). Water vapor permeability (WVP) and kinetic parameters for water uptake: water content at equilibrium ($M\infty$), diffusion coefficient (D) and sorption coefficient (S), of control film (CF) and cellulose nanofiber film (CNFF)

Film	WVP (10^{-11} g/m.s.Pa)	$M\infty$ (%)	D (10^{-8} mm^2/s)	S
CF	11.4 ± 1.1^a	3.28 ± 0.01^a	4.91 ± 0.3^a	0.83 ± 0.02^b
CNFF	20.23 ± 0.4^b	2.75 ± 0.05^a	2.68 ± 1.15^a	0.66 ± 0.02^a

CONCLUSION

Starch and cellulose nanofibers isolated from unripe bananas are potentially applicable as matrix and reinforcing material in biodegradable films, respectively. The banana starch matrix when incorporated with CNFs are able to produce a material with potential to be used as food packaging. The produced nanocomposites showed an improvement in mechanical and water barrier properties compared to starch film without CNF. The improvement in films performance can be explained by the chemical similarities between starch and cellulose, the nanometer size effect of the nanofibers and the hydrogen bonding interaction between the fillers and the matrix. These nanofibers offer potential for reinforcing biodegradable polymer composites, producing a material with add value. The nanocomposites produced in this work could find applications in the food packaging industry due their biocompatible, biodegradable, and non-toxic characteristics.

ACKNOWLEDGMENTS

The authors would like to acknowledge the financial support provided by Conselho Nacional de Desenvolvimento Científico e Tecnológico (CNPq) (477842/2011-9) and Coordenação de Aperfeiçoamento de Pessoal de Nível Superior (CAPES) (2952/2011). The authors would also like to acknowledge the Brazilian Nanotechnology National Laboratory (LNNano) for allocation of the TEM and AFM apparatus.

REFERENCES

1. S. Elanthikkal et al., *Carbohydrate Polymers* **80**, 852-859 (2010).
2. M.M. Andrade-Mahecha et. al. *Carbohydrate Polymers* **123**, 406-415 (2015).
3. F. M. Pelissari. *Journal of Colloid and Interface Science* **505**, 154–167 (2017).
4. O. Faruk et al. *Progress in Polymer Science* **37**, 1552-1596 (2012).
5. G. Siqueira. *BioResources* **5**, 727-740 (2010).
6. H. Tibolla et al. *LWT – Food Science and Technology* **59**, 1311-1318 (2014).
7. H. Tibolla et al. *Industrial Crops and Products* **95**, 664–674 (2017).
8. F.M Pelissari et al. *Food Hydrocolloids* **30**, 681-690 (2013).
9. ASTM – D882-02. In *Annual Book of ASTM Standards*. ASTM: Philadelphia, 2002.
10. ASTM – D644-99. In *Annual Book of ASTM Standards*. ASTM: Philadelphia, 1999.
11. N. Gontard et al. *Journal of Food Science* **57**, 190-195 (1992).
12. ASTM – E96-00. In *Annual Book of ASTM Standards*. ASTM: Philadelphia, 2000.
13. A. Dufresne et al. *Journal of Applied Polymer Science* **76**, 2080-2092 (2000).
14. M. O. Müller et al. *Food Hydrocolloids* **23**, 1328-1333 (2009).
15. Y. LU and W. X. Cao. *Mocromolecules Bioscience* **5**, 1101–1107 (2005).
16. M. D. Sanchez-garcia. *Carbohydrate Polymers* **71**, 235–244 (2008).

Montmorillonite supported Silver Nanoparticles based Antibacterial HDPE Nanocomposites: The future of commodity plastics

Roy Anasuya (1)*, Joshi Mangala (1), Butola B S (1)

(1) IIT Delhi - Delhi - India

In the present decade, increasing resistance of pathogenic microbes to drugs and bioactive organic compounds has led to extensive research on inorganic metal based nanoparticles as potential bacteriocidal materials in preparation of antimicrobial polymer nanocomposites. Although metal based nanoparticles like silver nanoparticles (Ag-NP) has established itself as a highly potent antimicrobial even at very low concentrations, instability stemming from high surface to volume ratio to form aggregates has been a barrier for its extensive application in fabrication of polymer nanocomposites. Thus, supports for Ag-NP to control its particle size has been under research attention to obtain uniform sized nanoparticles supported on carriers leading to high antimicrobial efficiency. In the present work, montmorillonite clay (Mt) has been utilised as a support for formation of silver nanoparticles. Mt clay was first acid-activated using mineral acid and further converted to silver-montmorillonite (Ag-Mt) nanohybrid complex using silver nitrate. TEM images of Ag-Mt confirmed deposition of Ag-NP (~10 nm) on surface of Mt sheets although no reduction medium was employed. Ag-Mt powder showed superior biocidal action against pathogenic bacteria at very low concentration (5 ppm). Ag-Mt was further melt-mixed in HDPE in a co-rotating twin-screw extruder with varying Ag-Mt concentration (from 1 wt% to 5 wt%). The developed HDPE nanocomposites were characterized using TEM and XRD which suggested exfoliated morphology at low concentration and intercalated morphology at higher concentrations. The nanocomposites exhibited exceptional antibacterial activities of >99.9% in colony counting method against both gram-positive and gram-negative bacteria, making it suitable for polyethylene commodity applications where infection control is a major issue e.g. kitchen dustbins, water storage tanks, crates, buckets etc.

Tailored dispersion of Zinc Oxide nanoparticles and hexagonal Boron Nitride in block copolymer nanocomposites for dielectric applications

Helal Emna [1]*, David Eric [1], Fréchette Michel [2], Demarquette Nicole Raymonde [1]

[1] École de Technologie Supérieure - Québec - Canada, [2] Institut de Recherche d'Hydro-Québec - Québec - Canada

Styrenic thermoplastic elastomers such as Poly[styrene-b-(ethylene-co-butylene)-b-styrene] (SEBS) constitute a special class of block copolymers composed of two styrene endblocks and a soft rubber midblock. Due to their immiscibility, the blocks self-assemble into well-ordered nanostructured morphologies that depend on several parameters such as their molecular weights, affinities and relative fractions. These organized morphologies have great potential to act as templates tailoring the dispersion and spatial distribution of nanoparticles with different shapes in order to achieve well designed nanocomposites with optimized functional properties. In this study, both Zinc Oxide (ZnO) nanospheres and exfoliated hexagonal boron nitride (h-BN) have been dispersed in SEBS grafted maleic anhydride (SEBS-MA) triblock copolymer in a trial to achieve thermally conductive electrically insulating nanocomposites, that benefit from potential synergetic effect of both nanoparticles. SEBS-MA/ZnO nanocomposites, studied beforehand, featured an interesting rheological behavior, even at low loading of ZnO, characterized by a consistent increase of dynamic moduli and reduced dependence on the frequency especially in the low values range indicating the formation of networks between ZnO nanoparticles and polymer chains. Moreover, improvements of breakdown strength, thermal conductivity and resistance to surface erosion were achieved. In a second step, h-BN particles were partially exfoliated and dispersed in SEBS-MA/ZnO mixture by means of high-shear mixing. The dispersion of these anisotropic fillers in the hybrid nanocomposite was enhanced compared to direct dispersion in neat SEBS-MA. The impact of this combination on the dielectric performance was investigated through the evaluation of the complex dielectric permittivity, breakdown strength and thermal conductivity in correlation with the rheological properties. Key words: block copolymer nanocomposite, tailored dispersion...

BENEFIT OF CLAY IN POLYMER NANOCOMPOSITE FOR BIOMEDICAL ENGENEERING: FROM CLAY PROPERTIES TO POLYMER COMPOSITE PERFORMANCE

Bastianini Maria [(1)]*, Sisani Michele [(1)], Petracci Annarita [(1)], Scatto Marco [(2)], Scopece Paolo [(2)], Patelli Alessandro [(3)]

[(1)] Prolabin & Tefarm - PG - Italy, [(2)] Nadir S.r.l - VE - Italy, [(3)] Department of Physics and Astronomy, Padova University - PD - Italy

Polymers are key building blocks in the development of smart material for biomedical applications, and a lot of polymers offer unique properties for specific applications. An increased matrix of material options is available through the use of polymer compounds. These compounds can incorporate performance-enhancing fillers, which provide properties not possible with ordinary polymers alone. An interesting and promising class of filler for polymeric composite are Layered Double Hydroxides (LDHs) LDHs are anionic clays that, due to the presence of exchangeable anions in the interlayer regions can be used as a host for antimicrobial, antioxidant, antibiotic anions which can be intercalated by ionic exchange process in the metal hydroxide sheets. So the organic guest anions can improve the compatibility of the inorganic layers with the polymeric matrix, and the mechanical, thermal, gas barrier properties of the composite. LDHs can also confers to polymer composite biological activity. A very interesting approach for the production of Bionanocomposite is used in our H2020 Project - FAST "Functionally Graded Additive Manufacturing Scaffolds by Hybrid Manufacturing" (NMP07, GA n. 685825), where processing parameters during Melt Compounding in Twin Screw Extruder, preparation method, feed composition, clay morphology, clay intercalation process were properly modulated until obtaining polymer clay Bionanocomposite with mechanical, reological, and biological properties useful for the preparation of scaffold prototypes for the bone regeneration using a 3D printer device (Bioscaffolder). In this work we have correlated morphological properties of LDH intercalated with antibiotic molecules, evaluated by XRD and SEM analyses with processing parameters of Twin Screw Extrusion in order to obtain a Bionanocomposite with intercalated/exfoliated morphology and with tuneable controlled release of antibiotic.

Bio-Inspired Polyethylene-based Composite reinforced by Thermoplastic Polyurethane (TPU) Fiber for Aerogel Production

Solmaz Karamikamkar[a], Chul B. Park[c*](reference author), and Hani E. Naguib[b]

[a]*RS210D, Mechanical and Industrial Engineering department, University of Toronto*
[b]*RS207A, Mechanical and Industrial Engineering department, University of Toronto*
[c*]*RS210A, Mechanical and Industrial Engineering department, University of Toronto*

Presenting author's email: solmazkk@mie.utoronto.ca
Reference author's email: park@mie.utoronto.ca

Abstract. A nanocomposite strategy for the selective inclusion of viscoelastic silica modified polyethylene (Si-PE) into thermoplastic polyurethane (TPU) fiber junctions is demonstrated to create effective stress-transfer pathways within three-dimensional (3-D) fiber-reinforced modified polyethylene-based composites with a phenomenal thermal insulation properties. Inspired by the bone architecture in the human body, large amounts of hard segments and small amounts of soft segments, the 3-D interconnected Si-PE w/25% TPU composite achieves synergistic strengthening. The composite is prepared by a wet chemical synthesis approach, a sol–gel reaction, with operation procedure of functional polyethylene while incorporating TPU fiber phase. The solution containing polyethylene precursor initiates network formation throughout the liquid to form a 3-D network (the gel) through chain polymerization. The obtained functional polyethylene-based gel reinforced with TPU fibers is then left to be aged stimulated by time and temperature for structure completion. By the end of aging process, the solvent is extracted from the gels through supercritical drying technique. The remained obtained solid structure is the Si-PE w/TPU composite nanocellular foam called aerogel. The resulting Si-PE w/25% TPU composite aerogel exhibited fully structural deformations and compressive mechanical strength (18 MPa at a compressive strain of 90%) at one compression cycle while showing an extremely great thermal stability up to 500 °C. Owing to the combination of excellent mechanical strength and thermal insulation properties, the Si-PE w/25% TPU composite can be used as thermal insulation material for strain sensors, showing very short response and load bearing with reliable sensitivity and extreme durability. Such bio-inspired architecture opens the door to fabricate new 3-D-multifunctional and mechanically durable nanocomposite for thermal insulation products and strong partially-flexible devices.

Keywords: Polyethylene, Thermoplastic Polyurethane, nanofiber, Thermal Insulation, Aerogel
PACS: 62.20 mj Brittleness

INTRODUCTION

Aerogels are ultralight porous solids made from a gel in which the liquid phase was replaced by air which was first produced by S. S. Kistler[1] in the early 1930's extracting the liquid phase from the gel at a very high temperature under applied pressure. They stand out by their very unusual properties such as a very low density, high porosity, high specific area, low thermal conductivity, low index of refraction and low dielectric constant. Unlike foams, aerogels have extremely small pores, on the nanometer scale with complex interconnectivity and branched structure of a few nanometer. This makes them 2-5 times more effective insulators than foams. Current foams made with Polyurethane (PU) and Polystyrene (PS) often have thermal conductivity of 0.03W/m.K. However; aerogels emerge as a super-insulator material (thermal conductivity <0.02W/m.K) with the highest heat barrier ever reported (0.012-0.018 W/m.K at atmospheric pressure, 0.004-0.014 W/m.K at vacuum pressure)[2]. These unique properties make them interesting in various industrial applications such as thermal insulation for windows, aerospace applications and many more[3–8]. However, they are not currently produced at industrial scale due to their weak mechanical properties being fragile which can break easily when touched, which restrain their use in many industrial fields. That is why much research has been done recently about the mechanical reinforcement of aerogels, with the aim to get strong and flexible materials, while keeping their outstanding properties. Different methods have been found to achieve this goal such as modifying the precursors of aerogels, polymerizing monomers during the aerogel's synthesis, or adding different kinds of fibers to reinforce their structure.

Polymer-based nanocomposites where nano-fibers are included as a dispersed phase in a small amount (less than 10%), have been reported to successfully reinforce the structure because of surface-area-to-volume-ratio effect[9–11]. In comparison to traditional understanding of mechanical nature of polymer-based nanocomposites, bio-based materials offer quite different strategy to design new material architecture. For example, cancellous bones (spongy bone) found at the ends of long bones and pelvic bones that are very porous structure with density of 0.03 to 0.12 g/cm3 and modulus of 0.04 to 1 GPa. This nature-created material, spongy bone contains 90% of hard ceramic combined with 10% of soft materials which gives it a unique structure for its special application in human body[12]. Such a composition is believed to resolve the complication of different tensions experienced by human body. Having been inspired by material mimicking spongy bone, strong enough nanocomposites obtained by the inclusion of TPU nanofibers with soft and hard segment in the polyethylene matrix is believed to demonstrate the achievement of synergistic toughness through the interfacial interaction between layer of nanofibers and matrix. However, simply including nanofibers in the polymeric matrix is not creating such a structure since polyethylene and TPU fibers may only interact through a weak van der walls interactions. To overcome this challenge, the attempt of polyethylene modification during its polymerization and further in its gelation to the final structure has been made. Herein, we demonstrate for the first-time bone-inspired design of three-dimensional tough structure that preserve the thermal insulation characteristics of aerogel.

EXPERIMENTAL

Process of TPU Electrospinning

A mixture of 15 wt% TPU (Lubrizol – Estane 58123) in THF (Tetrahydrofuran from Sigma-Aldrich, contains 250 ppm BHT as inhibitor, ACS reagent, ≥99.0%) and DMF (N,N-Dimethylformamide, ACS reagent, from ACP) has been made. The mixture of 15 wt% TPU and DMF/THF of 50:50 v% was prepared by mixing TPU in the solvent system at room temperature for 6 hours until a homogenous solution is resulted. TPU solution was electrospun under high voltage of 20 kV at room temperature with flow rate of 1 ml/hr and needle diameter of 1.37 mm. Porous fiber mat was obtained on the collector which was placed in 20 cm distance to the needle dispenser. The obtained fiber mat was then dried at 80 °C for 12 h in an oven. TPU electrospun nanofibers (90-210 nm) mat is prepared.

Sol-gel Chemistry

Functionalized polyethylene has been made from free-radical polymerization of VTMS (60 gr, vinyltrimethoxysilane, 98%, Sigma-Aldrich). VTMS was first mixed with 10 wt.% DTBP (6 gr, di-tert-butyl peroxide, 98%, sigma-Aldrich) in a glass reactor. The mixture was then heated up to 145 °C for 10 h under 2 bar of nitrogen environment. The resultant silane-functionalized polyethylene (Si-PE) was then cooled to room temperature and dissolved in dehydrated ethanol at 0.2 g/ml to give sample 1 (Si-PE). Sample 2 (Si-PE w/ TPU) has been made mixing 25wt% TPU fibers in the solution of Si-PE in ethanol at the density of 0.2 g/ml. Both solutions (Sample 1 and 2) were then gelled in a one-step sol-gel process by addition of ammonium hydroxide in a 1:1 molar ratio to Si portion where warmed in an oven at 40°C for 3 hours until the gels were formed. The gels were then bathed in pure dehydrated ethanol for another 4 hours before solvent exchange and supercritical drying step. The mechanism of the sol-gel process is shown on Fig.1.

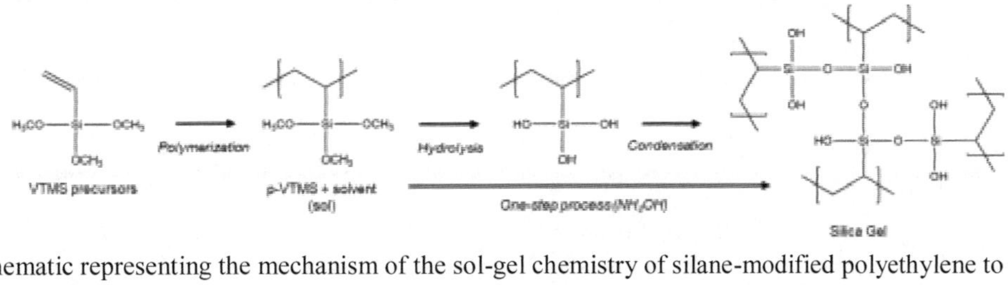

Figure 1. The schematic representing the mechanism of the sol-gel chemistry of silane-modified polyethylene to Si-PE aerogel.

Aerogel Production

Solvent exchange has been done by replacing the ethanol with CO2, in a chamber by placing them in a sealed chamber filled with ethanol and pressurized at 1500 psi (10.3 Mpa) by a high-pressure syringe pump (Fig. 2). At this pressure, the CO2 connected to the syringe pump becomes liquid and can diffuse in the gels and start replacing the initial solvent, ethanol. Then the chambers are regularly depressurized (80 mL every 2 h) to get rid of ethanol and letting CO2 replacing ethanol in the pores until all the CO2 has replaced the solvent in the gels (7 days). When there is no more ethanol coming out of the chamber, supercritical drying is proceeded to remove the CO2 from the gel by increasing the temperature from room temperature to 40°C. This way all the capillary forces will be canceled out to prevent creation of any crackes in the structure during drying.

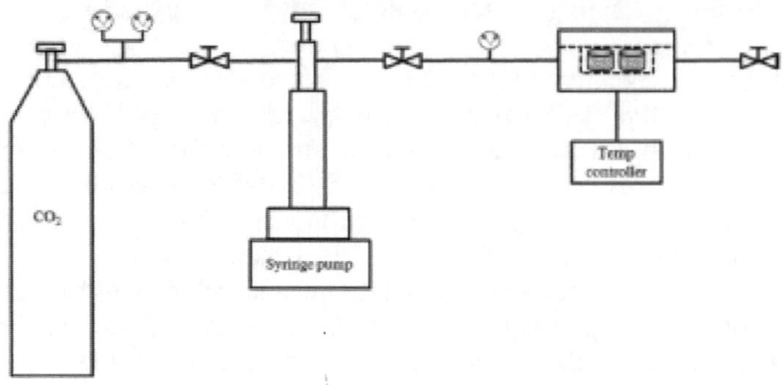

Figure 2. Schematic repressing the sealed chamber for solvent exchange and supercritical drying.

Characterization

The morphology of the fibers was characterized using Hitachi S-3000H scanning electron microscopy (SEM) imaging. The thermal conductivity/insolation of the aerogels is studied using Thermal Conductivity Analyzer (HotDisk) (Transient Plane Source, TPS2500). The compression Modulus (Y) of the aerogels is determined by a uniaxial compression test by applying various loads on the cylindrical aerogel samples. The mechanical characterization has been conducted using the aerogel experiencing compression in which the vertical length changes is measured during the compression load. The chemical structure of the synthesized material with and without fiber is studied using Fourier Transient Infrared (FTIR). NMR- cryoporometry is conducted to measure the pore size and its distribution.

RESULTS AND DISCUSSION

The factor that influences the final volume and density of the aerogel is the concentration of the precursors and fillers in the initial solution[13] with the assumption of no shrinkage in the structure during the gelation process and drying step. The aerogel network created during gelation step controls the microstructure and final material properties[13]. The bulk densities of the resultant aerogels with and without TPU fibers are calculated theoretically by their mass to volume ratios to be 0.2. However, after drying the samples experienced some percentages of shrinkage. The percentage of volume shrinkage is determined from the change in the volumes of the alcogel to aerogel as reported to be 21% for Si-PE w/ 25% TPU fiber aerogel (density of 0.08 g/cm3). **Table 1** summarizes the physical, and mechanical properties of both samples.

Table 1. Theoretical density, bulk density, gel time, shrinkage percentage, compression modulus of Si-PE aerogel and Si-PE w/TPU fibers aerogel

Sample	Theoretical Density	Bulk Density (ρ_b) (kg m-3)	Gel Time (hr)	Volume of Shrinkage (Vs%)	Compression Modulus (MPa)
Si-PE Aerogel	0.2	0.16	0.3	56	5
Si-PE W/25% TPU Aerogel	0.2	0.08	3	21	18

NMR cryoporometry is used to determine pore size distribution in the porous aerogel. This technique is since the melting point of a solid depends on the volume in which it is confined, via the following equation obtained from the Gibbs and Thompson equation[14]:

$$T_m\text{-}T_m(x) = \frac{k}{x}$$

where Tm is the melting point of the bulk solid, Tm(x) is the melting point of the solid in a pore of dimension x, and k is a constant characteristic of the solid. That means that a solid in a small pore will have a lower melting point than the bulk solid. To characterize a porous sample, it is imbibed with a liquid which is frozen, and then slowly warmed up, while the amount of melted liquid in it is determined at each temperature with nuclear magnetic resonance (NMR). The NMR peaks' intensity is proportional to the amount of liquid, which can thus be interpreted to obtain the pore size distribution of the material[14]. The peaks intensity is obtained using MestReNova software. As measured by NMR-cryoporometry, 17% of the pores are smaller than 17 nm while 78% of the pores are between 15 to 37 nm. Such a high percentage of nano-scale pores prevents gas conduction dramatically.

The morphology of aerogels is observed by scanning electronic microscope (SEM). The coating thickness as well as the quality/magnification of SEM pictures are highly sacrificed due to the brittle nature and extremely porous nature of the aerogel. As seen in **Fig. 3**, samples are presenting densely packed and crosslinked network of polymer chains which formed through the polymerization and subsequent gelation step.

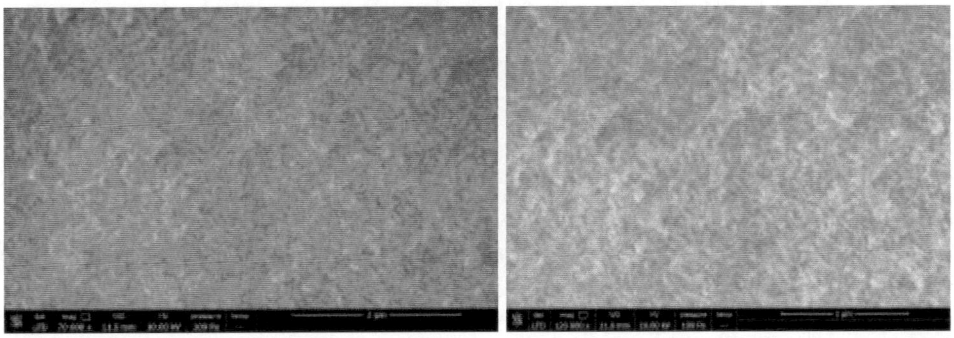

Figure 3. Representative SEM image of Si-PE aerogel. Left: 70,000 magnification, right: 100,000 magnification.

The FTIR spectroscopy is applied on both samples with and without TPU fibers. The FTIR spectra of non-reinforced and reinforced aerogels with TPU fiber in the underlying structure are presented in Fig. 4 to study any possible chemical changes in the structure. The main characteristic peaks of TPU, such as the C–H asymmetric and symmetric stretching of CH2 at 2948 cm-1, sp3 C-H stretch at around 2960 cm-1, the bonded C=O stretching of the urethane group at around 1734 cm-1, and the benzene ring at 1150 cm-1 were present at the same location after added to the Si-PE aerogel compared to pure TPU[15] showing that TPU fibers were successfully integrated into the aerogel backbone structure. The reduction in the peak height at around 1100 cm-1 suggests that the stretching vibration of C-O from functional silane groups are reduced meaning that most of them chemically interacted with fiber backbone.

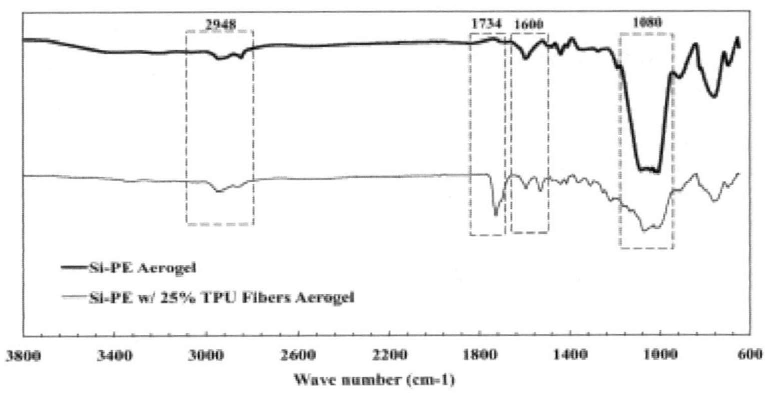

Figure 4. FTIR spectra of Si-PE aerogel and Si-PE w/ 25% TPU fiber aerogel.

The TGA test at the rate of 15 degree/min up to 700 °C is performed on both Si-PE aerogels with and without electrospun TPU fibers. As shown in **Fig. 5**, the addition of TPU fibers to the aerogel structure significantly increased the thermal resistance of the aerogel and caused the graphs to shift up noticeably. This observation along with increased degradation slope with addition of TPU fiber at around 500 °C, suggests that the addition of TPU fibers leads to more cross-linked polymer network structure at the nanoscale consequently resulting in stronger chemical interactions between the chains. The presence of TPU fiber improved the mechanical integrity of the aerogel by increasing the compression modulus from 7 MPa for Si-PE aerogel to 18 MPa for Si-PE w/TPU fiber aerogel at a compressive strain of 90% at one compression cycle.

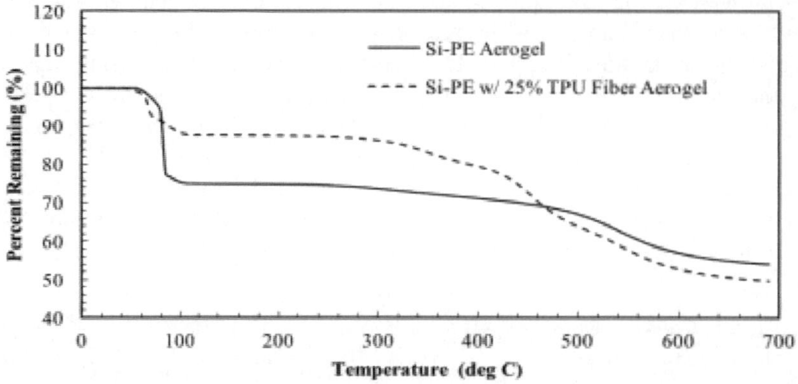

Figure 5. TGA graph of Si-PE aerogel (—) and Si-PE w/ 25% TPU fiber aerogel (---).

Conclusion

Silica-based polyethylene aerogel with and without electrospun thermoplastic polyurethane were produced by sol-gel chemistry and supercritical drying techniques. As compared together, the aerogel containing TPU fibers seemed to be strongly improved in terms of mechanical properties. The results presented in this paper suggests that inclusion of TPU fiber in the Si-functionalized polyethylene during lateral polymerization increased the density slightly yet improved the mechanical properties dramatically (compression modulus increase from 7 MPa to 18 MPa). There is more characterization carried out to elaborate on the physical and chemical changes in the structure backbone with inclusion of TPU fibers to discuss such a dramatic improvement in mechanical properties as atomic force microscopy (AFM) nanoindentation (AFM-NI). Multi-scale experimental studies are currently in progress to characterize the interfacial transition zone between aerogel matrix and the fibers using AFM-NI technique.

REFERENCES

1. S. S. Kistler, "Coherent expanded aerogels and jellies.," Nature (London, United Kingdom), vol. 127, p. 741, 1931.
2. Lu, X., Arduini-Schuster, M. C. & et al., Thermal Conductivity of Monolithic Organic Aerogels. Sci. Wash. 255, 971 (1992).
3. Cuce, E., Cuce, P. M., Wood, C. J. & Riffat, S. B. Toward aerogel based thermal superinsulation in buildings: A comprehensive review. Renew. Sustain. Energy Rev. 34, 273–299 (2014).
4. Bheekhun, N., Abu Talib, A. R. & Hassan, M. R. Aerogels in Aerospace: An Overview. Adv. Mater. Sci. Eng. 2013.
5. Koebel, M., Rigacci, A. & Achard, P. Aerogel-based thermal superinsulation: an overview. J. Sol-Gel Sci. Technol. 63, 315–339 (2012).
6. Baetens, R., Jelle, B. P. & Gustavsen, A. Aerogel insulation for building applications: A state-of-the-art review. Energy Build. 43, 761–769 (2011).
7. Hrubesh, L. W. Aerogel applications. J. Non-Cryst. Solids 225, 335–342 (1998).
8. Fricke, J. & Emmerling, A. Aerogels—Recent Progress in Production Techniques and Novel Applications. J. Sol-Gel Sci. Technol. 13, 299–303 (1998).
9. Suter, J. L., Groen, D. & Coveney*, P. V. Chemically Specific Multiscale Modeling of Clay–Polymer Nanocomposites Reveals Intercalation Dynamics, Tactoid Self-Assembly and Emergent Materials Properties. Adv. Mater. Deerfield Beach Fla 27, 966–984 (2015).
10. Baeza, G. P. et al. Network dynamics in nanofilled polymers. Nat. Commun. 7, ncomms11368 (2016).
11. Core–Shell Structured High-k Polymer Nanocomposites for Energy Storage and Dielectric Applications - Huang - 2014 - Advanced Materials - Wiley Online Library.
12. Bureau, M. N., Legoux, J.-G. & Denault, J. Implantable biomimetic prosthetic bone. (2006).
13. Hüsing, N. & Schubert, U. Aerogels—Airy Materials: Chemistry, Structure, and Properties. Angew. Chem. Int. Ed. 37, 22–45 (1998).
14. Strange, J. H., Rahman, M. & Smith, E. G. Characterization of porous solids by NMR. Phys. Rev. Lett. 71, 3589–3591 (1993).
15. Mi, H.-Y., Jing, X., Salick, M. R., Cordie, T. M. & Turng, L.-S. Carbon nanotube (CNT) and nanofibrillated cellulose (NFC) reinforcement effect on thermoplastic polyurethane (TPU) scaffolds fabricated via phase separation using dimethyl sulfoxide (DMSO) as solvent. J. Mech. Behav. Biomed. Mater. 62, 417–427 (2016).

Investigation of the mechanical properties of GNP/MWCNT reinforced PA66 hybrid nanocomposites

Saeed Doagou Rad[a,b,*], Aminul Islam[a,b], and Jacob Søndergaard Jensen [b]

[a] *Department of Mechanical Engineering, Technical University of Denmark Produktionstorvet, Building 427A, DK-2800 Kgs. Lyngby, Denmark*
[b] *Centre of Acoustic-Mechanical Micro Systems (CAMM), Technical University of Denmark Building 352, DK-2800 Kgs. Lyngby, Denmark*

sadora@mek.dtu.dk

Abstract. The multifunctional characteristics of nanocomposites have introduced novel possibilities for different industrial sectors. However, the stable and optimized production of polymeric nanocomposite components is challenging. This research investigates the mechanical behavior of thermoplastic based nanocomposites reinforced with two prominent nanofillers namely Multi Walled Carbon Nanotubes (MWCNT) and Graphene NanoPlatelets (GNP) manufactured through industrially viable methods. Three main groups of Polyamide (PA 66) based nano- and hybrid-composite specimens namely PA 66/MWCNT, PA 66/GNP, and PA 66/MWCNT/GNP are prepared. Different contents and mixture ratios of the nanofillers are incorporated in the polymeric matrix through the dilution process using a twin-screw extruder. The influence of the manufacturing parameters and content of the nanofillers on the mechanical properties of the nanocomposite specimens are investigated. Results show that the inclusion of either of the nanomaterials improves the elastic modulus and tensile strength of the nanocomposites significantly. Moreover, the combination of the two nanofillers in the nanocomposites is resulting into completely novel material properties which do not follow the linear combinations of the observed behaviors. In fact, the interaction between the two different nanofillers influences both dispersion state and mechanical properties. The mechanisms of modulation in the properties and dispersion states are also discussed using scanning electron microscopy and rheological investigations. The research provides an insight to manufacture tailored hybrid nanocomposites with the optimized mechanical properties.

Keywords: Nanocomposites, Mechanical properties, nano fillers, rheology.

INTRODUCTION

Novel nanomaterials and their corresponding composites have introduced new functionalities for different applications. Carbon nanotubes (CNTs) and Graphene nano Platelets (GnPs) due to presenting a variety of outstanding properties including very high mechanical (Elastic modulus ~1 TPa), electrical (~0.1 μΩcm), and thermal (~ 6000 W/mK) properties, etc. are being developed and utilized for many novel applications [1,2]. In order to exploit the advantages of these nanofillers in many applications, they should be effectively incorporated within a polymeric matrix. Several methods are being applied to mix the nanofillers with the polymeric matrix to produce the desired nanocomposites such as direct mixing, solution mixing, wet mixing, melt mixing, and in-situ polymerization [3]. In fact, attaining proper de-agglomeration and dispersion within the polymer matrix is crucial to reach the desired properties in the realm of nanocomposites [4]. Although, some mixing methods like solution mixing have shown promising results in the aspect of dispersion state of the nanofillers, they present some major drawbacks. Using large amounts of solvents, poor solubility of many polymers in the currently used solvents, and damaging nanofillers in addition to environmental issues make solution mixing not the preferable approach especially at the industrial scales [5]. In comparison, melt mixing provides several advantages such as creating considerably less pollutant residuals, mass production, high speed, and lower costs making it favored especially in the industrial scales. This process consists of melting the selected polymer at high temperatures and mixing it with the defined amount of nano additives through rotational and shear forces to reach the desirable content of the nanofillers within the composite. However, the quality of the products of this process depends on the parameters such as polymer characteristics, compatibility of the nanofiller and matrix, interface behavior, etc. [6]. Reaching to an acceptable state of dispersion in the composites reinforced with the CNTs and/or GnPs can be a challenging task due to their high interfacial energy. In fact, due to their very small dimensions and high aspect ratios, the attractive forces arising from Vander Waals and π-π interactions make de-agglomeration so difficult. Inclusion of the second nanofiller inside the microstructure of the nanocomposites have shown some improvement in the properties of the composites

[7–9]. It is speculated that the simultaneous presence of the two nanofillers can help homogenization of the whole microstructure through interfering in the attractive interfacial forces leading to enhanced interfaces.
The purpose of this study is to investigate of the influence of the addition of the second nanofiller on the dispersion state and the mechanical properties of the hybrid nanocomposites. Carbon nanotubes and Graphene interactions within the polymeric matrixes are investigated through their mechanical performances. Moreover, the underlying mechanisms are investigated through scanning electron microscopy and rheological analyses.

EXPERIMENTS

Materials

The selected matrix for the nano and hybrid composites is a Polyamide or Nylon (PA 6,6) from ALBIS Plastic GmbH (Tg = 54.8 oC, Tm = 262.9 °C). The MWCNTs used in this study are catalytic chemical vapor deposition produced thin MWCNTs (NC 7000 ™) by Nanocyl SA, Belgium, with the average aspect ratio of 67 (dave=10.4 nm). In addition, Graphene nanoplatelets with the average surface area of 500 m^2/g, and the size range of 100 nm to 1 μm (XGnP®) were also acquired from XGSciences, USA in an initially agglomerated state. PA 6,6 based masterbatch containing 15.0 wt. % of the aforementioned MWCNTs (according to the supplier Nanocyl SA) was also used in the manufacturing process.

Processing

A conical counter rotating twin-screw extruder (HAAKETM Rheomex CTW, Φ=31.8/20 mm rear/front, L=300 mm) was selected to perform the melt mixing process. Prior to feeding the materials to the extruder, the neat polymer, masterbatch and/or GNP nanofillers were dried, weighted, and hand-mixed in the sealed bags, respectively. Two series PA 66 based nanocomposites with the contents of 1.0, 3.0, and 6.0 wt. % of either MWCNT or GnP were produced. In addition, hybrid nanocomposites comprising different ratios MWCNTs and Graphene nanoplatelet were produced. The ratios of MWCNT to GNP within the polymeric composition of the hybrid nanocomposites were 1.0/2.0, 2.0/1.0, and 3.0/3.0 wt. %. During the blending process in the twin screw extruder, the temperature distribution was kept from 265 to 280 °C(feed section to die), with the average temperature of 275 °C.

Injection molding of the nanocomposite specimens was performed on Ferromatik, Milacron moulding machine following the instructions of ISO 294-1 standard. The geometry of the cavity of the mold was dog-bone shape which was designed based on ISO 527-2 2012 standard. At each experiment after reaching stability, 35 samples were acquired and stored in vacuum bags. Constant cooling time of 10 s was applied for all the moulding experiments.

Characterization

In order to characterize the mechanical properties of the injection-molded specimens, uniaxial tensile experiments were conducted according to ISO 527-1 2012 standard. From each series of specimens produced with the defined setting and content, 10 samples were selected randomly, and tested in the dry as molded state.

Furthermore, in order to study the dispersion state and other possible involved mechanisms in the behavior of the nanocomposites produced via different methods, scanning electron microscopy (SEM) (Quanta FEG 200 ESEM) was employed on the fractured surfaces of the nanocomposites using a charging method. Melt rheological properties of the nanocomposites were also acquired from experiments conducted using a Discovery HR-2 (TA Instruments). In order to prepare the 25 mm disc-shape samples, granulates of nanocomposites were compression molded between the preheated plates for 6 min. Subsequently, the oscillatory shear measurements were performed under nitrogen atmosphere, using 25 mm parallel plates with 1.0 mm gap in between. Each experiment was conducted on the three different samples using strains within the linear viscoelastic range, which had already been defined using strain amplitude sweeps.

RESULTS AND DISCUSSIONS

Mechanical Properties

Effect of Single Nanofiller Inclusion

In order to understand how the individual inclusion of the introduced nanofillers in the composites influences the mechanical properties of the nanocomposites, PA 66 based composites comprising different contents of either MWCNTs or XGnPs were produced and studied in the first step. Fig. 1 (a) and (b) show the Influence of the MWCNT and XGnP addition on the elastic moduli and tensile strengths of the nanocomposites, respectively. Results show that incorporation of either of the nanofillers in the polymeric matrix increases the elastic moduli consistently. In fact, addition of the 6.0 wt. % nanofillers in the nanocomposites lead to 45 and 41 % increase in the elastic moduli of the nanocomposites containing XGnPs, and MWCNTs, respectively. As it is evident from the Fig. 1 (a), the nanocomposites reinforced by Graphene Nano platelets were showing higher values in the similar contents compared to the ones reinforced with carbon nanotubes. In contrast, tensile strengths of nanocomposites reinforced with MWCNTs were higher in all the investigated nanofiller contents. It is also noteworthy to mention that in the both nanocomposite types, contents higher than 3.0 wt. % nanofiller inclusion resulted into lower tensile strengths indicating larger agglomeration sites and/or less control on the alignment along the flow.

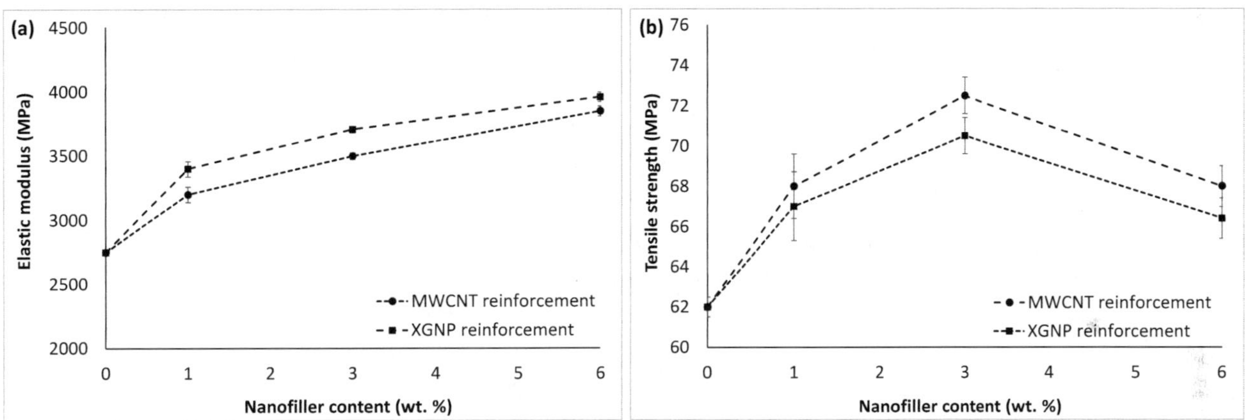

FIGURE 1. Influence of the MWCNT and XGnP content on the (a) Elastic moduli, and (b) tensile strengths of the corresponding nanocomposites.

Interaction between the Two Nanofillers

In the second step, PA 66 based nanocomposites containing both MWCNTs and XGnPs were produced and investigated. The reason behind this study was to investigate whether simultaneous presence of the nanofillers can result into improved dispersion and mechanical properties or not. Fig. 2 shows the comparison between the 3.0 wt. % nanofiller reinforced nanocomposites containing either or both of the nanofillers. Results show that replacing 1.0 wt. % carbon nanotubes with Graphene Nano platelets lead to notable increase in both elastic modulus and tensile strength compared to nanocomposites containing 3.0 wt. % of MWCNTs or XGnPs. However, replacing 1.0 wt. % Graphene Nano platelets with carbon nanotubes decreased the tensile strength values. It seems when the nanocomposite system is dominated by the carbon nanotubes, the system can benefit from the addition of the second nanofiller effectively.

In order to understand the interaction between the two nanofillers within the polymeric matrix better, nanocomposites containing 6.0 wt. % of the nanofillers were also studied (see Fig. 2(c) and (d)). Similar to the nanocomposites containing 3.0 wt. % nanofillers, the elastic modulus increased with introducing the second nanofiller in the nanocomposite system. However, the resulted tensile strength was less than when both nanocomposites were filled with just one type the nanofillers.

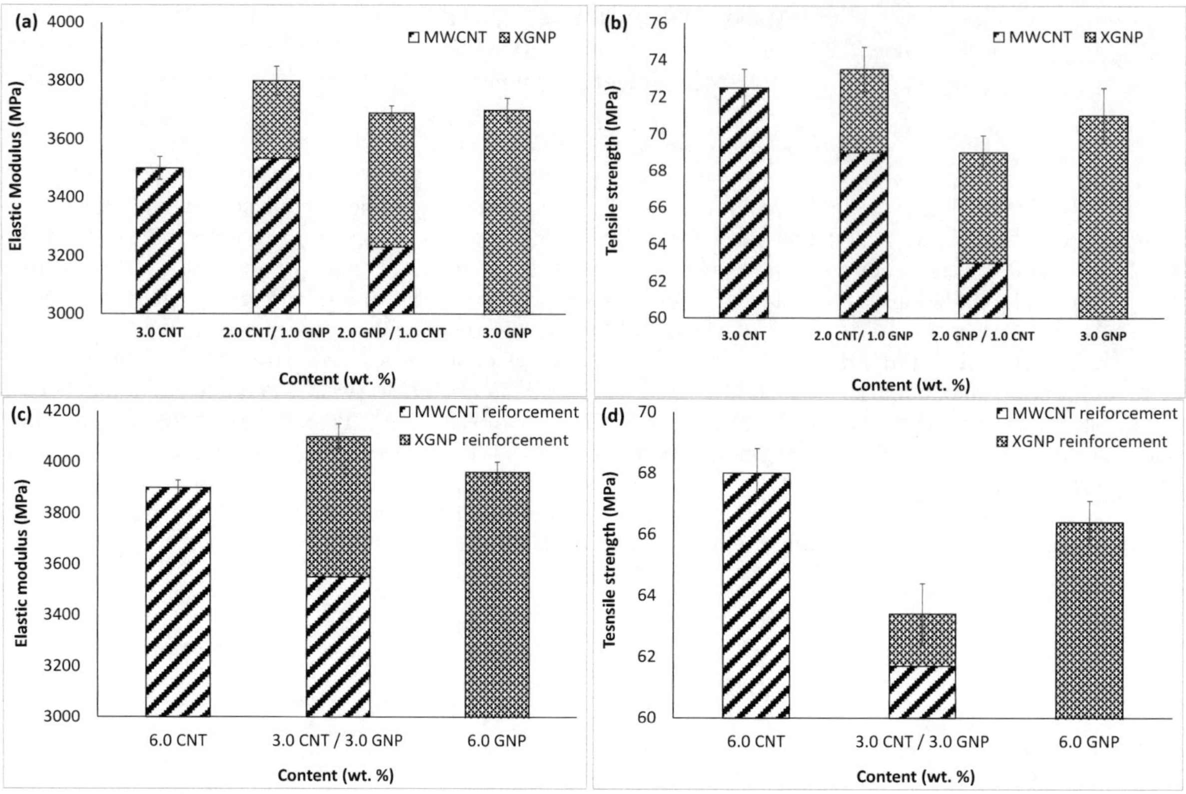

FIGURE 2. Mechanical properties of the respectively 3.0 and 6.0 wt. % nanofiller reinforced nanocomposites as the function of MWCNT / XGnP inclusion within the PA 66 matrix: (a), (c) elastic moduli, (b), (d) tensile strengths.

Rheological Investigations

Fig. 3 shows the influence of the type and content of the nanofillers within the polymeric matrix on the complex viscosity of the nanocomposites. The rheological behavior of the presented 3.0 and 6.0 wt. % nanofiller reinforced nanocomposites were investigated. As it can be seen from the figures, nanocomposites containing carbon nanotubes showed significantly higher viscosities compared to the nanocomposites containing Graphene Nano platelets. In the all nanocomposites, the Newtonian plateau was not present, and only the shear thinning is defining the melt behavior, while the viscosity decreases constantly with the increase of frequency.

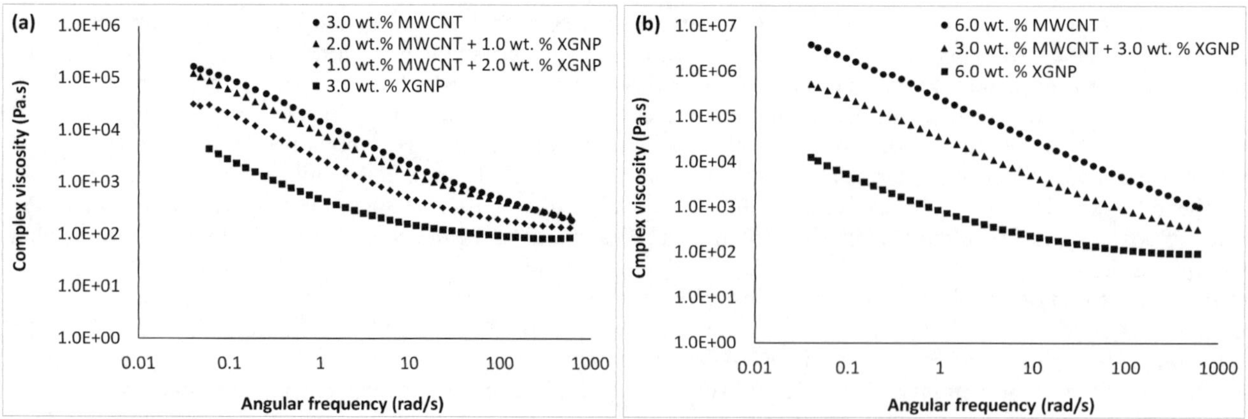

FIGURE 3. Complex viscosity of the (a) 3.0 and (b) 6.0 wt. % nanofiller reinforced nanocomposites as a function of MWCNT/ XGnP inclusion ratio and angular velocity.

Microstructure

Fig. 4 shows the SEM pictures of the nanocomposite microstructures. As it can be noticed from the pictures, in the both nanocomposites, acceptable states of dispersion have been achieved, indicating the successful mixing and injection molding process. An interesting observation was made in the Fig. 4(b), where the graphene nanoplatelets have been placed between the carbo nanotubes prohibiting the nanofillers to agglomerate additionally. Regarding the attained states of dispersion, one might assume that the nucleation influence of the carbon nanotubes in the composites might be reason for higher values of the tensile strengths.

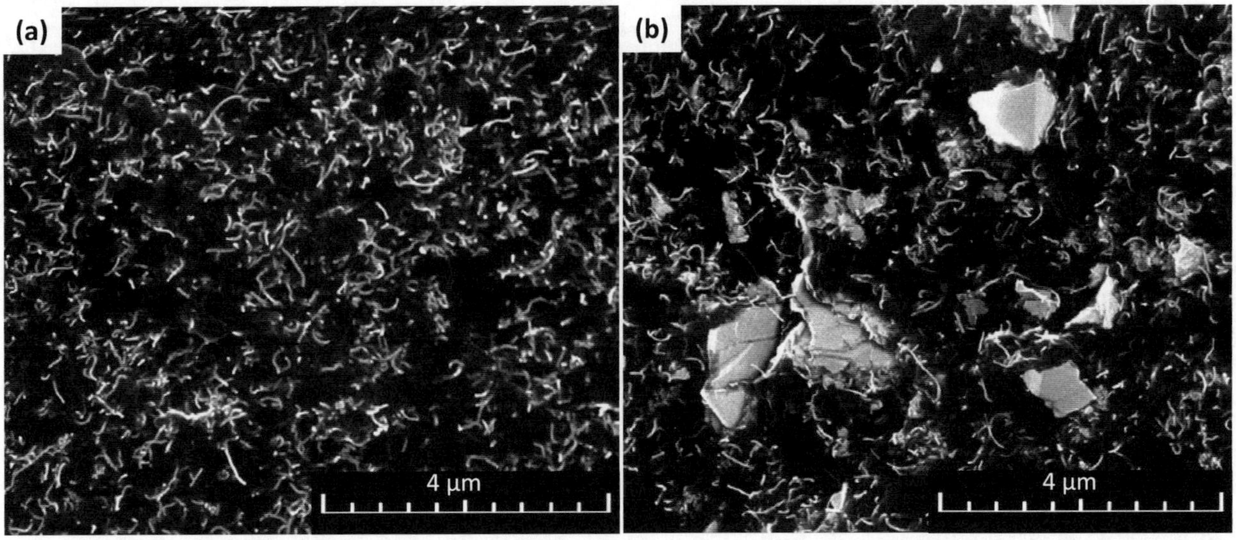

FIGURE 4. Scanning electron microscopy investigation of the (a) carbon nanotubes and (b) hybrid MWCNT/XGnP nanocomposites.

CONCLUSIONS

PA 66 based hybrid and nano composites containing different contents of carbon nanotubes and Graphene nano platelets were produced via melt blending and injection molding. Results showed that addition of the nanofillers increased the tensile strength and elastic moduli of the nanocomposites. Moreover, inclusion of the smaller percentage of the XGnPs in the MWCNTs reinforced composites additionally improved the mechanical properties. In fact, the novel hybrid composites showed enhanced mechanical properties compared to the nanocomposites with the similar content of just one type of nanofillers. The reasons behind the improvement were also investigated through rheological and electron microscopy tools.

REFERENCES

1. J. Hone, M. Whitney, C. Piskoti, A. Zettl, *Phys. Rev. B.* **59**, R2514 (1999).
2. M. Meyyappan, Carbon Nanotubes: Science and Applications, CRC Press, 2004.
3. P. M. Ajayan, L. S. Schadler, P. V Braun, Nanocomposite Science and Technology, Wiley, 2003.
4. E. Alishahi, S. Shadlou, S. Doagou-R, M. R. Ayatollahi, *Macromol. Mater. Eng.* **298** (2013).
5. E. Thostenson, C. Li, T. Chou, *Compos. Sci. Technol.* **65**, 491–516 (2005).
6. S. Doagou-Rad, A. Islam, J. S. Jensen, *Procedia CIRP.* **66**, 131–136 (2017).
7. J. Sumfleth, X. C. Adroher, K. Schulte, *J. Mater. Sci.* **44**, 3241 (2009).
8. M. H. Al-Saleh, *Synth. Met.* **209**, 41–46 (2015).
9. Ren P, Di Y, Zhang Q, Li L, Pang H, Li Z, *Macromol. Mater. Eng.* **297**, 437–443 (2012).

Reinforcing Effect of Cellulose Nanofibers on Biopolymers

Safdari Fatemeh[a], Carreau Pierre J.[a], Heuzey Marie-Claude[a], Kamal Musa R.[b]

[a] *Research Center for High Performance Polymer and Composite Systems (CREPEC), Chemical Engineering Department, Polytechnique Montreal, Montreal, Quebec H3T 1J4, Canada*

[b] *CREPEC, Chemical Engineering Department, McGill University, Montreal, Quebec H3A 0C5, Canada*

Abstract. The focus of this investigation has been to explore the potential use of cellulose nanofibers (CNFs) to enhance the properties of a bio-based/biodegradable (hydrophobic) polylactide (PLA). Initially, PLA/CNF biocomposites were prepared via a solution method, and the effects of CNFs on key properties of PLA were investigated. Subsequently, the compatibilization effects of poly(ethylene glycol) (PEG), a low molecular weight poly(ethylene oxide) (PEO), on the dispersion/distribution of the CNFs within the PLA matrix and on the rheological and mechanical properties of PLA/CNF biocomposites were examined. Scanning electron, atomic force and transmission electron microscopy confirmed that the CNFs were well dispersed in PLA. Moreover, improved dispersion/distribution of nanofibers was achieved using PEG as a compatibilizer. The complex viscosity and storage modulus of PLA increased substantially by incorporating CNFs. The storage modulus of a PLA/2 wt% CNF biocomposite, compatibilized with PEG, measured in dynamic mechanical thermal analysis at room temperature and 80 °C, increased by 42 and 553%, respectively, relative to the neat PLA. These results demonstrate that, using proper preparation methods, CNFs can improve the properties of hydrophobic matrices at very low contents.

Keywords: Cellulose nanofibers, polylactide, poly(ethylene glycol), composites.
PACS: 81

INTRODUCTION

Due to sustainability and environmental concerns it is desirable to use polymers derived from renewable resources as alternatives to petroleum-based polymers [1]. Polylactide (PLA) is probably the most promising biopolymer, owing to its interesting mechanical properties such as high modulus and tensile strength at room temperature, relatively good processability and moderate cost. However, PLA suffers from major drawbacks: slow crystallization, low heat resistance, and low melt strength. These drawbacks can be partly overcome by the incorporation in PLA of fillers such as cellulose nanofibers (CNFs). However, the challenge is to obtain a good dispersion and distribution of the hydrophilic fibers in the hydrophobic PLA. In our initial work [2], CNFs were incorporated into PLA via solvent casting to successfully improve the properties in different aspects without deteriorating the thermal stability and transparency of the matrix. In a subsequent study [3] PEG was used as a compatibilizer to further improve the dispersion/distribution of nanofibers within the PLA matrix, and obtain biocompatible and biodegradable PLA/CNF/PEG composites. Here we summarize our main findings.

EXPERIMENTAL

Experimental details can be found in [2, 3]. A polylactide (PLA) (Ingeo Biopolymer 3251D of NatureWorks) was used. A poly(ethylene glycol) (PEG) of a number-average molecular weight of 20,000 g/mol was purchased from Sigma-Aldrich Canada Co. The solvent, N,N-dimethylformamide (DMF), was used as received (anhydrous 99.8% from Sigma-Aldrich Canada Co. An aqueous suspension containing 2.3 wt% cellulose nanofibers (CNFs) was kindly provided by Prof. Mohini M. Sain of the University of Toronto.

The aqueous suspension of CNFs was freeze-dried for 48 h and the freeze-dried CNFs were mixed with an appropriate amount of DMF to form CNF suspensions containing 4-10 g of CNFs per L of DMF. The suspension was sonicated for 1 h using a water-bath sonicator (FS30 100 Watts Ultrasonic Cleaner of Fisher Scientific) and, then, stirred for 1 more hour using a magnetic stirrer at 400 rpm. Vacuum-dried PLA, at 80 °C for 24 h, was added to the suspension and stirred at 400 rpm, 70 °C and for 2 h until the PLA pellets were completely dissolved. Finally, the mixture was vacuum dried in an oven at 80 °C for 24 h, and, the product was then ground and kept for 48 h in a vacuum oven at 60 °C.

A masterbatch of CNFs and PEG was also prepared via an aqueous mixing method. The original CNF aqueous suspension was diluted with distilled water using a magnetic stirrer at 400 rpm for 1 h to form a 1 wt% CNF suspension and the desired amount of PEG was added for a weight ratio of CNFs to PEG 1:2, then, stirred for 1 more hour. Finally, the mixture was vacuum dried for 48 h at 50 °C to be used as the masterbatch of compatibilized CNFs for the preparation of composites following the same solution-preparation technique described above. The samples are identified based on their fiber and PEG contents, 0-5 and 0-4 wt%, respectively. For instance, PLA/2CNF/4PEG denotes a composite containing 2 wt% CNFs and 4 wt% PEG of the overall composite. Biocomposites were also prepared in the melt using an internal mixer, DDRV501 Brabender. However, these composites showed very low or no enhancements of their rheological and mechanical properties, indicative of a poor dispersion/distribution of the nanofibers in PLA. Hence, no results are reported here (for details refer to [3]).

RESULTS AND DISCUSSION

Scanning electron (SEM), atomic force (AFM) and transmission electron microscopic (TEM) images of PLA/2CNF and PLA/2CNF/4PEG are presented in Figure 1. The left images are for SEM, the centered ones for AFM and the right ones for TEM. In the upper images for PLA/2CNF (Figs. 1a-c) no large agglomerates are seen, but there are not many single fibers; however as shown by the lower images the quality of the nanofiber dispersion/distribution for composites prepared with PEG is considerably improved (Figs. 1d-f).

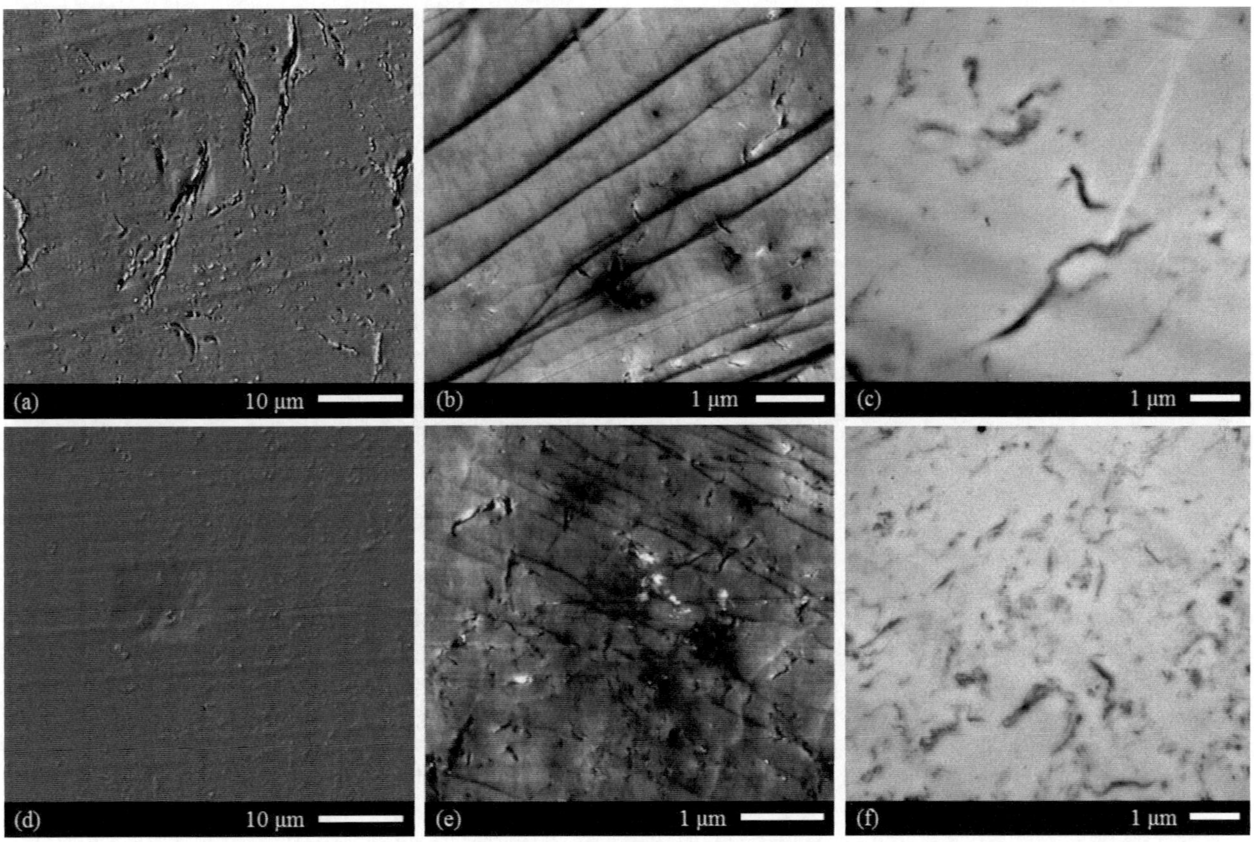

FIGURE 1. SEM (a, d), AFM (b, e) and TEM images (c, f) of PLA/2CNF (a-c) and PLA/2CNF/4PEG (d-f).

Figure 2 presents linear viscoelastic data in terms of the complex viscosity, η^*, and storage and loss moduli, G' and G'', as functions of frequency, ω, for all solution-produced samples. PLA exhibits a long plateau for η^* followed by shear thinning (Fig. 2a). Also terminal zones with slopes of two and one are observed for G' and G'' versus ω, respectively (Fig. 2b). All composites show a shear-thinning behavior without a plateau region for η^* (Fig. 2a) and a less frequency-dependency behavior for G' and G'' (Fig. 2b), especially at lower frequencies. Large increases in η^*, G' and G'' by one, four and one orders of magnitude, respectively, are observed for PLA/2CNF relative to the neat PLA. These results suggest a network formation, which consequently leads to a transition from liquid- to solid-like behavior. On the other hand, by introducing the masterbatch into PLA, i.e., PLA/2CNF/4PEG, the values for η^*, G'

and G'' of PLA/4PEG at low frequencies increase by two, six and two orders of magnitude, respectively. Also, Fig. 2b indicates that G' and G'' for PLA/2CNF/4PEG cross over at a frequency higher than that for PLA/2CNF. Interestingly, PLA/2CNF/4PEG exhibits a similar low-frequency complex viscosity behavior as the composite containing more CNFs, but without PEG, i.e., PLA/5CNF (Fig. 2a). In contrast, η^* of PLA/4PEG is lower than that of PLA due to the lower complex viscosity of PEG compared to PLA. Composites containing other particles such as cellulose nanocrystals [4, 5], organoclays [6] and carbon nanotubes [7] also showed similar rheological behavior.

The composites exhibit an apparent yield stress that can be quantified using the modified Carreau-Yasuda model [3]:

$$\eta^* = \sigma_y/\omega + \eta_0(1 + (\lambda\omega)^a)^{(n-1/a)} \tag{1}$$

where σ_y, a, λ, n and η_0 are the apparent yield stress, Yasuda parameter, time constant, flow index and zero-shear viscosity, respectively. Figure 2a shows that the model fits all the composite data very well.

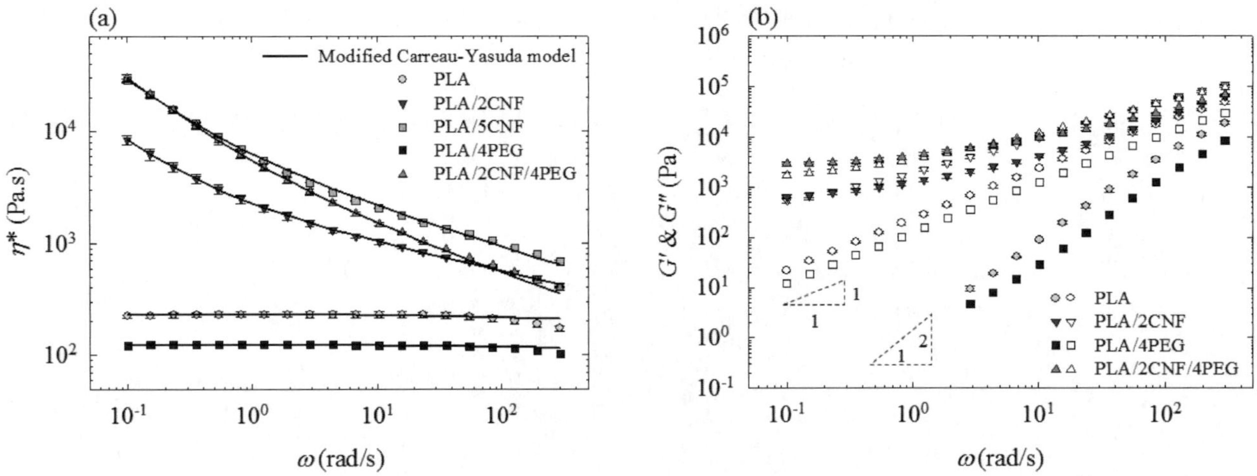

FIGURE 2. Complex viscosity (a), and storage and loss moduli (*filled* and *open symbols*, respectively) (b) vs. frequency at 175 °C and 0.05 strain amplitude for all solution-based samples. The solid lines in (a) represent the modified Carreau-Yasuda model fits, Eq. (1).

Figure 3 presents dynamic mechanical thermal analysis (DMTA) data in terms of the flexural storage modulus, E', and $\tan\delta$ for the samples prepared in solution as functions of temperature. The storage modulus of PLA increases over a wide range of temperatures by incorporating the nanofibers; the largest increases are obtained in the rubbery region when CNFs are added to the PLA via the masterbatch approach (Fig. 3a). The $\tan\delta$ peak (Fig. 3b), which is taken as the glass transition temperature, T_g, of PLA, is not affected significantly by the addition of the CNFs, compatibilizer nor via the use of the masterbatch; it increases from 59 °C for the neat PLA to 63 °C for PLA/2CNF. However, with the incorporation of either the CNFs or the compatibilizer within the PLA phase, the area under the $\tan\delta$ peak of PLA decreases and the decrease is more important for PLA/2CNF/4PEG. This indicates restricted PLA chain motion as a result of the good dispersion of the CNFs creating a large interfacial area and the effect of PEG on the crystallinity of PLA [3].

The storage modulus of PLA at 25 °C (1.71 GPa) increases up to 42% for PLA/2CNF/4PEG. Moreover, in the rubbery region, the storage modulus of PLA (3.9 MPa at 80 °C) increases by 132 and 553% for the samples containing 2 wt% CNFs without and with PEG, respectively (Fig. 3a). This shows that the reinforcement of the nanofibers is more effective when the matrix has a lower modulus in the rubbery region, and also confirms that a good state of dispersion is achieved using the compatibilizer; thus, PLA products can be exposed to higher temperatures.

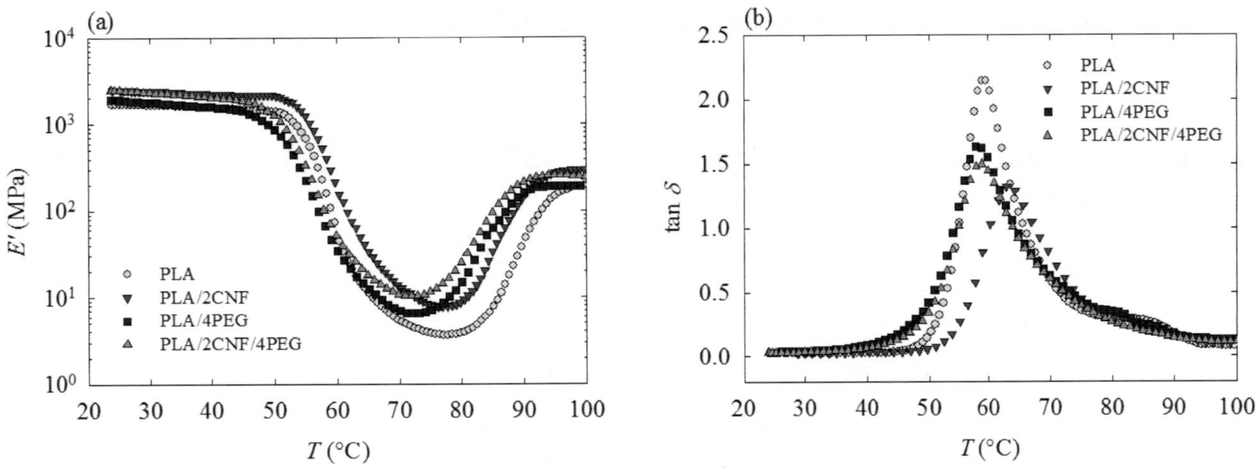

FIGURE 3. Flexural storage modulus (a) and tan δ (b) in DMTA of the solution-prepared samples.

CONCLUSION

Simple solution-based methods were utilized to prepare a CNF/PEG masterbatch and nanofiber-reinforced PLA composites in order to investigate the effects of PEG on the morphology, rheological, thermal, and mechanical properties of PLA/CNF composites. The results of SEM, AFM and TEM indicated a more uniform dispersion/distribution of the nanofibers in the PLA matrix when the CNFs were initially dispersed in PEG. Incorporation of CNFs caused substantial increases of the complex viscosity and storage and loss moduli of the PLA composites; an apparent yield stress was also observed. However, the effect of CNFs on the rheological properties was more pronounced when the masterbatch method was used, suggesting a better dispersion and distribution of CNFs. The composites prepared using melt processing hardly showed any effect on rheological and mechanical properties (results not shown). This confirms that dispersing highly-entangled CNFs in a non-polar matrix via direct mixing in the melt remains highly challenging.

The flexural storage modulus in DMTA was significantly enhanced for the composites in comparison with the neat PLA, especially in the rubbery zone; the storage modulus of PLA at 80 °C was improved by 132 and 553% for the samples containing 2 wt% CNFs without and with PEG, respectively. The results of this study point out to the efficiency of the solution preparation method and the substantial potential of PEG to further improve the properties of PLA/CNF biocomposites. This would promote the use of PLA in applications where its low heat resistance is of serious drawbacks.

ACKNOWLEDGMENTS

This work was supported by the Natural Sciences and Engineering Research Council of Canada (NSERC) through the Network for Innovative Plastic Materials and Manufacturing Processes (NIPMMP). The authors also wish to thank Professor Mohini M. Sain of the University of Toronto for providing the CNFs.

REFERENCES

1. J.M. Raquez, Y. Habibi, M. Murariu, P. Dubois P. *Prog. Polym. Sci.* **38**, 1504-1542 (2013).
2. F. Safdari, D. Bagheriasl D, P.J. Carreau, M.C.Heuzey, M.R. Kamal. Rheological, Mechanical, and Thermal Properties of Polylactide/Cellulose Nanofiber Biocomposites. *Polym. Compos*. doi: 10.1002/pc.24127 (2016).
3. F. Safdari, P.J. Carreau, M.C. Heuzey, M.R. Kamal. *Cellulose* 24, 2877-2893 (2017).
4. M.R. Kamal and V. Khoshkava. Carbohydr. Polym. **123,** 105-114 (2015).
5. D. Bagheriasl, P.J. Carreau, B. Riedl, C. Dubois, W.Y. Hamad. *Cellulose* **23**, 1885-1897 (2016).
6. A. Ghanbari, M.C. Heuzey, P.J. Carreau, M.T. Ton-That. *Rheol. Acta* **52**, 59-74(2013).
7. S. Abbasi, P.J. Carreau, A. Derdouri, M. Moan. *Rheol. Acta* **48**, 943-959 (2009).

Acetone detection using SERS in Poly (styrene) with Silver nanoparticles

marquez alfredo [(1)*]

[(1)] CIMAV - Chihuahua - Mexico

Acetone is an organic solvent; it is volatile and flammable and is a widely used solvent in the industry. According to the Occupational Safety and Health Administration (OSHA) the permissible exposure limit of acetone in air is 1000 ppm; above this value it can be harmful for humans. Many techniques have been used to measure acetone in air, like: Gas chromatography, Flame Ionization Detection, Ion Mobility Spectrometry and Mass Spectrometry to mention some2. These techniques have a number of limitations to be used in real time monitoring, because they require sample preparation, qualified personal and they are relatively expensive. Other option to very precisely measure the concentration of acetone in air is the Surface-Enhanced Raman Spectroscopy, a versatile molecular spectroscopic technique, which is sensitive, rapid, and accurate. In this present work we use Silver nanoparticles (AgNPs) on the surface of Polystyrene subtract (PS) for improving the Raman signal at low analite concentrations, silver nanoparticles act as Raman enhancers helping to detect an analite. This work presents the results of synthetizing Polystyrene/Silver nanoparticles (Ps/AgNPs) to be used as the sensitive section in a detector of acetone by Raman Spectroscopy.

Production of ecofriendly thermoplastic vulcanizate composites with high stretchability, high elasticity, and low electrical percolation thresholds

Kazemi Yasamin (1), Ramezani Kakroodi Adel (1), Wang Sai (1), Ameli Aboutaleb (2), Filleter Tobin (1), Park Chul B (1)*

(1) University of Toronto - Ontario - Canada, (2) Washington State University Tri-Cities - Washington - USA

Production of stretchable and conductive polymer composites has attracted the attention of the scientific community due to their high potential for various applications including smart clothing, stretchable circuits, and strain gauges. However, the successful production of these materials through cost-effective techniques has remained a challenge due to several issues. This work presents, for the first time, a practical and ecofriendly approach towards the production of ground tire rubber (GTR)-based stretchable and conductive composites. GTR, with an average particle size of < 50 m, was compounded with maleated polyethylene (MAPE) matrix and carbon nanotubes (CNTs) via melt blending. The pre-vulcanized state of the GTR enabled us to produce thermoplastic vulcanizates (TPVs) with high concentrations, 60 wt.%, of the rubbery phase while still maintaining the MAPE's continuous phase morphology. The GTR particles showed remarkable compatibility with the MAPE matrix. This behavior was shown to be a result of chemical bonding during the reactive extrusion of the MAPE and the GTR. The CNTs had an almost perfect level of dispersion within the MAPE matrix. This was ascribed to good MAPE/CNT compatibility caused by the interactions between the MAPE's maleic anhydride moiety and the CNTs. This, combined with the fact that the CNTs were only present in the MAPE phase (segregated blend morphology) led to the composites' very low electrical percolation thresholds. The composites showed excellent stretchabilities and elasticities due to the strong interfacial adhesion between the MAPE and the rubber particles. In addition, the composites showed excellent abilities to maintain their mechanical and electrical properties up to 1000 stretching cycles.

Improving thermal stability of polymers by using long single wall carbon nanotube as an antioxidant

Seisuke ATA

Advanced Institute of Science and Technology, 1-1-1 Higashi, Tusukuba, Ibaraki, 305-8565 Japan)
1Tel: +81-29-861-3767, Fax: +81-29-861- 4851, E-mail: ata-s@aist.go.jp

Abstract. Do not replace the word "abstract," but do replace the rest of this text. If you must insert a hard line break, please use Shift+Enter rather than just tapping your "Enter" key. You may want to print this page and refer to it as a style sample before you begin working on your paper. Super-growth single wall carbon nanotube (SG-SWNT), which was developed in AIST, offer several unique properties such as high aspect ratio, high purity, good dispersability, thermal stability and radical scavenging ability etc. Furthermore, the growth efficiency of SG-SWNT is superior to other SWNT and ton-scaled SG-SWNT will be provided in no distant future. We are developing polymer/SG-SWNT composite with excellent properties. Representative examples of development are shown below. 1,SG-SWNT/fluorinated rubber composite for high thermal durable shielding material. 2, SG-SWNT including CFRP to improve electrical conductivity and mechanical properties of CFRP. 3, SG-SWNT/carbon fiber/rubber composite for soft yet high thermal conductive material. In the presentation, I would like to introduce our developed manufacturing and applications of CNT/polymer composites.

Keywords: Carbon nanotube, composite, thermal resistivity, thermal conductivity
PACS: 61.48.De

Polymers, such as rubbers, thermal plastic elastomer, resins and engineering plastics etc., are widely used as an industrial material because of their excellent characteristics. However, polymers are weak for heat derived by thermal degradation of its main-polymer-chain by auto-oxidation reaction. The low-heat resistance of polymers limits their further wide-range of application. Therefore, improving thermal resistivity of polymers is one of the most important issue in the industrial field.

Thermal resistivity of polymers can be improved in several ways. Attempts to improve the thermal resistivity of polymers by changing and controlling the chemical structure of polymers have been carried out for a long time. However, as these attempts have been exhausted, it is difficult to achieve further improvement in thermal resistivity of polymers.

Another approach to improving the thermal resistivity of polymers is to add additives typified by antioxidants. By quenching and scavenging the radicals generated by thermal decomposition at high temperature, the antioxidant can greatly improve the heat resistance of the polymer material. However, the antioxidant has a problem that the heat resistance of the antioxidant itself is not so high. This is because the conventional antioxidant is low molecular weight, its function is lowered or decreased by heat denaturation at high temperature. Therefore, in order to further improve the thermal resistance of the polymer material, it is necessary to use an antioxidant having higher heat resistance.

Carbon nanotubes (CNT) is a material having a possibility to be used as an antioxidant which can be used under a high temperature environment. CNT has π-electrons on its surface, has high thermal resistance (~600°C in air) and is a high aspect ratio of fibrous material with a large specific surface area (~1200 m^2/g), so it has great potential as a high thermal resistive antioxidant.

However, on the other hand, as a weak point of CNT, it contains a metal (normally iron or its alloy) acting as a catalyst at synthesis process, which is a big problem for using CNTs as antioxidant. It is known that a metal catalyst functions as a decomposition catalyst of a polymer material.
Super growth single wall carbon nanotube[1] (SG-SWNT) is synthesized by CVD, it does not contain a metal catalyst. Furthermore, because of SG-SWNT is a single wall CNT, it is having a very large specific surface area more than 1000 m^2/g. Therefore, it is expected to be the CNT having the most excellent antioxidant among CNTs.
In this study, we report the effect of SG-SWNT as an antioxidant to improving thermal resistivity of polymers.

Polyetheretherketone (PEEK, 150P, Victrex) is known as high thermal resistive and high strength engineering plastics. For that reason, PEEK has attracted attention as a substitute for metallic materials. The deflection temperature under load of PEEK can be improved by composing carbon fiber or glass fiber. On the other hand, at 300 °C, thermal decomposition of PEEK starts, so the continuous use temperature is only about 250 °C.
Figure 1 shows the weight loss of PEEK under high temperature environment for 2hours. Pure PEEK (grey dots) was heavily weight reduced at 450 °C. On the other hand, the weight loss of CNT / PEEK composite material decreased. This result implies that CNT functions as a radical scavenger even at a very high temperature of 450 °C.
In the presentation, we will also report on the results of more detailed mechanisms.

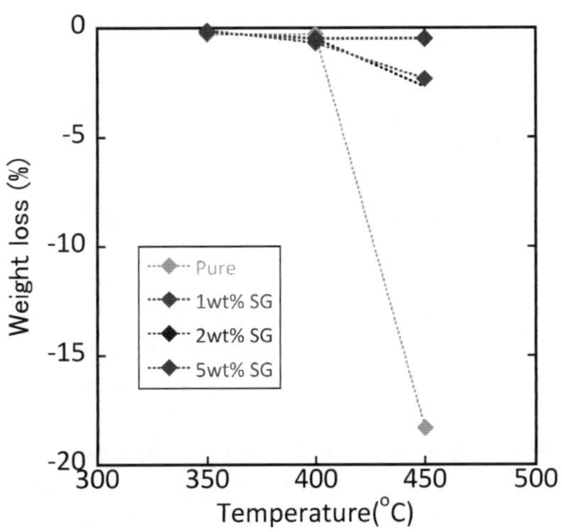

Figure1. Weight loss of PEEK and PEEK/SG-SWNT composites for 2hours.

References
1. Hata, K. *et al.* Water-assisted highly efficient synthesis of impurity-free single-walled carbon nanotubes. *Science* **306,** 1362–1364 (2004).

Dynamic Mechanical and Electrical Behavior of Polyvinylbutyral Reinforced with Graphene Nanoplatelets

Sérgio Henrique Pezzin[a], Jean Carlos Hoepfner[a], Márcio Rodrigo Loos[b]

[a]Santa Catarina State University – Center of Technological Sciences - UDESC
[b]Santa Catarina Federal University - UFSC

Abstract.. In this work, nanocomposites of PVB matrix reinforced with graphene nanoplatelets (GNP) and graphene oxide (GO) (PVB/GNP and PVB/GO) were synthesized 'in situ' with different concentrations. Nanoplatelets of GO and GNP were produced and characterized by X-ray diffraction and transmission electron microscopy analyses. The values of glass transition temperature had an increase of 10 ºC for the nanocomposite with 2.5 wt% GO. The dynamic mechanical analysis (DMA) showed that the sample with 2.5 wt% GO had the largest increase in storage modulus, which can be attributed to the interactions and better dispersion of the reinforcement in the PVB matrix. DMA results also showed the presence of two immiscible phases for all samples, one attributed to the PVB and the other to the PVA. Electrical impedance spectroscopy analysis showed that despite a homogeneous dispersion, no significant increases were observed in the values of electrical conductivity.

Keywords: Graphene nanoplatelets, polyvinylbutyral, nanocomposites, dynamic mechanical analysis
PACS: 81.07.-b; 81.05.ue; 81.05.Lg; 81.05.Qk; 81.70.Bt.

INTRODUCTION

Polyvinylbutyral (PVB) is a random copolymer obtained by the condensation reaction of polyvinylalcohol (PVA) with butyraldehyde (BU) in acid medium. PVB is used in making safety glass laminates, adhesives, coatings and paints due to their transparency combined with mechanical strength, flexibility and adhesiveness to metals and glass and resistance to weathering (rain, humidity, heat).[1] These properties make PVB suitable also for special applications such as solar cell and displays, or any optoelectronic system that requires transparency, mechanical strength, and resistance to natural events such as temperature variations.[2]

Graphene nanoplatelets (GNP) are multiple graphene layers stacked to form platelets. GNP is obtained from the oxidation of graphite and reduction of graphene oxide.[3,4] Graphene oxide (GO) consists of a 'decorated' graphene sheet with various oxygen groups, as carboxylic acids, hydroxyls and epoxy rings.[5,6] Both GNP and GO have high aspect ratio, and low density, and have been extensively studied to improve the characteristics of polymer matrices.[7,8,9]

There are few studies in the literature reporting PVB/graphene nanocomposites. A recent work showed that a PVB membrane reinforced with 6.0 wt% of silanized GNP can reach tensile strength and elongation at break of 35.6 MPa and 44.5%, respectively. Moreover, the incorporation of 3.0% of functionalized graphene enhanced the electrical conductivity of the material by seven orders of magnitude.[10]. Haijan and coworkers have synthesized PVB via condensation of PVA in aqueous solution with BU, achieving a degree of acetalization of 85 mol %. The nanocomposites, prepared by solution blending, showed increases in mechanical properties and thermal stability, when compared with neat PVB.[11]

In this work, we studied the influence of GNP and GO in thermal-mechanical properties of PVB nanocomposites and the influence of these nanoparticles in the synthesis of PVB. The dispersion and interfacial interaction of GNP and GO in PVB matrix were studied by dynamic mechanical analyses and correlated with infrared spectroscopy.

EXPERIMENTAL

Materials

GNP were supplied by Stream Chemicals, with thicknesses of approximately 6-8 nm and widths up to 25 µm. The graphite was purchased from Labsynth, with particle size <45 µm, 98% purity and 1.0% ash content. PVA with a molecular weight of 85.300 g mol^{-1} was provided by Vetec Chemistry and butyraldehyde 98% by Sigma-Aldrich.

Sulfuric acid 98% (Cinética), nitric acid 65% (Merck), hydrochloric acid 37% (Cinética), glacial acetic acid (Cinética), sodium lauryl sulphate (Cinética), sodium nitrate (Synth), potassium permanganate (Synth) and 35% hydrogen peroxide (Cinética) were used without prior treatments.

Production of graphene oxide

GO was produced from graphite using an adaptation of the Hummers method, according to another study from our research group.[12]

Synthesis and processing of nanocomposites

GNP and GO were dispersed by sonication (Sonics 750 W) in 50 mL of deionized water for 15 minutes. To this dispersion 14 g of PVA, 0.21 g of sodium lauryl sulfate, 140 ml of acetic acid, 90 ml water and a few drops of sulfuric acid were added. The system was kept in mechanical stirring in a ultrasonic bath at 10 °C for 120 minutes and then at 70 °C for 90 minutes. At the end of the reaction, water was added to precipitate the polymer. The polymer was filtered and washed with deionized water and a solution of NaOH 1 mol L^{-1} up to pH 7. The polymer was then dried in an oven at 50 °C until constant weight. For dynamic mechanical analysis (DMA) the nanocomposites were pressed in a hydraulic press under pressure of 2 Ton for 2 minutes at 160 ° C and then cooled for 7 minutes. For eletrical analysis, the nanocomposites were pressed in hydraulic press under pressure of 3 Ton for 5 minutes and cooled for 5 minutes. Table (1) shows de composition of nanocomposites.

TABLE (1). Compositions of the PVB nanocomposites

Sample	Nanoparticle	Concentration wt%
PVB/GO	Graphene oxide	0.5, 1.0, 2.5
PVB/GNP	Graphene nanoplatelets	0.5, 1.0, 2.5

CHARACTERIZATION

GO and GNP were characterized by X-ray diffractograms using a Shimadzu XRD 6000 diffractometer with a source of CuKα radiation. The analyses were performed with 2θ angle ranging from 5 to 60°. Transmission electron microscopy (TEM) was performed on a JEOL-2100. Fourier transform infrared spectroscopy (FTIR) spectra were collected with a Perkin Elmer Spectrum One B in the range of 4000 to 600 cm^{-1}, using resolution of 4 cm^{-1} and 32 scans. The mechanical and thermal properties of nanocomposites were analyzed by dynamic-mechanical analysis (DMA). Neat PVB and the nanocomposites were characterized in flexural mode. The measurements were recorded using a TA instrument model 2980 at a frequency of 2 Hz, in temperature range of 25 °C to 110 °C with heating rate of 5 °C min^{-1}. The conductivity analysis was performed by electrical impedance spectroscopy (Autolab PGSTAT20, EcoChemie).

RESULTS AND DISCUSSION

The X-ray diffractogram of GNP (FIGURE 1a) shows the presence of a characteristic peak at a 2θ angle of 26.8°, referring to a planar spacing (d_{002}) of 0.334 nm between the graphene plates.[13] In the case of GO, the peak at the angle 26.8 ° disappears and a high intensity peak at the 10.2° is observed, showing an increase in the d_{002} interplanar spacing to approximately 0.790 nm. This interplanar increase is due to the insertion of oxygen groups between the platelets of graphene.[14]

FIGURE 2 shows the FTIR spectra of PVA, PVB and nanocomposites. The bands at 1050 and 1150 cm^{-1} correspond to symmetric and asymmetric stretching of the C-O-C vibrations present in the butyral ring, while the band at 1740 cm^{-1} is attributed to the presence of acetal groups, confirming the conversion of PVA to PVB.[1,15] Nevertheless, it is observed that the PVB spectrum still has some bands related to PVA, as the OH stretch at 3490 cm^{-1}. The spectra observed for nanocomposites are similar to PVB, but the small changes indicate some differences on the degree of acetalization of PVB with the addition of GO.

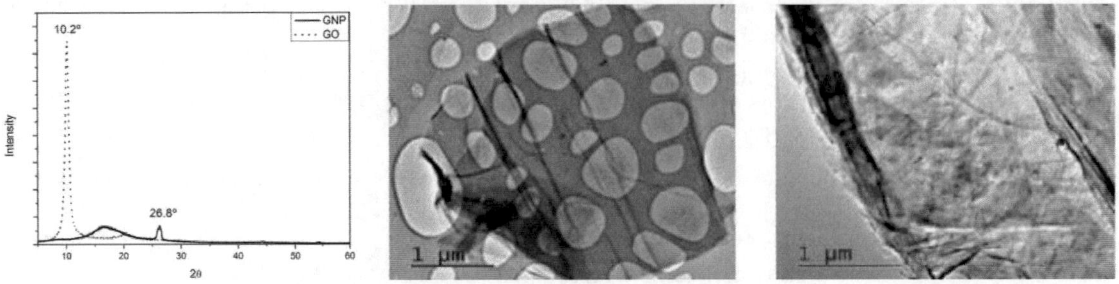

(a) **(b)** **(c)**

FIGURE 1. a) X-ray diffractograms of GO and GNP; b) TEM image of GO; c) TEM image of GNP.

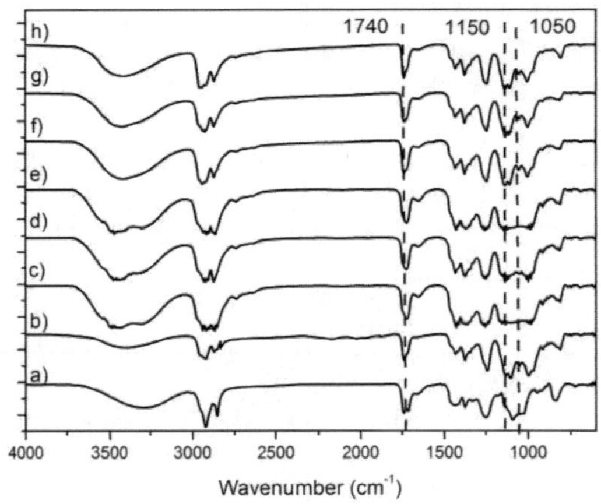

FIGURE 2. FTIR spectrum to a) PVA, b) PVB, c) PVB/GO 0.5 wt%, d) PVB/GO 1.0 wt%, e) PVB/GO 2.5 wt%, f) PVB/GNP 0.5 wt%, g) PVB/GNP 1.0 wt% and h) PVB/GNP 2.5 wt%

The behavior of the storage modulus with temperature for PVB and nanocomposites are observed in FIGURE 4a. The values are very similar to PVB, with the exception of the sample PVB/GO 2.5 wt% which presented a higher storage module at room temperature compared to the PVB.

PVB and PVB GO 2.5 wt% samples present a plateau up to the rubbery region, characteristic of rigid materials. All other samples do not present a plateau in the studied temperature range, which indicates that they are ductile materials, due to a lower degree of acetalization. Nevertheless, it is observed that the sample PVB GO 2.5 wt% has the highest storage modulus value of all the samples studied in this work, which can be related to the high concentration and totally exfoliated structure. The values of Tan δ for PVB and nanocomposites are plotted in FIGURE 5b. In the case of PVB, a peak is observed at 60 °C and another at 84 °C. The latter is related to PVA, whereas the peak of 60 ºC refers to PVB. As previously mentioned, the higher the degree of acetalization, the lower the Tg of PVB.[1] According to Saravanan and co-authors, the presence of shoulders in the Tan δ curve indicates a higher degree of acetylation.[16] For the PVB/GO 0.5 wt%, PVB/GO 1.0 wt% and PVB/GNP 1 wt% samples, it is observed a small decrease in the Tg values compared to PVB. This decrease can be associated a change in degree of acetalization. For PVB/GNP 0.5 wt% sample, an increase to 65 °C in the Tg value is observed.

For most of the samples, two peaks are observed, characterizing a two-phase immiscible system[17], corresponding to PVB (Tg = 60 °C) and PVA (Tg = 84 °C). For the PVB/GO 2.5 wt% a Tg value of 70 °C, with a shoulder at ca. 80 °C, is observed. This behavior indicates the formation of a partially miscible PVA/PVB system, probably due to a decrease in the degree of acetalization of PVB caused by the higher GO amount.

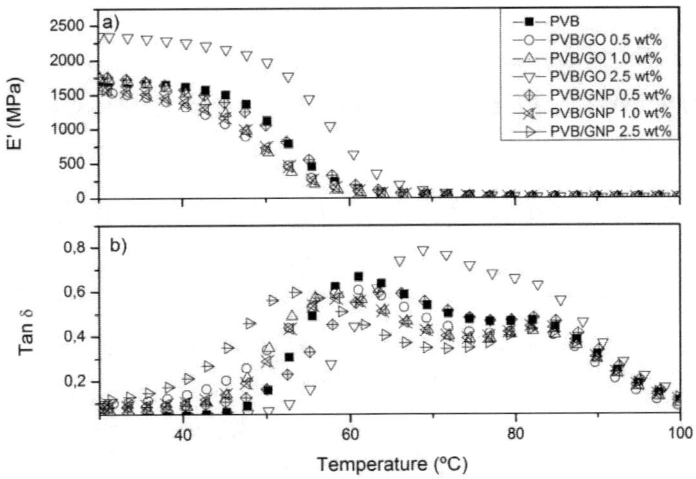

FIGURE 3.a) Storage modulus and b) values of Tan δ in a temperature range of 30 to 100 ° C for PVB and nanocomposites.

Electrical impedance spectroscopy analysis showed that despite a homogeneous dispersion, no significant increases were observed in the values of electrical conductivity. For all the samples an ohmic behavior was observed, exhibiting a linear increase of conductivity with frequency.

CONCLUSIONS

The synthesis of PVB nanocomposites with GO and GNP *in situ* was successfully carried out. Dynamic mechanical analysis showed the presence of two Tan δ peaks, indicating the presence of two immiscible phases for all samples, which can be related to different degrees of acetalization of PVB. The storage modulus of the matrix was significantly increased only with the addition of 2.5 wt% of GO. This result indicates that the *in situ* polymerization method contributed to improve the dispersion and interaction between GO and PVB. However, no electrical percolation was observed, despite the increase of an order of magnitude in the conductivity for all samples.

ACKNOWLEDGMENTS

The authors are grateful to FAPESC by financial support and for the scholarship to J.C. Hoepfner.

REFERENCES

1. Fernandez, M. D.; Fernandéz, M. J.; Hoces, P. *J. Appl. Polym. Sci.* **5**, 5007-5017 (2006).
2. Ding, L.; Liu, J.; Guo, X. J. *Qingdao Univer. Sci. Technol.* 31, 290-293 (2010).
3. Novoselov, K. S.; Geim, A. K.; Morozov, S. V.; Jiang D.; Zhang, Y.; Dubonos, S. V.; Grigorieva, I. V.; Firsov, A. A. *Science*, **306**, 666-669 (2004).
4. Geim, A. K.; Novoselov, K. S. *Nat. Mater.* **6**, 183-19 (2007).
5. Hummers, W. S.; Offeman, R. E. **J. Am. Chem. Soc**. **80**, 1339 (1958).
6. Park, S.; Ruoff, R. S.. *Nat. Nanotechnol*., **4**, 217-224 (2009).
7. Yu, A. P.; Ramesh, P.; Itkis, M. E.; Bekyarova, E.; Haddon, R.C. *J. Phys. Chem. C*, **21**, 7565-7569 (2001).
8. Vickery, J.L.; Patil, A .J.; Mann, S. *Adv. Mater*., **21**, 2180-2184 (2009).
10. Lei, D.; Ma, W.; Xiao, S. *Polym. Mater. Sci. Eng*., **30**, (2015).
11. Haijan, M.; Reisi, M. R.; Koohmareh, A. L.; Jam, A. R. Z. *J. Polym Res*., **19**, 9966 (2012).
12. da Silva, D. D.; dos Santos, W. F.; Pezzin, S. H. *Matéria*, **18**, 1260-1272 (2013).
13. Mcallister, M. J.; Li, J-L.; Adamson, D. H.; Schniepp, H. C.; Abdala, A. A.; Liu, J.; Herrera-alonso, M.; Milius, D. L.; Car, R., Prud'homme, R. K.; Aksay, I. A. *Chem. Mater*., **19**, 4396-4404 (2007).
14. Kuilla, T.; Bhadra, S.; Yao, D.; Kim, N. H.; Bose, S.; Lee, J. H. *Prog. Polym. Sci*., **35**, 1350-1375 (2010).
15. Yang, B.; Liu, R.; Huang, J.; Sun, H. *Ind. Eng. Chem. Res*., **52**, 7425-7431(2013).
16. Saravanan, S.; Gowda, K. A.; Varman, K. A.; Ramamurthy, P. C.; Madras, G. *Compos. Sci. Technol*., **117**, 417-427 (2015).
17. 30. Miles, I. S.; Zurek, A. *Polym. Eng. Sci*., **12**, 796-805 (1988).

Mouldability study of injected micro parts made of reinforced PA6 composites

Bellantone Vincenzo (1)*, Pagano Caludia (2), Surace Rossella (1), Fassi Irene (2)

(1) Institute of Industrial Technologies and Automation- National Research Council, ITIA-CNR - Bari - Italy, (2) Institute of Industrial Technologies and Automation- National Research Council, ITIA-CNR - Milano - Italy

Currently, the increasing interest in the study of polymer nanocomposites has been motivated by the potential of such materials to improve the properties of obtainable products significantly. Indeed, the combination of two materials (for example a polymer matrix and Multi-Wall Carbon Nanotubes- MWCNTs as nanofiller) is capable of merging the good characteristics of both and, as such, taking advantage to accomplish high design flexibility. Nonetheless, many of these polymer nanocomposites can be moulded into complex shapes. In this work, experimental and statistical studies have been performed to analyze the filling ability of a nanocomposite material, PA6/MWCNT, with filler fraction of 3wt%. The pristine material and the compound were processed via injection moulding process using different conditions defining their operating range. First, miniaturized dog bone have been prepared for tensile tests in order to confirm the reinforcement effect of the fillers. Then, samples with micro channels have been moulded in order to evaluate the material filling ability. It has been tested according to a Design of Experiment approach of the process parameters. As expected, the filler content affected the operative range of the micro injection process parameters, especially regarding the injection velocity. Nevertheless, a suitable choice of the process parameters allows to obtain good quality samples. Moreover, the most effective parameters on replication capability have been estimated at filler content variation evaluating a dimensional index defined as the differences between the injected parts and mould.

Preparation of Silica Nanotube/ PMMA Composites and their Physical Properties

Gwang-Yong Na[b], Jeong-Min Lee[a], Sang-Cheol Han[a], and Jeong-Cheol Kim[b]

[a]*Korea Innotech, Changwon-si, Gyongsangnamdo Korea,*
[b]*Energy and Applied Optics Team, Korea Institute of Industrial Technology, Gwangju 500-480, Korea*

Abstract. Silica nanotube, SNT, was used to prepare composite sheets of polymethyl methacrylate, PMMA, which showed improved flexural modulus than the virgin PMMA sheet. Surface of silica nanotube was treated by methacryloxypropyl trymethoxysilane, MPMA, to give better particle dispersion in the composite during melt mixing with PMMA through the twin screw extruder. Weight of MPMA in treated SNT was measured to be 7.5 percent of SNT by a thermogravimetric analyzer. The flexural modulus of SNT/PMMA composite was measured to be increased up to by 50% as the SNT content in the composite increased by 5%. Light transmittance decreased with SNT content in the composite, however, SNT with MPMA treatment showed better light transmittance than that without MPMA. 5% of MPMA treated SNT composite showed 80% of light transmittance, however, 5% of untreated SNT composite showed only 30%. TEM showed that the particle dispersion of MPMA treated SNT in the composite was better than without MPMA treatment. Breaking strength and strain decreased simultaneously with increasing SNT in the composite, but it did not look to be related to the treatment of MPMA on SNT.

Keywords: Nanocomposite, PMMA, Silica Nanotube, Flexural Modulus
PACS:

FLEXURAL MODULUS

The flexural modulus of PMMA/SNT nanocomposite was improved at 5wt% of SNT, while it was not at 1wt% of SNT. The average flexural modulus of the nanocomposites prepared using the solvent dispersion method containing 5 wt.% surface-modified SNT from ethanol-extraction (SEm5) was 6.5 GPa, and this showed an improvement of 50% compared to that of pure PMMA.. This high flexural modulus of the SEm5 sample is considered to have arisen from the surface modifier causing stronger interactions and unchanged SNT length with higher aspect ratio during the solvent dispersion method.

TABLE 1. Flexural Modulus of PMMA/SNT nanocomposite Samples from different Mixing Methods and Surface Treatment. Surface of SNT was treated by MPMA. (M=Melt mixing, Cn=Surfaced non-treated, S=Solvent Mixing, Em=Surface MPMA treated)

Samples	Flexural Modulus (GPa)	Specifications
V-PMMA	4.4	Virgin PMMA
MCn1	3.7	Melt mixing, SNT surfaced non treated, SNT 1wt%
MCn5	5.2	Melt mixing, SNT surfaced non treated, SNT 5wt%
SEm1	4.2	Solvent mixing, SNT surfaced MPMA treated, SNT 1wt%
SEm5	6.5	Solvent mixing, SNT surfaced MPMA treated, SNT 5wt%

LIGHT TRANSMITTANCE

The transmittance in the visible light region was found to be mostly reduced as a result of SNT addition. When the SNT content was 1 wt. %, the light transmittance of nanocomposites was in the range of 71.1 to 87.7%, which was not much lower than 88.5% of virgin PMMA. The light transmittance of nanocomposite by solvent dispersion method was higher than that by melt dispersion method. When SNT content became to 5 wt. %, it was found that the light transmittance of composite varied greatly according to the surface treatment of SNT.

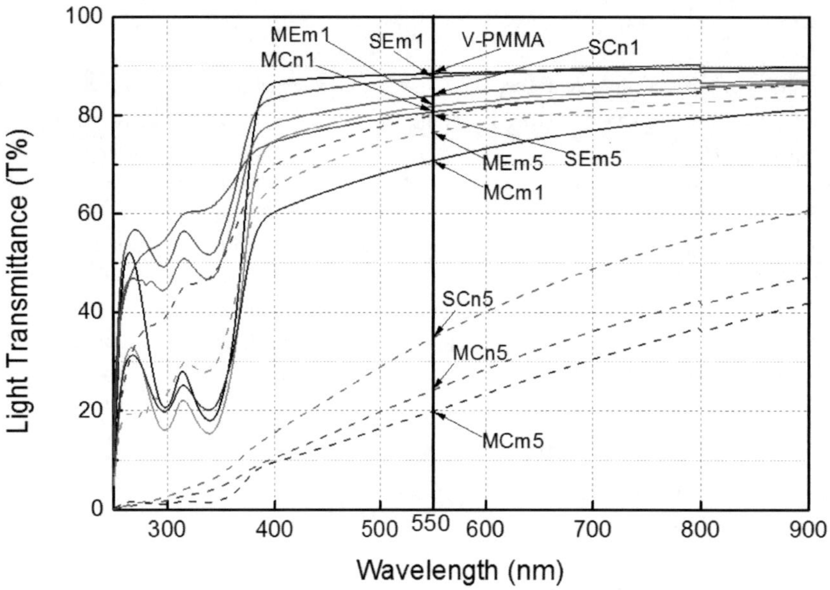

Figure 1. Light transmittance of PMMA/SNT nanocomposites at a wavelength of 550nm measured by UV-Visible. Surface modified samples showed higher light transmittance.

MECHANICAL PROPERTIES OF NANOCOMPOSITES

The strength and strain of the PMMA/SNT composites decreased with increasing SNT content. This might be from the inorganic additives acted as weak points in the composite matrix and adhesion instability existed at the organic-inorganic interface. It is also thought that the dispersion of the inorganic particles was not sufficiently high to increase the tensile strength of the composites. At the same amount of SNT, nanocomposites with surface-treated SNTs showed higher strength and elongation than those with SNT without surface treatment. However, tensile strength of composites did not appear to be correlated with dispersion method (solvent or melt).

TABLE 2. Tensile stress and strain of PMMA/SNT nanocomposite.

Sample	Tensile stress (MPa)	Tensile strain (%)
V-PMMA	85.84	9.54
MCn1	66.92	6.95
MCn5	43.25	4.21
SEm1	66.14	8.42
SEm5	45.68	5.05

CONCLUSIONS

PMMA/SNT nanocomposites were prepared by melt and solvent mixing methods. The effects of SNT surface modification, and contents of SNT in the composites were investigated by mechanical, and optical properties of PMMA/SNT composites. At the same SNT content, the composites with surface modified SNT showed higher strength and elongation than samples with SNT without surface treatment. Surface modification was found to be an important factor in the manufacture of composite. As a result of analysis of the dispersibility by TEM, it was found that the dispersibility was affected by the surface modification not by the method of manufacturing the composite (melt or solvent) or the template removal method. When the PMMA/SNT composite sheet was prepared by an appropriate method, the flexural modulus could be improved by about 50% compared to virgin PMMA sheet. Light transmittance could reached 80% which was close to that of virgin PMMA(88%).

ACKNOWLEDGMENTS

This work was supported by the Technology Innovation Program (No.10048243) funded By the Ministry of Trade, Industry & Energy (MOTIE, Korea)

REFERENCES

1. E.Pouget, E.Jujardin, A. Cavalier, A. Moreac, C. Valery, V. Artzner, T. Weiss, A. Renault, M. Patersnostre, and G. Artzner, Nature Mat., 6, 434, **(2007)**.
2. S. Han, and S. Park, Bull. Kor. Chem. Sci., 31, 3519 **(2010)**.
3. M.Mathew, K. Shenoy, and KS Ravishankar, Int. J. Exper. Dent. Sci., 3(1), 2 **(2014)**.
4. X. Song, X. Wang, H.Wang, W. Zhong, and Q. Du, Mat. Chem. Phys., *109*, 143 **(2007)**

Remarkably Enhanced Triboelectrification by Increasing the Conformable Contact and the Adhesion Energy with a Film-Covered Pillar Structure

Lee Ju Hyun (1), Yu Insang (1), Hyun Seung (1), Kim Jin Kon (1)*, Jeong Unyong (1)

(1) POSTECH - Gyeongbuk, Pohang - SouthKorea

Structural modification of the surface has been widely investigated in triboelectric energy generation to improve the power harvested. Large contact area, large surface area and good dielectric performance are the basic design principles needed for the structures, while energy harvesting from low-energy mechanical motions is desirable. This work suggests that a film-covered pillar structure made of polydimethylsiloxane (PDMS) improved its output performance by satisfying all the basic design parameters, especially, the conformable contact and easy deformation. The contact area and the adhesion energy of the film-covered pillar structure are quantitatively compared with those from the flat PDMS substrates and the pillar-only structures. We report that the film-covered pillar structure remarkably enhances the charge generation and increases the output voltage and the current density. We also emphasize the optimized thickness of the cover film for triboelectric performance. The triboelectric power generator operating by low-energy mechanical motions like a gentle finger-touch has been demonstrated as a possibility for human-based energy harvesting device in the future.

Graphene-based masterbatch obtained via modified Polyvinyl Alcohol liquid-shear exfoliation and its application in enhanced polymer composites

Simon Douglas [(1)*], Bischoff Eveline [(1)], Francisquetti Edson [(1)], Lavorgna Marino [(2)], Mauler Raquel Santos [(3)]

[(1)] IFRS - Farroupilha-RS - Brazil, [(2)] CNR - Napoli - Italy, [(3)] UFRGS - Porto Alegre-RS - Brazil

A simple and unexpensive method for the production of graphene-based masterbatch via polymer-assisted shear exfoliation of graphite in water was comprehensively investigated. In detail, a modified polyvinyl alcohol (mPVOH) characterized by surface energy comparable with that of graphene was used as surfactant for the production of graphene-like particles. These particles exhibit a higher output than that obtained by graphite liquid-exfoliation using common surfactants, along with a narrower size distribution. A mPVOH-masterbatch containing 4.38 wt% of graphene-like particles was produced, and directly used for the production of polymer nanocomposites by melt processing. Films prepared by blending the graphene-based masterbatch with an extrusion grade polyvinyl alcohol showed a 78% reduction in water permeability and a 48% increase in storage modulus when only 0.3wt% of graphene-like particles is added. Improved barrier properties were also observed for PLA and LDPE-based films, as well as an increment of about 520% in the storage modulus of PLA-based films. These results will open up a new pathway for using graphene-based masterbatch in the packaging industry.

PVDF-based dielectric composites with tunable coating of polydopamine on carbon nanotubes

Ji Xiaoying [1], Zhu Jiaming [1], Shen Jiabin [1]*

[1] Sichuan University - Sichuan Province - China

The pursuit of flexible dielectrics with high permittivity and low loss is favored by next-generation charge storage devices. Based on the micro-capacitor model, polymer/conductive filler dielectric (PCD) composites show the most potential in achieving high permittivity. However, a remarkable increase in dielectric loss generally limits their practical applications, due to inevitable contact between neighboring conductive particles. In this work, the carbon nanotubes (CNTs) were coated by polydopamine (PDA) through the self-polymerization of dopamine hydrochloride (DAH) monomer and the coating thickness could be controlled by changing the DAH/CNT ratios. When the coated CNTs were incorporated into the PVDF matrix, the electrical conductivity tended to be reduced with the thickening of PDA coatings, implying that undesired current leakage could be effectively suppressed. In addition, the effects of the coating thickness, loading contents and dispersion status of as-treated CNTs on dielectric properties of the composites were investigated. Results demonstrated that the composite filled with 0.2wt% coated CNTs (DAH/CNT weight ratio: 0.2) could obtain a high permittivity of 315 and a low loss factor of 0.35. Accordingly, it can be regarded as one of strategies to make the PCD composites achieve balanced dielectric behaviors.

ENHANCED THERMAL PROPERTIES OF HYDROGEL-CLAY NANOCOMPOSITES

Romina Ollier, Laura M. Sanchez*, Jimena S. Gonzalez and Vera A. Alvarez

Materiales Compuestos Termoplásticos (CoMP), INTEMA (CONICET).Facultad de Ingeniería (UNMdP). Av. Colón 10890, B7608FLC, Mar del Plata – ARGENTINA. TE: +54 0223 6260600.
**email: lsanchez@mdp.edu.ar*

Abstract. Hydrogels are one of the most widely employed materials in biological, medical and technological areas. They consist essentially on a low-cost and hydrophilic soft polymer, which forms a three-dimensional network and has the ability to absorb a large amount of water. In particular, the cryogels of polyvinylalcohol (PVA) constitute an eco-friendly alternative due to they can be obtained by a physical crosslinking method (through freezing/thawing cycles, F-T) thus avoiding the use of potentially harmful components. However, the use of hydrogels sometimes is restricted due to their specific properties. It is well-known that polymeric nanocomposites reinforced with clays show valuable improvements in their mechanical properties, heat resistance and gas permeability, regarding conventional micro and macrocompounds[1]. This is due to the large contact area between the matrix and the filler[2]. Composite hydrogels will be prepared employing PVA and bentonite (abundant and low cost smectite-type clay in Argentina) as raw materials through the previously optimized F-T crosslinking method. Different nanocomposite hydrogels will be obtained by varying the corresponding reactor feed. These materials have been recently chemically, morphologically and mechanically analyzed by our group[3]. Specifically, in the present work, thermal degradation behavior of PVA-bentonite hydrogels will be studied. Kinetic analytical models[4] will be applied to non-isothermal thermogravimetric (TGA) measurements trying to identify the possible mechanism for PVA degradation in the presence of bentonite.

Keywords: Hydrogel-clay nanocomposites; Thermal degradation; Modeling; Activation energy.

INTRODUCTION

Man-made composite materials could be considered as natural biocomposite materials-inspired. As examples, horn, bone and nacre are natural inorganic/organic nanocomposites with excellent mechanical properties[5]. The addition of both natural and synthetic nano-fillers to polymeric materials leads to have modifications on solvent, heat and impact resistance, transparency, electrical conductivity, molecular barriers, flame retardant and corrosion protection properties, mechanical strength and certain gases permeability of the composite material when compared with either the pure polymer or their conventional microcomposites, even at very low filler concentrations[6–18].

Durability, shelf lives, spectral stability, mechanical properties and life cycles of polymeric-based materials are ultimately governed by their thermal resistance due to all the above-mentioned properties will be gradually affected if the degradation process begins. As a result, controlling the material thermal resistance is a priority[19]. For this, a variety of fillers have been added to a number of polymeric matrix, including carbon, ceramics, clays, metals and metal oxides, among many other examples[12,17].

Polyvinyl alcohol (PVA) is a water soluble biodegradable polymer with a large extent of applications, usually considered as a good host synthetic material[3,6,8,20,21]. Several nanocomposites were prepared by employing PVA as matrix and sepiolite[6], silica[8,19], montomorillonite (MMT)[9,10,22,23], gold nanoparticles[24], $CoFe_2O_4$ magnetic nanoparticles[25], bentonite[3], magnetite nanoparticles[26], kaolinite[7], iron, cobalt and nickel sulfides[27] as fillers, among many other ones.

Clay materials constitute good raw materials since mostly of them are abundant and low-cost. Enhanced thermal stability temperatures for several PVA-clay nanocomposites have been reported[5,7,9,10,17,21]. The physical properties improvement achieved on the final nanocomposite product is strongly dependent on the clay mineral–polymer compatibility as well as on the corresponding processing conditions[16]. Three different morphological structures are mainly achievable: intercalated nanocomposites (polymer chains are inserted into a layered silicate structure in a crystallographically regular fashion), exfoliated or delaminated nanocomposites (the individual silicate layers are separated in the polymer matrix by average distances,) or flocculated (the hydroxylated edge–edge interactions of the silicate layers produce intercalated and stacked silicate layers flocculated to some extent) clay mineral particles[2]. It is important to remark that only exfoliated clay particles being part of the nanocomposite material could offer remarkable improved properties.

In the present work composite hydrogels prepared by employing PVA and bentonite as raw materials through the previously optimized freezing/thawing cycles (F-T) physical crosslinking method, thus avoiding the use of potentially harmful components. Particularly, bentonite is a smectite-type clay very abundant in Argentina characterized for having a moderate negative charge and good mechanical and chemical resistance[3]. Different nanocomposite hydrogels were obtained with bentonite contents from 1 to 5 wt. %. These materials have been recently chemically, morphologically and mechanically analyzed by our group[3]. Specifically, in the present work, the thermal degradation behavior of PVA-bentonite hydrogels will be deeply studied. The thermal degradation process will be modeled in order to explain the clay effect on it.

MATERIALS AND METHODS

Materials

Polyvinyl alcohol (PVA) was supplied by Sigma-Aldrich (Molecular weight of 89000-98000 g/mol, hydrolysis degree of 98-99%). Bentonite, supplied by Minarmco S.A. (Neuquén, Argentina), was used as received to prepare the nanocomposites.

Methods

Materials preparation

Materials were prepared according to authors previous reports[3]. Briefly, 10 g of PVA were dissolved in 100 mL of distilled water under constant stirring at 85 °C during 1h. After this, 1, 3 and 5 wt. % of bentonite (previously overnight swollen in water) was added and the stirring continued for 3h (samples named 1% bent, 3% bent and 5% bent, respectively). Then, solutions were placed in an ultrasonic bath (30 minutes). When the solutions reached the room temperature, PVA composite dispersions were cast onto anti-adherent containers. Samples were then frozen (-18 ºC, 1h, freezing process, F) and placed at room temperature (25 ºC, 1h, thawing process, T). The F-T procedure was repeated 3 times to obtain physical crosslinked hydrogels. PVA hydrogel (sample named PVA) samples were obtained in a similar procedure, without the addition of the nanoclay.

Materials characterization

All prepared materials have been recently chemically, morphologically and mechanically analyzed by our research group[3]. Briefly, Fourier transformed infrared spectroscopy (FTIR) spectra of samples were acquired with a Nicolet 6700 Thermo Scientific instrument equipped with a diamond ATR probe, over the range 400–4000 cm^{-1} from 32 coadded scans with 4 cm^{-1} resolution. X-Ray diffraction (XRD) tests were performed to characterize the clay dispersion inside the polymer matrix using an X-Pert Pro diffractometer, operating at 40 kV and 40 mA, with CuK_α radiation (λ = 0.154 nm). All samples were scanned in 2θ = 2 to 50 ° at a rate of 2 °/min. Field Emission Scanning Electron Microscopy (FESEM) was employed to analyze the pore morphology and sizes of hydrogels on a FESEM Zeiss Supra. Each sample was swollen, frozen, lyophilized, cryofractured and the coated with gold before testing. Thermogravimetric Analysis (TGA) measurements were carried out using a Shimadzu TGA-50 instrument at a set of heating rates: 5, 10, 15, 20 and 25 °C/min from room temperature to 800 °C under N_2 atmosphere in order to avoid the thermo-oxidative degradation. Degradation temperatures (Tp) were obtained from the maximum of the main loss-weight event in the derivative curves of TGA (DTGA).

RESULTS AND DISCUSSION

Materials Characterization

Morphological and Structural Characterization

From FTIR data the interaction between PVA matrix and bentonite can be studied. Concisely, there are shifts from 995 to 1040 cm^{-1} and from 3269 to 3287 cm^{-1} that were previously reported by our research group[3]. They could be attributed to the presence of hydrogen bonding between –OH groups of PVA and Si-O groups present in bentonite, allowing them to be considered as compatible one with the other one.

According to our previously reported XRD results[3], good dispersion of the filler inside the PVA matrix is observed for 1 and 3 wt.% bent, based on the fact that no characteristic clay diffraction peak appears at $2\theta = 2\text{-}10$ °.This absence is related to the possibility of having exfoliated or mixed exfoliated-intercalated silicate layers of the clay dispersed in PVA matrix[9]. In the case of nanocomposite with 5 wt.% bent sample, a minor increase in the interlamellar space in the clay could be inferred from the presence of a little peak around 4 °, which is associated with the polymer entrance in clay galleries[3].

Figures 1 show the FESEM micrographs of PVA and PVA-nanocomposites. As it can be seen, the introduction of clay inside the PVA hydrogel matrix produces a decrease on the pore size (Table 1)[3].

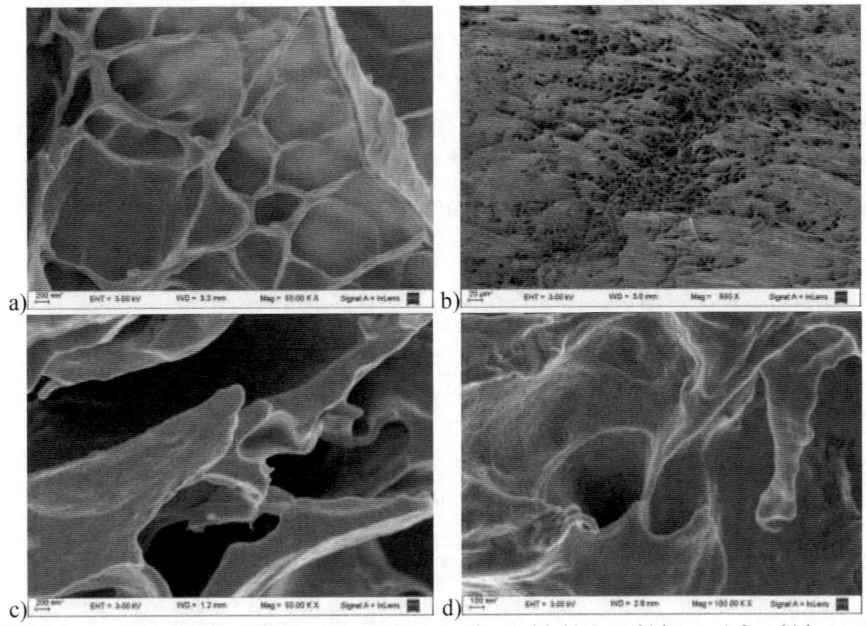

FIGURE 1. FESEM micrographs for a) PVA and nanocomposites with b) 1 wt.% bent, c) 3wt.% bent, d) 5 wt.% bent.

TABLE 1. Pore size determined from FESEM micrographs.

Sample	PVA	1 wt.% Bent	3 wt.% Bent	5 wt.% Bent
Porous size (μm)	8.5 ± 4.0	3.2 ± 1.4	4.0 ± 2.2	4.7 ± 2.3

Study of the Thermal Degradation Process

Thermal stability of PVA–bentonite systems has been investigated through the thermogravimetric analyses and the determination of certain characteristic temperatures. Figures 2 display the TGA and DTGA curves of PVA hydrogel matrix and nanocomposites obtained at a heating rate of 10 °C/min.

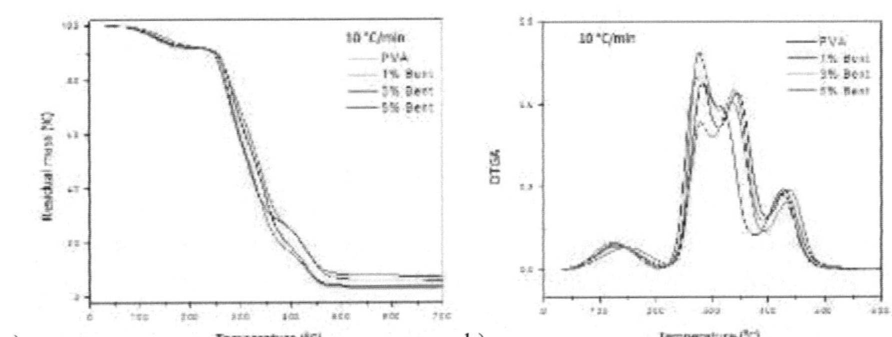

FIGURE 2. a) TGA and b) DTGA curves of PVA matrix and nanocomposites obtained at 10 °C/min.

The obtained curves show that the decomposition profiles of matrix and nanocomposites are characterized by three distinct peaks. The first peak, correlating to water evaporation, occurs between ambient temperature and 183 to 200 °C. A second peak at approximately 260 °C correlates to the first step degradation of polyvinyl alcohol-clay nanocomposite, while the third peak at around 440 °C is attributed to the second step decomposition. Although some researchers reported remarkable improvements on nanocomposites thermal stability by incorporating clays as nanofillers[7,9,10], some other ones observed no significant changes[21].

Bentonite may have two opposite functions in the thermal stability of nanocomposites: a barrier effect, which should improve the thermal stability or a catalytic effect on the degradation of the polymer matrix, which should decrease their thermal resistance. The clay acts as a heat barrier improving the overall thermal stability of the system. The presence of the filler restricts the polymer chain mobility and assists the char formation, during the thermal decomposition[2,3,25]. Besides, the silicate layers could hold accumulated heat that can act as a heat source to accelerate the decomposition process[28].

Modeling of the degradation process

One of the simplest models of the degradation process can be presented by an nth order kinetic equation:

$$\frac{d\alpha}{dt} = A.(1-\alpha)^n.\exp\left(\frac{-E_a}{R.T}\right) \tag{1}$$

were α is the average advancement of the degradation reaction, proportional to the material weight loss and varied between 0 and 1. Introducing the heating rate, β in Equation 1:

$$\frac{d\alpha}{(1-\alpha)^n} = \frac{A}{\beta}.\exp\left(\frac{-E_a}{R.T}\right).dT \tag{2}$$

This is the fundamental expression of the analytical methods used to calculate the kinetic parameters from TGA data. Several methods are commonly used to estimate the apparent activation energy. The proposed by Kissinger[18,29] is based on the following equation:

$$\frac{d(\ln(\beta/T_m^{\ 2}))}{d(1/T_m)} = \left(\frac{-E_a}{R}\right) \tag{3}$$

For the analysis, the whole degradation process was divided in two parts, each one associated to one of the degradation peaks. These peaks were separated by a Lorentzian multi-peak curve. The results of activation energy (from Kissinger method) of the first peak of each system are summarized in Table 2.

TABLE 2. Activation Energy of the First Peak of PVA and Nanocomposites Obtained from Kissinger's Method.

Sample	PVA	1 wt.% Bent	3 wt.% Bent	5 wt.% Bent
Ea (kJ)	107,38	82,55	147,74	146,19

It is possible to observe whereas for low clay content (1 wt.%) the apparent activation energy is reduced, the increase on the bentonite content produces an improvement on the thermal stability of the PVA hydrogel matrix. The effect at low bentonite content could be explained in terms of an easier elimination of the water and acetate groups from the PVA in the second step being facilitated by the incorporated clay[6]. However, at higher clay contents there are other factors, previously mentioned, that increase the thermal stability with respect to the neat PVA matrix. It seems that exist a clay content that acts as inflection point in the thermal properties. We are working on find this concentration and a deeply explanation of this behavior.

CONCLUSIONS

It was possible to prepare and characterize PVA hydrogels with different bentonite contents through a simple and low-cost methodology. The morphology characterization showed that the selected clay (bentonite without modifications) and the PVA matrix have interactions (as it was observed in FTIR spectra), having good dispersion (evidenced in XRD spectra) for all studied clay contents (it will be further confirmed by Transmission Electron Microscopy, TEM). Then, an initial characterization of the involved process in the PVA matrix and the nanocomposites thermal degradation showed three main degradation steps, the same for both, matrix and nanocomposites. From this, a first approximation to the process modeling was done by applying the Kissinger model: it was observed that whereas for low clay content (1 wt.%) the apparent activation energy is reduced, further increase on the bentonite content produces an improvement on the thermal stability of the PVA hydrogel matrix. Application of other models in order to analyze the effect of incorporate different clay contents on the thermal degradation of PVA nanocomposite hydrogels is being carried out. After that, all results and model's parameters will be correlated with the clay dispersion degree inside the PVA matrix.

ACKNOWLEDGMENTS

The authors acknowledge the financial support of CONICET (PIP00617), Universidad Nacional de Mar del Plata, ANPCyT (PICT 2014-3228 and PICT 2016-1905) and Fundación Bunge & Born.

REFERENCES

1. Haraguchi, K. *Curr. Opin. Solid State Mater. Sci.* **11,** 47–54 (2007).
2. Sinha Ray, S. and Okamoto, M. *Prog. Polym. Sci.* **28,** 1539–1641 (2003).
3. Gonzalez, J., Ponce, A. and Alvarez, V.. *Adv. Mater. Lett.* **7,** 979–985 (2016).
4. Gotor, F. J., Criado, J. M., Malek, J. and Koga, N. *J. Phys. Chem. A* **104,** 10777–10782 (2000).
5. Allison, P. G. *et al. J. Nanomater.* 9 pages (2015).
6. Alkan, M. & Benlikaya, R. *J. Appl. Polym. Sci.* **112,** 3764–3774 (2009).
7. Jia, X., Li, Y., Zhang, B., Cheng, Q. and Zhang, S. *Mater. Res. Bull.* **43,** 611–617 (2008).
8. Peng, Z., Kong, L. X. and Li, S. D. *J. Appl. Polym. Sci.* **96,** 1436–1442 (2005).
9. Yu, Y.-H., Lin, C.-Y., Yeh, J.-M. and Lin, W.-H. *Polymer (Guildf).* **44,** 3553–3560 (2003).
10. Ali, M. H. M., Kahder, M. M., Al-Saad, K. and Al-Meer, S. *Q. Science Connect* **1,** 1–9 (2013).
11. Lan, T. and Pinnavaia, T. J. *Chem. Mater.* **2219,** 2216–2219 (1994).
12. Sun, Y., Zhang, Z., Moon, K. S. and Wong, C. P. *J. Polym. Sci. Part B Polym. Phys.* **42,** 3849–3858 (2004).
13. Yasmin, A., Luo, J. J., Abot, J. L. and Daniel, I. M. *Compos. Sci. Technol.* **66,** 2415–2422 (2006).
14. Hsiue, G. H. O., Liu, Y. L. and Liao, H. H. A. O. *J. Polym. Sci. Part A Polym. Chem.* **39,** 986–996 (2001).
15. Ha, S. R., Ryu, S. H., Park, S. J. and Rhee, K. Y. *Mater. Sci. Eng. A* **448,** 264–268 (2007).
16. Guo, B., Jia, D. and Cai, C. *Eur. Polym. J.* **40,** 1743–1748 (2004).
17. Jose, T. *et al. Ind. Eng. Chem. Res.* **53,** 16820–16831 (2014).
18. Zhao, C. *et al.* **87,** 183–189 (2005).
19. Peng, Z. and Kong, L. X. *Polym. Degrad. Stab.* **92,** 1061–1071 (2007).
20. Guirguis, O. and Moselhey, M. *Nat. Sci.* **4,** 57–67 (2012).
21. Chang, J. H., Jang, T. G., Ihn, K. J., Lee, W. K. and Sur, G. S. *J. Appl. Polym. Sci.* **90,** 3208–3214 (2003).
22. Kokabi, M., Sirousazar, M. and Hassan, Z. M. *Eur. Polym. J.* **43,** 773–781 (2007).
23. Gaume, J. *et al. Sol. Energy Mater. Sol. Cells* **99,** 240–249 (2012).
24. Khanna, P. K. *et al. Mater. Chem. Phys.* **92,** 229–233 (2005).
25. Lopez, D., Cendoya, I., Torres, F., Tejada, J. and Mijangos, C. *J. Appl. Polym. Sci.* **82,** 3215–3222 (2001).
26. Goiti, E. *et al. Polym. Degrad. Stab.* **92,** 2198–2205 (2007).
27. Qian, X. F. *et al. J. Appl. Polym. Sci.* **82,** 2744–2749 (2001).
28. Sinha Ray, S. (2013).
29. Kissinger, H. E. *Anal Chem* **29,** 1702–1706 (1957).

Mechanical properties of polyamide with cellulose nanofiber (CNF) modified by Cardo material

Sato Hirokazu (1)*, Tokumitsu Katsuhisa (1), Yamada Masahiro (2), Sugimoto Masayuki (2)

(1) The University of Shiga Prefecture - Shiga Prefecture - Japan, (2) Osaka Gass Co.ltd - Osaka - Japan

In this study, mechanical and thermal properties of polyamide (PA) 6/66 with cellulose nano-fiber (CNF) treated by bisphenol fluorene diglycidil ether (BPFG), which is referred to as "B-CNF", was investigated. Although in PA/neat CNF composite the CNFs were aggregated, B-CNF can be dispersed homogeneously in PA matrix and the size of B-CNF was getting smaller than that of CNF. On dynamic mechanical measurement, the storage modulus (E') of PA/B-CNF was much higher than that of neat PA as well as PA/neat CNF, and the values of E' increased with B-CNF content especially at high temperature region. On the other hand, the peak intensity of tanδ of PA/B-CNF at the temperature region around Tg of PA was getting lower in comparison with the neat PA and PA/neat CNF, indicating that the molecular motion of PA was restricted by the presence of B-CNF because of the strong interaction between PA and B-CNF. From the result of thermal mechanical measurement, Coefficient of Thermal Expansion (CTE) of PA/B-CNF was getting lower remarkably than those of neat PA and PA/neat CNF. Moreover, on the basis of CTE results, we estimated the volume fraction of PA' phase, which interacted with B-CNF, by means of the "mixtures rule", and as a result, it was found that the volume fraction of PA' phase increased with the B-CNF contents and the values were much larger than those of PA/neat CNF. Furthermore, the volume fraction of PA phase, which was NOT interacted with CNF or B-CNF, had good correlation with the intensity of tanδ at Tg. These experimental results also supported our assumption mentioned above: the strong interaction existed between PA and B-CNF. Consequently, B-CNF was one of the most promising candidate to modify the mechanical properties of PA particularly at high temperature region.

Reinforced Resorcinol Formaldehyde Aerogel with Electrospun Polyacrylonitrile Nanofiber for Thermal Insulation Applications

Alshrah Mohammed [(1)]*, Naguib Hani [(1)], Park Chul B [(1)]

[(1)] University of Toronto - Ontario - Canada

A three-dimensional organic aerogel nano-network with high structural stability and strength was obtained by physically bonding polyacrylonitrile (PAN) nanofibers into a resorcinol-formaldehyde (RF) nano-network. The PAN nanofibers were produced by the electrospinning of PAN solution. The PAN fibers were then oxidized, to produce oxidized polyacrylonitrile (O-PAN) nanofibers. The two fibers were added separately into the RF nano-network to generate a new hybrid RF aerogel. The RF aggregates were co-assembled onto the surfaces of the electrospun PAN and O-PAN fibers during the gelation process. Which increased the RF structural stiffness, strength, and resilience. Further, graphene oxide (GO) sheets were added to the mixtures as a crosslinker between the fibers. With the GO sheets presence, the number of the RF aggregates on the surface of the fibers increased by an order of magnitude because of the hydrogen bonds formed between the amino groups on the fibers and the functional groups on the GO. The final RF aerogels with O-PAN nanofibers and GO sheets had double the strength and resilience and three times the stiffness compared to the traditional RF aerogels while maintaining their light weight characteristics and low thermal conductivity of 30 mW/m.K.

Study of PBAT films formulations with nanoclays and nano-TiO2 for packaging application

Calderaro Mayara Pollyane [(1)]*, Morales Ana Rita [(1)], Sarantópoulos Claire I. G. L. [(2)], Sanchez Elisabete M. S. [(3)]

[(1)] UNICAMP - São Paulo - Brazil, [(2)] CETEA - ITAL - São Paulo - Brazil, [(3)] no affiliation - São Paulo - Brazil

Study of PBAT films formulations with nanoclays and nano-TiO2 for packaging application The aim of this work was to evaluate the combination of nanofillers - organically modified montmorillonite (MMT), sepiolite (SEP) and nano titanium dioxide (TiO2) - on the tensile properties and light barrier of poly(butylene adipate-co-terephtalate) (PBAT). A 2^3 factorial rotatable central composite design (α=1.682) with 3 central points was conducted. Five concentrations (in mass) for each factor were used (0.0, 2.0, 4.5, 7.0 and 9.0 % for MMT); (0.0, 1.0, 2.5, 4.0 and 5.0 % for SEP); (0.0, 0.2, 0.7, 1.2 and 1.5 phr for TiO2). The clays concentrations were determined from percolation threshold: 4 % for sepiolite, considering the nanorod shape and 7 % for montmorillonite established from Nielsen model, assuming lamella shaped clays. Both methods require the clays aspect ratio, which were accessed by scanning electron microscopy (TEM). The films were prepared using a Drais mixer, a hot press and a blown film extruder. The light transmission percentage of ultraviolet (UV) and visible (Vis) radiation were measured. The mechanical properties – Young's modulus, maximum tensile strength and elongation at break - were accessed according to ASTM D882-12. Only nano-TiO2 factor was statistically significant on light transmission, for UV and Vis radiation. The effect values for TiO2 factor obtained by experimental design (DoE) analysis were negative indicating that the increase of its concentration favors the reduction on light transmission. Young's modulus showed an increase, especially for the sample with concentrations of 7 % MMT, 4 % SEP and TiO2 1.2 % for which the modulus increased around three times, and this increase could be associated to a good dispersion and reinforcement effect of the nanoclays in the polymer matrix. The nanofillers and its combinations showed statistically significant effetcs for this property by DoE analysis. The addition of nanofillers did not present statistical influence on the tensile strength and the elongation.

Fabrication and characterization of PVDF/PC blend-based multicomponent nanocomposites

Jhong Jian-Sheng [1], Chiu Fang-Chyou [1]*

[1] Chang Gung University - Taoyuan - Taiwan

PVDF/PC blend-based nanocomposites with CNT, GnP, and organo-MMT as nanofillers were successfully prepared. SEM results confirmed the selective localization of individual and hybrid fillers within the PC domains. Addition of CNT led to the formation of a quasi co-continuous PVDF-PC morphology. DSC data showed that organo-MMT, not CNT/GnP, facilitated PVDF crystallization in the composites. The rigidity of the PVDF/PC blend increased evidently after the formation of various nanocomposites. Rheological property measurements confirmed the achievement of a pseudo-network structure in the composites. Adding CNT increased the complex viscosity of the samples at a higher extent than that of adding GnP, and the viscosity values further increased with the co-existence of CNT/GnP and organo-MMT.

Water soluble metal/carbon/polymer nanocomposites for stretchable conductors

Jang Jaewong* [1], *Chee Sang-Soo* [2], *Ham Moon-Ho* [1]

[1] School of Materials Science and Engineering, Gwangju Institute of Science and Technology - Gwangju - SouthKorea, [2] School of Materials Science and Engineering, Gwangju Institute of Science and Technology - Gwangju - SouthKorea

With the development of wearable electronic, optoelectronic and sensing devices, stretchable conductors hold the promise for transforming the current technologies. There is substantial interest in realizing the stretchable conductors of polymer nanocomposites with metal/carbon nanomaterials. In this study, we report on a new water-based synthetic route for fabricating stretchable electrodes composed of Ag nanoparticles (NPs) and single-walled carbon nanotubes (SWNTs) embedded in a polymer matrix. The Ag NPs and SWNTs were uniformly dispersed in water with polymer surfactant, poly(acrylic acid). The mixture solution is subsequently incorporated into a polyurethane aqueous solution. With this aqueous approach, stretchable and conductive films were fabricated on polydimethylsiloxane substrate, with the conductivities of 600 and 120 S/cm in unstrained and 90% strain conditions, respectively.

Grafting of poly (methylmethacrylate) from Magnetic Nanoparticles using initiator coupling agent for controlled radical polymerization

Tichané Teddy [(1)]*, Blottière Benoît [(1)]

[(1)] Université Lyon - Rhône-Alpes - France

The development of composite materials with nanoparticles has been studied in recent years. But studies on the dispersion of magnetic nanoparticles in a polymer matrix are still difficult to achieve today. Our goal is to design this kind of material with good dispersion and high loading rates. We are particularly interested in nanoparticles dispersed in a matrix of PMMA with a mass loading rate of 1%. Poly(methylmethacrylate) is synthesised from the surface of magnetic nanoparticles by Reversible Addition-Fragmentation Chain Transfer (RAFT) using a silanic coupling agent with trithiocarbonate RAFT agent. First, magnetic nanoparticules were activate by nitric acid to create hydroxyl groups at their surface. Then, the RAFT agent is anchored to magnetic nanoparticles via condensation reactions. Thermogravimetric analysis and diffused reflection FT-IR were used to confirm and measure the grafting reaction. These results show the importance of the activation by nitric acid and a better grafting yield with collecting by solvent evaporation. The RAFT polymerisation of methyl methacrylate was carried out in the presence of AIBN in THF. The grafting density was determinated by thermogravimetric analysis, mechanic properties were measured by traction tests and molar mass and polydispersity index were caracterised by GPC analysis. First results show that composite material compared to matrix of PMMA is more elastic due to nanoparticules. PMMA has a deformation of 2% and up to 4% with nanoparticules. The next step is to improve mass loading rate to 5% and 10% and then to determinate magnetic properties of these materials.

Tailored dispersion of Zinc Oxide nanoparticles and hexagonal Boron Nitride in block copolymer nanocomposites for dielectric applications

Helal Emna [1]*, David Eric [1], Fréchette Michel [2], Demarquette Nicole Raymonde [1]

[1] École de Technologie Supérieure - Québec - Canada, [2] Institut de Recherche d'Hydro-Québec - Québec - Canada

Styrenic thermoplastic elastomers such as Poly[styrene-b-(ethylene-co-butylene)-b-styrene] (SEBS) constitute a special class of block copolymers composed of two styrene endblocks and a soft rubber midblock. Due to their immiscibility, the blocks self-assemble into well-ordered nanostructured morphologies that depend on several parameters such as their molecular weights, affinities and relative fractions. These organized morphologies have great potential to act as templates tailoring the dispersion and spatial distribution of nanoparticles with different shapes in order to achieve well designed nanocomposites with optimized functional properties. In this study, both Zinc Oxide (ZnO) nanospheres and exfoliated hexagonal boron nitride (h-BN) have been dispersed in SEBS grafted maleic anhydride (SEBS-MA) triblock copolymer in a trial to achieve thermally conductive electrically insulating nanocomposites, that benefit from potential synergetic effect of both nanoparticles. SEBS-MA/ZnO nanocomposites, studied beforehand, featured an interesting rheological behavior, even at low loading of ZnO, characterized by a consistent increase of dynamic moduli and reduced dependence on the frequency especially in the low values range indicating the formation of networks between ZnO nanoparticles and polymer chains. Moreover, improvements of breakdown strength, thermal conductivity and resistance to surface erosion were achieved. In a second step, h-BN particles were partially exfoliated and dispersed in SEBS-MA/ZnO mixture by means of high-shear mixing. The dispersion of these anisotropic fillers in the hybrid nanocomposite was enhanced compared to direct dispersion in neat SEBS-MA. The impact of this combination on the dielectric performance was investigated through the evaluation of the complex dielectric permittivity, breakdown strength and thermal conductivity in correlation with the rheological properties. Key words: block copolymer nanocomposite, tailored dispersion, network, dielectric properties...

Effect of MnO2 Architectures on Dielectric properties of PVDF/CNT Nanocomposites

Shayesteh Zeraati Ali [(1)]*, Sundararaj Uttandaraman [(1)]

[(1)] University of Calgary - Alberta - Canada

Poly (vinylidene fluoride) (PVDF) matrix hybrid nanocomposites incorporating MnO2 and Carbon nanotubes (CNT), were fabricated by melt mixing in a mini mixer followed by hot pressing. Two different hybrid architectures of CNT and MnO2 were prepared: (i) CNTs were uniformly coated by MnO2 flakes (MnO2@CNT); (ii) CNTs were mixed with MnO2 nanowires (CNT/MnO2NW) and both were used as nanofiller in PVDF matrix. Dielectric properties of fabricated nanocomposites were studied in X-band frequency (8.2-12.4 GHz). The conductive CNT increased the dielectric permittivity of the PVDF by serving as a nanocapacitor. Increasing CNT loading enhanced dielectric loss due to the formation of a conductive network. MnO2 addition hindered the conductive network formation and thereby affected dielectric properties. Adding MnO2@CNT decreased both dielectric permittivity and dielectric loss, while CNT/MnO2NW increased the dielectric permittivity and decreased dielectric loss. Details of the dielectric loss mechanisms for both architectures are presented in this manuscript. Besides achieving reliable dielectric properties, CNT/MnO2NW nanocomposites had high flexibility, making it useful for many flexible electronic and energy storage devices.

Mechanical and Thermal Properties of Polylactic Acid with Cellulose Nanofiber Modified by Cardo material (BPFG) or Bisphenol-A diglycidyl ether (BPAG) (II)

Tokumitsu Katsuhisa [1]*, Katsurada Shinya [1], Sato Hirokazu [1], Yamada Masahiro [2], Sugimoto Msayuki [3]

[1] University of Shiga Prefecture - Shiga - Japan, [2] Osaka Gas Co.Ltd - Osaka - Japan, [3] Osaka Gas Co. Ltd - Osaka - Japan

In this study, we prepared several cellulose nano-fiber (CNF) by Cardo material (BPFG), referred to as B-CNF, with various degrees of substitution to BPFG. Moreover, we also prepared the CNF treated by Bisphenol-A diglycidyl ether (BPAG), which does not have Cardo structure in its chemical structure, and determined the effect of the difference in chemical structure on their mechanical properties at high temperature. TEM observation revealed that both of the CNF treated with BPFG (B-CNF) and that treated with BPAG (BisA-CNF) dispersed homogeneously in the PLA. The storage moduli of B-CNF/PLA composites above Tg increased with the degree of BPFG substitution. The storage moduli of B-CNF/PLA and BisA-CNF/PLA with similar degree of substitution were about 5 times higher than that of neat PLA above Tg, and the modulus of B-CNF/PLA was slightly higher than that of BisA-CNF/PLA. Moreover, the peak intensity of tanδ of B-CNF/PLA and BisA-CNF/PLA were lower than that of neat PLA, and the intensity of tanδ of B-CNF/PLA was lower than that of BisA-CNF/PLA. These results suggest that the molecular motion of PLA was restricted by the presence of B-CNF and BisA-CNF because the interaction between PLA and B-CNF or BisA-CNF is much stronger than that in untreated CNF/PLA. Furthermore, B-CNF improved the mechanical properties of the composite more effectively than BisA-CNF for the PLA composite above Tg. From the result of Coefficient of Thermal Expansion (CTE) measurements, we estimated the volume fraction of PLA phase interacting with B-CNF and BisA-CNF by means of the "mixtures rule". As a result, it was found that the PLA phase interacting with of B-CNF/PLA is larger than that in BisA-CNF/PLA. This may be one of the factor which caused the difference in their mechanical properties of B-CNF/PLA and BisA-CNF/PLA. Finally, CTE measurements revealed that there is a strong relation between the intensity of tanδ at Tg and the volume fraction of the PLA interacting phase in PLA matrix.

Electrical conductivity of nanostructures of N-doped multiwall carbon nanotubes synthesized with different catalysts intertwined with polyaniline/polyamide 6 nanofibers

Bretas Rosário [1]*, Sundararaj Uttandaraman [2], Santos João Paulo Ferreira [3], Arjmand Mohammad [2], Melo Guilherme França [1], Chizari Kambiz [2]

[1] Federal University of São Carlos - São Paulo - Brazil, [2] University of Calgary - Alberta - Canada, [3] State University of Ponta Grossa - Paraná - Brazil

Semi conductive, porous nanostructures and with very high surface area constituted by polyamide 6 (PA6)/polyaniline (PAni) nanofibers coated with nitrogen-doped carbons nanotubes (N-CNTs) were obtained. The effect of the type of catalyst and temperature used in the N-CNT's synthesis, the structure and the final electrical conductivity of the N-CNTs and the nanostructures was studied. N-CNTs synthetized with three types of catalyst were used: Iron (Fe), Cobalt (Co) and Nickel (Ni). For the N-CNTs (Fe) the synthesis temperature was varied: 650°C, 750°C, 850°C and 950°C. The PA6/PAni nanofibers mats were produced using the electrospinning technique. For the coating with N-CNTs, the electrospun nanofiber mats were dipped in aqueous solution containing the N-CNTs dispersed and stabilized by a surfactant. The N-CNTs, the nanofiber mats and the nanostructures produced were morphologically characterized by scanning and transmission electronic microscopy (SEM and TEM). The N-CNTs were structurally characterized by Raman and x-ray photoelectron spectroscopy (XPS). The mass fraction of deposited N-CNTs was calculated by weighing. The electrical conductivity of N-CNTs pellets and nanostructures was measured by the four-point DC conductivity method. The results revealed that by varying the synthesis conditions, N-CNTs with different morphologies, structures and doping levels were obtained. The results showed, however, that such differences did not exert a large difference in the conductivity of compacted bulks of these N-CNTs. The conductivities of these bulks were practically of the same order of magnitude. However, for the nanostructures coated with N-CNTs, different behaviors were observed. The results suggest that higher aspect ratio of the N-CNTs, higher oxide content in the N-CNTs and the presence of PAni in the nanofibers contributed the interaction between N-CNTs and nanofibers resulting in the formation of a percolated network more easily.

Weather resistant thermoplastic polyurethane nanocomposite films

ADAK BAPAN [(1)]*, Joshi Mangala [(1)], Butola Bhupendra Singh [(1)]

[(1)] INDIAN INSTITUTE OF TECHNOLOGY DELHI - DELHI - India

In this study, the synergistic effect of different organic UV additives such as UV-absorber (Tinuvin 328), HALS (Tinuvin 770), antioxidant (Irganox 1076) and antiozonant (6PPD) as well as the effect of inorganic UV-Shielding nanopowder (nano-TiO2 and ZnO) on weather resistance property of thermoplastic polyurethane (TPU) films were studied. Ten polyurethane films were produced using these UV-additives in different combinations. The mixing of TPU and UV additives were done in a micro twin screw extruder, followed by preparation of film in compression molding. The accelerated weathering was carried out by exposing these films in a weather-o-meter up to 300h under UV-light plus raining cycle as per ISO 4892 method. The weather resistance performance was analyzed in term of change of UPF, tensile properties and yellowness index of the films before and after exposure. The chemical change before and after exposure were analyzed by FTIR. The surface morphology of the films (before and after exposure) was studied by SEM, where some crack was found in some films after 300h exposure. Incorporation of UV-shielding nanopower in TPU matrix resulted in improved tensile properties of the films. However, in presence of organic UV-additive with nanopowder, almost no improvement was observed in tensile properties, due to plasticizing effect of organic UV additives. It was observed that the organic UV absorber was quite efficient to increase UPF value initially and a good synergistic effect was found with HALS and antioxidant. However, there was an issue of sustainability due to continuous leaching out of organic UV additives. On the other hand, UPF value increased significantly with about 3wt% loading of UV-shielding nanopowder due to their UV absorbance plus UV reflectance property. However, only UV-shielding nanopowder was not effective to increase UPF value at lower concentration. Incorporation of organic UV-additives with UV-shielding nanopowder resulted in good weather resistance properties even at very low concentration about 1 wt% and also provided sustained effect. Similarly, the antiozonant showed good synergistic weather resistant effect with UV-shielding nanopowder.

The construction of specific nanostructure in epoxy thermosets via the self-organization of block copolymer

Heng Zhengguang (1), Chen Yang (2)*, Zou Huawei (2), Liang Mei (2)

(1) Polymer Research Institute of Sichuan University - State Key Laboratory of Polymer Materials Engineering - China, (2) Polymer Research Institute of Sichuan University - Chengdu - China

Over the past decades, materials scientists have aspired to exploit nature's assembly principles to create artificial materials. Self-organization of block copolymer in bulk and selective solvent has attracted considerable attention because it can yield ordered structures in a wide range of morphologies, including spheres, cylinders, vesicles, lamellae, bicontinuous structures, etc., which has been proved to have practical or potential applications in many fields. Epoxy thermosets can also be acted as selective solvent of block copolymer, however, the study on the construction of specific nanostructure in epoxy thermosets is rarely reported. It is well known that the macroscopic properties are closely related to microstructures of materials. The preparation of high performance and functional materials through the design of microstructure has become a new trend in the development of advanced composites. Thus, how to efficiently and accurately achieve required microstructure has become a very essential and urgent subject. In this work, different kinds of block copolymers (eg. polystyrene-poly(ε-caprolactone), polydimethylsiloxane- poly(ε-caprolactone), polydimethylsiloxane-poly(ε-caprolactone)-polystyrene, polystyrene-poly(ε-caprolactone)-polydimethylsiloxane-poly(ε-caprolactone)-polystyrene are synthesized. Various nanostructures, such as, spheres, "worm-like", lamellae, "flower-like", "core-shell", bicontinuous structures, tangled tubular structures, and so on, are effectively constructed in epoxy thermosets through appropriately designing the structure of block copolymer. The effects of component, composites, concentration of block copolymer on the morphologies and sizes of nanostructures are investigated in order to obtain the key factors in constructing specific nanostructure morphologies. Besides, the relationship between nanostructure and macroscopic performance of the epoxy composites is studied. It is believed that designing specific structures in epoxy resins through self-organization could provide new methods for the preparation of advanced functional epoxy composites.

Shape memory crosslinked polyethylene/POSS Nnnocomposites

Bicer Ezgi* [1]*, Kurt Gulsen* [1]*, Kodal Mehmet* [1]*, Ozkoc Guralp* [1]

[1] Kocaeli University Department of Chemical Engineering - Kocaeli - Turkey

Shape memory materials are capable tomemorize a temporary shape unless they are exposed to anexternal stimulus that triggers them to recover their permanent shape. The melting transition of crosslinked semi-crystalline networks can be employed to trigger shape recovery. Here, the secondary shapes are fixed by crystallization and the permanent shapes are established by chemical cross-linking. Compared with glassy materials, this class of materials is generally more compliant below the critical temperature, with a stiffness that is sensitive to the degree of crystallinity, and thus indirectly to the extent of cross-linking. Shape-recovery speeds are also faster for this first-order transition with an often sharper transition zone. One of the most widely used crosslinked shape memory polymers is polyethylene. Polyhedral oligomeric silsesquioxanes (POSSs) organic/inorganic hybrid materials are new generation nanofillers with the possibility of reinforcement and stabilization of the polymers. POSS with an empirical formula of (RSiO1.5)n (n is 8) has a □□Si–O– cage-like skeleton with organic sidegroups having reactive and/or nonreactive character. The chemistry of the side groups has a crucial importancebecause it controls the compatibility between POSS and polymers. Besides, the inorganic structure may provide polymers with molecular reinforcement, increased thermal stability, better flame resistance. In this study, the shape memory, rheological, thermal and mechanical properties of crosslinked polyethylene/POSS nanocomposites were investigated. The samples were prepared by melt compounding and subsequent crosslinking in hot-press. The effects of peroxide content, peroxide type and POSS content were examined.

Analysis of the flame retardancy and mechanical properties of HFFR compounds with incorporated Halloysite Nanotubes and Spherical Nano SiO2

Čermák Michal (1)*, Hamáček Aleš (2)

(1) University of West Bohemia, Silon, s.r.o. - Plana nad Luznici - Czech, (2) RICE, Electrical Engineering Faculty, University of West Bohemia - Pilsen - Czech

The topic of the fire retardancy of cable compounds has been gaining a great interest, nowadays. Unfortunately, it is due to the fact, that the safety in the event of fire has gradually decreased in the last few decades. Therefore, tightening of the fire retardancy requirements by introduction of a new Construction Products Regulation this year was a necessary step for ensuring the safety of newly constructed buildings within the EU. This situation has led to the new technological challenges, which in some aspects have not been fully solved. One of these challenges are the development of economically acceptable cable halogen-free polymer compounds with the high level of fire retardancy, long service life and wide processing window, which is from the customer side a highly expected. Therefore, the motivation of this article is to provide an investigation of mechanical and thermal responses of HM4 thermoplastic HFFR compounds of the 3rd generation – filled by Halloysite nanotubes (HNT) and Spherical nano amorphous SiO2 (SA-SiO2). Presented compounds were designed as a sheeting, suitable for 1kV applications with the operational temperature up to 90°C, fulfilling the requirements of VDE 0276-622, IEC 60502-1. Base formulations are consisted from: EVA, LLDPE, ATH (>60%), compatibilizers, HNT, SA-SiO2 and AOXs and were prepared on laboratory twin-extruder Leistritz. Results show a significant reduction of the heat release rate exceeding 40 % in comparison with the reference compound measured by Cone Calorimetry, improved char forming and increased LOI after the incorporation of 3% HNT into the formulation. Also, a strong synergism between HNT and SA-SiO2 has been observed in tested compounds in terms of self-extinguishing behaviour, peak heat release rate reduction, as well as, an improved mechanical properties. The reason for these phenomena can be explained by improved dispersion of all nanoparticles provided by SA-SiO2, which act as a ball bearings during the extrusion.

Determination of PSU Nanofibers Electrospinning Parameters via SEM Morphology Correlation

Daniel C. Silveira[a], Daniel M. Gil[a], Tiago T. S. Braga[a], Lília M. Guerrini[b] and Edson C. Botelho[a]

[a]*Department of Materials and Technology, São Paulo State University (UNESP), School of Engineering, Guaratinguetá, Brazil.*
[b]*Institute of Science and Technology, Federal University of São Paulo (UNIFESP), São José dos Campos, Brazil.*

danielsilveira.tech@gmail.com

Abstract. Electrospinning is regarded as one of the most versatile process for nanometer scale fibers generation from a variety of materials, including polymers and ceramics. The process consists in the application of intense electric field in a polymer solution, which by means of electrostatic forces is deposited in the form of a polymer non-woven layer of fibers. In this work the ideal parameters for the electrospinning process of polysulfone (PSU) nanofibers are determined, by means of correlation with morphological evaluations through scanning electron microscopy (SEM). The nanofibers were obtained by electrospinning of polymer solutions of PSU dissolved in N,N-Dimethylacetamide, considering several mass percentages for the polymer. The electrospinning process was applied with variations on the voltage (15, 18 and 20 kV), working distance (5, 10 and 15 cm) and polymer concentration (8, 12, 20, 26 and 28 %m/m). The correlation between the SEM of the obtained polymeric layers and the respective parameters applied during the electrospinning identified the best process voltage, working distance and polymer concentration. Thus, ideal parameters that favor the formation of nanofibers with greater homogeneity and aspect ratio (L/D), lower concentration of defects and decreased dispersion for the fibers diameters.

Keywords: electrospinning, polysulfone (PSU), N,N-Dimethylacetamide, scanning electron microscopy (SEM), process parameters.
PACS: 68, 81.

INTRODUCTION

Recently, polymer and composite nanofibers represent a class of developing nanostructures that show an extraordinary range of applications, such as filtration and water treatment [1], obtaining, converting and storing energy [2] and medical materials in the form of dressings and medicines [3]. The nanofibers can be obtained through different methods such as self-assembly, phase separation and electrospinning, mentioning some examples. [4]

Electrospinning process is very efficient due to its versatility, flexibility and ease of production. With their use, fibers with diameters ranging from nanometric to micrometric scale are produced when electrostatic forces are applied to polymer solutions or melts. The process consists in the application of high voltage on the polymer in liquid phase, leading to the formation of droplets. When the electrostatic repulsion exceeds the surface tension of the fluid, the droplet deforms in a conical shape (Taylor cone), leading to jet ejection of the polymer from the tip of the processing needle. The interaction between the electrostatic forces and the surface tension leads to the elongation of the generated fiber, leading to the evaporation of the solvent, the fiber then solidifies and is deposited in a grounded collector. The process causes the jet stream to be continuous, forming a fiber of reduced diameter, great length and uniform. [4,5]

The fibers obtained from electrospinning have properties that are determinant in their efficiency in the several areas of application cited, including large specific surface area, a higher aspect ratio and a better interconnectivity between pores [2,5]. However, by varying the processing parameters, such as the applied voltage, the polymer concentration, the working distance, the type of collector, the feed rate and the nozzle design, nanofibers with different morphologies and structures can be obtained. [4,5,6]

Among the diversity of materials that can be used in electrospinning, there is polysulfone: an amorphous thermoplastic polymer known for its high strength, high temperature stability, low creep, good electrical characteristics and resistant to many solvents and chemicals. Polysulfone (PSU) finds applications in several fields, such as: medical, electrical, electronic, aerospace, automotive, coating, among others. The organic solvent N,N-Dimethylacetamide (DMAc) is one of the substances in which the polysulfone is soluble, used for preparation of polymeric solution. [7,8]

Therefore, this work has the objective of analyzing and correlating the nanofibers obtained from the electrospinning of the polymer solution of PSU and DMAc, with the variation of the processing parameters, being the applied voltage, working distance and polymer concentration used. Aiming this objective, correlations between processing parameters and scanning electron microscopy (SEM) were used, determining the configuration of parameters that promotes the formation of higher quality nanofibers.

EXPERIMENTAL

Materials and Methods

For the polymeric solution preparation, Polysulfone polymer (PSU) was used in pellets and N,N-Dimethylacetamide (DMAc) as solvent. Several solutions were prepared with different concentrations: 8, 12, 20, 26 and 28% m/m of solute, in order to determine the range of nanofibers formation with the aid of scanning electron microscopy (SEM), the range has been identified between 26 and 28% m/m of solute. The dissolution of the polymer in the appropriate solvent was carried out in a water bath with constant stirring at 70 °C. For electrospinning, the assembled and used system is shown in Figure 1, which consists of a adjustable high voltage source connected to the tip of a syringe, in this case, a Hamilton's 20 ml type, positioned above a grounded cylindrical rotative collector with configurable rotation, adopting in this case 27 rpm. Several procedures were performed varying the processing parameters: applied tension (15, 18 and 20 kV) and working distance, the distance between the needle and the collector (5, 10 and 15 cm). The electrospinning was performed with controlled air humidity, in 50%, and room temperature.

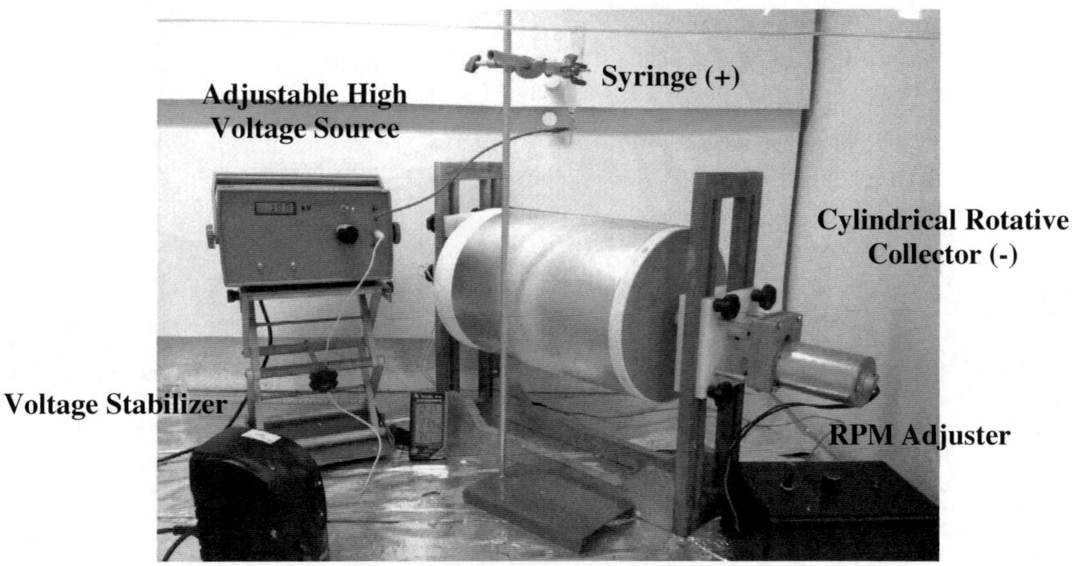

FIGURE 1. Electrospinning System.

A scanning electron microscope ZEISS brand, EVO series, model LS15, was used for the morphological analysis of the nanofiber blankets obtained from the electrospinning, available at the Laboratory of Images of the Department of Materials and Technology of the School of Engineering of Guaratinguetá, at São Paulo State University (UNESP), Brazil.

RESULTS

Processing Parameters and Nanofibers Quality Correlations

Polymer Concentration

The correlation between the applied electrospinning parameters and the quality of the nanofibers obtained was firstly based on different concentrations (%m/m) of solute used in the preparation of the polymer solution. The formation of nanofibers was confirmed by SEM.

At the beginning, between the concentrations of 20 and 24% m/m of PSU solute and with variation of the other processing parameters, tension and working distance, it was verified the existence of combinations in which nanofibers weren't formed. However, for polymeric solutions with a mass concentration of 26% m/m of PSU, the formation of nanofibers of higher or lower quality is observed, independently of the other parameters applied during the electrospinning process.

However, the upper limit, in terms of concentration, was determined as a result of the viscosity of the polymer solution, in other words, above a certain solute concentration, in this case 28% m/m, there was an obstruction of the needle nozzle used in the process, making electrospinning impossible. Thus, the ideal solute concentration range for electrospinning of polysulfone nanofibers was determined, between 26 and 28% m/m.

Voltage and Working Distance

Then, the analysis focused on the influence of the other parameters, voltage and working distance, on the quality of the nanofibers generated.

The SEM analysis of the 26% m/m blankets allowed concluding the intense influence of the processing parameters on the quality of the obtained fibers. It is observed that the increase of the working distance promotes the improvement of the nanofibers in all cases, mainly due to the fact of providing a greater distance, consequently a longer time for evaporation of the solvent until reaching the collector. However, the increase of the applied voltage promotes the improvement of the quality of the nanofibers until the voltage of 18 kV, since with 20 kV it is observed, for solutions of 26% m/m, the accumulation of the fibers in the form of beads.

The Figure 2 shows the SEM of the best blankets obtained at the concentration of 26% m/m. Emphasizing the configuration of 18 kV and 10 cm of working distance, since these samples present greater homogeneity of fiber diameter and fewer presence of defects. However, with presence of fibers in micrometric size.

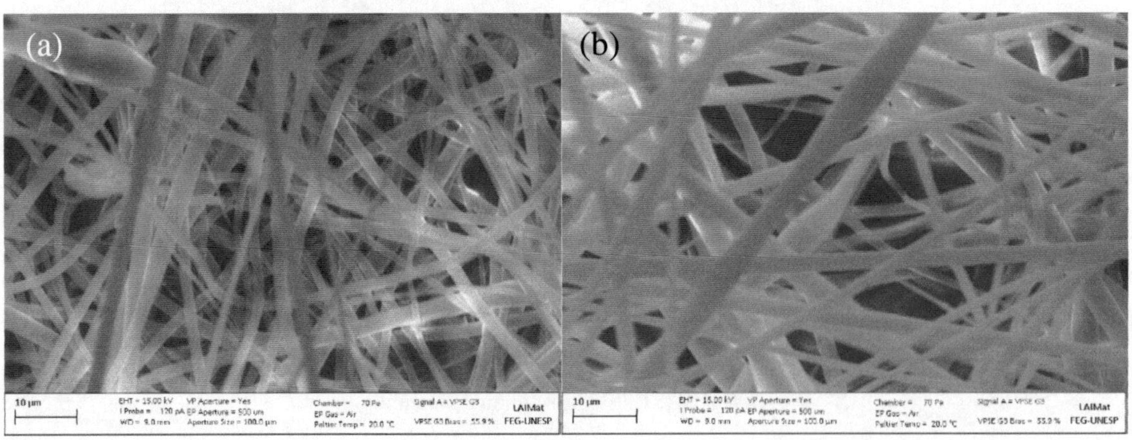

FIGURE 2. PSU fibers blanket SEM (3000x) and respective electrospinning parameters (a) 26%, 15 kV, 15 cm; (b) 26%, 18 kV, 10 cm.

The analysis of the SEM of the blankets obtained from polymer solutions with a concentration of 28% m/m show that the variation of the processing parameters presents a smaller influence on the quality of the processed fibers.

Nevertheless, with a more specific analysis, it is possible to determine the best combination of parameters to generate nanometric fibers and with greater homogeneity.

The improvement of the nanofibers obtained with the increase of the working distance was observed again, for the same reason, more time for the evaporation of the solvent. However, it was observed that the increase of the voltage promoted the improvement in the quality of the nanofibers up to the maximum voltage studied, 20 kV, different from the behavior observed in the 26% m/m concentration blankets, allowing to assume that the voltage could still be higher before it´s negative influence.

The Figure 3 shows the micrographs of the best blankets obtained at the concentration of 28% m/m. Emphasizing the configuration of 20 kV of tension and 10 cm of working distance due to the greater homogeneity of fiber diameter and fewer presence of defects.

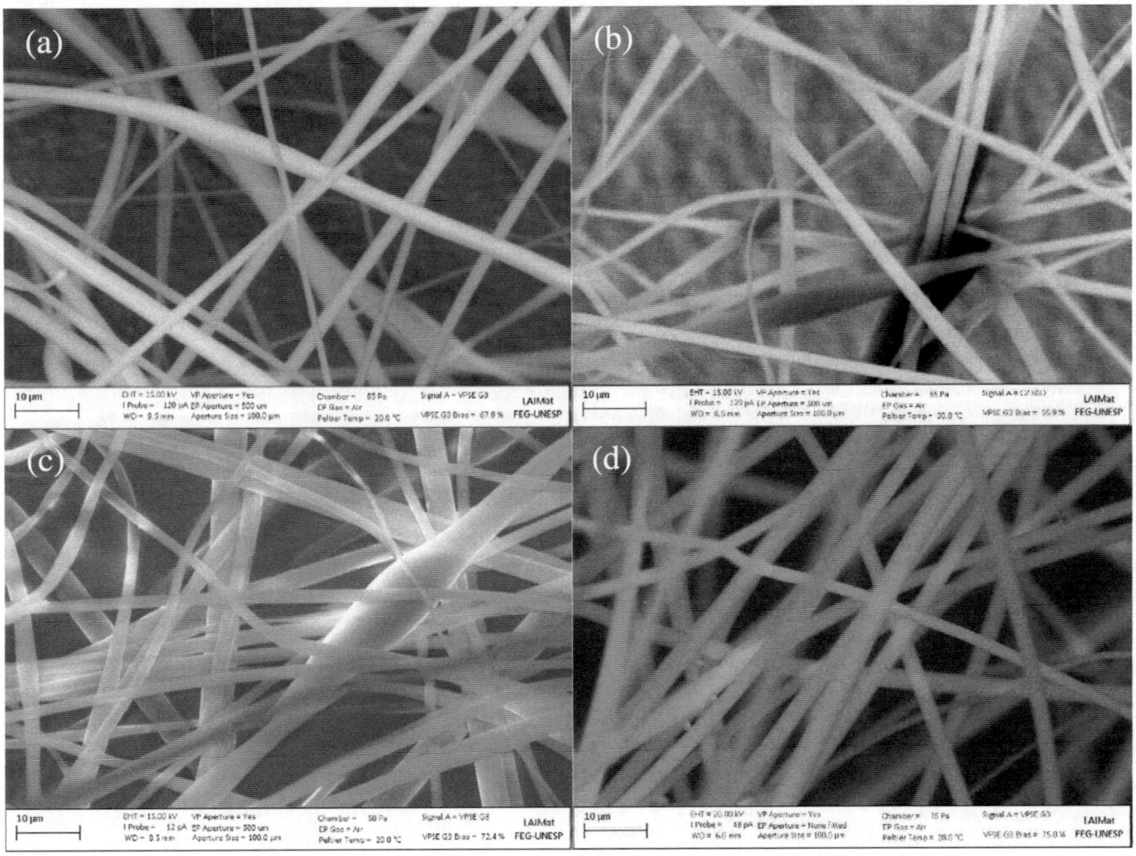

FIGURE 3. PSU fibers blanket SEM (3000x) and respective electrospinning parameters (a) 28%, 15 kV, 10 cm; (b) 28%, 18 kV, 15 cm; (c) 28%, 20 kV, 5 cm; (d) 28%, 20 kV, 10cm.

CONCLUSIONS

Polysulfone nanofibers were obtained by the electrospinning of polymer solutions of polysulfone dissolved in N-N-Dimethylacetamide, varying several processing parameters: solute concentration (8, 12, 20, 26 and 28% m / m), applied voltage (15, 18 and 20 kV) and working distance (5, 10 and 15 cm).

With the blankets generated from the electrospinning parameters combinations and analysis by scanning electron microscopy, it was possible to correlate the quality of the nanofibres obtained with each combination of parameters applied in the electrospinning processing of the material.

Nanofibers were only obtained at the concentrations of 26 and 28% m/m of PSU solute, for concentrations below this limit, micrometric fibers blankets were obtained, with higher diameter fibers and accumulation of polymeric droplets in the blankets. Concentrations above 28% m/m lead to increased viscosity of the polymer solution, making it impossible to electrospinning due to obstruction of the processing needle.

In general, the increase in working distance promoted the improvement of nanofibers, in some cases in a more discrete but still present form. However, the voltage variation showed a different behavior, according to the solute concentration, the increase of the voltage promoted the improvement of the fiber to a certain extent, negatively impacting with an even greater increase of tension (above 20 kV).

Therefore, it was possible to determine the best parameter configuration for both concentrations, for 26% m/m of solute, 18 kV of tension and 10 cm of working distance, and for 28% m/m of solute, 20 kV of tension and 15 cm of working distance. The solution with 28% m/m of solute is recommended for the obtaintion of nanofibers of higher homogeneity, lower concentration of defects, higher aspect ratio and greater constancy of diameters.

ACKNOWLEDGMENTS

The authors would like to thank CAPES, CNPq and the Image Laboratory (LAIMat) of São Paulo State University (UNESP/FEG), Brazil.

REFERENCES

1. R. J. Beck, Y. Zhao, H. Fong, T. J. Menkhaus, *Journal of Water Process Engineering Vol. 16,* Amsterdam: Elsevier, 2017 pp. 240-248.
2. X. Shi, W. Zhou, D. Ma, Q. Ma, D. Bridges, Y. Ma and A. Hu, *Journal of Nanomaterials Vol. 2015*, Cairo: Hindawi, 2015, pp. 1-20.
3. I. Rajzer, E. Menaszek, O. Castano. *Materials Science and Engineering C Vol. 77,* Amsterdam: Elsevier, 2017, pp. 493-499
4. P. S. Kumar, S. Jayaraman and G. Singh "Polymer and Composite Nanofiber: Electrospinning Parameters and Rheology properties" in *Rheology and Processing of Polymer Nanocomposites,* edited by S. Thomas, R. Muller and J. Abraham, New Jersey: John Wiley & Sons, 2016, pp. 329-333.
5. N. Bhardwaj, S. C. Kundu, *Biotechnology Advances Vol. 28*, Amsterdam: Elsevier, 2010, pp. 325-347.
6. A. Valizadeh, S. M. Farkhani, *IET Nanobiotechnology Vol. 8, Iss. 2*, Stevenage: The Institution of Engineering and Technology, 2014, pp. 83-92.
7. R. O. Ebewele, *Polymer Science and Technology*, Boca Raton: CRC Press, 2000.
8. D. M. Gil, Trabalho de Conclusão de Curso, Guaratinguetá: Universidade Estadual Paulista, 2016.

Modification of PP with CNT as painting agent: influence on performance of injected pieces

Rosales Caren [(1)]*

[(1)] Universidad Nacional Mar del Plata, Argentina - Buenos Aires - Argentina

From an industrial point of view, final plastic products should meet both certain structural requirements and adequate aesthetic quality. Manufactured pieces are commonly subjected to different surface finishing techniques to achieve desired aesthetic requirements, being the electrostatic painting the most efficient procedure. However, this technique cannot be directly applied in polymers due to their insulating properties. Plastic pieces need to be previously prepared by painting them with a primer, adding a further stage to the production cycle. To avoid this extra step, a painting agent could be added to the polymer to modify its electrical properties and make it suitable to be directly electrostatic painted. In this sense, carbon nanotubes (CNT) were proposed in this work as an electrostatic painting assistant for polypropylene (PP) injected pieces. The present work aims to evaluate how the addition of CNT to increase the panting transfer efficiency, influences the mechanical performance of intricate PP injected moldings.Then a complete characterization is presented, including mechanical properties and the morphology induced by processing. It was characterized at different locations (weld-line and injection point zones) of the moldings to assess the influence of in-homogeneities and flow pattern upon the dielectric behavior of the molded boxes and of double gated injection molded PP/CNT parts is presented. Finally, dielectric and electric properties and relaxation phenomena of the composite were investigated Differences in CNT orientation were found along the pieces: while a CNT orientation profile was observed along the thickness of the injected part induced by the molding thermo-mechanical field, CNT were seen randomly distributed in the weld line zone. These dissimilar orientations induced a difference of an order of magnitude in the electrical properties of the part that would result in distinct paint transfer efficiency in the same piece.

POLYETHYLENE-CELLULOSE COMPOSITES AND NANOCOMPOSITES. I. Effect of FIBER SIZE AND COMPOSITION

Maia Thiago Henrique Silveira [1], Calazans Marília [1], Lima Vitor [1], Moreira Francys Kley Vieira [1], Lucas Alessandra de Almeida [1]*

[1] University Federal of São Carlos - São Paulo - Brazil

The use of cellulose nanofibrils (CNF), conventional bleached micrometric cellulose pulp fibers (CF) and lignocellulose nanofibrils (LCNF) has been explored in a polyolefin aqueous dispersion based on ethylene copolymer modified with 6 wt.% of acrylic acid. The addition of 5, 10 and 30 wt% of fibers to neat PE act as reinforcement and increase the Young's modulus. The PE/LCNF composite exhibit a two-fold increase on Young's modulus of nanocomposite films containing 10 and 30 wt.%. The PE/CNF and PE/CF films are transparent to naked eye up to 10 wt%, while PE/LCNF are yellowish. The resultant composites showed high thermal stability on processing and application temperatures with only a slight reduction in the thermal degradation temperatures compared to the neat PE and a very low degree of water absorption.

Dissimilar Tendencies of Innovative Green Clay Organo-Modifier on the Final Properties of Poly(ε-caprolactone) Based Nanocomposites

D. Merino[a], L. N. Ludueña[a] and Vera A. Alvarez[a]

[a] Faculty of Engineering, Composite Materials Group, Research Institute of Materials Science and Technology (INTEMA), National University of Mar del Plata (UNMdP) - National Council of Scientific and Technical Research (CONICET), Colon, 10890 Mar del Plata, Argentina

Abstract. Composites based on poly(ε-caprolactone) (PCL) were prepared by melt blending the polymer with natural and modified bentonite. Soy lecithin (SL), a natural and nontoxic biosurfactant, was used as modifier agent. Three organoclays with different SL contents were employed as fillers and composites containing 1 and 2 wt% of each clay, were prepared. Thermal (by Thermogravimetric analysis, and differential scanning calorimetry), morphological (by X-ray diffraction (XRD) and melt rheology), mechanical (by tensile tests) and barrier properties (by means of water vapor permeability tests) of matrix and composites were studied as a function of clay type and content. Morphological analysis by XRD showed nanocomposites with intercalated-exfoliated structures while melt rheology suggests thermal degradation of PCL matrix catalyzed by SL bentonite modifier. Thermal and mechanical properties were consistent with this assumption, due to the slight increment in crystallinity percentage and detriment in the Young's modulus of the nanocomposites in comparison with the neat matrix. On the other hand, water vapor permeability of PCL significantly decreased in composites, as is expected for polymer/clay nanocomposites, favoring its potential application as food packaging. Thus, dissimilar tendencies were found in the final properties of the nanocomposites that were attributed to the matrix molecular weight degradation catalyzed by the novel green clay organo-modifier.

Keywords: Biodegradable polymer, Composite, Polyester degradation, Bentonite, Soy lecithin.

PACS: Replace this text with PACS numbers; choose from this list: http://www.aip.org/pacs/index.html

INTRODUCTION

The problem of litter requires an immediate solution. It not only results in visual pollution but also presents a threat to animals and results in a reduction of agricultural productivity [1]. Among the different options to deal with the problem of disposal of the used polymeric materials (biodegradation, incineration and recycling), biodegradation appears as the best solution to overcome this drawback [2].

Nowadays the packaging industry is by far the major user of plastics while food packaging is the largest growing sector within this market [1, 3]. Also, over 60% of post-consumer plastics waste is produced by households; most of it as single use packaging [1].

The goal of food packaging is to maintain food safety and minimize environmental impact [3]. Biodegradable plastics materials are considered as a viable alternative for conventional plastics [4]. However, to replace them, more versatile properties of the biodegradable plastics must be reached in order to cover full range of properties provided by the traditional ones [4].

Poly(ε-caprolactone) (PCL) is a linear semicrystalline biopolyester synthesized from fossil resources and produced by the ring opening polymerization of ε-caprolactone [5, 6]. It is a polymer with a strong hydrophobic nature, a degree of crystallinity around 45%, a glass-transition temperature around -60 °C and a melting point near 60 °C [6, 7]. It seems to be an important matrix for design of biodegradable nanocomposites because of its biocompatible and biodegradable properties [6]. Its potential uses are currently being examined in biodegradable packaging materials [8–11], in pharmaceutical controlled release systems [12–15] and other medical applications [15–18]. However, it has poor mechanical, barrier and thermal properties compared with conventional polymers [5, 6]. One way to improve these properties is to incorporate nano-sized fillers to the matrix. Nanocomposites based on PCL have been widely reported [19–23]. The most reported nano-fillers in poly(ε-caprolactone)-based materials are represented by organo-modified clays [3, 5, 7]. The main advantages of nanoclay particulates are easy availability, low cost, easy processability and good performance [3].

In order to improve polymer intercalation, clay minerals must be previously modified by the exchange of the native alkaline counterions by organic cations. The previous modification reduces their hydrophilic character, increasing the compatibility with the matrix [3, 5]. Primary, secondary, tertiary and quarternary alkylamonium cations [24, 25] or phosphonium cations [26, 27] having various substituents, have been widely employed as surface modifiers regardless the toxicity they introduce in the final product, especially when the material is designed for its use in food contact applications, cosmetic and biomedical industries [3, 5, 28, 29]. Among the employed modifiers found in bibliography, some natural compounds were used to modify bentonite, i.e. amino acids, proteins, nucleic acids, DNA and RNA. Most of these compounds were utilized as catalyst for biological reactions, drug delivery systems, enzyme immobilization, biosensing and biofunctional materials among others [30–32].

The aim of this work was to obtain a nontoxic and biodegradable material based on PCL and to study the effect of a novel green clay organo-modifier of the thermal, barrier, mechanical and morphological properties in order to analyze their use as a potential material for food packaging. We focus on the effect of modifier chemistry and structure on the main characteristics of the matrix and nanocomposites that are responsible of the final properties.

EXPERIMENTAL SECTION

Materials

Commercial grade PCL (CAPA FB 100) was supplied by Perstorp, with Mw = 100000 g/mol. The bentonite (Bent) was supplied by Minarmco S.A. (Argentina) and its cation exchange capacity (CEC) was 105 mequiv/100 gr, measured by the methylene blue method. It consisted predominantly of montmorillonite and contained quartz and feldspar as major impurities, as well as traces of gypsum and sepiolite as evidenced by X-ray diffraction (XRD) analysis [33]. Soy lecithin (SL) was provided by Melar S.A. (Argentina) and it was used as received.

Preparation of Soy Lecithin- Modified Clays

Soy lecithin-modified clays with different organic content were prepared by ionic exchange as reported in our previous work [34]. Briefly, 4 g of Bent were mixed with 1, 2 and 4 g of soy lecithin aqueous and acid solution and stirred at 75 ºC for 2 h. Then, they were washed (with an equimolar mixture of ethanol - distilled water) and lyophilized for 72 h. Samples were labeled A, B and C for the nanoclays prepared with 1, 2 and 4 g of soy lecithin, respectively.

Composites Preparation

PCL matrix nanocomposites sheets with 1 and 2 wt.% of each clay were prepared by melt-intercalation followed by compression-molding. An intensive Brabender type mixer with two counter-rotating roller rotors was used. Mixing temperature was 100 ºC; screws speed was 100 r.p.m. and mixing time was 10 min. Compression molding was carried out in a hydraulic press under 75 kg/cm2 for 10 minutes at 100 ºC. The thickness of the sheets was 1 mm. Nanocomposites sheets were identified by the following code: PCL-"clay content"% "clay type". For example, "PCL-1% Bent" is the nanocomposite sheet based on PCL matrix with 1wt.% of natural bentonite.

Characterization Methods

Diffraction patterns of organoclays and nanocomposites were obtained using a PANalytical X-Pert Pro diffractometer, operating at 40 kV and 40 mA, with Cu Kα radiation (λ=1.5406 Å), at room temperature. Organoclays were analyzed in the 2-10 º range at scanning speed of 2 º/min. The basal spacing distances (d_{001}) were calculated from the 2θ values using the Bragg's equation:

$$n\lambda = 2d \text{ sen } \theta \quad (1)$$

Where λ is the source wavelength, θ is the diffraction angle and d is the interlamellar distance. Nanocomposites were examined in the 2-40 º at the same conditions of organoclays.

Rheological tests were conducted in a Rheometric Scientific Ares rheometer under nitrogen atmosphere. Plate–plate geometry with a plate diameter of 25 mm was used. Samples were inserted and heated up to 80 °C. Low shear amplitude (2%) was used in order to avoid the destruction of any stabilized clay structure and work in the lineal viscoelastic regime. Data were taken for shear rates ($\gamma^{\cdot}$) in the range of 0.002 - 12.5 s^{-1}. The melt rheology curves were fitted to the power law expression in the range of 0.002 - 0.02 s^{-1} by the following equation:

$$\eta_{Rh} = A_{Rh}\dot{\gamma}^{(n_{Rh})} \quad (2)$$

Where the Rh subscript represents a rheology parameter, η_{Rh} is the complex viscosity, A_{Rh} is the pre-exponential factor, $\gamma^{\cdot}$ is the shear rate and n_{Rh} is the shear thinning exponent. In the double logarithmic plot, a linear zone at low shear rates (in our case up to 0.02 s-1) can be seen. The n_{Rh} parameter was calculated from the slope of this region [35].

Measurements of nanocomposites and PCL crystallinity percentage were conducted in a differential scanning calorimeter (DSC) TA instruments model Q2000, operating from 20 to 120 °C at a heating rate of 10 °C/min under nitrogen atmosphere. The degree of crystallinity of each sample, $X_C(\%)$, was calculated according to the following equation:

$$X_C(\%) = \frac{\Delta H_m}{w_{PCL}\, x\, \Delta H_{100}}\, x\, 100 \quad (3)$$

Where ΔH_m is the experimental melting enthalpy, w_{PCL}, is the PCL weight fraction and ΔH_{100} is the melting enthalpy of 100 % crystalline PCL (142 J/g [36]).

RESULTS AND DISCUSSION

Soy- Lecithin Modified Bentonite Characterization

The results of the characterization of soy-lecithin modified Bent were presented in our previous work [34]. Briefly, their main characteristics are shown in Table I.

It can be observed from the previous table that the organic content in the organoclays significantly increased with the initial amount of SL. However, the basal spacement shows a slow raising tendency as a function of SL content, probably because the additional amount of SL may be adsorbed on the clay sheets instead of the interlayer region [26, 36]. As it is known, thermal stability of organoclays is significantly influenced by the surfactant adsorbed on the external surfaces of the silicate layers. This phenomena can be observed from the thermal degradation temperatures ($T_{-10\%}$) that show a decrease in the thermal stability of organoclays as a function of the absorbed SL content [26]. Higher values of d_001 are mostly desired due to the possibility of easier intercalation of polymer chains between the silicate layers [21]. In this regard, there is a compromise between thermal stability and the interlayer spacing of the clays when using them as reinforcement of a polymer matrix. On the other hand, moisture absorption is another interesting parameter that can be related with the compatibility between the reinforcement and the polymer matrix. Lower moisture adsorption is desired for nanoclays used as reinforcement of PCL, which has a strong hydrophobic nature. Moisture adsorption tests (Table I) showed a huge reduction of this parameter for the organoclays when they are compared with raw bentonite. Also, it was found that the moisture adsorption slightly decrease as a function of the organic content increase. That can be attributed to the presence of minor amount of cations in the interlayer region of ionically-exchanged clays that are solvated by water molecules which interact most strongly with the compensation cations [36].

PCL-SL-Bent Nanocomposites

Bent can be intercalated or exfoliated depending on the increment in their basal spacement caused by interactions of polymer chains between the silicate layers [4, 5]. If an intercalated structure is found, the XRD pattern shows a shift on the 001 diffraction peak towards lower angle values, but if no more diffraction peaks are visible, we may be in the presence of an exfoliated or mixed intercalated-exfoliated structure [5, 37]. XRD curves of PCL, clays (Bent, A, B and C) and their nanocomposites are shown in **FIGURE 1**. Fig. 1a compares the position of the unmodified Bent diffraction peak (2θ = 6.69 °, d_{001} = 13.2 Å) with those for PCL-1% Bent (2θ = 6.01 °, d_{001} = 14.7 Å) and PCL - 2% Bent (2θ = 5.87 °, d_{001} = 15.03 Å). As can be seen, there is a shift of the Bent diffraction peak towards lower angles, indicating an increase in the interlaminar spacing of Bent in nanocomposites. Furthermore, this effect increases as a function of Bent content. This result suggests that intercalated nanocomposites were obtained [38]. **FIGURE 1b** again shows the appearance of an intercalated compound, due to the displacement of A diffraction peak (2θ = 5.88 °, d_{001} = 15.1 Å) to lower diffraction angles in the nanocomposites PCL-1% A (2θ = 5.19 °, d_{001} = 17.00 Å) and PCL-2% A (2θ = 5.11 °, d_{001} = 17.27 °). In the case of nanocomposites obtained with B, no significant changes were observed in the peak positions of B and B in PCL-1% B and PCL-2% B (**FIGURE 1c**). This suggest that micron-size particles of B are randomly oriented in the PCL matrix [39]. Finally, Fig.1d shows C diffraction peaks (2θ = 3.6 ° & 5.52 °, d_{001} = 24.5 & 16.0 Å) and C in PCL-1% C and PCL-2% C nanocomposites, in which peaks almost disappear because of its large widening and decreasing intensity, suggesting an intercalated-exfoliated structure [40].

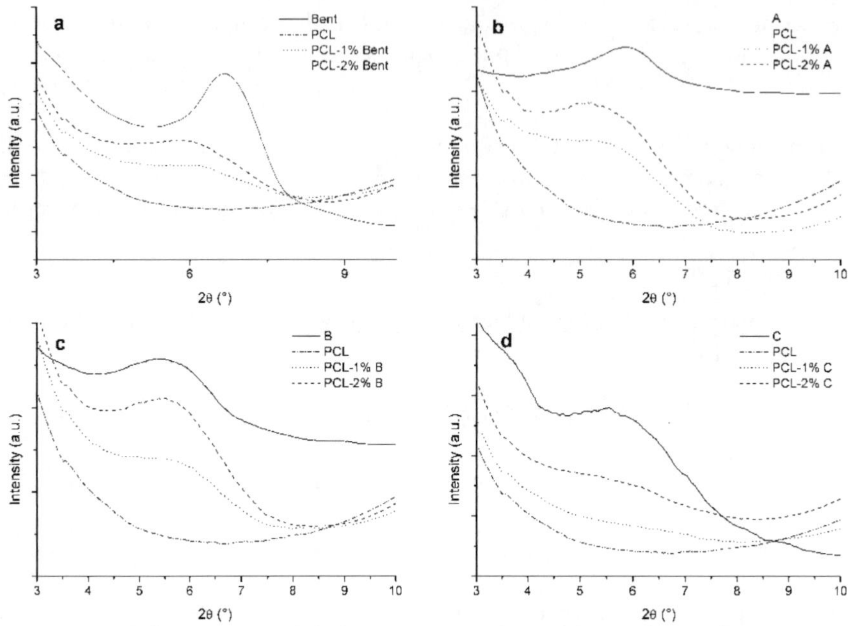

FIGURE 1 XRD patterns of clays, PCL and their nanocomposites with 1 and 2% of: **a**) raw Bent, **b**) organoclay A, **c**) organoclay B and **d**) organoclay C

Melt rheology can be also useful to compare the clay dispersion degree of polymer/clay nanocomposites. Several authors [34, 41–43] have found that the η_{Rh} parameter (see equation (2)) is higher as a function of the clay dispersion degree. Zhao et al. [41] have proposed η_{Rh} as a semi-quantitative measure of the clay dispersion degree of the sample. It must be noted that the average number of nanoplatelets per tactoid for a given nanocomposite cannot be calculated from η_{Rh}. Therefore, the measured rheology needs to be 'calibrated/correlated' with known nanocomposite morphology, previously characterized by the traditional techniques (XRD, TEM). It must be also taken into account that XRD and TEM analysis are performed in the solid phase while rheology in the molten one. Even so, good correlation between η_{Rh} studied by melt rheology and clay dispersion degree analyzed by XRD/TEM for different polymer/clay nanocomposites was found by several authors [34, 41, 43, 44]. **FIGURE 2** shows the log-log complex viscosity as a function of shear rate curves.

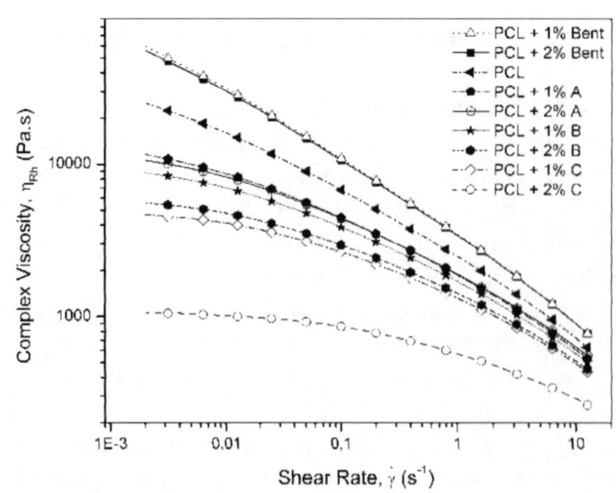

FIGURE 2.Double logarithmic plot of complex viscosity as a function of shear rate for the neat PCL and nanocomposite.

Lower η_{Rh} values for the nanocomposites with organo-modified clays are observed while the interlayer distance was higher for these materials. Therefore, the expected correlation between XRD and melt rheology was not observed. In addition, significant detriment of the viscosity of these nanocomposites is observed in comparison with the neat matrix (**FIGURE 2**) even when the incorporation of a rigid filler such as clays in a polymer matrix should increase the viscosity of the system. These unexpected results may be a consequence of polymer matrix degradation catalyzed by clay organo-modifiers. Lin et al. [45] studied the effect of catalytic degradation of molecular weight of PCL in solid state by the presence of a methanolic solution of alkylamines. They found that the primary amines possessed the ability to degrade the matrix by 50% after 8 days depending on the type of amine and the test conditions. Additionally, they found that the degradation rate increased with the length of the alkyl chain of the amine and that drugs with secondary and tertiary amino groups do not react directly with the PCL but catalytically accelerate its degradation. Similar results were reported by Cha Y. et al. and Shah S. et al. [46, 47], who studied the decrease of the molecular weight of PCL, PGLA (glycolic acid-co-DL-lactic acid) and PLLA (poly L-lactic acid) microspheres that housed drugs with amino functional groups.

Although there are not many publications that consider the effect of catalytic degradation of polyesters, this could be of practical importance for controlled drug release, because depending on the drug, the molecular weight and type of polyester could be regulated the time to release it. In the case of food packaging, the catalytic degradation of PCL can be a problem during processing and can lead to undesirable final mechanical and barrier properties of the final nanocomposite product. The balance between polymer matrix degradation and clay dispersion can produce lower η_{Rh} for better dispersed nanocomposites. In addition, the balance between polymer matrix degradation, clay dispersion degree and clay platelet content will define the final mechanical and barrier performance of the nanocomposites [48].

Percentages of crystallinity, Xc (%), and melting temperatures, T_m (°C), are introduced in **TABLE I**.

Table I. DSC test results for PCL matrix and nanocomposites.

Sample	T_m (°C)	X_C (%)
PCL	59.2	52.6
PCL - 1% Bent	59.5	54.1
PCL - 2% Bent	60.0	52.0
PCL - 1% A	59.0	59.8
PCL - 2% A	58.5	52.0
PCL - 1% B	58.5	52.2
PCL - 2% B	58.7	51.2
PCL - 1% C	59.3	56.7
PCL - 2% C	58.3	57.5

Pitt C. G. et al. [48] has reported that polymer chain degradation of PCL can be associated with an increase in their crystallinity. They claim that the degree of crystallinity of the PCL can increase by up to 40% by changing their molecular weight from 50,000 to 5000. It was observed that only PCL-1% A, PCL-1% C and PCL-2% C had a slightly increment in their percentage of crystallinity while the other samples analyzed have shown similar results than neat PCL. This small change in crystallinity can be related to the change in molecular weight of PCL in these nanocomposites, as suggested by rheology analysis. Regarding melting temperatures all values were similar to that for PCL and non-significant changes were observed.

CONCLUSIONS

This paper describes the preparation by melt blending of environmentally friendly compounds of PCL reinforced with natural and soy lecithin modified bentonite. X-ray characterization of the compounds suggested that PCL with Bent, A and B possessed an intercalated structure, while those obtained with C, an intercalated-exfoliated one. However, these results could not be directly confirmed by melt rheology. This characterization showed unexpected results that led to the belief that thermal degradation of the matrix occurs by catalytic action of the clay modifier (soy lecithin) due to the presence of amino groups that have the ability to break the ester bond present in the polymeric polyesters. The thermal analysis indicated that the compounds obtained with C had higher percentages of crystallinity than the pure PCL, which is in accordance with the alleged decrease in molecular weight. Finally, it can be concluded that clay organo-modifiers should be designed not only looking for improvements on the clay/matrix compatibility and clay interlayer distance but also analyzing the chemistry of the modifiers that may induce catalytic matrix molecular weight degradation and thus leading to unexpected tendencies on the final properties of the nanocomposites.

ACKNOWLEDGMENTS

Authors acknowledge the financial support of the National Research Council (CONICET) (PIP 00617), the National Agency for the Promotion of Science and Technology (ANPCyT) (FS Nano 004), and the University of Mar del Plata (UNMdP), Argentina.

REFERENCES

1. Scott G (1999) Polymers and the Environment.
2. Albertsson A, Huang S (1995) Degradable polymers, recycling, and plastics waste management.
3. Majeed K, Jawaid M, Hassan A, et al (2013) Potential materials for food packaging from nanoclay/natural fibres filled hybrid composites. Mater Des 46:391–410. doi: 10.1016/j.matdes.2012.10.044
4. Janigová I, Lednický F, Mošková DJ, Chodák I (2011) Nanocomposites with Biodegradable Polycaprolactone Matrix. Macromol Symp 301:1–8. doi: 10.1002/masy.201150301
5. Benali S, Dubois P (2012) PCL/Clay Nano-Biocomposites. In: Environ. Silic. Nano-Biocomposites. Springer, pp 119–142
6. Neppalli R, Causin V, Marega C, et al (2011) Structure, morphology, and biodegradability of poly (ε- caprolactone)- based nanocomposites. Polym Eng Sci 51:1489–1496.
7. Labidi S, Azema N, Perrin D, Lopez-Cuesta JM (2010) Organo-modified montmorillonite/poly(ε-caprolactone) nanocomposites prepared by melt intercalation in a twin-screw extruder. Polym Degrad Stab 95:382–388. doi: 10.1016/j.polymdegradstab.2009.11.013
8. Rhim J-W, Park H-M, Ha C-S (2013) Bio-nanocomposites for food packaging applications. Prog Polym Sci 38:1629–1652. doi: 10.1016/j.progpolymsci.2013.05.008
9. Siracusa V, Rocculi P, Romani S, Rosa MD (2008) Biodegradable polymers for food packaging: a review. Trends Food Sci Technol 19:634–643. doi: 10.1016/j.tifs.2008.07.003
10. Othman SH (2014) Bio-nanocomposite Materials for Food Packaging Applications: Types of Biopolymer and Nano-sized Filler. Agric Agric Sci Procedia 2:296–303. doi: 10.1016/j.aaspro.2014.11.042
11. Ortega-Toro R, Contreras J, Talens P, Chiralt. A (2015) Physical and structural properties and thermal behaviour of starch-poly(ε-caprolactone) blend films for food packaging. Food Packag Shelf Life 5:10–20. doi: 10.1016/j.fpsl.2015.04.001
12. Chawla JS, Amiji MM (2002) Biodegradable poly(e-caprolactone) nanoparticles for tumor-targeted delivery of tamoxifen. Int J Pharm 249:127–138. doi: 10.1016/S0378-5173(02)00483-0
13. Wei X, Gong C, Gou M, et al (2009) Biodegradable poly(??-caprolactone)-poly(ethylene glycol) copolymers as drug delivery system. Int J Pharm 381:1–18. doi: 10.1016/j.ijpharm.2009.07.033
14. Anzai R, Murakami Y (2015) Poly(ε-caprolactone) (PCL)–polymeric micelle hybrid sheets for the incorporation and release of hydrophilic proteins. Colloids Surfaces B Biointerfaces 127:292–299. doi: 10.1016/j.colsurfb.2015.01.028
15. Dash TK, Konkimalla VB (2012) Poly-є-caprolactone based formulations for drug delivery and tissue engineering: A review. J Control Release 158:15–33. doi: 10.1016/j.jconrel.2011.09.064
16. Jones DS, Djokic J, McCoy CP, Gorman SP (2002) Poly(ε-caprolactone) and poly(ε-caprolactone)-polyvinylpyrrolidone-iodine blends as ureteral biomaterials: characterisation of mechanical and surface properties, degradation and resistance to encrustation in vitro. Biomaterials 23:4449–4458. doi: 10.1016/S0142-9612(02)00158-8
17. Annabi N, Fathi A, Mithieux SM, et al (2011) Fabrication of porous PCL/elastin composite scaffolds for tissue engineering applications. J Supercrit Fluids 59:157–167. doi: 10.1016/j.supflu.2011.06.010
18. Ghasemi-Mobarakeh L, Prabhakaran MP, Morshed M, et al (2008) Electrospun poly(ε-caprolactone)/gelatin nanofibrous scaffolds for nerve tissue engineering. Biomaterials 29:4532–4539. doi: 10.1016/j.biomaterials.2008.08.007
19. Arslan M, Tasdelen MA, Uyar T, Yagci Y (2015) Poly(epsilon caprolactone)/clay nanocomposites via host–guest chemistry. Eur Polym J 71:259–267. doi: 10.1016/j.eurpolymj.2015.08.006
20. Ludueña LN, Alvarez V a., Vazquez A (2007) Processing and microstructure of PCL/clay nanocomposites. Mater Sci Eng A 460–461:121–129. doi: 10.1016/j.msea.2007.01.104
21. Ludueña LN, Kenny JM, Vázquez A, Alvarez VA (2011) Effect of clay organic modifier on the final performance of PCL/clay nanocomposites. Mater Sci Eng A 529:215–223. doi: 10.1016/j.msea.2011.09.020
22. Fukushima K, Tabuani D, Camino G (2009) Nanocomposites of PLA and PCL based on montmorillonite and sepiolite. Mater Sci Eng C 29:1433–1441. doi: 10.1016/j.msec.2008.11.005
23. Goffin a. L, Raquez JM, Duquesne E, et al (2011) Poly(??-caprolactone) based nanocomposites reinforced by surface-grafted cellulose nanowhiskers via extrusion processing: Morphology, rheology, and thermo-mechanical properties. Polymer (Guildf) 52:1532–1538. doi: 10.1016/j.polymer.2011.02.004

24. Huskić M, Žagar E, Žigon M (2012) The influence of a quaternary ammonium salt and MMT on the in situ intercalative polymerization of PMMA. Eur Polym J 48:1555–1560. doi: 10.1016/j.eurpolymj.2012.06.010
25. Mallakpour S, Moslemi S (2012) Dispersion of chiral amino acid organomodified Cloisite Na+ in poly (vinyl alcohol) matrix for designing of novel bionanocomposite films. Prog Org Coatings 74:8–13.
26. Calderon JU, Lennox B, Kamal MR (2008) Thermally stable phosphonium-montmorillonite organoclays. Appl Clay Sci 40:90–98. doi: 10.1016/j.clay.2007.08.004
27. Hedley CB, Yuan G, Theng Bkg (2007) Thermal analysis of montmorillonites modified with quaternary phosphonium and ammonium surfactants. Appl Clay Sci 35:180–188.
28. Sarkar B, Megharaj M, Shanmuganathan D, Naidu R (2013) Toxicity of organoclays to microbial processes and earthworm survival in soils. J Hazard Mater 261:783–800. doi: 10.1016/j.jhazmat.2012.11.061
29. Maisanaba S, Puerto M, Pichardo S, et al (2013) In vitro toxicological assessment of clays for their use in food packaging applications. 57:266–275.
30. Petra L, Billik P, Komadel P (2015) Preparation and characterization of hybrid materials consisting of high-energy ground montmorillonite and α-amino acids. Appl Clay Sci 115:174–178.
31. Yu WH, Li N, Tong DS, et al (2013) Adsorption of proteins and nucleic acids on clay minerals and their interactions: A review. Appl Clay Sci 80:443–452.
32. D'Amico DA, Ollier RP, Alvarez VA, et al (2014) Modification of bentonite by combination of reactions of acid-activation, silylation and ionic exchange. Appl Clay Sci 99:254–260. doi: http://dx.doi.org/10.1016/j.clay.2014.07.002
33. Merino D, Ollier R, Lanfranconi M, Alvarez V (2016) Preparation and characterization of soy lecithin-modified bentonites. Appl Clay Sci 127:17–22. doi: 10.1016/j.clay.2016.04.006
34. Wagener R, Reisinger TJG (2003) A rheological method to compare the degree of exfoliation of nanocomposites. Polymer (Guildf) 44:7513–7518. doi: 10.1016/j.polymer.2003.01.001
35. Tsuji H, Ishizaka T (2001) Porous Biodegradable Polyesters, 3. Preparation of Porous Poly (ε- caprolactone) Films from Blends by Selective Enzymatic Removal of Poly (L- lactide). Macromol Biosci 1:59–65.
36. Bergaya F, Jaber M, Lambert J-F (2012) Clays and clay minerals as layered nanofillers for (bio) polymers. In: Environ. Silic. Nano-Biocomposites. Springer, pp 41–75
37. Bellucci F, Fabiani D, Montanari GC, Testa L (2010) The processing of nanocomposites. In: Dielectr. Polym. Nanocomposites. Springer, pp 31–64
38. Zheng X, Wilkie CA (2003) Nanocomposites based on poly (ϵ-caprolactone) (PCL)/clay hybrid: polystyrene, high impact polystyrene, ABS, polypropylene and polyethylene. Polym Degrad Stab 82:441–450. doi: 10.1016/S0141-3910(03)00197-6
39. Lepoittevin B, Devalckenaere M, Pantoustier N, et al (2002) Poly (ε-caprolactone)/clay nanocomposites prepared by melt intercalation: mechanical, thermal and rheological properties. Polymer (Guildf) 43:4017–4023.
40. Di Y, Iannace S, Di Maio E, Nicolais L (2003) Nanocomposites by melt intercalation based on polycaprolactone and organoclay. J Polym Sci Part B Polym Phys 41:670–678. doi: 10.1002/polb.10420
41. Zhao J, Morgan AB, Harris JD (2005) Rheological characterization of polystyrene–clay nanocomposites to compare the degree of exfoliation and dispersion. Polymer (Guildf) 46:8641–8660. doi: 10.1016/j.polymer.2005.04.038
42. Krishnamoorti R, Ren J, Silva AS (2001) Shear response of layered silicate nanocomposites. J Chem Phys 114:4968. doi: 10.1063/1.1345908
43. Durmus A, Kasgoz A, Macosko CW (2007) Linear low density polyethylene (LLDPE)/clay nanocomposites. Part I: Structural characterization and quantifying clay dispersion by melt rheology. Polymer (Guildf) 48:4492–4502. doi: 10.1016/j.polymer.2007.05.074
44. VanderHart DL, Asano A, Gilman JW (2001) NMR Measurements Related to Clay-Dispersion Quality and Organic-Modifier Stability in Nylon-6/Clay Nanocomposites. Macromolecules 34:3819–3822. doi: 10.1021/ma002089z
45. Lin W, Flanagan DR, Linhardt RJ (1994) Accelerated Degradation of Poly(ε-caprolactone) by Organic Amines. Pharm Res 11:1030–1034. doi: 10.1023/A:1018943622498
46. Cha Y, Pitt CG (1989) The acceleration of degradation-controlled drug delivery from polyester microspheres. J Control Release 8:259–265. doi: 10.1016/0168-3659(89)90047-3
47. Shah SS, Cha Y, Pitt CG (1992) Poly (glycolic acid-co-dl-lactic acid): diffusion or degradation controlled drug delivery? J Control Release 18:261–270. doi: 10.1016/0168-3659(92)90171-M
48. Pitt CG, Zhong-wei G (1987) Modification of the rates of chain cleavage of poly(ϵ-caprolactone) and related polyesters in the solid state. J Control Release 4:283–292. doi: 10.1016/0168-3659(87)90020-4

Hydrogen storage capacity of nanocomposites of sulfonated poly(ether imide) and polypyrrole with sodium alanate and carbon nanotubes

Beatrice Cesar A. G. [(1)], Oliveira Amanda D. [(2)], Passador Fabio R. [(3)], Leiva Daniel R. [(1)], Pessan Luiz A. [(1)]*

[(1)] UFSCar - DEMa - São Paulo - Brazil, [(2)] UFPel - CDT - Rio Grande do Sul - Brazil, [(3)] UNIFESP - ICT - São Paulo - Brazil

The fabrication of porous materials with optimized and tailored structural and chemical properties is key to their use in numerous emergent applications such as storage of electrical energy in supercapacitors, hydrogen storage, catalyst supports in fuel cells, etc. Gas and energy storage are currently attracting much attention due to the increasing demand for energy and the environmental problems caused by society's continuing dependence on fossil fuels. Hydrogen is considered to be a clean, economical and safe renewable energy source that would be ideal to replace fossil fuels. However, it is easy to burn, which has the risk of fire and explosion. The problems of transportation and storage also restrict the application of hydrogen energy. This gas adsorbs at solid surfaces depending on the applied pressure and temperature. The development of materials for efficient hydrogen storage remains a challenge. Several studies have shown that narrow micropores are the most efficient for hydrogen storage. However, large surface area and high pore volume are also necessary to ensure high hydrogen uptake. The purpose of this work was to develop a hybrid porous materials consisting of sulfonated polyetherimide and polypyrrole blend with sodium alanate nanoparticles and carbon nanotubes. Dilute solutions with 30 wt% of hydride were first prepared under stirring at room temperature and the solutions were dried under vacuum. The hybrids were analyzed by differential scanning calorimetry, thermogravimetric analysis, transmission electron microscopy and hydrogen sorption measurements. The addition of sodium alanate changed the glass transition temperature of the hybrids when compared to the blend and the TEM images showed that simply physically mixture occurred between polymer and metallic nanoparticles. Hydrogen sorption tests showed an increase in the amount of hydrogen in the presence of higher carbon nanotubes amount.

The influence of clay slurry on the flexural strength, the conversion degree and the rheological properties of BIS-GMA/TEGDMA resin

Rivera-Gonzaga Alejandro (1)*, Cuevas-Suárez Carlos (1), Sánchez-Olivares Guadalupe (2), Calderas Fausto (2), Moreno Leonardo (3), Sánchez-Solís Antonio (3), Manero Octavio (3)

(1) Universidad Autónoma del Estado de Hidalgo - Hidalgo - Mexico, (2) CIATEC, A.C. - Guanajuato - Mexico, (3) Universidad Nacional Autónoma de México - Ciudad de México - Mexico

In this work nanocomposites were prepared using a mixture of Bis-phenol A glycidyl methacrylate and triethylene glycol dimethacrylate (Bis-GMA/TEGDMA, 70/30 w/w, Resin) and sodium montmorillonite (MMT) as inorganic filler in concentrations of 0.1, 0.2, 0.3 and 0.4 phr. Two groups were prepared: the first one by adding hydrated MMT to the resin (Resin-MMTh) and the second one by adding clay slurry to the resin (Resin-MMTs). The properties of these two groups were compared with the control group (Bis-GMA/TEGDMA without clay). The mechanical properties (flexural strength and flexural modulus) have been evaluated according to the established standards for dental materials application; the conversion degree of nanocomposites based on Bis-GMA/TEGDMA by FTIR; the state which the clay sheets are found in the polymer matrix has been studied by using XRD; the morphology of the nanocomposite has been analyzed with SEM and finally the interaction of the clay sheets with the polymeric matrix through rheological tests. For both groups (Resin-Na+Mth and Resin-Na+Mts) at a concentration of 0.3 phr and 0.2 phr, in the flexural strength tests, there were increases of 34% and 41% respectively, without significant variation in the flexural modulus. The values obtained for the conversion degree were higher with respect to the control at all concentrations and both groups except at a concentration of 0.4 phr for Resin-MMTh, this fact showed that there was no interference of the clay particles in the polymerization process. For a concentration of 0.3 phr for the two groups the results obtained by XRD showed that Bis-GMA/TEGDMA is able to penetrate the interlaminar space of the MMT causing a polymer intercalation in the clay sheets. Scanning electron microscopy exhibited a rough surface for the two groups containing 0.3 phr of clay, indicative of an electrostatic interaction between the nanoparticle and the polymer. Similarly, rheological tests also showed that there is an interaction betw

pps proceeding
Symposium: S09 - Polymer nanocomposites
Poster Presentation

EFFECT OF THE PRESENCE OF SILICON NANOPARTICLES IN THE COEFFICIENT OF LINEAR THERMAL EXPANSION OF MAIZE STARCH

Jiménez Calvo Mauricio (1), Reyes Samilpa Alicia (2)*, Chávez Gutiérrez Miguel (3), Hernández Martínez Ángel Ramón (4)

(1) Instituto Tecnologico de Oaxaca - Oaxaca de Juárez - Mexico, (2) CIIDIROaxaca - Oaxaca de Juárez - Mexico, (3) CIIDIR, Oaxaca, CONACYT - Instituto Politecnico Nacional - Oaxaca de Juárez - Mexico, (4) CFATA Juriquilla, Universidad Nacional Autonoma de Mexico - Queretaro - Mexico

The nanocomposite materials have great advantages compared to other plastics and composites, because when the properties of the polymers with a nano-sized reinforcement component are combined there is an improvement in their specific properties. In the present work, maize starch was used as the matrix of the nanocomposite, a natural biopolymer that is exploited for its biodegradability and low costs. The nanocomposite specimens were obtained by the extrusion-injection method. For material extrusion DSM Xplore MC-5 equipment was used at 90 °C in the three heating zones for 3 minutes at 100 RPM. To obtain specimens, the DSM Xplore IM5.5 injector was used. Previously, corn starch was wetted with 20% by weight water for 24 hours, 10% by weight glycerol was used as plasticizer. The addition of silica nanoparticles (Aerosil-vs Nis-silice) to the maize starch base structure was done to improve its mechanical and thermal properties. Different formulations were prepared by varying the amount of filler to be obtained, these being 2.5, 5 and 7.5% by weight. To evaluate the effects on the dimensional changes of the test specimens due to exposure to different temperatures, a thermomechanical analysis was performed to determine the linear coefficient of thermal expansion and the glass transition temperature in a TMA Q400, TA Instruments equipment under the standard ASTM E831-14.Mean values of thermal expansion coefficient were 0.0598 μ/mm °C for pure corn starch based plastic, and 0.0580, 0.1104 and 0.2196 μ/mm °C for 2.5%, 5% and 7.5% silica nanoparticles concentration on the corn starch base matrix. Glass transition temperature mean values were 55 °C for pure corn starch based plastic, and 51 °C, 45 °C and 39 °C for 2.5%, 5% and 7.5% silica nanoparticles concentrations, respectively.

Hydrocarbon Resins and compatibilizing agents to improve dispersion and mechanical properties in halloysite/ethylene–propylene copolymer nanocomposites

Bischoff Eveline [1]*, Simon Douglas [1], Francisquetti Edson [2], Liberman Susana [3], Mauler Raquel [4]

[1] IFRS - UFRGS - Farroupilha RS - Brazil, [2] IFRS - Farroupilha RS - Brazil, [3] UFRGS - Farroupilha RS - Brazil, [4] UFRGS - Porto Alegre RS - Brazil

The properties of polyolefin nanocomposites strongly depend on the dispersion level of the nanoparticles and the addition of dispersion agents has contributed to improvement in their performance. In this study, the morphology and the mechanical properties of heterophasic ethylene-propylene copolymer/halloysite nanocomposites were tailored by using two hydrogenated hydrocarbon resins: 90% and 100% hydrogenated and two compatibilizing agents: poly(propylene-g-maleic anhydride) and poly(ethylene-octene-g-maleic anhydride). The transmission electron microscopy indicated that the best dispersion of the halloysite nanotubes was achieved when hydrogenated hydrocarbon resins and poly(propylene-g-maleic anhydride) were used simultaneously. All nanocomposites showed an increase in mechanical stiffness and the most pronounced increase of 46% in the Young modulus was achieved with the system containing halloysite, poly(propylene-g-maleic anhydride) and the hydrocarbon resin with the higher degree of hydrogenation. Poly(ethylene-octene-g-maleic anhydride) caused the halloysite nanoparticles to concentrate preferentially in the rubber domains, wherein these hindered the crystallization of polypropylene and polyethylene chains, as showed by atomic force microscopy. In this case, the composite exhibited both high stiffness and improved toughness. These results highlight the key role of dispersion agents in promoting a good balance in the mechanical properties of resulting nanocomposites based on halloysite particles and heterophasic ethylene-propylene copolymers.

PREPARATION OF POLYMERIC NANOCOMPOSITES WITH GRAPHENE VIA MELT BLENDING AND THEIR RHEOLOGICAL EVALUATION

Iniestra-Galindo Maria Guadalupe (1)*, PEREZ-GONZALEZ JOSE (1), MARIN-SANTIBAÑEZ BENJAMIN MARCOS (2)

(1) INSTITUTO POLITECNICO NACIONAL, ESFM - CIUDAD DE MEXICO - Mexico, (2) INSTITUTO POLITECNICO NACIONAL, ESIQIE - CIUDAD DE MEXICO - Mexico

Polymers are used in many practical applications due to their versatility, low cost, easy processing and light weight. However, a drawback of polymers is their poor mechanical performance in structural applications. One way to overcome this limitation is by blending them with reinforcing materials to obtain a composite. Recently, the discovery and development of nanostructured materials as carbon nanotubes and graphene and their use as reinforcing materials in polymeric matrices have allowed the development of new polymeric nanocomposites materials with improved mechanical properties. One inexpensive method to prepare polymeric nanocomposites is the mixing and dispersion of the nanometer size particles in a molten polymeric matrix by using a batch mixer or twin screw extrusion. In this work, polymer nanocomposites of polypropylene with different concentrations of reduced graphene oxide, pure as well as with a functionalizing agent, were prepared and the rheological behavior under shear flow of the molten matrices as well as that of the nanocomposites was evaluated by using capillary rheometry and torsional flow. The changes in the rheological behavior of the nanocomposites as compared to the matrices are presented, and discussed in relation to the graphene contents, its functionalization and dispersion.

Biodegradable Starch-Based Nanocomposites Incorporated with Cellulose Nanofibers as a Reinforcing Agent

H. Tibolla[a], T. Sartori[a], F.M. Pelissari[b], F.C. Menegalli[a]

1 Department of Food Engineering, School of Food Engineering, University of Campinas, Campinas – SP, Brazil
2 Institute of Science and Technology, Food Engineering, University of Jequitinhonha and Mucuri, Diamantina – MG, Brazil

Abstract. The interest in biodegradable materials has motivated the use of natural resources capable to produce continuous matrices in order to develop composites. Natural fibers (cellulose source) of nanometer size had been studied as a reinforcing material when incorporated to a starch matrix, being an effective method to obtain high-performance bionanocomposites. Cellulose nanofibers (CNFs) were isolated from unripe banana peel using acid treatment (H_2SO_4 10%) and mechanical treament. Banana starch biodegradable films (CF) were produced by casting method and the effect of the CNFs incorporation to starch matrix (CNFF), was investigated. Scanning electron microscopy (SEM) presents topography and cross-section, showing that the structure of film was modified with CNFs incorporation. The starch matrix was added with CNFs, which are well dispersed in the matrix and did not agglomerate. But, it can be identified a non-homogeneous structure, suggesting that CNFF presents some open and more porous structure zones than the CF. The CF film showed similar density ($p>0.05$) and higher moisture and water solubility when compared with CNFF film ($p<0.05$). CNFF film presented a similar sorption coefficient and water content at equilibrium values compared with CF. With the CNFs incorporation the water diffusivity value was reduced however with no statistical significance ($p > 0.05$). The water vapor permeability decreased significantly ($p < 0.05$) when the starch matrix was reinforced with cellulose nanofibers. CNFF film was more rigid due to the greater tensile strength and Young's modulus. While films without CNFs were more flexible as a consequence of their higher elongation capacity. The nanocomposite has potential use in food packaging, since the use of CNFs can promote improvements on barrier and mechanical properties. Cellulose nanofibers isolated from agro-industrial residue (banana peels) offer the potential to reinforce composites of biodegradable polymers, producing a value-added material.

Keywords: cellulose, natural fibers, films, biodegradable polymers
PACS: S09-443

INTRODUCTION

Recently, considerable research effort has been undertaken to develop new materials with low environmental impact. In particular, composites and nanocomposites of biopolymers with natural fillers seem to be very promising for the development of environmentally friendly materials [1]. Biodegradable films, when used as packaging materials, can prevent the moisture migration, protect foodstuffs during transportation, avoiding mechanical damage and preserving their physical integrity. In addition, they function as a modified atmosphere package that promotes the conservation of sensory characteristics and prolongs the shelf life of products [2]. The starch-based composites have great commercial potential, however, they still present some weaknesses in the thermal, mechanical and barrier properties [3]. In the preparation of biodegradable films, the nanometric fibers of plants (cellulose source) have been exploited as an alternative to reinforce the polymer matrix, since they have many advantages offered when incorporated into composites, besides being renewable and biodegradable [4]. The development of the composites with improved properties made from polymers reinforced with natural nano-based filler material has received special attention in recent studies [5]. Because natural sources can be used as a substitute for synthetic materials, the use of agricultural waste is a trend in this area, besides of their annual renewability, it can be a valuable source of cellulose nanofibers. Since the banana peel is rich in cellulose, this material is potentially applicable as reinforcing component in high-performance composites, increasing its commercial value and providing a purpose for this byproduct [4].

Our previous studies showed that chemical treatment followed by high-pressure homogenization successfully isolates cellulose nanofibers from the banana peel [6]. The present work aimed at studying of the potential use of cellulose nanofibers obtained from banana peel by chemical treatments as reinforcing agents in starch polymer composites. The nanocomposites containing were produced and characterized with respect to their structure, mechanical and barrier properties.

MATERIALS AND METHODS

Nanocomposites Production

The banana starch isolation and the production of the biodegradable films were produced by the casting method according to the methodology adapted from Pelissari et al. [7]. The fruit, unripe banana from variety "Terra" (*Musa paradisiaca*), was obtained from the southeastern region of Brazil, and was not subjected to any postharvest treatment. CNFs were isolated from unripe banana peel by chemical treatments, according to the method described by Tibolla et al. [6]. Acid hydrolysis was performed with sulfuric acid (H_2SO_4 10%, v/v) followed by a homogenization at high pressure for 5 times (pressure in the first and second stages was 500 and 50 bar, respectively). Table 1 summarizes the characteristics of the CNFs (NM 10%) used in this study. The first step of the nanocomposites production involved the dispersion of CNFs suspension (5 g of nanofibers/100 g of starch) by magnetic stirring for 15 min, followed by homogenization with an Ultra-Turrax (5.000 rpm for 15 min). The CNFs suspension was added to a water solution of 4 % (w:w, dry basis) of banana starch previously stirred for 30 min. The mixture was homogenized by mechanical stirring for 30 min, followed by heating up to the process temperature (81 °C) under gentle stirring. Glycerol (25 g of glycerol/100 g of starch) was added at this point, and the solution was maintained at this temperature for 15 min. After that, the film-forming suspension (FFS) was sonicated for 10 min and 70.4 g of the FFS was poured onto acrylic plates (18 cm x 21 cm) to controlling the thickness of the nanocomposites. The films were dried in a chamber with air circulation under controlled temperature (54 °C) and relative humidity (RH) (48 %) for 4 h. A film without CNFs (control) was produced for comparison purposes. Before characterization, the films were conditioned in desiccators under 58% RH, at 25 °C, for 48 h. The produced films were designated as: CF = control film without CNFs and CNFF = films containing 5% (w/w starch) CNFs.

TABLE (1). Characteristics of the CNFs added to the films, according to Tibolla et al. [6].

CNFs	Length (L) (nm)	Diameter (d) (nm)	Aspect ratio (L\d)	Zeta potential (mV)	Crystallinity index (%)
	310.77	2.89	107.25	-53.2	66.4

Nanocomposites Characterization

Microstructure by Scanning Electron Microscopy (SEM)

The surface and cross-section of the nanocomposite films were analyzed by SEM for microstructure evaluation. The dried film samples were fractured with the help of liquid nitrogen, and small fragments were obtained. Samples of these fragments were fixed on aluminum stubs by means of a double-sided tape and were then coated with a layer of gold (Sputter Coater POLARON, model SC7620), to improve conductivity. The coated samples were viewed under a scanning electron microscope (LEO, model LEO 440i, Cambridge, England) operating at an acceleration voltage of 15 kV.

Thickness, Density and Mechanical Properties

Film thickness was measured using a digital micrometer (Digimatic Micrometer Series 293 MDC-Lite, Mitutoyo Corporation, Japan). The mean thickness of each film was determined from an average of 10 random measurements. To determine film density, 20 x 20 mm squares samples were cut and their thickness was measured (three random measurements). The film samples were dried at 105 °C for 24 h and weighted. The density of the films was calculated as the ratio between the film weight and film volume (film thickness x film area). The density experiments were accomplished in triplicate, and the data are reported as mean values.

The film mechanical properties were investigated using a texture analyzer (Stable Micro Systems, model TA.TXplus, Surrey, UK, England), according to the standard method D882-02 [8], by taking an average of six determinations for each film sample. The samples were cut into 20 x 80 mm strips by means of a scalpel. The initial grip separation and crosshead speed were set at 40 mm and 1.0 mm/s, respectively.

Moisture Content and Water-Soluble Fraction

The film moisture content was analyzed by gravimetry, in triplicate, by drying the samples at 105 °C for 24 h [9]. The film water-soluble fraction (S) was determined by employing the methodology described by Gontard et al. [10]. Three discs (diameter= 20mm) of each film were stored in a desiccator containing silica gel (~0% RH) for 48 h. The film samples were immersed into 50 mL of water containing sodium azide (0.2 g/L) at 25 °C for 24 h, under sporadic agitation. The solution containing the samples was filtered, and insoluble matter was dried at 105 °C for 24 h. Analyses were carried out in quadruplicate, and the film water-soluble fraction (*S*, %) was calculated by means of difference of the initial dry sample weight (g) and the dry resulting material weight (g).

Water Vapor Permeability (WVP)

WVP was determined by gravimetry [11], with modifications. The film samples (diameter = 0.06 m) was placed on the circular opening (area = 0.00196 m^2) of a permeation cell and was sealed with sealant ring, to ensure that humidity migration would occur only through the film. The interior of the cell was filled with saturated $MgCl^2$ saline solution (33% RH) and the system was stored in a desiccator containing saturated $NaNO^2$ solution at 25 °C (64% RH). The weight gain was monitored every 30 min, for 8 h. The analyses were conducted in triplicate and, WVP was calculated on the basis of Equation 2 and expressed in g/(m.s.Pa). Where: w/t is the slope of the line of weight gain (w) as a function of time (t) (g/s); δ is the mean sample thickness (m); A is the sample permeation area (m^2); and ΔP is the difference in water vapor pressure through the sample for pure water at 25 °C (Pa).

$$WVP = \left(w.t^{-1}\right)\left[\delta.\left(A.\Delta P\right)^{-1}\right] \quad (1)$$

Water Uptake (WU)

The water absorption kinetics was determined according to the method described by Dufresne et al. [12]. Films with the dimensions of 20 x 20 mm and, with known thickness, were stored in a desiccator containing silica gel (~0% RH) for 48 h. After this period, the films were weighted (w_0) and conditioned at 25 °C in hermetically sealed flasks containing Na_2SO_4 saturated solution, to ensure a RH of 95%. The samples were removed at desired intervals and weighted (w_t) until equilibrium state was reached. WU was calculated by means of difference of the sample weight after time "t" of exposure to 95% RH (g) and the initial weight of the sample (g).

Statistical analysis

One-way analysis of variance (ANOVA) and a Tukey test of multiple comparisons with a significance level of 5% ($p \leq 0.05$) were run using a Statistic software (StatSoft Inc, Tulsa, Oklahoma, USA).

RESULTS AND DISCUSSION

Nanocomposites characterization

The SEM images obtained from the surface and cross-section of the films CF and CNFF are shown in Figure 1. Incorporation of cellulose nanofibers significantly changed the microstructure of the nanocomposites (CNFF) as compared with the control film (CF). The nanocomposite presented a non-homogenous structure with irregular surface, while the control film displayed a more uniform and smooth surface. The cross-section images indicate that the control film exhibited a more dense and homogeneous polymeric structure as compared with the CNFF that presented small cracks and fissures. Results demonstrated that the nanoparticles adhered well to the polymer matrix and did not agglomerate significantly, indicating structural integrity.

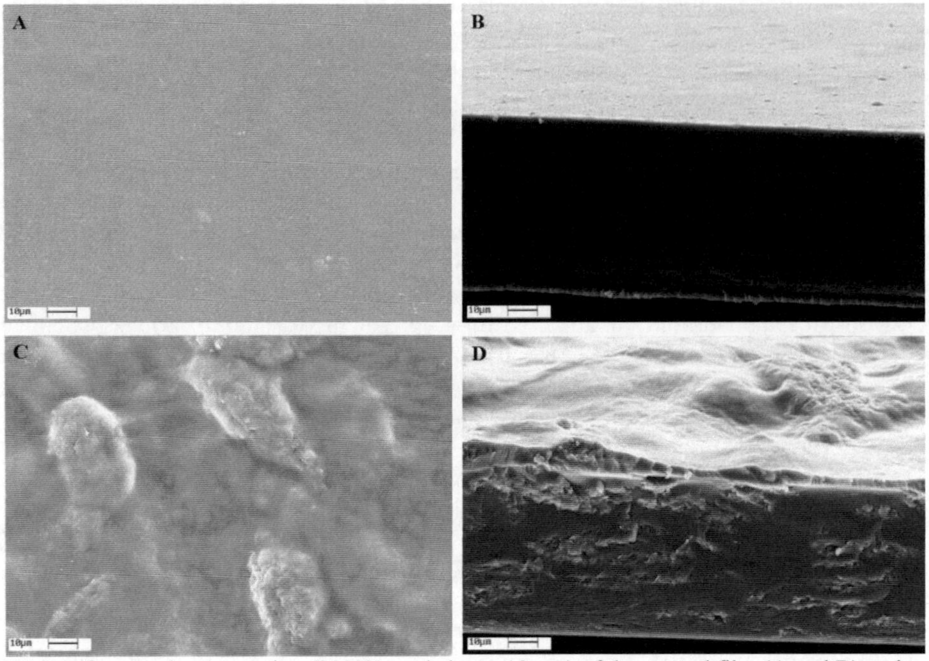

FIGURE 1. SEM surfaces and cross-section (2000X, scale bar = 10 μm) of the control film (A and B) and nanocomposite reinforced with CNFs (C and D).

The results of thickness, density, moisture and solubility in water of the produced films are presented in Table 2. The strict control of the film-forming during the casting procedure resulted in the formation of films with similar thickness values. The incorporation of cellulose caused a decrease in the density value of the CNFF in relation to CF, however there was no statistically significant difference ($p > 0.05$). SEM images (Figure 1) showed that the structure of the nanocomposites is more open, porous, and less dense than the structure of the control film.

The incorporation of CNFs also reduced the moisture content in nanocomposites, showing statistically significant difference ($p < 0.05$). This result can be explained by the lower water affinity of CNFs compared to the starch [13]. Because of the chemical composition of the starch (amylose and amylopectin), the produced films are considered hydrophilic and the CNFs incorporation into the starch matrix promote a reduction on water solubility ($p < 0.05$). This may be due to formation of the nanofibers network, which hinders water diffusion through the matrix, since strong hydrogen bonding interactions tend to stabilize the starch matrix when the nanocomposites are submitted to high humidity [10].

TABLE (2). Results of thickness, density, moisture and solubility in water of control film (CF) and nanocomposite film added with cellulose nanofibers (CNFF)

Film	Thickness (μm)	Density (g/cm^3)	Moisture (%)	Solubility in water (%)
CF	76.6 ± 0.1^a	1.10 ± 0.14^a	18.7 ± 0.1^b	31.6 ± 1.0^b
CNFF	78.3 ± 2.2^a	1.13 ± 0.01^a	15.1 ± 0.1^a	24.1 ± 0.8^a

Tensile strength, elongation at break and Young's modulus of the produced films are shown in Table 3. The tensile strength and Young's modulus of the nanocomposites increased with the addition of cellulose nanofibers, when compared with the control film ($p<0.05$), showing a tendency to form more rigid films. In this study, incorporation of the cellulose nanofibers affected the Young's modulus the most; this parameter increased >100% with the CNFs addition. This might be attributed to the formation of a fibrillar network within the polymer matrix and the interaction of the cellulose nanofibers through hydrogen bonds [14]. Besides, the high zeta potential of the nanofiber suspensions (-53.2 mV) preventing the formation of aggregates, giving more stable colloidal suspensions with individualized cellulose fibers, and promoting a strong network of nanofibers inside the material. The elongation at break indicates the flexibility and stretch of the film, which is determined at the point when the film breaks under tensile testing. Films without CNFs (CF) were flexible as a consequence of their higher elongation capacity. The lower elongation of the CNFF film might be due to the small size of the CNFs, that according to Pelissari et al. [4] increase the hardness and reduces the elongation at break of the films.

TABLE (3). Mechanical properties of control film (CF) and cellulose nanofiber film (CNFF)

Film	Tensile strength (MPa)	Elongation at break (%)	Young's modulus (MPa)
CF	8.9 ± 0.1^a	78.3 ± 6.9^b	283.7 ± 7.3^a
CNFF	16.7 ± 0.4^b	11.5 ± 2.5^a	$706.9 \pm 173.7.3^b$

Table 4 summarizes the results of WVP of films on a relative humidity gradient of 33%-64%, and the typical kinetic parameters for water uptake: water content at equilibrium ($M\infty$), diffusion coefficient (D), and sorption coefficient (S). The film incorporated with CNFs had lower WVP than the FC, showing statistical difference ($p < 0.05$). This may be due to difficulty of water vapor molecules in penetrating the cellulose crystalline region, since the tortuosity of the material was increased with the presence of the nanofibers [15]. These results are in accordance with the values of solubility in water.

The CNFF film presented lower water diffusivity (2.3×10^{-8} mm^2/s) than CF film (4.91×10^{-8} mm^2/s), but there was no statistical difference ($p > 0.05$). The water uptake at equilibrium (M_∞) and sorption coefficient (S) values were similar for both produced films ($p > 0.05$).

TABLE (4). Water vapor permeability (WVP) and kinetic parameters for water uptake: water content at equilibrium ($M\infty$), diffusion coefficient (D) and sorption coefficient (S), of control film (CF) and cellulose nanofiber film (CNFF)

Film	WVP (10^{-11} g/m.s.Pa)	$M\infty$ (%)	D (10^{-8} mm^2/s)	S
CF	11.4 ± 1.1^a	4.6 ± 0.1^a	4.91 ± 0.6^a	0.83 ± 0.02^a
CNFF	7.7 ± 0.2^b	4.3 ± 0.1^a	2.30 ± 0.1^a	0.83 ± 0.02^a

CONCLUSION

In general, the cellulose nanofibers improved the features of the starch-based material and are potentially applicable as reinforcing elements in a variety of polymer composites. The nanocomposites had a significantly improvement in their mechanical, barrier (water vapor permeability and solubility) properties compared to starch film without CNFs. The improvement of nanocomposite performance can be explained by the chemical similarities between starch and cellulose and the nanometer size effect of the fibers. This work presented a possible use of a material from an agricultural residue (high value material) as an innovative product for future applications in the food industry.

ACKNOWLEDGMENTS

The authors would like to acknowledge the financial support provided by Conselho Nacional de Desenvolvimento Científico e Tecnológico (CNPq) (477842/2011-9) and Coordenação de Aperfeiçoamento de Pessoal de Nível Superior (CAPES) (2952/2011). The authors would also like to acknowledge the Brazilian Nanotechnology National Laboratory (LNNano) for allocation of the TEM and AFM apparatus.

REFERENCES

1. A. Abdulkhani et al., *Polymer Testing* **35**, 73-79 (2014).
2. S. Mali et al. *Semina: Ciências Agrárias* **31**, 137-156 (2010).
3. N. Follain et al. *Journal of Membrane Science* **427**, 218-229 (2013).
4. F. M. Pelissari. *Journal of Colloid and Interface Science* **505**, 154–167 (2017).
5. M. R. de Moura et al. *Journal of Food Engineering* **104**, 154-160 (2011).
6. H. Tibolla et al. *Industrial Crops and Products* **95**, 664-674 (2017).
7. F.M Pelissari et al. *Food Hydrocolloids* **30**, 681-690 (2013).
8. ASTM – D882-02. In *Annual Book of ASTM Standards*. ASTM: Philadelphia, 2002.
9. ASTM – D644-99. In *Annual Book of ASTM Standards*. ASTM: Philadelphia, 1999.
10. N. Gontard et al. *Journal of Food Science* **57**, 190-195 (1992).
11. ASTM – E96-00. In *Annual Book of ASTM Standards*. ASTM: Philadelphia, 2000.
12. A. Dufresne et al. *Journal of Applied Polymer Science* **76**, 2080-2092 (2000).
13. M. O. Müller et al. *Food Hydrocolloids* **23**, 1328-1333 (2009).
14. W. Helbert, et al. Polymer Composites **17**, 604-611 (1996).
15. N. Lavoine et al. *Carbohydrate Polymers* **90**, 735-764 (2012).

Melt Mixed PP/MWCNT Composites: Influence of CNT Incorporation Strategy and Matrix Viscosity on Filler Dispersion and Electrical Resistivity

Petra Pötschke [1, a)], Beate Krause, [1, b)] Fanny Grugel [1,2]

[1]Leibniz-Institut für Polymerforschung Dresden e.V., Hohe Str. 6, 01069 Dresden, Germany.
[2] Technische Universität Dresden, 01062 Dresden, Germany

[a)]Corresponding author: poe@ipfdd.de
[b)]krause-beate@ipfdd.de

Abstract. Aiming answering the question, if masterbatch (MB) dilution is more effective on nanofiller dispersion than direct nanotube incorporation, small-scale melt mixing was performed for composites based on polypropylenes (PP) and 0.5 -7.5 wt.% multiwalled carbon nanotubes (MWCNT). Composites with 2 wt.% MWCNTs in 5 different PP types were compared when produced by both methods. After determination of the specific mechanical energy (SME) input the mixing time in direct incorporation was adapted to result in comparable SME values than in the masterbatch dilution approach. Interestingly, despite better dispersion when using the MB approach the electrical resistivity of samples with 2 wt.% MWCNT was higher than in direct incorporation, indicating a possibly more pronounced MWCNT shortening in this two-step procedure. In summary, this study on PP suggests that the masterbatch approach is very suitable for the dispersion of MWCNTs and results in advantages for nanotube dispersion, whereby the only slightly higher electrical resistivity may present an acceptable compromise between both properties.

INTRODUCTION

The state of dispersion of carbon nanotubes (CNTs) in melt mixed composites influences as well electrical as also mechanical properties. If nanotubes remain undispersed they cannot contribute in a suitable way to their network formation needed for electrical percolation. Remaining agglomerates also may cause problems in shaping processes and act as stress concentrators inducing earlier break under mechanical load. Thereby, melt mixed composites with CNTs can be produced by different approaches. On the one hand side, direct incorporation (DI) of CNTs is used in many scientific studies and influences of processing conditions on the state of dispersion and electrical and other properties have been studied. It was shown that the specific mechanical energy (SME) input significantly influences the state of dispersion. In general, with increasing SME the macrodispersion of the CNTs is improved [1-5]. The findings about the effect of SME on electrical resistivity are not consistent, with increasing SME decrease as well as increase was reported [1, 2, 5]. In addition it was found that CNTs are shortened during melt processing and that CNT length decreases with SME, which also has consequences on the electrical resistivity [4, 5]. However, the direct CNT incorporation method includes direct handling of the CNTs, which requires appropriate safety precautions and is not preferred in industrial applications. On the other side, the masterbatch (MB) dilution approach is quite common, especially in industrial applications. Here, the masterbatch having concentrations of CNTs between 10 and 25 wt% guaranties that the nanotubes are bound with the polymer and the handling is much easier. In the dilution process the masterbatch matrix should be miscible with the diluting polymer and the already existing nanotube network structure should be extended by incorporation of the additional polymer chains [6]. Due to the pre-existence of a CNT network and the two-step procedure better nanotube dispersion than in direct incorporation is expected. It is a disadvantage of the use of industrial produced masterbatches that in most cases the used grade of the polymer matrix and the nanotube type are not known. Thus, not in all cases good dispersion was achieved, as e.g. illustrated in a study of Mičušík et al. [7] where MB islands in different polypropylene matrices were found. In the rare published studies comparing DI with the MB approach better dispersion and lower electrical resistivity were observed for twin-screw extruded and compression molded MB samples [8]. However, in this study SME was not regarded so that a direct comparison is not appropriate.

Therefore, in this study polypropylene (PP) based composites with 0.5 -7.5 wt.% multiwalled carbon nanotubes (MWCNT) Nanocyl™ NC7000 were prepared by melt mixing using a DSM Xplore 15ccm co-rotating microcompounder. A comparison was made between the efficiency of MB dilution and direct nanotube incorporation on nanotube dispersion and electrical resistivity on three PP types differing in the molecular weight. In addition, two maleic anhydride (MA) modified PPs were used. To compare the two incorporation procedures, the specific mechanical energy (SME) during processing in the microcompounder was recorded and hold similar by adapting the mixing time in direct incorporation.

EXPERIMENTAL

Commercially available MWCNTs (Nanocyl™ NC7000 (Nanocyl S.A., Sambreville, Belgium) with a carbon purity of 90 %, an average diameter of 9.5 nm and an average length of 1.3 µm [9] were used. Two grades of a maleic anhydride (MA) modified polypropylene (PP), Orevac® PPC and Orevac® 18732 with different melt viscosities were selected. In addition three grades of unmodified polypropylene with similar viscosity behavior were used, namely Moplen HP501H, a PP matrix material which is used for CNT filled masterbatches from Hyperion Catalysis International Inc. (unknown trade name, marked as PPH) and Moplen HP400R. Some properties are given in Table 1.

TABLE 1. Properties of the five polymer matrices

trade name	notation	manufacturer	MFI [g/10 min]	MA-content* [wt.%]
Moplen HP501H	PP-h	LyondellBasell (Netherlands)	2.1	-
Orevac® PPC	PP-h-MA	Arkema Group (France)	2.0	0.14
PPH	PP-m	-	11.8	-
Orevac® 18732	PP-m-MA	Arkema Group (France)	6 – 10	0.13
Moplen HP400R	PP-l	LyondellBasell (Netherlands)	25	-

* according to [7]

For preparing composites containing 1 – 6 wt.% CNTs and the masterbatches (7.5wt.% CNTs) for the further diluting step, a DSM Xplore (The Netherlands) microcompounder with a filling volume of 15 cm^3 and co-rotating twin-screws was used. Polymer and MWCNTs were added simultaneously via hopper into the running compounder. After a mixing time of 5 min at 250 rpm and 210°C composite strands were extruded. The masterbatch dilution step was performed under the same conditions. To measure the electrical resistivity compression molded plates (thickness 0.5 mm, diameter 60 mm) were prepared with a press PW40EH (Paul-Otto Weber GmbH, Remshalden, Germany) at 210°C applying a pressure of 50 kN for 2 min.

Based on the force values recorded during the compounding, the specific mechanical energy SME was calculated using the following equation as introduced by Krause et al. [1]:

$$SME = \frac{2\pi Nr}{m}\int_{t_{Start}}^{t_{End}} Fdt \qquad (1)$$

where N is the rotation speed of the microcompounder (250 rpm), r is the distance from rotation axis (0.002 m), t is the mixing time (5 min) and F is the force, which is recorded. The SME used in direct incorporation is named SME_{direct}. The SME applied on the composites produced by the masterbatch dilution approach (SME_{MB}) was calculated by composing the energy part introduced in the corresponding masterbatch material mass used in the dilution and the energy part introduced during the diluting process:

$$SME_{MB} = \left(\frac{2\%}{7.5\%} SME_{MBP}\right) + SME_{MBD} \qquad (2)$$

where SME_{MBP} is the SME applied in the masterbatch production (corresponds to SME_{direct} for 7.5 wt.%) and SME_{MBD} the SME applied in the masterbatch dilution process from 7.5 wt.% to 2 wt.%.

In order to make the comparison between MB dilution and DI at similar mechanical energy inputs ($SMEs$), an adapted mixing time $t_{adapted}$ (applied in full minutes) was calculated using the following equations:

$$\frac{SME_{MB}}{SME_{direct}} = \frac{t_{adapted}}{t_{direct}}, \qquad t_{adapted} = \frac{SME_{MB} * t_{direct}}{SME_{direct}} \qquad (3), (4)$$

where t_{direct} is 5 min and the before calculated values of SME_{MB} and SME_{direct} were used.

Melt rheological measurements on the neat PPs were performed at 210°C under nitrogen atmosphere using an ARES oscillation rheometer (Rheometrics) with parallel plate geometry (diameter 25 mm, gap ca. 1 mm) in dynamic frequency sweeps (strain 10 %) with increasing and decreasing frequency (between 0.05 and 100 rad s^{-1}) and the latter sweep was used for interpretation. Fig. 1 left shows the obtained viscosity curves, illustrating the differences in viscosity and the comparability of PP-m and PP-h without and with MA.

The state of MWCNT macrodispersion was studied using transmission light microscopy (LM) (Olympus BH2 with a camera DP71 both from Olympus Deutschland GmbH, Hamburg, Germany) on thin sections of 10 µm thickness. These were prepared from granules with 2 wt.% MWCNTs using a Leica RM 2155 microtome (Leica Microsystems GmbH, Wetzlar, Germany) equipped with a hard metal knife at room temperature. The agglomerate area fraction A_A was calculated from LM images using 6 cuts, with a total area of ca. 3.2 mm^2, of various granules using the digital image processing software ImageJ Version 1.43g. The area of agglomerates A was divided by the total area of the image A_0. Considered were all agglomerate areas resulting in circle equivalent diameters > 1 µm.

Electrical volume resistivities > 10^7 Ohm·cm were measured applying a Keithley 8009 Resistivity Test Fixture based on ring electrodes combined with a Keithley electrometer 6517A (open symbols). For resistivities < 10^7 Ohm·cm but > 10^3 Ohm·cm the Keithley electrometer 6517A was combined with a 4 – point test fixture (gold contact wires with a distance of 16 mm between the source and 10 mm between the measuring electrodes, filled symbols). Therefore, strips (5 mm x 25 mm x 0.5 mm) were cut from the plates. Values lower than 10^3 Ohm·cm were measured using the 4 – point test fixture in combination with the electrometer DMM 2000 (filled symbols).

RESULTS

In a first step, percolation sets were performed applying the direct mixing approach with a constant mixing time of 5 min for all PP materials. Based on LM investigations (Fig. 2, 2 wt.%) it is found that dispersion improves as lower the matrix viscosity. In a study on polycarbonate by Kasaliwal et al. [10] better dispersion at higher matrix viscosity was reported when using constant mixing conditions but worse considering comparable stress values. This indicates the importance of melt infiltration into the primary agglomerates as first step of the dispersion process which is more pronounced at lower matrix viscosity. Also matrices having MA result at similar melt viscosity in better CNT dispersion (lower A_A) than unmodified PP indicating the compatibilizing effect of the anhydride groups. For the unmodified PPs the nanotube percolation threshold (Fig. 1 right) is lower at lower matrix viscosity. This is accordance with previous reports on PP but also on other polymer matrices [10, 11]. MA-grafting increases the CNT percolation threshold for PP with high viscosity but decreases it in case of middle viscous PP. Thus, the reasons for higher percolation threshold at higher matrix viscosity seem to be the worse dispersion combined with an expected more intensive nanotube shortening.

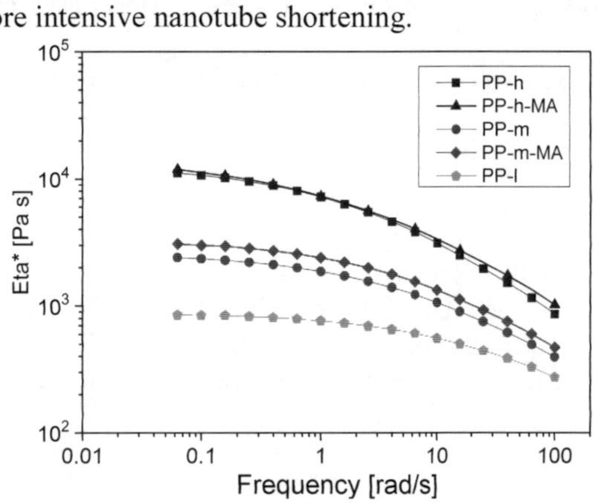

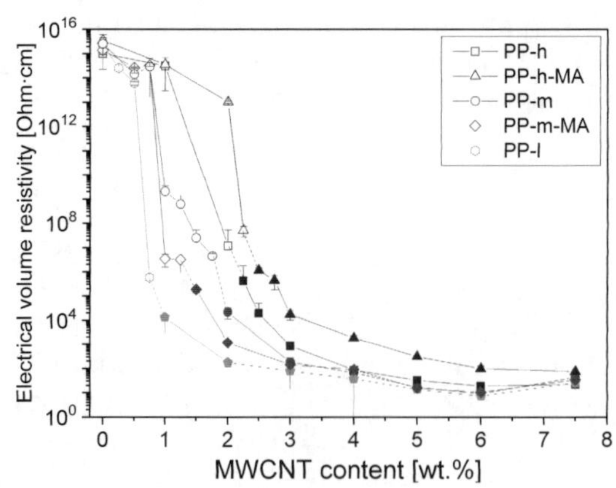

FIGURE 1. Left side: Complex viscosity |eta*| at 210°C of the polypropylene materials, right side: Electrical volume resistivity of the composites prepared by direct incorporation (5 min) of MWCNTs

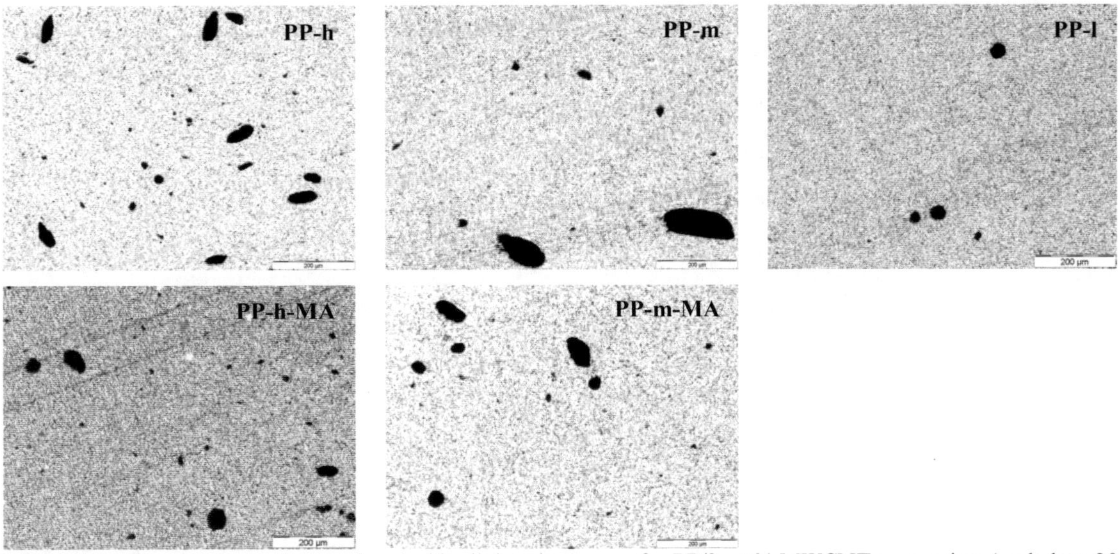

FIGURE 2. State of macro dispersion as observed by light microscopy for PP/2 wt.% MWCNT composites (scale bar: 200 μm)

TABLE 2. State of dispersion (agglomerate area ratio A_A) for composites at 2 wt.% MWCNT prepared using direct incorporation and masterbatch dilution (masterbatch with 7.5 wt.%, diluted to 2 wt.%); preparation at 210°C, 5 min, 250 rpm

Material	Direct incorporation A_A [%]	Direct incorporation SME [kWh/kg]	Masterbatch dilution A_A [%]	Masterbatch dilution SME [kWh/kg]
PP-h	3.9 ± 1.4%	1.44	1.2 ± 0.7%	1.69
PP-h-MA	2.1 ± 0.9%	1.45	0.8 ± 0.2%	1.80
PP-m	3.7 ± 2.2%	0.90	1.4 ± 1.0%	1.24
PP-m-MA	1.8 ± 0.7%	0.98	0.8 ± 0.2%	1.24
PP-l	1.6 ± 0.7%	0.73	0.5 ± 0.1%	0.93

TABLE 3. Calculated adapted mixing time, state of dispersion (agglomerate area ratio A_A) for composites with 2 wt.% CNT prepared using direct incorporation (210°C, adapted time, 250 rpm) and real SME measured under adapted conditions

Material	Adapted mixing time [min]	Direct incorporation adapted A_A [%]	Direct incorporation adapted SME [kWh/kg]
PP-h	6	2.9 ± 0.9%	1.62
PP-h-MA	7	2.1 ± 0.7%	1.90
PP-m	7	2.4 ± 1.0%	1.14
PP-m-MA	6	1.6 ± 0.3%	1.14
PP-l	8	1.0 ± 0.4%	1.05

In the second set, composites with 2 wt.% MWCNTs were produced using direct incorporation @ 5 min and masterbatch dilution as mentioned above. In all cases the two-step masterbatch dilution method resulted in significantly better macrodispersion independent of the PP viscosity and MA modification (see Table 2). However, when calculating the applied SME values also higher mechanical energies were applied. Thus, comparable SMEs were aimed by increasing the mixing time in direct incorporation (see Table 3). Even if this aim could not be completely achieved (still slight differences in SME between masterbatch diluting and direct incorporation with prolonged mixing time), some general tendencies can be observed. In all cases the MWCNT dispersion is better using the masterbatch dilution approach (Table 3, Fig. 3). Interestingly, despite better dispersion the electrical resistivity of samples with 2 wt.% MWCNT was higher when using the MB approach (Fig. 4 left), indicating a possibly more pronounced MWCNT shortening in this two-step procedure. The difference is highest for PP-h where also highest shortening may be expected. Furthermore, from Figure 4 (right) the general tendency of higher electrical resistivity at higher applied SME can be observed which corresponds to more pronounced CNT shortening with increasing SME [4, 11]. This tendency is independent on the preparation way of the composites.

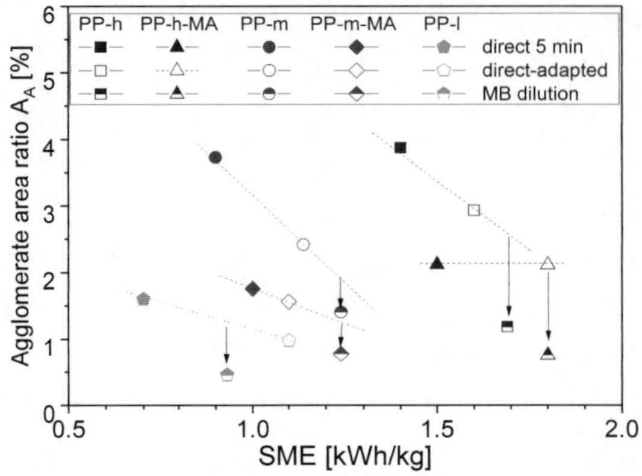

FIGURE 3. State of macro dispersion (A_A) vs. SME for PP/2 wt.% MWCNT prepared using different conditions

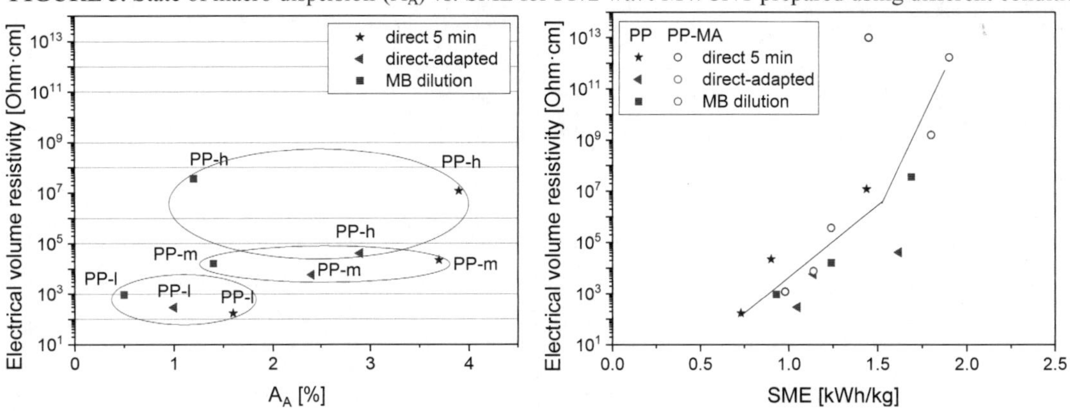

FIGURE 4. Electrical volume resistivity vs. state of macro dispersion (A_A) for unmodified PP/2 wt.% MWCNT composites (left); Electrical volume resistivity vs. SME for all PP/2 wt.% MWCNT, circles and lines are only for guiding the eyes.

SUMMARY

It could be shown on PP based composites with MWCNTs that, when performed with similar mechanical energy input, the masterbatch dilution technique resulted in better MWCNT macrodispersion than the direct incorporation. This finding was independent of the polypropylene matrix viscosity. In contrast, the electrical resistivity of compression molded samples with 2 wt.% MWCNT was slightly higher in case of low and middle viscous PP and significantly higher with high viscosity PP when using the masterbatch approach. In summary, this study on PP suggests that the masterbatch approach is very suitable for the dispersion of MWCNTs and results in advantages for nanotube dispersion in combination with good/acceptable electrical properties of melt mixed composites.

REFERENCES

1. B. Krause, P. Pötschke and L. Häußler, *Compos Sci Technol* **69** (10), 1505-1515 (2009).
2. G. R. Kasaliwal, S. Pegel, A. Göldel A, P. Pötschke P and G. Heinrich, *Polymer* **51,** 2708–2720 (2010).
3. T. Villmow, B. Kretzschmar and P. Pötschke, *Compos Sci Technol* **70**, 2045–2055 (2010)
4. B. Krause, M. Mende, G. Petzold, R. Boldt and P. Pötschke, *Chem Ing Tech* **84** (3), 263-271 (2012).
5. P. Pötschke, T. Villmow, B. Krause, *Polymer* **54,** 3071-3078 (2013).
6. G. Kasaliwal, T. Villmow, S. Pegel, P. Pötschke, chapter 4 in: *Polymer - carbon nanotube composites: Preparation, properties and applications*, edited by T McNally, P Pötschke (Woodhead Publishing Limited, 2011), pp. 92-132.
7. M. Mičušík, M. Omastová, I. Krupa, J. Prokeš, P. Pissis, E. Logakis, C. Pandis, P. Pötschke and J. Pionteck, *J Appl Polym Sci* **113** (4), 2536-2551 (2009).
8. M.T. Müller, B. Krause, B. Kretzschmar,and P. Pötschke, PPS Conference, **2017**, Dresden, Germany, S08-352.
9. B. Krause, R. Boldt and P. Pötschke, *Carbon* **49**, 1243-1247 (2011).
10. G. R. Kasaliwal, A. Göldel, P. Pötschke, G. Heinrich, *Polymer* **52** (4), 1027-1036 (2011).
11. R. Socher, B. Krause, M.T. Müller, R. Boldt and P. Pötschke, *Polymer* **53** (2), 495-504 (2012).

Polycaprolactone/Organoclay Biodegradable Nanocomposites: Effect of Different Clay Modifiers

Romina P. Ollier[a], Matias R. Lanfranconi[a], Vera A. Alvarez[a], Leandro N. Ludueña[a]

[a]Materiales Compuestos Termoplásticos (CoMP), Instituto de Investigaciones en Ciencia y Tecnología de Materiales (INTEMA), CONICET - Universidad Nacional de Mar del Plata (UNMdP). Av. Colón 10890, Mar del Plata, 7600, Argentina. E-mail: rominaollier@fi.mdp.edu.ar

Abstract. In this work, biodegradable nanocomposites based on polycaprolactone (PCL) reinforced with 2.5, 5.0 and 7.5 wt.% of two different clays, a commercial organo-clay (Cloisite 20A, C20A) and a laboratory modified bentonite with tributylhexadecyl phosphonium bromide (bTBHP), were prepared by melt intercalation followed by compression molding. The study contemplates the analysis of morphological (X-Ray Diffractometry, XRD, Scanning Electron Microscopy, SEM, and Transmission Electron Microscopy, TEM), rheological, thermal (Differential Scanning Calorimetry, DSC, and Thermogravimetrical Analysis, TGA) and mechanical properties (tensile tests), which are important properties for packaging applications.

In previous works, we concluded that higher clay dispersion degree inside the PCL matrix is expected when clays with large interlayer distance, strong hydrophobicity and strong processing stability are used. In the present work the opposite result was obtained. Although the phosphonium treated clay (bTBHP) showed the largest interlayer distance (d001), strongest hydrophobicity and the best processing stability, the clay dispersion degree inside PCL was worse than in the case of the alkylammonium treated clay (C20A). PCL/bTBHP nanocomposites showed weaker mechanical properties in comparison with PCL/C20A ones, which is in accordance with the morphological analysis. On the other hand, the thermal properties of the matrix were not substantially affected by clay incorporation in both nanocomposites

Keywords: nanocomposites, compatibility, biodegradable polymer, bentonite, montmorillonite.

INTRODUCTION

The development of environmental friendly, biodegradable, polymeric materials has attracted extensive interest in the last decades due to the pollution caused by the waste accumulation at the end of the life cycle of traditional polymer products. Polycaprolactone (PCL) is a synthetic aliphatic polyester which is fully biodegradable [1]. It has low glass transition (-60 °C) and melting (60 °C) temperatures [2]. Although PCL has been broadly used as matrix for the design of biodegradable nanocomposites for medical and packaging applications, in certain cases it is not competitive enough compared with conventional thermoplastics because some of its properties are limited [3]. The performance of this polymer can be greatly enhanced by the addition of a small amount (typically less than 10 wt.%) of nanometer-size inorganic fillers, such as layered silicates, producing nanocomposites [1, 4-6]. Smectite clays have increasingly attracted research interests for the preparation of polymer/clay nanocomposites in the past decade. Two important clays, due to environmental and economic relevance and mechanical and chemical resistance are bentonite (bent) and montmorillonite (MMT). These smectite clays of about 1 nm in thickness are the clays of choice for polymer nanocomposite production [6].

A key factor in the preparation of nanocomposites is to compatibilize the matrix and the filler, which will affect the nanostructure of the resulting material. In order to achieve the best properties, it is necessary to obtain a totally exfoliated structure, in which the individual clay platelets are completely and uniformly dispersed in the polymeric matrix. However, instead of fully exfoliated structures, intercalated structures (where the polymer chains are inserted between the platelets, increasing the interlayer distance, but without separating them completely) or a mixture of both, are generally achieved [7]. Hence, the characterization of the clay morphology (dispersion and distribution) inside the polymer matrix is crucial to understand these materials.

As most polymers are organophilic, the hydrophilic nature of clays hinders the exfoliation of the silicate layers. However, due to the rich intercalation and ion-exchange chemistry of bent and MMT, they can be organically modified in order to achieve a better compatibility with the matrix. The interlayer ions can be easily exchanged by organic ions, mainly ammonium or phosphonium cations having aliphatic or aromatic short or long alkyl chains, in order to produce an increment in the interlayer spacing (d_{001}) and reduce the hydrophilicity of the clay [8-11]. The resulting organo-modified clays are more compatible with hydrophobic polymers, which promotes interactions that facilitate the insertion of the polymer chains inside the clay galleries under defined processing conditions [8-11].

In previous works, we tried to explain the effect of several commercial organo-modified MMT [9,12-13] and a laboratory organo-modified bentonite [14] on the morphology and mechanical properties of PCL/clay nanocomposites prepared by melt intercalation and casting. The results suggested that the main characteristics responsible for the final morphology and properties of the PCL/clay nanocomposites are the hydrophobic/hydrophilic nature of the system, the initial basal spacing of the clay and the processing stability of the clay organo-modifiers.

The aim of this work was to characterize and compare morphological, thermal and mechanical properties of PCL/clay nanocomposites prepared by melt mixing. Two organically-modified clays were used: a commercially organo-MMT, modified with a quaternary ammonium salt (Cloisite 20A), and a bentonite containing a quaternary alkylphosphonium salt as modifier.

EXPERIMENTAL SECTION

Materials

PCL provided by Sigma Aldrich (Mn 80,000, Mn/Mw < 2, Glass Transition -60 °C, Melting Temperature 60 °C, Density 1.145 g/mL at 25 °C) was used as matrix. Cloisite® 20A (C20A, Southern Clay Products) and bentonite (bent, Minarmco S.A.) were used as fillers. The basal spacing of bent was 1.3 nm and its cation exchange capacity (CEC) was 0.939 meq.g-1 [29]. The bentonite was organo-modified with tributylhexadecylphosphonium bromide (TBHP) (Sigma Aldrich).

Samples Preparation

Modification of bentonite with TBHP [15]. Firstly, 2.5 g of bentonite were dispersed in 100 ml of deionised water. Then, an aqueous solution of the corresponding concentration of TBHP (0.9 times the CEC of the bentonite) was added to the suspension. The mixture was stirred for 4 hours at 70 ºC. After that, the suspension was filtered and washed several times with deionesed water until free of bromide. The wet organoclay, named bTBHP, was frozen for 24 hours and then lyophilized at 100 mTorr and −50 °C for 72 hours.

Preparation of nanocomposites. Neat polymer and nanocomposites with 2.5, 5.0 and 7.5 wt.% of each clay were prepared by melt-intercalation followed by compression-moulding. An intensive Brabender type mixer with two counter-rotating roller rotors was used. Mixing temperature was 100 ºC; speed of rotation was 150 r.p.m. and mixing time was 10 minutes. Compression moulding was carried out to prepare sheets in a hydraulic press for 10 minutes at 100 ºC. The thickness of the sheets was between 0.3-0.5 mm.

The samples will be referred as xbTBHP and xC20A, being x the wt.% content of bTBHP and C20A, respectively, in the nanocomposites.

Characterization techniques

X-Ray diffractometry (XRD) was performed using a PANalytical X'Pert PRO diffractometer equipped with CuKα radiation (λ=1.5406 Å) at a scanning speed of 1 °/min. The basal spacing distances (d_{001}) of the clays were calculated from the 2θ values using the Bragg's equation.

Thermogravimetric Analysis (TGA) measurements were carried out by using a TA Auto-MTGA Q500 Hi-Res, from 30 °C to 900 °C with a heating rate of 10 °C/min under air atmosphere.

Water absorption tests were carried out at 90 % relative humidity (RH). Samples were weighted at prefixed times and the absorption at each time was calculated as:

$$M_t(\%) = \frac{M_t - M_0}{M_0}.100 \tag{1}$$

where M_t is the mass of the sample at a time t and M_0 is the initial mass of the sample (dried).

Differential Scanning Calorimetry (DSC) measurements were performed in a TA Instruments model Q2000 calorimeter, from 25 to 100 ºC at a heating rate of 10 ºC/min under nitrogen atmosphere. The degree of crystallinity (Xcr) of each sample was calculated from the following equation:

$$X_{cr}(\%) = \frac{\Delta H_f}{w_{PCL} \times \Delta H_{100}} \times 100 \tag{2}$$

where ΔHf is the experimental heat of fusion, w_{PCL} is the PCL weight fraction and ΔH_{100} is the heat of fusion of 100 % crystalline PCL and its value is 136.1 J/g. The melting temperature (*Tm*) of each sample was also obtained.

Tensile tests were performed with a universal testing machine Instron 3369 at a constant crosshead speed of 50 mm/min. Samples were prepared according to the ASTM D882-91 standard.

Rheological tests were conducted in an Anton Paar rheometer under nitrogen atmosphere. Plate-plate geometry with a plate diameter of 25 mm was used. Samples were inserted and heated up to 80 °C. Low shear amplitude (2 %) was used in order to avoid the destruction of any stabilized clay structure and to work in the lineal viscoelastic regime. Data were taken for shear rates ($\dot{\gamma}$) in the range of 0.002 to10 s^{-1}. The melt rheology curves were fitted to the power law expression for $\dot{\gamma}$ in the range of 0.002-0.030 s^{-1}:

$$\eta = A_{Rh} \cdot \dot{\gamma}^{(n_{Rh})} \tag{3}$$

where the *Rh* subscript represents a rheology parameter, η is the complex viscosity; A_{Rh} a preexponential factor; $\dot{\gamma}$ the shear rate and *nRh* the shear thinning exponent. In the double logarithmic plot a linear zone at low shear rates ($\dot{\gamma}$ up to 0.030 s^{-1} for our both test conditions and polymer/clay system) can be seen. The n_{Rh} parameter was calculated from the slope of this region.

Scanning electron microscopy (SEM) measurements were conducted on a JEOL JSM-6460 LV equipment. The samples were fractured in liquid nitrogen.

Transmission Electron Microscopy (TEM) was performed at JEOL CX II using an acceleration voltage of 80 kV, to observe the dispersion of clay platelets within the polymer matrix.

RESULTS AND DISCUSSION

Physicochemical and structural characterization of the organoclays

Table 1 shows the chemical structure of both clay modifiers and summarizes the equilibrium water uptake percentage measured in a 90 % RH environment (M_{24h}), the interlaminar spacing (d_{001}) calculated by XRD, the organic content and the temperature for the maximum degradation rate (Tp) assessed by TGA.

TABLE 1. Characteristics of the organoclays used as reinforcement of PCL.

Clay	Chemical structure of the organic modifier	*Organic Content* (%)	d_{001} *(nm)*	M_{24h} *(%)*	*Tp*
bTBHP	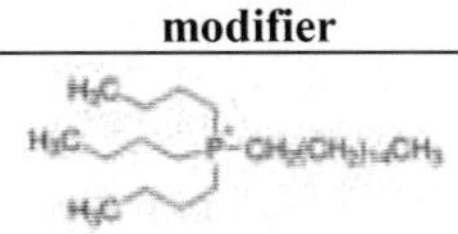	29	2.51	2.73	388
C20A	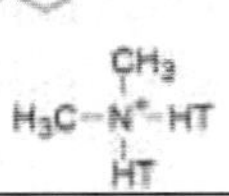	41	2.42	3.72	303

Regarding XRD characterization, C20A organoclay exhibits almost identical interlaminar spacing compared to the bTBHP organoclay. Hence, in both clays, the polymer chains could have enough space to penetrate between the clay platelets, which is important in order to obtain a well dispersed nanocomposite [26]. On the other hand, the bTBHP organoclay presents lower organic content than C20A, which suggests that the phosphonium salt was more effective improving the clay interlayer distance.

The moisture absorption after 24 hours at an environment with 90 % RH provides information about the bulk polarity of the organoclay, which is important to evaluate the compatibility with the host polymer matrix [26]. Taking into account that PCL is a hydrophobic polymer, both organoclays are good candidates for the preparation of PCL nanocomposites because of their low moisture content.

Another important parameter is the thermal stability of the organoclay. The thermal stability of alkylammonium treated organoclays, characterized by the Tp values, is in the range of processing temperatures of several traditional polymers. Phosphonium salts behave differently than their ammonium counterparts because of the greater steric tolerance of the phosphorus atom and the participation of its low-lying d-orbitals in the processes of making and breaking chemical bonds [10]. That is why bTBHP has a remarkable better thermal stability than C20A. In the present work, the processing temperature does not exceed 120 °C, thus, the thermal stability of both organoclays is adequate for the preparation of the PCL nanocomposites.

Characterization of PCL/organoclay nanocomposites

The XRD patterns of neat organoclays and PCL/clay nanocomposite samples are shown in **Fig. 1**. As it was previously mentioned, bTBHP and C20A show a 001 diffraction peak corresponding to an interlaminar spacing of 2.51 and 2.42 nm, respectively. All the prepared PCL nanocomposites containing clays with different chemical treatments, exhibited an increase of the d-value of the clay compared to the pure organoclays, which is an indication of intercalation of the polymer chains between the clay platelets [6]. This in turns indicates that regardless the organoclay concentration, PCL chains have intercalated into the layers of the bTBHP and C20A modified silicates [3]. The highest d_{001} value was observed at the highest clay content of each nanocomposite type. This result suggests that both clays do not agglomerate at 7.5 wt.% inside the PCL matrix in the processing conditions used in this work. The extent of intercalation cannot be quantified by this technique but it must be noted that the d_{001} values of C20A are 7%, 2% and 3% higher than bTBHP ones at 2.5 wt.%, 5.0 wt%, and 7.5 wt.% of clay content, respectively, suggesting highest clay dispersion degree for the C20A nanocomposites.

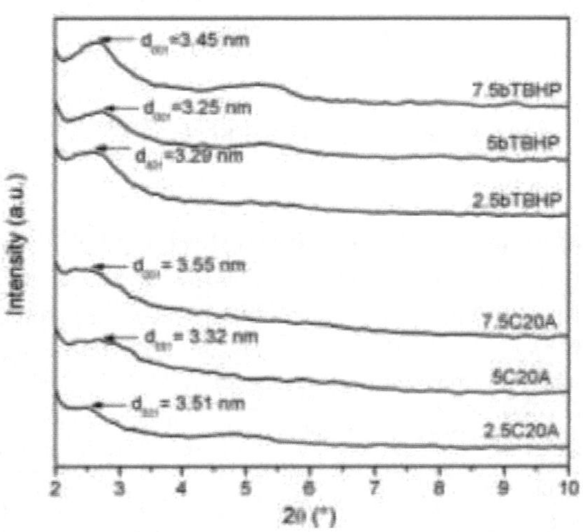

FIGURE 1. XRD spectra of PCL based nanocomposites.

Melt rheology can be useful to compare the clay dispersion degree of polymer/clay nanocomposites. This method has the advantages that the information obtained is representative of the whole sample instead of small zones of it and samples do not need to be specially conditioned [16]. Exfoliated or well-dispersed nanocomposites, including intercalated structures, increase the probability of edge to face and edge to edge interactions of the silicate platelets to occur, thus, it may build and mechanically stabilize mesoscale structures known as card-house [17]. These structures promote a pseudo-solid like rheological behavior at low shear rates which results in pronounced shear thinning behavior in that zone of the melt rheological curves. It has been reported that the n_{Rh} parameter (see eq.3) increases as a function of the clay dispersion degree [17]. **Fig. 2** shows the melt rheology curves and resumes the values of the n_{Rh} parameter of the neat matrix and their nanocomposites. All nanocomposites showed a pronounced shear thinning behavior at shear rates in the range of 0.002-0.030 s^{-1}. On the other hand, the obtained n_{Rh} values are in accordance with those reported in a previous work for similar polymer/clay systems [12]. The tendencies found for the n_{Rh} values are in accordance with the XRD results, the C20A nanocomposites displayed higher n_{Rh} values than the bTBHP ones at the same clay content.

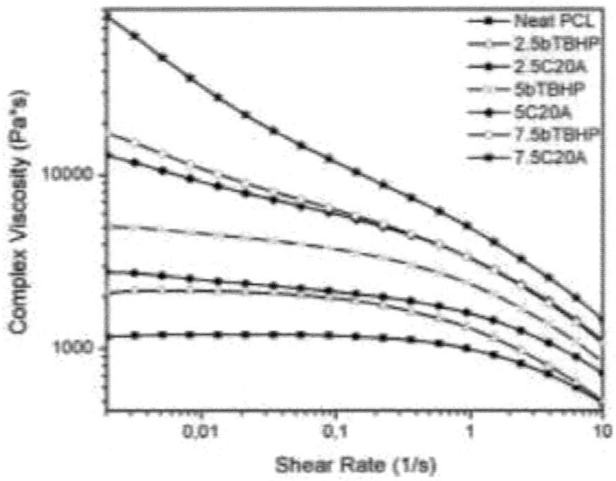

FIGURE 2. Melt rheology curves of neat PCL and PCL/clay nanocomposites.

Fig. 3 (a-c) shows the SEM/TEM micrographs of the pure PCL and their 5 wt.% nanocomposites. In the case of 5C20A, no evidence of clay agglomeration is observed. Small bumps with an average diameter of 200 ± 20 nm, whose can be associated to clay agglomerates, were found in the 5bTBHP nanocomposite (Fig. 3c).

FIGURE 3. SEM and TEM micrographs of the pure matrix and PCL/organoclay nanocomposites.

Better clay dispersion degree for the C20A nanocomposites than bTBHP ones was not expected since the latter showed the strongest hydrophobicity, largest initial basal spacing and the strongest thermal stability, which are factors that should promote clay dispersion degree inside the PCL hydrophobic matrix mixed by melt blending. This behavior can be attributed to the highest organic content of the C20A clay. Therefore, it can be concluded that the balance between basal spacing, hydrophobicity, thermal stability and organic content of the neat clays, and the processing route will control the final clay dispersion degree of the nanocomposites.

In **Table 2**, it can be noticed that both organoclays did not significantly affect the Tm of pure PCL, whereas the crystallinity degree of PCL slightly increased when the organoclays were added. This observation suggests that the presence of the clay platelets may act as nucleating sites for the PCL crystallization [3].

Moreover, from the observation of the $T_{d5\%}$ values in Table 2, it can be deduced that the thermal degradation of PCL is affected by the presence of both organoclays. The variations in the chemical composition of the organoclays was determinant for the degradation process observed in both composites. On the one hand, it is clear that the addition of quaternary alkylammonium modified clay within the PCL matrix, causes a reduction of the degradation temperature of the nanocomposites for all the clay loadings. This detrimental effect may be attributed to the fact that the layered silicates catalyze the PCL pyrolysis due to the presence of Lewis acidic sites formed by the degradation of the organic modifier [18]. On the other hand, in Fig. 5 b it can be noticed that the PCL/bTBHP nanocomposites exhibit slightly better thermal stability compared to pristine PCL. This can be attributed to the nature of the organic modifiers [10], which is in accordance with the thermal stability of clay alone reported in Table 1 and with the dispersion of each clay inside PCL matrix previously described in Figs.1 to 3.

TABLE 2. Thermal and mechanical properties of PCL matrix and nanocomposites with different clay contents.

Clay content (wt.%)	PCL/C20A							PCL/bTBHP						
	Xcr (%)	*Tm* (°C)	Residual mass (%)	$Td_{5\%}$ (°C)	E (MPa)	σ (MPa)	ε (%)	*Xcr* (%)	*Tm* (°C)	Residual mass (%)	$Td_{5\%}$ (°C)	E (MPa)	σ (MPa)	ε (%)
0	61.5	62.1	0.1	324	193.4 ± 6.0	18.2 ± 0.9	1070 ± 278	61.5	62.1	0.1	324	193.4 ± 6.0	18.2 ± 0.9	1070 ± 278
2.5	63.4	60.0	1.9	308	231.1 ± 7.7	18.6 ± 0.4	790 ± 119	65.7	61.0	1.6	334	260.5 ± 9.7	18.5 ± 0.3	582 ± 283
5.0	64.3	61.5	3.6	310	169.3 ± 14.8	16.8 ± 0.2	850 ± 166	65.0	60.7	3.5	333	286.5 ± 24.4	17.9 ± 1.8	231 ± 78
7.5	64.7	60.1	5.3	317	191.7 ± 7.5	15.3 ± 1.8	427 ± 178	63.6	61.0	5.3	323	312.8 ± 32.3	17.0 ± 0.7	705 ± 174

The tensile stress–strain curves for the pristine PCL and the nanocomposites are shown in **Fig. 4** a (for the PCL/C20A nanocomposites) and b (for the PCL/bTBHP nanocomposites), and the values of Young´s modulus (*E*), tensile strength (σ) and strain at break (ε) are given in **Table 2**. PCL/C20A nanocomposites showed higher Young´s modulus and tensile strength than PCL/bTBHP ones at the same clay content. The opposite trend was observed for the elongation at break. These results are a consequence of the improved clay dispersion degree obtained with the PCL/C20A nanocomposites; the correlation between clay dispersion degree and mechanical properties was widely demonstrated in the literature [6, 12]. The bTBHP organoclay showed the strongest hydrophobicity, the highest interlayer distance and the strongest processing stability, so better bTBHP dispersion inside the hydrophobic PCL matrix was expected but the opposite trend was observed; C20A showed better dispersion inside the PCL and hence improved mechanical properties. This result was attributed to the higher organic content of C20A.

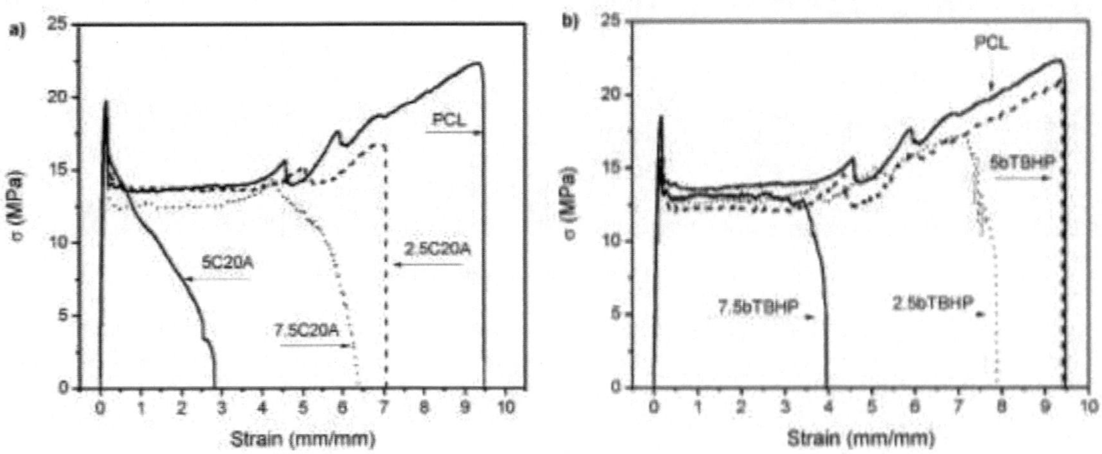

FIGURE 4. Tensile stress-strain curves of PCL matrix and nanocomposites with different contents of: (a) C20A; (b) bTBHP.

CONCLUSIONS

In the present work, the effect of the type of clay modifier on the structure and properties of polycaprolactone (PCL)/organoclay nanocomposites was studied. Two organoclays were compared: a commercially available montmorillonite organoclay, C20A and a laboratory modified bentonite bTBHP. The bTBHP organo-clay showed the strongest hydrophobicity, the highest interlayer distance and the strongest thermal stability. Even so, PCL/bTBHP nanocomposites displayed poor clay dispersion degree and hence lower Young´s modulus and tensile strength than the PCL/C20A ones. The Young´s modulus of the PCL increased 62% after the incorporation of 7.5 wt.% of C20A, while the tensile strength and elongation at break did not change significantly. The results showed that the hydrophobicity, the interlayer distance and the processing stability of the clays are not enough to predict polymer/clay compatibility. In this case, the higher organic content of C20A can be controlling the final clay dispersion degree. Therefore, a balance between basal spacing, hydrophobicity, thermal stability and organic content of the neat clays, including the processing route, will control the final clay dispersion degree of the nanocomposites.

ACKNOWLEDGMENTS

Authors acknowledge to National Scientific and Technical Research Council (CONICET) (PIP 00617), National Agency for the Promotion of Science and Technology (ANPCyT) (FS Nano 004), and the National University of Mar del Plata (UNMdP) (ING392/14) for the financial support.

REFERENCES

1. Kunioka, M.; Ninomiya, F. and Funabashi, M.; *Polym. Degrad. Stab.,* **92**, 1279 (2007).
2. Homminga, D.; Goderis, B.; Dolbnya, I. and Groeninckx G. *Polymer* **47**, 1620 (2006).
3. Yahiaoui, F.; Benhacine, F.; Ferfera-Harrar, H.; Habi, A.; Hadj-Hamou, A. and Grohens. Y.; *Polym. Bull.* **72**, 235 (2015).
4. Ludueña, L. N.; Alvarez, V. A. and Vazquez. A.; *Mater. Sci. Eng, A* **460-461**, 121 (2007).
5. Gorrasi, G.; Tortora, M.; Vittoria, V.; Pollet, E.; Lepoittevin, B.; Alexandre, M. and Dubois. P.; *Polymer,* **44**, 2271 (2003).
6. Alexandre, M. and Dubois. P.; *Mater. Sci. Eng. R Rep.* **28**, 1 (2000).
7. Fischer, H.; *Mater. Sci. Eng., C,* **23**, 763 (2003).
8. Picard, E.; Gauthier, H.; Gérard, J. F. and Espuche, E.; *J. Colloid Interface Sci.,* **307**, 364 (2007).
9. Xi, Y.; Frost, R. L. and He, H.; *J. Colloid Interface Sci.,* **305**, 150 (2007).
10. Xie, W.; Xie, R.; Pan, W. P.; Hunter, D.; Koene, B.; Tan and L. S. R.; Vaia. *Chem. Mater.,* **14**, 4837 (2002).
11. Bergaya, F.; Jaber, M. and Lambert, J.-F., in *Rubber-Clay Nanocomposites*; (Ed: Galimberti, M.), John Wiley & Sons, Inc., 2011, pp. 45.
12. Ludueña, L. N.; Kenny, J. M.; Vázquez, A. and Alvarez, V. A.; *Mater. Sci. Eng., A*, **529**, 215 (2011).
13. Ludueña, L. N.; Vázquez, A. and Alvarez, V. A., *J. Appl. Polym. Sci.*, **128**, 2648 (2013).
14. Ollier, R.; Vazquez, A. and Alvarez V.; in *Advances in Nanotechnology*; (Eds: Bartul, Z.; Trenor, J.), Nova Science Publishers; USA, 2011, 10, pp. 281.
15. D'Amico, D. A.; Ollier, R. P.; Alvarez, V. A., Schroeder, W. F. and Cyras, V. P.; *Appl. Clay Sci.* **99**, 254 (2014).
16. Ahmed, J.; Auras, R.; Kijchavengkul, T. and Varshney, S. K.; *J Food Eng.* **111**, 580 (2012).
17. Wagener, R. and Reisinger, T. J. G.; *Polymer* **44**, 7513 (2003).
18. Bellucci, F.; Camino, G.; Frache, A. and Sarra, A.; *Polym. Degrad. Stab.* **92**, 425 (2007).

Controlling the conductive network formation of polymer nanocomposites filled with nanorods through the electric field

Gao Yangyang (1)*

(1) Beijing University of Chemical Technology - Beijing - China

Facilitating the electric field to fabricate high performance polymer nanocomposites (PNCs) is always a great and promising strategy. In this work, the effect of the electric field on the conductive property of the PNCs is investigated by adopting coarse-grained molecular dynamics simulation. The translational and rotational diffusion of the nanorod gradually decreases with the increase of the nanorod aspect ratio; however, it is nearly independent of the nanorod volume fraction. Under the electric field, it exhibits a limited decrease, but more anisotropy of the translational diffusion of the nanorod. Meanwhile, the nanorods can not experience the random rotation, which is attributed to the electrostatic force. In addition, on one hand, the electrostatic force exerting on the nanorods induces the nanorod orientation, leading to the decrease of the homogeneous conductive probability; however, the electrostatic interaction induces the connection of the nanorod, resulting in the increase of the homogeneous conductive probability. Additionally, the directional conductive probability parallel to the electric field direction increases; however, the directional conductive probability perpendicular to the electric field direction shows a continuous decrease. Considering these two effects, the decrease or the increase of the homogeneous conductivity probability depends on the competition between the electrostatic force and the electrostatic interaction. The relationship among the anisotropy of the conductive probability, the nanorod volume fraction, the electrostatic force and the electrostatic interaction can be described quantitatively by an empirical formula. Another empirical formula is adopted to quantitatively describe the relationship among the percolation threshold, the electrostatic force and the electrostatic interaction. Meanwhile, the evolution process of the conductivity network structure is studied with or without the electric field. In summary, this work quan

Influence of Functionalization on Strength of CNT-epoxy nanocomposite

Billing Beant Kaur (1)*, Singh Narinder (1), Agnihotri Prabhat K (1)

(1) Indian Institute of Technology, Ropar - Punjab - India

Since the discovery of Carbon Nanotubes (CNTs) by Iijima in 1991, the outstanding electrical, thermal and mechanical properties have gained huge interest worldwide. To explore the mechanical properties, the growth of Carbon Nanotubes (CNTs) via Chemical Vapor Deposition method and its functionalization for evaluating their performance in polymer matrix was elucidated in this paper. To begin with, at temperature of 750 oC, CNT growth process was commenced. Further, to examine the strength of polymer matrix, acid functionalization was performed using acid mixture (H_2SO_4: HNO_3), which also results in removal of metal catalyst impurities. Moreover, acid treatment results in carboxyl group attachment onto the CNT surface. The COOH group gets attached at the defects, ends and at sidewalls of CNTs. In addition to this, carboxyl acid functionalized CNTs were reacted with amine resulting in amide functionalization. The functionalization improved the solubility of CNTs in various solvents. All the prepared samples were successfully characterized using SEM, EDX, FTIR, XRD, RAMAN and BET. To examine the mechanical properties of different samples, nano composites were formed. As grown CNTs, acid functionalized CNTs (CNT-COOH) and amide functionalized CNTs (CNT-CONH-NH_2) were dispersed in polymer for 25 minutes under sonication. Dispersed CNTs was cured by adding epoxy hardener and poured in a mould to prepare nano composites. All the 4 nano composite samples (including neat epoxy) were cut down in desired dimensions to perform 3- point bend test and micro hardness test for different nano composites to analyze the mechanical strength. All the results were compared and observed that the CNTs that were subjected to functionalization provide better reinforcement to epoxy matrix as compared to composite samples processed with grown CNTs.

Thermal and crystalline properties of polylactic acid-waxy maize starch nanocrystals nanocomposites prepared by melt processing

***Takkalkar Pooja Ankkush** (1)*, **Kao Nhol** (1), **Griffin Gregory** (1)*

(1) RMIT University - Victoria - Australia

Polylactic acid (PLA) based nanocomposites with waxy maize starch nanocrystals (WSNC) as nanofiller were melt-blended in a Haake Rheomix. WSNC were incorporated into PLA at three different loadings – 1 wt-%, 3 wt-% and 5 wt-% in PLA matrix. The morphological, thermal and crystalline properties of neat PLA and the prepared nanocomposites were investigated by scanning electron microscopy (SEM), thermo gravimetric analysis (TGA) and modulated differential scanning calorimetry (MDSC), respectively, to determine the effects of adding WSNC at different loadings in PLA. The microscopy study shows the effective dispersion of WSNC in PLA at lower loadings (0-3 wt-%) and agglomeration at 5 wt-% of WSNC. The thermal degradation studies confirmed that the nanocomposites and WSNC are very stable and safe to process in the actual processing temperature limits of PLA, 25-220 ºC. MDSC results show that there is no substantial variation in the glass transition temperature and the melting point of the nanocomposites. However, the cold crystallization temperature was reduced with the increase in WSNC loadings. This decline in cold crystallization temperature may be accredited to the nucleation function of WSNC to improve the crystallinity of the PLA matrix or to the greater interaction amongst the filler and the matrix.

(A2+B3) Approach Hyperbranched Polyester based Polyurethane-Urea (HBPU-Urea) and Polyurethane-Urea -Sulfur (HBPU-Urea-Sulfur) hybrid coating films: Brief Study of Hydrogen bonding interaction, morphology and mechanical properties

***JENA KISHORE KUMAR* (1)*, *ALHASSAN SAEED M* (1)**

(1) KUST THE PETROLEUM INSTITUTE - ABUDHABI - UAE

Hybrid coating based on hyper-branched polyurethane and elemental sulfur was synthesized by in-situ polycondensation and urethane reaction. The effect of Sulfur in the hydrogen bonding, thermo-mechanical properties and surface morphology of HBPU-Urea-Sulfur hybrid coating at 2wt. % concentration was the notable objective of this work. Deconvolution studies confirmed that more hydrogen bonding interaction happened in HBPU-Urea coating, thereby shifting the characteristic peak position of N–H and C=O groups to lower wavenumber values. HBPU-Urea coating exhibited superior mechanical properties compared to HBPU-Urea-Sulfur hybrid coating. With the incorporation of 2 wt. % sulfur, the resultant hybrid coating showed decreasing in tensile strength and young's modulus. Interestingly, the elongation at break percentage increased from 12% for HBPU-Urea to 18% for HBPU-Urea-Sulfur hybrid coating. Thermal stability of the coatings was examined by thermogravimetric analysis (TGA) depicted a 10°C decrease in thermal stability for 2 wt.% sulfur filled hybrid coating. DMTA measurements showed that the addition of sulfur nanoparticles decreased the glass transition temperature and crosslinking density. The HBPU-Urea-Sulfur hybrid coating displays a smooth surface because the size of the sulfur nanoparticles is reduced, leading to uniform dispersion and achieve good compatibility.

Effect of network formation on the rheological and mechanical properties of precipitated calcium carbonate/poly(carprolactone) composites

***Kim Jihwan* [1], *Ahn Kyung Hyun* [1]*, *Lee Seung Jong* [1]**

[1] Seoul National University - Seoul - Korea

For melt-compounded poly(caprolactone) (PCL) containing precipitated calcium carbonate (PCC) having high weight fractions Wpcc added up to 40 wt%, this study examined rheological and mechanical properties of PCC filled composites, with a focus being particularly placed on formation of PCC network structure in the Wpcc range between 30 and 40 wt%. For the composites containing Wpcc = 30 wt%, the linear viscoelastic measurements revealed only a marginal increase in low-frequency storage G' and loss moduli G" in comparison with those of the neat PCL matrix due only to filler effect of PCC. However, as Wpcc increased from 30 to 40 wt%, there were drastic increases in both low-frequency G' and G", exhibiting a liquid-solid transition resulted from the formation of the PCC network structure. This formation of the network structure was also observed from a scanning electron microscopy in which the image of Wpcc = 40 wt% revealed an interconnected PCC network. The PCC network formed in the composite having Wpcc = 40 wt% significantly enhanced the ultimate tensile strength and Young's modulus by 127.4 and 292.8 %, respectively, compared with those of the neat PCL matrix. In contrast, for the composite having Wpcc = 30 wt%, those changes in the mechanical properties were less significant (the tensile strength and the modulus were only increased by 117.7 and 163.8 %, respectively compared to the neat PCL matrix). The elongation at break was, however, decreased gradually to Wpcc = 30 wt%, and dropped suddenly with Wpcc = 40 wt%.

Synthesis and Characterization of Colored Polymer Nanoparticles for Electronic Paper Display

Nam Sang Yong [(1)]*, Yun Jun Seong [(1)], Kim Chul Am [(2)]

[(1)] Gyeongsang National University - Gyeongsangnamdo - Korea, [(2)] Electronics & Telecommunications Researech Institute - Daejeon - Korea

Recently, Several studies have shown achievement of Electronic Paper Display(EPD) and Colored polymer particles, and EPD has been of great interest for applications in information displays. EPD has many advantages such as high electrical efficiency, high flexibility and wide viewing angle. Colored polymer particles(PMMA) can be produced by dispersion polymerization and the surface of PMMA particles are modified by aminolysis. The manufacturing process of colored particles is, when the aprticles swell by the solvent, the dyes enter the particles. After the particles are dried, the dyes do not go out of the particle but stay. Finally, characteristics of the colored particles are as follows. The particle surface modification was confirmed by SEM, and the absorbance was confirmed using a UV/VIS spectrometer. Thermal properties were evaluated using TGA and DSC.

Development of corn starch-based films with added active and intelligent natural nano-fillers

Gutiérrez Tomy J. [(1)]***, Alvarez Vera A.*** [(1)*]

[(1)] Institute of Research in Materials Science and Technology - INTEMA - Buenos Aires - Argentina

Corn starch films with added active and intelligent (A&I) natural nano-fillers obtained from montmorillonite with the addition of blueberry extract were investigated. The objective of this study is focused on the development of A&I films to be applied in the food industry. The obtained films were characterized by thermogravimetric analysis (TGA), contact angle, atomic force microscopy (AFM), scanning electron microscopy (SEM), thickness, density, opacity, color parameters and response to pH changes. The results indicated that A&I natural nano-fillers had no response to pH changes within developed films, and that these A&I nano-fillers increased the phase separation in the starch thermoplastic material.

Expanding Applications Areas for Sheet Molding Compound (SMC) Through New Material Compositions and New Hybrid Processes

Hrymak, Andrew[a,b]; Bücheler, David[c]; Henning, Frank[b,c,d]; Ugresic, Vanja[b]; Hohberg, Martin[d]

[a] *University of Western Ontario, Chemical and Biochemical Engineering, London, Canada*
[b] *Fraunhofer Project Centre for Composites Research at Western University (FPC), London, Canada*
[c] *Fraunhofer Institute for Chemical Technology (ICT), Polymer Engineering, Pfinztal, Germany*
[d] *Karlsruhe Institute of Technology (KIT), Institute of Vehicle System Technology, Karlsruhe, Germany*

Abstract. Sheet Molding Compound (SMC) offers excellent characteristics in terms of complex part geometry, function integration, material utilization and productivity for an economical cost. Thus it is well known in industry and widely used. [1–7] However, to some extent new requirements have changed SMC processing. Examples for these new demands for versatility are low density, high stiffness, low emissions, use of recycled carbon fibers or heavy tows, co-molding of SMC and prepreg, and the need for computer aided design and process optimization. [8, 9] This paper provides an overview about recent developments in these fields. The focus is set on actual technology breakthroughs in the field of alternative resin systems, carbon fiber SMC, locally continuous-fiber reinforced SMC and computer aided engineering.

Keywords: Sheet Molding Compound, SMC, Tailored-SMC, LD-SMC, Advanced-SMC, CF-SMC, Heavy Tow
PACS: 81.05.Qk, 83.50.Uv, 88.30.mjwil

ALTERNATIVE RESIN SYSTEMS

SMC composites have developed over time to include many resin systems such as polyesters, vinyl esters, epoxy, and hybrid resins. Decades of formulating and compounding experience has enabled the reliable production of SMC components. A shift from conventional polyester and vinyl ester systems to the new resin systems, epoxies and polyurethanes has happened in response to the demands of industry for high performance and styrene free applications.

Epoxy

Epoxy shows high strength and low emissions and thus is an ideal resin when it comes to high performance but styrene free SMC. However, thickening of the material is a challenge and currently tackled by different resin suppliers [8, 10–12]. Hexion has developed a two component formulation which fits to the single process steps of SMC processing [13]. The resin *EP TRAC 06605* shows low initial viscosity, which can be matured as conventional UP based materials, and shows high flowability during compression-molding while having homogenous fiber transport. Furthermore, the semi-finished material can be stored at room temperature for at least 21 days.

Molding studies were performed at the Fraunhofer ICT in collaboration with Hexion to evaluate the mechanical performance of the system. After mixing of the resin with 23 phr of hardener (*EK TRAC 06608*) and 3 phr of mold release agent (*HA TRAC 06805*), the compounding is done by the help of a flat conveyor plant type *HM-LB-800* by *Schmidt & Heinzmann*. Hereby, the matrix gets reinforced by randomly distributed 12k carbon fibers (type *T700SC 12000 50C* by *Toray*) of 1 inch in length. Furthermore the material is heated up to around 80 °C while compounding to improve fiber wetting. After maturation, the material is cut and compression molded to plates of 456 mm x 456 mm x 3 mm, according to the molding conditions shown in TABLE 1.

TABLE 1: Molding conditions

Mold coverage	60 – 70	%
Closing velocity	1	mm/s
Final pressure	70	bar
Molding time	182	s
Molding temperature	150 – 155	°C

After molding, specimen preparation is done by water jet cutting. The fiber content was determined [14] to be 66.4 ± 4.1 wt.-%. The tensile properties were measured following DIN EN ISO 527-1. The material shows a tensile strength of 234.4 ± 34.4 MPa at a Young's Modulus of 40.4 ± 4.4 GPa. Bending properties were measured according to DIN EN ISO 14125-05. A flexural strength of 413.8 ± 76.8 MPa was achieved at a modulus of 34.2 ± 4.9 GPa. To investigate impact properties, tests according to ISO 179-1/2bU were performed delivering an impact strength of 110.8 ± 2.5 kJ/m².

The tensile properties as well as impact properties were compared to other commercially available carbon fiber reinforced molding compounds (see FIGURE 1). The manufactured material shows best in class tensile properties and is directly comparable to more expensive 3k reinforced materials. The impact strength is superior to the other EP-molding compounds considered.

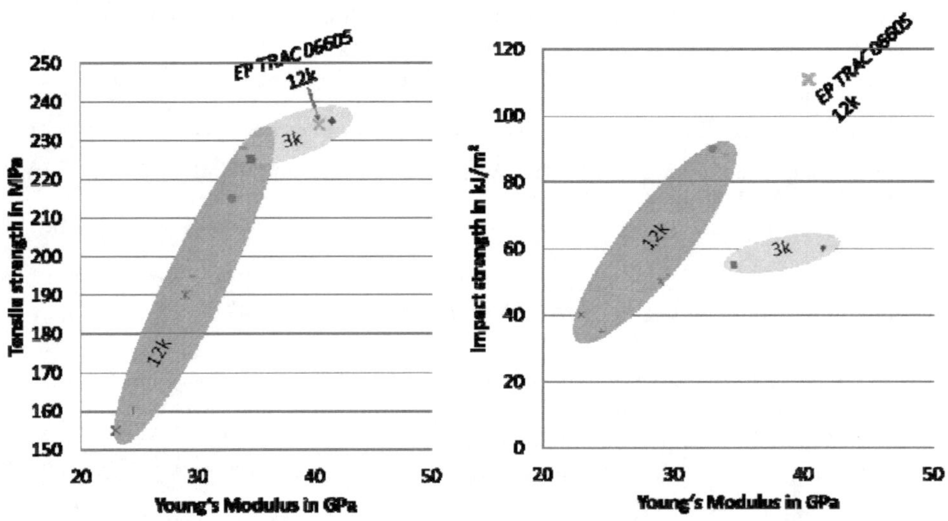

FIGURE 1: Manufactured material compared to other commercial available EP-Molding compounds
a) left: Tensile strength to Young's Modulus, b) right: Impact strength to Young's Modulus

Polyurethane

Polyurethane resins exhibit superior mechanical properties such as toughness and strength as compared to the conventional PE/VE resin systems used in SMC. The high reactivity of PU resin, which leads to short cycle times, is usually associated with a rapid increase in viscosity over time. The SMC process inherently requires stable viscosity build with initially low viscosity that ensures the wet-out of the fibres during compaction, followed by maturation and snap cure profile in the mold. This requirement has previously limited the use of PU systems in SMC. Developments of the VITROX resin system by Huntsman Polyurethanes has produced a PU system with decoupled viscosity and cure profiles. This major change in processing ability has made the implementation of PU resins in high volume composite manufacturing processes feasible, - including SMC.

PU SMC has been compounded on the direct sheet molding compound technology (D-SMC) where inline heating has been used to mature the SMC to a molding viscosity allowing for a continuous compounding and molding process. The advantage of continuous SMC process is that it gives near instantaneous feedback when compared to conventional SMC systems and limits the need for material storage and handling.

The reactivity of the PU resin systems challenges the shelf life for the semi-finished SMC. The shelf life of the matured PU SMC was found to be at least one hour without the need of freezing. The limits of the shelf life are yet to be fully investigated. Furthermore, it was found that after freezing the material behaved the same in terms of flowability. Studies were conducted at the Fraunhofer Project Centre in Composites Research at Western University in collaboration with Huntsman Polyurethanes, Johns Manville GmbH and ZOLTEK. These system are PU formulations with glass and carbon fiber and have been compared to VE and PE systems with similar fibre and filler content and processing.

TAILORED SHEET MOLDING COMPOUND

To further expand the application of SMC into highly loaded structural applications, it can be co-molded with prepreg material as shown in FIGURE 2. This approach is currently deeply investigated by the International Research and Training Group "Integrated engineering of continuous-discontinuous long fibre reinforced polymer structures" (DFG-GRK 2078) [15]. Besides investigations in processing and mechanical performance of the material, as described in [16] and [17], the integrated product development chain for this new material class is in focus. Key to this is a CAE chain allowing to optimize over the whole simulation chain.

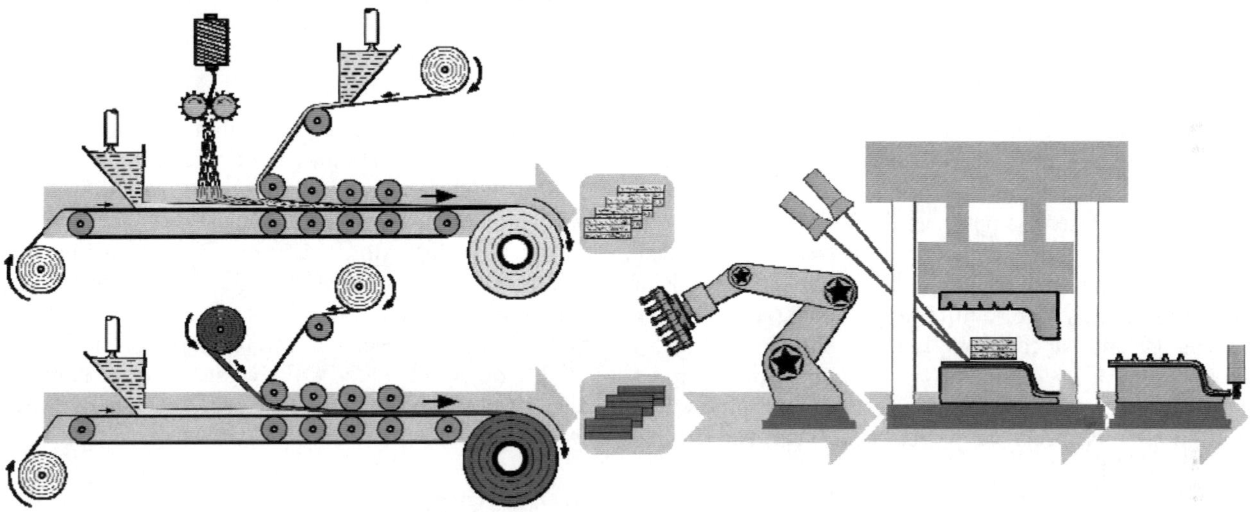

FIGURE 2: Concept of co-molding of SMC with prepreg

CAE-CHAIN

To utilize these new materials and material combinations in a product design, new simulation approaches are needed. For example, to consider the manufacturing effects in the structural simulation, not only the single simulation steps can be performed separately, but in a CAE-Chain to take into account results from previous simulation. Therefore, the combination of the steps (Design, Optimization, Process Simulation, Mapping and Structural Simulation) was implemented within the FAST CAE-Chain approach at the Institute of Vehicle System Technology (FAST). This chain was implemented for unidirectional fiber reinforced polymers [18] and for long fiber reinforced polymers [19]. The schematic procedure of the CAE-Chain for SMC is plotted in FIGURE 3. In both CAE-Chains, each step has criteria that must be fulfilled before results are passed to the next simulation step. Otherwise a process step internal optimization tries to fulfill the criteria by changing free parameter. If this is not possible or if the internal optimization fails, the previous step is done again with a varied parameter set.

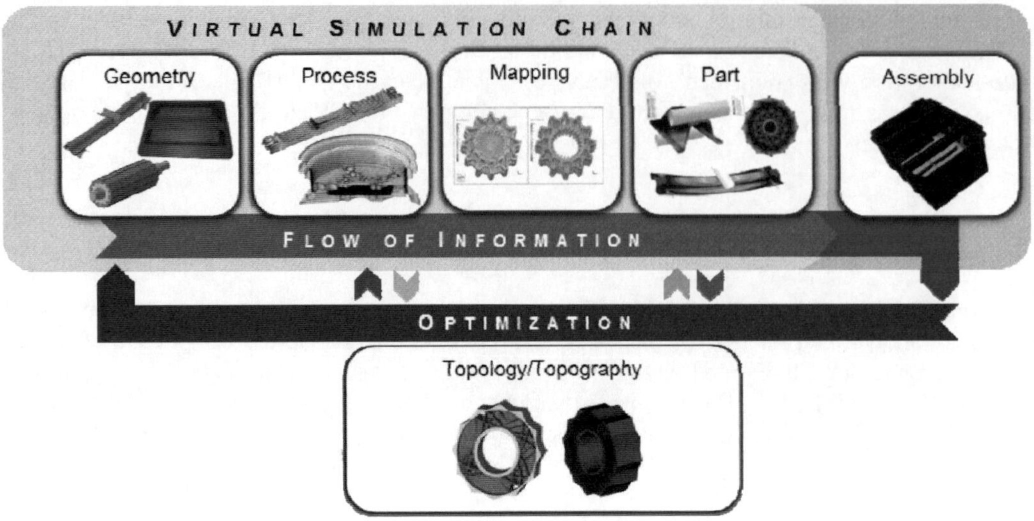

FIGURE 3: Steps of the KIT-FAST CAE-Chain [19]

The holistic long fiber FAST CAE-chain for SMC with local unidirectional reinforcements starts with a topology and shape optimization of the maximum available part space. This is done by combining isotropic and anisotropic methods to consider the fiber orientation within these optimizations. After deducing a CAD model from these first optimizations, a second optimization is done to identify the best position of local reinforcements. Therefore, a multi-objective optimization based on an evolutionary algorithm is used. This algorithm was implemented to find the minimal deflection or global part stiffness while using a minimum amount of unidirectional reinforcements. Furthermore, this algorithm considers manufacturing constraints as well as minimum or maximum reinforcement dimensions and the corresponding part stiffness [20, 21]. After weighting the objectives, one result out of the Pareto set (the sum of results in which it is impossible to make any single objective better without impairing another) is used for the design (see Figure 4 right). One of these intermediate results is plotted in Figure 4a. The final optimal configuration is processed for the further steps and therefore converted to a CAD file.

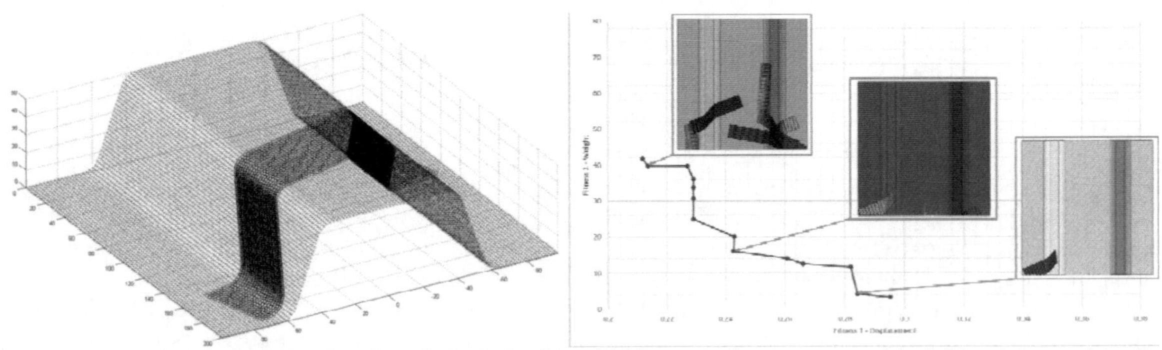

FIGURE 4: Evolutionary algorithm based patch-Optimization with kinematic draping approach(a, left) and the resulting Pareto front (b, right) [20]

The next step in the FAST CAE-Chain is process simulation. As commercial software is just using modified thermoplastic injection molding material models, which cannot distinguish shear and elongation viscosity, flow behavior, flow pattern and pressure distributions are not predicted fully [22]. Furthermore, it is not possible to predict large deflections of the local reinforcements during the molding process. For these reasons a self-written material model was implemented in Abaqus CEL (Coupled Eulerian Lagrangian), which can identify the resin rich thin lubrication layer at the mold surface with its shear dominated flow pattern and the relatively cold core region, which have an elongation flow behavior. Furthermore, the Folgar-Tucker model with the ARD-RSC extension and the IBOF closure approach was implemented to predict the fiber orientation distributions (see Figure 5). Due to the CEL approach, which contains a fluid-structure modelling, a prediction of the displacement of local reinforcements or part inserts can be done [23].

The process simulation tools are used to find the best positon for the initial charges and the best process parameter by doing a DOE study. Besides the main criteria mold filling, other criteria such as maximum press force and weld lines are considered to find the best setting. In the future, the criteria maximum displacement of the reinforcement during the process will be added.

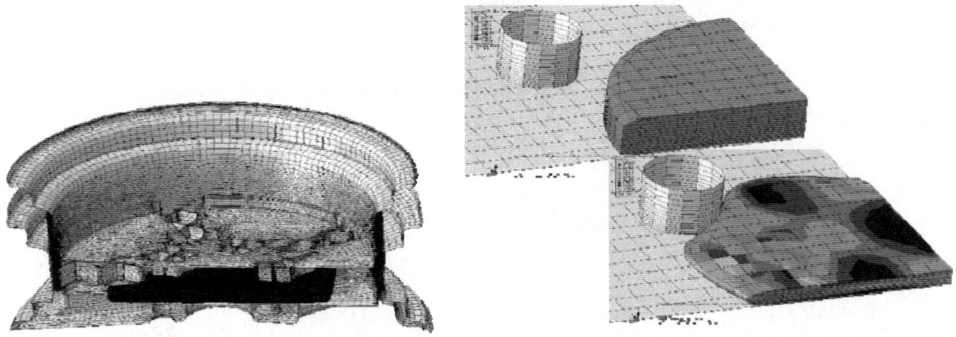

FIGURE 5: Process simulation of SMC: Model setup (a, left) and fiber orientation distribution (A_{xx}) evolution (b, right) [19]

The main idea of the CAE-Chain is to use results from previous simulations in the current study. To transfer element or node based results, the neutral vtk-format is generally used for the whole CAE-chain. By using this format, result information like fiber orientation distribution, temperature distribution or eigen-stresses are stored. To use these results in the next simulation steps the mapping library MpCCI MapLib [24] provided by the Fraunhofer SCAI is used to transfer the results from the process to the structural mesh.

Based on these results mapped to the structural mesh, linear elastic or more advanced material models can be used. For both approaches a clustering of similar fiber orientations can be done to reduce the number of material cards to speed up the simulation. For a fast first approximations these clusters can be homogenized purely elastic by using a Tandon-Weng approach. Depending on the SMC material composition, SMC shows a non-linear material behavior and a significant damage behavior for cyclic loads. To consider these behaviors, phenomenological materials models are developed and fitted to experimental results over a large data range (see

FIGURE6a) [25, 26]. The target of these simulations is to prove the concept under the given mechanical load cases. If the simulation results fails, a material substitution, a change of the process parameter or even a change of the geometry have to be conducted. Depending on the position of the correction in the CAE-chain, parts of the chain have to be redone.

As SMC shows relatively good mechanical properties even at higher temperatures (up to 140°C), thermal and thermo-mechanical simulations could be done during the structural simulation if needed (see

FIGURE6b). By considering the anisotropic thermal conductivity due to the fiber orientation distribution, the proof of the structural resistance under non-isothermal as well as the verification of thermal cooling concepts could be done.

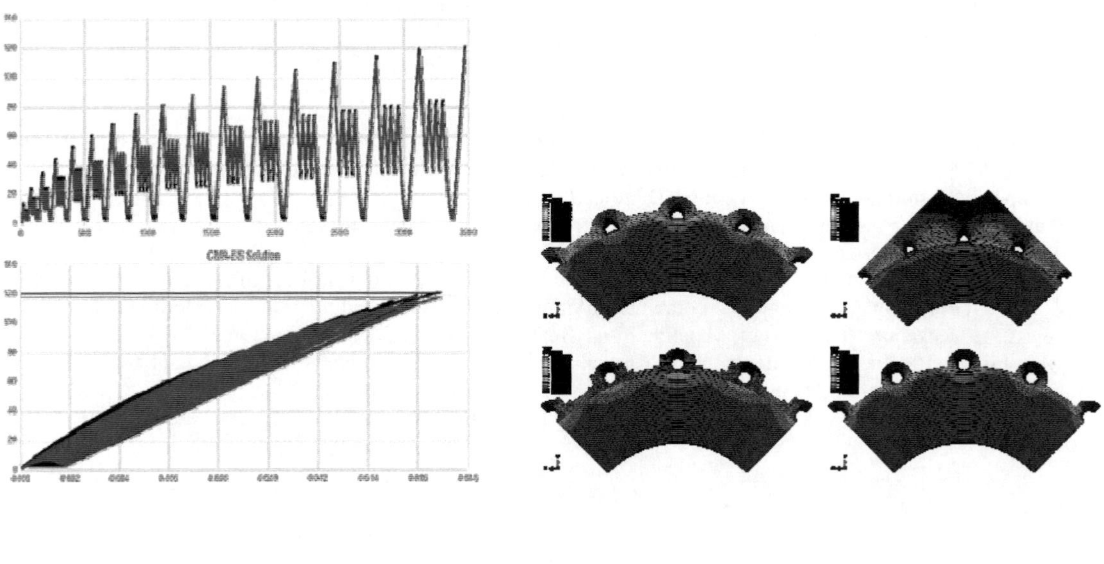

FIGURE 6: Development of material models to predict the viscos-elastic and the damage behavior (a,left). Thermal simulation of a thermoset electrical engine housing to predict the part stiffness under thermal loads (b,right) [19]

ACKNOWLEDGMENTS

The authors acknowledge the financial support by the German Research Foundation (DFG) within the International Research Training Group "Integrated engineering of continuous-discontinuous long fibre reinforced polymer structures" (GRK 2078).

REFERENCES

1. W. Assmann and E. Witten, *Handbuch Faserverbundkunststoffe - Composites: Grundlagen, Verarbeitung, Anwendungen,* 4th edn., Springer Fachmedien, Wiesbaden (2013).
2. G.W. Ehrenstein, *Faserverbund-Kunststoffe: Werkstoffe, Verarbeitung, Eigenschaften,* 2nd edn., Hanser, München [u.a.] (2006).
3. P. Elsner, P. Eyerer, and T. Hirth, *Kunststoffe: Eigenschaften und Anwendungen,* 8th edn., Springer, Dordrecht (2011).
4. F. Henning and E. Moeller, *Handbuch Leichtbau: Methoden, Werkstoffe, Fertigung*, Hanser, München (2011).
5. M. Biron, *Thermosets and composites: Material selection, applications, manufacturing, and cost analysis,* 2nd edn., Andrew, Elsevier, Oxford [u.a.] (2014).
6. M.M. Schwartz, Encyclopedia of materials, parts, and finishes, 2nd edn., CRC Press, Boca Raton (2002).
7. A. Wilson, Reinforced Plastics, 61, 100 (2017).
8. P. Malnati, CompositeWorld, 50 (2017).
9. K. Akiyama, Development of preforming process in PCM (Prepreg Compression Molding) technology, in 12th Annual Automotive Composites Conference and Exhibition 2012 (ACCE 2012) ; Troy, Michigan, USA, 11 - 13 September 2012, Annual Automotive Composites Conference and Exhibition and ACCE, Eds., Curran, Red Hook, NY (2013).
10. C. Yin, A. Kaffee, M. Henningsen, J. Zwecker, and L. Gehringer, Epoxy-resin composition for fiber-matrix semifinished products, US Patent Application 20160115283
11. Aditya Birla Chemicals (Thailand) Ltd., Automotive Applications: Epotec Epoxy Systems.
12. S. Black, Wait, wait, there's more: JEC 2017 (2017). http://www.compositesworld.com/blog/post/wait-wait-theres-more-jec-2017. Accessed 14 September 2017.
13. A. Schmidt, New epoxy systems enabling styrene-free high-performance SMC manufacturing, Paris (2017).
14. D. Bücheler, A. Kaiser, and F. Henning, Composites Part B: Engineering, 2016, 218 (2016).
15. L. Kärger, T. Böhlke, K.-A. Weidenmann, and F. Henning, Integrierte Entwicklung kontinuierlich-diskontinuierlich langfaserverstärkter Polymerstrukturen im Rahmen des internationalen Graduiertenkollegs GRK 2078, Hamburg (2016).
16. D. Bücheler and F. Henning, Hybrid Resin Improves Position and Alignment of Continuously Reinforced Prepreg During Compression Co-molding With Sheet Molding Compound, in 17th European Conference on Composite Materials (ECCM17), European Society for Composite Materials, Ed. (2016).
17. A. Trauth, D. Bücheler, K.A. Weidenmann, and F. Henning, Mechanical Properties of Unidirectional Continuous Carbon Fiber Reinforced Sheet Molding Compounds, in 17th European Conference on Composite Materials (ECCM17), European Society for Composite Materials, Ed. (2016).
18. L. Kärger, A. Bernath, F. Fritz, S. Galkin, D. Magagnato, A. Oeckerath, A. Schön, and F. Henning, Composite Structures, 132, 350 (2015).
19. M. Hohberg, L. Kärger, F. Henning, and A. Hrymak, "Process Simulation of Sheet Molding Compound (SMC) as key for the integrated Simulation Chain," in Simulation of Composites - Ready for Industry 4.0?, NAFEMS, Ed. (2016).
20. B. Fengler, L. Kärger, A. Hrymak, and F. Henning, Application of a Multi Objective Optimization approach for Continuous Fiber Tapes in Hybrid Composite Structures, Detroit, USA (2016).
21. B. Fengler, L. Kärger, and A. Hrymak, Influence of the tape number on the optimized structural performance of locally reinforced composite structures, Stuttgart, Germany (2017).
22. M. Hohberg, Process simulation of hybrid SMC structures with local continuous fiber reinforcements, Frankfurt (2015).
23. T. Müller and F. Henning, Simulation of combined forming and injection molding processes, Berlin (2015).
24. H. Spiess, A. Oeckerath, B. Landvogt, Mapping library manual version 2010.1 , Fraunhofer SCAI (2012).
25. M. Mrkonjić, U. Rayling, K. Weidenmann, L. Kärger, and F. Henning, Phenomenological Characterization and Macromechanical Modeling of Anisotropic, Non-linear Behavior of Sheet Molding Compounds (SMC), Karlsruhe (2015).
26. O. Saburow, J. Huether, R. Maertens, A. Trauth, Y. Kechaou, F. Henning, and K.A. Weidenmann, Procedia CIRP, 66, 265 (2017).

AUTHOR INDEX

AUTHOR INDEX

AUTHOR INDEX

AUTHOR INDEX

AUTHOR INDEX

AUTHOR INDEX

AUTHOR INDEX

AUTHOR INDEX

AUTHOR INDEX

AUTHOR INDEX

AUTHOR INDEX

AUTHOR INDEX

AUTHOR INDEX

Polymer Processing Society
810 E. 10th St.
Lawrence, KS 66044
USA

ISBN 978-1-5108-6330-9

CURRAN ASSOCIATES INC.
proceedings.com

33rd Polymer Processing Society Annual Meeting 2017 (PPS-33)

Cancun, Mexico
10 - 14 December 2017

Volume 2 of 2

33rd Polymer Processing Society Annual Meeting 2017 (PPS-33)

Cancun, Mexico
10 - 14 December 2017

Volume 2 of 2

ISBN: 978-1-5108-6330-9

Printed from e-media with permission by:

Curran Associates, Inc.
57 Morehouse Lane
Red Hook, NY 12571

Some format issues inherent in the e-media version may also appear in this print version.

Printed by Curran Associates, Inc. (2018)

For permission requests, please contact Polymer Processing Society
at the address below.

Polymer Processing Society
810 E. 10th St.
Lawrence, KS 66044
USA

Phone: +1-800-627-0326
+1-785-865-9403
Fax: +1-785-843-6153

pps@allenpress.com

Additional copies of this publication are available from:

Curran Associates, Inc.
57 Morehouse Lane
Red Hook, NY 12571 USA
Phone: 845-758-0400
Fax: 845-758-2633
Email: curran@proceedings.com
Web: www.proceedings.com

TABLE OF CONTENTS

VOLUME 1

S01 – EXTRUSION PROCESSES

S02 – INJECTION MOLDING

S03 – PROCESS MODELING

S04 – BLOW MOLDING, THERMOFORMING AND ROTOMOLDING

S05 - RECYCLING

S06 – FIBER, MEMBRANES AND BARRIER

S07 – MIXING AND COMPOUNDING

S08 – POLYMER BLENDS

S09 – POLYMER NANOCOMPOSITES

S10 - COMPOSITES

VOLUME 2

S11 – BIOPOLYMERS & NATURAL FIBERS

S12 – POLYMER FOAMS AND FILMS

S13 – RUBBER AND ELASTOMERS

S14 – RHEOLOGY AND RHEOMETRY

S15 - MORPHOLOGY

S16 – REACTIVE PROCESSING

S17 – ADVANCED PROCESSING TECHNOLOGY

S18 – AUTOMOTIVE AND MEDICAL

S19 – ADDITIVE TECHNOLOGY

Element Technology for High-cycle Molding of Continuous Fiber Reinforced Composites

Asami Nakai

Faculty of Engineering, Gifu University, 1-1 Yanagido, Gifu, 501-1193, Japan

Abstract. In recent years, continuous fiber reinforced thermoplastic composites are attracting a lot of attention in various area such as automobile industry and aircraft industry because of the improvement of the fuel efficiency and the capability of automobile and aircraft. Thermoplastic composite can be molded in high cycle compared to thermo-setting composite and it has also second workability and recyclability. However, the combination of continuous fiber and thermoplastic resin have disadvantage that impregnation of resin into fiber bundle is difficult since the melting viscosity of thermoplastic resin is high. In this study, in order to achieve high-cycle molding of continuous fiber reinforced thermoplastic composites, the element technologies were investigated in terms of 1) development of intermediate materials for processing, 2) optimization of fiber / resin interface, 3) development of high-cycle processing technology.

Keywords: Continuous fiber, High-cycle molding, Pultrusion molding, Hybrid molding
PACS: 81.05.Ni, 81.20.Hy, 83.50.Uv, 81.05.Qk

INTRODUCTION

In recent years, continuous fiber reinforced thermoplastic composites are attracting a lot of attention in various area such as automobile industry and aircraft industry because of the improvement of the fuel efficiency and the capability of automobile and aircraft. Thermoplastic composite can be molded in high cycle compared to thermo-setting composite and it has also second workability and recyclability. However, the combination of continuous fiber and thermoplastic resin have disadvantage that impregnation of resin into fiber bundle is difficult since the melting viscosity of thermoplastic resin is high. With increase in fiber length, mechanical properties are increased, on the other hand the processing ability was decreased such as cycle time, complex shape, and cost. Therefore, the high cycle molding with the use of high impregnated intermediate materials has been expected.

In order to achieve high-cycle molding of continuous fiber reinforced thermoplastic composites, the element technologies were investigated in terms of 1) development of intermediate materials for processing, 2) optimization of fiber / resin interface, 3) development of high-cycle processing technology. On account of the space, pultrusion molding and hybrid molding with compression molding for continuous fiber and injection molding for dis-continuous fiber were described.

PULTRUSION MOLDINGS

In one of the high cycle molding system for continuous fiber reinforced plastic with thermoplastic resin, there is pultrusion molding system. We have proposed one of the high cycle molding for FRTP with fibrous intermediate material and pultrusion systems by using braiding technology[1]. Braided fabrics fabricated with fibrous intermediate materials are pulled into the pultrusion molding die. Resin fiber in the braided fabric was melted and impregnated in the molding die and then melted resin was solidified. The moldings are continuously pulled out with pulling device.

Pultrusion molding system is composed of five devices such as mandrel having electric heater on the inside, preheater by far infrared ray, molding die, cooling system, drawing device. Schematic of pultrusion molding system are shown in Fig.1. Molding die is composed of two part such as taper part and straight part. The area of taper part is gradually decreases from the entrance to pressurize braided fabrics along with impregnation process. The area of straight is constant to acquire cooling time and arrange the shape of final products. The molding die can manufacture cylindrical composite material with a diameter of 18mm and an inside diameter of 15mm. In Fig.1, H1~H8 indicate electric heater which was inserted in top and bottom both ends of the die. This electric heater can control heating temperature independently each.

From instrumentation system for pultrusion molding, temperature history inside of braided fabric, pressure history in molding die, pultrusion force history were obtained. Impregnation state and bending properties were measured for composite pipes with various molding condition. From these results impregnation mechanism was clarified and molding condition for shortening molding time was investigated.

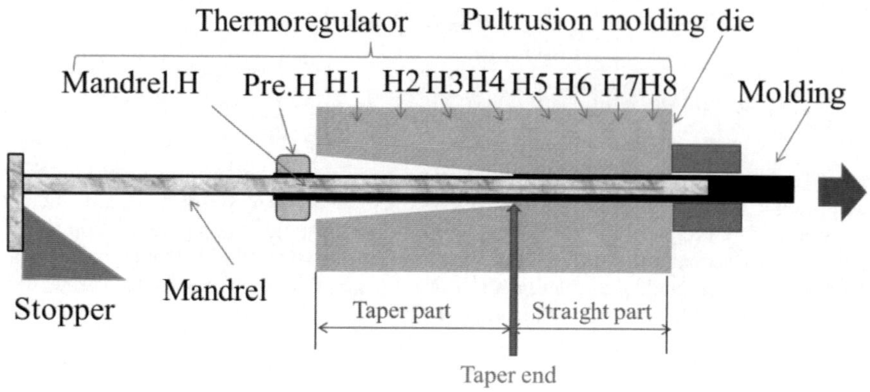

FIGURE 1. Schematic of pultrusion molding system.

Pultrusion molding system has issues that productivity is still lower for mass production such as automotive application. If the molding speed is increased, pultrusion force is extremely increased due to the friction between melted materials and moldings. Therefore, establishment of molding conditions which shortens molding time is demanded. The purpose of this study is to increase the pultrusion speed by decreasing the pultrusion force. In this study, the simple harmonic motion to the mandrel was applied during pultrusion molding.

The pressure was measured at the fixed point; the distance from entrance of molding die was 455 mm. When the material came to the pressure sensor positon, the pressure was gradually increased. And then, the mandrel was stopped by stopper, the pressure was rapidly decreased. After stopping of mandrel, material was slipped on mandrel and the pressure became steady state. After that, we have applied the simple harmonic motion. Figure 2 shows pressure history after stopping of mandrel by changing frequency of the vibration under the constant pultrusion speed. By applying the vibration, the pressure was decreased and reached constant value. By increasing pultrusion speed, the pressure was largely increased. The reason for the increase in pressure was that; the hating time was decreased with increase in pultrusion speed, and then melt viscosity became higher and increased friction between material and mold.

Figure 3 shows the pultrusion force during pultrusion molding. As same as pressure history, the vertical axis is the position of mandrel tip from the entrance of molding die. After stopping of mandrel, material was slipped on mandrel and the pultrusion force was drastically increased and became constant value. With increase in the frequency, the pultrusion force was drastically decreased from 1Hz to 2Hz and then increased again. In the case of higher pultrusion speed, the braided fabric was broken with increase in the pultrusion force and then the pultrusion force was decreased. The reason for increase in pultrusion force was resistance between material/molding die and material/mandrel. The resistance consists of friction force by molding pressure and force caused by the viscosity. The relative speed and the direction of braided fabric to mandrel were changed according to the frequency. Also mandrel was vibrated, the direction of the force was changed and the relative speed was also changed. If the direction of mandrel movement and braided fabric were same, the pressure was decreased with optimum frequency. If the direction of mandrel movement and braided fabric were opposite, the force caused by viscosity was increased with increase in frequency. This is the reason for optimum frequency to pultrusion force.

Consequently, the pultrusion speed was increased without increasing the pultrusion force. Also mechanical properties were improved since the interfacial properties were increased with increase in the frequency by the repeating applied pressure at the taper end.

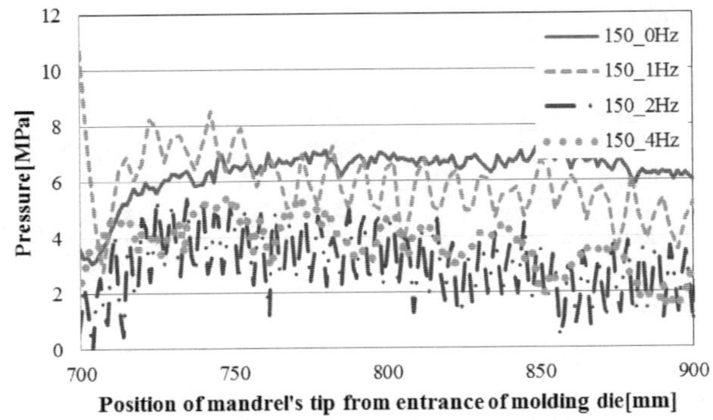

FIGURE 2. Pressure history.

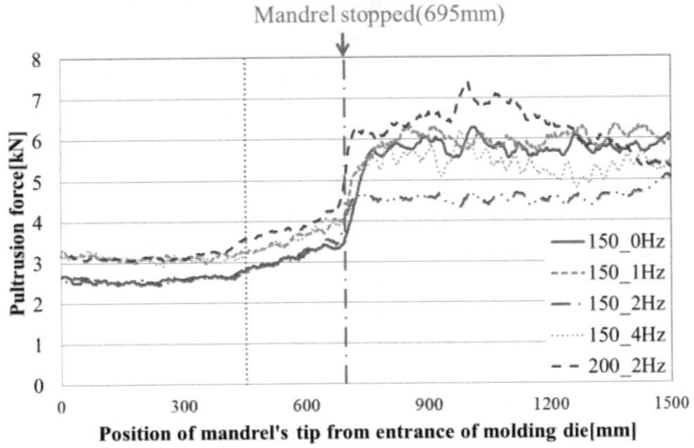

FIGURE 3. Pultrusion force history.

HYBRID MOLDINGS

The advantage for discontinuous fiber reinforced thermoplastic composite is to form a complicated shape in a short time. On the other hand, the continuous-FRTP has high mechanical properties, but its process ability is limited and it is difficult to produce complicated shapes such as ribs. Therefore, the hybrid molding method attracts attention. The hybrid molding method used in this research is a molding method combined the compression molding and the injection molding. Hybrid molding can take both advantage of continuous fiber and discontinuous fiber. The part of requiring strength is formed by compression molding of intermediate material with continuous fiber, and the complicated shape part is made by injection molding of discontinuous fiber. However, there is a problem with the strength of the joint part between the intermediate material part molded by compression molding and the part molded by injection molding. Joining characteristics are thought to be influenced by preheating conditions of c-FRTP, molding conditions such as temperature, pressure and speed of injection resin. Therefore, it was aimed to clarify the influence of the difference between the preheating condition for the intermediate material and the temperature of the injection resin on the mechanical properties of the molding and injection resin joint.

The molding process of the hybrid molding is shown in Figure 4. In hybrid molding, after an intermediate material such as prepreg consisting of continuous fiber and thermoplastic resin is inserted into a molding die, a preheating device is inserted into the die and heat the intermediate material. Thereafter, the preheating device is taken out, the mold is closed, pressurized, and the resin is injected. At this time, the injection resin and the intermediate material are welded by the heat from injection resin and the preheated intermediate material. After cooling, the molded product is taken out and molding is completed.

In this study, different wavelengths of near infrared and middle infrared preheating devices were used as preheating devices. A completely impregnated prepreg reinforced with a glass fiber woven fabric (TEPEX dynalite 104-RG600(X), Thickness 1.5mm, Bond Laminates) was used as an intermediate material. In the injection molding material, pellets in which the resin was polypropylene and the reinforced fibers was carbon fibers was used. The weight fraction of fiber (Wf) of the pellets was 30%, and the average fiber length was 150 μm.

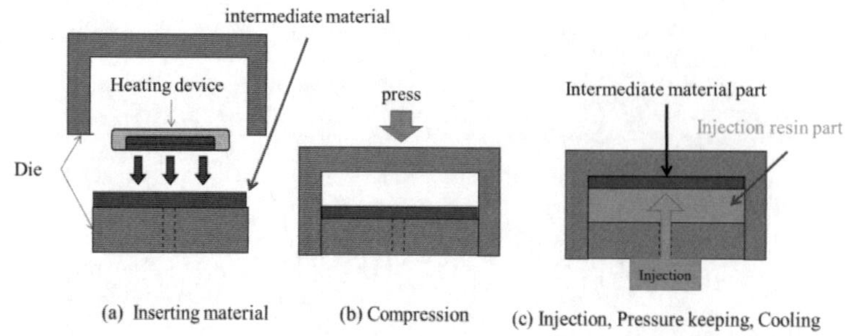

FIGURE 4. Schematic drawing of molding procedure of hybrid molding.

Figure 5 shows relationship between bending strength and preheating temperature an example of experimental results. When the preheating temperature is from 80 ℃ to 120 ℃, under the condition that the preheating condition and the injection resin temperature are equal, the bending strength becomes higher in the middle infrared radiation than in the near infrared radiation. The bending strength increased until the preheating temperature reached 100 ℃, and when it exceeded 120 ℃, the bending strength decreased. Bending strength showed the maximum value of 245 MPa under the condition of M100-240. In the same preheating condition, bending strength increased in higher injection temperature. In the non heated condition, the bending strength was about 110 MPa when the injection resin temperature was 200 ℃, but when the injection resin temperature was raised to 240 ℃, the bending strength became about 3 times. In Fig.8, the preheated specimen of M240-100 and M240-180 showed the largest and smallest bending strength from the load-displacement diagram. M240-180 fractured at displacement of 11 mm and the load suddenly decreased, but load reduction could not be seen clearly under other conditions.

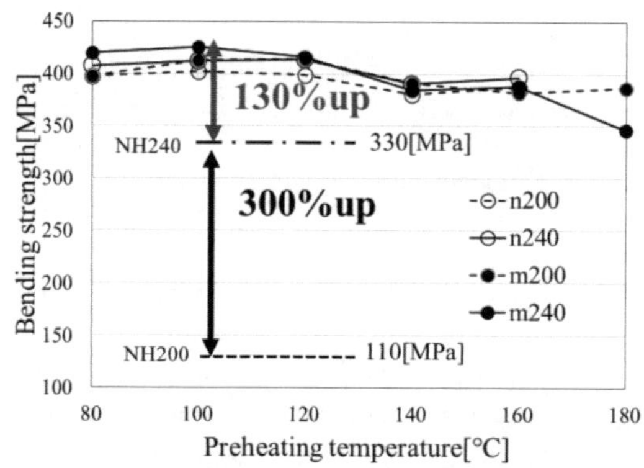

FIGURE 5. Schematic drawing of molding procedure of hybrid molding.

From the series of experiments, following results was obtained. The temperature rising rate of the middle infrared ray was higher than that of the near infrared ray. In addition, the maximum temperature was higher for middle infrared heating than for near infrared. The temperature just after the stop of heating is equal, the descent temperature in 10 seconds becomes larger in the middle infrared heating than in the near infrared heating. It is considered that the injection resin temperature and the preheating condition do not affect the elastic modulus. The bending strength increased until the preheating temperature reached 100 ℃, and when it exceeded 120 ℃, the bending strength decreased. It is considered that the flowability of the intermediate material resin increased due to the increase in the molding temperature and the fibers flowed toward the end of the molded part at the time of injection. The higher the molding temperature, the greater the disorder of the fiber, the more the distance between the layers and the lower the volume fraction of fiber. From these facts, it is considered that the bending strength decreased when the preheating temperature exceeded 120 ℃. From this study, middle infrared was found out to be the most suitable preheating device for hybrid molding, from its advantages that could preheat in a short time and improves mechanical property.

REFERENCES

1. Louis Laberge Lebel, Asami Nakai "Design and manufacturing of an L-shaped thermoplastic composite beam by braid-trusion", Composites Part A: Applied Science and Manufacturing, Vol. 43, No. 10, 1717-1729 (2012).

Polyamide/Graphite Composites with Improved Conductivity and Mechanical Properties

Osayuki Osazuwa[a], Marianna Kontopoulou[a,*] and Mark Gallerneault[b]

[a]*Department of Chemical Engineering, Queen's University Kingston Ontario K7L 3N6 Canada*
[b]*ALCERECO Inc., Grafoid Global Technology Centre, Kingston Ontario K7L 0E9 Canada*

Abstract. Blends of polyamide and an elastomer are prepared by melt compounding in the presence of high purity graphite particles. The surface energies of the blend components are evaluated, to assess their affinity with the graphite particles. The high-aspect ratio delaminated graphite particles impart conductivity and improved modulus to the polyamide matrix, with a percolation threshold of around 30 wt. %. Addition of the elastomeric impact modifier further alters the interfacial interactions and improves the impact properties. Microscopic characterization is used to interpret these results in terms of the dispersion of the high aspect ratio delaminated graphite platelets.

INTRODUCTION

Polymer composites containing conductive carbon-based fillers are used as electrical and thermal conductors and they have found applications in electrically conducting electrodes, antistatic coatings, and electromagnetic interference shielding [1-5]. These composites have attracted considerable research interest because of their superior properties high thermal and electrical conductivity, light weight, ease of processing, and corrosion resistance compared with metals [6, 7]. Graphite-based fillers are also added to polymer matrices to obtain composites with enhanced mechanical properties [8, 9].

Graphite is a well-known carbon-based mineral composed of individual graphene sheets, stacked to thicknesses in the micron range. As a result, it has low aspect ratio and therefore relatively high contents are needed to produce the desired level of enhancement in the electrical and thermal properties of the composites; this results in the deterioration of their mechanical properties like toughness and impact properties. Delamination of graphite to obtain graphite sheets with higher aspect ratios and specific surface area would be therefore beneficial [10, 11]. This can be achieved by separating the graphite stacks to thinner sheets, using processes such as thermal exfoliation of graphite oxide, followed by reduction [12, 13].

Polymer/graphite composites have been studied extensively to demonstrate the advantages of graphite exfoliation on composite properties [14]. Krupa *et al* investigated the physical properties of polyethylene/graphite composites produced by melt compounding and reported an electrical percolation concentration of 11 vol. % with a sharp decrease in elongation at break at increasing graphite content. Kim *et al.* employed melt compounding to improve the physical properties of nylon 6/graphite composites [15].

In this study, the *in-situ* delamination of graphite within a polyamide (PA) matrix is investigated by melt processing.The morphology and structure, flexural and impact properties, and electrical conductivity of the composites are reported. Furthermore, a maleated ethylene-copolymer (EOC-MA) is blended with the PA matrix as impact modifier and the resulting properties of the composites are characterized.

EXPERIMENTAL

Materials

Graphite (> 97% purity, thickness: 45 – 200 μm) was used as supplied by Alcereco (Kingston, ON). A polyamide, (PA) Zytel 151 with density 1.06 g/cm^3 from DuPont Canada and a maleated ethylene copolymer (Exxelor MDEX, density – 0.83g/cm^3, MFI – 7.5 g/ 10 min at 190 °C, MA content – 0.35 wt. %) from Exxon Mobil were used as matrix and impact modifier/toughening agent, respectively.

The (PA) had a melting point of 219.5°C and heat of fusion of 48.6 J/g, as measured by Differential Scanning Calorimetry (DSC). Thus the melt compounding temperature was set at 220 °C. At this temperature the shear viscosity was about 100 Pa.s.

Composites Processingand Characterization

Polymer composites were produced by melt mixing in a Haake Rheomix 600 QC mixer at a temperature of 220 °C and rotor speed of 100 rpm for 6 minutes. Samples for physical testing were compression molded at 230 °C for 2 minutes in a Carver press. Scanning Electron Microscopy (SEM) was carried out on cryo-fractured sample surfaces using a JEOL JMS-840A SEM. For blend morphology investigation, samples were cryo-fractured in liquid nitrogen and the elastomer phase was selectively etched using toluene at 70 °C for 2 hours. Electrical conductivity was measured using a Keithley 8009 Resistivity Test Fixture connected to a Keithley 6517B Electrometer/ High Resistance Meter under DC current at room temperature, for samples having a volume resistivity above 10^6 Ω cm. For samples exhibiting resistivity below 10^6 Ω cm, an Agilent 34401A 6 digit multimeter was used to measure the resistance [16, 17]. Flexural tests were performed on an Instron 3369 Universal tester, at a cross head speed of 1.3 mm/min using the 3-point bending method as per ASTM D790- 71 standard. Notched impact tests were performed on a SATEC Instron machine. The test method adopted was consistent with ASTM D256-78 method B.

RESULTS AND DISCUSSION

Table 1 summarizes the surface energy data obtained for graphite flakes, PA, elastomer and blends with varying elastomer content. Interfacial tension is minimized in composite systems with components having surface energies that are similar. Lower interfacial tension results in improved filler dispersion, due to better filler wetting by the matrix. The filler-matrix interaction is even better if there are functional groups in the matrix able to associate with the filler. PA and its blends are expected to have good interaction with graphite/graphite products, because their surface energies are relatively close, as shown in Table 1.

TABLE (1). Surface energy data for graphene, elastomer, PA and blends, determined at 25°C

Sample	Surface Energy (mJ/m^2)
Graphite	53.61
PA	43.44
EOC-MA	33.35
PA-10% EOC-MA	44.39
PA-30% EOC-MA	45.15

The SEM morphology of 20 wt. % graphite reinforced composite is presented in Fig. 1a. It is evident that some delamination is achieved in-situ during melt processing, as evidenced by the high aspect ratio multilayer graphite flakes (> 100 based on SEM images). Also, the fracture surfaces show that most of the flakes are embedded in the matrix, which is an indication of good filler-matrix interaction. Representative SEM images in Fig. 1b show that the PA/ EOC-MA blend compositions of 70/30 exhibited matrix/droplet morphology with average dispersed droplet sizes of about 0.2 μm. It is also interesting to note that the graphite flakes are well embedded the matrix, suggesting good interfacial adhesion.

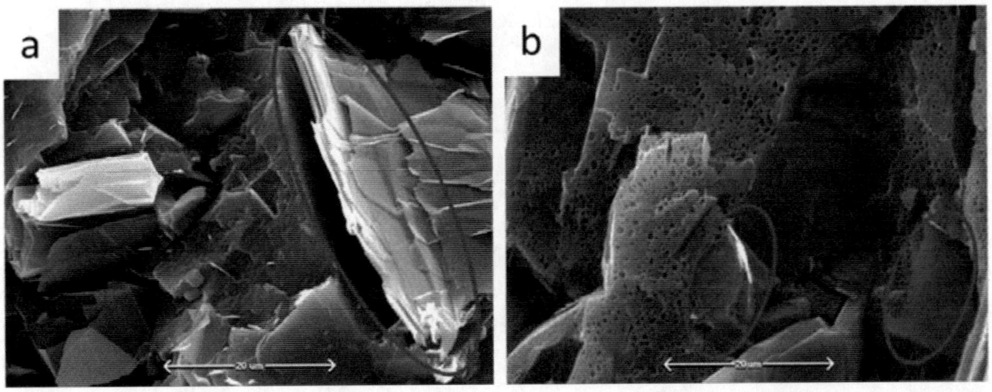

FIGURE 1. SEM images showing a) 30 wt. % PA/ graphite composites showing high aspect ratio multilayer flakes and b) blend morphology of PA/ 30% EOC-MA (etched surfaces – holes correspond to the dispersed phase)

The electrical conductivity data presented in Fig. 2 show that the composites have a percolation threshold of about 12 vol. % (30 wt. %). However, the PA composites are overall more conductive. The effects of elastomer content on the electrical conductivity of the composites are shown in Table 2; this may be attributed to adsorption of the EOC-MA onto the surface of the filler, thus forming an insulating layer [18].

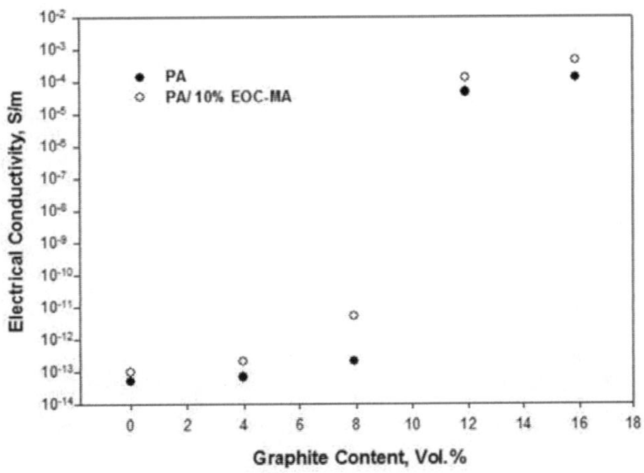

FIGURE 2. Electrical conductivity vs. graphite content for PA and PA/10EOC-MA/ graphite composites

TABLE (2). Effect of elastomer content on the electrical conductivity of PA/EOC-MA/ graphite composites, containing 40 wt% graphite

EOC-MA Content (wt. %)	Electrical Conductivity (S/m)
10	4.17E-04
20	3.33E-04
30	1.11E-06
40	4.42E-07

Fig. 3 shows that the flexural modulus of 40 wt. % graphite reinforced composite is nearly double that of the matrix; this a reflection on the good filler-matrix interaction observed in the SEM images of the composite fracture surface. Importantly, the impact strength was not affected significantly in the composite containing with 40 wt. % graphite. In Fig. 4, the elastomer toughened composites show a gradual reduction in flexural modulus with increasing elastomer content, due to the lower stiffness of the elastomer. However this drop is recovered in the composites containing 10 wt% EOC-MA and 40 wt% graphite.

Composites containing 30 wt% EOC-MA and 40 wt% graphite offered superior impact strength. These results open up an opportunity for a wide range of matrix/elastomer combinations with tailored properties to suit various applications.

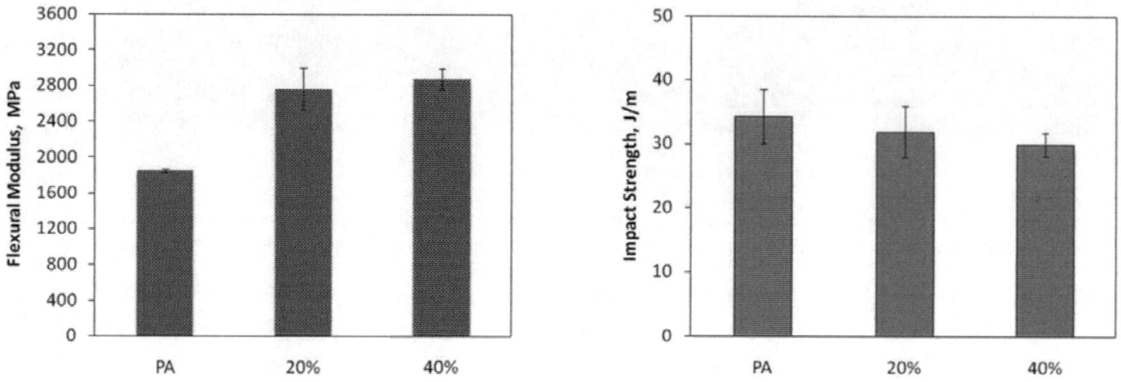

FIGURE 3. Flexural modulus and impact strength of PA/graphite composites

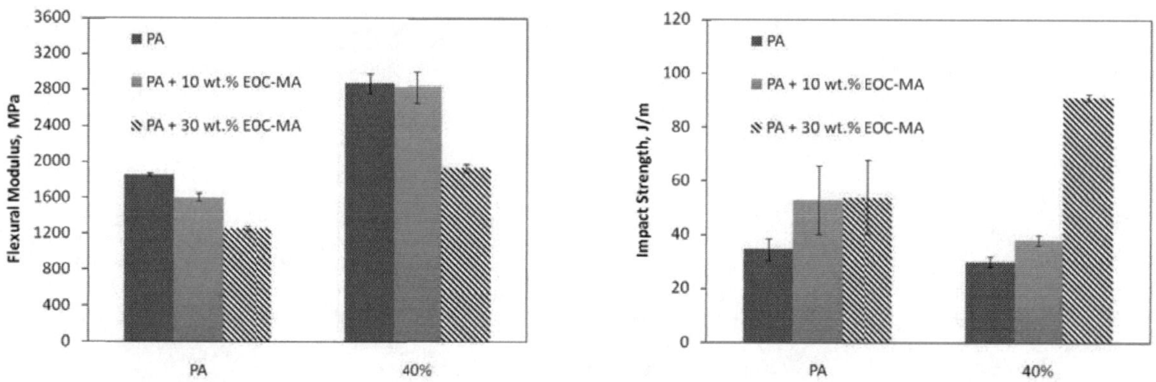

FIGURE 4. Effect of elastomer composition on the flexural modulus and impact strength of PA/graphite composites

CONCLUSIONS

Graphite had good compatibility between the matrix and filler, owing to the similar surface energies, which resulted in substantial enhancements in the mechanical properties of the composites. In the presence of the polyolefin elastomer, the interface between the graphite platelets and the polymer matrix was improved such that ternary composites with excellent impact and flexural properties were obtained, depending on the composition.

REFERENCES

1. A. Celzard, E. Mcrae, G. Furdin andJ. F. Mareche,*J. Phys. Condens. Mater* **9**, 2225–2237 (1997).
2. Q. Zheng, Y. Song, G. Wu, X. Yi,*J. Polym. Sci. Part B: Polym. Phys.* **39**, 2833–2842 (2001).
3. G. Yang,R. Tang andP. Xiao,*Polym. Compos* **18**, 477–483 (1997).
4. S. Paszkiewicz, A. Szymczyk, Z. Spitalky, M. Soccio, J. Mosnacek, T. A. Ezquerra and Z. Roslaniec,*J. Polym. Sci. Part B Polym. Phys.* **50**, 1645–1652 (2012).
5. H. Unal, K. Esmer andA. Mimaroglu, *J. Polym. Eng.* **33**, 351–355 (2013).
6. A. Kasgoz, D. Akın and A. Durmus, *Polym. Eng. Sci.* **52**, 2645–2653 (2012).
7. W. Zhou, C. Wang, T. Ai, K. Wu, F. Zhao and H. Gu,*Compos. A* **40**, 830–836 (2009).
8. G. Sun, X. Li, Y. Qu, X. Wang, H. Yan and Y. Zhang, *Materials Letters* **62**, 703-706 (2008).
9. A. Yasmin, J. Luo and I. M. Daniel, *Compos. Sci. and Technol.* **66**, 1179–1186 (2006).
10. D.D.L.Chung, *J. Mater. Sci. 37*, 1475–1489 (2002).
11. J. R. Potts, D. R. Dreyer,C. W. Bielawski and R. S. Ruoff, *Polymer 52*, 5–25 (2011).
12. S. Pei, and H.-M. Cheng, *Carbon* **50**, 3210–3228 (2012).
13. M. J. McAllister, J.-L. Li, D. H. Adamson, H. C. Schniepp, A. A. Abdala, J. Liu, M. Herrera-Alonso,D. L. Milius,R. Car, R. K. Prud'homme, *Chem. Mater.* **19**, 4396–4404 (2007).
14. R. Sengupta, M. Bhattacharya, S. Bandyopadhyay and A. K. Bhowmick, Prog. Polym. Sci. **36**, 638–670 (2011).
15. S. R. Kim, M. Poostforush, J. H. Kim, and S. G. Lee, *eXPRESSPolym. Lett.* **6**, 476–484 (2012).
16. A.A. Vasileiou, A. Docoslis, M. Kontopoulou, P. Xiang and Z. Ye, *Polymer* **54**, 5230–5240 (2013).
17. O. Osazuwa, M. Kontopoulou, P. Xiang, Z. Ye and A. Docoslis, *Carbon* **72**, 89-99 (2014).
18. O. Osazuwa, K. Petrie, M. Kontopoulou, P. Xiang, Z. Ye and A. Docoslis, *Compos. Sci. Technol.* **73,** 27-33 (2012).

Fabricating High-Quality Composite Laminates by Magnetic Compaction

Maya Pishvar, Mehrad Amirkhosravi, and M. Cengiz Altan*

School of Aerospace and Mechanical Engineering, University of Oklahoma
Norman, OK 73019

altan@ou.edu

Abstract. This paper demonstrates the feasibility of using high-power, permanent magnets in developing high consolidation pressure during a variety of fabrication methods for composite laminates. First, the salient features of this new manufacturing method, expected enhancement of the mechanical properties, and possible improvements in part quality are discussed. Then, the method of applying magnetic pressure over composite laminates with relative ease is presented. To validate this new method, a number of E-glass/epoxy and carbon/epoxy composite laminates were manufactured by utilizing Neodymium Iron Boron (NdFeB) magnets. Specific results will be given for: (i) wet lay-up/vacuum bag, (ii) VARTM, and (iii) oven curing (i.e. out-of-autoclave) processes, where the laminates fabricated with and without the magnetic compaction. The mechanical properties were shown to improve substantially, primarily due to increase in fiber volume fraction and decrease in process-induced voids. In most cases, both the flexural strength and modulus of the composites increased even as much as 60%, while the void content was reduced from 4-5% to less than 1%.

Keywords: Magnetic compaction, Mechanical properties, Open mold processes, Polymer-matrix composites
PACS: 81.05.Qk, 81.20.Hy, 81.70.Bt

INTRODUCTION

A number of manufacturing processes for composite laminates require applying consolidation pressure during cure to increase the fiber content and reduce process-induced defects such as voids, thus leading to higher mechanical properties. In open-mold processes, such as wet lay-up/vacuum bag (WLVB) and vacuum assisted resin transfer molding (VARTM), the consolidation pressure is limited to atmospheric pressure generated by the vacuum (i.e. around 0.1 MPa), while curing in an autoclave may utilize a much higher consolidation pressure, possibly up to 0.7 MPa. Thus, the laminates manufactured using these open-mold processes (i.e. WLVB and VARTM) have lower quality and mechanical performance compared to those fabricated in an autoclave. It is well-known that the composite parts made in an autoclave usually contain fiber volume fraction above 60% and void content less than 2% [1, 2]. For this reason, autoclave curing has been dominant in the industries where the highest mechanical properties and lower weight are required such as aerospace industry.

A few variants of WLVB and VARTM have been developed in which additional consolidation pressure is applied on the vacuum bag using an autoclave. For instance, Abraham et al. [3] applied a 1.2 MPa autoclave pressure on the WLVB assembly and found that plain weave epoxy laminates made by this method had a very high fiber volume fraction of 63.9% and a low void content of 1.57%. Recently, vacuum enhanced resin infusion technology (VERITy) has been proposed in which a 100 kPa autoclave pressure was applied on the vacuum bag after completing the resin infusion in VARTM [4-6]. The resulting unidirectional fabric composite laminates made by this method had the fiber volume fraction of up to 59% and void volume fraction of less than 0.6%. Nevertheless, both the high capital investment and operating costs limited the broader use of autoclaves to produce structural materials. Therefore, development of alternative methods to apply sufficiently high pressures similar to the autoclave pressure is particularly important in manufacturing high-quality composite laminates at a lower cost.

This paper explores the potential of using Neodymium Iron Boron (NdFeB) permanent magnets to apply sufficiently high consolidation pressure during cure of composite laminates. To validate this new method, a number of E-glass/epoxy and carbon/epoxy composite laminates were fabricated by applying magnetic compaction pressure on the lay-up prepared for (i) WLVB, (ii) VARTM, and (iii) oven curing. The fiber volume fraction, void volume fraction, and mechanical properties of the obtained composites were evaluated and compared with those of the samples made without applying external pressure. The experimental results showed that this method is very flexible, easy to implement, and effective in improving the composite parts quality. Even more importantly, by sliding the permanent magnets along the lay-up, it is possible to achieve much higher mechanical properties in large structural laminates.

EXPERIMENTAL DETAILS

In this study, two types of reinforcement materials and three different resins were used. The materials selected for WLVB process were randomly oriented chopped E-glass fabric (Fiberglast part #248) with EPON 862/EPIKURE 3300 (Miller-Stephenson Chemical Co.) epoxy resin system with a moderately low viscosity of 767 mPa s. For VARTM process, the same fabric and a different epoxy resin system (INF 114-INF 211 from PRO-SET) were used. The reason for the change was that the viscosity of the INF 114-INF 211 resin is very low 296 mPa s compared to EPON 862/EPIKURE 3300, so it is more suitable for infusion processes. For oven curing, the laminates were prepared from TENCATE IM7/EX-1522 prepreg consisting of plain weave carbon and an epoxy resin, having an initial resin content of 48 ± 3% by volume.

Neodymium Iron Boron (NdFeB) permanent magnets were used to apply a sufficiently high compaction pressure during cure of laminates. This method has been recently employed by Amirkhosravi et al. [7], and a number of applications of this method have been described in detail in Refs. [8, 9]. First, the typical vacuum bag lay-up was prepared for (i) wet lay-up/vacuum bag, (ii) VARTM, and (iii) oven curing (i.e. out-of-autoclave) on the magnetic bottom tool plate. Then, a set of NdFeB permanent magnets placed together in a square configuration on a magnetic steel plate and were positioned on the lay-up. To induce a large magnetic pressure on the WLVB and VARTM lay-up, high-grade (N52) NdFeB magnets with a maximum operating temperature of 80 °C and the energy product of 413.8 kJ/m^3 were chosen. However, the curing temperature of the IM7/EX-1522 prepreg is higher (177 °C), hence a different grade of magnets needs to be used. For this reason, N38EH NdFeB magnets with a maximum operating temperature of 200 °C and the energy product of 302.4 kJ/m^3 were utilized for applying pressure on the lay-up containing the prepreg materials. Moreover, the applied pressure by the permanent magnets reduces exponentially with increasing separation between the magnets and the bottom tool plate. Thus, the change of the generated pressure with respect to the vacuum bag lay-up thickness for an N52 and N38EH magnet was measured in two separate experiments using the mechanical testing machine (ComTen 705TN). It was found that for the WLVB and VARTM lay-up with the thicknesses of 3.7 mm and 1.8 mm, the N52 magnet can generate, 0.22 MPa and 0.35 MPa compaction pressure respectively. Also, for the oven curing lay-up with a thickness of 2.5 mm, the N38EH magnet can generate 0.23 MPa compaction pressure. It is also important to note that the magnetic pressure applied on the lay-up increases during cure of the laminates due to the decrease in the lay-up thickness.

LAMINATE FABRICATION AND CHARACTERIZATION

The WLVB lay-up was prepared by first laying the preforms on a tool plate. Each ply was saturated with resin using brushes and rollers, and a perforated release film, 0.3 mm thick caul plate, and breather were placed on top of the plies. Then, vacuum bag was placed over the lay-up and sealed around the mold, and a vacuum was applied. The curing process started at room temperature for 240 min, followed by demolding of the laminates and post-curing in an oven at 82 °C for 90 min and 150 °C for 90 min.

The VARTM lay-up was prepared by placing the preforms on the tool plate with a layer of release film on top, accompanied by a distribution media and peel ply on the inlet side. Inlet and outlet tubing were arranged on both sides of the fabric, and a vacuum bag covered the whole lay-up. Once the vacuum was drawn, the resin infusion was started. After full saturation of the preforms, the inlet was clamped. The cure cycle was 45 min at room temperature, followed by 8 h at 60 °C.

The oven curing lay-up consisted of the prepreg, perforated release film, breather plies, and 0.3 mm thick caul plate which were assembled on the tool plate. Then, the assembly was placed in an oven, and the vacuum was drawn. Finally, the cure cycle started with a heat ramp of 2.2 °C/min to 177 °C, continued for 180 min isothermal hold, and finished by cooling to room temperature at 3.5 °C/min rate.

To determine the effectiveness of applying magnetic pressure on the overall quality of WLVB, VARTM, and oven curing laminates, six fabrication scenarios, as shown in Table 1, were considered. For each fabrication scenario, two laminates were made to verify the results. In the first scenario (WLVB-6), 6-ply E-glass/epoxy laminates made with WLVB process without external pressure and they were considered as the reference WLVB laminates. In the second scenario (WLVB-6-M), the same material and lay-up sequence and geometry as WLVB-6 scenario were employed. However, once the vacuum was pulled, nine NdFeB, N52-2.54 cm×2.54 cm×1.27 cm thick magnets (magnetized through-the-thickness) were applied on the WLVB lay-up. The rest of the process was identical to the scenario WLVB-6. In the third scenario (VARTM-6), the reference 6-ply E-glass/epoxy VARTM laminates were manufactured without external pressure. The fourth scenario (VARTM-6-M) was identical to the third one except that twenty-five above-mentioned magnets were placed on the VARTM lay-up throughout the process: from start of the infusion to the full cure. To avoid the localized force on the laminates and to avoid indentation marks, a 0.3-mm thick caul plate was used between the magnets and the vacuum bag. In the fifth scenario (Oven-8), oven curing of the 8-ply carbon/epoxy prepreg under only vacuum was used as the reference. The sixth scenario (Oven-8-M) was identical to scenario Oven-8, except for the use of sixteen through-the-thickness magnetized NdFeB-N38EH 2.54 cm×2.54 cm×5.08 cm thick magnets to apply compaction pressure in the oven. As mentioned earlier in Ref. [2], in order to minimize the void content of the final parts, the magnetic pressure was applied on the lay-up once the minimum viscosity of resin was reached (i.e., at 55 min from the start of cure cycle and at a temperature of 145 °C).

The fiber volume fraction, void volume fraction, and flexural properties of the composites manufactured with and without magnetic compaction pressure were determined according to ASTM D3171-15, ASTM D2584-11, and ASTM D7264/D7264M-15.

TABLE (1). Six fabrication scenarios used in the manufacturing of fiber-reinforced epoxy matrix composites.

Fabrication scenario	Fiber	Resin	Fabrication process	Permanent magnet type
WLVB-6 [7]	Random mat E-glass	EPON 862 / EPIKURE 3300	6-py, WLVB, without external pressure	—
WLVB-6-M [7]	Random mat E-glass	EPON 862 / EPIKURE 3300	6-ply, WLVB, with magnetic pressure	1.27 cm thick N52
VARTM-6 [9]	Random mat E-glass	INF114 / INF211	6-ply, VARTM, without external pressure	—
VARTM-6-M [9]	Random mat E-glass	INF114 / INF211	6-ply, VARTM, with magnetic pressure	1.27 cm thick N52
Oven-8	IM7 carbon	EX-1522	8-ply, Oven curing, without external pressure	—
Oven-8-M	IM7 carbon	EX-1522	8-ply, Oven curing, with magnetic pressure	5.08 cm thick N38EH

RESULTS AND DISCUSSION

Average thickness, fiber volume fraction, and void volume fraction of the laminates

The average thickness, fiber volume fraction, and void volume fraction of the laminates fabricated under six manufacturing scenarios are presented in Table 2. In all scenarios, applying magnetic compaction pressure on the vacuum bag lay-up resulted in thinner laminates than those made without external pressure. The results show that the average thicknesses of 6-ply random mat/epoxy laminates fabricated by WLVB were reduced by 37% to 2 mm using magnetic compaction. Table 2 also represents that due to the magnetic pressure the thickness reduction in 6-ply random mat laminates fabricated by VARTM was 20% to 1.2 mm and in the 8-ply prepreg laminates made in the oven was 14% to 1.7 mm. Results also indicate that the magnetic compaction caused an increase in fiber volume fraction and a substantial decrease in void volume fraction of all scenarios compared to those made without external pressure. In WLVB laminates the fiber volume fraction increased by 55% to 27% and void volume fraction

TABLE (2). The average thickness, fiber volume fraction, and void volume fraction for the composite laminates manufactured by six different scenarios (n=6 for fiber and void volume fractions and n=35, 42, and 18 for thicknesses measurements of laminates made by WLVB, VARTM, and oven curing, 95% confidence intervals).

Fabrication scenario	Average thickness (mm)	Fiber volume fraction (%)	Void volume fraction (%)
WLVB-6	3.18 ± 0.02	17.3 ± 0.8	5.81 ± 1.24
WLVB-6-M	1.99 ± 0.04	26.9 ± 2.0	2.71 ± 0.36
VARTM-6	1.45 ± 0.02	45.7 ± 0.1	1.86 ± 0.72
VARTM-6-M	1.16 ± 0.01	51.8 ± 1.0	0.21 ± 0.09
Oven-8	1.96 ± 0.01	51.2 ± 0.7	4.81 ± 0.54
Oven-8-M	1.68 ± 0.03	60.4 ± 0.7	0.90 ± 0.16

decreased by 53% to 2.7%. Although the quality of WLVB laminates was improved significantly using magnetic compaction pressure, there is still a potential for further improvement by employing stronger magnets. In addition, the fiber volume fraction of VARTM laminates was increased by 13% to 52% while the void volume fraction was reduced to very low 0.2%. Indeed, applying magnetic pressure on the vacuum bag lay-up in the oven is very innovative because this method could be a possible low-cost alternative to autoclave curing. The results show that the magnetic compaction pressure in Oven-8-M samples could improve fiber volume fraction by 18% to 60% and decrease the void volume fraction from relatively high 4.8% to less than 1%. Therefore, all the results presented in Table 2 demonstrate that applying magnetic pressure is a simple but very effective method to improve the laminate quality made by a wide variety of fabrication techniques.

Flexural properties of the laminates

The flexural strength and modulus of the laminates fabricated by different manufacturing scenarios are shown in Figures 1 and 2. The void and fiber volume fractions of each scenario are also presented in Figures 1 and 2 to illustrate their influence on flexural properties. In all fabrication scenarios, flexural strength and modulus of the laminates made by magnetic compaction are considerably higher than those in the laminates made without external pressure due to the increased fiber volume fraction and decreased void content (see Figures 1 and 2). It is well-known that higher fiber volume fraction and lower void content result in higher mechanical properties of the composite laminates [10-12]. Figure 1 indicates that applying magnetic compaction pressure on the vacuum bag lay-up of the laminates made by WLVB, VARTM, and oven curing caused a substantial improvement in flexural strength by 60% to 254 MPa, by 39% to 365 MPa, and by 21% to 871 MPa, respectively. In addition, the flexural modulus of the fabricated laminates was improved significantly by 46% to 9.9 GPa in WLVB laminates, by 46% to 15.5 GPa in VARTM laminates, and by 14% to 47.6 GPa in the laminates made in the oven (shown in Figure 2).

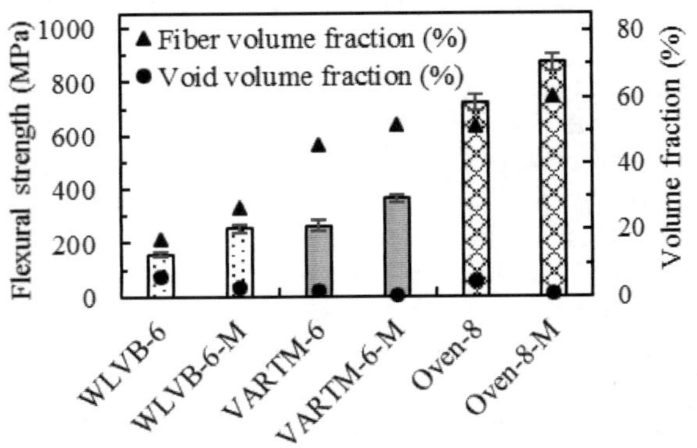

FIGURE 1. Flexural strength of composite laminates fabricated by six different scenarios. Fiber and void volume fractions for each scenario are also shown for reference. (n=7, 14, and 12 for laminates made by WLVB, VARTM, and oven, respectively, 95% confidence interval error bars)

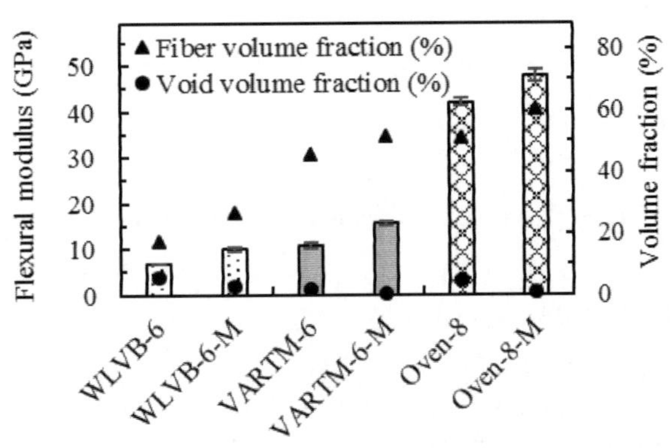

FIGURE 2. Flexural modulus of composite laminates fabricated by six different scenarios. Fiber and void volume fractions for each scenario are also shown for reference. (n=7, 14, and 12 for laminates made by WLVB, VARTM, and oven, respectively, 95% confidence interval error bars)

Interestingly, applying magnetic pressure caused higher improvement in flexural properties of the WLVB laminates compared to those in VARTM and oven curing laminates. This can be explained by the higher initial resin content of the WLVB laminates. When the magnetic pressure is applied, more resin can be squeezed out in WLVB compared to those in VARTM and oven curing. For this reason, applying magnetic pressure increased the fiber content of the WLVB laminates (≈55%) more than those in VARTM (≈13%) and oven curing laminates (≈18%).

The specific flexural properties (flexural properties divided by the bulk density) of the laminates made by six different scenarios are reported in Table 3 and are compared with those of the reference laminates. The specific flexural strength of the WLVB, VARTM, and oven curing laminates increased around 43%, 31%, and 14%, respectively with applying magnetic compaction pressure compared to those made without magnets. In addition, the specific flexural modulus is 29%, 38%, and 7% higher in WLVB-6-M, VARTM-6-M, and Oven-8-M laminates, respectively than those in the reference laminates. As a result, applying magnetic pressure in the fabrication of laminates has greatly increased their specific flexural properties, thus expanding their use in structural applications.

TABLE (3). The specific flexural properties of the laminates manufactured by six different scenarios and the percentage increase in specific flexural properties of the laminates made with magnets compared to those made without magnets

Fabrication scenario	Specific flexural strength (MPa/kg m^{-3})	Percentage increase in specific flexural strength (%)	Specific flexural modulus (GPa/kg m^{-3})	Percentage increase in specific flexural modulus (%)
WLVB-6	119	—	5.1	—
WLVB-6-M	170	43	6.6	29
VARTM-6	152	—	6.1	—
VARTM-6-M	199	31	8.4	38
Oven-8	462	—	26.9	—
Oven-8-M	528	14	28.9	7

CONCLUDING REMARKS

The feasibility of using permanent magnets to produce a high consolidation pressure during cure of composite laminates fabricated by (i) wet lay-up/vacuum bag, (ii) VARTM, and (iii) oven curing (i.e. out-of-autoclave) was presented. The E-glass/epoxy and carbon/epoxy composite laminates made under magnetic pressure exhibited higher fiber volume fraction and lower void content, as compared with those made without magnetic pressure. In WLVB, VARTM, and oven curing laminates, the fiber volume fraction was increased from 17% to 27%, from 46% to 52%, and from 51% to 60%, respectively. In addition, the high void volume fraction of 5.8% in WLVB was decreased to 2.7% where the magnetic consolidation pressure was utilized. Similarly, in both VARTM and oven curing laminates made under magnetic pressure, the void content was decreased from 2-5% to less than 1%. The flexural properties of the WLVB, VARTM, and oven curing laminates were dramatically improved even as much as 60%, by the incorporation of magnetic pressure. These results demonstrate that permanent magnets can generate sufficiently high compaction pressures that can have a significant effect on the quality and mechanical properties of laminates fabricated by wet lay-up/vacuum bag, VARTM, and oven curing processes.

REFERENCES

1. L. Grunenfelder and S. Nutt, *Compos. Sci. Technol.*, **70**, 2304-2309 (2010).
2. L. Liu, B.M. Zhang, D.F. Wang, and Z.J. Wu, *Compos. Struct.*, **73**, 303-309 (2006).
3. D. Abraham, S. Matthews, and R. McIlhagger, *Compos. Part A*, **29**, 795-801 (1998).
4. N. Gupta and R. Sundaram, *Compos. Part A*, **40**, 1065-1070 (2009)
5. K. K. Verma, B. Dinesh, K. Singh, K. M. Gaddikeri, and R. Sundaram, *Procedia Mater. Sci.*, **6**, 331-340 (2014).
6. K. K. Verma, B. Dinesh, K. Singh, K. M. Gaddikeri, V. Srinivasa, R. Kumar, and R. Sundaram, *J. Ind. Inst. Sci.*, **93**, 621-634 (2013)
7. M. Amirkhosravi, M. Pishvar, and M. C. Altan, *Compos. Part A*, **98**, 227-237 (2017).
8. M. Amirkhosravi, M. Pishvar, and M. C. Altan, ASC Conference Proceedings 32nd, American Society for Composites, West Lafayette, IN, 2017, ASC2017-317.
9. M. Pishvar, M. Amirkhosravi, and M. C. Altan, ASME Conference Proceedings 2017, International Mechanical Engineering Congress and Exposition, Tampa, FL, 2017, IMECE2017-72019.
10. S.F.M. de Almeida and Z.S.N. Neto, *Compos. Struct.*, **28**, 139-148 (1994).
11. P. Olivier, J. Cottu, and B. Ferret, *Composites*, **26**, 509-215 (1995).
12. D. Romanzini, A. Lavoratti, H. L. Ornaghi, S. C. Amico, and A. J. Zattera, *Mater. Des.*, **47**,9-15(2013).

Balance of Thermal Conductivity and Dielectric Strength of Ceramics Filler Filled Polymer Composites

Takushi Saito[a,*], Kosuke Soga[b], Tatsuya Kawaguchi[a] and Isao Satoh[a]

[a] *Department of Mechanical Engineering, Tokyo Institute of Technology, O-okayama, Meguro-ku, Tokyo, JAPAN*
[b] *Development Center, Hitachi Construction Machinery Co., Ltd., Kandatsu-machi, Tsuchiura-shi, Ibaraki, JAPAN*

Abstract. Polymer composite film containing ceramics filler used for electric power devices is desired to have good thermal conductivity and superior dielectric strength, simultaneously. However, the improvement of thermal conductivity and the maintenance of dielectric strength is usually conflicting to each other. Therefore, balanced condition of the thermal conductivity and the dielectric strength was investigated through a numerical simulation in this study. The filler shape was assumed as sphere. The influence of filler dispersion condition on the thermal conductivity and dielectric breakdown strength of the composite film was discussed according to the variation of thermal conductivity ratio and dielectric breakdown strength ratio between the polymer and filler. Based on the results obtained, an evaluation formula that comprehensively evaluated the performance of composite material was introduced. By using this formula, it was shown that the desirable dispersion condition depended on the material combination of filler and matrix polymer.

Keywords: Thermal conductivity, Dielectric breakdown strength, Polymer composite film

INTRODUCTION

Composite material consisting of ceramics filler and polymer matrix has a great potential of use for the electric power devices such as the electric motor and invertor, because the composite material has good electrical insulation and high thermal conductivity comparing with the original polymer material. It is generally known that the thermal conductivity of the composite material is varied with the filler content ratio and aggregation status. On the other hand, excessive increment of the filler content results in lowering the voltage of dielectric breakdown[1)]. This is caused by the fact that the dielectric breakdown voltage of the ceramics filler such as Boron Nitride and Aluminum Oxide, is usually lower than that of polymer matrix such as polyimide and polyamide.

Composite materials of this type are normally assumed to be used as a sheet to be inserted between a heat source and a heat sink. Therefore, a thick sheet having higher breakdown strength cannot be adopted because the thermal resistance increases. In other words, there is a trade-off relationship between thermal conductivity and dielectric breakdown strength, and it is not easy to determine optimum filler dispersion state, sheet thickness and filler concentration by trial and error approach. Although several studies[2, 3)] have been conducted to increase the thermal conductivity of the composite material and to improve the dielectric breakdown strength, discussions dealing with both physical properties at the same time are not sufficiently done.

In this study, three-dimensional numerical simulation of the influence of filler dispersion condition on the thermal conductivity and dielectric breakdown strength of the composite material was conducted, and an appropriate gap between fillers was discussed. In addition, an expression for comprehensively evaluating the performance of composite materials was introduced using thermal conductivity and dielectric breakdown strength as parameters. By using this formula, mapping of the desired aggregation/dispersion conditions that varies with the physical properties of the filler and polymer matrix was presented.

SIMULATION METHOD OF DIELECTRIC BREAKDOWN STRENGTH AND THERMAL CONDUCTIVITY

Method for Creating Pseudo Filler Distribution

The thermal conductivity and dielectric breakdown strength of the filler dispersed composites are influenced not only by the shape, physical properties and content ratio of filler but also by the arrangement of fillers inside the composites. Therefore, the pseudo filler distribution within the composite material was generated by using random numbers. The schematic procedure is shown in Fig. 1. First, the center position of a filler is determined by using random number, and the filler is arranged unless it interferes with other filler. If the filler interferes, the center position of the filler is recalculated until the filler does not interfere with the other filler. This procedure was repeated until the filler content ratio reached the prescribed value, then a virtual filler distribution for numerical simulation was obtained.

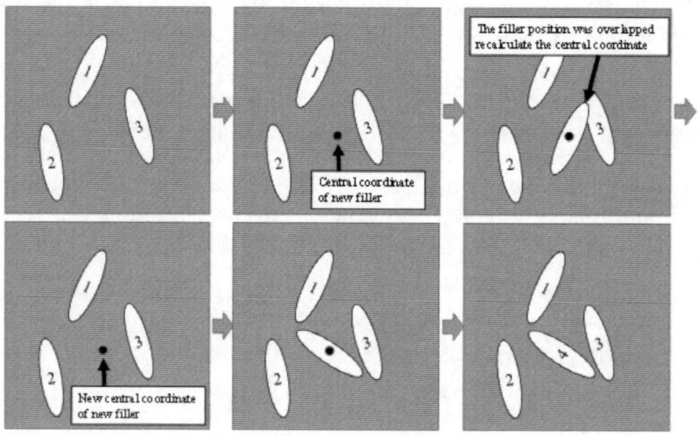

Figure 1 Schematic procedure of pseudo filler arrangement

Calculation of Thermal Conductivity

Conductive heat transfer inside the filler dispersed composite material was numerically simulated by a finite difference method, and the effective thermal conductivity of the calculation domain was estimated by the same method as the comparative method for the thermal conductivity measurement. As shown in Figure 2, a constant temperature difference between the upper and lower electrodes was maintained by giving Dirichlet boundary conditions to the xy plane at y=0 and Y. By calculating the temperature distribution in the composite material, the effective thermal conductivity of the composite material k_e was calculated from the average heat flux value passing through the xz plane at y=0. Equations (1) and (2) are the formulas of this procedure.

$$Q = \sum\left(-k\frac{\partial T}{\partial y}\Delta x \Delta z\right) \tag{1}$$

$$k_e = Q\frac{Y}{XZ\Delta T} \tag{2}$$

where Q, T, k mean thermal flow, temperature, and thermal conductivity. The coordinates consists of x, y, z, and X, Y, Z indicate the calculation domain size.

Calculation of Dielectric Breakdown Strength

The dielectric breakdown under DC load was modeled. In this study, a characteristic assumption was adopted that kept the electric field intensity constant at a region where the dielectric breakdown occurred. It was also assumed that dielectric breakdown in the calculation domain generated, when the local breakdown regions connected the upper and lower electrodes. The derivation of the characteristic dielectric constant in this study is summarized in Equation (3).

$$\varepsilon_{\text{new}} = \begin{cases} \varepsilon_0 & (E < E_{bd}) \\ \varepsilon_{\text{old}}\left(\dfrac{9E_{bd}+E}{10E_{bd}}\right) & (E > E_{bd}) \end{cases} \tag{3}$$

where ε, E, E_{bd} were dielectric constant, electric field intensity, and electric field intensity at dielectric breakdown. Subscripts of 0, old, new indicate initial, current step, next step, respectively. The voltage causing the dielectric breakdown was estimated by raising the DC load by 50 V at one step and repeating the above calculation procedure.

Evaluation Approach of Filler Aggregation Status

As the physical properties of the composite material, the dielectric breakdown strength E_{bdc} and the thermal conductivity k_c will be obtained by the numerical simulation. On the other hand, the dielectric breakdown voltage and thermal resistivity that are obtained in the actual measurement (Equations (4) and (5)) need to consider the sheet thickness d.

$$V_{bdc} = E_{bdc} \cdot d \tag{4}$$

$$R = \frac{d}{k_c} \tag{5}$$

Since these values include the sheet thickness d, a new evaluation method that did not include the sheet thickness was necessary for evaluating the performance of the composite sheet. Therefore, an evaluation number given by Equation (6) was introduced in this study.

$$EV = \frac{V_{bdc}}{R} = E_{bdc} \cdot k_c \tag{6}$$

As shown in the equation, EV is the product of dielectric breakdown strength and reciprocal of thermal resistance, which is increased by the improvement of overall performance of the composite sheet.

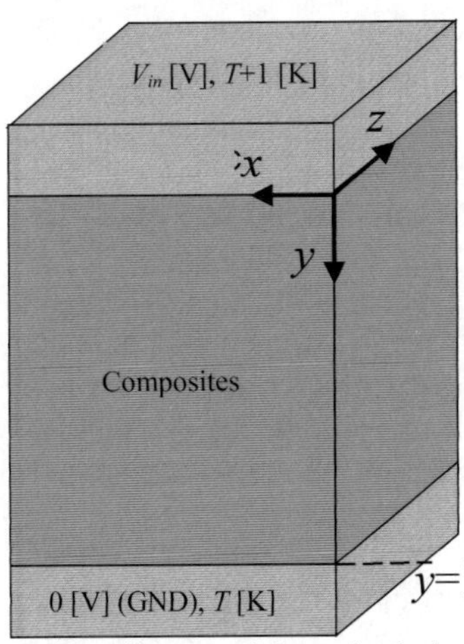

Figure 2 Coordinates of the calculation domain

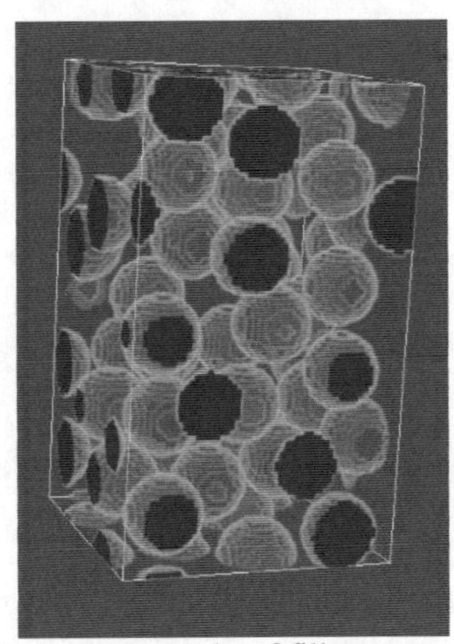

Figure 3 Example of filler arrangement

NUMERICAL SIMULATION MODEL

The shape of the filler used in this study was sphere with a diameter of 8 μm. In order to investigate the effect of filler gap on the thermal conductivity and dielectric breakdown strength, the minimum gap between the fillers was varied from 0 to 10% of the filler diameter. By assuming a sheet shape of filler dispersed composite material, the calculation domain had a rectangular shape having a thickness of 100 μm in the application direction of the voltage gap and the temperature gap and having a length of 60 μm in the in-plane direction (Figure 2.) Note that periodical boundary conditions were applied for xy and yz planes. The volume content of the filler was fixed at 30%. In the simulation of thermal conductivity, the specific thermal conductivity of the filler (k_f*) was varied in the range of 1 to 1000 with the thermal conductivity of polymer matrix as the denominator. On the other hand, the specific dielectric breakdown strength of the filler (E_{bdf}*) was varied in the range of 0.1 to 1 with the dielectric breakdown strength of polymer matrix as the denominator. An example of filler arrangement in the calculation domain is shown in Figure 3. Note that the polymer matrix in Figure 3 is transparent to clearly show the filler arrangement in the domain.

RESULTS AND DISCUSSION

Estimated Results of Thermal Conductivity

Figure 4 shows the effective thermal conductivity at the filler content of 30%. The results of agglomerated condition (gap value 0%) and dispersed condition (gap value 5, 10%) are shown in the figure. If the specific thermal conductivity of the filler was 10 or less, the distance between the fillers had little influence on the specific thermal conductivity of composite material. On the other hand, when the specific thermal conductivity of the filler exceeded 10, the increase rate in specific thermal conductivity of the agglomerated condition became larger than that of the dispersed condition. This result was caused by the fact that the effective conduction heat transfer in the bulk composite material was reduced by the increase of gap between each filler.

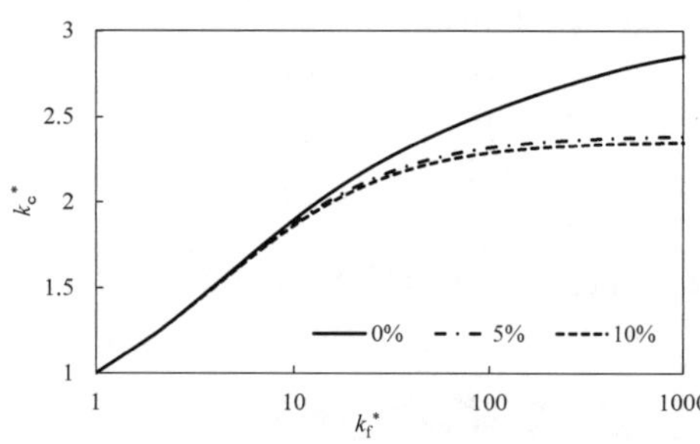

Figure 4 Effect of filler gap on thermal conductivity

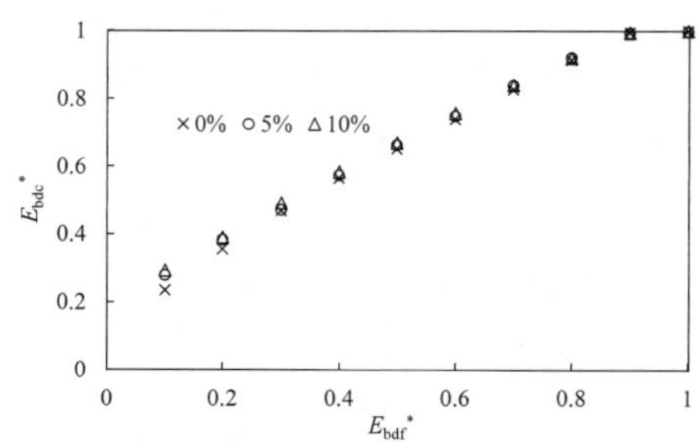

Figure 5 Effect of filler gap on dielectric breakdown strength

Estimated Results of Dielectric Breakdown Strength

Figure 5 shows the effect of filler gap on the dielectric breakdown strength. When the specific dielectric breakdown strength of the filler was 0.5 or less, the effective dielectric breakdown strength of the composite material improved with the increase of filler gap (dispersed condition). On the other hand, when the specific dielectric breakdown strength of the filler approached the breakdown strength of the matrix phase, the specific breakdown strength of the composite material became almost the same regardless of the dispersed condition.

Comprehensive Performance of Filler Dispersed Composite Sheet

Figure 6 shows the change of total evaluation number *EV** of 4 types of fillers having different intrinsic dielectric breakdown strength. Higher value of *EV** was better performance, basically. Although the comprehensive performance of the composite sheet was improved by ensuring a certain gap between the fillers when the thermal conductivity of the filler was low, the comprehensive performance was improved by decreasing the filler gap in case of high thermal conductivity. It was also shown that the effect of filler gap on *EV** varied with the dielectric breakdown strength of the filler. Furthermore, the desirable filler gap condition also varied with the specific dielectric breakdown strength of the filler. By using the total evaluation number introduced in this study, it becomes possible to estimate an appropriate filler gap based on the thermal conductivity and dielectric breakdown strength of the filler and matrix.

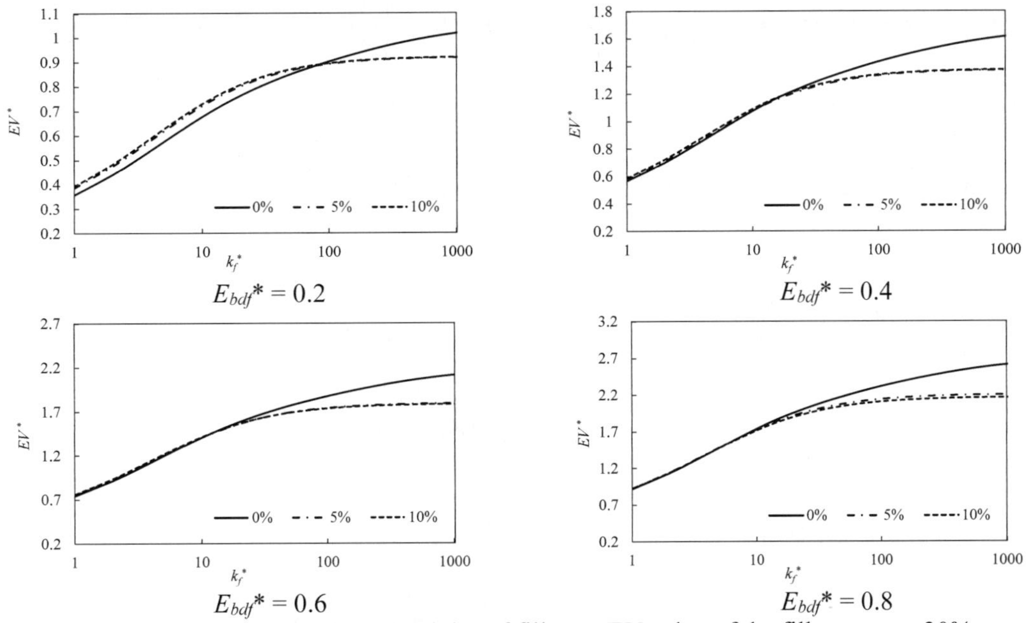

Figure 6 Effect of thermal conductivity of filler on EV value of the filler content 30%

Figure 7 shows the change of the intersection point of EV^* on the graph plane of the specific thermal conductivity and specific dielectric breakdown strength. By using this kind of chart map, an appropriate filler gap condition for better performance of the composite material was possible to determine from a comprehensive viewpoint.

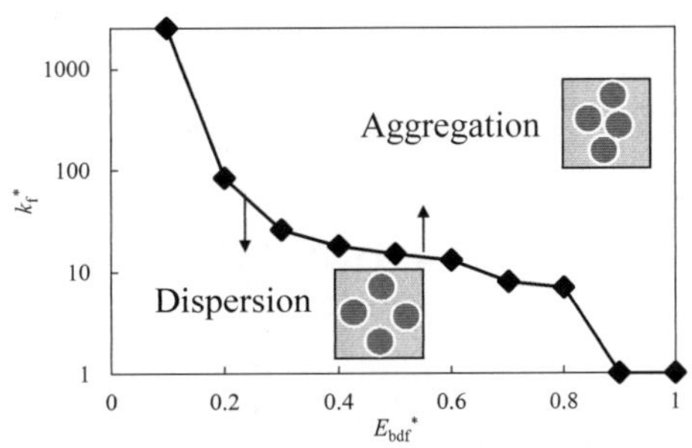

Figure 7 Chart map for the filler aggregation conditions

CONCLUSIONS

In this study, the effect of filler gap on the thermal conductivity and dielectric breakdown strength of the spherical filler dispersed composite material was investigated by numerical simulation. In order to comprehensively discuss the effect of filler gap on the thermal conductivity and dielectric breakdown strength, the evaluation number EV^* was introduced and the desirable filler dispersion state to improve the overall performance of the composite material was discussed. The model showed that the desirable aggregation condition of the filler was changed by the physical properties of the components of the composite material; that is, when the dielectric breakdown strength of the filler was much smaller than that of polymer matrix, dispersed filler condition was desirable. However, the agglomeration state of the filler was desirable if the dielectric breakdown strength of the filler was close to that of polymer matrix.

REFERENCES

1. Y. Agari, A. Ueda, M. Tanaka, S. Hagai, *Journal of Applied Polymer Science*, **40**, Issue 5-6, pp. 929-941 (1990)
2. K. Mimura, M. Masaki, Y. Nakamura, T. Nishimura, *IEICE Transactions on Electronics*, **J95-C**, No. 11, pp. 434-438 (2012)
3. A. Boudenne, L. Ibos, M. Fois, E. Gehin, J. C. Majeste, *Journal of Polymer Science Part B Polymer Physics*, **42**, Issue 4, pp. 722-732 (2004)

Potential supercapacitors made of polymer/perovskite composites

Andre Luis Marcomini (presenting author)[a], Jeferson A. Dias[a], Marcio R. Morelli[a]
Rosario E. S. Bretas[a*] (reference author)

[a]Department of Materials Engineering, Universidade Federal de São Carlos, Rodovia Washington Luiz, Km 235, PO Box 676, São Carlos, SP, 13565-905, Brazil.

Presenting author's e-mail:andre.marcomini@gmail.com
Reference author's email bretas@ufscar.br

Abstract. The objective of this project was to obtain high dielectric constant polymeric composites to be used as supercapacitors. The composites were formed by a polymeric matrix, polyvinylidene difluoride (PVDF) and perovskite particles of high dielectric constant ($Na_{1/3}Ca_{1/3}Bi_{1/3}Cu_3Ti_4O_{12}$). The perovskite was synthetized by solid-state reaction and added to the polymeric matrix by casting, followed by a melting mixing and injection molded processes. The masterbatches produced by casting were melted together with the matrix in a torque rheometer HaakeTM by RheomixTM ; subsequently each sample was molded in a Ray-RanTM Test Sample Injection Molding Press. After that, the potential application as supercapacitors was evaluated by measuring the dielectric properties in AC by impedance spectroscopy. The morphology of these composites was analyzed by scanning electron microscopy (SEM). The evaluation showed particle agglomeration and an improvement in the PVDF dielectric permittivity. The Maxwell Garnett equation (MG) and a new model (NM) were compared with the experimental data, showing that their predictions was far below the obtained ones.

Keywords: Supercapacitor, PVDF, perovskite, high-permittivity, composites

PACS: 77, 81

INTRODUCTION

In the last years, a high demand for materials that can be used as supercapacitors has increased exponentially, due to the high miniaturization of electronic devices. The notebooks, tablets and smartphones markets require that these devices be smaller and more functional [1]. These devices are made of a series of capacitive components and because the trend is to miniaturize still more, it is necessary also to obtain new capacitor materials capable of storing a high amount of energy per volume.

Historically, ceramic materials have been traditionally used for that application; however, in the last decades, it has been usual to add these ceramic materials as a powder to polymeric matrices [1-3], forming a composite with lower loss factor, lower fragility and endless molding possibilities; this type of composite, made of dielectric ceramic and dielectric polymer are called DD composite. In addition, it has been observed that the addition of conductive fillers below the percolation concentration allows the development of a composite also with good electrical energy storage capacity [2, 3]. In this type of composite, called conductive-dielectric composite or CD composite, the conductive filler acts as the electrode and the polymer between the conductive particles acts as the dielectric. In our project, we proposed to develop a DD composite, based on the work of Kum-Onsa et al [4] that synthesized for the first time an inorganic perovskite of chemical composition $Na_{1/3}Ca_{1/3}Bi_{1/3}Cu_3Ti_4O_{12}$ (NCBCTO). The dielectric constant of this material at 1 kHz was reported to be~25000; later [5], the authors developed a composite with polyvinylidene difluoride (PVDF) also with good dielectric properties. The incorporation of the NCBCTO in the PVDF matrix was made by powder mixing using a ball mill. In our work, we employed a different mixing route: casting followed by melt mixing and injection molding.

EXPERIMENTAL

Materials

Perovskite synthesis: Na_2CO_3 and $CaCO_3$ from Synth; Bi_2O_3, CuO and TiO_2 from Sigma Aldrich.

Polymeric matrix: PVDF Kynar® 1000HD from Arkema Inc.

Methods

The perovskite was synthesized via solid-state reaction; oxides and carbonates were mixed in a stoichiometric proportion and de-agglomerated in alcohol suspension in a ball mill during 24 h. The mixture was calcinated at 900°C during 6h; the perovskite phase was confirmed by wide angle x-rays diffraction (WAXS). The perovskite powder was mixed with the PVDF by solution. A PVDF solution in dimethylformamide (DMF) (0.2 g/ml) was first prepared by heating at 80°C during 1h; a perovskite suspension in DMF (0.25g/ml) prepared at 80°C during 1h was added to the PVDF solution , under magnetic stirring at 80° C during 3h. After this period, methanol (1 L) was added to the suspension to promote the composite precipitation. After precipitation, the material was filtered 4 times to obtain a precipitate without solvent. Each filtering produced a masterbatch with different perovskite concentration, this concentration increasing from the first masterbatch to the last one. Each masterbatch was then mixed with the PVDF in a torque rheometer Haake® from Rheomix® at 220°C and 120rpm during 8 min. The final concentrations (wt % perovskite) of the composites were: 1, 5, 10 and 20% , named PVDF_NCBCTO_1, PVDF_NCBCTO_5, PVDF_NCBCTO_10 and PVDF_NCBCTO_20, respectively. The pure PVDF matrix was also melt in the Haake rheometer at the same conditions that the composites and it was named just PVDF. Samples of the composites were prepared by transfer injection molding using a Ray-Ran™ Test Sample Injection Molding equipment. The morphological characterization was made by scanning electron microscopy (SEM) using a FEG-XL30 equipment from Philips. The dielectric properties were measured in an impedance analyzer Solartron-1260.

RESULTS AND DISCUSSION

Figure 1 shows some SEM micrographs of the composites.

FIGURE 1. (a) Composite PVDF_NCBCTO_1, (b) Composite PVDF_NCBCTO_20

The average particle size of the pure perovskite was 980 nm (not shown). From the figure it can be observed that the perovskite particles in the low concentration composite had an average particle size of 1533 nm and in the high concentration composite was between 46 nm and 4 μm; that means that the mixing with the PVDF promoted the agglomeration of the perovskite particles. The distribution of the agglomerates, however, was good.

Figure 2 shows the dielectric permittivity ε'(ω) of some of the samples , where ω is the frequency.

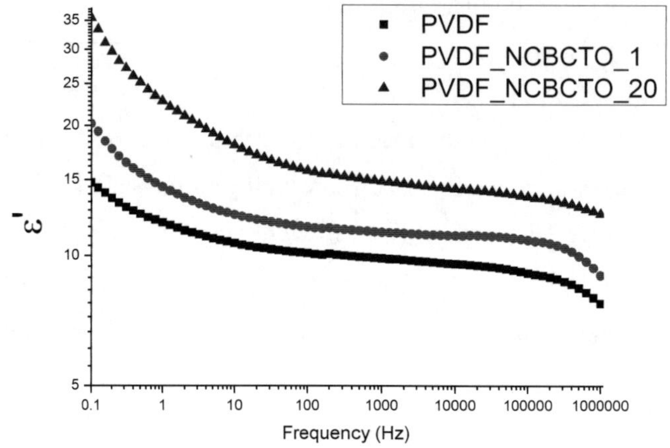

FIGURE 2. Dielectric permittivity ε'(ω) of some of the samples.

It can be observed that the dielectric permittivity decreased with the increase of the frequency for all the samples; however, between the frequency range of 100 and 10^5 Hz, ε' increased slightly with the increase in perovskite concentration from 10 (pure PVDF) to 15 (20% concentration), which configures an increase of 50% of the PVDF dielectric permittivity. It can also be observed that as the higher the perovskite concentration, the higher is the fall of the dielectric permittivity at low frequencies (0.1 -10 Hz), which is what it is found in the literature [5].

Figure 3 shows the dielectric loss (ε"(ω)) of the samples.

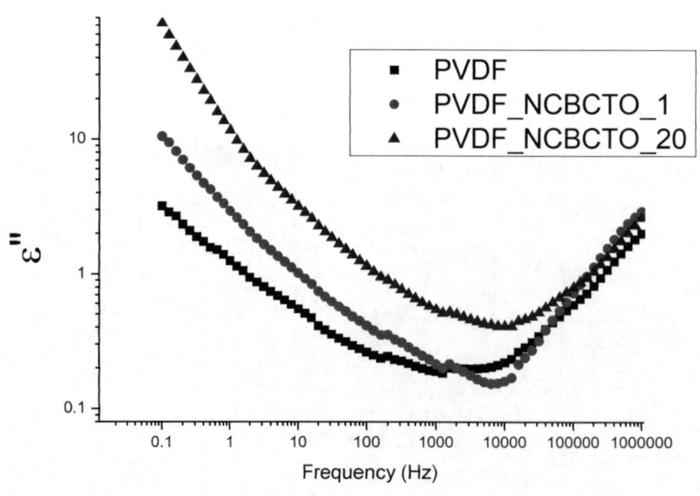

FIGURE 3. Dielectric loss ε"(ω) of the samples.

It can be observed that ε"(ω) decreased with the frequency up to approximately 10^4 Hz and then begin to increase again. As reported by Barber et al [6], the agglomeration of the ceramic particles can produce electronic conduction, which will lead to higher dielectric loss. Thus, the agglomeration of the perovskite nanoparticles constitutes a drawback of the composites. However, in general the dielectric loss of the samples was small. The values of dielectric permittivity and dielectric loss were within the values that the literature [3] establishes as good for a capacitor material.

To obtain the optimal perovskite concentration some DD composites theoretical models can be used. Thus, two of the more common models (the Maxwell-Garnet (MG) and the Jylhä et al (NM) [7] models) were used to predict this value for the composites. The values of the dielectric permittivity of the pure components were measured and used in the equations at each frequency. The experimental data was compared with the theoretical values as shown in Figure 4.

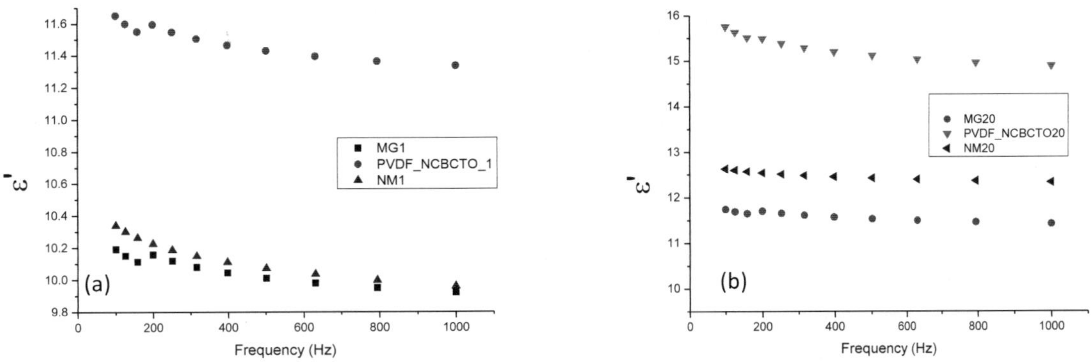

FIGURE 4 Dielectric permittivity as predicted by the MG and NM models as compared with the experimental data of: (a) Sample PVDF_NCBCTO_1 ; (b) Sample PVDF_NCBCTO_20

It can be observed that the prediction given by both models when the concentration is low (1%) is similar; however, when the concentration is higher (20%) the prediction of both models is not coincident. In both samples, however, the experimental data was higher than the predicted one. To observe the influence of the concentration on the permittivity, both equations were plotted as a function of the volumetric fraction of the perovskite. Figure 5 shows the result. It is observed that if the NM model is used, only above a volumetric fraction of 0.33 of perovskite, an increase in dielectric permittivity will be observed; for the MG model, only above a volumetric fraction of 0.90 of perovskite that some increase in dielectric permittivity can be obtained. The volumetric fraction of perovskite in the 1% composite was 0.0034 while in the 20% was 0.074, that is, well below the critical volumetric fraction of both models.

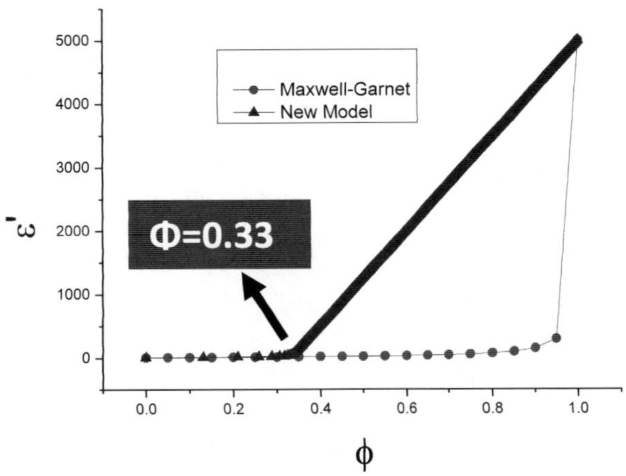

FIGURE 5 Dielectric permittivity as a function of volumetric fraction of perovskite predicted by the MG and NM models at 1kHz.

Kum-Onsa [5] obtained a value of dielectric permittivity at 1 kHz for a PVDF/NCBCTO of approximately the same volumetric fraction that our composite (0.074) of 16, which is very similar to the one obtained by us.

CONCLUSIONS

In this work, composites of PVDF with a novel high dielectric perovskite (NCBCTO) were produced by casting process, followed by melt mixing and injection molding. The SEM characterization showed, however, high particles agglomeration, which promoted an increase of dielectric loss at high frequencies. On the other hand, a 50 % of improvement on the dielectric permittivity value for a composite with 20 wt% of NCBCTO was obtained. Two models, the Maxwell-Garnett and a new model (NM) developed by Jylhä et al predicted values of this property well below the experimental ones.

ACKNOWLEDGMENTS

To CNPq and FAPESP (2015/2339-3) for the financial aid.

REFERENCES

1. Z. Dang , J. Yuan, J. Zha, T. Zhou, S. Li and G. Hud, *Prog. Mater Sci.* **57**, 660–723 (2012).
2. H. Liu, Y. Shen , Y. Song , C. Nan , Y. Lin , X. Yang, *Adv. Mater.* **23**, 5104–5108 (2011).
3. A. B. da Silva, M. Arjmand, U. Sundararaj and R.E.S.Bretas, *Polymer* **55**, 226-234 (2013).
4. P. Kum-Onsa, P. Thongbai, B. Putasaeng, T. Yamwong, S. Maensiri, *J. Eur. Ceram. Soc.* **35**, 1441–1447 (2015).
5. P. Kum-Onsa, P. Thongbai, S. Maensiri, P. Chindaprasirt, *J. Mater. Sci. : Mater. Electron.* **27,** 9650–9655 (2016).
6. P. Barber, S. Balasubramanian, Y. Anguchamy, S. Gong, A.Wibowo, H. Gao, H. J. Ploehn and H. C. Z. Loye, *Materials*, **2**, 1697-1733 (2009).
7. L. Jylhä and A. Sihvola, *J. Phys. D: Appl. Phys.* **40,** 4966-4973 (2007).

Obtaining Fiber Orientation Model Parameters Using Non-Lubricated Squeeze Flow

Gregory Lambert[a], Hongyu Chen[a], and Donald Baird[a]

[a]*Department of Chemical Engineering, Virginia Tech, 24061 Blacksburg, VA, USA*

Abstract. Although fiber reinforced thermoplastics have been a common material for several decades, accurately modeling the fiber orientation dynamics during injection molding remains an important challenge. An accurate description of the fiber orientation state is necessary to predict the mechanical properties (modulus, strength, etc.) of the injection molded part. All of the orientation models in common use rely extensively on empiricisms to account for inter-fiber contacts and confinement effects, among other variables. Naturally, this means every model comes with a set of empirical parameters that must be determined. Common industrial practice obtains these parameters by first injection molding a part, measuring the orientation state at some point in the part, and adjusting the empirical parameters until the orientation model fits the orientation data. Once this is done, the orientation model can predict the fiber orientation state throughout the rest of the part, and the part's mechanical properties can be predicted. This ultimately leads to a trial-and-error approach to part design with the orientation measurement and modeling being a major bottleneck. An experimental technique that could obtain orientation model parameters independent of the injection molding conditions would greatly expedite the design process. This work continues efforts to obtain orientation model parameters from rheological experiments. Here, a polypropylene/glass fiber composite is subjected to non-lubricated squeeze flow and an orientation evolution model is fit to measured orientation at one point in a sample. These parameters are then used in the model to predict fiber orientation at various points in a center-gated disk and results are compared against experimental data.

Keywords: Composites, Glass Fiber, Carbon Fiber, Squeeze Flow, Injection Molding, Rheology
PACS: 47.15.G, 81.05.Qk, 83

INTRODUCTION

Current automobile and aerospace design work relies extensively on fiber-reinforced thermoplastics. These composite materials combine the ease of processing associated with thermoplastics with the mechanical properties required for a variety of applications from plane wings to car engine brackets. Injection molding is a particularly useful method as it can be used to make parts with complex geometries, but the average fiber length is significantly reduced during injection molding, reducing the mechanical properties of the part. Furthermore, the fiber orientation distribution within the part is not uniform, nor is it easily controlled. These two variables, fiber length and fiber orientation, significantly affect the mechanical properties of the part. Their evolution during processing needs to be understood so that parts can be designed from first principles rather than through trial-and-error.

The orientation of a single fiber is readily described with a single vector aligned with the longest axis, but tracking thousands of individual fibers is computationally prohibitive. Using a Fokker-Planck equation to track the evolution of the orientation distribution function is feasible, but this requires solutions on two separate domains: the physical processing space and the spherical orientation space at each point in the physical space. To get around the computational demands, Advani and Tucker [1] suggested the even-ordered moments of the orientation distribution be tracked instead, with the first two defined as:

$$\boldsymbol{A} = \int \boldsymbol{pp}\psi(\boldsymbol{p},t)d\boldsymbol{p}, \tag{1}$$

$$\boldsymbol{A_4} = \int \boldsymbol{pppp}\psi(\boldsymbol{p},t)d\boldsymbol{p}. \tag{2}$$

in which $\psi(\mathbf{p},t)$ is the orientation distribution function, giving the probability of finding a fiber with orientation between $\mathbf{p}$ and $\mathbf{p}+d\mathbf{p}$.

Various models tying the orientation to the flow of the matrix can be found in the literature. Huynh [2] proposed the Strain Reduction Factor (SRF) model for rigid fibers:

$$\frac{DA}{Dt} = \alpha[\boldsymbol{W} \cdot \boldsymbol{A} - \boldsymbol{A} \cdot \boldsymbol{W} + \xi(\boldsymbol{D} \cdot \boldsymbol{A} + \boldsymbol{A} \cdot \boldsymbol{D} - 2\boldsymbol{A_4}{:}\boldsymbol{D}) + 2C_I D(\boldsymbol{I} - 3\boldsymbol{A})], \tag{3}$$

in which D/Dt is the substantial derivative, **W** is the vorticity tensor (defined as $\frac{1}{2}(\nabla \mathbf{u}^t - \nabla \mathbf{u})$), **D** is the rate-of-deformation tensor (defined as $\frac{1}{2}(\nabla \mathbf{u}^t + \nabla \mathbf{u})$), ξ is a shape parameter (defined as $(r^2-1)/(r^2+1)$ in which r is the aspect ratio of the fiber), D is the magnitude of the rate of strain tensor (defined as $[(1/2)\mathbf{D}{:}\mathbf{D}]^{1/2}$), and **I** is the unit tensor. The empirical parameters α and C_I account for fiber interactions by slowing the rate of orientation and randomizing the orientation state, respectively. The SRF model is not objective, and this was rectified later by Wang et al. [3] with the reduced strain closure model. However, later work [4] demonstrated that making the orientation model objective did not greatly improve the accuracy of the predictions of fiber orientation in injection molded parts. The non-objective SRF model is used in this study.

The SRF model uses empirical parameters, α and C_I, which must be determined if the models can be used to predict orientation during injection molding. Previous work [5-7] has obtained them by fitting to orientation data depicting the evolution of fiber orientation during startup of simple shear. Parameters obtained this way have been successful in the relatively simple center-gated disk (CGD) geometry [4,8], but make very poor predictions for more complicated geometries, such as an end-gated plaque [9]. The present study considers obtaining the empirical parameters by fitting to orientation data from non-lubricated squeeze flow through a rectangular channel. Unlike the startup of simple shear, this flow features both extensional and shear flow, similar to injection molding. Predictions using these parameters are compared with experimental data from a CGD as a first test of the method.

EXPERIMENTS

Raw material came in the form of 13 mm pellets of polypropylene (PP) with 50 wt% glass fiber from SABIC. Pellets were cut to 4 mm then fed into an injection molder (35E, Boy) with a 32 mm diameter screw. Nozzle purge was taken to create samples from non-lubricated squeeze flow (NLSF), and center-gated disks (CGDs) were molded as well. Nozzle purge material was generated using the same conditions as those used for molding the center-gated disks.

Disks (Figure 1) had a diameter of 11 mm (90% of full mold fill) and were 1.35 mm thick. Injection molding took place with a 10 bar back pressure, a screw rotational speed of 200 rpm, an injection speed of 6 mm/s, and disks were allowed to cool for 30 s. The barrel had five heating zones, and these were set to 200 °C, 215 °C, 220 °C, 220 °C, and 220 °C, while the mold was set to 80 °C. Average fiber length was measured by cutting out one-quarter of a disk, burning off the resin, collecting fibers from the center of the cut piece using a method proposed by Kunc et al. [10] with modification, scanning the captured fibers with a desktop scanner, and tracing the fibers with ImageJ (National Institutes of Health). The number and weight average fiber lengths were thus found to be 0.475 mm and 0.726 mm, respectively.

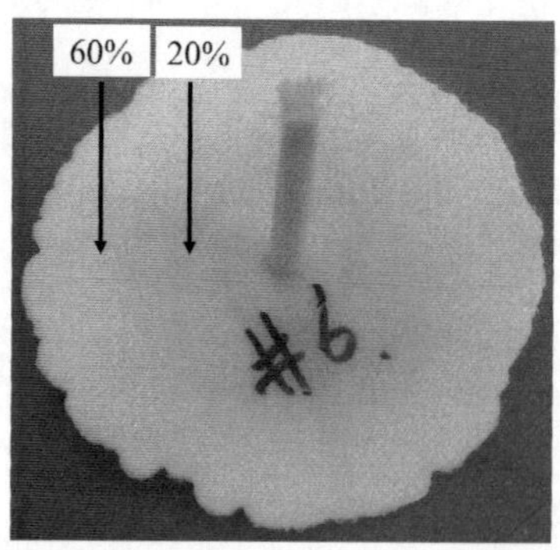

Figure 1. Example center-gated disk. Locations for orientation evolution profile are labeled.

Samples for NLSF were made by compression molding the nozzle purge at 200 °C and were 95.25 mm wide, 50.8 mm long, and 7.50 mm thick. Strands of the nozzle purge were laid down in the width direction to impart an initial fiber orientation state with orientation primarily in the width direction and very little thickness-direction orientation (i.e. a planar orientation state). Tests were run in a custom fixture built in-house (Figure 2a) and mounted on an Instron 5969 with an environmental chamber (3119-609, Instron) set to 200 °C. Samples were squeezed at a constant Hencky strain rate of -0.50 s^{-1} (note that the negative indicates compression) with a constant contact area between the sample and the top platen. The test occurred over 2 seconds, corresponding to a Hencky strain of 1.0.

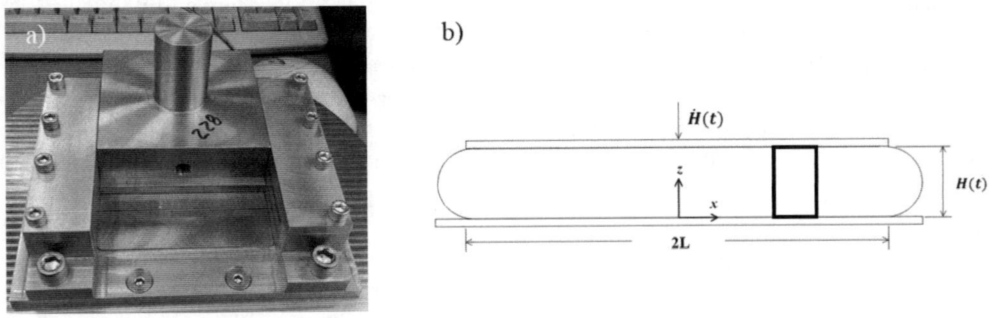

FIGURE 2. a) Fixture used for NLSF experiments. b) Schematic of the location of the orientation measurements (in the boxed region. This location is at 50% of the distance between the origin and the front of the sample.

Fiber orientation was measured with the method of ellipses, using techniques laid out by Velez-Garcia [11]. Orientation in the CGDs was measured through the thickness of the disks at radial location 20% and 60% along the radial length of the mold fill (Figure 1). Orientation in the NLSF samples was measured at a single point along the lateral center of the sample, halfway between the center and front edge of the sample (Figure 2b). Micrographs of the fiber orientation were analyzed using image analysis routines developed in-house using MATLab (Mathworks). Both the injection molding and non-lubricated squeeze flow were simulated using a finite element method described elsewhere [4]. The CGD simulation used the measured orientation state at 0% of the radial length as the initial orientation condition, while the NLSF simulation used an assumed initial orientation condition of $A_{11} = 0.15$, $A_{22} = 0.80$, and $A_{33} = 0.05$ (with all other components of **A** set to 0). Note that, in the NLSF simulation, the 1-direction refers to the flow direction, the 2-direction to the width direction, and the 3-direction to the thickness direction.

RESULTS AND DISCUSSION

Fiber orientation through the thickness of the non-lubricated squeeze flow (NLSF) sample at the location in Fig. 2b at 1.0 Hencky strain is shown in Fig. 3. Significant fiber reorientation occurred from an assumed initial condition of $A_{11} = 0.15$, $A_{22} = 0.80$, and $A_{33} = 0.05$. Note that the 1-direction corresponds with the flow-direction, the 2-direction with the width direction, and the 3-direction with the thickness direction. The strain reduction factor (SRF) model was fit to the data using the parameters $\alpha = 1.0$ and $C_I = 0.005$, which are similar to results from Lambert and Baird [12] and Lambert et al. [13] and suggest that extensional flow dominates the kinematics of NLSF.

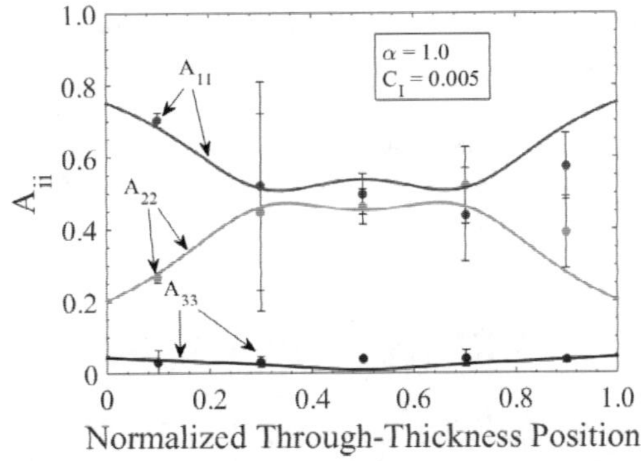

FIGURE 3. Measured and predicted fiber orientation through the thickness of the NLSF sample. Points represent experimental data (with error bars representing 95% confidence intervals) and solid lines represent predictions from the SRF model using $\alpha = 1.0$ and $C_I = 0.005$. The red points and lines correspond to A_{11} (flow-direction orientation), the green points and line correspond to A_{22} (width-direction orientation), and the blue points and line refer to A_{33} (thickness-direction orientation). Thickness position 0 corresponds with the bottom platen and thickness position 1 corresponds with the top platen.

These model parameters were subsequently used to predict fiber orientation through the thickness of a center-gated disk (CGD) at two locations, 20% and 60% along the radial length, and the predictions were compared against experimental orientation data, shown in Fig. 4. Predictions from the SRF model reasonably agree with the orientation profile towards the center of the disks, between z/H = -0.5 and 0.5. Considering that $\alpha = 1.0$ seems to correspond to extensional flow, the agreement towards the center of the disk suggests that the center of the disk is dominated by extensional flow in the tangential direction. However, the model fails to predict the profile near the top and bottom of the disk. The disagreement might be due to shear flow dominating the kinematics near the walls, but could also be due to solidification of the matrix at the walls, and the influence of the walls themselves on the fiber orientation dynamics.

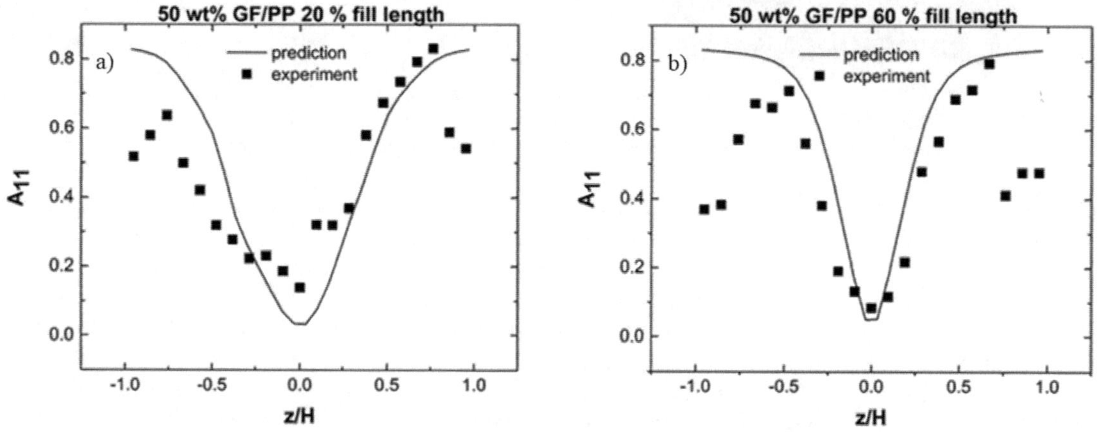

FIGURE 4. Measured and predicted fiber orientation through the thickness of the CGD at a) 20% and b) 60% of the fill length. Note that only the A_{11} component is shown, corresponding to flow-direction orientation. Points represent measured orientation and the solid line represents the prediction from the SRF model using $\alpha = 1.0$ and $C_I = 0.005$. The thickness position z/H = 1 corresponds with the top of the disk, and thickness position z/H = -1 corresponds with the bottom of the disk.

CONCLUSIONS

Non-lubricated squeeze flow (NLSF) was proposed as a method to obtain orientation model parameters because the corresponding kinematics feature both shear and extensional flow. The dual nature of the flow was posited to better reflect the kinematics experienced by fiber-polymer composites during injection molding and thus could be useful in obtaining orientation model parameters independent of injection molding. Based on previous findings that obtained and compared orientation model parameters from simple shear and planar extension [12], the model parameters obtained from NLSF in this study suggest that extensional kinematics dominate both NLSF and the core region of a center-gated disk (CGD) during injection molding. Furthermore, the model predictions were markedly worse near the walls of the CGD, where shear is expected to dominate the kinematics. This suggests that the strain reduction factor (i.e. α) could be a function of the velocity field. The NLSF geometry could be useful in further developing this relationship due to the simultaneous presence of shear and extensional flow.

ACKNOWLEDGMENTS

The authors thank Micahel Vaught for assisting in the design and fabrication of the NLSF fixture. The authors also thank Dr. Peter Wapperom of the Virginia Tech Mathematics department for running the NLSF simulations and fitting to the orientation data in those experiments. Finally, the authors also thank Junwei Luo and Emma Sparks for their assistance with the fiber length and orientation measurements. This work was funded in part by General Motors Co.

REFERENCES

1. S.G. Advani and C.L. Tucker III, *J. Rheol.* **31**, 751-784 (1987).
2. H.M. Huynh, "Improved Fiber Orientation Predictions for Injection-Molded Composites", Ph.D. Thesis, University of Illinois at Urbana-Champaign., 2001.
3. J. Wang, J.F. O'Gara, C.L. Tucker III, *J. Rheol.* **52**, 1179-1200 (2008).
4. S.M. Mazahir, G.M. Velez-Garcia, P. Wapperom, and D.G. Baird, *Composites Part A* **51**, 108-117 (2013).
5. A.P.R. Eberle, D.G. Baird, P. Wapperom, G.M. Velez-Garcia, *J. Rheol.* **53**, 685-705 (2009).
6. K. Ortman, D.G. Baird, P. Wapperom, and A. Whittington, *J. Rhol.* **56**, 955-981 (2012).
7. M.J. Cieslinski, D.G. Baird, and P. Wapperom, *J. Rheol.* **60**, 161-174 (2016).
8. K. Ortman, D.G. Baird, P. Wapperom, and A. Anning, *Polym. Compos.* **33**, 1360-1367 (2012).
9. K.J. Meyer, J.T. Hofmann, D.G. Baird, *Composites Part A* **62**, 77-86 (2014).
10. V. Kunc, B.J. Frame, B.N. Nghiep, C.L. Tucker III, G.M. Velez-Garcia, "Fiber length distribution for long glass and carbon fiber reinforced injection molded thermoplastics", in *Proceedings of the 7th Annual Society of Plastics Engineers Automotive Composites Conference and Exposition*, Troy, MI, 2007.
11. G.M. Velez-Garcia, "Experimental Evaluation and Simulations of Fiber Orientation in Injection Molding of Polymers Containing Short Glass Fibers", Ph.D. Thesis, Virginia Tech, 2012.
12. G.M. Lambert, D.G. Baird, *J. Man. Sci. Eng.* **139**, 031012-1 – 031012-7 (2017).
13. G.M. Lambert, P. Wapperom, D.G. Baird, *Phys. Fluids* (in press).

Compounding and Direct Compounding of Crimped Fiber Reinforced Thermoplastic Elastomers

Johannes Gattinger[a)], Raphael Köppl, Matthias Zeppenfeld

Technical University of Munich, Institute of Medical and Polymer Engineering
Boltzmannstr. 15, 85748 Garching, Germany

[a)]Corresponding author: johannes.gattinger@tum.de

Abstract. The use of wavy fibers as reinforcement elements accounts for higher strengths and preserves the flexibility of elastomers. For this purpose, the manufacturing of wavy short fiber reinforced thermoplastic elastomers by injection molding was investigated. Thermoplastic elastomer and crimped staple fibers were separately compounded and molded (SCM) or directly compounded in the injection molding machine (DCM). Comparative tension bars were manufactured with compounds derived from straight fiber (tow, same fiber type). Selected specimens were dissolved in cyclohexane and the fibers were characterized regarding fiber length and degree of crimp (quotient of the extended fiber length and the crimped fiber length). Tensile tests were performed to evaluate the mechanical properties. The examination of the resulting fiber length and shape in the injection molded parts showed that every processing step (compounding or injection molding) and every raw fiber, whether straight or crimped, produced wavy fibers in the resulting granules or parts.

Keywords: crimped fiber, soft composite, thermoplastic elastomer, compounding, injection molding
PACS: 81.05.Lg, 81.05.Qk, 83.80.Tc, 81.20.Hy

INTODUCTION

Reinforcing elastomers with wavy fibers gives us the opportunity to retain the elasticity and soft character of the elastomer, combined with an enhanced strength resulting from the fiber reinforcement. Such composites show a progressive J-shaped stress-strain relationship, with a region of low stiffness when the reinforcing fibers are still in a wavy pattern, and a region of high stiffness when the fibers are straightened due to elongation [4, 10]. Some biological tissues like tendons and ligaments show a similar mechanical behavior because the load-bearing elements in these tissues are collagen fibers that are crimped when in an unloaded state[3]. Iannace [8] and Ambrosio [1] investigated a tendon/ligament prosthesis of hydrogel polymer matrix reinforced with helically shaped poly(ethylene terephthalate) fibers. This composite prosthesis was manufactured by winding the fibers on a mandrel and subsequent overmolding (manual casting). Ramakrishna and Huang [6, 7, 11] investigated flexible composites of a polyurethane elastomer reinforced with knitted fabrics manufactured by compression molding. Bekisli [2] also manufactured flexible composites with knitted fabrics and a thermoplastic elastomer by compression molding. Jang [9] investigated a soft thin film composite with skin-like properties by overmolding (spin casting) a stereolithographically manufactured polyimide fiber network with silicone. Recently Hopmann [5] introduced a cruciate ligament prosthesis manufactured by winding poly(ethylene terephthalate) fibers on an injection-molded silicone cylinder and overmolding (injection molding) with silicone.

The aim of this study was to investigate whether it is possible to manufacture injection-molded parts out of a wavy, short, fiber reinforced thermoplastic elastomer that show such an introduced mechanical behavior with enabled high strain from the thermoplastic elastomer and enhanced strength through the fiber reinforcement.

MATERIALS AND METHODS

To manufacture injection molded parts of a crimped fiber reinforced thermoplastic elastomer (TPE), two different compounding techniques were investigated: separate compounding and subsequent injection molding (SCM) and direct compounding in the injection molding machine (DCM). For separate compounding, a co-rotating twin-screw compounder (Coperion ZSK 18 ML, Coperion GmbH, Germany) with a screw diameter of 18 mm, a screw length of 720 mm and a die diameter of 3 mm was used. The injection molding machine was an Arburg Alrounder 420 C 1300-

150/60 (ARBURG GmbH & Co KG, Germany) with a standard three-zone screw (diameter 25 mm, length 500 mm, nozzle diameter 2 mm). The mold was a two-cavity mold for tension bars in in accordance with ISO 527-2 Type 1A. The manufacturing parameters for compounding and injection molding are shown in Table 1.

The thermoplastic elastomer used was a soluble TPE-s based on hydrogenated styrene block copolymers (HTM8505/57, KRAIBURG TPE, Germany, 54 ShoreA). As reinforcing fibers oxidized polyacrylonitrile (PAN) fibers were used and crimped staple fibers (PANOX®C63-1.7/1.39-A110, SGL CARBON, Germany, tensile strength 220 MPa) were compared to straight fibers (tow, same fiber type). The crimped staple fibers had a cutting length of 63 mm, therefore the straight fiber tow was cut to the same length before processing.

TABLE 1. Parameters for Compounding and Injection Molding

Compounding Parameters		
Screw Rotation Speed	178 rpm	
Melt Temperature	240 °C	
Injection Molding Parameters	**Injection Molding (SCM)**	**Direct Compounding (DCM)**
Peripheral Screw Speed	200 mm/min	200 mm/min
Melt Temperature	240 °C	240 °C
Back Pressure	20 bar	100 bar
Injection Speed	50 m/s	50 m/s
Holding Pressure	200 bar, 10 s	200 bar, 10 s
Cooling Time	25 s	25 s
Mold Temperature	60 °C	60 °C

The staple fibers were not suitable for dosing with a standard feeder, therefore TPE granules and fibers were mixed with a fiber mass fraction of 10 % and fed directly in the feeding zone (dry-blend). This procedure was applied to compounding as well as direct compounding in the injection-molding machine and to both fiber types. To avoid air inclusions by DCM the back pressure was increased to 100 bar. For separate compounding and molding, the compounded extrudate was cut into 20 mm pellets. Table 2 gives a summary on the manufactured samples and performed tests.

TABLE 2. Manufactured Samples and Tests Performed

Processing	**Fiber Type**		**Tests**	
	crimped	**straight**	**Fiber Length and Shape**	**Tensile Tests**
Unprocessed Fiber (raw)	x		x	
Compounding (C)	x	x	x	
Compounding and Subsequent Injection Molding (SCM)	x	x	x	x
Direct Compounding in the Injection Molding Machine (DCM)	x	x	x	x

For the analysis of the fiber lengths and shapes, three samples (0.1 g) of every compounded material or injection molded parts were dissolved in cyclohexane (10 ml) and poured into petri dishes for analysis by light microscopy (Keyence VHX-500, magnification 100x). For the evaluation, the real fiber length l_1, the crimped fiber length l_2 and the crimping radii r of each fiber in 4 squares of 1600 µm edge length on each petri dish were measured (Figure 1). The degree of crimp is calculated as d = l1/ l2. The Mann–Whitney U test was performed to evaluate significant differences. The tensile tests were performed on a Zwick/Roell Z050 (Zwick GmbH & Co. KG, Ulm, Germany) at an extension rate of 40 mm/min.

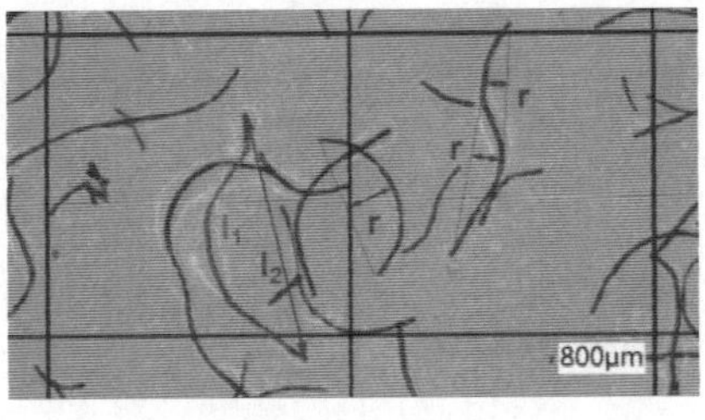

FIGURE 1. Evaluation of fiber lengths (real l_1/crimped l_2) by light microscopy (magnitude 100 x, grid edge 800 µm)

RESULTS AND DISCUSSION

The decrease in fiber length could be seen in every processing step (Figure 3). After compounding, the resulting fiber lengths in the granules are in a range (interquartile range) of 144 μm – 578 μm with a median at 294 μm. The subsequent injection molding step further reduced the fiber length by up to 40 % to a median of 180 μm and an interquartile range of 84 μm – 355 μm. Direct compounding in the injection molding machine (DCM) with straight raw fibers produced the highest resulting fiber lengths with a median at 325 μm and an interquartile range of 156 μm – 777 μm. DCM with crimped raw fibers produced shorter fibers with a median at 259 μm. The shape of the raw fiber had no influence on the resulting fiber lengths in compounding and injection molding. It is assumed that the aspect of dry-blending (mixing of the granules and dry fibers in the feeder) had a negative influence on the obtained fiber lengths due to fiber-screw interactions.

The examination of the fiber shape showed that the original crimp of the raw fibers could not be preserved because of fiber length degradation. Nevertheless, the fibers were properly crimped after processing although on another scale (Figure 2). Even the straight raw fibers were crimped after processing. Actually the degree of crimp after compounding was significantly ($p = 0.005$) higher (median 1.18) than the original degree of crimp of the raw crimped fibers (median 1.11). Subsequent injection molding slightly decreased the degree of crimp but in case of the crimped raw fibers this effect was not significant ($p = 0.4$). DCM increased the degree of crimp of the crimped raw fibers and crimped the straight raw fibers, but not as much as compounding did (Figure 3). The crimp radii were measured as a quantity for the change in scale. They decreased by 70 % - 80 % depending on the processing method. The reason for this crimping in the course of processing is seen in bending and heat setting of the fibers due to flow and shear conditions in the melt. The dwell time in the compounder was 2 – 4.5 minutes.

The mechanical behavior of the injection molded specimens was not the one that was intended . The tensile tests showed no J-shaped stress-strain function as reported in the literature [10] for endless wavy fiber reinforced elastomers (Figure 4). No difference was seen between the reinforced samples with crimped or straight raw fibers, which makes sense considering the even fiber shapes and lengths in the compounds and samples after processing. Three reasons are assumed for the non-existent J-shape: a too short fiber length, the fibers constrict themselves in straightening because of the nearly random distribution of the fibers after injection molding, and inappropriate mechanical properties of the matrix with its high stiffness in the beginning (due to a pronounced Mullins effect).

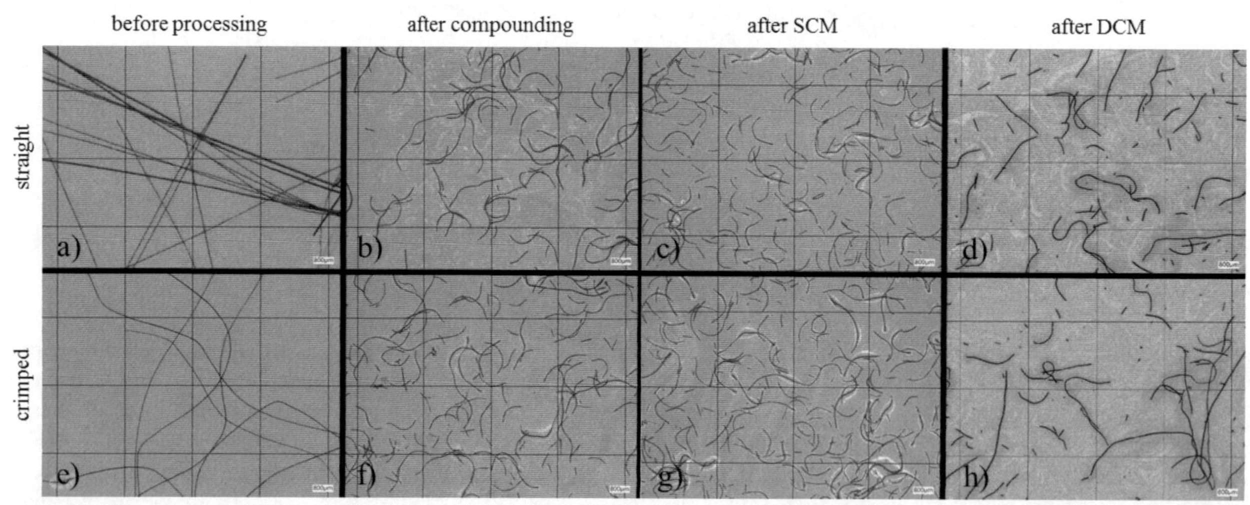

FIGURE 2. Light microscopy pictures of fibers before and after the different processing steps (magnitude 100 x, grid edge 800 μm). a) and e): straight and crimped fibers before processing. b) and f) dissolved samples after compounding with straight and crimped fibers. c) and g) dissolved samples after compounding and injection molding. d) and h) dissolved samples after direct compounding in the injection molding machine.

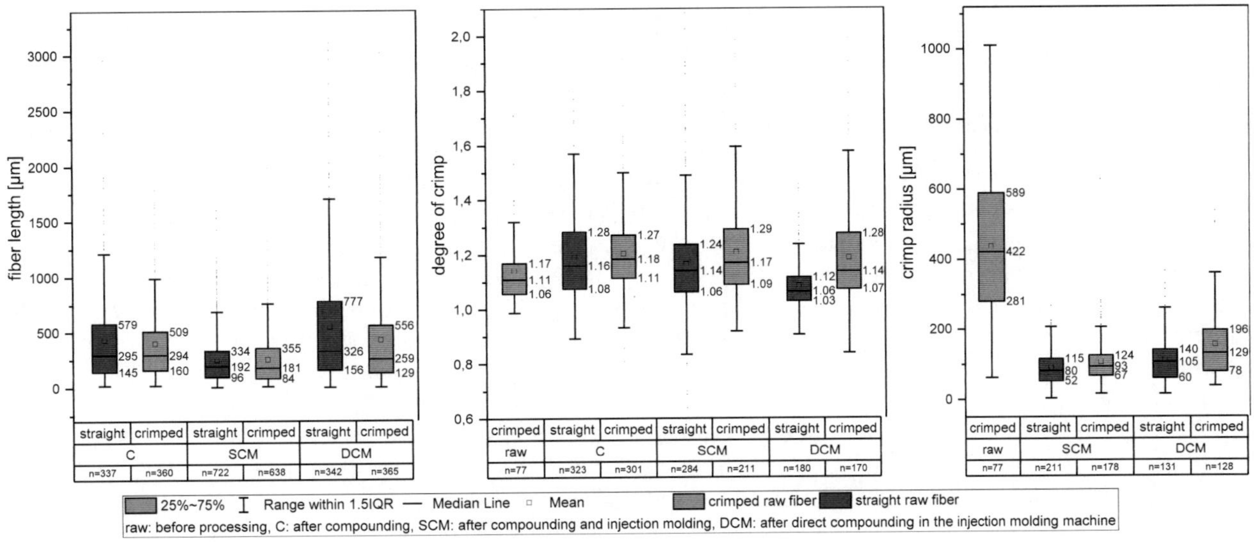

FIGURE 3. Fiber length and shape before and after processing by compounding, injection molding and direct compounding in the injection molding machine. After compounding, the median of the fiber length is at 294 µm regardless whether the raw fibers were straight or crimped. Subsequent injection molding further reduced the fiber length by up to 40 % to 180 µm (median). Direct compounding in the injection molding machine enabled the highest fiber lengths of 325 µm (median). The median of the degree of crimp is higher after compounding (1.18) than it is in raw crimped fibers (1.11). The straight raw fibers crimped by compounding then revealed a degree of crimp similar to that of crimped fibers after compounding. The subsequent injection molding caused no further change in the degree of crimp. Direct compounding (DCM) also affected the fiber shape but not as strongly as compounding. The crimp radius of the crimped fibers was significantly smaller after processing. The crimp radius of straight and crimped fibers were in the same range after processing.

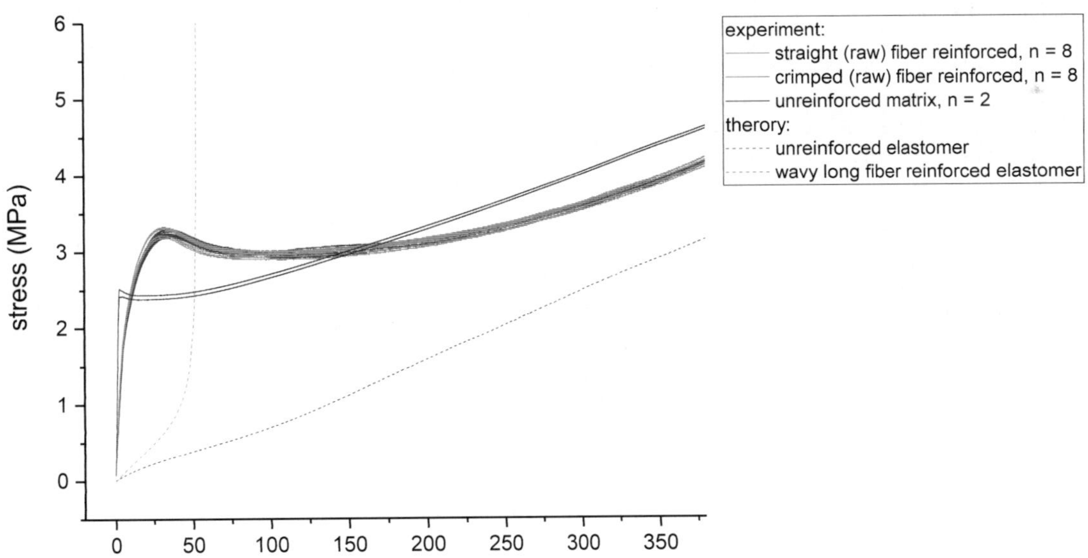

FIGURE 4. Stress-strain curves from tensile tests of reinforced and unreinforced TPE. The curves of the crimped and straight fiber reinforced specimens show no difference. The unreinforced specimens are stiffer than the reinforced ones. Dashed lines reflect the theory of the stress-strain relationship of wavy fiber reinforced elastomers as it is reported in the literature [10] for endless wavy fiber reinforce elastomers.

CONCLUSION

In conclusion, the manufacture of wavy short fiber reinforced thermoplastic elastomers by injection molding has been successful. Although the expected mechanical properties of the compound could not be reached and the influence of the fiber shape on the molded samples could not be investigated due to the occurrence of in-screw-crimping, thus preventing straight fibers for comparison in the resulting parts. The investigation of the fiber shape after compounding and injection molding showed that the fibers have a significant crimp in the granules and injection molded samples. It was found that thermoplastic fibers could be crimped by compounding or injection molding. This effect of in-screw-crimping may be explained with bending and heat setting of the fibers due to flow and shear conditions during processing. Further investigation has to be done to investigate the influence of a matrix with a much lower initial stiffness on the mechanical properties with the aim to obtain a progressive stress-strain curve based on the wavy reinforcement fibers.

ACKNOWLEDGMENTS

The support of KRAIBURG TPE GmbH & Co. KG is gratefully acknowledged.

REFERENCES

1. Ambrosio L, De Santis R, Iannace S, Netti PA, Nicolais L. Viscoelastic behavior of composite ligament prostheses. Journal of biomedical materials research. 1998;42(1):6-12.
2. Bekisli B, Nied HF. Flexible composites with knitted reinforcements. SPE Annual Technical Conference ANTEC; Boston: Society of Plastics Engineers; 2011. p. 578-83.
3. Chamberlain CS, Jr. RV. Ligament and Tendon. In: Murphy W, Black J, Hastings G, editors. Handbook of Biomaterial Properties. London: Chapman and Hall; 2016. p. 55-62.
4. Chou TW, Takahashi K. Non-linear elastic behaviour of flexible fibre composites. Composites. 1987;18(1):25-34.
5. Hopmann C, Gries T, Jockenhoevel S, Dornebusch H, Eggert F, Pishnamaz M. The Cruciate Ligament Implant of the Future? - Continuous Fiber-Reinforced Liquid Silicone Rubbers Create New Possibilities in Sports Medicine. Kunststoffe international. 2017;2017(6-7):18-21.
6. Huang ZM, Ramakrishna S, Dinner HP, Tay AAO. Characterization of a Knitted Fabric Reinforced Elastomer Composite. Journal of Reinforced Plastics and Composites. 1999;18(2):118–37.
7. Huang ZM, Ramakrishna S, Tay AAO. Modeling the stress/strain behavior of a knitted fabric-reinforced elastomer composite. Composites Science and Technology. 2000;60(5):671-91.
8. Iannace S, Sabatini G, Ambrosio L, Nicolais L. Mechanical behaviour of composite artificial tendons and ligaments. Biomaterials. 1995;16(9):675-80.
9. Jang KI, Chung HU, Xu S, Lee CH, Luan H, Jeong J, et al. Soft network composite materials with deterministic and bio-inspired designs. Nat Commun. 2015;6:6566.
10. Luo S-Y, Chou T-W. Finite Deformation and Nonlinear Elastic Behavior of Flexible Composites. Journal of Applied Mechanics. 1988;55(1):149-55.
11. Ramakrishna S, Ming Huang Z, Yew HM. Development of a novel flexible composite material. Journal of Materials Processing Technology. 1999;89–90:473-7.

Novel method for the measurement of fiber breakage during mold filling in injection molding of polymer composites

Pontes Antonio (1), Ludueña Leandro Nicolas (2)*

(1) IPC - UMinho - Guimaraes - Portugal, (2) INTEMA CONICET-UNMdP - Buenos Aires - Argentina

This work deals with the quantification of fiber breakage of thermoplastic polymers/fiber reinforced composites (TPFRC) during mold filling in injection molding. TPFRC have been developed over last 4 decades looking for higher mechanical performance for the production of lightweight parts. Injection molding is a processing technique used to produce high performance TPFRC parts with complex geometries at high production rates. Better mechanical performance of these materials is obtained maximizing their fiber length. For this reason, fiber breakage during mold filling should be prevented, which can be achieved by numerical simulations. The main limitation of the models used to predict fiber breakage in mold filling simulations is their experimental validation. Fiber breakage has only been experimentally quantified as a function of injection parameters such as flow rate or injection pressure but not as a function of flowing conditions (shear rate, residence time) in the injection molding range. The experimental setup designed in this work is able to isolate each one of the flowing parameters and geometrical/morphological features that cause fiber breakage during mold filling of injection molded polymer composites. It was done with the attachment of a capillary rheometer to the clamping unit of an injection molding machine. Polypropylene PP Domolen V20N filled with 20wt.% of glass fibers with 13.3µm and 516µm of number average diameter and length, respectively, was used as received for the injection molding trials. Four 4 capillaries with length 50, 100, 150 and 200mm and 4mm diameter were used. These capillary geometries made possible to work at shear rates in the injection molding range minimizing capillary wall/fibers contacts. The effect of 4 shear rates at constant residence time and 4 residence times at 4 different shear rates was analyzed. Al least 2000 fibers where measured at the initial and final 8mm of each strand collected from the capillaries.

Experimental Verification on Moldex3d Fiber Orientation Prediction in Injection-Molded Tensile Bar and Ribbed Plates

Huan-Chang Tseng and Dannick Deng

CoreTech System (Moldex3D) Co., Ltd., ChuPei City, Hsinchu County, 30265, Taiwan.

Abstract. The prediction of fiber orientation has become a standard calculation in injection-molding simulation of fiber-reinforced thermoplastic parts. Recently, commercially available injection molding software, Moldex3D, developed a new fiber orientation model, iARD-RPR. For simple parts such as plate and disk, several studies using iARD-RPR have shown good agreements between the predictions and experiments. This study extends the usage of iARD-RPR model to more complex shapes like a tensile bar and a ribbed plate. After comparing to the measured fiber orientation data, we successfully validate that adopting the iARD-RPR model can accurately predict fiber orientation in various kinds of injection molded parts, from simple geometry to complex ones, and provides a basis for further industrial applications.

Keywords: Fiber Orientation; Moldex3D; Injection Molding; iARD-RPR Orientation Model.

1. Introduction

In the automotive industry, fiber reinforced thermoplastic (FRT) composites have become widely used for manufacturing lightweight materials without decreasing safety and durability. Fiber orientation distribution causes anisotropy in material properties. An accurate orientation prediction is the prerequisite in a comprehensive analysis workflow, from injection molding simulation to structural analysis. Following the models (see [1] for an overview), Tseng et al. [2] proposed an new fiber orientation model, namely, iARD-RPR (improved Arbitrary Rotary Diffusion model and Retarding Principal Rate model) with three physically meaningful parameters. It is implemented in the commercial injection molding simulation software, Moldex3D and has been demonstrated to be able to predict short/long fiber orientation in several injection molded parts, such as plate and disk with different gate types. It is important to flexibly determine optimal parameters for material grades with different amounts and types of glass fibers [3]. Besides, the initial condition of the fiber orientation tensor at the runner entrance is given as isotropic random state. With the aid of 3D simulation, the fiber orientation is calculated directly without any auxiliary correction by experimental inlet condition. Nevertheless, a comprehensive study on the fiber orientation prediction from a simple geometry to a complex one with ribs which is closer to industrial applications has not yet been proposed. In this work, we firstly verify an optimal set of the iARD-RPR parameters on a rectangular plate. Then utilize the verified iARD-RPR parameters on a tensile bar and on two plates, one with a rib along flow direction (longitudinal ribbing design) and the other with a rib perpendicular to the flow direction (transverse ribbing design). Through this practice, we can therefore ascertain that an accurate prediction on fiber orientation in the broader applications of complex geometrical designs is feasible with the aid of the iARD-RPR parameter set.

2. Model of Fiber Orientation

A single fiber is regarded axisymmetric and rigid. The bond's unit vector **p** along its axis direction can describe fiber orientation. Advani and Tucker [4] defined the second order orientation tensor as:

$$\mathbf{A} = \oint \psi(\mathbf{p})\mathbf{pp}d\mathbf{p} \tag{1}$$

where $\psi(\mathbf{p})$ is the probability density distribution function over orientation space.

Tseng *et al.* [2] developed a new fiber orientation model to couple with Jeffery's hydrodynamic (HD) model, namely, the iARD-RPR model (known as Improved Anisotropic Rotary Diffusion model combined with Retarding Principal Rate model),

$$\dot{\mathbf{A}} = \dot{\mathbf{A}}_{\text{HD}} + \dot{\mathbf{A}}_{\text{iARD}}(C_I, C_M) + \dot{\mathbf{A}}_{\text{RPR}}(\alpha) \tag{2}$$

where $\dot{\mathbf{A}}$ represents the material derivative of **A**. Parameters C_I and C_M describe the fiber-fiber interaction and fiber-matrix interaction, while parameter α can slow down a response of fiber orientation. Details of the RPR and iARD models are described in reference [2].

According to these criteria, we can fine-tune three parameters to obtain an accurate orientation prediction, namely, determining an optimal parameter set. The shell-orientation layer depends on the iARD parameters (C_I and C_M), while the core region and skin layer are controlled by the RPR parameter (α). For the range of these three parameters, the best practices are: $0 < C_I < 0.1$; $0 < C_M < 1$; $0 < \alpha < 1$. When C_M and α all set to 0, iARD-RPR model goes back to Folgar-Tucker model [2].

3. Materials and Methods

First simulation-experimental validation was done on an injection molded rectangular plate. The rectangular plate with dimensions, 310 mm x 270 mm x 2 mm, designed by DSM is illustrated in **Figure 1**. The volume of the cavity is about 167 cm^3. For the processing conditions, the filling time and filling rate were 1.6 sec and 108 cc/sec, respectively. In addition, the mold and melt temperatures were set at 85°C and 270°C, respectively. At the point S2 fiber orientation is measured (location see **Figure 1**).The second simulation-experimental validation is done on a tensile bar which is essential for characterizing the mechanical behavior of a composite under loading. **Figure 2**, shows the dual cavity mould for producing (ISO5271A) tensile bars with a length of 170 mm and a thickness of 4 mm. For the processing conditions, the filling time and filling rate were 1.7 sec and 20 cc/sec, respectively.

We considered ribbed plate parts with two different ribbing directions, longitudinal and transverse. **Figure 3** shows the longitudinal ribbed plate (designed by DSM). The thickness of the plate with rib is 2 mm. For the processing conditions, the filling time and filling rate were about 1.0 sec and 30 cc/sec, respectively. Fiber orientations at region 1 and region 3 were measured and compared to the simulation predictions. With the same approach, the fiber orientation investigation on a transverse ribbed plate (**Figure 4**) was also performed.

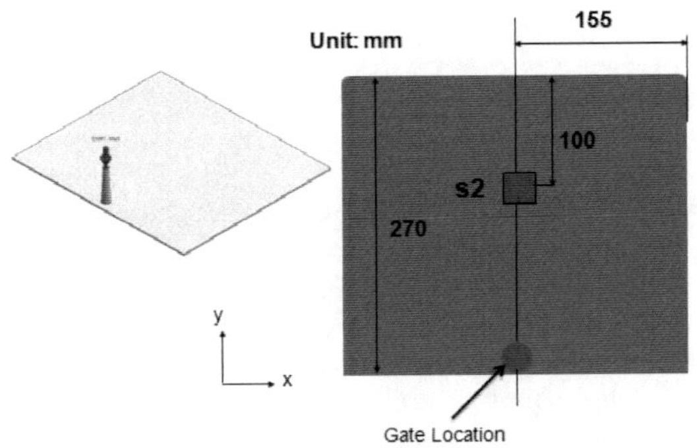

FIGURE 1. Geometry of DSM rectangular plate (310 mm x 270 mm x 2 mm) with one measurement location.

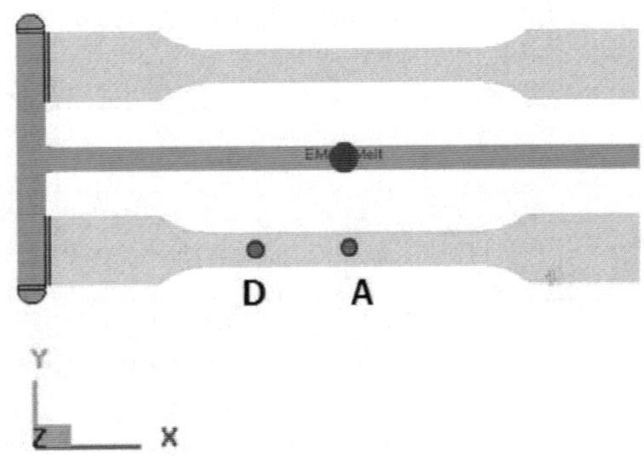

FIGURE 2. Geometry tensile bars with two measurement locations for glass fiber orientation.

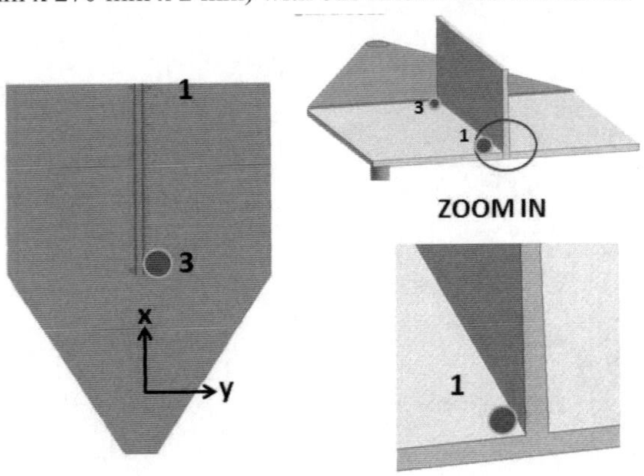

FIGURE 3. Longitudinal ribbed plate with two measurement locations.

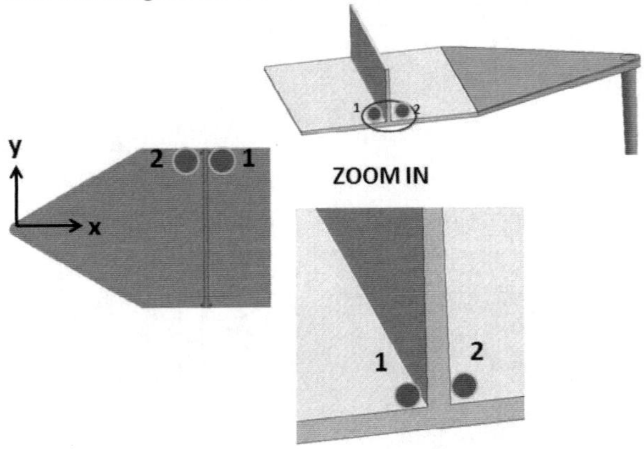

FIGURE 4. Transverse ribbed plate with two measurement locations.

One commercial DSM compound was used in this study: 30wt% short glass fiber-reinforced Polyamide6 (abbreviation: 30wt%-SGF/PA6). We adopted the R14-Solid version of Moldex3D to simulate the 3D filling stage in injection molding process for the DSM 30wt%-SGF/PA6 material. A hybrid mesh with 11 elements across the thickness was used for the parts considered. Shear viscosity flow curve of the 30wt%-SGF/PA6, used in Moldex3D [5] is shown in **Figure 5**. For the rectangular plate case, and the ribbed plate cases the fiber orientation was determined based on micro-CT data (a resolution of 2-3 microns is required). In-house developed software tools at DSM were used to determine the second order orientation tensor components. For the tensile bar, the fiber orientation was determined by an in-house image analysis facility developed at University of Leeds, details can be found in: [6]. Results are taken from an unpublished report [7].

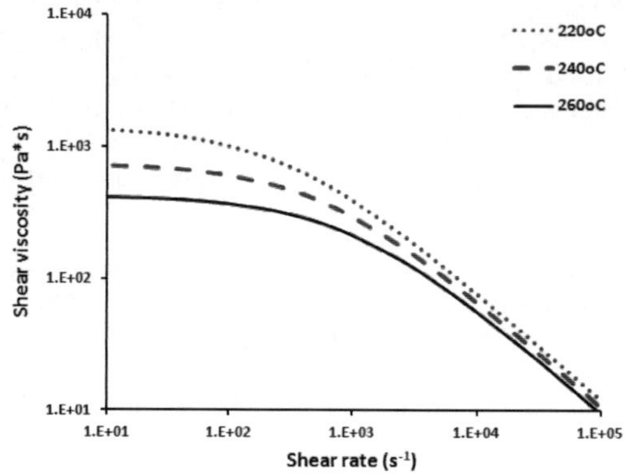

FIGURE 5. Viscosity flow curve of DSM 30wt%-SGF/PA6.

4. Results and Discussion

I. Rectangular plate and Tensile bar

Figure 6 shows the simulated the fiber orientation distributions through the thickness at S2, as compared with experimental data. The important orientation descriptors are the orientation tensor components: A11 is the flow direction along the *x* axis and A22 is the cross-flow direction along *y* axis. We firstly compare the results obtained by the iARD-RPR model to the one obtained by the Folgar-Tucker model. Dash lines are from the Folgar-Tucker model and Solid lines are from iARD-RPR models. For short fiber composite, parameter used in the Folgar-Tucker model is $C_I = 0.005$, and parameters used in the iARD-RPR model are: $C_I = 0.005$, $C_M = 0.1$, $\alpha = 0.7$. In **Figure 6**, we can see that the predicted fiber orientations from the core region, through shell, to the skin layer, the iARD-RPR model yields good agreement with the experimental data (symbols), while the Folgar-Tucker model fail to capture the orientation at the skin layer. Fibers in the shell layers are much aligned with the flow direction, characterized by the large values of A11 and the low values of A22; whereas the core layer is wide with low values of A11 and high values of A22.

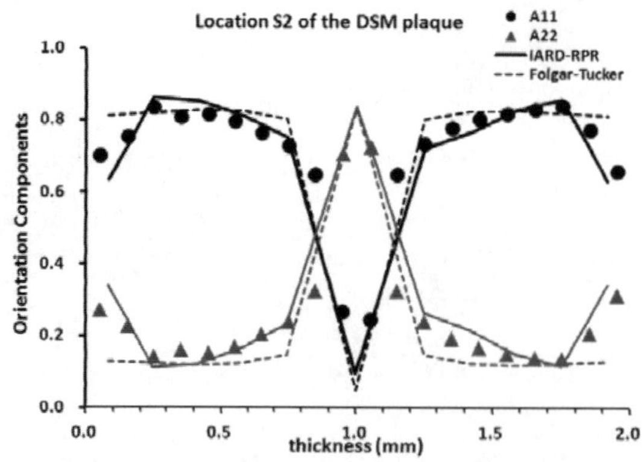

FIGURE 6. Comparison of the experimental and predicted data of orientation components through the thickness at **S2** of the DSM plaque.

II. Tensile bar

Simulation was conducted with the same iARD-RPR parameters as previous plate model. Fiber orientation were measured at region A and D of this tensile bar. **Figure 7** and **Figure 8** show the simulation-experimental comparison at region A and region D, respectively. A11 is the flow direction along the *x* axis and A22 is the cross-flow direction along *y* axis. From this comparison, we can see that the prediction highly matches the experimental result. Through the validation on the plate and the tensile bar, we can also conclude that the current iARD-RPR parameters ($C_I = 0.005$, $C_M = 0.1$, $\alpha = 0.7$) yield accurate fiber orientation on the geometries without drastic changes such as ribs or bosses.

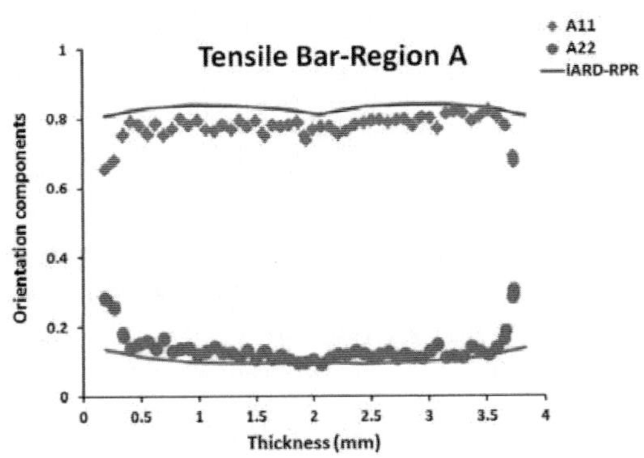

FIGURE 7. Comparison of the experimental and predicted data of orientation components through the thickness at **Region A** of the DSM tensile bar.

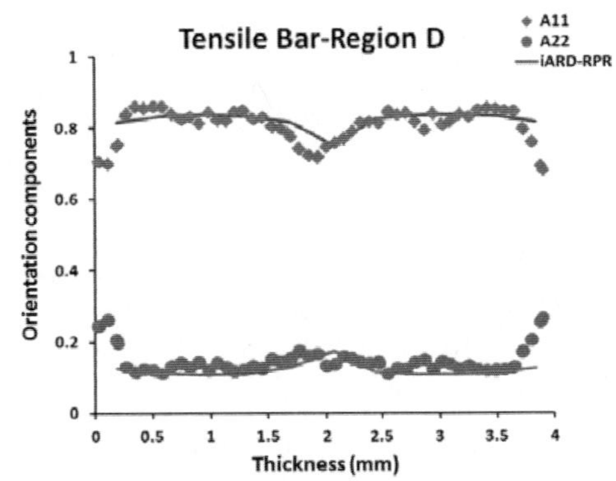

FIGURE 8. Comparison of the experimental and predicted data of orientation components through the thickness at **Region D** of the DSM tensile bar**.** locations.

III. Longitudinal and transverse ribbed plates

Simulation-experimental comparisons on fiber orientation distributions through the thickness at region 1 and region 3 of the longitudinal ribbed plate are shown in **Figure 9** and **Figure 10**, respectively. A11 is the flow direction along the *x* axis, A22 is the cross-flow direction along *y* axis and A33 is the thickness direction along *z* axis.For the orientation distribution plot of the ribbed plates, complex flow field occurring at the bottom of the rib might be the possible cause of the discrepancies between simulation results and experimental ones. However, if we take their average values for comparison, (see **Table 1**), good correlation between experimental results and simulated ones could still be observed (within 10% difference) which imply that for the macro-scale behavior, such as overall mechanical behaviors under loading, the predicted fiber orientation tensors could still provide some qualitative indications. With the same approach, the fiber orientation investigation on a transverse ribbed plate was performed, and the simulation-experimental comparisons are shown in **Figure 11** and **Figure 12**. In this case a better agreement with the experiments are presented (simulation avg. vs. experimental avg.) in **Table 1**.

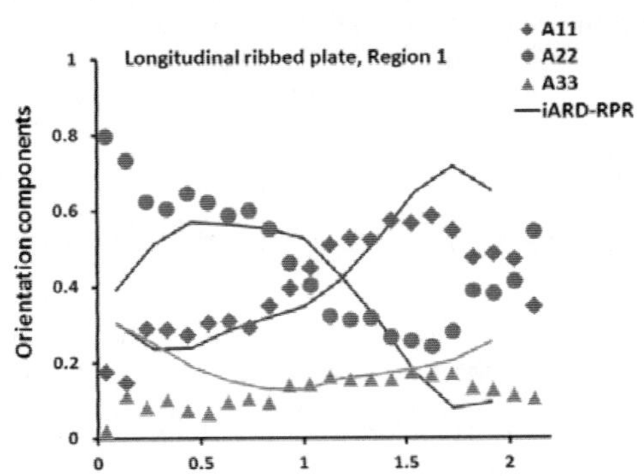

FIGURE 9. Comparison between the experimental and predicted data of orientation components through the thickness at **Region 1** of the longitudinal ribbed plate.

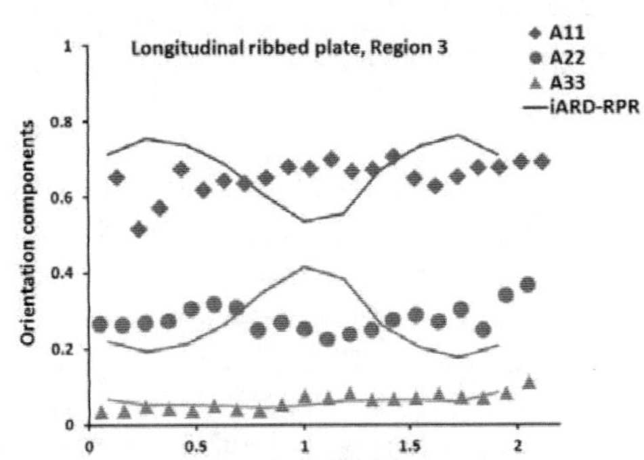

FIGURE 10. Comparison between the experimental and predicted data of orientation components through the thickness at **Region 3** of the longitudinal ribbed plate.

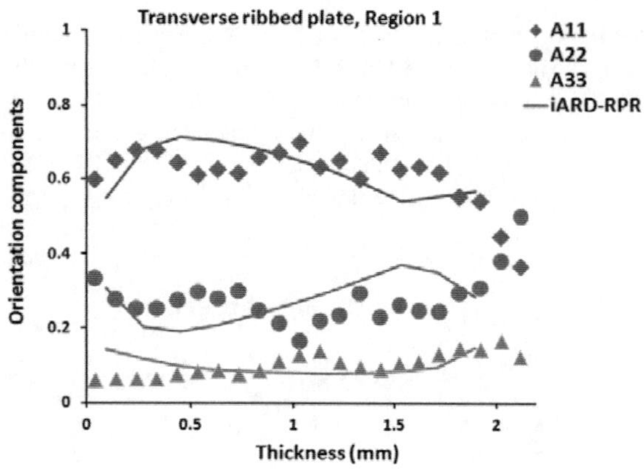

FIGURE 11. Comparison between the experimental and predicted data of orientation components through the thickness at **Region 1** of the transverse ribbed plate.

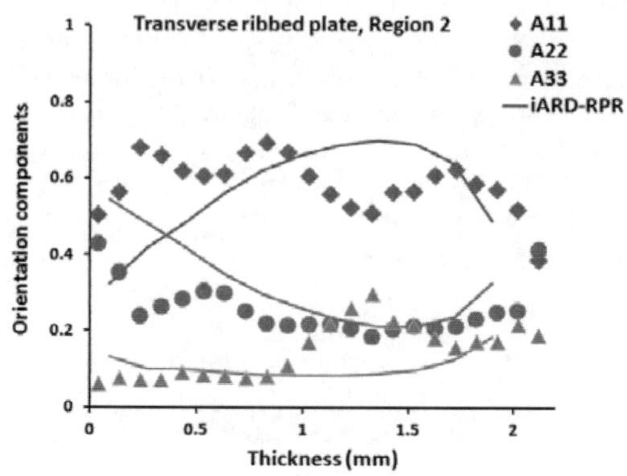

FIGURE 12. Comparison between the experimental and predicted data of orientation components through the thickness at **Region 2** of the transverse ribbed plate.

Table 1. Thickness-average value of experimental and predicted A_{11}. Note that the digit found in the bracket means the relative deviation percentage.

Location	Experiment	Prediction
Longitudinal-Rib #1	0.472	0.426 (-10%)
Longitudinal-Rib #3	0.656	0.679 (4%)
Transverse-Rib #1	0.616	0.624 (1%)
Transverse-Rib #2	0.589	0.570 (-3%)

5. Conclusion

In this work, we verified that using a default parameter set of the iARD-RPR model as the material parameters is able to provide good predictions on fiber orientation in different kinds of injection molded parts, from simple plate, tensile bar, to ribbed plates. This work also provides a basis for further applications such as parts with various changes in the direction of the flow, the inclusion of ribs and changes in thickness and holes. In the future, connecting the fiber orientation to the mechanical properties in order to obtain more reliable integrative simulations of structural analyses will be our next step.

Acknowledgments

Special thanks to DSM's Materials Science Center for providing the experimental data, models and molded parts used in the research project, as well as for their valuable suggestions and comments.

References

[1] Zheng, R., Tanner, R. I., and Fan, X.-J., Injection Molding: Integration of Theory and Modeling Methods, Springer, Berlin (2011).

[2] Tseng, H.-C., Chang, R.-Y., and Hsu, C.-H., Journal of Rheology, **60**, 215 (2016).

[3] Foss, P. H., Tseng, H.-C., Snawerdt, J., Chang, Y.-J., Yang, W.-H., and Hsu, C.-H., Polymer Composites, **35**, 671 (2014).

[4] Advani, S. G. and Tucker III, C. L., Journal of Rheology, **31**, 751 (1987).

[5] Moldex3D, User Manual and Material Database, CoreTech System, Hsinchu, Taiwan.

[6] Hine, P.J., Duckett, R.A.. Polymer Composites, **25**, 237 (2004).

[7] Hine, P.J., Fibre orientation report for DSM, **unpublished**, (2007).

Development of joining methods for highly filled graphite/PP composite based bipolar plates for fuel cells: adhesive joining and welding

P. Rzeczkowski[1,a)], M. Lucia[1,b)], A. Müller[1,c)], M. Facklam[2,d)], A. Cohnen[2,d)], P. Schäfer[2,d)], Ch. Hopmann[2,d)], T. Hickmann[3,e)], P. Pötschke[1,f)], B. Krause[1,g)]

[1] Leibniz-Institut für Polymerforschung Dresden e.V. (IPF), Hohe Straße 6, 01069 Dresden, Germany;
[2] Institute of Plastics Processing (IKV) at RWTH Aachen University, Seffenter Weg 201, 52074 Aachen, Germany;
[3] Eisenhuth GmbH & Co. KG, Friedrich Ebert Straße 203, 37520 Osterode am Harz, Germany;

[a)] rzeczkowski@ipfdd.de
[b)] lucia@ipfdd.de
[c)] mueller-anett@ipfdd.de
[d)] zentrale@ikv.rwth-aachen.de
[e)] T.Hickmann@Eisenhuth.de
[f)] poe@ipfdd.de
[g)] Corresponding author: krause-beate@ipfdd.de

Abstract: Novel material solutions for bipolar plates in fuel cells require adapted ways of joining and sealing technologies. Safe and life time enduring leak-tight contacts must be achieved by automatic processes using reasonable joint forces. A proper sealing should manage such challenges as good ageing properties, excellent leak-tightness, high thermal conductivity and low gas permeability. Hence in this work, adhesive bonding and welding are considered as suitable methods, which can fulfill the requirements mentioned above. Adhesive systems seem to be more easy to apply than conventional sealings (hand layed-up rubber gaskets), e.g. with automatic dispensers. Additionally, the properties of an adhesive joint can be enhanced by a process-specific surface pre-treatment. This work focuses on the characterization of adhesive systems and their joints with highly filled graphite composites. Mechanical properties of the joints were characterized through lap-shear tests.. The influence of ageing caused by humidity or acidic solvent at increased temperature on the bond line properties as well as neat adhesive was examined. The thermal conductivities of neat adhesives and through the entire joint were examined. In order to improve above conductivities, roughening, substrate pre-heating, post-curing and various contact pressure weights were applied. Plasma treatment was chosen as surface pre-treatment method for improving substrate's surface energy. An alternative to bonding is plastic welding, which does not require the use of sealants and adhesives. Based on former study of influences of filler content on the welding process using ultrasonic, hot plate or infrared welding, a welding method for joining the graphite compounds was derived.

INTRODUCTION

One of the major problems by assembling fuel cell stacks is ensuring tight sealing between both bipolar plates and preventing leakage of media from the inside of the cell. A proper sealing/joint of bipolar plates should exhibit sufficient mechanical properties, low gas permeability, resistance to humidity and increased temperature and provide high tightness. So far, O-rings or profile gaskets are mostly used for sealing in fuel cells, which require time-consuming hand lay-up and certain contact pressure for necessary tightness, which can be problematic for brittle, graphite-based bipolar plates. In those conventional sealings, gas feeding channels in the bipolar plates are still exposed to leakage [1, 2]. Therefore, there is a need to find a better sealing method in fuel cell stacks. Adhesive joining or welding seem to be suitable methods due to the possibility of automated processing and high tightness of the bonding between both components. Except of sealing, joints with good mechanical properties can be provided, so that the fuel cell stack can be assembled without necessary compression forces on bipolar plates, as it is needed in case of O-rings or gaskets.

Polypropylene (PP)/graphite-based composites seem to be promising materials for bipolar plates and are successfully applied in industry. However, PP is considered as a polymer with very poor adhesiveness due to its nonpolar properties. Therefore, a proper surface pre-treatment of every PP surface before sticking is necessary.

Corona or plasma treatments were found as most effective methods for improving polarity and adhesiveness of PP materials [3].

The production of material-locking plate joints using a plastic welding method offers the advantage that the assembly process is greatly simplified and the number of mounting elements and additional materials (sealing material or adhesives) can be reduced. In extensive welding tests, the welding behavior of conductive graphite compounds was analyzed using the ultrasonic, infrared, and heating element welding methods [4,5]. The welding quality of highly filled plastic composites is significantly influenced by the graphite content in the material as increasing graphite content reduces the polymer part available for welding. In addition, mechanical damping increases as well as the thermal diffusivity which all affects the weld formation. Thereby, the ultrasonic welding method is restricted by process limitations when joining composites highly filled with graphite. A major challenge for infrared and hot plate welding methods is the thermal conductivity of the composite materials. In order to establish a material-locking joint during the welding process, a defined melting of the material in the joining area is decisive. Due to of the heat flow into the parts to be joined, however, the needed welding temperature level is difficult to reach and to retain. Due to volumetric heating the entire joining part is heated up, which leads to deformation of the joined part wall when applying a joining force [4,5].

This work focuses on the development of bonding and welding methods, which will be suitable for an assembly of PP composite materials filled with high graphite contents used for bipolar plates in fuel cells, redox-flow batteries or housings in heat exchangers. The following properties of epoxy- and PUR based adhesive systems as well as adhesive joints were examined: mechanical properties (lap-shear test), thermal conductivity, gas permeability, and resistance to humidity and acidic solvents at increased temperature. Possible improvement of the properties of adhesive joints through surface pre-treatment (plasma, roughening, pre-heating of the surface) was preliminary tested. New idea of welding method for joining of graphite compounds will be presented.

EXPERIMENTAL PART

Materials

Three kinds of graphite based composites made based on PP/EPDM (70/30) blends with 65 wt.% (G65) or 80 wt.% (G80_1) expanded graphite and 80 wt.% (G80_2) synthetic graphite were used. The composites based on polypropylene (PP, Sabic PP579S) were melt mixed using the lab scale co-rotating twin-screw extruder Typ ZSK26Mc (Coperion) using a screw with a L/D ratio of 45. Compounding was performed at 200°C, a throughput of 10 kg/h and a rotation speed of 150 min^{-1} and compression molded to model bipolar plates with 200 mm x 220 mm x 3 mm. For lap-shear tests of adhesive joints aluminum samples (alloy 2024) were used.

Commercially available adhesive systems were selected based on requirements for fuel cell's and heat exchanger's working conditions:

- EP1 – 2K epoxy-based, filled with glass micro spheres
- EP2 – 2K epoxy-based, filled with glass micro spheres
- EP3 – 2K epoxy-based, in form of cast resin
- EP4 – 1K epoxy-based, filled with aluminum powder, thermal curing
- EP5 – 2K epoxy-based, low viscous
- PUR – 2K polyurethane-based.

In order to carry out welding tests, type 1BA tensile test rods were injection molded from a test compound consisting of the matrix material PP and the synthetic graphite (70 wt.%). Rectangular samples with the dimensions 15 x 10 x 4 mm^3 were prepared.

Test Methods

Lap-shear tests of single lap adhesive joints were carried out with the universal testing machine Zwick Roell 20 kN in accordance with DIN EN 1465. The samples were prepared with dimensions 100 x 25 x 3 mm^3 (graphite composites) and 100 x 25 x 1.6 mm^3 (aluminum) and stuck together with an overlapping length of 12.5 mm and a bondline thickness of 0.2~0.3 mm.

Thermal conductivity (TC, through-plane) was determined with the laser flash method on the device Netzsch LFA 447 at 25°C for neat adhesive resins and through the adhesive joints. Samples were pressed to cylindrical shape with diameter of 12.7 mm and thickness of 2 mm. Post-curing process was carried out at 70°C / 5hrs.

Dynamic contact angles were measured on the Data Physics OCA 35 XL device with two fluids: water and 1,5-pentanediol.

Ageing of single lap joints (aluminum substrate) was carried out in 100% relative humidity at 60°C over 10 days. Ageing of neat adhesives (same cylindrical samples as in thermal conductivity measurement) was performed in 25%-sulfuric acid solution at 60°C over 48hrs.

Graphite composite surfaces were either treated with the vacuum plasma device TePla 440-G (O_2, 10 cm^3/min, 60 sec, 0.23 mbar) or roughened with sandpaper P150, or annealed at 50°C / 5hrs.

FTIR spectroscopy of the cured adhesive surfaces was carried out with the FTIR microscope Hyperion 2000 with Vertex 7 and Raman spectroscopy on Raman Microscope WITec alpha 300R.

RESULTS

The results of the lap-shear tests with the aluminum substrate are shown in Fig. 1. Depending on the kind of adhesive, an increase or decrease of the lap-shear strength after ageing was observed. Almost all lap-joints fractured adhesively, except of EP1, where without ageing mixed fracture (partially adhesively, partially cohesively within adhesive layer) and after ageing adhesive fracture were observed. The adhesive EP5 exhibit no resistance against humidity at increased temperature and lost his mechanical properties after ageing. EP3 and EP5 have limited resistance against ageing and the decrease of lap-shear strength is around -25%. EP1 and EP4 seem to withstand conditions approximated to those in fuel cells with satisfying results, e.g. changes of lap-shear strength < ±10%. The increase of lap-shear strength of EP1 and EP2 could be caused by post-curing or softening of the adhesive during ageing process. Lap-shear tests conducted with graphite composite materials with EP1 show without and after ageing the same strength results namely fracture within the substrate material due to the tensile load. Hence, the adhesive joint exhibit higher strength than strength of substrate material and would be suitable for joining these graphite composites.

As summarized in Table 1, according to expectations neat adhesives exhibit low thermal conductivity (TC) in the range of 0.24 - 0.57 W/(m·K). The highest thermal conductivity has the adhesive EP4 (0.57 W/(m·K)) due to incorporated aluminum powder. The influence of post-curing on thermal conductivity of two adhesives (EP1 and EP3) was examined and only slight improvement was observed (+ 5-6%). Thermal conductivities measured through adhesive joints with graphite filled composites and several adhesives are also shown in Table 1. It was observed, that the thermal conductivities through the joint have values between the thermal conductivities of substrate and neat adhesive. Generally, higher thermal conductivity through the joint can be reached with adhesives having higher conductivity. The differences in thermal conductivity through the joints between adhesives EP3 and EP4 with the composite substrates G80_1 and G80_2 can be explained through the thicker adhesive layer when using the adhesive EP4. When increasing the pressing weight during the curing phase of the joints, the thermal conductivity through the joint improved as shown on the sample of G65. Joints made of adhesives with higher thermal conductivity and substrates with lower conductivity result in higher conductivity through the joint than joints made of adhesives with lower conductivity and substrates with higher conductivity.

The influence of roughening or pre-heating or both of the substrate surfaces on thermal conductivity was examined with the composite materials G65 and G80_2 combined with the adhesive EP3. When using G65 no improvement of thermal conductivity through the joint was observed after these treatments. This could be caused by the relatively high porosity of the composite substrate surface, so that roughening has not much contribution to the aimed increase of the contact surface area between the adhesive and the substrate. Roughening of the surface of the G80_2 composite indicated an improvement of thermal conductivity through the joint by 22%.

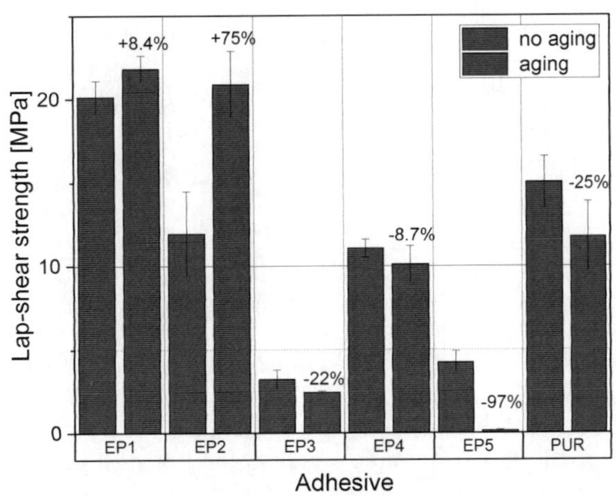

FIGURE 1. Results of lap-shear test with aluminum substrate without and after ageing (100% rel. humidity at 60°C, 10 days) [IPF].

In order to evaluate the changes on the composite materials by plasma treatment, dynamic contact angle measurements were performed on the untreated and plasma-treated surfaces of all graphite composites (Table 2). The untreated surfaces exhibit hydrophobic properties and insufficient wetting with water, e.g. contact angle of >90°. After vacuum plasma-treatment, an activated and more polar substrate surface was created and thus distinct improvement of wettability was observed, e.g. significant decreases of contact angles measured with both fluids.

The chemical resistance of the neat adhesive against 25%-sulfuric acid solution was tested. The adhesives EP1, EP2, and EP5 exhibit not sufficient chemical resistance. Significant changes on the sample surface (etching), in color, weight, dimensions and loss of the mechanical properties (softening) of the samples after ageing have been observed.

TABLE 1. Thermal conductivity (TC) of neat adhesives, substrate materials, substrate/adhesive/substrate (s/a/s) and pressing weight [IPF].

Material	Adhesive	Pressing weight [g]	Adhesive layer thickness [mm]	TC adhesive [W/(m·K)]	TC substrate [W/(m·K)]	TC through joint (s/a/s) [W/(m·K)]
G65	EP3	360	0.3	0.42	8.7	3.2 ±0.7
		750	0.2			4.4 ±0.3
		1200	0.2			2.5 ±1.1
G80_1		600	0.2		16.4	6.7 ±0.2
G80_2		600	0.2		15.7	6.3 ±0.5
G80_1	EP1	600	0.4	0.24	16.4	2.2 ±0.4
G80_2		600	0.4		15.7	2.3 ±0.4
G65	EP4	600	0.4	0.57	8.7	3.5 ±0.2
G80_1		600	0.4		16.4	4.3 ±0.5
G80_2		600	0.45		15.7	5.2 ±1.7

In the corresponding FTIR-spectra of these samples, distinct difference before and after chemical ageing occurred, which indicate decomposition or strong oxidation of the adhesives. Adhesive EP4 seems to withstand the chemical attack and EP3 and PUR exhibit limited resistance against sulfuric acid. In this case, no changes in weight or dimensions were found, but EP3 and PUR exhibited some color changes of the sample after ageing. In the FTIR spectra of EP3 and PUR bands indicating oxidation were observed. When using EP4 the FTIR-spectrum of the surfaces indicated somehow lower oxidation of this adhesive. In Raman spectroscopy of EP4, the spectrum of aluminum sulfate was found on the sample surface, which was created through the contact of aluminum powder with sulfuric acid

TABLE 2. Results of contact angle measurements with different fluids [IPF].

Material	Contact angle[°]			
	Water		1,5-pentanediol	
	untreated	plasma-treated	untreated	plasma-treated
G65	96.6 ±3.9	43.2 ±4.8	35.7 ±1.7	12.1 ±1.8
G80_1	94.3 ±3.3	49.5 ±0.5	33.3 ±2.7	12.3 ±1.5
G80_2	117.6 ±3.3		34.9 ±1.2	

Based on the results of the practical welding tests and material analyses, a concept for hot pressing of the highly filled graphite compounds at the IKV was derived and validated in an initial feasibility study. The thermal properties of the material can be used specifically for heat input by heating the joining area completely through contact with a heating element (see Fig. 2). At the same time, a joining force is applied so that the heating and joining phase do not run separately from each other as in infrared or hot plate welding method. The graphite filler prevents the melt from sticking to the heating element.

A constructive encapsulation of the melt is used to prevent undesired flow of the melt. As a further decisive process parameter compared to hot plate or infrared welding, the joining force can be significantly increased. Initial tests with graphite based polypropylene composite with 70 wt.% of synthetic graphite have shown promising results, although the fracture does not run along the joint interface in the case of mechanical failure. This indicates that the base material strength is reached. Further tests are required in order to be able to evaluate the hot pressing process. In this context, a variothermally temperature-controlled joining tool is developed, which allows controlled cooling of the melt after a certain joining time.

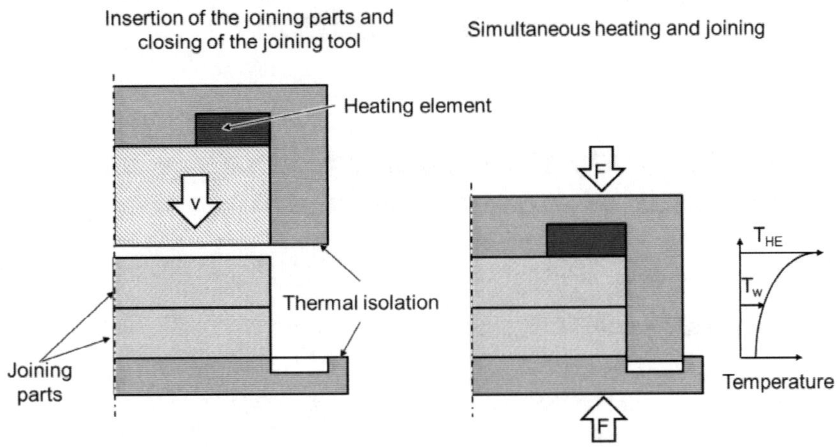

FIGURE 2. Schematic illustration of the welding of highly filled graphite compounds [IKV]

SUMMARY

Commercially available epoxy and PUR based adhesive systems were tested regarding the operating conditions in fuel cells, redox-flow batteries and heat exchangers. Depending on the material class of the adhesives, incorporated fillers and curing conditions, the adhesives exhibit different properties and can withstand differently the conditions in the named applications. Adhesive EP4 seems to fulfill all the application requirements due to sufficient mechanical properties and good resistance against humidity and chemicals at increased temperature. EP1 seems to be suitable for sealing and joining of bipolar plates in fuel cells and heat exchanger housings, where the durability in humid environment at increased temperature is required. The PUR based adhesive exhibits limited resistance against operating conditions in all applications and thus could be also considered as a suitable adhesive. Further investigation of all adhesives is needed.

Adhesives filled with conductive fillers (EP4) exhibit higher thermal conductivity than unfilled adhesive systems. The thermal conductivity through the joint is much lower than that of the substrate materials but significantly higher than that of neat adhesives. Joints with thicker adhesive layer show the tendency to lower thermal conductivity through the joint. It appears that the adhesive layer has a higher contribution to the thermal conductivity through the joint than the substrate material, taking into consideration the fact, that the adhesive layers have only thicknesses of 0.2~0.4 mm and the substrates of ~0.85 mm on each side of the joint. When increasing the pressing weight during the curing phase of the adhesive or roughening of the substrate surface, the thermal conductivity of the joints can be improved.

Plasma treatment of the substrate surfaces, which represents a fast and efficient method, was shown to improve wettability and polarity of the surfaces of the composites with high graphite contents, which is crucial for high quality adhesive joints.

Due to the high thermal conductivity of highly filled graphite composites, joining by means of conventional plastic welding methods is a major challenge. A material-adapted joining process seems to be promising in order to realize a high-quality weld seam. In order to assess the suitability, further investigations are necessary. For this purpose, a suitable test piece is used, which allows leak tests and bursting pressure tests. In order to optimize the process control, a variothermal temperature control system is integrated into the joining tool.

ACKNOWLEDGEMENT

The authors thank the German Bundesministerium für Bildung und Forschung (No. 01LY1512) for financial support of this work.

REFERENCES

1. K. Dilger, P. Beckhaus, Abschlussbericht zum IGF-Forschungsvorhaben Nr. 17.062 N/1 (2013).
2. J. Larminie, A. Dicks, *Fuel Cell Systems Explained* (John Wiley & Sons Ltd., West Sussex, 2003).
3. G. Habenicht, *Kleben: Grundlagen, Technologien, Anwendungen* (Springer-Verlag, Berlin, 2006).
4. Ch. Hopmann, M. Facklam, A. Cohnen, B. Krause, T. Hickmann, Joining Plastics **11**, 108-115 (2017).
5. T. Hickmann, C. Hopmann, M. Facklam, A. Cohnen, A.. Umdruck zur IKV Fachtagung Kunststoffe erfolgreich verbinden – Innovative Fügetechnologien für die Praxis. (IKV, Aachen, 2016).

A novel approach to study thermoplastic matrix materials' impregnation behavior of woven fibres

Längauer Manuel [1]*, Pree Maria [1], Zitzenbacher Gernot [1]

[1] University of Applied Sciences Upper Austria - Upper Austria - Austria

With increasing demand of thermoplastic composites for the automotive and aviation industry, thermoplastic matrix materials' impregnation behavior of woven fibres is of rising interest. Due to the temperature dependent flow behavior and the wide range of physical properties of thermoplastic melts, predictions of the wetting of the weft fibres by the melt are hard to make. This work focusses on a new attempt to study the impregnation behavior by investigating the wetting of fabrics by polymeric melts. The experiments were conducted in a Krüss drop shape analyzer (DSA30), equipped with a high-temperature chamber. Due to the fact that fabrics usually do not exhibit a flat surface which is additionally altered with increasing temperature depending on the thermal expansion of the material, a special sample holder had to be constructed to overcome those obstacles. The study includes different polymeric melts and a variety of woven fabrics. Polymeric pellets are placed on the preheated fabrics in the nitrogen-flooded high-temperature chamber and the time-dependent contact angle and the basic geometries as height and drop radius of the melt droplets are measured. The contact angle gives information about the wetting behavior, while the geometries can deliver data about the impregnation of the melt drops. The results show, that the impregnation behavior is strongly dependent on the area weight of the fabric and the kind of weave that is used. Furthermore, the temperature plays an important role in the wetting properties. An additional influence is derived from the interaction of the polymer with the fibres.

Confined carbon nanotube network lead to high dielectric constant and suppressed dielectric loss

Wu Kai (1), Fu Qiang (1)*

(1) Sichuan University - Sichuan - China

It is still a challenge to prepare dielectric materials with giant dielectric constant and low dielectric loss. In this study, we provide a new strategy to fabricate a segregated double network within PS matrix, where robust hexagonal boron nitride (h-BN) network serves as the insulating shell and is excluded from PS matrix to destroy the continuity of local multi-walled carbon nanotubes (MWCNT) network located in PS phase. To do this, firstly, PS/MWCNT material was simply fabricated by melt mixing PS with different volume content of MWCNT. Then the resultant composites were smashed to get micron-sized PS/MWCNT powders. Finally, different volume content of h-BN was coated on the PS/MWCNT powders under the intense mechanical mixing via ð-ð self-assembly. As a result, a high dielectric constant of 123 is achieved while a relatively low dielectric loss is also kept as 0.36. This suppressed dielectric loss is mainly ascribed to the dramatically reduced AC conductivity owing to the segregated effect of insulating h-BN layer. Besides, this special structure of segregated double network also leads to obviously enhanced thermal conductivity, arising from the high synergistic efficiency between segregated h-BN network and dense MWCNT network. We believe that these good comprehensive performances promise this structure to offer a simple and effective way to prepare high-performance dielectric materials with not only high dielectric constant and low dielectric loss, but also good capability of heat dissipation.

Innovative automated tape laying process – Manufacturing challenges and material properties

Nagl Benjamin Johann (1)*

(1) Upper Austria University of Applied Sciences - Upper Austria - Austria

BENJAMIN NAGL a), ROLAND HINTERHÖLZL a) WOLFGANG STADLBAUER a) a)University of Applied Science Upper Austria, Research and Development Ltd., Stelzhamerstraße 23, 4600 Wels, Austria Fibre reinforced composites are nowadays state of the art in many different industrial areas, like in aerospace and automotive applications as well as in wind power plants. Such composites typically consist of carbon or glass fibres as well as a thermosetting matrix. An emerging trend is to use thermoplastic polymers as the matrix material, as these exhibit some advantages, like low cycle time, thermoformability, recyclability and improved joining possibilities. Usually parts made of thermoplastic composites are either manufactured using forming or injection processes. In this work an alternative fabrication concept, compared to the ones already on the market, is used. The Set up consists of 3 robots and a deposition table which are working in coordination with each other. This technique can be classified as an automated tape laying process. In this work the influence of effects like the load distribution of the compaction roller and the absolute compaction force is investigated. The defects occurring during the manufacturing process like gaps, overlaps and bridging and their impact on the macroscopic material properties of the composites are also examined. The consolidated composites are investigated in respect to their morphology, as well as material degradation (via OIT and FTIR spectroscopy), their interlaminar properties (via mechanical testing) and their warpage due to residual stresses.

Study of the Influence of the mixing method on the Graphitization of phenolic resins for carbon-based refractories

C. G. Renda[a], A. P. Luz[a], R. Bertholdo[b], T. Venâncio[a] , V. C. Pandolfelli[a], A. A. Lucas[a]

[a] Federal University of São Carlos, Washington Luiz Highway, km 235, São Carlos, SP, 13565-905, Brazil

[b] Federal University of Alfenas, Institute of Science and Technology, José A. Vilela Highway, 11999, BR 267 Km 533, Poços de Caldas, MG, 37715-400, Brazil

Abstract. Phenolic resins (PR) are commonly applied as binders in MgO-C refractories bricks. Despite their good qualities, PR can be classified as non-graphitizing carbon sources. Aiming to improve the thermo-mechanical behavior of resin-bonded refractories products, there is a continuous search for alternative routes or additives (i.e. ferrocene) which can induce the graphitization of carbon derived from the phenolic resins at temperatures around 1000^{0}C . In this regard, the graphitization trend of two sorts of PR [a commercial novolak (Nv) and a modified synthesized novolak (Rc)] with and without 3 wt % of ferrocene (Fc) were evaluated. The effect of two different mixing procedures (ML = low energy and MH = high energy) were also analyzed. ML was carried out in a mechanical stirrer at 300 rpm for 20min + sonication during 10 min, whereas MH was performed at 2000 rpm for 30 seconds. The samples were heat treated as follow: 80^{0}C/2h, 100^{0}C/30min 150^0/30min, 200^{0}C or 400^{0}C/1h, followed by 500^{0}C/1h and 1000^{0}C/5h at a heating rate of 3^{0}C/min. The graphitization level (GL) was evaluated via XRD measurements. The Diffraction patterns were deconvoluted using Origin Peak Analyser tool, and the GL was calculated as the ratio between the area of the graphitic carbon peak and the total area of the patterns. The results indicated that the highest GL (~60%) were obtained for the Rc+3 wt% of Fc sample using the ML procedure, whereas GL ~50% was obtained when the same sample was stirred using the MH. FTIR showed Fc delayed the reaction for the Rc+3wt% (ML) added with NMR analysis where these samples have more methylene bridges (CH_2) probably giving enough time and mobility to have the highest GL. The NMR for the Nv+Fc resins were showed a chemical shift near to Fc (C aliphatic) and -CH_3, in the other hand, for the Rc+Fc resins, just the MH showed the methylol groups in PR structures. The investigations suggest that Rc+Fc+ML has more CH_2 and this mixture delay the reaction and improve the graphitization level.

Keywords: carbon, mixing, phenolic resin, resin-bonded refractories, reaction.
PACS: 81.05.U,47.51.+a, 45.70.Mg,81.05.Je,81.05.Mh, 82.30.-b, 82.33.-z, 82.35.-x

INTRODUCTION

The presence of carbon in carbon-containing refractories products provides higher thermal shock resistance and the corrosion resistance caused by siderurgical slag. Phenolic resins in these refractories can act as carbon source, as binder, or with the potential to graphitic carbon formation. Graphitic carbon diminishes the wettability to slag and improves the mechanical properties in low and high temperatures. The increased thermal conductivity of refractories induces to the metal layer formation or solidified slag into the refractories faces, protecting it from the degradation. All of these characteristics improve refractories shelf-life (1-3). Another advantage comes from the substitution to tar, which has been also traditionally used for health and sustainability reasons. Phenolic resins doesn´t need any additional water and they can form nanostructured phases in the presence of graphitizing agents. In a previous study, some parameters of the phenolic resins (two novolaks and a resole one) were evaluated, such as the cure temperature, and presence of firing and additives (ferrocene, boric acid, exfoliated graphite) with the goal to obtain the highest graphitization level. Our results had proposed that the ferrocene had induced better the graphitization level (3). Ferrocene is an organometallic compound which decomposes itself in Fe and Fe_3C nanoparticles, under high temperature, serving as sites for graphene sheets to organize themselves in a heterogeneous way (1,3). In order to understand the influence of ferrocene and mixing protocol on the graphitization of phenolic resin. Their chemical reactions with ferrocene during each heat treatment step is presented, based on results FTIR-ATR, ^{13}C-NMR, XRD and Raman Techniques.

Materials and Techniques

Two types of phenolic resins were used in this study: a commercial novolak (NV) and a modified synthesized novolak (Rc). Nv was Prefere 88510R, kindly donated by Dynea, Brazil, and Rc was synthetized with Phenol (99% purity, Proquímios, Brazil) and Formaldehyde (37% in solution, Proquimios, Brazil), in a 0.33 molar ratio, stirring in room temperature for 30 minutes, adding the catalyst acid HCl (0.67 volume ratio HCl:formaldehyde) under stirring for 150 minutes. Two phases were obtained being the bottom one the pre-polymer (4). The cure agent used was hexamethyltetramine (HMTA) (10wt%) and the graphitizing agent is ferrocene (150<d<70μm, 98% purity, Aldrich, USA)3wt%. Two mixing procedures for adding ferrocene to phenolic resin were used, ML or Low Energy and MH or High Energy. ML was with mechanical stirrer, with rotation speed of ~50 rpm for 20 minutes followed by ultrasonic bath for 10minutes. MH was conducted in a Vortex Mixer (StateMix VM-200, Canada), with a rotation speed ~2000 rpm for 30 seconds. The curing procedure comprised various intermediate heating steps (80^0C/2h, 100^0C/30min 150^0/30min and 200^0C/1h (Rc), followed by a thermal treatment at 500°C for 1h and at 1000°C for 5h (heating rate= 3^0C/min. For Nv was changed the step in 200^0C/1h for 400^0C/1h. The samples were stored the alumina crucibles (with lid), inside a sealed refractory box with reducing atmosphere during the samples' pyrolysis as proposed by Luz et al (3). For XRD analysis, a Siemens D5005, AXS equipment was used, with Cu-K α radiation [λ=1.54056 Å], nickel filter, using 40 mA, 40 mV and scanning step=0.032°xsec and 2θ=4 to 80 degrees. The graphitization level was evaluated by the relation of the graphitic carbon area by the total area of XRD diffraction pattern (or graphitic + non-graphitic carbon). The experiments were duplicated. After the cure, the samples were analyzed by the ^{13}C- solid state NMR in a Bruker Avance III-400 equipment, operating at a magnetic field of 9.4 T, CPTOSS-MAS pulse sequence was used, with a ^{13}C frequency of 100.57 MHz (400 MHz for ^{1}H nucleus). The NMR spectra of each sample was collected with approximately 2 hours, using a number of scans of about 2500. The reference standard for chemical shift calibration was adamantane (38.5 ppm for CH_2 signal). The recycle delay was 3 s, the contact time used was 3 ms and the spinning speed was 5 kHz. The Raman Spectrophotometer was a Labran HR - Horiba Jobin Yuon, with a sample area of 100μm^2. The equipament operates with He-Ne laser (632.8 nm) with a power of 17μw. The spectra were obtained for powder samples, liquids, viscoses or monoliths, for the range 200 a 2000 cm^{-1}, with accumulations of 2 spectra each 30 seconds. The bi-Lorentzian deconvolution was obtained by a software Origin. The FTIR spectrophotometer was an Agilent Technologies Cary 630 FT-IR. Each spectrum with 64 accumulations and a resolution of 4cm^{-1}. The spectra were obtained for powder samples, liquids, viscoses or monoliths, for the range 4000-650 cm^{-1}.

Results and Discussion

The XRD results are presented in Table i. According to Figure 1 the graphitization level was higher for synthesized modified novolak (Rc) than for commercial novolak (Nv). The findings corroborate with Luz et al (3). It was also possible to identify in the XRD results two phases Fe_3O_4 and Fe_2O_3.

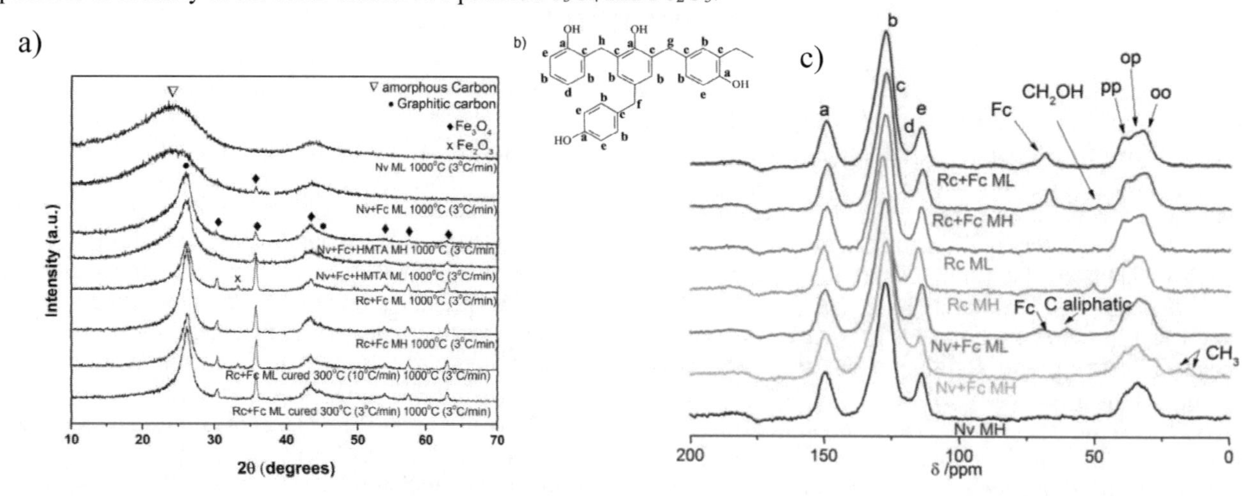

FIGURE 1. a) XRD Results, b) the structure model consider for ^{13}C-NMR attribution (5) and c) the ^{13}C-NMR spectra.

For ^{13}C-NMR results, the highlighted groups were assigned to the chemical shift hydroxy-substituted phenolic carbons and aromatic carbons (between 150-115ppm), 70ppm, 60 ppm, 50ppm, methylene bridges (between 40-30ppm), 20 ppm and 16 ppm. The chemical shift in 60 ppm, referred to aliphatic carbon appeared only in the Nv+Fc ML, probably due to the oxidation state of ferrocene having changed. The chemical shift in 50 ppm was attributed to CH_2OH (methylol groups) and it appeared in the Rc+Fc MH and Rc MH only. The ferrocene signal appeared in higher level in the Rc resins, as expected corroboring with the first work, because the GL was higher for the Rc resins (70 ppm). The chemical shift of pure ferrocene (as received) was detected in 70 ppm. The chemical shifts in 20 and 16 ppm can be attributed to the $-CH_3$ groups (only appeared for the Nv+Fc MH), probable the structure is similar as that founded in Omori et al (6). The ferrocene level, on the ML mixture, was higher to the Rc+Fc than Nv+Fc. The methylene bridges in the Nv+Fc resins in the *ortho-para* positions (35 ppm) is higher in ML than MH. The opposite occurred, for the Rc+Fc, but it is a small difference, less than 1%. Rc is higher than Rc+Fc for the ML mixture, in this chemical shift (35 ppm) and the opposite occurred when the MH mixture. For the *para-para* methylene bridges (40 ppm) the intensity of this chemical shifts Nv+Fc and Rc+Fc with MH procedure was smaller than ML. For the *ortho-ortho* (32 ppm) positions the Nv+Fc and the Rc+Fc presented the same behavior with MH smaller than the ML mixture. Comparing the methylene bridges only, the Nv had higher intensity than Nv+Fc both with the MH mixture and the Rc was higher intensity than the Rc+Fc both with the ML mixture. The ML mixture have more bridges than MH for Nv+Fc, Rc+Fc, and the Rc MH was higher than Rc ML and Rc MH is higher than Rc+Fc with the same mixture. The two samples with higher number of methylene bridges were Rc and Rc+Fc with ML mixture.

TABLE i) Graphitization Level (GL) evaluated by XRD

	GL (%)		GL (%)
Rc+Fc ML cured (10ºC/min) 1000ºC(3ºC/min)	53	Nv+Fc+HMTA ML 1000ºC (3ºC/min)	45
Rc+Fc ML cured (3ºC/min) 1000ºC(3ºC/min)	58	Nv+Fc+HMTA MH 1000ºC (3ºC/min)	44
Rc+Fc MH 1000ºC(3ºC/min)	53	Nv+ Fc ML 1000ºC (3ºC/min)	amorphous
Rc+Fc ML 1000ºC (3ºC/min)	56	Nv ML 1000ºC (3ºC/min)	amorphous

TABLE ii) structure model of phenolic resin consider for ^{13}C-NMR. Source: (5)

Description	Carbon	δ(ppm)	Description	Carbon	δ(ppm)
Hydroxy-substituted phenolic carbons	a	150	unsubstituted ortho aromatic carbons	e	115
unsubstituted meta aromatic carbons	b	130	para-para methylene bridge	f	40
methylene-substituted aromatic carbons	c	125	ortho-para methylene bridge	g	35
unsubstituted para aromatic carbons	d	120	ortho-ortho methylene bridge	h	30

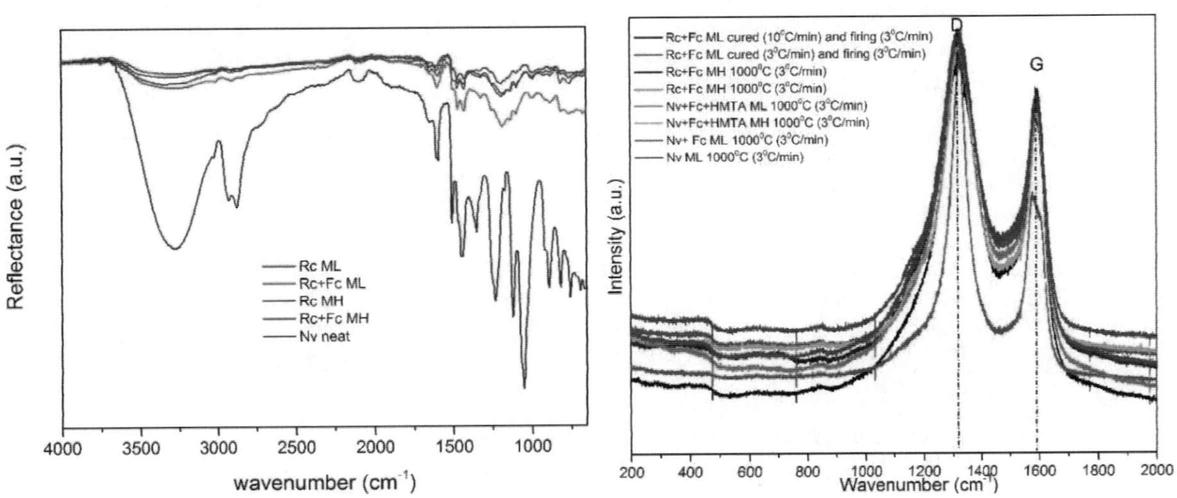

FIGURE 2. FT-IR and the Raman Results.

The FTIR results were normalized divided all peaks by the more intense peak spectra, and repeated twice. These results showed that Rc MH had more CH_2 (methylene bridges) and ML had more CH groups. The Rc+Fc had the inverse, ML with more methylol groups, in this case attributed to the delay in the reaction give more time to graphitization.

The Raman analysis was based in two main peaks, G band (around 1600cm^{-1}) was attributed to highly organized carbons and D band (around 1320cm^{-1}) was attributed to the disordered carbon. The lower ratio I_D/I_G represents the material with higher order of the carbons on the structure. The best ratio I_D/I_G was founded to Rc+Fc ML sample cured before firing, followed by the others from 2.50 to 2.77.

TABLE iii) The Ratio I_D/I_G of Raman Results

	I_D/I_G		I_D/I_G
Rc+Fc ML cured (10^{0}C/min) 1000^{0}C(3^{0}C/min)	2.59	Nv+Fc+HMTA ML 1000^{0}C (3^{0}C/min)	2.54
Rc+Fc ML cured (3^{0}C/min) 1000^{0}C(3^{0}C/min)	1.93	Nv+Fc+HMTA MH 1000^{0}C (3^{0}C/min)	2.50
Rc+Fc MH 1000^{0}C(3^{0}C/min)	2.65	Nv+ Fc ML 1000^{0}C (3^{0}C/min)	2.62
Rc+Fc ML 1000^{0}C (3^{0}C/min)	2.77	Nv ML 1000^{0}C (3^{0}C/min)	2.56

CONCLUSIONS

The results presented suggested an improvement of the capability of the carbon graphitization of phenolic resins, as confirmed by XRD, ^{13}C-NMR, FTIR and Raman results. The sample that seems more interesting is the Rc+Fc with ML mixture. The possibility of controlling the characteristics of the obtained materials creates the perspective of application on the refractories area.

ACKNOWLEDGMENTS

The authors are grateful to Magnesita Refratários S.A. for providing the raw materials used in this work and the Federation for International Refractory Research and Education (fire), the Laboratory of Nuclear Magnetic Resonance- Chemistry Department-UFSCar, DEMA-UFSCar, Ceramics Department-UFSCar, Unesp-Araraquara for partly supporting this study. The PPGCEM-UFSCAR and the CNPQ for the financial assistance to disseminate this research.

REFERENCES

1. C. S. BITENCOURT and V. C. PANDOLFELLI, "Refratários contendo carbono: propriedades, características e variáveis em sua composição," *Cerâmica*, vol. 59, no. 349, pp. 84–114, 2013.
2. C. S. BITENCOURT and V. C. PANDOLFELLI, "Resinas termofixas e a produção de refratários contendo carbono: base teórica e insights para futuros desenvolvimentos," *Cerâmica*, vol. 59, pp. 1–26, 2013.
3. P. Luz, C. G. Renda, A. A. Lucas, R. Bertholdo, C. G. Aneziris, and V. C. Pandolfelli, "Graphitization of phenolic resins for carbon-based refractories," *Ceram. Int.*, vol. 43, no. 11, pp. 8171–8182, 2017.
4. C. G. RENDA, "Estudo da Resina Fenólica FenolFormaldeído e sua Aplicação como Matriz na Preparação de Nanocompósitos Poliméricos contendo Nanotubos de Carbono," Universidade Federal de Alfenas - Unifal - Campus Poços de Caldas MG, 2015.
5. HATFIELD, G.R.; MACIEL, G.E. Solid-State NMR Study of the Hexamethylenetetramine Curing of Phenolic Resins, Macromolecules, 20 (1987) 606-615.
6. OMORI, M; NAGASHIMA, K; YAJIMA, S. Synthesis of Ferrocene-Phenol Resins and Liberation of Iron Particles by Pyrolysis. Bulletin of Chemical Society of Japan, vol.50, no. 5, pp. 1157-1160, 1977.
7. SILVERSTEIN, R. M.; WEBSTER, F. X.; KIEMLE, D. J., Identificação Espectrométrica de Compostos Orgânicos, 7.ed. Rio de Janeiro: [s.n.], 2006.

The Barrier Properties of Flake-Filled Composites

Papathanasiou Thanasis [1]*, Tsiantis A [1]

[1] University of Thessaly - Volos - Greece

We present the results of a direct numerical study to investigate the effect of orientation on the barrier properties of flake-filled composites. Over 5000 simulations have been conducted in two-dimensional, doubly-periodic unit cells, each containing up to 10,000 individual flake cross-sections. We consider long-flake systems (aspect ratio α=50, 100 and 1000) from the dilute ($\alpha\varphi$=0.01) and into the concentrated ($\alpha\varphi$=40) regime, where (φ) is the flake volume fraction. Two types of systems are considered. In the first, the flakes are randomly placed but uniformly oriented. In the second, the orientation angles assume random values in an interval [-ε,+ε], where 0<ε<π/2. At each level of (ε) and ($\alpha\varphi$), several realizations are generated. At each of those, the steady-state diffusion equation is solved and an effective diffusivity Deff calculated. For $\alpha\varphi$<1, the Barrier Improvement Factor (BIF) is a linear function of (ε) and a power function of ($\alpha\varphi$), with the exponent (~1.07) being independent of the extent of misorientation. In concentrated systems, we find that the power exponent decrease as (ε) increases; from 1.71 in the aligned system (ε=0) to ~0.9 in the fully random one (ε=π/2). We propose a scaling which incorporates the effects of both ($\alpha\varphi$) and (ε) on the BIF, resulting in a master curve for all ($\alpha\varphi$) and (ε). In unidirectional misaligned systems (angle (θ)), we find that the BIF reaches a plateau value. We derive analytical estimates for this value in terms of (θ) and ($\alpha\varphi$) and propose a model for the BIF which is capable of reproducing all computational results for ($\alpha\varphi$) up to 40 and 0<θ<π/2. Our results suggest that the anticipated improvement in barrier properties may not be realized if the flake orientations exhibit a significant scatter around the desired direction.

Investigation of Different Thermo-Rheological Measurement Methods for the Characterization of Material Aging

Michael Burgfeld[1, a)] and Johannes Wortberg[1]

[1]*University of Duisburg-Essen, Institute of Product Engineering, Engineering Design and Plastics Machinery, Lotharstraße 1, 47057 Duisburg, Germany*

[a)]michael.burgfeld@uni-due.de

Abstract. To investigate the influence of shear stress on material aging, different thermo-rheological measurement methods can be performed with a rotational rheometer. Thus, frequency and time sweep measurements conducted and subsequently compared. Through the use of frequency sweeps it is possible to investigate the shear thinning effect by various shear rates under a constant thermal stress. To determine time-dependent effects, which lead to a material aging, sequential frequency sweeps are conducted. Using time sweep measurements, the test sample is subjected to a constant thermal and mechanical stress over a defined time period. Afterwards, both methods are compared to find the best way to characterize the material aging. For this purpose, different materials and temperatures are investigated. This paper presents the results for a low-density polyethylene (LDPE) grade.

Keywords: rheology, viscosity, material aging, measurement method
PACS: 83.80.-k, 66.20-d

INTRODUCTION

During the processing of plastics, the material is subjected to thermal and mechanical stresses. In addition, a high residence time in stagnation areas and in the boundary layer causes material aging during the extrusion process. As a whole, the term "material aging" means the physical and chemical changes to plastics that can occur during the processing, storage, and use of the material over time. Material aging can be distinguished by internal and external causes. Internal causes of aging include chemical and physical aging processes that exert a major impact on the material behavior during the extrusion process [1]:

- Chemical aging processes: (irreversible) — Oxidation, degradation, post-condensation, post-polymerization
- Physical aging processes: (reversible) — Relaxation, post-crystallization

As a result of material change, product failure may occur because of optical and mechanical defects, caused by a change in the molecular structure of the material. This, in turn, induces a change in the viscosity. To prevent such failures, it is important to understand the material behavior during extrusion and to examine the aging process more closely. Besides the known influences on rheological properties of plastic melts such as shear rate, temperature, filler content, it is also necessary to investigate time-dependent changes.

The following presents two thermal-rheological measurement methods, using a rotational rheometer, which allow studying the influence of shear stress on the material aging process. Therefore, frequency-sweep and time-sweep measurements are conducted. Frequency-sweep measurements investigate the shear thinning effect by various shear rates under a constant thermal stress. To determine time-dependent changes, which lead to a material aging, sequential frequency sweeps are conducted. In this case, the same material sample is measured multiply times in succession. Alternatively, time-sweep measurement can be performed. This method enables analysis of material aging at a constant shear rate and temperature over a defined test time. To separately distinguish the thermal influence, the measurements are conducted at different temperatures. The applied material is a low-density polyethylene (LDPE, Sabic 2100N0).

EXPERIMENTS

A rotational rheometer with a plate/plate-system (diameter 25 mm) is used for the measurements. A possible material aging is indicated by the complex viscosity. If a change in the molecular structure of the material occurs, the measured viscosity values are altered as well. Thus, increasing viscosity values indicate a material build-up whereas decreasing values infer degradation. According to the Cox-Merz relation [2], the measured complex viscosity is equal to the viscosity. In addition, the angular speed is identical with the shear rate.

For preparing the test sample, the granules of the material are milled with a cryogenic vibration mill. Through this, a thermal stress of the material in generating the test sample is minimized because a dissipative heating is excluded. Furthermore, the comparability between the test samples is improved because of the same stress history. After milling the granules, the powder is pressed to form a test sample with a diameter of 25 mm and a height of 2 mm. Moreover, a test period of respectively 2 hours is defined for both measurement methods. To avoid chemical aging due to oxidation, the measurements are conducted under inert conditions (nitrogen atmosphere). In addition, all measurements are performed in the linear viscoelastic region (LVR). Thus, a deformation γ of 1% is used for a nondestructive testing of the specimen. This prevents an irreversible change in the material structure [3].

Frequency sweep

To reach the required test time, the frequency-sweep measurements were conducted 13 times in succession with a shear rate range from 0.01 s^{-1} to 100 s^{-1}. Every measurement curve has a measurement time of approximately 6 minutes in addition to 3 minutes of pause time. Furthermore, each measurement is conducted with the same test sample. As a result, it is possible to identify the time-dependent material aging which is apparent from the differences in the various measurement curves. This method has the advantage that a load duty cycle affects the material sample and reflects the material stress during the extrusion process. Furthermore, the material behavior is presented for a range of shear rates over the defined repetitions. The disadvantage here is that the different shear stresses may influence the material ageing process. Because of the acting load duty cycle, it is not possible to draw a clear conclusion about the influence at a defined shear rate.

Figure 1 displays the results of the measurements at 210 °C. For the sake of clarity, the first viscosity curve is plotted as well as every second curve. In this case, the material shows a general build-up that is independent of the temperature.

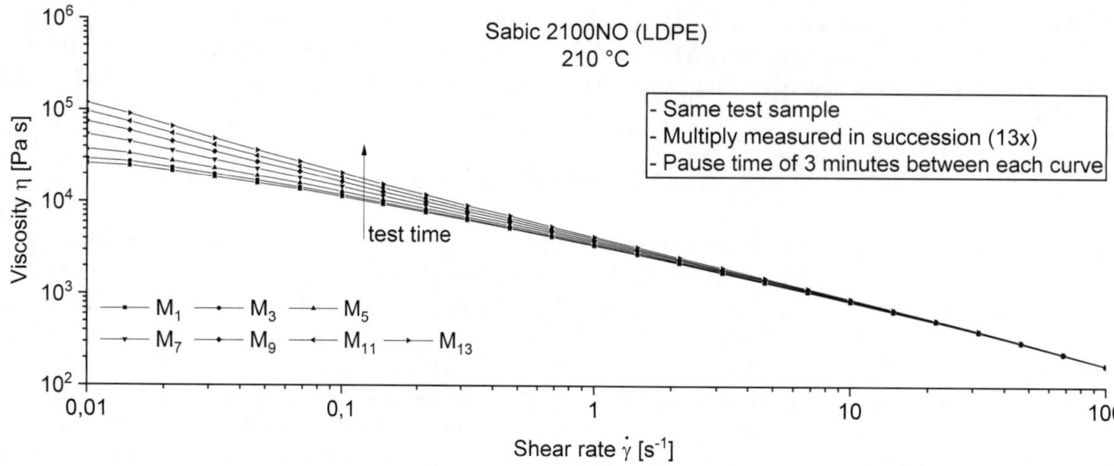

Figure 1. Sequential time-sweep measurements (210 ° – LDPE)

It is apparent that at higher shear rates, the effect of the build-up decreases. In the range of 0.01 s^{-1} to about 0.1 s^{-1}, changes at the molecular level can be distinguished well. In order to evaluate the acquired results, time-sweep measurements conducted in the next step.

Time sweep

For the time-sweep measurements, the material sample is subjected to a constant stress (thermal and mechanical) over a period of nearly 2 hours. In detail, the shear rates at 0.01 s^{-1}; 0.1 s^{-1}; 1 s^{-1}; 10 s^{-1}; 100 s^{-1} are investigated. Moreover, the detected material aging is presented by the viscosity values plotted as a function of test time. Every shear rate is measured three times with a new sample (S_x) because of deviations found between each measurement at the same shear rate due to inhomogeneities of the granules. Thus, the arithmetic mean (MV) is calculated and used for further investigations. Figure 2 shows an example of the deviations at a constant shear rate of 0.1 s^{-1} and a temperature of 210 °C. Here, the generation of test sample and experimental conditions are identical in all three tests. At the beginning of every measurement, a pause time of 3 minutes is used to reduce the stress inside the specimen.

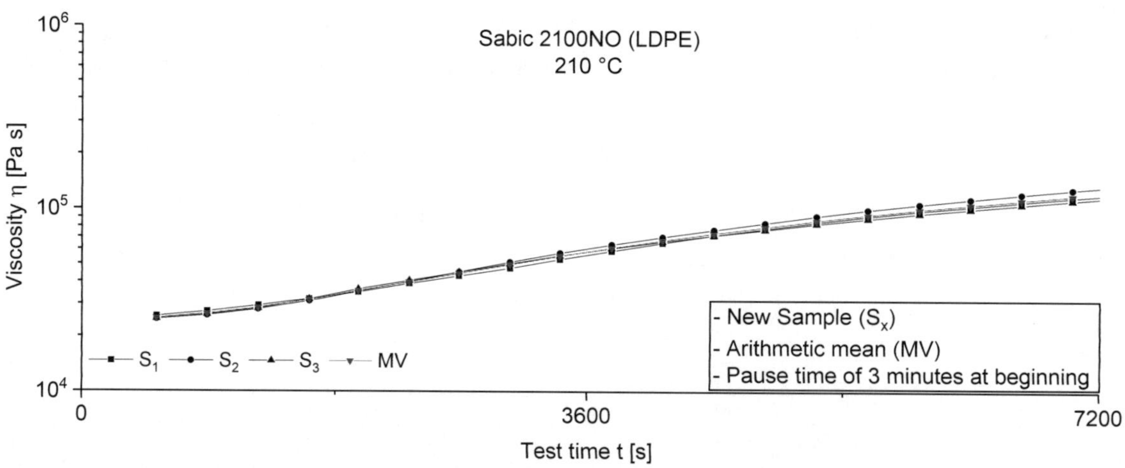

Figure 2. Deviation between time-sweep measurements at a constant shear rate of 0.1 s^{-1}. (210 °C – LDPE)

As the results of the sequential time sweep demonstrate, a change in the viscosity over the test time is only detectable at a shear rate range from 0.01 to 0.1 s^{-1}. The time-sweep measurements also reflect this finding (Fig. 3). This is attributed to the movement of the molecular chains. At higher shear rates, the molecular chain movement cannot follow the mechanical stress. By contrast, the molecular chain movement can follow the mechanical stress at lower shear rates [4]. Regarding the reptation model of de Gennes [5,6] and the later extended tube model by Doi and

Edwards [7], the molecular chain can leave its tube at lower shear rates. Accordingly, it is possible to measure changes at a molecular level on hand of the measured viscosity.

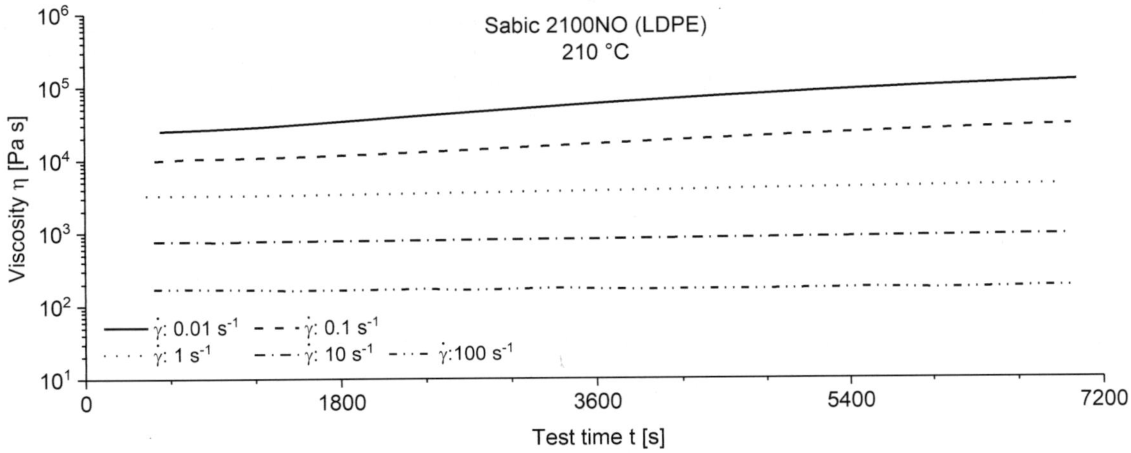

Figure 3. Viscosity as a function of test time at different shear rates (210 °C – LDPE)

Comparison of measurement methods

A 3D-diagram is used to compare the two measurement methods. Here, the three axes represent the shear rate (x-axis), test time (y-axis), and viscosity (z-axis), respectively. This 3D diagram enables one to visualize and distinguish the various qualitative results of both measurement methods. Figure 4 illustrates the measurement data for a temperature of 210 °C. Moreover, the shear rate of each time sweep is shown in the frequency sweep by means of highlighted points.

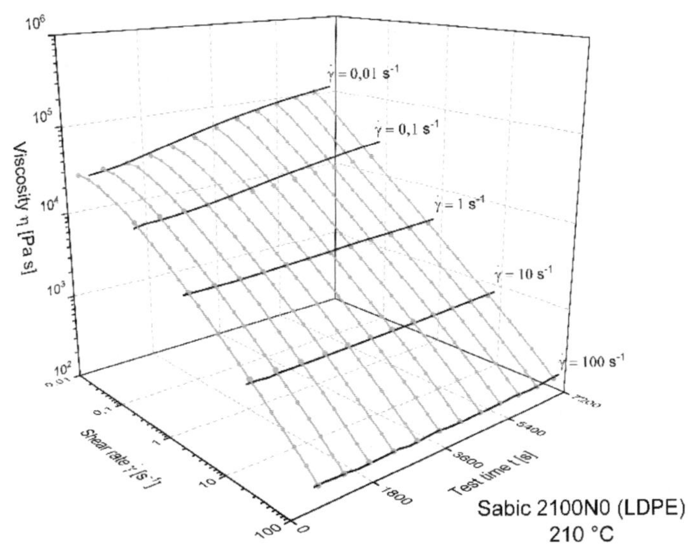

Figure 4. Comparison of frequency and time-sweep measurements (210 °C – LDPE)

As it can be seen from the figure, both measurement methods show good correlation with each other and similarly detect the material build-up at low shear rates (< 0.1 s^{-1}). Small differences between the curves at equal shear rate are attributed to the inhomogeneity of the granules and measurement inaccuracy. In studying material aging, it is important to measure the characteristic material behavior (build-up / degradation), whereby generating exact viscosity values is not necessary in this case.

CONCLUSION & OUTLOOK

As presented, both measurement methods show a good correlation with one another. To investigate the material aging, it is better to conduct frequency sweeps. In the same time period, a more detailed overview about the material aging by different shear rates can be obtained, enabling more extensive investigations. Viscosity values, however, allow only a conclusion to be drawn about the flowability of materials. On these grounds, it is advisable to consider the change in the storage and loss modulus. These two measured quantities give information about the structural character (gel or fluid) of the material.

ACKNOWLEDGMENTS

This publication was created within the scope of the research project "Thermo-rheological phenomena at wall boundary layers during color and material changes in plastics extrusion." (WO 302/46-1). We appreciate the funding by the German Research Foundation (Deutsche Forschungsgemeinschaft DFG).

REFERENCES

1. Ehrenstein G., Pongratz S.: *Resistance and Stability of Polymers,* Carl Hanser Verlag, Munich, 2013
2. Cox, W. P., Merz, E. H.: *Correlation of dynamic and steady flow viscosities,* Journal of Polymer Science. 118, 28, p. 619–622, 1958
3. Metzger, T.G.: *The Rheology Handbook,* Vincentz Network GmbH & Co. KG, 2011
4. Schwarzl, F.R.: *Polymermechanik,* Springer-Verlag, Berlin Heidelberg, 1990
5. De Gennes, P. G.: *Entangled polymers,* Physics Today, American Institute of Physics. 36 (6), p. 33, 1983
6. De Gennes, P. G.: *Reptation of a Polymer Chain in the Presence of Fixed Obstacles,* The Journal of Chemical Physics. American Institute of Physics. 55 (2): 572, 1971
7. Doi M., Edwards S.F.: *The theory of Polymer Dynamics,* Oxford University Press, 1986

QUANTITATIVE CHARACTERIZATION OF INTERDIFFUSION AT THE RESIN–RESIN AND RESIN–PREPREG INTERPHASES OF EPOXY SYSTEMS PROCESSED BY MODEL SQ-RTM

Ballout Wael (1)*, Van Velthem Pascal (1), Pardoen Thomas (2), Bailly Christian (1)

(1) IMCN-BSMA, Université Catholique de Louvain - Louvain-la-Neuve - Belgium, (2) IMMC, Material and process engineering, Université Catholique de Louvain - Louvain-la-Neuve - Belgium

The interdiffusion between a low and a high viscosity epoxy resin used in aerospace field was studied on model systems representing the novel composite manufacturing process called "Same Qualified-Resin Transfer Molding" (SQ-RTM). Neat resin model systems were first characterized after curing by Raman spectroscopy, energy dispersive X-ray microscopy, and nano-indentation, all methods reveal an interdiffusion distance of about 700–900 µm. Transmission electron microscopy further revealed a complex morphological gradient structure in the interdiffusion zone. The interdiffusion distance was about 800 µm in the absence of carbon fibers, reducing down to 500 µm when the viscous resin was replaced by the corresponding prepreg, due to the geometrical constraints imposed by the fibers. Moreover, some asymmetry was observed in the interdiffusion profile because of the viscosity difference between the resins. The results obtained on the model systems were found to match very well the interdiffusion profiles generated in real SQ-RTM composites processed with the same combination of resins.

Nature-Inspired Structures for Reduced Impact Load Transfer

Ahmed A.W. Anwer[a], Farooq Al Jahwari[d], and Hani E. Naguib[a,b,c]

[a]*Department of Mechanical and Industrial Engineering, University of Toronto*
[b]*Department of Materials Science and Engineering, University of Toronto*
[c]*Institute of Biomaterials and Biomedical Engineering, University of Toronto*
[d]*Department of Mechanical and Industrial Engineering, Sultan Qaboos University*

Abstract. In addition to withstanding structural stresses during an impact event, there is an increasing demand to reduce force transferred to the receiver/structure during an impact event. Although most material structures can absorb load without failing, these structures are not capable of reducing force transferred to adjacent structures and bodies. The high stiffness of most impact resistant material structures results in an isotropic high rate of load transfer, and the coupled effect of low damping properties of these material structures, results in a high load transfer. In this work, nature-inspired through-thickness reinforced composite structures were fabricated with graded lightweight material sheets. The proposed design will aim at mimicking capillary waves in water, for reducing load transfer. Polyurethane resins of different chemical compositions were used as matrix materials and ultrahigh molecular weight polyethylene braids were used for out-of-plane reinforcements of structures. Effects of fibrous reinforcement and material gradients on transferred loads were investigated with a customized drop weight set up. Correlation between transferred force and processing parameters, material gradients, and orientations will be presented.

Keywords: Out-of-plane reinforced composite structures, Impact properties, Impulse transfer

INTRODUCTION

During an impact scenario, the ability of a material to maintain structural stability and the impulse transferred to the adjacent face, are the two important defining characteristics for defining impact performance of a material. The impulse transferred to the adjacent face would address safety issues, since, during an impact, serious human fatalities and structural damage to the apparatus often occur as a result of the impulses transferred to the person/apparatus. Existing impact protection materials can absorb impact energy without taking structural damage. However, they transfer dangerously high impulses to the person/apparatus. This creates a critical issue to humans and equipment that use these protective materials, which in turn exposes these humans/equipment to dangerously high impulses. The focus of this work is on the impulse transferred to humans and equipment that use the proposed material as protective materials.

Many of the recent inspiration and development in impact resistant materials are based on designs that exist in nature. Impact resistant structures inside Mantis shrimps, Seashells, and bone are very good examples of impact resistant materials that exist in nature. All of the mentioned structures that exist in nature are structures that can deform under high stresses, while having a high stiffness as well. This is a unique property of these nature inspired structures, since toughness and stiffness of a material are normally inversely independent on each other. This inverse correlation exists, due to the inability of stiffer materials to flow and relax stresses during an impact. Hence, it is difficult to find materials that have a low density, high stiffness and high toughness. In order to get a low-density, high stiffness, and high toughness material, extrinsic reinforcement mechanisms were used. These extrinsic reinforcement mechanisms were combined with the intrinsic reinforcement effects, in an optimal manner, to develop a material with properties that exceed that of its individual components. Khayer and co-authors [1], concluded that the toughness in nacre-like structures is not because of the strong interfacial strength, rather the toughness of these nacre-like structures is because of ductility of these interfaces.

High toughness, high stiffness and low-density structures, found in nature, can be fabricated using composite structures. Composite structures have the advantage of combining individually weak components into a structure that outperforms each of the individual structures. However, in the combined structure, determining optimal strength of the interfaces is still a major problem. This is because weak interfaces generally cause delamination, which results in premature failure of structures. However, if delamination is controlled within the structure, then weak interfaces also provide a mechanism for impact energy dissipation, whereby interfacial shearing between faces causes dissipation of energy in the form of heat. Out-of-plane reinforcements prevents the issue of delamination, reduces the need for strong adhesion, and allows of energy dissipation by interfacial sliding between the different components. Out plane reinforcement also had the added benefit of providing extra reinforcement in the out-of-plane direction as well. Ching Quek and co-authors [3], E-glass reinforced polyurethane tubes. E-glass fibers, used for the pipe, were continuous fabrics and braids, since they wanted to study the energy absorption during compressive failure of different reinforcement types. During the compressive crushing, the fabric tubes failed due to lateral cracks, where as these lateral cracks were not observed in the braided tubes. In another study, bending behaviour of out-of-plane reinforced sandwich structures was tested by Dawood and co-authors [4]. It was found that the impact response of structures were significantly improved when compared to their non reinforced composites counterpart [5]. In addition, the out-of-plane reinforced composite structure showed limited delamination when compared to their non-reinforced counterparts. Potluri and co-authors also developed out of lane reinforced composite structure to address the issue of delamination, which is commonly found in composite structures. Sandwich structures with out-of-plane reinforcements showed and increase in the maximal cracking width and penetration depth by 67% and 4% respectively [7].

This study focuses on development, fabrication, and impact characterization of out-of-plane reinforced syntactic foamed structures. The structures were fabricated by incorporating strong fibres in a soft matrix, which improved structural stability and improved damping within the structure[8]. For the matrix, microspheres dispersed in a soft polyurethane matrix. This was done, since many researchers used polyurethane and its foam derivatives for enhanced energy impact energy applications [9, 10]. The fabricated composite structure comprised of three layers of 4mm thick polyurethane/microspheres, which were stacked one top of the other. Kevlar/carbon fiber mats were then embedded in between each of the stacked layers. The multiple polyurethane layers were held together by using polyurethane as an adhesive. Out-of-plane reinforcement was provided ultra high molecular weight polyethylene (UHMWPE) Braids. Reinforcement was provided in the form of stitches that ran along the edge of the composite structure. Preparation of the composite structure in such manner ensured that there was little delamination, while allowing for a global bending response of the material as well. In addition, the interfacial energy frictional energy loses, at the interface between the layers, allowed for greater damping in the composites structure as well. UHMWPE fibres were used in the composite structure due to their ability to provide structural strength while damping the impulses as well [11]. The composition of microspheres was varied through the thickness of the materials, since a previous study by the authors showed significant influence of gradient on impact energy absorption [12]. Other researchers also reported similar effect of the gradient on energy absorption [13]. Compressive and relaxation moduli of the fabricated structures were measured and correlated to the transferred impact load.

Material Structure and Processing

High tear resistant properties of polyurethane, along with its high damping properties were the reasons why polyurethane was selected for the matrix material. For his study, Polyurethane, grade 3360 from Fibre Glast, USA with a Shore A hardness of 60 was used for this study. For the reinforcement mats, a twill weave hybrid composite fabric of 50/50 carbon fiber/Kevlar composite mat was used. These composite mats functioned as the in-plane reinforcements for the composite structure. Out-of-plane reinforcements were provided using Ultrahigh molecular weight polyethylene (UHMWPE) braid with a diameter of 0.76mm.The grade of UHMWPE used, was the Spectra-200 from Twinline, USA. The reported tensile strength of these braids was 2048 MPa, which is about 5 times stronger than structural steel. The syntactic foams were fabricated by dispersing Expancel microspheres grade 980 DU 120 from Akzonobel, USA.

The polymeric microspheres were supplied as light yellow powder with a density of 14 kg/m^3 and with a microsphere diameter ranging from 25-40 μm. The microsphere includes a thermoplastic shell, which encapsulates a hydrocarbon liquid with a low boiling point. The wt% concentrations of microspheres considered in this study are microspheres 1%, 2% and 3%. These microspheres were uniformly dispersed in the matrix by sonication.

The fabrication process can be broadly categorized into two main phases. The phase includes, fabrication of 45 X 45 square polyurethane sheets. The square sheets were fabricated using 0%, 1%, and 2% wt% Expancel concentrations. The fabrication process is explained schematically in Figure 1. The polyurethane mixture was prepared by mixing 100:55 ratios of Part A: Part B. Uniform dispersion of microspheres was achieved by sonicating part A with the microspheres at 10% amplitude for 5 minutes using the s-4000 Sonicator from Misonix. The composite mixture was then cooled in an ice bath, to prevent setting of the composite mixture. The composite mixture was then manually mixed with part B for 1 minute. The mixture was then degassed for 7 minutes, at -30 in-Hg (-101592 Pa) in a vacuum chamber until no more air bubbles were observed. The mixture was finally poured slowly in the desired mold and cured at 50 °C for 12 hours.

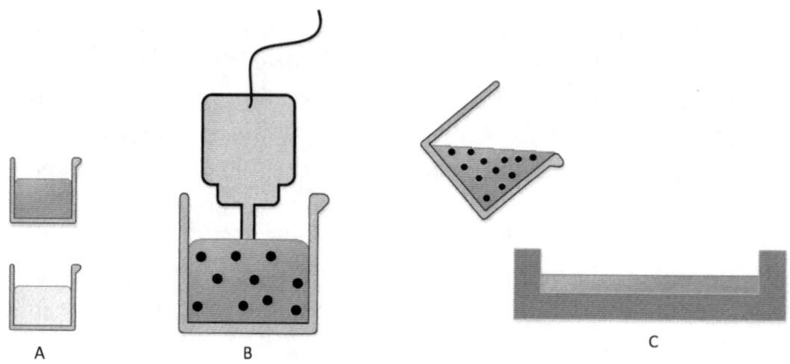

FIGURE 1. A. Weight amounts of resin and hardener based on weight ratio, B. Disperse microspheres into resin mixture using sonicator, C. Pour composite mixture into the desired mold.

In the second phase of the fabrication process, three layers of 4mm stacked on top of the other, and Kevlar/carbon fiber mats were placed in between the layers. Polyurethane was used as the wetting layer between the different fabric layers. UHMWPE braids were then used for out-plane-reinforcements as shown in the manner shown in Figure 2.

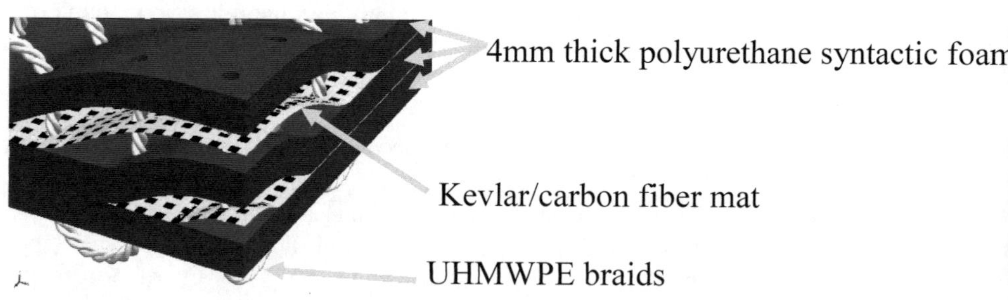

FIGURE 2. Schematic of Fabricated Sample

Experimental Setup and Testing

The experimental setup consisted of a load cell, on which the sample was placed. The sample was then impacted by predetermined weight, 1.1kg, upon which the load cell measured the force transferred to the side. The measurements were taken, using a high speed DAQ, and the maximum force, from the force vs. time was then recorded and reported.

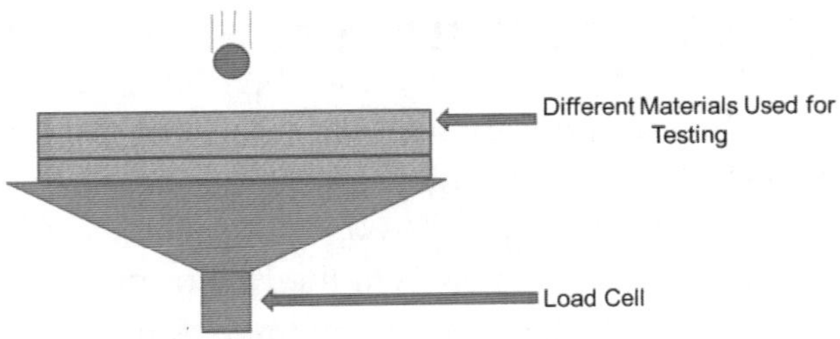

FIGURE 3. Experimental Setup Schematic

Results and Discussions

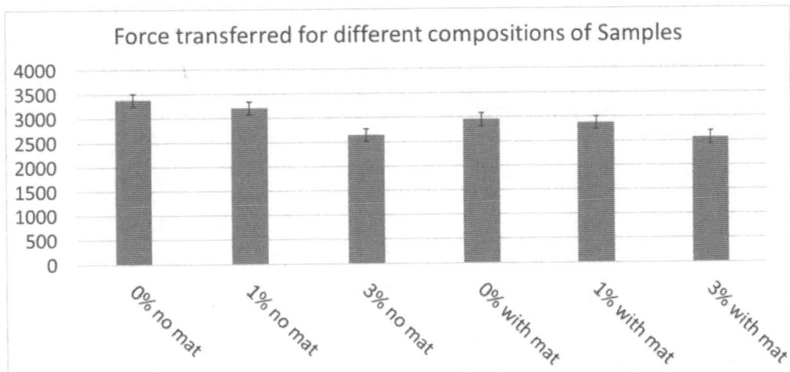

FIGURE 4. Results and Discussions for different fabricated samples

As can be seen from the results in figure 4, there is a considerable increase in the impact performance of the fabricated structures. As seen in the figure there is up to 15.38% decrease in force transferred to the adjacent face, by increasing the Expancel concentration from 0%wt to 3%wt. This is because of the increased impact energy dissipation by deformation of the microspheres within the sample, and because of friction losses at the polyurethane Expancel interface. In addition, from the figure it can be seen that there is up to 12.8% reduction in force transfer by addition of Kevlar/carbon fiber fabric in between the layers. This reduction in force transfer is due to the increased frictional loses at the polyurethane-fabric interface and because of the external reinforcement provided by the Kevlar/carbon fiber mats as well.

ACKNOWLEDGMENTS

The authors would like to acknowledge the following agencies for financial support: Ontario Graduate Scholarships of Ontario Canada, William L. Dunbar Memorial Scholarship, Natural Sciences and Engineering Research Council (NSERC) of Canada.

REFERENCES

1. A. Khayer Dastjerdi, R. Rabiei, and F. Barthelat, "The weak interfaces within tough natural composites: Experiments on three types of nacre," *Journal of the Mechanical Behavior of Biomedical Materials,* vol. 19, 2013, pp. 50-60.
2. R. G. Hoagland, J. P. Hirth, and A. Misra, "On the role of weak interfaces in blocking slip in nanoscale layered composites," *Philosophical Magazine,* vol. 86, 2006, pp. 3537-3558.
3. S. Ching Quek, A. M. Waas, J. Hoffman, and V. Agaram, "The crushing response of braided and CSM glass reinforced composite tubes," *Composite Structures,* vol. 52, 2001, pp. 103-112.
4. M. Dawood, E. Taylor, and S. Rizkalla, "Two-way bending behavior of 3-D GFRP sandwich panels with through-thickness fiber insertions," *Composite Structures,* vol. 92, 2010, pp. 950-963.
5. B. Lascoup, Z. Aboura, K. Khellil, and M. Benzeggagh, "Impact response of three-dimensional stitched sandwich composite," *Composite Structures,* vol. 92, 2010, pp. 347-353,.
6. P. Potluri, E. Kusak, and T. Y. Reddy, "Novel stitch-bonded sandwich composite structures," *Composite Structures,* vol. 59, 2003, pp. 251-259.
7. F. Xia and X.-q. Wu, "Study on impact properties of through-thickness stitched foam sandwich composites," *Composite Structures,* vol. 92, 2010, pp. 412-421.
8. P. Fratzl, I. Burgert, and H. S. Gupta, "On the role of interface polymers for the mechanics of natural polymeric composites," *Physical Chemistry Chemical Physics,* vol. 6, 2004, pp. 5575-5579.
9. G. Zhang, B. Wang, L. Ma, L. Wu, S. Pan, and J. Yang, "Energy absorption and low velocity impact response of polyurethane foam filled pyramidal lattice core sandwich panels," *Composite Structures,* vol. 108, 2014, pp. 304-310.
10. J. Zhou, X. Deng, Y. Yan, X. Chen, and Y. Liu, "Superelasticity and reversible energy absorption of polyurethane cellular structures with sand filler," *Composite Structures,* vol. 131, 2015, pp. 966-974.
11. T. K. Ćwik, L. Iannucci, P. Curtis, and D. Pope, "Investigation of the ballistic performance of ultra high molecular weight polyethylene composite panels," *Composite Structures,* vol. 149, 2016, pp. 197-212.
12. F. Al Jahwari, Y. Huang, H. E. Naguib, and J. Lo, "Relation of impact strength to the microstructure of functionally graded porous structures of acrylonitrile butadiene styrene (ABS) foamed by thermally activated microspheres," *Polymer,* vol. 98, 2016, pp. 270-281.
13. J. Zhou, Z. W. Guan, and W. J. Cantwell, "The impact response of graded foam sandwich structures," *Composite Structures,* vol. 97, 2013, pp. 370-377.

Flexible Mould Concept for the Individualised Production of CFRP Parts for the Automotive Repair Strategy of the Future

Christian Hopmann, Philipp Nicolas Wagner, Arne Boettcher, Kai Fischer

Institute of Plastics Processing (IKV) at RWTH Aachen University
Seffenter Weg 201, D-52074 Aachen, Germany

Abstract. Carbon fibre reinforced plastics (CFRP) are increasingly applied in high volume markets, such as the automotive industry. Due to the cost sensitivity of automotive products new repair strategies for such CFRP parts are required. The strategies need to enable a fast and economic repair of the damaged parts by restoring the mechanical properties with optimised material utilisation. At the same time, the suitability for the versatile geometries of automotive CFRP parts has to be assured. At the Institute of Plastics Processing (IKV) at RWTH Aachen University a highly integrated and sensor-aided repair strategy for CFRP vehicles has been developed. The strategy consists of multiple steps from the damage detection and damage evaluation to a repair patch based reconstruction of the vehicles in the repair workshop. All steps of the repair are connected through a industry 4.0 data-management system. One key element of the repair strategy is the centralised production and supply of pre-fabricated, individual repair patches for the repair of the damaged CFRP parts. For the individual production of these CFRP parts (repair patches), a flexible mould concept consisting of a pin-based support structure in combination with a diaphragm forming process has been developed. The pin-based support structure consists of 7,200 individually adjustable pins, a clamping solution for processing pressures up to 15 bar and a heating solution for a cavity temperature up to 140 °C. For a high complexity and geometric flexibility of the produced CFRP parts, standardised modular and additively manufactured elements can additionally be integrated. In this paper, results of feasibility analyses for the development and application of the mould concept are presented. Therefore, the functionality of the pin-based support structure for the CFRP repair patch production are evaluated on a specifically developed test bench. In conclusion the final mould concept including detailed solutions is presented.

Keywords: composites, production technology, mould technology, automotive repair
PACS: 61.41.+e; 81.05.Qk; 81.20.Hy

INTRODUCTION

Carbon fibre reinforced plastics (CFRP) are increasingly applied in high volume markets, such as the automotive industry due to the demand to reduce CO2 emissions [1]. This results in the challenge for the automotive industry and vehicle repair workshops to provide economic, product-independent repair methods for CFRP parts. Compared to metals the repair process of CFRP-based vehicles contains new challenges. CFRP components show a great variety of possible damages and the failure behaviour differs from metallic materials. While metals show visible deformation as consequence of a crash, damages are often not detectable on the surface of CFRP components [2, 3, 4]. For this reason reliable damage detection, a case-specific damage evaluation and a quality assured repair process are core elements for the establishment of sustainable CFRP-based vehicle concepts.

Therefore, a highly integrated and sensor-aided repair strategy for CFRP vehicles has been developed at the Institute of Plastics Processing (IKV) in Industry and the Skilled Crafts at RWTH Aachen University in close cooperation with six research institutes and around sixty industrial companies. New technologies are being developed for identifying and evaluating damages in CFRP, for enabling customized production of repair materials – so-called repair patches – within flexible manufacturing moulds, as well as the actual repair, including quality assurance. All individual steps of the repair strategy (damage detection, damage evaluation, repair patch production and repair) are connected through Industry 4.0 principles, where a centralized data management system ensures the flow of information to all involved service providers. Up to now, repair methods for automotive CFRP components have not yet been standardized across all brands. Up to now, specially trained staff is necessary for the repair of CFRP parts. One applied method are the concept of the so-called "Flying Doctor's" [5] by Lamborghini, Sant'Agata Bolognese, Italy, who are trained professionals for the damage detection, evaluation and the manual repair of specific vehicles at the location of the customer. In a different concept of BMW, Munich, Germany, selected repair centres are designated to perform the repair of CFRP-components based on the removal and replacement of predefined repair sections with spare-parts [6]. Compared to the prevailing methods of individual and manual repair, respectively of replacing large segments or whole CFRP structures, the possibility of a component and brand independent, automated and cost-efficient repair for large series vehicles is the objective of the new repair strategy by IKV and its partners.

One of the main challenges for the developed repair strategy is the high geometric complexity and large variety regarding materials and lay-up of structural parts in modern vehicles that have to be repaired after damage. Therefore, a production method for the automated manufacturing of highly individualised CFRP repair patches is required. For the individual production of these CFRP repair patches, a flexible mould concept has been developed and evaluated based on preliminary investigations [7]. The final mould design has been developed in close collaboration with FRIMO Sontra GmbH, Sontra, Germany. In this paper, results of feasibility analyses for the development and application of the mould concept on a specifically developed test bench are presented and the final mould concept including detailed solutions is described.

CONCEPT OF THE FLEXIBLE MOULD FOR THE INDIVIDUALISED REPAIR PATCH PRODUCTION

In principle, a large variety of mould technologies and development approaches is available for the production of individual CFRP parts. For the development of the flexible mould concept [7], different mould technologies were evaluated and combined regarding the specific requirements for the envisaged CFRP repair patch production. The requirements are result from the matrix materials (thermosets, thermoplastics) and the manufacturing process to be used (forming process to realise different part shapes in combination with a resin injection process for thermoset matrix systems) as well as the required repair patch geometry variants and complexities based on currently applied CFRP automotive parts and the required number of same-properties-parts. Based on a theoretical evaluation and combination of identified mould technologies, manufacturing trials were carried out in order to answer basic questions regarding evaluation of one- and two-sided mould concepts with regard to geometric accuracy of the produced parts, the homogeneity of the resin distribution and the formability of different laminate lay-ups [7].

Based on these results, the developed flexible mould concept consist of a temperature controlled, two sided pin-based support structure which is combined with a diaphragm forming process for an additional pressure application during forming, as depicted in Figure 1 (a). For a high complexity and geometric flexibility of the produced CFRP parts, standardised and additively manufactured modular elements can additionally be integrated as visualised in Figure 1 (b).

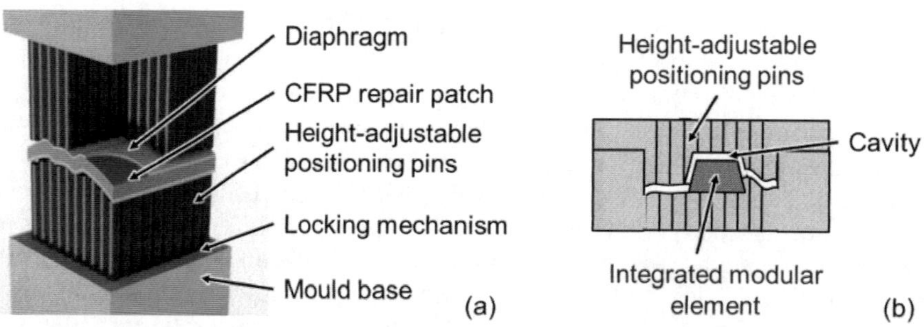

FIGURE 1. Schematic view of the flexible mould concept (a) and approach for combination with modular elements (b)

The required two sided mould cavity for the repair patch production is created by adaptively adjustable positioning pins. The positioning pins can be adjusted to the desired repair patch geometry in different ways: The use of modular templates, pattern parts or a passive milling-like process. The individual cavity geometry for repair patch production is prepared iteratively, by fastly pushing each pin down into the desired position with an industrial robot and then fixing each pin in the desired position during forming. Diaphragms are used for the transport of the lay-up of semi-finished products (thermosets / thermoplastic) into the cavity. The diaphragms also serve as protection of mould components from the matrix materials during processing and allow an additional air pressure application during forming. Modular metallic standard geometries as well as additively manufactured components can be used for local complex structures or steep flank angles which cannot be realised by the adjustment of the positioning pins. This increases the possible geometry complexity of the CFRP repair patches. The temperature control is realised by an indirect temperature control of the cavity via heating and cooling of the positioning pins.

DEVELOPED TEST BENCH AND RESULTS OF FEASIBILITY ANALYSES

In order to be able to process thermoset as well as thermoplastic CFRP materials and realise an accurate forming of complexly shaped parts with sufficient consolidation quality, the realisation of processing pressures of up to 15 bar and a cavity temperature at the top of the positioning pins of up to 140 °C were defined as requirements. The temperature requirement results due to the required curing temperature of the envisaged thermoset matrix materials. Regarding these two processing requirements, a feasible clamping solution as well as a heating solution for the temperature control is required.

Regarding the clamping solution, the precise adjustment and stable fixation of the mould cavity geometry during forming requires a locking mechanism of the positioning pins. Three main modes of operation should be considered. First, processing pressures of up to 15 bar need to be resisted during repair patch production, so that changes in the cavity's geometry during the forming process are prevented. Second, independent movements of the positioning pins during the height adjustment have to be avoided. Third, the positioning pins need to be resistance-free when the positioning pins are reset to their basic position after the repair patch production. The mentioned modes are made possible in the mould concept by using an elastomer block, in which each positioning pin is individually guided in a fit. In order to increase the clamping force and the associated locking of the pins in the desired position for repair patch production, the elastomer block experiences a volume contraction through a pressure plate and a circumferential cage. As a result, the fitting diameters are reduced and the pins are locked.

Regarding the temperature control, the main challenge lies in the homogeneous heating of each individually positioned positioning pin. In order to be able to heat all positioning pins at the same time, the heating is realised by a hot air supply which convectively heats the bottom part of the positioning pins. The heat is then transferred to the top of the pins which form the cavity by heat conduction in the pin material. The convective heating by hot air is combined with heated and cooled guiding plates and a surrounding insulation in order to reach the required cavity temperatures of up to 140 °C at the top of the positioning pins.

A small scale test bench was developed together by with FRIMO Sontra GmbH, Sontra, Germany, and used to validate the respective individual features of the mould concept for the developed clamping solution and the concept for the temperature control. The test bench was realised with 100 single positioning pins (diameter at the top of 3 mm) in order to evaluate the maximum pressure resistance of the positioning pins and the temperature at the top of the positioning pins for the chosen heating concept. The developed test bench is depicted in Figure 2.

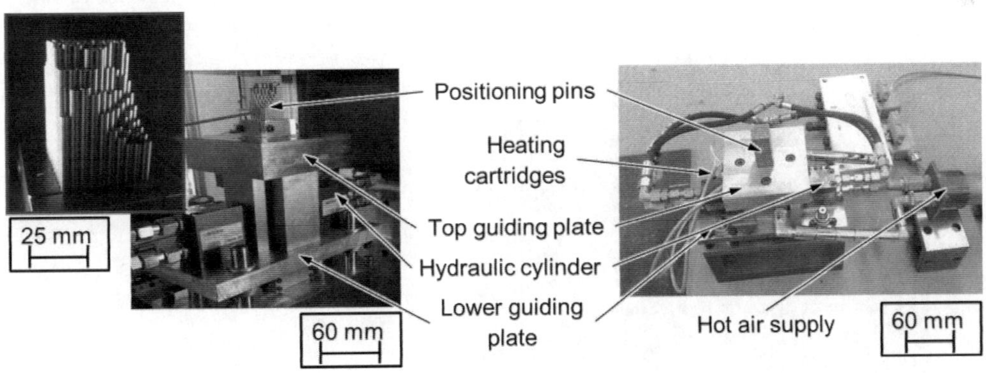

FIGURE 2. Developed test bench for feasibility analyses

For the analysis of the pressure resistance, the clamping of the positioning pins is realized via the volumetric contraction of the elastomer block (reduction of the fitting diameter of the pins) with an individual fit for each of the 100 positioning pins. The contraction of the elastomer block is realised over a guided pressure plate, also with an individual fit for each of the 100 positioning pins. A schematic view of the locking mechanism in the test bench is shown in Figure 3 (a). The results of the analysis of the pressure resistance of the clamped positioning pins against movement are depicted in Figure 3 (b). The force on the pressure plate is applied via a hydraulic cylinder. The results show, that the pressure resistance of the positioning pins exceeds the required processing pressures of up to 15 bar by a factor of 2 already at low applied hydraulic pressure 50 bar. The required force to reset the positioning pins to their basic position after the hydraulic pressure was released was determined to 20 N, allowing for a feasible resetting of the mould geometry after the repair patch production.

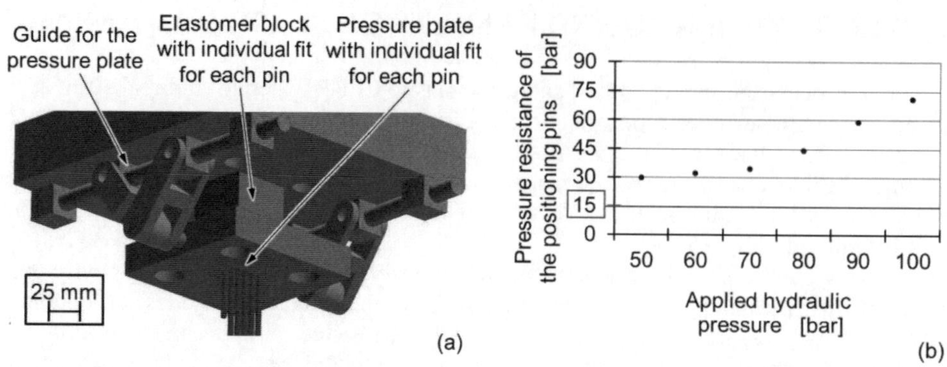

FIGURE 3. Locking mechanism in the test bench (a), pressure resistance of positioning pins (b)

For the analysis of the heating solution, the maximum achievable temperature at the top of the positioning pins was measured at the bottom mould side under convective heating of the positioning pins below the top guiding plate in the test bench with the hot air supply and additional heating of the top guiding plate itself by heating cartridges (see Figure 2). The analysis of the temperature at the top surface of the positioning pins which form the cavity show, that a maximum temperature of 126 °C can be realised. Due to this, an additional surrounding insulation of the positioning pins above the top guiding plate will be implemented in the final mould concept to reduce the heat loss over the pins' lateral surfaces and reach the required temperature at the top surface of the positioning pins of 140 °C.

FINAL MOULD CONCEPT

The analysis of the mould features in the test bench show, that the requirements for the flexible mould for the individual production of the CFRP repair patches can be fulfilled with the developed flexible mould concept. Therefore, the final mould design is based on a pin-based support structure with a two-sided cavity in combination with a diaphragm forming process. In order to realise a high complexity and geometric flexibility of the produced CFRP repair patches, standardised modular and additively manufactured elements can be integrated additionally to the flexible positioning of the positioning pins. An overview of the final mould design (bottom side) is depicted in Figure 4.

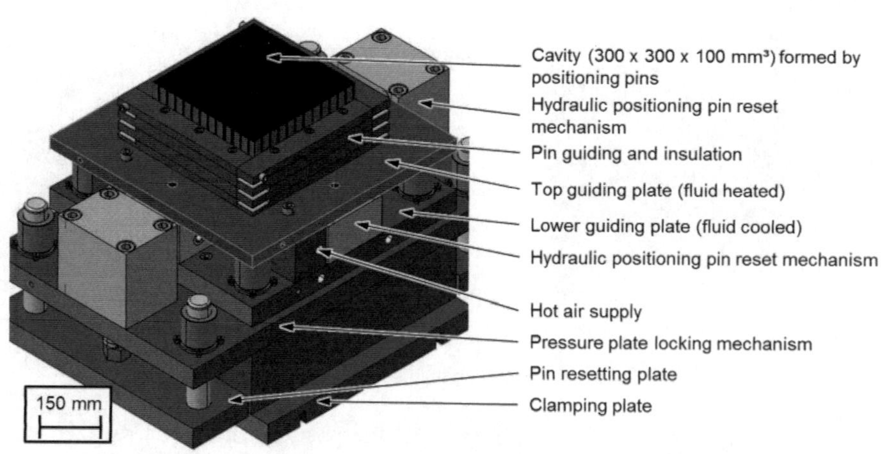

FIGURE 4. Final mould design (bottom side) for the production of individual repair patches

The pin-based support structure consists of 7,200 individually adjustable positioning pins (3,600 positioning pins per mould side, adapted diameter at the top of 5 mm), forming a cavity size of 300 x 300 x 100 mm³. Based on the results of the feasibility analyses, the clamping solution is realised by an elastomer block with a pressure plate, designed for processing pressures up to 15 bar and a locking as well as a reset mechanism for the release of the clamping force after part production. The heating of the mould respectively the temperature control is realised by a hot air supply in combination with heated and cooled guiding plates and an additional surrounding insulation in order to realise the required cavity temperatures up to 140 °C. In contrast to the test bench, the heating of the top guiding plate is realised by fluid temperature control instead of heating cartridges due to the larger cavity size.

CONCLUSION AND OUTLOOK

For the centralised production and supply of pre-fabricated, individual repair patches for the repair of CFRP parts, a developed flexible mould concept has been presented. The developed flexible mould concept consists of a temperature controlled, two sided pin-based support structure which is combined with a diaphragm forming process. For a high complexity and geometric flexibility of the produced CFRP parts, standardised and additively manufactured modular elements can additionally be integrated. The functionality of the pin-based support structure for the CFRP repair patch production was evaluated with regard to the clamping solution as well as the heating solution for the temperature control on a specifically developed test bench. The results showed, that the requirements for the individual production of the CFRP repair patches can be fulfilled with the developed final mould concept, where optimisations regarding the temperature control will be implemented. The presented final mould concept will be available at IKV in October 2017. Following the functional check of the mould, a test program is planned, wherein the suitability of the mould for the individual production of the CFRP repair patches is to be demonstrated. The experimental program includes the fabrication of different repair patch geometries of increasing degree of complexity from thermoset as well as thermoplastic laminate lay-ups.

ACKNOWLEDGMENTS

The research project 26 LN in context of the initiative "Leittechnologien für KMU" of the Forschungsvereinigung Kunststoffverarbeitung is sponsored as part of the "Industrielle Gemeinschaftsforschung und –entwicklung (IGF)" by the German Bundesministerium für Wirtschaft und Energie (BMWi) due to an enactment of the German Bundestag through the AiF. The Authors would like to extend their thanks to all organizations mentioned.

REFERENCES

1. Blunck, E.: 'Germany BMW's Sustainability Strategy of Evolution and Revolution towards a Circular Economy', in Anbumozhi, V. and J. Kim (eds.), *Towards a Circular Economy: Corporate Management and Policy Pathways.* ERIA Research Project Report 2014-44, 2016, pp.75-92.

2. Hyonny, K.: Impact Damage Formation on Composite Aircraft Structures. *Federal Aviation Administration, JAMS 2014, Technical Review Meeting, 14 March 2008*

3. Mabry, N. J.; Peters, K. J.; Seracino, R.: Depth Detection of Bond Defects in Multilayered Externally Bonded CFRP-to-Concrete Using Pulse Phase Thermography. *Journal of Composites for Construction* 19 (2015) 6, p. 04015002

4. Yashiro, S.; Ogi, K.; Yoshimura, A.; Sakaida, Y.: Characterization of high-velocity impact damage in CFRP laminates: Part II - prediction by smoothed particle hydronamics. *Composites Part A: Appl. Science and Manuf.g* 56 (2014) 1, pp. 308-318

5. Nothdurfter, S. K.: Oto, L.: Lightweight Construction in Automotive: The Lamborghini Aventador. *15th. European Conference on Composite Materials,* Venice, Italy, 2012

6. Kiebach, H.: BMW i3 – ein neues Fahrzeugkonzept Schadenerkennung und Instandsetzung am Elektrofahrzeug BMW i3. *Technische Information 01/2014 KTI GmbH & Co. KG,* 2017

7. Hopmann, C.; Wagner, P.N.; Fischer, K.; Böttcher, A.: Industry 4.0 repairs of automotive CFRP parts by individual repair patch production with geometrically adaptive manufacturing molds. *Proceedings* of *SAMPE Conference , Seattle, USA, 2017, pp. 816 - 829*

Characterizing the Biocompatibility and Conductivity of a Styrene- Isobutylene- Styrene (SIBS) and Carbon Black Pressure Sensor

Brigitte Morales[a], Alexandra Damley-Strnad[a], Gianluca D'Ippolito[b], Landon R. Grace[c]*

[a]*Department of Mechanical & Aerospace Engineering*
University of Miami, Coral Gables, FL, 33146, USA
[b]*Department of Biomedical Engineering*
University of Miami, Miami, FL, 33136, USA
[c]*Department of Mechanical & Aerospace Engineering*
North Carolina State University, Raleigh, NC, 27606, USA

bmm58@miami.edu
lgrace2@ncsu.edu

Abstract. The biocompatibility of a styrene-isobutylene-styrene (SIBS) and carbon black (CB) composite pressure sensor was investigated. SIBS is known for its excellent biocompatibility and lack of foreign body reaction. However, the effects of CB loading on the biocompatibility of the thermoplastic elastomer are unknown. The goal of this effort is to characterize the biocompatibility of implantable and/or external (stick-to-skin) pressure sensors. The SIBS/CB composites were manufactured through solution casting in toluene and a combination of high-shear mixing and ultrasonication. Cell toxicity was assessed for composites with 15, 20, and 25% CB by weight, as well as a neat sample and a control sample. Cell viability in the composites was determined using human bone marrow derived Mesenchymal-Isolated Adult Multilineage Inducible (MIAMI) cells. Cells were monitored to observe any toxicity effects of the composite scaffolds on the cells. After 7 days in culture, it was determined that the viability was >90%. None of the samples demonstrated any toxicity. A comparison across the different samples indicates that cells cultured with the 20% CB samples demonstrated as much expansion as the cell-only control, which was subsequently the largest amount of growth. It has previously been determined that the percolation threshold of CB/SIBS composite lies around 20% by weight CB. These results indicate that the optimum CB loading to achieve maximum pressure sensitivity for reinforced conductive polymer composites does not diminish the biocompatibility properties exhibited by SIBS. Consequently, SIBS-CB based composites are determined to be well-suited as the material for stick-to-skin eye pressure sensors.

Keywords: SIBS, Carbon Black, Biocompatibility, Cell Toxicity
PACS: 81.05.Qk

INTRODUCTION

Poly(Styrene-block-Isobutylene-block-Styrene) (SIBS) is a thermoplastic elastomer with tailorable mechanical properties due to its phase-separated structure. SIBS consists of ordered and separated blocks of soft isobutylene and hard polystyrene. The ability to customize the mechanical properties of SIBS by varying the styrene/isobutylene ratio without altering its composition offers versatility in its range of applications. SIBS has becoming an increasingly important material in the biomedical field due to its high degree of biocompatibility [1-5]. There is potential for the use of SIBS in medical applications where material property customization is required. The Food and Drug Administration approved the use of the material for ultra-long term vascular device applications [1, 4]. Some SIBS medical application examples are the TAXUS® drug-eluting coronary stent, a glaucoma-treating microshunt, and as alternatives to silicone breast implants [6-8]. Here, the target application is external (stick-to-skin) and implantable pressure sensors. Specifically, SIBS may be an excellent choice for stick-to-skin pressure sensors to mitigate pressure-based injuries such as decubitus ulcers (alternatively known as pressure ulcers). These injuries are extremely prevalent, and lead to serious complications and even death; US Joint Commission on Patient safety reports that more than 60,000 people die from pressure ulcer complications each year [9]. Apart from its biocompatibility, SIBS is an excellent choice for this application because of its ability to conform to complex anatomical surfaces, its commercial availability, and its ease of processing.

In order to use SIBS for pressure sensors, the electrical properties of the neat polymer must be enhanced by reinforcing the polymer with conductive fillers. In this study, carbon black (CB) nanoparticles were chosen due to their excellent adhesion to polymer chains compared to microspheres or microrods [10] and their wide availability with relatively low cost. Percolation theory predicts that at a certain concentration of CB filler, the electrical behavior of the composite matrix will be dramatically altered [11]. At this percolation threshold, electro-conductive channels are formed and "ideal" performance is achieved; i.e., large conductivity changes as a function of applied pressure. This is a consequence of the tunneling of electrons between isolated CB particles located sufficiently close together [12]. If the volume fraction of CB is too low, SIBS will remain an insulator but if it is too high, the composite will behave as a conductor even in the absence of external pressure. Previous studies have shown that the percolation threshold for SIBS/CB composites lies around 20% CB by weight [13]. Thus, in this study experiments have been run at concentrations slightly below, at, and slightly above the optimum CB loading percentage. Although the excellent biocompatibility of SIBS has been widely reported [1-5], it is necessary to determine whether the material retains these properties upon reinforcement with CB nanoparticles. The purpose of this study is to determine biocompatibility of SIBS/CB composite scaffolds of 0, 15, 20, and 25% CB loading by weight through a cell toxicity assay. This initial study attempts to assess the viability of SIBS/CB composites for external (stick-to-skin) pressure sensing applications.

EXPERIMENTAL

The SIBS pellets used in this experiment have a composition by weight of 22% styrene and 78% isobutylene. The block copolymer, trade name SIBSTAR™ 72T, was used in the as-received condition from Kaneka Corporation. Carbon black filler material with an average 32.2 nm particle size, trade name BLACK PEARLS® 2000, was also used as-received from Cabot Corporation. SIBS/CB particle composites were fabricated through a solvent casting procedure. SIBS pellets and CB particles were dissolved using 8.57% w/v toluene as the solvent. To ensure adequate dispersion of CB particles and achieve distribution homogeneity for the resulting composites, mixing was performed for a 6-hour period per sample using a high-shear mixer at 2000 revolutions per minute with 275 mL of toluene. Due to solvent evaporation, an additional 275 mL of toluene were added halfway through the process. Composite samples were then subjected to ultrasonication at 50 kHz for an additional 45 minutes with a QSonica to augment dispersion of filler in the soft polymer. Subsequently, solutions were poured into flat stainless-steel trays and left to cure inside a fume-hood for twenty-four hours. The fabricated conductive-filler polymer composites had the following compositions: 0%, 15%, 20%, and 25% CB loading by weight. Specimens produced were approximately 0.17 mm thick.

The SIBS/CB composites of various loadings, as well as the neat sample, were tested for cell viability. The cells utilized for this experiment were a sub-population of Mesenchymal Stem Cells, Marrow-Isolated Adult Multilineage Inducible (MIAMI) cells, derived from human bone marrow. Neat polymer and composite scaffolds were sterilized in 70% Ethanol for two hours. Each scaffold was rinsed three times with Phosphate-Buffered Saline (PBS). Scaffolds were then placed into MIAMI cells expansion medium consisting of the following: DMEM-low glucose, 3% FBS, 100 μM ascorbic acid, and a mix of fatty acids. Subsequently, samples were incubated in a tissue culture incubator for two hours. After two hours, 15k MIAMI cells were plated on top of the scaffolds and placed into a 24-well plate in expansion medium. Medium was changed after three days. During the incubation period, cells were monitored to observe any toxicity effect of the scaffolds on the cells. A control sample, consisting of MIAMI cells in expansion medium but in the absence of composite scaffolds, were also tested for cell viability. Following the 7-day incubation period, microscopy images were taken and SIBS/CB scaffolds were removed. To assess cell toxicity, cells were then trypsinized and counted.

RESULTS AND DISCUSSION

Table 1 shows the final cell count (after the 7-day incubation period) for well 1, well 2, the mean, and the standard deviation across each scaffold that was plated with cells. The cell count results shown in Table 1 are for the following samples: control, neat polymer, 15% CB, 20% CB, and 25% CB. Based on these results, it is clear that cells proliferated in each of the samples tested. It is expected that the cell count will increase under the control conditions (only cell expansion medium present). It is also expected that cells will proliferate in the neat polymer sample due to the excellent biocompatibility observed in SIBS. The results summarized in table 1 do not indicate any signs of cell toxicity in any of the scaffolds containing CB loadings.

TABLE 1. Summary of cell viability after seven days in culture

	Samples	Well 1	Well 2	Mean	SD
1	Control	41000	49000	45000	5656.854
2	Neat Polymer	42500	-	42500	-
3	15% CB	37500	48750	43125	7954.951
4	20% CB	40000	50500	45250	7424.621
5	25% CB	35000	45000	40000	7071.068

Figure 1 shows the microscopy images for each sample tested. Images were taken after the scaffolds had undergone a 7-day incubation period with MIAMI cells and expansion medium. Microscopy images show no signs of toxicity due to the presence of CB, regardless of loading percentage.

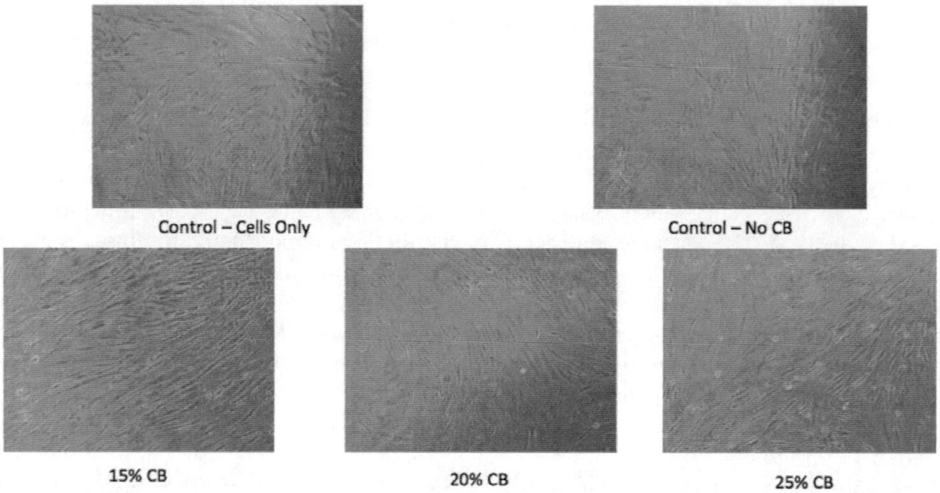

FIGURE 1. Microscopy images indicating no sign of toxicity from different concentrations of CB on Human Adult Stem Cells after 7 days

Figure 2 shows the initial and final cell count for each sample tested. It can clearly be seen that the most favorable conditions for cell proliferation were the control conditions and the 20% CB loading conditions. This is a favorable result to obtain because prior work on the relationship between conductivity and carbon black loading indicates the percolation threshold for conductivity in CB/SIBS composites and associated peak in sensor performance is at approximately the same loading percentage. Thus, fabrication of CB/SIBS composites using the percolation threshold necessary for optimum sensing performance is not likely to preclude the biocompatibility requirement for biomedical applications.

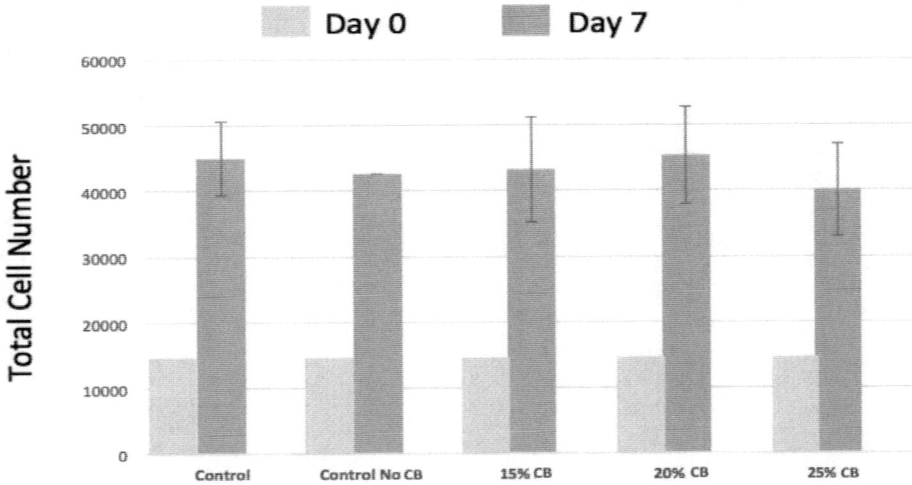

FIGURE 2. Total cell count for SIBS/CB specimens of various loadings at day 0 and day 7 of the incubation period

Figure 3 shows that the final average cell count for each scaffold incubated with MIAMI cells. All scaffold conditions were cultured with an initial 15,000 cell count. It was expected that the control group would achieve maximum cell proliferation. Data shows no significant variation in cell viability across the various loadings.

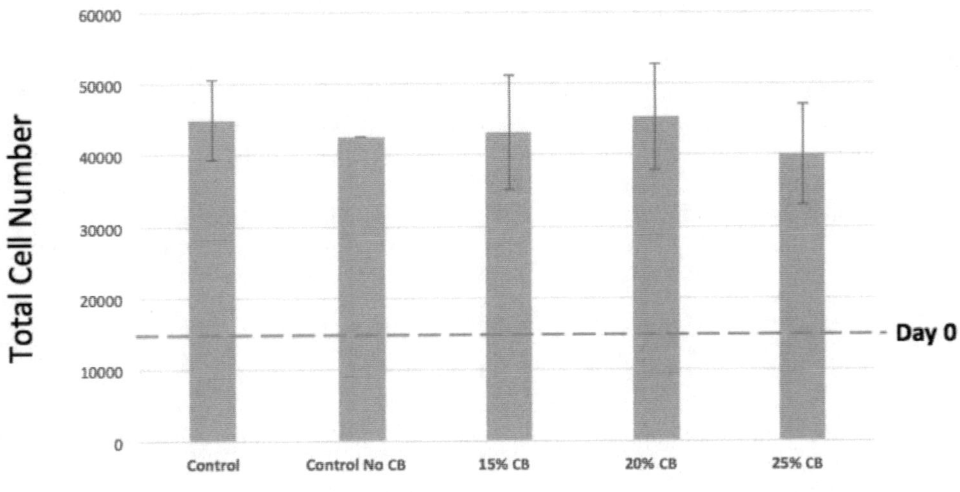

FIGURE 3. Final average cell count for control and SIBS/CB specimens of various loadings

Figure 4 shows a comparison in cell growth for each day of the 7-day incubation period across all of the samples tested. All of the loadings tested show a linear relationship for cell growth. The conditions with the steepest slope (fastest rate of growth) are the control condition and the 20% CB conditions Cell growth was observed as expected despite the presence of nano-filler. No cell toxicity was observed in any of the samples tested.

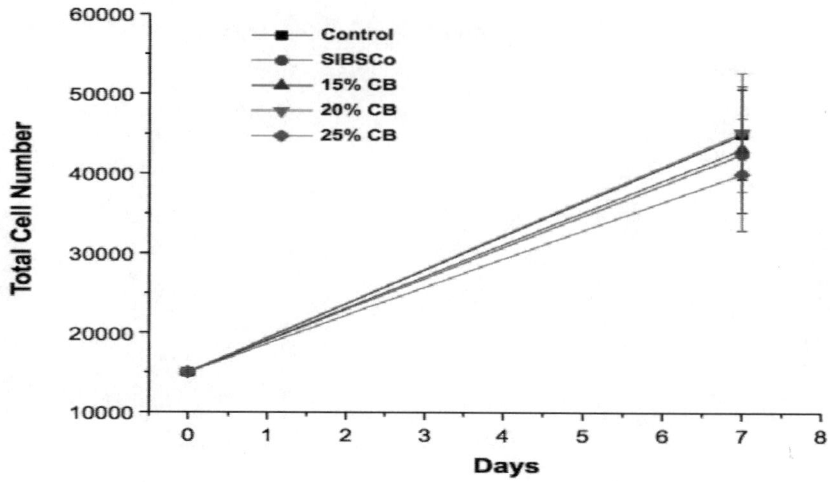

FIGURE 4. Comparison of cell count throughout a period of 7 days for the following samples: control (cells only), neat polymer, 15% CB, 20% CB, 25% CB

CONCLUSION

In this study, the effects on the biocompatibility of CB particle reinforced SIBS polymer were analyzed. MIAMI stem cells were incubated with CB/SIBS 72T scaffolding for a seven-day period. Cells were monitored and counted to ascertain any potential toxicity effects from the CB particles. Results indicate that none of the samples demonstrated any toxicity. Cell viability was >90%. From the data, it can be seen that the 20% CB scaffold exhibited cell growth similar to that of the control culture, though all loading percentages exhibited similar behavior. This likely indicates an effective, complete encapsulation of the CB particles. A comparison across various CB loadings indicates that despite higher CB loadings, cells continue to grow and proliferate as expected. These preliminary results indicate that the presence of CB particles, whether in low or higher quantities, may not alter the excellent biocompatibility observed in the neat polymer. Although these results do not limit the use of CB/SIBS composites for stick-to-skin pressure sensors, further sensitization and skin irritation testing is needed to definitively prove viability for external medical applications. Given the performance of SIBS/CB for detecting pressure and deformation, the results of this first-level biocompatibility tests suggest that SIBS/CB composites may be well-suited for biomedical sensing applications and that further immune response, genotoxicity, and carcinogenicity testing is justified.

REFERENCES

1. M. Boden, R. Richard, M.C. Schwarz, S. Kangas, B. Huibregtse, and J.J. Barry, *Journal of materials science. Materials in medicine* **20,** 1553-62 (2009).
2. M. El Fray, P. Prowans, J.E. Puskas, and V. Altstadt, *Biomacromolecules* **7,** 844-50 (2006).
3. S.L. Gallocher, A.F. Aguirre, V. Kasyanov, L. Pinchuk, and R.T. Schoephoerster, *Journal of biomedical materials research. Part B, Applied biomaterials* **79,** 325-34 (2006).
4. L. Pinchuk, G.J. Wilson, J.J. Barry, R.T. Schoephoerster, J.M. Parel, and J.P. Kennedy, *Biomaterials* **29,** 448-60 (2008).
5. F. Strickler, R. Richard, S. McFadden, J. Lindquist, M.C. Schwarz, R. Faust, G.J. Wilson, and M. Boden, *Journal of biomedical materials research. Part A* **92,** 773-82 (2010).
6. K.R. Kamath, J.J. Barry, and K.M. Miller, *Advanced drug delivery reviews* **58,** 412-36 (2006).
7. L.Pinchuk, I. Riss, J.F. Batlle, Y.P. Kato, J.B. Martin, E. Arrieta, P. Palmber, R.K. Parrish, B.A. Weber, Y. Kwon, and J.-M. Parel, *Regenerative Biomaterials* **3**, 137-42 (2016).
8. G.Teck Lim, S.A. Valente, C.R. Hart-Spicer, M.M Evanchp-Chapman, J.E. Puskas, W.I. Horne, and S.P. Schmidt, *Journal of the mechanical behavior of biomedical materials* **21**, 47-56 (2013).
9. E.A. Kruger, M. Pires, Y. Ngann, M. Sterling, and S. Rubayi, *The journal of spinal cord medicine* **36**, 572-85 (2013).
10. M. Knite, V. Teteris, A. Kiploka, and J. Kaupuzs, *Sensors and actuators* **110**, 142-49 (2004).
11. M. Knite, V. Teteris, B. Polyakov, and D. Erts, *Materials sciene and engineering* **19**, 15-19 (2002).
12. L. Wang, T. Ding, P. Wang, *IEEE Sensors journal* **9**, 1130-35 (2009).
13. B. Morales, M. Fittipaldi, A. Damley-Strnad, and L. Grace, "A biocompatible pressure sensor based on styrene-isobutylene-styrene and carbon black" in 32nd American Society of Composites Conference, Williamsburg, VA, Sept. 2016.

Enhanced Damping of Carbon Fiber-Reinforced Composites by Novel Liquid-Core Fibers

Leal A. Andrés (1), Heuberger Manfred (1), Hufenus Rudolf (1)*

(1) Empa - St. Gallen - Switzerland

We demonstrate the continuous production of liquid-core thermoplastic filaments in a stable melt-spinning process at the pilot plant scale, using an adapted bicomponent melt-spinning approach. Different polymers and liquids could successfully be combined over a wide range of core-sheath dimensions. The ability to produce a continuous liquid-core fiber (LCF) is attractive since post-filling of a hollow fiber with similar dimensions is not practical. We characterized the mechanical properties of the LCFs with particular attention to their damping properties. A LCF can exhibit significantly enhanced damping properties compared to plain polymeric filaments of same dimensions. Low frequency bending tests suggest that pressure propagation along the incompressible fluid core combines with damping properties of the polymer sheath at a distance. Significant damping factor increases observed at higher frequency suggest that this effect can be further optimized with suitable liquid/polymer combinations. This new type of filament could find future applications in the enhancement of the damping factor of fiber-reinforced lightweight structures. In a feasibility study we could improve the vibration damping characteristics of carbon fiber-reinforced epoxy composite materials. For this purpose, a novel fibrous grid composed of liquid-core thermoplastic filaments was introduced in the interlaminar regions of the composite. As a result, structural vibration damping of the carbon fiber-reinforced composite could be increased up to 4-fold at high frequencies.

In-Situ-Pultrusion – Investigation of the Viscosity of the Reacting PA6 under Process Conditions

Stefan Epple[a] and Christian Bonten[a]

[a] *Institut für Kunststofftechnik, University of Stuttgart, Pfaffenwaldring 32, D-70569, Stuttgart, Germany*

Abstract. The innovative combination of an anionic polymerisation from the ε-Caprolactam to polyamide 6 (PA6) with a continuous pultrusion process allows production of well impregnated continuous fiber reinforced thermoplastics with both high wall thickness and a high amount of fibers. Parts produced by such an in-situ-pultrusion are supposed e.g. to be used as local reinforcement of injection molded plastic parts. An important factor for the pultruded parts quality is to understand the physical and technical process-parameters during the pultrusion process. As there are a lot of differences between the chemical reactions in thermoset pultrusion processes and the in-situ-pultrusion process of continuous fiber reinforced thermoplastics, we investigated especially the behaviors of the reaction under process conditions. Mainly the determination of viscosity of PA6 over time during activated anionic polymerization was therefore analyzed using a modified rotational rheometer. In our experiments, we varied parameters such as temperature and catalyst concentration. Furthermore, the residual monomer content of the polyamide 6 was determined to receive information on a possible hydrolysis during polymerization in the rheometer. The detected viscosities were approximated by a continuously differentiable fit function with suitable parameters. One focus of this work lies in the determination of the specific moment when the PA6 reaction has proceeded so far that it can be demolded. We report on the influence of the tool length and the temperature profile of the zones of the tool on the materials quality and parameter and how these results can be implemented in a simulation tool for an in-situ-pultrusion-process.

Keywords: Composites, Injection molding, Thermoplastic Composites, Thermoplastic Pultrusion
PACS: 81.05.Ni

INTRODUCTION

In traditional fiber reinforced plastics, a thermosetting resin is used as matrix. These fiber reinforced thermosets have disadvantages, especially in batch production because e. g. the cycle times are too long, they are not weldable and poorly recyclable. In batch productions it is common to use fiber reinforced thermoplastics. Often, short fibers are used as reinforcement. By using glass fibers, mechanical properties increase, however, highly-stressed components needs continuous fiber reinforced plastics. The two major advantages of thermoplastic composites are that they have a higher impact resistance to comparable thermoset composites and can be welded and recycled. One objective of the current research is to produce a pultruded thermoplastic composite that could be heated and remolded to have complex shapes. This is not possible with thermosetting resins. It will also be possible to recycle the thermoplastic composites in contrast to thermoset composites. Thermosetting resins are of low viscosity and thereby can easily impregnate the reinforcing fibers before they react. In contrast, thermoplastics are usually of high viscosity, even in the molten state. It is not easy to impregnate reinforcing fibers. To impregnate reinforcing fibers with a thermoplastic matrix, plastic films and woven fabrics must be laminated. High pressure and high temperature are required. These fiber-reinforced thermoplastic laminates are called organo-sheets. Fiber mat or long glass fiber reinforced thermoplastics are also used to produce rugged parts. The problem of these parts is their limited shape variety. In-Situ-pultrusion is one way to produce thermoplastic FRP. Like in conventional pultrusion with a thermoset matrix, continuous fibers are impregnated with the monomeric precursor of the matrix material, which then reacts and forms the polymer. When using ε-caprolactam as a monomer, cast PA6 will be formed and act as the matrix [1]. Different to conventional pultrusion, the fibers cannot get impregnated by pulling them through a resin bath, as the surrounding humidity would disturb the chemical reaction. Thus, the reaction has to take place in an inert gas atmosphere inside the pultrusion die. The products from in-situ-pultrusion can e. g. be used as reinforcements for injection molded parts. One example is the addition of rip structures to bending beams composed of two FRP profiles via back injection molding. For this purpose, the profiles are inserted into an injection mold and overmolded with a thermoplastic. Thus, cycle times like in conventional injection molding processes are possible, combined with excellent mechanical properties and high component complexity. [2]

STATE OF THE ART

To bring the advantages of continuous fiber reinforced plastics also into parts with complex geometries, the Institut für Kunststofftechnik (IKT) put into effect a process based on the pultrusion process combined with reaction injection molding (RIM). In reaction injection molding, two components are mixed together. The mixture is then injected into the mold, where it reacts to the plastic. The RIM process offers the advantage of a very low viscosity of the mixture before the reaction starts. In this state, the two component mixture can impregnate the fibers very well.

It was the objective of the development to produce thermoplastic parts with a high amount of continuous glass fibers in various shapes and thicknesses. Adhesion between fibers and matrix and fiber coating were optimized [3].

Basic Principles of the In-Situ Pultrusion Process

Two principles are coupled in the In-Situ-pultrusion process. First, the two reactive components are mixed together. Then the mixture is injected into a mold. This process is often used to produce parts made of polyurethane. In pultrusion processes, fibers are pulled through a bath of thermosetting resin. Afterwards, the impregnated fibers are pulled through a heated die, where the resin reacts to the formed composite. A puller behind the die is pulling the material through the process at a consistent rate. Behind the puller, the parts are sawn off to the desired length. In current researches, the matrix of the composite was polyamide 6 (PA 6). That is why ε-caprolactam was used in the activated anionic polymerization process from monomer to polymer [4]. The resulting PA 6, also known as cast PA 6, has good properties like dimensional accuracy, low moisture adsorption and a good creep resistance. The In-Situ-pultrusion process used for the production of continuous glass fiber reinforced thermoplastics is schematically shown in **Figure 1**.

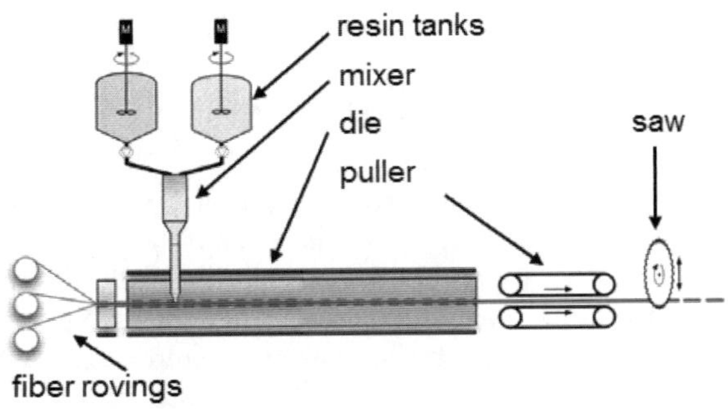

Figure 1: Scheme of the new RIM pultrusion process

The glass fibers are pulled through a pre-warming and sorting tool before they are entering the die. In the die, the fibers are impregnated with the monomer. In the following zones of the die, the monomer ε-caprolactam reacts to PA 6. The first part of the die is only heated while in the second part of the die, cooled or heated, depending on the process behaviour. This is important because in the process heat can be generated by the reaction which has to be removed from the die.

The puller and the saw are also shown in the scheme. Providing the mixtures of ε-caprolactam and catalyst as well as ε-caprolactam and activator in the correct way is very important for the process. Protective gas has to be used to reduce the influence of moisture on the reaction. [5]

Basic Principles of the Activated Anionic Polymerization of PA 6

The activated anionic polymerization of PA 6 can occur within minutes. In this polymerization, the monomer respectively the growing macromolecule is an anion. An activator to accelerate the catalytic reaction is also used in this system.

Especially the low temperatures below the melting temperature of PA 6 which are needed for this reaction (130 to 170 °C) allow its use in the present process.

Test Facility at IKT

At IKT, an In-Situ-pultrusion facility with two component units, puller, saw, die and mixer was installed.

It is important for the process that the material does not polymerize in the nozzle. The nozzle opening was therefore separated from the rest of the die and can be cooled. The die length was chosen in a way that in the times known from the preliminary studies, polymerization can take place in the die. It should be noted that the die should not be unnecessarily long. Otherwise the puller reaches its performance limits. On the other hand the die must not be too short because then the material would have too few time to polymerize. To design the In-Situ-pultrusion die (Figure 2), preliminary examinations were needed and the reaction behavior of ε-caprolactam under process conditions had to be described.

Figure 2: In-Situ-pultrusion facility

EXPERIMENTAL

To measure and to describe the reaction behavior from ε-caprolactam to PA6 under process conditions, the regularly used test tube experiments, which are well known from chemistry, are not useful, because they give results far away from process conditions. Especially influences like surrounding humidity, type of the used steel, shearing, and temperature profiles cannot be considered. That's why a modified rheometer test was developed. In **Figure 3**, the test is described. First, initiator (Akt) and catalyst (Kat) were poured in test tubes and heated under nitrogen atmosphere until they are liquid, which is from 80 °C. After that they are poured together in one test tube and are mixed by shaking the test tube. After that, 2 ml of the mixture were taken by using a syringe and then filled in the modified rheometer plate. The plate has an edge, because otherwise, the liquid ε-caprolactam would flow away. The rheometer has been closed and filled with nitrogen. Then the rheometer test startet and the viscosity was measured while the ε-caprolactam was reacting to PA6.

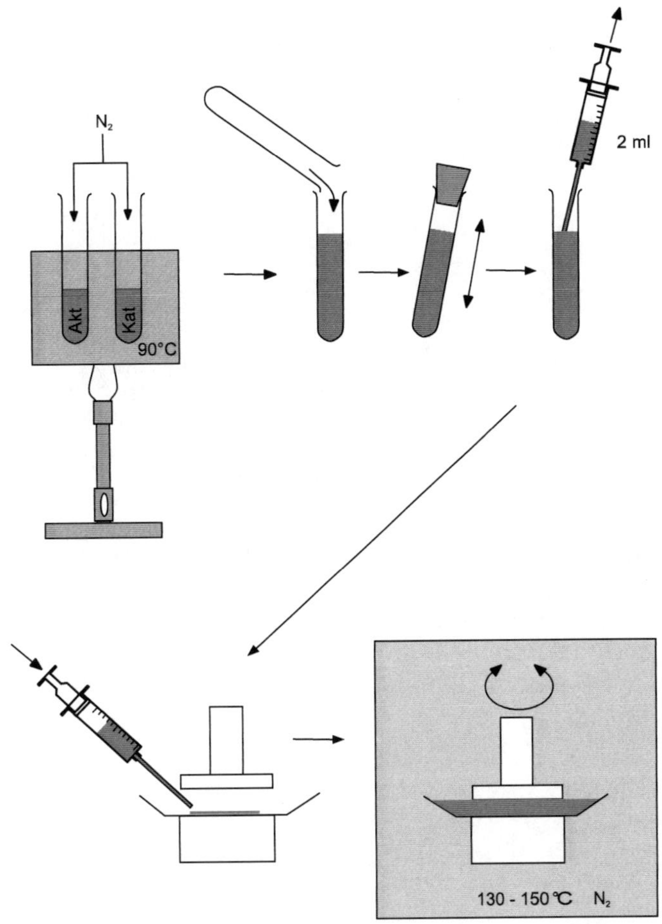

Figure 3: Modified rheometer test

RESULTS AND DISCUSSION

In all experiments, characteristic viscosities over the time were measured. At the start, when the ε-caprolactam was still liquid, the values were fluctuating. That's due to the fact that the transducer, used to measure the forces in the specimen was in this area of viscosity not accurate enough. At a certain point in time, a steep increase in the recorded values happened. The jump to higher viscosities is about five decades in viscosity and marks the time at which the material is stable in shape. After the jump, the viscosity increases continuously but at a reduced rate. At viscosities higher than 10^6 Pa·s, the measurements are broken as the maximum torque of the rheometer is exceeded and the measurement is terminated automatically. The measured values were investigated in the following with regard to the varied influencing parameters.

As a critical influencing factor during the measurements, the time between the mixing of the components and the start of the measurement was considered. In order to ensure that this has no influence on the results, these times were varied in preliminary tests under otherwise identical conditions. All three measurements in Figure 4 were carried out at a temperature of 130 ° C. and a catalyst concentration of 3.75%. The times between the mixing of the components and the start of the measurement were 45, 55 and 80 seconds. It has been shown that the results are equal independent from the time until the start of the reaction. At least three reference measurements were made for each test point.

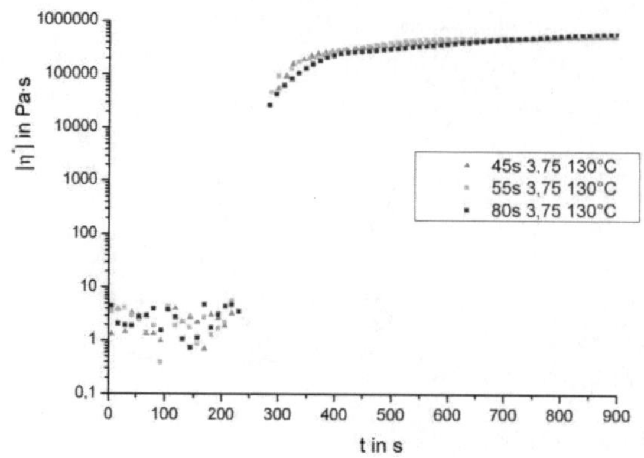

Figure 4: Variation of time until start of the rheometer tests

Under otherwise identical conditions, the temperature in the test chamber of the rheometer was varied in further experiments. Three temperatures, which are considered processing temperatures for the system, were used. It was found that by increasing the temperature by 20 K from 130 ° C. to 150 ° C., the duration to the start of the reaction could be shortened to less than half(**Figure 5**). The maximum torque of the rheometer was exceeded at the highest temperature (150 ° C.) in all experiments shortly after the start of the reaction and the measurements stopped.

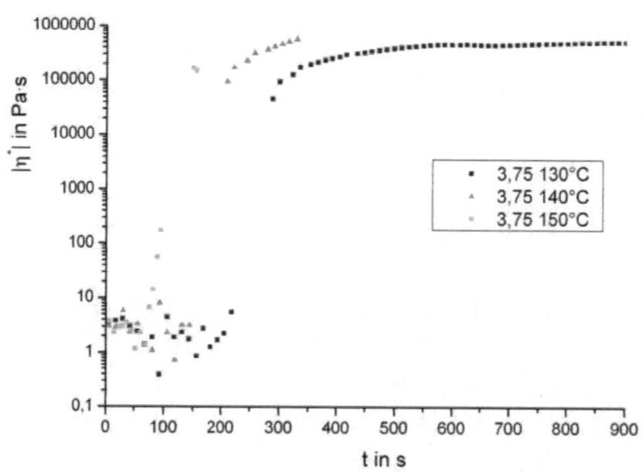

Figure 5: Variaton of the temperature

The ratio of initiator to catalyst during all experiments was 3 to 2. The concentration of the catalyst was varied in three stages. In preliminary tests, the catalyst concentration of 3.75 % proved to be a concentration with suitable reaction properties. From this concentration, two variations were made to a higher as well as a lower value. The concentration of the catalyst was increased to 4.5 %. The lower catalyst concentration was fixed at 3.0 %. The measurements clearly showed that a higher concentration of the catalyst has a significant influence on the reaction rate of the mixture. It can also be seen from Figure 6 that a reduction in the catalyst concentration from 3.75% to 3.0% does not lead to an extension of the time period until the beginning of the reaction at the 130 ° C.

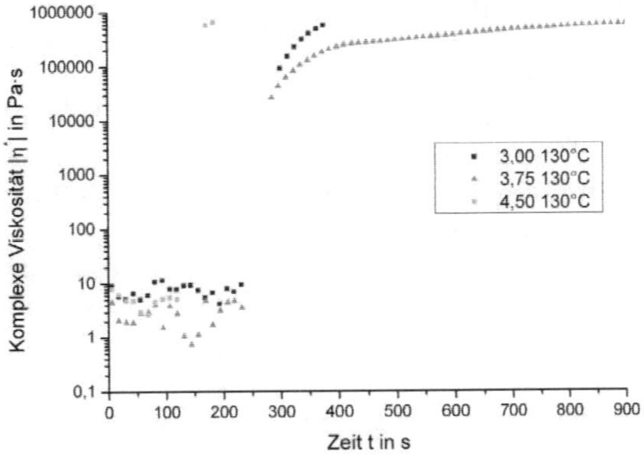

Figure 6: Variation of the catalyst concentration

CONCLUSION

For the desired process-related application of pultrusion, it is important to know at which moment the material is polymerized so far that it can leave the tool without losing its shape. If the polymerization has not progressed sufficiently, liquid material can pour out from the die. For manufacturing reasons, a short pultrusion tool is better. **Figure 7** shows the start times of the reaction over the temperature and the catalyst concentration. The shortest time to start the reaction (43 s) passes at high temperature (150 °C.) and high catalyst concentration (4.75 %). At a low temperature (130 °C) and a low catalyst concentration (3.0 %), it takes the longest time for the mixture to react (220 s).

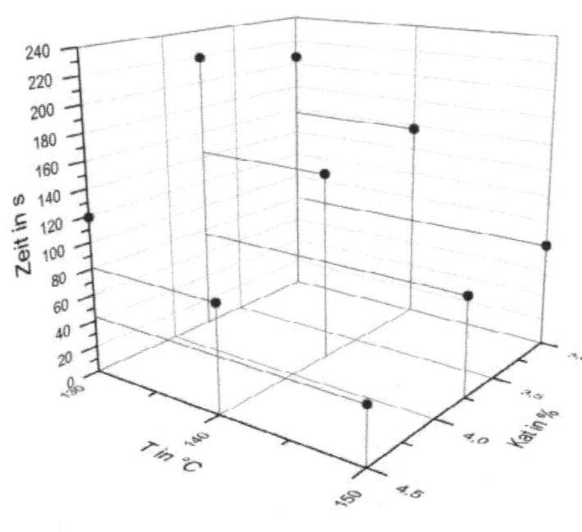

Figure 7: Times for start of the reaction

The samples with the catalyst concentrations of 3.0 % and 3.75 % behaved very similarly due to the dependence of the reaction start time on the temperature. This means that, in this concentration range, the reaction start time can be influenced solely by varying the temperature. This can bring advantages in the pultrusion process in particular if the metering of the catalyst as well as of the initiator is subject to slight fluctuations. The temperature then remains the only control variable in the process.

REFERENCES

1. GEIER, S. Optimierung von Steifigkeit/Zähigkeits-Eigenschaften nanoskaliger Polyamid 6-Verbundwerkstoffe durch Analyse von Struktur/Eigenschafts-Korrelationen. [S.l.: s.n.], 2011.

2. EPPLE, S. and C. BONTEN. Analyse der anionischen Polymerisation aus verarbeitungstechnischer Sicht. In: INSTITUT FÜR KUNSTSTOFFTECHNIK, Hg. Stuttgarter Kunststoffkolloquium.

3. ALEWELT, W., L. BOTTENBRUCH und G.W. BECKER. Polyamide. Neuausg. München: Hanser, 1998. Kunststoff-Handbuch Technische Thermoplaste. / hrsg. von Ludwig Bottenbruch ... ; 4. ISBN 3-446-16486-3.

4. LANG, K. Zur Erweiterung des Spektrums von Faser-Kunststoff-Verbunden mit thermoplastischer Matrix. Einsatz unidirektional faserverstärkter Halbzeuge in hoch belasteten Strukturbauteilen des Automobil- und Flugzeugbaus. Techn. Univ., Diss.--Darmstadt, 2012. Aachen: Shaker, 2012. Schriftenreihe Konstruktiver Leichtbau mit Faser-Kunststoff-Verbunden. ISBN 978-3-8440-1502-7.

5. EPPLE, S. and C. BONTEN. RIM-Pultrusion zur Herstellung endlosglasfaserverstärkter PA6-Halbzeuge. In: INSTITUT FÜR KUNSTSTOFFTECHNIK, Hg. Stuttgarter Kunststoffkolloquium.

Development of an Inline Monitoring System for Quality Assurance of Thermoplastic CFRP Tape Using Air-coupled Ultrasound

Wolfgang Essig[a] and Marc Kreutzbruck[a]

[a]*Institut für Kunststofftechnik, University of Stuttgart, Pfaffenwaldring 32, D-70569, Stuttgart, Germany*

Abstract. Besides the aerospace industry, fiber reinforced plastics have also spread towards many further applications such as automotive or civil engineering. Their superior strength and stiffness to mass ratio made them the number one material for achieving high performance. Especially continuous fiber reinforced plastics allow for the construction of structures which are custom tailored to their mechanical loads. The two main constituents of CFRP are carbon fibers and matrix. Two possibilities for matrix material exist: thermosetting and thermoplastic matrix. While thermosetting matrix may yield better properties, thermoplasticity opens a wide range of applications due to weldability and compatibility to e.g. thermoplastic injection molding materials. In this work an inline monitoring system was developed for quality assurance of continuous fiber reinforced, thermoplastic CFRP tapes using air-coupled ultrasound. This system is independent of tape thickness and tape width due to an adaption of the angle of incidence. By slanted incidence of the ultrasound on the tape surface, guided waves were excited in the material in fiber direction and perpendicular to the fiber direction. Artificial defects – fiber cuts, matrix cuts, circular holes, low velocity impacts from tool drop, and sharp bends – were produced. Experiments on a stationary tape showed good detectability of all artificial defects by guided waves. Also the effects of variation in material properties, fiber volume content and fiber matrix adhesion were examined, to allow for quality assessment. For validation of the monitoring system, an apparatus containing one endless loop of CFRP tape was constructed and built to simulate the pultrusion process. Several methods were investigated regarding the scanning path of single channel ultrasonic transducers in order to achieve 100 % testing of the tape. For final validation of the monitoring system a Probability of Detection (POD) analysis was conducted and analyzed.

Keywords: Nondestructive Testing, Ultrasonics, Air-coupled Ultrasound, CFRP, Tape.
PACS: *43.35.Zc, 81.70.Cv

INTRODUCTION

Whilst most prepregs are manufactured from carbon-fiber and thermosetting matrix, latest developments tend to replace the thermosetting with a thermoplastic matrix. The most common method of production is the pultrusion process, impregnating the fibers with molten thermoplast pellets. After the pultrusion process the carbon-fiber reinforced prepreg with thermoplastic matrix (CFR-Tape) is wound on a spool and ready for further processing [1].

For the analysis of the tapes air-coupled ultrasound testing was chosen because it is a well-known technique, it does not need any radiation shielding or coupling agent and offers a range of possible experimental set-ups [2]. Besides that, non-destructive testing may be part of an inline-quality inspection, just as demanded within the upcoming concept of "industry 4.0" [3]. Recent research was performed on the electromechanical response of polypropylene ferroelectret transducers [4], air-coupled cellular polypropylene transducers as mentioned in [5], as well as the resonance-free measuring and excitation of ultrasound [6].

SIMULATION OF ULTRASOUND PROPAGATION

Consisting of two materials, carbon fiber and polyamide 6, with quite different elastic properties the tape shows anisotropic material behavior similar to other composite materials. For the simulation of ultrasound propagation, the computer program "Disperse" [7], developed at the Imperial College London was used. With the program only accepting a single set of homogenized material parameters, the elastic properties of both materials have to be combined, applying Puck's and Jones' laws [7].

TABLE 1: Calculated phase velocities for f = 500 kHz

	v_{a0} acc. to Puck in m/s	v_{a0} acc. to Jones in m/s
longitudinal	888.3	980.6

TAPE SPECIMENS

The tape with a thickness of 0.18 mm was manufactured from UD carbon fibers and the widespread polyamide 6 with an intended fiber volume ratio of 50 %. In order to achieve comprehensive results, five different specimens with specific properties were prepared. Besides the detection of defects, a second task was to find a correlation between ultrasound data and the actual fiber volume ratio of the tape. The purpose of the first specimen, named TP1, was to serve as a standardized sample with specific defects to test different experimental set-ups. Its schematic representation is given in Figure 1. The second specimen (TP2) featured an incorrect fiber volume ratio over entire length, the third specimen (TP3) showed a transition from correct to incorrect fiber volume ratio with the fourth specimen (TP4) being a flawless specimen. The fifth and last specimen (TP5) was manufactured as a flawless continuous tape with a length of 2 m.

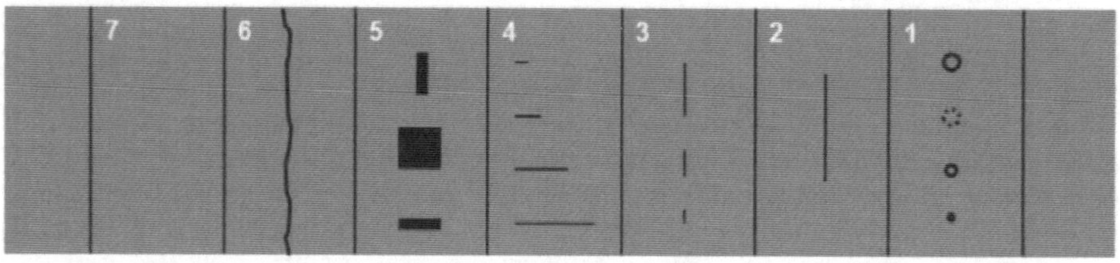

FIGURE 1: TP1, defect pattern

In the standardized specimen TP1, the first segment included holes with four different diameters, the second one featured one long cut perpendicular to the fiber orientation with a length of 40 mm. The third section displayed three cuts perpendicular to the fiber orientation with a length of 5 mm, 10 mm and 20 mm. The fourth segment showed four cuts parallel to the fiber orientation with a length of 5 mm, 10 mm, 20 mm and 30 mm, whereas the fifth segment featured three impact damages. The second to last segment included a crack perpendicular to the fiber orientation with the last section serving as a flawless reference zone.

ULTRASOUND TEST EQUIPMENT

For the experiments two ultrasound transducers manufactured by the Ultran Group were used. The first one, namely NCG500-S50, has an active area of 50 mm by 50 mm and served as the transmitter, whilst the other transducer, called NCG500-S25, with an active area of 25 mm by 25 mm was used as the receiver. Both transducers were driven by a Dasel Airscope TT, wired to a laptop which was used for the ultrasound user interface and for programming the control device of a scanning table. For testing the samples were mounted to an adjustable frame.

Measuring Methods

Two different alignment methods for the transducers have been tested within the analysis of TP1, revealing that single-sided ultrasound testing is unsuited for testing CFR-Tape. Consequently only two-sided alignment was applied to the other specimens with variation of transducer to surface angle from 27° up to 43° and variation of ultrasound coupling parallel or perpendicular to the fiber orientation.

Continuous Tape Test Rig

FIGURE 2: Experimental set-up for testing continuous tapes

A continuous tape test rig was developed in order to simulate the inline testing in the production process. The test rig was designed to accept specimens with a length of up to 1880 mm and to be run at speeds of up to 10 m min^{-1}. Aluminum profiles were used for the construction as well as acrylic glass elements. It features four cylinders directing the tape with one of them being driven by a stepper motor equipped with an encoder.

ULTRASOUND DATA EVALUATION

The data evaluation program based on Matlab was designed to function as a quality gate. After cutting off the interferences at the edges of the tape, the average value as well as the standard deviation for both the amplitude and the phase image is calculated. Moreover the relative standard deviation in the entire range of values is computed. In the program all pixels that are not in the range average value plus/minus standard deviation are marked as defect. After cleaning the image from very small defects using morphological processes, the ratio of area indicated per area examined is calculated. If the size of the indicated area exceeds a certain value, the tape also no longer matches the quality standards. Linking both methods, a good method for quality testing of CFR-Tapes was obtained. The whole process is shown in Figure 3.

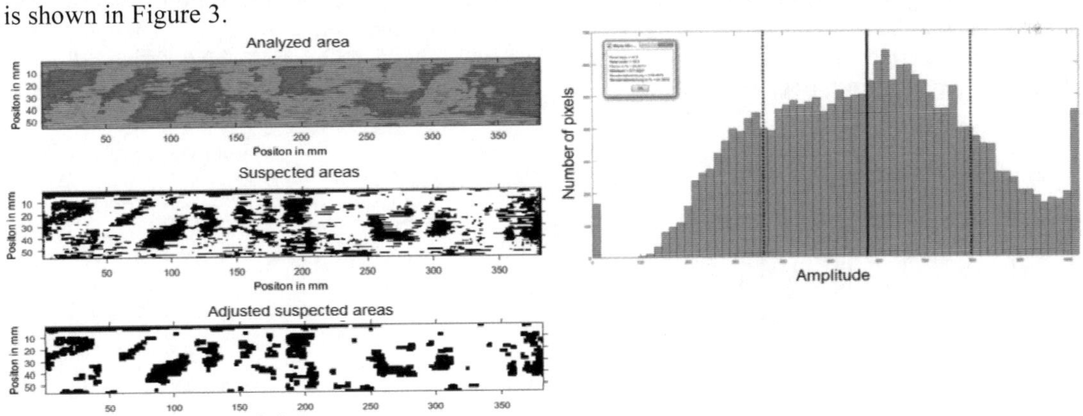

FIGURE 3: left: Amplitude image of TP2 with image processing shown to mark suspicious areas; right: Histogram of specimen TP2 with a separate window giving information about average value (solid line) and standard deviation (dashed lines) as well as suspected surface area.

Analysis of Fiber Volume Ratio

According to [8], it is possible to ascertain the fiber volume ratio by examining micrographs of the tape. After the microscopy, the micrographs are extracted from the surrounding microscope picture and the contrast of the image is increased. The fibers appear in white color whereas the matrix remains black. A program clears the image from smaller particles and calculates the total area of the cross-section, as well as the area of the fibers. With knowledge of both figures it is possible to give an estimate of the fiber volume ratio.

Maximum tape velocity

The maximum tape velocity was determined by observing the A-scan image of the specimen TP5 while increasing the tape velocity until gaps between two measuring points became visible on the user interface of the Airscope TT. Applying this method, the tape speed topped at 5.35 m min^{-1}. This limit is due to the used ultrasound system, it is not a physical limit.

RESULTS AND DISCUSSION

Artificial Defects and Decreasing Fiber Volume Ratio

Both, wave propagation parallel and perpendicular to the fiber orientation provided good data. As expected, defects with perpendicular orientation to the ultrasound wave propagation direction were detected more clearly than those oriented parallel to it. For the type of tape investigated the best results were gained with a transducer-to-surface angle of 27°, measured from the vertical direction. An example of an amplitude image of the specimen TP1 is given in Figure 4.

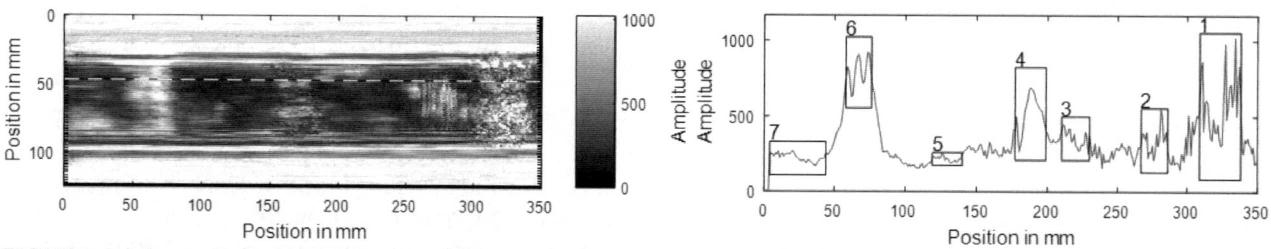

FIGURE 4: left: amplitude image of specimen TP1 measuring parallel to the fiber orientation; right: amplitude profile along line in c-scan

For the analysis of the tape TP3 with the transition from a good to a bad fiber volume ratio the tape was separated into eight sections and the amplitude values as well as the standard deviations were calculated for each sector. Within the flawless sections the amplitude values and the standard deviation remained steady. Regarding the sections displaying an incorrect fiber volume ratio, both the average values and the standard deviations varied as seen in Figure 5.

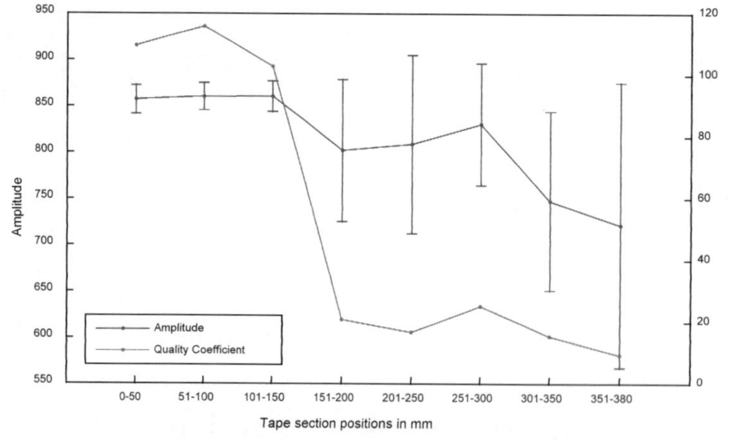

FIGURE 5: Specimen TP3; progress of the average amplitude value and standard deviation

With the amplitude values dropping and the standard deviation increasing for the sections with incorrect fiber volume ratio, the hypothesis about using the average values and the standard deviation as a quality gate was supported. This led to the definition of an empirical quality coefficient ***2mean/std***, which easily separates to flawless sections of the tape from those with incorrect fiber volume ratio.

Analysis of Fiber Volume Ratio and Simulated In-line Inspection

By overlaying a photograph of the second specimen with a transparent ultrasound image in an image processing software, it became apparent that the ultrasound image corresponds quite well with the visual pattern of incorrect fiber volume ratio.

Comparing the specimens with an incorrect fiber volume ratio to an optically flawless tape, the measurements showed exceptional variations. Consequently, no clear statement regarding a correlation can be made, although the calculated fiber volume ratio of the flawless specimen was measured in fact higher than the fiber volume ratio of the specimens with an obviously incorrect fiber volume ratio. According to [8], the method should be capable of exactly determining the fiber volume ratio, but further examination has to be done, in order to adjust the method to CFR-Tapes.

As the method of non-destructive testing with air coupled ultrasound should be evaluated as a quality assurance method, the process should be capable of adapting to different tape speeds. This fact was proven by running the tape at three different speeds. For measurements at 2.17 m min^{-1}, 2.97 m min^{-1} and 4.41 m min^{-1}, the values of the amplitude varied only by 0.4 % between measurements, whereas the values of the phase varied by 1.4 %. These results state clearly that the testing method results are independent of the tape production speed.

CONCLUSION AND FUTURE WORKS

Air coupled ultrasound is capable of being applied as a quality assurance element for thermoplastic CFR tapes. For future application it is essential to increase the spatial resolution by using focused or phased array transducers. Regarding the correlation between the fiber volume ratio and the measured ultrasound data, further examination needs to be performed.

REFERENCES

1. Lang, K. Zur Erweiterung des Spektrums von Faser-Kunststoff-Verbunden mit thermoplastischer Matrix. Einsatz industriell faserverstärkter Halbzeuge in hochbelasteten Strukturbauteilen des Automobil- und Flugzeugbaus. Ph.D. thesis. Darmstadt, 31. Jan. 2012.
2. Peters, J., V. Kommareddy, Z. Liu, D. F and D. Hsu. Non-Contact Inspection of Composites Using Air-Coupled Ultrasound. In: REVIEW OF PROGRESS IN QUANTITATIVE NONDESTRUCTIVE EVALUATION:Volume 22, S. pp. 973-980.
3. Hermann, M., T. Pentek und B. Otto. Design Principles for Industrie 4.0 Scenarios. A Literature Review. Dortmund.
4. Döring, J., V. Bovtun, M. Gaal, J. Bartusch, A. Erhard, M. Kreutzbruck and Y. Yakymenko. Piezoelectric and electrostrictive effects in ferroelectret ultrasonic transducers. In: Journal of Applied Physics, 2012, 112(8), S. 84505. Doi:10.1063/1.4759052
5. Mate Gaal, Elmar Dohse, Jürgen Bartusch, Enrico Köppe, Marc Kreutzbruck, Wolfgang Hillger and Jay Amos. Ultrasonic testing of adhesively bonded joints using air-coupled cellular polypropylene transducers.
6. Daschewski, M., M. Kreutzbruck, J. Prager, E. Dohse, M. Gaal and A. Harrer. Resonanzfreie Messung und Anregung von Ultraschall. In: tm - Technisches Messen, 2015, 82(3). Doi:10.1515/teme-2014-0020
7. Stellbrink, K.K.U. Micromechanics of composites. Composite properties of fiber and matrix constituents. Munich: Hanser, 1996. ISBN 1569902062.
8. George, A. Optimization of Resin Infusion Processing for Composite Materials. Simulation and Characterization Strategies. Dissertation. Stuttgart, 15. Mrz. 2011.
9. Blackstock, D.T. Fundamentals of physical acoustics. New York, NY: Wiley, 2000. A Wiley-Interscience publication. ISBN 978-0-471-31979-5.

Endowing the high efficiency thermally conductive and electrically insulating composites with excellent antistatic property through selective distribution of functional fillers in separate layers

Wu Hong* [1]

[1] Sichuan University - Sichuan - China

To fabricate thermally conductive and electrically insulating composites with excellent antistatic property is a huge challenge in the region of packaging materials of electronic devices, due to the contradiction between the electrical insulation and the antistatic property. In the present work, the peculiar multilayer structures with alternating high efficiency thermally, electrically conductive layers and thermally conductive, electrically insulating layers were constructed successfully by confining different functional fillers in separate layer spaces through a simple, one-step melt extrusion method. Such thermally conductive and electrically insulating composites possessed significant anisotropic electrical resistivities, and the in-plane electrical resistivities (parallel to the layer direction) were below 117 Ω•cm, while the through-plane electrical resistivities were over 5×1013 Ω•cm. Meanwhile, thermal conductivities of the composites were improved monotonously with the increase of the layer number, and reached as high as 1.45 W/(m×K) in the composite with 32 layers. In addition, tensile strengths and elongations at break of the composites were also enhanced due to the different deformation mechanism of separate layers. Furthermore, finite element analysis was also applied to give a deep insight into the enhancement mechanism of the thermally conductive efficiency, and the results indicated that high efficiency thermally, electrically conductive layers possessed positive synergetic and magnified effects on the heat dissipation. Therefore, the multilayer structure with alternating high efficiency thermally, electrically conductive layers and thermally conductive, electrically insulating layers can impart the composites with excellent comprehensive properties effectively, and it also sheds light on the design and fabrication of high performance materials for the application of thermal management or other fields.

pps proceeding
Symposium: S10 - Composites
Oral Presentation

Low percolation threshold in carbon fibre/epoxy composite containing carbon nanotubes for sensing of structural damage

Santos R. M. (1)*, Rocha J. (1), Silva M. (1), Rocha N. (1)

(1) INEGI - Porto - Portugal

Carbon fibre reinforced polymer (CFRP) composites comprise unidirectional layers or woven fibres susceptible to the mechanical failure through the formation and propagation of microcracks within the polymeric matrix, interfacial debonding, presence of voids, delamination, fibre breakage, among others. Conductive carbon nanoparticles, including carbon nanotubes (CNTs), graphite and its derivatives, have raised much enthusiasm in scientific and technological domains especially owing to their mechanical and electrical properties. Percolating nanotube network surrounding carbon fibres is expected to act as a sensor for detecting structural damages during cyclic loading deformation. However, the key challenge remains to achieve an appropriate dispersion during compounding and processing of composites, which strongly influences the electrical percolation threshold and the interface between the nanoparticle and the polymer. In this work, nanocomposites based on epoxy resin and containing different amounts of CNTs were prepared using a three-roll mill, at room temperature. The dispersion degree attained during compounding was adjusted by changing the gap width between rolls and rotational speed of the rolls. At each gap selected, the material was forced to pass through the feed roll at least five times. The influence of the compounding conditions (shear rate and residence time) on morphology, rheological properties and electrical conductivity of nanocomposites was investigated. Furthermore, pre-impregnated materials based on CFRP/CNT composite were prepared using a laboratorial drumwinder, under optimal processing conditions. Laminates produced in autoclave (at 120 °C, 3.5 bar for 2 hours) were characterized by mechanical experiments, with emphasis on intralaminar shear strength (ILSS), and the resulting improvements can be attributed to the enhanced dispersion and stronger interfaces promoted by covalent functionalization.

Carbon felt reinforced thermoplastic composite bipolar plates for PEMFC

Lee Ha Eun (1), Song Seung A (1), Kim Seong Su (2)*

(1) Chonbuk National University - Jeollabuk-do - Korea, (2) KAIST - Daejeon - Korea

Polymer electrolyte membrane fuel cells (PEMFCs) have high efficiency, low operating temperature and nearly zero environmental emissions. The bipolar plate is one of the main components in the PEMFC because it accounts for a great part in weight, volume and cost. Recently, carbon fiber reinforced thermoplastic composite bipolar plates for PEMFCs have been attracted owing to their simple fabrication method, high flexibility, excellent corrosion resistance, and damage resistance. The efficiency of the PEMFCs predominantly depends on the performances of the bipolar plates, however, the composite bipolar plates have high surface electrical contact resistance because of their resin-rich area. In this study, a carbon felt was used to fabricate the bipolar plate for PEMFCs. Also, sol-gel method and the double percolation concept were introduced to improve the impregnation state of the thermoplastic polymer and electrical conductivity of the composite plates respectively. The areal specific resistance, double percolation morphology, and impregnation state of the newly developed composite bipolar plates were measured by using four-point probe method, transmission electron microscopy analysis and scanning electron microscopy analysis respectively. The results of the experiments were compared with the Department of Energy's (DOE's) technical target value.

Material development and processing techniques for highly filled plastic compounds

M. Grundler[1, a)], M.Gillmann[1], P. Beckhaus[1] and A. Heinzel[1]

[1]*Zentrum für BrennstoffzellenTechnik GmbH,*
Carl-Benz-Str, 201, 47057 Duisburg, Germany

[a)]*m.grundler@zbt-duisburg.de*

Abstract. Beside the established use of reinforced plastics regarding lightweight designs, functionalized plastics in general are claiming more and more novel applications and do function as a metal replacement. Especially the so called "green technologies" like fuel cells or vanadium redox flow batteries require highly engineered plastics for many new needs e.g. a high level of thermal or electrical conductivity. To functionalize an intrinsic insolating plastic, high loadings of conductive fillers, like graphite and/or carbon blacks are necessary. The mixtures, the so called compounds, expose a significantly different behavior in processing and in their properties compared to the neat polymer. Thus the focus of this paper forms the development of electrically conductive thermoplastic compounds in the field of bipolar plates for vanadium redox flow batteries stacks which are processed by an innovative sheet extrusion processing technique for highly filled plastics.

INTRODUCTION

Bipolar plates (BPP) are an essential component of a vanadium redox flow battery (VRFB) stack guaranteeing electrical conductivity, physical separation of adjacent cells and distribution of electrolyte. Because of the required chemical resistance against corrosive electrolytes they are usually made of carbon based fillers like graphite and/or carbon black and polypropylene as binder material [1,2]. In order to achieve the necessary conductivity a high content (up to 75 vol%) of conductive filler material is needed. Such mixtures of a polymer and one or more fillers is called compound. The polymer functions as a binding matrix, which is mixed with the filler particles in a twin screw extrusion process. In a second step bipolar plates can be processed by conventional polymer processing methods like compression or injection molding [3]. But both methods are limited in the maximum producible size of bipolar plate because of among others high melt viscosity. Bipolar plates for new large-scale vanadium redox flow batteries with a single cell area of more than 2 m^2 cannot be produced by those techniques.

In series of investigations ZBT as part of a governmental funded project (BMWI; 03ET6050E) is actually developing the novel realization of bipolar plates with a size of approximately three square meters by an industrial and cost effective - which means continuous – sheet extrusion process. Compounds consisting of almost 90 wt% fillers and only 10 wt% flowable polymer require very special machining equipment, an exceptional understanding of the complete process chain and a very good material knowledge. Meanwhile, first approaches have led to quasi endless bipolar plates in 400 mm width, the results of electrical conductivity and good processability will be presented in the present paper.

ELECTRICAL CONDUCTIVITY AND RHEHOLOGY OF HIGHLY FILLED COMPOUNDS

Electrical properties of compounds change significantly by increase of filler amount. Figure 1a shows schematically the formation of conductive paths by conductive filler particles (coloured points) in an isolating polymer (white fields). By adding filler particles into the isolating polymer first contacting filler paths occur (1) but resistivity still stays nearly constant. Addition of more and more filler particles leads to a decrease of the resistivity by several dimensions (2). A further raise of filler content causes an additional small improval of resistivity (3) [3]. In general, compounds with a filling load in the range of 65 vol% – 75 vol% are expected to achieve an adequate conductivity for VRFB application.

With an increase of filler amount melt flowability decreases significantly. Compared to an unfilled polypropylene melt viscosity of a highly filled compound measured by capillary rheometer is many orders of magnitude higher (fig.1b).

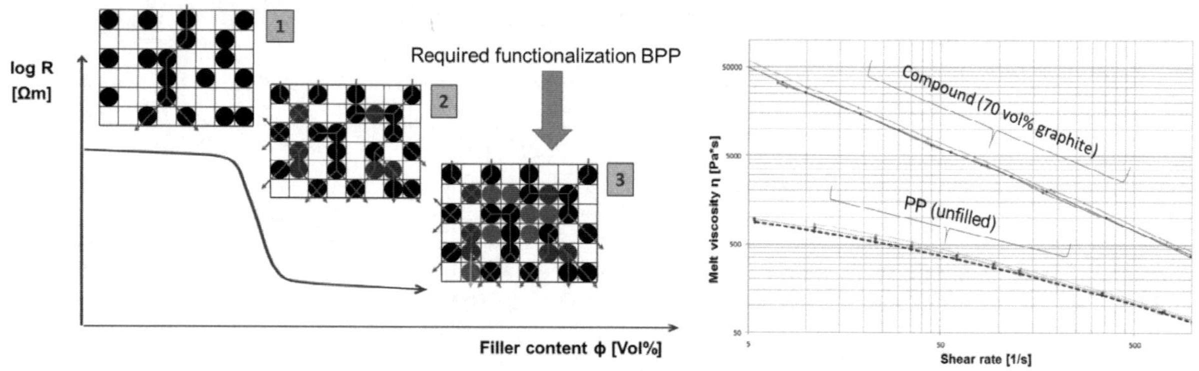

(a) (b)

FIGURE 1. (a) Electrical resistivity depending on filler content (reproduced from [3]), (b) Comparison of melt viscosity

SAMPLE PREPARATION

Production of electrically conductive plates took place in a sheet extrusion line (fig.2) consisting of a twin screw extruder (PTW25, Thermo Scientific), melt pump and a new innovative extrusion tool for highly filled compounds. To achieve a reproducible characterization of different compound materials all processing parameters were kept constant while production of plates with 7 mm thickness and 400 mm width. For each material formulation a plate of two meter length was produced.

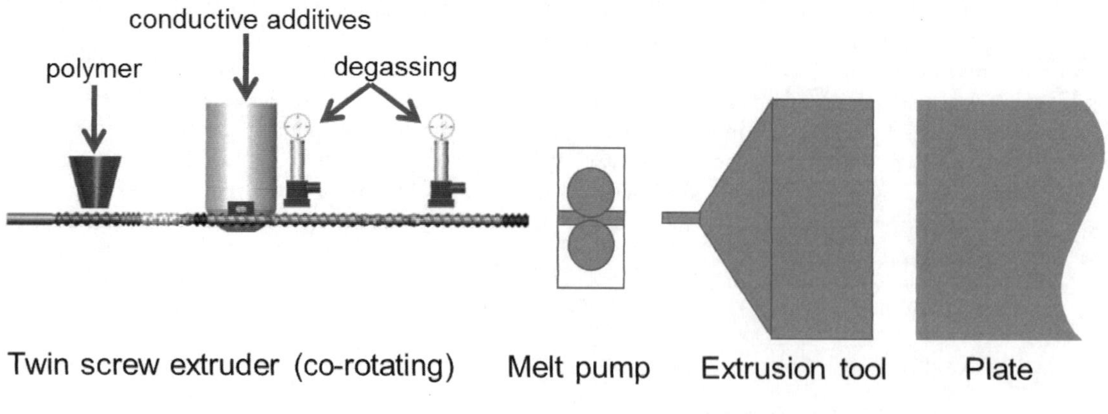

FIGURE 2. Extrusion line for highly filled bipolar plates (schematic representation)

Figure 3a shows the produced plate at the end of the extrusion tool. With an increased filler content production of homogeniously filled plates gets much more difficult and by filler contents > 75 vol% defects at the edges of the plate could be observed (fig. 3b). In a following step samples in dimension 400 mm x 50 mm were sawed out and milled on both sides to a plate thickness of 5 mm (fig 3c). The influence of different material formulations on the surface of a plate after the milling process can be seen in figure 3d.

(a) (b) (c) (d)

FIGURE 3. (a) Plate after extrusion tool, (b) good plate (top) and plate with defects at edge, (c) milled specimen for characterization, (d) surface after milling with different compound formulations

EXPERIMENTAL

Sheet extrusion of highly filled bipolar plates

For the first study four types of graphite with different particle size distribution (PSD) were used. All four graphites had a spherical morphology and a purity of > 99,5 % carbon. A method to describe the PSD of a graphite is to use D values. The D10, D50 and D90 are used to represent the midpoint and range of the particle sizes measured by sieving of a given graphite sample. After sieving a graphite an S-curve is created of cumulative mass retained against sieve mesh size, and calculating the intercepts for 10%, 50% and 90% mass [4]. In the study graphite with a D90 of 150 µm, 200 µm, 800 µm and 1250 µm were used. The PSD of a material can have an important effect on the characteristics of a bipolar plate.

In a first step the amount of filler was increased from 65 vol% to 75 vol%. With increasing filler level a rising trend of pressure at the inlet of the extrusion could be observed (Fig. 4a). A suitable indicator to describe the melt-viscosity of the compounds is the measured extrusion tool pressure during extrusion process. The higher the filling load the higher was the required pressure. Because of safety reasons and the possibility of damages of the extrusion toll a maximum pressure at extrusion tool inlet was set at 100 bar. That leads to the assumption that with increased filler load the viscosity of the compound rises significantly which in turn leads to a very challenging process.

In a second step volume resistivity was measured with a four pole measurement apparatus. The milled specimen (fig. 3b) were placed between two conductive sheet materials (gas diffusion layer) which are fixed on gold plated stainless steel stamps integrated in a hydraulic press. At a pressure of 2 Mpa a current of 500, 1000, 1500 and 2000 mA (Wenking HP 96-20) was applied and the resulting voltage recorded. The measured volume resistances were compared in figure 4b. It has been found, that with increasing filler content the volume resistivity decreases immensely while the process also becomes more difficult. With filler content of around 68 vol% the reference value of 0,055 Ohm*cm can be achieved. The combination of graphite and carbon black (CB) shows significantly lower resistivity compared to only graphite filled plates at same filler level. No clear trend of size of particles on extrusion tool pressure or resistivity of extruded plates could be found. Another essential point for extruded bipolar plates is the quality of surface after the milling process. As shown in fig. 3d plates with larger graphite particles have a rougher surface and show also defects after milling. Based on these results and within the focus of a good processability and low resistivity graphite G150 was chosen for further investigations.

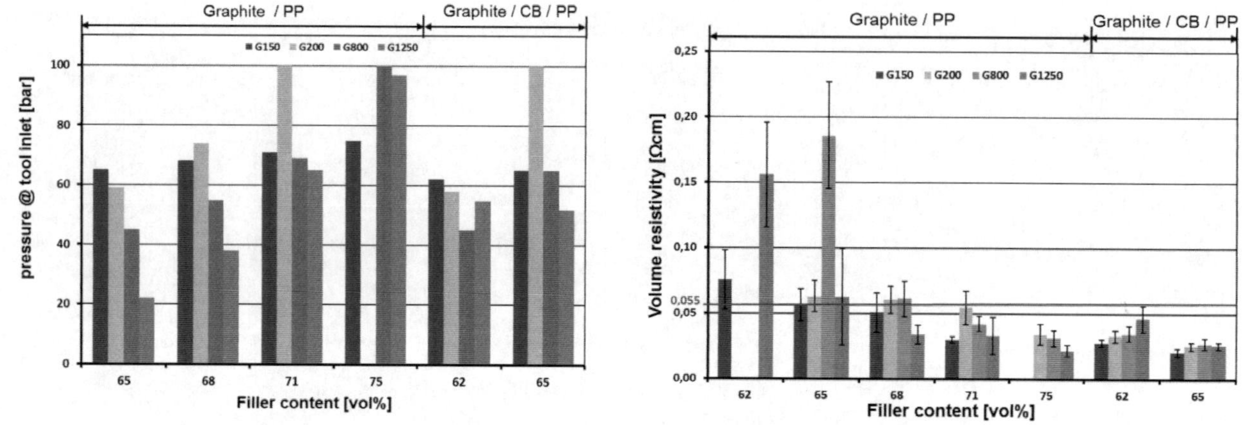

(a) (b)

FIGURE 4. (a) Pressure at extrusion tool inlet, (b) Volume resistivity of extruded bipolar plate

To improve performance of extruded plates in a second study the influence of polypropylene matrix was investigated. Figure 5a shows a table with the investigated kinds of polypropylene and their properties. Main difference is the melt flow rate which might have an effect on the plate performance after extrusion process. All material formulations were compounded and plates extruded under identical process conditions, with equal filling loads and graphite G150 as filler.

After plate extrusion and milling of specimen the volume resistivity was measured (fig. 5b). With increase of filler content again resistivity decreases for all types of polypropylene and reference resistivity value is achieved by around 68 vol%. Within this investigation no clear trend of influence of melt flow rate on volume resistivity could be observed. Because of lower resistivity at same filler content PP25 and PP100 were chosen for following extrusion tests.

Notation	PP 0,95	PP 1,8	PP 6,5	PP 25	PP 100
Melt flow rate (MFR) [g/10min] (230°C/2,16kg)	0,95	1,8	6,5	25	100
Tensile strength [MPa]	26	30	26	33	30
Strain [%]	> 500	> 50	> 500	> 500	10
Vicat softening temperature [°C] (A50 (50°C/h 10N))	150	130	144	153	154

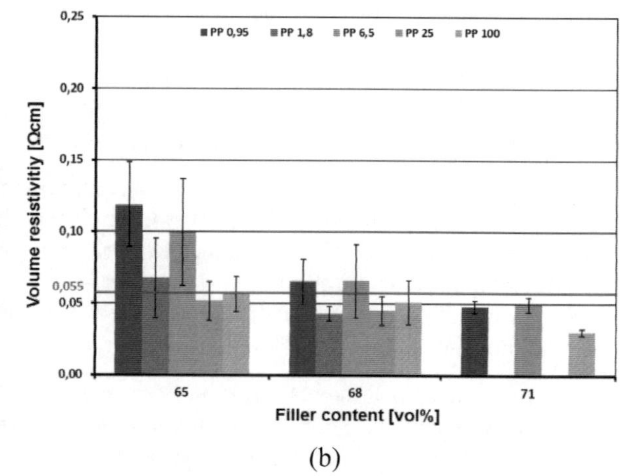

(a) (b)

FIGURE 5. (a) Tested polymers with technical properties, (b) Volume resistivity of extruded bipolar plate

CONCLUSION

The experiments have shown the potential of this novel processing technique for highly filled, conductive bipolar plates by a sheet extrusion process. In order to achieve a good processability and high electrical conductivities in extruded bipolar plates a high load of conductive fillers has to be incorporated. The reference value for electrical conductivity could be achieved in extruded plates with pure graphite and with a combination of graphite and carbon black in different types of polypropylene. Bipolar plate dimensions of 400 mm width and endless length could be successful produced what is much larger than today's commercially available injection or compression molded bipolar plates. A good surface quality after milling process could be realized with smaller graphite particles in the compound formulation.

ACKNOWLEDGMENTS

This work has been funded by the German „Bundesministerium für Wirtschaft und Energie“ through grant 03ET6050E.

REFERENCES

1. C. Minke, T. Hickmann, A. dos Santos, U. Kunz, T. Turek; *Cost and performance prospects for composite bipolar plates in fuel cells and redox flow batteries,* Journal of Power Sources 305 (2016), pp.182-190.
2. B. Satola, L. Komsiyska, G. Wittstock; *Bulk Aging of Graphite-Polypropylene Current Collectors Induced by Electrochemical Cycling in the Positive Electrolyte of Vanadium Redox Flow Batteries*, Journal of The Electrochemical Society, 164 (12) A2566-A2572 (2017).
3. T. Derieth; *Erweiterte Betrachtung der Perkolationstheorie an elektrisch leitfähigen Compounds für Bipolarplatten in PEM-Brennstoffzellen*, Phd thesis Universität Duisburg-Essen, 2013.
4. I. Klarica; *A Guide to D-values in Pharmaceutical Particle Characterisation,* https://www.innopharmalabs.com/tech/applications-and-processes/particle-size-distribution (August 2017)

Preparation, structure and properties of poly(vinyl alcohol)/shell powder composites

Chen Ning [(1)]*, Li Li [(1)], Wang Qi [(1)]

[(1)] State Key Lab of Polymer Materials Engineering, Polymer Research Institute, Sichuan University - Sichuan - China

Shells are a typical kind of organic/inorganic hybrid materials and has a special crossover layered structure, resulting in its better strength and toughness than calcium carbonate filler. In China, more than 70% marine culture production is shellfish, which exceeded 12 million tons. As a result, large amounts of shellfish shells are discarded around the coastal areas without efficient use, which not only wastes valuable renewable marine resources, but also leads to serious environmental pollution. Therefore, efficiently and environmental friendly recycling of discarded shells became a serious problem, but also a major challenge. In this paper, combining molecular complexation and plasticization technology with solid state shear milling (S3M) technology, the thermal processing of high-performance and biodegradable PVA/shell powder composites were prepared. The interactions between PVA and shell powder and their impacts on the structure and properties of PVA/shell powder composites were studied. The results showed that through molecular complexation and plasticization, PVA/shell powder composites had wide thermal processing window, which was 81℃ when shell powder content was 70wt%. PVA/shell powder composite was pseudoplastic fluid, and its viscosity met the need of traditional extrusion or injection molding. Shell powder were well dispersed in PVA matrix and the diameter could be decreased to 0.5μm due to the strong pulverization, dispersion, mixing and activation effects of S3M technology, resulting in better tensile strength than conventional thermal-processing method. Meanwhile, better interaction between shell powder and PVA further increased the thermal decomposition temperature of PVA. The shell powder reduced the number of free hydroxyl groups of PVA and improved water resistance of the composites. A small amount of organic matter in shell powder enhanced its interaction effect with PVA and the special organic/inorganic hybrid structure endowed PVA/shell powder composites with higher mechanical strength than PVA/CaCO3 composites. This work was supported by NSFC(51433006) and International Science & Technology Cooperation Program of China(2013DFG52300).

STUDY OF AMORPHOUS OXYNITRIDE SILICON COATING ON CF/PPS COMPOSITES

Cintra Maria Paula [(1)], Santos Alberto Lima [(1)*], Silva Patrícia [(2)], Ueda Mario [(3)], Lauke Bernd [(4)], Botelho Edson [(1)]

[(1)] School of Engineering, Universidade Estadual Paulista - UNESP - São Paulo - Brazil, [(2)] Associated Sensors Laboratory, INPE - São Paulo - Brazil, [(3)] Associated Plasma Laboratory, INPE - São Paulo - Brazil, [(4)] Leibniz Institute of Polymer Research - Saxony - Germany

This work deals with the processing and surface treatment of thermoplastic composites carbon fibers/Polyphenylene sulfide (CF/PPS) for aerospace applications. After the production of the thermoplastic composites trough hot compression molding, a silicon deposition was carried out on its surface using Plasma Enhanced Chemical Vapor Deposition (PECVD). Then, the Plasma Immersion Ion Implantation (PIII) process was carried out on the silicon coated samples to create an amorphous silicon nitride layer. Subsequently, characterization of the coated and uncoated samples was carried out by means of the following techniques: optical microscopy, atomic force microscopy, X-ray diffraction and X-ray excited photoelectron spectroscopy. By analyzing all the results obtained from the characterizations, it was concluded that it was generated a layer of amorphous silicon oxynitride on the CF/PPS composite without damaging it. This result is an innovative step concerning the application of this material in a space environment (low orbit).

EFFECT OF WATER ABSORPTION AND ACCELERATED AGEING ON MECHANICAL BEHAVIOR OF PLA BASED LAMINATES REINFORCED WITH WOVEN JUTE FIBERS

Russo Pietro (1)*, Abbate Mario (2), Simeoli Giorgio (3)

(1) IPCB-CNR - Pozzuoli Naples - Italy, (2) National Research Council of Italy - Napoli - Italy, (3) CRdC Tecnologie scarl - Naples - Italy

In the last decades, biocomposites involving biodegradable matrices and/or natural fibers have received a lot of attention to alleviate the large environmental impact related to the disposal of conventional fiber reinforced polymers widely used for several industrial applications. However, the intrinsic hygroscopic nature of the main constituents may adversely affect the service life of biocomposites. Indeed, the inevitable moisture absorption of the polymer matrix and/or of the included natural fibers gives rise to microstructural changes of the interphase resulting in formation of cracks and occurrence of debonding/delamination phenomena with worsening of the mechanical behavior and premature failure of manufacts. [1,2] In this work, 3.5 mm thick laminates [(0/90)4]s were prepared by alternating films of neat PLA with layers of a commercial jute fabric, both predried in a vacuum oven at 70 °C for 4 h, using the film stacking and compression molding techniques. Specimens of adequate shape and sizes for subsequent flexural and dynamic-mechanical tests were first dried in a vacuum oven at 50 °C until a constant weight is attained and, then, immersed in distilled water at constant temperatures of 30 °C and 70 °C. Weight gain results, collected at specific intervals of time, increasing as the sorption process gets closer to the equilibrium, were always averaged on at least three determinations conducted on as many conditioned specimens. Furthermore, the conditioning by an accelerated ageing protocol in a lab weathering camera set at 70 °C and 50% RH was also considered to verify the potential use of investigated laminates for interiors in automotive applications. Effects of water absorption and hygrothermal aging, periodically monitored by static and dynamic flexural tests performed in three-point bending mode, were supported by SEM morphological observation of crio-fractured sections highlighting structural modifications occurring at pre-defined conditioning time.

Phenoxy nanocomposite carriers for delivery of nanofillers in epoxy matrix for resin transfer molding (RTM)-manufactured composites

***van velthem Pascal* [1]*, *Ballout Wael* [1], *Pardoen Thomas* [1], *Bailly Christian* [1]**

[1] UCL - Brabant Wallon - Belgium

The use of phenoxy nanocomposite films as carriers of nanofillers involving multiwalled carbon nanotubes and nanoclays is successfully demonstrated for application in epoxy carbon fibers reinforced composites (CFRC) processed by RTM. Model studies on individual nanocomposite filaments embedded in epoxy precursors show that the nanofillers are passively transported by the interdiffusion gradient during heating over distance around 800 micrometers. A morphology gradient is generated after reaction induced phase separation and the nanofillers end up in the epoxy, despite their initial dispersion in the phenoxy. The proof of concept is extended to CFRC panels where nanocomposite phenoxy films are prepositioned between every odd carbon layer of the preform. Carbon nanotubes are filtered by the carbon fabrics, which limits their full diffusion and that of phenoxy through the preform. This has negative consequences on fracture toughness (GIc). For nanoclay, GIc is rather slightly improved although the origin is not fully clear.

ANALAYTICAL STUDY OF DISPERSION AND PROCESSING PARAMETERS ON MICROSCOPICALLY EVALUATED COMPOUNDED POLYCARBONATE

Alsadi Jamal [(1)*]

[(1)] Higher Colleges of Technology. HCT-ADMC - Abu-Dhabi - UAE

The objective of this paper is to evaluate the influence of variation processing parameters on the dispersion quality of polycarbonate compound. The influences of parameters, pigment size distribution, and morphology on the pigment dispersion were also studied. The information on pigment dispersion were obtained using digital optical microscopy (DOM), scanning electron microscopy (SEM), particle size analysis (PSA), and micro ct scanner -image analysis. The optimum objective of this study is to evaluate pigment dispersion in PCs composition, by understanding the relationship between the processing conditions, particle size distribution, and the colour difference output. This will help us in understanding additional dispersion issues that have a bearing on colour variations for different polycarbonate (PCs) compound grades.

SILARYLENE-CONTAINING POLY(AZINE)S AND POLY(AZOMETHINE)S. SYNTHESIS AND CHARACTERIZATION.

Terraza Claudio A. **(1)**, **_Tagle Luis H._** **(1)**, **_Tundidor-Camba Alain_** **(1)**, **_González-Henríquez Carmen M._** **(2)***, **_Sobarzo Patricio A._** **(1)**, **_Hauyon Rene A._** **(1)**, **_González Alexis F._** **(1)**, **_Ortiz Pablo A._** **(1)**

(1) RLOP, Organic Chemistry Department, Pontificia Universidad Católica de Chile - Santiago - Chile, (2) LNnMA, Chemistry Department, Universidad Tecnológica Metropolitana - Santiago - Chile

Recently, we have prepared aromatic dialdehydes and diamines containing a diphenylsilane unit as central element of their structure1. These monomers were adequately purified and characterized by spectroscopic techniques (NMR, IR) and elemental analysis. Later, using solution polycondensation conditions we synthesized two series of silylated poly(azomethine)s and poly(azine)s, which were also spectroscopically characterized. It is very common that the homologue polymers without silicon show low processability, with problems the solubility and elevated Tg values. In this sense, the presence of silicon atoms in the chains promote flexibility due to the sp3 hibridation of the heteroatom and the higher distance C-Si in relation to the C-C bond2-3. Likewise, the chains show a certain degree of polarity due to the difference of electronegativity between the Si and C atoms. These effects increase the solubility in organic polar solvent and decrease the Tg, maintaining the high decomposition thermal temperature (TDT). On the other hand, certain structural modifications developed on the structure monomers also are used to promote best processability. Thus, several research groups have incorporated asymmetric effects, flexibilizing units as oxyether unit1, and lateral groups of varied volume. In this work, we presented the synthetic results and some properties of poly(azine)s and poly(azomethine)s containing diphenylsilane units in their structure. Keywords: poly(azomethine)s, poly(azine)s, silarylene
References: 1. Claudio A. Terraza, Luis. H. Tagle, Alain Tundidor-Camba, Carmen M. González-Henríquez, Mauricio A. Sarabia-Vallejos and Deysma Coll. Synthesis and characterization of aromatic poly(ether-imide)s based on bis(4-(3,4-dicarboxyphenoxy)phenyl)-R,R-silane anhydrides (R = Me, Ph) – spontaneous formation of surface micropores from THF solutions. RSC Adv., 2016, 6, 49335–49347. DOI: 10.1039/c6ra00188b. 2. Bruma and B. Schulz. Silicon-containing aromatic polymers. J. Macromol. Sci., Polym. Rev., 2001,41, 1–40. 3. A. Terraza, L. H. Tagle, C. Contador, A. Tundidor-Camba and C. M. González-Henríquez. Silarylene-containing poly(ether-amide)s obtained by Yamazaki–Higashi phosphorylation technique: use of bis(4-(4-aminophenoxy)phenyl)ethylmethyl silane. Polym. Bull., 2014, 71,1101–1115. DOI: 10.1007/S00289-014-113-6.

Attaining good EMI Shielding with Reinforced Thermoplastic Compounds

Martins Luís C. [(1)], Pinto Eduardo B. [(1)], Pontes António J. [(1)]*

[(1)] Institute for Polymers and Composites I3N, University of Minho - Guimarães - Portugal

Engineering thermoplastics with electrical conductive properties have emerged significantly into the market of housings for electronic devices (e.g. computers, HMI or multimedia systems). These plastic composites have interesting Electromagnetic Shielding (EMI SE) properties and allow weight reduction, better handling, improved corrosion resistance, more design freedom, short production cycle and the adaptability to product requirements, which makes them an attractive alternative to traditional solutions using metallic sheets or hybrid products based on coating technologies or embedded thin metal foils. The use of reinforced thermoplastics compounds with electromagnetic shielding characteristics provided by electrically conductive fibers (Carbon or Stainless Steel) was investigated as an alternative to metal shielded housings for electronic devices. This paper exhibits experimental and numerical proceedings for shielding effectiveness analyses and product design pre-validation.

pps proceeding
Symposium: S10 - Composites
Poster Presentation

Anisotropic Composite Foams with Tailored Structural and Functional Responses

Sorrentino Luigi (1)*, D'Auria Marco (1), Pantani Roberto (2), Davino Daniele (3)

(1) Istituto per i Polimeri, Compositi e Biomateriali, Consiglio Nazionale delle Ricerche - Portici - Italy, (2) Dipartimento di Ingegneria Industriale, Università di Salerno - Salerno - Italy, (3) Dipartimento di Ingegneria, Università degli Studi del Sannio - Benevento - Italy

Foams reinforced with aligned particles are recently gaining increasing interest, confirming the potential of polymeric magneto-sensitive foams. Composite foams with anisotropic mechanical response were produced by assembling the reinforcement in anisotropic arrangements. Two polymeric matrices were used (a flexible and a rigid polyurethane) to show how the proposed approach can be effective on different systems. Magnetic micro-particles were embedded into the foam and their arrangement was induced by means of the application of a magnetic field (MF) during the foaming process. The particles aligned along the magnetic field lines, forming reinforcing chain-like structures. The presence of aligned micro-particles imparted an anisotropic mechanical and functional behavior along the particle alignment direction, thus strongly improving both stiffness and strength of the foam compared to randomly filled or unfilled systems. Functional properties, such as thermal conductivity and acoustic impedance were affected. Furthermore, the structural response of anisotropic composite foams prepared with a flexible matrix showed to be sensitive to the application of a magnetic field, thus exhibiting an active (smart) behaviour. Anisotropic composite foams not only showed relevant increments of both Young's modulus and compressive strength (up to 1000% with respect to samples with not aligned reinforcing particles) but also a potential active behavior due to mechanical relaxations induced by the application of a MF. In this paper we present lightweight composites based on PU matrix and magneto-sensitive micro-particles. These low density materials are characterized by both compressive and functional properties tunable with a MF during the foaming process and peculiar magneto-elastic behaviors.

Calculation of an effective viscosity to predict the flow behavior of randomly oriented discontinuous long-fiber reinforced thermoplastics (LFTs) in compression molding process

bhakta arka (1)*

(1) PhD Student University of Twente - Enschede Overijsel - Netherlands

Discontinuous Sheet Molding Compounds(thermoset SMCs) are being used in the aerospace industry for quite long. Discontinuous long fiber composites are easier to process than continuous fiber composites without much compromise on performance. They also offer a significant versatility from the ability to mold complex shapes in one shot. The interest here is on thermoplastics (TP) which generally have higher processing viscosity than thermosets. This highly viscous behavior with the densely packed fibers (especially for aerospace applications) affects the flow during processing. However, they are also favorable due to their recyclability, formability and weldability. This research attempts to smoothen the way for LFTs to enter the aerospace market by making advances in making their process modeling more predictable. Discontinuous long fiber reinforced thermoplastics generally come at a fixed length ranging from 3 mm to 48 mm as per process' needs. They are manually filled into a mold of known geometry and size before being taken through the compression molding consolidation cycle, where the material is treated with a synchronous (t) pressure (P) temperature (T) cycle before the part is demolded. Till now there have been few studies on processing variables of pressure, squeeze time and temperature during the compression molding process. Globally there is a knowledge gap in the processing analysis of these materials. This paper discusses a method to calculate the effective viscosity term of these materials to be later used in FEM modeling.Pressing experiments were conducted to analyze the flow depending on the duration of pressing. A square mold of 250 x 250 mm2 is used. Flow is restricted in one direction and the relative distance of the flow from the initial to its flowed position is measured. The pressing distances versus the duration are used to calculate the viscosity of the composite from existing squeeze flow equations, which is higher than pure polymer viscosity.

Microwave-assisted pultrusion process (MAP)

R. Emmerich[a], M. Graf[a], R. Wendel[a] and M. Reichmann[b]

[a] *Fraunhofer Institute Chemical Technology, Joseph-von-Fraunhofer-Str.7, D-76327 Pfinztal*

[b] *MUEGGE GMBH, Hochstrasse 4 – 6, D 64385 Reichelsheim*

Abstract. Pultrusion is a continuous and cost-effective process for the production of composite structural components with a constant cross-sectional area. This technological process could be made more effective by replacing conventional heaters by a high-frequency electromagnetic energy source characterized by fast, instantaneous, non-contact and volumetric heating. The state of the art in MAP processing is to replace the metal die at least partly by a microwave-transparent die or insert mostly made of ceramic, which is the main drawback and the main barrier to industrial uptake. This study presents a microwave module for accelerating the pultrusion process for (in principle) all tube geometries and shapes, without any use of microwave-transparent elements or inserts in the die. The module can be integrated into the pultrusion machine without any major modification of the die. It is also possible to retrofit existing pultrusion dies.

Keywords: Microwave, pultrusion
PACS: 81.05.Lg, polymers, 81.05.Qk

INTRODUCTION

Pultrusion is a continuous and cost-effective process for the production of composite structural components with a constant cross-sectional profile (Fig. 1). During pultrusion, the fiber reinforcements are saturated with the resin in a resin tank and then continuously pulled through a heated die by a puller. Inside the die the resin gradually cures and solidifies to form a composite part with the same cross-sectional profile as in the die. At the final stage a traveling cut-off saw cuts the composite profile to the desired length.

The curing time of the resin limits the pultrusion speed, which significantly influences the cost of product. High-frequency electromagnetic energy sources characterized by fast, instantaneous, non-contact and volumetric heating reduce the curing time of the resins and enable higher pultrusion speeds and shorter pultrusion dies. It should be noted that microwave heating [1] is currently successfully and widely used in various industrial curing processes [2, 3].

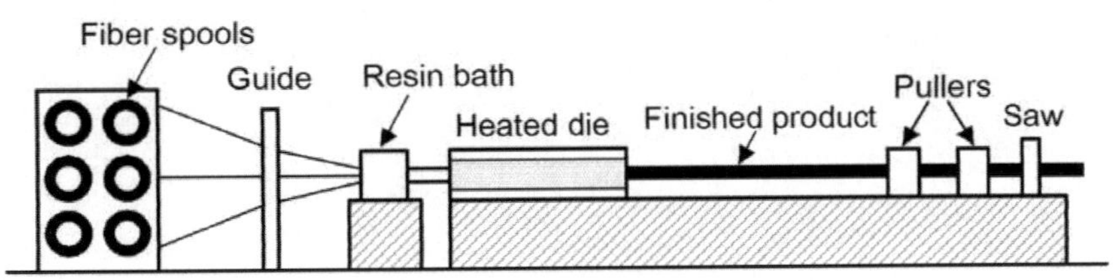

FIGURE 1. Pultrusion process

Microwaves are electromagnetic waves in the frequency range between 300 MHz and 300 GHz. Depending on the material, the applied microwaves are absorbed, transmitted or reflected. Metals in general have high electrical conductivity and consequently the microwaves are reflected. Microwaves heat material very fast, because they deposit contactless energy inside the material, independently of its thermal conductivity. Microwaves can be instantaneously controlled. For industrial application microwaves with a frequency of 915 MHz, 2.45 GHz or 5.8 GHz are used with a wave length of about 32 cm, 12 cm or 5.5 cm respectively.

Metheven et al. [4] described a MAP process in which the metal die was replaced by a microwave-transparent die and heating was applied in a so-called resonant or single mode cavities applicator. This approach has two principle drawbacks: the partial replacement of the metal die with ceramic and the limited possible diameters due to the cavity concept.

MICROWAVE MODULE

The objective was to develop a microwave module for a pultrusion die for tubes of any geometry and shape, for example circular, square or rectangular. The die for tube pultrusion consists of a hollow metal mold containing a metal core. The resin flows between the inner side of the metal mold and the core to fix the shape and geometry of the tube. The new approach is to use the die geometry for transmitting the microwave inside [5]. The microwave energy is directed from the beginning or end of the module into the core. The microwave travels along the core and is distributed inside along the metal mold. The resin there absorbs the microwave energy and is heated. The metal pultrusion die is completely retained and there is no need for any ceramic insert in the mold.

Figure 2 shows a sketch of the microwave module mounted at the end of the pultrusion die. The different colors indicate the different parts and the materials. The microwave module consists of a wave guide (grey) which is connected to the microwave generating part and a transfer module (brown). The length of the core of the die (green) is extended so that it continues into the transfer module, leaving a gap for the tube (blue). The metal mold (purple) ends on the outer side of the wave guide. The pultruded tube is already solid at the end of the pultrusion die and is guided through the transfer module.

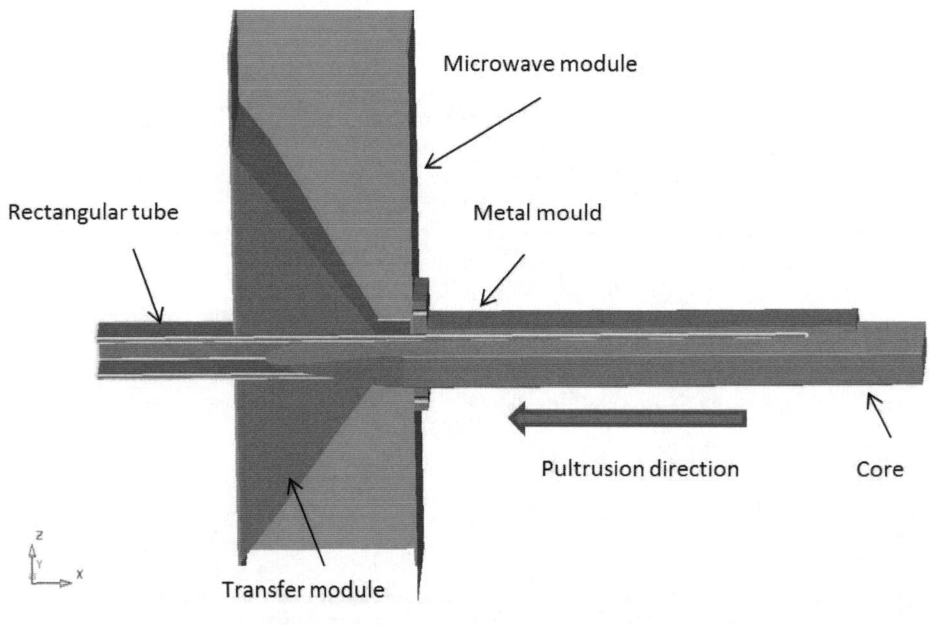

FIGURE 2. Schematic diagram of the electromagnetic model

To design the microwave module the microwave field in the module and in the resin-filled die is simulated with the FEM software package COMSOL Multiphysics. We decided to design the microwave module for a rectangular polyester glass fiber (POLRES 305BV + Unifilo 4800 tex) reinforced tube with the dimensions 80 mm length, 30 mm height and 3 mm wall thickness.

The simulation calculates the spatial electric field distribution, and consequently the dimensions and geometries for the design. Using the common approach of a harmonic oscillating electric field

$$\vec{E}(\vec{r},t) = \vec{E}(\vec{r}) \cdot e^{2\pi i f t}$$

Maxwell's equations could be written as following:

$$\nabla \times \nabla \times \vec{E}(\vec{r}) - \varepsilon_0 \mu_0 (2\pi f)^2 \varepsilon_r \vec{E}(\vec{r}) = \vec{0}$$

This complex valued equation is solved numerically for the amplitudes of the electric field with respect to the relative permittivity, which for loss dielectric materials like mixtures of glass fibers and polyester resin is a complex function of frequency, temperature and degree of curing.

$$\varepsilon_r(f,T,\alpha) = \varepsilon'(f,T,\alpha) - i \cdot \varepsilon''(f,T,\alpha)$$

$\vec{E}(\vec{r},t)$	harmonic oscillating electric field
t	time
$\vec{r}$	location vector
f	microwave frequency
ε_0	vacuum permittivity
μ_0	magnetic constant
ε_r	relative permittivity

The dielectric properties of the glass fiber/polyester resin composition were measured before with a cavity perturbation method [6] versus temperature, and mean values taken at a temperature of T=60 °C were used for the calculation of the electric field distribution.

Running through the optimization loop the geometry of the microwave module was adjusted to achieve a symmetric field distribution inside the die, and a minimized power reflection.

Figure 3 shows the microwave field distribution in top view in the die and the microwave module for the optimized geometry. Along the tube a standing wave pattern is generated, with symmetrically distributed high (red) and low (blue) microwave energy spots. While being pulled through the die the tube passes spots of high and low energy, resulting in a homogenous heating as an average. The field distribution on the side of the tube has the same spatial distribution.

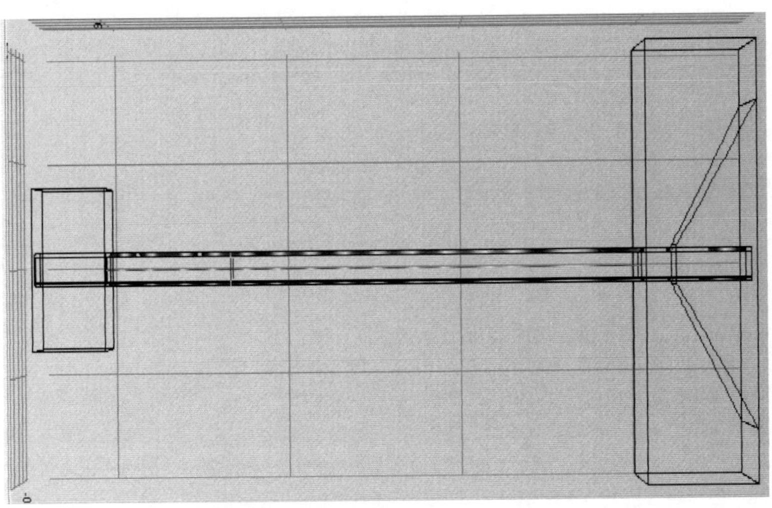

FIGURE 3. Field distribution top-view

DESIGN AND FIRST TEST OF THE MICROWAVE MODULE

On the basis of the numerical simulation results the microwave module was designed and built up. The module is mounted at the end of the pultrusion die (Fig. 4) and initial pultrusion tests are carried out. The tests show that the concept of the microwave module works. The microwave is radiated into the pultrusion die. The resin absorbs the radiation and cures considerably faster. This fast curing of the resin allows an increase of the pultrusion speed without any visible reduction in the quality of the pultruded tubes.

FIGURE 4. Mounted microwave module

CONCLUSION

With the developed microwave module, tubes can be pultruded more efficiently because of the higher possible pultrusion speed. The main feature of the concept is the microwave heating in the metal die without any major modifcations (e.g. no need for ceramic inserts). Further tests will be necessary to fully exploit the potential of the microwave module concept. In particular, it is necessary to determine the resins or polymers for which the microwave module will be most favourable.

ACKNOWLEDGMENTS

This project was funded by the European Union's Seventh Framework Programme (FP7/2007-2013) under Grant Agreement n° [609149]. Start day: September 1st, 2013. Duration: 42 Months. Web: **www.coaline.eu**

REFERENCES

1. A.C. Metaxas, R.J. Meredith, Industrial microwave heating, IEEE Power Engineering Series 4, Peter Peregrinus Limited. London, 1983.
2 E.T. Thostenson, T.-W. Chou, Microwave and conventional curing of thick-section thermoset composite laminates: experiment and simulation. Polymer Composites, 22 (2), 2001, 197-212.
3 L. Liu, S. Yi, L.S. Ong, K.S. Chian, S. Osiyemi, S.H. Lim, F. Su, Chemothermal modeling and finite-element analysis for microwave cure process of underfill in flip-chip packaging. IEEE Transactions on Electronics Packaging Manufacturing, 28
4 Methven,.J.M. et al, Materials Technology (UK). Vol. 14, no. 4, Dec. 1999, pp. 183-186.
5 O. Zinke, H. Brunswig Streifenleitungen in Hochfrequenztechnik 1, Springer Verlag 1995, pp.156-169
6 M. Sucher, J. Fox, *Handbook of microwave measurements, Vol. II, 3rd Edition, Polytechnic Press*. New York, London 1963.

Research of Molecular Orientation on Mechanical Properties of Biodegradable PLA Composites by Injection molding

Lee Chia-Cheng (1), Chen Chao-Chang Arthur (1)*, Cuo Chia-Hsing (1)

(1) National Taiwan University of Science and Technology, Mechanical Engineering Department - Taipei city - Taiwan

This study is to investigate the influence of injection molding process for mechanical properties of poly lactic acid (PLA) composite materials, and to observe the relation between molecular orientation and tensile strength. Hydroxyapatite (HA) is used as a filler material, and silane is selected as a couple agent for improving the interface of PLA and HA. Design of experiment is used to find the most important factors for tensile strength and impact strength, and to compare the Raman intensity ratio with tensile strength of specimens. The differential scanning calorimetry (DSC) and differential thermal analyzer (DTA) are used to measure the thermal properties of PLA and PLA composite materials. Results show that the injection velocity is the most significant factor for the tensile strength of both PLA and PLA composite materials. Tensile strength and Raman intensity ratio have been increased with increasing injection velocity. The packing pressure is the most important parameter for Young's modulus value of both materials. Impact strength of both material have not found significantly for current set-up of experiments. The degree of crystallinity of PLA specimen is increased with longer in-mold annealing time. The mechanical properties and thermal properties of PLA/S-HA are much better than those of PLA/HA. The toughness of PLA composites are lower than that of pure PLA. The mechanical properties and thermal properties of PLA/HA and PLA/S-HA content 5wt% are higher than those of content 10wt%. Result of this study can be applied on bio-medical prosthesis by injection molding of porous PLA composites. Keyword: Bio-degradable, Polylactic Acid, Hydroxyapatite, Mechanical Properties, In-mold Annealing.

Effects of surface modification of glass fiber on the performance of glass fiber reinforced poly(lactic acid) composites

Jing Mengfan (1), Fu Qiang (1)*

(1) Sichuan University - Sichuan - China

Considerable scientific and industrial interest have been attracted to tailor the properties of the environmental friendly polymers, such as poly(lactic acid) (PLA). Glass fiber (GF), as a quite popular reinforcement material for polymer, has not been given enough attention on the reinforcement of PLA. Considering poor wettability and adsorption of GF, silane coupling agents are commonly used to obtain surface modified GF to achieve effective interfacial bonding between fiber and matrix. Recently, a facile method using graphene oxide (GO) as a modifier for the surface treatment of GF has attract a lot of attention. In this work, a comparison study was carried to investigate the effect of the surface treatment on the reinforcement efficacy of GF reinforced PLA by using silane coupling agents (GF-S) and GO (GF-GO) as modifiers. The effect on crystallization behavior of the PLA matrix was conducted by differential scanning calorimetry. The mechanical performances and thermomechanical properties of the composites were evaluated by uniaxial tensile testing and dynamic mechanical analysis, respectively. It is found that the incorporation of GF-S significantly improves mechanical strength while kept relatively good toughness of the composites, while GF-GO exhibits excellent nucleation ability for PLA and could moderately increase the modulus of the composites. The degree of crystallinity, the interfacial interaction and the length of fiber altogether affect the mechanical performance of the samples.

Preparation and properties of polypropylene composites based on synthesized novel flame retardant

Amirova Lyaysan [1]*, Gohs Uwe [2], Müller Michael Thomas [2], Zschech Carsten [2]

[1] Leibniz-Institut für Polymerforschung Dresden e.V. - Dresden - Germany, [2] Leibniz-Institut für Polymerforschung Dresden e. V. - Dresden - Germany

Polypropylene (PP) is widely used in different fields owing to its mechanical properties, chemical resistance and low price. However, PP is a highly flammable material which releases smoke and poisonous gases during a fire because of the aliphatic hydrocarbon structure. So flame retardance is an important requirement for PP. A promising way to improve fire resistance is to use intumescent flame retardants (IFR) which have advantages such as high efficiency and production of low smoke and nontoxic gases, usually composed of three components: an acid source, a carbonization agent, and a blowing agent. In a widely used IFR system, ammonium polyphosphate (APP), pentaerythriol (PER) and melamine (MA) are employed. However, traditional IFR additives exhibit low flame retardant efficiency in PP. One method of obtaining more effective IFR is modification of APP with charring agent. Previously in our research group one-step modification of APP was successfully performed using allylamine. Synthesized allylamine polyphosphate (AAPP) showed remarkable flame retardancy efficiency as well as good crosslinking capability in PP composites. However the chain of APP modificator should be enough long to permit entanglements with the PP in the interphase and to increase mechanical properties of composites. Consequently, the aim of this work was one-step synthesis of novel long chain flame-retardant crosslinker based on APP. Using obtained flame-retardant PP composites have been prepared and for further improvement the thermal stability electron beam irradiation has been used to treat these composites. Possible crosslinking and flame-retardant mechanism of novel PP composites have been studied by thermogravimetric analysis coupled Fourier-transform infrared spectroscopy (TG-FTIR), thermal stability and fire behavior of these composites have been investigated by thermogravimetric analysis (TGA), limiting oxygen index (LOI), vertical burning test (UL-94), and cone calorimeter (CC) tests.

Determination of GF/Epoxy prepreg cure cycle based on 2^3 full factorial design for DSC in dynamic mode

Daniel C. Silveira[a], Michelle L. Costa[a], Messias B. Silva[b,c], Mirabel C. Rezende[d], Edson C. Botelho[a]

[a]*Department of Materials and Technology, São Paulo State University (UNESP), School of Engineering, Guaratinguetá, Brazil.*
[b]*Department of Production, São Paulo State University (UNESP), School of Engineering, Guaratinguetá, Brazil*
[c]*São Paulo University (USP), School of Engineering, Lorena (EEL), Brazil*
[d]*Institute of Science and Technology, Federal University of Sao Paulo (UNIFESP), São José dos Campos, Brazil*

danielsilveira.tech@gmail.com

Abstract. Pre-impregnated materials, such as epoxy resin/glass fiber fabric, have allowed the reduction of processing time of high structural performance composites, with contributions in the aerospace, automotive, construction, among other sectors. In this sense, understand the cure cycle of the material and its optimization becomes essential for structural performance improvement. In this work a 2^3 full factorial design is applied for differential scanning calorimetry (DSC) in dynamic mode. A total of 24 runs, 8 DSC experiments performed in triplicate, determined several temperatures for onset, peak and end of cure cycle during exothermic event. Therefore, low and high level of the factors nitrogen flow (20 and 100 mL/min), sample mass (1.3 and 7.2 mg) and heating rate (2 and 10 °C/min) were applied. Also the degradation temperatures were obtained during experiments and confirmed by means of thermogravimetric analyzes (TGA) in dynamic mode. Based on the onset and peak cure temperatures, where maximum rate of cure kinetics occurs, DSC experiments on isothermal mode were performed considering 3 different temperatures, allowing the precise determination, for each temperature, of required isotherm time for complete cure of the material. Based on these results, an optimized cure cycle were designed and performed on hot compression molding process, applying equal heating rates between thermal analysis and in situ processing of composites. Furthermore, the loss of mass of the material during the entire cure cycle was determined by means of isothermal and dynamic mode TGA, also considering the same parameters applied on hot compression molding process and thermal analysis. Therefore, this research collaborates for the optimization and better control of the curing cycle, with lower isothermal temperatures, consequently collaborating for a slower cure kinetics and greater homogeneity printed on the material, favoring better mechanical performance.

Keywords: GF/Epoxy prepreg, structural composites, 2^3 full factorial design, differential scanning calorimetry (DSC), thermogravimetric analysis (TGA), cure cycle.
PACS: 65, 81.

INTRODUCTION

The demand for productive processes of greater capacity and capability increases each year in the most varied sectors of the industry, including aeronautical, which reinforces the need to use pre-impregnated materials (prepregs). In general, prepregs are compositions of reinforcing fibers and polymer matrix impregnated in mass specific fraction, forming ready-made material for application in molding processes of high resolution parts with complex geometries [1].
The polymer matrices used during the impregnation may be thermoplastic or thermoset, with epoxy thermoset matrices widely used in the preparation of prepregs by means of the combination with glass, carbon, aramid fibers, among others, either in unidirectional reinforcement configurations or in the form of fabrics (woven) [2].
The parameters applied during the processing of the composite material, temperatures, pressures and time involved, control the curing kinetics of the material, determining the maximum rate of formation of crosslinked chemical bonds in the structure of the thermosetting matrix, which influences the mechanical properties of the structural composites obtained [3]. In this sense it becomes fundamental to understand the temperatures involved in the curing cycle of the same, in order to make it mechanically structural [4].
Curing reactions of thermosetting matrix systems and temperatures involved can be understood by thermoanalysis and rheology. Differential Scanning Calorimetry (DSC) can be applied in dynamic mode, allowing the determination of onset, peak and endset curing temperatures of the prepreg material. The times involved in the processing of the material can be determined, for specific temperatures applied during the hot compression molding cycle, by isothermal mode DSC [2, 4].
In this work, dynamic mode DSC were used to determine onset, peak and endset curing temperatures, as well as the degradation start temperature, which was later confirmed by thermogravimetric analysis (TGA). With this objective, 24

DSC evaluations were performed in dynamic mode, divided into 8 experiments performed in triplicate, based on a 2^3 full factorial experimental planning matrix, varying the factors nitrogen flow, sample mass and heating rate. Low and high levels were defined for the factors, respectively, nitrogen flow (20 and 100 mL/min), sample mass (1.3 and 7.2 mg) and heating rate (2 and 10 °C/min), applied during the DSC evaluations.

EXPERIMENTAL

Materials

The prepreg used was HexPly® F155, from Hexcel Co., with an advanced epoxy system of modified formulation and impregnated with plain weave type 1x1 glass fibers. Some of the properties of the epoxy matrix are: good adhesion and coverage for layup and assembly, specific mass of 1.335, glass transition temperature (T_g) of 121 °C, gel time of 4 to 10 min at 121 °C, characteristics of self-extinguishing flames, moisture absorption equilibrium at 9.4% and coefficient of thermal expansion of 5.27 x 10^{-5} cm/cm/°C. Still on the resin, it has fracture strength (K_{IC}) of 1.65 mPa√m, tensile modulus of 3.24 GPa, tensile strength of 80 MPa, specific deformation of 5.2% and is certified according to MIL-R-9300.

The fiberglass fabric of Hexcel is designated as 120-38"-F155, with fibers 450-1/2, aerial weight of 115 g/m^2 and volumetric fraction of fibers of 38% in dry condition. The fabric has tensile strength of 393 MPa, tensile modulus of 20.7 GPa, compressive strength of 444 MPa, all at 24 °C.

Differential exploratory calorimetry (DSC) and 2^3 full factorial planning

The DSC analyzes of the glass fiber/epoxy resin prepreg were performed on a TA Instruments equipment, model Q20, previously calibrated with indium. All analyzes were performed using hermetically sealed aluminum pan and $N_{2(g)}$ atmosphere with controlled flow. Based on the 2^3 full factorial *design of experiments* (DOE), 24 DSC scans were performed in dynamic mode, 8 experiments in triplicate, with low and high levels defined for each factor: nitrogen flow (20 and 100 mL/min), sample mass (1.3 and 7.2 mg) and heating rate (2 and 10 ° C/min).

Thermogravimetric analysis (TGA)

The thermogravimetric analysis were performed in dynamic mode using TGA/DTA 6200 equipment, model EXSTAR6000 from SII Nanotechnology. All analyzes were carried out using platinum pan and alumina as reference material. Based on recommendations of ASTM E2550-11, the oven atmosphere was controlled with $N_{2(g)}$ at 100 mL/min of flow rate, heating rate between 10 and 20 °C/min, specifically 7.64 °C/min equals to heating rate applied during hot compression molding process, and average sample mass equals to 10.40 mg. The program set-up was heating from 25 °C up to 900 °C and 3 min of isothermal step on the last temperature.

RESULTS

Figure 1 shows the dynamic mode DSC curves for each of the 8 experiments (replicas 2). Table 1 indicates the low (-) and high (+) levels of each factor, nitrogen flow, sample mass and heating rate, used in each experiment.

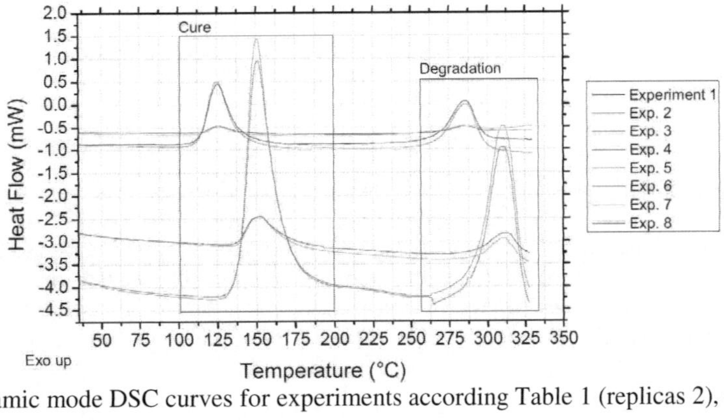

FIGURE 1. Dynamic mode DSC curves for experiments according Table 1 (replicas 2), exothermic events.

Figure 1 presents curves of heat flow (mW) as a function of temperature (°C), the main events, cure and degradation, are highlighted by squares. It is possible to verify the natural occurrence of the curing process with the increase in temperature, with peaks of degradation of the material occurring at higher temperatures, thus being displaced to the right of the figure, as expected.

Also in Table 1, are presented the values for onset cure temperature (Ti), peak cure temperature (Tp) and endset cure temperature (Tf), in addition to the degradation start temperature (Td) of the prepreg material. The 96 temperatures results are presented in specific positions of the planning matrix, allowing reading the respective experiments, replicas and factor levels used in the set-up of each DSC. It is noteworthy that the minimum and maximum values of each temperature involved in the curing cycle of the material are highlighted in red and green, respectively.

TABLE 1. 2^3 full factorial experimental planning matrix for dynamic mode DSC.

Exp.	Factors Level			Ti (°C)	Tp (°C)	Tf (°C)	Td (°C)
	A	B	C				
1	-	-	-	Replica 1 - 105.12 Replica 2 - 105.88 Replica 3 - 106.13	126.10 125.93 126.49	150.60 **148.58** 175.12	254.72 262.04 235.76
2	+	-	-	105.12 106.76 105.92	125.62 126.05 125.76	164.50 150.56 155.62	**222.88** 248.15 241.58
3	-	+	-	100.06 101.83 99.05	**124.67** 125.05 126.06	167.53 169.56 170.57	238.80 243.09 249.16
4	+	+	-	**98.97** 100.57 100.06	125.18 125.43 125.19	178.28 171.58 169.05	247.86 247.39 240.82
5	-	-	+	132.45 132.66 132.87	152.41 152.62 152.82	182.57 179.41 181.94	**283.23** 271.86 268.91
6	+	-	+	**132.90** 130.14 131.40	**153.36** 152.20 152.08	184.00 183.63 179.84	279.67 275.02 277.34
7	-	+	+	125.37 123.56 125.59	149.97 150.55 151.21	201.68 201.69 210.75	263.09 262.72 260.53
8	+	+	+	126.56 127.10 125.50	150.16 150.90 150.60	**223.22** 197.86 199.63	264.91 265.58 261.33

Regarding the initial temperature of degradation (Td), the highest temperature recorded during DSC was 283.2 °C. To confirm the degradation temperature of the material, thermogravimetric analysis (TGA) was performed in dynamic mode. The TGA curve is shown in Figure 2. For aeronautical application, the initial degradation temperature considered in this case is 286.9 °C, point where material presents 1% of mass loss. Also from Figure 2, the degradation of the prepreg occurred in only one main exothermic event, with peak degradation of the polymer structure at 311.2 °C and maximum thermal decomposition rate of 270.5 μg/min (DTG curve peak - blue). The residual mass is 74.3% at temperatures up to 880.8 ° C.

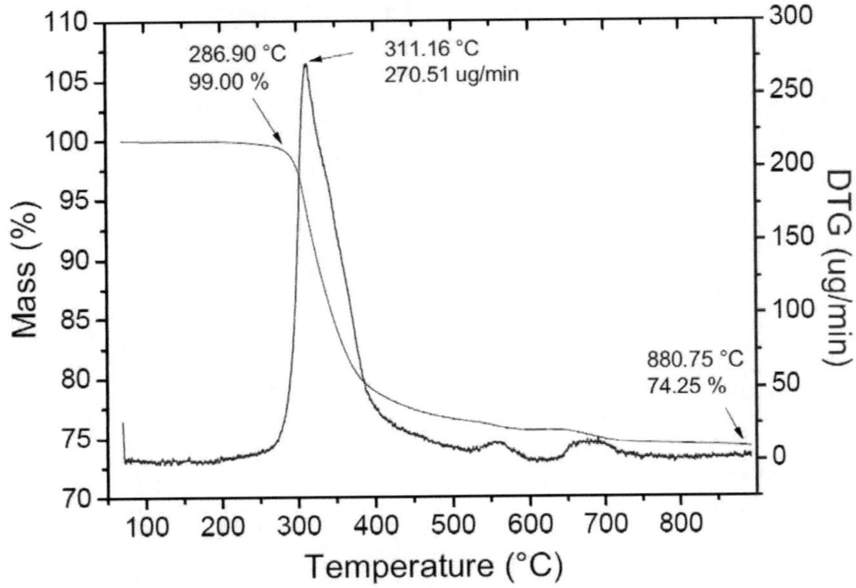

FIGURE 2. Dynamic mode TGA curve of the GF/epoxy resin prepreg in $N_{2(g)}$ atmosphere.

Based on the DSC curves in dynamic mode, three temperatures were chosen, in region of maximum cure kinetics (between Ti and Tp), for the DSC evaluations in isothermal mode, which allowed the determination of time required for the complete curing of the prepreg at each of the chosen temperatures. Figure 3 shows the isothermal curves obtained at the temperatures of 100, 115 and 130 °C. For 115 °C the required time for the total curing of the prepreg (complete exothermic event), considering the real heating rate applied in the hot compression molding press (7.64 °C/min), is 100 min.

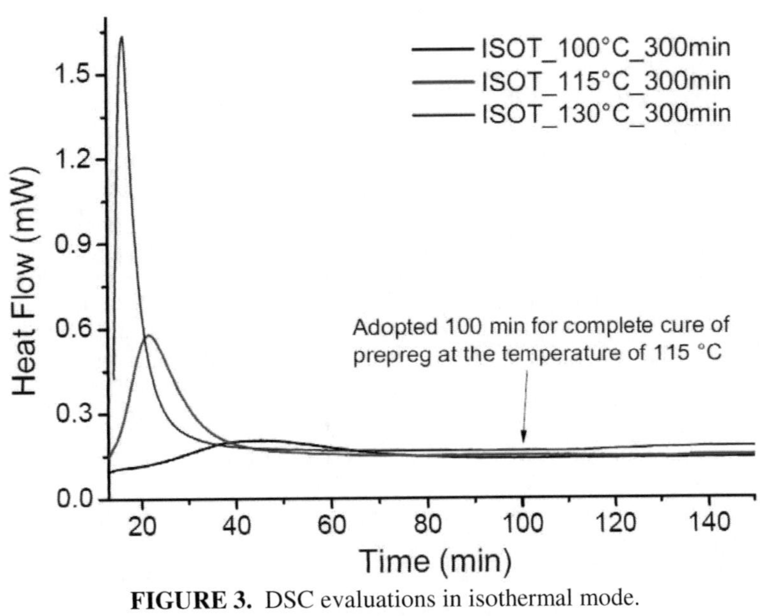

FIGURE 3. DSC evaluations in isothermal mode.

Based on all research exposed above the cure cycle for the GF/epoxy resin prepreg were proposed as presented in Figure 4. The molding process was performed at a heating rate of 7.6 °C/min up to 115 °C with an isothermal at this temperature for 100 min, the necessary parameters for complete exothermic reactions, secondary chemical bonds formation and cross-link of the polymeric matrix material. The heating rate was determined by linear regression based on measurements of temperatures every 1 min during 5 cycles at hot compression molding press, with controlled external temperature of 20 °C. The applied pressure was 0.64 MPa during the entire cure cycle.

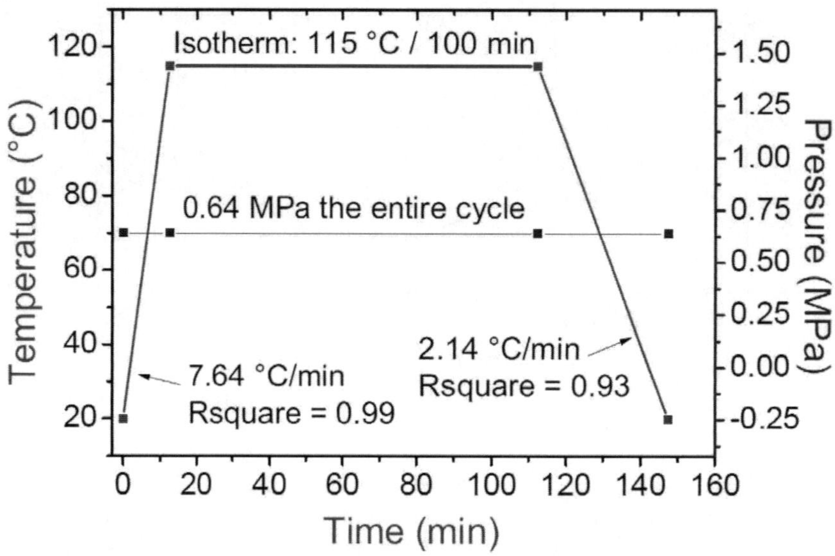

FIGURE 4. Proposed cure cycle for GF plain weave 1x1/epoxy resin matrix prepreg.

CONCLUSIONS

In this work the effect of the parameters applied in DSC analysis in dynamic mode on curing and degradation temperatures of GF/epoxy resin pre-impregnated material was carried out. Based on onset, peak and endset cure temperatures obtained by means of DSC dynamic mode experiments, performed according 2^3 full factorial *Design of Experiments* (DOE) a cure cycle was proposed for the complete cure and consequently secondary chemical bonds formation and cross-link of the polymeric matrix material during hot compression molding process. A total of 24 DSC scans were performed in dynamic mode, 8 experiments in triplicate, with low and high levels defined for each factor: nitrogen flow (20 and 100 mL/min), sample mass (1.3 and 7.2 mg) and heating rate (2 and 10 ° C/min). Degradation temperature were confirmed by means of dynamic mode TGA in nitrogen atmosphere, the initial degradation temperature considered is 286.9 °C, point where material presents 1% of mass loss. DSC isothermal mode at the temperatures of 100, 115 and 130 °C, located at maximum cure kinetics rate, allowed determination of required time for the total cure of the prepreg, considering the real heating rate applied in the hot compression molding press (7.64 °C/min).

ACKNOWLEDGMENTS

The authors would like to thank CNPq, CAPES, ALLTEC Composites and the Thermal Analysis Laboratory (LAT) of São Paulo State University (UNESP/FEG), Brazil.

REFERENCES

1. M. K. Hagnell, M. Åkermo, *Composites Part B: Engineering Vol. 79,* 2015, pp. 254-261.
2. M. L. Costa, J. M. de Paiva, E. C. Botelho, M. C. Rezende, *Polímeros: Ciência e Tecnologia Vol. 12, Iss. 3,* 2003, pp. 188-197.
3. S. Anandan, G. S. Dhaliwal, V. A. Samaranayake, K. Chandrashekhara, *Journal of Applied Polymer Science Vol. 133, Iss. 38,* 2016, pp. 1-7. DOI: 10.1002/app.43984.
4. D. C. Silveira, Masters Dissertation, Guaratinguetá/Brazil: São Paulo State University, 2016. https://repositorio.unesp.br/handle/11449/142858

Toughening of Epoxy Resin for the Development of a Carbon Fiber Based Tow Prepreg

Jeon Jinseok (1), Yun Chang Hyun (1), Park Yongmin (2), Park Nohyun (2), Nah Changwoon (1)*

(1) Chonbuk National University - Jeonju - SouthKorea, (2) EZ Composites Co. Ltd - Jeonju - SouthKorea

Pre-impregnated fiber with thermoset resins like epoxy is widely used for the manufacturing of pressure vessels for various applications. However, the inherent brittle nature of the epoxy resin is one of the limiting factors for using in these kinds of applications. Therefore, toughening of epoxy resin is very essential to make it a more convenient for the development of fiber based tow prepreg. This study focused on the toughening of a bisphenol-A type epoxy resin by two approaches. One is via conventional elastomeric toughening method with the use of a carboxyl-terminated butadiene-acrylonitrile (CTBN) rubber and the other one involves with the use of nano-diamond (ND) as filler. The concentration of CTBN was varied from 1 phr to a maximum of 30 phr and that of ND was varied from 1 phr to a maximum of 15 phr. The curing reaction was carried out using dicyandiamide (DICY) as the curative after proper mixing the epoxy resin with the toughening agents. The toughness of the epoxy resin was then measured using an Izod impact tester. It has been observed that the impact strength initially decreased up to 10 phr of CTBN and then increased as the content of CTBN increased. However, the incorporation of ND does not contribute much to improve the impact strength of the epoxy resin.

Study of the linear coefficient of thermal expansion of high density polyethylene matrix composite materials reinforced with agave bagasse

Pérez Ortega Fernando [(1)], Ruiz Jimenez Axel [(1)], Caballero Caballero Magdaleno [(1)], Chávez Gutiérrez Miguel [(2)]*

[(1)] Instituto Politecnico Nacional - Oaxaca - Mexico, [(2)] CONACYT - Instituto Politecnico Nacional - Oaxaca - Mexico

In recent years, there has been an increase in the use of environmentally friendly materials, which translates into greater environmental awareness among the population and a greater demand for these materials, in particular polymer matrix composite materials, use natural fibers to reduce their environmental impact, their density and improve their specific properties. In the present work, high density polyethylene (HDPE) matrix composites were prepared, which is a highly commercialized polymer. The composites were reinforced with agave fiber of the species Agustifolia Haw, from the bagasse generated by the mezcal production industry. The fiber was treated with a 5% NaOH solution before being introduced into the matrix, for improve the interfacial adhesion. It was tested with three different concentrations of the fiber: 5, 7.5 and 10% by weight. The composites were tested to evaluate their linear coefficient of thermal expansion at 35 ° C, using a thermomechanical analyzer TMA Q400, TA Instruments, according to the ASTM E831-14 standard. Regarding the matrix, a greater linear coefficient of thermal expansion was observed in the composites with 7.5 % of fiber; 7.7x10-2 and 8.3x10-2 µm mm-1 K-1, respectively. When the composite with 5% and 10% fiber content, it presented a behavior lower to that of the matrix, 6.7x10-2 and 7.3x10-2 µm mm-1 K-1, respectively. The increase in the thermal expansion could be attributed to the hydrophilic character of the fiber, which favors the dimensional changes, whereas the decrease in the thermal expansion can be due to a good adhesion between fiber and matrix, preventing the dilatation of the matrix. The composite with 5% fiber could be suitable for applications requiring greater dimensional stability.

Coir fibers reinforced Polylactic acid composites compatibilized by maleated PLA.

Takahashi Márcio Conti (1), Rigolin Talita Rocha (1), Kondo Drieli Litsue (1), Bettini Sílvia Helena Prado (1)*

(1) Universidade Federal de São Carlos - São Paulo - Brazil

The growing concerment about ambiental impacts of production and discard of consumer goods has fomented the developments in the field of renewable resources based and biodegradable polymers. Nevertheless, the applications of this type of materials is still limited by the high prices of the resins compared to the conventional ones. Polylactic acid (PLA) is an interesting polymer due to its attractive mechanical properties, similar to polystyrene. Aiming high mechanical performance and costs reduction, this study investigated the compatibilization of coir fibers reinforced PLA composites by addition of maleic anhydride grafed polymer (PLA-g-MA), prepared by reactive extrusion. Formulations with 10 and 30% wt of coir fibers, with (5 and 10 wt%) and without compatibilizer were evaluted. Maleated PLA samples were purified and the acidity was determinated by titration. Results revealed the introduction of acid groups in the chemical modification due to the grafting of maleic anhydride and the PLA degradation as well.Mechanical analysis were conducted by tensile and flexural tests and the results showed increasement in Young's modulus as the percent of coir fibers was higher. The compatibilizer didn't induce further improvement on this property. On the other hand, tensile strenght decreased with the addition of fibers. Although fracture surface images obtained by scanning electron microscopy (SEM) showed better interfacial adhesion between the components in the presence of compatibilizer, the tensile strenght was not modified on formulations with 10% of fibers. Formulations with 30% of fibers showed, to both percentages, 5% and 10%, of compatibilizer, significant reduction on tensile strenght. This effect is associated to higher moisture content due to hydrophilicity of the fibers, intensifying hydrolysis catalyzed by acid conditions during processing. Size exclusion cromatography (SEC) analysis proved that degradation was more pronounced on formulations with higher content of fibers, in the presence of maleic anhydride.

PEEK-TiO2 Compounds for Dental Applications

Koehler Valerie Laura [1]*

[1] Technical University of Munich - Bavaria - Germany

The high significance of functional teeth with aesthetic appeal offers a market for constant optimization of dental prosthesis and the applied materials. The use of special polymers increases as an alternative to established ceramic or metal materials. In comparison with ceramics, these materials are closer to match the mechanical properties of tooth and bone. Unlike metal they can be colored in natural tones. High stress and challenging environmental conditions in the mouth require a high-tech polymer as a matrix material and polyether ether ketone (PEEK, VESTAKEEP® 2000G, Evonik Industries AG, Germany) was chosen. For coloring the naturally brown PEEK, inorganic filler particles with a high refractive index have to be used due to the high processing temperatures of up to 400°C. The aim of this study was to investigate the influence of filler particles on the mechanical properties and the color of PEEK. Four different titanium dioxide pigments were compounded with PEEK at filler contents of 6 %, 10 %, and 20 %. Samples of each compound were produced by compression molding and tested by scaling the indentation hardness. The color was measured with and without gloss by determining the L*a*b*-values. The indentation hardness altered by the influence of the processing of the material and the filler type and content. The compounds showed a significantly brighter shade with increasing filler content. Even 6 % filler content sufficed to reach a tooth-like shade of color. Slight differences between the various pigments were detected. A comparison to the color of established dental materials will be made. Furthermore, the processing of samples with higher qualities by injection molding is conducted and even better mechanical properties are expected.

Improvement of electrical resistivity of highly filled graphite/PP composite based bipolar plates for fuel cells by addition of carbon black

B. Krause[1,a], P. Pötschke[1,b], T. Hickmann[2,c]

[1]Leibniz-Institut für Polymerforschung Dresden e.V. (IPF), Hohe Straße 6, 01069 Dresden, Germany
[2]Eisenhuth GmbH & Co. KG, Friedrich Ebert Straße 203, 37520 Osterode am Harz, German

[a)]Corresponding author: krause-beate@ipfdd.de
[b)]poe@ipfdd.de
[c)]T.Hickmann@Eisenhuth.de

Abstract. Novel material solutions for polymer based bipolar plates in fuel cells require adapted ways to develop suitable material compositions. The common pathway to develop materials with at the same time high electrical as well as thermal conductivity is the use of conductive graphite as filler with contents up to 80-85 wt.%. However, there is a need to develop recipes with maximized conductive behavior at lowest possible content of conductive filler to enhance the mechanical properties and allow good processability. In this study, composites based on polypropylene (PP) and different filler systems were melt-mixed using a lab scale co-rotating twin-screw extruder and compression molded to bipolar type plates. As fillers synthetic (G) or expanded (EG) graphites were incorporated. At the overall filler content of 80 wt% part of the graphite was replaced by highly conductive carbon black (CB, 2.5 wt.%, 5.0 wt.%). It was found that the addition of CB significantly reduced the electrical volume as well as the surface resistivity up to values of 0.12 Ω·cm or 4 mΩ/square, respectively. For the values of thermal conductivity the kind and particle size of the selected graphite was important. If expanded graphite was partially replaced by CB the thermal conductivity of PP/EG+CB composites decreased significantly. Otherwise, the combination of synthetic graphite and CB changed the thermal conductivity of PP composites only marginal at the same overall filler content. For both graphite types the filler with larger particle size resulted in higher thermal conductivity.

INTRODUCTION

Despite the established commercial polymer composite materials available for producing bipolar plates for fuel cell stacks with suitable conductive properties, there are still research activities needed to develop bipolar plates with maximized electrical and thermal conductivity [1-5]. The common pathway to develop such materials is to enhance the content of conductive fillers (e.g. synthetic graphite (G)) in the composite [6-10]. Other research activities focus on the use of expanded graphites (EGs) that have high intrinsic electrical and thermal conductivity. The approach of using bimodal filler systems based on graphite and carbon black (CB) seems to be promising. Dweiri et al. described for polypropylene (PP) composites containing constant 80 wt.% carbon based filler an increase of electrical conductivity if graphite was replaced by CB. The conductivity increased from 7 S/cm (0.14 Ohm·cm) for PP/G 20/80 wt.% up to 36 S/cm (0.03 Ohm·cm) for PP/G/CB 20/55/25 wt.% [10]. Király et al. [11] replaced PP by CB by holding the graphite content (vol.%) constant which resulted in an increase of electrical conductivity. At around 40 vol.% graphite, the CB addition leads to an increase from 0.1 S/cm (PP/CB/G 61.3/0/38.7 vol.%) up to 1 S/cm (PP/CB/G 57.9/2.5/39.6 vol.%) or 10 S/cm (PP/CB/G 50.7/7.8/41.6 vol.%). Derieth et al. [7] studied the influence of particle size of graphite on electrical conductivity whereas particle sizes varied between 5 and 20 μm. It was found that composites with smaller graphite particles tend to have lower conductivities than composites containing larger graphite particles.

In this study, composites based on polypropylene (PP) and carbon based filler systems of graphite and carbon black were melt mixed using a lab scale co-rotating twin screw extruder. The measurements of the electrical volume resistivity as well as thermal conductivity of the composites were performed on model bipolar plates compression molded under industrial conditions. The influence of combinations of graphite and carbon black in the composite on both conductive properties was investigated. Additionally, the fillers were characterized concerning their particle size distribution.

EXPERIMENTAL PART

Composites are melt-mixed using a lab scale co-rotating twin-screw extruder Zyp ZSK26Mc (Coperion) using a screw with a L/D ratio of 45 at 200°C, throughput of 10 kg/h and rotation speed of 150 min^{-1} . The final composites contained 14 wt.% polypropylene (PP, Sabic PP579S), 6 wt.% ethylene-propylene-diene-rubber (EPDM) for improved impact behavior and 80 wt.% filler. In a first extrusion step, the PP/EPDM blend was prepared having the mass ratio of 70 wt.% PP and 30 wt.% EPDM. In the second extrusion step the composites were extruded whereby graphite and carbon black were dosed by two side-feeders. For the mixtures of graphite and carbon black a part of the graphite was replaced by carbon black to remain the filler content constant at 80 wt.%.

As synthetic graphite Timcal Timrex® KS300-1250 and KS500 from Imerys Graphite & Carbon Switzerland Ltd., and as expanded graphite the types Sigratherm® GFG600 and GFG1000HD from SGL Group were used. As electrical conductive carbon black the type Ketjenblack® EC600JD from AkzoNobel Polymer Chemistry having a BET surface area of 1270 m^2/g and a primary particle radius of 34 nm were used.

The agglomerate size distribution of the graphite powders was determined by laser diffraction using a Helos/BF particle size analyzer coupled with a RODOS dry dispersion unit and ASPIROS micro dose module (Sympatec GmbH, Germany). The measurement range is 4.5–875 µm. From the distributions the characteristic particle diameters x_{10}, x_{50}, and x_{90} were calculated indicating that 10%, 50%, and 90% of the agglomerate particles are smaller than the given value. The mean value of particle size was calculated from three measurements.

The morphology of the graphite powders was observed by scanning electron microscopy (SEM) using the Ultra Plus from Carl Zeiss AG (Oberkochen, Germany). The powders were sputtered with 3 nm platinum.

The measurements of the electrical resistivity and thermal conductivity of the composites were performed on pressed model bipolar plates (20 cm x 22 cm). The pressing of the composites was done at 190-195°C and a pressure of 170 t. The electrical volume and surface resistivity was measured using Loresta-GP (Mitsubishi) combined with 4-pin-electrode ESP at different places on the plate. The mean value of electrical resistivity was calculated from five measurements. The thermal conductivity was determined by the laser flash method at 25 °C using the LFA 447 NanoFlash® (Netzsch, Germany). The mean value of thermal conductivity was calculated from two measurements.

RESULTS

The morphology of graphite and carbon black powders was characterized by measuring their particle size and the observation in a scanning electron microscope. The mean values of particle size are given in table 1. For the synthetic graphite KS300-1250 larger particles were measured compared to KS500. The particle size distribution of KS500 is much broader than for KS300-1250. For expanded graphite only the particle size of GFG600 could be determined and this material has smaller size than the synthetic graphites. The particles of GFG1000HD were too large to be measured in the used equipment. Therefore it was concluded that the particles are larger than the measuring range of 875 µm. As expected, the smallest particles size was found for carbon black EC600JD. However, the particle size distribution was bimodal having maxima at 15 µm and 215 µm.

TABLE 1. Particle size distribution of dry powders

sample	x_{10} [µm]	x_{50} [µm]	x_{90} [µm]	Kind of distribution
KS500	22.8	143.5	416.2	monomodal
KS300-1250	79.3	327.4	470.3	monomodal
GFG600	106.1	268.8	394.6	monomodal
GFG1000HD	Not measurable; all particle have a particle size larger than 875 µm			
EC600JD	5.6	65.0	256.2	Bimodal: maximum at 15 µm and 215 µm

Next to the particle size also the layer morphology plays a role for the dispersability of fillers with other fillers and with the polymer matrix. In Figure 2 exemplarily the morphology of synthetic and expanded graphite as well as carbon black is shown. The two kinds of graphite contain platelets which are much larger for expanded graphite. The carbon black material includes small spherical particles which are partially agglomerated. It is expected that the small carbon black particles can fill easily the place between the large platelets of graphite, leading to positive effects on the properties.

FIGURE 1. Scanning electron microscopy images of the carbon based fillers

The PP composite plates were characterized regarding thermal conductivity and electrical volume resistivity (Figure 2). For the interpretation of the influence of CB on the thermal conductivity and electrical resistivity of the composites it has to be noted that the PP composites containing only 2.5 or 5.0 wt.% CB have the same thermal conductivity as unfilled PP and are electrical non-conductive.

It was found that the replacement of 2.5 or 5.0 wt.% synthetic graphite by CB leads only to a marginal decrease of thermal conductivity from 12.5 to 11.5 W/(m·K) or 16 to 15 W/(m·K) for PP composites filled with KS500 or KS300-1250, respectively. As seen from these values, the graphite with larger particle size results in higher thermal conductivity values. In case of PP composites with expanded graphite the thermal conductivity decreases significantly if part of the EG was replaced by carbon black. For composites containing GFG1000 the mean values decreases significantly from 21.9 W/(m·K) to 13.2 (@ 2.5 wt.% CB) and 15.9 W/(m·K) (@ 5.0 wt.% CB). Also for PP/GFG600 the thermal conductivity decreases from 15.5 W/(m·K) to 11.4 and 13.2 W/(m·K) if 2.5 or 5.0 wt.% CB, respectively, were added. Again, the composites having EG with larger particle size show higher thermal conductivity values. Especially for PP/EG composites it seems that the CB addition damages the carbon network needed for the phonon transport responsible for thermal conduction. It can be concluded that especially the larger platelets of expanded graphite GFG1000 in combination with the small CB particles were not able to form a homogeneous filler network indicated by a significantly decrease of thermal conductivity to around 60%. However, the absolute values of thermal conductivity of PP/GFG1000+CB composites are higher than for PP/GFG600+CB composites.

Furthermore, the electrical volume and surface resistivities of the PP composites were measured and the values of volume resistivity are given in Figure 2. The largest reduction of resistivity values when adding CB was found for PP/KS500 composites. The replacement of synthetic graphite by carbon black leads to a decrease of volume resistivity from 0.42 to 0.21 Ohm·cm and of surface resistivity from 49 to 6-12 mOhm/square. The PP composite filled with KS300-1250 had with 0.31 Ohm·cm and 14 mOhm/square already lower starting resistivity values than PP/KS500. The addition of CB leads also to volume resistivities around 0.2 Ohm·cm and 6-9 mOhm/square being in the same range like the PP/KS500+CB composites.

In contrast, for PP/expanded graphite composites the tendencies of the influence of CB addition on electrical properties were different. The PP/expanded graphite (GFG600 or GFG1000HD) composites without carbon black have already low volume resistivities of 0.18 or 0.19 Ohm·cm and surface resistivities of 19 or 10 mOhm/square, which is in the range of resistivity achieved in PP/G composites if CB was added. If the PP/EG composites contain carbon black no clear tendency of volume resistivity could be detected. For the combination of 75 wt.% expanded graphite GFG600, 5 wt.% carbon black and 20 wt.% PP the lowest volume resistivity of this study was found, namely 0.12 Ohm·cm. For both kinds of expanded graphite the surface resistivity decreases up to 4-5 mOhm/square and is marginal lower compared to PP composites filled with synthetic graphite and carbon black.

It can be summarized that the replacement of graphite by carbon black leads to a significant decrease of electrical resistivities only for PP composites containing synthetic graphitess. However, the electrical resistivity values for

PP/EG composites were already lower than for PP/G composites with and without carbon black. Mixed carbon filler systems with CB improve electrical properties when the composites contain synthetic graphite and show only marginal increase when expanded graphite is added.

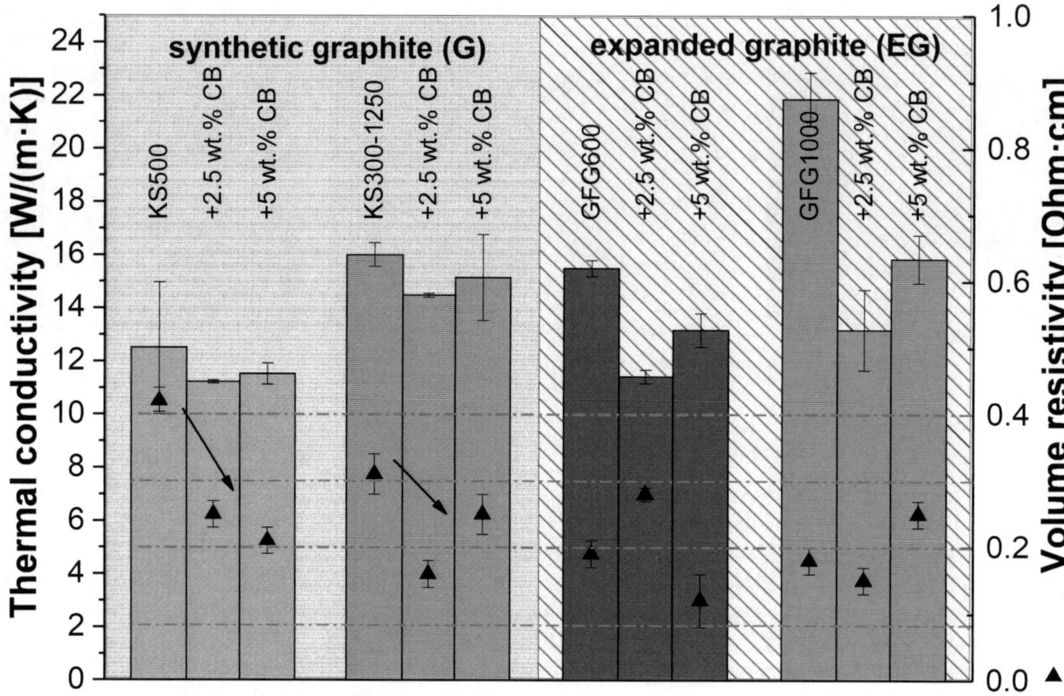

FIGURE 2. Thermal conductivity (bars) and electrical volume resistivity (triangles) of composites containing 80 wt.% (graphite and carbon black) and 20 wt.% PP/EPDM (70/30)

SUMMARY

In this study, composites based on polypropylene (PP)/ ethylene-propylene-diene-rubber (EPDM) (70/30) blends and different filler systems were melt mixed using a lab scale co-rotating twin-screw extruder. As filler synthetic or expanded graphites were incorporated. At the overall filler content of 80 wt.% a part of graphite was replaced by highly conductive carbon black (2.5 wt.%, 5.0 wt.%). The thermal conductivity as well as the electrical volume and surface resistivity of composites were studied on compression molded bipolar shape plates.

It was found that the replacement of graphite by carbon black significantly reduced the electrical volume as well as surface resistivity up to values of 0.12 Ohm·cm or 4 mOhm/square, respectively, (PP/EG GFG600/CB=20/75/5 wt.%) which correlates with values named in literature [7, 10, 11]. The influence of CB on the reduction of electrical resistivity was more pronounced for composites filled with synthetic graphite which showed higher starting values than PP/EG composites. The use of graphite having mean particle size around 300 μm seem to be the best choice to generate in PP composites high thermal conductivity and low electrical resistivity.

For the changes in thermal conductivity with CB replacement the kind of graphite was important. If expanded graphite was mixed with carbon black the thermal conductivity reduced to 60% of the value of PP/expanded graphite without CB. Otherwise, the combination of synthetic graphite and carbon black changed the thermal conductivity only marginal at the same overall filler content. In this comparison, the graphites with larger particle size showed significantly larger values than those with smaller sized graphite. This tendency was found for synthetic graphite as well as expanded graphite.

The use of graphite having mean particle size around 300 µm seems to be the best choice to generate high thermal conductivity and low electrical resistivity in PP composites.

ACKNOLEDGEMENT

The authors thank the German Federal Ministry of Education and Research (BMBF) for financial support within the project No. 01LY1512 and Mr. A. Cohnen (Institute of Plastics Processing (IKV) at RWTH Aachen University) for the preparation of the composites within this project. Furthermore, we thank Mr. U. Geißler, Mrs. S. Schwarz and Mrs. C. Steinbach for particle size analysis (all from IPF) and Mrs. M. Heber for SEM investigations.

REFERENCES

1. T. Hickmann, I. Kundler, U. Kunz, A. d. Santos, C. Hopmann, A. Cohnen, B. Krause and P. Pötschke, in *Fortschrittsberichte der Materialforschung und Werkstofftechnik / Bulletin of Materials Research and Engineering: 1. Niedersächsisches Symposium Materialtechnik*, edited by C. Z. f. Materialtechnik (Aachen, Germany: Shaker Verlag GmbH, 2015), Vol. 1, pp. 434-450.
2. C. Hopmann, C. Windeck, A. Cohnen, J. Onken, B. Krause, P. Pötschke and T. Hickmann, *AIP Conference Proceedings* **1779**, 030017 (2016).
3. C. Hopmann, M. Facklam, B. Krause and T. Hickmann, *Joining Plastics* **69**, 108-115 (2017).
4. B. Krause and P. Pötschke, *AIP Conference Proceedings* **1779**, 040003 (2016).
5. B. Krause, A. Cohnen, P. Pötschke, T. Hickmann, D. Koppler, B. Proksch, T. Kersting and C. Hopmann, *AIP Conference Proceedings* accepted (2017).
6. F. Mighri, M. A. Huneault and M. F. Champagne, *Polymer Engineering & Science* **44**, 1755-1765 (2004).
7. T. Derieth, G. Bandlamudi, P. Beckhaus, C. Kreuz, F. Mahlendorf and A. Heinzel, *Journal of New Materials for Electrochemical Systems* **11**, 21-29 (2008).
8. R. A. Antunes, M. C. L. de Oliveira, G. Ett and V. Ett, *Journal of Power Sources* **196**, 2945-2961 (2011).
9. R. Yeetsorn, M. W. Fowler and C. Tzoganakis, in *Nanocomposites with Unique Properties and Applications in Medicine and Industry*, edited by J. Cuppoletti (Rijeka: InTech, 2011), pp. Chapter 16.
10. R. Dweiri and J. Sahari, *Journal of Power Sources* **171**, 424-432 (2007).
11. A. Király and F. Ronkay, *Polymers and Polymer Composites* **21**, 93-100 (2013).

CONFINEMENT OF PLLA BY LAYER MULTIPLYING CO-EXTRUSION: EFFECT ON MICROSTRUCTURE AND ON GAS BARRIER PROPERTIES

GUINAULT Alain (1)*, NASSAR Samira (2), SOLLOGOUB Cyrille (1), DELPOUVE Nicolas (3), STOCLET Grégory (4), DOMENEK Sandra (2)

(1) Laboratoire PIMM UMR 8006 ENSAM CNRS CNAM - PARIS - France, (2) UMR 1145 Genial AgroParisTech INRA CNAM - Massy - France, (3) Normandie Univ, UNIROUEN Normandie, INSA Rouen, CNRS, Groupe de Physique des Matériaux - Rouen - France, (4) UMET, Université de Lille, Cité scientifique - Villeneuve d'Ascq - France

In recent years, much attention has been focused on biodegradable and biocompatible polymers, particularly from an environnemental point of view. Polylactide (PLA) is a good candidate for food packaging because it can be converted by conventional polymer processes, it has a good transparency like PET and good mechanical properties. However, one of its drawback is its poor gas barrier properties. To modulate the polymer barrier properties, many strategies can be developed like crystallization, biaxial streching, co-extrusion and more recently the crystallization under confinement [1]. The layer multiplying co-extrusion process can be used to confine polymer [2]. Starting with the co-extrusion of a two (A-B) or a three (A-B-A) layers assembly and by adjusting the number of multiplying elements, it is possible to create films with thousands of alternating nanometric layers In this work, PLLA was confined by poly(styrene) (PS) and poly(carbonate) (PC) chosen for their different chemical affinities towards PLA. Ten multiplying elements (2049 layers) were used in order to reach 20 to 80 nm continuous layers of PLLA Depending on the confining polymer and on the subsequent post-crystallization treatment conditions,gas barrier improvements were obtained compared to the annealed bulk PLLA. Thermal analysis and X-ray diffraction measurements showed that the crystal orientation and the quantity of a rigid amorphous phase fraction (RAF) could explain the results. In the case of PLLA/PC films, the compatibility between PLLA and PC was found to impact also the gas barrier properties. 1. Wang H, Keum JK, Hiltner A, Baer E. Macromolecules 2009; 42: 7055-7066 2. Ania F, Baltá-Calleja FJ, Flores A, Michler GH, Scholtyssek S, Khariwala D, Hiltner A, Baer E, Rong L, Hsiao BS. European Polymer Journal 2012; 48: 86-96

Biodegradation of Poly(lactic acid) Films during Composting Enhanced by Bioaugmentation

Auras Rafael [1]*, Castro-Aguirre Edgar [1], Selke Susan [1], Rubino Maria [1], Marsh Terence [2]

[1] School of Packaging, Michigan State University - Michigan - USA, [2] Department of Microbiology and Molecular Genetics, Michigan State University - Michigan - USA

Accumulation of plastics in landfills is still a major environmental concern. Biodegradable polymers provide an opportunity to divert plastic waste from landfills, with composting as disposal route. However, some biodegradable polymers, such as poly(lactic acid) (PLA), do not biodegrade as fast as other organic wastes during composting affecting their general acceptance in industrial composting facilities. Bioaugmentation, the addition of specific microbial strains, is a promising technique to accelerate the biodegradation of compostable plastics, so that they biodegrade in comparable time frames with other organic materials. In this work, we evaluated the effect of bioaugmentation on the biodegradation of PLA in simulated composting conditions. PLA films were produced and fully characterized. PLA-degrading strain was isolated with PLA as the sole carbon source at 58oC. Isolates were identified as Geobacillus using 16S rRNA gene sequencing, and further used to study the effect of bioaugmentation on the biodegradation rate of PLA in solid environments. The biotic and abiotic degradation was assessed in compost, inoculated vermiculite, and uninoculated vermiculite at 58°C and 50% RH by analysis of the evolved CO2 using an in-house built direct measurement respirometer. Size exclusion chromatography was also used to measure and to monitor the change in molecular weight of the PLA samples retrieved every week during the biodegradation test and correlated with the degradation kinetics. Additionally, microbial attachment on the surface of PLA of the isolated microbial strain and other microorganisms present in the compost was evaluated by a biofilm forming assay in wells incubated at 58°C. Bioaugmentation with Geobacillus increased the evolution of CO2 and accelerated the biodegradation phase of PLA when tested in compost and inoculated vermiculite with compost mixed culture. Bioaugmentation could be commercially used to accelerate the biodegradation of PLA in composting.

Facile, green approach for alkaline treatment of flax fibers

Fathi Babak (1)*, Elkoun Saïd (1), Robert Mathieu (1)

(1) Center for Innovation in Technological Ecodesign, Université de Sherbrooke - Quebec - Canada

Flax fiber, are relatively cheap, low density, biodegradable, and renewable material, which has the potential to replace the conventional composites' reinforcement like glass fiber. However, hydrophilicity of flax fibers along with the presence of waxy substances and pectin on their surface make them incompatible with most of the polymer matrices. Alkaline treatment, which is mostly done by sodium hydroxide (NaOH), is a common, low cost process to modify the surface of flax fibres. It removes non-cellulosic content of the fiber, increases the surface roughness, and disrupting the hydrogen bonding between hydroxyl groups. The main drawback in use of alkaline treatment lies in the fact that large amounts of hazardous chemicals are usually involved in the alkaline treatment that must be appropriately handled. This work aims to study the effect of a green surface treatment method, by using sodium bicarbonate, on properties of flax fibers. In particular, flax fiber's yarns were treated with a 10%w/w of sodium bicarbonate solution for different periods (30, 60 and 120 min), at 80±5 °C. Surface treated fibres were characterized through Fourier transform infrared spectroscopy and thermogravimetric analysis. In addition, the mechanical characterization of flax yarns was carried out through tensile tests. The results showed that 60 minutes is the optimum time for treating flax fiber to remove non-cellulosic content from the flax fiber's surface and achieve the proper mechanical properties.

Copolyesters with Tunable and Triggered Biodegradation for Controlled Release Applications

Sobkowicz Margaret J [(1)]*, Bi Siwen [(1)], Tan Bin [(1)]

[(1)] University of Massachusetts Lowell - Massachusetts - USA

Biobased and biodegradable polyesters are of interest for a range of controlled release functions including drug delivery, packaging, environmental and agricultural applications. Controlling the polymer crystallinity, hydrophobicity and surface area are typical strategies for tailoring degradation and release rates. This research compared the degradation rates of a set of compositionally variant copolyesters composed of 1,4-butanediol 1,6-hexamethylene diol, succinic acid and other biobased monomers in lipase solution. The kinetics of degradation were modeled depending on the lipase concentration, crystallinity and thermal properties of the copolymers. The chemical structure, thermal properties and crystallization behaviors of these polymers were characterized using nuclear magnetic resonance (NMR), differential scanning calorimetry (DSC) and wide-angle X-ray diffraction (WAXD), respectively. Through variation of the polymer architecture, early degradation rates of 5 wt% per day were achieved. Techniques for loading the copolyesters with agricultural fertilizers were also explored including solution mixing, melt compounding and encapsulation. The ability of lipase-catalyzed degradation to tune the release of active compounds was quantified through phosphate assay. This work has the potential to achieve more targeted delivery of nutrients and other essential chemicals and prevent environmentally deleterious overdosing.

Microinjection molded PLA parts with modulated degradation rates

A. De Meo[a], F. De Santis[a], V. Iozzino[a], R. Pantani[a]

[a] *Department of Industrial Engineering, University of Salerno, Via Giovanni Paolo II, 84084 Fisciano (SA), Italy*

**rpantani@unisa.it*

Abstract

In this work, an original and versatile system to dynamically control the local temperature of the cavity surfaces in microinjection molding was adopted to obtain 200μm thick PLA parts having a given distribution of crystalline and amorphous zones. The system allows to selectively change temperature of specific regions of the mold surface. This makes it possible not only to allow the complete filling of the mold even with thin cavities and highly viscous polymers, thanks to a high surface temperature during filling, but also to obtain, in the same microinjected part, regions presenting a higher degree of crystallinity and completely amorphous regions. The local morphology of the parts was investigated with different techniques (DSC, WAXS, microscopy) to understand the effect of the peculiar thermal history on the material. Given the well-known higher resistance of PLA crystalline regions to hydrolysis, the same molded parts naturally presents region with a faster degradation rate and regions with slower degradation rates. The degradation rates were assessed by immersing the samples in water at 58°C and measuring by GPC the molecular mass of samples taken from different regions of the parts. The preliminary results obtained are encouraging toward the possibility of effectively obtaining parts having selected regions with time modulated degradation rates.

INTRODUCTION

Poly(lactic acid), PLA, is a biodegradable polymer with extremely interesting properties [1]. In particular, the characteristics of being biodegradable, makes this polymer a candidate in many applications in which the solid material has to disappear after having played its role [2]. Obviously, each specific application can in principle require a given degradation rate, and furthermore it can be of practical interest to produce a part in which different regions present different degradation rates [3].

One of the most efficient ways to increase resistance to hydrolysis in PLA is by crystallinity control [4, 5]. Unfortunately, PLA is a slowly crystallizing material, with crystallization half times of the order of several minutes [6]. This means that producing a crystalline PLA part is almost impossible, except if a control of mold temperature is carried out. A dynamic control of surface temperature was recently developed at the University of Salerno [7]. The system is able to impose predetermined temperature histories with cooling and heating rates of the order of tens Celsius degrees per second. Furthermore, the dimension of the surfaces whose temperature can be controlled is about 4x5mm^2 and this makes this allows to apply this system to micro-injection molding.

MATERIALS AND METHODS

The PLA adopted in this work is produced by Natureworks with the trade name of 4032D. The D-Lactide content of this grade of PLA is about 2%, which makes it able to crystallize when cooled from the melt, even if with a low crystallization rate [8, 9].

For injection molding an HAAKE Minjet II by Thermo Scientific was used. This machine is equipped with a pneumatic piston to control the pressure during the molding. The mold presents a truncated cone shape with a diameter from 50 mm to 35 mm over a length of 90 mm. A cavity with rectangular shape and dimensions of 12 mm × 4 mm × 0.2 mm was adopted. The surface temperature control system employs 50 μm thick resistances as heating elements and attached near the surface of the cavity to cover the whole length. The resistances are supplied by a voltage of 12 V. The temperature control is implemented by thin T thermocouples located near the heating elements, close to the cavity surface. Two heating areas in sequence from the gate to the tip were adopted, as indicated in figure 1.

The molding tests were carried out by imposing an injection temperature of 220 °C, an injection pressure of 150 bar kept for 20 s and a maintaining pressure of 100 bar for 10 s. During this time, the mold surface temperature was kept at 160°C in all zones. Then, the surface temperature lowered a value below the glass transition temperature of the material (60°C). Without opening the mold, another isothermal step (crystallization step) was carried out heating up only the zone closer to the gate (Zone 1, in red in fig. 1) at a temperature of 105 °C. This temperature was kept constant for 1000 s. In these conditions the lowest crystallization time for 4032D is expected. Due to heat transfer, also the zone closer to the tip of the sample, although not directly heated, experiences an increase of temperature, which reaches about 70°C. The temperature measurements taken during the tests are reported in Fig. 2

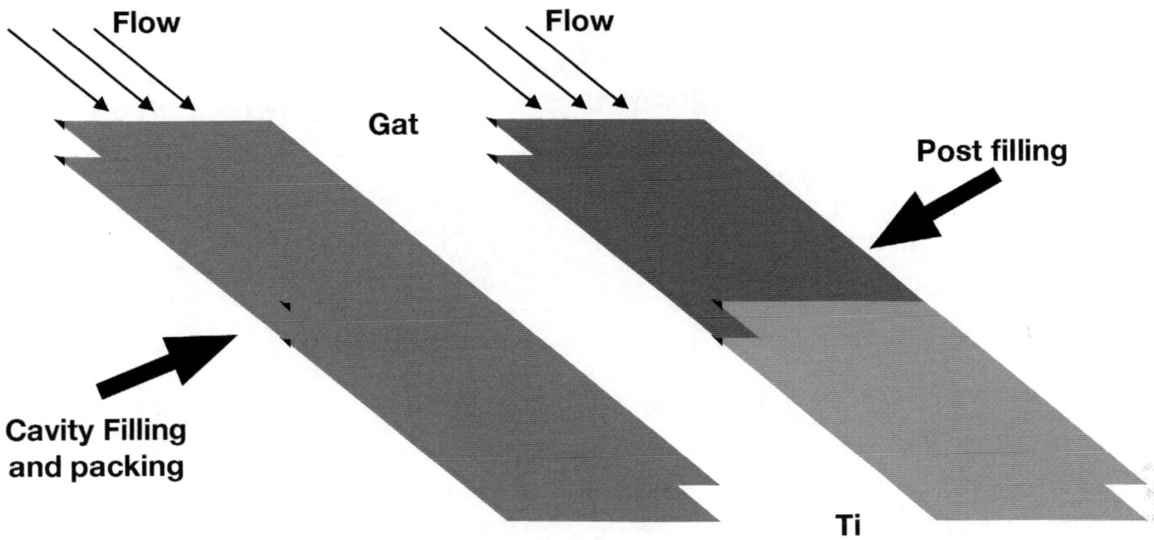

FIGURE 1 Schematics of temperature zones during and after cavity filling. The sample is 200μm thick, 12mm long and 4mm wide

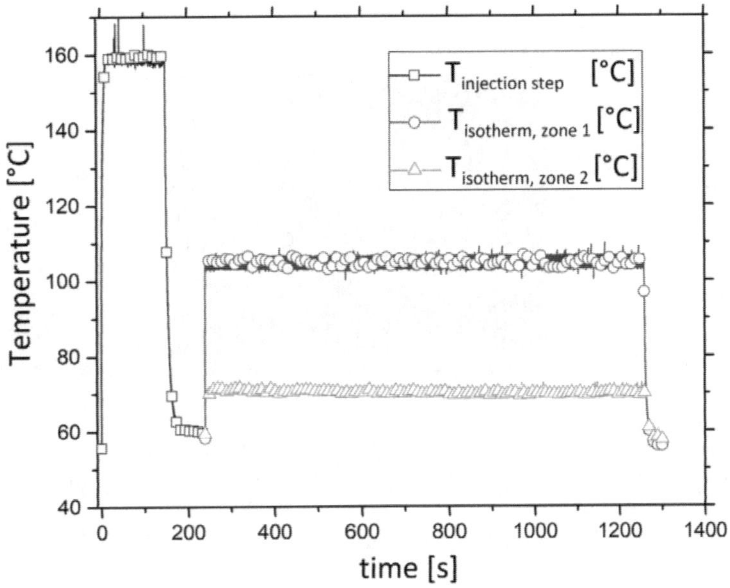

FIGURE 2 Mold surface temperature evolution during time. "Zone 1" refers to part of the sample closer to the gate. "Zone 2" refers to part of the sample closer to the tip

Hydrolysis tests were conducted according to the protocol reported in the literature [4]. In particular, the samples were kept in bidistilled water at 58°C. This temperature was selected because it is the temperature adopted for biodegradation tests according to ASTM and ISO standards. The mass amount of water was fixed in the ratio 500:1 with the mass of the dry sample. The water was changed every 24h.

In order to test the local fragility of the samples, some penetration tests were carried out by using a cylindrical tip with a diameter of 200μm. The tip was moved with an increasing force toward the surface of the sample at a rate of 0.2N/min, and until the breaking of the sample (identified by a sudden movement of the tip) was recorded.

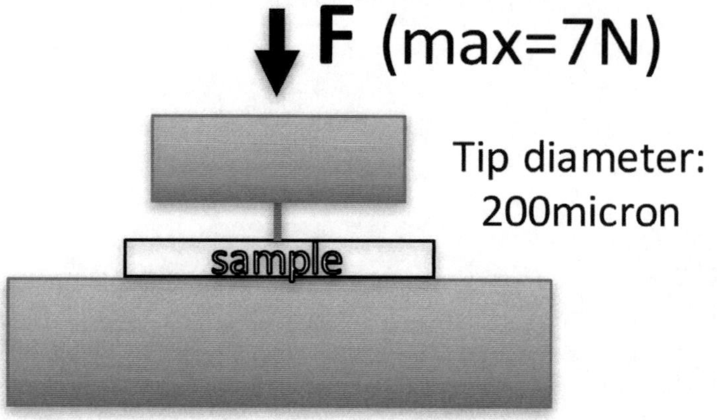

FIGURE 3 Schematics of the penetration tests

RESULTS

After the molding and post-molding step, the samples appeared as clearly made of two separate regions (Fig. 4): the one closer to the gate (named "Zone 1"), which was opaque, and the one closer to the tip (named "Zone 2"), which was transparent. An X-Ray analysis was carried out in the two zones, as also shown in fig 3. The Zone 1 presented a two-dimensional WAXD pattern typical of a crystalline sample, whereas the Zone 2 presented the halo typical of an amorphous sample. A DSC analysis confirmed a crystallinity degree of about 35% for the Zone 1 and a negligible crystallinity value (below 2%) for the Zone 2.

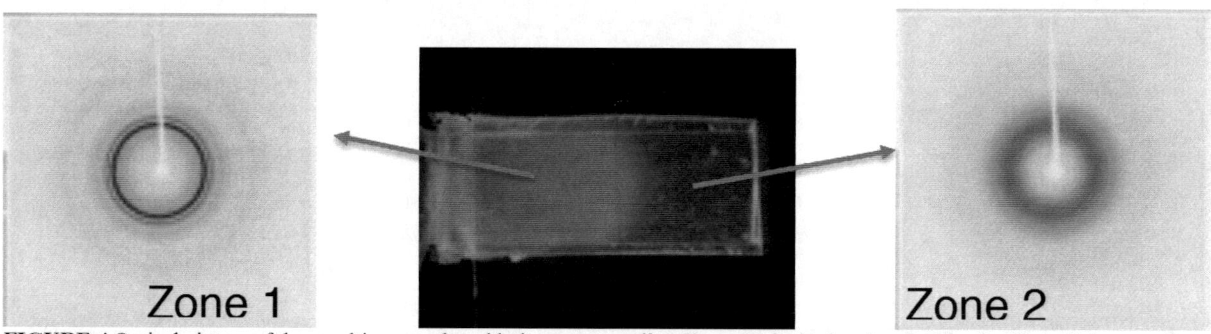

FIGURE 4 Optical picture of the resulting sample, with the corresponding X-ray analysis showing that the tip (Zone 2) is amorphous and the gate (Zone 1) is crystalline

After the molding and post-molding step, the samples appeared as clearly made of two separate regions (Fig. 4): the one closer to the gate (named "Zone 1"), which was opaque, and the one closer to the tip (named "Zone 2"), which was transparent. An X-Ray analysis was carried out in the two zones, as also shown in fig 4. The Zone 1 presented a two-dimensional WAXD pattern typical of a crystalline sample, whereas the Zone 2 presented the halo typical of an amorphous sample. A DSC analysis confirmed a crystallinity degree of about 35% for the Zone 1 and a negligible crystallinity value (below 2%) for the Zone 2.

As a preliminary analysis, the assessment of hydrolysis was conducted just by optical analysis. In fig. 5, images collected at different times (in days) are reported, considering both light transmission and reflection. It is clear that the two zones of the sample behave in a completely different way. In particular, the crystalline zone (Zone 1) remains essentially unchanged for the first 15 days, whereas the amorphous zone (Zone 2) changes from transparent to white opaque. For times longer than two weeks, some white stains appear in the samples. These stains grow in dimensions until the whole sample until after about 35 days the whole sample appears to be homogeneous. At a deeper investigation, the amorphous part of the samples starts to show some cracks which are visible at a larger magnification (Fig. 6). This phenomenon is not found in the crystalline region of the samples (Zone 1). This suggests that the crystalline region is, as expected, more stable with respect to hydrolysis.

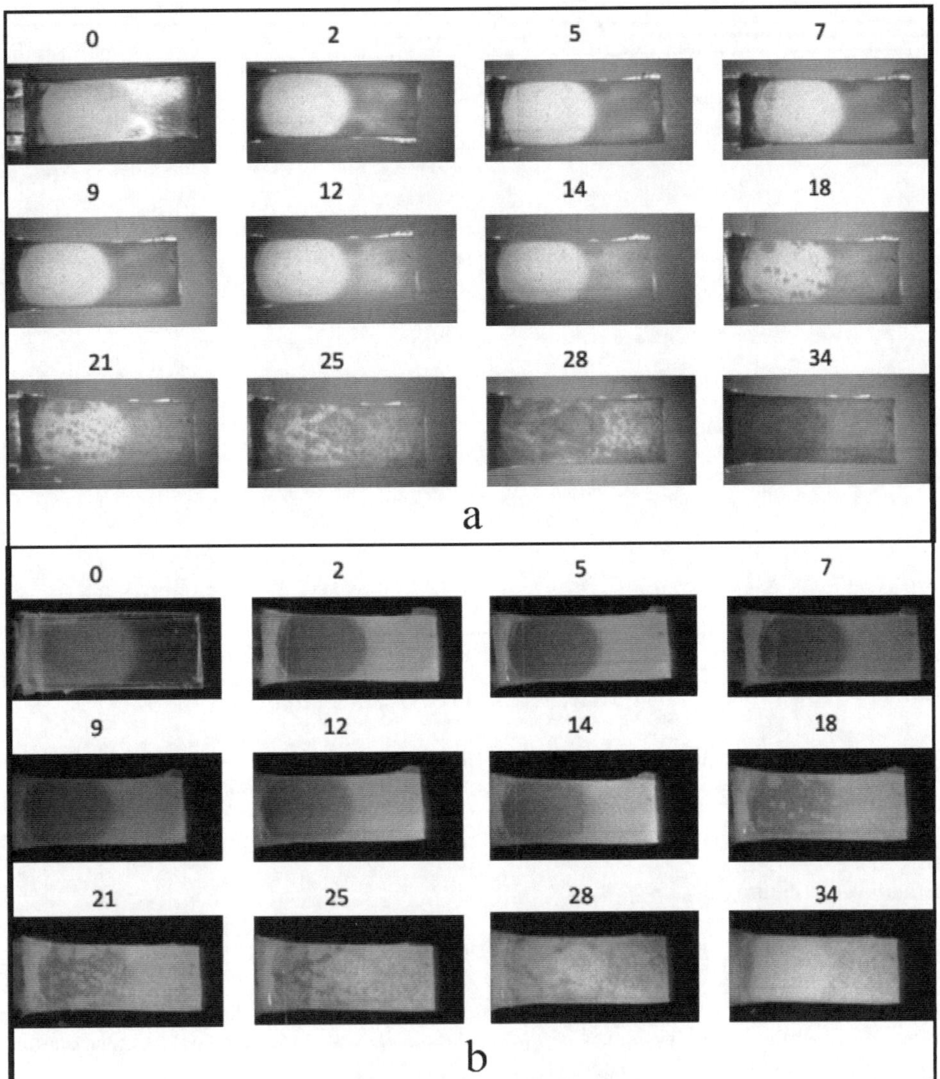

FIGURE 5 Evolution of samples' appearance during time in days. a) images in transmitted light; b) images in reflected light

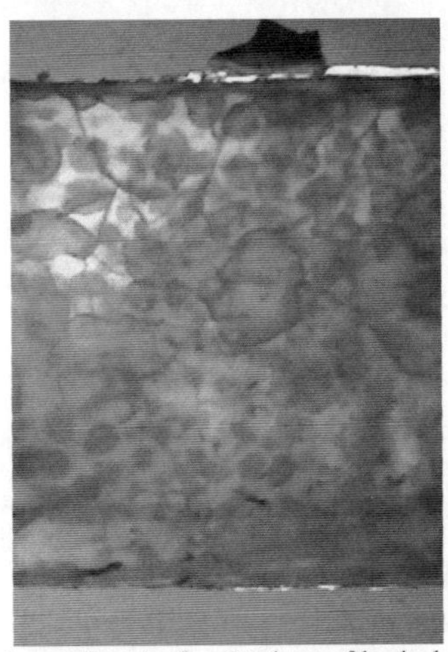

FIGURE 6 Larger magnification of the Zone 2 (amorphous) after 34 days of hydrolysis

Some results of the penetration tests are reported in Fig. 7. It can be noticed that the force needed to break the samples, although decreasing with the time of hydrolysis, is always larger for the crystalline part of the sample with respect the amorphous part.

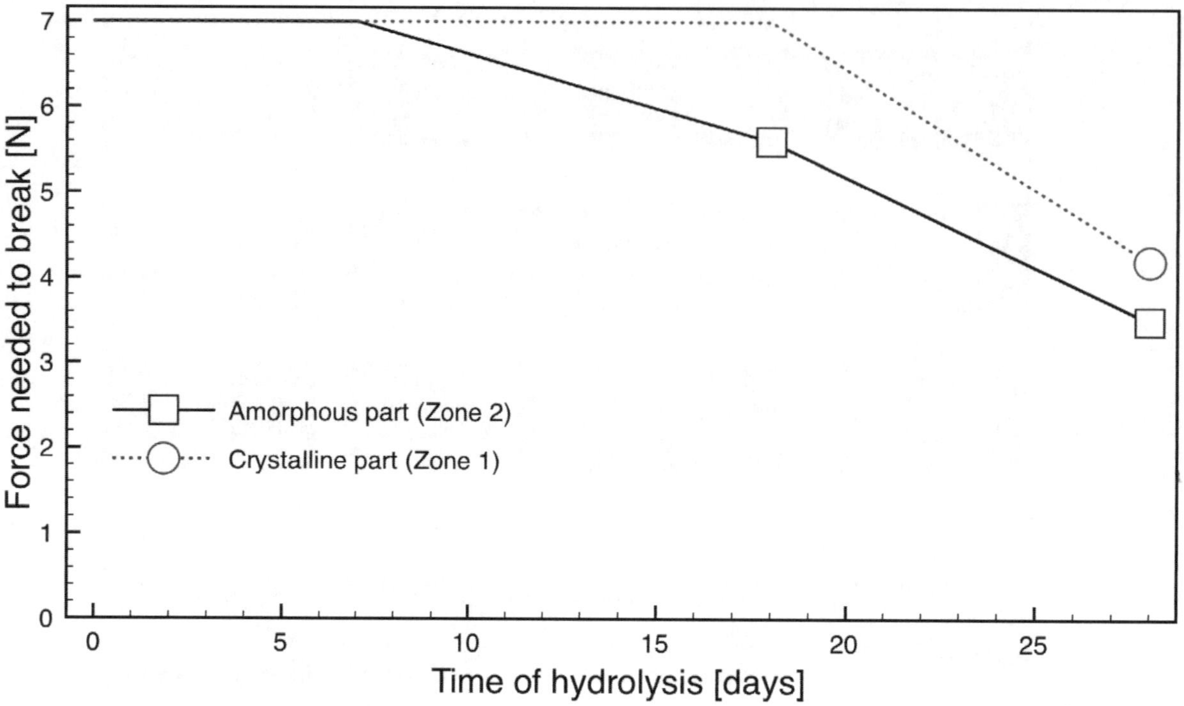

FIGURE 7 Results of penetration tests on some of the samples. The symbols are placed only for the cases in which breaking took place. A value of 7N means that the maximum force value was reached without sample breaking. In these cases the symbol is not reported on the lines

Further investigations are in progress.

CONCLUSIONS AND PERSPECTIVES

In this work, PLA micro injection molded parts were obtained by adopting a system to dynamically control the local temperature of the cavity surfaces. The system was applied in such a way to obtain the separate regions in the same molded parts: a crystalline and an amorphous zone.

The biphasic samples were subjected to hydrolysis tests at 58°C.

The preliminary results obtained are encouraging toward the possibility of effectively obtaining parts having selected regions with time modulated degradation rates.

The adopted system has multifarious applications, if one imagines that it gives the possibility of choosing the part of the samples which first breakes during the hydrolysis process.

REFERENCES

1. E. Castro-Aguirre, F. Iniguez-Franco, H. Samsudin, X. Fang and R. Auras, Adv Drug Deliv Rev **107**, 333-366 (2016).
2. Y. Chen, L. M. Geever, J. A. Killion, J. G. Lyons, C. L. Higginbotham and D. M. Devine, Taylor & Francis (2016).
3. R. R.-S. Jeancarlo, W. Orla, B. Claire-Helene, C. Dominique, P. Francesca, M. Eva, S. Alessandro, D. Frederic and K. Thomas, ACS Sustainable Chemistry & Engineering (2017).
4. G. Gorrasi and R. Pantani, Polymer Degradation and Stability **98** (5), 1006-1014 (2013).
5. F. Adam, A. Sushant and G. Rakesh, Journal of Applied Polymer Science **133** (43) (2016).
6. R. Pantani, F. De Santis, A. Sorrentino, F. De Maio and G. Titomanlio, Polymer Degradation And Stability **95**, 1148-1159 (2010).
7. F. De Santis and R. Pantani, Journal of Materials Processing Technology **237**, 1-11 (2016).
8. F. De Santis, R. Pantani and G. Titomanlio, Thermochimica Acta **522** (000294593900018), 128-134 (2011).
9. A. Jalali, M. A. Huneault and S. Elkoun, Journal of Materials Science **51** (16), 7768-7779 (2016).

pps proceeding
Symposium: S11 - Biopolymers & Natural Fibers
Oral Presentation

Enhancing the crystallization kinetics of thermoplastic biopolyesters through reactive modification

Najafi Chaloupli Naqi (1), Gopi Salim (1), Kontopoulou Marianna (1)*

(1) Queen's University - Ontario - Canada

Thermoplastic biopolyesters, such as polylactide (PLA) and polyhydroxyalkanoates (PHA) have attracted considerable attention in recent years, because of their biodegradability and compostability. They are also derived from renewable-based resources. A major obstacle that prevents their more widespread use that would allow replacement of conventional thermoplastics is their slow crystallization rate. This requires the addition of additives and plasticizers, which may not result in fully bio-based formulation. In this research we have implemented a reactive extrusion approach to alter the architecture of the linear backbone of PLA and elastomeric medium-chain-length PHAs. These modifications have resulted in significantly enhanced crystallization behaviour, as monitored by non-isothermal DSC experiments, including appearance of sharp exothermal crystallization peaks, shifting of the peak crystallization temperature to higher values, and increase in the overall crystallinity. Isothermal crystallization analysis has revealed substantial increases in the crystallization half-time, showing that the crystallization kinetics are significantly enhanced. Furthermore the altered crystalline structure results in mechanical property improvements.

Blends Based On Polyhydroxyalkanoates: An Efficient Way To Modulate Functional Properties

Amirouche Chikh[a,b], Idris Zembouai[a,b], Mustapha Kaci[b], Stéphane Bruzaud[a]

[a]*Université de Bretagne-Sud, Institut de Recherche Dupuy de Lôme, 56321 Lorient, France*
[b]*Université Abderrahmane Mira, Laboratoire des Matériaux Polymères Avancés, 06000 Bejaia, Algeria*

Abstract. In this work, different series of blends based on poly(3-hydroxybutyrate-*co*-3-hydroxyvalerate) (PHBV) were prepared by melt compounding. As a first step, PHBV were blended with polylactide (PLA) with different PHBV/PLA ratios, in the presence or not of a layered nanofiller, i.e. Cloisite 30B (C30B). Secondly, blends based on PHBV and poly(butylene succinate) (PBS) were also elaborated in the presence or not of sepiolite which is characterized by a needle-like shape. In both cases, the influence of a compatibilizer obtained by grafting of maleic anhydride (MA) on PHBV, i.e. PHBV-g-MA, was also studied. Regarding the morphology and the properties of PHBV-based blends, combined and synergistic effects were highlighted in the presence of both compatibilizer and filler.

Keywords: Blends, polyhydroxyalkanoates, elaboration, morphology.

INTRODUCTION

Polyhydroxylalkanoates (PHA) have been studied extensively for some years as replacements or substitutes for current mainstream petrochemical plastics **[1]**. PHA are microbial polyesters produced by numerous bacteria in nature as intracellular reserve of carbon or energy **[2-4]**. They are recognized as completely biosynthetic and biodegradable with zero toxic waste and recyclable into organic waste. They are also well known for presenting good biocompatibility, making them attractive as biomaterials with a wide array of properties that range from stiff thermoplastics to elastomers. A wide variety of different monomer compositions of PHA that provide different properties and functionalities has been described, as well as their future prospects for applications where renewable character, biodegradability or biocompatibility is required. Nevertheless, the processability and the brittleness of some PHA remains the major problem in applications **[5]**.

Many methods have been developed to increase the properties of biopolymers, such as random and block copolymerization, chemical modification or incorporation of numerous fillers, which modulate both the biodegradation rate and the mechanical properties of the final products. The preparation of bionanocomposites defined as a combination between a biopolymer and an inorganic nanofiller is a route to enhance some of the biodegradable polymers properties. Nanoclays (montmorillonite, hectorite, etc.) and lignocellulosic fillers (olive husk flour, wood flour, etc.) are classically used to improve biodegradable polymers stress and stiffness, reduce their gas/water vapor barrier properties, increase their thermal stability and modify their biodegradation rate **[6-8]**. The best results are generally obtained when the inorganic platelets are exfoliated within the polymer matrix which is a real challenge rarely achieved. Polymer blends have attracted more attention in the last decades, since physical blending is another effective and simple way to prepare biodegradable composites with different morphologies and physical characteristics. The physical and mechanical properties of blends can be easily modified by choosing the polymers and making the composition of the blend vary. Thus, one of the methods to improve the physical and mechanical properties of PHA is to blend PHA with other polymers, including biopolymers and conventional petroleum-based polymers **[9]**.

ELABORATION OF BLENDS

The part aims to study blend properties of biobased and biodegradable polymers of poly(3-hydroxybutyrate-co-3-hydroxyvalerate) (PHBV) and polylactide (PLA) prepared by melt mixing **[10]**. Blend compositions based on PHBV/PLA were investigated according to the following weight ratios, i.e. 100/0, 75/25, 50/50, 25/75 and 0/100 wt.%. The study showed through scanning electron microscopy (SEM) that blends of PHBV/PLA are not miscible (Fig. 1).

This is consistent with differential scanning calorimetry (DSC) data which indicate the presence of two distinct glass transition temperatures (T_g) and melting temperatures (T_m), attributed to the neat polymers, over all the range of blend compositions. Water and oxygen barrier properties of PHBV/PLA blends are significantly improved with increasing the PHBV content in the blend. Further, morphological analyzes indicated that increasing the PHBV content in the polymer blends results in increasing the PLA crystallinity due to the finely dispersed PHBV crystals acting as a filler and a nucleating agent for PLA. On the other hand, the addition of PLA to the blend results in a very impressive increase in the complex viscosity of PHBV. Moreover, the rheological data showed that, excluding the specific behavior of the neat polymers at low frequencies, i.e. less than 0.1 Hz, the complex viscosity of PHBV/PLA blends fits the mixing law well.

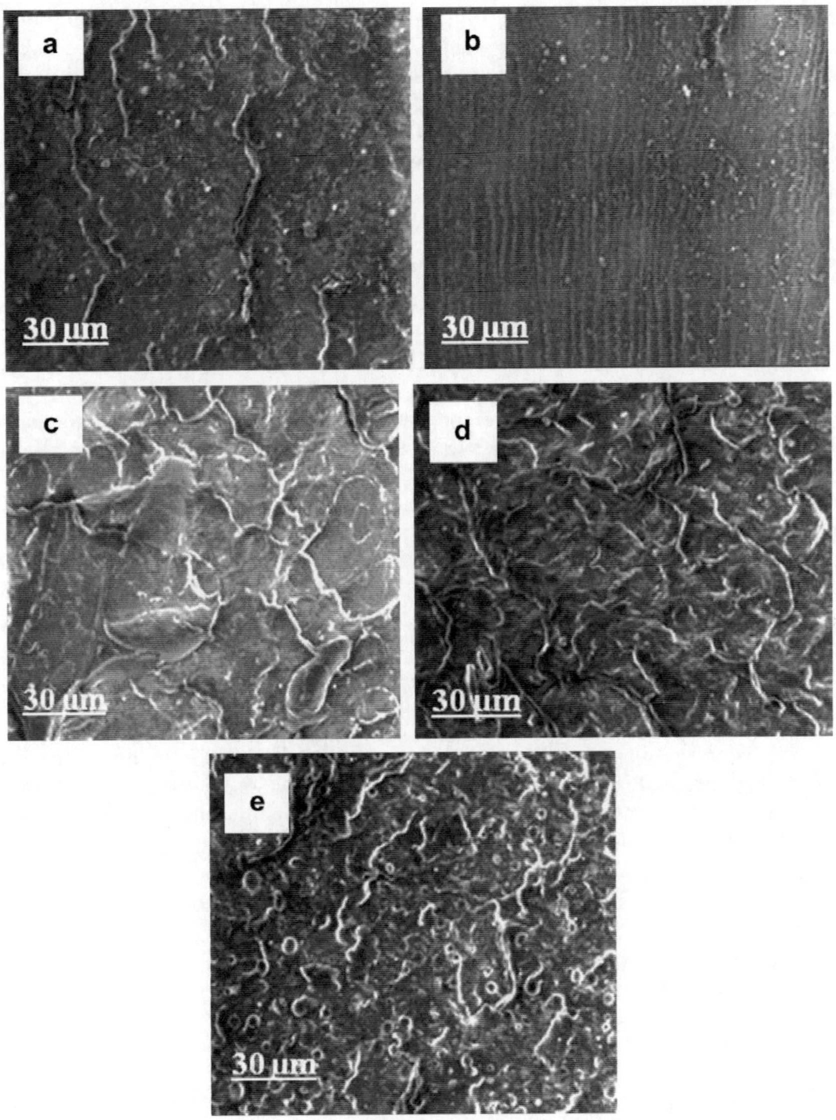

FIGURE 1. SEM micrographs (2000 x) of fracture surface of neat PHBV (a), neat PLA (b), PHBV/PLA 75/25 (c), PHBV/PLA 50/50 (d) and PHBV/PLA 25/75 (e) **[10]**.

Furthermore, DSC data show an improvement in crystallization of PLA due to the positive effect played by PHBV crystals on the nucleation of PLA. On the other hand, FTIR analysis indicates significant differences in compatibility and crystalline structures between PHBV and PLA polymers when blended. PHBV provides better water and oxygen barrier properties to PHBV/PLA blends acting as an efficient barrier promoter for PLA, even at quite low ratio. Barrier properties of PHBV/PLA blend would thus be a balance between PHBV/PLA ratio and phase distribution, as recently observed in the case of PHBV/PPC blends. The rheology of PHBV/PLA blends investigated in the dynamic mode reveals that PHBV exhibits a very impressive increase in complex viscosity with increasing the PLA content. In the PHBV/PLA blend and that at medium and high frequencies mixture, dynamic viscosity follows the mixing law.

It can be concluded that the production of blends based on PHBV and PLA can be an efficient and promising route to extend the applications as biodegradable materials, PHBV for instance, considering the possibility to finely tune their functional properties playing with the respective proportions of each component **[11]**.

Blends of PHBV and poly(butylene succinate) (PBS) with different PHBV/PBS weight ratios (100/0, 75/25, 50/50 and 0/100) were also elaborated by melt mixing **[12]**. Using an adequate solvent, the selective extraction of the different components of blends is usually carried out to describe in detail the morphology. PBS can be completely dissolved in dichloromethane at room temperature, which is without incidence on the PHBV solubility. The use of this solvent leading to a total dissolution of PBS phase in the PHBV/PBS blend enables us to calculate the continuity rate of each component and to observe the blend morphology using SEM. The morphological investigation of the different samples, in comparison with that of neat PHBV and neat PBS, pointed out that PHBV/PBS blends form a biphasic system over the whole composition range. This immiscibility between the both components of blends has been revealed using SEM analysis associated to selective solvent dissolution, which clearly highlighted the occurrence of a co-continuous morphology for the blend PHBV/PBS 50/50 while the other compositions of blends 75/25 led to a dispersed morphology.

INFLUENCE OF COMPATIBILIZER AND FILLER

To improve the miscibility between the two components of the blend, low amount of compatibilizing agent (5 wt.%), obtained by grafting maleic anhydride onto PHBV, was used **[13]**. When compared with the uncompatibilized blends, the compatibilizer presence induces a greater interfacial adhesion. The effect of Cloisite 30B (C30B) on the blend morphology and the blend properties was also investigated. The morphology of the different blends as well as the evolution of their material properties were discussed in terms of the nanoclay and compatibilizing agent contents. A synergistic effect of compatibilizer and C30B was highlighted leading to an improved miscibility of the two blend components. The resulting properties were correlated with the morphology observed for the different blends.

Low amount of compatibilizing agent (5 wt.%), obtained by grafting maleic anhydride onto PHBV, i.e. PHBV-g-MA, was also used for improving the miscibility between the two components of the PHBV/PBS blend. The incorporation of a fibrous filler as the sepiolite, easily dispersible in a polymer matrix, was also investigated. The morphology of the different blends as well as the evolution of their material properties were discussed in terms of the sepiolite and compatibilizing agent contents **[12]**. The dispersion of PBS in the PHBV matrix markedly became finer with incorporation of sepiolite and PHBV-g-MA, due to enhanced interactions between the components, highlighting a synergistic effect induced by the presence of both compatibilizer and sepiolite and leading to an improved miscibility of the two blend components (Fig. 2).

RECYCLING AND DEGRADATION

The effects of recycling cycles (up to 6 repeated cycles) on the structure and properties of neat PHBV, neat PLA, neat PBS, PHBV/PLA and PHBV/PBS blends (50/50 wt.%) elaborated by melt molding process were also investigated **[14,15]**. The changes in the chemical structure, the morphology and the thermal, mechanical and rheological properties of the recycled materials were studied. All the results showed that PHBV is the more sensitive component towards recycling due to thermomechanical degradation, but PLA as well as PBS play a stabilizing role on PHBV due to improved morphology. This better stability of the PHBV-based blends after multiple reprocessing cycles compared to that of neat PHBV is due probably to the co-continuous morphology which improves the material elasticity. A stabilizing effect of PLA or PBS on the recyclability of PHBV-based blends was highlighted and this may constitute a promising way in order to minimize the degradability of PHBV and to enhance its recyclability.

The gamma irradiation effects on morphology and properties of PHBV/PLA blends in presence of compatibilizer and C30B have been also studied. The different tests revealed that the organomodified clay, i.e. C30B, can act as a compatibilizer and compensate for the effect of irradiation on both thermal stability and fire retardancy of PHBV/PLA blends **[16]**. Then, electron beam radiation effects on properties and ecotoxicity of PHBV/PLA blends with and without C30B revealed the presence of C30B in the polymer blend has no influence on the nature of the degradation process. However, the good dispersion of C30B nanoparticles provides more stability to the molar mass and the thermal, mechanical and barrier properties of PHBV/PLA blend. At absorbed dose of 10 kGy, the irradiated samples are completely safe. Furthermore, ecotoxicity testing of both non irradiated and irradiated samples clearly showed no toxicity **[17]**.

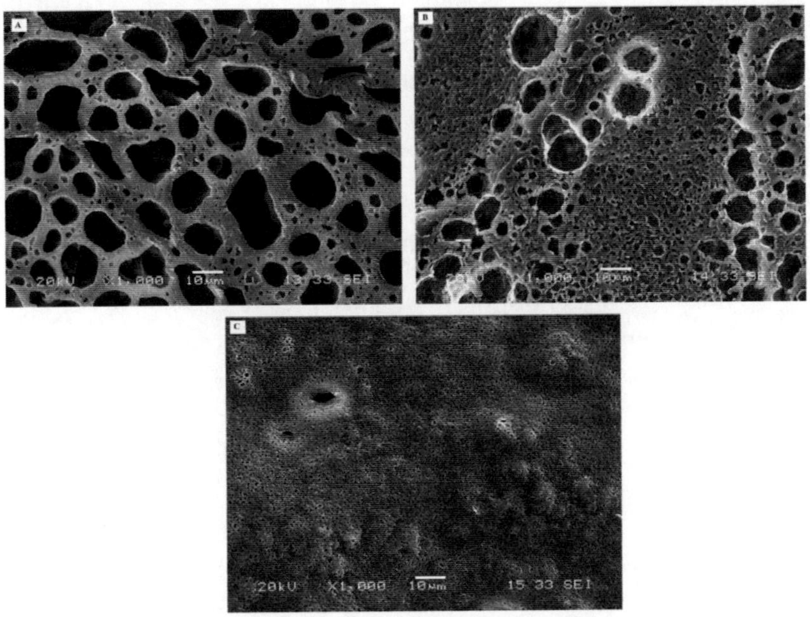

FIGURE 2. SEM micrographs of fracture surface of PHBV/PBS 75/25 (a), PHBV/PBS 75/25 with 5% of sepiolite (b) and PHBV/PBS 75/25 with 5% of sepiolite and 5% of PHBV-g-MA (c) **[12]**.

ACKNOWLEDGMENTS

The authors are grateful to CAMPUS FRANCE through the TASSILI program 12 MDU 871 and PROFAS B+ program for its financial support in this collaborative project.

REFERENCES

1. S. Bruzaud, in « Handbook of Polymer Nanocomposites: Processing, Performance and Application – Volume A: Layered Silicates », Edited by J.K. Pandey, R. Reddy, A.K. Mohanty and M. Misra, Springer-Verlag Berlin Heidelberg, **2014**, Chapter 1, pp 1-20.
2. S. Chardron, S. Bruzaud, B. Lignot, A. Elain, O. Sire, *Polym. Test.*, **2010**, *29*, 966-971.
3. A. Elain, A. Le Grand, Y.M. Corre, M. Le Fellic, N. Hachet, V. Le Tilly, P. Loulergue, J.L. Audic, S. Bruzaud, *Ind. Crops Prod.*, **2016**, *80*, 1-5.
4. A. Elain, M. Le Fellic, Y.M. Corre, V. Le Tilly, A. Legrand, J.L. Audic, S. Bruzaud, *World J. Microbiol. Biotechnol.*, **2015**, *31*, 1555-1563.
5. Y.M. Corre, S. Bruzaud, J.L. Audic, Y. Grohens, *Polym. Test.* **2012**, *31,* 226-235.
6. S. Bruzaud, A. Bourmaud, *Polym. Test.*, **2007**, *26*, 652-659.
7. L. Hassaini, M. Kaci, S. Bruzaud, A. Benhamida, I. Pillin, Y. Grohens, *J. Adhes. Sci. Technol.*, **2016**, *30*, 2061-2080.
8. L. Hassaini, M. Kaci, N. Touati, I. Pillin, A. Kervoelen, S. Bruzaud, *Polym. Test.*, **2017**, *59*, 430-440.
9. Y.M. Corre, S. Bruzaud, Y. Grohens, *Macromol. Mater. Eng.* **2013**, *298,* 1176-1183.
10. I. Zembouai, M. Kaci, S. Bruzaud, A. Benhamida, Y.M. Corre, Y. Grohens, *Polym. Test.* **2013**, *32,* 842-851.
11. I. Zembouai, S. Bruzaud, M. Kaci, A. Benhamida, Y.M. Corre, Y. Grohens, A. Taguet, J.M. Lopez-Cuesta, *J. Polym. Environ.* **2014**, *22,* 131-139.

12. A. Chikh, A. Benhamida, M. Kaci, I. Pillin, S. Bruzaud, *Polym. Test.* **2016**, *53*, 19-28.
13. I. Zembouai, S. Bruzaud, M. Kaci, A. Benhamida, Y.M. Corre, Y. Grohens, J.M. Lopez-Cuesta, *Polym. Eng. Sci.* **2014**, *54,* 2239-2251.
14. I. Zembouai, S. Bruzaud, M. Kaci, A. Benhamida, Y.M. Corre, Y. Grohens, *J. Polym. Environ.* **2014**, *22,* 449-459.
15. A. Chikh, A. Benhamida, M. Kaci, A. Bourmaud, S. Bruzaud, *Polym. Degrad. Stab.* **2017**, *142*, 234-243.
16. I. Zembouai, M. Kaci, S. Bruzaud, L. Dumazert, I. Pillin, A. Bourmaud, M. Mahlous. J.M. Lopez-Cuesta, Y. Grohens, *Polym. Test.* **2016**, *49,* 29-37.
17. I. Zembouai, M. Kaci, I. Pillin, J.L. Audic, S. Shayanfar, S.D. Pillai, S. Bruzaud, *Polym. Degrad. Stab.* **2016**, *132,* 117-126.

Biocomposites From Polyhydroxyalkanoate and Hemp Fibers: Effects of Processing and Compatibilizer on Structure/Performance Relationships

Francois Touchaleaume[a,b], Remi Auvergne[a], Sylvain Caillol[a], Sandrine Hoppe[c] and Helene Angellier-Coussy[b]

[a] Institut Charles Gerhardt, CNRS, ENSCM, Université de Montpellier - Montpellier - France,
[b]JRU IATE, Agropolymers Engineering & Emerging Technologies, INRA - Montpellier - France,
[d]University of Lorraine - CNRS, Reaction and Process Engineering Laboratory - Nancy - France,

Abstract. The objective of the present work was to develop fully bio-sourced and biodegradable composite materials based on polyhydroxybutyrate (PHB) and hemp fibers. It is well-known that the improvement of the final functional properties of composite materials is not only related to the filler content and intrinsic properties of constituents, but also to the filler dispersion state and to the properties at the filler/matrix interface. Hence, the lack of favourable interactions between the apolar PHB matrix and the polar hemp fibers has been overcome by adding a green compatibilizer which was previously synthetized using a solvent-free process, i.e. reactive extrusion. The modulation of the extrusion processing conditions (screw profile and speed) was tested in combination with the previous compatibilizing route in order to optimize the performances of the materials. The impact of processing and formulation on the structure and thermal properties of biocomposites (filler content of 20 wt%) has been discussed in relation to their mechanical properties and water vapour sensitivity. A strong nucleating effect provided by the fibers has been highlighted, making it unnecessary the recourse to mineral nucleating agents. Mechanical performances exhibited by the materials revealed a fiber-matrix load transfer, confirming the effectiveness of the prepared compatibilizer.

Keywords: polyhydroxyalkanoates, hemp, biocomposites, compatibilization, reactive extrusion, biodegradation
PACS: 64.70.pj , 65.90.+I , 68.37.Hk , 81.20.Hy , 81.20.Ka , 81.70.Bt , 81.70.Pg , 82.35.Lr , 81.05.Qk

INTRODUCTION

Bio-composite materials based on polyhydroxyalkanoates and lignocellulosic fibers constitute an answer to environmental and economic concerns, with a less important dependence to oil resources, a decreased environmental pollution, and specific new functionalities. One of the main bottlenecks of lignocellulosic fibers-polymer composites is the poor affinity between the fibers and the polymer resulting in limited tensile properties. To overcome this problem, one possible strategy is the introduction of amphiphilic compatibilizers [1-4]. If the use of compatibilizers is widely reported in literature, particularly by extrusion process [5], only a handful of papers depict the use of biobased compatibilizers[6], unliked biobased plasticizers [7]. Some authors proposed glycolysis of polylactide (PLA) with biobased polyols such as glycerol or sorbitol to yield compatibilizer oligomers to enhance compatibilization of PLA/starch composites [8]. Our team proposed a new strategy to synthesize biobased compatibilizers for PHB/hemp fiber composites. We intended to design oligoPHB with aromatic groups, which could develop affinity with hemp composition. Hence, PHB should be depolymerized with alcohol functions of biobased aromatic groups, for instance vanillin alcohol.

In addition to this chemical strategy that reduces the lack of affinity between the components, it is necessary to disperse homogeneously the fibers among the matrix for the final material to fully benefit from the introduction of fibers. For that purpose, processing conditions need to be optimized as largely reported in the literature, whatever the fiber type[9, 10].

The objective of the present work was to optimize the mechanical performance of PHB/hemp fibers biocomposites by investigating and combining two routes: 1/ extrusion processing conditions (screw profile and speed) and 2/ addition of biobased compatibilizers. The impact of processing and formulation on mechanical properties has been assessed, and was .discussed in relation to the structure of the materials.

EXPERIMENTAL

A commercial PHB was purchased from Metabolix, Germany (P(3HB-co-4HB), 5% 4-hydroxybutyrate, M2100 grade). Hemp fibers (retted, carded, 4mm long) were provided by Chanvriers de l'Est (France). Zinc acetate dihydrate (ZnAc2, 99%) and vanillic alcohol (VA, 98%) were purchased from Sigma-Aldrich. The compatibilizer was obtained by a transesterification of PHB by VA by a continuous reactive extrusion process. VA and Zinc acetate were previously powder-mixed and added into the extruder using a gravimetric feeder while PHB was introduced by using a volumetric feeder at a 3600 $g.h^{-1}$ feeding rate. Extruded material was collected at the rod die and air-cooled before pelletization.

PHB and compatibilizer pellets were dried for one night in a vacuum oven at 45°C and mixed in a plastic can before processing in a twin screw extruder equipped with 24 mm diameter screws and a L/D ratio of 28 (Haake Polylab, Thermo Scientific). Hemp fibers were dried at 80°C for 1 hour and added to the extruder by a lateral feeder directly into the melt. The temperature was set at 180°C all along the screw, the polymer/compatibilizer throughput was set at 3.6 $kg.h^{-1}$. Three screw speeds were chosen (from 250 to 750 rpm) and two screw profiles have been tested, a first one with several mixing zones and one reverse screw thread (Low Shear profile) and a high shear rate one with 5 reverse screw threads (High Shear profile). Extruded composites (with compatibilizer: PHB-C-H; without: PHB-H) were granulated and further injected as dog-bone samples with the aid of a DSM MC15 microcompounder and a DSM IM12 injection molding machine. Materials not containing fibers could not be processed along with this last step as injection molding requires crystallizing kinetics that those unfilled compositions did not show. No crystallization agent was employed.

Scanning electron microscopy (SEM) observations were performed on frozen broken samples using a scanning electron microscope (SEM S-4500, Hitachi, Japan) with an acceleration voltage of 2 kV and a detector for secondary electrons. The weight-average molar mass (Mw) were determined by size exclusion chromatography (SEC) on a Waters HPLC system equipped with a Waters 410 differential refractive-index detector, a 600 mm column PL gel MixC 5µm. The mobile phase was $CHCl_3$, its flow rate set at 0.8ml/min and toluene was used as flowmarker. The samples were prepared as $10mg.mL^{-1}$ solutions and when containing fibers, were filtered on a 45 µm-pores filter (VWR, filter paper, qualitative 413). Then, all samples were filtered using 0.2µm PTFE filters. A sample injection volume of 200µL was used in the GPC analysis and the calibration was performed with polystyrene standards.

Differential Scanning Calorimetry (DSC) measurements were performed with a calorimetric NETZSCH DSC 200 F3 calibrated with an indium standard. Samples of 10 ± 2 mg were placed in the measurement heating cell and were heated under inert atmosphere from −50 to 200 °C at a heating rate of 20 °C/min. Thermogravimetric analyses (TGA) were performed on a Q50 from TA Instrument, from RT to 550°C under a nitrogen flow (heating rate of 10°C/min). Mechanical tensile tests were performed on a Zwick metrology Z010 apparatus (owing to NF EN ISO 527-1 standard) equipped with a sensor of 500 N or 5 kN. Results are an average of measurements performed on at least five samples.

RESULTS AND DISCUSSION

Multi-Scale Structure of PHB/Hemp Fibers Biocomposites

The morphology of PHB-based composites was assessed by SEM observations of cryo-fractured surfaces of injected samples (figure 1). Dispersion of the hemp fibers among the PHB matrix appears homogeneous as can be noticed on the low magnification SEM picture (Figure 1, left). The absence of fiber cluster was an encouraging indicator for favored interactions between the matrix and the fibers. High magnification pictures (figure 1, right) even revealed a good fiber/matrix adhesion, also revealing the apparition of an interphase, initially not present in the case of non-compatibilized composites (pictures not shown here). This improved interface also pointed an efficient role played by the compatibilizer. However, all the compatibilized materials (PHB-C-H) exhibited quite similar dispersion states and interphases, whatever the processing conditions employed.

Although the composites prepared with the various processing conditions did not clearly differ in terms of fiber dispersion, their performances could also be linked to their macromolecular structure and molar mass.

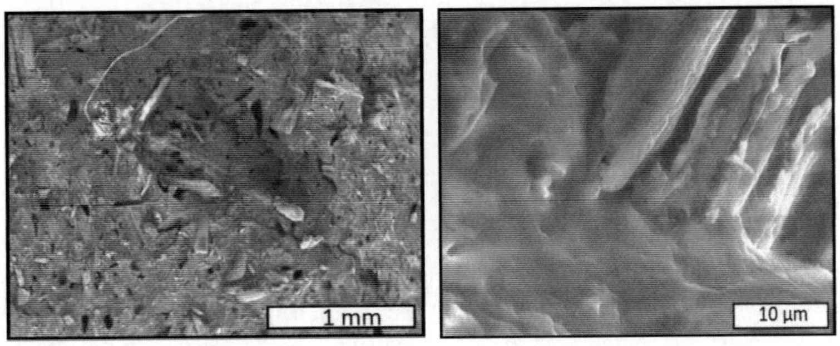

FIGURE 1. SEM pictures of PHB/hemp biocomposite containing 2.4w% compatibilizer (PHB-C-H)

In comparison with the pristine PHB matrix, all samples underwent a common reduction of the molar mass when the material was submitted to an extra processing step (table 1). More precisely, when comparing the two screw profiles at a similar screw speed, the Mw drop was stronger for the material processed with the high shear profile with the low shear one. The Mw reduction was even more pronounced when the screw speed increased (in spite of reduced residence times) and even more severe in presence of moisture (hydrolysis)[11]. When adding fibers, two phenomena contributed to worsen this Mw drop: 1/ PHB hydrolysis was favored by the presence of water, alcohols and/or acids produced during biomass thermal degradation [12, 13], 2/ melt shear viscosity was higher than without fiber, increasing then the internal warming during the process of thermally sensitive PHB. When the compatibilizer was introduced, PHB-C-H exhibited slightly lower Mw than the corresponding PHB-H as the processing conditions were similar to those of the compatibilizer synthesis (transesterification in reactive extrusion with catalyzer still in media) So the transesterification reaction went on during the compounding process.

DSC analyses (table 1) revealed that introducing hemp fibers drove to a plasticizing effect (presence of PHB oligomers confirmed by reduced Mw values) and a concomitant nucleating effect (Xc increased up to 39% and Tc shift up to 75°C in comparison with pure PHB). When the compatibilizer was used, this plasticizing effect was compensated by a macromolecular mobility reduction induced by improved fiber/matrix interactions and/or a better fiber dispersion. The thermal sensitization induced by the extrusion step was lower for the high-shear process ($T_{-5\%}$ reduced by 5°C for high-shear processed PHB compared to as-received PHB) than for the low-shear one (15°C decrease). This sensitization was associated to random chain scissions [14] (confirmed by Mw analyses) and creation of crotonate-ending PHB oligomers and crotonic acid [15].

TABLE 1 Thermal properties of PHB-based materials.

Screw profile	Screw speed	Sample	Mw (g/mol)	Tg (°C)	Tm(°C)	Xc(%)	Tc(°C)	$T_{-5\%}$ (°C)
-	-	PHB as received	150700	3,5	174	3,6	ind.	279,5
High Shear Profile	250 rpm	PHB	115600	2,5	171	16,9	0-50°C	274
	250 rpm	PHB-H	77200	2,5	171,5	40,0	76	261
	500 rpm	PHB-H	59900	-0,5	168,5	40,9	75	263,5
	750 rpm	PHB-H	47700	1,5	173	39,7	70	262
	250 rpm	PHB-C-H	74900	2,5	168	39,6	75,5	259
	500 rpm	PHB-C-H	52900	2	168	39,5	75	258
	750 rpm	PHB-C-H	43400	0,5	165	40,5	74	257
Low Shear Profile	250 rpm	PHB	130700	3	171	9,6	0-50°C	264,5
	250 rpm	PHB-H	122200	0	171	41,8	77,5	257,5
	500 rpm	PHB-H	90000	0,5	169,5	37,2	76	257,5
	750 rpm	PHB-H	63700	-1	167	37,8	76	254
	250 rpm	PHB-C-H	106500	1	170,5	40,7	76	258
	500 rpm	PHB-C-H	80600	3	169,5	40,0	73,5	258,5
	750 rpm	PHB-C-H	69200	1,5	170,5	41,0	76,5	258,5

The introduction of fibers induced a crystallinity increase concomitantly with a $T_{-5\%}$ reduction (lower drop for the low-shear process). This thermal sensitization was associated to a more degraded PHB matrix, to a material structure proner to thermal degradation (pores, holes and fiber clusters facilitated the transfer of degradation reagents and products) and to the inherent thermal instability of the hemp fibers. The introduction of compatibilizer slightly increased $T_{-5\%}$, in the case of low-shear processed materials, probably because of a better material structure. Fortunately, the degradation temperatures remained above the processing temperatures (180°C) for all formulations.

Mechanical Performances of PHB/Hemp Biocomposites

The effects induced by the processing conditions (screw speed and profile) were rather weak regarding the Young's modulus but more clearly defined on ultimate properties (table 2). Notably, stress at break decreased when the screw speed increased. This drop was even higher for the material processed with the high-shear profile. The ductility of the materials was reduced concomitantly. Indeed, when the processing conditions became more severe antagonist phenomena occurred on fiber dispersion (positive), possible degradation of fibers (negative) and degradation of polymers (negative)

TABLE 2 Mechanical performances of PHB-based materials.

Screw profile	Screw speed	Sample	Young modulus (MPa)	Stress at break (MPa)	Strain at break (%)
High-shear profile	250 rpm	PHB-H	336 ± 29	28,7 ± 0,8	4,7 ± 0,1
	500 rpm	PHB-H	334 ± 27	24,3 ± 1,6	3,7 ± 0,2
	750 rpm	PHB-H	304 ± 29	14,4 ± 0,7	2,6 ± 0,1
	250 rpm	PHB-C-H	395 ± 46	28,5 ± 0,9	4,4 ± 0,1
	500 rpm	PHB-C-H	308 ± 15	23,4 ± 1,3	3,6 ± 0,1
	750 rpm	PHB-C-H	351 ± 22	15,3 ± 0,9	2,6 ± 0,1
Low-shear profile	250 rpm	PHB-H	330 ± 25	33,9 ± 0,6	6,2 ± 0,2
	500 rpm	PHB-H	321 ± 21	32,7 ± 1,3	5,7 ± 0,2
	750 rpm	PHB-H	317 ± 10	29,5 ± 1,0	4,1 ± 0,0
	250 rpm	PHB-C-H	392 ± 51	31,6 ± 0,4	6,4 ± 0,4
	500 rpm	PHB-C-H	480 ± 16	29,8 ± 0,9	4,9 ± 0,2
	750 rpm	PHB-C-H	444 ± 43	25,3 ± 0,7	4,0 ± 0,1

The introduction of compatibilizer significantly increased the materials' stiffness, and this load-transfer was more noticeable for the materials that stayed in the melt state for a longer time (higher residence time for the low-shear process, enabling better compatibilization mechanisms). The fact that the stress at break was only slightly impacted by the addition of compatibilizer indicated that fiber/matrix interfacial adhesion could still be improved.

For the high-shear processed materials, the process effect deleted the influence of the compatibilizer while coupling fiber and matrix was an efficient way to tune the mechanical performances of the low-shear processed ones.

CONCLUSIONS

Polyhyroxyalkanoates/hemp fibers biocomposites have been successfully prepared by reactive extrusion. Structure analyses revealed a clear nucleating effect induced by the introduction of fibers, in association with a thermal sensitization of the polymer matrix. In the case of severe processing conditions, the fiber dispersion state was improved but degradation reactions occurred at the same time. For the materials processed under low-shear rate, mechanical performances were preserved and even improved when the compatibilizer was added. The original compatibilization route proposed here was confirmed and offers new opportunities to design biocomposites with tuned perfomances and low environmental impact.

ACKNOWLEDGMENTS

The authors would like to thank 3BCAR research network and Institut Carnot Chimie Balard for funding this study.

REFERENCES

1. Bledzki, A.K. and J. Gassan, *Progress in Polymer Science*, 1999. 24(2): p. 221-274.
2. Mohanty, A.K., M. Misra, and L.T. Drzal, *Composite Interfaces*, 2001. 8(5): p. 313-343.
3. Belgacem, M.N. and A. Gandini, *Composite Interfaces*, 2005. 12(1-2): p. 41-75.
4. Gandini, A. and M.N. Belgacem, Z. N.E., Editor. 2011, Woodhead Publishing: Cambridge. p. 3-42.
5. Reddy, J.P., M. Misra, and A. Mohanty, *Pure and Applied Chemistry*, 2013. 85(3): p. 521-532.
6. Yu, L., K. Dean, and L. Li, *Progress in Polymer Science*, 2006. 31(6): p. 576-602.
7. Bocque, M., et al., *Journal of Polymer Science Part a-Polymer Chemistry*, 2016. 54(1): p. 11-33.
8. Wu, D. and M. Hakkarainen, *European Polymer Journal*, 2015. 64: p. 126-137.
9. Faruk, O., et al., *Progress in Polymer Science*, 2012. 37(11): p. 1552-1596.
10. Gamon, G., P. Evon, and L. Rigal, *Industrial Crops and Products*, 2013. 46: p. 173-185.
11. Pachekoski, W.M., C. Dalmolin, and J.A.M. Agnelli, *Materials Research-Ibero-American Journal of Materials*, 2013. 16(2): p. 327-332.
12. Spitalsky, Z., et al., *Polymer Degradation and Stability*, 2006. 91(4): p. 856-861.
13. Berthet, M.A., et al., *Industrial Crops and Products*, 2015. 69: p. 110-122.
14. Nguyen, S., G.E. Yu, and R.H. Marchessault, *Biomacromolecules*, 2002. 3(1): p. 219-224.
15. Morikawa, H. and R.H. Marchessault, *Canadian Journal of Chemistry-Revue Canadienne De Chimie*, 1981. 59(15): p. 2306-2313.

Bacterial cellulose nanofiber mats as reinforcement for epoxy-anhydride systems

Yue Liang [(1)], Liu Fei [(2)], Gross Richard [(3)], Manas-Zloczower Ica [(1)]*

[(1)] Case Western Reserve University - Ohio - USA, [(2)] Rensselaer Polytechnic Institute - Ne York - USA, [(3)] Rensselaer Polytechnic Institute - New York - USA

Bacterial cellulose (BC) produced by extrusion through bacterial cell walls assemble and crystallize to form 3-D nanostructured matrices of ribbon-shaped 40-60 nm diameter. Since BC is known to have high chemical and mechanical stability, it was used in this work as a reinforcement matrix for an epoxy nanocomposite. The epoxy-anhydride resin was filled into cavities of the BC nanostructured matrix by impregnation under vacuum and then cured with a hot press. The resulting BC reinforced epoxy composites show significant improvement in mechanical strength and a decreased thermal-expansion coefficient. Moreover, the surface hydroxyl groups of the nanofibers can participate in the curing of the epoxy-anhydride system, offering good interfacial interactions between nanofibers and the epoxy matrix, contributing to an increase in the system glass transition temperature.

Degradation Profiles of Polylactide - A Comparative Study of Machine Learning Soft Sensors used for Inline Prediction

Konrad Mulrennan[1,a)], Marion McAfee[1], John Donovan[1], Leo Creedon[1], Thomas Smyth[1], Fraser Buchanan[2] and Mark Billham[2]

[1]*Institute of Technology Sligo, Ireland*
[2]*Queens University Belfast, United Kingdom*

[a)]Corresponding author: mulrennan.konrad@itsligo.ie

Abstract. Industries processing Polylactide (PLA), which include medical devices, pharmaceutical, food and waste packaging, face challenges in the melt processing of PLA due to its poor thermal stability which is influenced by processing temperatures and shearing. The characterisation of the processed products currently takes place offline in laboratory environments. Scrap rates of a PLA medical grade product can be high as there is no current inline method to identify whether or not these were within specification during production. The focus of this study is developing soft sensors which can infer the measurements of final properties from in process data. The PLA was processed in a laboratory scale twin screw extruder with a slit die under varying process conditions. The degradation characteristics of the extruded PLA were of interest. Developing a model of PLA degradation profiles will allow for the long term prediction of the performance of processed products. The metric used to characterise the degradation of samples was the samples molecular weight which was recorded at various time points during an accelerated degradation test to build the degradation profiles. The processing data, recorded by two pressure transducers in a slit die, is then used along with the observations of the degradation metrics to train a number of machine learning algorithms. These were Bagging, Random Forest and Neural Network models. The models performances were measured against each other by comparing the normalised root mean squared errors on the test data set. This work targets finding the optimal model and discusses the potential of implementing the model as an inline soft sensor.

Introduction

PLA has many degradation mechanisms which are determined by the factors, which catalyse the degradation process and also by the material properties [1]. During extrusion, PLA undergoes process induced degradation which has a significant effect on its final properties. It has been shown that complex reactions occur as a consequence of high temperatures experienced during processing [2]. The presence of moisture [3], residual or hydrolysed oligomers or monomers [4, 5] and the starting molecular weight of a PLA batch [6] further influence the level of degradation caused by processing.

Although best efforts are made to reduce variability in the degree of degradation resulting from processing, the final product's quality is assessed offline in laboratory environments. This means that manufacturers have no indication whether or not a product will be in or out of specification during the production run. There are no current inline monitoring technologies which can measure product properties of interest. The only knowledge available during processing is about the process itself rather than the product.

This work evaluates the use of machine learning algorithms, which were applied as soft sensors. These soft sensors can be used as inline quality assurance tools to bridge the gap between processing and lab assessment. A number of models have been developed using Bagging [7], Random Forest [8] and multi-layer Neural Networks [9] techniques. The models were then used to predict responses, using data which they have not been trained on, and the predictions of each of the models were compared to determine which approach best suits this problem.

Machine Learning Soft Sensors

In-process degradation of the material will have significant impact on the final mechanical properties of the PLA product as well as its degradation behaviour. Arguably, the degradation is best assessed by analysis of the molecular weight and/or degradation products present. Unfortunately it is not possible to do this kind of analysis in real-time on the production line. However, it is known that degradation will alter the viscosity of the PLA melt and this is turn will impact melt pressures and temperatures in the process. These relationships are complex and non-linear and cannot realistically be derived from first principles. Machine learning algorithms pose a potential solution due to their ability to identify relationships between input and output data sets and their ability to continue to auto adapt and update over time with further observations.

All of the models were developed using R (version 3.3.3) software as a back end and RStudio (version 1.0.143) as a front end. The Bagging and Random Forest models were built using the randomForest package (version 4.6-12). A detailed description of Bagging models and also Random Forest models can be found in [10]. The Neural Network models were built using the h2o package (version 3.10.5.3) in RStudio. The h2o package required the installation of the Java development kit (version JDK 8u144) and the Java run time environment (version JRE 8u144).

The original data set is split into a training set and a test set for the Bagging and Random Forest models. The split for the Bagging and Random Forest models is 80% training data and 20% test set. The Neural Network requires the original data set be split into training, validation and test sets. The split for the Neural Network models is 50% training data, 25% validation data and 25% test data. The test set in all cases contains data which the models will not see during the training or validation phase.

The Neural Networks were constructed using a feed forward architecture with multiple hidden layers. The networks were trained using backpropagation [9] along with the dropout technique [11] and maxout activation units [12]. The dropout technique trains a number of models by randomly dropping out input variables, hidden activation units and their connections. The final prediction is made by averaging the predictions from the ensemble of models trained. It should be noted that all of the models trained using the dropout technique share the same weights. The maxout activation unit is a piece-wise linear approximation function. The term maxout relates to choosing the maximum output from a number of linear feature extractors (each performs linear regression) as the input to the next layer. The output layer uses a linear activation function.

Experimental Set Up

The PLA grade used in this study is Ingeo™ Biopolymer 2003D and was obtained from NatureWorks LLC. This is a general purpose extrusion grade which is suitable for sheet production and can be processed on conventional extrusion equipment. The PLA was processed in a Haake Rheomex OS laboratory scale twin screw extruder. The extruder has 16mm co-rotating screws in a length to diameter ratio of 25:1. The material was fed from a volumetric feeder at a constant rate (20%) for all extrusion runs. The PLA resin was prepared for extrusion by undergoing cryogenic grinding into a powder form. Prior to processing the PLA was dried overnight at 65C to remove any residual moisture. There were twelve processing runs in total with varying process settings for the temperature profile of the barrel zones, adapter zone and die zone and the screw speed was varied also. The temperature in the barrel zones 1-3 remained constant for all processing runs. The temperature in barrel zones 4-5, the adapter and the die was varied in the range from 200-240C and the screw speed ranged from 200-400RPM. It should be noted that the data split for each of the modelling approaches is done equally among each of the processing conditions. This means there is an equal number of observations for each processing condition in the splits.

The PLA was extruded as a sheet through a slit die with a 130mm slit length, 2mm slit height and 39.25mm slit width. The slit die had two pressure transducers at a distance of 35mm apart along the length of the die. A slit die was used so that an inline estimate of the shear viscosity (η) could be developed. This approach was taken to investigate whether the η estimate would make a good predictor variable for the models by exploiting the relationship between η and the polymer's molecular weight i.e. a change in molecular weight will result in a change in η. [10] gives a detailed discussion on the slit die used in this study and the shear viscosity model. Post extrusion, the PLA underwent an accelerated degradation procedure and characterisation tests were carried out. The samples were tested at 0, 24, 72, 120 and 168 hours during the PLA accelerated degradation cycle. The accelerated degradation procedure degrades the PLA at approximately 27 times the normal rate. The samples were tested for mechanical properties, the percentage mass change and the molecular weight at each of the time points. [10] gives a detailed discussion on the accelerated degradation procedure and the mechanical and mass change post-processing characterisation of the PLA samples. This paper focuses on modelling the molecular weight degradation profile due to space restrictions. It can be noted that the models for the flexural strength and percentage mass change show similar performance.

Molecular Weight Analysis & Results

Gel Permeation Chromotography (GPC) analysis of the PLA samples molecular weight was carried out by high-performance liquid chromatography (HPLC). Samples were prepared in chloroform with a concentration of 2.5mg/mL and were injected at a flow rate of 0.8mL/min. The mobile phase was chloroform. The columns used were an Agilent ResiPore Guard precolumn 50 x 7.5mm and an Agilent ResiPore 300 x 7.5mm. The column was kept at a temperature of 30C. The HPLC system is a Thermo-Scientific Dionex Ultimate 3000 and the detector (Varian 385-LC) is an evaporative light-scattering detector. The detector had an evaporator temperature of 50C, a nebuliser temperature of 50C and a carrier flow rate of 1.4slm. The light source intensity was 10%. The system was calibrated using polystyrene standards 500, 2000, 3000, 10000, 20000, 30000, 70000, 150000 and 300,000Da. The injection volume of the samples was 10μL. Each extrusion processing condition had a single sample tested at each time point during the accelerated degradation cycle.

The molecular weight of the standard set is known and the retention time is recorded for each standard. The retention time refers to the amount of time which has passed from the sample being injected into the system until a peak is recorded. The samples molecular weight is assessed by measuring the retention time of each sample and fitting it to the standards curve. The analysis of the chromatograms indicate a clear shift in the retention time of the peaks of samples which were processed under the same conditions at each time point of the accelerated degradation cycle. The shift in peak retention time indicates that the molecular weight is reducing during the degradation cycle. All of the samples at 0, 24 and 72 hours had a single peak. A shoulder region develops for some of the samples at 120 hours of the degradation cycle. The other samples at this time point have two peaks. All of the samples at 168 hours have two peaks. This would indicate that the samples were beginning to break down into lower molecular weight components at these times. For all of the samples which have two peaks, an average molecular weight value has been used for modelling purposes.

Soft Sensor Results

The machine learning soft sensors were compared using a normalised root mean squared error (NRMSE) measurement. The RMSE is calculated in Equation 1 where y_i were the observed values from the unseen test data set, $\hat{y}_i$ were the predicted values using the unseen test data set and n is the number of observations/predictions. The unseen test data is used to get the predicted values from the Bagging models, Random Forest models and Neural Network models which were developed from the training data sets. The vector of observations and the vector of predictions were normalised ($\mu = 0$; $\sigma = 1$) to compare the models' performances at each of the degradation time points by having RMSE values on the same scale. The NRMSE is then calculated for each model at each time point during the PLA sample's degradation cycle. Figure 1 highlights the error profiles of the machine learning soft sensors when modelling the molecular weight.

$$RMSE = \sqrt{\frac{\sum(y_i - \hat{y}_i)^2}{n}} \tag{1}$$

An example of the degradation profiles for a single processing condition can be seen in Figures 2, 3 and 4. The profiles are created using an ensemble of Random Forest models (Figure 2), Bagging models (Figure 3) and Neural Network models (Figure 4). The focus in Figures 2, 3 and 4 are on a single processing condition (i.e. temperature in final zones = 200C, screw speed = 200RPM) so as not to overpopulate the graphs with all conditions. All of the processing conditions share similar results for the accuracy of the model predictions. The profiles highlight a mean prediction (black square) and one standard deviation (black bars either side of the square) to indicate the range of predictions made by the models using the unseen test data set. The observed data points are represented by black triangles. It is evident in Figures 2, 3 and 4 that the predicted values overlap the observed values. This indicates that all of the models perform very well.

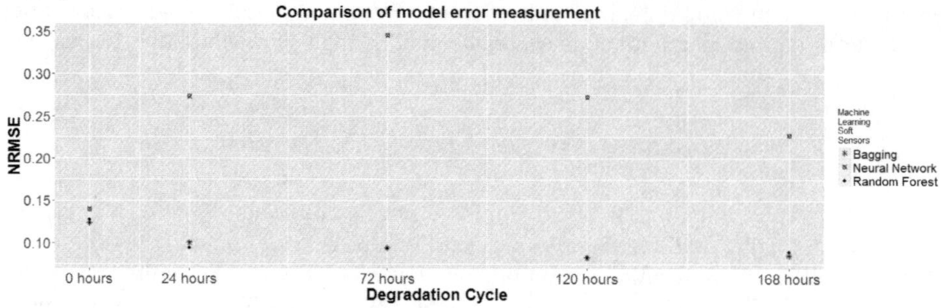

FIGURE 1. NRMSE

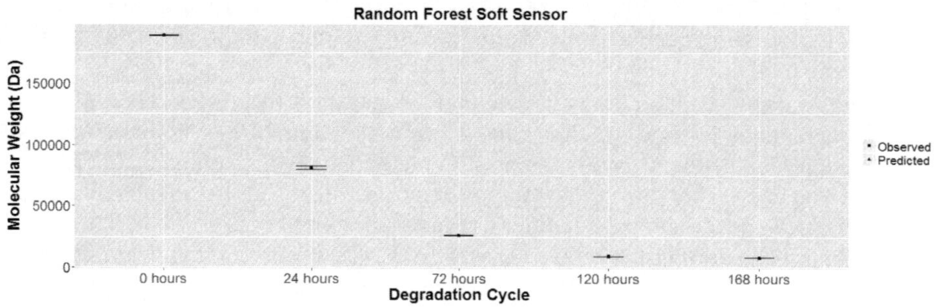

FIGURE 2. Random Forest Soft Sensor

Discussion & Conclusion

The development of the soft sensors proposed in this work can have a significant impact as they can provide immediate feedback during a production run. In the case of a medical device manufacturer, this can allow the early stopping or the adjustment of processing parameters if the soft sensors detect that the product is going out of specification. This would lead to reduced scrap rates and in turn reduced manufacturing costs. The technology could also potentially reduce the number of offline laboratory tests.

There will be a number of considerations for the choice of which machine learning soft sensor to use for any application. These primarily include the programmer's inherent knowledge of the models and the models suitability for a given application. Another consideration is the time required to train the models and find optimal parameters. When evaluating the models investigated here, there is a significant difference in time required to train and tune the Neural Networks compared to the Bagging and Random Forest approach's. The Neural Networks can take up to approximately 144 times longer to train, although this time differential is computer hardware related. Once the models were trained their predictions can be delivered near instantly in all cases.

Machine learning models have the ability to learn from experience over time. This is achieved by updating the model with new observations taken from production runs. This can lead to robust model predictions as the models can adapt to changing environmental conditions (e.g. ambient temperature, humidity levels), the change of a batch of material or wear on process hardware. The models learn the patterns in the data without being explicitly programmed to adapt to these changes.

All of the soft sensors described here have excellent performance, as can be viewed in Figures 2, 3 and 4, and highlight their potential for use in industry. It has been demonstrated that all of the models have accurately predicted the molecular weight degradation profile of PLA. It should be noted that these soft sensor models need to be validated on new data captured on a different day with different environmental conditions before they can be implemented in an industrial setting.

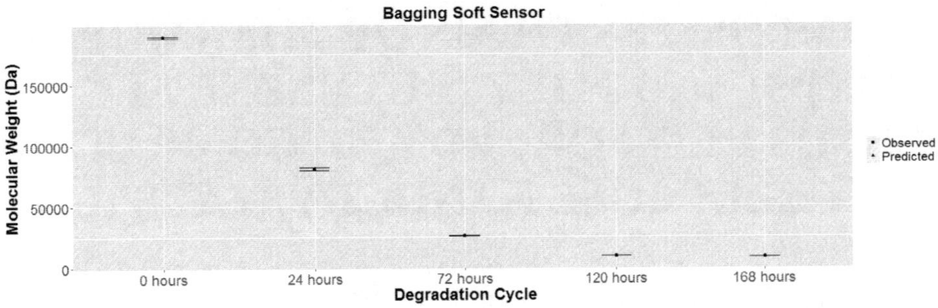

FIGURE 3. Bagging Soft Sensor

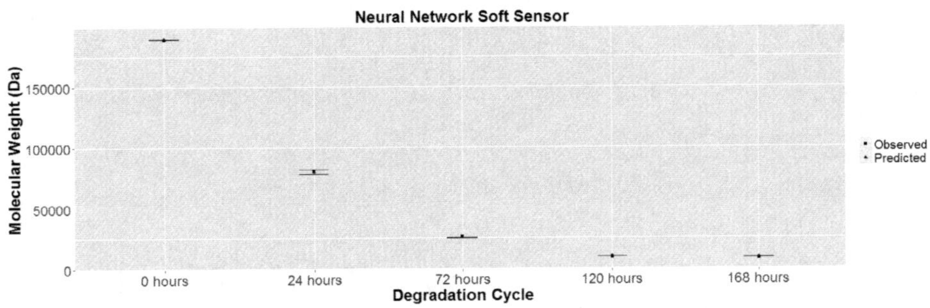

FIGURE 4. Neural Network Soft Sensor

Acknowledgements

The research leading to these results has received funding from the European Union's Seventh Framework Programme managed by REA-Research Executive Agency http://ec.europa.eu/research/rea (FP7/2007-2013) under grant agreement n605086FP7-SME-2012 and from the Institute of Technology Sligo Presidents Bursary Award 2012 and Research Capacity Building Fund 2016/17.

References

[1] Hideto Tsuji. "Hydrolytic degradation". In: *Poly(lactic acid) - Synthesis, Structures, Properties, Processing and Application*. Hoboken, NJ, USA: John Wiley Sons, Inc., 2010. Chap. 21, pp. 345–381.

[2] Hideto Tsuji, Ippei Fukui, Hiroyuki Daimon, and Koichi Fujie. In: *Polymer Degradation and Stability* **81** (2003), pp. 501–509.

[3] Yaming Wang, Bernd Steinhoff, Christoph Brinkmann, and Ingo Alig. In: *Polymer* **49** (2008), pp. 1257–1265.

[4] Rafael Auras, Bruce Harte, and Susan Selke. In: *Macromolecular bioscience* **4** (2004), pp. 835–64.

[5] Ville Ellä, Lila Nikkola, and Minna Kellomäki. In: *Journal of Applied Polymer Science* **119** (2010), pp. 2996–3003.

[6] K. Paakinaho, V. Ellä, S. Syrjälä, and M. Kellomäki. In: *Polymer Degradation and Stability* **94** (2009), pp. 438–442.

[7] Leo Breiman. In: *Machine Learning* **24** (1996), pp. 123–140.

[8] Leo Breiman. In: *Machine Learning* **45** (2001), pp. 5–32.

[9] David E. Rumelhart, Geoffrey E. Hinton, and Ronald J. Williams. In: *Nature* **323** (1986), pp. 533–536.

[10] Konrad Mulrennan, Marion McAfee, John Donovan, Leo Creedon, Fraser Buchanan, and Mark Billham. "Developing a soft sensor Random Forest model for the inline product characterization of Polylactide (PLA) in a twin screw melt extrusion process". In: *ANTEC Conference Proceedings*. Anaheim, CA, 2017, pp. 1024–1031.

[11] Nitish Srivastava, Geoffrey Hinton, Alex Krizhevsky, Ilya Sutskever, and Ruslan Salakhutdinov. In: *Journal of Machine Learning Research* **15** (2014), pp. 1929–1958.

[12] Ian Goodfellow, David Warde-Farley, Mehdi Mirza, Aaron Courville, and Yoshua Bengio. "Maxout Networks". In: *Proceedings of the 30th International Conference on Machine Learning*. Ed. by Sanjoy Dasgupta and David McAllester. Atlanta, Georgia, USA: PMLR, 2013, pp. 1319–1327.

NANOCOMPOSITES FILMS OF THERMOPLASTIC STARCH (TPS) AND POLY (BUTYLENE SUCCINATE) (PBS) BLENDS REINFORCED AND COMPATIBILIZER WITH CELLULOSE NANOCRYSTALS (CNCs)

Pedro Ivo Cunha Claro[a*], Thiago Henrique Siqueira Maia[a], Luiz Henrique Capparelli Mattoso[b], José Manoel Marconcini[b] and Alessandra Almeida de Lucas[c]

[a]Graduate Program in Materials Science and Engineering (PPG-CEM), Federal University of São Carlos, São Carlos, SP 13565-905, Brazil
[b]National Nanotechnology Laboratory for Agribusiness (LNNA), Embrapa Instrumentation, São Carlos, SP, 13560-970, Brazil
[c]Department of Materials Engineering - Federal University of Sao Carlos, São Carlos, SP, 13565-905, Brazil
**pedrocunhaclaro@ymail.com*

Abstract. Environmental problems and the lack of economic viability for packaging recycling are the motivators in the search for biodegradable materials. Biodegradable polymers alone do not exhibit satisfactory properties as thermoplastic starch (TPS) or have a high cost as poly (butylene succinate) (PBS), being necessary to develop blends and composites. This work aims to investigate the potential of using cellulose nanocrystals (CNCs) as a compatibilizer and as a reinforcer in PBS/TPS 70/30 w/w blends. The following films formulations were prepared by twin-screw extrusion and pressing: neat PBS; PBS/TPS 0 wt%; 2 wt%; 4 wt% and 6 wt% CNCs added in TPS phase. PBS/TPS 6 wt% presented compatibilization by CNCs due to adhesion of the two polymers after stretching (SEM images), elongation at break similar to neat PBS (~10%) and glass transition temperatures (Tgs) of PBS and TPS more closer on DMTA diagrams. The recrystallization peak at -96 °C on DSC cooling curve suggests that CNCs migrated to PBS/TPS interface region and became nucleating point in PBS phase. The blends with CNCs provided an effect on tensile strength relative to PBS/TPS without CNCs, indicating the CNCs potential use as reinforcement on PBS/TPS blends. Therefore, CNCs have the potential to reinforce and compatibilize polymer blends.

Keywords: Thermoplastic starch, poly(butylene succinate), cellulose nanocrystals, blends, compatibilizer, reinforcer.
PACS: 88.20.M-

INTRODUCTION

Friendly environmental polymers and its blends has received much attention, this kind of polymers are biodegradable and/or obtained from renewable resources (*1*).

PBS is synthesized by polycondensation of butanediol and succinic acid and it has good mechanical properties, good thermal stability (~350 °C), good processing properties and low melt temperature (~110 °C) (*2-4*). However, the price of PBS is still high when compared to polyolefins (*5*).

TPS consists of two major components, the linear amylose and the highly branched amylopectin (*6, 7*). It is economically feasible and can be produced by plastification process from a variety of native starch sources, including corn, wheat, rice, potato and cassava. However, neat TPS is fragile and hydrophilic (*6, 8, 9*). The mixture of PBS and TPS can balance the disadvantages of both biodegradable polymers (*10*).

The aim of this work was to investigate the incorporation of CNCs as reinforcement and compatibilizer in PBS/TPS blends. Mechanical, thermal and morphological properties are presented to prove the compatibilization and reinforcement of PBS/TPS blends with CNCs.

The glycerol was prepared with CNCs in various weight percentages using disperser homogenizer (ULTRA-TURRAX IKA T18). The cassava starch was mixed with 25 wt% glycerol before dried in an oven at 90°C for 12 h. The cellulose nanocrystals replaces part of starch in the blend. The TPS compositions were plasticized in a twin-screw extruder (Coperion 18 mm) at 300 rpm and the temperature profile of the feed channel to the matrix was 100 °C/105 °C/110 °C/115 °C/120 °C/130 °C/120 °C. The PBS/TPS 70/30 were mixed using a twin-screw extruder

(Baker & Perkins MP-19TC) at 100 rpm the temperature profile of the feed channel to the matrix was 135 °C/140 °C/145 °C/140 °C/135 °C. The nanocomposite blend films were obtained in a single screw extruder with planar die (AX Plastic) at 50 rpm the temperature profile of the feed channel to the matrix was 135 °C/145 °C/160 °C. Followed by pressing (Marconi MA 098/A) at 5 tons for one minute at a temperature of 140 °C. All the films compositions are showed in Table1.

TABLE 1. The films compositions

Films	CNCs [wt%] on TPS	TPS + CNCs [wt%]	PBS [wt%]
PBS/TPS 0%	-	30	70
PBS/TPS 2%	2	30	70
PBS/TPS 4%	4	30	70
PBS/TPS 6%	6	30	70

Results and Discussions

Figure 1a is shown PBS/TBS films with different concentrations of CNCs. The films maintained their transparency with increasing nanocrystals, making its viable the use as transparent packaging.

Figures 1b to h have shown micrographs of cellulose nanocrystal (CNC), the cryogenic fracture surface of films and the cryogenic fracture surface of films stretched.

The nanocrystals have 178 ± 59 nm length, 14 ± 5 nm diameter, and about 13 aspect ratio (length/diameter).

The TPS/PBS blend without CNCs showed dispersed TPS phase in PBS matrix, with voids throughout the matrix (indicated by yellow arrows in Fig. 1d). On the other hand, good dispersion of TPS with nanocrystals in PBS matrix is evidenced by the reduced voids observed (Fig. 1f, Fig. 1g and Fig. 1h). The cellulose nanocrystals were dispersed on the films and adhered also to the PBS phase on the blend up to 4 wt% CNCs (indicated by red arrows in Fig. 1f, Fig. 1g and Fig. 1h).

In Figure 1h was observed the TPS phase not adhered to the PBS phase with 4 wt% CNCs after stretching. In the other hand, TPS phase adhered to the PBS with 6 wt% CNCs after stretching. Therefore, Figure 1i indicates that the CNCs migrate to the interface region of TPS/PBS and anchor the two polymer phases for PBS/TPS 6% composition.

Table 2 shows the results of tensile strength, Young's Modulus and elongation at break of the films were obtained from the tensile tests.

TABLE 2. Tensile properties of neat PBS and PBS/TBS blends with CNCs in various weight percentages

Films	Tensile Strength [MPa]	Young's Modulus [GPa]	Elongation at Break [%]
Neat PBS	34.6 (±3.5)	0.59 (±0.04)	10.1 (±1.4)
PBS/TPS 0%	17.9 (±0.9)	0.51 (±0.04)	6.0 (±0.9)
PBS/TPS 2%	18.5 (±0.8)	0.55 (±0.03)	6.0 (±0.5)
PBS/TPS 4%	21.7 (±0.8)	0.70 (±0.04)	5.1 (±0.2)
PBS/TPS 6%	19.2 (±0.4)	0.43 (±0.01)	9.1 (±0.7)

The addition of 4 wt% of CNCs on TPS phase resulted in a maximum increase on tensile strength and Young's Modulus, and a decrease on elongation at break for the compositions with nanocrystals. The PBAT/TPS films with the addition of nanocrystals are shown also an increase in tensile strength, elastic modulus and elongation at break up to 6% wt of CNCs in relation to TPS phase on the blend without nanocrystals (*11*). The high concentrations of CNCs increase the possibility of forming a rigid continuous network linked cellulose nanoparticles with each other through hydrogen bonds which could increase the strength and hinder the elongation of the chains hamper stretching of the chains (*12, 13*).

PBS/TPS 6% composition showed a drop on tensile strength and elastic modulus with an increase of almost 100% in elongation at break compared to the composition PBS/TPS 4%. The fall on mechanical properties and increase on elongation at break is hard to be explained, other studies with cellulose nanocrystals also report this behavior, could be changing the crystallinity of the polymer (*14*). According to Figure 2h, CNCs migrate to the PBS/TPS interface and compatibilize these two phases, allowing PBS/TPS 6% to have an elongation at break similar to neat PBS (~10%).

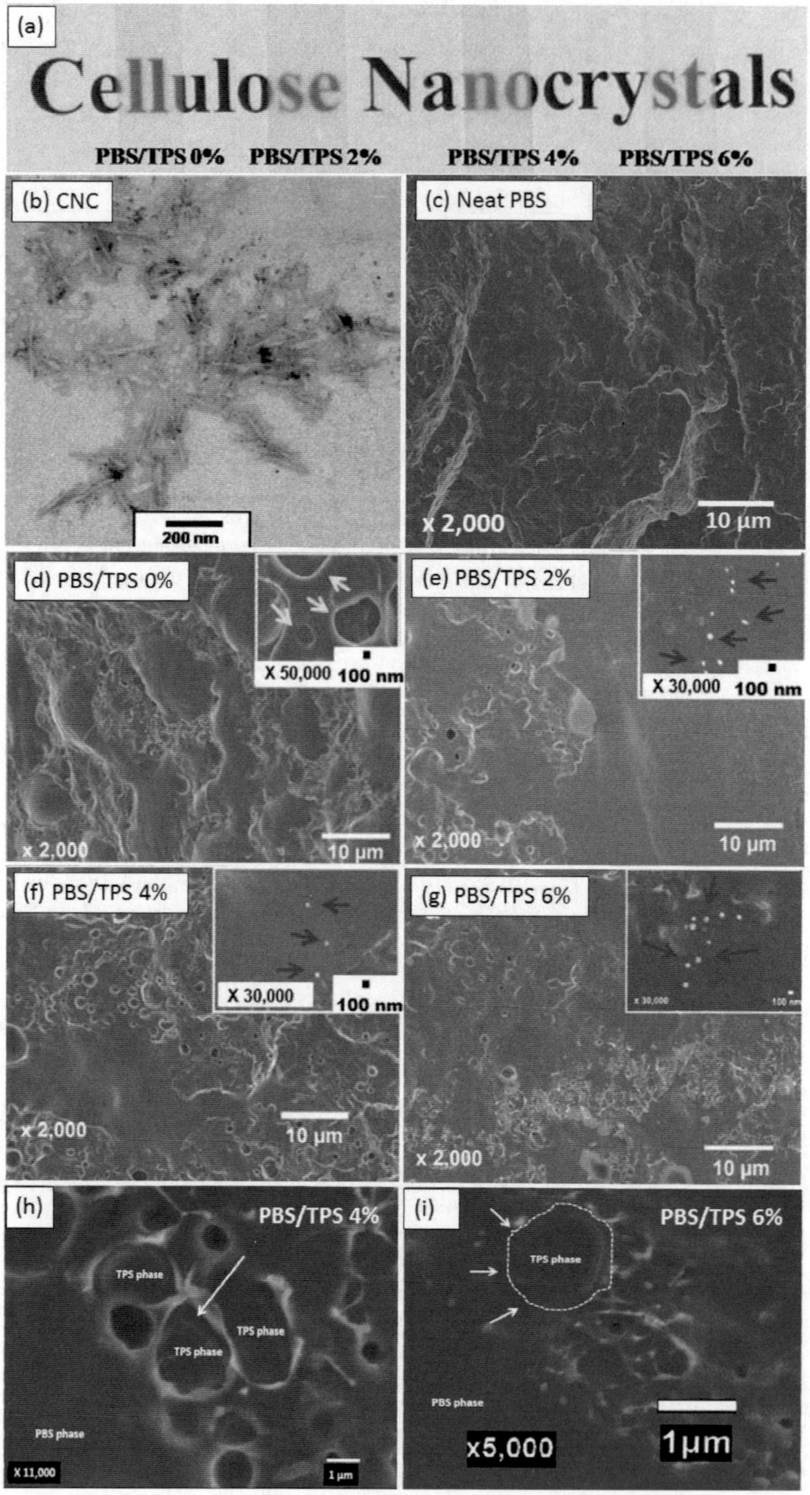

FIGURE 1. (a) PBS/TBS films with CNC in various weight percentages. Micrographs of: (b) CNC; (c) neat PBS; (d) PBS/TPS 0%; (e) PBS/TPS 2%; (f) PBS/TPS 4%; (g) PBS/TPS 6%; (h) PBS/TPS 4% strained and (i) PBS/TPS 6% strained.

Table 3 shows the values of transitions temperatures and crystallinity values relative to PBS phase from DSC, the values of factor tan δ peaks relative to TPS transitions temperatures by DMTA and the onset temperatures (OTT) values obtained by the first degradation on TGA curves.

TABLE 3. Values obtained from the analysis on TGA, DSC and DMTA curves

Films	TGA	DSC (PBS)							DMTA (TPS)	
	T_{onset}	Tg_1 (°C)	Tm_1 (°C)	Ic_1 (%)	Tc (°C)	Tg_2 (°C)	Tm_2 (°C)	Ic2 (%)	Tg (°C)	Td (°C)
Neat PBS	344	-31	107/117	58.8	88	-29	105/114	47.3	-	-
PBS/TPS 0%	286	-26	102/114	48.1	82	-28	102/113	47.0	26	48
PBS/TPS 2%	286	-31	102/115	48.4	82	-29	102/113	49.0	23	60
PBS/TPS 4%	285	-28	103/115	48.6	82	-27	102/113	50.8	26	60
PBS/TPS 6%	281	-31	103/116	49.2	83/96	-30	102/114	49.6	14	60

On cooling, TPS phase affected the crystallization kinetics of PBS phase, reducing the crystallization temperature of PBS. This reduction demonstrates the effect of TPS on PBS molecules. PBS/TPS 6% showed a further crystallization peak at 96 °C. This crystallization peak is an indication that CNCs migrated to PBS/TPS interface and began to act on the crystallization kinetics of PBS phase. This behavior confirms the increase on elongation at break near value of neat PBS, indicating a compatibilization of PBS/TPS as seen in Figure 1i. In Table 3, there were no significant changes in the index crystallinity (Ic) of PBS/TPS 6% in relation to PBS/TPS 4%. Therefore, it is not the crystallinity index that affected the mechanical properties between these two formulations, but the compatibility between PBS/TPS phases through the CNCs is the possible cause for change on mechanical properties. There was no change in the glass temperature (Tg) of PBS for either the first or the second heating with the addition of TPS and CNCs.

The temperature values estimated by the tan δ peak for Tg of TPS and the destructuring of recrystallized TPS phase (Td), shown in Table 3, indicate the effect of the CNCs and their migration to the PBS/TPS interface region for these values. The addition of CNCs increases the thermal stability of the crystals (Td) formed by the recrystallization of the TPS. The migration of the CNCs to the PBS/TPS interface region shifted Tg of TPS to the nearest Tg of PBS, indicating a possible compatibilization of these two polymers, to a certain degree of miscibility (*15*).

It is evident that the decrease in degradation temperature of the blend was caused by glycerol, TPS and CNCs (*16*). An interesting fact is the subtle change in PBS/TPS 6% curve in which a reduction in thermal degradation temperature compares the other blends compositions, proposing that compatibilization affects thermal stability.

Conclusions

The adhere of TPS phase on PBS phase for PBS/TPS 6% stretched observed on SEM micrographs, elongation at break similar of neat PBS and PBS/TPS 6% and glass transition temperatures nearest of PBS and TPS with 6 wt% CNCs than others compositions on DMTA diagrams prove the compatibilizing effect of CNCs in PBS/TPS blends. In PBS/TPS 6%, CNCs migrate to interface region of PBS/TPS, providing a recrystallization peak at -96 °C and became nucleating point in PBS phase, as observed on DSC cooling curve.

The presence of CNCs in blends provided an effect on tensile strength for all the blends films compared with PBS/TPS without nanocrystals, showing the potential of using CNCs as a reinforcement to PBS/TPS blends.

ACKNOWLEDGMENTS

The authors are grateful for the support provided by CAPES, CNPq, PPGCEM/UFSCar and Embrapa.

REFERENCES

1. S. Kalia, L. Avérous, *Biopolymers: biomedical and environmental applications*. John Wiley & Sons, 2011.
2. E. Yoo, S. I, *Journal of Polymer Science Part B: Polymer Physics* **37**, 1357-1366 (1999).

3. C. Ba, J. Yang, Q. Hao, X. Liu, A. Cao, *Biomacromolecules* **4**, 1827-1834 (2003).
4. R. Muthuraj, M. Misra, A. K. Mohanty, *Journal of Polymers and the Environment* **22**, 336-349 (2014).
5. D. Liu, Z. Qi, Y. Zhang, J. Xu, B. Guo, *Starch-Stärke* **67**, 802-809 (2015).
6. G. Moad, *Progress in Polymer Science* **36**, 218-237 (2011).
7. J. Gao, T. Vasanthan, R. Hoover, J. H. Li, *Starch-Starke* **64**, 313-325 (2012).
8. R. Zullo, S. Iannace, *Carbohydrate Polymers* **77**, 376-383 (2009).
9. A. Pereira *et al.*, *Materials Science and Engineering: C* **31**, 443-451 (2011).
10. J. Li, X. Luo, X. Lin, Y. Zhou, *Starch-Stärke* **65**, 831-839 (2013).
11. J. B. da Silva *et al.*, paper presented at the 12th International Conference on Biocomposites, Niagara Falls, Canada, 2012.
12. Y. Lu, L. Weng, X. Cao, *Carbohydrate polymers* **63**, 198-204 (2006).
13. G. Siqueira, H. Abdillahi, J. Bras, A. Dufresne, *Cellulose* **17**, 289-298 (2010).
14. K. Oksman, A. Mathew, D. Bondeson, I. Kvien, *Composites science and technology* **66**, 2776-2784 (2006).
15. M. Pang *et al.*, *Journal of applied polymer science* **107**, 2854-2860 (2008).
16. S. Mohanty, S. K. Nayak, *International Journal of Plastics Technology* **13**, 163-185 (2009).

OBTAINING A BIOPLASTIC BY EXTRUSION – INJECTION MOLDING METHOD FROM RESIDUES OF THE FISHERY INDUSTRY

Aragon-Lucero Irais[a], Hinojosa-Arango Gustavo[b], Felisberti Maria Isabel[c], Gutierrez Miguel Chávez[b]

[a] *Instituto Politécnico Nacional, CIIDIR-Oaxaca,Hornos 1003, Col. Noche Buena, Santa Cruz Xoxocotlán, 71230 Oaxaca. iaragonl1600@alumno.ipn.mx*

[b] *CONACYT-Instituto Politécnico Nacional, CIIDIR-Oaxaca, Hornos 1003, Col. NocheBuena, Santa Cruz Xoxocotlán, 71230 Oaxaca.*

[c] *Institute of Chemistry, University of Campinas. Box: 6154, 13083, Campinas S.P. Brazil.*

Abstract. Partial use of fish species targeted by the fishing industry is an economical and an environmental problem; however, the resulting bone and cartilage waste represents an important source of polymeric materials. This research explored the use of gelatin extracted from fish bones and processed by extrusion – injection molding, techniques that have not been used widely for biopolymers such as gelatins, for the development of a bioplastic to be used by the polymeric industry. Once obtained, the bioplastic was characterized by physic-chemical analyses showing a similar structure to the pig-skin gelatin according to the FT-IR analysis. DMA analysis showed that the Tg was 50 and 60ºC for samples obtained from marlin and shark species, respectively. The ratio of dissolution was less for the bioplastic of Marlin sample. The bioplastic obtained represents an alternative to the plastics from petroleum; in addition, the extrusion-injection technique allows obtaining a wide range of three-dimensional pieces.

Keywords: Gelatin fish, bioplastic, extrusion.
PACS: S11-357

INTRODUCTION

Plastics from petroleum are the main source of solid waste, which represents an important contribution to environmental pollution. Due to this problem, the industry and academia are interested in the study and obtaining of biodegradable materials as a substitute for conventional plastics. Recent research has reported the use of different natural sources to produce bioplastics such as proteins. Gelatin is a biopolymer derived from the hydrolysis of collagen that has been extensively studied for the production of plastic films and composites materials by casting, extrusion and blow processing [1]. However, there are few studies regarding its processing by the extrusion-injection molding method [2]. The objective of this work was to obtain a bioplastic material from fishery residues by means of the extrusion-injection molding method

EXPERIMENTAL

Extraction of fishes´gelatine

Gelatin was extracted from the discarged bones of indo-pacific blue marlin (*Makaira nigricans*) and cartilage from the silky shark (*Carcharhinus falciformis*). The residues were washed and cleaned to remove any remaining meat. The extraction was carried out with an acid-base treatment. The clean material was treated with a sequence of 2% NaOH, 2% H_2SO_4 and 1% citric acid. Each treatment took 45 min respectively, it was performed in duplicate and followed by a wash to pH = 7[1]. The final extraction was carried out with distilled water at 45°C for 24 hrs. The extract was filtered to remove impurities and dried in petri dishes in an oven at 60°C for 16 hrs [3]. After drying, the material was manually sprayed and stored in an airtight container at 25°C.

Bioplastic preparation

After extraction, gelatin was plasticized with glycerol to 10% by weight. The formulation was manually mixed for 10 min prior to extrusion and, lately processed at 120°C 100 rpm for 2 min, in a co-roting extrusor DSM Xplore, model MC-5. The material obtained in the extruder was injected with the following processing conditions: temperature of 120°C in the feed and 45°C in the mold, a pressure of 0.80 bar for 24 seconds in a molding injector DSM Xplore IM-5. The composites were characterized by FT-IR analyze, DMA analysis, Thermo-gravimetric analysis (TGA), Differential scanning calorimetry (DSC), and solubility test.

RESULTS AND DISCUSION

Gelatin characterization

FT-IR spectrum (Figure 1) gelatin extracted from species such as blue marlin and shark, present the same structure as commercial gelatin. We can observe the same functional groups as amide A, B, I, II and III. Principal peaks corresponding to amide A are showed in the $3250\ cm^{-1}$ region and in the region of $3000\ cm^{-1}$ to Amide B and Amide I appears in $1600\ cm^{-1}$, the amide II shows peaks in the region $1520\ cm^{-1}$ and the $1250\ cm^{-1}$ region, the peaks of amide are showed.

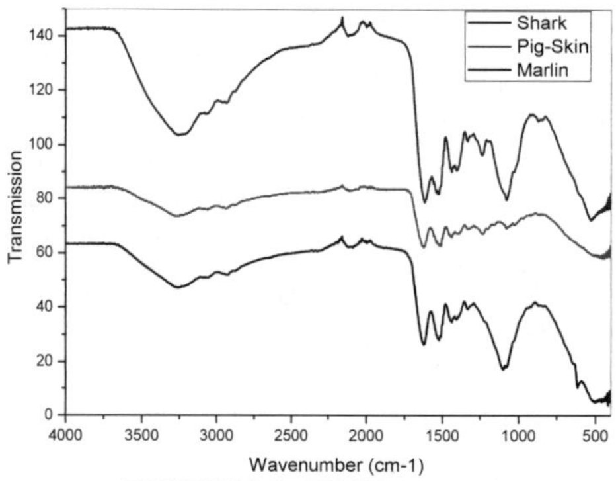

FIGURE 1. FT-IR spectrum

Bioplastic characterization

Figure 2 shows the dynamic mechanical properties of both samples; they have similar values of Tg, around 60°C, from storage and loss modulus. The bioplastic obtained from shark´s presents a fall at lower temperatures near Tg, most likely because the storage module is associated to the elastic characteristics of the material. Subsequently, the drop in the storage modulus for bioplastic of marlin is observed, probably because its Tg is at higher temperatures. The drop of the storage modulus near the Tg results from the softening of the material, that increases in the mobility of the polymer chains. The drop in the Loss modulus corresponding the β transition according to the amorphous fraction in the bioplastic.

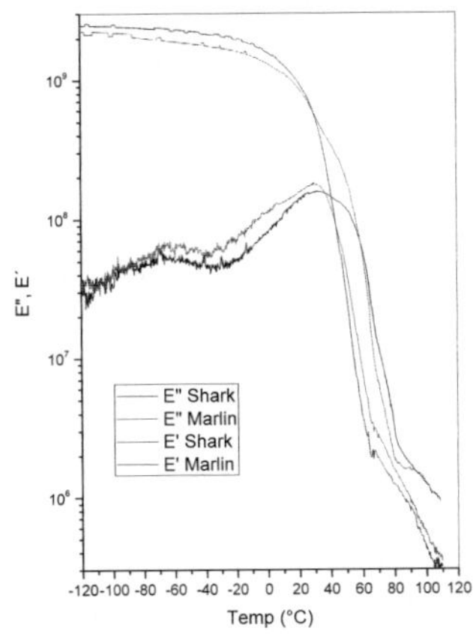

FIGURE 2. Dynamic mechanical properties Modulus E´ and E”

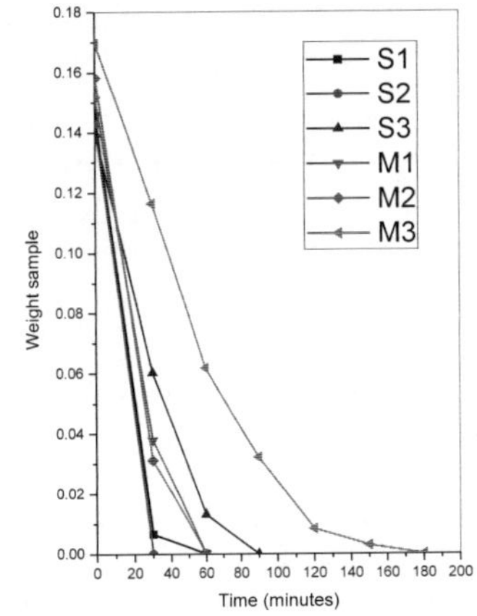

FIGURE 3. Dissolution time of Gelatin

Bioplastic from Marlin presented a better ratio of dissolution during 180 minutes (M3) in saline solution compared to the shark gelatin in the same solution, which was 90 minutes (Fig. 3). Samples placed in distilled water had a dissolution time of 30 minutes and drinking water had a dissolution time of 60 minutes for shark and marlin respectively, this can be associated with the hydrogen bridge bonds that are formed between the molecules from distilled and drinking water and the gelatin.

Samples experienced an initial weight loss of 7% at 200°C because of the presence of moisture in the sample (Fig. 4). A drop in the percentage of residual weight in 200 ° C and 400 ° C is associated with degradation of the material, DSC analyze show a drop near to 175 and 180ºC corresponding to melting temperature of bioplastic, in the DSC we can observed the same value of Tg, corresponding with the DMA analyzed showed in the Figure 5. This characteristics avoid processing the bioplastic by extrusion – injection moulding.

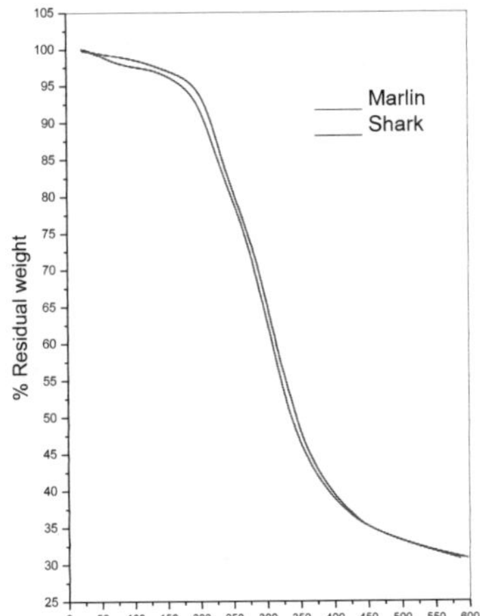

FIGURE 4.TGA results as weight loss curve

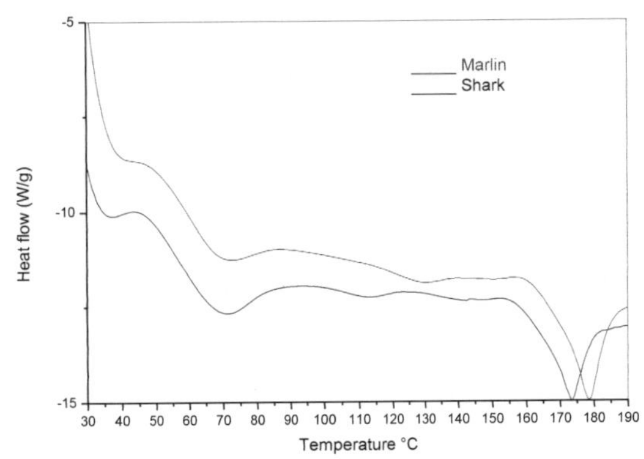

FIGURE 5.DSC first heating curve

CONCLUSION

Our methodology for the transformation of fisheries residues into bioplastic is an innovative approach for the integral use of targeted fish species. Gelatin samples obtained from blue marlin and shark cartilage bones were successfully prepared by the extrusion molding method. Shark´s Gelatin was the best option for the processing with a Tg of 50. The degradation Temperature for protein before and after extraction was 200 ° C according to TGA analysis; allowing its processing by the extrusion-injection method.

ACKNOWLEDGMENTS

Institute of chemistry (UNICAMP), CONACYT and IPN for the fellowship granted to Irais Aragón Lucero

REFERENCES

1. R. Jeya Shakila, E. Jeevithan, A. Varatharajakumar, G. Jeyasekaran, D. Sukumar. *Functional characterization of gelatin extracted from bones of red snapper and grouper in comparison with mammalian gelatin.* India. LWT - Food Science and Technology, 2012, pp. 30-36
2. A. Etxabide, K. de la Caba, P. Guerrero. *A novel approach to manufacture porous biocomposites using extrusion and injection moulding*. Spain. European Polymer Journal, 2016, pp. 324-333
3. J. M. Koli, S Basu, B. B. Nayak, S B. Patange, A.U. Pagarkar, V. Gudipatia. *Functional characteristics of gelatin extracted from skin and bone of Tiger-toothed croaker (Otolithes ruber) and Pink perch (Nemipterus japonicas*. India. Food and Bioproducts Processing, 2012, 555-562

PLA-Blends With Modified Starch Granules

Drying And Crosslinking Of Native Potato Starch As Well As Subsequent Blending With PLA In A One-Step Compounding Process

Johannes Fuchs[a], Maik Feldmann[a], Hans-Peter Heim[a]

Institute of Material Engineering, Polymer Engineering, University of Kassel, Germany

Keywords: Starch, Crosslinking, PLA, Compounding, Biopolymer, Bioplastic

Native starch is a suitable raw material for the synthesis of bioplastics (i.e., PLA), or as a filler for thermoplastic materials owing to its low costs. However, the compounding and injection molding processes must be modified to process native starches. Moreover, the hydrophilic properties necessitate a drying of the material before processing. The hydrophilic characteristics also restrict the number of fields in which starch blends can be applied. In order to make these materials able to compete with comparable materials, an enhancement of the mechanical properties is compulsory.

By means of newly developing, or tailoring the preparation and processing methods, it is possible to modify the starch, and, thus, increase the number of application fields in which it can be employed. Drying of the starch prior to the compounding process leads to significantly higher mechanical properties, already. Amongst others, Epichlorohydrin has been used to crosslink native starches. One disadvantage of Epichlorohydrin is its toxicity. Due to this handicap, starch-blends modified with Epichlorohydrin are limited in their areas of application (i.e., food packaging). Another option as reagent to modify native starch is nonhazardous Glycidylether. This cross-linking agent even shows a high reactivity at low temperatures using a catalyst. Up to now, several process steps are required to dry, modify and blend native starch with PLA. Due to time and cost expenditures it is necessary to improve the compounding process (fewer process steps) to make starch-blends competitive.

In the first instance, native potato starch Superior (Emsland) was modified in a two-step process. Therefore the starch was crosslinked using a laboratory mixer (MTI). Subsequently, the previously modified starch and PLA 3052D (NatureWorks) were compounded using a co-rotating twin screw extruder ZSE18 HPe (Leistritz). In the injection molding process, the blends were processed into samples. In order to determine the influence of the modification and process control on the hydrophobization and material properties of the starch blends, the materials were then examined regarding their thermomechanical and structural properties. Compared to conventional starch blends with moist or even pre-dried starch, the blends with modified starch made in the two-step process exhibited higher mechanical properties. Following, a novel one-step compounding process for drying and crosslinking the starch and blending with PLA was developed by modifying a customary co-rotating twin screw extruder.

The crosslinking of the native starch resulted in an increase in the mechanical properties (tensile strength, impact strength etc.) of the starch blends. Blends with a content of 50 weight percent starch show comparable mechanical properties to unfilled PLA. Furthermore, the blends with crosslinked starch display lower levels of moisture absorption according to blends with native starch.

ACKNOWLEDGMENTS

The authors would like to thank the "Fachagentur Nachwachsende Rohstoffe e.V. (FNR)", the "Hessen State Ministry of Higher Education, Research and the Arts" and the "Federal Ministry of Food and Agriculture (BMEL)" of the Federal Republic of Germany for their financial support for this investigation.

Chitosan and chitin based-Nanofiber Biosorbents for efficient removal of Zinc from wastewater

Botes Alicia [(1)], Sinha Ray Suprakas [(2)]*, Lutz Marietjie [(3)], Van Reenen Albert Johannes [(3)]

[(1)] University of Stellenbosch and CSIR - Gauteng - SouthAfrica, [(2)] National Centre for Nano-structured Materials CSIR - Gauteng - SouthAfrica, [(3)] Stellenbosch University - Western Cape - SouthAfrica

Heavy metal contamination is causing serious damage to the environment and posing dangerous health risks to humans1. It is the aim of this project to investigate the adsorption capacity of chitin derivatives such as chitosan and chitin nanowhiskers (chnw), with biodegradable and eco-friendly characteristics, for various heavy metals2,3. The investigation focuses on the possibility of using biopolymers created in our lab and comparing it to existing biosorbents such as pristine chitosan and also testing the behaviour of these biopolymers under varying pH and with varying heavy metal-ion concentration and types. Electrospinning was used to synthesise chnw/EVOH nanofibers and CTS nanofibers. The adsorption capacity for these biosorbents were then tested for Zn+2 metal cation using isotherm models from inductively couple plasma (ICP) analysis and EDX-SEM. The preparation and synthesis of the nanofibers and the nanowhiskers were characterized using transmission (TEM), scanning (SEM) and confocal microscopies as well as X-ray diffraction (XRD). The biopolymers synthesized in our lab showed an improved morphology and more uniform surface area compared to the pristine CTS. As expected, the pH of the solution influenced the adsorption capacity significantly for the biosorbents. Generally the adsorption capacity improved as the pH became more neutral, but there were important physiochemical properties for CTS nanofibers and chnw/EVOH nanofibers that affected the behaviour of these biosorbents at higher and lower pH. The concentration of the Zn salt also played an important role in the adsorption capacity (qe)4. The general trend showed that qe increased for higher heavy metal ion concentration. Chitosan nanofibers showed good adsorption at pH 5 and chnw/EVOH nanofibers showed good adsorption at pH 11. The maximum qe is 9.9 mg/g for CTS powder after 30 min and 4.45 mg/g for CTS nanofiber after 60 min at pH 5.

PBT AND BIO-BASED PTT REINFORCED WITH MAN-MADE CELLULOSE FIBERS

Nicole Gemmeke, Maik Feldmann, Hans-Peter Heim

University of Kassel

Keywords: polybutylene terephthalate, polytrimethylene terephthalate, man-made cellulose fiber, bio-based composites

Cellulosic fibers are based on natural resourced cellulose from softwood. To generate man-made cellulose fibers neat viscose go through a spin bath recovery before they were wound into bobbins. Those fibers have a higher degree of ductility and a lower density compared to glass fibers. The man-made cellulose fibers can be used for high performance tires, hoses and in composites as a natural filler for thermoplastic materials to increase the strength and the toughness and to reduce the weight of the compound at once. Due to their good properties, thermoplastic cellulose composites offer a high degree of crash-resistance, which is an important factor for the automotive industry. The challenge of compounding cellulosic fibers with technical polymers is to take care of the natural fibers when using high temperatures during the manufacturing process.

Technical polymers like PBT and bio-based PTT need high processing temperatures (>230°C). PBT is based on the petrochemical monomer butanediol while PTT is based on the biological monomer propanediol.

In the first instance, man-made cellulose fibers and PBT/PTT with different percentage by weight were compounded using a co-rotating twin screw extruder. The cellulose fibers with a length of 2 mm were used. Different additives and special screw designs were selected to protect the fibers. In the injection molding process, the blends were processed into samples. In order to determine the influence of different percentage by weight and material properties the cellulosic reinforced PBT/PTT compounds were examined regarding their thermomechanical and structural properties. Also the influence of different additives and screw designs were tested.

The investigation showed that the mechanical properties like tensile strength and Young´s modulus rise with increasing fiber content. The results of the notched Charpy tests revealed a very high level of impact strength with a fiber content of 30 percentage by weight. These samples reached a six times higher impact strength compared to neat PBT/PTT and also a higher level in comparison with standard reinforced materials like glass fibers. The heat deflection temperature (HDT-A) rises from 65 °C (PBT) to 150°C (PBT with 30 wt. % of man-made cellulose fibers).

Furthermore, with the possibilities regarding lightweight construction and energy absorption capacity the reinforced cellulosic fiber blends have a great potential for applications in vehicle construction. The results illustrate the potential usage and effectiveness of bio-based fibers as reinforcement in engineering polymers such as PBT and bio-based PTT.

ACKNOWLEDGMENTS

The authors would like to thank the “Fachagentur Nachwachsende Rohstoffe e.V. (FNR)”, the “Hessen State Ministry of Higher Education, Research and the Arts” and the “Federal Ministry of Food and Agriculture (BMEL)” of the Federal Republic of Germany for their financial support for this investigation.

Improving PCL's properties via morphological manipulation of biodegradable PCL/PLA compounds

Ramezani Kakroodi Adel (1), Kazemi Yasamin (1), Chu Raymond KM (1), Rodrigue Denis (2), Park Chul B (1)*

(1) University of Toronto - Ontario - Canada, (2) Laval University - Quebec - Canada

Improvement of the interfacial adhesion in blends of PCL and PLA has attracted the attention of the scientific community due to their complimentary mechanical properties. In this work, in situ nanofibrillation has been proposed as an ideal approach towards the green and economical production of PCL/PLA blends with an unparalleled level of compatibility of the blend components. Scanning electron microscopy showed that the transformation of the dispersed PLA phase into stretched nanofibrils drastically improved the blends' surface morphologies. Investigations on the blends' crystallization behaviors using polarized optical microscopy, differential scanning calorimetry, and wide angle X-ray scattering confirmed that the improvements in the nanofibrillar composites' (NFCs) interfacial adhesion were caused by an increase in the PCL's heterogeneous crystal nucleation on the nanofibrillated PLA domains. This behavior was ascribed to a promotion of the epitaxial crystallization of the PCL on the nanofibrillated PLA's crystals following a change in the orientation of the PLA's crystal lamellae. The PCL's tensile modulus and strength were significantly improved after the inclusion of even small quantities (as low as 5 wt.%) of the PLA nanofibrils. The tensile modulus of the NFC with 15 wt.% of the PLA was 192% higher than that of the neat PCL. Dynamic mechanical analyses and shear rheology measurements showed substantial improvements in the PCL's elastic behavior after nanofibrillation. This effect was shown to improve the PCL's tensile strength via delaying the yield point during tension. The NFCs also showed dramatically improved oxygen barrier properties due to the improvements in the PCL-PLA compatibility.

Fatigue behavior of short fiber bio-based composites for engineering applications

Maik Feldmann and Jan-Christoph Zarges

University of Kassel, Institut für Werkstofftechnik - Kunststofftechnik – Germany
Moenchebergstr. 3, 34125 Kassel
Tel.: +49 561 804-2837
E-Mail: feldmann@uni-kassel.de

Abstract. Bio-based plastics and their composites become more important in the last years. To provide these materials also in technical application, new aspects of stresses must be taken into account during characterization. Depending on the later application and besides other stresses, moisture and temperature is an important issue for most bio-based plastics and composites. In this study, different composites based on bio-based polyamide were manufactured and investigated. Besides fatigue tests with dry-state and conditioned (ISO 1110) specimens at constant load and increasing load, common mechanical properties were measured. The fatigue tests with constant load were performed with 100.000 cycles and the residual strength of the unbroken specimen was subsequently determined afterwards. It can be seen, that the compounds with cellulose fibers show similar breaking stress in the increasing load fatigue test compared to composites with glass fibers with the same fiber weight content. Similar results were observed with the conditioned specimen according to ISO 1110. At constant load, it can be observed, that the damage behavior of glass fiber and cellulose fiber composites differs. This leads to higher relative residual strengths at the subsequently performed tensile test. The reason for this behavior is probably the difference fiber-matrix adhesion and the different fiber properties.

Keywords: Biocomposites, Renewable Resources, Fatigue, Ageing, Injection Moulding, Mechanical Properties, Bio-based Polymers, Cellulose Fibers

Biocomposites : Innovative continuous processes for series production

***Ouali Ahmed-Amine* [1]*, *Rinberg Roman* [1], *Kroll Lothar* [1], *Nendel Wolfgang* [1], *Todorov Aleksandr* [1], *Cebulla Holger* [1]**

[1] Chemnitz University of Technology - Chemnitz - Germany

The development of innovative bio-based composites with high-efficient manufacturing processes is the purpose of the project C4 in the framework of the Excellence Cluster MERGE EXC 1075, funded by DFG (Deutsche Forschungsgemeinschaft). Natural fibers composite (NFC) prepregs have been recently investigated as a potential cost-efficient semi-finished product. By means of continuous production processes, prepregs rolls can be manufactured with unidirectional natural fibres (flax) fabrics as reinforcement and thermoplastic biopolymers films as matrix. The used natural fibre fabrics are made of high twisted yarns. For a better impregnation and higher stiffness properties, non-crimp fabrics with non-twisted yarns represent an appropriate solution. A second option is the use of low cost thermoplastic spunlace fabrics as a matrix material. With a higher permeability, those facilitate the release of humidity of natural fibers for impregnation. This can result in higher fiber ratio in NFC. With these materials, NFC prepregs processes are implemented by an innovative continuous impregnation machine system, which is composed of calender rolls and PTFE-coated metal belts, which permits a 6-meter-long material impregnation. A continuous drying of natural fibers with well-studied parameters is implemented directly before the impregnation step. With these material and process developments suitable for natural fibres, mechanical performances of NFC tend to be improved. From prepregs to finished parts can be implemented by discontinuous processes, with compression molding, or by continuous processes, with devices gathering several stages such as cutting, stacking, pre-heating and back injection molding. These innovative technologies can be incorporated into new concepts in several fields such as automotive industry or sport equipment. The mechanical properties profile of these biocomposites has been here studied in detail and has confirmed their high application potential.

Degradation of PLA During Long-Term Storage

Nikola Kocić[a], Marieluise Lang[a], Thomas Hochrein[a], Martin Bastian[a]

[a]*German Plastics Center SKZ*
97076 Wuerzburg, Germany

Abstract. The degradation of PLA proceeds through hydrolysis of the ester linkages in the polymer backbone. This leads to a significant reduction of molecular weight and thus to a deterioration of mechanical properties. However, up to now there are no studies of PLA-hydrolysis performed under practical conditions and long-term storage. In this work, the influence of molecular weight, pellet size, crystallinity, air humidity and temperature on the hydrolytic degradation of PLA was investigated. Two commercially available PLA types with different molecular weights were used in this study. Three different pellet sizes were prepared by compounding with a co-rotating twin-screw extruder. The pellets were annealed at defined temperatures so that crystallinity degrees in three different magnitudes were generated. Finally, all pellets were stored at different temperatures in the range of 10 to 50 °C and a relative humidity of 33 and 75 % for nine months. During this time, the viscosity of the samples was monitored with a rotational rheometer. A simple linear function was used to define the relationship between the weight average molecular weight and the zero viscosity in logarithmic form. By fitting the zero viscosity in the simplified kinetic equation, the degradation rate constants for each sample at each temperature were determined and used afterwards in the Arrhenius equation to calculate the activation energy of the hydrolytic degradation for each sample. Temperature and air humidity were identified as the parameters with the biggest impact on the hydrolytic degradation of PLA. The degradation rate constants and thus the rate of the hydrolytic degradation, increase significantly with increasing temperature and air humidity. Furthermore, small pellet size, high crystallinity and high molecular weight increase the activation energy and thus reduce the rate of PLA degradation during long-term storage.

Keywords: Biopolymers, Hydrolysis, Degradation
PACS: 81.05.Lg; 82.35.Lr; 82.35.Pq; 82.40.–g; 83.80.Sg; 89.20.Bb; 36.20.Cw

INTRODUCTION

Current research around the world on PLA modification is focused on producing high performance, biobased materials that can compete with conventional plastics in durable applications [1], [2], [3]. [4], [5]. In such case, the material has to be stable against degradation under the conditions of storage and use.

The hydrolytic degradation of PLA plays a key role in many applications and thus has been investigated in various contributions [6], [7], [8], [9], [10], [11] [12], [13]. In aqueous environments, PLA degrades by simple hydrolysis of the ester bonds [14]:

$$-COO- + H_2O = -COOH + -OH \qquad (1)$$

The rate of reaction (1) depends on various factors, which can be divided into two main groups: factors affecting the accessibility and factors affecting the reactivity of the ester groups of PLA [7]. To the first group belongs e.g. crystallinity of PLA as well as the parameters, which influence the mobility of PLA-macromolecules like e.g. molecular weight, presence of plasticizers and temperature. The hydrolytic chain cleavage proceeds preferentially in the amorphous regions, because water can intrude easily into these domains [12]. Therefore, most published works suggest that an increase in polymer crystallinity will decrease hydrolysis rate [8], [11]. Furthermore, lower molecular weights, presence of plasticizers as well as higher temperatures increase the mobility of the PLA-chains and therefore favor the hydrolysis [7], [13]. The hydrolysis of PLA is in the simplest case controlled by second order kinetics. Therefore, the rate of reaction (1) depends on the concentrations of the reactants, as follows [9], [10]:

$$\frac{d[COOH]}{dt} = k_2[-COO-][H_2O] \qquad (2)$$

In eq. (2) k_2 is the second order hydrolysis rate constant, while $[-COO-]$ and $[H_2O]$ are the ester and water concentrations, respectively. Substitution of $[COOH] = [-COO-]M_0 M_n^{-1}$ in eq. (2), where M_n is the number average molecular weight of PLA and M_0 is the molecular weight of the monomer and integration of eq. (2) results in [9]:

$$\frac{M_n^0}{M_n} = 1 + \frac{M_n^0 k_2 [H_2O]}{M_0} t \quad (3)$$

In eq. (3) M_n^0 and M_n are the number average molecular weight of PLA at the beginning of hydrolysis and at time t, respectively. Furthermore, it is well-known, that the hydrolysis of PLA can be catalyzed under acidic conditions [6]. Thus, the presence of acid increases the reactivity of PLA-ester groups, leading to a increased reaction rate. Also the end carboxylic groups of PLA can act as catalyst. The concentration of end carboxylic groups of PLA is proportional to 1/M at any time, where M is the molecular weight of the PLA chain [9]. Accordingly, the lower the M, the higher the concentration of carboxylic end-groups and thus the faster the hydrolytic degradation of PLA. In addition, the carboxylic groups newly formed as a product of hydrolysis can catalyze the reaction. If so, the rate of hydrolytic degradation is described by the third order rate equation [10]:

$$\frac{d[COOH]}{dt} = k_3 [-COO-][H_2O][COOH] \quad (4)$$

where k_3 is the third order hydrolysis rate constant. Substitution of $[COOH] = [-COO-]M_0 M_n^{-1}$ in eq. (4) and integration, results in [9]:

$$ln \frac{M_n^0}{M_n} = k_3 [-COO-][H_2O]t \quad (5)$$

After polymerization and compounding steps, PLA pellets are often stored for a longer period of time before final processing. During storage, the pellets are in contact with a certain amount of humidity from the air even in sealed begs, which causes the hydrolytic degradation of PLA. This can leads to a significant reduction of molecular weight and thus to a deterioration of mechanical properties, e.g. Young's modulus, tensile strength, elongation at break or impact strength, which are important for durable applications. In order to maintain the high quality and functionality of PLA over a long period of storage, some aspects of material storage have to be considered. However, up to now, hydrolysis of PLA was investigated only under intensified conditions in different solutios and over a short period of time. The main purpose of this study was to identify the critical factors affecting the hydrolytic degradation of PLA during material storage for the purpose of obtaining practical guidelines that can be used to minimize degradation during long term storage of PLA pellets. Therefore, the influence of molecular weight, pellet size, crystallinity, air humidity and temperature on the hydrolytic degradation of PLA was investigated. The experiments were conducted under practical conditions over a period of 9 months.

EXPERIMENTAL

Materials

Two semi-crystalline PLA-grades from NatureWorks, having different molecular weights and comparable contents of D-lactic acid of approx. 1.4 wt%, were used in this study: Ingeo 6251D, designated as PLA1, having a lower weight average molecular weight (M_w) of approx. 10^5 g/mol and Ingeo 4032D, having a higher M_w of approx. $1.6 \cdot 10^5$ g/mol labeled as PLA3. For the correlation between zero viscosity and M_w, an additional PLA grade Ingeo 3001D was used. This grade contains the same amount of D-lactic acid as 6251D and 4032D and the medium M_w of approx. $1.3 \cdot 10^5$ g/mol.

Sample Preparation and storage conditions

All compounds were prepared by melt compounding on a Leistritz (model ZSE 18 Hpe) corotating twin-screw extruder (L = 810 mm and D = 18 mm) with a felling temperature profile from 20 to 180 °C (die), screw speed of 400 min^{-1}, and throughput of 6 kg/h. The pelletizer speed was adjusted to obtain pellets with the same length and three different diameters, signed as G1 (smallest), G3 (medium) and G5 (largest) diameter. After compounding, the pellets were thoroughly dried at 40 °C for 10 h in a drying chamber to achieve a residual humidity of less than 300 ppm. Afterwards the pellets were annealed at 82 °C for 20 min and at 100 °C for 22 min to obtain different crystallinity, denoted as CM (medium crystallinity) and CH (high crystallinity), respectively. The samples without annealing were signed as CL (low crystallinity). After annealing, the pellets were stored in desiccators, which were tempered at 10, 20, 30, 40 and 50 °C. The desiccators were equipped with agitators and filled with magnesium or sodium chloride to achieve a uniform and constant humidity of 33 % and 75 %, respectively.

Sample Characterization

The crystallinity of the samples was measured after annealing and before storage using a nitrogen atmosphere in a Netzsch 204 F1 Phoenix differential scanning calorimeter. Indium, zinc, lead, bismuth, and tin standards were used for temperature calibration. Samples of 15 ± 0.5 mg were cut from the pellets and first cooled to 0 °C with a cooling rate of 10 K/min and afterwards heated to 200 °C with the same rate. Three measurements were performed for each sample and all calculated crystallinity-values are the averages obtained from the three measurements.

The surface quality of the pellets was visualized with a high-resolution, field emission scanning electron microscope Supra 40VP (Carl Zeiss). The pellets-dimensions were measured using optical light microscopy Axioskop 2 MAT (Carl Zeiss) equipped with AxioCam MRc. Ten measurements were performed for each sample and all diameter and length values are the averages obtained from the ten measurements.

Rheological measurements were carried out in regular time intervals over nine months on Haake Mars rotational rheometer from Thermo Fisher Scientific. The samples were dried for 4 h at 80 °C before each test to remove the moisture. The measurements were run with 20 mm parallel plate geometry and a gap of 1.0 mm at a temperature of 180 °C and a constant deformation of 0.01. The complex viscosity was monitored over angular velocity. Afterwards the zero shear viscosity values were determined as a complex viscosity at an angular velocity of 0.7 rad/s.

Weight average molecular weight of PLA, necessary to establish the correlation with molecular weight, was determined using a gel permeation chromatography (GPC) system equipped with four PSS SDV separating columns (5 µm bead size, pore size range from 100 – 10^5 Å) connected in series and a refractive index detector PSS-SECcurity 1260. The measurements were performed at 25 °C with THF as the mobile phase at a flow rate of 1.0 ml/min. Sample concentrations and injection volume were 3.0 g/ml and 100 µl, respectively. The data were analyzed using the software PSS-WinGPC UniChrom Version 8.1.

RESULTS AND DISCUSSION

Hydrolysis Kinetics

By neglecting the change of the polydispersity index ($PDI = M_w/M_n$) during the hydrolysis as well as by asuming the water and ester concentrations to be constant and by using the well-known linear relationship between the zero viscosity (η) and weight average molecular weight (M_w) in form of [15]:

$$log\,\eta = log\,K + \alpha\,log\,M_w \tag{6}$$

the eq. (3) and (5) change to:

$$\left(\frac{\eta^0}{\eta}\right)^{\frac{1}{\alpha}} = 1 + c_2 t \tag{7}$$

$$ln\frac{\eta^0}{\eta} = \alpha c_3 t \tag{8}$$

In equations (6) - (8) K is a constant at a given temperature [16] and the slope α contains information on the chain entanglement state [17], while η^0 and η are the measured zero viscosity before storage and after time t, respectively. The constants c_2 and c_3 are related to the second and third order rate constants k_2 and k_3, as follows:

$$c_2 = \frac{M_n^0 k_2 [H_2O]}{M_0} \tag{9}$$

$$c_3 = k_3[-COO-][H_2O] \tag{10}$$

If the left terms of the eq. (7) and (8) are plotted against t, straight lines with the slopes c_2 and αc_3 should be obtained, when the models represent the hydrolysis reaction adequately. Furthermore, the constants c_2 and c_3 can be used in the well-known logarithmic form of Arrhenius equation, as follows:

$$\ln c_2 = \ln \frac{AM_n^0[H_2O]}{M_0} - \left(\frac{E_a}{R}\right)T^{-1} \tag{11}$$

$$\ln c_3 = \ln \frac{A}{[-COO-][H_2O]} - \left(\frac{E_a}{R}\right)T^{-1} \tag{12}$$

In eq. (11) and (12) A is the pre-exponential factor, R is the universal gas constant and E_a is the activation energy, which is temperature independent and represents a minimum energy, required to start the hydrolytic degradation.

Experimental Results

The example of SEM and light microscopy pictures of the obtained pellets are presented in Fig. 1.

FIGURE 1. Scanning electron microscopy (a, b) and light microscopy (c, d) of the sample P1_G5 as example

As it can be seen from SEM-pictures (see Fig. 1, (a) and (b)), non-porous pellets of cylindrical shape, with a smooth surface were obtained. Afterwards, diameter and length of the pellets were measured by light microscopy, as presented in Fig. 1, (b) and (c). The determined values are presented in Fig. 2. All pellets possess the same length but different diameters, that can be classified in three categories: low diameter of approx. 1.5 mm, denoted as G1, medium diameter of approx. 1.8 mm, denoted as G3 and high diameter of approx. 2.6 mm, denoted as G5.

The DSC-thermograms of the three examples presented in Fig. 3 (a) display a shift in the signal baseline at about 65 °C, related to the glass transition. Furthermore, all three samples show similar endothermic melting peaks. Because of its unfavorable structure, PLA crystallizes very slowly. Therefore, the majority of the sample P1_G3_CL directly after compounding and without annealing remains amorphous. Nevertheless, the increased PLA chain motions during the DSC heating step allow a rearrangement of PLA chains of amorphous phase into the crystal structure during the so-called cold crystallization between approx. 108 and 127 °C (see Fig 3(a)). The crystals formed during the cold crystallization of the sample P1_G3_CL show a melting point between approx. 160 and 182 °C.

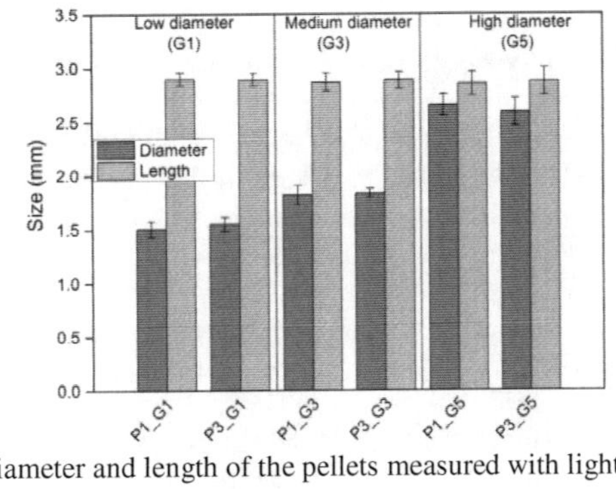

FIGURE 2. Diameter and length of the pellets measured with light microscopy

In contrast to that, the annealing of the sample P1_G3_CM (at 82 °C for 20 min) increases the amount of the crystalline phase. For higher portions of the crystalline phase formed during the annealing, less "crystallizable" material remains to crystallize during the cold crystallization. Therefore, the enthalpy of the cold crystallization (surface under the peak) of the sample P1_G3_CM is lower compared to that of the sample P1_G3_CL. Furthermore, the crystals formed during annealing can probably act as heterogeneous surface i.e. nucleation agents in the cold crystallization of the sample P1_G3_CM, thus leading to a shift of cold crystallization towards lower temperatures compared to P1_G3_CL. The sample annealed at 100 °C for 22 min has no cold crystallization peak. The enthalpy of fusion (ΔH_m) obtained from the heating step, the enthalpy of the cold crystallization (ΔH_{cc}) and the enthalpy of fusion of an infinitely thick crystal ΔH^0 = 93,6 J/g [18] were used to calculate the percentage crystallinity x_c as $100(\Delta H\text{-}\Delta H_{cc})/\Delta H^0$. The results are presented in Fig. 3 (b).

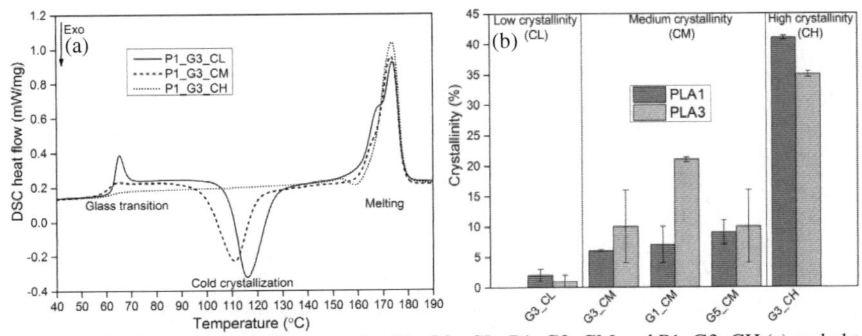

FIGURE 3. Examples of DSC-thermograms of the samples P1_G3_ CL, P1_G3_CM and P1_G3_CH (a) and obtained percentage crystallinity of the samples (b)

Accordingly, the samples were divided in three regions: low crystallinity (x_c = 1 - 2 %) medium crystallinity (x_c = 6 - 21 %) and high crystallinity (x_c = 35 - 41 %). Afterwards, the pellets were stored at five different temperatures in the range of 10 to 50 ºC and a relative humidity of either 33 % or 75 %. This results in a total of 75 samples which are summarized in Table (1).

The molecular weight distribution (MWD) of the sample P3_G3_CM before and after three months storage (at 50 °C/75 % r.h.) is presented as example in Fig. 4 (a). From this results different facts can be deduced. Firstly, the considerable hydrolytic degradation of PLA pellets occurs in material during the storage of 3 months at 50 °C and 75 % r.h. Consequentially, the molecular weight distribution curves are shifted toward lower molecular weights. Accordingly, the weight average molecular weight of the sample is drastically reduced from approx. $1.6 \cdot 10^5$ g/mol before to approx. $2.2 \cdot 10^4$ g/mol after the storage. Secondly, the shape of the distribution curves is changed after the storage. This can be explained by the fact, that the hydrolysis occurs predominantly in the amorphous phase of the polymer, because water can intrude easily into these domains [19]. Thus, the amorphous parts of the material degrade by random scission of its ester bonds along the polymer chain, while the macromolecules of the crystalline phase stay intact. This leads to a multimodal distributions of molecular weight [12], as presented in Fig. 4 (a). Furthermore, weight and number average molecular weights are used to calculate the polydispersion index (*PDI*) as

M_w/M_n before and after the storage. In spite of the slightly change of the shape of distribution curves, the change of *PDI* (see Fig.4(a)) is minor and thus can be neglected during development of eq. (7) and (8).

TABLE (1). Sample nomenclature and storage conditions

Sample name	M_W	Pellets diameter	Crystallinity	Rel. humidity (%)	Storage T (°C)
P1_G3_CL	Low	Medium	Low	33, 75	10, 20, 30 40, 50
P1_G3_CM	Low	Medium	Medium	33, 75	10, 20, 30 40, 50
P1_G3_CH	Low	Medium	High	33, 75	10, 20, 30 40, 50
P1_G1_CM	Low	Low	Medium	33, 75	10, 20, 30 40, 50
P1_G5_CM	Low	High	Medium	33, 75	10, 20, 30 40, 50
P3_G3_CL	High	Medium	Low	75	10, 20, 30 40, 50
P3_G3_CM	High	Medium	Medium	75	10, 20, 30 40, 50
P3_G3_CH	High	Medium	High	75	10, 20, 30 40, 50
P3_G1_CM	High	Low	Medium	75	10, 20, 30 40, 50
P3_G5_CM	High	High	Medium	75	10, 20, 30 40, 50

P1: Mw (low) = 100 000 g/mol G1: Pellets diameter (low) = 1.5 mm CL: Crystall. (low) = 1 - 2 %
P3: Mw (high) = 160 000 g/mol G3: Pellets diameter (medium) = 1.8 mm CM: Crystall. (medium) = 6 – 21 %
G5: Pellets diameter (high) = 2.6 mm CH: Crystall. (high) = 35 – 41 %

To establish the correlation between weight average molecular weight and zero viscosity, three PLA grades (see PLA1, 2 and 3 in experimental part) are processed on a twin screw extruder under different processing conditions. Accordingly, throughput, screw speed and temperature are systematically varied to obtain different levels of material degradation. For each obtained sample, the zero viscosity (η) and the weight average molecular weight (M_w) are determined and used to fit eq. (6). The results are presented in Fig. 4(b)

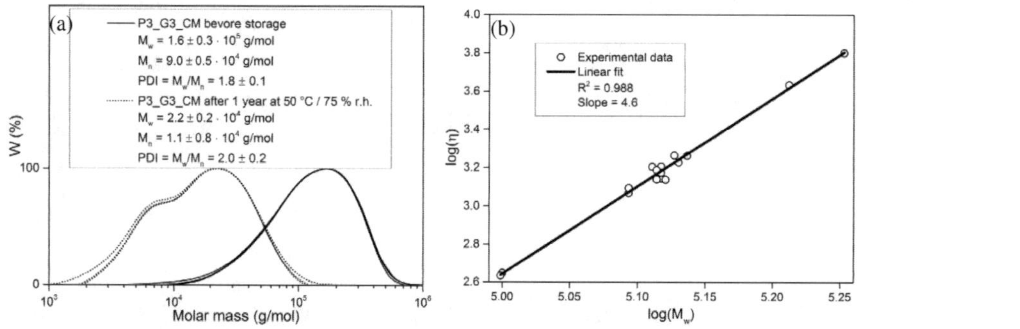

FIGURE 4. MWD of the sample P3_G3_CM before and after three months storage at 50 °C and 75 % r. h. (a) and the fitting of the experimental data (logη - logM_w) with eq. (6) (b)

It can be seen in Fig. 4(b) that experimental data fit well with eq. (6). Regarding this, a correlation coefficient R^2 of 0.988 was obtained as well as a slope of 4.6, which corresponds to parameter α in equation (6). This value is slightly higher, than the values reported in literature (e.g. 3.4 for monodisperse PS, polydisperse HDPE and PB [15], [17], from 3.5 to 3.7 for polydisperse PS [17], from 2.9 to 3.4 for BR [20] as well as 4.1 for LLDPE [21]).

The obtained parameter α was used afterwards to fit eq. (7) and (8) with experimental data. The results of the fitting for sample P3_G1_CM at different temperatures and rel. humidity of 75 % are presented in Fig. 5 as example. From the results presented in Fig. 5(a), it is obvious that a linear relationship does exist over the whole temperature range, with high linear regression correlation coefficients. This indicates that the hydrolytic degradation of PLA during the long term storage and at rel. humidity of 75 % obeys the second order rate kinetic given in eq. (7). In contrast to that, a satisfied linear correlation for the third order kinetic (see eq. (8)) was obtained only at a temperature of 40 and 50 °C, as shown in Fig. 5(b). Therefore, the auto accelerated degradation of PLA at 75 % r.h. can occur only at elevated temperatures, which may be the consequence of increased chain end mobility at temperature close to T_g (65 °C). Similar behavior was observed by Lyu et al [9]. Furthermore, in the case of lower air humidity of 33 %, good linear correlations were found only at a temperature of 50 °C. The determined constants c_2 in this case are by 2 orders of magnitude lower than the constants c_2 obtained at a higher air humidity of 75 % and at the same temperature.

The constants c_2 of the samples determined at air humidity of 75 % are summarized in table (2). As expected the constants c_2 differ by the orders of several magnitudes at different temperatures, which is the consequence of the strong temperature influence on the hydrolytic degradation of PLA during the long term storage of the pellets.

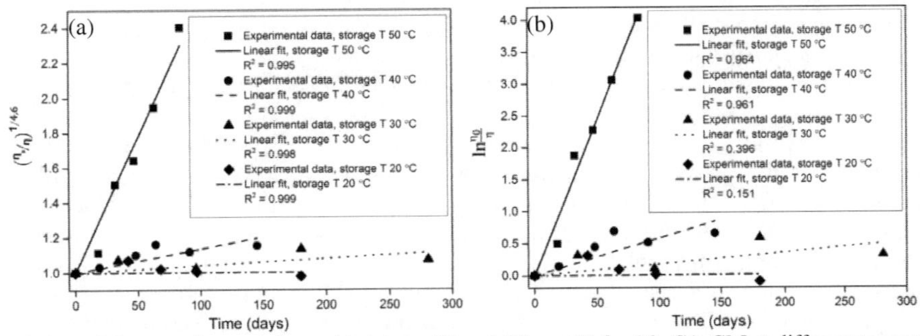

FIGURE 5. The fitting of the experimental data with (a) eq. (7) and (b) eq. (8) for P3_G1_CM at different temperatures and a rel. humidity of 75 %

TABLE (2). Experimentally determined constants c_2 from eq. (7) at different temperatures from 50 to 10 °C. The numbers in brackets show the correlation coefficients R^2.

Sample	$c_2/10^2$ (day^{-1}) 50 °C	$c_2/10^3$(day^{-1}) 40 °C	$c_2/10^4$(day^{-1}) 30 °C	$c_2/10^4$(day^{-1}) 20 °C	$c_2/10^5$ (day^{-1}) 10 °C
P1_G3_CL	1.27 (0.990)	1.99 (0.999)	5.39 (0.999)	1.43 (0.999)	6.34 (0.998)
P1_G3_CM	1.70 (0.995)	2.17 (0.994)	6.85 (0.999)	0.93 (0.999)	6.90 (0.986)
P1_G3_CH	1.11 (0.998)	1.81 (0.996)	3.53 (0.999)	0.75 (0.999)	3.29 (0.998)
P1_G1_CM	1.48 (0.995)	2.55 (0.999)	3.24 (0.998)	0.69 (0.999)	3.33 (0.997)
P1_G5_CM	1.54 (0.999)	2.17 (0.999)	7.84 (0.999)	3.70 (0.999)	17.1 (0.999)
P3_G3_CL	1.94 (0.994)	1.67 (0.999)	3,81 (0.997)	2.21 (0.999)	3.66 (0.999)
P3_G3_CM	1.63 (0.998)	1.50 (0.998)	2.46 (0.999)	0.71 (0.999)	2.84 (0.999)
P3_G3_CH	1.31 (0.996)	1.97 (0.998)	3.53 (0.999)	0.91 (0.998)	1.74 (0.997)
P3_G1_CM	1.54 (0.997)	1.35 (0.999)	4.03 (0.998)	0.18 (0.999)	0.16 (0.993)
P3_G5_CM	1.59 (0.994)	2.21 (0.995)	4.81 (0.999)	2.06 (0.999)	6.85 (0.999)

The values of c_2 listed in table (2) were fitted in eq. (11) and the results of the fitting are presented in Fig. 6(a). As can be seen, the experimentally obtained values of c_2 fit well with eq. (11). Therefore, the hydrolytic degradation of PLA during the long term storage at air humidity of 75 % obeys the Arrhenius equation. The activation energy values were determined from the slopes in Fig. 6(a) and presented in Figure 6(b).

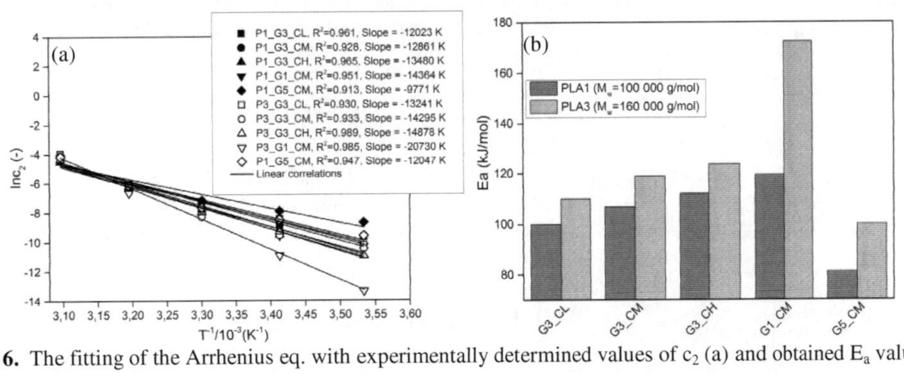

FIGURE 6. The fitting of the Arrhenius eq. with experimentally determined values of c_2 (a) and obtained E_a values (b)

The values of E_a between 81 and 172 kJ/mol determined in this work agree well with the values determined by Zhou et al. They obtained E_a by using a NaOH solution at pH 10.5 as the storage medium for PLA. The samples were stored in the temperature range between 50 and 70 °C and the degradation was monitored by measuring the changes of intrinsic viscosity [6]. Furthermore, similar E_a values were also obtained by Weir et al. [13]. These authors examined the hydrolytic degradation of PLA in phosphate-buffered saline solution in the temperature range between 37 and 70 °C in accordance with ISO 15814. The kinetic of the hydrolytic degradation of PLA was directly determined by measuring the changes of the sample molecular weight.

The higher E_a values obtained in this contribution for PLA3 compared to PLA1 (see Fig. 6(b)) clearly indicate that the hydrolytic degradation of PLA during the long term storage started with more difficulty for the PLA samples having higher molecular weight. This behavior can be explained by the fact, that longer macromolecules with higher molecular weights shows reduced mobility compared to the molecules with a lower molecular weight. Furthermore, it can be clearly seen, that the activation energy for the hydrolytic degradation of PLA during the long term storage depends very much on the pellets size. The greater the diameter of the pellets, the lower the activation energy and thus the faster the hydrolytic degradation of PLA. This appears in good agreement with the findings from Grizzi et al. and proposed diffusion-reaction model, predicting that the degradation of PLA occurs faster in the bulk of the material, than on its surface [22]. Thus, the pellets having a larger diameter possess a higher volume/surface ratio and therefore more "bulk" material available for the reaction with water molecules. This leads to the lower activation energy and accordingly to the faster hydrolytic degradation of pellets with a larger diameter. Furthermore, as expected, the increase of the PLA crystallinity leads to a higher value of activation energy and thus decreased hydrolysis reaction rate.

SUMMARY AND CONCLUSIONS

The PLA pellets having different molecular weights, crystallinities and pellets sizes were produced by melt compounding and subsequent annealing. Each of the samples was stored over a period of 9 months at 5 different temperatures over a range of 10 – 50 °C and air humidity of 33 or 75 %, respectively. The hydrolytic degradation of PLA pellets was monitored by measuring the zero viscosity over a time. The obtained values of zero viscosity were used in a simplified kinetic equation to find the temperature dependent reaction rate constants, which were afterwards applied together with related temperatures in Arrhenius equation. In this way, the material dependent activation energies for the hydrolytic degradation of PLA were determined. Several conclusions may be proposed:

- The method used in this work was shown to be applicable for the investigation of hydrolytic PLA-degradation during the long-term storage at 75 % r.h. over a temperature range of 10 - 50 °C
- Carboxylic acid end groups of PLA were shown to have no influence on the hydrolytic degradation for the storage temperatures from 10 to 30°C. Compared to that, at elevated temperatures of 40 and 50 °C the auto-accelerated degradation of PLA occurs probably simultaneously with non-accelerated hydrolysis..
- As expected, the temperature and humidity have the strongest influence on the hydrolytic degradation of PLA pellets during storage. At a lower air humidity of 33 %, PLA degradation has been detected only at a temperature of 50 °C and is far less pronounced compared to the degradation at 75 % r.h. at the same temperature
- A good agreement between the determined E_a values and the E_a values found in literature has been obtained
- Furthermore, it has been shown, that the reduction of pellet diameter, which can be achieved via increase of appropriate granulator speed, the increase of crystallinity as well as higher molecular weights have an increase of E_a as a consequence. Consequentially, PLA pellets more resident to hydrolytic degradation during long term storage are obtained.

ACKNOWLEDGMENTS

The presented results were obtained within the scope of the project „Competence Network for the Processing of Bioplastics" financially supported by the Federal Ministry of Food and Agriculture, Berlin, Germany via its project co-ordination agency FNR (Fachagentur Nachwachsende Rohstoffe e. V.). The financial support is gratefully acknowledged.

REFERENCES

[1] Krishnan, P. Pandey, S. Mohanty and S. K. Nayak, *Polymer-Plastics Technology and Engineering,* **55(15),** 1623-1652 (2016).

[2] J. Odent, J. Raquez and P. Dubois, „Highly Toughened Polylactide-Based Materials through Melt-Blending Techniques," in *Biodegradable Polyesters*, Berlin: Wiley-VCH Verlag, 2015.

[3] J. Zeng, K. Li and A. Du, *RSC Advances* **5(41)**, 32546– 32565 (2015).

[4] G. Kfoury, J. Raquez, F. Hassouna, J. Odent, V. Toniazzo, D. Ruch and P. Dubois, *Frontiers in Chemistry* **1(32),** 1-46 (2013).

[5] H. Liu and J. Zhang, *Journal of Polymer Science, Part B: Polymer Physics* **49(15)**, 1051– 1083 (2011).

[6] Q. Zhou and M. Xanthos, *Polymer Degradation and Stability* **93**, 1450-1459 (2008).

[7] X. Zhang, U. P. Wyss, D. Pichora and m. F. Goosen, *Journal of Bioactive and Compatible Polymers* **9**, 80-99 (1994).

[8] X. Zhang, M. Espiritu, A. Bilyk and l. Kurniawan, *Polymer Degradation and Stability* **93**, 1964-1970 (2008).

[9] S. P. Lyu, J. Schley, B. Loy, D. Lind, C. Hobot, R. Sparer and D. Untereker, *Biomacromolecules* **8**, 2301-2310 (2007).

[10] C. G. Pitt and Z. W. Gu, *Journal of Controlled Release* **4**, 283-292 (1987).

[11] S. H. Lee, S. H. Kim, Y. K. Han and Y. H. Kim, *Journal of Polymer Science: Part A: Polymer Chemistry* **39**, 973-985 (2001).

[12] H. Pistner, D. R. Bendix, J. Mühling and J. F. Reuther, *Biomaterials* **14(4)**, 291-298 (1993).

[13] N. A. Weir, F. J. Buchanan, J. F. Orr, D. F. Farrar and D. R. Dickson, *Proceedings of the Institution of mechanical Engineers, Part H: Journal of Engineering in Medicine* **218**, 321-330 (2004).

[14] R. Auras, L. T. Lim, S. E. Selke and H. Tsuji, *Poly(lactic acid): Synthesis, Structures, Properties, Processing, and Applications, Hoboken*, New Jersey: John Wiley & Sons, Inc., 2010, pp. 345-383.

[15] T. G. Fox and P. J. Flory, *Journal of Physical Chemistry* **55(2)**, 221-234 (1951).

[16] W. l. Peticolas and E. Menefee, *Journal of Chemical Physics* **35**, 951-954 (1961).

[17] D. Nichetti and I. Manas-Zloczower, *Journal of Rheology* **42(4)**, 951-969 (1998).

[18] R. Acioli-Moura and X. Sun, *Polymer Engineering and Science* **48(4)**, 829-836 (2008).

[19] E. W. Fisher, H. J. Sterzel and G. Wegner, *Kolloid-Z. und Z. Polymere* **251**, 980-990 (1973).

[20] R. H. Colby, L. j. Fetters and W. W. Graessley, *Macromolecules* **20**, 2226-2237 (1987).

[21] K. C. Seavey, Y. A. Liy and P. K. Neeraj, *Industrial and Engineering Chemistry Research* **42**, 5354-5362 (2003).

[22] A. Grizzi, H. Garreau, S. Li and M. Vert, *Biomaterials* **16**, 305-311 (1995).

Microstructure development and mechanical performance of thermoplastic starch/ polycaprolactone/ graphene

Javidi Zeinab (1), Nazockdast Hossein (1)*, Ghasemi Ismail (2)

(1) Amirkabir University of Technology - Tehran - Iran, (2) Iran Polymer and Petrochemical Institue - Tehran - Iran

In this work the effect of graphene nanoparticles on microstructure development of thermoplastic starch (TPS)/ polycaprolactone (PCL) binary blend with constant blend ratio prepared by melt compounding was studied by using combination of rheology, thermo-mechanical analysis and scanning electron microscopy. The nanocomposites samples varied in graphene concentration and the order of feeding. The results of frequency sweep measurements (G′ versus frequency) performed on TPS/PCL blend samples exhibited a solid like behavior at low frequency range similar to that of TPS matrix. The results of similar experiments on the graphene filled blend samples showed a graphene induced enhancement in storage modulus at low frequencies but not as much as that on TPS matrix. The nanofilled blend samples exhibited two completely different mechanical behavior (ductile or brittle) depending on the sequence of feeding. On the other hand, the SEM micrographs of the blend nanocomposites prepared by two different feeding routes showed some PCL agglomerates formed in the samples which were prepared by either direct feeding or using TPS masterbatch which was responsible for brittle behavior of these samples. Whereas the sample prepared by PCL masterbatch showed a ductile behavior similar to that unfilled blend sample.

PLA/TPU Shape Memory Bio-Polymer Blends

Dogan Sebnem (1), Ozkoc Guralp (1)*

(1) Kocaeli University Department of Chemical Engineering - Kocaeli - Turkey

Poly(lactic acid) (PLA)/thermoplastic polyurethane (TPU) blends were melt-mixed and compatibilized to investigate their biocompatibility, biodegradability and thermally induced shape memory properties. The blend compositions were PLA/TPU: 80/20 (20TPU) and PLA/TPU: 50/50 (50TPU). 1,4-phenylene diisocyanate (PDI) was used in order to compatibilize the components reactively. The PDI composition was 0.5, 1, 3% by weight. Biodegradability was assessed by enzymatic degradation tests. Biocompatibility was investigated through in-vitro cell-culture experiments. Shape memory tests exhibited that 20TPU blends have higher recovery ratio than that of 50TPU blends. It was observed that the shape recovery ratio was enhanced by the addition of PDI. The highest shape recovery ratio was obtained at 3%PDI in 20TPU blends. Enzymatic biodegradability tests showed that the increasing TPU content decreased the biodegradability of the blends. It was found that compatibilization slowed down the enzymatic degradation of PLA/TPU blends. In-vitro cell-culture experiments indicated that all blends were biocompatible, and no evidence of cytotoxicity was observed.

Controlled cavitation a key factor in enhancing the toughening effect in Polyhydroxybutyrate/Hydroxyapatite nanocomposites

Hadimani Imam Khasim Rajasaheb [(1)]*

[(1)] SABIC - Riyadh - SaudiArabia

This study reports the development of novel nanocomposite materials for bone replacements. Polyhydroxybutyrate was used as a matrix polymer and nanohydroxyapatite as a reinforcing osteoconductive filler. NanoHA was synthesized using micro emulsion technique with both spherical and needles like morphology. PHB with nanoHA were used for fabricating nanocomposite materials in the form of porous substrates, films and fibers. Optimum dispersion was achieved at 10 and 20 wt% of HA loading. Mechanical properties were evaluated for the synthesized nanocomposites on minimat tensile tester. It was observed that higher elongation at break was achieved with nanoneedles when compared with spherical HA nanoparticles with both 10 and 20 wt% loading. The toughness in polymer nanocomposites with semicrystalline polymer matrix is governed by several factors. a) particle diameter not too large and particle size distribution not too broad, b) interparticle distance (ID) smaller than a maximum IDmax and larger than a minimum distance IDmin, c) no strong, perfect interfacial bonding (interfacial strength between particles and matrix), so as to permit partial, localized debonding under load. Average agglomerates size of nanoneedle nancomposites were around 300-500 nm when compared to spherical nanocomposite systems which had 800-1200 nm. Fractured surfaces showed interesting deformation mechanisms for both the nanocomposites. Localized debonding and cavitation influenced by the nanoparticle geometry were the key factors, which enhanced toughness in the polymer matrix. Young's modulus obtained were also compared to simplified Halpin-Tsai models for proving the micromechanical deformation mechanisms in the nanocomposites.

Effect of adding carbon black on physcal structure of hydroxyapatite granules using the gelation of alginate droplet formation

effendi mochammad dachyar [(1)]*, ***gustiono dwi*** [(1)], ***riswoko asep*** [(1)]

[(1)] BPPT - banten - Indonesia

Calcium phosphate (Ca-P)-based ceramics of synthetic or biological origin can be used as an alternative to autogenous bone grafting. CaP granules are used for open defect of bone, where is not necessary a mechanical stability, in fact they are proposed for no-load bearing applications. The important feature in the physical structure of a synthetic ceramic is its porosity. Pore size, volume of porosity and interconnections between the pores are three crucial parameters. Pore structure is of great importance for osteoconduction.. The main objective of this study is to investigate the Effect of adding carbon black on physcal structure of hydroxyapatite granules of physical properties of droplet CaP. Accroding to that, certain carbon black (325 mesh) has added into composition of microspheres CaP-algintae gelation with ratio CaP : carbon black were 0.5, 1, and 1,5 weight ratio respectively. XRD was used to investigate of phases of sintered droplet. Scanning Electron Microscope (SEM) was used to inventigate microstructure of droplet.FTIR was used to ensure no existention of polymer after sintering process .BET was used to measure of prosity of sintered droplet

Sheet Extrusion of Novatein Thermoplastic Protein

Uitto Jussi Matias [1]*, Verbeek Casparus JR [1]

[1] University of Waikato - Waikato - NewZealand

Polymers from industrial byproducts and waste streams show great potential as a low-cost biomaterial resources. Novatein Thermoplastic Protein is made from a by-product of meat processing and can be processed into the plastic material using thermo-mechanical processes such as extrusion and injection moulding. However, its rheological characteristics are unique and make more advanced processing methods, such as sheet or film extrusion, difficult. The aim of this study was to investigate the processing behaviour of Novatein and to determine the relationship between its molecular structure, rheological properties and plasticization. Proteins are hygroscopic in nature, and after saturation with hydrophilic plasticisers, such as polyols, its hygroscopic character is modified, leading to secondary plasticization effects. In this study, a combination of different levels of water and polyols were used to understand the plasticization effect on the thermo-mechanical properties of Novatein. Furthermore, plasticisers affect a protein's secondary structure, which has been characterised using synchrotron-based FTIR. These changes are related to the rheological properties of Novatein, from which extensional viscosity, in particular, is drastically different to most commodity polymers. Novatein's extensional viscosity is significantly (100x) higher in comparison to some commodity polymers even though the shear viscosity values are at the same range. It was thought that this is the most significant factor preventing easy sheet extrusion of Novatein. Plasticisers have potential to improve the deformation properties up to 10 times better leading o significantly better performance with the sheet extrusion attempts.

Coffee chaff and spent coffee grounds as low-cost fillers in sustainable composite materials for consumer product uses

Wang Tao [(1)], MOHANTY Amar K. [(1)], Zarrinbakhsh Nima [(1)], Muthuraj Rajendran [(1)], Rodriguez-Uribe Arturo [(1)], Vivekanandhan Singaravelu [(2)], Misra Manju [(1)]*

[(1)] University of Guelph - Ontario - Canada, [(2)] University of Guelph - Guelph - Canada

This work explores the utilization of the organic wastes generated in the production and consumption of the world's most popular caffeinated drink, coffee, in biobased composites for food packaging uses. The roasting of green coffee beans gives coffee the characteristic aroma and brown color. The innermost skin of the coffee beans floats free during the roasting process and is commonly referred to as coffee chaff. Coffee chaff and the residue of coffee brewing, spent coffee grounds (SCGs), are major contributors of organic wastes and often end up in landfills. The first phase of this research is a comprehensive characterization of the properties of coffee chaff and spent coffee grounds. The compositional and elemental differences of the two materials were analyzed with respect to their implications on the materials' potential food-contact applications. Morphological studies showed that coffee chaff had a denser fibrous structure than spent coffee grounds. Thermal stabilities were investigated to evaluate the materials' suitability to be processed at high temperatures. In the second phase of the research, a compostable biocomposite filled with coffee chaff was successfully developed and adopted in a commercial packaging product. Maleic anhydride based compatibilizers were used to enhance the interaction between the coffee chaff and the polymer matrix through reactive extrusion. The utilization of the byproducts of food production in food packaging can have profound impact on reducing the environmental footprint of the industry.

Flow-induced crystallization of peroxyde-modified poly(lactic acid)

***Huneault Michel A.* [1]*, *Jalali Amirjalal* [1]**

[1] Université de Sherbrooke - Qc - Canada

Melt-Mixing of PLA with an organic peroxide is a means to increase the elasticity and melt viscosity of PLA, probably through branching reactions. The effect of shear and extensional flows on the crystallization of these branched poly(lactic acid), B-PLA, was investigated using rheometry. For weak shear flow, the samples were first maintained in the melt state to erase the thermal history and then rapidly cooled to a determined temperature and subjected to continuous shear flow varying from 0.01 to 0.1 s-1. For stronger shear flow, the samples were pre-sheared and then crystallization was monitored at the same temperature by observing the variation of storage modulus, G′, versus time, t. In elongational flow, the same thermal protocol was employed and transient viscosity was monitored until strain hardening occurred. It was found that branched PLA exhibited a shorter crystallization induction in shear but a slower crystal growth. In elongational, the flow had a stronger impact on B-PLA crystallization compared to shear flow. Thermal behavior and the microstructures of the sheared and stretched samples were studied by Differential Scanning Calorimetry, and Wide-angle X-ray Diffraction and will be discussed.

PREPARATION AND CHARACTERIZATION OF CROSSLINKED MUCILAGE OF NOPAL (Opuntia ficus indica) WITH EPICHLOROHYDRIN, POTENTIALLY ADSORBENT MATERIAL OF POLLUTANTS IN WASTEWATER

AVILA MORALES BARBARA BERENICE (1)*, CHAVEZ GUTIERREZ MIGUEL (2), SOTO CASTRO DELIA (3)

(1) INSTITUTO POLITÉCNICO NACIONAL, CIIDIR OAXACA - Oaxaca - Mexico, (2) CONACYT- INSTITUTO POLITÉCNICO NACIONAL, CIIDIR OAXACA - Oaxaca - Mexico, (3) CONACYT - INSTITUTO POLITÉCNICO NACIONAL, CIIDIR OAXACA - Oaxaca - Mexico

Dyes in wastewater are very difficult to treat contaminants because of their synthetic origin and the stability provided by their aromatic structures. For its degradation, it is necessary to consider its molecular structure, and the type, number and position of the molecular substituents to determine the possible interactions with some adsorbent. Biological materials can be effective in the decontamination of wastewater through biosorption such as alginates, agricultural residues, cellulosic materials, chitosan, bacteria, algae, cactus plants, xanthan and guar gum. The mucilage obtained from nopal (Opuntia ficus indica) is a biopolymer widely studied and has been used in Chile and Mexico as a water purifier and is known to reduce turbidity and hardness in the spring waters. On the other hand, the starch and its modified forms represent a promising class of adsorbent materials. In this work, nopal mucilage microcapsules were spray-dried on a Yamato model 311S equipment under inlet and outlet temperature conditions of 140 °C and 70 °C, respectively, at a pressure of 0.28 MPa and a flow rate of 7 mL/min. Cross-linking of the biopolymers with epichlorohydrin (0.1, 0.4 and 0.7 %P/P) was performed stirring the mixture at a set temperature and time in a salt bath. The temperature conditions (50, 60 and 70 °C), reaction time (8, 12 and 24 h) and pH of the NaOH (5.0 M) were varied to determine optimum cross-linking conditions. The morphological, mechanical and thermal characterization of the crosslinked materials was performed by the FTIR, SEM, TGA, DSC, TEM and TMA techniques.

Study of PBAT photostabilization using HALS and vitamin E additives in combination with carbon black, aiming application for mulch films

***Souza Patrícia Moraes Sinohara* [(1)]*, *Morales Ana Rita* [(1)], *Sanchez Elisabete Maria Saraiva* [(2)], *Mei Lucia Helena Innocentini* [(1)]**

[(1)] School of Chemical Engineering-UNICAMP - São Paulo - Brazil, [(2)] no affiliation - São Paulo - Brazil

This work evaluated the combinations of additives "carbon black and hindered amine light stabilizer (HALS)" and "carbon black and vitamin E" against photodegradation process of poly(butylene adipate-co-terephthalate) (PBAT), aiming application for mulch films. Two 22 factorial rotatable central composite designs (α=1.414) with 3 central points were conducted. Five different concentrations for each additive were used (0.1, 0.5, 1.0, 1.5 and 2.0% in mass). The films were prepared using a Drais mixer, a hot press and a blown film extruder. A sample without additives was also prepared. The formulations were exposed in an environmental simulator to radiation and condensation cycles. Analysis of light transmission and mechanical properties were conducted for all samples. The factor carbon black (linear) was statistically significant and the increase of its concentrations favored the reduction of light transmission values. The Young modulus of most formulations did not suffer statistically significant differences after aging. The pure polymer suffered the greater reduction for this property, probably due to a higher degree of chain scission suffered during aging test. The elongation and tensile strength results for the aged formulations with carbon black and vitamin E varied between 17-90% and 4-8 MPa, respectively. After aging, the PBAT films with carbon black and HALS presented elongation and tensile strength values in the range of 37-391% and 8-15 MPa, respectively. The results permit to infer that the phenolic stabilizer presented a lower efficiency than the HALS stabilizer when used in combination with carbon black. Considering the results obtained for light transmission and mechanical properties, the formulations with 1.5% concentration in mass of each additive were considered the most promising for application in mulch films.

Study of chain scission and crosslink due to photodegradation of PBAT films with carbon black in combination with vitamin E and HALS additives

Souza Patrícia Moraes Sinohara (1), Morales Ana Rita (1)*, Sanchez Elisabete Maria Saraiva (2), Mei Lucia Helena Innocentini (1)

(1) School of Chemical Engineering-UNICAMP - São Paulo - Brazil, (2) no affiliation - São Paulo - Brazil

The aim of this work was to evaluate the photodegradation of poly(butylene adipate-co-terephthalate) (PBAT) with different combinations of additives: "carbon black and hindered amine light stabilizer (HALS)" and "carbon black and vitamin E". Two 22 factorial rotatable central composite designs (α=1.414) with 3 central points were conducted. Five different concentrations for each additive were used (0.1, 0.5, 1.0, 1.5 and 2.0% in mass). The films were prepared using a Drais mixer, a hot press and a blown film extruder. A sample without additives was also prepared. The films were exposed in an environmental simulator to radiation and condensation cycles. The samples were analyzed by measurement of gel content and the gel structures were also observed by optical microscopy. The Chain Scission Distribution Function (CSDF) methodology was used to study chain scission and crosslink processes. Additionally, samples of pure PBAT were withdrawn from the environmental simulator in different periods in the first 36 h for quantification of the gel content. The gel content for all natural formulations with additives was null. After aging, the results varied from 42 to 62% for samples with carbon black and HALS and from 30 to 57% for samples with carbon black and vitamin E. The pure PBAT presented increasing of gel content during radiation cycles and decreasing during condensation cycles. This behavior was associated to the greater susceptibility of pure PBAT to the hydrolysis process. It was verified by CSDF curves that the chain scission was reduced in the presence of additives related to pure polymer, due to its protection action against photodegradation. The CSDF curves obtained for samples with carbon black and vitamin E indicated a higher chain scission in relation to the PBAT samples with carbon black and HALS. The HALS additive in combination with the carbon black was noted to be more efficient to minimize the PBAT chain scission.

Tribological Properties of Ureidosilane Treated Natural Fiber Reinforced Plant-Derived Polyamide1010 Biomass Composites

Yosuke Nishitani[1, a)], Natsuki Sugawara[1)], Kentaro Kawasaki[1)] and Tetsuto Kajiyama[2)]

[1]*Department of Mechanical Engineering, Faculty of Engineering, Kogakuin University, 2665-1 Nakano, Hachioji, Tokyo 1920015, Japan*
[2]*Jonan Branch, Tokyo Metropolitan Industrial Technology Research Institute, 1-20-20 Minamikamata, Ota, Tokyo 1440035, Japan*

[a)]Corresponding author: at13152@ns.kogakuin.ac.jp

Abstract. To develop new polymeric tribomaterials for mechanical sliding parts such as bearing, cum, gear, and seal, this study aimed to experimentally investigate the tribological properties of various natural fiber (NF) reinforced plant-derived polyamide 1010 (PA1010) biomass composites with surface treatment using the ureidosilane coupling agent. Four types of NF, namely hemp fiber (HF), jute fiber (JF), ramie fiber (RF) and sisal fiber (SF) were used as the reinforcement fibers. Various NFs were surface-treated by two types of surface treatment: (a) alkali treatment by sodium chloride ($NaClO_2$) solution and (b) surface treatment by ureidosilane coupling agents (A-1160). These NF/PA1010 biomass composites were extruded using a twin extruder, and injection-molded. Tribological properties such as frictional coefficient, specific wear rate and limiting *pv* value by ring-on-plate type sliding wear testing were evaluated. It was found that the frictional coefficient and the specific wear rate improved when filled with various NFs, however the results of tribological testing showed different tendencies according to the type of NF used. In particular, ramie fiber (RF) was the most effective natural fiber system for enhancing the wear properties of natural fiber reinforced plant-derived polyamide 1010 biomass composites. On the other hand, the limiting *pv* values of hemp fiber (HF) and sisal fiber (SF) filled composites were higher than those of jute fiber (JF) and ramie fiber (RF) ones.

INTRODUCTION

Recently, the use of natural fibers in polymeric composites are increasingly being considered in tribomaterials for mechanical sliding parts such as bearing, cum, gear, and seal [1-3]. To further enhance the eco-friendliness of materials, there is a strong need to use not only reinforcement fibers but also biopolymers obtained from plant-derived materials as the matrix polymer of these composites. Biomass polymer composites, which are used in both reinforcement fiber and matrix polymer based on renewable raw materials, are environmentally friendly to a large extent and have unique performances [4-6]. In previous studies, we conducted the development of new polymeric tribomaterias made of 100% inedible plant-derived materials [4-10]. In particular, we investigated the thermal, rheological, mechanical and tribological properties of hemp fiber (HF) reinforced plant-derived polyamide 1010 (PA1010) biomass composites (HF/PA1010). PA1010 is made by using sebacic acid and decamethylenediamine, which are obtained from plant-derived castor oil [11]. As castor oil is not used for food, there is no competition with human food production. It was found that the addition of HF, whose surface has been treated by alkali treatment or silane coupling agent, has strong influences on the thermal, rheological, mechanical and tribological properties of HF/PA1010 biomass composites [4-10]. The tribological properties of HF/PA1010 biomass composites improved by surface treatments such as alkali treatment and silane coupling agent. In particular, surface treatment by the combination of alkali treatment using sodium chlorite ($NaClO_2$) solution and ureidosilane coupling agent (A-1160) remarkably improved wear properties.

However, very few studies have been conducted on the effects of natural fibers other than HF on the tribological properties of all plant-derived polymer based biomass composites. To further enhance the tribological properties of these biomass composites, it is necessary to clarify the effects of the type of natural fibers on the tribological properties of these biomass composites. To develop a new polymeric tribomaterial made of only inedible plant-derived materials, this study aimed to experimentally investigate the tribological properties of various natural fibers-reinforced plant-derived polyamide 1010 biomass composites surface-treated using the ureidosilane coupling agent. In this study, four types of natural fibers such as hemp (HF), jute (JF), ramie (RF) and sisal fibers were used for the reinforcement fibers in the biomass composites.

EXPERIMENTAL

The materials used in this study were various natural fiber-reinforced plant-derived polyamide 1010 biomass composites surface-treated using ureidosilane coupling agent. Plant-derived polyamide 1010 (PA1010, Vestamid Terra DS16, Daicel Evonic Ltd., Japan) was used as the matrix polymer. Four types of natural fibers (NFs), namely hemp fiber (HF), jute fiber (JF), ramie fiber (RF), and sisal fiber (SF) were used as the reinforcement fibers. Various NFs were precut into 5 mm pieces and surfaced-treated by two types of surface treatment: (a) alkali treatment by sodium chlorite ($NaClO_2$) solution and (b) surface treatment by ureidosilane coupling agents [10]. Alkali treatment by $NaClO_2$ was employed as follows: 5% $NaClO_2$ solution was placed in a stainless beaker. Chopped NFs were then added to the beaker and stirred well. This was kept at room temperature for 4 h. The fibers were then washed thoroughly with water to remove the excess $NaClO_2$ sticking to the fibers. The alkali-treated fibers were dried in the air for 24 h and in a vacuum oven at 80 °C for 5 h. The surface treatment of NFs with the concentration of 1 wt.% ureidosilane (3-ureidopropyltrimethoxy silane, A-1160, Momentive Performance Materials Inc., Japan) was carried out in 0.5 wt % acetic acid aqueous solution where the pH of the solution was adjusted to 3.5 and stirred continuously for 15 min. Then, the fibers were immersed in the solution for 60 min. After treatment, fibers were removed from the solution and dried in air for 24 h and in a vacuum oven at 80 °C for 5 h. The volume fraction of fiber V_f was fixed with 20 vol % in this study.

Various NF/PA1010 biomass composites were extruded using a twin screw extruder and injection-molded [6, 12]. All the components were dried for 12 h at 80 °C in a vacuum oven beforehand until the moisture level was below 0.2%, then dry blended in a small bottle, and subsequently the melt was mixed at 85 rpm and 220 °C using a twin screw extruder (TEX-30HSS, Japan Steel Works, Ltd., Tokyo, Japan). After mixing, the extruded strands of various NF/PA1010 biomass composites were cut into 5 mm pieces using a pelletizer and dried again at 24 h at 80 °C in a vacuum oven. NF/PA1010 biomass composite samples of various shapes for various experiments were injection molded (NS20-A, Nissei Plastic Industrial, Nagano, Japan). The molding conditions were as follows: cylinder temperatures of 220 °C, mold (cavity) temperature of 30 °C, and the injection rate of 13 cm^3/s. To maintain the dry conditions of the samples in all the measurements, they were kept in accordance with JIS K 6920-2 for at least 24 h at 23 °C in desiccators after molding.

The tribological properties were measured using a ring-on-plate type sliding wear tester (EFM-3-EN, Orientec, Co. Ltd., Tokyo, Japan) [6, 12]. The tests were conducted using plate-shaped test specimens (30 mm × 30 mm × 3 mm) with the surface (R_a = 0.4 μm) finished by No. 240 and No. 800 polishing paper at room temperature under dry conditions in accordance with JIS K 7218a (tested by full contact between the test specimen and metal counterface). A carbon steel (S45C) ring (φ20 mm × φ25.6 mm × 15 mm, apparent contact area A_a = 2 cm^2) with the surface (R_a = 0.4 μm) finished with the same polishing paper was used as a metal counterpart. In this study, two types of tribological test were carried out: the constant normal load and constant sliding velocity test, and the limiting pv (pressure p × velocity v) value test by the step load method. First, the sliding wear test under constant normal and constant sliding velocity was conducted at the normal load P of 50 N (apparent contact pressure p_a = 0.25 MPa), the sliding velocity v of 0.2 m/s, and the sliding distance L of 600 m. Tribological properties were evaluated by the kinetic frictional coefficient μ and specific wear rate V_s. The kinetic frictional coefficient was calculated from the average value of the test distance range of 400 m to 600 m in which the frictional coefficient is stable. The specific wear rate was calculated by the weight of the test specimen before and after test. The sliding surfaces on the test specimen, wear debris, and counterface before and after tests, were observed using a SEM with platinum (Pt) sputter coated. Second, the limiting pv (pressure p × velocity v) value test by the step load method was conducted under a constant velocity v of 0.5 m/s at the initial normal load P_0 of 50 N (apparent contact pressure p_a = 0.25 MPa) and step load P of 25 N every 3 min until the test specimen fractured or melted. The step load just before the test specimen fractured or melted was defined as the limiting load P_{lim}. This limiting load was divided with the apparent contact area A_a = 2 cm^2, and the value obtained was taken as the apparent limiting contact pressure p. The limiting pv value was calculated by multiplying this limiting contact pressure and the test velocity v.

RESULTS AND DISCUSSION

Figure 1 shows the relationship between the specific wear rate V_s and the frictional coefficient μ of various natural fiber (NF) reinforced plant-derived polyamide 1010 (PA1010) biomass composites with the surface treatment by ureidosilane coupling agent by constant normal load P of 50N, constant sliding velocity v of 0.2m/s and sliding distance L of 600 m. The μ and V_s of pure plant-derived PA1010 (100%) improved when filled with various NFs, however the results of tribological testing show different tendencies according to the type of NFs. The μ of various NF/PA1010 biomass composites was found to decrease in the following order: PA1010 >> jute fiber (JF) > ramie fiber (RF) > hemp fiber (HF) = sisal fiber (SF). On the other hand, the V_s of various NF/PA1010 biomass composites showed a different tendency from that of μ, decreasing as follows: PA1010 >> HF > SF > JF > RF. This may be attributed to the change in the mode of wear mechanisms according to the types of natural fibers used caused by the difference in the self-lubricant properties of various natural fibers, the difference in the interface adhesion between NF and PA1010, and the difference in the fiber dispersion in these composites according to the type of fibers used. Thus, it is necessary to observe the formation of transfer films on the metallic counterface, wear debris, and worn surface of the materials in order to understand the mechanisms of tribological properties, because polymer composites are strongly influenced by their ability to form these [6,12]. Figure 2 presents the SEM photographs of the metallic counterface after sliding wear tests against various NF/PA1010 biomass composites: (a) HF/PA1010, (b) JF/PA1010, (c) RF/PA1010, and (d) SF/PA1010, respectively. The SEM observations of the counterface showed differences in the formation of transfer films on the metallic counterface according to the type of NFs. The thickness of transfer films on the counterface against RF/PA1010 biomass composites was thinner than that of other NF/PA1010 biomass composites. There were small scratches all over the surface, which can often be seen in constant surface roughness finish, and also some wear debris adhered on the counterface for RF/PA1010 biomass composites. These tendencies mentioned above are thought to affect the tribological properties of these biomass composites. However, it is necessary to further clarify the above mechanisms by carrying out detailed observations of the structural changes of frictional surface of various NF/PA1010 biomass composites such as the wear debris and worn surface. Their SEM photographs after sliding wear testing will be reported at the Conference presentation.

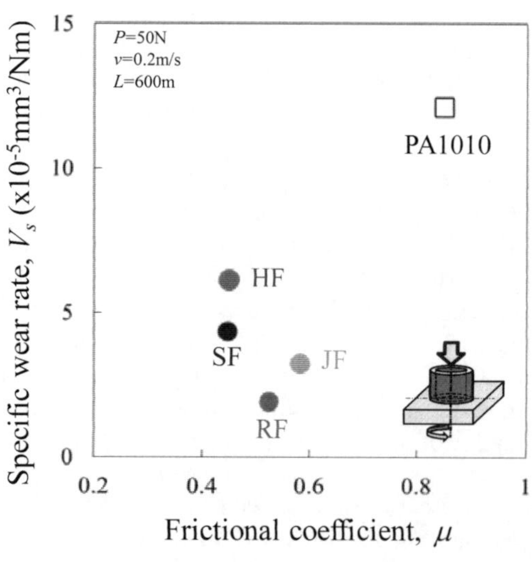

FIGURE 1. Relationship between specific wear rate and frictional coefficient of various NF/PA1010 biomass composites with surface treatment by ureidosilane coupling agent.

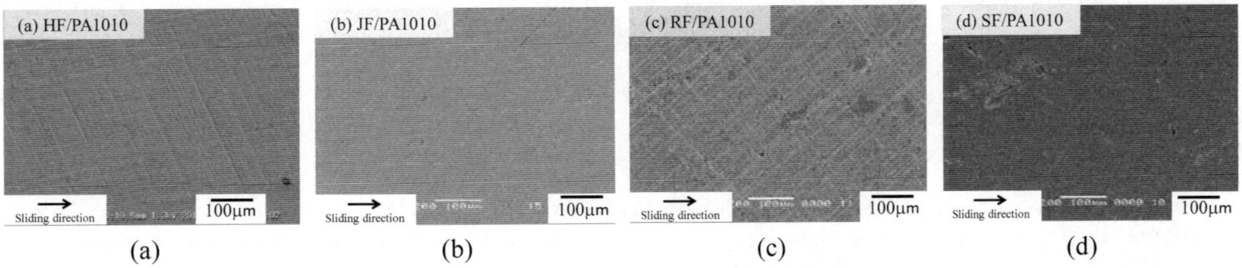

FIGURE 2. SEM photographs of metallic counterface after sliding wear tests of various NF/PA1010 biomass composites with surface treatment by ureidosilane coupling agent: (a) HF/PA1010, (b) JF/PA1010, (c) RF/PA1010 and (d) SF/PA1010.

Next, the limiting *pv* values test results by the step load method of various NF/PA1010 biomass composites, which is harsher than the constant load and speed test, are discussed [6]. Fig.3 shows the limiting *pv* value of various NF/PA1010 biomass composites by constant sliding velocity *v* of 0.5 m/s. The limiting *pv* value of pure plant-derived PA1010 (100%) improved when filled with various NFs. The limiting *pv* value of various NF/PA1010 biomass composites showed a different tendency from that of constant load and speed test as above earlier, increasing as follows: PA1010 < RF < JF < HF = SF. This is considered to be due to the fact that not only tribological properties but also mechanical properties, which help support load increases when filled with various NFs, are influenced by the improvement of the limiting *pv* values.

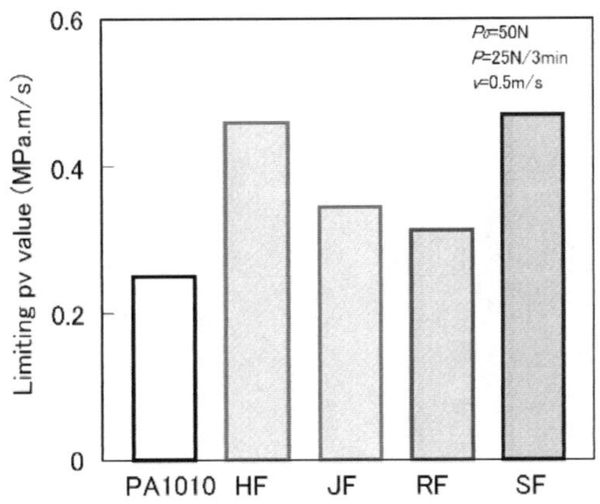

FIGURE 3. Limiting *pv* value of various NF/PA1010 biomass composites with the surface treatment by ureidosilane coupling agent.

CONCLUSION

This study aimed to experimentally investigate the tribological properties of various natural fiber (NF) reinforced plant-derived polyamide 1010 (PA1010) biomass composites whose surface has been treated using ureidosilane coupling agent. It was found that the frictional coefficient and the specific wear rate improved when filled with various NFs, however the results of tribological testing showed different tendencies according to the type of NFs. In particular, ramie fiber (RF) was the most effective natural fiber system for enhancing the wear properties of natural fiber reinforced plant-derived polyamide 1010 biomass composites. On the other hand, the limiting *pv* values of hemp fiber (HF) and sisal fiber (SF) filled composites were higher than those of jute fiber (JF) and ramie fiber (RF) ones.

ACKNOWLEDGEMENTS

This work was supported by JSPS KAKENHI Grant Number 16K06750. We would like to thank the Functional Microstructured Surfaces Research Center (FMS, MEXT, Japan) of Kogakuin University and the Project Research of the Research Institute for Science and Technology of Kogakuin University for funding this study.

REFERENCES

1. N. Chand and M. Fashim, in *Tribology of natural fiber polymer composites* (Woodhead Publishing Ltd., Cambridge, 2008), p. 84–196.
2. A. Shalwan and B.F. Yousif, Mater. Des. **48**, 14-24 (2013).
3. U. Nirmal, J. Hashim and M.M.H.M. Ahmad, Tribol. Int. **83**, 77–104 (2015).
4. Y. Nishitani, M. Hasumi and T. Kitano, AIP Conference Proceedings, **1664**, 060007 (2015).
5. Y. Nishitani, T. Yamanaka, T. Kajiyama and T. Kitano, in *Viscoelastic and Viscoplastic Materials*, edited by M. F. El-Amin (Intech, Rijeka, 2016), p. 53–79.
6. Y. Nishitani, T. Kajiyama and T. Yamanaka, Materials, **10**, 1040 (2017).
7. J. Mukaida, Y. Nishitani, and T. Kitano, AIP Conference Proceedings, **1664**, 060008 (2015).
8. Y. Nishitani, J. Mukaida, Y. Yamanka, T. Kajiyama and T. Kitano, AIP Conference Proceedings, **1713**, 120007 (2016).
9. Y. Nishitani, J. Mukaida, Y. Yamanka T. Kajiyama and T. Kitano, AIP Conference Proceedings, **1779**, 060004 (2016).
10. J. Mukaida, Y. Nishitani, Y. Yamanka, T. Kajiyama and T. Kitano, AIP Conference Proceedings, **1779**, 060005 (2016).
11. T. Shimizu, JETI, **59**, 71–83 (2011).
12. Y. Nishitani, Y. Yamada, C. Ishii, I. Sekiguchi and T. Kitano, Polym. Eng. Sci., **50**, 100–112 (2010).

Melt-processing, Moisture-resistant and Strength Retention Properties of Thermoplastic Starch and Glutaraldehyde Modified Thermoplastic Starch Resins with Supercritical CO_2-assisted Gelatinization

Xuan-long Peng[1], Da-wei Wang[1], Wei Wei[2], James Runt[3], Chi-Chuan Huang[4] and Jen-taut Yeh[1*, 3, 5]

[1] *Hubei Collaborative Innovation Center for Advanced Organic Chemical Materials Ministry of Education, Key Laboratory for the Green Preparation and Application of Functional Materials, Faculty of Materials Science and Engineering, Hubei University, Wuhan, China*

E-mail: jyeh@mail.ntust.edu.tw

[2] *Province Key Lab of Plasma Chemistry and Advanced Materials, School of Materials Science and Engineering, Wuhan Institute of Technology, Wuhan, China*

[3]*Department of Materials Science and Engineering, Penn State University, University Park, Pennsylvania, USA*

[4]*Department of Materials Engineering, TanTung University, Tainan, Taiwan*

[5]*Department of Materials Engineering, Kun Shan University, Tainan, Taiwan*

Abstract The ease of melt-processing, moisture-resistant and tensile retention properties of thermoplastic starch ($scCO_2TPS$) and glutaraldehyde (GA) modified TPS ($scCO_2TPS_{100}GA_x$) resins gelatinized and/or modified with the aid of supercritical carbon dioxide ($scCO_2$) fluid were first investigated and reported in the literature. The melt flow rate (MFR), initial and retention values of tensile strength (σ_f) and elongation at break (ε_f) of $scCO_2TPS$ and $scCO_2TPS_{100}GA_x$ resins prepared using $scCO_2$ fluid as the medium were considerably higher than those of TPS and $TPS_{100}GA_x$ resins, respectively, prepared using the conventional way. After maintaining at 20°C/50% RH for the same amounts of time, the moisture contents of conditioned $scCO_2TPS$ and $scCO_2TPS_{100}GA_x$ resins are considerably lower than those of the corresponding conditioned TPS and $TPS_{100}GA_x$ resins, respectively. In order to understand these drastically improved properties found for $scCO_2TPS$ and $scCO_2TPS_{100}GA_x$ specimens, the retrogradation effect, inter- or intra-molecular interaction present in starch molecules of TPS, $scCO_2TPS$, $TPS_{100}GA_x$ and $scCO_2TPS_{100}GA_x$ resins were investigated via Fourier transform infrared, dielectric relaxation, solid-state ^{13}C Nuclear Magnetic Resonance, thermal and wide angle X-ray diffraction analyses. Possible reasons accounting for the improvement in ease of melt-processing, moisture-resistant properties, initial and retention values of σ_f and ε_f found for $scCO_2TPS$ and $scCO_2TPS_{100}GA_x$ resins are proposed in this study.

Keywords: Thermoplastic starch, Intra-hydrogen bonding, Water proof, Strength retention.

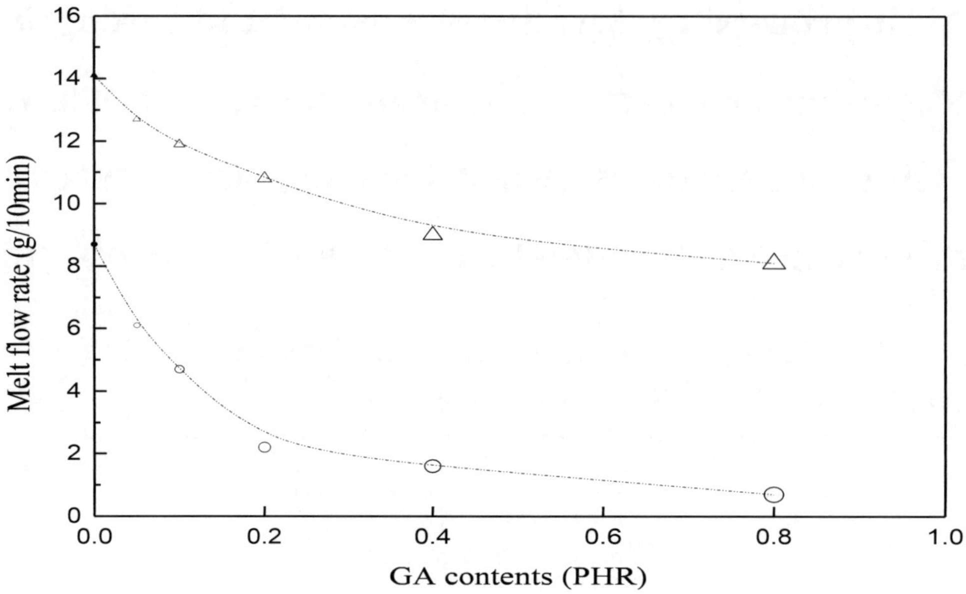

Figure 1 The melt flow rate values of TPS (●), $TPS_{100}GA_{0.05}$ (○), $TPS_{100}GA_{0.1}$ (○), $TPS_{100}GA_{0.2}$ (○), $TPS_{100}GA_{0.4}$ (○), $TPS_{100}GA_{0.8}$ (○), $scCO_2TPS$ (▲), $scCO_2TPS_{100}GA_{0.05}$ (△), $scCO_2TPS_{100}GA_{0.1}$ (△), $scCO_2TPS_{100}GA_{0.2}$ (△), $scCO_2TPS_{100}GA_{0.4}$ (△) and $scCO_2TPS_{100}GA_{0.8}$ (△) specimens.

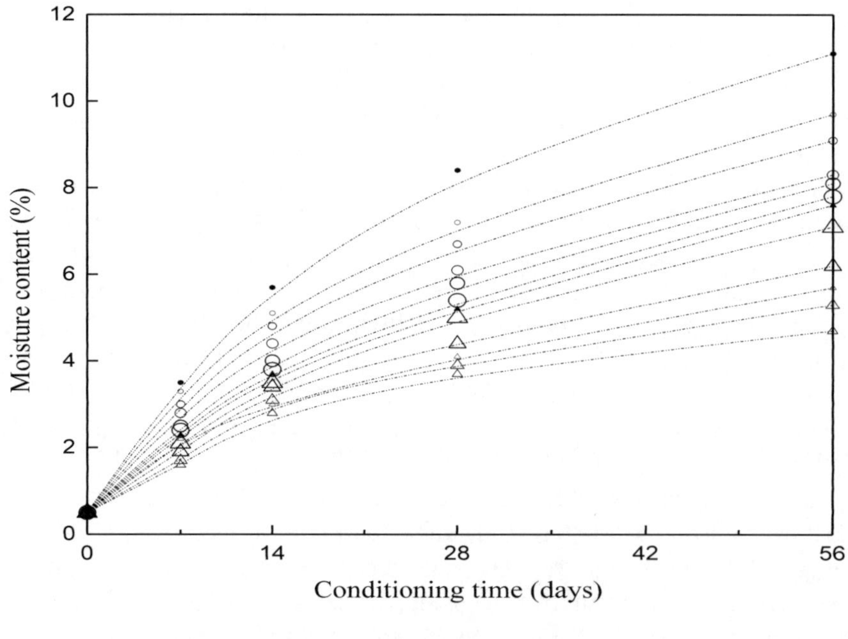

Figure 2 The moisture contents of initial and conditioned TPS (●), $TPS_{100}GA_{0.05}$ (○), $TPS_{100}GA_{0.1}$ (○), $TPS_{100}GA_{0.2}$ (○), $TPS_{100}GA_{0.4}$ (○), $TPS_{100}GA_{0.8}$ (○), $scCO_2TPS$ (▲), $scCO_2TPS_{100}GA_{0.05}$ (△), $scCO_2TPS_{100}GA_{0.1}$ (△), $scCO_2TPS_{100}GA_{0.2}$ (△), $scCO_2TPS_{100}GA_{0.4}$ (△) and $scCO_2TPS_{100}GA_{0.8}$ (△) specimens.

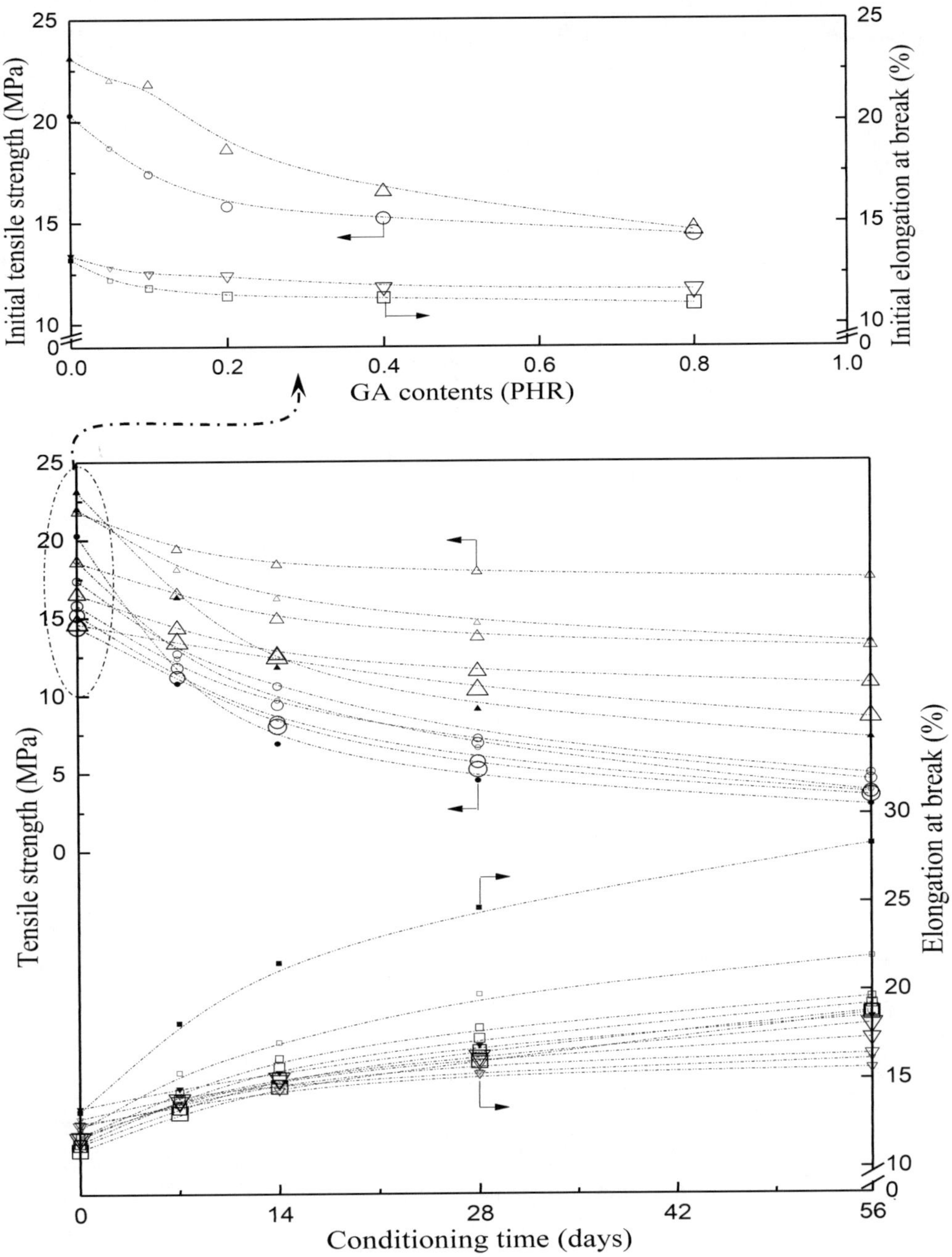

Figure 3 Tensile strength and elongation at break of initial and conditioned TPS (●, ■), $TPS_{100}GA_{0.05}$ (○, □), $TPS_{100}GA_{0.1}$ (○, □), $TPS_{100}GA_{0.2}$ (○, □), $TPS_{100}GA_{0.4}$ (○, □), $TPS_{100}GA_{0.8}$ (○, □), $scCO_2TPS$ (▲, ▼), $scCO_2TPS_{100}GA_{0.05}$ (△, ▽), $scCO_2TPS_{100}GA_{0.1}$ (△, ▽), $scCO_2TPS_{100}GA_{0.2}$ (△, ▽), $scCO_2TPS_{100}GA_{0.4}$ (△, ▽) and $scCO_2TPS_{100}GA_{0.8}$ (△, ▽) specimens conditioned at 20°C/50% RH for specified time periods.

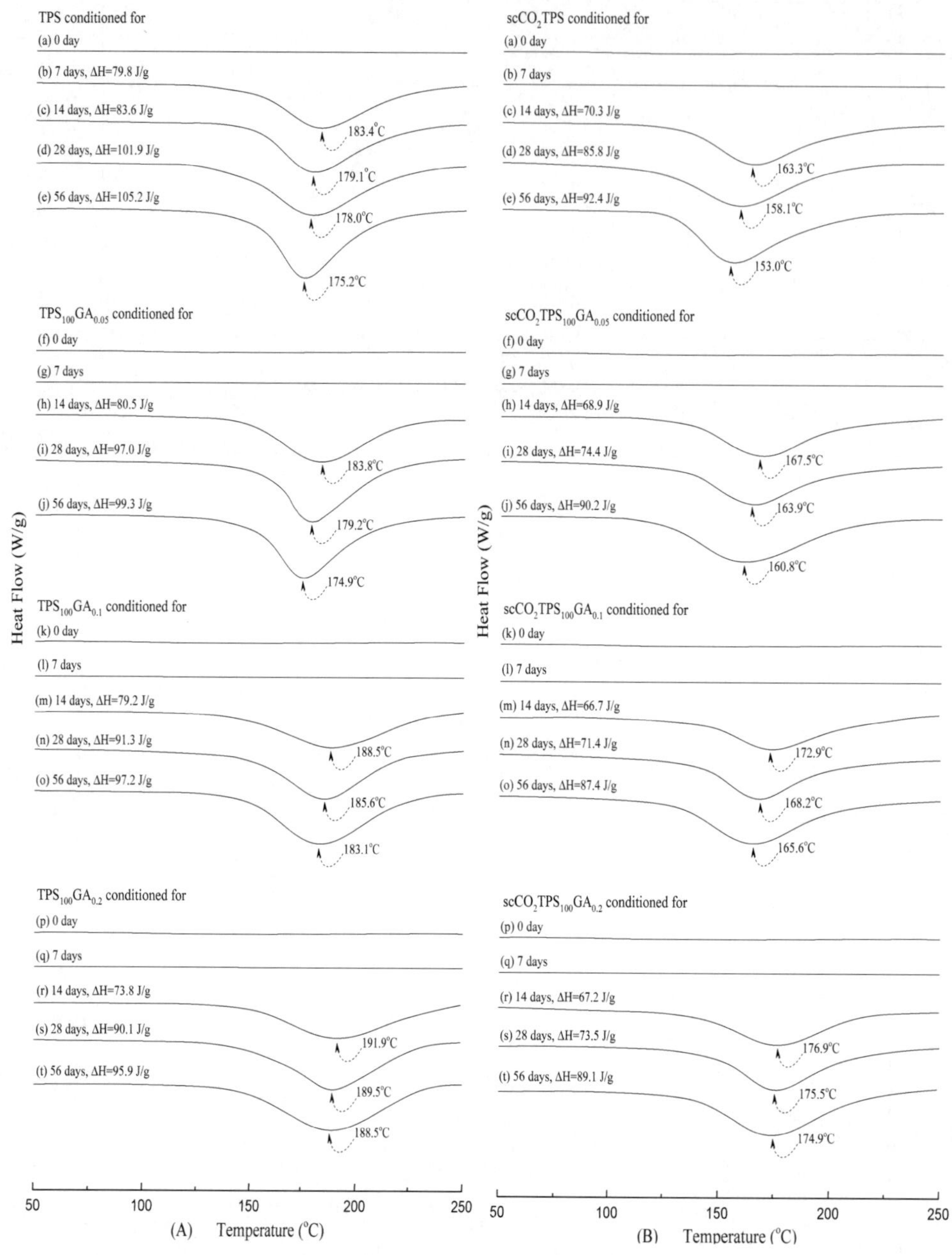

Figure 4 DSC thermograms of (A) TPS, $TPS_{100}GA_{0.05}$, $TPS_{100}GA_{0.1}$ and $TPS_{100}GA_{0.2}$, (B) $scCO_2TPS$, $scCO_2TPS_{100}GA_{0.05}$, $scCO_2TPS_{100}GA_{0.1}$ and $scCO_2TPS_{100}GA_{0.2}$ specimens conditioned at 20°C/50% RH for specified time periods.

REFERENCES

1. Harris, R., Jureller, S. H., Kerschner, J. L., Trzasko, P. T., Humphreys, R. W. R., 1999. U.S. Patent 5977348.
2. Muljana, H., Picchioni, F., Heeres, H. J., Janssen, L. P. B. M., 2011. Experimental and modeling studies on the solubility of sub- and supercritical carbon dioxide ($scCO_2$) in potato starch and derivatives. Polymer Engineering and Science 51, 28-36.
3. Yeh J. T., Hou Y. J., Cheng L. , Xu Y. J., Wang Y. Z., Yang L and Wang C. K., 2015, Water Proof and Strength Retention Properties of Thermoplastic Starch Based Biocomposites Modified with Glutaraldehyde, Carbohydrate Polymers, 127, 135.
4. Wang D. W., Xu Y. J., Li X., Huang C. M., Huang K. S., Wang C. K., and Yeh J. T., 2015, Mechanical Strength Retention and Water Proof Properties of Thermoplastic Starch Based Biocomposites Modified with Sodium Hexametaphosphate, Materials, 8, 3168.

Microencapsulation of bioactive compounds with Aloe Vera mucilage mixed with gelatin or maltodextrin by spray drying: Rheological evaluation and controlled release

Moreno Astorga Josué Joel (1), Medina Torres Luis (2)*, González Laredo Rubén Francisco (1), Nunez Ramirez Diola Marina (3), Gallegos Infante José Alberto (1), Minjares Fuentes José Rafael (4), Candelas Cadillo María Guadalupe (5), Calderas Fausto (6)

(1) Instituto Tecnológico de Durango - Durango, Durango - Mexico, (2) Universidad Nacional Autónoma de México - Ciudad de México - Mexico, (3) Facultad de Ciencias Quimicas, Universidad Juarez del Estado de Durango - Durango, Durango - Mexico, (4) University of the Balearic Islands - Palma de Mallorca - Spain, (5) Universidad Juárez del Estado de Durango - Durango, Gómez Palacio - Mexico, (6) Facultad de Estudios Superiores Zaragoza, UNAM - CDMX - Mexico

The stability of antioxidants can be improved with microencapsulation by spray drying. The wall material of the microcapsules protects the antioxidants compounds from oxygen, water or any condition that could affect their stability. Microencapsulations of gallic acid and curcumin was performed using mucilage blends of Aloe vera (Av), hydrolyzed collagen (gelatin) and maltodextrin as wall materials at feed flow of 1.5 L / H, inlet temperature of 170° C and spray rate of 27,500 rpm. The obtained microcapsules were characterized by scanning electron microscopy and infrared spectroscopy. Results show well-defined and undamaged granules presenting morphological and physicochemical characteristics ideal for encapsulating bioactive principles. In terms of rheology, the behavior in steady simple shear flow of the reconstituted powders was non-Newtonian with shear-thinning characteristics, whereas in linear oscillatory shear tests, it was observed that the viscous component (G'') predominated over the elastic one (G'). The microcapsules maintain high levels of antioxidant availability, indicating a minimum degradation during the drying process. The release performance for gallic acid showed more stable release profiles than the curcumin samples. Keywords: Microencapsulation, spray drying, bioactive principles, rheology, controlled release.

Cellulose fibers and nanofibrils/polypropylene composites obtained by solid state shear pulverization

Marcatto Gean Henrique* [1], *Dufresne Alain* [2], *Canto Leonardo Bresciani* [1], *Ruiz Lou* [3], *Lucas Alessandra DE ALMEIDA* [1]

[1] DEMa-UFSCar - SP - Brazil, [2] Pagora-INP-Grenoble - Grenoble - France, [3] University Joseph Fourier - Grenobe - France

This study investigated the use of solid-state shear pulverization (S3P) in a twin-screw extruder as an alternative processing technique for the preparation of polypropylene (PP) composites with three different cellulosic reinforcements. Unlike traditional S3P process, in which the samples must be processed at sub ambient temperature, the process was conducted at ambient temperature and at 100oC. The composites were prepared with 10, 20 and 30% of cellulose fiber (FC), cellulose nanofibrils (NFC) and lignocellulosic nanofibrils (NFLC). Polypropylene grafted with maleic anhydride was used as a compatibilizing agent. After S3P, composites were pelletized by melt-extrusion, injection molded as standard bars and characterized by tensile testing and impact resistance. The composite of PP containing 30% FC presented the best performance among the formulations studied, with increments of approximately 100% in the Young's modulus and 70% in the tensile strength at yield and 30% on impact resistance, in relation to the neat PP. Scanning electron microscopy, SEM, showed a bad dispersion of the fibers and nanofibrils, while the sample containing 30% of CF presented the best dispersion. A never drying approach was used for CNF and a better dispersion and distribution was observed by SEM.

Control of Hydrolytic Degradation of Poly(Lactic Acid) by Cast Extrusion Processing with Chain Extender

INIGUEZ-FRANCO FABIOLA (1)*, AURAS RAFAEL (2), RUBINO MARIA (2), SELKE SUSAN (2), DOLAN KIRK (3), SOTO-VALDEZ HERLINDA (4)

(1) School of Packaging, Michigan State University - MI - USA, (2) School of Packaging, Michigan State University - Michigan - USA, (3) Food Science and Health Nutrition, Michigan State University - Michigan - USA, (4) Centro de Investigación en Alimentación y Desarrollo - Sonora - Mexico

Poly(lactic acid) (PLA) is a bio-based and biodegradable polymer that has increasingly gained acceptance due to the global awareness of material sustainability and lower environmental footprint than plastics from fossil resources. The versatility of PLA has led to broad applications. Depending on the application, degradation of PLA can be an advantage or a disadvantage. In the case of applications for long service life, for instance as a packaging material, controlling the resistance to hydrolytic degradation is important since the properties of the container must remain intact during the shelf life of the product to maintain the quality of the product. Chain extenders have been used for thermal stabilization of PLA during processing, such as epoxy-acrylic additives that reconnects cleaved chains by increasing the molecular weight of the polymer. The aim of this work was to find the optimal conditions to process PLA film with high molecular weight and to study the effect of the chain extender on the hydrolytic degradation of PLA. PLA control (PLA-C) and PLA with Joncryl®4368 (1.5% wt) as the chain extender (PLA-J) were processed in a home cast film extruder using different residence times: 4, 8 and 16 min. Then, PLA-C and PLA-J with the highest molecular weight were exposed to hydrolysis in 50% ethanol solution at 80 °C, pH 11. PLA-J film processed with a residence time of 8 min had the highest molecular weight (Mn=272 ± 6.58 kDa) with a binomial distribution. The hydrolysis experiments showed that the rate of hydrolysis was ~3 times lower for PLA-J than PLA-C, being 0.065 ± 0.008 h-1 and 0.206 ± 0.014 h-1, respectively. PLA-J could be used to reduce the hydrolysis of PLA on water-ethanol solutions. This study provides a basic understanding of the effect of Joncryl®4368 on the hydrolytic degradation of PLA when it is exposed to water-ethanol solutions.

Bio-Based Polymers and Films from Wood Residues of the Chilean Sawmill Industry

Serrano Jesus (1)***, Clarke Alan*** (2)***, Briones Rodrigo*** (1)***, Martin Peter*** (2)*****, Cunningham Eoin*** (2)***, Toro Aedo Claudio*** (1)

(1) Centro de Investigacion de Polímeros Avanzados - Concepción - Chile, (2) Queen's University Belfast - Belfast - UnitedKingdom

The wood industry in the Bio-bío region of Chile generates large volumes of residual sawdust (50.000 m3/year), which are either burnt or dumped for no useful purpose. This material represents an available, abundant and low cost resource with strong potential for utilization as feedstock for polymer and biomaterial production. In this context, the physical/chemical/mechanical attributes of wood residues can be used as the basis of revenue streams to produce new added value bio-based materials in order to replace synthetic polymers. Liquefaction is regarded as an efficient way to convert the biomass into biopolyols with high content of reactive hydroxyl groups. Recently, we have successfully demonstrated the potential of several liquefied lignocellulosic residues (biopolyols) and starch (low cost biodegradable polymer material), to obtain improved thermoplastic starch films, which are innovative and potentially have high added value. The encouraging results obtained from our previous work with respect to the use of natural biopolyols for film production and their mechanical properties justify the interest to pursue the development of novel and low cost bio-based polymers. This paper will demonstrate the suitability of Radiata Pine sawdust (Chile's main forest plantation) to develop a natural biopolyol by the liquefaction process, and to produce bio-based films by extrusion processes.

Study on the Mechanical Properties of Sebacic Acid/Starch/ Polyurethane Blends

Chi-Yuan Huang [a] , Mei-Chuan Kuo [b] , Fang-Mei Chang [c] ,

Che-Wei Lin [d] and Hong-Kai Hsu [e]

Department of Materials Engineering, Tatung University, Taipei, Taiwan

[a] cyhuang@ttu.edu.tw, [b] kmc0901@gmail.com, [c] askjk1556@gmail.com ,

[d] rickylinth@gmail.com, [e] jackhsu8579@gmail.com

Abstract. The aim of this study is using sebacic acid(SA) to modify thermoplastic starch, and then polyurethane(TPU) and glycerol(GA) were added to made the blends. In this study, sebacic acid was dissolved in alcohol and mixed with glycerol modified starch, furthermore the polyurethane(TPU) was added to prepare the blends by using single-screw extruder. The effect of the content of sebacic acid(SA), SA treatment method and processing temperatures on the mechanical properties of the blends were investigated. FT-IR characteristic peak of C=O in ester was observed at 1730cm^{-1}. It proved that SA was reacted with glycerol by esterification and used to plasticise blends during extrusion. The hydrolysis, thermal stability and hydrophobic of blends could be improved. As thermoplastic polyurethane (TPU) added to the starch, the biodegradability and ductility of blends could be improved. In this study, the ductility, hydrophobic and green material were produced.

Keywords: Sebacic Acid, Starch, Polyurethane, Mechanical Properties

1. Introduction

At present, plastic in the industry and convenience products have a great demand. However, the plastic cannot be decomposed or reused, bringing a lot of ecological problems [1]. So the development of biodegradable thermoplastic starch can solve the growing ecological environment caused by plastic problems [2, 3]. As a result, many scholars have begun to study biodegradable composites. Over the past two decades, biodegradable polymers have had a considerable degree of basic research [4].

Starch is a rich, inexpensive, renewable natural polysaccharide that is easily biodegradable in the soil, so starch is widely used in different fields of polymer science [5]. But the starch used in biodegradable plastics often encounters some problems. For example, starch is hydrophilic, and this feature makes it difficult to blend starch and biodegradable plastics.

The intermolecular force of the starch causes the biodegradable compound to become brittle. Therefore, it is necessary to use a plasticizer to obtain the appropriate polymerization effect. Glycerol is added to the starch, which can increase the flexibility of the hydroxyl group and the interaction between molecules and molecules. Therefore, it is necessary to add glycerin to plasticize it [6, 7].

The modified starch has high fluidity and hydrophobicity, for example, hjdroxyl of anhydride modify starch so the modified starch has hydrophobicity. The organic acid esterification (from acetic acid to fatty acid) has good results in thermoplastic and hydrophobic materials. Long chain fatty acids are introduced into the starch to improve the thermoplastic properties and enhance the thermal stability of the starch. The introduced esterification functional groups improve the original hydrophilicity and gain enhanced or new thermal and mechanical properties.

2. Experiment

2.1 Compounding

First, sebacic acid (0.5, 2 g) was dissolved in alcohol (100 g) and then mixed with tapioca starch (250 g) in a water bath at 50℃, and after 3 hours of drying. Furthermore, above compound stirred with glycerol (75 g) for 15 minutes. After this step, the thermoplastic polyurethane (100 g) was added. The ratio of glycerol to tapioca starch (weight: weight) is about 30:100. Finally, the blend was placed in a single screw extruder (φ = 25 mm, L/D= 32). The four-stage temperature of the single screw extruder was 120, 125, 130, 125 ℃ / 130, 135, 140, 135℃ (from the feed zone to the end) and the screw speed remains constant at 20 rpm.

2.2 Tensile Strength Measurement

Hot pressing was carried out using a hot press (130 ° C) to prepare a test piece for tensile test. The tensile test was carried out according to the ASTM D638-99. The crosshead load is 100 kg and the test speed is 20 mm / min.

2.3 Fourier Transform Infrared Spectrometry

The different composition of material was subjected to FTIR spectroscopy in the range of 400-4000 cm^{-1} as KBr powder and the sample was mixed or blends film. FTIR spectroscopy was operated at a resolution of 4 cm^{-1} and accumulation 100 times in the range of 400-4000 cm^{-1}.

Table I：The abbreviations of various blends

sample	ST^a(g)	GA^b(g)	SA^c(g)	Al^d(g)	PU^e(g)	PT^f(℃)
STGT105			X	X	X	95/100/105/100
STGS20(X)T105			0.5	X	X	95/100/105/100
STGS20(X)T130			2.0	X	X	120/125/130/125
STGS20(O)T130	250	75	2.0	100	X	120/125/130/125
STGS20(X)T140			2.0	X	X	130/135/140/135
STGS20(O)T140			2.0	100	X	130/135/140/135
STGS05(X)T130			0.5	X	X	120/125/130/125
STGS05(O)T130			0.5	100	X	120/125/130/125
STGS20(O)UT130			2.0	100	100	120/125/130/125

a：starch

b：glycerol

c：sebacic acid

d：alcohol

e：polyurethane

f：processing temperature

3. Results and Discussion

3.1 FT-IR

Figure 1 shows the FT-IR spectra of various pure substance. It can be seen from the graph that the starch has an absorption peak at 1203 cm^{-1}, 1649 cm^{-1} and 1730 cm^{-1}. And the peak at 1203 cm^{-1} is C-O bond, 1203 cm^{-1} formation is due to the deformation of glycerol vibration[8]. And the peak at 1649 cm^{-1} and 1730 cm^{-1} is C = O bond, 1649 cm^{-1} formation is due to δ-OH bending, hydrophobic functional groups[9]. From the appearance of the above esterification peak, it can be concluded that the blend of this study is successfully esterified when prepared by single screw extruder, so it can effectively improve the surface hydrophobicity of the sample[10].

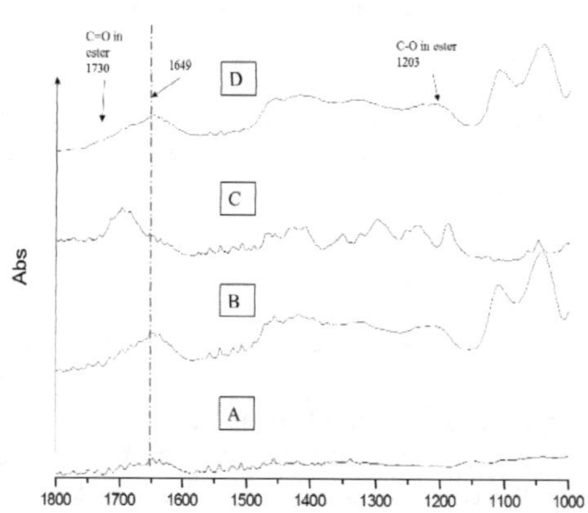

Figure 1 ：the FT-IR spectra of various pure substance（A）ST（B）GA（C）SA（D）GA-S.

3.2 Tensile strength

Figures 2 ~ 7 show the stress-strain curve of various blends. Figure 2 shows the stress-strain curves of STGT105 and STGS20(X)T105 blends. It shows that the tensile strength of STGT105 was about 12.8 MPa, elongation for 35%. As 2.0g SA was added, the tensile strength of STGS20 (X)T105 decreased to about 3 MPa, the elongation up to 220%. It was known that sebacic acid is a kind of long chain fatty acid. The elongation of the biodegradable material can be increased as the SA was added. In Figure 2, it was found that SA was a kind of good plasticizer for starch. However, the strength of STGS20 (X)T105 was just 3 MPa. It cannot use for the reason of low strength [11].

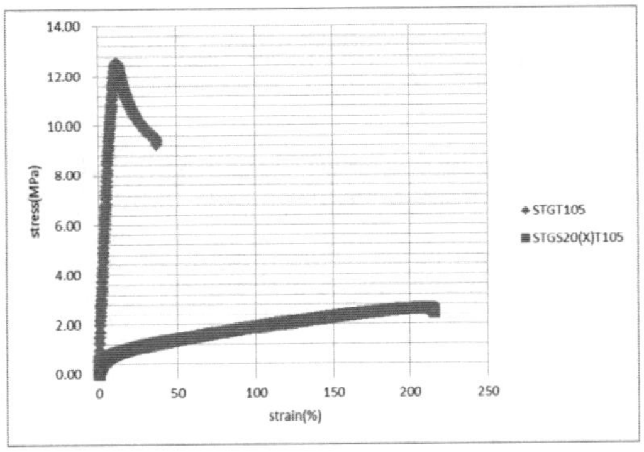

Figure 2 : Stress-Strain curves of various blends:(A)STGT105 ; (B)STGS20(X)T105

Figure 3 shows the Stress-Strain curves of STGT105, STGS20(X)T130, and STGS20(O)T130 blends. In Figure 3, it was found that the elongation of blends could be increased as the SA was homogeneous dispersed in starch. It was suggested that SA reacted with GA and homogeneous dispersed in starch and increased the plasticization efficiency.

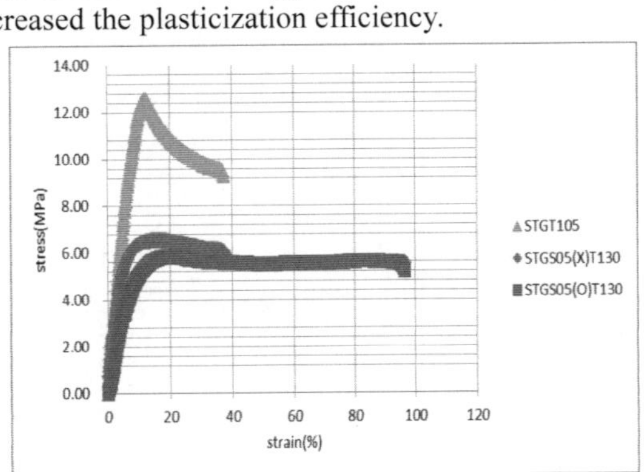

Figure 3 : Stress-Strain curves of various blends:(A)STGT105 ; (B)STGS20(X)T130 ; (C)STGS20(O)T130

Figure 4 shows the Stress-Strain curves of STGT105, ; STGS20(X)T140 and STGS20(O)T140 blends. In Figure 4, the tensile of STGS20(O)T140 blends was 7.8 MPa. The tensile of STGS20(X)T140 blends was 5.6 MPa. Figure 4 also shows that the elongation of STGS20(O)T140 blends was about 13%. The elongation of STGS20(X)T140 blends was 6 %. In Figure 4, it also shows that the elongation of blends could be increased as the SA was homogeneous dispersed in starch. Figure 4 also shows that the elongation of blends would be decreased as the processing temperature increased. It was because that the saccharification of starch was formed. Comparison the result of Figure 3 and Figure 4, the suitable processing temperature of the blends was 130℃ in this study.

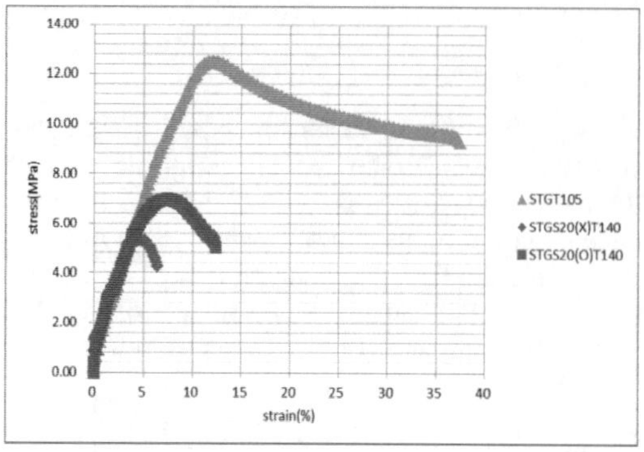

Figure 4：Stress-Strain curves of various blends:(A)STGT105；(B)STGS20(X)T140；(C)STGS20(O)T140

Figure 5 shows the Stress-Strain curves of STGT105, STGS05(O)T130 and STGS20(O)T130 blends. In Figure 5, the suitable amount addition of SA was 0.5g. Figure 6 shows the Stress-Strain curves of STGS20(O)T130 and STGS20(O)T140 blends. It shows that the suitable processing temperature of the blends was 130℃ in this study. This result coincides with those of Figure 3 and Figure 4.

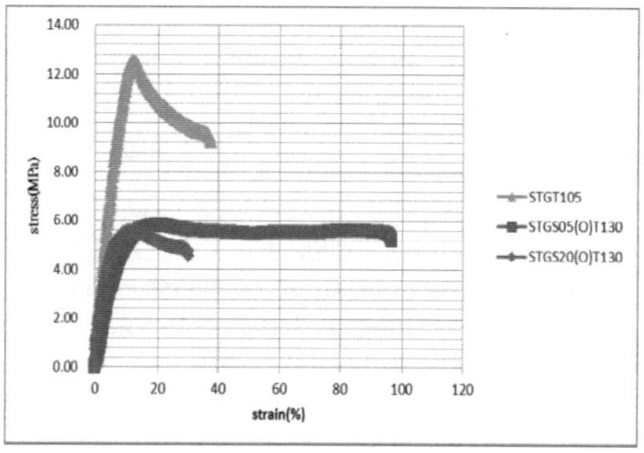

Figure 5：Stress-Strain curves of various blends:(A)STGT105；(B)STGS05(O)T130；(C)STGS20(O)T130

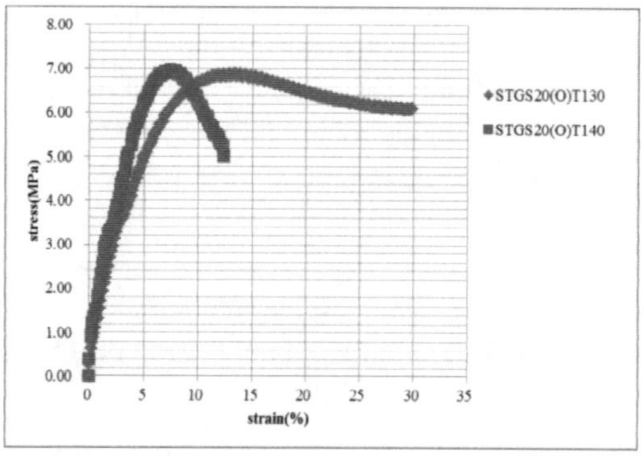

Figure 6：Stress-Strain curves of various blends:(A)STGS20(O)T130；(B)STGS20(O)T140

Figure 7 shows the Stress-Strain curves of STGT105, STGS20(O)T130 and STGS20(O)UT130 blends. Figure 7 shows that the elongation of blend could be increased as the PU was added. It was because that this PU possessed elastomer properties. For this reason the elongation blend could be increased.

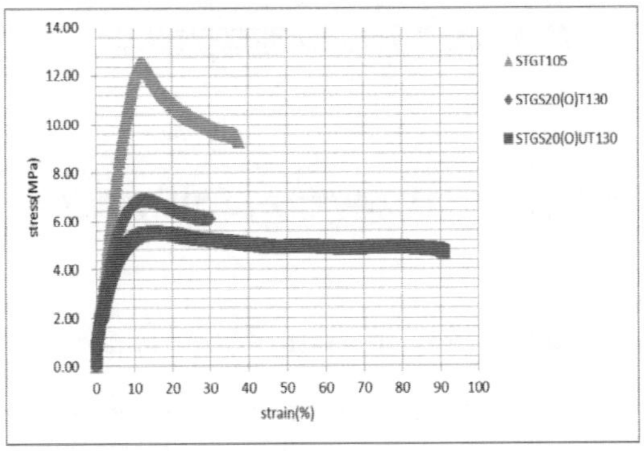

Figure 7 : Stress-Strain curves of various blends: (A)STGT105 ; (B)STGS20(O)T130 ; (C)STGS20(O)UT130

4. Conclusions

Since the added sebacic acid is a long chain fatty acid, it will produce esterification reaction with starch, which can increase the elongation of the blends. After the infiltration of alcohol, the starch and sebacic acid mixed more uniform, the elongation 0f blend will be more obvious to increase from 35% to 220%. As the processing temperature was high, the saccharification phenomenon of starch occurred. The blends become hard, brittle and lost its toughness property. In addition, the ductility of the biodegradable material can be greatly increased by the addition of the thermoplastic polyurethane, the stress of the STGS20 (O) UT130 is 5.6MPa, the strain is 90%, the original biodegradable properties are preserved with better elongation.

5. References

1. Szymanowski, H., et al. (2005). Surface and Coatings Technology 200(1-4): 539-543.
2. YU, L., Christie, G. J. Mater. Sci. 40 (2005) 111.
3. Kampeerapappun, P. P.; Duangdao, A. O.; Duanghathai, P. K.; Srikulkit, K., Carbohydr. Polym. 67 (2007) 155.
4. Nishat, N. and A. Malik (2012). Journal of Saudi Chemical Society.
5. Chi, H., et al. (2008). Food Chemistry 106(3): 923-928.
6. Park, H. M., Lee, W. K.; Park, C. Y., Cho, W. J., Ha, C. S., J. Mater. Sci. 38 (2003) 909.
7. Heydari, A., et al. (2013). Materials & Design 50: 954-961.
8. Yu, J. G.; Wang, N.; Ma, X. F., Starch/ Stärke 57 (2005) 494
9. Fang, J. M., Carbohydr. Polym. 47 (2002) 245.
10. Wei-Cheng Zhao, "Study on Mechanical Properties of Sebacic Acid, Maleic, Anhydride, Bacterial Cellulose and Exopolysaccharides Modified Thermoplastic Starch/Polyurethane Blends", Department of Materials Engineering, Tatung University, 2015.
11. Ching-Yi Hsiao, "Effect of Different Chain Length Acids(C6, C8, C10, C12)Modified by Methane Sulfonic Acid and Thermoplastic Starch/Glycerin Blends",Department of Materials Engineering, Tatung Unuversity, 2014.

Extraction of Pectin and Characterization of Pectin Films From Ripe Plantain Peel.

Obele Chizoba May [(1)]*, Ofoegbu Stanley Udochukwu [(2)], Awuzie Chioma Irene [(3)]

[(1)] Nnamdi Azikiwe University, Awka - Anambra State - Nigeria, [(2)] University of Aveiro, Campus Universitário de Santiago, 3810-193 Aveiro, Portugal - Aveiro - Portugal, [(3)] Federal Polytechnic Okoh - Anambra State - Nigeria

The extraction of pectin from plantain peel was achieved using acid extraction method. The yield was 40.5% while the moisture content was 87.56%. The degree of esterification was found to be 59.53% hence the pectin is categorized as high methoxyl pectin (DE≥ 50%). The equivalent weight was found to be low (1000g/ml) while the galacturonic acid content was 2.89mg/ml. The extracted pectin was used with polyvinyl alcohol (PVA) to synthesize a composite polymer film with and without glycerol as a plasticizer. Introduction of glycerol as plasticizer was observed to lead to improvements in mechanical properties of the composite film. The best mechanical properties of the polymer film; tensile strength, % elongation and abrasion resistance of 5.51MPa, 5.75% and 1.44 ×105 cm2 respectively were obtained with 10ml glycerol/1g pectin. The films were found to be water impermeable which is indicative of excellent barrier properties. Raman spectra showed peaks at 2930 cm1, 1750 cm-1 and 820cm-1 indicating CH- stretch, C=O ester group of pectin and COH ring in pectin respectively. Thermal analysis of the composite films indicates partial decomposition in a single stage between 220-270oC for plantain in Argon which is in line with that obtained for Citrus pectin in argon (200-230oC). Since the major component of the composite film, pectin is sourced from waste, the synthesized films are considered as a potential source of renewable and environmentally benign packaging materials for the food and packaging industry. Keywords: Pectin, Films, Packaging, Polysaccharide.

Enhanced Flame Retardancy of Polylactide/Jute Overmolded Composites

Elcheck Wis Abdulmounem [1], Atabek Savas Lemiye [2], Dogan Mehmet [2], Ozkoc Guralp [1]*

[1] Kocaeli University - Kocaeli - Turkey, [2] Erciyes University - Kayseri - Turkey

In recent years, the use of natural products, such as natural fibres and polymers, has gained attention in industrial applications due to their low environmental impact. The growth of natural material applications has contributed to the reduction of CO2 emission. The best match to obtain ecologically friendly composite materials is the combination of natural fibers with a biodegradable matrix. Poly(lactic acid) (PLA), a biodegradable, renewable, linear aliphatic polyester, is widely used to produce biodegradable composite materials. A new thermoplastic composite technology so called "organic hybrid composites" holds promise to obtain lightweight structural composite components for automotive parts. In this process, a thermoplastic material is over-molded on a continuous fiber based thermoplastic composite reinforcing sheet in an injection moulding process. The flammability of the automotive components is highly important. Phosphorus/nitrogen contained flame-retardants, especially the high content P/N compoıunds, exhibit good fire resistance and reduced toxic effects in PLA. In this work, a PLA compound of P/N based flame retardant was over-molded onto the jute continuous fiber reinforced PLA sheets to improve mechanical properties and flame-resistance of the composites simultaneously. Continuous jute fibre/PLA composite layers were prepared via film stacking method in a hot press. The over-moulded composites (OMCs) were prepared using an Xplore laboratory injection-moulding device. Prior to OMC manufacturing PLA was compounded with a P/N based flame retardant using an Xplore IM15. The mechanical, thermo-mechanical and morphological properties were investigated. Flame retardancy was examined via limiting oxygen index and UL 94 tests. The mechanical tests showed that flexural modulus and strength of OMCs was improved in comparison to neat PLA nearly without altering the density. The DMA results showed that the thermo-mechanical resistance of PLA was enhanced. SEM pict

Biobased thermoset polymer compounds, reinforced with natural fibers.

***Barajas Cervantes Alfonso* [(1)*], *Fernandes Felipe Cicaroni* [(2)], *Ortega Gudiño Pedro* [(1)], *Robledo Ortíz Jorge Ramón* [(1)], *Kirwan Kerry* [(2)], *Arellano Martínez Martín Rigoberto* [(1)]**

[(1)] University Center of Exact Sciences and Engineering, University of Guadalajara. - Jalisco - Mexico, [(2)] International Manufacturing Centre, University of Warwick - Coventry - UnitedKingdom, [(3)] University Center of Exact Sciences and Engineering, University of Guadalajara. - Jalisco - Mexico

The substitution of thermostable polymers derived from petroleum by those obtained from renewable resources has taken interest in the more demanding specialized industries mainly because it can be a strategy to eliminate the effect of the fluctuating price Among the various renewable sources that exist, the oils of vegetal origin are a raw material of vital importance for the industry of the chemical transformation due to its applications, low cost and availability. The versatility of vegetable oils lies in the unsaturated bonds responsible for the formation of crosslinked bonds during the formation of polymer Focusing on the research and development of these polymers could present an opportunity for growth for Mexico as it is among the top 10 countries producing vegetable oils. The aim is to obtain a thermoset polymer based only on renewable resources, however, particularly with epoxy resins based on natural oils, the structural characteristics of the triglycerides chain directly affect the structural properties of the material The objective of the present work was to analyze the viability of wheat germ oil as a precursor of epoxy resins synthesized from the Prileschajew epoxidation method to obtain thermostable polymers based, mix with DGEBA-based commercial resin in Concentrations of 5 and 10% wt as well as to see the influence on the mechanical properties when using a reinforcement mesh with natural fibers of jute and pine Preliminary results show low mechanical properties in the flexural and tensile tests for the thermoset polymer based solely on wheat germ oil compared to the commercial epoxy resin,Therefore, it is verified what was reported in the literature and proceeded to be mixed with the commercial epoxy resin based on DGEBA in concentrations of 5 and 10% As for the blends, the increase in the concentration of bio-based epoxy resin greatly decreases the mechanical properties, a phenomenon that can be diminished with the reinforcement with jute fibers

MANUFACTURE OF A PLA-PAPER BILAYER MATERIAL WITHOUT ADHESIVES

TRUJILLO MAGDALENA [(1)]*, MACIEL ALFREDO [(2)], RIVERA JOSUE [(3)]

[(1)] FACULTAD DE INGENEIRA UNIVERSIDAD NACIONAL AUTONOMA DE MEXICO - JURIQUILLA QUERETARO - Mexico, [(2)] INSTITUTO DE INVESTIGACIONES EN MATERIALES UNIVERSIDAD NACIONAL AUTONOMA DE MEXICO - CIUDAD DE MEXICO - Mexico, [(3)] FACULTAD DE INGENEIRA UNIVERSIDAD NACIONAL AUTONOMA DE MEXICO - CIUDAD DE MEXICO - Mexico

Nowadays, products based on polymeric materials are elements designed to satisfy a specific need. Taking advantage of polymer intrinsic characteristics and all existing manufacturing resources, benefits are achieved in material quality, low product costs, and use of innovative products. This paper shows the design and manufacture of a composite polymer material based on lactic acid polymer (PLA) and paper, which union is guaranteed by the mechanical anchorage between both components. lactic acid polymer-paper bond was obtained by controlling the PLA film surface at the extruder outflow using non-conventional cooling. Then, the mechanical anchorage between the paper and PLA film was obtained by heat and pressure. In order to evaluate the processed product properties, several adhesion tests were carried out using ASTM D-1876 standard, and the bilayer material rated up to 3 N/cm. Besides, tensile tests were carried out using ASTM D638-14 standard; as bilayer material as each layer was evaluated. Also, surface analysis on PLA film was implemented. (1) M. Ortenzi y L. Basilissi, «Evaluation of crystallinity and gas barrier properties of film obtained from PLA nanocomposites synthesized via in situ polymerization of L-latice with silane-modified nanosilca and montmorrillonite,» European Polymer Journal , nº 66, 2015. (2) E. Rudnik, Compostable polymer materials, Oxford: Elsevier, 2008.

Influence of Plasticizer and Clays on the Biodegradation of Poly(hydroxybutyrate-co-hydroxyvalerate)

Michele Theobald[a], Tales S. Daitx[b], Raquel S. Mauler[b], Janaina S. Crespo[c], and Larissa N. Carli[a,*]

[a]*Campus Blumenau, Universidade Federal de Santa Catarina, Blumenau, SC, Brazil.* *larissa.carli@ufsc.br
[b]*Instituto de Química, Universidade Federal do Rio Grande do Sul, Porto Alegre, RS, Brazil*
[c]*Área do Conhecimento de Ciências Exatas e Engenharias, Universidade de Caxias do Sul, Caxias do Sul, RS, Brazil*

Abstract. The aim of this work was to evaluate the influence of plasticizer (triethylcitrate – TEC) and two different clays (montmorillonite – Mt and halloysite – Hal) on the biodegradation of poly(hydroxybutyrate-co-hydroxyvalerate) (PHBV). The plasticizer was used alone or in combination with 3 wt% of clay in a concentration of 10 wt%. The compositions were prepared by melt processing and compression molding. The biodegradation in soil was analyzed by the weight loss during 90 days. The biodegraded samples were characterized by optical microscopy, differential scanning calorimetry, and molar mass determination. The samples containing TEC presented fragmentation after 30 days of exposure in soil. After 90 days, the weight loss of neat PHBV was higher than 50%, with a molar mass reduction of 25%, while the samples containing clays presented a weight loss between 20% and 30%. The degree of crystallinity did not change for the samples containing TEC, while neat PHBV, PHBV/Mt and PHBV/Hal presented a decrease with the time of exposure. Therefore, the additives affected the biodegradation of PHBV to different extents: the plasticizer alone favored the biodegradation process, whereas the addition of clays promoted a delay in the process, irrespective on the clay type. Similarly to the mechanical behavior, the combination of additives resulted in intermediate results, which allow the biodegradation of PHBV to be controlled for the use in the production of disposable products.

Keywords: Biodegradable polymers, Clay/nanoclay, Plasticizers, Biodegradation

INTRODUCTION

Biodegradable polymers from renewable sources are gaining increasing interest as alternatives to petroleum derived materials in the production of disposable packaging. Poly(hydroxybutyrate-co-hydroxyvalerate) (PHBV), a biodegradable polyester of polyhydroxyalkanoates family synthesized from bacteria, is a favorable option since its physical properties are similar to conventional thermoplastics [1].

The drawbacks that hinder the large scale application of PHBV, including high stiffness, brittleness, and thermal instability, may be overcome through the formation of nanocomposites. For PHBV, the most common nanoparticles used are based on clays and clay minerals [2]. In order to improve processability of PHBV, plasticizers are usually incorporated to the polymer matrix. Furthermore, these additives may improve flexibility and toughness of PHBV [3]. Therefore, the combined use of clay and plasticizers may promote a large scale application of PHBV by improving the processing characteristics as well as the physical, chemical, and mechanical properties of polymer through the synergistic effect between the components [4].

The biodegradation of a polymer can be affected by changes in the degree of crystallinity and by the formation of a tortuous path in the matrix by the incorporation of additives, among other aspects. However, this influence is not well established in the literature. In this context, the aim of this work was to evaluate the effect of plasticizer (triethylcitrate – TEC) and two different clays (montmorillonite – Mt and halloysite – Hal) on the biodegradation of PHBV.

EXPERIMENTAL

Materials

PHBV was supplied by Ningbo Tianan Biologic Material Co., Ltd. (ENMAT Y 1010). Commercial clays used were: bentonite, predominantly composed by sodium montmorillonite (Mt) (Argel T from Bentonit União Nordeste Ltda.) and halloysite (Hal) (Sigma-Aldrich). The plasticizer used was triethyl citrate (TEC) from Sigma-Aldrich.

Melt processing

Six samples were analyzed with varying combinations: neat PHBV, PHBV/TEC, PHBV/Hal, PHBV/TEC/Hal, PHBV/Mt, and PHBV/TEC/Mt, were the plasticizer and clay content were 10 wt% and 3 wt%, respectively. PHBV and clays were dried under vacuum at 80 °C for 6 h before use. The compositions were prepared in a Haake Rheomix internal mixer at 170 °C for PHBV and nanocomposites and 165 °C for plasticized samples, for 5 min at 100 rpm. The compositions were compression molded in a Carver hydraulic press at 190 °C for 3.5 min.

Biodegradation tests and characterization

Biodegradation tests were carried out on a simulated soil according to ASTM G 160-03 for 90 days. The samples, before and after the biodegradation test, were characterized by optical microscopy (OM), differential scanning calorimetry (DSC), and viscosimetry. Optical microscopy was performed in a Zeiss AxioLAb.A1 equipment. Crystallization characteristics were determined by DSC using a TA Instruments DSC Q-20. The samples were analyzed from 25 °C to 200 °C at a heating and cooling rate of 10 °C min^{-1} under nitrogen atmosphere. The measurements were performed in the first cooling and heating scan. The degree of crystallinity was determined using ΔH°_{m} = 146 J g^{-1} for PHBV [5]. The molar mass of neat PHBV before and after 30, 60, and 90 days of biodegradation was determined by intrinsic viscosity according to ASTM D 2857-95 using a Cannon Fenske Schott 513 00 no. 25 capillary viscometer. The intrinsic viscosity of the samples was estimated using the Solomon-Ciuta equation [6], and the viscosimetric molar mass was determined with the Mark-Houwink-Sakurada equation [7].

RESULTS AND DISCUSSION

The evolution of weight loss during the biodegradation test is shown in Figure 1. Neat PHBV and PHBV/TEC compositions presented the highest rate of biodegradation, achieving up to 50% of weight loss after 90 days of test. The addition of clays, combined or not with TEC, affected the biodegradation rate of the polymer and, generally, delayed the process, reaching ~30% of weight loss at the end of the period analyzed. In the literature, the effect of nanoparticles on the biodegradability of PHBV is still little investigated and is not completely defined. It is possible that the strong interaction of the clays with the matrix reduces the mobility of the polymer chains as well as the clay acts as a barrier for the diffusion of water and microorganisms inside the material [8].

Visually, PHBV/TEC presented a fragmentation after 30 days of exposure. From optical micrographs, it is clear that all the compositions containing TEC showed a rough and layered surface and the presence of holes after 60 days of exposure in soil. This is an indicative that the sample degradation occurred through a layer by layer erosion from the surface [9]. Additionally, this may point out that the plasticizer alone facilitates the biodegradation process. When combined with the clays, there are additional effects that inhibit the process.

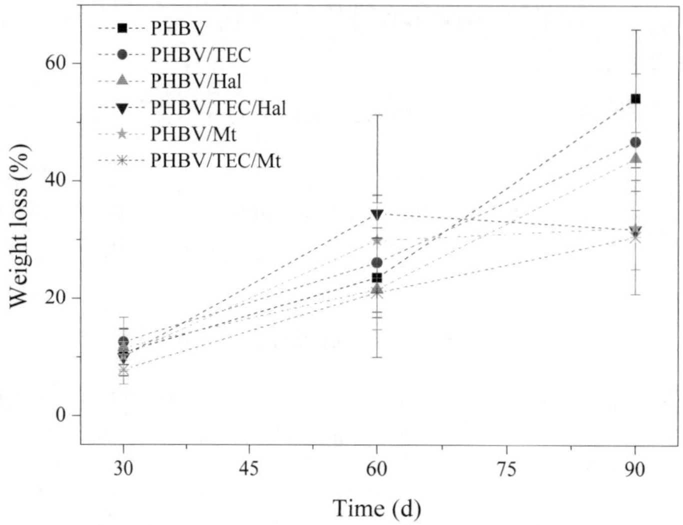

FIGURE 1. Weight loss evolution during the biodegradation test.

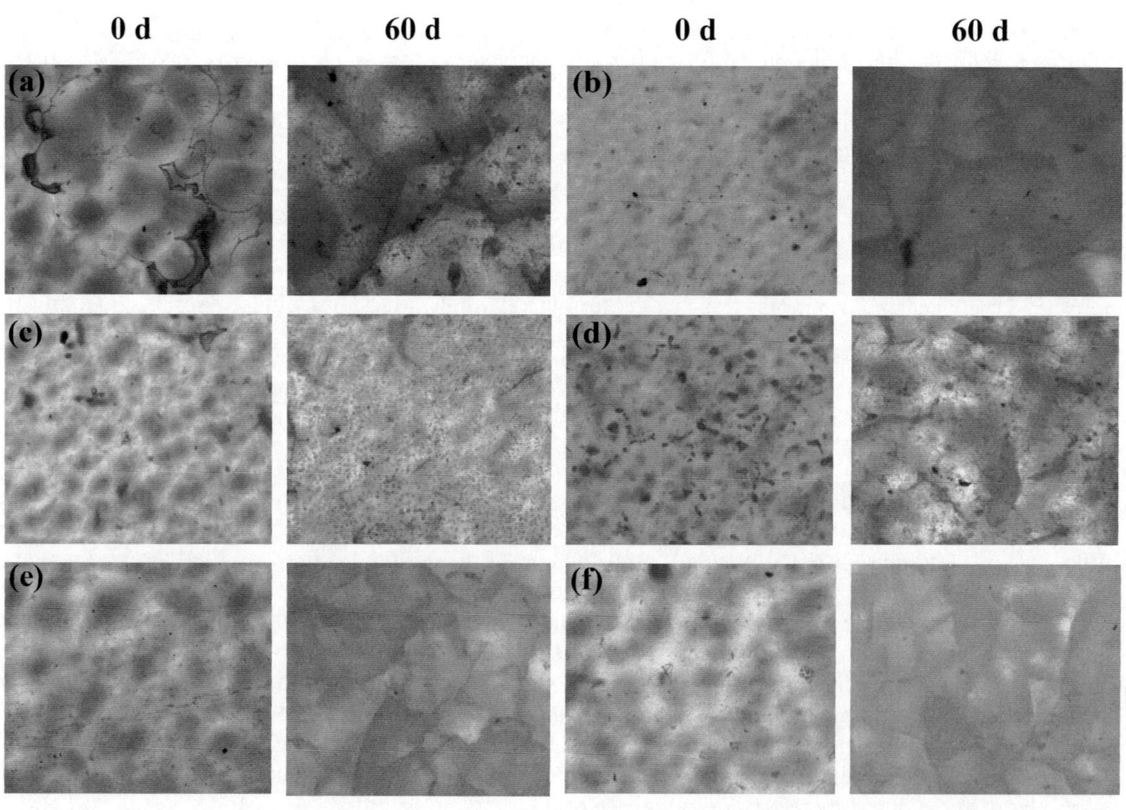

FIGURE 2. Optical micrographs of the samples before and after 60 days of biodegradation: (a) PHBV (b) PHBV/TEC (c) PHBV/Hal (d) PHBV/TEC/Hal (e) PHBV/Mt (f) PHBV/TEC/Mt.

The differences in the biodegradation rate may be associated to the structural changes that resulted from the attack of microorganisms [10]. Therefore, DSC analyses were performed to evaluate the variation on the degree of crystallinity (Xc) during the biodegradation test. Neat PHBV and PHBV nanocomposites (PHBV/Hal and PHBV/Mt) presented a decrease in Xc, while this value increased for PHBV/TEC sample. The increase in Xc may indicate a consumption of the amorphous phase during the biodegradation process [11]. For the compositions containing the combination plasticizer/clay, no noticeable variation was found, evidencing an intermediate behavior as a consequence of opposite effect of additives. This result suggest that the amorphous and crystalline phases degraded at the same rate with no change in Xc [9].

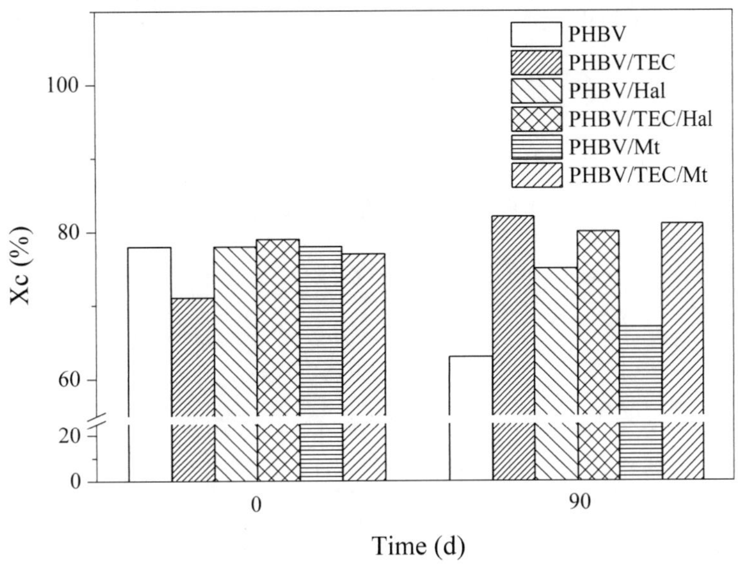

FIGURE 3. Variation in the degree of crystallinity before and after 90 days of biodegradation.

The results of molar mass for neat PHBV are shown in Figure 4. Comparing with the initial molar mass of PHBV (420000 g mol^{-1}), a drastic reduction during processing was observed, which refers to the high thermal instability of the PHBV. During 90 days of biodegradation in soil, this reduction was gradual, reaching 25% of the initial value at the end of the analyzed period.

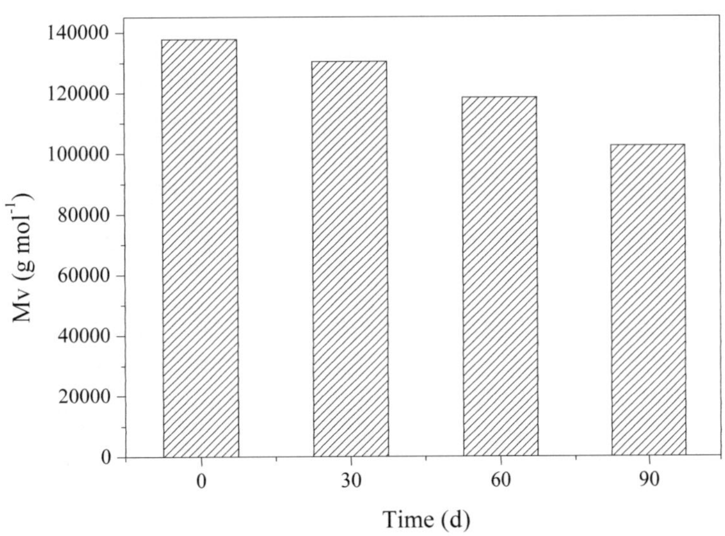

FIGURE 4. Viscosimetric molar mass variation during the biodegradation period for neat PHBV.

CONCLUSIONS

The presence of clay promoted a delay in the biodegradation process, while the addition of plasticizer resulted in an opposite effect. The samples containing the combination of plasticizer and clays maintained intermediate percentages of weight loss in the soil, as well as constant values of crystallinity. Therefore, it can be concluded that the clays, both Hal and Mt, were the additives that had the greatest influence on the biodegradation process. Similarly to the mechanical behavior, the combination of additives allows the biodegradation of PHBV to be controlled for the use in the production of disposable products.

ACKNOWLEDGMENTS

The authors are grateful to Conselho Nacional de Desenvolvimento Científico e Tecnológico (CNPq) for financial support.

REFERENCES

1. B. Laycock, P. Halley, S. Pratt, A. Werker and P. Lant *Prog. Polym. Sci.* **39**, 397-442 (2014).
2. S. S. Ray and M. Bousmina *Prog. Mater. Sci.* **50**, 962-1079 (2005).
3. M. Rahman and C. S. Brazel *Prog. Polym. Sci.* **29**, 1223-1248 (2004).
4. M. C. S. Corrêa, M. C. Branciforti, E. Pollet, J. A. M. Agnelli, P. A. P. Nascente and L. Avérous *J. Polym. Environ.* **20**, 283-290 (2012).
5. S. Gogolewski, M. Jovanovic, S. M. Perren, J. G. Dillon and M. K. Hughes *Polym. Degrad. Stab.* **40**, 313-322 (1993).
6. O. F. Solomon and I. Z. Ciuta *J. Appl. Polym. Sci.* **VI**, 683-686 (1962).
7. S. Akita, Y. Einaga, Y. Miyaki and H. Fujita *Macromolecules* **9**, 774-780 (1976).
8. S. Wang, C. Song, G. Chen, T. Guo, J. Liu, B. Zhang and S. Takeuchi *Polym. Degrad. Stab.* **87**, 69-76 (2005).
9. K. Iggui, N. L. Moigne, M. Kaci, S. Cambe, J.-R. Degorce-Dumas and A. Bergeret *Polym. Degrad. Stab.* **119**, 77-86 (2015).
10. D. G. Brunel, W. M. Pachekoski, C. Dalmolin and J. A. M. Agnelli *Mater. Res.* **17**, 1145-1156 (2014).
11. S. Kumar and P. Maiti *Polymer* **76**, 25-33 (2015).

The effect of processing conditions on the barrier properties of biodegradable polyesters

Gowman Alison (1), Wang Tao (1), MOHANTY Amar K. (1), Misra Manju (1)*

(1) University of Guelph - Ontario - Canada

The packaging industry today consumes ever-increasing amount of petroleum-based plastics. The practice is non-sustainable and raising strong environmental concerns. The development of renewable and biodegradable materials for the next-generation sustainable packaging requires these materials to meet the demand for gas and moisture barrier performances. In this work, the oxygen and water vapor permeabilities of two different grades of biobased polybutylene succinate (PBS) were compared and the differences were correlated with their physical properties such as the crystallization behavior. The effect of different processing techniques such as compression molding, film casting, and blow molding on the barrier performances of the PBS films was then studied. The knowledge gained can be used for tuning the material properties for practical applications.

ENCAPSULATION OF JIOTILLA DYES (CHIOTILLA ESCONTRIA) WITH MAIZE STARCH (ZEA MAYS) BY EXTRUSION METHOD

Gualberto-Rojas Miguel Angel [(1)], Soto-Castro Delia [(2)], Aragón-Lucero Irais [(3)], Santiago-Garcia Patricia Araceli [(3)], Chávez Gutiérrez Miguel [(2)]*

[(1)] Instituto Tecnologico de Oaxaca - Oaxaca - Mexico, [(2)] CONACYT - Instituto Politecnico Nacional - Oaxaca - Mexico, [(3)] Instituto Politecnico Nacional - Oaxaca - Mexico, [(4)] Instituto Tecnologico de Oaxaca - Oaxaca - Mexico

The jiotilla is the fruit of a cactus that grows wild in arid zones of the Mexican territory within the mixteca state of Oaxaca, produces fruits of red edible pulp of bittersweet flavor used in the elaboration of artesian snows and waters with flavors, nevertheless it is a product underutilized in the food industry for being a seasonal fruit that occurs in the months March to May. In the present work, jiotilla dyes were encapsulated by means of extrusion with maize starch, for which the pulp of the jiotilla was mixed with nopal mucilage in three different concentrations (4: 100, 4: 150, 4: 200, g pulp / ml mucilage) and then dried the samples by means of Yamato model ADL 3115 spray dryer at temperature of 140 °C at the inlet and 70 °C at the outlet and at a feed flow of 6.88 ml / min. Obtaining powders with different shades of pink, of hanging of the concentration. Subsequently, the encapsulation by the extrusion method was carried out in a DSM Xplore model MC.5 conical twin screw extruder, at 90 °C and 2 minutes residence time, using a 54% starch mixture, water at 23%, 13% coconut oil and 10% dye, all by weight. Water and oil were used as plasticizers for starch. The moisture content for the product obtained during the mucilage extraction, the microencapsulation by spray and after the encapsulation process by extrusion, were determined for analysis by UV-vis spectrometry and to determine the loss of color in the samples. Observing a better color retention for the sample with 4 g of pulp per 200 of mucilage, due to the protective effect of the mucilage, when the samples were submitted to higher temperatures respect to room temperature.

Investigation the printability of LDPE/ starch/nanoclay bio composite

ataeefard maryam (1)*, Salehi Mohammad Mahdi (2)

(1) Institute for Color Science and Technology - tehran - Iran, (2) and Science Technology Division, Research Institute of Petroleum Industry - tehran - Iran

The mechanical properties of biopolymer-based food packaging can potentially be enhanced by the use of layered silicates (nanoclay) to produce nanocomposites. In this study, starch/clay. /LDPE nanocomposites were synthesized by a melt extrusion method. Closite 15A were chosen for the nanocomposite preparation. The structures of the hybrids were characterized by X-ray diffraction (XRD). Nanocomposite films showed higher tensile strength and better printability properties. To find the best combinations of starch, the effects of starch content (10, 20, 50 weight% LDPE) were investigated. The mechanical properties and printability of nanocomposite films vary significantly with different starch amount

Influence of surface modification of eucalyptus cellulose nanofibrils on the crystallization behavior of PHBV bionanocomposites

Uieda Beatriz [1], Beatrice Cesar A. G. [2], Lucas Alessandra A. [2]*

[1] UFSCar - PPGCEM - São Paulo - Brazil, [2] UFSCar - DEMa - São Paulo - Brazil

The use of cellulose nanofibrils (CNF) as reinforcements in polymers strongly increased in the last years, due to its high elastic modulus and tensile strength, low permeability to oxygen, allied to its low density, low cost, renewability and abundance. Cellulose is highly hydrophilic while most polymeric matrices are hydrophobic, turning the dispersion of nanofibrils expressive challenge during the development of these materials. Many attempts deal with the chemical modification of cellulose nanostructures in order to avoid irreversible agglomeration after drying. There are few studies in literature that show cellulose nanofibrils acting as effective nucleating agents for some biopolymers. The purpose of this study was to develop nanocomposites of a biodegradable poly(hydroxybutyrate-co-hydroxyvalerate) (PHBV) matrix reinforced with eucalyptus cellulose nanofibrils through melt processing. The CNF were mixed in aqueous solution in the presence of different physical modifiers, such as: poly(ethylene oxide) (PEO), ethylene copolymer modified with 6 wt% of acrylic acid (PEAA) and maleic anhydride (MA), and then freeze-dried and milled for feeding with PHBV in a torque rheometer. The influence of CNF before and after the surface modification on the crystallinity index and crystallization rate of PHBV, as well as the thermal, rheological and morphological properties of the nanocomposites were evaluated. From those analysis, it can be concluded that the CNF acted as a nucleating agent for PHBV, accelerating its crystallization. The thermal stability of the PHBV changed according to the type of surface modification of CNF, as revealed by thermogravimetry.

Studies on physical and chemical properties of the novel bio-carbon obtained from pyrolysis of bio-oil

Arnold Stefanie (1), Rodriguez-Uribe Arturo (2), Mohanty Amar (2), Misra Manju (2)*

(1) University of Hohenheim - Stuttgart - Germany, (2) University of Guelph - Ontario - Canada

This presentation describes the conversion of bio-oil into bio-carbon by using pyrolysis. The effects of two different temperatures (600 and 900°C) on the morphology, elemental composition, thermal stability, functionality, as well as the thermal and electrical conductivity of the produced bio char were investigated. Temperatures of pyrolysis of 900°C were found to have a direct impact in the morphology of the synthesized bio-carbon.

Design of high performance biomaterial with varying sizes of biocarbon particles

Codou Amandine [(1)], Misra Manju [(1)], MOHANTY Amar [(1)]*

[(1)] University of Guelph - Ontario - Canada

This study consisted to addition micrometric biocarbon particles into engineering plastic based blends. A range of particle sizes varying from ~ 300 µm to ~3 µm was used and melt compounded using a twin-screw extruder and then examined through a variety of approaches. Rheology allowed highlighting the relaxation mechanisms involved in the melt to complement the dynamic mechanical analysis used in the glassy, the rubbery and the melt states. The excellent particles repartition, verified by scanning electron microscopy, was impacting significantly the mechanical properties of the system considered, measured by means of tensile, flexural and impact tests. Moreover, the benefit of lower particle sizes on all mechanical properties was highlighted.

Microencapsulation by spray drying of gallic acid with mucilage of Aloe vera barbadensis Miller

Medina Torres Luis [(1)]*, Herrera-Valencia Edtson E. [(2)], González-Laredo Rubén Francisco [(3)], Nuñez Ramirez Diola Marina [(4)], Bernad-Bernad María Josefa [(1)], Moreno-Astorga Josué Joel [(3)], Calderas Fausto [(5)]

[(1)] Facultad de Química, Universidad Nacional Autónoma de México - Mexico City - Mexico, [(2)] Facultad de Estudios Superiores Zaragoza, UNAM - CDMX - Mexico, [(3)] TecNM-Instituto Tecnológico de Durango - Durango, Dgo. - Mexico, [(4)] Facultad de Ciencias Quimicas, Universidad Juarez del Estado de Durango - Durango, Dgo. - Mexico, [(5)] Facultad de Estudios Superiores Zaragoza, UNAM - CDMX - Mexico

The microencapsulation by spray drying (SD) of gallic acid (GA) with mucilage of Aloe vera (Av) as an alternate encapsulating agent was studied. Microstructure of granules was observed by microscopy and described without damages for both Av and MD capsules. The particle size distribution from the Av mucilage samples exhibited a quasi-modal distribution unlike the GA samples, which showed a bimodal distribution. Samples of Av and MD, containing GA were characterized by infrared spectroscopy and characteristic absorption peaks of the active compounds were observed. Gallic acid peaks were not observed in the expected wave number ranges of the spectrum, indicating a possible encapsulation of GA. In the antioxidant capacity evaluation, results showed that encapsulated GA retains its capacity due to a short contact time in the SD (<5 s). The behavior in steady simple shear flow of the reconstituted mucilage samples evidenced non-Newtonian shear-thinning characteristics. In linear oscillatory shear tests, it was observed that the viscous component (G”) predominated over the elastic counterpart (G'). Comparing the release profiles of the micro particles obtained from the SD, a difference is observed in the maximum release time for the GA, as well in its release yield percentage; being greater in both cases for the microcapsules based on the Av mucilage.

Miscibility and Thermo-Mechanical Study of a Biocomposite Blend Utilizing PP, PLA, and PHBV with a Sustainable Biocarbon Filler

Snowdon Michael R. (1)*, MOHANTY Amar K. (1), Misra Manju (1)

(1) University of Guelph - Ontario - Canada

This project targets the performance of biocomposites containing greater amounts of biobased biodegradable plastics in polypropylene (PP) through varying weight ratio optimization of poly(lactic acid) (PLA) and poly(hydroxybutyrate-co-hydroxyvalerate) (PHBV) with while improving impact, enhancing the thermal utilization range, and minimizing mechanical performance loss. This focus matches with the current aim of many composite polymer processing practices that are driven on utilizing sustainable products. To add further incentive into this PP blend, the inclusion of a biobased carbon alternative to talc and/or carbon black is used for offsetting thermal property loss, reducing polymer requirements and providing a less dense filler. In this study, the incorporation of biopolymers PLA and PHBV were added to a polypropylene matrix to increase the biobased content of the blend to 30% without causing mechanical performance loss in flexural and impact properties. Further replacement of the polymer component with a sustainable biocarbon filler was utilized to increase the bio-content to 37% and enhance the thermal stability of the composite material by reducing the coefficient of linear thermal expansion by 50% and retaining a heat deflection temperature above 115°C, matching the neat PP, which was not obtainable with a talc filler. The miscibility and phase morphology of the ternary blends were analyzed with experimental and theoretical calculations including solubility parameters and interfacial adhesion to address the observed results. This research is beneficial in demonstrating the utilization of under-valued biocarbon as a filler in polymeric components, along with expanding upon the limited research into multicomponent polymeric composites and their blending characteristics.

Influence of Maleic Anhydride Treatment on the Tribological Properties of Hemp Fiber Reinforced Plant-Derived Polyamide 1010 Biomass Composites

Tetsuto Kajiyama[1, a], Toshiyuki Yamanaka[2], Kentaro Kawasaki[3], Natsuki Sugawara[3] and Yosuke Nishitani[3]

[1]*Jonan Branch, Tokyo Metropolitan Industrial Technology Research Institute, 1-20-20, Minamikamata, Ota, Tokyo 144-0035, Japan*
[2]*Administration Planning Section, Tokyo Metropolitan Industrial Technology Research Institute, 2-4-10, Aomi, Koto, Tokyo 135-0064, Japan*
[3]*Department of Mechanical Engineering, Faculty of Engineering, Kogakuin University, 2665-1 Nakano, Hachioji, Tokyo 1920015, Japan*

[a]Corresponding author: kajiyama.tetsuto@iri-tokyo.jp

Abstract. The need for environment-friendly alternatives to traditional petroleum-derived polymers has attracted much recent interest in biomass composites, which are made from carbon-neutral raw materials. This study aims to investigate the frictional and wear properties of hemp fiber (HF) reinforced plant-derived polyamide 1010 (PA1010) biomass composites for developing new engineering materials such as those used in mechanical sliding parts (tribomaterials) based on inedible plant-derived materials. PA1010 was made from sebacic acid and decamethylenediamine, which are obtained from plant-derived castor oil. HF was surface-treated by four methods: (a) alkali treatment by sodium hydroxide solution and surface treatment by ureidosilane coupling agent, (b) alkali treatment by sodium hydroxide solution and surface treatment by maleic anhydride, (c) alkali treatment by sodium chlorite solution and surface treatment by ureidosilane coupling agent, and (d) alkali treatment by sodium chlorite solution and surface treatment by maleic anhydride. These biomass composites were extruded using a twin extruder and injection molded. Frictional and wear properties were evaluated using ring-on-plate-type sliding wear testing. The surface of these biomass composites was observed using a scanning electron microscope. In particular, treatment using maleic anhydride was the most effective surface treatment for the improvement of the tribological properties of these biomass composites. These results may be attributed to the change in the mode of wear mechanism by natural fiber and the surface treatment due to the good dispersion of fibers in the composites.

INTRODUCTION

The need for environment-friendly alternatives to traditional petroleum-derived polymers has attracted much recent interest in biomass composites, which are made from carbon-neutral raw materials[1), 2)]. However, their low performances compared to traditional petroleum polymers have limited their applications; in addition, the supply of raw materials is unstable because many biopolymers are made from edible biomass such as corn. To solve these problems, new engineering materials based on inedible plants-derived materials are strongly required. In the previous studies undertaken by the authors, the effect of surface treatment on the mechanical and tribological properties of hemp fiber (HF) reinforced polyamide 1010 (PA1010) composites and the blends of these composites and plants-derived thermoplastic elastomers (bio-TPE) such as polyamide 11 elastomer (PA11E) and thermoplastic polyurethane elastomer (TPU)[3)-8)] have been studied. In those studies, the mechanical and tribological properties of these composites have been found to have improved with the filling of HFs, their surface treatment by alkali and silane coupling agent, and the addition of bio-TPE. Nevertheless, to achieve better performance in all plant-derived polymer-based biomass composites, the clarification of the effect of alkali treatment (mercerization) of natural fiber such as HF on the mechanical properties of all plant-derived biomass-polymer composites is necessary. Alkali treatment is a chemical treatment of natural fiber; it is most commonly used to reinforce thermoplastics and thermosets[9), 10)]. The important modification resulting from alkali treatment is the disruption of hydrogen bonding in the network structure, thereby increasing surface roughness. This treatment removes a certain amount of lignin, hemicellulose, wax, and oils that cover the external surface of the fiber cell wall, and depolymerizes cellulose and exposes the short length crystallites. Therefore, a strong effect can be expected from the treatment. To develop new engineering materials and tribomaterials based on plant-derived materials, the influence of the types of alkali treatment on the mechanical and tribological properties of HF reinforced plant-derived PA1010 (HF/PA1010) biomass composites has been studied experimentally. The effect of the type of alkali treatment on the mechanical and tribological properties of the composites was found to differ for each material property[11)-13)]. In the present study, the influence of maleic anhydride and ureidosilane coupling agent on the mechanical and tribological properties of HF/PA1010 biomass composites is studied.

EXPERIMENTAL

The material used in this study was HF/PA1010 biomass composite. PA1010 (Vestamid Terra DS16, Daicel-Evonic Ltd., Japan) was used as the matrix polymer. PA1010 was made from sebacic acid and decamethylenediamine, which are obtained from plant-derived castor oil[14)]. HF (ϕ 50–100 μm, Hemp Levo Ink., Japan) was used as a reinforcement fiber. The volume fraction of fiber V_f was fixed as 20 vol.%. HFs were previously cut into 5-mm-long pieces and were surface-treated via four methods (Figure 1): (a) alkali treatment by sodium hydroxide solution and surface treatment by ureidosilane coupling agent (3-ureidopropyltrimethoxy silane, A-1160, Momentive Performance Material Inc., USA) (HF-NaOH+A1160), (b) alkali treatment by sodium hydroxide solution and surface treatment by maleic anhydride (MA) (HF-NaOH+MA), (c) alkali treatment by sodium chlorite solution and surface treatment by ureidosilane coupling agent (HF-$NaClO_2$+A1160), and (d) alkali treatment by sodium chlorite solution and surface treatment by maleic anhydride (HF-$NaClO_2$+MA). Two types of aqueous solutions such as NaOH and $NaClO_2$ solutions were used as mercerization agents in the alkali treatment. Alkali treatment by NaOH and $NaClO_2$ was performed as follows: a 5 wt.% solution of alkali agents was taken in a stainless steel beaker and the chopped HFs were added into the beaker and stirred well. This arrangement was kept at room temperature for 4 h. The fibers were then washed thoroughly with water to remove the excess alkali agents sticking to the fibers. The alkali treated fibers were dried in air for 24 h and subsequently in a vacuum oven at 80°C for 5 h. The surface treatment of HFs with 1 wt.% ureidosilane coupling agent was conducted in 0.5 wt.% of acetic acid aqueous solution for 15 min. Then, the fibers were immersed in the solution for 60 min. The surface-treated HFs were removed from the solution and kept in air for 24 h and then in a vacuum oven at 80°C for 5 h. The surface treatment of HFs with 1 wt.% maleic anhydride was carried out for 15 for 60 min. The surface-treated HFs were removed from the solution and kept in air for 24 h and then in a vacuum oven at 80°C for 5 h.

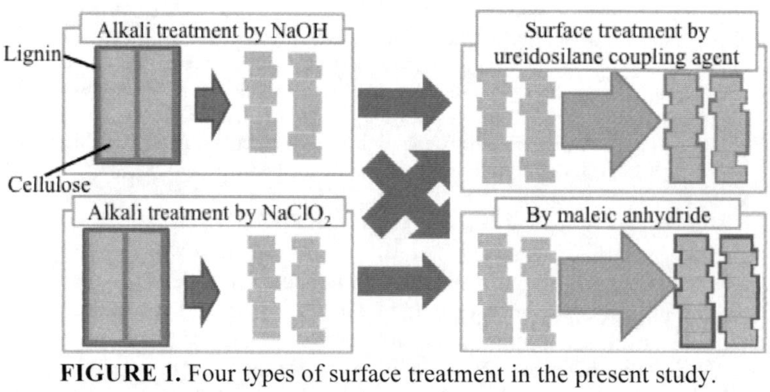

FIGURE 1. Four types of surface treatment in the present study.

All the components that were dried in the vacuum oven were dry blended in a small plastic bottle and subsequently melt-mixed at 85 rpm and 220°C on a twin-screw extruder (TEX-30, Japan Steel Works, Ltd., Japan). After mixing, the extruded strands of various HF/PA1010 biomass composites were cut using a pelletizer and were dried again at 80°C for 12 h in a vacuum oven. The samples of different shapes were injection molded (NS20-A, Nissei Plastic Industrial, Japan) for testing their mechanical and tribological properties. The molding conditions were as follows: cylinder temperature = 220°C, mold (cavity) temperature = 30°C, and injection rate = 13 cm^3/s. Furthermore, mechanical and tribological properties for these samples were evaluated. Tensile testing was conducted with dog-bone samples on a Toyo Seiki testing machine V-10 in accordance to the JIS K 7161 at a cross-head speed of 50 mm/min. Three-point bending test was conducted with coupon specimens on the V-10 in accordance with the JIS K 7171 and at a cross-head speed of 2 mm/min.

Tribological properties were measured via a ring-on-plate-type sliding wear tester (EFM-3-EN, A&D Co. Ltd., Japan) under dry condition. A carbon steel (S45C) ring with its surface finished by No.240 and No.800 polishing paper was used as a counter-piece. Two types of tribological testing were performed using a constant normal load (P = 50 N). The surfaces of the samples were observed via a scanning electron microscope (SEM, JSM-6360LA. JEOL Ltd., Japan).

RESULTS AND DISCUSSION

First, the influence of the different alkali treatments on the tensile and bending properties of HF/PA1010 biomass composites is discussed. Figure 2 shows the mechanical properties of various surface-treated HF/PA1010 biomass composites. In Figure 2(a), the tensile strength σ_t is plotted against the tensile modulus E_t for pure PA1010 (100%) and HF/PA1010 biomass composites for the four surface treatment methods. σ_t and E_t of PA1010 improved with the filling of HF and their surface treatment via alkali treatments using ureidosilane coupling agent or maleic anhydride. σ_t increased in the following order: HF-NaOH+MA < HF-NaOH+A1160 = HF-$NaClO_2$+A1160 = HF-$NaClO_2$+MA. E_t increased in the following order: HF-NaOH+MA < HF-NaOH-S3 < HF-$NaClO_2$-S3 = HF-$NaClO_2$+MA. These results may be attributed to the improvement of the interphase adhesion and interaction between HF and the matrix polymer. The interaction between HFs and polymer matrix was considered to have improved after the surface treatment by $NaClO_2$ and ureidosilane coupling agent or maleic anhydride. The bending strength σ_b of various PA1010 composites is plotted against the bending modulus E_b in Figure 2(b). σ_b and E_b of HF-S/PA1010 composites improved with the filling of HF and surface treatment; σ_b increased in the following order: HF-NaOH+A1160 < HF-NaOH+MA < HF-$NaClO_2$+MA < HF-$NaClO_2$+A1160, and E_b increased in the following order: HF-$NaClO_2$+MA < HF-NaOH+A1160 < HF-$NaClO_2$+A1160 = HF-NaOH+MA. Thus, the influence of the method of alkali treatment used on the mechanical properties differed for each property. This difference was assumed to be due to the differences in the mode of testing tensile and bending properties such as the testing speed, direction of deformation, and fracture mechanism. The observation of the internal structure of various surface-treated HF/PA1010 biomass composites is necessary for understanding the relation between the mechanical properties and the interphase adhesion between HF and PA1010.

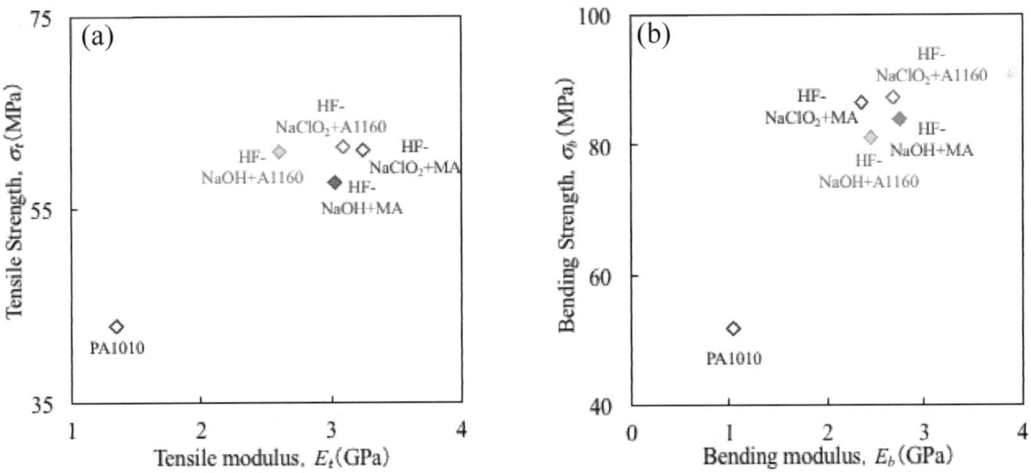

FIGURE 2. Mechanical properties of various polyamide 1010 (PA1010) composites. (a): Relation between the tensile strength and the tensile modulus of various PA1010 composites and (b): Relation between the tensile strength and the tensile modulus of various PA1010 composites.

Figure 3 shows the SEM images of the cryogenically fractured surface of various surface-treated HF/PA1010 biomass composites: (a) HF-NaOH+A1160, (b) HF-NaOH+MA, (c) HF-$NaClO_2$+A1160, and (d) HF-$NaClO_2$+MA. The morphologies of the composite surfaces treated by both alkali treatment and ureidosilane coupling agent or maleic anhydride show good interaction between fiber and matrix polymer; the fiber does not leave any voids on the fracture surface. These results could be attributed to the compatibility between the alkali treatments and the ureidosilane coupling agent or maleic anhydride for improving the mechanical properties of HF/PA1010 biomass composites. The observations from these SEM images agree with the mechanical properties such as tensile and bending strengths and moduli mentioned above.

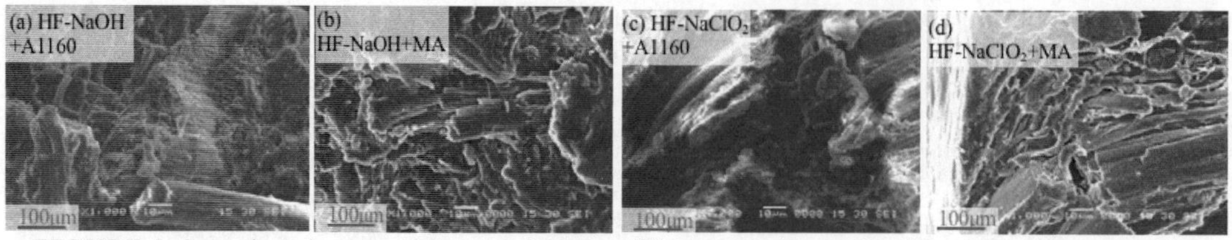

FIGURE 3. Scanning electron microscope images of fracture surfaces of various PA1010 composites.

Second, the tribological properties of various surface-treated HF/PA1010 biomass composites using a ring-on-plate-type sliding wear tester are discussed. The relation between the specific wear rate *Vs* and frictional coefficient μ of the composites by constant-normal-load testing is shown in Figure 4. μ of pure PA1010 improves with the filling of HF and their surface treatment by alkali treatments and ureidosilane coupling agent or maleic anhydride. The μ of various surface-treated composites decrease in the following order: PA1010 < HF-$NaClO_2$+A1160 < HF-NaOH+A1160 = HF-$NaClO_2$+MA < HF-NaOH+MA. Conversely, *Vs* is highly influenced by the types of alkali treatments with surface treatment by ureidosilane coupling agent or maleic anhydride. The *Vs* of various surface-treated HF/PA1010 biomass composites behaved differently from the trend exhibited by the μ, and it decreased in the following order: PA1010 < HF-NaOH+A1160 < HF-$NaClO_2$+A1160 < HF-NaOH+MA < HF-$NaClO_2$+MA. In particular, the surface treatment using maleic anhydride had a more pronounced effect on the specific wear rate of HF/PA1010 biomass composites than on the frictional coefficient.

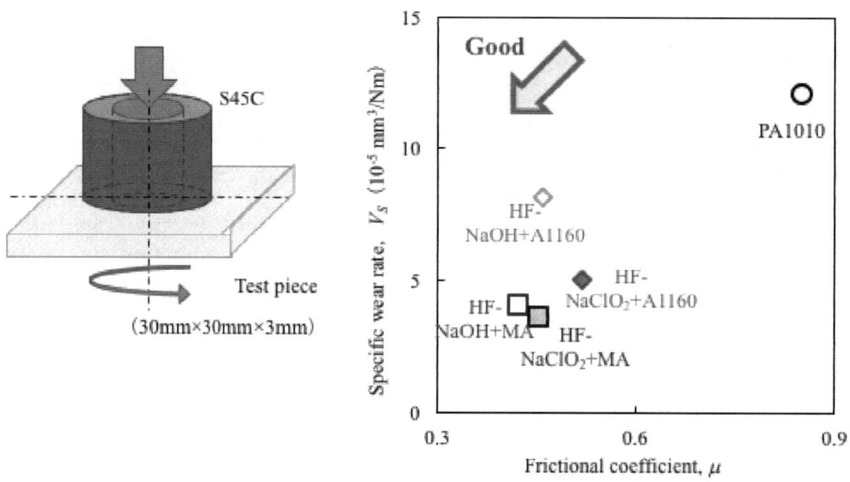

FIGURE 4. Relation between V_s and μ of various PA1010 composites.

CONCLUSION

The aim of this study was to experimentally investigate the influence of different types of alkali treatment on the mechanical properties of HF/PA1010 biomass composites. Such an investigation is important for developing new and improved engineering materials, specifically tribomaterials, from inedible plant-derived substances. From this study, we found that the effect of the type of alkali treatment on the mechanical properties of the composites was different for each property. In general, the mechanical properties are improved with alkali treatment by NaOH and $NaClO_2$ along with surface treatment via ureidosilane coupling agent or maleic anhydride. Tribological properties as measured by a constant-normal-load test for HF/PA1010 biomass composites showed improvement when the material was subject to surface treatment via ureidosilane coupling agent or maleic anhydride. Experimental results indicated that the surface treatment using maleic anhydride was the most effective method for improving the tribological properties of biomass composites such as those tested in this study.

ACKNOWLEDGMENTS

This work was supported by JSPS KAKENHI Grant Number 16K06750. We would like to thank the Functional Microstructured Surfaces Research Center (FMS, MEXT, JAPAN) of Kogakuin University and Project Research of the Research Institute for Science and Technology of Kogakuin University for funding this study.

REFERENCES

1. A. C. Wibowo, A. K. Mohanty, M. Misra and L. T. Drzal, Ind. Eng. Chem. Res. 43, 4883-4888 (2004).
2. L. Petersson and K. Oksman, Compos. Sci. Technol. 66, 2187-2196 (2006).
3. M. Hasumi, Y. Nishitani and T. Kitano, The proceedings of Seikei-kako symposia'11, Akita, 491-492 (2011) (in Japanese).
4. M. Hasumi, Y. Nishitani and T. Kitano, The proceedings of Tribology Conference Autumn'12, Muroran, 211-212 (2012) (in Japanese).
5. M. Hasumi, Y. Nishitani and T. Kitano, The proceedings of the Polymer Processing Society 28th Annual Meeting (PPS-28), Pattaya, P-07-324 (2012).
6. Y. Nishitani, M. Hasumi and T. Kitano, AIP Conf. Proc. 1664, 060007 (2015).
7. J. Mukaida, Y. Nishitani and T. Kitano, AIP Conf. Proc. 1664, 060008 (2015).
8. Y. Nishitani, J. Mukaida, T. Yamanaka, T. Kajiyama and T. Kitano, The proceedings of the 31st International Conference of the Polymer Processing Society (PPS-31), Jeju, Korea, 1116 (2015).
9. N. Chand and M. Fahin, "Physical treatment," in Tribology of natural fiber polymer composites, (Woodhead Publishing, Cambridge, 2008), PP.22-23.
10. T. Shimiztu; JETI 59, 71-83 (2011) (in Japanese).
11. J. Mukaida, Y. Nishitani, Toshiyuki Yamanaka, Tetsuto Kajiyama and T. Kitano, AIP Conf. Proc. 1779, 060005 (2016).
12. Y. Nishitani, Jun Mukaida, Toshiyuki Yamanaka, Tetsuto Kajiyama and T. Kitano, AIP Conf. Proc. 1779, 060004 (2016).
13. Y. Nishitani, Tetsuto Kajiyama and T. Yamanaka, Materials. 10, 1040 (2017).
14. A. K. Bledzki and J. Gassan, Prog. Polym. Sci. 24, 221-274 (1999).

PLA/GO composites: Effect of the addition of GO in the thermomechanical properties

MENDOZA MONICA (1)*, Estrada Ivan A. (1), Flores Sergio G. (1), López Erika I. (1), Zaragoza Armando (1), Vega Alejandro (1)

(1) CIMAV - CHIHUAHUA - Mexico

This study is focused on investigate the effect of the inclusion of graphite oxide (GO) into a poly-lactide (PLA) matrix on the thermal, rheological and mechanical properties of the composite. The incorporation of GO fillers into PLA was carried out by means of an internal mixing chamber. Four formulations with GO concentrations of 0, 0.1, 0.5 and 1 wt% were prepared. The mixing temperature was 190°C. Thermal stability, thermo mechanical behaviour and crystallinity of the composites were evaluated by means of TGA, DMA and DSC respectivelly. It was found that degradation temperature (T5%), cristallinity grade (Xc %) and the elastic modulus (E') of the composites were modified the addition of GO

Preparation and Characterization of Chitosan and Chitosan/Montmorillonite Beads for the Removal of Methyl Violet From Aqueous Solution

Marion I. Buse[a], Tales S. Daitx[b], Raquel S. Mauler[b], Dagoberto O. Silva[c], Josiel B. Domingos[c] and Larissa N. Carli[a,*]

[a]*Campus Blumenau, Universidade Federal de Santa Catarina, Blumenau, SC, Brazil. *larissa.carli@ufsc.br*
[b]*Instituto de Química, Universidade Federal do Rio Grande do Sul, Porto Alegre, RS, Brazil*
[c]*Laboratório de Catálise Biomimética, Universidade Federal de Santa Catarina, Florianópolis, SC, Brazil*

Abstract. The aim of this work was to prepare and characterize chitosan (Ch) and chitosan/montmorillonite (Ch/Mt) beads for the adsorption of methyl violet 10B (MV) from aqueous solution. The beads were prepared by dissolution of chitosan in acetic acid and coagulation of the gel by dropping into an alkaline coagulating mixture of $H_2O/CH_3OH/NaOH$. Different compositions were prepared varying the Mt content (Ch/Mt3, Ch/Mt10 and Ch/Mt50, with 3, 10, and 50 wt% Mt respectively). For comparison, beads of neat chitosan (Ch/Mt0) were also prepared and characterized. The gel formed nearly spherical beads with approximately 2-3 mm diameter. SEM results indicated a rough surface of neat chitosan and a good distribution and adhesion of the particles in the Ch matrix for the composition containing 10 wt% and 50 wt% Mt. For the adsorption tests, the distinct forms of chitosan were evaluated, including as received chitosan (Ch). The adsorption tests were performed in 25 mL aqueous solution of MV 50 mg L^{-1} for 4 h, using dry and wet beads. The determination of the removal efficiency (%) was carried out by UV-Vis absorption spectra using the maximum absorbance of MV at 583 nm. Compositions of up to 10 wt% Mt presented a good stability during the test period. The Ch/Mt10 wet gel beads also presented the higher adsorption capacity, thus indicating the potential of this material for the application in the industrial wastewater treatment.

Keywords: Chitosan, Clay/Nanoclay, Morphology, Adsorption

INTRODUCTION

Dyes are extensively used in industrial fields such as textiles, paper, plastic and cosmetics, resulting in a large amount of wastewater. This is a source of pollution to aquatic systems and problems to human health [1]. Several processes for removal of dyes from wastewater have been developed, including adsorption, biological treatment, chemical oxidation, coagulation, and membrane filtration. Among these, adsorption has become the most popular one since it includes advantages such as high efficiency, low energy requirements, and the use of low cost adsorbents [2].

Chitosan is a biopolymer derived from the alkaline deacetylation of chitin (40-98%), which is the second most abundant polysaccharide after cellulose [3]. The chemical structure of chitosan is composed by glucosamine and N-acetyl-glucosamine [4], with OH functional groups at C-3 and C-6, and NH_2 group at C-2 position. The presence of such groups results in a high affinity for the removal of contaminants such as heavy metals and dyes [5].

The combination of chitosan and nanoparticles may result in a material with enhanced gel strength, improved swelling properties, low production cost, and high adsorption capacity [6,7]. Montmorillonite (Mt) is a layered clay mineral with a thickness of 1 nm that can interact with a variety of organic molecules through several interaction mechanisms [8]. The aim of this work was to develop economical, non-toxic, and readily available adsorbents from chitosan/montmorillonite. The removal efficiency of such adsorbents was evaluated for methyl violet 10B (MV) dye. For the composition that presented the best results, optimized conditions were achieved through the variation of contact time and adsorbent mass.

EXPERIMENTAL

Preparation and characterization of chitosan beads

The preparation of chitosan beads was based in the method of Pal et al. [9]. A solution of chitosan (Sigma-Aldrich) 1.2% (w/v) in acetic acid 7% (w/v) was prepared and dropped into an alkaline coagulation mixture of H_2O:CH_3OH:NaOH (4:5:1 w/w). The chitosan gel was coagulated to nearly spherical beads and rinsed with distilled water till neutralization. Chitosan/montmorillonite beads were prepared following the same procedure as described above, varying the content of Mt: Ch/Mt3, Ch/Mt10 and Ch/Mt50, with 3, 10, and 50 wt% Mt, respectively. In this case, the clay (bentonite, predominantly composed by sodium montmorillonite (Mt) (Argel T from Bentonit União Nordeste Ltda.)) was previously dispersed in acetic acid solution for 1 h and then mixed with the chitosan solution for 1 h. The beads were characterized after drying in an oven at 70 °C for 4 h. The morphology of the chitosan was analyzed by scanning electron microscopy (SEM) in a Jeol JSM 6390 at 10 kV. In the adsorption tests, both dry and swollen beads were evaluated.

Adsorption tests

Initially, five different types of chitosan were evaluated by the efficiency in the removal of MV: as received chitosan, chitosan beads (Ch), and chitosan/montmorillonite beads (Ch/Mt3, Ch/Mt10 and Ch/Mt50). In the adsorption tests, 25 mL of a 50 mg L^{-1} MV solution was used. The adsorption process was optimized for the composition that presented the best results, varying the contact time (30 min to 420 min) and the initial adsorbent mass (0.0125 g to 0.14 g, dry weight) under continuous stirring at 80 rpm in a MarqLabor BM/DR orbital shaker. The dye removal was quantified by molecular absorption in the visible region at a wavelength of 583 nm using a Spectramax Plus 384 UV-Vis spectrophotometer (Molecular Devices) from the calibration curve obtained at concentrations of 2.0 to 14.0 mg L^{-1}.

RESULTS AND DISCUSSION

Characterization of chitosan beads

Figure 1 shows the beads of neat chitosan (Ch) and Ch/Mt50 composition. It is possible to observe that both swollen beads present an uniform size varying from 2-3 mm.

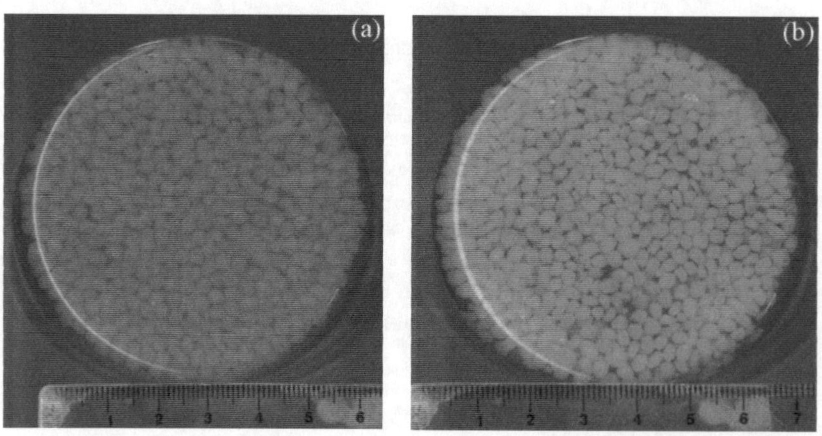

FIGURE 1. Images of (a) neat chitosan and (b) Ch/Mt50 beads.

SEM image in Figure 2a indicated the smooth surface of neat chitosan beads. By contrast, the incorporation of Mt resulted in an increased roughness. It is also possible to visualize the presence of Mt aggregates with a poor adhesion with the polymer matrix for the composition containing 3 wt%. This can be attributed to the high interaction among the nanoparticles and re-aggregation during the coagulation process. For higher contents of Mt, SEM images evidenced a more homogeneous distribution and adhesion of the nanoparticles, as shown in Figures 2c and 2d. The changes in viscosity of chitosan solution may have favored the mixture and consequently the interaction among the phases. For the Ch/Mt50, a more irregular surface was observed due to the high amount of clay.

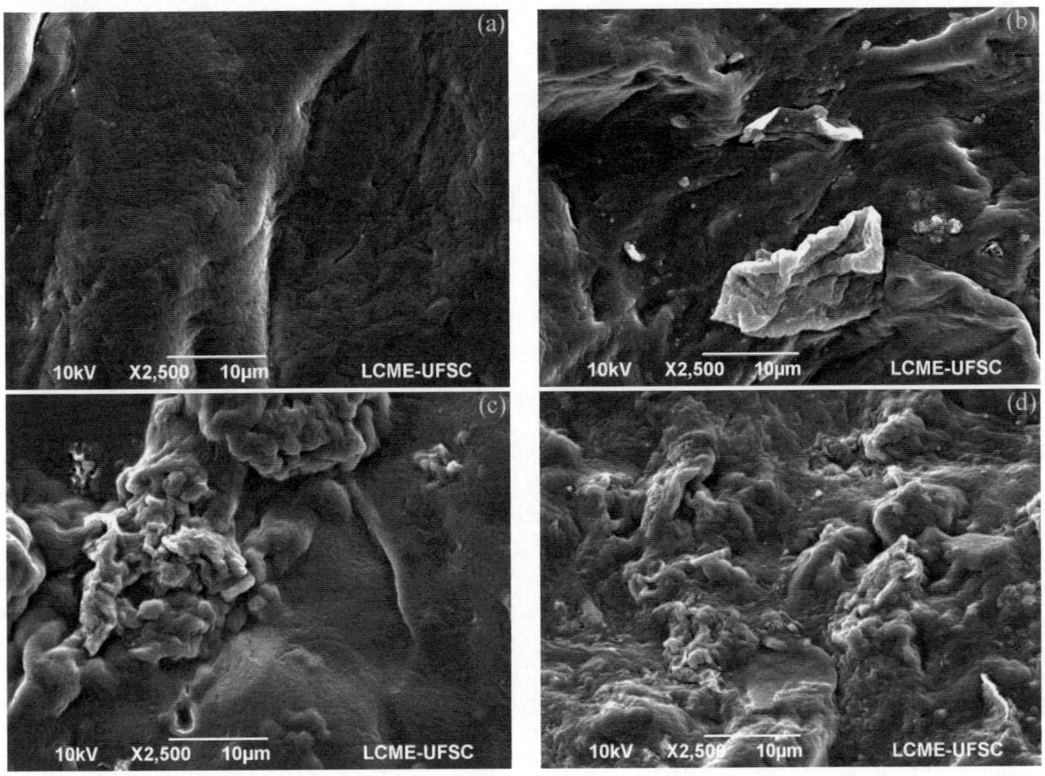

FIGURE 2. SEM micrographs of chitosan beads: (a) Ch (b) Ch/Mt3 (c) Ch/Mt10, and (d) Ch/Mt50.

Adsorption tests

Preliminary tests (Table 1) showed a high efficiency of removal of MV for the dried beads of Ch/Mt50. However, this composition presented a low mechanical strength during the test. Comparable efficiency values were observed for the swollen Ch/Mt10 composition. The higher adsorption capacity of swollen beads may be related to the ease of diffusion of the dye into the material. Therefore, Ch/Mt10 was selected for the optimization of the process.

TABLE 1. Removal efficiency of MV for different chitosan compositions.

Sample	Removal efficiency (%)
Ch (as received)	5.2
Ch	10.4
Ch/Mt3	6.2
Ch/Mt10	13.6
Ch/Mt50	32.4
Ch (swollen)	21.8
Ch/Mt10 (swollen)	37.6

Experimental conditions: 25 mL MV solution 50 mg L^{-1}, 0.1 g adsorbent, 240 min, room temperature.

The removal efficiency of MV as a function of contact time and adsorbent mass is shown in Figure 3. The maximum dye removal occurs at 180 min (Figure 3a), reaching approximately 97% of efficiency, and remains constant for longer periods. The removal efficiency increases with the amount of adsorbent (Figure 3b) since more sites are available for the adsorption. However, there is a smaller amount of dye being adsorbed by mass of adsorbent [2]. Therefore, the ratio considered ideal for the practical application was 0.1 g since it presents high removal efficiency without the use of an excess of adsorbent.

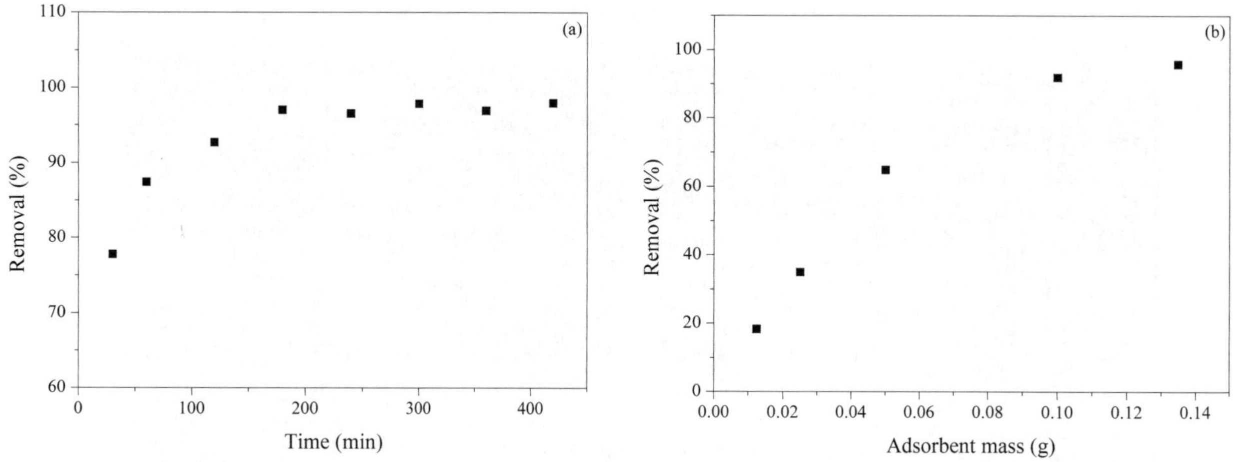

FIGURE 3. Influence of (a) contact time for 0.15 g of adsorbent and (b) mass of adsorbent for 180 min on the removal of MV.

CONCLUSIONS

In this study, chitosan/montmorillonite beads were successfully prepared and characterized. The experimental parameters were of great influence on the adsorption process, which presented a higher efficiency with 0.1 g of adsorbent, reaching up to 92% of removal in 180 min. The results of the composition Ch/Mt10 highlight the potential of this material for application in the removal of dyes in aqueous medium through adsorption.

ACKNOWLEDGMENTS

The authors are grateful to Conselho Nacional de Desenvolvimento Científico e Tecnológico (CNPq) for financial support and to Laboratório Central de Microscopia Eletrônica (LCME) from UFSC for SEM analyses.

REFERENCES

1. J. Duan, R. Liu, T. Chen, B. Zhang and J. Liu *Desalination* **293**, 46-52 (2012).
2. R. Liu, B. Zhang, D. Mei, H. Zhang and J. Liu **268**, 111-116 (2011).
3. G. Crini and P.-M. Badot *Prog. Polym. Sci.* **33**, 399-447 (2008).
4. I. Corazzari, R. Nisticò, F. Turci, M. G. Faga, F. Franzoso, S. Tabasso and G. Magnacca *Polym. Degrad. Stab.* **112**, 1 (2015).
5. M. Vakili, M. Rafatullah, B. Salamatinia, A. Z. Abdullah, M. H. Ibrahim, K. B. Tan, Z. Gholami and P. Amouzgar *Carbohydr. Polym.* **113**, 115-130 (2014).
6. D. Chen, W. Li, Y. Wu, Q. Zhu, Z. Lu and G. Du *Chem. Eng. J.* **221**, 8-15 (2013).
7. R. T. D. Silva, P. Pasbakhsh, K. L. Goh, S.-P. Chai and H. Ismail *Polym. Test.* **32**, 265-271 (2013).
8. N. H. Kim, S. V. Malhotra and M. Xanthos *Microporous Mesoporous Mater.* **96**, 29-35 (2006).
9. A. Pal, S. Pan and S. Saha **217**, 426-434 (2013).

Effect of additives on the performance of poly(trimethylene terephtalate), PTT and PTT/biochar composites

Torres Simonet [(1)], Myllytie Petri [(2)], Misra Manju [(2)*], MOHANTY Amar K. [(2)]

[(1)] CEDENNA, Universidad de Santiago de Chile, USACH - Casilla 40, Santiago C.P. 33 - Chile, [(2)] University of Guelph - Ontario - Canada

Partly bio-based thermoplastic engineering polyester poly (trimethylene terephthalate) and biocarbon were used in this work to study the performance of PTT and biochar composite with additive. Neat poly (trimethylene terephthalate) (PTT) and composites containing 20% biochar (BC) were processed with compounding-molding equipment DSM Xplore from DSM. Neat PTT and PTT containing 20% of biochar in the absence and presence of chain extender were processed at selected conditions. Mechanical properties, thermal and rheological properties were tested in this work. Electrical and thermal properties were also analyzed for neat PTT and PTT/biochar composites. Tensile modulus and tensile strength increase from PTT and PTT/biochar (80/20), the same behavior when added chain extender. Rheological behavior of the poly(trimethylene terephthalate) (PTT) and PTT/biochar composites with and without chain extender (CE) was investigated. Frequency sweep experiment (100 to 0.1 rad/s) at 1 % strain rate was used to study the rheological behavior of the samples. With the addition of 0.4% CE to neat PTT similar trend was observed but with increased value of complex viscosity. Composites with biochar particles exhibited Newtonian behavior at lower frequencies and shear thinning behavior at higher frequencies. Complex viscosity of PTT /biochar (80/20) increased from 300 Pa.s to 850 Pa.s with the addition of 0.4% CE. Such increase in viscosity is an indication of the chain extension effect which preserved the molecular chains of PTT from thermal degradation.

Films with Functional Properties for Packaging

***Ajji Abdellah* [(1)*]**

[(1)] Polytechnique Montreal - Quebec - Canada

During the last decades, the food, pharmaceutical and many other industries have seen several changes in packaging technology and applications because of new consumer demands and market trends. These drivers can be summarized as requirements for high quality, freshness and extended shelf-life of products. On the other hand, public demand and awareness for food safety has become a significant concern. This prompted the public awareness to question food quality in stores and technological solutions that could prevent contamination and/or alert consumers, which may provide better public protection. In the past, most of the studies were directed to the improvement of structural properties of films and multilayer structures (mainly mechanical: strength, tear, toughness etc...) and much less interest was directed to functionality such as barrier, antibacterial/antimicrobial, oxygen scavenging, detection, biological interactions etc. In this study, the results of recent efforts directed towards addressing those concerns will be presented and discussed. In particular, films with antibacterial properties, structures that can detect food degradation by-products and some oxygen scavenging compounds and formulations will be presented and their results discussed.

pps proceeding
Symposium: S12 - Polymer foams and films
Keynote Presentation

The role of elasticity in cell wall breaking

Di Maio Ernesto [(1)*]

[(1)] University of Naples Federico II - Naples - Italy

The breaking dynamics of the thin film separating a bubble from bulk air has been studied since several centuries. When a hole is induced in a bubble made by a Newtonian fluid, a creep due to the surface tension works provoking the hole opening. The kinematics of the hole opening is tricky, but easily understandable, with viscous forces and inertia involved in the process. However, when the liquid entrapping air is of viscoelastic origin, unusual behavior in rupture is displayed. Elasticity is indeed an important parameter, both during the bubble floating and in the hole opening: viscoelastic fluids can, in fact, store elastic energy (which can be released during rupture) that, in turn, strongly affect the breaking dynamics. The mechanism of viscoelastic bubble breaking is unknown and only very few experimental observations have been recently reported in the literature. In this contribution, the retraction of a circular viscoelastic sheet with a hole in the middle is studied by finite element simulations. The interplay between inertial, viscous, surface and elastic forces on the hole evolution is investigated. The simulation results are compared with experiments where the breaking dynamics of a viscoelastic bubble is thoroughly tracked through a fast camera.

pps proceeding
Symposium: S12 - Polymer foams and films
Oral Presentation

Inverse Mold Design To Accommodate Manufacturing Warpage in Chemically Blown PMDI Foams

Long Kevin N [(1)]*, Roberts Christine C [(1)], Soehnel Melissa [(1)], Troyer Kevin [(1)], Rao Rekha R. [(1)]

[(1)] Sandia National Laboratories - New Mexico - USA

We simulate the warpage of chemicaly blow pMDI (polyurethane) foam components (10-40 pounds per cubic foot) from manufacturing processes. We deploy a component-scale "cradle-to-grave" finite element simulation framework (developed in previous work) to consider foaming and filling, secondary cure, vitrification, cool down, demolding, physical aging, and depressurization to predict component warpage from birth to three years. Initially, spatial varyings in density are generated during the filling process and locked in when the polymerization process reaches gelation (captured both by the model and as seen experimentally). Using this as an initial condition, viscoelastic cure and thermal shrinkage are calculated on removal from the mold and cool down. The pore depressurization for a chemically blown foamed is then used to predict aging shrinkage and mass loss over time. Based on this shape change, we then warp the mold by the combined net predicted shrinkage. Components following the same process (but with the new mold) show very little warpage compared with the target shape, which demonstrates that foamed components can be designed through this method with minimal net warpage. Long term measurements show that viscoelastic aging and pore depressurization are not sufficient to describe continued warpage months after our PMDI foam components are made leaving open the discussion for other mechanism (physical and chemical) of warpage during foam aging. We discuss varying levels of model complexity and their associated predicted responses during manufacturing as well as a variety of manufacturing sensitivity studies on the warpage of exemplar components. *Sandia National Laboratories is a multi-program laboratory managed and operated by Sandia Corporation, a wholly owned subsidiary of Lockheed Martin Corporation, for the U.S. Department of Energy's National Nuclear Security Administration under contract DE-AC04-94AL85000.

Preparation of Rigid Polyurethane Foams as Inner Wetted Thermal Insulation

Ugis Cabulis[a], Vladimir Yakushin[a] and Wolfgang P.P. Fischer[b]

[a]Latvian State Institute of Wood Chemistry, Dzerbenes 27, LV-1006, Riga, Latvia
[b]ArianeGroup GmbH, Airbus Allee 1, 28199, Bremen, Germany

Abstract. This paper provides an overview about the recently performed development of rigid polyurethane (PUR) foams on a cryogenic insulation for liquefied hydrogen (LH2) tanks application and gives information on the characterization test campaigns performed so far. Since this specific insulation is foreseen to be applied on inner surfaces of the propellant tanks it is called Inner Wetted Thermal Insulation (IWTI). The IWTI obtained by spraying method has been especially designed (with apparent core density 85 ± 2 kg/m^3) for an optimum chemistry and cell structure morphology in order to achieve superior mechanical and thermal performance under these exceptional conditions. Studies of the foams` tension and compression properties at 295 and 77 K were carried out. The obtained safety coefficient of cryogenic insulation material reached 3.7.

Keywords: Polyurethane, Foams, Morphology, Cryogenic characteristics
PACS: 85.35.-x; 66.70.Hk

INTRODUCTION

So far, different methods of hydrogen storage for ground and aerospace vehicles have been proposed and investigated in different studies [1-4], and it has been shown that the most favorable type of tank system depends highly on the kind of vehicle and operation [5]. When using hydrogen as an aviation fuel, whether at subsonic or hypersonic speeds, lightweight, insulated cryogenic tanks are a key technical enabler as the low density of H_2, even when stored as a liquid, leads to a tank volume that is about 4 times higher than that of an equivalent kerosene fuel in spite of the fuel mass reduction [6, 7]. The high temperatures encountered at hypersonic speeds make tank construction potentially even more challenging as the insulation system has to be designed to withstand a higher thermal load for a comparatively long time.

Since the development of the vacuum flask by Sir James Dewar up today the vacuum is still the primary player in most insulation schemes. The use of the vacuum has helped to minimize the effects of gas heat conduction and convection. Naturally the degree of the vacuum within the construction space has a direct impact on the overall efficiency of the insulation characteristics. At low temperatures in a high vacuum, radiation is the dominant heat transfer mechanism. For the same reason within a vacuum, a number of solid materials have been introduced to reduce the radiation heat transfer within the system. These include expanded perlite, silica aerogel, diatomaceous earth, lampblack, glass and phenolic microspheres, and many others [8]. This approach is relatively inexpensive and has demonstrated excellent results.

Rigid polyurethane (PUR) foams have outstanding properties, and some of them are successfully used as cryogenic insulation. Also in the future, PUR foams and other plastic foams are regarded as the main material for liquefied gas tank cryogenic insulation in different projects of ground, air and space vehicles. There are a whole range of examples of the use of plastic foams as cryogenic insulation, although manufacturers and developers of insulation not always provide information on the chemical nature and properties of the used plastic foams. PUR foams possess many advantageous properties, such as low thermal conductivity, light weight, low water absorption/permeability, and dimensional stability. In particular, PUR foams do not change in volume along temperature gradients, in other words, they exhibit significant size stability [9, 10].

In the present article, work is carried out on the development of PUR foam material for inner wetted cryogenic insulation (IWTI) of a liquefied hydrogen (LH2) tank. The main advantage of PUR foams in comparison with other thermal insulation materials is possibility to cover this material on the complicated shape metal surfaces by spraying method. The IWTI obtained by spraying method has been especially designed (with apparent core density 85 ± 2 kg/m^3) for an optimum chemistry and cell morphology in order to achieve superior mechanical and thermal performance under these exceptional conditions.

MATERIALS AND METHODS

PUR foam composition was developed in the frame of commercial contract with ArianeGroup GmbH, and designed for insulation of common bulkhead between of LH2 and liquefied oxygen (LOX) tanks in re-ignitable cryogenic upper-stage (**FIGURE 1**) of space launchers [11, 12]. The current baseline is to apply this material as spray-on foam with all the advantages concerning simpler processing and cost reduction. The nature of the material allows this even for large structures where current solutions are very limited (glued as single tiles) and tend to form unacceptable cracks during the tank filling process. For application on the inner tank structure the insulation has to provide sufficient barrier properties in view of permeation of LH2 (or He) molecules.

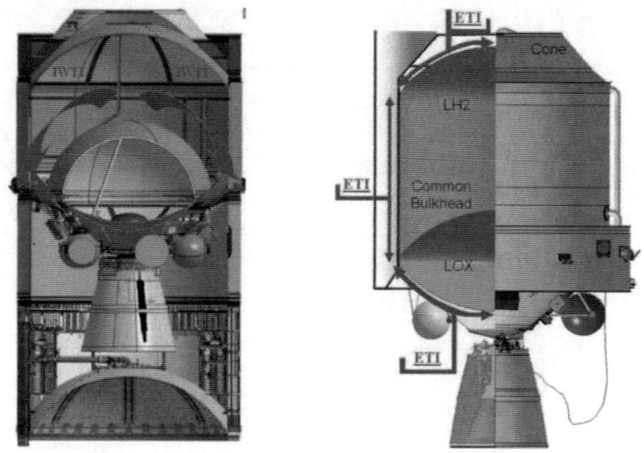

FIGURE 1. Locations of IWTI and external thermal insulation (ETI) on current LH2 & LOX Tank Configurations for Upper Stage

Manufacturing of SOFI foams samples were done with the high pressure spraying machine Glass Craft. For producing PUR foams by spraying method, the following parameters were controlled and fixed: A-component (polyol) temperature, B-component (polyisocyanate) temperature, A and B components' volume ratios, pressure at A and B components' supply lines and room temperature, relative air moisture, cream time, time after which the following layer was sprayed, metal sheet temperature. The used PU foam composition is calculated and optimised using approach for development of cryogenic insulation described in [13] and composition is given in **TABLE 1**.

TABLE 1. Composition of PUR foams for IWTI, parts by weight (pbw)

Component	pbw
A-component	
Polyether polyol	30
Polyester polyol	35
Chain extender	35
Surfactant	2
Catalyst package	3 - 4
Physical blowing agent	10 - 12
B-component	
Polyisocyanate	156

For manufacturing of samples for verification tests, spray-applied polyurethane foam panels were produced. The GlassCraft high pressure variable ratio machine MH VR and the spray gun Probler P2 Elite with a round type nozzle was used for PUR foam spraying. The output range of the unit was about 5 kg/min.

A Static Materials Testing Machine Z010 TN (Zwick GmbH & Co) with 1 kN force cell and the basic program testXpert II was used for foam testing at 295 K. Compression tests were performed according to ISO 844-2014 in two directions: parallel (Z) and perpendicular (X) to foam rise. Samples 30×30×30 mm were used.

Tensile tests were performed in perpendicular (X) direction. Dumb-bell samples with the length 140 mm, thickness 10 mm and width 40 (25) mm according to ISO 1926-2009 were used. For strain measurement on the gauge length 40 mm, a non-contact extensometer ME-46 Full Image VideoXtens was used.

The testing of the foams' mechanical properties at 77 K (liquefied nitrogen (LN2) media) was carried out according to the methods described in [11]. In this case, a Static Materials Testing Machine Z100 (Zwick GmbH & Co. KG)

with 1 kN force cell and the basic program testXpert V11.02 was used. Testing speed was 10% h/min (where h was the specimen height) like in ISO 844:2004.

Average thermal expansion coefficient in the temperature range 77-295 K was calculated from the contraction of the samples immersed in liquid nitrogen for 60 min. For precise measurement of the average thermal expansion coefficient in the temperature range of 113-295 K, a TMA PT1600 thermomechanical analyzer (LINSEIS GmbH) were used.

Thermal conductivity coefficient (λ) was determined using a Linseis HFM 200 thermal analyzer. The size of samples was 200x200x50 mm.

RESULTS AND DISCUSSIONS

The obtained PUR foams characterized by fine cell structure, which confirmed by SEM images and smooth thermal expansion (**FIGURE 2**) and basic physical – mechanical characteristics (**TABLE 2**). The SEM investigations of structure revealed that foams are nearly isotropic in the central part of sprayed IWTI foam layer. Near external and internal skins foams are anisotropic, with finer cell size and polymer struts oriented mainly parallel to the rise direction.

(a) 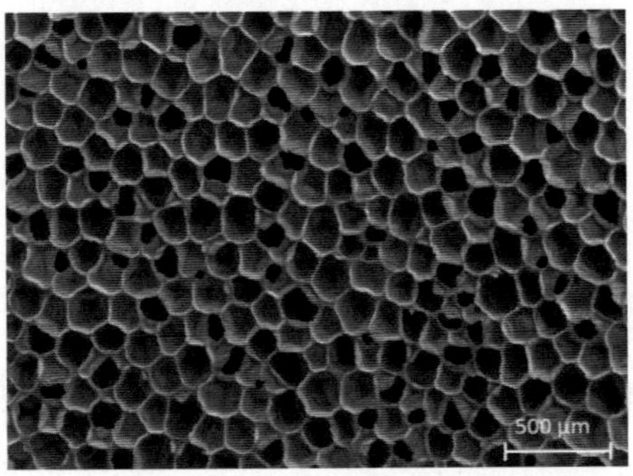

(b) 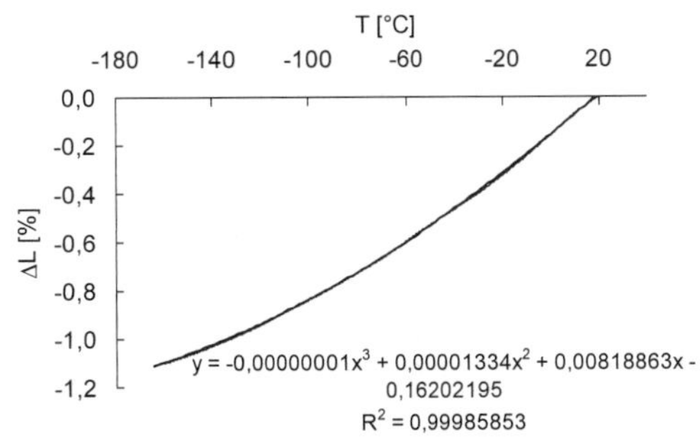

FIGURE 2(a) SEM image of central sprayed IWTI foam part; **(b)** Coefficient of thermal expansion of the IWTI foams

TABLE 2. Properties of the IWTI foam CF93 at 295 K and 77 K

Characteristics	at temperature	
	295 K	**77 K**
Density, kg/m^3	85 ± 2	
Compressive strength *Z*, MPa	0.68 ± 0.02	1.94 ± 0.06
Compressive strength *X*, MPa	0.59 ± 0.01	1.57 ± 0.07
Compressive modulus *Z*, MPa	21.6 ± 0.5	32.9 ± 1.0
Compressive modulus *X*, MPa	14.7 ± 0.4	20.5± 0.6
Tensile strength *X*, MPa	0.98 ± 0.06	1.48 ± 0.11
Elongation, %	11.5 ± 0.8	4.07 ± 0.13
$\alpha_X \times 10^6$ (295 K-77 K), K^{-1}	50.4 ± 0.7	

The IWTI possibility to withstand thermal strains can be characterized by safety coefficient K_S, which can be calculated from following equation:

$$K_S = \frac{\varepsilon_{77}}{\Delta l_{77-295}} = \frac{\varepsilon_{77}}{\alpha_x \cdot \Delta T} \quad (1)$$

where:

ε_{77} – PUR foam elongation at 77K perpendicular to foaming direction, %;

$\Delta l_{77\text{-}295}$ – the contraction of PUR foam perpendicular to foaming direction freezing it from 295 K to 77 K, %;

α_x – coefficient of thermal expansion;

ΔT – temperatures difference.

The PUR foams contraction $\Delta l_{77\text{-}295}$ were measured at fast chilling process in LN2 medium, similar to conditions when the tank will be filled with LH2 or LOX. In all cases the CTE were calculated using the contraction characteristics at temperature difference 218 K.

Safety coefficient of the obtained IWTI foam material is 3.7. Cryogenic insulation materials characterizes by safety coefficient above 3 [13], it means, that elongation at break 3 folds or more increase shrinkage at thermal shock.

Due to the very fine and closed cell structure, the IWTI material has a particularly low coefficient of thermal conductivity at room temperature (RT) and at cryogenic temperatures (**FIGURE 3**). Characteristic decrease of λ in long term storage for foams obtained with physical blowing agent are very slow and plateau value reaches at 22.2 W/m·K.

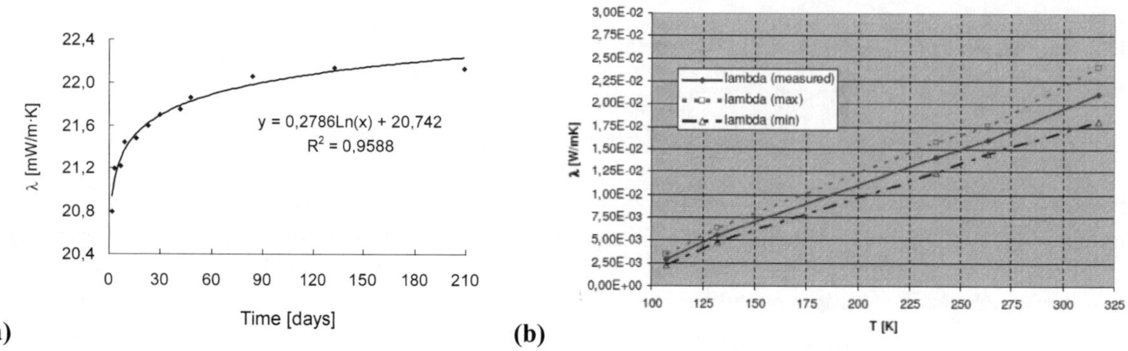

(a) **(b)**

FIGURE 3. Coefficient of thermal conductivity **(a)** changes of λ depending from time at RT; **(b)** changes of λ depending from temperature [12].

SUMMARY

This paper presents the results achieved so far in the current development of IWTI. For the chemical composition and morphological structure of IWTI are very satisfying so that no further development is needed from current point of view. This is confirmed by testing and measurement of major performance as well other properties. Characterization of IWTI from RT to LN2 range is also mostly completed and results are better than those expected. A method for IWTI safety coefficient (K_s) calculation, has been developed. For given composition the K_s is 3.7, and it confirm, that developed material can withstand cryogenic shock. Characterization for LH2 conditions has been commenced. First results on compatibility to LH2 and measurement of most important performance parameters (thermal conductivity) are very positive.

ACKNOWLEDGMENTS

The research leading to these results has received funding from the Future Launcher Preparatory Program project financed by ArianeGroup GmbH; Contract No. FLPP-RIBRE-CON-0007.

REFERENCES

1. K. Pehr, P. Sauermann, O. Traeger, M. Bracha, *Int. J. Hydrogen Energy* **26**, 777-782 (2001).
2. S. Aceves, G. Berry, J. Martinez-Frias, F. Espinosa-Loza, *Int. J. Hydrogen Energy* **31**, 2274-2283 (2006).
3. R. Ahluwalia and J. Peng, *Int. J. Hydrogen Energy* **33**, 4622-4633 (2008).
4. S. Aceves, F. Espinosa-Loza, E. Ledesma-Orozco, T. Ross, A. Weisberg, T. Brunner, *Int. J. Hydrogen Energy* **35**, 1219-1226 (2010).
5. A. Westenberger, "Liquid hydrogen fuelled aircraft-system analysis (CRYOPLANE)", Final Technical Report (Publishable version). Airbus Deutschland GmbH, 2003.
6. D. Verstraete, P. Hendrick, P. Pilidis, K. Ramsden, *Int. J. Hydrogen Energy* **35**, 11085-11089 (2010).
7. D. Verstraete, D. Long, *Int. J. Hydrogen Energy* **38**, 14824-14831 (2013).
8. L. Adams, *Cryogenic Techn. J. of the Cryogenic Soc. of America*, **1**, 249-251, (1965).
9. A. Bahadori, *Thermal insulation handbook for the oil, gas, and petrochemical industries*, Oxford: Gulf Professional Publishing; 2014.
10. S. Desai, I.M. Thakore, B.D. Sarawade, S. Devi, *Eur. Polym. J.* **36**, 711-725 (2000).
11. W.P.P. Fischer, U. Stirna, V. Yakushin, U. Cabulis, "Cryogenic Insulation for LOX and LH2-Tank Application", in 40th Int. Conf. on Environmental Systems (ICES), Barcelona, Spain (2010).
12. W.P.P. Fischer, "Development of Cryogenic Insulations for Launcher Upper Stages", in 44th Int. Conf. on Environmental Systems (ICES), Tuscon, Arizona, USA (2014).
13. U. Stirna, I. Beverte, V. Yakushin, U. Cabulis, *J. of Cell. Plast.* **47**, 337-355 (2011).

Predicting the final microstructure of polyurethane foams using direct numerical simulations

***Mitrias Christos* [(1)*], *Jaensson Nick O.* [(1)], *Hulsen Martien A.* [(1)], *Anderson Patrick D.* [(1)]**

[(1)] Technische Universiteit Eindhoven - Eindhoven - Netherlands

The mechanical, thermal and electrical properties of polyurethane foams are governed by both the microstructure and the properties of the matrix material. The microstructure arises during the foaming process, in which growing gas bubbles are created by blowing agents in a viscous liquid. Important parameters that determine the material properties of the foam, such as bubble size and wall thickness between the bubbles, are difficult to control a-priori, due to the large range of parameters that play a role (viscosity of the liquid, surface tension between the bubble and the liquid, etc.). We have developed a novel direct numerical simulation technique with which we are able to predict the morphology of the foam. Moreover, we can then use this morphology in simulations of solid mechanical testing, and determine the mechanical properties of the foam. We describe the foam by a representative volume element (RVE) containing a number of randomly distributed bubbles. By applying periodic boundary conditions, the RVE can describe the bulk behaviour of the foam, while remaining computationally tractable. We assume that Brownian motion and inertia can be neglected and that the fluid of the matrix is Newtonian and incompressible. An internal pressure on the bubble interface is enforced, thus making the bubbles grow. The resulting equations are discretized in space using the 3D finite element method. Adaptive meshing is employed, to ensure the results remain accurate in the thin layers between the bubbles. The solidification process is simulated by increasing the viscosity of the liquid as the simulation progresses. The final mesh of the fluid foam, after solidification has been reached, can then be used directly as a representative volume element in solid mechanics simulations. We will present 3D direct numerical simulations of the foaming process that allow us to predict the microstructure of the foam and thus control the material properties of the final product.

Rigid PUR Foam Impact Absorption Material Obtained from Sustainable Resources

Mikelis Kirpluks[a], Elena Cischino[b], Ugis Cabulis[a] and Janis Andersons[a]

[a] *Polymer department, Latvian State Institute of Wood Chemistry; 27 Dzerbenes str., LV-1006, Riga, Latvia;*
[b]*Pininfarina S.p.A. Via Nazionale 30, Cambiano 10020, Italy*

Abstract. Polyols from different sustainable resources have been widely studied as one of the raw materials for polyurethane (PUR) material production in recent years. The renewable materials have been proven as a viable alternative for petrochemical products from their properties perspective. This study describes the development of rigid PUR foam impact absorption material that was used in electric vehicle crashboxes and crossbeam system. It was a part of EU FP7 Project EVolution that's goal was to demonstrate the sustainable production of a full electric 600 kg vehicle – NidoEV. This electric car was designed by a company Pininfarina S.p.A. Lightweight materials developed in the project were used to reduce the weight of Body in White which is the heaviest vehicle element, representing about 40% of total vehicle weight. The PUR foams were obtained from polyols synthesized from sustainable resources, like rape seed oil, tall oil and recycled polyethene terephthalate. Properties, such as tensile, compression, impact, compression at high strain rates, thermal stability and morphology, were tested for rigid PUR foams.The high strain rate compression tests were done by modifying puncture impact testing equipment to obtain stress-strain curves of foam material at 100-300 s^{-1} strain rates. The obtained data were used by project partners to develop finite element model of crossbeam front subsystem impact behaviour into a solid wall. Proposed concept was also realized at a demonstrator level and rigid wall crash tests of the developed front subsystem were performed. Obtained results showed that such concept is a viable option for vehicle impact absorption front subsystem and sustainable materials are a suitable alternative for petrochemicals. Final weight reduction was ~50% in respect to the steel baseline, which shows this concepts potential application in automotive industry.

Keywords: Polyurethane Foam, Sustainable Polyols, Foam Injection moulding, Impact absorption
PACS: 81.05.Lg ; 82.35.Pq ; 81.05.Qk ; 83.80.Jx ;

INTRODUCTION

Global trends toward CO_2 reduction and resource efficiency have significantly increased the importance of lightweight materials over the last years. The European Commission (EC) has set severe targets for average new car CO_2 emissions of 95 g/km by 2020, and by 2030 EC aims to reduce CO_2 emissions down to 75 g/km [1].

Weight reduction of the vehicle directly decreases the energy consumption and subsequently reduces the CO_2 emissions. Additional reduction of the CO_2 emissions could be achieved by the use of materials that are obtained from the renewable or recycled feedstock, like polyurethane (PUR) foams obtained from vegetable oils [2].

The weight reduction is particularly important for electric vehicles whose broader implementation is crucial to reduce the CO_2 emissions. Currently, the battery costs and the expected range of the vehicle are the most limiting factors for the slow rate of Full Electric Vehicle (FEVs) adoption for a larger consumer base. As a rough estimation, considering the battery costs for the year 2015, each slot of vehicle weight reduction of 50 kg implies a battery cost decreasing of about 500 €/car. This cost reduction can compensate the additional cost due to light weighting, contributing in market penetration of EVs [3].

The Body in White (BiW) is the heaviest vehicle element, representing about 40% of total vehicle weight, hence the implementation of lightweight measures here appears effective. The project EVolution stands for "The Electric Vehicle revOLUTION enabled by advanced materials highly hybridized into lightweight components for easy integration and dismantling providing a reduced life cycle cost logic". Project EVolution finished in November 2016 and was a result of a cooperation between 24 partners from 11 different European Union (EU) countries, with the goal to demonstrate the sustainable production of a full electric 600 kg vehicle. EVolution project was principally based on Pininfarina Nido concept which is fully electric and is an A-segment car conceived as a multifunctional rolling chassis plus a non-structural upper frame [4]. The Nido FEV and the Nido BiW concept is shown in Figure 1.

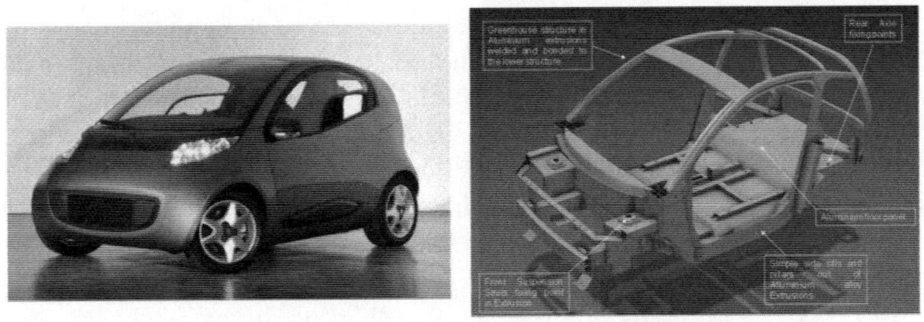

FIGURE 1. Nido concept (left) and Nido BiW (right)

EVolution project was focused on some specific car body areas. Different demonstrators were designed using lightweight materials to achieve 50% weight saving over existing steel equivalent. Selected demonstrators were: the underbody, the side door, the front crossbeam, the structural node (shotgun system) and the front mechanical subframe. Previous fully aluminium BiW of the Nido was redesigned to integrate the demonstrators, following the approach to lightweight everywhere reinforcing only where it is necessary while accomplishing structural performances, crashworthiness included.

Latvian State Institute of Wood Chemistry (IWC) was one of the partners cooperating in EVolution project to design novel rigid PUR foam material for impact absorption used in the front crossbeam of the vehicle. Rigid PUR foams were developed from sustainable resources, polyols obtained from vegetable oils and recycled polyethylene terephthalate (PET). Rigid PUR foams were developed in a broad range of the apparent densities (50-600 kg/m^3) and their mechanical properties were investigated to select the material that is the best suited for the core of the impact absorption element. The performance of the designed concept in vehicle crash was evaluated using finite element modelling and subsequent crash test of the developed front crossbeam. To develop a good model, the data of the material mechanical deformation behaviour at high strain rates have to be aquired. The high strain rate compression tests were carried out by modifying puncture impact testing equipment to obtain stress-strain curves of foam material at 100-300 s^{-1} strain rates.

EXPERIMENTAL

Rigid PUR Foam Development

IWC studied different sustainable raw materials for the rigid PUR foam development, like rapeseed (canola) oil, tall oil and recycled PET. Mentioned feedstock have to be transformed into polyols before they can be used for PUR material development. High density (~200 kg/m^3) rigid PUR foams were developed from the most perspective "green" polyols and their mechanical and thermal properties compared to select most perspective raw material for further development of the lightweight components. Initial test showed that rigid PUR foams produced from polyol based on recycled PET delivered the most stable properties, highest compression strength and compression modulus of 2.6±0.1 MPa and 74.5±3.7 MPa, respectively and most uniform cell structure (cell diameter 123±21 μm; anisotropy coefficient 1.00±0.02) [5]. Polyol synthesized from recycled PET was selected to develop series of rigid PUR foams with different apparent densities ~50-600 kg/m^3 to select material best suited for the impact absorption. The compression strength and compression modulus of developed rigid PUR foams is shown in Figure 2.

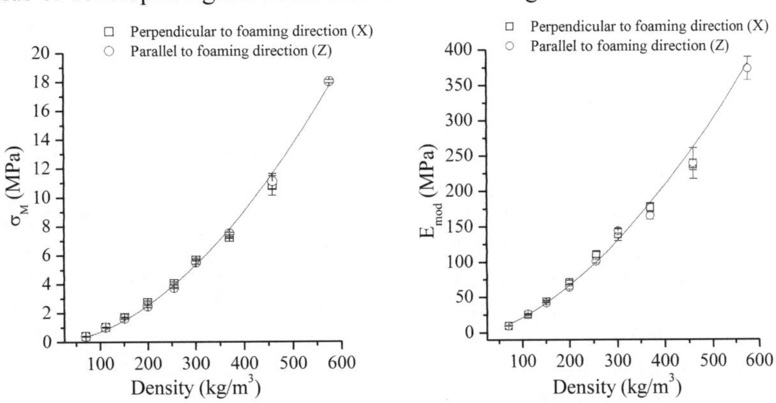

FIGURE 2. Compression strength and compression modulus of rigid PUR foams obtained from recycled PET

Rigid PUR foams were obtained in a closed type stainless steel mould to simulate a reaction injection moulding manufacturing process. It was possible to obtain foams with isotropic cell structure (Figure 3. a). due to the high pressure during the rigid PUR foam curing process. Developed rigid PUR foam formulations showed stable and predictable properties in all of the proposed density range. The apparent density of the foams was varied by changing the amount of blowing agent used in the foam formulation.

Rigid PUR foam testing at high strain rates

The mechanical characteristics of foams are strongly affected by their density, therefore, the optimum foam density, providing the intended mechanical response at a minimum foam weight, can be determined by employing the FEM modeling for crash simulation. As input data for such simulations, compressive stress-strain diagrams of foams at different densities and loading rates are utilized.

The mechanical properties of the rigid PUR foams usually are tested at quasi-static strain rates of 0.002-0.5 s^{-1}. To optimize the mechanical properties of the rigid PUR foam and to develop finite element model of the front crossbeam demonstrator, it is necessary to obtain stress/strain compression response data at high strain rates 100-300 s^{-1}. Typical equipment for such measurement is Split-Hopkinson Pressure Bar Test apparatus. Unfortunately, it was not possible to develop a testing procedure for the rigid PUR foams. It was not possible to obtain reproducible results using Split-Hopkinson Pressure Bar due to the high amount of noise signals resulting from the travelling pressure wave on the interface between bar and sample.

Instead, the stress/strain curve data at high strain rates were obtained by modifying CEAST 9340 Drop Tower (Instron) equipment. Usually this equipment is used to test puncture impact behaviour according to ISO 6603-2 standard. PUR foam cylinders were compressed at a wide range of strain rates (quasi-static testing: 0.002-0.5 s^{-1}; high strain rate testing: 100-300 s^{-1}) up to 80% of the sample deformation. The testing rig is shown in Figure 3. b) where the compression of the PUR foam sample is achieved by the impact of the sample against a steel plate. Obtained stress-strain curves are shown in Figure 3. c) and other data was used in finite element modelling which identified rigid PUR foams with density 70 kg/m^3 and 100 kg/m^3 as the optimal ones to fill the core of crossbeam and crash boxes respectively.

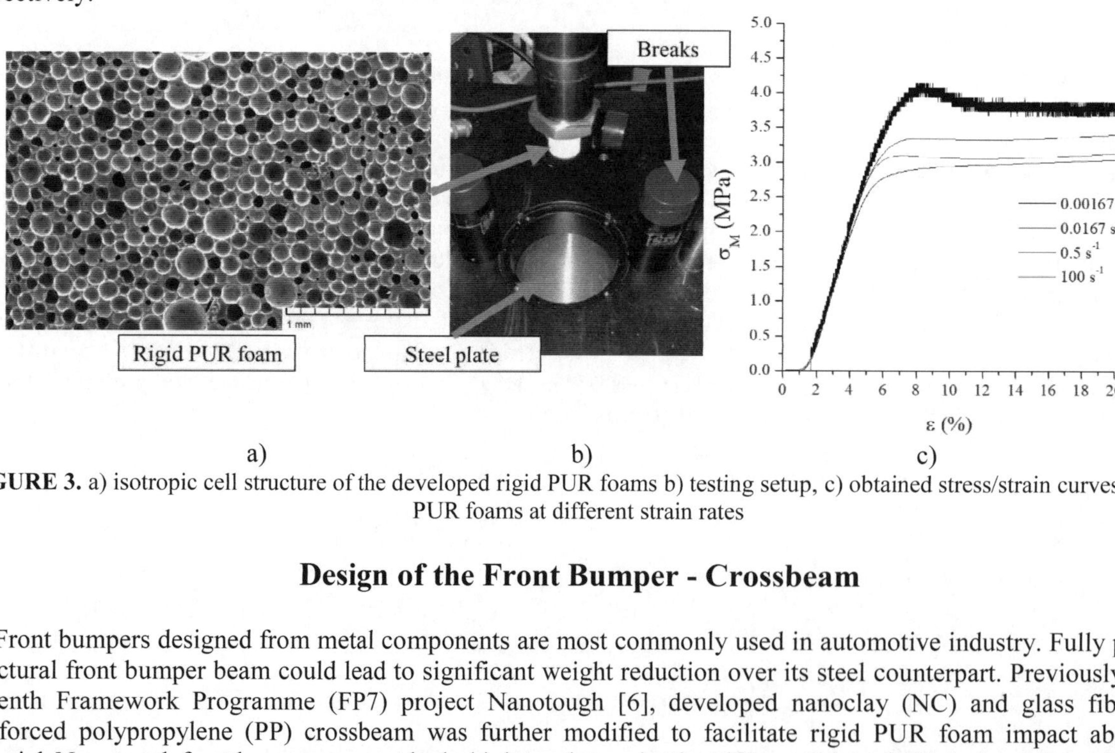

FIGURE 3. a) isotropic cell structure of the developed rigid PUR foams b) testing setup, c) obtained stress/strain curves of rigid PUR foams at different strain rates

Design of the Front Bumper - Crossbeam

Front bumpers designed from metal components are most commonly used in automotive industry. Fully polymer structural front bumper beam could lead to significant weight reduction over its steel counterpart. Previously, in EU Seventh Framework Programme (FP7) project Nanotough [6], developed nanoclay (NC) and glass fiber (GF) reinforced polypropylene (PP) crossbeam was further modified to facilitate rigid PUR foam impact absorption material. Nanotough front bumper concept had a high toughness, but its stiffness did not fulfil the impact requirements, so additional impact absorption material was introduced. The proposed design was analyzed using ECE94 Full Front Crash simulations with the support of Pininfarina.

The new beam design consisted of two components, front and rear beam shell members and a core of rigid PUR foams. Additional crash boxes made from NC and GF reinforced PP filled with rigid PUR foams were developed and the whole system was tested for impact absorption. The proposed front crossbeam design is shown in Figure 4.

The target weight saving was set 50% over existing steel equivalent, which was achieved. The crash test results underlined the potentiality of this solution in terms of raw materials, processes and performances.

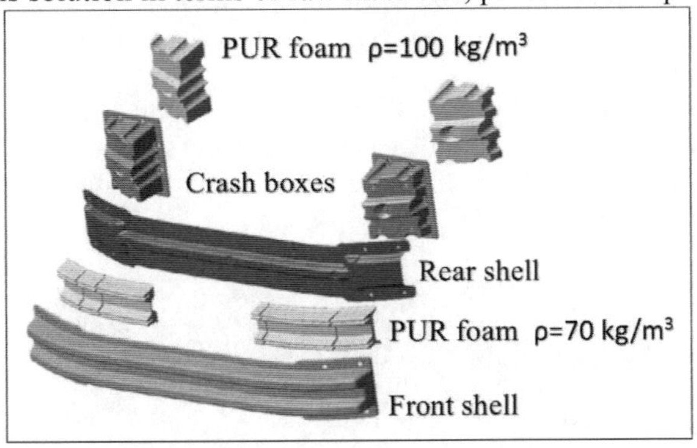

FIGURE 4. Evolution front crossbeam demonstrator concept

The manufacturing process

Starting from the background knowledge and the results obtained within the Nanotough project, partner ICECHIM together with partners from Alborg University proposed a new polymer material composition as a combination of two key solutions from nanotechnology and advanced materials field [6, 7]. Therefore, the composition of the new polymer material was specially tailored using a special nanosilicate additive (PP with 30-50% GF) which was used by partner FPK-BATZ. The nanosilicate was embedded into a PP matrix to obtain a smart masterbatch. The masterbatch composition and the optimum ratio between components enabled its use directly in the extrusion-injection process of the FPK-BATZ products. An increase of 300% of impact resistance was observed when masterbatch was used in comparison to virgin PP. The stiffness and mechanical strength did not decrease due to a synergistically effect between the components [8].

FPK-BATZ produced the PP parts (front shell, rear shell and crashboxes) using NC and GF reinforced PP masterbatch by injection moulding. Different foam feeding tests were carried out by IWC on part geometry, in order to define the filling strategy and the PP/PUR adhesion. A series of metallic inserts were introduced into all parts to improve joint strength into plastic parts and to avoid creep or stress relaxation.

Front and rear shell members were bonded together with a Polyurethane-based two component adhesive, after a plasma treatment of the surfaces, then the crossbeam and the crashboxes were bolted together, obtaining the complete demonstrator shown in Figure 5.

The produced demonstrator was subjected to a physical crash test based on ALLIANZ protocol. The tests were focused on analyzing impact resistance of the new structural component. Results were promising and showed that proposed design has similar impact absorption properties than the steel front bumper counterpart.

FIGURE 5. Evolution Front Bumper Crossmember demonstrator prototype

SUMMARY

Proposed design elements of Nido demonstrate an interesting lightweight solution that could be implemented in modern vehicles. Proposed BiW of the Nido is estimated to be 113 kg versus 160 kg of the initial full aluminum baseline at the start of the project and more than 50 % lighter than the steel counterpart. An finite element analysis of the virtual full vehicle indicated a good structural behavior, considering EU standards of crash homologation and global static and dynamic performances.

A technological mock-up representing the full vehicle environment was conceived and manufactured by Pininfarina to demonstrate the part assemblies and showcase the project solution as shown in Figure 6. The production costs together with savings made from lightweight solution implementation are in line for a medium vehicle production volumes (≥ 30,000 units/year).

FIGURE 6. The technological mock-up with integrated demonstrators

The project demonstrated that it is possible to implement sustainable materials where high performance requirements are demanded. Rigid PUR foam formulations obtained from recycled PET showed high stability of properties and good mechanical properties. Recycled PET polyols could be used to replace aromatic polyester polyols obtained from petrochemical materials.

ACKNOWLEDGMENT

The research leading to these results has received funding from the EU FP7 (2007-2013) under grant agreement no. 314744 in frame of the project EVolution.

REFERENCES

1. Regulation (EU) No 333/2014 of the European Parliament and of the Council of 11 March 2014
2. R. K. Helling and D. Russell, *Green Chem,* **11**(3), 380-389 (2009).
3. S. Faßbender, Investigation of the Trade-off Between Lightweight and Battery Cost for an Aluminium-intensive Electric Vehicle, FKA, Report 106330
4. E-light, GA n. 266284, theme GC-SST.2010.7-5, full title: "Advanced Structural Light-Weight Architectures for Electric Vehicles"
5. M. Kirpluks, U. Cabulis, A. Ivdre, M. Kuranska, M. Zieleniewska and M. Auguscik, *J. Renew. Mater.* **4**(1), 86–100 (2016).
6. Nanotough. GA n. 213436, full title. "Toughened Hybrid Nanocomposites for High Performance Application".
7. C. G. Sanporean, Z. Vuluga, C. Radovici, D. M. Panaitescu, M. Iorga, J. deClaville Christiansen and A. Mosca, *RSC Adv.*, **4**, 6573-6579 (2014).
8. E. Cischino, Z. Vuluga, C. E. Ezeiza, I. L. Benito, E. Mangino, J. de Claville Christiansen, C.-G. Sanporean, F. Di Paolo, M. Kirpluks, and P. Cabulis, *Procedia CIRP* **66**, 300–305 (2017).

Performance and Biodegradability of Polymer Mulch Films after Field Ageing

Touchaleaume François[a], Angellier-Coussy Hélène[a], César Guy[b], Raffard Grégory[a], Gontard Nathalie[a], Gastaldi Emmanuelle[a]

[a]*JRU IATE, Agropolymers Engineering & Emerging Technologies, INRA - Montpellier - France,*
[b] *SERPBIO, University Bretagne Sud - Lorient - France*

Abstract. In the aim of resolving environmental key issues such as irreversible soil pollution by non-biodegradable and non-recoverable polyethylene (PE) fragments, several biodegradable films based on poly(butylene adipate-co-terephtalate) (PBAT/PLA, PBAT/PPC and PBAT/Starch) have been tested as mulch films in real field conditions. The impact of outdoor ageing on film's mechanical and functional properties was investigated both on films exposed at the soil surface and buried into the soil in order to assess the respective contribution of the main related ageing factors i.e. UV radiations and microorganisms. The early loss of integrity of the biodegradable materials occurred concurrently and was attributed to the crosslinking of polymer chains induced by UV radiations. Considering that the major part of the three materials studied is made of PBAT, even if probably from various origins and containing different aliphatic/aromatic ratios, this suggested that the nature of the other polymer constituting the blend would not have a significant impact on the ageing mechanism of the material. The first signs of biodegradation were noticed on the soil-facing surface of films 18 months after installation without previous burying step. Biodegradation analyses conducted by respirometry in a compost medium indicated that field ageing had a low impact on the kinetics of mineralization and led to similar percentages according to whether the materials had previously been aged or not.

Keywords: polymer, biodegradable, PBAT, field ageing, UV radiations
PACS: 64.70.pj , 65.90.+I , 81.70.Bt , 82.35.Lr , 89.20.Bb , 68.90.+g

INTRODUCTION

The introduction and growing use of plastics in agriculture as mulch films, tunnels, drip irrigation tubes, string, clips, etc. has led to greatly improve crop production while decreasing the dependency toward climatic conditions and limited water resources. As a result, mulch films consumption has greatly increased during the last 10 years

The main benefits of mulch films for crops are: an increase of the soil temperature and moisture retention, a reduction of weed pressure and pests infestation that lead to earlier and higher crop yields. However, most of these materials are made of (PE), which cannot be totally removed from the field after the harvest, then fragment into non-biodegradable microplastics and finally contaminate soils. PE long lasting durability is inappropriate for applications in which plastic films are required for short time. For this reason materials designed to biodegrade with tunable lifetimes are becoming important for such agricultural applications.

Then, although the degradation of those biodegradable materials is beneficial for their after-use management, it comes with a progressive decrease of the functional properties of the film [1-3]. Consequently, making the mulch film degradation kinetics match the plants' needs requires mastering the material behavior in its environment by identifying the degrading mechanisms. Concerning usual PE mulch films, the main ageing factors have been identified as light [4-6], heat [4, 6, 7] and/or water [6, 8, 9]. Among them, the UV radiations are considered as the most detrimental one. Moreover, in addition to those ageing factors which refer to polyolefin, biodegradable films undergo biotic degradation mechanisms involving soil bacteria [10], fungi [11] and related microbial enzymes [12].

Poly(butylene adipate-co-terephthalate) (PBAT) is an oil-based polyester displaying similarities with polyethylene, both in processability and mechanical performances. In addition to this, PBAT is biodegradable in soil [13, 14] making it a possible substitute to PE in mulching applications. A way to tune PBAT performances while reducing its environmental impact is to blend it with another biodegradable polymer [15, 16]. Poly(lactic acid) (PLA), poly(propylene carbonate) (PPC) and Starch exhibit relevant processability and functional properties combined with environmental advantages.

The current work aims at providing a better understanding of the mechanisms involved in the ageing of biodegradable mulch films subjected to field conditions in order to assess the respective contribution of UV radiations and microorganisms. The impact of field ageing on their mechanical performances and ultimate biodegradability was assessed. Those films were either exposed to physical factors (temperature and UV radiations) when laid at the soil surface or subjected to biological factors (soil microorganisms) when buried into the soil. In both cases, the morphological and structural changes occurring in the materials during their ageing were investigated.

EXPERIMENTAL

Three biodegradable materials were tested as mulch films in real field conditions: Mater-Bi CF04P made of PBAT/Starch (BioBag, Askim, Norway) containing 37wt% bio-based carbon (as determined according to ASTM D6866), Bioflex F2110 made of PBAT/PLA in a 70/30 ratio (FKuR, Willich, Germany) and a PBAT/PPC compound in a 80/20 ratio (Hangzhou Xinfu Pharmaceutical, Zheijang, China). For comparison, a conventional PE film (Agripolyane, Saint Chamont, France) was also used. All the materials were obtained by blow-extrusion of 40 μm thick films. They were three-layered and each extruder was feed with the same composition. 5wt% carbon black was used as a colour additive using master batches prepared with the same blends as the matrices. Screw diameters were 65, 65 and 85 mm with L/D ratios of 30 for the three extruders. Total throughputs were 140 kg/h and blow-up ratios set on 5,6. Die gaps were 1mm thick.

The mulch films were installed mechanically in two vineyards located in France (43°12 N-02°36 E). Two full-scale experiments were launched at one-year interval. The mulch films were fixed onto wood frames (size 20 x 60 cm) installed at the soil surface beside the vineyard and periodically collected. Samples of varying ageing degrees were submitted to tensile tests. Tensile tests were performed on a Zwick Z2.5/TN1S (Germany) tensile machine. The gel content of the film samples was measured according to ASTM D2765 [17] using DMSO (in which thermoplastic starch is soluble) and chloroform (in which PLA, PPC and PBAT are soluble) as consecutive solvents.

Respirometric tests were conducted in compost medium at 58°C to assess the biodegradability of the materials before and after 18 months spent in field. Details of the testing conditions are available elsewhere [19].

RESULTS AND DISCUSSION

During the field test, the first signs of material loss of integrity (i.e. formation of cracks and tears of the film) were observed about 5 months after their installation, on all the biodegradable mulch films. In contrast, PE film did not undergo any visual degradation. Interestingly, comparing the two field trials launched ate one-year interval revealed that the films broke not after spending the same time in field but after receiving the same amount of cumulated radiations, which was around 260-270 kJ/cm². Observations of the sun-facing side of the mulch films (SEM pictures not reported here) did not reveal any significant change even after 18 months of field exposition. In contrast, the soil-facing surfaces became progressively rough whatever the material considered. The most altered one was made of PBAT/Starch and exhibited numerous holes. Concerning the parts of the films buried in soil, observations revealed the presence of small holes on the PBAT/Starch film 5 months after burying whereas the surface of the films made of PBAT/PPC and PBAT/PLA appeared almost intact. Those holes were ascribed to the disappearance of some starch granules since their diameters were comparable with the ones of original starch particles [39].

Impact of UV Radiations on the Mechanical Properties of Above-Soil Mulch Films

Effects of the microstructural modifications on the mechanical performances were assessed by tensile tests. All initial films exhibited a ductile behavior (figure 1), with elongation at break values higher than 300%. During ageing, PE film maintained an elongation at break higher than 400% whatever the amount of radiations received. On the contrary, biodegradable films first maintained their initial ductility for radiations lower than 100 kJ/cm² and strongly embrittled afterwards. Indeed, for radiations higher than around 150 kJ/cm² (i.e. after around 2 months in field), PBAT/PLA, PBAT/PPC and PBAT/Starch films broke with elongations lower than 100%.

This transition of the failure mode, from ductile to brittle, can be explained by the photochemical degradation of the biodegradable materials. Indeed, PBAT-based materials underwent Norrish reactions [18] when exposed to solar radiations, because of the macromolecular structure and the high density of ester linkages which favored the absorption of UV radiations. Those chemical reactions drove to main chain scissions and/or crosslinking. Such modifications occurred in the amorphous phase of the material, leading the material to reduce its molecular mobility and eventually adopt some thermosetting behavior.

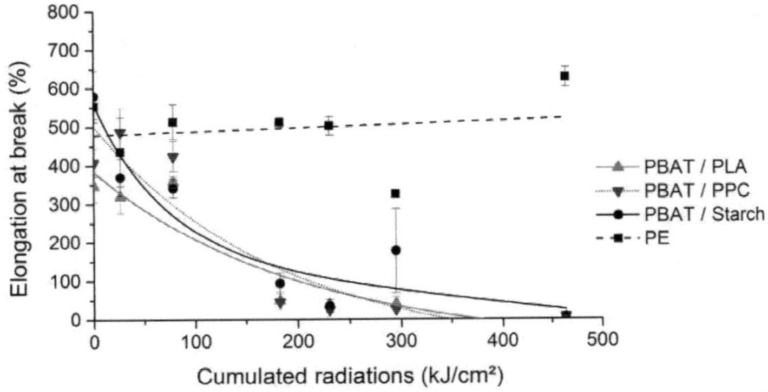

FIGURE 1. Elongation at break of mulch films in transverse mode as a function of cumulated UV radiations (symbols are experimental points and lines are provided to help reading)

Table 1 presents the gel content of the PBAT-based films exposed to various amounts of solar radiations. As expected, the crosslinked part of the biodegradable films increased with solar radiations uptake. Whatever the material's composition, the gel content was initially rather low (lower than 0.5 wt%) and increased after UV exposition. However, PLA is not crosslinkable in such conditions as already demonstrated by Stloukal [19] and Koo et al. [20] and Tao et al. [21] reported that PPC was hardly crosslinkable [22] as no gel fraction was observed without any crosslinking initiator for PPC submitted to 12 kJ/m² irradiation. Other works demonstrated that the main requirements for starch crosslinking are the use of a crosslinking agent [23, 24] or a sensitizer [25] that were not present here. Moreover, the fast consumption of starch granules by microorganisms during ageing increased the relative polyester part in PBAT/Starch film. at the expense of Starch part. By the way, the initial bio-based part of the Mater-Bi used in this study (CF04P grade) has been estimated around 37 wt%. Hence, in the present study, as no crosslinking agent had been incorporated into the film compositions, the gel contents of the three materials were attributed to the PBAT phase, which relatively increased during ageing in the case of PBAT/Starch.

Microscopically, the aged materials contained dense clusters made of 3D covalently bonded chains. Under stretching, those crosslinked chains could not unfold and then prevented the materials from plastic deformation and limited their ultimate properties. The UV-induced chain cleavage also drove to molecular weight reduction, which also contributed to lessen the ultimate properties.

TABLE 1. Gel content of biodegradable mulch films submitted to various irradiation levels

Time in field (months)	Cumulated radiations (kJ/cm²)	Gel content (w%)		
		PBAT / PLA	PBAT / PPC	PBAT / Starch
0	0	0,2 ± 0,9	0,5 ± 0,9	0,4 ± 0,3
5	295	2,6 ± 0,9	2,7 ± 0,7	4,5 ± 0,9
18	953	3,9 ± 0,8	3,0 ± 0,6	7,5 ± 1,1

Impact of Field-Ageing on the Biodegradability of Above-Soil Mulch Films

Biodegradation of field aged films containing notable crosslinked parts have been evaluated in comparison with the corresponding unaged materials in compost conditions (figure 2). Results indicated that field ageing had few impact on the biodegradability of the three materials since the percentages of mineralization reached after 75 days were not significantly different whether the materials had previously been aged or not. Concerning the PBAT/Starch film, a 10% decrease in the percentage of mineralization was noticed for the aged film compared to the un-aged one at the end of the experiment. A two-steps mechanism was proposed: microorganisms first consumed starch granules and then the hydrolysis of the PBAT phase started. This last step required more time in the case of the aged film due to the crosslinked part. Moreover, the starch part was lower for the aged films than for the unaged one. Differences in crosslinked part and starch content drove to the distinct biodegradation patterns. In contrast, PBAT/PPC and PBAT/PLA films previously field-aged or not led to the same percentage of biodegradation but exhibited changes in the kinetics of mineralization during the first part of the assay. The rate of biodegradation slowed down for aged PBAT/PPC film whereas aged PBAT/PLA degraded faster as compared to un-aged ones. In the case of PLA, such a lag-phase at the beginning of the biodegradation process was attributed to the need for the ester bonds to be cleaved hydrolytically to shorter molecular fragments before assimilation by microorganisms and ultimate mineralization [26]. This lag-phase was reduced for aged PBAT/PLA films as chain scissions already started during UV-exposition.

Polymer photo-ageing could either improve or decrease the biodegradable character of aliphatic-aromatic (co)polyester depending on the prevalence of aliphatic or aromatic features in the chemical structure. Actually, the effect depends on the balance between chain scission and crosslinking mechanisms undergone by the material during UV-weathering. In any case, the biodegradable character of the three materials could not be questioned even after undergoing photo ageing in field conditions during 18 months

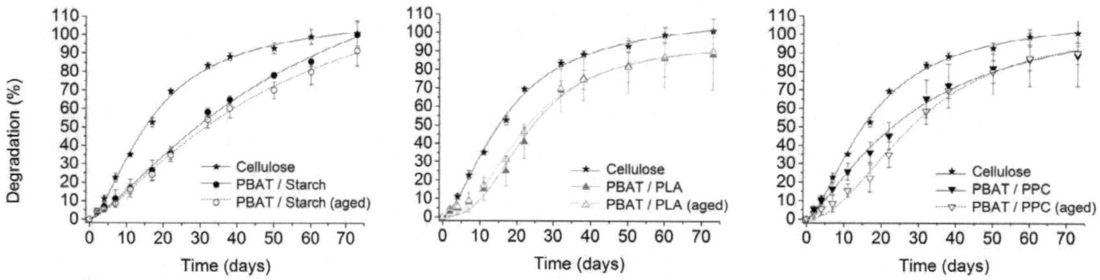

FIGURE 2. Kinetics of biodegradation in compost conditions of the three PBAT-based mulch films before and after field-ageing. (Symbols: experimental data; lines: Hill model fitting curves with r^2>0.99)

CONCLUSION

Three black biodegradable films based on PBAT-blends of different compositions (PBAT/PLA, PBAT/PPC and PBAT/Starch) were tested for vine mulching in real field conditions. The morphological analysis of the films revealed that their degradation state depended on the location of the studied material, the soil-buried and the above-soil parts undergoing different ageing mechanisms. The fact that the soil-facing surface exhibited holes 18 months after installation suggested that the biodegradation process could occur above the soil even without a preliminary burying step. However, whatever their composition, UV radiations led to a concomitant breakage of the mulch films which occurred after cumulating 260-270 kJ/cm² radiations as confirmed by the two consecutive and consistent assays. Beyond this threshold, the photochemical modifications undergone by the material were too high to sustain its integrity. The PBAT parts were shown to drive the overall behavior of the materials by developing UV-induced crosslinked clusters, which embrittled the films. In addition to this drop of mechanical performances, the presence of gel content also modified the biodegradation kinetics patterns of the PBAT-based films. However, the biodegradable nature of the materials was not altered by UV-ageing phenomena.

ACKNOWLEDGMENTS

This study has been carried out in the frame of the AGROBIOFILM project with financial support from the European Union 7th Framework Program (FP7/2007-2013) under the grant agreement number 262257. The authors gratefully acknowledge Olivier Mandeville (Château Vaissiere) for kindly providing vineyard and Carlos Rodrigues and Joao Pardao from Silvex (Portugal) for valuable contribution in discussion about plastic film extrusion.

REFERENCES

1. Scarascia-Mugnozza, G., et al., *Mechanical properties decay and morphological behaviour of biodegradable films for agricultural mulching in real scale experiment.* Polymer Degradation and Stability, 2006. **91**(11): p. 2801-2808.

2. Briassoulis, D., *Analysis of the mechanical and degradation performances of optimised agricultural biodegradable films.* Polymer Degradation and Stability, 2007. **92**(6): p. 1115-1132.
3. Bilck, A.P., M.V.E. Grossmann, and F. Yamashita, *Biodegradable mulch films for strawberry production.* Polymer Testing, 2010. **29**(4): p. 471-476.
4. Corrales, T., et al., *Photooxidative and thermal degradation of polyethylenes: interrelationship by chemiluminescence, thermal gravimetric analysis and FTIR data.* Journal of Photochemistry and Photobiology a-Chemistry, 2002. **147**(3): p. 213-224.
5. Stark, N.M. and L.M. Matuana, *Surface chemistry changes of weathered HDPE/wood-flour composites studied by XPS and FTIR spectroscopy.* Polymer Degradation and Stability, 2004. **86**(1): p. 1-9.
6. Dehbi, A., et al., *Degradation of Thermomechanical Performance and Lifetime Estimation of Multilayer Greenhouse Polyethylene Films Under Simulated Climatic Conditions.* Polymer Engineering and Science, 2015. **55**(2): p. 287-298.
7. Gugumus, F., *Re-examination of the thermal oxidation reactions of polymers - 2. Thermal oxidation of polyethylene.* Polymer Degradation and Stability, 2002. **76**(2): p. 329-340.
8. Massey, S., et al., *Action of water in the degradation of low-density polyethylene studied by X-ray photoelectron spectroscopy.* Express Polymer Letters, 2007. **1**(8): p. 506-511.
9. Montes, J.C., et al., *Ageing of polyethylene at raised temperature in contact with chlorinated sanitary hot water. Part I - Chemical aspects.* Polymer Degradation and Stability, 2012. **97**(2): p. 149-157.
10. Oceguera-Cervantes, A., et al., *Characterization of the polyurethanolytic activity of two Alicycliphilus sp strains able to degrade polyurethane and N-methylpyrrolidone.* Applied and Environmental Microbiology, 2007. **73**(19): p. 6214-6223.
11. Kessler, F., et al., *Biodegradation improvement of poly(3-hydroxy-butyrate) films by entomopathogenic fungi and UV-assisted surface functionalization.* Journal of Photochemistry and Photobiology B-Biology, 2014. **130**: p. 57-67.
12. Bhardwaj, H., R. Gupta, and A. Tiwari, *Communities of Microbial Enzymes Associated with Biodegradation of Plastics.* Journal of Polymers and the Environment, 2013. **21**(2): p. 575-579.
13. Saadi, Z., et al., *Fungal Degradation of Poly(Butylene Adipate-Co-Terephthalate) in Soil and in Compost.* Journal of Polymers and the Environment, 2013. **21**(4): p. 893-901.
14. Muthuraj, R., M. Misra, and A.K. Mohanty, *Hydrolytic degradation of biodegradable polyesters under simulated environmental conditions.* Journal of Applied Polymer Science, 2015. **132**(27).
15. Briassoulis, D., *Mechanical behaviour of biodegradable agricultural films under real field conditions.* Polymer Degradation and Stability, 2006. **91**(6): p. 1256-1272.
16. Kasirajan, S. and M. Ngouajio, *Polyethylene and biodegradable mulches for agricultural applications: a review.* Agronomy for Sustainable Development, 2012. **32**(2): p. 501-529.
17. ASTM, *ASTM D2765 - 11 - Standard Test Methods for Determination of Gel Content and Swell Ratio of Crosslinked Ethylene Plastics*. 2011.
18. Albertsson, A., *Biodegradation of polymers*, in *Handbook of polymer degradation.* 1992: New York. p. 345-363.
19. Stloukal, P., et al., *Assessment of the interrelation between photooxidation and biodegradation of selected polyesters after artificial weathering.* Chemosphere, 2012. **88**(10): p. 1214-1219.
20. Koo, G.H. and J. Jang, *Preparation of melting-free poly(lactic acid) by amorphous and crystal crosslinking under UV irradiation.* Journal of Applied Polymer Science, 2013. **127**(6): p. 4515-4523.
21. Tao, Y.H., et al., *Crosslinkable poly(propylene carbonate): High-yield synthesis and performance improvement.* Journal of Polymer Science Part a-Polymer Chemistry, 2006. **44**(18): p. 5329-5336.
22. An, J., et al., *A novel method to improve the thermal stability of poly(propylene carbonate).* Polymer Chemistry, 2014. **5**(14): p. 4245-4250.
23. Niazi, M.B.K. and A.A. Broekhuis, *Surface photo-crosslinking of plasticized thermoplastic starch films.* European Polymer Journal, 2015. **64**: p. 229-243.
24. Zhou, J., et al., *Surface photo-crosslinking of corn starch sheets.* Carbohydrate Polymers, 2008. **74**(3).
25. Delville, J., et al., *Solid state photocrosslinked starch based films: a new family of homogeneous modified starches.* Carbohydrate Polymers, 2002. **49**(1): p. 71-81.
26. Mueller, R.J., *Biological degradation of synthetic polyesters - Enzymes as potential catalysts for polyester recycling.* Process Biochemistry, 2006. **41**(10): p. 2124-2128.

Controlling cellular structure and surface quality in foam injection molding by an innovative core-back mold

Tromm Mike [1]*, Heim Hans-Peter [1]

[1] University of Kassel - Hessen - Germany

High-pressure foam injection molding in combination with core-back is a special procedure to achieve high density reductions or high expansion ratios and can lead to an improved and tailored cellular structure. Within the last years there is an increasing interest of industry in this technology. Beyond the technical possibilities of the standard core-back method, a new unique injection mold was built for investigations in high-pressure foam injection molding in context with core-back and local-core back method. Inside the closed mold the cavity volume can be locally expanded after volumetrically injection of the melt containing blowing agent. In doing so, injection molded parts with local foam structures (up to an expansion ratio of 500%) can be created while adjacent areas remain non-expanded and nearly unfoamed. Thus, a specific manipulation of the structure at different positions of the part is possible. The key features of the new mold are a controllable hydraulic cylinder, enabling a precise definition of speed, pressure and movement profile, a changeable core-back insert for foaming different geometries as well a variotherm mold temperature control via rapid heating ceramic inserts to achieve high surface qualities. Thus, the foam structure and the surface quality can be influenced in a wide range. The mold contributes to find out the possibilities and restrictions of geometrical design and influencing the cellular structure by using core-back and local core back process. The interworking of the special components of the mold and first results are presented. By the help of design of experiments especially the specific core-back parameters were varied and their influence on the final foam structure was analyzed.

Effect of the particle diameter of the chemical foaming agent on the foaming process and the cellular structure of one-shot compression molded polyethylene foams

Junichiro Tateishi[a], Tsuyoshi Nishiwaki[a]
Sathish K. Sukumaran[b], Masataka Sugimoto[b]

[a] *Institute of Sport Science, ASICS, Hyogo 651-2271, Japan*
[b] *Graduate School of Science and Engineering, Yamagata University, Yamagata 992-8510, Japan*

junichiro.tateishi@asics.com

Abstract. Effect of the particle diameter and the particle content of a chemical foaming agent on the foaming process and the cellular structure of one-shot compression molded polyethylene foams are investigated. As indicated by the time evolution of the cavity pressure, the thermal decomposition of ADCA accelerated with decreasing particle diameter of ADCA. The cell size in the foam decreased with a decrease in the particle diameter of the ADCA. The cell number density in the unfoamed samples (N_F) was proportional to the ADCA particle number density in the unfoamed samples (N_C), with a proportionality constant less than 1. A model for the foaming process where the ADCA particles nucleate the cells and the nucleated cells coalesce can offer an explanation for the observed proportionality between N_F and N_C and the proportionality constant being less than 1.

Keywords: chemical foaming, cell structure, polymer foam
PACS: 83.50.Uv

INTRODUCTION

Chemically cross-linked polyethylene foams by the one-shot chemical foaming process have been widely used for shock attenuation parts of shoe soles. A typical one-shot chemical foaming process involves several steps. First, polymer compounds including the chemical foaming agent (CFA), the CFA activator and the polymer cross-linker are filled into the mold at a certain temperature. Next, cross-linking reaction of the polymer and thermal decomposition reaction of the CFA proceed simultaneously, while suppressing expansion of compounds by the mold clamping pressure. Finally, as the mold clamping pressure is released, the compounds immediately expand, and are cooled. Then cross-linked polymer foams are obtained. Concerning the one-shot chemical foaming process, the forming mechanism of the cell structure has not been revealed yet.

In this study, the effect of varying the particle diameter and the particle content of a CFA on the foaming process and the cellular structure of one-shot compression molded polyethylene foams are investigated. Investigation for foaming process was accomplished using a custom-built mold with the capability to measure the instantaneous pressure in the cavity.

EXPERIMENTAL

Commercial metallocene linear low-density polyethylene (LLDPE, PRIME POLYMER, Evolue SP1540; melting temperature 113°C; MFR 3.8 g/10 min, 2.16 kg, 190°C; density 0.913 g/cm3) was used as the base polymer. Azodicarbonamide (ADCA, Eiwa Chemical, Vinyfor, density 1.65 g/cm3) of two different particle diameters (labeled ADCA-L and ADCA-S) were used as the CFA. The median particle diameters are 30 and 6 μm respectively. Sample compositions are shown in TABLE 1. ADCA contents were 2, 6 and 10 phr. The DCP (crosslinker) content was kept constant at 1.2 phr. ZnO (CFA activator) content was maintained at 50% of the weight of ADCA in all of the investigated samples. A mold cavity is 100 mm in length and width, and 5 mm in depth with a taper angle of 45° at the edges. The pressure sensor (Kistler Japan, 6172AAA) is attached to the bottom of the cavity at the center. In this study, the time evolution of the thermal decomposition of the ADCA during the foaming process was monitored by measuring the temporal variation of the cavity pressure. Molding temperature and time were 175 °C and 1200 s, respectively. It was ensured that the sample did not leak out of the mold cavity during the foaming process. Foamed samples were stored at 23 ± 3 °C for 1 week before their cellular structure was characterized.

TABLE 1 Sample compositions

Code	L2	L6	L10	S2	S6	S10
LLDPE	100	100	100	100	100	100
DCP	1.2	1.2	1.2	1.2	1.2	1.2
ADCA-L	2	6	10			
ADCA-S				2	6	10
ZnO	1	3	5	1	3	5

unit:phr

The cell number density, N_F, (cells/cm^3) in the unfoamed sample was calculated as follows [1]:

$$N_F = \left(\frac{n_F}{A_F}\right)^{\frac{3}{2}} \times E_R \qquad (1),$$

where, n_F is the number of cells in the area, A_F and E_R is volume expansion ratio. In the following, N_F is used as a proxy for the number of cell nuclei in the unfoamed samples. The ADCA particle number density in the unfoamed samples, N_C, (particles/cm^3) was calculated as follows:

$$N_C = \left(\frac{n_C}{A_C}\right)^{\frac{3}{2}} \qquad (2),$$

where, n_C is the number of ADCA particles in the area, A_C.

RESULTS AND DISCUSSION

FIGURE 1 shows the time evolution of the cavity pressure for two particle diameters and three additive contents of ADCA. The time dependence of the pressure change is s-shaped for all of the samples. The s-shape suggests that, after an initial induction phase, the ADCA rapidly decomposes leading to a correspondingly rapid pressure increase. The pressure increase at intermediate times started earlier, its rate increased and the saturation of the cavity pressure occurred at earlier times as the ADCA particle diameter decreased. This suggests that, the finer the ADCA particle and consequently, higher the surface area per unit volume, the faster is the rate of decomposition [2].

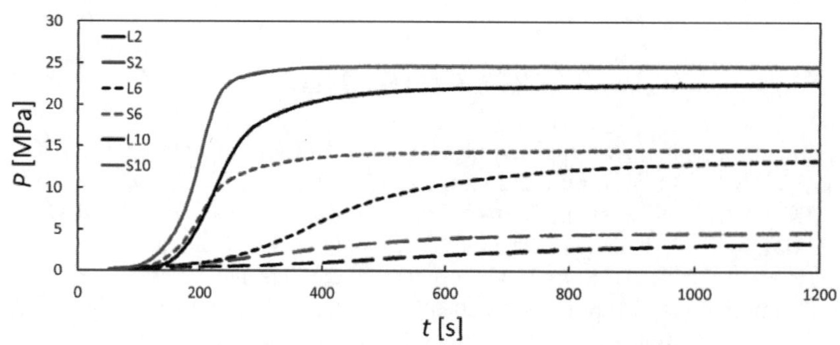

FIGURE 1 Time evolution of cavity pressure with different particle diameters and contents of ADCA.

FIGURE 2(a-f) displays SEM micrographs of the foamed samples with different particle diameters and contents of ADCA. As can be seen from the micrographs, all of the foamed samples exhibit closed-cell structure, and a decrease in the particle diameter of the ADCA is accompanied by a decrease in the cell diameter of the foamed samples. In addition, it is germane to note here that a more careful consideration of the relationship between the ADCA particle diameter distribution and the corresponding cell diameter distribution might be able to shed further light on the foaming mechanism and the final cell structure. This will be pursued in future work.

Both the average cell diameter and the volume expansion ratio of different foams are plotted against ADCA content in FIGURE 3. It should be noted that expansion ratio linearly increases with ADCA content increment even though cell diameter keeps almost constant level depending on ADCA particle size. This relation indicates that the increment of expansion ratio is not caused by increment of not cell size, but increment of cell number. This suggestion is also in agreement with visualization studies [3, 4] using uncrosslinked low-density polyethylene with very low CFA content. The visualization studies also reported that cell growth occurred around each CFA particle when the concentration of the gas generated by the CFA was sufficient to cause saturation in certain regions of the polymer melt. Combining this observation with the results in FIGURE 3 suggests that, at a fixed ADCA content, smaller average diameter ADCA particles could generate larger numbers of cell nuclei when compared to larger average diameter ADCA particles.

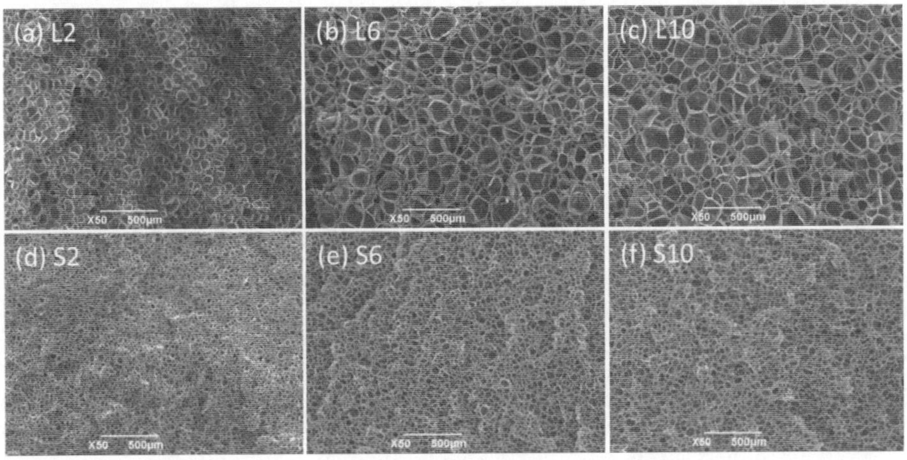

FIGURE 2 SEM micrographs of the foamed samples, (a)L2, (b)L6, (c)L10, (d)S2, (e)S6 and (f)S10.

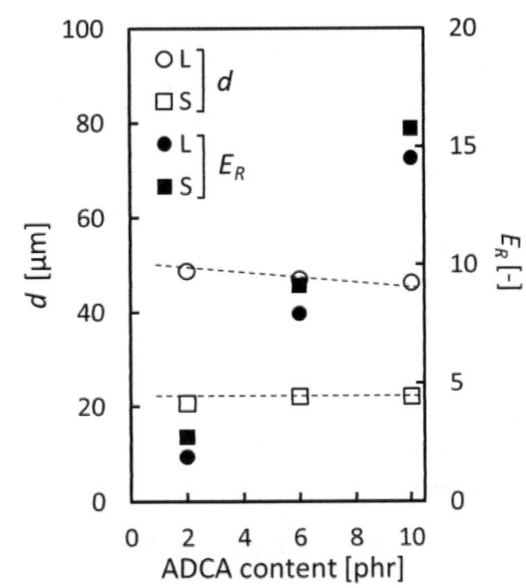

FIGURE 3 Average cell diameter d and volume expansion ratio E_R plotted against ADCA content for with different particle diameters of ADCA.

To further discuss the quantitative relationship between cell structure and the characteristics of ADCA particle diameter and number, the dependence of N_F on N_C, the number density of the ADCA particles, was investigated. Note that, as already mentioned, N_F is to be a proxy for the number density of cell nuclei in the unfoamed samples in this study. In FIGURE 4, N_F is plotted against N_C for (a)ADCA-L and (b)ADCA-S. The black lines correspond to a slope of 1. N_F increases in proportion to N_C but with slope less than 1. Moreover, slope decreases with decreasing average particle diameter of ADCA. The slopes are 0.38 and 0.09 for ADCA-L and ADCA-S respectively. If each ADCA particle nucleates a cell and all of the nucleated cells survive in the final foam, then N_F must equal N_C. However, for all of the samples, $N_F < N_C$. One way the proportionality of N_F on N_C but $N_F < N_C$ can be understood is by hypothesizing that while each ADCA particle nucleates a cell, only a certain fraction of the cells survive in the final foam and that the fraction of the cells that survive depends on the diameter of the particle nucleating the cell.

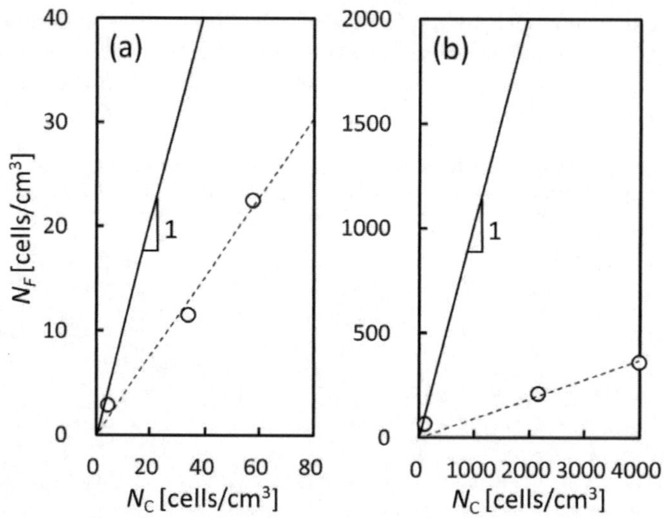

FIGURE 4 Dependence of the cell number density in the unfoamed sample, N_F, on the ADCA particle number density, N_C, for large, (A)ADCA-L and small, (C)ADCA-S particles.

CONCLUSIONS

The effect of varying the particle diameter and particle content of the chemical foaming agent ADCA on the foaming process and the cellular structure of one-shot compression molded polyethylene foams were investigated. Using a purpose-built mold cavity with a built-in pressure sensor enabled us to monitor the time evolution of the cavity pressure during foaming. A summary of the findings and the conclusions drawn are as follows:

- As indicated by the rate of increase of the cavity pressure, the rate of thermal decomposition of the ADCA increases with decreasing average particle diameter of the ADCA.
- The cell size in the final foam decreases with a decrease in the particle diameter of the ADCA.
- The cell number density in the foamed samples, N_F, is proportional to the ADCA particle number density, N_C. When combined with past work, this suggests that the ADCA particles serve as cell nucleation sites during the foaming process.
- The proportionality constant between N_F and N_C was less than 1 for all of the investigated samples and decreased with decreasing average ADCA particle diameter. This suggests that, even if each ADCA particle nucleates a cell, not all of the nucleated cells survive in the final foam.

REFERENCES

1. Naguib HE, Park CB, Panzer U, Reichelt N. *Polymer Engineering and Science*, 2002, 42, 1481-1492
2. Quinn S. *Plastics Additives & Compounding*, 2001, 3, 16-21
3. Pop-iliev R, Dong N, Xu D, Park CB. *Advances in Polymer Technology*, 2007, 26, 213-222
4. Leung SN, Wong A, Guo Q, Park CB, Zong J. *Chemical Engineering Science*, 2009, 64, 4899-4907

CO_2 as a Physical Blowing Agent in Polyurethane Foams

Maria Rosaria Di Caprio[a], Ernesto Di Maio[a*], Sara Cavalca[b], Vanni Parenti[b], Pellegrino Musto[c] and Salvatore Iannace[d]

[a]Dipartimento di Ingegneria Chimica, dei Materiali e della Produzione Industriale, University of Naples Federico II, P.leTecchio 80, 80125 Naples, Italy,
[b]Dow Italia s.r.l, Polyurethanes R&D, via Carpi 29, 42015 Correggio, Italy,
[c]Institute for Polymers, Composites and Biomaterials, National Research Council, Via Campi Flegrei, 34, 80078 Pozzuoli (Na), Italy
[d]Institute for Macromolecular Studies, National Research Council, Via Edoardo Bassini, 15, 20133 Milan, Italy

Abstract. The aim of the present study was the achievement of thermosetting polyurethane foams by a batch foaming process that uses CO_2 and N_2 as physical blowing agents. In particular, to address the recent interest in combining the gas (physical) foaming with the classical (chemical) polyurethane foaming, a novel instrumented pressure vessel was designed for investigating: i) sorption under high gas pressure on the two, separate, components of the polyurethane foams, ii) synthesis under high gas pressure after the two components mixing, and iii) foaming upon release of the pressure. Results revealed a significant effect of sorbed CO_2 on the polyurethane synthesis and the need for the design of a new chemistry to exploit the use of physical foaming on thermosetting polyurethanes.

Keywords: Polyurethane, CO_2,sorption, synthesis, foaming, FT-NIR
PACS: 82.35.Lr, 82.70.Rr

INTRODUCTION

In the field of thermosetting polyurethane foams (PUs) obtained by the curing reaction between polyol and isocyanate, the selection and the amount of the blowing agents is very important to the final cellular structure and properties of the foam. Since the beginning of the history of PUs, CO_2, derived from the blowing reaction of isocyanate with water, is used as blowing agent. In the late 1950s, large PUs production also utilized chlorofluorocarbons (CFCs) and hydro-chlorofluorocarbons (HCFCs) as PBAs, then replaced by hydrocarbons (HCs) (e.g. pentane) because of their negative environmental impact. However, although cheap and with zero ozone depletion potential (ODP), HCs use was limited by flammability concerns. In this context, CO_2 is eco-friendly and safe, with zero ODP and the lowest GWP among known blowing agents.[1]

In order to use CO_2 as a physical blowing agent (in a totally different manner to what has been done previously, with CO_2 resulting from the reaction of isocyanate with water) knowledge of the behavior of the whole system in the presence of high pressure CO_2 is fundamental. In particular, the following are needed: 1) the physical properties of polymer/gas solutions, namely solubility, diffusivity, specific volume and interfacial tension of the polyol- and isocyanate-CO_2 solutions in contact with the external, high pressure CO_2 phase; 2) the effect of the CO_2 on the reactions involving the different chemicals in the polyurethane synthesis; 3) the way the reactive system behaves when CO_2 is rapidly released to induce foaming.

We herein report our approach to the problem, with the design of new experimental apparatus. We report, in particular, measurements at 35°C of solubility, mutual diffusivity, specific volume and interfacial tension of polyol/CO_2 solution at CO_2 pressures up to 80 bar and isocyanate/CO_2 solution up to 55 bar, obtained by a coupled gravimetry-Axisymmetric Drop Shape Analysis (ADSA).[2-4] Foaming experiments, conducted at 35°C and various CO_2 pressures, with a novel instrumented pressure vessel capable of retrieving data on sorption, synthesis and foaming are described. This pressure vessel makes use of a sample holder with a flexible impeller that allows to keep polyol and PMDI separated during the sorption stage and to mix the components to start the PU synthesis.

Furthermore, it makes use of a sapphire window and FT-NIR spectroscopy to monitor sorption and the synthesis. The effect of CO_2 on the reaction has been evaluated at 35°C and 40bar.

MATERIALS AND METHODS

Materials

A formulated polyol and a polymeric methylene diphenyl di-isocyanate (PMDI) were supplied by DOW Italia S.r.l. (Correggio, RE, Italy) and used "as received". High purity grade CO_2 was supplied by SOL (Naples, Italy).

Sorption Measurements of CO_2 in Polyol and PMDI

Coupled sorption-ADSA experiments were conducted by using magnetic suspension balance (MSB) (Rubotherm Prazisionsmesstechnik GmbH, Germany), equipped with a high temperature/high pressure (HT-HP) view cell. The simultaneous measurement was made possible by adopting a custom-designed measuring device, which consists of a crucible containing the polymer/CO_2 solution (for weight monitoring) and a Teflon rod to which the polymer/CO_2 solution drop is stuck (for ADSA) (see Figure 1a). Drop changes in volume and shape were observed through two optical quality windows, by using an adjustable high resolution CCD camera (BV-7105H, Appro), equipped with a modular zoom lens system (Zoom6000, Navitar). The CCD camera is connected to a computer, and a commercial software (FTA32 Video 2.0, First Ten Angstroms) is used to analyze drop profile.[2-4] Furthermore, in order to achieving the optimal threshold background for digitizing the drop image, a uniform bright background was be provided by light emitting diodes.

The data flow adopted for the elaboration of the data acquired during the coupled sorption-ADSA measurement is illustrated in Figure 1b.

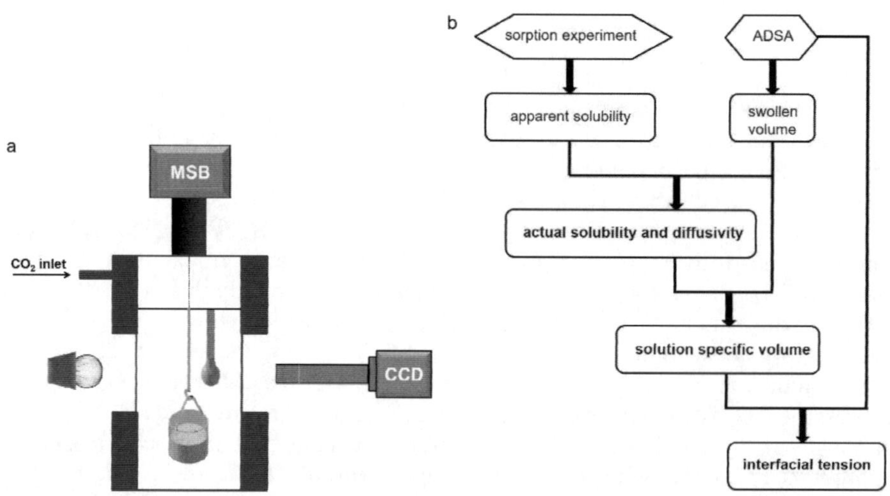

FIGURE 1. (a) Schematic illustration of the experiment and (b) data flow used in the coupled sorption-ADSA measurement.

Sorption and ADSA experiments were carried out by isothermal pressure increments up to 80 bar for polyol while up to 55 bar for PMDI, at 35°C, after having placed both the crucible containing the polymer and the rod with the polymer drop in the HT–HP view cell. In detail, sorption measurements were performed by step-wise increments of the gas pressure (500 kPa steps ca.), after the attainment of equilibrium sorption in the previous step.

Concurrently, during each pressure step, image acquisition of the pendant drop was performed every 10 min. and drop volume was calculated by using the commercial software.

PU Foaming with CO_2

In Figure 2, the steps of foaming experiment are shown. Inside the pressure vessel, polyol and PMDI, separated by a flexible impeller, are placed (1), pressurized with CO_2 at 35°C until saturation (2) and mixed (3). Finally, at any stage of the curing reaction physical foaming occurs due to pressure release (4).

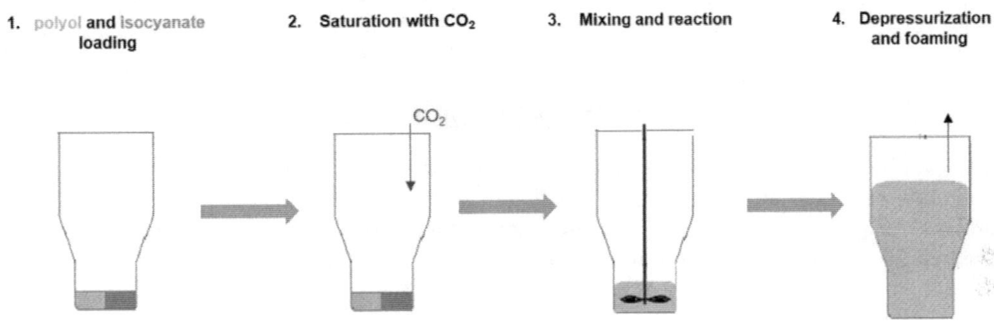

FIGURE 2. Steps of foaming experiment.

Different foaming tests will be carried out, by changing process parameters as sorption time in order to evaluate its influence on the PU morphology.

Spectroscopic Measurements

The curing reaction, at 35°C and atmospheric pressure, was monitoring by using a fiber optic reflectance probe, placed beneath the sampler holder contained the two reactants, connected to Frontier™ NIR spectrometer (Perkin Elmer Inc., Waltham, MA, USA) equipped with equipped with a NIR source, a CaF_2 beam splitter and a INGAAS detector.

During the cure, spectra were recorded automatically at regular time interval of 3s for 10 minutes using Perkin Elmer TimeBase software, in the spectral range 4000-10000cm^{-1} with a resolution of 16 cm^{-1} and 4 scans just after mixing.

RESULTS

Sorption Measurements of CO_2 in Polyol and PMDI

Sorption measurements for the polyol/CO_2 solution show quite a large CO_2 solubilization (17%wt ca. at 80 bar and 35°C), in turn responsible for a moderate swelling of the polymer and an extensive effect on the interfacial tension, which reaches vanishing values at the highest investigated pressure. Mutual diffusivity goes from 1e-6 to 1e-5 cm^2/s in the whole experimental range. The same study conducted on the PMDI at 35°C and CO_2 pressures up to 50 bar has shown for the sorption isotherm a maximum of about 6%wt, with mutual diffusivity of 1e-6 cm^2/s ca. and a minor swelling of the polymer/gas solution with respect to the neat polymer, while the interfacial tension reaches the value of 15 mN/m ca. In both sorption measurements, for higher pressure it has been observed that low molecular weight components of polyol and PMDI are solubilized in CO_2, as confirmed by Gel Permeation Chromatography analysis.

PU foaming with CO_2

In Figure 3 PUs obtained at 35°C and 40 bar of CO_2 pressure by changing sorption time (or CO_2 amount) are shown. Curing time was fixed at 1 minute and mixing at 250 rpm for 10s. It is possible to note that by increasing sorption time, PU change its morphology from an inhomogeneous to more homogeneous one until to become a pulverized product for high sorption time indicating a CO_2 effect on the system.

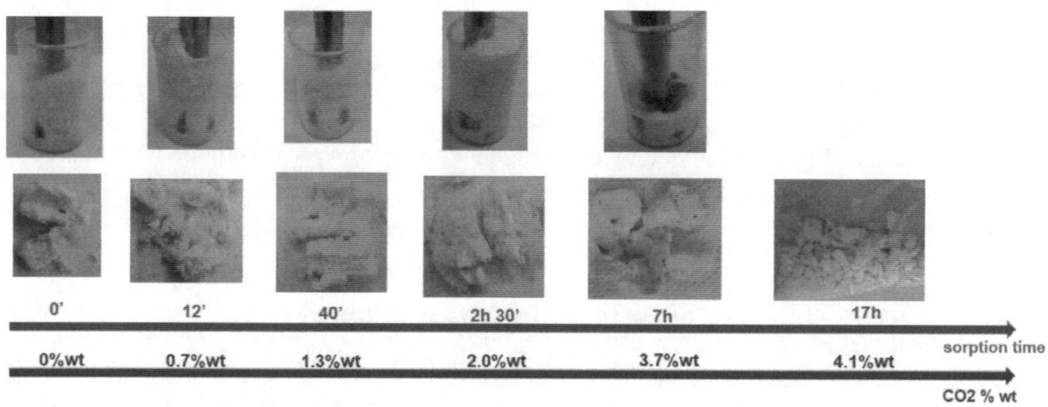

Figure 3. PUs obtained at different sorption times (or CO_2 amount).

Spectroscopic Measurements

NIR spectra of polyol and PMDI at room temperature are shown in Figure 4 a and b, with their characteristic absorption peaks.

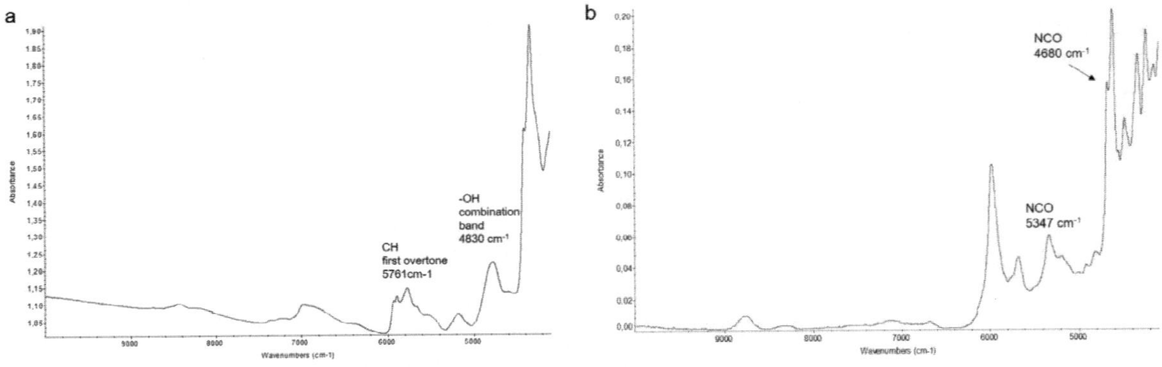

Figure 4. NIR spectra of polyol (a) and PMDI (b) at room temperature.

In Figure 5 spectra collected at the beginning (blue line) and the end (red line) of the curing reaction, at room temperature and atmospheric pressure, are reported. From these spectra is possible to follow the NCO band decrease and the NH band (of urethane group) formation and increase.[5,6]

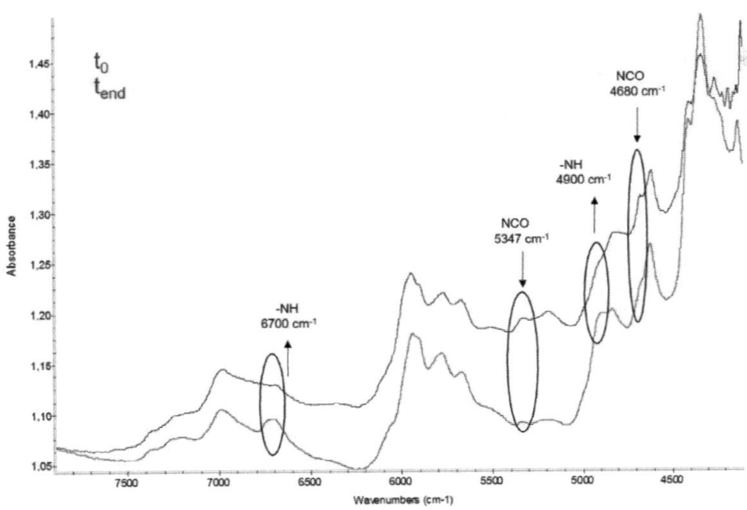

Figure 5. NIR spectra collected at the beginning t_0 (blue line) and the end t_{end} (red line) of the cure.

CONCLUSIONS

The use of CO_2 as PBA to obtain PUs on lab scale was investigated. Solubility, mutual diffusivity, specific volume and interfacial tension of polyol/CO_2 solution at CO_2 pressures up to 80 bar and PMDI/CO_2 solution up to 55 bar, were measured by using a fully-experimental, coupled gravimetry-Axisymmetric Drop Shape Analysis at 35°C. These data were used to produce PUs in a pressure vessel, at 35°C and 40bar of CO_2 pressure, by changing the sorption time. It was noticed an influence of CO_2 on the final product, in fact by increasing sorption time PU morphology improves until an optimum and then it changes resulting in a pulverized product. FT-NIR spectra were recorded to monitor the curing reaction after mixing of the two reactants at room temperature and atmospheric pressure.

ACKNOWLEDGMENTS

The financial support of the European Union's (LIFE13-EN/IT/001238 project, http://ec.europa.eu/environment/life ; www.dow.com/k12) is gratefully acknowledged.

REFERENCES

1. S. N. Singh, "Chapter 2" in *Blowing Agents for Polyurethane Foams*, edited by Rapra Technology Ltd, 2002, pp. 6-7
2. M. G. Pastore Carbone, E. Di Maio, S. Iannace and G. Mensitieri, *Polym. Test.* **30**, 303-309 (2011).
3. M. G. Pastore Carbone, E. Di Maio, G. Scherillo, G. Mensitieri and S. Iannace, *J. Supercrit. Fluids* **67**, 131-138 (2012).
4. M. R. Di Caprio, G. Dal Poggetto, M. G. Pastore Carbone, E. Di Maio, S. Cavalca, V. Parenti and S. Iannace, *Fluid Phase Equilib.* **425**, 342-350 (2016).
5. J. Dupuy, S. Benali, A.Maazouz, G. Lachenal and D. Bertrand, *Macromol. Symp.* **184**, 249-260 (2002).
6. F. A. De Thomas, J. W. Hall and L. S. Monfre, *Talanta*, **41**, 425-431 (1994).

pps proceeding
Symposium: S12 - Polymer foams and films
Oral Presentation

Investigation of Gas/SCF Diffusivity and Solubility in Polymers Using Dielectric Properties

Yao Selina X. [(1)], **Romero Diez Sandra** [(1)], **Park Hee Eon** [(1)], **Lee Patrick C.** [(1)*]

[(1)] Department of Mechanical Engineering Univ of Vermont - VT - USA

The attributes of a gas/SCF in polymers benefited many polymer processing applications (light-weight structural components, thermal insulation panels, and food packaging containers). For the applications involving gas/SCF infused polymer systems, the gas/SCF solubility and diffusivity are important parameters. In this research, we are investigating the relationship between gas/SCF solubility and diffusivity in three amorphous polymers (polystyrene, polycarbonate, cyclic olefin polymer) and their dielectric constants. In our preliminary study, the amorphous polymer films were saturated with supercritical CO2 at 5.5 MPa and 25 °C for 24 hours in a pressure chamber and removed from the chamber for gas desorption (unsteady diffusion) experiments. The capacitance values of the samples were recorded with an Inductance, Capacitance and Resistance (LCR) meter over time. These values were used to calculate dielectric constants. In parallel, CO2 weight percentages were measured by a sensitive scale which were used to calculate experimental diffusivity coefficients. It was demonstrated that the trend of dielectric constant was similar to that of the CO2 weight percentage change during gas desorption. Experimental diffusivity coefficients from our previous work agree well with literature data. To be more relevant to foaming applications, which usually involve high pressures and high temperatures, we modified a pressure chamber with a custom-made dielectric property measurement fixture. This system allows capacitance value measurements at high pressures and temperatures with short saturation time. The effect of polymer swelling (film thickness change) during the experiments at these conditions was estimated by Sanchez and Lacombe equation of state (SL-EOS). The trend of dielectric constant with a high-pressure gas into the polymer was consistent with estimated solubility results in the literature.

Trade-Offs and Interplay between Fluid Handling & Mechanics in Polymeric Foams for Absorbent Products

Carla Vito (1)*

(1) Procter & Gamble - OHIO - USA

This talk focuses on trade-offs and presents the existing open questions on the interplay between fluid handling and mechanics in polymeric foams. Polymeric foams are quite versatile materials for application as absorbent materials. Conventional porous media structures utilized as acquisition and storage layers in the hygiene product industry show a known trade-off between driving force for fluid acquisition (capillary suction) and inverse of resistance to the flow (permeability). This is due to flow into porous material strongly depending on structure geometry and particularly specific surface area (SSA). Capillary suction increases as the SSA increases, because more surfaces are available to wick fluid, while permeability decreases because the fluid path becomes inherently more complex and tortuous, as prescribed by statistical percolation theories. Conversely, whenever the SSA is low, then resistance to flow is reduced at the expenses of capillary suction. Polymeric foams can be engineered to optimize such trade-off and work quite effectively as acquisition layers or storage cores. While the above trade-off has been understood, polymeric foams in absorbent articles also present another trade-off: the interplay between fluid handling properties and mechanical characteristics. In particular, depending on their structure and polymer chemistry, foams are susceptible to external stresses that could significantly modify the resulting fluid handling properties. More specifically, foams relevant for the absorbent industry tend to easily undergo compressions within the range of forces they experience while in use (due to softness requirements), resulting in a compressed, modified structure with reduced permeability and storage capacity. The set of ideal properties for any absorbent article comprising polymeric foams must be well understood to be able to correctly formulate the problem statement behind these trade-offs and help develop improved polymeric foam structures.

Functionalized porous polymers with hydrated ferric oxide nanoparticles for arsenic removal from water

Zowada Ryan [1], Foudazi Reza [1]*

[1] Department of Chemical & Materials Engineering, New Mexico State University - New Mexico - USA

The aim of this work is to assess the capabilities of functionalized high internal phase emulsion (HIPE) templated polymers (abbreviated as polyHIPEs) to remove arsenic ions from drinking water. PolyHIPEs were synthesized using hydrophilic monomer and cross-linker. The monoliths were then functionalized with iron(III) oxide nanoparticles which reside in the cross-linked matrix enhancing the arsenic ions sorption. The morphology of polyHIPEs was investigated using scanning and transmission electron microscopies. The polyHIPEs were found to have a water uptake capacity of more than 4000 wt.%. 1 gram of the functionalized polyHIPEs has the potential to remove arsenate from 2.70 L of contaminated water in 50 ppb concentration.

Development of Ultralight Isotactic Polypropylene (iPP)/ Cellulose Nanofibers (CNF) Foams Using a Core-Back Foam Injection Molding

Long Wang[*,a], Shota Ishihara[a], Yuta Hikima[a], Masahiro Ohshima[**,a], Takafumi Sekiguchi[b], Akihiro Sato[b] and Hiroyuki Yano[c]

[a]*Department of Chemical Engineering, Kyoto University, Kyoto 615-8510, Japan*
[b]*New Business Development Division, SEIKO PMC Corp., Chiba 267-0056, Japan*
[c]*Research Institute for Sustainable Humano-sphere, Kyoto University, Kyoto 611-0011, Japan*

Abstract. In this work, nanocomposite foams based on isotactic polypropylene (iPP) and cellulose nanofibers (CNFs) were prepared using a core-back foam injection molding (FIM) technique with nitrogen (N_2) as the physical blowing agent. To increase the dispersion of CNFs in PP matrix, hydrophobic-modified CNFs, which were prepared by surface modification using alkenyl succinic anhydride (ASA), was adopted here. Different rheological curves such as complex viscosity and storage modulus highlighted the reinforcing potential of CNFs on the viscoelastic properties of iPP. Unprecedentedly, an ultrahigh expansion ratio of 18-fold PP/CNF nanocomposite foams were fabricated using a FIM with core-back operation for the first time in the foaming world. In contrast, the achievable highest expansion ratio obtained for pure iPP foams was only 10-fold, indicating that the incorporation of CNFs improved the foamability of PP significantly. Using transmission electron microscopy, we revealed that the CNFs were well-dispersed and aligned along the cell wall in the core-back direction, which could help the thin cell wall withstand the elongational force from breaking. Interestingly, the hybrid shish-kebab and classic shish-kebab were simultaneously formed in the cell walls, which were ascribed to the strong elongational flow induced during the core-back manipulation. The highly-oriented structures were further corroborated by the two-dimensional wide-angle X-ray diffraction (2D WAXD) and small-angle X-ray scattering (SAXS) characterizations.

Keywords: Cellulose Nanofibers, Foam Injection Molding, High Expansion Foam, Shish-Kebab
PACS: 82.35.-x, 61.41.+e, 61.20.Lc, 61.25.Hq, 81.05.Rm

INTRODUCTION

Cellulose is the most abundant, renewable, green, and environmentally friendly biopolymer in nature [1]. There has been increasing attention on the use of cellulose nanoparticles as a reinforcement in polymer systems due to its low weight, interesting mechanical and thermal properties, and broad capacity for chemical-modification. Cellulose nanocrystals (CNCs) and cellulose nanofibers (CNFs) are two mainly kinds of cellulose nanoparticles (CNPs). Compared with CNCs, CNFs have a much larger aspect ratio, higher surface area, and stronger percolation network, which usually result in their exhibiting stronger reinforcing effects than CNCs. However, one main challenge of CNFs as a reinforcement is associated with its strong hydrophilic nature, which limits its application in hydrophobic polymers [1,2]. In addition, few attempts have been conducted to fabricate CNFs-containing polymer nanocomposites through a melt mixing technique, which is a more environmentally and industrially favorable method [2,3].

Microcellular foam injection molding (FIM), as one of the most important ways of manufacturing foamed polymer parts in industrial applications, can be used to fabricate foams with complex, three-dimensional geometries [4]. Additionally, the injection foamed plastic parts display an improvement in their mechanical properties and a higher acoustic and thermal insulation. Due to the low void fraction (5–30%) in the conventional FIM approach, FIM techniques with core-back operation have been proposed, which can significantly increase the void fraction of the injection-molded foam products [5]. As a biorenewable nanofiller, CNFs show promise as a bubble nucleating agent for the polymer foams in the FIM process. However, there are few applications of CNFs in polymer foams, and they are mainly limited to the batch foaming process [6]. In this study, we have fabricated PP/CNF nanocomposite foams using FIM with a core-back operation with nitrogen (N_2) as the physical blowing agent. The cellular structures, distribution of CNFs, and the hierarchical crystalline structure of such composites will be studied.

[*]*E-main address*: wangl.kevin@cheme.kyoto-u.ac.jp (Long Wang) ohshima@cheme.kyoto-u.ac.jp (Masahiro Ohshima)

EXPERIMENTAL

Materials

Isotactic polypropylene (F133A) with a high tacticity of 97 % was provided by Prime Polymer Corp., Tokyo, Japan. It has a weight-averaged molecular weight (M_w) of 379 kg/mol and a melt flow rate (MFI) of 3.0 g/10 min (230 °C/2.16 kg). Nitrogen with a purity of 99.9 % was provided by Izumi Sangyo, Kyoto, Japan and was used as the physical blowing agent.

Cellulose nanofibers (CNFs) were prepared from needle-leaf bleached kraft pulp using the surface modification method [2,3]. Briefly, needle-leaf bleached kraft pulp, which is the source of CNFs, was mechanically disintegrated via a bead mill, which could provide sufficient CNFs to melt the compounding. The obtained CNFs were then surface modified by alkenyl succinic anhydride, and their hydrophobicity was tuned by controlling the degree of substitution (DS). In this study, CNFs with a DS of approximately 0.4 were used and showed a uniform dispersion within the resin matrix. The PP/CNF masterbatch with 17 wt % CNFs was pre-prepared by blending with the pure iPP using a twin-screw extrusion process (KZW15-TW; Technovel Corp., Osaka, Japan), and the detailed fabrication information can be found in previous studies [2,3].

Foam Injection Molding with Core-back Operation

A 35-ton clamping force FIM machine (35EL III-F, Japan Steel Work, Ltd, Hiroshima, Japan), with a 20-mm diameter screw equipped and a gas injection unit (Trexel SCF system SII TRJ-10-A-MPD, Trexel Inc., Wilmington, MA, U.S.A.), was used to conduct the foam injection molding (FIM) experiments. For the PP/CNF nanocomposites, the masterbatch was diluted by the pure iPP, and a fixed CNF content of 5 wt % was used. N_2 at a small level of 0.2 % was used. The mold was a box-type cavity with a gate of 1.5 mm in diameter. The cavity dimension is 70 mm in length, 50 mm in width and 1 mm in the original thickness. To obtain a high expansion ratio, the FIM technique with a core-back operation was applied. Herein, a maximum core-back distance of 17 mm was used to achieve the highest expansion ratio of 18-fold, and a fixed core-back rate of 20 mm/s was used. The processing parameters used for the FIM experiments are listed in Table 1.

TABLE (1). Processing conditions for foam injection experiments

Parameters	Values
Melt temperature (°C)	190 (iPP), 210 (PP/CNF)
Mold temperature (°C)	40
Cooling water (°C)	20
Injection speed (mm/s)	70
Injection pressure (MPa)	180
Dwelling time (s)	2.8 (iPP), 3.0 (PP/CNF)
Shot size (mm)	35
Packing pressure (MPa)	40
Core-back rate (mm/s)	20
Core-back distance (mm)	1, 4, 6, 9,14, 17
N_2 content (%)	0.2

Characterization of Materials and Foams

The dynamic rheological properties were measured using a rheometer (AREX-G2, TA Instruments). The oscillation frequency sweep tests were conducted at 180 °C at a strain of 1 % in the linear viscoelastic regime and the temperature sweep experiments were measured at a cooling rate of 2 °C/min after the polymer melting at a temperature of 220 °C.

The foamed specimens were cryogenically fractured in liquid nitrogen. After the surfaces of the specimens were covered with a thin layer of gold, the foam morphology was observed by a scanning electron microscopy (Tiny-SEM Mighty-8, Technex), at an accelerating voltage of 17 kV. Image analysis was performed using ImageJ Software. A gas pycnometer (AccuPycII, Shimadzu, Kyoto, Japan) was used to characterize the open cell content (OCC) of the foamed samples. The volume measured by the gas pycnometer excluded the samples' open cell volume. The TEM images were obtained using an FEI Tecnai G2 Sphera TEM, with a field emission gun (FEG) at 200 kV.

To observe the crystalline morphology and orientation, the scanning microbeam WAXD measurements for the foamed samples were conducted at the beamline BL15U1 of Shanghai Synchrotron Radiation Facility (SSRF, Shanghai, China). WAXD patterns were then collected using an X-ray CCD detector (model SX165, Rayonix Co. Ltd., USA). Two-dimensional small-angle X-ray scattering (SAXS) measurements were also performed on the beamline BL16B1 of the Shanghai Synchrotron Radiation Facility (SSRF, Shanghai, China). An X-ray CCD detector (Model Mar165) with a resolution of 2048 × 2048 pixels was used to collect the SAXS images.

RESULTS AND DISCUSSION

Oscillatory Shear Flow Properties

Linear viscoelastic properties are very sensitive to the structural change in the materials. Figure 1 shows the results of temperature sweep of the complex viscosities as a function of temperature for iPP and PP/CNF composites. As observed in Figure 1, there is a drastic increase in the complex viscosity for each sample, which is caused by PP crystallization as the melt temperature is reduced. At higher melt temperatures (before the crystallization occurs), the incorporation of CNFs clearly increased the complex viscosities of PP and reinforced the PP melt. On the other hand, because the inflection point of the curves is regarded as the crystallization temperature, it can be considered that the crystallization temperature and complex viscosities increased with the incorporation of CNF. This indicates that the modified CNFs could act as a crystal nucleating agent for PP and promote the crystallization process.

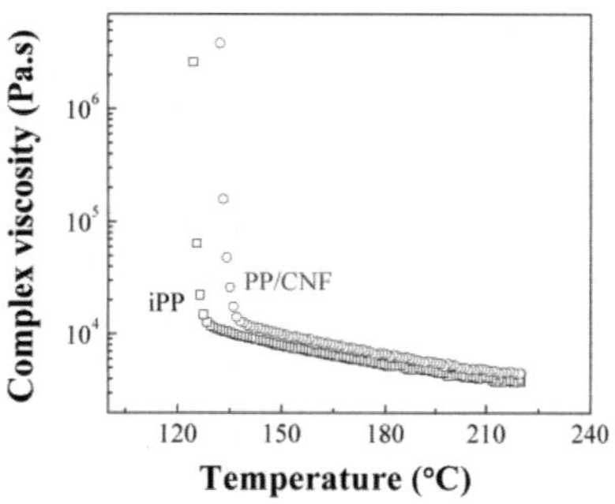

FIGURE 1. Complex viscosity of iPP and LCBPP samples.

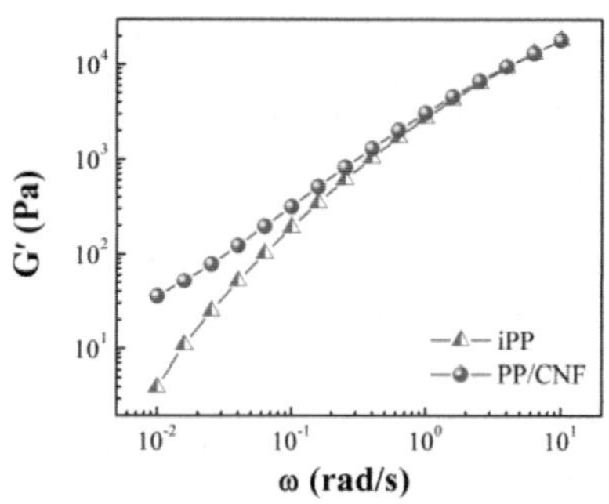

FIGURE 2. Storage modulus of iPP and LCBPP samples.

Figures 2 shows the changes in the storage modulus (G') as functions of the frequency. It was observed that at low frequency, the G' of the PP/CNF composites clearly had higher values relative to the neat iPP. The iPP melt exhibited a liquid-like behavior at low frequency, where the slope of G' was 1.85. The small differences from the theoretical values of 2 could be due to the relatively wide molecular weight distribution of the commercial PP resin. The slope of the G' curve against the frequency for the PP/CNF composites at the low-frequency region was 0.90. This weak dependence of the storage modulus on the frequency of CNFs is a characteristic of solid-like behavior, which indicates that a network-like structure is likely to be generated in the PP/CNF composites.

Cell Structures

Figure 3 shows the SEM images of the iPP and PP/CNF composite foams prepared at various expansion ratio. It was observed that 2-fold expansion foams maintained a spherical shape while much smaller cell sizes were obtained in PP/CNF composites. As Figure 3a shows, cells in the 5-fold expansion iPP foams became elliptical and deformed along the core-back direction, which was attributed to the induced extensional stress during the mold-opening operation. With an increase of expansion ratio, cells were further deformed and oriented along the core-back direction, and thus we can observe that the cells were highly stretched for the 10-fold iPP foams. Compared with the iPP foams, a similar tendency of cell development was observed in the PP/CNF composite foams but with much smaller deformed cells. At the expansion ratio of 10-fold, cells in the PP/CNF composites were severely deformed and regularly aligned in the core-back direction.

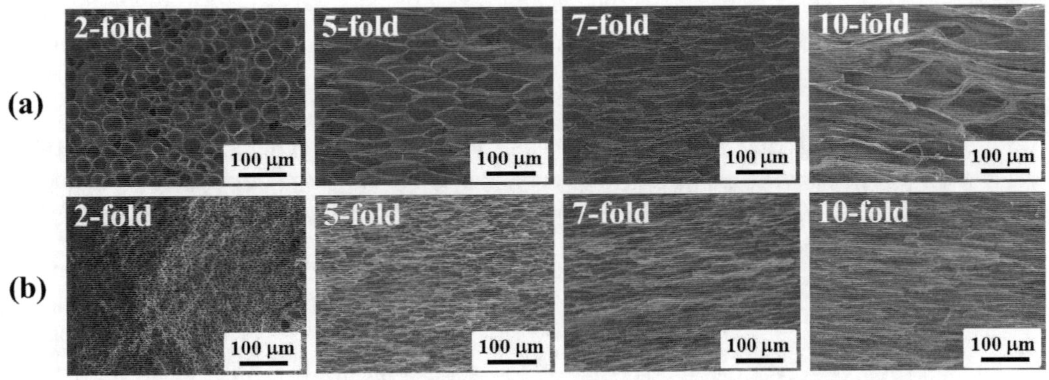

FIGURE 3. SEM images of the (a) iPP and (b) PP/CNF composite foams viewed parallel to the core-back direction.

Furthermore, microcellular foams with an ultrahigh expansion ratio of 18-fold (Figure 4) were obtained for the PP/CNF composites, while the highest expansion ratio achieved for the iPP foams was only 10-fold. Figure 4a revealed that the cell walls were highly deformed and oriented along the core-back direction in a way of fibrillary structures. As shown in Figure 4b, a highly interconnected cellular structure was observed for the 18-fold PP/CNF composites foams with a very high open cell content (OCC) of 93%. Figure 5 shows the TEM images of the iPP and PP/CNF nanocomposites foams. As displayed in Figure 5a, it revealed that the highly-regularly aligned lamellar (kebab) were attached to the surface of the CNFs, indicating the formation of hybrid shish-kebab in the PP/CNF composites. Besides, opposite to the adjacent CNFs areas, a great deal of well-regularly aligned lamellar simultaneous occurred. Similarly, we can observe regularly aligned lamellar in the pure iPP foams in the core-back direction (Figure 5b).

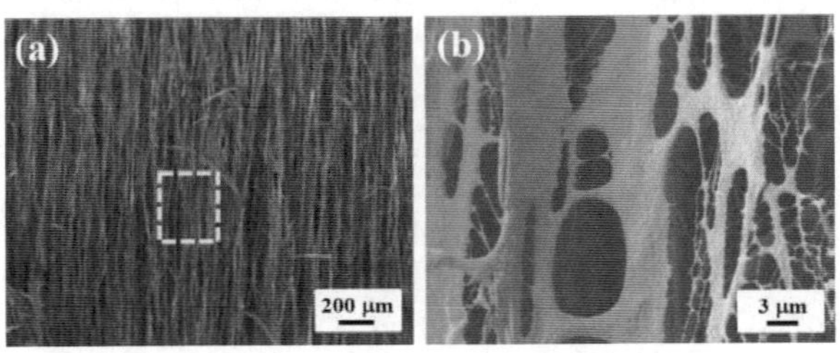

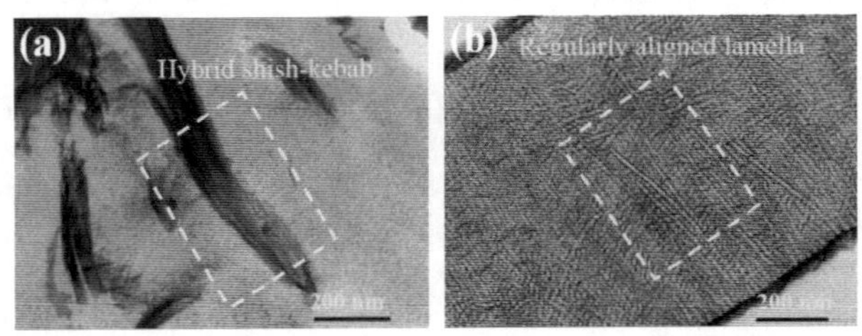

FIGURE 4. SEM image of the 18-fold PP/CNF composite. **FIGURE 5.** TEM of (a) 18-fold PP/CNF and (b) 10-fold iPP foams.

Crystalline Structures

Figure 6a displays the representative 2D WAXD diffraction patterns from the core layer of injection-molded foams observed along the core-back direction. At a low expansion ratio of 2-fold, mainly isotropic diffraction circles were observed for the neat iPP foams. While the diffraction circles of the iPP foams transformed into arclike diffraction at a higher expansion ratio of 10-fold, which manifested that a higher degree of orientation was achieved for high expansion FIM foams. This might be ascribed to the formation of shish-kebab structure in the high expansion foams. Similar 2D WAXD patterns was obtained for the 10-fold PP/CNF composites. Besides, with a further increase of core-back distance, a much narrower arclike diffraction refection was achieved for the 18-fold PP/CNF composites foams, which might be caused by the increasing amounts of hybrid shish-kebab and classic shish-kebab superstructure because of enhanced flow field.

Figure 6b shows the representative SAXS patterns for the iPP and PP/CNF composites. Generally, for the 10-fold and 18-fold foams, they exhibit two bulb-shape lobes and two triangular streaks in the meridional direction and equatorial direction, respectively, which further confirm the existence of shish-kebab structures. Compared with the 10-fold iPP foams, the equatorial diffraction intensity of 10-fold PP/CNF composites foams was clearly increased, indicating a large size or high-volume fraction of shish, which might be due to the role of CNFs as a shish. In contrast, no scattering reflection could be detected in the 2-fold iPP foams, which shows the randomly distributed crystalline lamellae in the core layer due to the lack of or very weak flow field. Therefore, combing the TEM results with the

WAXD and SAXS patterns, it is reasonably confirmed that the formation of shish-kebab structures in core layer of the high expansion foamed iPP and PP/CNF composites, which was ascribed to the strong elongational flow field induced during the core-back operation.

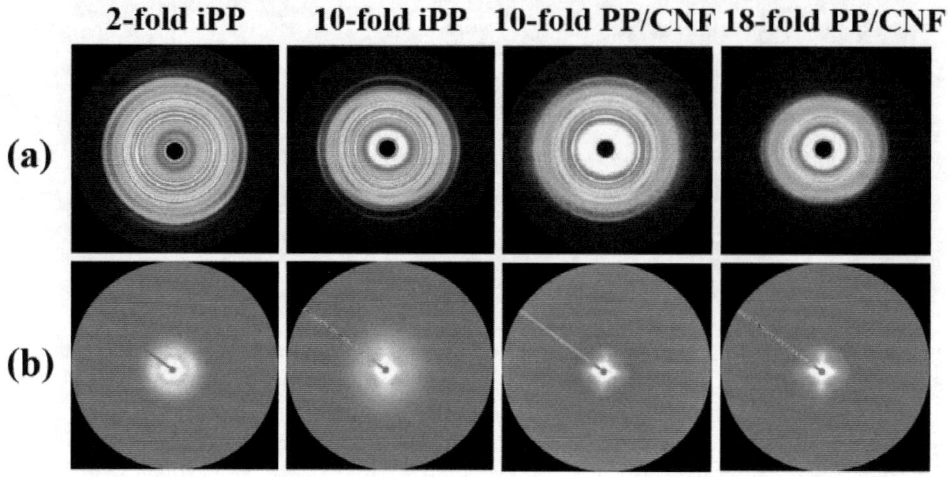

FIGURE 6. Typical (a) 2D WAXD and (b) SAXS patterns in the core layer for the iPP and PP/CNF composites samples.

CONCLUSIONS

In this study, ultrahigh expansion microcellular PP/CNF foams were successfully fabricated using a foam injection molding machine with core-back operation. The rheological results revealed that the CNFs could increase the crystallization temperature of PP as well as enhance its melt strength. Consequently, the presence of CNFs remarkably improved the foaming behavior of the iPP and led to smaller cell sizes. This was ascribed to the promotion of the crystallization induced by the CNFs, the improvement of the melt strength and the bubble nucleating effects of the CNFs. For the iPP, the achievable highest expansion ratio was only 10-fold. However, an ultrahigh expansion ratio of 18-fold was achieved for the PP/CNF nanocomposite. The TEM results shown that the CNFs were aligned along the cell wall in the core-back direction, which could help the thin cell wall withstand the elongational force from breaking. Interestingly, the hybrid shish-kebab and classic shish-kebab were simultaneously achieved in injection-molded PP/CNFs composites, which were further evidenced from the 2D WAXD and SAXS results.

ACKNOWLEDGMENTS

The research was conducted in association with the Special Priority Research Areas (White Biotechnology) of the Advanced Low Carbon Technology Research and Development Program (ALCA) of the Japan Science and Technology Agency.

REFERENCES

1. R. J. Moon, A. Martini, J. Nairn, J. Simonsen, J. Youngblood, *Chem. Soc. Rev.* **40**, 3941-3994 (2011).
2. A. Sato, D. Kabusaki, H. Okumura, T. Nakatani, F. Nakatsubo, H. Yano, *Composites, Part A* **83**, 72–79 (2016).
3. L. Wang, M. Ando, M. Kubota, S. Ishihara, Y. Hikima, M. Ohshima, T. Sekiguchi, A. Sato, H. Yano, *Composites, Part A* **98** 166-173 (2017).
4. R. Miyamoto, S. Yasuhara, H. Shikuma and M. Ohshima, *Polym. Eng. Sci.* **54**, 2075–2085 (2014).
5. L. Wang, S. Ishihara, M. Ando, A. Minato, Y. Hikima, M. Ohshima, *Ind. Eng. Chem. Res.* **55** 11970-11982 (2016).
6. W. D. Ding, T. Kuboki, A. Wong, C. B. Park, M. Sain, *RSC Adv.* **5** 91544-91557 (2015).

Processing-structure relationship in high-pressure foam injection molding of polypropylene/carbon dioxide system

Vahid Shaayegan[a], Chongda Wang[a], Franco Costa[b], Sejin Han[c], Chul B. Park[a]*

[1] *Department of Mechanical and industrial Engineering, University of Toronto, Toronto, Canada*
[2] *Autodesk, Inc. 259-261 Colchester Rd., Kilsyth, VIC. 3137, Australia*
[3] *Autodesk, Inc. 2353 North Triphammer Road, Ithaca, NY 14850, USA*

* *Corresponding author: park@mie.utoronto.ca*

Abstract. Development and formation of cellular structure in foam injection molding (FIM) of polypropylene (PP)/carbon dioxide (CO_2) was studied. The PP was provided by Braskem (H521; MFR 3.6 g/10min at 230 °C/2.16 Kg) and CO_2 was provided from Linde Gas Canada. A 50-ton Arburg Allrounder injection molding machine was employed to inject melt/gas mixture into a rectangular mold cavity, 135 mm × 111 mm × 3.2 mm, fed by a fan gate. Carbon dioxide in supercritical condition was injected and mixed with PP melt using MuCell® technology (Trexel, Inc., Woburn, Massachusetts). Talc, as nucleating agent, was mixed with PP to promote the cell nucleation. For this purpose, a Leistritz co-rotating twin-screw extruder was used. The mold-opening technology (MO) was used to achieve foams with high void fractions.

Full-shot FIM+MO experiments were performed with and without using a packing pressure [1]. With no packing used, *i.e.,* injection of a full-shot melt/gas mixture alone, the cell structure after mold opening was characterized by irregular shaped and very large cells and voids due to the presence (and coalescence) of un-dissolved cells which had nucleated during mold-filling stage [2]. This scenario was more noticeable in PP + talc foams. Hence, a high-pressure protocol based on the application of melt-packing pressure was employed to increase and maintain the mold cavity pressure. By using a proper packing pressure, premature nucleated cells from the mold-filling stage were first re-dissolved back into the melt. Then, cell nucleation occurred right at the mold-opening instant, increasing the obtainable pressure-drop rate for cell nucleation. In addition, the absence of premature cells enhanced the structure uniformity [1,2].

The addition of talc had two major roles in high-pressure FIM+MO of PP/CO_2 system. This enhanced the cell nucleation through heterogeneous nucleation mechanism, resulting in much finer, more uniform, and higher cell density in obtained expanded foams [3]. In addition, a much higher void fraction was achieved in the presence of the talc. In fact, the PP composite demonstrated higher expansion during foaming, and thus, the mold-opening distance could be increase twice of that used for foaming the pure PP. We should note that the crystallization of PP had an evident influence on the foaming behaviour of both pure and composite PP such that ultra-fine cell structures were observed in areas which were cooled faster and crystallized during the foaming process [4, 5].

To conclude, it was observed that the mold cavity pressure plays a crucial role in both cell nucleation and the uniformity of the final foam cellular structure. Packing pressure, as an important processing parameter, can be used to control the cavity pressure. By controlling the cavity pressure, ultra low-density foams with expansions up to 9 fold for the foamed layer could be fabricated. In addition, the presence of talc enabled us to achieve very fine cell structure with high void fractions. The crystallization of PP resulted in a graded cell structure in injection molded foam parts.

Keywords: High-pressure Foam Injection Molding, Mold Opening; Packing Pressure; Polypropylene.

REFERENCES

1. V. Shaayegan, G. Wang and C.B. Park, "*Euro. Polym. J.*" **76,** 2-13 (2016).
2. V. Shaayegan, C. Wang, F. Costa, S. Han and C.B. Park, "*Euro. Polym. J.*" **92**, 314-325 (2017).
3. V. Shaayegan, G. Wang and C.B. Park, "*Chem. Eng. Sci.*" **155,** 27-37 (2016).
3. J. Wang, W. Zhu, H. Zhang and C. B. Park" *Chem. Eng. Sci.*" **75,** 390-399 (2012).
5. K. Taki, D. Kitano and M. Ohshima, "*Ind. Eng. Chem. Res.*" **50**, 3247-3252 (2011).

Utilization of Crystals in the Foaming Technologies

Park Chul B [1]*

[1] University of Toronto - Ontario - Canada

This plenary speech will explain the details of the modern foaming technologies that use crystals to control foam processing. First, the effects of crystals on cell nucleation and cell growth will be discussed. Both growing and developed crystals can significantly affect the foaming process. Next, I will discuss the gas's effect on crystallization kinetics. We found that the crystallization temperature decreased because of the gas's plasticizing effect through increased dilation of the polymer. A high-pressure DSC analysis showed that the dissolved gas significantly promoted crystal nucleation. But crystal growth was reduced because of the increased entanglement of the molecules. Consequently, the final crystallinities were often eventually decreased. I will also explain the nanofibrils' effect on the crystallization kinetics of polymer/nanofibril composites, both with and without gas. The dissolved gas and the nanofibrils dispersed in the polymer matrix acted synergistically on the crystallization behavior, especially for PLA with slow crystallization kinetics. I will show how crystallization is controlled and how crystals affect the foaming behaviors in the following processes. 1. Bead foaming: I will show how crystals can be formed before foaming, and the effects of those crystals on the cell morphology and on the bead-to-bead sintering behaviors. 2. Extrusion foaming: An in-situ visualization tool was used to demonstrate how crystals can form instantly in the extensional and shear field, even at a temperature 13oC higher than Tm, with a view to controlling the foaming behavior. Such strain-induced crystallization was quantified and correlated to the resultant cell structure. 3. Foam injection molding: High-pressure foam injection molding with a full shot, for re-dissolving all the earlier nucleated cells, is utilized. In this technology, foaming can be induced by mold opening, after promoting crystallization with packing first.

Numerical Simulation of Polyurethane Chemical Foaming Process

Li-Yang Chang, Yuan-Jung Chang and Chia-Hsiang Hsu[a]

[a]*CoreTech System (Moldex3D) Co., Ltd., Chupei City, Hsinchu Country, 302, Taiwan*

Abstract. Polyurethane (PU) foam molding has become a common molding process in chemical foaming molding simulation of foamed parts. In this work we present the recent development of three-dimensional water blown polyurethane foam injection molding. Reaction kinetic model, density model and viscosity model are applied to describe the complex polyurethane foam system in the simulation study. In this work, the melt front time and the part density are predicted. It will be helpful to understand the mold filling behavior in the foam polyurethane

Keywords: Polyurethane Foam Molding, Foaming Kinetics

INTRODUCTION

Polyurethane (PU) foam molding is the common molding process in chemical foaming molding. Polyurethane foam can be used in the automotive industry, refrigeration industry, footwear industry and medical industry. The polyurethane resin is mainly composed of polyol or polyether, which contains OH group, and isocyanate. The molecule chain can be grown to form polymer chains by the reaction and polymerize to the cross-linked network structure. If add water as a blowing agent in the raw material, isocyanate reacts with water to produce CO_2 and finally produce a porous foamed polyurethane. The basic process of polyurethane foam molding is mixing the polyol, isocyanate, and water and inject into the mold cavity. In the injection stage, cavity usually not be completely filled, and then fill the remaining space by foaming expansion. In this process, the polyurethane release the carbon dioxide gas due to the chemical reaction, and the viscosity of the polyurethane also increase due to the crosslinking reaction. At the same time, the exothermic effect caused by the chemical reaction will increase the temperature of the mold, and further release the carbon dioxide into the polyurethane at the relatively high temperature. The foaming process is complete while the mold cavity is full filled with polyurethane foam or the polyurethane is fully cured. The challenge of the polyurethane foam molding process is how to fill the mold cavity with less material without short shot. Short shot occur when the cavity filled with less material, or the foaming gas amount is insufficient, or polyurethane curing rate is too fast. However, large amounts of waste material occur when inject too much material and product too much gas. With simulation of polyurethane foaming, it can more accurately predict the filling behavior of polyurethane and optimize the injected amount of materials.

METHODS

During polyurethane foaming process, the following chemical reaction occurs when mixing the raw materials (isocyanate, polyol, and water) [1]:

Gelation reaction: The chemical reaction between isocyanate and polyol produce urethane bond. The polymerization reaction of urethane bond proceeds to produce the polyurethane polymer

$$\underset{\text{isocyanate}}{\text{R–NCO}} + \underset{\text{polyol}}{\text{R}'\text{–OH}} \longrightarrow \underset{\text{urethane bond}}{\text{R–NH–CO–O–R}'}$$

Blowing reaction: In the same time, the chemical reaction between isocyanate and water produce carbon dioxide gas

$$\underset{\text{isocyanate}}{\text{2R–NCO}} + \underset{\text{water}}{H_2O} \longrightarrow \underset{\text{urea bond}}{\text{R–NH–CO–NH–R}} + \underset{\text{carbon dioxide}}{CO_2}$$

Under the assumption that less amount of the water concentration in the system, the conversion in the chemical kinetics of polyurethane can be described as [1]

$$\frac{d\alpha}{dt} = A\exp\left(-\frac{E}{R_g T}\right)(1-\alpha)(C_{c,0} - 2C_{w,0}\alpha_f - C_{p,0}\alpha) \tag{1}$$

$$\frac{d\alpha_f}{dt} = A_f \exp\left(-\frac{E_f}{R_g T}\right)(1-\alpha_f) \tag{2}$$

Heat contribution from the curing and foaming reaction are described as

$$Q_{rxn} = C_{p,0}\frac{D\alpha}{Dt}\Delta H_p + C_{w,0}\frac{D\alpha_f}{Dt}\Delta H_w \tag{3}$$

The viscosity model used in simulation is Cross Castro Macosko model

$$\eta = \frac{\eta_0\left(\frac{\alpha_g}{\alpha_g - \alpha}\right)^{c_1 + c_2\alpha}}{1 + \left(\frac{\eta_0\dot{\gamma}}{\tau^*}\right)}, \eta_0 = A\exp\left(\frac{T_b}{T}\right) \tag{4}$$

Assuming that the gas obeys to the ideal gas law, the mixing density of the foamed polyurethane can be described by the following equation [2]

$$\rho = \frac{m_p + m_g}{V_p + V_g} = \frac{1}{\frac{C_{w0}\alpha_f R_g T}{P\rho_p} + \frac{1}{\rho_p}} \tag{5}$$

Eqs. (1)-(5) are coupled with continuity equation, momentum equation and energy equation to describe the whole system.

In the experimental method, Sandia report [3] shows the typical front-tracking video at an oven temperature of 30°C, as showed in **FIGURE 1**. Foam started expanding in such a short time that it was difficult to capture the initial time data, even at this relatively low temperature where the reactions are slower. The height evolution of a foam column with time is observed to estimate the rate of gas formation, and assume that the increase in volume is solely due to the production of CO_2. This method is useful since it predicts the apparent CO_2 generation that goes into volume generation. It also hypothesizes that the blowing reaction could produce pressurized bubbles, especially between gelation and verification of the continuous phase. Therefore, volume change, temperature and pressure are measured. The number of moles of CO_2 is then estimated through the ideal gas law.

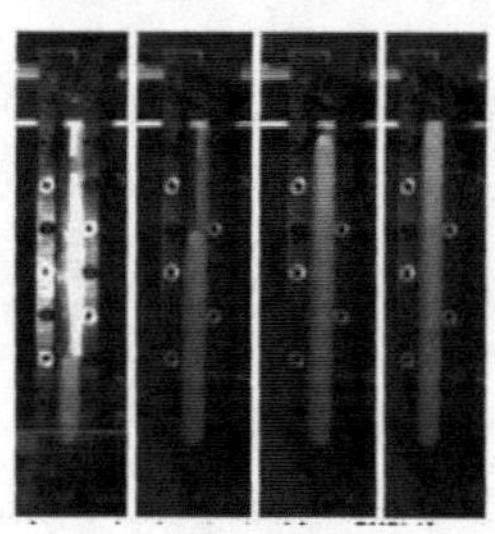

FIGURE 2: Video frames showing structural foam expanding in a bar mold. [3]

RESULTS AND DISCUSSIONS

The part of interesting is a steering wheel part. The part volume is about 933 cc. The initial concentration of polyol, isocyanate, and water are 3045, 4604, 780 mole/m^3 respectively. In the process conditions, the partial fill volume percentage is 35% in 3.5 second. The mold temperature is 50 °C, inlet stream at 25 °C.

FIGURE 2 shows the melt front of foamed polyurethane. First, the polyurethane melt will begin filling along the bottom of the cavity because of gravity. When the filling reaches at 35% of the cavity, CO_2 gas will be released by the polyurethane reaction, and the polymer viscosity will increase due to the gelation reaction. Meanwhile, the reaction heat of the exothermic reaction will be released to increase the temperature of the cavity. As a result, more and more CO_2 gases are released to the polymer melt under a relatively high temperature condition. Eventually, polyurethane foam will completely fill the cavity.

(a) (b)

FIGURE 2: Melt front at (a) 35% partial filled and (b) at end of foaming

FIGURE 3 shows the density distribution of foamed polyurethane. Density distribution reveals the expansion effect during foaming process. When the foaming conversion is higher, the more CO_2 gas is produced. It causes the larger volume of polyurethane foam, and the lighter density.

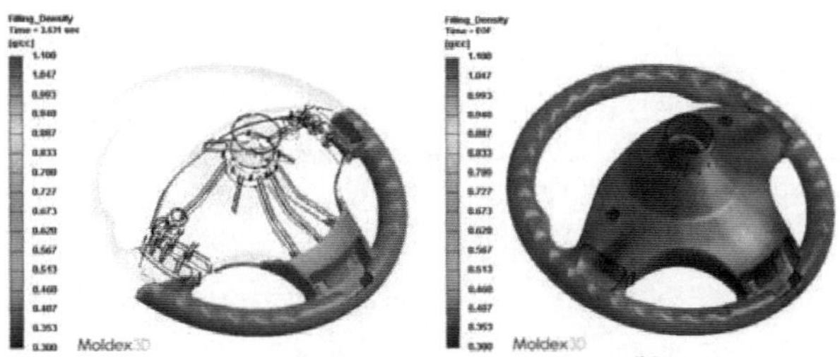

(a) (b)

FIGURE 3: Density distribution at (a) 35% partial filled and (b) at end of foaming.

FIGURE 4 shows the gas volume fraction of foamed polyurethane during foaming process. Foam initiation time is delay since isocyanate and water are not fully mixing to produce CO_2 at time before five seconds. As time increases, the higher the foaming conversion rate, the more the amount of gas produces. Finally, gas volume fraction stops increasing when the cavity is fully-filled.

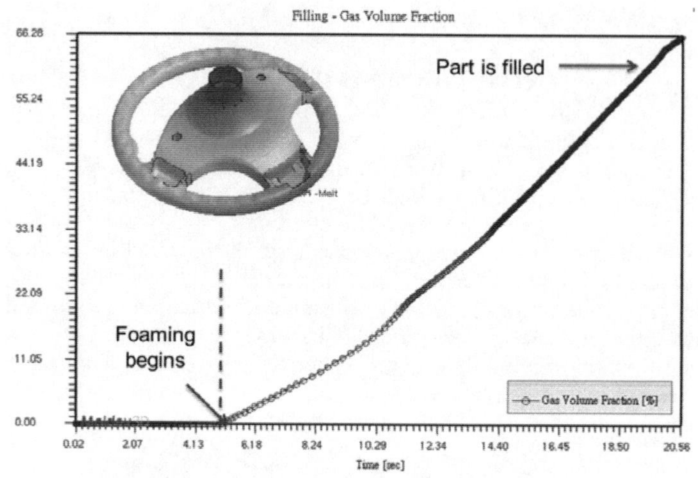

FIGURE 4: Gas volume fraction during foaming process.

CONCLUSIONS

In summary, with Moldex3D's new simulation feature of polyurethane foaming, it can fully understand the dynamic behavior of the foam polyurethane material in both filling and foaming stage. More importantly, the in-depth analysis of polyurethane foaming enables users to avoid trial-and-error which subsequently leads to substantial time and production cost reductions

REFERENCES

1. P. Ferkl et al., *Che. Eng. Sci.*, **148**, 55-64 (2016)
2. S. Baser and D. Khakhar, *Polym. Eng. Sci.*, **34**, 642-649 (1994)
3. R. Rao, et al., *A kinetic approach to modeling the manufacture of high density structural foam: Foaming and polymerization*, Sandia National Laboratories, 2015

Investigation on foamability of polypropylene with varying amorphous content

Banerjee Debjyoti [1], Dutta Anindya [1], Kapur G. S. [2], Ghosh Anup K. [1]*

[1] Centre for Polymer Science and Engineering, Indian Institute of Technology Delhi - Delhi - India, [2] Research & Development Centre, Indian Oil Corporation Ltd. - Haryana - India

The objective of this investigation was to recognize the fundamental insights on foamability of polypropylene with varying amorphous content were looked upon. Three types of impact copolymers, one beta nucleated homo-polymer and one random copolymer of polypropylene were chosen as the raw materials. Tensile and impact samples of all PP were foamed at varying foaming temperatures maintain the same temperature during gas saturation of samples. The saturation pressure, saturation time and depressurization rate were constant. The microscopic morphology of foams was characterized by SEM and density measurements were performed to identify the void fraction and volume expansion ratio calculation. The mechanical properties of the foamed samples were measured and compared with the corresponding unfoamed samples. On a parallel basis, the effect of foaming conditions (both the contribution of saturation temperature as well as carbon dioxide) on the nature of crystals formation and their corresponding changes were analyzed in detail by comparing the melting point, percentage crystallinity of unfoamed, foamed and unfoamed annealed samples. The change in thermo-mechanical behaviour of the polymer in terms of the molecular architecture before and after foaming was studied by DMA. The Foams of impact copolymers showed a very good balance between strength and density. DSC measurements showed a significant increase in melting point of the foams as compared to unfoamed polymers, whereas thermo-mechanical analysis indicated unique changes for impact copolymers.

Development and Characterization of Piezoelectric PP-Cellular Films

Mohebbi Abolfazl [(1)], Rodrigue Denis [(1)], Mighri Frej [(1)*], Ajji Abdellah [(2)]

[(1)] Laval University - Quebec - Canada, [(2)] Polytechnique Montreal - Quebec - Canada

In this work, we showed how cellular polypropylene (PP) films of controlled morphologies and good piezoelectric properties were developed by using twin-screw extrusion/calendaring. It was observed that the use of supercritical nitrogen (N_2) as a physical blowing agent led to thin cellular PP films of around 500 microns with a controlled cellular morphology. In addition, the use of calcium carbonate ($CaCO_3$) as nucleating agent substantially improved the cell size and cell density of the developed cellular films. The optimization of the extrusion parameters, such as screw configuration and temperature profile, as well as post-extrusion conditions such as calendaring temperature and rotational speed, led to a stretched eye-like cellular shape of uniform cell-size distribution. This particular morphology led to higher storage and loss moduli in the extrusion/calendaring (longitudinal) direction than for the transverse direction. Particularly, higher cell aspect ratio led to lower Young's modulus and consequently a higher quasi-static piezoelectric d_{33} coefficient of around 550 pC/N. In addition, we used the dynamic mechanical analysis (DMA) as a simple and rapid technique to predict the piezoelectric behavior of the cellular films. In this method, two new parameters that we called anisotropic moduli ratios (AMR1 and AMR2) were proposed to quantify the film cellular structure. They respectively correspond to the ratio of the storage and loss moduli in the longitudinal (L) direction over those in the transverse (T) direction (AMR1 = EŒ(L)/EŒ(T) and AMR2 = E(L)/E(T)). The results showed that film samples with more elongated eye-like cells (higher cell aspect ratios) had higher AMR1 and AMR2 values. The variation of the d_{33} coefficient with respect to AMR1 and AMR2 was also studied and the corresponding results showed that d_{33} increases with increasing both AMR1 and AMR2.

Energy-Efficient Filtration of Organic Microdroplets through Reusable Adsorbent Foam

Cherukupally Pavani [(1)]*, Bilton Amy M. [(1)], Park Chul B. [(1)]

[(1)] University of Toronto - Ontario - Canada

Removal of micro-oil droplets from industrial effluents is a global challenge. These droplets are generated in various industrial processes, oil spills, and petrochemical effluents. For example, Canada's oil sands industry alone has generated over 1 billion m3 of dissolved organic contaminated water that caused severe damage to the ecosystem. Polymeric foams are promising solution for this application due to their low cost, effectiveness, and multifunctional properties. To date, numerous bioinspired superwetting foams have been reported for surface oil removal, but not for the challenging to remove micro-oil droplets. Herein, we report an interesting new application of microcellular foams as an adsorbent for removal of aqueous micropollutants from water. For the adsorption removal of neutral and polar micro-oil droplets, a polyester polyurethane foam constituting both charged sites and hydrophobic sites was applied. Due to foam's high porosity (~98%) and micro pores (200-1100 μm) the pressure drop across the filter were measured to be only less than 1% of an atmosphere. The foam filter was able to remove complex crude oil micro-droplets with over 99% efficiency to below acceptable EPA standards at low energy inputs. Furthermore, the foam was able to be regenerated both by compression and solvent backwash cycles. The adsorbed crude oil was also recovered. Due to production of clean water, multiple usage of the foam, and recovery of lost oil product, the foam-based filtration is a sustainable system. This work could open new pathways for development of foam-based advanced water technologies and minimize environmental impacts.

pps proceeding
Symposium: S12 - Polymer foams and films
Poster Presentation

The effects of processing procedure on melting behavior and cell size of expanded polypropylene bead via batch-foaming process

***Lyu Mingfu* [(1)]*, *Xu Yaohui* [(1)], *Guo Peng* [(1)], *Zhang Shijun* [(1)]**

[(1)] Beijing Research Institute of Chemical Industry - Beijing - China

The expanded polypropylene beads with microcellular structure were prepared by an autoclave-based batch foaming process using CO2 as the foaming agent. Herein, polypropylene micropellets (diameter: 0.6-1.0 mm) were obtained from three kinds of polypropylene resin by underwater micro-pelletizer system. The results from differential scanning calorimetry indicated that the double melting peaks of beads moved to higher temperature with increasing saturation temperature and pressure. The kind of comonomer did not exert noticeable influence on the double melting behaviors. The X-ray diffraction of expanded polypropylene bead revealed that characteristic peaks of a-type crystal did not change compared with that of original polypropylene micropellets. The cell size of the foam was influenced greatly by the processing procedure. The average cell size could be prepared from 30μm to 300μm. The relationship between cell morphology and saturation temperature/ pressure of expanded polypropylene beads are discussed preliminarily by scanning electron microscope.

ANTIBACTERIAL WRINKLED PATTERNS BASED ON FUNCTIONAL HYDROGEL THIN FILMS.

Vera-Alarcón A. M. (1), González-Henríquez C. M. (1), Alfaro-Cerda P. J. (1), Sagredo-Oyarce D. H. (1)*, Sarabia-Vallejos M. A. (2), Rodríguez-Hernández J. (3)

(1) Universidad Tecnológica Metropolitana - Santiago - Chile, (2) Escuela de Ingeniería, Pontificia Universidad Católica de Chile - Santiago - Chile, (3) Instituto de Ciencia y Tecnología de Polímeros – Consejo Superior de Investigaciones Científicas - Madrid - Spain

Hydrogels are polymers based on hydrophilic three-dimensional networks that possess remarkable physico-chemical characteristics which resemble the extracellular matrix. These are used for a wide range of applications in regenerative medicine, such as scaffold or transdermal patches for tissue engineering [1]. Hydrogel properties are strongly dependent on the monomers molecular structure, the preparation methods and their crosslinking degree. Extern stimulus, such as temperature or UV radiation, can generate surfaces instabilities into two layer systems with different mechanical properties between them. These differences produce intern mechanical stress, leading to the generation of forces perpendicular to the hydrogel surface, which resembles into surface deformations producing finally the so called wrinkled patterns [2]. Our research group is focused on develop sophisticated smart-hydrogels with tunable physicochemical properties, with particular emphasis on materials composed by photothermosensitive polymers with rough topographic shapes and morphologies that can be modulated in the micrometric scale, these materials could be used for control bacterial adhesion or cellular proliferation [3]. In this work, hydrogels were formed from a mixture of hydroxyehtyl methacrylate (HEMA, hydrophilic monomer), N,N-diethylaminoethyl methacrylate (DEAEMA, hydrophobic monomer) and poly(ethylene glycol) diacrylate (PEGDA575, crosslinking agent). These films were supported over a glass surface functionalized with a monolayer of (trimethoxysilyl) propyl methacrylate (TSM) in order to improve their adherence. The monomer influence over the hydrogel characteristics was analyzed by preparing DEAEMA / HEMA mixtures with variable molar composition. The obtained hydrogels were treated with methyl iodide (CH3I) with the finality of quaternize the amine groups for obtain bactericide surface. Antibacterial and adherence test will be performed to the hydrogel layers by using gram positive and negative strains. Keywords: Smart Hydrogels, Wrinkle Patterns, Antibacterial Surfaces References: [1] C.M. González-Henríquez, M.A. Sarabia-Vallejos, J. Rodriguez-Hernandez, Advances in the Fabrication of Antimicrobial Hydrogels for Biomedical Applications, Materials (Basel). 10 (2017) 1–23. doi:10.3390/ma10030232. [2] A.K. Gaharwar, N.A. Peppas, A. Khademhosseini, Nanocomposite hydrogels for biomedical applications, Biotechnol. Bioeng. 111 (2014) 441–453. doi:10.1002/bit.25160. [3] C.M. González-Henríquez, D.H. Sagredo-Oyarce, M.A. Sarabia, J. Rodríguez-Hernández, Fabrication of micro and sub-micrometer wrinkled hydrogel surfaces through thermal and photocrosslinking processes, Polymer (Guildf). 101 (2016) 24–33. doi:10.1016/j.polymer.2016.08.051.

Experimental Study on Bubble Removal and Reduction of Surface Roughness during Preformed Phosphor Film Forming Using the Mold

Jeong Yeon Park[a], Dongearn Kim[a], Young Bae Ko[a], Young Moo Heo[a] and Gil Sang Yoon[a, #]

[a]*Molds & Dies Technology R&D Group, Korea Institute of Industrial Technology (156, Gaetbeol-ro, Yeonsu-Gu, Incheon, 21999, Republic of KOREA)*
#: seviaygs@kitech.re.kr

Abstract. This research attempts to identify a means of reducing the surface roughness of phosphor films and removing bubbles within phosphor films during the thermal curing process without the use of vacuum processing. Slurries combining 3 types of silicone resins of different viscosities and 2 types of yellow phosphors were applied in molds to form tensile specimens. During this process, the compression pressures of the top side of the mold were changed to cure the tensile specimen in the heat pressure chamber. By analyzing the density and bubbles of the molded products, the bubble removal tendencies of the tensile specimens molded under each of the conditions were analyzed. Finally, bubble removal tendency, the surface roughness and the degree of deformation of the cured phosphor film were compared when the phosphor film was formed with each different compression pressure value. The results of this process indicated that regardless of silicone resin viscosity, diameter of phosphor, and phosphor weight percentage in the slurry, the maximum bubble removal efficiency was achieved at a compression pressure of 0.4 MPa. However, through actual phosphor film molding, it was found that a compression pressure of 0.2 MPa was needed to inhibit phosphor film deformation. On the other hand, material features that influenced bubble removal efficiency were found to be silicone resin viscosities and phosphor weight percentage within a slurry. Phosphor diameters were indirectly found to not have a large effect.

Keywords: white LED (Light Emitting Diode), bubble removal tendency of phosphor film, compression pressure, the viscosity of silicone resin, the diameter of phosphor, phosphor weight percent
PACS: Replace this text with PACS numbers; choose from this list: http://www.aip.org/pacs/index.html

INTRODUCTION

Due to its fast response, environmental friendliness, high luminance, long lifespan, and low electricity consumption compared to incandescent lights, white LEDs using phosphor converted LED (PC-LED) are currently being used in next generation lighting systems [1-2]. Although there are several methods of manufacturing white LEDs using PC-LED, the method of arranging yellow luminescent layers on top of blue or UV LED chips to create white light is the simplest and most cost-effective approach [3].

In general, white LEDs are manufactured by dispensing a mixed substance of phosphors and transparent resins (hereinafter, phosphor slurry) on top of an LED chip using screw pumps. During the thermal curing process, precipitation of the phosphor occurs. This results in the thickness of the luminescent layer becoming uneven, and further results in the imbalanced color coordination of each of the LED lights. Also, to charge the reflection cup of LED package, relatively large amounts of transparent resins and phosphors are needed [4]. On the other hand, if a luminescent layer of consistent thickness can be formed on top of an LED light emitting area, the above-mentioned disadvantage can be solved. Especially, if the mold is used during the phosphor film forming process, the phosphor film can be manufactured in the desired shape by changing the shape of the mold.

It is reported that bubbles are generated during the process of applying the phosphor slurry in the mold, and remain on the surface or inside the phosphor film. These bubbles result in a light characteristic imbalance among LED lights due to the scattering of light from the LED light emitting areas. For this reason, the amount of residual bubbles within the phosphor film must be minimized in order to achieve balanced light characteristics for each LED display. But thus far, there has been a lack of research regarding phosphor slurry regarding the removal of air when foreign substances such as fillers are used within liquid state polymers.

This study aims to derive a method for reducing the surface roughness of the phosphor film and eliminating bubbles inside the phosphor film without the vacuum process during the forming process. First, the curing process temperatures needed to mold the phosphor film were identified through a heat analysis of the 3 types of silicone resins having different viscosities. By applying different diameters and weight percentages of each of the 3 types of silicone resins to create the mixed phosphor slurry, tensile specimens were fabricated by changing the compression pressures applied to the top surfaces of each of the phosphor slurry molds. Using the densities and surface roughness measured from each of the tensile pieces that were completely cured, bubble removal efficiencies based on silicone resin viscosity, phosphor size, and phosphor weight percentage were compared. Finally, by measuring the bubble sizes and surface roughness of each of the phosphor films molded with different compression pressures applied to its top surface during the molding process, the compression pressure at which maximum bubble removal efficiencies are achieved while preventing phosphor film deformation was identified.

EXPERIMENT

To compare the air removal tendencies of each viscosity of silicone resin, 3 types of phenyl silicone resin as shown in Table 1 were prepared. Resin 1(350 cP) was purchased from Shin-Etsu and Resins 2(2,500 cP) and 3(7,150 cP) were purchased from Dow-Corning.

All samples from sample 1 to sample 3 were dosed into the Differential Scanning Calorimeter (DSC Q-100, TA Instruments, USA) in a mixed state with the pre-polymer (A) and curing agent (B), after which heat flow at a temperature increase rate of 10 ℃/min was measured and the temperature at the end point of curing was set as the curing process temperature for samples 1 through 3. Figure (1) indicates the heat flow results of each silicone resin according to changes in temperature. The curing process temperature was 120 ℃ for Resin 1 and 2 and 160℃ for Resin 3.

To compare air removal tendencies based on diameter, it is prepared two kinds of phosphor. One is silicate based yellow phosphor having an average diameter of 15 μm (hereinafter, phosphor 1, Y4453, Intematix, USA) and the other is Yttrium Aluminum Garnet (YAG) based yellow phosphor (hereinafter, phosphor 2, QMK58/F-U1, U-VIX Corporation, Japan) having an average diameter of 4 μm. By mixing the 3 types of silicone resins and two types of yellow phosphors as shown in Table (1), 5 types of phosphor slurries were produced.

TABLE (1). 3 types of phosphor slurry to be used in the Experiment.

Silicone	Curing condition (in compression chamber)	Phosphor 1 40w% included	Phosphor 2 40w% included	Phosphor 1 80w% included
Resin 1	120 ℃ / 20 min	Slurry 1	x	x
Resin 2	120 ℃ / 20 min	Slurry 2	Slurry 4	Slurry 5
Resin 3	160 ℃ / 20 min	Slurry 3	x	x

By stirring each of the 5 types of phosphor slurries at 2,000 rpm for 2 minutes, air bubbles were intentionally generated, and the air bubble generated phosphor slurries were dispensed on the tension specimen (ASTM D648) shape molds. After placing the molds in the heat compression chamber, After placing the molds in the heat compression chamber, it is performed the first heat curing of the phosphor slurry in the mold with applying the compression pressure 0 / 0.2 / 0.4 / 0.6 MPa onto the top surface of the molds respectively. Thereafter, specimens that had been subjected to 1 hour of post curing in an oven set at 150℃ (1) had their densities measured using the electronic densimeter (MD300S, MIRAGE, Japan) to confirm residual porosity within the tensile specimen, after which (2) the maximum bubble sizes of the tensile specimens were measured using a video microscope (VMS-305EG, EG Tech) to compare the bubble removal tendencies according to different silicone resin viscosities, phosphor sizes and the compression pressure values.

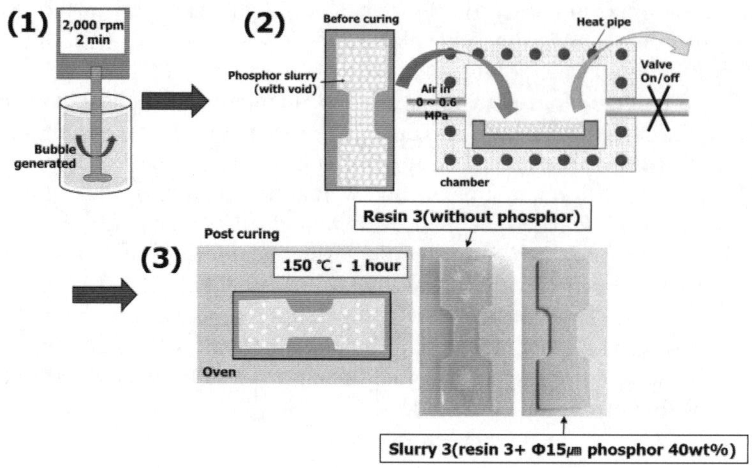

FIGURE 2. Thermo-cure process of phosphor slurry.

RESULTS AND CONCLUSION

Figure (3-a) compares the formed bubble sizes of each type of cured phosphor slurry that was not subject to compression pressures on the top side of the mold. In cases where phosphor was not included, residual bubble sizes were found to increase in the cured tensile specimen as the viscosity of the silicone resin increased. Regardless of the type of silicone resin, when mixed with phosphors, the diameters of the bubbles were found to be reduced by approximately 400 ㎛ or more compared to when they were not mixed with phosphors. This was estimated to be due to the destruction of bubbles within the phosphor slurry because of the weight of the phosphor.

Figure (3-b) presents the increase in density of the molded tensile specimen according to the compression pressures applied to the top surface of the mold in silicone resins not mixed with phosphors. Density increase rates can be calculated by 『dividing the tensile specimen density subjected to a specified compression pressure value by the tensile specimen density subjected to 0 MPa of compression pressure』. For the tensile specimen based on the highly viscous resin, resin 3 (7,150 CP), the tensile specimen density was found to increase as the compression pressure within the chamber increased. This implied that higher compression pressures in the chamber within the range of 0~0.6 MPa resulted in higher bubble removal efficiency. On the other hand, in the specimens with comparatively lower viscosity resins, resin 1 (350 cP) and resin 2 (2,500 cP), the application of compression pressure of 0.4 MPa was found to result in a lower density than the application of compression pressure of 0.6MPa. This was considered to be due to compression pressure in the chamber being higher than the vaporization pressure of the silicone resin needed for bubbles to escape from the silicone resin during the curing process, resulting in bubbles becoming trapped in the tensile specimen.

Figure (3-c) compares the density increase rates of the tensile specimens according to increases in compression pressures during the molding of the tensile specimens applied with slurries 1~3. With the exception of the tensile specimen applied with slurry 1 (350 cP Resin 1 + Φ15 ㎛ 40 wt %) and compression pressures of 0.6 MPa resulting in lower density, it was found that applications of higher compression pressures resulted in higher densities of the tensile specimen. In other words, generally, it was found that the application of higher compression pressures resulted in higher efficiency of bubble removal from the tensile specimen. The efficiency of bubble removal from the tensile specimen applied with the slurry mixed the phosphor (40 wt %) from (b) and (c) of Figure 3 was found to be 100 %p or more higher than in the tensile specimen applied only silicone resin. Thus, the application of compression pressures to the top surface of the mold was considered to help the removal of bubbles when molding the phosphor film by applying phosphor slurries.

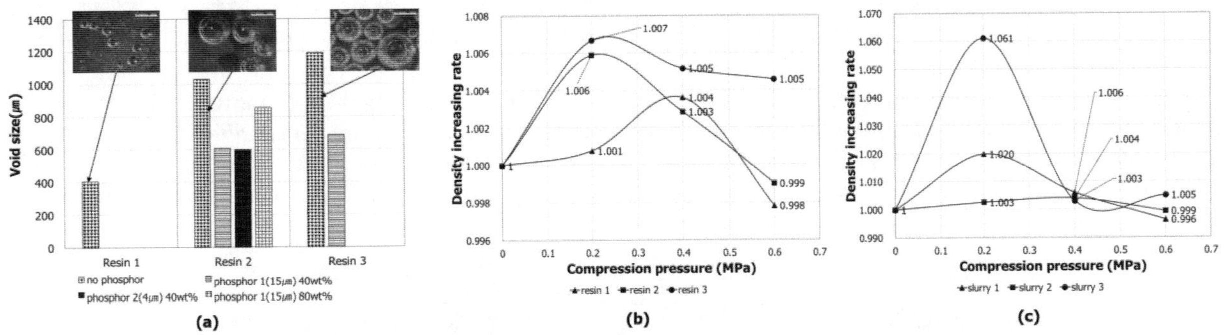

Figure 3. (a) Void size of tensile specimens after the thermal curing process (0.0 MPa) (b) Comparison of tensile specimen densi ty increase rates based on compression pressure and silicone resin viscosity, with phosphor 1 not applied (c) Comparison of tensil e specimen density increase rates based on compression pressure and silicone resin viscosity when 40 wt% phosphor 1 is applied.

Figure (4) presents a graph that compares the density of tensile specimen applied with different diameters of phosphor (b) and the density of tensile specimens applied with different mass weight percentages of phosphors within the phosphor slurry using resin 2 having a viscosity of 2,500 cP. As with the results shown in Figure (3), regardless of the phosphor diameters or mass weight percentages within the phosphor slurry, density rates were found to be maximized at a compression pressure of 0.4 MPa. The largest increase in density in Figure (4-a) was shown in the case mixed 80 wt% of phosphor and the resin 2(the density value was 0.024 g/㎤ increased). The density increase rates of cases not mixed with phosphors or mixed with 40 wt% (slurry 2) were found to be at a level of 0.01 g/㎤. On the other hand, as shown in Figure (4-b), regardless of the mixture of phosphors or diameter sizes, the maximum density increase rate according to increases in compression pressures was found to be 0.01 g/㎤. Meanwhile, in Figure (3) and Figure (4) we can see that it was indirectly observed that the material properties that influenced bubble removal efficiencies were silicone resin viscosities and phosphor content within a slurry; phosphor diameters did not have a large effect.

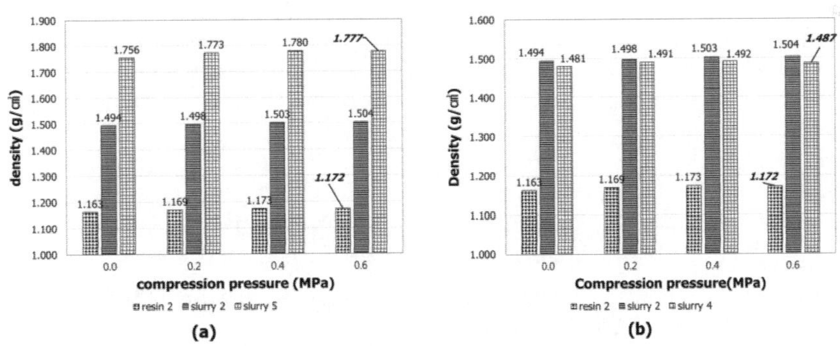

Figure 4. (A) Comparison of bubble removal tendencies of tensile specimens according to increases in phosphor mass weight pe rcentages (B) Comparison of bubble removal tendencies of tensile specimens according to changes in phosphor diameters.

Finally, a phosphor film having a thickness of 0.2 mm and a width x length = 90 mm x 90 mm applied with slurry 1 (2,500 cP silicone resin + phosphor 1 40 wt%) was molded. After applying screen printer to dispense slurry 1 within the mold, the phosphor film was thermally cured under the same conditions applied to mold the tensile specimen (120 ℃, 20 minutes). The comparison of results of bubble formations within the phosphor films at different compression pressures are as shown in Figure 5. For cases in which compression pressures were not applied to the mold's top surface, as in Figure (5-a), bubbles were discharged during the curing process on a micro level, resulting in an uneven phosphor film surface upon completion of curing. Measurement of surface roughness of the molded phosphor films using a laser scanning microscope indicated Ra=5.08 μm, Rz=42.3 μm, and Rmax=39.3 μm when compression pressures were not applied. But, in Figure (5-b), which shows the case in which a compression pressure of 0.2 MPa was applied to the top surface of the mold, bubbles were removed during the curing process and a phosphor film having an even surface was molded. Surface roughness of the phosphor film in this case was found to be Ra=2.47

μm, Rz=26.6 μm, and Rmax=28.7 μm, which indicated a decrease in surface roughness by approximately 50% compared to cases in which no compression pressure was applied.

On the other hand, when applying compression pressures of 0.4 MPa or higher as shown in Figure 5 (c-d), intumescence of the phosphor film was occurred due to a higher compression pressure being applied to the top surface of the mold than the bubble vaporization pressure of the bubbles escaping from the phosphor slurry during the curing process, which caused bubbles to remain at the bottom surface of the phosphor film in contact with the mold.

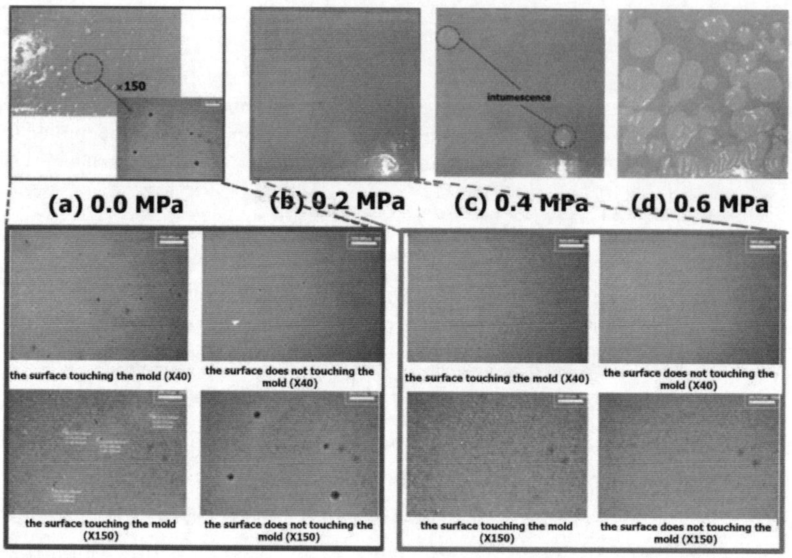

Figure 5. Comparison of deformation and air bubble formation tendencies on phosphor films according to compression pressure sizes (Applied to slurry 1)

Through the results above, the optimal compression pressure to be applied to the top surface of the mold during the manufacture of phosphor films was found to be 0.2 MPa. By additionally applying air vacuuming processes or only applying compression pressures during the thermal curing of the phosphor film, the time spent on the process of removing air bubbles from the mold of phosphor films is expected to be largely reduced.

ACKNOWLEDGEMENT

This study has been conducted with the support of the Korea Institute of Industrial Technology as "Research equipment coordination project(kitech UR-17-0036)".

REFERENCE

1. X. Luo, W. Xiong, T. Cheng and S. Liu, 2009, *IET Optoelectron*, **3**(5), 225-232(2009).
2. T. Egawa, B. Zhang, and H. Ishikawa, *IEEE Electr Device l*, **26**(3), 169-171(2005)
3. K. H. Lee and S. W. R. Lee, "Process Development for Yellow Phosphor Coating on Blue Light Emitting Diodes (LEDs) for White Light Illumination," Proc. 8th Electronic Packaging Technology Conference (EPTC), 2006, Singapore, pp. 379-384.
4. K.C. Lee, D.G. Kim, S.M. Kim, H.S. Oh, and J.H. Baek, "The Influence of Phosphor Sedimentation on the White LEDs with Different Structure Chip", Asia Communications and Photonics Conference and Exhibition, Proceedings of SPIE, 2011, Shanghai, China, paper 83120E

Foam injection molding of magneto sensitive polymeric parts

Volpe Valentina [(1)], Marco D Auria [(2)], Sorrentino Luigi [(3)], Davino Daniele [(2)], Pantani Roberto [(1)]*

[(1)] University of Salerno - Salerno - Italy, [(2)] University of Sannio - benevento - Italy, [(3)] IPCB - CNR - Napoli - Italy

The purpose of this work is obtaining magneto sensitive foamed parts made by iron microparticles distributed in a flexible polymer matrix. The material adopted is a compound of a thermoplastic elastomer with 20% by weight of iron microparticles. In particular, a masterbatch containing 50% by weight of iron microparticles was obtained by melt compounding in a twin screw extruder and was then diluted with the pure polymer directly in the hopper of the injection molding machine. A conventional injection molding machine was adapted to carry out the microcellular injection molding process. The cylinder of the injection molding machine was modified to host a system for controlling the quantity of gas injected during the batching step. The screw was modified introducing a mixing section in order to improve the dissolution of the gas inside the melt. An aluminum mold was designed and built in order to host an electro-magnet able to apply a magnetic field (up to 0.2 Tesla) during the extrudate (!?) injection and foaming. The samples obtained by foam injection molding achieved a density reduction up to 40% when compared with the corresponding non-foamed samples. In the case of samples obtained in the presence of magnetic field, the iron particles aligned in the direction of the field and organize into fibrillar structures of variable length. The mechanical tests have shown that the samples in which the iron particles are aligned in the direction of the magnetic field have a higher modulus compared to samples in which the particles are randomly dispersed. Furthermore, it was verified that the magnetic field aligned particles induces an anisotropic structural reinforcement. The research activity was supported by the Italian Ministry of Education and Research (prin2012jwpmn9)

In-situ Visualization of Crystal Nucleation and Growth Behaviors of Various Polymers under High Pressure Gases

Ducharme Taylor [(1)], Park Heon E. [(1)], Lee Patrick C. [(1)]*

[(1)] Department of Mechanical Engineering Univ of Vermont - VT - USA

A new visualization system was designed and built which is capable of capturing crystallization kinetics of polymers under high-pressure gases. This pressure cell has a viewing capability and can operate at high pressures up to 25MPa, temperatures up to 230oC, with shearing capability up to 300 s-1. Upon saturation, the polymer/gas sample was cooled to various temperatures above the crystallization temperature as determined by a high-pressure DSC. Polar optical microscopy (POM) coupled with a charged coupled device (CCD) camera was used to capture the evolution of crystal nucleation and growth with high spatial and temporal resolution. After the stress-free case, this process was repeated for various shear rates to see the shearing effect on crystallization. Some studies have used temperature controlled static or extensional view cells to study crystallization of polymer/gas systems, however there are very few publications for polymer crystallization with shear deformation and gas absorption. Many industrial polymer processes such as foam extrusion involve shear deformation under a high-pressure gas, and thus the effect of shear deformation on the crystallization should be thoroughly understood. The result from the study effectively decouples the shearing effect on the polymer melt from the complex shear and extensional flows common in foam extrusion processes. This study will show the effects of shear and the effect of various absorbed gases on the crystallization behavior. We will use the results of this study to tune the crystallization kinetics to produce desirable crystalline morphologies in polymers.

MORPHOMETRIC ALTERATIONS OF SURFACE WRINKLED PATTERNS USING FUNCTIONAL MONOMERS.

González-Henríquez Carmen M. (1)*, Sagredo-Oyarce D. H. (1), Vera-Alarcón A. M. (1), Sarabia-Vallejos M. A. (2), Rodríguez-Hernández J. (3)

(1) Universidad Tecnológica Metropolitana - Santiago - Chile, (2) Escuela de Ingeniería, Pontificia Universidad Católica de Chile - Santiago - Chile, (3) Instituto de Ciencia y Tecnología de Polímeros, Consejo Superior de Investigaciones Científicas - Madrid - Spain

Wrinkled patterns can be generated via the creation of certain surface instabilities over elastomeric materials conforming a multilayer structure by applying an elastic deformation – residual stress or mechanical compression – between top and underlying layers (elastic deformable foundation and a rigid top layer) [1]. In this context, we explored the formation of functional wrinkled hydrogel surfaces by finely tuning the synthesis and polymerization conditions using two different types of monomers. Hydroxyehtyl methacrylate (HEMA, hydrophilic monomer), trifluoroethyl methacrylate (TFMA, hydrophobic monomer) and poly(ethylene glycol) diacrylate (PEGDA575, crosslinking agent) were used as base for create thin layers with variable surface characteristics via a two-step polymerization. These changes permit to modulate wrinkled pattern morphologies according to the application needed [2]. The wettability of the sample was altered via the variation of the composition ratio between the monomers HEMA and TFMA. The samples were first thermo-polymerized in order to acquire a proper viscosity for been deposited as thin films via spin-coating technique, then were exposed to UV light with the finality to conclude the polymerization. The samples were characterized via field emission scanning electron microscopy (FE-SEM), atomic force microscopy (AFM), force spectroscopy, confocal microscopy (profilometer), contact angle, Raman spectroscopy (1-D, 2-D and depth-profile). The results show that a thin rigid layer of hydrogel (~ 1 μm) is formed on the top of the material due to the UV polymerization. The morphology of the wrinkled patterns is altered by the mechanical properties of the underneath layers, which can be modulated according to the ratio between the monomers; different structurations such as wrinkles, folds, creases and crumples can be observed. The monomer ratio also alters the wettability of the surface, property that is reflected in the variation on the contact angle, showing an increase in the hydrophobicity when the amount of TFMA increase in the reaction mixture. Keywords: Control of wrinkle patterns, Functional hydrogels, Deswelling process References: [1] K. Efimenko, M. Rackaitis, E. Manias, A. Vaziri, L. Mahadevan, J. Genzer, Nested self-similar wrinkling patterns in skins., Nat. Mater. 4 (2005) 293–7. doi:10.1038/nmat1342. [2] C.M. González-Henríquez, D.H. Sagredo-Oyarce, M.A. Sarabia, J. Rodríguez-Hernández, Fabrication of micro and sub-micrometer wrinkled hydrogel surfaces through thermal and photocrosslinking processes, Polymer (Guildf). 101 (2016) 24–33. doi:10.1016/j.polymer.2016.08.051.

ANTIFOULING PATTERNED HYDROGEL SURFACES.

González-Henríquez Carmen M. (1), Vera-Alarcón A. M. (1), Sagredo-Oyarce D. H. (1), Sarabia-Vallejos M. A. (2)*, Terraza C. A. (3), Rodríguez-Hernández J. (4)

(1) Universidad Tecnológica Metropolitana - Santiago - Chile, (2) Escuela de Ingeniería, Pontificia Universidad Católica de Chile - Santiago - Chile, (3) Departamento de Química Orgánica, Pontificia Universidad Católica de Chile - Santiago - Chile, (4) Instituto de Ciencia y Tecnología de Polímeros – Consejo Superior de Investigaciones Científicas - Madrid - Spain

A major issue in the use of biomaterials in hospitals is the microorganism adhesion to their surface. In this context, the focus of scientists turned to the development of coatings capable of resisting bacterial colonization and that can be placed on the surfaces of medical devices. There are several strategies to either reduce or prevent bacterial adhesion [1]. They basically can be divided into two different methodologies: the first type includes strategies that involve chemical modification in the material which generate antifouling or antimicrobial activity. The second type refers to those methodologies that resort to the formation of micro/nanostructures at the biomaterial surface. In this work, we aim to generate functional wrinkled patterns surfaces that take advantage of these two strategies. Hydrogel thin films based on functional monomers – hydroxyehtyl methacrylate (HEMA, hydrophilic monomer) and trifluoroethyl methacrylate (TFMA, hydrophobic monomer) – were thermo- and photo-polymerized in a two-step polymerization with the finality to form homogeneous and high-coverage layers via spin coating deposition. The ratio between these two monomers will be changed in order to determinate which produce the surface with lowest bacterial adherence. The variation in their chemical composition also generates changes in the mechanical properties of the layers, producing alterations in the structuration and morphometry of the wrinkled patterns [2]. The alterations in surface topography will be monitored via field emission scanning electron microscopy (FE-SEM) and atomic force microscopy (AFM). We expect that these micro-wrinkled patterns also improve antifouling properties of the surface. Keywords: Antifouling Surfaces, Wrinkled Patterns, Smart Hydrogels References: [1] J. Rodríguez-Hernández, Antimicrobial/Antifouling Surfaces Obtained by Surface Modification, in: Polym. against Microorg., Springer International Publishing, Cham, 2017: pp. 95–123. doi:10.1007/978-3-319-47961-3_5. [2] C.M. González-Henríquez, D.H. Sagredo-Oyarce, M.A. Sarabia, J. Rodríguez-Hernández, Fabrication of micro and sub-micrometer wrinkled hydrogel surfaces through thermal and photocrosslinking processes, Polymer (Guildf). 101 (2016) 24–33. doi:10.1016/j.polymer.2016.08.051.

Preparation and characterization of PVDF-diluent materials for thermally induced phase separation

Nam Sang Yong [(1)]***, Kim Se Jong*** [(1)]***, Lee Jeong Woo*** [(1)]***, Jang Jae Young*** [(2)]

[(1)] Gyeongsang National University - Gyeongsangnamdo - Korea, [(2)] Pure envitech Co., Ltd. - Gyeonggido - Korea

In this study, polyvinylidene fluoride (PVDF) membrane with excellent mechanical properties and chemical resistance was prepared and characterized for the application of water treatment. Dibutyl-phthalate (DBP) was used as a diluent for making a membranes through temperature induced phase separation (TIPS) method, and the crystallization temperature, melting point, cloud point and SEM image were observed with different ratio of diluent in polymer/diluent mixture. The crystallization temperature and melting point increased proportionally with the content of polymer, while the cloud point temperature decreased. Finally, it was confirmed that stable membrane could be manufactured at a polymer content of 62 wt% and a temperature 125 C using the phase diagram of PVDF/DBP mixtures with temperatures.

Progress in multi-scaled structure and related properties of elastomer nanocomposites explored by molecular dynamics simulation and experimental approaches

Zhang Liqun* [1]

[1] Beijing University of Chemical Technology - Beijing - China

In this talk I will systematically present some important simulated results of elastomer nanocomposites (ENCs) achieved via molecular dynamics simulation. First, we studied the dispersion and aggregation behavior of bare nanoparticles(NPs) with different geometries such as spherical, sheet-like and rod-like on the molecular scale. Second, we probed the translational and relaxation dynamics at the chain and segmental length scales of the interfacial regions, hoping to elucidate whether "glassy polymer layers" exist around NPs. Third, we simulated the enhancement of the stress-strain and fracture toughness induced by NPs, providing a molecular reinforcing mechanism. Fourth, the famous "Payne effect", namely the decrease of the storage modulus as a function of the strain amplitude was examined, uncovering the underlying reason responsible for this non-linear behavior, and meanwhile how the introduced carbon nano-springs can effectively reduce the dynamic hysteresis of ENCs is illustrated. Fifth, through simulation synthesis approach, we put forward a new and achievable approach to design and prepare a nanoparticle chemical network, with the NPs acting as "giant cross-linkers" or netpoints to chemically connect the dual end-groups of each polymer chain to form a network. We find this new network structure possesses excellent static and dynamic mechanical properties, highlighting a ultralow dynamic hysteresis loss tailored for green automobile tires. Computer simulation is shown to have the capability to obtain some fundamental understanding of ENCs, in hopes of providing some design basis and principles for synthesizing and fabricating multi-functional and high performance ENCs. In the meanwhile, I will show the preparation, structure-property relation and their industrialization of various high performance elastomer nanocomposites filled with clay, fibrillar silicate, silica, carbon nanotubes and graphene.

Toward Replacement of Petroleum Oils by Modified Bio-based Oils in Elastomers

***Isayev Avraam I.* [(1)]*, *Li Jiaxi* [(1)], *Ren Xiaofeng* [(1)], *Soucek Mark D.* [(1)]**

[(1)] The University of Akron - OH - USA

Nonrenewable petroleum-based plasticizers are widely used to lower the viscosity of rubber compounds, improve processability and decrease cost. However, some popular petroleum-based plasticizers, such as aromatic oils, are considered to be carcinogenic due to the high polycyclic aromatic hydrocarbon (PAH) content. Therefore, renewable bio-based oil such as soybean oil (SO) is recently emerging as sustainable replacement for petroleum-based plasticizers in rubbers. This is done by modification of SO through the reaction with dienes including dicyclopentadiene (DCPD) converting the C=C double bonds in the SO into cycloaliphatic groups of higher reactivity. The effect of incorporation of modified SO in styrene-butadiene rubber (SBR), chloroprene rubber (CR) and butyl rubber (IIR) containing carbon black (CB), precipitated silica and their mixture and also different curing systems including sulfur, metal oxide and phenolic resin is investigated. Rheological, thermal, dynamic, mechanical and aging properties of rubber compounds containing modified SO are compared with those of virgin SO and a petroleum-based naphthenic oil (NO). Results showed that SO and modified SO can react with various additives such as curatives and silane coupling agent during the curing process and change various properties. Rubbers containing modified SO showed improved elongation at break and tensile strength, cure rate and aging resistance than those with NO and SO. Both SO and modified SO provided rubbers with better thermal stability and lower glass transition temperature than NO. Rubbers containing modified SO also showed dynamic properties suitable for manufacturing tires with better wet traction and lower rolling resistance, and also better abrasion resistance simultaneously compared with those with NO. It is shown that modified SO is a good environmentally friendly replacement for petroleum-based rubber plasticizers in rubber compounds along with improved performance of rubber products an

pps proceeding
Symposium: S13 - Rubber and elastomers
Oral Presentation

Analysis of the axial temperature profile of differently stored rubber compounds in the screw antechamber during rubber injection molding

Hutterer Thomas [1]*, Berger Gerald R. [2], Friesenbichler Walter [2]

[1] Polymer Competence Center Leoben GmbH - Steiermark - Austria, [2] Montanuniversitaet Leoben - Steiermark - Austria

For injection molding of rubbers, the bulk material temperature at cavity entrance of rubber compounds is critical to the subsequent curing reaction. Currently, all commercial process simulations employ the initial material temperature as constant. However, the actual bulk temperature can greatly differ from the desired set point, as it depends on shear energy input by the screw, back pressure and energy flow by heat conduction. Thus, a strong axial (and radial) temperature profile of the material in the screw antechamber can occur. In this work, the axial temperature profile is characterized, and models for its dependence on key machine parameters are derived. To do so, a Central Composite Design of Experiments (CCD), where the effects of dosing speed, back pressure and cylinder set temperature are taken into account, is conducted and analyzed. The injection molding equipment consists of a rubber injection molding machine, and a specially designed test set-up to measure the bulk temperature. This device is mounted on the moveable clamping plate of the injection molding machine, which facilitates the exact lateral positioning of a thermocouple pin in the screw antechamber. The investigated material is and carbon-black-filled NBR rubber compound. Two batches of this compound, differing in their storage prehistory, are used for the characterization. A temperature inhomogeneity of several degrees centigrade was measured for both compounds. By means of statistical analysis, it was possible to derive key dosing factors and to quantify their influence on the axial temperature. The deviation of the mean bulk temperature as well as the shape of the axial temperature profile will be discussed. Finally, the differences between the two NBR batches due to storage effects will be demonstrated.

Hard segment influented on dielectric properties of polyurethane based polycarbonate diol

Manuspiya Hathaikarn [(1)*]

(1) Chulalongkorn University - Bangkok - Thailand

Polyurethane (PU) is interesting to study the electrical properties because it can be tuned by controlling hard/soft segments. In addition, structures can response to the electric field due to the dipole moment in hard segment. This research prepared polyurethane via solution polymerization using N,N-dimethylacetamine (DMAc) as a solvent and varying molar ratio of methylene diphenyl diisocyanate(MDI) and ethylene glycol, using the two different structures of polycarbonate diol(PCD), Eternacoll®UH200 and Eternacoll®PH100. FTIR result shows disappearance of the NCO stretching band at 2273 cm-1 revealed the reaction of –OH with –NCO to form PU. Peaks at 1540 cm-1, 1705 cm-1 and 3300 cm-1 were observed and attributed to NH-bending, carbonyl stretching and NH-stretching from the urethane linkage respectively. DSC results showed that miscible of hard/soft segment decreased when the molecular weight of polycarbonate diol increased. Subsequently, dielectric constant and dielectric loss were obtained from LCR meter at 10-106 Hz and revealed that the higher percent hard segment content, the higher of dielectric constant. This work can represent polyurethane based PCD, which provides its flexibility and manufactures easily for being utilized in actuator applications

An all atom molecular dynamic simulation and experimental study on damping properties of the hindered phenol/NBR composites

*YANG Ruining (a), ZOU Hua (b), ZHU Jing (b), ZHAO Xiuying (b), WU Sizhu (b)**

a. Center of Advanced Elastomer Materials, Beijing University of Chemical Technology, Beijing 100029, P.R. China

b. Key Laboratory of Beijing City on Preparation and Processing of Novel Polymer Materials, Beijing 100029, P.R. China

Abstract. Three types of neat hindered phenol/NBR composites were prepared and the inherent damping mechanism was explored through a combined an all atom molecular dynamic simulation and experimental study. With a series of studies on hydrogen bond (H-bond) and damping performance carried out, it is found that the effective damping peak area from DMA experiments could be quantitatively described by the level of the intermolecular (H-bond). With the combined study method, the role of H-bond in damping materials has been understood more clearly.

Keywords: **composites, damping materials, MD simulation, H-bond**

MD Simulation Strategies and Experimental Results

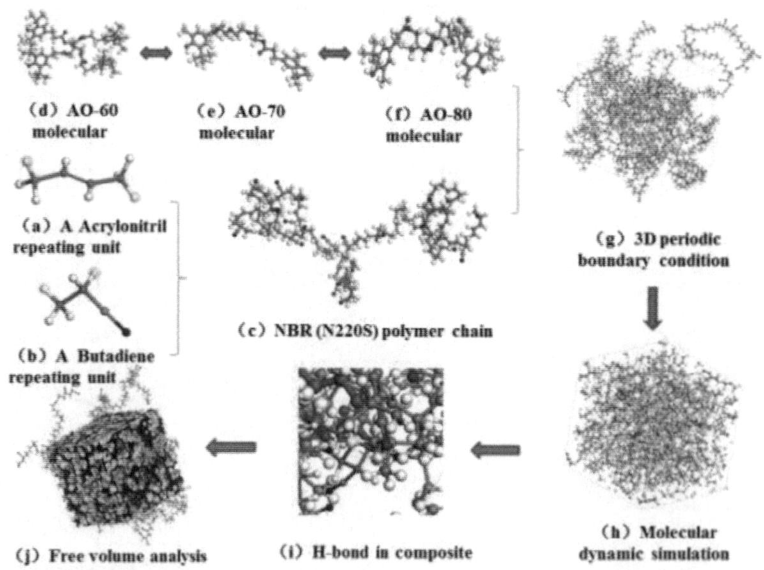

FIGURE 1 . Models for MD simulation of three neat systems (the grey sphere represents C atom, red sphere represents O atom, white sphere represents H atom, blue sphere represents N atom).

The construction procedures of the amorphous cells are shown in Fig.1. Firstly, according to N220S (a kind of NBR with an acrylonitrile mass fraction of 41%), the butadiene repeat units (Fig. 1(a)) and acrylonitrile repeat units (Fig. 1(b)) were random copolymerized as a NBR polymer chain with corresponding mass fraction. Secondly, the 3D periodic cells (Fig. 1(g)) were constructed containing four NBR chains (Fig. 1(c)) and a certain amount of the specific hindered phenol AO-60 (Fig. 1(d)), AO-70 (Fig. 1(e)) or AO-80 (Fig. 1(f)). Subsequently, the cells were optimized by Forcite modulus to equilibrate the conformation. Annealing process was set with the annealing cycle at a temperature ranged from 198 K to 498 K for 200 ps for eliminating the internal stress. Finally, isothermal-isochoric (NVT) as well as isothermal-isobaric (NPT) steps was configured at 298 K with a simulation time about 1000 ps to obtain energy optimized cells (Fig. 1(h)), and the configurations were saved for every 5 ps in these procedures. The resulting cells were used to calculate the type and quantity of H-bonds (Fig. 1(i)). Fractional free volume (FFV) (Fig. 1(j)) as well as binding energy ($E_{binding}$) were also available from the resulting cells.

Number and Types of H-bonds in Three Systems

The quantities of intermolecular H-bonds at the similar mass ratio in three systems were shown in Fig. 2(a). It is found that the intermolecular H-bonds increased linearly with the addition of hindered phenol in AO-70/NBR and AO-80/NBR composites from the following histogram, but the increasing trend in AO-60/NBR composite is not as same as the other two systems. Compared with the intermolecular H-bonds, another type of H-bonds shown in (Fig. 2(b)) was really tough to form in all of these composites. Only few intramolecular H-bonds have formed in the first few components among these three systems and the phenomenon still exists in spite of the increasing filler. The intermolecular H-bonds are easier to form than intramolecular H-bonds in hindered phenol/NBR composites according to the statistical results.

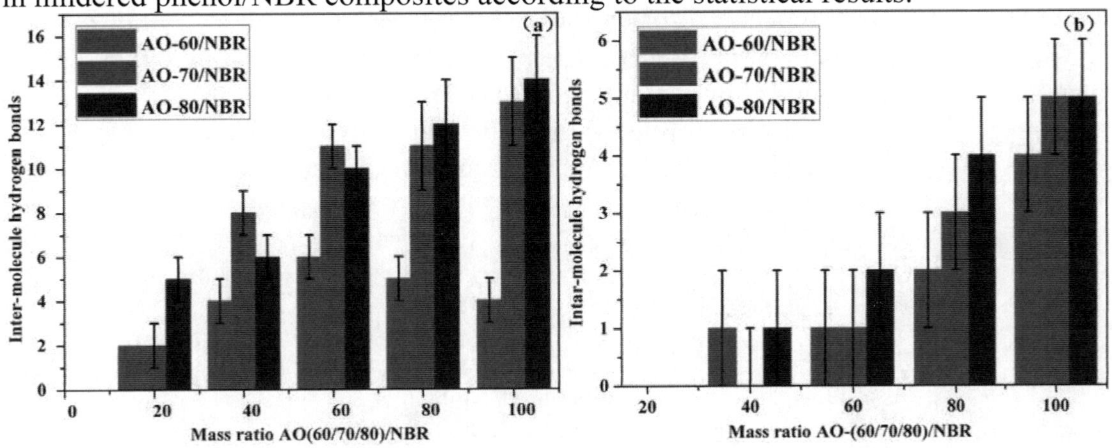

FIGURE.2 (a) Number of intermolecular H-bonds in similar mass ratio at 3 composites, (b) number of intramolecular H-bonds in similar mass ratio at 3 composites.

Binding Energy of Three Systems

Binding energy ($E_{binding}$) which describes the tightness of the components in composition is defined as the negative value of the interaction energy (E_{inter}). Specific value of $E_{binding}$ in three systems can be calculated by following equation:

$$E_{binding} = - E_{inter} = - (E_{total} - E_{(AO\text{-}60/AO\text{-}70/AO\text{-}80)} - E_{NBR})$$

As shown in Fig. 3, the $E_{binding}$ increased with the addition of hindered phenol in AO-70/NBR and AO-80/NBR systems, and the increase indicates stronger intermolecular interactions between hindered phenol and NBR matrix. The most dramatic increase gets arise in AO-80/NBR system and the interactions reach the top at 97 phr, which correspond to the largest number of H-bonds.

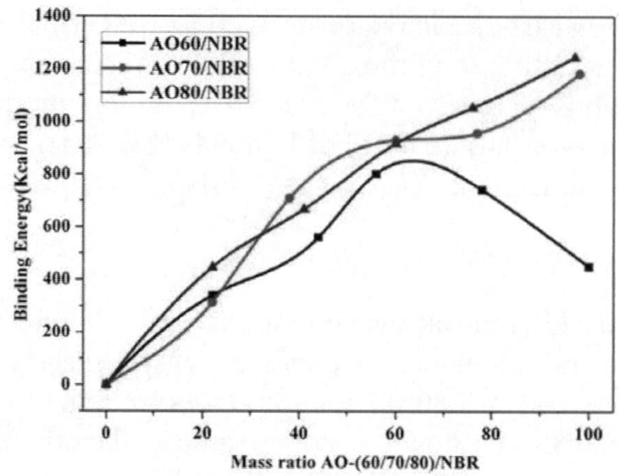

FIGURE. 3 binding energy measured by MD simulation at similar mass ratio in three systems.

Dynamic Mechanical Properties of Three Systems

Fig. 4(b-c) has shown the tanδ of similar mass ratio in these different composites. In AO-70/NBR and AO-80/NBR composites, the tanδ peak shifts from -15.1 °C to 12.9 °C and 22.6 °C, respectively. The value of tanδ also gradually increased with the addition of hindered phenol, which may owe to the intense H-bonds interactions between AO-70 or AO-80 and NBR matrix. Enhanced damping performance is also performed in AO-60/NBR composites, but when the content of AO-60 is 56 phr, the interactions between AO-60 and NBR reach the maximum value.

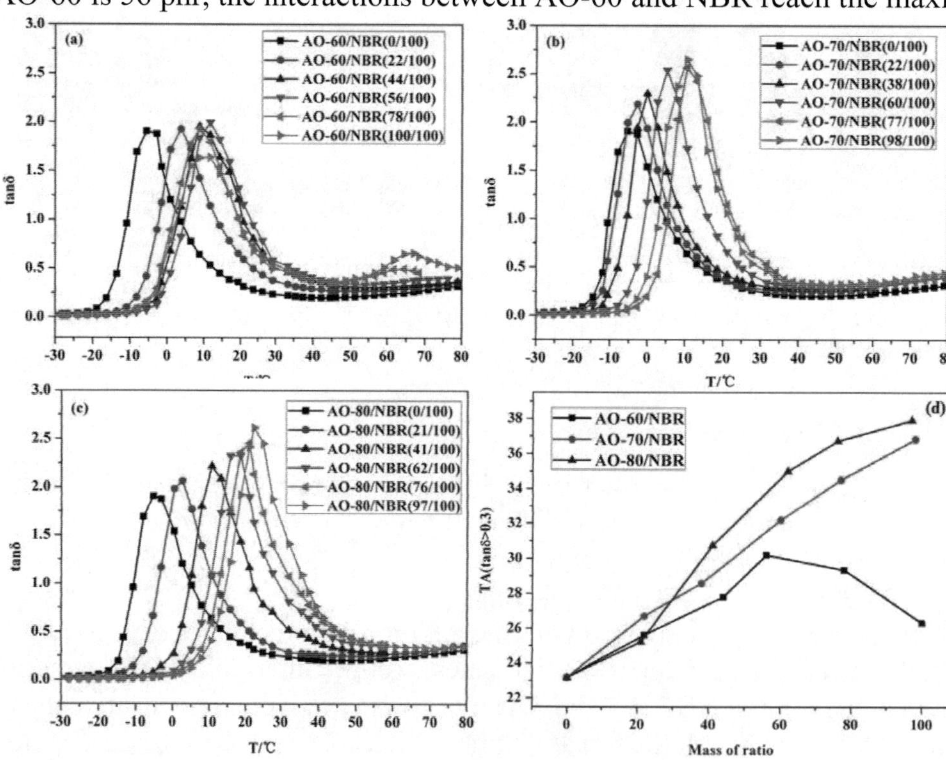

FIGURE. 4 Temperature dependence of loss factor (tanδ) for (a) AO-60/NBR composites, (b) AO-70/NBR composites, (c) AO-80/NBR composites and (d) the value of TA in the similar mass ratio of different composites.

To better compare the damping properties of three systems, the peak area (TA) at tanδ> 0.3 in the curve of tanδ versus temperature have been summarized in Fig. 4(d). AO-80/NBR composites have the best damping performance among three systems at the same temperatures. Furthermore, TA of all the three systems is consistent with the binding energy from MD simulation results. And For further illustrate the relationship, the linear regression line of Ebinding and TA was made in Fig.5. It can be found from the image that the value of TA increased with the Ebinding and an linear correlation coefficient R^2 of 0.92 was obtained (without phase separation points in AO-60/NBR composite), demonstrating that the intermolecular H-bonds do improve damping performance of materials.

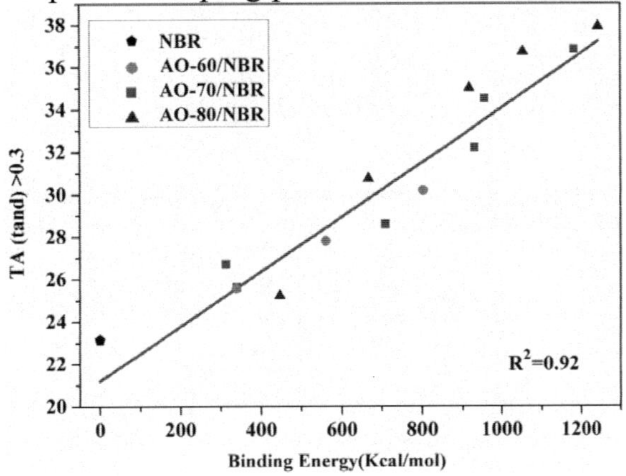

Fig.5 Linear regression coefficient of binging energy and the value of TA (without separation point).

Conclusions

In this article, with a combination of an all atom molecular dynamic simulation and experimental study, we have analyzed the microstructure and damping mechanism in three neat systems including AO-60/NBR, AO-70/NBR and AO-80/NBR composites. All atom molecular dynamic simulation results have proved that the intermolecular H-bonds with the largest number were the main type among three composites. DMA results indicated that the intense intermolecular H-bonds interaction was formed on account of the added hindered phenol. The TA can be quantitatively described by the level of the intermolecular $E_{binding}$, which put forward a clearer relationship for future study. Besides, not only all atom molecular dynamic simulation but also experimental results have revealed that AO-80 was the optimum hindered phenol for NBR systems.

ACKNOWLEDGMENTS

The authors gratefully acknowledge the National Natural Science Foundation of China (Grant No. 51473012, 51320105012 and 51603236).

REFERENCES

1.Qiao B, Zhao X, Yue D, Zhang L, Wu S. *Journal of Materials Chemistry* 2012, 22 (24):12339.
2. Song M, Zhao X, Li Y, Hu S, Zhang L, Wu S *Rsc Advances* 2014, 4 (13):6719-6729.
3. Xu K, Zhang F, Zhang X, Hu Q, Wu H, Guo S *Journal of Materials Chemistry A* 2014, 2 (22):8545.
4.Yang DW, Zhao XY, Chan T, Zhang LQ, Wu SZ *Journal of Materials Science* 2016, 51 (12):5760-5774.
5.Zhao X, Zhang G, Lu F, Zhang L, Wu S *RSC Advance* 2016, 6 (89):85994-86005.

Structure and Properties of Silicone Rubber/Styrene–Butadiene Rubber Composites with In-situ Interface Coupling by Thiol-ene Click Reaction

Sun Zheng[a], Wu Youping[a]*, Huang Qiang[b], Zhang Liqun[a], and Wang Youzhi[b]

[a]*Beijing University of Chemical Technology - Beijing - China*
[b] *Chengdu Guibao Science and Technology Co., Ltd - Sichuan - China*

Abstract. The low surface energy of silicone rubber (SiR) makes it difficult to blend SiR with conventional rubbers such as solution styrene–butadiene rubber (SSBR) and natural rubber. In this study, to enhance the interface interaction between SiR and SSBR, trimethylolpropane tris(3-mercaptopropionate) (TMPMP) was carefully chosen to couple SiR and SSBR by in-situ thiol-ene click reactions between mercapto groups of TMPMP and vinyl groups of SiR and SSBR. The reaction of TMPMP with SiR and SSBR was characterized via torque–time curves, element analysis, and the dissolving–swelling test. To overcome the low reactivity of TMPMP with SiR containing low-vinyl-content, a two-step strategy was proposed to prepare low-vinyl-content SiR/SSBR composites, where TMPMP reacted with SiR first. Transmission electron microscopy indicated that the two-step strategy reduced the phase domain size of the blend. The corresponding composite exhibited higher mechanical properties and lower hysteresis at 60 °C. The results would extend the applications of SiR.

Keywords: Rubber, Silicone, Reactive processing, Compatibilizer.

INTRODUCTION

Silicone rubber (SiR) has extremely low glass translation temperature (Tg), great resilience and outstanding heat resistance due to the high flexibility and high bond energy of its Si–O–Si backbone.[1] However, the weak mechanical strength and incompatibility with other conventional rubbers greatly limit SiR replacing the petrochemical rubbers.

Solution polymerized styrene-butadiene rubber (SSBR) is tailored synthesized to achieve high wet-skid resistance and low rolling resistance when used in passenger tires tread.[2] In this work, we are attempting to apply SiR replacing the high flexible but petrochemical cis-butadiene rubber to reduce the rolling resistance and abrasion in tread compounds.

The key to form an ideal SiR/SSBR blend is to enhance the interface interaction. In-situ interface coupling is an efficient way, but it also requires very high reactivity to form the copolymer during mixing.[3] The thiol-ene click reactions have the qualities of high efficiency and selectivity, and insensitivity to water and oxygen,[4,5] which are appropriate to the in-situ interface coupling.

In this paper, we used the trifunctional thiol, trimethylolpropane tris(3-mercaptopropionate) (TMPMP), to couple the chains of SiR with SSBR at their interface. The in-situ reaction was carried out during mixing process, which makes it suitable to rubber processing. Considering the different reactivity of TMPMP with the SiR and SSBR, a two-step strategy was carried out to make the TMPMP react with the SiR in first step. The results can help to extend the applications of SiR in tread compounds.

EXPERIMENTAL

Materials

Commercial SiR 110-2 (vinyl content of 0.15 mol%) was purchased from Chengrand Research Institute of Chemical Industry Co., Ltd., China. Commercial SSBR 2466 (styrene content of 23.5 wt% and vinyl content of 41.3 wt%) was purchased from Taiwan Synthetic Rubber Co., Ltd. TMPMP was the product of Sigma-Aldrich Co., LLC. Fumed silica AEROSIL 200 and precipitated silica ULTRASIL VN3 were purchased from Evonik Industries AG. All the other ingredients were commercially available industrial products.

Characterization of the Reactivity of TMPMP with SiR and SSBR

To investigate the reactivity of TMPMP with SiR and SSBR, a serial of mixtures was mixed as formulas in Table 1 on two-roll mill at room temperature. Torque-time curves of the mixtures were measured by MR-C3 rotorless rheometer (Beijing Ruidayuchen instrument Co., Ltd., China) at 170°C.

Table (1). Formulas of mixtures for the characterization of the reactivity of TMPMP with SiR and SSBR

Ingredients	Contents of Mixtures (phr[a])					
	SSBR T=0 D=0.1	SSBR T=1 D=0	SSBR T=1 D=0.1	SiR T=0 D=0.1	SiR T=1 D=0	SiR T=1 D=0.1
SiR	100	100	100	0	0	0
SSBR	0	0	0	100	100	100
Precipitated silica	30	30	30	30	30	30
TMPMP	0	1	1	0	1	1
DBPMH[b]	0.1	0	0.1	0.1	0	0.1

[a] phr: parts per hundred parts of rubbers

[b]DBPMH:2,5-bis(tert-butylperoxy)-2,5-dimethylhexane

Characterization of Interface Coupling Reaction of TMPMP Modified SiR (T-SiR)/SSBR Mixture

SiR, TMPMP and DBPMH were first mixed in mass ratio of 100/4/0.3 on two-roll mill at room temperature, then the mixture was heated in 170°C for 15 min to form the T-SiR, which was slightly crosslinked by the side reactions. After that, the T-SiR was extracted by tetrahydrofuran for 24 hours to remove the unreacted TMPMP. Then the T-SiR gel was dried in vacuum oven at 50°C for 24 hours. At last the dried T-SiR was mixed with SSBR in mass ratio of 2/8 on two-roll mill at room temperature to form T-SiR/SSBR mixture.

The sulfur element concentration of the dried T-SiR sample was measured by vario EL cube elemental analysis instrument (Elementar Analysensysteme GmbH).

Torque-time curves of the T-SiR/SSBR mixture and pure SSBR were measured by MR-C3 rotorless rheometer (Beijing Ruidayuchen instrument Co., Ltd., China) at 170°C to investigate the interface coupling reaction.

Preparation of Silica/SiR Composites

Silica/SiR masterbatches were prepared for matching the viscosity of SSBR. Formulas of SiR masterbatches were shown in Table 2. The masterbatches were prepared on two-roll mill at room temperature. After the ingredients were well-dispersed, the masterbatch of the two-step was heated in 170°C for 15 min to make the TMPMP adequately react with the SiR. The masterbatches of the blank and the one-step without heat treatment were also prepared.

Table (2). Formulas of silica/SiR masterbatches

Ingredients	Contents of Masterbatches (phr)		
	Blank	One-step	Two-step
SiR	20	20	20
Fumed silica	6	6	6
TMPMP	0	0.8	0.8
DBPMH	0	0	0.06

The compounds were mixed in three stages as follows. First, the masterbatches contained 20 phr of SiR, 54phr of silica, 4.8 phr of TESPT, 1 phr of SA, and 2.5 phr of hydroxyl silicone oil were dispersed in 80 phr of SSBR on a two-roll mill at room temperature. Second, the compounds were sheared for 8 min on another two-roll mill at 150°C to make the TMPMP react with the rubbers and the TESPT take the coupling reactions. Last, the compounds were cooled down to room temperature and 1.3 phr of DBPMH, 1.5 phr of 4010NA and 1.5 phr of RD were well dispersed in the compounds. The compounds were vulcanized at 170°C for 16 min.

Characterization and Properties Measurement of SiR/SSBR Composites

The vulcanization characteristics of the compounds were measured by an MR-C3 rotorless rheometer (Beijing Ruidayuchen instrument Co., Ltd., China) at 170°C.

The microstructure morphology of the composites was investigated by a Tecnai G220 transmission electron microscope (TEM) (FEI Co., USA) under accelerating voltage of 200 kV. The samples for TEM observations were cut by an ultra-microtome under liquid nitrogen cooling. The OsO4 stained samples were used to study the rubber phase separation, which was observed by the contrast of vinyl group concentration between SiR and SSBR.

The tensile properties were measured by an electronic tensile machine (Shenzhen SANS Test Machine Co., Ltd., China) at a crosshead speed of 500 mm/min, according to standards ISO 37: 2011. The shore A hardness was measured according to ISO 868: 2003.

The dynamic mechanical properties were measured by a VA 3000 dynamic mechanical analyzer (DMA) (Metravib Co., Ltd., France). The specimens were tested under the tensile mode, strain amplitude of 0.1%, frequency of 10 Hz, and temperature range from -40°C to 80°C at a heating rate of 3°C/min.

RESULTS AND DISCUSSION

Reactivity of TMPMP with SSBR and SiR

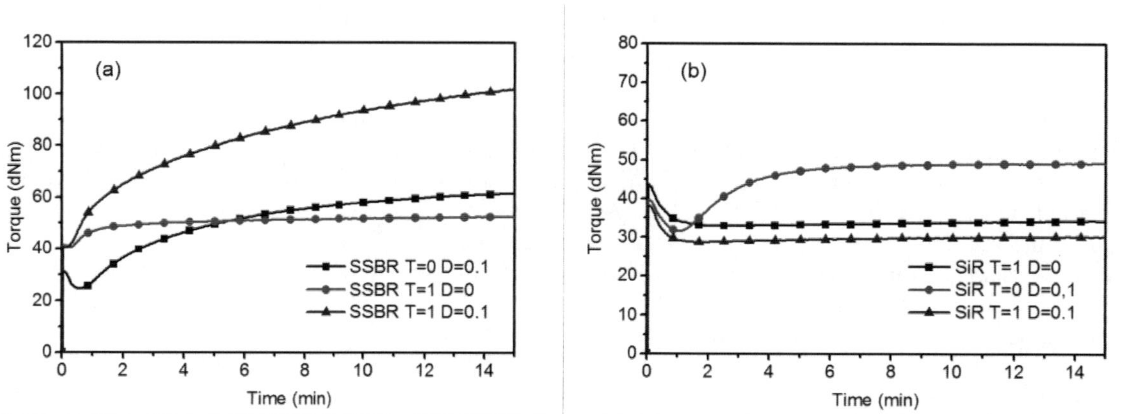

FIGURE 1. Torque vs. time curves of SSBR (a) and SiR (b) with TMPMP and DBPMH independent or synergistic.

The thiol-ene click reactions can be triggered by heat or UV-light. We used DBPMH as the thermal initiator to trigger the reaction of TMPMP with the vinyl groups of SSBR or SiR. Meanwhile, DBPMH can vulcanize the rubber independently. Considering the difference in vinyl concentration of SSBR and SiR, we measured the torque vs. time of SSBR or SiR with TMPMP and DBPMH independently or synergistically. The curves were shown in Figure 1. For SSBR in Figure 1 (a), 0.1 phr of DBPMH crosslinked SSBR evidently. And TMPMP could significantly crosslink SSBR without initiator. This was due to the high reactivity of the trifunctional thiol in TMPMP and the high vinyl content of SSBR (41.3 wt%). When TMPMP and DBPMH were used together in SSBR, a synergistic effect was observed. For SiR in Figure 1 (b), 0.1 phr DBPMH could crosslink SiR, whereas TMPMP could not significantly crosslink the SiR. This was due to the low vinyl concentration of SiR (0.15 mol%). Theoretically, when TMPMP was excessive, each of the vinyl group could react with one TMPMP molecule, with the redundant mercapto groups as side groups of the silicone molecules, forming T-SiR. Moreover, when TMPMP and DBPMH were used together in SiR, the torque showed no significant rising, indicating that the vinyl groups mostly reacted with TMPMP rather than formed significant crosslink. Meanwhile, the side reactions made the T-SiR highly swelled rather than dissolved during extraction, although the slight crosslinking was not evident in torque rising in Figure 1 (b).

Characterization of Interface Coupling Reaction of TMPMP Modified SiR (T-SiR)/SSBR Mixture

To confirm the mercapto groups of TMPMP were successfully reacted with the vinyl groups of SiR, the slightly crosslinked T-SiR was extracted by tetrahydrofuran for 24 hours to remove the unreacted TMPMP. The elemental analysis showed that the sulfur concentration of T-SiR was 0.20 wt%, which converted into mole ratio between

TMPMP and original vinyl groups was 1.02:1, indicating that the vinyl groups were fully reacted and T-SiR was successfully formed with dangled mercapto groups.

To confirm the mercapto groups dangled on T-SiR could react with the vinyl groups of SSBR, the extracted T-SiR was mixed with SSBR and the torque vs. time curves of the mixture and pure SSBR were shown in Figure 2. Contrasted to the constant torque of pure SSBR, the torque of T-SiR/SSBR mixture increased, indicating that the interface coupling reaction between T-SiR and SSBR could successfully take place.

Based on the above results, we proposed to improve the interface interaction between SSBR and SiR as follow, and the schematic diagram of the two-step interface coupling was shown in Figure 3. The first step was mixing excessive TMPMP with SiR, the mole ratio of TMPMP molecule to vinyl groups was 3.2:1, and making TMPMP fully reacted with SiR under 170°C to form T-SiR; the second step was mixing T-SiR with SSBR, and then the dangled mercapto groups of T-SiR in-situ reacted with SSBR under shear and heat, forming the chemical bonds between SiR and SSBR. For comparison, a one-step strategy blend, where the TMPMP were added after the blending of SiR and SSBR, was also prepare as a reference.

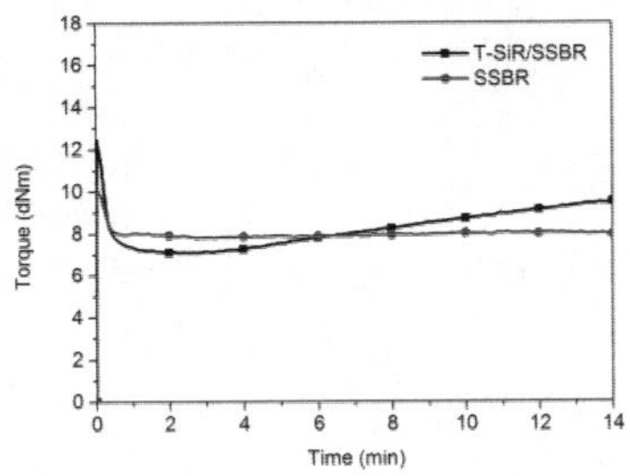

FIGURE 2. Torque vs. time curves of the T-SiR/SSBR mixture and the pure SSBR.

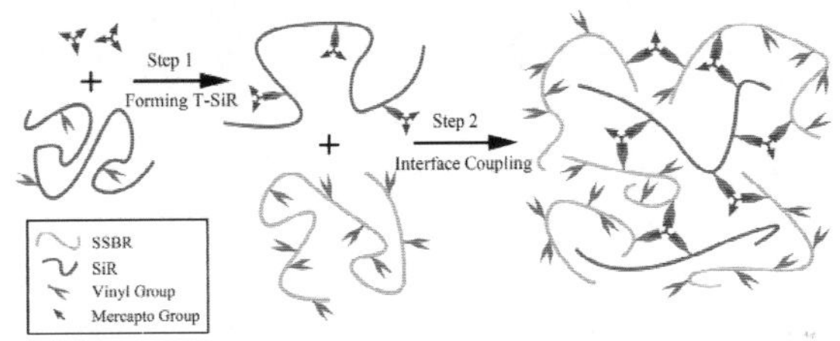

FIGURE 3. Schematic diagram of the two-step interface coupling reactions.

Static Mechanical Properties of the Composites

Table (3). Static mechanical properties of the three composites

Samples	Blank	One-step	Two-step
Tensile strength (MPa)	14.7	14.1	15.6
Elongation at break (%)	368	264	290
Modulus at 100% (MPa)	1.9	2.6	2.5
Modulus at 200% (MPa)	5.5	8.4	7.8
Shore A Hardness	63	63	64

The static mechanical properties of the three composites were shown in Table 3. Compared with the blank composite, both of the one-step and the two-step composites showed an increased modulus, and the one-step composite showed a higher rise in modulus. This was consistent with the torque results of the compounds. Whereas the tensile strength and elongation at break of the one-step composite decreased, indicating the increasing of stress concentration, which was introduced by the tendency of TMPMP to crosslink the SSBR rather than couple the SiR with the SSBR. However, the two-step composite showed an increased tensile strength and tensile modulus, indicating that the interface interaction between SiR and SSBR were successfully enhanced by the coupling reaction.

Dynamic Mechanical Properties of the Composites

The viscoelastic properties of the composites were measured to investigate the interface interaction of the blends. The temperature sweep was carried out and the curves were shown in Figure 4. The glass transition peak of SSBR showed no significant differences among the composites, suggesting that the thermodynamic immiscibility of SiR and SSBR could hardly be changed by the interface coupling. However, the one-step and the two-step composites had lower tan δ at high temperature than the blank composite. This phenomenon was ascribed to the stronger interface interaction, which decreased the interface friction. The dynamic mechanical properties could also reflect the potential of this interface coupling technique in tire industry. The two-step composite showed lower tan δ at 60°C and higher tan δ at 0°C than the blank, which correlated to lower rolling resistance and higher wet-skid resistance when the compound was used as treads.

Microstructure Morphology of the Composites

The phase separation morphology is an important factor in determining the properties of the polymer blends. TEM was used to evaluate the phase separation morphology of the blends. The SSBR in the blends was stained by OsO4 to make a contrast with the unstained SiR. Figure 5 (a) and (b) showed the morphology of the blend without interface coupling. The phase separation was clear and the domain scale was over 100 nm, suggesting the low surface energy of SiR led to the aggregation of SiR. Figure 5 (c) and (d) showed the morphology of the two-step composite. The SiR domain was smaller and dispersed better in SSBR matrix. The misty interface suggested that the SiR molecules were coupled with the SSBR molecules during the in-situ interface coupling process. The in-situ formed copolymers of SSBR and SiR also played the role in reducing the surface tension, and then the domain size. From macroscopic view, Figure 6 showed the backside illuminated image under the same lighting and exposure condition. Consistent to Figure 5, the two-step composite slice was more transparent than the blank, suggesting the better dispersion of SiR.

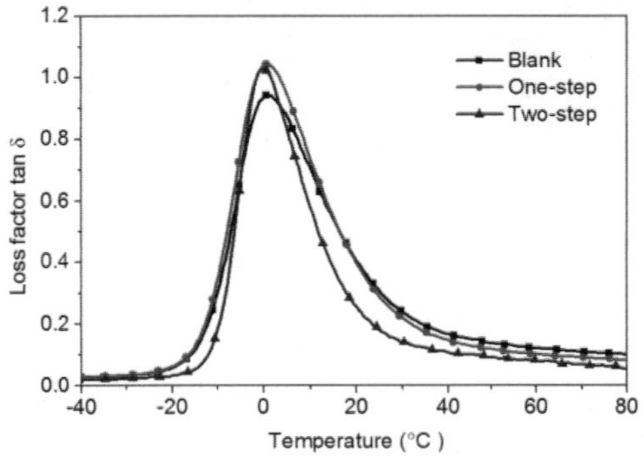

FIGURE 4. Loss factor tan δ vs. temperature curves of the composites.

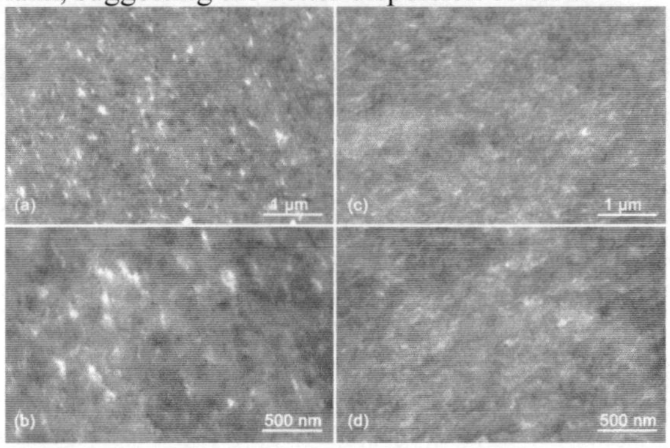

FIGURE 5. The stained microstructure morphology images of SiR/SSBR blends: (a) and (b) blank; (c) and (d) two-step interface coupling strategy.

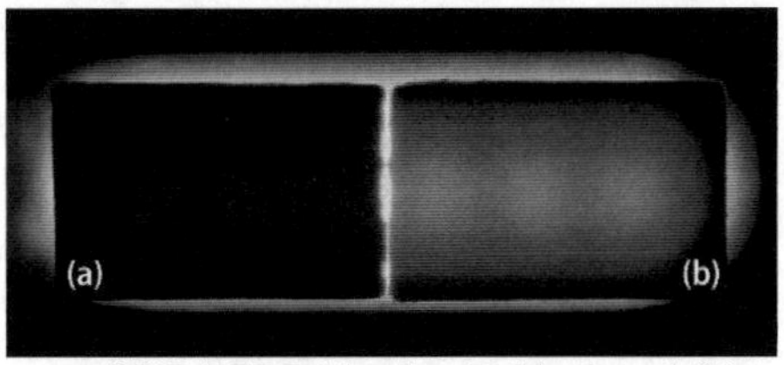

Figure 6. Backside illuminated images of SiR/SSBR blends: (a) blank and (b) two-step interface coupling strategy.

CONCLUSIONS

We used the trifunctional thiol TMPMP to couple the molecules of SSBR and SiR at interface through thiol-ene click reactions during the high-temperature mixing. Because of the rare concentration of vinyl groups in SiR, we proposed a two-step interface coupling strategy, which made the TMPMP react with SiR first. By adding excessive TMPMP into SiR, the mercapto modified SiR was successfully synthesized in the first step. Then the dangled mercapto groups reacted with SSBR in the second step. The phase separation morphology confirmed that the in-situ coupling could reduce the domain size of SiR. The hysteresis tan δ of the composites was also reduced due to the lower interface friction. This in-situ interface coupling technique would make the SiR wildly used and partly replacing the petrochemical rubber in tire industry and other fields.

REFERENCES

1. İ.Yilgör and J.E.Mcgrath, *Adv. Polym. Sci.* **86**, 1 (2006).
2. X. Liu and S. Zhao, *Polymer*, **55**, 1964 (2014).
3. P. Lucas and J.J. Robin, *Adv. Polym. Sci.* **209**, 111 (2007).
4. C. E. Hoyle and C. N. Bowman, *Angew. Chem. Int. Ed.* **49**, 1540 (2010).
5. M. J. Kade and D. J. Burke, *Polym. Sci. Part A Polym. Chem.* **48**, 743 (2010).

A comparison of melt and solution mixing on the dispersion of carbon nanotubes in a styrene–Butadiene–styrene block copolymer

Santos João (1)*, Melo Guilherme Henrique França (2), Gonçalvez Andre (2), Eiras José (2), Bretas Rosário Elida Suman (2)

(1) Universidade Estadual de Ponta Grossa - Paraná - Brazil, (2) Universidade Federal de São Carlos - São Paulo - Brazil

Multi-walled carbon nanotube (MWCNT)/ styrene–butadiene–styrene (SBS) nanocomposites were prepared by two frequently used approaches, melt and solution mixing. SBS is a thermoplastic elastomer which can self-assembly into well-defined ordered nanostructures. Taking this capability into account, block copolymer materials have long been recognized as scaffolds and used to patter nanoparticles to construct hybrid structures. Therefore, the combination of electrical conductive MWCNTs and SBS in a well-mixed nanocomposite with stable morphology may lead to a material with new and interesting properties for electronic devices. To obtain the equilibrium surface morphology, the SBS film was prepared by solvent casting annealing for 168 h (seven days) at 110 C. The surface morphologies of the SBS specimens were examined by non-contact mode atomic force microscopy (AFM). To produce the nanocomposites by solution the MWCNTs were mixed with SBS using toluene as a solvent. Melt mixed nanocomposites were made using an internal mixer (Haake rheometer, model Rheomix 600p) operating at 150°C, 100 rpm for 7 min. The nanocomposites were characterized by impedance to obtain the AC electrical conductivity, by stress strain tests to obtain the elastic modulus, elongation and tensile at break, and the dispersion of the MWCNT was evaluated by scanning electron microscopy (SEM). The results showed that nanocomposites prepared by solution casting showed a higher electrical conductivity than the melted. It was found that the dispersion state of MWCNTs was dependent not only on the concentration of MWCNTs but also the mixing method. Melt mixing led to a much better dispersion of MWCNTs than solution mixing which decresed the conductivity.

Influence Of The Injection Molding Process On The Peeling Resistance Of Thermoplastic / Silicone Composites

Dipl.-Ing. Christof Schlitt[a], Dipl.-Ing. Ralf-Urs Giesen[a]
and Prof. Dr.-Ing. Hans-Peter Heim[a*] (reference author)

[a] *Polymer Application Center UNIpace; Institute of Material Engineering- Plastic Engineering, University of Kassel; Mönchebergstraße 3, 34125 Kassel, Germany*

christof.schlitt@gast.uni-kassel.de
heim@uni-kassel.de

By means of using diverse materials, various properties were able to be integrated in according areas of multicomponent parts. For example, in these applications, there are hard components that provide the required stability and soft components that function as sealing or shock absorbing elements.
These processes are well-established for the production of thermoplastic components. One of the most common applications are probably tooth brushes, which are made of a hard, solid body that is overmolded with a softer thermoplastic material for a better grip. [1]
When it comes to technical applications that have higher requirements for the employed materials, thermoplastic elastomers (TPEs) reach their limits. Silicones are much better suited for such requirements. They are characterized by highly reversible elastomeric behavior, good temperature resistance (long-term 180°C, short-term 250°C), and good flexibility at temperatures as low as -50°C, as well as a high resistance against chemicals, weather conditions and aging. Furthermore, silicones are harmless from a physiological point of view, which makes them very interesting for medical applications. [2, 3] Due to the varying morphological material characteristics, the creation of a strong bond between the different components is still challenging. Different theories exist that attempt to explain how these combinations could work.
To investigate on how the injection molding process and its parameter settings influence the bond between the different components a modular tool was designed for the production of multicomponent specimens. In this tool, a thermoplastic plate is molded on the one side, and is then automatically transferred by a robot system to be overmolded with silicone on the other side. The approach presented in the VDI 2019 guideline "Testing the Adhesion of Thermoplastic Elastomers (TPE) on Injection-Molded Substrates" was chosen for testing the material composites, which consisted of thermoplastics and LSR. [4]
Using an all-electric Arburg Allrounder 370A multicomponent injection molding machine composite samples were produced. In this study the thermoplastic plates were made of polyamide 6 (Ultramid B3K by BASF) and a self bonding liquid silicone rubber (Elastosil LR 3070 by Wacker) was used for overmolding. In the study the parameters for the thermoplastic injection molding process as well as for the liquid injection molding process (LIM) were varied.
It was found out that the melt temperature, the mold temperature and the holding pressure of the thermoplastic process as well as the injection speed of the silicone process significantly influences the resulting resistance against peeling of the specimens.
Only by varying the process parameters the resulting bonding strength can be influenced by up to 30%.

REFERENCES

1. van Onna, E.: Innovative Materialien und Überzüge für Innenräume. Birkhäuser 2003
2. Silicone Rubber - The Elastomer Of Choice For Medical Applications, Briehn, C., 2011
3. Clarson, S. J. (Hrsg.): Silicon. Springer-Verlag 2009
4. VDI 2019; Oktober 2014. *Testing the adhesion of thermoplastic elastomers (TPE) on substrates*

Quantitatively Studying the Interphase of Rubber Nanocomposites by a New Mode of AFM

Ming Tian[a,b,c*], Guangyu Chu[c], Chunmeng Miao, Tong Mi, Yanyan Guo,

Yuxing Feng[c], Yonglai Lu[a,c], Nanying Ning[a,c] and Liqun Zhang[a,b,c]

[a]State Key Lab of Organic-Inorganic Composites, Beijing University of Chemical Technology, Beijing 100029, China.

[b]Beijing Advanced Innovation Center for Soft Matter Science and Engineering, Beijing University of Chemical Technology, Beijing 100029, China.

[c]Key Laboratory of Carbon Fiber and Functional Polymers, Ministry of Education, Beijing University of Chemical Technology, Beijing 10029, China.

Abstract. The interphase between nano-filler and rubber matrix, also known as "bound rubber (BR)" play an important role in the properties of rubber nanocomposites. Herein, we developed a new method to quantitatively identify the interphase of silica/rubber nanocomposites and carbon nanotubes (CNTs)/ rubber composites at nanoscale resolution by using peak force quantitative nanomechanical mapping (PF-QNM) mode of Atomic Force Microscope. The modulus and adhesion distribution around the fillers of the composites at nanoscale resolution was measured, and accordingly the interfacial thickness between rubber and fillers with or without silane coupling agent modification was tested according to the gradient change in nanomechanical properties from matrix to fillers. We further studied the effect of physical and chemical charateristic of fillers and rubber macromolecules on the interfacial thickness of rubber nanocomposites. Interestingly, the double layer structure of interphase was directly identified in silica/rubber nanocomposites based on PF-QNM images and the corresponding force-distance curves, and was further evidenced by high resolution transmission electron microscopy. This study provides a new method to quantitative characterize the interface between filler and elastomer matrix at nanoscale resolution.

Keywords: interface, polymer, nanocomposites, AFM, microstructure characterization

INTRODUCTION

Rubber is a kind of high molecular weight polymer with very flexible macromolecules, it is widely employed in various applications such as tires, conveyor belts, hoses and sealing[1]. The reinforcement of rubber with suitable nanoscale fillers is of significant importance because of its low mechanical strength.

SSBR has attracted a lot of attention owing to its advantages of wear-resisting, cold resistant property, low heat generation, high resilience, faster cure rate and so on[2-4]. According to the report, comparing with other tire, the tire made of SSBR has a great improvement in the wet resistance, the rolling resistance and the wearing-resisting, which has a great contribution to the environment.

Silica is widely used to improve the static and dynamic mechanical properties of rubber, such as wet skid resistance and rolling resistance[5, 6]. Furthermore, silica is also independent of oil resources and environmentally friendly filler used as a replacement for carbon black[7, 8]. As a result, compound with the usual "green" tire tread SSBR, silica/SSBR nanocomposites has been considered as promising materials for "green" tire in recent years[9, 10].

Carbon nanotubes (CNT), having the unique one dimension carbon structure[11], possess excellent mechanical properties , thermal properties[12,13] and unique electromagnetic properties and optical properties[14, 15], good chemical stability[16] which can greatly improve the performance of the composite materials.

It is well known that the homogeneous dispersion of the filler and the strong interfacial interaction between the filler and the polymer matrix can effectively improve the mechanical, electrical, and thermal performances of the composites[17-20]. Due to the enormous surface to volume ratio of the nanoparticles, the polymer close to the interface constitutes a significant fraction of the material, and its behavior significantly affect or even dominate the properties of the composite. So interfacial properties is an traditional but important issue that attracted the interest of many researchers[21-23].

In recent years, atomic force microscope (AFM) has demonstrated a versatility in the characterization of nanostructured materials[24-26]. It enables us to investigate topological and mechanical properties at the same point and the same time. Dong Wang et al[27]. use a procedure together with AFM force volume measurements to investigate the intermediate modulus region formed around CNTs, and demonstrated the existence of interaction between CNTs and NR matrix. However, there is almost none system research about the influence of filler properties on interfacial layer through direct or nanomechanical characterization.

In our study, we explored the method using AFM with Peak force Quantitative Nanomechanical Mapping(PF-QNM) mode to measure the interfacial layer thickness of silica/SSBR and CNTs/SSBR nanocomposites, and then systematically discussed the influence of physical and chemical charateristic of fillers on the interfacial layer.

EXPERIMENTAL

Materials

Solution styrene-butadiene rubber (SSBR-T2003) composed of 75% butadiene with a 13.5% vinyl content, and 25% styrene, was provided by Shanghai Gaoqiao Petrochemical Co., Ltd. (China), and its Mooney viscosity ML(1+4) 100 °C was 33. The precipitated silica of Ultrasil VN3 (average diameter is about 15nm, BET specific surface area is 168 m2/g) was obtained from Degussa Chemical Industry Co., Ltd. (China). Silica coupling agent g-(methacryloxy) proxyltrimethoxysilane (KH570) and bis-(3-(triethoxysilyl)-propyl)-tetrasulfide (Si69) were purchased from Nanjing Crompton Shuguang Organosilicon Specialties Co., Ltd. (China). All other materials were commercially available.

MWCNTs were purchased from Nanjing XFNANO Materials Tech Co.,Ltd. SSBR (Shanghai Gaoqiao Co. ltd.T2003) was used as the rubber matrix. The parameters of the MWCNTs are listed in TABLE 1.

TABLE (1) The parameters of the MWCNTs provided by XFNANO Materials Tech Co.,Ltd

Name	Length (um)	ID(nm)	OD(nm)	Purity	SSA(m^2/g)
1-MWCNT	0.5-2um	5-10nm	10-20nm	>95%	>200
2-MWCNT	0.5-2um	5-10nm	20-30nm	>95%	>110
3-MWCNT	0.5-2um	5-12nm	30-50nm	>95%	>60

Preparation of Composites

For the SiO2/SSBR composites, 100 phr (wt%) SSBR and 5 phr silica were first mixed on a two-roll mill (Zhanjiang Machinery Factory, China) at room temperature. The gap between the two rolls was adjusted to ~0.5 mm for high-shear blending, which was followed by the addition of (0.5 phr) vulcanizer, DCP. The obtained compounds were sheeted into slab of 2 mm, and vulcanized at 150 °C in 15 MPa for the optimum cure time (t90) in a standard mould to obtain the rubber composite. The same procedure was used to prepare modified SiO2/SSBR composites.

The basic formulas are listed in Table2. The SSBR was placed on the two –roll mill and rolled upon themselves. Then the MWCNT and DCP were mixed with the SSBR by order. The raw materials were cured by compression moulding at 150C°. Then the MWCNT/SSBR composites were obtained.

Characterization

The morphology and microstructure of the SiO2/SSBR composites were characterized by a scanning electron microscope (SEM, S4800, Hitachi Co., Japan). The samples used in SEM examinations were first tensile fractured and coated with gold.

All AFM experiments were performed using a commercial AFM system(Multimode series with a NanoScope 8 controller, Bruker) under ambient condition. Type SNL-10 probe is used for nanomechanical mapping. We used the spring constant of 0.35 N/m, which was appropriate to deform rubbery regions. The cantilever is made from Si3N4, and the tip is made of Si. We operated force mapping measurement with a peakforce setpoint range from 500pN to 2nN. The AFM nanomechanical mapping technique was employed, namely, 256*256 approaching force-distance curves were obtained over a two-dimensional area. The sheet 2mm sample was cut by a Leica EMFC7 ultramicrotome at -80 °C to obtain a flat surface for AFM observation.

A high resolution transmission electron microscope ((HRTEM, JEM-3010, JEOL, Hitachi, Japan) was used to observe the bound rubber surrounding the silica particle. The uncrosslinked compounds were immersed into cyclohexane solvent. After stirring for 12 h, the undissolving systems were centrifuged and washed with cyclohexane for three times. Samples for HRTEM observation were prepared by dissolving the products in deionized water to a concentration of 0.01 mg/ml, then dropping the suspension on microgrids.

The viscoelastic properties of the composites were characterized with a VA3000 DMTA (01 dB-Metravib, France) in the tension mode. The temperature was from -100 °C to 80 °C, the heating rate was 3 °C min-1, the frequency was 1 Hz, and the strain amplitude was 0.1%.

An optical contact angle meter (Kruss DSA 100, Germany) was used to measure the dynamic contact angle (θ). The surface energy and the thermodynamic parameters can be calculated from the contact angle data.

The static mechanical properties of the composites were measured according to ASTM D412 by a CMT4204 Electrical Tensile Tester (Shenzhen SANS Test Machine Co., China). For tensile tests, the dumb-bell shaped specimens were stretched until break at a crosshead rate of 500 mm min-1 according to a China standard GB/T528-1998. The tensile strength, elongation and stress at 100%, and 300% strains were the average values of five specimens.

RESULTS AND DISCUSSION

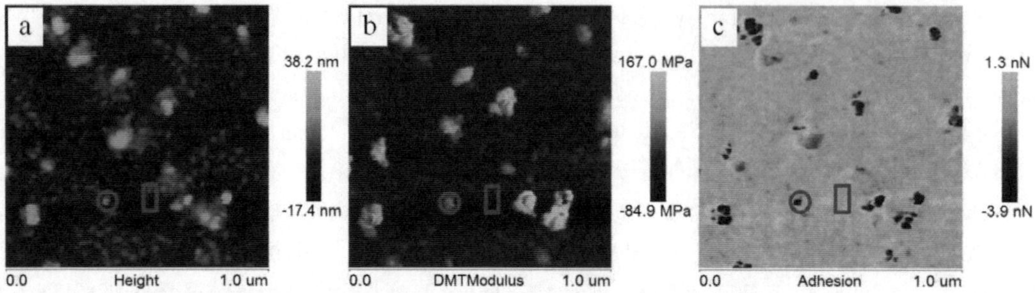

FIGURE 1. AFM images of SiO2/SSBR nanocomposites (a) height image (nm) (b) elastic modulus image (MPa) (c) adhesion image (nN). Scan size is 2 um.

FIGURE 1 shows simultaneously generated the height, elastic modulus and adhesion maps of SiO2/SSBR nanocomposites as measured via Peak Force Quantitative Nanomechanical Mapping mode AFM. We select the appropriate particles shown as FIGURE 2(a) to characterize the interfacial thickness more accurately, excluding silica aggregates and isolated silica embedding by rubber, which height above polished surface is greater than the particle radius. The adhesion-length curve is obtained by drawing a transverse line through the selected filler inFIGURE 2(c). the area have adhesion gradient is interfacial layer, and the length from high to low adhesive energy could be the thickness of the interface.

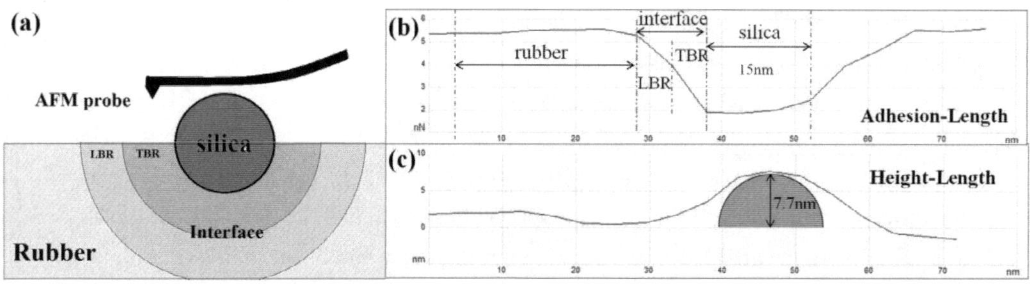

FIGURE 2. The measurement method of interfacial layer thickness (a) the appropriate particle (b) Adhesion-Length image (c) Height-Length image

What's more, based on the double layer structure, the interfacial layer can divide into two different segments, tightly bound rubber (TBR) immediately surrounding the silica and loosely bound rubber (LBR) of intermediate mobility. the nanomechanical properties of LBR change slower than TBR, corresponding to the two different adhesion gradient part in Fig.2(b).

TABLE (2) Grafting degree of silica with different modifier types and contents (%)

Sample	unmodified	Si69-2	Si69-6	Si69-10	Si69-20	Si69-40	KH570-40
grafting degree(%)	0	0.88	1.29	2.32	5.10	6.06	6.31

In order to explore the effect of modifier types on interfacial thickness, we prepared unmodified-SiO2/SSBR, 40% KH570-SiO2/SSBR and 40% Si69-SiO2/SSBR composites. The influence of modifier grafting degree is ignored because it is similar measured by TGA [24], as shown in TABLE 2. Using the method mentioned above, we can see that the interfacial thickness increased after modification, owing to the improvement of the activity of filler surface and the filler-rubber compatibility.as shown in TABLE 3. we obtained the total interface thickness are 8.3±0.2 nm, 8.5±0.1 nm and 17.2±0.3 nm, respectively. In order to verify the established method, we followed a series of tests to verify it, such as TEM and DMA. The reliability of the above methods is well verified.

TABLE (3) Comparision of the interfacial layer thickness for SiO2/SSBR nanocomposites

Sample	TBR(nm)	LBR(nm)
Co-unmodified	3.9±0.11	8.3±0.2
Co-KH570	4.0±0.09	8.5±0.1
Co-Si69	3.9±0.28	17.2±0.3

Quantitatively Studying the Interphase of CNTs/SSBR Nanocomposites

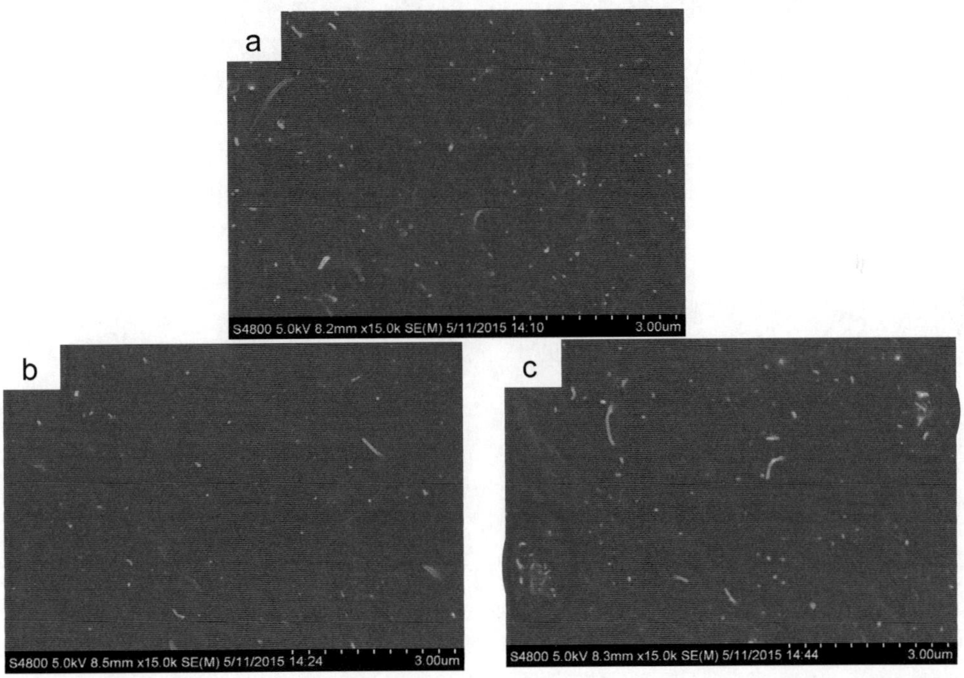

FIGURE 3. SEM micrographs of the fracture surfaces of MWCNT/SSBR(a) MWCNT-1/SSBR (b) MWCNT-2/SSBR(c)MWCNT-3/SSBR, All of the magnification is 15 k.

The SEM is a direct way to character the interface between the rubber and the nanoparticles. The fracture surface can show the adhesive state as a whole base.The SEM micrographs of the fracture surfaces of the three types of MWCNT/SSBR is presented in FIGURE 3.The micrographs presented that MWCNT-1/SSBR has the most homogeneous dispersion, followed by MWCNT-2/SSBR. However, the micrograph of MWCNT-3/SSBR can be seen the obvious agglomeration(especially in ellipse). Additionally, the MWCNT-1/SSBR exhibits the coarse fracture surface.

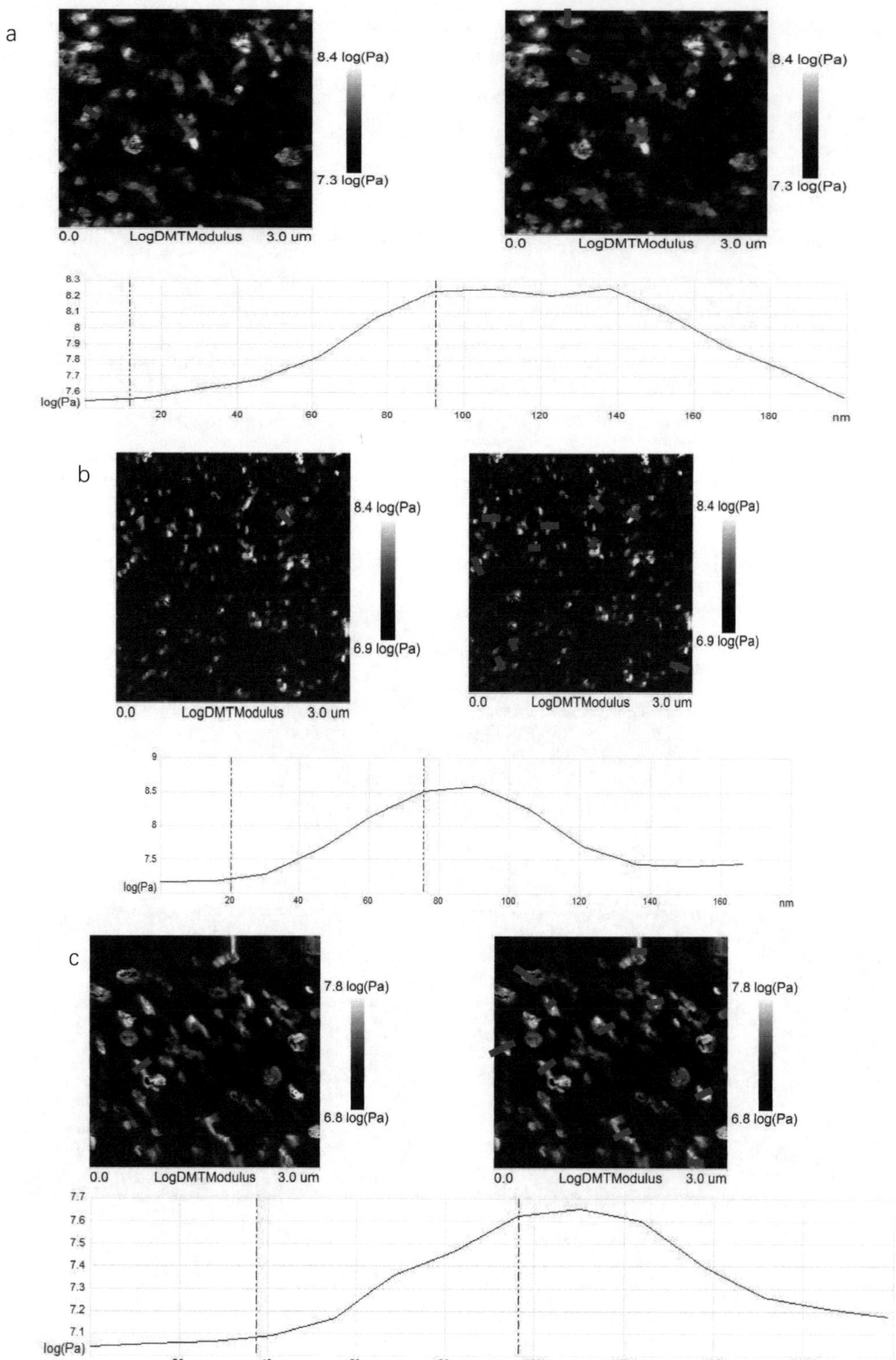

FIGURE 4. the AFM images of the height and the Modulus (a) 1-MWCNT/SSBR (b) 2-MWCNT/SSBR (c) 3-MWCNT/SSBR

In this section, we use peking-force AFM which can record one force curve as the feedback signal of image to regulating the tip force tip pressed on the sample. The image of height, Young modulus, adhensive energy can be obtained at the same time. FIGURE 4 shows the height image and the LogYonug image of the three types MWCNT/SSBR. The height image shows that CNT has uniform dispersion in the composites. Corresponding to the image of the LogYoung modulus, there are two distinct area, the high LogYoung modulus area and the low LogYoung modulus area. The structure of the high LogYoung modulus is probably assigned to the MWCNT and the low LogYoung modulus is probably assigned to the SSBR matrix. We found that, as the specific surface area of the carbon nanotube rises, the thickness between the interfaces is also steadily increasing. so we propose a hypothesis: the stronger the interaction is, the stronger the entanglement is, the more thick the thickness of the interface is.

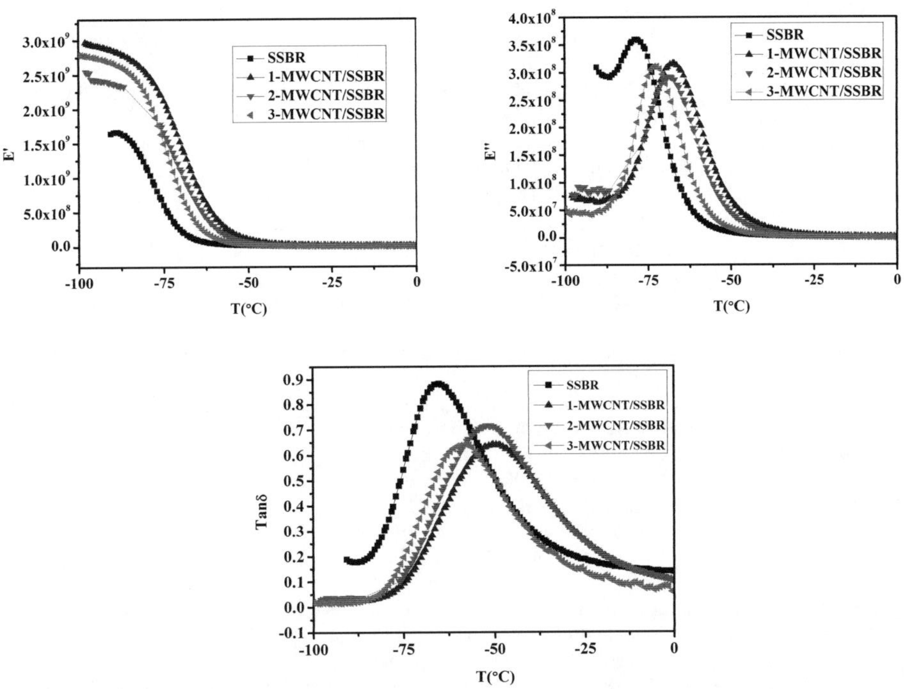

FIGURE 5. Temperature-dependence of storage modulus, loss modulus, and tanδ of the composites.

FIGURE 5 shows the emperature-dependence of storage modulus, loss modulus, and tanδ of the three composites. FIGURE 5(a) shows that the filled composites have higher storage modulus than the unfilled composites especially in the low temperature. It indicates the stored energy of the viscoelastic materials is in the process of the elastic deformation and represents the resilience. The MWCNT-1/SSBR has the best resilience owing to the largest SSA which can contributes to strengthen the interaction on the interface and then form the best network structure. FIGURE 5(b) shows the loss modulus of the composites. It indicates that the the three filled composites have tiny differences. However, there are some differences. This is mainly because that the MWCNT-1/SSBR has the strong interaction.

CONCLUSION

We have developed a new method to quantitatively identify the interphase of silica/rubber nanocomposites and carbon nanotubes (CNTs)/ rubber composites at nanoscale resolution by using PF-QNM mode of AFM. The double layer structure of interphase was directly identified in silica/rubber nanocomposites based on PF-QNM images and the corresponding force-distance curves, and was further evidenced by high resolution transmission electron microscopy. We further studied the effect of physical and chemical charateristic of fillers and rubber macromolecules on the interfacial thickness of rubber nanocomposites. This study provides a new method to quantitative characterize the interface between filler and elastomer matrix at nanoscale resolution.

REFERENCES

1. Kohls, D. and G. Beaucage, Rational design of reinforced rubber. Current opinion in solid state and materials science, 2002. 6(3): p. 183-194.
2. Liu, X.G., et al., Synergistic effect of nanofiller geometry and shear flow on the morphology evolution in SSBR/LPI/filler ternary system. Polymer, 2015. 72: p. 193-201.
3. Wang, L., et al., Study on the structure and properties of SSBR with large-volume functional groups at the end of chains. Polymer, 2010. 51(9): p. 2084-2090.
4. Liu, X., et al., Synergistic effect of nanofiller geometry and shear flow on the morphology evolution in SSBR/LPI/filler ternary system. Polymer, 2015. 72: p. 193-201.
5. Pan, X.-D., Impact of reinforcing filler on the dynamic moduli of elastomer compounds under shear deformation in relation to wet sliding friction. Rheologica acta, 2005. 44(4): p. 379-395.
6. Le Gal, A., X. Yang, and M. Klüppel, Evaluation of sliding friction and contact mechanics of elastomers based on dynamic-mechanical analysis. The Journal of chemical physics, 2005. 123(1): p. 014704.
7. Gerspacher, M., et al., High frequency viscoelasticity of carbon black filled compounds. Rubber chemistry and technology, 1996. 69(5): p. 786-800.
8. Wang, M.-J., New developments in carbon black dispersion. KGK. Kautschuk, Gummi, Kunststoffe, 2005. 58(12): p. 626-637.
9. Uhrlandt, S. and A. Blume, Silica in green tyres-Processes, products, properties. Kautschuk Gummi Kunststoffe, 2001. 54(10): p. 520-520.
10. Guy, L., et al., New insights in the dynamic properties of precipitated silica filled rubber using a new high surface silica. KGK. Kautschuk, Gummi, Kunststoffe, 2009. 62(7-8): p. 383-391.
11. Kuo, D.H. and M.Y. Su, The effects of hydrogen and temperature on the growth and microstructure of carbon nanotubes obtained by the Fe(CO)(5) gas-phase-catalytic chemical vapor deposition. Surface & Coatings Technology, 2007. 201(22-23): p. 9172-9178.
12. Hameed, A., et al., Thermal and Mechanical Properties of Carbon Nanotube/Epoxy Nanocomposites Reinforced With Pristine and Functionalized Multiwalled Carbon Nanotubes. Polymer Composites, 2015. 36(10): p. 1891-1898.
13. Pu, Z., K. Jia, and X. Liu, Covalent grafting of a-CNTs on copper phthalocyanine for the preparation of PEN nanocomposites with high dielectric constant and high thermal stability. Journal of Materials Science-Materials in Electronics, 2015. 26(11): p. 8922-8932.
14. Kim, M.-S., et al., Synergistic Effects of Carbon Nanotubes and Exfoliated Graphite Nanoplatelets for Electromagnetic Interference Shielding and Soundproofing. Journal of Applied Polymer Science, 2013. 130(6): p. 3947-3951.
15. Theilmann, P., et al., Superior electromagnetic interference shielding and dielectric properties of carbon nanotube composites through the use of high aspect ratio CNTs and three-roll milling. Organic Electronics, 2013. 14(6): p. 1531-1537.
16. Qu, M., et al., Nanoscale visualization and multiscale mechanical implications of bound rubber interphases in rubber–carbon black nanocomposites. Soft Matter, 2011. 7(3): p. 1066-1077.
17. Li, C., et al., Reinforcement and deformation behaviors of polyvinyl alcohol/graphene/montmorillonite clay composites. Composites Science and Technology, 2015. 118: p. 1-8.
18. Mensah, B., et al., Preparation and properties of acrylonitrile–butadiene rubber–graphene nanocomposites. Journal of Applied Polymer Science, 2015. 132(36).
19. Bokobza, L., Enhanced electrical and mechanical properties of multiwall carbon nanotube rubber composites. Polymers for Advanced Technologies, 2012. 23(12): p. 1543-1549.
20. Bokobza, L., Multiwall carbon nanotube elastomeric composites: A review. Polymer, 2007. 48(17): p. 4907-4920.
21. Zhong, B., et al., Reinforcement and reinforcing mechanism of styrene–butadiene rubber by antioxidant-modified silica. Composites Part A: Applied Science and Manufacturing, 2015. 78: p. 303-310.
22. Liu, J., et al., Surface modification of silica and its compounding with polydimethylsiloxane matrix: interaction of modified silica filler with PDMS. Iranian Polymer Journal, 2012. 21(9): p. 583-589.
23. Liu, J., et al., Nanoparticle dispersion and aggregation in polymer nanocomposites: insights from molecular dynamics simulation. Langmuir, 2011. 27(12): p. 7926-7933.

24. Johnson, D. and N. Hilal, Characterisation and quantification of membrane surface properties using atomic force microscopy: A comprehensive review. Desalination, 2015. 356: p. 149-164.
25. Schön, P., et al., Quantitative mapping of elastic moduli at the nanoscale in phase separated polyurethanes by AFM. European Polymer Journal, 2011. 47(4): p. 692-698.
26. Matos, C.F., F. Galembeck, and A.J. Zarbin, Multifunctional materials based on iron/iron oxide-filled carbon nanotubes/natural rubber composites. Carbon, 2012. 50(12): p. 4685-4695.
27. Wang, D., et al., Visualization of nanomechanical mapping on polymer nanocomposites by AFM force measurement. Polymer, 2010. 51(12): p. 2455-2459.

Modified glass powder as filler in rubber mixtures

Hans-Joachim Radusch[a,b*], Melanie Keller[a], André Wutzler[b], Uwe Ferner[c], Hans-Jürgen Voß[c]

[a]*University Halle-Wittenberg, Polymer Technology, 06099 Halle, Germany*
[b]*Polymer Service GmbH Merseburg, 06217 Merseburg, Germany*
[c]*TROVOtech GmbH, 06766 Bitterfeld-Wolfen, Germany*

hans-joachim.radusch@iw.uni-halle.de

Abstract. Rubber goods are characterized by a huge variety in a lot of applications that becomes possible both by the variety in the recipe of the vulcanization system and by the filler inserted into the rubber mixture. Mainly carbon black, silica, talk or nanofillers like carbon nanotubes, graphene or nanoclay have been used successfully in technical rubber goods and tires. A novel approach of rubber filling was developed as foamed borosilicate glass particles (TP) were inserted into the rubber mixture. Foamed borosilicate glass was generated by extrusion of glass melt using a specific foaming agent and subsequent cooling and milling of the glass foam. Thus, glass particles with a size of 2-18 μm were produced and applied. Additionally, the procedure allowed a modification of the surface of glass particles. Unmodified as well as zinc (ZnTP) and N-cyclohexyl-2-benzothiazole sulfenamide (CBS) modified glass powders were used in SSBR, BR and SSBR/BR blends. The investigation was targeted to the characterization of the influence of glass powder on vulcanization behavior and final properties of the vulcanizates. The comparative discussion of the influence of unmodified and modified borosilicate glass particles on vulcanization behavior and final elastomer properties shows that borosilicate glass particles are generally suitable as carrier of functionalizing groups for elastomer modification. Here, the substitution of environmental unfriendly zinc oxide (ZnO) as vulcanization accelerator in rubber mixtures by Zn modified glass powder became specific interest. Insertion of Zn modified glass powder resulted in the reduction of incubation time and cross-linking period during cross-linking process as well as reduction of loss factor at 60°C. Furthermore, the content of Zn modified glass powder had a significant influence on the properties of the vulcanized samples investigated, expressed by the increased ultimate mechanical properties tensile strength and elongation at break.
Keywords: **rubber, filler, glass powder, vulcanization, mechanical properties**

POROUS GLASS POWDER AS FILLER IN RUBBER MIXTURES

The application properties of rubber can be influenced in a broad manner by blending, variation of crosslinking system, insertion of several processing aids, plasticizers and stabilizers as well as especially by the addition of fillers. Fillers influence both the processing and vulcanization behavior and the final properties of the vulcanized rubber too. In practice of elastomer technology, concerning the fillers, carbon black has still the highest industrial importance, but the market part of silica increases mainly due to the continuous development of the silane technology. It is well known that carbon black and also silica as so called active fillers does not act as reinforcing agents only, but they influence also the viscosity of the mixture and the crosslinking behavior significantly. Consequently, the processing behavior and the final properties are impressed definitively. Generally, the modification of elastomers by means of carbon black or silica bases on the filler-filler as well as filler-polymer interactions, but also the interactions between filler and crosslinking agents as well as other processing or vulcanizing aids like plasticizers or stabilizers have to be taken into consideration.

A new approach in the field of elastomers is the application of porous glass particles as filler, whereat the knowledge of the influence of porous glass powder on the processing and vulcanization behavior as well as on the resulting application properties of rubber mixtures is rather small currently. Therefore, the present research deals with the investigation of the influence of type, chemical composition and morphological performance of the glass powder on the reaction behavior of the crosslinking agents, the processing behavior and final application properties.

The fabrication of the porous glass powder particles was realized via extrusion of glass foam with subsequent cooling and milling of the foam followed by particle classification [1]. The glass particles have an average particle size of some micrometers (Fig. 2a, b) and an irregular shape and surface (Fig. 2c). The procedure allowed a modification of the surface of glass particles. Unmodified as well as zinc and N-cyclohexyl-2-benzothiazole sulfen-

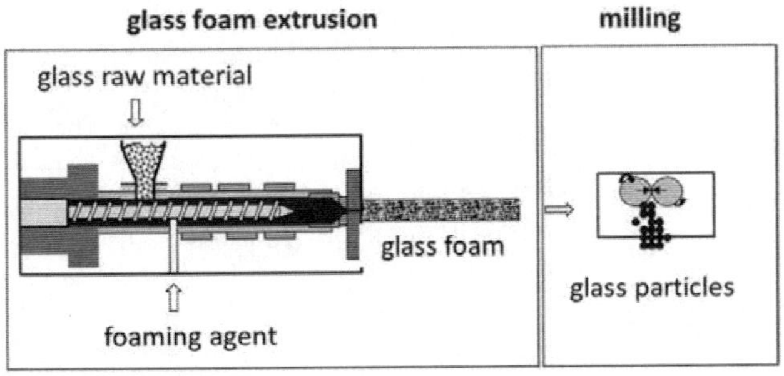

FIGURE 1. Scheme of the production of porous glass-foam particles

amide (CBS) modified glass particles were produced. Volume frequency distribution and cumulative particle volume of the borosilicate glass particles (TP and ZnTP) determined by Laser scattering are shown in figure 2.The figure displays the bi- or multimodal particle distribution of TP and ZnTP, respectively. It is assumed that the local maxima at 45 and 140 µm at ZnTP are caused by agglomeration of the modified particles.

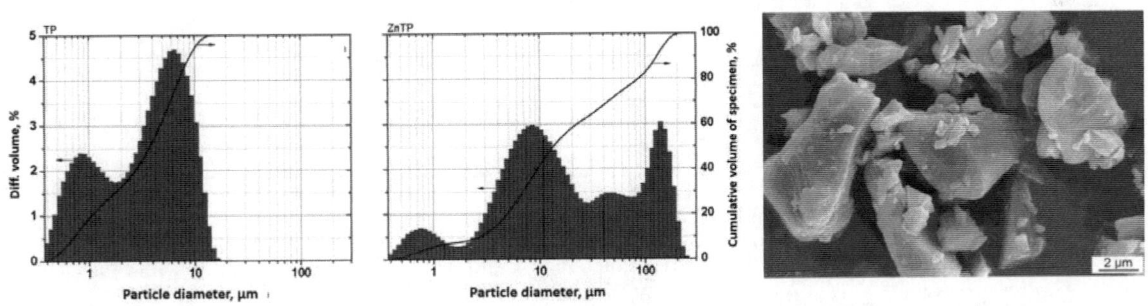

a) b) c)

FIGURE 2. Particle distribution of glass foam particles TP (a) und ZnTP (b) as well as SEM micrograph of ZnTP (c)

Figure 2c shows clearly the polyhedral shape of the particles. The application of zinc-modified borosilicate glass (ZnTP) for the generation of the specific glass filler has been the specific goal of the research. Using zinc-modified borosilicate glass opens up the opportunity for a target-oriented manipulation of the vulcanization process. The question has been of special interest whether zinc oxide, which is commonly used as vulcanizing activator, can be substituted by the ZnTP. Zinc oxide was identified to be toxic for organisms in water [2]. Furthermore, the investigation of suitability of glass-powder particles as carrier for several functionalities – especially for vulcanization accelerators – has been of serious interest.

SOLUTION STYRENE-BUTADIENE RUBBER- AND POLYBUTADIENE RUBBER-GLASS POWDER COMPOUNDS

Vulcanization behavior of SSBR and BR with ZnTP

Both the unmodified as well as the zinc and N-cyclohexyl-2-benzothiazole sulfenamide modified glass powders were used in SSBR, BR and SSBR/BR blends. The recipes of the SSBR and BR single mixtures are displayed in Table 1. The rubber mixtures were produced by means of the Brabender Plasticorder PL 2000 with temperature and time regime adapted to the composition of the rubber mixtures [3].

The influence of ZnTP on the vulcanization behavior of the used rubbers was characterized by means of the vulcameter Scarabaeus SIS-V50 (TA Instruments). The measurements were realized at a frequency of 1.67 Hz in the temperature range of 150-190°C. Incubation time t_i, conversion time t_{90}, torque moment ΔM and activation energy E_A were determined.

TABLE 1. Composition of SSBR and BR based compounds (single rubber mixtures) with ZnTP

Component [phr]	incl. activators (standard) [phr]	without activators	ZnTP [phr]
SSBR or BR	100		
S	1.4		
CBS	1.5		
StA	1	0	0
ZnO	2.5	0	0
ZnTP	0	0	2.5 / 5 / 10

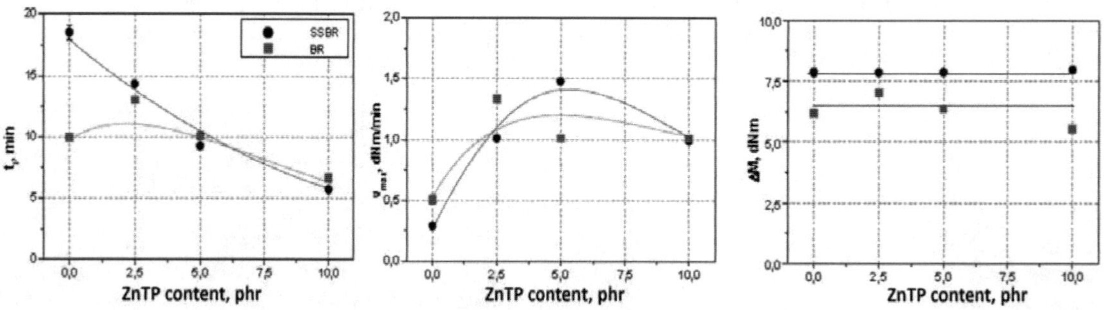

a) b) c)

FIGURE 3: Incubation time t_i, maximum reaction rate v_{max} and torque moment ΔM in dependence on ZnTP content for SSBS and BR

In case of SSBR the incubation time decrease with the content of ZnTP, while in case of BR at low ZnTP concentration the incubation time increases but decreases too at higher ZnTP content. The reaction rate has a maximum at 5 phr ZnTP. It is assumed that ZnTP promotes the scission reaction of the vulcanizing agent CBS, hence the incubation time decreases and the reaction rate increases. The torque moment seems to be unchanged, i.e. the glass particles do not have a negative influence on the viscosity of the system.

Vulcanization behavior and properties of SSBR/BR blends with ZnTP

SSBR/BR blends were produced with a component ratio of 60/40. The blend composition corresponds to table 2. The "adapted" activator content means a ratio of ZnO and Stearic acid (StA) that corresponds to the content of Zn-ions of the ZnTP content in the blend.

TABLE 2. Composition of SSBR/BR blends with ZnTP

Component [phr]	incl. activators (standard) [phr]	without activators	"adapted" activator	ZnTP [phr]
SSBR		60		
BR		40		
S		1.4		
CBS		1.5		
StA	1	0	2.17	0
ZnO	2.5	0	0.29	0
ZnTP	0	0	0	2.5 / 5 / 7 / 10

The blends produced by means of Brabender Plasticorder PL 2000 are heterogeneous with island-matrix morphology as displayed in figure 4a. Although the heterogeneous character appears in the AFM image, the mechanical loss factor tan δ from dynamic-mechanical analysis (Fig. 4b) shows only one broad peak. It is assumed that co-vulcanization between the polymer components happened, which results in such a single broad peak.

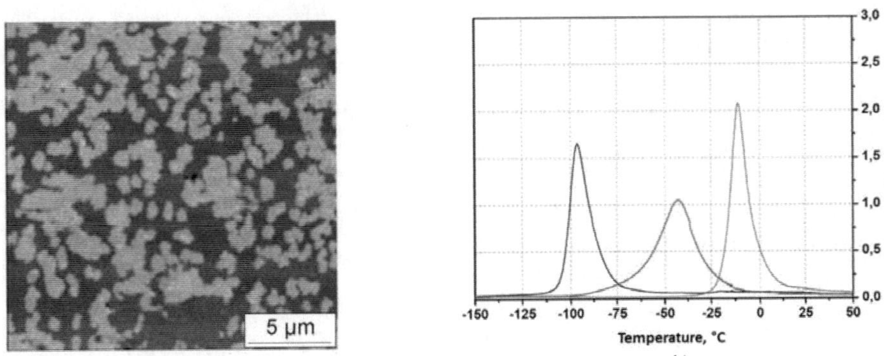

a) b)

FIGURE 4. Phase morphology of unfilled SSBR/BR blend. (a) AFM, phase image. (b) tan δ in dependence on temperature.

Crosslinking isothermes and reaction rates of the SSBR/BR blends with different composition were determined again at 160 °C (Fig. 5). The curves demonstrate that the activators strongly influence the vulcanizing behavior of the SSBR/BR blends. Both the development of the torque moment as a measure for the crosslinking degree and the resulting reaction rates show high differences caused by the respective activator system.

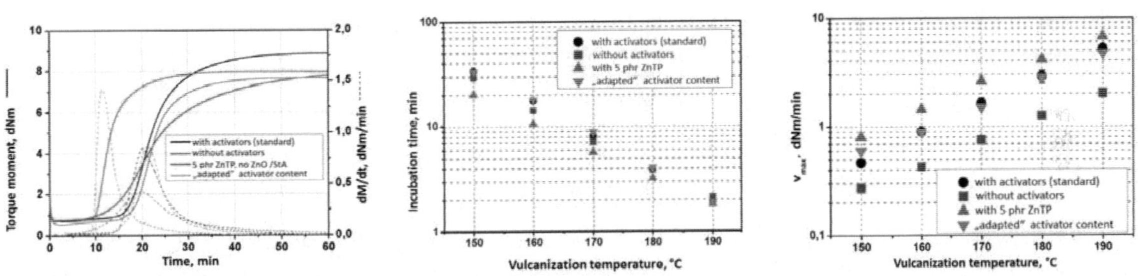

a) b) c)

FIGURE 5. Crosslinking isothermes at 160°C (a), incubation time (b) and reaction rate (c) of SSBR/BR blends 60/40 with and without activators as well as with 5 phr ZnTP.

Figure 5 demonstrates that the substitution of the activators by ZnTP, i.e. absence of the activators ZnO and StA, results in a significant decrease of the incubation time (Fig 5b) as well as in the increase of the reaction rate (Fig 5c) in comparison to the standard system. That corresponds to the behavior of the single rubber systems (Fig. 3) and confirms the assumption that ZnTP promotes the scission reaction of the vulcanizing agent CBS. The same crosslinking mechanism is running in the co-vulcanizing SSBR/BR blend too. Furthermore, figure 5 shows that the "adapted" content of activator does not result in a significant difference to the incubation times and reaction rates of

Table 3. Relative activation energy based on the incubation time

Activator system	Relative activation energy E_A^{ti}, %
with activators (standard)	100 *(=118 kJ/mol)*
without activators	91
with "adapted" activator content	97
5 phr ZnTP	83

the standard system at all temperatures investigated. The determination of the activation energies (Table 3) showed that the insertion of activators results in an increase of this parameter in comparison to the activator-free or ZnTP-containing blends.

The stress-strain behavior of the rubber compounds was characterized by means of a Zwick tensile tester Z005 at room temperature and the testing rate 200 mm/min using specimen typ S2. The stress-strain behavior of SSBR/BR blends with different activators and in dependence on ZnTP content is represented in figure 6. The stress-strain curve of the 5 phr ZnTP activated SSBR/BR blend runs below the curve of the standard mixture, but reaches higher ultimate stress and strain values (Fig.6a). While the stress values σ_{50} and σ_{100} are only slightly reduced, the ultimate stress gains with increasing ZnTP content (Fig. 6b). Simultaneously, the elongation at break increases in the same manner (Fig. 6c).

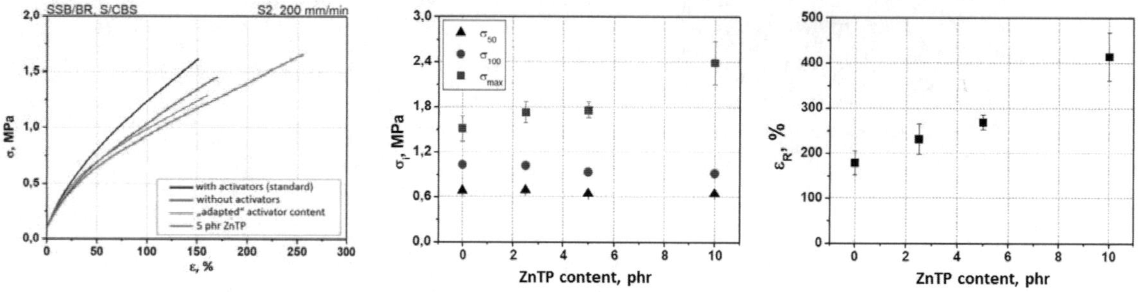

a) b) c)

FIGURE 6. Stress-strain behavior of SSBR/BR blends with and without activators, with „adapted" activator content and with 5 phr ZnTP (a) as well as stress values at different elongation (b) and elongation at break (c) in dependence on ZnTP content.

The results show that the ZnTP content has a significant influence on the mechanical properties of the vulcanized SSBR/BR blends. The reason for this specific mechanical behavior is seen in the reduced crosslinking degree with increasing ZnTP content, that is supported by the Mooney-Rivlin parameter C1, which was calculated on the basis of the stress-strain curves [3].

CONCLUSIONS

Functionalized glass powder ZnTP could be identified as an interesting filler for synthetic rubber and rubber blends. The investigation showed that both the vulcanization behavior and the mechanical properties could be advantageously influenced under the condition of an optimization of the entire vulcanizing system. For SSBR/BR blends could be demonstrated that the substitution or partial substitution of ZnO and StA by ZnTP is possible under consideration of the reduction of mechanical stress values σ_{50} and σ_{100} but an increase of the ultimate stress and elongation at break that is caused by a reduced crosslinking degree of the rubber. Finally, the application of zinc modified borosilicate glass particles as filler in rubber is an opportunity to influence the vulcanization process and application properties of rubber mixtures in targeted manner.

REFERENCES

1. U. Ferner, H. U. Lugner, Patent WO 2012113525 A1 (2011), Doped porous, amorphous glass particles made of continuously produced glass foam
2. Commission Directive 2004/73/EC of 29 April 2004
3. M. Keller, Influence of functionalized borosilicate glass particles on vulcanization behavior and properties of SSBR/BR composites, Ph.D. Thesis, Martin Luther University Halle-Wittenberg, Halle 2015

Preparation of EPDM seals with high hardness and low compression set used for PEM fuel cells through multilayered co-extrusion system

Xia Lichao [1], Shen Liyuan [1], Wu Hong [1]*, Guo Shaoyun [1]

[1] Polymer Research Institute of Sichuan University - Sichuan - China

Seal gaskets are widely used in Proton Electrolyte Membrane (PEM) fuel cells to keep the reactant gases and liquid within their respective regions. High hardness and low compression set are both essential to seal performance of gaskets. However, the compression set usually increases with the increase of the hardness. Therefore, in this paper, EPDM seal materials with soft and hard alternating multilayered structure are fabricated through the multilayered co-extrusion system. The structure and morphology of multilayered and blended EPDM vulcanizates were observed by Scanning electron microscopy (SEM). The results of hardness, compression set and compression stress relaxation test show that multilayered vulcanizates have lower compression set but higher hardness than blended one. The compression set of multilayered vulcanizates decreases with the increase of the layer number, indicating better sealing capacity and longer working life. The results of aging test indicate that the mechanical properties of EPDM seal materials decrease with time, but multilayered vulcanizates still have lower compression set than blended one after aging.

Multi-Component Injection Molding of Polycarbonate-LSR-Composites with Integrated Surface Activation Created by UVC Irradiation

Hartung Michael [1]*, Giesen Ralf-Urs [1], Heim Hans-Peter [1]

[1] University of Kassel - Kassel - Germany

Liquid silicone rubber with modified adhesive properties enables multi-component thermoplast-silicone composites to be injection molded without requiring mechanical linkage. In most cases, the chemical structure of the in situ bonding agent is only capable of connecting to chemically compatible thermoplasts, such as PBT and PA. When dealing with PC, adhesion with these LSR types is only possible if previous surface modifications have been performed, i.e., plasma-, corona- or flame-treatments. Unfortunately, these intricate processes also come at a high cost, and the complexity of integration into a multi-component injection molding process is an evident disadvantage. Another possibility to achieve the adhesion of LSR on polycarbonate is by means of irradiating the thermoplastic surface with UVC light. The effect is very powerful as long as the irradiation dose is high enough. Using a common mercury vapor arc lamp and optimal conditions, irradiation periods under five seconds generate a strong bond with cohesive failure within the silicone rubber. In comparison to other studies in literature, which investigated the optimization of liquid coatings, the UVC irradiation doses required to activate the surface of PC to achieve bonding to LSR are much lower. In this study, the effects, correlations, and limits of the UVC light treatment applied to PC-LSR composites were evaluated using experimental designs. The VDI guideline 2019 was used to determine the adhesive strengths of the composites. It became clear that surface activation works for different polycarbonate and LSR types. The most important aspect is the amount of UVC irradiation needed on the thermoplastic surface. Surprisingly, the production conditions in injection molding play a minor role. The surface pre-treatment of polycarbonate with UVC light is, unlike other treatment processes, stable long-term and resistant against various secondary treatment processes, such as sterilization, heat, or storage in media.

High Performance ZnO-graphene/Chloroprene Rubber Composites

Shipeng Wen, Zongchao Xu, Wencai Wang, Liqun Zhang, Li Liu*

Beijing Engineering Research Centre of Advanced Elastomers, Beijing University of Chemical Technology, Beijing 100029, China.

Abstract. In this paper, ZnO-graphene oxide (ZG) hybrids were synthesized using a simple chemical and deposition method and subsequently ZG/chloroprene rubber (CR) composites were obtained by direct mechanical blending. The structure and morphology of the ZG hybrids were characterized by FTIR, Raman, XRD and SEM spectroscopy. The results showed that ZnO particles were uniformly dispersed on the surface of GO sheets. In addition, with the incorporation of ZG hybrids into CR matrix, the mechanical properties of CR nanocomposites were significantly higher than CR gum. These results demonstrated that ZnO particles on the surface of GO sheets not only inhibited the aggregation of GO layers but also acted as an efficient cure agent in the vulcanization process. ZG hybrid as an ideal functional hybrid filler opens up a new opportunity for the fabrication of high performance CR nanocomposites.

Keywords: ZnO-graphene hybrids, Chloroprene rubber, Mechanical properties.
PACS: 81.05.Lg

INTRODUCTION

In recent years, graphene and its derivatives have attracted a great deal of attention because of their unique structure, physical and chemical properties[1]. In particular, the application of graphene in the rubber composites has attracted great interest[2]. In the previous study, graphene and its derivatives have been proven to be useful fillers which can significantly improve the vulcanization performance[3], mechanical properties[4], dynamic properties[5], fatigue performance[6] of rubber composites. However, the dispersion of graphene in the matrix and the interfacial interactions between graphene and the matrix are still the two major challenges due to the van der Waals force in the interlayers of graphene. The simple but effective ways is expected to be explore to prepare the graphene/rubber composites with a good graphene dispersion.

Zinc oxide (ZnO) is an inorganic metal oxide, which is widely used in rubber composites. It can be used as rubber activator and vulcanization agent. However, ZnO compounds have been classified as toxic chemicals. The European Union has claimed that its application in rubber technology should be controlled and reduced[7]. It is necessary to control the content of ZnO in the manufacturing process of rubber products as low as possible in the future. Therefore, improving the efficiency of ZnO particle in rubber matrix is an effective way.

In this study, ZnO particles was decorated on the surface of the graphene sheets, and added the ZG hybrids into CR matrix to obtain the ZG/CR composites. As a result, the ZG hybrids not only can be used as a reinforcing agent but also a vulcanizing agent at the same time. In the rubber matrix, the ZnO particles prevented the agglomeration of the graphene sheets and enhanced the interfacial interface between rubber and graphene.

EXPERIMENTAL

Materials

Graphene oxide was purchased from Jinxi nanomaterials Technology Co., Ltd, Chloroprene rubber (CR210) was produced by Lanxess Chemical Industry Co., Ltd. Zinc acetate ($Zn(CH_3COO)_2 \cdot 2H_2O$), sodium hydroxide (NaOH) were all provided by Beijing Chemical Factory. All other rubber ingredients were commercially available.

Preparation of ZnO-GO hybrids

In the first step, 0.0614 mol $Zn(CH_3COO)_2 \cdot 2H_2O$ was dissolve in 100 ml deionized water and then added into the GO suspension. Secondly, the pH of the solution was adjusted to 10 by adding sodium hydroxide suspension and then the solution was stirred for 2h at 80°C. Finally, the solution was filtered for three times with ethanol and water, respectively. The filter products were dried in vacuum at 50 °C. The ZnO nanoparticles was also synthesized in a similar way as a reference sample.

Preparation of ZnO-GO/CR composites

ZnO-GO/CR composites was fabricated in the open two-row mixer with blending CR and ZnO-GO powders with different GO contents. The formula is as follows: CR 100, SA 1, RD 2, 4010 2, DM 1, MgO 4, ZnO-GO variable.

RESULTS AND DISCUSSION

Structural characterization of ZnO-graphene hybrids

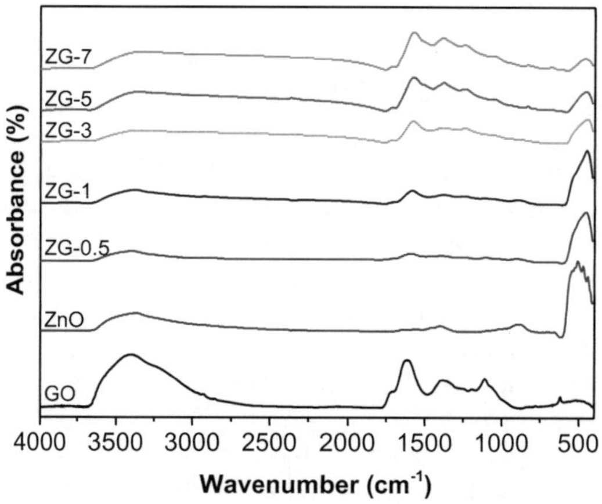

FIGURE 1. FTIR spectra of GO, ZnO and ZnO-GO hybrids

FTIR spectra was used to confirm the change of the chemical structure of ZG hybrids. The peaks at 1109 cm^{-1}, 1386 cm^{-1}, 1612 cm^{-1}, 1715 cm^{-1}, 3397 cm^{-1} of GO were assigned to C-O-C stretching vibrations, O-H deformation vibrations, C=C stretching mode of sp2 carbon backbones, C=O stretching vibrations and O-H stretching vibrations, respectively. As for ZnO, the strong peak at 497cm^{-1} was ascribed to Zn-O stretching vibration[8, 9]. However, for the ZG hybrids, it was founded an obvious characteristic peak around 1600cm^{-1} and 450cm^{-1} appeared, which were corresponding to the stretching vibrations of C=C in GO sheets and the Zn-O stretching vibration.

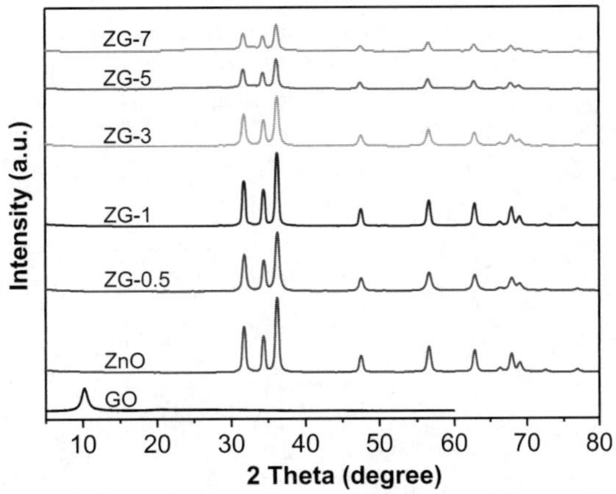

FIGURE 2. XRD patterns of GO, ZnO and ZnO-GO hybrids

XRD patterns of GO, ZnO and ZnO-GO hybrids were shown in Fig. 2. Obvious diffraction peak at $2\theta=10.2^{o}$ was associated with the 001 reflection of GO sheets. The diffraction peaks in the XRD pattern of ZnO nanoparticles were consistent with the hexagonal phase wurtzite ZnO[10]. The diffraction peaks of ZG hybrids were similar to those of hexagonal ZnO. However, the (001) diffraction peak at 10.2 ° of GO disappeared, which suggested that the regular layered structure of GO had been destroyed. The exfoliated GO layers were formed due to the growth of ZnO nanocrystals between the GO layers.

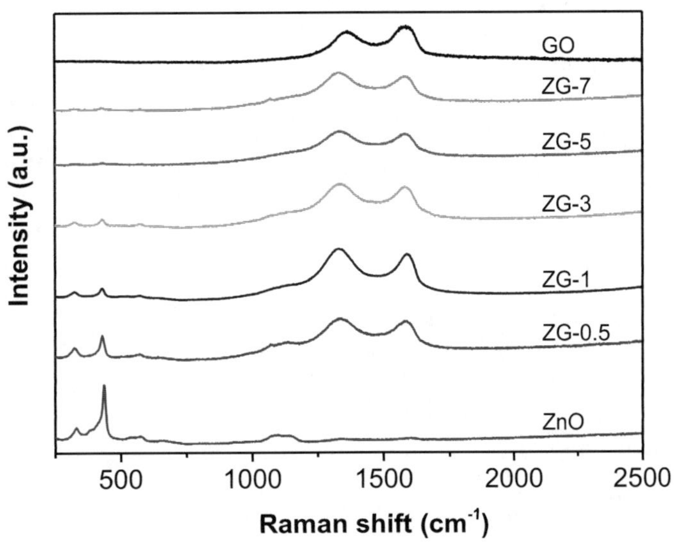

FIGURE 3. Raman spectra of GO, ZnO and ZnO-GO hybrids

In curve for pure ZnO, the peaks at 329cm^{-1} and 1143cm^{-1} were the characteristic peaks for 2E1 scattering modes. The peak at 437cm^{-1} was attributed to the E2 scattering mode. In curves for ZG hybrids, the intensity of the two peaks decreased gradually as the contents of GO increased. In curves for ZG hybrids, the G band at around 1580cm^{-1} stemed from the bond stretching of sp2 carbon atoms in GO lattice, and the D band at 1353cm^{-1} originated from the disorder of GO sheets. The intensity ratio of the D band and G band (I_D/I_G) is usually used to reflect the defects of graphene and its derivatives. Comparison with GO, the intensity of I_D/I_G increased in ZG hybrids, which was attributed to the ZnO absorbed on the surface of GO sheets decreased the average size of sp2 domains during the chemical deposition stage. It was noted that the peak of E2 scattering mode in ZG hybrids shifted to 429 cm^{-1} compared with the pure ZnO, suggesting that there was strong interaction between the ZnO particle and GO sheets[11].

FIGURE 4. SEM picture of GO (a), ZnO (b), ZG-1 (c), ZG-5 (d)

Figure 4 shows the SEM images of GO, ZnO and ZG hybrids. In Fig. 4. (a), it can be seen that the GO sheets aggregated due to the strong van der waals forces between interlayers. In Fig. 4. (c) and (d) , GO layers were uniformly decorated by ZnO particles, which can prevent the stacking of GO sheets.

Vulcanization performance of ZnO-GO/CR nanocomposites

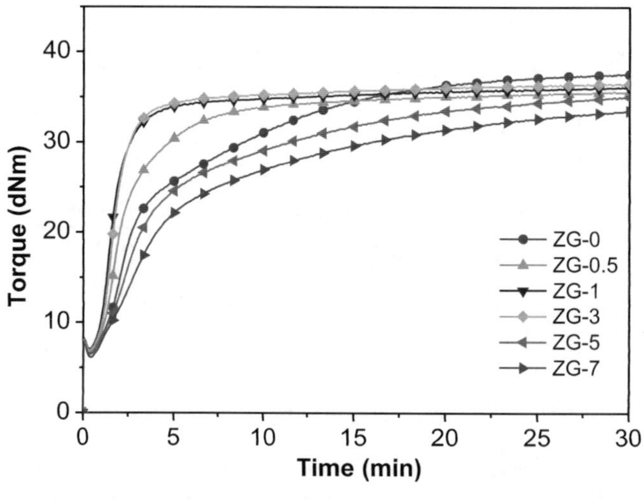

FIGURE 5. Vulcanization curves of CR composites

The influence of the ZG hybrids on the vulcanization properties was studied. When the GO content was below 3 phr, the composites with ZG hybrids showed high efficiency with the increase of GO content. However, the curing efficiency of composites decreased when the GO content was above 3 phr. The decreased efficiency was attributed to the facts that ZnO particles easily wrapped by the GO sheets at the high GO content.

Mechanical properties of ZnO-GO/CR nanocomposites

TABLE 1. Mechanical properties of CR composites

Sample	ZG-0	ZG-0.5	ZG-1	ZG-3	ZG-5	ZG-7
Hardness (Shore A)	40	40	39	39	38	38
Modulus at 100% strain (Mpa)	1.0	0.8	0.9	0.8	0.8	0.8
Modulus at 300% strain (Mpa)	1.5	1.6	2.1	2.1	1.5	1.6
Tensile strength (Mpa)	8.1	9.3	9.0	10.7	15.0	14.0
Elongation at break (%)	783	622	642	703	857	911

The static mechanical properties of the ZG/CR composites are shown in Table 1. Compared with the CR composites without GO, the tensile strength was significantly improved by introducing the ZG hybrids.

CONCLUSION

The ZG hybrids were synthesized via a chemical and deposition route. The results showed that the ZnO particles was decorated onto the surface of GO sheets. The ZnO particles significantly prevent the stacking of GO sheets. When compared with pure ZnO, the ZG hybrids exhibits efficient vulcanization efficiency. The tensile strength CR composites filled with ZG-5 improved 85% compared with the CR without GO.

REFERENCES

1. G. AK and N. KS, *Nature materials* **6**, 183 (2007).
2. H. Kim, A. A. Abdala and C. W. Macosko, *Macromolecules* **43**, 6515-6530 (2010).
3. J. Wu, W. Xing, G. Huang, H. Li, M. Tang, S. Wu and Y. Liu, *Polymer* **54**, 3314-3323 (2013).
4. Y. Mao, S. Wen, Y. Chen, F. Zhang, P. Panine, T. W. Chan, L. Zhang, Y. Liang and L. Liu, *Scientific reports* **3** (2013).
5. H. Kang, Y. Tang, L. Yao, F. Yang, Q. Fang and D. Hui, *Composites Part B: Engineering* **112**, 1-7 (2017).
6. S. Zhang, L. Zheng, D. Liu, Z. Xu, L. Zhang, L. Liu and S. Wen, *RSC Advances* **7**, 40813-40818 (2017).
7. B. Panampilly and S. Thomas, *Polymer Engineering & Science* **53**, 1337-1346 (2013).
8. Y. Lin, Y. Chen, Z. Zeng, J. Zhu, Y. Wei, F. Li and L. Liu, *Composites Part A Applied Science & Manufacturing* **70**, 35-44 (2015).
9. J. Wu, X. Shen, J. Lei, K. Wang and K. Chen, *Applied Surface Science* **256**, 2826-2830 (2010).
10. B. Li and H. Cao, *Journal of Materials Chemistry* **21**, 3346-3349 (2011).
11. R. Beura and P. Thangadurai, *Journal of Physics & Chemistry of Solids* **102**, 168-177 (2017).

Synthesis of Functionalized Carbon Quantum Dots and Rh/CDs Nanoparticles Application in Rubber

Zhongyu Yi, Jie Zhang, Qingqing Zhang, Liqun Zhang, Dongmei Yue*

State Key Laboratory of Organic-Inorganic Composites, Beijing University of Chemical Technology, Beijing 100029, China.

*e-mail: yuedm@mail.buct.edu.cn

Abstract. Carbon quantum dots（CDs）are recently attracted wide attention because of their strong fluorescence. Most of the current researches focus on the luminescent properties of carbon dots and their applications in biological area[1-4].In this paper, N-dope carbon quantum dots were synthesized by microwave assisted pyrolysis, and subsequent surface vinylation. Functionalized carbon quantum dots were prepared by free radical polymerization, finding its multistimuli-responsive property. In addition, rhodium nanoparticles has been developed by a new convenient method, in which carbon dots (CDs) were used both as stabilizing and reducing agent by heating Rh^{3+} with CDs for at 120 °C. The hydrogenation activities of Rh/CDs composited nanoparticles demonstrated that the nanoparticles had highly catalytic activity in the hydrogenation reaction of NBR. Also, the presence of CDs could improve the fluorescent properties of hydrogenated NBR.

Keywords: Carbon quantum dots, Catalyst, Hydrogenated NBR

1. INTRODUCTION

Graphene and carbon nano tubes as carbon materials have been widely used in nature. Recently, new carbon materials, CDs have attracted great interests because of their unique functions, chemical sensing, drug delivery, photocatalyst, low toxicity and bioimaging [5-7]. This nature assigns CDs both the role of electron acceptors and electron donors since there are many electrons and holes on the surface of CDs. Thus CDs can serve either as oxidizing agent or reducing agent in some cases [8]. Additionally, with numerous hydroxyl, amido or carboxyl on their surface, CDs can coordinate with metal ions and stabilize metal nanoparticles [9].In this paper, functionalized carbon quantum dots were prepared by free radical polymerization, finding its multistimuli-responsive property. In addition, a new convenient method has been developed to synthesize rhodium nanoparticles, in which carbon dots (CDs) were used both as stabilizing and reducing agents, reducing Rh^{3+} to Rh^0. Rh/CDs nanoparticles had highly catalytic activity in the NBR hydrogenation. Also, the presence of CDs could improve the fluorescent properties of hydrogenated NBR.

2. EXPERIMENT

2.1 Material

Chitosan (Sigma Aldrich Co. Ltd), acetic acid (analytical grade, Beijing Chemical Company), hydrazine hydrate (reagent grade,Sigma Aldrich Co. Ltd), ascorbic acid (reagent grade, Sigma Aldrich Co. Ltd), $RhCl_3 \cdot 3H_2O$ (99.9%, Shanghai Jiuling Chemical Co. Ltd., China), HTBN (Zibo Qilong Chemical Industry), dimethylbenzene (analytical grade, Beijing Chemical Company) were used as received and without further purification. Dialysis membrane (1000 Da) was purchased from Beijing Biodee Biotechnology Co., Ltd (China) and boiled 20 min before using.

2.2 Rh/CDs Nanoparticles and NBR Hydrogenation

CDs were dissolved in deionized water and then the $RhCl_3$ aqueous solution was added to the CDs aqueous solution for 2 h. The resulting mixture was added into a polytetrafluoroethylene vessel and reacted at 120 ◦ C for 1 h. After reaction, the mixture was centrifuged and dialyzed with water for 24 h. Finally, the Rh/CDs nanoparticles were obtained under reduced pressure at 55°C for 12 h. Then, Rh/CDs nanoparticles were dispersed in dimethyl- benzene. The catalyst solution was added into NBR solution for 4 h.

2.3 Characterization

The fluorescence test was carried out on an F-7000 fluorescence Spectrophotometer. The UV–vis spectra were recorded on a Jasco J-810 spectropolarimeter. Transmission electron microscopy (TEM) images were obtained with an JEOL JEM-3010 instrument. A JEOL JEM-3010 instrument was employed to obtain the high revolution transmission electron microscopy (HRTEM) images. X-ray diffraction (XRD) analyses were carried out on a Shimadzu XRD-6000 diffractometer. The energy dispersive spectrometry (EDS) was performed with a GENESIS 307 system. 1 H nuclear magnetic resonance (NMR) spectrawere were recorded on a Bruker AV400 spectrometer.

3. RESULTS AND DISCUSSION

3.1 Preparation of Functionalized Carbon Quantum Dots

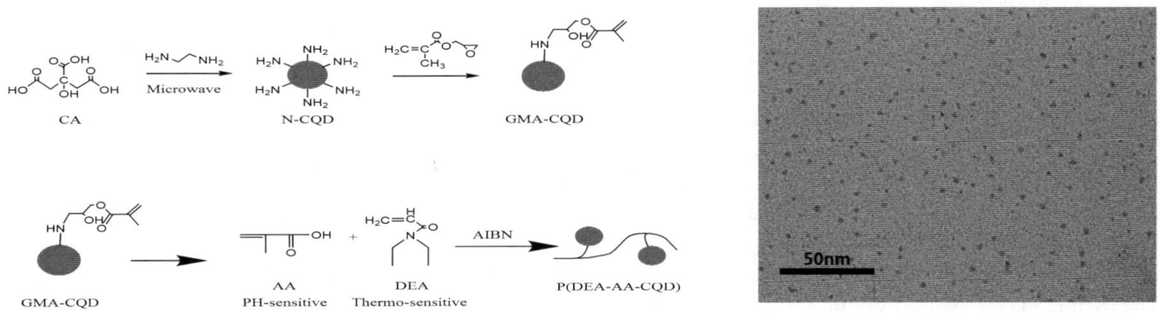

FIGURE 1 Synthetic Route of N-CQDs, GMA-CQDs, and FCQDs and TEM Images of N-doped CQDs

In the production of N-CQDs via microwave assisted pyrolysis .The diameter of the N-CQDs is about 5nm(Figure 1) with a 78% QY. The functionalized Carbon Quantum (FCQD) shows the different emission intensity in different temperature.The emission intensity decrease with the increase of temperature. The transmittance also decrease with the increase of temperature. The FCQD shows the different emission intensity in different PH. The emission intensity reach the highest point at PH=11. FCQD shows PH-sensitive property. FCQD's cloud point decrease with the increasing PH and With the increase of concentration of NaCl, the cloud point increase ether. The FCQD shows the themo-sensitive Property. The FCQD shows Multistimuli-Responsive property.

3.2 Rh/CDs Nanoparticles and Catalytic Hydrogenation

In this part, the CDs were synthesized by chitosan used as carbon source under microwave irradiation with a yield of 35% and the average size 9.6 nm. FTIR spectra indicated that the surface of CDs had lots of electron-rich groups that could coordinate with metal ion and stabilize metal nanoparticles. The fluorescent property of CDs display a bright blue fluorescence under ultraviolet light with a quantum efficiency of about 6.65%.

Herein, Rh/CDs nanoparticles named as "Rh-CD-c" was prepared by a new method, in which CDs were used both as stabilizing and reducing agent. By comparing, two kinds of Rh/CDs nanoparticles named as "Rh-CD-a" and "Rh-CD-b" were obtained by hydrazine hydrate and ascorbic as reducing agent under the same condition. The XRD patterns (**FIGURE 2.**) of the Rh/CDs nanoparticles of Rh-CD-b and Rh-CD-c exhibits diffraction peaks in the positions of 41.1°, 47.8°, 69.8°and 84.4°, which representing the (111), (200), (220), and (311) planes of Rh^0 nanoparticles. By comparing XRD curves, it can be seen that Rh-CD-c and Rh-CD-b exhibits sharper peaks than Rh-CD-a, indicating better crystallization properties of Rh nanoparticles in those two nanoparticles. The element compositions of the series of Rh/CDs nanoparticles were analysed by EDS, indicating that Rh^0 were formed using CDs both as stabilizing and reducing agent. Based on the above results, Rh-CD-c showed optimal performances with higher Rh content, excellent crystallization properties, smaller Rh sizes and adaptive fluorescent properties, and is possilbe to have the best catalytic activities in the following hydrogation reaction.

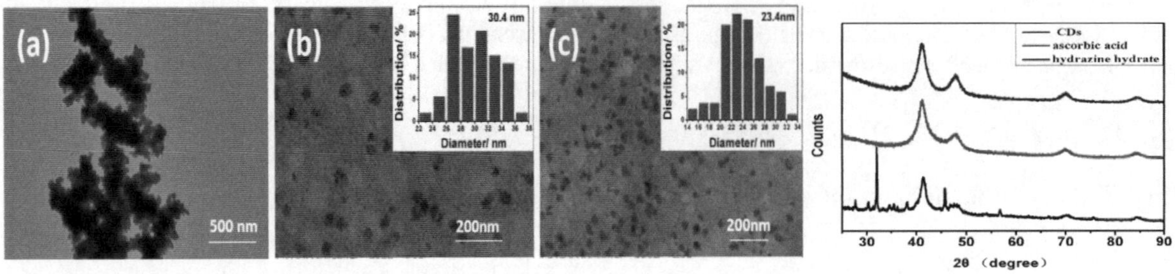

FIGURE 2. (a) TEM image of Rh-CD-a; (b) TEM images of Rh-CD-b (particle size distribution inset); (c) TEM images of Rh-CD-c (particle size distribution inset); (d) XRD spectra of Rh-CD-a, Rh-CD-b and Rh-CD-c

To further evaluate the hydrogenation activity of Rh/CDs nanoparticles, NBR was chosen as the reaction substrate. Rh/CDs nanoparticle (Rh-CD-a, Rh-CD-b, Rh-CD-c) as catalysts were applied in the catalytic hydrogenation of NBR. The NBR macromolecule and H_2 molecules were absorbed on the catalytic active centers and then reacted with each other. H_2 pressure, agitation speed, catalyst content and temperature are important to the progress of hydrogenation. H_2 pressure, agitation speed, catalyst content and temperature were examined separately, finding that hydrogenation degree of NBR was able to achieve above 80% at the H_2 pressure (4 MPa), agitation speed (700 r/min), mass ratio of [Cat.]/[NBR] (1.5wt%) and temperature (80 ℃) by Rh-CD-c as catalyst.

3.3 Fluorescent Properties of Hydrogenated HTBN

HTBN is fluorescent rubbers and show good fluorescent properties before hydrogenation. The curve (a) is the emission spectrum of HTBN before hydrogenation and the peak at 362 nm can be attributed to the conjugated structures in backbones (as shown in **FIGURE 3**). The fluorescent effects is attributed to the conjugated structures in backbones, which are advantageous to the electron excitation. Normally, the conjugated structures are gradually destroyed with the increase of HD, thus the fluorescence intensities gradually decrease until it completely disappear (as shown in curve (e)). However, CDs improved the fluorescent properties of hydrogenation products (as shown in curve (c)) in **FIGURE 3**.

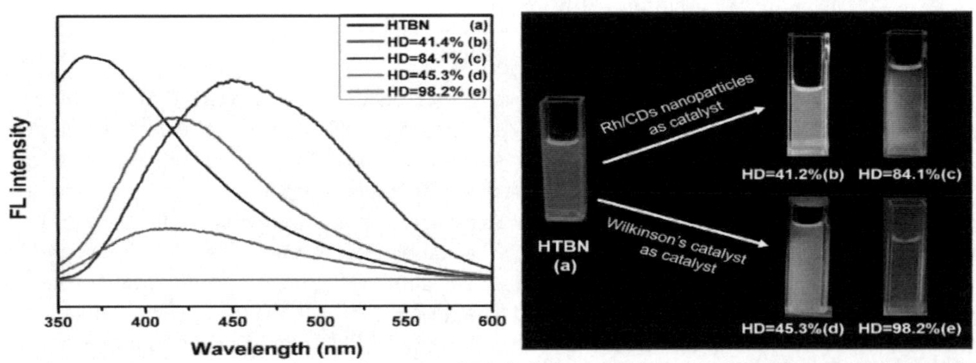

FIGURE 3. Fluorescence Emission Spectra and Digital Photographs of the Hydrogenated NBR under UV Light

ACKNOWLEDGMENTS

The financial support of the National Natural Science Foundation of China under Grant no. 20774009 is gratefully acknowledged.

REFERENCES

1.Haitao Li, Zhenhui Kang, Yang Liu,Shuit-Tong Lee, J. Mater. Chem, 22, 24230,2012
2.Youfu Wang ,Aiguo Hu,J. Mater. Chem. C, 2,6921,2014
3.Lu Cheng,Yongmao Li,Xinyun Zhai,Bing Xu,Zhiqiang Cao, Wenguang Liu ,ACS Appl. Mater. Interfaces , 6, 20487,2014.

4.Bin Chen, Jiachun Feng, J. Phys. Chem. C , 119, 7865,2015
5. D. Mosconi, D. Mazzier, S. Silvestrini, A. Privitera, C., ACS Nano 9 , 4156,2015.
6. P. Chen, Z.Y. Wang, S.F. Zong, D. Zhu, H. Chen,, Biosens. Bioelectron. 75, 446,2016.
7.K.A.S. Fernando, S. Sahu, Y. Liu, W.K. Lewis, E.A. Guliants, ACS Appl. Mater. Interfaces 7,8363,2015.
8.S. Han, H. Zhang, Y.J. Xie, L.L. Liu, C.F. Shan, Appl. Surf. Sci. 328,368,2015.
9.L.M. Shen, M.L. Chen, L.L. Hu, X.W. Chen, J.H. Wang, Langmuir 29,16135,2013.

The Path from the Rubber Strip to the Dosed Compound: A Way full of Challenges

Kerschbaumer Roman Christopher [(1)]*, Lechner Bernhard [(1)], Graninger Georg [(1)], Friesenbichler Walter [(2)]

[(1)] Polymer Competence Center Leoben GmbH - Styria - Austria, [(2)] Injection Molding of Polymers, Montanuniversitaet Leoben - Styria - Austria

INTRODUCTION In rubber processing, the material temperature and distribution significantly influences the reduction of the incubation time. Thus, it is substantial to know (1) the flow behavior in the screw channel as well as (2) the temperature distribution of the dosed rubber compound prior to injection to be able to ensure the processability of the used compound. METHODS The apparatus used in this work was a specially designed screw test stand consisting of a plasticizing cylinder and a downstream throttle valve. The design of the plasticizing cylinder and the temperature control is identical in comparison to a vertical Maplan rubber injection-molding machine. In addition, it is equipped with pressure and temperature sensors and offers the ability to remove the screw after curing of the compound in the plasticizing cylinder. Furthermore, the throttle valve offers the ability to measure the temperature distribution of the dosed material as well as to adjust the back pressure. To determine the temperature distribution of the dosed material it is necessary to (1) heat up the plasticizing cylinder to the testing temperature, adjust (2) the back pressure and (3) screw rotational speed and (4) perform the measurement under steady state conditions. The characterization of the flow behavior in the screw channel was performed as followed: after reaching the steady state conditions (1) the two colored rubber compound was cured in the plasticizing cylinder, (2) removed and (3) sliced on defined positions perpendicular to the screw channel to visualize the cross-flow. CONCLUSION Interactions of screw rotational speed and back pressure influences the formation of the cross-flow in the screw channel. Furthermore, a significant temperature rise of the dosed rubber compound could be observed.

Results of Mullins Experiment to compare stress softening of different types of liquid silicone rubber

Svenja Marl[a], Ralf-Urs Giesen[a] and Hans-Peter Heim[a]

[a]*Institute of Material Engineering, Polymer Engineering, University of Kassel, Germany*

Keywords: Liquid Silicone Rubber, Mullins Experiment,

Stress softening during the Mullins experiment (hysteresis tensile test) is a phenomenon of filled and crystalline rubber respectively. Previous publications presented different theories to explain the effect at the molecular level. These theories include for example bond rupture, molecule slipping, filler rupture and disentanglements. But the mechanism of the Mullins effect is not solved finally. Moreover the literature reports mostly one material with a model fit. Publications which compare different materials are missing. The question remains how to compare results of different materials.

Liquid silicone rubber (LSR) is a material with a broad range of mechanical properties. These greatly depend on the amount of filler, the degree of curing and the type of curing agent. The hysteresis stress-elongation curve is different for each type of LSR caused by these factors (Figure 1 a). The aim is to find a characteristic value which allows to compare different rubbers.

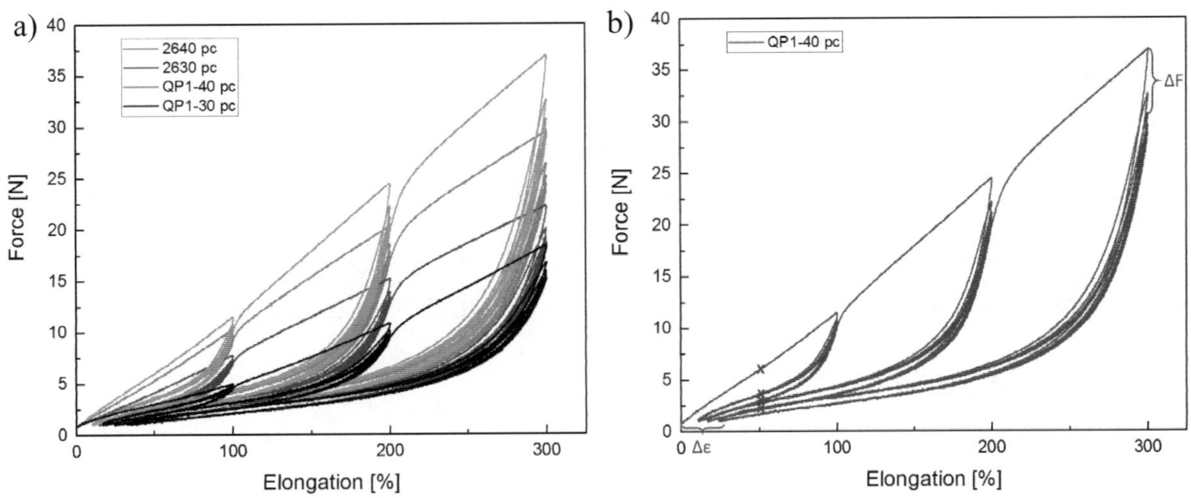

FIGURE 1. a) Comparison of four different types of LSR, b) Possible values to characterize the hysteresis stress-elongation curve of the LSR

One possible value is the loss of force at the maximum elongation ΔF and another the elastic recovery Δε. It is also possible to compare the area which is a value for the irreversible damage. Another possibility is to observe the decrease of force at one defined elongation (e.g. 50%) (Figure 1 b).

Preparation of Thermo-Reversible Elastomers by Catalyzed Ring-Opening and Diels-Alder Reaction via One-Step Melt Blending: Toward Industrialization

Nanying Ning*, *Zhanbin Feng, Jing Hu, Hongchi Tian, Hongli Zuo, Liqun Zhang, and Ming Tian*

State Key Laboratory of Organic-Inorganic Composites, Beijing University of Chemical Technology, Beijing 100029

Abstract. Traditional rubbers commonly use either sulfur or peroxide to form an irreversible chemical crosslinks, which make it difficult to be recycled and repeatedly used, so that it will produce a large amount of waste rubbers. So it is important for us to prepare thermo-reversible elastomers (TRE). The preparation of TRE are generally carried out in solution, which require solvent and long reaction time. These disadvantages limit the large scale preparation. In this study, we used epoxidized natural rubber (ENR) as the matrix and designed a new method to prepare TRE by Diels-Alder (DA) reaction through one-step melt blending for the first time. Furfurylamine (FA) was first grafted onto ENR and then bismaleimide (BMI) was added to crosslink with FA grafted ENR via DA reaction, which can be easily realized during melt blending. The mechanical properties of TRE could be tailored by the ratio of FA to BMI. Furthermore, the dynamic cross-linked ENR had good thermo-reversibility and self-healing behaviors. This new technique is promising in industry for rubber recycling and large-scale preparation of TRE.

.

Keywords: thermo-reversible elastomers, Diels-Alder reaction, epoxidized natural rubber (ENR), one-step melt blending, rubber recycling

PACS: Replace this text with PACS numbers; choose from this list: http://www.aip.org/pacs/index.html

RESULTS AND DISCUSSION

FA modified ENR by RO reaction

ENR-FA was prepared by RO reaction. The structure of ENR-FA was evidenced by using ^{1}H-NMR, and the results are shown in **Figure 1**(A). Compared with the ^{1}H-NMR spectrum of unmodified ENR, a new characteristic peak at $\delta = 1.45$ ppm (proton b), assigned to the proton of amine group connecting to the furan ring, occurs on the spectrum of ENR-FA (Figure 1(A)). The area of the peak located between 2.70 and 2.90 ppm (proton a), representing the proton connecting to the carbon atom of epoxy groups on the main chain, decreases after the RO reaction. These ^{1}H-NMR results indicate that furan groups have been successfully grafted onto ENR.

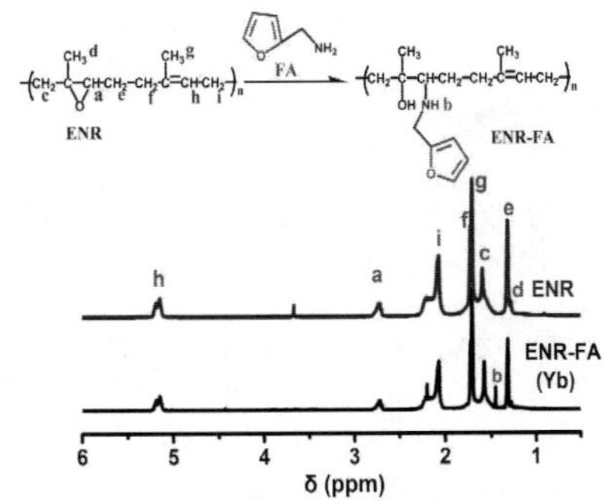

FIGURE 1. ^{1}H-NMR spectra of ENR, ENR-FA with Yb

Diels-Alder Cross-Link of ENR-FA and BMI.

As the unmodified ENR can be directly cross-linked by BMI at about 170 °C, here, we carried out a comparative study on DA crosslinking reaction at lower temperature for ENR-FA-BMI (DA crosslinking of ENR-FA and BMI) and unmodified ENR-BMI (crosslinking of ENR and BMI) and we would take ENR-0.125BMI and ENR-0.125FA-0.125BMI for example. The torque and temperature as a function of time of ENR-FA and ENR-FA-BMI are shown in **Figure 2**. When BMI was added into Haake rheomixer and the temperature rose to about 120°C, the torque of ENR-BMI decreased, indicating that the crosslinking reaction of unmodified ENR did not occur, whereas the torque of ENR-FA-BMI increased, indicating the occurrence of crosslinking reaction, as shown in Figure 2(B). 10 minutes later, the torque of ENR-FA-BMI remained the same indicating that the DA cross-linking reaction between ENR-FA and BMI was finished.

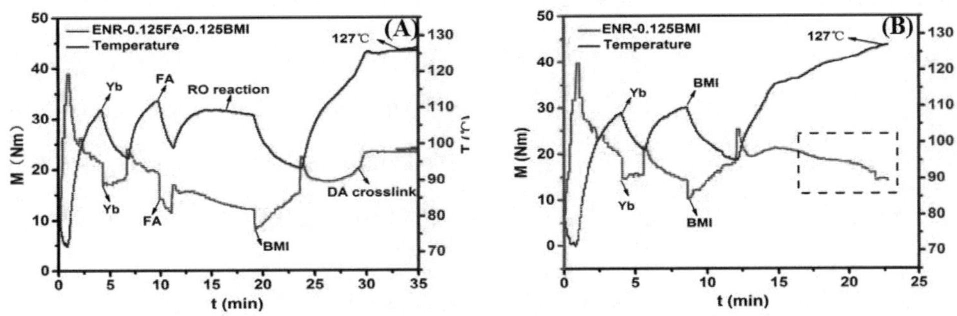

FIGURE 2. Haake rheomixer curves of (A) ENR-0.125BMI and (B) ENR-0.125FA-0.125BMI,

The thermo-reversibility of TRENR

The thermo-reversibility of TRENR was first demonstrated by the variation of solubility under different temperatures. **Figure 3**(A) shows the dissolving behavior of DA cross-linked sample (taking the sample ENR-0.125FA-0.125BMI as an example) in DMF. Clearly, DA cross-linked ENR can only swell at 30°C in DMF, as the temperature increases up to 170 °C, the cross-linked ENR gradually dissolves in DMF to form solution after 1h, indicating the decrosslinking (de-DA) reaction. The sample precipitated at the bottom of the bottle as the solution was cooled down to room temperature, and dissolved again at 170 °C, demonstrating that the cross-linked ENR has thermal reversible dynamic cross-linked network.

DSC was further carried out on the cross-linked sample to verify the thermo-reversibility and the thermo-reversible temperature range for the retro DA (rDA) reaction due to de-DA reaction and re-crosslinking reaction. As shown in Figure 3(B), a single exothermic peak occurs at about 130 °C, indicating the occurrence of re-crosslinking reaction. In addition, a single endothermic peak at about 150°C is observed, representing the occurrence of de-DA reaction and the rupture of DA cross-link network.

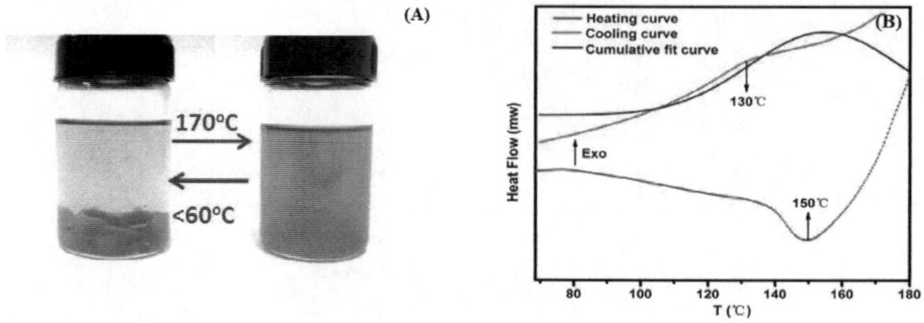

FIGURE 3. (A) the dissolving of DA cross-linked ENR , de-cross-linked ENR and re-crosslinked ENR in DMF. (B) DSC curve for re-DA and de-DA reaction of cross-linked ENR.

ACKNOWLEDGMENTS

We would like to express our sincere thanks to National Natural Science Fundation of China (Grant No. 51525301 and 51521062) for financial supports.

REFERENCES

1 T. Tambouratzis, D. Karalekas, N. Moustakas, A Methodological Study for Optimizing Material Selection in Sustainable Product Design. J. Ind. Ecol. 18 (2014) 508−516.
2 A. Gandini, A. J. D. Silvestre, D. Coelho, Reversible click chemistry at the service of macromolecular materials. Polym. Chem. 2 (2011) 1713-1719.
3 H. Nandivada, X. Jiang, J. Lahann, Click Chemistry: Versatility and Control in the Hands of Materials Scientists. Adv. Mater. 19 (2007) 2197–2208.
4 X. Chen, M. A. Dam, K. Ono, A. Mal, H. Shen, S. R. Nutt, K. Sheran, F. A. Wudl, Thermally Remendable Cross-Linked Polymeric Material. Science 295 (2002) 1698−1702.
5 A. Gandini, The furan/maleimide Diels–Alder reaction: A versatile click–unclick tool in macromolecular synthesis. Prog. Polym. Sci. 38 (2013) 1-29.
6 E. Goiti, M. B. Huglin, J. M. Rego, Thermal Breakdown by the Retro Diels–Alder Reaction of Crosslinking in Poly[styrene-co-(furfuryl methacrylate)]. Macromol. Rapid Commun. 24 (2003) 692-696.
7 R. Gheneim, C. Perez-Berumen, A. Gandini, Diels−Alder Reactions with Novel Polymeric Dienes and Dienophiles: Synthesis of Reversibly Cross-Linked Elastomers. Macromolecules 35 (2002) 7246-7253.

pps proceeding
Symposium: S13 - Rubber and elastomers
Poster Presentation

Flow instabilities of silica filled SBR compounds

Honda Shinichiro [1]*, Kawamura Takanobu [1], Tada Toshio [1]

[1] Sumitomo Rubber Industries, LTD - Kobe, Hyogo - Japan

Effect of temperature and shear rate on appearances of extruded silica filled styrene butadiene rubber compound was investigated by capillary rheometer. It was revealed that in low shear rate region flow instability appeared as shark-skin, and the types of flow instability changed to melt fracture with increasing shear rate. The shear viscosity with Bagley correction and the extensional viscosity calculated by Cogswell's method were also investigated. The shear viscosity with Bagley correction didn't relate to the melt fracture occurrence. The extensional viscosity calculated by Cogswell's method, however, got rising with increasing extensional rate, and then, shark-skin switched to melt fracture. This switching phenomenon indicates that melt fracture occurrence of rubber compound is due to strain-hardening by extensional flow in entrance of capillary die and then wall slip phenomenon also happens.

Impact of the Filler Condition on Spatial Thermal Conductivity Fluctuations of an Injection Molded Rubber Part

Kerschbaumer Roman Christopher (1)*, Lechner Bernhard (1), Tscharnuter Daniel (1), Wanner Andrea (1), Friesenbichler Walter (2)

(1) Polymer Competence Center Leoben GmbH - Styria - Austria, (2) Injection Molding of Polymers, Montanuniversitaet Leoben - Styria - Austria

INTRODUCTION For the injection molding simulation of rubber parts, the thermal conductivity of the rubber material is mainly implemented as a function of temperature. This work shows that the thermal conductivity is also significantly dependent on the local filler condition. MATERIALS AND METHODS The investigation was carried out with an injection molded part (100x100x15 mm) of an industrial SBR rubber compound. The part was separated by cutting into several layers in order to be able to determine the local thermal conductivity. A sample with a 50 mm diameter was taken from each layer and the thermal conductivity was determined by means of a stationary method. RESULTS AND DISCUSSION Differences of up to 50 % in thermal conductivity could be observed between the surface and middle layer. Furthermore, an almost symmetrical distribution of the results could be found with the highest thermal conductivity occurring in the center of the part. This distribution is due to (1) the flow profile during injection and (2) the resulting filler condition. CONCLUSION Significant differences of the thermal conductivity between the surface and middle layer of an injection molded rubber part could be observed. Accordingly, the thermal conductivity is not only a function of temperature but also depending on the filler condition.

Study on the Remove Protein from Natural Rubber Latex and Its Application

Nghia Phan Trung [1]***_, Hang Trinh Thi_** [1]**_, Thuong Nghiem Thi_** [1]**_, Ha Nguyen Thu_** [1]**_, Yamamot Yoshimasa_** [2]**_, Kawahara Seiichi_** [3]

[1] Hanoi University of Science and Technology - Hanoi - Vietnam, [2] Tokyo National College of Technology - Tokyo - Japan, [3] Nagaoka University of Technology - Nagaoka - Japan

The low protein content of natural rubber was prepared in latex stage by the novel procedure to remove all proteins from natural rubber with either urea or acetic acid or ammonium sulfate and a polar organic solvent as a cacatyst in the presence of sodium dodecyl sulfonate (SDS) surfactant. The condition for the removal of proteins was investigated in terms of concentration of the solvents, and repeating times for washing latex with centrifuge. Acetone and anionic surfactant were found to be effective for the removal of the proteins. Remaining proteins assessed by the amount of nitrogen in the sample was determined by Kjeldahl method. The results showed in the presence of acetone, wt% of nitrogen decreased from 0.02 wt% to 0.005 wt%, i.e. removal efficiency increased 4-fold compared to not using acetone. In addition, the study also found that when using various amount of acetone, different removal efficiencies were obtained. Optimal concentration of 0.25% by weight of acetone and nitrogen content in the sample was removed to 0.005 wt% after three centrifugation. The removal of proteins from natural rubber was confirmed via Fourier Transform Infrared (FT-IR) and Nuclear Magnetic Resonance (NMR) spectroscopy.

The rheological significance of sagging in extensional measurements

Pasquino Rossana **[(1)]**, ***Grizzuti Nino*** **[(1)*]**

[(1)] University of Naples Federico II - Napoli - Italy

Sagging is a well known, undesired phenomenon in polymer processing that depends on processing as well as material parameters. Among the latter, viscosity and viscoelasticity play the most important role. Sagging is also relevant in extensional rheometry, as the collapse of the fiber- or sheet-shaped polymer sample can compromise the measurement. In this work, we analyze the sagging phenomenon taking place in the Sentmanat Extensional Rheometer (SER). We performed experiments with Polstyrenes (PS) of different molecular weight and at different temperatures. Films of PS were placed in the SER fixture and time evolution of the sagging effect was quantitatively measured by a photographic time-lapse technique. The results are interpreted in terms of existing theories (Stephenson, et al., Polym. Eng. Sci 1999, Giacomin, et al., Polym. Eng. Sci 2010). Criteria for a correct, sagging-free measurement in the SER fixture are also discussed.

An innovative triborheometry fixture / design to study skin simulant materials

Medina Dora I [(1)]*, Elmoumni Aadil [(2)], McKinley Gareth H [(3)]

[(1)] Massachusetts Institute of Technology, Cambridge MA, USA - Tecnologico de Monterrey - Estado de Mexico - Mexico, [(2)] TA Instrumetns - Waters LLC - New Castle, DE - USA, [(3)] Massachusetts Institute of Technology - Cambridge MA - USA

The surfaces characteristics of the skin are of much interest in the scientific community and only a few methods have been developed to study their frictional properties. This research describes an innovative triborheometry fixture / design to study the skin surface with special significance on the measurement of coefficient of friction at different angular velocities and pressures. The study includes different types of skin which in different fields and applications are used to mimic and/or replace human skin. Skin substitutes demand structural and mechanical features for optimal performance which are directly associated with the surface properties. We propose to obtain an insight into the fluid mechanics of different solid/liquid systems such as skin simulant and substitutes / fluids.

Unusual viscoelastic behaviour of plasticized cellulose acetate in the melt state

Dreux Xavier [(1)]*, Majesté Jean-Charles [(1)], Carrot Christian [(1)], Argoud Alexandra [(2)], Vergelati Caroll [(2)]

[(1)] Laboratoire d'Ingénierie des Matériaux Polymères - Rhône-Alpes - France, [(2)] Solvay in Axel'One - Rhône-Alpes - France

This present work deals with the rheological properties of cellulose acetate which is a biobased and biodegradable polymer manufactured from cellulose. The hydroxyl groups of this abundant organic resource are partially converted to acetyl groups. Depending on the average number of acetyl groups in the anhydroglucose unit of cellulose, the degree of substitution (DS) modify the polymer properties. In order to facilitate its forming process, cellulose acetate is usually blended with plasticizers. As the conventional industrial processes need rheological information for different flow conditions, viscoelastic behaviour of a plasticized cellulose acetate with a degree of substitution of 2,45 was then investigated by different rheology tests. In this study the plasticizer contents are low. Measurements in linear viscoelasticity regime were carried out for different plasticizer contents and some unexpected behaviours have been highlighted. Compared to conventional thermoplastic polymers, the structure and dynamics of the macromolecular chains of cellulose acetate seems to be more complex than that of common synthetic polymers. The plasticizer content and its nature do not have any significant impact on the molecular dynamics of the chains. The results show that both relaxation times and apparent molecular weight between entanglements are not affected by plasticizer content whatever its nature. These results suggests that the viscoelastic behaviour of cellulose acetate chains is no more controlled by entanglements but rather by another correlation length given by strong interactions between chains which are not directly affected by the plasticizer. These interactions assign unusual scaling behaviour upon plasticizer content and temperature.

Contribution of Inter-Molecular Cross-Correlations in Viscoelasticity of Entangled Polymers

Yuichi Masubuchi

Nagoya University, Nagoya 4648603, Japan

Abstract. Although not frequently discussed, there exist considerable contributions from inter-molecular cross-correlations in viscoelasticity of entangled polymers. In this study, molecular simulations were performed and the viscoelastic response was decomposed into the contributions from self-correlation in the same molecule and cross-correlation between different molecules in various systems. For monodisperse linear polymers, the contribution from cross-correlation is mildly time dependent and it gradually increases up to 40% of the relaxation modulus. This behavior is almost universal for monodisperse systems with different molecular weights, and even for branch polymers. The universality of the cross-correlation is also observed for bidisperse systems in which long and short chains are mixed given that the amount of mixing and the molecular weights of long and short chains are not largely different. Although the results mentioned above suggest that the effects of cross-correlation could be adsorbed by a tuning of unit of modulus, the cross-correlation contribution is of further complication for some systems. In particular, for bidisperse melts in which the amount of long chains is relatively small and the long chain molecular weight is significantly larger than the short chain, the time development of the cross-correlation contribution becomes different from the universal behavior, showing bumps reflecting the relaxation time of short and long chains. The results above suggest that the cross-correlation is of significance and to be separately considered from constraint release, though it is neglected in single chain model

Keywords: **Enter Keywords here.**
PACS: 83.10.Mj, 83.10.Kn, 83.10.Rs

INTRODUCTION

Viscoelasticity of entangled polymers has been explained in the molecular picture called tube model[1], in which the motion of entangled polymers is replaced by the single chain dynamics confined in a tube. The molecular models based on the tube picture have attained the remarkable success[2,3], as long as the model incorporates the significant relaxation mechanisms, i.e., reptation, contour length fluctuation (CLF), and constraint release (CR).

However, there still remain unsolved issues for the modeling. One of such problems is cross-correlation between different chains. In the tube picture, dynamical correlations between different chains are not considered explicitly, but incorporated in a mean-field way in the context of CR. Thus, by construction, the relaxation modulus is calculated from the stress of each chain. On the other hand, there have been reported considerable contributions from cross-correlations between different chains. For instance, Cao and Likhtman[4] extracted the cross-correlation in the bead-spring molecular simulations to report that the contribution is up to 40% of the relaxation modulus. Although their study is the first report to quantify the ratio in relaxation modulus, there have been earlier studies reporting consistent results, not only via simulations[5] but also from experimental side[6–8].

The nature of cross-correlation has not been revealed. Cao and Likhtman[4] reported that the amount of cross-correlation in relaxation modulus is mildly time-dependent. It increases over time to a steady value with a certain characteristic time, which is sufficiently larger than that for the glassy modes. The ratio of cross-correlation contribution to the total relaxation modulus is virtually insensitive to the molecular weight and its distribution. These important findings suggest that the cross-correlation contribution might be embedded into the stress-optical coefficient in the single chain models. In other words, a tuning for the entanglement molecular weight and/or the plateau modulus would be sufficient to take account of the cross-correlation for the tube models. Such a strategy is intuitive, in a sense that the cross-correlation contribution would be embedded into the parameters of the coarse-grained elements, as long as it is generated by local interactions between atoms. But it should be further examined if the cross-correlation is solely affected by the local interactions, and if the universality of cross-correlation is observed for the systems that were not examined by Cao and Likhtman[4]. Indeed, owing to the computational costs, their results are restricted to scarcely entangled an unentangled linear polymers.

In this study, the inter-chain cross-correlation was examined by multi-chain slip-link simulations for various entangled systems. In spite of the fact that there are no explicit two-body interactions considered in the model, the cross-correlation is observed. The cross-correlation seems thus generated by the force balance around the entanglements. The results also revealed that the cross-correlation is fairly universal with respect to the molecular weight distribution, unless the contrast between molecular weights and volume fractions is strong. The universality was observed for long-chain branching as well. These results lend support the argument, in which the effects of cross-correlation can be embedded into the coarse-grained unit. Details are shown below.

MODEL AND SIMULATIONS

The primitive chain network model[9] was used for the simulations. In the model, the entangled polymers are replaced by consecutive segments. Each segment is assumed to carry a molecular weight comparable to the entanglement molecular weight. The chains are bundled by slip-links in pair to form a network. Dynamics of the system is thus described by the kinetics of slip-link position, sliding motion of the chain through the slip-links, and creation/destruction of the slip-links at the chain ends. The slip-link position obeys the Langevin-type equation of motion incorporating the force balance among drag force, segment tensions, osmotic force and random force. The chain sliding is expressed by the rate equation for the number of monomers assigned to each segment. In the equation, the same force balance with the equation for slip-link position is considered along the chain. The creation/destruction of the slip-links take place at each chain end owing to the chain sliding. If the chain protrudes from the slip-link at the end, a new slip-link is created to hook another segment randomly chosen from the surroundings. Vice versa, the end slip-link is removed when the chain slides off.

The model construction mentioned above is virtually the same with that considered in the tube models with the full-chain description[10–12] apart from a few distinctive features. One of them is the explicit pairing of entanglement between different chains. In the tube models, even though the two-body assumption is widely used, the pairing is considered in mean-field ways such as the double reptation, Rouse-CR and dynamic tube dilation. The explicit pairing is free from possible artifacts in mean-field approaches. The other feature of the model is the force balance around entanglements. There are other slip-link models, in which the explicit pairing is employed[13]. But such models assume that the slip-link position (i.e., the position of entanglement) is immobile in space. This immobile assumption of entanglements is essential for the single-chain models, though there exist a few models that allow some fluctuations for entanglements. Nevertheless, the force balance around entanglement is not considered in the other entanglement-based models.

The model parameters are unit of length, unit of time and unit of energy. The unit length is chosen as the average segment length. The unit time is the diffusion time of slip-links. The unit energy is $k_B T$. These parameters are determined via the fitting of rheological data. With the appropriate parameters, the simulation can reproduce linear and non-linear rheology of various entangled polymer systems, as reported earlier[14–21].

Note that the relaxation modulus here is discussed in terms of the Green-Kubo formula, in which the relaxation modulus $G(t)$ is expressed by the auto-correlation function of shear stress $\sigma(t)$ as written below.

$$G(t) = \frac{V}{kT}\,\sigma(t)\sigma(0) \tag{1}$$

Because the shear stress is the sum of the stress from each molecule, $G(t)$ can be decomposed into self-correlation terms from same molecule and cross-correlation terms between different molecules.

$$G(t) = \frac{V}{kT}\left(\sum_i \sigma_i(t)\sigma_i(0) + \sum\sum_{i\neq j} \sigma_i(t)\sigma_j(0)\right) = G_A(t) + G_C(t) \tag{2}$$

Here, $G_A(t)$ and $G_C(t)$ are the contributions from self-correlation and cross-correlation, respectively. Note that $G_C(t)$ is zero given that the stress from each molecule is independent of each other. In other words, if the molecules move independently as assumed in the single chain models, $G_C(t)$ is zero. For the normalization purpose, the inter-molecular correlation was quantified by the coupling parameter $\kappa(t)$ defined as[4]

$$\kappa(t) \equiv G_C(t)/G(t) \tag{3}$$

RESULTS AND DISCUSSION

Figure 1 shows the viscoelasticity for a H-branch polystyrene melt (left) and a bidisperse polystyrene melt (right) obtained from simulated $G(t)$, in comparison to the literature data[22,23]. As reported earlier, the viscoelasticity is semi-quantitatively reproduced, demonstrating that the simulated $G(t)$ is in fair agreement with the experiments. On the basis of this agreement, $G(t)$ was decomposed into $G_A(t)$ and $G_C(t)$, and $\kappa(t)$ was obtained for the analysis of cross-correlations as shown below.

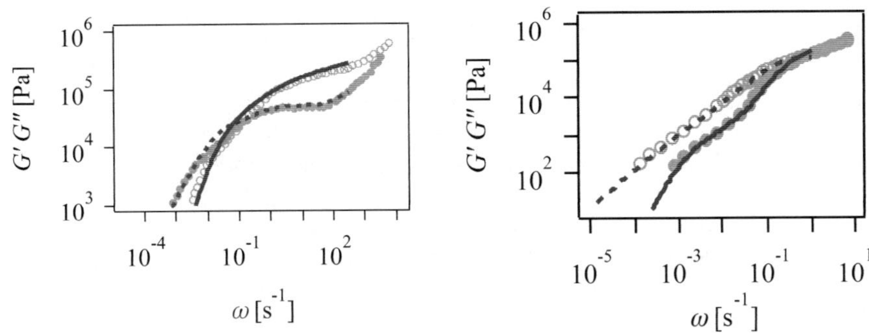

FIGURE 1. Linear viscoelasticity of H-branch PS (left) and bidisperse linear PS (right). The simulation results and the experimental data[22,23] are shown by curves and symbols. The molecular weight of H-branch PS is 10^5 for each arm and bridge segment. For the bidisperse PS, the molecular weights are 390k and 52k for long and short components. The long chain amount is 4%. The reference temperature was 169.5 °C (left) and 130 °C (right), respectively,

Figure 2 shows the coupling parameter for star and H-branch polymers in comparison to the linear polymer (left), and for binary mixtures of linear polymers (right). For the systems examined, $G(t)$ is in good agreement with the experimental data for polystyrene melts in the literatures[22–24], as shown Fig 1 for example.

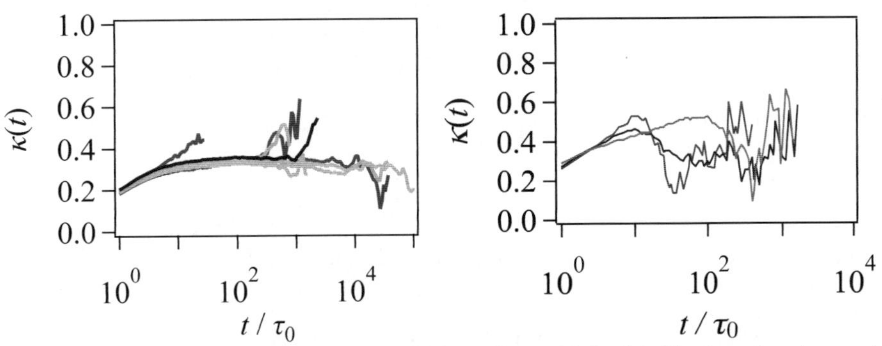

FIGURE 2. Coupling parameter for branch PS (left) and bidispese linear PS (right). The branch polymers include H (red) and 4-arm and 6-arm stars (green) with various molecular weights ranging from 100k to 1200k. Black curve indicates the linear polymer result. For the bidisperse blends, the molecular weights of short and long chains and the long chain fraction are (390k, 52k, 4%), (390k, 52k, 14%) and (390k, 103k, 14%), shown in red, blue and green, respectively,

It is remarkable that $\kappa(t)$ is not zero. Indeed, it mildly increases over time and saturates around 0.3. This behavior is consistent with that observed for the bead-spring simulations[4]. Note, however, that the characteristic time of $\kappa(t)$ is much larger and the plateau value is somewhat smaller than those for the bead-spring simulations. The discrepancy is apparently due to the difference in the inter-molecular interactions. In our model the glassy behavior is not considered and the density fluctuation is large. Further cross-correlation would appear if the density fluctuation is suppressed and the motional coupling according to the incompressibility is enhanced. Indeed, another set of simulations for a similar multi-chain model without the motion of slip-links[25], $\kappa(t)$ becomes smaller than that observed here.

For the branch polymers (Fig 2, left panel), it seems that the cross-correlation is essentially universal with respect to the branching structure. Namely, the star polymer results (green) and the H-polymer results (red) overlaps with the linear polymer result (black). There are deviations around the terminal times owing to insufficient data statistics for the calculation of correlation functions. The result might be not intuitive if one may argue the effects of intra-molecular entanglements among the chains that belong to same molecule. Such intra-molecular and inter-chain correlations certainly exist and actually observed in the simulations. However, the contribution is much smaller than that from the inter-molecular interactions, owing to the fact that the intra-molecular entanglements are rather minor.

For binary blends Fig 2, right panel), the cross-correlation apparently behaves in a manner that depends on the blending condition[26]. Indeed, there seem a bump, in which the peak exists at a certain characteristic time comparable to the relaxation time of the short chain in the system. This bumping behavior is not observed given that the difference of molecular weights between long and short chains is not large or the blend ratio is close to 50%.

CONCLUSIONS

The cross-correlation between different molecules in entangled polymers was investigated via the multi-chain slip-link simulations. In spite of the lack of explicit two-body interactions in the model, there exists the considerable contribution of cross-correlation in the relaxation modulus. Owing to the nature of the cross-correlation, which is mildly time-dependent and universal with respect to the long chain branching, the effects would be embedded into the tube models by a tuning of parameters, such as the plateau modulus. However, for bidisperse blends in which the contrast in the molecular weights and the blend ratio is strong, the universality is violated. This observation would suggest a possible source of difficulties in the development of single-chain models, in which the cross-correlation is neglected.

ACKNOWLEDGMENTS

This work is supported in part by Grant-in-Aid for Scientific Research (A) (17H01152) from JSPS and by Council for Science, Technology and Innovation, Cross-ministerial Strategic Innovation Promotion Program, "Structural Materials for Innovation" from JST.

REFERENCES

1. M. Doi and S.F. Edwards, *The Theory of Polymer Dynamics* (Clarendon press, Oxford, 1986).
2. Y. Masubuchi, *Annu. Rev. Chem. Biomol. Eng.* **5**, 11 (2014).
3. Y. Masubuchi, in *Ref. Modul. Mater. Sci. Mater. Eng.* (Elsevier, 2016), pp. 1–7.
4. J. Cao and A.E. Likhtman, *Phys. Rev. Lett.* **104**, 207801 (2010).
5. J. Gao and J.H. Weiner, *J. Chem. Phys.* **91**, 3168 (1989).
6. J. A. Kornfield, G.G. Fuller, and D.S. Pearson, *Macromolecules* **22**, 1334 (1989).
7. C.M. Ylitalo, J. A Kornfield, G.G. Fuller, and D.S. Pearson, *Macromolecules* **24**, 749 (1991).
8. M. Doi, D. Pearson, J. Kornfield, and G. Fuller, *Macromolecules* **22**, 1488 (1989).
9. Y. Masubuchi, J.-I.I. Takimoto, K. Koyama, G. Ianniruberto, G. Marrucci, and F. Greco, *J. Chem. Phys.* **115**, 4387 (2001).
10. C.C. Hua and J. Schieber, *J. Rheol.* **42**, 477 (1998).
11. J.D. Schieber, J. Neergaard, and S. Gupta, *J. Rheol.* **47**, 213 (2003).
12. A.E. Likhtman and T.C.B. McLeish, *Macromolecules* **35**, 6332 (2002).
13. M. Doi and J. -i. Takimoto, *Philos. Trans. R. Soc. London Ser. A-Mathematical Phys. Eng. Sci.* **361**, 641 (2003).
14. Y. Masubuchi, G. Ianniruberto, F. Greco, and G. Marrucci, *J. Chem. Phys.* **119**, 6925 (2003).
15. Y. Masubuchi, G. Ianniruberto, F. Greco, and G. Marrucci, *Model. Simul. Mater. Sci. Eng.* **12**, S91 (2004).
16. Y. Masubuchi, H. Watanabe, G. Ianniruberto, F. Greco, and G. Marrucci, *Macromolecules* **41**, 8275 (2008).
17. Y. Masubuchi, T. Yaoita, Y. Matsumiya, and H. Watanabe, *J. Chem. Phys.* **134**, 194905 (2011).
18. Y. Masubuchi, Y. Matsumiya, H. Watanabe, S. Shiromoto, M. Tsutsubuchi, and Y. Togawa, *Rheol. Acta* **51**, 1 (2012).
19. Y. Masubuchi, Y. Matsumiya, H. Watanabe, G. Marrucci, and G. Ianniruberto, *Macromolecules* **47**, 3511 (2014).
20. Y. Masubuchi, Y. Matsumiya, and H. Watanabe, *Macromolecules* **47**, 6768 (2014).
21. T. Yaoita, T. Isaki, Y. Masubuchi, H. Watanabe, G. Ianniruberto, and G. Marrucci, *Macromolecules* **45**, 2773 (2012).
22. J. Roovers, *Macromolecules* **17**, 1196 (1984).
23. J.K. Nielsen, H.K. Rasmussen, O. Hassager, and G.H. McKinley, *J. Rheol.* **50**, 453 (2006).
24. W.W. Graessley and J. Roovers, *Macromolecules* **12**, 959 (1979).
25. Y. Masubuchi, A. Pandey, and Y. Amamoto, *J. Soc. Rheol. Jpn* (2017).
26. Y. Masubuchi and Y. Amamoto, *Macromolecules* **49**, 9258 (2016).

Effects of short glass fibre orientation on rheological properties of Nylon-6 composites

Bhattacharya Sati N [1]*, Perumal Deverpiran Vishak [1], Gupta Rahul K [1], Costa Franco S [2]

[1] RMIT University - Victoria - Australia, [2] Autodesk Australia Pty Ltd - Victoria - Australia

In order to understand the effects of short glass fibre orientation on the rheology of polymer melts (non-Newtonian medium), experimental testing on compression moulded Nylon-6 samples with 30% weight short glass fibres (PA6-30GF) was carried out. Unfilled Nylon-6 (PA6) was also tested to provide a basis for comparison. Steady shear experiments were performed using a parallel plate rheometer and PA6-30GF samples displayed a viscosity overshoot before reaching steady state viscosity as a function of strain units. Unfilled PA-6 did not display any viscosity overshoot. The viscosity overshoot has been attributed to the reorientation of fibres from an initial orientation to a steady state orientation. Based on this behaviour, samples subjected to different strain units were rapidly cooled within the rheometer to preserve the fibre orientation. X-Ray Computed Tomography(X-CT) was used to obtain 3-dimesional fibre orientation and fibre concentration of the sheared samples. The orientation tensor value corresponding to the shearing direction increased as a function of strain units as more fibres were aligned by the applied force. In order to validate the experimental fibre orientation data, a comparison was performed with simulations from Folgar-Tucker based Reduced Strain Closure (RSC) orientation model. The predicted fibre orientation results showed reasonable agreement with experimental data. Estimated interaction coefficient obtained from the simulation for the Nylon-6 glass fibre system was compared with those reported in literature for Polypropylene and High-Density Polyethylene (HDPE) composites. Using this data, a non-Newtonian viscosity model incorporating effects of fibre orientation is being developed for improving composite flow predictions.

Viscoelastic Effects on the Roll-Separating Force and Power Input on the Calendering Process of a Non-Newtonian Fluid

Muñoz Sanchez Jaime Ernesto (1)*

(1) IPN ESIME Azcapotzalco - Mexico City - Mexico

This study deals with the calendering process in viscoelastic sheets of finite initial thickness. The rheological constitutive equations of the fluid under consideration follow the Simplified–Phan–Thien–Tanner (SPTT) fluid model. The corresponding governing equations of mass and momentum balance, based on lubrication theory, are solved for the velocity and pressure fields using numerical techniques. When the above variables were obtained, we continued determining quantities of engineering interest such as the roll-separating force and the power transmitted to the fluid by the rolls. Is our interest to determine the influence of viscoelastic effects (represented by the Weissenberg number Wi) on such quantities.

Rheology of Polymer in Couette Flow Condition

***Ema Masuma Khatoon* [(1)]*, *Armstrong Connor* [(1)], *Bigio David* [(1)]**

[(1)] Department of Mechanical Engineering, University of Maryland - Maryland - USA

Flow analysis of polymer with particle suspension affecting on microstructure and rheology remains one of the challenging problems in colloidal science. In this paper, polymers containing soft beads were used in a flat plate rotational rheometer at constant stresses. A simple rheological model of a non-Newtonian fluid has been evaluated to explain the time-dependent viscosity of polymer. Viscosity bifurcation of the polymer has been also explained in terms of rheological parameters at constant applied stress. Hence, the rheological behavior of soft beads into polymer carrier has been explained by Couette flow.

Rheological fingerprints of self-assembled block copolymers in oil/water mixtures

Qavi Sahar [(1)], Foudazi Reza [(1)*]

[(1)] Department of Chemical & Materials Engineering, New Mexico State University - New Mexico - USA

Amphiphilic molecules self-assemble in water/oil mixtures and form mesomorphic structures, known as mesophases. The self-assembled mesophases are used in a broad range of applications such as drug delivery, templating synthesis, and nanostructure production. In this work, mixtures of water, oil, and poly(ethylene oxide)-b-poly(propylene oxide)-b-poly(ethylene oxide) copolymers, known as Pluronics, are prepared in hexagonal and lamellar structures. The rheological fingerprints of such mesophases are studied through flow curve,frequency sweep, and large-amplitude oscillatory shear (LAOS) measurements.

pps proceeding
Symposium: S14 - Rheology and rheometry
Oral Presentation

Rheological Behavior of Polyolefin/In-situ Shrinking Microfiber Composites in Foaming Process

Kim Eric S. [1], ***Lee Heon E.*** [1], ***Lee Patrick C.*** [1]*

[1] Department of Mechanical Engineering Univ of Vermont - VT - USA

Polyolefin is widely used in many processing applications such as extrusion and injection molding. However, linear polyolefins generally do not show strain hardening behavior that is an essential characteristic for various processes involving extension such as foaming, fiber spinning, and film forming. Even though branched polyolefins have been widely used in industry, the cost of producing polyolefins with controlled branching structure is still relatively high compared to the base resins. In this study, unlike chemically branched polyolefins of which branching is difficult to control to suit the desirable characteristics, a smart microfiber blending technology would be able to tune the optimum degree of strain hardening behavior more cost efficiently. These smart microfibers can induce controlled pre-stressing or post-tensioning in the matrix by responding to external stimuli such as temperature variation. This study describes the effects of blending polyolefins with active microfibers on their rheological behavior, and thus final foam morphologies. This new technology includes the design and fabrication of in-situ shrinking microfibers to improve extensional behavior for foaming applications. The modification of the rheological properties by the microfiber shrinkage is discussed. The extensional viscosity measurements are described in terms of strain-hardening behaviors in polyolefin composites containing shrinking microfibers, and in neat linear polyolefins. Final foam properties resulting from these structures are also presented.

Influence of Shearing Load on the Material Aging Process of Polyethylene

Michael Burgfeld[1, a)] and Johannes Wortberg[1,b)]

[1] *University of Duisburg-Essen, Institute of Product Engineering, Engineering Design and Plastics Machinery, Lotharstraße 1, 47057 Duisburg, Germany*

[a)] michael.burgfeld@uni-due.de

Abstract. A long residence time of plastic melt in extrusion processes leads to a change in material viscosity. This change is based on reactions at a molecular level. To predict the material aging and analyze the material behavior and the potential occurrence of defects in the product, it is necessary to extend the knowledge of the influence of shearing load on material aging. Thus, a low-density polyethylene investigated. As a criterion for the detection of material aging, the complex viscosity measured by a rotational rheometer is applied. Hence, frequency- and time-sweep measurements are performed with different shear rates at constant temperature.

KEYWORDS: rheology, viscosity, material aging, shear load influence
PACS: 83.80.-k, 83.60.F.g, 66.20.-d

INTRODUCTION

Mechanical and physical properties of plastics show a dependence on time, temperature, strains, and external influences because of their macromolecular structure and weak binding forces [1]. This leads to chemical and physical changes during the extrusion process that, in turn, affect material properties such as the viscosity. These processes, called "material aging", can be distinguished as internal and external causes of aging. In the present case, internal causes of aging are important, due to their major impact on the material behavior during the extrusion process. Material aging can be further divided into chemical and physical aging processes [1]. Chemical aging processes are irreversible and include processes such as oxidation, degradation, post-condensation, and post-polymerization. By contrast, physical aging processes, such as relaxation and post-crystallization effects, are reversible. A detailed description of different influencing factors leading to material aging can be found in [1, 2].

In the following, the impact of a mechanical load on material aging is shown. Nevertheless, its effects on material aging has not been fully determined, and only a few explanatory approaches exist. On the one hand, a deformation of the bond angles and distance causes a mechanical failure in the polymer chain due to a mechanical degradation [3]. On the other hand, a high mechanical load can lead to chain scission and an accelerated thermal-oxidative degradation [1, 4]. During the extrusion of polypropylene at 230 °C, the mechanical load leads to the creation of alkyl radicals [5, 6].

This paper presents the effect of the mechanical load during the extrusion process. For this purpose, the viscosity is measured at different test conditions.

EXPERIMENTS

To investigate the influence of the mechanical load in terms of shearing on the material aging of a LDPE (Sabic 2100N0) specimen, time-sweep measurements are conducted. The material is tested using a rotational rheometer with a plate/plate-system (25 mm) at a constant stress (thermal and mechanical) over a defined test period of nearly 2 hours. Furthermore, the measurements are conducted under inert conditions (nitrogen atmosphere) to avoid a chemical aging process based on oxidation. To prevent an irreversible change in the material, the measurements are performed in the linear viscoelastic region (LVR) by using a deformation γ of 1 % for nondestructive testing [7]. A possible material aging is detected by the measured complex viscosity, which is plotted against the test time. In the following, the complex viscosity η^* and the angular speed ω are denoted as viscosity η and shear rate $\dot{\gamma}$ in accordance with the Cox-Merz relation [8]. At the beginning of every measurement, a pause time of 3 minutes is used to reduce the specimen's stresses.

In detail, shear rates at 0.01 s^{-1}; 0.1 s^{-1}; 1 s^{-1}; 10 s^{-1}; 100 s^{-1} investigated. Each setting is used three times with a new specimen to compensate for the material's deviations. Afterwards, the arithmetic mean is calculated and applied for further comparisons. Figure 1 shows the results of the time-sweep measurements at different shear rates at a temperature of 210 °C. Conspicuously, in a range of 0.01 s^{-1} to 0.1 s^{-1}, the viscosity shows a strong increase. At higher shear rates, however, the measured viscosity values are nearly constant over the test time.

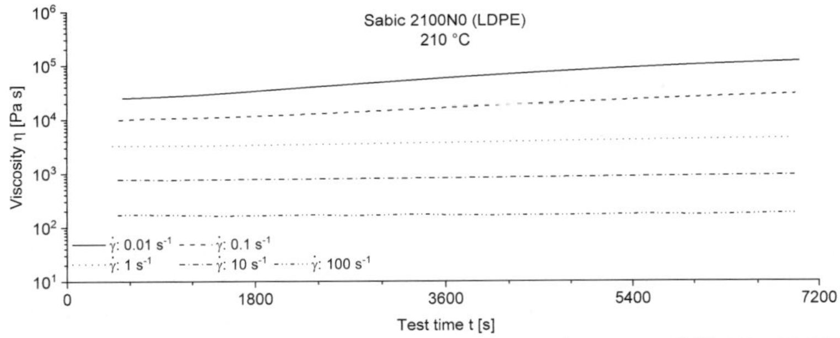

Figure 1. Acquisition of the material aging for different shear rates (210 °C – LDPE)

A possible explanation for this trend is the fact that the molecular chain movement can follow the mechanical stress best at lower shear rates [9]. Regarding the reptation model of de Gennes [10, 11] and the later extended tube model by Doi and Edwards [12], molecular chain can leave its tube at lower shear rates. This process can be measured and identified by an increase in viscosity over test time. To obtain a more detailed insight into the influence of the shear rate on the material aging, the same test sample is measured many times at different shear rates. This also ensures that the method is capable of detecting material aging at higher shear rates. The measurement time is set to nearly 2 hours at a test temperature of 210 °C, a setting which is already used in the other time-sweep measurements. First, the specimen is tested at a shear rate of 100 s^{-1} (0h/1h) for 1 hour and thereafter at 0.1 s^{-1} (+1h/1h). Furthermore, the result of this experiment is compared with those attained at a constant shear rate (0.1 s^{-1}, 100 s^{-1}). The following figure depicts the results of both measurements. The dashed curves present the results at a constant shear rate over the entire measurement time, while the solid curves present the results of the multiple measurements of the test sample.

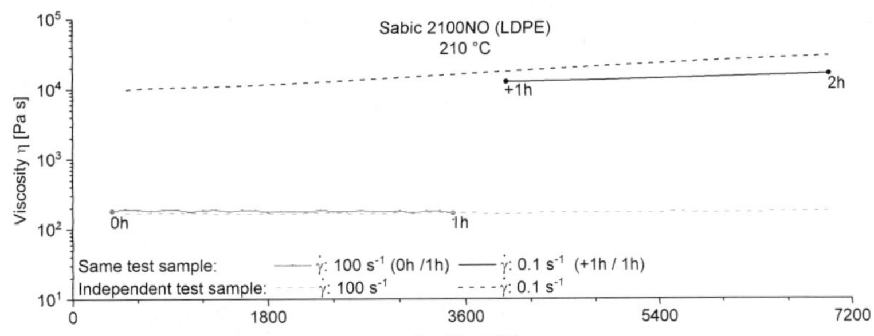

Figure 2. Investigation of the shear load on the material aging

As can be seen in Figure 2, both curves of the viscosity, which are measured at 100 s^{-1} are identical and do not exhibit a material change. To help draw a conclusion about the possibility of a material aging during this time, the measurement at 0.1 s^{-1} is used. If a material change already occurs at 100 s^{-1}, the measured viscosity at 0.1 s^{-1} (constant line) must be at the same level as the measurement result at a constant shear rate of 0.1 s^{-1} (dashed line) after 1 hour. A further criterion to evaluate a potential material aging at shear rates of 100 s^{-1} is the gradient of the viscosity curve, which reflects the reaction kinetic and speed. As it can be seen, both criteria show that the reaction only starts at the beginning of the measurement at 0.1 s^{-1}. Both measurements are compared in detail in Figure 3. Moreover, to enable a better comparison of both measurements, the time of the multiply measured specimen postponed for 1hour.

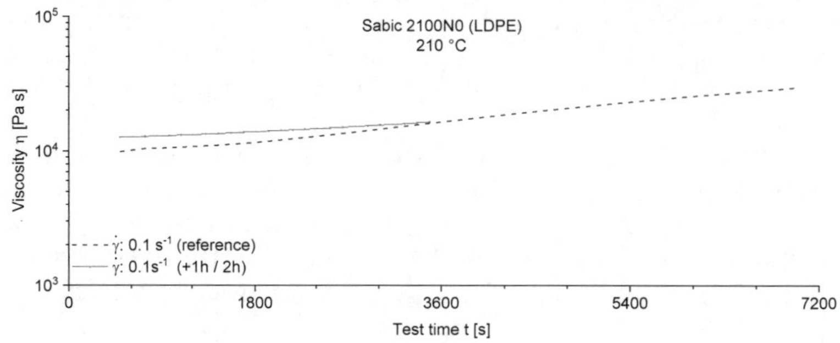

Figure 3. Comparison between the measurements at 0.1 s^{-1}

As Figure 3 illustrates, both curves show the same characteristics regarding the change in viscosity as a function of measurement time. The variation between them is attributed to the inhomogeneity of the granules, thereby leading to different viscosity values at the beginning of the measurement. Generally, it can be noted that higher shear rates decelerate the reaction speed or inhibit material aging. To evaluate this, multiple frequency sweeps are conducted with the same test time of the time-sweep measurements of nearly 2 hours. For this purpose, the same specimen is measured 13 times in succession at a shear rate ranging from 0.01 s^{-1} to 100 s^{-1}. Every measurement curve reflects a measurement time of approximately 6 minutes in addition to 3 minutes pause time at the beginning of the measurement. This conforms to the first measurement settings. For the second measurement setting, the same test sample is tested 20 times in succession at the same shear rate range as in configuration 1. Moreover, the pause time is skipped, leading to a constant shearing of the material sample at load duty cycle. The results of the two measurements are depicted in Figure 4.

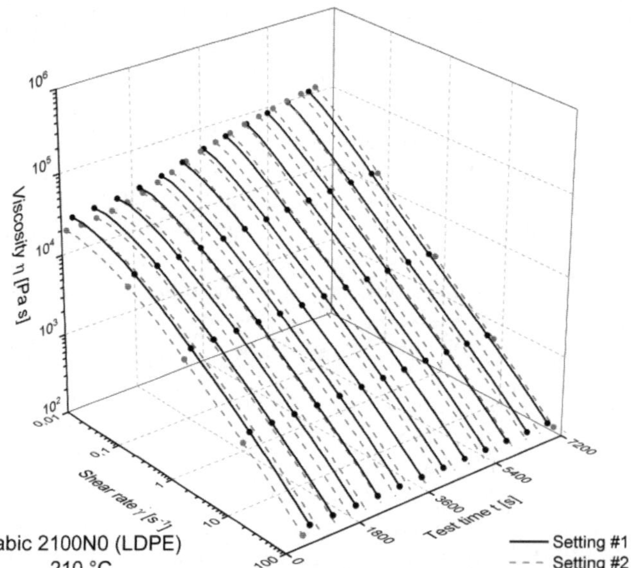

Figure 4. Influence of the load cycle duty on the material aging

As shown in the figure, the measurement curves show an increase in the viscosity in a range of 0.01 s^{-1} to 0.1 s^{-1}. At higher shear rates ($\dot{\gamma} > 0.1\, s^{-1}$), a nearly constant viscosity is measured over the test time. Thus, in principle, frequency sweeps can represent the influence of the shear rate on the material aging process. However, the difference between the two-measurement setting of the frequency sweeps is obvious. As a criterion for the evaluation of the material change, the viscosity value at a shear rate of 0.01 s^{-1} is applied. At setting 2, the viscosity tends to slightly increase over test time compared to setting 1. This can be explained by higher shear loads, which prevent the reaction kinetic. After nearly a 1-hour test time, both viscosity curves are nearly equal in value. This can be explained by the storage (G′) and loss (G′′) modulus, which are used to calculate the complex viscosity according to Equation 1

$$|\eta^*| = \sqrt{\left(\frac{G'}{\omega}\right)^2 + \left(\frac{G''}{\omega}\right)^2} \qquad \text{Eq. (1)}$$

A change in the intersection point ($\dot{\gamma}_{co}$) between the storage and loss modulus allow an interpretation of the molecular change and indicates a change in the structural character. If the storage module is greater than the loss module (G′ > G′′), the polymer structure is inflexible and limited in the movement of its molecular chains. Furthermore, the material changes from having a liquid- to a gel-character [7]. The following figure exemplarily describes the course of both moduli.

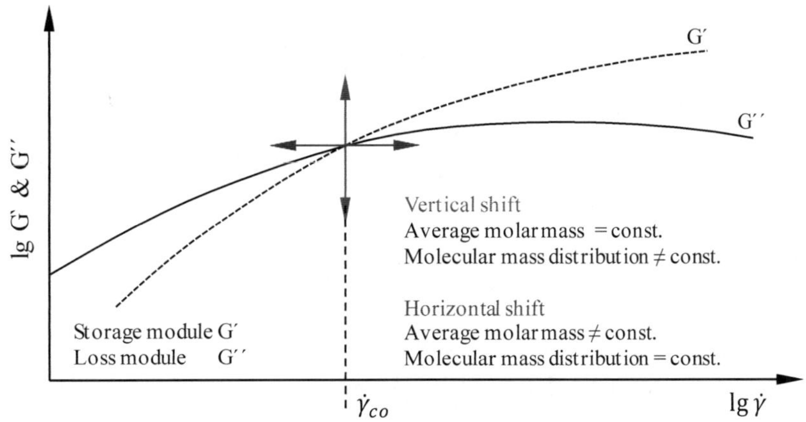

Figure 5. Schematic course of the storage and loss modulus

Consequently, a flow on the molecular level is not possible. Thus, it is conceivable that the radicals and reactive parts cannot find reactants. Figure 6 below depicts the change in the intersection point (IP) for both measurement settings over the test time. As it can be seen, the intersection point moves to lower shear rates. This means that detection of the material aging is probably impossible below the lowest shear rate of 0.01 s^{-1}. It has to be expected that the intersection point moves on to lower shear rates and leads to a constant measured viscosity at both settings at a shear rate of 0.01 s^{-1}.

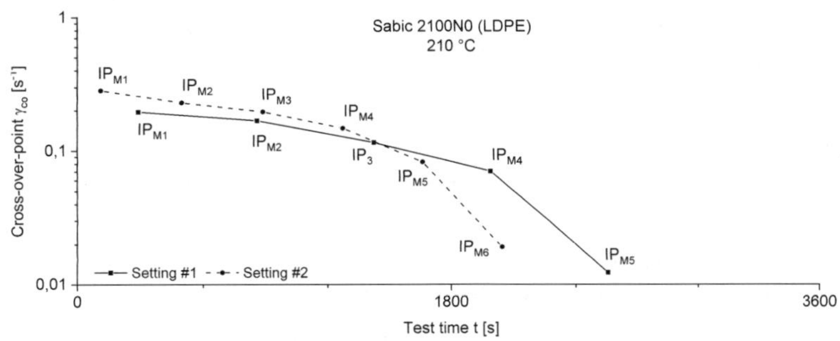

Figure 6. Change in the intersection point ($\dot{\gamma}_{co}$) over the test time

CONCLUSION & OUTLOOK

As presented, the shear rate affects the material aging. Higher shear rates lead to a suppression of the material aging and consequently to a nearly constant viscosity over measurement time. When the material is exposed to shear rates in a range of 0.01 s^{-1} to 0.1 s^{-1}, reactions do take place. With regard to the extrusion process, this means that material aging takes place in stagnation areas or in the boundary layer due to slow flow velocities. This implies that for color -or material change processes, the required time for alteration increases, because the displacement of the old material with the new one is more difficult. Furthermore, it must be considered that the measurements are made in the linear viscoelastic region. This excludes degradation of the material as a result of the mechanical stress on the basis of the measurement principle.

During the extrusion process, additional factors have an impact on the material aging. A possible scission of molecular chains due to the mechanical stress is difficult to prove. During extrusion, a higher mechanical stress in terms of shear rate leads to a higher dissipation that might cause an increase in temperature. Accordingly, it is not possible to draw conclusions about the influence, which leads to a material aging between the acting thermal and mechanical loads such as a Measuring Mixers or the in [13] presented system. Consequently, the results of this investigation should be cross checked with another measurement systems. Furthermore, the material aging in stagnation zones should be analyzed. Therefore, the material should be exposed to a defined temperature over a defined time. Subsequently, the material aging ought to be measured by a time sweep with the rotational rheometer at a shear rate of 0.01 s^{-1}. At least the investigation of the shear rate's effect on the material aging should be tested with other materials, such as HDPE, EVOH, PA.

ACKNOWLEDGMENTS

This publication was created within the scope of the research project "Thermo-rheological phenomena at wall boundary layers during color and material changes in plastics extrusion." (WO 302/46-1). We appreciate the funding by the German Research Foundation (Deutsche Forschungsgemeinschaft DFG).

REFERENCES

1. Ehrenstein G., Pongratz S.: *Resistance and Stability of Polymers,* Carl Hanser Verlag, Munich, 2013
2. Pongratz, S.: *Alterung von Kunststoffen während der Verarbeitung und Gebrauch*, PhD Thesis, University Erlangen-Nürnberg, 2000
3. Gutman, E.M.: *Mechanochemistry of Solid Surfaces*, World Scientific Publ., Singapore, New Jersey, London, Hon Kong, 1984
4. Pavlov, N.N., Kudrjavceva, G.A., Najdenova, G.P.: *Änderung von Deformations- und Festigkeitskennwerten von Konstruktionsthermoplasten durch Alterung im freien und deformierten Zustand*, Plaste und Kautschuk, 38, 11-12, p. 385-386, 1991
5. Allen, N.S., Harrison, M.J. et. al: *Thermal and photo-chemical degradation of nylon 6,6 polymer: Part III-Influence of iron and metal deactivators*, Polymer Degradation and Stability, 23, p. 165-174, 1989
6. Hepperle, J.: *Schädigungsmechanismen bei Polymeren, Polymeraufbereiteung 2002 – technischer Fortschritt zur Steigerung von Leistung und Produktqualität*, VDI Verlag, Düsseldorf, 2002
7 Metzger, T.G.: *The Rheology Handbook,* Vincentz Network GmbH & Co. KG, 2011
8 Cox, W. P., Merz, E. H.: *Correlation of dynamic and steady flow viscosities,* Journal of Polymer Science. 118, 28, p. 619–622, 1958
9. Schwarzl, F.R.: *Polymermechanik,* Springer-Verlag, Berlin Heidelberg, 1990
10. De Gennes, P. G.: *Entangled polymers,* Physics Today, American Institute of Physics. 36 (6), p. 33, 1983
11. De Gennes, P. G.*: Reptation of a Polymer Chain in the Presence of Fixed Obstacles*, The Journal of Chemical Physics. American Institute of Physics. 55 (2): 572, 1971
12. Doi M., Edwards S.F.: *The theory of Polymer Dynamics,* Oxford University Press, 1986
13. Burgfeld, M., Wortberg, J.: Development of a new measurement concept for the process-oriented characterization of rheological material properties, Dresden, 2017

Immobilization of shear thickening fluids in energy absorbing structures

Wierzbicki Lukasz (1)*, Leonowicz Marcin (1)

(1) Warsaw University of Technology Faculty of Materials Science and Engineering - Mazowieckie - Poland

Shear Thickening Fluids (STFs) are non-Newtonian fluids - viscosity of which increases with growing share rate. STFs can be classified as smart materials, due to their ability of predictable response to the external stimuli. Smart materials are able to combine functions of a sensor, a processor and an actuator.This feature can easily be adapted in energy absorbing systems.The studies revealed that the silica particles, dispersed in polypropylene glycol, demonstrate shear-thickening properties. During typical utilization the STFs will be characterized by low viscosity, while in the case of sudden strike, the viscosity of STFs will abruptly increase - up to solid-like properties.Thereby, STFs are considered as a component of various protective structures, such as: liquid armour [1], sport protectors, bumpers or dumpers [2,3], etc. Main challenge, related to common application of STFs in everyday life, relies on development of effective methods of their immobilization. Several methods of immobilization were tested: sealing STF in polyethylene (PE) and ultra high molecular weight polyethylene (UHMWPE) bags, impregnation of foams sealed in UHMWPE, placement inside various silicone structures, encapsulation in gelatine capsules and placement inside spacer fabrics and saturation of flexible polyurethane foams. The energy absorbing tests performed by dropping an impactor with energy of 5J allowed to identify significant differences in the energy absorbing properties of created structures. Also, the properties such as weight, surface density, thickness, flexibility and mass fraction of STF in crafted composites were compared.

[1] Yang Hai-lin, RuanJian-ming, Zhou Zhong-cheng, ZouJian-pe, Wu Qui-mei, Xie Yuan-yan, Shear thickening rheological response of PCC/PEG suspensions, J. Cent. South Univ. Technol. (2009) 16:0926 [2] M. Leonowicz, J. Kozłowska, Ł. Wierzbicki, Rheological Fluids for Energy Absorbing Systems, Applied Mechanics and Materials, (2013), 440, 13 [3] Lee Y S, Wetzel E D, Wagner N J. The ballistic impact characteristics of Kevlar woven fabrics impregnated with a colloidal shear thickening fluid, Journal of Materials Science, (2003), 38: 2825

The Melt Rheology of Poly(ethylene oxide) Powder Mixtures of Varying Initial Molecular Weight Distribution Subject to Non-Oxidative Thermal Degradation

Mansfield Craig Daniel [(1)*], ***Baird Donald G*** [(1)], ***Pino Christopher J*** [(2)]

[(1)] Virginia Polytechnic Institute and State University - Virginia - USA, [(2)] Innovative BioTherapies - Michigan - USA

The chemical kinetics of the non-oxidative thermal degradation of poly(ethylene oxide) (PEO) are well characterized in the literature. However, the rheological behavior of PEO subject to these conditions has not previously been addressed. The rheology of melt material streams plays an important role in predicting the optimal performance of melt processes, such as the co-extrusion process. Connecting the rheology of degrading PEO to the governing kinetics is of fundamental importance to predicting composition and rheological behavior as a function of processing conditions. In this work, the rheological behavior of melts formed from PEO powder mixtures of varying molecular weight was characterized using a combination of dynamic small amplitude oscillatory shear (SAOS) tests and inception of steady shear tests. SAOS tests were used to eliminate shear induced mixing, which was studied separately with inception of shear tests. The SAOS test scheme used allowed for isolation of thermal degradation effects during data analysis. The effect of degradation on the rheology of PEO will be discussed. In addition, a model for non-oxidative thermal degradation connecting the observed rheological behavior to the governing chemical kinetics will be introduced.

Long-branched Structures of Polybutadienes Analyzed by Rheology

Shuichi Maeda[a], Koji Shiba[b] and Michinori Suzuki[b]

[a]*Advanced Science and Innovational Research Center, Organization for Research Initiatives, Yamaguchi University, 2-16-1, Tokiwadai, Ube, Yamaguchi 755-8611, Japan*
maeshu1@yamaguchi-u.ac.jp
[b]*Ube Industries, Ltd., 8-1, Goi-Minamikaigan, Ichihara, Chiba 290-0045, Japan*

Linear and non-linear viscoelastic properties of five polybutadiene(PB) series (Y1, Y2, V, CoB, CoL) synthesized by three different catalysts and two processes were measured to investigate their molecular structures, especially long-branched structures in this study. PB solution (10%, 30%, 50%) samples were prepared by using liquid paraffin as solvent. Storage moduli, G' and loss moduli, G'' for all samples were corrected to $G'c^{-2}$ and $G''c^{-2}$ regarding their concentration (c). The relationships between $G'c^{-2}$ and $G''c^{-2}$ corrected by concentration for PB solutions and melts strongly depend on whether the PB has long branches in the molecule or not. The value of $G'c^{-2}$ at $G''c^{-2}$ = 20,000 Pa of PB having long-branched structure has not only increased with concentration, but it got larger with more complicated long-branched structure. On the other hand, the value of $G'c^{-2}$ at $G''c^{-2}$ = 20,000 Pa for PBs having linear structure was not dependent on concentration and was the same for all PB solutions and melts. From these measurements, the order on complex long-branched structures is Y1 > CoB > CoL > Y2 = V. From the measurement of the large deformation stress relaxation of 30% PB solutions, the order of alpha values in damping function is Y1 < CoB < CoL = Y2 = V. These results were essentially the same as the case of comparing the linear viscoelastic functions of the PB solutions. By the rheological method of this study, it is possible to obtain knowledge concerning the long-branched structure of PBs.

Keywords: **Polybutadiene / Long-branched structure / Linear and non-linear viscoelastic properties /**

Fine Tuning of a Polymer Melts' Rheological Properties

Youngwook P. Seo, Sehyun Kim,[+] Yongsok Seo*
RIAM, Dept. Materials Science & Engineering, Seoul National University, Kwanakro 1, Kwanakku, Seoul, Korea
[+] Polymer Processing Team, LG Techcenter, Daejon, Korea

Abstract. Some of Polymer melts are viscous fluids without the appearance of significant elastic property though their molar masses are high enough to show it. Poly(ε-caprolactam) (nylon 6) is one of them. In this study, we modulated the molar mass of nylon 6 by adding a linking agent that enables the chain entanglements in different sheet folds. The presence of branched molecules as well as intra-sheet molecules results in latent entanglements between chains. A remarkable transition was seen at the optimum molar mass of nylon 6. At this molar mass, the viscoelastic behavior crossed from viscous relaxation which is the characteristics of most polyamide melts to a predominantly elastic and rubber-like response.

PACS: 83.85.Jn Viscosity measurements

INTRODUCTION

Nylon 6 (poly-ε-caprolactam) is a thermoplastic resin that has found a wide range of applications in many sectors since it combines high strength and rigidity with good thermal and wear resistance, and has good resistance to chemical agents and flame retardancy.[1,2] At the same time, it has quite distinctive rheological property; its viscosity follows Newtonian fluid behavior (a wide plateau value) over a wide range of shear rates (or oscillating frequencies). [2,3] Further, the viscosity in the Newtonian fluid region is quite low which means this resin does not have that appropriate rheological properties for applications in fields that require a high dynamic melt viscosity in the absence of (or with weak) shear forces and a lower melt viscosity when subjected to these forces. [2] Although the low melt viscosity of nylon 6 has been ascribed to its low molar mass, it is typically much greater than its entanglement molecular weight (M_e).[4] If the molar mass of a polymer is greater than M_e, its melt viscosity is expected to be high at low shear rates due to the entanglements of polymer chains.[5] This rheological behavior of nylon 6 has been addressed in many studies, but its cause of low viscosity and wide Newtonian fluid behavior has not been clearly elucidated yet.[5] In this paper, we clarify the reason of the peculiar rheological properties of nylon 6 melt and pursue the origin of elastic properties of polymer melts. For the first time, we show a high shear thinning melt viscosity of neat nylon 6 melt. The unexpected high melt viscosity as well as highly solid-like behavior (melt elasticity) due to the emergence of chain entanglements are of significant importance in the design and applications.[19]

EXPERIMENTAL SECTION

4,4'-Di(2,3-epoxypropyloxy) phenyl benzoate (DEPPB) was synthesized with following the previously reported procedure.[19] (1) 4,4'-dihydroxy phenyl benzoate) (DHPB) was synthesized first. 4-hydroxybenzoic acid (60g, 0.43 mol) and NaOH (38g, 0.96mol) were added to 900mL of water and then stirred for 10 min at 5°C. Ethylchloroformate (54g, 0.5 mol) was added dropwise into the solution and then stirred for 10 min. 2N HCl solution (500mL) was poured into the solution and the resulting white precipitate was filtered and washed with water several times. The solid product was recrystallized from acetone and white needle-like crystals were obtained. These crystals and a few drops of DMF (N,N-dimethyl formamide) were poured into thionyl chloride and boiled for 1h. The excess thionyl chloride was evaporated under vacuum at 10^{-2} mmHg. The product was dissolved in 300mL of CH_2Cl_2 and the solution was added dropwise to a THF (Tetrahydrofuran) solution containing hydroquinone (55g, 0.5 mol) and pyridine (40mL, 0.5 mol).

After reaction for 12 h at room temperature, the solution was precipitated in 2N NaOH solution. After stirring for 2 h, the ethanol was evaporated, the product was washed with water, and then recrystallized from methylcellulose acetate. The yield was 68% (67g). (2) A mixture of DHPB (33g, 0.2mol), 3-bromopropene (26mL, 0.30 mol) and K_2CO_3(70g, 0.5 mol) was added to 400mL of acetone and then boiled for 24 h. The solid was filtered and acetone was evaporated for 24 h. The remaining solid was washed successively with 5% Na_2CO_3 solution, excess water and 200 mL of cold ethanol. After drying, a white powder was obtained, which was recrystallized from acetonitrile / isopropanol (1:1). The yield was 74% (46g). (3) Diallyl monomer (31g, 0.1mol) from step 2 was oxidized with 3-chloroperoxybenzoic acid and the product was recrystallized from acetonitrille / isopropanol (1:1). The yield was 75% (25g). the melting temperature (T_m) of the product was found to be 119°C. Chemical structure was verified using ^1H NMR and FT-IR.

FT-IR spectra were obtained using a Bruker 200 spectrometer (IF 66) with an average of 200 scans at a resolution of 4 cm^{-1}. ^1H-NMR spectra were recorded with a Varian Gemini 600 by using $CDCl_3$ as the solvent. Tetrasilane was used as the internal standard material. XRD patterns for the prepared specimens were recorded on a Rigaku D/ max 3C diffractometer using filtered Cu $K\alpha_1$ radiation ($\lambda = 0.15406$ nm).The diffractometer was operated at 40 kV and 40 mA. The XRD data were collected from 10 to 35° (2θ) in fixed time mode with a step interval of 0.02° (2θ). Nylon 6 was a Kolon product (KN171). The weight average molar mass was found to be $8.5x10^4$g/mol with a polydispersity index of 3.5. The nylon 6 pellets and DEPPB powder were dried in a vacuum oven at 100°C and 80°C respectively for 24 h. The DEPPB powder was then premixed in a container with dried nylon 6 pellets at a predetermined weight ratio. Finally, the mixture was blended in a twin screw extruder (PRISM) at 280°C. The rheological properties were measured at 250°C with a UDS200 (Physica, Germany) rheometer on which a 25-mm diameter cone and plate were mounted. The frequency range was set at 0.1-500 rad/s, and the applied strain was 5%. Before the measurement, the samples were prepared with a compression molder at 280°C. The measurements were carried out under a nitrogen atmosphere. The viscosity molar mass was calculated by the Mark-Houwink equation of $[\eta] = k [M]^a$ with k = 0.0226 and a = 0.82 where [η] is the intrinsic viscosity which was measured using an Ubelhode viscometer at 25°C. [13] 85% formic acid was used as a solvent.

RESULTS and DISCUSSION

Like other polymers such as polyethylene, polyoxymethylene, and poly(ethylene terephthalate), nylon 6 forms lamella sheets consisting of folded chains.[6] According to Li and Goddard's theoretical study, these folding sheets contain adjacent re-entry chain folding.[7] A polymer chain in one lamellae can extend out of and re-enter the same lamellae or exit the lamella, or extend out of the lamella and connect through an amorphous region to an adjacent lamella. However, the energy constraints that arise due to hydrogen bonding between folded chains favor intramolecular rather than intermolecular hydrogen bonding.[7] Welch and Muthukumar[8] concluded from their simulation study that: "Lamellae thickening is a highly cooperative process requiring the mobility of all chains in the crystals" whereby the formation of hydrogen bonds within each sheet is enthalpically favored. Since intramolecular hydrogen bonds are favored over intermolecular hydrogen bonds, both energetically and kinetically, the lowest enthalpy for a folded molecule with intramolecular hydrogen bonds arises for a 2D sheet with a constant length between folds. Two such sheets can pack to form a 3D structure in which the inter molecular interactions are dominated by van der Waals (non-Coulombic) interactions because hydrogen bonds are only present within the sheets.[9] Because of this 3D structure formed by the stacking of 2D sheets, there are few entanglements between the molecules of nylon 6 even though its molar mass is much greater than that of entanglement (M_e). The melt viscosity value is quite low and shows a wide Newtonian flow plateau due to the lack of chain entanglements.[3] In contrast, vinyl-type polymers such as polypropylene and polyethylene easily form folded sheets with inter-sheet molecules because of the absence of any energy constraints, such as hydrogen bonding in nylon 6.[10] When vinyl polymers are subjected to shear forces in the melt state, many entanglements arise between molecules, which results in a high dynamic viscosity. [11] If we can provide some entanglements between chains, the rheological behavior of nylon 6 may be significantly varied. We carried out some experiments in which the increase of nylon 6 molar mass and formation of inter-sheet bonds for chain entanglements were assisted by the addition of a chain linking agent (DEPPB, Scheme1). This chain linking agent contains two reacting end groups (epoxy groups) that can react with the amine end groups of nylon 6 chains. Thus it can form a covalent bond with a nylon 6 molecule at each end, to double the molar mass of nylon 6.[12] Viscosity molar mass increased with the addition of the linking agent. It reaches the maximum when the molar ratio of the reactive functional groups (amine and epoxy) is close to 1, but decreases with more addition of the linking agent. When more epoxy groups than the equimolar reacting amount were added (the amount of functional groups are mismatched), only one of two epoxy groups on the same linking agent molecule would react with an amine end group of nylon 6 due to the lack of amine groups; the excess epoxy molecules then form dangling groups attached to nylon 6 molecules rather than bridging two nylon 6 molecules. The molar mass increase due to the linking-agent is then lessened.

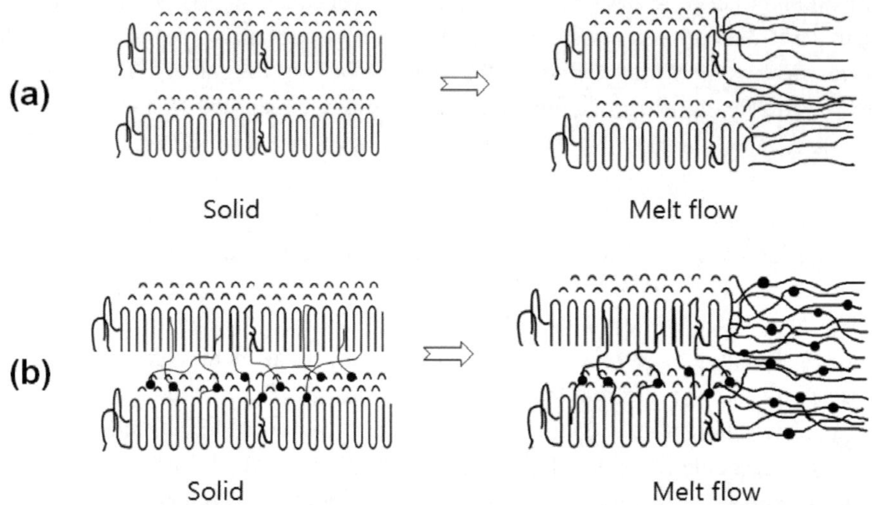

Figure 1. Schematic diagrams of nylon 6 melt behavior. (a) For neat nylon 6: (b) Modified nylon 6 contains some inter-sheet molecules due to the linking agent. (shown as filled circles) The inter-sheet molecules become entangled when subjected to shear stress in the melt state, so modified nylon 6 exhibits a high melt viscosity.

The presence of the linking agent hampers the inclusion of whole molecules in a single sheet (Figure 1): the increased molar mass and the inclusion of bulky groups signify that whole molecules are not easily incorporated into a single folding sheet. The XRD data of some samples showed that there is no structural change of the crystal lattice. The pair of peaks at $2\theta = 20.5^{o}$ and 24° are distinctive features of the α phase in nylon 6. They are the reflections from the (200) and the (002) plane.[10] The degree of crystallinity determined using peak fitting was *ca*.39.5% for nylon 6 and others indicates they have almost the same crystallinity. This implicates that addition of the linking agent inducing chain entanglements does not significantly affect or change the crystalline structure. The half of the molar mass augmented chains can more easily participate in the chain folding of different sheets. [7] Also, intra-sheet fold-loop can easily slide with each other in the melt flow rather than forms entanglements. These latent entanglements between inter-sheet prevent the free slide of molecules in the melt state. This outcome is verified with the rheological property measurements.[13]

The melt viscosities of modified nylon 6 with the frequency shows that there is a broad Newtonian plateau of low viscosity for neat nylon 6 at frequencies below 100 rad/s; above this frequency nylon 6 exhibits weak shear-thinning. Modified nylon 6 reveals a different trend. Modified nylon 6 with ca. 0.16 wt% linking agent (approximate ratio between reacted amine and epoxy groups of 1:0.2) exhibits Newtonian behavior up to a frequency of 10 rad/s, above which it exhibits a shear-thinning behavior. More addition of the linking agent induces higher melt viscosity at low frequencies. The absolute value of the melt viscosity,η, of modified nylon 6 becomes significantly higher than that of neat nylon 6 with the linking agent contents. The physical picture to imagine, as the molar mass of nylon 6 increases with the addition of the linking agent, is a progressive chain entanglements of the molecules in different folding sheets that otherwise flows like a Newtonian fluid (Figure 1). Notably, the viscosity exhibits a very strong shear thinning over the wider frequency range with more linking agent, compared to the near constant viscosity for the nylon 6 melt.

The sample having equimolar reacted amount of the epoxy group of the linking agent and the amine group of nylon 6 (1:1 ratio) exhibits the highest viscosity molar mass and the highest melt viscosity. Surprisingly, there is no Newtonian plateau for this sample but only strong shear-thinning behavior throughout the whole frequency range! Within our knowledge, this is the first time to report that nylon 6 melt shows only shear thinning behavior without the appearance of the Newtonian plateau! The phenomenon is believed to reflect the chain *entanglements*.[11] Similar rheological behavior was frequently observed in various nano-composites containing carbon nanotubes, in which neat matrix polymers show little frequency dependence whereas the nano-composites show strong shear-thinning effect when those composites have filler nanotubes more than the rheological threshold.[19] The nanotubes influence the polymer relaxation dynamics at length scales longer than the entanglement distance. By the same token, we expect chain coupling between nylon 6 molecules by the linking agent to influence the chains mobility and entanglements.

In the modified nylon 6 melts, more coupling between nylon 6 molecules restrain polymer motion to a greater extent, especially at lower frequency to show higher melt viscosity. On flowing, when the chain with the lower folder is pulled out of the lamellar, it will try to pull out also the chains in the upper fold. Interactions between the two chain segments can give rise to actual entanglements (Figure 1) when the involved chain segments are drawn out of the lamellar. [12, 13] When excessive linking agent is added, the molar mass increase is then lessened as are the entanglements between chains. Thus the melt viscosity also decreases because of less entanglements between chains.

CONCLUSIONS

The chain entanglements of modified nylon 6 molecules result in the restriction of nylon 6 chain relaxation as well as of short range dynamics (the local motions of the nylon 6 chains). As the molar mass increases, the dynamic modulus approaches the plateau modulus, which is independent of the overall molar mass of the polymer.[5] When excess linking agent is added, however, the modulus decreases again because of the attenuation of molar mass enhancement and chain entanglements as mentioned before.

ACKNOWLEDGMENTS

This work was supported by MIKE through Original Materials Research Program (RIAM IAC 1410) and by NRF through BK 21 Plus (Materials Science & Engineering Program).

REFERENCES

(1) Hatfield G. R.; Glans, J.H.; Hammond, W.B. (1990 *Macromolecules* 23, 1654-1658
(2) Palmer, R.J. (2003) "Polyamides Plastics" in *Encyclopedia of Polymer Science and Technology*, Vol. 3, 3rd Edn. ed. H.F. Mark, (Wiley Interscience, New York) pp 618-643
(3) Russo, G.M.; Simon, G.P.; Incarnato, L. *Macromolecules* **2006**, 39, 3855-3864
(4) Wu, S., *J. Polym. Sci. B. Polym. Phys*. **1989**,27, 723-741
(5) Doi, M. ; Edwards, S.F. , *The Theory of Polymer Dynamics* (Clarendon Press, Oxford) Chap.7, pp 222-236, **1986**
(6) Cooper, S.J.; Atkins,E.O., Hill, M.J. *Macromolecules* **1998**, 31, 5032-5042
(7) Li, Y.; Goddard III W.A., *Macromolecules* **2002**, 35, 8440-8455
(8) Welch, P.; Muthukumar, M. *Phys. Rev. Lett,* **2001,** 87, pp 218302,1-4
(9) Li, Y.; Goddard III W.A., *Macromolecules* **2003**, 36, 900-907
(10) Laun, H.M., *Rheolo.Acta* **1978**,17, 1-16
(11) Ottani, S.; Porter, R.S., *Macromol. Rapid. Commun.*, **1995**, 16, 813-819
(12) Tanner, R.I. in *Engineering Rheology* (Oxford University Press, New York) Chap.1 pp19-22, **2000**
(13) Seo, Y. submitted to Langmuir 2017

Time-resolved mechanical spectroscopy as a technique to study the influence of thermal history on thermo-oxidative degradation of polyethylene and its blends with polyamide.

Reza Salehiyan[a] and Suprakas Sinha Ray [a, b]

[a]*DST-CSIR National Centre for Nanostructured Materials, Council for Scientific and Industrial Research, Pretoria 0001, South Africa*

[b]*Department of Applied Chemistry, University of Johannesburg, Doorfontein 2028, Johannesburg, South Africa*

Abstract. Thermo-oxidative degradation of linear low density polyethylene (LLDPE) processed at different temperatures (180°C, 220°C, 250°C and 260°C) and their corresponding blends with polyamide (PA6) were studied using time-resolved mechanical spectroscopy (TRMS) technique. The elastic moduli of the LLDPE and LLDPE/PA6 blends increased regardless of the thermal histories. Changing the processing temperatures during the extrusion led to different viscoelastic behaviors for the LLDPE samples where sharper increase in moduli was observed at shorter times in case of samples processed at lower temperatures. Moreover, measuring temperature was found to be influencing the rheological results where the higher measuring temperatures caused the more pronounced growth in moduli. Later on, blending these LLDPE's having different thermal histories with PA6 at 260°C did not show any difference in rheological responses of LLDPE/PA6 where all blends behaved similarly irrespective of the processing temperatures of the LLDPE's. It is known that branching/cross-linking is a considerable phenomenon during the oxidation of the PE and PA6 at high temperatures. Therefore, the growth in moduli of the LLDPE and LLDPE/PA6 was suggested to be due to cross-linking reactions as a result of thermo-oxidative degradation. Further, it was suggested that TRMS is a more accurate technique to measure frequency dependent viscoelastic properties of such transient polymers than the conventional frequency sweep tests where longer experimental time related to the low frequency regions could surpass the degradation time of the material. Therefore, extrapolating the moduli at different scanning frequencies during TRMS to obtain the moduli at time-zero could be the more reliable way to acquire the frequency dependent viscoelastic responses. In addition, isochronal recording of the moduli at different frequencies during TRMS was found to be a useful tool to understand the degree of cross-linking.

Keywords: Thermo-oxidative degradation, Time-resolved rheometry, LLDPE, PA, blends.
PACS: 83.80.Tc Polymer Blends (Rheology), 83.50.Xa Materials Processing (Rheology)

INTRODUCTION

It is well documented that both polymers are prone to thermal degradation when certain conditions are met, e.g., specific temperature ranges, oxygen content, and shear intensities [1, 2]. During their degradation, polymers may undergo chain scission, chain branching (crosslinking), or both simultaneously as competitive reactions, depending on the conditions and extent of the process. Alternatively, the degree of degradation (chain scission or branching) could affect physical properties such as the viscosity and final thermal stability of the products. Therefore, it is important to study the effect of the thermo-oxidation process for polymers to develop materials with superior performance under optimized processing conditions. Among the different rheological testing methods, time-resolved rheology has greater potential for investigating the polymer degradation at the melt state. Time-resolved rheometry is a technique to sequentially acquiring the frequency dependencies of materials (e.g., transient polymers) by probing the dynamic mechanical properties of such materials within their linear viscoelastic regions (constant strain amplitude) over a specific measuring time and at a fixed frequency [3, 6]. Conventional frequency sweep tests are very common to study the dynamic of polymers with respect to their molecular structures and morphologies as they probe polymers within their linear viscoelastic regions where molecular structures do not change during the tests. Low frequency results are richer in information as they are associated with longer relaxation times related to the larger portions of the materials. However, the longer required experimental time related to the low frequency regions could surpass the degradation time of the testing materials. Thus, the long relaxation time associated to the structure of the polymer might not be accurate as degradation has already interfered with the structure and hence the properties. Therefore, time-resolved rheometry is a more reliable alternative to the conventional frequency sweep tests since nonlinear effects from large deformations are excluded, and the viscoelastic properties of polymers can be investigated as a function of time at a fixed frequency rather than being observed in frequency-dependant tests where the polymer is concurrently subjected to both frequency changes and degradation phenomena [5, 6]. The aim of this study was to observe the effect of the thermal history on the kinetics of the thermo-oxidative degradation of linear low density polyethylene (LLDPE) samples and their corresponding blends with PA via a time-resolved rheology technique. Different LLDPE samples were processed at different temperatures from 180 °C to 260 °C in an extruder. Then, in the next step, these differently processed LLDPE samples passed through another extrusion to be blended with PA at 260 °C. This research may help us to understand the degradation kinetics of blends in the melt state.

Experimental

Materials

The polymers used in this study were commercial products. LLDPE, designated as HF120, was purchased from Sasol, South Africa. According to the supplier, it has a melt flow index (190 °C/2.16 kg) = 1.0 g/10 min, density = 0.920 g/m^3, and recommended melt-processing temperature of 190 °C. Nylon 6 (PA6), which was designated as UBE nylon 1030B, was purchased from UBE industries, Ltd. According to the supplier, it has a density = 1.14 g/m^3 and melt range temperature = 215–225 °C. Its measured melt flow index (250 °C/2.16 kg) was found to be 4.56 g/10 min.

Blend Preparation

Neat LLDPE and an 80 wt% pre-processed LLDPE/20 wt% neat PA6 blend were processed using a TE-30 co-rotating twin screw extruder (Nanjing ONLY Extrusion Machinery Co. Ltd), with an L/D of 40. Four neat LLDPE samples were processed at different temperatures:

(1) Processed at 180 °C; the following zone temperatures were used: 120, 160, 180, 180, 180, 180, 180, 180, and 180 °C; the die temperature was 180 °C; and the screw speed was 200 rpm.

(2) Processed at 220 °C; the following zone temperatures were used: 120, 180, 220, 220, 220, 220, 220, 220, and 220 °C; the die temperature was 220 °C; and the screw speed was 200 rpm.

(3) Processed at 250 °C; the following zone temperatures were used: 120, 220, 250, 250, 250, 250, 250, 250, and 250 °C; the die temperature was 250 °C; and the screw speed was 200 rpm.

(4) Processed at 260 °C; the following zone temperatures were used: 120, 220, 260, 260, 260, 260, 260, 260, and 260 °C; die temperature was 260 °C; and the screw speed was 200 rpm.

Blends of 80 wt% pre-processed LLDPE samples and neat 20 wt% neat PA were processed using the following zone temperatures: 120, 200, 260, 260, 260, 260, 250, 250, and 245 °C, with a die temperature of 240 °C and screw speed of 200 rpm. During processing we did not find any change in residence time and color of the blend samples. The extrudate was further moulded using an injection moulding machine (model: ENGEL, e-mac 50) to produce samples for rheological measurements. The following temperature zones were used: 220, 230, 235, 240, and 36 °C (mould temperature).

Rheological Characterization

Rheological analyses were conducted using a Physica MCR501 rheometer (Anton Paar Austria) equipped with 25 mm parallel plates. Time resolved mechanical spectroscopy (TRMS) measurements were carried out under an air atmosphere and a strain amplitude of 0.5% (linear region) within a 1 h period. The selected frequency for the melt-processed LLDPE samples was 0.1 rad/s, whereas various LLDPE/PA blend samples were probed at different frequencies of 0.1, 0.3, 1, 3, 6.28, 10, 30, and 60 rad/s to cover the low, intermediate, and high frequency regions. The time evolution rheological properties of the LLDPE were determined at different temperatures of 190, 220, and 240 °C, whereas the temperature was fixed at 240 °C for the LLDPE/PA blends, which was equal to their mixing temperature during injection moulding. Frequency sweep tests were performed under an air atmosphere and a strain amplitude of 0.5%, with frequencies varying from 0.1 to 100 rad/s.

Result and Discussion

The LLDPE processed at higher temperatures had higher moduli than those processed at lower temperatures in the following descending order LLDPE: (260 °C) > LLDPE (250 °C) >LLDPE (220 °C) >LLDPE (180 °C) > neat LLDPE. In other words, there had already been some alterations in the molecular architecture during processing, leading to preliminary branching stages, where the LLDPE samples processed at higher temperatures had higher degrees of branching. This in turn resulted in the higher modulus during TRMS. Alternatively, it is worth noting that increasing the measuring temperature ramped up the modulus growth rate. This confirms the temperature dependency of the cross-linking reactions in LLDPE samples, which led to a gelation behaviour when G' (t) intersected G" (t) within a few minutes of the measurements.

Interestingly the blends show the same response regardless of the thermal history of the LLDPE samples used in the blends. This suggests that when the LLDPE underwent the second mixing process with PA at 260 °C, its molecular structure was balanced and blending them with the PA eliminated the previous processing temperature effects. Identical morphologies of the blends with different LLDPEs, confirming that the thermal history of the LLDPE during processing has no influence during the second mixing process of blending with PA. Hence, the LLDPE (180 °C)/PA blend was selected to investigate the kinetics of the thermo-oxidative degradation and will be referred to simply the LLDPE/PA blend.

A more reliable way to monitor the degree of cross-linking due to the thermo-oxidative degradation was to record moduli at different times during the multiple-frequency TRMS tests at different frequencies (vertical lines in Figure. 1).

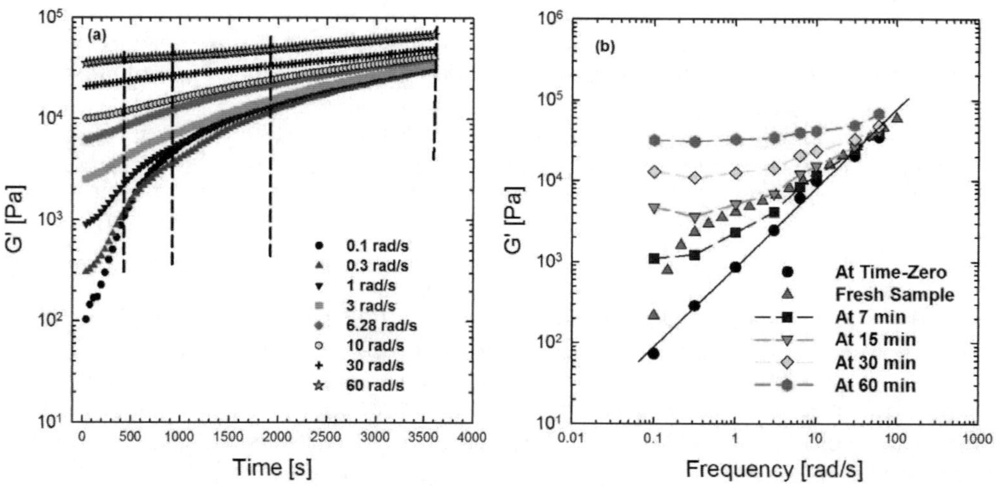

FIGURE 1. (a) TRMS results for elastic (storage) moduli $G'(t)$ of the (80/20) LLDPE/PA blend at different frequencies, with fixed strain amplitude of 0.5% and temperature of 240 °C under air atmosphere. The dashed vertical lines indicate the times selected (7, 15, 30, and 60 min) to collect the isochronal moduli of the blend at different frequencies shown in (b).

Figure 1b. shows the master curves of the isochronal elastic moduli of the blend recorded at different times during the TRMS course, as indicated by the dashed vertical lines in Figure 1a., along with the frequency sweep results for a fresh sample to monitor the extent of cross-linking from the thermo-oxidative degradation process. The modulus of the blend extrapolated at time zero exhibits a power-law dependency at the different frequencies, indicating a terminal behaviour. On the other hand, it shows that if a conventional frequency sweep test is performed, the results are completely different because the molecular structure is still changing during the frequency sweep test as a result of rapid degradation. It can be seen that at low frequencies, the modulus increases rapidly, and then the rate of change slows down, but with higher values than those obtained at time zero. A frequency sweep test (0.1~100 rad/s) takes about 12 min (720 s) and it is clear that the significant growth in moduli occurs at early times within 500 s.

This once again implies that a frequency sweep test may not be a suitable characterization method for such rapidly degrading materials. Further, a non-terminal behaviour (plateau modulus) at lower frequencies appeared after different periods of exposure to air under the TRMS test. This plateau modulus reached higher frequencies as the exposure time increased because the isochronal modulus of the blend at 60 min behaved almost like a plateau throughout the frequency range. The non-terminal behaviour was indicative of a long relaxation mechanism for the blend associated with an increase in molar mass or cross-linking reactions during the exposure time to the air atmosphere under the TRMS test. Thus, it can be reiterated that obtaining isochronal moduli at different frequencies from TRMS tests is a more accurate method for determining the degree of cross-linking due to thermo-oxidative degradation of the polymers.

CONCLUSION

Thermo-oxidative degradation of LLDPE samples with different thermal histories and their corresponding blends was investigated using a time-resolved rheometry technique, where the time evolution rheological properties of the blends were probed using scans at different frequencies. It was found that the processing temperature had significant effects on the viscoelastic behaviour and degradation of the LLDPE samples. However, it did not have any influence on the viscoelastic and morphological properties of the LLDPE/PA blends. The thermo-oxidative degradation led to discoloration at the edges of the samples where they were in direct contact with air. This in turn caused a significant upturn shift in the elastic modulus, especially at shorter times. It was concluded that dominant cross-linking reactions were the reason for the increase in the modulus and discoloration of the blends upon the thermo-oxidative degradation. It was shown that the extrapolation of data points for such transient materials from TRMS tests at different frequencies was a more accurate method for acquiring the rheological properties as a function of the frequency compared to conducting conventional frequency sweep tests, which suffer from molecular architectural alterations due to rapid degradation mechanisms. Finally, it was suggested that the isochronal collection of data at different frequencies was a more convenient way to observe the degree of network formation as a result of degradation at different exposure times. Although different mechanism routes were previously proposed for the degradation reactions (chain scission and cross-linking) of individual polymers (PEs and PA), the reactions at the interphase between LLDPE and PA are unknown and could be investigated in further studies.

ACKNOWLEDGMENTS

The authors would like to thank the Council for Scientific and Industrial Research (HGER20S) and Department of Science and Technology (HGERA8X) for financial support.

REFERENCES

1. E. M. Hoang, N. S. Allen, C. M. Liauw, E. Fontan and P. Lafuente. *Polym. Degrad. Stabil.* **91**, 1363-1372 (2006).
2. W. Dong and P. Gijsman. *Polym. Degrad. Stabil.* **95,** 1054-1062 (2010).
3. M. Kruse and M. H. Wagner. *Rheol. Acta.* **55,** 789-800 (2016).
4. M. Mours and H. H. Winter. *Rheol. Acta.* **33**, 385-397 (1994).
5. G. Filippone, S. C. Carroccio, R. Mendichi, L. Gioiella and N.Tz. Dintcheva. *Polymer* **72,** 134-141 (2015).
6. G. Filippone, S. C. Carroccio, G. Curcuruto, E. Passaglia and C. Gambarotti. *Polymer* **73**, 102-110 (2015).

Effect of morphology and rheology on Nylon 6 – glass fibre composites

Gupta Rahul K [1]*, Perumal Deverpiran Vishak [1], Bhattacharya Sati N [1], Costa Franco S [2]

[1] RMIT University - Victoria - Australia, [2] Autodesk Australia Pty Ltd - Victori - Australia

In an attempt to understand the influence of fillers on the rheology of composite melts, oscillatory shear experiments were carried out on Nylon-6 composites containing short glass fibres (length <1mm) of different weight fractions: 0% (PA-6), 15% (PA6-15GF) and 30% (PA6-30GF). Oscillatory time sweep experiments carried out within the Linear Viscoelastic Region (LVR) using a parallel plate rheometer showed an increase in complex viscosity as a function of time for filled systems. PA6-30GF samples exhibited a greater increase in viscosity than PA6-15GF samples while unfilled PA-6 samples showed a constant viscosity. While an increase in fibre orientation usually results in the reduction of complex viscosity, these results suggest that other phenomenon such as fibre migration and flow induced crystallisation aided by fillers could be occurring. Differential Scanning Calorimetry (DSC) was used to characterise the crystallisation behaviour of the samples. X-Ray Computed Tomography (X-CT) was employed to measure 3-dimensional fibre orientation and local fibre concentration in the fibre filled samples at a resolution of 2.5µm. The research work aims to gain a better understanding of how specific flow mechanisms impact rheological properties of high fibre content composites which are commonly used in industrial processes.

pps proceeding
Symposium: S14 - Rheology and rheometry
Oral Presentation

Transient rheology and 3-dimensional fibre orientation evolution in Nylon-6 short glass fibre composites

Perumal Deverpiran Vishak (1)*, Gupta Rahul K (1), Bhattacharya Sati N (1), Costa Franco S (2)

(1) RMIT University - Victoria - Australia, (2) Autodesk Australia Pty Ltd - Victoria - Australia

In an attempt to relate transient rheology to 3-dimensional fibre orientation, experimental testing on compression moulded Nylon-6 samples with 30% weight short glass fibres (PA6-30GF) and unfilled PA-6 were carried out. The application of steady shear on PA6-30GF samples resulted in a viscosity overshoot before reaching steady state values. Unfilled PA-6 did not display any viscosity overshoot. Observation of the viscosity overshoot has been attributed to the reorientation of fibres from an initial orientation to a steady state orientation. Samples subjected to different strain units were then obtained from the rheometer to preserve the fibre orientation. X-Ray Computed Tomography(X-CT) was used to obtain 3-dimesional fibre orientation and fibre concentration. As a means of obtaining a more repeatable initial fibre orientation, rheological and morphological data from injection moulded PA6-30GF samples were also obtained for comparison. With higher applied strains, the orientation of fibres in the shearing direction was observed to increase while the orientation in the radial and thickness directions were observed to decrease. In order to validate the experimental fibre orientation data, a comparison was performed with simulations from Folgar-Tucker and Reduced Strain Closure (RSC) orientation models. The predicted fibre orientation results showed agreement with experimental data after empirical fitting. Using this data, a constitutive stress equation incorporating effects of fibre orientation is being developed for improving composite flow predictions.

Rheological Behavior of Cellulose Nanofibers from Cassava Peel Obtained by Different Methods

Aline Czaikoski[a] and Florencia Cecilia Menegalli[a]

[a]*Department of Food Engineering, School of Food Engineering, University of Campinas, Campinas, São Paulo, Brazil.*

Abstract. To identify the rheological properties of cellulose nanofibers suspensions is of high importance to understand the interaction of nanofibers with other components and to evaluate the possible use in materials and process as coating, thickening and extrusion. Thus, the main objective of this study was to determine the rheological behaviors of aqueous suspensions with different cellulose nanofibers extracted from cassava peel. To isolate the nanofibers, a combination of pre treatments with acid hydrolysis or catalytic oxidation with TEMPO and ultrasonic disintegration were used. It has been observed that the nanofibers presented network structure and diameters on the nanometer scale. The results of the oscillatory tests showed that all the nanofibers produced presented a gel-like behavior. The nanofibers obtained by acid hydrolysis presented higher gel strength than those produced by TEMPO-mediated oxidation. The sonication treatment increased the storage and loss modulus. Meanwhile, in the steady state tests, it was found that the nanofibers obtained by acid hydrolysis presented a thixotropic character and a viscosity profile with three regions (shear thinning - Newtonian plateau - shear thinning). It can be observed that the nanofibers produced can be used as reinforcement material in nanocomposites materials.

Keywords: viscosity; rheology; mechanical resistance; nanoparticles; biodegradable packaging.
PACS: S14-454

INTRODUCTION

Cellulose is a structural polymer of plant cell walls and is the most abundant biopolymer on earth. In addition to its biodegradability and renewability, the cellulosic fibers in nano dimensions have technological characteristics favorable for use in packaging, such as low thermal expansion, high aspect ratio (length in relation to diameter), strengthening effect, good mechanical and optical properties[1]. Recently due to the favorable technological characteristics properties of nanofibers that increased interest in production and characterization of cellulose nanofibers, it is recognize that their rheological behavior may play an important role in much process[2].

Cassava (Manihot esculenta) is a root crop largely grown in Brazil that generates a high amount of wastes as peels and residual bagasse when it is used in the production of cassava starch or minimum processed tubers for final consume. Then, with the aim of reduce environmental problems and to generate products with higher added value, this study related the morphological characterization with the rheological behaviors of cellulose nanofibers produced by different methods of extracting, for the correct applications.

MATERIAL AND METHODS

The peels (inner, peel and bark) of cassava were obtained in the region of Campinas - Brazil, during the harvest of January 2016. The peels were properly classified and washed in running water. Then they were sanitized with sodium hypochlorite solution (250 ppm) for 10 min and dried in a forced convection oven at 50 ° C for 48 h. Next the peelings were cut and ground on a high-performance professional blender LT-2.0 Super Skymsen from Metallurgical Siemsen Ltda (Santa Catarina, Brazil). Before going through the processes for obtaining the cellulose nanofibres, the cassava peels underwent a pre treatment for the removal of the amorphous components, according to the methodology of Leite, Zanon & Menegalli[3].

Acid hydrolysis, TEMPO-mediated oxidation and Physical treatment

The acid hydrolysis was carried out according to the methodology Leite, Zanon & Menegalli[3] with sulfuric acid concentration was 30 %. Already the process TEMPO-mediated oxidation was carried out using the method of Saito et al. (2007)[4], for twenty-five minutes. Half the suspensions obtained by acid hydrolysis and TEMPO-mediated oxidation underwent sonication and the other half did not. An Ultrasonic Disruptor / Sonicator (QR 750 W Ultranique from ECO-SONICSP, Brazil) was used for approximately 20 minutes with an ultrasonic power of 300 W.

Cellulose nanofibers characterization

The morphology of cellulose nanofibers were annualized using transmission electron microscopy (TEM). TEM images were captured with a Zeiss CEM902 microscope equipped with an EELS analyzer; the accelerating voltage was 15 kV. The surface charge were determined using the Zetasizer model Nano ZS from MALVERN Instruments Ltd., U.K by the dynamic light scattering technique at a detection angle of 173 °. AFM images were acquired on a Microscope Park Systems, model NX-10 (Suwon, Korea) equipped with Si Nano sensor probes manufactured with constant spring of 42 $N m^{-1}$; the resonance frequency was about 320 kHz. The acquired images were treated in the software GWYDDION, version 2.4. The average diameter and length of nanofibers were measured using the same software for calculating the aspect ratio.

Rheological characterization

The suspensions of nanofibers were diluted in distilled water to give different concentrations of nanofibers (1.0%, 1.4% and 1.8% w / w). The rheological behavior of the CNFs suspensions was studied using a stress-controlled rheometer MCR 301 (Anton Paar, Austria) equipped with cone and plate geometry with 6 cm of diameter, gap of 0.208 mm and angle of 2 °. The oscillation strain sweeps from 0.1 to 10% at constant angular frequency of 1 rad/s were carried out to define the linear viscoelastic range. Then the oscillation frequency sweeps in the range of 0.01–10 rad/s. The strain (%) for each samples were performed. After, the mechanical spectra of the various suspensions were determined by frequency sweep between 0.01 Hz to 10 Hz, the constant strain in the linear viscoelasticity. All the measurements were carried out at controlled temperature of 25 °C.

RESULTS AND DISCUSSION

The viscoelastic properties of the CNF suspensions prepared with different methods are illustrate in Figure 1. The cellulose nanofibers obtained by the acid hydrolysis presented the storage modulus (G ') and loss modulus (G'') with less dependence of the angular frequency (ω), than those obtained by catalytic oxidation. All the CNF suspensions exhibit gel-like properties with G'>G''. However, the nanofibers isolated by acid hydrolysis showed G' larger than those isolated by TEMPO-mediated oxidation. This may have occurred because CNF obtained by acid hydrolysis present more negative charges on their surfaces than CNF obtained by TEMPO-mediated oxidation. For the zeta potential of the nanofibers obtained by acid hydrolysis (-49 mV) is higher than that of the nanofibers obtained catalytic oxidation (-42 mV). These negative charges raise repulsion forces between the nanofibers, causing a better dispersion, and immobilize water molecules near the fibers, thus increasing the elastic character of the suspensions[5, 6].

The sonication was also responsible for the increase of the storage modules of the CNF. The gel-like behavior of nanofibres is also obtained by physical interaction between the fibers, which are the interlacing[6]. Mishra et al.[7] reports that the sonication treatment causes a higher fibrillation of the nanofibers, and this can be observed in Figure 2. This better dispersion or fibrillation caused by the ultrasonic treatment increased the surface area, which facilitated the physical interactions between the nanofibers causing the entanglement, thus increasing their gel strength. The range of average aspect ratio (L/D) of the cellulose nanofibers was of 242-371.

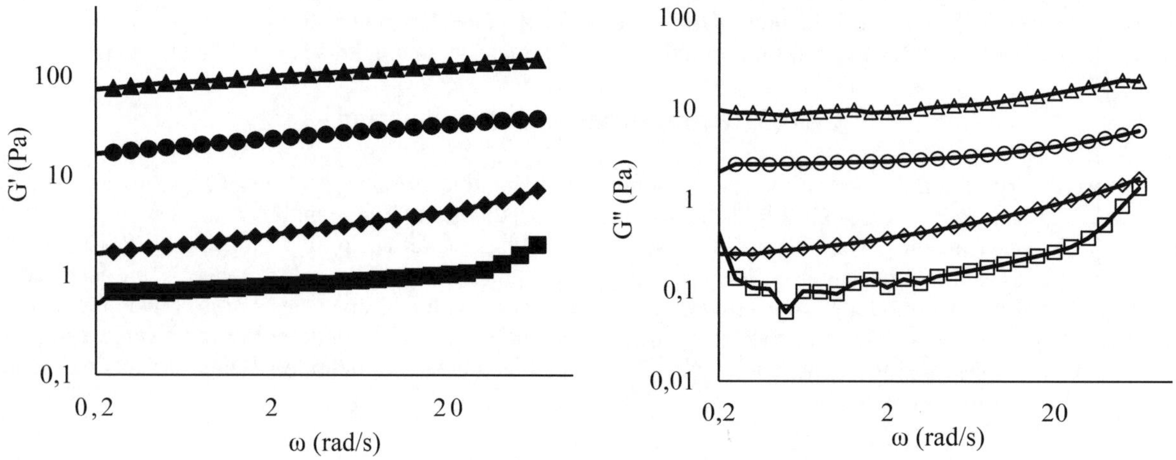

FIGURE 1. The storage modulus (G') and the loss modulus (G'') of suspensions with 1.4 % (w/w) CNF as a function of frequency for different CNF suspensions: NTOws (♦); NTOwos (■); NAHws (▲); NAHwos (●). NTOws (◊); NTOwos (□); NAHws (Δ) and NAHwos (○).Where: "N" refers to nanofibers, "TO" refers to TEMPO-mediated oxidation, "AH" refers to acid hydrolysis, "ws" refers to nanofiber with sonication and "wos" refers to nanofiber without sonication.

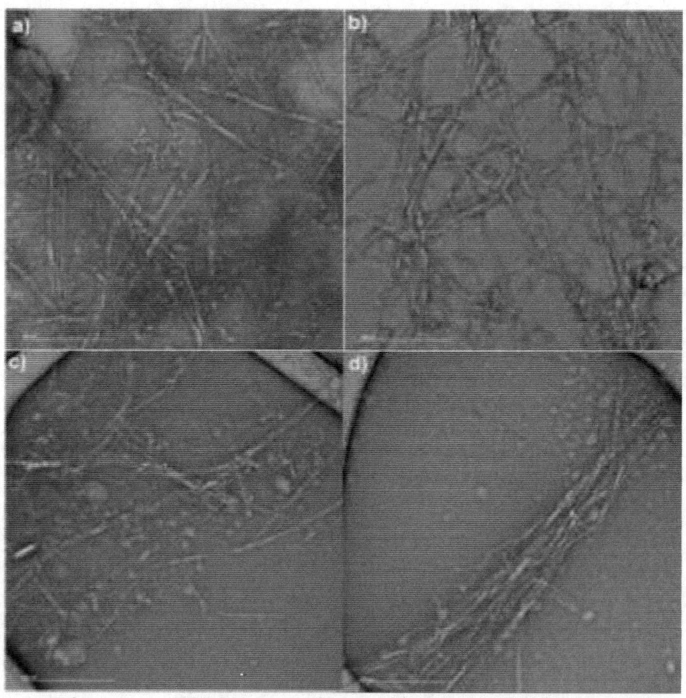

FIGURE 2. Transmission electron microscopy (TEM) images for: a) NAHws; b) NAHwos; c) NTOws and d) NTOwos. Where: "N" refers to nanofibers, "TO" refer to TEMPO-mediated oxidation, "AH" refers to acid hydrolysis, "ws" refers to nanofiber with sonication and "wos" refers to nanofiber without sonication.

All the studied NFC suspensions exhibit a shear-thinning behavior, as seen from the flow curves in Figure 3. The viscosities of CNF obtained by acid hydrolysis decreased in shear rates <10 s^{-1}, a shear thinning behavior. At shear rates between 10 s^{-1} and 25 s^{-1}, the viscosities became constant, forming Newtonian plateaus. Above these rates, the viscosities decreased again (shear thinning). However, the nanofibers obtained by catalytic oxidation presented only the shear thinning behavior along the shear rates used. The similar behavior of CNF obtained by acid hydrolysis was reported in the other studies. Iotti et al.[8] found this viscosity behavior for microfibrillated cellulose. They believed that the formation of the Newtonian plateau was due to an induction of the formation of another structure by the shear. However, Karppinen et al.[9] also working with suspensions of microfibrillated cellulose and using images captured by a transparent outer cylinder in a concentric cylinder sensor, suggested that a shear-induced phase separation in the shear range where the Newtonian plate appeared. Nechyporchuk, Belgacem & Pignon[2] also reported that this phenomenon occurred due to phase separation suspensions of microfibrillated cellulose. Bettaieb et al.[10] attributed this behavior in the flow of nanofibers to slippage of the suspension over the sensor, although the authors used rough sensors.

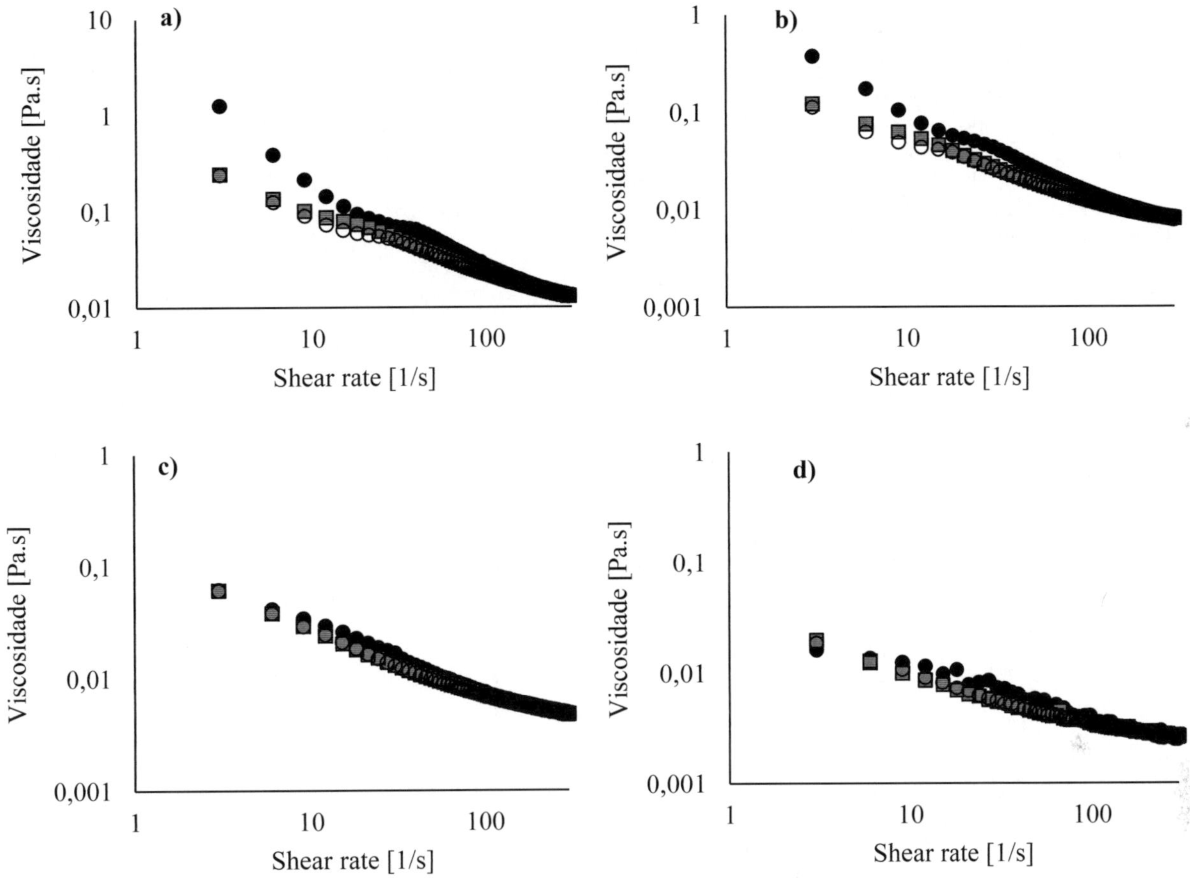

FIGURE 3. Viscosity of suspensions with 1.4 % (w/w) CNF as a function of shear rate for: a) NAHws; b) NAHwos; c) NTOws and d) NTOwos. (●) ramp 1; (□) ramp 2 and (○) ramp 3. Where: "N" refers to nanofibers, "TO" refer to TEMPO-mediated oxidation, "AH" refers to acid hydrolysis, "ws" refers to nanofiber with sonication and "wos" refers to nanofiber without sonication.

CONCLUSION

This work demonstrated that the rheological studies revels to be an efficient method to choose the better treatment to isolate cellulose nanofibers and that the agricultural residue cassava peel has great potential for the isolation cellulose nanofibers. All the cellulose nanofibres obtained in this study presented behavior gel-like. The acid hydrolysis produced nanofibers with higher gel strength than the nanofibers obtained by TEMPO-mediated oxidation, because the acid treatment was more efficient in introducing surface negative charges. The sonication process used also interfered in the gel strength of these materials, since the energy storage modulus of the aqueous suspensions of the nanofibers showed an increase after this process. With the flow curves, it was possible to observe that the cellulose suspensions present inhomogeneous flow at low shear rates.

ACKNOWLEDGMENTS

The authors would like to acknowledge the financial support provided by Coordenação de Aperfeiçoamento de Pessoal de Nível Superior (CAPES) and Conselho Nacional de Desenvolvimento Científico e Tecnológico (CNPq)

(131520/2015-6). The authors would also like to acknowledge the Brazilian Nanotechnology National Laboratory (LNNano) for allocation of the equipments.

REFERENCES

1. H. A. Khalil, Y. Davoudpour, M. N. Islam, A. Mustapha, K. Sudesh, R. Dungani and M. Jawaid. *Production and modification of nanofibrillated cellulose using various mechanical processes: a review*. Carbohydrate polymers, 2014, v. 99, pp. 649-665.
2. O. Nechyporchuk, M. N. Belgacem and F. Pignon. *Rheological properties of micro-/nanofibrillated cellulose suspensions: wall-slip and shear banding phenomena*. Carbohydrate polymers, 2014, v. 112, pp. 432-439.
3. A. L. M. P. Leite, C. D. Zanon and F. C. Menegalli. *Isolation and characterization of cellulose nanofibers from cassava root bagasse and peelings*. Carbohydrate Polymers, 2017, v. 157, pp. 962-970.
4. T. Saito, S. Kimura, Y. Nishiyama and A. Isogai. *Cellulose nanofibers prepared by TEMPO-mediated oxidation of native cellulose*. Biomacromolecules, 2007, v. 8, pp. 2485-2491.
5. K. Benhamou, A. Dufresne, A. Magnin, G. Mortha and H. Kaddami. *Control of size and viscoelastic properties of nanofibrillated cellulose from palm tree by varying the TEMPO-mediated oxidation time*. Carbohydrate Polymers, 2014, v. 99, pp. 74 – 83.
6. M.-C. Li, Q. Wu, K. Song, S. Lee, Y. Qing and Y. Wu. *Cellulose Nanoparticles: Structure-Morphology-Rheology Relationships*. Acs Sustainable Chemistry & Engineering, 2015, v. 3, pp. 821-832.
7. S. P. Mishra, A. S. Manent, B. Chabot and C. Daneault. *Production of nanocellulose from native cellulose–various options utilizing ultrasound*. BioResources, 2011, v. 7, pp. 0422-0436.
8. M. Iotti, O. W. Gregersen, S. Moe and M. Lenes. *Rheological studies of microfibrillar cellulose water dispersions*. J Polym Environ, 2011, v. 19, pp. 137–145.
9. A. Karppinen, T. Saarinen, J. Salmela, A. Laukkanen, M. Nuopponen and J. Seppa¨la¨. *Flocculation of microfibrillated cellulose in shear flow*. Cellulose, 2012, v. 19, pp.1807–1819.
10. F. Bettaieb, O. Nechyporchuk, R. Khiari, M. F. Mhenni, A. Dufresne and M. N. Belgacem. *Effect of the oxidation treatment on the production of cellulose nanofiber suspensions from Posidonia oceanica: the rheological aspect*. Carbohydrate polymers, 2015, v. 134, pp. 664-672.

Comparative Investigations on Different Shear Thickening Suspensions for Protective Applications

Mishra Swarna [(1)], ***Kumari Neetu*** [(1)], ***Pattanayek Sudip*** [(1)], ***Ghosh Anup K.*** [(1)]*

[(1)] Center of Polymer Science and Engineering, Indian Institute of Technology Delhi - Delhi - India

Although suspensions comprising of nano or micro sized particles in a Newtonian liquid medium are ubiquitous, certain criteria has to be met for a suspension to be called as shear thickening fluids. They function post a critical particle concentration depending on the nature of particle, suspending medium, additives etc. Even with the development of scaling relations and predictive models for rheological parameters, the contribution of forces of interaction (responsible for their flow behavior) present in a suspension is still ambiguous. In the current study we consider non-colloidal (cornstarch) and colloidal (silica) particles in a Newtonian fluid (water, PEG) as comparative studies correlating with their rheological properties. However, with the help of rheology (steady, non-linear dynamic), we present the occurrence of shear thinning in colloidal particle suspensions and compare the forces of interaction. For 50 wt% of cornstarch in water we observed a continuous shear thickening with a critical shear rate of 21.5 s-1. The viscosity rises from 0.026 to 0.239 Pa.s post critical shear rate upto 1000 s-1. However, with the addition of PEG 400 to water in 1:1 ratio at same cornstarch concentration, a shear thinning behavior is observed with high yield stress. Similarly, fumed silica in PEG 400 at 19 wt% particle concentrations and surface functionalized silica in PEG 400 at 62 wt% particle concentrations also showed a shear thinning behavior with high yield stress. The dynamic rheology of all the suspensions under study however showed high elastic modulus in the frequency range of 0.1 to 100 s-1. The amplitude strain under which linear viscoelastic behavior observed is less than 0.1% for all the suspensions. Further, we also present impact behavior and energy dissipation analysis for shear thickening suspensions using Abaqus. By combining the experimental and simulation results, we provide a route where one can choose the suspensions for suitable applications.

Numerical simulation of complex fluids under simple and oscillatory shear flow.

Gómez-López Aldo,[a] Ferrer-López Víctor H.,[b] Mil-Martínez Rubén,[b] Chávez-Castellanos Ángel E.[c] and Vargas-Aguilar René O.[d]

[a]*Facultad de Ingeniería, Universidad Nacional Autónoma de México, Ciudad de México, 04510, Mexico*
[b]*ESIME-Zacatenco, Instituto Poliécnico Nacional, Ciudad de México, 07738, Mexico*
[c]*Departamento de Ingeniería Química, Facultad de Química, Universidad Nacional Autónoma de México, Ciudad de México, 04510, Mexico*
[d]*ESIME Azcapotzalco, Instituto Politécnico Nacional, Av. De las Granjas 682, Col. Santa Catarina, Del. Azcapotzalco, Ciudad de México, 02250, Mexico*

Abstract. In this work, a multiscale approach of kinetic models and transient network dynamics in dilute regime to model the rheology of complex fluids is presented. The Brownian Configuration Fields (BCF) method is applied to kinetic models and different microstates representing the physical interaction between segments. The evolution of the microstates includes a function that represents a variable limiting extension length of the mesoscopic structure. Numerical results show the influence of microscopic and macroscopic parameters on the transient and steady rheological characterization of complex fluids.

Keywords: Complex fluids, Mesoscopic scale, Numerical Simulation.

INTRODUCTION

The micro-macro technics are versatile tools to model complex fluids with microstructure evolution under flow. The macroscopic behavior depends on the microstructure evolution and on the required energy to form or destroy complex or simple structures, full details can be found in [5-7,10]. In this work, the numerical simulation of simple and oscillatory shear flow of a complex fluid confined between to plates (see Fig. 1) are presented.

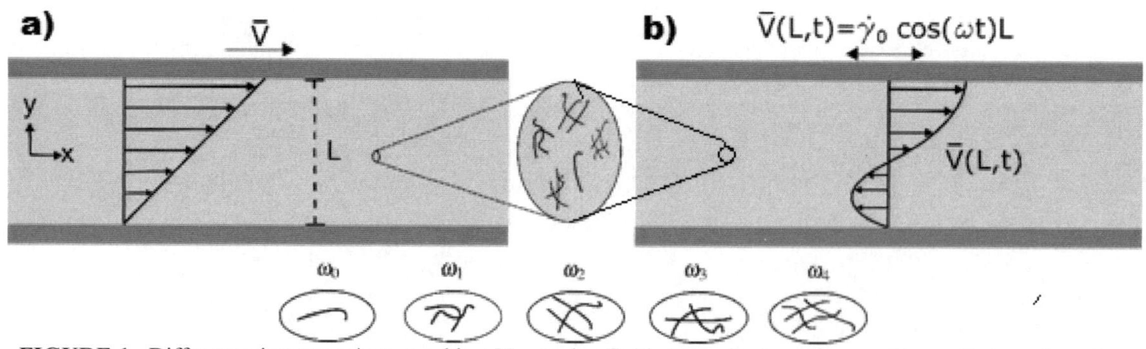

FIGURE 1. Different microstates immersed in a Newtonian fluid a) simple shear flow and b) oscillatory shear flow.

MATHEMATICAL MODELING

Governing Equations

To apply the BCF method, the equations to be solved correspond to the momentum (1) and configurational space connector vector (2). It is implicit that the continuity equation for an incompressible fluid must be satisfied.

$$\mathrm{Re}\frac{\partial \underline{v}}{dt}=\beta\frac{\partial^2 \underline{v}}{\partial \underline{x}^2}+\frac{\partial \underline{\tau}_p}{\partial \underline{x}} \tag{1}$$

$$d\underline{Q}=\left(-\underline{v}\cdot\nabla\underline{Q}+\underline{\kappa}\cdot\underline{Q}-\frac{1}{2}\underline{F}(\underline{Q})\right)dt+d\underline{W} \tag{2}$$

with $\beta=\eta_s/\eta_0$ and $\underline{F}=\underline{Q}/[1-Q^2/b]$ representing the viscosity ratio and Warner spring law, respectively. Finally, the non-Newtonian contribution to the extra stress is evaluated using the Kramers expression:

$$\underline{\tau}_p=\alpha_{b,d}(1-\beta)\left(\langle \underline{Q}\otimes\underline{F}(\underline{Q})\rangle-\underline{I}\right) \tag{3}$$

where $\alpha_{b,d}=1$ for Hookean materials $(b\to\infty)$ and $\alpha_{b,d}=(b+d+2)/b$ for FENE. The dimensional analysis of equations (1) and (2) was stablished by using the following dimensionless variables

$$\underline{x}=\frac{\underline{x}^*}{L},\underline{v}=\frac{\underline{v}^*}{U},t=\frac{t^*}{\lambda},\underline{\underline{\tau}}=\frac{\lambda\underline{\underline{\tau}}^*}{\eta_0},b=\frac{HQ_c^2}{kT},Q_c=\sqrt{\frac{kT}{H}}$$

Kinetic Equations

When microstructure evolution is considered, the following kinetic equations are used to describe generation and destruction of the microstates. The usage of these equations is well described in Rincón et al. (2005).

$$\frac{dC_0}{dt}=-AC_0^p+mBC_n \tag{4}$$

$$\frac{dC_n}{dt}=\frac{1}{m}AC_0^p-BC_n \tag{5}$$

Numerical Procedure

The numerical solution of equations (1) to (5) was performed using a finite difference discretization and applying suitable boundary conditions. Particularly, the solution of equation (2) is carried out by applying Öttinger's predictor corrector scheme (Öttinger, 1996; Vargas et al., 2009; Ferrer et al. 2017). The numerical algorithm consists in the following steps:

1. Set initial and boundary conditions for velocity vector and microstate concentrations.
2. Generate a family of random numbers and assign them to each connector vector to stablish their initial equilibrium configuration when the fluid is at rest.
3. Solve momentum equation and used the updated velocity gradient to compute viscous dissipation.
4. Solve kinetic equations using the viscous dissipation computed in the previous step, then update the maximum extension length of the dumbbells.
5. Solve equation (2) by applying the Öttinger predictor-corrector scheme, then use Kramers expression to compute the non-Newtonian contribution to the extra stress.
6. Repeat from step 2 until steady state is achieved.

RESULTS AND DISCUSSION

Transient Network Dynamics under Simple Shear Flow

For comparison, three different constitutive models are used (figure 2). These models are the Oldroyd-B, FENE-P and FENE-M models. The Oldroyd-B is used without a retardation time. The FENE-P and FENE-M are completely equivalent in steady state as shown in figure 2 (d) and they follow different paths during transient state as shown in figures (b) and (c). Since FENE-M is more stable than FENE-P, another simulation with a higher maximum extension length, b=900, is used. This simulation is more proximate to the FENE behavior as expected from theory.

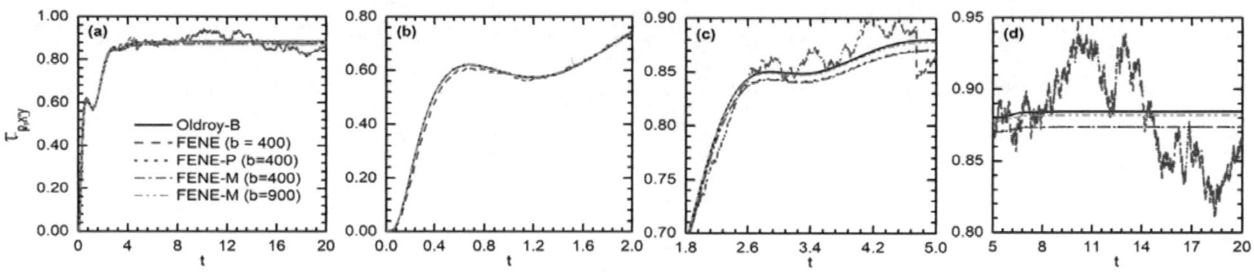

FIGURE 2. Polymeric stress evolution using different models.

In figure 3, the apparent viscosities predicted by the present model are compared with experimental results. In this case, microstructure evolution promotes different equilibrium values in the power law zone for a fluid with $\beta = 1000$ and different kinetic constants representing a weakly structure network (A=0.01, B=1) and a highly structure network (A=0.1, B=0.01). Although the numerical results do not match at all with the experimental data, this figure suggests that adjusting some model parameters is sufficient to improve the predictions.

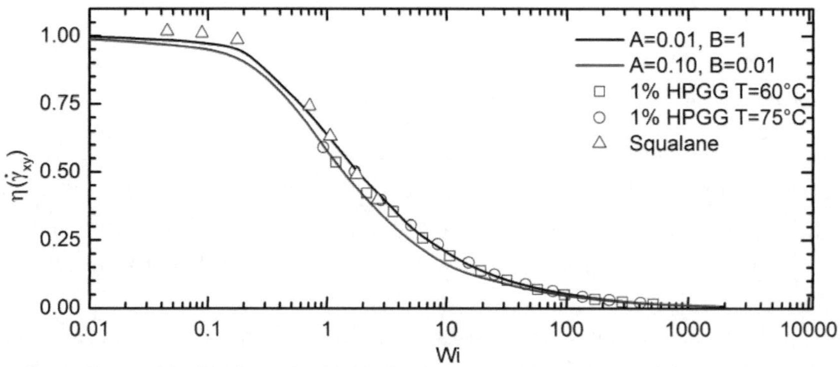

FIGURE 3. Comparison of a weakly (black) and a highly (red) structured networks with experimental results [Zhang et al., 2007; Bair et al., 2014].

The shear stress growth coefficient at different Wi and different β are plotted in figure 4. The insets show the velocity transient evolution with their characteristic viscoelastic effects. At low Wi, the noise from Brownian motion is evident, but at high Wi its influence diminishes due to molecular alignment. Let thixotropy be represented by microstructure kinetics, then it is evident that thixotropy is related to the occurrence of multiple overshoots in the shear stress growth coefficient curves.

Oscillatory Shear Flow for a Complex Fluid

In Figure 5 are shown the velocity profiles at two instances (t=0.1 and t=10) and its corresponding microstructure distribution. At early times for a strong structured network, only the ω_4 microstructure is presented, this structure is destroyed to form different basic structures and for longer times, the structure ω_0 dominates close to the upper plate because this region is where the viscous dissipation is higher.

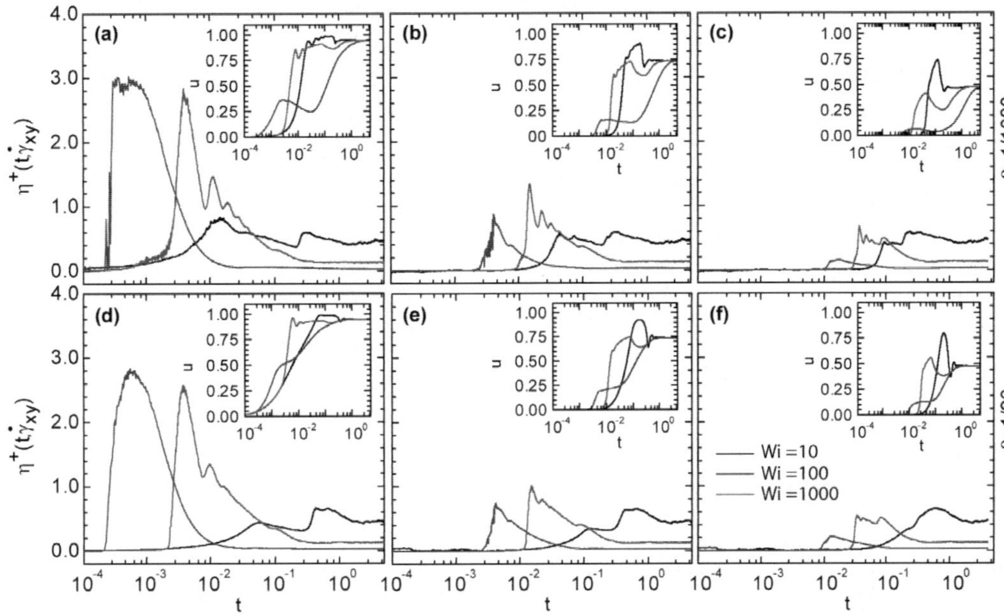

FIGURE 4. Shear stress growth coefficient of a weakly structured network at different heights. a) and d) y=0.95, b) and e) y=0.75, c) and f) y=0.5. The insets show the velocity evolution.

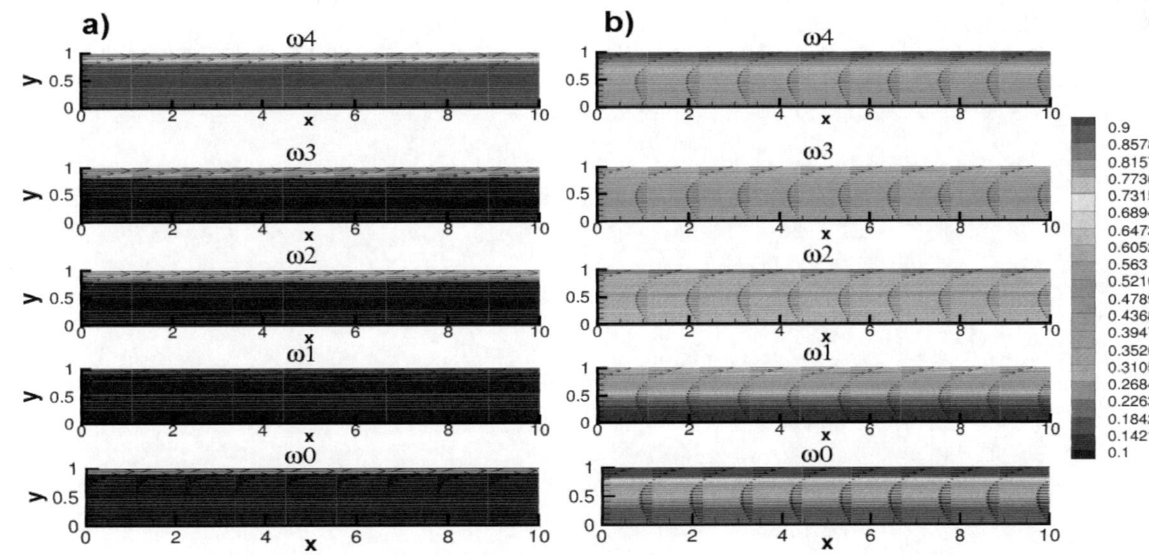

FIGURE 5. Microstates evolution and velocity vector at a) t=0.1 and b) t=10

CONCLUSIONS

In this work, the numerical simulation of a complex fluid confined between two plates is presented. The simple and oscillatory shear flow using a micro-macro technique are shown, with this technique the principal macroscopic and microscopic variables of a complex fluid can be analyzed.

ACKNOWLEDGMENTS

A. Gómez and V. H. Ferrer would like to acknowledge the financial support from the National Council of Science and Technology (CONACyT) of Mexico. R. O. Vargas appreciates support from project SIP-IPN 20171482.

REFERENCES

1. M.S. Green and A. Tobolsky, A new approach to the theory of relaxing polymeric media. *J. Chemical Physics*.1946, 14, 80-89.
2. A.S. Lodge, A network theory of flow birefringence and stress in concentrated polymer solutions. *Trans. Faraday Soc.* 1956, 52, 120.
3. M. Yamamoto, The visco-elastic properties of network structure I. General formalism *J. Phys. Soc. Jpn*. 1956 11, 413.
4. M.D. Chilcott and J. M. Rallison, Creeping flow of dilute solution past cylinder and sphere *J. Non-Newt. Fluid Mech.*, 1988, 29, 381-432.
5. E. Rincón, A. E. Chávez, R. Herrera, O. Manero, Rheological modelling of complex fluids: A transient network model with microstates. *J. Non-Newt. Fluid Mech.* 2005, 131, 64-77.
6. H.C. Öttinger: Stochastic Processes in Polymeric Fluids: Tools and examples for Developing Simulation Algorithms. Springer-Verlag, Berlin, 1996.
7. Ferrer, V. H., A. Gómez, J. A. Ortega, O. Manero, E. Rincón, F. López-Serrano and R. O. Vargas (2017). Modeling of complex fluids using micro-macro approach with transient network dynamics.
8. Rheologica Acta 56(5), 445–459.Li-Ming Zhang, Tao Kong, and Peter S Hui. Semi-dilute solutions of hydroxypropyl guar gum: Viscosity behaviour and thixotropic properties. Journal of the Science of Food and Agriculture, 87(4):684-688, 2007.
9. Scott Bair, Tsuyoshi Yamaguchi, Ludwig Brouwer, Hubert Schwarze, Philippe Vergne, and Gerhard Poll. Oscillatory and steady shear viscosity: The cox-merz rule, superposition, and application to ehl friction. Tribology International, 79:126-131, 2014..
10. R.O. Vargas, O. Manero, and T.N. Phillips. Viscoelastic flow past confined objects using a micro-macro approach. Rheol. Acta, 48:373-395, 2009.

Dual-slit rheo-optical die for the characterization of process sensitive melts

Teixeira Paulo (1), Ferrás Luís (1), Hilliou Loic (1), Covas Jose Antonio (1)*

(1) University of Minho - Guimaraes - Portugal

Instrumented slit dies coupled to single or twin screw extruders have been used by many researchers to measure shear viscosity and normal-stress differences and perform rheo-optical experiments, namely small angle light scattering, birefringence, polarized optical microscopy and turbidimetry. Generally, generation of a range of shear rates implies operating the extruder with varying screw speeds or feed rates. However, this changes the thermomechanical environment inside the extruder and, consequently, may affect the characteristics of the material at the inlet of the die, thus putting at risk the validity of the rheo-optical characterization. To avoid this problem, the authors developed a dual-slit die, which is an adaptation of the twin-slit concept proposed earlier [1,2]. It consists of two flow channels, each fitted with an output control valve at its inlet. By handling simultaneously the two valves, shear rate in the measuring channel can be changed within a range while maintaining constant the total pressure drop. A few constructive and measurement issues are discussed. Experimental data for a few representative polymer systems are presented. [1] P. Springer, R. S. Broadkey, R.E. Lynn, Development of an extrusion rheometer suitable for on-line rheological measurements, Polym Eng Sci, 15, 583 (1975) [2] B. Vergnes, G. Della Valle, J. Tayeb, A specific slit die rheometer for extruded starchy products. Design, validation and application to maize starch, Rheol Acta, 32, 465 (1993)

TRIBORHEOLOGICAL STUDY OF SYSTEMS SOLID–LIQUID USING LIQUID CRYSTAL POLYMERS AS LUBRICANTS

TAFOYA NORA A. **_(1)_**, **_MEDINA DORA I._** **_(2)*_**

(1) IPN-ESIQIE - CIUDAD DE MÉXICO - Mexico, (2) TECNOLÓGICO DE MONTERREY - ESTADO DE MÉXICO - Mexico

This work is concerned with the study of the triborheological properties of solid-liquid systems. The research and development of more efficient lubricants is of fundamental technological and economic importance. The material´s wear implies great economic losses derived from the dissipation of energy by friction (about half the energy consumed in the mechanical systems), corrective maintenance of machinery and replacement of metallic parts. The triborheological study was realized with a Discovery Hybrid Rheometer, model DHR-3 of TA Instruments, using surfaces (plates) of different materials (steel, copper, aluminum and ceramic), and liquid crystal polymers (PLC) as lubricants and lubricant additives. The results showed ultra-low frictional coefficients and wear rates with respect to other conventional lubricants. This is due to PLC molecules have a great ability to orient themselves on the sliding surfaces and to form stable adsorbed films, thus avoiding direct contact which allows almost frictionless slippage. The efforts of this research are part of a complete characterization of the triborheological behavior of solid/ lubricating liquid systems, in order to generate innovative solutions regarding the maintenance of machinery and friction pairs in general, related to the current and future alloys subject to the effects of friction.

Static Yield Stress of a Magnetorheological Fluid Containing Pickering Emulsion Polymerized Fe_3O_4/Polystyrene Composite Particles

Sangsok Han, Junsok Choi, Sehyun Kim,[+] Yongsok Seo*
RIAM, Dept. Materials Science & Engineering, Seoul National University, Kwanakro 1, Kwanakku, Seoul, Korea
[+] Polymer Processing Team, LG Techcenter, Daejon, Korea

Abstract. Magnetorheological (MR) fluids are a type of smart material with rheological properties that may be controlled through mesostructural transformations. MR fluids form solid-like fibril structures along the magnetic field direction upon application of a magnetic field due to magneto-polarization of particles suspended in the inert medium. A reverse structural transition occurs upon removal of the applied field. The structural changes occur very quickly, on the order of milliseconds. Recent applications have increased the demand for better performing MR materials with good long-term stability. A variety of industrial MR materials have been developed and tested in experimental and theoretical studies. This article briefly describes our recent progress in the search for advanced MR fluid materials and novel approaches to analyzing the MR flow behavior. Because modelling and analysis are inevitable steps toward optimizing a material design, a novel macroscale structural model was devised to distinguish between the static yield stress and the dynamic yield stress and to describe the flow behavior under a broad range of shear rates. Several routes to improving the stability and efficiency of the MR fluids are summarized.

PACS: 83.85.Jn Viscosity measurements

INTRODUCTION

Magnetorheological (MR) fluids comprising suspensions of fine susceptible particles in a magnetically insulating fluid are a type of smart material because application of a magnetic field can transform the fluid from a suspension to a solid-like fibril structures in which the fibrils are aligned along the magnetic field direction due to magneto-polarization among the suspended particles.[1] A reverse structural transition occurs once the applied magnetic field is removed. The structural changes proceed very quickly, on the order of milliseconds. The aligned structural changes formed in response to a magnetic field tune the rheological properties of the MR fluid over three to four orders of magnitude. The chain-like structure formed under the external magnetic field provides the suspension with a yield stress as the fluid begins to flow. The rheological properties of MR fluids vary with the application of a magnetic field, yielding a non-Newtonian viscoplastic flow behaviors. The rheological properties of MR fluids can be fine-tuned by varying the magnetic field intensity for use in a variety of material applications. [2] MR fluids have attracted significant interest over the past two decades, and their promising features have driven many scientific studies and industrial applications. MR fluid applications can be found in automobile devices, for example, in linear dampers, rotor dampers, rotary brakes, as well as in other diverse industry products, such as directional control valves, hybrid actuators, and the haptic devices. [2] MR fluids have been used in medical applications, for example in novel cancer therapeutic methods. MR fluids can inhibit the blood supply to a tumor or enable drug delivery systems. [2] On a larger scale, MR suspension systems are present in China's Dong Ting Lake Bridge to counteract vibrations caused by sudden gusts of wind. [3,4] The same principle has been applied to stabilize buildings against earthquakes. More applications can be found in our recent report. [5]

EXPERIMENTS

Synthesis of Polystyrene/ Fe_3O_4 Particles. Submicron-sized PS/Fe_3O_4 particles were synthesized by Pickering-emulsion polymerization method using Fe_3O_4 nanoparticles as a surfactant and a stabilizer. Magnetic Fe_3O_4 nanoparticles (5g) (particle size of 50-100 nm and density of 4.87 g/cm^3, Sigma- Aldrich) were added to 200 mL of deionized water and then ultrasonicated for several minutes. After ultrasonication, styrene monomer (9 and 13.5g) (above 99% purity , Daejung , Korea) including the radical initiator , 2,2'-Azobis(2-methyl-propionitrile) solution (15ml) (AIBN solution , Sigma-Aldrich) was poured into Fe_3O_4 suspension and agitated vigorously using a homogenizer until the styrene layer was well-dispersed in the aqueous phase as emulsified droplets. The emulsified suspension was placed into a 250 mL round-bottomed three-neck flask. The reactor was sealed with a rubber septum, purged with nitrogen gas, and the mixture was stirred with a mechanical stirrer. The system was heated to 70°C and maintained for 24 h. The final product was washed with methanol and distilled water to remove the excess initiator, monomer and dried in a vacuum oven for a day.

Characterization Methods. The densities of Fe_3O_4, PS / Fe_3O_4, and PSF / Fe_3O_4 particles were measured using a helium pycnometer (AccuPyc 1330, Micromeritics Instrument Corporation, Norcross, GA). The morphology of the synthesized PS / Fe_3O_4 particles was observed by scanning electron microscopy (SEM ; SUPRA 55VP , Carl Zeiss , Germany) and transmission electron microscopy (TEM ; JEM1010 , JEOL , JAPAN). The crystal structures of the samples were examined by X-ray diffraction (XRD; D8-Advance, BRUKER MILLER Co., USA). The magnetic characteristics were examined using a vibrating sample magnetometer (VSM; Model 7370 , Lake Shore Cryotronics, Westerville, OH) with a range -5000 ~ 5000Oe. The rheological properties of the MR fluids were measured using a commercial rotational rheometer (Physica MCR300, Stuttgart, Germany) equipped with a magnetic generator (Physica MRD 180). A parallel-plate measuring system with a diameter of 20mm was employed at a gap distance of 1 mm. Silicon oil (KF-96 , 10cS , Shin Etsu , Japan) was used as a medium oil and a particle concentration was 25 vol%. The sedimentation stability of Fe_3O_4, PS / Fe_3O_4, and foamed PSF / Fe_3O_4 particle suspensions was analyzed by using a Turbiscan (Classic MA2000, Formulation, France).

RESULTS and DISCUSSION

Although MR fluids have made substantial progress toward commercialization, long term stability has been a crucial factor limiting its use in large devices and huge structures such as buildings, bridges and dams. Density mismatch between the particles and the carrier liquid causes particle sedimentation that worsens the MR efficiency over time. Other factors such as the poor redispersibility of aggregated or settled particles, weak chemical degradation resistance also affect the long term stabilities of MR devices. The stabilities of MR fluids may be improved via several approaches, including the introduction of polymer coatings or passive layers on magnetic particles, the use of viscoplastic medium as a carrier liquid, the addition of additives, fillers or surfactants, the use of viscoplastic medium as a carrier liquid, and the use of non-spherical particles, nano-fibers or bidisperse MR suspensions. [3,4] Alleviating the density mismatch is the most feasible method to improve the stabilities of MR fluids because this technique reduces the particle sedimentation speed.

Depending on the required yield stress values, two different approaches were tested ; 1) attaching the magnetic particles onto the surface of a light polymer particle or an inorganic particle when small field stress values are desirable for small devices or microfluidics, [5] and 2) coating the magnetic particles with lighter materials when large yield stress values are required for huge structural applications. [1,5] The first approach takes an advantage of Pickering emulsion polymerization method, [3,4] in which the solid inorganic particles, such as the magnetic nanoparticles, are used as a surfactant instead of an organic surfactant. Since the main body is composed of polymeric molecules, the density of the entire particle is much lower than that of the metal particle, thereby reducing the density mismatch. [5] The second approach, a core-shell method, involves wrapping up the magnetic metal particles with a coating layer that can decrease the particle density, enhance the durability of the particles, and improve the surface free energy of magnetic particles, thereby yielding superior compatibility (a higher wettability) with the carrier liquid. [1]

A magnetic material that can be used in Pickering emulsion polymerization is magnetite (Fe_3O_4). Because the magnetization saturation value of Fe_3O_4 is 61 emu/g, twice the value of Fe_2O_3 (30 emu/g), PS/Fe_3O_4 particle suspension can provide a better performance (higher yield stress) if its surfactant activity is as good as that of Fe_2O_3. We applied the same method to prepare a PS/Fe_3O_4 suspension and investigated its MR fluid performance. The fabricated PS/Fe_3O_4 particles displayed an even lower density of 2.11 g/cm^3 compared to the value of PS/Fe_2O_3.[3] The shear stress of the PS/Fe_3O_4 suspensions. The PS/Fe_3O_4 suspension exhibits less stress variation with the shear rate than the PS/Fe_2O_3 suspension. [30] This was ascribable to stronger interactions among the Fe_3O_4 particles due to a larger magnetization compared to that in Fe_2O_3. Surprisingly, the yield stress value of the PS/Fe_3O_4 suspension was much larger than the corresponding value obtained from the PS/Fe_2O_3 suspension, indicating a strong magnetic interaction between Fe_3O_4 particles adsorbed particles onto the surface of the PS core. These results

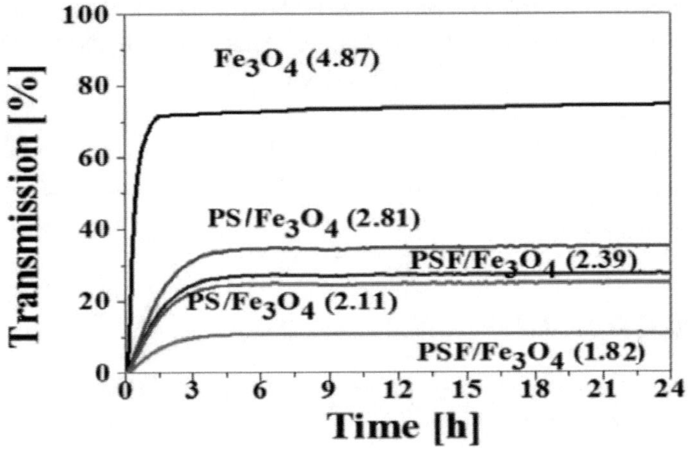

Figure 1. Transmission [%] curve as a function of time [h] for pure Fe_3O_4, PS / Fe_3O_4 (2.81), PSF / Fe_3O_4 (2.39), PS / Fe_3O_4 (2.11) and PSF / Fe_3O_4 (1.82) particles for 24h.

Table 2. Static yield stress and dynamic yield stress

	Yield stress (Pa)					
	M = 86 (kA/m)			M = 343 (kA/m)		
	Static Exp (τ_{sy})	Dynamic (τ_{dy})	Static SS (τ_{sy})	Static Exp (τ_{sy})	Dynamic (τ_{dy})	Static SS (τ_{sy})
Fe_3O_4	970	920	970	3730	3740	3800
PS / Fe_3O_4(2.81)	405	400	410	1590	1550	1610
PSF / Fe_3O_4(2.39)	230	225	230	925	910	930
PS / Fe_3O_4(2.11)	355	350	360	1385	1350	1400
PSF / Fe_3O_4(1.82)	200	190	200	790	770	800

verified that surfactant activity of Fe_3O_4 was stronger than that of Fe_2O_3; however, the suspension stability was slightly lower than that of PS/Fe_2O_3 due to the higher density of PS/Fe_3O_4 (2.8 g/cm^3) compared to PS/Fe_2O_3. The average PS/Fe_3O_4 particle size was 350nm, whereas that of the PS/Fe_2O_3 particle was less than 270nm. Larger PS/Fe_3O_4 particles featured a larger surface area and more surfactant adsorption. Because the sedimentation velocity of suspended particles settling is proportional to the density mismatch, a lower particle density may improve the stability.

With the goal of reducing particle density, foamed PS/Fe_3O_4, particles were fabricat PS/Fe_3O_4 ed using asupercritical carbon dioxide fluid (sc CO_2) as a physical foaming agent. The PS core was foamed to provide a porous structure that reduced the density (2.83 g/cm^3) and enhanced the suspension stability. A 15% density reduction provided 33% improvement in stability. [3]

The PS core size could be controlled by tuning the initial quantity of styrene used in the emulsion. Larger PS cores provided lower particle densities. Larger PS core particles (~600nm) were also synthesized having a density of 2.32 g/cm^3 The produced PS/Fe_3O_4 particles were foamed to decrease the particle density even further, to 2.11 g/cm^3.

CONCLUSIONS

The Pickering –emulsion polymerization method using Fe_3O_4 nanoparticles as a surfactant enables easy preparation of core-shell structured PS/ Fe_3O_4 particles. The interfacial activity of Fe_3O_4 nanoparticle was found out to be stronger than Fe_2O_3 particles. The suspensions of PS/ Fe_3O_4 particles provides much better MR performance than PS/Fe_2O_3 particle suspension. On the other hand, the PS/Fe_3O_4 particle suspension has the edge of stability compared to CI/PS suspension because of the reduced density mismatch with the liquid medium and rougher surface. Foaming the PS core provides extra stability; the suspension of foamed PS/Fe_3O_4 (ρ =1.82 g/cm^3) showed remarkably improved stability (~10% more stable compared to unfoamed PS/Fe_3O_4 suspension (ρ =2.11 g/cm^3) and ~62% compared to bare Fe_3O_4 particle (ρ = 4.87 g/cm^3) suspension).

ACKNOWLEDGMENTS

This work was supported by MIKE through Original Materials Research Program (RIAM IAC 1410) and by NRF through BK 21 Plus (Materials Science & Engineering Program).

REFERENCES

(1) Y, P, Seo.; H, J, Choi.; J, R, Lee.; Y, Seo. *Colloids Surf. A* **2014**, *457*, 363-367.
(2) Seo, Y. P.; Seo, Y. *Langmuir* **2012**, *28* (6), 3077-3084.
(3) Seo, Y.P.; Choi, H. J.; Seo, Y. *Soft Matter* **2012**, *8* , 4659
(4) Chuah , W. ; Zhang , W. ; Choi , H. J.; Seo , Y. *Macromolecules* **2015** , 48 , 7311
(5) Seo, Y.P.; Choi, H. J.; Seo, Y. *ACS Appl. Mat.Interf.* **2017** submitted.

Rheo-PIV measurements of a yield-stress fluid in Couette flow

Medina-Bañuelos Esteban F. **[1]***, **Marín-Santibáñez Benjamín M.** **[1]**, **Pérez-González José** **[2]**

[1] Instituto Politécnico Nacional ESIQIE SEPI - Ciudad de México - Mexico, [2] Instituto Politécnico Nacional ESFM - Ciudad de México - Mexico

Yield-stress fluids are fluids that do not flow unless a critical shear stress, τ_y, called the yield stress, is overcome. This sort of materials behave as elastic solids for shear stress values below the yield one and may exhibit a range of behaviors under flow for shear stresses beyond, among them, the shear thinning behavior. In rheology, hydrogels have been used as yield-stress model fluids due to their low cost, lack of toxicity and easy preparation. Thus, the flow kinematics of a yield-stress fluid, consisting of a polyacrylic acid hydrogel (Carbopol® 940) at a concentration of 0.12 wt.% and temperature of 25 °C, in a Couette rheometer with three different gaps, namely, 1, 4.65 and 28.5 mm, respectively, was analyzed in this work via the particle image velocimetry technique (reo-PIV). The use of the Rheo-PIV technique allowed for a reliable description of the rheological behavior of the hydrogel and evidenced flow characteristics that had not been reported. For shear stresses below the yield one, the velocity profiles were consistent with a rigid solid movement under rotation and simultaneously corroborated the existence of the yield stress. For shear stresses higher than the yield one, the velocity profiles exhibited two regions with different behavior inside the gap, one with flow and another without flow, which were separated by a critical position that determined the yield stress and was dependent on the applied shear stress. On the other hand, slip was observed on both cylinders of the Couette cells for all the flow conditions analyzed in this work and the magnitude of the slip velocity increased along with the shear stress applied. Finally, the flow behavior of the hydrogel was properly described by the Herschel-Bulkley model and it was shown that the experimental velocity profiles agree very well with those obtained by the numerical solution of the conservation equations for the Couette flow.

Molecular Motion of Poly(methyl methacrylate) Containing Metal Salt

A. Ito[a*], R. Maeno[a], M. Soga[a], and M. Yamaguchi[a]

[a]*School of Materials Science, Japan Advanced Institute of Science and Technology, Japan – asae.ito@jaist.ac.jp*

Abstract. The enhancement of glass transition temperature (T_g) of amorphous plastics has been required to expand the service temperature for the practical use. In this study, the T_g enhancement by the addition of two kinds of lithium salts, i.e., lithium trifluorosulfonate ($LiCF_3SO_3$) and lithium bromide (LiBr), was investigated using poly(methyl methacrylate) (PMMA) as a host polymer. The addition of $LiCF_3SO_3$ had a great influence on T_g enhancement of PMMA. Once sample films were exposed to high humidity condition, however, PMMA containing $LiCF_3SO_3$ absorbed a large amount of water that act as a plasticizer. As a result, T_g shifted to a lower temperature, which restricts industrial applications. In contrast, T_g was slightly affected by the absorbed water for PMMA containing LiBr. This is attributed to the difference in the strength of the intermolecular interaction between PMMA and salts. Moreover, it was revealed that the distribution of relaxation time becomes broad by the LiBr addition. Since LiBr has larger binding energy between the cation and anion than $LiCF_3SO_3$, it is expected to cause the lower degree of dissociation. Lower career density and lower ionic conductivity due to a small amount of dissociated lithium cation would be responsible for broader distribution of ion-dipole interaction between lithium cation and PMMA and thereby the broader relaxation time distribution.

Keywords: Poly(methyl methacrylate); Glass transition; Ion-dipole interaction; Relaxation
PACS: 62.20.Dc; 81.40.Jj;

INTRODUCTION

Amorphous plastics are widely utilized in various applications these days and thus inevitable in our daily life. Considering the practical application and processing of amorphous plastics, the enhancement of glass transition temperature (T_g) has been always required to expand the service temperature. Therefore, in past decades, various methods to increase T_g have been proposed, such as decrease in the number of chain ends, in other words, increase in the molecular weight [1–4], addition of a miscible polymer possessing high T_g [5,6], and suppression of molecular motion by introducing chemical/physical crosslink points [7–11]. Among crosslinking structures, hydrogen bonding [10,11], π-π stacking [12], and electrostatic interaction [13] are also involved besides the chemical crosslinking generally employed for rubber.

Recently, our research group found that the addition of lithium trifluoromethanesulfonate ($LiCF_3SO_3$) can affect T_g enhancement of poly(methyl methacrylate) (PMMA) [14]. Although the metal salt addition was known to cause the T_g enhancement of an ionomer [15–17], this should be noted because PMMA does not contain any acid/base functions like carboxyl group. In these days, the relationship between ion-dipole interaction in polymer-metal salt composites and the T_g enhancement has been studied in the field of solid electrolyte [18, 19]. However, there have been few reports on T_g enhancement for conventional glassy plastics. Therefore, the mechanism has been barely revealed. Considering this situation, we decided to focus on the T_g enhancement of glassy plastics using metal salts. The ion-dipole interaction between lithium cation that is dissociated in PMMA and oxygen atoms in the carbonyl group is related to the reduced segmental motion in the glass-to-rubber transition region, leading to the enhancement of T_g [14]. Although this phenomenon must have a great impact on the industry, the hygroscopic property and the effect of water absorption on T_g should be clarified prior to applications because most salts absorbed a lot of water.

In this paper, we investigated the effect of a metal salt addition on T_g under dry and humid conditions for PMMA containing $LiCF_3SO_3$ and lithium bromide (LiBr).

EXPERIMENTAL

Materials

The polymer used in this study was a general poly(methyl methacrylate) (PMMA) (Mitsubishi Chemical Corp., Acrypet VH). The number- and weight- average molecular weights were determined by a size exclusion chromatography (Tosoh Corp., HLC-8020) using chloroform as a solvent, and they were Mn = 58,000 and Mw = 120,000, respectively. Two types of lithium salts, i.e., lithium bromide (LiBr > 99.0 %, Tokyo Chemical Industry Co., Ltd.) and lithium trifluoromethanesulfonate ($LiCF_3SO_3$, Kanto Chemical Co., Ltd.), were utilized for experiments without any purification.

Sample Preparation

PMMA and a metal salt were completely dissolved in mixture solvent of dichloromethane and methanol in 9 to 1 weight ratio and stirred for 1 h. The molar ratio of lithium to carbonyl group in PMMA was fixed to 0.07, which were 10 wt.% of $LiCF_3SO_3$ and 6 wt.% of LiBr in the blends. The solutions were casted in a container and dried at 160 °C for 30 h in vacuum. After dryed under a vacuum condition, all samples were molded into films whose thichness was 350 μm using a compression-molding machine. After heated at 200 °C for 10 min, the samples were quenched at 25 °C using the compression-molding machine. Some sample films were used for the measurements immediately after compression-molding. Moreover, the other films were kept at 50 °C and 50 %RH in a temperature-and-humidity controlled chamber prior to the measurements.

Measurements

Thermal properties were measured by a differential scanning calorimeter (DSC, PerkinElmer, DSC 8500) with a heating rate of 10 °C/min from room temperature. Approximately 10 mg of samples were sealed in an aluminum pan.

The temperature dependence of oscillatory tensile moduli, i.e., storage modulus E' and loss modulus E'', was measured from 25 °C to 180 °C by a dynamic mechanical analyzer (UBM, Rheogel-E4000). The frequency and heating rate of this measurement were 10 Hz and 2 °C/min, respectively.

RESULTS AND DISCUSSIONS

It was found from the DSC heating curves of the dried samples, i.e., the samples immediately after compression-molding, that the glass-to-rubber transition were clearly detected at least for pure PMMA and PMMA/$LiCF_3SO_3$. T_g's of pure PMMA and PMMA/$LiCF_3SO_3$ were estimated to be 108 °C and 123 °C, thus, T_g was enhanced by the $LiCF_3SO_3$ addition. In the case of PMMA/LiBr, the transition seemed to occur in a broad range of temperature.

The temperature dependences of dynamic tensile moduli are shown in **FIGURE 1** for PMMA/$LiCF_3SO_3$ and PMMA/LiBr. The data of pure PMMA is also shown in the figures as gray symbols for references. As described in the previous report [14], T_g was enhanced by the addition of $LiCF_3SO_3$. Furthermore, E' at high temperature in the glassy region is high and insensitive to the temperature. Accordingly, the E'' peak ascribed to T_g is located at high temperature with a large maximum value.

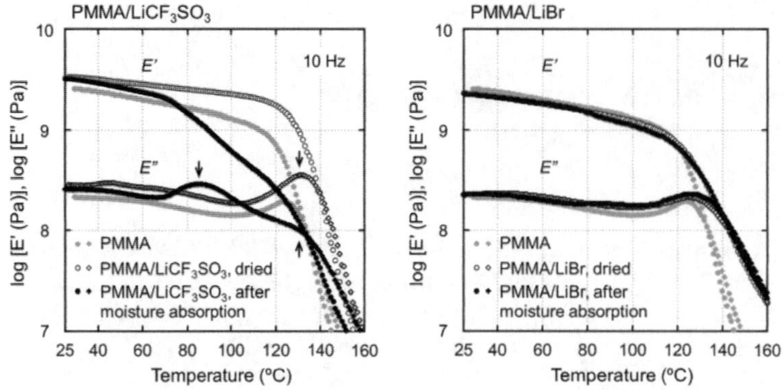

FIGURE 1. Temperature dependence of the dynamic tensile modulus at 10 Hz for (a) PMMA/LiCF$_3$SO$_3$ and (b) PMMA/LiBr. (gray symbols) pure PMMA, (open symbols) dried blend films and (closed symbols) moisture-absorbed (stored in a temperature-and-humidity controled chamber at 50 °C and 50% RH) blend films.

The ion-dipole interaction between lithium cation and carbonyl group in PMMA is responsible for the T_g enhancement. However, after being stored for 5 days in a temperature-and-humidity controlled chamber at 50 ℃ and 50% RH, *E'* decreases with temperature prominently even at room temperature. Furthermore, there is a broad ambiguous peak in the *E"* curve at 85 °C, as indicated as an arrow in the figure, besides the peak at 130 °C. The peak at the low temperature would be attributed to T_g of the region containing a large amount of water, as described later. In other words, water molecules act as a plasticizer [20-22]. As a result, *E'* of PMMA/LiCF$_3$SO$_3$ after water absorption was greatly lower than that of pure PMMA in the wide range of temperature (< 120 °C). Moreover, the glassy and transition regions cannot be defined clearly. This result demonstrates that the LiCF$_3$SO$_3$ addition reduces the service temperature of PMMA especially at high temperatures after the water absorption.

In the case of PMMA/LiBr (**FIGURE 1** (b)), PMMA/LiBr also shows higher T_g than pure PMMA. However, the dynamic mechanical properties are different from those of PMMA/LiCF$_3$SO$_3$. In the glassy region, *E'* decreases slightly as temperature increase as similar to the pure PMMA. Therefore, the temperature dependence of *E'* in the glassy region is prominent for PMMA/LiBr as compared with PMMA/LiCF$_3$SO$_3$. Beyond 110 °C, *E'* of PMMA/LiBr is higher than that of pure PMMA, which is obvious in the high temperature region. As a result, the *E"* peak becomes broad, suggesting the broad distribution of relaxation modes. These results correspond with the DSC heating curves. The broad distribution of relaxation modes indicates that a part of ion-dipole interactions in the PMMA/LiBr system is significantly strong, which is responsible for a prolonged relaxation mechanism. Finally, it should be noted that the dynamic mechanical properties for PMMA/LiBr are barely affected by the water absorption. This is a very important phenomenon because this technique can be useful in industry.

The growth curves of the sample weight, measured by a chemical balance, at 50 °C and 50 %RH are shown in **FIGURE 2**.

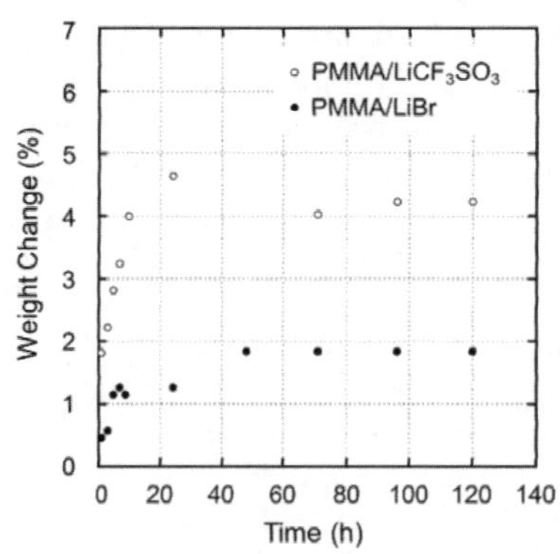

FIGURE 2. Growth curves of the weight change for (open symbols) PMMA/LiCF$_3$SO$_3$ and (closed symbols) PMMA/LiBr films stored at 50 °C and 50 %RH.

As seen in the figure, the water adsorption occurs immediately (< 10 hrs) and becomes a constant value, i.e., equilibrium moisture content. The value for PMMA with $LiCF_3SO_3$ is approximately 4 %. On the other hand, that of PMMA containing LiBr is only 2 %. Although the value is larger than that of pure PMMA (approximately 1 %), the result demonstrates that the hygroscopic property depends on the species of a salt. The difference in water content between the two types of salts, which affects T_g enhancement, provides a fundamental question on interaction between a salt and PMMA. The dissociation and ionization of the salts in PMMA are explained using the binding energy based on the coulomb-coulomb interaction, which is defined by the equation below [23];

$$\varpi(r) = \frac{Q_1 Q_2}{4\pi\varepsilon_0 \varepsilon r}.$$

where Q_1 and Q_2 are charges, ε (= 3.6) and ε_0 are the dielectric constants of medium and vacuum [24, 25], and r is the distance between cation and anion. Here, we can assume that r is equal to the sum of cation and anion radii (where a_+ is the cation radius [26] and a_- is the anion radius [27] in **TABLE a)** Based on this equation, a salt having a large value of the sum of ion radius, shown in **TABLE a)** shows the dissociation easily.

The coulomb interaction energies of ionic crystals in PMMA are estimated as shown in the table. Since the sum of the ion radius of $LiCF_3SO_3$ is relatively large because of the large ionic radius of triflate anion, the dissociation of $LiCF_3SO_3$ occurs more frequently than that of LiBr in the same solvent. Furthermore, small ions tend to be easily solvated by surrounding molecules because of their higher charge density. Therefore, lithium cations are strongly solvated by PMMA molecules. In contrast, marked solvation is not expected for triflate ions. Considering both dissociation and solvation states, LiBr is expected to show the strong association state with PMMA molecules. Because of such tight bond with LiBr, the motion of PMMA segments would be restricted even at high temperature, providing a prolonged relaxation mechanism.

TABLE a). Ionic radii of lithium cation and triflate anion, and bromide anion from references [26, 27]. $\varpi(r)$ shows calculated binding energy considering the coulomb-coulomb interaction between cation and anion in ionic crystals in a medium (PMMA).

	a_+ (nm)	a_- (nm)	$\varpi(r)$ (J)
$LiCF_3SO_3$	0.073	0.230	-2.1×10^{-19}
LiBr	0.073	0.196	-2.4×10^{-19}

CONCLUSIONS

Thermal and dynamic mechanical properties were studied for PMMA with lithium salts, i.e., $LiCF_3SO_3$ and LiBr. Both salts have a capability to enhance T_g of PMMA due to ion-dipole interaction between lithium cation and carbonyl group of PMMA, with a marked T_g enhancement of $LiCF_3SO_3$. However, the salt addition enhances the water absorption. Because water molecules play the role on a plasticizer, which is more predominant for $LiCF_3SO_3$, T_g was depressed with low moduli even in the glassy region in the case of PMMA/$LiCF_3SO_3$. On the other hand, the plasticization by water absorption was barely detected for the blend of LiBr. The strong interaction between PMMA and LiBr with undissociated condition, leading to a prolonged relaxation mode, is responsible for the difference from the results observed in PMMA/$LiCF_3SO_3$. Although LiBr is a good candidate to modify the heat resistance of PMMA, it still needs to be explored to find an appropriate metal salt for PMMA to enhance T_g without hydrophylic propertiy.

ACKNOWLEDGMENTS

The authors would like to express their sincere gratitude to Mitsubishi Chemical Corporation for their valuable suggestions and the kind supply of the sample employed in this study. A part of this study was supported by COI program "Construction of next-generation infrastructure system using innovative materials" – Realization of safe and secure society that can coexist with the Earth for centuries – Supported by Japan Science and Technology Agency (JST).

REFERENCES

1. J. M. G. Cowie, *Euro. Polym. J.*, **11**, 297-300 (1975).
2. K. Tadlaoui, Y. Pietrasanta, A. Michel, V. Verney, *Polymer*, **32**, 2234-2237 (1991).
3. K. L. Wooley, C. J. Hawker, J. M. Pochan, J. M. J. Frechet, *Macromolecules*, **26**, 1514-1519 (1993).
4. J. A. Forrest, K. Dalnoki-Veress, J. R. Dutcher, *Phys. Rev. E*, **56**, 5705-5716 (1997).
5. P. C. Painter, J. F. Graf, M. M. Coleman, *Macromolecules*, **24**, 5630-5638 (1991).
6. T. P. Lodge, T. C. B. McLeish, *Macromolecules*, **33**, 5278-5284 (2000).
7. A. Shefer, M. Gottlieb, *Macromolcules*, **25**, 4036-4042 (1992).
8. H. Liu, S. Zheng, *Macromol Rapid Comm.*, **26**, 196-200 (2005).
9. B. D. Fairbanks, T. F. Scott, C. J. Kloxin, K. S. Anseth, C. N. Bowman, *Macromolcules*, **42**, 211-217 (2009).
10. T. Y. Juang, J. K. Liu, C. C. Chang, S. M. Shau, M. H. Tsai, S. A. Dai, W. C. Su, C. H. Lin, R. J. Jeng, *J. Polym. Res.*, **18**, 1169-1176 (2011).
11. Y. Shangguan, J. Yang, Q. Zheng, *RSC Adv.*, **7**, 15978-15985 (2017).
12. B. P. Donat, A. Viallat, J. F. Blachot, C. Lombard, *Adv. Matter.*, **18**, 1401-1405 (2006).
13. S. Yang, H. Fan, Y. Jiao, Z. Cai, P. Zhang, Y. Li, *Comp. Sci. Technol.*, **138**, 161-168 (2017).
14. A. Miyagawa, V. Ayerdurai, S. Nobukawa, M. Yamaguchi, *J. Polym. Sci. B Polym Phys.*, **54**, 2388-2394 (2016).
15. A. Eisenberg, J. S. Kim, *Introduction to ionomers*, New York: Wiley, 1998.
16. J. S. Kim, R. J. Jackman, A. Eisenberg, *Macromolecules*, **27**, 2789-2803 (1994).
17. M. Hara, A. Eisenberg, *Macromolecules*, **17**, 1335-1340 (1984).
18. M. Watanabe, S. Nagano, K. Sanui, N. Ogata, *Solid State Ion.*, **18&19**, 338-342 (1986).
19. M. Watanabe, M. Itoh, K. Sanui, N. Ogata, *Macromolcules*, **20**, 569-573 (1987).
20. J. Zhou, J. P. Lucas, *Polymer*, **40**, 5513-5522 (1999).
21. J. Shen, C. C. Chen, J. A. Sauer, *Polymer*, **26**, 511 -518(1985).
22. L. S. A. Smith, V. Schmitz, *Polymer*, **29**, 1871-1878 (1988).
23. J. N. Israelachvili, *Intermolecular and Surface Forces*: Third Ed.; Amsterdam: Elsevier, 2011, pp. 44-45.
24. J. Brandup, E. H. Immergut, E. A. Grulke, *Polymer Handbook*: Fourth Ed.; Hoboken: Wiley Interscience, 1999, Chapter v, p. 87.
25. P. Atkins, J. D. Paula, *ATKINS' PHYSICAL CHEMISTRY*: Eighth Ed.; Oxford: Oxford University Press, 2006, p. 110.
26. The Chemical Society of Japan, *Handbook: Pure Chemistry*: Fifth Ed.; Tokyo: Maruzen, 2004, p. 887. (in Japanese)
27. R. Custelcean, B. A. Moyer, *Eur. J. Inorg. Chem.*, **10**, 1321-1340 (2007).

Mathematical modelling, Rheology and Transport Phenomena of the pulsatile flow system of a viscoelastic biological structural fluid: Human Blood with different cholesterol levels

Calderas Fausto (1)*, Sanchez Hugo (2)

(1) CIATEC, A.C. - Distrito Federal - Mexico, (2) CIATEC - Guanajuato - Mexico

In this work, the rectilinear flow of a complex fluid (human blood) under a pulsating time-dependent pressure gradient is analyzed. A first approximation of the real case of blood flowing in a vein is described. The pressure gradient simulates the pumping work of the heart while the flow geometry (circular tube) is assumed rigid, smooth and cylindrical. The rheological behavior of blood with different cholesterol levels is modelled using the Generalized Bautista-Manero-Puig (GBMP) constitutive equation. The coupling between flow, structural parameters and diffusion naturally arise in this model, which has been derived from the Extended Irreversible Thermodynamic (EIT) formalism. In shear flow, the system shows shear thinning, shear thickening, yield stress, thixotropy rheopexy and shear banding flow respectively. According to the analytical solution, a flow enhancement is predicted to first order which represents the optimum pumping work of the heart which governs the flow of blood in the entire body. This work is a contribution to the understanding of the complex rheology involved in the discontinuous pressure driven flow of blood in the human body. The effect of different flow conditions on the fluid structure are analyzed through momentum and mass transfer equations.

Influence of copolymer architecture on generation of defects in reactive multilayer coextrusion (for Keynote lecture)

Bondon Arnaud [1], LAMNAWAR Khalid [1], Maazouz Abderrahim [1]*

[1] Université de Lyon, INSA de Lyon, IMP UMR CNRS 5223 - Villeurbanne - France

Several polymers can be combined in one multilayer structure by reactive coextrusion. Tie-layers are often used to compatibilize adjacent layers and may reduce or suppress interfacial instabilities and defects in multilayer coextrusion flow. However, a new additional defect defined as "grainy" defect can be observed. In our best of knowledge, no study in literature has been dedicated to understand its origin. The phenomena are quite complex due to the coupling of the effects of flow and the physico-chemical mechanisms at the interface. The aim of this work is to understand the relationship between the instabilities and defects encountered in multilayer coextruded films and the role of the copolymer formed in-situ between tie and barrier layers. Polyamide 6 (PA6) and ethylene-vinyl alcohol copolymer (EVOH) were used as barrier layers sandwiched in polypropylene (PP) with or without tie-layer based on polypropylene grafted maleic anhydride (PP-g-MA). Influence of process parameters and nature of the polymer pair on the generation of "grainy" defect has been assessed and related to the rheological and the physico-chemical properties of layers. These experiments showed that this defect appeared mainly in the compatibilized EVOH system and could be distinguished from the usual coextrusion instabilities. Interfacial properties between tie and barrier layers have been investigated. Shear stress relaxation experiments have been carried out on reactive tie/barrier bilayers. Due to the interphase generated in-situ, the relaxation behavior was altered by extending the relaxation time. Investigation of interfacial morphology highlighted that the copolymer architecture significantly affected the interface/interphase development and interface roughness. Hence, relationships between relaxation process, interfacial morphology and copolymer structure were correlated with the generation of grainy defects in coextrusion.

Highly viscous polymeric foam flowing through an orifice

Ching-Hsien Chen*, Bart Hallmark and John Frank Davidson

Department of Chemical Engineering and Biotechnology, University of Cambridge
Philippa Fawcett Drive, Cambridge CB3 0AS

* *chc57@cam.ac.uk*

Abstract. The rapid movement of gas-liquid foams through narrow contractions is commercially challenging and technically important, with wide industrial applications, including the flow and moulding of aerated products, e.g. ice cream and shaving foam, foam-sclerotherapy for varicose veins, and oil-well cementing when foam-cement slurries are pumped into narrow annuli. This study investigates the flow of wet foams with a polymeric liquid phase, which has a high viscosity and complex rheology, using a Multi-Pass Rheometer. The in-situ visualization of foam flowing through an orifice, recorded by high-speed camera, reveals rich and complex behaviour and interaction between bubbles. The bubble behaviour in the polymer melt is qualitatively compared with that observed in a highly viscous Newtonian liquid.

Keywords: Polymeric foam, Bubbly liquid, Multi-Pass Rheometer
PACS: 47.50.-d

1. Introduction

Thermoplastic foam is of importance due to its wide applications in various areas: (i) where impact strength is required, (ii) for thermal and electrical insulation, and (iii) for lightweight material-saving structures. The processes for producing polymeric foam range from using polymer with dissolved blowing agents in batch mode processes to using expandable polymer beads containing blowing agents for injection moulding and hot-melt extrusion in continuous foaming processes. The cellular structures of the foams are determined by the effects of polymer properties, the operating conditions such as pressure, temperature, and shear flow, and the foaming mechanisms involving bubble nucleation, bubble growth, and cell coalescence.

The visualization of polymer foaming is useful for studying foaming mechanisms. Many researchers have worked on the in-situ observation of bubble nucleation and growth either in batch foaming or continuous foaming, and directly observe the foaming process during injection moulding or extrusion[1-7]. Experiments have also been conducted to study bubble nucleation in a shear flow field, revealing several kinds of nucleation mechanisms[8-10].

The Multi-Pass Rheometer (MPR) is a twin piston rheometer invented in the early 1990s by Professor Malcolm Mackley at the University of Cambridge. Since then, the MPR has found useful applications in a wide range of areas[11] such as (i) using the MPR as a capillary rheometer for measuring rheological properties of Newtonian and Non-Newtonian liquids, (ii) studies of multiphase fluids, (iii) flow induced birefringence processing, (iv) flow induced crystallisation, (v) polymer melt instability, and (vi) polymer foaming. These studies demonstrated the flexibility of the MPR as a research tool which can be modified for new experiments. The MPR has also been used for studying bubble nucleation and growth in polystyrene melt containing pentane gas[12]. In this paper the MPR has been used to study the flow of highly viscous polymeric foam and to make observation of bubble behaviours as the foam passes through a narrow orifice.

2. Experimental apparatus and procedures

The experiments were performed by using a Multi-Pass Rheometer equipped with an optical module, where a flow chamber is formed by two parallel quartz windows with stainless inserts between the windows. By changing the shape of the inserts, a flow passage with a chosen geometric pattern is obtained; in this work an orifice plate and a contraction-expansion channel were used. Both temperature and pressure were measured by a pair of pressure transducers each with an integrated thermocouple, installed on both ends of the chamber. A separate thermocouple was also placed at the centre of the flow chamber.

The experimental setup is shown in figure 1. A high speed CMOS camera (Balser acA1300-200uc) with LED light source (Thorlabs MCWHLP1) backlighting was used to obtain the bubble images. The white box shown in figure 1 indicates the field of view where the sequential images were recorded. The blue area represents the twin pistons that can generate fluid upflow or downflow through the flow passage; the pink area represents the loaded test liquid. The test liquids were polybutene oil (Brookfield viscosity standard liquid B73000, 73.0 Pa s at 25 °C) and expandable polystyrene (INEOS Styrenics EPS-FR), e.g. polystyrene beads containing pentane. A rotational rheometer (TA instruments ARES), equipped with 20.0 mm diameter parallel plates, was used to measure the apparent viscosities of both liquids; results are shown in figure 2. The expandable polystyrene beads were degassed by first being loaded into a mould and pressed to a round disc by means of a heated press; some of the gas is released, but some remains in the disc as bubbles. By repeating the same process several times to release the gas, a pentane-free polystyrene disc can be obtained for testing the apparent viscosity of polystyrene melt. The flow curve shown in figure 2 demonstrates the polybutene has a constant apparent viscosity of 67 Pa s, showing that it is a Newtonian liquid. The deviation from the reported 73 Pa s is probably due to temperature effects. The polystyrene melt is shear-thinning, with a zero-shear rate viscosity of about 1250 Pa s

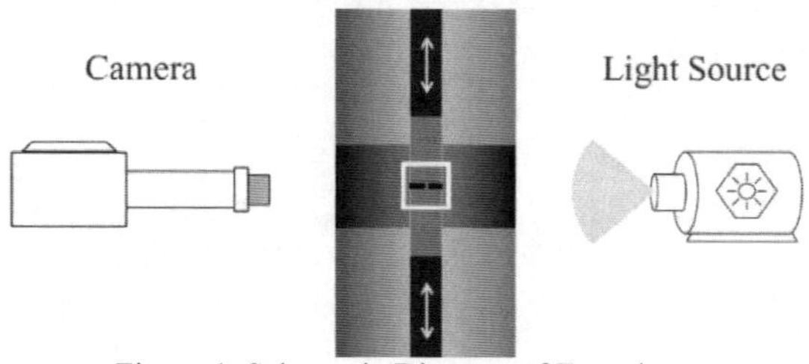

Figure 1. Schematic Diagram of Experiment

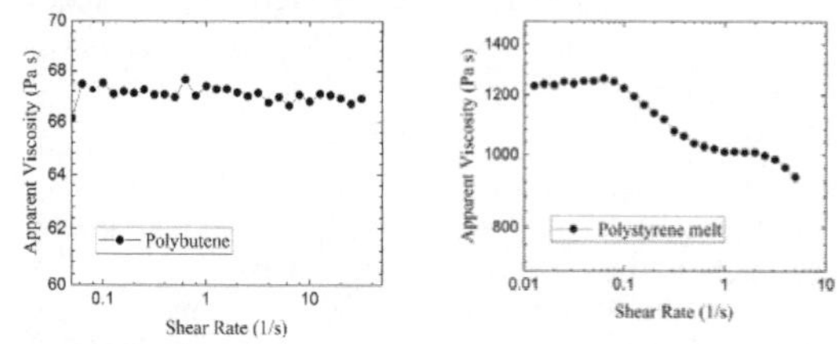

Figure 2. Plots of apparent viscosity as a function of shear rate

The steps of the experimental procedure are as follows:

(i) Assembling the flow chamber by selecting different quartz windows and stainless inserts to obtain the desired geometry of flow passage.

(ii) Loading and preparing the test materials.

The polybutene oil is first premixed with air bubbles which were entrained by hand stirring using a spatula; the bubbly liquid is then injected into the flow chamber using a syringe. Due to the high viscosity of the continuous phase, bubbles are stably suspended within the oil during the experiment; the effects of buoyancy and Ostwald ripening are observed to be relatively slow, hence the position and size distribution of bubbles change negligibly during an experiment.

A different loading protocol is required for bubbly polystyrene melt[12]. The beads, each bead containing dissolved gas, are first loaded into the flow chamber at room temperature and then heated to around 80 °C where they soften without gas release. Then, the beads are further compacted by hand to leave space for additional beads to be added. This compaction and addition process is repeated several times. Finally, the flow chamber is sealed with the upper piston, and the temperature increased to over 110 °C, giving rising to bubble formation, resulting in even bubble distribution within the flow chamber.

A typical transformation of the polystyrene beads containing pentane from solid through to bubbly liquid is shown in figure 3. Initially, the beads appear as dark round spheres, each bead identifiable at room temperature in photo A, loaded into a contracting-expanding flow geometry; then the beads are heated through photos (B), (C) and (D) where they swell as they reach 110 °C until finally forming a bubbly liquid, pentane bubbles in polystyrene liquid, at 160 °C, see photo (E). Increasing the pressure of the melt causes the bubbles to dissolve, so the volume fraction of bubbles diminishes, see photos F, G, and H. Finally, a single phase polymer melt is achieved at photo I.

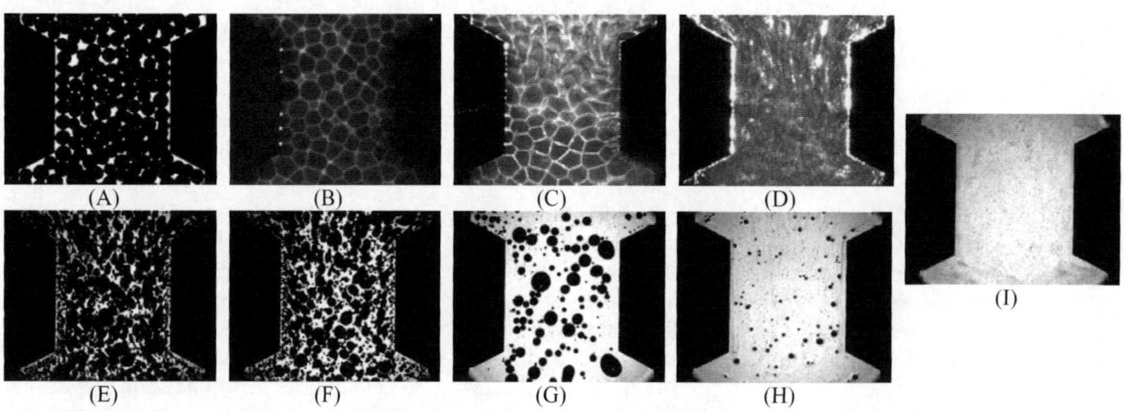

Figure 3. Sequential images illustrating the transformations of the beads from solid through to bubbly liquid.

(iii)Performing experiments.

The upper and lower pistons can be controlled, either individually or together, to generate desired pressure and flow rate within the chamber. Piston velocities ranging from 0.01 up to 10.0 mm/s were used. During a given experiment, images of the bubble motion were recorded using the high speed camera and the differential pressure across the flow geometry was measured.

3. Results and Discussion

Experiments were carried out to visualize the bubbly polymer melt flow. The width of the flow chamber was chosen as 10.0 mm; from this chamber, the flow channel abruptly reduced to a 1.4 mm orifice opening with a length of 1.5 mm. The depth, normal to the flow, was 1.0 mm to reduce overlapping of the bubbles. A typical flow of expandable polystyrene is shown in figure 4; the bubbly polymer melt flowing downward through an orifice and at rest are shown in the left and right photographs, respectively. In these images, bubbles appear as black shadows. The bubbles upstream of the orifice were elongated along the flow direction towards the orifice; whereas downstream the bubbles tended to coalesce into bigger bubbles, giving big dark cavities; these bubbles were stretched perpendicular to the flow direction outward from the orifice. When the flow stopped, all the deformed bubbles gradually returned to spherical shape due to a balance between the viscous and surface tension forces.

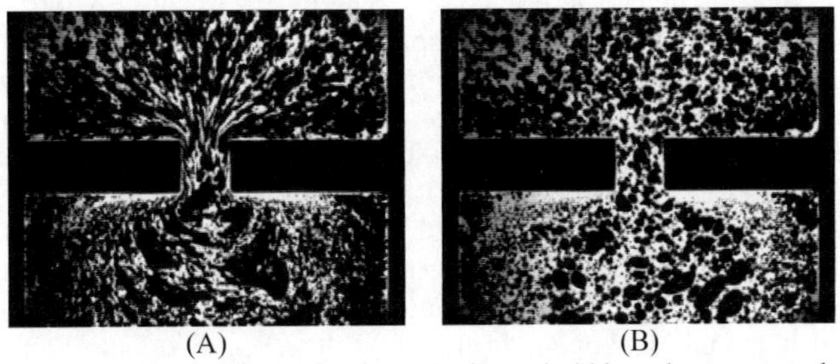

Figure 4. Typical flow of polymeric foam, bubbly polystyrene melt (A) flowing downwards through an orifice and (B) at rest.

In a system of reduced bubble density, the behaviour of individual bubbles and the interaction between bubbles can be observed more clearly. By carefully manipulating the pressure within the flow chamber, a desired volume fraction of pentane gas in the polystyrene melt can be obtained. Upflow of bubbly polystyrene melt through an orifice, with low bubble volume fraction, is shown in figure 5. The polydisperse pentane bubbles were sparsely and randomly distributed within the flow chamber. In the flowing polymeric melt, the initially spherical bubbles were deformed and elongated as they moved towards the orifice; significant coalescence of the bubbles occurred near the orifice and downstream of the orifice. The dynamics of bubble deformation upstream of the orifice is similar to that observed in a previous study[10]. At a time of 6.25 sec, bigger bubbles with distorted boundaries were observed downstream; here the bubbles are stretched laterally like those observed in dense polymer foam systems. From 12.5 sec, after the start of the experiment, the liquid motion slowed and came to a full stop at 43.25 sec; all the deformed bubbles subsequently relaxed back to a spherical shape.

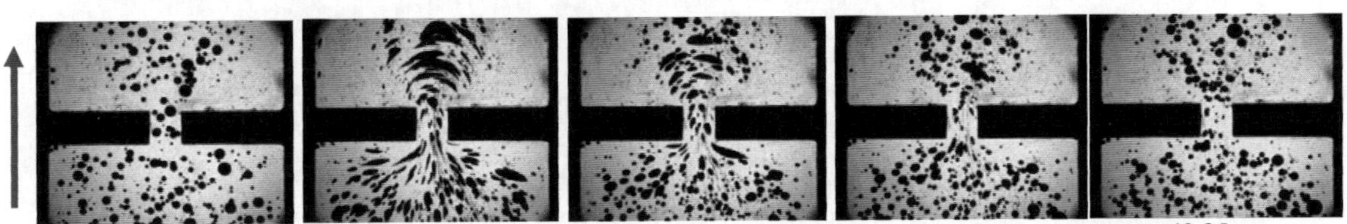

Figure 5. Pentane gas bubbles in polystyrene melt, non-Newtonian, flowing upward through orifice. Motion slows from 12.5 Sec and stops at 43.25 Sec.

In order to understand the behaviour of the bubbles in non-Newtonian liquid, it is helpful to examine the dynamic behaviour of bubbles in a Newtonian liquid for comparison. Air bubbles in polybutene oil, with a bubble volume fraction similar to that used in the non-Newtonian liquid, are shown flowing through the orifice in figure 6. The bubbly liquid started from rest; the bubbles were deformed and elongated as they approached the orifice as observed in the non-Newtonian liquid. The motion slowed down from 10.0 sec and stopped at 17.5 sec where the spherical shape of bubbles is shown in figure 6. However, coalescence of bubbles was not observed, even for the bigger bubbles; in Figure 6 at 5.0 sec two larger bubbles are seen without coalescence. Bubble coalescence appears to be suppressed in this viscous Newtonian liquid. The bigger bubbles in the Newtonian liquid flowing through the orifice were usually stretched, downstream, into a smooth crescent-shaped bubble with very sharp tips. This is different from the distorted bubble shapes observed under similar conditions in non-Newtonian liquid. The formation of the crescent bubble shape is thought to be due to the radial flow of liquid so that the front of the bubble moves more slowly than the back. The sharp tips of each bubble is preserved from breakup due to the surface tension force being less than viscous force. A separate paper is currently in preparation to discuss these crescent-shaped bubbles in more detail.

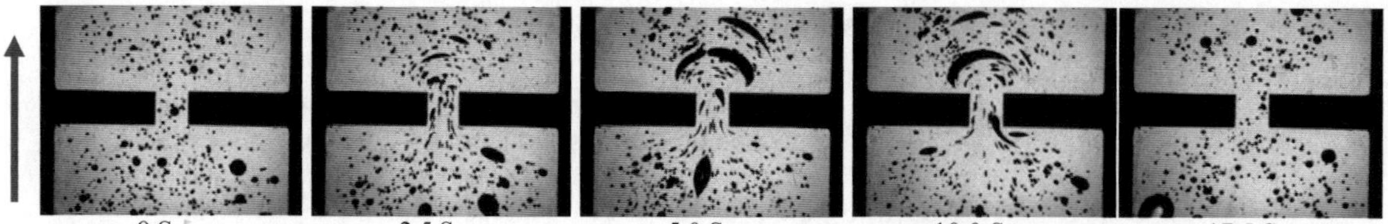
0 Sec 2.5 Sec 5.0 Sec 10.0 Sec 17.5 Sec

Figure 6. Air bubbles in Polybutene liquid (Newtonian liquid) flowing through Orifice. Motion slows after 10.0 Sec, and stops before 17.5 Sec.

By further increasing the pressure of the fluid, most of the bubbles dissolved, leaving a single bubble or a few bubbles in the polymer melt. This greatly facilitated the study of bubble dynamics in non-Newtonian liquid. Figure 7 shows breakup and coalescence of bubbles in polystyrene melt during the upflow through the orifice, which are not observed in the Newtonian case. Initially, four large spherical bubbles moved from the stationary position; the bubbles were stretched and distorted after passing through the orifice, at the time 1.5 sec. The bubbles were each split up into a major bubble and a satellite bubble due to bubble distortion under the radial flow. Subsequently, there is coalescence of those major bubbles at 2.0 sec. At 2.5 sec, a big bubble is seen flowing towards the orifice, bubble breakup was again observed, giving major and satellite bubbles at 3.5 sec. From 7.0 sec, the flow slowed and the bubbles stopped at 15.0 sec, restoring to spherical shape.

This case shows flow induced bubble breakup and coalescence in non-Newtonian liquid. The mechanism of bubble coalescence, observed in figure 7, can be explained from previous studies on the coalescence between bubbles rising under buoyancy in non-Newtonian liquids[13, 14]. The memory effect of the residual stress in a shear thinning viscoelastic liquid holds for a certain time in the wake of the leading bubble, causing the local viscosity to decrease, which induces acceleration of the following bubble. After coalescence, the resultant bigger bubble will shear the liquid more strongly, accelerating the following bubbles; thus, there is a dynamical competition between the creation and relaxation of stresses due to the consecutive passage of bubbles. When the bubble density is further increased to a certain threshold, the interaction between bubbles is no longer regularly periodic; chaotic coalescence dominates [14, 15].

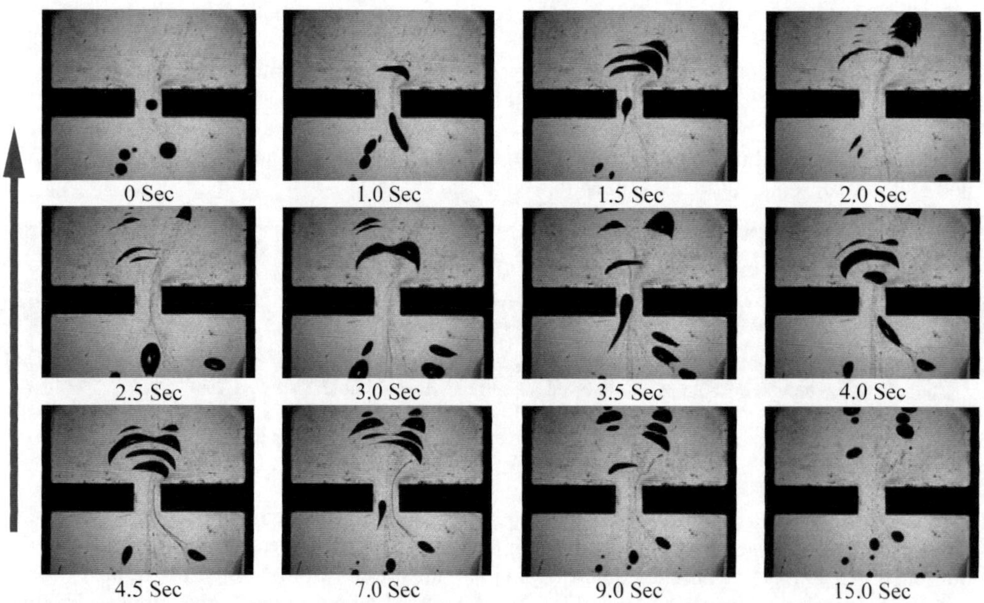

Figure 7. Bubble coalescence and breakup in polystyrene melt, non-Newtonian. Steady upward flow, starting from stationary, at 0 Sec. From 7.0 sec, the flow slows and stops at 15.0 sec.

In figure 8, a 45° contraction-expansion channel with both ends of 10.0 mm width reducing down to a flow chamber with both 6.0 mm in width and length was used to demonstrate this flow-induced coalescence in the polystyrene melt. The channel depth was 1.0 mm, normal to the flow direction. The bubbly polystyrene melt was cyclically moved up and down with a 3.0 sec stop at the end of each moving. In the first downflow, row A in Figure 8, initially spherical bubbles starting from stationary were deformed; significant coalescence of bubbles greatly reduced the bubble number density in the channel. After a 3.0 sec stop, the flow was reversed, row B in figure 8. A large gas cavity and fewer large bubbles passing through the channel were observed. In the subsequent flow cycles, C and D, the large gas cavity became bigger with fewer large bubbles left in the channel.

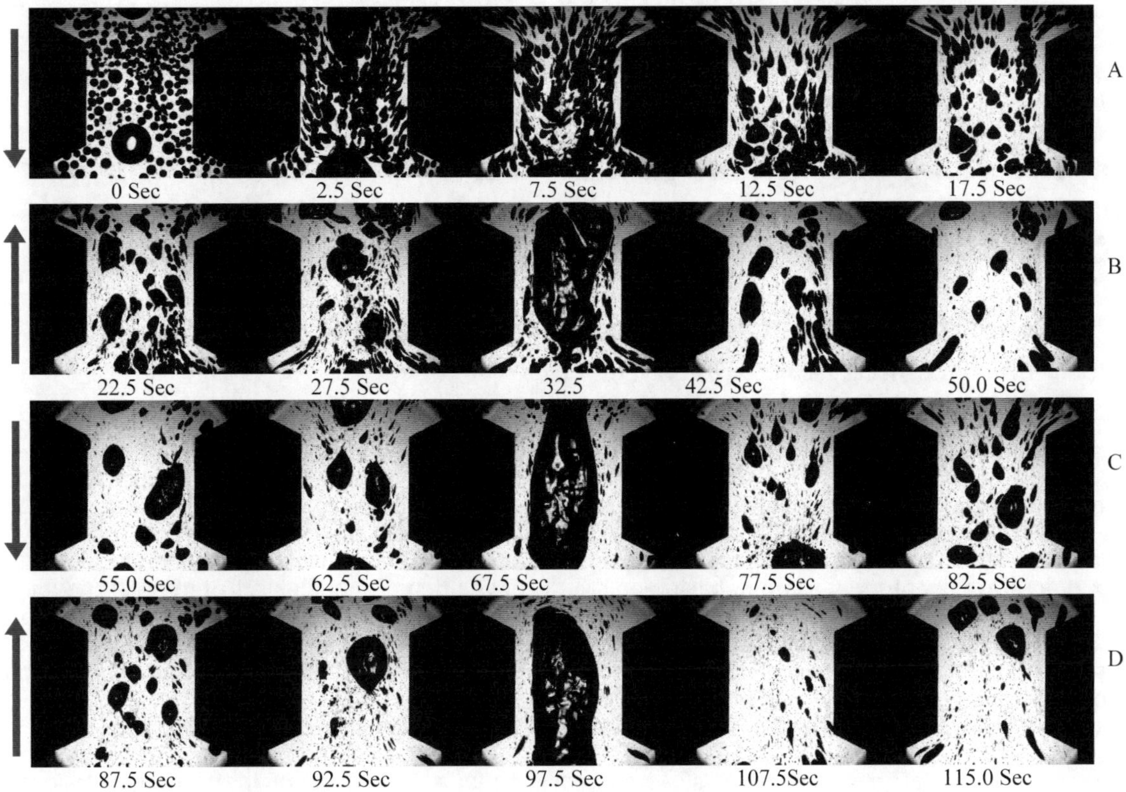

Figure 8. Pentane bubbles in polystyrene melt (Non-Newtonian) flowing cyclically through 45 degree contraction-expansion channel. The arrows denote the upflow or downflow.

Conclusions

Experiments used a Multi-pass rheometer to study the flow of a bubbly polystyrene melt flowing through an orifice or a 45° contraction-expansion channel. By controlling the pressure within the flow chamber, a system of reduced bubble density can be obtained to study the dynamics of single bubble or the interaction between bubbles in polymer melt. The dynamics of bubbles in polystyrene melt, a non-Newtonian liquid, is qualitatively compared with that observed in polybutene oil, a Newtonian liquid, showing great differences of bubble deformation, instability in bubble pinch-off, and flow-induced coalescence, especially downstream of the orifice.

Acknowledgements

The authors thank for the financial support of EPSRC grant, EP/N00230X/1.

References

1. K. Taki, T. Nakayama, T. Yatsuzuka and M. Ohshima, *Journal of Cellular Plastics*, 2003, **39**, 155-169.
2. C. D. Han and C. A. Villamizar, *Polymer Engineering & Science*, 1978, **18**, 687-698.
3. C. A. Villamizar and C. D. Han, *Polymer Engineering & Science*, 1978, **18**, 699-710.
4. A. Ahmadzai, A. H. Behravesh, M. T. Sarabi and P. Shahi, *Journal of Cellular Plastics*, 2014, **50**, 279-300.
5. V. Shaayegan, L. H. Mark, A. Tabatabaei and C. B. Park, *Express Polymer Letters*, 2016, **10**, 462-469.
6. Q. P. Guo, J. Wang, C. B. Park and M. Ohshima, *Industrial & Engineering Chemistry Research*, 2006, **45**, 6153-6161.
7. C. D. Han and H. J. Yoo, *Polymer Engineering and Science*, 1981, **21**, 518-533.
8. J. H. Han and C. D. Han, *Polymer Engineering & Science*, 1988, **28**, 1616-1627.
9. A. Wong and C. B. Park, *Polymer Testing*, 2012, **31**, 417-424.
10. H. J. Yoo and C. D. Han, *Polymer Engineering & Science*, 1981, **21**, 69-75.
11. M. R. Mackley and D. G. Hassell, *Journal of Non-Newtonian Fluid Mechanics*, 2011, **166**, 421-456.
12. T. R. Tuladhar and M. R. Mackley, *Chemical Engineering Science*, 2004, **59**, 5997-6014.
13. H. A. Z. Li, *Chemical Engineering Science*, 1999, **54**, 2247-2254.
14. H. Z. Li, X. Frank, D. Funfschilling and Y. Mouline, *Chemical Engineering Science*, 2001, **56**, 6419-6425.
15. H. Z. Li, Y. Mouline, L. Choplin and N. Midoux, *International Journal of Multiphase Flow*, 1997, **23**, 713-723.

New methodology for phase identification of modulus mapping images of macromolecular complex systems collected by peak force QNM AFM

Lu Yonglai [(1)*]***, Zhang Xi*** [(1)]***, Sun Shuquan*** [(1)]***, Zhang Liqun*** [(1)]

[(1)] Beijing University of Chemical Technology - Beijing - China

Peak-Force Quantitative Nano Mechanic (QNM) AFM is an edge-cutting scanning probe microscope and can be used to quantitatively investigate mechanic properties of materials in the nanoscale. With force mapping technique, the different phases could be distinguished according to their different mechanic features. For some macromolecular complex systems, however, the phase contrast of their force mapping images is fuzzy, resulting in that the phase identification might very depend on personal judgment of the researchers and some tiny microstructure could not be detected. In order to resolve this problem, we developed a new methodology for objective and accurate phase identification, basing on the deconvolution of modulus distribution of the force mapping images. In consideration with the fact that different phase exhibits different elastic modulus with normal distribution, the modulus distribution curve the image of the macromolecular complex system was fitted into several Gauss peaks; and accordingly how many different phases in the complex phases can be identified. The peak position and peak area reflect the characteristics modulus and relative fraction of corresponding phase. According to the deconvolution result, the modulus mapping images was recolored so that the size, shape and space distribution of identified phases can be demonstrated visually. This new method was used to analyze modulus mapping images of several complex systems, including functionalized SSBR/performance resin, isoprene rubber under stretching deformation, and carbon nanotube reinforced NR nanocomposites. Some hidden but valuable microstructural information were disclosed, facilitating deep understanding structure-property relationship.

Revealing the Polymorphism of Isotactic Polypropylene Transcrystallinity Using Gradient Temperature Field

Zhao Jing (1), Wang Ke (2)*, Fu Qiang (1)

(1) Sichuan university - Sichuan province - China, (2) Sichuan university - Sichaun province - China

The shear-induced isotactic polypropylene (iPP) trancrystallinity (TC) results in two crystal form, known as α-cylindrites and β-cylindrites. In this study, the trancrystallization of iPP induced by carbon fibers pulling under gradient temperature field (g-T field) was systematically investigated. A combined effect of temperature and shear rate on the formation of β-iPP cylindrites can be identified by stepwise adjust pulling rate under continuous temperature field. The results demonstrated that the trancrystallization depended strongly on the both carbon fiber pulling rate and the temperatures. It was found that the β-iPP cylindrites can only be produced under the condition of fiber pulling rate higher than 100 μm/s. Moreover, the trancrystallization varied from α-iPP cylindrites to mixed α-iPP and β-iPP cylindrites and finally pure β-iPP cylindrites along the g-T field from 150 °C to 130 °C. It was demonstrated that there is a highest temperature of 145 °C for the formation of β-iPP cylindrites, which is indenpendent on the pulling rate. Such temperature (145 °C) is highly possible as a starting temperature for ocurrance of β-to-α transition. So the appearance of β polymorph in iPP TC is essentially dominated by the thermodynamic stability of β-form. The findings in our study are meaningful for the manipulation of polymorphic structure of iPP trancrystallization.

Roughness Characteristics in Random Superhydrophobic Surfaces

Mead Joey [(1)]*, Nahum Tehila [(1)], Dodiuk Hanna [(2)], Kenig Samuel [(2)], Panwar Artee [(1)], Barry Carol [(1)]

[(1)] University of Massachusetts Lowell - MA - USA, [(2)] Shenkar College of Engineering Design and Art - Ramat Gan - Israel

To investigate the role of roughness in superhydrophobic coatings a variety of superhydrophobic and non-superhydrophobic surfaces were synthesized using various polymer binders, nanosilica particles and fluoro chemistry on both glass and polycarbonate substrates. The roughness of the coatings was measured by profilometry and atomic force microscopy (AFM) and analyzed by a variety of statistical methods. The skewness and kurtosis for the surfaces indicated that all coated samples, both superhydrophobic and non-superhydrophobic, had a random Gaussian roughness distribution. The skewness and kurtosis values, however, were not sufficient to determine superhydrophobicity, hence, other analysis methods had to be used to assess superhydrophobicity. It was concluded that the power spectral density function (PSDF) of the coatings was an effective tool to study the roughness needed to predict superhydrophobicity. Both RMS (root mean square) – the vertical scale of roughness and ACL (auto correlation length) – the horizontal scale of roughness that were extracted from the PSDF could predict superhydrophobicity. The ACL for all superhydrophobic formulations indicated that tightly packed asperities are essential for superhydrophobicity. In addition, the RMS roughness in nanoscale should be above 100 nm with ACL below 350 nm.

Layer breakup in nanolayer coextrusion

Bironeau Adrien (1), Guinault Alain (1), Salez Thomas (2), Miquelard-Garnier Guillaume (1), Sollogoub Cyrille (1)*

(1) PIMM - Paris - France, (2) Gulliver - Paris - France

Nanolayer coextrusion is an innovative process allowing to create films composed of thousands of layers of alternating polymers with individual layer thicknesses below 100 nm [1]. It has been observed that below a certain layer thickness, the layers tend to lose their integrity, i.e. they break spontaneously during the process. This study focuses on polystyrene (PS) / poly(methyl methacrylate) (PMMA) nanolayered films: different films have been fabricated with the aim to reach individual layer thicknesses as small as possible, varying the number of layers, the mass composition of both components and the final total thickness of the film. Films are characterized by atomic force microscopy (AFM) and a statistical analysis is used to determine the distribution in layer thicknesses and the continuity of layers [2]. For the PS/PMMA nanolayered systems, results point out the existence of a critical layer thickness around 10 nm, below which the layers break up. Possible mechanisms for critical layer thickness are examined and discussed. We propose this beak up phenomenon is due to small interfacial perturbations that are amplified by (van der Waals) disjoining forces.

Modulus enhancement of polycarbonate by additives

Takumi Sako[a] and Yamaguchi Masayuki[a]

[a]School of Materials Science, Japan Advanced Institute of Science and Technology (JAIST)
1-1 Asahidai, Nomi, Ishikawa, 923-1211 Japan
E-mail: m_yama@jaist.ac.jp

Abstract. Polycarbonate (PC) is widely employed in industrial fields because of its excellent transparency, mechanical toughness and heat resistance. However, the surface haedness is low as compared with other plastic glasses, leading to the poor anti-scratch property. Therefore, the modulus enhancement is required to widen its applications. We demonstrates new methods to enhance the modulus of PC without losing its advantages by blending a specific metal salt or poly(methyl methacrylate) with low molecular weight (L-PMMA). The modulus enhancement is achieved by a small amount of lithium perchlorate ($LiClO_4$) without losing its transparency, mechanical toughness and heat resistance, which is significantly different from a previously reported antiplasticized system. The electrostatic interaction between PC chain and Li cation is responsible for the phenomenon. The modulus enhancement is also detected by adding L-PMMA. L-PMMA can be miscible with PC even at the processing temperature and it can enhance the modulus of PC.

Keywords: Polycarbonate, Antiplasticizer, Poly(methyl methacrylate)
PACS: 83.80.Tc, 82.35.Lr

INTRODUCTION

Polycarbonate (PC) is expected as a substitution of inorganic glass because PC has marked transparency, heat resistance and mechanical toughness. The substitution is required especially in the automobile industry to reduce the weight of vehicles. However, the rigidity, i.e., modulus, of PC is low compared with that of inorganic glass. The modulus enhancement of PC without losing its transparency is desired to widen the application.

One of the well-known methods to improve the modulus of PC without losing its transparency is the addition of an antiplasticizer.[1)] It is believed that the antiplasticizers reduce the free volume fraction of polymers, which is responsible for the modulus enhancement.[2)] Despite of the marked modulus enhancement, the antiplasticizers are not employed in industry due to the decrease of the glass-transition temperature (T_g) and mechanical toughness.[3)]

In this research, we demonstrate two novel methods to improve the modulus of PC. The first method is the addition of a metal salt. It has been reported that some specific metal salts ionize in polymers with polar group such as poly(ethylene oxide) (PEO)[4)] and poly(propylene oxide) (PPO).[5)] Presumably, the ionized anion and/or cation show electrostatic interaction with polymer chains and affects the modulus of a polymer.

The second method is the addition of low-molecular-weight poly(methyl methacrylate) (L-PMMA), which has high modulus and miscibility with PC. Generally, the PC/PMMA blend is opaque due to the light scattering originated from phase-separated structure.[6)] However, our recent study indicates that a miscible PC/PMMA blend can be obtained by using L-PMMA[7)].

EXPERIMENTAL

Materials and Sample preparation

A commercially available bisphenol A type PC was used in this study. The number- and weight-average molecular weights, determined by size exclusion chromatography (HLC-8020, Tosoh) with polystyrene standards, were M_n=28,000 and M_w=46,000, respectively. Lithium perchlorate ($LiClO_4$) was purchased as a trihydrate form ($LiClO_4$ $3H_2O$). Low-molecular-weight PMMA (L-PMMA) was also used as an additive, which was prepared by Mitsubishi Rayon. The number- and weight-average molecular weights were M_n=8,900 and M_w=15,000.

PC and $LiClO_4$ were mixed in tetrahydrofuran (THF). The content of $LiClO_4$ was 5.0 wt.% in PC. The mixed solution was cast onto a glass plate and dried at room temperature for 5h under vacuum. The obtained films were subsequently dried at 120 °C for 30 h under vacuum to remove remaining solvent.

PC and L-PMMA were blended by melt-mixing method. Melt-mixing was performed in a 30 cm^3 batch-type internal mixer (Labo-Plastmill, Toyoseiki) at 240 °C for 3 min. The content of L-PMMA was 10 wt.%. The blend was compressed into a flat sheet with thickness of 0.3 mm. The sheet was subsequently cooled in another compression-molding machine with a cooling unit at 25 °C.

Measurements

The temperature dependence of oscillatory tensile modulus, storage modulus (*E'*) and loss modulus (*E"*), was measured from 25 to 180 °C by using a dynamic mechanical analyzer (E4000, UBM). The frequency and heating rate were 10 Hz and 2 °C/min, respectively. Rectangular specimens, in which the width is 5 mm and the length is 10mm, were cut out from the obtained sheets.

Tensile tests were performed using a uniaxial tensile machine (S1000-DVE, UBM) at room temperature. The films were cut into dumb-bell shape with an initial gage length of 8 mm. The tests were performed 10 times and the average value was calculated.

RESULTS AND DISCUSSION

Blend with $LiClO_4$

The temperature dependence of tensile storage modulus (*E'*) and loss modulus (*E"*) for the solution-cast films of PC and PC/$LiClO_4$ (5%) is shown in Figure 1. As shown in the figure, *E'* at room temperature for PC/$LiClO_4$ is higher than that for PC. The *E'* values at 25 °C for PC and PC/$LiClO_4$ (5%) are 2.00 and 2.69 GPa, respectively. The T_g of PC, determined by the peak temperature of *E"*, does not decrease by the addition of $LiClO_4$.

Figure 2 shows the stress-strain curves for PC and PC/$LiClO_4$ (5%). The Young's modulus is enhanced by the addition of $LiClO_4$, corresponding to the results of dynamic mechanical measurements. Moreover, it should be noted that the $LiClO_4$ addition does not affect the strain at break significantly. Since the stress level is higher than that of the pure PC in the whole experimental region without losing the strain at break, the mechanical toughness, is not reduced by the $LiClO_4$ addition. This behavior is completely different from the previously reported antiplasticized systems. Presumably, electrostatic interaction between ionized lithium and carbonyl fraction in the polymer chain is responsible for the mechanical properties.

Although this system is difficult to apply for thermal processing due to the crystallization of the salt, we achieved the modulus enhancement of PC without losing its transparency, heat resistance and mechanical toughness.

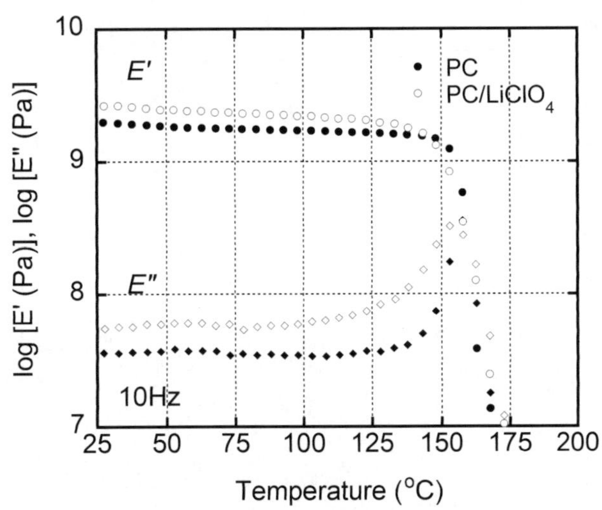

FIGURE 1. Temperature dependence of tensile storage modulus *E'* and loss modulus *E"* at 10 Hz for (filled symbols) PC and (open symbols) PC/LiClO4 (5%)

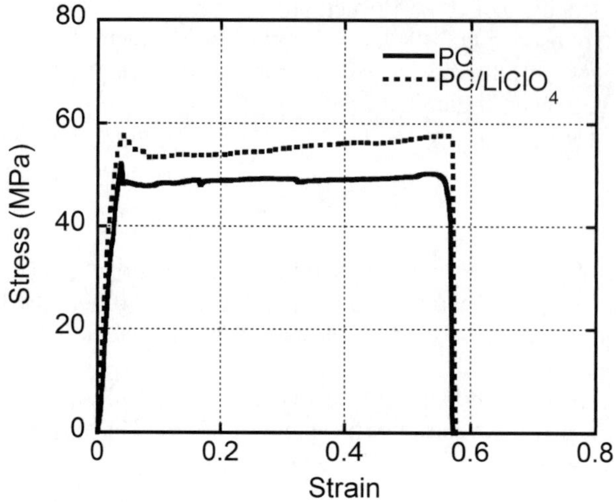

FIGURE 2. Stress-strain curves for solution-cast films of (solid line) PC and (dotted line) PC/$LiClO_4$ (5%)

Blend with L-PMMA

The temperature dependence of E' and E'' for the compressed sheets of PC and PC/L-PMMA (10%) is shown in Figure 2. The blend also shows a single T_g and keeps its transparency. This result indicates that the blend is miscible even at the processing temperature. The T_g of the blend slightly decreased due to the miscibility with L-PMMA. The T_g for PC and PC/L-PMMA (10%) were 161 and 151 °C, respectively. However, E' at room temperature is increased by the addition of L-PMMA. The E' values at 25 °C for PC and PC/L-PMMA (10%) are 1.90 and 2.13 GPa, respectively. It was revealed that L-PMMA also works as an antiplasticizer for PC.

Figure 4 shows stress-strain curves for the compression-molded sheets of PC and PC/L-PMMA (10%). Although the strain at break slightly decreased by the addition of L-PMMA, the Young's modulus and yield stress increased significantly.

This method is easy to apply for industrial products because transparent products can be obtained easily by conventional processing operations.

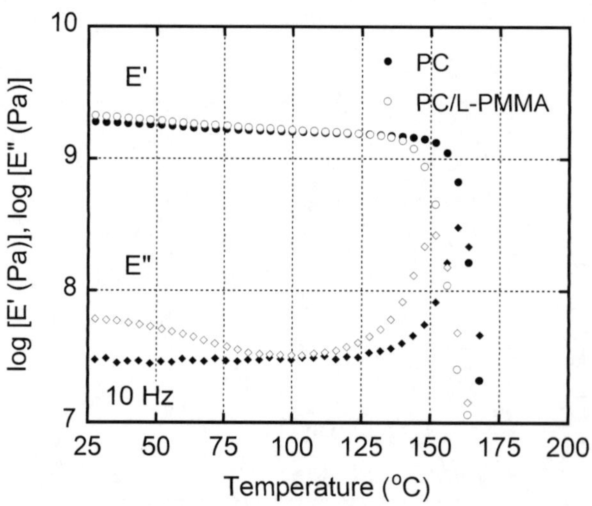

FIGURE 3. Temperature dependence of tensile storage modulus E' and loss modulus E'' at 10 Hz for (filled symbols) PC, and (open symbols) PC/L-PMMA (10%)

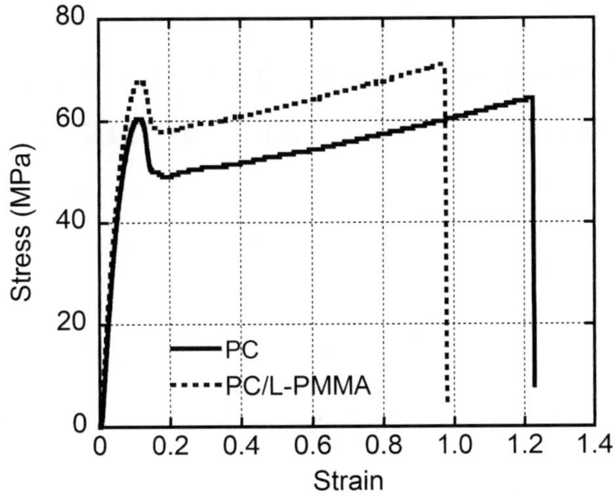

FIGURE 4. Stress-strain curves for compression-molded sheets of (solid line) PC and (dotted line) PC/L-PMMA (10%)

CONCLUSION

We demonstrated two novel methods to enhance the modulus of PC. The addition of $LiClO_4$ is effective for the enhancement of mechanical properties of PC. The addition of L-PMMA also enhances the modulus of PC without losing its transparency. This method is available in the industry because it can be applicable for conventional processing operations.

REFERENCES

1. W. J. Jackson and J. R. Caldwell, *J. Appl. Polym. Sci.* **11**, 211-226 (1967)
2. R. E. Cais, M. Nozomi, M. Kawai and A. Miyake, *Macromolecules* **25**, 4588-4596 (1992)
3. M. Ueda, *Polym. Eng. Sci.* **44**, 1877-1884 (2004)
4. K. Yoshida, H. Manabe, Y. Takahashi and T. Furukawa, *Electrochim. Acta* **57**, 139-146 (2011)
5. T. Furukawa, Y. Mukasa, T. Suzuki and K. Kano, *J. Polym. Sci. B: Polym. Phys.* **40**, 613-622 (2002)
6. M. Nishimoto, H. Keskkula and D. R. Paul, *Polymer* **32**, 272-278 (1991)
7. T. Sako, S. Nobukawa and M. Yamaguchi, *Polym.J.* **47**, 576-579 (2015)

Investigation of morphology and structure formation in Li-Ion Battery Electrodes during drying

Prof. Dr. Wilhelm Schabel

Co-autors: Dr. P. Scharfer, S. Jaiser, J. Kumberg

Karlsruhe Institute of Technology
Thin Film Technology Group, Institute of Thermal Process Engineering (TVT-TFT)
Institute of Micro Structure Technology (IMT-TFT)

Wilhelm.Schabel@kit.edu

Keywords: Processing, Casting. Coating and Drying, Thin Film Process Technology

ABSTRACT

Battery technologies are known to be key technologies for e-mobilty and stationary applications. There are still many developments needed before cheaper and larger storage systems will be more affordable. On one hand battery manufacturers have to increase the energy density and cycling capability and on the other hand processing costs must decrease in future. To overcome the current limitations and increase production costs, a better understanding of the production processes and limitations are essential. Goals are increase the production speed of electrode processing, i.e. improving coating and subsequent drying processes in the next decades.

State-of-the-art anodes in lithium-ion batteries are typically composed of porous graphite layers that are coated on a thin, current-collecting copper foil. Their properties strongly depend on the complex microstructure, which is defined by the pore structure and the distribution of functional additives throughout the film. Polymeric binder is added in small amounts in order to provide the film with mechanical integrity, while carbon black constitutes a conductivity enhancing agent. The microstructure essentially evolves during solvent removal, which makes drying a paramount, yet insufficiently understood processing step.

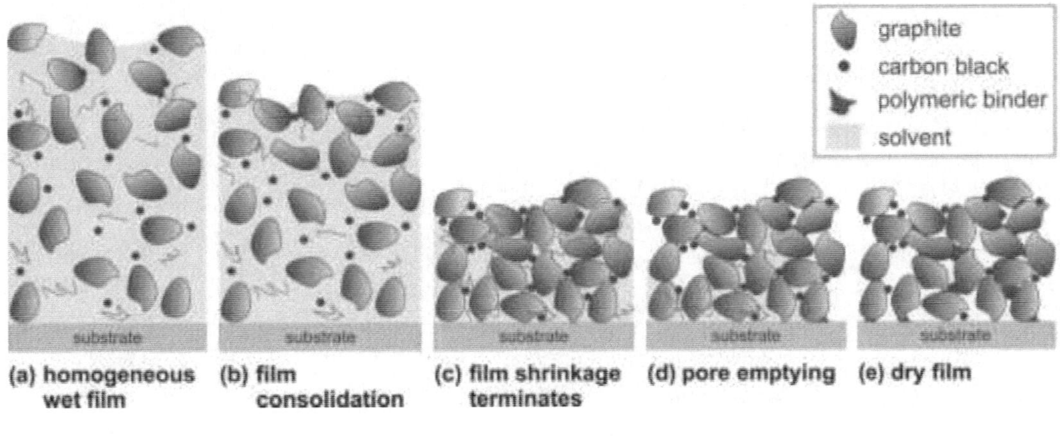

Figure 1 **Illustration of the film formation process.** *Schematic illustration of the film formation of particulate electrode films during drying, reference [1-4]*

(a) Solvent evaporation induces film shrinkage of the initially homogeneous wet layer.

(b) Graphite particles approach each other as solvent evaporates.

(c) Film shrinkage terminates, as the graphite particles obtain their most dense packing.

(d) further solvent evaporation causes the liquid phase to recede into the electrode structure

(e) Dry electrode.

Figure taken from: S. Jaiser; J. Schmatz; J. Kumberg; J. Klaver; P. Scharfer; W. Schabel; Microstructure Formation of Lithium-Ion Battery Electrodes during Drying – An Ex-Situ Study using Cryogenic Broad Ion Beam Slope-Cutting and Scanning Electron Microscopy (Cryo-BIB-SEM). J. Power Sources 2017, 345, 97-107

In the present work, different experimental approaches are introduced that allows for optical evaluation of microsture during drying to determine the onset of pore emptying and component migration as a function of defined process parameters. It could be shown that the drying play a key role with regard to pore emptying and component distribution and electrode properties.

REFERENCES

1. S. Jaiser; J. Schmatz; J. Kumberg; J. Klaver; P. Scharfer; W. Schabel; Microstructure Formation of Lithium-Ion Battery Electrodes during Drying – An Ex-Situ Study using Cryogenic Broad Ion Beam Slope-Cutting and Scanning Electron Microscopy (Cryo-BIB-SEM). *J. Power Sources* 2017, 345, 97-107

2. S. Jaiser; L. Funk; M. Baunach; P. Scharfer; W. Schabel; Experimental Investigation into Battery Electrode Surfaces: The Distribution of Surficial Liquid and the Emptying of Pores during Drying. *J. Colloid Interface Sci.* 2017, 494, 22-31

3. M. Baunach; S. Jaiser; P. Scharfer; W. Schabel,
Local heat transfer characteristics of a slot nozzle array for batch drying of thin films under industrial process conditions. *J. Coating. Tech. Res.* 2015, 12 (5), 915-920

4. J. Kaiser; V. Wenzel; H. Nirschl; B. Bitsch; N. Willenbacher; M. Baunach; S. Jaiser; M. Schmitt; P. Scharfer; W. Schabel,
Prozess- und Produktentwicklung von Elektroden für Lithiumionenzellen. *Chemie Ingenieur Technik* 2014, 86, 695-706

Acknowledgements:

We would like to thank to DFG for funding research on these drying topics

Crystallization behavior of polymeric materials with alternating multilayer structure and its influence on mechanical performances

Guo Shaoyun (1)*, ***Yang Shuo*** (1), ***Luo Shanshan*** (1), ***Shen Jiabin*** (1), ***Li Jiang*** (1)

(1) Sichuan University - Sichuan Province - China

Hierarchical structure is one of preferred choices in the nature and has been widely mimicked for practical applications in the past decades. As a kind of efficient technology, layer-multiplying coextrusion has been applied to fabricate polymeric materials with alternating multilayer structure by combining an assembly of layer multiplying elements (LMEs) with two extruders. With increasing the number of LMEs, the layers tended to be multiplied and the thickness of each layer could be reduced from micron to nanometer. Herein, the crystallization behavior of semi-crystalline polymers in the multilayer structure was investigated. Studies revealed that the confined layer interfaces and layer spaces induced different crystal morphologies and types from those in bulk systems. In order to clearly clarify the crystallization mechanisms, the influence of layer structure, the type of nucleating agents and the processing conditions were considered. Besides, the distinct crystalline structure possessed the ability to resist the mechanical deformation and destruction, which provided a non-additive strategy to fabricate polymeric materials with balanced performances.

Isothermal crystallization of polypropylene/ expanded graphite microplatelets composites: Graphite size and volume fraction

BEUGUEL Quentin [(1)], MONGE Gabriel [(1)], PEUVREL-DISDIER Edith [(1)], VERGNES Bruno [(1)], HAUDIN Jean-Marc [(1)], BOYER Severine A.E. [(1)]*

[(1)] CEMEF UMR CNRS 7635 - MINES ParisTech PSL - France

Interest in graphite fillers has grown since the separation of graphene from graphite by micromechanical cleavage [1]. More especially, one of the interests is to understand the influence of graphite microplatelets, with different size, on the crystallization behavior of a polymer matrix, after elaboration by melt mixing. The selected matrix is a neat polypropylene (PP), referenced as Moplen HP400R (LyondellBasell). The selected expanded graphite particles are KNG-180, KNG-150 [2] and KNG-G5 [3] (Knano®). Composites with three volume fractions of graphite varying from 0.3 to 2 % are concerned. The dispersion state of the particles in the composites is characterized at different scales using SEM and TEM; and crystallization using optical microscopy, DSC and X-ray diffraction. The discussions deal with the strong kinetics enhance of crystallization, at a given volume fraction, in the presence of respectively KNG-150, KNG-G5 and KNG-180. The expanded graphite particles are found to have a flake shape with a micrometric size. KNGs act as nucleating agent and induce an epitaxial growth of alpha crystalline phase of PP, but related to the difference in stacking dimension of graphite particles [4]. Radial growths are also observed, randomly located as classically reported. The coarsest and ordered KNG favors PP transcrystallinity, in such a way that (040) plane of PP is parallel to (002) plane of graphite platelets. Acknowledgement Expanded graphite particles were kindly provided by the CNRS research group project n°3661 'GDR PolyNano'. [1] K.S. Novoselov, Science 306 (2004) 666 [2] G. Chen, C. Wu, W. Weng, D. Wu, W. Yan, Polymer 44 (2003) 1781 [3] W. Zhao, F. Wu, H. Wu, G. Chen, J. Nanomater. (2010) 528235 [4] P. Sowiński, E. Piorkowska, S.A.E. Boyer, J.-M. Haudin, K. Zapala, Colloid Polym. Sci. 293 (2015) 665

Layered compounds as active fillers of biopolymeric nanocomposites for 3D printed functionally graded scaffolds for bone tissue regeneration

Sisani Michele (1)*, Bastianini Maria (1), Scatto Marco (2), Sanchez Alberto De Andres (3), Villanueva Sara Diez (3), Egizabal Ainhoa Luzuriaga (4)

(1) Prolabin & Tefarm S.r.l. - Perugia - Italy, (2) Nadir S.r.l. - Campus Scientifico Università Ca' Foscari Venezia - Mestre Venezia - Italy, (3) Tecnalia - Sustainable Construction Division - Azpeitia - Spain, (4) Tecnalia - Health Division - Azpeitia - Spain

In bone regeneration, autografts are considered the gold standard used for medical procedures, but several complications might occur for patients treated with autologous tissue due to the limited availability and also possible tissue morbidity. Scaffolds produced by Additive Manufacturing (AM) technologies allows the control of the pore architecture. This results in 3D matrices with controlled mechanical properties matching those of the tissue to be replaced. The use of polymeric nanocomposites with bio-functionalised layered fillers in a 3D printing device is a smart and effective solution for the production of scaffolds for bone regeneration. Biocompatible layered fillers, like hydrotalcites and zirconium phosphate, can be modified through the intercalation of active species with biocidal activity and then dispersed in a proper bioresorbable polymer using the melt blending technique. The exfoliation of these layered fillers into the polymer allows the preparation of bionanocomposites with better mechanical, cell differentiation and drug delivery properties. In this work MgAl-hydrotalcites and zirconium phosphate intercalated with high loading (up to 54 wt%) with ciprofloxacin and gentamicin antibiotics respectively were successfully employed as active fillers of PEOT/PBT copolymer, already used for clinical applications in orthopedic surgery. The fine dispersion of the fillers, confirmed by XRD and SEM/TEM analyses, was obtained using a twin screw extruder in the melt blending process that allows a good exfoliation/intercalation of the layered compounds into the polymeric composite materials. Composites with different loading of active fillers (from 5 to 20 wt%) were prepared and characterized in order to determine the mechanical (tensile, flexural and compression test). The evaluation of the biocidal activity on Staphylococcus epidermidis (Gram +) and Pseudomonas aeruginosa (Gram -) strains was also carried out by means of a Dose Response screening assay.

Exploring the role of nanoparticle geometry on the morphological evolution of cylinderical Polystyrene-b-poly(ethylene-butylene)-b-polystyrene (SEBS) triblock copolymer

N.Hasanabadi[1,2], M.Lattuada[2], B. Ranjbar[1], H.Nazockdast[1]*

[1]*Department of Polymer Engineering & Color Technology, Amirkabir University of Technology, No. 424, Hafez Ave., P.O. Box 15875-4413, Tehran, Iran*

[2]*Adolphe Merkle Institute,* University of Fribourg, *Chemin des Verdiers 4 CH-1700 Fribourg,* Switzerland

Abstract: Block copolymer(BCP) nanocomposites systems have attracted considerable technical and scientific interests. However, some aspects of these composite nanostructured materials including the impact of nanoparticle(NPs) geometry has not been comprehensively studied. The objective of present study is to examine the role nanoparticle geometry (Carbon Nanotube/ graphene nanosheet) on the microstructural development of BCP hybrids based on cylindrical forming polystyrene-b-poly(ethylene-butylene)-b-polystyrene (SEBS) by means of rheological techniques as well as small angle x-ray scattering(SAXS). SAXS experiment was utilized to probe the structural morphology of neat and also hybrid BCP samples. As observed in the SAXS pattern of neat BCP, the Bragg peak position ratios q_n/q_1 are equal to $1,\sqrt{3},\sqrt{7}$, indicating that it consists of cylindrical domains spatially arranged in a 2-D hexagonal lattice. The obtained results for hybrid samples confirmed no microstructural changes and/or disruption of BCP ordering upon incorporation of NPs. The results of temperature sweep experiments implied an enhancing effect of NPs on increasing the microphase separation temperature and accelerating its kinetic mainly due to the confinement of BCP segments. Such confinement leads to the stretching of polymer chains and reducing the number of chain conformations, resulting to the domination of enthalpic energy over entropic one and microphase separated structure is formed at higher temperatures. Graphene nanosheets showed to have more significant influence on microphase separation due to the construction of a more constrained space as compared to CNTs.
Moreover, DMTA results indicated that BCP samples exhibited three peaks associated to the damping of EB (ethylene-buthylene) domains, interphase region and PS domains. Incorporation of nanoparticles leads to the promotion of BCP microphase separation as confirmed by reducing the interphase region area.

Keywords: block copolymers, Multi-Walled Carbon Nanotubes, Graphene nanosheets, microstructure

INTRODUCTION

Block copolymers (BCPs) represent a subject of broad current research due to their ability to infuse the dissimilar properties of homopolymers into a single material by spontaneous self-assembly at the molecular level. Ordered structures observed in block copolymers include spheres, cylinders, gyroids and lamella [1]. The ultimate morphology of block copolymers depends on the self-assembly phenomenon, which is governed by thermodynamic incompatibility, chain connectivity and relative length of blocks [2–8]. In order to further improve BCP properties, various studies have investigated the incorporation of nanoparticles into block copolymers as a highly promising method to tailor the material's properties and enhance their mechanical, electrical and thermal properties.

Carbon nanotubes (CNTs) and grapheme nanosheets, because of their superb physical and mechanical characteristics have been used as one of the most useful building blocks candidates for producing new generation of composite materials for a variety of applications [9-10].

The objective of the present study is to examine the effect of CNT and grapheme nanosheets on the microstructural evolution of cylindrical forming SEBS block copolymers by employing a combination of rheological and small angle x-ray scattering(SAXS).

EXPERIMENTAL SECTION

Materials

A commercial SEBS block copolymer with 30 wt% PS content (SEBS G1652; solution viscosity at 25 ^{0}C and 25 wt% in toluene: 1800 cps) was kindly supplied by Kraton Corporation.

Multi-Walled Carbon Nanotubes (MWCNTs) with the trade name of Nanocyl NC 700 (average diameter of 9.5 nm, average length of 1.5 μm and purity of 90%) and graphene nanoplates, N002-PDR, from Angstron Materials (thickness <1 nm, average diameter <10 μm, average aspect ratio of about 4000, surface area of 400–800 m^2 g^{-1} and comprised of stacks of 1-3 monolayer graphene sheets) were used as the dispersed phase.

Preparation of Nanocomposites

All nanocomposite samples varying in MWCNT or graphene content were prepared using the melt-compounding method in a laboratory internal mixer, 60 cm^3, equipped with a banbury-type rotor design (Brabender Plasticorder W5) at 220 °C and rotor speed of 80 rpm for a total mixing time of 15 min.

Small-angle X-ray scattering (SAXS)

The morphology of block copolymer and their nanocomposites samples were evaluated by SAXS. The SAXS spectra were recorded using a NanoMax-IQ camera (Rigaku Innovative Technologies, Auburn Hills, MI USA). The samples were kept under vacuum at room temperature during the measurements. The scattering spectra are presented as a function of momentum transfer, defined by $q = 4\pi\lambda-1 \sin(\theta/2)$ (where θ is the scattering angle and $\lambda = 0.1524$ nm is the photon wavelength).

Rheological measurements

To study microphase separation of samples, isochronal dynamic temperature sweep tests were carried out at a strain of 1% and ω=1 s^{-1} and a cooling rate of 1 °C min^{-1} under nitrogen purge to avoid oxidative degradation.

Dynamic mechanical thermal analysis (DMTA)

DMTA was recorded by a Diamond DMA Perkin Elmer instrument in tensile mode on samples of 20 × 11 × 1 mm^3 cut from compression molded sheets. Thermograms were measured at frequency of 1 Hz and a heating rate of 5 °C min^{-1} over temperature range from -100 to 140 °C.

RESULTS AND DISCUSSION

SAXS was used to obtain structural information on the BCP nanocomposite samples. Their SAXS profiles are shown in Figure 1. Bragg peaks appear at relative q-positions of 1, $\sqrt{3}$ and $\sqrt{7}$, indicating all samples exhibit a morphology of hexagonally packed cylinders[11].

From angular averaged SAXS profiles, it can be noticed that no changes were observed in the feature of the SAXS patterns upon incorporation of CNT or graphene nanoparticles, thus suggesting the BCPs retain their original structure and addition of nanoparticles could not disrupt the BCP ordering.

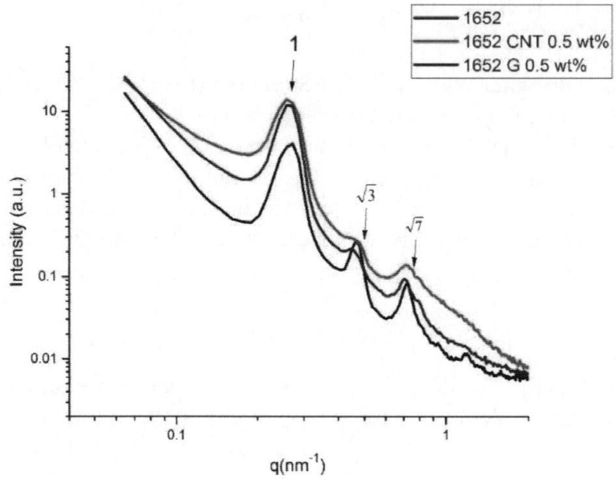

FIGURE 1. 1D SAXS plots for SEBS G1652 neat BCPs and their nanocomposite samples containing 0.5 wt% nanofillers.

It has been demonstrated that rheology can be employed to probe the microstructural evolution of partially miscible blends and block copolymers. Microstructural developement temperature of BCP can be estimated through a distinct change in the slope of the storage modulus curve vs. temperature, as a result of microphase separation-induced elastic (solid-like) response[12]. The results of temperature sweep experiment performed on neat SEBS G1652 and its nanocomposite samples are shown in Figure 2.

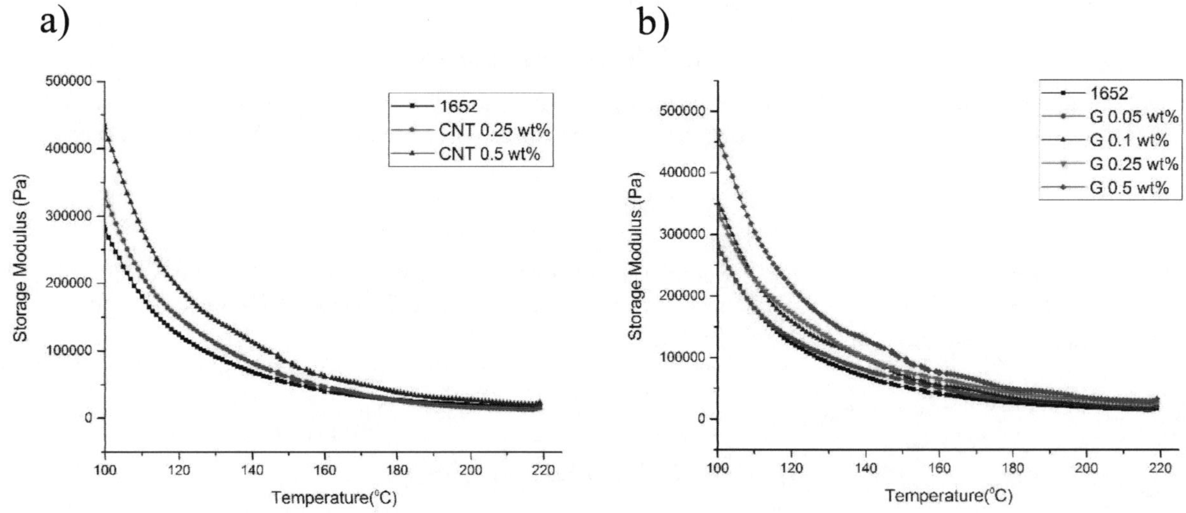

FIGURE 2. Storage modulus versus temperature for SEBS G1652 nanocomposite samples containing (a) MWCNT, (b) graphene nanosheets.

It could be observed that for both CNT and graphene filled samples, the addition of nanoparticles promotes the microstructural development temperature, and also that the presence of graphene has a greater influence on increasing the kinetics and the temperature of microstructural development temperature compared to that of CNT. This can be explained in terms of greater extent of PS segments confinement in graphene interlayer space, which provides higher potential for PS domain formation in graphene filled sample compared to CNT. Such a confinement of PS segments leads to stretching of polymer chains and reducing the number of chain conformations. Hence, the enthalpic interactions begin to dominate and a microphase separated structure is formed. A similar behavior has also been observed on the crystalline behavior of nanocomposite samples containing CNT and graphene nanoplates [13-14].

Figure 3 exhibits the results of damping factor vs. temperature for neat BCP and the nanocomposite samples. The samples display three peaks at around -30 °C, 40-50 °C and 100-120 °C which can be assigned to the glass transition of ethylene-butylene (EB) domains, interphase regions and PS domains., This interphase is a transitional region between adjacent pure PS and EB domains where the chain segments of both blocks coexist.

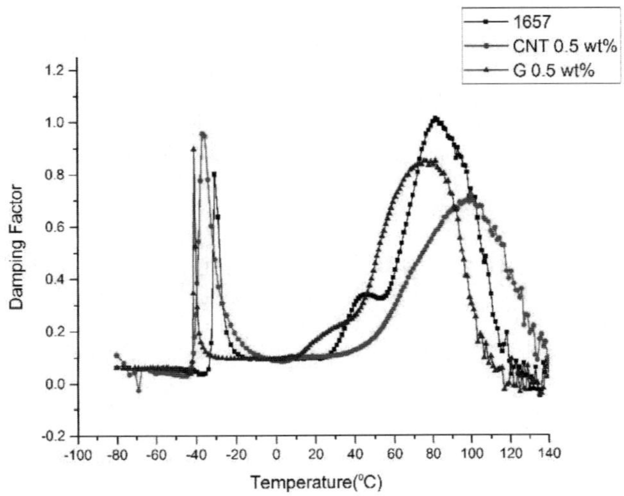

FIGURE 3. tan δ versus temperature for SEBS G1652 block copolymers nanocomposite samples containing 0.5 wt% nanofillers.

Preferential interaction of nanoparticles with PS segments leads to increasing the segmental incompatibility of BCP. This increased interfacial interaction can also result in decreasing the entropic energy of the system through slowing down the segmental motion. These two effects promote microphase separation. Improved microphase separation drives more PS segments to be arranged in PS domains instead of interphase region. Therefore, PS chain segmental movement would be more restrained. In this case, the area of tan(δ) of both interphase region and PS domains would be decreased due to lower content of PS segments in interphase region and also restriction of segments in PS domains, respectively.

CONCLUSIONS

Block copolymer-based materials containing nanoparticles has drawn much attention in the literature. This work aims at using CNTs and graphene nanosheets in cylindrical forming SEBS block copolymer. SAXS spectra confirm that the microstructure of the BCP is not affected by the presence of the nanoparticles. Temperature sweep measurements indicate that the addition of nanoparticles enhances the microstructural development temperature and accelerates its kinetic, mainly due to the confinement of BCP segments. This effect is more pronounced in the case of graphene nanosheets due to the providing of more confined geometry in comparison to CNTs. Additionally, the DMTA results indicat a promotion of the BCP microphase separation by incorporation of nanoparticles following the same effect observed for thermal annealing.

REFERENCES

1. B.H. Kim, J.Y. Kim, S.O. Kim, S.Y. Park, P.B. Messersmith, I.S. Choi, R. Langer, D.G. Anderson, H. Lee, S.O. Kim, E.L. Thomas, Y.S. Jung, Directed self-assembly of block copolymers for universal nanopatterning, Soft Matter. 9 (2013) 2780.
2. G.H. Fredrickson, F.S. Bates, Dynamics of Block Copolymers: Theory and Experiment, Annu. Rev. Mater. Sci. 26 (1996) 501–550.
3. F.S. Bates, G.H. Fredrickson, Block Copolymers—Designer Soft Materials, Phys. Today. 52 (1999) 32–38.
4. Y. Mai, A. Eisenberg, R.J.M. Nolte, J.C.M. van Hest, S.P. Armes, A.J. Ryan, A.L. Lewis, G. Battaglia, S. Tan, A. Brisson, V. Dupuis, O. Sandre, S. Lecommandoux, Self-assembly of block copolymers, Chem. Soc. Rev. 41 (2012) 5969.
5. S.B. Darling, Directing the self-assembly of block copolymers, Prog. Polym. Sci. 32 (2007) 1152–1204.
6. J.K. Kim, S.Y. Yang, Y. Lee, Y. Kim, Functional nanomaterials based on block copolymer self-assembly, Prog. Polym. Sci. 35 (2010) 1325–1349.
7. T. P. Lodge, "Block Copolymers: Past Successes and Future Challenges," *Macromol. Chem. Phys.*, vol. 204, no. 2, pp. 265–273, Feb. 2003.
8. A. Horechyy, "Ordered Structures from Nanoparticles/Block Copolymer Hybrids: Ex-situ Approaches toward Binary and Ternary Nanocomposites.", Leibniz-Institut für Polymerforschung Dresden PhD thesis, 2007.
9. H. Garate, M.L. Fascio, I. Mondragon, N.B. D'Accorso, S. Goyanes, Surfactant-aided dispersion of polystyrene-functionalized carbon nanotubes in a nanostructured poly(styrene-b-isoprene-b-styrene) block copolymer, Polymer. 52 (2011) 2214–2220.
10. G. Mittal, V. Dhand, K.Y. Rhee, S.-J. Park, W.R. Lee, A review on carbon nanotubes and graphene as fillers in reinforced polymer nanocomposites, J. Ind. Eng. Chem. 21 (2015) 11–25.
11. I.W. Hamley, V. Castelletto, Small-angle scattering of block copolymers: in the melt, solution and crystal states, Prog. Polym. Sci. 29 (2004) 909–948.
12. G.P. Kar, A. Bharati, P. Xavier, G. Madras, S. Bose, B.A. Wolf, Q. Tran-cong, T. Chang, I.C. Sanchez, The key role of polymer grafted nanoparticles in the phase miscibility of an LCST mixture, Phys. Chem. Chem. Phys. 17 (2015) 868–877.
13. S. Xin, Y. Li, H. Zhao, Y. Bian, W. Li, C. Han, Q. Dong, Z. Ning, L. Dong, Confinement crystallization of poly(l-lactide) induced by multiwalled carbon nanotubes and graphene nanosheets, J. Therm. Anal. Calorim. 122 (2015) 379–391.
14. S. Aoyama, Y.T. Park, T. Ougizawa, C.W. Macosko, Melt crystallization of poly(ethylene terephthalate): Comparing addition of graphene vs. carbon nanotubes, Polymer (Guildf). 55 (2014) 2077–2085.

The Influence of Shear Flow and Specific Work on the Peak Time of Crystallization and Resulting Microstructure of Polyamide 6,6

Rhoades Alicyn M [(1)]*, Gohn Anne M [(1)], Seo Jiho [(2)], Androsch Rene [(3)], Colby Ralph H [(2)]

[(1)] Penn State Behrend - PA - USA, [(2)] Penn State University - PA - USA, [(3)] Martin Luther University - Saxony Anhalt - Germany

The excellent physical properties of Polyamide 6,6, (PA66) result from the cooperative interactions within the semi-crystalline microstructure of the polymer. Several factors drive the development of the microstructure on cooling, including melting temperature and the amount of shear to which the melt is subject prior to solidification. These two factors contribute to the amount of specific work (MPa) introduced to the melt. In this study, samples were prepared using rotational rheometry to introduce a known amount of specific work to each sample. The range of specific work studied spanned from 6.25 MPa to 200 MPa. After shearing, the polymer was allowed to solidify and was then sectioned for subsequent thermal analysis. Crystallization studies using fast scanning calorimetry (FSC) were designed to isolate the peak time of crystallization for several previously-sheared samples of PA66 in five-degree increments across the temperature range of 85°C – 220°C. In the high-temperature range, peak time of crystallization decreased with increasing specific work, indicating flow-induced effects on the nucleation process for PA66. To a lesser extent, accelerated crystallization was also observed for the low-temperature crystallization processes. Interestingly, the temperature at which PA66 transitions from the high-temperature to low-temperature crystallization mechanism shifted to lower temperatures with increasing specific work. ESEM imaging of PA66 subject to strategic isothermal/specific work combinations reveals a shift in microstructure as a function of specific work and isothermal crystallization temperature.

Temperature-dependent Transitions of Block Copolymer Confined in Nanolayer Assemblies

Tan Xiao *(1)**

(1) Sichuan University - SiChuan - China

Block copolymers (BCPs) are capable of self-assembling into periodic nanostructures, making them fascinating candidates in applications of nanotechnologies. When placed into a spatially confined environment, BCPs can self-assemble into novel morphologies deviated from bulk. And the phase behavior become more complex and sensitive to subtle changes of confining conditions. Therefore, in this paper we report a new method to investigate temperature-dependent phase behavior of BCPs under confinement by preparing nanolayer assemblies via layer-multiplying coextrusion technology. The transitions are characterized with in-situ small-angle X-ray scattering (SAXS), rheological measurements, and transmission electron microscopy (TEM). Our studies show that the transitions and stability of microdomains have a strong dependence on the confinement factors, including layer thickness and surface affinity, with shifts of transition temperatures.

Routes Towards The Multiscale Simulation of Filled Polymers: The Strain-Bridging Algorithm

Clemens Holzer[1, a)], Ali Gooneie[2, b)],

[1]*Chair of Polymer Processing, Department of Polymer Engineering and Science, Montanuniversitaet Leoben, Otto Gloeckel-Strasse 2, Leoben 8700, Austria*

[2]*Laboratory of Advanced Fibers, Empa, Swiss Federal Laboratories for Materials Science and Technology, Lerchenfeldstrasse 5, CH-9014 St. Gallen, Switzerland*

[a)]Corresponding author: clemens.holzer@unileoben.ac.at

[b)]ali.gooneie@empa.ch

Abstract. In this manuscript, we investigate the possibilities and challenges to simulate morphological evolutions in filled polymers under applied flows. Various mesoscopic examples from polymers containing layered silicates, carbon nanotubes, and cellulose nanocrystals are discussed. In particular, dissipative particle dynamics (DPD) is highlighted as a promising tool to attack the existing challenges. The orientation of anisometric particles under shear flow is simulated at various flow conditions. The results are shown to be in good agreement with previous experiments. Furthermore, self-assembly of nanorods/nanotubes is manifested as a function of their length and content under thermodynamic conditions. Molecular dynamics (MD) is also represented as an underlying simulation strategy to provide DPD parameters. Moreover, the value of MD in atomistic simulations of filled polymers is emphasized by an example from highly-filled polypropylene with Fe_2O_3 microparticles. The conformation of polymer chains under confinement is discussed in terms of their bond orientation, segmental orientation, and chain size. Eventually, we discuss the most promising routes to couple MD, DPD, and finite element method (FEM)/finite volume method (FVM) methods to uncover a multiscale framework for the simulation of such filled systems. In this context, the strain-bridging algorithm is revisited using which one can setup a multiscale simulation of the formation of oriented microstructures of anisometric nanoparticles in polymers under shear deformations.

INTRODUCTION

For many years, the microstructure of complex polymeric materials such as immiscible blends, nanocomposites, highly-filled composites of fibers, carbon black, ceramics, and metal powders has been investigated and proven to determine the final properties of such advanced materials. Experiments have been conducted and theories developed in order to provide a better understanding of the phenomena involved in the microstructural evolutions during processing. Various complications have surfaced ever since depending on the system specific physics which can immensely influence the materials behavior before, during, and after the flow. To give a few examples, we start by the adsorption of polymer chains on the surface of microparticles in highly-filled polypropylene/Fe_2O_3 materials. The extend of the adsorption as well as the local configurations of the chains in the vicinity of Fe_2O_3 is undoubtedly a key parameter which dictates the binding and/or debinding during a sintering process of polymer/metal composites. In another class of materials, the anisometric particles such as layered silicates have shown to rotate and orient when imposed to external flows. Such orientations could provide anisometric properties in the final part if it is frozen before relaxation occurs. Therefore, it is important to understand the underlying physics of particle orientation in flow. At the same time, other anisometric nanoparticles, such as cellulose nanocrystals (CNCs) and carbon nanotubes (CNTs) have shown to undergo certain self-assembly patterns in thermodynamic equilibrium with a kinetics which critically depends on their length distribution. Finally, on a totally different length scale, the development of velocity, pressure, and temperature fields in the extruder, in extrusion dies, and in injection molds has shown to be strongly affected by a number of phenomena including wall slippage, discontinuities in the flow, and viscoelastic behaviors. These phenomena are all dependent on the underlying microscopic and atomistic physics of such complex materials.

Computer simulation methods such as molecular dynamics, Monte-Carlo, lattice-Boltzmann, finite volumes, etc., have served well in many research areas to tackle such problems and produce reliable predictions. Over the past decade, a variation of molecular dynamics method namely dissipative particle dynamics (DPD) has been developed and incorporated to model dynamic systems [1]. DPD is a coarse-grained particle-based method which simplifies the molecular structures into interacting beads. Such an approach makes it possible to access longer time scales while still preserving significant molecular details [2]. In this work, we aim to show some of the mentioned problems with solutions from MD and DPD and later discuss how one can link the microscopic data of such simulations to the macroscopic FEM/FVM models. To achieve the latter, the orientation of layered silicate nanoparticles in a polymer is monitored during the flow in a strain-bridging algorithm. In order to preserve the brevity, we encourage the interested reader to refer to our previous papers which outline the details of all numerical simulations explained in this manuscript [3].

MODEL SYSTEMS, METHODS AND RESULTS

Local configuration of polypropylene chains on Fe_2O_3 substrate. Utilizing atomistic MD simulations, the local ordering and properties of polypropylene chains sandwiched between Fe_2O_3 surfaces were studied, see figure 1a [4]. The density profile of this system was calculated and showed an adsorbed polymer layer on the metal oxide surface with a thickness of ~7 Å, see figure 1b. The density was similar to that of unconfined polypropylene at distances beyond 20 Å. The local configuration analyses revealed that at the interfacial region, the backbone and side chain C–C bonds of the chains orient parallel and normal to the solid surface, respectively. Furthermore, the data shows that the chains are compressed in the normal direction to the substrates in the interfacial regions to some extent. Moreover, the local pressure tensor is anisotropic in the vicinity of the solid surfaces of Fe_2O_3 indicating the influence of the confinement on the buildup imbalance of normal and tangential pressures, see figure 1c. This imbalance allowed for the calculation of the interfacial tension γ. These results confirm the applicability of polypropylene as a backbone component in a metal injection molding feedstock containing ferritic materials due to the strong adsorption of the polymer at the surface of Fe_2O_3 substrates. This notion is particularly important since the backbone polymer should hold the metal particles in place during the shaping process, and while other components are being removed during the debinding step. This guarantees that the geometry of the molded part will be maintained and sintering can be done to obtain a metal part with complex shape.

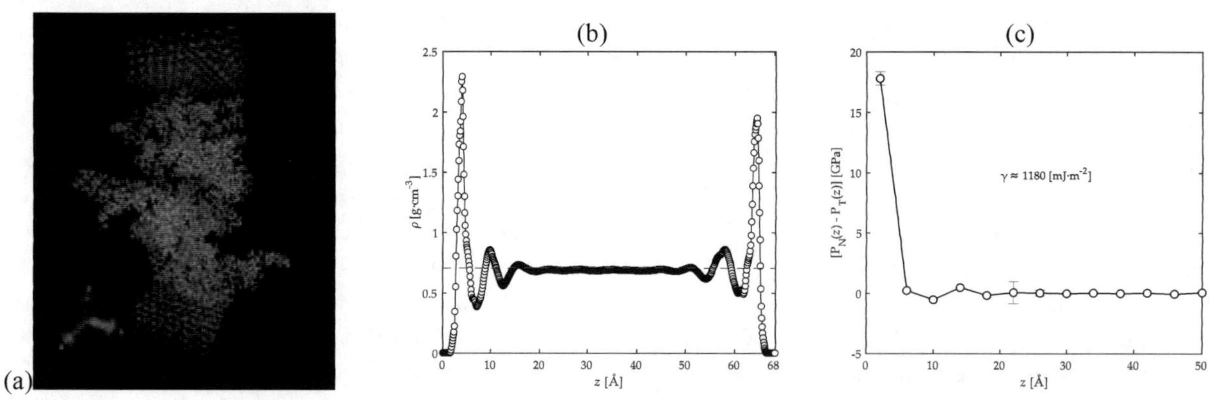

FIGURE 1. (a) The polypropylene chains between two Fe_2O_3 substrates. Atoms are: Red: O, White: H, Cyan: Fe/C. (b) Local mass density profile, ρ, across the film thickness. (c) The profile of the difference of normal (P_N (z)) and tangential pressures (P_T (z)) across the film thickness. For further details, refer to Gooneie et al. [4] Reprinted from Gooneie et al. under the terms of the Creative Commons Attribution License.

Orientation of layered silicate nanoparticles under shear. The orientation of the layered silicate was studied by applying different shearing directions in a DPD framework [5,6]. Figure 2 demonstrates the orientation of the particle at different simulation steps for various shearing directions. It is obvious that the layers rearrange themselves so that their surfaces would be normal to the velocity gradient direction regardless of the applied flow direction. Understanding the orientation kinetics helps to define possible directional properties in the final material.

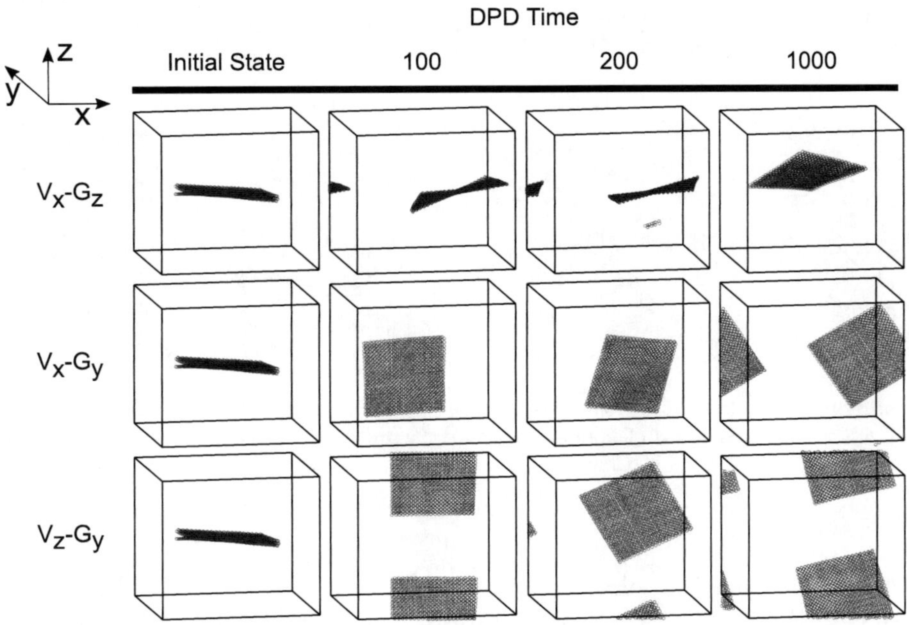

FIGURE 2. The orientation of a 3-layered silicate particle with DPD steps. The applied shear directions are shown for each row by V_i and G_j as the applied velocity in *i*-direction and applied velocity gradient in *j*-direction, respectively. Reprinted from Gooneie et al. [5] Copyright 2017, with permission from John Wiley & Sons Inc.

Self-assembly of bidisperse nanorods/nanotubes in equilibrium. Polymer-based materials reinforced by nanorods/nanotubes have drawn considerable attention of the scientific community in the last few decades. The aspect ratio of rod-like nanofillers is an important factor that governs the percolation threshold and therewith the properties of nanocomposites formed with a polymer of interest. Sample calculations using DPD clearly shows the temporal developments of equilibrium microstructures, i.e. self-assembly, of such dispersions in nanocomposites [7]. These self-assembled bundles only form under neutral and/or unfavorable dispersion parameters, see figure 3. Similar self-assemblies have also been observed in DPD models of CNTs at the interface of immiscible polymer blends [8]. It appears that in the bidisperse mixtures self-assembly kinetics is substantially slowed down particularly in the initial stages, suggesting the spatial interferences and providing evidence for a synergistic reinforcement. The findings are particularly relevant to materials with an inherent length distribution, for instance cellulose nanocrystals (CNCs).

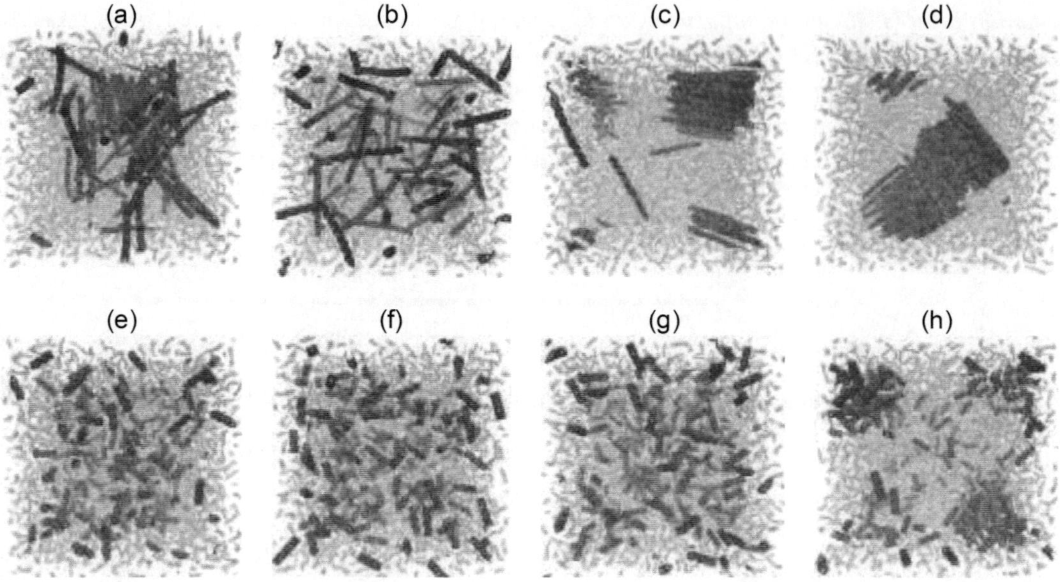

FIGURE 3. The equilibrium microstructures of nanorods/nanotubes in polymer for long (a to d) and short (e to h) nanotubes: (a, e) the initial microstructure, (b, f) system with favorable dispersion parameters, (c, g) system with neutral dispersion parameters, and (d, h) system with unfavorable dispersion parameters. The long and short nanotubes are shown in blue and red, respectively. The polymers are presented in green in all parts. Reprinted from Gooneie et al. [7] Copyright 2017, with permission from Elsevier.

Strain-bridging of orientation data in a multiscale framework. The orientation data of DPD models (e.g. figure 2) can be incorporated to build an upscaling method for the mesoscopic orientation patterns to the macroscopic flows [3,6]. This upscaling method was tested successfully against the available standard orientation models from the literature. In this strategy, the trajectories of the orientation process of weakly-interacting layered silicates are parametrized as a function of the applied shear strain instead of the time, based on the experiments which propose strain-dependent rather than time-dependent structural evolutions in such non-Brownian materials, see figure 4. Benefitting from the notion that the orientation kinetics is simply the rate of change with respect to strain rather than time, the applied strain was selected to pass the orientation parameters to an upper scale through a simple combination of affine and nonaffine deformations. This combination was pictured in its simplest form to be a random mixing of DPD unit cells (simulating nonaffine deformations) in a larger cell which distributes an affine deformation over the unit cells. It was noted that this strategy could be used to perform multiscale simulations of orientation process provided that the unit cells represent a precise description of the interactions between the components. A comparison of this methodology with the strain reduction factor model showed the success of the multiscale simulation of the evolution of orientation parameters against the applied shear strain.

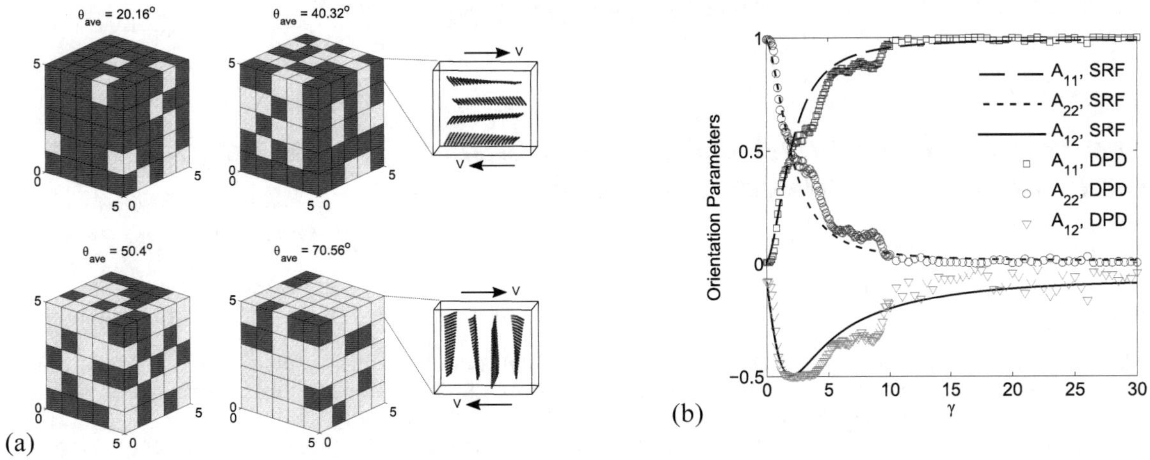

FIGURE 4. (a) Examples of construction of a large cell for the upper scale simulation benefitting from a random mixing of unit cells resulting in various average initial orientation angles, θ_{ave}. The initial configurations of the unit cells before the flow starts are also given. (b)The orientation process defined by the orientation parameters as a function of the shear strain, γ. The results are derived from DPD models and strain reduction factor (SRF) model. Reprinted from Gooneie et al. [6] Copyright 2017, with permission from John Wiley & Sons Inc.

CONCLUSION

As it was shown, atomistic and microscopic methods such as MD and DPD offer great potential to simulate the complex systems with high relevance for the industrial applications. The data of such detailed simulations can be further transferred to macroscopic simulation techniques including FEM and FVM in order to investigate the effect of the processing parameters on the microscopic microstructures and vice versa. The strain-bridging algorithm is proposed here for non-Brownian materials such as polymers to be a possible route for such a communication between the scales. Indeed, further investigations are required on a variety of systems in order to understand its benefits and shortcomings.

REFERENCES

1. R.D. Groot and P.B. Warren, J. Chem. Phys. **107**, 4423–4435 (1997).
2. A. Gooneie, S. Schuschnigg, and C. Holzer, Macromol. Theory Simulations **25**, 170–186 (2016).
3. A. Gooneie, S. Schuschnigg, and C. Holzer, Polymers (Basel) **9**, 16 (2017).
4. A. Gooneie, J. Gonzalez-Gutierrez, and C. Holzer, Polymers (Basel) **8**, 361 (2016).
5. A. Gooneie, S. Schuschnigg, and C. Holzer, Macromol. Theory Simulations **25**, 85–98 (2016).
6. A. Gooneie, S. Schuschnigg, and C. Holzer, Macromol. Theory Simulations **25**, 287–302 (2016).
7. A. Gooneie, J. Sapkota, A. Shirole, and C. Holzer, Polymer (Guildf) **118**, 236–248 (2017).
8. A. Gooneie and C. Holzer, Polymer (Guildf) **125**, 90–101 (2017).

Bio-inspired Highly Sensitive and Reversible Mechanochromisms Based on a Polymer Bilayer Structure

Sun Luyi **(1)***

(1) University of Connecticut - CT - USA

A series of scalable, simple, and low cost elastomer based bilayer systems with a rigid surface layer of polymer nanocomposite were designed and prepared. The bilayer systems exhibit different surface morphologies and strain-responsive properties by forming micro-cracks or wrinkles on the top surface. Although cracks and wrinkles are typically considered as defects and mechanical instabilities, respectively, to be avoided in most cases, proper control of the formation and evolution of micro-cracks and wrinkles can lead to many novel functionalities. With the assistance of modeling, we created a series of highly sensitive and reversible mechanochromisms. Upon uniaxially stretching and releasing, as well as additional stimuli if needed, the devices can reversibly switch the topography between flattening surface and rough morphology with longitudinal micro-cracks or wrinkles, resulting in striking optical changes, including changes in transparency, luminescence, and coloration, promising for widespread application.

Tailoring Structure of Epoxy/Polyethersulfone Composites for Thermal Conductivity Enhancement

Yucai Shen, Yuwei Zhang, Tingwei Wang

Nanjing Tech University, Nanjing, Jiangsu, China, 210009

Abstract. Epoxy/polyethersulfone (PES) blends were prepared without using solvent by dissolving the PES in the curing agent. Morphological structure of the blends during the reaction-induced phase separation were studied. At various PES loadings, different morphologies of the blends were obtained: sea-island structure formed with the low PES loading, nodular structure formed with the high PES loading, and interesting layered structure or capsule-like structure as a whole formed in the medium PES loading where the outer layers consisted mostly of the epoxy-rich phase. The evolution of reaction-induced phase separation process and final morphology of various Epoxy/PES blends were investigated. It was suggested that at certain PES loading, bicontinuous structure formed and grew in size at the early stage of phase separation, then the PES-rich network became shrinking, and epoxy-rich domains eventually flew out to be outer layer. With the addition of graphite nanoplatelet in the optimized epoxy/PES composites, an enhancement of thermal conductivity was obtained due to the formation of thermally conductive network.

Keywords: Epoxy, reaction-induced phase separation, thermal conductivity
PACS: 66.10.cd

INTRODUCTION

Reaction-induced phase separation (RIPS) has been extensively studied and applied in polymer blending and modification. The formation of various morphologies through two different matrixes can greatly change the properties of materials. The structural evolutions and the forming mechanisms are also very complicated. Therefore, it is very important to study the detailed structures and monitor the changes of the structures during the RIPS. Pascault et al. [1] studied the structural evolution of the PEI-modified epoxy system, and it was found that the mixture showed the sea-island structure, the co-continuous phase structure and the inversion phase structure with the different PEI contents. Yu et al. [2, 3] demonstrated that viscoelastic phase separation exists in thermoplastics (TP)-modified epoxy system experimentally. Han et al. [4] found a unique three-layer structure in the polyimide-modified epoxy blend firstly. Further, it was found this unique three-layer structure existed in many systems with different kinds of TP, different curing temperatures and different thicknesses.

Thermally conductive polymeric materials have strong potential as thermal management materials due to their facile processing, light weight and low cost. Recently, a lot of efforts have been devoted to attaining the high TC at low filler content through designing a continuous filler-network in composite. The most common method is designing a segregated structure [5~8]. However, this strategy is limited to thermoplastic resins.

In this study, morphological structures and rheological behaviors of the blends in RIPS were studied by optical microscope and electron microscope combined with rheometer. With the addition of graphite nanoplatelet in the epoxy/PES composites, an enhancement of thermal conductivity was obtained due to the formation of thermally conductive network, and the mechanism was also discussed.

EXPERIMENTAL

To avoid the negative effects of solvent, PES was dissolved in the curing agent at a mild temperature, which effectively avoided the effects of the residual solvent on the interfacial structures of components. Different contents of PES (10 wt.%, 20 wt.%, 30 wt.% of epoxy) were mixed with the curing agent at 90 °C under mechanical stirring until complete dissolution. A certain amount of epoxy (DGEBA : MTHPA = 5 : 4, by weight radio) was added and stirred vigorously at the same temperature for 1 h. The mixtures were degassed in a vacuum at 90 °C for 30 min. Homogeneous mixtures with different PES concentrations were obtained. Similarly, for epoxy composites with GNPs, different content of GNPs were mixed with the curing agent at 90 °C for 30 min. The compositions were cured at 145°C for 4 h.

The microstructures of the cured epoxy blends and dispersibility of the GNPs filler were evaluated in both a JSM-6510 scanning electron microscope (SEM, Jeol, Japan) and metallographic microscope (Axio Observer A1m, Carl Zeiss Jena, Germany). Thermal conductivity was measured by a Hot Disk thermal analyzer (TPS-2500), based upon the transient plane source (TPS) method

RESULTS AND DISCUSSION

The morphologies of composites at various epoxy/PES weight ratios (epoxy/PES = 10/1, 10/2, 10/3) were characterized using OM (Fig. 1). At 10 wt%, PES forms a large number of small domains (Fig. 1a); at 20 wt%, a co-continuous phase structure is observed (Fig. 1b); and at 30 wt%, phase inversion occurs (i.e., epoxy forms small phase domains throughout the PES matrix) (Fig. 1c).

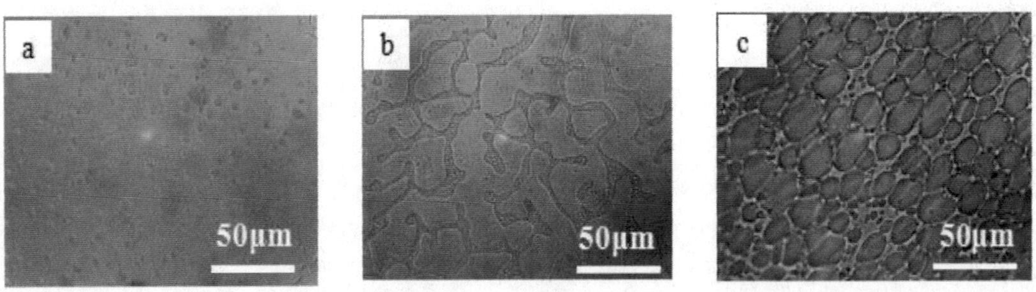

FIGURE 1. OM images of epoxy/PES composites with different PES contents: (a) 10 wt% PES; (b) 20 wt%; (c) 30 wt%.

Fig. 2 shows the SEM micrographs of fractured surfaces with different PES contents. At the low PES content (10 wt.%), PES was distributed uniformly in epoxy matrix in the form of solitary particles with a size distribution of 1~3 μm (Fig. 2a). At the high PES content (30 wt.%), PES-rich phase formed a continuous network structure (Fig. 2c). Epoxy was embedded in the PES network with irregular spherical shape. The sizes of the disperse epoxy domains were between 4 and 10 μm. When the content of PES was medium (20 wt.%), the whole sample showed a three-layer structure (Fig. 2b). The structure of middle layer was similar to structure of the blend with 30 wt.% PES, and the structure of upper and lower layers were similar to structure of the blend with 10 wt.% PES. The thickness of the middle layer was 50~60 μm, and the thickness of the upper and lower layers were both around 30 μm.

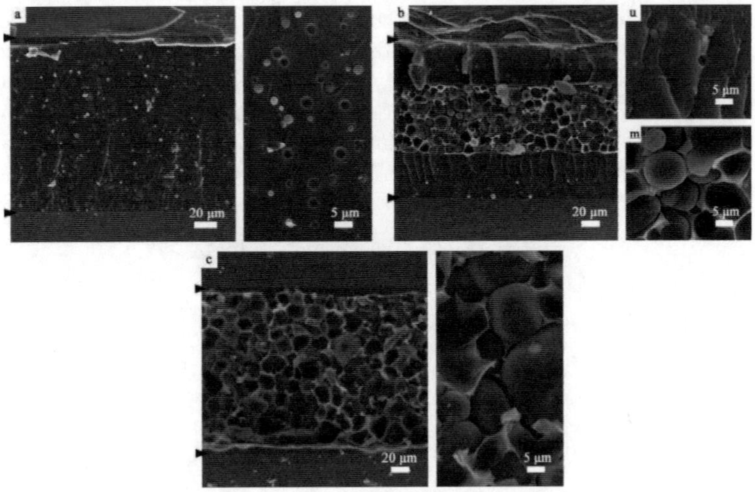

FIGURE 2. SEM micrographs of fractured surfaces at different PES content: (a) 10 wt.%, (b) 20 wt.% and (c) 30 wt.%. The location of the glass surfaces was indicated by arrows. Right pictures in each panel are at higher magnifications. (u) and (m) represent the upper and middle parts of the sample with 20 wt.%, respectively.

Fig. 3 shows the storage modulus (G') and loss modulus (G") of the system during the RIPS at 145 °C for various contents of PES. Neglecting the experimental uncertainties and resolution differences from different instruments, development of the morphologies obtained in the optical microscope was consistent with phase separation evolution detected by rotational rheometer. The modulus varied together with the evolution of complex phase separation, and complex viscosity profiles showed an exponential growth with curing reaction.

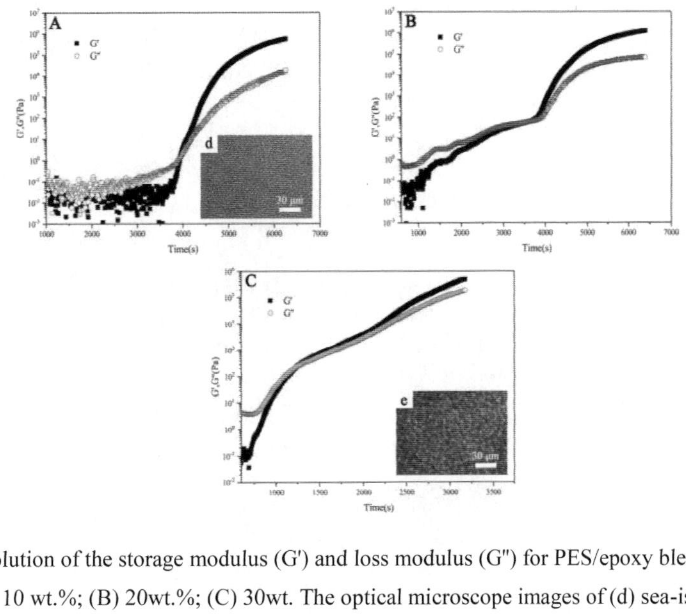

FIGURE 3. Time evolution of the storage modulus (G') and loss modulus (G") for PES/epoxy blends with different PES weight fractions: (A) 10 wt.%; (B) 20wt.%; (C) 30wt. The optical microscope images of (d) sea-island structure and (e) inversion phase structure.

Fig. 4 shows the thermal conductivities of epoxy/GNPs and epoxy/PES/GNPs composites with different amounts of GNPs. Significant thermal conductivity enhancement was observed for epoxy/PES/GNPs composites as GNPs content increased from 7 wt% to 10 wt%. This demonstrates that 7 ~10 wt% of GNPs is a critical enhancement window for epoxy/PES/GNPs composites. When the GNPs content is below 7 wt%, thermal conductivity increased slightly with increasing GNPs. With further increase in GNPs content, thermal conductivity in epoxy/PES/GNPs composites increases remarkably, while the thermal conductivity of epoxy/GNPs composites is almost linear with the GNPs content. This could be attributed to the formation of a thermally conductive network initiated by the reaction-induced phase separation.

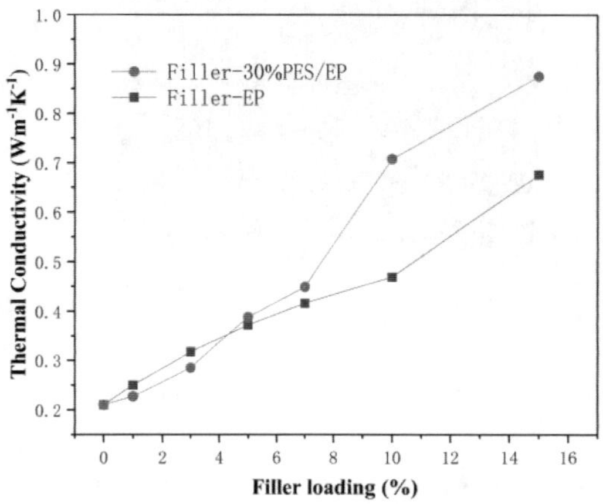

FIGURE 4. Thermal conductivities of epoxy/PES/GNPs and epoxy/GNPs composites as a function of the GNPs content.

ACKNOWLEDGMENTS

This work was supported by the Natural Science Foundation of Jiangsu Province, the Priority Academic Program Development of Jiangsu Higher Education Institutions (PAPD) and National Natural Science Fund of China (NSFC).

REFERENCES

1. E. Girard-Reydet, H. Sautereau, J.P. Pascault, *Polymer* **40**, 1677-1687 (1999)
2. G. Shen, Z. Hu, Z. Liu, R. Wen, X. Tang, Y. Yu, *Rsc Adv.* **6**, 34120-34130 (2016)
3. X. Tang, X. Zhong, G. Zhan, Y. Yu, H. Zhang, *Int. J. Polym. Sci.* **2013**, 331-350 (2013)
4. F. Chen, X. Wang, X. Zhao, J. Liu, S. Yang, C.C. Han, *Macromol. Rapid Comm.* **29**, 74-79 (2008)
5. X. Wang, P. Wu, *Acs Appl Mater Inter.* **9**, 19934-19944 (2017)
6. K. Wu, C. Lei, R. Huang, W. Yang, S. Chai, C. Geng, F. Chen, Q. Fu, *ACS Appl Mater Inter.* **9**, 7637-7647 (2017)
7. Y. Jiang, Y. Liu, P. Min, G. Sui, *Compos Sci Technol.* **144**, 63-69 (2017)
8. F.E. Alam, W. Dai, M. Yang, S. Du, X. Li, J. Yu, N. Jiang, C. Lin, *J Mater Chem A.* **5**, 6164-6169 (2017)

Exploiting a Simple Strategy for Tunable Multi-shape Memory Polymers

Ji Xiaoying (1), Shen Jiabin (1)*, Guo Shaoyun (1)

(1) Sichuan University - Sichuan Province - China

Shape memory polymers can be fixed into a temporary shape and subsequently recover to its permanent shape when triggered by an external stimulus, typically heating. Recently, increasing attention have been focused on developing tunable triple- or even multi-shape memory polymers(MSMPs), which are believed to have a great potential for a wide range of applications, such as smart devices, sensors, actuators and biomedical products. Herein, we fabricated a tunable MSMPs with alternating multilayered structure consisting of poly (vinylidene fluoride) (PVDF) and poly (methyl methacrylate) (PMMA) via layer-assembly coextrusion technology. The number of layers was multiplied by combining an assembly of layer-multiplying elements with two extrudes, and induced the simultaneous increase of the shape fixing (Rf) and recovery ratios (Rr). For the specimen having 1024 layers, both of the Rf and Rr reached as high as above 90%, which is much better than the conventional PVDF/PMMA blends. In addition, numerous layer interfaces formed in the 1024-layer system caused a broad glass transition range (about 110℃), due to the interfacial diffusion between PVDF and PMMA layers. Consequently, the 1024-layer system exhibited good triple- or even quadruple-shape memory effect. The results achieved in this work reveal that the multilayered assembly provides a simple and efficient strategy to fabricate tunable MSMPs. Keywords: multi-shape memory polymers, multilayer structure, interface

Development of micro-nano polymer structures with cellular porous architectures with energy absorption properties under the action of mechanical loading

***Tapia Monica* [(1)], *Medina Dora I* [(1)*]**

[(1)] ITESM CEM - México - Mexico

3D printing is one of the new polymer processing technologies that has transformed the field of manufacturing for product development with complex geometries. The Additive Manufacturing (AM) is a process that consists in production of parts from a 3D digital design by successive deposition of layers of material. One of the printing methods of this type is stereolithography (SLA) which by the deposition of thin layers while being cured with UV forms the part from a specific design. A great capability of additive manufacturing (AM) is the production of cellular capillary geometries on a scale required in applications such as personal protection equipment (PPE). Thermoplastic elastomers (TPEs) are flexible polymers, which are compatible with the additive manufacturing process. Thermoplastic polyurethane (TPU) is a material that mitigates impact damage, one of its special applications is in sports safety gear, like the inside of American football players' helmets. The evaluation of the mechanical behavior of capillary cell geometries with thermoplastic polyurethane (TPU) material denotes the damping capacity of this novel structure made with additive manufacturing (AM) technology. The characterization and development of this novel structure has a promising application in personal protection equipment (PPE), as in devices for hip protection in elderly people and also bicycle helmets.

Thermodynamic Properties and Crystallization Kinetics of Isotactic Polypropylene under Pressure

Felice De Santis[1,2, a)] and Roberto Pantani[1, b)]

[1]*Department of Industrial Engineering, University of Salerno, Fisciano - SA - Italy*
[2]*Department of Pharmacy, University of Salerno, Fisciano - SA – Italy*

[a)]Corresponding author: fedesantis@unisa.it
[b)]rpantani@unisa.it

Abstract. The effect of pressure on the processing and, in particular, on the crystallization kinetics of a polymeric material, is often ignored because of the difficulties involved in the experimental measurements. As a consequence, crystallization behaviors were rarely described by using a model which takes into account the effect of pressure on the crystallization. In this study, experimental data on the dependence of the specific volume on temperature and pressure of isotactic polypropylene using a high-pressure dilatometer up to 150 MPa are reported and discussed. In the complex polymorphism of isotactic polypropylene, the main way to obtain γ structure for a high molecular weight homopolymer is to perform crystallization under high pressure. From the subsequent heating step after the isothermal crystallization, the melting temperatures at the same pressures have been determined to assess the pressure effect on the equilibrium melting temperature. The crystallization times of isothermal experiments have been analyzed.

INTRODUCTION

Isotactic polypropylene (i-PP) is one of the most investigated polymers because of its industrial importance. Depending on thermal treatments, molecular mass, and degree of tacticity, when crystallized from the melt, the i-PP chains can organize in several spatial arrangements giving rise to different crystalline polymorphs (α-monoclinic, β-hexagonal, and γ-triclinic) and a mesomorphic phase. Furthermore the effect of the flow, often relevant in the processing, also affects crystallization both for the final phases (microstructure) and macrostructure. Consequently, the crystallization behavior of i-PP is very complex. Numerous groups have aimed their research on a better understanding of the different crystal modifications.

The existence of the γ crystalline phase of isotactic polypropylene was reported soon after the elucidation of the more common α phase. The γ-phase is usually associated with the α-phase, although individual single crystals of the γ-phase have also been observed. The γ-modification may be formed in degraded, low molecular weight isotactic polypropylene and in samples crystallized under high pressures.

Specifically, about 50 years ago, it was established that the ratio between the γ -iPP and the α -iPP polymorphs grows with increasing pressure.

In polymer processing, the specific volume plays a major role. It is influenced by processing conditions like temperature, pressure, and flow history and it determines shrinkage, which expresses itself by dimensional (in)stability.

EXPERIMENTAL

Material

The resin adopted for the experiments performed in this work is a commercial iPP[1-4] (tradename T30G, M_w = 376 kg mol^{-1}, polydispersity index M_w/M_n of 6.7) supplied by Montell (Ferrara, Italy), now LyondellBasell Industries.

Dilatometry

The experiments were performed with a PVT dilatometer, Gnomix-PVT apparatus (GNOMIX Inc., Boulder, CO, USA), developed by Zoller et al.[5] using the confining-fluid technique. The sample is enclosed in a rigid sample chamber, and it is submerged into a fluid (mercury). The cell is closed using a flexible wall or bellows. The bellows are used to apply hydrostatic pressure to the fluid and polymer by reducing the sample chamber volume, and sensing the cumulative volume change of fluid and polymer. The absolute specific volume of the polymer can be obtained by correcting the relative volume difference with the specific volume of the confining fluid. Both pressure and temperature can be varied. The advantages of this technique are:

- pressure is purely hydrostatic as the sample is surrounded by the confining fluid in both a melted and solid state;
- there is an absence of leakage and friction.

The disadvantage is:

- the volumetric changes measured are not that of the polymeric sample only.

The PVT data, most of them obtained on Gnomix equipment, are currently used in the simulation of injection molding, i.e. by Moldflow. The PVT instrument has some advantages compared to a capillary rheometer if accurate numbers are desired (0.0001 cm^3/g)[6].

The Gnomix manufacturer suggests three “modes” to perform experimental tests:

- in the isobaric mode (IBA) the sample will be heated or cooled at a defined rate and a defined pressure.
- in the isothermal standard (ITS) mode, after reaching thermal equilibrium, the pressure is raised from 10 to 200 MPa and all data are recorded in steps of 10 MPa
- in a special (data acquisition, DAQ) mode it is possible to follow changes in the specific volume at constant pressure and constant temperature, which may be caused by a phase transition.

In particular, the special DAQ mode is adopted in this work as summarized in Table 1.

Isobaric isothermal crystallization

To erase the previous thermomechanical history the sample in the PVT apparatus was heated to 240°C and held at this temperature for 10 minutes. In order to study melt isothermal crystallization, the sample was cooled from 240°C directly to isothermal test temperature (crystallization temperature), with a cooling rate of 3 K/min. This experimental protocol is just the same modus operandi usually adopted in calorimetry, in which thermal effects linked to crystallization are detected. Obviously in dilatometry, during the isothermal step, the volumetric changes related to crystallization, i.e. specific volume decrease, are measured under pressure.

The cooling rate of 3 K/min before the isothermal step is substantially restricted by the total heat capacity of the system and the sample size, as in calorimetry or nanocalorimetry[2] it could be of the order of 10 K/min or 1000 K/s, respectively. As consequence, this cooling rate limits the minimum reachable temperature of the isothermal step, without evident previous crystallization during the cooling step. The temperatures of the isothermal step, using the special DAQ mode at various pressure, are summarized in Table 1.

TABLE (1). Crystallization temperatures of experimental PVT isotherms.

P [MPa]	10	100	150
Temperature [°C]	127.1	152.0	162.5
	131.9	154.8	164.2
	133.6	155.5	167.8
	136.6	157.6	169.5
	138.6	159.6	
	141.3	162.5	
		167.3	
		169.3	

Furthermore, as known by the previous literature[7], the pressure increase has the effect to accelerate crystallization rate of iPP and thus the higher is the pressure the greater is the minimum reachable temperature of the isothermal step in Table 1, without evident previous crystallization during the cooling step.

An example of an isothermal experimental test is reported in Figure 1: during the cooling, the specific volume change is almost linear with the temperature until the isothermal step temperature is reached.

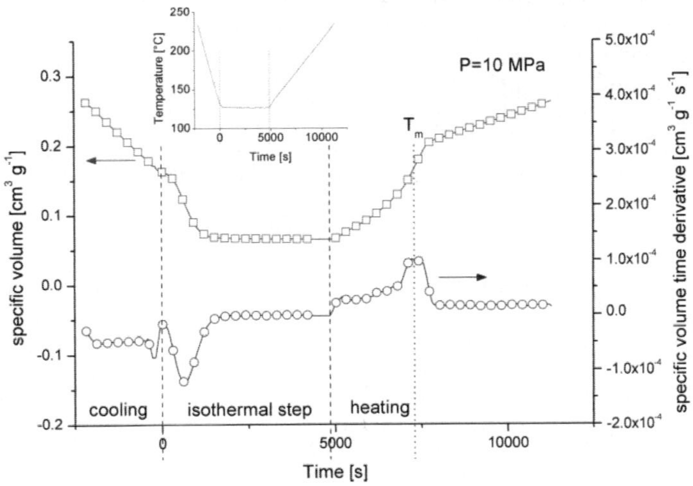

FIGURE 1. Experimental data of specific volume, *v*, during an isobaric crystallization test (on the left axis), and resultant time derivative (on the right axis) to evaluate melting temperature. In the inset the experimental protocol: temperature evolution during the test.

During the isothermal step, the specific volume shows an inverse sigmoid curve related to the crystallization. After the isothermal step, the sample is heated again to 240 °C, at a heating rate of 1 K/min, to explore the corresponding melting behavior. The peak of the time derivative of the specific volume permits to evaluate the melting temperature during the heating. After the complete melting, the specific volume change is again linear with the temperature during the heating to 240 °C.

The temperature experimental protocol is shown in the inset in Figure 1, for the isothermal test at 127.1 °C.

RESULTS AND DISCUSSION

From the analysis of the specific volume evolution during the isothermal step is possible to evaluate crystallization half-time for each test, performed changing the temperature and pressure of the PvT experiments.

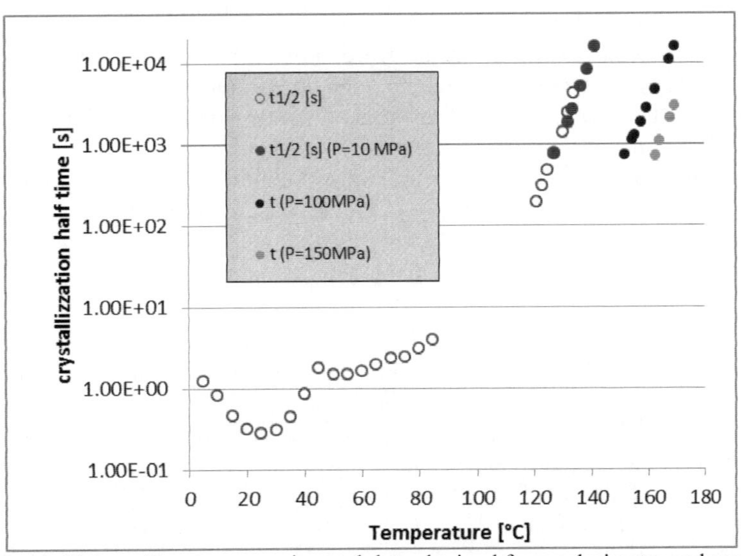

FIGURE 2. Crystallization half-time: literature experimental data obtained from calorimetry and nanocalorimetry[2] (open symbol) and data from the analysis of isobaric isothermal crystallization PvT tests (full symbol) with different pressures

The crystallization half-times are reported in Fig. 2: the data obtained from low pressure PvT experiments, at 10 MPa, compares well with isothermal calorimetry results taken from literature.

Furthermore for the tests at higher pressures, i.e. at 100 and 150 MPa, a clear effect on crystallization is reported and crystallization rate is improved by the effect of pressure. The analysis of these data is still in progress, in order to take into account the influence of pressure in a crystallization kinetics model.

From the subsequent heating step after the isothermal crystallization, the melting temperatures at the same pressures have been determined to assess the pressure effect on the equilibrium melting temperature.

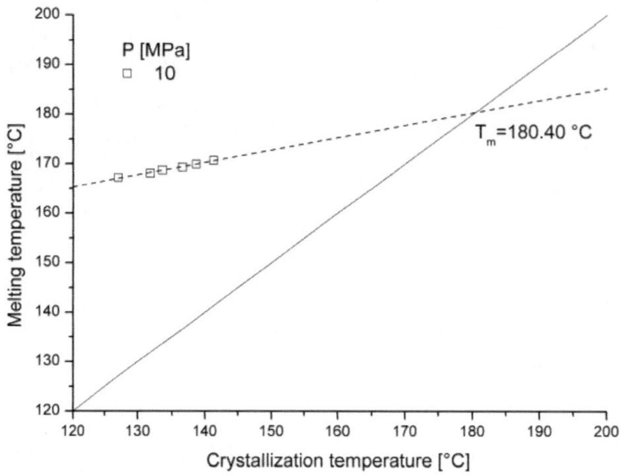

FIGURE 3. Hoffman-Weeks plot for the pressure of 10 MPa

The effect of the pressure on the equilibrium melting temperature was analyzed by means of Hoffman-Weeks plots of crystallization and melting. The results are summarized in Figure 3, for the pressure of 10 MPa, evaluating the corresponding equilibrium melting temperature, with a linear extrapolation, of 180.4°C. Actually, this is value is lower than equilibrium melting temperature of 194°C, reported for this resin at ambient pressure but estimated through the non-linear Hoffman—Weeks analysis[4, 8]. On the other hand a higher equilibrium melting temperature, if compared to the one at ambient pressure, could be expected at higher pressure[7].

CONCLUSIONS

The crystallization was affected by temperature, pressure, and shear stress. In particular, the change of temperature and pressure is an important factor governing the crystallization progress during injection molding.

Isobaric isothermal crystallization test were performed in a PvT apparatus, following the evolution of specific volume during the crystallization and subsequent melting.

The investigation of these experiments allowed to evaluate the effect of pressure on crystallization kinetics. A further analysis should descript the crystallization kinetics model taking into account the effect of pressure.

REFERENCES

1. F. De Santis, R. Pantani and G. Titomanlio, Polymer **90**, 102-110 (2016).
2. F. De Santis, S. Adamovsky, G. Titomanlio and C. Schick, Macromolecules **40** (25), 9026-9031 (2007).
3. F. De Santis, S. Adamovsky, G. Titomanlio and C. Schick, Macromolecules **39** (7), 2562-2567 (2006).
4. I. Coccorullo, R. Pantani and G. Titomanlio, International Polymer Processing **20** (2), 186-190 (2005).
5. P. Zoller, P. Bolli, V. Pahud and H. Ackermann, Review of Scientific Instruments **47** (8), 948-952 (1976).
6. P. Zoller and Y. A. Fakhreddine, Thermochimica acta **238**, 397-415 (1994).
7. C. Angelloz, R. Fulchiron, A. Douillard, B. Chabert, R. Fillit, A. Vautrin and L. David, Macromolecules **33** (11), 4138-4145 (2000).
8. P. Juhász, J. Varga, K. Belina and H. Marand, Journal of thermal analysis and calorimetry **69** (2), 561-574 (2002).

Fast scanning calorimetry of as-received polymer samples

Schawe Juergen E.K. *[1]**

[1] Mettler-Toledo GmbH - Switzerland - Switzerland

Because of reorganization at slow heating, conventional DSC is not an approvable technique to characterize the melting behavior of polymer samples with thermo-mechanical history. For characterization of samples as received higher scanning rates should be used to reduce reorganization effects. Fast Scanning Calorimetry (FSC) offers sufficient high heating rates, however the heat contact between sensor and sample is not good enough for reproducible measurements. Therefore a contact medium (oil) is recommended for such measurements. On the example of isotactic polypropylene (iPP) a technique is introduced to analyses polymer samples as received by a METTLER TOLEDO Flash DSC 1 at heating rates of 1000 K/s.

Isothermal beta-crystal growth in polypropylene nucleated with calcium pimelate: Growth under pressure with CRISTAPRESS

***BOYER SEVERINE A.E.* [1]**, *SOWINSKI Przemyslaw* [2], *PIORKOWSKA Ewa* [2], *HAUDIN Jean-Marc* [1]**

[1] CEMEF UMR CNRS 7635 - MINES ParisTech PSL - France, [2] CMMS - Polish Academy of Sciences - Poland

In our research works dealing with polymer cristallization in severe conditions, recently we focused on high-pressure induced gamma-form nucleation in nucleated isotactic polypropylene (iPP). Alpha- and beta-nucleating agents (NAs) were selected. Gamma-form nucleation was investigated in isothermal and nonisothermal conditions [1,2]. The present work offers a new focus on crystal growth under hydrostatic pressure. It refers to an original optical high-pressure cell, named CRISTAPRESS, which has been especially designed to investigate liquid-solid transitions under elevated pressure [3]. The growth of the beta-form of iPP beta-nucleated with calcium pimelate (CaPim) was studied. The first results show the ability of CRISTAPRESS to follow in-situ crystallization in the beta-form in quiescent and isothermal mode explored in the ranges 10-50 MPa and 127-145 °C. CaPim nucleated efficiently the beta-form. The beta-crystal growth rate, depending on temperature and pressure, was successfully analyzed. Acknowledgement PAN/CNRS Collaborative Project n°179990 'High pressure crystallization and structure of polymers based nanocomposites'. CNRS, ARMINES, MINES ParisTech, Conseil Régional de Provence-Alpes-Côte d'Azur, and Conseil Général des Alpes-Maritimes for financial supports in the development of the «CRISTAPRESS» project. [1] Sowiński P., Piorkowska E., Boyer S.A.E., Haudin J.-M., Zapala K., Colloid Polym. Sci. 293 (2015) 665 [2] Sowiński P., Piorkowska E., Boyer S.A.E., Haudin J.-M., Eur. Polym. J. 85 (2016) 564 [3] Boyer S.A.E., Fournier F.E.J, Gandin Ch.-A., Haudin J.-M., Rev. Sci. Instrum. 85 (2014) 013906 / 029902

Controlling the Length of Branched Chains on Polypropylene Using Tetraethyl Thiuram Disulfide

Aramesh Hossein[a], Nazockdast Hossein[b]

[a] *Department of Polymer Engineering, Amirkabir University of Technology, Mahshahr Campus, P.O. Box 415, Mahshahr, Iran*

[b] *Department of Polymer Engineering & Color Technology, Amirkabir University of Technology, No. 424, Hafez Ave., P.O. Box 15875-4413, Tehran, Iran*

Abstract. In this work parameters affecting the efficiency and the length of branched chains on polypropylene using melt free radical grafting in a modular and multifed twin screw extruder were investigated. The concentration of polyfunctional monomer (trimethylol propane triacrylate (TMPTA)) and co-agent (tetarethylthiuram disulfide (TETDS)) in the presence of dibenzoyl peroxide (BPO) were among parameters studied. The melt viscoelastic results together with thermal analysis and fourier transform infrared spectroscopy (FTIR) results showed that TETDS could result in higher efficiency and longer chain branch length onto PP backbone. This could be explained in terms of great capability of TETDS in minimizing the β-scission of PP backbone and restrain the homopolymerization of TMPTA. The results of frequency sweep experiment (G′ versus ω) performed on the long chain branched sample exhibited a pronounced positive deviation from rosian viscoelastic behavior at low frequency range. This was considered as a major of degree and the length of long chain branching reaction. This was supported by the results of FTIR and thermal analysis.

Keywords: Polypropylene, Polyfunctional monomer, Co-agent, Melt free radical grafting, Long chain branch.

INTRODUCTION

Isotactic polypropylene (iPP) has become one of the most widely used commercial polymers. However poor melt strength of polypropylene limits its use in applications such as thermoforming, foaming, extrusion coating and blow molding [1-4]. One of the effective approaches to improve the melt strength of PP is to introduce long chain branches (LCB) onto PP backbone [5,6].

There are several methods for preparing long chain branched polypropylene (LCB-PP), including in situ polymerization in a reactor, high-energy irradiation and melt free radical grafting [7-9]. In the third method, the primary radicals (RO•) generated by the thermal decomposition of peroxide would abstract preferentially the tertiary hydrogen atoms from PP backbone, forming tertiary macroradicals which can further undergo β-scission or react with PP and polyfunctional monomer. The LCB-PP is formed through recombining macroradicals whatever directly or indirectly via polyfunctional monomer. However, two undesirable side reactions, β-scission of macroradicals and the homopolymerization of polyfunctional monomers, offten occur during preparing LCB-PP. Prolonging the life time of macroradicals by forming reversible radicals by using thiuram disulfide [2], copper N,N-dimethyldithiocarbamate [8] and different zinc dithiocarbamates [10] or stabilizing macroradicals by changing radical intermediate states by using styrene [11] and furan derivatives [12] are efficient routes to reduce β-scission and improve grafting reaction.

In this work, our efforts focused on using co-agent (tetarethylthiuram disulfide (TETDS)) in controlling the length of branched chains on polypropylene by melt free radical grafting in the presence of benzoyl peroxide (BPO) and trimethylol propane triacrylate (TMPTA). The length of branched chains was evaluated by using melt linear viscoelastic measurements in conjunction with fourier transform infrared spectroscopy (FTIR) and thermal analysis.

EXPERIMENTAL SECTION

Materials

Additive-free isotactic polypropylene powder (HP525J) was obtained from Jam Polypropylene Co., Iran. The melt flow index (MFI) was 3 g/10 min measured at 2.16 kg and 230 °C. The PP powder was stabilized by the addition of 0.1 wt % Iraganox 1010 (Ciba, Switzerland) antioxidant. Benzoyl peroxide (BPO) and co-agent tetraethylthiuram disulfide (TETDS) were supplied by Merck, and polyfunctional monomer trimethylol propane triacrylate (TMPTA) was purchased from Aldrich. All compounds were used as received.

Reactive Extrusion process

A Plsati-Corder Lab-Station intermeshing corotating twin-screw extruder equipped with a devolatilization zone and required controlling devices (Plsati-Corder Lab-Station, Brabender® GmbH & Co. KG, Duisburg, Germany) was used to prepare the long chain branched (modified) samples. The temperature was set to 160 °C for the first zone and 210 °C for the die (6 heating zones).

The formulations of the modified polypropylenes are listed in Table 1. In order to obtain fine dispersion of BPO, TMPTA, TETDS and Irganox 1010 in PP powder, these were dissolved in 50 ml acetone, and resulting solution was added to PP powder and stirred for 10 minutes. At last, the mixture was placed in a ventilating cabinet at room temperature until the acetone volatilized completely.

TABLE (1). Formulation and abbreviation of samples.

Sample	BPO (wt %)	TETDS (wt %)	TMPTA (wt %)
PP	0.000	0.000	0.000
BPP1	0.030	0.025	0.660
BPP2	0.030	0.025	1.320
BPP2-T	0.030	0.000	1.320

FTIR Analysis

The modified samples were dissolved in xylene for 2 h at 140 °C, and then the solutions were charged into acetone at room temperature. Unreacted TMPTA monomer and copolymerized TMPTA remained soluble, while PP and PP-g-TMPTA precipitated out. PP and PP-g-TMPTA were separated by filtration, and then were dried at 60 °C for 24 h under vacuum. The purified samples were pressed into films at 200 °C, 12 MPa with the same thickness about 0.1 mm and then were analyzed using a FTIR spectrometer (SPG9100G, Bomem, Canada).

Rheological Characterization in Shear Flow

The rheological measurements in shear flow were conducted using a stress-controlled rheometer (Physica MCR 300, Anton Paar, Graz, Austria) in plate-plate geometry. Cylindrical samples of a diameter of 25 mm and a thickness of 1 mm were prepared with a hot press at a temperature of 180 °C under nitrogen atmosphere. Small-amplitude oscillatory shear (SAOS) was performed in the frequency range 0.01- 100 rad/s at 180 ºC under nitrogen atmosphere. The deformation amplitude applied was set to 1% for all measurements, which was proven to be in the linear viscoelastic regime.

RESULTS AND DISCUSSION

FTIR spectra of the purified samples (BPP1, BPP2 and BPP2-T) are shown in Figure 1. The three modified samples showed a peak around 1735 cm^{-1} corresponding the stretching vibration of the carbonyl group of the ester in the grafted TMPTA molecule, which indicates that TMPTA was grafted onto PP backbone. However as can be seen the intensity of this peak is increased by increasing TMPTA concentration. Nevertheless the increase in intensity of this peak does not necessarily imply the increase in the length of branched chains.

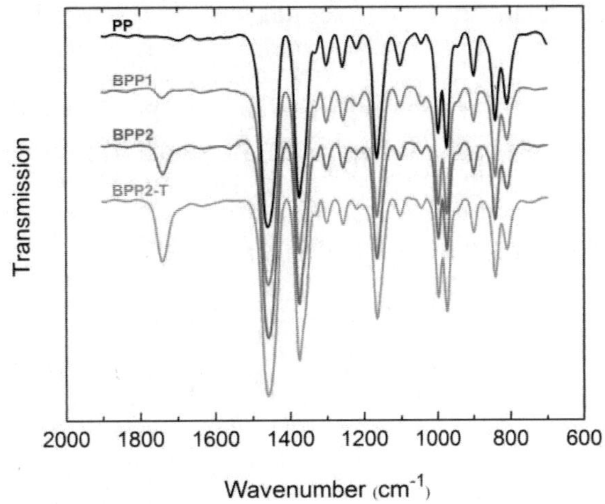

FIGURE 1. FTIR spectra (700-1900 cm^{-1}) of the plain PP and modified PP samples after purification.

Figure 2 shows the results of complex viscosity (η^*) and storage modulus (G′) as a function of angular frequency obtained for the plain PP and modified PP samples. As it can be observed in Figure 2a, the modified PP samples exhibit a highly enhanced viscosity along with shear thining at low frequency range compared to the plain PP which showes an almost Newtonian behavior at the same frequency range. Similar results have been reported by other research groups. It has recognize that the presence of very low amounts of LCB can change the zero-shear viscosity (η_0) and the degree of shear thinning, as compared to the linear polymers with similar molecular weight [7].

By comparing the results of Figure 2a, one main note is that BPP2 which was prepared by using higher TMPTA concentration and the same amount of TETDS as that of BPP1 exhibits higher viscosity and stronger shear thining. Whereas little increase in visco

sity was observed for BPP2-T which was preparing by using the same TMPTA concentration but without using TETDS. The increase in the TMPTA concentration above 1.32 wt % was found to have little effect on increasing the viscosity. These results suggest that presence of TETDS is necessary to obtain long chain branched PP. These results could also be explained in terms of variation in storage modulus values and changing its slope at low frequency range (Figure 2b). The greater increase in storage modulus and its deeper slope can be considered as indications of increase in the length of branched chains and degree of long chain branching.

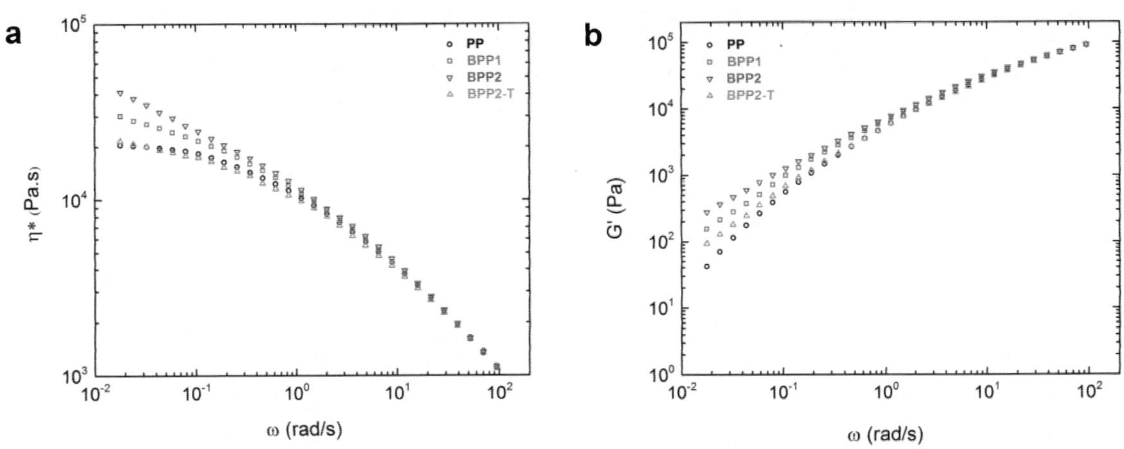

FIGURE 2. (a) Complex viscosity versus angular frequency and (b) storage modulus versus angular frequency of the plain PP and modified PP samples.

More insight into understanding the extent of chain branching reaction can be obtain by considering the Cole-Cole plot shown in Figure 3a. The strong tail observed for BPP2 is a clear evidence of presence of molecules with high elasticity, which may be attributed to increased the molecular entanglement between the long branched chains. Accordingly, the number of long branched chains formed in BPP1 is much lower than that of BPP2. More importantly, results of Figure 3a clearly evidence the absence of long branched chains in BPP2-T and therefore necessity of using TETDS.

The weighted relaxation spectrum of the plain PP and modified PP samples are shown in Figure 3b. As can be seen the plain PP used in this work is a bimodal type PP. Interestingly these results show two characteristic peaks corresponding to two distinct relaxation time range (0.1 to 0.7 and 1 to 11). The highly intensified peak appeared at the higher relaxation time range is a clear evidence for formation of LCB-PP molecules, while the first peak represents the increase in number of molecules with short relaxation time.

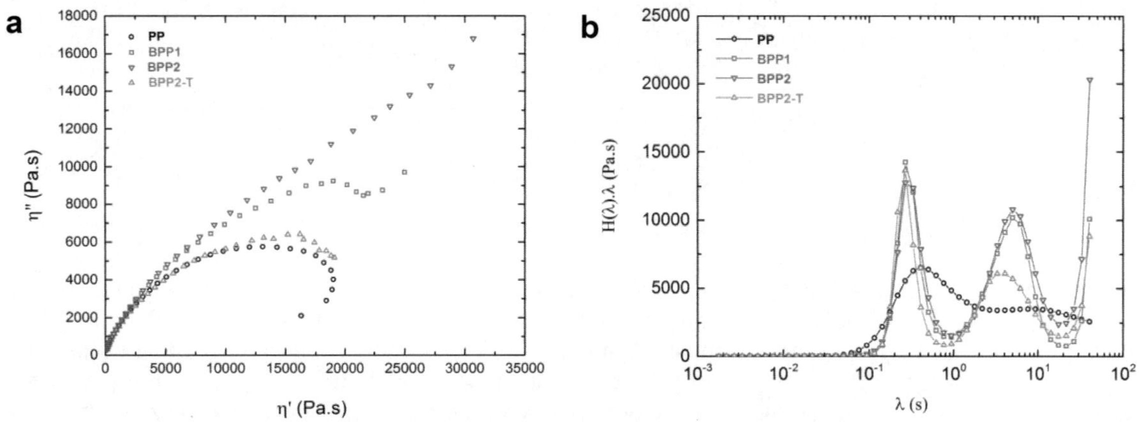

FIGURE 3. (a) Cole-Cole plots and (b) weighted relaxation spectrum of the plain PP and modified PP samples.

From the rheological results discussed above and considering the FTIR spectra of these samples (Figure 1) it can be realized that the increase in the intensity of the peak around 1735 cm^{-1} does not necessarily imply the increase in the length of branched chains. Thus the increase in the length of branched chains and total number of long branched chains can be monitored by employing the rheological results in conjunction with FTIR spectra.

CONCLUSIONS

It was demonstrated that the long chain branching reaction of PP can be monitored by using the melt viscoelastic measurements in conjunction with FTIR. It was found that in order to produce the LCB-PP with acceptable branched length, it is necessarily to use an appropriate amount of TETDS. Finally, comparing of our results with those reported by other research groups revealed that more efficient long chain branching reaction of PP can be achieved by using a bimodal PP.

REFERENCES

1. C.J. Tsenoglou, A.D. Gotsis, Rheological characterization of long chain branching in a melt of evolving molecular architecture, Macromolecules 34(14) (2001) 4685-4687.
2. D. Graebling, Synthesis of branched polypropylene by a reactive extrusion process, Macromolecules 35(12) (2002) 4602-4610.
3. W. Zhao, G. Wu, Q. Yang, Controlling the transition of long-and short-chain branching polypropylene, Polymer-Plastics Technology and Engineering 51(7) (2012) 716-723.
4. B. Lu, T. Chung, Synthesis of long chain branched polypropylene with relatively well-defined molecular structure, Macromolecules 32(25) (1999) 8678-8680.
5. W. Zhao, Y. Huang, X. Liao, Q. Yang, The molecular structure characteristics of long chain branched polypropylene and its effects on non-isothermal crystallization and mechanical properties, Polymer 54(4) (2013) 1455-1462.
6. W. Weng, W. Hu, A.H. Dekmezian, C.J. Ruff, Long chain branched isotactic polypropylene, Macromolecules 35(10) (2002) 3838-3843.
7. J. Tian, W. Yu, C. Zhou, The preparation and rheology characterization of long chain branching polypropylene, Polymer 47(23) (2006) 7962-7969.
8. Z. Zhang, H. Xing, J. Qiu, Z. Jiang, H. Yu, X. Du, Y. Wang, L. Ma, T. Tang, Controlling melt reactions during preparing long chain branched polypropylene using copper N, N-dimethyldithiocarbamate, Polymer 51(7) (2010) 1593-1598.
9. D. Auhl, J. Stange, H. Münstedt, B. Krause, D. Voigt, A. Lederer, U. Lappan, K. Lunkwitz, Long-chain branched polypropylenes by electron beam irradiation and their rheological properties, Macromolecules 37(25) (2004) 9465-9472.
10. Z. Zhang, D. Wan, Y. An, F. Liu, H. Xing, L. Wang, Z. Jiang, T. Tang, Controlling the evolution of polypropylene microstructure during melt free radical modification using zinc dithiocarbamate with different N-substituted groups, Polymer degradation and stability 96(4) (2011) 653-659.
11. H. Cartier, G.H. Hu, Styrene-assisted free radical grafting of glycidyl methacrylate onto polyethylene in the melt, Journal of Polymer Science Part A: Polymer Chemistry 36(15) (1998) 2763-2774.
12. S. Coiai, E. Passaglia, M. Aglietto, F. Ciardelli, Control of degradation reactions during radical functionalization of polypropylene in the melt, Macromolecules 37(22) (2004) 8414-8423.

INFLUENCE of WATER during MANUFACTURING of APA6 in the THERMOPLASTIC RTM PROCESS

Rainer Wendel[a], Bernd Thoma[a] and Frank Henning[a,b]

a Department of Polymer Engineering, Fraunhofer Institute for Chemical Technology (ICT), Joseph-von-Fraunhofer Straße 7, 76327 Pfinztal, Germany

b Institute for Vehicle Systems Technology, Karlsruhe Institute of Technology (KIT), Kaiserstraße 12, 76131 Karlsruhe, Germany

Abstract. The thermoplastic resin transfer molding (T-RTM) process has the potential for high-volume production of high performance fiber-reinforced components in the automotive industry. The fast anionic polymerization of APA6 and the development of robust injection equipment lead to short cycle times. The sensitivity against water of the reactive APA6 is well known. All resin components from the manufacturer are dried to achieve water content below 200 ppm. In T-RTM high-performance composite manufacturing fibers, binder systems or core materials are needed. During the manufacturing (preforming, handling, infiltration) of these materials it is necessary to ensure that the total amount of water is small enough to achieve good polymerization. Previous studies established the influence of water for non-activated polymerization. However, there is no knowledge about the influence of water during T-RTM manufacturing using fast-reacting activated polymerization. In this paper, different concentrations of catalyst are investigated to analyze the influence of water in the T-RTM process during curing of the resin. Residual monomer content and viscosity number was measured to analyze the conversion of manufactured plates. To characterize the impact on the mechanical properties, tensile tests were performed. The results show the possibility to monitor the anionic polymerization during the manufacturing by observing the trend of the cavity pressures. In addition, the influence of water on the conversion and the resulting mechanical properties can be compensated by the concentration of the catalyst.

Keywords: thermoplastic resin transfer molding (T-RTM), cast polyamide 6 (cast PA6), anionic polyamide 6 (APA6), anionic polymerization, moisture, ε-caprolactam, thermoplastics
PACS: 81.05.Lg

INTRODUCTION

Plastics are increasingly prevalent in lightweight applications. The advantages they offer over thermosets include their good impact properties, toughness, the possibility of welding and their reusability with secondary manufacturing processes [1,2,3]. The high viscosity of plastics prevents the infiltration of continuous fibrous reinforcements. The manufacturing of reactive plastics provides the advantage of fast infiltration of fabrics with high fiber contents due to the very low viscosity of the monomer melt. The anionic polymerization of ε-Caprolactam to APA 6 has been investigated extensively due to the good chemical conversion and the fast polymerization. The use of APA6 in the resin transfer molding process offers a new possibility to manufacture continuous-fiber-reinforced composites with complex geometries.

During the last few years extensive research and development was carried out on the so-called thermoplastic resin transfer molding process. Some demonstrators were also developed for structural parts in automotive applications, such as a b-pillar and roof frame [4,5]. Abisset-Chavanne et al. developed a simulation model for reactive T-RTM processes to predict the mold filling. In this context they considered the thermo-chemico-mechanical couplings. They showed that the crucial parameter is the evolution of resin viscosity with the polymerization degree for the filling [6]. Schell et al. studied process variants like compression RTM for low viscosity thermoplastics to avoid fiber washing and to reduce cycle time [7]. Flame retardant additives were developed for the manufacturing of APA6 in T-RTM by Höhne et al [8]. Verrey and co-workers conducted a cost study of automotive body-in-white structures for thermoplastic and thermoset RTM [9]. Ehleben et al. showed the potential to use T-RTM for the new, large-scale production of an automotive application like the b-pillar. The good drapability of the dry preforms is an advantage compared with forming technologies, which are limited by geometrical issues [10].

However, there was skepticism about manufacturing in product lines, due to the sensitive influence of water during the anionic polymerization. Reimschüssel et al. and Bernat et al. have analyzed the influence of water on non-activated polymerization on a lab scale [11,12]. Even the resin suppliers recommend the manufacturing of reactive monomer under dry and oxygen-free conditions, because the oxygen in water and the air hinders the reaction. A

robust and reproducible process with well-known constraints is essential for production. All resin components from the manufacturers are dried to achieve water contents of less than 200 ppm. In T-RTM high-performance composite manufacturing, fibers [13,14,15,16], binder systems or core materials are therefore needed. During T-RTM manufacturing it is important to ensure that the total amount of water in these materials in each process step (preforming, handling, infiltration) is small enough to achieve good polymerization. In this study the influence of water on fast, activated polymerization is investigated in the T-RTM manufacturing process.
For manufacturing using the T-RTM processes it is important to understand the influence of water on the anionic polymerization and to know threshold values for the manufacturing. In this study the influence of water for different concentrations of activator and catalyst is investigated during the T-RTM manufacturing process. The impact on the mechanical properties is also described.

EXPERIMENTAL PROCEDURES

For the manufacturing of APA6 the monomer ε-Caprolactam, the activator Brüggolen C20 and the catalyst Brüggolen C10 from Brüggemann Chemical were used. The ratio of 1:2 of activator and catalyst was kept constant due to the bi-functionality of the activator. The following concentrations of activator and catalyst were used:

TABLE 1. Formulations with different concentration of activator and catalyst

Formulation	Concentration Activator C20	Concentration Catalyst C10
F1	1%	2%
F2	1,5%	3%
F3	2%	4%
F4	2,5%	5%

Unreinforced APA6 plates (900x550x2mm³) are manufactured via T-RTM processing. The manufacturing equipment consists of a low-pressure T-RTM injection machine from ATP, a press Type 630 DYL from Dieffenbacher and a plate mold. The resin components ε-Caprolactam, activator and catalyst were stored in the vessels of the injection machine at 110°C. Defined amounts of water are added in the vessels. After 5 minutes mixing the next trials are performed.
For the tensile test measurements, an inspect table 50kN tester (Hegewald & Peschke) with a 50 kN-measuring box and a test speed of 1 mm/min were used. The tests were carried out according to DIN EN ISO 527-1 standard. The samples for the tensile test were dried for 72 h at 80 °C to achieve constant weight. The residual monomer content of the samples was extracted with water and measured with a HPLC laboratory device Agilent 1100 Series. The viscosity number was investigated with an ubbelohde viscometer according to DIN 51562. The measurements were carried out at 20°C. The solvent of the polymer was 95% sulfuric acid.

RESULTS AND DISCUSSION

The anionic polymerization is monitored during the T-RTM manufacturing process by measuring the cavity pressures during the injection and curing of the reactive components in the mold. First the cavity pressure of different formulations is tested with 0% water (Figure 1) to understand the influence of the concentration of activator and catalyst on the cavity pressure.
The manufacturing T-RTM process can be divided into 3 phases: Vacuum application, injection and curing. At the beginning of the T-RTM process, a vacuum is applied for 30 seconds after closing the mold. The injection is then started, and the cavity pressures increase as the cavity is filled. After the injection is completed the injection point in the mold is closed and the clamping force of the press is increased. The cavity pressures increase up to their maximum. Subsequently the cavity pressures decrease and reach a plateau at about 50 bar. The shape of this plateau depends on the amount of activator and catalyst. Afterwards, the pressures drop to 0 bar. The increase of activator and catalyst leads to a drop in the cavity pressure after 50 bar. The time until the pressures drop to 0 bar is much shorter.
The cavity pressures show a regressive trend after applying the dwell pressure. This decrease is caused by shrinkage which takes place due to polymerization and crystallization during the anionic polymerization [17]. A probable reason for the fast decrease after the plateau is that the material becomes solid and loses contact with the pressure sensor. The higher the concentrations of activator and catalyst, the faster the conversion of the monomers [18,2].

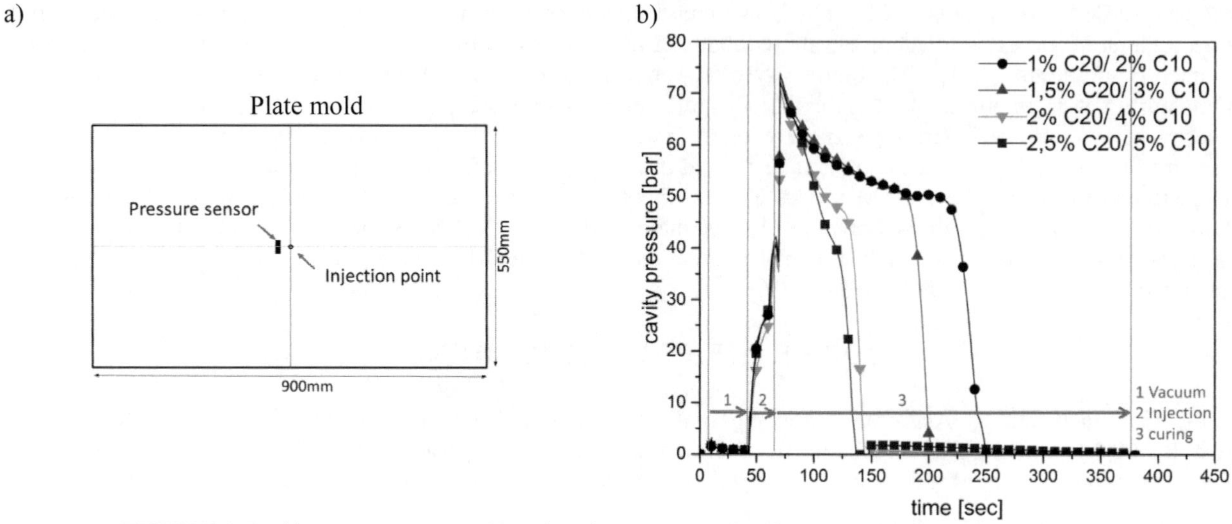

FIGURE 1. Position pressure sensor (a) and cavity pressures of different formulations with 0% water (b)

For the characterization of the anionic polymerization during the T-RTM manufacturing process, the water content in the reactive melt was increased gradually by 0.02%. For all formulations it can be seen that the addition of water slows the regressive decrease of the cavity pressures and strengthens the plateau before the pressure falls away sharply. With increasing water content it takes longer for the pressures to drop to a minimum. This decrease did not take place until the mold opened. The reaction was not finished. The part could therefore not be demolded. This could be seen for all formulations. 0.02% water added to F1 is enough to prevent curing of the melt within 300 seconds curing time (Figure 2).

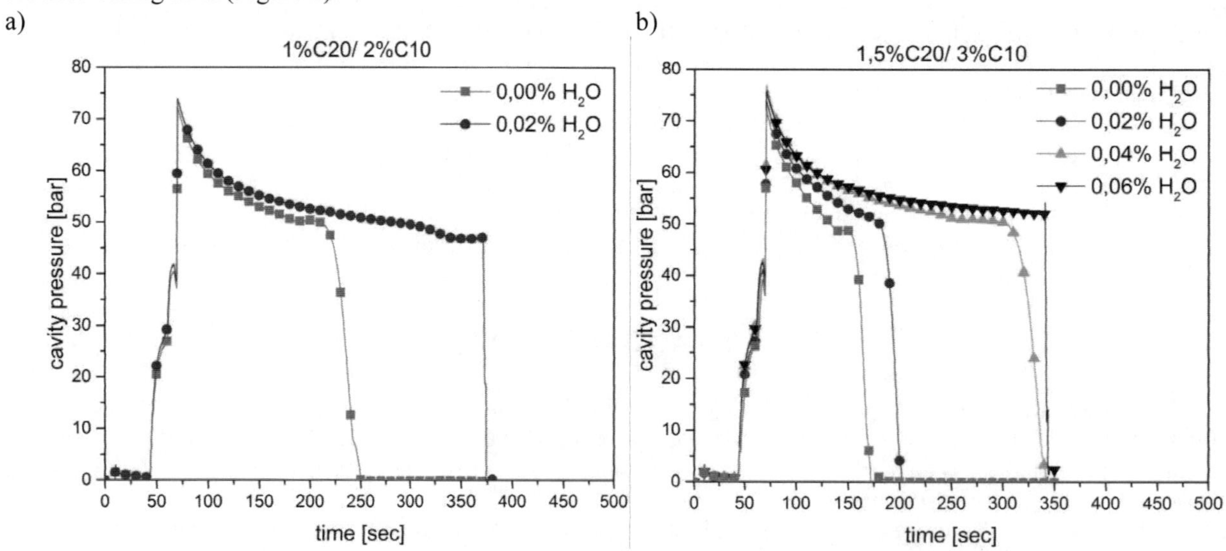

FIGURE 2: Cavity pressures during curing of formulation 1 (a) and formulation 2 (b)

With the extension of the curing time, and consequently the later opening of the mold, the curing takes place. This can be seen in the brown curve (0.02% H_2O). At 1.5% activator and 3% catalyst the demolding with 0.04% water is possible. The pressure trends show a rapid decrease of the pressure shortly before opening the mold. At a water content of 0.06% there is no decrease after achieving the plateau. No polymerization occurred until the mold opened.

The manufacturing of solid plates with F3 is possible for water contents up to 0.06% (Figure 3a). A further increase of water up to 0.08% prevents the polymerization for a curing time of 300 seconds until the mold opens. The formulation with 2.5% activator and 5% catalyst shows similar trends (Figure 3b). The manufacturing of cured plates is possible for water contents of up to 0.08%. At 0.1% water, the area in the cavity at the pressure sensor is not cured.

The comparison of the cavity pressures of the different formulations of activator and catalyst reveals a significant difference. The lower the activator/catalyst concentration, the less water can be added before the anionic polymerization of the reactive melt stops. The presence of water initiates a side reaction [19]. The catalyst is consumed. There are fewer anions available for the reaction. By extending the closing time of the press the T-RTM process can be adjusted to allow the melt to cure and to demold the part. Time spent cleaning non-reacted material from the mold can be saved. At small amounts of activator and catalyst the cavity pressure needs more time to drop. The shrinkage of the material during reaction is slower due to the lower conversion rate of the reaction.

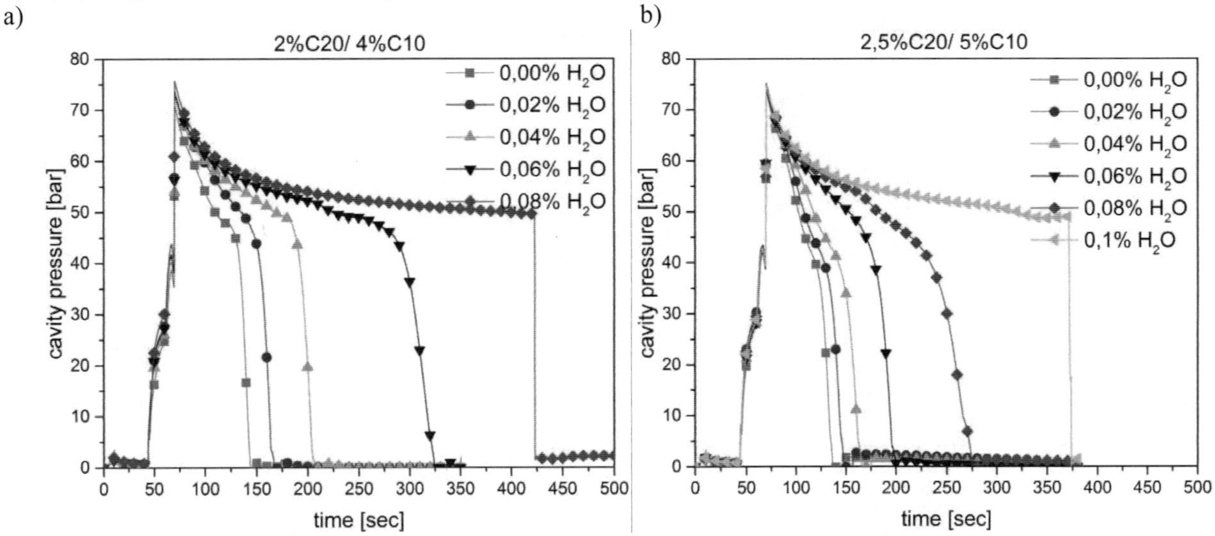

FIGURE 3: Cavity pressures during curing of formulation 3 (a) and formulation 4 (b)

The mechanical properties were measured in tensile tests. The mechanical testing was only possible with polymerized samples which could be demolded as solid plates after the curing time of 300 seconds.

In Figure 4 the tensile properties of the manufactured plates are shown with different water contents during curing. The reduction of the activator and catalyst increases the tensile stiffness of the samples with 0% water content. The influence of the concentration of activator/catalyst on the tensile strength is negligible. The addition of water influences the stiffness and strength of the cured material. At small concentrations of activator and catalyst the reaction is extremely sensitive. Small amounts of water (0.02%) lead to a significant decrease in stiffness and strength. The decrease of stiffness and strength at higher concentrations of activator and catalyst therefore occurs only at higher concentrations of water.

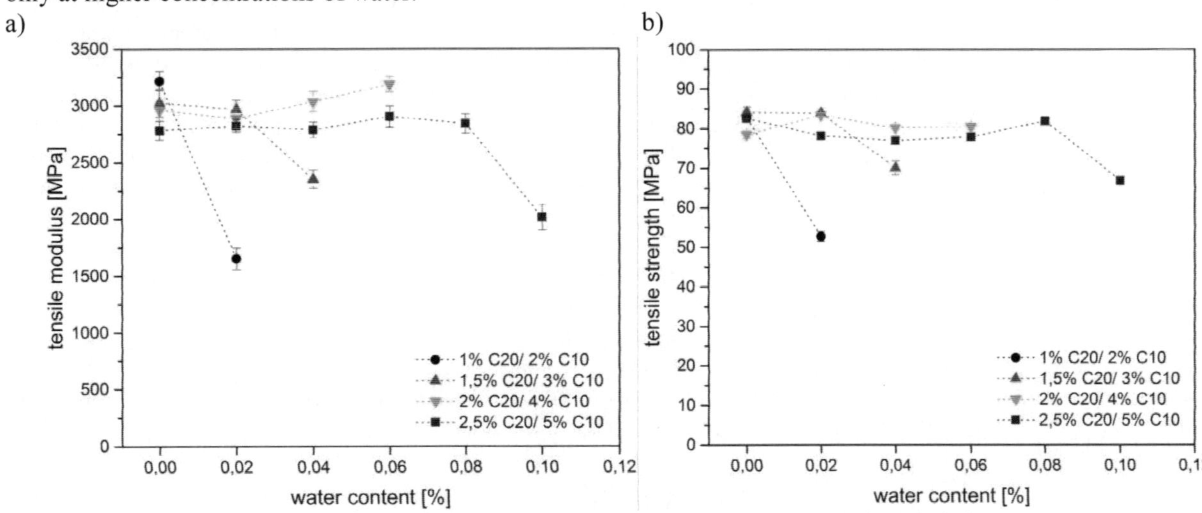

FIGURE 4: Influence of water on tensile modulus (a) and strength (b)

Zhang et al. investigated methods to measure the conversion rate [20]. The residual monomer content (RC) and the viscosity number (VZ) is used in this study as an indicator for the conversion of the anionic polymerization. The RC

indicates the monomer content which is not converted during the reaction. The VZ is directly related to the average molecular weight, and can be used as one indicator for the rate of the polymerization [21].
In Figure 5 the RC of the formulations are shown. A RC of 1% builds the threshold value for good polymerization. F1 with only 1% activator and 2% catalyst is extremely sensitive to water. The RC of F1 increases from 0.6% up to 12 % on the addition of 0.02% water. For F2, 3 and 4, with higher concentrations of activator and catalyst this increase of RC fails to appear at water contents of 0.02%. F2 shows a higher RC of 1.8% with 0.04% water. The increase in RC at F3 starts at 0.06%. Due to the buffering of F4 the increase occurs at 0.1% water.

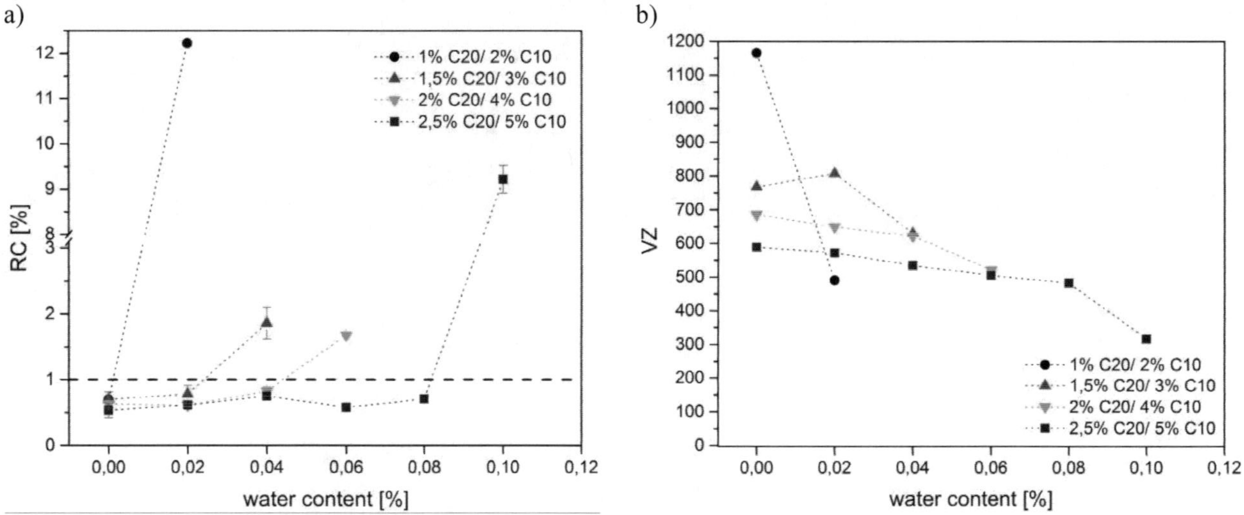

FIGURE 5: Influence of water on residual monomer content (a) and viscosity number (b)

The VZ offers valuable clues about the conversion of the anionic polymerization. The results with 0% water show a significant difference between the individual formulations. As expected, the concentration of the activator and catalyst influences the VZ. At low concentrations of activator and catalyst the VZ is much higher than at higher concentrations of activator and catalyst [Figure 5]. The graph only shows the average because the standard deviation of the three measurements is negligible (about 1 %). The VZ decreases when water is added. This effect can be seen for every formulation. The decrease of VZ when water is added depends on the formulation. Low concentrations are more sensitive to water. At low concentrations of activator and catalyst the decrease on adding water occurs much faster than at higher concentrations.
The influence of activator and catalyst on the mechanical properties can be seen in the tensile tests with 0% water. F1 with the lowest content of activator and catalyst shows the best properties and the highest molecular weight.
The increase in the amount of residual monomer at low concentrations of activator and catalyst starts at lower amounts of water. The water molecules deactivate the reactive groups. As more reactive groups are deactivated, the RC increases. With increased concentration of the activator and catalyst the reactive system is tougher against water or humidity because more anions are available for the reaction. However, the influence of water or humidity is still a major determining factor which cannot be neglected. It is important to ensure that the amount of water in the system (machine, resin components, fibers, cores) does not exceed the threshold values to ensure reproducible manufacturing. The VZ of anionically manufactured PA6 is higher than that of hydrolytically manufactured PA6. Measurements of hydrolytically manufactured PA6 (Ultramid B3K) showed a VZ of 150. The higher VZ of anionically manufactured samples can be attributed to the higher molecular weight. The molecular chains break during cleaning of the polycondensated materials due to the high temperatures, solvents and high shear forces. The molecular weight is reduced to lower the melt viscosity to enable manufacturing [22].
The results of RC are well in line with the results of VZ. A quality fall-off takes place at the same concentration of water. F1 shows this quality fall-off at 0.02%, F2 at 0.04%, F3 at 0.06% and F4 at 0.1%.

CONCLUSION

In this study, fast activated anionic polymerization was investigated for the T-RTM manufacturing process. The influence of water on the curing time was demonstrated. The presence of water does not stop the polymerization, but it slows the reaction by deactivating the reactive components.

The results show the possibility of monitoring the rate of anionic polymerization during manufacturing by observing the trend of the cavity pressures. The absorption and the delivery of water in semi-finished products like preforms will be carried out in future studies. Necessary pretreatments will be deviated with the results of these studies.

ACKNOWLEDGEMENTS

We thank Ronny Hanich (Fraunhofer ICT) for performing many of the measurements presented, and Matthias Stricker (Fraunhofer ICT) for the tests with the ubbelohde viscosimeter, Uwe Podwojewski (Fraunhofer ICT) for the tensile tests and Anja Kaiser (Fraunhofer ICT) for the residual monomer measurements. We also thank Cahit Arik (Fraunhofer ICT) for performing the T-RTM process, and Brüggemann Chemical for providing the materials.

REFERENCES

1 D. Stavrov; H.E.N. Bersee, "Resistance welding of thermoplastic composites-an overview", Compos. A: Appl. Sci. Manuf.2005, 36, 39.

2 P. Hodge, "Cyclodepolymerization as a method for the synthesis of macrocyclic oligomers", Reactive & Functional Polymers 80 (2014) 21-32

3 B. R. Formisano; C. Bonten; "Recycling von Gusspolyamid-Abfällen", WAK Journal of Plastics Technology 2016

4 M. Bitterlich; M. Ehleben; A. Wollny; P. Desbois; J. Renkl; S. Schmidhuber, "Tailored to Reactive Polyamide 6", Carl Hanser Verlag, Munich, Kunststoffe International 3/2014

5 S. Schmidhuber; E. Fries; P. Zimmermann, "It couldn´t be more hybrid", Carl Hanser Verlag, Munich, Kunststoffe international 1-2/2017

6 E. Abisset-Chavanne; F. Chinesta, "Toward an optimisation of the reactive resin transfer molding process: thermo-chemico-mechanical coupled simulations", Int J Mater Forming. DOI: 10.1007/s12289-013-1124-0.

7 J.S.U. Schell; H. Aimé; S. Comas-Cardona; Ch. Binetruy; A. Lemasçon; M.F. Lacrampe ; M. Deléglise-Lagardère, ITHEC 2014, MESSE BREMEN

8 C.-C. Höhne; R. Wendel; B. Käbisch; T. Anders; F. Henning; E. Kroke; "Hexaphenoxycyclotriphosphazene as FR for CFR anionic PA6 via T-RTM: a study of mechanical and thermal properties" Fire Mater, 2016

9 J. Verrey; M.D. Wakeman; V. Michaud; J.-A.E. Manson, "Manufacturing cost comparison of thermoplastic and thermoset RTM for an automotive floor pan", Composites: Part A 37 (2006) 9–22

10 M. Ehleben; M. Bitterlich; O. Täger, "New industrial-scale production processes for thermoplastic fibre reinforced polymer parts", VDI Conference 2015

11 H. K. Reimschüssel, "Nylon 6 Chemistry and Mechanisms" J. POLYMER SCI.: Macromolecular Reviews, Vol. 12,65-139 (1977) by John.Wiley & Sons, Inc

12 P. Bernat; O. Hladká; M. Fišmanová; J. Roda; J. Brožek: "Polymerization of lactams. 98: Influence of water on the non-activated polymerization of e-caprolactam" European Polymer Journal 44 (2008) 32–41

13 R. Hoto; J. Andres; B. Cabillic; L. Gascon; J.A. Garcia, "Optimization of mechanical properties of basalt woven/APA-6 composite parts by means of velocity control". 18th international conference on composite materials

14 S. Pillay; U. K. Vaidya; G. M. Janowski, "Liquid Molding of Carbon Fabric-reinforced Nylon Matrix Composite Laminates Journal of Thermoplastic Composite Materials", Vol. 18 (2005) 509

15 A. Alfonso; R. Hoto; J. Andrés; J.A. García, "Industrial viability of APA-6 matrix-based basalt fiber reinforced composites" 18th international conference on composite materials

16 Chun Yan, Hongzhou Li, Xiaoqing Zhang, Yingdan Zhu, Xinyu Fan, Liping, "Preparation and properties of continuous glass fiber reinforced anionic polyamide-6 thermoplastic composite"s Materials and Design 46 (2013) 688-695

17 P.P. Parlevliet; H.E.N. Bersee, "Measurement of polymerization shrinkage of matrices for liquid moulding"

18 N. Barhoumi; A. Maazouz; M. Jaziri; R. Abdelhedi, Polyamide from lactams by reactive rotational molding via anionic ring-opening polymerization Polymer Letters Vol.7, No.1 (2013) 76–87

19 K. Ueda; K. Yamada; M. Nakai; T. Matsuda; M. Hosoda; K. Tai, "Synthesis of high molecular weight nylon 6 by anionic polymerization of e-caprolactam" Polym J 1996;28:446–51

20 C.-L. Zhang; L.-F. Feng; G.-H. Hu, "Anionic polymerization of lactams: A comparative study on various methods of measuring the conversion of ε-caprolactam to polyamide 6" Journal of Applied Polymer Science, Vol. 101, 1972–1981 (2006) Wiley Periodicals, Inc.

21 K. Ueda; M. Nakai; K. Hattori; K. Yamada; K. Tai, Kobunshi Ronbunshu 1997, 56, 401

22 K. van Rijswijk; H.E.N. Bersee, "Reactive processing of textile fiber-reinforced thermoplastic composites – An overview" Composites: Part A 38 (2007) 666–681

Relationship between structures and properties of modified polyamide by reactive extrusion

Auclerc Mathilde [(1)*]

[(1)] hutchinson IMP laboratory - Villeurbanne - France

Lighten vehicles replacing metals by plastic materials is one of the most important challenge of the automobile industry. With this in mind, researches are pushed to develop new polymeric materials presenting enhanced properties. One strategy is to modify common polymers such as polyamide. In that frame, chemical reactions carried out by reactive extrusion can answer to this challenge but the main difficulty in that case is to elucidate the new complex structures created. In our study, to scrutinize the reactive medium obtained by the reactive modification of a polyamide by a trifunctional derivative, the coupling of several physico-chemical analyses was necessary. The correlation between the chemical structures and physical properties can be evaluated respectively by i) Liquid NMR, Matrix Assisted Laser Desorption Ionisation/Time Of Fligh (MALDI-TOF) and size-exclusion chromatography (SEC) analyses and ii) X-ray crystallography, Small Angle X-rays Scattering (SAXS) and Wide-angle X-ray scattering (WAXS). From these information it is possible to evidence both grafting and polyamide chain scissions and to highlight the diminution of crystalline and amorphous lamellar thickness while the degree of crystallinity slightly increased. In conclusion, these coupling experiments demonstrate the possibility to modify the amorphous phase mobility by controlling the polyamide chemical modification through reactive extrusion.

Modified talcum as crosslinking agent for polyolefins

A. Uray[a,b], G. Riess[a], T. Lucyshyn[b], C. Holzer[b] and W. Kern[a]

[a] *Montanuniversitaet Leoben, Department Polymer Engineering and Science, Chair of Chemistry of Polymeric Materials, Leoben, Austria*

[b] *Montanuniversitaet Leoben, Department Polymer Engineering and Science, Chair of Polymer Processing, Leoben, Austria*

corresponding and presenting author: anna.uray@unileoben.ac.at

Abstract. The aim of this study is to investigate the modification of talcum and its support as crosslinking agent for polyolefins. To improve the crosslinking of polyolefin chains a thermal initiator is linked to the talcum surface. During the extrusion process the initiator decomposes and radicals are formed. These radicals will start and support crosslinking between the polymer chains. To produce crosslinked polymers, two steps were necessary. The first step was the modification of the mineral filler. For that purpose alkoxysilanes were grafted onto the talcum surface. After that, the initiator was linked to the silanes grafted onto the talcum surface. To prove the active molecular components on the particles, the silanized and modified talcums were characterized with different measurement methods such as x-ray photoelectron spectroscopy, infrared spectroscopy and thermogravimetric measurement. The second step were the crosslinking experiments with the modified talcum in polyolefins. Compounds of polyolefins and modified talcum were kneaded, plates were compression molded and analyzed with a plate/plate rheometer regarding their crosslinking behavior.

Keywords: crosslinking, talcum, thermal initiator
PACS: 61.25.hp

INTRODUCTION

Polyolefins are the most used polymers for different applications. For example, polyethylene is used in packaging industries, insulation components, medical applications, household items and many more. However, the application of polyethylene is limited by temperature. Therefore, polyethylene is often crosslinked to use it in applications at higher temperatures [1].

For crosslinking polymers, an initiator is necessary to get free radicals at the polymer chains. Different methods are existing to crosslink polyethylene. Electron beam irradiation, for example, is a method to produce radicals without any additives [2]. A popular method is the use of peroxide. In this method, the peroxide forms radicals which eliminate hydrogen atoms to create carbon-carbon links between the polymer chains [3]. Another procedure is to combine peroxide with silane, especially alkoxysilanes. The peroxide decomposes, forms radicals, which eliminate hydrogen atoms of the polymer chains and then the silane groups are grafted to the polymer chains and form linkages between the chains [1, 3, 4].

The purpose of this study is to modify a typical inorganic filler, talcum, which is frequently present in polyolefins as an active filler (for improved mechanical properties or as nucleating agent [5]), as an additional crosslinking agent to further improve the crosslinked network. To achieve that, we linked a thermal initiator 4,4′-azobis(4-cyanovaleric acid) to the talcum surface. This initiator is also used in some other studies for grafting of polymers on solid surfaces [6, 7, 8, 9]. For the linkage of the initiator to the talcum surface three steps were necessary and were performed according to the study of Gert Boven et al. [9].

MATERIALS

In this study, a standard talcum grade by Mondo Minerals B.V., was used. For the silanization alkoxysilane (3-aminopropyl)trimethoxysilane and as solvent toluene were used. Phosphorus pentachloride, dichloromethane, hexane, the thermal initiator 4,4′-azobis(4-cyanovaleric acid) (ABCA), triethylamine and ethanol were utilized in the second and third step of the modification. All chemicals were purchased from Sigma Aldrich and used without any purification. For the crosslinking experiments, a low density polyethylene (LDPE) with a melt flow rate (MFR) of 0.65 g/10 min (at 230 °C/2.16 kg) and an ethylene octane copolymer (EOC) with an MFR of 0.5 g/10 min (at

230 °C/2.16 kg) were used. For crosslinking 3-Methacryloxypropyltrimethoxysilane from Wacker Chemie AG and dicumyl peroxide from Sigma Aldrich were utilized.

EXPERIMENTAL

Modification of talcum

The first step in this study was the silanization of talcum. For that purpose, alkoxysilanes were grafted onto the talcum surface. The second step was the chlorination of the thermal initiator for an easier linkage between the functional groups of the aminosilanes and the chlorine. The last step was the modification of the silanized talcum: the thermal initiator was linked to the functional groups of the silane grafted onto the talcum surface. All steps were performed corresponding to the study of Gert Boven et al. [9]. Figure 1 shows the modified talcum with the linked thermal initiator to the alkoxysilanes.

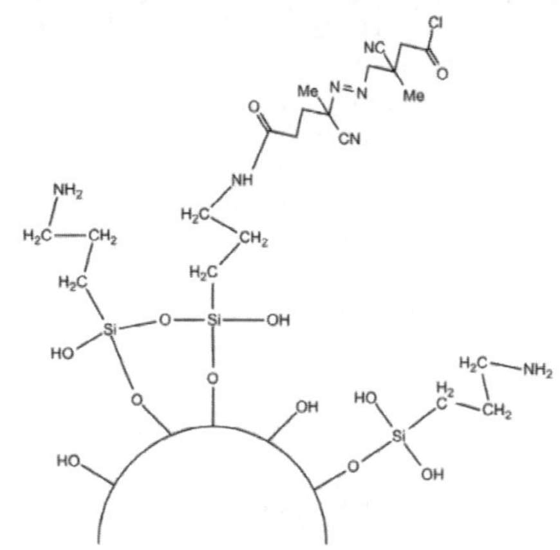

FIGURE 1. Modified talcum

For the evidence of the molecular components on the talcum particles, after each step the silanized and modified talcum was characterized with x-ray photoelectron spectroscopy (Thermo Fisher Scientific K-alpha, Vienna, Austria), infrared spectroscopy (Vertex 70, Bruker Optics GmbH, Ettlingen, Germany) and thermogravimetric measurement (TGA/DSC 1, Mettler Toledo GmbH, Vienna, Austria).

Crosslinking experiments

For the crosslinking experiments, polymer blends with a ratio of 80/20 of LDPE and EOC with 4 phr silane, 0.5 phr dicumyl peroxide and with and without modified talcum were produced in a kneader (Haake Rheocord 3000/600, Thermo Fisher Scientific GmbH, Karlsruhe, Germany). The used material formulations are shown in table 1. The compression molded plates (produced on a vacuum press Type P200PV, Dr. Collin GmbH, Ebersberg, Germany) were stored at 60 °C for a defined time and rheologically analyzed using a plate/plate rheometer (Physica MCR501, Anton Paar GmbH, Graz, Austria).

TABLE (1). Used material formulations.

LDPE/EOC (ratio)	Silane (phr)	Dicumyl peroxide (phr)	Modified talcum (wt.-%)
80/20	-	-	-
80/20	4	0.5	1.6
80/20	4	0.5	-

Results and Discussion

Before the modified talcum was used as crosslinking agent the silanized and modified talcum was characterized. To determine the alkoxysilanes on the talcum surface, three different measurement methods were used. Thermogravimetric measurement is a method which determines the weight loss dependent on the temperature. The

weight loss of the silanized talcum was about 3 % and about 2.5 % of the pure talcum. The drop of the pure talcum curve is because of the dissociation of the H_2O molecules from the surface. The decomposition of the organic groups of the silane components leads to a higher weight loss of the silanized talcum.
Another method to prove the success of the silanization is the infrared spectroscopy. It was possible to determine the stretching vibrations of the C-N groups of the aminosilane (1320 cm^{-1}) [10].
The third method that was used in this study is the x-ray photoelectron spectroscopy. The results are shown in table 2. This method investigates the element composition on the surface of the pure and the silanized talcum. Since magnesium (Mg) only occurs in talcum this element is used as reference (Mg/R). "Spot" is the result of the measurement. The increase of the silicon value is one evidence for the presence of silane on the talcum surface. This increase is caused by the silicon components of the aminosilane. The detection of nitrogen on the talcum surface can also be seen as an evidence of a successful silanization.

TABLE (2). Results of the x-ray photoelectron spectroscopy of the pure and silanized talcum.

Elements	Pure talcum		Silanized talcum	
	Spot (atom%)	Mg/R	Spot (atom%)	Mg/R
Magnesium	19.22	1	11.39	1
Silicon	22.71	1.18	20.89	1.83
Nitrogen	-	-	3.71	0.33

We used infrared spectroscopy also to investigate the modified talcum. We detected a peak at around 1660 cm^{-1} which indicates the carbonyl groups of the amides [9]. These carbonyl groups arise because of the linkage between the thermal initiator and the NH_2 groups of the aminosilanes. The carbonyl chlorine bonds and the carbonyl groups of the carboxylic acid chloride bonds of the chlorinated initiator were not detected. This is another indication for the linkage of the initiator with the silane components.
The x-ray photoelectron spectroscopy results of the modified talcum compared to the silanized talcum are shown in table 3. The results of the modified talcum show a lower amount of magnesium. This indicates the presence of other components at the surface of talcum. The increase of the nitrogen value can be explained due to the N=N groups of the thermal initiator. It was also possible to determine chlorine which is another proof for the linkage of the initiator.

TABLE (3). Results of the x-ray photoelectron spectroscopy of the silanized and modified talcum.

Elements	Silanized talcum		Modified talcum	
	Spot (atom%)	Mg/R	Spot (atom%)	Mg/R
Magnesium	11.39	1	9.83	1
Silicon	20.89	1.83	19.61	1.99
Nitrogen	3.71	0.33	5.18	0.53
Chlorine	-	-	1	0.1

After the evidence of the linkage of the thermal initiator to the talcum surface, the polymer blends were kneaded and analyzed with a plate/plate rheometer using compression molded samples. The complex viscosity (η), loss modulus (G") and storage modulus (G') were determined for the crosslinked and non-crosslinked polymer blends. Figure 2 shows the complex viscosity of the crosslinked polymer blends. Because of the grafting reaction of the polymer chains the complex viscosity increased at a lower frequency region and shows no longer a Newtonian plateau compared to the non-crosslinked blend [11]. In figure 2 it can be seen that the blend with the modified talcum shows a higher complex viscosity, at lower angular frequencies, than the blends without the filler. These results indicate that talcum modified with a thermal initiator acts as an additional crosslinking agent and leads to a better crosslinking network.

By measuring the loss and storage moduli the crosslinking network can also be determined (figure 3). The crosslinked sample shows no cross over point of the loss and storage moduli compared to the non-crosslinked samples. This indicates a crosslinking network [12].

Figure 3 shows also a higher storage modulus for the crosslinked blend at low frequencies (ω). This effect can be explained by the restricted movement of the chains in the crosslinked network [13].
Figure 4 shows the comparison of the storage and loss moduli of the sample with and without the use of the modified talcum. At low angular frequencies the sample with modified talcum shows a higher storage modulus than

without. This indicates a better crosslinked network because of the additionally formed radicals of the initiator linked to the talcum surface. Details of the results are shown in table 4.

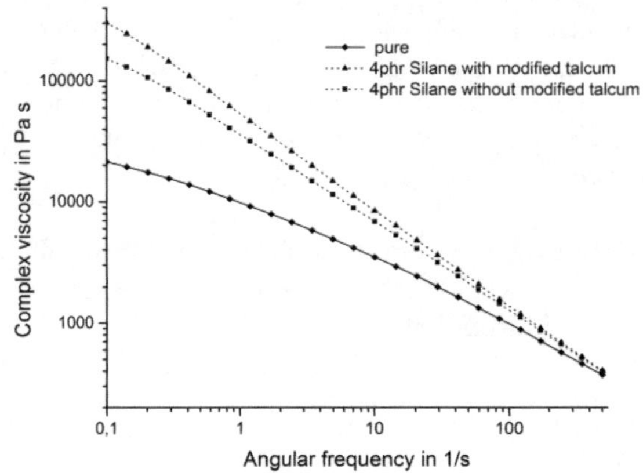

FIGURE 2. Complex viscosity of non-crosslinked and crosslinked polymer blends

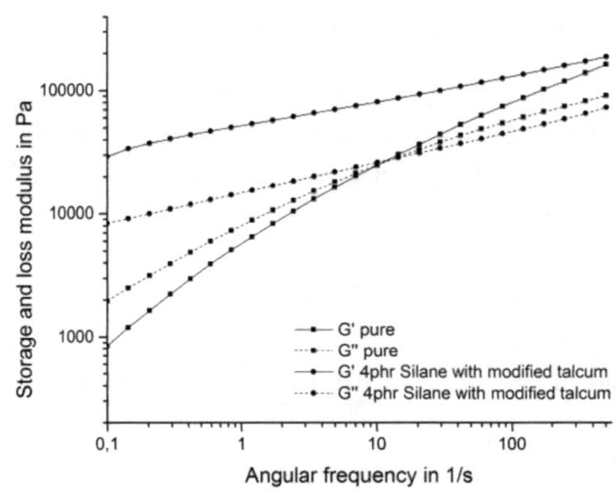

FIGURE 3. Storage and loss moduli (G' and G") of non-crosslinked and crosslinked polymer blends

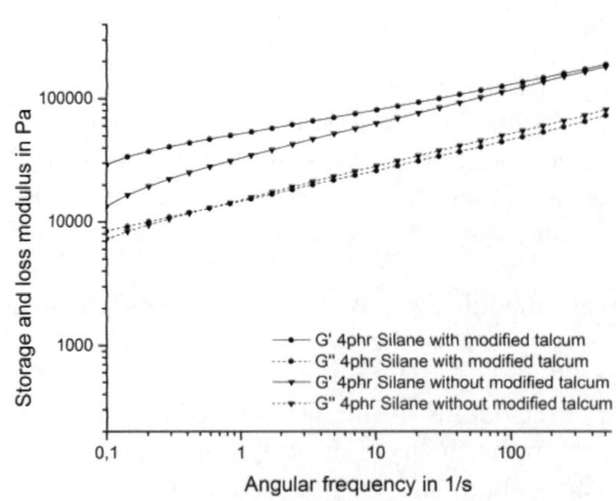

FIGURE 4. Storage and loss moduli (G' and G") of crosslinked polymer blends with and without modified talcum

TABLE (4). Results of the complex viscosity and storage modulus (G') at 0.1 s^{-1}.

Sample	Complex viscosity (kPa s)	G' (kPa)
pure	21.4	0.844
4phr Silan without modified talcum	152	13.4
4phr Silan with modified talcum	304	29.2

Conclusions

In this study, we investigated the use of talcum with a modified surface as additional crosslinking agent. The first step was the modification of talcum with a thermal initiator. Using different measurement methods, we were able to prove the linkage of the alkoxysilanes and the initiator to the talcum surface.

After the modification of the talcum, polymer blends with silane, dicumyl peroxide and with/without the modified talcum were produced and stored at 60 °C for a defined time. Afterwards the samples were analyzed using a plate/plate rheometer. The results of the complex viscosity showed an increase at a lower frequency region for the crosslinked samples compared to the pure polyethylene and an even higher increase for the sample with the modified talcum.

The curves of the loss and storage moduli of the crosslinked blends showed no cross over point, which is an evidence for crosslinked polymer chains. The results at low angular frequencies also showed a higher storage modulus for the sample with the modified talcum due to the additionally formed radicals.

These results indicate that the formed radicals of the thermal initiator linked to the talcum supported the crosslinking process and led to a better crosslinking network. Therefore, the modification of talcum with an initiator is a possibility to support and improve the crosslinking process for polyolefins.

ACKNOWLEDGMENTS

The research work of this paper was performed at the department Polymer Engineering and Science, Chair of Chemistry of Polymeric Materials and Chair of Polymer Processing (Montanuniversitaet Leoben) within a framework of the Austrian Research Promotion Agency (FFG - Bridge No. 853291) and the financial support of Steinbacher Dämmstoff GmbH.

REFERENCES

1. Tamboli, S. M.; Mhaske, S. T.; Kale, D. D., *Indian J. Chem. Technol.* **11**, 853–864 (2004).
2. Berger, K., Keimel, C., Helfer, E., Haar, B., Eder, A., Mattausch, H., Riess, G., Kern, W. *J. Appl. Polym. Sci.* **134**, DOI: 10.1002/APP.44968 (2017).
3. Morshedian, J., Hoseinpour, P., *Iran. Polym. J.* **18**, 103–128 (2009).
4. Berger, K., Keimel, C., Helfer, E., Haar, B., Eder, A., Mattausch, H., Riess, G., *J. Appl. Polym. Sci.* **134**, DOI: 10.1002/APP.45374 (2017).
5. Saechtling Kunststoff Taschenbuch, 30. Ausgabe, Carl Hanser Verlag, (2015).
6. Müller T. K. H., Cao P., Ewert S., Wohlgemuth J., Liu H., Willett T. C., Theodosiou E., Thomas O. R. T., Franzreb M., *J. Chromatogr. A* **1285**, 97-109 (2013).
7. Zimmermann J.: "Polyol- und Azlacton-Makromonomere für Netzwerksysteme, neue Werkstoffe und biomedizinische Anwendungen", Ph.D. Thesis, Fakultät für Chemie und Pharmazie der Albert-Ludwigs-Universität Freiburg i. Br., (2001).
8. Golze S.: "Oberflächengebundene Polymermonolagen für die Herstellung von DNA-Chips", Ph.D. Thesis, Chemie und Pharmazie der Johannes Gutenberg-Universität Mainz, (2001).
9. Boven G., Oosterling M. L. C. M., Ger Challa, Schouten, A. J., *Polymer* **31**, 2377-2383 (1990).
10. Rafael A. Bini., *Journal of Magnetism and Magnetic Materials* **324**, 534-539 (2012).
11. Sharif-Pakdaman, A., Morshedian, J. & Jahani, Y., *J. Appl. Polym. Sci.* **125**, 305–313 (2012).
12. Mezger, T. G. The Rheology Handbook: For Users of Rotational and Oscillatory Rheometers. (Vincentz Network GmbH & Co KG, 2006).
13. Liu, H., Fang, Z., Peng, M., Shen, L. & Wang, Y., *Radiat. Phys. Chem.* **78**, 922–926 (2009).

IN SITU POLYMERIZATION OF STYRENE INTO A PMMA MATRIX BY USING AN EXTENSIONAL FLOW MIXING DEVICE: A NEW APPROACH TO ELABORATE POLYMER BLENDS

***Bouquey Michel* [(1)]*, *Pierrot François Xavier* [(1)], *Ibarra Gomez Rigoberto* [(1)], *Muller René* [(1)], *Serra Christophe* [(1)]**

[(1)] University of strasbourg_ICS - Strasbourg - France

Different approaches have been used nowadays to elaborate new and functional plastic compounds. In a practical and economic way, a wide range of bulk properties for polymeric materials is still obtained by direct blending of polymers. Nevertheless, the inherent downstream direction of the melt blending process by starting out from polymers as raw materials makes quite difficult the targeting or tuning of desired morphologies, thus, final properties of blends. Different strategies other than direct melt mixing have been continuously evolving to prepare innovative polymer blends. Among the most interesting ones are those based primarily on the in situ polymerization approach applied to polymer processing, which is the frame of the present work to be detailed below. In this, unlike the conventional melt blending, one or more monomers in addition to polymeric components may be present in the starting mixture. On this respect our strategy introduces a new brand mixing device, RMX®, in order to highly enhance the dispersion capacity thanks to a contraction flow type mixer capable of generate strong extensional flow fields. We present our first experiments, PMMA/styrene systems leading to PMMA/PS blends, to valid the feasibility of our technic which, in one step, combines the in situ polymerization approach along with extensional mixing with the aim of preparing innovative polymer blends. By using this method, a fine and controlled balance between formulation design, desired morphologies and mixing efficiency is intended to be attained.

In-situ reduction of graphene oxide for conductive graphene bionanocomposite extrusion

Lo Re Giada [1]*, Mianehrow Hanieh [1], Olsson Richard [1], Berglund Lars [1]

[1] KTH - Royal Institute of Technology - Stockholm - Sweden

Extrusion of polymeric materials is the most used processing technique by volume of material in the world and examples of potential future use for large-scale continuous extrusion could include preparation of non-linear and linearly conductive cable accessories, piezoelectric components and emerging filaments for 3D printing wherein more complex geometries and applications may be foreseen. Moreover, conductive (bio)polymeric nanocomposites has emerged as a relatively new and promising material within extrusion, but still requires more advances in the field. Graphene as a conductive nanomaterial has been widely used in different polymeric systems to make conductive polymeric nanocomposites. However, because of strong π-π interactions between graphene sheets, it is difficult to achieve homogeneous dispersions of graphene in the resultant nanocomposite. An efficient way to overcome this obstacle is to start from graphene oxide and reduce it to get electrical conductivity and suitable dispersion at the same time. Here, the challenge is to control the dispersion of the nano sheets in the matrix. By reaching sufficient concentration of graphene sheets, which may provide an interconnected carbon network, the exploiting of conductivity in (bio)polymeric nanocomposites may provide new opportunities while maintaining low percolation threshold and high mechanical properties of the composites. Different reductants have been used for reduction of graphene oxide to graphene. Among them, recently L-Ascorbic acid was demonstrated to be a green and efficient reductant for graphene oxide in (bio)nanocomposites [1]. This work contributes to the efforts aiming a sustainable and scalable approach to in-situ reduce graphene oxide and process in one-step continuous fashion (bio)nanocomposite with high mechanical properties. Reference: [1] H. Nassira et al. Adv Mater.2016

pps proceeding
Symposium: S16 - Reactive processing
Oral Presentation

SUSTAINABLE ENZYMATIC COPOLYMERISATION OF BIODEGRADABLE POLYESTERS BY REACTIVE EXTRUSION

Lo Re Giada [(1)]*, Gioia Claudio [(1)], Berglund Lars [(1)]

[(1)] KTH - Royal Institute of Technology - Stockholm - Sweden

In the last decades, both industrial and academic research are interested on enzymes as a green alternative to metallorganic catalytic systems. Enzymes have shown ability to achieve high molecular weight in the synthesis of biodegradable polyesters, together with easier remouval/deactivation, making these emergent class of polymers more attractive for special uses such in biomedical or enviromental friendly applications. A further significant step toward fully sustainable materials can be designed by coupling the enzymatic catalysis together with a solvent-free continous reactive polymerization that can produce, in a high efficient fashion, end-user products directly from the monomers [1]. In particular, copolyesters synthesized using this pathway could broaden the application field of biodegradable and/or environmentally friendly polymers as new matrices or compatibilizers [2]. A solvent-free enzyme-catalyzed synthesis of fully bio-degradable poly(3-hydroxybutyrate-co-butylene succinate)(PBS)-poly(caprolactone)(PCL) copolyesters is here reported for the fist time. A one-pot process from 1,4-butanediol, dimethyl succinate and ▯-caprolactone or preformed PCL copolyesters were produced in bulk with Candida Antarctica lipase B (CALB) as catalyst. The synthetic composition, kinetic and monomers sequences were studied and optimized in bulk in order to transfer the copolymerization to a micro-compounder. References: 1.Spinella, S., Ganesh, M., Lo Re, G., Zhang, S., Raquez, J. M., Dubois, P., & Gross, R. A. (2015). Green Chemistry 2015, 17(8), 4146-4150. 2.Todd, R., Tempelaar, S., Lo Re, G., Spinella, S., McCallum, S. A., Gross, R. A. & Dubois, P. ACS Macro Letters 2015, 4(4), 408-411.

An Advanced Kinetic Modeling for Reactive Polymer Processing

Julio C. Hernández-Ortiz[a], Paul H. M. Van Steenberge[a], Jan N. E. Duchateau[b], Carolina Toloza [b], Fons Schreurs [b], Marie-Françoise Reyniers[a], Guy B. Marin[a], and Dagmar R. D'hooge[a,c]

[a]Ghent University, Laboratory for Chemical Technology (LCT), B-9052 Gent, Belgium
[b]SABIC, PO Box 319 / 6160 AH Geleen, The Netherlands
[c]Ghent University, Centre for Textile Science and Engineering, Technologiepark 907, B-9052 Gent, Belgium

Abstract. An advanced novel kinetic Monte Carlo (*k*MC) methodology is developed, allowing the description of all the microstructural properties and the complete reaction event distribution of complex individual architectures formed during free radical induced grafting (FRIG) of vinyl monomer onto polyolefin chains at the micro-scale, *i.e.* for given, bulk concentrations and temperature. The methodology is constructed in a modular fashion. The first module permits the calculation of important properties, such as the overall monomer conversion, the grafting selectivity, the grafting yield, and the chain length distribution (CLD) of the macrospecies. The average grafting density, the average chain length of grafts, and the average crosslinking density can also be computed. The second module tracks for the first time the reaction event of every individual chain and links this information to the macroscopic properties. The CLD of grafted chains and the bivariate copolymer composition-CLD can therefore be retrieved. Furthermore, the reaction event distributions can be obtained, showing the region of the polyolefin CLD where certain reaction predominantly acts. The effect of depropagation of the vinyl monomer on the kinetics of grafting is evaluated. Additionally, the effect of diffusional limitations on the grafting kinetics is assessed. The kinetic model can be extended with reactor models to account for phase segregation and imperfect macromixing.

Keywords: Reactive processing, Modeling, Polymerization, Post-polymerization functionalization.
PACS: 81. Materials science

INTRODUCTION

Post polymerization functionalization (PPF) is an advantageous reactive polymer processing technique that allows to incorporate new reactive groups into the original polymer. The PPF of polyolefins, one of the most widely produced polymers, is mainly achieved via free radical induced grafting (FRIG)[1]. Despite being a mature technology, there are still several issues to be addressed, such as the grafting efficiency and control of side reactions[2] and an improved mechanistic knowledge.[3] Typically, this process involves the treatment of molten polyolefin chains with a source of free radicals and a vinyl monomer. During this process, the primary free radicals abstract hydrogens from the backbone of the polyolefin, creating mid-chain macroradicals (MCRs). These MCRs can add to the double bond of the vinyl monomer, creating grafting points through a grafting "from" process. However, the action of the primary radicals is not selective and they can also add directly to the double bonds of the vinyl monomer, generating homopolymer radicals. Nevertheless, the latter radicals can still contribute to further functionalization if they recombine with MCRs, generating grafting "to" points. The identification of the optimal process conditions that help increase the level of grafting and reduce the extent of homopolymerization becomes crucial for the better performance of this type of reactive processing. Kinetic modeling can serve as a tool to achieve a detailed understanding of the synthesis-structure relationship and to optimize the process conditions.

METHODOLOGY

In this contribution, a novel kinetic Monte Carlo methodology is presented,[4] allowing for the first time to describe all relevant reaction characteristics and microstructural properties during FRIG processes. The methodology is composed by two modules. The first module (Module A) permits the calculation of important properties, such as the overall monomer conversion, the grafting selectivity, the grafting yield, and the chain length distribution (CLD) of the macrospecies. The average grafting density, the average chain length of grafts, and the average crosslinking density are also computed. The second module (Module B) tracks the reaction event of every individual chain and links this information to the macroscopic properties. Thus, the CLD of grafted chains and the bivariate copolymer composition chain length distribution (CLD) can be retrieved. Furthermore, the reaction event distributions can be obtained, showing the region of dominance with respect to the polyolefin CLD. These two modules can be run in parallel (**FIGURE 1**). For modeling the hydrogen abstraction reaction events, a mass-weighted CLD for the polyolefin is employed for the *k*MC sampling. Depropagation of the vinyl monomer is also considered, due a reaction temperate close to the ceiling temperature. An effective forward propagation rate coefficient (k_p^{eff}) has been used to account for depropagation steps, according to **EQUATION (1)**, where k_p represents the intrinsic rate coefficient of propagation, K_{eq} stands for the equilibrium coefficient of depropagation, and $[M]_{eq}$ indicates the monomer concentration at the equilibrium.

$$k_p^{eff} = k_p\left(1-\frac{K_{eq}}{[M]}\right) = k_p\left(1-\frac{[M]_{eq}}{[M]}\right) \tag{1}$$

The kinetic model also accounts for diffusional limitations on the micro-scale, based on the "parallel" encounter pair model to evaluate the apparent rate coefficients.[5] The self-diffusion coefficients are calculated using the free-volume theory.[6] Under practical conditions the kinetic model also needs to be combined with a reactor model, which is currently under investigation.

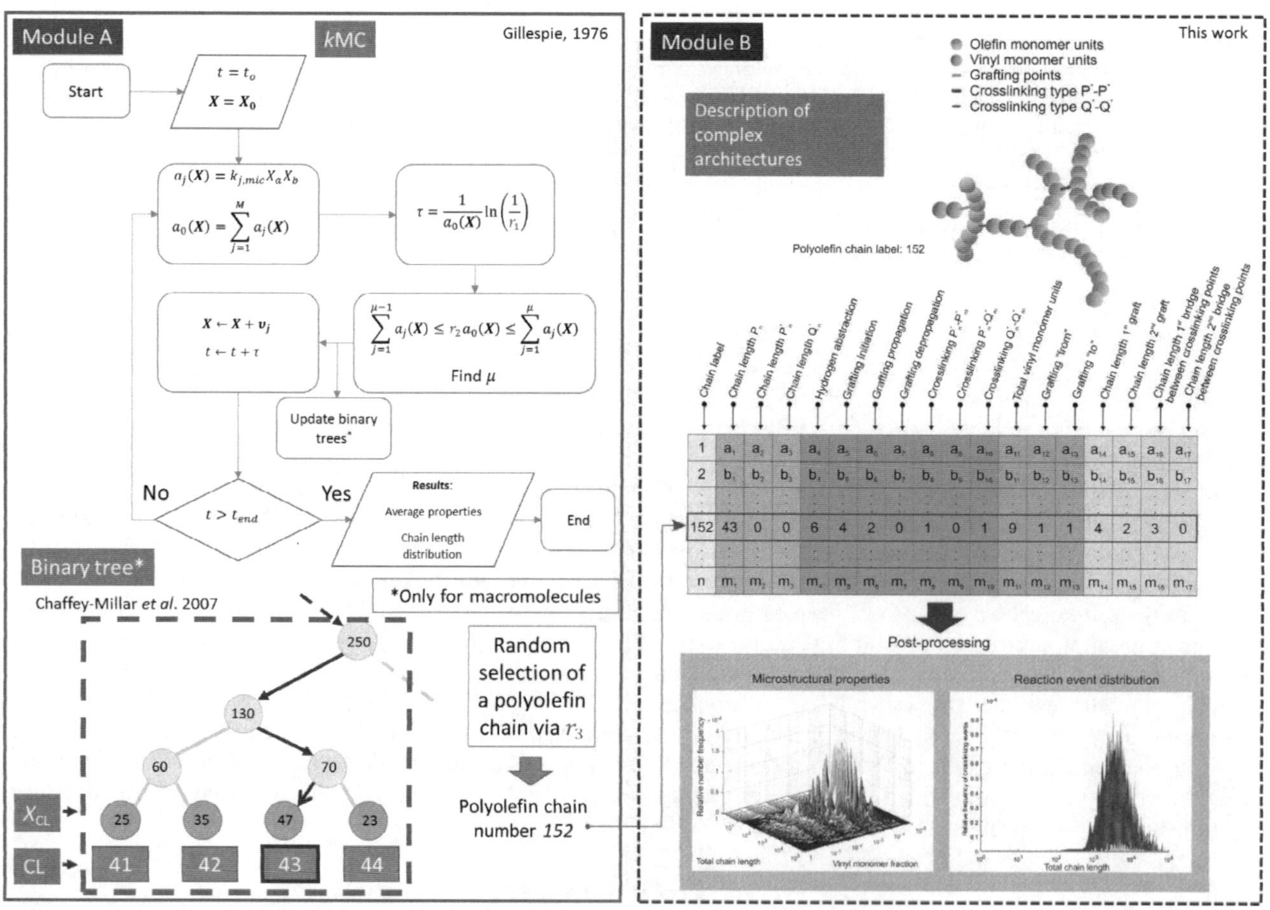

FIGURE 1. Conceptual framework to obtain the average characteristics during FRIG (left; Module A) and to additionally track the microstructure of individual complex macromolecules and their complete reaction event history (right; module B).

RESULTS AND DISCUSSIONS

The kinetic model has been applied at the micro-scale to investigate the FRIG of a general vinyl monomer onto polyethylene (PE) under melt conditions initiated by a monofunctional initiator, assuming isothermal and homogeneous conditions. The initial concentrations for the reactants are summarized in **TABLE 1**. Under those conditions and considering a control volume for the *k*MC simulations equal to 4.0×10^{-15} liter, the initial number of polyolefin macromolecules in the calculations is 3.3×10^{7}, the initial number of monomer molecules is 1.8×10^{9}, and the initial number of molecules of initiator 2.1×10^{7}. The CLD of virgin PE is presented in **FIGURE 2**, both in number and mass basis. In what follows, a distinction is made between simulations results ignoring and accounting for diffusional limitations. The former are known as intrinsic simulations and the latter are based on apparent rate coefficients. These results are also key upon a further model refinement accounting for macromixing.

TABLE (1). Initial concentration of chemical species.

Chemical species	Concentration (mol L^{-1})
Polyolefin (PE)	1.37×10^{-2}
Vinyl monomer	7.39×10^{-1}
Monofunctional initiator	8.76×10^{-3}

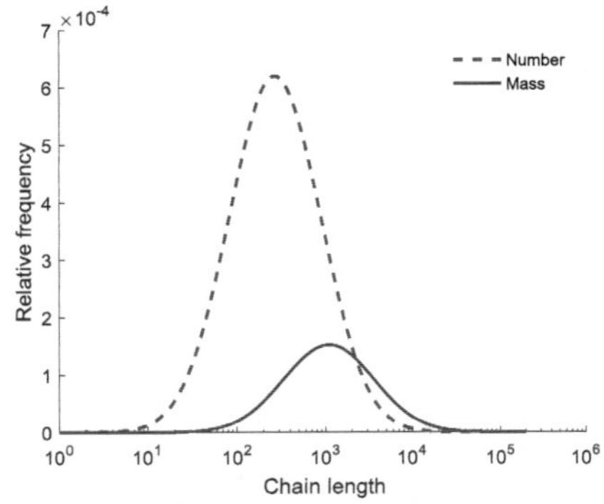

FIGURE 2. Number and mass-CLD for the starting polyolefin for the grafting process, based on a log-normal distribution with mass-average chain length of 8810 and dispersity (*Đ*) of 3.99.

Intrinsic Simulation Results

In order to analyze the general trends for the kinetics of FRIG, typical values for intrinsic rate coefficients have been selected[7], except for the rate coefficient of hydrogen abstraction (k_H) and the equilibrium coefficient of depropagation (K_{eq}), for which a parametric sensitivity analysis was carried out. The outcome for monomer conversion versus reaction time, grafting selectivity, grafting yield and average chain length of grafts as a function of k_H and K_{eq} is presented in **FIGURE 3**. From **FIGURE 3a** it can be observed that K_{eq} has a strong impact on the profile of monomer conversion, reaching a plateau at lower values of conversion if the K_{eq} values are higher. The plateau indicates the onset of the equilibrium of propagation/depropagation. On the other hand, both k_H and K_{eq} play and important role on the values of grafting selectivity, grafting yield and average chain length of grafts, as shown in **FIGURE 3b-d**. In general, higher values of k_H promotes more grafting with respect to homopolymerization, leading to higher values for grafting selectivity, grafting yield, and average chain length of grafts. Higher K_{eq} values lead to higher values of grafting selectivity and yield, but shorter grafted chains due to a more pronounced depropagation. The application of Module B allows calculating the reaction event distribution. **FIGURE 3e** presents the relative frequency of hydrogen abstraction as a function of the chain length (CL) of polyolefin.

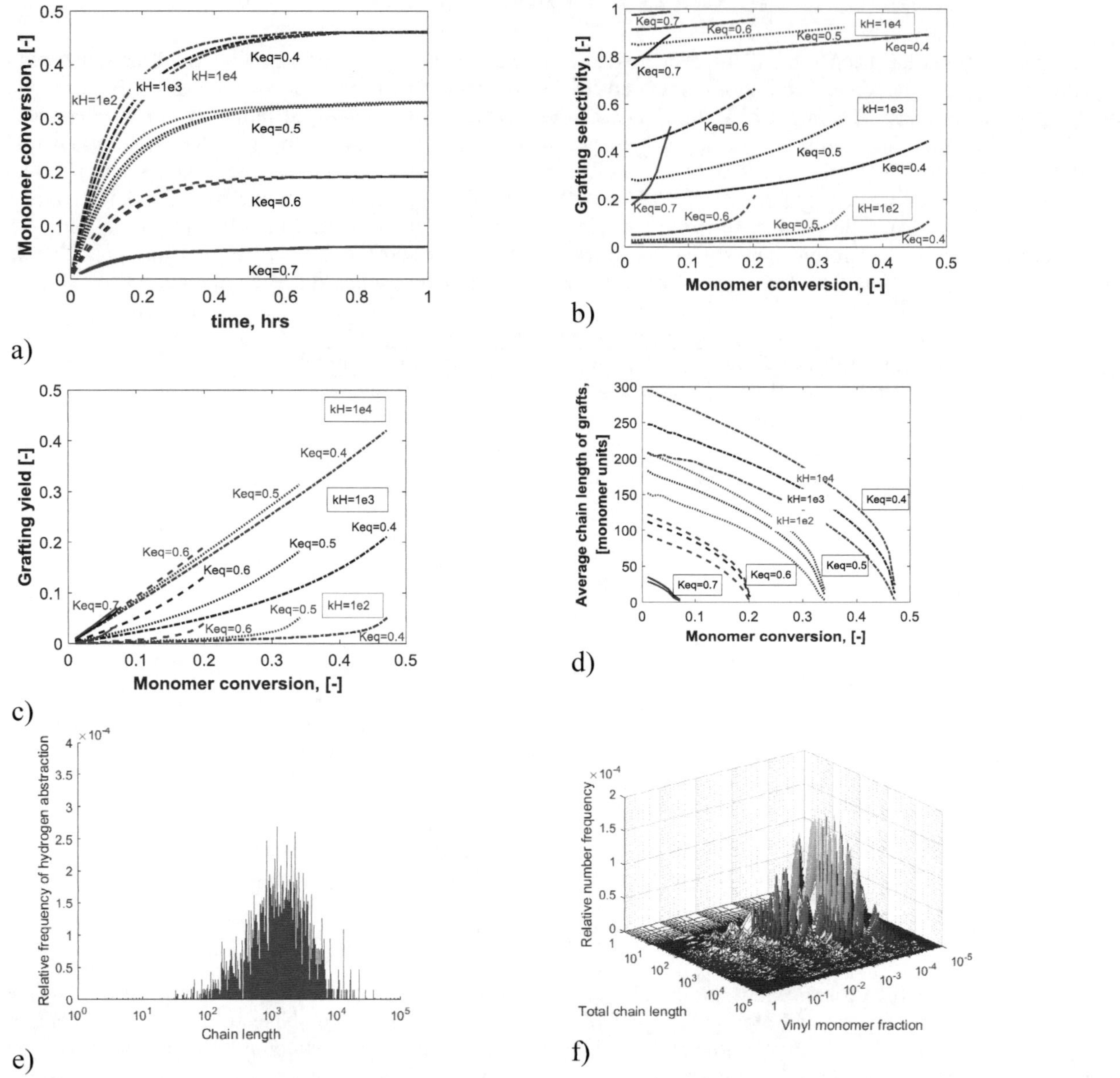

FIGURE 3. General intrinsic trends. Module A was employed to carry out a parametric sensitivity analysis for a) monomer conversion versus time; b) grafting selectivity; c) grafting yield and d) average chain length of grafts, varying the values for the kinetic parameter of hydrogen abstraction and depropagation equilibrium. Module B was employed for calculating e) the reaction event distribution for hydrogen abstraction, and f) the bivariate copolymer composition-CLD, both at monomer conversion of 0.47, and k_H=1.0×10^2 L·mol^{-1}·s^{-1} and K_{eq}=0.4 mol·L^{-1} (same units for legends)

It can be observed that this reaction predominantly occurs in polyolefin chains with CL around 10^3, which corresponds with the maximum of the mass-CLD of the polyolefin, as shown in **FIGURE 2**. This is in agreement with the mass-CL dependency of the hydrogen abstraction. **FIGURE 3f** presents the bivariate copolymer composition-CLD obtained with Module B. This figure shows that most of the functionalized polyolefin chains possess a level of vinyl monomer less than 1% and a CL of around 10^3.

Simulations Accounting for Diffusional Limitations

The decrease in the apparent value of the kinetic rate coefficients due to the limited mobility of the chemical species in the molten polyolefin is now taken into account. It can be noted in **FIGURE 4** that diffusional limitations have a strong impact on the profile of monomer conversion, increasing dramatically the reaction rate as a consequence of the reduced termination, which also brings about a higher CL for the grafts, a higher functionalization selectivity and a lower crosslinking density.

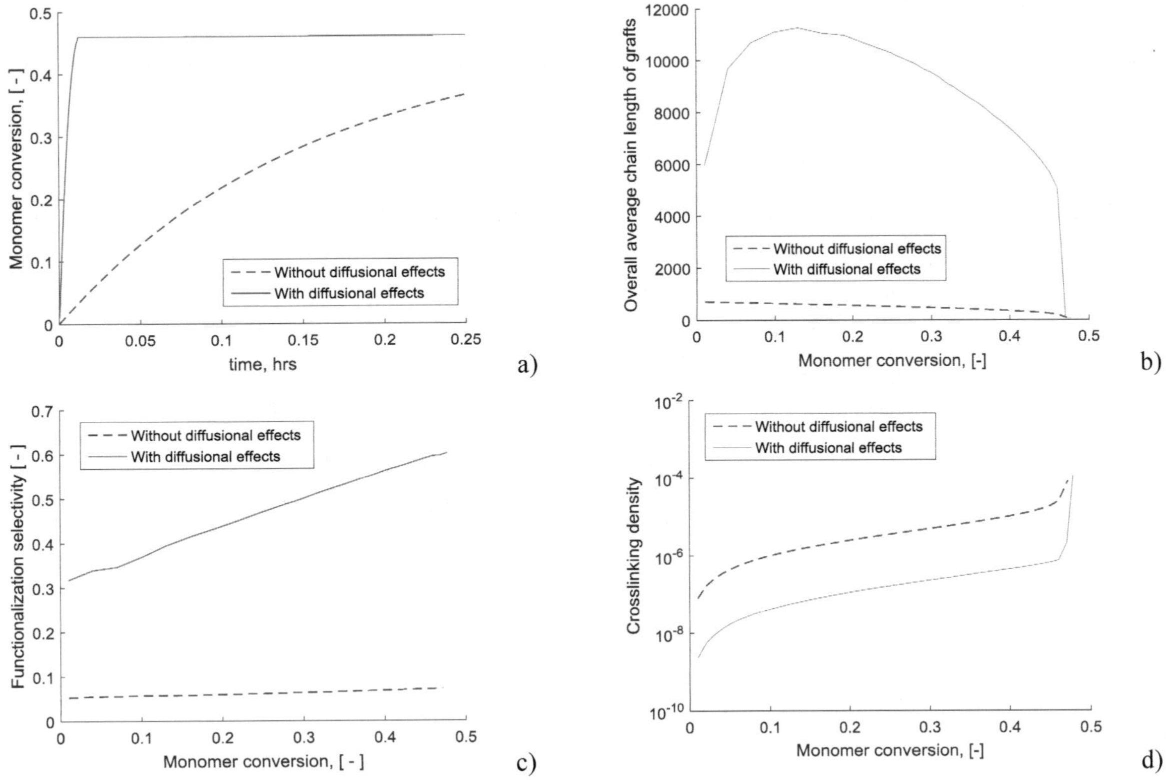

FIGURE 4. Calculated profiles for a) monomer conversion versus time, b) average chain length of grafts, c) functionalization selectivity, and c) crosslinking density during FRIG of vinyl monomer onto PE, as obtained with Module A, ignoring and accounting for diffusional limitations.

ACKNOWLEDGMENTS

P.H.M.V.S. and D.R.D. acknowledge the Fund for Scientific Research Flanders FWO through a postdoctoral fellowship.

REFERENCES

1. N.M.G Franssen, J.N.H Reek, B. de Bruin, *Chem. Soc. Rev.* **42**, 5809-5832 (2013).
2. T.C. Chung, "Functionalization by Post-polymerization Processes" in *Functionalization of Polyolefins*, edited by T.C. Chung, London: Academic Press, 2002. pp. 65-78.
3. M. van Duin *Macromol. Symp.* **202**,1-10 (2003).
4. J.C. Hernández-Ortiz, P.H.M. Van Steenberge, J.N.E. Duchateau, K. Remerie, C. Toloza, A.L. Vaz, F. Schreurs, M.-F. Reyniers, G.B. Marin, D.R. D'hooge *AIChE J.* (2017) doi:10.1002/aic.15842.
5. D.R. D'hooge, M.-F. Reyniers G.B. Marin, *Macromol. React. Eng.* **7**, 362-379 (2013).
6. J.S. Vrentas, J.L. Duda, M.K. Lau, *J. Appl. Polym. Sci.* **27**, 3987-3997 (1982).
7. G. Moad and D.H. Solomon, *The Chemistry of Radical Polymerization*, Oxford: Elsevier Science, 2005.

pps proceeding
Symposium: S16 - Reactive processing
Oral Presentation

Influence of molecular weight and crystalline structure on rheological behavior of controlled-rheology-polypropylene using viscosity modifiers

Hadimani Imam Khasim Rajasaheb [1]*, Siddiqui Inaamul Haq [1], Al Mohanna Mohammad Hamad [1]

[1] SABIC - Riyadh - SaudiArabia

Results of experimental studies of free-radical-induced degradation of polypropylene (PP) in the melt phase are presented. Experiments were carried out in a co-rotating twin-screw extruder and single screw extruder. The initiator, luperox 101 and trigonox 301 were used as a radical generator. Concentration of the peroxide was in the range 0.03–0.14 %wt. Melt flow index, mechanical, thermal and rheological properties were measured to see the effect of initiator concentration, temperature (200 and 230 °C) and screw rpm (70 and 100 rpm). Initiator concentration of the peroxides and initial melt flow of PP was the most significant variable. Melt flow index increased with initiator concentration. Reaction temperature and screw rpm had no significant effect on the melt flow index of extrudate. This was attributed to the minimum residence time in the extruder being sufficient for all degradation reactions to be completed. Degraded resins had more Newtonian behavior than the original resin. Viscosity reduced with peroxide concentration. Melting point of degraded resin remained relatively constant at different levels of peroxide but the crystalline fraction increased slightly. No significant change of the mechanical properties was seen up to 0.1 % wt of peroxide. Beyond 0.1 %wt elongation at break and impact strength reduced considerably. The rheological PDI determined by Mod Sep method were compared to MWD obtained by GPC.

REACTIVE EXTRUSION of POLYAMIDE-6 USING A NOVEL CHAIN EXTENDER

Tuna Basak [(1)], Benkreira Hadj [(1)]*

[(1)] University of Bradford - Yorkshire - UnitedKingdom

REACTIVE EXTRUSION of POLYAMIDE-6 USING A NOVEL CHAIN EXTENDER Basak Tuna, Hadj Benkreira* R & KT Centre in Advanced Materials Engineering, University of Bradford, UK *Presenting and Corresponding Author: H.Benkreira@bradford.ac.uk Polyamide-6 (PA-6) is an important engineering thermoplastic having improved mechanical properties, processability and chemical resistance, attributes which enable it to be used in textile, packaging and automotive industry. PA-6 requires high processing temperatures during melt compounding and undergoes thermal degradation which leads to a decrease in molecular weight and thus a reduction in mechanical and rheological properties. Solid state polymerisation (SSP) has been used over many years to enhance molecular weight of polycondensates. It is however an added expensive processing step. Chain extension during extrusion is a better cheaper alternative, especially when small amount (1%) of the chain extender are required. In this study a novel multi-functional chain extender, Joncryl® ADR 3400, was used. The experiments were performed in APV twin screw extruder (L/D=25) at different temperature profiles and screw speeds and with different amount of Joncryl in an attempt to optimise the chain extension reaction. Rheological data in the LVR region were used to assess the change of structure resulting from the chain extension reaction. This was complemented by mechanical and thermal properties measurements of the solid extrudate. In addition, experiments on a smaller scale were performed using a Haake Minilab Microcompounder and Xplore 15 ml Microcompounder to assess the suitability of these minimixers for such studies. In the presentation, the experimental methods and data will be critically presented to guide practical consideration of the technique. References Loontjens, J.A. (2005), Ph.D. Thesis. Eindhoven University of Technology. Najafi, N., Heuzey, M.C., Carreau, P.J. and Adams, P.M. (2012), Polymer Degradation and Stability 97.

CHEMICAL MODIFICATION OF POLY(LACTIC ACID) WITH VINYL MONOMERS BY REACTIVE PROCESSING

Rigolin Talita Rocha (1)*

(1) Universidade Federal de São Carlos - UFSCar - São Paulo - Brazil

Modification of poly(lactic acid) (PLA) with different monomers, maleic anhydride (MA) and glycidyl methacrylate (GMA), in the presence of peroxide, was conducted in a Haake torque rheometer according to a 22 factorial design in which the monomer and peroxide concentrations were varied. The modified poly(lactic acid) was characterized by size-exclusion chromatography, titration of acid groups and differential scanning calorimetry. Results showed that the PLA modification with maleic anhydride was seen to be accompanied by increase in acidity percentage and reduction in molecular weight, whereas the glycidyl methacrylate led to an opposite result. Torque curves demonstrated the same behavior, with increase and reduction of torque on samples with GMA and MA, respectively.

Properties of Films Made from Plantain Flour/Poly(ε-caprolactone) Blends Under Conditions of Reactive Extrusion using Zirconium Octanoate as a Catalyst

Tomy J. Gutiérrez[a] and Vera A. Alvarez[a]

[a]*Grupo de Materiales Compuestos Termoplásticos (CoMP), Instituto de Investigaciones en Ciencia y Tecnología de Materiales (INTEMA), Facultad de Ingeniería, Universidad Nacional de Mar del Plata (UNMdP) y Consejo Nacional de Investigaciones Científicas y Técnicas (CONICET), Colón 10850, B7608FLC, Mar del Plata, Argentina.*

Abstract. Plantain flour (*Musa ssp*., group AAB, sub-group clone Harton)/poly(ε-caprolactone) (PCL) blends, containing glycerol as a plasticizer, were prepared by reactive extrusion (REx) in a twin-screw extruder using zirconium octanoate ($Zr(Oct)_4$) as a catalyst, followed by thermo-compression molding for film development. The films were then characterized in terms of their: infrared (FTIR) spectra, water solubility, thermogravimetric (TGA) curves, differential scanning calorimetry (DSC) thermograms, and X-ray diffraction (XDR) diffractograms, as well as their microstructural, mechanical and antimicrobial properties in order to 1) compare the effects of PCLs with two different molecular weights (Mw) on the characteristics of the plantain flour/PCL blends, and 2) determine whether using $Zr(Oct)_4$ in the production of active composite polymer materials improves their properties. The plantain flour/PCL blends were all developed successfully. The higher Mw PCL gave more hydrophobic and thermally stable films with improved mechanical properties. The addition of the $Zr(Oct)_4$ catalyst to the plantain flour/PCL blends also resulted in films with similar characteristics to those described above, due to the cross-linking of the polymers. In addition, the films containing the catalyst showed antimicrobial activity against *Escherichia coli* O157:H7 and *Staphylococcus aureus* indicating a dual effect of $Zr(Oct)_4$, and making it an attractive alternative for the development of active films.

Keywords: Active films; Antimicrobial properties; Cross-linking; Poly(ε-caprolactone); Starch.
PACS: S16-266

1. INTRODUCTION

Starch-based films have been proposed as an environmentally friendly alternative to traditional food packaging materials (1). This is because they are biodegradable and thus reduce, in part, the problems caused by synthetic plastics obtained from the oil industry. Recent studies have reported excellent properties of edible films made from plantain flour (2, 3). Nevertheless, these materials have some drawbacks in particular their hydrophilic nature that makes them highly sensitive to water. For this reason in recent years several alternatives have been proposed to improve their properties. Among the most promising of these is the development of starch-based films blended with biopolyesters in the polyhydroxyalkanoate family (PHAs) (4).

Poly(ε-caprolactone) (PCL) is a hydrophobic aliphatic polyester that may be obtained by chemical synthesis either from crude oil or renewable resources such as polysaccharides. PCL is a thermoplastic, biodegradable, biocompatible and semi-crystalline polymer with a very low viscosity, glass transition temperature (approx. -60 °C) and melting point (58-60 °C). In addition, PCL is a fully biodegradable and easily processable synthetic material that can be used with conventional melt processing equipment. In fact, this material is already used in the food packaging and biomedical fields (4).

Among the advantages of using PCL are that: 1) it is a hydrophobic material with respect to other biopolymers such as proteins and therefore is not prone to plasticization or swelling, 2) it shows excellent stretchability and low water vapor permeability and 3) PCL-based films show good water resistance (4).

Nevertheless, biopolyesters and polysaccharides are thermodynamically immiscible, i.e. they show a lack of polymeric compatibility owing to their different polarities. Starch-PCL blends obtained by simple mixing are thus not adequate for film preparation as their low affinity leads to phase separation. This inhibits the adhesion of the polymeric interfaces to each other resulting in poor film properties (4). For example, Gutiérrez and Alvarez (4) observed phase separation in sagu flour (*Canna edulis* Kerr)/PCL and sagu starch/PCL blends, with the former showing the greater separation. However, the use of starch-rich flours as raw materials for edible and/or biodegradable film development has generated great interest, since their high performance and low cost compared to pure starch makes them a promising alternative for the production of polymer matrices that could replace synthetic polymers obtained from oil.

Some of the disadvantages of starch/PCL blends could be solved by preparing them using reactive extrusion (REx). REx combines mass and heat transport operations which, together with simultaneous chemical reactions taking place inside the extruder, modify the properties of the existing polymers or generate new ones. The extruder can thus be considered as a reactor, where polymers can be modified in a single step while they are being processed. This procedure is increasingly becoming recognized as a powerful technique to develop and manufacture a wide variety of novel polymeric materials in a highly efficient and flexible way. The combination of chemical reactions and transport phenomena in an extruder provides an invaluable window of opportunity for the compatibilization of synthetic resins and biopolymers such as starch, or the compatibilization of synthetic resins and natural fillers under current industrial conditions (5, 6).

With this in mind, several catalysts have been used in conjunction with REx to generate cross-linking reactions within polymer blends especially those incorporating thermoplastic starch (TPS) (6). The use of catalysts with REx is based on the fact that the temperatures and pressures used to process starch-based biodegradable composites activate the catalyst, thereby generating an increase in the reaction rate, i.e. the kinetics of the cross-linking reaction are favored. In this way, the compatibility of starch-based polymer blends can be improved by a process that is scalable industrially.

In particular, the use of octanoates as catalysts for food packaging production has generated great interest especially since some octanoates, for example tin octanoate ($Sn(Oct)_2$), are already widely used (7) and have been approved as food additives by the U.S. Food and Drug Administration (8). According to Raquez et al. (9) the most probable reaction mechanism involves the direct catalytic action of $Sn(Oct)_2$, whereby the catalyst first activates the monomer forming a donor-acceptor complex, which then participates directly in the first propagation step. $Sn(Oct)_2$ is thereafter liberated in every ensuing propagation step. This means that Sn^{2+} atoms are not covalently bound to the polymer at any stage during polymerization. Another likely mechanism, the "active chain-end" mechanism, was proposed by Penczek et al. (10). This involves the *in-situ* formation of Sn-alkoxide bonds at chain-ends as observed by MALDI-TOF and fully confirmed by kinetic studies (7). Thus, through a rapid exchange equilibrium $Sn(Oct)_2$, and most probably any other covalent metal carboxylate, is first converted by reacting with protic compounds (ROH) into tin (or other metal) alkoxides which then act as active centers for polymerization. Polymerization itself involves a "coordination insertion" mechanism similar to those reported for covalent metal alkoxides and dialkylaluminum alkoxides.

Recently our research group has used zinc octanoate ($Zn(Oct)_2$) as a catalyst in corn starch/polystyrene (PS) blends under REx conditions, observing an improvement in the compatibility between the bio- and synthetic polymers with the use of the catalyst. In addition, the $Zn(Oct)_2$ showed antimicrobial activity. Unfortunately, however, this had no effect on the developed materials possibly because its migration to the environment was compromised by the cross-linked starch/PS network (6). With all this in mind we feel that it is worth examining the potential dual function (catalyst and antimicrobial agent) of another octanoate, zirconium octanoate ($Zr(Oct)_4$), for the production of active eco-friendly materials for food packaging. It is worth noting that there is currently no record in the literature of the use of $Zr(Oct)_4$ in food packaging systems.

In addition, most of the studies found in the literature related to starch materials have focused on films obtained by the casting method, whereas processing methods such as blown film extrusion, compression, or injection molding are less reported. Solvent casting is the most common method employed for small-scale biopolymer film preparation and involves solubilization, casting, and drying steps. Despite this being a good and adequate technique on a laboratory scale, it is considered to be a high energy-consuming procedure and thus not appropriate for large-scale production (11).

Procedures for the efficient manufacture of high quality, biodegradable films are thus still required by industry. With this in mind, the scaling-up of processing methods using equipment designed for synthetic polymers is indispensable, and extrusion, blowing, injection and thermo-compression are all viable alternatives due to their energy efficiency and high productivity. In particular, extrusion followed by thermo-compression is useful as a processing method because of its simplicity. These techniques could thus be used to provide feasible ways of developing biomaterials that are both economical and environmentally friendly (11).

The aims of this study were thus to 1) compare the effects of using low and high molecular weight (Mw) PCL on the properties of plantain flour/PCL blends and 2) determine whether the use of $Zr(Oct)_4$ can improve the properties of active composite polymer materials.

2. EXPERIMENTAL

2.1. Materials

Plantain flour (*Musa ssp.*, group AAB, sub-group clone Harton) was obtained by the method described by Pacheco (12). The plantains were purchased at a local market in Caracas, Venezuela, and had a degree of maturation of 1 according to the scale put forward by Loesecke (13), i.e. unripe fruit. The plantain flour obtained was characterized in terms of its moisture content (9.4%), total proteins (2.64%), crude fat (0.36%), ash (2.41%), crude fiber (0.297%), total carbohydrates (84.9%), starch content (85%), amylose content (24%) and average molecular weight (Mw) (70 x 10^6 g/mol) (14). Poly(ε-caprolactone) (PCL) pellets with molar weights (Mw) of 10,000 and 80,000 g/mol were purchased from Aldrich (product codes: 440744-500G and 440752-500G, respectively) and used to prepare the plantain flour/PCL blends. The catalyst (zirconium 2-ethylhexanoate [IUPAC name] - zirconium octanoate [trade name] - $Zr(C_8H_{15}O_2)_4$ [molecular formula] - $Zr(Oct)_4$ [condensed molecular formula]; density = 0.870 g/mL at 25 °C; and boiling point = 167-178 °C at 760 mmHg) was kindly provided by Ghion, Laboratorios Químicos, S.R.L. (Mar del Plata, Argentina) (approx. 12% w/w of Zr). Glycerol (density = 1.26 g/mL; boiling point = 290 °C) (Aurum, Mar del Plata, Argentina) was employed as the plasticizer for film formation.

2.2. Extrusion and Film Formation

Films were prepared from plantain flour, and plantain flour/PCL blends at a 50:50 ratio (152.5 g of starch and 152.5 g of PCL), plus 67.5 g of glycerol. To some of the samples 10 mL of the catalyst were then also added. The blends were homogenized manually before being introduced into the extruder. The instrument used was a twin-screw extruder with six heating zones and two feed zones. The first feed zone is in the first heating step and the second in the third heating step. The extrusion profile selected was 50/70/90/100/100/100 °C. Films were obtained by compressing at 130 °C for 15 min. at 100 bars and then cooling to 30 °C. Five film systems were obtained as follows: plantain flour (*TPPF*); plantain flour + PCL (Mw = 10,000 g/mol) (*TPPF/PCL(10,000)*); plantain flour + PCL (Mw = 10,000 g/mol) + catalyst (*TPPF/PCL(10,000)+CAT*); plantain flour + PCL (Mw = 80,000 g/mol) (*TPPF/PCL(80,000)*); and plantain flour + PCL (Mw = 80,000 g/mol) + catalyst (*TPPF/PCL(80,000)+CAT*). The resulting materials were conditioned with a saturated NaBr solution (a_w ~0.575 at 25 °C) for seven days prior to each test.

2.3. Characterization of the Films

2.3.1. Attenuated Total Reflectance Fourier Transform Infrared Spectroscopy (ATR/FTIR)

The IR spectra of the films were determined using a Nicolet 8700 infrared spectrometer (FTIR) (Thermo Scientific Instrument Corporation, Madison, Wisconsin, USA) equipped with a diamond at an incident angle of 45° to collect the infrared spectra of the film samples. The spectra were recorded in Attenuated Total Reflectance mode (ATR) between 700 and 4000 cm^{-1}, using 40 scans at a resolution of 4 cm^{-1}. Each sample was scanned three times observing good reproducibility.

2.3.2. Water Solubility (WS)

Water solubility (WS) was determined according to the method described by Gutiérrez et al. (15). Film pieces were cut (2 × 2 cm) and the initial weights registered (w_i). The initial dry matter content of each film system was determined by drying them to a constant weight in an oven (Memmert, Germany) at 105 °C. To determine the weight of dry matter not solubilized in water the film pieces were immersed in 50 mL of distilled water and stored for 24 h at 25 °C. After this, they were filtered through previously desiccated and weighed filter paper, and the undissolved films obtained dried again to a constant weight (w_f) in the same oven at 105 °C for 24 h. The solubility of each film system was then determined as follows:

$$\text{WS}(\%) = (w_i - w_f) \times w_i^{-1} \times 100 \qquad (1)$$

where w_i and w_f are the initial and final weights of each sample, respectively. Results were reported as % water solubility ± standard deviation (SD) from three measurements per film system.

2.3.3. Thermogravimetric Analysis (TGA)

Thermogravimetric tests were carried out with a thermal analyzer (Model TGA Q500, Hüllhorst, Germany). Samples were heated at a constant rate of 10 °C/min from room temperature up to 500 °C under a nitrogen flow of 30 mL/min. Film weights were in the range of 13-17 mg. Analyses were performed in triplicate to ensure repeatability.

2.3.4. Differential Scanning Calorimetry (DSC)

A DSC-50 differential thermal analyzer (Shimadzu Corporation, Oregon, USA) was used to determine the melting temperatures (T_m) of the different films. The temperature and heat flux of the instrument were previously calibrated using indium and zinc. About 5 mg of each sample was placed in hermetically sealed aluminum pans and heated from room temperature up to 110 °C at a scanning rate of 10 °C/min. under a nitrogen atmosphere. Changes of phase or state and the corresponding enthalpies were determined from the melting peaks of the DSC thermograms (16). The crystallinity percentage values of each sample were then calculated according to the following equation (17):

$$\text{Crystallinity (\%)} = \left(\Delta H_m / w_{PCL} \times \Delta H_{100} \right) \times 100 \tag{2}$$

where ΔH_m is the experimental melting enthalpy, wPCL is the PCL weight fraction and ΔH_{100} is the melting enthalpy of 100% crystalline PCL (142 J/g) (18). Three samples were analyzed for each film system. The crystallinity percentage values ± SD were reported from three samples for each film system.

2.3.5. X-ray Diffraction (XRD)

X-ray diffractograms of the different films were obtained with a PAN analytical X'Pert PRO diffractometer (Netherlands) equipped with a monochromatic Cu K_α radiation source (λ = 1.5406 Å) operating at a voltage of 40 kV and current 40 mA at a scanning rate of 1° per min. The scanning region of the samples was in a 2θ range between 3° and 33°. The distances between the planes of the crystals *d* (Å) were calculated from the diffraction angles (°) measured from the X-ray diffractograms according to Bragg's law:

$$n\lambda = 2d\sin(\theta) \tag{3}$$

where λ is the Cu K_α radiation wavelength and n is the order of reflection. For the calculations, *n* was taken as 1. The thicknesses of the samples on the slides were ~1.12 mm. The crystallinity percentage values of each film were also calculated taking into account the main peaks (main *d*-spacings) as described by Hermans & Weidinger (19) by the following equation:

$$\text{Crystallinity (\%)} = \left(A_c / A_c + A_a \right) \times 100 \tag{4}$$

where A_c is the crystalline area, A_a is the amorphous area, and $A_c + A_a$ is the total area under each of the curves of the different X-ray diffractograms. Percentage crystallinity was determined to verify the results obtained by the DSC calculations.

2.3.6. Scanning Electron Microscopy (SEM)

Both the surfaces and the cryo-fractured surfaces of the thermo-compressed films were analyzed using a JEOL JSM-6460 LV scanning electron microscope. Films were cryo-fractured by immersion in liquid nitrogen. For both analyses the films were mounted on bronze stubs and sputter-coated (Sputter coater SPI Module, Santa Clara, CA, USA) with a thin layer of gold for 35 s.

2.3.7. Uniaxial Tensile Tests

The mechanical properties of the films were determined from samples cut in a bone-shape with an effective area of ~28.6 mm long × ~5.5 mm width and cross-sectional area ~ 6.5 mm^2 (the exact value differed slightly between samples). Film samples (10 per formulation) were mounted and clamped with tensile grips (A/TG model) in an INSTRON 4467 machine to generate the force-distance curves following the ISO 527-2 (20) norm. The samples were stretched at a constant speed of 0.01 mm/s at 25 °C until they broke.

The overall stress acting on each sample during tension was expressed as the so-called true tensile strength (σ). This is the force normal to the film cross-section F(*N*) divided by the initial area A_0 (m^2) of the sample (21):

$$\sigma = {}^{F}/_{A_o} \tag{5}$$

The overall strain (ε) of each film was expressed as a percentage of the initial height as follows:

$$\varepsilon = {}^{L}/_{L_o} \times 100\% \tag{6}$$

where L_o and L (mm) are the initial and final longitudes of each film before and after deformation, respectively.

The mechanical properties at break: maximum stress (σ_m) and strain at break (ε_b) were obtained from the stress-distance curves. For this, the curves were transformed into stress-strain curves as outlined in ISO 527-2 (20). Young's modulus (*E*) was then determined from the linear regression slope of the stress-strain curves, and toughness (*T*) was calculated from the area under the stress-strain curves.

2.3.8. Antimicrobial Activity

Antimicrobial activity tests of the films were carried out using the agar diffusion method according to Gutiérrez and Alvarez (6). Films were cut into circular-shaped discs (diameter 12 mm) and hydrated with 1 mL of sterile water. They were then placed on Mueller Hinton agar plates (Merck, Darmstadt, Germany) previously seeded with 0.1 mL of inoculum containing approximately 10^5 - 10^6 CFU/mL each of the two pathogenic bacteria (one Gram-negative and the other Gram-positive) tested: *Escherichia coli* O157:H7 (32158, American Type Culture Collection) and *Staphylococcus aureus*, respectively, both of which were provided by the Centro de Referencia de Lactobacilos (CERELA, Tucumán, Argentina). The zone of inhibition assay on solid media was used for determining the antimicrobial effect of each film. Plates were incubated for 24 h at 37 °C and then examined to study the inhibitory effect. The total area was used to evaluate the antimicrobial potential of the films by exactly measuring the diameter of the inhibitory zones surrounding the wells as well as the contact area of these with the agar surface. The films were then evaluated according to Ponce et al. (22) who classified the susceptibility of microorganisms to a given inhibitory agent based on the diameters of inhibition halos as: not sensitive, sensitive, very sensitive and extremely sensitive. Experiments were done in triplicate on two separate experimental runs.

2.4. Statistical Analysis

The experimental data was analyzed using OriginPro 8 (Version 8.5, Northampton, USA) software. The results were expressed as means ± SD, and the one-way ANOVA procedure followed by Duncan's multiple range tests was adopted to determine significant differences ($p < 0.05$) between treatments.

3. RESULTS AND DISCUSSION

3.1. Attenuated Total Reflectance Fourier Transform Infrared Spectroscopy (*ATR/FTIR*)

Fig. 1A shows the ATR/FTIR spectra of the different films studied over the whole absorption range. A significant absorption peak at around 3309 cm^{-1} associated with the stretching of the OH groups belonging to the starch, glycerol and water and the stretching vibration of the C-O groups was observed in all the systems studied (1). Gutiérrez (24) suggested that materials with more free OH groups vibrate more easily and thus tend to present wider bands with higher absorbance. It can thus be inferred from Fig. 1B that the number of OH groups in each of the film formulations decreased in the following order: TPPF > TPPF/PCL(10,000) > TPPF/PCL(10,000)+CAT > TPPF/PCL(80,000) > TPPF/PCL(80,000)+CAT. Shi et al. (25) proposed another method to compare the number of available OH groups in these types of systems: the ratio between the band intensities at 3300 cm^{-1} (I_{3300}) and 1149 cm^{-1} (I_{1149}) (A_{3300}/A_{1149} ratio). This was 1.03, 0.54, 0.50, 0.32 and 0.27 for the TPPF, TPPF/PCL(10,000), TPPF/PCL(10,000)+CAT, TPPF/PCL(80,000) and TPPF/PCL(80,000)+CAT films, respectively. Both methods thus gave the same results. As expected, the films containing poly(ε-caprolactone) (PCL) had fewer OH groups than those without PCL. The fewer the number of OH groups the more hydrophobic the material, thus showing that the incorporation of this hydrophobic polymer reduced the hydrophilicity of the flour/PCL blends. The films with the high molecular weight (M_w) PCL contained the fewest free OH groups, particularly those with the added catalyst (zirconium octanoate - $Zr(Oct)_4$). In the latter case this was possibly due to the cross-linking of the starchy matrix and the PCL. Similar results have been reported by Gutiérrez and Alvarez (6) for films developed from corn starch/polystyrene (PS) blends using zinc octanoate ($Zn(Oct)_2$) under reactive extrusion (REx) conditions.

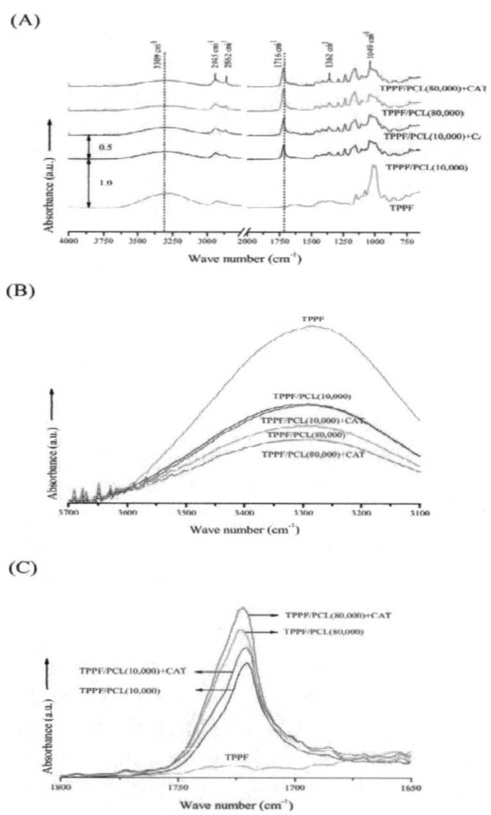

FIGURE 1. Panel A- FTIR Spectra of the Different Films Studied in All the Absorption Range. Panel B- FTIR Spectra in the Absorption Range Corresponding to C-O Group (OH Stretching) of the Different Films Studied. Panel C- FTIR Spectra in the Absorption Range Corresponding to C=O Group. Film Systems: plantain flour (TPPF), plantain flour + PCL (M_w = 10,000 g/mol) (TPPF/PCL(10,000)), plantain flour + PCL (M_w = 10,000 g/mol) + catalyst (TPPF/PCL(10,000)+CAT), plantain flour + PCL (M_w = 80,000 g/mol) (TPPF/PCL(80,000)) and plantain flour + PCL (M_w = 80,000 g/mol) + catalyst (TPPF/PCL(80,000)+CAT).

The band located at 1716 cm^{-1} (Fig. 1A) in the PCL-containing films represents the C=O groups of the esters in the PCL, thus confirming their presence. An increase in the intensity of this band can be seen in Fig. 1C for the films containing the catalyst (TPPF/PCL(10,000)+CAT and TPPF/PCL(80,000)+CAT) compared to the analogous films prepared without the catalyst (TPPF/PCL(10,000) and TPPF/PCL(80,000)). This could be related to the formation of new C=O groups in the esters formed between the plantain flour and the PCL during cross-linking. Interestingly, there is a more significant increase in the intensity of the band located at 1716 cm^{-1} in the TPPF/PCL(80,000)+CAT films than the TPPF/PCL(10,000)+CAT films, thus suggesting a higher degree of cross-linking between the plantain flour and the higher M_w PCL in the presence of the catalyst.

Other peaks located at 2862 and 2945 cm^{-1} are associated with C-H stretch vibrations (5) and are characteristic of polymeric matrix materials, and the bands observed between 1362 cm^{-1} and 1497 cm^{-1} are assigned to C-O angular deformations (6). Specifically, the band detected at ~1049 cm^{-1} in all the films developed is associated with -C-O-C-glycosidic bonds (2).

3.2. Water Solubility (WS)

Table i shows the water solubility values at 25 °C for the different films studied. As can be observed all of the films made from the plantain flour/PCL blends, with and without the catalyst, were significantly less water soluble ($p \leq 0.05$) than the films prepared from plantain flour alone (TPPF). These results were expected as the hydrophilic nature of a starchy matrix should decrease when blended with a hydrophobic polymer such as PCL. In addition, the incorporation of the high Mw PCL in the plantain flour/PCL blends resulted in a significant decrease in the water solubility ($p \leq 0.05$) of the TPPF/PCL(80,000) and TPPF/PCL(80,000)+CAT films compared to the films made with the low M_w PCL blends (TPPF/PCL(10,000) and TPPF/PCL(10,000)+CAT). According to the results obtained from the FTIR spectra, this is associated with the lower number of OH groups available in these blends. The water solubility of starch films thus provides an indication of their integrity in an aqueous medium, such that higher solubility values indicate a lower water resistance (7). To summarize: blends made with the high M_w PCL gave films with improved water resistance. These results agree with the FTIR spectra analyses.

TABLE i). Water Solubility (*WS*) and Crystallinity Percentages of the Films Studied.

Parameter	TPPF	TPPF/PCL(10,000)	TPPF/PCL(10,000)+CAT	TPPF/PCL(80,000)	TPPF/PCL(80,000)+CAT
WS (%)	30.8 ± 0.5^{c}	26.3 ± 0.3^{b}	27.1 ± 0.5^{b}	25.1 ± 0.4^{a}	25.6 ± 0.3^{a}
Crystallinity (%)	$< 1^{a}$	36 ± 1^{b}	49 ± 1^{c}	35 ± 1^{b}	62 ± 1^{d}

Equal letters in the same row indicate no statistically significant differences ($p \leq 0.05$). Film systems: plantain flour (TPPF), plantain flour + PCL (M_w = 10,000 g/mol) (TPPF/PCL(10,000)), plantain flour + PCL (M_w = 10,000 g/mol) + catalyst (TPPF/PCL(10,000)+CAT), plantain flour + PCL (M_w = 80,000 g/mol) (TPPF/PCL(80,000)) and plantain flour + PCL (M_w = 80,000 g/mol) + catalyst (TPPF/PCL(80,000)+CAT).

Likewise, the results obtained from the FTIR spectra suggest that the incorporation of the catalyst into the polymer blends (TPPF/PCL(10,000)+CAT and TPPF/PCL(80,000)+CAT) leads to a decrease in the number of free OH groups. However, the water solubility of the films with and without the catalyst, (TPPF/PCL(10,000)+CAT and TPPF/PCL(80,000)+CAT) *vs* (TPPF/PCL(10,000) and TPPF/PCL(80,000), respectively) were not statistically different ($p \geq 0.05$). We cannot thus confirm yet that the presence of the catalyst caused a reduction in the number of free OH groups in the films studied. Nevertheless, other studies have reported that the addition of a catalyst under REx conditions does decrease the water solubility of films made from starch/synthetic polymer blends (6).

3.3. Thermogravimetric Analysis (*TGA*)

The TGA curves for the systems under study are given in Fig. 2. This assay was conducted in order to analyze the thermal stability of the five types of film prepared. According to Pelissari et al. (2) and Gutiérrez and Álvarez (14) the thermal decomposition of glycerol-starch-containing films occurs in three main stages. The first stage of degradation, at around 100 °C, sees the onset of weight loss associated with moisture loss. During the second stage, between 160 and 290 °C, the glycerol-rich phase, which also contains starch, evaporates. Finally, in the third stage at temperatures above 330 °C, the degradation of the partially decomposed starch is completed (10).

The first fact worth noting from the TGA curves is the double weight loss shown by the TPPF film. This is related to an obvious phase separation caused by the low compatibility of the flour-glycerol blend. Similar results have been reported by Gutiérrez et al. (3) and Gutiérrez and Álvarez (14) for films also prepared from plantain flour, but obtained by casting. The phase separation observed in the TPPF film suggests that the plasticizer (glycerol) is free, and is thus more easily solubilized in water resulting in a more soluble system. This agrees with the results obtained from the FTIR spectra: the TPPF film had the highest A_{3300}/A_{1149} ratio indicating a higher number of available OH groups.

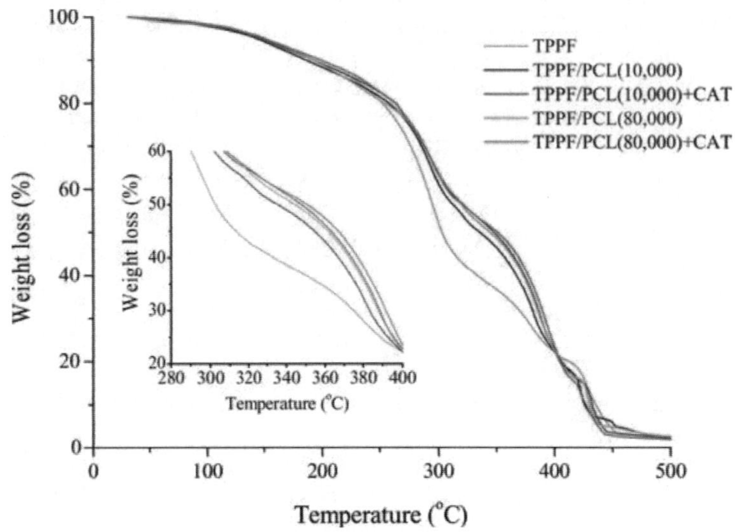

FIGURE 2. TGA Curves of the Films Studied: plantain flour (TPPF), plantain flour + PCL (M_w = 10,000 g/mol) (TPPF/PCL(10,000)), plantain flour + PCL (M_w = 10,000 g/mol) + catalyst (TPPF/PCL(10,000)+CAT), plantain flour + PCL (M_w = 80,000 g/mol) (TPPF/PCL(80,000)) and plantain flour + PCL (M_w = 80,000 g/mol) + catalyst (TPPF/PCL(80,000)+CAT).

On the other hand, only a low degree of phase separation was observed for the films made from the plantain flour/PCL blends, which was far less obvious than that of the TPPF films. These results were not expected as it is widely known that the polymers used, i.e. starch/PCL, are immiscible. Nonetheless, it came as a welcome surprise and may be because the lipids contained in the chemical composition of the plantain flour acted as compatibilizing agents between the starch contained in the flour and the PCL. This would also explain the decrease in the water solubility observed for the flour/PCL film systems.

Another important result to highlight is that the films made with the high Mw PCL (TPPF/PCL(80,000) and TPPF/PCL(80,000)+CAT blends) showed higher degradation temperatures than those made with the low M_w PCL blends (TPPF/PCL(10,000) and TPPF/PCL(10,000)+CAT) (Fig. 2). It is well known that polymer systems with higher M_w tend to have higher thermal resistance. This is in line with the results obtained. In addition, using a high M_w PCL in the presence of a catalyst under REx conditions increases the degree of the cross-linking reactions, demonstrated by a further increase in the thermal degradation temperature of these carbohydrate-based polymers. The results discussed so far thus provide evidence of the cross-linking of the polymeric matrices in the catalyst ($Zr(Oct)_4$)-containing films (TPPF/PCL(10,000)+CAT and TPPF/PCL(80,000)+CAT) processed under REx conditions. Similar results were reported by Gutiérrez and Álvarez (14) for plantain flour based films using aloe vera gel as a cross-linking agent.

A positive effect of the use of the catalyst under REx conditions to produce the flour/PLC blends was a slight tendency to avoid phase separation, possibly related to the cross-linking between the polymers used.

3.4. Differential Scanning Calorimetry (*DSC*)

Fig. 3 shows the DSC curves for the developed films. According to Labet and Thielemans (30) the melting temperature (T_m) of PCL is between 56 and 65 °C. The endothermic peaks of the film systems prepared from plantain flour/PCL blends were thus consistent with the T_m of the crystalline fraction of the PCL.

As a general rule any structural feature that reduces the mobility of the polymer chains or the free volume will increase the T_m (13). The increase in the Tm of the systems containing the high M_w PCL (TPPF/PCL(80,000) and TPPF/PCL(80,000)+CAT) can thus be attributed to the lower mobility and rearrangement of the macromolecules of this PCL. Gutiérrez and Alvarez (6) concluded that the increase they found in the T_m of film systems made from starch/PS was associated with cross-linking between the polymeric matrices when $Zn(Oct)_2$ was used as the catalyst under REx conditions. The slight increase in the T_m of the TPPF/PCL(80,000)+CAT films compared to the TPPF/PCL(80,000) films can thus be attributed to the cross-linking between the plantain flour and the PCL, triggered by the use of the catalyst. Thus the reduction in the number of available OH groups in these systems (see section 3.1) was brought about by the cross-linking reactions of starch chains that occur through these hydroxyl groups, which act as nucleophiles in the addition reaction on the double bond of the caprolactone monomer (6). Nevertheless, no significant differences were observed in the T_m of the TPPF/PCL(10,000)+CAT films compared to the TPPF/PCL(10,000) films.

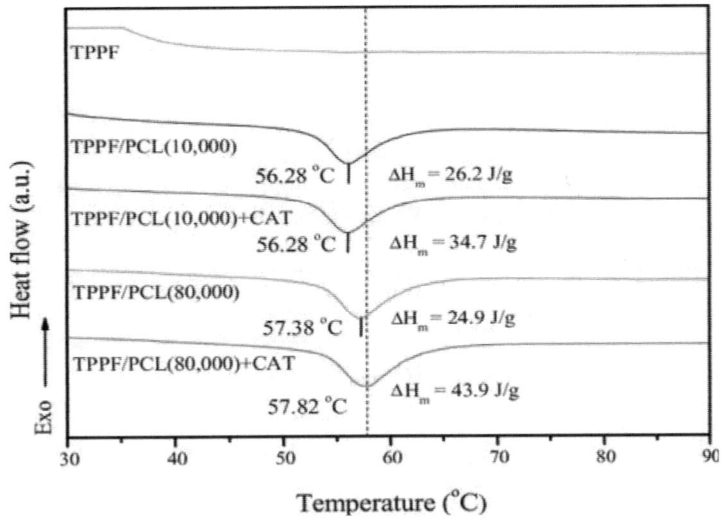

FIGURE 3. Heating Thermograms of the Films Studied: plantain flour (TPPF), plantain flour + PCL (M_w = 10,000 g/mol) (TPPF/PCL(10,000)), plantain flour + PCL (M_w = 10,000 g/mol) + catalyst (TPPF/PCL(10,000)+CAT), plantain flour + PCL (M_w = 80,000 g/mol) (TPPF/PCL(80,000)) and plantain flour + PCL (M_w = 80,000 g/mol) + catalyst (TPPF/PCL(80,000)+CAT).

In addition, higher melting enthalpy values (ΔH_m) values have been correlated with the strengthening of inter and intramolecular interactions between starch-starch chains and starch-synthetic polymer chains (6, 24, 32). According to Bertuzzi et al. (33) the cross-linking reactions between the polymers cause the starch molecules to align leading to an increase in the growth rate of the crystalline phase in these semi-crystalline biomaterials. Thus the higher ΔH_m values of the catalyst-containing films (TPPF/PCL(10,000)+CAT and TPPF/PCL(80,000)+CAT) compared to those of the films prepared without the catalyst (TPPF/PCL(10,000) and TPPF/PCL(80,000)) confirm that cross-linking has occurred between the plantain flour and the PCL. This agrees with the rest of the results discussed above.

The crystallinity percentages of all the systems were calculated from the ΔH_m values following the methodology proposed by Merino et al. (17) (Table i). The TPPF films were almost completely amorphous (< 1% crystallinity), whilst the films made from flour plantain/PCL blends were significantly more crystalline ($p \leq 0.05$). Obviously these results are due to the semi-crystalline nature of PCL as reported in the literature (4). It is worth noting that no statistically significant differences in percent crystallinity ($p \geq 0.05$) were observed between the TPPF/PCL(10,000) and TPPF/PCL(80,000) films. However, the films containing the catalyst (TPPF/PCL(10,000)+CAT and TPPF/PCL(80,000)+CAT) were significantly more crystalline ($p \leq 0.05$) than the films prepared without it (TPPF/PCL(10,000) and TPPF/PCL(80,000)). This confirms that it was the action of the catalyst that generated the cross-linking reactions between the polymer matrices under REx conditions. As mentioned previously, the cross-linking reactions caused the polymers to align in parallel leading to a rise in the crystallinity of these systems (18). This had a positive effect on the films, since cross-linking inhibits the retrogradation of these types of materials resulting in improved mechanical properties (19). An even more significant increase ($p \leq 0.05$) in the percent crystallinity was obtained by incorporating a higher M_w PCL in the presence of the catalyst (TPPF/PCL(80,000)+CAT film) suggesting that increasing the M_w of the PCL generated a higher rate of reactivity in the catalyst compared to the TPPF/PCL(10,000)+CAT film. This may be due to an increase in the number of active sites on the PCL leading to an increase in the reaction (cross-linking) rate. Thus it seems that the use of higher M_w PCL can result in a higher degree of cross-linking.

3.5. X-ray Diffraction (*XRD*)

Fig. 4A shows the X-ray diffractograms and percent crystallinities of the films studied. It can be observed that the results obtained by XRD agree with those of the DSC. Thus, in general, both the methodologies used to determine percent crystallinity (XRD and Merino et al. (17) – DSC - Hermans and Weidinger (19)) gave the same results. Specifically, the TPPF film shows a spectrum typical of an amorphous material, in line with the crystallinity percentage values calculated.

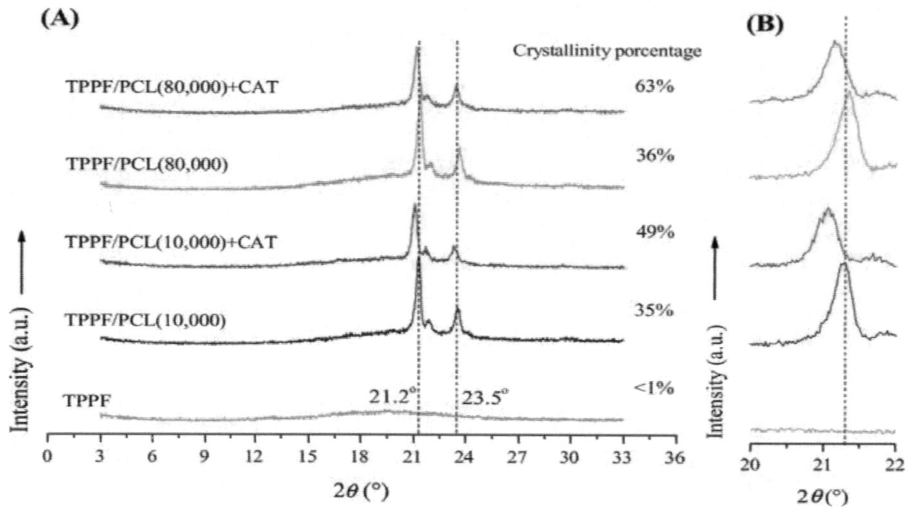

FIGURE 4. Panel A- X-ray Diffraction Diffractograms in the Range of 3-33° (2θ) and Crystallinity Percentages of the Film Systems Studied. Panel B- X-ray Diffraction Diffractograms in the Range of 20-22° (2θ) of the Film Systems. Film Systems: plantain flour (TPPF), plantain flour + PCL (M_w = 10,000 g/mol) (TPPF/PCL(10,000)), plantain flour + PCL (M_w = 10,000 g/mol) + catalyst (TPPF/PCL(10,000)+CAT), plantain flour + PCL (M_w = 80,000 g/mol) (TPPF/PCL(80,000)) and plantain flour + PCL (M_w = 80,000 g/mol) + catalyst (TPPF/PCL(80,000)+CAT).

The films made with the PCL showed the characteristic peaks of the crystallinity of this polymer at 2θ = 21.2°, 21.8° and 23.5°, corresponding to the *d*-spacings ≅ 4.2 Å, 4.1 Å, and 3.8 Å, respectively (21). Similar results were reported by Ortega-Toro et al. (37) for starch-based films with different starch-PCL ratios.

On the other hand, it is evident from Fig. 4B that the peak at 2θ = 21.2° exhibits a slight displacement towards lower 2θ values in the films containing the catalyst (TPPF/PCL(10,000)+CAT and TPPF/PCL(80,000)+CAT) indicating an increase in the molecular spacing of the polymer chains.

3.6. Scanning Electron Microscopy (*SEM*)

Fig. 5 shows the SEM images of the thermo-compressed and cryo-fractured surfaces of the different films. The thermo-compressed surfaces were very smooth and flat except for the films containing the catalyst (TPPF/PCL(10,000)+CAT and TPPF/PCL(80,000)+CAT) which had some cracks.

The cryo-fractured surfaces of the films derived from the plantain flour/PCL blends were very heterogeneous, demonstrating the immiscibility of the polymers (plantain flour and PCL). This is consistent with the TGA results. Similar structures were reported by Ortega-Toro et al. (37) and Gutiérrez and Alvarez (6) for films prepared from starch/PCL blends in different proportions, and for films made from sagu starch and flour/PCL blends, respectively.

Finally, none of the surfaces of the films examined showed structures similar to starch granules. We can thus assume that the processing conditions selected produced the complete gelatinization of the starch.

3.7. Uniaxial Tensile Tests

The stress-strain curves of the developed films are shown in Fig. 6A. The mechanical behavior of the films made from the plantain flour and low M_w PCL blends (TPPF/PCL(10,000) and TPPF/PCL(10,000)+CAT) could not be assessed as they were too fragile (see Fig. 6B and C). The results outlined above are also consistent with the TGA results which indicated a higher thermal resistance in films made from plantain flour/high M_w PCL blends (TPPF/PCL(80,000) and TPPF/PCL(80,000)+CAT). It thus seems that the improvements observed in the mechanical properties of the high M_w PCL systems are related to the greater molecular interactions that occur within them. Cheung et al. (38) reported similar results for high M_w polyethylene glycol (PEG) films.

With regard to the other films, a small linear elastic zone followed by a non-linear zone was observed until break point in the TPPF film, and in the films containing the high M_w PCL (TPPF/PCL(80,000) and TPPF/PCL(80,000)+CAT) a creep zone was observed. The presence of a creep zone indicates a change in mechanical behavior from an elastic to a plastic zone, and was irreversible from 0.59% elongation in the TPPF/PCL(80,000) film and 0.54% elongation in the TPPF/PCL(80,000)+CAT film. In other words, theoretically, after elastic deformation these materials undergo a volume change.

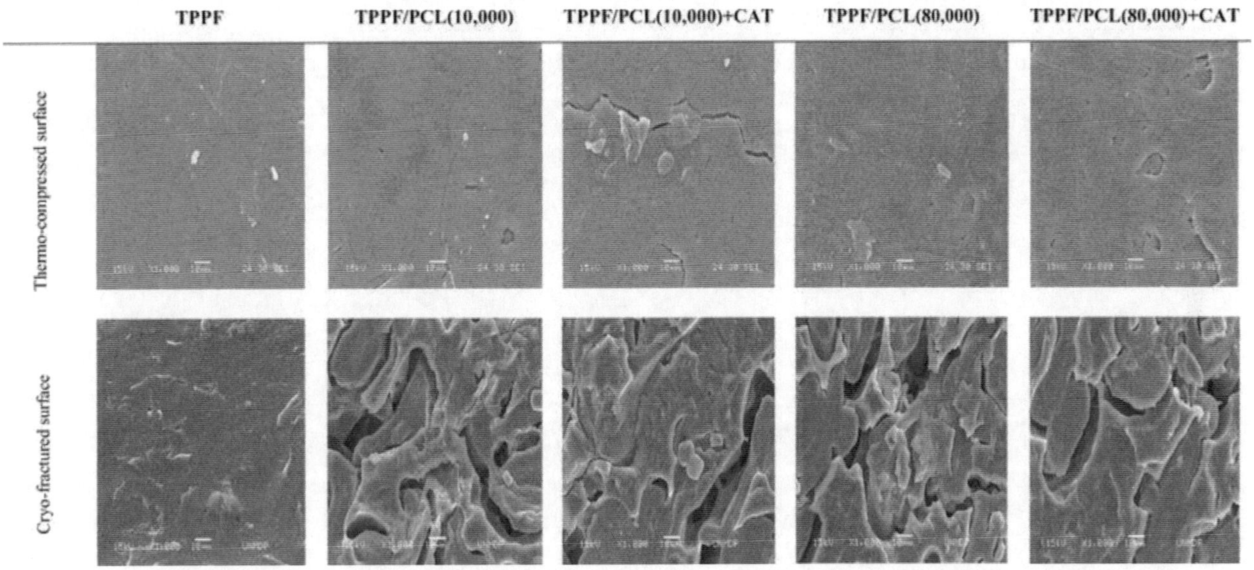

FIGURE 5. SEM Micrographs of the Thermo-Compressed and the Cryo-Fractured Surfaces of the Films Studied: plantain flour (TPPF), plantain flour + PCL (M_w = 10,000 g/mol) (TPPF/PCL(10,000)), plantain flour + PCL (M_w = 10,000 g/mol) + catalyst (TPPF/PCL(10,000)+CAT), plantain flour + PCL (M_w = 80,000 g/mol) (TPPF/PCL(80,000)) and plantain flour + PCL (M_w = 80,000 g/mol) + catalyst (TPPF/PCL(80,000)+CAT).

The TPPF/PCL(80,000)+CAT film had a smaller creep zone than the TPPF/PCL(80,000) film, a result similar to that reported by Gutiérrez and Alvarez (6) for films made from starch/PS blends using $Zn(Oct)_2$ as a catalyst under REx conditions. It is well known that after the creep zone, Hooke's Law for elastic, linear and isotropic materials is not met. This means that normal deformation does not occur anywhere in the solid, nor along any of the normal stresses in the orthogonal directions, i.e. the principal directions of the stress matrix do not coincide with the principal direction of the matrix of deformation (6).

In addition, the positive slope of the curve in the plastic deformation range for the films containing the catalyst (TPPF/PCL(80,000)+CAT) implies that hardening occurred until the sample finally broke. This correlates with the assertion that the polymer chains in these films were aligned parallel to the load (4) and fits well with the crystallinity results obtained by both DSC and XRD for this system (TPPF/PCL(80,000)+CAT) as compared to the analogous films prepared without the use of the catalyst (TPPF/PCL(80,000)). Once again, the results obtained confirm the cross-linking of the plantain flour with the PCL under REx conditions using $Zr(Oct)_4$ as a catalyst.

It is important to note, however, that the mechanical properties observed for the TPPF/PCL(80,000) and TPPF/PCL(80,000)+CAT films would signify the failure of these materials in a kinematic system. They are thus not appropriate for food packaging applications where the duration of the stress, the applied load and temperature are important considerations.

Generally the Young's modulus and maximum stress values of films made from 50/50 flour/PCL blends significantly increase ($p \leq 0.05$) compared to films that do not contain PCL independently of whether they are processed by blending or extrusion (Table ii). The incorporation of PCL in itself thus seems to have a positive effect on the mechanical properties of film systems. However, higher values of these properties were obtained here (plantain flour/PCL blends processed by Rex) than in a previous study conducted by our research group using sagu flour/PCL blends processed by blending (4).

TABLE ii). Parameters of the uniaxial tensile strength tests: Young's modulus (E), maximum stress (σ_m), strain at break (ε_b) and toughness (T).

Material	E (MPa)	σ_m (MPa)	ε_b (%)	T (x10^3) (J/m^3)
TPPF	0.78 ± 0.06^a	1.1 ± 0.1^a	4 ± 1^a	3 ± 1^a
TPPF/PCL(10,000)	N.d.	N.d.	N.d.	N.d.
TPPF/PCL(10,000)+CAT	N.d.	N.d.	N.d.	N.d.
TPPF/PCL(80,000)	1.35 ± 0.04^b	2.1 ± 0.9^b	2.9 ± 0.6^a	4 ± 2^a
TPPF/PCL(80,000)+CAT	1.38 ± 0.02^b	3.4 ± 0.1^c	5.4 ± 0.3^b	12 ± 1^b

Equal letters in the same column indicate no statistically significant difference ($p \leq 0.05$). Film systems: plantain flour (TPPF), plantain flour + PCL (M_w = 10,000 g/mol) (TPPF/PCL(10,000)), plantain flour + PCL (M_w = 10,000 g/mol) + catalyst (TPPF/PCL(10,000)+CAT), plantain flour + PCL (M_w = 80,000 g/mol) (TPPF/PCL(80,000)) and plantain flour + PCL (M_w = 80,000 g/mol) + catalyst (TPPF/PCL(80,000)+CAT). Not determined: N.d.

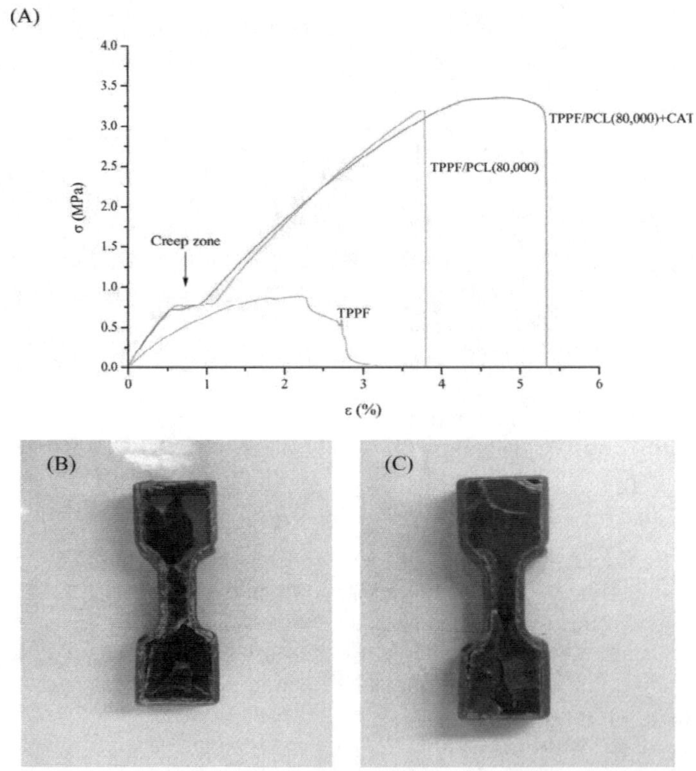

FIGURE 6. Panel A- Stress (σ)-Strain (ε) Curves of the Films Studied: plantain flour (TPPF), plantain flour + PCL (M_w = 80,000 g/mol) (TPPF/PCL(80,000)) and plantain flour + PCL (M_w = 80,000 g/mol) + catalyst (TPPF/PCL(80,000)+CAT). Panel B- Appearance of the Film Prepared from Plantain Flour + PCL (M_w = 10,000 g/mol) (TPPF/PCL(10,000)). Panel C- Appearance of the Film Prepared from Plantain Flour + PCL (M_w = 10,000 g/mol) + Catalyst (TPPF/PCL(10,000)+CAT).

The films that contained the catalyst (TPPF/PCL(80,000)+CAT) also showed improved mechanical properties compared to the films that did not (TPPF/PCL(80,000)) ($p \leq 0.05$) (Table ii). This fits well with the proposed cross-linking of the catalyst-containing blends. Similar behavior was observed by Gutiérrez and González (40) for films derived from plantain flour cross-linked with *aloe vera* gel obtained by casting.

Finally, the highest toughness value was obtained for the TPPF/PCL(80,000)+CAT film. According to Gutiérrez and Alvarez (4) this could be very useful, as these films could absorb more energy thus minimizing damage caused by knocks to foods during transport and storage. Nevertheless, all the toughness values obtained here are at least twice as low as those reported for films made from corn starch/PS blends using $Zn(Oct)_2$ under REx conditions (6).

3.8. Antimicrobial Activity

The antimicrobial activity of all the film systems was evaluated in order to establish whether they had potential as active packaging materials. As expected, the films prepared from plantain flour/PCL blends (TPPF/PCL(10,000) and TPPF/PCL(80,000)) showed no antimicrobial activity. In contrast, the films containing Zr(Oct)$_4$ (TPPF/PCL(80,000)+CAT and TPPF/PCL(10,000)+CAT) inhibited the growth of the two pathogenic microorganisms evaluated (*Escherichia coli* O157:H7 and *Staphylococcus aureus*) (see Fig. 7), which is promising for the development of eco-friendly and active materials by REx. The dual effect of Zr(Oct)$_4$ (as a catalyst and an antimicrobial agent) is particularly interesting. This means that after generating cross-linking reactions between the polymers it remains within the polymer structure where it acts as an antimicrobial additive, making it an attractive alternative in the development of active food packaging. It is worth noting that the antimicrobial effect exhibited by Zr(Oct)$_4$ was effective against both the Gram-negative and the Gram-positive microorganisms evaluated indicating a broad antimicrobial spectrum. Both films containing the catalyst (TPPF/PCL(10,000)+CAT and TPPF/PCL(80,000)+CAT) showed the same degree of antimicrobial activity, with the susceptibility of the two microorganisms classified as "sensitive" according to the criteria used by Ponce et al. (22). Conflicting results, however, were obtained by our research group using very low amount of Zn(Oct)$_2$ (approx. 1.90%) in starch/PS blends under REx conditions (6). This could be due to three factors: 1) the catalyst content used in this study was 1.73 times greater (approx. 3.3%) than that employed in the study previously carried out using Zn(Oct)$_2$ (6), 2) the concentration of the cation of the catalyst used here (Zr(Oct)$_4$) was at least twice that of the Zn(Oct)$_2$. It is well known that the anti-microbial effect depends on the concentration and type of cation of the organic salts typically used as antimicrobials, since this affects the oxidation-reduction potential for microbiological development (8), and 3) The matrices obtained in this study permitted a greater mobility of the catalyst within the polymer structure, thereby enabling its migration into the environment where it could act against the microorganisms. As regards this last point, it is worth remembering that the molecular spacing of the polymer chains was greater in the catalyst-containing films than in the films without the catalyst (see section 3.5). This could explain the lower steric hindrance which permitted the catalyst to migrate towards the medium.

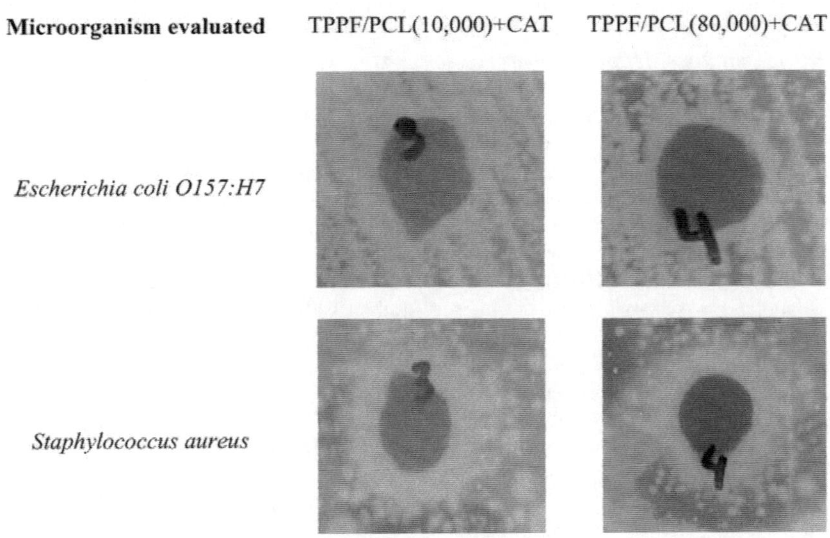

FIGURE 7. Antimicrobial Activity of the Films Studied: plantain flour + PCL (M_w = 10,000 g/mol) + catalyst (TPPF/PCL(10,000)+CAT) and plantain flour + PCL (M_w = 80,000 g/mol) + catalyst (TPPF/PCL(80,000)+CAT).

4. CONCLUSIONS

Films made from plantain flour/poly(ε-caprolactone) (PCL) blends were obtained under reactive extrusion (REx) conditions using zirconium octanoate ($Zr(Oct)_4$) as a catalyst. Films containing the catalyst showed cross-linking between the plantain flour and the PCL. The catalyst-containing films also showed antimicrobial activity against both the Gram-positive and Gram-negative pathogenic microorganisms studied (*E. coli* O157:H7 and *S. aureus*) with both microorganisms being "sensitive" to their presence. This demonstrates that active eco-friendly films with improved properties can be obtained under REx conditions. The addition of PCL to the plantain flour gave films with improved properties compared to the films without this polymer. The use of high molecular weight (M_w) PCL in the blends reduced water sensitivity, whilst improving the thermal resistance and mechanical properties of the film systems.

ACKNOWLEDGMENTS

The authors would like to thank the Consejo Nacional de Investigaciones Científicas y Técnicas (CONICET) (Postdoctoral fellowship internal PDTS-Resolution 2417), Universidad Nacional de Mar del Plata (UNMdP) for financial support, and Dr. Mirian Carmona-Rodríguez.

REFERENCES

1. T. J. Gutiérrez, R. Guzmán, C. Medina Jaramillo, and L. Famá, Effect of beet flour on films made from biological macromolecules: Native and modified plantain flour. *Int. J. Biol. Macromol.* **82**, 395–403. (2016).
2. F. M. Pelissari, M. M. Andrade-Mahecha, P. J. do A Sobral, and F. C. Menegalli, Comparative study on the properties of flour and starch films of plantain bananas (*Musa paradisiaca*). *Food Hydrocoll.* **30**(2), 681–690. (2013).
3. T. J. Gutiérrez, J. Suniaga, A. Monsalve, and N. L. García, Influence of beet flour on the relationship surface-properties of edible and intelligent films made from native and modified plantain flour. *Food Hydrocoll.* **54**, 234–244. (2016).
4. T. J. Gutiérrez, and V. A. Alvarez, Films made by blending poly(ε-caprolactone) with starch and flour from sagu rhizome grown at the Venezuelan amazons. *J. Polym. Environ.* **25**(3), 701–716. (2017).
5. T. J. Gutiérrez, M. P. Guarás, and V. A. Alvarez, "Reactive extrusion for the production of starch-based biopackaging", in: Biopackaging, edited by M. A. Masuelli, Miami: CRC Press, Taylor & Francis Group. (2017).
6. T. J. Gutiérrez, and V. A. Alvarez, Properties of native and oxidized corn starch/polystyrene blends under conditions of reactive extrusion using zinc octanoate as a catalyst. *React. Funct. Polym.* **112**, 33–44. (2017).
7. A. Kowalski, A. Duda, and S .Penczek, Kinetics and mechanism of cyclic esters polymerization initiated with tin(II) octoate, 1. Polymerization of ε-caprolactone. *Macromol. Rapid. Commun.* **19**(11), 567–572. (1998).
8. FDA. Indirect additives used in food contact substances [Internet]. Code of Federal Regulations Title 21 175.105 and 175.300. 2016. Available from: https://www.accessdata.fda.gov/scripts/fdcc/index.cfm?set=IndirectAdditives&id=STANNOUSETHYLHEXANOATE
9. J.- M. Raquez, R. Narayan, and P. Dubois, Recent advances in reactive extrusion processing of biodegradable polymer-based compositions. *Macromol. Mater. Eng.* **293**(6), 447–470. (2008).
10. S. Penczek, A. Duda, A. Kowalski, J. Libiszowski, K. Majerska, and T. Biela, On the mechanism of polymerization of cyclic esters induced by tin(II) octoate. *Macromol. Symp.* **157**(1), 61–70. (2000).
11. T. J. Gutiérrez, and V. A. Alvarez, Cellulosic materials as natural fillers in starch-containing matrix-based films: a review. *Polym. Bull.* **74**(6), 2401–2430. (2017).
12. E. Pacheco, Evaluación nutricional de sopas deshidratadas a base de harina de plátano verde. Digestibilidad in vitro de almidón. *Acta Científica Venez.* **52**(4), 278–282. (2001).
13. H. V. Loesecke, Bananas. New York Intersci. 189. (1950).
14. T. J. Gutiérrez, and K. Álvarez, Physico-chemical properties and in vitro digestibility of edible films made from plantain flour with added Aloe vera gel. *J. Funct. Foods* **26**, 750–762. (2016).
15. T. J. Gutiérrez, N. J. Morales, E. Pérez, M. S. Tapia, and L. Famá, Physico-chemical properties of edible films derived from native and phosphated cush-cush yam and cassava starches. *Food Packag. Shelf Life* **3**, 1–8. (2015).
16. C. G. Biliaderis, A. Lazaridou, and I. Arvanitoyannis, Glass transition and physical properties of polyol-plasticised pullulan–starch blends at low moisture. *Carbohydr. Polym.* **40**(1), 29–47. (1999).
17. D. Merino, L. N. Ludueña, and V. A. Alvarez, Dissimilar tendencies of innovative green clay organo-modifier on the final properties of poly(ε-caprolactone) based nanocomposites. *J. Polym. Environ.* 1–12. (2017).
18. H. Tsuji, and T. Ishizaka, Porous biodegradable polyesters, 3. preparation of porous poly(ε-caprolactone) films from blends by selective enzymatic removal of poly(L-lactide). *Macromol. Biosci.* **1**(2), 59–65. (2001).
19. P. H. Hermans, and A. Weidinger, On the determination of the crystalline fraction of polyethylenes from X-ray diffraction. *Macromol. Chem. Phys.* **44**(1), 24–36. (1961).

20. ISO 527-2. Determination of tensile properties of plastics. 2012. Available from: https://www.iso.org/obp/ui/#iso:std:56046:en
21. D. D. Hamann, J. Zhang, C. R. Daubert, E. A. Foegeding, and K. C. Diehl, Analysis of compression, tension and torsion for testing food gel fracture properties. *J. Texture Stud.* **37**(6), 620–639. (2006).
22. A. G. Ponce, R. Fritz, C. del Valle, and S. I. Roura. Antimicrobial activity of essential oils on the native microflora of organic Swiss chard. *LWT - Food Sci. Technol.* **36**(7), 679–684. (2003).
23. V. A. Pereira, I. N. Q. de Arruda, and R. Stefani, Active chitosan/PVA films with anthocyanins from Brassica oleraceae (Red Cabbage) as time–temperature indicators for application in intelligent food packaging. *Food Hydrocoll.* **43**, 180–8. (2015).
24. T. J. Gutiérrez, Effects of exposure to pulsed light on molecular aspects of edible films made from cassava and taro starch. *Innov. Food Sci. Emerg. Technol.* **41**, 387–396. (2017).
25. R. Shi, Z. Zhang, Q. Liu, Y. Han, L. Zhang, D. Chen, et al., Characterization of citric acid/glycerol co-plasticized thermoplastic starch prepared by melt blending. *Carbohydr. Polym.* **69**(4), 748–55. (2007).
26. S. Mathew, M. Brahmakumar, and T. E. Abraham, Microstructural imaging and characterization of the mechanical, chemical, thermal, and swelling properties of starch–chitosan blend films. *Biopolymers* **82**(2), 176–187. (2006).
27. L. C. Batista Reis, C. Oliveira de Souza, J. B. Alves da Silva, A. C. Martins, I. Larroza Nunes, and J. I. Druzian. Active biocomposites of cassava starch: The effect of yerba mate extract and mango pulp as antioxidant additives on the properties and the stability of a packaged product. *Food Bioprod. Process.* **94**, 382–91. (2015).
28. C. A. Romero-Bastida, L. A. Bello-Pérez, M. A. García, M. N. Martino, J. Solorza-Feria, and N. E. Zaritzky. Physicochemical and microstructural characterization of films prepared by thermal and cold gelatinization from non-conventional sources of starches. *Carbohydr. Polym.* **60**(2), 235–244. (2005).
29. M. L. Sanyang, S. M. Sapuan, M. Jawaid, M. R. Ishak, and J. Sahari, Effect of plasticizer type and concentration on tensile, thermal and barrier properties of biodegradable films based on sugar palm (*Arenga pinnata*) starch. *Polymers* (Basel). **7**(6), 1106–1124. (2015).
30. M. Labet, and W. Thielemans, Synthesis of polycaprolactone: a review. *Chem. Soc. Rev.* **38**(12), 3484–3504. (2009).
31. M. Mitrus, Glass transition temperature of thermoplastic starches. *Int. Agrophysics.* **19**(3), 237–41. (2005).
32. S. Mali, M. V. E. Grossmann, M. A. Garcia, M. N. Martino, and N. E.Zaritzky, Microstructural characterization of yam starch films. *Carbohydr. Polym.* **50**(4), 379–386. (2002).
33. M. A. Bertuzzi, E. F. Castro Vidaurre, M. Armada, and J. C. Gottifredi, Water vapor permeability of edible starch based films. *J. Food Eng.* **80**(3), 972–978. (2007).
34. T. J. Gutiérrez, M. S. Tapia, E. Pérez, and L. Famá, Edible films based on native and phosphated 80:20 waxy:normal corn starch. *Starch/Stärke* **67**(1–2), 90–97. (2015).
35. Y. V. García-Tejeda, C. López-González, J. P. Pérez-Orozco, R. Rendón-Villalobos, A. Jiménez-Pérez, E. Flores-Huicochea, et al., Physicochemical and mechanical properties of extruded laminates from native and oxidized banana starch during storage. *LWT - Food Sci. Technol.* **54**(2), 447–455. (2013).
36. M. D. Ninago, O. V. López, M. M. S. Lencina, M. A. García, N. A. Andreucetti, A. E. Ciolino, et al. Enhancement of thermoplastic starch final properties by blending with poly(ε-caprolactone). *Carbohydr. Polym.* **134**, 205–212. (2015).
37. R. Ortega-Toro, J. Contreras, P. Talens, and A. Chiralt, Physical and structural properties and thermal behaviour of starch-poly(ε-caprolactone) blend films for food packaging. *Food Packag. Shelf Life* **5**, 10–20. (2015).
38. H.- Y. Cheung, Lau K.- T., T.- P. Lu, and D. Hui, A critical review on polymer-based bio-engineered materials for scaffold development. *Compos. Part B Eng.* **38**(3), 291–300. (2007).
39. M. Berruezo, L. N. Ludueña, E. Rodriguez, and V. A. Alvarez, Preparation and characterization of polystyrene/starch blends for packaging applications. *J. Plast. Film Sheeting.* **30**(2), 141–161. (2014).
40. T. J. Gutiérrez, and G. González, Effect of cross-linking with Aloe vera gel on surface and physicochemical properties of edible films made from plantain flour. *Food Biophys.* **12**(1), 11–22. (2017).
41. J. M. Jay. "Intrinsic and extrinsic parameters of foods that affect microbial growth". in: Modern Food Microbiology, edited by J. M. Jay. Fifth. Boston, MA: Springer US. p. 38–66. (1995).

CHEMICAL MODIFICATION OF POLY(LACTIC ACID) WITH VINYL MONOMERS BY REACTIVE PROCESSING

Rigolin Talita Rocha [1], Costa Lidiane Cristina [2], Bettini Silvia Helena Prado [1]*

[1] Universidade Federal de São Carlos - UFSCar - São Paulo - Brazil, [2] Universidade Federal de São Carlos - UFSCar - São Paulo - Brazil

Modification of poly(lactic acid) (PLA) with different monomers, maleic anhydride (MA) and glycidyl methacrylate (GMA), in the presence of peroxide, was conducted in a Haake torque rheometer according to a 22 factorial design in which the monomer and peroxide concentrations were varied. The modified poly(lactic acid) was characterized by size-exclusion chromatography, titration of acid groups and differential scanning calorimetry. Results showed that the PLA modification with maleic anhydride was seen to be accompanied by increase in acidity percentage and reduction in molecular weight, whereas the glycidyl methacrylate led to an opposite result. Torque curves demonstrated the same behavior, with increase and reduction of torque on samples with GMA and MA, respectively.

Nano-engineered Polymer Surfaces with Enhanced Mechanical and Thermal Properties Obtained by Nanoimprint Method

Paritat Muanchan[a], Takashi Kyotani[b] and Hiroshi Ito[a]

[a] *Research Center for GREEN Materials and Advanced Processing (GMAP) & Graduate School of Organic Materials Science, Yamagata University, Japan*
[b] *Institute of Multidisciplinary Research for Advanced Materials, Tohoku University, Japan*

Abstract. In this study, Vertically Aligned Composite Nanostructures (VACNs) of Polystyrene–Graphene Nanoplatelets (PS–GNPs) (1.0–5.0 wt/wt%) with 100 nm–diameter were precisely replicated by thermal nanoimprint using Anodic Aluminum Oxide (AAO) template. Implications of nanofiller orientation on nanostructured properties including physical, chemical, thermal, and mechanical properties were investigated. Imprinted nanostructures with the length of 10–70 μm were fabricated, resulting the enhancement of surface wetting ability in the water contact angle from 87±3° of the flat film increases to 132±2°. The increase in surface friction coefficient, surface durability, surface modulus and hardness were respectively obtained. The glass transition temperature (T_g) of VACNs was shifted toward about ~1.0 to ~4.0 °C as compared with their bulk composites because of the immobilization the polymer chain owing to nanoconfinement and surface interfacial interaction effects. Interestingly, it was found that thermal conductivity of VACNs became higher than their composite films due to the one-dimensional property in the parallel orientation of GNPs nanofilled can be controlled.

Keywords: Composite Nanostructures; Surface and Interfaces; Nanoimprint Lithography.
PACS: 81.16. –c

INTRODUCTION

Nano–engineered surfaces inspired by nature of gecko feet and cicada wings were widely mimicked to replicate the artificial nanostructures on various polymer films [1]. These natural surfaces often provide multifunctional properties and advanced material surfaces such as adhesive, hydrophobic surfaces, bio–medical applications, nanosensing materials, scaffold materials and energy storages [2]. In order to precise fabricate these nanostructures, templating technique using Anodic Aluminum Oxide (AAO) combined with melt wetting and solvent wetting methods are extensively proposed. Especially, most of these methods are mainly focused on supramolecular chemistry and polymerization for materials development due to the interfacial interaction effect under nanoconfinement conditions within AAO template. However, it still doesn't applicable for the mass production and environmental friendly because of the long running process time and the harsh chemical used of these methods. Therefore, many types of research are focusing on the rapid fabrication processes which have a good benefit for industrial scale.

Since the last decade, nanoimprint lithography has been presented as the high class and high–resolution technology for the fabrication of micro–/nano– structures on material surfaces [3]. It is very successful in both of academic area and industrial area to utilize nanoimprint process combined with AAO template in order to fabricate the ultra–fine nanostructures of plastic materials. This technique combines nanotextured design, reproducibility, high–precision with material versatility, inexpensive, large area and high productivity. The variety of plastic materials and with many sacrificial processes have been explored in this field. However, until now, the low potential of practical uses of these nanostructures due the poor in surface mechanical stability and surface durability still has not received much attention and studies. To solve these problems, polymer nanocomposites based on carbon materials especially for Carbon Nanotubes (CNTs) and Graphene Nanoplatelets (GNPs) have much a great interested. It is well known that the polymer nanocomposites with excellent properties of thermal properties, electrical properties, mechanical properties, and chemical resistances were successfully prepared by using these carbon nanofilled materials.

In this present work, Vertically Aligned Composite Nanostructures (VACNs) were precisely produced by thermal nanoimprint process using AAO template. The model material of PS–GNPs was used. Surface properties such as physical, chemical, thermal and mechanical properties were experimentally evaluated including wetting ability, friction coefficient and durability, surface modulus and hardness. The effect of GNPs nanofilled content and its orientation on the change in glass transition temperature (T_g) and thermal conductivity were respectively reported.

EXPERIMENTAL

Materials

Commercial grade PS (T_g = 87 °C, GPPS 679; PS Japan Corp.) and GNPs (6–8 nm–thickness, G0499, Tokyo Chemical Industry Co., Ltd.) were used in this study. For the preparation of polymer nanocomposites, in order to disperse the GNPs throughout the PS matrix, a solution mixing method was employed using Dimethyl Formamide (DMF). In this study, PS–GNPs 1.0–5.0 wt/wt% were prepared. The polymer and polymer nanocomposites film were prepared with thethickness of 500 µm using hot press with a vacuum system (IMC–11FA; Imoto Machinery Co. Ltd.).

Fabrication of Nanostructures by Thermal Nanoimprint

1D polymer nanostructure arrays were fabricated using thermal nanoimprinting machine (Izumi Tech., Japan) with AAO templates, the experimental procedure is portrayed in Figure 1. The polymer films were placed onto AAO with pore diameter of 100 nm and insert into the thermal nanoimprint machine. The imprinting conditions with the temperature of 120–180 °C, pressure of 5.0 MPa, and compressed time of 30 min were performed. After imprinting, AAO was removed by dissolution in 4.0 M NaOH solution for 24 h, followed by water dissolving at 60 °C and drying process respectively.

Characterization

The surface morphological of AAO templates and polymer nanostructures were observed using scanning electron microscopy (SEM, JSM–6510; JEOL/TM–1000; Hitachi High–Technologies Corp.) and field emission scanning electron microscopy (FE–SEM, SU–8000; Hitachi High–Technologies Corp.).

A commercial contact angle meter (DM 500; Kyowa Interface Science Co. Ltd., Japan) was used to measure the surface interaction between molten polymers/AAO surfaces and water droplets/polymer surfaces.

The surface chemical property of polymer and polymer nanocomposite films were analyzed by attenuated total reflection (ATR, Mapping Mode), IR Affinity–1s & Automatic Infrared Microscope AIM–8800 (Shimadzu). The contents of GNPs within imprinted nanostructures were evaluated by thermal gravimetric analysis (TGA, TGA Q50; TA Instruments Inc.).

The surface mechanical at macroscale was evaluated by friction and wear tester (EFM–III–F; Orientec). The friction test conditions using applied force 20–40 N with rotational speed 10 rpm for 30 s were evaluated in this study. All of the tests were performed at the ambient condition at 25 °C and relative humidity of 35 %. The surface mechanical at nanoscale was also evaluated by Agilent nanoindenter G200 (Agilent Technologies Inc.).

The T_g of polymers/polymer nanocomposites and their nanostructures were analyzed using modulated differential scanning calorimetry (M–DSC, DSC Q200; TA Instruments Inc.). Thermal conductivity (λ) was simply calculated by equation (1) as following

$$\lambda = \alpha\rho C_P \tag{1}$$

where α is thermal diffusivity (ai-Phase Mobile 1; ai-Phase Co., Ltd), ρ is density (AD–1653; A&D Co., Ltd), and is the specific heat capacity (from DSC analysis).

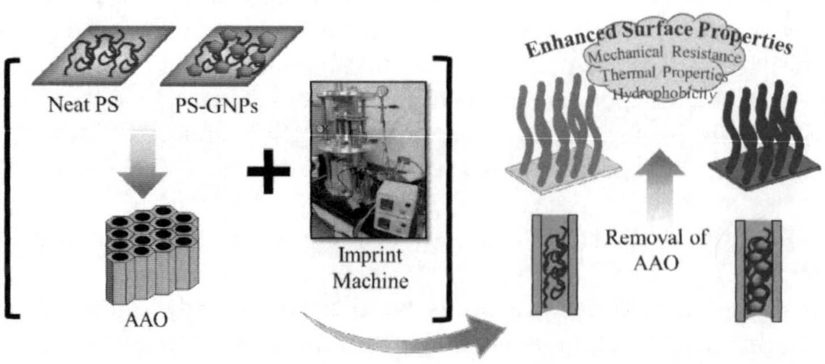

FIGURE 1. Overview of the fabrication of nanostructures by thermal nanoimprint using AAO template.

RESULTS AND DISCUSSION

Fabrication of Nanostructures by Thermal Nanoimprint

SEM images (top view) of VACN of PS–GNPs 5.0 wt/wt% obtained by thermal nanoimprint using AAO template are shown in Figure 2 (right). For the each of SEM images, nanostructure arrays of PS–GNPs 5.0 wt/wt% was produced. It has been noted that the crucially important point for performing the thermal nanoimprint is that the imprinting temperatures must be above T_g and/or T_m of polymers for sufficient polymer chain mobilities, also adequate for polymer flowing ability at nanoscale within AAO cavities which were driven by imprinting pressures [3]. In Figure 2 (left), VANs of PS and VACNs of PS–GNPs (1.0–5.0 wt/wt%) with the length of 10–70 μm were obtained.

The results revealed that the flowing ability was increased due to the increase of imprinting temperature with the reduction of polymer viscosity but it was reduced when increases the content of GNPs. The increase in GNPs content induced flowing resistant of polymer within AAO nanochannel because of the molecular interaction between GNPs and PS that the orientation of PS was changed due to the confinement effect and attractive surfaces can restrict the PS chain mobility. Moreover, the flowing resistance can be caused by geometric and the dispersion of the nanoplatelets.

Surface Properties

Surface Chemical Properties

The dispersion of GNPs within PS nanotextured was studied by ATR–FTIR method in order to confirm the presence of GNPs on the obtained nanostructure surfaces. The presence of GNPs was proven by the reduction of 3 peak point intensities (753, 1451–1493 and 2991–3023 cm^{-1}) of the IR spectrums, indicating the GNPs has appeared on the PS nanostructure surfaces. These three points of spectrum are assigned to the characteristic peaks of benzene rings in the PS skeleton. This present work also performs the TGA in order to measure remained GNPs within imprinted nanostructures. The remained GNPs within PS of the imprinted nanostructures were evaluated by TGA.

Surface Wetting Properties

For the flat films, the contact angles of water droplets on PS and PS–GNPs materials with the values of 87.7° and 87.9° were respectively observed. The imprinted nanostructures on their films present the enhanced wetting ability with the higher contact angle than their flat films were obtained and became to 132.4° and 133.1° for the neat PS and PS–GNPs respectively. The presence of GNPs with the low content of 5.0 wt/wt% has not significant effect on the interfacial surface tension between the surface of the water and composite materials.

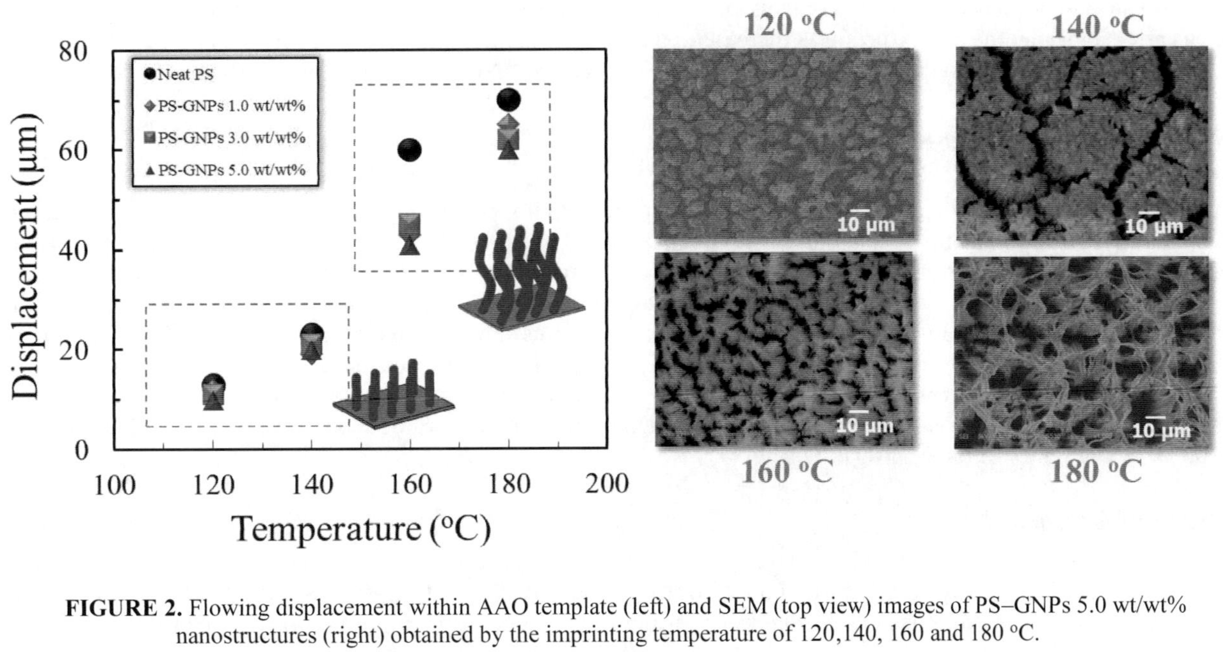

FIGURE 2. Flowing displacement within AAO template (left) and SEM (top view) images of PS–GNPs 5.0 wt/wt% nanostructures (right) obtained by the imprinting temperature of 120,140, 160 and 180 °C.

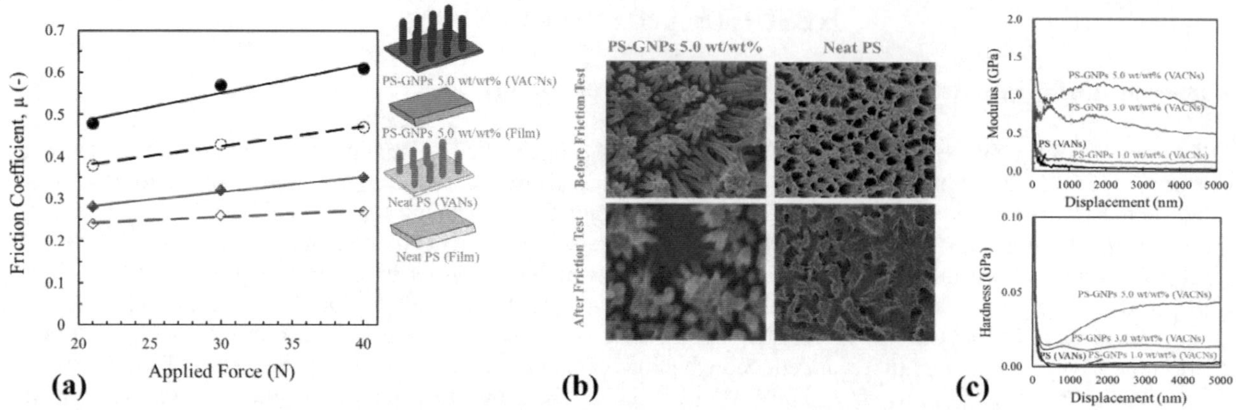

FIGURE 3. Surface mechanical characteristics in term of (a) friction coefficient (b) surface durability and (c) nanoindentation test results.

Surface Mechanical Properties

The friction coefficient of material surfaces are showed in Figure 3 (a). The flat surfaces of neat PS and PS–GNPs 5.0 wt/wt% display a relatively low friction coefficient between 0.24–0.27 and 0.28–0.35 while all of the nanotextured surfaces of those materials have the higher between 0.38–0.47 and 0.48–0.61 respectively. The replicated nanostructures on the material surfaces can be developed the surface roughness, resulting in increased the friction coefficient of surfaces. This nanotextured enables enhanced energy dissipation through molecular ploughing that enhances the friction coefficient becomes higher. Moreover, the reinforcement of GNPs enhances the material stiffness, resulting in the surface mechanical resistance and drag force at the surfaces between the composite materials and friction test material (stainless steel) were increased.

The surface durability of the imprinted surfaces was observed by SEM, comparing the nanostructured surfaces before and after friction test. Figure 3 (b) shows the SEM images (top view) of the surfaces of the samples for both in term of before and after friction test. This observation reveals that the destruction of the PS VANs was clearly observed. The neat PS nanostructures as the vertically aligned structures transformed to the randomly in the horizontal direction owing to the applied force during the friction test. In contrast, the PS–GNPs still remained the vertically aligned nanostructures because of the improvement of mechanical resistance of polymer by adding GNPs. However, it was observed that the broken nanostructures for a few area (less than 5 %) due to the brittle of composite material.

The surface modulus and surface hardness at nanoscale were precisely evaluated by nanoindentation test. Figure 3 (c) presents the influence of GNPs contents on the improvement of surface mechanical properties. The surface hardness of imprinted nanostructures was increased up to 11 folds and the surface hardness was also enhanced up to 4 folds, comparing between the neat PS nanostructures and the PS–GNPs 5.0 wt/wt% nanostructures.

Glass Transition Temperature

It has no change in the T_g (imprinted)–T_g (bulk) for the neat PS. However, the increase in the gap of T_g (imprinted)–T_g (bulk) up to +4 °C when increasing the content of GNPs (see Figure 4 (left)). The increased T_g (imprinted)–T_g (bulk) due to the confinement effect and the surface interfacial interaction effect can be promoted and became dominant effect by the increasing content of GNPs within nanocylidrical pore. The chain mobility of polymer was not only restricted by the GNPs but also by the applied imprinting pressure with the high content of GNPs. With the nanoimprinting process, after flowing into the pores the polymer chains were not stretched in the direction of flow but were compressed in a direction orthogonal to the flowing due to confinement under applied pressure [4]. Another reports on the increase in T_g (imprinted)–T_g (bulk) was caused by the rapid cooling process that promotes the restriction of polymer chain mobilities at AAO pore wall.

Thermal Conductivity

Figure 4 (right) displays the thermal conductivity of the neat PS and PS–GNPs 5.0 wt/wt% for both of 2D film structures and 1D of vertically aligned nanostructures, calculating from equation (1). Both of 2D and 1D structures revealed the enhancement of the thermal conductivity can be obtained due to the increase in GNPs content. The higher in thermal conductivity of the imprinted nanostructures compared to their bulk films was observed for all of GNPs contents due to the 1D structure has the higher heat transfer in the parallel direction which presents the directional transportation of the thermal energy. In the case of 2D materials, the heat transfer has ineffectiveness in the perpendicular direction that influences the reduction of thermal conductivity was observed. In addition, the dispersion and the orientation effect of GNPs in 1D structures owing to the shear induced orientation has the resulted in the improvement of the thermal conductivity [5]. In addition, the confinement with applied imprinting pressure at nanoscale effects to molecular orientation, polymer chain distances and thermal capacity of imprinted nanostructures that promoted thermal conductivity was increased. In general, heat is transferred more easily along the chains than in the transverse direction [5] hence the thermal conductivity can be significantly higher in the direction which polymer chains are preferentially oriented.

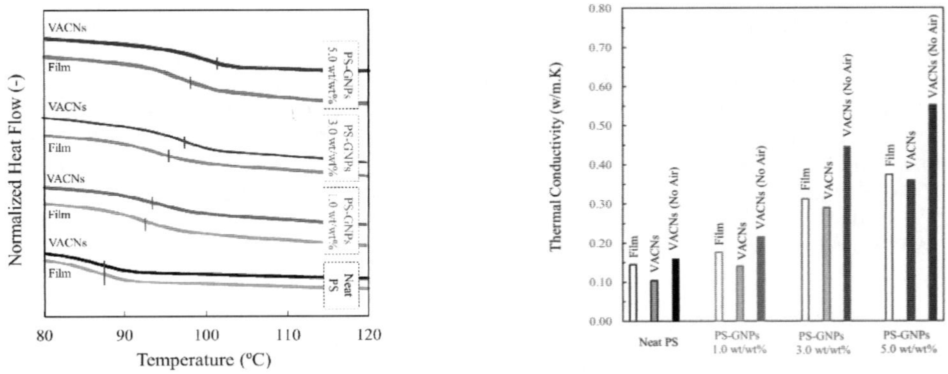

FIGURE 4. DSC thermographs (left) and results of thermal conductivity measurement (right).

CONCLUSION

In summary, VACNs of PS–GNPs were successfully prepared by thermal nanoimprint using AAO template. The nanostructures with 100 nm–diameter and 10–70 μm was replicated depending up on imprinting conditions and materials precursors. The replicated nanostructures with enhanced the surface wetting ability with the contact angle of water droplet for both of neat PS and PS–GNPs 5.0 wt/wt% than their flat film was obtained owing to the roughness of its surfaces. However, both of materials found that the quite similar wetting ability with the water contact angle of 132.4° and 133.1°, also the contact angle of water with 87.7° and 87.9° for the flat film surface of the neat PS and PS–GNPs respectively. The enhanced surface mechanical properties in both of macroscale and nanoscale were found including the increases in friction coefficient, surface durability, surface modulus and surface hardness. The confinement effect at nanoscale and the surface interfacial interaction between PS– GNPs –AAO have the resulted in the increased glass transition temperature of composite nanostructures as compared with their bulk film. The orientation of nanofilled can be controlled with the enhanced thermal conductivity.

ACKNOWLEDGMENTS

The authors gratefully acknowledge the support of the Japan Government (MEXT) and the Japan Society for the Promotion of Science (JSPS) in cooperation with the Innovative Flex Course for Frontier Organic Material Systems (iFront) of Yamagata University to promote full doctoral scholarship for Mr. Paritat Muanchan. This research was financially supported by JSPS Grant–in–Aid for Scientific Research (C), project ID 16K06740.

REFERENCES

1. J. J. Hernández, M. A. Monclús, I. Navarro-Baena, F. Viela, J. M. Molina-Aldareguia and I. Rodríguez, *Sci. Rep.* **7**, 43450 (1–10) (2017).
2. C. Mijangos, R. Hernández and J. Martín, *Prog. Polym. Sci.* **54–55**, 148–182 (2016).
3. P. Muanchan, S. Suzuki, T. Kyotani and H. Ito, *Polym. Eng. Sci.* **57**, 214–223 (2017).
4. J. Chen, L. Li, D. Zhou, X. Wang and G. Xue, *Phys. Rev. E* **92**, 032306 (1–8) (2015).
5. H. Chen, V. V. Ginzburg, J. Yang, Y. Yang, W. Liu, Y. Huang, L. Du and B. Chen, *Prog. Polym. Sci.* **59**, 41–85 (2016).

Spray Processing of Polymers and Complex Fluids: PARC's Filament Extension Atomizer Technology

Unidad Jerome [1]*, Karatay Elif [1], Neelakantan Ravi [1], Jose Arun [1], Johnson David [1]

[1] Palo Alto Research Center - California - USA

Polymers and similar complex fluids are notoriously difficult to spray process due to their strain hardening behavior in extension. In most cases, strain hardening leads to the formation of stable filaments that resist break-up into droplets. PARC's filament extension atomizer (FEA) is an innovative solution that addresses this problem. FEA takes advantage of the strain hardening behavior of polymers and elastocapillary thinning of the filaments in the high throughput generation of droplets with a narrow size distribution. This technology could be disruptive and could enable the creation of new platforms with a wide range of applications including additive manufacturing. It could also change the formulation paradigm for consumer and specialty products intended for spraying. In this contribution, we will present a few case studies on different materials and applications where FEA can be applied and how it can change the way we think about spray processing.

3D Printed Microfibrillated Polyamide/Polylactide Composites

Aliheidari Nahal [1], Hohimer Cameron [1], Ramezani Kakroodi Adel [2], Ameli Amir [1]*, Park Chul B. Caleb [2]

[1] Washington State University-Tri-Cities - Washington - USA, [2] University of Toronto - Ontario - Canada

Fused deposition modeling (FDM) as the most common type of additive manufacturing (3D printing) is experiencing rapid growth. FDM employs thermoplastic-based materials to create complex and three-dimensional objects layer by layer. To expand the applications of FDM for end-use structural and functional products, novel printable materials with enhanced properties must be developed. Polylactide (PLA) has been well established as one of the most common feedstock materials in FDM 3D printing. However, PLA suffers from brittleness, narrow service window, and relatively poor mechanical properties. In this work, we report a printable PLA-based filament reinforced with microfibrillated polyamide-6 (PA-6) that results in printed structures that are stronger, stiffer, and tougher, compared to those made of pure PLA. The microfibrillated PA-6/PLA fibers were produced by melt mixing-hot stretching process. The filament feedstock was then successfully prepared using a single extruder process. The morphology, thermal, and mechanical properties of the microfibrillated composites were investigated before and after printing. The results reveal that the fibril structure could be maintained after filament fabrication and printing processes and generated samples with higher tensile modulus, tensile strength, and strain-at-break. The results of this work are highly applicable in developing high-performance printed materials.

Preparation and 3D printing of polymer-based micro/nano functional composites

Wang Qi [(1)]*, *Chen Ning* [(1)], *Xia Hesheng* [(1)], *Lu Canhui* [(1)], *Zhang Jie* [(2)]

[(1)] State Key Lab of Polymer Materials Engineering, Polymer Research Institute, Sichuan University - Sichuan - China, [(2)] State Key Lab of Polymer Materials Engineering, College of Polymer Science and Engineering SCU - Sichuan - China

Polymer is one of the most suitable materials used for 3D printing due to its unique structure and properties. However, the polymers used in 3D printing at present are most single-component polymer, such as PA12, PLA, ABS, which can create any complicated objects that can't be achieved by conventional polymer processing, but it can hardly meet the electrical, magnetic, biomedical and other functional requirement for high-tech fields. So it need novel polymer based micro/nano composites with high performances and multi-functions and realize their 3D printing. In this paper, the polymer-based micro/nano functional composites suitable for Selective Laser Sintering (SLS) and Fused Deposition Modeling (FDM) processing were successfully prepared using solid state shear milling, ultrasonic irradiation, intermolecular complexation and other techniques, e. g. PA11/BaTiO3 piezoelectric powders, PU/CNTs conductive powders, PVA/PLA/HA biomedical filaments, PP/PE/CNTs high thermal conductivity filaments, et al. The novel technique for large-scale production of spherical polymer-based micro/nano composite powders was established, and the new conical screw extruded 3D printer was invented. The SLS processing of functional polymer-based micro/nano composite powders, such as PA11/BaTiO3, PU/CNTs and FDM processing of functional composite filaments, such as PVA/PLA/HA, PP/PE/CNTs , were successfully achieved. The low-dimensional fabrication, adding material layer upon layer, processing forming without mould, and complex solid-liquid-solid phase transition process in 3D printing process were investigated. The functional parts with complex geometric structure which cannot be manufactured by conventional processing methods were fabricated by optimizing 3D printing process, such as PA11/BaTiO3 piezoelectric device, flexible PU/CNTs sensor, individualized mandible model and cooling fans with good thermal conductivity, and so on, which makes a breakthrough in manufacturing complex shape products that can't be processed by traditional processing method and the functional products that can't be processed by current 3D printing. This work was supported by NSFC(51433006).

Polymerized Processing of Polypropylene-Based Alloys, Blends, and Nanocomposites

Jin-Yong Dong[a, b]

[a]*CAS Key Laboratory of Engineering Plastics, Institute of Chemistry, Chinese Academy of Sciences, Beijing 100190, China*

[b]*University of Chinese Academy of Sciences, Beijing 100049, China*

Abstract. Over the years, we have been focusing our research on developing synthetic approaches to polypropylene (PP)-based alloys, blends, and nanocomposites. In view of its inert chemical nature and easily degradable polymer chain at high temperature, PP is indeed not a polymer that can enjoy to the fullest of the convenient and most industrially desired processing art of melt-extrusion. PP is inherently immiscible and incompatible with most other polymers and materials, which raises two disturbing uncertainties, one being that the efficacy of mixing is merely dependant on how good the tool that is used for mixing is and there is no telling if the mixed systems have reached their climax in mixing effectiveness, the other being that the mixed systems even in their final stage are not at a thermodynamically stable state and any additional melt operations will make their multi-phasic structures fluctuate and consequently change, often aggravate, the ultimate material properties. It is thus our mission to solve all these processing dilemmas of PP by taking advantages of the powerful Ziegler-Natta and metallocene/post-metallocene catalysis for propylene polymerization that has shown an exceptional ability to be utilized flexibly to generate hetero-phasic copolymers including alloys, blends, and nanocomposites. With the so-named "polymerized processing technology" developed, we aim at new generations of hetero-phasic copolymers of PP with still alloy, blend, and nanocomposite appearances however with unprecedented structural and morphological contents. Most importantly, these new hetero-phasic PP are expected of controllable structural stability that will be allowed structural derivations during further processing however only to controlled degrees for bettering the polymer properties and performances.

Keywords: Polypropylene, Polymerized processing, Alloy, Blend, Nanocomposite.
PACS: 81.05.Lg

INTRODUCTION

Polyolefins, including polypropylene (PP, usually isotactic) and polyethylene (PE), are the most important family of synthetic polymers with extremely wide application range that shows no signs of diminishing with time. PP, in particular, with a successful transformation from being a long-time commodity polymer to an emerging new engineering plastic, thanks to the many advanced modification techniques, is being increasingly used in areas traditionally the strongholds of high-priced engineering plastics.[1, 2] In fact, with low cost, low weight, high durability, and high mechanical and thermal resistance, PP has been deemed one of the most promising polymer materials in the 21st century.[3]

There are two major ways to PP modification, which work on different structure hierarchical levels fulfilling different needs. On the chain structure level, polymer chain rearrangement and compositional enrichment may serve to, among others, increase PP's melt strength and make it polar-compatible; on the hyper-chain level, on the other hand, alloying, blending, and nanocompositing are highly facile ways rendering PP with property improvements in almost all-round respects.[4] Nevertheless, in view of its inert chemical nature and easily degradable polymer chain at high temperature, PP is indeed not a polymer that can enjoy to the fullest of the convenient and most industrially desired processing art of melt-extrusion. PP is inherently immiscible and incompatible with most other polymers and materials, which raises two disturbing uncertainties, one being that the efficacy of mixing is merely dependant on how good the tool that is used for mixing is and there is no telling if the mixed systems have reached their climax in mixing effectiveness, the other being that the mixed systems even in their final stage are not at a thermodynamically stable state and any additional melt operations will make their multi-phasic structures fluctuate and consequently change, often aggravate, the ultimate material properties. It is thus our mission to solve all these processing dilemmas of PP by taking advantages of the powerful Ziegler-Natta and metallocene/post-metallocene catalysis for propylene polymerization that has shown an exceptional ability to be utilized flexibly to generate hetero-phasic copolymers including alloys, blends, and nanocomposites. With the so-named "polymerized processing technology" developed, we aim at new generations of hetero-phasic copolymers of PP with still alloy, blend, and nanocomposite appearances however with unprecedented structural and morphological contents. Most importantly, these new hetero-phasic PP are expected of controllable structural stability that will be allowed structural derivations during further processing however only to controlled degrees for bettering the polymer properties and performances. In this paper, we present our latest results on polymerized processing of PP alloys, blends, and nanocomposites.

POLYMERIZED PROCESSING OF PP ALLOYS

One of the most prominent challenges in polymerized processing of PP alloys is the control of physical growth pattern of the incoming rubbery EPR. As a matter of fact, as far as the important EPR physical growth is concerned, the current PP in-reactor alloy technique is still more of a spontaneous polymerization process, far from being controllable. The PP alloy polymerized processing is composed of a first step of propylene homopolymerization, producing PP matrix in the form of hard particles, and a second step of ethylene/propylene random copolymerization in which alloy is formed by soft EPR being generated inside the preformed PP particles. Unlike PP, EPR is a random copolymer whose chain structure is unable or at least with difficulty to crystallize at the increased temperature of polymerization. In consequence, during its formation, EPR poses as polymer liquid with considerable flowability, making it unable to stay with the solid catalyst fragments from which it is originated. Rather, it will flow freely inside the pore network of the PP particles, the consequences of which are, one, EPR phase is discretely segregated from rather than homogeneously alloyed with PP matrix, two, there is a composition cap on EPR depending on the pore volume of the preformed PP particles, beyond which the liquid-like EPR will be no longer confined inside PP particle and exceed to the surface causing polymer particle lumping as well as reactor fouling, which are major failures in industrial PP polymerization practice and strictly prohibited from occurring whatsoever the reactor setup is. In fact, though of reactor origin, commercial grades of PP in-reactor alloys such as hiPP are known to be inhomogeneous in EPR dispersion, for which it is very important and absolutely necessary to carefully control the followed melt-processing operations for ensuring a fine EPR dispersion morphology in the final fabricated articles. And it is also common to add additional EPR or EPDM to hiPP in order to compensate the inadequacy in EPR content of the reactor-grade hiPP.

The newly developed polymerized processing technique is able to address this issue, which eventually leads to new PP in-reactor alloy materials that are much different from traditional hiPP and TPO in both structural and property forefronts.[5] Since the source of the concerned problem is the excessive flowability of EPR, the essence of the new technique is to reduce its flowability, by means of simultaneous cross-linking (Figure 1). Thus, after the first step of propylene homopolymerization has completed, with homogeneous dispersion of catalyst fragments in the PP matrix, the second step of ethylene/propylene copolymerization will be carried out involving a cross-linking reaction that ensures simultaneous crosslinking of EPR simultaneous with its formation. With sufficient degree of cross-linking, EPR will, instead of flowing away from the mother catalyst fragments, stay with them forming discretely dispersed particles. Further formation of EPR will only add to the volume of the EPR particles without entering the adjacent pores of the PP particles. In this pattern, without being randomly squeezed into the pore network and segregated from the PP matrix, controlled growth of EPR will be achieved, which is in turn expected to solve both the phase morphology inhomogeneity and EPR composition cap issues that currently limits the PP alloy polymerized processing technology.

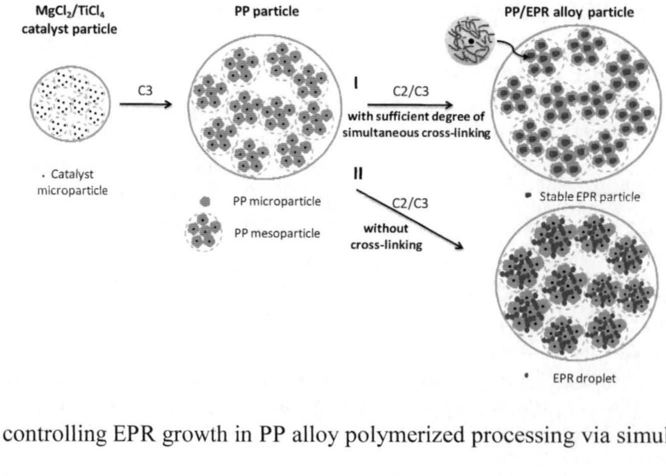

FIGURE 1. Schematic of controlling EPR growth in PP alloy polymerized processing via simultaneous chain cross-linking.

POLYMERIZED-PROCESSING OF PP BLENDS

In this part, we demonstrate a first example of precision polyolefin nanoalloys, of which polypropylene (PP) is the matrix and poly(ε-caprolactone) (PCL) the nanometric dispersed phase.[6] We will show, as illustrated in Figure 2, PCL the dispersed phase that is thermodynamically immiscible with any polyolefins including PP will not only be nanometrically dispersed in the alien PP matrix but its morphology (including domain size and distribution) will be precisely controlled and stable against polymer chain motion during melting that usually leads to morphology uncertainty in otherwise conventional immiscible polymer alloys/blends. The chemistry relies largely on the pre-preparation of PP/MWCNTs nanocomposite that disperses the high-aspect ratio, high-stiffness nanofiller of MWCNTs in PP, which lays the very foundation for the subsequent elaborate alloying with PCL the exotic polymer. As schemed in Figure 1, being grafted on and along the length of individual MWCNTs, which are both rigid and discretely dispersed, PCL is reserved a well-ordered pattern for phase structure formation. The dispersion discreteness and structural rigidity of MWCNTs will prevent the grafted PCL chains on different nanotubes as well as those on the same nanotubes but distant from each other from interwining. Without these inter- and intra-nanotube interventions, phase formation of PCL on each individual nanotube will be strictly molecular weight-dependant: the lower the molecular weight of PCL, the smaller the length range of nanotube within which the grafted PCL chains will be engaged in one domain formation, and the smaller the formed domain size and the more the domain number, and vice versa. Each of the dispersed MWCNTs is repeated the same PCL phase formation process, which together constitute the overall PP/PCL nanoalloy formation. Thus, by controlling its molecular weight from low to high and containing the molecular weight distribution to be relatively narrow, PCL will form different-sized (from small to large) and -numbered (from more to less) nanodomains in alloys with PP. Domain coalescence, a general phenomenon and a problematic issue causing morphology instability in conventional immiscible polymer alloys/blends especially in those of polyolefins, is here of insignificance since confinement by MWCNTs will greatly reduce the chain mobility of PCL. Moreover, PCL grafting will stabilize the dispersion of MWCNTs by both reducing their extremely large surface energy and annihilating the strong π -π interaction in between individual nanotubes, which are otherwise the two main reasons for MWCNTs' re-aggregation in their polymer nanocomposites. In combination, precision PP/PCL nanoalloys with PCL dispersion are in reasonable expectation.

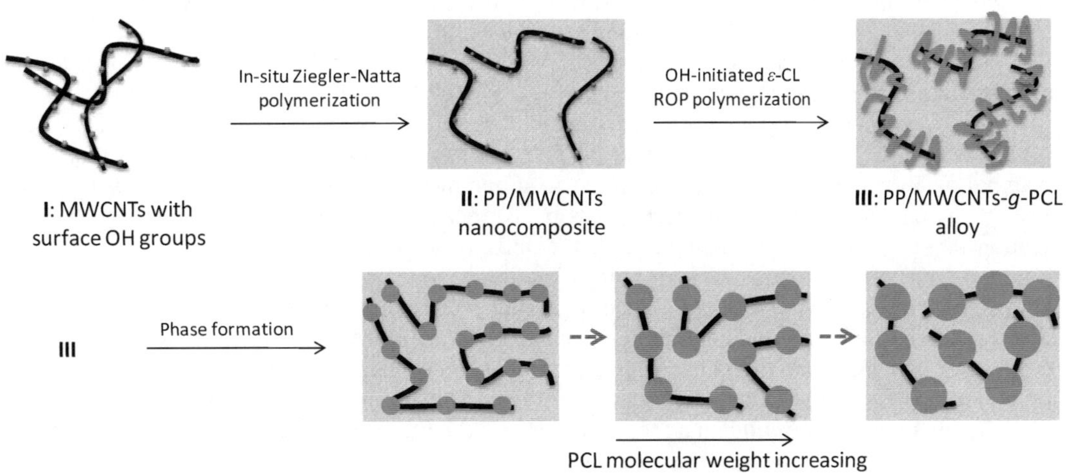

FIGURE 2. Schematic of fabrication of precision polyolefin nanoalloys with PP as the matrix and PCL as the dispersed phase regulated by MWCNTs.

POLYMERIZED-PROCESSING OF PP NANOCOMPOSITES

For polymerized processing of PP nanocomposites, for the moment with most chemistry issues having been largely resolved, the major concern is that it remains a premature technology locked in the laboratory, which, as pretty as it is, has yet to be tested and accepted by industry to unveil its true beauty. The difficulty in the wayis a lack of control of the produced polymer particle granule morphology, which, in most cases, is observed as amorphous, sometimes flake-like powder that causes severe reactor fouling and is destined to be harmful to industrial polymerization process. After all the in-reactor nanocompositing of PP in essence is a propylene polymerization process, for which to be industrially viable, the formed polymer particles are required at least a controlled granular morphology to keep from reactor fouling and other process problems.

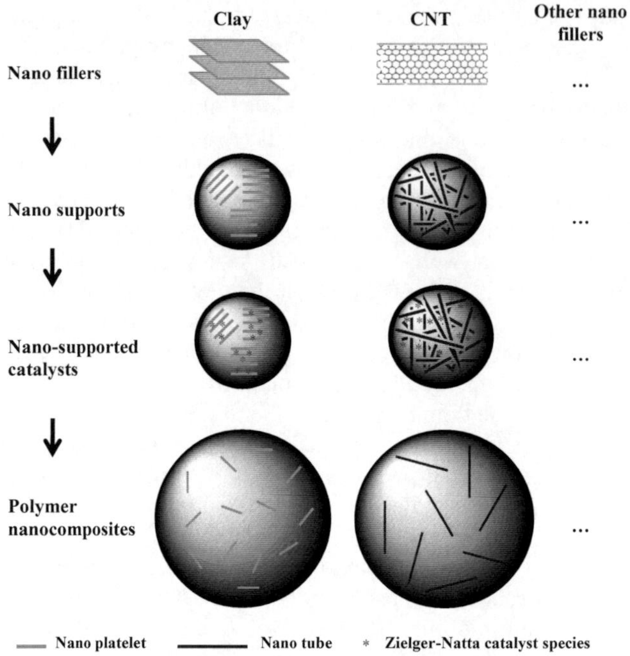

FIGURE 3. Nanofiller support-fabrication strategy for in-reactor nanocomposite of PP.

We have met this challenge by a nanofiller support-fabrication strategy, which follows the classical particle growth mechanism of heterogeneous propylene polymerization.[7] It is well known that a typical heterogeneous catalyst, e.g., $MgCl_2$-supported $TiCl_4$, is of a controlled granular particle morphology, with $MgCl_2$ crystal containing surface $TiCl_4$ being the basic structural unit (the primary particle). Polymerization takes place on each and every primary catalyst particle, driving fragmentation and dispersion of the $MgCl_2$support in the formed PP. The whole of the polymer particle growth is actually a particle-swelling process where all the primary particles, although driven apart, stay inside one and the same catalyst/polymer particle. The final polymer particle, with a much larger size, mimics the original catalyst particle in morphology (catalyst/polymer particle morphology replication). Based on this mechanism and assuming individual nanoparticles as primary catalyst particles, we have laid out this nanofiller support-fabrication strategy, which would proceed in the following manner (Figure 3). First and foremost, the nanofillers are granulated into a controlled morphology similar to that of a typical heterogeneous catalyst, with desired particle shape, size, and pore structure. Then transition metal components ($MgCl_2/TiCl_4$, metallocene and non-metallocene, etc.) are anchored on the granulated nanofillers with each individual nanofiller being the true carrier of the catalyst components. With the nanofillers having become a catalytically active heterogeneous catalyst, its propylene polymerization will follow the regular particle growth mechanism: The particle whole of the nanocomposite will remain granular replicating morphology from the catalyst, while the individual nanoparticles are driven apart inside the polymer granule forming nano-dispersed structure.

ACKNOWLEDGMENTS

Financial support from NSFC is gratefully acknowledged.

REFERENCES

1. B. Wang, Y. Long, Y. Li, Y. Men, Y. Li, *Polymer* **61**,108-114 (2015).
2. B. Zhang, Q. Dong, Z. Fu, Z. Fan, *Polymer* **55**, 4865-4872 (2015).

3. E. Passaglia, S. Coiai, S. Augier, *Prog. Polym. Sci.* **34**, 911-947 (2009).
4. P. Galli, *Prog. Polym. Sci.* **19**:959-974 (1994).
5. J. Shi and J. Dong, *Polymer* **85**, 10-18 (2016).
6. Y. Liu, N. Wang, J. Dong, *Macromol. Rapid Commun.* **36**: 1971-1978 (2015)
7. Y. Qin, N. Wang, Y. Zhou, J. Dong, *Macromol. Rapid Commun.* **32**: 1052-1059 (2011)

Mechanical Properties of Solid Films of Deoxyribonucleic Acid

Hisao Matsuno[a,b] Yuma Morimitsu,[a] Noboru Ohta,[c] Hiroshi Sekiguchi,[c] Atsushi Takahara[d] and Keiji Tanaka[a,b]

[a] *Department of Applied Chemistry, Kyushu University, 744, Motooka, Nishi-ku, Fukuoka, Japan 819-0395*
[b] *International Institute for Carbon-Neutral Energy Research, Kyushu University, 744, Motooka, Nishi-ku, Fukuoka, Japan 819-0395*
[c] *Japan Synchrotron Radiation Research Institute (JASRI), 1-1-1, Kouto, Sayo-cho, Sayo-gun, Hyogo, Japan 679-5198*
[d] *Institute for Materials Chemistry and Engineering, Kyushu University, 744, Motooka, Nishi-ku, Fukuoka, Japan 819-0395*

Abstract. Promoting green innovation to establish a worldwide low-carbon society is an urgent priority. Effective utilization of natural resources excluding fossil fuels is one of the efforts that will contribute to the resolution of this issue. Here, we show that solid-state films made from deoxyribonucleic acid (DNA) can be used as a structural material. The great advantage of the DNA films over the ones made from synthetic polymers is that the mechanical properties are controllable, from glassy to rubbery, via semicrystalline by simply regulating the water content in the film. Why and how such unique mechanical properties can be manifested by the DNA films is determined from detail structural analyses using Fourier-transform infrared spectroscopy and wide-angle X-ray diffraction measurements. With an increasing water content, the conformation of double-stranded DNA (dsDNA) was changed from the A-form in an amorphous state to the B-form in a partially packed one. dsDNA in the B-form became densely packed as the film was stretched. In addition, hydrated ionic liquid as a non-volatile substance was introduced into DNA solid films to maintain the tensile properties for longer time.

Keywords: Deoxyribonucleic Acid (DNA), Solid film, Mechanical properties, Thermal molecular Motion
PACS: 62.20.-x

INTRODUCTION

Natural polymers have been revisited as a material which may contribute to a low-carbon society. However, the utilization of such natural polymers is quite limited even now. Previously, we have revealed that double-stranded deoxyribonucleic acid (DNA) is one of promising resources as a structural material having peculiar tensile properties.[1] A self-supported film of DNA prepared by a solvent-casting method can be regulated from glassy to rubbery state via semicrystalline-like by its water content. However, there is a problem that the water evaporation causes changing the tensile properties of the films in a short time. To overcome this problem, in this study, we focused on hydrated ionic liquid as a non-volatile substance to maintain the tensile properties of DNA films for a long time. Tensile properties, aggregation structure, and thermal molecular motion of DNA films blended with hydrated ionic liquid were studied.

EXPERIMENTAL PROCEDURES

As a material, commercially-available DNA purified from fish sperm was used. As hydrated ionic liquid, choline dihydrogen phosphate (CDP) was used. FIGURE 1 shows chemical structure of CDP. DNA/CDP blend solutions with a weight ratio of 80/20, 74/26 and 70/30 were prepared by mixing aqueous solutions of DNA and CDP. Then, the DNA/CDP blend films were prepared by a simple solvent-casting method. Hereafter, these are referred to D/C20, D/C26 and D/C30 films. FIGURE 2 shows a typical DNA/CDP film. The mechanical properties of the films were carried out on a standard tensile testing machine at room temperature with a crosshead speed of 10 mm/min. The aggregation states of DNA molecules in the films were examined by wide-angle X-ray diffraction

(WAXD) measurements, which were performed at the BL40B2 beamline of SPring-8 (JASRI). The thermal molecular motion of DNA molecules in the films was measured by dynamic mechanical analysis (DMA). The measurements were carried out at temperatures ranging from 125 to 373 K.

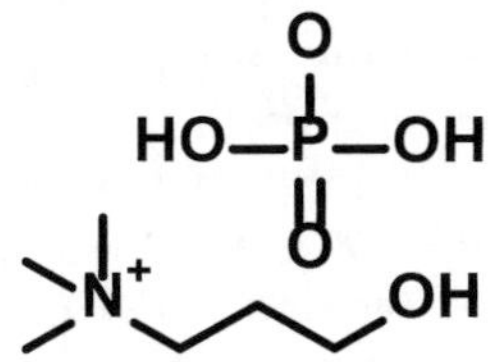

FIGURE 1. Chemical structure of CDP.

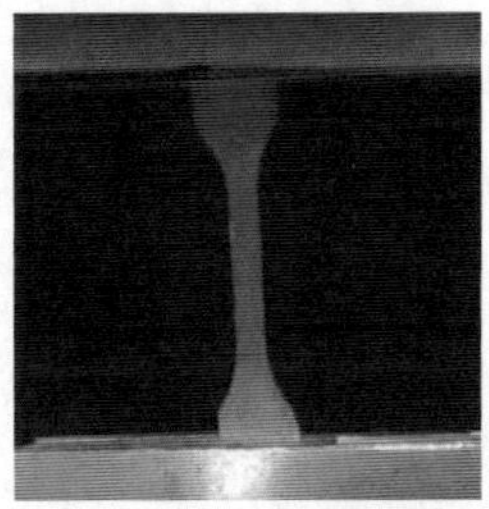

FIGURE 2. Appearance of a DNA/CDP blend film.

RESULTS AND DISCUSSION

Mechanical Properties of DNA/CDP Solid Films

FIGURE 3 shows stress-strain (S-S) curves for DNA/CDP solid films. The S-S curve strongly depended on the CDP content in the films. The D/C20 film behaved like a glassy polymer with a relatively higher Young's modulus (E) and a yield point. As the CDP content increased, the shape of S-S curves changed to ductility-like or rubbery-like polymers. In the case of the D/C26 film, the strain softening after the yield point at a strain (ε) of approximately 0.1 as well as the strain hardening beyond ε of 1.2 were observed. As the CDP content further increased to 30 wt%, the yield stress more decreased and the clear strain hardening like elastomers was found. Overall, the breaking point and breaking energy increased, and E and yield stress decreased with increasing CDP amount in the film. And also, these tensile properties were maintained for more than several hours because hydrated CDP could be hardly evaporated.

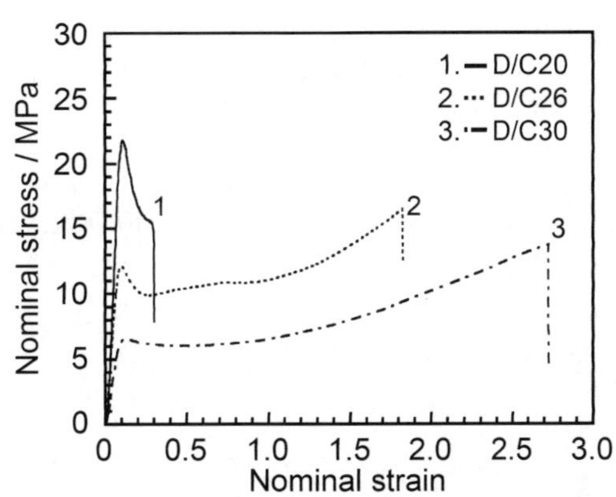

FIGURE 3. Stress−strain curves for DNA/CDP films.

Aggregation Structure of DNA/CDP Solid Films

FIGURE 4 shows two-dimensional WAXD patterns for a D/C30 film acquired at ε of 0 (unstretched), 1.5 and 2.0. Two diffraction rings corresponding to a distance of 0.34 and 2.4 nm were observed for the unstretched film. The distance between base pairs of double-stranded DNA in a B-form, which is the most typical conformation of DNA, is 0.34 nm. Thus, it can be claimed that DNA chains in the D/C30 film maintained the B-form conformation even after blending with CDP. On the other hand, the distance of 2.4 nm was larger than the diameter of 2.0 nm in the B-form. This larger periodicity was explained in terms of loosely-packed double strands of DNA interacted with hydrated CDP in the film. After stretching the D/C30 film to ε of 1.5, the diffraction arcs corresponding to the distance of 0.34 and 2.2 nm were observed along the equatorial direction, namely parallel to the stretching direction, and the meridian direction, respectively. This means that DNA strands with CDP were orientated parallel to the stretching directions and the distance between DNA strands decreased due to the stretching. The WAXD pattern for the D/C30 film obtained at ε of 2.0 was almost the same to the one at ε =1.5. Combined the S-S curves discussed above, it seems likely that structural relaxation of DNA strands occurred up to ε of 1.5.

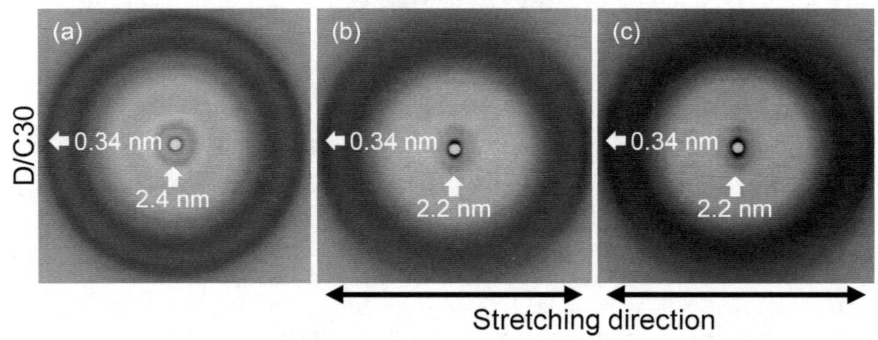

FIGURE 4. Two-dimensional WAXD patterns for a D/C30 film. The strain (ε) was set to be (a) 0, (b) 1.5 and (c) 2.0, respectively.

Thermal Molecular Motion of DNA/CDP Solid Films

FIGURE 5 shows temperature (T) dependence of dynamic storage and loss moduli (E' and E'') for a D/C30 film. On the temperature versus E'' plot, two peaks were observed at approximately 250 and 180 K. Since the peak temperatures were dependent on the frequency (f), the peaks can be assigned to molecular motion with different size scales, namely relaxation processes, at these temperatures. According to the established practice of DMA, these peaks are named as α and β, starting from the higher temperature. Based on the relationships between ln f and the inverse of T for the E'' maxima, the apparent activation energy (ΔH) for the α- and β-relaxation processes was calculated to be 187±7 kJ·mol^{-1} and 111±4 kJ·mol^{-1}, respectively. In our previous study, it was claimed that the E'' peaks at approximately 230 and 180 K for a DNA film without CDP were assigned to the cooperative motion with water molecules (ΔH of 131±3 kJ·mol^{-1}) and the B_I to B_{II} conformational transition derived from changes in the phosphate backbone (ΔH of 44±1 kJ·mol^{-1}), respectively.[2] That is, ΔH values increased with introducing CDP in the solid film. This can be explained in terms of a cooperative motion of DNA with CDP and water.

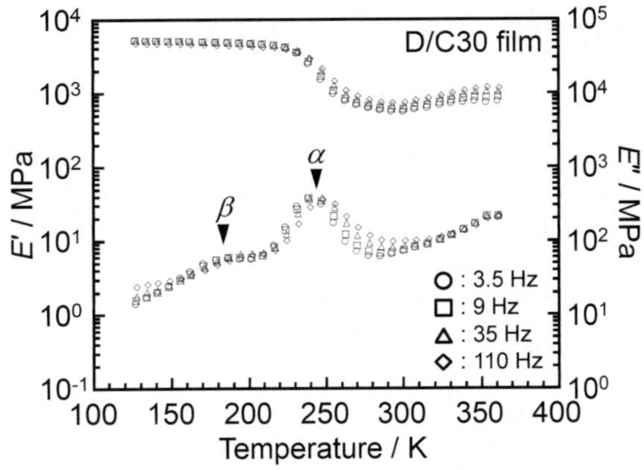

FIGURE 5. Temperature dependence of E' and E'' for a D/C30 film.

CONCLUSIONS

To suppress the water evaporation which causes changing the tensile properties of the solid films of DNA in a short time, we here used hydrated ionic liquid, CDP. The tensile properties of DNA/CDP films could be maintained for a time being longer than several hours. As CDP content increased, DNA/CDP film showed rubbery-like tensile properties. After stretching the D/C30 film, DNA strands with CDP were orientated parallel to the stretching directions and the distance between DNA strands decreased due to the stretching.

ACKNOWLEDGMENTS

This research was partly supported by JSPS KAKENHI, Grant-in-Aids for Scientific Research (C) (15K05633) to HM. The WAXD measurements were carried out at BL40B2 at SPring-8 with the proposal number 2016A1113, 2017A1522 and 2017B1090.

REFERENCES

1. Jie Zhan, Hisao Matsuno, Hiroyasu Masunaga, Hiroki Ogawa, Keiji Tanaka, *NPG Asia Mater.*, **6**, e92 (2014).
2. Hisao Matsuno, Junichi Nakahara, Keiji Tanaka, *Biomacromolecules*, **12**, 173 (2011).

Processing of a Laminated Composite Part by Microwave Heating

Hermine Tertrais[a] , Chady Ghnatios[b], Anaïs Barasinski[a] and Francisco Chinesta[c]

[a]*GeM Institute – Ecole Centrale de Nantes*
1 Rue de la Noé, 44321 Nantes, France
[b]*Notre-Dame University*
Louaize, Zouk Mosbeh, Lebanon
[c]*ICI Institute – Ecole Centrale de Nantes*
1 Rue de la Noé, 44321 Nantes, France

Abstract. Microwave (MW) heating relies on volumetric heat sources due to induction and dielectric phenomena. The main drawback of this technology today is that the complex physics involved in the conversion of electromagnetic energy to thermal energy is not fully understood or even controlled. To overcome this issue, we propose the following work which consist in modelling and simulate the propagation of the electromagnetic field in a laminated composite. The material is composed of carbon fibers and resin matrix. The full simulation from electromagnetics to thermal aspects will provide us with a better understanding of the process and its parameters.

Keywords: Composite processing, microwaves, PGD
PACS: 41, 44

INTRODUCTION

Composite parts tend to represent an increasing volume of production in transport industries (aeronautic and automotive). This is due to their combination of high mechanical properties and low mass. Yet, one disadvantage still remains: their long cycle time.

Conventional processing methods for producing polymer composite parts usually involve the application of heat to the material by convection or conductive heating through elements: surface heat transfers rule the process. On the other hand, microwave (MW) technology relies on volumetric heating: thermal energy is transferred through electromagnetic fields to materials that can absorb it at specific frequencies (in the case of MW heating: 2.45 GHz). Volumetric heating enables better process temperature control [1] and less overall energy use, which can results in shorter processing cycles [2]. Furthermore, comparable mechanical properties are shown between parts made with the MW technology and parts made with a traditional curing system (autoclave in the case of [3]). These virtues of the MW technology have attracted interest in developing the method and adopting it for the production of thermoset as well as thermoplastic composite materials.

The main drawback of this technology today is that the complex physics involved in the conversion of electromagnetic energy to thermal energy (heating) is not entirely understood and controlled.

The principal objective of this work is to model the interactions of the MW field with a laminate part, made of a sequence of prepreg composite layers (here, carbon fibers impregnated of a polymeric matrix), at macro and meso scales, that is to say at the scale of the part and also the scale of the ply. These models will simulate the way electromagnetic energy is propagated within the material volume and the various interfaces, then converted to thermal energy. The main challenge concerns the high-resolution description of the electromagnetic and thermal fields in a composite laminate, that involve plies whose characteristic in-plane dimension is orders of magnitude higher than the ones related to the thickness (typical aspect ratio are of tens of thousands). In that situation the use of in-plane-out-of-plane separated representations within the Proper Generalized Decomposition –PGD- framework seems an appealing and valuable route for solving 3D models, very rich in both, the in-plane and the out-of-plane directions, while ensuring a computational complexity of standard 2D models [4- 6].

PROCESS DESCRIPTION

One kind of oven cavity used for such a process is a Hephaistos© one (**FIGURE 1a**). A metallic bench is placed in the oven and the composite part (Carbon Fiber Reinforced Polymer material) is displayed on it within a mold made of a material transparent to microwave but coated in its interior surface by an absorbent layer. (**FIGURE 1**).

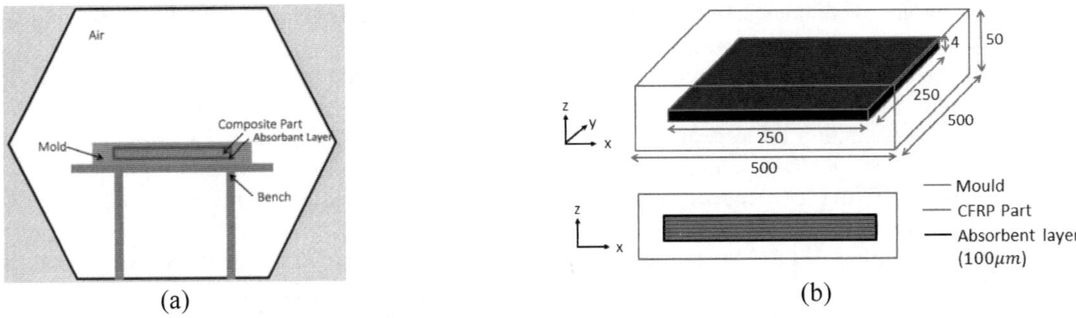

FIGURE 1. Schematic representation of the process (a) and zoom of the mold and composite part with its characteristic dimensions in mm (b)

MODELLING THE MICROWAVE HEATING

Electromagnetic Model

A coupled electromagnetic and thermal model is proposed in order to simulate the emerging process o microwave heating for composite material. The physical model for electromagnetism consists of the well-known Maxwell equations, which after some manipulations and assuming a harmonic field, are reduced to:

$$\nabla \times \frac{1}{\mu} \nabla \times \boldsymbol{E} = \gamma^2 \boldsymbol{E} \text{ with } \gamma^2 = i\omega(\sigma + i\omega\varepsilon) \quad \text{in } \Omega \subset \mathbb{R}^3 \tag{1}$$

$$\boldsymbol{n} \times \boldsymbol{E} = \boldsymbol{E}^t \quad \text{in } \partial\Omega \tag{2}$$

where μ, ε and σ and respectively the permeability, the permittivity and the electrical conductivity, all three of them depending on the considered material and $\omega = 2\pi f$ with f the frequency. $\boldsymbol{n}$ refers to the unit outwards vector defined on the domain boundary and $\boldsymbol{E}^t$ is the prescribed electric field (assumed known) on the domain boundary.

However, it is known [7] that this weak from version of the above strong formulation produces spurious solutions. Even if (1) ensures the verification of the Gauss equation $\nabla \cdot (\varepsilon \boldsymbol{E}) = 0$. its discrete version after approximating the different fields implied in the weak form does not ensure the fulfillment of the Gauss equation. Therefore, a regularization is needed to avoid spurious solutions. The regularized form is then:

$$\nabla \times \frac{1}{\mu} \nabla \times \boldsymbol{E} - \bar{\varepsilon} \nabla \left(\frac{1}{\bar{\varepsilon}\varepsilon\mu} \nabla \cdot (\varepsilon \boldsymbol{E})\right) = \gamma^2 \boldsymbol{E} \tag{3}$$

whose associated weak form (with Dirichlet BC applying on the whole domain boundary) is:

$$\int_\Omega \frac{1}{\mu} (\nabla \times \boldsymbol{E}) \cdot (\nabla \times \overline{\boldsymbol{E}}^*) d\Omega - \int_\Omega \gamma^2 \boldsymbol{E} \cdot \overline{\boldsymbol{E}}^* \, d\Omega + \int_\Omega \frac{1}{\bar{\varepsilon}\varepsilon\mu} \big(\nabla \cdot (\varepsilon \boldsymbol{E})\big)\big(\nabla \cdot (\varepsilon \overline{\boldsymbol{E}}^*)\big) \, d\Omega - \int_{\partial\Omega} \frac{1}{\bar{\varepsilon}\varepsilon\mu} \big(\nabla \cdot (\varepsilon \boldsymbol{E})\big) \cdot (\mathbf{n} \cdot (\bar{\varepsilon} \overline{\boldsymbol{E}}^*)) d\Gamma = 0 \tag{4}$$

In-plane-out-of-plane Separated Representation

As the composite part considered in the process is a plate, its characteristic in-plane dimension is orders of magnitude higher than the ones related to the thickness as shown on **FIGURE 1b**. In that situation, the use of the in-plane-out-of-plane separated representation within the Proper Generalized Decomposition (PGD) framework allows writing the electric field in the 3D separated form:

$$\boldsymbol{E}(x,y,z) \approx \sum_{i=1}^{N} \boldsymbol{X}_i(x,y) \circ \boldsymbol{Z}_i(z) \tag{5}$$

where the bullet ∘ denotes the Hadamar product. Thus, the electric field is expressed as a finite sum of functional couples involving a function depending on the in-plane coordinates (x,y) and the other involving the coordinate related to the thickness (z). Thus, the 3D solution is obtained from a number (in the order of N) of 2D problems involving the in-plane coordinates and the same number of 1D problems involving the thickness.

Such a separated representation seems an appealing and valuable route for solving 3D models, very rich in both, the in-plane and the out-of-plane directions, while ensuring a computational complexity of standard 2D models [4]. This separated representation is especially interesting for addressing domains in which at least one of its characteristic dimensions is much smaller than the others, and the solution exhibits significant richness in the thickness direction implying the necessity of using a fine enough representation (mesh) to capture the transmission conditions at the ply interfaces. When using standard mesh-based discretization the use of a fine mesh in the thickness direction within a 3D discretization usually implies extremely fine meshes in the whole domain, involving a prohibitive number of degrees of freedom. The use of separated representations makes independent the in-plane and the thickness approximations and the use of different discretization techniques for solving the associated problems related to the in-plane and the thickness coordinates is possible.

Thermal Model

From the electromagnetic solution, the thermal source term Q can be determined and the thermal model defined and solved. The heat conduction model inside the laminate allows calculating the temperature field at each position and time $T(x,y,z,t)$ by solving the heat equation:

$$\rho C_p \frac{\partial T}{\partial t} = \nabla \cdot (\boldsymbol{K} \cdot \nabla T) + Q(x,y,z,t) \tag{6}$$

where $\boldsymbol{K}$ represents the thermal conductivity tensor defined at the ply level and which in the case of unidirectional reinforcement or fabrics becomes anisotropic justifying its tensorial character.

Again, the too different in-plane and thickness characteristic lengths, motivates the use of a separated representation of the temperature field, which now writes:

$$T(x,y,z,t) \approx \sum_{i=1}^{M} X_i(x,y) Z_i(z) \Theta_i(t) \tag{7}$$

TEST CASE

The teste case described on the geometry description on **FIGURE 1b** is simulated in the following paragraph. The mesh is composed of 60x60 elements in the plane (x,y) and 3000 elements in the out-of-plane direction. The composite material is composed of a stack of plies described by the sequence $[(0°, 90°)_{10}]_{sym}$. The material properties are given in **TABLE 1.**

TABLE 1 Dielectric Material properties used in the simulation. For anisotropic materials, tensor is reduced to its diagonal [11-component, 22-compoenet, 33-component].

Dielectric property	0°-oriented CFRP ply	90°-oriented CFRP ply	Mold	Absorbent layer
Relative permittivity	[13 1 1]	[1 13 1]	2	1
Relative permeability	[1 1 1]	[1 1 1]	1	1
Electrical conductivity (S/m)	[5000 10 10]	[10 5000 10]	0	10000

The boundary conditions applied are:

$$\boldsymbol{n} \times \boldsymbol{E} = \begin{pmatrix} \cos(2\pi n x) + \cos(2\pi n y) \\ \cos(2\pi n x) \\ \cos(2\pi n y) \end{pmatrix} \text{ with } n\text{=10} \tag{8}$$

The results of the simulation are depicted on **FIGURE 2** and **FIGURE 3**. We can see that the electric field is down to zero when arriving on the absorbent layer. The discontinuity observed comes from the interface conditions which needed to be enforced in the simulation tool. The temperature distribution in the thickness shows that the process allows us to heat the part of about 200°C in 30s. A future concern would be to optimize the process such as the heating would as much homogeneous as possible to ensure part quality.

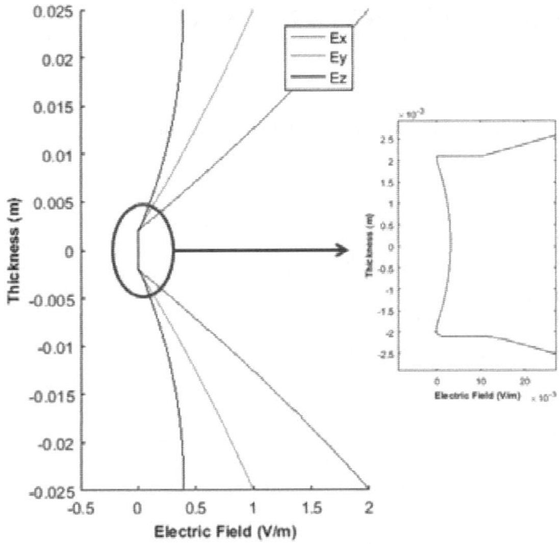

FIGURE 2. Electric field distribution along the line (x_{middle},y_{middle},z) with zoom at the composite interface.

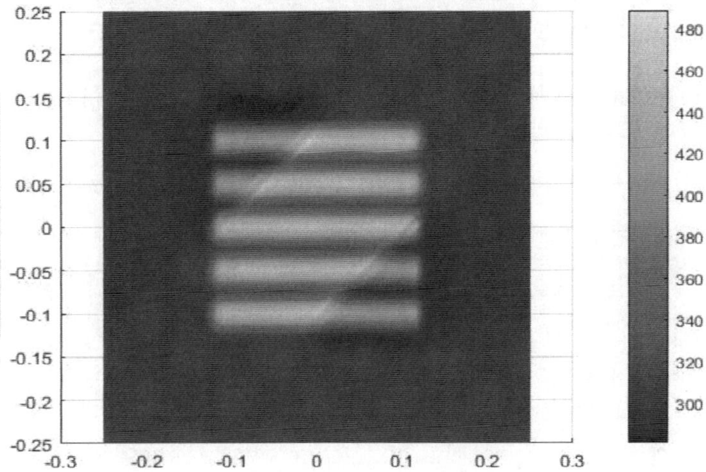

FIGURE 3. Temperature field (in K) in the plane (x,y,z_{middle}) showing the inmpact of the electric field distribution on the heating

CONCLUSION

The modelling of the electromagnetic heating of a CFRP plate has been presented in the paper. First steps of the simulation tool have been detailed and show that the simulation approach chosen (in-plane-out-of-plane representation within the PGD framework) ensures good results and fast computation. The coupling between electromagnetism and thermic physics has also been presented. Further simulation work is to develop the simulation tool to take into account the plies scale in laminates and use the tool in an optimization approach of the process.

ACKNOWLEDGMENTS

This work has been done in the framework of the SIMUTOOL project (H2020).

REFERENCES

1. G. Hug, "Behavior analysis of carbon-epoxy laminates under high-speed loading: manufacture of the same materials by means of microwave curing for comparison" Ph.D. thesis, Arts et Métiers ParisTech, 2005.
2. R. Youssof, M.K. Aroua, A. Nesbitt and R.J. Day, "Curing of polymeric composites using microwave resin transfer moulding (RTM)", in *Journal of Engineering Science and Technology*, **2**(2), 51-163
3. M. Kwak, P. Robinson, A. Bismarck and R. Wise, "Curing of Composite Materials Using the Recently Developed Hephaistos Microwave", in 18th International Conference on Composite Materials, 21-26 (2011)
4. B. Bognet, A. Leygue, F. Chinesta, A. Poitou and F. Bordeu, "Advanced simulation of models defined in plate geometries: 3D solutions with 2D computational complexity", in *Computer Meth. in Applied Mechnanics and Engineering*, **201**, 1-12 (2012)
5. F. Chinesta, A. Leygue, B. Bognet, Ch. Ghnatios, F. Poulhaon, F. Bordeu, A. Barasinski, A. Poitou, S. Chatel and S. Maison-Le-Poec, "First steps towards an advanced simulation of composites manufacturing by automated tape placement" in International Journal of Material Forming, 7(1), 81-92 (2014)
6. Ch. Ghnatios, F. Chinesta and Ch. Binetruy, "3D Modeling of squeeze flows occurring in composite laminates", in *International Journal of Material Forming*, **8**(1), 73-83 (2015)
7. R. Otin, "Regularized Maxwell Equations and Nodal Finite Elements for Electromagnetic Field Computations" in *Electromagnetics*, **30**(1-2), 190-204 (2010)

Advanced simulation of Automated Tape Placement : Effect of reinforced resins on consolidation

LEON Angel [1], PEREZ Marta [1], BARASINSKI Anaïs [1]*, Abisset-Chavanne Emmanuelle [2], CHINESTA Francisco [3]

[1] GeM - ECN - Nantes - France, [2] ICI - ECN - Nantes - France, [3] ESI Chair - ECN - . - France

Automated tape placement is a promising composite forming process based on the fusion bonding of a prepreg tape on a substrate. By heating the interface between the incoming tape and substrate, and applying pressure, a laminate part, out of autoclave, can be obtained in one step, making the process at- tractive for industry. In order to reach in-situ consolidation with this process, the material has to undergo several steps: heating, consolidation, and cooling. During these three stages, different physical phenomena occur, the important point is that all these phenomena take place at material interfaces and are governed by temperature and pressure. One critical aspect concerns the consolidation step, when the bonding occurs, this part has already been treated in many previous work, but the output descriptors (either rectangular, fractal,) are not satisfying from our point of view as they are not enough discriminant, ie almost different surfaces can be reduced with the same descriptors but obviously will not present the same evolution of Dic - degree of intimate contact- when put together under a thermo-mechanical field. In previous work, we proposed a novel approach to improve the understanding of the consolidation step, we first proposed to model the surface as a fractal which parameters are extracted from a prepeg experimental characterisation, the surface was then compressed and thermal exchanges at the microscales computed. An improvement of the fractal modelisation is now proposed by a wavelet description of the surface in order to work on a real surface topology and thermal fields are computed at a microscale at the interface. Indeed today many possibilies exists to modify more or less easily the material parameters, e.g. by adding some particules in the matrix. In that work, it is proposed to look at the influence of the thermal properties, and viscosity of the matrix, induced by the addition particules in the matrix, on the consolidation evolution.

Controlling Surface Wettability by Microstructure and Surface Energy Manipulation on Three Dimensional Substrates for Versatile Oil/Water Separation

Hao-Yang Mi[a, b], Xin Jing[a, b], Han-Xiong Huang[b], Lih-Sheng Turng[a]*

[a] *Department of Mechanical Engineering, University of Wisconsin–Madison, Madison, WI 53706, USA*
[b] *Department of Industrial Equipment and Control Engineering, South China University of Technology, Guangzhou, 510640, China*

Abstract. Superwettable materials have gained tremendous attention due to their special wetting abilities. The key to obtaining and tuning superwettability is to precisely control the interfacial microstructures and surface energies of materials. Herein, we propose a novel approach to controlling the superwettability of three-dimensional (3D) foams. The surface microstructure was manipulated by the layer-by-layer covalent grafting of multidimensional nanoparticles (e.g. silica, carbon nanotubes, and graphene oxide), and the surface energy was tailored by grafting chemicals with different functional groups. This grafting approach improved the mechanical performance, reduced particle loading, and prevented particle disassociation, thereby increasing the absorption capacity and durability of the functionalized foams. More importantly, superhydrophobic/superoleophilic foams were obtained after heptanol grafting. They showed water contact angles of 153° in air and 158° in oil, an absorption capacity 113 times their weight gain, and a remarkable flux of 25 L m-2 s-1 when separating oil from water driven by gravity. Polydopamine grafting resulted in superhydrophilic/ underwater superoleophobic foams which had an oil contact angle of 152° under water and a high flux of 15 L m-2 s-1 when separating water from oil. Thus, this study offers not only intelligent materials for versatile oil/water separation, but also a profound approach for engineering high-performance superwettable materials.

Keywords: Superhydrophobic, underwater superoleophobic, oil/water separation, microstructure

INTRODUCTION

Oil spills are one of the main causes of water pollution[1]. Developing materials and products for effective oil recycling and water remediation is an urgent need. Fabrication of high-performance superabsorbent and oil/water separation materials has been attracting much attention from both academic and industrial communities. Materials that have superwettability show promising performance in the selective separation of oil and water due to their special wetting phenomenon. To date, two types of superwettable materials have been developed for gravity-driven oil/water separation. Superhydrophobic/ superoleophilic materials that allow oil penetration but repel water can separate heavy oil from water by gravity.[2,3] Likewise, superhydrophilic/underwater superoleophobic materials that enable water transport but prevent oil flow can separate water from light oil by gravity.[4,5]

So far, there are a vast number of studies on particulate-modified superhydrophobic materials since hierarchical nanostructures can be easily generated by loading nanoparticles.[6] The nanoparticles are normally loaded onto substrates by dip coating,[7] spraying,[8] *in situ* deposition,[9] or by using a commercial binder.[10] However, a particular issue has arisen in practical applications for materials fabricated using these methods. That is the particles tend to fall off and contaminate the recovered oil during usage due to the lack of strong chemical bonds. To resolve this problem, in this study, we proposed a layer-by-layer chemical grafting method to create strong covalent bonds between the substrate and the particles. Recently, it was reported that combining multidimensional particles was an effective way to improve the surface roughness of the substrate.[11] Therefore, in this study, 0D silica nanoparticles, 1D functionalized CNTs (fCNTs), and 2D GO were used to engineer the surface microstructure of melamine foam (MF). These three multidimensional particles are collectively called GSC particles for simplicity. Different chemical groups were grafted to produce materials for versatile oil/water separation.

FABRICATION OF MULTIFUNCTIONAL FOAMS:

The overall fabrication process of multifunctional foam is illustrated in Figure 1 and mainly includes two steps. The first step involves microstructure manipulation. MF was modified by hexamethylene diisocyanate (HDI) in dimethylformamide (DMF), followed by dipping into silica, fCNT, and a GO solution, and subsequent curing. Then the foam was sonicated in DMF to remove unreacted particles and vacuum dried. This procedure was repeated to achieve layer-by-layer grafting. The second step involves surface energy manipulation. Particle-grafted MF (GSCMF) was modified with isocyanate groups by reacting with HDI, then reacted with heptanol in DMF to graft long carbon chains, which yielded the superhydrophobic MF (HepGSCMF). On the other hand, GSCMF was also subjected to a dopamine (DA) coating in aqueous solution with pH = 8.5 to enhance the hydrophilicity. Dopamine was able to self-polymerize and coated GSCMF to result in superhydrophilic MF (DAGSCMF). In this process, since all reactions happen on the foam surface, the reactant solutions can be reused until all particles are consumed, which significantly reduces cost compared with other batch-to-batch modification methods.

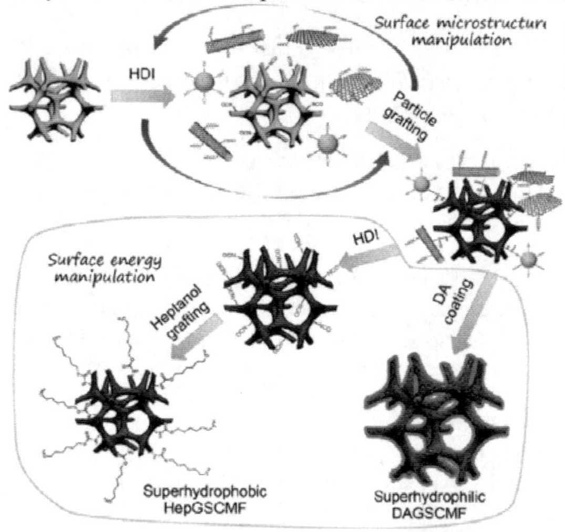

FIGURE 1. Schematic illustration of MF surface microstructure and surface energy manipulation procedures to produce superhydrophobic HepGSCMF and superhydrophilic DAGSCMF.

Characterization

Fourier Transform Infrared Spectroscopy (FTIR): FTIR spectra were recorded in transmittance mode using a Bruker Tensor 27 spectrometer (Thermo Scientific) in the range of 4000–600 cm^{-1} with a resolution of 4 cm^{-1}.

X-ray Photoelectron Spectroscopy (XPS): XPS measurements were performed on an X-ray photoelectron spectrometer with a focused, monochromatic K-alpha X-ray source and a monoatomic/cluster ion gun (Thermo Scientific). Survey scans and C1s core-level scans were recorded.

Scanning Electron Microscopy (SEM): SEM analysis was performed on a fully digital LEO GEMINI 1530 SEM (Zeiss, Germany) at a voltage of 3 kV. Samples were sputtered with a thin layer of gold prior imaging.

Brunauer–Emmett–Teller (BET) Surface Area: The BET surface area was measured using a Gemini VII 2390 surface area analyzer (Micromeritics Instrument Corp.). Specimens were dried under vacuum for 2 h at 100 °C and saturated with N_2 prior to measurement.

Contact Angle: The contact angle measurement was performed at room temperature with a Dataphysics OCA 15 optical contact angle measuring system using the sessile drop method. A water bath and hexane bath were used for the underwater and under-oil measurements.

Compression Tests: Compression tests were carried out using an Instron universal testing machine equipped with a 50 N load cell. All functionalized MF specimens were compressed to 70% strain with a crosshead speed of 1 mm min^{-1}. The compressive modulus and strength were recorded. At least five specimens for each group were tested. Cyclic compression tests of GSC-4L were performed in two ways. First, different strains (10%, 20%, 40%, and 60%) were set as the termination of compression to investigate its response to different strains. Second, the specimen was loaded and unloaded for 500 cycles with a fixed strain of 50% to investigate its long term recoverability.

Absorption Capacity: Specimens were dried under vacuum at 100 °C for 30 min, then saturated in hexane or chloroform and weighed again. The absorption capacity (AC) is calculated using Equation (1).

$$AC = \left(\frac{W_{s1} - W_0}{W_0} \right) \times 100\% \quad (1)$$

The parameter W_0 is the weight of the original specimen, W_{s1} is the weight of the saturated specimen.

Flux: The flux of HepGSCMF, which was measured using chloroform and water, was used for DAGSCMF. Briefly, 60 mL of test liquid was loaded in the setup shown in Figure 4 and a thin Teflon film was placed on top of the foam to block the liquid. The film was quickly removed and the time used for all liquid to pass through was recorded. The flux was calculated using Equation (2). The test was repeated 10 times on the same sample.

$$F = \frac{V}{A \times t} \tag{2}$$

RESULTS AND DISCUSSION

The covalent bond was created not only between MF and nanoparticles, but also among GSC particles because of their abundant–OH and –COOH reactive groups. To reveal this, three particles were reacted in HDI and observed using transmission electron microscopy (TEM). Tight connections between silica and GO, GO and fCNT, fCNT and silica, and among GO, silica, and fCNT were observed (Figure 2a-c). The bridging among particles enabled multiple layer coating. In addition, Fourier transform infrared spectroscopy (FTIR) spectra confirmed the presence of C=O, N–H, and C–N bonds from urethane and amide bonds in GSCMF, thus indicating the successful grafting of particles via covalent bonds (Figure 3).

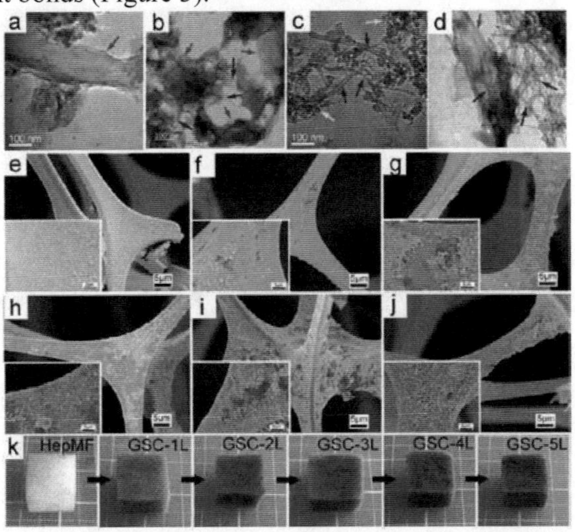

FIGURE 2. TEM images of a) GO grafted with silica, b) GO grafted with fCNT, c) silica grafted with fCNT, d) GO and silica grafted with fCNT. Arrows indicate GO, silica, and fCNT particles. SEM images of e) HepMF, f) GSC-1L, g) GSC-2L, h) GSC-3L, i) GSC-4L, and j) GSC-5L. k) Photographs of corresponding functionalized foams.

Scanning electron microscopy (SEM) was used to investigate the morphology of hydrophobic foams. MF grafted with only heptanol showed a relatively smooth surface (Figure 2e). When grafted with multidimensional GSC particles, the surface became rougher with the increase of grafting layers (Figure 2f-j). The particle aggregates were able to be distinguished when there were less than four grafted layers, but it became indistinguishable after five layers of grafting (GSC-5L), which may be due to the simultaneous increase of HDI content. The photographs of corresponding foams (Figure 2k) clearly showed an increase of darkness as the GSC grafting layers increased.

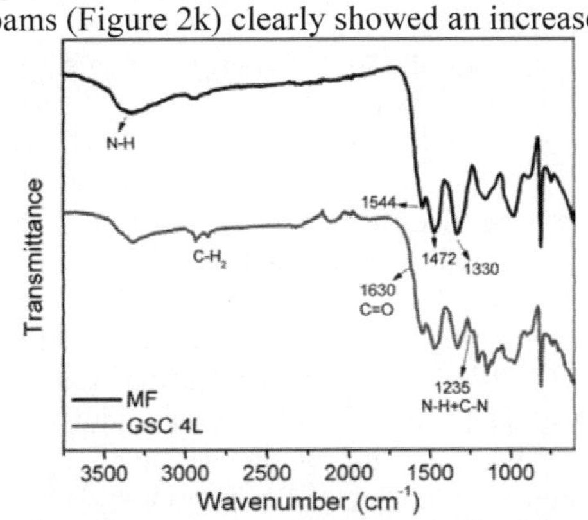

FIGURE 3. FTIR spectra of MF and GSCMF with four layers of coating (GSC-4L).

The density and surface area are important bulk characteristics of absorbent materials. According to Figure 4a, the bulk density increased with an increase in grafting layers, while the surface area showed a different trend where GSC-2L was the highest. This is because the surface area was increase by introducing nanoparticles initially, but it decreased when the weight of the loaded particles was too high. The water contact angle (WCA) of heptanol-modified MF (HepMF) was 130.2°, and it increased to 152.6° for GSC-4L indicating superhydrophobicity (Figure

4b). In order to investigate the sustainability of the grafting, specimens were either thermally treated at 200 °C or immersed in liquid nitrogen for up to 4 h (Figure 4c and d). All specimens showed a reduction of WCA of less than 3°, suggesting that the modification was highly thermally stable. In the compression test, the compressive modulus and strength of modified MFs were significantly improved after GSC grafting (Figure 4f). They also showed excellent recoverability (Figure 4e), which is an important property for absorbent materials to achieve effective oil recovery. When compressed to higher strains, the hysteresis loop of GSC-4L became larger along with a significant increase in strength (Figure 4g). Statistical data showed an increase of hysteresis loss from 23.4% to 54.5% when strain was increased from 10% to 60%, while GSC-4L still maintained a high recovery rate over 99.2%. In 500 cycles of loading and unloading, GSC-4L showed a small reduction in strength and hysteresis loss, while a recovery rate of 96.3% was still achieved, even after 500 cycles (Figure 4h). These results elaborate on the extraordinary mechanical and chemical stability of the modified foams.

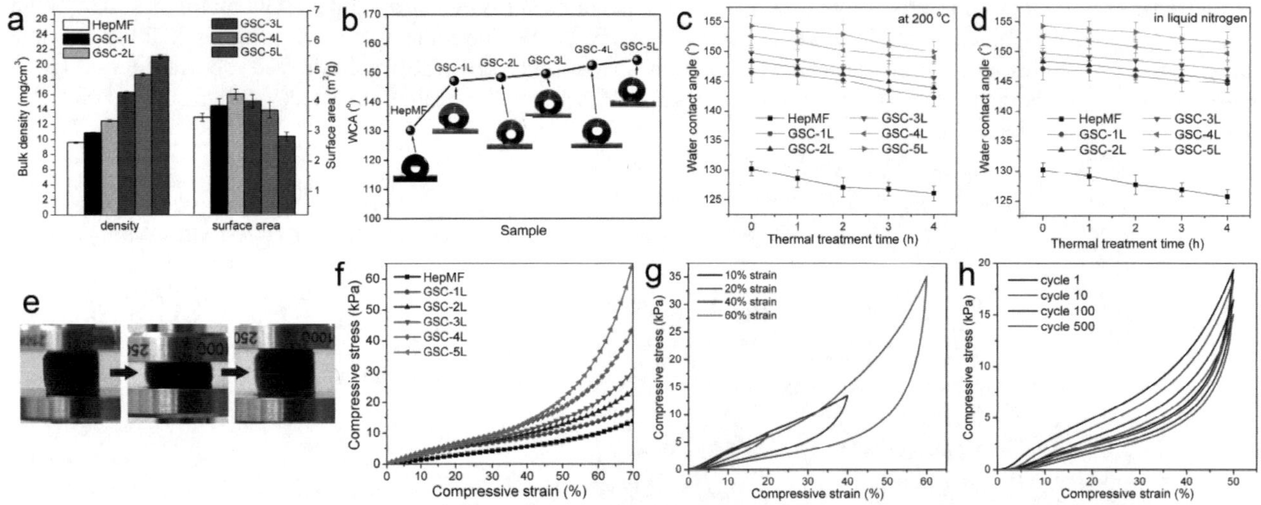

FIGURE 4. a) Bulk density and surface area, and b) water contact angle of functionalized foams. Water contact angle change c) at 200 °C and d) in liquid N_2 for up to 4 h. e) Photographs demonstrating high recoverability. f) Compression test stress vs. strain curves. Cyclical compression test g) at different strains and h) for 500 cycles.

In practical applications, people often have to face the trade-off between absorption capacity and separation efficiency when choosing absorbent materials. Achieving high capacity and absorption efficiency with minimum particle loading is a major direction in superabsorbent materials research. Because GSC-4L is superhydrophobic (WCA > 150°) and has relatively high absorption capacity (11.3~46.3 times weight gain), it was used as absorbent and separation materials in the following study and is denominated as HepGSCMF in the following sections.

To demonstrate the durability of HepGSCMF when used in various applications, it was subjected to repeated mechanical squeezing, ultrasonication, and sand paper abrasion. As shown in Figure 5a, HepGSCMF demonstrated excellent tolerance to squeezing and ultrasonication with a WCA exceeding 150° after 500 cycles of squeezing and 60 min ultrasonication. After abrasion using sand paper for 10 m, the WCA was still as high as 146°, revealing HepGSCMF's high tolerance to abrasion. In addition to neutral water (pH = 7.0), HepGSCMF showed a high contact angle (CA) to aqueous solutions, with a pH ranging from 1 to 13.5, and its tolerance to acidic solutions seemed slightly higher than to basic solution (Figure 5b). Figure 5c shows the relative absorption efficiency (RAE) results of HepGSCMF, which indicates the sustainability of the absorption ability in multiple cycles. The results showed that the decrease of RAE was less than 7% after 10 cycles of usage for both hexane and chloroform. In addition, the mass of the specimen remained almost the same, thus indicating negligible particle loss. The remarkable robustness of HepGSDMF emphasized the profound effects of the strong bonding between substrate and nanoparticles. As a demonstration of application of HepGSCMF as a superabsorbent material (Figure 5d), it was used to absorb heavy oil under water and light oil atop water. HepGSCMF was able to selectively absorb oils from water rapidly, and the absorbed oil can be simply recycled by squeezing or distillation. Figure 5e indicates that HepGSCMF was able to absorb various oils and organic solvents, and HepGSCMF had a higher capacity than most particulate-modified commercial foams and some carbon-based aerogels. Taking its excellent durability into consideration, HepGSCMF is an excellent, durable superabsorbent material for high efficiency, selective oil recycling applications.

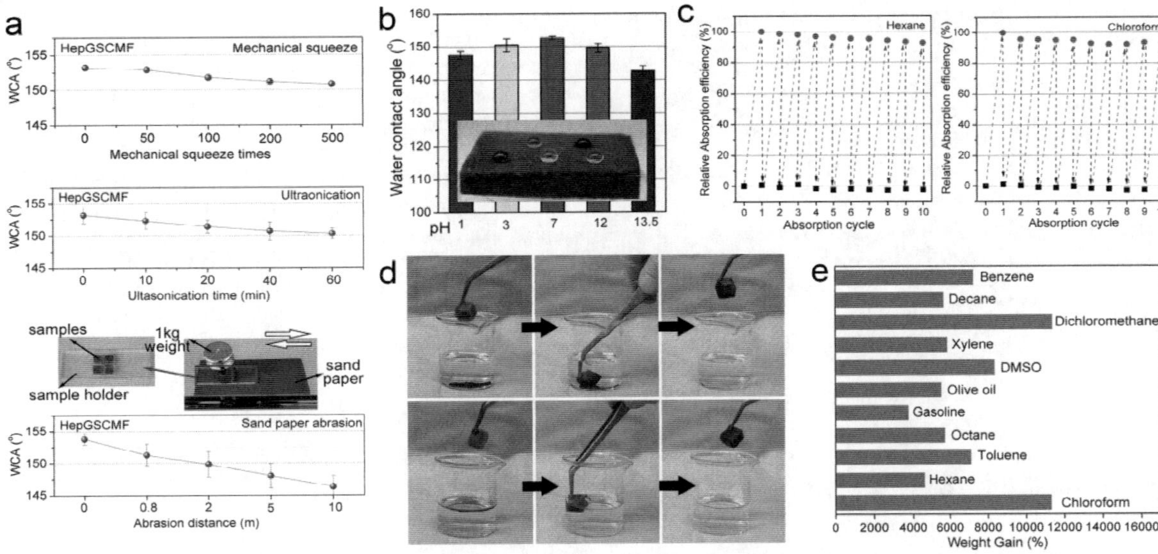

FIGURE 5. a) WCA of HepGSCMF after being subjected to 500 cycles of mechanical squeezing, ultrasonication up to 60 min, and sand paper abrasion up to 10 m. b) The WCA of HepGSCMF to aqueous solutions with a pH ranging from 1 to 13.5. c) Relative absorption efficiency of HepGSCMF to hexane and chloroform over 10 cycles. e) Demonstration of absorption of heavy oil under water and light oil atop water. e) Absorption capacity of HepGSCMF to various oils.

In addition to its superabsorption properties, HepGSCMF can be used for oil/water separation. As shown in Figure 6a, heavy oil (chloroform) was able to flow through thick water layers and be separated by HepGSCMF with high efficiency, as indicated by the microscope image of separated chloroform. In this study, superhydrophilic/underwater superoleophobic foams (DAGSCMF) was also fabricated by simply coating the GSC-grafted MF with dopamine. A pre-wet DAGSCMF was used for water/oil separation as shown in Figure 6c and. Similarly, water was able to flow through light oil (hexane) and high purity water was reclaimed. HepGSCMF and DAGSCMF had the same surface microstructure but different surface chemistries which resulted in their distinctly different surface superwettabilities.

Conventional composite boundaries for superhydrophobic surfaces involve solid–liquid–gas interfaces, which are governed by equation 3. The air trapped beneath the liquid (air pocket) is believed to be the reason preventing liquid infiltration.

$$\cos\theta^{*} = \varphi(\cos\theta + 1) - 1 \tag{3}$$

However, the composite boundary changes to the solid–oil–water interfaces in the case of separation.Thus, Equation 3 should still be applicable, but θ needs to be replaced with CA in liquid (i.e. $\theta_{o\text{-}w}$) instead of air. Because oil and water are generally immiscible, θo-w is typically a large value. For example, HepGSCMF has a WCA of 152.6°, but the WCA increases to 158.4° when it is submerged in hexane, while its CA to chloroform is 0° in air (superoleophilic) (Figure 6e). Since HepGSCMF is superhydrophobic to water, no pre-wetting is needed for separation. As illustrated in Figure 6b, the water layer is supported by air and oil pockets. The case is more complicated for DAGSCMF. DAGSCMF is both superhydrophilic and superoleophilic in air. In order to block the oil flow, the air in the foam needs to be replaced with water, which has a high θo-w. Figure 6f showed that DAGSCMF was able to absorb water and showed a WCA of 0° in air, and its CA to chloroform was 151.3° when measured underwater indicating under water superoleophobicity. Therefore, the pre-wetted DAGSCMF was able to prevent the flow of oil thanks to the water pockets (Figure 6d). In addition to its versatile separation attributes, these modified foams showed excellent flux (over 25 L m^{-2} s^{-1} for HepGSCMF and over 15 L m^{-2} s^{-1} for DAGSCMF), and separation efficiency (over 99% initial efficiency). The high flux was attributed to the low particle loading content, which led to large pore sizes. The high separation efficiency and property stability were attributed to the superwettability and strong bonding between substrate and nanoparticles.

CONCLUSIONS

A novel approach is proposed to manipulate surface wettability of 3D substrates to produce superhydrophobic/ superoleophilic materials and superhydrophilic/ underwater superoleophobic materials. The approach involves layer-by-layer covalent grafting of multidimensional particles that manipulate surface microstructure, and the grafting of

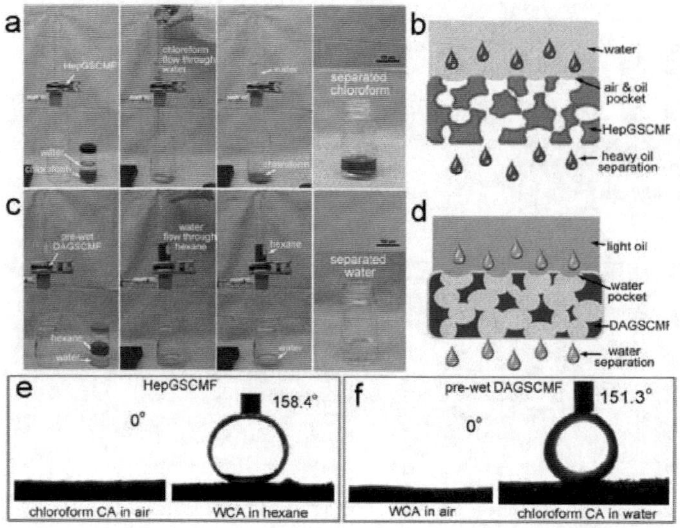

FIGURE 6. a) Demonstration of heavy oil separation from water using HepGSCMF. b) Mechanism of oil/water separation using superhydrophobic/superoleophilic material. c) Demonstration of water separation from light oil using DAGSCMF. d) Mechanism of water/oil separation using superhydrophilic/underwater superoleophobic material. e) Chloroform CA in air and WCA in hexane of HepGSCMF. f) WCA in air and chloroform CA in water of pre-wet DAGSCMF.

chemicals with different functional groups. Covalent grafting of particles improved mechanical performance, reduced necessary particle loading content, and prevented particle disassociation, thereby increasing durability, absorption capacity and performance stability, and prevented contamination of recovered oil or water. Changing the material surface energy resulted in intelligent separation materials that were capable of separating heavy oil from water and water from light oil with unprecedented flux and separation efficiency. Grafting fluoro chains on the surface further enhanced superhydrophobicity and endowed amphiphobicity. Besides the outstanding performance of the materials, this study elucidates the fundamental theory in directing the design of advanced functional materials for a host of applications and providing guidelines for engineering robust surfaces with superwettability.

ACKNOWLEDGEMENTS

The authors would like to acknowledge the financial support of the Office of the Vice Chancellor for Research and Graduate Education, the College of Engineering, and the Wisconsin Institute for Discovery (WID) at the University of Wisconsin–Madison, and the National Natural Science Foundation of China (51603075; 21604026).

REFERENCES

1. Aurell, J.; Gullett, B. K., Environ Sci Technol 44, 9431 2010.
2. Wen, L. P.; Tian, Y.; Jiang, L., Angew Chem Int Edit 54, 3387 2015.
3. Du, R.; Feng, Q. L.; Ren, H. Y.; Zhao, Q. C.; Gao, X.; Zhang, J., J Mater Chem A 4, 938 2016.
4. Gao, X. F.; Xu, L. P.; Xue, Z. X.; Feng, L.; Peng, J. T.; Wen, Y. Q.; Wang, S. T.; Zhang, X. J., Adv Mater 26, 1771 2014.
5. Rana, M.; Chen, J. T.; Yang, S. D.; Ma, P. C., Adv Mater Interfaces 32016.
6. Cao, N.; Yang, B.; Barras, A.; Szunerits, S.; Boukherroub, R., Chem Eng J 307, 319 2017.
7. Zhu, X. T.; Zhang, Z. Z.; Ge, B.; Men, X. H.; Zhou, X. Y.; Xue, Q. J., J Colloid Interf Sci 432, 105 2014.
8. Sasaki, K.; Tenjimbayashi, M.; Manabe, K.; Shiratori, S., ACS Appl. Mater. Interfaces 8, 651 2016.
9. Du, R.; Zhao, Q. C.; Li, P.; Ren, H. Y.; Gao, X.; Zhang, J., ACS Appl. Mater. Interfaces 8, 1025 2016.
10. Brown, P. S.; Bhushan, B., Scientific Reports 52015.
11. Du, R.; Gao, X.; Feng, Q. L.; Zhao, Q. C.; Li, P.; Deng, S. B.; Shi, L. R.; Zhang, J., Adv Mater 28, 936 2016.

Preparation of High-Performance Polylactide Fibers via Manipulating Crystallization with Fibrillar Nucleating Agent

Zhang Huixian (1), Bai Hongwei (1), Fu Qiang (1)*

(1) Sichuan university - Sichaun - China

As an environment-friendly alternative to conventional synthetic polymers, bio-based and biodegradable polylactide (PLA) displays huge application potential in the textile industry owing to its inspiring elastic recovery, moisture regain, and flammability. However, the inefficient tensile strength and dimensional stability make PLA fibers unable to be used in various practical applications. In this work, a facile and effective strategy is demonstrated to improve these properties via simultaneously manipulating crystallinity and lamellae orientation with the aid of a highly active nucleating agent (NA). The NA can be completely dissolved in PLA melt and self-organizes into fine fibrils upon cooling. Under the strong elongational flow field in the melt spinning, the NA fibrils tend to align along the flow direction and subsequently act as nucleation templates to induce the rapid growth of kebab-like PLA lamellae perpendicular to their long axis, finally forming highly orientated crystal structure in melt-spun PLA fibers. As a result, the crystallization manipulation imparts the PLA fibers with superior mechanical strength and heat resistance. Compared with neat PLA fibers, a significant increase of 78% in tensile strength and a dramatic decrease of 1069% in boiling water shrinkage are obtained in the nucleated fibers. This work suggests a promising avenue toward the design and development of high-performance PLA fibers by using fibrillar nucleating agent as a nucleation template to manipulate crystallization in the melt spinning process.

Enhance PLA for 3D Printing

Lanfranconi Matias[a], Salinas Tobias[a], Torres Nicolini Andres[b], Alvarez Vera[b]

[a]*Nairotech Desarrollo e Innovación, Mar del Plata, Argentina.*
[b]*CoMP INTEMA, Mar del Plata, Argentina*

Abstract. Polylactid acid (PLA) is the most common polymer and easy to use for 3D printing. The two major problems that PLA has are its fragility and its low heat resistance (heat deflection temperature, HDT, around 55 °C). In this work, we improve both characteristics by increasing the quantity and proportion of crystallinity. To achieve this we add different nucleation agents: submicron talc and nanoclay (commercial and own lab made), and we compare is effect over the properties.

Keywords: Crystallisation, Polylactic acid, Bioplastic, Additive Manufacturing.

INTRODUCTION

Polylactic acid (PLA) is a commercially available compostable biobased material. It is a transparent and rigid material with a relatively low crystallization rate, relatively easy to process, and this makes it a widely used material for 3D printing. Unfortunately the range of application of PLA is severely limited because of its low heat deflection temperature (around 55–60°C). Increasing the crystallinity of the PLA the crystalline phase maintain the mechanical properties, thus, this paper focus on increasing the crystallinity by adding nucleation agents, Increasing the crystallization speed of PLA is thus desired [1]. Also, its fragility can be enhanced by increasing the quantity and proportion of crystallinity, with more and finer crystals [2].

In processes such as 3D printing, the orientation is limited and the cooling rate not possible to control, is difficult to develop high proportion of crystallinity with fine grains, and thus formulation changes are required. Researchers have studied the crystallization behaviour [3–6] of PLA; and crystallization kinetics [7-9]. In order to increase crystallinity we considered one of the possible routes, add a nucleating agent that will lower the surface free energy barrier towards nucleation and thus initiates crystallization at higher temperature upon cooling.

Talc is a widely used nucleating agent. It was shown that talc nucleates the crystallization of polymers, and in the case of PLA, it is shown that the crystallization half-time can be reduced by more than one order of magnitude to less than 1 min when 1% talc is added [10].

Also nanoclays have been proved that can increase the mechanical behavior of polymer materiales, nad also increase HDT of PLA [11].

In this paper we investigated the effect of microtalc and nanotalc, bentonite clay and nanoclays, on the crystalline content and its effect over HDT and fragility of the polymer over 3D printer samples.

EXPERIMENTAL

Materials

For this work we use PLA Ingeo 4043D from Naturworks (MFR 6g/10min, Notched Izod Impact 16 J/m, HDT 55 °C). The additives we use are: Bentonite (BENT) from Minarmco S.A.; Dellite 67G, 72T and 43B from Laviosa Chimica Mineraria; Hallgreen IM-8830 Ether from Hallstar (ETHER); ultrafine talc Jefine 3CA (NTALC) and coarse talc Luzenac 1445 (MTALC) from Imerys.

Sample Preparation

The material has been prepared at two scales. As a first approach we use an intensive mixer, two screw counter rotating bulk mixer (bravender type) of 25ml capacity for the preparation of the samples. The used temperature was 190 °C during 7 minutes at a speed of 50 and 100 rpm.

After initials characterization, some systems have been selected and we upscale the method of preparation of the sample. As a second step we use a twin screw extruder (D25mm, L/D36) for the material preparation. PLA it was added from the primary entry and the additives from a lateral secondary entry at half of length of the extruder. The material has been pelletized.

To measure mechanical properties, sheets have been prepared using a parallel plate press at 190 °C and 75 Kg/cm^2 of pressure.

Characterization Techniques

Differential Scanning Calorimetry (DSC) measurements were performed in a TA Instruments model Q2000 calorimeter, from 25 to 200 °C at a heating rate of 10 °C/min under nitrogen atmosphere. The degree of crystallinity (X_{cr}) of each sample was calculated from the following equation:

$$X_{cr}(\%) = \frac{\Delta H_f}{w_{PCL} \times \Delta H_{100}} \times 100$$

where ΔH_f is the experimental heat of fusion, w is the PLA weight fraction and ΔH_{100} is the melting enthalpy of the 100% crystalline PLA and its value is 93.0 J/g [12,13]. The melting temperature (Tm) of each sample was also obtained.

Impact test were performed in a Ceast Fractovis gravity falling drop with a mass of 23.5Kg and a velocity of 1m/s. Impact resistance was calculated and compared.

Heat deflection temperature was tested using a TA DMA Q800, from 25 to 200 °C at a heating rate of 2 °C/min.

RESULTS AND DISCUSSION

NanoClay Studies

Table 1 shows the different nanocomposites that have been mixed.

TABLE 1. PLA/Clay Nanocomposites.

Composite	Clay	*Clay Content* (%)	*Mix velocity (rpm)*
PLA3%D43B50rpm	Dellite 43B	3	50
PLA3%D43B100rpm	Dellite 43B	3	100
PLA3%D67G50rpm	Dellite 67G	3	50
PLA3%D67G100rpm	Dellite 67G	3	100
PLA3%D72T50rpm	Dellite 72T	3	50
PLA3%D72T100rpm	Dellite 72T	3	100
PLA4%D43B50rpm	Dellite 43B	4	50
PLA4%D43B100rpm	Dellite 43B	4	100
PLA4%D67G50rpm	Dellite 67G	4	50

PLA4%D67G100rpm	Dellite 67G	4	100
PLA4%D72T50rpm	Dellite 72T	4	50
PLA4%D72T100rpm	Dellite 72T	4	100

The following photos, **Figures 1** and **2**, shows the color of the composite as they came out from the mixer.

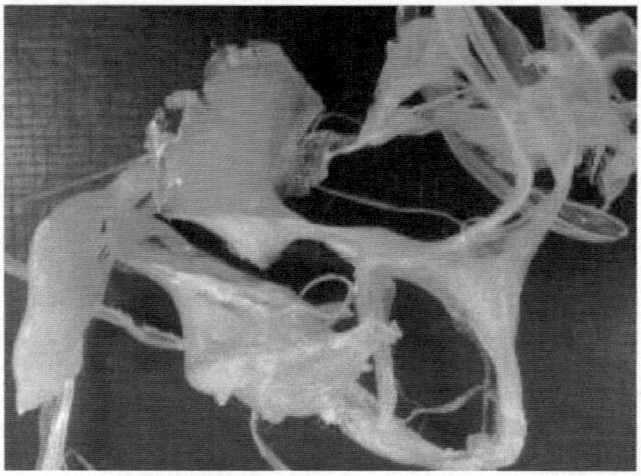

FIGURE 1. PLA3% D43B50rpm

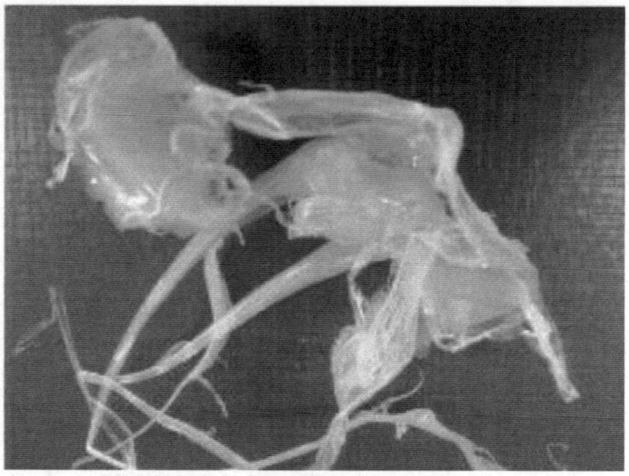

FIGURE 2. PLA4% D72T100rpm

All the composites were successfully mix with the conditions described. The look homogeneous at naked eye. XRD experiments are carried out.

DSC experiments were carried out and no significant increase on the crystallinity of the PLA has been found. As an example, DSC results of net PLA and one of the nanocomposites are presented on Figures 3 and 4:

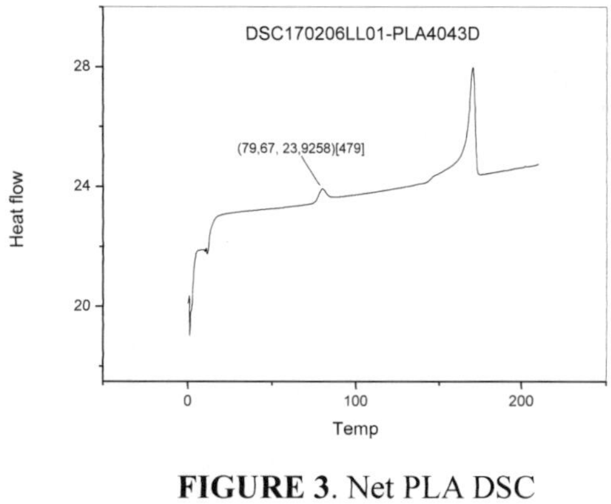

FIGURE 3. Net PLA DSC

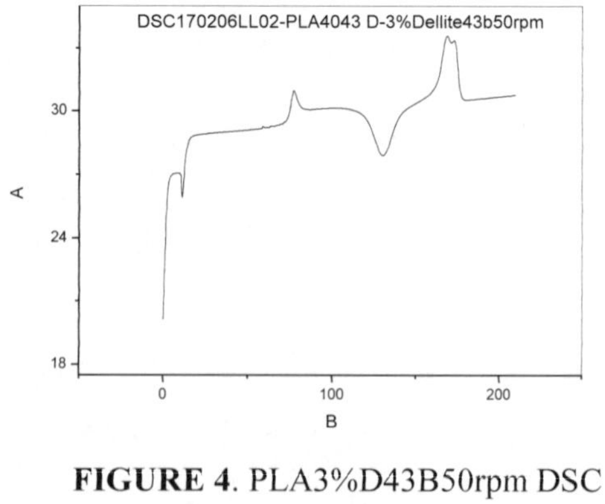

FIGURE 4. PLA3%D43B50rpm DSC

CONCLUSIONS

In this work, the effect of different nanoclays, concentration and processing conditions, over crystallinity have been studied. For the continuation of this work it is necessary to complete the calorimetry studies, and perform the

DRX, DMA and Impact experiments over the samples. Also new samples are being processing with the use of different talcs, and the incorporation of an ether impact improver. The same studies are being done.

ACKNOWLEDGMENTS

Authors acknowledge to National Scientific and Technical Research Council (CONICET) (PIP 00617), National Agency for the Promotion of Science and Technology (ANPCyT) (FS Nano 004), and the National University of Mar del Plata (UNMdP) (ING392/14) for the financial support.

REFERENCES

1. Shen L., Haufe J., Patel M. K.: Product overview and market projection of emerging biobased plastics: PRO-BIP 2009. Final Report, Utrecht University (2009).
2. William G. Perkins: Polymer toughness and impact resistance. Polymer Engineer and Science, 39, 2445–2460 (1999).
3. Nijenhuis A. J., Grijpma D. W., Pennings A. J.: Highly crystalline as-polymerized poly(L-lactide). Polymer Bulletin, 26, 71–77 (1991).
4. Kalb B., Pennings A. J.: General crystallization behaviour of poly(L-lactic acid). Polymer, 21, 607–612 (1980).
5. Cohn D., Younes, H., Marom G.: Amorphous and crystalline morphologies in glycolic acid and lactic acid polymers. Polymer, 28, 2018–2022 (1987).
6. Migliaresi C., de Lollis A., Fambri L., Cohn D.: The effect of thermal history on the crystallinity of different molecular weight PLLA biodegradable polymers. Clinical Materials, 8, 111–118 (1991).
7. Urbanovici E., Schneider H. A., Brizzolara D., Cantow H. J.: Isothermal melt crystallization kinetics of poly(Llactic acid). Journal of Thermal Analysis and Calorimetry, 47, 931–939 (1996).
8. Tábi T., Sajó I. E., Szabó F., Luyt A. S., Kovács J. G.: Crystalline structure of annealed polylactic acid and its relation to processing. Express Polymer Letters, 4, 659–668 (2010).
9. Kolstad J. J.: Crystallization kinetics of poly(L-lactideco-meso-lactide). Journal of Applied Polymer Science, 62, 1079–1091 (1996).
10. Ke T., Sun X.: Melting behavior and crystallization kinetics of starch and poly(lactic acid) composites. Journal of Applied Polymer Science, 89, 1203–1210 (2003).
11. Suprakas Sinha Ray, Kazunobu Yamada, Masami Okamoto, and Kazue Ueda. Polylactide-Layered Silicate Nanocomposite: A novel Biodegradble Material. American Chemical Society, NanoLetters, 2, 1093-1096 (2002).
22. Turner J. F., Riga A., O'Connor A., Zhang J., Collis J.: Characterization of drawn and undrawn poly-L-lactide films by differential scanning calorimetry. Journal of Thermal Analysis and Calorimetry, 75, 257–268 (2004). DOI:
13. Rayner J. H., Brown G.: The crystal structure of talc. Clays and Clay Minerals, 21, 103–114 (1973). DOI:

Examination of the causal relation between the processing parameters in a direct impregnation device and the fiber wetting of glass- fibers

Stephan Lehmann, Ivonne Jahn and Peter Michel

Fraunhofer Institut for Microstructure of Systems and Materials IMWS
Walter-Hülse-Straße 1
06120 Halle (Saale)
Germany

Abstract. In this work the process of the production of unidirectional tapes (UD-Tapes) is observed. UD-Tapes are made of a thermoplastic matrix and reinforcing fibers. The process is a kind of pultrusion process and is used for an economical production of the tapes. For good properties of the composites the wetting and the fiber distribution in the tapes is crucial. In this work two different effects considered. The first one is the fiber distribution over the tape width and the second one is the fiber wetting of the glass fibers. Therefore a theoretical approach of the fiber wetting was done by Darcy´s Law and the occurring physical effects. These results then compared to the theoretical approach. In this work the differences in the spreading of glass fibers and the corresponding fiber wetting is shown.

Keywords: Composites, Glass Fibers, Pultrusion, Thermoplastics
PACS: 81.05.Qk

THE DIRECT IMPREGNATION PROCESS

The direct impregnation process is a high efficient process to process thermoplastic fiber reinforced composites. In Figure 1 the processing is shown in an abstracted way. As seen the process consists of three different parts, which are important for the treatment. First one is the supply of the fibers, with the pay- off creel, with the fiber bobbins, and the fiber spreading. At the pay off- creel the tension in fibers can be adjusted. The spreading unit is used for a pre-spreading of the fibers. The spreading is induced by the tension in the fibers and the radii of the spreading bars.

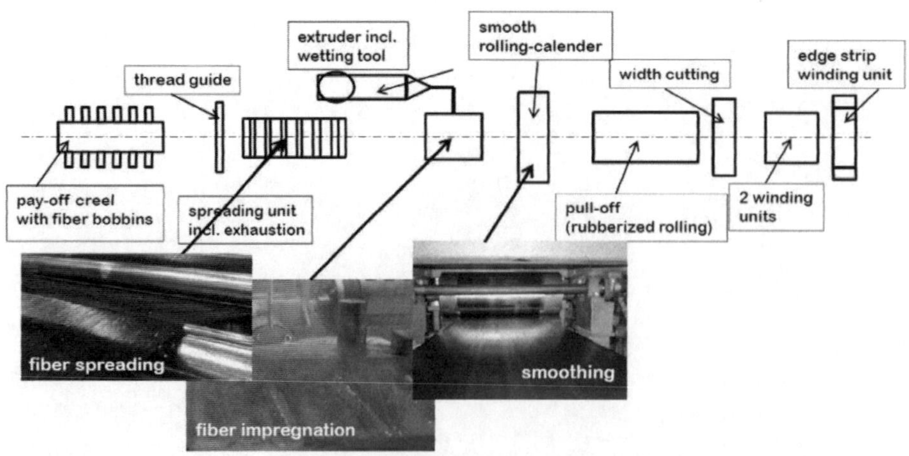

Figure 1: The impregnation process for the production of thermoplastic reinforced composites with a unidirectional reinforcement

The main part of the plant consists of the extruder and the impregnation device. The extruder melts the thermoplastic granule and provides it to the impregnation device. In the device the fibers are impregnated by the matrix due to pressure gradients in the fiber material caused by impregnation pins and the fiber tension. The third part of the plant consists of a calender, for the additional smoothing and cooling of the tape, and a pull off unit.

Theoretical assumptions for the impregnation process

For the impregnation of the fibers with matrix 5 different equations are necessary, to get an estimation of the time and the degree of impregnation [1, 2, 3, 5, 6]. The Euler-Eytelwein Equation (1) where F_1 is the occurring force at the pin, F_0 is the force prior to the pin, μ the friction coefficient between the surface of the impregnation device and α angle of impregnation.

$$F_1 = F_0 \cdot e^{\mu\alpha} \quad (1)$$

In this equation the exponential increase of the force over the impregnation angle can be shown. This parameter is crucial for the estimation of the pressure gradient (2) in the fibers. In the equation p is the pressure in the fibers and A is the area of the surface of the impregnation pins.

$$\Delta p = \frac{F_1}{A} \quad (2)$$

The pressure on the fibers has also a direct impact on the spreading of the fibers and the homogenization of the tape. Challenging is the prediction of the spreading under different forces respective pressures.

$$K = \frac{r_f^{\,2} \cdot (1 - v_f)^3}{4 \cdot \kappa \cdot v_f^{\,2}}$$

Equation above is the Carman-Kozeny equation. It is used for the prediction of the permeability of porous materials in the transversal direction. In this case the glass fibers are the porous material. For the calculation the radius of the fibers r_f, the filament volume fraction inside the fiber bundle v_f and the Kozeny-Constant κ, which is estimated with 17.9 [7].

$$\Delta t = \frac{\eta \cdot z^2}{2 \cdot K \cdot \Delta p} \quad (4)$$

For the impregnation time equation 4 is used. The viscosity η of the polymer and the impregnation length are shown in this equation. As seen the impregnation time has a linear correlation to the viscosity of the polymer, the pressure gradient and the permeability of the fibers. A quadratic impact on the time has the impregnation length z.

The radius of the fibers in equation 3 can be calculated by the equation 5, where n is the number of filaments in the fiber bundle, r_{fil} the radius of the single filament and p_d the packaging density of the fibers in the bundle.

$$r_f = \sqrt{\frac{n \cdot r_{fil}}{p_d}} \quad (5)$$

Calculation of the theoretical impregnation time

In Figure 2 an impregnation device is shown. The important geometrical boundary condition is the impregnation angle. In this device are in sum the impregnation angle is 700°. The radius of the pins is 28 mm.

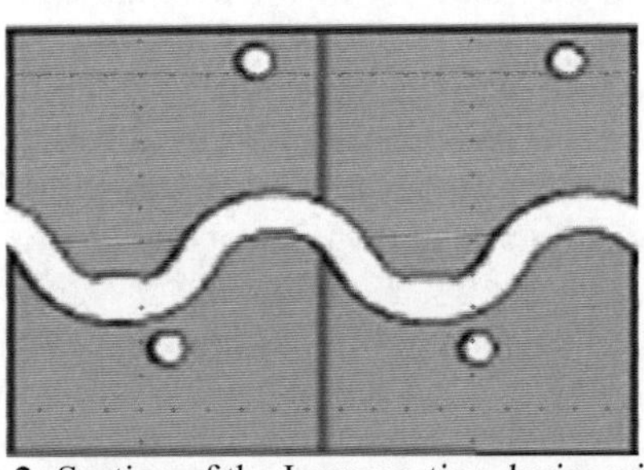

Figure 2: Section of the Impregnation device with pins

Due to this angle the maximal impregnation time can be calculated, by knowing the number of impregnation pins and the haul of speed (8.5 m/min) of the winding unit.

$$t = \frac{n_{pin} \cdot \frac{\alpha}{360°} \cdot 2 \cdot \pi \cdot r_{pin}}{v} \quad (6)$$

Equation 6 estimates the impregnation time t by the geometrical factors of the number of pins n_{pin} and the radius of the pins r_{pin}. The haul off speed v corresponds direct with the impregnation time. With the boundary conditions of the device the calculated impregnation time in the device is 2.41s. This is just the time at the pins, where the impregnation happens, due to a pressure gradient in the filaments. Between the pins there is no impregnation of the fiber bundles, because there are just working capillary effects, which are very slow in comparison to the pressure induced impregnation.

For the experiments three different types of glass fibers used, see Table 1. The used Polypropylene is a Borealis Copolymer with a MFI of 100. The used fibers have differences in the linear weight and the diameter, which corresponds direct with the needed time for the impregnation.

Table 1: Properties of the glass fibers

Name of the fiber	Linear weight [tex]	Diameter of the filament [µm]
E – Glass	1 200	17
S- Glass	1 650	9
ECR- Glass	2 400	17

For the calculation of the of the theoretical impregnation times, some estimations are needed. First of all the Kozeny Constant is 17.9 and the packaging density p_d of the filaments in the fiber bundle is 0.77 [8]. The fiber volume content in the fiber is 0.99, regarding to additives on the filaments (sizing e.g.). In Table 2 the results of the calculation are shown.

Table 2: Impregnation conditions for the glass fibers

Name of the fiber	Radius of the fiber bundle r_f [µm]	Permeability of the fibers K [m^2]	Average pressure in the device p [bar]
E – Glass	213	$6.5 \cdot 10^{-12}$	15
S- Glass	340	$1.6 \cdot 10^{-11}$	7.5
ECR- Glass	298	$1.27 \cdot 10^{-11}$	9.7

As shown there are some differences in the permeability of the glass roving. The E- Glass fibers have a lower permeability with the power of ten in comparison to the other two, what was not expectable by taking account on the fiber diameter which is for both fibers the same. Surprisingly the S- Glass and ECR- type have almost the same permeability, even though the diameter of the filaments divers. For every kind of filaments the pressure differs, due to the difference in the number of bobbins. The force that holds back one bobbin is about 1N. This means this force has to be multiplied by the number of bobbins. In the tape a fiber volume content of 40 % should be realized, so a different number of bobbins are needed for the impregnation process. That means for the E- glass fibers 80 bobbins needed, for the S-Glass fibers 80 bobbins and for the ECR-glass 60 bobbins. So this differences leads to changes in the average pressure in the device. For the calculation of the average pressure a friction coefficient between the surface of the device (steel) and the glass fibers of $\mu = 0.5$ was estimated.

In Figure 3 is the needed time for the full wetting of the fibers shown. The fibers of E-Glass and S-Glass have almost an equal impregnation time. There are differences in permeability (see Table 2) and in the thickness of the fiber bet to. In the equation 4 the thickness of the tape has a quadratic relationship to the impregnation time. Due to this the result in the impregnation time is the same. The fiber bundles of ECR Glass have the highest needed impregnation time. By comparing the needed impregnation time for the fibers and the residence time of the fibers in the impregnation zones of the device, the fibers should be theoretically impregnated 100%. The calculations made under the assumption of a uniform distribution of the filaments and no additives on the fibers that reduce the permeability, because they acts like an adhesive.

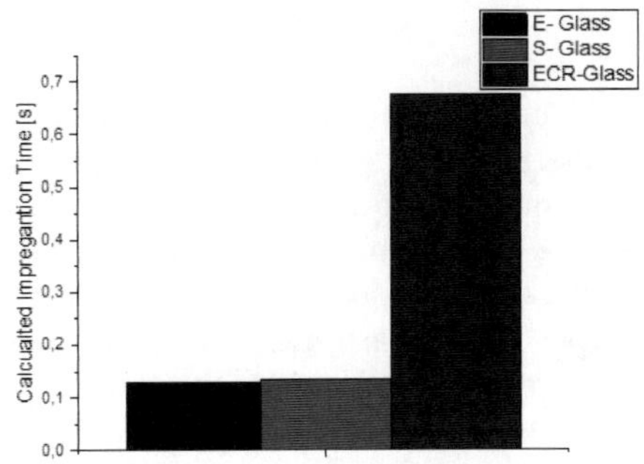

Figure 3: Calculated Impregnation Time for the different Types of Fibers

Experiments for the degree of impregnation regarding to the fiber distribution in the tapes

The fiber distribution gives a prediction of the thickness of the tapes at the respective position. The tapes have been cut into slices with a width of 50 mm. The slices have been pyrolysed by heat treatment to get information about the fiber mass and void content in the tape (DIN EN 2564 [9]). In Figure 4 (left) the distribution of the fibers in the tape are shown. For all three types of fibers an identical distribution can be seen. The accumulation of the fibers is skewed to the left in the tape, what means the highest fiber mass content is there and more fibers layered about each other. This influences the degree of impregnation of the tapes, due to higher impregnation lengths. For the degree of impregnation the distribution of the void content has to be the same, by taking into account the theory, that the impregnation length has a square influence.

The distribution of the fiber mass contents corresponds with the void content in tapes (Figure 4 (right)). It can be shown, that there are differences in the fiber wetting in comparison to the various types of fibers. The void content in the types varies from 0.3% to 5 % for the S- glass, from 0.2 % to 8% for the ECR type and from 1.8 % to 7 % for the E-glass. The distribution of the void content shows the same course as the distribution of the fiber mass content. The differences in standard deviation in the void content differ a lot between the different types of fibers in comparison to the fiber mass content. This shows the impact of the fiber wetting is not just by the fiber distribution, but in the kind of used fibers. For the S-glass the impregnation is the best, but it has the same fiber distribution than the other two types.

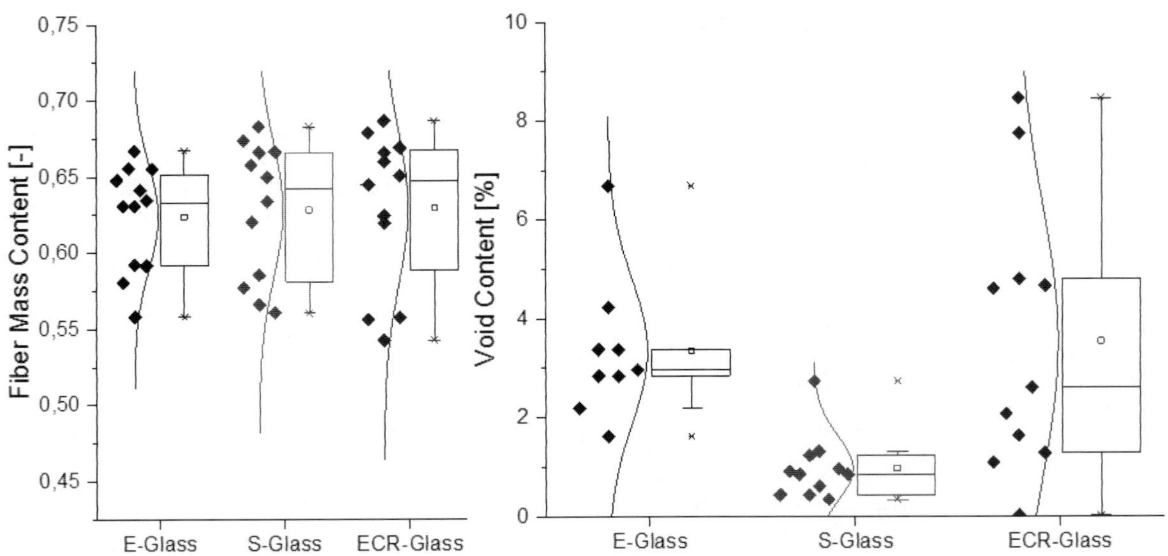

Figure 4: Distribution of the fiber mass content (left) and void content (right) in the UD-tapes

Conclusion

As shown in the articles before the calculation of the impregnation time is possible to get estimation, whether it takes seconds or longer to impregnate the fiber up to 100 %. For an accurate prediction in the direct impregnation process it is necessary to take other parameters into account. The calculated impregnation time for the fibers has been much smaller than the residence time of the fibers at the impregnation pins. It was shown that the theoretical and the processing time differ much. This could be seen best in the void content of the tape. Theoretical the fibers have to be full impregnated, but two types of the glass fibers (E- and ECR-Glass) are not full impregnated. To get more reliable times it is necessary to adjust the rule of Darcy and the corresponding equations. Especially the calculation of the permeability contains errors, because the dominant impregnation is transversal to the tape and not longitudinal. Therefore a new method has to be evaluated.

On the other hand it is a material specific problem. For the calculation just the geometric parameters of the fibers used, but this is not sufficient. For the inter phase strength it is necessary to treat the fibers with a sizing. This sizing is acting like an adhesive, what lowers the permeability.

The third difficulty to forecast the impregnation time is induced by the processing. As shown in the upper article there is not a homogeneous fiber bet. Due to this there are differences in the impregnation length that influences the needed impregnation time.

Impregnation by the use of S-glass can be predicted in a very reliable way. As shown the void content of the S-Glass tape is on a very low scale. Due to this the impregnation time fulfills the requirements of the processing of S-glass fiber reinforced tapes.

For more reliable result in the forecast of the impregnation process, for all types of fibers, it is immanent important to get into account of the fiber material (sizing, geometric parameters, height of the fiber bet), the process (fiber distribution) and a better understanding of the transversal permeability of the fibers.

ACKNOWLEDGMENTS

The authors are thankful to the “Investitionsbank Sachsen Anhalt” and “European Regional Development Fund” who support this project. The name of the project, where these results has been evaluated, is “Entwicklung und Erprobung einer thermoplastisch gebundenen Blattfeder” with the project number 1604/00020.

REFERENCES

1. P. J. Bates, “Melt Impregnation of a Glass Roving in a Thermoplastic Ppultrusion Compounding Process”, Doctor of Engineering Thesis, McGill University, 1993
2. T. Weustink, “Development of a Rapid Impregnation Device”, PhD Thesis, Technische Universiteit Delft, 2008
3. P. J. Bates et al, “Pressure Built up During Melt Impregnation” Composite Science and Technologie 62 2002, pp 379 – 384
4. D. Shou et al, “Longitudinal Permeability Determination of Dual-Scale Fibrous Materials”, Composites Part A 68 2015, pp 42 – 46
5. M. F. Foley and J. W. Gillespie, “Modeling the Effect of Fiber Diameter and Fiber Bundle Count on Tow Impregnation during Liquid Molding Process”, Journal of Composite Materials Vol. 39,
6. 178K. Lang, “Zur Erweiterung des Spektrums von Faser-Kunststoff-Verbunden mit thermoplastischer Matrix- Einsatz unidirektional faserverstärkter Halbzeuge in hoch belasteten Strukturbauteielen des Automobil- und Flugzeugbaus”, Dr.-Ing. Thesis, 2012
7. 2005P. J. Bates et al, “Compaction and Transverse Permeability of Glass Rovings”, Applied Composite Materials 8: 2001, pp 163 –
8. F. Fodor,” The Densest Packing of 19 Congruent Circles in a Circle”, Geometrica Dedicata 74, 1999, pp 139 - 145
9. DIN EN ISO 2564, “Kohlenstofffaser-Laminate – Bestimmung der Faser-, Harz- und Porenanteile, Deutsche Fassung EN 2564, 1998

Experimental Study on the Processing Conditions of In-Mold Coating for Injection-Molded Plates

Kim Dongmi [(1)], Lee Hosang [(1)]*

[(1)] Korea National University of Transportation - Chungbuk - Korea

In-Mold Coating(IMC) is novel coating method which achieves simultaneous injection molding and surface coating. In this paper, a mold for in-mold coating experiments was fabricated, which consisted of an overflow area, a coating channel, and sealing edges. The coating material used was two-component polyurethane with good transparency, high glossy, durability, and scratch resistance. The effects of the processing conditions of the in-mold coating on the bubbles and mixing characteristics in the injection molded plates were investigated. The bubbles that appeared over the substrate surface decreased, as the injected weight of the coating material increased. On the other hand, the bubbles increased with an increase of mass-flow rate. When the discharged flow rate was low, inhomogeneous surfaces appeared due to the poor mixing between polyol and isocyanate. It was found that the nozzle pressure of more than 180 bar was required to achieve uniform mixing.

Dual delivery of active anti-bactericidal agents and bone morphogenetic protein using biodegradable sheath-core-structured drug-eluting nanofibers

Liu Shih-Jung ***(1)*, Hsu Yung-Heng (2)***

(1) Chang Gung University - Tao-Yuan - Taiwan, (2) Chang Gung Memorial Hospital - Tao-Yuan - Taiwan

In this study, we developed biodegradable sheath-core-structured drug-eluting nanofibers for sustainable delivery of antibiotics (vancomycin and ceftazidime) and recombinant human bone morphogenetic protein (rhBMP-2) via electrospinning. To prepare the biodegradable sheath-core nanofibers, we first prepared solutions of poly(D,L)-lactide-co-glycolide (PLGA), vancomycin, and ceftazidime in 1,1,1,3,3,3-hexafluoro-2-propanol (HFIP) and rhBMP-2 in phosphate-buffered solution. The PLGA/antibiotics and rhBMP-2 solutions were then fed into two different capillary tubes controlled by two independent pumps for coaxial electrospinning. The electrospun nanofiber morphology was observed under a scanning electron microscope. We further characterized the in vitro antibiotic release from the nanofibers via high-performance liquid chromatography and that of rhBMP-2 via enzyme-linked immunosorbent assay and alkaline phosphatase activity.

Damage Of PE-HD Induced By Liquid Media – FNCT Testing And Imaging Analysis

Schilling Markus [(1)]*, Böhning Martin [(1)], Niebergall Ute [(1)]

[(1)] Bundesanstalt für Materialforschung und -prüfung - BAM - Berlin - Germany

Due to the increasing demand for utilization and improvement of energy efficient materials, especially concerning requirements in lightweight construction and design, polyolefin materials are used extensively and in an increasingly broad range of applications. Although loaded with stresses under yield stress, preformed voids, inhomogeneities or notches can possibly lead to damage and unexpected failure induced by liquid media in polyolefin materials. For PE-HD, slow crack growth (SCG) as well as environmental stress cracking (ESC) are relevant damage mechanisms. SCG appears without influence of a surrounding medium or in "inert" media whereas ESC occurs in "active" media, which have decisive influence on damage mechanism and time to failure. To characterize the inherent resistance of the material against those damage mechanisms, the Full-Notch Creep Test (FNCT) is widely used. Therefore, the FNCT is of paramount importance particularly for the assessment of pipe and blow molding types of PE-HD – especially for automotive but also for most other high-performance applications. In this study, the FNCT – usually applied as a standardized testing method (ISO 16770) using a few universal model liquid media – is extended by investigations of selected relevant PE-HD materials with a variety of properties also in real media. Mainly topical fuels, customary in the market such as diesel and biodiesel are examined. Especially the influence of temperature, and the ESC behavior of PE-HD in media that are sorbed to a significant extent, are addressed. The investigations were performed using a novel FNCT-device with 12 individual sub-stations, each equipped with individual electronic stress and temperature control and continuous online monitoring of the specimen elongation. Additional to acquired results concerning time to failure and elongation behavior, imaging techniques, such as light microscopy (LM), laser scanning microscopy (LSM) and X-ray computed tomography (CT scan) were utilized to obtain data concerning crack propagation and media migration. Correlations of elongation behavior and data derived from imaging analysis such as fracture plane roughness were obtained.

Fiber-Metal Laminates with Pins Deposited by CMT PIN

Iaroslav Skhabovskyi[a], Natassia Lona Batista[b,c], Cesar Augusto Damato[d], Ruham Pablo Reis[a], Edson Cocchieri Botelho[b], Américo Scotti[a,e]

[a]Laprosolda - Center for Research and Development of Welding Processes, Faculty of Mechanical Engineering, Universidade Federal de Uberlândia (UFU), Uberlândia, Brazil; [b]Materials and Technology Department, School of Engineering, Universidade Estadual Paulista (UNESP), Guaratinguetá, Brazil; [c]Structures and Composite Materials Laboratory, Department of Mechanical Engineering, McGill University, Montreal, Canada; [d]ALLTEC Materiais Compostos, São José dos Campos, Brazil; [e]Production Technology West, University West, Trollhättan, Sweden

Abstract. The increasing demand for high-performance-lightweight structures has strongly stimulated the evolution of fiber-metal laminates. The adhesion between metal and composite is key to improve the effectiveness of this combination. Specifically aiming at improving the performance of composite-metal joints, CMT welding has been successfully considered for hybrid joining approaches that couple metal pin anchoring with adhesive bonding. However, this hybrid joining concept has not been extended to fiber-metal laminate panels yet. Thus, this work aims to employ metal pins deposited by CMT PIN to reinforce fiber-metal laminate panels. The methodological approach included panels with different pin deposition patterns that were compared with panels without pins in terms of energy dissipation during drop-weight testing, impact damage characterization and finally in terms of damage tolerance by means of compression and buckling tests after impact. Concerning impact energy dissipation, the pins did not make the fiber-metal laminate panels more brittle and the change in deposition pattern did not show any significant effect on the capacity they have to dissipate impact energy. Overall, regardless of the deposition pattern, the pins increased the maximum permanent deformation of the panels after impact, especially if less spaced. The in-plane damage areas were larger in the panels with pins, corroborating the damage depth profile results and their connection with impact energy dissipation. The compression and buckling tests indicated that the pins have potential to improve damage tolerance of fiber-metal laminate panels, since a less-catastrophic behavior was observed. Pin anchoring seems to retard propagation of debonding between metal sheets and composite and hold back delamination within the composite part.

Keywords: Fiber-metal laminate panels; Pin anchoring; CMT PIN; Composite-metal hybrid joining.
PACS: 81.05.Qk; 81.05.Bx.

1) INTRODUCTION

The increasing demand for high-performance-lightweight structures has strongly stimulated the development of fiber-metal laminates (FMLs). Cortes and Cantwell [1] describe FMLs as hybrid composite structures based on thin sheets of metal alloys and plies of fiber-reinforced polymeric materials. Vlot [2] states that FMLs combine the advantages of metal materials and fiber reinforced matrix systems. Sinmazçelik et al. [3] justify this combination by considering that metals are isotropic materials, have high bearing strength and impact resistance and are easy to repair, while full composites generally have high strength and stiffness and excellent fatigue characteristics. Botelho et al. [4] state additionally that composite materials provide excellent corrosion resistance. Despite exhibiting low weights, according to Seydel and Chang [5], composite materials are very susceptive to impact damage. On the other hand, metals are in general heavier, but might have excellent impact resistance. The combination of these materials, and then of these properties, has been early highlighted in the development of FMLs. As per Vlot [6], impact damage zones of FMLs are smaller than those found in fiber-reinforced composites alone. In summary, FMLs are characterized by a balance of low structural weight and high strength and stiffness, when compared with metals, including aerospace grade aluminum alloys.

Ucsnik et al. [7] showed that the use of pins deposited by CMT PIN on the metal side improved the strength of metal-composite joints, when compared to simply bonded ones. Later, Ucsnik et al. [8] and Stelzer et al. [9] confirmed that the pins might be applied to improve the performance of composite-composite joints. They used thin metal sheets with pins on both sides, like inserts, between composite lap joints. According to Graham et al. [10], advanced hybrid joining (adhesive bonding with CMT PIN anchorage) was able to consistently outperform the co-cured (adhesive bonding only) specimens in terms of strength and energy absorption at quasi-static and high loading rates, as well as in aspects of damage tolerance after drop-weight testing, environmental durability and mechanical fatigue performance. The CMT (Cold-Metal Transfer) PIN is based on its parent process called CMT, in which metal wire is continuously fed and an arc (protected by gas) is open between the wire tip and the material to be welded, as in all gas metal arc welding (GMAW) versions. The CMT PIN embarks a specific program for realizing pin depositions. After short-circuiting, it causes the wire to stay leaning against the weld pool, with no arc, allowing cooling and consequent welding of the wire to the base metal. Shortly thereafter, a low current is imposed in such a way to heat and soften the welded wire close to its mid length (between the torch and the work piece), but not enough to break it and open the arc, as it is controlled by the CMT PIN software. With the reversing motion started, the consequent soft wire is subjected to a tensile force and is broken (similarly to a tensile test), leaving a pin deposited on the base metal. By changing some of the deposition parameters, it is possible to form pins of different sizes and geometries, as explained in a Fronius [11] literature source.

The CMT PIN process seems to have found its main application in enhancing the performance of composite-metal advanced hybrid joints. However, its potential might be also applied to FML panels, as the authors of this work are proposing; arrays of pins on the metal sides would act as an agent of anchorage for the composite sides throughout the panels, contributing to the performance along with the adhesive bonding between the composite and metal sheets. Thus, this work aims to employ metal pins deposited by CMT PIN on metal surfaces used as layers of FMLs. The methodological approach includes comparing miniaturized FML panels with different pin deposition patterns to a reference (without pins) FML panel and to a panel consisting only of composite, in terms of energy dissipation during drop-weight testing, impact damage extent and compression and buckling tests after impact.

2) FABRICATION OF MINIATURIZED PANELS

The miniaturized FML panels reinforced with metal pins were produced as illustrated in Fig. 1. The FML panels without pins were produced in the same way, with exception of the pin deposition, whereas the panels consisting only of composite were produced only by laying-up of prepregs, followed by pressing and curing. The metals parts were composed of AISI 430 stainless steel sheets (200×80×0.4 mm). A Fronius TransPuls Synergic 5000 power source, a VR7000-CMT wire feeder, a PullMig CMT torch and a RCU 5000i remote control unit, with DB0875 data base selected (synergic line CrNi 19 9 PIN), were used for pin deposition. A 6-axis welding robot was used to move the torch according to the deposition pattern. An AWS ER309L filler wire with nominal diameter of 1 mm was used as pin material and Ar with 4 % of CO_2 gas at a flow rate of 8 L/min for protecting the deposition. The torch was always maintained perpendicular to the metal parts. The metal surfaces were all homogeneously wiped with acetone wetted cloth. Other input parameters, such as contact-tip to work-piece distance (5 mm), metal sheet temperature (room temperature ≈ 27 °C) and pin deposition sequence were kept unchanged. The CMT PIN process was then parameterized to get minimum-height (considering the thickness of the FMLs) and small ball-head pins. Ucsnik et al. [8] has shown that the ball-head shape presents potential in metal-composite hybrid joining. Pins of 2.50 ± 0.06 mm in height and 1.40 ± 0.03 mm in head diameter were produced, resulting in an average pin weight of 0.023 ± 0.002 g.

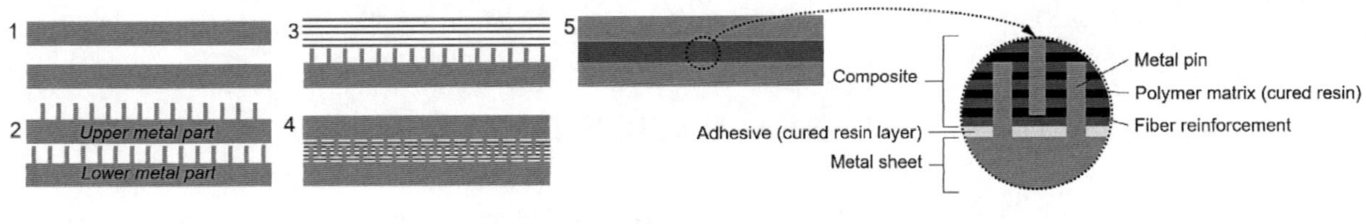

FIGURE 1. Sequence of fabrication of the miniaturized FML panels reinforced with metal pins, where: 1 - cleaning of metal parts; 2 - deposition of pins; 3 - laying-up of prepregs on the lower metal part; 4 - placing of the upper metal part, pressing and curing; 5 - FML panel

After pin deposition, all metal parts (with and without pins) went through ultrasonic cleaning by immersion in acetone for eight minutes to remove grease, welding fumes and other unwanted substances, which could impair the adherence of composite layers. For composite production, prepregs made of glass fiber (8-Harness Satin Weave) and epoxy resin from Hexcel Corporation® (product data 7781-38"-F155) were utilized. The prepreg layers were stacked on top of the lower metal parts (with or without pins), always aligning the warp yarns with their length and, consequently, the filling yarns with their width. Next, the upper metal parts were placed aligned to the prepreg layers and lower metal parts. All prepreg handling was executed in a white room wearing gloves, avoiding to the maximum the contamination. All miniaturized panels were wrapped with a thin film of polyamide (release agent) and then processed in groups of three in a CARVER® CMG100H-15-C hot-curing press, with curing cycle as recommended by the prepreg manufacturer. After curing, the excess of composite material at the edges of each panel was removed with a band saw followed by belt sanding, reaching panels of 200×80×4 mm.

Three types of miniaturized panels were made: 1) composite only; 2) conventional FML panel (without pins), i.e., metal-composite-metal (M-C-M); 3) FML panel with pins, i.e., pined metal-composite-pined metal (MPin-C-MPin). The MPin-C-MPin type was produced with 10 and 20 mm of spacing between the pins with two deposition patterns (Fig. 2(a) and Fig. 2(b)). The pins of the upper metal parts were deposited with an adequate displacement in relation to the pins of the lower metal parts, avoiding contact between the upper and lower pins and providing homogenous pin distribution within the composite. Each of the resulting six miniaturized panels was replicated once. An example of a cross section of a FML panels with pins is shown in Fig. 2(c).

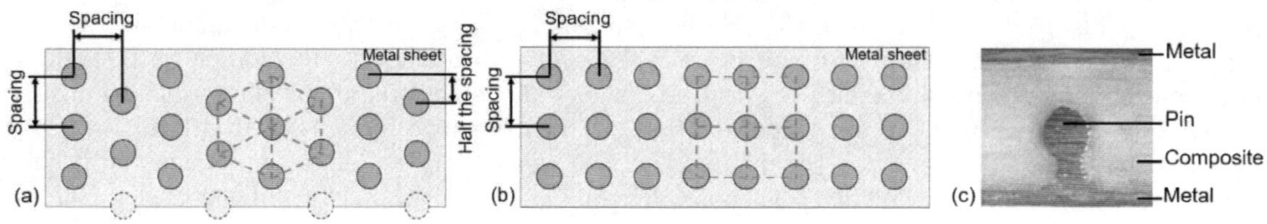

FIGURE 2. Illustration of deposition patterns of MPin-C-MPin panels, where: (a) Hexagonal; (b) Squared and (c) Cross section of pin inside the composite

As shown in Table 1, an attempt to keep the same thickness (approximately 4 mm) for all panels was made by having a different number of prepreg layers per type of panel. Pin density was estimated for each pertinent case, by considering the respective number of pins divided by the panel surface area (320 cm^2). The FML panels with higher density of pins were slightly thicker because of the consequent larger total volume of pins pushed into the composite parts. However, concerning mass density, usually crucial for FMLs, the presence of pins has only marginal effect, regardless of pin density. The overall average mass density of FML panels with pins was 3.1 ± 0.1 g/cm^3, the same as in the case of M-C-M (no pins). Thus, the pins, despite in large number, seem to have their potential to add weight counterbalanced by the mass/volume of composite replaced by their own mass/volume.

TABLE (1). Characteristics of the miniaturized FML panels

Panel type	Pins per panel	Prepreg layers per panel	Actual thickness (mm)	Pin density (pin/cm^2)	Mass density (g/cm^3)
Composite only	0	22	4.24 ± 0.00	0	2.1
M-C-M (no pins)	0	19	4.34 ± 0.00	0	3.1
MPin-C-MPin Hexagonal 10 mm	285	16	4.29 ± 0.00	0.89	3.1
MPin-C-MPin Hexagonal 20 mm	63	16	3.93 ± 0.01	0.20	3.2
MPin-C-MPin Squared 10 mm	293	16	4.49 ± 0.00	0.92	3.0
MPin-C-MPin Squared 20 mm	67	16	3.98 ± 0.01	0.21	3.2

3) EVALUATION TESTS

3.1) Drop-Weight Testing

The FML panels reinforced with pins and the reference panels were submitted to impact damage by drop-weight testing to verify whether the pins would have positive or deleterious effects on the FML panels and their capacity to absorb impact energy. Based on the ASTM D7136 standard [12], a rig to impose free fall (from around 1.810 mm of height) of a constant mass (2.024 kg) over the miniaturized panel surface was devised. This mass was composed of a 17-mm-spherical head made of hard steel attached to a plain carbon steel cylinder (50 mm of diameter and 130 mm of length). The rig included a latching device for ensuring no mass bouncing (single impact). The free fall height aimed a potential energy according to the thickness of each panel and a level sufficient for causing apparent damage at impact, which resulted in 8.5 J per each millimeter of panel thickness (gravitational acceleration considered as 9.81 m/s^2). High-speed images (2,000 frames per second) were used for determination of the falling/raising mass velocities immediately before/after impact (Fig. 3(a)), based on displacements (visualized from frame to frame) of its spherical head lower surface and respective time lapses. The velocities involved in the impacts, in turn, were used to calculate the impact and return energies (Fig. 3(b) and Fig. 3(c)).

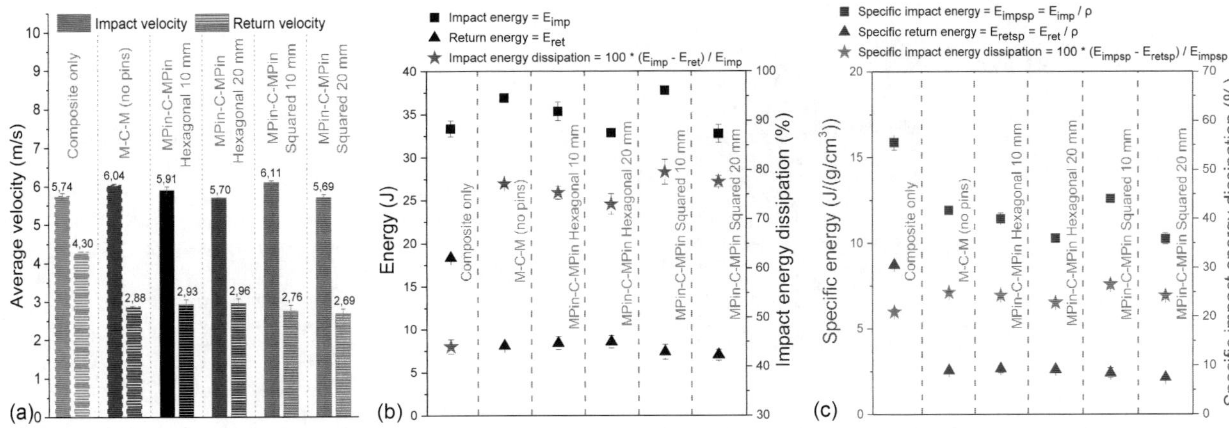

FIGURE 3. (a) Average impact and return velocities for each panel; (b) Average impact and return energies and consequent energy dissipation during drop-weight testing; (c) Average specific impact and specific return energies and consequent specific energy dissipation during drop-weight testing; (ρ = panel mass density)

3.2) Impact Damage Characterization

Due to the intrinsic difficulties related to panel damage assessment, the damage extent after drop-weight testing was estimated by two nondestructive testing methods; 3D digital probing, to delineate damage depth profile, and infrared thermography, to estimate the damage area. The damage profile was determined by 3D scanning the upper and lower surfaces of the panels. The average maximum permanent deformation of each panel is shown in Fig. 4(a). The in-plane damage extent, hereinafter called damage area, of the panels after impact was estimated by infrared thermography using an active pulsed transient transmission technique. All panels were also inspected before impact for as-manufactured integrity check. Fig. 4(b) shows the average estimated damage areas at different sampling times after heating start, being the damage areas taken as the average maximum indications verified in each case.

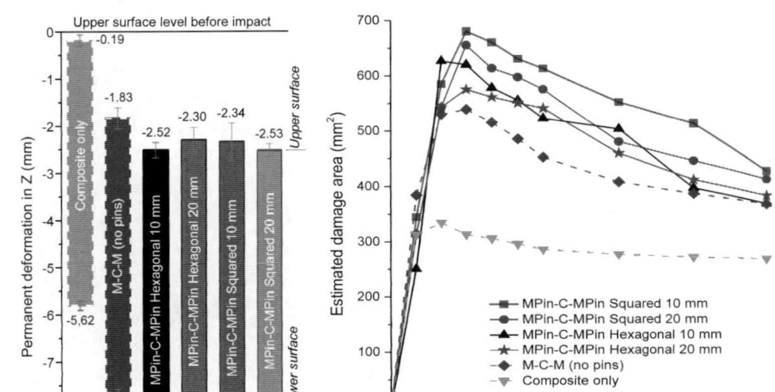

FIGURE 4. (a) Maximum permanent deformation in each panel; (b) Damage areas at different sampling times after heating start

3.3) Compression and Buckling Tests after Impact – Damage Tolerance

All types of panels were subjected to compression (Fig. 5(a)) and buckling (Fig. 6(a)) tests after impact (drop-weight test) as initial investigation on the influence of pins on damage tolerance. For each type of miniaturized panels fabricated, one was tested for compression and another for buckling. An electromechanical universal tensile test machine was used. Upper and lower supports for the panels were designed and built to allow proper alignment and fixture for both tests. In accordance with ASTM D7137 standard [13], the testing speed was always set at 1.25 mm/min. Fig. 5 (b) and Fig. 6(b) shows, respectively, the evolution of load *versus* axial displacement in compression and buckling tests of all panel types after impact damaging.

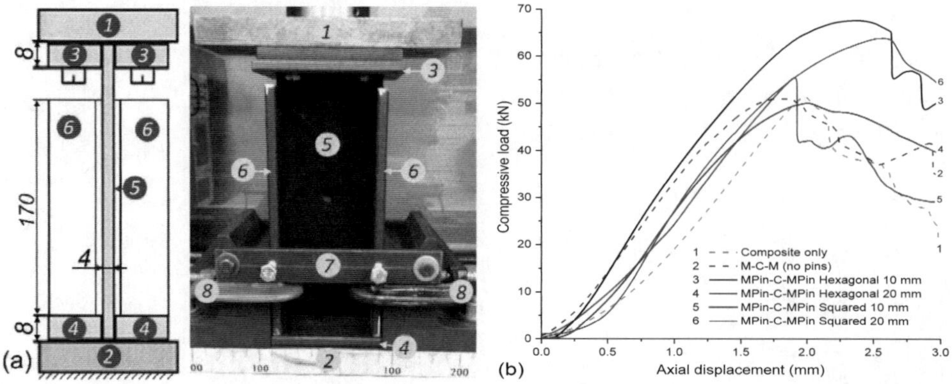

FIGURE 5. (a) Schematic side view of the compression set up and image of the miniaturized panel accordingly assembled in the testing machine, where: where: 1 - moving upper head; 2 - fixed lower head; 3 - upper support (two parts); 4 - lower support (two parts); 5 - panel; 6 - support fixture side plates; 7 - auxiliary holding bar; 8 - auxiliary clamps (dimensions in mm); (a) Evolution of load *versus* axial displacement in compression tests of all types of panels after impact damaging

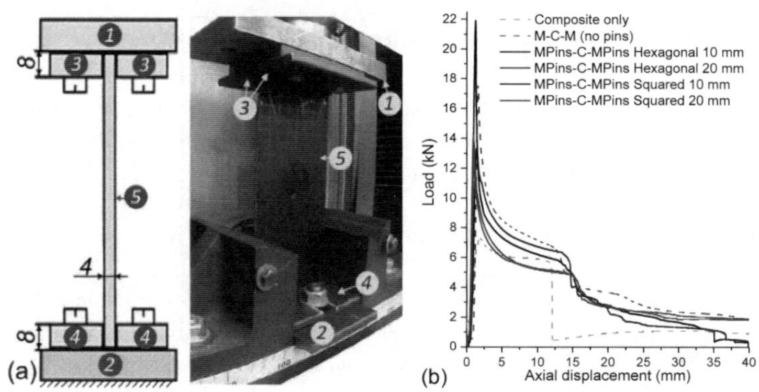

FIGURE 6. (a) Schematic side view of the buckling set up and image of the miniaturized panel accordingly assembled in the testing machine; where: 1 - moving upper head; 2 - fixed lower head; 3 - upper support (two parts); 4 - lower support (two parts); 5 - panel; 6 - support fixture side plates; 7 - auxiliary holding bar; 8 - auxiliary clamps (dimensions in mm); (b) Evolution of load *versus* axial displacement in buckling tests of all types of panels after impact damaging

4) CONCLUSIONS

• In terms of impact energy dissipation, all FML panels with pins exhibited similar performance, generally equivalent to the conventional FML panel (M-C-M (no pins)), but significantly higher than that achieved with the panel consisting only of composite. Besides, the addition of weight by the pins as anchorages does not penalize the capacity of the panels to dissipate impact energy, as all panel types dissipated similar levels of specific energy during impact. Therefore, the pins did not make the FML panels more brittle and the change in their deposition pattern did not show any significant effect concerning the capacity of the panels to dissipate impact energy;

• Concerning the damage depth profiles caused by drop-weight testing, all FML panels with pins suffered damages similar to that of the conventional FML panel (M-C-M (no pins)), but not as concentrated as verified in the panel consisting only of composite. Overall, regardless of the deposition pattern, the pins increased the maximum permanent deformation after impact, especially if less spaced. The in-plane damage extents, estimated by infrared thermography, were larger for the FML panels with pins, intermediate for the conventional FML panel (without pins) and much smaller for the composite panel, corroborating the impact energy dissipation results;

• Regarding damage tolerance, the FML panels with pins exhibited a less catastrophic trend, i.e., achieving significantly higher loads at longer axial displacements in compression testing, even with previous larger damages. However, in buckling testing after impact, the less catastrophic trend is not so evident, but the pins tend to retard the debonding propagation between metal sheets and composite, for anchoring them to each other. The pins also would hold back the delamination spreading in the composite, for clamping the layers together between the metal sheets and ball-heads of the pins.

ACKNOWLEDGMENTS

The authors acknowledge the Center for Research and Development of Welding Processes (Laprosolda) at UFU and the Department of Materials and Technology at UNESP-Guaratinguetá, for providing laboratorial infrastructure and support, and ALLTEC Materiais Compostos, for providing prepreg materials. They also acknowledge the Brazilian agencies CNPq (Procs. 302863/2016-8 and 303224/2016-9), CAPES (PROEX 0309/2015), and FAPEMIG (Proc. 11304), for research grants and scholarships.

REFERENCES

1. P. Cortes and W.J. Cantwell, "The prediction of tensile failure in titanium-based thermoplastic fibre–metal laminates", *Composite Science Technology*, 66 (2006), pp. 2306–2316.
2. A. Vlot, "Impact properties of fibre metal laminates", *Compos Eng.*, 3 (1993), pp. 911–927.
3. T. Sinmazçelik, E. Avcu, M.Ö. Bora, O. Çoban, "A review: Fibre metal laminates, background, bonding types and applied test methods", *Materials and Design*, 32(7) (2011), pp. 3671–3685.
4. E.C. Botelho, R.A. Silva, L.C. Pardini, M.C. Rezende, "A review on the development and properties of continuous fiber/epoxy/aluminum hybrid composites for aircraft structures", *Mater Res.*, 9(3) (2006), pp. 247–256.
5. R. Seydel and F. Chang, "Impact identification of stiffed composite panels: i. system development", *Smart Materials and Structures*, 10(2) (2001), pp. 354–369.
6. A. Vlot, "Impact loading on fibre metal laminates", *Int. J. Impact Engng.*, 18(3) (1996), pp. 291–307.
7. S. Ucsnik, M. Scheerer, S. Zaremba, H.D. Pahr, "Experimental investigation of a novel hybrid metal-composite joining technology", *Composites Part A: Applied Science and Manufacturing*, 41(3) (2010), pp. 369 – 374.
8. S. Ucsnik, S. Stelzer, H. Sehrschön, G. Sieglhuberc, "Composite to composite joint with lightweight metal reinforcement for enhanced damage tolerance" in *ECCM16 - 16th European conference on composite materials*, 22-26 Jun, 2014, Seville, Spain.
9. S. Stelzer, S. Ucsnik, G. Pinter, "Fatigue behaviour of composite–composite joints reinforced with cold metal transfer welded pins", *International Journal of Fatigue*, 81 (2015), pp. 37–47.
10. D.P. Graham, A. Rezai, D. Baker, P.A. Smith, J.F. Watts, "The development and scalability of a high strength, damage tolerant, hybrid joining scheme for composite-metal structures", *Composites Part A: Applied Science and Manufacturing*, 64 (2014), pp. 11–24.
11. Fronius, *Current welding practice: CMT technology*, first edition, Vol 11, DVS Media GmbH, Dusseldorf, 2014.
12. ASTM D7136/D7136M-15 "Standard Test Method for Measuring the Damage Resistance of a Fiber-Reinforced Polymer Matrix Composite to a Drop-Weight Impact Event".
13. ASTM D7137/D7137M-12 "Standard Test Method for Compressive Residual Strength Properties of Damaged Polymer Matrix Composite Plates".

Hybridized Nylon-based Biocarbon Composites for Lightweight, Durable Automotive Applications

Andrew Anstey[a], Amandine Codou[a], Manjusri Misra[a,b], Amar K. Mohanty[a,b]

[a] *Bioproducts Discovery and Development Centre, Crop Science Building, University of Guelph, 50 Stone Road East, Guelph, Ontario, Canada*
[b] *School of Engineering, Thornbrough Building, University of Guelph, 50 Stone Road East, Guelph, Ontario, Canada*

Abstract

With increasingly stringent modern regulations on vehicle fuel economy, lightweighting has become a key aspect in the design of new automobiles and automotive materials. This research aims to produce lightweight biocomposite materials that can replace conventional nylon composites that are utilized in under-the-hood applications. Compatibilized blends of nylon 6 (PA6), poly(lactic acid) (PLA), and polypropylene (PP) were developed to match the mechanical properties of nylon 6, with improved hydrophobicity, increased bio-based content, and reduced density. The compatibilized blend was utilized to produce biocomposites with a hybridized reinforcement of biocarbon and recycled carbon fibre. Finally, thermal stabilizing additives were incorporated into the biocomposite to improve its long-term thermal stability. The blends and biocomposites were investigated using scanning electron microscopy (SEM) to study their morphology and polymer phase distribution. Mechanical testing was conducted to determine the tensile and flexural strength and modulus of the materials. Accelerated heat aging studies were conducted to analyze the long-term thermal stability of the biocomposites.

The study yielded nylon-based biocomposite materials with high strength and stiffness that match the industrial requirements for the automotive components. The generated biocomposites not only meet these requirements, but also weigh up to 15% less than conventional nylon/glass fibre/talc composites and contain up to 34% bio-based content by weight. The incorporation of biocarbon additionally serves to reduce both the cost of the material and the environmental impact due to that fact that it is a carbon-neutral material. The biocomposites developed in this research have excellent potential for providing a more sustainable alternative for nylon automotive components.

Keywords: Nylon, biocarbon, durability, lightweight, automotive, composites.
PACS: 81.05.Qk

INTRODUCTION

Since the introduction of fuel economy regulations in the United States in the 1970s, automotive manufacturers have strived to improved fuel economy through increasingly lightweight components. One of the major revolutions in this area has been the replacement of metal parts with lighter plastics and plastic composites. For "under-the-hood" components such as air intake manifolds, covers, reservoirs, shielding and tanks, polyamides (PA) such as nylon 6 (PA6) and nylon 6,6 (PA66) are the plastic of choice [1]. Due to their strength and excellent resistance to chemicals and heat, nylons are ideal materials for these applications.

Nylon resins can be easily used to create injection moulded parts with much lower weight than metal components. PA6 and PA66 have densities around 60% lower than that of aluminum. In order to provide adequate strength and stiffness, under-the-hood components are typically made using fibre- and mineral-filled nylon composite resins. Glass fibres are sized to improve their adhesion to the nylon, and contribute substantially to tensile strength and modulus. Mineral fillers also contribute to stiffness, but most importantly improve the dimensional stability of the resin and reduce warpage after moulding. Dimensional stability is especially improved by mineral fillers with small particle size and low aspect ratio, and is a crucial aspect for manufacturing components with high consistency and low tolerances [2]. Hybridization of glass fibre and mineral reinforcement is often employed to provide resins with high strength and stiffness, as well as excellent dimensional stability.

TABLE 1. Typical properties of commercial PA6 composite resins. Properties compiled from publicly available technical data sheets from manufacturers [3], [4].

	Manufacturer	Material name	Reinforcement	Tensile strength (MPa)	Young's modulus (GPa)	Density (g/cm^3)
Neat polymer	BASF	Ultramid B3K	None	85	3.0	1.13
	DuPont	Zytel 7301	None	80	2.9	1.13
Mineral reinforced	BASF	Ultramid B3M6	30% mineral	75	4.5	1.36
	DuPont	Minlon 73M30	30% mineral	82	4.8	1.35
Glass fibre reinforced	BASF	Ultramid B3EG6	30% glass fibre	185	9.5	1.36
	DuPont	Zytel 73G30	30% glass fibre	180	9.5	1.36
Hybrid reinforcement	BASF	Ultramid B3GM35 Q643	40% mineral and glass fibre	120	8.5	1.48
	DuPont	Minlon 73GM40	40% mineral and glass fibre	140	8.3	1.46

From TABLE 1, it is evident that while conventional reinforcement strategies are highly effective in improving the tensile strength and modulus of PA6, this comes at the cost of increased density and therefore increased weight. Furthermore, there is a high environmental cost associated with the production of glass fibre for composites. Every kilogram of glass fiber reinforcement requires 48 MJ of energy to manufacture, and generates 2 kilograms of CO_2 [5]. In this study, nylon composites reinforced with biocarbon and post-industrial recycled carbon fibre (RCF) were investigated to meet targeted specifications for commercial applications.

Biocarbon is a carbon-rich solid product created through the pyrolysis of biomass at high temperatures in an oxygen limited environment. Biocarbon is targeted as a particulate filler due to its relatively low density (1.35 – 1.40 g/cm^3) compared to mineral fillers such as talc (2.6 – 2.8 g/cm^3). In addition, biocarbon has a low environmental impact, with net zero or net negative generation CO_2 due to its sequestration of carbon in a stable, solid form [6]. RCF consists of short fibres reclaimed from carbon fibre scraps generated during industrial processes. Post-industrial recycling of RCF diverts it from industrial waste streams and generates new opportunities for value-added applications of the material. Both biocarbon and RCF are targeted due to their potential for providing lightweight reinforcement to PA6, at a lower cost and reduced environmental impact compared to conventional methods.

In a similar fashion to glass fibre/mineral reinforcement, hybridized reinforcement with particulate biocarbon and RCF was used to provide both improved strength/stiffness and dimensional stability. Improvement in heat deflection temperature (HDT) through the addition of biochar has been demonstrated for multiple engineering polymer, including polyamides. Myllytie et al. observed an increase in HDT of over 80°C in PTT through the addition of 20 wt.% biocarbon, resulting in composites with HDT similar to commercial mineral filled PTT composites [7]. Ogunsona et al. demonstrated that the HDT of PA6,10 can be increased from 150°C to over 185°C through the addition of 20 wt.% biocarbon, with the greatest improvement achieved when the biocarbon particle size was reduced below 63 μm [8].

Further lightweighting of PA6 resins was targeted through blending with polypropylene (PP). Though not as strong or stiff as PA6, PP is much less dense (0.94 g/cm^3 compared to 1.13 g/cm^3) and is very hydrophobic, which can counteract the issue of moisture absorption in PA6 systems. Furthermore, properly compatibilized blends of PA6 and PP are known to demonstrate improved impact strength [9]–[12]. A reactive extrusion scheme with maleic anhydride grafted PP (MA-g-PP) was used to compatibilize PA6 and PP. The addition of poly(lactic acid) (PLA) was also investigated to improve the mechanical properties of the polymer blend. PLA is a bio-based and biodegradable polymer with excellent stiffness, and high tensile and flexural strength.

EXPERIMENTAL

Materials

The PA6 used in this study was Ultramid B27E, provided by BASF (Germany). The PP used was PP 1120H, provided by Pinnacle Polymers (USA). The PLA was Ingeo 3251D, provided by NatureWorks LLC (USA). Fusabond P353 was used as the MA-g-PP compatibilizer, and was supplied by DuPont (USA). The biocarbon used in the study was provided by Competitive Green Technologies (Canada), and was pyrolyzed between 630-690°C. A proprietary copper salt-based heat stabilizer was used to improve thermal durability.

Processing

Neat polymers and composite blends were processed in a Leistritz (Germany) Micro-27 co-rotating twin-screw extruder at 250°C and 100 RPM, with an approximate average residence time of around 90 seconds, and subsequently pelletized and dried. Dried pellets were used to create injection moulded testing samples using an Arburg (Germany) AllRounder 77-ton injection moulding machine, with an injection temperature of 250°C and a mould temperature of 40°C.

Characterization

Tensile and flexural properties were evaluated using an Instron (USA) 3382 Universal Testing machine. Tensile testing was completed following ASTM D638–14 using a crosshead speed of 50mm/min, while flexural testing was conducted following ASTM D790–17. Charpy impact strength was conducted on notched samples following ASTM D6110–10. Microscopic analysis of polymer blends was conducted with a Phenom ProX scanning electron microscope (SEM) (Netherlands) on samples sputter-coated in gold with a Cressington 108 (UK). Samples were aged in an EB-10 heat aging oven from Qualitest (USA) for 1000 hours at 140°C with an average air change rate of 7 changes per hour.

RESULTS AND DISCUSSION

Mechanical Properties

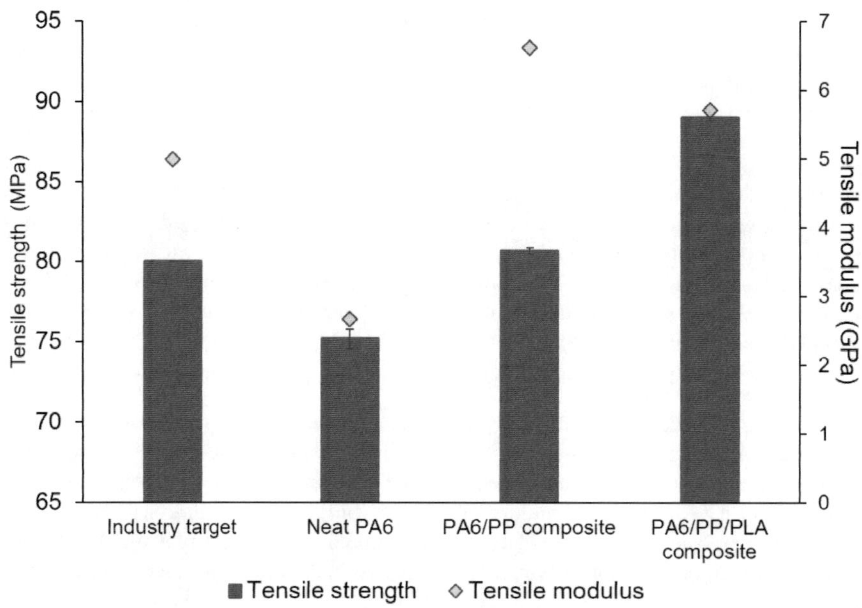

FIGURE 1: Tensile properties of hybridized PA6 composites with 27 wt. % biocarbon (BC) and 5 wt.% recycled carbon fibre (RCF), as compared to industrial target specifications (PA6 reinforced with 10% glass fibre and 20% mineral filler).

The mechanical properties of hybridized biocarbon-RCF composites were investigated and compared to target performance specified by an industrial partner for an under-the-hood component. As seen in Figure 1, the goal was to achieve materials with tensile strength and modulus greater than 80 MPa and 5.0 GPa, respectively, with a Charpy impact strength of 0.5 kJ/m^2. Typical materials used for the targeted part were reinforced with 10% glass fibre and 20% mineral filler by weight, with a density between 1.32 and 1.45 g/cm^3. The impact requirement of the targeted component was relatively low, and easily achieved in all composites. PA6/PP blend composites demonstrated impact strengths just slightly lower than that of neat PA6 (5.7 vs. 6.0 kJ/m^2, respectively). While the addition of particulate fillers such as biocarbon are detrimental to impact strength in PA6 composites, reactive blending with PP improved the toughness sufficiently to counteract this effect. The impact strength was decreased with the incorporation of PLA into the polymer matrix system, which is expected due to the relative brittleness of PLA. However, at 3.7 kJ/m^2 the PA6/PP/PLA blend still greatly exceeded the target impact strength.

Reinforcement with biocarbon and RCF improved the tensile strength and modulus of PA6 to exceed the targeted specifications. The blend of PA6/PP with hybridized reinforcement provided the necessary tensile strength, and exceeded the targeted modulus by over 30%. The incorporation of PLA served to increase the tensile strength further to 89.0 MPa, but also resulted in a lower modulus and impact strength than the PA6/PP blend. Very importantly, both of these hybrid composites were significantly lighter than conventional glass fibre/mineral reinforced composites. With a value of ~1.20 g/cm^3, the biocarbon composite density was reduced by 9 % of the lower bound and 17 % of the upper bound of the density requirement.

Blend Morphology

Using SEM, drastic changes in morphology were observed as a result of the blend compatibilization through reactive extrusion with MA-g-PP, as shown in Figure 2. Reactive extrusion of PA6/polyolefin blends with maleated polyolefin additives is a well-established compatibilization route [13], [12], [9]. In Figure 2A, the poor miscibility between PA6 and PP is evident by the large droplets of PP embedded in the PA6 matrix, with clear gaps at the polymer-polymer interface due to a lack of adhesion. Large voids are created by the pullout of poorly adhered PP domains during impact fracture. This is starkly contrasted by the morphology observed in Figure 2B, wherein the addition of MA-g-PP has substantially improved the miscibility of PP and PA6. The droplet size of dispersed PP is reduced to below 1 micron in the compatibilized blend, which results in the superior toughness and strength required for the application [10].

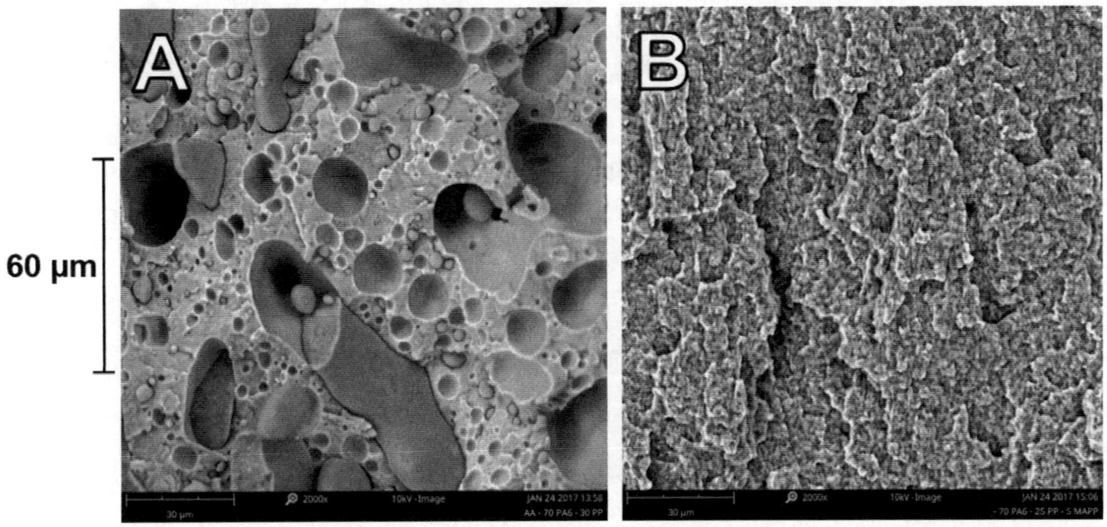

FIGURE 2. Comparison of SEM micrographs at 2000x magnification of impact fractured samples (A) uncompatibilized PA6/PP blend and (B) PA6/PP blend compatibilized with MA-g-PP.

Durability in Accelerated Heat Aging

Accelerated heat aging was used to simulate the thermal exposure of components in under-the-hood applications. This is a crucial aspect to consider when designing components that will undergo repeated thermal loading at a wide range of ambient temperatures during the life cycle of an automobile. Both the PA6/PP and PA6/PP/PLA composites shown in Table 2 were subjected to 1000 hours of aging at 140°C. Parallel blends of these composites were prepared with 0.5 phr of a copper salt-based stabilizer. The tensile strengths of each of these four composites are shown in Figure 3; all composites were prepared with 27% biocarbon and 5% RCF by weight. In both of the non-stabilized blends, the tensile strength is severely reduced by more than 50% after 1000 hours of aging. The incorporation of the stabilizer drastically improved the strength retention of the PA6/PP blend, which lost less than 1% of its original tensile strength after aging. While the PA6/PP/PLA blends started with a higher tensile strength, they were weakened by 37% despite the addition of a heat stabilizer. PLA has excellent tensile strength, which improves the initial performance of this blend, but its susceptibility to oxidative degradation negatively impacts the blend after thermal aging.

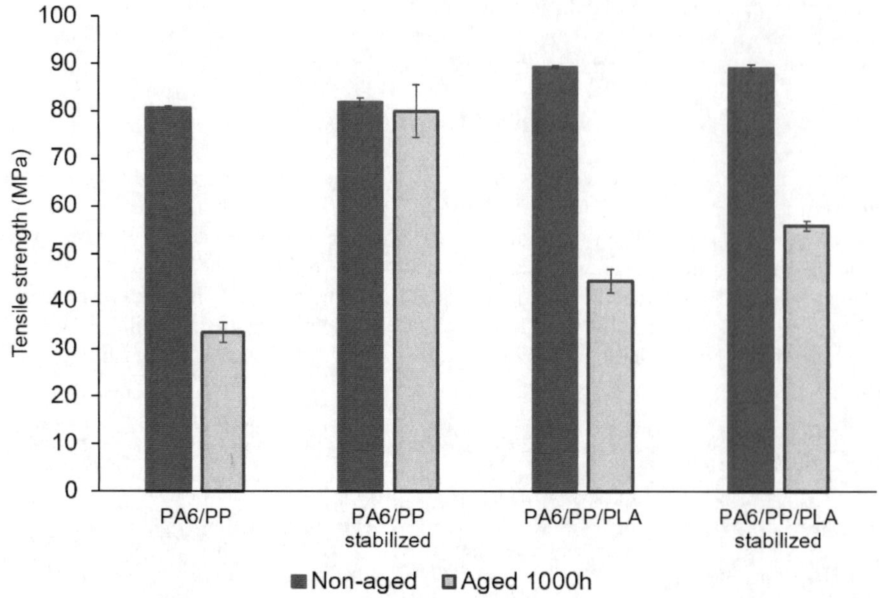

FIGURE 3. Tensile strength retention of PA6/PP and PA6/PP/PLA composites before and after 1000 hours of aging at 140°C. All composites contain 27% biocarbon and 5% RCF by weight; stabilized composites contain 0.5 phr of a copper salt-based stabilizer.

CONCLUSION

Through hybridization of reinforcement, PA6 composites with high bio-based content, excellent durability, and reduced density were prepared. These composites were able to exceed automotive industry specifications, while providing a weight reduction of up to 17%. Weight reduction was accomplished through the incorporation of biocarbon and RCF as hybrid reinforcements, as well as through reactive blending of PA6 with PP.

ACKNOWLEDGEMENTS

This research is financially supported by the Ontario Ministry of Agriculture, Food and Rural Affairs (OMAFRA), Canada/University of Guelph-Bioeconomy for Industrial Uses Research Program Theme (Project # 030176); the Natural Sciences and Engineering Research Council (NSERC), Canada–Discovery Grants (Project # 401111) and Agriculture and Agri-Food Canada and Competitive Green Technologies (CGTech) through AgriInnovation Program (Project # TBD); Natural Sciences and Engineering Research Council, Canada for the Discovery Grants (Project #401111); and the Ontario Research Fund, Research Excellence Program; Round-7 (ORF-RE07) from the Ontario Ministry of Research and Innovation (MRI), currently known as the Ontario Ministry of Research, Innovation and Science (MRIS) (Project # 052665 and #052644).

REFERENCES

1. E. Carlson and K. Nelson, "Nylon Under the Hood: A History of Innovation," Automot. Eng., vol. 104, pp. 84–89, 1996.
2. P. J. Yoon, T. D. Fornes, and D. R. Paul, "Thermal expansion behavior of nylon 6 nanocomposites," Polymer, vol. 43, no. 25, pp. 6727–6741, Jan. 2002.
3. DuPont, "DuPont Material Data Centre," 2017. [Online]. Available: http://dupont.materialdatacenter.com/profiler/1WeV0/. [Accessed: 25-Sep-2017].
4. BASF, "PlasticsPortal: Ultramid Polyamide (nylon)," 2017. [Online]. Available: http://iwww.plasticsportal.com/products/ultramid.html. [Accessed: 25-Sep-2017].
5. S. V. Joshi, L. T. Drzal, A. K. Mohanty, and S. Arora, "Are natural fiber composites environmentally superior to glass fiber reinforced composites?," Compos. Part A Appl. Sci. Manuf., vol. 35, no. 3, pp. 371–376, Mar. 2004.
6. K. G. Roberts, B. A. Gloy, S. Joseph, N. R. Scott, and J. Lehmann, "Life Cycle Assessment of Biochar Systems: Estimating the Energetic, Economic, and Climate Change Potential," Environ. Sci. Technol., vol. 44, no. 2, pp. 827–833, Jan. 2010.
7. P. Myllytie, M. Misra, and A. K. Mohanty, "Carbonized Lignin as Sustainable Filler in Biobased Poly(trimethylene terephthalate) Polymer for Injection Molding Applications," ACS Sustain. Chem. Eng., vol. 4, no. 1, pp. 102–110, Jan.

2016.
8. E. O. Ogunsona, M. Misra, and A. K. Mohanty, "Sustainable Biocomposites from Biobased Polyamide 6,10 and Pyrolyzed Biocarbon from Miscanthus Fibers," J. Appl. Polym. Sci. Rev., 2016.
9. A. González-Montiel, H. Keskkula, and D. R. Paul, "Morphology of nylon 6/polypropylene blends compatibilized with maleated polypropylene," J. Polym. Sci. Part B Polym. Phys., vol. 33, no. 12, pp. 1751–1767, 1995.
10. A. González-Montiel, H. Keskkula, and D. R. Paul, "Impact-modified nylon 6/polypropylene blends: 1. Morphology-property relationships," Polymer, vol. 36, no. 24, pp. 4587–4603, Jan. 1995.
11. A. Tedesco, R. V Barbosa, S. M. B. Nachtigall, and R. S. Mauler, "Comparative study of PP-MA and PP-GMA as compatibilizing agents on polypropylene/nylon 6 blends," Polym. Test., vol. 21, pp. 11–15, 2002.
12. P. Agrawal, S. I. Oliveira, E. M. Araújo, and T. J. A. Melo, "Effect of different polypropylenes and compatibilizers on the rheological, mechanical and morphological properties of nylon 6/PP blends," J. Mater. Sci., vol. 42, no. 13, pp. 5007–5012, Jun. 2007.
13. T. Huber, M. Misra, and A. K. Mohanty, "Mechanical properties of compatibilized nylon 6/polypropylene blends; studies of the interfacial behavior through an emulsion model," J. Appl. Polym. Sci., vol. 131, no. 18, Sep. 2014.

Injection molded carbonaceous hybrid bio-composites for the automotive industry

Arturo Rodriguez-Uribe[a], Singaravelu Vivekanandhan[b], Manjusri Misra[a,c], Amar K. Mohanty[a, c]

[a]Bioproducts Discovery and Development Centre, Department of Plant Agriculture, Crop Science Building, University of Guelph, Guelph, N1G 2W1, Ontario, Canada, [b]Sustainable Materials and Nanotechnology Lab, Department of Physics, VHNSN College, Virudhunagar, 626001, Tamilnadu, India, [c]School of Engineering, Thornbrough Building, University of Guelph, Guelph, N1G 2W1, Ontario, Canada

Abstract.

Automotive plastics are generally loaded with high amounts of mineral fillers or synthetic fibers as methods of property enhancement. These traditional fillers are of very high density and results in the overall weight increase of the vehicles. As an alternative, natural fibers of renewable origin and lightweight nature have been explored for the automotive applications. However, the physicochemical properties of natural fibers or bio-fillers, which includes lower thermal stability and higher moisture absorption, hinders their utilization in commercial products. Hybridization of fillers/ reinforcements can be used to improve the mechanical and thermal properties of bio-based composites. The thermal stability of biomass can be improved by adopting the various pyrolysis strategies. The introduction of pyrolyzed biomass (carbonaceous products or biocarbon) improves the thermal stability of the composites, therefore creating wider processing windows. By using hybridization, it is possible to keep relatively high loadings of natural fibers in the composites while achieving balanced mechanical properties. Some of these mechanical properties include stiffness, strength as well as impact resistance. Furthermore, hybridization of biocarbon positively impacts moldability, which further improves the processability of the composites by reducing the cycling time during the injection molding of the parts. Hybridization of biocarbon or carbonaceous biomasses also helps in the reduction of synthetic additives such as carbon black. The technology used to produce composite blends can be easily scaled to industrial practices since the current technology used for plastics is also used for bio-composites. These production technologies include extrusion and injection molding. However, there are subtle differences that must be taken into account in order to successfully process these bio-based composites which are discussed here.

Keywords: Extrusion technologies, injection molding, bio-fillers, bio-carbon, bio-materials, bio-composites

INTRODUCTION

Natural fibres can be obtained from forests, energy crops, agricultural staples, or agricultural residues in more than 1 billion tons annually. Plant fibers are widely investigated for strengthening of synthetic plastics including thermoplastic and thermoset polymers. The properties of plastics reinforced with natural fibers have a similar trend when compared with glass fiber reinforced composites, and therefore represent an important commercial alternative for the fiber-plastic composites industry.

Fiber reinforced plastic composites can be produced by a variety of techniques including thermoforming, vacuum forming, compression molding, extrusion molding, and injection molding. Plant fibers can act as partial or complete substitutes for glass fibers in many applications. Apart from that, the cost of plant fibers is substantially less than the compared to glass fibers. For example, plant fibers cost approximately $US 0.20-0.25/lb while general purpose E-glass fiber cost approximately $US 2.00/lb. Natural fibers/fillers obtained from energy crops (e.g., perennial grasses like *Miscanthus*, switch grass etc.) cost around $0.10/lb. Similarly, agri-residue natural fibers cost approximately $0.10 /lb as well. However, the utilization of various plant fibers in composite materials may experience performance limitations due to lower processing temperatures and potential reduced compatibility with the plastic matrix material during the extrusion and/or injection molding processes. Thus, there is a need for composite materials, particularly reinforced with plant fibers and/or natural fillers, which possess improved chemical, physical, and/or mechanical properties (1).

In order to address the issues related to the thermal stability of natural fibres, an alternative solution is to convert the fibers into bio-carbon. Bio-carbon is a pyrolysed biomass that can be used as a practical and efficient reinforcement with any polymeric matrix including high melting temperature polymers such as nylon and PTT. Biocarbon is also an alternative to synthetic carbon black and can act as colorant or filler or reinforcing phase. The injection molding of materials containing bio-carbon can be radically improved by using bio-carbon. Since the particle size of bio-carbon can be manipulated, biocarbon based composites can facilitate the injection molding of bio-based composites. Since there is an increasing desire for effective stiffness and strength balance in automotive composites, the hybridization of natural fiber and biocarbon as reinforcement has been explored (2).

COMPOSITION OF HYBRID COMPOSITES

The possibilities of production of composites are practically limit-less. Composites may contain one or more sources of biomass including natural fibers and/or fillers and synthetic fibers and/or fillers, in addition to one or more sources of synthetic plastics and additives including impact modifiers, compatibilizers, fillers, phenolic antioxidants, etc. The possibility of sourcing natural fibers and or biochar from grasses to woods are almost unlimited, due to the substantial generation of byproducts such as saw dust, industrial and municipal wastes. Furthermore, materials treated under different pyrolysis conditions may vary from torrefied materials or bio-carbons and therefore possess different degrees of functionalities (2).

EXTRUSION AND INJECTION MOLDING

A wide range of formulations were explored, which include blends of commodity polymers such as: polypropylene, impact polypropylene, homopolymer polypropylene, rubbers or functionalized terpolymers, functionalized polyolefin maleic anhydride, functionalized homopolymer/impact polypropylene maleic anhydride (in situ prepared for commercially availability), formulations of reinforcements with natural fiber *Miscanthus*, as well as the biocarbon produced from *Miscanthus* fiber or other sources. In situ polymerization was performed by employing maleic anhydride (MAH) as an oxidizer in the presence of an initiator (mostly peroxides).

Optimization of suitable formulations for product development, i.e. utilization of materials in minimal amounts, was performed with a micro extruder and injection molding machine. The use of a micro extruder resulted in the successful dispersion of filler into the matrix material due to the enclosed environment and recirculation capabilities of the machine. However, the use of extruders at industrial scale possesses an additional challenge to achieve good dispersion of fiber throughout the matrix material. Large scale production uses an extruder with the basic configuration shown in Fig 1. The extruder has two inserts for feeding materials (polymers in the first feeder and fiber in the second feeder). The processing conditions can be performed between 180 and 220°C for optimal results. The main variables involved during extrusion processing include: the machine and screw configuration, temperature, screw speed, feed rate, and most importantly, the melt temperature at the die. The control of the melt temperature is crucial in the processing of biomaterials or temperature sensitive polymers. The feeding protocol is of imperative since different materials having different physical properties may run at different speeds at the time of feeding. Calibration of different feeders is therefore of extreme importance.

Typically, the variables discussed and analyzed during the injection molding process are the pressure, holding time, holding pressure, cooling time, cycle time, mold temperature, melt temperature, and pressure in the hydraulic system. A general diagram of an injection molding apparatus is shown in Figure 2. Gate size is a commonly overlooked factor that it is of extreme importance when analyzing issues related to processing. In particular, gate size is important when considering materials with significantly large particle size of 2 mm or bigger such as natural or synthetic fibers. In particular, biofillers such as biocarbon or natural fibers can be used to substitute heavy synthetic fibers/fillers leading to a reduction in the density of the final product.

OUTCOMES

Overall, the hybridization of fillers/ reinforcements, which may consist in the combination of different materials from synthetic and biobased sources, can facilitate the effective engineering of new advanced materials through various processing (extrusion/injection molding). Furthermore, it can also be used as a methodology to enhance the performance, reduce the density, and improve the aesthetics of bio-composites. The added advantage of this hybridization process is its fulfilment for overall industrial requirements.

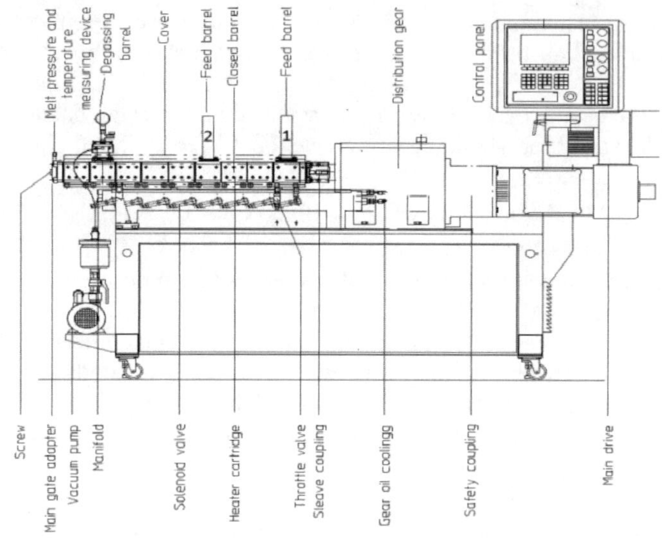

FIGURE 1. Schematic representation of twin screw extruder used for the fabrication of the biocomposites (obtained from the manual of the Leistritz extruder, type MIC/GL-480 convertible from co- to counter –rotating).

Injection molding is a technology used for

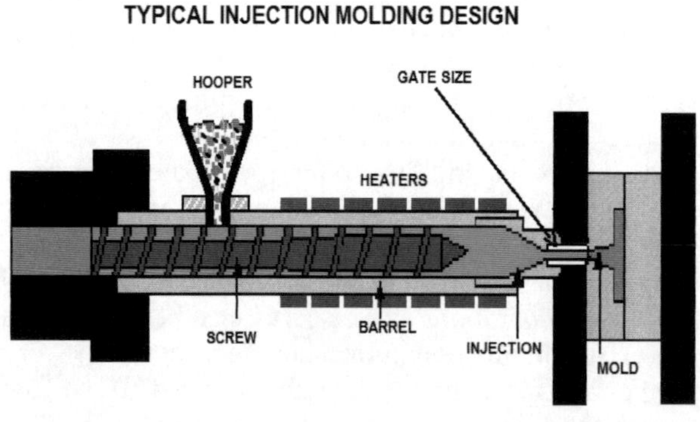

FIGURE 2. General representation of an injection molding unit

ACKNOWLEDGMENTS

The authors gratefully acknowledge the financial support of the Agriculture and Agri-Food Canada and Competitive Green Technologies (CGtech) through AgriInnovation Program (Project # 052882 and 051910); the Ontario Ministry of Agriculture, Food, and Rural Affairs - University of Guelph Bioeconomy Industrial uses Research Program (Project # 030055); Natural Sciences and Engineering Research Council, Canada for the Discovery Grants (Project #401111 and 400320); and the Ontario Research Fund, Research Excellence Program; Round-7 (ORF-RE07) from the Ontario Ministry of Research and Innovation (MRI), currently known as the Ontario Ministry of Research, Innovation and Science (MRIS) (Project # 052665 and #052644) to carry out this research.

REFERENCES

1. Amar Mohanty, ManjusriMisra, Arturo Rodriguez-Uribe, SingaravelyVivekanandhan. Hybrid sustainable composites and methods of making and using thereof.CA2924867A1, US20160229997 (Mar 26, 2015).

2. Amar Mohanty, ManjusriMisra, Atul Bali, Arturo Rodriguez-Uribe. Renewable replacements for carbon black in composites and methods of making and using thereof.CA2945688A1, US20170107334 (Sep 17, 2015).

Orthopaedic compression screws formed from body-temperature reverting, shape-memory polymers

Thomson Brain [(1)]*, Thompson Glen [(1)], Vgenopoulos Dimitrios [(1)], Nair Karthik [(1)], Howell Ken [(1)], Martyn Mike [(1)], Coates Philip [(1)]

[(1)] Bradford University - Yorkshire - UnitedKingdom

Orthopaedic compression screws promote fracture-repair by drawing-together and stabilising adjacent bone fragments. Shape memory polymers (SMP) are smart materials that switch from one macroscopic shape to another following the application of external 'triggering' stimuli (e.g. by shrinking in length when heated above Tg). SMP compression screws that shorten in length following surgical implantation (i.e. display body temperature reversion) could potentially pull bone fragments together and thereby facilitate fracture healing. The biocompatible, bioresorbable polymer polylactic acid (Purasorb 7038; PLDLA) shows shape memory properties but the Tg of unmodified PLDLA is too high for body temperature reversion. Accordingly we have used the plasticiser Tributyl O-acetylcitrate (3.5% v/w; TAC) to reduce PLDLA's Tg and produce a plasticised shape memory material (TAC-PLDLA) that is dimensionally stable at -20°C but which shows shape memory reversion when incubated under physiological conditions (i.e. saline at 37°C). To produce this material, PLDLA and TAC were blended by melt extrusion, pelletized and extruded into 5 mm rods. Shape memory properties were imparted by drawing the rods at 85C, followed by rapid cooling to 4C (draw ratio 4.7, final diameter 2.3 mm). TAC-PLDLA is likely to be biocompatible: TAC is considered 'Not hazardous' according to regulation (EC) No. 1272/2008 whilst preliminary cell biological studies show that TAC-PLDLA supports cell adhesion. A prototype TAC-PLDLA orthopaedic compression screw was designed based upon knowledge of existing metal compression screws and clinical requirements. Aspects of this design (e.g. thread profile) were optimised in silico. Shape memory polymers can be moulded by selective heating and partial reversion within a mould. Subsections of a candidate mould for the proposed screw were therefore designed in silico, and 3D printed in metal. Ongoing work aims to find the optimise moulding, reversion and mechanical strength

pps proceeding
Symposium: S18 - Automotive and medical applications
Oral Presentation

Mechanical and Thermal Behaviors of Polyketone Thermoplastic Materials

WEON J.-I. [(1)*]***, CHOI B. I.*** [(2)]***, PARK S. P.*** [(2)]***, KIM Y. S.*** [(2)]***, YOU S.*** [(2)]***, LEE J. S.*** [(3)]***, KIM S. H.*** [(3)]

[(1)] Department of Safety Engineering, Dongguk University - Gyeongju, Gyeongbuk - SouthKorea, [(2)] Korea Testing & Research Institute - Gwacheon, Gyeonggi-do - SouthKorea, [(3)] Hyosung Corporation - Anyang, Gyunggi-do - SouthKorea

Polyketone polymers with different olefin units have been recently commercialized and mass-produced. The polyketone is a high performance thermoplastic polymer and unique engineering plastic with the back bone of only carbons. In this study, the injection-molded test specimens were prepared and various processing methods and industrial application were also discussed. Consequentially, the mechanical and thermal behaviors of polyketone materials were investigated. In addition, the development of international standards for designation system and specification of polyketone will be addressed.

Photocatalytically active TiO_2 – Silicone Rubber

Theresa Fischer[a)], Markus Ahrens[b)] and Markus Eblenkamp[c)]

Institute of Medical and Polymer Engineering, Technical University of Munich
Boltzmannstraße 15, 85748 Garching, Germany

[a)]Corresponding author: theresa.fischer@tum.de
[b)]markus.ahrens@tum.de
[c)]markus.eblenkamp@tum.de

Abstract. Bacteria are present in all areas of everyday life and can lead to dangerous infections. Surfaces with antimicrobial properties are used to reduce the risk of infection in sanitary facilities and hospitals. Apart from the addition of biocides or antibiotic agents to synthetic materials, recent research show, that it is possible to use the semiconductor titanium dioxide (TiO_2) to generate antibacterial surfaces. The photocatalytically active TiO_2 leads to the development of reactive oxygen species (ROS) that are able to kill bacteria. The aim of this research is to generate silicone rubbers with a photocatalytically active surface that has the possibility to lead to an antimicrobial property by incorporating the additive TiO_2 into the bulk material. AEROXIDE® TiO_2 P25 was used. As matrix material representatives of the three silicone rubber classes were tested. A room-temperature-vulcanization silicone rubber (RTV), a liquid silicone rubber (LSR) and a high-temperature-vulcanization silicone rubber (HTV) were examined. Compounds with various concentrations of TiO_2 were produced and the influence on the cross-linking reaction was investigated using differential scanning calorimetry (DSC). The photocatalytic effect on the surface of the modified material was studied via contact angle measurements and methylene blue tests. The DSC analysis showed that titanium dioxide inhibits the curing of the additivated RTV and LSR. The incorporation of TiO_2 in the HTV matrix material up to 15 wt% was possible, with an extension of the cross-linking time. The investigation of the HTV siivone rubbers considering the photocatalytic effect yield different results. While the expected photo-induced hydrophilic effect could not be verified for the modified HTV silicone rubbers. However the methylene blue trial showed a high significant discoloration of the titanium dioxide addtivtated HTV polymer and proves that a photocatalytic active HTV silicone rubber surface can be produced using the additive titanium dioxide.

Keywords: silicone, titanium dioxide, photocatalyst, curing density
PACS: 81.05.Hd, 81.05.Lg, 81.20.Wk, 82.35.Np

INTRODUCTION

Microbial infections are currently a primary cause of human death. Consequently, multiple antimicrobial solutions have been developed in order to kill or inhibit the growth of microorganisms in an environment with a high risk of infection. Disinfectants such as hypochlorite, hydrogen peroxide, or quaternary ammonium compounds are repeatedly applied and investigated, but their short-term effect and environmental safety limit their use. So antimicrobial polymers arise as promising materials for long-term effective infection prevention at sanitary facilities and hospitals [3,7]. Recent research [6] shows, that titanium dioxide (TiO_2) is a promising antimicrobial agent for polymers. The antibacterial activity of TiO_2 is based on its photocatalytic effect, its ability to induce reactive oxygen species (ROS) in the presence of oxygen and water under UV-irradiation. These ROS cause oxidative degradation of biological structures followed by microorganism death [1].

The aim of the present research is to use TiO_2 to generate a photocatalytically active silicone rubber surface. Therefore, the production of titanium dioxide - silicone rubber compounds was investigated. The examination of the influence of the additive, and its amount, on the cross linking process was performed by using the method of differential scanning calorimetry (DSC). Furthermore, the photocatalytic activity of the surfaces of the produced specimens was analyzed using contact angle measurement and discoloration of methylene blue.

MATERIALS AND METHODS

A nanoscale TiO_2 type was used to reach an antimicrobial and self-cleaning surface: AEROXIDE® TiO2 P25 (Evonik Industries AG, Germany). As polymer matrix three different silicone rubbers were compounded with different concentrations of TiO_2. One representative per silicone rubber class was selected: Silpuran® 2430 AB (RTV), Elastosil® 3003/60 AB (LSR), and Silpuran® 8020/40 (HTV) (Wacker Chemie AG, Germany). In figure 1 an overview of the used materials and its compounds is given.

TABLE (1). Overview of used materials and compositions of the tested compounds

Polymer Matrix	Silicone rubber class	TiO_2 content in wt%
Silpuran® 2430 AB	RTV	0, 5, 10, 15
Elastosil® 3003/60 AB	LSR	0, 5, 10, 15
Silpuran® 8020/40	HTV	0, 5, 10, 15

Sample Preparation

The TiO_2 was incorporated into the RTV and LSR using a mixer. Before the samples were formed by compression molding (Heating press P 300 P, Dr. Collin GmbH, Germany), the compounds were stored in a desiccator for 10 minutes to reduce failures of the surface caused by air residues in the material. The HTV was processed by using a roll-mill (Bench-Type Roll-Mill W 100 T, Dr. Collin GmbH, Germany) to incorporate the TiO_2 into the silicone rubber. In order to simplify the HTV compound production, the additive was solved in acetone. Subsequently the uncured material was formed to test specimens using compressing molding. Table 2 shows the parameters. The samples (figure 1) of each silicone rubber classes were tempered for 4 hours at 200 °C.

TABLE (2). Overview of used parameters for the pressing using the P 300 P

Polymer Matrix	Time [min]	Pressure [bar]	Temperature [°C]
Silpuran® 2430 AB	60	50	200
Elastosil® 3003/60 AB	10	50	200
Silpuran® 8020/40	25	50	170

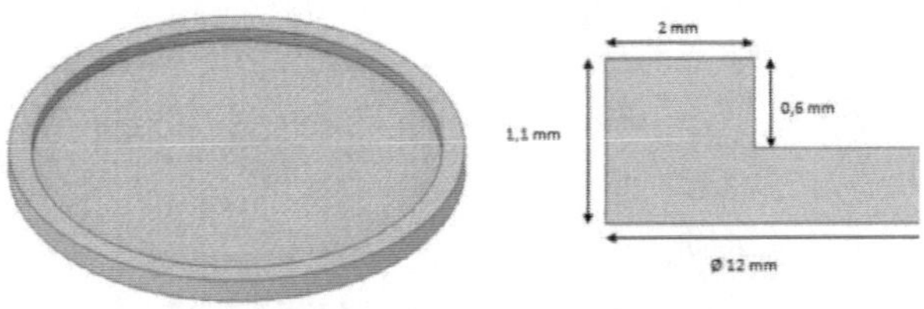

FIGURE 1. Produced specimens

Curing Process Investigation

Bogner et al. [2] showed that differential scanning calorimetry (DSC) can be used to study the crosslinking density of silicone rubbers over time and temperature. By measuring the heat transfer of the material at a specific temperature profile, statements about phase conversion can be made. The investigation of the curing process in this work was based on the standard DIN EN ISO 1357-5. Samples of the various uncured compounds (7.5 mg ± 2.5 mg) were weighed into aluminum pans and analyzed immediately in the DSC chamber. The DSC chamber, continuously flushed with helium to avoid side reactions, was cooled down to -180 °C at the beginning of the analysis and then heated up to 200 °C (HTV to 170 °C) with a constant heating rate of 20 °C/min. The heat flow of the silicone rubber samples was measured and the curing density α can be calculated at time K with formula (1).

$$\alpha_K = \frac{\Delta H_K}{\Delta H_{tot}} \cdot 100 \tag{1}$$

The reaction enthalpy ΔH_K at a specific time K and the total enthalpy of reaction ΔH_{tot} can be computed by integration of the heat transfer over time with the help of the trapezoidal rule. The curing density increases with time and reaches the end value of 100 % asymptotically.

Photocatalytic Effect Examination

The photocatalytic effect of the test samples was investigated by using contact angle measurements (photo-induced hydrophilic effect) and the discoloration of methylene blue.

During photocatalytic activation, oxygen vacancies occur in the titanium dioxide structure where water molecules can bind and lead to an absorption of OH groups that lead to a higher free surface energy [4]. So the contact angle measurement can be used to measure the impact of the photocatalytic effect on the conformation of the test specimens. Therefore contact angles were measured using the DSA 25 contact angle measuring system (Krüss GmbH, Germany) before and after an irradiation with simulated sunlight (250 – 800 nm) for 48 hours. The hydrophilic properties were studied with 2 µl drops of distilled water. The contact angle was measured 5 seconds after wetting.

Based on the standard DIN 52980, the second method to confirm the photocatalytic effect is to measure the discoloration of methylene blue. Hydroxyl radicals as part of the ROS decompose methylene blue to leucomethylene blue [5]. The samples were stored in methylene blue solution during irradiation with simulated sunlight for 240 minutes until the methylene blue is dismantled. The difference of the optical density of the solution was measured every 20 minutes using a UV-VIS-spectrometer (Multiskan FC, Thermo Scientific, Germany).

Statistical analyses were performed for both methods. The level of significance is stated by: * $p < 0.05$, ** $p < 0.01$, *** $p < 0.001$.

RESULTS AND DISCUSSION

Compounds of the silicone rubbers and TiO_2 were successfully produced with 5, 10 and 15 wt% by mixing and using a roll mill for HTV. Because of its high specific surface nanostructured titanium dioxide tends to agglomerate [8] which complicates the homogenous distribution of the additive in the matrix material and can yield to inconsistences in the following analysis.

Initinally the curing process of the materials was investigated. The results of the DSC investigation of the HTV titanium dioxide additivated silicone rubbers can be seen in figure 2.

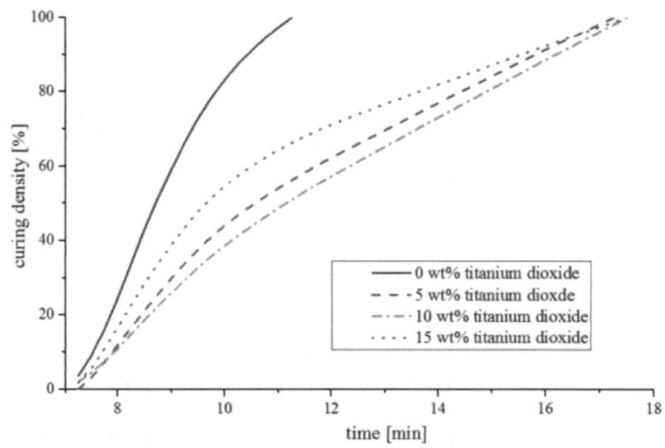

FIGURE 2. Curing density of various additivated HTV silicone rubbers over time.

As figure 2 shows, the curing time of HTV silicone rubber increases with higher amount of titanium dioxide. While pure HTV is completely crosslinked after 11.25 minutes, the HTV with 5 wt% titanium dioxide takes 17.25 minutes to reach a curing density of 100 %. Materials with 10 and 15 wt% incorporated TiO_2 are cured after 17.5 minutes. The DSC investigation of the LSR material shows different results. The crosslinking time of the pure LSR does not extend with an increasing amount of titanium dioxide, but the analysis of the various compounds yield differences in the total reaction enthalpy ΔH_{tot}. The enthalpy ΔH_{tot} decreases from 4.1 J/g for the pure material to 3.5 J/g for 15 wt% TiO_2 additivated LSR. This indicates that the reaction of the crosslinking process of the modified material was not completed. The curing process of the modified RTV silicone rubbers could not be measured by using DSC, even at low amounts of titanium dioxide. While the crosslinking of the pure material starts already after 2.5 minutes, the DSC analysis of the additivated RTV material could not detect any reaction enthalpy ΔH_{tot} . These results point out that the additive titanium dioxide inhibits the platinum catalyst of the silicone rubbers and weakens or even prevents the curing process of silicone rubbers (RTV and LSR). Further analysis showed that the curing of the additivated RTV and LSR could be completed by adding additional platinum catalyst.

Considering the results of the DSC analysis, further investigations concerning a photocatalytic active silicone rubber surface were only carried out with the HTV material. In figure 3 the results of the contact angle measurement before and after irradiation with simulated sunlight can be seen. Taking into account the standard deviation, the additivated samples show very little difference in their measured values. The contact angles before and after irradiation with UV-light are in a hydrophobic range of 109 ° ± 12 °. The expected hydrophilicity on the surface of the specimens as a result of the photocatalytic effect could not be shown, except the 5 wt% TiO_2 showed a decrease of its value after an irradiation of 48 hours.

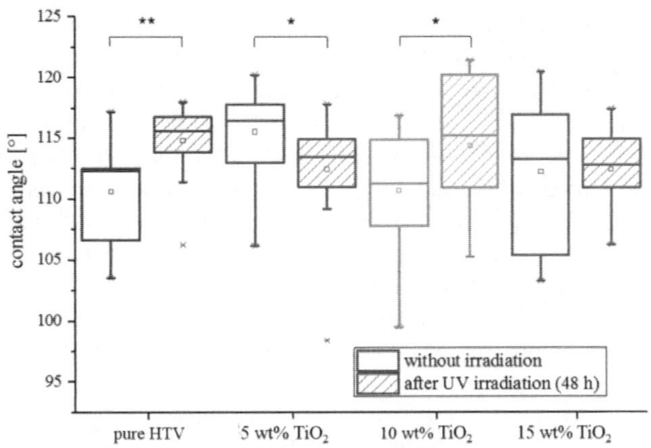

FIGURE 3. Results of the contact angle measurements.

In case of the pure HTV and the 10 wt% TiO_2 additivated material, the contact angle increased significantly to a more hydrophobic surface. This could result from the effect of the hydrophobic recovery of silicone rubbers and the roughening of the surface during an UV irradiation. The hydrophobic recovery causes a change in location of the hydrophilic functional groups on the surface and the hydrophobic functional groups in the bulk [10]. Furthermore, hydrophobic surfaces show an increase of the contact angle if the roughness of the surface increases [9]. During UV irradiation, induced ROS are able to react with the polymer chains, damage the polymer structure and surface and can lead to a roughening of the surface that increases the contact angle.

In order to determine the photocatalytically induced ROS on the sample surface, the methylene blue trial is used. The percentage dismantling of methylene blue into leucomethylene blue during UV irradiation of the HTV test specimens is shown in figure 4.

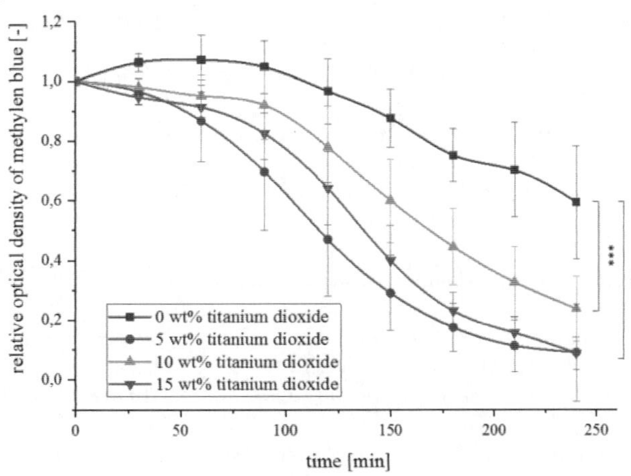

FIGURE 4. Discoloration of methylene during UV irradiation for 240 minutes.

Figure 4 demonstrates that photocatalytically active HTV silicone rubber surface can be produced using the additive titanium dioxide. The methylene blue trial yields a highly significant discoloration of the three modified HTV silicone rubbers compared to the pure material. After 240 minutes, the silicone rubber with 10 wt% TiO_2 shows a value of 23 ± 13 %, while the 5 and 15 wt% additivated silicone rubbers reaches 8 ± 10 %. These results indicate that the modified materials induce ROS and consequently hydroxyl radicals to dismantle the color indicator.

Further investigations should involve an improved incorporation by preventing the agglomeration of the additive. So inconsistences in the contact angle measurement or in the methylene blue trial can be avoided. Furthermore a plasma-etching technique will be investigated to increase the titanium dioxide content on the surface of the test specimens in order to get a more active photocatalytic HTV silicone rubber surface.

CONCLUSION

In this research compounds of HTV silicone rubber and TiO_2 were successfully produced with 5, 10 and 15 wt%. The DSC analysis showed that it was not possible to produce titanium dioxide additivated RTV and LSR silicone rubbers, only the TiO_2 HTV compounds crosslinked to a curing density of 100 %. The crosslinking investigation of the HTV silicone rubbers demonstrated that the curing time increases by adding titanium dioxide to the matrix material. Considering the DSC results, further investigations are only done on modified HTV material.

The expected photo-induced hydrophilic effect could not be verified for the modified HTV silicone rubbers. The hydrophobic recovery of the hydrophobic silicone rubber and the roughening caused by an UV irradiation is a possible explanation of these results. The existence of photocatalytically induced ROS was verfied by the methylene blue trial. This analysis showed a highly significant discoloration of the titanium dioxide addtivtated HTV polymer and proves that a photocatalytically active HTV silicone rubber surface can be produced using the additive titanium dioxide. Further investigations will proof if the photocatalytically active surface hast the possibility to lead to an antimicrobial property.

ACKNOWLEDGMENTS

The authors greatfully acknowledge Evonik Resources Efficiency GmbH and Wacker Chemie AG for supporting this research with their material. The presented research is funded by the AiF. We would like to extend our thanks to AiF.

REFERENCES

1. S. Banerjee, J. Gopal, P. Muraleedharan, A.K. Tyagi and B. Raj, *Physics and chemistry of photocatalytic titanium dioxide: Visualization of bactericidal activity using atomic force microscopy*. Current Science, 2006. **90**(10): p. 1378-1383.
2. R.H. Bogner, J.-C. Liu and Y.W. Chien, *Kinetics of curing and physicomechanical behavior of silicone membrane*, Journal of Controlled Release, 1990. 14(1): p.11-19.
3. L.E. Fisher, A.L. Hook, W. Ashraf, A. Yousef, D.A. Barrett, D.J. Scurr, X. Chen, E.F. Smith, M. Fay, C.C.J. Parmenter, R. Parkinson and R. Baysotn, *Biomaterial modification of urinary catheters with antimicrobials to give long-term broadspectrum antibiofilm activity.* Journal of Controlled Release, 2015. **215**(1): p. 57-64.
4. A. Fujishima, X. Zhang, D.A. Tryk, *TiO_2 photocatalysis and related surface phenomena*. Surface science reports, 2008. **63**(12): p. 515-582.
5. A. Ghauch, A.M. Tuqan, N. Kibbi and S. Geryes, *Methylene blue discoloration by heated persulfate in aqueous solution.* Chemical Engineering Journal, 2012. **213**(1): p. 259-271.
6. T. Huppmann, S. Yatsenko, L. Leonhardt, E. Krampe, I. Radovanovic, M. Bastian, E. Wintermantel, *Antimicrobial polymers – The antibacterial effect of photoactivated nano titanium dioxide polymer composites*. AIP Conference Proceedings, 2014. **1593**(1): p. 440-443.
7. A. Jain, L.S. Duvvuri, S. Farah, N. Beyth, A. Domb and W. Khan, *Antimicrobial Polymers*. Advanced Healthcare Materials, 2014. **3**(12): p. 1969-1985.
8. J. Jinag, G. Oberdörster and P. Biswas, *Characterization of size, surface charge, and agglomeration state of nanoparticle dispersions for toxicological studies,* Journal of Nanoparticle Research, 2009. 11(1): p. 77-89.
9. M. Miwa and A. Nakajima, *Effects of the Surface Roughness on Sliding Angles of Water Droplets on Superhydrophobic Surfaces*, Langmuir, 2000. **16**(13): p.5754-5760.
10. M.J. Owen, P.J. Smith, *Plasma treatment of Polydimethylsiloxane,* Journal of Adhesion Science and Technology, 2012. **8**(10): p. 1063-1075

3D Printing of Custom Fabricated Textured Flexible Fiber Composites

Ali Anwer[a], Hui (Vina) Ye[a] and Hani E. Naguib[a,b,c]

[a]*Department of Mechanical and Industrial Engineering*
University of Toronto
5 Kings College Rd., Toronto, M5S3G8, Canada
[b]*Department of Materials Science and Engineering,*
184 College Rd.., Toronto, M5S3E4, Canada
University of Toronto
[c]*Institute of Biomaterials and Biomedical Engineering*
University of Toronto
164 College St., Toronto, M5S3G9, Canada
University of Toronto

Abstract. his study conducts a parametric study of using custom fabricated flexible 3D printer composite filament that can be used as "ink" to be fed into a Fused Deposition modeling printer to investigate the fiber alignment during the printing process. Different volume fractions of flexible thermoplastic polyurethane carbon fiber-composites have been fabricated and the effect of fiber alignment during the printing process was elucidated by conducting tensile testing. The surface was then textured using fiber-deboning and pullout process to expose the fibrous phase, and the surface morphology was characterized for roughness.

Keywords: 3D Printing, Flexible Fiber composite, Carbon Fiber, Fiber Alignment, Fused Deposition Modelling
PACS: Replace this text with PACS numbers; choose from this list: http://www.aip.org/pacs/index.html

INTRODUCTION

Fiber alignment is becoming an increasingly crucial component in composite design. Controlling the orientation would significantly impact the mechanical and surface properties such as hydrophobicity[1–4]. In addition to this, texturing of fiber composites using fiber pullout, based on internal fiber orientation, has become increasingly popular for created advanced functional surfaces. Fused Deposition Modelling, an Additive manufacturing technique, is a new field that has emerged for rapidly prototyping composites with controlled properties [5]. There has been research in the where fiber orientation can be controlled using the additive manufacturing and inherently effect the mechanical properties of the composite, however very little research effort has been done for flexible reinforced thermoplastic composites [6] . In a study by Compton et al., they investigated the tensile strength of the developed epoxy based SiC composite [6]. When depositing the fiber composite during the 3D printing process, it caused the fibers to align in the direction of extrusion and found that the modulus was 10X higher than that of pure epoxy. Similarly, this fiber alignment process could be extended to other thermoplastic fused deposition modelling printers. In such FDM printers, the fibers within the molten material align themselves in the extrusion direction, thereby imparting uni/bi-directional improvement in the mechanical properties [7]. However, most of the previous studies on using FDM for fiber composites has been conducted on stiffer polymers such as PLA and ABS [7]. This gives rise to the potential for using fiber alignment in additive manufacturing for several different applications in the future to create more functionalized surface, one of which is surface texturing applications. Very few studies have been conducted on using fiber alignment on flexible fiber composites for surface texturing application.

This study details the uni-directional improvement in mechanical properties. Other studies use off-the shelf 3D printer filaments such as Poly lactic acid (PLA) infused with carbon fiber (CF) [8]. However, in this study we have developed a fabrication technique for manufacturing and 3D printing flexible carbon fiber composite with uni-directional improved mechanical properties. Briefly, Thermoplastic Plastic Polyurethane (TPU) masterbatch batch with 10 wt% Multi-wall Carbon-nanotubes (MWCNT) was used with varying carbon fiber fraction from 0 to 4vol %. Tensile samples and rectangular samples were printed in 45° And 90° orientation to study the effect of fiber alignment during the printing process. Following fabrication of these samples, the composite surfaces were textured using fiber debonding and pullout methods detailed in our previous study to prove fiber orientation in the printed samples [9].

This was achieved by measuring the surface roughness of the textured composite surface. In this study, we employ a technique that does not require complex post texturing processes such as plasma etching, chemical etching and photolithography to create a functional textured surface [10–12].

METHODOLOGY

The fabrication of the 3D printed composite was performed using the following procedure (**Error! Reference source not found.**a). Initially, a master batch of TPU (Lubrizol Estane ETE 55DT3) with 10wt% MWCNT (cheaptubes) was compounded using a melt at 250° C with varying carbon fiber fraction of 0 vol%, 1 vol%, 2 vol% and 4 vol% using a using a DSM Xplore 15 (DSM, Netherlands) lab scale mixer at 200°C, with a residence time of 10 min, and a screw rpm of 50. MWCNT was used to provide sufficient stiffness of the filament for use in the FDM printer without causing clogging of the extrusion nozzle on the FDM printer. Having a masterbatch of constant 10wt% MWCNT with TPU allowed the gears in the 3d printer that feed the polymer into the nozzle, to have sufficient grip and not cause slippage of the softer TPU filament being pushed down. The uniformly blended mix was then allowed to exit a 2.5mm diameter exit nozzle and air cooled to form a uniformly blended hybrid composite mixed filament. This filament was then pelletized in a pelletizer, and re-fed through a single screw extruder (Filastruder 2.0, USA). This single screw extruder was kept at 210°C, and had an exit filament diameter of 1.75mm. As the filament exited the end of the extruder, it was simultaneously air-cooled and allowed to pass through an infrared dynamic positioning sensor as it was wrapped around a spool mounted on the filament winder (Filawinder 2.0, USA). This sensor allowed the filament winder to increase/ decrease its speed (depending on the position of the filament) as the composite filament was exiting the extruder nozzle. Following fabrication of this hybrid composite filament on to a spool, the spool was then mounted onto an FDM dual nozzle 3d Printer (E3D Online- Big Box Dual V1.1). A 0.6mm nozzle was used and the extruder was kept at 255 °C to prevent clogging of the nozzle. The bed temperature of the printer was set to 105 °C for sufficient bonding between the bed and the composite during the print and did not allow the base layer to shift during printing. The infill percentage was set to 100%, and the X/Y movement was set to 1500 mm/min, and the Z to 1000 mm/min to ensure sufficient resolution of printing. The g code of the print file was modified, so that the composite filaments could be laid out onto the bed 45° And 90° during the print to verify fiber alignment during the printing process schematized in 1b. A standard tensile sample with thickness (1.5mm) was printed based on ASTM D639 standard was printed to see the effect of fiber orientation and alignment during the print process on the tensile modulus and strength. To further investigate the effect of fiber alignment and different fiber orientations, a rectangular sample with 56mm X 13mm X 1.5mm dimensions was printed for comparing the storage modulus of the different printed composites. A schematic of the fabrication process can be seen in **Error! Reference source not found.**a. These rectangular samples were then cut to expose the fibrous phase from the surface, thereby texturing the surface. This process of fiber de-bonding and pull-out for texturing the composite surface has been elucidated in our previous studies [9,13]

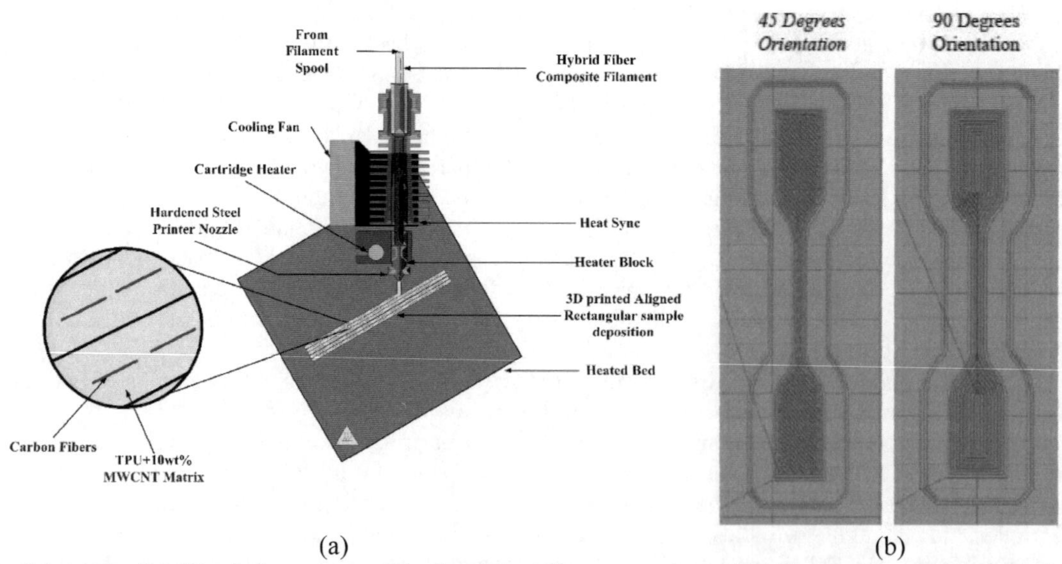

(a) (b)

FIGURE 1. Schematic of a) 3D printing process of flexible carbon fiber composite and (b) filament layout orientation for tensile testing.

RESULTS AND DISCUSSION

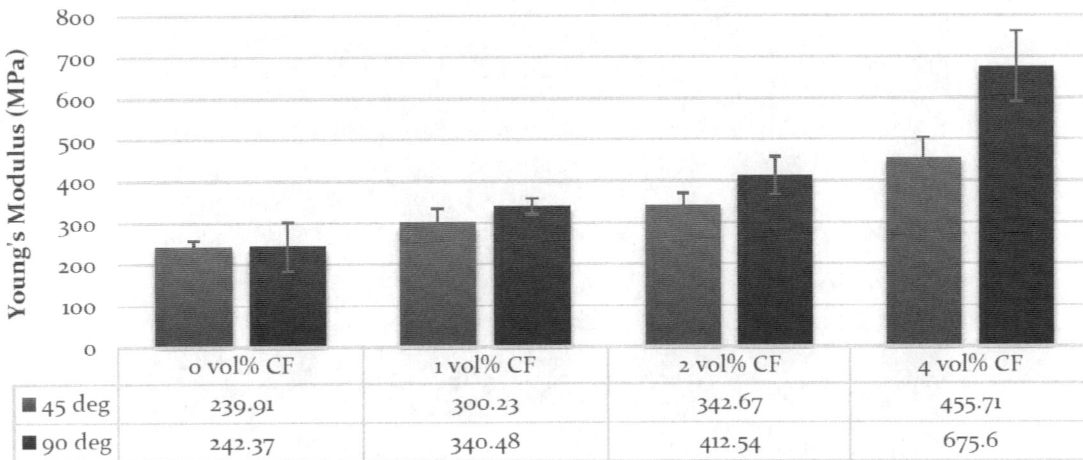

FIGURE 2. Tensile Modulus comparison of varying carbon volume fraction (0-4 vol%) in 10wt% MWCNT+TPU

Tensile Testing was conducted at 5mm/min on an INSTRON 5848 to elucidate the fiber alignment during the 3D printing process of this hybrid flexible carbon fiber composite material. As evident from Figure 2, we can see that the modulus is steadily increasing from 0 vol% to 4 vol%. At lower vol% of CF, the difference between the 45° and 90° composite is not significant because this developed fabrication technique for 3D printing aligned fiber composite does not align the nanoscopic nanotubes, but rather the microscopic carbon fibers only. Hence, at higher vol%, we can see a much higher % improvement in the tensile modulus of >30% due to greater number of carbon fibers are oriented in the direction of loading that carry higher loads. To further elucidate the fiber alignment of the flexible fiber composite, the surface was textured using fiber debonding and pullout using a linear surface profilometer Mitutoyo SJ 210 [9]. The textured composite surface was then evaluated using surface roughness. Through measurement of the surface roughness as seen in Figure 3, we can see that the 45° printed composite shows a lower surface roughness value as compared to the 90° composites when compared within the same vol%. This is because the surface profilometer does not traverse the entire length of the carbon fiber during its linear motion, thereby the linear z movement is reduced, indicated by the lower surface roughness values. Figure 4, shows an SEM image of 1 vol% CF+(10wt%MWCNT+TPU) composite with 45° oriented fibers indicated by the arrows

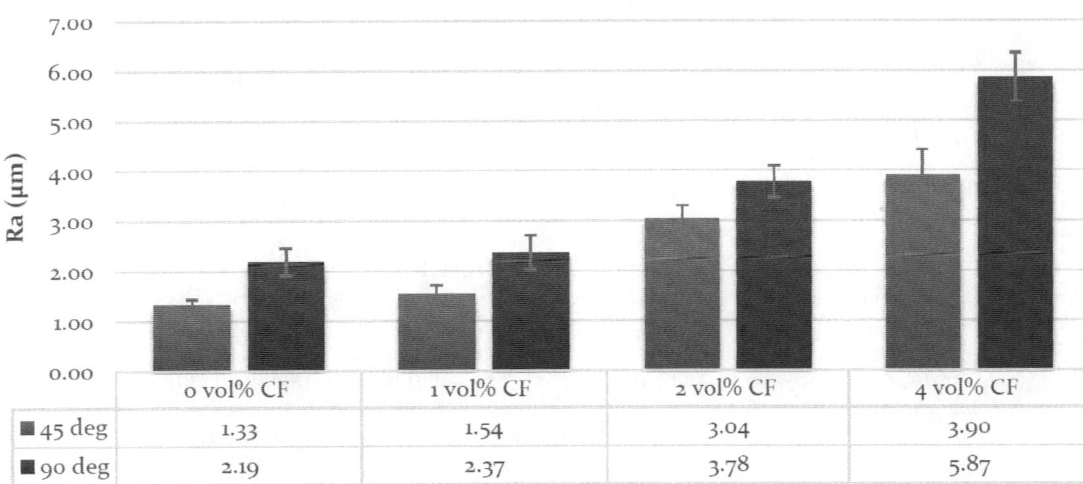

FIGURE 3. Surface Roughness comparison of varying carbon volume fraction (0-4 vol%) in 10wt% MWCNT+TPU

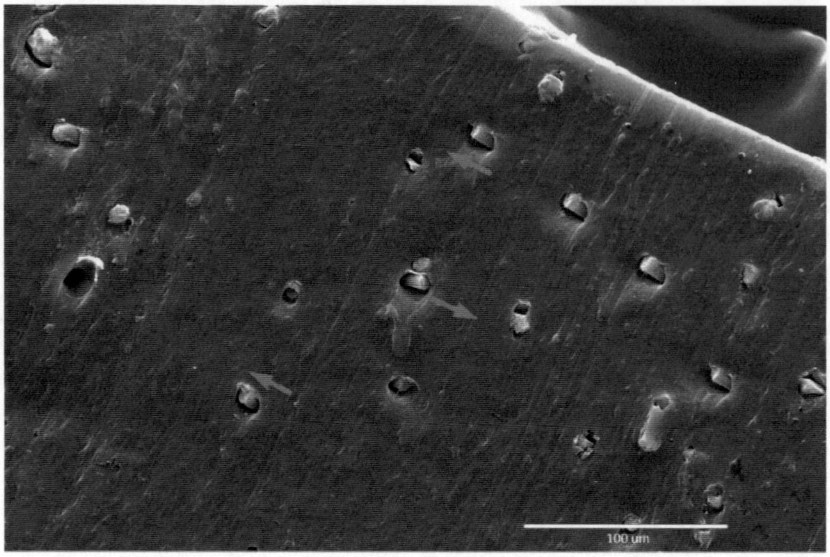

FIGURE 4. SEM Image of 1vol% 45° composite showing alternating 45° layer orientation of the protruding fibers

CONCLUSION

In this study, we have developed a fabrication technique for manufacturing flexible carbon fiber composite for use in Fused Deposition Modelling 3D printing. Tensile testing was conducted to prove a 34% improvement for the 90° as compared to the 45° composite in tensile modulus during filament deposition from the 3D printer nozzle. To further elucidate the fiber alignment in this flexible carbon fiber composite, roughness testing was conducted that showed an increase in the surface roughness for the 90° layout fiber composites when compared within the same vol%. By developing this novel methodology for manufacturing flexible fiber composite filaments and controlling the fiber orientation during the FDM printing process, we were able to impart directional properties to the composite, making this suitable for several applications.

ACKNOWLEDGMENTS

We would also like to acknowledge funding from Natural Sciences and Engineering Council (NSERC) of Canada.

REFERENCES

[1] R. Matsuzaki, M. Ueda, M. Namiki, T.-K. Jeong, H. Asahara, K. Horiguchi, T. Nakamura, A. Todoroki, Y. Hirano, Three-dimensional printing of continuous-fiber composites by in-nozzle impregnation, Sci. Rep. 6 (2016).

[2] J.M. Holzwarth, P.X. Ma, Biomimetic nanofibrous scaffolds for bone tissue engineering, Biomaterials. 32 (2011) 9622–9629.

[3] J. Visser, F.P.W. Melchels, J.E. Jeon, E.M. Van Bussel, L.S. Kimpton, H.M. Byrne, W.J.A. Dhert, P.D. Dalton, D.W. Hutmacher, J. Malda, Reinforcement of hydrogels using three-dimensionally printed microfibres, Nat. Commun. 6 (2015).

[4] Z. Quan, A. Wu, M. Keefe, X. Qin, J. Yu, J. Suhr, J.-H. Byun, B.-S. Kim, T.-W. Chou, Additive manufacturing of multi-directional preforms for composites: opportunities and challenges, Mater. Today. 18 (2015) 503–512.

[5] F. Ning, W. Cong, J. Qiu, J. Wei, S. Wang, Additive manufacturing of carbon fiber reinforced thermoplastic composites using fused deposition modeling, Compos. Part B Eng. 80 (2015) 369–378.

[6] B.G. Compton, J.A. Lewis, 3D-printing of lightweight cellular composites, Adv. Mater. 26 (2014) 5930–5935. doi:10.1002/adma.201401804.

[7] H.L. Tekinalp, V. Kunc, G.M. Velez-Garcia, C.E. Duty, L.J. Love, A.K. Naskar, C.A. Blue, S. Ozcan, Highly oriented carbon fiber–polymer composites via additive manufacturing, Compos. Sci. Technol. 105 (2014) 144–150.

[8] R.T.L. Ferreira, I.C. Amatte, T.A. Dutra, D. Bürger, Experimental characterization and micrography of 3D printed PLA and PLA reinforced with short carbon fibers, Compos. Part B Eng. 124 (2017) 88–100.

[9] R. Rizvi, A. Anwer, G. Fernie, T. Dutta, H. Naguib, Multifunctional Textured Surfaces with Enhanced Friction and

Hydrophobic Behaviors Produced by Fiber Debonding and Pullout, ACS Appl. Mater. Interfaces. 8 (2016) 29818–29826. doi:10.1021/acsami.6b11497.
[10] B.K. Lee, I. Baek, B. Lee, Y. Kim, W.I. Jang, Y.S. Yoon, H.Y. Yu, Fabrication of a Large-Area Hierarchical Structure Array by Combining Replica Molding and Atmospheric Pressure Plasma Etching, Adv. Mater. Interfaces. 2 (2015).
[11] T. Wang, G. Liu, J. Kong, C. Wong, Fabrication and Hydrophobic Properties of Column-Array Silicon Using Wood-Structured Silver-Assisted Chemical Etching, Adv. Mater. Interfaces. (2016).
[12] M. Laurenti, A. Verna, M. Fontana, S. Stassi, G. Canavese, S.L. Marasso, V. Cauda, How Micropatterning and Surface Functionalization Affect the Wetting Behavior of ZnO Nanostructured Surfaces, Adv. Mater. Interfaces. (2016).
[13] R. Rizvi, H. Naguib, G. Fernie, T. Dutta, High friction on ice provided by elastomeric fiber composites with textured surfaces, Appl. Phys. Lett. 106 (2015) 111601.

Optimization of Freeforming process for amorphous and semi-crystalline thermoplastics

Charlon Sebastien [1], Lacrampe Marie-France [1], Soulestin Jérémie Jérémie [1]*

[1] IMT Lille Douai - Nord - France

Freeformer from Arburg offers a new technology for additive manufacturing of polymers that enables to use standard pellets as raw material and gives new opportunities to manufacture parts of high quality thanks to its nozzle technology. The main advantage is the possibility to use any material commercially available. However, if printing with qualified materials (ABS, TPU, supporting material, PS) is as easy as inkjet printing, the process has to be optimized before getting parts of good quality in other cases. In the same way that for regular polymer processing technologies like injection molding, optimization of the processing parameters and the choice of appropriate grade is needed. The Freeformer technology is based on two injection molding units that enables to melt the standard granulate and to feed the printing head. The discharge unit featuring a pulsed nozzle closure generates small (down to 200 µm) molten polymer droplets to build, layer- by-layer, three-dimensional parts. Thanks to the presence of two discharge units, parts of complex geometries using water soluble supporting material or parts combining hard and soft materials are easily feasible. Influence of the processing parameters on the microstructure (using SEM, figure 1a and b) of the samples will be presented and its influence on resulting properties (mechanical properties, dimensional accuracy, surface roughness,...) will be illustrated in the case of ABS. The influence of the type of material on the resulting properties of the parts after process optimization will be demonstrated by comparing ABS with other materials like polyolefins, SEBS, EVA,... In the case of soft materials comparison with injection molding will confirm that Freeformer technology is able to manufacture parts with mechanical properties as good as injected parts and even better.

pps proceeding
Symposium: S19 - Additive technology
Oral Presentation

Tailoring properties of Additively Manufactured Parts Using Off-Axis Printing

Rudolph Natalie [(1)]*, van Hulle Luke [(1)]

[(1)] University of Wisconsin-Madison - Wisconsin - USA

In the last few years, Fused Filament Fabrication (FFF) - often referred to as 3D printing or Fused Deposition Modeling (FDM®) - has become the fastest growing AM technology. FFF generates 3D parts by melting thermoplastic polymer filaments in a heated nozzle to form beads and fill out a 2D layers. These layers are stacked up to build the parts. This sequential part build causes an anisotropy within and between layers. The mechanical strength is strongest in bead direction and weakest between layers. The resulting anisotropy in FFF can be used by preferentially orienting beads in load direction and thus tailor the part properties. To fully control all print parameters a custom python™ program was developed and programed. In addition to traditional user adjustable print parameters, custom infill patterns, distance between beads, and amount of material extruded along a bead can be set. All of these parameters can be independently set for every layer, and infill pattern can also be set for sub-regions in a layer. In addition to analyzing the anisotropy as a function of bead orientation, build orientation and layer print time in tensile test specimen, a C-bracket was used as a demonstrator to apply the gained knowledge to a more complex geometry. The infill angle was determined using Pareto-based topology optimization and was compared to the default infill settings. This is the first step towards a tool pathing algorithm that prioritizes by strength requirements rather than simple geometrical constraints. To take full advantage of the anisotropy and either improve performance or further reduce weight, off-axis printing was investigated. Mounted on a 6-axis robot, the heated print platform is moving in any plane and direction under the heated printer nozzle to build up the part. This enables unrestricted design freedom in all three dimensions and significant reduction of support material. The advantages and disadvantages of this novel printer will be discussed.

Experimental Studies on Liquid Additive Mixing in a Hyperbolic Die to Incorporate Liquids in Polymer Based Additive Manufacturing

Aditya Sangli[a], Marcelo Arispe-Guzman[a], Connor Armstrong[a] and David Bigio[a]

[a]*Department of Mechanical Engineering, University of Maryland, College Park, MD 20740, USA*
asangli@umd.edu

Abstract. Additive manufacturing methods currently use solid fillers in the polymer matrix to make composites. Many researchers have studied the effect of die design on solid fillers like glass and carbon fibers but there is no research on incorporating immiscible liquid additives which can lead to additive manufacturing of solid parts with a controllable immiscible liquid morphology internal to the part. To accomplish this, experiments must be done to determine the effect of die design on the deformation induced on an immiscible droplet in the flow field. In this research, the feasibility of using shear flows and extensional flows to deform a droplet is first studied. A channel with walls based on a hyperbolic equation can impose a constant stretch rate at the center and makes it easier to model the flow. Applying this principle, a converging channel is built, filled with Silicone oil (Matrix liquid) and subject to hyperbolic flows. The deformation seen by a droplet of immiscible Castor oil is studied for various positions of injection using high speed imagery. It was observed that droplets injected at the center of the channel did not see any significant stretch but the ones injected closer to the walls stretched and reached the affine state. Simulations studies were performed using ANSYS Fluent to find the velocity gradients along each streamline where the droplet was injected. This data, combined with the droplet widths measured from the images, was used to track the Capillary number changes for a given droplet. It was observed that droplets with a higher initial Capillary number saw a steeper reduction than droplets with lower initial Capillary numbers. Further, close examination of the droplet dimensions from each image showed that the droplet width reduced and reached an asymptote in the channel. These results aid in understanding liquid filler behavior in a polymer flow and helps in incorporating liquids in polymer based additive manufacturing.

Keywords: Additive Manufacturing, Extensional Flow, Hyperbolic Channel.

INTRODUCTION

Liquid additives which are immiscible in a polymer undergo certain changes in morphology as the matrix polymer is subjected to a velocity. Incorporating immiscible liquid additives will lead to a unique additive manufacturing process in which liquid-filled parts can be manufactured. The liquid morphology internal to the part, namely droplet distribution and tube lengths, can be predicted based upon the operating conditions and controllable by flow conditions seen by the matrix polymer.

The behavior of immiscible droplets in a deforming matrix has been widely studied in experimental devices like the four-roll mill. Shear and extensional deformation can be imposed on the droplet in the four-roll mill but they are stagnation flows in nature since the center of the droplet remains motionless. The four-roll mill also does not have any practical significance and there is a need to develop a die which can deform a droplet in a flow. Production of shear flows present experimental difficulties and has a lower magnitude compared to an extensional flow, which can be produced in a converging die.

The die used in this research had converging walls based on a hyperbolic equation which has the potential to deform the suspended droplet in a near constant extensional rate. Achieving a constant extensional rate is a unique feature since the extensional rate in converging channels with straight walls, used by many other researchers, is not constant in the center. The die was filled with a fluid and an immiscible droplet was subject to the converging flow to observe the changes in the droplet morphology. In order to see the behavior of the droplet along different streamlines, the droplet was injected at offset positions and the changes were observed. Simulation data was used in conjunction with the experimental results to calculate the Capillary number associated with the droplet as it moved along the channel. This research is an attempt to understand the behavior of an immiscible droplet in an industrially significant process. The hyperbolic converging die can be scaled down and incorporated into polymer based additive manufacturing machines to dynamically produce liquid-filled solid parts.

BACKGROUND

The deformation of an immiscible droplet in a matrix was studied by Taylor [1] who invented the four-roll mill and subjected the droplet to shear and extensional flows. He found a theoretical expression for deformation as a function of fluid properties like viscosity and interfacial tension. In the four-roll mill, the center of the droplet remains stationary while the ends are deformed depending upon the type of flow. This stagnation deformation of the droplet is a characteristic of the four-roll mill and the device has been in popular use by researchers trying to predict immiscible droplet behavior.

Grace [2] used the four-roll mill and found the characteristic time taken by the droplet to reach certain dimensions in shear and extensional flows. He identified the term "Capillary number" associated with the droplet deformation and plotted the critical Capillary number with respect to viscosity ratios. Critical Capillary number is referred to as the state wherein no stable droplet shape can exist at the applied deformation rate which is kept constant. This definition is not based on the state when droplets breakup. The critical Capillary number plots were useful in predicting dispersion of color concentrates in plastics industries.

Bentley and Leal [3] did a careful study and produced the critical Capillary number curve for a variety of fluid flows. In their results, they comment upon the critical Capillary number being around 0.4 for an extensional flow in a system with viscosity ratio of one. The Capillary number vs. viscosity ratio plots have been accepted as the standard for shear and extensional flows in the velocity field produced by a four-roll mill. However, non-stagnant extensional flows can be produced with polymer flow in converging channels. The velocity field in such channels is different than the field in a four-roll mill. Further, the boundary conditions on the droplet in a flow are unlike that in the four-roll mill which is basically a stagnation type extensional flow. It is thus worthy to explore the behavior of a droplet in extensional flows that can be produced in converging channels.

A foray in this area was done by Chin and Han [4] who used converging dies with linear bounding walls and subject a droplet in the center of the channel in the flow of both Newtonian and Non-Newtonian fluid matrices. When a fluid flows between converging walls, an immiscible droplet in the center of the field will experience an extensional deformation because of the converging action of the streamlines. The rate of deformation tensor has components that impose a slight shear on the sides but is largely an extensional flow. However, the extensional rate, which can be calculated from the rate of deformation tensor, is not a constant along the centerline of the matrix fluid flow. There is a need to explore different wall profiles which can impose a near constant extensional rate at the center.

Oliveira et al., [5] did numerical simulations on microfluidic dies which have a rectangular cross section and the walls are shaped based upon a hyperbolic equation. They proved that the extensional rate can be made near constant in such dies and this fact used by Mulligan and Rothstein [6] to develop hyperbolic dies. They provide wall equations and the equation for average extensional rate seen by a droplet in such a die. However, the average extensional rate is not the right measure along streamlines offset from the center. In such cases, the net strain rate must be calculated.

In this research, the hyperbolic equations are chosen and a unique die is developed for polymer processing applications. This is an effort to develop a die that is fundamentally different than the four-roll mill and model the behavior of droplets in the flow. Droplets are injected into the channel at different locations and the behavior is observed. The conditions at which the droplet shape becomes affine as well as certain interesting morphologies during the transition have been illuminated.

EXPERIMENTAL SETUP

The hyperbolic die which was developed has been shown in Figure 1. It has a rectangular cross section with constant depth of 20 *mm*. The initial section of 70 *mm* is for the motion of the plunger which exerts a uniform velocity on the fluid. The converging portion has a length of 40 *mm* along which the width reduces hyperbolically based on Equation 1.

$$y = \frac{600}{3x + 40} \qquad (1)$$

In this converging section, the extensional rate at the center of the channel, for different flow rates, is given by the Equation 2 [6].

$$\dot{e} = \frac{2Q}{wLh} \tag{2}$$

Where *'Q'* is the flow rate in the channel, *'w'* is the half width of the channel in the downstream, *'L'* is the length of the converging zone, and *'h'* is the constant depth of the channel. Thus, the average extensional rate is only affected by the flow rate in the channel. However, this equation is not valid for other streamlines and necessitates the calculation of the strain rate from the rate of deformation tensor.

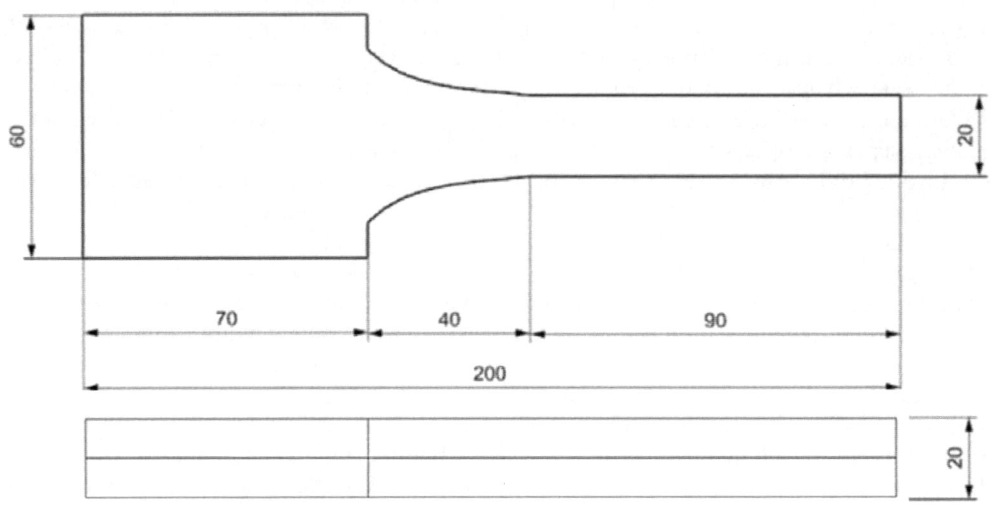

FIGURE 1. Design of the Hyperbolic die. Streamwise direction is from left to right. All dimensions in *mm*.

SIMULATION STUDIES

Computational Fluid Dynamics (CFD) simulation was carried out on the die in ANSYS Fluent, Academic research, Release 17.2 for the channel geometry. For an inlet speed of 34 *mm/s*, the change in the streamwise component of velocity (*u*) along the centerline is shown in Figure 2.

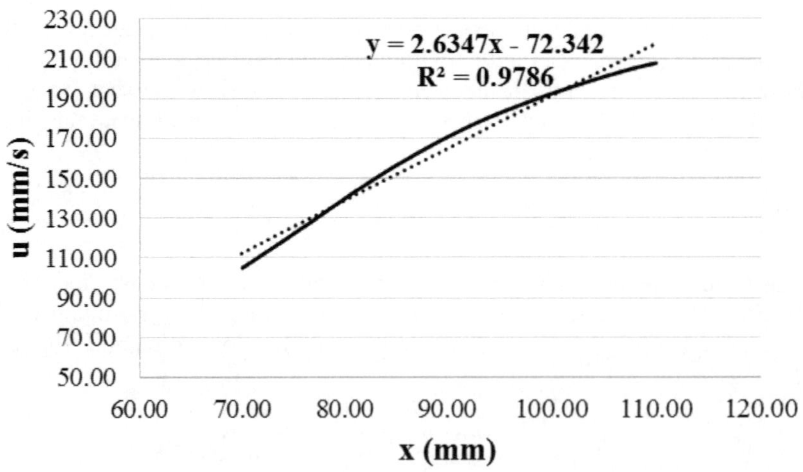

FIGURE 2. Variation of the streamwise component of velocity in the converging zone for an inlet speed of 34 *mm/s*. The converging portion begins at 70 *mm* and ends at 110 *mm*.

It can be seen from the graph that the variation in velocity is a linear trend. The slope of the line gives *du/dx* which is the extensional rate along the line. Since the slope is constant, the value can be assumed to be the extensional rate along the centerline for the given inlet speed.

Simulations were carried out along other straight-line paths in the channel where a droplet can be deployed. These are depicted in Figure 3.

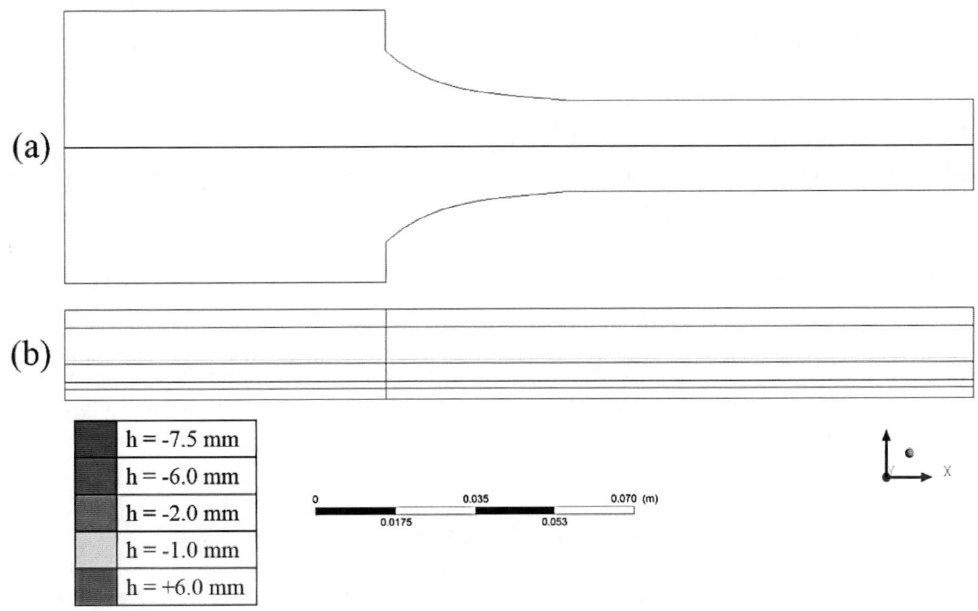

FIGURE 3. Depiction of different droplet injection positions. *h* refers to the distance from the centerline.

A sample set of results is discussed for the line $h = -7.5$ *mm*. Simulation results showed that the net strain rate had to be considered in offset deployment cases instead of just *du/dx*. This was because other terms in the rate of deformation tensor (the off-diagonal terms) had a significant influence apart from *du/dx*. Table 1 shows the magnitude of strain rate along the line $h = -7.5$ *mm*

TABLE 1. Velocity gradients along line $h = -7.5$ *mm*. Strain rate, shown in the last column, is the second invariant of the rate of deformation tensor.

Position along *x* (*mm*)	***du/dx* (s^{-1})**	***du/dz* (s^{-1})**	***dv/dy* (s^{-1})**	***dw/dz* (s^{-1})**	**Strain Rate (s^{-1})**
70.00	1.53	15.92	-1.48	-0.03	15.99
74.53	1.68	18.27	-1.65	-0.03	18.35
80.56	1.68	21.56	-1.66	-0.03	21.62
85.15	1.53	23.92	-1.51	-0.02	23.97
90.82	1.25	26.41	-1.24	-0.02	26.44
95.58	1.04	28.10	-1.03	-0.02	28.12
100.40	0.92	29.54	-0.90	-0.02	29.55
105.19	0.82	30.80	-0.81	-0.02	30.81
110.00	0.51	31.81	-0.51	0.00	31.81

The magnitude of the strain rate, which is the second invariant of the rate of deformation tensor, is much higher than in the center deployment case, which is a case of pure extension. Experiments had to be conducted to observe the droplet morphological changes along these lines.

EXPERIMENTAL METHODS

The fluids used in the study were Castor oil and Silicone oil, as the droplet and the matrix respectively. The viscosity of Castor oil was 600 *cSt* and Silicone oil was 1000 *cSt* with the interfacial tension between them being

0.004 *N/m*. Droplets of Castor oil were subject both in the center of the channel as well as in an offset position as per Table 2 and Table 3.

TABLE 2. Center deployment trials.

Trial number	Plunger speed (*mm/s*)	Initial droplet size (*mm*)
1	34.08	3.10
2	42.02	2.77
3	49.96	3.10
4	57.90	2.61
5	65.84	2.52
6	73.78	2.65

TABLE 3. Offset deployment trials.

Plunger speed (*mm/s*)	Distance from center '*h*' (*mm*)	Initial droplet size (*mm*)
	-1.0	3.17
	-2.0	3.58
34.08	-6.0	3.39
	+6.0	3.73
	-7.5	3.41

A high-speed camera (Nikon D7100) was used to capture images as the droplets moved down the channel. The images were processed in Photoshop, and MATLAB was used to measure the width of the droplet at different stages in the channel which formed the basis of calculating the Capillary number. The image processing also yielded certain unique droplet shapes in the offset deployment experiments and is discussed further.

RESULTS AND DISCUSSION

A Based upon the experimental conditions in Table 2 and Table 3, droplets were injected in the center of the channel and along different streamlines offset from the center. The results from the experiments are elaborated.

Center Deployment Trials

A series of images showing the droplet deformation when subjected to the center of the channel is shown in Figure 4.

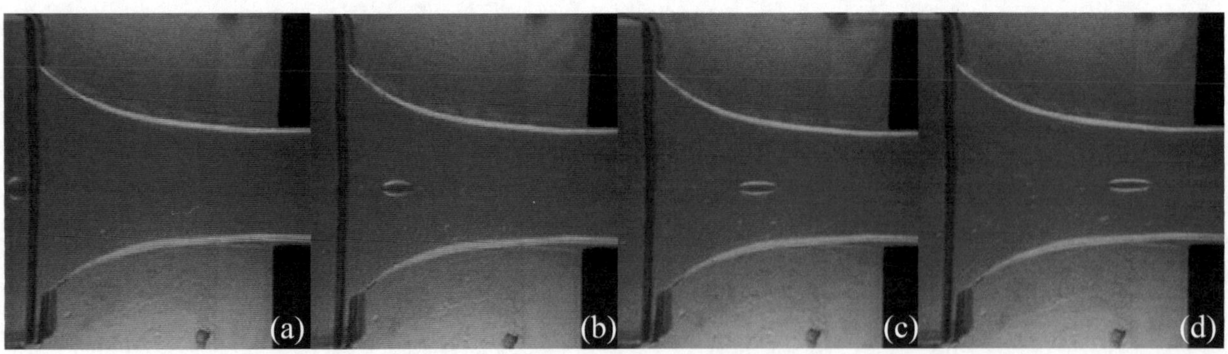

FIGURE 4. Castor oil droplet deployed in the center of the channel deforming with the flow (Trial 1)

The deformation is symmetric with very little evidence of asymmetry. By processing the collected images, the droplet morphological changes are plotted as it moves down the channel. Figure 5 shows a sample image processing result for one of the numerous experimental trials.

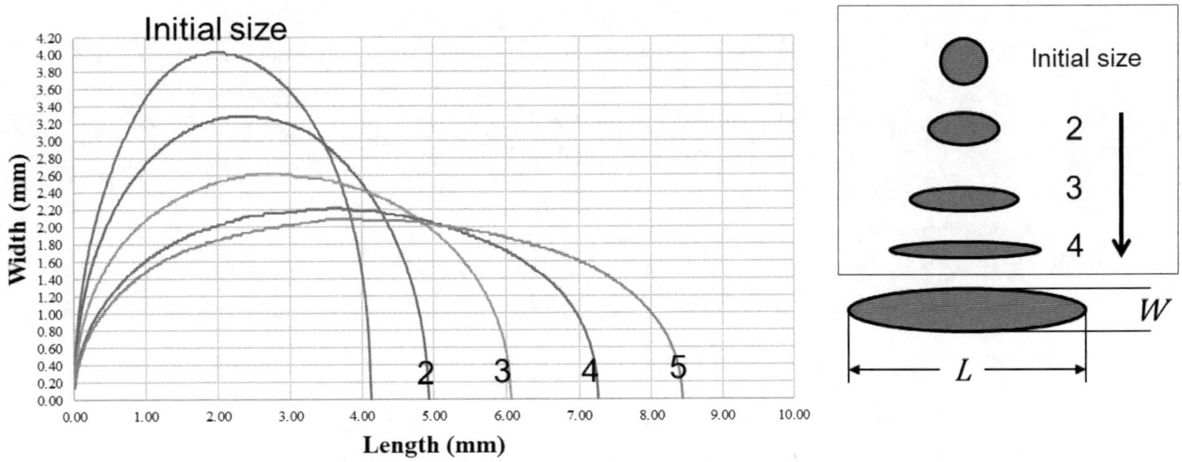

FIGURE 5. Droplet morphology, in a center deployment trial, extracted from the high-speed images.

From the calculated widths, the Capillary number can be calculated and is plotted for all the trials in Figure 6.

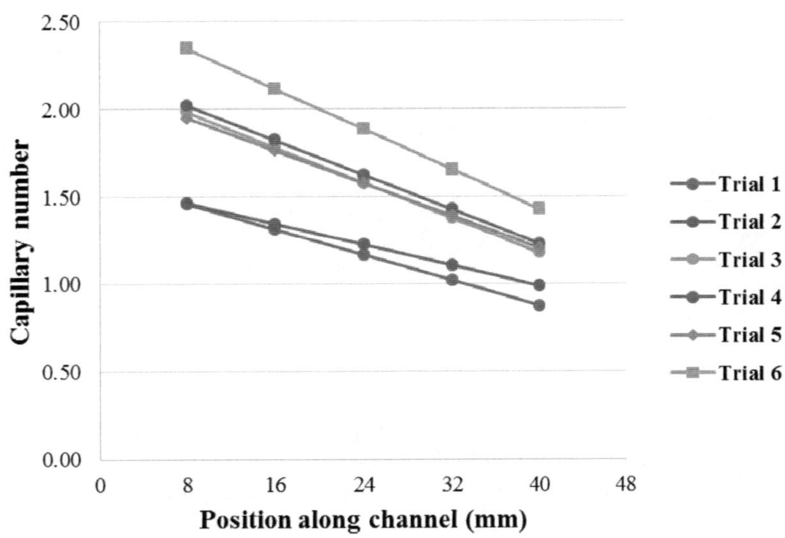

FIGURE 6. Reduction in Capillary number of droplets in the center deployment experiments.

A linear fit was done on the data and it can be seen that droplets with a higher initial Capillary number reduce at a higher rate than droplets with lower initial Capillary number. However, the strength of the extensional rate is inadequate to stretch the droplet to an affine state.

Offset Deployment Trials

A series of images showing the droplet deformation when deployed at an offset position is shown in Figure 7. The flow is from left to right as seen in the images.

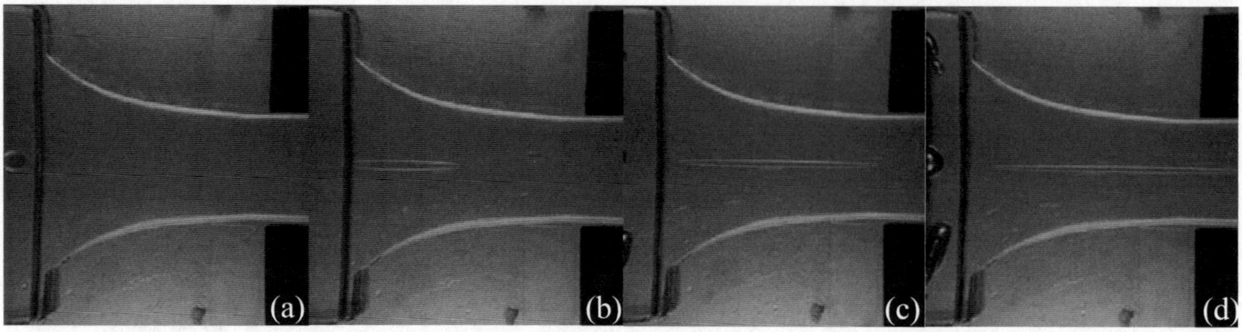

FIGURE 7. Castor oil droplet deployed at an offset position (h = -7.5 *mm*) in the hyperbolic channel.

A large degree of deformation can be seen for the same initial speed as the center deployment trials. Plotting the droplet morphology after analyzing data from the image processing, Figure 8 depicts the change.

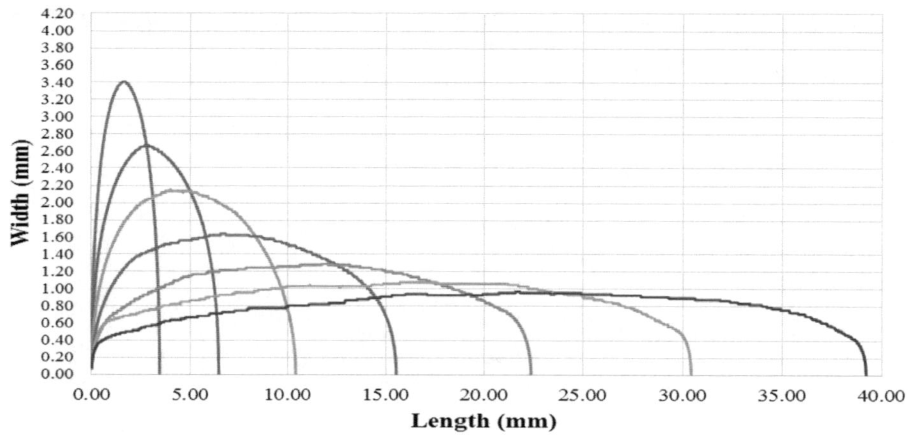

FIGURE 8. Droplet morphology as it deforms along an offset position of h = -7.5 *mm*.

There is a significant deviation from the symmetric profile at each instant of time, with the width reaching an asymptote. Further, pointed leading and trailing ends can be observed which have a width lesser than the width at the center. Change in Capillary number for each of the trails is shown in Figure 9.

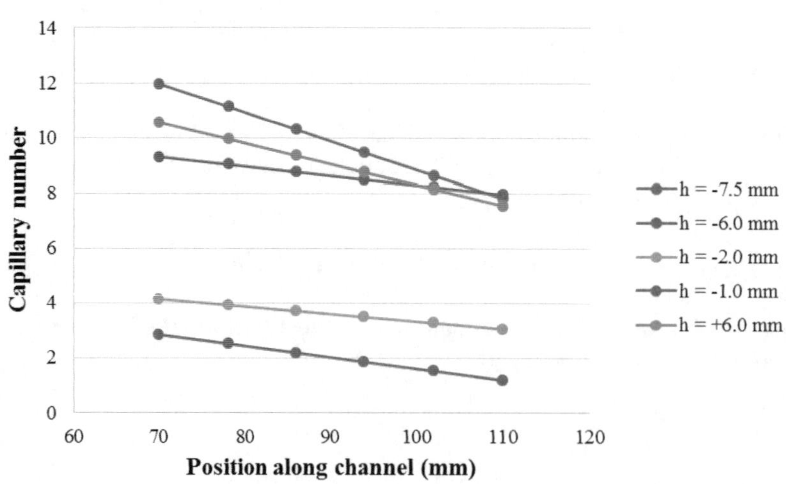

FIGURE 9. Reduction in Capillary number of droplets in the offset deployment experiments.

The significantly higher deformation can be explained since some components in the rate of deformation tensor were orders of magnitude higher than others as seen in Table 1. In particular, *du/dz* was very significant in these cases which represents a shear term in the channel geometry. On top of the extensional effect due to the converging walls, this shear component dominated the deformation dynamics of the experiment. The affine state was attained in trials where the droplet was injected closer to the walls and the width of the droplet reduced as square of the inverse.

CONCLUSION

In the hyperbolic channel, the deformation of an immiscible droplet was explored for different droplet injection positions. When the droplet was deployed in the center of the channel, the channel geometry imposed a constant extensional rate on it as it traveled through the channel. For the designed geometry, the droplet did not reach the affine state in any of the trials and exhibited a symmetric deformation. Exploring other droplet deployment positions, when the droplets were injected at an offset position from the centerline, significant deformation was observed with the droplets reaching the affine state.

Simulation results confirmed that there was a strong shear effect which raised the net strain rate to a value which was an order of magnitude higher than what was seen along the centerline. This high strain rate is the cause for the attainment of the affine state with an asymmetric deformation of the droplet. Further, the droplets had pointed ends with lower width than the center. New channels with stronger contractions and longer lengths have to be fabricated in order to draw conclusions about the critical Capillary number in the device.

This research was an attempt to understand the behavior of droplets in a hyperbolic die subjected to laminar flows. Currently, the four-roll mill experiments are regarded as a standard in the field of droplet deformation but the flow fields produced in four roll mills are very typical stagnation flows. This research attempted to extend the droplet deformation experiment into real, industrially applicable flows. The hyperbolic die used in this research was also unique in the sense that the extensional rate can be kept constant at a given speed along the center. The effect of die on the deformation dynamics of a droplet will help in incorporating the die for additive manufacturing applications.

ACKNOWLEDGMENTS

The authors acknowledge the guidance of Dr. David Bigio who was the primary research advisor and thank colleagues at the Advanced Manufacturing Lab in the University of Maryland, College Park.

REFERENCES

1. G. I. Taylor, *Proc. R. Soc. Lond. Ser. Contain. Pap. Math. Phys. Character* **146**, 501-523 (1934)
2. H. P. Grace, *Chem. Eng. Commun.* **14**, 225–277 (1982).
3. B. J. Bentley and L. G. Leal, *J. Fluid Mech.* **167**, 241-283 (1986).
4. H. B. Chin and C. D. Han, *J. Rheol.* **23**, 557-590 (1979)
5. M. S. N. Oliveira, M. A. Alves, F. T. Pinho, and G. H. McKinley, *Exp. Fluids* **43**, 437-451 (2007)
6. M. K. Mulligan and J. P. Rothstein, *Phys. Fluids* **23**, 022004 (2011)

Development of a new Polyamide 6 Material for Laser Sintering Process

Andreas Wegner[a,b*]

[a]*University of Duisburg-Essen, Chair for Manufacturing Technology, Lotharstraße 1, 47057 Duisburg, Germany*
[b]*AM Polymer Research UG, Bismarckstraße 120, 47057 Duisburg, Germany*
andreas.wegner@uni-due.de

Abstract. Laser sintering of polymers gets more and more importance for small series production. However, there is only a little number of materials available for the process. In most cases parts are still build up using polyamide 11 or polyamide 12 which are both not the most common polyamide materials in traditional plastic processing. Therefore, laser sintered parts sometimes do not fulfil the required part properties for intended applications. Especially temperature stability or mechanical properties are sometimes not sufficient. Some new developments on the market should close this gap. However, some of these new materials are difficult to process or do not have the expected properties. Oftentimes there is insufficient knowledge on the processing of alternative materials in laser sintering. Additionally, material characteristics are sometimes not optimal for processing and do not fulfil the high requirements of the complex processing conditions in laser sintering. Fundamental knowledge is needed to improve the understanding on essential material characteristics for the laser sintering process.

Aim of the presented study was to qualify a new polyamide 6 material for the laser sintering process. Within the study different material grades as well as a commercial product were compared in order to analyse influences on the processing and the part properties. Problems of qualifying new materials are discussed using some examples while all material grades were processed successfully. Furthermore, the processing conditions as well as the part properties (part density, mechanical properties) of the used materials are discussed. Correlations between energy input and part properties are established. Therefore, enhanced knowledge on the processing of new materials in laser sintering is established.

Keywords: Laser Sintering, New Materials, Polyamide 6, Powder properties, Mechanical Properties.
PACS: 62.20.de, 62.20fg, 62.20fk, 81.05.Lg, 81.20.Ev, 83.50.Uv, 83.80.Sg,

INTRODUCTION

Increasing competition, decreasing product life cycles, the wish for customized products and a shortage of resources cause the need for innovative manufacturing techniques for small series production [1]. Going beyond the stage of Rapid Prototyping on to Rapid Manufacturing, Additive Manufacturing offers possibilities for small series production of customized products and an increased freedom of design due to the lack of tools [2]. The laser sintering of plastic parts is, aside from beam melting of metal parts, one of only two AM-processes which have the capability to be used for Rapid Manufacturing in the near future [3]. In laser sintering parts are built up layer by layer. The machine produces the parts by repeating four stages for each layer: Firstly, the platform descends by the thickness of one layer. Secondly, powder is spread across the build platform by a leveling roller or coater. Thirdly, the layer is preheated to a temperature close to the material's melting point by a radiant heater. Then, a CO_2-laser beam melts the powder by tracing the actual cross section line after line using a scanner system. These steps are repeated until the parts are completed. [4]

Laser sintering has reached a high technical level within the past two decades. However, there is only a little number of materials available for the process. In about 90 to 95 % of all cases parts are build up using polyamide 12 (PA 12) or polyamide 11 [5, 6]. Alternative materials like standard polymers for injection molding like polypropylene or polyamide 6 are seldom used. Reasons for that are high prices, a restricted availability, poor mechanical part properties, a more difficult processing or an insufficient understanding of the processing. The problems with other materials result from the very complex processing conditions in laser sintering with high requirements on the material characteristics. Therefore, it is very complicated to develop new materials with good processing behavior, good mechanical properties as well as a sufficient ductility of the parts. Within this area, fundamental knowledge was established at the chair for manufacturing technology within the last years. Aim of the presented study is to qualify a new polyamide 6 material for the laser sintering process.

STATE OF THE ART

Technical literature contains several papers with experiments on new materials. Rietzel analyzed in [7] polyoxymethylene (POM), polypropylene (PP), polybutylene terephthalate (PBT) and polyethylene (PE). Fiedler performed different experiments with PP in [7, 8]. Goodrigde studied UHMWPE in [5]. Fraunhofer Umsicht and RPM evaluated the processing of PBT in [10, 11]. However, none of these studies led to new commercial materials.

Reasons for that are the very complex processing conditions in laser sintering. These result in high requirements on the powder materials as discussed in [7, 12, 13]. Material should have a wide processing window between the beginning of melting processes when heated up and the start of the crystallization process when cooled down again. Crystallization speed should be as low as possible in order to avoid warpage. The melt of the materials should have a suitable rheology and surface tension in order to form a flat film after laser exposure. The used powders need to have good flowing properties and preferable round particle shape in order to allow good powder spreading during the process. In addition to that, powders should have high bulk density and a suitable powder size distribution for the laser sintering process. Furthermore, the material should have a high absorption of the CO2 laser beam wavelength of 10.6 μm.

Own research of the author concentrated in the last years on the development of a polyethylene, a polypropylene as well as an elastomeric material for laser sintering process [14-17]. All three were commercialized in the meantime.

Other studies focus different points of the material development. Rietzel studied in [7] the thermal and powder properties, the processing conditions as well as the mechanical properties of different materials like PE, POM, PBT or PP. The analyzed PE and PP material was produced by precipitation while the POM and PBT powder was made by cryogenic grinding. The named powders had a percentaged bulk density fewer than 40 % oftentimes even under 35 % while standard laser sintering PA 12 has values over 40 %. Also the powder flowability showed worse values for the PP, PE and POM materials compared to PA 12 while the flowability of PBT is only a little bit lower than for PA 12.

There are only few studies on the development of polyamide 6, while two commercial products from Solvay and BASF are available on the market.

A first study on polyamide 6 was performed by Kim and Creasy in [18] in 2002. They studied the processing of neat PA6 and a compound produced out of that PA6 and 5 wt.-% clay nanoparticles in the laser sintering process. Both used powder materials were produced by milling. They did different material characterizations like DSC, DMA, MFI in order to analyze the suitability for the laser sintering process [18, 19]. All tests were done in comparison with the unfilled PA6 material. They produced one layer parts using a DTM Sinterstation 2000. However, the made SEM pictures showed a high porosity. Additionally, in [19] the sintering behavior and the resulting density of the neat material and the compound were investigated in dependency of temperature and time using a sintering cell in an oven. Maximum density was found with 0.96 g/cm³ for the neat material.

Another study on laser sintering of polyamide 6 was performed by Wahab et al. [20, 21] using a commercially PA6 powder with a particle size of 15 to 20 microns. Basing on that material they produced another PA6 powder as well as PA6 compounds using on the one hand zirconium nanoparticles and on the other hand an organoclay. All these powders were produced by dissolution of the PA6 and spray drying having a medium powder size of 30 microns. The materials were analyzed by DSC. Additionally, laser sintering tests were performed using an experimental machine setup. The produced specimen were tested regarding their mechanical properties and the part density. Density of all tested materials is very low and lies under 0.85 g/cm³. Highest tensile strength was found for the original material with 50 MPa while the value of the spray dried materials is much lower. However, maximum elongation is less than 1 mm. No information on the conditioning state of the specimen is given in that paper.

Salmoria et al. tried to produce PA6/PA12 blends using laser sintering process [22]. They used a commercial PA6 powder with a medium particle size of 150 microns and an EOS PA12 powder. The tested the neat materials as well as blends with 20, 50 and 80 wt.-% PA6. Specimen were produced using a self-developed laser sintering machine using laser powers of maximum 3.3 W. They chose a low preheating temperature of only 120 to 140 °C to produce the specimen using the powder blends. DMA as well as tensile tests were performed with the different materials. Results showed for the neat PA6 a maximum stress of 60 MPa and a maximum strain of 11 %. However, elastic modulus is only 110 MPa. The blend and PA12 show similar low modulus values. These are much lower than standard values for PA6 or PA12. A reason might be the porosity found in SEM micrographs. The DMA measurements showed higher values of about of up to 1000 MPa. Additionally, they performed some fatigue and creep tests with the materials. However, processing was not optimal which is proved by the reported uncommon values for the mechanical properties of the commercial PA12 material.

There is only one study using a commercial PA6 laser sintering powder. Lohn and Schmid used in [23] the Solvay Sinterline powder in order to validate the performance of a self-developed laser sintering machine. In addition to that, a new PA613 polymer developed by Evonik was tested for comparison reasons. Both materials were also compared to the performance of a standard laser sintering PA12 produced by EOS. Powder shape as well as particle size distribution and melting behavior were analyzed. The Solvay PA6 show a rough powder which seemed to be produced by milling. The powder size distribution of the materials is similar. However, the PA6 powder had a sharp buckle at 80 microns which indicates an additional sieving process. The parts were produced using different energy densities. It was found that processing of the PA6 needs very high energy input of over 0.67 J/mm³ while for the PA613 energy densities between 0.35 and 0.47 J/mm³ are sufficient to produce parts with highest tensile strength of 57 MPa. Values for the Solvay PA6 are higher exceeding 65 MPa.

State of the art gives much information on the optimal characteristics of materials for laser sintering. There are also several studies on the development of new materials. However, it seems very difficult to transfer the given requirements to the development of powders. Therefore, parts show low mechanical properties or materials are too expensive. Within the work presented here experiments on the development of a new polyamide 6 laser sintering material are performed. Therefore different PA6 grades were processed in the laser sintering process. Powder properties as well as part properties were studied and compared to a commercial PA6 laser sintering material showing some potential of improvement.

EXPERIMENTAL

Based on the aforementioned state of the art, the analyses were planned. Unfilled Solvay Sinterline Technyl PA6 powder was chosen as commercial reference product. Three different unmodified PA6 granule grades of different type were used for the experiments. The materials were converted to powder by ROWAK AG an expert in producing fine powder fractions with good bulk properties and flowability. Powder properties were further improved using a suitable flow agent. For all material grades the percentaged bulk density (bulk density/ material density) was determined according to DIN EN ISO 60. Furthermore, the Hausner-ratio (tap density/ bulk density) was determined measuring additionally the tap density according to VDI 3405 Part 1.1. The quality of the virgin powder as well as powder properties of overflow and partcake powder was measured. Additionally, the powder properties are compared to those of some other commercial laser sintering materials like EOS PA2200 or Solvay Sinterline Technyl.

After characterization of material and powder properties, the different PA6 powders were tested in the laser sintering process. The materials were processed in virgin state on a 3D-Systems Vanguard HiQ HS using a build space reduction. Variation of parameters was done by changing volume energy density. Laser power of the fill scan was set at different levels between 20 and 40 W for the experiments while hatch distance, scan speed and outline laser power was held on a constant level. Energy density is often used to describe correlations in laser sintering. This value for the energy input was introduced by Nelson (1993) and extended by Starr et al. (2011) and Kaddar (2010) with the layer thickness to define the volume energy density. It is defined as the ratio of fill laser power divided by beam speed, scan spacing and the layer thickness. For the PA6 tests energy density was varied on five steps between 0.26 and 0.52 J/mm³. Processing temperatures were chosen based on the specific melting temperatures of the materials.

Tensile specimens according to DIN EN ISO 3167 and density cubes were produced for testing. Tensile bars were produced in x-direction. The tensile tests in compliance with DIN EN ISO 527-1 and DIN EN ISO 10350-1 were performed using a Zwick Z020 M (MutiXtense) and a testing speed of 5 mm/min. The tensile specimen were stored under vacuum at 80 °C for three weeks before testing. Therefore part properties are given for the dry state. Part's density was measured for cubes using the Archimedes method according to DIN 1183. Properties of the different grades are compared with the commercial reference material Solvay Sinterline.

RESULTS AND DISCUSSION

The powder properties were measured according to the methods described in experiments chapter. The properties of the three PA6 grades are studied considering also different powder states. The results for percentaged bulk density as well as the Hausner-ratio are summarized in Figure 1 and Figure 2 using values for commercial Solvay Sinterline Technyl PA6 and EOS PA2200 PA 12 laser sintering materials as reference.

Figure 1 shows the percentaged bulk density of the materials. The commercial reference materials Solvay Sinterline and EOS PA2200 have a bulk density of 42 % respectively 40 % in the virgin state. All tested experimental PA6 grades show significant higher values than the reference materials of over 44.4 %. Due to that, also packing density during the process might be higher. The highest value is achieved for grade three. This material shows a bulk density of over 52 % which is one of the highest values ever reported for laser sintering materials. Similar effects are found for the different powder states. The material from the overflow bins after processing shows for all materials only a little reduction of bulk density. Values decrease only little by fewer than 2 %. Highest decrease is found for grade 3. However, relative bulk density is still over 50 % and still higher than all other values. Bigger differences are achieved for the partcake material recycled from the build envelope. Especially for the Solvay material bulk density drops significantly on values of 29 % which is 15 % lower than for the virgin respectively overflow state. However, partcake for this material was very hard due to a little bit too high temperature. All experimental grades do not show such a big decrease. Bulk density of grade 1 and 2 decreases by 6 % on values of about 38 %. Lowest decrease and highest values are again achieved for grade 3. Relative bulk density of partcake materials is still over 49 %. Therefore the bulk density of the recycled material lies even over the value of the virgin state of all other materials. Bulk density reduces only by 3 % compared to the virgin state.

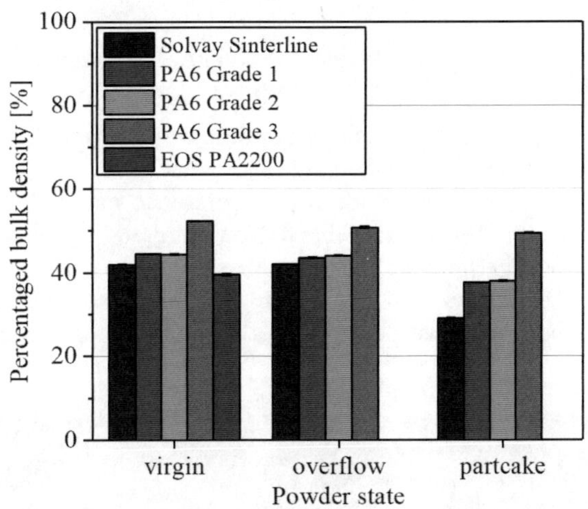

FIGURE 1. Percentaged bulk density of the different PA6 grades and the PA6 and PA12 reference material in dependence of the powder state

Figure 2 summarizes the Hausner-ratios of the different analyzed materials and powder states. Effects are similar to the results of bulk density. The virgin reference materials has values of 1.19 (Solvay Sinterline) and 1.15 (EOS PA2200). A Hausner-ratio of less than 1.25 stands for good powder flowability. A value between 1.25 and 1.4 represents reduced powder flowability while a ratio over 1.4 stands for no flowabilty. Therefore, especially the EOS material shows a very good flowability. Hausner ratio of virgin grade 1 and 2 powder is comparable to the Solvay PA6 having Hausner ratios of about 1.2. Again, grade 3 shows much better value of 1.12 which is even better than the value for the EOS material. The overflow state causes only little change of the powder flowability. Only the Hausner ratio of grade three increases to 1.14. However, this value is again still better than the values of all other materials. Bigger effects are achieved for the partcake material. In contrast to the overflow state, grade 3 shows only a very little further increase on 1.15. The Hausner ratio of all other PA6 materials increases significantly. The value for grade 1 and grade 2 increases to about 1.25. Therefore, the material is at the limit to reduced powder flowability. Hausner ratio of the Solvay Sinterline partcake material rises to 1.41. Therefore, the material shows per definition no flowabilty. Reason might be again a little too high temperature during processing. In summary, all three studied experimental PA6 grades show good powder properties comparable or even better than some commercial laser sintering materials. Results show that powder properties and especially bulk density is much higher than for many other experimental powders for the laser sintering process. Especially grade 3 shows superior powder properties exceeding also commercial laser sintering materials significantly.

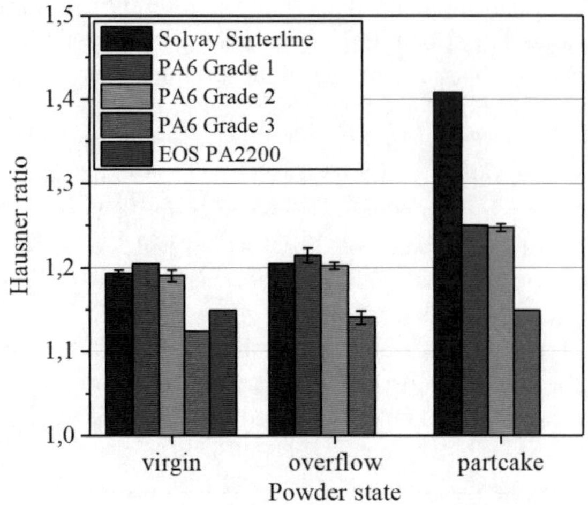

FIGURE 2. Hausner ratio of the different PA6 grades and the PA6 and PA12 reference material in dependence of the powder state

The powder properties of the different materials show good or suitable values for laser sintering. Therefore, no problems in powder spreading occur during the processing of the materials. Different specimen were produced out of all materials successfully, Figure 3. The parts made by the reference material Solvay Sinterline have a slightly grey color. In contrast to that, the experimental grade have a beige (grade 1), brown beige (grade 2) or yellowish (grade 3) color. Most parts have good geometrical shape with sharp edges. However, if energy input is too high over-sintering occurred for all processed materials. Warpage is a well-known phenomenon in laser sintering of polyamide 6. Especially the tensile bars show warpage. This effect is found for all materials also for the commercial reference material.

Solvay Sinterline PA6 grade 1 PA6 grade 2 PA6 grade 3

FIGURE 3. Density cubes produced with the different PA6 grades and the PA6 reference material

After manufacturing of the specimen, the density and the mechanical properties of the different PA6 grades as well as the PA6 reference material were analyzed. Within the study the influence of energy input on different part properties was analyzed. Therefore, correlations between energy density and part density as well as tensile properties are established.

All tested materials showed good powder properties suitable for laser sintering. However, the processing of the materials do not led in all cases to good part properties. The results for the part density measurement are summarized in Figure 4. The density of the reference material Solvay Sinterline lies between 0.93 and 1.09 g/cm³. It decreases with an increase of energy density due to over-sintering. Best values of over 1.07 g/cm³ are achieved for energy density between 0.26 and 0.39 J/mm³. Therefore, optimal energy input is much lower than reported in [23]. Reference value for the experimental PA6 grades is therefore 1.07 g/cm³. Grade 1 shows the poorest values. No values over 0.89 g/cm³ could be measured for that grade. Values slightly increase for higher energy densities. However, no density value can be given for 0.52 J/mm³ due to wrong parameter settings during processing. Both other PA6 grades show significantly better part densities. For grade 2 highest part density is found at 1.05 g/cm³ for energy densities of 0.33 and 0.39 J/mm³. Therefore, values are little lower than for the reference material. Highest density of grade 3 is achieved for an energy input of maximum 0.39 J/mm³. Highest part density is 1.08 g/cm³ which is equal to the Solvay reference. Grade 2 and 3 show both a decrease of part density with an increase of volume energy density. However, the decrease is less strong than for the Solvay material. Reason for the decrease might be on the one hand over-sintering of the parts and on the other hand a beginning degradation or oxidation of the material due to too high energy input. Especially the two highest energy levels are very high compared to laser sintering processing of polyamide 12. Results prove that the production of parts having high part density is possible when using grade 2 and 3.

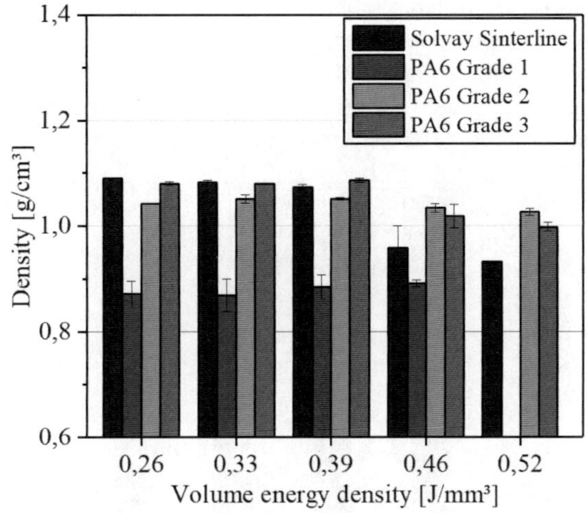

FIGURE 4. Part density of the different PA6 grades and the PA6 reference material as a function of energy density

Young's modulus shows similar correlations to the energy density, Figure 5. For the reference material no value can be reported for the highest energy density of 0.52 g/cm³. The parts show too strong over-sintering effects. Therefore, no measurement of part properties was possible. Highest modulus values of 3350 N/mm² are achieved at the two lowest ene rgy densities of 0.26 and 0.33 J/mm³. Values again decrease for higher energy input. The experimental grades show in some cases significant lower young's modulus values. For grade 1 maximum values of 1926 N/mm² are achieved which is over 40 % lower than the reference. For grade 1 the found increase of modulus values with an increase of energy input does not lead to better values. Young's modulus values for grade two are significant higher. Values range from 2860 to 3060 N/mm² while highest values are achieved for 0.26 J/mm³. The maximum value is only about 9 % lower than the reference. Again, a decrease of values is found when energy input is increased. The same effect is also found for grade 3. However, young's modulus for grade 3 is comparable to the reference resulting also in values of maximum 3350 N/mm². Again the highest modulus values are achieved for the two lowest energy density levels of 0.26 and 0.33 J/mm³. In summary, grade 3 results again in the best property values comparable to the reference material.

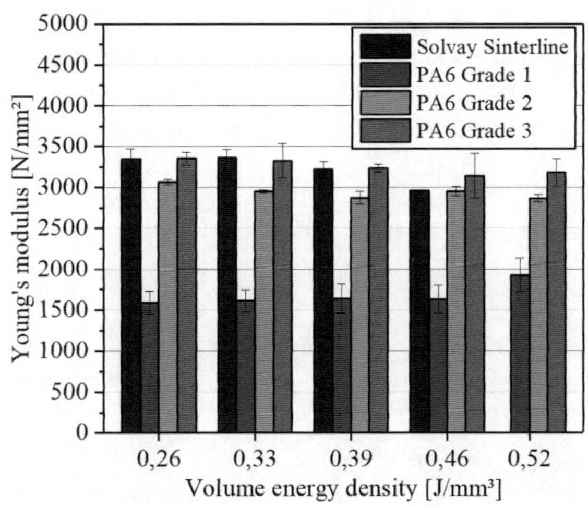

FIGURE 5. Young's modulus of the different PA6 grades and the PA6 reference material as a function of energy density

Figure 6 summarizes the values measured for the elongation at break (EAB). Results are very different to the properties considered before. The Solvay material has an EAB of 2.1 to 2.9 % while highest values are found for the lowest energy input. In contrast to part density and young's modulus, grade 2 and 3 lead to significantly lower EAB values compared to the reference. All measured EAB values are in the range of 0.93 to 1.3 %. Volume energy density has also only little influence on the part properties. Therefore, parts made by grade 2 or 3 PA6 material are very brittle and not suitable for application. No analyses are made yet on the conditioned state. Maybe the materials show then lower difference to the reference. However, this had to be studied in the future. In contrast to the analyses before, grade 1 shows the best elongation at break of all test materials. Values range from 2.5 to 3.9 %. Maximum values are achieved for energy densities of up to 0.46 J/mm³ being nearly constant up to that level. EAB is therefore about 34 % higher than for the commercial reference material.

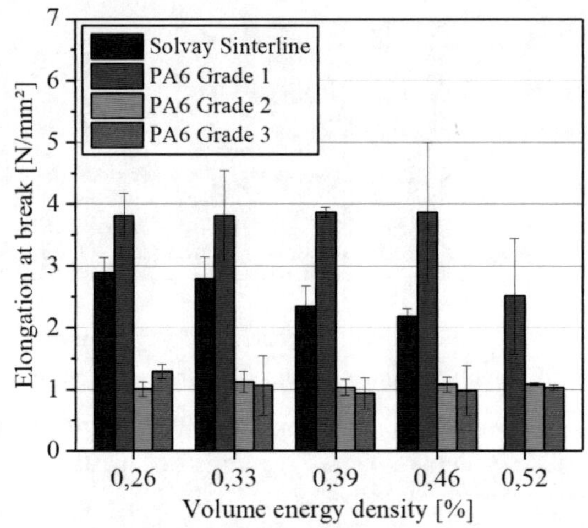

FIGURE 6. Elongation at break of the different PA6 grades and the PA6 reference material as a function of energy density

CONCLUSIONS

Within the presented study the powder properties as well as the processing behavior and part properties of three self-developed experimental PA6 laser sintering materials were studied. It was found that the produced powder have better powder properties than some commercially available powders. The processing of all grades was successful resulting in parts of good geometrical shape and for some mechanical properties even in good values. However, no one of the self-developed PA6 grades lead in total to better properties than the commercial laser sintering PA6 material produced by Solvay. But, results prove that powder and part properties can be potentially improved by further investigations on PA6. However, further restriction of the tested experimental grades is the tensile strength. No material exceeds 41 N/mm² in the analyzed dry state while the reference reaches values of over 75 N/mm². Reasons for these effects were not found yet and have to be clarified in further investigations.

Future work should improve the processing conditions and part properties of the studied experimental materials. Aim is PA6 materials which show similar part properties like in injection molding and having a sufficient ductility for use of parts in serial applications.

REFERENCES

1. E. Abele and G. Reinhart, *Zukunft der Produktion - Herausforderungen, Forschungsfelder, Chancen*, München: Carl Hanser Verlag, 2011.
2. R. Hague et al,: Proc. Inst. Mech. Eng., Part C: J. Mech. Eng. Sci., **217/1**, 25-30 (2004).
3. N. N., *VDI 3405 Additive Fertigungsverfahren – Grundlagen, Begriffe, Verfahrensbeschreibungen, 2014.*
4. A. Gebhardt, *Generative Fertigungsverfahren – Rapid Prototyping – Rapid Tooling – Rapid Manufacturing*, München: Carl Hanser Verlag, 2007.
5. R.D. Goodridge et al., *Prog. Mat. Sci.*, **57/2**, 229-267 (2012).
6. M. Schmid, *Selektives Lasersintern (SLS) mit Kunststoffen – Technologie, Prozesse und Werkstoffe*, München: Carl Hanser Verlag, 2015.

7. D. Rietzel, "Werkstoffverhalten und Prozessanalyse beim Laser-Sintern von Thermoplasten", Ph. D. Thesis, Friedrich-Alexander-University of Erlangen-Nürnberg, 2011.
8. L. Fiedler, L.O. Garcia Correa, H.-J. Radusch, *Zeitschrift Kunststofftechnik*, **3-4**, 1-14 (2007).
9. L. Fiedler, A. Hähndel, A. Wutzler, J. Gerken and H.-J. Radusch, "Development of new Polypropylene based Blends for Laser Sintering" in: Proceedings of the Polymer Processing Society 24th Annual Meeting PPS-24, Salerno (Italien), Polymer Processing Society, 2008, pp. 203-207.
10. J. Gerken, "Verdichtetes Kohlendioxid als Prozessadditiv zur Herstellung polymerer und mikronisierte Nanokomposite", Final report "nanocrosser" BMBF-Projekt 03X0009 E rpm GmbH, 2009.
11. J. Bertling and C. Eloo, C., "Verdichtetes Kohlendioxid als Prozessadditiv zur Herstellung polymerer und mikronisierte Nanokomposite", Final report "nanocrosser" BMBF-Projekt 03X0009 A-F Fraunhofer UMSICHT, 2009.
12. G. Alscher, "Das Verhalten teilkristalliner Thermoplaste beim Lasersintern", Ph. D. Thesis, University of Essen, 2000.
13. M. Schmid, A. Amado and K. Wegener, "Polymer Powders for Selective Laser Sintering (SLS)" in: Proceedings of the Polymer Processing Society 30th Annual Meeting PPS-30, Cleveland, Ohio (USA), Polymer Processing Society, 2014, pp. S15-377.
14. A. Wegner, *Physics Procedia*, **83**, 1003-1012 (2016).
15. A. Wegner and T. Ünlü, "Powder Life Cycle Analyses for a new Polypropylene Laser Sintering Material" in: Proceedings of the 27th International Solid Freeform Fabrication Symposium (SSF 2016), The University of Texas at Austin, 2016, pp. 834-846.
16. A. Wegner, "New Materials for Laser Sintering: Processing conditions of Polyethylene and Polyoxymethylene" in: Proceedings of the Fraunhofer Direct Digital Manufacturing Conference 2016, Berlin, 2016.
17. A. Wegner and T. Ünlü, "Adjustment of Part Properties for an Elastomeric Laser Sintering Material" in: Proceedings of the 28th International Solid Freeform Fabrication Symposium (SSF 2017), The University of Texas at Austin, 2017.
18. J.H. Kim, and T.S. Creasy, " The SLS characteristics of 1.8wt% clay nanoparticle/Nylon 6 composite" in: Solid Freeform Fabrication Proceedings, Austin, Texas, 2002, pp. 224 - 230.
19. J. Kim and T. S. Creasy, "Measurement of Sintering Characteristics of Clay-Reinforced Polyamide 6 Nanocomposite" in: Proceedings of the Antec Conference 2004, Chicago, Society of Plastics Engineers, 2004, pp. 1325-1329.
20. M.S. Wahab, K.W. Dalgarno and R.F. Cochrane, "Selective Laser Sintering of Polymer Nanocomposites" in: Proceedings of the 18th International Solid Freeform Fabrication Symposium (SFF 2007), The University of Texas at Austin, 2007, pp. 358-366.
21. M.S. Wahab, et al., "Development of Polymer Nanocomposites for Rapid Prototyping Process" in: Proceedings of the World Congress on Engineering 2009 Vol. II, London UK, WCE 2009, 2009.
22. G. V. Salmoria, et al., *Polymer Testing*, **31**, 411-416 (2012).
23. J. Lohn and H.-J. Schmid, "Aufbau einer Lasersinter-Versuchsanlage und Verarbeitung neuer Materialien am Beispiel von Polyamid 6 und Polyamid 613" in: Proceedings of the 14th Rapid.Tech Conference, 2017, pp. 312-326.

Strategies for Laser Sintering of PA12/polymer composites

Strobbe Dieter [(1)]*, ***Van Puyvelde Peter*** [(1)], ***Kruth Jean-Pierre*** [(1)]

[(1)] KU Leuven - Leuven - Belgium

Laser sintering (LS) is an Additive Manufacturing technique, using a polymer powder feedstock and the heat of a laser beam, to build up complex 3D parts from a CAD file in a layer-wise manner. Currently, the commercially available powders for LS are limited to a few grades, of which 95% is in the polyamide family. This limits the possibilities of LS for production of parts for structural applications. In the development of new polymer powders for this technique, the industrial standard polyamide-12 grade (PA12) is reinforced by other materials (glass beads, carbon tubes, aluminum flakes). This yields parts with modified properties, without adapting the processing window. However, these composites often show bad interfacial adhesion and high residual porosity, deteriorating mechanical properties. Another strategy is to blend another polymer with PA12 by mixing their respective powders. This results in parts with properties intermediate between PA12 and the in-blended polymer and significant changes to processing. In this work, the addition of two polymer powders (an aramide grade and a polyamide-4,6 grade) to PA12 is studied. These two powders have melting temperatures significantly above the processing window of PA12 and standard Laser Sintering machines, but chemistry similar to PA12. It is shown that using the standard processing window, composite parts with a PA12 matrix and reinforcing polymer phase with good interfacial adhesion are produced. This opens up new perspectives for production of polymers with advanced mechanical properties, without costly changes to machines and processing.

Micro-FDM Process Capability and post-processing effects on mechanical properties

Claudia Pagano[a], Vito Basile[b], Francesco Modica[b] and Irene Fassi[a]

[a] *Institute of Industrial Technologies and Automation - CNR, Via Corti 12, 20133 Milan, Italy*
[b] *Institute of Industrial Technologies and Automation - CNR, Via Lembo 38/F, 70124 Bari, Italy*

Abstract. The mechanical behaviour of specimens fabricated using FDM machine has been thoroughly studied and several works have been presented. However, they are focused on few materials, in particular ABS, while very few papers analysed the mechanical properties of FDM samples made by PLA filament. Even though ABS is well known for its superior mechanical properties, some applications might require materials with other properties, such as PLA. Therefore, a deeper investigation of the effect of the process on its properties is needed. In this study, at first, the influence of the main FDM process parameters, such as raster angle, extrusion width, air gap and extrusion temperature, on mechanical performance of PLA samples has been evaluated. The mechanical tests have been carried out on miniaturized tensile specimens focusing on the tensile test main properties: ultimate tensile stress, Young modulus and elongation at break. Moreover, an experimental campaign has been carried out on the effect of a thermal post-processing, in order to evaluate the effects of the treatment and its relation with the process parameters on the mechanical properties of the specimens.

Keywords: Fused deposition modelling, additive manufacturing, tensile test, annealing.
PACS: 81.05.Lg, 81.70.Bt , 81.40.Gh, 61.72.Cc, 89.20.Bb.

1. INTRODUCTION

Among the Additive Manufacturing (AM) technologies, Fused Deposition Modelling (FDM) is becoming more and more popular, mainly due to the lost cost of investment required and its easiness to use. It allows the manufacturing of 3D products with a great variety of geometry and in short time. For these very reasons, it began as a process for creating prototypes, but it is now becoming largely used for the production of end-use parts. However, it present a few limitations, such as the material range and the anisotropy of the products. Indeed, the product are built up layer by layer, extruding a molten thermoplastic filament, which creates each layer following a specified toolpath. The thermal fusion of the polymer allows the bond between the layers and across the adjacent beads, but some voids are present and the anisotropic structure affects the properties of the part. Therefore, in order to manufacture end-use parts with suitable mechanical properties, not only is important the choice of the material, but also a correct process control, which is often more influent than the material [1]. Several works have been presented in the literature in order to study and improve the mechanical properties of FDM products, including the control of the process parameters during fabrication and the development of materials with superior properties and compatible with the technology [2, 3]. However, these works studied the properties of Acrylonitrile Butadiene Styrene (ABS), one of the most common polymer used to obtain tough parts with acceptable strength in FDM and very few considered other materials [4, 5, 6]. Therefore, considering the peculiar properties of Polylactic acid (PLA), namely stiffness and environmental impact, this paper is focused on the effects of the process parameters on the tensile tests of miniaturized samples made of PLA. Miniaturization can have a high impact on several industrial fields - such as biomedical, electronics, optics and high-precision micro mechanical industry - and the advantages of AM technologies could help the spreading of microproducts. Moreover, in order to study the effect of temperature history on FDM specimens a thermal post-treatment (annealing) has been investigated in relation with the process parameters. The effects of heat treatments on mechanical properties of PLA has been widely investigated in last years [7]. Tensile modulus and heat deflection temperature (HDT) are increased by annealing at 100°C. The effects of thermal post-process was investigated also with respect to the chemical composition of the PLA. Additives change the native crystallinity of the material and its sensitivity to annealing, however, no matter the chemical composition of PLA (reinforcement additives) the tensile modulus increases while the elongation at break decreases with thermal post-processing (4h@100°C). Tensile strength does not exhibit a specific trend, it can be increased or decrease depending on material composition. The cause of this improved mechanical properties is referred to an increase of crystallinity and secondary bonds [8, 9, 10]. However, they are mostly applied on the injection moulding process, and there is a lack of research works on the combination of the PLA material with FDM and the effects of annealing on mechanical properties at different process conditions. More insight are requested into this specific technological niche, in order to investigate whether this could be a solution to some of the limitations of the fabrication technology.

After the description of material and methods in section two, in the third section, the focus has been given on FDM micro-part production capability and process parameter optimization; while in the fourth section, an investigation on heat treatment and process parameter effects on mechanical properties is presented.

2. MATERIAL AND METHODS

A first design of experiment *(DoE)* has been developed focusing on the influence on ultimate tensile stress (*UTS*), elongation at break (*EB%*) and Young modulus (*E*) of four of the most influent parameters according to the literature and preliminary studies, i.e. raster angle (*RA*), air gap (*AG*), extrusion width (*EW*) and extrusion temperature (*ET*). The first two parameters define the filling of the layer in terms of orientation and distance between two depositions in the raster. The other parameters (layer thickness, extruder speed, the temperature of the build platform and the filament cooling) have been fixed for all the tests. Preliminary tests have been performed in order to guarantee the achievement of the values of A*G* and *EW* by calibration of the parameter extrusion multiplier, which controls the ratio of extrusion and feed speeds. The DoE has been conceived in which several process parameters, summarized in table 1, have been combined to obtain a variety of process conditions. Each of the 24 conditions has been replied 3 times and 3 miniaturized dog-bone specimens have been fabricated at each replication.

TABLE 1. Process parameter values.

Factor	Unit	Low level	Central level	High level
Air Gap (*AG*)	mm	-0,05	-	0
Extrusion Width (*EW*)	mm	0,5	-	0,6
Extrusion Temperature (*ET*)	°C	200	-	230
Raster angle (*RA*)	degrees	0°/0°	+45°/-45°	0°/90°
Layer thickness	mm		0,1	
Extrusion speed	mm/min		2400	
Bed Temperature	°C		50	
Cooling	-		OFF	

In order to define a miniaturized geometry for mechanical test specimens, several geometries have been proposed in recent years, downscaling the models of the standards ISO 527 [11] and ASTM D638 [12] or creating original geometries [13, 14]. The dimensions of the proposed geometry is reduced enough to comply with the definition of micro-component [15], and, at the same time, adequate to perform tensile mechanical tests.

The specimen geometry has been defined based on ISO527-5B. However, in order to avoid incomplete filling of the samples, the specimen width (*A*) is modified taking into account the variation of Air Gap and Extrusion Width with the aim of deposit 8 beads for all the layers with a raster angle of 0°. Figure 1 shows the nominal width of the specimen when *AG* and *EW* are varied and where a negative value of *AG* means that an overlap of adjacent depositions occur.

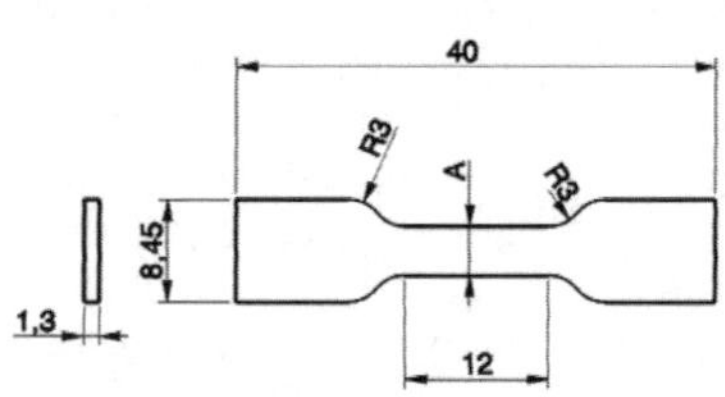

Extrusion Width (EW) mm	Air Gap (AG) mm	Specimen Width (A) mm
0.5	0	4.00
0.5	-0.05	3.65
0.6	0	4.80
0.6	-0.05	4.45

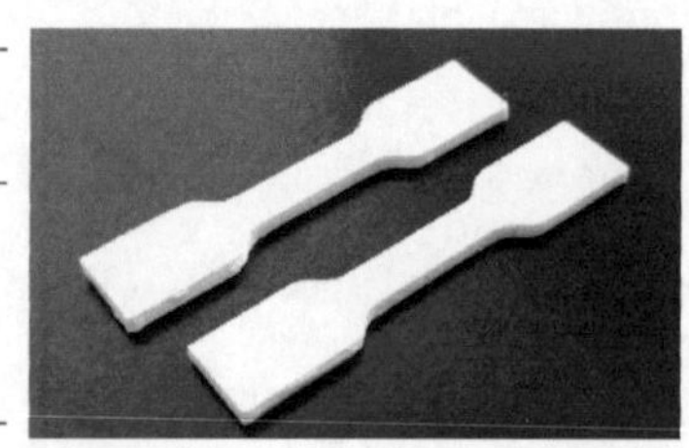

FIGURE 1. Mini dog-bone geometry.

The specimens are fabricated in PLA (PlastInk) with filament diameter of 1.75 mm, density of 1.24 gr/cm^3, colour white and a nominal decomposition temperature of 250 °C. The FDM samples have been manufactured using a GIMAX3D S2 machine equipped with a double extrusion head, heated build platform, nozzle diameter of 0.4 mm, with a resolution of 40 μm for X and Y axes and a resolution of 10 μm for Z axis. The part program has been generated using Simplify3D CAM software. The specimen width (*A*) has been measured for each sample by optical microscope (Zeiss Discovery V.20) while the thickness has been measured with a Mitutoyo digital caliper with a nominal resolution of 0.01 mm. The tensile tests on FDM specimens have been carried out using a Shimadzu EZ-S equipped with a 500 N load cell and crosshead speed 1 mm/min, according with ISO527 standards.

A second DoE has been executed in order to identify the effects of annealing on mechanical properties. For this purpose, the extrusion temperature and raster angle have been varied while air gap and extrusion width have been fixed to 0 and 0.5 mm respectively. The levels chosen for the two factors are reported in table 2. The other process parameters are equal to the values set in the previous experimental setup (table 1). Each FDM fabrication run consists of 3 samples replied for three times (replications).

TABLE 2. Process parameter values.

Factor	Unit	Low level	Central level	High level
Extrusion Temperature (ET)	°C	200	-	230
Raster angle (RA)	degrees	0°/0°	+45°/-45°	0°/90°
Air Gap (AG)	mm		0	
Extrusion width (EW)	mm		0.5	

For each DoE an analysis of variance (ANOVA) has been carried out based on the following variables: UTS, EB% and E, checking that the ANOVA hypotheses of normality distribution, equal variances, and independence of residuals were fully satisfied.

3. PLA MECHANICAL PROPERTIES AND FDM CAPABILITY

The statistical analysis shows that the raster angle is highly significant ($P < 0.001$) for UTS and EB%, while EW is highly significant for E. The extrusion temperature is statistically significant ($P < 0.05$) for EB% and E. Moreover, the interaction between raster angle and extrusion width (RA*EW) is statistically significant for UTS with a P value of 0.025, RA*ET has a P value of 0.007 for EB and EW*AG is statistically significant ($P = 0.049$) for E.

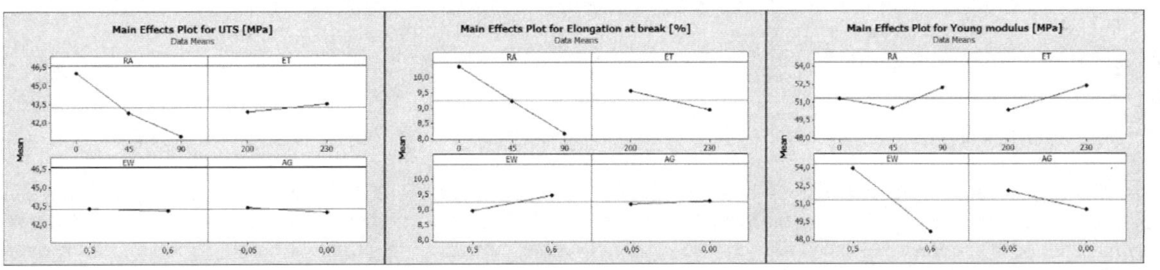

FIGURE 2. Main effects plots of UTS, Elongation at break (*EB%*) and Young modulus (*E*).

FIGURE 2 reports the main effects plots for UTS, EB% and E. Raster angle has a similar trend on UTS and EB%. The minimum value is obtained with the lowest level of RA (0°/90°), for which the samples are weakened by the one other layer where the infill orientation is perpendicular to the force during the test. The highest result is for samples where the beads are deposited along the direction of the tensile test (RA = 0°/0°) as expected. Lower extrusion width leads to a higher Young modulus, as well as higher extrusion temperature for which a better coalescence between the beads can be achieved.

4. EFFECTS OF ANNEALING ON PLA MECHANICAL PROPERTIES.

In this section the effect of a thermal treatment on the PLA samples is investigated. The samples are fabricated following a reduced DoE of the section 3 where extrusion temperature and raster angle have been varied while air gap and extrusion width have been fixed to 0 and 0.5 mm respectively.

The thermal treatments adopted is characterized by a trapezoidal profile showed in figure 3, where an annealing temperature (T_a) of 80 ± 1 °C is kept constant for a time interval (t_a) of 10 minutes. The rising time (t_r) and falling time (t_f) are set to 40 minutes in order to have slow heating and cooling phases (± 2 °C/min). The annealing temperature is higher (+35%) than the glass transition temperature (T_g) of PLA, which is around 60 °C. The crystallization and bonding processes is promoted along the period when the temperature is higher than the T_g and consequently an extended annealing time ($t_{a\text{-}ext}$) of 30 minutes should be considered.

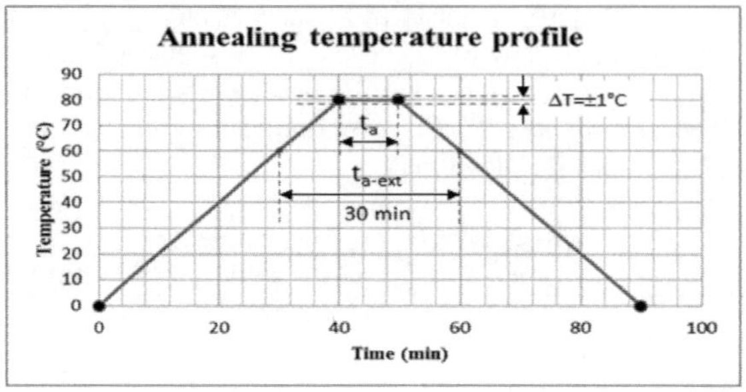

FIGURE 3. Annealing temperature profile (right).

The ANOVA shows that for *UTS* and *EB%* raster angle and its interaction with the temperature are statistically significant ($P < 0.001$), while the extrusion temperature is statistically significant ($P = 0.001$) for the Young modulus. As expected, the main effect plots in figure 4 show a trend similar to the case of the specimens that were not annealed (figure 2), but a comparison, as reported in the figure 5, can highlight the effect of the annealing.

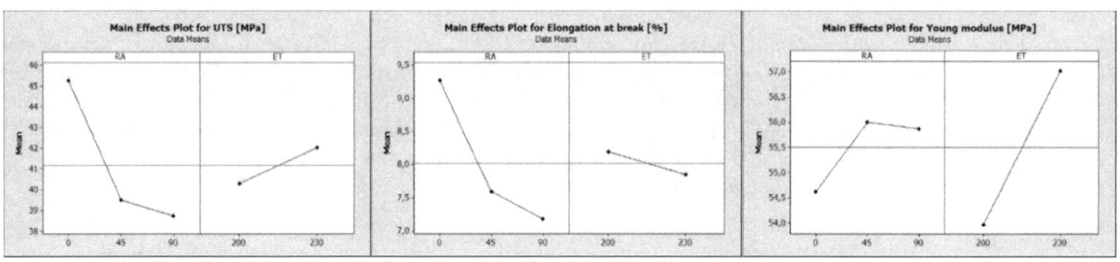

FIGURE 4. Main effects plots of *UTS* and Elongation at break and Young modulus for the annealed samples.

While the annealing adopted does not statistically influence the UTS, it is resulted statistically influent on Elongation at break ($P = 0.001$) and on Young modulus ($P = 0.005$). In particular, it causes a decrease of EB% and an increase of E in agreement with the literature [7]. Concerning the Young modulus a smaller dispersion of the data has been also observed. It must be underline that the annealing increases the side bonds between adjacent beads within a layer and between layers. Consequently, this effect produces an enhancement of the Young Modulus for non-zero raster angle. On contrary, side bonding has lower influence for zero value of raster angle since the beads are oriented along the load application direction and their bonds are less important. This effect can be clearly seen in figure 5, where the values of the Young modulus of the samples annealed is much higher for specimens with raster angle ±45° than for samples with beads at 0°/0°.

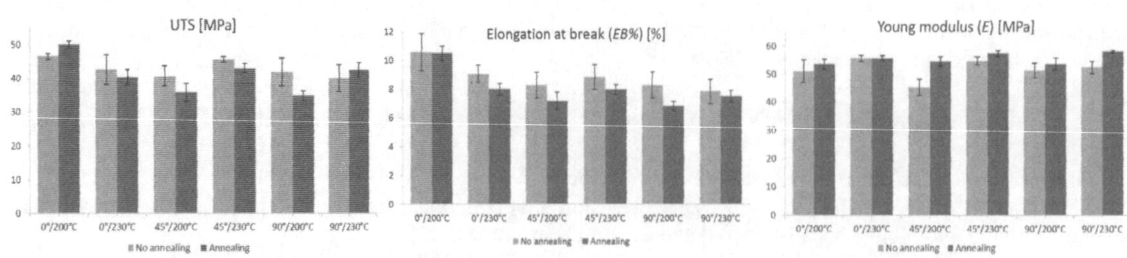

FIGURE 5. Histograms of *UTS*, *EB%* and *E* for different process parameters and annealing.

In addition, results show that the lower the extrusion temperature, the higher the gain of stiffness achieved with annealing. This latter fact can be easily explained since low extrusion temperatures result in lower bonding (by surface melting) between beads during the fabrication.

5. CONCLUSIONS.

In this paper, fused deposition modelling (FDM) technology has been analysed in terms of process capability for micro-component manufacturing and mechanical properties. In addition, post-process heat treatment (annealing) has been applied to PLA specimens in order to investigate its effects on mechanical properties at different process conditions. The experimentations have been carried out on micro dog-bones samples designed on the base of ISO527-5B.

The orientation of infill (raster angle) affects toughness and ductility of PLA samples since the more oriented along the stress direction the infill is, the higher the UTS and elongation at break are. On contrary no effect has been found on Young modulus (stiffness). This latter property is significantly improved by the extrusion temperature. Higher values of extrusion width result in lower values of Young modulus but no effect is found on UTS and EB%. No statistical significance has been found for the air gap parameters.

These results entail that it is possible to improve mechanical properties of the PLA products through an adequate choice of raster angle, temperature and extrusion width but an accurate study of the operating load conditions should be performed in order to carefully evaluate the stress states.

Finally, the effect of the annealing on tensile mechanical properties has been investigated in conjunction with process parameters. The results confirmed that the annealing improves the Young modulus, worsens the EB% and has no effect on UTS. Moreover, the improvement on Young modulus is more significant for samples with non-zero raster angle. This result should be more investigated, but preliminary it can be explained with a bonding process between beads which promote cross-linking and improved stiffness. In addition, results show that the lower the extrusion temperature, the higher the gain of stiffness achieved with annealing. This latter fact can be easily explained since low extrusion temperatures result in lower bonding (by surface melting) between beads during the fabrication.

ACKNOWLEDGMENTS

The authors would like to thank Dr. Eng. Lara Rebaioli for the helpful insights into the statistical analyses.

REFERENCES

1. O. Ivanova, et al., Rapid Prototyping Journal, 19 (5), 353-364 (2013)
2. O.A. Mohamed et al., Adv. Manuf. 3, 42-53 (2015).
3. M. Dawoud et al., Journal of manufacturing Processes 21, 39-45 (2016).
4. M. S. Hossain et al., Int. Solid Freeform Fabrication Symposium: An Additive Manuf. Conf. Austin, 2013, 380-392 (2013).
5. A. Bagsik, Proceedings of ANTEC 2011, Boston (2011).
6. O. Lužanin et al., Journal for Technology of Plasticity, 39 (1), 49-58 (2014).
7. J. Wootthikanokkah et Al., Journal of applied polymer science, DOI: 10.1002/APP.38715 (2013).
8. Y. Srithep, P. Nealey, L.S. Turng, Polymer engineering and science, 580-588, DOI: 10.1002/pen.23304 (2013).
9. A.M. Harris, E. C. Lee, Journal of applied polymer science, Vol. 107, 2246-2255 DOI: 10.1002/APP.27261 (2008).
10. L.S. Turng, Y. Srihep, Society of Polymer Engineers Plastic Research online, DOI: 10.2417/spepro.005392 (2014).
11. E. Haberstroh et al., Macromolecular Materials and Engineering, 287 (12), 881-888 (2002).
12. C. H. Wu et al, Polymer Engineering and Science, 45 (7), 1021-1030 (2005).
13. T. Koch et al., Macromolecular Symposia WILEY-VCH Verlag GmbH, 181 (1), 499-506 (2002).
14. L. Xie et al., Microsystem Technologies, 15 (6), 913-917 (2009).
15. J. Giboz et al., Journal of Micromechanics and Microengineering 17, 96 (2007).

Generation of High performance Polyphenylene Sulfide-Thermotropic Liquid Crystalline Polymer Composite Filaments for Use in Fused Filament Fabrication

Mubashir Q. Ansari[a] and Donald G. Baird[a,b]

[a]*Department of Chemical Engineering, Virginia Polytechnic Institute and State University, Blacksburg, VA 24061-0211, USA*
[b]*Macromolecular and Interfaces Institute, Virginia Polytechnic Institute and State University, Blacksburg, VA 24061-0211, USA*

Abstract. In this work, a novel dual extrusion process was used to fabricate high performance continuously-reinforced composite filaments for use in Fused Filament Fabrication (FFF). The processing conditions required to reinforce polyphenylene sulfide (PPS) using a commercially available thermotropic liquid crystalline polymer (TLCP), composed of terephthalic acid (TA), 4-hydroxybenzoic acid (HBA), hydroquinone (HQ) and hydroquinone derivatives (HQ-derivatives) were established. The TLCPs possess high mechanical properties arising from the stearic hindrance of their rodlike monomers. The TLCP of interest melts at 320 °C but can stay molten up to around 35 °C below its melting temperature when cooled from 360 °C, exhibiting a phenomenon known as supercooling behavior. However, the melt stability was found to vary drastically with the temperature. At 290 and 295 °C, the sample begins to solidify in 43 and 317 seconds, respectively. However, at 300 °C and above, the melt is stable for more than 3600 seconds. Isothermal time sweeps performed on PPS indicate an unstable melt complex viscosity above 320 °C. Shear step strain tests performed on the TLCP indicate that the TLCP should be heated to at least 360 °C to melt all the residual crystallites. Due to the difference in processing temperatures, the filaments are generated using a unique process in which the higher melting point TLCP is spun within the matrix to form an in situ reinforced filament. The resins are plasticated in two different extruders at different temperatures and then the stream of the TLCP is injected into the stream of PPS. The composite stream is passed through a series of static mixers which subdivide the streams into finer streams, and then the composite stream is drawn to impart molecular orientation to the TLCP phase. The filaments so produced are processed in FFF by melting the PPS without melting the TLCP. The properties of the plaques generated from these strands will be reported.

Keywords: Dual Extrusion Technology, Thermotropic Liquid Crystalline Polymer, Fused Filament Fabrication.
PACS: 81.05.Qk, 83.85.Cg

INTRODUCTION

Additive manufacturing (AM) has recently attracted immense attention both in industry as well as academia, due to its advantages, such as the ability to fabricate lighter parts due to innovative designs, less waste generation, cost reduction and lesser lead time in product development [1, 2]. The push to adopt AM by the auto and aerospace industries comes from the intense pressure on these industries to improve fuel efficiency which significantly reduces the cost per mile of transportation and also the greenhouse gas emissions [3].

The composites, due to low density, have an edge over metals which has prompted escalating replacement of metals in the automobile industry [4].On the other hand, the potential of composites has not been utilized to the required extent in AM, primarily due to the availability of composites that are compatible with the AM technologies and also due to the inability of the available technologies to handle composites in general [5].

Fused Filament Fabrication (FFF), also known as Fused Deposition Modeling (FDM) is an extrusion based AM technology, which uses filaments as feedstock to fabricate objects [6]. The mechanical properties of the filaments generated using pure matrices are inadequate for printing load-bearing components [7]. The mechanical properties of the printed parts are found to be even lower than those of the filaments due to the introduction of multiple weld lines during the printing process [8, 9]. Researchers have directed their attention towards improving the mechanical properties of the printed parts by improving the weld strength [10] and also improving the filament properties by using reinforcement fibers. The processing method utilized to reinforce the filaments using extruders is known to break the length of the fibers, thereby diminishing the reinforcing potential of the fibers [9]. Although researchers have separately impregnated the polymer during the printing process to obtain continuous reinforcements, these are done at the cost of modifications in the FFF machines [11, 12].

Another solution to obtain longer fibers is to use in-situ composites of thermotropic liquid crystalline polymers (TLCP) as the process of fibril generation takes place during the processing stage [13]. TLCP's are composed of rigid rod-like monomers, which under extensional flow conditions are highly aligned and form long fibrils giving significantly improved tensile properties [13, 14]. Other advantages are the low viscosity of the blend, minimal wear on the processing equipment and low density [15]. Because most TLCP's are processed above the degradation temperature of the matrix polymers, dual extrusion technology developed by Baird and Sukhadia [13, 16], solves the problem. This effort is to use PPS and a TLCP to obtain high performance composite filaments using dual extrusion technology, discussed elsewhere in detail [13], for application in FFF.

EXPERIMENTAL

Materials

The PPS used in this work, Ryton®, is a commercial grade and was generously supplied by Solvay. The TLCP used, HX3000, was supplied by Dupont and is believed to be synthesized from different ratios of terephthalic acid (TA), 4-hydroxybenzoic acid (HBA), hydroquinone (HQ) and hydroquinone derivatives (HQ-derivatives. The melting temperatures of the PPS and HX3000 are 285 and 320 °C, respectively.

Rheological Measurements

All the rheological tests were carried out using an ARES-G2, a TA instrument manufactured rheometer. These tests were carried out in a nitrogen atmosphere. 25 mm parallel plate fixtures were used for testing PPS whereas 25 mm cone and plate fixtures were used for testing HX3000. Wherever it has not been stated, the strain was kept at 5% and the angular frequency used was 1 radian/sec. SAOS tests in the isothermal conditions, referred henceforth to as isothermal time sweeps, were conducted on PPS at 290, 300, 310, 320, 330 and 340 °C to establish the maximum temperature below which degradation is minimal. The complex viscosity as a function of time was tracked for an hour. HX3000 was subjected to SAOS tests in the isothermal conditions at 290, 295, 300, 305 and 310 °C, referred henceforth to as isothermal time sweeps. The sample was equilibrated at 360 °C for 3 minutes, following which the sample was cooled at 5 °C/min to the desired temperature in the supercooled regime where isothermal time sweeps were carried out. HX3000 was also subjected to shear step strain tests at 340, 350, 360 and 370 °C to identify the maximum temperature that HX3000 should be equilibrated at to melt all the residual crystallites. Shear step strain tests were also conducted on HX3000 at 290, 300, 310, and 320 °C to identify the minimum temperature below which HX3000 should not be cooled to for avoiding the formation of crystallites. PPS and HX3000 were dried at 120 °C in a vacuum oven prior to testing.

Post Processing Using FFF and Compression Molding

Filaments generated using dual extrusion technology were cut 3″ long and compression molded by laying the strands parallel to each other in a square mold. HX300/PPS (67.54 wt./wt.%) was compression molded for 5 minutes at 290, 300, 310 and 320 °C. The molded plaques were cut into strips to assess tensile properties in the alignment direction to understand the effect of temperature on tensile properties using an Instron Mechanical Tester (model 4204). A RoVa3D 5 Extruder FDM 3D Printer was used to print tensile testing specimens in the raster printing pattern.

RESULTS

Rheology

As seen in Figure 1(a), PPS exhibits unstable complex viscosity above 320 °C due to chain scission reactions which is in agreement with the study conducted by Peters and Still [17]. Thus, to process PPS, the temperature should not be taken above 320 °C. Any degradation leads to gaseous products, eventually forming bubbles in the extrudate.

The results for the isothermal time sweeps performed on HX3000 is shown in Figure 1(b). The crossover of the loss and storage modulus, which represents the onset of crystallization of the TLCP's in the supercooled state [18], was found to occur in 43 and 317 seconds at 290 and 295 °C, respectively. At temperatures above 300 °C, the samples did not solidify in 1 hour. Figure 2(a) and 2(b) show the results for shear step strain tests carried out on HX3000 at temperatures above and below the melting temperature of HX3000, respectively. Done [19] ascribed the tails in the relaxation curves to the presence of some structures in the melt, most probably the presence of higher melting crystallites which do not allow the relaxation modulus to drop to zero. Figure 2(a) suggests that HX3000 should be heated to at least 360 C to melt all the residual crystallites. The presence of tails in Figure 2(b) can be seen at temperatures equal to and below 300 °C, indicating that the temperature should not be dropped below 310 °C.

Dual extrusion of PPS and HX3000.

Using the rheological information obtained for PPS and HX3000, processing temperatures selected to carry out dual extrusion are shown in the schematic in Figure 3(a). T_R, T_M, T_{RP}, T_{MP}, T_D, are the reinforcement extruder, matrix extruder, reinforcement pump, matrix pump and the die temperatures, respectively. T_N and T_S represent the temperatures at the 28-port injection nozzle and, at the 4-element KOCH and 4-element Kenics mixers, respectively. The extrudate was drawn, cooled in a water bath and taken up on a winder. Figure 7b shows the filaments obtained.

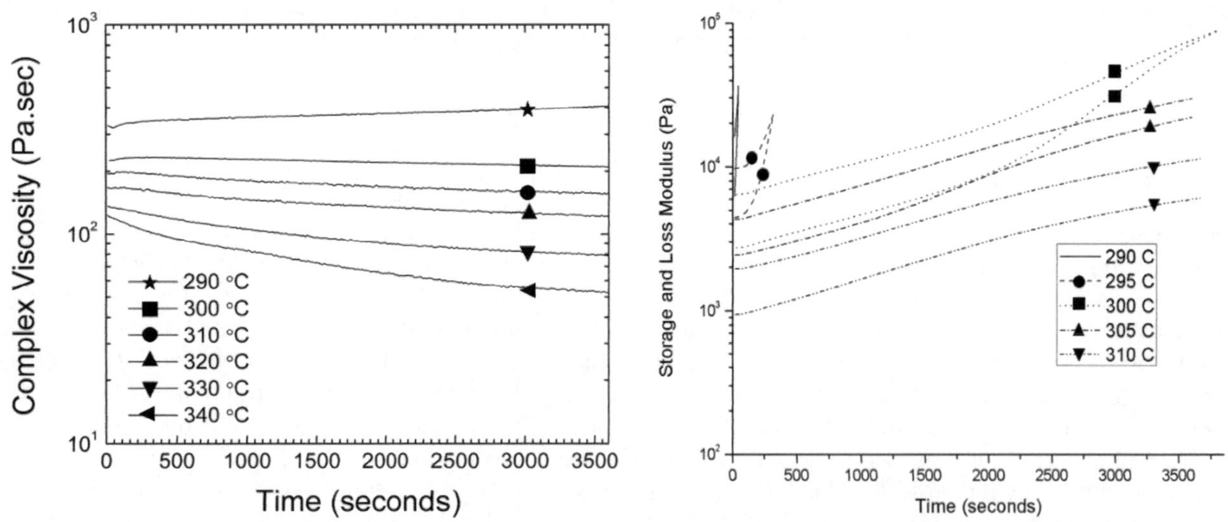

(a) (b)

Figure 1: (a) Isothermal time sweeps on PPS in nitrogen atmosphere (b) Cooling temperature ramps on HX3000. The top and bottom curves at each temperature represent the loss and the storage modulus.

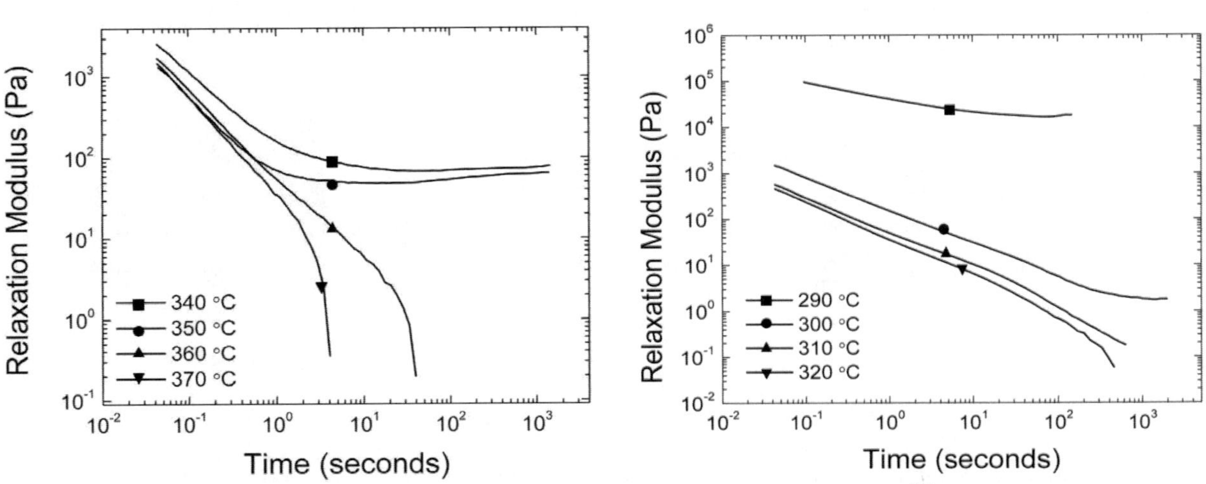

(a) (b)

Figure 2: Shear step strain performed on HX3000 (a) above and (b) below the melting temperature of HX3000.

Postprocessing using FFF.

HX3000/PPS (67.54 wt./wt.%) was used to determine printability of the filaments. The specimens seen in Figure 3(b) were printed at 310 °C, which is below the melting temperature of HX3000. It was observed that the excessive pressure by the feeding rollers on the filament leads to fiber breakage and eventually flow stoppage. An important consideration to process continuously reinforced filaments is to match length per unit time of the filament fed into the liquefier with the scan speeds in x, y and z directions. This is to avoid breakage or kinking of the reinforcing fibrils.

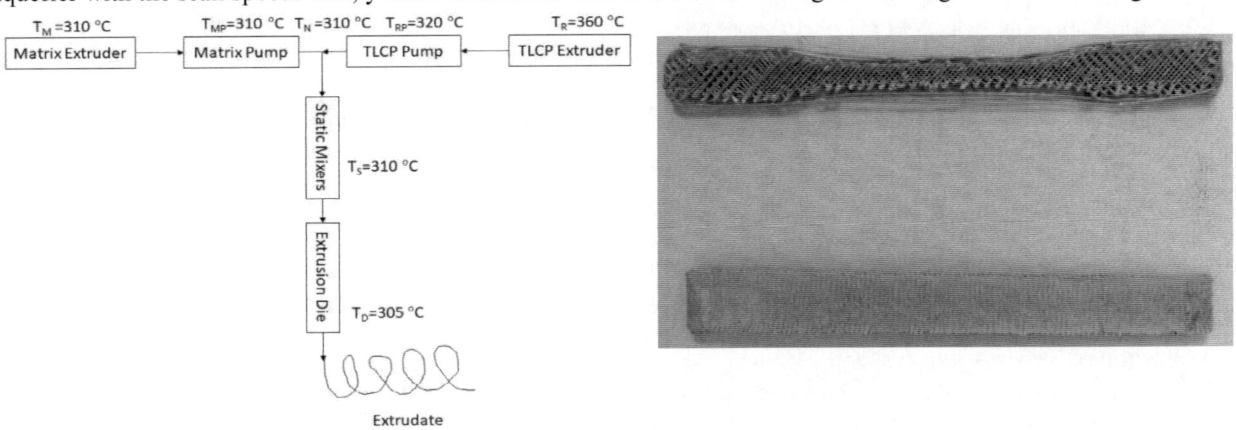

(a) (b)

Figure 3: (a) Schematic of the temperature profile in the dual extrusion system (b) Printed composite specimens.

Tensile testing

The tensile strength and modulus obtained in the alignment direction for HX3000/PPS (67.54 wt./wt.%) are shown in Figure 4(a) and 4(b), respectively. Significant improvement in modulus can be seen. Increase in the strength could not be obtained and needs further investigation.

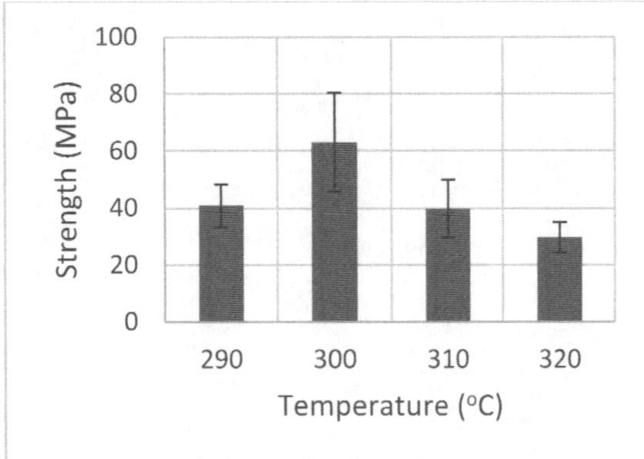

(a)

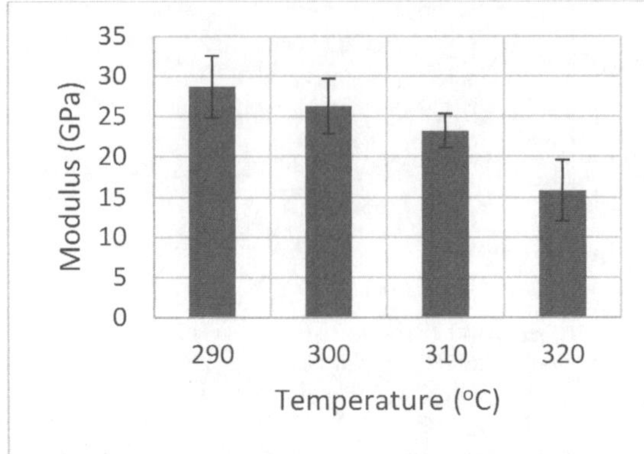

(b)

Figure 4: (a) Tensile strength and (b) modulus of HX3000/PPS (67.54 wt./wt.%) as a function of molding temperature in the alignment direction. Error bars represent 95% confidence intervals.

CONCLUSION

Processing parameters determined using rheology allowed one to successfully blend PPS and HX3000 using the dual extrusion system. Improvement in modulus was obtained on tensile testing the compression molded specimens. Moreover, the composite filament was successfully printable below the melting temperature of HX3000 using FFF.

REFERENCES

1. B. Lyons, The Bridge 44 (3) (2014).
2. S. H. Huang, P. Liu, A. Mokasdar and L. Hou, The International Journal of Advanced Manufacturing Technology, 1-13 (2013).
3. B. P. Conner, G. P. Manogharan, A. N. Martof, L. M. Rodomsky, C. M. Rodomsky, D. C. Jordan and J. W. Limperos, Additive Manufacturing 1, 64-76 (2014).
4. G. Cole and A. Sherman, Materials characterization 35 (1), 3-9 (1995).
5. I. J. Petrick and T. W. Simpson, Research-Technology Management 56 (6), 12-16 (2013).
6. I. Gibson, D. Rosen and B. Stucker, *Additive manufacturing technologies: 3D printing, rapid prototyping, and direct digital manufacturing*. (Springer, 2014).
7. F. Ning, W. Cong, J. Qiu, J. Wei and S. Wang, Composites Part B: Engineering 80, 369-378 (2015).
8. S.-H. Ahn, M. Montero, D. Odell, S. Roundy and P. K. Wright, Rapid prototyping journal 8 (4), 248-257 (2002).
9. H. L. Tekinalp, V. Kunc, G. M. Velez-Garcia, C. E. Duty, L. J. Love, A. K. Naskar, C. A. Blue and S. Ozcan, Composites Science and Technology 105, 144-150 (2014).
10. N. P. Levenhagen and M. D. Dadmun, Polymer 122, 232-241 (2017).
11. N. Li, Y. Li and S. Liu, Journal of Materials Processing Technology 238, 218-225 (2016).
12. G. W. Melenka, B. K. Cheung, J. S. Schofield, M. R. Dawson and J. P. Carey, Composite Structures 153, 866-875 (2016).
13. A. Handlos and D. Baird, Journal of Macromolecular Science, Part C: Polymer Reviews 35 (2), 183-238 (1995).
14. R. W. Gray IV, D. G. Baird and J. Helge Bøhn, Rapid Prototyping Journal 4 (1), 14-25 (1998).
15. M. Mcleod and D. Baird, Journal of Elastomers & Plastics 31 (2), 143-161 (1999).
16. D. G. Baird and A. Sukhadia, (Google Patents, 1993).
17. O. A. Peters and R. H. Still, Polymer degradation and stability 42 (1), 41-48 (1993).
18. C. Qian and D. G. Baird, Polymer 102, 63-72 (2016).
19. D. Done and D. G. Baird, Journal of Rheology 34 (5), 749-762 (1990).

Prediction of Porosity Dissolution in SLS Process

LY Hai-Bang [(1)], MONTEIRO Eric [(1)], DAL Morgan [(1)], REGNIER Gilles [(1)]*

[(1)] PIMM-ENSAM-CNRS-CNAM - PARIS - France

The polymer grains coalesce faster at surface than deeper in the powder bed because of higher laser radiation absorption at powder surface. Then gas porosities are embedded in the melted polymer layer, whose thickness can reach a few hundred microns. Due to the pressure induced by polymer surface tension around the porosity, the gas slowly diffuses through polymer melt enabling porosities to be reduced or dissolved. This phenomenon has been studied by several researchers considering a gas porosity in an infinite medium, but porosities in SLS process are too close to the surface to considerate an infinite medium. Therefore, to assess porosity resorption kinetic, we first considered the dissolution of an air porosity embedded in a melted polymer sphere. The melted polymer was considered as Newtonian and gas diffusion was supposed to obey a Fick law. At polymer/gas interface, the concentration of gas in the polymer was supposed to follow Henry Law. The problem was first solved in 1D thanks to a finite element method, flow and diffusion was fully coupled in an implicit scheme. Then, a 2D simulation was performed by considering a porosity close to a plane surface. We first find that a ratio of about ten between the diameter of the melted polymer sphere and the one of the porosity is equivalent to porosity in an infinite medium. We show that the influence of porosity dissolution time versus diameter, surface tension and diffusion coefficient obeys to a power law. The influence of the porosity distance to the surface is determined. The viscosity has a small and linear influence on dissolution time, which means that the time scale of diffusion is larger than the one of the flow. We confirm the strong influence of the initial gas concentration in polymer medium, which decreases with the thickness of melted polymer. A last, it seems important to take into account the advection in the flow simulation, it decreases significantly the time of porosity dissolution.

Maximisation of intra- and inter-layer strengths in parts produced by fused filament fabrication

Spoerk Martin [1]*, Arbeiter Florian [2], Cajner Hrvoje [3], Sapkota Janak [1], Holzer Clemens [1]

[1] Polymer Processing, Montanuniversitaet Leoben - Steiermark - Austria, [2] Materials Science and Testing of Polymers, Montanuniversitaet Leoben - Steiermark - Austria, [3] Faculty of Mechanical Engineering and Naval Architecture, University of Zagreb - Zagreb - Croatia

Extrusion-based additive manufacturing, also known as fused filament fabrication (FFF) represents an increasingly popular additive manufacturing technology. Its major drawback of yielding unsatisfactory mechanical properties due to the large number of weld lines still prevents many industry sectors from relying on this technique. The present work aims at tackling this issue by maximising the intra- and inter-layer cohesion between adjacent strands within one printing layer and between two consecutive layers, respectively. A detailed parametric study is presented, in which the most significant factors regarding the printing of commercially available polylactic acid (PLA) were analysed and optimised. The influence of the printing temperature, flow rate, layer thickness, and the type of layer-overlap was studied by means of two characterisation techniques. An adapted tensile test setup, addressing the strands' intra-layer cohesion, was complemented by the fracture mechanical technique of the double cantilever beam (DCB) test, investigating the inter-layer cohesion. The DCB test, which is novel for additively manufactured parts, revealed beneficial settings for deliberately augmenting inter-layer strengths. Morphology analyses of the cross-sections and the fracture surfaces, as well as rheological studies, confirmed the results for both characterisation techniques. In sum, the present work describes a pathway and the experimental design to fabricate 3D-printed parts with optimised mechanical properties similar to those of homogeneous compression-moulded samples.

Novel Process Routes towards the Production of Spherical Polymer Powders for Selective Laser Sintering

Jochen Schmidt, Maximilian Dechet, Juan Gomez Bonilla, Stephanie Kloos, Karl-Ernst Wirth, Wolfgang Peukert

Friedrich-Alexander-Universität Erlangen-Nürnberg, Institute of Particle Technology, Cauerstr. 4, D-91058 Erlangen, Germany

Abstract. Additive manufacturing processes allow the production of individualized and complex parts without the use of molds or tools. Especially powder and beam-based AM processes like selective laser sintering (SLS) of polymers are promising, although the choice of commercial SLS powders is very limited. To further extend the applications of SLS, new and cheap materials with good processability must be supplied. Within this contribution two innovative methods for production of spherical polymer micro particles for SLS are discussed: The first approach is a process chain applicable to a wide range of polymers consisting of wet grinding in a stirred media mill followed by rounding of the obtained irregular-shaped particles in a heated downer reactor and subsequent dry coating of the spherical particles for improvement of powder flowability. It will be outlined for polybutylene terephthalate (PBT) – polycarbonate (PC) and glass-filled PBT systems. The second approach is melt emulsification with subsequent dry coating. In this process a polymer melt is emulsified by elongational and shear stresses in a rotor stator device. The method is exemplified for polyethylene waxes and polypropylene (PP). Both approaches allow for production of spherical polymer micro particles. The effect of flowability, shape and bulk density on the powders' processabilities in SLS is assessed.

Keywords: polymer powder, additive manufacturing, melt emulsification, rounding
PACS: 81.05.Lg; 82.35.Lr; 81.20.Wk; 47.55.-t; 81.15.-z

INTRODUCTION

Additive manufacturing methods allow the tool-free production of complex structures. Especially powder-based generative manufacturing processes like selective laser sintering (SLS) seem promising for individualized serial production. With the transition from rapid prototyping to additive manufacturing, i.e. the application of generative manufacturing processes in serial production, the requirements on mechanical part properties, process stability and robustness are constantly rising [1–6]. Besides optimal settings of the SLS process parameters, c.f. for example beam power, beam deflection velocity or building chamber temperature, powders of good flowability and improved bulk density are a prerequisite for obtaining dense parts of proper dimensions [2,4]. Device quality, processability and particle properties are directly connected. Especially the powder flowability is crucial: Powders of good flowability allow for homogeneous powder deposition and frequently show a higher bulk density than powders of inferior flowability, which is advantageous for building of dense devices. Currently the spectrum of commercially available laser sintering powders is limited to polyamides (PA), polystyrene (PS), polypropylene (PP), thermoplastic elastomers (TPE) and poly ether ether ketone (PEEK) are available [2,4,5], whereby PA-based systems make up more than 90% of the market share Polymer powders produced by cryogenic dry grinding [7] are available, too, however they often show poor processability due to the bad flowability of the typical irregular shaped particles obtained by comminution processes [8]. Two innovative methods for the production of spherically shaped polymer microparticles for additive manufacturing are discussed in this contribution: In the first approach, a process chain consisting of wet grinding [8] with subsequent rounding of the irregular particles in a heated downer reactor [9] and dry coating [10] is applied. Both, the change of the particle shape and the increase of surface roughness (dry coating) improve the powder flowability. The second process, melt emulsification [11], allows to obtain spherical particles in a single step. In this work the effect of particle shape and surface roughness on powder properties will be outlined exemplarily.

METHODS

Laser Diffraction Particle Sizing

Particle size distributions of the product particles were determined by laser diffraction particle sizing using a Mastersizer 2000 (Malvern) equipped with a Hydro 2000S wet dispersion unit. To obtain stable dispersions appropriate stabilization has been assured.

Scanning Electron Microscopy (SEM)

Shape and morphology of the polymer particles was assessed by electron microscopy A Gemini Ultra 55 (Zeiss) device equipped with a through-the-hole detector was used at an acceleration voltage of 1 to 1.2 kV.

Particle Sizing

Static laser diffraction particle sizing was applied to measure particle size distributions using a Mastersizer 2000 / Hydro 2000S (Malvern). A small amount of dried PA11 powder was suspended in deionized water. Small quantities of sodium dodecyl sulphate solution (SDS, 98% (Merck)) and ethanol (96%, denatured with 1% MEK (VWR)) have been added to assure dispersion stability. Prior to measurements, ultrasonication of the dispersions has been performed. Measurements were conducted under constant stirring and ultrasonication.

Powder Flowability

Flowability of the products was assessed by measuring the tensile strength of the powders using a tensile strength tester. For details on the procedure please refer to [9,12,13].

RESULTS

Process Chain for Spherical Glass-filled PBT Particles

Wet Grinding

Wet grinding of glass-filled PBT (Ultradur 4520 HS, BASF) has been performed in the batch stirred media mill PE5 (Netzsch). A three-disc stirrer was used and the grinding chamber was lined with silicon nitride ceramics. A mass fraction of about 18.5 % feed material was applied, the solvent was ethanol. A stress energy SE [14] of 1.9 mJ was chosen by using grinding media made of Yttria-stabilized zirconia (YTZ, density ρ_{GM} = 6,050 kg m^{-3}) of 2.0 mm diameter (d_{GM}) and operating the mill at a stirrer tip speed of v_{tip} = 6.3 m s^{-1} corresponding to a stirrer rotational frequency of 800 min^{-1}. Approximately 14.3 kg grinding beads were filled into the grinding chamber. The process temperature was (20 +/- 2) °C. The product particles were separated from the solvent by centrifugation and dried prior to rounding.

The effect of the volume fraction of the filler (glass) (on the grinding kinetics at constant stressing conditions represented by the temporal evolution of the product particle size $x_{50,3}$ is shown in Figure 1. For comparison, the grinding kinetics of non-filled PBT are given as well (blue asterisks) The composite material is not strengthened by the filler, on the contrary a weakening effect was observed: in wet comminution smaller product particle sizes and faster grinding kinetics are observed for larger filling degrees.

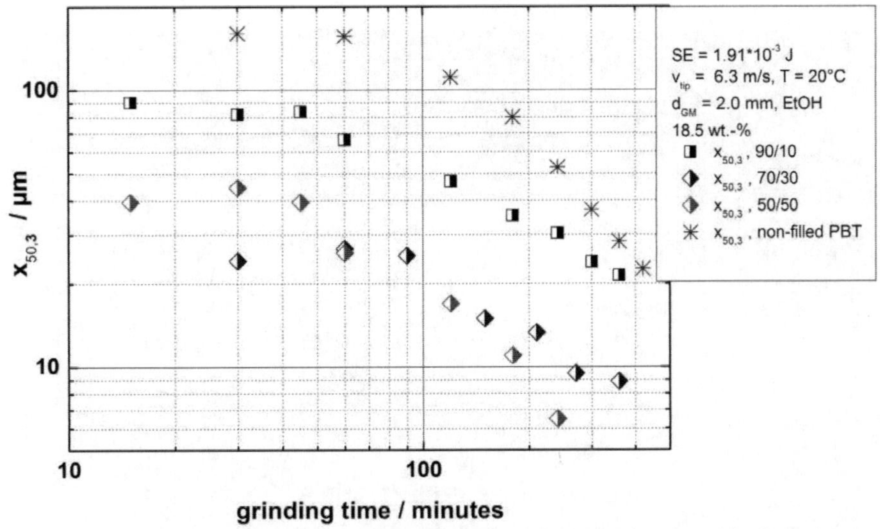

FIGURE 1. Product particle size x50,3 during wet grinding of glass-filled PBT in ethanol in dependency on process time: influence of polymer-to-filler volume ratio.

Rounding

The rounding of the particles obtained by wet comminution was performed in a heated downer reactor. In the reactor the particles are molten, rounded due to the action of surface tension and finally solidified. For details on the experimental setup and the choice of appropriate process parameters, please see [9]. Heating is realized by a nine stage heating system (Thermal Technology) and the particles are introduced into the heated downer as an aerosol. For aerosol generation a Venturi disperser was used. The carrier gas was nitrogen (purity 5.0, Linde) and the aerosol gas flow was surrounded by a sheath gas flow (nitrogen) to minimize particle-particle and particle-wall contacts. The applied temperature profile is given Figure 2. The aerosol and sheath gas flow enter the reactor at its top section. The total gas flow is heated to the process temperature necessary for rounding then step by step lowered to allow solidification. Rounded particles are collected at the bottom of the reactor.

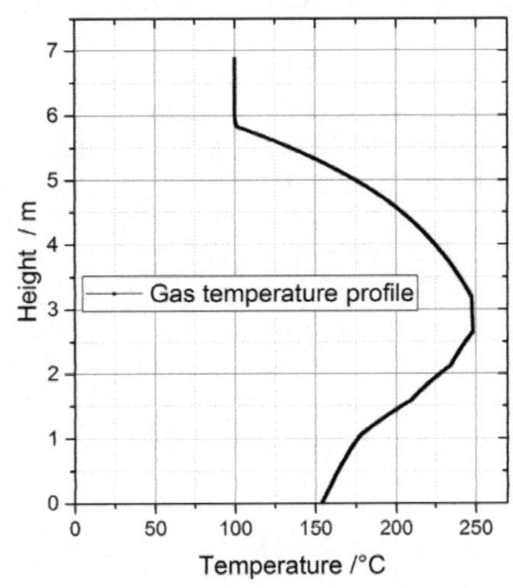

FIGURE 2. Temperature profile used for rounding of PBT-glass comminution products.

SEM images of PBT-glass comminution products and the obtained rounded particles are shown in Figure 3: the comminution product consists of irregular shaped, plate-like particles (Figure 3, left). Disintegration of the compositeand breakage of filler particles occurs. Rounding of the comminution product was possible up to a PBT-glass-ratio of 70/30.

FIGURE 3. SEM images of the comminuted (left) and rounded (right) PBT-glass composite powders: 90/10 (top), 70/30 (bottom).

Influence of Shape on Powder Tensile Strength

The tensile strength of comminuted and rounded powders was (13.7 +/- 1.6) Pa and (4.5 +/- 1.4) Pa, respectively. The comminuted product is cohesive. The rounding step leads to substantial improvement of flowability which should be reflected by better processability. A small tensile strength corresponds to good powder flowability. Further improvement of flowability is realized by dry particle coating [9-11].

Melt Emulsification of Polypropylene

Melt emulsification of polymers was performed for polypropylene PP 520P (Sabic). Hexadecane (Alfa Aesar) was the continuous phase and sodium dioctyl sulfosuccinate (AOT; Acros Organics) was the emulsifier and stabilizer. A scheme of the plant used is given in Figure 4. All pipes and the stirred tank 'B-1' are double-walled and can be heated with heat-transfer oil. The stirred tank 'B-2' is double-walled as well and may be water-cooled. First, the continuous phase, the stabilizer and the polymer granules were introduced into the stirred tank 'B-1' and the whole device was flushed and pressurized with nitrogen (purity 5.0, Linde) as inert gas. Then the device was heated above the melting temperature of the polymer and a pre-emulsion was formed using the stirrer in tank 'B-1'. The hot pre-emulsion was fed by a gear pump into the rotor-stator device where the fine emulsion is formed by intensive shear and elongational stress. The product fine emulsion is transferred into the (cooled) stirred tank 'B-2'. After setting the pressure of the device to ambient conditions the polymer particle suspension may be collected by opening the bottom valve of the stirred tank. PP was melt-emulsified in the aforementioned system at a dispersed phase concentration of 5 wt.-% at 190°C. After a process time of 30 minutes spherical particles sized $x_{50,3}$ = 22.1 µm were obtained. After spray drying and dry coating with fumed silica the powders could be successfully applied to produce thin layer specimen in SLS. The process is outlined in [11]; please refer to this work for further details.

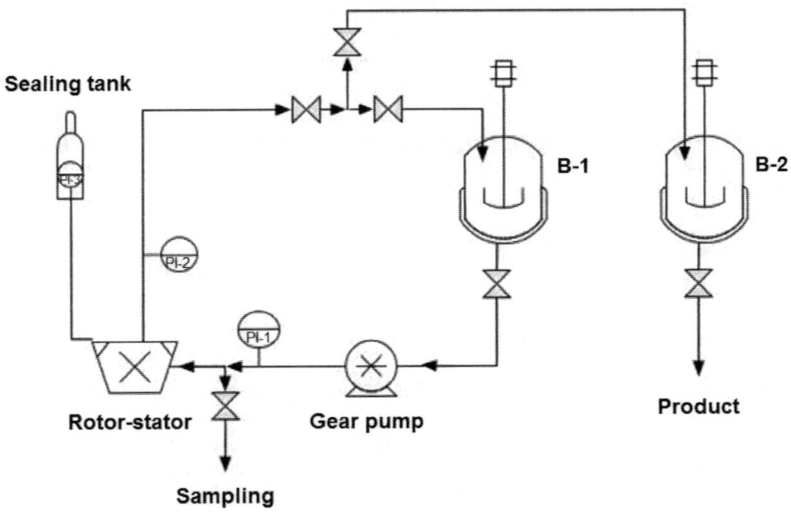

FIGURE 5. Scheme of the melt emulsification plant.

CONCLUSIONS

Two innovative approaches [9,11] for production of spherical polymer micro particles of good flowability for SLS have been discussed: a process chain consisting of a wet comminution step followed by rounding of the grinding product and dry coating has been exemplified for glass-filled PBT, the second approach, melt emulsification with subsequent dry coating for PP. Both scalable approaches allow for powders of improved flowability and may be applied to a variety of polymers and might be transferred to the plant scale.

ACKNOWLEDGMENTS

This study has been supported by Deutsche Forschungsgemeinschaft (DFG) in the framework of the collaborative research center CRC 814 'Additive Manufacturing' (projects A1, A2, A3). Financial support is gratefully acknowledged.

REFERENCES

1. J.P. Kruth, X. Wang, T. Laoui, L. Froyen, *Assembly Automation* **23,** 357–371 (2003).
2. R.D. Goodridge, C.J. Tuck, R. Hague, *Progress in Materials Science* **57**, 229–267 (2012)..
3. A. Gebhardt, *Generative Fertigungsverfahren: Additive manufacturing und 3D-Drucken für Prototyping - Tooling - Produktion*, 4th ed., Hanser, München, 2013.
4. T.T. Wohlers, *Wohlers Report 2016*, Fort Collins, Col.: Wohlers Associates, 2016.
5. M. Schmid, *Selektives Lasersintern (SLS) mit Kunststoffen: Technologie, Prozesse und Werkstoffe*, Hanser, 2015.
6. G. Witt, *Neue Entwicklungen in der Additiven Fertigung*, Springer Berlin Heidelberg, 2016.
7. M. Wilczek, J. Bertling, D. Hintemann, *International Journal of Mineral Processing* **74**, S425-S434 (2004)..
8. J. Schmidt, M. Plata, S. Tröger, W. Peukert, *Powder Technology* **228**, 84–90 (2012)..
9. J. Schmidt, M. Sachs, C. Blümel, B. Winzer, F. Toni, K.-E. Wirth, W. Peukert, *Powder Technology* **261**, 78–86 (2014).
10. C. Blümel, M. Sachs, T. Laumer, B. Winzer, J. Schmidt, M. Schmidt, W. Peukert, K.-E. Wirth, *Rapid Prototyp. J.* **21**, 697–704 (2015).
11. S. Fanselow, S.E. Emamjomeh, K.-E. Wirth, J. Schmidt, W. Peukert,, *Chemical Engineering Science* **141**, 282–292 (2016).
12. A. Schweiger, I. Zimmermann, *Powder Technol.* **101,** 1–9 (1998).
13. K. Meyer, I. Zimmermann, *Powder Technol.* **139,** 40–54 (2004).
14. A. Kwade, Powder Technology **105**, 382–388 (1999).

The effect of processing parameters on the dimensional characteristics of SLS parts

Pontes Antonio J. (1)*, Lima Andre (1), Freitas Luis (1), Sousa Ana (1), Barros Patricia (1), Sampaio Alvaro M. (1)

(1) Institute for Polymers and Composites,I3N - University of Minho - Guimarães - Portugal

The Selective Laser Sintering (SLS) is an Additive Manufacturing technique (AM) in which parts are created layer by layer, for example using polyamide (PA) powder, directly from 3D CAD data. The dimensional characteristics (including resolution and accuracy) of different features can vary from AM process to process and also with part orientation, thickness, and other characteristics of the part and the build. This paper presents the ongoing studies done on the SLS process, focussing on the effect of processing parameters on the dimensional characteristics of SLS parts. In this paper, a set of benchmark parts are designed, fabricated, and measured to build up design rules and dimensional data to use in the development and manufacturing of AM parts. The benchmark features were characterized experimentally and includes circular hole and circular bosses resolution versus part thickness, and thin wall resolution. All feature resolution experiments were conducted in six orientations, from parallel to orthogonal to the build plate from facing forward to backward, from upwards to downwards and two tilted versions at 45º in different directions. The information obtained by the preliminary studies was applied to an automotive part, revealing that it is possible to predict which is the right processing parameters to produce a SLS part with the final required dimensions.

Modification of polyamide powders and influence on their processability in Selective Laser Sintering

Ruggi David (1)*

(1) INSA - Villeurbanne - France

The SLS process involves almost exclusively semi-crystalline thermoplastic polymer powders. In principle, any semi-crystalline polymer might be used, but the actual range of SLS processable polymers is limited by the complex, and not yet well understood, interactions between the physico-chemical polymer properties in the powder state and the process conditions. As a consequence, the present knowledge and technological boundaries lead to the use of polyamides mainly. Our study focuses on the polyamide type: PA 6 and PA 12. Polyamides are subject to crystal perfection and Solid State Polymerization while they are maintained at high temperature, such as those encountered during the SLS fabrication procedure. The aforementioned phenomena limit the recyclability of the powder and the newly-created part quality, due to increasing viscosity and powder cohesiveness. The present work addresses the evolution of the properties of polyamide powders designed for the SLS process. Based on the SSP phenomena described in the literature , , one can control the crystallinity perfection, the molecular mass and the powder flowability at high temperature by appropriate powder treatment. Indeed, by varying the thermal treatment in different atmospheres, significant evolutions can be observed regarding the molecular weight, melting and crystalline features, and surface aspect. The study of the modifications of the powder/polymer properties provides better understanding of the behavior of such powders in SLS machines, and suggests that a better control over the powder ageing could be beneficial to processability and recyclability. Vouyiouka, Karakatsani, and Papaspyrides, "Solid State Polymerization." Wiloth, "Über Den Mechanismus Und Die Kinetik Der ε-Caprolactam-Polymerisation in Gegenwart von Wasser. 15. Mitt."

.Characterization of additive manufacturing processes for polymer micro parts productions using direct light processing (DLP) method

A. Davoudinejad, D. B. Pedersen, G. Tosello

Department of Mechanical Engineering, Technical University of Denmark, Building 427A, Produktionstorvet, 2800 Kgs. Lyngby, Denmark

Abstract. The process capability of additive manufacturing (AM) for direct production of miniaturized polymer components with micro features is analyzed in this work. The consideration of the minimum printable feature size and obtainable tolerances of AM process is a critical step to establish a process chains for the production of parts with micro scale features. A specifically designed direct light processing (DLP) AM machine suitable for precision printing has been used. A test part is designed having features with different sizes and aspect ratios in order to evaluate the DLP AM machine capability to fabricate polymer micro scale features geometries. Four different factors are evaluated for the AM process analysis: printing layer thickness, exposure time, film thickness and geometry. The process optimization of the workpiece quality features is carried out to highlight potential and challenges of the micro AM process.

INTRODUCTION

Direct AM of components is significant for different industrial applications to produce the end components having all the properties of marketable products. AM processes are thus characterized by a complex parameter optimization that must be done individually depending on the complexity or feature size of the parts [1]. In order to evaluate the performance of AM machines different test parts were used to investigate how accurate parts were printed with different complexity [2] in the macro scale. Another study investigated the resolution and repeatability of AM processes with voxel sizes at the micro scale to provide clear information about the resolution of the 3D printer that produced them [3].

This study evaluate the experimental printing parameters with two different test parts are designed having features with different sizes and aspect ratios. The results reveals the number of printed feature and the smallest printable size for parts and the height variation with this AM method.

METHOD

Part Design

The test part was designed in a way to cover various requirements in terms of the shape and size of the features. Two different shape was considered as a box and cylinder with a specific distance of 250 μm between each other and 3:4 aspect ratio for the lateral size and 1:2 for the height. Figure 1 shows the parts design in two different geometries. The base of the part was 12x12x2 mm2. The features raised the maximum height of the test part to 2 mm. The geometries were designed to observe the accuracy of micro print features in different shapes.

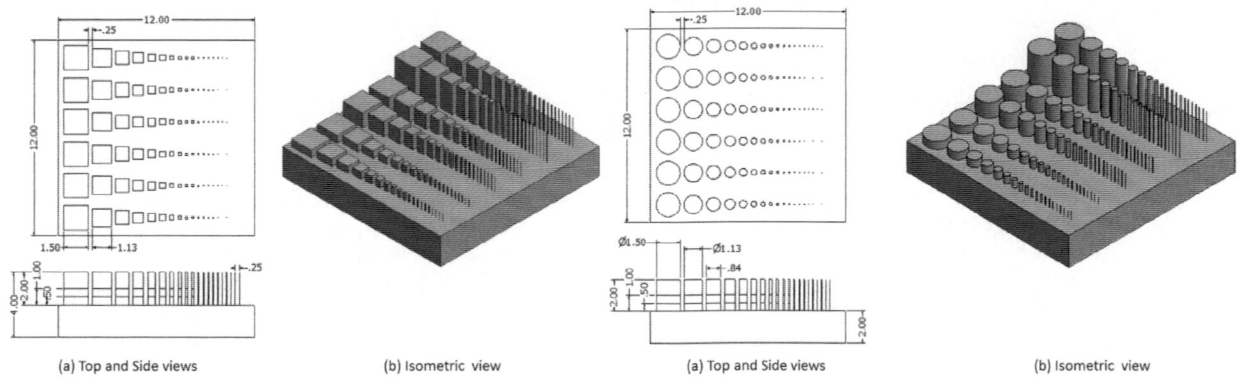

FIGURE 1. Drawing of the test part (a) Top& side views and (b) Isometric view (dimensions in mm)

Printing and Measurement Procedure

In vat photopolymerization method liquid photopolymer in a vat is selectively cured by light activated polymerization [4]. In this study, Direct Light Processing (DLP) was applied by using a light projector to solidify liquid photopolymer. By changing the light pattern and vertical position of the workpiece, the favorite geometry is build up layer by layer. The 3D printing machine for precision printing with the level of accuracy down to 1 μm resolution in the z-stage that has been developed, built and validated at the AM laboratory at the Technical University of Denmark (DTU) as shown in Figure 2 [5].

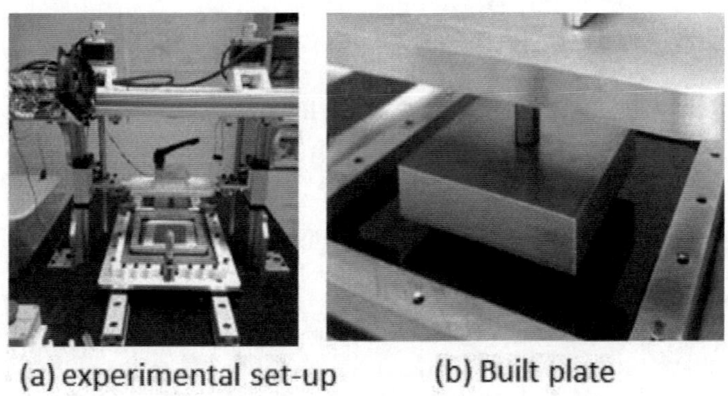

FIGURE 2. Experimental set-up 3D printing

In order to design the experimental plan, and decide on the significant factors to evaluate and values of the different levels some preliminary experiments were carried out to certain optimal values [6][7]. The printing parameters were selected as film thickness, layer thickness, exposure time and the sample geometries. A full factorial experiment 24 was applied to optimize an array of important printing parameters in order to improve the expected printing quality. Table 1 shows the experimental conditions, four different factors and two levels. The total number 16 samples were printed and printing time for each sample takes about45 minutes.

TABLE (1) Experimental conditions

Experiment number	Layer Thickness	Film Thickness	Exposure Time	Geometry
	μm	μm	s	
1	16	100	4	Square
2	18	100	4	Square
3	16	100	4.5	Square
4	18	100	4.5	Square
5	16	100	4	Circle
6	18	100	4	Circle
7	16	100	4.5	Circle
8	18	100	4.5	Circle
9	16	254	4	Square
10	18	254	4	Square
11	16	254	4.5	Square
12	18	254	4.5	Square
13	16	254	4	Circle
14	18	254	4	Circle
15	16	254	4.5	Circle
16	18	254	4.5	Circle

Printing included some manual steps such as setting the built plate and reference points, adjusting the parameters, afterwards post processing cleaning the printed sample with isopropanol in order to remove residual resin. The last step is to place the sample for curing in the UV oven, where it is exposed to UV light for 60 minutes.

Regarding the measurement different area on the samples were measured and investigated in order compare and measure the results in an effective way. Consequently, the height of pillars for both geometries, the width of the square in both *X* and *Y* directions and diameter of the cylinder were evaluated. For the measurements, the Olympus LEXT electronic microscope, and for analysis a scanning probe image-processing software was employed for the purpose (SPIP). For observing, the total printed intact pillars the Zeiss microscope was used for inspection of the samples.

RESULTS

The initial inspection in terms of the samples distortion reveals that the high aspect ratio printed features tended to collapse at 2 mm height. Figure 3 shows the printed samples with different geometries. The higher magnification of the parts in the red box area shows how height variation affect the sample with similar size. At 0.5 mm height, all printed features were in the correct shape (perpendicular to the base) however, they tend to be distorted at 1 mm height. Then more features were printed in the lower height (0.5 mm) about eleven features with the smallest size of 84 μm. However, at 2 mm height about 8-7 features were printed with the smallest size of 266 μm and 200 μm. Figure 4 presents the number of printed features in different geometries at various parameters at 0.5 mm height. The sloid fill graph are the square and the gradient fill are the cylinder. In terms of the geometry, it was reveals higher number of features were printed with cylindrical shape due to the limitation of the pixel at the corner of the square.

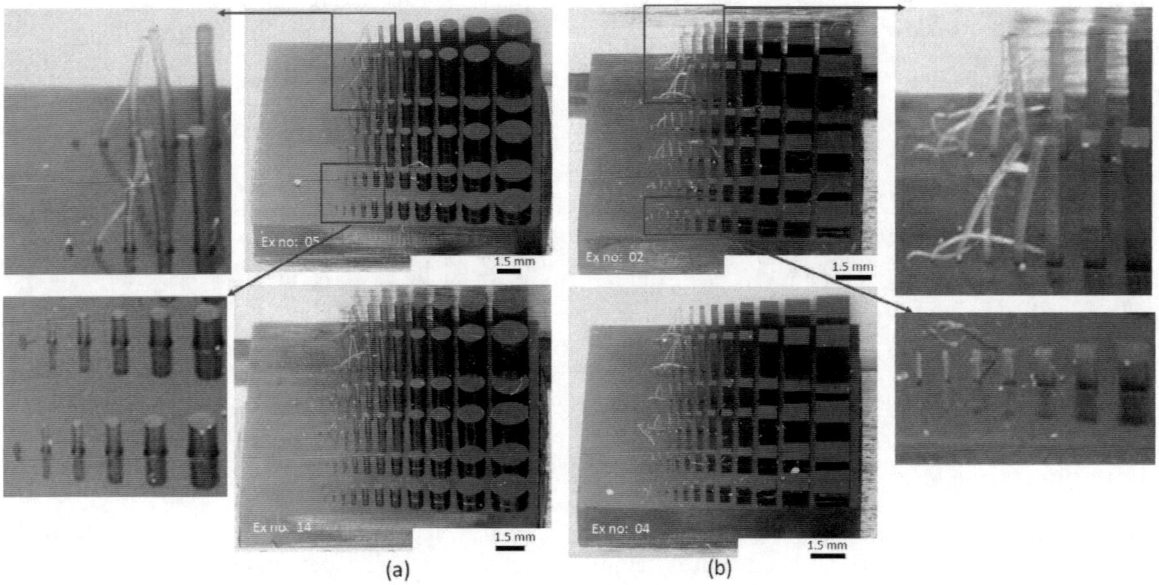

FIGURE 3 the printed samples (a) cylindrical (b) cube.

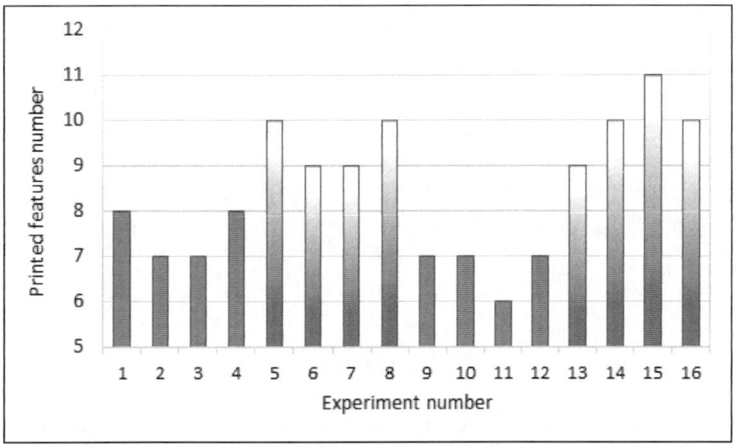

FIGURE 4 Number of printed features, sloid fill square and the gradient fill cylinder

Regarding the height measurement, design of experiments (DOE) analysis was carried out to illustrate the printing parameters contribution. Figure 5 shows the main effect plot and the height measurement at 500 μm features. The main effect plot (Figure 5 (a)) shows the influence of the factors on the height variation. The film thickness significantly affect the height variation and printed features with the thicker film tends to have higher size. This might be due to the less distortion of the thicker film in the vat for solidification of each layer. Then the geometry of the printed features and the layer thickness influenced the height of the features.

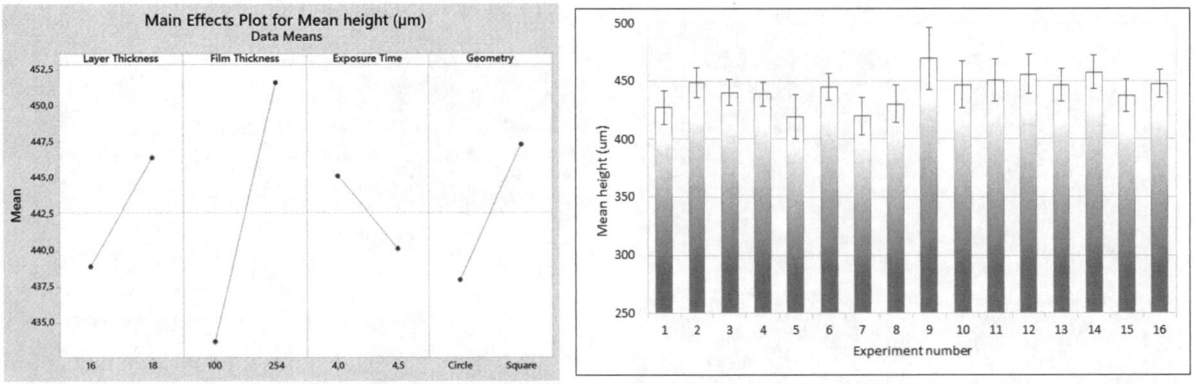

FIGURE 5 (a) main effects plot (b) height measurements.

CONCLUSION

The paper presents the evaluation of AM vat photopolymerization method in production of miniaturized polymer components. Two test parts were designed with different geometry, sizes and aspect ratios. Four different printing parameters as layer thickness, film thickness, exposure time and geometry were selected. The most significant factor for the printed features height was the film thickness and with thicker film 254 µm higher features printed. In terms of the smallest printable feature, the geometry affect the results, with the cylindrical shape more features were printed than square shape. It was observed that features with high aspect ratios tend to be distorted with 200-150 µm at 2 mm height however, at 05 mm 84 µm features size was printed.

ACKNOWLEDGEMENT

The research leading to these results has received funding from the People Programme (Marie Curie Actions) of the European Union's Seventh Framework Programme (FP7/2007-2013) under REA grant agreement no. 609405 (COFUNDPostdocDTU).

REFERENCES

[1] I. Gibson, D. W. D. W. Rosen, and B. Stucker, Additive Manufacturing Technologies: Rapid Prototyping to Direct Digital Manufacturing, vol. 54. 2009.

[2] D. Dlmitrov and K. Schrevel, "An lnvestlgatlon of the Capablltty Proflle of the Three Dlrnenslonal Prlntlng Process m ," no. 1, pp. 1–4.

[3] M. K. Thompson and M. Mischkot, "Design of Test Parts to Characterize Micro Additive Manufacturing Processes," Procedia CIRP, vol. 34, pp. 223–228, 2015.

[4] I. Cooperstein, M. Layani, and S. Magdassi, "3D printing of porous structures by UV-curable O/W emulsion for fabrication of conductive objects," J. Mater. Chem. C, vol. 3, no. 9, pp. 2040–2044, 2015.

[5] A. R. Jørgensen, "Design and development of an improved direct light processing (DLP) platform for precision additive manufacturing," Technical university of denmark (DTU), 2015.

[6] A. Davoudinejad, D. B. Pedersen, and G. Tosello, "Evaluation of polymer micro parts produced by additive manufacturing processes by using vat photopolymerization method," 2017, no. International Conference of the European Society for Precision Engineering and Nanotechnology.

[7] A. Davoudinejad, M. M. Ribo, D. B. Pedersen, G. Tosello, and A. Islam, "Biological features produced by additive manufacturing processes using vat photopolymerization method," no. Euspen conference proceeding micro/nano manufacturing November 2017, pp. 3–5, 2017.

Additive Manufacturing for Development of Polymeric Ron-Resch-Like Origami Structure for Damping Applications

Kshad Mohamed Ali Emhmed [(1)], Naguib Hani E. [(1)*]

[(1)] University of Toronto - Ontario - Canada

Additive manufacturing was used to study Ron-Resch-Like origami sandwich cores structures, which can be used for damping and the energy dissipation applications. Ron-Resch origami pattern is characterized by its easy way to form flat or curved shape, and its capacity to distribute the incoming forces and dissipate energy. This behavior achieved from the reentrant of the structure elements of the origami tessellation that fold when loaded and unfold when stretched. The main concern of the study is the effectiveness of the Ron-Resch origami structures to dissipate the impact energy. The design utilizes fused deposition modeling of polymeric material that allows flexible design, dimensional stability, and good surface quality. Compression and impact tests were conducted and then the results have been validated and discussed.

Using the Simplified Viscoelastic Fluid Model to Predict Extrudate Swell and Fiber Orientation in Fused Filament Fabrication Nozzle Flow

Z. Wang[a] and D.E. Smith[a]

[a]*Department of Mechanical Engineering, Baylor University, Waco, TX 76798)*

Abstract. Short carbon fibers suspended in the polymer feedstock enhances the mechanical performance of products produced with the Fused Filament Fabrication (FFF) Additive Manufacturing (AM) process. As the melted short fiber polymer composite is extruded and deposited on a moving platform, the velocity gradients within the polymer melt flow orientate the fibers, and the final orientation has a direct effect on the mechanical properties of printed beads. This paper numerically simulates FFF nozzle flow, including the extrudate material beyond the nozzle exit, using different rheology models. Particularly, we employ the Simplified Viscoelastic Model (SVM) in ANSYS Polyflow to evaluate the polymer melt velocity field and die swell. Fiber orientation tensors are computed along streamlines using the Orthotropic Closure and Folgar-Tucker isotropic rotary diffusion. The predictions indicate that the SVM yields reasonable die swell prediction with much less computational cost than traditional viscoelastic models. Orientation tensor values calculated from the flow results of the viscoelastic rheology models yields lower principal alignment in printed beads than those computed with Generalized Newtonian Fluids models.

Keywords: Fused Filament Fabrication, Simplified Viscoelastic Model, Die Swell, Fiber Orientation Tensor.
PACS: Replace this text with PACS numbers; choose from this list: http://www.aip.org/pacs/index.html

INTRODUCTION

Large scale polymer deposition Additive Manufacturing (AM) has recently experienced significant advances with applications that included automotive chassis, housing, and composite tooling. One example of large scale deposition is the Oak Ridge National Lab Big Area Additve Manufacturing (BAAM) system which has a build volume of 6m x 2.4m x 1.8m, capable of processing upwards of 100 lbs/hr of polymer through extruder nozzles having exit diameters up to 0.5 inches. Extrudate swell occurs in BAAM and other extrusion-based polymer processes as seen with the Strangpresse (Strangpresse, Inc, Youngstown, OH) AM extruder nozzle appearing in Figure 1(a). Prior results showed that the sudden change in the flow boundary at the nozzle exit results in a notable variation in fiber orientation downstream.

Beraudo, et al. [1] introduced a finite-element-based method to investigate the extrudate swell of LLDPE and LDPE melts using the Phan-Thien-Tanner (PTT) Model [2] for a 2D axisymmetric capillary die flow and the results shown a good agreement with related experimental data under low shear. Ganvir, et al. [3] proposed an Arbitrary Lagrangian Eulerian (ALE) algorithm for solving the die swell problem for both the steady state and transient problems. Their formulation behaved well at high shear rate but was unstable under low shear. Alternatively, Limtrakarn, et al. [4] employed the Simplified Viscoelastic Model in ANSYS Polyflow (ANSYS, Inc., Canonsburg, PA) with the implementation of the Discrete Elastic Viscous Stress Splitting (DEVSS) to predict the die swell of a 3D circular die flow for LDPE. They achieved a good agreement with the experimental data. Clemeur, et al. [5] also found that the SVM is a cost-effective tool to model the viscoelastic fluid flow in comparison with conventional viscoelastic fluid flow models (i.e. the Phan-Thien-Tanner model).

This paper presents a numerical approach to evaluate fiber orientation in a large scale FFF extruder nozzle and the post-nozzle extrudate, including a prediction of the die swell at the nozzle exit. Several rheology models are included in this study, and the Simplified Viscoelastic Model [6] is specifically employed in order to find a cost-effective solution for viscoelastic die swell and fiber orientation for the nozzle flow problem. The decoupled formation used in Heller, et al. [7] is applied in computing the fiber orientation of the flow domain.

NOZZLE FLOW MODELING

The simulations in this research are based on the large scale AM Strangpresse Model-19 extruder appearing in Figure 1. Figure 1(a) shows the extrusion of neat ABS polymer from the nozzle and important geometrical dimensions of the Strangpresse Model 19 large scale AM extruder nozzle appear in the Figure 1(b). Our extruder nozzle flow finite element model is based on this design, and also includes a 20-mm strand of polymer extrudate beyond the nozzle exit to simulate the effect of die swell. Due to the axisymmetry of the nozzle geometry and assumed flow, we consider only half of the flow domain using a 2-D axisymmetric model to reduce computational expense. Note that this axisymmetric assumption ignores any swirling motion of the flow that may result from the extruder screw. The boundary conditions for the flow domain are shown in Figure 2(a), where Q is the volumetric flow rate, F_s is the tangential force, F_n is the normal force, v_s is the tangential velocity, v_n is the normal velocity, **v** is the velocity vector at the free surface, and **n** is normal vector to the free surface [6]. The finite element domain is discretized into 20,382 nodes and 19,866 elements using the 4-node axisymmetric quadrilateral element. The mesh in the convergence zone and the nozzle exit is shown in Figure 2-(b). Additionally, we assume an isothermal polymer melt flow for simplicity.

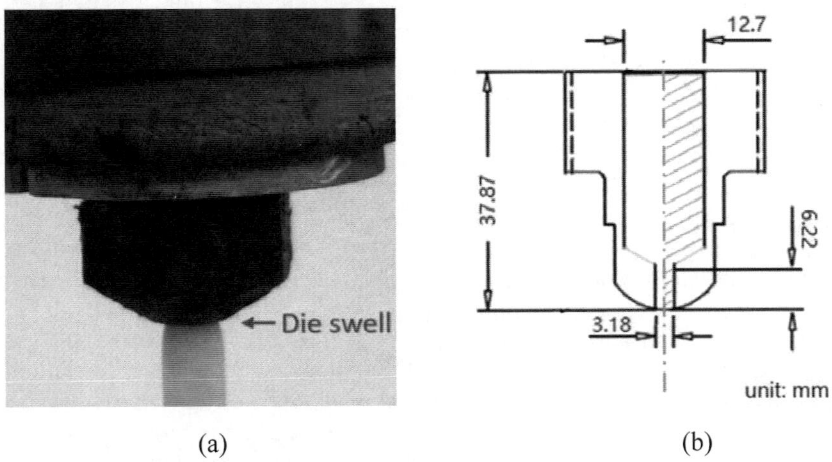

FIGURE 1. (a) Neat ABS exhibit obvious die swell as is extruded throught the Model-19 extruder nozzle. (b) Essential geometric dimensions of the nozzle.

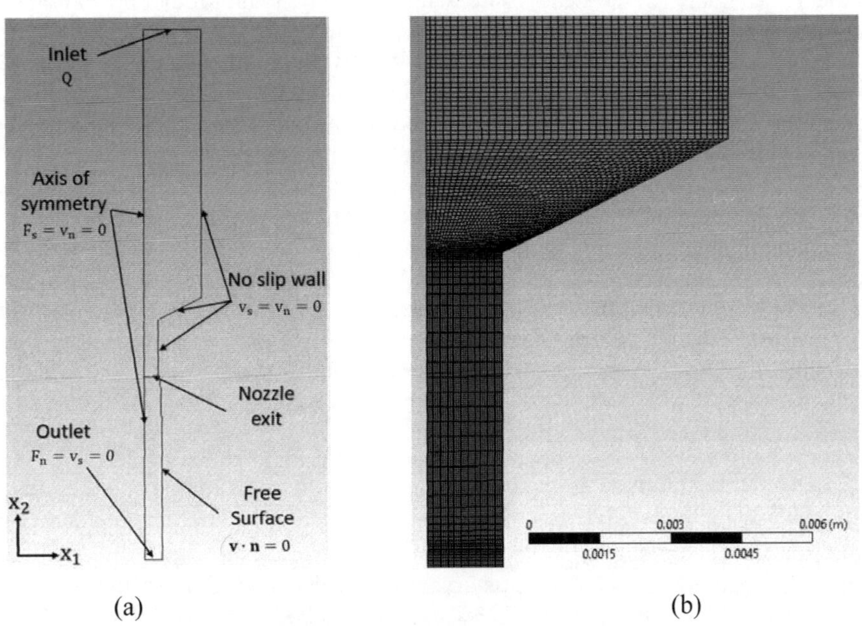

FIGURE 2. (a) Boundary conditions for the flow domain (b) Mesh of nozzle convergent zone as well as the nozzle exit.

RHEOLOGY MODELS

Carbon fiber filled ABS is commonly used in large scale AM polymer deposition, however, comparable die swell studies using ABS have yet been found. Therefore, we have chosen to use LLDPE properties from Ganvir, et al. [3] in this study to assess the validity of our viscoelastic flow modeling approach given that data exist in the literature for this comparison. To include shear thinning rheology behavior in our flow simulations, the Generalized Newtonian Fluid (GNF) models are applied (Power law, Bird-Carreau law, and Cross law, see their constitutive equations in [8]). We also consider a differential-type viscoelastic model (Phan-Thien-Tanner model [2]) which is employed to capture elastic effects of the polymer melt. The Simplified Viscoelastic Model (SVM) [6] is implemented as well in order to find a computationally more efficient approach for die swell prediction.

EXTRUDATE SWELL

The melt flow simulation was primarily performed for the large scale deposition nozzle flow using the PTT viscoelastic fluid model. The ANSYS-Polyflow PTT model was employed using the LLDPE parameters appearing in Ganvir, et al. [3]. Before simulating the FFF nozzle with the PTT fluid, we first considered the 2D axisymmetric capillary flow domain shown in Figure 3(a) taken from Ganvir, et al. [3] to validate our PTT flow simulations. Note that the lower 14.2 mm of the flow model is used to capture the die swell. In this model, the die length to diameter ratio (L/D) is 19.2 and the volumetric flow rate is about 10 mm^3/s. It can be seen that the predicted free surface shown in Figure 3(b) has a good agreement with the published literature. Once our modeling approach was validated with literature data above, we applied the PTT rheology model to the FFF nozzle flow domain shown in Figure 2(a) to obtain the velocity field within the melt flow and the predicted die swell. The flow rate for the FFF nozzle flow is also kept at 10 mm^3/s. Besides the PTT model, a several GNF rheology models are also employed to evaluate the effect of shear thinning behavior, exclusively. The parameter swell ratio is used to assess the amount of die swell predicted in the polymer extrudate for each fluid rheology model. The swell ratio used here is the ratio of the difference between the steady state swell flow diameter and the nozzle exit diameter to the nozzle exit diameter.

Figure 4(a) shows the comparison among predicted die swell profiles. Heller, et al. [7] calculated a 13% die swell when modeling a Newtonian fluid exiting a small scale FFF nozzle. Our Newtonian fluid die swell results that appear in Figure 4(a) yield a similar value as shown. Die swell predictions using GNF models generally yield lower die swell ratios as compared with Newtonian fluid results. In detail, the Power Law [8] gives a 1.8% swell ratio, Bird-Carreau Law and Cross Law [8] give similar results, which are about 8.8%. Furthermore, the PTT model yields a higher die swell than the GNF models due to the included consideration of the elastic effect. Note, the PTT model's 22.4% swell ratio is computed based on the swelled diameter at the 20 mm position. Moreover, the swell profile of the PTT model does not reach the steady state within the first 20 mm of the free extrudate while the results of Newtonian Fluid model and GNF models reach steady state at roughly the first 2 mm. The above analysis indicates that the elastic effects of the viscoelastic fluid flow significantly enhance the extrudate swell phenomenon even with considering the decreasing effect of the shear thinning behavior.

The SVM is also employed to our nozzle flow domain. To find the optimum λ and Ψ to predict die swell, various combinations of λ and Ψ values were simulated. The parameter fitting procedure is referenced from section 6.5.4 of Polymat User Guide [9] and the best-fitted prediction is shown in Figure 4-(b). Although the simplified-model-predicted swell ratio at the 20 mm matches the PTT result, the development of the predicted swell diameter does not show a good agreement with that of the PTT model. However, the computational costs shown in TABLE (1) indicates a significant drop by implementing the simplified model instead of the PTT model. Additionally, the modeling experiences also inform us that the simplified model requires far less efforts to find a convergent iteration scheme than that of the PTT model, which inherently consists of highly non-linear terms in the constitutive equations (c.f. [2]). Finally, the flow kinematics solved using the simplified model (λ_m=0.52, Ψ=0.3) will be used in further study of fiber orientation.

TABLE (1). Computational Cost of using Different Rheology Models

Model Name	PTT Model	Simplified Viscoelastic Modele
CPU time for a typical simulation run (sec.)	~95,000	~300

Note: The computer used for this simulation is assembled with the processor: Intel(R) Xeon(R) CPU E5-2640 v4 @ 2.40 GHz 2.40 GHz (2 processors) and the installed memory (RAM): 256 GB.

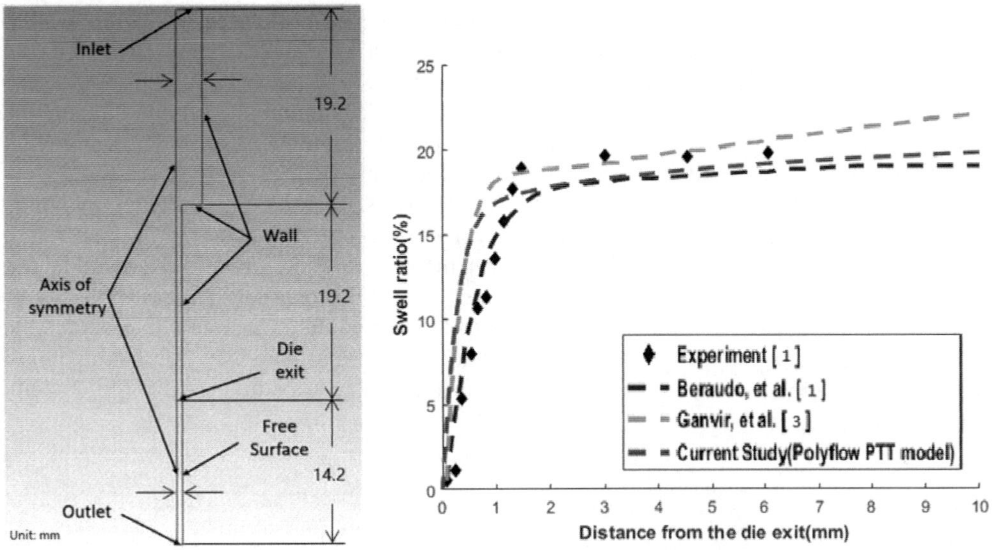

(a) (b)

FIGURE 3. (a) Representation of the capillary die used in the validation study (b) Extrudate swell predictions for the die flow domain.

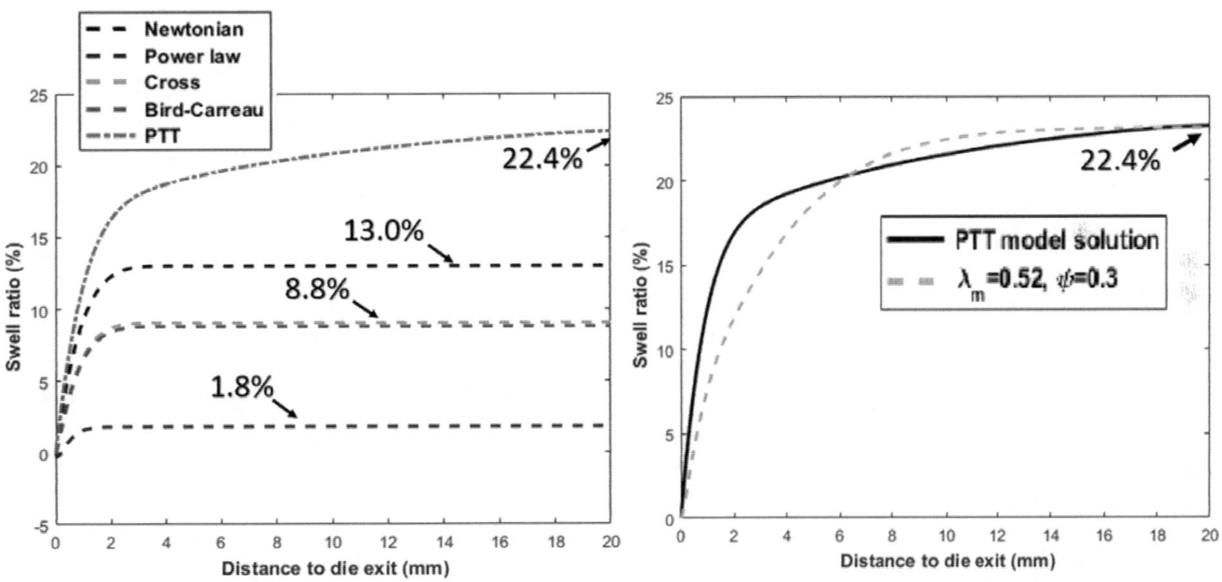

(a) (b)

FIGURE 4. Extrudate swell predictions of the PTT model, NF model, and GNF models.

FIBER ORIENTATION

The flow-induced fiber orientation pattern is determined using the decoupled fiber orientation tensor formulation [7], which consists of the Advani-Tucker [10] fiber orientation tensors ($\boldsymbol{A}_{ij}$) and the Folgar-Tucker isotropic rotary diffusion [11]. Additionally, the Orthotropic Closure (ORT) [12] is used to compute the 4th order fiber orientation tensor from $\boldsymbol{A}_{ij}$. Specifically, one diagonal component of $\boldsymbol{A}$ having a value of one represents full alignment in a given direction (see $\boldsymbol{A}_{ij}^2$ in Figure 5 shows a quasi-full alignment along x_2 aixs), and values of 1/3 on all diagonal components represent the case of a uniformly random orientations (see $\boldsymbol{A}_{ij}^1$ in Figure 5). Recall, in Figure 2(a), we define the extrusion direction of the melt is along x_2 direction, thus, we plot the A_{22} component of the fiber orientation tensor in Figure 5 since it indicates how well the fibers align along the principal extrusion direction. Note that the solution shown in Figure 5 is computed based on the flow kinematics solved using the PTT rheology model. In our solution, we assume the initial fiber orientation state at the nozzle entrance is uniformly random, and we note that other initial conditions on $\boldsymbol{A}$ (at the inlet) will likely result in different fiber orientation solutions within the nozzle geometry. Additionally, other model parameters in the solution are set as they were in Heller, et al. [7].

A relative comparison is given in Figure 6, including the steady state values of the A_{22} component computed using the Newtonian fluid model, GNF models, the simplified model (λ_m=0.52, Ψ=0.3), and the PTT model. Among the comparison, it is shown that the PTT model kinematics yields the lowest steady state principal fiber orientation alignment, on average. The variation of the principal fiber orientation decrease initially and recovers gradually as moving from the boundary toward to the center of the nozzle. In particular, the simplified model yields similar prediction of A_{22} values at the shear-dominated boundary region, but its results varies significantly from the those of the PTT model in the inner region of the flow domain. Recall, the relaxation time of the simplified model is expressed by a single value, λ_m.

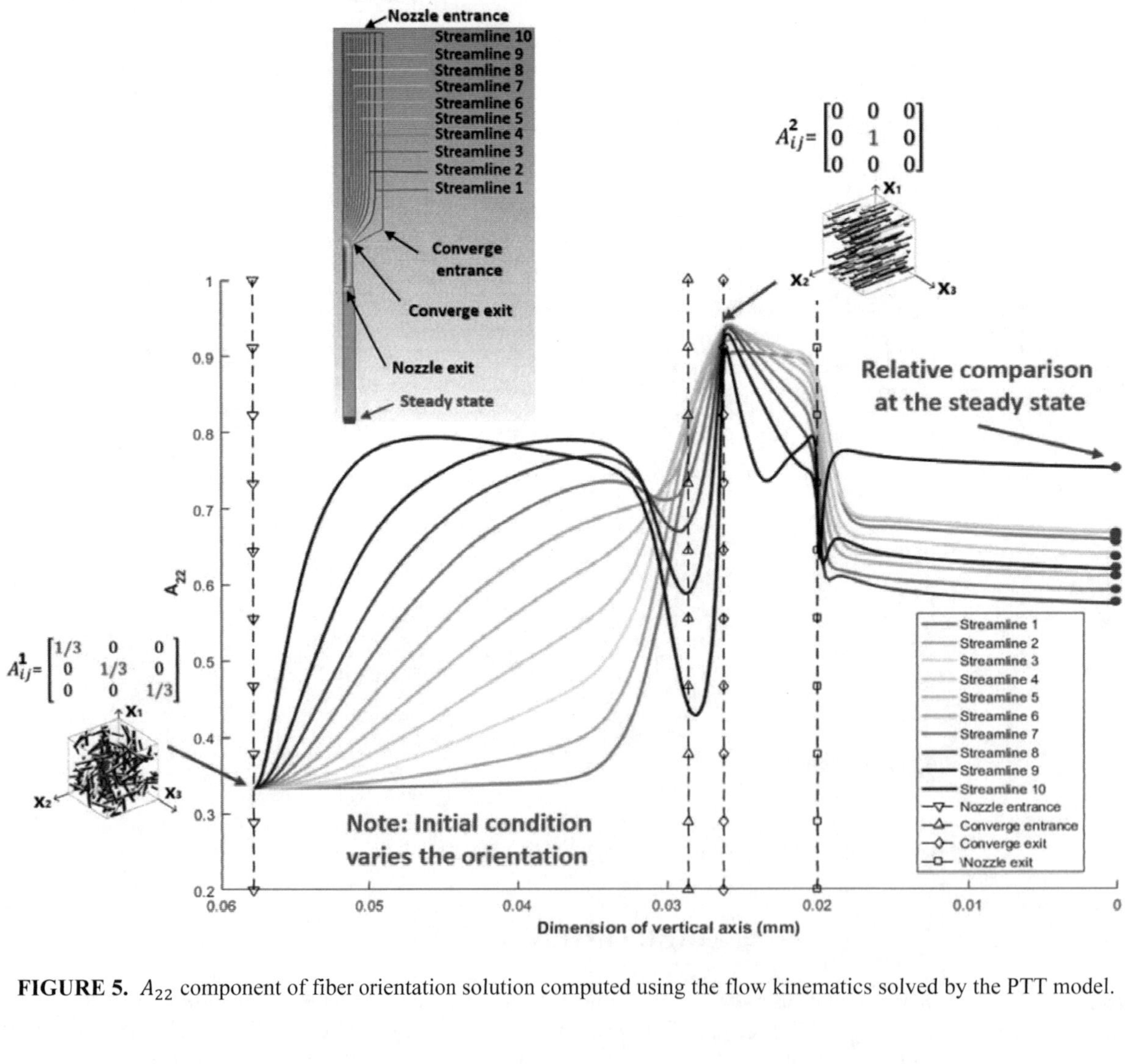

FIGURE 5. A_{22} component of fiber orientation solution computed using the flow kinematics solved by the PTT model.

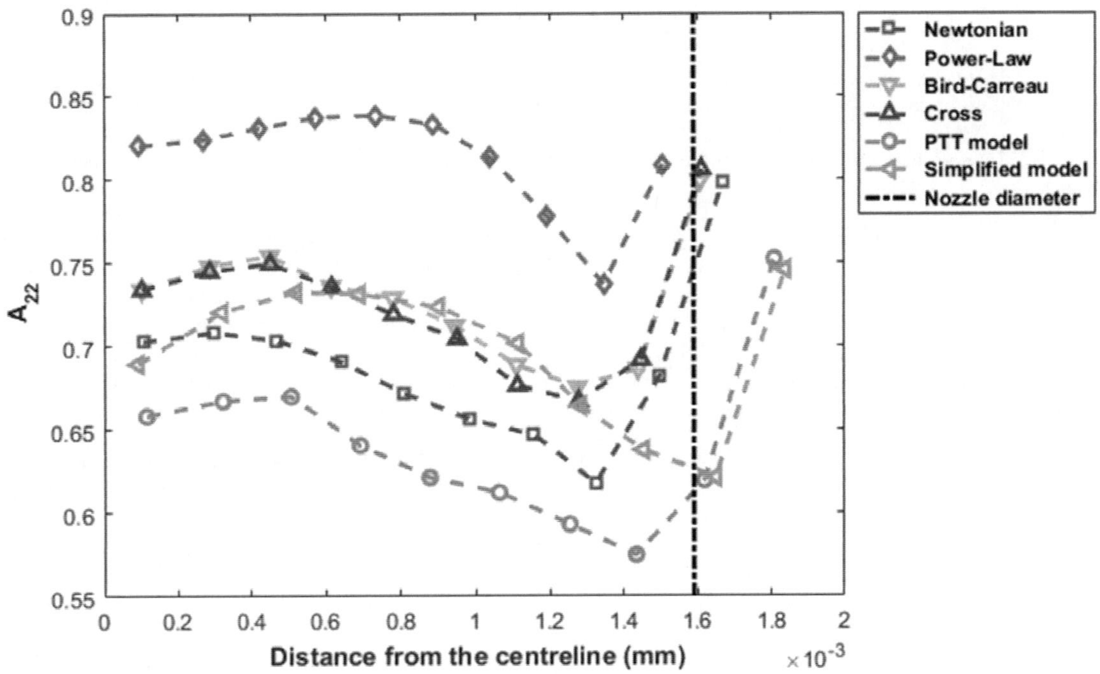

FIGURE 6. Comparison between the steady state A_{22} component solved by different flow kinematic results.

CONCLUSION

A few conclusions have been reached based on the simulations presented in this paper. The shear thinning behavior of polymer melts reduces die swell effect but the viscoelastic rheology of polymer melts greatly enhances the die swell. The Simplified Viscoelastic Model (SVM) [6] can be used to predict die swell for the FFF application but empirical adjustment on parameters of relaxation time and weighting factor is required.

Based on the fiber orientation prediction of the decoupled model, it is observed that the highest principal alignment state occurs at the streamline closest to the boundary, and shows a concave down variation trend as moving from boundary to the centerline. Furthermore, the PTT model yields lower principal fiber alignment results than results computed by other investigated rheology models (i.e. Newtonian Fluid model, GNF models). Specifically, the steady state prediction of the Simplified Viscoelastic Model (λ_m=0.52, Ψ=0.3) shows a good agreement with that of the PTT model at the shear-dominated boundary region.

FUTURE WORK

Future work will include simulations with materials and flow rates that are typical of large scale polymer deposition additive manufacturing. This includes ABS and carbon fiber filled ABS materials, and flow rates in the 10 to 15 lbs per hour range. Also, a shear rate dependent expression of the relaxation time spectrum is needed when applying the Simplified Viscoelastic Model to predict die swell. Finally, the initial assumption of the fiber orientation prediction at the nozzle entrance should take swirling effects from the screw into account and a coupled fiber orientation and flow velocity analysis is also needed to more accurately account for influence of the suspended fibers on melt viscosity.

ACKNOWLEDGMENTS

The authors would like to thank the Oak Ridge National Lab (RAMP-UP 40001455134) as well as Baylor University for funding this research project. Also, we would like to thank the Strangpresse corporation for donating their Model-19 extruder to our study.

REFERENCES

1. Béraudo, C., Fortin, A., Coupez, T., Demay, Y., Vergnes, B., Agassant, J.F., A finite element method for computing the flow of multi-mode viscoelastic fluids: comparison with experiments. Journal of non-newtonian fluid mechanics. 1998 Feb 28;75(1):1-23.
2. Thien, N.P., Tanner, R.I., A new constitutive equation derived from network theory. Journal of Non-Newtonian Fluid Mechanics. 1977 Jul 1;2(4):353-65.
3. Ganvir, V., Lele, A., Thaokar, R., Gautham, B.P., Prediction of extrudate swell in polymer melt extrusion using an Arbitrary Lagrangian Eulerian (ALE) based finite element method. Journal of Non-Newtonian Fluid Mechanics. 2009 Jan 31;156(1):21-8.
4. Limtrakarn W, Pratumwal Y, Krunate J, Prahsarn C, Phompan W, Sooksomsong T, Klinsukhon W. Circular Die Swell Evaluation of LDPE Using Simplified Viscoelastic Model. King Mongkut's University of Technology North Bangkok International Journal of Applied Science and Technology. 2013 Dec 24;6(3):59-68.
5. Clemeur N, Rutgers RP, Debbaut B. Numerical simulation of abrupt contraction flows using the Double Convected Pom–Pom model. Journal of non-Newtonian fluid mechanics. 2004 Feb 20;117(2):193-209.
6. ANSYS POLYFLOW User's Guide, Ansys Inc., 2013.
7. Heller, B., Smith, D.E., Jack, D.A., The Effects of Extrudate Swell, Nozzle Shape, and the Nozzle Convergence Zone on Fiber Orientation in Fused Deposition Modeling Nozzle Flow. InAmerican Society of Composites-30th Technical Conference 2015 Nov 2.
8. Tadmor, Z., Gogos, C.G., Principles of polymer processing. John Wiley & Sons; 2006.
9. ANSYS POLYMAT User's Guide, Ansys Inc., 2013.
10. Advani S.G., Tucker, C.L., The use of tensors to describe and predict fiber orientation in short fiber composites. Journal of rheology. 1987 Nov;31(8):751-84.
11. Phelps, J.H., Tucker, C.L., An anisotropic rotary diffusion model for fiber orientation in short-and long-fiber thermoplastics. Journal of Non-Newtonian Fluid Mechanics. 2009 Feb 28;156(3):165-76.
12. Verweyst B.E., Tucker C.L., Fiber suspensions in complex geometries: flow/orientation coupling. The Canadian Journal of Chemical Engineering. 2002 Dec 1;80(6):1093-1106.

pps proceeding
Symposium: S19 - Additive technology
Oral Presentation

3D printing with directionally thermally conductive composite materials by fused filament fabrication

Mulholland Tom [(1)]*, Rudolph Natalie [(1)]

[(1)] University of Wisconsin - Madison - Wisconsin - USA

Fused filament fabrication (FFF) is an additive manufacturing (AM) technique that is often considered a rapid prototyping technology. However, the ability to use composite materials, a distinct advantage of FFF over other AM methods, will help reorient the technique as a true manufacturing technology. Thermally conductive thermoplastic composites have been successful and competitive in traditional plastics manufacturing techniques, such as extrusion and injection molding, and the intrinsic geometric freedom of AM adds to their applicability. Composites of polyamide 6 (PA6) and copper or carbon fiber fillers are examined for their mechanical , thermal, and rheological properties. The rheology of the melts provides insights into the printability of the materials, although the filler geometry and content can contribute to manufacturing issues such as clogging. The filler orientation is investigated for its effect on the anisotropic thermal diffusivity, using the laser flash analysis technique. The anisotropy can be advantageous in FFF, since the thermal diffusivity and conductivity can be several times higher in the extrusion direction. However, the lower diffusivity perpendicular to extrusion is also a disadvantage for certain features, such as thin walls. These issues are explored with respect to the larger aim of the project, which is the manufacturing of air-cooled heat exchangers. The filler orientation is affected by processing and print parameters, which is examined using micro computed tomography scans.

Silicone formulations for additive manufacturing

Crouillere Marie* (1)

(1) IMP@INSA UMR 5223 - Lyon - France

Marie CROUILLERE, Claire BARRES, François GANACHAUD "Ingénierie des matériaux polymères", UMR 5223, INSA Lyon, France Key words: FDM, silicone, structure, rheology Additive manufacturing is a rapidly growing technical area within which a wide variety of methods have been developed. These new processes provide a higher degree of geometrical freedom when building specific objects for small series production or for custom-made use. The benefits against traditional processes are reductions of time, of cost and of material consumption, thanks notably to the suppression of specific equipment (e.g. moulds). Nowadays different types of materials are processed in additive manufacturing, like thermoplastic polymers (PLA, ABS, PA), metals or ceramics. However crosslinkable elastomeric materials like silicones have hardly been considered although they are widely used in several industries. This study focuses on 3D printing of new silicone formulations, in a view to creating new structures of light densities [1][2]. Material extrusion (or fused deposition modelling, FDM) is used as an additive manufacturing process in which material is selectively dispensed through a nozzle. This method, commonly used with thermoplastic materials, requires a number of modifications and improvements with silicone formulations. In particular, the network, rheology and curing time have been studied to come up with a formulation that both fits perfectly with additive manufacturing and generates good mechanical properties of the finished product. [1] E. B. Duoss, T. Weisgraber, K. Hearon et al., "Three-Dimensional Printing of Elastomeric, Cellular Architectures with Negative Stiffness," Adv. Funct. Mater., vol. 24, no. 31, pp. 4905–4913, Aug, 2014. [2] A. Maiti, W. Small, J. Lewicki et al., "3D printed cellular solid outperforms traditional stochastic foam in long-term mechanical response," Nature, vol. 6, no. April, 2016.

Experimental setup for in situ visualization studies of laser sintering of polymer particles

Hejmady Prakhyat (1)*, Cardinaels Ruth M (1), van Breemen Lambert C. A. (1), Anderson Patrick D (1)

(1) Eindhoven University of Technology - Noord-Brabant - Netherlands

Products manufactured by Selective Laser Sintering suffer from low mechanical stability and reproducibility [1]. This is mainly caused by an incomplete sintering of the polymer particles, resulting in significant remaining porosity after sintering, as well as limited interlayer adhesion. Hence, more insight in the effects of the different processing conditions and material characteristics on the final product morphology is required. Therefore it is important to understand the sintering of the particles during the process and to utilize this knowledge in material processing. The sintering of two polymer particles has been studied in literature by hot stage microscopy [2], but this technique does not simulate the conditions in an SLS machine. Therefore a dedicated experimental setup, which incorporates the main features of a 3D SLS printing device and at the same time allows in-situ visualization of the sintering dynamics by means of optical microscopy and/or X-rays is developed. A visible light laser is used to achieve good spatial resolution and control over the supplied energy. Energy absorption of the polymers containing colored dye was quantified by means of UV-VIS spectroscopy. Initial thermal characterization by DSC and TGA enabled us to define the stable sintering region for the particles. Subsequently, experiments were conducted using pairs of polymer particles inside a temperature controlled chamber. The polymer particles were subjected to a known amount of laser energy with the ultimate objective to study the sintering dynamics, which is a crucial stage in SLS. The dynamics of the growth of the neck between both particles is compared to available models for sintering kinetics of molten polymers. References: [1] Zarringhalam, Hadi, et al. Materials Science and Engineering: A 435 (2006): 172-180. [2] Berretta et al. Journal of Material Science 51.10 (2016): 4778 - 4794.

Special Binder Systems for the Use with Metal Powders for Highly Filled Filaments for Fused Filament Fabrication

Gonzalez-Gutierrez Joamin (1)*, Kukla Christian (2), Arbeiter Florian (3), Holzer Clemens (1)

(1) Montanuniversitaet Leoben, Chair of Polymer Processing - Leoben, Styria - Austria, (2) Montanuniversitaet Leoben, Industrial Liaison - Leoben, Styria - Austria, (3) Montanuniversitaet Leoben, Chair of Polymer Testing and Characterization - Leoben, Styria - Austria

Fused filament fabrication (FFF) is an extrusion-based additive manufacturing process, which is very popular for the fabrication of polymeric parts with complex geometry. FFF can be used as an economical alternative for the production of metal parts, too, by using filaments with a volume content of metal powder greater than 50 vol%. The additive manufacturing process must be followed by a debinding step and a sintering step. The addition of metal particles to matrices usually used in FFF makes the filaments brittle and non-flexible; therefore the extrusion process during FFF is hindered. In order to overcome the brittleness a special polymeric binder system had to be prepared, consisting of flexible and stiff components. With this matrix and 55 vol% of three different metal powders – stainless steel, titanium and a magnetic alloy of NdFeB – filaments were produced and their tensile properties were tested. The printing trials, performed on a conventional FFF machine, proved that all of those materials were still printable even though their tensile properties were very different. The printed parts were debound with a solvent and after sintering metallic parts were obtained.

pps proceeding
Symposium: S19 - Additive technology
Poster Presentation

Assessing Selective Laser Melting in Extrusion Blow Molding

Gross Heinz [(1)], Vidal João Paulo Oliveira [(2)], Nobrega Joao Miguel [(3)]*

[(1)] Dr. Groß Kunststoff-Verfahrenstechnik - Roßdorf - Germany, [(2)] University of Minho - Guimarães - Portugal, [(3)] University of Minho - Guimarães - Germany

Nowadays, with the availability of powerful numerical codes processing tool designers are able to conceive, theoretically, high performance tools. However, in several situations, conventional machining methods limit significantly the geometries that can be manufactured, either due to technical or budget limitations. With the advent of additive manufacturing techniques, the boundaries of the machinable geometries are being pushed forward. In this work the authors present a case study, comprising the design of an blow molding extrusion die, that combines numerical modelling tools and a manufacturing approach comprising Selective Laser Melting (SLM), which aims the reduction of time required to change materials during production. The results obtained, in what concerns to the decrease of the time required to change the materials, clearly evidence the advantages of using numerical modeling tools to aid the design of processing equipment and the potential of employing additive manufacturing approaches, as SLM, to machine polymer processing equipment.

An Integrated Biodegradable 3D-Printed Cage and Nanofibrous Drug-Eluting Membrane for Treating Metaphyseal Comminuted Fractures

***Liu Shih-Jung* [1]*, *Chou Ying-Chao* [2]**

[1] Chang Gung University - Tao-Yuan - Taiwan, [2] Chang Gung Memorial Hospital - Tao-Yuan - Taiwan

A biodegradable polylactide (PLA) truncated pyramid cage was designed and manufactured using a fused deposition modeling 3D printer, while drug-eluting poly(D,L)-lactide-co-glycolide (PLGA) nanofibers were produced via electrospinning. The mechanical properties of printed bone cages were measured. The morphology of electrospun nanofibers was characterized by scanning electron microscopy. The in vitro and in vivo release characteristics of pharmaceuticals from the bone cages were investigated. A bacterial inhibition test was also performed to determine the relative activity of the released antibiotics. In addition, the efficacy of biodegradable cages as bone grafts for repairing bone defects was investigated by using the clinical outcomes of radiographic bone healing conditions and final leg length comparisons.

Formation of Spherical Micron-Sized Polyamide Particles for Additive Manufacturing via Liquid-Liquid Phase Separation

Maximilian A. Dechet, Stephanie Kloos, Wolfgang Peukert and Jochen Schmidt

Friedrich-Alexander-Universität Erlangen-Nürnberg, Institute of Particle Technology, Cauerstr. 4, D-91058 Erlangen, Germany

Abstract. Selective laser sintering (SLS) is an additive manufacturing (AM) process that yields excellent part qualities with good mechanical properties. The SLS process employs micron-sized polymer particles, which are selectively fused by a laser. While there seem to be hardly any boundaries regarding design, there are quite some restrictions concerning the variety of commercially available SLS materials. With a market share of roughly 90%, polyamide 12 (PA12) is currently the most widely used polymer material for SLS. In order to increase the understanding of the precipitation based manufacturing process for polyamide powders, production of polyamide 11 powders via precipitation in ethanol was investigated. The polymer particles need to be optimized regarding size, shape, flowability and packing characteristics. In this contribution particle manufacturing via liquid-liquid phase separation is discussed. Bulk polymer material of polyamide 11 is directly converted to spherical micron sized particles in a single unit operation. The produced particles are characterized regarding their size and morphology. Furthermore, also structural characteristics, crystallinity and flowability are analyzed and the product properties are correlated with process parameters. We discuss the potential of this process for the production of novel, easy-to-handle polymer particle systems for SLS.

Keywords: polymer powder, precipitation, additive manufacturing, selective laser sintering
PACS: 81.30.Mh, 64.70.dg. 81.05.Lg

INTRODUCTION

Additive manufacturing (AM) technologies, commonly called 3D-printing, offer new and innovative manufacturing possibilities. The variety of technologies and the fields of application of parts made thereof is steadily increasing. Especially for polymers, a wide variety of technologies with comparatively low investment costs exist. Among them are fused deposition modeling, stereolithography or binder jetting, rendering a large selection of solid to flexible parts readily available, particularly prototype devices and decorative parts. Only few polymer-based AM technologies are feasible, if functional resistant parts, e.g. parts under static or dynamic stress, are required. Such qualities are almost exclusively obtainable from powder-based technologies like selective laser sintering (SLS) [1,2]. In the SLS process, a layer of powder is applied onto a movable part bed and a locally confined energy input via laser is melting the powder exactly where the part contours are. The part bed is retracted by the height of one powder layer and a new powder layer is delivered, followed by another laser exposure phase. In this manner, a part is built layer by layer from a powder bed [3]. To ensure high part qualities, the powders must meet several properties like high flowability, narrow size distribution and semi-crystallinity with a sufficient 'thermal sintering window' (c.f. temperature difference between melting and crystallization) [4]. More than 90% of the currently used SLS powder material is based on polyamides, especially PA12, with only two commercial producers (Arkema and Evonik Industries) of these powders. One of them is using a direct emulsion polymerization route to obtain particles with high sphericity and narrow size distribution [5]. The other one applies a process, in which bulk PA is dissolved under pressure and high temperature. Upon cooling, the PA precipitates and crystallizes in the form of potato-shaped particles with narrow size distribution [6]. It is very likely, that a thermally-induced liquid-liquid phase separation precedes the precipitation, causing the morphology of semi-coalesced crystalline polyamide particles [7,8]. To increase the understanding of this process and investigate its feasibility for the production of SLS-grade polymer powders, we used polyamide 11 bulk material and manufactured spherical micron-sized particles via thermally-induced precipitation from ethanol.

METHODS

PA11 Precipitation

Dissolution and subsequent precipitation of PA11 bulk material (Rilsan, Arkema) was performed in autoclaves equipped with PTFE-liners. The feed material and the ethanol (99.5%, denatured with 1% MEK (VWR)) were weighed directly in the PTFE-liners. Size of the batches was always 40 g and the PA11 concentration was varied between 3 wt.% and 20 wt.%. In some experiments flow aid is added to the system. In this case, fumed silica (Aerosil, Evonik) was added to the PTFE-liners prior to starting the experiment. All chemicals were used without further purification. To ensure proper mixing during the process, a magnetic stirring bar was added to the PTFE-liners before the liners were closed and fixed in the autoclaves. The autoclaves were put in a metal heating block located on hot plates (MR-Hei-Tec, Heidolph). The system was heated under constant stirring to 210 °C, while the temperature was measured via a thermocouple (type-K) located in the heating block. After the temperature was reached, stirring and heating were stopped and the autoclaves were allowed to cool down. The autoclaves were opened at approximately 50 °C and the product suspension was transferred to a suction filtration device for particle separation. Subsequently, the particles were vacuum dried at 40°C and 500 mbar. After these processing steps, white and odorless fine powders with apparently good flowability were obtained.

Scanning Electron Microscopy

Analysis of shape and morphology of the polymer particles was performed using a Gemini Ultra 55 (Carl Zeiss AG). The electron microscope was operated using the SE2 detector and an acceleration voltage of 1.0 kV. SEM samples were prepared by diluting a small fraction of the product particle suspension in pure ethanol and applying it on a silicon wafer.

X-Ray Diffraction

Structural characteristics of the PA11 particles were analyzed by means of powder X-ray diffraction (XRD) using a Bruker AXS D8 Advance diffractometer. Cu K_α radiation (λ = 154.06 pm) was used and the device was equipped with a VANTEC-1 detector and a Ni filter. The X-ray diffractograms were recorded in Bragg-Brentano geometry in the range $20° \leq 2\theta \leq 80°$. The step width was 0.014° and the accumulation time was 1 s for each step.

Particle Sizing

Static laser diffraction particle sizing was applied to measure particle size distributions using a Mastersizer 2000 / Hydro 2000S (Malvern). A small amount of dried PA11 powder was suspended in deionized water. Small quantities of sodium dodecyl sulphate solution (SDS, 98% (Merck)) and ethanol (96%, denatured with 1% MEK (VWR)) have been added to assure dispersion stability. Prior to measurements, ultrasonication of the dispersions has been performed. Measurements were conducted under constant stirring and ultrasonication.

Powder Flowability

The flowability of the PA11 powders was investigated with a modified Zimmermann tensile strength tester [9,10]. The measurement is based on adhesion forces between adjacent powder layers in a nearly uncompacted powder bed. The applied load to the bed was 153 Pa in all experiments. Generally, low tensile strength values indicate high flowability.

RESULTS

Particle Size and Morphology

The particle size distribution measurements, depicted in Figure 1, show a clear trend towards larger particle sizes with increasing PA11 concentration. While the mean particle size for the sample with 3 wt.% polymer is in the range of 38 µm, it increases to 55 µm for 8 wt.%, 60 µm for 10 wt.% and finally 123 µm for 20 wt.%. The distribution width, or span, given as $((x_{90,3}-x_{10,3})/x_{50,3})$, changes from 1.04 for 3 wt.%, to 1.14 for 8 wt.% and 1.13 for 10 wt.%. For the experiment with 20 wt.% PA11 the span is 1.3, i.e. a somewhat broader particle size distribution is found. The size distributions of the manufactured powders up to 10 wt.% are very narrow, indicated by the low spans. In all of the powders a fine fraction (sub 10 µm particles) was observed, with the sample with 15 wt.% showing an exceedingly high amount of fine particles. These fine fractions have negative effects on the SLS-process and should be removed, e.g. by sieving or air-classifiers. The particle size distributions show, except for the 20 wt.% experiment, no significant coarse fraction (> 200 µm) and suitable PA11 powders in terms of mean particle size and distribution width were obtained being comparable to commercial SLS powders [4].

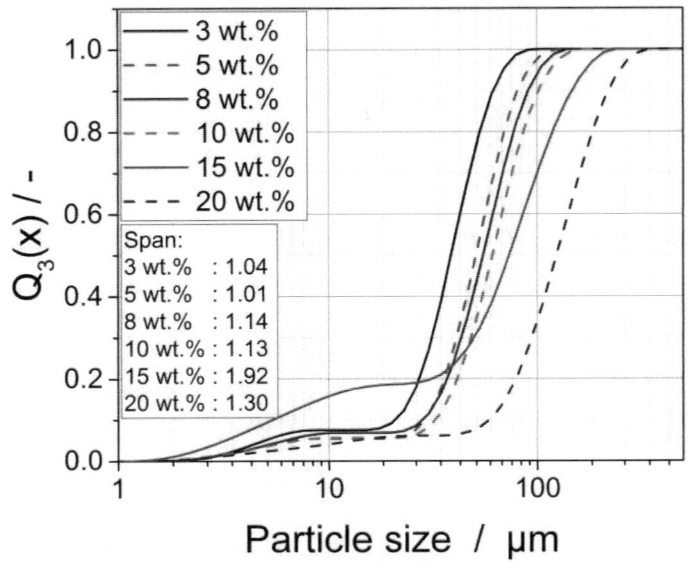

FIGURE 1. Particle size distributions of manufactured PA11 particles precipitated from batches with different polymer concentrations.

Next to the size distribution, the morphology of the particles governs the powder's SLS processablity. Spherical and dense particles are needed to obtain high packing densities and high bulk densities with low compressibility. Figure 2 depicts SEM images of produced PA11 particles at different polymer concentration. The spherical shape of the particles is clearly observable and the particle sizes correlate well with the ones measured via laser diffraction sizing. The surface of the particles is rough and rugged, which might be caused by a lamellar spherulitic crystallization starting from the center of the spherical particles with the ends of the lamellas directed at the particle surface [11,12].

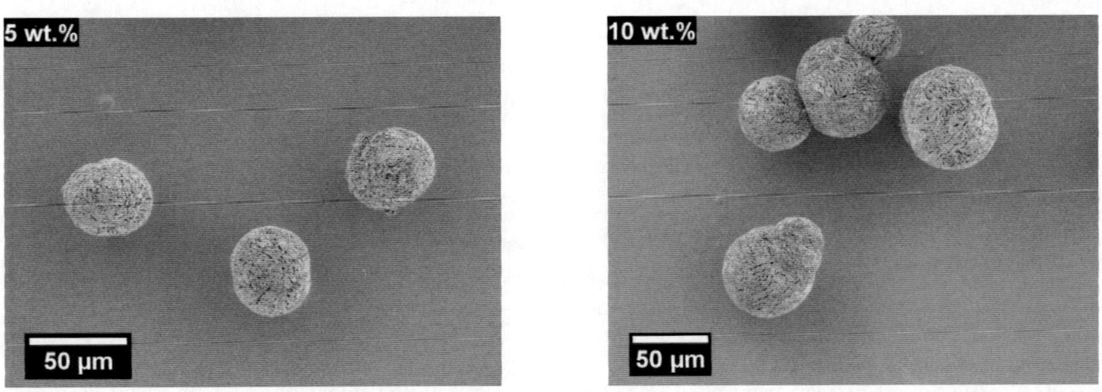

FIGURE 2. SEM images of precipitated PA11 particles. Left: Image of particles prepared from a 5 wt.% batch. Right: SEM image of PA11 particles from a 10 wt.% batch.

X-Ray Diffraction

X-ray diffractograms of the powders produced from 15 wt.% and 20 wt.% PA11 precipitated from ethanol show the typical PA11 diffraction pattern, visible in Figure 3: the (200) reflection at 20° and (010) + (210) reflection at 23.5° being characteristic for the α-form is observed [13]. The narrow reflection breadths indicate high crystallinity of the particles.

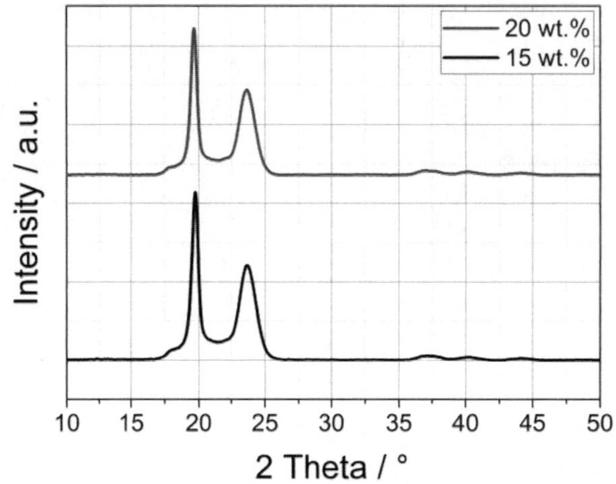

FIGURE 3. X-Ray diffractograms of manufactured PA11 powders showing the typical diffraction patterns of semi-crystalline PA11

Tensile Strength

The tensile strength measurements depicted in Figure 4 show, that the produced powders exhibit good flowability, but a trend towards higher tensile strength and worse flowability with increasing polymer concentration is observable. A general approach to enhance the flowability of powders is dry coating of the particles with e.g. fumed silica nanoparticles [14]. In a similar approach, hydrophobic silica was added in the precipitation process with the expectation that the fumed silica will adhere to the particles during precipitation and crystallization.

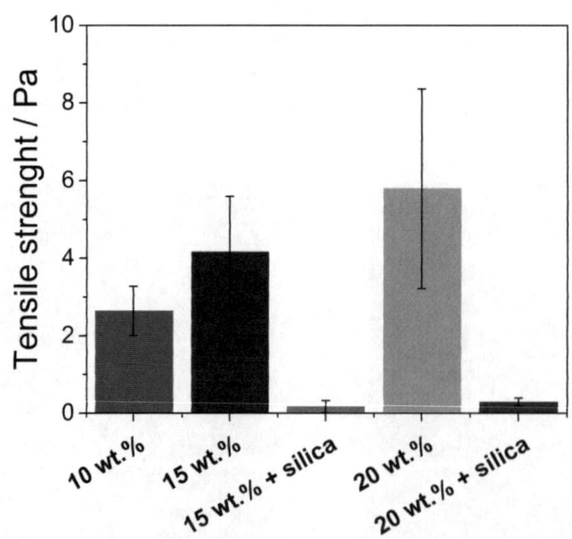

FIGURE 4. Tensile strength measurements of different PA11 powders with and without the addition of fumed silica to the precipitation process. Lower tensile strength values indicate higher flowability.

In Figure 5 it is clearly visible, that the particle surface is coated with fumed silica nanoparticles. The tensile strength measurements of the coated samples with values below 1 Pa indicate an extremely high flowability. Thus, the in-situ coating during precipitation was proved to be successful.

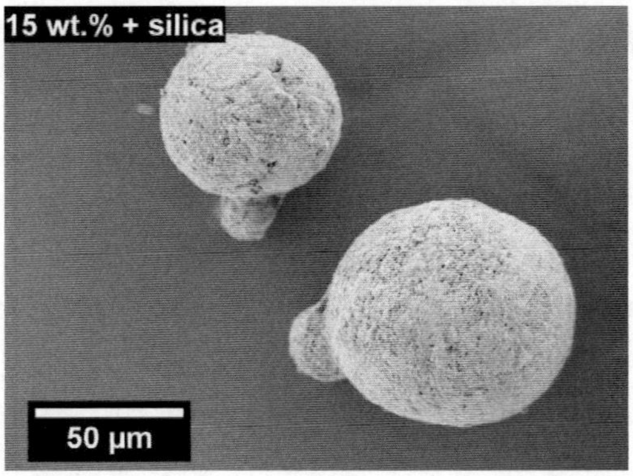

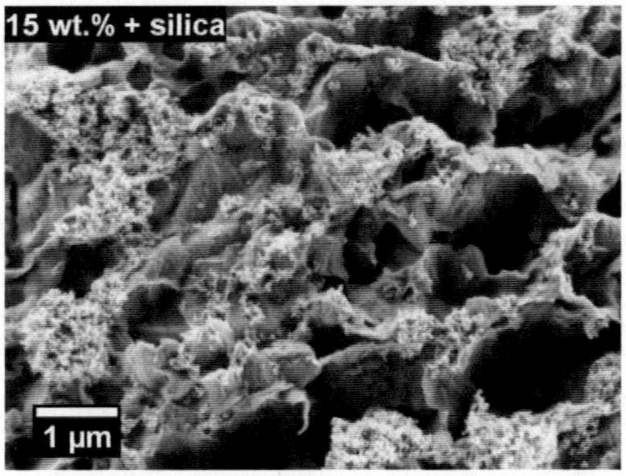

FIGURE 5. SEM images of particles precipitated from 15 wt.% PA11 with added fumed silica nanoparticles. Left: Overview image indicating the spherical shape of the particles. Right: Detailed image of the particle surface with high magnification, showing the adhesion of the fumed silica nanoparticles on the PA11 particle surface.

CONCLUSIONS

In this study we were able to produce PA11 powders via a thermally-induced precipitation process for the use in SLS processes. We investigated the influence of the polymer concentration on the powder properties and could show the advantages of an in-situ coating during the precipitation process. It is expected, that similar processes can be transferred to other polymer systems, opening new possibilities to broaden the field of available SLS materials.

ACKNOWLEDGMENTS

This study has been supported by Deutsche Forschungsgemeinschaft (DFG) in the framework of the collaborative research center CRC 814 'Additive Manufacturing' (projects A1 & A3). Financial support is gratefully acknowledged.

REFERENCES

1. T.T. Wohlers, *Wohlers Report 2016*, Fort Collins, Col.: Wohlers Associates, 2016.
2. I. Gibson, D. Rosen, B. Stucker, *Additive manufacturing technologies, 2. ed.*, New York, NY: Springer, 2015.
3. M. Schmid, *Selektives Lasersintern (SLS) mit Kunststoffen: Technologie, Prozesse und Werkstoffe*, Hanser, 2015.
4. M. Schmid, A. Amado, K. Wegener, *Polymer powders for selective laser sintering (SLS)*, Proc. 30th Int. Conf. Polym. Process. Soc. (2015).
5. K. Loyen, H. Senff, F.-X. Pauly, EP20050290448 (11 February 2009)
6. F.E. Baumann, N. Wilczok, DE19708946, (10 September 1998)
7. P. van de Witte, P.J. Dijkstra, J.W.A. van den Berg, J. Feijen, *J. Memb. Sci.* **117**, 1–31 (1996).
8. S.-J. Wang, J.-Y. Liu, L.-Q. Chu, H. Zou, S.-J. Zhang, C.-J. Wu, *J. Polym. Sci. Part B Polym. Phys.* **55**, 320–329 (2017).
9. A. Schweiger, I. Zimmermann, *Powder Technol.* **101**, 1–9 (1998).
10. K. Meyer, I. Zimmermann, *Powder Technol.* **139**, 40–54 (2004).
11. P. Schaaf, B. Lotz, *Polymer* **28**, 193–200 (1987).
12. L. Aerts, M. Kunz, H. Berghmans, R. Koningsveld, *Die Makromol. Chemie.* **194**, 2697–2712 (1993).
13. P. Ricou, E. Pinel, N. Juhasz, *Adv. X-Ray Anal.* **48**, 170–175 (2005).
14. C. Blümel, M. Sachs, T. Laumer, B. Winzer, J. Schmidt, M. Schmidt, W. Peukert, K.-E. Wirth, *Rapid Prototyp. J.* **21**, 697–704 (2015).

3D Printed Silicon Rubber Cellular Architectures: Rheological Control, Structures and Properties

Geng Chengzhen [(1)]*, Yu Fengmei [(1)], Lu Ai [(1)], Zhao Xiuli [(1)]

[(1)] Institute of Chemical Materials, China Academy of Engineering Physics - Sichuan - China

Creating lightweight materials with desired mechanical properties has long been an engineering pursuit, and the rise of three-dimensional (3D) printing techniques has made the fabrication of lightweight, porous materials with tailored architectures not only possible but also feasible. A variety of cellular materials based on plastics, metals and ceramics with engineered mechanical responses were created 1-3, but 3D printing of cellular rubbers is still a challenge. Here, a 3D direct ink writing technique was adopted to print a viscoelastic silicon rubber/nano silica system into orderly arranged sub-millimeter struts. By controlling the size, content and interface of nano silica, the particle-particle interaction and particle-polymer interaction could be varied, so the rheological behaviour of silicon rubber/nano silica composite could be tailored, and a viscoelastic system with a solid-like rheological behaviour ($G' > G''$) at low oscillation stress and a liquid-like rheological behaviour ($G'' > G'$) at high oscillation stress could be obtained. In this way, the system is printable and 3D-printed silicon cellular architectures could be prepared after thermal curing. By manipulating the specially designed architectures of the printed 3D patterns, materials with unique stress-strain behaviour could be obtained. Besides, the printed solids show lower compressive set compared to tradition silicon rubber foams with the same density, which is beneficial to their long-term services. This work not only paves a feasible way to control the cellular structure of elastomers in the sub-millimeter scale, but also suggests the ability to tailor mechanical properties of cellular materials via micro-patterned design, which may find potential applications in emerging fields or replacement of traditional foams.

Evaluation of surface roughness and geometrical characteristic of additive manufacturing inserts for precision injection moulding

A. Davoudinejad, A. Charalambis, D. B. Pedersen, G. Tosello

Department of Mechanical Engineering, Technical University of Denmark, Building 427A, Produktionstorvet, 2800 Kgs. Lyngby, Denmark

alidav@mek.dtu.dk

Abstract. In order to use additive manufacturing (AM) processes for producing complex feature geometries and benefiting from the possibility of manufacturing large series by using medium/high production techniques, the repeatability of the AM process and accuracy assessments is a critical aspect to be considered in the polymer insert printing and it's integration into the injection moulding process chains. This study investigates the AM tool inserts and injection moulded parts in terms of surface roughness and geometrical accuracy prior and after injection moulding. The vat photopolymerization method was used to produce additively manufactured tool inserts. The insert was designed in a way to involve several miniaturized features with different shapes in order to evaluate the process repeatability over a large variety of geometries, dimensions and aspect ratios.

INTRODUCTION

Different processes are applied for manufacturing microinjection moulding tool inserts depending on the number of parts needed. The tool fabrication for injection molding can be categorized in four main different manufacturing schemes which two use silicon substrates and two use other materials as the substrate (polymers or metals). Two schemes utilize subtractive micromachining (reactive ion etching, mechanical micromachining or laser micromachining) and two schemes involve additive micromachining (UV-lithography combined with electroforming) [1].

Additive manufacturing (AM) processes with fundamental abilities to fabricate three dimensional parts with high aspect ratio is thus becoming a standard method in part production and rapid tooling applications [2]. Polymer materials with low melting point or with liquid state are widely used in AM industry because of low weight, low cost and processing flexibility. Different AM methods have been employed for printing the polymers composite materials which some of the well-recognized techniques to fabricate polymer parts are as follow; Stereolithography (SL) [3], Selective laser sintering (SLS®) [4], Direct Laser Writing (DLW) [5], Vat photopolymerization [6] and etc. The selection of proper AM processes is depend on different factors such as the ease of manufacturing, speed, fabrication cost, part design, material, processes limitations, surface finish and dimensional accuracy of the part.

This study will focus on the AM process of the IM inserts to evaluate the injected parts with different printed inserts. The inserts were directly printed with vat photopolymerization process then the inserts were used for IM process and the repeatability of the parts were evaluated.

EXPERIMENTAL PROCEDURE

Mold inserts fabrication

The mould inserts were produced in the photopolymer by vat photopolymerization method in which liquid photopolymer in a vat is selectively cured by light activated polymerization layer by layer [7]. A high precision industrial 3D-printing system Digital Light Processing DLP-LED Stereolithography was used with a 50 µm pixel detail in X and Y directions and the precision of ± 50 µm in X, Y, and Z directions. Due to the rather larger build, envelop of the printing 96×54×150 mm (X/Y/Z) most of the inserts were printed in the same batch. The test part is designed in a way to evaluate the ability to reproduce specific geometries. It is a flat cuboid (19 x 19 x 3 mm3) with rounded corners and includes both micro and macro features: On top of the cuboid, there are two heart-shaped artefacts and two more cuboids (5.30x 2.67x1.17 mm3) with eight cylinders (0.83 mm diameter) on the top surface of each as shown in Fig. 1.

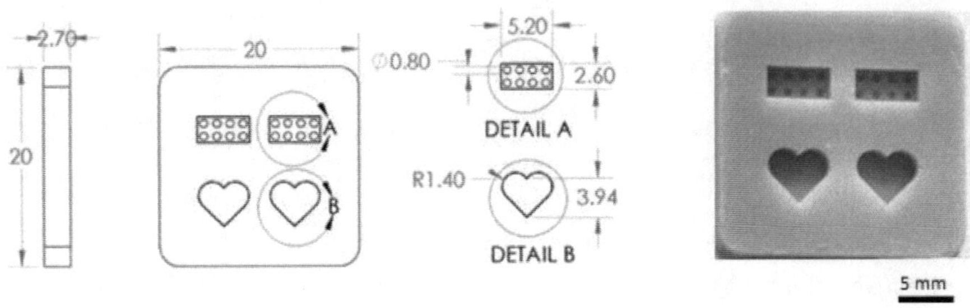

FIGURE 1. (a) Drawing of insert and details (dimensions in mm) (b) printed insert

Injection Moulding

The IM was carried out on an Arburg (370A 600-70), 60 tonne moulding machine. Injection moulding material with a higher melting point was used. Acrylonitrile butadiene styrene (ABS) Terluran GP-35 was chosen for its ease of processing and injection temperature of 220-260°C. The material was dried for 4 hours at 80°C in a HELIOS WINsystem Micro D dryer. The IM is carried out with standard settings for the ABS. The 200 bar packing pressure and 4 s packing time were kept constant in different conditions. In order to collect the exact manufactured parts in the initial of the moulding process for each insert only one cavity with AM insert was used for each moulding out of four cavities. Table 1 shows the IM parameters applied for each insert.

TABLE 1 MIM experimental parameters

Insert	T melt (°C)	Injection speed (mm/s)	T mold (°C)
A	220	80	25
B	260	40	50
C	220	80	50
D	220	40	25

Measurements and Analysis

The inserts were measured prior to the IM experiments in selected positions within the focus areas are systematically considered, to carry out similar measurements on all inserts as show in if Figure 2. The measurement results of the heart shape is presented in this paper. The inserts measurements were conducted with infinite focus Alicona (3D optical microscope). The areal roughness (Sa) measurements were conducted under 100x magnification with exposure time about 1-2 ms. Roughness of the insert cavity surfaces are measured at specific points. Each heart cavity is measured at three locations. Regarding the IM parts, similar measurements were carried out on the five initial injected parts to find out the repeatability of the process without considering the wear of the inserts.

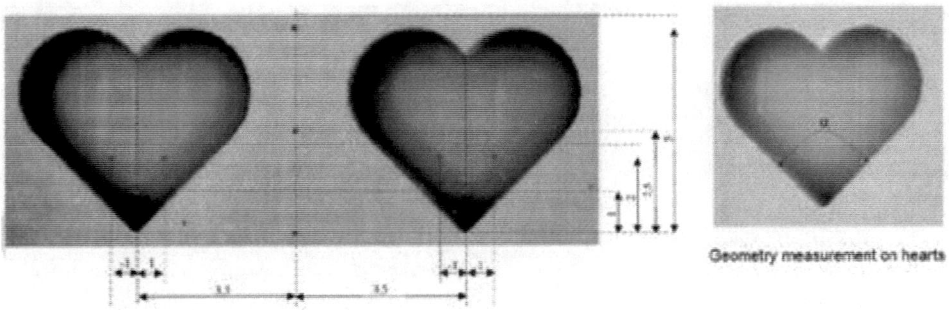

FIGURE 2. Diameters of the brick knobs (Indicated with withe circles) and roughness measurements zone of bricks (top) and heart (bottom, left, on outer surface between and inside holes); angle measurement on hearts (bottom, right).

RESULTS

The replication of the injection moulding parts were continued until the AM inserts were broken. The inserts tool life were different, the cracks were appeared at various area of the inserts. The inserts after number of shots started with a crack and then the crack propagated and resulted in the total breakage. Most of the runs reached at least 30 cycles. Figure 3 shows insert crack and final damage for one of the insert with lowest number of shots. The crack resulted in the damage of the top left corner of the left brick and the experiment was stopped as total failure prevents further use of the insert.

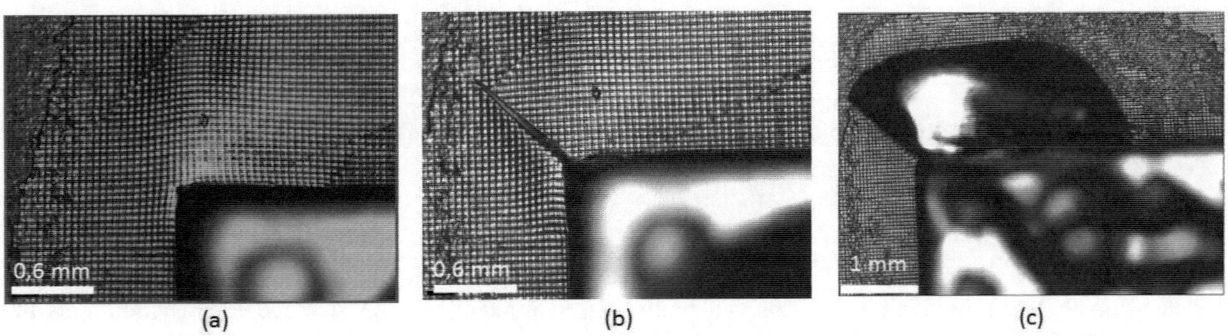

FIGURE 3 Crack propagation and final damage (a) new insert (b) after 30 shots (c) after 38 shots.

Figure 4 presents the surface roughness results obtained at the bottom of hearts on the inserts (Figure 2) prior to the IM experiments and the five initial injection moulded parts on the top of the hearts. The results reveals slight difference surface roughness at the left and right hearts in both 3D printed inserts and IM parts. The variation in terms of inserts might be due to the non-uniform quality of the 3D printer vat that replicate the insert with the same un-even quality. Concerning the IM parts, the final quality was affected by the surface of the insert and the IM parameters as well. The best surface was achieved with the first batch of the IM parts that shows the lowest surface roughness of the insert and the combination of the IM parameters. Figure 5 shows the height variation of the inserts and the IM parts. The height differences reveals a comparable variation between the inserts and the parts. There is a decreasing tendency in the IM parts with the same inserts, which is due to the shrinkage in the IM process, affected by material and IM parameters.

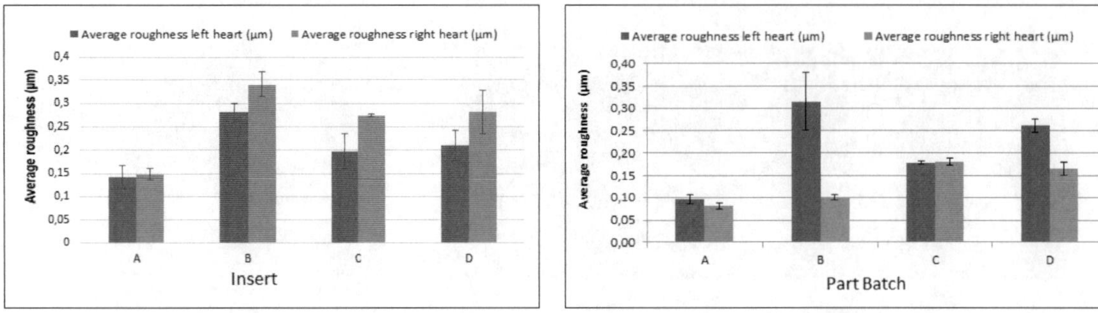

(a) (b)

FIGURE 4 Areal surface roughness in the hearts (a) 3D printed inserts (b) IM parts

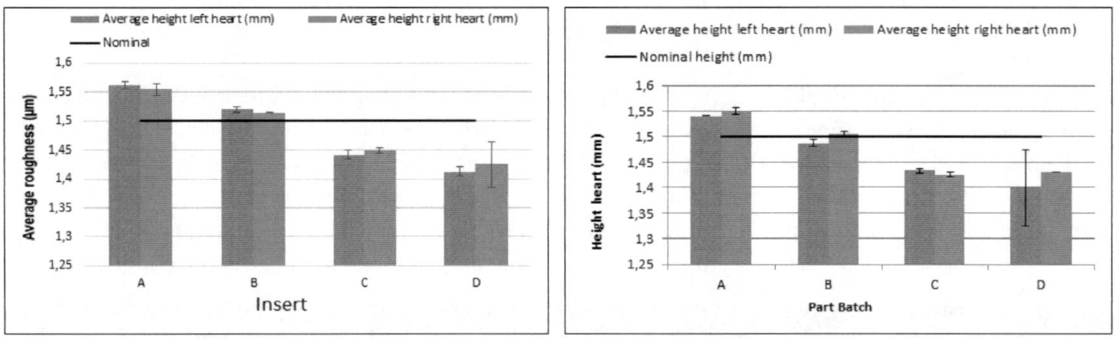

(a) (b)

FIGURE 5 the height variation of the hearts (a) 3D printed inserts (b) IM parts

CONCLUSION

This study investigated the capability of the soft tooling insert produced by AM method for injection moulding components. The inserts were manufactured with vat photopolymerization method. The defect was appeared as a crack on the inserts and then the crack propagated to break the insert. The surface roughness was slightly difference on each side of the inserts and IM parts. The variation of the parts affected by IM parameters as well. In terms of height, similar trend was observed with a little variation due to the shrinkage and IM parameters.

ACKNOWLEDGMENTS

The research leading to these results has received funding from the People Programme (Marie Curie Actions) of the European Union's Seventh Framework Programme (FP7/2007-2013) under REA grant agreement no. 609405 (COFUNDPostdocDTU).

REFERENCES

[1] G. Bissacco, H. N. Hansen, P. T. Tang, and J. Fugl, "Precision manufacturing methods of inserts for injection molding of microfluidic systems . Technical University of Denmark," pp. 1–7.

[2] A. AHMAD, S. DARMOUL, W. AMEEN, M. H. ABIDI, and A. M. AL-AHMARI, "Rapid Prototyping for Assembly Training and Validation," IFAC-PapersOnLine, vol. 48, no. 3, pp. 412–417, 2015.

[3] F. P. W. Melchels, J. Feijen, and D. W. Grijpma, "A review on stereolithography and its applications in

biomedical engineering," Biomaterials, vol. 31, no. 24, pp. 6121–6130, 2010.

[4] K. Subramanian, N. K. Vail, J. W. Barlow, and H. Marcus, "Selective laser sintering of alumina with polymer binders," Rapid Prototyp. J., vol. 1, no. 2, pp. 24–35, 1995.

[5] A. Selimis, V. Mironov, and M. Farsari, "Microelectronic Engineering Direct laser writing : Principles and materials for scaffold 3D printing," Microelectron. Eng., vol. 132, pp. 83–89, 2015.

[6] I. Cooperstein, M. Layani, and S. Magdassi, "3D printing of porous structures by UV-curable O/W emulsion for fabrication of conductive objects," J. Mater. Chem. C, vol. 3, no. 9, pp. 2040–2044, 2015.

[7] A. Davoudinejad, D. B. Pedersen, and G. Tosello, "Evaluation of polymer micro parts produced by additive manufacturing processes by using vat photopolymerization method," 2017, no. International Conference of the European Society for Precision Engineering and Nanotechnology.

Novel Fused Filament Fabrication Procedure and Printhead to Process Thermoplastic Elastomers

Johannes Gattinger[a)] and Stefan Fischer[b)]

Technical University of Munich, Institute of Medical and Polymer Engineering
Boltzmannstr. 15, 85748 Garching, Germany

[a)]Corresponding author: johannes.gattinger@tum.de
[b)]st.fischer@tum.de

Abstract. In the medical sector, flexible material parts have widespread applications in e.g. orthosis or surgical instruments. Another trend in the medical sector is to produce medical devices individually for every patient or surgeon. To date, only a few rapid manufacturing technologies are able to handle flexible materials. Especially thermoplasts with low hardness are very difficult to handle. Therefore, research is needed on printheads for additive manufacturing technologies that are able to fabricate highly flexible materials. In this study, a printhead for fused filament fabrication was developed that can handle thermoplastic elastomers with low hardness. A first prototype was constructed and tested using a flexible material with a hardness of 47 shore A.

Keywords: fused filament fabrication, thermoplastic elastomer, fused deposition modeling, 3D printing, hot end

INTRODUCTION

Elastic material properties are needed for several medical device applications. Beside seals, tubes or solid joints, the possibility to produce individualized medical devices is a major area of research. Additive Manufacturing (AM) is a technology, which fits perfectly in the field of producing such parts. However, the handling of flexible material leads to some problems due to its material properties. Current FFF printers are restricted in the range of printable thermoplastic elastomers (TPE). Commercially available printers are not able to process TPE with low hardness due to the established extrusion principle. Common soft TPE filaments have a hardness of down to 80 – 90 Shore A. TPE filaments with a hardness below that have not been reliably processable so far. Current FFF printers work on the principle of pushing a thermoplastic filament into the printhead where the filament is melted, and the polymer melt is deposited in lines for each layer onto the build plate using a fine nozzle [3, 5]. Soft TPE filaments cannot be pushed this way from the extruder in the printhead because it buckles or bends in the inliner or knots in excess space [4]. Besides FFF, there are a few other AM technologies that can handle flexible materials. Selective Laser Sintering (SLS) is a process, which is also capable of molding thermoplastic elastomers. During SLS, a thermoplastic powder is fused by a laser in each layer. The advantage of this technology is the fast laser positioning system – mainly implemented by mirrors – which results in increased print speed compared to FFF. Furthermore, the non-cured powder is commonly used as a support structure, which facilitates the manufacture of overhanging structures. Currently, materials with a hardness of 60 Shore A can be processed with SLS [1]. However, compared to FFF, SLS is a cost-intensive rapid manufacturing technology due to its laser module.

Another AM technology that can handle flexible material down to 30 Shore A is inkjet technology. During this process, a photopolymer is sprayed with many fine nozzles onto a build plate and is cured afterward by a radiation source in each layer [6]. Similar to SLS, inkjet technology is very cost-intensive. Another issue worth mentioning is biocompatibility. For medical devices, it is very important that the polymer is well cured so that there are no monomers inside after the building process. Thus, compared to thermoplastic polymers, the handling of photocuring polymers is more critical.

The ACEO® printer is based on this technology, and it is capable of building parts using silicone with a hardness of 30 Shore A [2]. Silicone is a commonly used material in the medical sector. Compared to FFF, the technology is much more expensive and so far, the manufacturer does not sell machines but offers a print service of parts.

Due to the issues mentioned, the purpose of this study was to develop a cost-effective FFF machine, which can handle thermoplastic elastomers with low hardness for medical applications.

NEW PRINTHEAD FEEDING CONCEPT

To counter the mentioned problems of the buckling and bending of soft TPE filaments, a novel FFF process, filament and printhead were developed. The concept of this FFF process is shown in Figure 1. In this process, a fiber reinforced TPE filament is used whereby the fiber roving is removed in the printhead and only the pure TPE is deposited to build the part. The reinforcement leads to a much stiffer filament, which prevents buckling or knotting while feeding the printhead. Furthermore, during the extraction of the roving, a pulling force can be applied on the roving and therefore, the filament can be pulled in the printhead, which prevents the filament from buckling and leads to a pressure on the remaining TPE in the printhead to squeeze out the melt through the nozzle. A standard extruder can be used to support the feeding of the printhead in addition to the pulling-in force on the fiber roving.

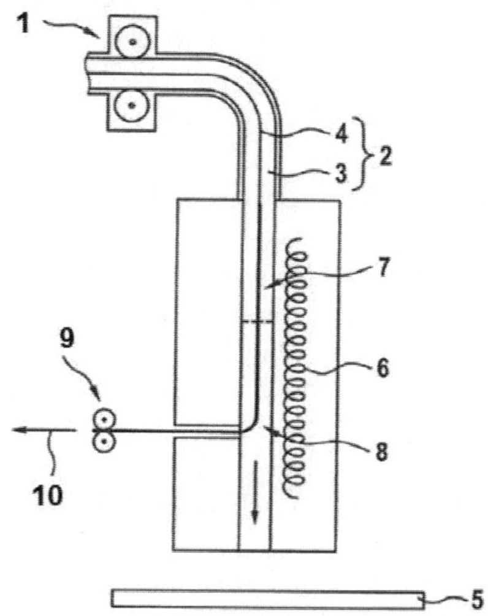

FIGURE 1. Novel printhead feeding concept. 1) Extruder, 2) Filament, 3) TPE print material, 4) Fiber roving, 5) Build plate, 6) Heater, 7) Unmolten filament, 8) TPE melt, 9) Pull-off drive, 10) Pull-off force

PROTOTYPE AND RESULTS

A prototype of such a printhead was developed and manufactured (Figure 2). Experiments to prove the feasibility of such a printing principle were executed with a fiber reinforced TPE filament, which was manufactured by extrusion with a cross-head pressure coating die. The mantle of the filament was a TPE-s, based on hydrogenated styrene block copolymers with a hardness of 47 Shore A (THERMOLAST® K TF5THT, KRAIBURG TPE GmbH & Co. KG, Waldkraiburg, Germany). The reinforcing core was a Twaron® 1000 aramid roving (1680 dtex, Teijin Aramid, Arnhem, The Netherlands). The resulting filament had a diameter of 1.7 mm.

In a first experiment, it could be shown that it was possible to push out TPE melt through the nozzle by simply pulling off the aramid roving (Figure 2B). A problem at this stage was the pulling-in of the filament. Figure 2C shows the knotting of the TPE at the inlet that occurred through detaching the fiber core and the mantle. Two main reasons for this pulling-in malfunction are presumed: low core-mantle adhesion and the absence of cooling of the inlet region. The combination of this TPE and aramid fiber led to very low fiber-matrix adhesion, which is beneficial (but not necessary) for the separation of the core and the mantle in the hot end but supports knotting of the filament on the inlet through a higher frictional force between the mantle and the inliner than the adhesion force between the core and mantle. This problem should be solved in future research through another fiber-matrix combination or suitable bonding agent. By not cooling the inlet region, it is expected that the filament is molten in the whole inliner up to the inlet. In the region of molten TPE, a force transmission from the fiber core to the mantle is not possible and therefore, proper pulling-in is excessively difficult or virtually impossible if the TPE becomes molten instantly at the inlet. This problem can easily be solved by cooling the inlet region.

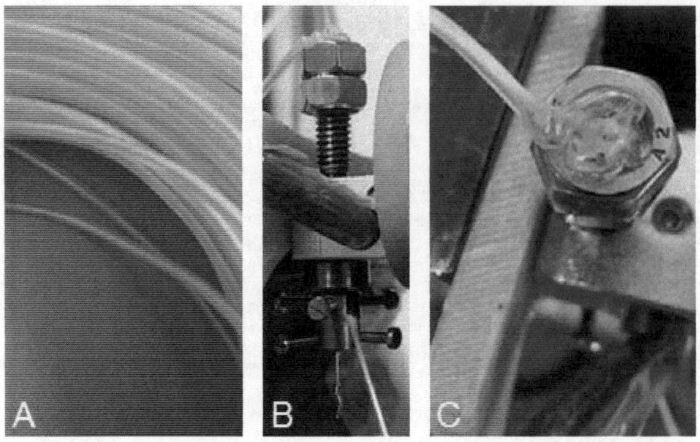

FIGURE 2. A: TPE filament with aramid core. B: Prototype hot end with pushed out TPE through a pull-off force on the aramid roving. C: Knotted TPE from the filament at the inlet.

CONCLUSION

The first results were promising and showed that such an FFF process is in principle possible, though, at this stage, the printer and filament have to be developed further. Future research must address the problem of pulling in the filament at the inlet. Therefore, the core-mantle adhesion of the filament has to be increased and the inlet region of the inliner has to be cooled, for example by a water-cooled cuff. Furthermore, the printhead can be extended with an extruder that helps push in the filament.

ACKNOWLEDGMENTS

The support of AM Medical/Kumos is gratefully acknowledged.

REFERENCES

1. 3D Systems. DuraForm TPU Elastomer Datasheet. 3D Systems; 2016.
2. Aceo. Aceo 3D Printer: Aceo; 2017 [18.09.2017]. Available from: https://www.aceo3d.com/.
3. Cooper K. Rapid prototyping technology: selection and application: CRC press; 2001.
4. Elkins K, Nordby H, Janak C, Gray RW, Helge J, Baird DG. Soft Elastomers for Fused Deposition Modeling: Virginia Polytechnic Institute and State University; 1997.
5. Gebhardt A, Hötter J-S. Additive Manufacturing - 3D Printing for Prototyping and Manufacturing: Carl Hanser Verlag; 2016.
6. Stratasys. Agilus 30 Polyjet Rubber-Like Material Datasheet. Stratasys; 2017.

Fundamental Investigations for a Combined Forming and Joining Process of Thermoplastic Composite Sheets and Additively Manufactured Structures

Christian Hopmann[a], Markus Hildebrandt[a], Ruggero Guisti[b], Max Ophuels[a], Arne Boettcher[a], Kai Fischer[a], Giovanni Lucchetta[b]

[a]Institute of Plastics Processing (IKV) at RWTH Aachen University
Seffenter Weg 201, D-52074 Aachen, Germany

[b]Department of Industrial Engineering at University of Padova
Via Venezia, 1, I-35131 Padova PD, Italy

Abstract. The production of lightweight composite products in large series is increasingly influenced by the demand for diversification and individualisation of the products. The demand for customised, functionalised components at low prices causes challenges for the manufacturers in the fields of component development, process flexibility and part costs. For functionalised thermoplastic composite (TPC) components, the common manufacturing route is forming and back injection moulding of TPC sheets. However, the production of individualised products as well as small, variable series in this manufacturing route is economically challenging due to the cost-intensive mould technology.
In order to increase process flexibility and extend the range of application for TPC components, an innovative combined forming and joining process is developed at the Institute of Plastics Processing (IKV) at RWTH Aachen University within the BMBF funded project LightFlex together with eleven partners from industry and science. The idea is to individually produce functionalised TPC components by forming and directly joining of continuous fibre reinforced, load optimised TPC sheets to additively manufactured functional structures (e.g. ribs, fastening elements) in one process step. Therein, the functional structure is preheated and the melted TPC sheet is formed onto the structure using a one sided diaphragm based forming process. One challenge for this combined forming and joining is the preheating of the functional structure in order to achieve a sufficient bonding quality. In this paper, initial results of fundamental process investigations regarding the challenges for the combined forming and joining between the TPC sheet and the additively manufactured structure are presented.

Keywords: Additive Manufacturing, Thermoplastic Composites, Diaphragm-Forming, Joining, Laser
PACS: 61.41.+e; 81.05.Qk; 81.20.Hy; 81.20.Vj

INTRODUCTION

Fibre reinforced thermoplastics, also named thermoplastic composites (TPC), are increasingly implemented in aircraft and automotive lightweight applications [1]. The production of such lightweight composite products is increasingly influenced by the demand for individualisation and customisation at low prices [2, 3]. To address this demand, thermoplastic composites offer a high potential as they offer the possibility to freely combine fibre and matrix materials as well as the mechanical properties e.g. by targeted adaption of the fibre orientation [4]. However, the production of individualised TPC components, as well as small, variable series is economically challenging due to the cost-intensive mould technology required in the common manufacturing route of forming and back injection moulding of TPC sheets [5].

Therefore, the development of novel production technologies to increase process flexibility and extend the range of applications for TPC components to individualised, functionally integrated TPC components is in the foreground of current research activities at the Institute for Plastics Processing (IKV) in Industry and Skilled Crafts at RWTH within the BMBF funded project LightFlex together with eleven partners from industry and science. In order to achieve the highest possible degree of individualisation and functionalisation at low tooling costs, the goal is to form customised TPC sheets and join them with additively manufactured functional structures in one process step [6]. In the best case, a cost-intensive mould is completely omitted and the functional structure itself is used to define the forming geometry. In the presented approach, the forming process is carried out with the aid of a single-sided diaphragm and air pressure. In order to do so, the high-temperature diaphragm process [7] developed at IKV is modified for this purpose. The key questions arise regarding how the functional structures must be heated up in order to achieve a sufficient bonding quality, which temperature level is necessary and which bonding strengths can be achieved between TPC sheet and functional structure. This paper presents the general process concept and discusses various heating concepts.

APPROACH FOR THE DEVELOPMENT OF THE COMBINED FORMING AND JOINING PROCESS

The approach of the combined forming and joining process developed at IKV is to combine the forming and joining of load-optimised TPC sheets with additively manufactured thermoplastic functional structures in a single process step [6, 8]. For the application in small series or for prototype production, complex mould adaptations for the change of the component geometry or the applied materials have to be avoided to enable a high degree of flexibility and a cost-efficient indiviualised production. For this purpose, the TPC sheets are formed directly onto the functional structure in a positive forming process by air pressure via a diaphragm after heating. Due to this approach, low-cost materials (e. g. aluminum or block resin moulds from temperature-resistant resins) can be used for the supporting structure which assures the stability of the functional structure during the process. Figure 1 illustrates the concept of the combined forming and joining process. Preliminary investigations have shown that heating of only the TPC sheet is not sufficient to successfully connect the joining partners with each other. Since heating the TPC sheet with infrared radiation is the established process, the question is whether heating the functional structure via infrared radiation as well enables a good bonding quality.

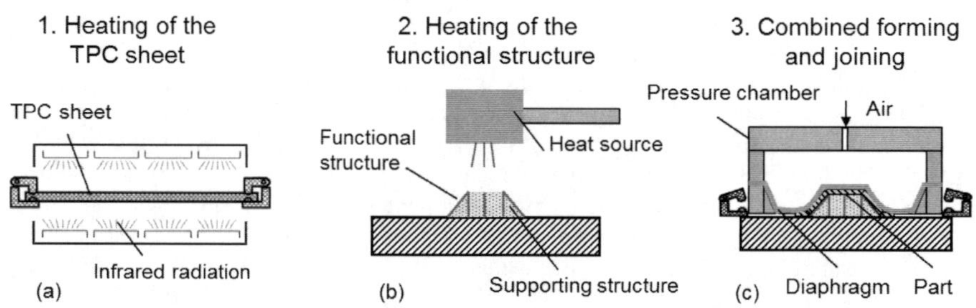

FIGURE 1. Process setup for the combined forming and joining process – a) heating of the TPC sheet b) heating of the additively manufactured structure c) combined forming and joining process

EXPERIMENTAL

Material

All investigation described in this paper are conducted with woven glass fibre reinforced polyamide TPC sheets (TEPEX® Dynalite 102, fibre volume content 47 %, 2x2 twill weave) with a thickness of 2 mm provided by Bond Laminates, Brilon, Germany. The melting point is specified by the manufacturer to 220 °C. The functional structures are made of made of PA 2200 and produced by the Adam Opel AG, Rüsselsheim, Germany by selective laser sintering (SLS) on a EOS SINT 760. Both the material and the SLS machine are fabricated by the EOS GmbH, Krailing, Germany. The nominal density of the structures after sintering is 930 kg/m³, while the melting point of the material is 176 °C. The functional structures are fabricated as T-joint ribs with the dimensions 50.5 mm x 20 mm x 4 mm. The production direction was carried out in layers along the longest component edge.

Process Technology

The TPC sheet is heated between two infrared radiator fields with a nominal power of 30 kW from KRELUS AG, Oberntfelden, Switzerland, each with a distance of 175 mm to the TPC sheet. The surface temperature is recorded on the top and bottom side of the TPC sheet with a pyrometer type Raytek MI13 of the company Fluke Process Instruments, Berlin, Germany. In order to ensure good formability of the TPC sheet, the TPC sheet is heated up above melting temperature to 250 °C. When this temperature is detected by the pyrometers, the emitter fields are switched off and the TPC sheet is positioned above the functional structure in the forming station via a shuttle.

The heating of the functional structures is carried out by four ceramic radiators of the company Elstein-Werk M. Steinmetz GmbH & Co. KG, Northeim, Germany with a maximum rated power of 1 kW. Each ceramic radiator is equipped with a thermocouple type K and is controlled separately. The control is carried out by an external controller of the type Z1293 from HASCO Hasenclever GmbH + Co KG, Lüdenscheid, Germany. The heating is carried out at the joining zone on the top side of the functional structure. Since there is no possibility of inline temperature measurement on the surface of the functional structure, the process is controlled by the heating time. A radiator temperature of 300 °C is specified for all process variations. In order to ensure safe handling and stabilisation during the joining process, the functional structure is held in a preheated support structure made of aluminium. For the joining tests, the heating time of the functional structures manufactured and the joining pressure are varied. Table 1 summarises the process variations with the used parameters. For each process variation seven parts are produced.

TABLE (1). Investigated process variations and used parameters

Process variation	Heating time [s]	Joining pressure [bar]
Both low level	600	2.5
Long heating time	900	2.5
High pressure level	600	3.5
Both high level	900	3.5

Methods

The bonding strength is determined by means of T-joint tensile tests. The tests are carried out on a floor standing machine type Z150 from Zwick GmbH & Co. KG, Ulm, Germany, with a test speed of 2 mm/min. The TPC sheet is clamped between two metal plates and held in place during the test. The functional structure is clamped between the upper movable fixture. The test setup is illustrated in Figure 2. The ultimate tensile strength (UTS) is chosen as the evaluation criterion for the bonding strength of the TPC sheet and the functional structure.

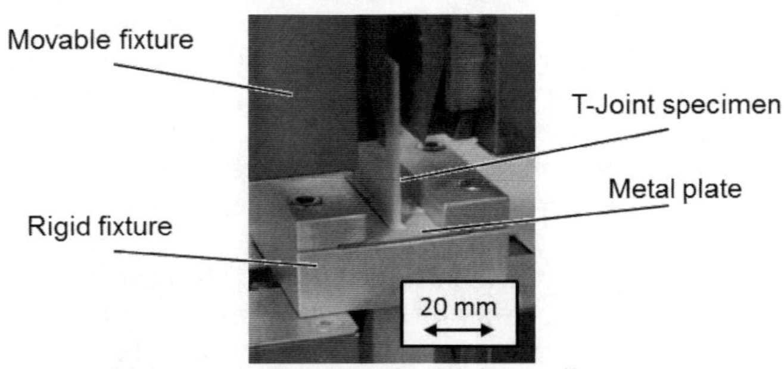

FIGURE 2. Test setup for T-joint tensile test

RESULTS AND DISCUSSION

Figure 3 shows the UTS for the different process variations. With the short heating time of 600 s and a joining pressure of 2.5 bar, a UTS of approx. 1.3 MPa is achieved. An increase in joining pressure causes the bond strength to drop to approx. 0.7 MPa. During the heating period, a melting bath is produced, but an increase in the joining pressure leads to the melt being almost completely displaced from the joining zone, which results in a reduction of the UTS. With an increased heating time of 900 s, a reverse effect can be observed. In this case, the UTS increases from approx. 1.1 MPa to 1.5 MPa by increasing the joining pressure, because the extended heating time results in a larger melting bed. However, it is noticeable that no sufficient bonding strengths are achieved at any parameter set for the applied infrared heating of both materials. Due to the long heating times of 600 seconds and more, significant heat conduction effects in the functional structure occur. Not only is the surface of the functional structure heated up, but also the entire component. This means that the dimensional stability of the functional structure is insufficient for the joining. During the joining process, only a fraction of the deformation resistance counteracts the joining pressure, which means that the functional structure is deformed and the joining pressure in the joining zone is not sufficient for a stable, substance-locking connection.

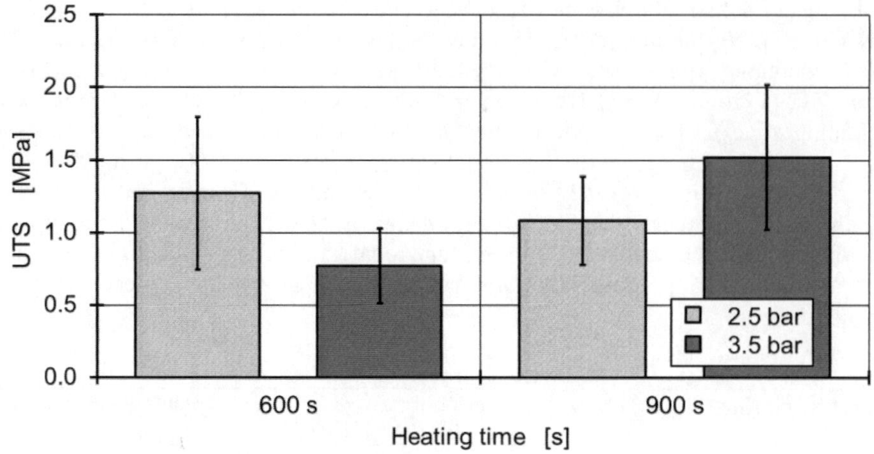

FIGURE 3. Ultimate tensile strengths with T-joint tensile test

To conclude, the approach of infrared heating of the additively manufactured functional structure is not feasible. Instead, it is necessary to concentrate the heating locally on the surface of the joining zone and to significantly reduce the heating time. In order to meet these requirements, a concept for the heating of the functional structure with a laser scanning system has been developed.

INTEGRATION OF THE LASER SCANNING SYSTEM INTO THE TPC FORMING FACILITY

The laser unit used is a diode laser of the type LDM 1000-40 from Laserline GmbH, Muehlheim-Kaerlich, Germany, which has a wavelength of 980 nm and a nominal power of 1 kW. The laser is coupled with a 3d-scanning head of the type Elephant from Arges GmbH, Wackersdorf, Germany, in order to melt the surface of differently shaped three-dimensional functional structures. In order to ensure a flexible design of the additively manufactured functional structure and at the same time to ensure homogeneous heating and melting of its surfaces at different heights a digital focus shift of the laser is accessible. With the used laser system a usable working height of 403 mm to 645 mm at the distance of the decoupling at the laser scanning head, which corresponds to a maximum structure height of 140 mm is possible. The maximum projected area of the functional structures that can be heated is 400 x 400 mm².

For the integration of the laser scanning system into the existing TPC forming facility and control technology at IKV, engineering requirements such as integration in the control system and aspects of laser safety as well as process criteria such as the fixation of the melted functional structure and the process speed had to be considered. Especially the fixation and support of the functional structure is essential for stable and reproducible process control. Free-standing sections may show sink marks during heating due to a lack of dimensional stability of the functional structure. To avoid a handling of the heated functional structure, integrating the laser into the TPC forming facility's forming and joining area would be the simplest option, since the functional structure would not have to be moved during the process. From an economic point of view, however, this approach would result in additional non-productive times in the process. The main process would have to be interrupted until the heating of the functional structure with the laser is completed. Additionally, the working area of the laser, in this approach the forming and joining area, would have to be covered in a laser-safe manner. To enable the transfer of the heated TPC sheet into this area, at least two doors would have to be inserted into the laser protection chamber, which must be opened when the handling unit is moved. In order to circumvent these circumstances and the resulting non-productive times, an integration concept was chosen in which a laser protection chamber is implemented as an external, self-contained system and the heated functional structure is handled and transferred to the forming and joining area after heating together with the supporting structure. The transfer is carried out on a plate with defined positioning devices in the forming and joining station. The concept of the laser protection chamber was developed in cooperation with Arges GmbH, Wackersdorf, Germany and has been put into operation at IKV. The TPC forming facility with the integrated laser protection chamber and scanning system is illustrated in Figure 4. The laser and the laser protection chamber have been put into operation at IKV. Investigations regarding the bonding strength with this promising approach will be presented in the future.

FIGURE 4. TPC facility with integrated laser scanning system (a), infrared heating station for TPC sheet (b), additively manufactured functional structure (c)

CONCLUSION AND OUTLOOK

In this paper, an approach for a combined forming and joining process for load-optimised TPC sheets and additively manufactured functional structures has been described. The heating of the functional structure is one of the key elements for a successful combined forming and joining process. The investigations have shown that heating with infrared radiation does not lead to the desired bonding quality, as the entire functional structure is heated and the dimensional stability is lost. In order to guarantee a defined local heating of the joining zone, a concept is presented to integrate a laser scanning system into the TPC forming facility. In further investigations, the presented system technology is used to determine and investigate the processing parameters (e.g. laser power for melting the functional structure, joining pressure) for different material combinations and the geometric possibilities and limitations of the process (design of functional structure, part complexity) as well as the resulting joining strengths. Furthermore, interactions between different process settings between laser parameters and forming parameters are to be investigated.

ACKNOWLEDGMENTS

The investigations set out in this report received financial support from the *German Bundesministerium für Bildung und Forschung* (No. 03XP0013B), to whom we extend our thanks.

REFERENCES

1. Ch. Hopmann, M. Hildebrandt, R. Bouffier and K. Fischer, „Inline-Impregnation – Individualized productionof thermoplastic continuous fiber reinforced composite parts", Proceedings of the SAMPE 2015 Conference. Baltimore, MD, USA, 2015
2. B. Ebel and M. Hofer, *Automotive Management*, Berlin Heidelberg, Germany: Springer, 2014
3. Ch. Hopmann, M. Hildebrandt, H. Janssen, Ch. Beste and D. Werner, „ Individualised production of thermoplastic composite parts – Combining additive manufacturing and thermoforming" Proceedings of ITHEC 2016 Conference, Bremen, Germany 11.10.2016
4. P. Bocr, „Maßgeschneiderte thermoplastische Einleger in Kombination mit Spritzgießen", Proceedings of 1st International Composite Congress (ICC). Stuttgart, Germany, 2015
5. H. Janssen, „Effiziente Fertigung von funktionsintegrierten FVK-Bauteilen aus belastungs- und verschnittoptimierten Organoblechen", Umdruck zur Veranstaltung des VDI Wissensforums. Nürnberg,Germany, 2015
6. Ch. Hopmann, M. Hildebrandt, H. Janssen, Ch. Beste and D. Werner, „Individualisierte Produktion funktionalisierter thermoplastischer FVK Bauteile", Umdruck zur 5. VDI-Fachkonferenz Fügen im Automobilbau, Bremen, Germany, 2016
7. Ch. Hopmann, W. Michaeli and R. Schoeldgen: "Advanced Diaphragm Forming Process for high temperature applications", Proceedings of the SAMPE 2011 Conference. Long Beach, USA, 23.05.2011.
8. Ch. Hopmann, M. Hildebrand, H. Janssen, Ch. Beste and D. Werner, „LightFlex - TP-FVK-Halbzeuge auf Basis von UD-Tapes individuell herstellen, verarbeiten und fügen", Umdruck IKV Fachtagung Großserie für TP-FVK-Bauteile – Herstellung und effiziente Verarbeitung thermoplastischer Halbzeuge, Aachen, Germany, 2015

Comparation of two methodologies to determine the cation exchange capacity on a Brazilian Sodium Bentonite Clay

Pelinson Lucas Gustavo (1), Rivas Guilherme Guedes (1), Ferreira Henderson Luiz (1), Torin Rondes Ferreira da Silva (1), Maia Anderson (1)*

(1) Faculdade de Tecnologia de Mauá - São Paulo - Brazil

It was analyzed through this project the effectivity of theorganofilization process on a Brazilian sodium bentonite clay, using the quaternary ammonium salt Dodigen 1611 by cation exchange in aqueous dispersion in a laboratory scale. The organofilization process main goal is increase the distance between the clay lamellae area, improving it's nanocomposites aplications. The contact's area increase and the organofilic property make this material extremely useful in the polymers industry. Three pre-experiments were required before the organofilization, a purification using the Hydrogen Peroxide Method and the cation exchange capacity using the Blue Methylene Index Method and a high precision instrumental analysis by Atomic Absorption Spectroscopy. The final and the pure product were caractherized by X ray diffraction, showing a variation between the galleries with the intercalation of organic cations, by thermogravimetric analysis to evaluate the tenor of organic cations and the thermal stability, which was expected according to the literature and by scanning electronic microscopy to evaluate the morphology of the particles.

pps proceeding
Symposium: S20 - Soft matter applications
Keynote Presentation

Magnetic Rotational Spectroscopy with Nanorods for Rheological Characterization of Nanoliter Droplets and Micrometer Thick Films

Kornev Konstantin *(1)**

(1) Clemson University - South Carolina - USA

Magnetic nanorods are attractive materials enabling assembly, ordering, control, and reconfiguration of different magnetic lattices within milliseconds. MilliTesla magnetic fields are sufficient to manipulate with these nanorods. In this talk, a few examples of applications of magnetic nanorods will be shown. First we will discuss the remote controlled rotation of magnetic nanorods providing a new nanoscale tool to probe different properties of liquids and solids at the micrometer scale. The current challenge of the in situ characterization of minute amounts of fluids rapidly changing their rheological properties will be addressed. Basic theory of Magnetic Rotational Spectroscopy for rotation of a magnetic nanorod in non-Newtonian fluids will be introduced and specific features of rotating nanorods will be explained. Then we will experimentally demonstrate that magnetic rotational spectroscopy provides rich physicochemical information about the gelation processes of polymers and sols. Recent progress in understanding of the basic physical principles of the nanorod alignment in thin films by external magnetic field and characterization of these films by magnetic rotational spectroscopy will be presented.

PVA-Borax-Curcumin Hydrogel: Correlation between viscoelastic properties and drug release rate

Rezvan Gelareh [1], Pircheraghi Gholamreza [1]*, Bagheri Reza [1]

[1] Sharif University of Technology - Tehran - Iran

In the present work, polyvinyl alcohol (PVA) hydrogel physically crosslinked using disodium tetraborate (Borax) have been prepared at different PVA: Borax ratios. To study the drug release rate as a function of viscoelastic properties of prepared hydrogels, fixed amount of curcumin encapsulated in pluronic F127 micelle were loaded to hydrogels and drug release rates were investigated in PBS solution using UV-Vis spectroscopy at two temperatures, i.e. 25 and 37°C. Interactions between PVA chains and Borax and curcumin molecules were studied by means of FTIR spectroscopy, while viscoelastic properties of prepared hydrogels were measured in the oscillatory shear mode at linear viscoelastic region. It was observed that with increasing the Borax: PVA ratio the low frequencies storage modulus shows an increasing-decreasing trend. The highest storage modulus and complex viscosity were achieved in a hydrogel sample prepared by PVA: borax ratio of 4: 1. While, Curcumin release rate shows a non-fickian power law behavior at studied temperatures. Interestingly it was observed that there is a linear relationship between the power law exponent of drug release rate, n, and viscoelastic parameter of hydrogels, R, which is defined as crossover frequency divided by plateau storage modulus. These results showed it could be possible to predict drug release rate from viscoelastic properties of curcumin loaded PB hydrogel.

Cholesteric Film Formation Flows of Dilute Collagen Solutions

***Aguilar Gutierrez Oscar F.* [(1)*], *Rey Alejandro D.* [(1)]**

[(1)] McGill University - Quebec - Canada

Dilute collagen solutions have been used as precursors of highly ordered cholesteric thin films obtained through a multi-step process whose first stage is a complex inhomogeneous flow deposition. The purpose of this initial stage is to pre-align the mesogens and form the paranematic phase that allows a more systematic control of the desired helical axis to be formed in later stages. We present the theory and simulations of this initial flow steps by means of the Landau - de Gennes nematodynamics equations where different flow types are explored using the four roll mill kinematics allowing to evaluate the microstructure response of the precursor solution. We provide a phase diagram indicating the full state of orientation as a function of flow type and intensity. Based on the film formation process requirements, we provide the operational region to which the solutions should be submitted in order to produce highly ordered homogeneous chiral films.

pps proceeding
Symposium: S20 - Soft matter applications
Oral Presentation

Electrorheological model based on Liquid Crystals Membranes With Applications to Outer Hair Cells

Herrera Valencia Edtson Emilio [(1)]*, Rey Alejandro D [(2)]

[(1)] McGill - Quebec - Canada, [(2)] McGill university - Montreal, Quebec - Canada

Abstract Background: Liquid crystal flexoelectric actuation uses an imposed electric field to create membrane bending and it is used by the Outer Hair Cells (OHC) located in the inner ear, whose role is to amplify sound through generation of mechanical power. Oscillations in the OHC membranes create periodic viscoelastic flows in the contacting fluid media. A key objective of this work on flexoelectric actuation relevant to OHC is to find the relations and impact of the electro-mechanical properties of the membrane, the rheological properties of the viscoelastic Jeffreys media, and the frequency response of the generated mechanical power output. Methods: The model developed and used in this work is based on the integration of: (i) the flexoelectric membrane shape equation applied to a circular membrane attached to the inner surface of a circular capillary, and (ii) the coupled capillary flow of contacting viscoelastic phases, which are characterized by the Jeffreys constitutive equation with different material conditions. The membrane flexoelectric oscillations drive periodic viscoelastic capillary flows, as in OHCs. Results: By applying the Fourier transform formalism to the governing equation, an analytical expression for the transfer function, associated to the average curvature and volumetric rate flow as a function of the electrical field were found. Assuming small Deborah number, the average membrane curvature or volumetric rate flow as a function of the electrical field can be expressed as an second order differential equation which depend on the material properties in the system. Neglecting the inertial mechanisms, the system can be expressed in terms of a spring and dashpots such as a mechanical system. Taking into account the inertial mechanisms, the power spectrum shows several resonance peaks in the average membrane curvature and volumetric flow rate. When the inertia is neglected, the system shows a non-monotonically behavior in the power spectr

Thermoresponsive Behavior of Temperature-Sensitive-Gel(TSG)/Polymer Composites

Jin Gong[1, 2]*, Eiichi Hosaka[2], Hiroshi Ito[2]*,
Yoshikazu Shibata[3], Dai Nakanishi[3] and Shin-ichiro Ishihara[3]

[1] *Department of Mechanical Systems Engineering, Graduate School of Science and Engineering, Yamagata University, 4-3-16 Jonan, Yonezawa, Yamagata, 992-8510, Japan*
[2] *Department of Polymer Science & Engineering, Graduate School of Organic Materials Science, Yamagata University, 4-3-16 Jonan, Yonezawa, Yamagata 992-8510, Japan*
[3] *KOHJIN Film & Chemicals Co., Ltd, 1-1 Koukokumachi, Yatsushiro, Kumamoto, 866-8686, Japan*
** Tel: +81-238-26-3135, Fax: +81-238-26-3248, E-mail: jingong@yz.yamagata-u.ac.jp;*
** Tel & Fax: +81-238-26- 3081, E-mail: ihiroshi@yz.yamagata-u.ac.jp*

Abstract. Temperature sensitive gel (TSGs) were widely used in many fields such as medical, robotics, MEMS and also in daily life. In this paper, we designed TSGs with lower transfer temperature at around 60 in order to expand their further application. We discussed the synthesis of TSG, the preparation of TSG/polymer composites, and the thermoresponsive behavior of these composites. Differential scanning calorimetry (DSC), dynamic mechanical analysis (DMA), thermal mechanical analysis (TMA), scanning electric microphotography (SEM) and tensile mechanical testing were performed to evaluate thermal and mechanical properties, and thermoresponsive behavior regarding to clarify the thermo response mechanism of TSG/polymer composites.

Keywords: Temperature sensitive gel, Polymer composites, Thermoresposive behavior.
PACS: 81.05.QK

INTRODUCTION

Temperature sensitive gel (TSGs) exhibit physical changes in response to changes in temperature. Many kinds of TSGs were developed currently in our group, and all these TSGs show high crystallinity due to their high crystalline side chains. Phase change resulting from the melting of side-chain crystals by increasing the thermal energy, result in a changing behavior in physical properties, especially in viscosity, flowability and mechanical strength. TSGs also have good strength and toughness because the side-chain crystals with controlled length give TSGs a good balance of mechanical properties. TSGs are expected to use in many fields such as medical, robot as well as daily life [1, 2]. TSGs change their modulus at transfer temperature [3, 4]. In this paper, we report a new TSG having a lower transfer temperature at around 60℃ to open up boarder-application possibilities of TSGs as functional filler for developing new functional films.

Functional films are widely used in fields such as food packaging, electronics, energy sector and daily necessities. There has been a growing demand for functional films recently. TSG sheets are synthesized from monomers with vinyl group through bulk photopolymeriazation, then ground into fine particles mechanically. TSG particles are added as functional fillers into polymer as Linear Low Density Polyethylene (LLDPE) to develop a new air and moisture permeable membrane. The TSG/polymer functional films were prepared and their thermal, mechanical properties and thermoresponsive behavior were characterized.

EXPERIMENTAL

Materials

Two monomers with vinyl group (VM1 and VM2) were used as monomers. N, N'-methylenebisacrylamide (MBAA) used as a cross-linker, and benzophenone (BP) used as a UV polymerization initiator were obtained from Wako Pure Chemical Industries ltd., Japan. LLDPE were supplied by KOHJIN Film & Chemicals Co., Ltd. All the above chemicals are used as received.

Synthesis of TSGs

The synthesis of TSG is described here as an example. Into a 50mL snap vial were placed monomer VM1, monomer VM2 and cross-linker MBAA. The mixture was stirred for 30min and then initiator BP was added. The nitrogen bubbling was performed during the above process to get rid of the oxygen which prevents the radical polymerization. Then the solution was poured into self-made mold and irradiated under UV lamp for 9~20 hours at 30℃. TSG sheet was obtained after removing the mold. Other TSGs were also prepared in the similar mannar.

Preparation of TSG/LLDPE films

TSG/LLDPE(10/90, wt/wt) composites with different TSG composition were prepared by blending with 10wt% of TSG and 90wt% of LLDPE using twin-screw extruder (LABO PLASTOMILL 4C150, TOYOSEIKI). The processing conditions were the following: screw speed 60rpm, feeder speed 10rpm, screw temperature (from feeder side) 190℃, 200℃, 200℃. In order to prepare the polymer composite films, the obtained TSG/LLDPE were molded into films with hot press machine (Imoto Machinery Co., Ltd.). The hot press conditions were the following: mold size 120mm×80mm×0.24mm; heating process: temperature 190℃, melting time 2min, pressure 8MPa; cooling process: cooling time 1min, pressure 4 MPa.

Characterization

TSG particles were prepared by freeze-grinding of TSG sheet using dancing mill (ALM90DM, Nitto Kagaku Co. Ltd.) and liquid nitrogen. The size of TSG particles was measured from SEM images (Miniscope TM-1000, Hitachi Ltd.). TSG particles were sputter-coated with platinum-palladium for SEM analyses. To obtain the mean size and size distribution, at least 100-150 particles were selected randomly, and measurements were performed for statistical analysis.

Thermal properties of prepared TSG and TSG/LLDPE were studied by use of differential scanning calorimetry (DSC) (Q-200, TA instruments), thermal gravimetric analysis (TGA) (Q-50, TA instruments) and tensile test and thermal mechanical analysis (TMA) (Q-400, TA instruments). DSC measurement conditions were the following: heating rate 5℃/min, temperature range of 10~70℃. TGA of TSG particles was undergone at the heating rate of 10℃/min and temperature range of 30~600℃. TMA was performed under the following conditions: measurement mode film/fiber, force 0.02N, heating rate 5℃/min, cooling process force 0.00N, cooling rate 2℃/min, temperature range 10~70℃. Test pieces for TMA with 20mm in length, 4mm in width and 250μm in thickness were obtained by punching of the film.

Tensile test was carried out by versatile tension-testing machine (STROGRAPH VGS-1-E, Toyo Seiki Seisaku-sho, Ltd.). Tension test conditions was as follow: tension speed 300mm/min, distance between chucks 15mm. Test pieces were obtained by punching of the film. The size of the test pieces is 15mm in length, 2mm in width and around 250μm in thickness.

RESULTS AND DISCUSSION

The size, morphology and thermal properties of TSG particles

The TSG white sheets were synthesized via bulk photopolymerization at 30℃ as shown in FIGURE 1 (a). TSG particles as shown in FIGURE 1 (b) were obtained by freeze-grinding with dancing mill and liquid nitrogen. It was

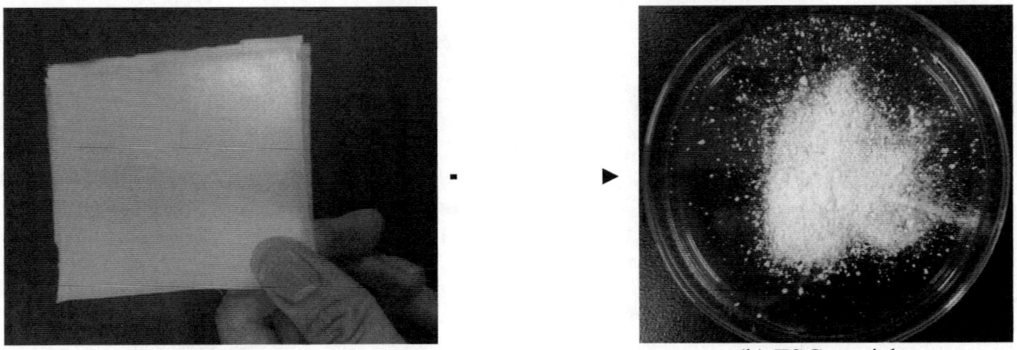

(a) TSG sheet (b) TSG particles

FIGURE 1. TSG white sheet synthesized via bulk photopolymerization (a) and TSG particles prepared by freeze-grinding with dancing mill and liquid nitrogen (b).

FIGURE 2. SEM micrographs for TSG particles.
The size of TSG particles is less than 300μm, and their mean size is nearly 100μm.

found that the size of TSG particles is less than 300μm, and their mean size is nearly 100μm from the SEM image (FIGURE 2). The TSG particles are angular but their shapes are relatively irregular. The DSC results (FIGURE 3) showed TSG particles have high crystallinity and the melting temperature of TSG particles is around 58℃. The TGA results (FIGURE 4) showed the start and end decomposition temperatures of TSG particles were 236℃ and 427℃, respectively, indicating TSG particles have enough thermal resistance to use as fillers.

The thermal and mechanical properties of TSG/LLDPE composites

The thermal properties of TSG/LLDPE composites were investigated using DSC and TMA. The DSC and TMA evaluation of TSG and LLDPE were also performed for comparison. FIGURE 3 and 4 shows the DSC and TMA curves of TSG particles, LLDPE and TSG/LLDPE composites, respectively. Peak of TSG almost disappeared in TSG/LLDPE composites and only very small peak at around 55℃ that seems derived from TSG appeared. The heat flow of TSG/LLDPE composites showed almost the same behavior as that of LLDPE. According to the TMA results, the TSG showed quite large dimension change in the temperature range of 40~65℃, and its total dimensional change was over 2000μm. TSG/LLDPE composites containing 10wt% TSG also showed larger dimension change than LLDPE due to the thermal expansion effect of TSG.

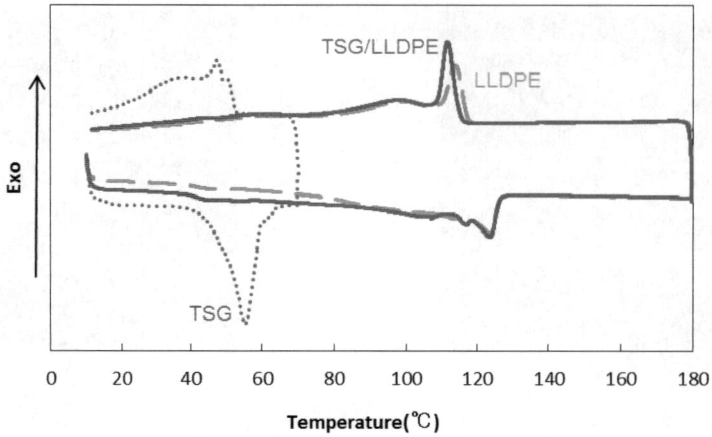

FIGURE 3. DSC curves of TSG particles, LLDPE and TSG/LLDPE composites.

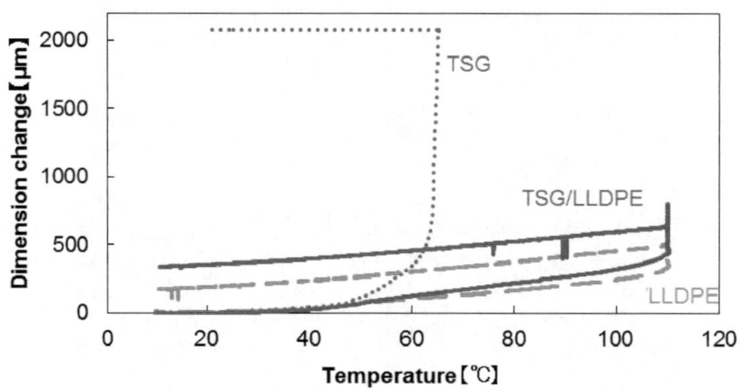

FIGURE 4. TMA results of TSG, LLDPE and TSG/LLDPE composites.

The loss tangent (tan δ) of LLDPE and TSG/LLDPE measured from TMA test were shown in FIGURE 5. It can be seen clearly that peak at the low-temperature side shifted to the high-temperature side, although the peak temperature could not be measured exactly probably due to the crystallinity falling of TSG crystals during the preparation processing of TSG/LLDPE composites.

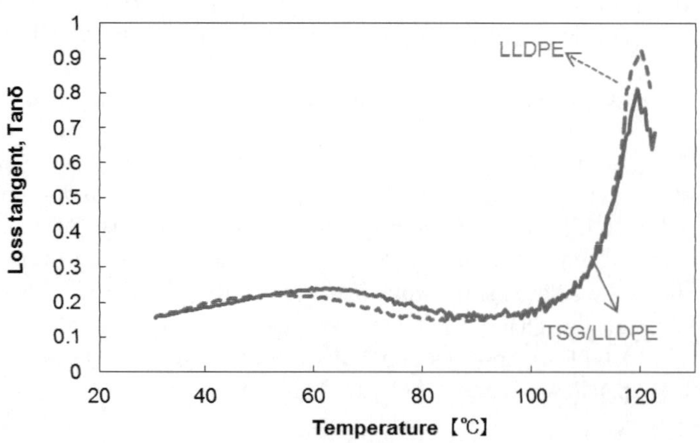

FIGURE 5. Loss tangent (Tan δ) of LLDPE and TSG/LLDPE measured from TMA test.

The stress-strain curves of tension testing of TSG, LLDPE and TSG/LLDPE were displayed in FIGURE 6, and values of tensile modulus, tensile strength and elongation were listed in TABLE 1. Modulus of TSG/LLDPE was significantly improved by almost 3.4 times from 127 MPa to 426 MPa comparing with virgin LLDPE. The reason of modulus increase is supposed that the orientation of TSG and LLDPE occured at compounding and hot press process, but further studies are needed to make the reason clear. The elongation at break of TSG/LLDPE composite films increased greatly as compared to that of TSG and did not decrease largely as compared to that of LLDPE because elongation propety is mainly decided by LLDPE matrix polymer.

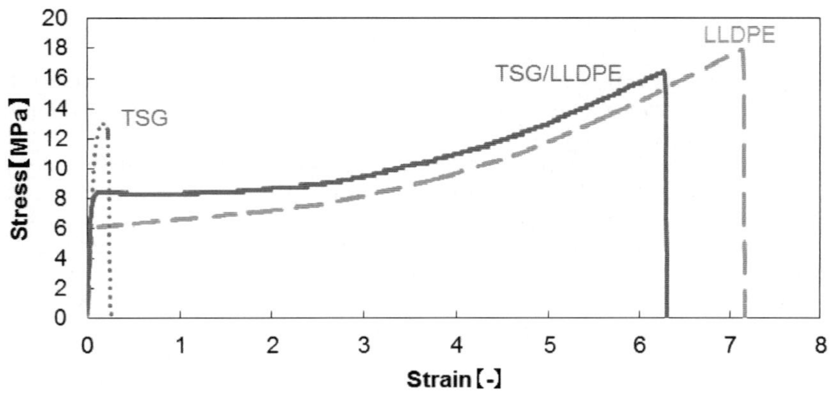

FIGURE 6. Stress-Strain curves of TSG, LLDPE and TSG/LLDPE measured on tension-testing machine.

TABLE 1. Values of tensile modulus, tensile strength and elongation.

Sample	Tensile modulus (MPa)	Tensile strength (MPa)	Elongation (%)
TSG	196.1	13.0	71.3
LLDPE	127.1	17.9	716.5
TSG/LLDPE	426.0	16.2	673.7

CONCLUSION

New TSGs having a lower transfer temperature of around 58℃ were synthesized successfully. These TSGs show high crystallinity which results in their good strength and toughness because the side crystalline chains were introduced into the TSG's structure with controlled the length of side chains. The developed TSG were dispersed in the matrix polymer LLDPE to produce new functional films. The TSG/LLDPE functional films were prepared and their thermal and mechanical properties were characterized. The melting peak of TSG crystals almost disapeareed in TSG/LLDPE composites. Adding 10wt% TSG into LLDPE had light influence on thermal properties of TSG/LLDPE. Modulus of TSG/LLDPE composite was significantly improved by almost 3.4 times from 127 MPa to 426 MPa as compared to that of virgin LLDPE. The elongation at break of TSG/LLDPE composite films increased greatly as compared to that of TSG and did not decrease largely as compared to that of LLDPE.

REFERENCES

1. S. R. Sershen, S. L. Westcott, N. J. Halas, J. L. West, *J. Biomed. Mater. Res. Part A* **2000**, 51(3): 293-298.
2. Y. M. Kwon and S. W. Kim, "Thermosensitive Biodegradable Hydrogels for the Delivery of Therapeutic Agents", in *Drugs and the Pharmaceutical Sciences*, edited by G. S. Kwon, Boca Raton, FL, Taylor & Francis Group, **2005**, pp. 251-274.
3. J. Gong, H. Furukawa, *Expected Materials for the Future* **2013**, 13(1): 5-7.
4. S. Harada, R. Hidema, J. Gong, H. Furukawa, *Chemistry Letters* **2012**, 41: 1047-1049.

Molecular simulation for the prediction of plasticizer efficiency and stability in a polymer matrix

Li Dongyang [(1)], Panchal Kushal [(1)], Xi Li [(1)]*

[(1)] Department of Chemical Engineering, McMaster University - Ontario - Canada

Despite the widespread use of plasticizers in the polymer industry, the mechanism of plasticization remains poorly understood. Industrial selection of plasticizers largely relies on semi-empirical rules and even for successful candidates, it is often difficult to answer why they outperform other molecules in terms of effectively reducing the glass transition temperature and elastic modulus of the material. In addition to plasticization efficiency, plasticizer migration is an issue that is causing increasing concern. Plasticizers that are not chemically bonded to the host polymer often migrate out of the matrix, either evaporating into the atmosphere or being absorbed into any liquids or other polymers in contact with the plastic. The loss of plasticizers over time not only causes the deterioration of material properties, it is also an environmental and safety concern. In particular, it is recently found that phthalates, the most consumed family of plasticizers on the market, are hazardous to human health. We use molecular simulation at both the full-atom and coarse-grained levels to understand the plasticization mechanism. From this knowledge, a number of molecular descriptors will be identified for a semi-quantitative prediction of the efficiency of any plasticizer based on its chemical structure. In addition, predictions will also be made on the thermodynamic compatibility between the plasticizer and its host polymer, as well as the diffusion coefficient of the plasticizer in the latter; these will help us estimate the tendency and speed of plasticizer migration. The end goal is to develop a computational protocol for screening plasticizers and finding high-performance and stable alternatives to phthalates.

Fourier transform applied to the description of flexoelectric membranes immersed in viscoelastic Burgers fluids.

Vizcaya Mendoza Ricardo* (1), *Herrera Valencia Edtson Emilio* (2)

(1) UNAM - Ciudad de México - Canada, (2) UNAM - Ciudad de México - Mexico

Membrane flexoelectricity is an electromechanical coupling effect between the membrane average curvature and macroscopic electric polarization, which is essential to the physiology of hearing. Flexoelectric actuation uses an imposed electric field to create membrane bending and is used by the Outer Hair Cells (OHCs) located in the inner ear. Motivated by the functioning of the OHC, in this paper we model the small amplitude oscillatory dynamics of a tethered circular membrane immersed in Burgers viscoelastic media driven by a small amplitude harmonic electric field of arbitrary frequency. The model for membrane average curvature dynamics as a function of the electric field dynamics is second order in both inputs and outputs and maps into the classical mechanical Burgers solid model. The three dimensional material space that characterizes the inertia, viscosity, and elasticity of the viscoelastic fluid/flexoelectric membrane material system is defined and used to classify and characterize the frequency response of the electro-mechanical system. The frequency response is characteristic of a second order dynamical system with a second order input and can display a single resonant peak in the total power. The amplitude, frequency and width of the power peak, of relevance to the functioning of OHC is dependent on the inertia emerging from the contacting viscoelastic phases and the ratio between the membrane elasticity and the elasticity of contacting liquids. The integrated flexoelectric/viscoelastic model and the novel findings contribute to the ongoing quest for a fundamental understanding of the functioning of outer hair cells (OHCs)

Effect of alginate oxidation condition on properties of alginate-based nanocomposite hydrogels

Emami Zahra Alsadat [1], Ehsani Morteza [1]*, Zandi Mojgan [1], Foudazi Reza [2]

[1] Iran Polymer and Petrochemical Insittue - Tehran - Iran, [2] Department of Chemical Engineering, New Mexico State University - NEW MEXICO - USA

In the present study, nanocomposite hydrogels composed of alginate-gelatin-nanohydroxyapatite were fabricated using oxidized alginates. We investigated the effect of oxidation conditions on the properties of nanocomposite hydrogels. Alginate was oxidized by sodium metaperiodate at different initial solution concentrations (1, 2 and 5%), while the ratio of sodium metaperiodate to number of alginate's repetitive units was fixed at 60 mol% for all samples. To prepare nanocomposite hydrogels, nanohydroxyapatite, synthesized by precipitation method, was added to the alginate and gelatin solution. The results showed that higher concentration of alginate solution during oxidation step resulted in higher degree of oxidation but lower yield of oxidized alginate. Rheological results obtained at oscillatory shear mode and linear viscoelastic region, revealed the lower storage modulus and complex viscosity in hydrogel samples prepared using highly oxidized alginate. In other words, the initial concentration of alginate solution strongly affects the oxidization reaction yield and microstructure of oxidized alginate which is reflected in rheological behavior.

ON THE PULSATING FLOW BEHAVIOR OF A BIOLOGICAL FLUID: HUMAN BLOOD

Herrera Valencia Edtson Emilio [1]*, Calderas Fausto [2], Sánchez-Villavicencio Mayra Luz [3], Medina-Torres Luis [4], Manero Octavio [5], Varela Martinez Aldo Fernando [6], velasco Cesar [7]

[1] UNAM - Ciudad de México - Mexico, [2] CIATEC, A. C. - León, Guanajuato. - Mexico, [3] UAM-I - Ciudad de México - Mexico, [4] Facultad de Química, UNAM - Ciudad de México - Mexico, [5] IIM-UNAM - Ciudad de México - Mexico, [6] Facultad de Estudios Superiores Zaragoza UNAM - Mexico City - Mexico, [7] UNAM - MEXICO CITY - Mexico

Abstract: In this work, the rectilinear flow of a complex fluid (human blood) under a pulsating time-dependent pressure gradient is analyzed. A first approximation of the real case of blood flowing in a vein is described. The pressure gradient which is assumed that is in the same order of the magnitude that would be expected with the human blood circulatory system, simulates the pumping work of the heart while the flow geometry (circular tube) is assumed rigid, smooth and cylindrical. The rheological behavior of blood with different cholesterol levels is modelled using the Bautista-Manero-Puig (BMP) constitutive equation. According to the analytical solution, a flow enhancement is predicted to first order which represents the optimum pumping work of the heart which governs the flow of blood in the entire body. This work is a contribution to the understanding of the complex rheology involved in the discontinuous pressure driven flow of blood in the human body. Hemorheology, Analytical Solutions, Pulsatile flow, Stochastic Noise, Complex Fluid, BMP Constitutive Equation

Scaling features of the tribology of polymer brushes of increasing grafting density around the mushroom – to – brush transition

Mayoral Villa Estela [1]*, Gama Goicochea Armando [2], Klapp Escribano Jaime [1]

[1] ININ - Estado de Mexico - Mexico, [2] ITESE - Estado de Mexico - Mexico

Non equilibrium coarse – grained, dissipative particle dynamics simulations of complex fluids, made up of polymer brushes tethered to planar surfaces immersed in a solvent yield non monotonic behavior of the friction coefficient as a function of the polymer grating density on the substrates (Gamma), while the viscosity shows a monotonically increasing dependence on Gamma. This effect is shown to be independent of the degree of polymerization, N,and the size of the system. It arises from the composition and the structure of the first particle layer adjacent to each surface that results from the confinement of the fluid. Whenever such layers are made up of as close a proportion of polymer beads to solvent particles as there are in the fluid, the friction coefficient shows a minimum, while for disparate proportions the friction coefficient grows. At the mushroom to brush transition (MBT) the viscosity scales with an exponent that depends on the characteristic exponent of the MBT (6/5) and the solvent quality exponent (nu = 0.5, for theta solvent), but it is independent of the polymerization degree (N). On the other hand, the friction coefficient at the MBT scales as $\mu \sim N6/5$, while the grafting density at the MBT scales as $\Gamma \sim N-6/5$ when friction is minimal, in agreement with previous scaling theories. We argue these aspects are the result of cooperative phenomena that have important implications for the understanding of biological brushes and the design of microfluidics devices, among other applications of current academic and industrial interest.

AUTHOR INDEX

AUTHOR INDEX

AUTHOR INDEX

AUTHOR INDEX

AUTHOR INDEX

AUTHOR INDEX

AUTHOR INDEX

AUTHOR INDEX

AUTHOR INDEX

AUTHOR INDEX

AUTHOR INDEX

AUTHOR INDEX

AUTHOR INDEX

Polymer Processing Society
810 E. 10th St.
Lawrence, KS 66044
USA

CURRAN ASSOCIATES INC.
proceedings.com

ISBN 978-1-5108-6330-9